Periodic Table of the Elements

Main Groups

1 IA	2 IIA	13 IIIA	14 IVA	15 VA	16 VIA	17 VIIA	18 VIIIA	Shells
1 H							2 He	K
3 Li	4 Be	5 B	6 C	7 N	8 O	9 F	10 Ne	K–L
11 Na	12 Mg	13 Al	14 Si	15 P	16 S	17 Cl	18 Ar	K–L–M
19 K	20 Ca	31 Ga	32 Ge	33 As	34 Se	35 Br	36 Kr	–L–M–N
37 Rb	38 Sr	49 In	50 Sn	51 Sb	52 Te	53 I	54 Xe	–M–N–O
55 Cs	56 Ba	81 Tl	82 Pb	83 Bi	84 Po	85 At	86 Rn	–N–O–P
87 Fr	88 Ra							–O–P–Q

Subgroups

3 IIIB	4 IVB	5 VB	6 VIB	7 VIIB	8 VIII (1)	9 VIII (2)	10 VIII (3)	11 IB	12 IIB	Shells
21 Sc	22 Ti	23 V	24 Cr	25 Mn	26 Fe	27 Co	28 Ni	29 Cu	30 Zn	–L–M–N
39 Y	40 Zr	41 Nb	42 Mo	43 Tc	44 Ru	45 Rh	46 Pd	47 Ag	48 Cd	–M–N–O
57 La	72 Hf	73 Ta	74 W	75 Re	76 Os	77 Ir	78 Pt	79 Au	80 Hg	–N–O–P
89 Ac	104 Rf	105 Db	106 Sg	107 Bh	108 Hs	109 Mt	110 Ds	111 Rg	112	–O–P–Q

Lanthanides (Shells –N–O–P)

58 Ce	59 Pr	60 Nd	61 Pm	62 Sm	63 Eu	64 Gd	65 Tb	66 Dy	67 Ho	68 Er	69 Tm	70 Yb	71 Lu

Actinides (Shells –O–P–Q)

90 Th	91 Pa	92 U	93 Np	94 Pu	95 Am	96 Cm	97 Bk	98 Cf	99 Es	100 Fm	101 Md	102 No	103 Lr

Springer Handbook
of Condensed Matter and Materials Data

Springer Handbook provides a concise compilation of approved key information on methods of research, general principles, and functional relationships in physics and engineering. The world's leading experts in the fields of physics and engineering will be assigned by one or several renowned editors to write the chapters comprising each volume. The content is selected by these experts from Springer sources (books, journals, online content) and other systematic and approved recent publications of physical and technical information.

The volumes will be designed to be useful as readable desk reference book to give a fast and comprehensive overview and easy retrieval of essential reliable key information, including tables, graphs, and bibliographies. References to extensive sources are provided.

Springer Handbook
of Condensed Matter and Materials Data

W. Martienssen and H. Warlimont (Eds.)

With 1025 Figures and 914 Tables

Professor Werner Martienssen
Universität Frankfurt
Fachbereich 13, Physik
Robert-Mayer-Str. 2–4
60325 Frankfurt
Germany

Professor Hans Warlimont
IWF Dresden
Helmholzstr. 20
01069 Dresden
Germany

Library of Congress Control Number: 2004116537

ISBN 3-540-44376-2
Spinger Berlin Heidelberg New York

This work is subject to copyright. All rights reserved, whether the whole or part of the material is concerned, specifically the rights of translation, reprinting, reuse of illustrations, recitation, broadcasting, reproduction on microfilm or in any other way, and storage in data banks. Duplication of this publication or parts thereof is permitted only under the provisions of the German Copyright Law of September, 9, 1965, in its current version, and permission for use must always be obtained from Springer. Violations are liable for prosecution under the German Copyright Law.

Springer is a part of Springer Science+Business Media

springeronline.com

©Springer Berlin Heidelberg 2005
Printed in Germany

The use of designations, trademarks, etc. in this publication does not imply, even in the absence of a specific statement, that such names are exempt from the relevant protective laws and regulations and therefore free for general use.

Product liability: The publisher cannot guarantee the accuracy of any information about dosage and application contained in this book. In every individual case the user must check such information by consulting the relevant literature.

Production and typesetting: LE-TeX GbR, Leipzig
Handbook coordinator: Dr. W. Skolaut, Heidelberg
Typography, layout and illustrations: schreiberVIS, Seeheim
Cover design: eStudio Calamar Steinen, Barcelona
Cover production: *design&production* GmbH, Heidelberg
Printing and binding: Stürtz GmbH, Würzburg

Printed on acid-free paper
SPIN 10678245   62/3141/   5 4 3 2 1 0

Preface

The Springer Handbook of Condensed Matter and Materials Data is the realization of a new concept in reference literature, which combines introductory and explanatory texts with a compilation of selected data and functional relationships from the fields of solid-state physics and materials in a single volume. The data have been extracted from various specialized and more comprehensive data sources, in particular the Landolt–Börnstein data collection, as well as more recent publications. This Handbook is designed to be used as a desktop reference book for fast and easy finding of essential information and reliable key data. References to more extensive data sources are provided in each section. The main users of this new Handbook are envisaged to be students, scientists, engineers, and other knowledge-seeking persons interested and engaged in the fields of solid-state sciences and materials technologies.

The editors have striven to find authors for the individual sections who were experienced in the full breadth of their subject field and ready to provide succinct accounts in the form of both descriptive text and representative data. It goes without saying that the sections represent the individual approaches of the authors to their subject and their understanding of this task. Accordingly, the sections vary somewhat in character. While some editorial influence was exercised, the flexibility that we have shown is deliberate. The editors are grateful to all of the authors for their readiness to provide a contribution, and to cooperate in delivering their manuscripts and by accepting essentially all alterations which the editors requested to achieve a reasonably coherent presentation.

An onerous task such as this could not have been completed without encouragement and support from the publisher. Springer has entrusted us with this novel project, and Dr. Hubertus von Riedesel has been a persistent but patient reminder and promoter of our work throughout. Dr. Rainer Poerschke has accompanied and helped the editors constantly with his professional attitude and very personable style during the process of developing the concept, soliciting authors, and dealing with technical matters. In the later stages, Dr. Werner Skolaut became a relentless and hard-working member of our team with his painstaking contribution to technically editing the authors' manuscripts and linking the editors' work with the copy editing and production of the book.

Prof. Werner Martienssen

Prof. Hans Warlimont

We should also like to thank our families for having graciously tolerated the many hours we have spent in working on this publication.

We hope that the users of this Handbook, whose needs we have tried to anticipate, will find it helpful and informative. In view of the novelty of the approach and any possible inadvertent deficiencies which this first edition may contain, we shall be grateful for any criticisms and suggestions which could help to improve subsequent editions so that they will serve the expectations of the users even better and more completely.

September 2004
Frankfurt am Main, Dresden

Werner Martienssen,
Hans Warlimont

List of Authors

Wolf Assmus
Johann Wolfgang Goethe-University
Physics Department
Robert-Mayer-Str. 2 – 4
60054 Frankfurt am Main, Germany
e-mail: *assmus@physik.uni-frankfurt.de*

Stefan Brühne
Johann Wolfgang Goethe-University
Physics Department
Robert-Mayer-Str. 2 – 4
60054 Frankfurt am Main, Germany
e-mail: *bruehne@physik.uni-frankfurt.de*

Fabrice Charra
Commissariat à l'Énergie Atomique, Saclay
Département de Recherche sur l'État Condensé,
les Atomes et les Molécules
DRECAM-SPCSI, Centre d'Études de Saclay
91191 Gif-sur-Yvette, France
e-mail: *fabrice.charra@cea.fr*

Gianfranco Chiarotti
University of Rome "Tor Vergata"
Department of Physics
Via della Ricerca Scientifica 1
00133 Roma, Italy
e-mail: *chiarotti@roma2.infn.it*

Claus Fischer
Formerly Institute of Solid State and Materials
Research (IFW)
Georg-Schumann-Str. 20
01187 Dresden, Germany
e-mail: *A_C.FischerDD@t-online.de*

Günter Fuchs
Leibniz Institute for Solid State and Materials
Research (IFW) Dresden
Magnetism and Superconductivity in the Institute
of Metallic Materials
Helmholtzstraße 20
01171 Dresden, Germany
e-mail: *fuchs@ifw-dresden.de*

Frank Goodwin
International Lead Zinc Research Organization, Inc.
PO BOX 12036
Research Triangle Parc, NC 27709, USA
e-mail: *fgoodwin@ilzro.org*

Susana Gota-Goldmann
Commissariat à l'Energie Atomique (CEA)
Direction de la Recherche Technologique (DRT)
Centre de Fontenay aux Roses BP 6
92265 Fontenay aux Roses Cédex, France
e-mail: *susana.gota-goldmann@cea.fr*

Sivaraman Guruswamy
University of Utah
Metallurgical Engineering
135 South 1460 East RM 412
Salt Lake City, UT 84112-0114, USA
e-mail: *sguruswa@mines.utah.edu*

Gagik G. Gurzadyan
Technical University of Munich
Institute for Physical and Theoretical Chemistry
Lichtenbergstrasse 4
85748 Garching, Germany
e-mail: *gurzadyan@ch.tum.de*

Hideki Harada
High Tech Association Ltd.
3730-30 Higashikaya, Fukaya,Saitama, Japan
e-mail: *khb16457@nifty.com*

Bernhard Holzapfel
Leibniz Institute for Solid State and Materials
Research Dresden – Institute of Metallic Materials
Superconducting Materials
Helmholtzstr. 20
01069 Dresden, Germany
e-mail: *B.Holzapfel@ifw-dresden.de*

List of Authors

Karl U. Kainer
GKSS Research Center Geesthacht
Institute for Materials Research
Max-Planck-Str. 1
21502 Geesthacht, Germany
e-mail: *karl.kainer@gkss.de*

Catrin Kammer
METALL – Intl. Journal for Metallurgy
Kielsche Straße 43B
38642 Goslar, Germany
e-mail: *Kammer@metall-news.com*

Wolfram Knabl
Plansee AG
Technology Center
6600 Reutte, Austria
e-mail: *wolfram.knabl@plansee.com*

Alfred Koethe
Leibniz-Institut für Festkörper- und Werkstoffforschung
Institut für Metallische Werkstoffe (retired)
Lessingstrasse 11
01099 Dresden, Germany
e-mail: *alfred.koethe@web.de*

Dieter Krause
Schott AG
Research and Technology-Development
PO BOX 2480
55014 Mainz, Germany
e-mail: *dieter.krause@schott.com*

Manfred D. Lechner
Universität Osnabrück
Institut für Chemie – Physikalische Chemie
Barbarastraße 7
46069 Osnabrück, Germany
e-mail: *lechner@uni-osnabrueck.de*

Gerhard Leichtfried
Plansee AG
Technology Center
6600 Reutte, Austria
e-mail: *gerhard.leichtfried@plansee.com*

Werner Martienssen
Universität Frankfurt/Main
Physikalisches Institut
Robert-Mayer-Strasse 2 – 4
60054 Frankfurt/Main, Germany
e-mail: *Martienssen@Physik.uni-frankfurt.de*

Toshio Mitsui
Osaka University
Nakasuji-Yamate, 3-6-24
665-0875 Takarazuka, Japan
e-mail: *t-mitsui@jttk.zaq.ne.jp*

Manfred Müller
Dresden University of Technology
Institute of Materials Science
Leubnitzer Straße 26
01069 Dresden, Germany
e-mail: *m.mueller33@t-online.de*

Sergei Pestov
Moscow State Academy of Fine Chemical Technology
Department of Inorganic Chemistry
Vernadsky pr., 86
119571 Moscow, Russia
e-mail: *pestovsm@yandex.ru*

Günther Schlamp
Metallgesellschaft Ffm and Degussa Demetron (retired)
Stettinerstr. 25
61449 Steinbach/Ts, Germany
e-mail:

Barbara Schüpp-Niewa
Leibniz-Institute for Solid State and Materials Research Dresden
Institute for Metallic Materials
Helmholtzstraße 20
01069 Dresden, Germany
e-mail: *b.schuepp@ifw-dresden.de*

Roland Stickler
University of Vienna
Department of Chemistry
Währingerstr. 42
1090 Vienna, Austria
e-mail: *roland.stickler@univie.ac.at*

Pancho Tzankov
Max Born Institute for Nonlinear Optics
and Short Pulse Spectroscopy
Max-Born-Str. 2A
12489 Berlin, Germany
e-mail: *tzankov@mbi-berlin.de*

Volkmar Vill
University of Hamburg
Department of Chemistry,
Institute of Organic Chemistry
Martin-Luther-King-Platz 6
20146 Hamburg, Germany
e-mail: *vill@chemie.uni-hamburg.de*

Hans Warlimont
DSL Dresden Material-Innovation GmbH
Helmholtzstrasse 20
01069 Dresden, Germany
e-mail: *warlimont@ifw-dresden.de*

List of Authors

Contents

List of Abbreviations XV

Part 1 General Tables

1 The Fundamental Constants
Werner Martienssen 3
1.1 What are the Fundamental Constants
       and Who Takes Care of Them? 3
1.2 The CODATA Recommended Values of the Fundamental Constants 4
References 9

2 The International System of Units (SI), Physical Quantities, and Their Dimensions
Werner Martienssen 11
2.1 The International System of Units (SI) 11
2.2 Physical Quantities 12
2.3 The SI Base Units 13
2.4 The SI Derived Units 16
2.5 Decimal Multiples and Submultiples of SI Units 19
2.6 Units Outside the SI 20
2.7 Some Energy Equivalents 24
References 25

3 Rudiments of Crystallography
Wolf Assmus, Stefan Brühne 27
3.1 Crystalline Materials 28
3.2 Disorder 38
3.3 Amorphous Materials 39
3.4 Methods for Investigating Crystallographic Structure 39
References 41

Part 2 The Elements

1 The Elements
Werner Martienssen 45
1.1 Introduction 45
1.2 Description of Properties Tabulated 46
1.3 Sources 49
1.4 Tables of the Elements in Different Orders 49
1.5 Data 54
References 158

Part 3 Classes of Materials

1 Metals
Frank Goodwin, Sivaraman Guruswamy, Karl U. Kainer, Catrin Kammer, Wolfram Knabl, Alfred Koethe, Gerhard Leichtfried, Günther Schlamp, Roland Stickler, Hans Warlimont 161
- 1.1 Magnesium and Magnesium Alloys 162
- 1.2 Aluminium and Aluminium Alloys 171
- 1.3 Titanium and Titanium Alloys 206
- 1.4 Zirconium and Zirconium Alloys 217
- 1.5 Iron and Steels 221
- 1.6 Cobalt and Cobalt Alloys 272
- 1.7 Nickel and Nickel Alloys 279
- 1.8 Copper and Copper Alloys 296
- 1.9 Refractory Metals and Alloys 303
- 1.10 Noble Metals and Noble Metal Alloys 329
- 1.11 Lead and Lead Alloys 407
- References 422

2 Ceramics
Hans Warlimont 431
- 2.1 Traditional Ceramics and Cements 432
- 2.2 Silicate Ceramics 433
- 2.3 Refractory Ceramics 437
- 2.4 Oxide Ceramics 437
- 2.5 Non-Oxide Ceramics 451
- References 476

3 Polymers
Manfred D. Lechner 477
- 3.1 Structural Units of Polymers 480
- 3.2 Abbreviations 482
- 3.3 Tables and Figures 483
- References 522

4 Glasses
Dieter Krause 523
- 4.1 Properties of Glasses – General Comments 526
- 4.2 Composition and Properties of Glasses 527
- 4.3 Flat Glass and Hollowware 528
- 4.4 Technical Specialty Glasses 530
- 4.5 Optical Glasses 543
- 4.6 Vitreous Silica 556
- 4.7 Glass-Ceramics 558
- 4.8 Glasses for Miscellaneous Applications 559
- References 572

Part 4 Functional Materials

1 Semiconductors
Werner Martienssen 575
- 1.1 Group IV Semiconductors and IV–IV Compounds 578
- 1.2 III–V Compounds 604
- 1.3 II–VI Compounds 652
- References 691

2 Superconductors
Claus Fischer, Günter Fuchs, Bernhard Holzapfel, Barbara Schüpp-Niewa, Hans Warlimont 695
- 2.1 Metallic Superconductors 696
- 2.2 Non-Metallic Superconductors 711
- References 749

3 Magnetic Materials
Hideki Harada, Manfred Müller, Hans Warlimont 755
- 3.1 Basic Magnetic Properties 755
- 3.2 Soft Magnetic Alloys 758
- 3.3 Hard Magnetic Alloys 794
- 3.4 Magnetic Oxides 811
- References 814

4 Dielectrics and Electrooptics
Gagik G. Gurzadyan, Pancho Tzankov 817
- 4.1 Dielectric Materials: Low-Frequency Properties 822
- 4.2 Optical Materials: High-Frequency Properties 824
- 4.3 Guidelines for Use of Tables 826
- 4.4 Tables of Numerical Data for Dielectrics and Electrooptics 828
- References 890

5 Ferroelectrics and Antiferroelectrics
Toshio Mitsui 903
- 5.1 Definition of Ferroelectrics and Antiferroelectrics 903
- 5.2 Survey of Research on Ferroelectrics 904
- 5.3 Classification of Ferroelectrics 906
- 5.4 Physical Properties of 43 Representative Ferroelectrics 912
- References 936

Part 5 Special Structures

1 Liquid Crystals
Sergei Pestov, Volkmar Vill 941
- 1.1 Liquid Crystalline State 941
- 1.2 Physical Properties of the Most Common Liquid Crystalline Substances 946

1.3　Physical Properties of Some Liquid Crystalline Mixtures 975
　　　References 977

2　**The Physics of Solid Surfaces**
　　Gianfranco Chiarotti 979
　　2.1　The Structure of Ideal Surfaces 979
　　2.2　Surface Reconstruction and Relaxation 986
　　2.3　Electronic Structure of Surfaces 996
　　2.4　Surface Phonons 1012
　　2.5　The Space Charge Layer at the Surface of a Semiconductor 1020
　　2.6　Most Frequently Used Acronyms 1026
　　References 1029

3　**Mesoscopic and Nanostructured Materials**
　　Fabrice Charra, Susana Gota-Goldmann 1031
　　3.1　Introduction and Survey 1031
　　3.2　Electronic Structure and Spectroscopy 1035
　　3.3　Electromagnetic Confinement 1044
　　3.4　Magnetic Nanostructures 1048
　　3.5　Preparation Techniques 1063
　　References 1066

Acknowledgements 1073
About the Authors 1075
Detailed Contents 1081
Subject Index 1091

List of Abbreviations

2D-BZ	2-dimensional Brillouin zone
2P-PES	2-photon photoemission spectroscopy

A

AES	Auger electron spectroscopy
AFM	atomic force microscope
AISI	American Iron and Steel Institute
APS	appearance potential spectroscopy
ARUPS	angle-resolved ultraviolet photoemission spectroscopy
ARXPS	angle-resolved X-ray photoemission spectroscopy
ASTM	American Society for Testing and Materials
ATR	attenuated total reflection

B

BBZ	bulk Brillouin zone
BIPM	Bureau International des Poids et Mesures
BZ	Brillouin zone

C

CB	conduction band
CBM	conduction band minimum
CISS	collision ion scattering spectroscopy
CITS	current imaging tunneling spectroscopy
CMOS	complementary metal–oxide–semiconductor
CODATA	Committee on Data for Science and Technology
CVD	chemical vapour deposition

D

DFB	distributed-feedback
DFG	difference frequency generation
DOS	density of states
DSC	differential scanning calorimetry
DTA	differential thermal analysis

E

EB	electron-beam melting
ECS	electron capture spectroscopy
EELS	electron-energy loss spectroscopy
ELEED	elastic low-energy electron diffraction
ESD	electron-stimulated desorption
EXAFS	extended X-ray absorption fine structure

F

FEM	field emission microscope/microscopy
FIM	field ion microscope/microscopy

G

GMR	giant magnetoresistance

H

HAS	helium atom scattering
HATOF	helium atom time-of-flight spectroscopy
HB	Brinell hardness number
HEED	high-energy electron diffraction
HEIS	high-energy ion scattering/high-energy ion scattering spectroscopy
HK	Knoop hardness
HOPG	highly oriented pyrolytic graphite
HPDC	high-pressure die casting
HR-EELS	high-resolution electron energy loss spectroscopy
HR-LEED	high-resolution LEED
HR-RHEED	high-resolution RHEED
HREELS	high-resolution electron energy loss spectroscopy
HRTEM	high-resolution transition electron microscopy
HT	high temperature
HTSC	high-temperature superconductor
HV	Vicker's Hardness

I

IACS	International Annealed Copper Standard
IB	ion bombardment
IBAD	ion-beam-assisted deposition
ICISS	impact ion scattering spectroscopy
ICSU	International Council of the Scientific Unions
IPE	inverse photoemission
IPES	inverse photoemission spectroscopy
ISO	International Organization for Standardization
ISS	ion scattering spectroscopy
IUPAC	International Union of Pure and Applied Chemistry

J

JDOS	joint density of states

K

KRIPES	K-resolved inverse photoelectron spectroscopy

L

LAPW	linearized augmented-plane-wave method
LB	Langmuir–Blodgett
LCM	liquid crystal material
LCP	liquid crystal polymer
LCs	liquid crystals
LDA	local-density approximation
LDOS	local density of states
LEED	low-energy electron diffraction
LEIS	low-energy ion scattering/low-energy ion scattering spectroscopy
LPE	liquid phase epitaxy

M

MBE	molecular-beam epitaxy
MD	molecular dynamics
MEED	medium-energy electron diffraction
MEIS	medium-energy ion scattering/medium-energy ion scattering spectroscopy
MFM	magnetic force microscopy
ML	monolayer
MOCVD	metal-organic chemical vapor deposition
MOKE	magneto-optical Kerr effect
MOSFET	MOS field-effect transistor
MQW	multiple quantum well

N

NICISS	neutral impact collision ion scattering spectroscopy
NIMs	National Institutes for Metrology

O

OPO	optical parametric oscillation

P

PDS	photothermal displacement spectroscopy
PED	photoelectron diffraction
PES	photoemission spectroscopy
PLAP	pulsed laser atom probe
PLD	pulsed laser deposition
PSZ	stabilized zirconia
PZT	piezoelectric material

R

RAS	reflectance anisotropy spectroscopy
RE	rare earth
REM	reflection electron microscope/microscopy
RHEED	reflection high-energy electron diffraction
RIE	reactive ion etching
RPA	random-phase approximation
RT	room temperature
RTP	room temperaure and standard pressure

S

SAM	self-assembled monolayer
SAM	scanning Auger microscope/microscopy
SARS	scattering and recoiling ion spectroscopy
SAW	surface acoustic wave
SBZ	surface Brillouin zone
SCLS	surface core level shift
SDR	surface differential reflectivity
SEM	scanning electron microscope
SEXAFS	surface-sensitive EXAFS
SFG	sum frequency generation
SH	second harmonic
SHG	second-harmonic generation
SI	Système International d'Unités
SIMS	secondary-ion mass spectroscopy
SNR	signal-to-noise ratio
SPARPES	spin polarized angle-resolved photoemission spectroscopy
SPIPES	spin-polarized inverse photoemission spectroscopy
SPLEED	spin-polarized
SPV	surface photovoltage spectroscopy
SQUIDS	superconducting quantum interference devices
SS	surface state
STM	scanning tunneling microscope/microscopy
STS	scanning tunneling spectroscopy
SXRD	surface X-ray diffraction

T

TAFF	thermally activated flux flow
TEM	transmission electron microscope/microscopy
TFT	thin-film transistor
TMR	tunnel magnetoresistance
TMT	thermomechanical treatment
TOF	time of flight
TOM	torsion oscillation magnetometry

TRS	truncation rod scattering	**V**	
TTT	time-temperature-transformation	VBM	valence band maximum
U		VLEED	very low-energy electron diffraction
UHV	ultra-high vacuum	**X**	
UPS	ultraviolet photoemission spectroscopy	XPS	X-ray photoemission spectroscopy
UV	ultraviolet		

Part 1 General Tables

1 **The Fundamental Constants**
   Werner Martienssen, Frankfurt/Main, Germany

2 **The International System of Units (SI), Physical Quantities, and Their Dimensions**
   Werner Martienssen, Frankfurt/Main, Germany

3 **Rudiments of Crystallography**
   Wolf Assmus, Frankfurt am Main, Germany
   Stefan Brühne, Frankfurt am Main, Germany

1.1. The Fundamental Constants

In the quantitative description of physical phenomena and physical relationships, we find constant parameters which appear to be independent of the scale of the phenomena, independent of the place where the phenomena happen, and independent of the time when the phenomena is observed. These parameters are called fundamental constants. In Sect. 1.1.1, we give a qualitative description of these basic parameters and explain how "recommended values" for the numerical values of the fundamental constants are found. In Sect. 1.1.2, we present tables of the most recently determined recommended numerical values for a large number of those fundamental constants which play a role in solid-state physics and chemistry and in materials science.

1.1.1   What are the Fundamental Constants and Who Takes Care of Them? 3
1.1.2   The CODATA Recommended Values of the Fundamental Constants 4
    1.1.2.1   The Most Frequently Used Fundamental Constants 4
    1.1.2.2   Detailed Lists of the Fundamental Constants in Different Fields of Application   5
    1.1.2.3   Constants from Atomic Physics and Particle Physics 7
References 9

1.1.1 What are the Fundamental Constants and Who Takes Care of Them?

The fundamental constants are constant parameters in the laws of nature. They determine the size and strength of the phenomena in the natural and technological worlds. We conclude from observation that the numerical values of the fundamental constants are independent of space and time; at least, we can say that if there is any dependence of the fundamental constants on space and time, then this dependence must be an extremely weak one. Also, we observe that the numerical values are independent of the scale of the phenomena observed; for example, they seem to be the same in astrophysics and in atomic physics. In addition, the numerical values are quite independent of the environmental conditions. So we have confidence in the idea that the numerical values of the fundamental constants form a set of numbers which are the same everywhere in the world, and which have been the same in the past and will be the same in the future. Whereas the properties of all material objects in nature are more or less subject to continuous change, the fundamental constants seem to represent a constituent of the world which is absolutely permanent.

On the basis of this expected invariance of the fundamental constants in space and time, it appears reasonable to relate the units of measurement for physical quantities to fundamental constants as far as possible. This would guarantee that also the units of measurement become independent of space and time and of environmental conditions. Within the frame work of the International System of Units (Système International d'Unités, abbreviated to SI), the International Committee for Weights and Measures (Comité International des Poids et Mesures, CIPM) has succeeded in relating a large number of units of measurement for physical quantities to the numerical values of selected fundamental constants; however, several units for physical quantities are still represented by prototypes. For example, the unit of length 1 meter, is defined as the distance light travels in vacuum during a fixed time; so the unit of length is related to the fundamental constant c, i.e. the speed of light, and the unit of time, 1 second. The unit of mass, 1 kilogram, however, is still represented by a prototype, the mass of a metal cylinder made of a platinum–iridium alloy, which is carefully stored at the International Office for Weights and Measures (Bureau International des Poids et Mesures, BIPM), at Sèvres near Paris. In a few years, however, it might become possible also to relate the unit of mass to one or more fundamental constants.

The fundamental constants play an important role in basic physics as well as in applied physics and technology; in fact, they have a key function in the development of a system of reproducible and unchanging units for physical quantities. Nevertheless, there is, at present, no theory which would allow us to calculate the numerical values of the fundamental constants. Therefore, national institutes for metrology (NIMs), together with research institutes and university laboratories, are making efforts worldwide to determine the fundamental constants experimentally with the greatest possible accuracy and reliability. This, of course, is a continuous process, with hundreds of new publications every year.

The Committee on Data for Science and Technology (CODATA), established in 1966 as an interdisciplinary, international committee of the International Council of the Scientific Unions (ICSU), has taken the responsibility for improving the quality, reliability, processing, management, and accessibility of data of importance to science and technology. The CODATA task group on fundamental constants, established in 1969, has taken on the job of periodically providing the scientific and technological community with a self-consistent set of internationally recommended values of the fundamental constants based on all relevant data available at given points in time.

What is the meaning of "recommended values" of the fundamental constants?

Many fundamental constants are not independent of one another; they are related to one another by equations which allow one to calculate a numerical value for one particular constant from the numerical values of other constants. In consequence, the numerical value of a constant can be determined either by measuring it directly or by calculating it from the measured values of other constants related to it. In addition, there are usually several different experimental methods for measuring the value of any particular fundamental constant. This allows one to compute an adjustment on the basis of a least-squares fit to the whole set of experimental data in order to determine a set of best-fitting fundamental constants from the large set of all experimental data. Such an adjustment is done today about every four years by the CODATA task group mentioned above. The resulting set of best-fit values is then called the "CODATA recommended values of the fundamental constants" based on the adjustment of the appropriate year.

The tables in Sect. 1.1.2 show the CODATA recommended values of the fundamental constants of science and technology based on the 2002 adjustment. This adjustment takes into account all data that became available before 31 December 2002. A detailed description of the adjustment has been published by *Mohr* and *Taylor* of the National Institute of Standards and Technology, Gaithersburg, in [1.1].

The editors of this Handbook, H. Warlimont and W. Martienssen, would like to express their sincere thanks to Mohr and Taylor for kindly putting their data at our disposal prior to publication in Reviews of Modern Physics. These data first became available in December 2003 on the web site of the NIST fundamental constants data center [1.2].

1.1.2 The CODATA Recommended Values of the Fundamental Constants

1.1.2.1 The Most Frequently Used Fundamental Constants

Tables 1.1-1 – 1.1-9 list the CODATA recommended values of the fundamental constants based on the 2002 adjustment.

Table 1.1-1 Brief list of the most frequently used fundamental constants

Quantity	Symbol and relation	Numerical value	Units	Relative standard uncertainty
Speed of light in vacuum	c	299 792 458	m/s	Fixed by definition
Magnetic constant	$\mu_0 = 4\pi \times 10^{-7}$	$12.566370614\ldots \times 10^{-7}$	N/A^2	Fixed by definition
Electric constant	$\varepsilon_0 = 1/(\mu_0 c^2)$	$8.854187817\ldots \times 10^{-12}$	F/m	Fixed by definition
Newtonian constant of gravitation	G	$6.6742(10) \times 10^{-11}$	m^3/(kg s^2)	1.5×10^{-4}
Planck constant	h	$4.13566743(35) \times 10^{-15}$	eV s	8.5×10^{-8}
Reduced Planck constant	$\hbar = h/2\pi$	$6.58211915(56) \times 10^{-16}$	eV s	8.5×10^{-8}
Elementary charge	e	$1.60217653(14) \times 10^{-19}$	C	8.5×10^{-8}

Table 1.1-1 Brief list of the most frequently used fundamental constants, cont.

Quantity	Symbol and relation	Numerical value	Units	Relative standard uncertainty
Fine-structure constant	$\alpha = (1/4\pi\varepsilon_0)(e^2/\hbar c)$	$7.297352568(24) \times 10^{-3}$		3.3×10^{-9}
Magnetic flux quantum	$\Phi_0 = h/2e$	$2.06783372(18) \times 10^{-15}$	Wb	8.5×10^{-8}
Conductance quantum	$G_0 = 2e^2/h$	$7.748091733(26) \times 10^{-5}$	S	3.3×10^{-9}
Rydberg constant	$R_\infty = \alpha^2 m_e c/2h$	$10\,973\,731.568525(73)$	1/m	6.6×10^{-12}
Electron mass	m_e	$9.1093826(16) \times 10^{-31}$	kg	1.7×10^{-7}
Proton mass	m_p	$1.67262171(29) \times 10^{-27}$	kg	1.7×10^{-7}
Proton–electron mass ratio	m_p/m_e	$1836.15267261(85)$		4.6×10^{-10}
Avogadro number	N_A, L	$6.0221415(10) \times 10^{23}$		1.7×10^{-7}
Faraday constant	$F = N_A e$	$96\,485.3383(83)$	C	8.6×10^{-8}
Molar gas constant	R	$8.314472(15)$	J/K	1.7×10^{-6}
Boltzmann constant	$k = R/N_A$	$1.3806505(24) \times 10^{-23}$	J/K	1.8×10^{-6}
		$8.617343(15) \times 10^{-5}$	eV/K	1.8×10^{-6}
Stefan–Boltzmann constant	$\sigma = (\pi^2/60)(k^4/(\hbar^3 c^2))$	$5.670400(40) \times 10^{-8}$	W/(m² K⁴)	7.0×10^{-6}

1.1.2.2 Detailed Lists of the Fundamental Constants in Different Fields of Application

Table 1.1-2 Universal constants

Quantity	Symbol and relation	Numerical value	Units	Relative standard uncertainty
Speed of light in vacuum	c	$299\,792\,458$	m/s	Fixed by definition
Magnetic constant	$\mu_0 = 4\pi \times 10^{-7}$	$12.566370614\ldots \times 10^{-7}$	N/A²	Fixed by definition
Electric constant	$\varepsilon_0 = 1/(\mu_0 c^2)$	$8.854187817\ldots \times 10^{-12}$	F/m	Fixed by definition
Characteristic impedance of vacuum	$Z_0 = (\mu_0/\varepsilon_0)^{1/2} = \mu_0 c$	$376.730313461\ldots$	Ω	Fixed by definition
Newtonian constant of gravitation	G	$6.6742(10) \times 10^{-11}$	m³/(kg s²)	1.5×10^{-4}
Reduced Planck constant	$G/\hbar c$	$6.7087(10) \times 10^{-39}$	(GeV/c²)⁻²	1.5×10^{-4}
Planck constant	h	$6.6260693(11) \times 10^{-34}$	J s	1.7×10^{-7}
		$4.13566743(35) \times 10^{-15}$	eV s	8.5×10^{-8}
(Ratio)	$\hbar = h/2\pi$	$1.05457168(18) \times 10^{-34}$	J s	1.7×10^{-7}
		$6.58211915(56) \times 10^{-16}$	eV s	8.5×10^{-8}
(Product)	$\hbar c$	$197.326968(17)$	MeV fm	8.5×10^{-8}
(Product)	$c_1 = 2\pi h c^2$	$3.74177138(64) \times 10^{-16}$	W m²	1.7×10^{-7}
(Product)	$(1/\pi)c_1 = 2hc^2$	$1.19104282(20) \times 10^{-16}$	W m²/sr	1.7×10^{-7}
(Product)	$c_2 = h(c/k)$	$1.4387752(25) \times 10^{-2}$	m K	1.7×10^{-6}
Stefan–Boltzmann constant	$\sigma = (\pi^2/60)(k^4/(\hbar^3 c^2))$	$5.670400(40) \times 10^{-8}$	W/(m² K⁴)	7.0×10^{-6}
Wien displacement law constant	$b = \lambda_{max} T = c_2/4.965114231$	$2.8977685(51) \times 10^{-3}$	m K	1.7×10^{-6}
Planck mass	$m_P = (\hbar c/G)^{1/2}$	$2.17645(16) \times 10^{-8}$	kg	7.5×10^{-5}
Planck temperature	$T_P = (1/k)(\hbar c^5/G)^{1/2}$	$1.41679(11) \times 10^{32}$	K	7.5×10^{-5}
Planck length	$l_P = \hbar/m_P c$ $= (\hbar G/c^3)^{1/2}$	$1.61624(12) \times 10^{-35}$	m	7.5×10^{-5}
Planck time	$t_P = l_P/c = (\hbar G/c^5)^{1/2}$	$5.39121(40) \times 10^{-44}$	s	7.5×10^{-5}

Table 1.1-3 Electromagnetic constants

Quantity	Symbol and relation	Numerical value	Units	Relative standard uncertainty
Elementary charge	e	$1.60217653(14) \times 10^{-19}$	C	8.5×10^{-8}
(Ratio)	e/h	$2.41798940(21) \times 10^{14}$	A/J	8.5×10^{-8}
Fine-structure constant	$\alpha = (1/4\pi\varepsilon_0)(e^2/\hbar c)$	$7.297352568(24) \times 10^{-3}$		3.3×10^{-9}
Inverse fine-structure constant	$1/\alpha$	$137.03599911(46)$		3.3×10^{-9}
Magnetic flux quantum	$\Phi_0 = h/2e$	$2.06783372(18) \times 10^{-15}$	Wb	8.5×10^{-8}
Conductance quantum	$G_0 = 2e^2/h$	$7.748091733(26) \times 10^{-5}$	S	3.3×10^{-9}
Inverse of conductance quantum	$1/G_0$	$12\,906.403725(43)$	Ω	3.3×10^{-9}
Josephson constant[a]	$K_J = 2e/h$	$483\,597.879(41) \times 10^9$	Hz/V	8.5×10^{-8}
von Klitzing constant[b]	$R_K = h/e^2 = \mu_0 c/2\alpha$	$25\,812.807449(86)$	Ω	3.3×10^{-9}
Bohr magneton	$\mu_B = e\hbar/2m_e$	$927.400949(80) \times 10^{-26}$	J/T	8.6×10^{-8}
		$5.788381804(39) \times 10^{-5}$	eV/T	6.7×10^{-9}
(Ratio)	μ_B/h	$13.9962458(12) \times 10^9$	Hz/T	8.6×10^{-8}
(Ratio)	μ_B/hc	$46.6864507(40)$	1/(m T)	8.6×10^{-8}
(Ratio)	μ_B/k	$0.6717131(12)$	K/T	1.8×10^{-6}
Nuclear magneton	$\mu_N = e\hbar/2m_p$	$5.05078343(43) \times 10^{-27}$	J/T	8.6×10^{-8}
		$3.152451259(21) \times 10^{-8}$	eV/T	6.7×10^{-9}
(Ratio)	μ_N/h	$7.62259371(65)$	MHz/T	8.6×10^{-8}
(Ratio)	μ_N/hc	$2.54262358(22) \times 10^{-2}$	1/(m T)	8.6×10^{-8}
(Ratio)	μ_N/k	$3.6582637(64) \times 10^{-4}$	K/T	1.8×10^{-6}

[a] See Table 1.2-16 for the conventional value adopted internationally for realizing representations of the volt using the Josephson effect.
[b] See Table 1.2-16 for the conventional value adopted internationally for realizing representations of the ohm using the quantum Hall effect.

Table 1.1-4 Thermodynamic constants

Quantity	Symbol and relation	Numerical value	Units	Relative standard uncertainty
Avogadro number	N_A, L	$6.0221415(10) \times 10^{23}$		1.7×10^{-7}
Atomic mass constant	$u = (1/12)m(^{12}C)$ $= (1/N_A) \times 10^{-3}$ kg	$1.66053886(28) \times 10^{-27}$	kg	1.7×10^{-7}
Energy equivalent of atomic mass constant	$m_u c^2$	$1.49241790(26) \times 10^{-10}$	J	1.7×10^{-7}
		$931.494043(80)$	MeV	8.6×10^{-8}
Faraday constant	$F = N_A e$	$96\,485.3383(83)$	C	8.6×10^{-8}
Molar Planck constant	$N_A h$	$3.990312716(27) \times 10^{-10}$	J s	6.7×10^{-9}
(Product)	$N_A hc$	$0.11962656572(80)$	J m	6.7×10^{-9}
Molar gas constant	R	$8.314472(15)$	J/K	1.7×10^{-6}
Boltzmann constant	$k = R/N_A$	$1.3806505(24) \times 10^{-23}$	J/K	1.8×10^{-6}
		$8.617343(15) \times 10^{-5}$	eV/K	1.8×10^{-6}
(Ratio)	k/h	$2.0836644(36) \times 10^{10}$	Hz/K	1.7×10^{-6}
(Ratio)	k/hc	$69.50356(12)$	1/(m K)	1.7×10^{-6}
Molar volume of ideal gas at STP	$V_m = RT/p$ at $T = 273.15$ K and $p = 101.325$ kPa	$22.413996(39) \times 10^{-3}$	m^3	1.7×10^{-6}
Loschmidt constant	$n_0 = N_A/V_m$	$2.6867773(47) \times 10^{25}$	1/m^3	1.8×10^{-6}
Stefan–Boltzmann constant	$\sigma = (\pi^2/60)(k^4/(\hbar^3 c^2))$	$5.670400(40) \times 10^{-8}$	W/(m^2 K^4)	7.0×10^{-6}
Wien displacement law constant	$b = \lambda_{max} T = c_2/4.965114231$	$2.8977685(51) \times 10^{-3}$	m K	1.7×10^{-6}

1.1.2.3 Constants from Atomic Physics and Particle Physics

Table 1.1-5 Constants from atomic physics

Quantity	Symbol and relation	Numerical value	Units	Relative standard uncertainty
Rydberg constant	$R_\infty = \alpha^2 m_e c / 2h$	10 973 731.568525(73)	1/m	6.6×10^{-12}
(Product)	$R_\infty c$	$3.289841960360(22) \times 10^{15}$	Hz	6.6×10^{-12}
(Product)	$R_\infty h c$	$2.17987209(37) \times 10^{-18}$	J	1.7×10^{-7}
		13.6056923(12)	eV	8.5×10^{-8}
Bohr radius	$a_0 = \alpha / 4\pi R_\infty$ $= 4\pi\varepsilon_0 \hbar^2 / m_e e^2$	$0.5291772108(18) \times 10^{-10}$	m	3.3×10^{-9}
Hartree energy	$E_H = e^2 / 4\pi\varepsilon_0 a_0$ $= 2R_\infty h c = \alpha^2 m_e c^2$	$4.35974417(75) \times 10^{-18}$	J	1.7×10^{-7}
		27.2113845(23)	eV	8.5×10^{-8}
Quantum of circulation	$h/2m_e$	$3.636947550(24) \times 10^{-4}$	m²/s	6.7×10^{-9}
(Product)	h/m_e	$7.273895101(48) \times 10^{-4}$	m²/s	6.7×10^{-9}

Table 1.1-6 Properties of the electron

Quantity	Symbol and relation	Numerical value	Units	Relative standard uncertainty
Electron mass	m_e	$9.1093826(16) \times 10^{-31}$	kg	1.7×10^{-7}
		$5.4857990945(24) \times 10^{-4}$	u	4.4×10^{-10}
Energy equivalent of electron mass	$m_e c^2$	$8.1871047(14) \times 10^{-14}$	J	1.7×10^{-7}
		0.510998918(44)	MeV	8.6×10^{-8}
Electron–proton mass ratio	m_e/m_p	$5.4461702173(25) \times 10^{-4}$		4.6×10^{-10}
Electron–neutron mass ratio	m_e/m_n	$5.4386734481(38) \times 10^{-4}$		7.0×10^{-10}
Electron–muon mass ratio	m_e/m_μ	$4.83633167(13) \times 10^{-3}$		2.6×10^{-8}
Electron molar mass	$M(e) = N_A m_e$	$5.4857990945(24) \times 10^{-7}$	kg	4.4×10^{-10}
Charge-to-mass ratio	$-e/m_e$	$-1.75882012(15) \times 10^{11}$	C/kg	8.6×10^{-8}
Compton wavelength	$\lambda_C = h/m_e c$	$2.426310238(16) \times 10^{-12}$	m	6.7×10^{-9}
(Ratio)	$\lambda_C/2\pi = \alpha a_0 = \alpha^2/(4\pi R_\infty)$	$386.1592678(26) \times 10^{-15}$	m	6.7×10^{-9}
Classical electron radius	$r_e = \alpha^2 a_0$	$2.817940325(28) \times 10^{-15}$	m	1.0×10^{-8}
Thomson cross section	$\sigma_e = (8\pi/3) r_e^2$	$0.665245873(13) \times 10^{-28}$	m²	2.0×10^{-8}
Magnetic moment	μ_e	$-928.476412(80) \times 10^{-26}$	J/T	8.6×10^{-8}
Ratio of magnetic moment to Bohr magneton	μ_e/μ_B	$-1.0011596521859(38)$		3.8×10^{-12}
Ratio of magnetic moment to nuclear magneton	μ_e/μ_N	$-1838.28197107(85)$		4.6×10^{-10}
Ratio of magnetic moment to proton magnetic moment	μ_e/μ_p	$-658.2106862(66)$		1.0×10^{-8}
Ratio of magnetic moment to neutron magnetic moment	μ_e/μ_n	960.92050(23)		2.4×10^{-7}
Electron magnetic-moment anomaly	$a_e = |\mu_e|/(\mu_B - 1)$	$1.1596521859(38) \times 10^{-3}$		3.2×10^{-9}
g-factor	$g_e = -2(1+a_e)$	$-2.0023193043718(75)$		3.8×10^{-12}
Gyromagnetic ratio	$\gamma_e = 2|\mu_e|/\hbar$	$1.76085974(15) \times 10^{11}$	1/(s T)	8.6×10^{-8}
(Ratio)	$\gamma_e/2\pi$	28 024.9532(24)	MHz/T	8.6×10^{-8}

Table 1.1-7 Properties of the proton

Quantity	Symbol and relation	Numerical value	Units	Relative standard uncertainty
Proton mass	m_p	$1.67262171(29) \times 10^{-27}$	kg	1.7×10^{-7}
		$1.00727646688(13)$	u	1.3×10^{-10}
Energy equivalent of proton mass	$m_p c^2$	$1.50327743(26) \times 10^{-10}$	J	1.7×10^{-7}
		$938.272029(80)$	MeV	8.6×10^{-8}
Proton–electron mass ratio	m_p/m_e	$1836.15267261(85)$		4.6×10^{-10}
Proton–neutron mass ratio	m_p/m_n	$0.99862347872(58)$		5.8×10^{-10}
Proton molar mass	$M(p) = N_A m_p$	$1.00727646688(13) \times 10^{-3}$	kg	1.3×10^{-10}
Charge-to-mass ratio	e/m_p	$9.57883376(82) \times 10^7$	C/kg	8.6×10^{-8}
Compton wavelength	$\lambda_{C,p} = h/m_p c$	$1.3214098555(88) \times 10^{-15}$	m	6.7×10^{-9}
(Ratio)	$(1/2\pi)\lambda_{C,p}$	$0.2103089104(14) \times 10^{-15}$	m	6.7×10^{-9}
rms charge radius	R_p	$0.8750(68) \times 10^{-15}$	m	7.8×10^{-3}
Magnetic moment	μ_p	$1.41060671(12) \times 10^{-26}$	J/T	8.7×10^{-8}
Ratio of magnetic moment to Bohr magneton	μ_p/μ_B	$1.521032206(15) \times 10^{-3}$		1.0×10^{-8}
Ratio of magnetic moment to nuclear magneton	μ_p/μ_N	$2.792847351(28)$		1.0×10^{-8}
Ratio of magnetic moment to neutron magnetic moment	μ_p/μ_n	$-1.45989805(34)$		2.4×10^{-7}
g-factor	$g_p = 2\mu_p/\mu_N$	$5.585694701(56)$		1.0×10^{-8}
Gyromagnetic ratio	$\gamma_p = 2\mu_p/\hbar$	$2.67522205(23) \times 10^8$	1/(s T)	8.6×10^{-8}
(Ratio)	$(1/2\pi)\gamma_p$	$42.5774813(37)$	MHz/T	8.6×10^{-8}

Table 1.1-8 Properties of the neutron

Quantity	Symbol and relation	Numerical value	Units	Relative standard uncertainty
Neutron mass	m_n	$1.67492728(29) \times 10^{-27}$	kg	1.7×10^{-7}
		$1.00866491560(55)$	u	5.5×10^{-10}
Energy equivalent	$m_n c^2$	$1.50534957(26) \times 10^{-10}$	J	1.7×10^{-7}
		$939.565360(81)$	MeV	8.6×10^{-8}
Neutron–electron mass ratio	m_n/m_e	$1838.6836598(13)$		7.0×10^{-10}
Neutron–proton mass ratio	m_n/m_p	$1.00137841870(58)$		5.8×10^{-10}
Molar mass	$M(n) = N_A m_n$	$1.00866491560(55) \times 10^{-3}$	kg	5.5×10^{-10}
Compton wavelength	$\lambda_{C,n} = h/(m_n c)$	$1.3195909067(88) \times 10^{-15}$	m	6.7×10^{-9}
(Ratio)	$(1/2\pi)\lambda_{C,n}$	$0.2100194157(14) \times 10^{-15}$	m	6.7×10^{-9}
Magnetic moment	μ_n	$-0.96623645(24) \times 10^{-26}$	J/T	2.5×10^{-7}
Ratio of magnetic moment to Bohr magneton	μ_n/μ_B	$-1.04187563(25) \times 10^{-3}$		2.4×10^{-7}
Ratio of magnetic moment to nuclear magneton	μ_n/μ_N	$-1.91304273(45)$		2.4×10^{-7}
Ratio of magnetic moment to electron magnetic moment	μ_n/μ_e	$1.04066882(25) \times 10^{-3}$		2.4×10^{-7}
Ratio of magnetic moment to proton magnetic moment	μ_n/μ_p	$-0.68497934(16)$		2.4×10^{-7}
g-factor	$g_n = 2\mu_n/\mu_N$	$-3.82608546(90)$		2.4×10^{-7}
Gyromagnetic ratio	$\gamma_n = 2|\mu_n|/\hbar$	$1.83247183(46) \times 10^8$	1/(s T)	2.5×10^{-7}
(Ratio)	$(1/2\pi)\gamma_n$	$29.1646950(73)$	MHz/T	2.5×10^{-7}

Table 1.1-9 Properties of the alpha particle

Quantity	Symbol and relation	Numerical value	Units	Relative standard uncertainty
Alpha particle mass[a]	m_α	6.6446565(11) × 10^{-27} 4.001506179149(56)	kg u	1.7 × 10^{-7} 1.4 × 10^{-11}
Energy equivalent of alpha particle mass	$m_\alpha c^2$	5.9719194(10) × 10^{-10} 3727.37917(32)	J MeV	1.7 × 10^{-7} 8.6 × 10^{-8}
Ratio of alpha particle mass to electron mass	m_α/m_e	7294.2995363(32)		4.4 × 10^{-10}
Ratio of alpha particle mass to proton mass	m_α/m_p	3.97259968907(52)		1.3 × 10^{-10}
Alpha particle molar mass	$M(\alpha) = N_A m_\alpha$	4.001506179149(56) × 10^{-3}	kg/mol	1.4 × 10^{-11}

[a] The mass of the alpha particle in units of the atomic mass unit u is given by $m_\alpha = A_r(\alpha)$ u; in words, the alpha particle mass is given by the relative atomic mass $A_r(\alpha)$ of the alpha particle, multiplied by the atomic mass unit u

References

1.1 P. J. Mohr, B. N. Taylor: CODATA recommended values of the fundamental physical constants, Rev. Mod. Phys. (2004) (in press)

1.2 NIST Physics Laboratory: Web pages of the Fundamental Constants Data Center, http://physics.nist.gov/constants

1.2. The International System of Units (SI), Physical Quantities, and Their Dimensions

In this chapter, we introduce the International System of Units (SI) on the basis of the SI brochure "Le Système international d'unités (SI)" [2.1], supplemented by [2.2]. We give a short review of how the SI was worked out and who is responsible for the further development of the system. Following the above-mentioned publications, we explain the concepts of base physical quantities and derived physical quantities on which the SI is founded, and present a detailed description of the SI base units and of a large selection of SI derived units. We also discuss a number of non-SI units which still are in use, especially in some specialized fields. A table (Table 1.2-17) presenting the values of various energy equivalents closes the section.

1.2.1	The International System of Units (SI)	11
1.2.2	Physical Quantities	12
1.2.3	The SI Base Units	13
1.2.3.1	Unit of Length: the Meter	13
1.2.3.2	Unit of Mass: the Kilogram	14
1.2.3.3	Unit of Time: the Second	14
1.2.3.4	Unit of Electric Current: the Ampere	14
1.2.3.5	Unit of (Thermodynamic) Temperature: the Kelvin	14
1.2.3.6	Unit of Amount of Substance: the Mole	14
1.2.3.7	Unit of Luminous Intensity: the Candela	15
1.2.4	The SI Derived Units	16
1.2.5	Decimal Multiples and Submultiples of SI Units	19
1.2.6	Units Outside the SI	20
1.2.6.1	Units Used with the SI	20
1.2.6.2	Other Non-SI Units	20
1.2.7	Some Energy Equivalents	24
	References	25

1.2.1 The International System of Units (SI)

All data in this handbook are given in the International System of Units (Système International d'Unités), abbreviated internationally to SI, which is the modern metric system of measurement and is acknowledged worldwide. The system of SI units was introduced by the General Conference of Weights and Measures (Conférence Générale des Poids et Mesures), abbreviated internationally to CGPM, in 1960. The system not only is used in science, but also is dominant in technology, industrial production, and international commerce and trade.

Who takes care of this system of SI units? The Bureau International des Poids et Mesures (BIPM), which has its headquarters in Sèvres near Paris, has taken on a commitment to ensure worldwide unification of physical measurements. Its function is thus to:

- establish fundamental standards and scales for the measurement of the principal physical quantities and maintain the international prototypes;
- carry out comparison of national and international standards;
- ensure the coordination of the corresponding measuring techniques;
- carry out and coordinate measurements of the fundamental physical constants relevant to those activities.

The BIPM operates under the exclusive supervision of the Comité International des Poids et Mesures (CIPM), which itself comes under the authority of the

Conférence Générale des Poids et Mesures and reports to it on the work accomplished by the BIPM. The BIPM itself was set up by the Convention du Mètre signed in Paris in 1875 by 17 states during the final session of the Conference on the Meter. The convention was amended in 1921.

Delegates from all member states of the Convention du Mètre attend the Conférence Générale, which, at present, meets every four years. The function of these meetings is to:

- discuss and initiate the arrangements required to ensure the propagation and improvement of the International System of Units
- confirm the results of new fundamental metrological determinations and confirm various scientific resolutions with international scope
- take all major decisions concerning the finance, organization, and development of the BIPM.

The CIPM has 18 members, each from a different state; at present, it meets every year. The officers of this committee present an annual report on the administrative and financial position of the BIPM to the governments of the member states of the Convention du Mètre. The principal task of the CIPM is to ensure worldwide uniformity in units of measurement. It does this by direct action or by submitting proposals to the CGPM.

The BIPM publishes monographs on special metrological subjects and the brochure *Le Système international d'unités (SI)* [2.1, 2], which is periodically updated and in which all decisions and recommendations concerning units are collected together.

The scientific work of the BIPM is published in the open scientific literature, and an annual list of publications appears in the *Procès-Verbaux* of the CIPM.

Since 1965, *Metrologica*, an international journal published under the auspices of the CIPM, has printed articles dealing with scientific metrology, improvements in methods of measurements, and work on standards and units, as well as reports concerning the activities, decisions, and recommendations of the various bodies created under the Convention du Mètre.

1.2.2 Physical Quantities

Physical quantities are tools which allow us to specify and quantify the properties of physical objects and to model the events, phenomena, and patterns of behavior of objects in nature and in technology. The system of physical quantities used with the SI units is dealt by Technical Committee 12 of the International Organization for Standardization (ISO/TC 12). Since 1955, ISO/TC 12 has published a series of international standards on quantities and their units, in which the use of SI units is strongly recommended.

How are Physical Quantities Defined?

It turns out that it is possible to divide the system of all known physical quantities into two groups:

- a small number of *base quantities*;
- a much larger number of other quantities, which are called *derived quantities*.

The derived quantities are introduced into physics unambiguously by a defining equation in terms of the base quantities; the relationships between the derived quantities and the base quantities are expressed in a series of equations, which contain a good deal of our knowledge of physics but are used in this system as the defining equations for new physical quantities. One might say that, in this system, physics is described in the rather low-dimensional space of a small number of base quantities.

Base quantities, on the other hand, cannot be introduced by a defining equation; they cannot be traced back to other quantities; this is what we mean by calling them "base". How can base quantities then be introduced unambiguously into physics at all?

Base physical quantities are introduced into physics in three steps:

- We borrow the qualitative meaning of the word for a base quantity from the meaning of the corresponding word in everyday language.
- We specify this meaning by indicating an appropriate method for measuring the quantity. For example, length is measured by a measuring rule, and time is measured by a clock.
- We fix a unit for this quantity, which allows us to communicate the result of a measurement. Length, for example, is measured in meters; time is measured in seconds.

On the basis of these three steps, it is expected that everyone will understand what is meant when the name of a base quantity is mentioned.

In fact, the number of base quantities chosen and the selection of the quantities which are considered as base quantities are a matter of expediency; in different fields and applications of physics, it might well be expedient to use different numbers of base quantities and different selections of base quantities. It should be kept in mind, however, that the number and selection of base quantities are only a matter of different representations of physics; physics itself is not affected by the choice that is made.

Today, many scientists therefore prefer to use the *conventional system* recommended by ISO. This system uses seven base quantities, which are selected according to the seven base units of the SI system. Table 1.2-1 shows the recommended names, symbols, and measuring devices for the conventional seven base quantities.

Table 1.2-1 The ISO recommended base quantities

Name of quantity	Symbol	Measured by
Length	l	A measuring rule
Time	t	A clock
Mass	m	A balance
Electrical current	I	A balance
Temperature	T	A thermometer
Particle number	N	Counting
Luminous intensity	I_v	A photometer

All other physical quantities can then be defined as derived quantities; this means they can be defined by equations in terms of the seven base quantities. Within this "conventional system", the set of all defining equations for the derived physical quantities also defines the units for the derived quantities in terms of the units of the base quantities. This is the great advantage of the "conventional system".

The quantity velocity v, for example, is defined by the equation

$$v = \frac{dl}{dt}.$$

In this way, velocity is traced back to the two base quantities length l and time t. On the right-hand side of this equation, we have a differential of length dl divided by a differential of time dt. The algebraic combination of the base quantities in the defining equation for a derived quantity is called the *dimensions* of the derived quantity. So velocity has the dimensions length/time, acceleration has the dimensions length/time squared, and so on.

Data for a physical quantity are always given as a product of a number (the *numerical value* of the physical quantity) and a unit in which the quantity has been measured.

1.2.3 The SI Base Units

Formal definitions of the seven SI base units have to be approved by the CGPM. The first such definition was approved in 1889. These definitions have been modified, however, from time to time as techniques of measurement have evolved and allowed more accurate realizations of the base units. Table 1.2-2 summarizes the present status of the SI base units and their symbols.

In the following, the current definitions of the base units adopted by the CGPM are shown in detail, together with some explanatory notes. Related decisions which clarify these definitions but are not formally part of them are also shown indented, but in a font of normal weight.

1.2.3.1 Unit of Length: the Meter

The unit of length, the meter, was defined in the first CGPM approval in 1889 by an international prototype: the length of a bar made of a platinum–iridium alloy defined a length of 1 m. In 1960 this definition was replaced

Table 1.2-2 The seven SI base units and their symbols

Base quantity	Symbol for quantity	Unit	Symbol for unit
Length	l	metre	m
Time	t	second	s
Mass	m	kilogram	kg
Electrical current	I	ampere	A
Temperature	T	kelvin	K
Particle number	n	mole	mol
Luminous intensity	I_v	candela	cd

by a definition based upon a wavelength of krypton-86 radiation. Since 1983 (17th CGPM), the meter has been defined as:

> *The metre is the length of the path travelled by light in vacuum during a time interval of 1/299 792 458 of a second.*

As a result of this definition, the fundamental constant "speed of light in vacuum c" is fixed at exactly 299 792 458 m/s.

1.2.3.2 Unit of Mass: the Kilogram

Since the first CGPM in 1889, the unit of mass, the kilogram, has been defined by an international prototype, a metal block made of a platinum–iridium alloy, kept at the BIPM at Sèvres. The relevant declaration was modified slightly at the third CGPM in 1901 to confirm that:

> The kilogram is the unit of mass; it is equal to the mass of the international prototype of the kilogram.

1.2.3.3 Unit of Time: the Second

The unit of time, the second, was originally considered to be the fraction 1/86 400 of the mean solar day. Measurements, however, showed that irregularities in the rotation of the Earth could not be taken into account by theory, and these irregularities have the effect that this definition does not allow the required accuracy to be achieved. The same turned out to be true for other definitions based on astronomical data. Experimental work, however, had already shown that an atomic standard of time interval, based on a transition between two energy levels of an atom or a molecule, could be realized and reproduced much more precisely. Therefore, the 13th CGPM (1967–1968) replaced the definition of the second by:

> The second is the duration of 9 192 631 770 periods of the radiation corresponding to the transition between the two hyperfine levels of the ground state of the caesium 133 atom.

At its 1997 meeting, the CIPM affirmed that:

> This definition refers to a caesium atom at rest at a temperature of 0 K.

This note was intended to make it clear that the definition of the SI second is based on a Cs atom unperturbed by black-body radiation, that is, in an environment whose temperature is 0 K.

1.2.3.4 Unit of Electric Current: the Ampere

"International" electrical units for current and resistance were introduced by the International Electrical Congress in Chicago as early as in 1893 and were confirmed by an international conference in London in 1908. They were replaced by an "absolute" definition of the ampere as the unit for electric current at the 9th CGPM in 1948, which stated:

> The ampere is that constant current which, if maintained in two straight parallel conductors of infinite length, of negligible circular cross-section, and placed 1 meter apart in vacuum, would produce between these conductors a force equal to 2×10^{-7} newton per meter of length.

As a result of this definition, the fundamental constant "magnetic field constant μ_0" (also known as the permeability of free space) is fixed at exactly $4\pi \times 10^{-7}\,\text{N/A}^2$.

1.2.3.5 Unit of (Thermodynamic) Temperature: the Kelvin

The definition of the unit of (thermodynamic) temperature was given in substance by the 10th CGPM in 1954, which selected the triple point of water as the fundamental fixed point and assigned to it the temperature 273.16 K, so defining the unit. After smaller amendments, made at the 13th CGPM in 1967–1968, the definition of the unit of (thermodynamic) temperature reads

> The kelvin, unit of thermodynamic temperature, is the fraction 1/273.16 of the thermodynamic temperature of the triple point of water.

Because of the way temperature scales used to be defined, it remains common practice to express a thermodynamic temperature, symbol T, in terms of its difference from the reference temperature $T_0 = 273.15$ K, the ice point. This temperature difference is called the Celsius temperature, symbol t, and is defined by the equation $t = T - T_0$. The unit of Celsius temperature is the degree Celsius, symbol °C, which is, by definition, equal in magnitude to the kelvin. A difference or interval of temperature may therefore be expressed either in kelvin or in degrees Celsius.

The numerical value of a Celsius temperature t expressed in degrees Celsius is given by $t(°C) = T(K) - 273.15$.

1.2.3.6 Unit of Amount of Substance: the Mole

The "amount of substance" of a sample is understood as a measure of the number of elementary entities (for example atoms or molecules) that the sample consists of. Owing to the fact that on macroscopic scales this number cannot be counted directly in most cases, one has to relate this quantity "amount of substance" to a more easily measurable quantity, the mass of a sample of that substance.

On the basis of an agreement between the International Union of Pure and Applied Physics (IUPAP) and the International Union of Pure and Applied Chemistry (IUPAC) in 1959/1960, physicists and chemists have ever since agreed to assign, by definition, the value 12, exactly, to the relative atomic mass (formerly called "atomic weight") of the isotope of carbon with mass number 12 (carbon-12, ^{12}C). The scale of the masses of all other atoms and isotopes based on this agreement has been called, since then, the scale of relative atomic masses.

It remains to define the unit of the "amount of substance" in terms of the mass of the corresponding amount of the substance. This is done by fixing the mass of a particular amount of carbon-12; by international agreement, this mass has been fixed at 0.012 kg. The corresponding unit of the quantity "amount of substance" has been given the name "mole" (symbol mol).

On the basis of proposals by IUPAC, IUPAP, and ISO, the CIPM formulated a definition of the mole in 1967 and confirmed it in 1969. This definition was adopted by the 14th CGPM in 1971 in two statements:

> 1. *The mole is the amount of substance of a system which contains as many elementary entities as there are atoms in 0.012 kilogram of carbon-12; its symbol is "mol".*
> 2. *When the mole is used, the elementary entities must be specified and may be atoms, molecules, ions, electrons, other particles, or specified groups of such particles.*

In 1980, the CIPM approved the report of the Comité Consultatif des Unités (CCU), which specified that:

> In this definition, it is understood that unbound atoms of carbon-12, at rest and in their ground state, are referred to.

1.2.3.7 Unit of Luminous Intensity: the Candela

The base unit candela allows one to establish a quantitative relation between radiometric and photometric measurements of light intensities. In physics and chemistry, the intensities of radiation fields of various natures are normally determined by radiometry; in visual optics, in lighting engineering, and in the physiology of the visual system, however, it is necessary to assess the intensity of the radiation field by photometric means.

There are, in fact, three different ways to quantify the intensity of a radiation beam. One way is to measure the "radiant intensity" I_e, defined as the radiant flux $\Delta \Phi_e$ per unit solid angle $\Delta \Omega$ of the beam. The subscript "e" stands for "energetic". Here, the radiant flux Φ_e is defined as the energy of the radiation per unit time, and is accordingly measured in units of watts (W). The radiant intensity I_e, therefore, has the dimensions of energy per time per solid angle, and is measured in the derived unit "watt per steradian" (W/sr).

Another way to quantify the intensity of a beam of radiation is to measure the "particle intensity" I_p, which is defined as the particle flux Φ_p divided by the solid angle $\Delta \Omega$ of the beam. The subscript "p" stands for "particle". The particle flux Φ_p itself is measured by counting the number of particles per unit time in the beam; in the case of a light beam, for example, the particles are photons. The corresponding SI unit for the particle flux is seconds^{-1} (1/s). The quantity particle intensity I_p therefore has the dimensions of number per time per solid angle; the corresponding derived SI unit for the particle intensity I_p is "seconds^{-1} times steradian^{-1}" (1/(s sr)).

In addition to these two radiometric assessments of the beam intensity, for beams of visible light there is a third possibility, which is to quantify the intensity of the beam by the intensity of visual perception by the human eye. Physical quantities connected with this physiological type of assessment are called photometric quantities, in contrast to the two radiometric quantities described above. In photometry, the intensity of the beam is called the "luminous intensity" I_v. The subscript "v" stands for "visual". The luminous intensity I_v is an ISO recommended base quantity; the corresponding SI base unit is the candela (cd). The luminous flux Φ_v is determined as the product of the luminous intensity and the solid angle. Its dimensions therefore are luminous intensity times solid angle, so that the SI unit of the luminous flux Φ_v turns out to be "candela times steradian" (cd sr). A derived unit, the lumen (lm), such that 1 lm = 1 cd sr, has been introduced for this product.

Table 1.2-3 summarizes the names, definitions, and SI units for the most frequently used radiometric and photometric quantities in radiation physics.

The history of the base unit candela is as follows. Before 1948, the units for photometric measurements were be based on flame or incandescent-filament standards. They were replaced initially by the "new candle" based on the luminance of a Planckian radiator (a blackbody radiator) at the temperature of freezing platinum. This modification was ratified in 1948 by the 9th CGPM, which also adopted the new international name for the base unit of luminous intensity, the candela, and its symbol cd. The 13th CGPM gave an amended version of the 1948 definition in 1967.

Table 1.2-3 Radiometric and photometric quantities in radiation physics

Quantity	Symbol and definition	Dimensions	SI unit	Symbol for unit
Radiant flux	$\Phi_e = \Delta E/\Delta t$ [a]	Power = energy/time	watt	$W = J/s$
Particle flux, activity	$\Phi_p = \Delta N_p/\Delta t$	1/time	$second^{-1}$	$1/s$
Luminous flux	$\Phi_v = I_v \Omega$ [b]	Luminous intensity times solid angle	lumen	$lm = cd\,sr$
Radiant intensity	$I_e = \Delta\Phi_e/\Delta\Omega$	Power/solid angle	watt/steradian	W/sr
Particle intensity	$I_p = \Delta\Phi_p/\Delta\Omega$	(Time times solid angle)$^{-1}$	(second times steradian)$^{-1}$	$1/(s\,sr)$
Luminous intensity	I_v, basic quantity	Luminous intensity	candela	cd
Radiance [c]	$L_e = \Delta I_e(\varphi)/[\Delta A_1 g(\varphi)]$ [d]	Power per source area and solid angle	watt/(meter^2 times steradian)	$W/(m^2\,sr) = kg/(s^3\,sr)$
Particle radiance [c]	$L_p = \Delta I_p(\varphi)/[\Delta A_1 g(\varphi)]$ [d]	(Time times area times solid angle)$^{-1}$	1/(second times meter^2 times steradian)	$1/(s\,m^2\,sr)$
Luminance [c]	$L_v = \Delta I_v(\varphi)/[\Delta A_1 g(\varphi)]$ [d]	Luminous intensity/source area	candela/meter^2	cd/m^2
Irradiance	$E_e = \Delta\Phi_e/\Delta A_2$ [e]	Power/area	watt/meter^2	W/m^2
Particle irradiance	$E_p = \Delta\Phi_p/\Delta A_2$ [e]	Number of particles per (time times area)	1/(second times meter^2)	$1/(s\,m^2)$
Illuminance	$E_v = \Delta\Phi_v/\Delta A_2$ [e]	Luminous flux per area	lux = lumen/meter^2	$lx = lm/m^2 = cd\,sr/m^2$

[a] The symbol E stands for the radiant energy (see Table 1.2-5).
[b] I_v stands for the luminous intensity, and Ω stands for the solid angle (see Table 1.2-5).
[c] The radiance L_e, particle radiance L_p, and luminance L_v are important characteristic properties of sources, not radiation fields. For a blackbody source, the radiance L_e, for example, is dependent only on the frequency of the radiation and the temperature of the black body. The dependence is given by Planck's radiation law. In optical imaging, the radiance L_e of an object turns out to show an invariant property. In correct imaging, the image always radiates with the same radiance L_e as the object, independent of the magnification.
[d] φ is the angle between the direction of the beam axis and the direction perpendicular to the source area; A_1 indicates the area of the source; and $g(\varphi)$ is the directional characteristic of the source.
[e] A_2 indicates the irradiated area or the area of the detector.

Because of experimental difficulties in realizing a Planckian radiator at high temperatures and because of new possibilities in the measurement of optical radiation power, the 16th CGPM in 1979 adopted a new definition of the candela as follows:

The candela is the luminous intensity, in a given direction, of a source that emits monochromatic radiation of frequency 540×10^{12} hertz and that has a radiant intensity in that direction of 1/683 watt per steradian.

1.2.4 The SI Derived Units

The SI derived units are the SI units for derived physical quantities. In accordance with the defining equations for derived physical quantities in terms of the base physical quantities, the units for derived quantities can be expressed as products or ratios of the units for the base quantities. Table 1.2-4 shows some examples of SI derived units in terms of SI base units.

For convenience, certain derived units, which are listed in Table 1.2-5, have been given special names and symbols. Among these, the last four entries in Table 1.2-5 are of particular note, since they were accepted by the 15th (1975), 16th (1979), and 21st (1999) CGPMs specifically with a view to safeguarding human health.

In Tables 1.2-5 and 1.2-6, the final column shows how the SI units concerned may be expressed in terms of SI base units. In this column, factors such as m^0 and kg^0, etc., which are equal to 1, are not shown explicitly.

The special names and symbols for derived units listed in Table 1.2-5 may themselves be used to express other derived units: Table 1.2-6 shows some examples. The special names and symbols provide a compact form for the expression of units which are used frequently.

Table 1.2-4 Examples of SI derived units (for derived physical quantities) in terms of base units

Derived quantity	Defining equation	Name of SI derived unit	Symbol for unit
Area	$A = l_1 l_2$	square meter	m^2
Volume	$V = l_1 l_2 l_3$	cubic meter	m^3
Velocity	$v = dl/dt$	meter per second	m/s
Acceleration	$a = d^2 l/dt^2$	meter per second squared	m/s^2
Angular momentum	$L = \Theta \omega$	meter squared kilogram/second	m^2 kg/s
Wavenumber	$k = 2\pi/\lambda$	reciprocal meter	1/m
Density	$\varrho = m/V$	kilogram per cubic meter	kg/m^3
Concentration (of amount of substance)	Concentration = amount/V	mole per cubic meter	mol/m^3
Current density	$j = I/A$	ampere per square meter	A/m^2
Magnetic exciting field	$H = I/l$	ampere per meter	A/m
Radiance (of a radiation source)	$L_e = \Delta I_e(\varphi)/[\Delta A_1 g(\varphi)]$ [a]	watt per (square meter × steradian)	$W/(m^2 \, sr) = kg/(s^3 \, sr)$
Luminance (of a light source)	$L_v = \Delta I_v(\varphi)/[\Delta A_1 g(\varphi)]$ [b]	candela per square meter	cd/m^2
Refractive index	$n = c_{mat}/c$	(number one)	1

[a] φ is the angle between the direction of the beam axis and the direction perpendicular to the source area; $I_e(\varphi)$ is the radiant intensity emitted in the direction φ; A_1 is the radiating area of the source; and $g(\varphi)$ is the directional characteristic of the source.

[b] φ is the angle between the direction of the beam axis and the direction perpendicular to the source area; $I_v(\varphi)$ is the luminous intensity emitted in the direction φ; A_1 is the radiating area of the light source; and $g(\varphi)$ is the directional characteristic of the source.

Table 1.2-5 SI derived units with special names and symbols

Derived quantity		SI derived unit			
Name	Symbol	Name	Symbol	Expressed in terms of other SI units	Expressed in terms of SI base units
Plane angle	$\alpha, \Delta\alpha$	radian [a]	rad		m/m = 1 [b]
Solid angle	$\Omega, \Delta\Omega$	steradian [a]	sr [c]		$m^2/m^2 = 1$ [b]
Frequency	ν	hertz	Hz		1/s
Force	F	newton	N		$m \, kg/s^2$
Pressure, stress	P	pascal	Pa	N/m^2	$(1/m) \, kg/s^2$
Energy, work, quantity of heat	E, A, Q	joule	J	N m	$m^2 \, kg/s^2$
Power, radiant flux	P, Φ_e	watt	W	J/s	$m^2 \, kg/s^3$
Electric charge, quantity of electricity	q, e	coulomb	C		A s
Electric potential difference, electromotive force	V	volt	V	W/A	$(1/A) \, m^2 \, kg/s^3$
Capacitance	C	farad	F	C/V	$A^2 \, (1/(m^2 \, kg)) \, s^4$
Electrical resistance	R	ohm	Ω	V/A	$(1/A^2) \, m^2 \, kg/s^3$
Electrical conductance	$1/R$	siemens	S	A/V	$A^2 \, (1/(m^2 \, kg)) s^3$
Magnetic flux	Φ	weber	Wb	V s	$(1/A) \, m^2 \, kg(1/s^2)$
Magnetic field strength	B	tesla	T	Wb/m^2	$(1/A) \, kg/s^2$
Inductance	L	henry	H	Wb/A	$(1/A^2) \, m^2 \, kg/s^2$
Celsius temperature	t	degree Celsius	°C		K; $T/K = t/°C + 273.15$
Luminous flux	Φ_v	lumen	lm	cd sr [c]	$(m^2/m^2) \, cd = cd$

Table 1.2–5 SI derived units with special names and symbols, cont.

Derived quantity		SI derived unit			
Name	Symbol	Name	Symbol	Expressed in terms of other SI units	Expressed in terms of SI base units
Illuminance	$E_v = \Delta\Phi_v/\Delta A$	lux	lx	lm/m²	(m²/m⁴) cd = cd/m²
Activity (referred to a radionuclide)	A	becquerel	Bq		1/s
Absorbed dose	D	gray	Gy	J/kg	m²/s²
Dose equivalent	H	sievert	Sv	J/kg	m²/s²
Catalytic activity		katal	kat		(1/s) mol

[a] The units radian and steradian may be used with advantage in expressions for derived units to distinguish between quantities of different nature but the same dimensions. Some examples of their use in forming derived units are given in Tables 1.2-5 and 1.2-6.
[b] In practice, the symbols rad and sr are used where appropriate, but the derived unit "1" is generally omitted in combination with a numerical value.
[c] In photometry, the name steradian and the symbol sr are frequently retained in expressions for units.

Table 1.2–6 Examples of SI derived units whose names and symbols include SI derived units with special names and symbols

Derived quantity		SI derived unit		
Name	Symbol	Name	Symbol	Expressed in terms of SI base units
Dynamic viscosity	η	pascal second	Pa s	(1/m) kg/s
Moment of force	M	newton meter	N m	m² kg/s²
Surface tension	σ	newton per meter	N/m	kg/s²
Angular velocity	ω	radian per second	rad/s	m/(m s) = 1/s
Angular acceleration	$d\omega/dt$	radian per second squared	rad/s²	m/(m s²) = 1/s²
Heat flux density	q_{th}	watt per square meter	W/m²	kg/s³
Heat capacity, entropy	C, S	joule per kelvin	J/K	m² kg/(s² K)
Specific heat capacity, specific entropy	C_{mass}, S_{mass}	joule per (kilogram kelvin)	J/(kg K)	m²/(s² K)
Specific energy		joule per kilogram	J/kg	m²/s²
Energy density	w	joule per cubic meter	J/m³	(1/m) kg/s²
Thermal conductivity	λ	watt per (meter kelvin)	W/(m K)	m kg/(s³ K)
Electric charge density	ρ	coulomb per cubic meter	C/m³	(1/m³) s A
Electric field strength	E	volt per meter	V/m	m kg/(s³ A)
Exciting electric field[b]	D	coulomb per square meter	C/m²	(1/m²) s A
Molar energy	E_{mol}	joule per mole	J/mol	m² kg/(s² mol)
Molar heat capacity, molar entropy	C_{mol}, S_{mol}	joule per (mole kelvin)	J/(mol K)	m² kg/(s² K mol)
Exposure (X- and γ-rays)		coulomb per kilogram	C/kg	(1/kg) s A
Absorbed dose rate	dD/dt	gray per second	Gy/s	m²/s³
Radiant intensity	I_ε	watt per steradian	W/sr	(m⁴/m²) kg/s³ = m² kg/s³
Radiance[a]	L_e	watt per (square meter steradian)	W/(m² sr)	(m²/m²) kg/s³ = kg/s³
Catalytic (activity) concentration		katal per cubic meter	kat/m³	(1/(m³ s)) mol

[a] The radiance is a property of the source of the radiation, not of the radiation field (see footnote c to Table 1.2-3).
[b] also called "electric flux density"

A derived unit can often be expressed in several different ways by combining the names of base units with special names for derived units. This, however, is an algebraic freedom whose use should be limited by common-sense physical considerations. The joule, for example, may formally be written "newton meter" or even "kilogram meter squared per second squared", but in a given situation some forms may be more helpful than others.

In practice, with certain quantities, preference is given to the use of certain special unit names or combinations of unit names, in order to facilitate making a distinction between different quantities that have the same dimensions. For example, the SI unit of frequency is called the hertz rather than the reciprocal second, and the SI unit of angular velocity is called the radian per second rather than the reciprocal second (in this case, retaining the word "radian" emphasizes that the angular velocity is equal to 2π times the rotational frequency). Similarly, the SI unit of moment of force is called the newton meter rather than the joule.

In the field of ionizing radiation, the SI unit of activity is called the becquerel rather than the reciprocal second, and the SI units of absorbed dose and dose equivalent are called the gray and the sievert, respectively, rather than the joule per kilogram. In the field of catalysis, the SI unit of catalytic activity is called the katal rather than the mole per second. The special names becquerel, gray, sievert, and katal were specifically introduced because of the dangers to human health which might arise from mistakes involving the units reciprocal second, joule per kilogram, and mole per second.

1.2.5 Decimal Multiples and Submultiples of SI Units

The 11th CGPM adopted, in 1960, a series of prefixes and prefix symbols for forming the names and symbols of the decimal multiples and submultiples of SI units ranging from 10^{12} to 10^{-12}. Prefixes for 10^{-15} and 10^{-18} were added by the 12th CGPM in 1964, and for 10^{15} and 10^{18} by the 15th CGPM in 1975. The 19th CGPM extended the scale in 1991 from 10^{-24} to 10^{24}. Table 1.2-7 lists all approved prefixes and symbols.

Table 1.2-7 SI prefixes and their symbols

Factor	Name	Symbol
10^{24}	yotta	Y
10^{21}	zeta	Z
10^{18}	exa	E
10^{15}	peta	P
10^{12}	tera	T
10^{9}	giga	G
10^{6}	mega	M
10^{3}	kilo	k
10^{2}	hecto	h
10^{1}	deca	da
10^{-1}	deci	d
10^{-2}	centi	c
10^{-3}	milli	m
10^{-6}	micro	µ
10^{-9}	nano	n
10^{-12}	pico	p
10^{-15}	femto	f
10^{-18}	atto	a
10^{-21}	zepto	z
10^{-24}	yocto	y

1.2.6 Units Outside the SI

The SI base units and SI derived units, including those with special names, have the important advantage of forming a coherent set, with the effect that unit conversions are not required when one is inserting particular values for quantities into equations involving quantities.

Nonetheless, it is recognized that some non-SI units still appear widely in the scientific, technical, and commercial literature, and some will probably continue to be used for many years. Other non-SI units, such as the units of time, are so widely used in everyday life and are so deeply embedded in the history and culture of human beings that they will continue to be used for the foreseeable future. For these reasons, some of the more important non-SI units are listed.

1.2.6.1 Units Used with the SI

In 1996 the CIPM agreed upon a categorization of the units used with the SI into three groups: units accepted for use with the SI, units accepted for use with the SI whose values are obtained experimentally, and other units currently accepted for use with the SI to satisfy the needs of special interests. The three groups are listed in Tables 1.2-8 – 1.2-10.

Table 1.2-9 lists three non-SI units accepted for use with the SI, whose values expressed in SI units must be obtained by experiment and are therefore not known exactly. Their values are given with their combined standard uncertainties, which apply to the last two digits, shown in parentheses. These units are in common use in certain specialized fields.

Table 1.2-10 lists some other non-SI units which are currently accepted for use with the SI to satisfy the needs of commercial, legal, and specialized scientific interests. These units should be defined in relation to SI units in every document in which they are used. Their use is not encouraged.

1.2.6.2 Other Non-SI Units

Certain other non-SI units are still occasionally used. Some are important for the interpretation of older scientific texts. These are listed in Tables 1.2-11 – 1.2-16, but their use is not encouraged.

Table 1.2-8 Non-SI units accepted for use with the International System

Name	Symbol	Value in SI units
minute	min	$1\,\text{min} = 60\,\text{s}$
hour	h	$1\,\text{h} = 60\,\text{min} = 3600\,\text{s}$
day	d	$1\,\text{d} = 24\,\text{h} = 86\,400\,\text{s}$
degree[a]	°	$1° = (\pi/180)\,\text{rad}$
minute of arc	′	$1′ = (1/60)° = (\pi/10\,800)\,\text{rad}$
second of arc	″	$1″ = (1/60)′ = (\pi/648\,000)\,\text{rad}$
litre[b]	l, L	$1\,\text{l} = 1\,\text{dm}^3 = 10^{-3}\,\text{m}^3$
tonne[c]	t	$1\,\text{t} = 10^3\,\text{kg}$

[a] ISO 31 recommends that the degree be subdivided decimally rather than using the minute and second.
[b] Unfortunately, printers from all over the world seem not to be willing to admit that in some texts it would be very helpful to have distinguishable symbols for "the number 1" and "the letter l". Giving up any further discussion, the 16th CGPM therefore decided in 1979 that the symbol L should also be adopted to indicate the unit litre in order to avoid the risk of confusion between "the number 1" and "the letter l".
[c] This unit is also called the "metric ton" in some countries.

Table 1.2-9 Non-SI units accepted for use with the International System, whose values in SI units are obtained experimentally

Unit	Definition	Symbol	Value in SI units
Electron volt[a]	[b]	eV	$1\,\text{eV} = 1.60217653(14) \times 10^{-19}\,\text{J}$
Unified atomic mass unit[a]	[c]	u	$1\,\text{u} = 1.66053886(28) \times 10^{-27}\,\text{kg}$
Astronomical unit[d]	[e]	ua	$1\,\text{ua} = 1.49597870691(30) \times 10^{11}\,\text{m}$

[a] For the electron volt and the unified atomic mass unit, the values are quoted from the CODATA recommended values 2002 (see Chapt. 1.1).
[b] The electron volt is the kinetic energy acquired by an electron in passing through a potential difference of 1 V in vacuum.
[c] The unified atomic mass unit is equal to 1/12 of the mass of an unbound atom of the nuclide ^{12}C, at rest and in its ground state. In the field of biochemistry, the unified atomic mass unit is also called the dalton, symbol Da.
[d] The value given for the astronomical unit is quoted from the IERS Convention (1996).
[e] The astronomical unit is a unit of length approximately equal to the mean Earth–Sun distance. Its value is such that, when it is used to describe the motion of bodies in the solar system, the heliocentric gravitational constant is $(0.01720209895)^2\,\text{ua}^3/\text{d}^2$.

Table 1.2-10 Other non-SI units currently accepted for use with the International System

Unit	Symbol	Value in SI units
nautical mile[a]		1 nautical mile = 1852 m
knot		1 knot = 1 nautical mile per hour = (1852/3600) m/s
are[b]	a	$1\,a = 1\,dam^2 = 10^2\,m^2$
hectare[b]	ha	$1\,ha = 1\,hm^2 = 10^4\,m^2$
Bar[c]	bar	$1\,bar = 0.1\,MPa = 100\,kPa = 1000\,hPa = 10^5\,Pa$
angstrom	Å	$1\,\text{Å} = 0.1\,nm = 10^{-10}\,m$
barn[d]	b	$1\,b = 100\,fm^2 = 10^{-28}\,m^2$

[a] The nautical mile is a special unit employed for marine and aerial navigation to express distance. The conventional value given above was adopted by the First International Extraordinary Hydrographic Conference, Monaco, 1929, under the name "international nautical mile". As yet there is no internationally agreed symbol. This unit was originally chosen because one nautical mile on the surface of the Earth subtends approximately one minute of arc at the center.
[b] The units are and hectare and their symbols were adopted by the CIPM in 1879 and are used to express areas of land.
[c] The bar and its symbol were included in Resolution 7 of the 9th CGPM (1948).
[d] The barn is a special unit employed in nuclear physics to express effective cross sections.

Table 1.2-11 deals with the relationship between CGS units and SI units, and lists those CGS units that were assigned special names. In the field of mechanics, the CGS system of units was built upon three quantities and their corresponding base units: the centimeter, the gram, and the second. In the field of electricity and magnetism, units were expressed in terms of these three base units. Because this can be done in different ways, this led to the establishment of several different systems, for example the CGS electrostatic system, the CGS electromagnetic system, and the CGS Gaussian system. In those three systems, the system of quantities used and the corresponding system of defining equations for the derived quantities differ from those used with SI units.

Table 1.2-11 Derived CGS units with special names

Unit	Symbol	Value in SI units
erg[a]	erg	$1\,erg = 10^{-7}\,J$
dyne[a]	dyn	$1\,dyn = 10^{-5}\,N$
poise[a]	P	$1\,P = 1\,dyn\,s/cm^2 = 0.1\,Pa\,s$
stokes	St	$1\,St = 1\,cm^2/s = 10^{-4}\,m^2/s$
gauss[b]	G	$1\,G \equiv 10^{-4}\,T$
oersted[b]	Oe	$1\,Oe \equiv (1000/4\pi)\,A/m$
maxwell[b]	Mx	$1\,Mx \equiv 10^{-8}\,Wb$
stilb[a]	sb	$1\,sb = 1\,cd/cm^2 = 10^4\,cd/m^2$
phot	ph	$1\,ph = 10^4\,lx$
gal[c]	Gal	$1\,Gal = 1\,cm/s^2 = 10^{-2}\,m/s^2$

[a] This unit and its symbol were included in Resolution 7 of the 9th CGPM (1948).
[b] This unit is part of the "electromagnetic" three-dimensional CGS system and cannot strictly be compared with the corresponding unit of the International System, which has four dimensions if only mechanical and electrical quantities are considered. For this reason, this unit is linked to the SI unit using the mathematical symbol for "equivalent to" ($\equiv$) here.
[c] The gal is a special unit employed in geodesy and geophysics to express the acceleration due to gravity.

Table 1.2-12 deals with the *natural units*, which are based directly on fundamental constants or combinations of fundamental constants. Like the CGS system, this system is based on mechanical quantities only. The numerical values in SI units are given here according to the 2002 CODATA adjustment.

Table 1.2-13 presents numerical values in SI units for some of the most frequently used *atomic units* (a.u.), again based on the 2002 CODATA adjustment.

Table 1.2-14 presents numerical values in SI units (based on the 2002 CODATA adjustment) for some X-ray-related quantities used in crystallography.

Table 1.2-12 Natural units (n.u.)

Unit	Symbol and definition	Value in SI units
n.u. of velocity: speed of light in vacuum	c	299 792 458 m/s
n.u. of action: reduced Planck constant	$\hbar = h/2\pi$	$1.05457168(18) \times 10^{-34}\,J\,s$ $6.58211915(56) \times 10^{-16}\,eV\,s$
n.u. of mass: electron mass	m_e	$9.1093826(16) \times 10^{-31}\,kg$
n.u. of energy	$m_e c^2$	$8.1871047(14) \times 10^{-14}\,J$ $0.510998918(44)\,MeV$
n.u. of momentum	$m_e c$	$2.73092419(47) \times 10^{-22}\,kg\,m/s$ $0.510998918(44)\,MeV/c$
n.u. of length	$\lambdabar_C = \hbar/m_e c$	$386.1592678(26) \times 10^{-15}\,m$
n.u. of time	$\hbar/m_e c^2$	$1.2880886677(86) \times 10^{-21}\,s$

Table 1.2-13 Atomic units (a.u.)

Unit	Symbol and definition	Value in SI units
a.u. of charge: elementary charge	e	$1.60217653(14) \times 10^{-19}$ C
a.u. of mass: electron mass	m_e	$9.1093826(16) \times 10^{-31}$ kg
a.u. of action: reduced Planck constant	$\hbar = h/2\pi$	$1.05457168(18) \times 10^{-34}$ J s
a.u. of length, 1 bohr: Bohr radius	$a_0 = \alpha/(4\pi R_\infty)$	$0.5291772108(18) \times 10^{-10}$ m
a.u. of energy, 1 hartree: Hartree energy[a]	E_H	$4.35974417(75) \times 10^{-18}$ J
a.u. of time	$\hbar/E_H$	$2.418884326505(16) \times 10^{-17}$ s
a.u. of force	E_H/a_0	$8.2387225(14) \times 10^{-8}$ N
a.u. of velocity	$\alpha c = a_0 E_H/\hbar$	$2.1876912633(73) \times 10^6$ m/s
a.u. of momentum	$\hbar/a_0$	$1.99285166(34) \times 10^{-24}$ kg m/s
a.u. of current	$e E_H/\hbar$	$6.62361782(57) \times 10^{-3}$ A
a.u. of charge density	e/a_0^3	$1.081202317(93) \times 10^{12}$ C/m^3
a.u. of electric potential	E_H/e	$27.2113845(23)$ V
a.u. of electric field	$E_H/(ea_0)$	$5.14220642(44) \times 10^{11}$ V/m
a.u. of electric dipole moment	ea_0	$8.47835309(73) \times 10^{-30}$ C m
a.u. of electric polarizability	$e^2 a_0^2/E_H$	$1.648777274(16) \times 10^{-41}$ C^2 m^2/J
a.u. of magnetic field B	$\hbar/(ea_0^2)$	$2.35051742(20) \times 10^5$ T
a.u. of magnetic dipole moment ($2\mu_B$)	$2\mu_B = \hbar e/m_e$	$1.85480190(16) \times 10^{-23}$ J/T
a.u. of magnetizability	$e^2 a_0^2/m_e$	$7.89103660(13) \times 10^{-29}$ J/T^2
a.u. of permittivity	$e^2/(a_0 E_H)$	Fixed by definition as: $10^7/c^2 = 1.112650056\ldots 10^{-10}$ F/m

[a] The Hartree energy is defined as $E_H = e^2/(4\pi\varepsilon_0 a_0) = 2R_\infty hc = \alpha^2 m_e c^2$.

Table 1.2-14 Units of some special X-ray-related quantities

Unit	Definition	Symbol	Value in SI units
Cu X unit	$\lambda(CuK\alpha_1)/1537.400$	xu(CuKα_1)	$1.00207710(29) \times 10^{-13}$ m
Mo X unit	$\lambda(MoK\alpha_1)/707.831$	xu(MoKα_1)	$1.00209966(53) \times 10^{-13}$ m
angstrom star	$\lambda(WK\alpha_1)/0.2090100$	Å*	$1.00001509(90) \times 10^{-10}$ m
Lattice parameter[a] of Si (in vacuum, at 22.5 °C)		a	$543.102122(20) \times 10^{-12}$ m
(220) lattice spacing of Si (in vacuum, at 22.5 °C)	$d_{220} = a/\sqrt{8}$	d_{220}	$192.0155965(70) \times 10^{-12}$ m
Molar volume of Si (in vacuum, at 22.5 °C)	$V_m(Si) = N_A a^3/8$	$V_m(Si)$	$12.0588382(24) \times 10^{-6}$ m^3/mol

[a] This is the lattice parameter (unit cell edge length) of an ideal single crystal of naturally occurring silicon free from impurities and imperfections, and is deduced from measurements on extremely pure, nearly perfect single crystals of Si by correcting for the effects of impurities.

Table 1.2-15 lists some other units which are common in older texts. For current texts, it should be noted that if these units are used, the advantages of the SI are lost. The relation of these units to SI units should be specified in every document in which they are used.

For some selected quantities, there exists an international agreement that the numerical values of these quantities measured in SI units are fixed at the values given in Table 1.2-16.

Table 1.2-15 Examples of other non-SI units

Unit	Symbol	Value in SI units
curie[a]	Ci	$1\,\text{Ci} = 3.7 \times 10^{10}\,\text{Bq}$
röntgen[b]	R	$1\,\text{R} = 2.58 \times 10^{-4}\,\text{C/kg}$
rad[c,d]	rad	$1\,\text{rad} = 1\,\text{cGy} = 10^{-2}\,\text{Gy}$
rem[d,e]	rem	$1\,\text{rem} = 1\,\text{cSv} = 10^{-2}\,\text{Sv}$
X unit[f]		$1\,\text{X unit} \cong 1.002 \times 10^{-4}\,\text{nm}$
gamma[d]	γ	$1\,\gamma = 1\,\text{nT} = 10^{-9}\,\text{T}$
jansky	Jy	$1\,\text{Jy} = 10^{-26}\,\text{W/(m}^2\,\text{Hz)}$
fermi[d]		$1\,\text{fermi} = 1\,\text{fm} = 10^{-15}\,\text{m}$
metric carat[g]		$1\,\text{metric carat} = 200\,\text{mg} = 2 \times 10^{-4}\,\text{kg}$
torr	Torr	$1\,\text{Torr} = (101\,325/760)\,\text{Pa}$
standard atmosphere	atm[h]	$1\,\text{atm} = 101\,325\,\text{Pa}$
calorie	cal	[i]
micron[j]	μ	$1\,\mu = 1\,\mu\text{m} = 10^{-6}\,\text{m}$

[a] The curie is a special unit employed in nuclear physics to express the activity of radionuclides.
[b] The röntgen is a special unit employed to express exposure to X-ray or γ radiation.
[c] The rad is a special unit employed to express absorbed dose of ionizing radiation. When there is a risk of confusion with the symbol for the radian, rd may be used as the symbol for $10^{-2}\,\text{Gy}$.
[e] The rem is a special unit used in radioprotection to express dose equivalent.
[f] The X unit was employed to express wavelengths of X-rays. Its relationship to SI units is an approximate one.
[d] Note that this non-SI unit is exactly equivalent to an SI unit with an appropriate submultiple prefix.
[g] The metric carat was adopted by the 4th CGPM in 1907 for commercial dealings in diamonds, pearls, and precious stones.
[h] Resolution 4 of the 10th CGPM, 1954. The designation "standard atmosphere" for a reference pressure of 101 325 Pa is still acceptable.
[i] Several "calories" have been in use:
  - the 15 °C calorie: $1\,\text{cal}_{15} = 4.1855\,\text{J}$ (value adopted by the CIPM in 1950);
  - the IT (International Table) calorie: $1\,\text{cal}_{\text{IT}} = 4.1868\,\text{J}$ (5th International Conference on the Properties of Steam, London, 1956);
  - the thermochemical calorie: $1\,\text{cal}_{\text{th}} = 4.184\,\text{J}$.
[j] The micron and its symbol, adopted by the CIPM in 1879 and repeated in Resolution 7 of the 9th CGPM (1948), were abolished by the 13th CGPM (1967–1968).

Table 1.2-16 Internationally adopted numerical values for selected quantities

Quantity	Symbol	Numerical value	Unit
Relative atomic mass[a] of ^{12}C	$A_\text{r}(^{12}\text{C})$	12	
Molar mass constant	M_u	1×10^{-3}	kg/mol
Molar mass of ^{12}C	$M(^{12}\text{C})$	12×10^{-3}	kg/mol
Conventional value of the Josephson constant[b]	$K_{\text{J-90}}$	483 597.9	GHz/V
Conventional value of the von Klitzing constant[c]	$R_{\text{K-90}}$	25 812.807	Ω
Standard atmosphere		101 325	Pa
Standard acceleration of free fall[d]	g_n	9.80665	m/s^2

[a] The relative atomic mass $A_\text{r}(X)$ of a particle X with mass $m(X)$ is defined by $A_\text{r}(X) = m(X)/m_\text{u}$, where $m_\text{u} = m(^{12}\text{C})/12 = M_\text{u}/N_\text{A} = 1\,\text{u}$ is the atomic mass constant, M_u is the molar mass constant, N_A is the Avogadro number, and u is the (unified) atomic mass unit. Thus the mass of a particle X is $m(X) = A_\text{r}(X)\,\text{u}$ and the molar mass of X is $M(X) = A_\text{r}(X)M_\text{u}$.
[b] This is the value adopted internationally for realizing representations of the volt using the Josephson effect.
[c] This is the value adopted internationally for realizing representations of the ohm using the quantum Hall effect.
[d] The value given was adopted by the 3rd General Conference on Weights and Measures (CGPM), 1903, and was the conventional value used to calculate the now obsolete unit kilogram force.

1.2.7 Some Energy Equivalents

In science and technology, energy is measured in many different units. Different units are used depending on the field of application, but owing to the different possible forms of the energy concerned, it is possible also to express the energy in terms of other quantities. All forms of the energy, however, are quantitatively related to one another and are therefore considered as being equivalent. Some of the most important equivalence relations are

$$E = eU = mc^2 = hc/\lambda = h\nu = kT \, .$$

These equations tell us that a given energy E, which is usually measured either in units of joule (J) or units of the Hartree energy ($E_H = 1$ hartree), can also be specified by giving a voltage U, a mass m, a wavelength λ, a frequency ν, or a temperature T. These equations contain, in addition to those variables, only well-known fundamental constants.

Table 1.2-17 gives the values of the energy equivalents of the joule and the hartree and for the SI units corresponding to the five quantities U, m, λ, ν, and T. The equivalents have been calculated on the basis of the 2002 CODATA adjustment of the values of the constants.

Table 1.2-17 Energy equivalents, expressed in the units joule (J), hartree (E_H), volt (V), kilogram (kg), (unified) atomic mass unit (u), reciprocal meter (m^{-1}), hertz (Hz), and kelvin (K)

Energy	Joule	Hartree	Volt	Kilogram
1 J	$(1\,\text{J}) = 1\,\text{J}$	$(1\,\text{J}) = 2.29371257(39) \times 10^{17} \, E_H$	$(1\,\text{J}) = 6.24150947(53) \times 10^{18}\,\text{eV}$	$(1\,\text{J})/c^2 = 1.112650056 \times 10^{-17}\,\text{kg}$
1 E_H	$(1\,E_H) = 4.35974417(75) \times 10^{-18}\,\text{J}$	$(1\,E_H) = 1\,E_H$	$(1\,E_H) = 27.2113845(23)\,\text{eV}$	$(1\,E_H)/c^2 = 4.85086960(83) \times 10^{-35}\,\text{kg}$
1 eV	$(1\,\text{eV}) = 1.60217653(14) \times 10^{-19}\,\text{J}$	$(1\,\text{eV}) = 3.67493245(31) \times 10^{-2}\,E_H$	$(1\,\text{eV}) = 1\,\text{eV}$	$(1\,\text{eV})/c^2 = 1.78266181(15) \times 10^{-36}\,\text{kg}$
1 kg	$(1\,\text{kg})\,c^2 = 8.987551787 \times 10^{16}\,\text{J}$	$(1\,\text{kg})\,c^2 = 2.06148605(35) \times 10^{34}\,E_H$	$(1\,\text{kg})\,c^2 = 5.60958896(48) \times 10^{35}\,\text{eV}$	$(1\,\text{kg}) = 1\,\text{kg}$
1 u	$(1\,\text{u})\,c^2 = 1.49241790(26) \times 10^{-10}\,\text{J}$	$(1\,\text{u})\,c^2 = 3.423177686(23) \times 10^7\,E_H$	$(1\,\text{u})\,c^2 = 931.494043(80) \times 10^6\,\text{eV}$	$(1\,\text{u}) = 1.66053886(28) \times 10^{-27}\,\text{kg}$
1 m^{-1}	$(1\,\text{m}^{-1})\,hc = 1.98644561(34) \times 10^{-25}\,\text{J}$	$(1\,\text{m}^{-1})\,hc = 4.556335252760(30) \times 10^{-8}\,E_H$	$(1\,\text{m}^{-1})\,hc = 1.23984191(11) \times 10^{-6}\,\text{eV}$	$(1\,\text{m}^{-1})\,h/c = 2.21021881(38) \times 10^{-42}\,\text{kg}$
1 Hz	$(1\,\text{Hz})\,h = 6.6260693(11) \times 10^{-34}\,\text{J}$	$(1\,\text{Hz})\,h = 1.519829846006(10) \times 10^{-16}\,E_H$	$(1\,\text{Hz})\,h = 4.13566743(35) \times 10^{-15}\,\text{eV}$	$(1\,\text{Hz})\,h/c^2 = 7.3724964(13) \times 10^{-51}\,\text{kg}$
1 K	$(1\,\text{K})\,k = 1.3806505(24) \times 10^{-23}\,\text{J}$	$(1\,\text{K})\,k = 3.1668153(55) \times 10^{-6}\,E_H$	$(1\,\text{K})\,k = 8.617343(15) \times 10^{-5}\,\text{eV}$	$(1\,\text{K})\,k/c^2 = 1.5361808(27) \times 10^{-40}\,\text{kg}$

Energy	Atomic mass unit	Reciprocal meter	Hertz	Kelvin
1 J	$(1\,\text{J})/c^2 = 6.7005361(11) \times 10^9\,\text{u}$	$(1\,\text{J})/hc = 5.03411720(86) \times 10^{24}\,\text{m}^{-1}$	$(1\,\text{J})/h = 1.50919037(26) \times 10^{33}\,\text{Hz}$	$(1\,\text{J})/k = 7.242963(13) \times 10^{22}\,\text{K}$
1 E_H	$(1\,E_H)/c^2 = 2.921262323(19) \times 10^{-8}\,\text{u}$	$(1\,E_H)/hc = 2.194746313705(15) \times 10^7\,\text{m}^{-1}$	$(1\,E_H)/h = 6.579683920721(44) \times 10^{15}\,\text{Hz}$	$(1\,E_H)/k = 3.1577465(55) \times 10^5\,\text{K}$
1 eV	$(1\,\text{eV})/c^2 = 1.073544171(92) \times 10^{-9}\,\text{u}$	$(1\,\text{eV})/hc = 8.06554445(69) \times 10^5\,\text{m}^{-1}$	$(1\,\text{eV})/h = 2.41798940(21) \times 10^{14}\,\text{Hz}$	$(1\,\text{eV})/k = 1.1604505(20) \times 10^4\,\text{K}$
1 kg	$(1\,\text{kg}) = 6.0221415(10) \times 10^{26}\,\text{u}$	$(1\,\text{kg})\,c/h = 4.52443891(77) \times 10^{41}\,\text{m}^{-1}$	$(1\,\text{kg})\,c^2/h = 1.35639266(23) \times 10^{50}\,\text{Hz}$	$(1\,\text{kg})\,c^2/k = 6.509650(11) \times 10^{39}\,\text{K}$
1 u	$(1\,\text{u}) = 1\,\text{u}$	$(1\,\text{u})\,c/h = 7.513006608(50) \times 10^{14}\,\text{m}^{-1}$	$(1\,\text{u})\,c^2/h = 2.252342718(15) \times 10^{23}\,\text{Hz}$	$(1\,\text{u})\,c^2/k = 1.0809527(19) \times 10^{13}\,\text{K}$
1 m^{-1}	$(1\,\text{m}^{-1})\,h/c = 1.3310250506(89) \times 10^{-15}\,\text{u}$	$(1\,\text{m}^{-1}) = 1\,\text{m}^{-1}$	$(1\,\text{m}^{-1})\,c = 299\,792\,458\,\text{Hz}$	$(1\,\text{m}^{-1})\,hc/k = 1.4387752(25) \times 10^{-2}\,\text{K}$
1 Hz	$(1\,\text{Hz})\,h/c^2 = 4.439821667(30) \times 10^{-24}\,\text{u}$	$(1\,\text{Hz})/c = 3.335640951 \times 10^{-9}\,\text{m}^{-1}$	$(1\,\text{Hz}) = 1\,\text{Hz}$	$(1\,\text{Hz})\,h/k = 4.7992374(84) \times 10^{-11}\,\text{K}$
1 K	$(1\,\text{K})\,k/c^2 = 9.251098(16) \times 10^{-14}\,\text{u}$	$(1\,\text{K})\,k/hc = 69.50356(12)\,\text{m}^{-1}$	$(1\,\text{K})\,k/h = 2.0836644(36) \times 10^{10}\,\text{Hz}$	$(1\,\text{K}) = 1\,\text{K}$

References

2.1 Bureau International des Poids et Mesures: *Le système international d'unités*, 7th edn. (Bureau International des Poids et Mesures, Sèvres 1998)

2.2 Organisation Intergouvernementale de la Convention du Mètre: *The International System of Units (SI), Addenda and Corrigenda to the 7th Edition* (Bureau International des Poids et Mesures, Sèvres 2000)

1.3. Rudiments of Crystallography

Crystallography deals basically with the question "Where are the atoms in solids?" The purpose of this section is to introduce briefly the basics of modern crystallography. The focus is on the description of periodic solids, which represent the major proportion of condensed matter. A coherent introduction to the formalism required to do this is given, and the basic concepts and technical terms are briefly explained. Paying attention to recent developments in materials research, we treat aperiodic, disordered, and amorphous materials as well. Consequently, besides the conventional three-dimensional (3-D) descriptions, the higher-dimensional crystallographic approach is outlined, and so is the atomic pair distribution function used to describe local phenomena. The section is concluded by touching on the basics of diffraction methods, the most powerful tool kit used by experimentalists dealing with structure at the atomic level in the solid state.

1.3.1	Crystalline Materials	28
	1.3.1.1 Periodic Materials	28
	1.3.1.2 Aperiodic Materials	33
1.3.2	Disorder	38
1.3.3	Amorphous Materials	39
1.3.4	Methods for Investigating Crystallographic Structure	39
References		41

The structure of a solid material is very important, because the physical properties are closely related to the structure. In most cases solids are crystalline: they may consist of one single crystal, or be polycrystalline, consisting of many tiny single crystals in different orientations. All periodic crystals have a perfect translational symmetry. This leads to selection rules, which are very useful for the understanding of the physical properties of solids. Therefore, most textbooks on solid-state physics begin with some chapters on symmetry and structure. Today we know that other solids, which have no translational symmetry, also exist. These are amorphous materials, which have little order (in most cases restricted to the short-range arrangement of the atoms), and aperiodic crystals, which show perfect long-range order, but no periodicity – at least in 3-D space. In this chapter of the book, the basic concepts of crystallography – how the space of a solid can be filled with atoms – are briefly discussed. Readers who want to inform themselves in more detail about crystallography are referred to the classic textbooks [3.1–5].

Many crystalline materials, especially minerals and gems, were described more than 2000 years ago. The regular form of crystals and the existence of facets, which have fixed angles between them, gave rise to a belief that crystals were formed by a regular repetition of tiny, identical building blocks. After the discovery of X-rays by Röntgen, Laue investigated crystals in 1912 using these X-rays and detected interference effects caused by the periodic array of atoms. One year later, Bragg determined the crystal structures of alkali halides by X-ray diffraction.

Today we know that a crystal is a 3-D array of atoms or molecules, with various types of long-range order. A more modern definition is that all materials which show sharp diffraction peaks are crystalline. In this sense, aperiodic or quasicrystalline materials, as well as periodic materials, are crystals. A real crystal is never a perfect arrangement. Defects in the form of vacancies, dislocations, impurities, and other imperfections are often very important for the physical properties of a crystal. This aspect has been largely neglected in classical crystallography but is becoming more and more a topic of modern crystallographic investigations [3.6, 7].

As indicated in Table 1.3-1, condensed matter can be classified as either crystalline or amorphous. Both of these states and their formal subdivisions will be discussed in the following. The terms "matter", "structure", and "material" always refer to single-phase solids.

Table 1.3-1 Classification of solids

Condensed matter (solids)			
Crystalline materials			Amorphous materials
Periodic structures	Aperiodic structures		
	Modulated structures	Composite structures	Quasicrystals

1.3.1 Crystalline Materials

1.3.1.1 Periodic Materials

Lattice Concept

A periodic crystal is described by two entities, the *lattice* and the *basis*. The (translational) lattice is a perfect geometrical array of points. All lattice points are equivalent and have identical surroundings. This lattice is defined by three fundamental translation vectors a, b, c. Starting from an arbitrarily chosen origin of the lattice, any other lattice point can be reached by a translation vector r that satisfies

$$r = ua + vb + wc \,,$$

where u, v, and w are arbitrary integers.

The lattice is an abstract mathematical construction; the description of the crystal is completed by attaching a set of atoms – the basis – to each lattice point. Therefore the crystal structure is formed by a lattice and a basis (see Fig. 1.3-1).

The parallelepiped that is defined by the axes a, b, c is called a *primitive cell* if this cell has the smallest volume out of all possible cells. It contains one lattice point per cell only (Fig. 1.3-2a). This cell is a type of unit cell which fills the space of the crystal completely under the application of the translation operations of the lattice, i.e. movements along the vectors r.

Conventionally, the smallest cell with the highest symmetry is chosen. Crystal lattices can be transformed into themselves by translation along the fundamental vectors a, b, c, but also by other symmetry operations. It can be shown that only onefold (rotation angle $\varphi = 2\pi/1$), twofold ($2\pi/2$), threefold ($2\pi/3$), fourfold ($2\pi/4$), and sixfold ($2\pi/6$) rotation axes are permissible. Other rotational axes cannot exist in a lattice, because they would violate the translational symmetry. For example, it is not possible to fill the space completely with a fivefold ($2\pi/5$) array of regular pentagons. Additionally, mirror planes and centers of inversion may exist. The restriction to high-symmetry cells may also lead to what is known as centering. Figure 1.3-2b illustrates a 2-D case. The centering types in 3-D are listed in Table 1.3-2.

Planes and Directions in Lattices

If one peers through a 3-D lattice from various angles, an infinity of equidistant planes can be seen. The position and orientation of such a crystal plane are determined by three points. It is easy to describe a plane if all three points lie on crystal axes (i.e. the directions of unit cell vectors); in this case only the intercepts need to be used. It is common to use *Miller indices* to de-

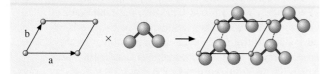

Fig. 1.3-1 A periodic crystal can be described as a convolution of a mathematical point lattice with a basis (set of atoms). *Open circles*, mathematical points; *filled circles*, atoms

Table 1.3-2 Centering types for 3-D crystallographic unit cells

Symbol	Description	Points per unit cell
P	No centering (primitive)	1
I	Body-centered (*innenzentriert*)	2
F	All-face-centered	4
S; A, B, C in specific cases	One-face-centered (*seitenzentriert*); (b, c), (a, c), and (a, b), respectively, in specific cases	2
R	Hexagonal cell, rhombohedrally centered	3

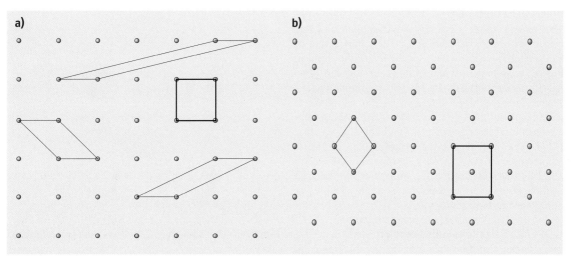

Fig. 1.3-2a,b Possible primitive and centered cells in 2-D lattices. *Open circles* denote mathematical points. (**a**) In this lattice, the conventional cell is the *bold square cell* because of its highest symmetry, $4mm$. (**b**) Here, convention prefers 90° angles: a centered cell of symmetry $2mm$ is chosen. It contains two lattice points and is twice the area of the primitive cell

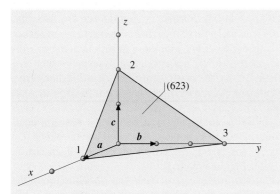

Fig. 1.3-3 Miller indices: the intercepts of the (623) plane with the coordinate axes

scribe lattice planes. These indices are determined as follows:

1. For the plane of interest, determine the intercepts x, y, z of the crystal axes $\boldsymbol{a}, \boldsymbol{b}, \boldsymbol{c}$.
2. Express the intercepts in terms of the basic vectors $\boldsymbol{a}, \boldsymbol{b}, \boldsymbol{c}$ of the unit cell, i.e. as x/a, y/b, z/c (where $a = |\boldsymbol{a}|, \ldots$).
3. Form the reciprocals $a/x, b/y, c/z$.
4. Reduce this set to the smallest integers h, k, l. The result is written (hkl).

The distance from the origin to the plane (hkl) inside the unit cell is the interplanar spacing d_{hkl}. Negative intercepts, leading to negative Miller indices, are written as $\bar{h}$. Figure 1.3-3 shows a (623) plane and its construction.

A direction in a crystal is given as a set of three integers in square brackets $[uvw]$; u, v, and w correspond to the above definition of the translation vector $\boldsymbol{r}$. A direction in a cubic crystal can be described also by Miller indices, as a plane can be defined by its normal. The indices of a direction are expressed as the smallest integers which have the same ratio as the components of a vector (expressed in terms of the axis vectors $\boldsymbol{a}, \boldsymbol{b}, \boldsymbol{c}$) in that direction. Thus the sets of integers 1, 1, 1 and 3, 3, 3 represent the same direction in a crystal, but the indices of the direction are [111] and not [333]. To give another example, the x axis of an orthogonal x, y, z coordinate system has Miller indices [100]; the plane perpendicular to this direction has indices (100).

For all crystals, except for the hexagonal system, the Miller indices are given in a three-digit system in the form (hkl). However, for the hexagonal system, it is common to use four digits $(hkil)$. The four-digit hexagonal indices are based on a coordinate system containing four axes. Three axes lie in the basal plane of the hexagon, crossing at angles of 120°: $\boldsymbol{a}, \boldsymbol{b}$, and $-(\boldsymbol{a}+\boldsymbol{b})$. As the third vector in the basal plane can be expressed in terms of $\boldsymbol{a}$ and $\boldsymbol{b}$, the index can be expressed in terms of h and k: $i = -(h+k)$. The fourth axis is the $\boldsymbol{c}$ axis normal to the basal plane.

Crystal Morphology

The regular facets of a crystal are planes of the type described above. Here, the lattice architecture of the crystal is visible macroscopically at the surface. Figure 1.3-4 shows some surfaces of a cubic crystal. If the crystal had the shape or morphology of a cube, this would be described by the set of facets $\{(100), (010), (001), (\bar{1}00), (0\bar{1}0), (00\bar{1})\}$. An octahedron would be described by $\{(111), (\bar{1}11), (1\bar{1}1), (11\bar{1}), (\bar{1}\bar{1}\bar{1}), (1\bar{1}\bar{1}), (\bar{1}1\bar{1}), (\bar{1}\bar{1}1)\}$. The morphology of a crystalline material may be of technological interest (in relation to the bulk density, flow properties, etc.) and can be influenced in various ways, for example by additives during the crystallization process.

The 32 Crystallographic Point Groups

The symmetry of the space surrounding a lattice point can be described by the point group, which is a set of symmetry elements acting on the lattice. The crystallographic symbols for the symmetry elements of point groups compatible with a translational lattice are the rotation axes 1, 2, 3, 4, and 6, mirror planes m, and the center of inversion $\bar{1}$. Figure 1.3-5 illustrates, as an example, the point group $2/m$. The "2" denotes a twofold axis perpendicular ("/") to a mirror plane "m". Note that this combination of 2 and m implies, or generates automatically, an inversion center $\bar{1}$. We have used the Hermann–Mauguin notation here; however, point groups of isolated molecules are more often denoted by the Schoenflies symbols. For a translation list, see Table 1.3-3.

No crystal can have a higher point group symmetry than the point group of its lattice, called the *holohedry*. In accordance with the various rotational symmetries,

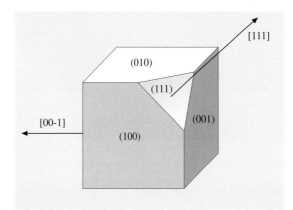

Fig. 1.3-4 Some crystal planes and directions in a cubic crystal, and their Miller indices

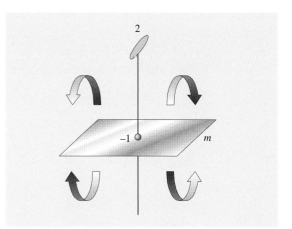

Fig. 1.3-5 The point group $2/m$ (C_{2h}). Any object in space can be rotated by $\varphi = 2\pi/2$ around the twofold rotational axis 2 and reflected by the perpendicular mirror plane m, generating identical copies. The inversion center $\bar{1}$ is implied by the coupling of 2 and m

there are seven crystal systems (see Table 1.3-3), and the seven holohedries are $\bar{1}$, $2/m$, mmm, $4/mmm$, $\bar{3}m$, $6/mmm$, and $m\bar{3}m$. Other, less symmetric, point groups are also compatible with these lattices, leading to a total number of 32 crystallographic point groups (see Table 1.3-4). A lower symmetry than the holohedry can be introduced by a less symmetric basis in the unit cell.

Since $\bar{3}m$ and $6/mmm$ are included in the same point lattice, they are sometimes subsumed into the hexagonal crystal family. So there are seven crystal systems but six crystal families. Note further that rhombohedral symmetry is a special case of centering (R-centering) of the trigonal crystal system and offers two equivalent possibilities for selecting the cell parameters: hexagonal or rhombohedral axes (see Table 1.3-4 again).

It can be shown that in 3-D there are 14 different periodic ways of arranging identical points. These 14 3-D periodic point lattices are called the (translational) Bravais lattices and are shown in Fig. 1.3-6. Table 1.3-4 presents data related to some of the crystallographic terms used here. The 1-D and 2-D space groups can be classified analogously but are omitted here.

The 230 Crystallographic Space Groups

Owing to the 3-D translational periodicity, symmetry operations other than point group operations are possible in addition: these are *glide planes* and *screw axes*. A glide plane couples a mirror operation and a translational shift. The symbols for glide planes

Table 1.3-3 The 32 crystallographic point groups: translation list from the Hermann–Mauguin to the Schoenflies notation

Crystal system	Hermann–Mauguin symbol	Schoenflies symbol	Crystal system	Hermann–Mauguin symbol	Schoenflies symbol
Triclinic	1	C_1	Trigonal	3	C_3
	$\bar{1}$	C_i		$\bar{3}$	C_{3i}
Monoclinic	2	C_2		32	D_3
	m	C_s		$3m$	C_{3v}
	$2/m$	C_{2h}		$\bar{3}m$	D_{3d}
Orthorhombic	222	D_2	Hexagonal	6	C_6
	$mm2$	C_{2v}		$\bar{6}$	C_{3h}
	mmm	D_{2h}		$6/m$	C_{6h}
Tetragonal	4	C_4		622	D_6
	$\bar{4}$	S_4		$6mm$	C_{6v}
	$4/m$	C_{4h}		$\bar{6}2m$	D_{3h}
	422	D_4		$6/mmm$	D_{6h}
	$4mm$	C_{4v}	Cubic	23	T
	$\bar{4}2m$	D_{2d}		$m\bar{3}$	T_h
	$4/mmm$	D_{4h}		432	O
				$\bar{4}3m$	T_d
				$m\bar{3}m$	O_h

are a, b, and c for translations along the lattice vectors $\boldsymbol{a}, \boldsymbol{b}$, and $\boldsymbol{c}$, respectively, and n and d for some special lattice vector combinations. A screw axis is always parallel to a rotational axis. The symbols are $2_1, 3_1, 3_2, 4_1, 4_2, 4_3, 6_1, 6_2, 6_3, 6_4$, and 6_5, where, for example, 6_3 means a rotation through an angle $\varphi = 2\pi/6$ followed by a translation of $3/6 (= 1/2)$ of a full translational period along the sixfold axis.

Thus the combination of 3-D translational and point symmetry operations leads to an infinite number of sets of symmetry operations. Mathematically, each of these sets forms a group, and they are called *space groups*. It can be shown that all possible periodic crystals can be described by only 230 space groups. These 230 space groups are described in tables, for example the *International Tables for Crystallography* [3.8].

In this formalism, a conventional space group symbol reflects the symmetry elements, arranged in the order of standardized *blickrichtungen* (symmetry directions). We shall confine ourselves here to explain one instructive example: $P4_2/mcm$, space group number 132 [3.8]. The full space group symbol is $P\,4_2/m\,2/c\,2/m$. The mean-

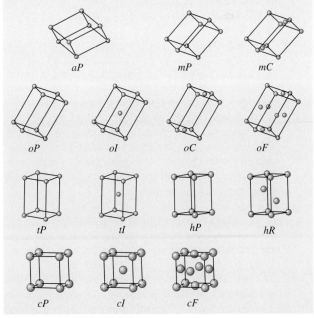

Fig. 1.3-6 The 14 Bravais lattices

Table 1.3-4 Crystal families, crystal systems, crystallographic point groups, conventional coordinate systems, and Bravais lattices in three dimensions. Lattice point symmetries (holohedries) are given in **bold**

Crystal family	Symbol	Crystal system	Crystallographic point groups	No. of space groups	Conventional coordinate system		Bravais lattice (Pearson symbol)
					Restrictions on cell parameters	Parameters to be determined	
Triclinic (anorthic)	a	Triclinic	1, **$\bar{1}$**	2	None	$a, b, c,$ α, β, γ	aP
monoclinic	m	Monoclinic	2, m, **2/m**	13	Setting with b unique: $\alpha = \gamma = 90°$	$a, b, c,$ β	mP, mS (mC, mA, mI)
					Setting with c unique: $\alpha = \beta = 90°$	$a, b, c,$ γ	mP, mS (mA, mB, mI)
orthorhombic	o	Orthorhombic	222, mm2, **mmm**	59	$\alpha = \beta = \gamma = 90°$	a, b, c	oP, oS, (oC, oA, oB) oI, oF
tetragonal	t	Tetragonal	4, $\bar{4}$, 4/m, 422, 4mm, $\bar{4}2m$, **4/mmm**	68	$a = b$ $\alpha = \beta = \gamma = 90°$	a, c	tP, tI
hexagonal	h	Trigonal	3, $\bar{3}$, 32, 3m, **$\bar{3}m$**	18	$a = b,$ $\alpha = \beta = 90°$ $\gamma = 120°$ (hexagonal axes)	a, c	hP
				7	$a = b = c$ $\alpha = \beta = \gamma \neq 90°$ (rhombohedral axes)	a, α	hR
		Hexagonal	6, $\bar{6}$, 6/m, 622, 6mm, $\bar{6}2m$, **6/mmm**	27	$a = b,$ $\alpha = \beta = 90°$ $\gamma = 120°$	a, c	hP
cubic	c	Cubic	23, $m\bar{3}$, 432, $\bar{4}3m$, **$m\bar{3}m$**	36	$a = b = c$ $\alpha = \beta = \gamma = 90°$	a	cP, cI, cF

ing of the symbols is the following: P denotes a primitive Bravais lattice. It belongs to the tetragonal crystal system indicated by 4. Along the first standard *blickrichtung* [001] there is a 4_2 screw axis with a perpendicular mirror plane m. Along [100] there is a twofold rotation axis, named 2, with a perpendicular glide plane c parallel to c. Third, along [110] there is a twofold rotation axis 2, with a perpendicular mirror plane m.

Decoration of the Lattice with the Basis

At this point we have to recall that in a real crystal structure we have not only the lattice, but also the basis. In [3.8], there are standardized sets of general and special positions (i.e. coordinates x, y, z) within the unit cell (Wyckoff positions). An atom placed in a general position is transformed into more than one atom by the action of all symmetry operators of the respective space group. Special positions are located on special points which are mapped onto themselves by one or more symmetry operations – for example a position in a mirror plane or exactly on a rotational axis. Reference [3.8] also provides information about symmetry relations between individual space groups (group–subgroup relations). These are often useful for describing relationships between crystal structures and for describing phase transitions of materials.

The use of the space group allows us to further reduce the basis to the *asymmetric unit*: this is the minimal set of atoms that needs to be given so that the whole crystal structure can be generated via the symmetry of the space group. This represents the main power of a crystallographically correct description of a material: just some 10 parameters are sufficient to describe an ensemble of some 10^{23} atoms.

Thus, a crystallographically periodic structure of a material is unambiguously characterized by

- the cell parameters;
- the space group;
- the coordinates of the atoms (and their chemical type) in the asymmetric unit;
- the occupation and thermal displacement factors of the atoms in the asymmetric unit.

For an example, the reader is referred to the crystallographic description of the spinel structure of $MgAl_2O_4$ given below under the heading "Structure Types".

To complete the information on space group symmetries given here, periodic magnetic materials should also be mentioned. Magnetic materials contain magnetic moments carried by atoms in certain positions in the unit cell. If we take into account the magnetic moments in the description of the structure, the classification by space groups (the 230 "gray" groups, described above) has to be extended to 1651 the "black and white", or Shubnikov, groups [3.9]. A magnetic periodic structure is then characterized by

- the crystallographic structure;
- the Shubnikov group;
- the cell parameters of the magnetic unit cell;
- the coordinates of the atoms carrying magnetic moments (the asymmetric unit in the magnetic unit cell);
- the magnitude and direction of the magnetic moments on these atoms.

Structure Types

It is useful to classify the crystal structures of materials by the assignment of *structure types*. The structure type is based on a representative crystal structure, the parameters of which describe the essential crystallographic features of other materials of the same type. As an example, we consider the structure of the spinel oxides AB_2O_4. The generic structure type is $MgAl_2O_4$, cF56. The *Pearson symbol*, here cF56, denotes the cubic crystal family and a face-centered Bravais lattice with 56 atoms per unit cell (see Table 1.3-5 and also the last column in Table 1.3-4).

Regarding the free parameters, for example a, the notation 8.174(1) in Table 1.3-5 means 8.174 ± 0.001. The chemical formula and the unit cell contents can easily be calculated from the site multiplicities (given by the Wyckoff positions) and the occupancies. So can the (crystallographic) density, using the appropriate atomic masses.

There is a huge variety of other materials belonging to the same structure type as in this example. The only parameters that differ (slightly) are the numerical value of a, the types of atoms in the positions, the numerical value of the parameter x for Wyckoff position 32e, and the occupancies. Thus, for example, the crystal structure of the iron sulfide Fe_3S_4 can be characterized in its essential features via the information that it belongs to the same structure type.

1.3.1.2 Aperiodic Materials

In addition to the crystalline periodic state of matter, a class of materials exists that lacks 3-D translational symmetry and is called *aperiodic*. Aperiodic materials cannot be described by any of the 230 space groups mentioned above. Nevertheless, they show another type of long-range order and are therefore included in the term "crystal". This notion of long-range order is the major feature that distinguishes crystals from amorphous materials. Three types of aperiodic order may

Table 1.3-5 Complete crystallographic parameter set for $MgAl_2O_4$, spinel structure type

Material	$MgAl_2O_4$				
Structure type	$MgAl_2O_4$, spinel				
Pearson symbol	cF56				
Space group	$Fd\bar{3}m$ (No. 227)				
a (Å)	8.174(1)				

Atom	Wyckoff position	x	y	z	Occupancy
Mg	8a	0	0	0	1.0
Al	16d	5/8	5/8	5/8	1.0
O	32e	0.3863(2)	x	x	1.0

be distinguished, namely modulated structures, composite structures, and quasicrystals. All aperiodic solids exhibit an essentially discrete diffraction pattern and can be described as atomic structures obtained from a 3-D section of a n-dimensional (n-D) ($n > 3$) periodic structure.

Modulated Structures

In a modulated structure, periodic deviations of the atomic parameters from a reference or basic structure are present. The basic structure can be understood as a periodic structure as described above. Periodic deviations of one or several of the following atomic parameters are superimposed on this basic structure:

- atomic coordinates;
- occupancy factors;
- thermal displacement factors;
- orientations of magnetic moments.

Let the period of the basic structure be a and the modulation wavelength be λ; the ratio a/λ may be (1) a rational or (2) an irrational number (Fig. 1.3-7). In case (1), the structure is commensurately modulated; we observe a qa superstructure, where $q = 1/\lambda$. This superstructure is periodic. In case (2), the structure is incommensurately modulated. Of course, the experimental distinction between the two cases is limited by the finite experimental resolution. q may be a function of external variables such as temperature, pressure, or chemical composition, i.e. $q = f(T, p, X)$, and may adopt a rational value to result in a commensurate "lock-in" structure. On the other hand, an incommensurate charge-density wave may exist; this can be moved through a basic crystal without changing the internal energy U of the crystal.

When a 1-D basic structure and its modulation function are combined in a 2-D hyperspace $R = R^{\text{parallel}} \oplus R^{\text{perpendicular}}$, periodicity on a 2-D lattice results. The real atoms are generated by the intersection of the 1-D physical (external, parallel) space R^{parallel} with the hyperatoms in the complementary 1-D internal space $R^{\text{perpendicular}}$. In the case of a modulated structure, the hyperatoms have the shape of the sinusoidal modulation function in $R^{\text{perpendicular}}$.

Figure 1.3-8 illustrates this construction. We have to choose a basis (a_1, a_2) in R where the slope of a_1 with respect to R^{parallel} corresponds to the length of the modulation λ.

It is clear that real atomic structures are always manifestations of matter in 3-D real, physical space. The cutting of the 2-D hyperspace to obtain real 1-D atoms illustrated in Fig. 1.3-8 may serve as an instructive basic example of the concept of higher-dimensional (n-D, $n > 3$) crystallography. The concept is also called a superspace description; it applies to all aperiodic structures and provides a convenient finite set of variables that can be used to compute the positions of all atoms in the real 3-D structure.

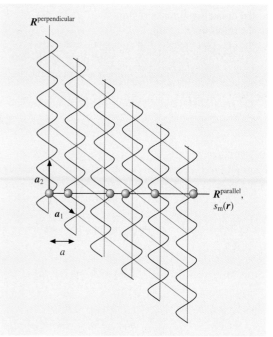

Fig. 1.3-8 2-D hyperspace description of the example of Fig. 1.3-7. The basis of the hyperspace $R = R^{\text{parallel}} \oplus R^{\text{perpendicular}}$ is (a_1, a_2); the slope of a_1 with respect to R^{parallel} is proportional to λ. Atoms of the modulated structure $s_{\text{m}}(r)$ occur in the physical space R^{parallel} and are represented by *circles*

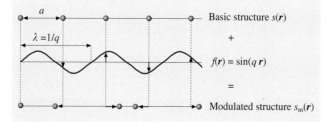

Fig. 1.3-7 A 1-D modulated structure $s_{\text{m}}(r)$ can be described as a sum of a basic structure $s(r)$ and a modulation function $f(r)$ of its atomic coordinates. If a/λ is irrational, the structure is incommensurately modulated. *Circles* denote atoms

The modulation may occur in one, two, or three directions of the basic structure, yielding 1-D, 2-D, or 3-D modulated structures. If we introduce one additional dimension per modulation vector (the direction r that the modulation corresponding to λ runs along), these structures can be described as periodic in 4-D, 5-D, or 6-D superspace, respectively.

Composite Structures

Composite crystals are crystalline structures that consist of two or more periodic substructures, each one having its own 3-D periodicity to a first approximation. The symmetry of each of these subsystems is characterized by one of the 230 space groups. However, owing to their mutual interaction, the true structure consists of a collection of incommensurately modulated subsystems. All known composite structures to date have at least one lattice direction in common and consist of a maximum of three substructures. There are three main classes:

- channel structures;
- columnar packings;
- layer packings.

These composite structures are also known as intergrowth or host–guest structures. Figure 1.3-9 illustrates an example of a host with channels along a, in which atoms of the substructure with a periodicity λa reside as a guest.

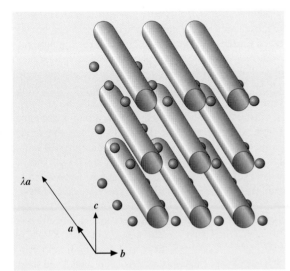

Fig. 1.3-9 Host–guest channel structure. The guest atoms reside in channels parallel to a, with a periodicity λa

The higher-dimensional n-D formalism ($n > 3$) used to describe composite structures is essentially the same as that which applies to modulated structures.

Quasicrystals

Quasicrystals represent the third type of aperiodic materials. Quasiperiodicity may occur in one, two, or three dimensions of physical space and is associated with special irrational numbers such as the golden mean $\tau = (1+\sqrt{5})/2$, and $\xi = 2+\sqrt{3}$. The most remarkable feature of quasicrystals is the appearance of noncrystallographic point group symmetries in their diffraction patterns, such as $8/mmm$, $10/mmm$, $12/mmm$, and $2/m\bar{3}\bar{5}$. The golden mean is related to fivefold symmetry via the relation $\tau = 2\cos(\pi/5)$; τ can be considered as the "most irrational" number, since it is the irrational number that has the worst approximation by a truncated continued fraction,

$$\tau = 1 + \cfrac{1}{1+\cfrac{1}{1+\cfrac{1}{1+\cfrac{1}{1+\cfrac{1}{1+\cdots}}}}} \ .$$

This might be a reason for the stability of quasiperiodic systems where τ plays a role. A prominent 1-D example is the Fibonacci sequence, an aperiodic chain of short and long segments S and L with lengths S and L, where the relations $L/S = \tau$ and $L + S = \tau L$ hold. A Fibonacci chain can be constructed by the simple substitution or inflation rule $L \to LS$ and $S \to L$ (Table 1.3-6, Fig. 1.3-10). Materials quasiperiodically modulated in 1-D along one direction may occur. Again, their structures are readily described using the superspace formalism as above.

The Fibonacci sequence can be used to explain the idea of a periodic rational approximant. If the sequence ...LSLLSLSLS... represents a quasicrystal, then the

Table 1.3-6 Generation of the Fibonacci sequence using the inflation rule $L \to LS$ and $S \to L$. The ratio F_{n+1}/F_n tends towards τ for $n \to \infty$. F_n is a Fibonacci number; $F_{n+1} = F_n + F_{n-1}$. The sequence starts with $F_0 = 0$, $F_1 = 1$

Sequence	n	F_{n+1}/F_n
L	1	$1/1 = 1$
LS	2	$2/1 = 2$
LSL	3	$3/2 = 1.5$
LSLLS	4	$5/3 = 1.66666\ldots$
LSLLSLSL	5	$8/5 = 1.6$
...		
...LSLLSLSLS...	∞	$\tau = 1.61803\ldots$

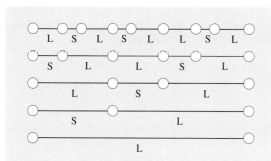

Fig. 1.3-10 1-D Fibonacci sequence. Moving downwards corresponds to an inflation of the self-similar chains, and moving upwards corresponds to a deflation

periodic sequence . . .LSLSLSLSLS. . ., consisting only of the word LS, is its 2/1 approximant (Table 1.3-6). In real systems, such approximants often exist as large-unit-cell (periodic!) structures with atomic arrangements locally very similar to those in the corresponding quasicrystal. When described in terms of superspace, they would result via cutting with a rational slope, in the above example 2/1 = 2, instead of $\tau = 1.6180\ldots$.

To date, all known 2-D quasiperiodic materials exhibit noncrystallographic diffraction symmetries of $8/mmm$, $10/mmm$, or $12/mmm$. The structures of these materials are called octagonal, decagonal, and dodecagonal structures, respectively. Quasiperiodicity is present only in planes stacked along a perpendicular periodic direction. To index the lattice points in a plane, four basis vectors a_1, a_2, a_3, a_4 are needed; a fifth one, a_5, describes the periodic direction. Thus a 5-D hypercrystal is appropriate for describing the solid periodically. In an analogous way to the 230 3-D space groups, the 5-D superspace groups (e.g. $P10_5/mmc$) provide

- the multiplicity and Wyckoff positions;
- the site symmetry;
- the coordinates of the hyperatoms.

Again, the quasiperiodic structure in 3-D can be obtained from an intersection with the external space.

On the atomic scale, these quasicrystals consist of units of some 100 atoms, called clusters. These clusters, of point symmetry $8/mmm$, $10/mmm$, or $12/mmm$

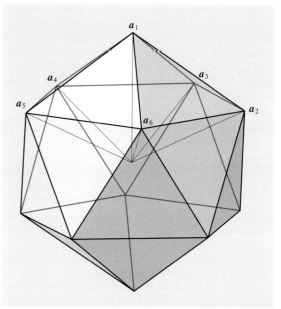

Fig. 1.3-11 Unit vectors $a_1, \ldots, a_6$ of an icosahedral lattice

(or less), are fused, may interpenetrate partially, and can be considered to decorate quasiperiodic tilings. In a diffraction experiment, their superposition leads to an overall noncrystallographic symmetry. There are a number of different tilings that show such noncrystallographic symmetries. Figure 1.3-12 depicts four of them, as examples of the octagonal, decagonal, and dodecagonal cases.

Icosahedral quasicrystals are also known. In 3-D, the icosahedral diffraction symmetry $2/m\bar{3}\bar{5}$ can be observed for these quasicrystals. Their diffraction patterns can be indexed using six integers, leading to a 6-D superspace description (see Fig. 1.3-11). On the atomic scale in 3-D, in physical space, clusters of some 100 atoms are arranged on the nodes of 3-D icosahedral tilings; the clusters have an icosahedral point group symmetry or less, partially interpenetrate, and generate an overall symmetry $2/m\bar{3}\bar{5}$. Many of their structures are still waiting to be determined completely. Figure 1.3-13 shows the two golden rhombohedra and the four Danzer tetrahedra that can be used to tile 3-D space icosahedrally.

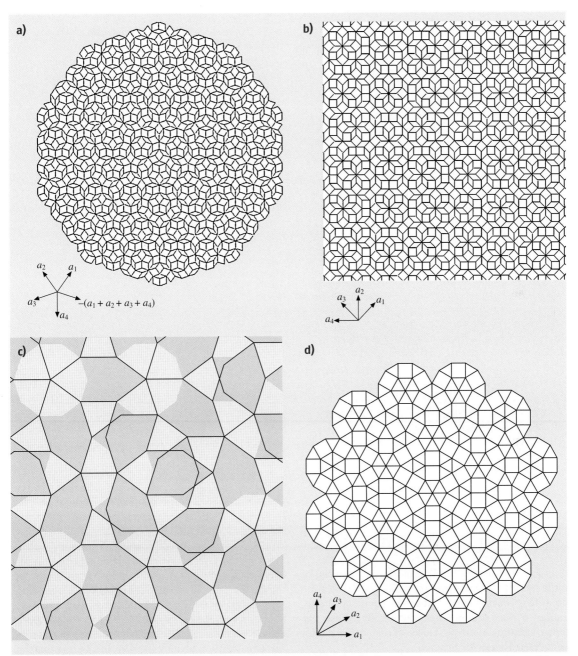

Fig. 1.3-12a–d Some 2-D quasiperiodic tilings; the corresponding four basis vectors $a_1, \ldots, a_4$ are shown. Linear combinations of $r = \sum_i u_i a_i$ reach all lattice points. (**a**) Penrose tiling with local symmetry $5mm$ and diffraction symmetry $10mm$, (**b**) octagonal tiling with diffraction symmetry $8mm$, (**c**) Gummelt tiling with diffraction symmetry $10mm$, and (**d**) dodecagonal Stampfli-type tiling with diffraction symmetry $12mm$

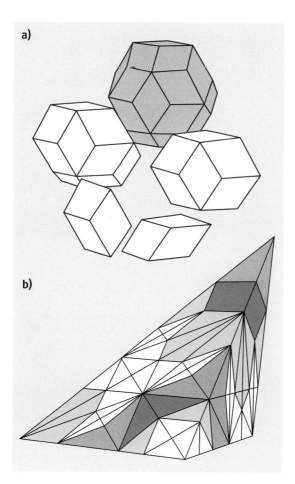

Fig. 1.3-13 Icosahedral tilings. (**a**) The two golden rhombohedra (*bottom*) can be used to form icosahedral objects (the rhombic triacontahedron with point symmetry $m\bar{3}\bar{5}$ shown in *gray*). (**b**) Danzer's {ABCK} tiling: three inflation steps for prototile A

1.3.2 Disorder

In between the ideal crystalline and the purely amorphous states, most real crystals contain degrees of disorder. Two types of statistical disorder have to be distinguished: chemical disorder and displacive disorder (Fig. 1.3-14). Statistical disorder contributes to the entropy S of the solid and is manifested by diffuse scattering in diffraction experiments. It may occur in both periodic and aperiodic materials.

Chemical Disorder
Chemical disorder is observed, for example, in the case of solid solutions, say of B in A, or $A_{1-x}B_x$ for short. Here, an average crystal structure exists. On the crystallographic atomic positions, different atomic species (the chemical elements A and B) are distributed randomly. Generally, the cell parameter a varies with x. For $x = 0$ or 1, the pure end member is present. A linear variation of $a(x)$ is predicted by Vegard's law. On the atomic scale, however, differences in the local structure, are present owing to the different contacts A–A, B–B, and A–B. These differences are usually represented by enlarged displacement factors, but can be investigated by analyzing the pair distribution function $G(r)$. $G(r)$ represents the probability of finding any atom at a distance r from any other atom relative to an average density. Chemical disorder can also occur on only one or a few of the crystallographically different atomic positions (e.g. $A(X_{1-x}Y_x)_2$). This type of disorder is often intrinsic to a material and may be temperature-dependent.

Displacive Disorder

The displacive type of disorder can be introduced by the presence of voids or vacancies in the structure or may exist for other reasons. Vacancies can be an important feature of a material: for example, they may leading to ionic conductivity or influence the mechanical properties.

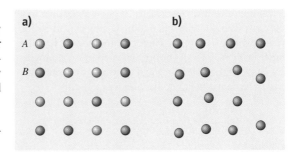

Fig. 1.3-14 Schematic sketch of (**a**) chemical and (**b**) displacive disorder

1.3.3 Amorphous Materials

The second large group of condensed matter is classified as the amorphous or glassy state. No long-range order is observed. The atoms are more or less statistically distributed in space, but a certain short-range order is present.

This short-range order is reflected in the certain average coordination numbers or average coordination geometries. If there are strong (covalent) interactions between neighboring atoms, similar basic units may occur, which are in turn oriented randomly with respect to each other. The SiO_4 tetrahedron in silicate glasses is a well-known example. In an X-ray diffraction experiment on an amorphous solid, only isotropic diffuse scattering is observed. From this information, the radial atomic pair distribution function (Fig. 1.3-15) can be obtained. This function $G(r)$ can be interpreted as the probability of finding any atom at a distance r from any other atom relative to an average density.

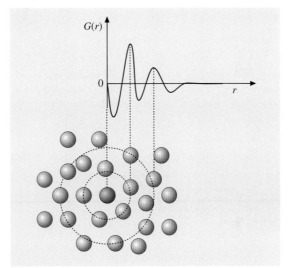

Fig. 1.3-15 Radial atomic pair distribution function $G(r)$ of an amorphous material. Its shape can be deduced from diffuse scattering

1.3.4 Methods for Investigating Crystallographic Structure

So far, we have been dealing with the formal description of solids. To conclude this chapter, the tool kit that an experimentalist needs to obtain structural information about a material in front of him/her will be briefly described.

The major technique used to derive the atomic structure of solids is the diffraction method. To obtain the most comprehensive information about a solid, other techniques besides may be used to complement a model based on diffraction data. These techniques include scanning electron microscopy (SEM), wavelength-dispersive analysis of X-rays (WDX), energy-dispersive analysis of X-rays (EDX), extended X-ray atomic fine-structure analysis (EXAFS), transmission electron microscopy (TEM), high-resolution transmission electron microscopy (HRTEM), differential thermal analysis (DTA), and a number of other methods.

For diffraction experiments, three types of radiation with a wavelength λ of the order of magnitude of interatomic distances are used: X-rays, electrons, and neutrons. The shortest interatomic distances in solids are a few times 10^{-10} m. Therefore the non-SI unit the ångström (1 Å = 10^{-10} m) is often used in crystallography. In the case of electrons and neutrons, their energies have to be converted to de Broglie wavelengths:

$$\lambda = h/mv \,,$$

$$\lambda(\text{Å}) = 0.28/\sqrt{E(\text{eV})} \,.$$

Figure 1.3-16 compares the energies and wavelengths of the three types of radiation.

From wave optics, it is known that radiation of wavelength λ is diffracted by a grid of spacing d. If we take a 3-D crystal lattice as such a grid, we expect diffraction maxima to occur at angles 2θ, given by the Bragg equation (Fig. 1.3-17)

$$\lambda = 2d_{hkl} \sin\theta_{hkl} \,.$$

For the aperiodic (n-D periodic crystal) case, d_{hkl} has to be replaced by $d_{h_1 h_2 \ldots h_i \ldots h_n}$. To give a simple 3-D example, for the determination of the cell parameter a in the cubic case, the Bragg equation can be rewritten in the form

$$(Q/2\pi)^2 = 4 \sin^2\theta_{hkl}/\lambda^2 = (h^2 + k^2 + l^2)/a^2 \,.$$

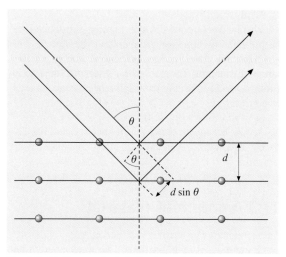

Fig. 1.3-17 Geometrical derivation of the Bragg equation $n\lambda = 2d \sin\theta$. n can be set to 1 when it is included in a higher-order hkl

Thus the crystal lattice is determined by a set of θ_{hkl}. In the case of X-rays and neutrons, information about the atomic structure is contained in the set of diffraction intensities I_{hkl}. Here we have $I_{hkl} = F_{hkl}^2$ where F_{hkl} are the structure factors.

To reconstruct the matter distribution $\rho(xyz)$ inside a unit cell of volume V, the crystallographic phase problem has to be solved. Once the phase factor ϕ for each hkl is known, the crystal structure is solved.

$$\rho(xyz) = 1/V$$
$$\times \sum\sum\sum_{\text{all } h,k,l} |F| \cos[2\pi(hx + ky + lz) - \phi] \,.$$

Non-Bragg diffraction intensities $I(Q)$ and therefore a normalized structure function $S(Q)$ can be obtained, for example, from an X-ray or neutron powder diffractogram. The sine Fourier transform of $S(Q)$ yields a normalized radial atomic pair distribution function $G(r)$:

$$G(r) = (2/\pi) \int_0^\infty Q[S(Q) - 1] \sin(Qr)\, dQ \,.$$

For measurements at high Q, the 1-D function $G(r)$ contains detailed information about the local structure. This function therefore resolves, for example, disorder or vacancy distributions in a material. The method can be applied to 3-D diffuse scattering distributions as well and thus can include angular information with respect to r.

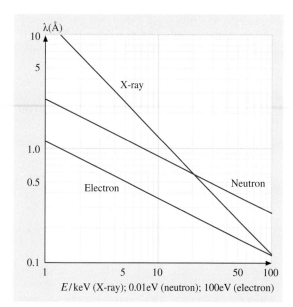

Fig. 1.3-16 Wavelengths λ in Å and particle energies E for X-ray photons (energies in keV), neutrons (energies in 0.01 eV), and electrons (energies in 100 eV)

X-rays

X-rays can be produced in the laboratory using a conventional X-ray tube. Depending on the anode material, wavelengths λ from 0.56 Å (Ag Kα) to 2.29 Å (Cr Kα) can be generated. Filtered or monochromatized radiation is usually used to collect diffraction data, from either single crystals or polycrystalline fine powders. A continuous X-ray spectrum, obtained from a tungsten anode, for example, is used to obtain Laue images to check the quality, orientation, and symmetry of single crystals.

X-rays with a higher intensity, a tunable energy, a narrower distribution, and higher brilliance are provided by synchrotron radiation facilities.

X-rays interact with the electrons in a structure and therefore provide information about the electron density distribution – mainly about the electrons near the atomic cores.

Neutron Diffraction

Neutrons, generated in a nuclear reactor, are useful for complementing X-ray diffraction information. They interact with the atomic nuclei, and with the magnetic moments of unpaired electrons if they are present in a structure. Hydrogen atoms, which are difficult to locate using X-rays (the contain one electron, if at all, near the proton), give a far better contrast in neutron diffraction experiments. The exact positions of atomic nuclei permit "X minus N" structure determinations, so that the location of valence electrons can be made observable. Furthermore, the magnetic structure of a material can be determined.

Electron Diffraction

The third type of radiation which can be used for diffraction purposes is an electron beam; this is usually done in combination with TEM or HRTEM. Because electrons have only a short penetration distance – electrons, being charged particles, interact strongly with the material – electron diffraction is mainly used for thin crystallites, surfaces, and thin films. In the TEM mode, domains and other features on the nanometer scale are visible. Nevertheless, crystallographic parameters such as unit cell dimensions, and symmetry and space group information can be obtained from selected areas.

In some cases, information about, for example, stacking faults or superstructures obtained from an electron diffraction experiment may lead to a revised, detailed crystal structure model that is "truer" than the model which was originally deduced from X-ray diffraction data. If only small crystals of a material are available, crystallographic models obtained from unit cell and symmetry information can be simulated and then adapted to fit HRTEM results.

The descriptions above provide the equipment needed to understand the structure of solid matter on the atomic scale. The concepts of crystallography, the technical terms, and the language used in this framework have been presented. The complementarities of the various experimental methods used to extract coherent, comprehensive information from a sample of material have been outlined. The "rudiments" presented here, however, should be understood only as a first step into the fascinating field of the atomic structure of condensed matter.

References

3.1 L.V. Azaroff: *Elements of X-Ray Crystallography* (McGraw-Hill, New York 1968)
3.2 J. Pickworth Glusker, K.N. Trueblood: *Crystal Structure Analysis – A Primer* (Oxford Univ. Press, Oxford 1985)
3.3 E.R. Wölfel: *Theorie und Praxis der Röntgenstrukturanalyse* (Vieweg, Braunschweig 1987)
3.4 W. Kleber, H.-J. Bautsch, J. Bohm: *Einführung in die Kristallographie* (Verlag Technik, Berlin 1998)
3.5 C. Giacovazzo (Ed.): *Fundamentals of Crystallography*, IUCr Texts on Crystallography (Oxford Univ. Press., Oxford 1992)
3.6 C. Janot: *Quasicrystals – A Primer* (Oxford Univ. Press, Oxford 1992)
3.7 S.J.L. Billinge, T. Egami: *Underneath the Bragg Peaks: Structural Analysis of Complex Materials* (Elsevier, Amsterdam 2003)
3.8 T. Hahn (Ed.): *International Tables for Crystallography*, Vol. A (Kluwer, Dordrecht 1992)
3.9 A.V. Shubnikov, N.V. Belov: *Colored Symmetry* (Pergamon Press, Oxford 1964)

Part 2 The Elements

1 The Elements
Werner Martienssen, Frankfurt/Main, Germany

44

2.1. The Elements

This section provides tables of the physical and physicochemical properties of the elements. Emphasis is given to properties of the elements in the condensed state. The tables are structured according to the Periodic Table of the elements. Most of the tables deal with the properties of elements of one particular group (column) of the Periodic Table. Only the elements of the first period (hydrogen and helium), the lanthanides, and the actinides are arranged according to the periods (rows) of the Periodic Table. This synoptic representation is intended to provide an immediate overview of the trends in the data for chemically related elements.

2.1.1	**Introduction**		45
	2.1.1.1	How to Use This Section	45
2.1.2	**Description of Properties Tabulated**		46
	2.1.2.1	Parts A of the Tables	46
	2.1.2.2	Parts B of the Tables	46
	2.1.2.3	Parts C of the Tables	48
	2.1.2.4	Parts D of the Tables	49
2.1.3	**Sources**		49
2.1.4	**Tables of the Elements in Different Orders**		49
2.1.5	**Data**		54
	2.1.5.1	Elements of the First Period	54
	2.1.5.2	Elements of the Main Groups and Subgroup I to IV	59
	2.1.5.3	Elements of the Main Groups and Subgroup V to VIII	98
	2.1.5.4	Elements of the Lanthanides Period	142
	2.1.5.5	Elements of the Actinides Period	151
References			158

2.1.1 Introduction

2.1.1.1 How to Use This Section

To find properties of a specific element or group of elements, start from one of Tables 2.1-1 – 2.1-5 and proceed in one of the following ways:

1. If you know the *name* of the element, refer to Table 2.1-1, where an alphabetical list of the elements is given, together with the numbers of the pages where the properties of these elements will be found.
2. If you know the *chemical symbol* of the element, refer to Table 2.1-2, where an alphabetical list of element symbols is given, together with the numbers of the pages where the properties of the corresponding elements will be found.
3. If you know the *atomic number* Z of the element, refer to Table 2.1-3, where a list of the elements in order of atomic number is given, together with the numbers of the pages where the properties of these elements will be found.
4. If you know the *group of the Periodic Table* that contains the element of interest, refer to Table 2.1-4, which gives the numbers of the pages where the properties of the elements of each group will be found.
5. If you wish to look up the element in the *Periodic Table*, refer to Table 2.1-5, where the element symbol and the atomic number will be found. Then use Table 2.1-2 or Table 2.1-3 to find the numbers of the pages where the properties of the element of interest are tabulated.
6. Alternatively you can also find the name and the chemical symbol of the element you are looking for in the alphabetic index at the end of the volume. The index again will give you the first number of those pages on which

you can find the properties of the element described.

The data-tables corresponding to the Periodes and Groups of the Periodic Table are subdivided in the following way:

A. Atomic, ionic, and molecular properties.
B. Materials data:
(a) Crystallographic properties.
(b) Mechanical properties.
(c) Thermal and thermodynamic properties.
(d) Electronic, electromagnetic, and optical properties.
C. Allotropic and high-pressure modifications.
D. Ionic radii.

2.1.2 Description of Properties Tabulated

2.1.2.1 Parts A of the Tables

The properties tabulated in parts A of the tables concern the atomic, ionic, and molecular properties of the elements:

- The relative atomic mass, or atomic weight, A.
- The abundance in the lithosphere and in the sea.
- The atomic radius: the radius r_{cov} for single covalent bonding (after Pauling), the radius r_{met} for metallic bonding with a coordination number of 12 (after Pauling), the radius r_{vdW} for van der Waals bonding (after Bondi), and, for some elements, the radius r_{os} of the outer-shell orbital are given.
- The completely and partially occupied electron shells in the atom.
- The symbol for the electronic ground state.
- The electronic configuration.
- The oxidation states.
- The electron affinity.
- The electronegativity X_A (after Allred and Rochow).
- The first, second, third, and fourth ionization energies and the standard electrode potential E^0.
- The internuclear distance in the molecule.
- The dissociation energy of the molecule.

2.1.2.2 Parts B of the Tables

Parts B of the tables contain data on the macroscopic properties of the elements. Most of the data concern the condensed phases. If not indicated otherwise, the data in this section apply to the standard state of the element, that is, they are valid at standard temperature and pressure (STP, i.e. $T = 298.15\,\text{K}$ and $p = 100\,\text{kPa} = 1\,\text{bar}$). For those elements which are stable in the gas phase at STP, data are given for the macroscopic properties in the gas phase.

The quantities describing the physical and physicochemical properties of materials can be divided into two classes. The first class contains all those quantities which are not directly connected with external (generalized) forces, these quantities have well-defined values even in the absence of external forces. Some examples are the electronic ground-state configuration of the atom, the coordination number in the crystallized state and the surface tension in the liquid state. The second class contains those quantities which describe the response of the material to externally applied (generalized) forces F. Such a force might be a mechanical stress field, an electric or magnetic field, a field gradient, or a temperature gradient. The response of the material to the external force might be observed via a suitable observable O, such as a mechanical strain, an electric current density, a dielectric polarization, a magnetization, or a heat current density. Assuming homogeneous conditions, the dependence of the observable O on the force F can be used to define material-specific parameters χ, which are also called physical properties of the material. Some examples are the elastic moduli or compliance constants, the electrical conductivity, the dielectric constant, the magnetic susceptibility, and the thermal conductivity.

In the *linear-response* regime, that is, under weak external forces F, these parameters χ are considered as being independent of the strength of the forces. The dependence of an observable O on a force F is then the simple proportionality

$$O = \chi F. \tag{1.1}$$

For strong external fields, the dependence of the response on the strength of the forces can be expressed by a power expansion in the forces, which then – in addition to the linear parameters χ – defines *nonlinear field-dependent* materials properties $\chi^{(nl)}(F)$, where

$$\chi^{(nl)}(F) = \chi + \chi^{(1)} F + \ldots. \tag{1.2}$$

In general, the class of materials properties that describe the response to externally applied forces have *tensor* character. The *rank* of the property tensor χ depends on the rank of the external force F and that of the observable O considered. In the case of Ohm's law, $j = \sigma E$, in which the current density j and the electric field strength E are vectors, the conductivity tensor σ is of rank 2; in the case of the generalized Hooke's law, $\boldsymbol{\varepsilon} = \boldsymbol{s}\boldsymbol{\sigma}$, the strain tensor $\boldsymbol{\varepsilon}$ and the stress tensor $\boldsymbol{\sigma}$ both are of rank 2, so that the elastic compliance tensor $\boldsymbol{s}$ is of rank 4. A vector can be considered as a tensor of rank 1, and a scalar, correspondingly, as a tensor of rank 0. A second-rank tensor, such as the electrical conductivity σ, in general has nine components in three-dimensional space; a tensor of rank n in general has 3^n components in three-dimensional space. Symmetry, however, of both the underlying crystal lattice and the physical phenomenon (for example, action = reaction), may reduce the number of independent nonvanishing components in the tensor. The tensor components reflect the crystal symmetry by being invariant under those orthogonal transformations which are elements of the point group of the crystal. In cubic crystals, for example, physical properties described by tensors of rank 2 are characterized by only one nonvanishing tensor component. Therefore cubic crystals are isotropic with respect to their electrical conductivity, their heat conductivity, and their dielectric properties.

Subdivisions B(a) of the Tables

These parts deal with the *crystallographic properties*. Here you will find the crystal system and the Bravais lattice in which the element is stable in its standard state; the structure type in which the element crystallizes; the lattice constants $a, b, c, \alpha, \beta, \gamma$ (symmetry reduces the number of independent lattice constants); the space group; the Schoenflies symbol; the *Strukturbericht* type; the Pearson symbol; the number A of atoms per cell; the coordination number; and the shortest interatomic distance between atoms in the solid state and in the liquid state.

Basic concepts of crystallography are explained in Chapt. 1.3.

Subdivisions B(b) of the Tables

These parts cover the *mechanical properties*. At the top of the table, you will find the density of the material in the solid state (ϱ_s) and in the liquid state (ϱ_l), and the molar volume V_{mol} in the solid state. Here, one mole is the amount of substance which contains as many elementary particles (atoms or molecules) as there are atoms in 0.012 kg of the carbon isotope with a relative atomic mass of 12. This number of particles is called *Avogadro's number* and is approximately equal to 6.022×10^{23}. The next three rows present the viscosity η, the surface tension, and its temperature dependence, in the liquid state. The next properties are the coefficient of linear thermal expansion α and the sound velocity, both in the solid and in the liquid state. A number of quantities are tabulated for the presentation of the elastic properties. For isotropic materials, we list the volume compressibility $\kappa = -(1/V)(dV/dP)$, and in some cases also its reciprocal value, the bulk modulus (or compression modulus); the elastic modulus (or Young's modulus) E; the shear modulus G; and the Poisson number (or Poisson's ratio) μ. Hooke's law, which expresses the linear relation between the strain ε and the stress σ in terms of Young's modulus, reads $\sigma = E\varepsilon$. For monocrystalline materials, the components of the elastic compliance tensor $\boldsymbol{s}$ and the components of the elastic stiffness tensor $\boldsymbol{c}$ are given. The elastic compliance tensor $\boldsymbol{s}$ and the elastic stiffness tensor $\boldsymbol{c}$ are both defined by the generalized forms of Hooke's law, $\boldsymbol{\sigma} = \boldsymbol{c}\boldsymbol{\varepsilon}$ and $\boldsymbol{\varepsilon} = \boldsymbol{s}\boldsymbol{\sigma}$. At the end of the list, the tensile strength, the Vickers hardness, and the Mohs hardness are given for some elements.

Subdivisions B(c) of the Tables

The *thermal and thermodynamic properties* are tabulated in these subdivisions of the tables. The properties tabulated are:

- The thermal conductivity λ.
- The molar heat capacity at constant pressure, c_p.
- The standard entropy S^0, that is, the molar entropy of the element at 298.15 K and 100 kPa.
- The enthalpy difference $H_{298} - H_0$, that is, the difference between the molar enthalpies of the element at 298.15 K and at 0 K.
- The melting temperature T_m.
- The molar enthalpy change ΔH_m and molar entropy change ΔS_m at the melting temperature.
- The relative volume change $\Delta V_m = (V_l - V_s)/V_l$ on melting.
- The boiling temperature T_b.
- The molar enthalpy change ΔH_b of boiling, and, for some elements, the molar enthalpy of sublimation.

In addition, the critical temperature T_c, the critical pressure p_c, the critical density ϱ_c, the triple-point temperature T_{tr}, and the triple-point pressure p_{tr} are given for some elements. For the element helium, the table also contains data for the λ point, at which liquid helium passes from the normal-fluid phase helium I (above the λ point) to the superfluid phase helium II (below the λ point), for ^{4}He and ^{3}He.

Throughout Sect. 2.1, temperature is measured in units of kelvin (K), the unit of thermodynamic temperature. 1 K is defined as the fraction 1/273.16 of the thermodynamic temperature of the triple point of water. To convert data given in kelvin into degrees Celsius (°C), the following equation can be used:

$$T(°C) = (T(K) - 273.15\,K)(°C/K).$$

This can be expressed in words as follows: the Celsius scale is shifted towards higher temperatures by 273.15 K relative to the kelvin scale, such that the temperature 273.15 K becomes 0 °C and the temperature 0 K becomes -273.15 °C. To convert data given in kelvin into degrees Fahrenheit (°F), the following equation can be used:

$$T(°F) = (9/5)(T(K) - 273.15\,K)(°F/K) + 32\,°F.$$

This can be expressed approximately in words as follows: the Fahrenheit scale is shifted relative to the kelvin scale and also differs by a scaling so that its degrees are smaller than those of the kelvin scale by nearly a factor of 2.

Subdivisions B(d) of the Tables

These subdivisions of the tables present data on the *electronic, electromagnetic, and optical properties* of the elements. Data are given for the following:

- The electrical resistivity ρ_s in the solid state, and its temperature and pressure dependence.
- The electrical resistivity ρ_l in the liquid state, and the resistivity ratio ρ_l/ρ_s at the melting temperature.
- The critical temperature T_{cr} and critical field strength H_{cr} for superconductivity.
- The electronic band gap ΔE.
- The Hall coefficient R, together with the range of magnetic field strength B over which it was measured.
- The thermoelectric coefficient.
- The electronic work function.
- The thermal work function.
- The intrinsic charge carrier concentration.
- The electron and hole mobilities.
- The static dielectric constant ε of the element in the solid state, and in some cases also in the liquid state.
- The molar magnetic susceptibility χ_{mol} and the mass magnetic susceptibility χ_{mass} of the element in the solid state, and in some cases also in the liquid state. The susceptibilities are given in the definitions of both the SI system and the cgs system (see below).
- The refractive index n in the solid and liquid states.

The magnetic susceptibility is the parameter that describes the response of the material to an externally applied magnetic field H, as measured by the observable magnetization M, in the linear regime, via $M = \chi H$. Three different forms of the term "magnetization" are in use, depending on the specific application: first, the volume magnetization M_{vol}, equal to the magnetic dipole moment divided by the volume of the sample; second, the molar magnetization, or magnetization related to the number of particles, M_{mol}, equal to the magnetic dipole moment divided by the number of particles measures in moles; and third, the mass magnetization M_{mass}, equal to the magnetic dipole moment divided by the mass of the sample. Correspondingly, there are three different magnetic susceptibilities. The *volume susceptibility* χ_{vol} is a dimensionless number because in this case M and H are both measured in the same units, namely A/m in the SI system and gauss in the cgs system. The dimensionless character of χ_{vol} might be the reason why, in physics textbooks, mostly only this susceptibility is mentioned. The other two susceptibilities, the *molar susceptibility* χ_{mol} and the *mass susceptibility* χ_{mass}, are more useful for practical applications. In both the SI system and the cgs system, the molar susceptibility is measured in units of cm^3/mol, and the mass susceptibility is measured in units of cm^3/g. In this Handbook, data are given for the molar and mass susceptibilities.

Although susceptibilities have the same dimensions in the SI and cgs systems, the numerical values in the cgs system are smaller than those in the SI system by a factor of 4π. This is due to the different definitions of the quantities dipole moment and magnetization in the two systems. The difference can be seen most clearly in the general relations between the magnetization M and the field strengths B and H in the two systems. In the SI system, this relation reads $B = \mu_0(H + M)$, whereas in the cgs system, it reads $B = H + 4\pi M$. Because of this difference, the magnetic-susceptibility data in Sect. 2.1.5 are given for both the SI and the cgs definitions.

2.1.2.3 Parts C of the Tables

Parts C of the tables present crystallographic data for allotropic and high-pressure modifications of the elements. The left-hand columns contain data for allotropic modifications that are stable at a pressure of 100 kPa over the temperature ranges indicated, and the right-hand columns contain data for modifications stable at higher pressures as indicated. The modifications stable at 100 kPa are denoted by Greek letters in front of the chemical symbol of the element (normally starting with α for the modification stable over the lowest temperature range), and the high-pressure modifications are denoted by Roman numerals after the chemical symbol. In these parts of the tables, "RT" stands for "room temperature", and "RTP" stands for "room temperature and standard pressure", i.e. 100 kPa.

2.1.2.4 Parts D of the Tables

Parts D of the tables contain data on *ionic radii* determined from crystal structures. The first row lists the elements, and the second row lists the positive and negative ions for which data are given. The remaining rows give the ionic radii of these ions for the most common coordination numbers.

2.1.3 Sources

Most of the data presented here have been taken from Landolt–Börnstein [1.1]. Additional data have been taken from the D'Ans-Lax series [1.2] and the *CRC Handbook of Chemistry and Physics* [1.3].

2.1.4 Tables of the Elements in Different Orders

Table 2.1-1 The elements ordered by their names

Element	Symbol	Atomic Number	Page	Element	Symbol	Atomic Number	Page	Element	Symbol	Atomic Number	Page
Actinium	Ac	89	84	Gold	Au	79	65	Praseodymium	Pr	59	142
Aluminium	Al	13	78	Hafnium	Hf	72	94	Promethium	Pm	61	142
Americium	Am	95	151	Hassium	Hs	108	131	Protactinium	Pa	91	151
Antimony	Sb	51	98	Helium	He	2	54	Radium	Ra	88	68
Argon	Ar	18	128	Holmium	Ho	67	142	Radon	Rn	86	128
Arsenic	As	33	98	Hydrogen	H	1	54	Rhenium	Re	75	124
Astatine	At	85	118	Indium	In	49	78	Rhodium	Rh	45	135
Barium	Ba	56	68	Iodine	I	53	118	Roentgenium	Rg	111	
Berkelium	Bk	97	151	Iridium	Ir	77	135	Rubidium	Rb	37	59
Beryllium	Be	4	68	Iron	Fe	26	131	Ruthenium	Ru	44	131
Bismuth	Bi	83	98	Krypton	Kr	36	128	Rutherfordium	Rf	104	94
Bohrium	Bh	107	124	Lanthanum	La	57	84	Samarium	Sm	62	142
Boron	B	5	78	Lawrencium	Lr	103	151	Scandium	Sc	21	84
Bromine	Br	35	118	Lead	Pb	82	88	Seaborgium	Sg	106	114
Cadmium	Cd	48	73	Lithium	Li	3	59	Selenium	Se	34	108
Calcium	Ca	20	68	Lutetium	Lu	71	142	Silicon	Si	14	88
Californium	Cf	98	151	Magnesium	Mg	12	68	Silver	Ag	47	65
Carbon	C	6	88	Manganese	Mn	25	124	Sodium	Na	11	59
Cerium	Ce	58	142	Meitnerium	Mt	109	135	Strontium	Sr	38	68
Cesium	Cs	55	59	Mendelevium	Md	101	151	Sulfur	S	16	108
Chlorine	Cl	17	118	Mercury	Hg	80	73	Tantalum	Ta	73	105
Chromium	Cr	24	114	Molybdenum	Mo	42	114	Technetium	Tc	43	124
Cobalt	Co	27	135	Neodymium	Nd	60	142	Tellurium	Te	52	108
Copper	Cu	29	65	Neon	Ne	10	128	Terbium	Tb	65	142
Curium	Cm	96	151	Neptunium	Np	93	151	Thallium	Tl	81	78
Darmstadtium	Ds	110	139	Nickel	Ni	28	139	Thorium	Th	90	151
Dubnium	Db	105	105	Niobium	Nb	41	105	Thulium	Tm	69	142
Dysprosium	Dy	66	142	Nitrogen	N	7	98	Tin	Sn	50	88
Einsteinium	Es	99	151	Nobelium	No	102	151	Titanium	Ti	22	94
Erbium	Er	68	142	Osmium	Os	76	131	Tungsten	W	74	114
Europium	Eu	63	142	Oxygen	O	8	108	Uranium	U	92	151

Table 2.1-1 The elements ordered by their names, cont.

Element	Symbol	Atomic Number	Page	Element	Symbol	Atomic Number	Page	Element	Symbol	Atomic Number	Page
Fermium	Fm	100	151	Palladium	Pd	46	139	Vanadium	V	23	105
Fluorine	F	9	118	Phosphorus	P	15	98	Xenon	Xe	54	128
Francium	Fr	87	59	Platinum	Pt	78	139	Ytterbium	Yb	70	142
Gadolinium	Gd	64	142	Plutonium	Pu	94	151	Yttrium	Y	39	84
Gallium	Ga	31	78	Polonium	Po	84	108	Zinc	Zn	30	73
Germanium	Ge	32	88	Potassium	K	19	59	Zirconium	Zr	40	94

[a] See Tungsten.

Table 2.1-2 The elements ordered by their chemical symbols

Element	Symbol	Atomic Number	Page	Element	Symbol	Atomic Number	Page	Element	Symbol	Atomic Number	Page
Actinium	Ac	89	84	Gadolinium	Gd	64	142	Polonium	Po	84	108
Silver	Ag	47	65	Germanium	Ge	32	88	Praseodymium	Pr	59	142
Aluminium	Al	13	78	Hydrogen	H	1	54	Platinum	Pt	78	139
Americium	Am	95	151	Helium	He	2	54	Plutonium	Pu	94	151
Argon	Ar	18	128	Mercury	Hg	80	73	Radium	Ra	88	68
Arsenic	As	33	98	Hafnium	Hf	72	94	Rubidium	Rb	37	59
Astatine	At	85	118	Holmium	Ho	67	142	Rhenium	Re	75	124
Gold	Au	79	65	Hassium	Hs	108	131	Rutherfordium	Rf	104	94
Boron	B	5	78	Iodine	I	53	118	Roentgenium	Rg	111	
Barium	Ba	56	68	Indium	In	49	78	Rhodium	Rh	45	135
Beryllium	Be	4	68	Iridium	Ir	77	135	Radon	Rn	86	128
Bohrium	Bh	107	124	Potassium	K	19	59	Ruthenium	Ru	44	131
Bismuth	Bi	83	98	Krypton	Kr	36	128	Sulfur	S	16	108
Berkelium	Bk	97	151	Lanthanum	La	57	84	Antimony	Sb	51	98
Bromine	Br	35	118	Lithium	Li	3	59	Scandium	Sc	21	84
Carbon	C	6	88	Lawrencium	Lr	103	151	Selenium	Se	34	108
Calcium	Ca	20	68	Lutetium	Lu	71	142	Seaborgium	Sg	106	114
Cadmium	Cd	48	73	Mendelevium	Md	101	151	Silicon	Si	14	88
Cerium	Ce	58	142	Magnesium	Mg	12	68	Samarium	Sm	62	142
Californium	Cf	98	151	Manganese	Mn	25	124	Tin	Sn	50	88
Chlorine	Cl	17	118	Molybdenum	Mo	42	114	Strontium	Sr	38	68
Curium	Cm	96	151	Meitnerium	Mt	109	135	Tantalum	Ta	73	105
Cobalt	Co	27	135	Nitrogen	N	7	98	Terbium	Tb	65	142
Chromium	Cr	24	114	Sodium	Na	11	59	Technetium	Tc	43	124
Cesium	Cs	55	59	Niobium	Nb	41	105	Tellurium	Te	52	108
Copper	Cu	29	65	Neodymium	Nd	60	142	Thorium	Th	90	151
Dubnium	Db	105	105	Neon	Ne	10	128	Titanium	Ti	22	94
Darmstadtium	Ds	110	139	Nickel	Ni	28	139	Thallium	Tl	81	78
Dysprosium	Dy	66	142	Nobelium	No	102	151	Thulium	Tm	69	142
Erbium	Er	68	142	Neptunium	Np	93	151	Uranium	U	92	151
Einsteinium	Es	99	151	Oxygen	O	8	108	Vanadium	V	23	105
Europium	Eu	63	142	Osmium	Os	76	131	Tungsten	W	74	114
Fluorine	F	9	118	Phosphorus	P	15	98	Xenon	Xe	54	128
Iron	Fe	26	131	Protactinium	Pa	91	151	Yttrium	Y	39	84
Fermium	Fm	100	151	Lead	Pb	82	88	Ytterbium	Yb	70	142
Francium	Fr	87	59	Palladium	Pd	46	139	Zinc	Zn	30	73
Gallium	Ga	31	78	Promethium	Pm	61	142	Zirconium	Zr	40	94

Table 2.1-3 The elements ordered by their atomic numbers

Element	Symbol	Atomic Number	Page	Element	Symbol	Atomic Number	Page	Element	Symbol	Atomic Number	Page
Hydrogen	H	1	54	Strontium	Sr	38	68	Rhenium	Re	75	124
Helium	He	2	54	Yttrium	Y	39	84	Osmium	Os	76	131
Lithium	Li	3	59	Zirconium	Zr	40	94	Iridium	Ir	77	135
Beryllium	Be	4	68	Niobium	Nb	41	105	Platinum	Pt	78	139
Boron	B	5	78	Molybdenum	Mo	42	114	Gold	Au	79	65
Carbon	C	6	88	Technetium	Tc	43	124	Mercury	Hg	80	73
Nitrogen	N	7	98	Ruthenium	Ru	44	131	Thallium	Tl	81	78
Oxygen	O	8	108	Rhodium	Rh	45	135	Lead	Pb	82	88
Fluorine	F	9	118	Palladium	Pd	46	139	Bismuth	Bi	83	98
Neon	Ne	10	128	Silver	Ag	47	65	Polonium	Po	84	108
Sodium	Na	11	59	Cadmium	Cd	48	73	Astatine	At	85	118
Magnesium	Mg	12	68	Indium	In	49	78	Radon	Rn	86	128
Aluminium	Al	13	78	Tin	Sn	50	88	Francium	Fr	87	59
Silicon	Si	14	88	Antimony	Sb	51	98	Radium	Ra	88	68
Phosphorus	P	15	98	Tellurium	Te	52	108	Actinium	Ac	89	84
Sulfur	S	16	108	Iodine	I	53	118	Thorium	Th	90	151
Chlorine	Cl	17	118	Xenon	Xe	54	128	Protactinium	Pa	91	151
Argon	Ar	18	128	Cesium	Cs	55	59	Uranium	U	92	151
Potassium	K	19	59	Barium	Ba	56	68	Neptunium	Np	93	151
Calcium	Ca	20	68	Lanthanum	La	57	84	Plutonium	Pu	94	151
Scandium	Sc	21	84	Cerium	Ce	58	142	Americium	Am	95	151
Titanium	Ti	22	94	Praseodymium	Pr	59	142	Curium	Cm	96	151
Vanadium	V	23	105	Neodymium	Nd	60	142	Berkelium	Bk	97	151
Chromium	Cr	24	114	Promethium	Pm	61	142	Californium	Cf	98	151
Manganese	Mn	25	124	Samarium	Sm	62	142	Einsteinium	Es	99	151
Iron	Fe	26	131	Europium	Eu	63	142	Fermium	Fm	100	151
Cobalt	Co	27	135	Gadolinium	Gd	64	142	Mendelevium	Md	101	151
Nickel	Ni	28	139	Terbium	Tb	65	142	Nobelium	No	102	151
Copper	Cu	29	65	Dysprosium	Dy	66	142	Lawrencium	Lr	103	151
Zinc	Zn	30	73	Holmium	Ho	67	142	Rutherfordium	Rf	104	94
Gallium	Ga	31	78	Erbium	Er	68	142	Dubnium	Db	105	105
Germanium	Ge	32	88	Thulium	Tm	69	142	Seaborgium	Sg	106	114
Arsenic	As	33	98	Ytterbium	Yb	70	142	Bohrium	Bh	107	124
Selenium	Se	34	108	Lutetium	Lu	71	142	Hassium	Hs	108	131
Bromine	Br	35	118	Hafnium	Hf	72	94	Meitnerium	Mt	109	135
Krypton	Kr	36	128	Tantalum	Ta	73	105	Darmstadtium	Ds	110	139
Rubidium	Rb	37	59	Tungsten	W	74	114	Roentgenium	Rg	111	

Table 2.1-4 The elements ordered according to the Periodic Table

						Page
Elements of the first period						54
1 Hydrogen	1 Deuterium	1 Tritium	2 Helium-4	2 Helium-3		
Elements of Group IA						59
3 Lithium	11 Sodium	19 Potassium	37 Rubidium	55 Cesium	87 Francium	
Elements of Group IB						65
29 Copper	47 Silver	79 Gold	111 Roentgenium			
Elements of Group IIA						68
4 Beryllium	12 Magnesium	20 Calcium	38 Strontium	56 Barium	88 Radium	
Elements of Group IIB						73
30 Zinc	48 Cadmium	80 Mercury				
Elements of Group IIIA						78
5 Boron	13 Aluminium	31 Gallium	49 Indium	81 Thallium		
Elements of Group IIIB						84
21 Scandium	39 Yttrium	57 Lanthanum	89 Actinium			
Elements of Group IVA						88
6 Carbon	14 Silicon	32 Germanium	50 Tin	82 Lead		
Elements of Group IVB						94
22 Titanium	40 Zirconium	72 Hafnium	104 Rutherfordium			
Elements of Group VA						98
7 Nitrogen	15 Phosphorus	33 Arsenic	51 Antimony	83 Bismuth		
Elements of Group VB						105
23 Vanadium	41 Niobium	73 Tantalum	105 Dubnium			
Elements of Group VIA						108
8 Oxygen	16 Sulfur	34 Selenium	52 Tellurium	84 Polonium		
Elements of Group VIB						114
24 Chromium	42 Molybdenum	74 Tungsten	106 Seaborgium			
Elements of Group VIIA						118
9 Fluorine	17 Chlorine	35 Bromine	53 Iodine	85 Astatine		
Elements of Group VIIB						124
25 Manganese	43 Technetium	75 Rhenium	107 Bohrium			
Elements of Group VIIIA						128
10 Neon	18 Argon	36 Krypton	54 Xenon	86 Radon		
Elements of Group VIII(1)						131
26 Iron	44 Ruthenium	76 Osmium	108 Hassium			
Elements of Group VIII(2)						135
27 Cobalt	45 Rhodium	77 Iridium	109 Meitnerium			
Elements of Group VIII(3)						139
28 Nickel	46 Palladium	78 Platinum	110 Darmstadtium			
Lanthanides						142
58 Cerium	59 Praseodymium	60 Neodymium	61 Promethium	62 Samarium	63 Europium	64 Gadolinium
65 Terbium	66 Dysprosium	67 Holmium	68 Erbium	69 Thulium	70 Ytterbium	71 Lutetium
Actinides						151
90 Thorium	91 Protactinium	92 Uranium	93 Neptunium	94 Plutonium	95 Americium	96 Curium
97 Berkelium	98 Californium	99 Einsteinium	100 Fermium	101 Mendelevium	102 Nobelium	103 Lawrencium

Table 2.1-5 Periodic Table of the elements

Periodic Table of the Elements

Legend: 18 / VIIIA (IUPAC Notation / CAS Notation); 2 / He (Atomic Number / Element Symbol); Unstable Nuclei.

Main Groups

1 IA	2 IIA	13 IIIA	14 IVA	15 VA	16 VIA	17 VIIA	18 VIIIA	Shells
1 H							2 He	K
3 Li	4 Be	5 B	6 C	7 N	8 O	9 F	10 Ne	K–L
11 Na	12 Mg	13 Al	14 Si	15 P	16 S	17 Cl	18 Ar	K–L–M
19 K	20 Ca	31 Ga	32 Ge	33 As	34 Se	35 Br	36 Kr	–L–M–N
37 Rb	38 Sr	49 In	50 Sn	51 Sb	52 Te	53 I	54 Xe	–M–N–O
55 Cs	56 Ba	81 Tl	82 Pb	83 Bi	84 Po	85 At	86 Rn	–N–O–P
87 Fr	88 Ra							–O–P–Q

Subgroups

3 IIIB	4 IVB	5 VB	6 VIB	7 VIIB	8 VIII (1)	9 VIII (2)	10 VIII (3)	11 IB	12 IIB	Shells
21 Sc	22 Ti	23 V	24 Cr	25 Mn	26 Fe	27 Co	28 Ni	29 Cu	30 Zn	–L–M–N
39 Y	40 Zr	41 Nb	42 Mo	43 Tc	44 Ru	45 Rh	46 Pd	47 Ag	48 Cd	–M–N–O
57 La	72 Hf	73 Ta	74 W	75 Re	76 Os	77 Ir	78 Pt	79 Au	80 Hg	–N–O–P
89 Ac	104 Rf	105 Db	106 Sg	107 Bh	108 Hs	109 Mt	110 Ds	111 Rg	112	–O–P–Q

Lanthanides (Shells –N–O–P)

58 Ce	59 Pr	60 Nd	61 Pm	62 Sm	63 Eu	64 Gd	65 Tb	66 Dy	67 Ho	68 Er	69 Tm	70 Yb	71 Lu

Actinides (Shells –O–P–Q)

90 Th	91 Pa	92 U	93 Np	94 Pu	95 Am	96 Cm	97 Bk	98 Cf	99 Es	100 Fm	101 Md	102 No	103 Lr

2.1.5 Data

2.1.5.1 Elements of the First Period

Table 2.1-6A Elements of the first period (hydrogen and helium). Atomic, ionic, and molecular properties

	Hydrogen			Helium			
Element name							
Special name	Hydrogen	Deuterium	Tritium	Helium 4	Helium 3		
Chemical symbol	H	^{2}H or D	^{3}H or T	^{4}He	^{3}He		
Atomic number Z	1	1	1	2	2		
						Units	**Remarks**
Characteristics			Radioactive				
Half-life			12.32			y	
Relative atomic mass A (atomic weight)	1.00794(7)	2.01408	3.01605	4.002602(2)			
Abundance in lithosphere	1400×10^{-6}	1.4×10^{-4}		4.2×10^{-7}			Mass ratio
Abundance in sea	110×10^{-3}						Mass ratio
Atomic radius r_{cov}	30					pm	Covalent radius
Atomic radius r_{met}				140		pm	Metallic radius, $CN^a = 12$
Atomic radius r_{vdW}	140			140		pm	van der Waals radius
Electron shells	K	K	K	K	K		
Electronic ground state	$^2S_{1/2}$	$^2S_{1/2}$	$^2S_{1/2}$	$^1S_0$	$^1S_0$		
Electronic configuration	$1s^1$	$1s^1$	$1s^1$	$1s^2$	$1s^2$		
Oxidation states	1−, 1+						
Electron affinity	0.754			0.19		eV	
Electronegativity χ_A	2.20			3.20			Allred and Rochow
1st ionization energy	13.59844			24.58741		eV	
2nd ionization energy				54.41778		eV	
Standard electrode potential E^0	0.00000					V	Reaction $2H^+ + 2e^- = H_2$
Molecular form in gaseous state	H_2						
Internuclear distance in molecule	74.166					pm	
Dissociation energy	4.475					eV	Extrapolated to $T = 0\,K$

^{a} Coordination number.

Table 2.1-6B(a) Elements of the first period (hydrogen and helium). Crystallographic properties (for allotropic and high-pressure modifications, see Table 2.1-6C)

Element name	Hydrogen			Helium		Units	Remarks
Special name	Hydrogen	Deuterium	Tritium	Helium 4	Helium 3		
Chemical symbol	H	^{2}H or D	^{3}H or T	^{4}He	^{3}He		
Atomic number Z	1	1	1	2	2		
State	H$_2$ at 4.2 K			At 1.5 K			
Crystal system, Bravais lattice	hex			hex			
Structure type	Mg			Mg			
Lattice constant a	377.1			357.7		pm	
Lattice constant c	615.6			584.2			
Space group	$P6_3/mmc$			$P6_3/mmc$			
Schoenflies symbol	D_{6h}^4			D_{6h}^4			
Strukturbericht type	A3			AQ3			
Pearson symbol	hP2			hP2			
Number A of atoms per cell	2			2			
Coordination number	12			12			
Shortest interatomic distance, solid				357		pm	

Table 2.1–6B(b) Elements of the fist period (hydrogen and helium). Mechanical properties

Element name	Hydrogen			Helium			
Special name	Hydrogen	Deuterium	Tritium	Helium 4	Helium 3		
Chemical symbol	H	^{2}H or D	^{3}H or T	^{4}He	^{3}He		
Atomic number Z	1	1	1	2	2	Units	Remarks
State	H_2	D_2					
Density ϱ, liquid		0.162	0.260			g/cm^3	Near T_m
Density ϱ, gas	0.0899×10^{-3}	0.0032^a	0.0031^a	0.1785×10^{-3}		g/cm^3	At 273 K^b
Molar volume V_{mol}, gas	13.26			32.07		cm^3/mol	
Viscosity η, gas	8.67					mPa s	At 293 K, 101 kPa
Sound velocity, gas	1237			969 (STP)		m/s	STP
Sound velocity, liquid	1340					m/s	At T_m, 12 MHz
Elastic modulus E	0.2			0.1		GPa	Solid state, estimated
Elastic compliance s_{11}	2930 (4.2 K)	1400 (4.2 K)		$+226 \times 10^3$ (1.6 K)	$+202 \times 10^3$ (0.4 K)	1/TPa	
Elastic compliance s_{33}	2000 (4.2 K)	995 (4.2 K)					
Elastic compliance s_{44}	9090 (4.2 K)	4350 (4.2 K)		$+46 \times 10^3$ (1.6 K)	$+108 \times 10^3$ (0.4 K)	1/TPa	
Elastic compliance s_{12}	-1240 (4.2 K)	-490 (4.2 K)		-107×10^3 (1.6 K)	-91.6×10^3 (0.4 K)	1/TPa	
Elastic compliance s_{13}	-166 (4.2 K)	-81 (4.2 K)					
Elastic stiffness c_{11}	0.042 (4.2 K)	0.82 (4.2 K)		$+31.1 \times 10^{-3}$ (1.6 K)	$+20.0 \times 10^{-3}$ (0.4 K)	GPa	
Elastic stiffness c_{33}	0.51 (4.2 K)	1.02 (4.2 K)					
Elastic stiffness c_{44}	0.11 (4.2 K)	0.23 (4.2 K)		$+21.7 \times 10^{-3}$ (1.6 K)	$+9.2 \times 10^{-3}$ (0.4 K)	GPa	
Elastic stiffness c_{12}	0.18 (4.2 K)	0.29 (4.2 K)		$+28.1 \times 10^{-3}$ (1.6 K)	$+16.4 \times 10^{-3}$ (0.4 K)	GPa	
Elastic stiffness c_{13}	0.05 (4.2 K)						
Solubility in water α_W	0.0178^b			0.0086			

^{a} At 25 K.
^{b} 101.3 kPa H_2 pressure.

Table 2.1–6B(c) Elements of the first period (hydrogen and helium). Thermal and thermodynamic properties

Element name	Hydrogen			Helium			
Special name	Hydrogen	Deuterium	Tritium	Helium 4	Helium 3		
Chemical symbol	H	^{2}H or D	^{3}H or T	^{4}He	^{3}He		
Atomic number Z	1	1	1	2	2		
						Units	Remarks
State	H$_2$ gas			He gas			
Thermal conductivity λ	0.171			143.0×10^{-3}		W/(m K)	At 300 K
Molar heat capacity c_p	14.418			20.786		J/(mol K)	At 298 K
Standard entropy S^0	130.680			126.152		J/(mol K)	At 298 K and 100 kPa
Enthalpy difference $H_{298} - H_0$	8.4684			6.1965		kJ/mol	At 298 K
Melting temperature T_m	13.81	18.65	20.65	0.95[a]		K	
Enthalpy change ΔH_m	0.12			0.021		kJ/mol	
Boiling temperature T_b	20.30	23.65	25.05	4.215	3.2	K	
Enthalpy change ΔH_b	0.46			0.082		kJ/mol	
Sublimation enthalpy	0.376			0.060		kJ/mol	At 0 K
Critical temperature T_c	33.2	38.55	39.95	5.23		K	
Critical pressure p_c	1.297	1.645		0.229		MPa	
Critical density ϱ_c	0.03102			69.3×10^{-3}		g/cm^3	
Triple-point temperature T_{tr}	14.0			[b]		K	
Triple-point pressure p_{tr}	7.2					kPa	
λ point				2.184	0.0027	K	

[a] At 2.6 MPa.
[b] He does not have a triple point.

Table 2.1-6B(d) Elements of the first period (hydrogen and helium). Electronic, electromagnetic, and optical properties

Element name	Hydrogen			Helium			
Special name	Hydrogen	Deuterium	Tritium	Helium 4	Helium 3		
Chemical symbol	H	^{2}H or D	^{3}H or T	^{4}He	^{3}He		
Atomic number Z	1	1	1	2	2		
						Units	Remarks
State	H$_2$ gas			He gas			
Characteristics	Light gas			Noble gas			
Dielectric constant ($\varepsilon - 1$), gas	$+264 \times 10^{-6}$			$+68 \times 10^{-6}$			At 273 K
Dielectric constant ε, liquid	1.225			1.048 (3.15 K)			At 20.30 K
Dielectric constant ε, solid							
Molar magnetic susceptibility χ_{mol}, gas (SI system)	-50.1×10^{-6}			-25.4×10^{-6}		cm^3/mol	At 295 K
Molar magnetic susceptibility χ_{mol}, gas (cgs system)	-3.99×10^{-6}			-2.02×10^{-6}		cm^3/mol	At 295 K
Mass magnetic susceptibility χ_{mass}, liquid (SI system)	-68.4×10^{-6}					cm^3/mol	At 20.3 K
Mass magnetic susceptibility χ_{mass}, liquid (cgs system)	-5.44×10^{-6}					cm^3/mol	At 20.3 K
Mass magnetic susceptibility χ_{mass}, gas (SI system)	-25×10^{-6}			-5.9×10^{-6}		cm^3/g	At 295 K
Refractive index ($n - 1$), gas	132×10^{-6}			36×10^{-6}			At 273.15 K, 101×10^5 Pa, $\lambda = 589.3$ nm
Refractive index n, liquid	1.112			1.026 (3.7 K)			$\lambda = 589.3$ nm

Table 2.1-6C Elements of the first period (hydrogen and helium). Allotropic and high-pressure modifications

Element	Hydrogen		Helium			
Modification	α-H (H$_2$)	β-H (H$_2$)	α-He	β-He	γ-He	Units
Crystal system, Bravais lattice	cub, fc	hex, cp	hex, cp	cub, fc	cub, bc	
Structure type	Cu	Mg	Mg	Cu	W	
Lattice constant a	533.4	377.1	357.7	4240	4110	pm
Lattice constant c		615.6	584.2			pm
Space group	$Fm\bar{3}m$	$P6_3/mmc$	$P6_3/mmc$	$Fm\bar{3}m$	$Im\bar{3}m$	
Schoenflies symbol	O_h^5	D_{6h}^4	D_{6h}^4	O_h^5	O_h^9	
Strukturbericht type	A1	A3	A3	A1	A2	
Pearson symbol	cF4	hP2	hP2	cF4	cI2	
Number A of atoms per cell	4	2	2	4	2	
Coordination number	12	12	12	12	8	
Shortest interatomic distance, solid			357	300	356	pm
Range of stability	< 1.25 K	< 13.81 K	< 0.95 K	1.6 K; 0.125 GPa	1.73 K; 0.03 GPa	

2.1.5.2 Elements of the Main Groups and Subgroup I to IV

Table 2.1-7A Elements of Group IA (CAS notation), or Group 1 (new IUPAC notation). Atomic, ionic, and molecular properties (see Table 2.1-7D for ionic radii)

Element name	Lithium	Sodium	Potassium	Rubidium	Cesium	Francium	Units	Remarks
Chemical symbol	Li	Na	K	Rb	Cs	Fr		
Atomic number Z	3	11	19	37	55	87		
Characteristics						Radioactive		
Relative atomic mass A (atomic weight)	6.941(2)	22.989770(2)	39.0983(1)	85.4678(3)	132.90545(2)	[223]		
Abundance in lithosphere	65×10^{-6}	$28\,300 \times 10^{-6}$	$25\,900 \times 10^{-6}$	280×10^{-6}	3.2×10^{-6}			Mass ratio
Abundance in sea	0.18×10^{-6}	$10\,770 \times 10^{-6}$	380×10^{-6}	0.12×10^{-6}	4×10^{-10}			Mass ratio
Atomic radius r_{cov}	123	157	203	216	253		pm	Covalent radius
Atomic radius r_{met}	156	192	238	250	272		pm	Metallic radius, CN = 12
Atomic radius r_{os}	158.6	171.3	216.2	228.7	251.8	244.7	pm	Outer-shell orbital radius
Atomic radius r_{vdW}	180	230	280	244	262		pm	van der Waals radius
Electron shells	KL	KLM	–LMN	–MNO	–NOP	–OPQ		
Electronic ground state	$^2S_{1/2}$	$^2S_{1/2}$	$^2S_{1/2}$	$^2S_{1/2}$	$^2S_{1/2}$	$^2S_{1/2}$		
Electronic configuration	[He]2s^1	[Ne]3s^1	[Ar]4s^1	[Kr]5s^1	[Xe]6s^1	[Rn]7s^1		
Oxidation states	1+	1+	1+	1+	1+	1+		
Electron affinity	0.618	0.548	0.501	0.486	0.472	0.46	eV	
Electronegativity χ_A	0.97	1.01	0.91	0.89	0.86	(0.86)		Allred and Rochow
1st ionization energy	5.39172	5.13908	4.34066	4.17713	3.89390	4.0727	eV	
2nd ionization energy	75.64018	47.2864	31.63	27.285	23.15745		eV	
3rd ionization energy	122.45429	71.6200	45.806	40			eV	
4th ionization energy		98.91	60.91	52.6			eV	
Standard electrode potential E^0	–3.040	–2.71	–2.931	–2.98	–2.92		V	Reaction type $Li^+ + e^- = Li$

Table 2.1–7B(a) Elements of Group IA (CAS notation), or Group 1 (new IUPAC notation). Crystallographic properties (see Table 2.1-7C for allotropic and high-pressure modifications)

Element name	Lithium	Sodium	Potassium	Rubidium	Cesium	Francium		
Chemical symbol	Li	Na	K	Rb	Cs	Fr		
Atomic number Z	3	11	19	37	55	87		
							Units	Remarks
Modification	β-Li	β-Na			Cs-I			
Crystal system, Bravais lattice	cub, bc	cub, bc	cub, bc	cub, bc	cub, bc			
Structure type	W	W	W	W	W			
Lattice constant a	350.93	420.96	532.1	570.3	614.1		pm	
Space group	$Im\bar{3}m$	$Im\bar{3}m$	$Im\bar{3}m$	$Im\bar{3}m$	$Im\bar{3}m$			
Schoenflies symbol	O_h^9	O_h^9	O_h^9	O_h^9	O_h^9			
Strukturbericht type	A2	A2	A2	A2	A2			
Pearson symbol	cI2	cI2	cI2	cI2	cI2			
Number A of atoms per cell	2	2	2	2	2			
Coordination number	8	8	8	8	8			
Shortest interatomic distance, solid	303	371	462	487	524		pm	
Shortest interatomic distance, liquid				497 (313 K)			pm	

Table 2.1-7B(b) Elements of Group IA (CAS notation), or Group 1 (new IUPAC notation). Mechanical properties

Element name	Lithium	Sodium	Potassium	Rubidium	Cesium	Francium	Units	Remarks
Chemical symbol	Li	Na	K	Rb	Cs	Fr		
Atomic number Z	3	11	19	37	55	87		
Density ϱ, solid	0.532	0.970	0.862	1.532	1.87	2.410	g/cm^3	
Density ϱ, liquid	0.508 (453 K)	0.927	0.827	1.470	1.847		g/cm^3	Near T_m
Molar volume V_{mol}	13.00	23.68	45.36	55.79	70.96	9.25	cm^3/mol	
Viscosity η, liquid	0.566 (473 K)	0.565 (416 K)	0.64	0.52			mPa s	
Surface tension, liquid	0.396	0.193	0.116	0.092	0.060		N/m	
Temperature coefficient		-0.05×10^{-3}	-0.06×10^{-3}		-0.046×10^{-3}		N/(m K)	
Coefficient of linear thermal expansion α	56×10^{-6}	70.6×10^{-6}	83×10^{-6}	90×10^{-6}	97×10^{-6}		1/K	At 298 K
Sound velocity, liquid		2395 (371 K)					m/s	At T_m
Sound velocity, solid, transverse	2820	1620	1230	1260	590		m/s	At 12 MHz
Sound velocity, solid, longitudinal	6030	3310	2600	1430	1090		m/s	
Compressibility κ	8.93×10^{-5}	13.4×10^{-5}	23.7×10^{-5}	33.0×10^{-5}	0.75×10^{-5}		1/MPa	Volume compressibility
Elastic modulus E	11.5	6.80	3.52	2.35	1.69		GPa	
Shear modulus G	4.24	2.94	1.28	0.91	0.65		GPa	
Poisson number μ	0.36	0.34	0.35	0.29	0.30			
Elastic compliance s_{11}	315	549	1339	1330	1190 (280 K)		1/TPa	
Elastic compliance s_{44}	104	233	526	625	690 (280 K)		1/TPa	
Elastic compliance s_{12}	-144	-250	-620	-600	-450 (280 K)		1/TPa	
Elastic stiffness c_{11}	13.4	7.59	3.69	2.96	1.60 (280 K)		GPa	
Elastic stiffness c_{44}	9.6	4.30	1.90	1.60	1.44 (280 K)		GPa	
Elastic stiffness c_{12}	11.3	6.33	3.18	2.44	0.99 (280 K)		GPa	
Tensile strength	0.6						MPa	
Vickers hardness				0.37	0.15			
Mohs hardness	0.6							

Table 2.1-7B(c) Elements of Group IA (CAS notation), or Group 1 (new IUPAC notation). Thermal and thermodynamic properties

Element name	Lithium	Sodium	Potassium	Rubidium	Cesium	Francium	Units	Remarks
Chemical symbol	Li	Na	K	Rb	Cs	Fr		
Atomic number Z	3	11	19	37	55	87		
Thermal conductivity λ	84.7	141	102.4	58.2	35.9		W/(m K)	At 300 K
Molar heat capacity c_p	24.77	28.24	29.58	31.062	32.18		J/(mol K)	At 298 K
Standard entropy S^0	29.120	51.300	64.680	76.776	85.230	101.00	J/(mol K)	At 298 K and 100 kPa
Enthalpy difference $H_{298} - H_0$	4.6320	6.4600	7.0880	7.4890	7.7110	10.000	kJ/mol	
Melting temperature T_m	453.69	370.87	336.86	312.47	301.59	300	K	
Enthalpy change ΔH_m	3.0000	2.5970	2.3208	2.1924	2.0960		kJ/mol	
Entropy change ΔS_m	6.612	7.002	6.890	7.016	6.950		J/(mol K)	
Relative volume change ΔV_m	0.0151	0.027	0.0291	0.0228	0.0263			$(V_l - V_s)/V_l$ at T_m
Boiling temperature T_b	1620	1156	1040	970	947	950 (estimated)	K	
Enthalpy change ΔH_b	147.7	99.2	79.1	75.7	67.7		kJ/mol	
Critical temperature T_c		2500	2280		2010		K	
Critical pressure p_c		25.3	16.1		11.3		MPa	
Critical density ϱ_c		0.210	0.190		0.410		g/cm^3	

Table 2.1-7B(d) Elements of Group IA (CAS notation), or Group 1 (new IUPAC notation). Electronic, electromagnetic, and optical properties

Element name Chemical symbol Atomic number Z	Lithium Li 3	Sodium Na 11	Potassium K 19	Rubidium Rb 37	Cesium Cs 55	Francium Fr 87	Units	Remarks
Characteristics	Very reactive metal	Reactive metal	Soft, reactive metal	Soft, reactive metal	Alkali metal	Alkali metal		
Electrical resistivity ρ_s	85.5	42	61	116	188		nΩ m	Solid, at 293 K
Temperature coefficient	48.9×10^{-4}	54.6×10^{-4}	67.3×10^{-4}	63.7×10^{-4}	50.3×10^{-4}		1/K	Solid
Pressure coefficient	-2.1×10^{-9}	-38.3×10^{-9}	-69.7×10^{-9}	-62.9×10^{-9}	0.5×10^{-9}		1/hPa	Solid
Electrical resistivity ρ_l	240 at T_m		129.7	220	367		nΩ m	Liquid
Resistivity ratio at T_m	1.68	1.44	1.56	1.612	1.66			ρ_l/ρ_s at T_m
Hall coefficient R^a	-1.70×10^{-10}	-2.1×10^{-10}	-4.2×10^{-10}	-5.92×10^{-10}	-7.8×10^{-10}		m^3/(A s)	At 300 K
Thermoelectric coefficient	14.37		12	-8.26	0.2		μV/K	
Electronic work function	2.28	2.75	2.30	2.05	1.94		V	
Thermal work function	2.39		2.15	2.13	1.87		V	
Molar magnetic susceptibility χ_{mol}, solid (SI)	178×10^{-6}	201×10^{-6}	261×10^{-6}	214×10^{-6}	364×10^{-6}		cm^3/mol	At 295 K
Molar magnetic susceptibility χ_{mol}, solid (cgs)	14.2×10^{-6}	16×10^{-6}	20.8×10^{-6}	17×10^{-6}	29×10^{-6}		cm^3/mol	At 295 K
Mass magnetic susceptibility χ_{mass}, solid (SI)	25.6×10^{-6}	8.8×10^{-6}	6.7×10^{-6}	2.49×10^{-6}	2.8×10^{-6}		cm^3/g	At 295 K
Mass magnetic susceptibility χ_{mass}, liquid (SI)					2.6×10^{-6}		cm^3/g	
Refractive index n, solid		4.22	0.024 (134 nm)					$\lambda = 589.3$ nm
Refractive index n, liquid		0.0045						$\lambda = 589.3$ nm

^{a} $B = 1-3$ T

Table 2.1-7C Elements of Group IA (CAS notation), or Group 1 (new IUPAC notation). Allotropic and high-pressure modifications

Element	Lithium			Sodium		Cesium			Units
Modification	α-Li	β-Li	γ-Li	α-Na	β-Na	Cs-I	Cs-II	Cs-III	
Crystal system, Bravais lattice	hex, cp	cub, bc	cub, fc	hex, cp	cub, bc	cub, bc	cub, fc	cub, fc	
Structure type	Mg	W	Cu	Mg	W	W	Cu	Cu	
Lattice constant a	311.1	309.3		376.7	420.96	614.1	598.4	580.0	pm
Lattice constant c	509.3			615.4					
Space group	$P6_3/mmc$	$Im\bar{3}m$	$Fm\bar{3}m$	$P6_3/mmc$	$Im\bar{3}m$	$Im\bar{3}m$	$Fm\bar{3}m$	$Fm\bar{3}m$	
Schoenflies symbol	D_{6h}^4	O_h^9	O_h^5	D_{6h}^4	O_h^9	O_h^9	O_h^5	O_h^5	
Strukturbericht type	A3	A2	A1	A3	A2	A2	A1	A1	
Pearson symbol	hP2	cI2	cF4	hP2	cI2	cI2	cF4	cF4	
Number A of atoms per cell	2	2	4	2	2	2	4	4	
Coordination number	6+6	8	12	12	8	8	12	12	
Shortest interatomic distance, solid	310	303	310	377	371	524	457	410	pm
Range of stability	< 72 K	RT	< 36 K	RT	RTP	> 2.37 GPa	> 4.22 GPa		

Table 2.1-7D Elements of Group IA (CAS notation), or Group 1 (new IUPAC notation). Ionic radii (determined from crystal structures)

Element	Lithium	Sodium	Potassium	Rubidium	Cesium	Francium	Units
Ion	Li$^+$	Na$^+$	K$^+$	Rb$^+$	Cs$^+$	Fr$^+$	
Coordination number							
4	59	99	137				pm
6	76	102	138	152	167	180	pm
8	92	118	151	161	174		pm
9		124					pm
10				166	181		pm
12		139	164	172	188		pm

Table 2.1-8A Elements of Group IB (CAS notation), or Group 11 (new IUPAC notation). Atomic, ionic, and molecular properties (see Table 2.1-8D for ionic radii)

Element name	Copper	Silver	Gold		
Chemical symbol	Cu	Ag	Au		
Atomic number Z	29	47	79		
				Units	Remarks
Relative atomic mass A_r (atomic weight)	63.546(3)	107.8682(2)	196.96655(2)		
Abundance in lithosphere	70×10^{-6}	2.0×10^{-6}	3×10^{-9}		Mass ratio
Abundance in sea	5×10^{-10}	4×10^{-11}	4×10^{-12}		Mass ratio
Atomic radius r_{cov}	117	134	134	pm	Covalent radius
Atomic radius r_{met}	128	144	144	pm	Metallic radius, CN = 12
Atomic radius r_{vdW}	140	170	170	pm	van der Waals radius
Electron shells	–LMN	–MNO	–NOP		
Electronic ground state	$^2S_{1/2}$	$^2S_{1/2}$	$^2S_{1/2}$		
Electronic configuration	$[Ar]3d^{10}4s^1$	$[Kr]4d^{10}5s^1$	$[Xe]4f^{14}5d^{10}6s^1$		
Oxidation states	2+, 1+	1+	3+, 1+		
Electron affinity	1.24	1.30	2.309	eV	
Electronegativity χ_A	1.75	1.42	(1.42)		Allred and Rochow
1st ionization energy	7.72638	7.57624	9.22567	eV	
2nd ionization energy	20.29240	21.49	20.5	eV	
3rd ionization energy	36.841	34.83		eV	
4th ionization energy	57.38			eV	
Standard electrode potential E^0	+0.521	+0.7996	+1.692	V	Reaction type $Cu^+ + e^- = Cu$
	+0.342			V	Reaction type $Cu^{2+} + 2e^- = Cu$
	+0.153			V	Reaction type $Cu^{2+} + e^- = Cu^+$
			+1.498	V	Reaction type $Au^{3+} + 3e^- = Au$
			+1.401	V	Reaction type $Au^{3+} + 2e^- = Au^+$

Table 2.1-8B(a) Elements of Group IB (CAS notation), or Group 11 (new IUPAC notation). Crystallographic properties

Element name	Copper	Silver	Gold		
Chemical symbol	Cu	Ag	Au		
Atomic number Z	29	47	79		
				Units	Remarks
Crystal system, Bravais lattice	cub, fc	cub, fc	cub, fc		
Structure type	Cu	Cu	Cu		
Lattice constant a	361.49	408.61	407.84	pm	At 298 K
Space group	$Fm\bar{3}m$	$Fm\bar{3}m$	$Fm\bar{3}m$		
Schoenflies symbol	O_h^5	O_h^5	O_h^5		
Strukturbericht type	A1	A1	A1		
Pearson symbol	cF4	cF4	cF4		
Number A of atoms per cell	4	4	4		
Coordination number	12	12	12		
Shortest interatomic distance, solid	255.6	288	288	pm	At 293 K
Shortest interatomic distance, liquid	257 (1363 K)			pm	

Table 2.1-8B(b) Elements of Group IB (CAS notation), or Group 11 (new IUPAC notation). Mechanical properties

Element name	Copper	Silver	Gold		
Chemical symbol	Cu	Ag	Au		
Atomic number Z	29	47	79		
				Units	Remarks
Density ϱ, solid	8.960	10.50	19.30	g/cm^3	
Density ϱ, liquid	8.000	9.345	17.300	g/cm^3	
Molar volume V_{mol}	7.09	10.27	10.19	cm^3/mol	
Viscosity η, liquid	3.36	3.62	5.38	mPa s	At T_m
Surface tension, liquid	1.300	0.923	1.128	N/m	
Temperature coefficient	-0.18×10^{-3}	-0.13×10^{-3}	-0.10×10^{-3}	N/(m K)	
Coefficient of linear thermal expansion α	16.5×10^{-6}	19.2×10^{-6}	14.16×10^{-6}	1/K	At 298 K
Sound velocity, solid, transverse	2300	1690	1190	m/s	
Sound velocity, solid, longitudinal	4760	3640	3280	m/s	
Compressibility κ	0.702×10^{-5}	0.95×10^{-5}	0.563×10^{-5}	1/MPa	Volume compressibility
Elastic modulus E	128	80.0	78.0	GPa	
Shear modulus G	46.8	29.5	26.0	GPa	
Poisson number μ	0.34	0.38	0.42		
Elastic compliance s_{11}	15.0	23.0	23.4	1/TPa	
Elastic compliance s_{44}	13.3	22.0	23.8	1/TPa	
Elastic compliance s_{12}	-6.3	-9.8	-10.8	1/TPa	
Elastic stiffness c_{11}	169	122	191	GPa	
Elastic stiffness c_{44}	75.3	45.5	42.2	GPa	
Elastic stiffness c_{12}	122	92	162	GPa	
Tensile strength	209	125	110	MPa	At 293 K
Vickers hardness	369	251	216		At 293 K

Table 2.1-8B(c) Elements of Group IB (CAS notation), or Group 11 (new IUPAC notation). Thermal and thermodynamic properties

Element name	Copper	Silver	Gold		
Chemical symbol	Cu	Ag	Au		
Atomic number Z	29	47	79		
				Units	Remarks
Thermal conductivity λ	1401	429	317	W/(m K)	At 300 K
Molar heat capacity c_p	24.443	25.36	25.42	J/(mol K)	At 298 K
Standard entropy S^0	33.150	42.551	47.488	J/(mol K)	At 298 K and 100 kPa
Enthalpy difference $H_{298} - H_0$	5.0040	5.7446	6.0166	kJ/mol	
Melting temperature T_m	1357.77	1234.93	1337.33	K	
Enthalpy change ΔH_m	13.263	11.297	12.552	kJ/mol	
Entropy change ΔS_m	9.768	9.148	9.386	J/(mol K)	
Relative volume change	$+0.0415$	0.038	0.051		$(V_l - V_s)/V_l$ at T_m
Boiling temperature T_b	2843	2436	3130	K	
Enthalpy change ΔH_b	300.7	250.6	334.4	kJ/mol	

Table 2.1-8B(d) Elements of Group IB (CAS notation), or Group 11 (new IUPAC notation). Electronic, electromagnetic, and optical properties. There is no Table 2.1-8C, because no allotropic or high-pressure modifications are known

Element name	Copper	Silver	Gold		
Chemical symbol	Cu	Ag	Au		
Atomic number Z	29	47	79		
				Units	Remarks
Characteristics	Metal	Noble metal	Noble metal		
Electrical resistivity ρ_s	16.78	147	20.5	nΩ m	At 293 K
Temperature coefficient	43.8×10^{-4}	43×10^{-4}	40.2×10^{-4}	1/K	
Pressure coefficient	-1.86×10^{-9}	-3.38×10^{-9}	-2.93×10^{-9}	1/hPa	
Electrical resistivity ρ_l	21.5		312	nΩ m	
Resistivity ratio	2.07		2.28		ρ_l/ρ_s at T_m
Hall coefficient R	-0.536×10^{-10}	-0.84×10^{-10}	-0.704×10^{-10}	m^3/(A s)	B = 0.3–2.2 T, 298 K
Thermoelectric coefficient	1.72	1.42	1.72	μV/K	
Electronic work function	4.65	4.26	5.1	V	
Thermal work function	4.39	4.31	4.25	V	
Molar magnetic susceptibility χ_{mol}, solid (SI)	-68.6×10^{-6}	-245×10^{-6}	-352×10^{-6}	cm^3/mol	At 295 K
Molar magnetic susceptibility χ_{mol}, solid (cgs)	-5.46×10^{-6}	-19.5×10^{-6}	-28×10^{-6}	cm^3/mol	At 295 K
Mass magnetic susceptibility χ_{mass}, solid (SI)	-1.081×10^{-6}	-2.27×10^{-6}	-1.78×10^{-6}	cm^3/g	At 295 K
Mass magnetic susceptibility χ_{mass}, liquid (SI)	-1.2×10^{-6}	-2.83×10^{-6}	-2.16×10^{-6}	cm^3/g	

Table 2.1-8D Elements of Group IB (CAS notation), or Group 11 (new IUPAC notation). Ionic radii (determined from crystal structures)

Element	Copper		Silver		Gold		
Ion	Cu$^+$	Cu^{2+}	Ag$^+$	Ag^{2+}	Au$^+$	Au^{3+}	
Coordination number							Units
2	46						pm
4	60	57	100	79		64	pm
6	77	73	115	94	137	85	pm
8			128				pm

Table 2.1-9A Elements of Group IIA (CAS notation), or Group 2 (new IUPAC notation). Atomic, ionic, and molecular properties (see Table 2.1-9D for ionic radii)

Element name	Beryllium	Magnesium	Calcium	Strontium	Barium	Radium	Units	Remarks
Chemical symbol	Be	Mg	Ca	Sr	Ba	Ra		
Atomic number Z	4	12	20	38	56	88		
Characteristics								
Relative atomic mass A (atomic weight)	9.012182(3)	24.3050(6)	40.078(4)	87.62(1)	137.327(7)	[226]		Radioactive
Abundance in lithosphere	6×10^{-6}	20900×10^{-6}	26300×10^{-6}	150×10^{-6}	430×10^{-6}			Mass ratio
Abundance in sea	6×10^{-13}	1290×10^{-6}	412×10^{-6}	8.0×10^{-6}	2×10^{-9}			Mass ratio
Atomic radius r_{cov}	89	136	174	191	198		pm	Covalent radius
Atomic radius r_{met}	112	160	197	215	224	230	pm	Metallic radius, CN = 12
Atomic radius r_{vdW}		170					pm	van der Waals radius
Electron shells	KL	KLM	–LMN	–MNO	–NOP	–OPQ		
Electronic ground state	$^1S_0$	$^1S_0$	$^1S_0$	$^1S_0$	$^1S_0$	$^1S_0$		
Electronic configuration	[He]2s²	[Ne]2s²	[Ar]4s²	[Kr]5s²	[Xe]6s²	[Rn]7s²		
Oxidation states	2+	2+	2+	2+	2+	2+		
Electron affinity	Not stable	Not stable	0.0246	0.048	0.15		eV	
Electronegativity χ_A	1.47	1.23	1.04	0.99	0.97	(0.97)		Allred and Rochow
1st ionization energy	9.32263	7.64624	6.11316	5.69484	5.21170	5.27892	eV	
2nd ionization energy	18.21116	15.03528	11.87172	11.03013	10.00390	10.14716	eV	
3rd ionization energy	153.89661	80.1437	50.9131	42.89			eV	
4th ionization energy	217.71865	109.2655	67.27	57			eV	
Standard electrode potential E^0	–1.847	–2.372	–2.868	–2.89	–2.90		V	Reaction type $Be^{++} + 2e^- = Be$

Table 2.1–9B(a) Elements of Group IIA (CAS notation), or Group 2 (new IUPAC notation). Crystallographic properties (see Table 2.1-9C for allotropic and high-pressure modifications)

Element name	Beryllium	Magnesium	Calcium	Strontium	Barium	Radium	Units	Remarks
Chemical symbol	**Be**	**Mg**	**Ca**	**Sr**	**Ba**	**Ra**		
Atomic number Z	**4**	**12**	**20**	**38**	**56**	**88**		
Modification	α-Be		α-Ca	α-Sr				
Crystal system, Bravais lattice	hex	hex	cub, fc	cub, fc	cub, bc	cub, bc		
Structure type	Mg	Mg	Cu	Cu	W	W		
Lattice constant a	228.57	320.93	558.84	608.4	502.3	514.8	pm	
Lattice constant c	358.39	521.07					pm	
Space group	$P6_3/mmc$	$P6_3/mmc$	$Fm\bar{3}m$	$Fm\bar{3}m$	$Im\bar{3}m$	$Im\bar{3}m$		
Schoenflies symbol	D_{6h}^4	D_{6h}^4	O_h^5	O_h^5	O_h^9	O_h^9		
Strukturbericht type	A3	A3	A1	A1	A2	A2		
Pearson symbol	hP2	hP2	cF4	cF4	cI2	cI2		
Number A of atoms per cell	2	2	4	4	2	2		
Coordination number	12	6+6	12	12	8	8		
Shortest interatomic distance, solid	222	319	393	430	434	446	pm	

Table 2.1–9B(b) Elements of Group IIA (CAS notation), or Group 2 (new IUPAC notation). Mechanical properties

Element name Chemical symbol Atomic number Z	Beryllium Be 4	Magnesium Mg 12	Calcium Ca 20	Strontium Sr 38	Barium Ba 56	Radium Ra 88	Units	Remarks
Density ϱ, solid	1.85	1.74	1.55	2.60	3.50	5.000	g/cm^3	
Density ϱ, liquid	1.420 (1770 K)	1.590	1.365	2.375	3.325		g/cm^3	
Molar volume V_{mol}	4.88	13.98	25.86	34.50	38.21	45.2	cm^3/mol	
Viscosity η, liquid		1.23	1.06				mPa s	
Surface tension γ, liquid Temperature coefficient	1.1 (1770 K)	0.563 (954 K)	0.361 -0.068×10^3	0.303	0.276 -0.095	0.45	N/m N/(m K)	
Coefficient of linear thermal expansion α	11.5×10^{-6}	26.1×10^{-6}	22×10^{-6}	23×10^{-6}	20.7×10^{-6}	20.2×10^{-6}	1/K	
Sound velocity, solid, transverse	8330	3170	2210	1520	1160		m/s	
Sound velocity, solid, longitudinal	12 720	5700	4180	2780	2080		m/s	
Compressibility κ	0.765×10^{-5}	2.88×10^{-5}	5.73×10^{-5} [a]	7.97×10^{-5}	10.0×10^{-5}		1/MPa	Volume compressibility
Elastic modulus E	286	44.4	19.6 [a]	15.7	12.8	13.2 [b]	GPa	
Shear modulus G	133	16.9	7.85 [a]	6.03	4.84		GPa	
Poisson number μ	0.12	0.28	0.31 [a]	0.28	0.28			
Elastic compliance s_{11}	3.45	22.0	104	218	157		1/TPa	
Elastic compliance s_{33}	2.87	19.7					1/TPa	
Elastic compliance s_{44}	6.16	60.9	71	135	105		1/TPa	
Elastic compliance s_{12}	-0.28	-7.8	-42	-90	-61		1/TPa	
Elastic compliance s_{13}	-0.05	-5.0					1/TPa	
Elastic stiffness c_{11}	292	59.3	22.8	10.94	12.6		GPa	
Elastic stiffness c_{33}	349	61.5					GPa	
Elastic stiffness c_{44}	163	16.4	14	7.41	9.5		GPa	
Elastic stiffness c_{12}	24	25.7	16.0	7.69	8.0		GPa	
Elastic stiffness c_{13}	6	21.4					GPa	
Tensile strength	228–352	90					GPa	
Vickers hardness	1670		130	140	42			
Brinell hardness	1060–1300	300–500						At 293 K

[a] At 348 K.
[b] Estimated.

Table 2.1-9B(c) Elements of Group IIA (CAS notation), or Group 2 (new IUPAC notation). Thermal and thermodynamic properties

Element name Chemical symbol Atomic number Z	Beryllium Be 4	Magnesium Mg 12	Calcium Ca 20	Strontium Sr 38	Barium Ba 56	Radium Ra 88	Units	Remarks
Thermal conductivity λ	200	171	200	35.3	18.4	18.6	W/(m K)	At 298 K
Molar heat capacity c_p	16.44	24.895	25.940	26.4	28.09		J/(mol K)	At 298 K
Standard entropy S^0	9.500	32.671	41.588	55.694	62.500	69.000	J/(mol K)	At 298 K and 100 kPa
Enthalpy difference $H_{298} - H_0$	1.9500	4.9957	5.7360	6.5680	6.9100	7.2000	kJ/mol	
Melting temperature T_m	1560.0	923.00	1115.0	1050.0	1000.0	969.00	K	
Enthalpy change ΔH_m	7.8950	8.4768	8.5395	7.4310	7.1190	7.7000	kJ/mol	
Entropy change ΔS_m	5.061	9.184	7.659	7.077	7.119	7.946	J/(mol K)	
Relative volume change ΔV_m		0.042	0.047					$(V_l - V_s)/V_l$ at T_m
Boiling temperature T_b	2741	1366	1773	1685	2118		K	
Enthalpy change ΔH_b	291.58	127.4	153.6	137.19	141.5	136.7	kJ/mol	

Table 2.1-9B(d) Elements of Group IIA (CAS notation), or Group 2 (new IUPAC notation). Electronic, electromagnetic, and optical properties

Element name Chemical symbol Atomic number Z	Beryllium Be 4	Magnesium Mg 12	Calcium Ca 20	Strontium Sr 38	Barium Ba 56	Radium Ra 88	Units	Remarks
Characteristics	Light metal	Light metal	Metal	Soft metal	Soft metal	Metal		
Electrical resistivity ρ_s	28	39.4	31.6	303	500		nΩ m	At RT
Temperature coefficient	90.0×10^{-4}	41.2×10^{-4}	41.7×10^{-4}	38.2×10^{-4}	64.9×10^{-4}		1/K	
Pressure coefficient	-1.6×10^{-9}	-4.7×10^{-9}	15.2×10^{-9}	5.56×10^{-9}	-3.0×10^{-9}		1/hPa	
Electrical resistivity ρ_l	27.4	1.78					nΩ m	
Resistivity ratio								ρ_l/ρ_s at T_m
Hall coefficient R	2.4×10^{-10}	-0.83×10^{-10}	-8.2				m^3/(A s)	At 300 K, $B = 1$ T
Thermoelectric coefficient		-0.4					μV/K	
Electronic work function	4.98	3.97	2.87	2.74	2.56		V	
Thermal work function	3.37	3.46	2.76	2.35	2.29		V	
Molar magnetic susceptibility χ_{mol}, solid (SI)	-113×10^{-6}	165×10^{-6}	503×10^{-6}	1.16×10^{-3}	259×10^{-6}		cm^3/mol	At 295 K
Molar magnetic susceptibility χ_{mol}, solid (cgs)	-9.0×10^{-6}	13.1×10^{-6}	40.0×10^{-6}	92.0×10^{-6}	20.6×10^{-6}		cm^3/mol	At 295 K
Mass magnetic susceptibility χ_{mass}, solid (SI)	-13×10^{-6}	6.8×10^{-6}	14×10^{-6}	13.2×10^{-6}	1.9×10^{-6}		cm^3/g	At 295 K
Refractive index n, solid		0.37						$\lambda = 589$ nm

Table 2.1-9C Elements of Group IIA (CAS notation), or Group 2 (new IUPAC notation). Allotropic and high-pressure modifications

Element	Beryllium		Calcium		Strontium			Units
Modification	α-Be	β-Be	α-Ca	γ-Ca	α-Sr	β-Sr	γ-Sr	Sr-II	
Crystal system, Bravais lattice	hex, cp	cub, bc	cub, fc	cub, bc	cub, fc	hex, cp	cub, bc	cub, bc	
Structure type	Mg	W	Cu	W	Cu	Mg	W	W	
Lattice constant a	228.57	255.15	558.84	448.0	608.4	428	487	443.7	pm
Lattice constant c	358.39					705			pm
Space group	$P6_3/mmc$	$Im\bar{3}m$	$Fm\bar{3}m$	$Im\bar{3}m$	$Fm\bar{3}m$	$P6_3/mmc$	$Im\bar{3}m$	$Im\bar{3}m$	
Schoenflies symbol	D_{6h}^4	O_h^9	O_h^5	O_h^9	O_h^5	D_{6h}^4	O_h^9	O_h^9	
Strukturbericht type	A3	A2	A1	A2	A1	A3	A2	A2	
Pearson symbol	hP2	cI2	cF4	cI2	cF4	hP2	cI2	cI2	
Number A of atoms per cell	2	2	4	2	4	2	2	2	
Coordination number	12	8	12	8	12	12	8	8	
Shortest interatomic distance, solid	222	221	393	388	430		422		pm
Range of stability	RT	> 1523 K	RT	> 1010 K	RT	> 486 K	> 878 K	> 3.5 GPa	

Table 2.1-9D Elements of Group IIA (CAS notation), or Group 2 (new IUPAC notation). Ionic radii (determined from crystal structures)

Element	Beryllium	Magnesium	Calcium	Strontium	Barium	Radium	
Ion	Be^{2+}	Mg^{2+}	Ca^{2+}	Sr^{2+}	Ba^{2+}	Ra^{2+}	Units
Coordination number							
4	27	57					pm
6	45	72	100	118	135		pm
8		89	112	126	142	148	pm
10			123	136			pm
12			134	144	161	170	pm

Table 2.1-10A Elements of Group IIB (CAS notation), or Group 12 (new IUPAC notation). Atomic, ionic, and molecular properties (see Table 2.1-10D for ionic radii)

Element name	Zinc	Cadmium	Mercury		
Chemical symbol	Zn	Cd	Hg		
Atomic number Z	30	48	80		
				Units	Remarks
Relative atomic mass A (atomic weight)	65.39(2)	112.411(8)	200.59(2)		
Abundance in lithosphere	80×10^{-6}	0.18×10^{-6}	0.5×10^{-6}		Mass ratio
Abundance in sea	4.9×10^{-9}	1×10^{-10}	3×10^{-11}		Mass ratio
Atomic radius r_{cov}	125	141	144	pm	Covalent radius
Atomic radius r_{met}	134	149	162	pm	Metallic radius, CN = 12
Atomic radius r_{vdW}	140	160	150	pm	van der Waals radius
Electron shells	–LMN	–MNO	–NOP		
Electronic ground state	$^1S_0$	$^1S_0$	$^1S_0$		
Electronic configuration	$[Ar]3d^{10}4s^2$	$[Kr]4d^{10}5s^2$	$[Xe]4f^{14}5d^{10}6s^2$		
Oxidation states	2+	2+	2+, 1+		
Electron affinity	Not stable	Not stable	Not stable	eV	
Electronegativity χ_A	1.66	1.46	(1.44)		Allred and Rochow
1st ionization energy	9.39405	8.99367	10.43750	eV	
2nd ionization energy	17.96440	16.90832	18.756	eV	
3rd ionization energ	39.723	37.48	34.2	eV	
4th ionization energy	59.4			eV	
Standard electrode potential E^0			+0.7973	V	Reaction type $Hg^{2+} + 2e^- = 2Hg$
	−0.7618	−0.403		V	Reaction type $Cu^{2+} + 2e^- = Cu$

Table 2.1-10B(a) Elements of Group IIB (CAS notation), or Group 12 (new IUPAC notation). Crystallographic properties (see Table 2.1-10 for allotropic and high-pressure modifications)

Element name	Zinc	Cadmium	Mercury		
Chemical symbol	Zn	Cd	Hg		
Atomic number Z	30	48	80		
				Units	Remarks
Modification			α-Hg		
Crystal system, Bravais lattice	hex	hex	trig, R		
Structure type	Mg	Mg	α-Hg		
Lattice constant a	266.44	297.88	300.5 (225 K)	pm	At 298 K
Lattice constant c	494.94	561.67		pm	At 298 K
Lattice angle α			70.53 (225 K)	deg	
Space group	$P6_3/mmc$	$P6_3/mmc$	$R\bar{3}m$		
Schoenflies symbol	D_{6h}^4	D_{6h}^4	D_{3d}^5		
Strukturbericht type	A3	A3	A 10		
Pearson symbol	hP2	hP2	hR1		
Number A of atoms per cell	2	2	1		
Coordination number	6+6	6+6	6+6		
Shortest interatomic distance, solid	266	297	299 (at 234 K)	pm	At 293 K

Table 2.1–10B(b) Elements of Group IIB (CAS notation), or Group 12 (new IUPAC notation). Mechanical properties

Element name	Zinc	Cadmium	Mercury		
Chemical symbol	Zn	Cd	Hg		
Atomic number Z	30	48	80		
				Units	Remarks
Density ϱ, solid	7.14	8.65		g/cm^3	
Density ϱ, liquid	6.570	8.02	13.53 (25 °C)	g/cm^3	
Molar volume V_{mol}	9.17	13.00	14.81	cm^3/mol	
Viscosity η, liquid	2.95	2.29	1.55	mPa s	At T_m
Surface tension, liquid	0.816	0.564	0.476	N/m	
Temperature coefficient	0.25×10^{-3}	0.39×10^{-3}	-0.20×10^{-3}	N/(m K)	
Coefficient of linear thermal expansion α	25.0×10^{-6}	29.8×10^{-6}	18.1	1/K	At 298.15 K
Sound velocity, liquid	2790	2200[a]	1451	m/s	12 MHz[a]
Sound velocity, solid, transverse	2290	1690		m/s	
Sound velocity, solid, longitudinal	3890	2980		m/s	
Compressibility κ	1.65×10^{-5}	2.14×10^{-5}	3.77×10^{-5}	1/MPa	Volume compressibility
Elastic modulus E	92.7	62.3	25	GPa	
Shear modulus G	34.3	24.5		GPa	
Poisson number μ	0.29	0.30			
Elastic compliance s_{11}	8.22	12.4	154 (83 K)	1/TPa	
Elastic compliance s_{33}	27.7	34.6	45 (83 K)	1/TPa	
Elastic compliance s_{44}	25.3	53.1	151 (83 K)	1/TPa	
Elastic compliance s_{12}	0.60	-1.2	-119 (83 K)	1/TPa	
Elastic compliance s_{13}	-7.0	-9.1	-21 (83 K)	1/TPa	
Elastic compliance s_{14}			-100 (83 K)	1/TPa	
Elastic stiffness c_{11}	165	114.1	36.0 (83 K)	GPa	
Elastic stiffness c_{33}	61.8	49.9	50.5 (83 K)	GPa	
Elastic stiffness c_{44}	39.6	19.0	12.9 (83 K)	GPa	
Elastic stiffness c_{12}	31.1	41.0	28.9 (83 K)	GPa	
Elastic stiffness c_{13}	50.0	40.3	30.3 (83 K)	GPa	
Elastic stiffness c_{14}			4.7 (83 K)	GPa	
Tensile strength	20–40	71		MPa	293 K
Brinell hardness	280–330	180–230			

[a] At 594 K

Table 2.1–10B(c) Elements of Group IIB (CAS notation), or Group 12 (new IUPAC notation). Thermal and thermodynamic properties

Element name	Zinc	Cadmium	Mercury		
Chemical symbol	Zn	Cd	Hg		
Atomic number Z	30	48	80		
				Units	Remarks
State	Solid	Solid	Liquid		
Thermal conductivity λ	121	96.8	8.34	W/(m K)	At 300 K
Molar heat capacity c_p	25.44	25.98	27.983	J/(mol K)	At 298 K
Standard entropy S^0	41.631	51.800	75.900	J/(mol K)	At 298 K and 100 kPa
Enthalpy difference $H_{298} - H_0$	5.6570	6.2470	9.3420	kJ/mol	
Melting temperature T_m	692.68	594.22	234.32	K	
Enthalpy change ΔH_m	7.3220	6.1923	2.2953	kJ/mol	
Entropy change ΔS_m	10.571	10.421	9.796	J/(mol K)	
Relative volume change	0.0730	0.0474	0.037		$(V_l - V_s)/V_l$ at T_m
Boiling temperature T_b	1180	1040	629	K	
Enthalpy change ΔH_b	115.3	97.40	59.2	kJ/mol	
Critical temperature T_c			1750	K	
Critical pressure p_c			167	MPa	
Critical density ϱ_c			5.700	g/cm^3	

Table 2.1–10B(d) Elements of Group IIB (CAS notation), or Group 12 (new IUPAC notation). Electronic, electromagnetic, and optical properties

Element name	Zinc	Cadmium	Mercury		
Chemical symbol	Zn	Cd	Hg		
Atomic number Z	30	48	80		
				Units	Remarks
Characteristics	Ductile metal	Soft metal	Noble metal		
Electrical resistivity ρ_s	54.3	68		nΩ m	At 293 K
Temperature coefficient	41.7×10^{-4}	46.2×10^{-4}		1/K	
Pressure coefficient	-6.3×10^{-9}	-7.32×10^{-9}		1/hPa	
Electrical resistivity ρ_l	326	337	958 (298 K)	nΩ m	
Resistivity ratio	2.1	1.97	3.74–4.94		ρ_l/ρ_s at T_m
Superconducting critical temperature T_{crit}	0.88	0.55	4.15 K	K	α-Hg
Superconducting critical field H_{crit}	53	30	412	Oe	α-Hg
Hall coefficient R	0.63×10^{-10}	0.589×10^{-10}	-0.73×10^{-10}	m^3/(A s)	$B = 0.3$–2.2 T 298 K
Thermoelectric coefficient	2.9	2.8	-3.4	μV/K	
Electronic work function	4.22	4.04		V	
Thermal work function	3.74	3.92		V	
Molar magnetic susceptibility χ_{mol}, solid (SI)	-115×10^{-6}	-248×10^{-6}	-303×10^{-6} (234 K)	cm^3/mol	At 295 K
Molar magnetic susceptibility χ_{mol}, solid (cgs)	-9.15×10^{-6}	-19.7×10^{-6}	-24.1×10^{-6} (234 K)	cm^3/mol	At 295 K
Molar magnetic susceptibility χ_{mol}, liquid (SI)			-421×10^{-6}	cm^3/mol	At 295 K
Molar magnetic susceptibility χ_{mol}, liquid (cgs)			-33.5×10^{-6}	cm^3/mol	At 295 K
Mass magnetic susceptibility χ_{mass}, solid (SI)	-1.38×10^{-6}	-2.21×10^{-6}		cm^3/g	At 295 K
Mass magnetic susceptibility χ_{mass}, liquid (SI)	-1.32×10^{-6}	-1.83×10^{-6}		cm^3/g	
Refractive index $(n-1)$, gas			$+1882 \times 10^{-6}$ (Hg$_2$ vapor)		
Refractive index n, solid	1.19 ($\lambda = 550$ nm)	1.8 (578 nm)			

Table 2.1-10C Elements of Group IIB (CAS notation), or Group 12 (new IUPAC notation). Allotropic and high-pressure modifications

Element	Mercury		
Modification	α-Hg	β-Hg	
			Units
Crystal system, Bravais lattice	trig, R	tetr	
Structure type	α-Hg	In	
Lattice constant a	300.5 (225 K)	399.5	pm
Lattice constant c		282.5	pm
Lattice angle α	70.53 (225 K)		
Space group	$R\bar{3}m$	$I4/mmm$	
Schoenflies symbol	D_{3d}^5	D_{4h}^{17}	
Strukturbericht type	A10		
Pearson symbol	hR1	tI2	
Number A of atoms per cell	1	2	
Coordination number	6+6	2+8	
Shortest interatomic distance, solid	299	283	pm
Range of stability	< 234.2 K	77 K, high pressure	

Table 2.1-10D Elements of Group IIB (CAS notation), or Group 12 (new IUPAC notation). Ionic radii (determined from crystal structures)

Element	Zinc	Cadmium	Mercury		
Ion	Zn^{2+}	Cd^{2+}	Hg^+	Hg^{2+}	
Coordination number					Units
2				69	pm
4	60	78		96	pm
6	74	95	119	102	pm
8	90	110		114	pm
12		131			pm

Table 2.1-11A Elements of Group IIIA (CAS notation), or Group 13 (new IUPAC notation). Atomic, ionic, and molecular properties (see Table 2.1-11D for ionic radii)

Element name Chemical symbol Atomic number Z	Boron B 5	Aluminium Al 13	Gallium Ga 31	Indium In 49	Thallium Tl 81	Units	Remarks
Relative atomic mass A (atomic weight)	10.811(7)	26.981538(2)	69.723(1)	114.818(3)	204.3833(2)		
Abundance in lithosphere	10×10^{-6}	$81\,300 \times 10^{-6}$	15×10^{-6}	0.06×10^{-6}	0.3×10^{-6}		Mass ratio
Abundance in sea	4.4×10^{-6}	0.002×10^{-6}	3×10^{-11}	1×10^{-13}	1×10^{-11}		Mass ratio
Atomic radius r_{cov}	81	125	125	150	155	pm	Covalent radius
Atomic radius r_{met}	89	143	153	167	170	pm	Metallic radius, CN = 12
Atomic radius r_{vdW}		205	190	190	200	pm	van der Waals radius
Electron shells	KL	KLM	–LMN	–MNO	–NOP		
Electronic ground state	$^2P_{1/2}$	$^2P_{1/2}$	$^2P_{1/2}$	$^2P_{1/2}$	$^2P_{1/2}$		
Electronic configuration	[He]$2s^22p^1$	[Ne]$3s^23p^1$	[Ar]$3d^{10}4s^24p^1$	[Kr]$4d^{10}5s^25p^1$	[Xe]$4f^{14}5d^{10}6s^26p^1$		
Oxidation states	3+	3+	3+	3+	3+, 1+		
Electron affinity	0.277	0.441	0.3	0.3	0.2	eV	
Electronegativity χ_A	2.01	1.47	1.82	1.49	1.44		Allred and Rochow
1st ionization energy	8.29803	5.98577	5.99930	5.78636	6.10829	eV	
2nd ionization energy	25.15484	18.82856	20.5142	18.8698	20.428	eV	
3rd ionization energy	37.93064	28.44765	30.71	28.03	29.83	eV	
4th ionization energy	259.37521	119.992	64	54		eV	
Standard electrode potential E^0		-1.662	-0.560	-0.338	-0.336	V	Reaction type $Tl^+ + e^- = Tl$
						V	Reaction type $Al^{3+} + 3e^- = Al$
					$+1.252$	V	Reaction type $Tl^{3+} + 2e^- = Tl^+$

Table 2.1-11B(a) Elements of Group IIIA (CAS notation), or Group 13 (new IUPAC notation). Crystallographic properties (see Table 2.1-11C for allotropic and high-pressure modifications)

Element name	Boron	Aluminium	Gallium	Indium	Thallium		
Chemical symbol	B	Al	Ga	In	Tl		
Atomic number Z	5	13	31	49	81		
						Units	Remarks
Modification	β-B	α-Al	α-Ga					
Crystal system, Bravais lattice	trig, R	cub, fc	orth, C	tetr, I	hex		
Structure type	β-B	Cu	α-Ga	In	Mg		
Lattice constant a		361.49	451.92	459.90	345.63	pm	
Lattice constant b			765.86			pm	
Lattice constant c			452.58	494.70	552.63	pm	
Space group	$R\bar{3}m$	$Fm\bar{3}m$	$Cmca$	$I4/mmm$	$P6_3/mmc$		
Schoenflies symbol	D_{3d}^5	O_h^5	D_{2h}^{18}	D_{4h}^{17}	D_{6h}^4		
Strukturbericht type		A1	A11	A6	A3		
Pearson symbol	hR105	cF4	oC8	tI2	hP2		
Number A of atoms per cell	105	4	8	2	2		
Coordination number		12	1+2+2+2	4+8			
Shortest interatomic distance, solid	162–192	286	247	325		pm	

Table 2.1–11B(b) Elements of Group IIIA (CAS notation), or Group 13 (new IUPAC notation). Mechanical properties

Element name	Boron	Aluminium	Gallium	Indium	Thallium	Units	Remarks
Chemical symbol	B	Al	Ga	In	Tl		
Atomic number Z	5	13	31	49	81		
Density ϱ, solid	2.46	2.70	5.91	7.31	11.85	g/cm^3	
Density ϱ, liquid		2.39	6.200	6.990	11.29	g/cm^3	At T_m
Molar volume V_{mol}	4.62	10.00	11.81	15.71	17.24	cm^3/mol	
Viscosity η, liquid		1.38	1.70	1.65		mPa s	At T_m
Surface tension γ, liquid		0.860	0.718	0.556	0.447	N/m	At T_m
Temperature coefficient		-0.135×10^{-3}		-0.09×10^{-3}	-0.07×10^{-3}	N/(m K)	
Coefficient of linear thermal expansion α	5×10^{-6}	23.03×10^{-6}	18.3×10^{-6}	33×10^{-6}	28×10^{-6}	1/K	
Sound velocity, liquid		4650	2740	2215		m/s	
Sound velocity, solid, transverse		3130	750	710	480	m/s	
Sound velocity, solid, longitudinal		6360	3030	2460	1630	m/s	
Compressibility κ	0.539×10^{-5}	1.33×10^{-5}	1.96×10^{-5}	2.70×10^{-5}	3.41×10^{-5}	1/MPa	Volume compressibility
Elastic modulus E	178[a]	70.2	9.81	10.6	7.89	GPa	
Shear modulus G		27.8	6.67	3.68	2.67	GPa	
Poisson number μ		0.34	0.47	0.45	0.45		
Elastic compliance s_{11}		16.0	12.2	148.8	104	1/TPa	
Elastic compliance s_{22}			14.0			1/TPa	
Elastic compliance s_{33}			8.49	196.2	31.1	1/TPa	
Elastic compliance s_{44}		35.3	28.6	153.7	139	1/TPa	
Elastic compliance s_{55}			23.9			1/TPa	
Elastic compliance s_{66}			24.8	83.2		1/TPa	
Elastic compliance s_{12}		-5.8	-4.4	-46.0	-83	1/TPa	
Elastic compliance s_{13}			-1.7	-94.5	-11.6	1/TPa	
Elastic compliance s_{23}			-2.4			1/TPa	

[a] Estimated.

Table 2.1-11B(b) Elements of Group IIIA (CAS notation), or Group 13 (new IUPAC notation). Mechanical properties, cont.

Element name	Boron	Aluminium	Gallium	Indium	Thallium		
Chemical symbol	B	Al	Ga	In	Tl		
Atomic number Z	5	13	31	49	81		
						Units	Remarks
Elastic stiffness c_{11}	467	108	100	45.1	41.9	GPa	
Elastic stiffness c_{22}			90.2			GPa	
Elastic stiffness c_{33}	473		135	44.6	54.9	GPa	
Elastic stiffness c_{44}	198	28.3	35.0	6.51	7.20	GPa	
Elastic stiffness c_{55}			41.8			GPa	
Elastic stiffness c_{66}			40.3	12.0		GPa	
Elastic stiffness c_{12}	241	62	37	40.0	36.6	GPa	
Elastic stiffness c_{13}			33	41.0	29.9	GPa	
Elastic stiffness c_{23}			31			GPa	
Elastic stiffness c_{14}	15.1					GPa	
Tensile strength	16–24[a]	90–100			8.9	MPa	
Vickers hardness	49 000	167			9		At 293 K
Brinell hardness				0.9			
Mohs hardness	9.3		1.5–2.5				

[a] Amorphous.

Table 2.1-11B(c) Elements of Group IIIA (CAS notation), or Group 13 (new IUPAC notation). Thermal and thermodynamic properties

Element name	Boron	Aluminium	Gallium	Indium	Thallium		
Chemical symbol	B	Al	Ga	In	Tl		
Atomic number Z	5	13	31	49	81		
						Units	Remarks
State	Crystalline						
Thermal conductivity λ	27.0	237	33.5	81.6	46.1	W/(m K)	
Thermal conductivity (liquid) λ_l		90				W/(m K)	At T_m
Molar heat capacity c_p	11.20	24.392	26.15	26.732	26.32	J/(mol K)	At 298 K
Standard entropy S^0	5.900	28.300	40.727	57.650	64.300	J/(mol K)	At 298 K and 100 kPa
Enthalpy difference $H_{298} - H_0$	1.2220	4.5400	5.5720	6.6100	6.832	kJ/mol	
Melting temperature T_m	2348.00	933.47	302.91	429.75	577.00	K	
Enthalpy change ΔH_m	50.200	10.7110	5.5898	3.2830	4.1422	kJ/mol	
Entropy change ΔS_m	21.380	11.474	18.454	7.639	7.179	J/(mol K)	
Volume change ΔV_m		0.065	−0.034	0.025	0.0323		$(V_l - V_s)/V_l$ at T_m
Boiling temperature T_b	4138	2790	2478	2346	1746	K	
Enthalpy change ΔH_b	480.5	294.0	258.7	231.45	164.1	kJ/mol	

Table 2.1-11B(d) Elements of Group IIIA (CAS notation), or Group 13 (new IUPAC notation). Electronic, electromagnetic, and optical properties

Element name Chemical symbol Atomic number Z	Boron B 5	Aluminium Al 13	Gallium Ga 31	Indium In 49	Thallium Tl 81	Units	Remarks
Characteristics	Semiconductor	Light metal	Soft metal	Soft metal	Soft metal, toxic		
Electrical resistivity ρ_s	6500×10^9 [a]	25.0	136	80.0	150	$n\Omega\,m$	[a] At 300 K
Temperature coefficient		46×10^{-4}	39.6×10^{-4}	49.0×10^{-4}	51.7×10^{-4}	1/K	
Pressure coefficient		-4.06×10^{-9}	-2.47×10^{-9}	-12.2×10^{-9}	-3.4×10^{-9}	1/hPa	
Electrical resistivity ρ_l		200	258	331	740	$n\Omega\,m$	
Resistivity ratio at T_m		1.64	1.9	2.18			ρ_l/ρ_s at T_m
Superconducting critical temperature T_{cr}		1.2	1.09	3.4	2.4	K	
Superconducting critical field H_{cr}		99	51	293	171	Oe	
Electronic band gap ΔE	1.5					eV	
Electronic work function	4.79	4.28	4.35	4.08	4.05	V	
Thermal work function	5.71	3.74	4.12	4.0	3.76	V	
Intrinsic charge carrier concentration	$+5 \times 10^{14}$					$1/cm^3$	At 430 K
Electron mobility	1					$cm^2/(V\,s)$	
Hole mobility	55					$cm^2/(V\,s)$	
Hall coefficient R		-0.343×10^{-10}	-0.63×10^{-10}	-0.24×10^{-10}	0.240×10^{-10}	$m^3/(A\,s)$	$B=1.0-1.3\,T$ 300 K
Thermoelectric coefficient		-0.6		2.4	0.4	$\mu V/K$	
Dielectric constant ε, solid	13 (0.5 MHz)						
Molar magnetic susceptibility χ_{mol}, solid (SI)	-84.2×10^{-6}	207×10^{-6}	-271×10^{-6}	-128×10^{-6}	-628×10^{-6}	cm^3/mol	At 295 K
Molar magnetic susceptibility χ_{mol}, solid (cgs)	-6.7×10^{-6}	16.5×10^{-6}	-21.6×10^{-6}	-10.2×10^{-6}	-50.0×10^{-6}	cm^3/mol	At 295 K
Mass magnetic susceptibility χ_{mass}, solid (SI)		$+7.9 \times 10^{-6}$	-3.9×10^{-6}	-1.4×10^{-6}	-3.13×10^{-6}	cm^3/g	At 295 K
Mass magnetic susceptibility χ_{mass}, liquid (SI)					-1.72×10^{-6}	cm^3/g	At T_m
Refractive index n, solid	3.2 ($\lambda = 1\,\mu m$)						

[a] At 300 K.

Table 2.1-11C Elements of Group IIIA (CAS notation), or Group 13 (new IUPAC notation). Allotropic and high-pressure modifications

Element	Aluminium		Gallium			Thallium			Units
Modification	α-Al	β-Al	α-Ga	β-Ga	γ-Ga	α-Tl	β-Tl	γ-Tl	
Crystal system, Bravais lattice	cub, fc	hex, cp	orth, C	tetr	orth	hex, cp	cub, bc	cub, fc	
Structure type	Cu	Mg	α-Ga	In	γ-Ga	Mg	W	Cu	
Lattice constant a	404.96	269.3	451.92	280.8	1059.3	345.63	387.9		pm
Lattice constant b			765.86		1352.3				pm
Lattice constant c		439.8	452.58	445.8	520.3	552.63			pm
Space group	$Fm\bar{3}m$	$P6_3/mmc$	$Cmca$	$I4/mmm$	$Cmcm$	$P6_3/mmc$	$Im\bar{3}m$	$Fm\bar{3}m$	
Schoenflies symbol	O_h^5	D_{6h}^4	D_{2h}^{18}		D_{2h}^{17}	D_{6h}^4	O_h^9	O_h^5	
Strukturbericht type	A1	A3	A11	A6		A3	A2	A1	
Pearson symbol	cF4	hP2	oC8	tI2	oC40	hP2	cI2	cF4	
Number A of atoms per cell	4	2	8	2	40	2	2	4	
Coordination number	12	12	1+2+2+2	4+8			8	12	
Shortest interatomic distance, solid	286		247	281	260–308		336		
Range of stability	RTP	> 20.5 GPa	RTP	> 1.2 GPa	220 K, > 3.0 GPa	RTP	> 503 K	High pressure	

Table 2.1-11D Elements of Group IIIA (CAS notation), or Group 13 (new IUPAC notation). Ionic radii (determined from crystal structures)

Element	Aluminium	Gallium	Indium	Thallium		Units
Ion	Al^{3+}	Ga^{3+}	In^{3+}	Tl$^+$	Tl^{3+}	
Coordination number						
4	39	47	62		75	pm
5	48					pm
6	54	62	80	150	89	pm
8				159	98	pm
12				170		pm

Table 2.1-12A Elements of Group IIIB (CAS notation), or Group 3 (new IUPAC notation). Atomic, ionic, and molecular properties (see Table 2.1-12D for ionic radii)

Element name	Scandium	Yttrium	Lanthanum	Actinium		
Chemical symbol	Sc	Y	La	Ac		
Atomic number Z	21	39	57	89		
Characteristics				Radioactive	Units	Remarks
Relative atomic mass A (atomic weight)	44.955910(8)	88.90585(2)	138.9055(2)	[227]		
Abundance in lithosphere	5×10^{-6}	28.1×10^{-6}	18.3×10^{-6}			Mass ratio
Abundance in sea	6×10^{-13}	3×10^{-12}	3×10^{-12}			Mass ratio
Atomic radius r_{cov}	144	162	169		pm	Covalent radius
Atomic radius r_{met}	166	178	187	188	pm	Metallic radius, CN = 12
Electron shells	–LMN	–MNO	–NOP	–OPQ		
Electronic ground state	$^2D_{3/2}$	$^2D_{3/2}$	$^2D_{3/2}$	$^2D_{3/2}$		
Electronic configuration	[Ar]$3d^1 4s^2$	[Kr]$4d^1 5s^2$	[Xe]$5d^1 6s^2$	[Rn]$6d^1 7s^2$		
Oxidation states	3+	3+	3+	3+		
Electron affinity	0.188	0.307	0.5		eV	
Electronegativity χ_A	1.20	1.11	1.08	(1.00)		Allred and Rochow
1st ionization energy	6.56144	6.217	5.5770	5.17	eV	
2nd ionization energy	12.79967	12.24	11.060	12.1	eV	
3rd ionization energy	24.75666	20.52	19.1773		eV	
4th ionization energy	73.4894	60.597	49.95		eV	
Standard electrode potential E^0			−2.522		V	Reaction type La^{3+} + 3e$^-$ = La

Table 2.1-12B(a) Elements of Group IIIB (CAS notation), or Group 3 (new IUPAC notation). Crystallographic properties (see Table 2.1-12C for allotropic and high-pressure modifications)

Element name	Scandium	Yttrium	Lanthanum	Actinium		
Chemical symbol	Sc	Y	La	Ac		
Atomic number Z	21	39	57	89		
					Units	Remarks
Modification	α-Sc	α-Y	α-La			
Crystal system, Bravais lattice	hex	hex	hex			
Structure type	Mg	Mg	α-La			
Lattice constant a	330.88	364.82	377.40		pm	
Lattice constant c	526.80	573.18	1217.1		pm	
Space group	$P6_3/mmc$	$P6_3/mmc$	$P6_3/mmc$			
Schoenflies symbol	D_{6h}^4	D_{6h}^4	D_{6h}^4			
Strukturbericht type	A3	A3	A3'			
Pearson symbol	hP2	hP2	hP4			
Number A of atoms per cell	2	2	4			
Coordination number	12	6+6	12			
Shortest interatomic distance, solid	166		364		pm	

Table 2.1-12B(b) Elements of Group IIIB (CAS notation), or Group 3 (new IUPAC notation). Mechanical properties

Element name	Scandium	Yttrium	Lanthanum	Actinium		
Chemical symbol	Sc	Y	La	Ac		
Atomic number Z	21	39	57	89		
					Units	Remarks
Density ϱ, solid	2.989	4.50	6.70	10.07	g/cm^3	
Molar volume V_{mol}	15.04	19.89	22.60		cm^3/mol	
Surface tension, liquid	0.9	0.9	0.71		N/m	
Coefficient of linear thermal expansion α	10.0×10^{-6}	10.6×10^{-6}	4.9×10^{-6}		1/K	
Sound velocity, solid, transverse		2420	1540		m/s	
Sound velocity, solid, longitudinal		4280	2770		m/s	
Compressibility κ	2.22×10^{-5}	2.62×10^{-5}	3.96×10^{-5}		1/MPa	Volume compressibility
Elastic modulus E	75.2	66.3	39.2	25[a]	GPa	
Shear modulus G	29.7	25.5	14.9		GPa	
Poisson number μ	0.28	0.27	0.28			
Elastic compliance s_{11}	12.5	15.4	51.7[b]		1/TPa	
Elastic compliance s_{33}	10.6	14.4			1/TPa	
Elastic compliance s_{44}	36.1	41.1	55.7[b]		1/TPa	
Elastic compliance s_{12}	−4.3	−5.1	−19.2[b]		1/TPa	
Elastic compliance s_{13}	−2.2	−2.7			1/TPa	
Elastic stiffness c_{11}	99.3	77.9	34.5[b]		GPa	
Elastic stiffness c_{33}	107	76.9			GPa	
Elastic stiffness c_{44}	27.7	24.3	18.0[b]		GPa	
Elastic stiffness c_{12}	39.7	29.2	20.4[b]		GPa	
Elastic stiffness c_{13}	29.4	20			GPa	
Tensile strength	256	250–380			MPa	
Vickers hardness	350	40	491			

[a] Estimated.
[b] For lanthanum in its metastable fcc phase at room temperature.

Table 2.1–12B(c) Elements of Group IIIB (CAS notation), or Group 3 (new IUPAC notation). Thermal and thermodynamic properties

Element name	Scandium	Yttrium	Lanthanum	Actinium		
Chemical symbol	Sc	Y	La	Ac		
Atomic number Z	21	39	57	89		
					Units	Remarks
Thermal conductivity λ	15.8	17.2	13.5		W/(m K)	
Molar heat capacity c_p	25.52	26.53	27.11	27.2	J/(mol K)	At 298 K
Standard entropy S^0	34.644	44.788	56.902	62.000	J/(mol K)	At 298 K and 100 kPa
Enthalpy difference $H_{298} - H_0$	5.2174	5.9835	6.6651	6.7000	kJ/mol	
Melting temperature T_m	1814.00	1795.15	1193.00	1323.00	K	
Enthalpy change ΔH_m	14.0959	11.3942	6.1965	12.000	kJ/mol	
Entropy change ΔS_m	7.771	6.347	5.194	9.070	J/(mol K)	
Relative volume change ΔV_m			0.006			$(V_l - V_s)/V_l$ at T_m
Boiling temperature T_b	3104	3611	3730	3473	K	
Enthalpy change ΔH_b	314.2	363.3	413.7	293	kJ/mol	

Table 2.1-12B(d) Elements of Group IIIB (CAS notation), or Group 3 (new IUPAC notation). Electronic, electromagnetic, and optical properties

Element name	Scandium	Yttrium	Lanthanum	Actinium		
Chemical symbol	Sc	Y	La	Ac		
Atomic number Z	21	39	57	89		
Characteristics	Soft metal	Reactive metal	Very reactive metal		Units	Remarks
Electrical resistivity ρ_s	505	550	540		$n\Omega\,m$	At RT
Temperature coefficient	28.2×10^{-4}	27.1×10^{-4}	21.8×10^{-4}		1/K	
Pressure coefficient			-1.7×10^{-9}		1/hPa	
Electrical resistivity ρ_l			1350		$n\Omega\,m$	
Superconducting critical temperature T_{crit}			5.0		K	
Superconducting critical field H_{crit}						At 293 K, $B = 0.5$–1.0 T
Hall coefficient R	-0.67×10^{-10}	-0.770×10^{-10}	-0.8×10^{-10}		$m^3/(A\,s)$	
Thermoelectric coefficient	-3.6	2.2			$\mu V/K$	
Electronic work function	3.5	3.1	3.5		V	
Thermal work function	3.23	3.07	3.3		V	
Molar magnetic susceptibility χ_{mol}, solid (SI)	3710×10^{-6}	2359×10^{-6}	$+1205 \times 10^{-6}$		cm^3/mol	At 295 K
Molar magnetic susceptibility χ_{mol}, solid (cgs)	295×10^{-6}	188×10^{-6}	$+95.9 \times 10^{-6}$		cm^3/mol	At 295 K
Mass magnetic susceptibility χ_{mass}, solid (SI)	88×10^{-6}	27.0×10^{-6}	$+11 \times 10^{-6}$		cm^3/g	

Table 2.1-12C Elements of Group IIIB (CAS notation), or Group 3 (new IUPAC notation). Allotropic and high-pressure modifications

Element	Scandium		Yttrium		Lanthanum				Units
Modification	α-Sc	β-Sc	α-Y	β-Y	α-La	β-La	γ-La	β'-La	
Crystal system, Bravais lattice	hex, cp	cub, bc	hex, cp	cub, bc	hex	cub, fc	cub, bc	cub, fc	
Structure type	Mg	W	Mg	W	α-La	Cu	W	Cu	
Lattice constant a	330.88		364.82		377.40	530.45	426.5	517	pm
Lattice constant c	526.80		573.18		1217.1				pm
Space group	$P6_3/mmc$	$Im\bar{3}m$	$P6_3/mmc$	$Im\bar{3}m$	$P6_3/mmc$	$Fm\bar{3}m$	$Im\bar{3}m$	$Fm\bar{3}m$	
Schoenflies symbol	D_{6h}^4	O_h^9	D_{6h}^4	O_h^9	D_{6h}^4	O_h^5	O_h^9	O_h^5	
Strukturbericht type	A3	A2	A3	A2	A3'	A1	A2	A1	
Pearson symbol	hP2	cI2	hP2	cI2	hP4	cF4	cI2	cF4	
Number A of atoms per cell	2	2	2	2	4	4	2	4	
Coordination number	12	8	6 + 6	8	12	12	8	12	
Shortest interatomic distance, solid			356		374	375	369		pm
Range of stability	RT	> 1607 K	RT	> 1752 K	RTP	> 613 K	> 1141 K	> 2.0 GPa	

Table 2.1-12D Elements of Group IIIB (CAS notation), or Group 3 (new IUPAC notation). Ionic radii (determined from crystal structures)

Element	Scandium	Yttrium	Lanthanum	Actinium	
Ion	Sc^{3+}	Y^{3+}	La^{3+}	Ac^{3+}	
Coordination number					Units
6	75	90	103		pm
8	87	102	116	112	pm
9		108			pm
10			127		pm
12			136		pm

Table 2.1-13A Elements of Group IVA (CAS notation), or Group 14 (new IUPAC notation). Atomic, ionic, and molecular properties (see Table 2.1-13D for ionic radii)

Element name	Carbon	Silicon	Germanium	Tin	Lead		
Chemical symbol	C	Si	Ge	Sn	Pb		
Atomic number Z	6	14	32	50	82	Units	Remarks
Relative atomic mass A (atomic weight)	12.0107(8)	28.0855(3)	72.61(2)	118.710(7)	207.2(1)		
Abundance in lithosphere	320×10^{-6}	$277\,200 \times 10^{-6}$	7×10^{-6}	40×10^{-6}	16×10^{-6}		Mass ratio
Abundance in sea	28×10^{-6}	2×10^{-6}	5×10^{-11}	1×10^{-11}	3×10^{-11}		Mass ratio
Atomic radius r_{cov}	77	117	122	140	154	pm	Covalent radius
Atomic radius r_{met}	91	132	137	220	175	pm	Metallic radius, CN = 12
Atomic radius r_{vdW}	170	210	234	158	200	pm	van der Waals radius
Electron shells	KL	KLM	–LMN	–MNO	–NOP		
Electronic ground state	$^3P_0$	$^3P_0$	$^3P_0$	$^3P_0$	$^3P_0$		
Electronic configuration	$1s^22s^22p^2$	$[Ne]3s^23p^2$	$[Ar]3d^{10}4s^24p^2$	$[Kr]4d^{10}5s^25p^2$	$[Xe]4f^{14}5d^{10}6s^26p^2$		
Oxidation states	4+, 4−, 2+	4+	4+	4+, 2+	4+, 2+		
Electron affinity	1.26	1.39	1.23	1.11	0.364	eV	
Electronegativity χ_A	2.50	1.74	2.02	1.72	1.55		Allred and Rochow
1st ionization energy	11.26030	8.15169	7.900	7.34381	7.41666	eV	
2nd ionization energy	24.38332	16.34585	15.93462	14.63225	15.0322	eV	
3rd ionization energy	47.8878	33.49302	34.2241	30.50260	31.9373	eV	
4th ionization energy	64.4939	45.14181	45.7131	40.73502	42.32	eV	
5th ionization energy	392.087	166.767	93.5	72.28	68.8	eV	
Standard electrode potential E^0				−0.137	−0.126	V	Reaction type $Sn^{2+} + 2e^- = Sn$
				+0.151		V	Reaction type $Sn^{4+} + 2e^- = Sn^{2+}$

Table 2.1-13B(a) Elements of Group IVA (CAS notation), or Group 14 (new IUPAC notation). Crystallographic properties (see Table 2.1-13C for allotropic and high-pressure modifications)

Element name	Carbon		Silicon	Germanium	Tin	Lead	Units	Remarks
Chemical symbol	C		Si	Ge	Sn	Pb		
Atomic number Z	6		14	32	50	82		
Modification	Diamond	Graphite			β-Sn,[a] white tin, Sn I			
Crystal system, Bravais lattice	cub	hex	cub, fc	cub, fc	tetr, I	cub, fc		
Structure type	Diamond	C	Diamond	Diamond	β-Sn	Cu		
Lattice constant a	356.71	246.12	543.102 (22.5 °C)	565.9(1) RT	581.97 (300 K)	495.02	pm	
Lattice constant c		670.90			317.49		pm	
Space group	$Fd3m$	$P6_3/mmc$	$Fd3m$	$Fd3m$	$I4_1/amd$	$Fm3m$		
Schoenflies symbol	O_h^7	D_{6v}^4	O_h^7	O_h^7	D_{4h}^{19}	O_h^5		
Strukturbericht type	A9	A9	A4	A4	A5	A1		
Pearson symbol		hP4	cF8	cF8	tI4	cF4		
Number A of atoms per cell	8	4	8	8	4	4		
Coordination number	4	3	4	4	4	12		
Shortest interatomic distance, solid	154.45	142.10	235	244	302	349	pm	Graphite, within layers
		335.45					pm	Graphite, between layers
Range of stability	RT, > 60 GPa	RTP						

[a] At ambient pressure Sn crystallizes in the diamond structure (gray tin, α-Sn) and below 17 °C in the β-tin structure (white tin, β-Sn, Sn-I) at room temperature. If it is alloyed with In or Hg, the simple hexagonal γ-Sn structure is observed

Table 2.1–13B(b) Elements of Group IVA (CAS notation), or Group 14 (new IUPAC notation). Mechanical properties

Element name Chemical symbol Atomic number Z	Carbon C 6		Silicon Si 14	Germanium Ge 32	Tin Sn 50	Lead Pb 82	Units	Remarks
Modification	Diamond	Graphite			β-Sn (white)			
Density ϱ, solid	3.513	2.266	2.33	5.32	7.30	11.4	g/cm^3	At 293 K
Density ϱ, liquid			2.525	5.500	6.978	10.678	g/cm^3	
Molar volume V_{mol}	3.42	5.3	12.06	13.64	16.24	18.26	cm^3/mol	
Viscosity η, liquid			2.0		2.71	1.67	mPa s	
Surface tension γ, liquid			0.735	0.650	0.545	0.470	N/m	
Temperature coefficient			-0.5×10^{-3}	-0.20×10^{-3}	-0.075×10^{-3}	-0.26×10^{-3}	N/(m K)	
Coefficient of linear thermal expansion α	1.06×10^{-6}	1.9×10^{-6} a	2.56×10^{-6}	5.57×10^{-6}	21.2×10^{-6}	29.1×10^{-6}	1/K	At 293 K
Sound velocity, liquid			5845		2270	1790	m/s	At T_m 12 MHz
Sound velocity, solid, transverse				2420	1650	710	m/s	At 293 K
Sound velocity, solid, longitudinal	11 220	3450	8433	4580	3300	2050	m/s	
Compressibility κ	2.25×10^{-6}	1.56×10^{-6}	10.2×10^{-6}	13.4×10^{-6}	18.3×10^{-6}	23.7×10^{-6}	1/MPa	Volume compressibility
Bulk modulus B_0	444		98.0	74.9	56.6	15.8	GPa	At 295 K
Elastic modulus E	545		112	79.9	52.9	5.54	GPa	
Shear modulus G			80.5	29.6	19.9		GPa	
Poisson number μ				0.34	0.33	0.44		
Elastic compliance s_{11}	0.98		7.73	9.73	41.6	93.7	1/TPa	
Elastic compliance s_{33}	27.5				14.9		1/TPa	
Elastic compliance s_{44}	1.732	250	12.7	14.9	45.6	68.0	1/TPa	
Elastic compliance s_{66}					42.8		1/TPa	
Elastic compliance s_{12}	-0.0987	-0.16	-2.15	-2.64	31.2	-43.0	1/TPa	
Elastic compliance s_{13}		-0.33			4.6		1/TPa	
Elastic stiffness c_{11}	1079	1060	165.6	129	72.30	48.8	GPa	At 300 K
Elastic stiffness c_{33}		36.5			88.40		GPa	At 300 K
Elastic stiffness c_{44}	578	4	79.6	67.1	22.03	14.8	GPa	At 300 K
Elastic stiffness c_{66}					24.00		GPa	At 300 K
Elastic stiffness c_{12}	124.5	180	63.9	48.3	59.40	41.4	GPa	At 300 K
Elastic stiffness c_{13}		15			35.78		GPa	At 300 K
Tensile strength			690			700	MPa	
Vickers hardness			2350			39		
Mohs hardness	10							

a At 298 K, parallel to layer planes. The corresponding value perpendicular to the layer planes is 2.9×10^{-6}/K

Table 2.1-13B(c) Elements of Group IVA (CAS notation), or Group 14 (new IUPAC notation). Thermal and thermodynamic properties

Element name	Carbon		Silicon	Germanium	Tin	Lead	Units	Remarks
Chemical symbol	C		Si	Ge	Sn	Pb		
Atomic number Z	6		14	32	50	82		
Modification	Diamond	Graphite			β-Sn (white)			
Thermal conductivity λ	1000–2320	5.7[a]	83.7	58.6	66.6	35.2	W/(m K)	At 300 K
		1960[b]					W/(m K)	At 300 K
Molar heat capacity c_p	6.11	8.519	20.00	33.347	27.17	26.51	J/(mol K)	At 298 K
Standard entropy S^0	2.360	5.742	18.810	31.090	51.180	64.800	J/(mol K)	At 298 K and 100 kPa
Enthalpy difference $H_{298} - H_0$	0.5188	1.0540	3.2170	4.6360	6.3230	6.8700	kJ/mol	
Melting temperature T_m		4765.30	1687.00	1211.0	505.08	600.61	K	
Enthalpy change ΔH_m		117.3690	50.208	36.9447	7.01940	4.7739	kJ/mol	
Entropy change ΔS_m		24.630	29.762	30.498	14.243	7.948	J/(mol K)	
Relative volume change ΔV_m			−0.10	−0.054	0.028	0.032		$(V_l - V_s)/V_l$ at T_m
Boiling temperature T_b		3915	3505	3107	2876	2019	K	
Enthalpy change ΔH_b		710.9	383.3	331	295.8	177.58	kJ/mol	
Transformation temperature	1900–2100						K	Transforms to graphite

[a] Perpendicular to layer planes.
[b] Parallel to layer planes.

Table 2.1-13B(d) Elements of Group IVA (CAS notation), or Group 14 (new IUPAC notation). Electronic, electromagnetic, and optical properties

Element name	Carbon	Silicon	Germanium	Tin	Lead	Units	Remarks
Chemical symbol	C	Si	Ge	Sn	Pb		
Atomic number Z	6	14	32	50	82		
Modification	Diamond Graphite			β-Sn(white)			
Characteristics	Very hard insulator Soft conductor	Hard semicondor	Semicondor	Ductile metal	Soft metal		
Electrical resistivity ρ_s	10^{11} 1.4×10^{-5}		0.45	110×10^{-9}	192×10^{-9}	Ω m	At 293 K
Temperature coefficient				46.5×10^{-4}	42.8×10^{-4}	1/K	
Pressure coefficient				-9.2×10^{-9}	-12.5×10^{-9}	1/hPa	
Electrical resistivity ρ_l			710			nΩ m	
Resistivity ratio			0.071		1.94		ρ_l^*, ρ_s at T_m
Superconducting critical temperature T_{crit}				3.72	7.2	K	
Superconducting critical field H_{crit}				309	803	Oe	
Hall coefficient R [a]	-487×10^{-10}	-100	0.1	0.041×10^{-10}	0.09×10^{-10}	m^3/(A s)	At 300 K
Thermoelectric coefficient	11.06		302.5	0.1	-0.1	μV/K	
Electronic band gap ΔE	5.4	1.107	0.6642			eV	At 300 K
Temperature dependence		-2.3×10^{-4}				eV/K	At 300 K
Electronic work function	4.81	4.95	5.0	4.42	4.25	V	
Thermal work function	4.00	4.1	4.56	4.11	3.83	V	
Electron mobility	1800	1900	3800			cm^2/(V s)	
Hole mobility	1400	480	1820			cm^2/(V s)	
Dielectric constant ε, static, solid	5.68	11.7[b]	16.0				
Dielectric constant ε, high-frequency, solid	5.9(1)	12.0	16	24			
Molar magnetic susceptibility χ_{mol}, solid [c] (SI)	-74.1×10^{-6}	-39.2×10^{-6}	-146×10^{-6}	-470×10^{-6}	-289×10^{-6}	cm^3/mol	At 295 K
Molar magnetic susceptibility χ_{mol}, solid [c] (cgs)	-5.88×10^{-6}	-3.12×10^{-6}	-11.6×10^{-6}	-37.4×10^{-6}	-23×10^{-6}	cm^3/mol	At 295 K
Mass magnetic susceptibility χ_{mass}, solid (SI)	-6.17×10^{-6}	-1.8×10^{-6}	-1.328×10^{-6}	-3.3×10^{-6}	-1.39×10^{-6}	cm^3/g	At 295 K
Mass magnetic susceptibility χ_{mass}, liquid (SI)				-4.4×10^{-6}		cm^3/g	
Refractive index n, solid	2.4173	4.24	4.00 (25 μm)	1.0	2.01		$\lambda = 589$ nm
Refractive index n, liquid				1.7			

[a] $B = 1.0$–1.6 T.
[b] At 11 K and 1 MHz.
[c] The values for Sn apply to gray Sn.

Table 2.1-13C Elements of Group IVA (CAS notation), or Group 14 (new IUPAC notation). Allotropic and high-pressure modifications

Element	Silicon				Germanium				Units
Modification	α-Si	β-Si	γ-Si	δ-Si	α-Ge	β-Ge	γ-Ge	δ-Ge	
Crystal system, Bravais lattice	cub, fc	tetr	cub	hex	cub, fc	tetr	tetr	cub, bc	
Structure type	Diamond	β-Sn		α-La	Diamond	β-Sn		γ-Si	
Lattice constant a	543.06	468.6	636	380	565.74	488.4	593	692	pm
Lattice constant c		258.5		628		269.2	698		pm
Space group	$Fd\bar{3}m$	$I4_1/amd$	$Im\bar{3}m$	$P6_3/mmc$	$Fd\bar{3}m$	$I4_1/amd$	$P4_32_12$	$Im\bar{3}m$	
Schoenflies symbol	O_h^7	D_{4h}^{19}	O_h^9	D_{6v}^4	O_h^7	D_{4h}^{19}	D_4^8		
Strukturbericht type	A4	A5		A3′	A4	A5			
Pearson symbol	cF8	tI4	cI16	hP4	cF8	tI4	tP12	cI16	
Number A of atoms per cell	8	4		4	8	4	12	16	
Coordination number	4	4+2			4	4+2	4+2+2		
Shortest interatomic distance, solid	235	243			245	253	249		
Range of stability	RTP	> 9.5 GPa	> 16.0 GPa	[a]	RTP	> 12.0 GPa	[b]	> 12.0 GPa	

[a] Decompressed β-Si.
[b] Decompressed β-Ge.

Table 2.1-13C Elements of Group IVA (CAS notation), or Group 14 (new IUPAC notation). Allotropic and high-pressure modifications, cont.

Element	Tin			Lead		Units
Modification	α-Sn (gray tin)	β-Sn (white tin)	γ-Sn	Pb I	Pb II	
Crystal system, Bravais lattice	cub, fc	tetr, I	tetr	cub, fc	hex, cp	
Structure type	Diamond	β-Sn	In	Cu	Mg	
Lattice constant a	648.92	583.16	370	495.02	326.5	pm
Lattice constant c		318.15	337		538.7	pm
Space group	$Fd\bar{3}m$	$I4_1/amd$		$Fm\bar{3}m$	$P6_3/mmc$	
Schoenflies symbol	O_h^7	D_{4h}^{19}		O_h^5	D_{6h}^4	
Strukturbericht type	A4	A5		A1	A3	
Pearson symbol	cF8	tI4	tI2	cF4	hP2	
Number A of atoms per cell	8	4	2	4	2	
Coordination number	4			12		
Shortest interatomic distance, solid	281	302		349		
Range of stability	< 291 K	RT	> 9 GPa	RTP	> 10.3 GPa	

Table 2.1–13D Elements of Group IVA (CAS notation), or Group 14 (new IUPAC notation). Ionic radii (determined from crystal structures)

Element	Carbon	Silicon	Germanium		Tin	Lead		
Ion	C^{4+}	Si^{4+}	Ge^{2+}	Ge^{4+}	Sn^{4+}	Pb^{2+}	Pb^{4+}	Units
Coordination number								
4	15	26		39	55		65	pm
6	16	40	73	53	69	119	78	pm
8					81	129	94	pm
10						140		
12						149		pm

Table 2.1–14A Elements of Group IVB (CAS notation), or Group 4 (new IUPAC notation). Atomic, ionic, and molecular properties (see Table 2.1-14D for ionic radii)

Element name	Titanium	Zirconium	Hafnium	Rutherfordium	Units	Remarks
Chemical symbol	Ti	Zr	Hf	Rf		
Atomic number Z	22	40	72	104		
Characteristics				Radioactive		
Relative atomic mass A (atomic weight)	47.867(1)	91.224(2)	178.49(2)	[261]		
Abundance in lithosphere	4400×10^{-6}	220×10^{-6}	4.5×10^{-6}			Mass ratio
Abundance in sea	1×10^{-9}	3×10^{-11}	7×10^{-12}			Mass ratio
Atomic radius r_{cov}	132	145	144		pm	Covalent radius
Atomic radius r_{met}	147	160	156		pm	Metallic radius, CN = 12
Electron shells	–LMN	–MNO	–NOP	–OPQ		
Electronic ground state	$^3F_2$	$^3F_2$	$^3F_2$	$^3F_2$		
Electronic configuration	[Ar]$3d^24s^2$	[Kr]$4d^25s^2$	[Xe]$4f^{14}5d^26s^2$			
Oxidation states	4+, 3+	4+	4+			
Electron affinity	0.079	0.426	near 0		eV	
Electronegativity χ_A	1.32	1.22	(1.23)			Allred and Rochow
1st ionization energy	6.8282	6.63390	6.82507		eV	
2nd ionization energy	13.5755	13.13	14.9		eV	
3rd ionization energy	27.4917	22.99	23.3		eV	
4th ionization energy	43.2672	34.34	33.33		eV	
Standard electrode potential E^0	–1.630	–1.553[a]			V	Reaction type $Ti^{2+} + 2e^- = Ti$
	–0.368				V	Reaction type $Ti^{3+} + e^- = Ti^{2+}$
	–0.04				V	Reaction type $Ti^{4+} + e^- = Ti^{3+}$

[a] Reaction type $ZrO_2 + 4H^+ + 4e^- = Zr + 2H_2O$.

Table 2.1-14B(a) Elements of Group IVB (CAS notation), or Group 4 (new IUPAC notation). Crystallographic properties (see Table 2.1-14C for allotropic and high-pressure modifications)

Element name	Titanium	Zirconium	Hafnium	Rutherfordium			Units	Remarks
Chemical symbol	Ti	Zr	Hf	Rf				
Atomic number Z	22	40	72	104				
Modification	α-Ti	α-Zr	α-Hf					
Crystal system, Bravais lattice	hex	hex	hex					
Structure type	Mg	Mg	Mg					
Lattice constant a	295.03	323.17	319.46				pm	
Lattice constant c	468.36	514.76	505.11				pm	
Space group	$P6_3/mmc$	$P6_3/mmc$	$P6_3/mmc$					
Schoenflies symbol	D_{6h}^4	D_{6h}^4	D_{6h}^4					
Strukturbericht type	A3	A3	A3					
Pearson symbol	hP2	hP2	hP2					
Number A of atoms per cell	2	2	2					
Coordination number	12	6+6	12					
Shortest interatomic distance, solid	291		318				pm	

Table 2.1-14B(b) Elements of Group IVB (CAS notation), or Group 4 (new IUPAC notation). Mechanical properties

Element name	Titanium	Zirconium	Hafnium	Rutherfordium			Units	Remarks
Chemical symbol	Ti	Zr	Hf	Rf				
Atomic number Z	22	40	72	104				
Density ϱ, solid	4.50	6.49	13.10				g/cm^3	
Density ϱ, liquid	4.110	5.80	12.00				g/cm^3	
Molar volume V_{mol}	10.55	14.02	13.41				cm^3/mol	
Surface tension, liquid	1.65	1.48	1.63				N/m	Near T_m
Temperature coefficient	0.26×10^{-3}	-0.2×10^{-3}	-0.21×10^{-3}				N/(m K)	
Coefficient of linear thermal expansion α	8.35×10^{-6}	5.78×10^{-6}	5.9×10^{-6}				1/K	
Sound velocity, solid, transverse	2920	1950	2000				m/s	
Sound velocity, solid, longitudinal	6260	4360	3671				m/s	
Compressibility κ	0.779×10^{-5}	1.08×10^{-5}	0.80×10^{-5}				1/MPa	Volume compressibility
Elastic modulus E	102	68.0	138				GPa	
Shear modulus G	37.3	24.8	53.0				GPa	
Poisson number μ	0.35	0.37	0.29					
Elastic compliance s_{11}	9.69	10.1	7.16				1/TPa	
Elastic compliance s_{33}	6.86	8.0	6.13				1/TPa	
Elastic compliance s_{44}	21.5	30.1	18.0				1/TPa	
Elastic compliance s_{12}	-4.71	-4.0	-2.48				1/TPa	
Elastic compliance s_{13}	-1.82	-2.4	-1.57				1/TPa	
Elastic stiffness c_{11}	160	144	181				GPa	
Elastic stiffness c_{33}	181	166	197				GPa	
Elastic stiffness c_{44}	46.5	33.4	55.7				GPa	
Elastic stiffness c_{12}	90	74	77				GPa	
Elastic stiffness c_{13}	66	67	66				GPa	
Tensile strength	235	150–450	400				MPa	
Vickers hardness	2000–3500	903	1760					

Table 2.1-14B(c) Elements of Group IVB (CAS notation), or Group 4 (new IUPAC notation). Thermal and thermodynamic properties

Element name	Titanium	Zirconium	Hafnium	Rutherfordium		
Chemical symbol	Ti	Zr	Hf	Rf		
Atomic number Z	22	40	72	104		
					Units	Remarks
Modification		α-Zr				
Thermal conductivity λ	21	22.7	23.0		W/(m K)	
Molar heat capacity c_p	25.02	25.36	25.3		J/(mol K)	At 298 K
Standard entropy S^0	30.720	39.181	43.560		J/(mol K)	At 298 K and 100 kPa
Enthalpy difference $H_{298} - H_0$	4.8240	5.5663	5.8450		kJ/mol	
Melting temperature T_m	1941.00	2127.85	2506.00		K	
Enthalpy change ΔH_m	14.1460	20.9978	27.1960		kJ/mol	
Entropy change ΔS_m	7.288	9.868	10.852		J/(mol K)	
Volume change ΔV_m						$(V_l - V_s)/V_l$ at T_m
Boiling temperature T_b	3631	4203	4963		K	
Enthalpy change ΔH_b	410.0	561.3	575.5		kJ/mol	

Table 2.1-14B(d) Elements of Group IVB (CAS notation), or Group 4 (new IUPAC notation). Electronic, electromagnetic, and optical properties

Element name	Titanium	Zirconium	Hafnium	Rutherfordium		
Chemical symbol	Ti	Zr	Hf	Rf		
Atomic number Z	22	40	72	104		
					Units	Remarks
Modification		α-Zr				
Characteristics	Hard, light metal	Resistant metal	Metal			
Electrical resistivity ρ_s	390	410	296		nΩ m	At 293 K
Temperature coefficient	54.6×10^{-4}	44.0×10^{-4}	44×10^{-4}		1/K	
Pressure coefficient	-1.118×10^{-9}	0.33×10^{-9}	-0.87×10^{-9}		1/hPa	
Superconducting critical temperature T_{crit}	0.39	0.55	0.35		K	
Superconducting critical field H_{crit}	100	47			Oe	
Hall coefficient R	-1.2×10^{-10}	0.212×10^{-10}	0.43×10^{-10}		m^3/(A s)	At 300 K, $B = 0.4$–2.8 T
Electronic work function	4.31	4.05	3.9		V	
Thermal work function	4.16	4.12	3.53		V	
Molar magnetic susceptibility χ_{mol}, solid (SI)	1898×10^{-6}	1508×10^{-6}	892×10^{-6}		cm^3/mol	At 295 K
Molar magnetic susceptibility χ_{mol}, solid (cgs)	151×10^{-6}	120×10^{-6}	71×10^{-6}		cm^3/mol	At 295 K
Mass magnetic susceptibility χ_{mass}, solid (SI)	40.1×10^{-6}	16.8×10^{-6}	5.3×10^{-6}		cm^3/g	At 295 K

Table 2.1-14C Elements of Group IVB (CAS notation), or Group 4 (new IUPAC notation). Allotropic and high-pressure modifications

Element	Titanium		Zirconium		Hafnium			
Modification	α-Ti	β-Ti	α-Zr	β-Zr	α-Hf	β-Hf	Hf-II	
								Units
Crystal system, Bravais lattice	hex, cp	cub, bc	hex, cp	cub, bc	hex, cp	cub, bc	hex	
Structure type	Mg	W	Mg	W	Mg	W		
Lattice constant a	295.03	330.65	323.17	360.9	319.46	361.0		pm
Lattice constant c	468.36		514.76		505.11			pm
Space group	$P6_3/mmc$	$Im\bar{3}m$	$P6_3/mmc$	$Im\bar{3}m$	$P6_3/mmc$	$Im\bar{3}m$		
Schoenflies symbol	D_{6h}^4	O_h^9	D_{6h}^4	O_h^9	D_{6h}^4	O_h^9		
Strukturbericht type	A3	A2	A3	A2	A3	A2		
Pearson symbol	hP2	cI2	hP2	cI2	hP2	cI2		
Number A of atoms per cell	2	2	2	2	2	2		
Coordination number	12	8	6+6	8	12	8		
Shortest interatomic distance, solid	291	286		313	318	313		
Range of stability	RT	>1173 K	RT	>1138 K	RTP	>2268 K	>38.8 GPa	

Table 2.1-14D Elements of Group IVB (CAS notation), or Group 4 (new IUPAC notation). Ionic radii (determined from crystal structures)

Element	Titanium			Zirconium	Hafnium	
Ion	Ti^{2+}	Ti^{3+}	Ti^{4+}	Zr^{4+}	Hf^{4+}	
Coordination number						Units
4			42	59	58	pm
6	86	67	61	72	71	pm
8			74	84	83	pm
9				89		pm

2.1.5.3 Elements of the Main Groups and Subgroup V to VIII

Table 2.1-15A Elements of Group VA (CAS notation), or Group 15 (new IUPAC notation). Atomic, ionic, and molecular properties (see Table 2.1-15D for ionic radii)

Element name	Nitrogen	Phosphorus	Arsenic	Antimony	Bismuth	Units	Remarks
Chemical symbol	N	P	As	Sb	Bi		
Atomic number Z	7	15	33	51	83		
Relative atomic mass A (atomic weight)	14.00674(7)	30.973761(2)	74.92160(2)	121.760(1)	208.98038(2)		
Abundance in lithosphere	20×10^{-6}	1200×10^{-6}	5×10^{-6}	0.2×10^{-6}	0.2×10^{-6}		Mass ratio
Abundance in sea	150×10^{-6}	0.06×10^{-6}	3.7×10^{-9}	2.4×10^{-10}	2×10^{-11}		Mass ratio
Atomic radius r_{cov}	70	110	121	141	146	pm	Covalent radius
Atomic radius r_{met}	92	128	139	159	182	pm	Metallic radius, CN = 12
Atomic radius r_{vdW}	155	190	200	220	240	pm	van der Waals radius
Electron shells	KL	KLM	–LMN	–MNO	–NOP		
Electronic ground state	$^4S_{3/2}$	$^4S_{3/2}$	$^4S_{3/2}$	$^4S_{3/2}$	$^4S_{3/2}$		
Electronic configuration	[He]$2s^2 2p^3$	[Ne]$3s^2 3p^3$	[Ar]$3d^{10} 4s^2 4p^3$	[Kr]$4d^{10} 5s^2 5p^3$	[Xe]$4f^{14} 5d^{10} 6s^2 6p^3$		
Oxidation states	5+, 3+, 3–	5+, 3+, 3–	5+, 3+, 3–	5+, 3+, 3–	5+, 3+		
Electron affinity	Not stable	0.747	0.81	1.046	0.946	eV	
Electronegativity χ_A	3.07	2.06	2.20	1.82	(1.67)		Allred and Rochow
1st ionization energy	14.53414	10.48669	9.8152	8.64	7.289	eV	
2nd ionization energy	29.6013	19.7694	18.633	16.53051	16.69	eV	
3rd ionization energy	47.44924	30.2027	28.351	25.3	25.56	eV	
4th ionization energy	77.4735	51.4439	50.13	44.2	45.3	eV	
5th ionization energy	97.8902	65.0251	62.63	56	56.0	eV	
6th ionization energy	552.0718	220.421	127.6	108	88.3	eV	

Table 2.1–15B(a) Elements of Group VA (CAS notation), or Group 15 (new IUPAC notation). Crystallographic properties (see Table 2.1-15C for allotropic and high-pressure modifications)

Element name	Nitrogen	Phosphorus	Arsenic	Antimony	Bismuth		
Chemical symbol	N	P	As	Sb	Bi		
Atomic number Z	7	15	33	51	83		
						Units	Remarks
State	N_2, 4.2 K	P, black					
Crystal system, Bravais lattice	cub, sc	orth, C	trig, R	trig, R	trig, R		
Structure type	α-N	P, black	α-As	α-As	α-As		
Lattice constant a	565.9	331.36	413.20	450.65	474.60	pm	
Lattice constant b		1047.8				pm	
Lattice constant c		43 763				pm	
Lattice angle α			54.12	57.11	57.23	deg	
Space group	$Pa3$	$Cmca$	$R\bar{3}m$	$R\bar{3}m$	$R\bar{3}m$		
Schoenflies symbol	T_h^6	D_{2h}^{18}	D_{3d}^5	D_{3d}^5	D_{3d}^5		
Strukturbericht type		A11	A7	A7	A7		
Pearson symbol	cP8	oC8	hR2	hR2	hR2		
Number A of atoms per cell	4×2	8	2	2	2		
Coordination number		2+1	3	3	3		
Shortest interatomic distance, solid		222	252	291	307	pm	

Table 2.1-15B(b) Elements of Group VA (CAS notation), or Group 15 (new IUPAC notation). Mechanical properties

Element name	Nitrogen	Phosphorus	Arsenic	Antimony	Bismuth		
Chemical symbol	N	P	As	Sb	Bi		
Atomic number Z	7	15	33	51	83	Units	Remarks
State	N_2 gas	P, black					
Density ϱ, solid		2.690	5.72	6.68	9.80	g/cm^3	
Density ϱ, liquid					10.05	g/cm^3	At 273 K
Density ϱ, gas	1.2506×10^{-3}					g/cm^3	At 323 K
Molar volume V_{mol}	13.65		12.95	18.20	21.44	cm^3/mol	
Viscosity η, gas	18.9					μPa s	
Viscosity η, liquid				1.50	1.65	mPa s	Near T_m
Surface tension γ, liquid				0.384	0.376	N/m	
Coefficient of linear thermal expansion α			4.7×10^{-6}	8.5×10^{-6}	13.4×10^{-6}	1/K	
Sound velocity, gas	336.9					m/s	
Sound velocity, liquid	929 (70 K)			1800	1635	m/s	
Sound velocity, solid, transverse					1140	m/s	
Sound velocity, solid, longitudinal				3140	2298	m/s	
Compressibility κ				2.6×10^{-5}	2.86×10^{-5}	1/MPa	Volume compressibility
Elastic modulus E	1.2 ^{a,b}	30.4 ^{b}	22	54.4	34.0	GPa	
Shear modulus G				20.6	12.8	GPa	
Poisson number μ				0.25	0.33		
Elastic compliance s_{11}			46.71	16.1	26.0	1/TPa	
Elastic compliance s_{33}			202.9	29.9	42.0	1/TPa	
Elastic compliance s_{44}			44.91	38.9	114	1/TPa	
Elastic compliance s_{12}			36.94	−6.1	−7.9	1/TPa	
Elastic compliance s_{13}			−88.2	−6.2	−11.9	1/TPa	
Elastic compliance s_{14}			1.80	−12.3	−21.2	1/TPa	

^{a} Solid.
^{b} Estimated.

Table 2.1-15B(b) Elements of Group VA (CAS notation), or Group 15 (new IUPAC notation). Mechanical properties, cont.

Element name	Nitrogen	Phosphorus	Arsenic	Antimony	Bismuth		
Chemical symbol	N	P	As	Sb	Bi		
Atomic number Z	7	15	33	51	83	Units	Remarks
State	N_2 gas	P, black					
Elastic stiffness c_{11}		73.9	123.6	101	63.4	GPa	
Elastic stiffness c_{22}		277				GPa	
Elastic stiffness c_{33}	53.7	53.7	59.1	44.9	37.9	GPa	
Elastic stiffness c_{44}		15.6	22.6	39.5	11.5	GPa	
Elastic stiffness c_{55}		11.5				GPa	
Elastic stiffness c_{66}		56.7				GPa	
Elastic stiffness c_{12}			19.7	32.2	24.5	GPa	
Elastic stiffness c_{13}			62.3	27.6	24.9	GPa	
Elastic stiffness c_{14}			−4.16	21.8	7.3	GPa	
Brinell hardness					200		
Mohs hardness			3.5				
Solubility in water α_W, 293 K	0.01557						α_W = vol (gas)/vol (water)

Table 2.1–15B(c) Elements of Group VA (CAS notation), or Group 15 (new IUPAC notation). Thermal and thermodynamic properties

Element name	Nitrogen	Phosphorus	Arsenic	Antimony	Bismuth		
Chemical symbol	N	P	As	Sb	Bi		
Atomic number Z	7	15	33	51	83		
						Units	Remarks
State	N_2 gas	P, white [a]	α-As				
Thermal conductivity λ	0.02598		50.0	25.9	7.87	W/(m K)	
Molar heat capacity c_p	14.560	23.824	24.65	25.23	25.52	J/(mol K)	At 298 K
Standard entropy S^0	191.611	41.090	35.689	45.522	56.735	J/(mol K)	At 298 K and 100 kPa
Enthalpy difference $H_{298} - H_0$	8.6692	5.3600	5.1170	5.8702	6.4266	kJ/mol	
Melting temperature T_m	63.1458	317.30	1090.00	903.78	544.55	K	
Enthalpy change ΔH_m	0.720	0.6590	24.4429	19.8740	11.2968	kJ/mol	
Entropy change ΔS_m		2.077	22.425	21.990	20.745	J/(mol K)	
Volume change ΔV_m			0.10	−0.008	−0.33		$(V_l - V_s)/V_l$ at T_m
Boiling temperature T_b	77.35	550		1860	1837	K	
Enthalpy change ΔH_b	5.577	51.9	34.8	165.8	174.1	kJ/mol	
Critical temperature T_c	126.25		1089			K	
Critical pressure p_c	3.40		36			MPa	
Critical density ϱ_c	0.311					g/cm^3	
Triple-point temperature T_{tr}	63.14					K	
Triple-point pressure p_{tr}	12.5					kPa	

[a] In Landolt–Börnstein, Group IV, Vol. 19A, Part 1 [1.4], the white form of phosphorus has been chosen as the reference phase for all phosphides because the more stable red form is difficult to characterize.

Table 2.1-15B(d) Elements of Group VA (CAS notation), or Group 15 (new IUPAC notation). Electronic, electromagnetic, and optical properties

Element name	Nitrogen	Phosphorus	Arsenic	Antimony	Bismuth	Units	Remarks
Chemical symbol	N	P	As	Sb	Bi		
Atomic number Z	7	15	33	51	83		
State	N_2 gas	P, black					
Characteristics	Gas	Semiconductor	Semiconductor	Semimetal	Brittle metal		
Electrical resistivity ρ_s			260	370	1068	$n\Omega\,m$	Solid, at RT
Temperature coefficient			42×10^{-4}	51.1×10^{-4}	45.4×10^{-4}	1/K	
Pressure coefficient				6.0×10^{-9}	15.2×10^{-9}	1/hPa	
Electrical resistivty ρ_l				1135	1280	$n\Omega\,m$	Liquid
Resistivity ratio				0.61	0.43		ρ_l/ρ_s at T_m
Electronic band gap ΔE		0.57	1.14			eV	At 300 K
Temperature coefficient		8×10^{-4}				eV/K	
Electronic work function			4.79	4.56	4.36	V	
Thermal work function			5.71	4.08	4.28	V	
Electron mobility		220				$cm^2/(V\,s)$	
Hole mobility		350				$cm^2/(V\,s)$	
Hall coefficient R			0.45×10^{-7}	0.27×10^{-7}	-6.33×10^{-7}	$m^3/(A\,s)$	$B = 0.4-1.0\,T$
Thermoelectric coefficient				35	-70	$\mu V/K$	
Dielectric constant $(\varepsilon - 1)$, gas	580×10^{-6}						At 293 K
Dielectric constant ε, liquid	1.45						At 74.8 K
Dielectric constant ε, solid			11.2 (optical)				
Molar magnetic suscepti-bility χ_{mol}, (SI)	-151×10^{-6}	a,b	-70.4×10^{-6} c	-1244×10^{-6}	-3520×10^{-6}	cm^3/mol	At 295 K
Molar magnetic suscepti-bility χ_{mol}, (cgs)	-12.0×10^{-6}	a,b	-5.60×10^{-6} c	-99×10^{-6}	-280.1×10^{-6}	cm^3/mol	At 295 K
Mass magnetic suscepti-bility χ_{mass}, (SI)	-5.4×10^{-6}		-3.9×10^{-6}	-10×10^{-6}	-16.8×10^{-6}	cm^3/g	At 295 K
Refractive index $(n-1)$, gas	297×10^{-6}						589.3 nm
Refractive index n, liquid	1.929 (78 K)						589.3 nm
Refractive index n, solid			3.35 (0.8 μm)				

[a] White P: molar magnetic susceptibility: -26.7×10^{-6} (cgs) and -335×10^{-6} (SI) cm^3/mol.
[b] Red P: molar magnetic susceptibility: -20.8×10^{-6} (cgs) and -261×10^{-6} (SI) cm^3/mol.
[c] Yellow As: molar magnetic susceptibility: -292×10^{-6} (SI) and -23.2×10^{-6} (cgs) cm^3/mol.

Table 2.1-15C Elements of Group VA (CAS notation), or Group 15 (new IUPAC notation). Allotropic and high-pressure modifications

Element	Nitrogen			Arsenic		Antimony				Units
Modification	α-N	β-N	γ-N	α-As	ε-As	Sb-I	Sb-II	Sb-III	Sb-IV	
Crystal system, Bravais lattice	cub, P	hex	tetr	trig, R	orth	trig, R	cub	hex, cp	mon	
Structure type	α-N	La		As	α-Ga	α-As		Mg		
Lattice constant a	565.9	404.6	395.7	413.20	362	450.65	299.2	337.6	556	pm
Lattice constant b					1085				404	pm
Lattice constant c		662.9	510.1		448			534.1	422	pm
Lattice angle α				54.12		57.11				deg
Lattice angle β									86.0	deg
Space group	$Pa\bar{3}$	$P6_3/mmc$	$P4_2/mnm$	$R\bar{3}m$	$Cmca$	$R\bar{3}m$	$Pm\bar{3}m$	$P6_3/mmc$		
Schoenflies symbol	T_h^6	D_{6h}^4	D_{4h}^{12}	D_{3d}^5	D_{2h}^{18}	D_{3d}^5	O_h^1	D_{6h}^4		
Strukturbericht type		A9		A7	A11	A7		A3		
Pearson symbol	cP8	hP4	tP4	hR2	oC8	hR2	cP1	hP2		
Number A of atoms per cell	4×2	4	4	2	8	2	1	2		
Coordination number		3		3		3				
Shortest interatomic distance, solid				252		291				pm
Range of stability	< 20 K	> 35.6 K	20 K, > 3.3 GPa		> 721 K	RTP	> 5.0 GPa	> 7.5 GPa	14.0 GPa	

Table 2.1-15C Elements of Group VA (CAS notation), or Group 15 (new IUPAC notation). Allotropic and high-pressure modifications, cont.

Element	Bismuth					Units
Modification	α-Bi	β-Bi	γ-Bi	δ-Bi	ε-Bi	ζ-Bi	
Crystal system, Bravais lattice	trig, R	mon	mon			cub, bc	
Structure type	α-As					W	
Lattice constant a	474.60		605			3800	pm
Lattice constant b			420				pm
Lattice constant c			465				pm
Lattice angle α	57.23						deg
Space group	$R\bar{3}m$	$C2/m$				$Im\bar{3}m$	
Schoenflies symbol	D_{3d}^5	C_{2h}^3				O_h^9	
Strukturbericht type	A7					A2	
Pearson symbol	hR2	mC4	mP3			cI2	
Number A of atoms per cell	2	4	3			2	
Coordination number	3						
Shortest interatomic distance, solid	307						pm
Range of stability	RTP	> 0.28 GPa	> 3.0 GPa	> 4.3 GPa	> 6.5 GPa	> 9.0 GPa	

Table 2.1-15D Elements of Group VA (CAS notation), or Group 15 (new IUPAC notation). Ionic radii (determined from crystal structures)

Element	Nitrogen		Phosphorus	Arsenic		Antimony		Bismuth		
Ion	N^{3+}	N^{5+}	P^{5+}	As^{3+}	As^{5+}	Sb^{3+}	Sb^{5+}	Bi^{3+}	Bi^{5+}	Units
Coordination number										
4		13	17		34					pm
5										pm
6	16		38	58	46	76	60	96	76	pm
8								103		pm
								117		

Table 2.1-16A Elements of Group VB (CAS notation), or Group 5 (new IUPAC notation). Atomic, ionic, and molecular properties (see Table 2.1-16D for ionic radii)

Element name	Vanadium	Niobium	Tantalum	Dubnium	Units	Remarks
Chemical symbol	V	Nb	Ta	Db		
Atomic number Z	23	41	73	105		
Relative atomic mass A (atomic weight)	50.9415(1)	92.90638(2)	180.9479(1)	[262]		
Abundance in lithosphere	150×10^{-6}	20×10^{-6}	2.1×10^{-6}			Mass ratio
Abundance in sea	2.5×10^{-9}	1×10^{-11}	2×10^{-12}			Mass ratio
Atomic radius r_{cov}	122	134	134		pm	Covalent radius
Atomic radius r_{met}	135	147	147		pm	Metallic radius, CN = 12
Electron shells	–LMN	–MNO	–NOP	–OPQ		
Electronic ground state	$^4F_{3/2}$	$^6D_{1/2}$	$^4F_{3/2}$			
Electronic configuration	$[Ar]3d^34s^2$	$[Kr]4d^45s^1$	$[Xe]4f^{14}5d^36s^2$			
Oxidation states	5+, 4+, 3+, 2+	5+, 3+	5+			
Electron affinity	0.525	0.893	0.322		eV	
Electronegativity χ_A	1.45	1.23	(1.33)			Allred and Rochow
1st ionization energy	6.7463	6.75885	7.89		eV	
2nd ionization energy	14.66	14.32			eV	
3rd ionization energy	29.311	25.04			eV	
4th ionization energy	46.709	38.3			eV	
5th ionization energy	65.282	50.55			eV	
Standard electrode potential E^0	−1.175				V	Reaction type $V^{2+} + 2e^- = V$
	−0.255				V	Reaction type $V^{3+} + e^- = V^{2+}$

Table 2.1-16B(a) Elements of Group VB (CAS notation), or Group 5 (new IUPAC notation). Crystallographic properties

Element name	Vanadium	Niobium	Tantalum	Dubnium		
Chemical symbol	V	Nb	Ta	Db		
Atomic number Z	23	41	73	105		
					Units	Remarks
Crystal system, Bravais lattice	cub, bc	cub, bc	cub, bc			
Structure type	W	W	W			
Lattice constant a	302.38	330.07	330.31		pm	
Space group	$Im\bar{3}m$	$Im\bar{3}m$	$Im\bar{3}m$			
Schoenflies symbol	O_h^9	O_h^9	O_h^9			
Strukturbericht type	A2	A2	A2			
Pearson symbol	cI2	cI2	cI2			
Number A of atoms per cell	2	2	2			
Coordination number	8	8	8			
Shortest interatomic distance, solid	263	285	285		pm	

Table 2.1-16B(b) Elements of Group VB (CAS notation), or Group 5 (new IUPAC notation). Mechanical properties

Element name	Vanadium	Niobium	Tantalum	Dubnium		
Chemical symbol	V	Nb	Ta	Db		
Atomic number Z	23	41	73	105		
					Units	Remarks
Density ϱ, solid	5.80	8.35	16.60		g/cm^3	
Density ϱ, liquid	5.55	7.830	15,00		g/cm^3	
Molar volume V_{mol}	8.34	10.84	10.87		cm^3/mol	
Surface tension, liquid	1.95	2.0	2.15		N/m	
Temperature coefficient	0.3×10^{-3}	-0.24×10^{-3}	-0.25×10^{-3}		N/(m K)	
Coefficient of linear thermal expansion α	8.3×10^{-6}	7.34×10^{-6}	6.64×10^{-6}		1/K	300 K
Sound velocity, solid, transverse	2780	2100	2900		m/s	
Sound velocity, solid, longitudinal	6000	4900	4100		m/s	
Compressibility κ	0.63×10^{-5}	0.56×10^{-5}	0.465×10^{-5}		1/MPa	Volume compressibility
Elastic modulus E	127	104	185		GPa	
Shear modulus G	46.6	59.5	64.7		GPa	
Poisson number μ	0.36	0.38	0.35			
Elastic compliance s_{11}	6.75	6.56	6.89		1/TPa	
Elastic compliance s_{44}	23.2	35.2	12.1		1/TPa	
Elastic compliance s_{12}	-2.31	-2.29	-2.58		1/TPa	
Elastic stiffness c_{11}	230	245	264		GPa	
Elastic stiffness c_{44}	43.1	28.4	82.6		GPa	
Elastic stiffness c_{12}	120	132	158		GPa	
Vickers hardness	630	70–250 (HV 10)	80–300 (HV 10)			At 293 K

Table 2.1-16B(c) Elements of Group VB (CAS notation), or Group 5 (new IUPAC notation). Thermal and thermodynamic properties

Element name	Vanadium	Niobium	Tantalum	Dubnium	Units	Remarks
Chemical symbol	V	Nb	Ta	Db		
Atomic number Z	23	41	73	105		
Thermal conductivity λ	30.7	59.0	60.7		W/(m K)	At 300 K
Molar heat capacity c_p	24.90	24.69	25.30		J/(mol K)	At 298 K
Standard entropy S^0	30.890	36.270	41.472		J/(mol K)	At 298 K and 100 kPa
Enthalpy difference $H_{298} - H_0$	4.5070	5.2200	5.6819		kJ/mol	
Melting temperature T_m	2183.00	2750.00	3290.00		K	
Enthalpy change ΔH_m	21.5000	30.0000	36.5682		kJ/mol	
Entropy change ΔS_m	9.849	10.909	11.115		J/(mol K)	
Relative volume change ΔV_m						$(V_l - V_s)/V_l$ at T_m
Boiling temperature T_b	3690	5017	5778		K	
Enthalpy change ΔH_b	451.8	683.2	743.1		kJ/mol	

Table 2.1-16B(d) Elements of Group VB (CAS notation), or Group 5 (new IUPAC notation). Electronic, electromagnetic, and optical properties. There is no Table 2.1-16C, because no allotropic or high-pressure modifications are known

Element name	Vanadium	Niobium	Tantalum	Dubnium	Units	Remarks
Chemical symbol	V	Nb	Ta	Db		
Atomic number Z	23	41	73	105		
Electrical resistivity ρ_s	248	152	125		nΩ m	At RT
Temperature coefficient	39.0×10^{-4}	25.8×10^{-4}	38.2×10^{-4}		1/K	
Pressure coefficient	-1.6×10^{-9}	-1.37×10^{-9}	-1.62×10^{-9}		1/hPa	
Superconducting critical temperature T_{crit}	5.3	9.13	4.49		K	
Superconducting critical field H_{crit}	1020	1980	830		Oe	
Hall coefficient R	0.82×10^{-10}	0.88×10^{-10}	1.01×10^{-10}		m^3/(A s)	At 273 K, $B = 0.5$–2.9 T
Thermoelectric coefficient			-5.0		μV/K	
Electronic work function	4.3	4.3	4.3		V	
Thermal work function	4.09	3.99	4.25		V	
Molar magnetic susceptibility, solid χ_{mol}, (SI)	3581×10^{-6}	2614×10^{-6}	1935×10^{-6}		cm^3/mol	At 295 K
Molar magnetic susceptibility, solid χ_{mol}, (cgs)	285×10^{-6}	208×10^{-6}	154×10^{-6}		cm^3/mol	At 295 K
Mass magnetic susceptibility, solid χ_{mass}, (SI)	62.8×10^{-6}	27.6×10^{-6}	10.7×10^{-6}		cm^3/g	At 295 K

Table 2.1-16D Elements of Group VB (CAS notation), or Group 5 (new IUPAC notation). Ionic radii (determined from crystal structures)

Element	Vanadium			Niobium			Tantalum			
Ion	V^{2+}	V^{3+}	V^{4+}	V^{5+}	Nb^{3+}	Nb^{4+}	Nb^{5+}	Ta^{3+}	Ta^{4+}	Ta^{5+}	Units
Coordination number											
4				36			48				pm
5			53	46							pm
6	79	64	58	54	72	68	64	72	68	64	pm
8			72		79		74				pm

Table 2.1-17A Elements of Group VIA (CAS notation), or Group 16 (new IUPAP notation). Atomic, ionic and molecular properties (see Table 2.1-17D for ionic radii)

Element name	Oxygen	Sulfur	Selenium	Tellurium	Polonium	Units	Remarks
Chemical symbol	O	S	Se	Te	Po		
Atomic number Z	8	16	34	52	84		
Characteristics					Radioactive		
Relative atomic mass A (atomic weight)	15.9994(3)	32.066(6)	78.96(3)	127.60(3)	[209]		Mass ratio
Abundance in lithosphere	$464\,000 \times 10^{-6}$	520×10^{-6}	0.09×10^{-6}	2×10^{-9}			Mass ratio
Abundance in sea	$880\,000 \times 10^{-6}$	905×10^{-6}	2×10^{-10}				
Atomic radius r_{cov}	66	104	117	137	146	pm	Covalent radius
Atomic radius r_{met}		127	140	160	176	pm	Metallic radius, CN = 12
Atomic radius r_{vdW}	150	185	200	220			van der Waals radius
Electron shells	KL	KLM	–LMN	–MNO	–NOP		
Electronic ground state	$^3P_2$	$^3P_2$	$^3P_2$	$^3P_2$	$^3P_2$		
Electronic configuration	[He]$2s^2 2p^4$	[Ne]$3s^2 3p^4$	[Ar]$3d^{10}4s^2 4p^4$	[Kr]$4d^{10}5s^2 5p^4$	[Xe]$4f^{14}5d^{10}6s^2 6p^4$		
Oxidation states	2–	6+, 4+, 2+, 2–	6+, 4+, 2+, 2–	6+, 4+, 2+, 2–	4+, 2+		
Electron affinity	1.46	2.08	2.02	1.97	1.9	eV	Reaction type $O + e^- = O^-$
	–8.75	–5.51				eV	Reaction type $O^- + e^- = O^{2-}$
Electronegativity χ_A	3.50	2.44	2.48	2.01	(1.76)		Allred and Rochow
1st ionization energy	13.61806	10.36001	9.75238	9.0096	8.41671	eV	
2nd ionization energy	35.11730	23.3379	21.19	18.6		eV	
3rd ionization energy	54.9355	34.79	30.8204	27.96		eV	
4th ionization energy	77.41353	47.222	42.9450	37.41		eV	
Standard electrode potential E^0	–0.476	–0.924		–1.143	+0.56	V	Reaction type $Te^{2-} = Te - 2e^-$
						V	Reaction type $Po^{3+} + 3e^- = Po$
				+0.568		V	Reaction type $Te^{4+} + 4e^- = Te$

Table 2.1-17B(a) Elements of Group VIA (CAS notation), or Group 16 (new IUPAC notation). Crystallographic properties (see Table 2.1-17C for allotropic and high-pressure modifications)

Element name	Oxygen	Sulfur	Selenium	Tellurium	Polonium		
Chemical symbol	O	S	Se	Te	Po		
Atomic number Z	8	16	34	52	84		
						Units	Remarks
State	α-O, $T < 23$ K	S_8, α-S	Gray Se, Se chains		α-Po		
Crystal system, Bravais lattice	mon, C	orth, F	hex	hex	cub, sc		
Structure type	α-O	α-S	γ-Se	γ-Se	α-Po		
Lattice constant a	540.3	1046.4	436.55	445.61	336.6	pm	
Lattice constant b	342.9	1286.60				pm	
Lattice constant c	508.6	2448.60	495.76	592.71		pm	
Lattice angle γ	132.53					deg	
Space group	$C2/m$	$Fddd$	$P3_121$	$P3_121$	$Pm3m$		
Schoenflies symbol	C_{2h}^3	D_{2h}^{24}	D_3^4	D_3^4	O_h^1		
Strukturbericht type		A16	A8	A8	A_h		
Pearson symbol	mC4	oF128	hP3	hP3	cP1		
Number A of atoms per cell	2×2	128	3	3	1		
Coordination number		2	2	2+4	6		
Shortest interatomic distance, solid		204	237	283	337	pm	

Table 2.1–17B(b) Elements of Group VIA (CAS notation), or Group 16 (new IUPAC notation). Mechanical properties

Element name	Oxygen	Sulfur	Selenium	Tellurium	Polonium		
Chemical symbol	O	S	Se	Te	Po		
Atomic number Z	8	16	34	52	84		
						Units	Remarks
State	O_2 gas	α-S	Gray Se				
Density ϱ, solid		2.037	4.79	6.24	9.40	g/cm^3	At 293 K
Density ϱ, liquid		1.819	3.990	5.797		g/cm^3	
Density ϱ, gas	1.429×10^{-3}					g/cm^3	
Molar volume V_{mol}	8.00	15.49	16.48	20.45	22.4	cm^3/mol	
Viscosity η, gas	19.5					µPa s	
Viscosity η, liquid		11.5	1260			mPa s	At T_m
Surface tension, liquid		0.061	0.106	0.186		N/m	At T_m
Coefficient of linear thermal expansion α		74.33×10^{-6}	36.9×10^{-6}	16.75×10^{-6}	23×10^{-6}	1/K	
Sound velocity, gas	336.95 (70 K)					m/s	
Sound velocity, liquid	1079 (70 K)					m/s	
Compressibility κ		13.0×10^{-5}	11.6×10^{-5}	4.8×10^{-5}		1/MPa	Volume compressibilty
Elastic modulus E		17.8 [a]	58.0	47.1	26 [a]	GPa	
Shear modulus G			6.46	16.7		GPa	
Poisson number μ			0.45	0.23			
Elastic compliance s_{11}	1280 [b]	74.6				1/TPa	
Elastic compliance s_{22}		111				1/TPa	
Elastic compliance s_{33}		75.4				1/TPa	
Elastic compliance s_{44}	3640 [b]	121				1/TPa	
Elastic compliance s_{55}		234				1/TPa	
Elastic compliance s_{66}		229				1/TPa	
Elastic compliance s_{12}	−570 [b]	−13.1				1/TPa	
Elastic compliance s_{13}		−7.1				1/TPa	
Elastic compliance s_{23}		−45.8				1/TPa	
Elastic stiffness c_{11}	2.60 [b]	14.22				GPa	
Elastic stiffness c_{22}		12.68				GPa	
Elastic stiffness c_{33}		18.30				GPa	
Elastic stiffness c_{44}	0.275 [b]	8.27				GPa	
Elastic stiffness c_{55}		4.28				GPa	
Elastic stiffness c_{66}		4.37				GPa	
Elastic stiffness c_{12}	2.06 [b]	2.99				GPa	
Elastic stiffness c_{13}		3.14				GPa	
Elastic stiffness c_{23}		7.95				GPa	
Tensile strength				10.8–12.25		MPa	
Brinell hardness				250			
Mohs hardness			2.0				
Solubility in water α_W [c]	0.0310						At 293 K and 1013 hPa

[a] Estimated.   [b] γ-Oxygen, $T = 54.4$ K.   [c] α_W = vol (gas)/vol (water).

Table 2.1-17B(c) Elements of Group VIA (CAS notation), or Group 16 (new IUPAC notation). Thermal and thermodynamic properties

Element name	Oxygen	Sulfur	Selenium	Tellurium	Polonium	Units	Remarks
Chemical symbol	O	S	Se	Te	Po		
Atomic number Z	8	16	34	52	84		
State	O_2 gas	α-S	α-Se	α-Te	α-Po		
Thermal conductivity λ	0.0245	0.269	2.48	1.7	20	W/(m K)	At STP
Molar heat capacity c_p	14.690	22.70	25.04	25.73		J/(mol K)	At 298 K
Standard entropy S^0	205.147	32.070	41.966	49.221	62.000	J/(mol K)	At 298 K and 100 kPa
Enthalpy difference $H_{298} - H_0$	8.6800	4.4120	5.5145	6.0800	6.700	kJ/mol	
Melting temperature T_m	54.361	388.36	494	722.66	527.00	K	
Enthalpy change ΔH_m	0.444	1.7210	6.6944	17.3760	10.000	kJ/mol	
Entropy change ΔS_m		4.431	13.551	24.045	18.975	J/(mol K)	
Relative volume change ΔV_m		0.515	+0.168	0.05			$(V_l - V_s)/V_l$ at T_m
Boiling temperature T_b	90.18	882	958	1261	1335	K	
Enthalpy change ΔH_b	6.2	9.62	90	104.6	100.8	kJ/mol	
Critical temperature T_c	154.58		1863			K	
Critical pressure p_c	5.4		38			MPa	
Critical density ϱ_c	0.419					g/cm^3	
Triple-point temperature T_{tr}	54.4					K	
Triple-point pressure p_{tr}	1.52					hPa	

Table 2.1-17B(d) Elements of Group VIA (CAS notation), or Group 16 (new IUPAC notation). Electronic, electromagnetic, and optical properties

Element name	Oxygen	Sulfur	Selenium	Tellurium	Polonium	Units	Remarks
Chemical symbol	O	S	Se	Te	Po		
Atomic number Z	8	16	34	52	84		
State	O_2 gas						
Characteristics		Solid insulator	Gray Se Semiconductor	Semiconductor	Volatile metal		
Electrical resistivity ρ_s			100	1–50		$M\Omega\,m$	At RT
Electrical resistivity ρ_l			20	$6.0\,\mu\Omega\,m$		$M\Omega\,m$	At 298 K
Resistivity ratio at T_m			1.0	0.048–0.091			
Hall coefficient R				0.24×10^{-10}		$m^3/(A\,s)$	At 298 K
Thermoelectric coefficient				400		$\mu V/K$	
Electronic band gap ΔE		3.6	1.79	0.33		eV	
Temperature coefficient		-6.8×10^{-4}	-9×10^{-4}			eV/K	
Electronic work function			5.9	4.95		V	
Thermal work function			4.72	4.73		V	
Electron mobility				1100		$cm^2/(V\,s)$	At 375 K
Hole mobility				560		$cm^2/(V\,s)$	At 815 K
Dielectric constant $(\varepsilon - 1)$, gas	525×10^{-6}						
Dielectric constant ε, liquid	1.505						
Dielectric constant ε, solid			8.5 ($\lambda = 3.3$ cm)	5.0 $\parallel c$; 2.2 $\perp c$			
Molar magnetic susceptibility χ_{mol}(SI)	$43\,341 \times 10^{-6}$ [a]	-195×10^{-6}	-314×10^{-6}	-478×10^{-6}		cm^3/mol	At 295 K
Molar magnetic susceptibility χ_{mol} (cgs)	3449×10^{-6} [b]	-15.5×10^{-6}	-25×10^{-6}	-38×10^{-6}		cm^3/mol	At 295 K
Mass magnetic susceptibility χ_{mass} (SI)	1.34×10^{-6} [c]	-6.09×10^{-6}	-4.0×10^{-6}	-3.9×10^{-6}		cm^3/g	At 295 K
Mass magnetic susceptibility χ_{mass}, liquid (SI)				-0.6×10^{-6}		cm^3/g	
Refractive index $(n-1)$, gas	270.6×10^{-6}						$\lambda = 589.3$ nm
Refractive index n, liquid	1.221 (92 K)						$\lambda = 589.3$ nm
Refractive index n, solid			4.0				$\lambda = 589.3$ nm

[a] Liquid O_2, 90 K, 96 748 cm^3/mol; solid O_2, 54 K, 128 177 cm^3/mol. [b] Liquid O_2, 90 K, 7699 cm^3/mol; solid O_2, 54 K, 10 200 cm^3/mol.
[c] At 280 K.

Further remarks

Liquid O_3 has a molar magnetic susceptibility of 84.2×10^{-6} cm^3/mol (SI) and 6.7×10^{-6} cm^3/mol (cgs).
The values given for the molar magnetic susceptibility χ_{mol} of sulfur apply to rhombic sulfur. The corresponding values for monoclinic sulfur are -187×10^{-6} cm^3/mol (SI) and -14.9×10^{-6} cm^3/mol (cgs).

Table 2.1–17C Elements of Group VIA (CAS notation), or Group 16 (new IUPAC notation). Allotropic and high-pressure modifications

Element	Oxygen			Selenium			
Modification	α-O	β-O	γ-O	γ-Se	α-Se	β-Se	
							Units
Crystal system, Bravais lattice	mon, C	trig, R	cub, P	hex	mon	mon	
Structure type	α-O	α-As	γ-O	γ-Se			
Lattice constant a	540.3	421.0	683	436.55	905.4	1501.8	pm
Lattice constant b	342.9				908.3	1471.3	pm
Lattice constant c	508.6			495.76	233.6	887.9	pm
Lattice angle α		46.27					deg
Lattice angle γ	132.53				90.82	93.6	deg
Space group	$C2/m$	$R\bar{3}m$	$Pm\bar{3}m$	$P3_121$	$P2_1/m$	$P2_1/b$	
Schoenflies symbol	C_{2h}^3	D_{3d}^5	O_h^3	D_3^4	C_{2h}^2	C_{2h}^5	
Strukturbericht type		A7	A15	A8			
Pearson symbol	mC4	hR2	cP16	hP3		mP32	
Number A of atoms per cell	2×2	2	16	3		32	
Coordination number				2	2	2	
Shortest interatomic distance, solid				237	233–235	233–236	
Range of stability	< 23 K	> 23.9 K	> 43.6 K	RT	RT	RT	
Characteristics				Gray, Se chains	Red, Se₈ rings;	Red, Se₈ rings	

Table 2.1–17C Elements of Group VIA, or Group 16. Allotropic and high-pressure modifications, cont.

Element	Tellurium			Polonium		
Modification	α-Te	β-Te	γ-Te	α-Po	β-Po	
						Units
Crystal system, Bravais lattice	hex	trig, R	trig, R	cub, P	trig, R	
Structure type	γ-Se	α-As	α-Hg	α-Po	α-Hg	
Lattice constant a	445.61	469	300.2	336.6	337.3	pm
Lattice constant c	592.71					pm
Lattice angle α		53.30	103.3		98.08	deg
Space group	$P3_121$	$R\bar{3}m$	$R\bar{3}m$	$Pm\bar{3}m$	$R\bar{3}m$	
Schoenflies symbol	D_3^4	D_{3d}^5	D_{3d}^5	O_h^1	D_{3d}^5	
Strukturbericht type	A8	A7	A10	A_h	A10	
Pearson symbol	hP3	hR2	hR1	cP1	hR1	
Number A of atoms per cell	3	2	1	1	1	
Coordination number	2+4			6	6	
Shortest interatomic distance, solid	283			337	337	
Range of stability	RTP	> 2 GPa	> 7.0 GPa	RTP	> 327 K	

Table 2.1–17D Elements of Group VIA (CAS notation), or Group 16 (new IUPAC notation). Ionic radii (determined from crystal structures)

Element	Oxygen	Sulfur			Selenium			Tellurium			Polonium	
Ion	O^{2-}	S^{2-}	S^{4+}	S^{6+}	Se^{2-}	Se^{4+}	Se^{6+}	Te^{2-}	Te^{4+}	Te^{6+}	Po^{4+}	
Coordination number												Units
2	121											pm
4				12		28			66	43		pm
6	140	184	37	29	198	50	42	221	97	56	97	pm
8	142											pm

Table 2.1-18A Elements of Group VIB (CAS notation), or Group 6 (new IUPAC notation). Atomic, ionic, and molecular properties (see Table 2.1-18D for ionic radii)

Element name	Chromium	Molybdenum	Tungsten	Seaborgium	Units	Remarks
Chemical symbol	Cr	Mo	W	Sg		
Atomic number Z	24	42	74	106		
Characteristics				Radioactive		
Relative atomic mass A (atomic weight)	51.9961(6)	95.94(1)	183.84(1)			
Abundance in lithosphere	200×10^{-6}	2.3×10^{-6}	1×10^{-6}			Mass ratio
Abundance in sea	3×10^{-10}	0.01×10^{-6}	1×10^{-10}			Mass ratio
Atomic radius r_{cov}	118	130	130		pm	Covalent radius
Atomic radius r_{met}	129	140	141		pm	Metallic radius, CN = 12
Electron shells	–LMN	–MNO	–NOP	–OPQ		
Electronic ground state	$^7S_3$	$^7S_3$	$^5D_0$			
Electronic configuration	[Ar]3d^{5}4s^1	[Kr]4d^{5}5s^1	[Xe]4f^{14}5d^{4}6s^2			
Oxidation states	6+, 3+, 2+	6+, 5+, 4+, 3+, 2+, 1+, 2–	6+, 5+, 4+, 3+, 2+, 1–, 2–			
Electron affinity	0.666	0.748	0.815		eV	
Electronegativity χ_A	1.56	1.30	(1.40)			Allred and Rochow
1st ionization energy	6.76664	7.09243	7.98		eV	
2nd ionization energy	16.4857	16.16			eV	
3rd ionization energy	30.96	27.13			eV	
4th ionization energy	49.16	46.4			eV	
Standard electrode potential E^0	–0.913	–0.2			V	Reaction type $Cr^{2+} + 2e^- = Cr$
	–0.744				V	Reaction type $Cr^{3+} + 3e^- = Cr$
	–0.407				V	Reaction type $Cr^{3+} + e^- = Cr^{2+}$

Table 2.1–18B(a) Elements of Group VIB (CAS notation), or Group 6 (new IUPAC notation). Crystallographic properties (see Table 2.1-18C for allotropic and high-pressure modifications)

Element name	Chromium	Molybdenum	Tungsten	Seaborgium		
Chemical symbol	Cr	Mo	W	Sg		
Atomic number Z	24	42	74	106		
					Units	Remarks
Crystal system, Bravais lattice	cub, bc	cub, bc	cub, bc			
Structure type	W	W	W			
Lattice constant a	288.47	314.70	316.51		pm	
Space group	$Im\bar{3}m$	$Im\bar{3}m$	$Im\bar{3}m$			
Schoenflies symbol	O_h^9	O_h^9	O_h^9			
Strukturbericht type	A2	A2	A2			
Pearson symbol	cI2	cI2	cI2			
Number A of atoms per cell	2	2	2			
Coordination number	8	8	8			
Shortest interatomic distance, solid	249	272	273		pm	

Table 2.1–18B(b) Elements of Group VIB (CAS notation), or Group 6 (new IUPAC notation). Mechanical properties

Element name	Chromium	Molybdenum	Tungsten	Seaborgium		
Chemical symbol	Cr	Mo	W	Sg		
Atomic number Z	24	42	74	106		
					Units	Remarks
Density ϱ, solid	7.19	10.220	19.30		g/cm^3	
Density ϱ, liquid	6.460		17.60		g/cm^3	
Molar volume V_{mol}	7.23	9.39	9.53		cm^3/mol	
Surface tension, liquid	1.6	2.25	2.31		N/m	
Temperature coefficient		-0.3×10^{-3}	-0.29×10^{-3}		N/(m K)	
Coefficient of linear thermal expansion α	6.2×10^{-6}	5.35×10^{-6}	4.31×10^{-6}		1/K	300 K
Sound velocity, solid, transverse	3980	3350	2870		m/s	
Sound velocity, solid, longitudinal	6850	6250	5180		m/s	
Compressibility κ	0.78×10^{-5}	0.338×10^{-5}	0.28×10^{-5}		1/MPa	Volume compressibility
Elastic modulus E	145	330	407		GPa	
Shear modulus G	71.6	123	152		GPa	
Poisson number μ	0.31	0.31	0.28			
Elastic compliance s_{11}	3.05	2.63	2.45		1/TPa	
Elastic compliance s_{44}	9.98	9.20	6.24		1/TPa	
Elastic compliance s_{12}	-0.49	-0.68	-0.69		1/TPa	
Elastic stiffness c_{11}	348	465	523		GPa	
Elastic stiffness c_{44}	100.0	109	160		GPa	
Elastic stiffness c_{12}	67	163	203		GPa	
Tensile strength	Strongly dependent on microstructure					
Vickers hardness	1060	160–400 (HV 10)	360–600 (HV 30)			

Table 2.1-18B(c) Elements of Group VIB (CAS notation), or Group 6 (new IUPAC notation). Thermal and thermodynamic properties

Element name	Chromium	Molybdenum	Tungsten	Seaborgium		Units	Remarks
Chemical symbol	Cr	Mo	W	Sg			
Atomic number Z	24	42	74	106			
Thermal conductivity λ	93.7	142	164			W/(m K)	At 293 K
Molar heat capacity c_p	23.44	23.932	24.27			J/(mol K)	At 298 K
Standard entropy S^0	23.543	28.560	32.618			J/(mol K)	At 298 K and 100 kPa
Enthalpy difference $H_{298} - H_0$	4.0500	4.5890	4.9700			kJ/mol	
Melting temperature T_m	2180.00	2893	3693			K	
Enthalpy change ΔH_m	21.0040	37.4798	52.3137			kJ/mol	
Entropy change ΔS_m	9.635	12.942	14.158			J/(mol K)	
Relative volume change ΔV_m						$(V_l - V_s)/V_l$	at T_m
Boiling temperature T_b	2952	4952	5828			K	
Enthalpy chnage ΔH_b	344.3	582.2	806.8			kJ/mol	
Critical temperature T_c		11 000				K	
Critical pressure p_c		540				MPa	
Critical density ϱ_c		2.630				g/cm^3	

Table 2.1-18B(d) Elements of Group VIB (CAS notation), or Group 6 (new IUPAC notation). Electronic, electromagnetic, and optical properties

Element name	Chromium	Molybdenum	Tungsten	Seaborgium		Units	Remarks
Chemical symbol	Cr	Mo	W	Sg			
Atomic number Z	24	42	74	106			
Electrical resistivity ρ_s	127	52	54.9			nΩ m	At 293 K
Temperature coefficient	30.1×10^{-4}	43.3×10^{-4}	51×10^{-4}			1/K	
Pressure coefficient	-17.3×10^{-9}	-1.29×10^{-9}	-1.333×10^{-9}			1/hPa	
Superconducting critical temperature T_{crit}		0.92	0.005			K	
Superconducting critical field H_{crit}		98				Oe	
Hall coefficient R	3.63×10^{-10}	1.26×10^{-10}	0.856×10^{-10}			m^3/(A s)	$B = 0.5$–2.0 T, $T = 293$ K
Thermoelectric coefficient		5.9	1.5			μV/K	
Electronic work function	4.5	4.39	4.54			eV	
Thermal work function	4.6	4.26	4.50			V	
Molar magnetic susceptibility χ_{mol} (SI)	2099×10^{-6}	905×10^{-6}	666×10^{-6}			cm^3/mol	At 295 K
Molar magnetic susceptibility χ_{mol} (cgs)	167×10^{-6}	72×10^{-6}	53×10^{-6}			cm^3/mol	At 295 K
Mass magnetic susceptibility χ_{mass} (SI)	44×10^{-6}	12×10^{-6}	4.0×10^{-6}			cm^3/g	At 295 K

Table 2.1-18C Elements of Group VIB (CAS notation), or Group 6 (new IUPAC notation). Allotropic and high-pressure modifications

Element	Chromium		
Modification	α-Cr	α′-Cr	
			Units
Crystal system, Bravais lattice	cub, bc	cub, bc	
Structure type	W	W	
Lattice constant a	288.47	288.2	pm
Space group	$Im\bar{3}m$	$Im\bar{3}m$	
Schoenflies symbol	O_h^9	O_h^9	
Strukturbericht type	A2	A2	
Pearson symbol	cI2	cI2	
Number A of atoms per cell	2	2	
Coordination number	8	8	
Shortest interatomic distance, solid	249		
Range of stability	RTP	High pressure	

Table 2.1-18D Elements of Group VIB (CAS notation), or Group 6 (new IUPAC notation). Ionic radii (determined from crystal structures)

Element	Chromium				Molybdenum				Tungsten			
Ion	Cr^{2+}	Cr^{3+}	Cr^{4+}	Cr^{6+}	Mo^{3+}	Mo^{4+}	Mo^{5+}	Mo^{6+}	W^{4+}	W^{5+}	W^{6+}	
Coordination number												Units
4			41	26			46	41			42	pm
5											51	pm
6	73	62	55	44	69	65	61	59	66	62	60	pm
7								73				pm

Table 2.1-19A Elements of Group VIIA (CAS notation), or Group 17 (new IUPAC notation). Atomic, ionic, and molecular properties (see Table 2.1-19D for ionic radii)

Element name	Fluorine	Chlorine	Bromine	Iodine	Astatine	Units	Remarks
Chemical symbol	F	Cl	Br	I	At		
Atomic number Z	9	17	35	53	85		
Characterization					Radioactive		
Relative atomic mass A (atomic weight)	18.9984032(5)	35.4527(9)	79.904(1)	126.90447(3)	[210]		Mass ratio
Abundance in lithosphere	800×10^{-6}	480×10^{-6}	2.5×10^{-6}	0.3×10^{-6}			Mass ratio
Abundance in sea	1.3×10^{-6}	$18\,800 \times 10^{-6}$	67×10^{-6}	0.06×10^{-6}			Mass ratio
Atomic radius r_{cov}	64	99	114	133	145	pm	Covalent radius
Atomic radius r_{vdW}	150–160	175	200	210		pm	van der Waals radius
Electron shells	KL	KLM	–LMN	–MNO	–NOP		
Electronic ground state	$^2P_{3/2}$	$^2P_{3/2}$	$^2P_{3/2}$	$^2P_{3/2}$	$^2P_{3/2}$		
Electronic configuration	$[He]2s^2 2p^5$	$[Ne]3s^2 3p^5$	$[Ar]3d^{10}4s^2 4p^5$	$[Kr]4d^{10}5s^2 5p^5$	$[Xe]4f^{14}5d^{10}6s^2 6p^5$		
Oxidation states	1–	1+, 3+, 5+, 7+, 1–, 3–, 5–, 7–					
Electron affinity	3.40	3.61	3.36	3.06	2.8	eV	Reaction type $F + e^- = F^-$
Electronegativity χ_A	4.10	2.83	2.74	2.21	(1.90)		Allred and Rochow
1st ionization energy	17.42282	12.96764	11.81381	10.45126		eV	
2nd ionization energy	34.97082	23.814	21.8	19.1313		eV	
3rd ionization energy	62.7084	39.61	36	33		eV	
4th ionization energy	87.1398	53.4652	47.3			eV	
Standard electrode potential E^0	+2.866	+1.358	+1.066	+0.536		V	Reaction type $2Cl^- = Cl_2 + 2e^-$
Molecular form in gaseous state	F_2	Cl_2	Br_2	I_2			
Internuclear distance in molecule		198.8				pm	
Dissociation energy	1.5417	2.475				eV	Extrapolated to $T = 0\,K$

Table 2.1-19B(a) Elements of Group VIIA (CAS notation), or Group 17 (new IUPAC notation). Crystallographic properties (see Table 2.1-19C for allotropic and high-pressure modifications)

Element name	Fluorine	Chlorine	Bromine	Iodine	Astatine		
Chemical symbol	F	Cl	Br	I	At		
Atomic number Z	9	17	35	53	85		
						Units	Remarks
State	F_2, < 45.6 K	Cl_2, 113 K	Br_2, 123 K	I_2			
Crystal system, Bravais lattice	mon, C	orth, C	orth, C	orth, C			
Structure type				Ga			
Lattice constant a	550	624	668	726.8		pm	
Lattice constant b	328	448	449	479.7		pm	
Lattice constant c	728	826	874	979.7		pm	
Lattice angle β	102.17					deg	
Space group	$C2/m$	$Cmca$	$Cmca$	$Cmca$			
Schoenflies symbol	C_{2h}^3	D_{2h}^{18}	D_{2h}^{18}	D_{2h}^{18}			
Strukturbericht type	C34	A11	A11	A11			
Pearson symbol	mC6	oC8	oC8	oC8			
Number A of atoms per cell	6	2×4	2×4	2×4			
Coordination number	1	1	1	1			
Shortest interatomic distance, solid	149	198.0	227	269		pm	

Table 2.1–19B(b) Elements of Group VIIA (CAS notation), or Group 17 (new IUPAC notation). Mechanical properties

Element name	Fluorine	Chlorine	Bromine	Iodine	Astatine		
Chemical symbol	F	Cl	Br	I	At		
Atomic number Z	9	17	35	53	85		
						Units	Remarks
State	F_2, gas	Cl_2, gas	Br_2, crystalline	I_2, crystalline			
Density ϱ, solid				4.92		g/cm^3	
Density ϱ, liquid			3.119 (293 K)			g/cm^3	
Density ϱ, gas	1.696×10^{-3}	3.17×10^{-3}				g/cm^3	At 273 K, 1 bar
Molar volume V_{mol}	18.05	17.46	19.73	25.74		cm^3/mol	
Viscosity η, gas	209.3×10^2					µPa s	
Viscosity η, liquid			0.916 (299 K)	2.27		mPa s	
Surface tension, liquid			0.0441 (286 K)	0.0557		N/m	
Sound velocity, gas	336 (375 K)	206				m/s	STP
Elastic compliance s_{11}				328		1/TPa	
Elastic compliance s_{22}				103		1/TPa	
Elastic compliance s_{33}				132		1/TPa	
Elastic compliance s_{44}				303		1/TPa	
Elastic compliance s_{55}				67.7		1/TPa	
Elastic compliance s_{66}				170		1/TPa	
Elastic compliance s_{12}				−97		1/TPa	
Elastic compliance s_{13}				−173		1/TPa	
Elastic compliance s_{23}				49		1/TPa	
Elastic stiffness c_{11}			11.5			GPa	
Elastic stiffness c_{22}			13.5			GPa	
Elastic stiffness c_{33}			25.0			GPa	
Elastic stiffness c_{44}			3.30			GPa	
Elastic stiffness c_{55}			14.8			GPa	
Elastic stiffness c_{66}			5.88			GPa	
Elastic stiffness c_{12}			4.50			GPa	
Elastic stiffness c_{13}			13.5			GPa	
Elastic stiffness c_{23}			0.93			GPa	

Table 2.1-19B(c) Elements of Group VIIA (CAS notation), or Group 17 (new IUPAC notation). Thermal and thermodynamic properties

Element name	Fluorine	Chlorine	Bromine	Iodine	Astatine	Units	Remarks
Chemical symbol	F	Cl	Br	I	At		
Atomic number Z	9	17	35	53	85		
State	F_2, gas	Cl_2, gas	Br_2, liquid	I_2, crystalline	At_2, crystalline		
Thermal conductivity λ	2.43×10^{-2}	9.3×10^{-3}		0.4	1.7	W/(m K)	STP
Molar heat capacity c_p	15.66	16.974	37.84	27.21		J/(mol K)	At 298 K
Standard entropy S^0	202.789	223.079	152.210	116.139	54.000	J/(mol K)	At 298 K and 100 kPa
Enthalpy difference $H_{298} - H_0$	8.8250	9.1810		13.1963	13.4000	kJ/mol	
Melting temperature T_m	53.48	172.18	265.90	386.75	575	K	
Enthalpy change ΔH_m		6.41	10.8	15.5172	23.8	kJ/mol	
Entropy change ΔS_m				40.122	40.0	J/(mol K)	
Volume change ΔV_m							$(V_l - V_s)/V_l$ at T_m
Boiling temperature T_b	84.95	239.1	332.3	458.4		K	
Enthalpy change ΔH_b		20.40	29.56	41.96	91	kJ/mol	
Critical temperature T_c	144.3	417		819		K	
Critical pressure p_c	5.22	7.98				MPa	
Critical density ϱ_c	0.630	0.573				g/cm^3	
Triple-point temperature T_{tr}	55	162				K	
Triple-point pressure p_{tr}	0.221	1.39				kPa	
Solubility in water[a]		4.610					At 273 K and 101 kPa

[a] m^3 gas/m^3 water.

Table 2.1-19B(d) Elements of Group VIIA (CAS notation), or Group 17 (new IUPAC notation). Electronic, electromagnetic, and optical properties

Element name	Fluorine	Chlorine	Bromine	Iodine	Astatine	Units	Remarks
Chemical symbol	F	Cl	Br	I	At		
Atomic number Z	9	17	35	53	85		
State	F_2, gas	Cl_2, gas	Br_2, liquid	I_2, crystalline			
Characteristics	Yellow gas, very reactive	Yellow-green gas	Liquid halogen	Solid semiconductor			
Electronic work function				2.8		eV	
Dielectric constant ε, liquid	1.517 (83.2 K)	2.15 (213 K)					
Molar magnetic susceptibility χ_{mol}, gas (SI)			−924			cm^3/mol	
Molar magnetic susceptibility χ_{mol}, gas (cgs)		−508 × 10^{-6}	−73.5 × 10^{-6}			cm^3/mol	
Molar magnetic susceptibility χ_{mol}, liquid (SI)			−709			cm^3/mol	
Molar magnetic susceptibility χ_{mol}, liquid (cgs)		−40.4 × 10^{-6}	−56.4 × 10^{-6}			cm^3/mol	
Molar magnetic susceptibility χ_{mol}, solid (SI)				−1131 × 10^{-6}		cm^3/mol	
Molar magnetic susceptibility χ_{mol}, solid (cgs)		−7.2 × 10^{-6}	−11.1 × 10^{-6}	−90 × 10^{-6}		cm^3/mol	
Mass magnetic susceptibility χ_{mass}, solid (SI)				−4.40 × 10^{-6}		cm^3/g	
Refractive index $(n − 1)$, gas [a]	206 × 10^{-6}	773 × 10^{-6}					At 273 K and 101 kPa
Refractive index n, liquid [a]		1.367 [b]	1.659				

[a] $\lambda = 589$ nm. [b] At 92 K, $\varrho = 1.330$ g/cm^3.

Table 2.1-19C Elements of Group VIIA (CAS notation), or Group 17 (new IUPAC notation). Allotropic and high-pressure modifications

Element	Fluorine		
Modification	α-F (F$_2$)	β-F	
			Units
Crystal system, Bravais lattice	mon, C	cub, P	
Structure type		γ-O	
Lattice constant a	550	667	pm
Lattice constant b	328		pm
Lattice constant c	728		pm
Lattice angle γ	102.17		deg
Space group	$C2/m$	$Pm\bar{3}m$	
Schoenflies symbol	C_{2h}^3	O_h^3	
Strukturbericht type	C34	A15	
Pearson symbol	mC6	cP16	
Number A of atoms per cell	6	16	
Coordination number	1		
Shortest interatomic distance, solid	149		pm
Range of stability	< 45.6 K	> 45.6 K	

Table 2.1-19D Elements of Group VIIA (CAS notation), or Group 17 (new IUPAC notation). Ionic radii (determined from crystal structures)

Element	Fluorine		Chlorine			Bromine			Iodine			
Ion	F$^-$	F^{7+}	Cl$^-$	Cl^{5+}	Cl^{7+}	Br$^-$	Br^{5+}	Br^{7+}	I$^-$	I^{5+}	I^{7+}	
Coordination number												Units
3				12			31			44		pm
3					8			25			42	pm
6	133	8	181			196		39	220	95	53	pm

Table 2.1-20A Elements of Group VIIB (CAS notation), or Group 7 (new IUPAC notation). Atomic, ionic, and molecular properties (see Table 2.1-20D for ionic radii)

Element name	Manganese	Technetium	Rhenium	Bohrium	Units	Remarks
Chemical symbol	Mn	Tc	Re	Bh		
Atomic number Z	25	43	75	107		
Characteristics		Radioactive				
Relative atomic mass A (atomic weight)	54.938049(9)	[98]	186.207(1)			
Abundance in lithosphere	1000×10^{-6}					Mass ratio
Abundance in sea	2×10^{-10}		1×10^{-9}			Mass ratio
Atomic radius r_{cov}	118	127	128		pm	Covalent radius
Atomic radius r_{met}	137	137	137		pm	Metallic radius, CN = 12
Electron shells	–LMN	–MNO	–NOP	–OPQ		
Electronic ground state	$^6S_{5/2}$	$^6S_{5/2}$	$^6S_{5/2}$			
Electronic configuration	[Ar]3d^{5}4s^2	[Kr]4d^{5}5s^2	[Xe]4f^{14}5d^{5}6s^2			
Oxidation states	7+, 6+, 4+, 3+, 2+	7+	7+, 6+, 4+, 2+, 1–			
Electron affinity	Not stable	0.55	0.15		eV	Reaction type Tc + e$^-$ = Tc$^-$
Electronegativity χ_A	1.60	1.36	(1.46)			Allred and Rochow
1st ionization energy	7.43402	7.28	7.88		eV	
2nd ionization energy	15.63999	15.26			eV	
3rd ionization energy	33.668	29.54			eV	
4th ionization energy	51.2				eV	
Standard electrode potential E^0	–1.185				V	Reaction type Mn^{2+} + 2e$^-$ = Mn

Table 2.1-20B(a) Elements of Group VIIB (CAS notation), or Group 7 (new IUPAC notation). Crystallographic properties (see Table 2.1-20C for allotropic and high-pressure modifications)

Element name	Manganese	Technetium	Rhenium	Bohrium		
Chemical symbol	Mn	Tc	Re	Bh		
Atomic number Z	25	43	75	107		
					Units	Remarks
State	α-Mn					
Crystal system, Bravais lattice	cub, bc	hex	hex			
Structure type	α-Mn	Mg	Mg			
Lattice constant a	892.19	273.8	276.08		pm	
Lattice constant c		439.4	445.80		pm	
Space group	$Im\bar{3}m$	$P6_3/mmc$	$P6_3/mmc$			
Schoenflies symbol	T_d^3	D_{6h}^4	D_{6h}^4			
Strukturbericht type	A12	A3	A3			
Pearson symbol	cI58	hP2	hP2			
Number A of atoms per cell	58	2	2			
Coordination number		12	12			
Shortest interatomic distance, solid	226–293	274			pm	

Table 2.1-20B(b) Elements of Group VIIB (CAS notation), or Group 7 (new IUPAC notation). Mechanical properties

Element name	Manganese	Technetium	Rhenium	Bohrium		
Chemical symbol	Mn	Tc	Re	Bh		
Atomic number Z	25	43	75	107		
					Units	Remarks
Density ϱ, solid	7.470	11.50	21.00		g/cm^3	
Density ϱ, liquid	6.430		18.80		g/cm^3	
Molar volume V_{mol}	7.38	8.6	8.86		cm^3/mol	
Surface tension, liquid	1.10		2.65		N/m	
Temperature coefficient	0		-0.34×10^{-3}		N/(m K)	
Coefficient of linear thermal expansion α	23×10^{-6}	8.06×10^{-6}	6.63×10^{-6}		1/K	
Sound velocity, solid, transverse	3280	50.6	2930		m/s	
Sound velocity, solid, longitudinal	5560	3270	5360		m/s	
Compressibility κ	0.716×10^{-5}	3.22×10^{-5}	0.264×10^{-5}		1/MPa	Volume compressibility
Elastic modulus E	196	407	520		GPa	At 293 K
Shear modulus G	79.4	162	180		GPa	
Poisson number μ	0.24	0.26	0.26			
Elastic compliance s_{11}		3.2	2.11		1/TPa	
Elastic compliance s_{33}		2.9	1.70		1/TPa	
Elastic compliance s_{44}		5.7	6.21		1/TPa	
Elastic compliance s_{12}		-1.1	-0.80		1/TPa	
Elastic compliance s_{13}		-0.9	-0.40		1/TPa	
Elastic stiffness c_{11}		433	616		GPa	
Elastic stiffness c_{33}		470	683		GPa	
Elastic stiffness c_{44}		177	161		GPa	
Elastic stiffness c_{12}		199	273		GPa	
Elastic stiffness c_{13}		199	206		GPa	
Tensile strength		0.40–0.74	1.16		GPa	
Vickers hardness	9.81	1.510	2.45–8.00			

Table 2.1–20B(c) Elements of Group VIIB (CAS notation), or Group 7 (new IUPAC notation). Thermal and thermodynamic properties

Element name	Manganese	Technetium	Rhenium	Bohrium		
Chemical symbol	Mn	Tc	Re	Bh		
Atomic number Z	25	43	75	107		
					Units	Remarks
State	α-Mn					
Thermal conductivity λ	29.7	185	71.2		W/(m K)	
Molar heat capacity c_p	26.28		25.31		J/(mol K)	At 298 K
Standard entropy S^0	32.220	32.985	36.482		J/(mol K)	At 298 K and 100 kPa
Enthalpy difference $H_{298} - H_0$	4.9957		5.3330		kJ/mol	
Melting temperature T_m	1519.00	2430.01	3458.00		K	
Transition	δ–liquid	α–liquid	α–liquid			
Enthalpy change ΔH_m	12.9089	33.2912	34.0750		kJ/mol	
Entropy change ΔS_m	8.498	13.700	9.854		J/(mol K)	
Relative volume change ΔV_m	0.017					$(V_l - V_s)/V_l$ at T_m
Boiling temperature T_b	2335	4538	5869		K	
Enthalpy change ΔH_b	226.7	592.9	714.8		kJ/mol	
Critical temperature T_c			2090		K	
Critical pressure p_c			14.5		MPa	
Critical density ϱ_c			0.320		g/cm^3	

Table 2.1–20B(d) Elements of Group VIIB (CAS notation), or Group 7 (new IUPAC notation). Electronic, electromagnetic, and optical properties

Element name	Manganese	Technetium	Rhenium	Bohrium		
Chemical symbol	Mn	Tc	Re	Bh		
Atomic number Z	25	43	75	107		
					Units	Remarks
Characteristics	Brittle metal	Metal	Metal			
Electrical resistivity ρ_s	1380	1510	172		nΩ m	At RT
Temperature coefficient	5.0×10^{-4}		44.8×10^{-4}		1/K	
Pressure coefficient	-3.54×10^{-9}				1/hPa	
Electrical resistivity ρ_l	400				nΩ m	
Resistivity ratio at T_m	0.61					
Superconducting critical temperature T_{crit}			1.70		K	
Superconducting critical field H_{crit}			198		Oe	
Hall coefficient R	0.84×10^{-10}		3.15×10^{-10}		m^3/(A s)	At 297 K, $B = 0.5$–5.0 T
Electronic work function	4.08		About 5.0		V	
Thermal work function	3.91		4.96		V	
Molar magnetic susceptibility χ_{mol} (SI)	6421×10^{-6}	1445×10^{-6}	842×10^{-6}		cm^3/mol	At 295 K
Molar magnetic susceptibility χ_{mol} (cgs)	511×10^{-6}	115×10^{-6}	67×10^{-6}		cm^3/mol	At 295 K
Mass magnetic susceptibility χ_{mass} (SI)	121×10^{-6}	31×10^{-6}	4.56×10^{-6}		cm^3/g	At 295 K

Table 2.1-20C Elements of Group VIIB (CAS notation), or Group 7 (new IUPAC notation). Allotropic and high-pressure modifications

Element	Manganese				
Modification	α-Mn	β-Mn	γ-Mn	δ-Mn	
					Units
Crystal system, Bravais lattice	cub, bc	cub, P	cub, fc	cub, bc	
Structure type	α-Mn	β-Mn	Cu	W	
Lattice constant a	892.19	631.52	386.24	308.06	pm
Lattice constant c					pm
Space group	$Im\bar{3}m$	$P4_132$	$Fm\bar{3}m$	$Im\bar{3}m$	
Schoenflies symbol	T_d^3	O^1	O_h^5	O_h^9	
Strukturbericht type	A12	A13	A1	A2	
Pearson symbol	cI58	cP20	cF4	cI2	
Number A of atoms per cell	58	20	4	2	
Coordination number			12	8	
Shortest interatomic distance, solid	226–293		273	267	pm
Range of stability	RTP	> 1000 K	> 1368 K	> 1408 K	

Table 2.1-20D Elements of Group VIIB (CAS notation), or Group 7 (new IUPAC notation). Ionic radii (determined from crystal structures)

Element	Manganese						Technetium	Rhenium				
Ion	Mn^{2+}	Mn^{3+}	Mn^{4+}	Mn^{5+}	Mn^{6+}	Mn^{7+}	Tc^{4+}	Re^{4+}	Re^{5+}	Re^{6+}	Re^{7+}	
Coordination number												Units
4	66		39	33	26	25					38	pm
6	83	58	53				65	63	58	55		pm
8	96											pm

Table 2.1–21A Elements of Group VIIIA (CAS notation), or Group 18 (new IUPAC notation). Atomic, ionic, and molecular properties

Element name Chemical symbol Atomic number Z	Neon Ne 10	Argon Ar 18	Krypton Kr 36	Xenon Xe 54	Radon Rn 86	Units	Remarks
Characteristics					Radioactive		
Relative atomic mass A (atomic weight)	20.1797(6)	39.948(1)	83.80(1)	131.29(2)	[222]		
Atomic radius r_{cov}	(69)	(97)	110	130		pm	Covalent radius
Atomic radius r_{met}	154	188	202	216	240	pm	Metallic radius, CN = 12
Atomic radius r_{vdW}	160	188	200	220	240	pm	van der Waals radius
Electron shells	KL	KLM	–LMN	–MNO	–NOP		
Electronic ground state	$^1S_0$	$^1S_0$	$^1S_0$	$^1S_0$	$^1S_0$		
Electronic configuration	[He]2s²2p⁶	[Ne]3s²3p⁶	[Ar]3d¹⁰4s²4p⁶	[Kr]4d¹⁰5s²5p⁶	[Xe]4f¹⁴5d¹⁰6s²6p⁶		
Electron affinity	Not stable	Not stable	Not stable	Not stable	Not stable	eV	Reaction type $He + e^- = He^-$
Electronegativity χ_A	5.10	3.30	3.10	2.40	(2.06)		Allred and Rochow
1st ionization energy	21.56454	15.75962	13.99961	12.12987	10.74850	eV	
2nd ionization energy	40.96328	27.62967	24.35985	21.20979		eV	
3rd ionization energy	63.45	40.74	36.950	32.1230		eV	
4th ionization energy	97.12	59.81	52.5			eV	

Table 2.1–21B(a) Elements of Group VIIIA (CAS notation), or Group 18 (new IUPAC notation). Crystallographic properties (see Table 2.1-21C for allotropic and high-pressure modifications)

Element name Chemical symbol Atomic number Z	Neon Ne 10	Argon Ar 18	Krypton Kr 36	Xenon Xe 54	Radon Rn 86	Units	Remarks
State	At 4.2 K	At 4.2 K	At 4.2 K	At 4.2 K			
Crystal system, Bravais lattice	cub, fc	cub, fc	cub, fc	cub, fc			
Structure type	Cu	Cu	Cu	Cu			
Lattice constant a	446.22	531.2	564.59	613.2		pm	
Space group	$Fm\bar{3}m$	$Fm\bar{3}m$	$Fm\bar{3}m$	$Fm\bar{3}m$			
Schoenflies symbol	O_h^5	O_h^5	O_h^5	O_h^5			
Strukturbericht type	A1	A1	A1	A1			
Pearson symbol	cF4	cF4	cF4	cF4			
Number A of atoms per cell	4	4	4	4			
Coordination number	12	12	12	12			
Shortest interatomic distance, solid	320	383	405	441		pm	

Table 2.1–21B(b) Elements of Group VIIIA (CAS notation), or Group 18 (new IUPAC notation). Mechanical properties

Element name	Neon	Argon	Krypton	Xenon	Radon		
Chemical symbol	Ne	Ar	Kr	Xe	Rn		
Atomic number Z	10	18	36	54	86		
Characteristics						Units	Remarks
Density ϱ, solid	Noble gas	Noble gas	Noble gas	Noble gas	Noble gas	g/cm³	
Density ϱ, gas		1.736 (40 K)				g/cm³	
Molar volume V_{mol}	0.8994×10^{-3}		3.7493×10^{-3}	5.8971×10^{-3}	9.73×10^{-3}	g/cm³	At 273 K
	13.97		29.68	37.09	50.5	cm³/mol	
Viscosity η, gas	29.8	21	23.4	21.2		µPa s	At 293 K
Sound velocity, gas	461	308	213	168		m/s	At STP
Sound velocity, liquid		855 (85 K)				m/s	
Elastic modulus E	1.0	1.6	1.8			GPa	Low temperature, estimated values
Elastic compliance s_{11}	1020 (4.7 K)	593 (80 K)	618 (115 K)	690 (160.5 K)		1/TPa	
Elastic compliance s_{44}	1000 (4.7 K)	1073 (80 K)	744 (115 K)	708 (160.5 K)		1/TPa	
Elastic compliance s_{12}	−370 (4.7 K)	−205 (80 K)	−226 (115 K)	−271 (160.5 K)		1/TPa	
Elastic stiffness c_{11}	1.69 (4.7 K)	2.77 (80 K)	2.85 (115 K)	2.93 (160.5 K)		GPa	
Elastic stiffness c_{44}	1.00 (4.7 K)	0.98 (80 K)	1.35 (115 K)	1.41 (160.5 K)		GPa	
Elastic stiffness c_{12}	0.97 (4.7 K)	1.37 (80 K)	1.60 (115 K)	1.89 (160.5 K)		GPa	
Solubility in water α_W	0.010	0.0340	0.059	0.108			At 293 K, $\alpha_W = \mathrm{vol(gas)/vol(water)}$

Table 2.1–21B(c) Elements of Group VIIIA (CAS notation), or Group 18 (new IUPAC notation). Thermal and thermodynamic properties

Element name	Neon	Argon	Krypton	Xenon	Radon		
Chemical symbol	Ne	Ar	Kr	Xe	Rn		
Atomic number Z	10	18	36	54	86		
State	Ne gas	Ar gas	Kr gas	Xe gas	Rn gas	Units	Remarks
Thermal conductivity λ	49.3×10^{-3}	18.0×10^{-3}	9.49×10^{-3}	5.1×10^{-3}	3.64×10^{-3}	W/(m K)	At 298.15 K
Molar heat capacity c_p	20.786	20.87	20.786	20.744	20.786	J/(mol K)	At 298.15 K and 100 kPa
Standard entropy S^0	146.328	154.842	164.085	169.575	176.234	J/(mol K)	At 298.15 K
Enthalpy difference $H_{298} - H_0$	6.1965	6.1965	6.1965	6.1970	6.1970	kJ/mol	
Melting temperature T_m	24.563	83.8	115.765	161.391	202	K	
Transition	α–liquid	α–liquid		α–liquid	α–liquid		
Enthalpy change ΔH_m	27.0	1.21	1.64	3.10	2.7	kJ/mol	
Boiling temperature T_b	27.10	87.30	119.80	165.03	211	K	
Enthalpy change ΔH_b	0.324	6.3	9.05	12.65	18.1	kJ/mol	
Critical temperature T_c	44.0	150.75	209.4	289.74		K	
Critical pressure p_c	2.75	4.86	5.50	5.840		MPa	
Critical density ϱ_c	0.4835	0.307	0.9085	1.105		g/cm³	
Triple-point temperature T_{tr}	24.5	83.85	115.95	161.25		K	
Triple-point pressure p_{tr}	43.3	68.75	73.19	81.6		kPa	

Table 2.1-21B(d) Elements of Group VIIIA (CAS notation), or Group 18 (new IUPAC notation). Electronic, electromagnetic, and optical properties

Element name	Neon	Argon	Krypton	Xenon	Radon	Units	Remarks
Chemical symbol	Ne	Ar	Kr	Xe	Rn		
Atomic number Z	10	18	36	54	86		
State	Ne gas	Ar gas	Kr gas	Xe gas	Rn gas		At STP
Dielectric constant $\varepsilon - 1$, gas	130×10^{-6}	545×10^{-6}	7×10^{-6}	1238×10^{-6}			At 295 K
Dielectric constant ε, liquid		1.516 (89 K)					
Molar magnetic susceptibility χ_{mol}, gas (SI)	-87.5×10^{-6}	-243×10^{-6}	-364×10^{-6}	-572×10^{-6}		cm^3/mol	At 295 K
Molar magnetic susceptibility χ_{mol}, gas (cgs)	-6.96×10^{-6}	-19.3×10^{-6}	-29.0×10^{-6}	-45.5×10^{-6}		cm^3/mol	At 295 K
Mass magnetic susceptibility χ_{mass}, gas (SI)	-4.2×10^{-6}	-6.16×10^{-6}	-4.32×10^{-6}	-4.20×10^{-6}		cm^3/g	At 293 K
Refractive index $(n - 1)$, gas	7.25×10^{-6}	281×10^{-6}		706×10^{-6}			$\lambda = 589.3$ nm
Refractive index n, liquid		1.233 (84 K)					$\lambda = 589.3$ nm

Table 2.1-21C Elements of Group VIIIA (CAS notation), or Group 18 (new IUPAC notation). Allotropic and high-pressure modifications. There is no Table 2.1-21D, because no data on ionic radii are available

Element	Argon		Units
Modification	α-Ar	β-Ar	
Crystal system, Bravais lattice	cub, fc	hex, cp	
Structure type	Cu	Mg	
Lattice constant a	531.2	376.0	pm
Lattice constant c		614.1	pm
Space group	$Fm\bar{3}m$	$P6_3/mmc$	
Schoenflies symbol	O_h^5	D_{6h}^4	
Strukturbericht type	A1	A3	
Pearson symbol	cF4	hP2	
Number A of atoms per cell	4	2	
Coordination number	12	12	
Shortest interatomic distance, solid	383		pm
Range of stability	< 83.8 K	> 83.8 K	

Table 2.1-22A Elements of Group VIII(1) (CAS notation), or Group 8 (new IUPAC notation). Atomic, ionic, and molecular properties (see Table 2.1-22D for ionic radii)

Element name	Iron	Ruthenium	Osmium	Hassium		
Chemical symbol	Fe	Ru	Os	Hs		
Atomic number Z	26	44	76	108		
Quantity					Units	Remarks
Relative atomic mass A (atomic weight)	55.845(2)	101.07(2)	190.23(3)			
Abundance in lithosphere	$50\,000 \times 10^{-6}$					Mass ratio
Abundance in sea	2×10^{-9}					Mass ratio
Atomic radius r_{cov}	116	125	126		pm	Covalent radius
Atomic radius r_{met}	126	132.5	134		pm	Metallic radius, CN = 12
Electron shells	-LMN	-MNO	-NOP	-OPQ		
Electronic ground state	$^5D_4$	$^5F_5$	$^5D_4$			
Electronic configuration	[Ar]$3d^6 4s^2$	[Kr]$4d^7 5s^1$	[Xe]$4f^{14} 5d^6 6s^2$			
Oxidation states	3+, 2+	8+, 6+, 4+, 3+, 2+	8+, 6+, 4+, 3+, 2+			
Electron affinity	0.151	1.05	1.1		eV	
Electronegativity χ_A	1.64	1.42	(1.52)			Allred and Rochow
1st ionization energy	7.9024	7.36050	8.7		eV	
2nd ionization energy	16.1878	16.76			eV	
3rd ionization energy	30.652	28.47			eV	
4th ionization energy	54.8				eV	
5th ionization energy	75.0				eV	
Standard electrode potential E^0	−0.447				V	Reaction type $Fe^{2+} + 2e^- = Fe$
	−0.037				V	Reaction type $Fe^{3+} + 3e^- = Fe$
	+0.771				V	Reaction type $Fe^{3+} + e^- = Fe^{2+}$

Table 2.1-22B(a) Elements of Group VIII(1) (CAS notation), or Group 8 (new IUPAC notation). Crystallographic properties (see Table 2.1-22C for allotropic and high-pressure modifications)

Element name	Iron	Ruthenium	Osmium	Hassium		
Chemical symbol	Fe	Ru	Os	Hs		
Atomic number Z	26	44	76	108		
					Units	Remarks
Modification	α-Fe					
Crystal system, Bravais lattice	cub, bc	hex	hex			
Structure type	W	Mg	Mg			
Lattice constant a	286.65	270.53	273.48		pm	
Lattice constant c		428.14	439.13		pm	
Space group	$Im\bar{3}m$	$P6_3/mmc$	$P6_3/mmc$			
Schoenflies symbol	O_h^9	D_{6h}^4	D_{6h}^4			
Strukturbericht type	A2	A3	A3			
Pearson symbol	cI2	hP2	hP2			
Number A of atoms per cell	2	2	2			
Coordination number	8	6+6	6+6			
Shortest interatomic distance, solid	248	265	267		pm	

Table 2.1-22B(b) Elements of Group VIII(1) (CAS notation), or Group 8 (new IUPAC notation). Mechanical properties

Element name	Iron	Ruthenium	Osmium	Hassium		
Chemical symbol	Fe	Ru	Os	Hs		
Atomic number Z	26	44	76	108		
					Units	Remarks
Density ϱ, solid	7.86	12.20	22.4		g/cm^3	
Density ϱ, liquid	7.020	10.90	20.100		g/cm^3	
Molar volume V_{mol}	7.09	8.14	8.43		cm^3/mol	
Viscosity η, liquid	5.53				mPa s	
Surface tension, liquid	1.65	2.25	2.5		N/m	
Temperature coefficient		-0.31×10^{-3}	-0.33×10^{-3}		N/(m K)	
Coefficient of linear thermal expansion α	12.3×10^{-6}	9.1×10^{-6}	6.1×10^{-6}		1/K	
Sound velocity, solid, transverse	3220	3740	3340		m/s	
Sound velocity, solid, longitudinal	5920	6530	5480		m/s	
Compressibility κ	0.56×10^{-5}	0.331×10^{-5}	0.261×10^{-5}		1/MPa	Volume compressibility
Elastic modulus E	211	432	559		GPa	
Shear modulus G	80.4	173	223		GPa	
Poisson number μ	0.29	0.25	0.25			
Elastic compliance s_{11}	7.67	2.09			1/TPa	
Elastic compliance s_{33}		1.82			1/TPa	
Elastic compliance s_{44}	8.57	5.53			1/TPa	
Elastic compliance s_{12}	-2.83	-0.58			1/TPa	
Elastic compliance s_{13}		-0.41			1/TPa	
Elastic stiffness c_{11}	230	563			GPa	
Elastic stiffness c_{33}		624			GPa	
Elastic stiffness c_{44}	117	181			GPa	
Elastic stiffness c_{12}	135	188			GPa	
Elastic stiffness c_{13}		168			GPa	
Tensile strength	193–206 [a]	540			MPa	
Vickers hardness	608 [a]	$2\text{–}5 \times 10^3$	800			

[a] Strongly dependent on microstructure

Table 2.1-22B(c) Elements of Group VIII(1) (CAS notation), or Group 8 (new IUPAC notation). Thermal and thermodynamic properties

Element name	Iron	Ruthenium	Osmium	Hassium		
Chemical symbol	Fe	Ru	Os	Hs		
Atomic number Z	26	44	76	108		
					Units	Remarks
Modification	α-Fe					
Thermal conductivity λ	80.2	117	87.6		W/(m K)	
Molar heat capacity c_p	25.10	24.06	24.7		J/(mol K)	At 298.2 K
Standard entropy S^0	27.280	28.614	32.635		J/(mol K)	At 298.15 K and 100 kPa
Enthalpy difference $H_{298} - H_0$	4.4890	4.6024			kJ/mol	At 298.15 K
Melting temperature T_m	1811.0	2607.0	3306.0		K	
Transition	δ–liquid	α–liquid	δ–liquid			
Enthalpy change ΔH_m	13.8060	38.5890	57.8550		kJ/mol	
Entropy change ΔS_m	7.623	14.802	17.500		J/(mol K)	
Relative volume change ΔV_m	0.034					$(V_l - V_s)/V_l$ at T_m
Boiling temperature T_b	3139	4423	5285		K	
Enthalpy change ΔH_b	349.6	595.5	746		kJ/mol	

Table 2.1-22B(d) Elements of Group VIII(1) (CAS notation), or Group 8 (new IUPAC notation). Electronic, electromagnetic, and optical properties

Element name	Iron	Ruthenium	Osmium	Hassium		
Chemical symbol	Fe	Ru	Os	Hs		
Atomic number Z	26	44	76	108		
					Units	Remarks
Characteristics	Soft metal	Metal	Brittle metal			
Electrical resistivity ρ_s	89	76	81		nΩ m	At RT
Temperature coefficient	65.1×10^{-4}	45.8×10^{-4}	42×10^{-4}		1/K	
Pressure coefficient	-2.34×10^{-9}	-2.48×10^{-9}			1/hPa	
Superconducting critical temperature T_{crit}		0.49	0.66		K	
Superconducting critical field H_{crit}		66	65		Oe	
Hall coefficient R	8×10^{-10}	2.2×10^{-10}			m^3/(A s)	At 300 K, $B = 4-5$ T
Thermoelectric coefficient	-51.34				μV/K	
Electronic work function	4.70	4.71			V	
Thermal work function	4.50	4.73	4.83		V	
Molar magnetic susceptibility χ_{mol} (SI)	Ferromagnetic	490	138		cm^3/mol	At 295 K
Molar magnetic susceptibility χ_{mol} (cgs)	Ferromagnetic	39	11		cm^3/mol	At 295 K
Mass magnetic susceptibility χ_{mass} (SI)	Ferromagnetic	5.37×10^{-6}	0.65×10^{-6}		cm^3/g	At 295 K

Table 2.1–22C Elements of Group VIII(1) (CAS notation), or Group 8 (new IUPAC notation). Allotropic and high-pressure modifications

Element	Iron				
Modification	α-Fe	γ-Fe	δ-Fe	ε-Fe	
					Units
Crystal system, Bravais lattice	cub, bc	cub, fc	cub, bc	hex, cp	
Structure type	W	Cu	W	Mg	
Lattice constant a	286.65	364.67	291.35	248.5	pm
Lattice constant c				399.0	pm
Space group	$Im\bar{3}m$	$Fm\bar{3}m$	$Im\bar{3}m$	$P6_3/mmc$	
Schoenflies symbol	O_h^9	O_h^5	O_h^9	D_{6h}^4	
Strukturbericht type	A2	A1	A2	A3	
Pearson symbol	cI2	cF4	cI2	hP2	
Number A of atoms per cell	2	4	2	2	
Coordination number	8	12	8	6+6	
Shortest interatomic distance, solid	248	258	254	241	pm
Range of stability	RTP	> 1183 K	> 1663 K	> 13.0 GPa	

Table 2.1–22D Elements of Group VIII(1) (CAS notation), or Group 8 (new IUPAC notation). Ionic radii (determined from crystal structures)

Element	Iron		Ruthenium					Osmium				
Ion	Fe^{2+}	Fe^{3+}	Ru^{3+}	Ru^{4+}	Ru^{5+}	Ru^{7+}	Ru^{8+}	Os^{4+}	Os^{5+}	Os^{6+}	Os^{8+}	
Coordination number												Units
4	63	49				38	36				39	pm
6	61	55	68	62	57			63	58	55		pm
8	92	78										pm

Table 2.1-23A Elements of Group VIII(2) (CAS notation), or Group 9 (new IUPAC notation). Atomic, ionic, and molecular properties (see Table 2.1-23D for ionic radii)

Element name	Cobalt	Rhodium	Iridium	Meitnerium	Units	Remarks
Chemical symbol	Co	Rh	Ir	Mt		
Atomic number Z	27	45	77	109		
Relative atomic mass A (atomic weight)	58.933200(9)	102.90550(2)	192.217(3)			
Abundance in lithosphere	40×10^{-6}	0.001×10^{-6}	1×10^{-9}			Mass ratio
Abundance in sea	5×10^{-11}					Mass ratio
Atomic radius r_{cov}	116	125	127		pm	Covalent radius
Atomic radius r_{met}	125	134.5	136		pm	Metallic radius, CN = 12
Electron shells	–LMN	–MNO	–NOP	–OPQ		
Electronic ground state	$^4F_{9/2}$	$^4F_{9/2}$	$^4F_{9/2}$			
Electronic configuration	[Ar]3d^{7}4s^2	[Kr]4d^{8}5s^1	[Xe]4f^{14}5d^{7}6s^2			
Oxidation states	3+, 2+	4+, 3+, 2+	6+, 4+, 3+, 2+			
Electron affinity	0.662	1.14	1.57		eV	
Electronegativity χ_A	1.75	1.35	(1.44)			Allred and Rochow
1st ionization energy	7.8810	7.45890	9.1		eV	
2nd ionization energy	17.083	18.08			eV	
3rd ionization energy	33.50	31.06			eV	
4th ionization energy	51.3				eV	
Standard electrode potential E^0	−0.28	+0.6	+1.156		V	Reaction type Co^{2+} + 2e$^-$ = Co
					V	Reaction type Ir^{3+} + 3e$^-$ = Ir

Table 2.1-23B(a) Elements of Group VIII(2) (CAS notation), or Group 9 (new IUPAC notation). Crystallographic properties (see Table 2.1-23C for allotropic and high-pressure modifications)

Element name	Cobalt	Rhodium	Iridium	Meitnerium		
Chemical symbol	Co	Rh	Ir	Mt		
Atomic number Z	27	45	77	109		
					Units	Remarks
Modifaction	ε-Co					
Crystal system, Bravais lattice	hex, cp	cub, fc	cub, fc			
Structure type	Mg	Cu	Cu			
Lattice constant a	250.71	280.32	383.91		pm	
Lattice constant c	406.94				pm	
Space group	$P6_3/mmc$	$Fm\overline{3}m$	$Fm\overline{3}m$			
Schoenflies symbol	D_{6h}^4	O_h^5	O_h^5			
Strukturbericht type	A3	A1	A1			
Pearson symbol	hP2	cF4	cF4			
Number A of atoms per cell	2	4	4			
Coordination number	6+6	12	12			
Shortest interatomic distance, solid	250	269	271		pm	

Table 2.1-23B(b) Elements of Group VIII(2) (CAS notation), or Group 9 (new IUPAC notation). Mechanical properties

Element name	Cobalt	Rhodium	Iridium	Meitnerium		
Chemical symbol	Co	Rh	Ir	Mt		
Atomic number Z	27	45	77	109		
					Units	Remarks
Modification	ε-Co					
Density ϱ, solid	8.90	12.40	22.50		g/cm^3	
Density ϱ, liquid	7.670	10.800	20.000		g/cm^3	
Molar volume V_{mol}	6.62	8.29	8.57		cm^3/mol	
Viscosity η, liquid	4.8				mPa s	Near T_m
Coefficient of linear thermal expansion α	13.36×10^{-6}	8.40×10^{-6}	6.8×10^{-6}		1/K	
Surface tension, liquid	1.520	1.97	2.250		N/m	
Temperature coefficient	-0.92×10^{-3}	-0.3×10^{-3}	-0.31×10^{-3}		N/(m K)	
Sound velocity, solid, transverse	3000	3470	3050		m/s	
Sound velocity, solid, longitudinal	5730	6190	5380		m/s	
Compressibility κ	0.525×10^{-5}	0.350×10^{-5}	0.258×10^{-5}		1/MPa	Volume compressibility
Elastic modulus E	204	379	528		GPa	
Shear modulus G	77.3	149	209		GPa	
Poisson number μ	0.32	0.27	0.26			
Elastic compliance s_{11}	5.11	3.46	2.28		1/TPa	
Elastic compliance s_{33}	3.69				1/TPa	
Elastic compliance s_{44}	14.1	5.43	3.90		1/TPa	
Elastic compliance s_{12}	-2.37	-1.10	-0.67		1/TPa	
Elastic compliance s_{13}	-0.94				1/TPa	
Elastic stiffness c_{11}	295	413	580		GPa	
Elastic stiffness c_{33}	335				GPa	
Elastic stiffness c_{44}	71.0	184	256		GPa	
Elastic stiffness c_{12}	159	194	242		GPa	
Elastic stiffness c_{13}	111				GPa	
Tensile strength	255	951	623×10^3		MPa	
Vickers hardness	1043	1246	1760			

Table 2.1–23B(c) Elements of Group VIII(2) (CAS notation), or Group 9 (new IUPAC notation). Thermal and thermodynamic properties

Element name	Cobalt	Rhodium	Iridium	Meitnerium		
Chemical symbol	Co	Rh	Ir	Mt		
Atomic number Z	27	45	77	109		
					Units	Remarks
Modification	ε-Co					
Thermal conductivity λ	100	150	147		W/(m K)	
Molar heat capacity c_p	24.811	24.98	24.98		J/(mol K)	At 298 K
Standard entropy S^0	30.040	31.556	35.505		J/(mol K)	At 298 K and 100 kPa
Enthalpy difference $H_{298} - H_0$	4.7655		5.2677		kJ/mol	
Melting temperature T_m	1768.0	2237.0	2719.0		K	
Transition	α–liquid	α–liquid	α–liquid			
Enthalpy change ΔH_m	16.200	26.5935	41.124		kJ/mol	
Entropy change ΔS_m	9.163	11.888	15.125		J/(mol K)	
Relative volume change ΔV_m		0.12			1	$(V_l - V_s)/V_l$ at T_m
Boiling temperature T_b	3184	3970	4701		K	
Enthalpy change ΔH_b	376.6	493.3	604.1		kJ/mol	

Table 2.1–23B(d) Elements of Group VIII(2) (CAS notation), or Group 9 (new IUPAC notation). Electronic, electromagnetic, and optical properties

Element name	Cobalt	Rhodium	Iridium	Meitnerium		
Chemical symbol	Co	Rh	Ir	Mt		
Atomic number Z	27	45	77	109		
					Units	Remarks
Modification	α-Co					
Characteristics	Hard metal	Metal	Brittle metal			
Electrical resistivity ρ_s	56	43.0	47		nΩ m	At RT
Temperature coefficient	6.04×10^{-3}	4.62×10^{-3}	4.11×10^{-3}		1/K	
Pressure coefficient	-0.904×10^{-9}	-1.62×10^{-9}	-1.37×10^{-9}		1/hPa	
Electrical resistivity ρ_l	1020				nΩ m	
Resistivity ratio at T_m	1.05					
Superconducting critical temperature T_{crit}			0.14		K	
Superconducting critical field H_{crit}			77		Oe	
Hall coefficient R	360×10^{-12}	50.5×10^{-12}	31.8×10^{-12}		m^3/(A s)	At 300 K, $B = 4.5$–5.0 T
Thermoelectronic coefficient	17.5	1.0	1.2		μV/K	
Electronic work function	4.97	4.98			V	
Thermal work function	4.37	4.68	5.03		V	
Molar magnetic susceptibility χ_{mol} (SI)	Ferromagnetic	1282×10^{-6}	314×10^{-6}		cm^3/mol	At 295 K
Molar magnetic susceptibility χ_{mol} (cgs)	Ferromagnetic	102×10^{-6}	25×10^{-6}		cm^3/mol	At 295 K
Mass magnetic susceptibility χ_{mass} (SI)	Ferromagnetic	13.6×10^{-6}	1.67×10^{-6}		cm^3/g	At 295 K

Table 2.1-23C Elements of Group VIII(2) (CAS notation), or Group 9 (new IUPAC notation). Allotropic and high-pressure modifications

Element	Cobalt		
Modification	ε-Co	α-Co	
			Units
Crystal system, Bravais lattice	hex, cp	cub, fc	
Structure type	Mg	Cu	
Lattice constant a	250.71	354.45	pm
Lattice constant c	406.94		pm
Space group	$P6_3/mmc$	$Fm\overline{3}m$	
Schoenflies symbol	D_{6h}^4	O_h^5	
Strukturbericht type	A3	A1	
Pearson symbol	hP2	cF4	
Number A of atoms per cell	2	4	
Coordination number	6+6	12	
Shortest interatomic distance, solid	250	251	pm
Range of stability	RTP	> 661 K	

Table 2.1-23D Elements of Group VIII(2) (CAS notation), or Group 9 (new IUPAC notation). Ionic radii (determined from crystal structures)

Element	Cobalt		Rhodium			Iridium			
Ion	Co^{2+}	Co^{3+}	Rh^{3+}	Rh^{4+}	Rh^{5+}	Ir^{3+}	Ir^{4+}	Ir^{5+}	
Coordination number									Units
4	56								pm
6	65	55	67	60	55	68	63	57	pm
8	90								pm

Table 2.1–24A Elements of Group VIII(3) (CAS notation) or Group 10 (new IUPAC notation). Atomic, ionic, and molecular properties (see Table 2.1-24D for ionic radii)

Element name Chemical symbol Atomic number Z	Nickel Ni 28	Palladium Pd 46	Platinum Pt 78	Uun 110	Units	Remarks
Relative atomic mass A (atomic weight)	58.6934(2)	106.42(1)	195.078(2)			Mass ratio
Abundance in lithosphere	100×10^{-6}	0.01×10^{-6}	5×10^{-9}			Mass ratio
Abundance in sea	1.7×10^{-9}					
Atomic radius r_{cov}	115	128	130		pm	Covalent radius
Atomic radius r_{met}	125	138	137		pm	Metallic radius, CN = 12
Atomic radius r_{vdW}	160	160	170–180		pm	van der Waals radius
Electron shells	–LMN	–MNO	–NOP	–OPQ		
Electronic ground state	$^3F_4$	$^1S_0$	$^3D_3$			
Electronic configuration	$[Ar]3d^84s^2$	$[Kr]4d^{10}$	$[Xe]4f^{14}5d^96s^1$			
Oxidation states	3+, 2+	4+, 2+	4+, 2+			
Electron affinity	1.16	0.562	2.13		eV	
Electronegativity χ_A	1.75	1.35	(1.44)			Allred and Rochow
1st ionization energy	7.6398	8.3369	9.0		eV	
2nd ionization energy	18.16884	19.43	18.563		eV	
3rd ionization energy	35.19	32.93			eV	
4th ionization energy	54.9				eV	
Standard electrode potential E^0	–0.257	+0.951	+1.118		V	Reaction type $Ni^{2+} + 2e^- = Ni$

Remark for first row noted: Reaction type $Ni + e^- = Ni^-$

Table 2.1–24B(a) Elements of Group VIII(3) (CAS notation) or Group 10 (new IUPAC notation). Crystallographic properties

Element name Chemical symbol Atomic number Z	Nickel Ni 28	Palladium Pd 46	Platinum Pt 78	Uun 110	Units	Remarks
Crystal system, Bravais lattice	cub, fc	cub, fc	cub, fc			
Structure type	Cu	Cu	Cu			
Lattice constant a	352.41	389.01	392.33		pm	
Space group	$Fm\bar{3}m$	$Fm\bar{3}m$	$Fm\bar{3}m$			
Schoenflies symbol	O_h^5	O_h^5	O_h^5			
Strukturbericht type	A1	A1	A1			
Pearson symbol	cF4	cF4	cF4			
Number A of atoms per cell	4	4	4			
Coordination number	12	12	12			
Shortest interatomic distance, solid	249	274	277		pm	

Table 2.1–24B(b) Elements of Group VIII(3) (CAS notation), or Group 10 (new IUPAC notation). Mechanical properties

Element name	Nickel	Palladium	Platinum			
Chemical symbol	Ni	Pd	Pt	Uun		
Atomic number Z	28	46	78	110		
					Units	Remarks
Density ϱ, solid	8.90	12.00	21.40		g/cm^3	
Density ϱ, liquid	7.910	10.700	19.700		g/cm^3	
Molar volume V_{mol}	6.59	8.85	9.10		cm^3/mol	
Viscosity η, liquid	5.0				mPa s	
Surface tension, liquid	1.725	1.50	1.866		N/m	
Temperature coefficient	-0.98×10^{-3}	-0.22×10^{-3}	-0.17×10^{-3}		N/(m K)	
Coefficient of linear thermal expansion α	13.3×10^{-6}	11.2×10^{-6}	9.0×10^{-6}		1/K	
Sound velocity, solid, transverse	3080	1900	1690		m/s	
Sound velocity, solid, longitudinal	5810	4540	4080		m/s	
Compressibility κ	0.513×10^{-5}	0.505×10^{-5}	0.351×10^{-5}		1/MPa	Volume compressibility
Elastic modulus E	220	121	170		GPa	
Shear modulus G	78.5	43.5	60.9		GPa	
Poisson number μ	0.31	0.39	0.39			
Elastic compliance s_{11}	7.67	13.7	7.35		1/TPa	
Elastic compliance s_{44}	8.23	14.1	13.1		1/TPa	
Elastic compliance s_{12}	-2.93	-6.0	-3.08		1/TPa	
Elastic stiffness c_{11}	247	221	347		GPa	
Elastic stiffness c_{44}	122	70.8	76.5		GPa	
Elastic stiffness c_{12}	153	171	251		GPa	
Tensile strength	317		134		MPa	At 293 K
Vickers hardness	640	461	560			At 293 K

Table 2.1–24B(c) Elements of Group VIII(3) (CAS notation), or Group 10 (new IUPAC notation). Thermal and thermodynamic properties

Element name	Nickel	Palladium	Platinum			
Chemical symbol	Ni	Pd	Pt	Uun		
Atomic number Z	28	46	78	110		
					Units	Remarks
Thermal conductivity λ	83	71.8	71.6		W/(m K)	
Molar heat capacity c_p	26.07	25.98	25.85		J/(mol K)	At 298 K
Standard entropy S^0	29.796	37.823	41.631		J/(mol K)	At 298 K and 100 kPa
Enthalpy difference $H_{298} - H_0$	4.7870	5.4685	5.7237		kJ/mol	
Melting temperature T_m	1728.30	1828.0	2041.50		K	
Transition	α–liquid	α–liquid	α–liquid			
Enthalpy change ΔH_m	17.4798	16.736	22.1750		kJ/mol	
Entropy change ΔS_m	10.114	9.155	10.862		J/(mol K)	
Relative volume change ΔV_m						$(V_l - V_s)/V_l$ at T_m
Boiling temperature T_b	3157	3237	4100		K	
Enthalpy change ΔH_b	369.24	357.6	509.8		kJ/mol	

Table 2.1-24B(d) Elements of Group VIII(3) (CAS notation), or Group 10 (new IUPAC notation). Electronic, electromagnetic, and optical properties. There is no Table 2.1-24C, because no allotropic or high-pressure modifications are known

Element name	Nickel	Palladium	Platinum			
Chemical symbol	Ni	Pd	Pt	Uun		
Atomic number Z	28	46	78	110		
					Units	Remarks
Characteristics	Ductile metal	Ductile metal	Ductile metal			
Electrical resistivity ρ_s	59.0	101	98.1		nΩ m	At RT
Temperature coefficient	69.2×10^{-4}	37.7×10^{-4}	39.6×10^{-4}		1/K	
Pressure coefficient	1.82×10^{-9}	-2.1×10^{-9}	-1.88×10^{-9}		1/hPa	
Electrical resistivity ρ_l	850				nΩ m	
Resistivity ratio at T_m	1.3					
Hall coefficient R	-60×10^{-12}	-86×10^{-12}	-24.4×10^{-12}		m^3/(A s)	At 298 K, $B = 0.3$–4.6 T
Thermoelectronic coefficient	-18	-9.54	-3.50		μV/K	
Electronic work function	5.15	5.12	5.65		V	
Thermal work function	4.60	4.99	5.30		V	
Molar magnetic susceptibility χ_{mol} (SI)	Ferromagnetic	6786×10^{-6}	2425×10^{-6}		cm^3/mol	At 295 K
Molar magnetic susceptibility χ_{mol} (cgs)	Ferromagnetic	540×10^{-6}	193×10^{-6}		cm^3/mol	At 295 K
Mass magnetic susceptibility χ_{mass} (SI)	Ferromagnetic	67.0×10^{-6}	13.0×10^{-6}		cm^3/g	At 295 K

Table 2.1-24D Elements of Group VIII(3) (CAS notation), or Group 10 (new IUPAC notation). Ionic radii (determined from crystal structures)

Element	Nickel		Palladium			Platinum		
Ion	Ni^{2+}	Ni^{3+}	Pd^{2+}	Pd^{3+}	Pd^{4+}	Pt^{2+}	Pt^{4+}	
Coordination number								Units
4	49		64			60		pm
6	69	56	86	76	62	80	63	pm

2.1.5.4 Elements of the Lanthanides Period

Table 2.1-25A Lanthanides. Atomic, ionic, and molecular properties (see Table 2.1-25D for ionic radii)

Element name Chemical symbol Atomic number Z	Cerium Ce 58	Praseodymium Pr 59	Neodymium Nd 60	Promethium Pm 61	Samarium Sm 62	Europium Eu 63	Gadolinium Gd 64	Units
Characteristics				Radioactive				
Relative atomic mass A (atomic weight)	140.116(1)	140.90765(2)	144.24(3)	[145]	150.36(3)	151.964(1)	157.25(3)	
Abundance in lithosphere [a]	41.6×10^{-6}	5.5×10^{-6}	23.9×10^{-6}		6.5×10^{-6}	1.1×10^{-6}	6.4×10^{-6}	
Abundance in sea [a]	1×10^{-12}	6×10^{-13}	3×10^{-12}		5×10^{-14}	1×10^{-14}	7×10^{-13}	
Atomic radius r_{cov}	165	165	164	165	162	185	161	pm
Atomic radius r_{met}	182	183	182	181	180	200	179	pm
Electron shells	–NOP	–NOP	–NOP	–NOP	–NOP	–NOP	–NOP	
Electronic ground state	$^3H_4$	$^4I_{9/2}$	$^5I_4$	$^6H_{5/2}$	$^7F_0$	$^8S_{7/2}$	$^9D_2$	
Electronic configuration	$[Xe]4f^15d^16s^2$	$[Xe]4f^36s^2$	$[Xe]4f^46s^2$	$[Xe]4f^56s^2$	$[Xe]4f^66s^2$	$[Xe]4f^76s^2$	$[Xe]4f^75d^16s^2$	
Oxidation states	4+, 3+	4+, 3+	3+	3+	3+, 2+	3+, 2+	3+	
Electronegativity χ_A [b]	(1.08)	(1.07)	(1.07)	(1.07)	(1.07)	(1.01)	1.11	
1st ionization energy	5.5387	5.464	5.5250	5.55	5.6437	5.6704	6.1500	eV
2nd ionization energy	10.85	10.55	10.73	10.90	11.07	11.241	12.09	eV
3rd ionization energy	20.198	21.624	22.1	22.3	23.4	24.92	20.63	eV
4th ionization energy	36.758	38.98	40.4	41.1	41.4	42.7	44.0	eV

[a] Mass ratio.   [b] According to Allred and Rochow.

Element name Chemical symbol Atomic number Z	Terbium Tb 65	Dysprosium Dy 66	Holmium Ho 67	Erbium Er 68	Thulium Tm 69	Ytterbium Yb 70	Lutetium Lu 71	Units
Relative atomic mass A (atomic weight)	158.92534(2)	162.50(3)	164.93032(2)	167.26(3)	168.93421(2)	173.04(3)	174.967(1)	
Abundance in lithosphere [a]	1×10^{-6}	4.5×10^{-6}	1.2×10^{-6}	2.5×10^{-6}	0.2×10^{-6}	2.7×10^{-6}	0.8×10^{-6}	
Abundance in sea [a]	1×10^{-13}	9×10^{-13}	2×10^{-13}	8×10^{-13}	2×10^{-13}	8×10^{-13}	2×10^{-13}	
Atomic radius r_{cov}	159	159	158	157	156	174	156	pm
Atomic radius r_{met}	176	175	174	173	172	194	172	pm
Electron shells	–NOP	–NOP	–NOP	–NOP	–NOP	–NOP	–NOP	
Electronic ground state	$^6H_{15/2}$	$^5I_8$	$^4I_{15/2}$	$^3H_6$	$^2F_{7/2}$	$^1S_0$	$^2D_{3/2}$	
Electronic configuration	$[Xe]4f^96s^2$	$[Xe]4f^{10}6s^2$	$[Xe]4f^{11}6s^2$	$[Xe]4f^{12}6s^2$	$[Xe]4f^{13}6s^2$	$[Xe]4f^{14}6s^2$	$[Xe]4f^{14}5d^16s^2$	
Oxidation states	4+, 3+	3+	3+	3+	3+, 2+	3+, 2+	3+	
Electronegativity χ_A [b]	(1.10)	(1.10)	(1.10)	(1.11)	(1.11)	(1.06)	(1.14)	
1st ionization energy	5.8639	5.9389	6.0216	6.1078	6.18431	6.25416	5.42585	eV
2nd ionization energy	11.52	11.67	11.80	11.93	12.05	12.1761	13.9	eV
3rd ionization energy	21.91	22.8	22.84	22.74	23.68	25.05	20.9594	eV
4th ionization energy	39.79	41.4	42.5	42.7	42.7	43.56	45.25	eV

[a] Mass ratio.   [b] According to Allred and Rochow.

Table 2.1-25B(a) Lanthanides. Crystallographic properties (see Table 2.1-25C for allotropic and high-pressure modifications)

Element name	Cerium	Praseodymium	Neodymium	Promethium	Samarium	Europium	Gadolinium	Units
Chemical symbol	Ce	Pr	Nd	Pm	Sm	Eu	Gd	
Atomic number Z	58	59	60	61	62	63	64	
Modification					α-Sm			
Crystal system, Bravais lattice	cub, fc	hex	hex	hex	hex	cub, bc	hex	
Structure type	Cu	α-La	α-La	α-La	Sm	W	Mg	
Lattice constant a	516.10	367.21	365.82	365	362.90	458.27	363.36	pm
Lattice constant c		1183.26	1179.66	1165	2620.7		578.10	pm
Space group	$Fm\bar{3}m$	$P6_3/mmc$	$P6_3/mmc$	$P6_3/mmc$		$Im\bar{3}m$	$P6_3/mmc$	
Schoenflies symbol	O_h^5	D_{6h}^4	D_{6h}^4	D_{6h}^4		O_h^9	D_{6h}^4	
Strukturbericht type	A1	A3′	A3′	A3′		A2	A3	
Pearson symbol	cF4	hP4	hP4	hP4		cI2	hP2	
Number A of atoms per cell	4	4	4	4		2	2	
Coordination number	12	12	12	12			12	
Shortest interatomic distance, solid	364		364	362	397		360	pm

Element name	Terbium	Dysprosium	Holmium	Erbium	Thulium	Ytterbium	Lutetium	Units
Chemical symbol	Tb	Dy	Ho	Er	Tm	Yb	Lu	
Atomic number Z	65	66	67	68	69	70	71	
Crystal system, Bravais lattice	hex	hex	hex	hex	hex	cub, fc	hex	
Structure type	Mg	Mg	Mg	Mg	Mg	Cu	Mg	
Lattice constant a	360.55	359.15	357.78	355.92	353.75	548.48	350.52	pm
Lattice constant c	569.66	565.01	561.78	558.50	555.40		554.94	pm
Space group	$P6_3/mmc$	$P6_3/mmc$	$P6_3/mmc$	$P6_3/mmc$	$P6_3/mmc$	$Fm\bar{3}m$	$P6_3/mmc$	
Schoenflies symbol	D_{6h}^4	D_{6h}^4	D_{6h}^4	D_{6h}^4	D_{6h}^4	O_h^5	D_{6h}^4	
Strukturbericht type	A3	A3	A3	A3	A3	A1	A3	
Pearson symbol	hP2	hP2	hP2	hP2	hP2	cF4	hP2	
Number A of atoms per cell	2	2	2	2	2	4	2	
Coordination number	12	12	12	12	12	12	12	
Shortest interatomic distance, solid	357	355	353	351	349	388	347	pm

Table 2.1–25B(b) Lanthanides. Mechanical properties

Element name	Cerium[a]	Praseodymium	Neodymium	Promethium	Samarium	Europium	Gadolinium	Units
Chemical symbol	Ce	Pr	Nd	Pm	Sm	Eu	Gd	
Atomic number Z	58	59	60	61	62	63	64	
Density ϱ, solid	6.78	6.77	7.00	6.48	7.54	5.26	7.89	g/cm^3
Density ϱ, liquid		6.609						g/cm^3
Molar volume V_{mol}	17.00	20.80	20.59	20.1	20.00	28.29	19.90	cm^3/mol
Viscosity η, liquid		0.431						mPa s
Surface tension, liquid	0.72	0.7	0.688	0.65	0.6	0.450	0.816	N/m
Coefficient of linear thermal expansion α	6.3×10^{-6}	6.79×10^{-6}	9.6×10^{-6}	11×10^{-6}	10.4×10^{-6}	32×10^{-6}	9.4×10^{-6}	1/K
Sound velocity, solid, transverse	1230	1410	1440		1290		1680	m/s
Sound velocity, solid, longitudinal	3060	2660	2720		2700		2950	m/s
Volume compressibility κ	5.06×10^{-5}	3.15×10^{-5}	2.94×10^{-5}	2.9×10^{-5}	3.27×10^{-5}	8.13×10^{-5}	2.47×10^{-5}	1/MPa
Elastic modulus E	44.1	35.2	37.9	46	53.9	18.2	56.3	GPa
Shear modulus G	12.1	13.2	14.3	18	12.7	7.8	22.4	GPa
Poisson number μ	0.25	0.31	0.31	0.28	0.35	0.17	0.26	
Elastic compliance s_{11}	62.8	26.6	23.7				18.0	1/TPa
Elastic compliance s_{33}		19.3	18.5				16.1	1/TPa
Elastic compliance s_{44}	57.8	73.6	66.5				48.1	1/TPa
Elastic compliance s_{12}	−22.3	−11.3	−9.5				−5.7	1/TPa
Elastic compliance s_{13}		−3.8	−3.9				−3.6	1/TPa
Elastic stiffness c_{11}	26.0	49.4	54.8				67.8	GPa
Elastic stiffness c_{33}		57.4	60.9				71.2	GPa
Elastic stiffness c_{44}	17.3	13.6	15.0				20.8	GPa
Elastic stiffness c_{12}	14.3	23.0	24.6				25.6	GPa
Elastic stiffness c_{13}		14.3	16.6				20.7	GPa
Tensile strength	117	169	169	170	157		122	MPa
Vickers hardness	275	400	348		412	167	510–638	

[a] The data for the elastic behavior of cerium concern γ-cerium.

Table 2.1–25B(b) Lanthanides. Mechanical properties, cont.

Element name	Terbium	Dysprosium	Holmium	Erbium	Thulium	Ytterbium	Lutetium	
Chemical symbol	Tb	Dy	Ho	Er	Tm	Yb	Lu	
Atomic number Z	65	66	67	68	69	70	71	Units
Density ϱ, solid	8.27	8.54	8.80	9.05	9.33	6.98	9.840	g/cm^3
Density ϱ, liquid						6.292		g/cm^3
Molar volume V_{mol}	19.31	19.00	18.75	18.44	18.12	24.84	17.78	cm^3/mol
Viscosity η, liquid						0.424		mPa s
Surface tension, liquid	0.65	0.650	0.650	0.620	0.62	0.85		N/m
Coefficient of linear thermal expansion α	7.0×10^{-6}	10.0×10^{-6}	11.2×10^{-6}	9.2×10^{-6}	13.3×10^{-6}	25.0×10^{-6}	8.12×10^{-6}	1/K
Sound velocity, solid, transverse	1060	1720	1740	1810		1000		m/s
Sound velocity, solid, longitudinal	2920	2960	3040	3080		1820		m/s
Volume compressibility κ	2.40×10^{-5}	2.50×10^{-5}	2.42×10^{-5}	2.34×10^{-5}	2.42×10^{-5}	7.24×10^{-5}	2.33×10^{-5}	1/MPa
Elastic modulus E	57.5	63.1	67.1	73.3	74.0	18.4	68.4	GPa
Shear modulus G	22.8	25.5	26.7	29.6	30.4	7.16	27.1	GPa
Poisson number μ	0.26	0.24	0.26	0.24	0.27	0.28	0.27	
Elastic compliance s_{11}	17.4	16.0	15.3	14.1		89.2	14.3	1/TPa
Elastic compliance s_{33}	15.6	14.5	14.0	13.2			14.8	1/TPa
Elastic compliance s_{44}	46.0	41.2	38.6	36.4		56.4	37.3	1/TPa
Elastic compliance s_{12}	−5.2	−4.6	−4.3	−4.2		−31.9	−4.2	1/TPa
Elastic compliance s_{13}	−3.6	−3.2	−2.9	−2.6			−3.5	1/TPa
Elastic stiffness c_{11}	69.2	74.0	76.5	84.1		18.6	86.2	GPa
Elastic stiffness c_{33}	74.4	78.6	79.6	84.7			80.9	GPa
Elastic stiffness c_{44}	21.8	24.3	25.9	27.4		17.7	26.8	GPa
Elastic stiffness c_{12}	25.0	25.5	25.6	29.4		10.4	32.0	GPa
Elastic stiffness c_{13}	21.8	21.8	21.0	22.6			28.0	GPa
Tensile strength	About 122		About 132	139	About 140	72.5	139	MPa
Vickers hardness	863	544	481	589	520	206	1160	

Table 2.1–25B(c) Lanthanides. Thermal and thermodynamic properties

Element name	Cerium	Praseodymium	Neodymium	Promethium	Samarium	Europium	Gadolinium	Units	Remarks
Chemical symbol	Ce	Pr	Nd	Pm	Sm	Eu	Gd		
Atomic number Z	58	59	60	61	62	63	64		
Thermal conductivity λ	11.4	12.5	16.5	17.9	13.3	13.9	10.6	W/(m K)	At 298.2 K
Molar heat capacity c_p	26.4	27.20	27.45		29.54	27.6	37.02	J/(mol K)	At 298.15 K and 100 kPa
Standard entropy S^0	69.454	73.931	71.086	72.00	69.496	80.793	68.089	J/(mol K)	At 298.15 K
Enthalpy difference $H_{298} - H_0$	7.2802	7.4182	7.1337	7.3000	7.5730	8.0040	9.0876	kJ/mol	
Melting temperature T_m	1072.0	1204.0	1289.0	1315.0	1345.0	1095.0	1586.15	K	
Transition	δ–liquid	β–liquid	β–liquid	β–liquid	γ–liquid	α–liquid	β–liquid		
Enthalpy change ΔH_m	5.4601	6.8869	7.1421	7.7000	8.6190	9.2132	9.6680	kJ/mol	
Entropy change ΔS_m	5.093	5.720	5.541	5.856	6.408	8.414	6.095	J/(mol K)	
Relative volume change ΔV_m	0.011	0.02	0.09		0.036	0.048	0.02		$(V_l - V_s)/V_l$ at T_m
Boiling temperature T_b	3699	3785	3341	3785	2064	1870	3569	K	
Enthalpy change ΔH_b	414.2	296.8	273.0		166.4	144.7	359.4	kJ/mol	

Table 2.1–25B(c) Lanthanides. Thermal and thermodynamic properties, cont.

Element name	Terbium	Dysprosium	Holmium	Erbium	Thulium	Ytterbium	Lutetium	Units	Remarks
Chemical symbol	Tb	Dy	Ho	Er	Tm	Yb	Lu		
Atomic number Z	65	66	67	68	69	70	71		
Thermal conductivity λ	11.1	10.7	16.2	14	16.8	38.5	16.4	W/(m K)	At 298.2 K
Molar heat capacity c_p	28.91	28.16	27.15	28.12	27.03	26.74	26.86	J/(mol K)	At 298.15 K and 100 kPa
Standard entropy S^0	73.8	74.956	75.019	73.178	74.015	59.831	50.961	J/(mol K)	At 298.15 K
Enthalpy difference $H_{298} - H_0$	9.4266	8.8659	7.9956	7.3923	7.3973	6.7111	6.3890	kJ/mol	
Melting temperature T_m	1632	1685.15	1745.0	1802.0	1818.0	1097	1936.0	K	
Transition	β–liquid	β–liquid	β–liquid	α–liquid	α–liquid	γ–liquid	α–liquid		
Enthalpy change ΔH_m	10.1504	11.3505	11.7570	19.9033	16.8406	7.6567	18.6481	kJ/mol	
Entropy change ΔS_m	6.220	6.736	6.738	11.045	9.263	6.980	9.632	J/(mol K)	
Relative volume change ΔV_m	0.031	0.45	0.074	0.09	0.069	0.051	0.036		$(V_l - V_s)/V_l$ at T_m
Boiling temperature T_b	3496	2835	2968	3136	2220	1467	3668	K	
Enthalpy change ΔH_b	330.9	230	242.50	261.4	190.7	128.83	355.9	kJ/mol	

Table 2.1–25B(d) Lanthanides. Electronic, electromagnetic, and optical properties

Element name	Cerium	Praseodymium	Neodymium	Promethium	Samarium	Europium	Gadolinium	Units	Remarks
Chemical symbol	Ce	Pr	Nd	Pm	Sm	Eu	Gd		
Atomic number Z	58	59	60	61	62	63	64		
Characteristics	Metal	Ductile reactive metal	Reactive metal	Metal	Metal	Soft metal very reactive	Soft metal		
Electrical resistivity ρ_s	730	650	610	500	914	890	1260	$n\Omega\,m$	At RT
Temperature coefficient	9.7×10^{-4}	17.1×10^{-4}	21.3×10^{-4}	28×10^{-4}	14.8×10^{-4}	81.3×10^{-4}	17.6×10^{-4}	$1/K$	
Pressure coefficient	-45.2×10^{-9}	-0.4×10^{-9}	-1.5×10^{-9}		-3.57×10^{-9}		-4.5×10^{-9}	$1/hPa$	
Electrical resistivity ρ_l		1130	1550			2440	1950	$n\Omega\,m$	
Hall coefficient R	1.92×10^{-10}	0.709×10^{-10}	0.971×10^{-10}		-0.2×10^{-10}		0.95×10^{-10}	$m^3/(A\,s)$	At 293 K, $B < 1$ T
Thermoelectric coefficient	4.39							$\mu V/K$	
Electronic work function	2.88						3.1	V	
Thermal work function	2.6	2.7	3.3		2.7	2.5	3.07	V	
Molar magnetic susceptibility χ_{mol} (SI)	31.4×10^{-3}	69.5×10^{-3}	74.5×10^{-3}		16.1×10^{-3}	388×10^{-3}	2.324	cm^3/mol	At 295 K
Molar magnetic susceptibility χ_{mol} (cgs)	2.5×10^{-3}	5.53×10^{-3}	5.93×10^{-3}		1.28×10^{-3}	30.9×10^{-3}	185×10^{-3}	cm^3/mol	At 295 K
Mass magnetic susceptibility χ_{mass} (SI)	217×10^{-6}	474×10^{-6}	490×10^{-6}		106×10^{-6}	2.81×10^{-3}	14.9×10^{-3}	cm^3/g	At 295 K

Element name	Terbium	Dysprosium	Holmium	Erbium	Thulium	Ytterbium	Lutetium	Units	Remarks
Chemical symbol	Tb	Dy	Ho	Er	Tm	Yb	Lu		
Atomic number Z	65	66	67	68	69	70	71		
Characteristics	Ductile metal	Metal	Soft metal	Soft metal	Ductile metal	Ductile metal	Metal		
Electrical resistivity ρ_s	1130	890	814	810	670	250	540	$n\Omega\,m$	At RT
Temperature coefficient	11.9×10^{-4}	11.9×10^{-4}	17.1×10^{-4}	20.1×10^{-4}	19.5×10^{-4}	13×10^{-4}	24×10^{-4}	$1/K$	
Pressure coefficient		-2.3×10^{-9}	-2.2×10^{-9}	-27×10^{-9}	-2.6×10^{-9}	9.7×10^{-9}	-1.31×10^{-9}	$1/hPa$	
Electrical resistivity ρ_l	1930	2100	2210	2260		1080		$n\Omega\,m$	
Hall coefficient R				-0.34×10^{-10}		-0.53×10^{-10}		$m^3/(A\,s)$	At 293 K, $B < 1$ T
Electronic work function	3.4	3.09	3.09	3.12	3.12	2.50	3.3	V	
Thermal work function	$2.136\ ^a$	$1.232\ ^a$					3.14	V	
Molar magnetic susceptibility χ_{mol} (SI)	$170 \times 10^{-3}\ ^a$	$98 \times 10^{-3}\ ^a$	916×10^{-3}	603×10^{-3}	329×10^{-3}	842×10^{-6}	2.30×10^{-3}	cm^3/mol	At 295 K
Molar magnetic susceptibility χ_{mol} (cgs)	15.3×10^{-3}	8.00×10^{-3}	72.9×10^{-3}	48×10^{-3}	26.2×10^{-3}	67×10^{-6}	183×10^{-6}	cm^3/mol	At 295 K
Mass magnetic susceptibility χ_{mass} (SI)			5.49×10^{-3}	3.33×10^{-3}	1.90×10^{-3}	4.9×10^{-6}	13×10^{-6}	cm^3/g	At 295 K

[a] The molar magnetic susceptibility is given for Tb and Dy for the α-phase [1.3].

Table 2.1-25C Lanthanides. Allotropic and high-pressure modifications

Element	Cerium					Praseodymium			
Modification	α-Ce	β-Ce	γ-Ce	α'-Ce	Ce-III	α-Pr	β-Pr	γ-Pr	
									Units
Crystal system, Bravais lattice	cub, fc	hex	cub, fc	cub, fc	orth	hex	cub, bc	cub, fc	
Structure type	Cu	α-La	Cu	Cu	α-U	α-La	W	Cu	
Lattice constant a	516.10	367.3		482		367.21	413	488	pm
Lattice constant c		1180.2				1183.26			pm
Space group	$Fm\bar{3}m$	$P6_3/mmc$	$Fm\bar{3}m$	$Fm\bar{3}m$	$Cmcm$	$P6_3/mmc$	$Im\bar{3}m$	$Fm\bar{3}m$	
Schoenflies symbol	O_h^5	D_{6h}^4	O_h^5	O_h^5	D_{2h}^{17}	D_{6h}^4	O_h^9	O_h^5	
Strukturbericht type	A1	A3′	A1	A1	A20	A3′	A2	A1	
Pearson symbol	cF4	hP4	cF4	cF4	oC4	hP4	cI2	cF4	
Number A of atoms per cell	4	4	4	4	4	4	2	4	
Coordination number	12	12	12	12	2+2 +4+4	12	8	12	
Shortest interatomic distance, solid	364					358		345	pm
Range of stability	RTP	> 263 K	< 95 K	> 1.5 GPa	5.1 GPa	RTP	> 1094 K	> 4.0 GPa	

Element	Neodymium			Promethium		
Modification	α-Nd	β-Nd	γ-Nd	α-Pm	β-Pm	
						Units
Crystal system, Bravais lattice	hex	cub, bc	cub, fc	hex	cub, bc	
Structure type	α-La	W	Cu	α-La	W	
Lattice constant a	365.82	413	480	365		pm
Lattice constant c	1179.66			1165		pm
Space group	$P6_3/mmc$	$Im\bar{3}m$	$Fm\bar{3}m$	$P6_3/mmc$	$Im\bar{3}m$	
Schoenflies symbol	D_{6h}^4	O_h^9	O_h^5	D_{6h}^4	O_h^9	
Strukturbericht type	A3′	A2	A1	A3′	A2	
Pearson symbol	hP4	cI2	cF4	hP4	cI2	
Number A of atoms per cell	4	2	4	4	2	
Coordination number	12	8	12	12	8	
Shortest interatomic distance, solid	364	358	339	362		pm
Range of stability	RTP	> 1135 K	> 5.0 GPa	RTP	> 1163 K	

Element	Samarium			Gadolinium			
Modification	α-Sm	β-Sm	γ-Sm	α-Gd	β-Gd	γ-Gd	
							Units
Crystal system, Bravais lattice	trig	cub, bc	hex	hex, cp	cub, bc	trig	
Structure type	α-Sm	W	α-La	Mg	W	α-Sm	
Lattice constant a	362.90		361.8	363.36	406	361	pm
Lattice constant c	2620.7		1166	578.10		2603	pm
Space group	$R\bar{3}m$	$Im\bar{3}m$	$P6_3/mmc$	$P6_3/mmc$	$Im\bar{3}m$	$R\bar{3}m$	
Schoenflies symbol	D_{3d}^5	O_h^9	D_{6h}^4	D_{6h}^4	O_h^9	D_{3d}^5	
Strukturbericht type	A7	A2	A3′	A3	A2	A7	
Pearson symbol	hR3	cI2	hP4	hP2	cI2	hR3	
Number A of atoms per cell	3	2	4	2	2	3	
Coordination number	12	8	12	12	8		
Shortest interatomic distance, solid	361	353	360	360	351		pm
Range of stability	RTP	> 1190 K	> 4.0 GPa	RTP	> 1535 K	> 3.0 GPa	

Table 2.1–25C Lanthanides. Allotropic and high-pressure modifications, cont.

Element	Terbium			Dysprosium				
Modification	α-Tb	β-Tb	Tb-II	α-Dy	β-Dy	α′-Dy	γ-Dy	
								Units
Crystal system, Bravais lattice	hex, cp	cub, bc	trig	hex, cp	cub, bc	orth	trig	
Structure type	Mg	W	α-Sm	Mg	W		α-Sm	
Lattice constant a	360.55		341	359.15		359.5	343.6	pm
Lattice constant b						618.4		pm
Lattice constant c	569.66		2450	565.01		567.8	2483.0	pm
Space group	$P6_3/mmc$	$Im\bar{3}m$	$R\bar{3}m$	$P6_3/mmc$	$Im\bar{3}m$	$Cmcm$	$R\bar{3}m$	
Schoenflies symbol	D_{6h}^4	O_h^9	D_{3d}^5	D_{6h}^4	O_h^9	D_{2h}^{17}	D_{3d}^5	
Strukturbericht type	A3	A2	A7	A3	A2		A7	
Pearson symbol	hP2	cI2	hR3	hP2	cI2	oC4	hR3	
Number A of atoms per cell	2	2	3	2	2	4	3	
Coordination number	12			12	8			
Shortest interatomic distance, solid	357			355	345			pm
Range of stability	RTP	>1589 K	>6.0 GPa	RTP	>1243 K	<86 K	>7.5 GPa	

Element	Holmium			Erbium		Thulium			
Modification	α-Ho	β-Ho	γ-Ho	α-Er	β-Er	α-Tm	β-Tm	Tm-II	
									Units
Crystal system, Bravais lattice	hex, cp	cub, bc	trig	hex, cp	cub, bc	hex, cp	cub, bc	trig	
Structure type	Mg	W	α-Sm	Mg	W	Mg	W	α-Sm	
Lattice constant a	357.78		334	355.92		353.75			pm
Lattice constant c	561.78		2450	558.50		555.40			pm
Space group	$P6_3/mmc$	$Im\bar{3}m$	$R\bar{3}m$	$P6_3/mmc$	$Im\bar{3}m$	$P6_3/mmc$	$Im\bar{3}m$	$R\bar{3}m$	
Schoenflies symbol	D_{6h}^4	O_h^9	D_{3d}^5	D_{6h}^4	O_h^9	D_{6h}^4	O_h^9	D_{3d}^5	
Strukturbericht type	A3	A2	A7	A3	A2	A3	A2		
Pearson symbol	hP2	cI2	hR3	hP2	cI2	hP2	cI2	hR3	
Number A of atoms per cell	2	2	3	2	2	2	2	3	
Coordination number	12	8		12	8	12	8		
Shortest interatomic distance, solid	353	343		351		349			pm
Range of stability	RTP	High temperature	>4.0 GPa	RTP	High temperature	RTP	High temperature	>6.0 GPa	

Element	Ytterbium			Lutetium			
Modification	α-Yb	β-Yb	γ-Yb	α-Lu	β-Lu	Lu-II	
							Units
Crystal system, Bravais lattice	cub, fc	cub, bc	hex, cp	hex, cp	cub, bc	trig	
Structure type	Cu	W	Mg	Mg	W	α-Sm	
Lattice constant a	548.48	444	387.99	350.52			pm
Lattice constant c			638.59	554.94			pm
Space group	$Fm\bar{3}m$	$Im\bar{3}m$	$P6_3/mmc$	$P6_3/mmc$	$Im\bar{3}m$	$R\bar{3}m$	
Schoenflies symbol	O_h^5	O_h^9	D_{6h}^4	D_{6h}^4	O_h^9	D_{3d}^5	
Strukturbericht type	A1	A2	A3	A3	A2		
Pearson symbol	cF4	cI2	hP2	hP2	cI2	hR3	
Number A of atoms per cell	4	2	2	2	2	3	
Coordination number	12	8	6+6	12	8	12	
Shortest interatomic distance, solid	388	385	388	347	385		pm
Range of stability	RTP	>1005 K	<270 K	RTP	>1005 K	>23 GPa	

Table 2.1–25D Lanthanides. Ionic radii (determined from crystal structures)

Element	Cerium		Praseodymium		Neodymium	Promethium	Samarium		Europium		Gadolinium	
Ion	Ce^{3+}	Ce^{4+}	Pr^{3+}	Pr^{4+}	Nd^{3+}	Pm^{3+}	Sm^{2+}	Sm^{3+}	Eu^{2+}	Eu^{3+}	Gd^{3+}	
Coordination number												Units
6	101	87	99	85	98		119	96	117	95	94	pm
8	114	97	113	96	112	109	127	108	125	107	105	pm
9					116							pm
10	125	107							135			pm
12	134	114			127			124				pm

Element	Terbium		Dysprosium		Erbium	Thulium		Ytterbium		Lutetium	
Ion	Tb^{3+}	Tb^{4+}	Dy^{2+}	Dy^{3+}	Er^{3+}	Tm^{2+}	Tm^{3+}	Yb^{2+}	Yb^{3+}	Lu^{3+}	
Coordination number											Units
6	92	76	107	91	89	101	88	102		86	pm
7						109					pm
8	104	88	119	103	100		99	114	99	97	pm
9									104		pm

2.1.5.5 Elements of the Actinides Period

Table 2.1-26A Actinides. Atomic, ionic, and molecular properties (see Table 2.1-26D for ionic radii)

Element name	Thorium	Protactinium	Uranium	Neptunium	Plutonium	Americium	Curium	Units
Chemical symbol	Th	Pa	U	Np	Pu	Am	Cm	
Atomic number Z	90	91	92	93	94	95	96	
Characteristics	Radioactive	Radioactive	Radioactive	Radioactive	Radioactive	Radioactive	Radioactive	
Relative atomic mass A (atomic weight)	232.0381(1)	231.03588(2)	238.0289(1)	[237]	[244]	[243]	[247]	
Abundance in lithosphere [a]	11.5×10^{-6}		4×10^{-6}					
Atomic radius r_{cov}	165		142					pm
Atomic radius r_{met}	180	164	154	150	164	173	174	pm
Electron shells	–OPQ	–OPQ	–OPQ	–OPQ	–OPQ	–OPQ	–OPQ	
Electronic ground state	$^3F_2$	$^4K_{11/2}$	$^5L_6$	$^6L_{11/2}$	$^7F_0$	$^8S_{7/2}$	$^9D_2$	
Electronic configuration	[Rn]$6d^27s^2$	[Rn]$5f^26d^17s^2$	[Rn]$5f^36d^17s^2$	[Rn]$5f^46d^17s^2$	[Rn]$5f^67s^2$	[Rn]$5f^77s^2$	[Rn]$5f^76d^17s^2$	
Oxidation states	4+	5+, 4+	6+, 5+, 4+, 3+	6+, 5+, 4+, 3+	6+, 5+, 4+, 3+	6+, 5+, 4+, 3+	3+	
Electronegativity χ_A [b]	(1.11)	(1.14)	(1.22)	(1.22)	(1.22)	(1.2)	(1.2)	
1st ionization energy	6.08	5.89	6.19405	6.2657	6.06	5.993	6.02	eV
2nd ionization energy	11.5							eV
3rd ionization energy	20.0							eV
4th ionization energy	28.8							eV
Standard electrode potential E^0 [c]	−1.899							V

[a] Mass ratio. [b] According to Allred and Rochow. [c] Reaction type Th^{4+} + 4e$^-$ = Th

Element name	Berkelium	Californium	Einsteinium	Fermium	Mendelevium	Nobelium	Lawrencium	Units
Chemical symbol	Bk	Cf	Es	Fm	Md	No	Lr	
Atomic number Z	97	98	99	100	101	102	103	
Characteristics	Radioactive	Radioactive	Radioactive	Radioactive	Radioactive	Radioactive	Radioactive	
Relative atomic mass A (atomic weight)	[247]	[251]	[252]	[257]	[258]	[259]	[262]	
Atomic radius r_{met}	170	186	186					pm
Electron shells	–OPQ	–OPQ	–OPQ	–OPQ	–OPQ	–OPQ	–OPQ	
Electronic ground state	$^6H_{15/2}$	$^5I_8$	$^5I_{15/2}$	$^3H_6$	$^2F_{7/2}$	$^1S_0$	$^2D_{5/2}$	
Electronic configuration	[Rn]$5f^97s^2$	[Rn]$5f^{10}7s^2$	[Rn]$5f^{11}7s^2$	[Rn]$5f^{12}7s^2$	[Rn]$5f^{13}7s^2$	[Rn]$5f^{14}7s^2$	[Rn]$5f^{14}6d^17s^2$	
Oxidation states	4+, 3+	3+					3+	
Electronegativity χ_A [a]	(1.2)	(1.2)	(1.2)	(1.2)	(1.2)			
1st ionization energy	6.23	6.30	6.42	6.50	6.58	6.65		eV

[a] According to Allred and Rochow.

Table 2.1-26B(a) Actinides. Crystallographic properties (see Table 2.1-26C for allotropic and high-pressure modifications)

Element name	Thorium	Protactinium	Uranium	Neptunium	Plutonium	Americium	Curium	Units
Chemical symbol	Th	Pa	U	Np	Pu	Am	Cu	
Atomic number Z	90	91	92	93	94	95	96	
Modification			α-U		α-Pu			
Crystal system, Bravais lattice	cub, fc	tetr, I	orth, C	orth, P	mon, P	hex	hex	
Structure type	Cu	α-Pa	α-U	α-Np	α-Pu	α-La	α-La	
Lattice constant a	508.51	394.5	285.38	666.3	618.3	346.8	349.6	pm
Lattice constant b			586.80	472.3	482.2			pm
Lattice constant c		324.2	495.57	488.7	1096.8	1124.1	1133.1	pm
Lattice angle γ					101.78			deg
Space group	$Fm3m$	$I4/mmm$	$Cmcm$	$Pnma$	$P2_1/m$	$P6_3/mmc$	$P6_3/mmc$	
Schoenflies symbol	O_h^5	D_{4h}^{17}	D_{2h}^{17}	D_{2h}^{16}	C_{2h}^2	D_{6h}^4	D_{6h}^4	
Strukturbericht type	A1	A6	A20	A_c		A3'	A3'	
Pearson symbol	cF4	tI2	oC4	oP8	mP16	hP4	hP4	
Number A of atoms per cell	4	2	4	8	16	4	4	
Coordination number	12	8+2	2+2+4+4	8+6		6+6	6+6	
Shortest interatomic distance, solid	360	321	276	259–335	257–278	347	349	pm

Element name	Berkelium	Californium	Einsteinium	Fermium	Mendelevium	Nobelium	Lawrencium	Units
Chemical symbol	Bk	Cf	Es	Fm	Md	No	Lr	
Atomic number Z	97	98	99	100	101	102	103	
Crystal system, Bravais lattice	hex	hex	hex					
Structure type	α-La	α-La	α-La					
Lattice constant a	341.6							pm
Lattice constant c	1106.9							pm
Space group	$P6_3/mmc$	$P6_3/mmc$	$P6_3/mmc$					
Schoenflies symbol	D_{6h}^4	D_{6h}^4	D_{6h}^4					
Strukturbericht type	A3'	A3'	A3'					
Pearson symbol	hP4	hP4	hP4					
Number A of atoms per cell	4	4	4					

Table 2.1–26B(b) Actinides. Mechanical properties

Element name	Thorium	Protactinium	Uranium	Neptunium	Plutonium	Americium	Curium	Units
Chemical symbol	Th	Pa	U	Np	Pu	Am	Cu	
Atomic number Z	90	91	92	93	94	95	96	
Modification					α-Pu			
Density ϱ, solid	11.70	15.40	18.90	20.40	19.80	13.60	13.510	g/cm^3
Density ϱ, liquid	10.35		17.907		16.623			g/cm^3
Molar volume V_{mol}	19.80	15.0	12.56	11.71	12.3	17.78	18.3	cm^3/mol
Viscosity η, liquid					7.4			mPa s
Surface tension, liquid	1.05		1.53		0.55			N/m
Temperature coefficient			-0.14×10^{-3}					N/(m K)
Coefficient of linear thermal expansion α	12.5×10^{-6}	7.3×10^{-6}	12.6×10^{-6}	27.5×10^{-6}	55×10^{-6}			1/K
Sound velocity, solid, transverse	1630		1940					m/s
Sound velocity, solid, longitudinal	2850		3370					m/s
Volume compressibility κ	1.86×10^{-5}		0.785×10^{-5}					1/MPa
Elastic modulus E	78.3	76	177	68	92.7			GPa
Shear modulus G	30.8		70.6		45			GPa
Poisson number μ	0.26		0.25		0.18			
Elastic compliance s_{11}	27.4		4.91					1/TPa
Elastic compliance s_{22}			6.73					1/TPa
Elastic compliance s_{33}			4.79					1/TPa
Elastic compliance s_{44}	22.0		8.04					1/TPa
Elastic compliance s_{55}			13.6					1/TPa
Elastic compliance s_{66}			13.4					1/TPa
Elastic compliance s_{12}	-10.9		-1.19					1/TPa
Elastic compliance s_{13}			0.08					1/TPa
Elastic compliance s_{23}			-2.61					1/TPa
Elastic stiffness c_{11}	77.0		215					GPa
Elastic stiffness c_{22}			199					GPa
Elastic stiffness c_{33}			267					GPa
Elastic stiffness c_{44}	45.5		124					GPa
Elastic stiffness c_{55}			73.4					GPa
Elastic stiffness c_{66}	50.9		74.3					GPa
Elastic stiffness c_{12}			46					GPa
Elastic stiffness c_{13}			22					GPa
Elastic stiffness c_{23}			108					GPa
Tensile strength	0.219		0.585		0.525			GPa
Vickers hardness	294–687		1960		2500–2800			

Table 2.1-26B(b) Actinides. Mechanical properties, cont.

Element name	Berkelium	Californium	Einsteinium	Fermium	Mendelevium	Nobelium	Lawrencium	Units
Chemical symbol	Bk	Cf	Es	Fm	Md	No	Lr	
Atomic number Z	97	98	99	100	101	102	103	
Density ϱ, solid	14.790	9.310	8.840					g/cm^3
Molar volume V_{mol}	16.70	26.96	28.5					cm^3/mol

Table 2.1-26B(c) Actinides. Thermal and thermodynamic properties

Element name	Thorium	Protactinium	Uranium	Neptunium	Plutonium	Americium	Curium	Units	Remarks
Chemical symbol	Th	Pa	U	Np	Pu	Am	Cm		
Atomic number Z	90	91	92	93	94	95	96		
Modification	α-Th	α-Pa	α-U	α-Np	α-Pu				
Thermal conductivity λ	77.0	About 47	22.0	6.3	6.74	About 10	About 10	W/(m K)	At 298 K
Molar heat capacity c_p	27.32	(27.61)	27.66	29.62	32.84	25.9	(27.70)	J/(mol K)	At 298.15 K and 100 kPa
Standard entropy S^0	51.800	51.882	50.200	50.459	54.461	55.396	71.965	J/(mol K)	
Enthalpy difference $H_{298} - H_0$	6.3500	6.4392	6.3640	6.6065	6.9023	6.4070	6.1337	kJ/mol	
Melting temperature T_m	2022.99	1844.78	1408.0	917.0	913.0	1449.0	1618.0	K	
Transition	β–liquid	β–liquid	γ–liquid	γ–liquid	ε–liquid	γ–liquid	β–liquid		
Enthalpy change ΔH_m	13.8072	12.3412	9.1420	3.1986	2.8240	14.3930	14.6440	kJ/mol	
Entropy change ΔS_m	6.825	6.690	6.493	3.488	3.093	9.933	9.051	J/(mol K)	
Relative volume change ΔV_m	0.05		0.022						$(V_l - V_s)/V_l$ at T_m
Boiling temperature T_b	5061		4407		3503	2880		K	
Enthalpy change ΔH_b	514.1	481	464.1	423.4	260.0	238.5	395.7	kJ/mol	

Element name	Berkelium	Californium	Einsteinium	Fermium	Mendelevium	Nobelium	Lawrencium	Units	Remarks
Chemical symbol	Bk	Cf	Es	Fm	Md	No	Lr		
Atomic number Z	97	98	99	100	101	102	103		
Thermal conductivity λ	About 10	About 10	About 10					W/(m K)	
Standard entropy S^0	(76.15)	80.542	89.471	87.236				J/(mol K)	At 298.15 K and 100 kPa
Melting temperature T_m	1256	1213	1133.0	1800	1100	1100	1900	K	
Transition			β–liquid						
Enthalpy change ΔH_m			9.4056	1.02				kJ/mol	
Entropy change ΔS_m			8.302					J/(mol K)	
Enthalpy of boiling ΔH_b				3.26				kJ/mol	

Table 2.1-26B(d) Actinides. Electronic, electromagnetic, and optical properties

Element name	Thorium	Protactinium	Uranium	Neptunium	Plutonium	Americium	Curium		Units	Remarks
Chemical symbol	Th	Pa	U	Np	Pu	Am	Cm			
Atomic number Z	90	91	92	93	94	95	96			
Characteristics	Soft metal	Toxic metal	Ductile metal	Reactive metal	Metal, very toxic	Metal	Metal, very reactive			
Electrical resistivity ρ_s	147	177	280		1414	680	860		$n\Omega\,m$	At RT
Temperature coefficient	27.5×10^{-4}		28.2×10^{-4}			33×10^{-4}			1/K	
Pressure coefficient	-3.4×10^{-9}								1/Pa	
Superconducting critical temperature T_{crit}	1.37								K	
Superconducting critical field H_{crit}	162								Oe	
Hall coefficient R [a]	-1.2×10^{-10}		0.34×10^{-10}		0.69×10^{-6}				$m^3/(A\,s)$	
Thermoelectronic coefficient			5.0						$\mu V/K$	
Electronic work function	3.67		3.47						V	
Thermal work function	3.42	4.8	3.47						V	
Molar magnetic susceptibility χ_{mol} (SI)	1.22×10^{-3}	3.48×10^{-3}	5.14×10^{-3}	7.23×10^{-3}	6.60×10^{-3}				cm^3/mol	
Molar magnetic susceptibility χ_{mol} (cgs)	97×10^{-6}	277×10^{-6}	409×10^{-6}	575×10^{-6}	525×10^{-6}				cm^3/mol	
Mass magnetic susceptibility χ_{mol} (SI)	7.2×10^{-6}		21.6×10^{-6}		31.7×10^{-6}	50×10^{-6}			cm^3/g	

[a] At 298 K, $B = 0.3$–0.7 T.

Element name	Berkelium	Californium	Einsteinium	Fermium	Mendelevium	Nobelium	Lawrencium
Chemical symbol	Bk	Cf	Es	Fm	Md	No	Lr
Atomic number Z	97	98	99	100	101	102	103
Characteristics	Metal	Metal	Metal		Metal		

Table 2.1–26C Actinides. Allotropic and high-pressure modifications

Element	Thorium		Protactinium		Uranium				
Modification	α-Th	β-Th	α-Pa	β-Pa	α-U	β-U	γ-U		Units
Crystal system, Bravais lattice	cub, fc	cub, bc	tetr, I	cub, bc	orth, C	tetr	cub, bc		
Structure type	Cu	W	In	W	α-U		W		
Lattice constant a	508.51	411	394.5		285.38	1075.9	352.4		pm
Lattice constant b					586.80				pm
Lattice constant c			324.2		495.57	565.4			pm
Space group	$Fm\bar{3}m$	$Im\bar{3}m$	$I4/mmm$	$Im\bar{3}m$	$Cmcm$	$P4_2/mmm$	$Im\bar{3}m$		
Schoenflies symbol	O_h^5	O_h^9	D_{4h}^{17}	O_h^9	D_{2h}^{17}	C_{2h}^{14}	O_h^9		
Strukturbericht type	A1	A2	A6	A2	A20		A2		
Pearson symbol	cF4	cI2	tI2	cI2	oC4	tP30	cI2		
Number A of atoms per cell	4	2	2	2	4	30	2		
Coordination number	12	8	8+2	2+2+4+4	2+2+4+4	12	8		
Shortest interatomic distance, solid	360	356	321		276	287–353	347		pm
Range of stability	RTP	>1673 K	RTP	>1443 K	RTP	>935 K	>1045 K		

Element	Neptunium			Plutonium						
Modification	α-Np	β-Np	γ-Np	α-Pu	β-Pu	γ-Pu	δ-Pu	δ′-Pu	ε-Pu	Units
Crystal system, Bravais lattice	orth, P	tetr	cub, bc	mon, P	mon	orth	cub, fc	tetr	cub, bc	
Structure type	α-Np		W				Cu	In	W	
Lattice constant a	666.3	489.6	352	618.3	928.4	315.87	463.71	332.61	570.3	pm
Lattice constant b	472.3			482.2	1046.3	576.82				pm
Lattice constant c	488.7	338.7		1096.8	785.9	1016.2		446.30		pm
Lattice angle γ				101.78	92.13					deg
Space group	$Pnma$	$P4_22_12$	$Im\bar{3}m$	$P2_1/m$	$I2/m$	$Fddd$	$Fm\bar{3}m$	$I4/mmm$	$Im\bar{3}m$	
Schoenflies symbol	D_{2h}^{16}	D_{4h}^7	O_h^9	C_{2h}^2	C_{2h}^3	D_{2h}^{24}	O_h^5	D_{4h}^{17}	O_h^9	
Strukturbericht type	A_c	A_d	A2				A1	A6	A2	
Pearson symbol	oP8	tP4	cI2	mP16	mI34	oF8	cF4	tI2	cI2	
Number A of atoms per cell	8	4	2	16	34	8	4	2	2	
Coordination number	8+6	8+6	8+6			4+2+4	12	4+8	8+6	
Shortest interatomic distance, solid	260–336	275–356	305	257–278	259–310	303	328	333	315	pm
Range of stability	RTP	>553 K	>850 K	RTP	>395 K	>508 K	>592 K	>726 K	>744 K	

Table 2.1-26C Actinides. Allotropic and high-pressure modifications, cont.

Element	Americium			Curium		
Modification	α-Am	β-Am	γ-Am	α-Cm	β-Cm	
						Units
Crystal system, Bravais lattice	hex	cub, fc	orth	hex	cub, fc	
Structure type	α-La	Cu	α-U	α-La	Cu	
Lattice constant a	346.8	489.4	306.3	349.6	438.1	pm
Lattice constant b			596.8			pm
Lattice constant c	1124.1		516.9	1133.1		pm
Space group	$P6_3/mmc$	$Fm\bar{3}m$	$Cmcm$	$P6_3/mmc$	$Fm\bar{3}m$	
Schoenflies symbol	D_{6h}^4	O_h^5	D_{2h}^{17}	D_{6h}^4	O_h^5	
Strukturbericht type	A3′	A1	A20	A3′	A1	
Pearson symbol	hP4	cF4	oC4	hP4	cF4	
Number A of atoms per cell	4	4	4	4	4	
Coordination number	6+6	12		6+6	12	
Shortest interatomic distance, solid	347	346		350	310	pm
Range of stability	RTP	> 878 K	> 15.0 GPa	RTP	> 1449 K	
Element	Berkelium		Californium		Einsteinium		
Modification	α-Bk	β-Bk	α-Cf	β-Cf	α-Es	β-Es	
							Units
Crystal system, Bravais lattice	hex	cub, fc	hex	cub, fc	hex	cub, fc	
Structure type	α-La	Cu	α-La	Cu	α-La	Cu	
Lattice constant a	341.6	499.7					pm
Lattice constant c	1106.9						pm
Space group	$P6_3/mmc$	$Fm\bar{3}m$	$P6_3/mmc$	$Fm\bar{3}m$	$P6_3/mmc$	$Fm\bar{3}m$	
Schoenflies symbol	D_{6h}^4	O_h^5	D_{6h}^4	O_h^5	D_{6h}^4	O_h^5	
Strukturbericht type	A3′	A1	A3′	A1	A3′	A1	
Pearson symbol	hP4	cF4	hP4	cF4	hP4	cF4	
Number A of atoms per cell	4	4	4	4	4	4	
Coordination number		12		12			
Shortest interatomic distance, solid		353		406			pm
Range of stability	RTP	> 1183 K		> 1213 K	RTP	> 1093 K	

Table 2.1-26D Actinides. Ionic radii (determined from crystal structures)

Element	Thorium	Protactinium			Uranium				Neptunium				
Ion	Th^{4+}	Pa^{3+}	Pa^{4+}	Pa^{5+}	U^{3+}	U^{4+}	U^{5+}	U^{6+}	Np^{3+}	Np^{4+}	Np^{5+}	Np^{6+}	
Coordination number													Units
2								45					pm
4								52					pm
6	94	104	90	78	103	89	76	73	101	87	75	72	pm
8	105					100		86					pm
10	113												pm
12	121					117							pm

Table 2.1-26D Actinides. Ionic radii (determined from crystal structures), cont.

Element	Plutonium				Americium		Curium		
Ion	Pu^{3+}	Pu^{4+}	Pu^{5+}	Pu^{6+}	Am^{3+}	Am^{4+}	Cm^{3+}	Cm^{4+}	
Coordination number									Units
6	100	86	74	71	98	85	97	85	pm
8					109	95		95	pm

Element	Berkelium		Californium		
Ion	Bk^{3+}	Bk^{4+}	Cf^{3+}	Cf^{4+}	
Coordination number					Units
6	96	83	95	82	pm
8		93		92	pm

References

1.1 W. Martienssen (Ed.): *Numerical Data and Functional Relationships in Science and Technology*, Landolt–Börnstein, New Series III and IV (Springer, Berlin, Heidelberg 1970–2003)

1.2 R. Blachnik (Ed.): *Elemente, anorganische Verbindungen und Materialien, Minerale*, D'Ans-Lax, Taschenbuch für Chemiker und Physiker, Vol. 3, 4th edn. (Springer, Berlin, Heidelberg 1998)

1.3 D. R. Lide (Ed.): *CRC Handbook of Chemistry and Physics*, 80th edn. (CRC Press, Boca Raton 1999)

1.4 Lehrstuhl für Werkstoffchemie, T.H. Aachen: *Thermodynamic Properties of Inorganic Materials*, Landolt–Börnstein, New Series IV/19 (Springer, Berlin, Heidelberg 1999)

Part 3 Classes of Materials

1. **Metals**
   Frank Goodwin, Research Triangle Parc, USA
   Sivaraman Guruswamy, Salt Lake City, USA
   Karl U. Kainer, Geesthacht, Germany
   Catrin Kammer, Goslar, Germany
   Wolfram Knabl, Reutte, Austria
   Alfred Koethe, Dresden, Germany
   Gerhard Leichtfried, Reutte, Austria
   Günther Schlamp, Steinbach/Ts, Germany
   Roland Stickler, Vienna, Austria
   Hans Warlimont, Dresden, Germany

2. **Ceramics**
   Hans Warlimont, Dresden, Germany

3. **Polymers**
   Manfred D. Lechner, Osnabrück, Germany

4. **Glasses**
   Dieter Krause, Mainz, Germany

3.1. Metals

Whereas the fundamental properties of all metallic elements are covered systematically and comprehensively in Chapt. 2.1, this section chapter treats those metals that are applied as base and alloying elements of metallic materials. According to common usage, the section is subdivided into treatments of metallic materials based on a single elements (Mg, Al, Ti, Zr, Fe, Co, Ni, Cu, Pb), and treatments of groups of metals with common dominating features (refractory metals, noble metals). The term metal is used indiscriminately for pure metals and for multicomponent metallic materials, i. e., alloys.

The properties of metallic materials depend sensitively not only on their chemical composition and on the electronic and crystal structure of the phases formed, but also to a large degree on their microstructure including the kind and distribution of lattice defects. The phase composition and microstructure of metallic materials are strongly dependent, in turn, on the thermal and mechanical treatments, which are applied under well-controlled conditions to achieve the desired properties. Accordingly, the production of metallic semifinished products and final parts on one hand, and the properties in the final state on the other hand, are usually intricately linked. This also applies to metallic materials treated in other chapters (Chapt. 4.2 on superconductors and Chapt. 4.3 on magnetic materials), as well as to the majority of other inorganic and organic materials.

According to the complexity of the interrelations between fundamental (intrinsic) and microstructure-dependent (extrinsic) properties of metallic materials, this section provides a substantial amount of explanatory text. By the same token, the data given are mostly typical examples indicating characteristic ranges of properties achievable rather than providing complete listings. More comprehensive databases are indicated by way of reference.

3.1.1	**Magnesium and Magnesium Alloys**	162
	3.1.1.1 Magnesium Alloys	163
	3.1.1.2 Melting and Casting Practices, Heat Treatment	168
	3.1.1.3 Joining	169
	3.1.1.4 Corrosion Behavior	169
	3.1.1.5 Recent Developments	170
3.1.2	**Aluminium and Aluminium Alloys**	171
	3.1.2.1 Introduction	171
	3.1.2.2 Production of Aluminium	171
	3.1.2.3 Properties of Pure Al	172
	3.1.2.4 Aluminium Alloy Phase Diagrams	174
	3.1.2.5 Classification of Aluminium Alloys	179
	3.1.2.6 Structure and Basic Mechanical Properties of Wrought Work-Hardenable Aluminium Alloys	180
	3.1.2.7 Structure and Basic Mechanical Properties of Wrought Age-Hardenable Aluminium Alloys	182
	3.1.2.8 Structure and Basic Mechanical Properties of Aluminium Casting Alloys	184
	3.1.2.9 Technical Properties of Aluminium Alloys	186
	3.1.2.10 Thermal and Mechanical Treatment	194
	3.1.2.11 Corrosion Behavior of Aluminium	202
3.1.3	**Titanium and Titanium Alloys**	206
	3.1.3.1 Commercially Pure Grades of Ti and Low-Alloy Ti Materials	207
	3.1.3.2 Ti-Based Alloys	208
	3.1.3.3 Intermetallic Ti–Al Materials	209
	3.1.3.4 TiNi Shape-Memory Alloys	216
3.1.4	**Zirconium and Zirconium Alloys**	217
	3.1.4.1 Technically-Pure and Low-Alloy Zirconium Materials	217
	3.1.4.2 Zirconium Alloys in Nuclear Applications	218
	3.1.4.3 Zirconium-Based Bulk Glassy Alloys	218

3.1.5	**Iron and Steels**......... 221		3.1.8.3	Brasses............... 298	
	3.1.5.1	Phase Relations and Phase Transformations....... 222	3.1.8.4	Bronzes............... 298	
	3.1.5.2	Carbon and Low-Alloy Steels.... 227	3.1.8.5	Copper–Nickel and Copper–Nickel–Zinc Alloys........ 300	
	3.1.5.3	High-Strength Low-Alloy Steels 240			
	3.1.5.4	Stainless Steels....... 240	3.1.9	**Refractory Metals and Alloys**........ 303	
	3.1.5.5	Heat-Resistant Steels........ 257		3.1.9.1	Physical Properties...... 306
	3.1.5.6	Tool Steels........... 262		3.1.9.2	Chemical Properties..... 308
	3.1.5.7	Cast Irons........... 268		3.1.9.3	Recrystallization Behavior..... 311
3.1.6	**Cobalt and Cobalt Alloys**.......... 272			3.1.9.4	Mechanical Properties...... 314
	3.1.6.1	Co-Based Alloys........ 272	3.1.10	**Noble Metals and Noble Metal Alloys**.... 329	
	3.1.6.2	Co-Based Hard-Facing Alloys and Related Materials........ 274		3.1.10.1	Silver and Silver Alloys......... 330
				3.1.10.2	Gold and Gold Alloys........ 347
	3.1.6.3	Co-Based Heat-Resistant Alloys, Superalloys......... 274		3.1.10.3	Platinum Group Metals and Alloys........ 363
	3.1.6.4	Co-Based Corrosion-Resistant Alloys........... 276		3.1.10.4	Rhodium, Iridium, Rhutenium, Osmium, and their Alloys....... 386
	3.1.6.5	Co-Based Surgical Implant Alloys........ 277	3.1.11	**Lead and Lead Alloys**......... 407	
	3.1.6.6	Cemented Carbides........ 277		3.1.11.1	Pure Grades of Lead....... 407
3.1.7	**Nickel and Nickel Alloys**......... 279			3.1.11.2	Pb–Sb Alloys......... 411
	3.1.7.1	Commercially Pure and Low-Alloy Nickels...... 279		3.1.11.3	Pb–Sn Alloys......... 414
				3.1.11.4	Pb–Ca Alloys......... 416
	3.1.7.2	Highly Alloyed Ni-Based Materials........ 279		3.1.11.5	Pb–Bi Alloys......... 419
				3.1.11.6	Pb–Ag Alloys......... 420
	3.1.7.3	Ni-Based Superalloys........ 284		3.1.11.7	Pb–Cu, Pb–Te, and Pb–Cu–Te Alloys....... 421
	3.1.7.4	Ni Plating............ 288			
3.1.8	**Copper and Copper Alloys**........ 296			3.1.11.8	Pb–As Alloys......... 421
	3.1.8.1	Unalloyed Coppers...... 296		3.1.11.9	Lead Cable Sheathing Alloys..... 421
	3.1.8.2	High Copper Alloys........ 297		3.1.11.10	Other Lead Alloys....... 421
			References............ 422		

3.1.1 Magnesium and Magnesium Alloys

Magnesium is the lightest structural metal with a density of $1.74\,\mathrm{g\,cm^{-3}}$. It is produced by two basic processes. One is the electrolysis of fused anhydrous magnesium chloride ($MgCl_2$) derived from magnesite, brine, or seawater, and recently from serpentine ores. The other one is the thermal reduction of magnesium oxide (MgO) by ferrosilicon derived from carbonate ores [1.1]. The use of primary Mg is shown in Fig. 3.1-1. Only one third is used for structural parts, mainly for castings, while the major amount of Mg is still used as an alloying element in Al alloys.

Pure Mg is rarely used for structural applications due to its poor mechanical properties. Therefore, Al and Zn have been introduced as major alloying elements for high-pressure die casting alloys. The main reason for using magnesium alloys is to lower the weigh load, predominantly in transportation industries. The weight reduction amounts to about 30% compared to Al alloys and to about 75% compared to steel. Magnesium has a hexagonal close-packed crystal structure (hcp). Therefore its deformation properties are poor. At room temperature slip occurs on the basal plane $\{0001\}$ in the $\langle 11\bar{2}0 \rangle$ direction only. In addition to slip on basal planes the deformation by twinning is also possible, the twinning system is $\{10\bar{1}2\}\langle 10\bar{1}1 \rangle$. Twin formation leads to different behavior of Mg alloys under tensile and compressive load. With increasing temperature above 200–225 °C, prismatic slip planes are activated in addition. This is of main importance for the processing wrought magnesium alloys.

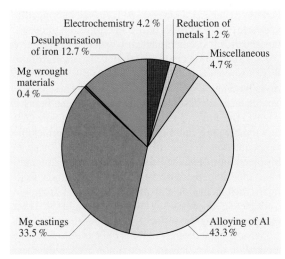

Fig. 3.1-1 World consumption of Mg in 2001 (329 480 t) [1.2]

Alloying elements used in Mg alloys and their abbreviations applied in the designation of Mg materials are listed in Table 3.1-1. General data, maximum solubility data, and compounds formed in major binary magnesium alloy systems are given in Table 3.1-2 [1.4]. Since Al is one of the most important alloying elements, Fig. 3.1-2 [1.3] shows an example of an Al–Mg binary alloy system. The effect of alloying elements in Mg materials is given in Table 3.1-3 [1.3,4].

3.1.1.1 Magnesium Alloys

The major alloying elements are manganese, aluminium, zinc, zirconium, silicon, thorium, and rare earth metals (E). At present E elements are the most promising candidates for magnesium alloys, with high temperature stability as well as improved corrosion behavior. E metals are forming stable intermetallic compounds at high temperature and therefore they decrease castability. Aluminium and zinc are introduced mainly to

Table 3.1-1 Alloying elements and abbreviations used for Mg materials [1.3]

A	C	E	K	L	M	Q	S	W	Z
Al	Cu	Rare earth	Zr	Li	Mn	Ag	Si	Y	Zn

Table 3.1-2 Solubility data and intermetallic phases in binary magnesium alloys [1.4]

Solute element	Maximum solubility (wt%)	(at%)	Adjacent intermetallic phase	Melting point of intermetallic phase (°C)	Type of equilibrium
Lithium	5.5	17.0	–	–	Eutectic
Aluminium	12.7	11.8	$Mg_{17}Al_{12}$	402	Eutectic
Silver	15.0	3.8	Mg_3Ag	492	Eutectic
Yttrium	12.5	3.75	$Mg_{24}Y_5$	620	Eutectic
Zinc	6.2	2.4	$MgZn$	347	Eutectic
Neodymium	≈ 3	≈ 1	$Mg_{41}Nd_5$	560	Eutectic
Zirconium	3.8	1.0	Zr	1855	Peritectic
Manganese	2.2	1.0	Mn	1245	Peritectic
Thorium	4.75	0.52	$Mg_{23}Th_6$	772	Eutectic
Cerium	0.5	0.1	$Mg_{12}Ge$	611	Eutectic
Indium	53.2	19.4	Mg_3In	484	Peritectic
Thallium	60.5	15.4	Mg_5Tl_2	413	Eutectic
Scandium	≈ 24.5	≈ 15	$MgSc$	–	Peritectic
Lead	41.9	7.75	Mg_2Pb	538	Eutectic
Thulium	31.8	6.3	$Mg_{24}Tm_6$	645	Eutectic
Terbium	24.0	4.6	$Mg_{24}Tb_5$	–	Eutectic
Tin	14.5	3.35	Mg_2Sn	770	Eutectic
Gallium	8.4	3.1	Mg_5Ga_2	456	Eutectic
Bismuth	8.9	1.1	Mg_3Bi_2	821	Eutectic
Calcium	1.35	0.82	Mg_2Ca	714	Eutectic
Samarium	≈ 6.4	≈ 1.0	$Mg_{6.2}Sm$	–	Eutectic

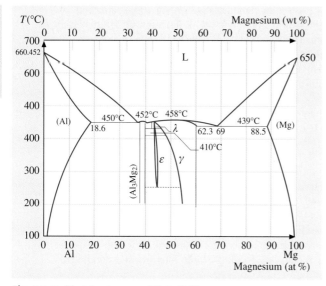

Fig. 3.1-2 Al–Mg phase equilibria [1.3]

increase castability. However, Al forms an intermetallic phase $Mg_{17}Al_{12}$ which is brittle and limits ductility and, thus, the use of Al-containing alloys such as AZ91 above 120 °C. Manganese has been introduced to increase ductility. It is replacing Zn in several cast alloys. Moreover, Mn is binding Fe in intermetallic phases. These can be separated before casting and therefore Mn can be used to purify the alloys. Silicon was also thought to increase high temperature stability by forming the intermetallic compound Mg_2Si. Due to the typical needle shape of this precipitate (so-called chinese script microstructure) it has only limited use because it acts like a notch, leading to crack formation at higher stress levels. Zirconium is not used in Al-containing alloys but it is added to other alloys acting as grain refiner.

Wrought magnesium alloys are not used widely. The application of AZ31 is most common for forgings, extrusions, as well as for sheetmetal. There is an increasing demand for the use of wrought magnesium alloys, especially in the automotive industries, in the form of sheet

Table 3.1-3 General effects of alloying elements in magnesium materials [1.3–5]

Series	Alloying elements	Melting and casting behavior	Mechanical and technological properties
AZ	Al, Zn	Improve castability; tendency to microporosity; increase fluidity of the melt; refine weak grain	Solid solution hardener; precipitation hardening at low temperatures (< 120 °C); improve strength at ambient temperatures; tendency to brittleness and hot shortness unless Zr is refined
QE	Ag, rare earths	Improve castability; reduce microporosity	Solid solution and precipitation hardening at ambient and elevated temperatures; improve elevated temperature tensile and creep properties in the presence of rare earth metals
AM	Al, Mn	Improve castability; tendency to microporosity; control of Fe content by precipitating Fe–Mn compound; refinement of precipitates	Solid solution hardener; precipitation hardening at low temperatures (< 120 °C); increase creep resistivity
AE	Al, rare earth	Improve castability; reduce microporosity	Solid solution and precipitation hardening at ambient and elevated temperatures; improve elevated temperature tensile and creep properties; increase creep resistivity
AS	Al, Si	Tendency to microporosity; decreased castability; formation of stable silicide alloying elements; compatible with Al, Zn, and Ag; refine week grain	Solid solution hardener, precipitation hardening at low temperatures (< 120 °C); improved creep properties
WE	Y, rare earths	Grain refining effect; reduce microporosity	Improve elevated temperature tensile and creep properties; solid solution and precipitation hardening at ambient and elevated temperatures; improve elevated temperature tensile and creep properties

material. Therefore, a number of processes, which are introduced for the processing of aluminium and steel are under research to find out the economic boundary conditions for the processing of magnesium alloys. Beside the investigations on processing routes, there is ongoing research on alloy development since AZ31 does not meet the given aims in most requirements. The compositions of selected cast and wrought magnesium alloys are given in Tables 3.1-4 and 3.1-5. The mechanical properties of the cast and wrought magnesium alloys are listed in Tables 3.1-6 and 3.1-7 [1.3, 4].

The system to denote Mg alloys is generally showing major alloying elements and their content in wt%. The first two letters are key letters and are used for the major alloying elements (Table 3.1-1). The two letters are followed by a number, which represents the nominal composition of the major alloying elements in wt%. The number indicates the content rounded to

Table 3.1-4 Nominal composition of selected cast Mg alloys [1.3, 4]

ASTM designation	Nominal composition (wt%)										
	Al	Zn	Mn	Si	Cu	Zr	RE (MM)	RE (Nd)	Th	Y	Ag
AZ63	6	3	0.3								
AZ81	8	0.5	0.3								
AZ91	9.5	0.5	0.3								
AM50	5		0.3								
AM20	2		0.5								
AS41	4		0.3	1							
AS21	2		0.4	1							
ZK51		4.5				0.7					
ZK61		6				0.7					
ZE41		4.2				0.7	1.3				
ZC63		6	0.5		3						
EZ33		2.7				0.7	3.2				
HK31						0.7			3.2		
HZ32		2.2				0.7			3.2		
QE22						0.7		2.5			2.5
QH21						0.7		1	1		2.5
WE54						0.5		3.25		5.1	
WE43						0.5		3.25		4	

Table 3.1-5 Nominal composition of selected wrought Mg alloys [1.3, 4]

ASTM designation	Nominal composition (wt%)							
	Al	Zn	Mn	Zr	Th	Cu	Li	
M1				1.5				
AZ31	3	1	0.3 (0.20 min.)					
AZ61	6.5	1	0.3 (0.15 min.)					
AZ80	8.5	0.5	0.2 (0.12 min.)					
ZM 21		2	1					
ZMC711		6.5	0.75			1.25		
LA141	1.2		0.15 (min.)				14	
ZK61		6		0.8				
HK31					0.7	3.2		
HM21			0.8		2			
HZ11		0.6		0.6	0.8			

Table 3.1-6 Typical tensile properties and characteristics of selected cast Mg alloys [1.3–5]

ASTM designation	Condition	Tensile properties			Characteristics
		0.2% proof stress (MPa)	Tensile strength (MPa)	Elongation to fracture (%)	
AZ63	As-sand cast	75	180	4	Good room-temperature strength and ductility
	T6	110	230	3	
AZ81	As-sand cast	80	140	3	Tough, leak-tight casting with 0.0015 Be, used for pressure die casting
	T4	80	220	5	
AZ91	As-sand cast	95	135	2	General-purpose alloy used for sand and die casting
	T4	80	230	4	
	T6	120	200	3	
	As-chill cast	100	170	2	
	T4	80	215	5	
	T6	120	215	2	
AM50	As-die cast	125	200	7	High-pressure die casting
AM20	As-die cast	105	135	10	Good ductility and impact strength
AS41	As-die cast	135	225	4.5	Good creep properties up to 150 °C
AS21	As-die cast	110	170	4	Good creep properties up to 150 °C
ZK51	T5	140	253	5	Sand casting, good room-temperature strength and ductility
ZK61	T5	175	275	5	As for ZK51
ZE41	T5	135	180	2	Sand casting, good room-temperature strength, improved castability
ZC 63	T6	145	240	5	Pressure-tight casting, good elevated temperature strength, weldable
EZ33	Sand cast T5	95	140	3	Good castability, pressure tight, weldable, creep resistant up to 250 °C
	Chill cast T5	100	155	3	
HK31	Sand cast T6	90	185	4	Sand casting, good castability, weldable, creep resistant up to 350 °C
HZ32	Sand or chill cast T5	90	185	4	As for HK31
QE22	Sand or chill cast T6	185	240	2	Pressure tight and weldable, high proof stress up to 250 °C
QH21	As-sand cast T6	185	240	2	Pressure tight, weldable, good creep resistance and stress-proof to 300 °C
WE54	T6	200	285	4	High strength at room and elevated temperatures, good corrosion resistance
WE43	T6	190	250	7	Weldable

the nearest digit for the range of that element. For variation within this range suffix letters A, B, C, etc., are used to indicate the stage of development of the alloy, and X is used for alloys which are still experimental. For example, AZ91D indicates that the two major alloying elements are 9 wt% Al and 1 wt% Zn. The letter D signifies that it is the fourth stage in development. The designations of the tempers are identical to those used for Al alloys and are given in Table 3.1-8 [1.3].

Table 3.1-7 Typical tensile properties and characteristics of selected wrought Mg alloys [1.3–5]. (For temper designations see Table 3.1-8)

ASTM designation	Condition	Tensile properties			Characteristics
		0.2% proof stress (MPa)	Tensile strength (MPa)	Elongation to fracture (%)	
M1	Sheet, plate F	70	200	4	Low to medium strength alloy,
	Extrusion F	130	230	4	weldable, corrosion-resistant
	Forgings F	105	200	4	
AZ31	Sheet, plate O	120	240	11	Medium-strength alloy
	H24	160	250	6	weldable, good formability
	Extrusion F	130	230	4	
	Forging F	105	200	4	
AZ61	Extrusion F	105	260	7	High-strength alloy,
	Forging F	160	275	7	weldable
AZ80	Forging T6	200	290	6	High-strength alloy
ZM 21	Sheet, plate O	120	240	11	Medium-strength alloy,
	H24	165	250	6	good formability, good
	Extrusions	155	235	8	damping capacity
	Forgings	125	200	9	
ZMC711	Extrusions T6	300	325	3	High-strength alloy
LA141	Sheet, plate T7	95	115	10	Ultra-light weight (S.G. 1.35)
ZK61	Extrusion F	210	185	6	High strength alloy
	T5	240	305	4	
	Forging T5	160	275	7	
HK31	Sheet, plate H24	170	230	4	High-creep resistance to 350 °C, weldable
	Extrusion T5	180	255	4	
HM21	Sheet, plate T8	135	215	6	High-creep resistance to 350 °C,
	T81	180	255	4	weldable after short time exposure to 425 °C
	Forging T5	175	225	3	
HZ11	Extrusion F	120	215	7	Creep resistance to 350 °C, weldable

Table 3.1-8 Temper designations [1.3]

	General Designations		
F	As fabricated.		
O	Annealed, recrystallized (Wrought products only).		
H	Strain-hardened.		
T	Thermally treated to produce stable tempers other than F, O, or H.		
W	Solution heat-treated (unstable temper).		
Subdivisions of H		**Subdivisions of T**	
H1, Plus one or more digits	Strained only	T2	Annealed (Cast products only)
H2, Plus one or more digits	Strain-hardened and then partially annealed.	T3	Solution heat-treated and cold-worked
H3, Plus one or more digits	Strain-hardened and then stabilized.	T4	Solution heat-treated
		T5	Artificially aged only
		T6	Solution heat-treated and artificially aged
		T7	Solution heat-treated and stabilized
		T8	Solution heat-treated, cold-worked and artificially aged
		T9	Solution heat-treated, artificially aged, and cold-worked
		T10	Artificially aged and cold-worked

3.1.1.2 Melting and Casting Practices, Heat Treatment

Due to the relatively low melting temperature of Mg, casting is the preferred processing route of Mg materials. The addition of Al and Zn are known to improve castability. All processing operations such as melting, alloying, refining, and the cleaning of melts can be carried out in plain carbon steel crucibles. The use of protective gases or fluxes is a recommended practice during melting. Fluxes consist of salts such as KCl, $MgCl_2$, $BaCl_2$, and $CaCl_2$ [1.6]. Gases are used in different combinations, also forming a stable film during melting. Combinations of argon and/or nitrogen with additions of SF_6 or SO_2 are in use. The application of vacuum during melting cannot be recommended due to the high vapor pressure of liquid Mg at low pressure.

During melting, a control of impurities is also necessary. The elements Fe, Ni, Co, and Cu are critical due to their effect of decreasing the corrosion resistance. The Fe content can be controlled by adding Mn, forming intermetallic compounds which are settling at the bottom of the melting crucible. But in general the amount of these critical elements has to be controlled during the primary production of Mg itself. Fortunately Cu is not part of any feedstock material used for primary production. Therefore only master alloys or alloying elements for producing the diverse alloys have to be controlled for Co and also for Fe, which is a major impurity in a number of master alloys and alloying elements.

Any casting process can be used to manufacture parts from Mg alloys. But in view of the low liquidus temperatures, the broad melting ranges, and the excellent fluidity of Mg casting alloys, high-pressure die casting (HPDC) is used most frequently at present. Mainly alloys from the AZ or AM series are used in HPDC but alloys from the AS or AE series are also useable when cast with cold chamber HPDC machines. Alloys from the QE or WE series are not suitable for HPDC but for sand casting or permanent mold casting.

The mechanical properties of the Mg alloys can be improved by heat treatment. Mainly T4, T5, and T6 heat treatments are in use. A T4 treatment means dissolution of precipitates. In general, a T4 treatment is followed by an artificial aging (T6). Stress relieving can also be applied to the cast alloys. Major variables which affect the heat-treatments are section size and heating time, annealing time and temperature, and the protective atmosphere. Welded parts made of Mg alloys can be stress relieved. In general, the heat treatments can be applied to castings with the exception of HPDCs. It is not recommended to heat-treat HPDCs materials because of the entrapment of gases during casting. The absorption of gases can be avoided by applying vacuum during casting. In this case all heat treatments can be applied as well. Table 3.1-9 shows Mg alloys with their respective useful heat treatments.

3.1.1.3 Joining

Brazing and soldering are not praticable due to corrosion problems and the formation of brittle phases. Similar problems occur during welding processes such as metal inert gas welding, laser welding, or even electron beam welding. As with brazing or soldering, the formation of brittle phases leads to failures, drastically decreasing the reliability of welded joints. Riveting and mechanical joining using screws are well introduced. Especially when the joint is supported with adhesives for sealing and for adding additional forces for bonding, riveting and screwing can still be viewed as state of the art in the joining of Mg alloys. Mg alloys are seldom used as rivets or screws due to their limited mechanical properties. But it was proven that Al alloys (5052, 5056, or 6061) for rivets and surface coated steel for screws can be used to fasten Mg alloys due to similarity in corrosion potentials. Moreover, the use of inert washers or layers to minimize contact between different materials is recommended [1.3, 7]. Due to the increasing demand for Mg wrought materials, especially in sheet form, new joining processes are under investigation. These processes are friction stir welding or clinching, often combined with the use of adhesives. While a certain degree of ductility (A > 12%) is necessary in clinching processes, friction stir welding appears to be independent of that requirement. Experiments have shown that even cast magnesium alloys can be joined using this process.

3.1.1.4 Corrosion Behavior

In general only poor corrosion behavior is attributed to Mg and its alloys. This is mainly true when Mg is in contact with other metals and alloys in accordance with the electrochemical potential series of metals. Therefore, special care has to be taken to avoid conductive contact between Mg alloys and other metals with different electrochemical potentials. When joining Mg parts with other materials, insulating washers, surface-coated screws or nonconducting films and surface coatings can be applied. When investigating a Mg

Table 3.1-9 Heat treatments applied to magnesium alloys [1.3]

Alloy	Heat Treatment			
Cast Alloy	F	T4	T5	T6
AM100A		×	×	×
AZ63A		×	×	×
AZ81A		×		
AZ91C		×		×
AZ92A		×		×
EZ33A			×	
EQ21A				×
QE22A				×
WE43A				×
WE54A				×
ZC63A				×
ZE41A			×	
ZE63A				×
ZK51A			×	
ZK61A		×		×
Wrought Alloys				
AZ80A			×	
ZC71A	×		×	×
ZK60A			×	

alloy by itself the material shows similar behavior as plain cabon steels, even in different environments. The main reason for an increase in corrosion rate could be found in the presence of impurities such as Fe, Ni, Co, and Cu. In order to improve the stand-alone corrosion properties these impurities have to be controlled regarding their content. While Co will normally not pose a problem since it is not part of any material used for the primary production of Mg, the content of the other elements needs to be controlled from the beginning in the production of primary Mg. According to the influence of Fe, Ni, and Cu, limits have been defined for high purity alloys which are mainly used today in the production of Mg parts (Fe/Mn < 0.0232, Cu < 0.04 wt%, Ni < 0.005 wt% for AZ91 hp). But it should be mentioned that these limits depend on the contents of other alloying elements as well as the casting processes applied [1.3, 8]. Figure 3.1-3 shows the influence of alloying elements on the rate of corrosion [1.7].

3.1.1.5 Recent Developments

Squeeze casting and semi-solid metal processing (SSMP) are recent developments in casting technology by which high quality castings can be produced. Almost all Mg alloys can be subjected to SSMP if they have a broad melting interval [1.3]. First trials using AZ91 led also to the development of different processes which are making use of processing in the semi-solid state. Recently Thixomolding® was introduced, a technique similar to processing techniques for polymers. Thixomolding is in use mainly for the production of housings for handheld devices such as laptops or cameras. It is under research to investigate the capability of this technology for processing alloys which are showing limited castability mainly in HPDC. Another process using the semi-solid behavior of some magnesium alloys is thixocasting which is still a topic of research, not of production. The New RheoCast process (NRC) which is well introduced in Al industries can, also, be applied to casting semi-solid materials based on Mg.

Beyond monolithic alloys the investigation of Mg matrix composites is still in progress. They can be processed using either the ingot metallurgy route or the powder metallurgy route. There are promising candidates that may be introduced to applications but the production routes are quite expensive and, moreover, the corrosion behavior is not well known.

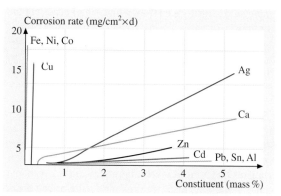

Fig. 3.1-3 Influence of alloying elements on the corrosion rate of Mg [1.7]

Apart from cast or powder metallurgical products, the use of wrought Mg materials is thought to increase in the future. To support the demands stemming mainly from the automotive industry the total production route is under investigation right now. This means that the developments of new wrought magnesium alloys as well as the development of suitable deformation processes in combination with appropriate joining techniques are underway. These studies are accompanied by developing modeling and simulation tools to ensure high quality and accurate prediction of properties and lifetime behavior of magnesium-based materials.

Controlling corrosion is the most pressing problem under research at the moment. Since Mg alloys are about to be widely used in several applications, the corrosion problem, i.e., mainly contact corrosion in combination with other metals and alloys is receiving more and more attention. Therefore, the influence of alloying elements, mainly in combination with each other, is investigated intensively. Aside from alloying, different kinds of surface protection such as surface layers and coatings are under investigation. While castings are normally showing sufficient volume to withstand corrosion attack, this is not the case for magnesium sheet materials even for a long time.

3.1.2 Aluminium and Aluminium Alloys

3.1.2.1 Introduction

Aluminium alloys are the second most widely used metallic materials after steels. Comprehensive treatments and data of Al-based materials are given in [1.9, 10]. Their most important properties are: *low density* ($2.7\,\text{g/cm}^{-3}$); alloying can result in significant further density reductions in Al-Li alloys; the low density can lead to significant energy savings, especially in applications in transportation; *good mechanical properties* offering optimum tensile strength ranging from 60 to 530 MPa; *good workability* permitting most varied shapes to be produced; *good castability* with a variety of casting techniques: sand, mould, die-casting; *good machinability*; *ease of joining* using all commonly applied techniques; *comparatively high corrosion resistance* thanks to the spontaneous formation of a strongly-adherent passivating surface film in air; *different surface treatments* are applicable; *high electrical and thermal conductivity*, especially of unalloyed aluminium; *good optical properties* depending on the degree of purity; *non-magnetic*; *low absorption cross section* for thermal neutrons; *non-combustible*, not causing sparking; *no health risk* is associated with the use of aluminium and its alloys; excellent *recycling* properties.

3.1.2.2 Production of Aluminium

The production of aluminium is based on the electrolysis of molten alumina Al_2O_3 using the Hall–Herault process. Alumina is extracted in the Bayer process form the bauxite ore which contains 20 to 30 wt% Al. In 2002 the main producers of the ore are Australia (28%), Guinea (20%), Brazil (14%), Jamaica (7%), India (4%), and Guyana (3%) [1.9].

After milling, the ore is first broken down using an Al-containing NaOH solution which is seeded to precipitate aluminium hydroxide $Al(OH)_3$. This is dehydrated at about $1100\,°C$ according to $2Al(OH)_3 \rightarrow Al_2O_3 + 3H_2O$.

An electrolytic reduction is carried out in a cell as shown in Fig. 3.1-4. Cryolithe Na_3AlF_6 is used as an additive to decrease the high melting point of pure Al_2O_3 (about $2050\,°C$) since the two compounds form a eutectic near 10 wt% of Al_2O_3 in Na_3AlF_6, Fig. 3.1-5 [1.11, 12]. The electrolysis is carried out in a cell lined with carbon, which serves as the cathode. Carbon anodes are suspended from above the cell into the electrolyte. Two main reactions occur: $2Al^{3+} + 6e^- \rightarrow 2Al$, and oxygen ions react with the carbon of the cathode to form CO_2 (consumption of the cathode). A primary aluminium smelter requires on average 13 to 14 kW h per kg of Al. The main power used is hydro-electric (52.5% in 2001 [1.13]). In recent decades

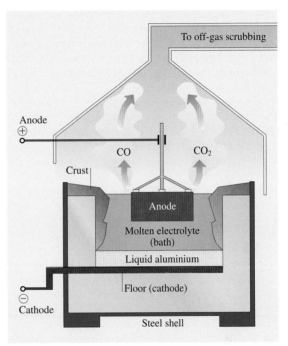

Fig. 3.1-4 Schematic representation of an electrolytic cell for Al winning [1.9]

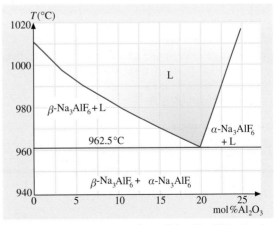

Fig. 3.1-5 Phase diagram of cryolithe Na_3AlF_6-alumina Al_2O_3 [1.9]

secondary production, i. e., recycling of scrap and waste material, has reached 30% of the total production, with high regional variations. Up to 95% less energy is needed to produce Al via secondary production.

3.1.2.3 Properties of Pure Al

Aluminium can be classified as "unalloyed," "pure," or "refined," depending on its degree of purity. The Al from conventional electrolysis is 99.5 to 99.9 wt% pure. Higher purity is produced by "triple-layer refining electrolysis" that can reach ≥ 99.99 wt% purity. The latter grade is used, for example, in the electronics sector.

Physical Properties

The physical properties of pure Al are given in Chapt. 2.1, Table 2.1-11. The temperature dependence of its density is shown in Fig. 3.1-6. The contraction of 6.5% upon solidification corresponds to an increase in density from $2.37\,\mathrm{g\,cm^{-3}}$ in the liquid state to $2.55\,\mathrm{g\,cm^{-3}}$ in the solid state. The temperature dependence of the coefficient of thermal expansion is given in Table 3.1-10.

The specific heat in the solid state increases with temperature from 720 kJ at $-100\,°C$, to 900 at $20\,°C$, and 1110 at $500\,°C$. At its melting point Al has a specific heat of 1220 (solid) and 1040 (liquid). In the liquid state the specific heat rises further, e.g., to 1060 kJ at $800\,°C$. Aluminium has a high reflectivity for light, heat, and for electromagnetic radiation. The Youngs's modulus E of aluminium materials is usually taken to be 70 GPa; it varies between 60 and 78 GPa depending on alloy composition. The shear modulus G varies between 22 and 28 GPa, the value for refined aluminium is 25.0 GPa. Poisson's ratio ν varies between 0.32 and 0.40 (0.35 for refined aluminium).

Unalloyed aluminium has an electrical conductivity of 34 to $38\,\mathrm{m\,\Omega^{-1}\,mm^{-2}}$. Due to this relatively high conductivity, a large fraction of unalloyed Al and Al−Mg−Si alloys are used for electrical conductors. The temperature dependence of the electrical conductivity depends on alloying additions, impurities, and microstructure (Figs. 3.1-7 – 3.1-9). Superconductivity occurs at $T_c = 1.2\,\mathrm{K}$ in refined grades of aluminium (≥ 99.99 wt% Al). The electrical conductivity of high-purity aluminium is still high at 4.2 K.

Mechanical Properties

The ultimate tensile strength of pure aluminium increases markedly with increasing amounts of alloying or impurity additions, as shown in Fig. 3.1-10. Unalloyed aluminium is soft (tensile strength 10–30 MPa, Table 3.1-11) and, like all fcc metals, shows a low rate of work hardening.

Chemical Properties

Aluminium, as a relatively reactive element, is a very strong base, as shown by its position in the electrochemical series and its low standard potential ($V_H = -1.66\,\mathrm{V}$). Therefore, it is not possible to obtain the element from aqueous solutions by electrolysis. Similarly, it is not possible to obtain aluminium by a carbother-

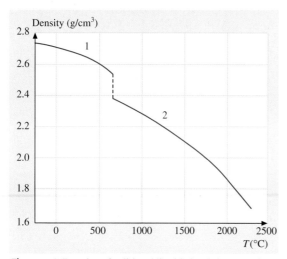

Fig. 3.1-6 Density of solid and liquid aluminium as a function of temperature [1.9]

Table 3.1-10 Coefficient of thermal expansion of Al 99.99 as a function of temperature [1.9]

Temperature range (°C)	Average linear coefficient of thermal expansion ($10^{-6}/\mathrm{K}^{-1}$)
(−200) – 20	18.0
(−200) – 20	21.0
20–100	23.6
20–200	24.5
20–400	26.4
20–600	28.5

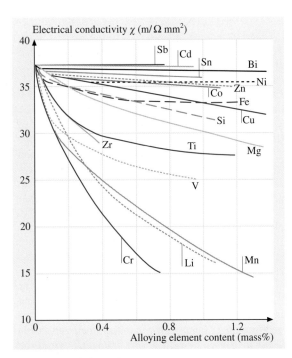

Fig. 3.1-8 Electrical conductivity of as-cast binary aluminium alloys (containing larger amounts of the alloying additions) as a function of concentration of the alloying element [1.14, 15]

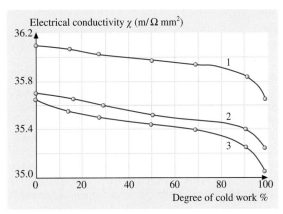

Fig. 3.1-7 Electrical conductivity of as-cast binary alloys based on high-purity aluminium as a function of the alloying element concentration [1.14, 15]

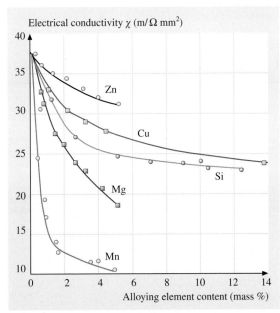

Fig. 3.1-9 Electrical conductivity of unalloyed Al (0.21 wt% Fe; 0.11 wt% Si) as a function of the degree of cold work and the degree of supersaturation and/or intermediate annealing temperature [1.14]; (1) intermediate annealing temperature 350 °C; (2) intermediate annealing temperature 500 °C; (3) as extruded

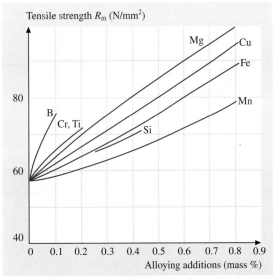

Fig. 3.1-10 Effect of small additions of alloying elements or impurities on the ultimate tensile strength of aluminium of high-purity (99.98 wt% Al, soft condition, 5 h, 360 °C) [1.14]

Table 3.1-11 Typical mechanical properties of refined aluminium (Al 99.98) [1.9]

Condition	Proof stress $R_{p0.2}$ (MPa)	Ultimate tensile strength R_m (MPa)	Elongation at fracture A_{10} (%)	Brinell hardness number HB
Soft	10–25	39–49	30–45	15
Hard	69–98	88–118	1–3	25

mic reaction. In chemical compounds, aluminium is positively charged and trivalent (Al^{3+}). It reacts readily with hydrochloric acid and caustic soda, but less readily with sulfuric acid; dilute sulfuric acid does not attack aluminium. It is not attacked by cold nitric acid at any concentration, and hardly so when heated. The reaction with sodium hydroxide is given by: $2Al + 2NaOH + 3H_2O \rightarrow 2Na[Al(OH)_4] + 3H_2$.

Aluminium-based materials are non-flammable. Even turnings and chippings do not ignite. Extremely fine aluminium particles can undergo spontaneous combustion and thus cause explosions. The heat of formation of the aluminium oxide Al_2O_3 is about $1590\,\text{kJ mol}^{-1}$, making aluminium a very effective deoxidiser for the steel industry and in metalothermic metal reduction processes (aluminothermy; e.g., $3V_2O_5 + 10Al \rightarrow 6V + 5Al_2O_3$ and aluminothermic welding, "thermit process" $3Fe_3O_4 + 8Al \rightarrow 9Fe + 4Al_2O_3$).

Important aluminium compounds include aluminium oxide Al_2O_3, which is commonly called alumina (in powder form) or corundum (in coarse crystalline structure), and aluminium hydroxide $Al(OH)_3$ ("hydrated alumina," usually extracted from bauxite in the Bayer process).

3.1.2.4 Aluminium Alloy Phase Diagrams

The properties of aluminium strongly depend on the concentration of alloying additions and impurities. Even the low residual contents of Fe and Si in unalloyed aluminium (Al99 to Al99.9) have a marked effect.

The main alloying elements of Al materials are Cu, Si, Mg, and Zn while Mn, Fe, Cr, and Ti are frequently present in small quantities, either as impurities or additives. Ni, Co, Ag, Li, Sn, Pb, and Bi are added to produce special alloys. Be, B, Na, Sr, and Sb may be added as important trace elements. All of these elements affect the structure and thus the properties of an alloy. The compositions of the more important aluminium materials are discussed below, using the relevant phase diagram. All alloying components are completely soluble in liquid aluminium if the temperature is sufficiently high. However, these elements have only limited solubility in solid solution. Continuous solid solubility does not occur in any of the alloy systems of Al.

Aluminium-rich solid solutions are often formed and are referred to as α-phase, α_{Al}-phase, or α-Al solid solution. Most of the phases occurring in equilibrium with α-Al are hard. They consist of elements or intermetallic compounds such as Al_2Cu, Al_8Mg_5, Al_6Mn, Al_3Fe, and AlLi.

Binary Al-Based Systems

Aluminium–Copper. Al–Cu forms a simple eutectic system in the range from 0 to 53 wt% Cu, as shown in Fig. 3.1-11. The α-Al solid solution and the intermetallic compound Al_2Cu (θ phase) are in equilibrium. At intermediate temperatures, metastable transition phases may form and precipitate from the supersaturated solid solution. These metastable phases may be characterised according to their crystal structure, the nature of the phase boundary they form, and their size:

- From room temperature up to $\approx 150\,°C$ the coherent Cu-rich Guinier-Preston zones I (GP I phase) form; they are only one to two {001} layers thick and have a highly strained, coherent phase boundary with the α-Al matrix phase.

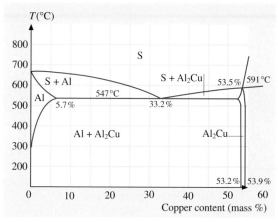

Fig. 3.1-11 Al–Al_2Cu section of the Al–Cu system [1.16]

- At ≈ 80 to ≈ 200 °C the GP II phase, also called θ'' phase, forms; it has a superlattice structure of the Al fcc lattice and has a coherent interface also, but the particle size grows larger than that of GP I.
- Above ≈ 150 °C the θ' phase forms; it is only partially coherent.
- Above ≈ 300 °C the incoherent, stable θ-phase Al_2Cu is formed (over-ageing).

These phase transformations are decisive for the precipitation-hardening behavior of Al–Cu-based technical Al alloys.

Aluminium–Silicon. Al–Si forms a simple eutectic system (Fig. 3.1-12). At room temperature the solubility of Si in Al is negligible. Thanks to the good casting properties of the Al–Si eutectic, this alloy system is the basis of a major part of the Al-based casting alloys. However, when slowly cooled (e.g., in sand casting), a degenerate form of eutectic, microstructure may occur. Instead of the desired fine eutectic array, the alloy develops a structure that is characterised by larger, plate-like primary Si crystals, leading to very brittle behavior. This degenerate behavior can be suppressed by the addition of small amounts of Na, Sb, or Sr to the melt at about 720 to 780 °C. This "modification" causes a lowering of the eutectic temperature and a shift in concentration of the eutectic point, as indicated in Fig. 3.1-12; the extent of the shift is dependent on the rate of solidification.

Aluminium–Iron. Figure 3.1-13 shows the Al-rich part of this system. The solubility of Fe in Al is very low. In the range shown Al_3Fe is formed by a peritectic reaction at 1160 °C. The eutectic between the Al phase and Al_3Fe crystallises in a degenerate manner to form brittle needles of Al_3Fe. The formation of Al_3Fe needles is also occurring in Al–Fe–Si alloys such as commercially (available) pure aluminium (Fig. 3.1-14).

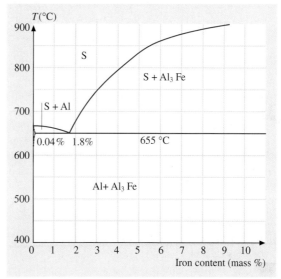

Fig. 3.1-13 Al–Fe system up to 10 wt% Fe

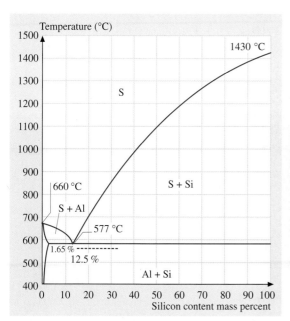

Fig. 3.1-12 Al–Si system; the *dotted line* shows the extent to which alloys can be supercooled

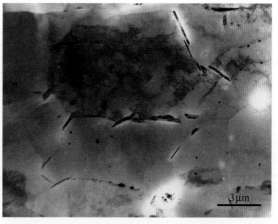

Fig. 3.1-14 Al_3Fe needles formed after annealing commercially pure aluminium sheet (Al99.5) for 65 h at 590 °C, 800X [1.17]

Table 3.1-12 Solubility of some elements in aluminium solid solutions [1.16, 18]

Element	Temperature of eutectic (E) or peritectic (P) eqilibrium °C	Phase in equilibrium with Al solid solution	T_E or T_P (wt%)	Solubility (wt%)			
				500 °C	400 °C	300 °C	200 °C
Cu	547 (E)	Al_2Cu	5.7	4.4	1.6	0.6	0.2
Fe	655 (E)	Al_3Fe	~0.04	0.005	<0.001		
Li	602 (E)	AlLi	4.7	2.8	2.0	1.5	1.0
Mg	450 (E)	Al_8Mg_5	17.4	~12.0	12.2	6.6	3.5
Mn	657 (E)	Al_6Mn	1.82	0.36	0.17	0.02	
Si	577 (E)	Si	1.65	0.8	0.3	0.07	0.01
Zn	275 Eutectoid equilibrium	Zn	31.6				14.5

Aluminium–Lithium. The Al-rich part of this system is shown in Fig. 3.1-15. The Al(Li) solid solution and the η (Al_3Li) phase are in equilibrium. Al–Li-based alloys are used for their low density.

Aluminium–Magnesium. This system is shown in Fig. 3.1-16. The solid phases are the α-Al solid solution and the β-Al_8Mg_5 intermetallic compound. The high solubility of Mg in Al can be put to practical use for solid solution hardening. The β phase precipitates preferentially at grain boundaries, forming a continuous network at the grain boundaries of the Al-rich α solid solution, e.g., after slow cooling from temperatures above 400 °C, especially in alloys containing more than 3 wt% Mg. Precipitates of the β phase are less noble electrochemically than the α phase and are thus subject to preferential attack by corrosive media. Accordingly the disadvantage of Al–Mg alloys is their potentially high susceptibility to intergranular corrosion.

Aluminium–Manganese. The Al-rich part of this system includes the intermetallic phases Al_4Mn (above 710 °C) and Al_6Mn (Fig. 3.1-17). The α-Al solid solution and the Al_6Mn phase form a eutectic (Fig. 3.1-17). The solubility of Mn in Al at room temperature is negligibly small. In hypereutectic Al–Mn alloys, pre-

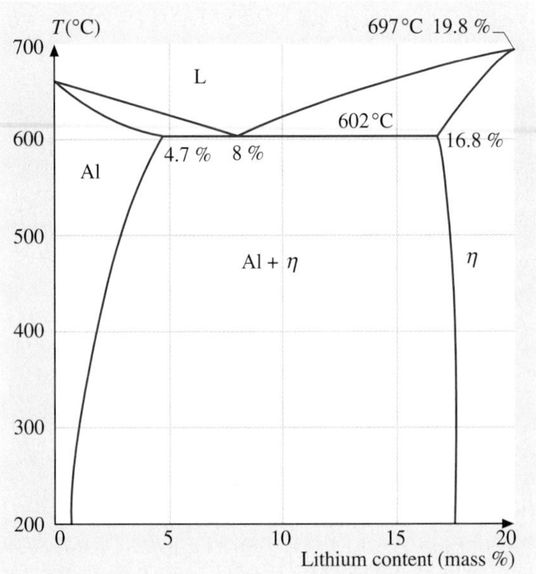

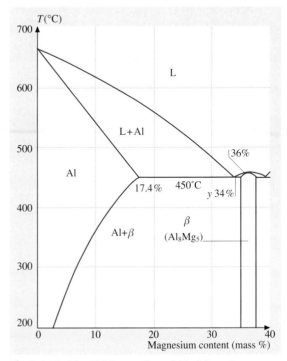

Fig. 3.1-15 Al–Li system for up to 20 wt% Li [1.18]

Fig. 3.1-16 Al–Al_8Mg_5 section of Al–Mg system

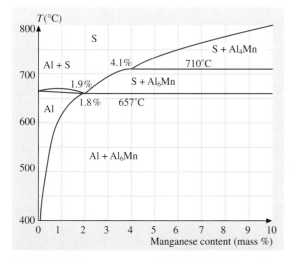

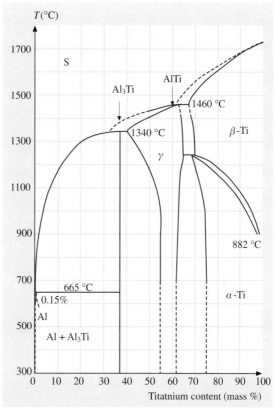

Fig. 3.1-18 Al–Ti system

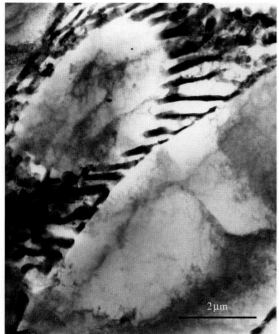

Fig. 3.1-17 (a) Al–Mn equilibrium system up to 10 wt% Mn. **(b)** Microstructure of an as-cast Al–1 wt% Mn alloy (TEM micrograph, 7000X) [1.17]

cipitation of primary, bar-shaped Al_2Mn crystals occurs in the eutectic structure. These have a marked embrittling effect. Therefore, the maximum Mn content is limited to 2 wt% in commercial Al–Mn alloys.

Aluminium–Titanium. Intermetallic phases with a significantly higher melting point (dissociation tempera- ture) than Al exist in this system: Al_3Ti (1340 °C) and AlTi (1460 °C) (Fig. 3.1-18). The structure of Al-rich Al–Ti alloys consists of Al_3Ti and an Al-rich solid solution.

Aluminium–Zinc. Zn is highly soluble in Al. The eutectic point occurs at Zn-rich concentration (Fig. 3.1-19). The liquidus and solidus temperatures depend markedly on temperature.

Ternary Al-Based Systems

Most technical Al alloys contain more than two components because of the presence of impuritiy and alloying elements.

Aluminium–Iron–Silicon. Figure 3.1-20 shows the Al-rich corner as a section for 0.5 wt% Fe. This concentration range is of particular interest for commercially pure, unalloyed Al. Apart from the solid solution and the AlFe and Si phases there are two ternary phases, i.e., α-$Al_{12}Fe_3Si$ and β-$Al_9Fe_2Si_2$. The exact

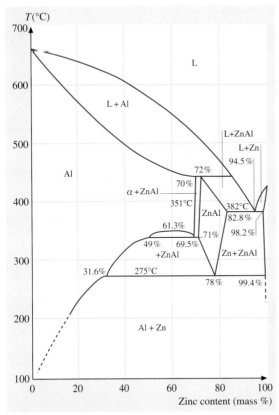

Fig. 3.1-19 Al–Zn system

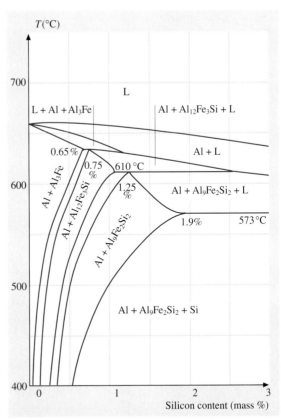

Fig. 3.1-20 Section through the Al–Fe–Si phase diagram at 0.5 wt% Fe [1.11]

compositions are subject to discussion and there are corresponding differences in the description of the solidification and precipitation processes. At the level of Fe and Si contents found in commercially pure aluminium (Al + Si ≤ 1 wt%), α-Al$_{12}$Fe$_3$Si will form as a result of the transformation of the α-solid solution and AlFe with decreasing temperature, as shown in Fig. 3.1-14. If the temperature further decreases, precipitation of β-Al$_9$Fe$_2$Si$_2$ and even Si will occur. Due to the low rate of diffusion at low temperatures it is possible that all four phases coexist with the Al-rich solid solution phase. The solid solution becomes supersaturated in Fe and Si at the cooling rates used in industrial practice. Moreover, non-equilibrium ternary phases form, often locally at the grain boundaries of the as-cast microstructure. Fe and Si are in solution inside the grains. For a given cooling rate, the maximum Fe solubility decreases with increasing Si content of the alloy, whereas the Si solubility is independent of Fe content [1.19–21]. Thermodynamically, such non-equilibrium microstructures are very stable. The primary non-equilibrium ternary phases only decompose to the secondary equilibrium phases Al$_3$Fe and Si at about 600 °C [1.22].

Aluminium–Magnesium–Silicon. Figures 3.1-21a and 3.1-21b show the Al-rich corner as a quasi-binary section Al–Mg$_2$Si (Mg/Si ratio 1:0.58). It divides the system into the two simple ternary eutectic systems shown. The fine-grained microstructure of Al–Mg–Si alloys consists of α-Al(Mg,Si) and numerous intermetallic compounds, such as Mg$_2$Si (forming a characteristic particle shape called "Chinese script"), Al$_6$Mn, and Al$_3$Fe. Mg$_2$Si has an important effect on properties. Its solid solubility in the aluminium matrix is temperature-dependent and, thus, leads to hardening effects, which are exploited in technical alloys. A coarse network of intermetallic phases impairs forming behavior, but annealing before further processing produces finely-dispersed precipitates. These improve workability but are, also, effective in retarding recrystallization.

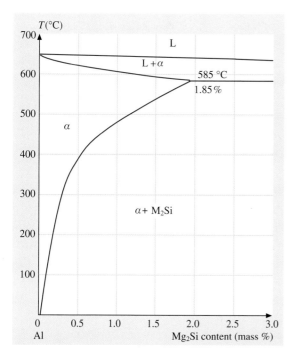

Aluminium–Copper–Magnesium. In addition to the two binary phases, Al_8Mg_5 (β) and Al_2Cu (θ), there are two ternary phases, Al_2CuMg (S) and Al_6Mg_4Cu (T), in equilibrium with the Al-rich solid solution. The microstructure of castings shows various ternary eutectics which, in addition to Al_2Cu, also contain Mg_2Si, resulting from Si as an impurity, and AlCuMg [1.9].

Aluminium–Copper–Lithium. In addition to binary phases between all three elements, three ternary phases, Al_7Cu_4Li, Al_2CuLi, and Al_5Li_3Cu, occur in equilibrium with the Al-rich solid solution in the aluminium-rich corner of this system.

Aluminium–Zinc–Magnesium. The Al-rich corner consists of two binary phases, Al_8Mg_5 and $MgZn_2$, and a ternary phase T with a nominal composition $Al_2Mg_2Zn_3$ having a wide range of homogeneity. The $Al-MgZn_2$ and $Al-T$ sections may be regarded as quasi-binary systems with eutectic temperatures at 475 and 489 °C, respectively. The $Al-Al_8Mg_5-T$ range constitutes a ternary eutectic, $T_E = 450$ °C. In the $Al-T-Zn$ range there is a four-phase reaction in which the T phase transforms to Mg_2Zn. At high Zn contents, Mg_2Zn_{11} transforms to $MgZn_2$ in another four-phase reaction at 365 °C. $MgZn_2$ subsequently solidifies eutectically at 343 °C together with Al and Zn. There is evidence of eutectic solidification in the as-cast structure of these alloys. If ternary Al–Zn–Mg alloys are solution-treated at temperatures above 450 °C they, consist of a homogeneous α phase solid solution which is supersaturated with respect to one phase at least, corresponding to its composition [1.9].

3.1.2.5 Classification of Aluminium Alloys

Technical aluminium alloys are subdivided first into the two main groups of cast and wrought alloys (Fig. 3.1-22). Typically, the alloying content of casting alloys is 10 to 12 wt%. This is significantly higher than the value for wrought alloys, most of which contain only a total of 1 to 2 wt% alloying elements; their content may be as high as 6 or even 8 wt% in individual cases.

Aluminium alloys are further subdivided, depending on whether or not an alloy can be hardened by the addition of alloying elements, as is the case with

- Precipitation-hardenable alloys which can be strengthened by aging and
- Non-precipitation-hardenable alloys which can be strengthened by work-hardening only.

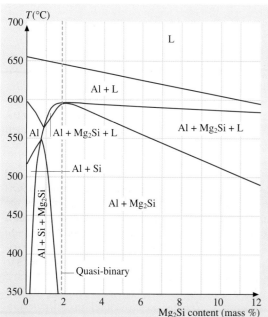

Fig. 3.1-21a,b Al–Mg–Si system (**a**) Quasi-binary section Al–Mg_2Si; (**b**) Section at a constant Si content of 1 wt% [1.11]

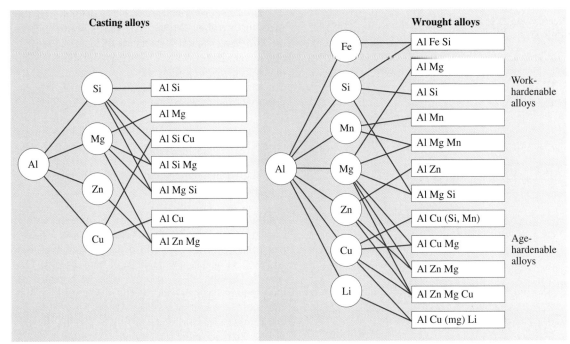

Fig. 3.1-22 Schematic array of cast and wrought aluminium alloys. The numbers are given according to European standardization. In the case of wrought alloys the numbers are the same as those used in North-American standardization

The addition of further alloying elements will always cause hardening, but not all elements have the same hardening effect. Hardening will also depend on whether the solute atoms are present in solid solution or as particles. Alloy hardening can be divided into

- Solid-solution hardening (as with non-precipitation hardening, work-hardenable alloys) and
- Hardening due to elements that are initially in solid solution and are precipitating as second phases (as is the case with age-hardenable alloys).

It should be noted that age-hardenable alloys can be strengthened by the use of a suitable heat treatment whereas the same heat treatment of alloys which are not age-hardenable leads to a loss in strength.

3.1.2.6 Structure and Basic Mechanical Properties of Wrought Work-Hardenable Aluminium Alloys

Al–Fe–Si and Unalloyed Aluminium (1xxx)

Al–Fe–Si contains about 0.6 wt% Fe and 0.8 wt% Si that has been added deliberately. The properties of Al–Fe–Si alloys, and of unalloyed aluminium, are strongly influenced by the elements which are in solid solution and the binary and higher phases that form. Increasing amounts of alloying additions lead to a marked increase in strength but there is a decrease in electrical conductivity since transition elements have a high effective scattering power for electrons.

Wrought Al–Mn (3xxx)

Manganese additions increase the strength of unalloyed aluminium (Fig. 3.1-23). The chemical resistance is not impaired. These alloys have very good forming properties, when the Mn content is below the maximum solubility of Mn in the Al-rich α-phase, i.e., practically below 1.5 wt%. At higher Mn content, brittle Al_6Mn crystals form and impair workability.

If Al–Mn alloys are rapidly solidified as in continuous casting, considerable supersaturation of Mn occurs. Fe reduces the solubility for Mn and promotes its precipitation in the form of multicomponent phases. Fe is often added to counteract supersaturation and for an increase of the tensile strength (Fig. 3.1-24). Depending on the amount of precipitation, Mn can inhibit recrystallization.

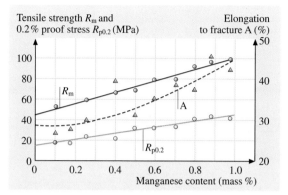

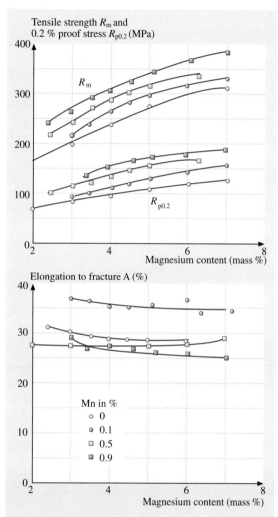

Fig. 3.1-23 Effect of manganese additions on the strength of aluminium; alloys based on 99.5 wt% Al, quenched from 565 °C, sheet samples, 1.6 mm thick [1.23]

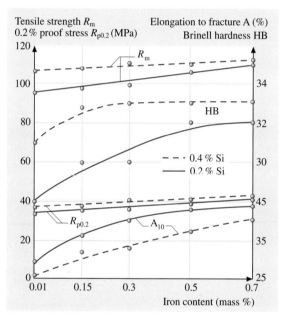

Fig. 3.1-24 Effect of iron and silicon on the strength of an Al–Mn alloy containing 1.2 wt% Mn; soft condition, sheet sample [1.24]

Wrought Al–Si (4xxx)
Aside from 1 to 12.5 wt% Si, wrought Al–Si alloys contain other elements such as Mg, Fe, Mn or Cu.

Wrought Al–Mg and Al–Mg–Mn (5xxx)
The two non-age-hardening alloy systems Al–Mg and Al–Mg–Mn cover the entire compositional range from 0.5 to 5.5 wt% Mg, 0 to 1.1 wt% Mg, and 0 to 0.35 wt% Cr. Alloys with more than 5.6 wt% Mg are of no significance as wrought alloys.

In Al–Mg alloys both tensile strength and 0.2% proof stress increase with increasing Mg content,

Fig. 3.1-25 Effect of Mn content on the mechanical properties of Al–Mg alloys [1.23]

whereas elongation shows a steady decrease up to about 3 wt% Mg, beyond which it increases again slightly. Embrittlement does not occur at low temperatures. The solubility of Mg in the Al-rich solid solution decreases rapidly with decreasing temperature. Thus most Al–Mg alloys are effectively supersaturated at room temperature. This is of practical significance in alloys containing more than 3 wt% Mg, where precipitation of the β-Al_8Mg_5 can occur, especially after prior cold working. This is not associated with any beneficial increases in strength but leads to deterioration in the corrosion resistance by intergranular corrosion (Fig. 3.1-16). It is necessary to avoid the formation of continuous networks of β particles at the grain boundaries. Individual β particles can be obtained by a globulization treatment (200–250 °C), followed by slow cooling.

Another means to increase strength is the addition of Mn. Al–Mn–Mg alloys show an additional increase in tensile strength, which is significantly higher than for binary Al–Mn alloys (Fig. 3.1-26). The alloys show good toughness and can be used at low temperatures. If the Mn content exceeds 0.6 wt%, the recrystallization temperature can be increased to such an extent that recrystallization does not occur during extrusion. With extruded sections, there is an increase in tensile strength and proof stress in the longitudinal direction termed "press effect."

3.1.2.7 Structure and Basic Mechanical Properties of Wrought Age-Hardenable Aluminium Alloys

Wrought Al–Cu–Mg and Al–Cu–Si–Mn (2xxx) Alloys

The 2xxx series alloys usually contain 3.5 to 5.5 wt% Cu and additions of Mg, Si and Mn and residual Fe. They show a significant increase in tensile strength by aging (310 to 440 MPa). Natural (room temperature) or artificial aging (at elevated temperature) may be preferable, depending on the alloy composition (Fig. 3.1-26). Additions of up to 1.5 wt% Mg increase both tensile strength and proof stress. At these Mg contents, the Al-rich solid solution is in equilibrium with the ternary Al$_2$CuMg phase which, together with the θ phase (Al$_2$Cu), is responsible for hardening. Mn additions increase the strength (Fig. 3.1-27) and can also cause a press effect. For achieving good ductility, the Mn content is limited to ≈ 1 wt%.

Wrought Al–Mg–Si (6xxx)

Al–Mg–Si alloys are the most widely used wrought age-hardenable alloys. Hardening is attributable to the formation of the Mg$_2$Si phase. The compositional range of interest is 0.30 to 1.5 wt% Mg, 0.20 to 1.6 wt% Si, up to 1.0 wt% Mn, and up to 0.35 wt% Cr. This is equivalent to about 0.40 to 1.6 vol% Mg$_2$Si and varying amounts of free Si and/or Mg. Some alloys are close to the pseudo-binary section Al–Mg$_2$Si, in others there is a significant excess of Si. This affects the maximum

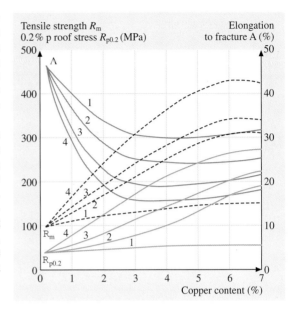

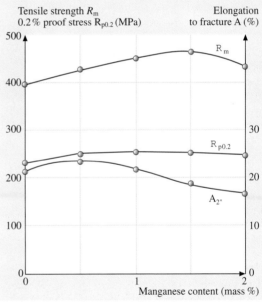

Fig. 3.1-26a,b Behavior of wrought Al–Cu-alloys (**a**) Effect of copper on the strength of binary Al–Cu alloys; base material 99.95 wt% Al, 1.6 mm thick sheet [1.23]. (1) Soft annealed; (2) Solution annealed and quenched; (3) Naturally aged; (4) Artificially aged. (**b**) Effect of Mn on the strength of an Al–Cu–Mg alloy with approx. 4 wt% Cu and 0.5 wt% Mg; naturally aged [1.23]

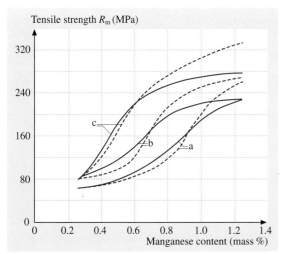

Fig. 3.1-27 Effect of Mg$_2$Si on tensile strength [1.9] (1) stoichiometric composition, (2) 0.3 wt% excess Mg, (3) 0.3 wt% excess Si. The *solid line* refers to alloys that were quenched + immediate ageing at 160 °C, the *dotted line* represents data for alloys that were quenched + intermediate ageing for 24 h at 20 °C + ageing at 160 °C

attainable strength values because of the resulting variations in the amount of Mg$_2$Si, as shown in Fig. 3.1-28 for the quenched and artificially aged condition. Additions of 0.2 to 1.0 wt% Mn lead to an increase in the notch impact toughness of Al–Mg–Si alloys (Fig. 3.1-29), and affect the recrystallization behavior. Additions of Cr cause similar effects.

Wrought Al–Zn–Mg- and Al–Zn–Mg–Cu Alloys (7xxx)

Additions of Zn to Al lead to an insignificant increase in strength. Combined additions of Zn and Mg cause age hardening and thus an increase in strength (Fig. 3.1-28). The sum of the Zn and Mg contents is limited to about 6–7% because of the risk of stress corrosion cracking at higher levels in Cu-free Al–Zn–Mg alloys; this results in medium strength. Zr, Mn, and Cr are added to reduce the tendency to recrystallize. If suitably heat-treated, such alloys have adequate corrosion resistance. It is important that cooling after solution treatment is not too rapid and that the proper precipitation treatment, usually step ageing, should be used. The alloys are of interest for welded structures because of their low quench

Fig. 3.1-29 Effect of Mn and Fe on the notch impact toughness of AlMgSi1 alloy 1 wt% Si and 0.75 wt% Mg; artificially aged, flat bars, 60 × 10 mm^2 [1.24]

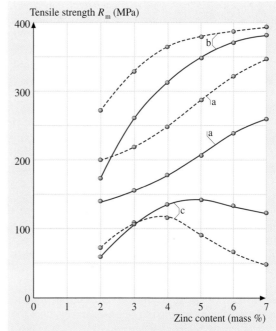

Fig. 3.1-28 Effect of Zn and Mg on the strength and aging effect of Al–Zn–Mg alloys [1.24] (1) Solution treatment at 450 °C and quenching. (2) Solution treatment at 450 °C and quenching plus natural aging for 3 months. (3) Ageing effect i.e., the increase in strength attributable to natural ageing; difference between (1) and (2)

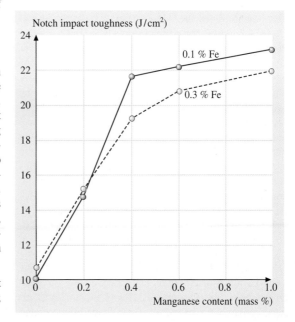

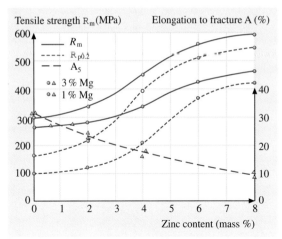

Fig. 3.1-30 Effect of zinc on the tensile properties of Al–Zn–Mg–Cu alloys; solution treated at 460 °C, quenched, aged for 12 h at 135 °C, 1.5 wt% Cu [1.23]

sensitivity. The low-strength heat-affected zone formed during welding is restored to full hardness without the need for renewed solution treatment. Al–Zn–Mg alloys usually contain 0.1 to 0.2 wt% Zr and some Ti in order to improve their weldability and resistance to stress corrosion cracking. Cu additions are avoided, despite their favourable effect on stress corrosion cracking, because they increase susceptibility to weld cracking.

Al–Zn–Mg–Cu Alloys

The addition of 0.5 to 2.0 wt% Cu strengthens Al–Zn–Mg alloys. Cu also reduces the tendency for stress corrosion cracking so that the upper limit for (Zn + Mg) can be increased to 9 wt%, provided additions of Cr are also made. The Zn/Mg ratio should preferably lie between 2 and 3. Al–Zn–Mg–Cu alloys can be aged both naturally and artificially. They attain the highest strength levels of all aluminium alloys. The actual hardening mechanism is attributable to Mg and Zn. Cu increases the rate of aging and acts as a nucleus for the hardening phases. Figure 3.1-30 shows some properties as a function of the Zn content at constant Cu content and two different Mg contents.

3.1.2.8 Structure and Basic Mechanical Properties of Aluminium Casting Alloys

Al–Si Casting Alloys

Due to the Al–Si eutectic (Fig. 3.1-12), these alloys, containing 5 to ≤ 20 wt% Si, have good casting

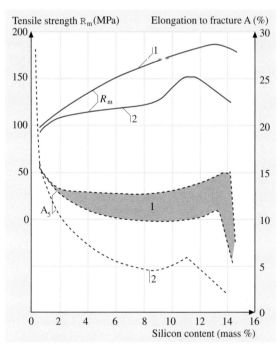

Fig. 3.1-31 Effect of Si on the strength and ductility of Al–Si casting alloys; modified and unmodified sand castings. [1.9] (1) modified; (2) unmodified

properties. The tensile strength increases with the Si content (Fig. 3.1-31). Apart from Si other elements may be added, e.g., for the modification of the eutectic. Cu is present in residual amounts and impairs chemical resistance if levels exceed 0.05 wt%. Additions of about 1 wt% Cu yield an increase in solid solution hardening and thus reduce the tendency for smearing during machining. Residual Fe reduces sticking tendency, but leads to the formation of β-AlFeSi needles which reduce strength and ductility. Therefore the Fe content must be limited. An addition of Mn has a favorable effect on the sticking tendency. Mn leads to the formation of a quaternary phase; but it poses no problem because of its globular shape. "Piston alloys" have hypereutectic compositions of up to 25 wt% Si. During solidification, primary Si crystals are formed which increase wear strength and reduce thermal expansion.

Al–Si–Mg Casting Alloys

These alloys contain about 5 wt%, 7 wt%, or 10 wt% Si and between 0.3 and 0.5 wt% Mg. The optimum amount of Mg decreases with increasing Si content. Mg causes high strength and a moderate ductility depending on the temper (Fig. 3.1-32). The alloys can be age hardened both naturally and artificially. Fe leads to a reduction in sticking tendency, but also to a drastic reduction in ductility.

Al–Mg Casting Alloys

These alloys contain 3–12 wt% Mg. Strength increases with increasing Mg content (Fig. 3.1-33). Above about 7 wt% Mg, the alloy has to be heat-treated to homogenise the structure and thus obtain good tensile properties. With Mg contents of up to 5 wt%, Si additions of up to 1 wt% are possible, and these lead mainly to improvements in the casting properties. The addition of Si causes hardening due to the formation of Mg_2Si.

Al–Zn–Mg Casting Alloys

They contain 4 to 7 wt% Zn and 0.3 to 0.7 wt% Mg. They can be naturally or artificially age hardened in the as-cast condition, without the need for prior solution treatment. Mg has a significant effect on the mechanical properties in the aged condition (Fig. 3.1-34). If the Mg content is limited, Al–Zn–Mg casting alloys can exhibit particularly good elongation to fracture.

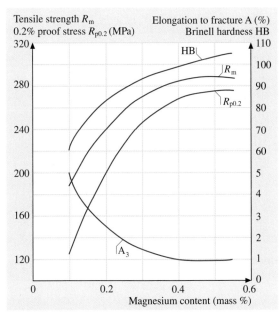

Fig. 3.1-32 Effect of Mg on the mechanical properties of G-AlSi10Mg casting alloy (with 9.5 wt% Si, 0.45 wt% Fe and 0.3 wt% Mn), sand cast and artificially aged [1.25–27]

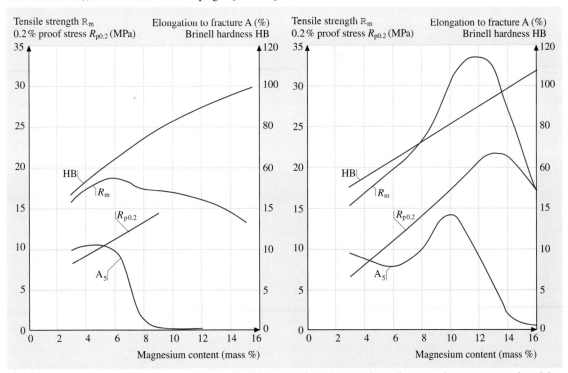

Fig. 3.1-33a,b Effect of Mg on the mechanical properties of Al–Mg casting alloys, sand cast, untreated and homogenised [1.9]. (**a**) As-cast condition. (**b**) Homogenized

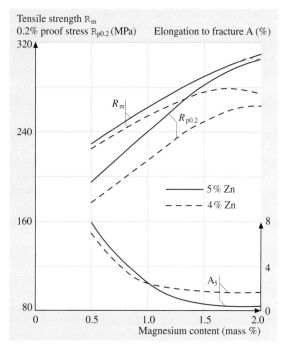

Fig. 3.1-34 Effect of Mg on the strength of cast Al–Zn–Mg alloys; sand casting, artificially aged [1.9]

3.1.2.9 Technical Properties of Aluminium Alloys

The applications of aluminium and its alloys are depending on the properties. In most cases, mechanical properties are an important criterion for assessing the suitability of an Al alloy for a specific application. Other properties, such as electrical conductivity or corrosion resistance, may also be included in the assessment process.

Mechanical Properties

Hardness. The Brinell hardness number HB ranges from 15 for unalloyed Al in the soft temper to about 140 for an artificially aged Al−Zn−Mg−1.5 wt%Cu alloy.

Tensile Strength. Figures 3.1-35 and 3.1-36 indicate the typical levels of strength attainable with wrought and cast aluminium alloys, respectively. There will be a range of tensile strengths in specific alloying systems because of possible additional increases in strength due to hardening by cold work or precipitation. The different alloying elements cause differing degrees of strengthening.

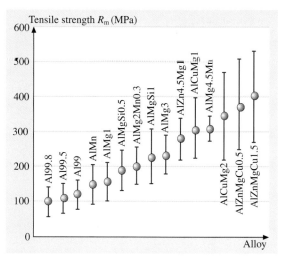

Fig. 3.1-35 Ranges of tensile strength for some important wrought aluminium alloys [1.9]

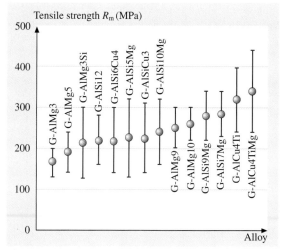

Fig. 3.1-36 Ranges of tensile strength for some important aluminium casting alloys [1.9]

Strength at Elevated Temperatures. An example of the temperature and time dependence of the different mechanical properties is given in Fig. 3.1-37. With regard to the resistance to softening the materials can be classified as follows, depending on their temper:

- Wrought alloys in the soft temper and as-cast non-age-hardenable casting alloys are all practically thermally stable.
- For cold worked wrought alloys, partially annealed to obtain an intermediate temper, the increase in

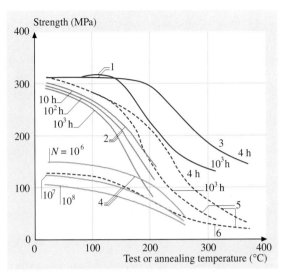

Fig. 3.1-37 Properties of 6061-T6 at elevated temperatures; 1.0 wt% Mg, 0.6 wt% Si, 0.3 wt% Cu, 0.25 wt% Cr, artificially aged [1.9]. (1) tensile strength R_m at 20 °C, (2) stress rupture strength, (3) duration of exposure, (4) fatigue strength, (5) tensile strength at test temperature, (6) solution treated

strength due to forming is diminished with increasing temperature and exposure time.
- Artificially aged alloys do not undergo any permanent changes on annealing up to approaching the ageing temperature. At higher temperatures, aging continues from the point where it was interrupted during the aging process (Figs. 3.1-38 and 3.1-39).
- Naturally aged alloys usually exhibit increased hardening when exposed to higher temperatures as a result of artificial ageing. At higher temperatures, the behavior is similar to that of artificially aged alloys.

High Temperature Mechanical Properties in Short-Term Tests. In materials that are not thermally stable, there will be an effect due to irreversible changes in properties. Its magnitude depends on the temperature and duration of exposure, Figs. 3.1-40 and 3.1-41.

Creep Behavior. Creep of Al alloys starts to play an important role at temperatures above 100 to 150 °C (Figs. 3.1-42 and 3.1-43). Material, amount of cold work

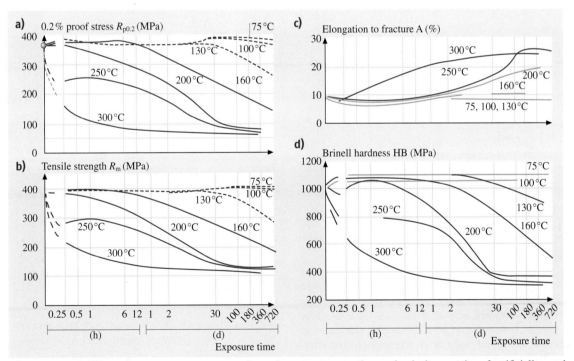

Fig. 3.1-38a–d Effect of temporary exposure to elevated temperature on the mechanical properties of artificially aged AlSi1MgMn [6082] at 20 °C [1.9]. (a) tensile strength; (b) 0.2% proofstress $R_{P0.2}$; (c) elongation to fracture A; (d) Brinell hardness

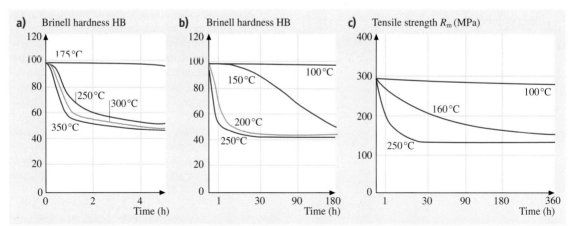

Fig. 3.1-39a–c Effect of temporary exposure to elevated temperature on the hardness and tensile strength at 20 °C of cast G-AlSi10Mg [43000, AlSi10Mg(a)] alloy, artificially aged [1.28]. (**a**) 0–4 h; (**b**) 1–180 d; (**c**) 1–360 d

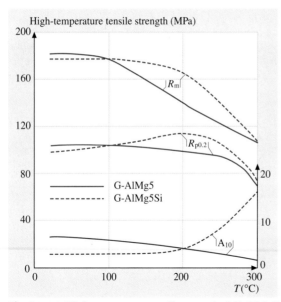

Fig. 3.1-40 High-temperature tensile strength of G-AlMg5 [51300] and G-AlMg5Si [51400] casting alloys, sand cast, after 30 min prior exposure to the test temperature [1.29]

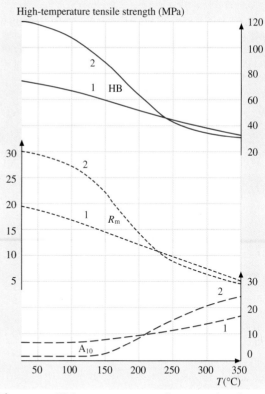

Fig. 3.1-41 High-temperature tensile strength of casting alloys [1.28]. (1) As-cast G-AlSi12 alloy [44300, AlSi12(Fe)]. (2) G-AlSi10Mg alloy [43000, AlSi10Mg(a)] after preheating for 8 days at the test temperature

and degree of age-hardening can affect creep behavior in various ways. With non-age-hardenable alloys, the effect of cold working is more pronounced at low temperatures, below about 150 °C. At higher temperatures, the behavior rapidly approaches that of the softened material. Thus, more highly alloyed materials in the soft temper may exhibit better creep properties than cold worked alloys that are less highly alloyed. Artificially

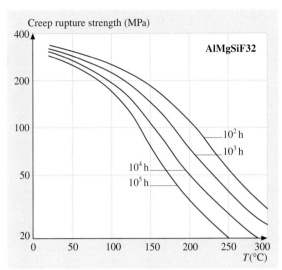

Fig. 3.1-42 Creep rupture behavior of 6082, AlSi1MgMn, T6 alloy, strain-hardened, min. tensile strength 320 MPa; extruded rods, 25 mm diameter [1.30, 31]

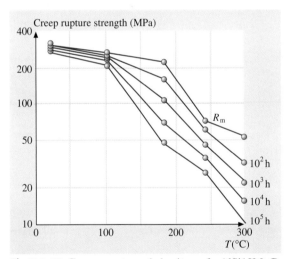

Fig. 3.1-43 Creep rupture behavior of AlSi10MgCa [43000] casting alloy, artificially aged, 100 h prior exposure to the test temperature [1.30, 31]

aged alloys should only be exposed for prolonged periods to temperatures that are significantly lower than the temperature of artificial ageing, otherwise over-ageing will lead to a complete loss of strength.

Mechanical Properties at Low Temperatures. Based on its fcc crystal structure, Al and its alloys show neither a rapid increase in yield stress nor a rapid decrease in fracture toughness with decreasing temperature (Figs. 3.1-44 – 3.1-46). Tests carried out on Al–Mn, Al–Mg, Al–Mg–Si, Al–Cu–Mg, and Al–Zn–Mg wrought alloys at −268 °C showed that the values of elongation to fracture at extremely low temperatures were even higher than at room temperature. In most cases, tensile strength and 0.2% proof stress increased weakly with decreasing temperature. Casting alloys behave similar to wrought alloys at low temperatures. The behavior will depend on composition, temper (especially size, shape and distribution of precipitates), and on the casting process used (Fig. 3.1-47).

Fatigue. There is a marked effect of composition, heat treatment, and method of processing on fatigue strength. Solid solution hardening, cold work, and age hardening all lead to an increase in fatigue strength. For a given alloy composition, extruded sections usually have a higher fatigue strength than sheet or forgings. Fine grains are generally beneficial whereas coarse grains and coarse intermetallic phase particles can lead to a reduction in fatigue strength. Often there is an increase in fatigue strength with decreasing sample thickness, especially with bending stresses. Moreover, the effect of roughness or surface defects in thin specimens is usually less than that in thicker samples. With wrought aluminium alloys, there is a marked difference between age-hardenable alloys and non-age-hardenable alloys. It manifests itself in the shape of the S–N curve (Fig. 3.1-47), which is almost horizontal after about 10^6 cycles for non-age-hardenable alloys and after about 10^8 cycles for age-hardenable alloys. Fig-

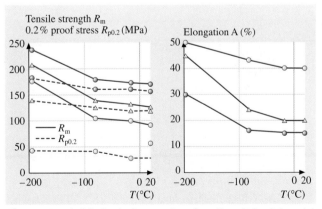

Fig. 3.1-44a,b Mechanical properties of unalloyed aluminium AA 1100 (Al99) at low temperatures [1.9]. (**a**) Tensile strength R_m; (**b**) Elongation A

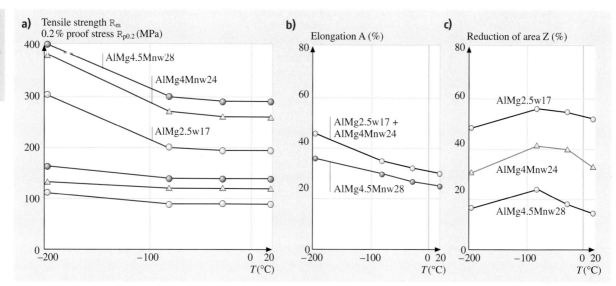

Fig. 3.1-45a–c Mechanical properties at low temperatures for some Al–Mg and Al–Mg–Mn alloys in the soft temper [1.9]. (**a**) Tensile strength R_m and 0.2 proof stress $R_{p0.2}$; (**b**) Elongation A_2''; (**c**) reduction of area AlMg4Mn ($\approx$ 5086, AlMg4), W24 ~O/H111; AlMg2.5 ($\approx$ 5052, AlMg2.5), W17 ~O/H111; AlMg4.5Mn ($\approx$ 5083, AlMg4.5Mn), W28 ~O/H111

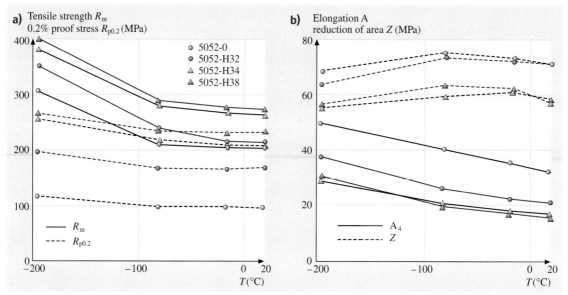

Fig. 3.1-46a,b Effect of cold working on the tensile properties of AlMg2.5 (5052) at low temperatures [1.9]. (**a**) Tensile strength and 0.2 proof stress; (**b**) Elongation and reduction of area

ure 3.1-48 illustrates that the choice of 10^8 cycles as the ultimate number of stress cycles is regarded as adequate. Figure 3.1-49 shows fatigue curves for a number of casting alloys.

Technological Properties

Abrasion Resistance. The wear resistance of Al alloys is low, especially in the absence of lubricants. There is no relation between hardness, strength and

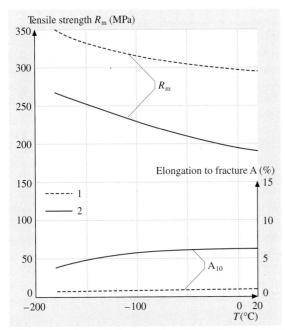

Fig. 3.1-47 Mechanical properties of artificially aged G-AlSi12 and G-AlSi10Mg casting alloys at low temperatures [1.28]. The *solid line* represents G-AlSi10Mg, artificially aged; the *dotted line* represents G-AlSi12, as-cast

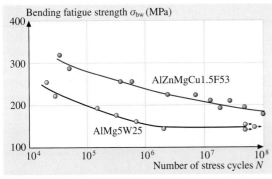

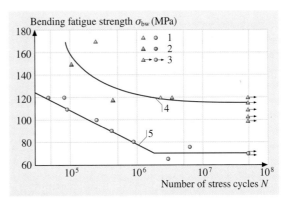

Fig. 3.1-49 Bending fatigue strength of two aluminium pressure die-casting alloys; flat samples with cast skin [1.9]. (1) Fracture; (2) Fracture, with evidence of defect; (3) No fracture; (4) GD-AlSi6Cu3; (not in EN); (5) GD-AlSi12 [≈ 44300, AlSi12(Fe)]

abrasion resistance. Under suitable conditions of lubrication, aluminium alloys can be safely used where they will encounter friction, as shown by their widespread use in the fabrication of pistons and sliding bearings. Wear can be drastically reduced by the suitable surface treatments.

Sheet Formability. Typical values for deep-drawing indices for commonly used sheet materials are shown in Fig. 3.1-50, whereas typical values of strain-hardening exponent n and degree of anisotropy r are given in Table 3.1-13. The r value is strongly dependent on the manufacturing process in particular, as is the case with all texture-dependent properties. In general, materials with high n and r values are deep-drawable.

Fig. 3.1-48 Typical S-N curves for reverse bending tests on an age-hardenable and a non-age-hardenable alloy [1.9], AlZnMgCu1.5 F53 (≈ 7075, AlZn5.5MgCu, T651) and AlMg5 (≈ 5019, AlMg5), W25

Table 3.1-13 Typical values of strain-hardening exponent n and degree of anisotropy r for some aluminium-base materials based on data from various sources [1.9]; n.a. – not available

Material, Alloy		Temper	n	r
Al99.5	1050A	O/H111	0.25	0.62
AlMn1	3103	O/H111	0.15	n.a.
AlMg3	5754	O/H111	0.20	0.64
AlMg3	5754	H14	0.16	0.75
AlMg3	5754	H18	0.12	1.10
AlMg2Mn0.8	5049	O/H111	0.20	0.66
AlMg2Mn0.8	5049	H112	0.20	0.92
AlMg4.5Mn	5063	O/H111	0.15	n.a.

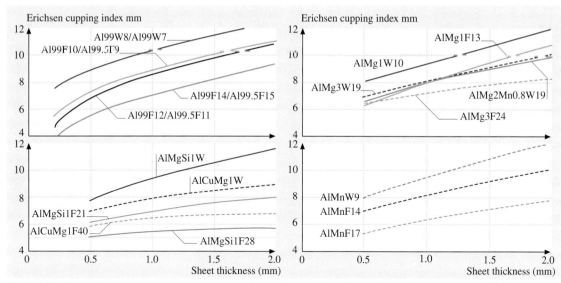

Fig. 3.1-50 Dependence of deep-drawing index on sheet thickness for different aluminium sheet alloys [1.9]

Machinability. There are difficulties to classify aluminium alloys according to their machinability. In general, soft wrought alloys perform worse than other materials regardless of the machining conditions because of chip shape. Harder wrought alloys perform somewhat better and special machining alloys and casting alloys perform best of all. In Si-containing alloys, including wrought alloys, there is a marked increase in tool wear with increasing Si content if the relatively hard Si is present in its elemental form.

Physical Properties
The most important physical properties of pure Al are listed in Chapt. 2.1,Table 3.1-11 Characteristic ranges of property variations with the amount of impurities or alloying elements are compiled in Table 3.1-14.

Coefficient of Thermal Expansion. Values of the average linear coefficient of thermal expansion for some Al-based materials at temperatures of practical significance are given in Table 3.1-15. Alloying leads to

Table 3.1-14 Examples for physical properties of aluminium alloys [1.9][a]

Designation EN	EN	former designation	Density (g cm^{-3})	Freezing range [b] (°C)	Electrical conductivity (m Ω^{-1} mm^{-2})	Thermal conductivity (W m^{-1} K^{-1})	Coefficient of linear expansion (10^{-6}/K)
				Wrought alloys			
1098	Al99.98	Al99.98R	2.70	660	37.6	232	23.6
1050A	Al99.5	Al99.5	2.70	646–657	34–36	210–220	23.5
1200	Al99	Al99	2.71	644–657	33–34	205–210	23.5
8011A	AlFeSi(A)	AlFeSi	2.71	640–655	34–35	210–220	23.5
3103	AlMn1	AlMn	2.73	645–655	22–28	160–200	23.5
3003	AlMn1Cu	AlMnCu	2.73	643–654	23–29	160–200	23.2
3105	AlMn0.5Mg0.5	AlMn0.5Mg0.5	2.71	635–654	25–27	180–190	23.2
3004	AlMn1Mg1	AlMnlMg1	2.72	629–654	23–25	160–190	23.2
5005A	AlMn1(C)	AlMg1	2.69	630–650	23–31	160–220	23.6
5052	AlMg2.5	AlMg2.5	2.68	607–649	19–21	130–150	23.8
5754	AlMg3	AlMg3	2.66	610–640	20–23	140–160	23.9
5019	AlMg5	AlMg5	2.64	575–630	15–19	110–140	24.1
5049	AlMg2Mn0.8	AlMg2Mn0.8	2.71	620–650	20–25	140–180	23.7

Table 3.1-14 Examples for physical properties of aluminium alloys [1.9][a], cont.

Designation EN	EN former designation		Density (g cm^{-3})	Freezing range [b] (°C)	Electrical conductivity (m Ω^{-1} mm^{-2})	Thermal conductivity (W m^{-1} K^{-1})	Coefficient of linear expansion (10^{-6}/K)
5454	AlMg2.7Mn	AlMg2.7Mn	2.68	602–646	19–21	130–150	23.6
5086	AlMg4	AlMg4Mn	2.66	585–641	17–19	120–140	23.8
5083	AlMg4.5Mn0.7	AlMg4.5Mn	2.66	574–638	16–19	110–140	24.2
6060	AlMgSi	AlMgSi0.5	2.70	585–650	28–34	200–220	23.4
6181	AlSi1Mg0.8	AlMgSi0.8	2.70	585–650	24–32	170–220	23.4
6082	AlSi1MgMn	AlMgSi1	2.70	585–650	24–32	170–220	23.4
6012	AlMgSiPb	AlMgSiPb	2.75	585–650	24–32	170–220	23.4
2011	AlCu6BiPb	AlCuBiPb	2.82	535–640	22–26	160–180	23.1
2007	AlCu4PbMgMn	AlCuMgPb	2.85	507–650	18–22	130–160	23
2117	AlCu2.5Mg0.5	AlCu2.5Mg0.5	2.74	554–650	21–25	150–180	23.8
2017A	AlCu4MgSi(A)	AlCuMg1	2.80	512–650	18–28	130–200	23.0
2024	AlCu4Mg1	AlCuMg2	2.78	505–640	18–21	130–150	22.9
2014	AlCu4SiMg	AlCuSiMn	2.80	507–638	20–29	140–200	22.8
7020	AlZn4.5Mg1	AlZn4.5Mg1	2.77	600–650	19–23	130–160	23.1
7022	AlZn5Mg3Cu	AlZnMgCu0.5	2.78	485–640	19–23	130–160	23.6
7075	AlZn5.5MgCu	AlZnMgCu1.5	2.80	480–640	19–23	130–160	23.4
Casting alloys							
44100	AlSi12(b)	G-AlSi12	2.65	575–585	17–27	120–190	20
47000	AlSi12(Cu)	G-AlSi12(Cu)	2.65	570–585	16–23	110–160	20
43000	AlSi10Mg(a)	G-AlSi10Mg	2.65	575–620	17–26	120–180	20
43200	AlSi10Mg(Cu)	G-AlSi10Mg(Cu)	2.65	570–620	16–20	110–140	20
42000	AlSi7Mg	G-AlSi7Mg	2.70	550–625	21–32	150–220	22
51000	AlMg3(b)	G-AlMg3	2.70	580–650	16–24	110–170	23
51300	AlMg5	G-AlMg5	2.60	560–630	15–22	100–160	23
51400	AlMg5(Si)	G-AlMg5Si	2.60	560–630	16–21	110–150	23
51200	AlMg9	GD-AlMg9	2.60	510–620	11–15	80–110	24
46200	AlSi8Cu3	G-AlSi8Cu3	2.75	510–610	14–18	100–130	22
45000	AlSi6Cu4	G-AlSi6Cu4	2.75	510–620	15–18	110–130	22
21100	AlCu4Ti	G-AlCu4Ti	2.75	550–640	16–20	110–140	23
21000	AlCu4MgTi	G-AlCu4TiMg	2.75	540–640	16–20	110–140	23

[a] The actual values will depend on the material composition within the permitted range; electrical and thermal conductivity will also depend on the material structure
[b] Where a eutectic structure is expected to form because of segregation, this will solidify at the lowest temperature given

Table 3.1-15 Linear coefficient of thermal expansion of some Al alloys for different ranges of temperature [1.9]

Designation (according to EN 573.3 and EN 1706)		Average coefficient of linear expansion (10^{-6} K^{-1}), for the range			
		−50 to 20 °C	20 to 100 °C	20 to 200 °C	20 to 300 °C
1098	Al99.98	21.8	23.6	24.5	25.5
1050A	Al99.5	21.7	23.5	24.4	25.4
3003	AlMn1Cu	21.5	23.2	24.1	25.2
3004	AlMn1Mg1	21.5	23.2	24.1	25.1
5005A	AlMg1(C)	21.8	23.6	24.5	25.5
5019	AlMg5	22.5	24.1	25.1	26.1
5474	AlMg3Mn	21.9	23.7	24.6	25,6
6060	AlMgSi0.5	21.8	23.4	24.5	25,6
2011	AlCu6BiPb	21.4	23.1	24.0	25,0
44100	AlSi12(b)	–	20.0	21.0	22.0

only small changes in the coefficient of thermal expansion (Tables 3.1-14 and 3.1-15). Si additions with 1.2% reduction in the coefficient of thermal expansion for each wt% Si have the greatest effect of the alloying elements commonly used. This effect is applied to the manufacture of piston alloys.

Specific Heat. The specific heat of aluminium in the solid state increases continuously from 0 at 0 K to a maximum at the melting point. This is only the case with alloys when there are no solid state reactions. The effect of alloying elements in solution is not very marked.

Elastic Properties. The Young's modulus of Al and its alloys is usually taken to be about 70 GPa. Values given in the literature for aluminium of all grades of purity and aluminium alloys range from about 60 GPa to 78 GPa. Unalloyed and low-alloyed materials are found in the lower part of the range. The age-hardenable alloys are in the middle to upper part. The Young's modulus is dependent on texture, because the elastic anisotropy as expressed by the ratio of the elastic moduli for single crystals with [111] and [100] orientations is 1.17. Furthermore, there is a marked effect of cold working and test temperature (Fig. 3.1-51).

Electrical Conductivity. The electrical conductivity is influenced largely by alloying or impurities and the

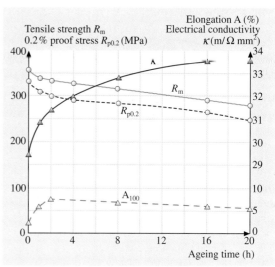

Fig. 3.1-52 Changes in the mechanical properties and electrical conductivity of electrical grade E-AlMgSi ($\approx$ 6101) alloy wire during artificial ageing at 160 °C; solution treated at 525 °C, water quenched, naturally aged for 14 days, 95% cold-worked and then artificially aged [1.14]

structure. Elements present in solid solution lead to a greater reduction in electrical conductivity than precipitates (Figs. 3.1-7 and 3.1-8). By use of a suitable combination of heat treatment and cold working, it is possible to obtain microstructures with an adequate combination of tensile strength and electrical conductivity. This is shown in Fig. 3.1-52, using E-AlMgSi wire as an example.

Behavior in Magnetic Fields. Aluminium is weakly paramagnetic. The specific susceptibility of Al and its alloys is about 7.7×10^{-9} m^3 kg^{-1} at room temperature.

Nuclear Properties. Thanks to its small absorption cross section for thermal neutrons, aluminium is often used in reactor components requiring low neutron absorption.

Optical Properties. The integral reflection is almost independent of the technical brightening process applied, but it depends on the purity of the metal. It is 84–85% for high-purity aluminium Al99.99 and 83–84% for unalloyed aluminium Al99.9. Higher amounts of diffuse reflection can be obtained by mechanical or chemical pre-treatment, such as sand blasting or strong pickling before brightening.

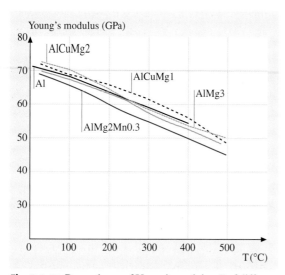

Fig. 3.1-51 Dependence of Young's modulus E of different Al alloys on temperature; determined using 9 mm thick plate [1.32]

3.1.2.10 Thermal and Mechanical Treatment

Thermal and mechanical treatments have a great influence on the properties of aluminium and its alloys. Moreover these treatments influence some physical properties, for example the electrical conductivity. Thus these treatments are used to obtain a material with an optimum of properties required.

Work Hardening

Strengthening by work hardening can be achieved by both cold- and warm working. Figures 3.1-53 – 3.1-56 illustrate the effect of cold working. During hot working, strengthening and softening processes occur simultaneously; with aluminium-based materials, this will be above about 150–200 °C. Therefore hardness decreases with increasing hot-working temperature (Fig. 3.1-57). The formation of subgrains occurs in the microstructure; their size increases with increasing rolling temperature. Alloying atoms, such as Mg, impede growth or coarsening of the subgrains.

Thermal Softening

Figures 3.1-57a and 3.1-57b show typical softening curves for aluminium alloys in three characteristic stages: recovery, recrystallization, and grain-growth. In the case of the untreated continuously cast and rolled strip shown in Fig. 3.1-57a, recrystallization will occur at temperatures between about 260 and 290 °C. In Fig. 3.1-57b, recrystallization starts after about half an hour and is complete after about an hour. The

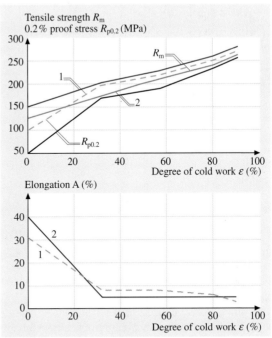

Fig. 3.1-54 Typical mechanical properties of (1) continuously cast and rolled strip and (2) strip produced conventionally by hot-rolling and then cold-rolling cast billets of AlMn1 (3103) [1.34]

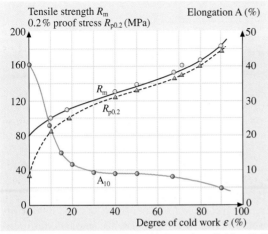

Fig. 3.1-53 Hardening of Al99.5 strip (0.15 wt% Si, 0.28 wt% Fe) after recrystallization annealing and subsequent cold-rolling [1.33]

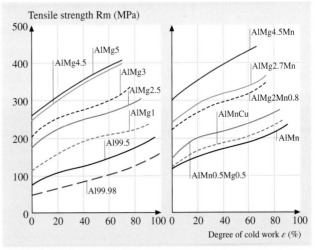

Fig. 3.1-55 Tensile strength of various wrought non-age-hardenable aluminium alloys as a function of the degree of cold work [1.33]: AlMg4.5 – 5082; AlMg5 – 5019; AlMg3 – 5754; AlMg2.5 – 5052; AlMg1 – 5005; Al99.5 – 1050A; Al99.8 – 1080°; AlMg4.5Mn – 5083; AlMg2.7Mn – 5454; AlMg2Mn0.8 – 5049; AlMn0.5Mg0.5 – 3105; AlMn – 3103; AlMnCu – 3003

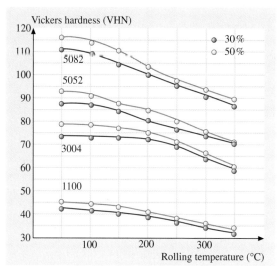

Fig. 3.1-56 Effect of rolling temperature on hardness; specimens quenched after hot-rolling: Al99.0Cu – 1100; AlMn1Mg2 – 3004; AlMg2.5 – 5052; AlMg4.5 – 5082

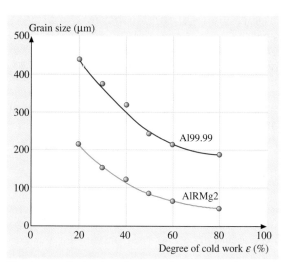

Fig. 3.1-58 Grain size after complete recrystallization of Al99.99 and AlRMg2 as a function of degree of cold work (annealed at 350 to 415 °C until complete recrystallization) [1.9]

softening curves depend not only on the alloy composition but also on the degree of prior cold working to a marked extent. Further factors are the content of alloying and impurity elements, annealing time, heating rate, the microstructure prior to deformation and prior thermomechanical treatment, which can also include the casting process used to produce the starting material.

The softening due to recovery or partial recrystallization is very important when producing semi-finished products of medium hardness (e.g., half hard). The relationship between the degree of cold work and grain size is apparent in Fig. 3.1-58. If the degree of cold work is below a critical value, no recrystallization will occur. This threshold depends on material and prior thermo-

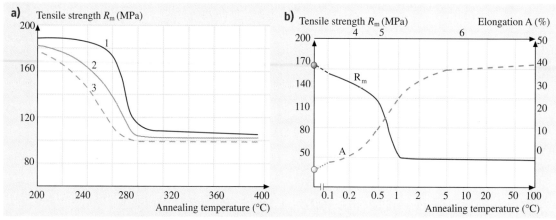

Fig. 3.1-57a,b Typical softening curves for cold-rolled Al99.5 [1.35, 36]. (**a**) Recrystallization after different initial tempers to 90% cold work and subsequent annealing for 1 h at different temperatures: (1) Continuously cast and rolled strip, subsequently cold-worked without an intermediate annealing treatment; (2) as 1 but with an intermediate annealing treatment for 1 h at 580 °C; (3) Strip conventionally produced using permanent mould casting and hot-rolling. (**b**) Continuously cast and rolled strip, cold-worked 90% and then annealed at 320 °C for time shown: (4) Recovery; (5) Recrystallization; (6) Grain growth

mechanical treatment, and is about 2–15%. If the degree of cold work is near the critical value, coarse grains can form during recrystallization, as clearly shown in the three-dimensional diagram in Fig. 3.1-59.

Other elements can affect the recrystallization temperature and the grain size after recrystallization. Mg does not have a marked effect (Fig. 3.1-60), while Mn, Fe, Cr, Ti, V, and Zr increase the recrystallization temperature (Fig. 3.1-61 delaying effect = plateau).

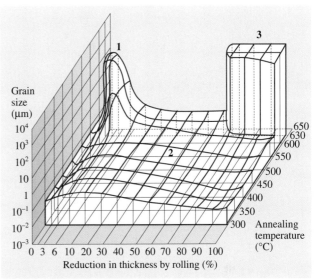

Fig. 3.1-59 Recrystallization diagram showing grain size of Al99.6 (1060) as a function of cold work and annealing for 2 h at the temperatures shown [1.9]: (1) Coarse grains in the region of critical degree of cold work (primary recrystallization); (2) Area of primary recrystallization with fine and medium-sized grains, (3) Coarse grains as a result of secondary recrystallization

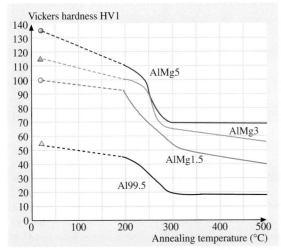

Fig. 3.1-60 Effect of Al–Mg alloy composition on the softening behavior; the material was 90% cold-worked and annealed for 1 h at temperature shown [1.36]

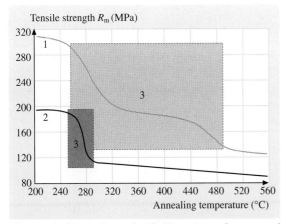

Fig. 3.1-61 Comparison of softening curves for cast and hot-rolled Al99.5 and AlMn1 strip; material cold-worked 90% and annealed for 1 h at the temperature shown [1.36]: (1) Cast and hot-rolled AlMn1 (3103) strip; (2) Cast and hot-rolled Al99.5 (1050A) strip; (3) Recrystallization range

This is dependent on the amount and distribution of these elements, i.e., whether they are present as precipitates or in supersaturated solid solution, and/or whether they form intermetallic phases. In Fig. 3.1-62 it is clearly apparent that solution treatment reduces the degree of supersaturation and that the range over which tensile strength and 0.2% proof stress decrease rapidly with temperature is thus shifted to significantly lower temperatures. Interactions between the various elements present can also affect recrystallization behavior.

Soft Annealing, Stabilization

With cold-worked materials, soft annealing consists of recrystallization annealing. The duration of the treatment, degree of cold work, and intermediate annealing treatments need to be selected with grain size in mind. In age-hardenable alloys, soft annealing permits most of the supersaturated components to precipitate out in coarse form or coherent or partially coherent phases to transform to incoherent stable phases. Soft annealing of casting alloys involves annealing the as-cast structure at 350 to 450 °C. With age-hardenable alloys, this is also possible using a solution treatment followed by furnace cooling

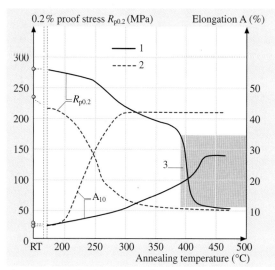

Fig. 3.1-62 Softening curves for continuously cast and rolled AlMn1 strip after annealing for 4 h at elevated temperatures [1.37,38]: (1) Untreated material, as-cast condition; (2) Homogenised at 540 °C after cold working; (3) Coarse grains

Homogenization

This treatment is used for the elimination of residual casting stresses and segregation, for dissolution of eutectoid components at grain boundaries, or for producing a more uniform precipitation of supersaturated elements (e.g., Mn and Fe). Additionally, the elements which are responsible for hardening in age-hardenable alloys are taken into solid solution. A homogenization is carried out at temperatures which are composition-dependent and close to the solidus temperature of the alloy in question. Long exposure times are required, typically about 10–12 hr but possibly even longer, depending on the relevant diffusion coefficients and microstructure.

Aging

Age hardening treatments require the three steps of solution treatment, quenching, and aging (Fig. 3.1-63). The purpose of solution treatment is to produce a homogenous α-Al solid solution. The annealing temperature is determined by the relevant phase. It should be as high as possible in order to avoid excessively long annealing times, but lower than the solidus of α and the melting point of the lowest melting phase (Table 3.1-16). It should be noted that segregation effects tend to displace the effective phase boundaries to lower temperatures and incipient melting may occur. In addition, the effect of the selected solution treatment temperature on strength levels attained after aging will vary with alloy (Figs. 3.1-64 and 3.1-65). The annealing time depends

or nonforced air cooling. Al–Mg alloys with more than 4 wt% Mg may have to be stabilized to produce a structure that is not susceptible to intergranular corrosion.

Stress-Relieving

The temperatures used for thermal stress relief are relatively low, i.e., at the lower end of the recovery range, or even lower, and between 200 and 300 °C for non-age-hardenable alloy castings, otherwise there will be an unacceptably large loss of strength.

Fig. 3.1-64 Effect of solution treatment temperature on the tensile strength of wrought aluminium alloys [1.9]: AlZnMgCu1.5 ($\approx$ 7075, AlZn5.5MgCu), 24 h at 120 °C; AlCuMg1 [$\approx$ 2017A, AlCu4MgSi(A)], 5 days at room temperature; AlZn4.5Mg1 7020, ($\approx$ AlZn4.5Mhg1), 1 month at room temperature; AlMgSi1 ($\approx$ 6082, AlSi1MgMn), 16 h at 160 °C ▶

Table 3.1-16 Conditions for aging treatments [a]

Alloy	Annealing Temperature [b] (°C)	Quenching medium	Natural aging time (d)	Artificial aging Temperature (°C)	Time (h)
E-AlMgSi	525–540	Water	5–8	155–190	4–16
AlMgSi0.5	525–540	Air/water	5–8	155–190	4–16
AlMgSi1	525–540	Water/air	5–8	155–190	4–16
AlMg1SiCu	525–540	Water/air	5–8	155–190	4–16
AlCuBiPb	515–525	Water up to 65 °C	5–8	165–185	8–16
AlCuMg1	495–505	Water	5–8	[d]	4[d]
AlCuMg2	495–505	Water		180–195[d]	16–24[d]
AlZn4.5Mg	460–485	Air At least	90	I 90–100/ II 140–160[c]	I 8–12/ II 16–24[c]

[a] Recommendations only; exact specification as per agreement with semi-finishing plant
[b] Metal temperature
[c] Stages I and II of step annealing
[d] Usually only naturally aged

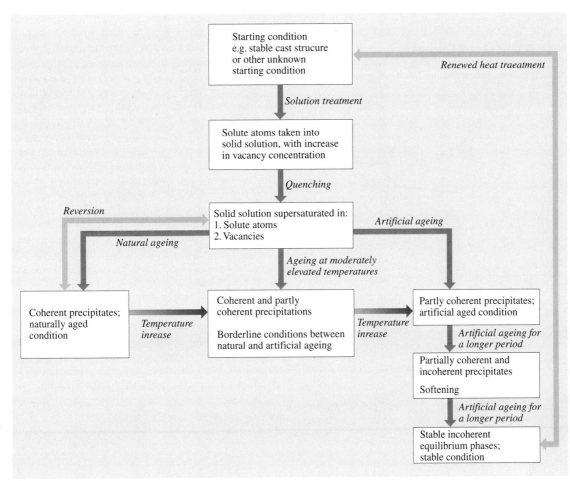

Fig. 3.1-63 Schematic of the age hardening process

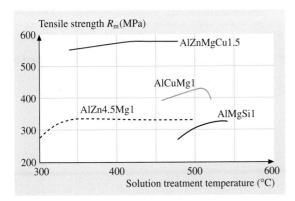

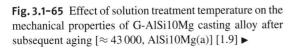

Fig. 3.1-65 Effect of solution treatment temperature on the mechanical properties of G-AlSi10Mg casting alloy after subsequent aging [≈ 43 000, AlSi10Mg(a)] [1.9] ▶

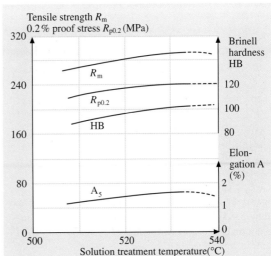

mainly on the initial condition of the semi-finished material, type of semi-finished product, and the wall thickness (Figs. 3.1-66 and 3.1-67).

During quenching, it is important to pass through the temperature range around 200 °C as quickly as possible, in order to prevent premature precipitation of the elements in supersaturated solid solution.

Some alloys require high cooling rates, such as those obtained by quenching in water. For others, especially with thinner sections, cooling in a forced draft of air or a mist spray suffices. In this case, quenching takes place immediately after hot working or even directly in the extrusion press. With most alloys it is important that the annealed material is rapidly transferred to the quenching bath as any delay (incipient cooling) can have a detrimental effect on strength and corrosion resistance.

After the first two stages of the hardening process, the solid solution will be supersaturated in both vacancies and solute atoms. It will tend to attain equilibrium conditions by precipitation of the supersaturated solute atoms. As the process is both temperature- and time-dependent, precipitation is basically possible either by natural or artificial aging, i. e., at room or elevated temperature respectively.

Aging at room temperature, or slightly elevated temperature, is accompanied by the formation of metastable Guinier–Preston zones (GP zones). These metastable phases increase the strength by differing amounts depending on their type, size, volume fraction, and distribution. With fully coherent precipitates, the crystal lattice is strongly strained locally, resulting in a marked increase in hardness and strength, while ductility and electrical conductivity decreases (Figs. 3.1-68 and 3.1-69).

At higher aging temperatures of about 100 °C to 200 °C, metastable phases form that are partially coherent. The local coherency strain is less pronounced

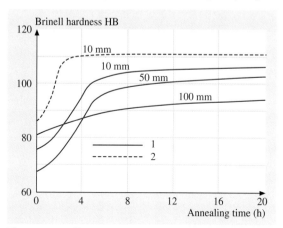

Fig. 3.1-66 Effect of annealing time on the hardness of artificially aged G-AlSi10Mg casting alloy for different casting techniques and specimen diameters [1.9]: (1) Sand cast; (2) Permanent mould cast

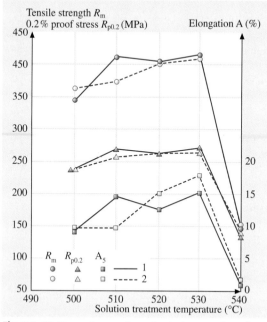

Fig. 3.1-67 Dependence of mechanical properties of naturally aged G-AlCu4TiMg casting alloy on the maximum solution treatment temperature: (1) Step annealing; (2) Single step annealing

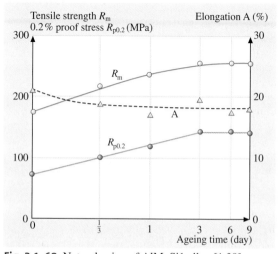

Fig. 3.1-68 Natural aging of AlMgSi1 alloy [1.39]

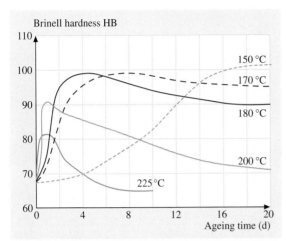

Fig. 3.1-69 Artificial aging of G-AlSi10Mg casting alloy [1.9]

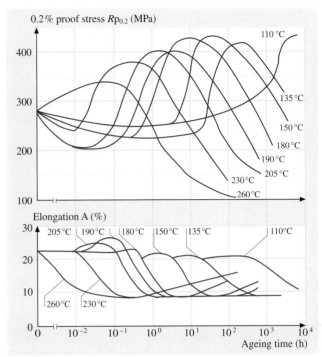

Fig. 3.1-70 Artificial aging curves for AlCuSiMn alloy [1.9]

because the stresses are partly reduced by the formation of interfacial dislocations. This would lead to a lower increase in strength in principle. However, since the particle size of the metastable phases formed in this range is often larger than that of the coherent phases which result from natural aging, there is a marked increase in strength (Figs. 3.1-70 and 3.1-71).

Higher temperatures and longer aging times lead to the formation of incoherent equilibrium phases (e.g., Al_2Cu, Mg_2Si, and $MgZn_2$) and there is a reduction in the hardening effect. This is called over-aging.

Effects of Plastic Deformation on Age-Hardening Behavior

For wrought age-hardenable aluminium alloys, the combination of heat treatments with hot and/or cold working (thermo-mechanical treatment) is of great practical significance. It can be a method to obtain a better combination of mechanical properties, such as moderate ductility and higher strength. The effects of thermomechanical treatments can be attributed to the fact that all processes occurring during heat treatment are influenced by the concentration of defects introduced during working. It is not necessary to carry out both steps simultaneously; the heat treatment can be carried out after forming [1.40]. As far as precipitation is concerned, lattice defects facilitate diffusion and at the same time act as nucleation sites. Thus, by deforming a material by a certain amount *prior* to aging and thereby obtaining the desired defect concentration, one can influence the size, quantity and distribution of the precipitates that form subsequently. It is important, however, that the final ageing treatment is carried out at a temperature below the recrystallization temperature, as the lattice defect concentration that is generated is fully effective only then.

Controlled hot working, often in combination with additional heat treatments, is another form of thermomechanical treatment and is used to improve fracture toughness, creep strength and fatigue strength. One tries to obtain a suitable recrystallized grain size and an optimum distribution of lattice defects and precipitates. Such treatments are very often used with high strength Al−Zn−Mg−Cu alloys. They usually consist of a series of solution treatments with controlled cooling and hot-rolling under well defined conditions. Cold working carried out between the various stages of a step-aging treatment is also referred to as thermomechanical treatment. Examples are shown in Fig. 3.1-72 and Fig. 3.1-73.

Simultaneous Softening and Precipitation

One of the factors that influence the softening of an alloy is the possible supersaturation by alloying of impurity elements. Supersaturation can be achieved deliberately, e.g., by solution treatment and subsequent quenching, or, as is often the case, it may be the result of rapid cool-

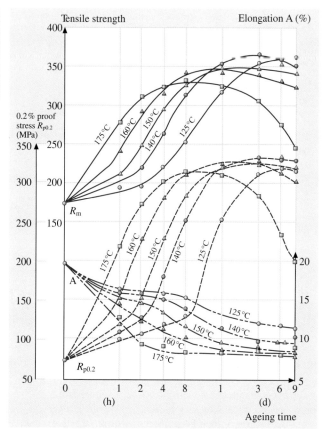

Fig. 3.1-71 Artificial aging curves for AlMgSi1 alloy after solution treatment at 520 °C and water quenching

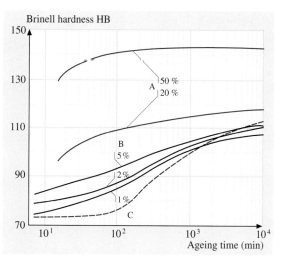

Fig. 3.1-72 Effect of cold work on the natural aging of AlCuMg1 (2017A) [1.9]; A – Rolled; B – Stretched; C – Without cold work

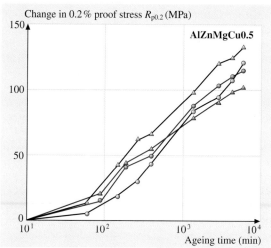

Fig. 3.1-73 Changes in the 0.2% proof stress during natural aging of 0.5 mm diameter wire specimens AlZnMgCu0.5 (7022) alloys after different degrees of cold working. The alloys were first solution-treated at 490 °C and then immediately subjected to cold working as indicated below [1.41]

ing, e.g., of castings. If such a supersaturated material is cold-worked and then aged, it reveals markedly different softening behavior because solute atom clusters or precipitates are formed during recovery at lattice defects in the deformed or recovered structure which pin lattice defects and the formation of solute atom clusters or of precipitates during recrystallization at the recrystallization front, which pin down the recrystallization front. There is a slowing down of the softening process as a result of these pinning effects (Figs. 3.1-74a and 3.1-74b).

The Stage III area in Fig. 3.1-74b is of particular interest, as by definition this is a thermomechanical treatment because the lattice defects introduced by working are fully effective during the formation of the precipitates. Dislocations, which are present in large quantities, act as nuclei for precipitation and this results in the formation of numerous finely-dispersed precipitates. These impair subsequent recrystallization at every stage – the

formation of recrystallization nuclei, the moving of boundaries and the subsequent grain growth. These effects increase the thermal resistance and delay the loss of strength that occurs after prolonged exposure because of grain growth. The Figs. 3.1-77 to 3.1-79 show some examples of TTT-Diagrams for unalloyed aluminium and for alloys.

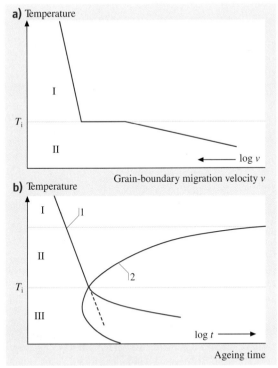

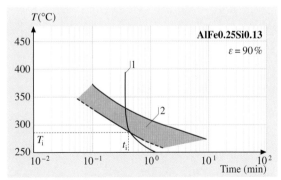

Fig. 3.1-75 TTT diagram for Al-0.25 wt% Fe-0.13 wt% Si alloy Hunter-Engineering cast strip after 90% cold work [1.35]. (1) First signs of precipitation; (2) Recrystallization zone

Fig. 3.1-74a,b Effect of segregation and precipitation on recrystallization (grain-boundary migration): (1) Start of recrystallization; (2) Start of precipitation. (**a**) Basic effects of solute atom clusters on the grain boundary migration velocity [1.42]. Region I, high temperatures: The grain boundary migration velocity is only weakly dependent on temperature because it is unaffected by the clustering ($\rightarrow$ high velocity). Region II: Below T_i, the mobility of the boundaries is reduced because atom clusters have to be dragged along by the boundary, and the diffusion rate of these atoms determines the migration velocity of the boundary ($\rightarrow$ low velocity). (**b**) Interrelationship between recrystallization and precipitation as TTT (time-temperature-transformation) diagram [1.43]. Region I: Recrystallization as a homogenous stable solid solution, unaffected by precipitation processes, i.e., at these temperatures there is no longer any supersaturation. Region II: Recrystallization in a supersaturated solid solution; the number of potential precipitation nuclei (dislocations) decreases. Result: Fewer but larger precipitations. Region III: Precipitation prior to recrystallization, or simultaneously (T_i) the precipitates delay recrystallization significantly

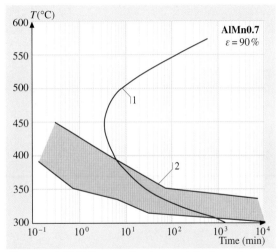

Fig. 3.1-76 TTT diagram showing precipitation and recrystallization in 90% cold-rolled, high-purity AlMn0.7 alloy, strip produced by permanent mould casting, hot-rolling, and then cold-rolling [1.44]. (1) Start of precipitation; (2) Recrystallization zone

Fig. 3.1-77 TTT diagram for Al-0.5 wt%Fe-0.15 wt% Si alloy Harvey cast strip after 80% cold work [1.45] (1) Start of precipitation; (2) Recrystallization zone

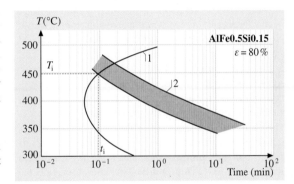

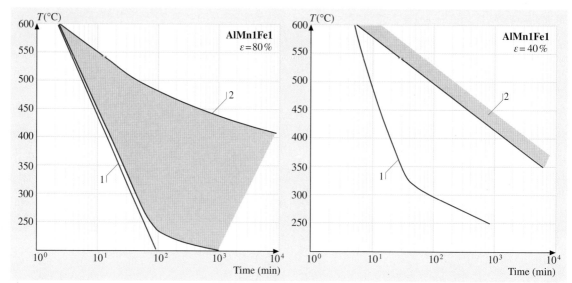

Fig. 3.1-79a,b TTT diagrams for precipitation and recrystallization in commercially pure AlMn1Fe1 (0.16 wt% Si), Hunter-Engineering cast strip after varying degrees of cold-rolling [1.49]. (1) Start of precipitation; (2) Recrystallization zone. **(a)** 90%. **(b)** 40%

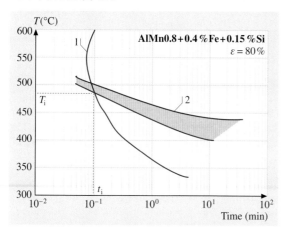

Fig. 3.1-78 TTT diagram showing precipitation and recrystallization in commercially pure AlMn0.8 alloy strip (0.4 wt% Fe and 0.15 wt% Si) produced by strip casting and 85% cold-rolled [1.46–48]. (1) Start of precipitation; (2) Recrystallization zone

3.1.2.11 Corrosion Behavior of Aluminium

From a thermodynamic point of view, aluminium would have to react with water to form hydrogen. However, Al and Al alloys have proven to be very corrosion-resistant in a wide range of practical applications. This corrosion resistance is attributable to the reaction of Al with oxygen or water vapor and the formation of a thin but compact natural oxide film when it is exposed to air, i.e. Al is passivated. In contrast to the oxide layers formed on many other materials, this oxide layer is strongly adherent and thus protects the underlying metal against further oxidation. This property explains the good resistance of aluminium when exposed to the weather or a large number of organic and inorganic substances. The corrosion resistance can be increased further by various surface treatments.

Aluminium and all standardized aluminium alloys are non-toxic. Aluminium products are easy to clean, can be sterilized and meet all hygienic and antitoxic requirements.

Surface Layers

Aluminium and Al alloys react with oxygen and water vapor in the air to produce a thin, compact surface oxide film which protects the underlying metal from further attack (Fig. 3.1-80). The surface layer contains mainly amorphous Al_2O_3 in several layers. The so-called barrier layer has an extremely low conductivity for electrons and ions and thus acts as an insulator in any interfacial electrochemical reactions. It thus affords effective protection against corrosion. If mechanical damage of the protective layer occurs, or if the layer is removed by pickling, it re-forms immediately. Aluminium and Al alloys thus exhibit good corrosion resistance to chemicals, seawater, and the weather.

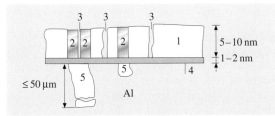

Fig. 3.1-80 Schematic representation of the structure of the oxide film formed on unalloyed aluminium in dry air; the total thickness is typically 0.005 to 0.02 mm [1.50]. Al = Aluminium; 1 = Surface layer; 2 = Mixed oxides; 3 = Pores; 4 = Barrier layer; 5 = Heterogeneous components

The oxide film which forms on bare aluminium in dry air at room temperature grows to a thickness of a few µm in a few minutes. It then grows to about two or three times this thickness in a few days at a continuously decreasing rate, so-called self-protection. Higher temperatures, such as those during heat treatments, accelerate the rate of growth of the natural oxide layer and lead to the growth of thicker films (Fig. 3.1-81). In moist air, the oxide films grow rapidly at first but then more slowly, and they are markedly thicker than the films formed in dry air.

The composition of the atmosphere has a significant effect on the behavior of the oxide layer. The aggressiveness of the atmosphere is particularly dependent on the amounts of sulfur dioxide, sulfur trioxide, dust, soot, and salts present. Rain water hitting the surface and running off flushes away these substances and thus reduces their influence.

Tap water or natural waters cause growth of the outer layer on top of the barrier layer of the oxide. The growth will depend on the alloy composition, the nature of the water and temperature. In aggressive waters, especially those containing chlorides and heavy metals, pitting corrosion can occur if oxygen from the air or other oxidizing media are introduced into the water. Traces of Cu (from copper piping or fittings containing copper) are particularly aggressive. Copper ions enter the oxide layer on the aluminium surface via defects and precipitate out as metallic Cu. Copper then acts as a cathode in the resulting local galvanic element such that Al is dissolved anodically.

Corrosion

If Al is exposed to acids or bases these dissolve the oxide film. The pH value of the electrolyte strongly influences corrosion in aqueous media. The protective film on aluminium is practically insoluble in the pH range from 4.5

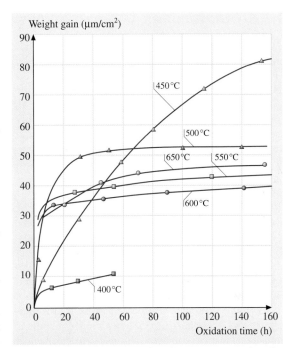

Fig. 3.1-81 Growth of the oxide film on super-purity aluminium in dry oxygen during the first 160 h [1.22]

to 8.5, which explains why aluminium is usually only used in this range.

Aluminium alloys have heterogeneous microstructural components such as intermetallic phases and resulting oxides in the surface and barrier layers. This explains why unalloyed aluminium and aluminium alloys have lower corrosion resistance than high-purity aluminium.

Apart from the effects of alloying and impurities, there are some other factors affecting corrosion, for example, changes in microstructure by thermal or mechanical treatments and ensuing changes of the surface condition.

Corrosion protection covers any measure aimed at modifying a corrosion system in order to mitigate corrosion damage. This can involve influencing the properties of the metal or the corrosive medium, or separating the metal from the medium by the use of protective layers. One can differentiate between active and passive measures. Passive measures, such as the use of organic polymer coatings (paints), will provide temporary protection, the level of which will depend on the nature, thickness, and quality of the layer. Active measures, such as the use of sacrificial magnesium or zinc anodes, offer long-term protection.

3.1.3 Titanium and Titanium Alloys

Titanium and its alloys are used as technical materials mainly because of the low density ($\varrho = 4.5\,\mathrm{g\,cm^{-3}}$) of Ti at technically useful levels of mechanical properties, and the formation of a passivating, protective oxide layer in air, which leads to a pronounced stability in corrosive media and at elevated temperatures. Further useful properties to be noted are its paramagnetic behavior, low temperature ductility, low thermal conductivity ($\kappa = 21\,\mathrm{W\,m^{-1}\,K^{-1}}$), low thermal expansion coefficient ($\lambda = 8.9 \times 10^{-6}\,\mathrm{K^{-1}}$), and its biocompatibility which is essentially due to its passivating oxide layer.

Five groups of materials based on Ti may be distinguished [1.51–53]: commercially pure (i. e., commercially available) Ti (cp-Ti), low-alloy Ti materials, Ti-base alloys, intermetallic Ti-Al materials, and highly alloyed functional materials: TiNi shape memory alloys, Nb–Ti superconducting materials (Sect. 4.2.1), and Ti-Fe-Mn materials for hydrogen storage.

Titanium undergoes a structural phase transformation at 882 °C. Like in steels, this transformation is crucial for the microstructural design and the mechanical properties of Ti-based alloys. The low temperature phase α-Ti has an almost close packed hexagonal (A3) structure that is somewhat compressed along the c axis. Its lattice parameters are $c = 0.4679$ nm and $a = 0.2951$ nm at room temperature. The high-temperature phase β-Ti has a body-centered cubic (A2) structure. The lattice parameter of β-Ti at room temperature can be obtained as $a = 0.3269$ nm by extrapolation from alloy solid solutions.

The binary phase diagrams of Ti [1.54] indicate that several interstitial components such as O, N, C, and H form extended solid solutions with α-Ti. As an example Fig. 3.1-82 shows the Ti-rich part of the Ti–O phase diagram. The high local lattice strains caused by the interstitial atoms lead to pronounced solid solution strengthening which is exploited, in particular, to harden commercially pure Ti by O additions. On average, the interstitials lead to a pronounced elongation of both the a and c axes of the α-Ti lattice, as shown in Fig. 3.1-83.

Figure 3.1-84 shows the influence of the concentration of residual impurity elements O, N, C, Fe, and Si on the increase of hardness HB.

Commercially pure grades of Ti, cp-Ti, are produced via the reduction of TiCl$_4$ by Mg (Kroll process). The product is Ti sponge. Table 3.1-17 gives typical concentration levels of the main impurity components and corresponding hardness values of the Ti sponge. Impu-

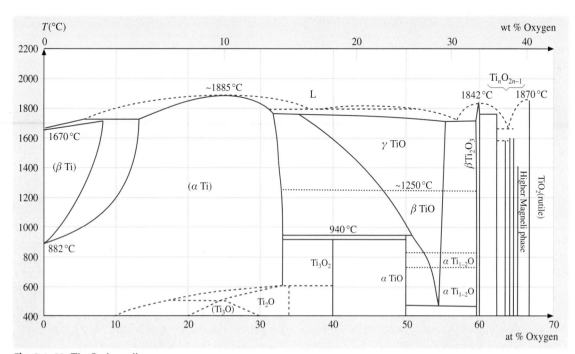

Fig. 3.1-82 Ti–O phase diagram

Table 3.1-17 Chemical composition of titanium sponge [1.53]

Reduction process/ element	C	N	O	H	Fe	Mg	Na	Cl	Maximum hardness (HB)
				(wt%)					
Magnesium	0.006	0.003	0.045	0.002	0.02	0.03		0.10	95
	0.008	0.005	0.05	0.002	0.04			0.09	100
	0.008	0.005	0.06	0.002	0.05	0.03		0.08	110
	0.008	0.005	0.07	0.002	0.06	0.03		0.09	120
Sodium electrolysis	0.009	0.004	0.07	0.01	0.016		0.08	0.13	120
	0.003	0.002	0.01	0.003	0.001			0.08	60
	0.013	0.004	0.03	0.004	0.004			0.09	75
	0.018	0.004	0.04	0.004	0.020			0.16	90

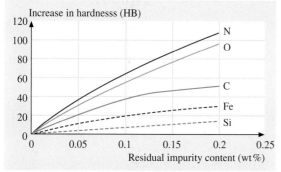

Fig. 3.1-84 Increase of hardness HB of Ti as a function of concentration of residual impurities O, N, C, Fe, and Si

rity levels resulting from earlier reduction processes by Na and electrolysis are also listed. The concentration of the residual content of the reducing elements Mg, Na, and their chlorides is lowered by the subsequent melting processes. High purity levels (5N) of Ti can be obtained by the iodide reduction process. These grades are used for electronic devices.

3.1.3.1 Commercially Pure Grades of Ti and Low-Alloy Ti Materials

Materials designated as commercially pure grades of Ti are interstitial solid solutions of O, N, C, and H in Ti. Oxygen is the only element added deliberately for solid solution strengthening. The other interstitial solutes are impurities resulting from the production process as indicated. Table 3.1-18 shows the chemical compositions and mechanical properties of commercially pure grades of Ti. The temperature dependence of their mechanical properties is shown in Fig. 3.1-85. Low-alloy Ti materials containing Pd, Ru, and Ni + Mo, respectively, are providing increased corrosion resistance at

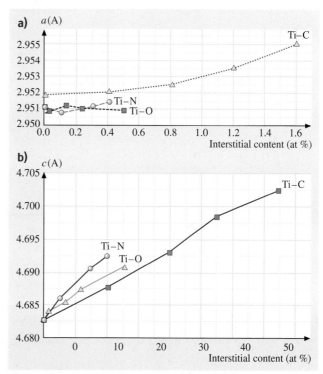

Fig. 3.1-83a,b Lattice parameters a (**a**) and c (**b**) of α-Ti as a function of interstitial content

identical levels of mechanical properties as those of the corresponding cp-Ti grades. The compositions and mechanical properties of typical alloys are also shown in Table 3.1-18. All commercially pure and low-alloy Ti materials may be strengthened by cold work. The tensile strength R_m is about doubled by cold work, which amounts to 80 to 90% reduction in area, and the fracture strain A_5 and the reduction in area are about halved.

Table 3.1-18 Chemical composition (maximum contents) and mechanical properties of commercially pure and low-alloy grades of Ti

O wt%	Tensile strength R_m (MPa)	Yield strength $R_{p0.2}$ (MPa)	Fracture strain A_{10} (%)	Standard grade[a] cp	Standard grade[a] low alloyed
0.12	290–410	> 180	> 30	Grade 1	Pd: grade 11
0.18	390–540	> 250	> 22	Grade 2	Pd: grade 7 Ru: grade 27
0.25	460–590	> 320	> 18	Grade 3	Ru: grade 26
0.35	540–740	> 390	> 16	Grade 4	
0.25	> 480	> 345	> 18		Ni+Mo: grade 12

[a] ASTM B265, ed. 2001; N_{max}: 0.03 wt%; C_{max}: 0.08 wt%; H_{max}: 0.015 wt%

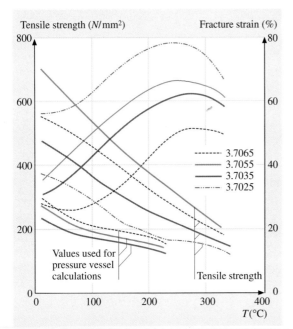

Fig. 3.1-85 High-temperature tensile strength and fracture strain of titanium, and values used for titanium pressure vessel calculations (105 h)

Most applications of cp-Ti and low-alloy Ti materials are based on their high corrosion resistance, such as components in chemical plants, heat exchangers, offshore technology, seawater desalination plants, Ni winning, electroplating plants, medical applications such as heart pacer casings, surgical implants including ear implants [1.55], and automotive applications. Applications of unalloyed Ti in architecture and jewellery are based on coloring the surface by oxide growth upon heat treatment in air. Unalloyed Ti grades are, also, applied at elevated temperature because of their favorable mechanical properties and oxidation resistance at high temperature. Figure 3.1-86 shows the creep date for grade 4 Ti.

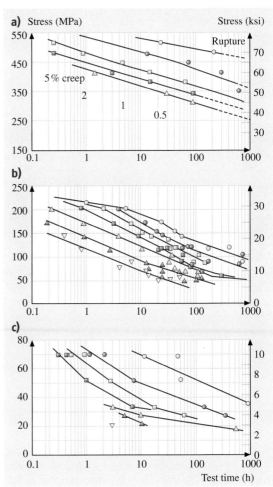

Fig. 3.1-86a–c Creep behavior of commercially pure Ti grade 4, mill annealed with a minimum yield stress of 480 MPa [1.56]. **(a)** 25 °C; **(b)** 425 °C; **(c)** 540 °C

3.1.3.2 Ti-Based Alloys

Alloying elements are added to Ti to improve its mechanical properties. Based on the phase transformation of pure Ti, alloying elements influence the phase equilibria and, thus, the transformations and the resulting microstructural states. The phase composition of the microstructure may be varied from pure α through ($\alpha + \beta$) to pure β, depending on the alloy content [1.54]. The α phase is stabilized by O, N, C, and Al. The β phase range is expanded by H, V, Mo, Fe, Cr, Cu, Pd, and Si. High solubility in α and β is exhibited by Zr and Sn. The alloy variants are characterized by the phases present in the annealed state at room temperature, as shown in Table 3.1-19. Alloys consisting partially or fully of the β phase are more easily deformed because of the larger number of slip systems in the bcc structure of the β phase compared to those in the hcp α phase. The alloying elements provide, also, solid solution strengthening. A further increase in yield strength may be attained by quenching and aging of β-phase alloys which leads to the coherent precipitation of the α and ω phases. However, the strength increase is moderate (10–20%) compared to that caused by precipitation hardening effects in Al and Cu-based alloys.

Based on their high specific strength, Ti alloys are widely used in aerospace applications, furthermore for high-speed moving parts, corrosion resistant pressure vessels and pipes, and for sports gear. The high-temperature strength has been increased by alloying for elevated temperature applications, as shown in Table 3.1-20. The densely packed hcp structure of the α phase is more creep-resistant than the more open bcc structure of the β phase. Accordingly the creep strength of the α- and near-α-phase alloys is clearly higher than that of $\alpha + \beta$-phase alloys, as shown in Fig. 3.1-87.

Titanium alloys may be processed like stainless steel. This leads to a cost effective production of a wide range of semi-finished and finished products and parts. If the tendency to oxidation and welding above approximately 350 °C and the low thermal conductivity are taken into account, parts can be manufactured from Ti alloys, quite similar to manufacturing from stainless steels. A wide variety of working, joining, and coating processes is well established.

Table 3.1-19 Chemical composition and mechanical properties of Ti-base alloys at room temperature (minimum values)

Alloy composition [a]	Alloy type	Tensile strength R_m (MPa)	Yield strength $R_{p0.2}$ (MPa)	Density ϱ (g/cm³)	Young's modulus E (GPa)	Main property	Standard grade [b]
Ti5Al2.5Sn	α	830	780	4.48	110	High strength	
Ti6Al2Sn4Zr2MoSi	near α	900	830	4.54	114	High-temperature strength	3.7145
Ti6Al5Zr0.5MoSi	near α	950	880	4.45	125	High-temperature strength	3.7155
Ti5.8Al4Sn3.5Zr0.7Nb 0.5Mo0.2Si0.05C	near α	1030	910	4.55	120	High-temperature strength	
Ti6Al4V	$\alpha + \beta$	900	830	4.43	114	High strength	
Ti4Al4Mo2Sn	$\alpha + \beta$	1100	960	4.60	114	High strength	3.7185
Ti6Al6V2Sn	$\alpha + \beta$	1030	970	4.54	116	High strength	3.7175
Ti10V2Fe3Al	near β	1250	1100	4.65	103	High strength	
Ti15V3Cr3Sn3Al	β	1000	965	4.76	103	High strength; cold formability	
Ti3Al8V6Cr4Zr4Mo	β	1170	1100	4.82	103	High corrosion; resistance	
Ti15Mo3Nb3AlSi	β	1030	965	4.94	96	High corrosion; resistance	

[a] Figure before chemical symbol denotes nominal wt%
[b] according to DIN 17851, ASTM B 265 ed. 2001

Table 3.1-20 Upper temperature limit for Ti alloys developed for elevated-temperature applications

Alloy	Alloy type	Year of introduction	Useful maximum °C
Ti-6Al-4V (Ti-64)	$\alpha+\beta$	1954	300
Ti-4Al-2Sn-4Mo-0.5Si (IMI-550)	$\alpha+\beta$	1956	400
Ti-8Al-1Mo-1V (Ti-811)	near-α	1961	425
Ti-2Al-11Sn-5Zr-1Mo-0.2Si (IMI-679)	near-α	1961	450
Ti-6Al-2Sn-4Zr-6Mo (Ti-6246)	$\alpha+\beta$	1966	450
Ti-6Al-2Sn-4Zr-2Mo (Ti-6242)	near-α	1967	450
Ti-3Al-6Sn-4Zr-0.5Mo-0.5Si (Hylite 65)	near-α	1967	520
Ti-6Al-5Zr-0.5Mo-0.25Si (IMI-685)	near-α	1969	520
Ti-5Al-5Sn-2Zr-2Mo-0.2Si (Ti5522S)	near-α or $\alpha+\beta$	1972	520
Ti-6Al-2Sn-1.5Zr-1Mo-0.1Si-0.3Bi (Ti-11)	near-α	1972	540
Ti-6Al-2Sn-4Zr-2Mo-0.1Si (Ti-6242S)	near-α	1974	520
Ti-5Al-5Sn-2Zr-4Mo-0.1Si (Ti-5524S)	near-α or $\alpha+\beta$	1976	500
Ti-5.5Al-3.5Sn-3Zr-0.3Mo-1Nb-0.3Si (IMI-829)	near-α	1976	580
Ti-5.5Al-4Sn-4Zr-0.3Mo-1Nb-0.5Si-0.06C (IMI-834)	near-α	1984	590
Ti-6Al-2.75Sn-4Zr-0.4Mo-0.45Si (Ti-1100)	near-α	1987	590
Ti-15Mo-3Al-2.75Nb-0.25Si (Beta-21S)	β	1988	590

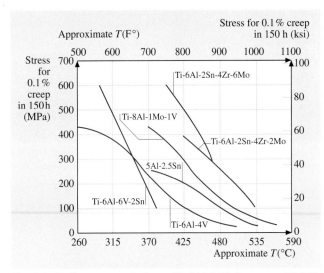

Fig. 3.1-87 Creep strength for α (Ti-5Al-2.5Sn), near-α (Ti-8Al-1Mo-1V and Ti-6Al-2Sn-4Zr-2Mo), and $\alpha+\beta$ (Ti-6Al-4V and Ti-6Al-2Sn-4Zr-6Mo) alloys

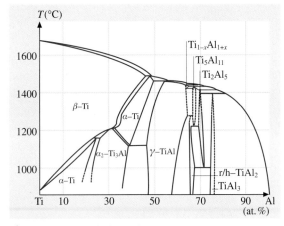

Fig. 3.1-88 Ti–Al phase diagram

Various versions of the binary Ti–Al phase diagram are available [1.54] and are still under discussion because of conflicting results. The phase diagram shown in Fig. 3.1-88 incorporates recent results.

3.1.3.3 Intermetallic Ti–Al Materials

The intermetallic compounds Ti_3Al and $TiAl$ are studied for high-temperature materials developments. Extensive accounts [1.52, 57] are the sources of the data presented here. In Table 3.1-21 the property ranges of Ti_3Al- and $TiAl$-based alloys are compared to those of conventional Ti alloys and Ni-based superalloys.

Ti_3Al-Based Alloys

Various Ti_3Al-based alloys have been developed with niobium as a major alloying element and further components for obtaining an optimised balance of strength, formability, toughness, and oxidation resistance. The alloys are two-phase or three-phase. Current Ti_3Al-based alloys with engineering significance are listed in Table 3.1-23.

Table 3.1-21 Properties of alloys based on the titanium aluminides Ti_3Al and TiAl [1.50] and of conventional titanium alloys and nickel based superalloys

Property	Ti-based alloys	Ti_3Al-based intermetallic materials	TiAl-based intermetallic materials	Ni-based superalloys
Structure	A3/A2	DO_{19}/A2/B2	$L1_0$/DO_{19}	A1/$L1_2$
Density (g/cm^3)	4.5–4.6	4.1–4.7	3.7–3.9	7.9–9.1
Thermal conductivity (W/m K)	21	7	22	11
Young's modulus at room temperature (GN/m^2)	95–115	100–145	160–180	195–220
Yield strength at room temperature (MN/m^2)	380–1150	700–990	400–650	250–1310
Tensile strength at room temperature (MN/m^2)	480–1200	800–1140	450–800	620–1620
Temperature limit due to creep (°C)	600	760	1000	1090
Temperature limit due to oxidation (°C)	600	650	900	1090
Tensile strain to fracture at room temperature (%)	10–25	2–26	1–4	3–50
Tensile strain to fracture at high temperature (%)	12–50	10–20	10–60	8–125
Fracture toughness K_{Ic} at room temperature (MN/m$^{3/2}$)	High	13–42	10–20	25

Table 3.1-22 Crystal structure data

Phase designation	Composition	Strukturbericht designation (prototype)
α-Ti(Al)	0–45 at.% Al	A3 (Mg)
β-Ti(Al)	0–47.5 at.% Al	A2 (W)
β_1		B2 (CsCl)
α_2-Ti_3Al	22–39 at.% Al	DO_{19} (Ni_3Sn)
γ-TiAl	48–69.5 at.% Al	$L1_0$ (AuCu)
O	Ti_2AlNb	
ω	Ti_4Al_3Nb	$B8_2$

Table 3.1-23 Major Ti_3Al-based alloys [1.50]

Alloy composition (at.%)	Phases	Designation
Ti-24Al-11Nb	$\alpha_2 + \beta$	24-11
Ti-25Al-11Nb	$\alpha_2 + \beta$	25-11
Ti-25Al-8Nb-2Mo-2Ta	$\alpha_2 + \beta$	8-2-2
Ti-25Al-16Nb	$\alpha_2 + \beta + O$	25-16
Ti-25Al-17Nb	$\alpha_2 + \beta + O$	25-17
Ti-27Al-15Nb	$\alpha_2 + \beta + O$	27-15
Ti-27Al-15Nb-1Mo	$\alpha_2 + \beta + O$	27-15-1
Ti-22Al-17Nb-1Mo	$\alpha_2 + \beta + O$	22-17-1
Ti-25Al-17Nb-1Mo	$\alpha_2 + \beta + O$	25-17-1
Ti-25Al-10Nb-3V-1Mo	$\alpha_2 + \beta + O$	10-3-1
Ti-25Al-24Nb	$O + \beta$	25-24
Ti-22Al-27Nb	$O + \beta$	22-27
Ti-30Al-20Nb	$O + \omega$	30-20 (1986)
SCS/6Ti-24Al-11Nb	$\alpha_2 + \beta$ + SiC	SCS-6/24-11

Alloy 24-11 and alloy 10-3-1 have already been produced on a production mill scale.

Mechanical Properties. The elastic properties of selected alloys are presented in Table 3.1-24. Strength, ductility, and toughness of the Ti_3Al-based alloys are sensitive functions of both composition and microstructure which are controlled by prior processing. Characteristic data are shown in Table 3.1-25 for various alloys. Different property data for the same alloy composition indicate the effect of different prior thermomechanical treatments. The effect of microstructure on toughness is exemplified by Table 3.1-26. Particularly high strengths can be obtained in intermetallic Ti_3Al-based matrix composites.

Table 3.1-26 Effect of microstructure variation on fracture toughness K_{Ic} at room temperature for a Ti-25Al-10Nb-3V-1Mo alloy

Microstructure	K_{Ic} [MPa m$^{1/2}$]
Coarse globular	11
Fine globular	10
Coarse bimodal	11
Fine bimodal	14
Fine lamellar	8
Coarse lamellar	50

Table 3.1-24 Elastic moduli E, G, K, and Poisson's ratio v at various temperatures T for selected simple Ti_3Al-based alloys

Alloy (at.%)	T (°C)	E (GPa)	G (GPa)	v	K (GPa)
Ti-25.5Al	27	144	56.2	0.283	
Ti-25.5Al	−270	151	59.0	0.280	
Ti-25Al-10Nb-3V-1Mo (coarse globular)	23	72			
Ti-25Al-10Nb-3V-1Mo (fine lamellar)	23	120			
Ti-25Al-10Nb-3V-1Mo (coarse lamellar)	23	159			
Ti-26.6Al-4.9Nb	26–883	134.47−0.0482T	53.56−0.0180T	0.257−0.00004T	
Ti-27.4Al-20.3Nb	25	128.8	48.8	0.32	118.4
Ti-22Al-23Nb/Ultra-SCS	25	201 ± 8			
Ti-22Al-23Nb/Ultra-SCS	649	173			
Ti-22Al-23Nb/Ultra-SCS	760	135 ± 8			

Table 3.1-25 Selected tensile data for yield stress R_p, fracture stress R_m, total elongation A, fracture toughness K_{Ic} at room temperature, and creep time to rupture t_r at 650 °C and 380 MPa for various Ti_3Al-based alloys

Alloy (at.%)	R_p (MPa)	R_m (MPa)	A (%)	K_{Ic} (MPa m$^{1/2}$)	t_r (h)
Ti-25Al	538	538	0.3		
Ti-24Al-11Nb	787	824	0.7		44.7
Ti-24Al-11Nb	510		2.0	20.7	
Ti-24Al-11Nb	761	967	4.8		
Ti-24Al-14Nb	831	977	2.1		59.5
Ti-24Al-14Nb	790		3.3	15.8	60.4
Ti-15Al-22.5Nb	860	963	6.7	42.3	0.9
Ti-25Al-10Nb-3V-1Mo	825	1042	2.2	13.5	360
Ti-25Al-10Nb-3V-1Mo	823	950	0.8		
Ti-25Al-10Nb-3V-1Mo	745	907	1.1		
Ti-25Al-10Nb-3V-1Mo	759	963	2.6		
Ti-25Al-10Nb-3V-1Mo	942	1097	2.7		
Ti-24.5Al-17Nb	952	1010	5.8	28.3	62
Ti-24.5Al-17Nb	705	940	10.0		
Ti-25Al-17Nb-1Mo	989	1133	3.4	20.9	476
Ti-22Al-23Nb	863	1077	5.6		
Ti-22Al-27Nb	1000		5.0	30.0	
Ti-22Al-20Nb-5V	900	1161	18.8		
Ti-22Al-20Nb-5V	1092	1308	8.8		

Chemical Properties. The oxidation resistance of the Ti_3Al-based alloys is higher than that of conventional Ti alloys with TiO_2 formation, but lower than that of Al_2O_3-forming alloys. The rate controlling mechanisms are complex and change with atmosphere, temperature, and time. Oxidation is moderate with porous scales up to 800 °C whereas spallation occurs at higher temperatures. The addition of Nb to the Ti_3Al-based alloys is beneficial to the oxidation resistance as well as nitrogen in the atmosphere [1.58]. Table 3.1-27 presents characteristic data for oxidation at 800 °C.

It should be noted that H is easily dissolved in Ti_3Al-based alloys, leading to embrittlement and stress corrosion cracking. Furthermore, the thermal stability of intermetallic Ti_3Al-based matrix composites – in particular SCS-6 SiC/Ti-25Al-10Nb-3V-1Mo – is affected by chemical reactions between the matrix and the strengthening phase.

TiAl-Based Alloys

Alloys based on γ-TiAl, also called gamma titanium aluminides, excel due to their high strength per unit density. These alloys contain the α_2-phase Ti_3Al as a second phase and are further alloyed with other elements for property optimization. The composition range is Ti-(45–48)Al-(0–2)(Cr, Mn, V)-(0–5)(Nb, Ta, W)-(0–2)(Si, B, Fe, N) (at.%). Further components such as Hf, Sn, C, and/or 0.8–7 vol.% TiB_2 are added for dispersion strengthening [1.59]. Table 3.1-28 shows the standard microstructures and Table 3.1-29 lists the alloys of primary interest.

Various jet engine and car engine components have been identified for application of TiAl-based alloys. It is particularly noteworthy that large sheets, which are suitable for superplastic forming and joining by diffusion bonding, can be produced from these intrinsically brittle intermetallic compound alloys.

Physical Properties. Some thermal data for single crystalline and polycrystalline TiAl alloys are shown in Tables 3.1-30 and 3.1-31.

Mechanical Properties. Elastic properties of TiAl-based materials are compiled in Table 3.1-32. Strength, ductility, and toughness of the TiAl-based alloys are sensitive functions of both composition and microstructure which is controlled by prior processing [1.59]. Characteristic data are shown in Table 3.1-33 for various alloys. Different property data for the same alloy composition indicate the effect of different prior thermo-mechanical treatments. It is noted that TiAl-based alloys are prone to hydrogen/environmental embrittlement depending on the amount of α_2 from Ti_3Al [1.60]. Creep and fatigue data are available [1.59, 61].

Table 3.1-27 Characteristic oxidation data for various Ti_3Al-based alloys at 800 °C. Apparent parabolic rate constant k_p and apparent activation energy Q [1.58]

Alloy (at.%)	Atmosphere	Test duration (h)	k_p (10^{-12} g^2 cm^{-4} s^{-1})	Q (kJ mol^{-1})
Ti-25Al	O_2	9	19	289
Ti-25Al-11Nb	O_2	4	2.0	330
Ti-24Al-15Nb	O_2	6	2.2	329
Ti-24Al-15Nb	20% O_2 + 80% N_2	5	0.25	274
Ti-24Al-10Nb-3V-1Mo	20% O_2 + 80% N_2	15	1150	217
Ti-25Al-10Nb-3V-1Mo	Air	24	2.4	248

Table 3.1-28 Standard microstructures of TiAl-based alloys [1.59]

Type	Phase distribution
Near-gamma (NG)	γ grains + α_2 particles
Duplex (DP)	γ grains + α_2 plates or particles
Nearly lamellar (NL)	Lamellar γ + γ grains
Fully lamellar (FL)	Lamellar γ + residual γ grains on grain boundaries
Modified NL (MNL)	Lamellar γ + fine γ grains
Modified FL (MFL)	Lamellar γ

Table 3.1-29 Important TiAl-based alloys with microstructures according to Table 3.1-28 [1.59]

Composition (at.%)	Microstructure type	Designation
Ti-48Al-1V-0.3C 0.2O	DP	48-1-(0.3C)
Ti-48Al-1V-0.2C-0.14O	DP/NL	48-1-(0.2C)
Ti-48Al-2Cr-2Nb	DP, NL, DP/FL	48-2-2
Ti-47Al-1Cr-1V-2.6Nb	DP/FL	G1
Ti-45Al-1.6Mn	NL	Sumitomo
Ti-47.3Al-0.7V-1.5Fe-0.7B	–	IHI Alloy
Ti-47Al-2W-0.5Si	DP	ABB Alloy
Ti-47Al-2Mn-2Nb-0.8TiB$_2$	NL + TiB$_2$	47XD
Ti-45Al-2Mn-2Nb-0.8TiB$_2$	MFL + TiB$_2$	45XD
Ti-46.2Al-xCr-y(Ta,Nb)	NL	GE Alloy 204b
Ti-47Al-2Nb-2Cr-1Ta	DP	Ti-47Al-2Nb-2Cr-1Ta
Ti-46Al-4Nb-1W	NL	Alloy 7
Ti-46.5Al-2Cr-3Nb-0.2W	DP/MFL	Alloy K5
Ti-48Al-2Cr	NG, DP, FL	Ti-48Al-2Cr
Ti-47Al-1.5Nb-1Cr-1Mn-0.7Si-0.5B	MFL	γ-TAB
Ti-46.5Al-4(Cr, Nb, Ta, B)	MFL	γ-Met

Table 3.1-30 Temperature dependence of the thermal expansion coefficient α for single crystalline TiAl

Alloy (at.%)	T (°C)	$\alpha_{[100]}$ (K^{-1})	$\alpha_{[001]}$ (K^{-1})
Ti-56Al	25–500	$9.77 \times 10^{-6} + 4.46 \times 10^{-9}(T+273)$	$9.26 \times 10^{-6} + 3.76 \times 10^{-9}(T+273)$

Table 3.1-31 Thermal expansion coefficient α, thermal conductivity λ and specific heat c_p at various temperatures T for polycrystalline TiAl alloys [1.62]

Alloy (at.%)	T (°C)	α (K^{-1})	λ (W K^{-1} m^{-1})	c_p (J g^{-1} K^{-1})
Ti-50Al	27	11×10^{-6}	11	0.4
Ti-50Al	850	13×10^{-6}	19	0.5
Ti-48Al-2Cr	100	9.1×10^{-6}	–	–
Ti-48Al-2Cr	700	11.8×10^{-6}	–	–

Table 3.1-32 Young's modulus E, shear modulus G, Poisson's ratio ν and bulk modulus K for single-phase TiAl and a TiAl composite

Alloy (at.%)	T (°C)	E (GPa)	G (GPa)	ν	K (GPa)
Ti-49.4Al	25–935	$173.59 - 0.0342T$	$70.39 - 0.0141T$	0.234	
Ti-50Al	25	185	75.7		110
Ti-50Al	−273	190	78.3		112
Ti-47Al-2V-7 vol.%TiB$_2$	27	180			
Ti-47Al-2V-7 vol.%TiB$_2$	647	162			

Table 3.1-33 Selected tensile data for yield stress R_p, fracture stress R_m, total elongation A, toughness K_{Ic} at room temperature (RT) or higher temperature T for the TiAl-based alloys of Table 3.1-29 with microstructures of Table 3.1-28 [1.59, 61]

Alloy designation	Alloy condition, processing, microstructure	T [°C]	R_p (MPa)	R_m (MPa)	A (%)	K_{Ic} (MPa m$^{1/2}$)
48-1-(0.3C)	Forging + HT	RT	392	406	1.4	12.3
		760	320	470	10.8	–
	Casting-duplex	RT	490	–	–	24.3
48-1-(0.2C)	Forging + HT	RT	480	530	1.5	–
		815	360	450	720	–
48-2-2	Casting + HIP + HT	RT	331	413	2.3	20–30
		760	310	430	–	–
	Extrusion + HT: DP/FL	RT	480/454	–	3.1/0.5	–
		870	330/350	–	53/19	–
	PM extrusion + HT	RT	510	597	2.9	–
		700	421	581	5.2	–
G1	Forging + HT: DP/FL	RT	480/330	548/383	2.3/0.8	12/30-36
		600	383/–	507/–	3.1/–	16/–
		800	324/290	492/378	55/1.5	–/40-70
Sumitomo	Reactive sintering	RT	465	566	1.4	–
		800	370	540	14	–
IHI Alloy	Casting	RT	–	520	0.6	–
		800	–	424	40	–
ABB Alloy	Casting + HT	RT	425	520	1.0	22
		760	350	460	2.5	–
47XD	Casting + HIP + HT	RT	402	482	1.5	15–16
		760	344	458	–	–
45XD	Casting + HIP + HT	RT	550–590	670–720	1.5	15–19
		760	415	510	19	–
GE Alloy 204b	Casting + HIP + HT	RT	442	575	1.5	34.5
		840	381	549	12.2	–
Ti-47Al-2Nb-2Cr-1Ta	Casting + HIP + HT	RT	430	515	1.0	–
		870	334	403	14.6	–
Alloy 7	Extrusion + HT	RT	648	717	1.6	–
		760	517	692	–	–
Alloy K5	Forging + HT: DP/MFL	RT	462/473	579/557	2.8/1.2	11/20-22
		870	–/362	–/485	–/12.0	–
Ti-48Al-2Cr	Casting + HIP + HT	RT	390	405	1.5	–
		700	340	395	2.7	–
γ-TAB	Casting + HIP + HT	RT	475	–	1.6	–
		700	385	–	6.0	–

Chemical Properties. The oxidation behavior of TiAl alloys is complex as both Al_2O_3 and/or TiO_2 are formed at parabolic rate constants ranging from 3×10^{-13} to 3×10^{-9} g^2 cm^{-4} s^{-1} at 950 °C, depending not only on alloy composition, atmosphere, and temperature, but also on alloy surface quality. The oxidation rate is higher in air than in pure oxygen. Alloying additions of Nb, W, Ta, or Si generally improve the oxidation resistance. Additions of V and Mo promote Al_2O_3 formation above 900 °C, which is beneficial, but accelerates scale growth at lower temperatures, i. e., at possible service temperatures. Small additions of Cr reduce the oxidation resistance whereas larger amounts improve it. Small additions of P or Cl have been found to increase the oxidation resistance. Some characteristic data are shown in Table 3.1-34.

Table 3.1-34 Characteristic oxidation data for some TiAl-based alloys of Table 3.1-29. Apparent parabolic rate constant k_p at 800 °C in air

Alloy designation	Test duration (h)	k_p (10^{-12} g^2 cm^{-4} s^{-1})
48-2-2	500	2
Alloy K5	1000	0.5
Alloy 7 (5 at.% Nb)	1000	0.1

Usually γ-TiAl-based alloys contain α_2-Ti$_3$Al as a second phase and are, therefore, subject to hydrogen uptake and hydrogen/environmental embrittlement depending on the amount of α_2-Ti$_3$Al.

3.1.3.4 TiNi Shape-Memory Alloys

The shape-memory effect is based on martensitic transformations of intermetallic phases with a high degree of reversibility [1.63]. The alloy TiNi is the most advanced and has widespread shape memory. The present state of development and applications is summarized in [1.63, 64]. The high-temperature TiNi phase, having a cubic B2 (CsCl type) structure, is transformed martensitically, i.e., by a combination of transformation shears, into a low-temperature phase with a monoclinic DO$_{19}$ structure upon cooling or upon applying a stress and strain. The transformation can be associated with a macroscopic shape change. The transformation and the associated shape change are reversed on heating (shape-memory effect) or on releasing the stress and strain isothermally (superelasticity). The range of transformation start temperatures M_s varies between -200 °C and 110 °C depending on the alloy composition, as shown in Fig. 3.1-89.

Further reversible, diffusionless, but non-martensitic transformation phenomena occur in the TiNi high temperature phase on approaching the temperature range of martensitic transformation. Microscopically, these transformation phenomena are associated with phonon softening. Crystallographically, they are observable as localised rhombohedral distortions of the B2 structure. Furthermore, these localized displacements appear to be stabilized by annealing. Annealing treatments in the temperature range of 300 to 800 °C are increasing the M_s temperature. Figure 3.1-90 shows a characteristic example.

The shape-memory effect is a complex function of composition, martensite start temperature M_s, stress, strain, microstructure, texture, and aging treatment. It consists essentially of a reversible transformation strain and the associated macroscopic shape change. At low numbers of transformation cycles (e.g., up to 100), the

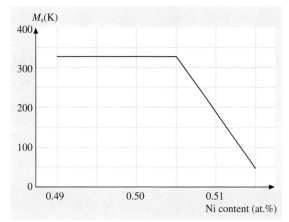

Fig. 3.1-89 Martensite start temperature M_s of the TiNi phase as a function of Ni content

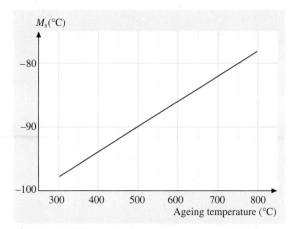

Fig. 3.1-90 Effect of aging temperature on the M_s temperature of an Ni$_{47.3}$Ti$_{43.8}$Nb$_{8.9}$ alloy

reversible transformation strain decreases from a typical initial value 8% to about 6%; even after $\geq 100\,000$ transformation cycles it stays at a technically useful level of $\cong 2\%$. A considerable number and variety of technical and medical applications of the shape-memory effect and of the closely related effect of superelasticity have been developed [1.65, 66].

3.1.4 Zirconium and Zirconium Alloys

Zirconium is used – similarly to titanium but less extensively so – as a special structural material of high corrosion resistance due to its highly stable protective oxide layer. This oxide layer forms in air at ambient temperature spontaneously and can be increased in thickness and stability by heat treatment in air. More extensive treatments are found in [1.51, 67]. Technically-pure and low-alloy Zr materials are especially used in applications requiring structural parts of high chemical stability. The processing is closely analogous to that of Ti. Another main application of Zr is in high-purity Zr alloys used as cladding and structural materials in nuclear applications due to the low thermal neutron capture cross section of Zr. A third group of Zr-based materials with application potential is due to the fact that some Zr-based alloys form the amorphous state easily upon cooling at a comparatively low rate. They are suitable for the production of bulk glassy alloys.

At 1138 K, Zr undergoes a structural phase transformation from low-temperature, hexagonal close-packed α-Zr to high-temperature, body-centered cubic β-Zr. Alloying elements such as Nb stabilize the β phase. On slow cooling and annealing of the β phase, a metastable ω phase can be obtained which forms by coherent precipitation.

3.1.4.1 Technically-Pure and Low-Alloy Zirconium Materials

Only two zirconium materials are standardised: Zr 702 which is essentially technically-pure Zr and consists of the α phase only, and Zr 705 which contains 2 to 3 wt% Nb, has an $\alpha + \beta$ microstructure and a higher yield stress. Their composition ranges and mechanical properties are listed in Tables 3.1-35 and 3.1-36. Since Hf is usually contained in the Zr ore zircon ($ZrSiO_4$) and is difficult to separate, the Zr materials are specified to contain Hf up to 4.5 wt%. The presence of Hf in Zr alloys up to this level of composition does not affect the mechanical and corrosion properties significantly.

The Zr–Nb phase diagram (Fig. 3.1-91) indicates that $\alpha + \beta$ microstructures can be obtained at the 2–3 wt% Nb level. This leads to both substitutional solid solution strengthening and dual phase strengthening effects. Oxygen has a high interstitial solubility in α-Zr. Fig. 3.1-92 shows the Zr–O phase diagram. It is inter-

Table 3.1-35 Composition of technically-pure and low-alloy Zr materials

Common designation	Zr 702	Zr 705
ASTM/UNS designation	R 60702	R 60705
Alloying, elements (wt%)		
Zr+Hf, min	99.2	95.5
Hf, max	4.5	4.5
Fe+Cr, max	0.20	0.20
Nb	–	2.0–3.0
O, max	0.16	0.18
H, max	0.005	0.005
N, max	0.025	0.025
C, max	0.05	0.05

Table 3.1-36 Mechanical properties of technically-pure and low-alloy Zr materials

| Alloy | Property | Test temperature (°C) | | | | |
		20	95	150	260	370
Zr 702	$R_{p0.2}$ (MPa)	321	267	196	129	82
	R_m (MPa)	468	364	304	201	157
	A (%)	28.9	31.5	42.5	49.0	44.1
Zr 705	$R_{p0.2}$ (MPa)	506		272	196	
	R_m (MPa)	615		389	326	
	A (%)	18.8		31.7	28.9	

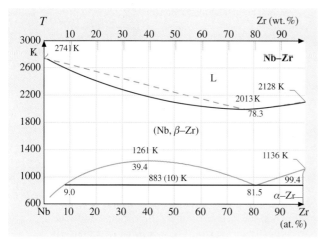

Fig. 3.1-91 Zr–Nb phase diagram

3.1.4.2 Zirconium Alloys in Nuclear Applications

Based on its low cross section for thermal neutrons and its high corrosion resistance in water, Zr is the preferred material for structural parts and cladding in pressurized and boiling water nuclear reactors. An extensive account of these materials and of their behavior under irradiation is given in [1.68]. Designations and composition ranges of the 4 standardized alloys are listed in Table 3.1-37. The addition of Sn gives rise to a strengthening contribution and increases corrosion resistance by reducing the deleterious effect of nitrogen. The metals Fe, Cr, and Ni, which are highly soluble in the β phase, have very low solubility in α-Zr (maximum solubility: Fe 120 ppm, Cr 200 ppm). This is the basis of a heat treatment involving a high-temperature solution annealing and a low-temperature precipitation treatment. The resulting precipitates are the Laves phases $Zr_2(Ni, Fe)$ and $Zr(Cr, Fe)_2$. In their presence the corrosion resistance is increased by reducing forms of localized corrosion. A specified O content serves to adjust the yield stress due to interstitial solid solution strengthening. Hafnium is removed and its upper limit is specified in Zr alloys for nuclear applications because of its high cross section for thermal neutrons.

esting to note that the O atoms undergo ordering on the interstitial lattice sites of the α-Zr solid solution above about 10 at.% O. These interstitially ordered phases are termed suboxides. The interstitial O content leads to a strong interstitial solid solution strengthening effect: about 0.01 wt% O increases the yield stress by about 150 MPa.

Extensive data and references are available on the corrosion behavior of Zr in many types of media and under various conditions of exposure [1.67].

The irradiation effects on the mechanical properties are significant. The interstitial atoms and vacancies resulting from irradiation-induced atomic displacements give rise to the formation of dislocation loops of interstitial and vacancy type. These dislocation loops act as obstacles to slip dislocations and lead to an increase in yield stress and decrease in elongation upon fracture as a function of dose, as shown in Fig. 3.1-93. The effect saturates at about 10^{24} nm^{-2}.

3.1.4.3 Zirconium-Based Bulk Glassy Alloys

Zirconium-based alloys were among the first non-noble metal-based alloys found to solidify in the amorphous state upon cooling from the melt at comparatively low rates R, such as 10 K s^{-1} [1.69, 70]. The term bulk glassy alloys refers to the fact that this solidification behavior permits us to obtain bulky parts with an amorphous structure by conventional casting procedures, e.g., in rod form, up to 30 mm. Data are given in [1.71, 72].

The tendency to solidification in the amorphous state is due to a low rate of nucleation of the equilibrium phase(s) at a particular alloy composition. An empirical method to determine the ease of glass formation has proved to be an evaluation of the glass-transition tem-

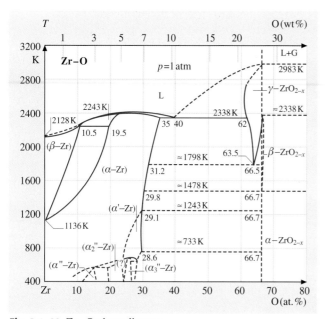

Fig. 3.1-92 Zr–O phase diagram

Table 3.1-37 Designations and compositions of the standardized Zr alloys for nuclear applications [1.68]

Common designation ASTM/UNS designation	Zircaloy-2 R 60802	Zircaloy-4 R 60804	Zr–Nb R 60901	Zr–Nb R 60904
Alloying elements (wt%)				
Sn	1.2–1.7	1.2–1.7	–	–
Fe	0.07–0.2	0.18–0.24	–	–
Cr	0.05–0.15	0.07–0.13	–	–
Ni	0.03–0.08	–	–	–
Nb	–	–	2.4–2.8	2.5–2.8
O		0.010–0.014 (t.b.s.)[a]	0.09–0.13	(t.b.s.)[a]
Impurity elements, maximum permissible content (wt ppm)				
Al	75	75	75	75
C	270	270	270	150
Cu	50	50	50	50
Hf	100	100	100	50
Mn	50	50	50	50
Mo	50	50	50	50
Ni	–	80	80	65
Si	120	120	120	120
Ti	50	50	50	50
W	100	100	100	100

[a] t.b.s. – to be specified

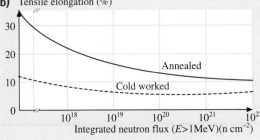

Fig. 3.1-93 Irradiation effects on mechanical properties of a zircaloy [1.68]. YS – yield stress; UTS – ultimate tensile strength. *Solid lines*: irradiation in the annealed state. *Dashed lines*: irradiation in the 10% cold worked state

perature T_g and the crystallization temperature T_x, which can be determined by thermal analysis of the amorphous alloy in question upon heating. For a great number of alloys which form the amorphous state upon cooling, it has been shown that the critical cooling rate R_c, i.e., lowest rate which is sufficient for complete glass formation, decreases strongly with two parameters:

- The relative glass-transition temperature $T_{rg} = T_g/T_{l/el}$, where $T_{l/el}$ is the liquidus temperature, and
- The magnitude of the temperature range $\Delta T_x = T_x - T_g$.

Figures 3.1-94 – 3.1-97 show characteristic data for Zr alloys which have been investigated in a wide range of compositions.

It is impossible to date to derive from first principles which alloy compositions are prone to easy glass formation. It has been found empirically that multi-component alloys with components of significantly different ionic radii are suitable candidates in principle. Examples of Zr-based systems that show bulk glassy solidification behavior are listed in Table 3.1-38.

Empirical findings have been subjected to a systematic treatment involving a mismatch entropy term S_σ and the enthalpy of mixing ΔH [1.73]. Such re-

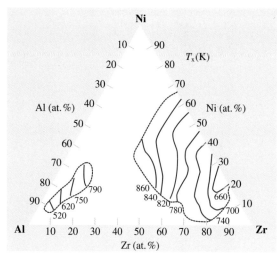

Fig. 3.1-94 Dependence of crystallization temperature T_x on composition in amorphous Zr–Al–Ni alloys [1.72]

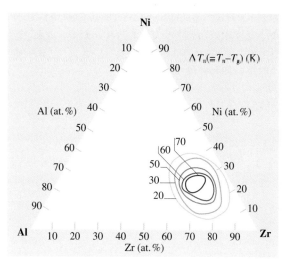

Fig. 3.1-96 Dependence of ΔT_x on composition in amorphous Zr–Al–Ni alloys [1.72]

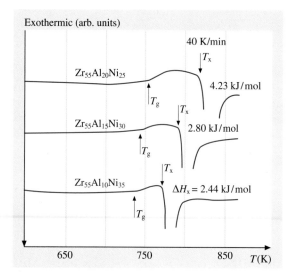

Fig. 3.1-95 Dependence of ΔT_x on composition as determined by differential thermal analysis of amorphous Zr–Al–Ni alloys [1.72]

Table 3.1-38 Zr-based alloy systems exhibiting bulk glassy behavior

Alloy system	Reference
Zr–Al–TM	[1.72]
Zr–Al–Ni–Cu–(Ti, Nb, Pd)	[1.72]
Zr–Al–Co	[1.74]
Zr–Ti–Ni–Cu–(Be)	[1.75]
Zr–Cu–Ni–(Al, Ti, Ta)	[1.76]

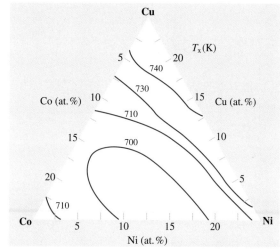

Fig. 3.1-97 Dependence of T_x on composition in amorphous $Zr_{65}Al_{7.5}Cu_{2.5}(Co_{1-x-y}Ni_xCu_y)_{25}$ alloys [1.72]

lations are based on systematic investigations of the effects of individual alloying elements on the characteristic properties T_g, T_x, and ΔT_x, as shown in Fig. 3.1-98 for a characteristic example.

There are no tabulated data of the properties of Zr-based bulk glassy alloys available yet. A typical set of data is given in Table 3.1-39 [1.75].

Table 3.1-39 Properties of amorphous $Zr_{41.2}Ti_{13.8}Cu_{12.5}Ni_{10}Be_{22.5}$

E (GPa)	G (GPa)	ν	Elastic strain limit %	$R_{p0.2}$ (GPa)	HV	ϱ g/cm³	$R_{p0.2}/\varrho$ (GPa cm³ g⁻¹)	A_e^a (%)	A_c^b (%)	K_{1c} (MN m^{1/2})
90±1	33±1	0.354	2.0–2.2	1.9±0.05	590	5.9	0.32	~1	1–20	20–40

^a Plastic strain to failure in tension
^b Plastic strain to failure in compression

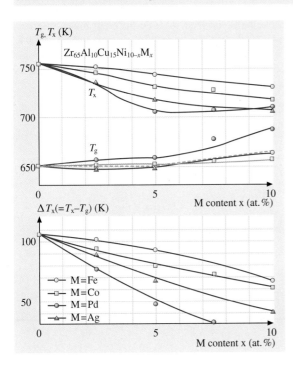

Fig. 3.1-98 Variation of T_g, T_x, and ΔT_x with M content for melt-spun $Zr_{65}Al_{10}Cu_{15}Ni_{10-x}M_x$ (M = Fe, Co, Pd, Ag) amorphous alloys

3.1.5 Iron and Steels

Iron is technically the most versatile and economically the most important base metal for a great variety of structural and magnetic materials, most of which are called steels. With increasing temperature, iron undergoes two structural phase transitions,

α (bcc) $\leftrightarrow \gamma$ (fcc) at 911 °C

and

γ (fcc) $\leftrightarrow \delta$ (bcc) at 1392 °C,

and a magnetic phase transition,

$\alpha_{\text{ferromagnetic}} \leftrightarrow \alpha_{\text{paramagnetic}}$ at 769 °C,

at ambient pressure. At elevated pressure levels, Fe forms a third structural phase ε (hcp). These phase transitions, their variation upon alloying, and the concomitant phase transformations are the thermodynamic, structural and microstructural basis for the unique variety of iron-based alloys and of their properties [1.52, 77–80].

The wide variety of standards for steels which have developed from national standards and efforts of international standardization are compiled in [1.81]. In the present section we are using different standard designations as provided by the sources used. In the following sections, the SAE (Society of Automotive Engineers), AISI (American Iron and Steel Institute), and UNS (Unified Numbering System) designations are the dominating ones.

3.1.5.1 Phase Relations and Phase Transformations

Iron–Carbon Alloys

The most frequent alloying element of iron is carbon. The Fe–C phase diagram (Fig. 3.1-99) shows the important metastable phase equilibria involving the metastable carbide Fe_3C, called cementite, in dashed lines, whereas the stable equilibria with graphite C are shown in solid lines. The formation of Fe_3C predominates in most carbon and low-alloy steels because the activation energy of its nucleation is considerably lower than that of graphite. At higher carbon contents (2.5–4.0 wt% C) and in the presence of Si (1.0–3.0 wt% Si), graphite formation is favored. This is the basis of alloying and microstructure of gray cast iron (see Sect. 3.1.5.7).

The Fe–C phase forms interstitial solid solutions of α- and γ-Fe. The solid solution phase of α-Fe is called ferrite, the solid solution phase based on γ-Fe is called austenite in the binary Fe–C system. These terms for the solid solutions phases of α- and γ-Fe are applied to all other Fe-based alloy systems as well. Since phase transformations induced by cooling from the austenite phase field play a major role to induce particular microstructures and properties, some resulting microstructures have also been given particular terms and form the basis of the nomenclature in steels. Cooling of Fe–C alloys from the austenite phase field can lead to three different phase transformations below the eutectoid temperature of 1009 K (736 °C). Their kinetics of formation depends on composition and cooling rate. The transformation products are:

- Pearlite, a lamellar product of ferrite and cementite. It is formed by a discontinuous (pearlitic) transformation, i.e., both phases are formed side by side in the reaction front. The lamellar spacing decreases with decreasing temperature of formation. The maximum rate of formation occurs at about 500 °C.

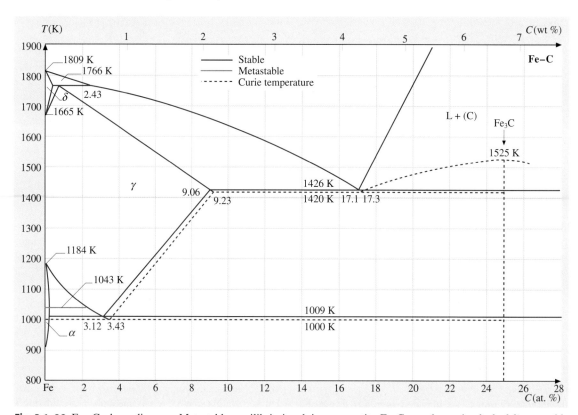

Fig. 3.1-99 Fe–C phase diagram. Metastable equilibria involving cementite Fe_3C are shown in *dashed lines*, stable equilibria with graphite C are shown in *solid lines* [1.82] (*dotted lines* – Curie temperature)

- Bainite, a plate- or spearhead-shaped product consisting of a ferrite matrix in which carbide particles are dispersed. The bainitic transformation mechanism depends sensitively on alloy composition and the temperature of transformation, yielding essentially two microstructural variants. A somewhat coarser transformation product formed at about 450 °C is called upper bainite and a finer transformation product formed at about 350 °C is termed lower bainite.
- Martensite, a plate-shaped product formed by a diffusionless, athermal transformation. Thermodynamically it is a metastable ferrite, designated as α' and supersaturated in carbon. But by the displacive mechanism of the transformation, the distribution of the C atoms in the martensite lattice is anisotropic such that is has a body-centered tetragonal crystal structure and its c and a parameters vary with the C content accordingly (Fig. 3.1-100). The temperature below which martensite begins to form upon queching is termed martensite start temperature M_s and depends strongly on the C concentration (Fig. 3.1-101). M_f designates the temperature at which the transformation is complete. In order to promote the diffusionless martensitic transformation, the diffusion-dependent transformations to pearlite and bainite have to be suppressed by rapid cooling, usually termed quenching.

Since martensite formation is used as a main hardening mechanism in steels, the hardenability is a main concern of alloy design and consequence of alloy composition. The lower the rate of formation of the

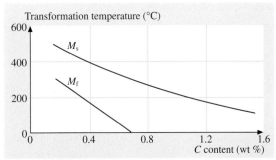

Fig. 3.1-101 Concentration dependence of the martensite transformation temperatures. M_s – martensite start; M_f – martensite finish, i.e., austenite is transformed completely

diffusion-dependent transformations, the higher is the fraction of martensite formed upon cooling from the austenite range, i. e., the hardenability (see Sect. 3.1.5.2). The rates of pearlite and bainite formation are reduced by alloying with carbon and all substitutional alloying elements except by Co. But the decrease of M_s with increasing alloy content has, also, to be taken into account.

Subsequent heat treatment of the phases formed is termed annealing with regard to ferrite and bainite, and tempering with regard to martensite. These heat treatments play a major role in optimizing the microstructure to obtain specific properties. Upon subsequent heat treatment the transformation products listed above undergo the following reactions:

- Pearlite is coarsened by the transition of the cementite lamellae into spherical particles, thus reducing the interfacial free energy per unit volume. The process is called spheroidization and consequently the resulting microstructural constituent is termed spheroidite.
- Bainite is coarsened as well both by recovery of the ferrite plates and by coarsening of the carbide particles.
- Martensite is essentially transformed into bcc ferrite by the precipitation of carbide particles during tempering. The tempering treatment usually leads to the precipitation of metastable carbides from the martensite phase. Different metastable carbides may be formed depending on alloy composition (including substitutional alloying elements), temperature, and time of annealing. A compilation of all metastable carbides occurring in Fe−C(−X) alloys is given in Table 3.1-40.

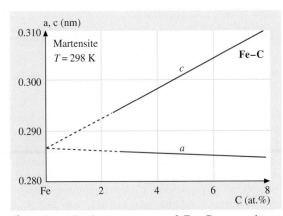

Fig. 3.1-100 Lattice parameters of Fe−C martensite as a function of composition [1.82]

Table 3.1-40 Metastable and stable carbide phases occurring in the Fe−C(−X) alloy system [1.82]

Phase	Structure	Type	a (nm)	b (nm)	c (nm)
Fe_4C	cub		0.3878		
Fe_3C	orth	Fe_3C	0.50889	0.67433	0.452353
$\varepsilon\text{-}Fe_3C$	hex		0.273		0.433
$Fe_{2-3}C$	hex		0.4767		0.4354
Fe_5C_2	mon	Mn_5C_2	1.1563	0.4573 $\beta = 97.73°$	0.5058
Fe_7C_3	hex	Th_7Fe_3	0.6882		0.4540
$Fe_{20}C_9$	orth		0.9061	1.5695	0.7937

Heat Treatments

The heat treatments referred to above need to be specified rather succinctly such that they can be correlated with the ensuing microstructures and properties. Furthermore, the specifications of heat treatments require taking the cross section and form of the part to be heat treated into account (at least if the cross sections get larger than, say, 0.5 mm). The finite thermal conductivity and the heat capacity of the material will cause any temperature change applied to the surface to occur at a decreasing rate with increasing depth in the heat-treated part. Thus, not only the time and temperature of an isothermal treatment but also the rate of cooling or the rate of heating are common parameters to be specified. Beyond those referred to above, the following treatments are widely applied to steels:

Austenitizing. Heating to and holding in the range of the austenite phase is commonly the first stage of transformation heat treatments. The higher the austenitizing temperature, the more lattice defects such as dislocations and grain boundaries are annihilated. This lowers the rate of nucleation of subsequent phase transformations.

Soft Annealing. This term is used for heat treatment of hardenable steels containing $\geq 0.4\,\text{wt\%}\,C$ at temperatures closely below the eutectoid temperature for a duration of $\leq 100\,\text{h}$. It results in a microstructure of coarse grained, ductile ferrite, and coarsened cementite.

Normalizing. This heat treatment is applied to obtain a uniform, fine-grained microstructure. The first step consists of heating the metal rapidly to, and holding it at a temperature $30–50\,\text{K}$ above the $(\alpha + \gamma)/\gamma$ phase boundary (also referred to as the A_{c3} line) for hypo-eutectoid steels, and heating rapidly to and holding at about $50\,\text{K}$ above the eutectoid temperature (also referred to as A_{c1} line). This step results in the formation of a fairly fine-grained austenite and ferrite structure in the hypo-eutectoid and in a fine-grained austenite with coagulated grain boundary cementite in the hyper-eutectoid compositions. Upon cooling, the austenite transforms into pearlite and this microstructural state has a favorable combination of strength, ductility, and machinability.

Substitutional Iron−Based Alloys

For the phase diagrams with substitutional alloying components shown in later sections, a major aspect pertaining to both the binary alloys shown and the steels alloyed with these components is whether the α or the γ phase of Fe is stabilized. i.e., which phase field is expanded or contracted upon alloying.

Fe−Ni. Figure 3.1-102 shows the Fe−Ni phase equilibria indicating that Ni stabilizes the fcc γ phase. If Fe-rich alloys are quenched from the γ phase field they transform martensitically to bcc α' martensite. The transformation temperatures are shown in Fig. 3.1-103. The Fe−Ni phase diagram is particularly relevant for the controlled thermal expansion and constant-modulus alloys as well as for the soft magnetic Fe−Ni based materials at higher Ni contents. These, in turn, derive their magnetic properties in part from the occurrence of the superlattice phase $FeNi_3$.

Nickel is added to Fe−C alloys to increase the hardenability and to increase the yield strength of ferrite by solid solution hardening.

Fe−Mn. Figure 3.1-104 shows the Fe−Mn phase equilibria indicating that Mn is stabilizing the fcc γ phase similar to Ni. It should be noted that quenching Fe-rich alloys from the γ-phase field leads to two different martensitic transformations which may result in a bcc structure (α' martensite) or an hcp structure (ε' martensite). The transformation temperatures are shown in Fig. 3.1-105. The martensitic transformation can also be induced by deformation. This property is exploited

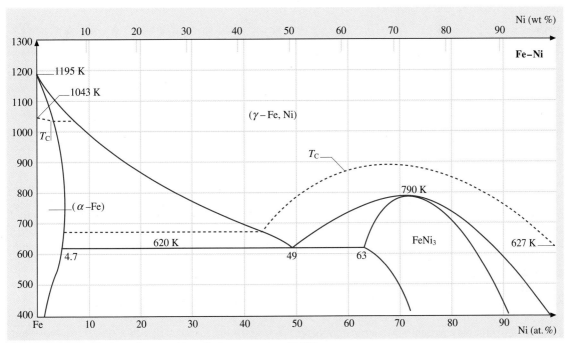

Fig. 3.1-102 Fe−Ni phase diagram. T_C − Curie temperature [1.82]

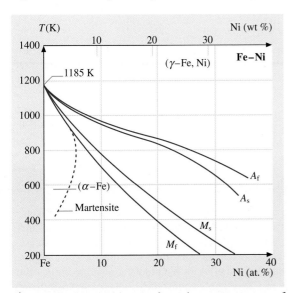

Fig. 3.1-103 Martensitic transformation temperatures of Fe-rich Fe−Ni alloys. The reverse transformation is characterized by the A_s (austenite start) and A_f (austenite finish) temperatures [1.82]

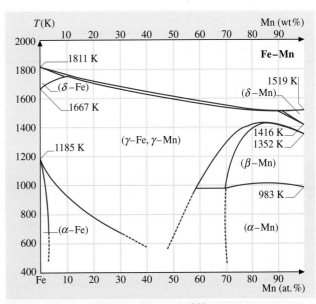

Fig. 3.1-104 Fe−Mn phase diagram [1.82]

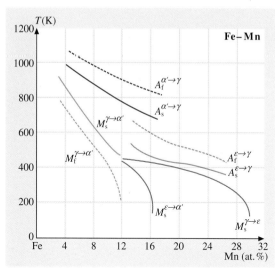

Fig. 3.1-105 Martensitic transformation temperatures of Fe-rich Fe–Mn alloys. The superscripts indicate the transforming phases [1.82]

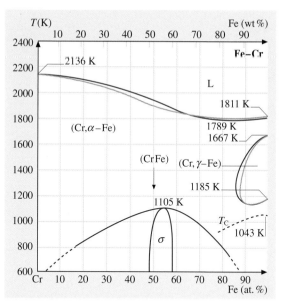

Fig. 3.1-106 Fe–Cr phase diagram [1.82]

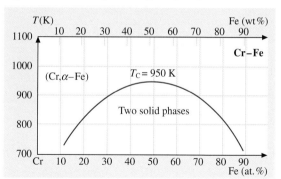

Fig. 3.1-107 Metastable miscibility gap in the Fe–Cr alloy system [1.82]

in the design of wear-resistant steels (Hatfield steel: 12 wt% Mn, 1 wt% C).

Manganese is contained in practically all commercial steels because it is used for deoxidation of the melt. Typical contents are 0.3–0.9 wt% Mn. Manganese increases the hardenability of steels and contributes moderately to the yield strength by solid solution hardening.

Fe–Cr. The Fe–Cr phase diagram, Fig. 3.1-106, is the prototype of the case of an iron-based system with an α-phase stabilizing component. Chromium is the most important alloying element of corrosion resistant, ferritic stainless steels and ferritic heat-resistant steels. If α-Fe–Cr alloys are quenched from above 1105 K and subsequently annealed, they decompose according to a metastable miscibility gap shown in Fig. 3.1-107. This decomposition reaction can cause severe embrittlement which is called "475 °C-embrittlement" in ferritic chromium steels. Embrittlement can also occur upon formation of the σ phase.

In carbon steels, Cr is added to increase corrosion and oxidation resistance because it promotes the formation of stable passivating and protective oxide layers. Moreover, Cr is a strong carbide former which modifies and delays the formation of pearlite and bainite, thus increasing the hardenability. In heat-resistant steels Cr contributes to the high-temperature yield strength.

Fe–Si. The phase diagram Fe–Si, Fig. 3.1-108, shows that Si is a strong ferrite former. The main application of binary Fe–Si alloys is in the form of steels with ≤ 3.5 wt% Si which have an optimum combination of high magnetic moment, low magnetostriction, and low magnetocrystalline anisotropy such that they are the ideal material for high induction and low magnetic power loss applications such as power transformers. Data are given in Sect. 4.3.2.3.

In Fe–C steelmaking, Si is one of the principal deoxidizers. It may amount to 0.05–0.3 wt% Si in the steel depending on the deoxidizing treatment and the amount of other deoxidants used. At these levels of concentration Si contributes only moderately to the strength of ferrite and causes no significant loss of ductility.

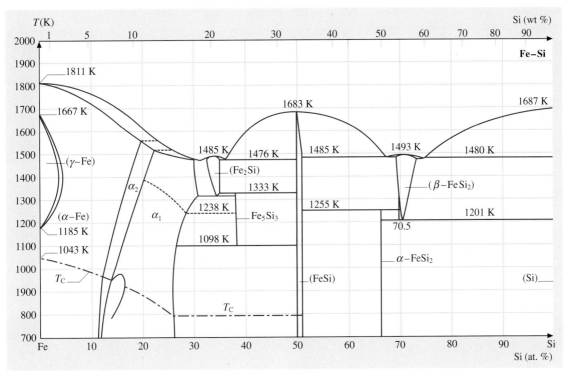

Fig. 3.1-108 Fe–Si phase diagram [1.82]

3.1.5.2 Carbon and Low-Alloy Steels

The largest group of steels produced both by number of variants and by volume is that of carbon and low-alloy steels. It is characterized by the fact that most of the phase relations and phase transformations may be referred to the binary Fe–C phase diagram or comparatively small deviations from it. These steels are treated extensively in [1.80].

Compositions and Properties of Carbon Steels

According to the effect of carbon concentration on the phases formed and on their properties, Fig. 3.1-109 shows the variation of the effective average mechanical properties of as-rolled 25-mm bars of plain carbon steels as an approximate survey of the typical concentration dependence.

Carbon steels are defined as containing up to 1 wt% C and a total of 2 wt% alloying elements. Apart from the deoxidizing alloying elements Mn and Si, two impurity elements are always present in carbon steels: phosphorous and sulfur. Phosphorus increases strength and hardness significantly by solid solution hardening, but severely decreases ductility and toughness. Only in exceptional cases may P be added deliberately to increase machinability and corrosion resistance. Sulfur has essentially no effect on the strength properties since it is practically insoluble in ferrite. However, it decreases the ductility and fracture toughness. But S is added deliberately along with an increased Mn content to promote the formation of MnS. This compound is formed in small particles which are comparatively soft and serve as effective chip breakers in free-cutting steel grades, thus increasing machinability. On the basis of these effects of the most common alloying and impurity elements, carbon steel compositions are specified as listed in Table 3.1-41 and free-cutting carbon steel compositions are specified as listed in Table 3.1-42.

A survey of the alloying elements used and of the ranges of composition applied in carbon and low-alloy steels may be gained from the SAE–AISI system of designations for carbon and alloy steels listed in Table 3.1-43. Extensive cross references to other standards may be found in [1.81].

Table 3.1-41 Standard carbon steel compositions applicable to semi-finished products for forging, hot-rolled and cold-finished bars, wire rods, and seamless tubing [1.80]. Selected grades

Designation		Cast or heat chemical ranges and limits[a] (wt%)			
UN number	SAE-AISI number	C	Mn	P_{max}	S_{max}
G10050	1005	0.06 max	0.35 max	0.040	0.050
G10100	1010	0.08–0.13	0.30–0.60	0.040	0.050
G10200	1020	0.18–0.23	0.30–0.60	0.040	0.050
G10300	1030	0.28–0.34	0.60–0.90	0.040	0.050
G10400	1040	0.37–0.44	0.60–0.90	0.040	0.050
G10500	1050	0.48–0.55	0.60–0.90	0.040	0.050
G10600	1060	0.55–0.65	0.60–0.90	0.040	0.050
G10700	1070	0.65–0.75	0.60–0.90	0.040	0.050
G10800	1080	0.75–0.88	0.60–0.90	0.040	0.050
G10900	1090	0.85–0.98	0.60–0.90	0.040	0.050

[a] When silicon ranges or limits are required for bar and semifinished products, the following ranges are commonly used: 0.10% max; 0.10 to 0.20%; 0.15 to 0.35%; 0.20 to 0.40%; or 0.30 to 0.60%. For rods the following ranges are commonly used: 0.10 max; 0.07–0.15%; 0.10–0.20%; 0.15–0.35%; 0.20–0.40%; and 0.30–0.60%. Steels listed in this table can be produced with additions of leaf or boron. Leaded steels typically contain 0.15–0.40% Pb and are identified by inserting the letter L in the designation (10L45); boron steels can be expected to contain 0.0005–0.003% B and are identified by inserting the letter B in the desingnation (10B46)

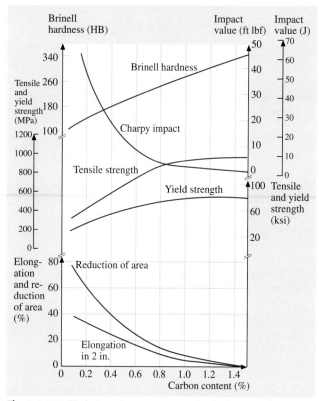

Fig. 3.1-109 Variations in average mechanical properties of as-rolled 25-mm-diam. bars of plain carbon steels as a function of carbon content [1.80]

Table 3.1-42 Standard free-cutting (re-sulfurized) carbon steel compositions applicable to semi-finished products for forging, hot-rolled and cold-finished bars, and seamless tubing [1.80]

Designation UN number	SAE-AISI number	Cast or heat chemical ranges and limits[a] (wt%)			
		C	Mn	P_{max}	S
G11080	1108	0.08–0.13	0.85–0.98	0.040	0.08–0.13
G11100	1110	0.08–0.13	0.30–0.60	0.040	0.08–0.13
G11170	1117	0.14–0.20	1.00–1.30	0.040	0.08–0.13
G11180	1118	0.14–0.20	1.30–1.60	0.040	0.08–0.13
G11370	1137	0.32–0.39	1.35–1.65	0.040	0.08–0.13
G11390	1139	0.35–0.43	1.35–1.65	0.040	0.13–0.20
G11400	1140	0.37–0.44	0.70–1.00	0.040	0.08–0.13
G11410	1141	0.37–0.45	1.35–1.65	0.040	0.08–0.13
G11440	1144	0.40–0.48	1.35–1.65	0.040	0.24–0.33
G11460	1146	0.42–0.49	0.70–1.00	0.040	0.08–0.13
G11510	1151	0.48–0.55	0.70–1.00	0.040	0.08–0.13

[a] When lead ranges or limits are required or when silicon or limits are required for bars or semifinished products, the values in Table 5 apply. For rods, the following ranges and limits for silicon are commonly used: up to SAE 1110 inclusive, 0.10% max; SAE 1117 and over, 0.10%, 0.10–0.20%, or 0.15–0.35%

Table 3.1-43 SAE–AISI system of designations for carbon and low-alloy steels [1.80]

Numerals and digits	Type of steel and nominal alloy content (wt%)
Carbon steels	
10xx[a]	Plain carbon (Mn 100 max)
11xx	Resulfurized
12xx	Resulfurized and rephosphorized
15xx	Plain carbon (max Mn range: 1.00–1.65)
Manganese steels	
13xx	Mn 1.75
Nickel steels	
23xx	Ni 3.50
25xx	Ni 5.00
Nickel–chromium steels	
31xx	Ni 1.25; Cr 0.65 and 0.80
32xx	Ni 1.75; Cr 1.07
33xx	Ni 3.50; Cr 1.50 and 1.57
34xx	Ni 3.00; Cr 0.77
Molybdenum steels	
40xx	Mo 0.20 and 0.25
44xx	Mo 0.40 and 0.52
Chromium–molybdenum steels	
41xx	Cr 0.50, 0.80, and 0.95; Mo 0.12, 0.20, 0.25, and 0.30
Nickel–chromium–molybdenum steels	
43xx	Ni 1.82; Cr 0.50 and 0.80; Mo 0.25
43BVxx	Ni 1.82; Cr 0.50; Mo 0.12 and 0.25; V 0.03 min
47xx	Ni 1.05; Cr 0.45; Mo 0.20 and 0.35
81xx	Ni 0.30; Cr 0.40; Mo 0.12
86xx	Ni 0.55; Cr 0.50; Mo 0.20

Table 3.1-43 SAE–AISI system of designations for carbon and low-alloy steels [1.80], cont.

Numerals and digits	Type of steel and nominal alloy content (wt%)
87xx	Ni 0.55; Cr 0.50; Mo 0.25
88xx	Ni 0.55; Cr 0.50; Mo 0.35
93xx	Ni 3.25; Cr 1.20; Mo 0.12
94xx	Ni 0.45; Cr 0.40; Mo 0.12
97xx	Ni 0.55; Cr 0.20; Mo 0.20
98xx	Ni 1.00; Cr 0.80; Mo 0.25
Nickel–molybdenum steels	
46xx	Ni 0.85 and 1.82; Mo 0.20 and 0.25
48xx	Ni 3.50; Mo 0.25
Chromium steels	
50xx	Cr 0.27, 0.40, 0.50, and 0.65
51xx	Cr 0.80, 0.87, 0.92, 0.95, 1.00, and 1.05
50xxx	Cr 0.50; C 1.00 min
51xxx	Cr 1.02; C 1.00 min
52xxx	Cr 1.45; C 1.00 min
Chromium–vanadium steels	
61xx	Cr 0.60, 0.80, and 0.95; V 0.10 and 0.15 min
Tungsten–chromium steels	
72xx	W 1.75; Cr 0.75
Silicon–manganese steels	
92xx	Si 1.40 and 2.00; Mn 0.65, 0.82, and 0.85; Cr 0 and 0.65
Boron steels	
xxBxx	B denotes boron steel
Leaded steels	
xxLxx	L denotes leaded steel
Vanadium steels	
xxVxx	V denotes vanadium steel

[a] The xx in the last two digits of these designations indicates that the carbon content (in hundredths of a weight percent) is to be inserted

Turning to the mechanical properties, it should be emphasized that the microstructure has a decisive influence on the properties of all steels. Therefore the composition and the prior thermal, mechanical, or thermomechanical treatments which determine the phase transformations and ensuing microstructural state of a steel will always have to be taken into account. Accordingly, tabulated property data will invariably be given with reference to mechanical and thermal treatments applied. The terms used and their specific definitions are outlined in Sect. 3.1.5.1.

In plain carbon steels the C content and microstructure are determining the mechanical properties. Manganese is providing moderate solid solution strengthening and increases the hardenability. The properties of plain carbon steels are also affected by the other common residual elements Si, P, and S. Furthermore, the gasses O, N, and H and their reaction products may play a role. Their content depends largely on the melting, deoxidizing and pouring practice. While Fig. 3.1-109 illustrates the general effect of C content on the mechanical properties if the austenite grain size and transformation microstructure are held essentially constant. Tables 3.1-44 and 3.1-45 list the mechanical properties of representative carbon and low alloy steels in specified states as a function of deformation and heat treatment.

Table 3.1-44 Mechanical properties of selected carbon and low-alloy steels in the hot-rolled, normalized, and annealed conditions [1.80]

AISI No.[a]	Treatment	Austenitizing temperature (°C)	Tensile strength (MPa)	Yield strength (MPa)	Elongation (%)	Reduction in area (%)	Hardness (HB)
1015	As-rolled	–	420.6	313.7	39.0	61.0	126
	Normalized	925	424.0	324.1	37.0	69.6	121
	Annealed	870	386.1	284.4	37.0	69.7	111
1020	As-rolled	–	448.2	330.9	36.0	59.0	143
	Normalized	870	441.3	346.5	35.8	67.9	131
	Annealed	870	394.7	294.8	36.5	66.0	111
1022	As-rolled	–	503.3	358.5	35.0	67.0	149
	Normalized	925	482.6	358.5	34.0	67.5	143
	Annealed	870	429.2	317.2	35.0	63.6	137
1030	As-rolled	–	551.6	344.7	32.0	57.0	179
	Normalized	925	520.6	344.7	32.0	60.8	149
	Annealed	845	463.7	341.3	31.2	57.9	126
1040	As-rolled	–	620.5	413.7	25.0	50.0	201
	Normalized	900	589.5	374.0	28.0	54.9	170
	Annealed	790	518.8	353.4	30.2	57.2	149
1050	As-rolled	–	723.9	413.7	20.0	40.0	229
	Normalized	900	748.1	427.5	20.0	39.4	217
	Annealed	790	636.0	365.4	23.7	39.9	187
1060	As-rolled	–	813.6	482.6	17.0	34.0	241
	Normalized	900	775.7	420.6	18.0	37.2	229
	Annealed	790	625.7	372.3	22.5	38.2	179
1080	As-rolled	–	965.3	586.1	12.0	17.0	293
	Normalized	900	1010.1	524.0	11.0	20.6	293
	Annealed	790	615.4	375.8	24.7	45.0	174
1095	As-rolled	–	965.3	572.3	9.0	18.0	293
	Normalized	900	1013.5	499.9	9.5	13.5	293
	Annealed	790	656.7	379.2	13.0	20.6	192
1117	As-rolled	–	486.8	305.4	33.0	63.0	143
	Normalized	900	467.1	303.4	33.5	63.8	137
	Annealed	855	429.5	279.2	32.8	58.0	121
1118	As-rolled	–	521.2	316.5	32.0	70.0	149
	Normalized	925	477.8	319.2	33.5	65.9	143
	Annealed	790	450.2	284.8	34.5	66.8	131
1137	As-rolled	–	627.4	379.2	28.0	61.0	192
	Normalized	900	668.8	396.4	22.5	48.5	197
	Annealed	790	584.7	344.7	26.8	53.9	174
1141	As-rolled	–	675.7	358.5	22.0	38.0	192
	Normalized	900	706.7	405.4	22.7	55.5	201
	Annealed	815	598.5	353.0	25.5	49.3	163
1144	As-rolled	–	703.3	420.6	21.0	41.0	212
	Normalized	900	667.4	399.9	21.0	40.4	197
	Annealed	790	584.7	346.8	24.8	41.3	167
1340	Normalized	870	836.3	558.5	22.0	62.9	248
	Annealed	800	703.3	436.4	25.5	57.3	207
3140	Normalized	870	891.5	599.8	19.7	57.3	262
	Annealed	815	689.5	422.6	24.5	50.8	197

Table 3.1-44 Mechanical properties of selected carbon and low-alloy steels [1.80], cont.

AISI No.[a]	Treatment	Austenitizing temperature (°C)	Tensile strength (MPa)	Yield strength (MPa)	Elongation (%)	Reduction in area (%)	Hardness (HB)
4130	Normalized	870	668.8	436.4	25.5	59.5	197
	Annealed	865	560.5	360.6	28.2	55.6	156
4140	Normalized	870	1020.4	655.0	17.7	46.8	302
	Annealed	815	655.0	417.1	25.7	56.9	197
4150	Normalized	870	1154.9	734.3	11.7	30.8	321
	Annealed	815	729.5	379.2	20.2	40.2	197
4320	Normalized	895	792.9	464.0	20.8	50.7	235
	Annealed	850	579.2	609.5	29.0	58.4	163
4340	Normalized	870	1279.0	861.8	12.2	36.3	363
	Annealed	810	744.6	472.3	22.0	49.9	217
4620	Normalized	900	574.3	366.1	29.0	66.7	174
	Annealed	855	512.3	372.3	31.3	60.3	149
4820	Normalized	860	75.0	484.7	24.0	59.2	229
	Annealed	815	681.2	464.0	22.3	58.8	197
5140	Normalized	870	792.9	472.3	22.7	59.2	229
	Annealed	830	572.3	293.0	28.6	57.3	167
5150	Normalized	870	870.8	529.0	20.7	58.7	255
	Annealed	825	675.7	357.1	22.0	43.7	197
5160	Normalized	855	957.0	530.9	17.5	44.8	269
	Annealed	815	722.6	275.8	17.2	30.6	197
6150	Normalized	815	722.6	275.8	17.2	30.6	197
	Annealed	815	667.4	412.3	23.0	48.4	197
8620	Normalized	915	632.9	357.1	26.3	59.7	183
	Annealed	870	536.4	385.4	31.3	62.1	149
8630	Normalized	870	650.2	429.5	23.5	53.5	187
	Annealed	845	564.0	372.3	29.0	58.9	156
8650	Normalized	870	1023.9	688.1	14.0	40.4	302
	Annealed	795	715.7	386.1	22.5	46.4	212
8740	Normalized	870	929.4	606.7	16.0	47.9	269
	Annealed	815	695.0	415.8	22.2	46.4	201
9255	Normalized	900	932.9	579.2	19.7	43.4	269
	Annealed	845	774.3	486.1	21.7	41.1	229
9310	Normalized	890	906.7	570.9	18.8	58.1	269
	Annealed	845	820.5	439.9	17.3	42.1	241

[a] All grades are fine-grained except for those in the 1100 series, which are coarse-grained. Heat-treated specimens were oil quenched unless otherwise indicated

Table 3.1-45 Mechanical properties of selected carbon and alloy steels in the quenched-and-tempered condition, heat treated as 25 mm rounds [1.80]

AISI No.[a]	Tempering temperature (°C)	Tensile strength (MPa)	Yield strength (MPa)	Elongation (%)	Reduction in area (%)	Hardness HB
1030[b]	205	848	648	17	47	495
	315	800	621	19	53	401
	425	731	579	23	60	302
	540	669	517	28	65	255
	650	586	441	32	70	207
1040[b]	205	896	662	16	45	514
	315	889	648	18	52	444
	425	841	634	21	57	352
	540	779	593	23	61	269
	650	669	496	28	68	201
1040	205	779	593	19	48	262
	315	779	593	20	53	255
	425	758	552	21	54	241
	540	717	490	26	57	212
	650	634	434	29	65	192
1050[b]	205	1124	807	9	27	514
	315	1089	793	13	36	444
	425	1000	758	19	48	375
	540	862	655	23	58	293
	650	717	538	28	65	235
1050	205	–	–	–	–	–
	315	979	724	14	47	321
	425	938	655	20	50	277
	540	876	579	23	53	262
	650	738	469	29	60	223
1060	205	1103	779	13	40	321
	315	1103	779	13	40	321
	425	1076	765	14	41	311
	540	965	669	17	45	277
	650	800	524	23	54	229
1080	205	1310	979	12	35	388
	315	1303	979	12	35	388
	425	1289	951	13	36	375
	540	1131	807	16	40	321
	650	889	600	21	50	255
1095[b]	205	1489	1048	10	31	601
	315	1462	1034	11	33	534
	425	1372	958	13	35	388
	540	1138	758	15	40	293
	650	841	586	20	47	235
1095	205	1289	827	10	30	401
	315	1262	813	10	30	375
	425	1213	772	12	32	363
	540	1089	676	15	37	321
	650	896	552	21	47	269

Table 3.1-45 Mechanical properties of selected carbon and alloy steels, cont.

AISI No.[a]	Tempering temperature (°C)	Tensile strength (MPa)	Yield strength (MPa)	Elongation (%)	Reduction in area (%)	Hardness (HB)
1137	205	1082	938	5	22	352
	315	986	841	10	33	285
	425	876	731	15	48	262
	540	758	607	24	62	229
	650	655	483	28	69	197
1137[b]	205	1496	1165	5	17	415
	315	1372	1124	9	25	375
	425	1103	986	14	40	311
	540	827	724	19	60	262
	650	648	531	25	69	187
1141	205	1634	1213	6	17	461
	315	1462	1282	9	32	415
	425	1165	1034	12	47	331
	540	896	765	18	57	262
	650	710	593	23	62	217
1144	205	876	627	17	36	277
	315	869	621	17	40	262
	425	848	607	18	42	248
	540	807	572	20	46	235
	650	724	503	23	55	217
1330[b]	205	1600	1455	9	39	459
	315	1427	1282	9	44	402
	425	1158	1034	15	53	335
	540	876	772	18	60	263
	650	731	572	23	63	216
1340	205	1806	1593	11	35	505
	315	1586	1420	12	43	453
	425	1262	1151	14	51	375
	540	965	827	14	58	295
	650	800	621	–	66	252
4037	205	1027	758	6	38	310
	315	951	765	14	53	295
	425	876	731	20	60	270
	540	793	655	23	63	247
	650	696	421	29	60	220
4042	205	1800	1662	12	37	516
	315	1613	1455	13	42	455
	425	1289	1172	15	51	380
	540	986	883	20	59	300
	650	793	689	28	66	238
4130[b]	205	1627	1462	10	41	467
	315	1496	1379	11	43	435
	425	1282	1193	13	49	380
	540	1034	910	17	57	315
	650	814	703	22	64	245

Table 3.1-45 Mechanical properties of selected carbon and alloy steels, cont.

AISI No.[a]	Tempering temperature (°C)	Tensile strength (MPa)	Yield strength (MPa)	Elongation (%)	Reduction in area (%)	Hardness (HB)
4140	205	1772	1641	8	38	510
	315	1551	1434	9	43	445
	425	1248	1138	13	49	370
	540	951	834	18	58	258
	650	758	655	22	63	230
4150	205	1931	1724	10	39	530
	315	1765	1593	10	40	495
	425	1517	1379	12	45	440
	540	1207	1103	15	52	370
	650	958	841	19	60	290
4340	205	1875	1675	10	38	520
	315	1724	1586	10	40	486
	425	1469	1365	10	44	430
	540	1172	1076	13	51	360
	650	965	855	19	60	280
5046	205	1744	1407	9	25	482
	315	1413	1158	10	37	401
	425	1138	931	13	50	336
	540	938	765	18	61	282
	650	786	655	24	66	235
50B46	205	–	–	–	–	560
	315	1779	1620	10	37	505
	425	1393	1248	13	47	405
	540	1082	979	17	51	322
	650	883	793	22	60	273
50B60	205	–	–	–	–	600
	315	1882	1772	8	32	525
	425	1510	1386	11	34	435
	540	1124	1000	15	38	350
	650	896	779	19	50	290
5130	205	1613	1517	10	40	475
	315	1496	1407	10	46	440
	425	1275	1207	12	51	379
	540	1034	938	15	56	305
	650	793	689	20	63	245
5140	205	1793	1641	9	38	490
	315	1579	1448	10	43	450
	425	1310	1172	13	50	365
	540	1000	862	17	58	280
	650	758	662	25	66	235
5150	205	1944	1731	5	37	525
	315	1737	1586	6	40	475
	425	1448	1310	9	47	410
	540	1124	1034	15	54	340
	650	807	814	20	60	270

Table 3.1-45 Mechanical properties of selected carbon and alloy steels, cont.

AISI No.[a]	Tempering temperature (°C)	Tensile strength (MPa)	Yield strength (MPa)	Elongation (%)	Reduction in area (%)	Hardness (HB)
5160	205	2220	1793	4	10	627
	315	1999	1772	9	30	555
	425	1606	1462	10	37	461
	540	1165	1041	12	47	341
	650	896	800	20	56	269
51B60	205	–	–	–	–	600
	315	–	–	–	–	540
	425	1634	1489	11	36	460
	540	1207	1103	15	44	355
	650	965	869	20	47	290
6150	205	1931	1689	8	38	538
	315	1724	1572	8	39	483
	425	1434	1331	10	43	420
	540	1158	1069	13	50	345
	650	945	841	17	58	282
81B45	205	2034	1724	10	33	550
	315	1765	1572	8	42	475
	425	1407	1310	11	48	405
	540	1103	1027	16	53	338
	650	896	793	20	55	280
8630	205	1641	1503	9	38	465
	315	1482	1392	10	42	430
	425	1276	1172	13	47	375
	540	1034	896	17	54	310
	650	772	689	23	63	240
8640	205	1862	1669	10	40	505
	315	1655	1517	10	41	460
	425	1379	1296	12	45	400
	540	1103	1034	16	54	340
	650	896	800	20	62	280
86B45	205	1979	1641	9	31	525
	315	1696	1551	9	40	475
	425	1379	1317	11	41	395
	540	1103	1034	15	49	335
	650	903	876	19	58	280
8650	205	1937	1675	10	38	525
	315	1724	1551	10	40	490
	425	1448	1324	12	45	420
	540	1172	1055	15	51	340
	650	965	827	20	58	280
8660	205	–	–	–	–	580
	315	–	–	–	–	535
	425	1634	1551	13	37	460
	540	1310	1213	17	46	370
	650	1068	951	20	53	315

Table 3.1-45 Mechanical properties of selected carbon and alloy steels, cont.

AISI No.[a]	Tempering temperature (°C)	Tensile strength (MPa)	Yield strength (MPa)	Elongation (%)	Reduction in area (%)	Hardness (HB)
8740	205	1999	1655	10	41	578
	315	1717	1551	11	46	495
	425	1434	1358	13	50	415
	540	1207	1138	15	55	363
	650	986	903	20	60	302
9255	205	2103	2048	1	3	601
	315	1937	1793	4	10	578
	425	1606	1489	8	22	477
	540	1255	1103	15	32	352
	650	993	814	20	42	285
9260	205	–	–	–	–	600
	315	–	–	–	–	540
	425	1758	1503	8	24	470
	540	1324	1131	12	30	390
	650	979	814	20	43	295
94B30	205	1724	1551	12	46	475
	315	1600	1420	12	49	445
	425	1344	1207	13	57	382
	540	1000	931	16	65	307
	650	827	724	21	69	250

[a] All grades are fine-grained except for those in the 1100 series, which are coarse-grained. Heat-treated specimens were oil quenched unless otherwise indicated
[b] Water quenched

Hardenability

Hardening of steels is the heat treatment consisting of heating to the range of austenite, cooling in water, oil, or air, and subsequent tempering. The term hardenability refers to the suitability of a steel to be transformed partially or completely from austenite to martensite to a specified depth below the free surface of a workpiece when cooled under specified conditions. This definition reflects that the term hardenability does not only refer to the magnitude of hardness which can be attained for a particular steel, but it relates the extent of hardening achievable to the macroscopic or local cooling rate or isothermal holding time in the transformation range and, thus, to the mechanisms, kinetic phase transformations, and their effects on the mechanical properties.

The amount of martensite formed upon cooling is a function of C content and total steel composition. This behavior is due to two decisive factors of influence:

- The temperature range of transformation of austenite to martensite depends on the C content, as shown in Fig. 3.1-101 of Sect. 3.1.5.1, and
- The diffusion-dependent transformations of austenite (to ferrite, pearlite, and bainite) compete with the martensitic transformation such that the volume fraction available for the latter will decrease as the volume transformed by the former increases. This transformation kinetics of the diffusional phase transformations is strongly dependent on alloy composition.

The lower the cooling rate may be while still permitting one to obtain a high fraction of martensite, the higher the hardenability of steel. This can best be tested by varying the cooling rate and analyzing the resulting microstructure and its hardness (and other mechanical properties). The method used most widely is an ingeniously simple testing procedure, the Jominy end-quench test illustrated in Fig. 3.1-110. The mater-

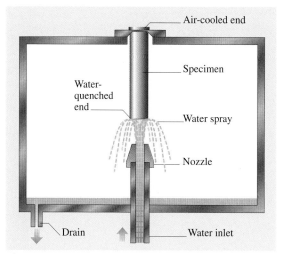

Fig. 3.1-110 Jominy end-quench apparatus

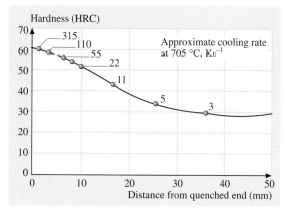

Fig. 3.1-111 Plot of end-quench hardenability data of an AISI 8650 steel (0.49 wt% C, 0.98 wt% Mn, 0.29 wt% Si, 0.59 wt% Ni, 0.47 wt% Cr, 0.19 wt% Mo) [1.80]

ial to be tested is normalized and a test bar 100 mm long and 25 mm in diameter is machined. This specimen is austenitized and transferred to the fixture (shown in Fig. 3.1-110) holding it vertically 13 mm above the nozzle through which water is directed against the bottom face of the specimen. While the bottom end is quenched by water, the top end is slowly cooled in air and intermediate cooling rates occur at intermediate positions. After the test, hardness readings and microstructural analyses may be taken along the bar. They can be correlated to the pre-determined approximate cooling rate at a given temperature, as shown for a characteristic example in Fig. 3.1-111.

The hardenability of numerous commonly used steels has been characterised quantitatively by extensive investigations of their transformation behavior as a function of temperature and time. These investigations have been carried out in two modes of heat treatment:

- Quenching from the austenite range to a temperature in the transformation range at which the specimen is held isothermally in a salt or a lead bath for different times, and finally quenched to room temperature to be investigated regarding the transformation products by microstructural and supplementary measurements. The resulting plots of the beginning of formation of the different transformation products (sometimes including the fractional amounts and end of transformation) are termed time-temperature-transformation (TTT) or isothermal transformation (IT) diagrams. An example is shown as Fig. 3.1-112.
- Cooling from the austenite range through the transformation range at different cooling rates, and investigating the temperature of onset of transformation of the different products for different cooling rates by microstructural and supplementary measurements, e.g., dilatometry. The resulting plot is called a continuous-cooling-transformation (CCT) diagram. An example is shown in Fig. 3.1-113. The critical cooling rate means the lowest rate for which a fully martensitic state can be obtained.

Apart from empirical determinations of these transformation diagrams, methods of prediction based on nucleation theory and phenomenological growth theory using the Johnson–Mehl–Avrami equation have been devised to estimate TTT diagrams [1.83].

The hardenability increases with increasing carbon and metallic alloy element concentration (with the exception of Co). The transformation kinetics and ensuing hardenability properties are documented extensively in compilations of TTT and CCT diagrams such as [1.84].

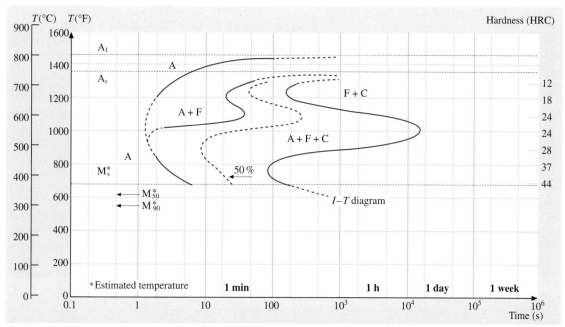

Fig. 3.1-112 Time-temperature-transformation (TTT) diagram of a 4130 grade low-alloy steel. A – austenite; F – ferrite; C – cementite; M – Martensite (the suffix indicates the amount of martensite formed in vol.%)

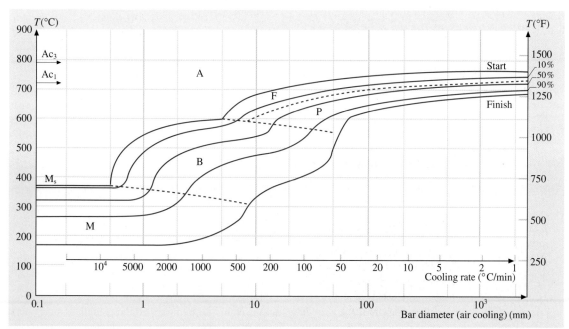

Fig. 3.1-113 Continuous-cooling-transformation (CCT) diagram for a 4130 grade low-alloy steel. Ac_3 and Ac_1 signify the temperatures of the $\gamma/(\gamma + \alpha)$ and eutectoid reation, respectively. A – austenite, F – ferrite, B – bainite, P – pearlite, M – martensite. The cooling rate is measured at 705 °C. The calculated critical cooling rate is 143 K/s [1.80]

3.1.5.3 High-Strength Low-Alloy Steels

High-strength low-alloy (HSLA) steels are designed to provide higher mechanical property values and/or higher resistance to atmospheric corrosion than conventional low-alloy steels of comparable level of alloy content. Higher yield stress is achieved by adding ≤ 0.1 wt% N, Nb, V, Ti, and/or Zr (micro-alloying) which form carbide or carbonitrite precipitates, and by special, closely-controlled processing which yields mostly fine-grained microstructures.

HSLA steels contain 0.05 to 0.25 wt% C, ≤ 2 wt% Mn and mainly Cr, Ni, Mo, and Cu as further alloying elements. Their yield stress is in the range ≥ 275 MPa. They are primarily hot-rolled into usual wrought product forms and commonly delivered in the as-rolled condition.

The particular processing methods of HSLA steels include [1.80]:

- *Controlled rolling* of micro-alloyed, precipitation hardening variants to obtain fine equiaxial and/or highly deformed, pancake-shaped austenite grains. During cooling these austenite grains transform into fine ferrite grains, providing an optimum combination of high yield strength and ductility.
- *Accelerated cooling* of controlled-rolled steels to enhance the formation of fine ferrite grains.
- *Quenching of steels containing* ≤ 0.08 wt% C such that acicular ferrite or low-carbon bainite is formed. This microstructural state provides an excellent combination of high yield strengths of 275 to 690 MPa, ductility, formability, and weldability.
- *Normalizing of V-alloyed steel*, thus increasing yield strength and ductility.
- *Intercritical annealing*, i. e., annealing in the $\gamma + \alpha$ phase field to obtain a dual-phase microstructure which, after cooling, consists of martensite islands dispersed in a ferrite matrix. This microstructure exhibits a somewhat lower yield strength but a high rate of work-hardening, providing a better combination of tensile strength, ductility, and formability than conventional HSLA steels.

HSLA steels include numerous standard and proprietary grades designed to provide specific desirable combinations of properties such as strength, ductility, weldability, and atmospheric corrosion resistance. Table 3.1-46 lists characteristic compositions and Table 3.1-47 lists mechanical properties of these characteristic variants.

In view of the multitude of compositional and processing variants of HSLA steels it is useful to have a summarizing overview as provided in Table 3.1-48.

3.1.5.4 Stainless Steels

Stainless steels are treated extensively in [1.79]. Compared to carbon or low-alloy steels, they are characterized by an increased resistance against corrosion in aggressive media. The corrosion resistance is achieved basically by an alloy content of at least $11-12$ wt% Cr. This content is required to form a dense, pore-free protective surface layer consisting mainly of chromium oxides and hydroxides. The corrosion resistance can be further increased by additional alloying with elements such as Ni, Mo, W, Mn, Si, Cu, Co, Al, or N.

Since chromium has a high affinity to carbon, the formation of chromium carbides may reduce the local concentration of Cr in solution and thus deteriorate the corrosion resistance. This can be avoided by

- Low carbon content of the steel.
- Suitable heat treatment.
- Bonding of carbon by other elements with higher carbon affinity, such as Ti and Nb, so-called stabilization.

Similar effects can be attained by the formation of chromium nitrides. Thus, in addition to the chemical composition of the steel, its corrosion properties are strongly influenced by its heat treatment condition.

Depending on the intended field of application, the corrosion resistance of the steel must often be combined with other useful properties such as high strength or hardness, high temperature strength, good formability, low temperature fracture toughness, weldability or machinability. However, since optimization of one property is generally only possible at the expense of others, the property spectrum of a stainless steel is often the result of a compromise. Consequently, a large number of steel grades has been developed to meet different property requirements. Some steels have been developed just for a single application. The following section gives typical representatives of the various types of stainless steel grades.

Depending on alloy composition and cooling conditions from elevated temperature, stainless steels may occur in different types of microstructure: ferritic (bcc), austenitic (fcc), martensitic, or mixtures of two or all three of these phases. The bcc structure is promoted by the ferrite forming elements Cr, Mo, W, Ti, V, Nb, Al, and Si, whereas the fcc structure is promoted by the austenite

Table 3.1-46 Compositional limits of HSLA steel grades according to ASTM standards [1.80]

ASTM specification[a]	Type or grade	UNS designation	Heat compositional limits (wt%)[b]									
			C	Mn	P	S	Si	Cr	Ni	Cu	V	Other
A 242	Type 1	K11510	0.15	1.00	0.45	0.05	–	–	–	0.20 min	–	–
A 572	Grade 42	–	0.21	1.35[c]	0.04	0.05	0.30[c]	–	–	0.20 min[d]	–	e
	Grade 50	–	0.23	1.35[c]	0.04	0.05	0.30[c]	–	–	0.20 min[d]	–	e
	Grade 60	–	0.26	1.35[c]	0.04	0.05	0.30	–	–	0.20 min[d]	–	e
	Grade 65	–	0.23[c]	1.65[c]	0.04	0.05	0.30	–	–	0.20 min[d]	–	e
A 588	Grade A	K11430	0.10–0.19	0.90–1.25	0.04	0.05	0.15–0.30	0.40–0.65	–	0.25–0.40	0.02–0.10	–
	Grade B	K12043	0.20	0.75–1.25	0.04	0.05	0.15–0.30	0.40–0.70	0.25–0.50	0.20–0.40	0.01–0.10	–
	Grade C	K11538	0.15	0.80–1.35	0.04	0.05	0.15–0.30	0.30–0.50	0.25–0.50	0.20–0.50	0.01–0.10	–
	Grade D	K11552	0.10–0.20	0.75–1.25	0.04	0.05	0.50–0.90	0.50–0.90	–	0.30	–	0.04 Nb, 0.05–0.35 Zr
	Grade K	–	0.17	0.50–1.20	0.04	0.05	0.25–0.50	0.40–0.70	0.40	0.30–0.50	–	0.10 Mo, 0.005–0.05 Nb
A 606	–	–	0.22	1.25	–	0.05	–	–	–	–	–	e
A 607	Grade 45	–	0.22	1.35	0.04	0.05	–	–	–	0.20 min[d]	–	e
	Grade 50	–	0.23	1.35	0.04	0.05	–	–	–	0.20 min[d]	–	e
	Grade 55	–	0.25	1.35	0.04	0.05	–	–	–	0.20 min[d]	–	e
	Grade 60	–	0.26	1.50	0.04	0.05	–	–	–	0.20 min[d]	–	e
	Grade 65	–	0.26	1.50	0.04	0.05	–	–	–	0.20 min[d]	–	e
	Grade 70	–	0.26	1.65	0.04	0.05	–	–	–	0.20 min[d]	–	e
A 618	Grade Ia	–	0.15	1.00	0.15	0.05	–	–	–	0.20 min	–	–
	Grade Ib	–	0.20	1.35	0.04	0.05	–	–	–	0.20 min[f]	–	–
	Grade II	K12609	0.22	0.85–1.25	0.04	0.05	0.30	–	–	–	0.02 min	–
	Grade III	K12700	0.23	1.35	0.04	0.05	0.30	–	–	–	0.02 min	0.005 Nb min[g]
A 633	Grade A	K01802	0.18	1.00–1.35	0.04	0.05	0.15–0.30	–	–	–	–	0.05 Nb
	Grade C	K12000	0.20	1.15–1.50	0.04	0.05	0.15–0.50	–	–	–	–	0.05–0.05 Nb
	Grade D	K02003	0.20	0.70–1.60[c]	0.04	0.05	0.15–0.50	0.25	0.25	0.35	–	0.08 Mo
	Grade E	K12202	0.22	1.15–1.50	0.04	0.05	0.15–0.50	–	–	–	0.04–0.11	0.01–0.05 Nb[d], 0.01–0.03 N
A 656	Type 3	–	0.18	1.65	0.025	0.035	0.60	–	–	–	0.08	0.020 N, 0.005–0.15 Nb
	Type 7	–	0.18	1.65	0.025	0.035	0.60	–	–	–	0.005–0.15	0.020 N, 0.005–0.10 Nb

Table 3.1-46 Compositional limits of HSLA steel grades according to ASTM standards [1.80], cont.

ASTM specification[a]	Type or grade	UNS designation	Heat compositional limits (wt%)[b]								
			C	Mn	P	S	Si	Cr	Ni	Cu	V	Other
A 690	–	K12249	0.22	0.60–0.90	0.08–0.15	0.05	0.10	–	0.40–0.75	0.50 min	–	–
A 709	Grade 50, type 1	–	0.23	1.35	0.04	0.05	0.40	–	–	–	–	0.005–0.05 Nb
	Grade 50, type 2	–	0.23	1.35	0.04	0.05	0.40	–	–	–	0.01–0.15	–
	Grade 50, type 3	–	0.23	1.35	0.04	0.05	0.40	–	–	–	[h]	0.05 Nb max
	Grade 50, type 4	–	0.23	1.35	0.04	0.05	0.40	–	–	–	[i]	0.015 Nb max
A 715	–	–	0.15	1.65	0.025	0.035	–	–	–	–	Added as necessary	Ti, Nb added as necessary
A 808	–	–	0.12	1.65	0.04	0.05 max or 0.010 max	0.15–0.50	–	–	–	0.10	0.02–0.10 Nb, V + Nb = 0.15 max
A 812	65	–	0.23	1.40	0.035	0.04	0.15–0.50[j]	–	–	–	V + Nb = 0.02–0.15	0.05 Nb max
	80	–	0.23	1.50	0.035	0.04	0.15–0.50	0.35	–	–	V + Nb = 0.02–0.15	0.05 Nb max
A 841	–	–	0.20	[k]	0.030	0.030	0.15–0.50	0.25	0.25	0.35	0.06	0.08 Mo, 0.03 Nb, 0.02 Al total
A 871	–	–	0.20	1.50	0.04	0.05	0.90	0.90	1.25	1.00	0.10	0.25 Mo, 0.15 Zr, 0.05 Nb, 0.05 Ti

[a] For characteristics and intended uses, see Table 3.1-48; for mechanical properties, see Table 3.1-47.
[b] If a single value is shown, it is a maximum unless otherwise stated.
[c] Values may vary, or minimum value may exist, depending on product size and mill form.
[d] Optional or when specified
[e] May purchased as type 1 (0.005–0.05 Nb), type 2 (0.01–0.15 V), type 3 (0.05 Nb, max, plus 0.02–0.15 V) or type 4 (0.015 N, max, plus V ≥ 4 N).
[f] If chromium and silicon are each 0.50% min, the copper minimum does not apply.
[g] May be substituted for all or part of V.
[h] Niobium plus vanadium, 0.02 to 0.15%.
[i] Nitrogen with vanadium content of 0.015% (max) with a minimum vanadium-to-nitrogen ratio of 4:1.
[j] When silicon-killed steel is specified.
[k] For plate under 40 mm (1.5 in.), manganese contents are 0.70 to 1.35% or up to 1.60% if carbon equivalents do not exceed 0.47%. For plate thicker than 40 mm (1 to 5 in.), ASTM A 841 specifies manganese contents of 1.00 to 1.60%

Table 3.1-47 Tensile properties of HSLA steel grades specified in ASTM standards [1.80]

ASTM specification [a]	Type, grade or condition	Product thickness [b] (mm)	Minimum tensile strength [c] (MPa)	Minimum yield strength [c] (MPa)	Minimum elongation (%) [c] in 200 mm	in 50 mm	Bend radius [c] Longitudinal	Transverse
A242	Type 1	20	480	345	18	–	–	–
		20–40	460	315	18	21	–	–
		40–100	435	290	18	21	–	–
A572	Grade 42	150	415	290	20	24	d	–
	Grade 50	100	450	345	18	21	d	–
	Grade 60	32	520	415	16	18	d	–
	Grade 65	32	550	450	15	17	d	–
A588	Grades A–K	100	485	345	18	21	d	–
		100–125	460	315	–	21	d	–
		125–200	435	290	–	21	d	–
A606	Hot rolled	sheet	480	345	–	22	t	2t – 3t
	Hot rolled and annealed or normalized	sheet	450	310	–	22	t	2t – 3t
	Cold rolled	sheet	450	310	–	22	t	2t – 3t
A607	Grade 45	sheet	410	310	–	22–25	t	1.5t
	Grade 50	sheet	450	345	–	20–22	t	1.5t
	Grade 55	sheet	480	380	–	18–20	1.5t	2t
	Grade 60	sheet	520	415	–	16–18	2t	3t
	Grade 65	sheet	550	450	–	15–16	2.5t	3.5t
	Grade 70	sheet	590	485	–	14	3t	4t
A618	Ia, Ib, II	19	485	345	19	22	t – 2t	–
	Ia, Ib, II, III	19–38	460	315	18	22	t – 2t	–
A633	A	100	430–570	290	18	23	d	–
	C, D	65	485–620	345	18	23	d	–
	C, D	65–100	450–590	315	18	23	d	–
	E	100	550–690	415	18	23	d	–
	E	100–150	515–655	380	18	23	d	–
A656	50	50	415	345	20	–	d	–
	60	40	485	415	17	–	d	–
	70	25	550	485	14	–	d	–
	80	20	620	550	12	–	d	–
A690	–	100	485	345	18	–	2t	–
A709	50	100	450	345	18	21	–	–
	50W	100	450	345	18	21	–	–
A715	Grade 50	sheet	415	345	–	22–24	0	t
	Grade 60	sheet	485	415	–	20–22	0	t
	Grade 70	sheet	550	485	–	18–20	t	1.5t
	Grade 80	sheet	620	550	–	16–18	t	1.5t
A808	–	40	450	345	18	22	–	–
		40–50	450	315	18	22	–	–
		50–65	415	290	18	22	–	–

Table 3.1-47 Tensile properties of HSLA steel grades specified in ASTM standards [1.80], cont.

ASTM specification[a]	Type, grade or condition	Product thickness[b] (mm)	Minimum tensile strength[c] (MPa)	Minimum yield strength[c] (MPa)	Minimum elongation (%)[c] in 200 mm	in 50 mm	Bend radius[c] Longitudinal	Transverse
A812	65	sheet	585	450	–	13–15	–	–
	80	sheet	690	550	–	11–13	–	–
A841	–	65	485–620	345	18	22	–	–
		65–100	450–585	310	18	22	–	–
A871	60, as-hot-rolled	5–35	520	415	16	18	–	–
	65, as-hot-rolled	5–20	550	450	15	17	–	–

[a] For characteristics and intended uses, see Table 3.1-48; for specified composition limits, see Table 3.1-46
[b] Maximum product thickness exept when a range is given. No thickness are specified for sheet products.
[c] May vary with product size and mill form
[d] Optional supplementary requierement given in ASTM A6

Table 3.1-48 Summary of characteristics and uses of HSLA steels according to ASTM standards [1.80]

ASTM specification[a]	Title	Alloying elements[b]	Avalible mill forms	Special characteristics	Intended uses
A 242	High-strength low-alloy structural steel	Cr, Cu, N, Ni, Si, Ti, V, Zr	Plate, bar, and shapes ≤ 100 mm in thickness	Atmospheric-corrosion resistance four times of carbon steel	Structural members in welded, bolted, or riveted construction
A 572	High-strength low-alloy niobium-vanadium steels of structural quality	Nb, V, N	Plate, bar, and sheet piling ≤ 150 mm in thickness	Yield strength of 290 to 450 MPa in six grades	Welded, bolted, or riveted structures, but many bolted or riveted bridges and buildings
A 588	High-strength low-alloy structural steel with 345 MPa minimum yield point ≤ 100 mm in thickness	Nb, V, Cr, Ni, Mo, Cu, Si, Ti, Zr	Plate, bar, and shapes ≤ 200 mm in thickness	Atmospheric-corrosion resistance four times of carbon steel; nine grades of similar stregth	Welded, bolted, or riveted structures, but primarily welded bridges and buildings in which weight savings or added durability is important
A 606	Steel sheet and strip hot-rolled steel and cold-rolled, high-strength low-alloy with improved corrosion resistance	Not specified	Hot-rolled and cold-rolled sheet and strip	Atmospheric-corrosion twice that of carbon steel (type 2) or four times of carbon steel (type 4)	Structural and miscellaneous purposes for which weight savings or added durability is important
A 607	Steel sheet and strip hot-rolled steel and cold-rolled, high-strength low-alloy niobium and/or vanadium	Nb, V, N, Cu	Hot-rolled and cold-rolled sheet and strip	Atmospheric-corrosion twice that of carbon steel, but only when copper content is specified; yield strength of 310 to 485 MPa in six grades	Structural and miscellaneous purposes for which greater strength or weight savings is important
A 618	Hot formed welded and seamless high-strength low-alloy structural tubing	Nb, V, Si, Cu	Square, rectangular round and special-shape structural welded or seamless tubing	Three grades of similar yield strength; may be purchased with atmospheric-corrosion resistance twice that of carbon steel	General structural purposes include welded, bolted or riveted bridges and buildings

Table 3.1-48 Summary of characteristics and uses of HSLA steels according to ASTM standards [1.80], cont.

ASTM specification[a]	Title	Alloying elements[b]	Avalible mill forms	Special characteristics	Intended uses
A 633	Normalized high-strength low-alloy structural steel	Nb, V, Cr, Ni, Mo, Cu, N, Si	Plate, bar, and shapes ≤ 150 mm in thickness	Enhanced notch toughness; yield strength of 290 to 415 MPa in five grades	Welded, bolted, or riveted structures for service at temperatures at or above −45 °C
A 656	High strength low-alloy, hot rolled structural vanadium-aluminium-nitrogen and titanium-aluminium steels	V, Al, N, Ti, Si	Plate, normally ≤ 16 mm in thickness	Yield strength of 552 MPa	Truck frames, brackets, crane booms, rail cars and other applications for witch weight savings is important
A 690	High-strength low-alloy steel H-piles and sheet piling	Ni, Cu, Si	Structural-quality H-pills and sheet piling	Corrosion resistance two to three times greater than that of carbon steel in the splash zone of marine marine structures	Dock walls, sea walls Bulkheads, excavations and similar structures exposed to seawater
A 709, grade 50 and 50W	Structural steel	V, Nb, N, Cr, Ni, Mo	All structural-shape groups and plate ≤ 100 mm in thickness	Minimum yield strength of 345 MPa, Grade 50W is a weathering steel	Bridges
A 714	High-strength low-alloy welded and seamless steel pipe	V, NI, Cr, Mo, Cu, Nb	Pipe with nominal pipe size diameters of 13 to 660 mm	Minimum yield strength of ≤ 345 MPa and corrosion resistance two or four times that of carbon steel	Piping
A 715	Steel sheet and strip hot-rolled. high-strength low-alloy with improved formability	Nb, V, Cr, Mo, N, Ti, Zr, B	Hot-rolled sheet and strip	Improved formability[c] compared to a A 606 and A 607; yield strength of 345 to 550 MPa in four grades	Structural and miscellaneous applications for which high strength, weight savings, improved formability and good weldability are important
A 808	High-strength low-alloy steel with improved notch toughness	V, Nb	Hot-rolled plate ≤ 65 mm in thickness	Charpy V-notch impact energies of 40–60 J (40–60 ft lfb) at −45 °C	Railway tank cars
A 812	High-strength low-alloy steels	V, Nb	Steel sheet in coil form	Yield strength of 450–550 MPa	Welded layered pressure vesels
A 841	Plate produced by themomechanical controlled processes	V, Nb, Cr Mo, Ni	Plates ≤ 100 mm in thickness	Yield strength of 310–345 MPa	Welded pressure vessels
A 847	Cold formed welded and seamless high-strength low-alloy structural rubing with improved atmospheric corrosion resistance	Cu, Cr, Ni, Si, V, Ti, Zr, Nb	Welded rubing with maximum periphery of 1625 mm and wall thickness of 16 mm or seamless tubing with maximum periphery of 810 mm and wall thickness of 13 mm	Minimum yield strength ≤ 345 MPa with atmospherieric-corrosion twice that of carbon steel	Round, square, or specially shaped structural tubing for welded, riveted or bolted construction of bridges and buildings

Table 3.1-48 Summary of characteristics and uses of HSLA steels according to ASTM standards [1.80], cont.

ASTM specification[a]	Title	Alloying elements[b]	Available mill forms	Special characteristics	Intended uses
A 860	High-strength butt-welding fittings of wrought high-strength low-alloy steel	Cu, Cr, Ni, Mo, V, Nb, Ti	Normalized or quenched-and-tempered wrought fittings	Minimum yield strength ≤ 485 MPa	High-pressure gas and oil transmission lines
A 871	High-strength low-alloy steel with atmospheric corrosion resistance	V, Nb, Ti Cu, Mo, Cr	As-rolled plate ≤ 35 mm in thickness	Atmosperic-corrosion resistance four times that of carbon structural steel	Tubular structures and poles

[a] For grades and mechanical properties, see Table 3.1-47.
[b] In addition to carbon, manganese, phosphorus, and sulfur. A given grade may contain one or more of the listed elements, but not necessarily all of them; for specified compositional limits, see Table 3.1-46.
[c] Obtained by producing killed steel, made to fine grain practice, and with microalloying elements such as niobium, vanadium titanium, and zirconium in the composition.

forming elements C, N, Ni, Mn, Cu, and Co. All alloying elements suppress the austenite to martensite transformation by reducing the M_s temperature, so that the steel may remain fcc at and below room temperature at a sufficiently high alloy content. For a rough estimate of the structural components of a stainless steel as a function of alloy composition, the so-called Schaeffler diagram (initially determined by M. Strauss and E. Maurer in 1920) can be used (see Fig. 3.1-114 [1.77]). It relates the equivalent Cr and Ni content to the observed fractions of martensite, austenite, and ferrite. One of the formulas of the Ni and Cr equivalent used most frequently is that of Schneider [1.85] (in wt%):

$$Cr_{equ} = Cr + 2\,Si + 1.5\,Mo + 5\,V + 5.5\,Al + 1.5\,Ti + 0.7\,W$$

and

$$Ni_{equ} = Ni + Co + 0.5\,Mn + 0.3\,Cu + 25\,N + 30\,C\,.$$

Ferritic Chromium Steels

High chromium (≥ 18 wt% Cr) and low carbon concentrations result in a fully ferritic structure of the steels at all temperatures, i.e., with a bcc delta ferrite structure and no phase transformations. Therefore these steels cannot be strengthened by quenching and tempering. The possibility to increase strength by cold deformation is limited since it decreases ductility and toughness. Steels containing < 18 wt% Cr form some austenite during heating which can be transformed into martensite by fast cooling, thus strengthening the steel. A tempering treatment just below Ar_1 results in a mixture of δ-ferrite, α-ferrite, and carbides formed from tempered martensite.

In general the toughness of the conventional stainless ferritic chromium steels is not very high, the impact transition temperature often being at or above room temperature and reaching $100\,°C$ after welding (see curve (e) in Fig. 3.1-115). This is due to (i) the tendency of the ferritic stainless steels to pronounced grain coarsing on heat treatments, leading to a relatively large grain size (which cannot be refined by phase transformation) and (ii) the precipitation of chromium carbides at the grain boundaries. Both effects render the steel rather brittle after welding. Furthermore, precipitation of chromium carbides at the grain boundaries causes susceptibility to intergranular corrosion due to the formation of zones with local chromium depletion along the grain boundaries. Some improvement can be achieved

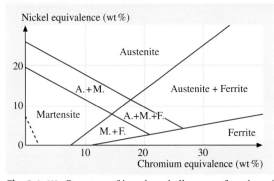

Fig. 3.1-114 Structure of iron-based alloys as a function of the concentrations of the chromium and nickel equivalent elements. A: austenite (face centered cubic); F: ferrite (body centered cubic); M: martensite (tetragonal-body centered cubic) [1.77]

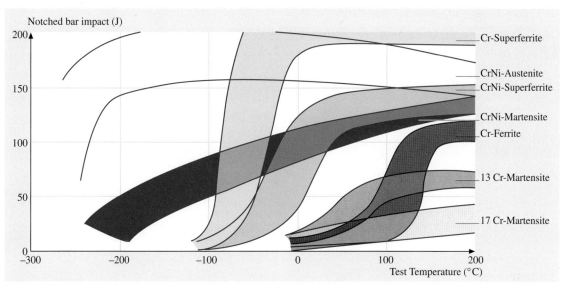

Fig. 3.1-115 Notched bar impact energy as a function of test temperature of various types of stainless steels (DVM samples: dimension $10 \times 10 \times 55$ mm^3, notch depth 3 mm, notch root radius 1 mm)

(i) by quenching or a suitable diffusion heat treatment for equilibration of the chromium distribution, and (ii) by a stabilization of the steels with small amounts of Ti and/or Nb which are added to bind the carbon and nitrogen in more stable compounds. In unstabilized steels the carbon content is generally restricted to ≤ 0.1 wt%. Up to 2 wt% Mo improves the corrosion resistance of ferritic chromium steels, especially in chloride-containing media.

In Table 3.1-49 the chemical composition of a number of ferritic stainless steels is given. The most common representative is the grade X6Cr17/AISI 430. Table 3.1-50 presents some information on hot deformation and recommended heat treatment parameters. Tables 3.1-51 and 3.1-52 show typical data of the mechanical and physical properties, respectively, and Table 3.1-53 reviews the weldability of these steels.

Good ductility and toughness at sub-zero temperatures can be achieved when the carbon and nitrogen content in ferritic steels is reduced to very low concentrations (below 100 ppm) (see curve (a) in Fig. 3.1-115). With higher alloy concentrations on the order of 26–30 wt% Cr, up to 2 wt% Mo, and up to 4 wt% Ni, these so-called superferritic steels exhibit excellent corrosion properties, i.e., high resistance against transcrystalline stress corrosion cracking and chloride-induced pitting corrosion, as well as intercrystalline and general corrosion. However, in order to maintain the low C and N contents even after welding, an effective inert gas shielding is required during welding as it is known for welding of Ti. Obviously, the high purity requirements which have to be observed during melting, hot forming and welding, and some embrittling effects which occur in high Cr steels after lengthy high temperature exposure (475° embrittlement and σ-phase formation) have precluded as yet a wider application of these steels, in spite of their attractive properties. In addition to their good corrosion resistance these steels possess a higher thermal conductivity than austenitic stainless steels, which is of special interest in heat exchanger applications, and a lower rate of work hardening on cold deformation.

Table 3.1-49 Chemical composition of ferritic stainless steels

Grade no. (EN 10088)	Steel designation	ASTM A 276/ AISI grade	Chemical composition (wt%)							
			C	Si	Mn	S	P	Cr	Mo	Others
1.4016	X6Cr17	430	≤0.08	≤1	≤1	≤0.03		15.5–17.5		
1.4104	X12CrMoS17	430F	0.10–0.17	≤1.5	≤1	0.15–0.35	≤0.06	15.5–17.5	0.2–0.6	
1.4105	X4CrMoS18		≤0.06	≤1.5	≤1	0.15–0.35	≤0.06	16.5–18.5	0.2–0.6	
1.4509	X2CrTiNb18	441	≤0.03	≤1	≤1	≤0.015		17.5–18.5		Ti 0.1–0.6 Nb $9x$%C+ 0.3–1.0
1.4510	X3CrTi17	430Ti	≤0.08	≤1	≤1	≤0.03	≤0.045	16–18		Ti ≥ $7x$%C up to 1.20%
1.4511	X3CrNb17		≤0.08	≤1	≤1	≤0.03	≤0.045	16–18		Nb ≥ $12x$%C up to 1.20%
1.4113	X6Cr Mo17-1	434	≤0.08	≤1	≤1	≤0.03	≤0.045	16–18	0.9–1.3	
1.4520	X2CrTi15		≤0.015	≤0.5	≤0.5	≤0.020	≤0.025	14–16		Ti 0.25–0.40
1.4521	X2CrMoTi18-2	444	≤0.025	≤1	≤1	≤0.03	≤0.045	17–19	1.8–2.3	Ti ≥ $7x$%(C+N) up to 0.8, (C+N) ≤ 0.04
	X20Cr20	442	≤0.2	≤1	≤1	≤0.03	≤0.04	18–23		
1.3810	X20Cr25	446	≤0.25	≤0.5–2.0	≤0.5	≤0.03	≤0.04	23–27		
[a]	X1CrMo26-1		0.002	0.1	0.3	0.015	0.01	26	1	N 0.006
[a]	X1CrMo29-4		0.003	0.1	0.04	0.01	0.01	29.5	4	
[a]	X1CrMoNi29-4-2		0.002	0.1	0.1	0.01	0.01	29.5	4	Ni 2.2
[a]	X2CrMoNiTi25-4-4		0.012	0.04	0.3	0.006		25	4	Ni 4, Ti 0.4

[a] Typical values

Table 3.1-50 Heat treatment conditions of ferritic stainless steels

Grade no.	Rolling and forging (°C)	Soft annealing		
		Temperature (°C)	Time (min)	Cooling
1.4016	1100–800	750–850	20–30	Air/water
1.4104	1100–800	750–850	120–180	Air/furnace
1.4105	1100–800	750–850		Air/water
1.4510	1100–800	750–850	20–30	Air/water
1.4511	1050–750	750–850	20–30	Air/water
1.4113	1050–750	750–850	20–30	Air/water
1.4521	1150–750	750–900		Air/water
X1CrMo26-1	1150–750	750–900		Air/water
X1CrMo29-4	1100–800	750–800	120–360	Air/furnace
X1CrMoNi29-4-2	1150–750	750–800	15–30	Air
X2CrMoNiTi25-4-4	1100–800	730–780	120–360	Air/furnace

Table 3.1-51 Mechanical properties of ferritic stainless steels

Grade no.	Heat treatment condition	Tensile properties of flat products ≤ 25 mm in thickness		
		Min. yield strength or 0.2% proof strength (MPa)	Ultimate tensile strength (MPa)	Min. fracture elongation A_5 (%)
1.4016	Annealed	270	450–600	20
1.4104	Annealed	300	540–740	16
1.4105	Annealed	270	450–650	20
1.4510	Annealed	270	450–600	20
1.4511	Annealed	250	450–600	20
1.4113	Annealed	260	480–630	20
1.4521	Annealed	320	450–650	20
X1CrMo26-1	Annealed	275[a]	450[a]	22[a]
X1CrMo29-4	Annealed	415[a]	550[a]	20[a]
X1CrMoNi29-4-2	Annealed	415[a]	550[a]	20[a]
X2CrMoNiTi25-4-4	Annealed	550[a]	650[a]	20[a]

[a] Typical values

Table 3.1-52 Physical properties of ferritic stainless steels

Grade no.	Mean thermal expansion coefficient between 20 °C and T (°C) in 10^{-6} K^{-1}				Density (kg/dm³)	Thermal conductivity at 20 °C (W/K m)	Specific heat at 20 °C (J/g K)	Electrical resistvity at 20 °C (Ω mm²/m)	Modulus of elasticity (kN/mm²)	Magnetizable
	100	200	300	400						
1.4016	10.0	10.0	10.5	10.5	7.7	25	0.46	0.60	220	yes
1.4104	10.0	10.5	10.5	10.5	7.7	25	0.46	0.70	216	yes
1.4105	10.0	10.5	10.5	10.5	7.7	25	0.46	0.70	216	yes
1.4510	10.0	10.0	10.5	10.5	7.7	25	0.46	0.60	220	yes
1.4511	10.0	10.0	10.5	10.5	7.7	25	0.46	0.60	220	yes
1.4113	10.0	10.0	10.5	10.5	7.7	25	0.46	0.70	216	yes
1.4521	10.4	10.8	11.2	11.6	7.7	15	0.43	0.80	220	yes

Table 3.1-53 Weldability of ferritic stainless steels

Grade no.	Weldable	Welding method				Preheating (°C)	After-treatment	
		SAW/MIG/TIG welding	Arc welding	Resistance welding	Autogenous welding		Annealing	at T (°C)
1.4016	Yes	+	+	+	(+)	200	(+)	700
1.4104	No	–	–	–	–	–	–	–
1.4105	No	–	–	–	–	–	–	–
1.4510	Yes	+	+	+	–	200	(+)	750
1.4511	Yes	+	+	+	–	200	(+)	750
1.4113	Yes	+	+	–	–	–	+	750
1.4521	Yes	+	–	–	–			
X2CrMoTi29-4	Yes	+	–	–	–			
X1CrMo26-1	Yes	+	–	–	–			
X1CrMo29-4	Yes	+	–	–	–			
X1CrMoNi29-4-2	Yes	+	–	–	–			

Martensitic and Martensitic-Ferritic Chromium Steels

Steels with 11.5 to 18 wt% Cr and up to 1.2 wt% C are austenitic at high temperatures and can be transformed into martensite by fast cooling. Due to the high chromium content, the hardenability of these steels is relatively high, so a fully martensitic transformation can be achieved even with relatively large cross sections and moderate cooling rates. They are normally used for applications requiring a combination of high hardness with good corrosion and wear resistance. With carbon contents below about 0.35 wt% C the steels are hypo-eutectic, and some ferrite will be present after heat treatment. They can be austenitized at $\geq 960\,°C$. Steels with ≥ 0.40 wt% C and 13 wt% Cr are already hyper-eutectic and will contain some undissolved primary carbides after quenching. To dissolve more carbon, these steels are usually quenched from a higher austenitization temperature around 1050 °C. The steels cannot be hardened if the carbon concentration is below 0.12 wt% C and the chromium content ≥ 16 wt% Cr.

In some grades a small content of 0.5–1 wt% Mo (and sometimes some W, V, or Nb) is used to increase the tempering resistance, i.e., to retain higher hardness on tempering by means of precipitation reactions. Increased S contents or Se additions improve the machinability of the steels. About 1–2.5 wt% Ni together with reduced carbon content is applied if hardenabilty of higher cross sections and an improved weldability are required.

The hardness after quenching depends on the carbon content, as illustrated in Fig. 3.1-116 for a 12 wt% Cr steel [1.77]. In the as-quenched state the steels are hard but very brittle. Thus a tempering treatment is necessary to adjust toughness and strength to the level required for a specific application. For high hardness, annealing temperatures of about 100–300 °C are applied, whereas increases in ductility and toughness require tempering at or above 650 °C.

The chemical composition for a number of martensitic Cr steels is given in Table 3.1-54. Tables 3.1-55 – 3.1-58 review data on recommended heat treatment conditions, mechanical properties, physical properties, and weldability, respectively.

It should be noted that a different way to produce stainless steels with high hardness is by precipitation hardening. Such steels have a low carbon content and contain in addition to chromium a few wt% of Ni and Cu. The hardening is caused by Cu precipitates. Others use precipitation hardening by intermetallic phases such as NiTi, TiAl, or NiAl.

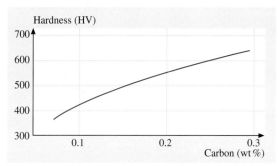

Fig. 3.1-116 As-quenched hardness as a function of the carbon content of a martensitic 12 wt% Cr steel

Table 3.1-54 Chemical composition of martensitic and martensitic-ferritic chromium steels

Grade no. (EN 10088)	Steel designation	ASTM A 276/ AISI grade	Chemical composition (wt%)							
			C	Si	Mn	P	S	Cr	Mo	Ni
1.4006	X12Cr13	410	0.08–0.12	≤1.0	≤1.0	≤0.045	≤0.030	12.0–14.0	–	–
1.4005	X12CrS13	416	≤0.15	≤1.0	≤1.0	0.15–0.25	≤0.045	12.0–13.0	–	–
1.4021	X20Cr13	420	0.17–0.25	≤1.0	≤1.0	≤0.045	≤0.030	12.0–14.0	–	–
1.4028	X30Cr13	420	0.28–0.35	≤1.0	≤1.0	≤0.045	≤0.030	12.0–14.0	–	–
1.4104	X12CrMoS17-2	430F	0.10–0.17	≤1.0	≤1.5	≤0.060	0.15–0.35	15.5–17.5	0.2–0.6	
1.4057	X20CrNi17-2	431	0.14–0.23	≤1.0	≤1.0	≤0.045	≤0.030	15.5–17.5	–	1.50–2.50
1.4109	X70CrMo15	440A	0.60–0.75	≤1.0	≤1.0	≤0.045	≤0.030	13.0–15.0	0.50–0.60	–
1.4125	X105CrMo17	440C	0.95–1.20	≤1.0	≤1.0	≤0.045	≤0.030	16.0–18.0	0.40–0.80	

Table 3.1-55 Heat treatment conditions of martensitic and martensitic-ferritic chromium steels

Grade no.	Rolling and forging T (°C)	Soft annealing T (°C)	Time (min)	Cooling	Quenching T (°C)	Medium	Hardness HRC ca.	Annealing T (°C)
1.4006	1100–800	750–800	120–360	Air/furnace	950–1000	Air/oil	31	780–680
1.4005	1150–750	750–800	15–30	Air	950–1000	Oil	31	700–600
1.4021	1100–800	730–780	120–360	Air/furnace	980–1030	Air/oil	47	750–650
1.4028	1100–800	730–780		Air/furnace	980–1030	Air/oil		740–640
1.4104	1100–800	750–850	120–180	Air/furnace	980–1030	Air/oil	27	650–550
1.4057	1100–800	650–750	180–240	Air/furnace	980–1030	Air/oil	47	720–620
1.4109	1100–900	790–840	120–360	Furnace	1020–1060	Oil	59	200–150
1.4125	1100–900	800–850	160–240	Furnace	1000–1050	Oil	61	300–100

Table 3.1-56 Mechanical properties of martensitic and martensitic-ferritic chromium steels

Grade no.	Heat treatment condition	Tensile properties of flat products ≤25 mm thickness			CVN impact energy at room temperature (J) longitudinal/ transversal	Min. yield strength or 0.2% proof strength at T (°C) in MPa			
		Min. yield strength or 0.2% proof strength (MPa)	Ultimate tensile strength (MPa)	Fracture elongation A_5 (%) long./transv.		100	200	300	400
1.4006	Annealed	250	450–650	20/15	–/–	235	225	220	195
	Quenched and tempered	420	600–800	16/12	–/–	420	400	365	305
1.4005	Quenched and tempered	440	590–780	12/–	–/–	–	–	–	–
1.4021	Annealed	–	≤740	–	–/–				
	Quenched and tempered	450	650–800	15/11	30/–	420	400	365	305
		550	750–950	13/10	25/–				
1.4028	Annealed	–	≤780	–	–/–				
	Quenched and tempered	600	800–1000	–	11/–				
1.4104	Annealed	300	540–740	–	–/–				
	Quenched and tempered	450	640–840	–	–/–				
1.4057	Annealed	–	≤950	–	–/–				
	Quenched and tempered	550	750–950	14/10	20/–	495	460	430	345

Table 3.1-57 Physical properties of martensitic and martensitic-ferritic chromium steels

Grade no.	Mean thermal expansion coefficient between 20 °C and T (°C) in 10^{-6} K^{-1}				Density (kg/dm³)	Thermal conductivity at 20 °C (W/K m)	Specific heat at 20 °C (J/g K)	Electrical resistvity at 20 °C (Ω mm²/m)	Modulus of elasticity (kN/mm²)	Magnetizable
	100	200	300	400						
1.4006	10.5	11.0	11.5	12.0	7.7	30	0.46	0.60	216	yes
1.4005	10.5	11.0	11.5	12.0	7.7	30	0.46	0.60	216	yes
1.4021	10.0	10.0	10.5	10.5	7.7	30	0.46	0.60	216	yes
1.4028	10.5	11.0	11.5	12.0	7.7	30	0.46	0.65	220	yes
1.4104	10.0	10.5	10.5	10.5	7.7	25	0.46	0.70	216	yes
1.4057	10.0	10.5	11.0	11.0	7.7	25	0.46	0.70	216	yes
1.4109	10.5	11.0	11.0	11.5	7.7	30	0.46	0.65	210	yes
1.4125	10.4	10.8	11.2	11.6	7.7	15	0.43	0.80	220	yes

Table 3.1-58 Weldability of martensitic and martensitic-ferritic chromium steels

Grade no.	Weldable	Welding method				Pre-heating (°C)	After-treatment			Q+T anew
		SAW/MIG TIG welding	Arc welding	Resistance welding	Autogenous welding		Annealing	at T (°C)		
1.4006	Yes	+	+	+	(+)	250	+	750		−
1.4005	No	−	−	−	−	−	−	−		−
1.4021	Condit.	+	+	−	−	350	+	720		(+)
1.4028	Yes	−	+	−	−	−	+	720		−
1.4104	No	−	−	−	−	−	−	−		−
1.4057	Condit.	+	+	−	−	200	+	700		(+)
1.4109	No	−	−	−	−	−	−	−		−
1.4125	No	−	−	−	−	−	−	−		−

Austenitic Stainless Steels

By adding of austenite forming elements, mainly Ni, the range of stability of the fcc phase is extended down to and below room temperature. The favorable combinations of ductility, toughness, hot and cold formability, weldability, and corrosion resistance have made the austenitic CrNi steels by far the most important and popular stainless steels. The most widespread representatives are the steel grades AISI 304 (X5CrNi18-10, 1.4301) and AISI 316 (X5CrNiMo17-12-2, 1.4401). The austenitic steels are applied in the solution-annealed (at 1000–1100 °C) and fast-cooled state which yields a microstructure that is free of carbide precipitates and has a homogeneous distribution of the alloying elements necessary for good corrosion resistance.

Compared to the bcc ferrite phase the fcc austenite phase is characterized by a higher solubility but a lower diffusivity of almost all alloying elements. The first fact allows the production of single phase fcc alloys with a broad composition spectrum. This permits adjustment of the properties of the steel to specific requirements of corrosion and oxidation resistance, cold and hot strength etc. The low diffusivity makes precipitation processes rather sluggish.

The austenite of the quenched steels can be unstable and can transform into martensite as a result of cold deformation or cooling to sub-zero temperatures, especially if these steels are relatively weakly-alloyed. This transformation leads to increased strength and reduced ductility. Martensite can be detected by magnetization measurements because the austenite is paramagnetic whereas the martensite is ferromagnetic. Since the fcc austenitic steels do not exhibit the ductile-brittle transition characteristic of ferritic alloys, the austenitic steels stay sufficiently ductile even at liquid He temperature (4.2 K) and are thus preferred materials for cryogenic applications. Their yield strength of 200–250 MPa is lower than that of the ferritic steels, but due to their pronounced work hardening by cold deformation (making extensive cold-forming operations difficult and often requiring intermediate soft annealing heat treatments), they have a higher tensile strength and fracture elongation. A pronounced solid solution strengthening is possible, especially by higher concentrations of nitrogen (up to about 0.4 wt% N). The nitrogen-alloyed CrNi steels are characterized by particularly favorable combinations of mechanical and corrosion properties.

As mentioned above, after fast cooling from $\geq 1000\,°C$, the austenitic stainless steels are free of carbide precipitates. But since steels with ≥ 0.05 wt% C are already carbon-supersaturated at temperatures below $900\,°C$, holding the material at temperatures between about 400 and $900\,°C$ will lead to precipitation of chromium carbides (mainly $M_{23}C_6$ type), preferentially at the grain boundaries. This will result in an increased susceptibility to intergranular corrosion due to local chromium depletion. The means to avoid this undesirable effect are the same as for the ferritic steels: bonding of the carbon (and nitrogen) atoms by an overstoichiometric alloying with Nb, Ta, or Ti (so-called stabilization of the steels), or the reduction of the carbon content to below 0.03 wt% C resulting in the development of the extra low carbon (ELC) steels. However, in strongly oxidative media (such as concentrated nitric acid) even steels that have been sufficiently stabilized by Ti may exhibit intergranular corrosion due to selective dissolution of the TiC. Under less severe corrosion conditions, the Ti stabilized steels are as stable as the Nb stabilized grades but at somewhat lower costs.

Chemical compositions, heat treatment conditions, physical properties, and hints to weldability of austenitic stainless steels are presented in Tables 3.1-59 – 3.1-63.

Table 3.1-59 Heat treatment conditions of austenitic stainless steels

Grade no.	Rolling and forging temperature (°C)	Cooling	Quenching temperature (°C)	Cooling
1.4541	1150–750	Air	1020–1100	Air, > 2 mm water
1.4401				
1.4402				
1.4406				
1.4436				
1.4571				
1.4580				
1.4301	1150–750	Air	1000–1080	Air, > 2 mm water
1.4303				
1.4305				
1.4306				
1.4311				
1.4550	1150–750	Air	1050–1150	Air, > 2 mm water
1.4429	1150–750	Air	1040–1120	Air, > 2 mm water
1.4438				
1.4439				
1.4435	1150–750	Air, below 600 °C furnace	1020–1100	Air, > 2 mm water

Table 3.1-60 Chemical composition of austenitic stainless steels

Grade no. (EN 10088)	Steel designation	ASTM A 276/ AISI grade	Chemical composition (wt%)								
			C	Si	Mn	P	S	Cr	Ni	Mo	Others
1.4310	X12CrNi17-7	301	≤0.12	≤1.5	≤2.0	≤0.045	0.030	16.0–18.0	6.0–9.0		
1.4301	X5CrNi18-10	304	≤0.07	≤1.0	≤2.0	≤0.045	≤0.030	17.0–19.0	8.5–10.5		
1.4303	X5CrNi18-12	305, 308	≤0.07	≤1.0	≤2.0	≤0.045	≤0.045	17.0–19.0	11.0–13.0		
1.4305	X10CrNiS18-9	303	≤0.12	≤1.0	≤2.0	≤0.060	0.15–0.35	17.0–19.0	8.0–10.0		
1.4306	X2CrNi19-11	304L	≤0.03	≤1.0	≤2.0	≤0.045	≤0.030	18.0–20.0	10.0–12.5		
1.4311	X2CrNiN18-10	304LN	≤0.03	≤1.0	≤2.0	≤0.060	0.15–0.35	17.0–19.0	8.5–11.5		N 0.12–0.22
1.4541	X6CrNiTi18-10	321	≤0.08	≤1.0	≤2.0	≤0.045	≤0.030	17.0–19.0	9.0–12.0		Ti 5x%C up to 0.80
1.4550	X6CrNiNb18-10	347	≤0.08	≤1.0	≤2.0	≤0.045	≤0.030	17.0–19.0	9.0–12.0		Nb 10x%C up to 1.0
1.4401	X5CrNiMo17-12-2	316	≤0.07	≤1.0	≤2.0	≤0.045	≤0.030	16.5–18.5	10.5–13.5	2.0–2.5	
1.4404	X2CrNiMo17-13-2	316L	≤0.03	≤1.0	≤2.0	≤0.045	≤0.030	16.5–18.5	11.0–14.0	2.0–2.5	
1.4406	X2CrNiMoN17-12-2	316LN	≤0.03	≤1.0	≤2.0	≤0.045	≤0.030	16.5–18.5	10.5–13.5	2.0–2.5	N 0.12–0.22
1.4571	X6CrNiMoTi17-12-2	316Ti	≤0.08	≤1.0	≤2.0	≤0.045	≤0.030	16.5–18.5	10.5–13.5	2.0–2.5	Ti 5x%C up to 0.80
1.4580	X6CrNiMoNb17-12-2	316Cb	≤0.08	≤1.0	≤2.0	≤0.045	≤0.030	16.5–18.5	10.5–13.5	2.0–2.5	Nb 10x%C up to 1.0
1.4429	X2CrNiMoN17-13-3	316LN	≤0.03	≤1.0	≤2.0	≤0.045	≤0.025	16.5–18.5	11.5–14.5	2.5–3.0	N 0.14–0.22
1.4435	X2CrNiMo18-14-3	316L 317L	≤0.03	≤1.0	≤2.0	≤0.045	≤0.025	17.0–18.5	12.5–15.0	2.5–3.0	
1.4436	X5CrNiMo17-13-3	316	≤0.07	≤1.0	≤2.0	≤0.045	≤0.025	16.5–18.5	11.0–14.0	2.5–3.0	
1.4438	X2CrNiMo18-16-4	317L	≤0.03	≤1.0	≤2.0	≤0.045	≤0.025	17.5–19.5	14.0–17.0	3.0–4.0	
1.4439	X2CrNiMoN17-13-5	F48	≤0.03	≤1.0	≤2.0	≤0.045	≤0.025	16.5–18.5	12.5–14.5	4.0–5.0	N 0.12–0.22
		201	≤0.15	≤1.0	5.5–7.5	≤0.06	≤0.03	16.8–18.0	3.5–5.5		N 0.25
		202	≤0.15	≤1.0	7.5–10.0	≤0.06	≤0.03	17.0–19.0	4.0–6.0		N 0.25
(1.4828)	(X15CrNiSi20-12)	309	≤0.20	≤1.0	≤2.0	≤0.045	≤0.03	22.0–24.0	12.0–15.0		
(1.4845)	(X12CrNi25-21)	310	≤0.25	≤1.5	≤2.0	≤0.045	≤0.03	24.0–26.0	19.0–22.0		

Table 3.1-61 Mechanical properties of austenitic stainless steels

Grade no.	Heat treatment condition	Tensile properties of flat products ≤ 25 mm thickness			Min. CVN impact energy at room temperarture (J)	Min. yield strength or 0.2% proof strength at T (°C) in MPa				
		Min. yield strength or 0.2% proof strength (MPa)	Ultimate tensile strength (MPa)	Min. fracture elongation A_5 (%)		100	200	300	400	500
1.4310	Quenched	260	600–950	35	105	–	–	–	–	–
1.4301	Quenched	195	500–700	45	85	157	127	110	98	92
1.4303	Quenched	185	490–690	45	85	155	127	110	98	92
1.4305	Quenched	195	500–700	35	–	–	–	–	–	–
1.4306	Quenched	180	460–680	45	85	147	118	100	89	81
1.4311	Quenched	270	550–760	40	85	205	157	136	125	119
1.4541	Quenched	200	500–730	40	85	176	157	136	125	119
1.4550	Quenched	205	510–740	40	85	177	157	136	125	119
1.4401	Quenched	205	510–710	40	85	177	147	127	115	110
1.4404	Quenched	190	490–690	40	85	166	137	118	108	100
1.4406	Quenched	280	580–800	40	85	211	167	145	135	129
1.4571	Quenched	210	500–730	35	85	185	167	145	135	129
1.4580	Quenched	215	510–740	35	85	186	167	145	135	129
1.4429	Quenched	295	580–800	40	85	225	178	155	145	138
1.4435	Quenched	190	490–690	35	85	166	137	118	108	100
1.4436	Quenched	205	510–710	40	85	177	147	127	115	110
1.4438	Quenched	195	490–690	35	85	172	147	127	115	110
1.4439	Quenched	285	580–800	35	85	225	185	165	150	–
AISI 201		370[a]	1125[a]	57[a]						
AISI 202		359[a]	1110[a]	55[a]						
AISI 309		334[a]	691[a]	48[a]						
AISI 310		319[a]	678[a]	50[a]						

[a] Typical values

Table 3.1-62 Physical properties of austenitic stainless steels

Grade no.	Mean thermal expansion coefficient between 20 °C and T (°C) in 10^{-6} K^{-1}							Density (kg/dm³)	Thermal conductivity at 20 °C (W/K m)	Specific heat at 20 °C (J/g K)	Electrical resistivity at 20 °C (Ω mm²/m)	Modulus of elasticity at 20 °C (kN/mm²)	Magnetizable
	100	200	300	400	500	600	700	800						
1.4310	16.4	16.9	17.4	17.8	18.2				8.0	15	0.45	0.80	198	No
1.4301	16.0	17.0	17.0	18.0	18.0	18.5	18.5	19.0	7.9	15	0.50	0.73	200	No
1.4303	16.0	17.0	17.0	18.0	18.0				7.9	15	0.50	0.73	200	No
1.4305 1.4306 1.4311 1.4541 1.4550	16.0	17.0	17.0	18.0	18.0	18.5	18.5	19.0	7.9	15	0.50	0.73	200	No

Table 3.1-62 Physical properties of austenitic stainless steels, cont.

Grade no.	Mean thermal expansion coefficient between 20 °C and T (°C) in 10^{-6} K^{-1}							Density (kg/dm^3)	Thermal conductivity at 20 °C (W/K m)	Specific heat at 20 °C (J/g K)	Electrical resistivity at 20 °C (Ω mm^2/m)	Modulus of elasticity at 20 °C (kN/mm^2)	Magnetizable
	100	200	300	400	500	600	700	800					
1.4401 1.4404 1.4406 1.4435 1.4436 1.4571 1.4580	16.5	17.5	17.5	18.5	18.5	19.0	19.5	19.5	7.98	15	0.50	0.75	200	No
1.4429	16.5	17.5	17.5	18.5	18.5				7.98	15	0.50	0.75	200	No
1.4438	16.5	17.5	18.0	18.5	19.0	19.0	19.5	19.5	8.0	14	0.50	0.85	200	No
1.4439	16.5	17.5	17.5	18.5	18.5				8.02	14	0.50	0.85	200	No

Table 3.1-63 Weldability of austenitic stainless steels welding methods not in parentheses are to be preferred

Grade no.	Weldable	Welding method SAW/MIG/TIG welding	Arc welding	Resistance welding	Autogenous welding	Preheating (°C)	After-treatment
1.4310	Yes	+	+	+	+	−	−
1.4301 1.4303 1.4306 1.4438 1.4541 1.4550 1.4401 1.4404 1.4406 1.4571 1.4580	Yes	+	+	+	(+)	−	−
1.4305	No	−	−	−	−	−	−
1.4311 1.4429 1.4435 1.4436 1.4439	Yes	+	+	+	−	−	−

Duplex Stainless Steels

Duplex steels have a mixed structure of ferrite and austenite. They contain the ferrite-forming elements Cr and Mo at levels of 20–29 wt% Cr and up to 4 wt% Mo, respectively, and the austenite-forming elements of about 4–9 wt% Ni and up to 0.3 wt% N. Typical examples are presented in Table 3.1-64. These steels solidify as δ-ferrite which will partly transform into austenite upon cooling. Thus, the phase fractions of ferrite and austenite depend not only on the chemical composition, but also on the annealing and cooling conditions.

The advantages of the duplex steels compared to the austenitic steels are a substantially higher yield and tensile strength (cf. Table 3.1-66) and a better resistance against stress corrosion cracking. A few data on heat treatment conditions, physical properties, and weldability are given in Tables 3.1-65, 3.1-67, and 3.1-68. In comparison with ferritic stainless steels, the duplex steels have a better weldability, a higher low temperature toughness, and a lower susceptibility to general and intergranular corrosion. With respect to optimum toughness, the δ-ferrite content of duplex steels should be below 60 vol%. The influence of annealing temperature and cooling conditions on the ferrite content of a steel with 0.05 wt% C, 25 wt% Cr, 8 wt% Ni, 2.5 wt% Mo, and 1.5 wt% Cu is illustrated in Table 3.1-69 [1.78]. During the partial phase transformation of ferrite into austenite below 1350 °C, a redistribution of the alloying elements occurs with the ferrite-forming elements Cr, Mo, and Ti enriched in the α-phase and the austenite-forming elements C, N, Ni, and Mn enriched in the γ-phase. The reduced C content in the α-phase delays the formation of chromium carbides. The relatively high Mo and N contents ensure good resistance against pitting corrosion.

The formation of the brittle Cr-rich σ-phase, which can form in the temperature range of 1000–500 °C and can lead to a drastic reduction of toughness and corrosion resistance, requires special attention. Therefore, duplex steels are usually water quenched from a solution treatment at 1100–1150 °C.

Table 3.1-64 Chemical composition of duplex stainless steels

Grade no. (EN 10088)	Steel designation	ASTM A 276/ AISI grade	Chemical composition (wt%)								
			C	Si	Mn	S	P	Cr	Ni	Mo	N
1.4460	X4CrNiMoN27-5-2	329	≤0.05	≤1	≤2	≤0.03	≤0.045	25.0–28.0	4.5–6.0	1.3–2.0	0.05–0.20
1.4462	X2CrNiMoN22-5-3	F51	≤0.03	≤1	≤2	≤0.02	≤0.03	21.0–23.0	4.5–6.5	2.5–3.5	0.08–0.20

Table 3.1-65 Heat treatment conditions of duplex stainless steels

Grade no.	Rolling and forging temperature (°C)	Cooling	Quenching temperature (°C)	Cooling
1.4460	1150–900	Air	1020–1100	Air, water
1.4462	1150–900	Air	1020–1100	Air, water

Table 3.1-66 Mechanical properties of duplex stainless steels

Grade no.	Heat treatment condition	Tensile properties of flat products ≤ 25 mm thickness			Min. CVN impact energy at room temp. (J)	Min. yield strength or 0.2% proof strength at T (°C) in MPa				
		Min. yield strength or 0.2% proof strength (MPa)	Ultimate tensile strength (MPa)	Min. fracture elongation A_5 (%)		100	200	300	400	500
1.4460	Quenched	450	600–800	20	55	360	310	–	–	–
1.4462	Quenched	450	640–900	30	120	360	310	280	–	–

Table 3.1-67 Physical properties of duplex stainless steels

Grade no.	Mean thermal expansion coefficient between 20 °C and T (°C) in 10^{-6} K^{-1}							Density (kg/dm^3)	Thermal conductivity at 20 °C (W/K m)	Specific heat at 20 °C (J/g)	Electrical resistivity at 20 °C (Ω mm^2/m)	Modulus of elasticity at 20 °C (kN/mm^2)	Magnetizable
	100	200	300	400	500	600	700	800					
1.4460	12.0	12.5	13.0	13.5	–	–	–	–	7.8	15	0.45	0.80	200	Yes
1.4462	12.0	12.5	13.0	–	–	–	–	–	7.8	15	0.45	0.80	200	Yes

Table 3.1-68 Weldability of duplex stainless steels

Grade no.	Weldable	Welding method SAW/MIG/TIG welding	Arc welding	Resistance welding	Autogenous welding	Preheating (°C)	Aftertreatment
1.4460	Yes	+	+	+	+	–	1020 °C
1.4462	Yes	+	+	+	–	–	–

Table 3.1-69 Influence of annealing temperature and cooling conditions on the ferrite content of a steel with 0.05 wt% C, 25 wt% Cr, 8 wt% Ni, 2.5 wt% Mo and 1.5 wt% Cu [1.10]

Annealing temperature (°C) (holding time 15 min)	Ferrite content (vol.%) after	
	water quenching	air cooling
1350	93.0	78.8
1300	70.2	61.8
1250	43.5	37.5
1150	35.7	34.2
1050	24.0	23.7
1000	7.6	7.6

3.1.5.5 Heat-Resistant Steels

Heat-resistant steels are treated extensively in [1.52], creep data are compiled in [1.86]. Steels are considered heat-resistant if they possess–in addition to good mechanical properties at ambient temperature–special resistance against short or long term exposure to hot gases, combustion products and melts of metals or salts at temperatures above about 550 °C where non- or low-alloyed steels are no longer applicable due to extensive scaling and creep.

Thus, heat-resistant steels are characterized by a combination of good high temperature strength, scaling resistance, a sufficient hot and cold formabilty, and weldability. They are sufficiently stable against embrittling processes at the high application temperatures. The resistance against scaling and hot gas corrosion is affected by the formation of a protective dense, pore-free, and tightly adherent oxide layer at the surface. The main alloying elements leading to such an oxide layer are Cr, Al, and Si. The oxidation and scaling resistance increases with increasing Cr content between about 6 and 25 wt% Cr. Higher Cr concentrations do not lead to further improvement. Additions of up to 2 wt% Al and up to 3 wt% Si enhance the effect of Cr. Small additions of rare earth metals, e.g., of Ce, can improve the adherence and the ductility of the oxide layer. In order to keep the protective oxide layer intact during temperature changes, the steels should exhibit low volume changes and, if possible, no phase transformations during heating and cooling. Consequently, there are two main groups of heat resistant steels: ferritic and austenitic steels, both showing no phase transformations.

The *ferritic* Cr (or Cr−Al−Si) steels are less expensive but have a lower creep strength above 800 °C. They may suffer from three embrittling mechanisms:

- The "475 °C embrittlement" due to decomposition in the metastable miscibility gap of the Fe−Cr solid solution, occuring between about 350 and 550 °C at Cr contents above 15 wt% Cr (Fig. 3.1-107);
- The formation of the brittle intermetallic FeCr σ-phase at temperatures between about 550 and 900 °C (Sect. 3.1.5.1, Fig. 3.1-106);
- Grain coarsening at temperatures above about 900 °C.

However, these embrittling mechanisms will not impair the behavior at high operating temperatures if taken into account properly, but will deteriorate the toughness after cooling to room temperature. Heating to temperatures above the range of occurrence of the embrittling phases followed by sufficiently fast cooling will suppress the embrittling effects.

The Cr and Ni alloyed *austenitic* steels possess a higher temperature strength and better ductility, toughness, and weldability. The susceptibility to embrittling effects is considerably lower. At Ni contents above 30 wt% Ni, they are outside the stability region of the brittle σ-phase.

The properties of the *ferritic-austenitic* steels lie between those of the ferritic and austenitic steels. They are characterized by a higher fracture toughness, cold formability, high temperature strength, and weldability than the fully ferritic grades, and by a higher chemical resistance in sulphurous gases than the austenitic grades.

It is obvious that the scaling resistance of the heat-resisting steels will be detrimentally influenced by any other corrosion mechanism which may be destroying the oxide layer, e.g., by chemical reactions with other metal oxides, chlorine, or chlorides. Thus in general, the heat resistance cannot be characterized by a single test method or measuring parameter but will depend on the specific environmental conditions.

In Table 3.1-70 chemical compositions of the most important grades of heat resistant steels are presented [1.87]. Table 3.1-71 contains some information about recommended temperature ranges for heat treatment and hot forming. In Tables 3.1-72 and 3.1-73 the mechanical properties at room temperature and at high temperatures are listed, respectively. Table 3.1-74 shows some physical properties. The ferritic and ferritic-austenitic steels are magnetisable while the austenitic grades are nonmagnetic. Qualitative data on the high temperature behavior in special gas atmospheres are given in Table 3.1-75. In carburizing atmospheres carbon can diffuse into the steel, reacting with the chromium to form chromium carbides which can lead to embrittlement and reduced scaling resistance due to chromium depletion in the matrix. Higher Ni and Si contents reduce the carburization susceptibility. In sulfur-containing atmospheres, which contain the sulfur mostly in the form of SO_2 or H_2S, the formation of sulfides at the surface may inhibit the formation of the protecting oxide layer. Under oxidizing conditions this process will proceed rather slowly, but under reducing conditions the pick-up of sulfur occurs very

Table 3.1-70 Chemical composition[a] of heat-resistant steels according to SEW [1.87]

Grade[a]	Steel designation	ASTM/ AISI grade	C	Si	Mn	P	S	Al	Cr	Ni	Others
Ferritic steels											
1.4713	X10CrAlSi7	–	≤0.12	0.5–1.0	≤1	≤0.04	≤0.03	0.5–1.0	6.0–8.0	–	–
1.4720	X7CrTi12	–	≤0.08	≤1.0	≤1	≤0.04	≤0.03	–	10.5–12.5	–	Ti ≥ 6×wt% C up to 1.0
1.4724	X10CrAlSi13	405	≤0.12	0.7–1.4	≤1	≤0.04	≤0.03	0.7–1.2	12.0–14.0	–	–
1.4742	X10CrAlSi18	430	≤0.12	0.7–1.4	≤1	≤0.04	≤0.03	0.7–1.2	17.0–19.0	–	–
1.4762	X10CrAlSi25	446	≤0.12	0.7–1.4	≤1	≤0.04	≤0.03	1.2–1.7	23.0–26.0	–	–
Ferritic-austenitic steels											
1.4821	X15CrNiSi25-4	327	0.10–0.20	0.8–1.5	2.0	≤0.04	≤0.03	–	24.0–27.0	3.5–5.5	–
Austenitic steels											
1.4878	X10CrNiTi18-10	321	≤0.12	≤1	≤2.0	≤0.045	≤0.03	–	17.0–19.0	9.0–12.0	Ti ≥ 4×wt% C up to 0.8
1.4828	X15CrNiSi20-12	309	≤0.20	1.5–2.5	≤2.0	≤0.045	≤0.03	–	19.0–21.0	11.0–13.0	–
1.4833	X12CrNi23-12	309S	≤0.08	≤1	≤2.0	≤0.045	≤0.03	–	21.0–23.0	12.0–15.0	–
1.4845	X8CrNi25-21-12	310S	≤0.15	≤0.75	≤2.0	≤0.045	≤0.03	–	24.0–26.0	19.0–22.0	–
1.4841	X15CrNiSi25-20	310/314	≤0.20	1.5–2.5	≤2.0	≤0.045	≤0.03	–	24.0–26.0	19.0–22.0	–
1.4864	X12NiCrSi36-18	330	≤0.15	1.0–2.0	≤2.0	≤0.030	≤0.02	–	15.0–17.0	33.0–37.0	–
1.4876	X10NiCrAlTi32-21	B163	≤0.12	≤1	≤2.0	0.030	≤0.02	0.15–0.6	19.0–23.0	30.0–34.0	Ti 0.15–0.6

[a] According to SEW [1.87]

Table 3.1-71 Recommended conditions for heat treatment and hot forming of heat-resistant steels

Grade[a]	Hot forming temperature (°C)	Soft annealing (°C) Cooling in air (water)	Quenching temperature (°C) Cooling in water (air)	Limiting scaling temperature in air (°C)
Ferritic steels				
1.4713	1100–750	750–800	–	620
1.4720	1050–750	750–850	–	800
1.4724	1100–750	800–850	–	850
1.4742	1100–750	800–850	–	1000
1.4762	1100–750	800–850	–	1150
Ferritic–austenitic steels				
1.4821	1150–800	–	1000–1050	1100
Austenitic steels				
1.4878	1150–800	–	1020–1070	850
1.4828	1150–800	–	1050–1100	1000
1.4833	1150–900	–	1050–1100	1000
1.4845	1150–800	–	1050–1100	1050
1.4841	1150–800	–	1050–1100	1150
1.4864	1150–800	–	1050–1100	1100
1.4876	1150–800	900–980 (recrystallization annealing)	1100–1150 (solution annealing)	1100

[a] According to SEW [1.87]

Table 3.1-72 Mechanical properties of heat-resistant steels at 20 °C

Grade[a]	Heat treatment condition	Hardness (HB) max.	0.2% proof stress (MPa) min.	Ultimate tensile strength (MPa)	Elongation $L_0 = 5d_0$ min. (%) long.	transv.
Ferritic steels						
1.4713	Annealed	192	220	420–620	20	15
1.4720	Annealed	179	210	400–600	25	20
1.4724	Annealed	192	250	450–650	15	11
1.4742	Annealed	212	270	500–700	12	9
1.4762	Annealed	223	280	520–720	10	7
Ferritic–austenitic steels						
1.4821	Quenched	235	400	600–850	16	12
Austenitic steels						
1.4878	Quenched	192	210	500–750	40	30
1.4828	Quenched	223	230	500–750	30	22
1.4833	Quenched	192	210	500–750	35	26
1.4845	Quenched	192	210	500–750	35	26
1.4841	Quenched	223	230	550–800	30	22
1.4864	Quenched	223	230	550–800	30	22
1.4876	Recryst. annealed	192	210	500–750	30	22
	Solution annealed	192	170	450–700	30	22

[a] According to SEW [1.87]

Table 3.1-73 Long-term mechanical properties of heat-resistant steels at high temperatures; average values of scatter bands

Grade[a]	Temperature (°C)	1% creep limit (MPa) at $t =$		Creep rupture strength (MPa) at $t =$		
		1000 h	10 000 h	1000 h	10 000 h	100 000 h
Ferritic and ferritic-austenitic steels						
1.4713	500	80	50	160	100	55
1.4720	600	27.5	17.5	55	35	20
1.4724	700	8.5	4.7	17	9.5	5
1.4742	800	3.7	2.1	7.5	4.3	2.3
1.4762	900	1.8	1.0	3.6	1.9	1.0
1.4821						
Austenitic steels						
1.4878	600	110	85	185	115	65
	700	45	30	80	45	22
	800	15	10	35	20	10
1.4828	600	120	80	190	120	65
1.4833	700	50	25	75	36	16
	800	20	10	35	18	7.5
	900	8	4	15	8.5	3.0
1.4845	600	150	105	230	160	80
1.4841	700	53	37	80	40	18
	800	23	12	35	18	7
	900	10	5.7	15	8.5	3.0
1.4864	600	105	80	180	125	75
	700	50	35	75	45	25
	800	25	15	35	20	7
	900	12	5	15	8	3
1.4876 (solution annealed)	600	130	90	200	152	114
	700	70	40	90	68	47
	800	30	15	45	30	19
	900	13	5	20	11	4

[a] According to SEW [1.87]

Table 3.1-74 Physical properties of heat-resistant steels

Grade[a]	Density at 20 °C (g cm^{-3})	Average coefficient of thermal expansion between 20 °C and T (°C) ($\times 10^{-6}$ K^{-1})						Thermal conductivity at T (°C) (W m^{-1} K^{-1})		Specific heat at 20 °C (J g^{-1} K^{-1})	Specific electrical resistivity at 20 °C (Ω mm^2 m^{-1})
		200	400	600	800	1000	1200	20	500		
Ferritic steels											
1.4713	7.7	11.5	12.0	12.5	13.0	–	–	23	25	0.45	0.70
1.4720	7.7	11.0	12.0	12.5	13.0	–	–	25	28	0.45	0.60
1.4724	7.7	11.0	11.5	12.0	12.5	13.5	–	21	23	0.45	0.90
1.4742	7.7	10.5	11.5	12.0	12.5	13.5	–	19	25	0.45	0.95
1.4762	7.7	10.5	11.5	12.0	12.5	13.5	15.0	17	23	0.45	1.10
Ferritic–austenitic steels											
1.4821	7.7	13.0	13.5	14.0	14.5	15.0	15.5	17	23	0.50	0.90

Table 3.1-74 Physical properties of heat-resistant steels, cont.

Grade [a]	Density at 20 °C (g cm^{-3})	Average coefficient of thermal expansion between 20 °C and T (°C) ($\times 10^{-6}$ K^{-1})						Thermal conductivity at T (°C) (W m^{-1} K^{-1})		Specific heat at 20 °C (J g^{-1} K^{-1})	Specific electrical resistivity at 20 °C (Ω mm^2 m^{-1})
		200	400	600	800	1000	1200	20	500		
Austenitic steels											
1.4878	7.9	17.0	18.0	18.5	19.0	–	–	15	21	0.50	0.75
1.4828	7.9	16.5	17.5	18.0	18.5	19.5	–	15	21	0.50	0.85
1.4833	7.9	16.0	17.5	18.0	18.5	19.5	–	15	19	0.50	0.80
1.4845	7.9	15.5	17.0	17.5	18.0	19.0	–	14	19	0.50	0.85
1.4841	7.9	15.5	17.0	17.5	18.0	19.0	19.5	15	19	0.50	0.90
1.4864	8.0	15.0	16.0	17.0	17.5	18.5	–	13	19	0.50	1.00
1.4876	8.0	15.0	16.0	17.0	17.5	18.5	–	12	19	0.50	1.00

[a] According to SEW [1.87]

Table 3.1-75 Resistance of heat-resistant steels in various media

Grade [a]	Resistance to Carburization	Sulphurous gases Oxidizing	Sulphurous gases Reducing	Nitrogenous and low-oxygen gases	Maximum operating temperature in air (°C)
Ferritic steels					
1.4713	Medium	Very high	Medium	Low	800
1.4762	Medium	Very high	High	Low	1150
Ferritic–austenitic steels					
1.4821	Medium	High	Medium	Medium	1100
Austenitic steels					
1.4878	Low	Medium	Low	High	850
1.4828	Low	Medium	Low	High	1000
1.4841	Low	Medium	Low	High	1150
1.4864	High	Medium	Low	High	1100

[a] According to SEW [1.87]

fast. This is especially true with Ni-alloyed steels due to the formation at about 650 °C of a low-melting Ni/NiS eutectic. Thus under such conditions the ferritic steels are more stable than the austenitic grades. In sulfur-containing atmospheres the maximum service temperatures will be about 100 to 200 °C lower than in air.

The heat-resistant steels are weldable by the usual processes, with arc welding preferred over gas fusion welding. For the ferritic steels, the tendency to grain coarsening in the heat affected zone has to be kept in mind. The application of austenitic filler metals will lead to better mechanical properties of the weld connection than those of the base metal (however, with respect to the scaling resistance, different thermal expansions of the ferritic and austenitic materials may be a problem). Filler materials should be at least as highly alloyed as the base metal. In sulfurizing atmospheres it is advisable to use ferritic electrodes for the cap passes only in order to ensure a tough weld. Post-weld heat treatments are generally not necessary.

3.1.5.6 Tool Steels

Tool steels are the largest group of materials used to make tools for cutting, forming, or otherwise shaping a material into a part or component. An extensive account is given in [1.88]. Other major groups of tool materials are cemented carbides (Sect. 3.1.6.6), and ceramics including diamond (Chapt. 3.2).

The most commonly used materials are wrought tool steels, which are either carbon, alloy, or high-speed steels capable of being hardened by quenching and tempering to hardness levels ≤ 70 HRC. High-speed tool steels are so named because of their suitability to machine materials at high cutting speed. Other steels used for metalworking applications include steels produced by powder metallurgy, medium-carbon alloy steels, high-carbon martensitic stainless steels, and maraging steels.

Wrought tool steels are essentially hardenable alloy steels with relatively high contents of the carbide forming elements Cr, Mo, W, and V. If the steels are quenched and tempered, the dependence of their hardness on tempering temperature indicates the level of hardening achieved as well as its temperature stability, (Fig. 3.1-117). The rate of effective softening at tempering temperatures up to about 300 °C is mainly due to the competing effects of recovery and the precipitation of iron carbides (Table 3.1-40). The hardening at higher temperatures is associated with the precipitation of alloy carbides which can form at elevated temperatures only because of their high melting points and transformation kinetics. They give rise to a second maximum on isothermal tempering curves, as curves 3 and 4 in Fig. 3.1-117, which is referred to as secondary hardening.

The alloy carbide phases which precipitate and give rise to secondary hardening are listed in Table 3.1-76.

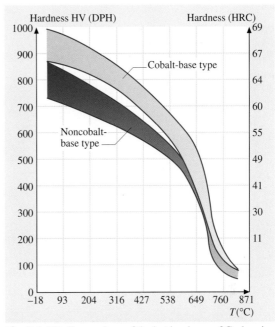

Fig. 3.1-118 Comparison of the hot hardness of Co-bearing versus non-Co bearing high-speed tool steels [1.88]

Co is an alloying element which raises the high temperature stability of tool steels by raising their melting temperature. Figure 3.1-118 shows this effect by a comparison of the hardness vs. temperature data for non-Co-based and Co-based high-speed tool steels.

Table 3.1-77 gives composition ranges for the tool steels most commonly used. According to the AISI classification, each group of similar composition and properties is given a capital letter, somewhat related to the major alloying element. Thus, high-speed steels are classified by M for molybdenum and T for tungsten. Within each group individual types are assigned code numbers.

The basic properties of tool steels that determine their performance in service are hardness, wear

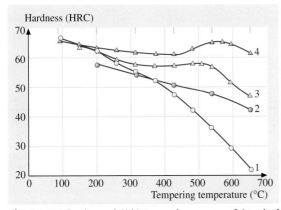

Fig. 3.1-117 Isothermal (1 h) tempering curves of 4 typical tool steels. Curves 1 and 2: softening of AISI grade W (water-hardening) and O (oil-hardening) steels; Curves 3 and 4: softening and secondary hardening of AISI grade A2 (air-hardening medium alloy) and M2 (Mo high-speed) steels [1.88]

Table 3.1-76 Alloy carbides occurring in tool steels

Type of carbide	Prototype	Lattice type	Occurrence, composition [a]
M_7C_3	Cr_7C_3	Hexagonal	In Cr alloy steels M = **Cr**
$M_{23}C_6$	$Cr_{23}C_6$	Face-centered cubic	In high-Cr steels M = **Cr**, Fe, W, Mo
M_6C	W_6C	Face-centered cubic	M = **W, Mo**, Cr, V, Co
M_2C	W_2C	Hexagonal	M = **W, Mo**, Cr
MC	VC	Face-centered cubic	VC

[a] Bold letters indicate major components

Table 3.1-77 Composition ranges of principle types of tool steels according to AISI and UNS classifications [1.88]

Designation AISI	UNS No.	Composition[b] (wt%) C	Mn	Si	Cr	Ni	Mo	W	V	Co
Molybdenum high-speed steels										
M1	T11301	0.78–0.88	0.15–0.40	0.20–0.50	3.50–4.00	0.30 max	8.20–9.20	1.40–2.10	1.00–1.35	–
M2	T11302	0.78–0.88; 0.95–1.05	0.15–0.40	0.20–0.45	3.75–4.50	0.30 max	4.50–5.50	5.50–6.75	1.75–2.20	–
M3, class 1	T11313	1.00–1.10	0.15–0.40	0.20–0.45	3.75–4.50	0.30 max	4.75–6.50	5.00–6.75	2.25–2.75	–
M3, class 2	T11323	1.15–1.25	0.15–0.40	0.20–0.45	3.75–4.50	0.30 max	4.75–6.50	5.00–6.75	2.75–3.75	–
M4	T11304	1.25–1.40	0.15–0.40	0.20–0.45	3.75–4.75	0.30 max	4.25–5.50	5.25–6.50	3.75–4.50	–
M7	T11307	0.97–1.05	0.15–0.40	0.20–0.55	3.50–4.00	0.30 max	8.20–9.20	1.40–2.10	1.75–2.25	–
M10	T11310	0.84–0.94; 0.95–1.05	0.10–0.40	0.20–0.45	3.75–4.50	0.30 max	7.75–8.50	–	1.80–2.20	–
M30	T11330	0.75–0.85	0.15–0.40	0.20–0.45	3.50–4.25	0.30 max	7.75–9.00	1.30–2.30	1.00–1.40	4.50–5.50
M33	T11333	0.85–0.92	0.15–0.40	0.15–0.50	3.50–4.00	0.30 max	9.00–10.00	1.30–2.10	1.00–1.35	7.75–8.75
M34	T11334	0.85–0.92	0.15–0.40	0.20–0.45	3.50–4.00	0.30 max	7.75–9.20	1.40–2.10	1.90–2.30	7.75–8.75
M35	T11335	0.82–0.88	0.15–0.40	0.20–0.45	3.75–4.50	0.30 max	4.50–5.50	5.50–6.75	1.75–2.20	4.50–5.50
M36	T11336	0.80–0.90	0.15–0.40	0.20–0.45	3.75–4.50	0.30 max	4.50–5.50	5.50–6.50	1.75–2.25	7.75–8.75
M41	T11341	1.05–1.15	0.20–0.60	0.15–0.50	3.75–4.50	0.30 max	3.25–4.25	6.25–7.00	1.75–2.25	4.75–5.75
M42	T11342	1.05–1.15	0.15–0.40	0.15–0.65	3.50–4.25	0.30 max	9.00–10.00	1.15–1.85	0.95–1.35	7.75–8.75
M43	T11343	1.15–1.25	0.20–0.40	0.15–0.65	3.50–4.25	0.30 max	7.50–8.50	2.25–3.00	1.50–1.75	7.75–8.75
M44	T11344	1.10–1.20	0.20–0.40	0.30–0.55	4.00–4.75	0.30 max	6.00–7.00	5.00–5.75	1.85–2.20	11.00–12.25
M46	T11346	1.22–1.30	0.20–0.40	0.40–0.65	3.70–4.20	0.30 max	8.00–8.50	1.90–2.20	3.00–3.30	7.80–8.80
M47	T11347	1.05–1.15	0.15–0.40	0.20–0.45	3.50–4.00	0.30 max	9.25–10.00	1.30–1.80	1.15–1.35	4.75–5.25
M48	T11348	1.42–1.52	0.15–0.40	0.15–0.40	3.50–4.00	0.30 max	4.75–5.50	9.50–10.50	2.75–3.25	8.00–10.00
M62	T11362	1.25–1.35	0.15–0.40	0.15–0.40	3.50–4.00	0.30 max	10.00–11.00	5.75–6.50	1.80–2.10	–
Tungsten high-speed steels										
T1	T12001	0.65–0.80	0.10–0.40	0.20–0.40	3.75–4.50	0.30 max	–	17.25–18.75	0.90–1.30	–
T2	T12002	0.80–0.90	0.20–0.40	0.20–0.40	3.75–4.50	0.30 max	1.0 max	17.50–19.00	1.80–2.40	–
T4	T12004	0.70–0.80	0.10–0.40	0.20–0.40	3.75–4.50	0.30 max	0.40–1.00	17.50–19.00	0.80–1.20	4.25–5.75
T5	T12005	0.75–0.85	0.20–0.40	0.20–0.40	3.75–5.00	0.30 max	0.50–1.25	17.50–19.00	1.80–2.40	7.00–9.50
T6	T12006	0.75–0.85	0.20–0.40	0.20–0.40	4.00–4.75	0.30 max	0.40–1.00	18.50–21.00	1.50–2.10	11.00–13.00
T8	T12008	0.75–0.85	0.20–0.40	0.20–0.40	3.75–4.50	0.30 max	0.40–1.00	13.25–14.75	1.80–2.40	4.25–5.75
T15	T12015	1.50–1.60	0.15–0.40	0.15–0.40	3.75–5.00	0.30 max	1.00 max	11.75–13.00	4.50–5.25	4.75–5.25
Intermediate high-speed steels										
M50	T11350	0.78–0.88	0.15–0.45	0.20–0.60	3.75–4.50	0.30 max	3.90–4.75	–	0.80–1.25	–
M52	T11352	0.85–0.95	0.15–0.45	0.20–0.60	3.50–4.30	0.30 max	4.00–4.90	0.75–1.50	1.65–2.25	–

Table 3.1-77 Composition ranges of principle types of tool steels according to AISI and UNS classifications [1.88], cont.

Designation AISI	UNS No.	Composition[b] (wt%) C	Mn	Si	Cr	Ni	Mo	W	V	Co
Chromium hot-worked steels										
H10	T20810	0.35–0.45	0.25–0.70	0.80–1.20	3.00–3.75	0.30 max	2.00–3.00	–	0.25–0.75	–
H11	T20811	0.33–0.43	0.20–0.50	0.80–1.20	4.75–5.50	0.30 max	1.10–1.60	–	0.30–0.60	–
H12	T20812	0.30–0.40	0.20–0.50	0.80–1.20	4.75–5.50	0.30 max	1.25–1.75	1.00–1.70	0.50 max	–
H13	T20813	0.32–0.45	0.20–0.50	0.80–1.20	4.75–5.50	0.30 max	1.10–1.75	–	0.80–1.20	–
H14	T20814	0.35–0.45	0.20–0.50	0.80–1.20	4.75–5.50	0.30 max	–	4.00–5.25	–	–
H19	T20819	0.32–0.45	0.20–0.50	0.20–0.50	4.00–4.75	0.30 max	0.30–0.55	3.75–4.50	1.75–2.20	4.00–4.50
Tungsten hot-worked steels										
H21	T20821	0.26–0.36	0.15–0.40	0.15–0.50	3.00–3.75	0.30 max	–	8.50–10.00	0.30–0.60	–
H22	T20822	0.30–0.40	0.15–0.40	0.15–0.40	1.75–3.75	0.30 max	–	10.00–11.75	0.25–0.50	–
H23	T20823	0.25–0.45	0.15–0.40	0.15–0.60	11.00–12.75	0.30 max	–	11.00–12.75	0.75–1.25	–
H24	T20824	0.42–0.53	0.15–0.40	0.15–0.40	2.50–3.50	0.30 max	–	14.00–16.00	0.40–0.60	–
H25	T20825	0.22–0.32	0.15–0.40	0.15–0.40	3.75–4.50	0.30 max	–	14.00–16.00	0.40–0.60	–
H26	T20826	0.45–0.55[b]	0.15–0.40	0.15–0.40	3.75–4.50	0.30 max	–	17.25–19.00	0.75–1.25	–
Molybdenum hot-worked steels										
H42	T20842	0.55–0.70[b]	0.15–0.40	–	3.75–4.50	0.30 max	4.50–5.50	5.50–6.75	1.75–2.20	–
Air-hardening, medium-alloy, cold-worked steels										
A2	T30102	0.95–1.05	1.00 max	0.50 max	4.75–5.50	0.30 max	0.90–1.40	–	0.15–0.50	–
A3	T30103	1.20–1.30	0.40–0.60	0.50 max	4.75–5.50	0.30 max	0.90–1.40	–	0.80–1.40	–
A4	T30104	0.95–1.05	1.80–2.20	0.50 max	0.90–2.20	0.30 max	0.90–1.40	–	–	–
A6	T30106	0.65–0.75	1.80–2.50	0.50 max	0.90–1.20	0.30 max	0.90–1.40	–	–	–
A7	T30107	2.00–2.85	0.80 max	0.50 max	5.00–5.75	0.30 max	0.90–1.40	0.50–1.50	3.90–5.15	–
A8	T30108	0.50–0.60	0.50 max	0.75–1.10	4.75–5.50	0.30 max	1.15–1.65	1.00–1.50	–	–
A9	T30109	0.45–0.55	0.50 max	0.95–1.15	4.75–5.50	1.25–1.75	1.30–1.80	–	0.80–1.40	–
A10	T30110	1.25–1.50[b]	1.60–2.10	1.00–1.50	–	1.55–2.05	1.25–1.75	–	–	–
High-carbon, high-chromium, cold-worked steels										
D2	T30402	1.40–1.60	0.60 max	0.60 max	11.00–13.00	0.30 max	0.70–1.20	–	1.10 max	–
D3	T30403	2.00–2.35	0.60 max	0.60 max	11.00–13.50	0.30 max	–	1.00 max	1.00 max	–
D4	T30404	2.05–2.40	0.60 max	0.60 max	11.00–13.00	0.30 max	0.70–1.20	–	1.00 max	–
D5	T30405	1.40–1.60	0.60 max	0.60 max	11.00–13.00	0.30 max	0.70–1.20	–	1.00 max	2.50–3.50
D7	T30407	2.15–2.50	0.60 max	0.60 max	11.50–13.50	0.30 max	0.70–1.20	–	3.80–4.40	–

Table 3.1-77 Composition ranges of principle types of tool steels according to AISI and UNS classifications [1.88], cont.

Designation AISI	UNS No.	Composition[b] (wt%) C	Mn	Si	Cr	Ni	Mo	W	V	Co
Oil-hardening cold-worked steels										
O1	T31501	0.85–1.00	1.00–1.40	0.50 max	0.40–0.60	0.30 max	–	0.40–0.60	0.30 max	–
O2	T31502	0.85–0.95	1.40–1.80	0.50 max	0.50 max	0.30 max	0.30 max	–	0.30 max	–
O6	T31506	1.25–1.55[c]	0.30–1.10	0.55–1.50	0.30 max	0.30 max	0.20–0.30	–	–	–
O7	T31507	1.10–1.30	1.00 max	0.60 max	0.35–0.85	0.30 max	0.30 max	1.00–2.00	0.40 max	–
Shock-resisting steels										
S1	T41901	0.40–0.55	0.10–0.40	0.15–1.20	1.00–1.80	0.30 max	0.50 max	1.50–3.00	0.15–0.30	–
S2	T41902	0.40–0.55	0.30–0.50	0.90–1.20	–	0.30 max	0.30–0.60	–	0.50 max	–
S5	T41905	0.50–0.65	0.60–1.00	1.75–2.25	0.50 max	–	0.20–1.35	–	0.35 max	–
S6	T41906	0.40–0.50	1.20–1.50	2.00–2.50	1.20–1.50	–	0.30–0.50	–	0.20–0.40	–
S7	T41907	0.45–0.55	0.20–0.90	0.20–1.00	3.00–3.50	–	1.30–1.80	–	0.20–0.30[d]	–
Low-alloy special-purpose tool steels										
L2	T61202	0.45–1.00[b]	0.10–0.90	0.50 max	0.70–1.20	–	0.25 max	–	0.10–0.30	–
L6	T61206	0.65–0.75	0.25–0.80	0.50 max	0.60–1.20	1.25–2.00	0.50 max	–	0.20–0.30[d]	–
Low-carbon mold steels										
P2	T51602	0.10 max	0.10–0.40	0.10–0.40	0.75–1.25	0.10–0.50	0.15–0.40	–	–	–
P3	T51603	0.10 max	0.20–0.60	0.40 max	0.40–0.75	1.00–1.50	–	–	–	–
P4	T51604	0.12 max	0.20–0.60	0.10–0.40	4.00–5.25	–	0.40–1.00	–	–	–
P5	T51605	0.10 max	0.20–0.60	0.40 max	2.00–2.50	0.35 max	–	–	–	–
P6	T51606	0.05–0.15	0.35–0.70	0.10–0.40	1.25–1.75	3.25–3.75	–	–	–	–
P20	T51620	0.28–0.40	0.60–1.00	0.20–0.80	1.40–2.00	–	0.30–0.55	–	–	–
P21	T51621	0.18–0.22	0.20–0.40	0.20–0.40	0.50 max	3.90–4.25	–	–	0.15–0.25	1.05–1.25Al
Water-hardening tool steels										
W1	T72301	0.70–1.50[e]	0.10–0.40	0.10–0.40	0.15 max	0.20 max	0.10 max	0.15 max	0.10 max	–
W2	T72302	0.85–1.50[e]	0.10–0.40	0.10–0.40	0.15 max	0.20 max	0.10 max	0.15 max	0.15–0.35	–
W3	T72303	1.05–1.15	0.10–0.40	0.10–0.40	0.40–0.60	0.20 max	0.10 max	0.15 max	0.10 max	–

[a] All steels except group W contain 0.25 max Cu, 0.03 max P, and 0.03 max S; group W steels contain 0.20 max Cu, 0.025 max P, and 0.025 max S. Where specified, sulfur may be increased to 0.06 to 0.15% to improve machinability of group A, D, H, M and T steels.
[b] Available in several carbon ranges.
[c] Contains free graphite in the microstructure.
[d] Optional.
[e] Specified carbon ranges are designated by suffix numbers.

resistance, ductility and fracture toughness, and in many applications stability against softening at elevated temperatures. Characteristic mechanical properties at room temperature as a function of hardening treatment are listed for group L and group S steels in Table 3.1-78.

Table 3.1-78 Mechanical properties of group L and group S tool steels at room temperature as a function of hardening treatment [1.88]

Type	Condition	Tensile strength (MPa)	0.2% yield strength (MPa)	Elongation[a] (%)	Reduction in area (%)	Hardness (HRC)
L2	Annealed	710	510	25	50	96 HRB
	Oil quenched from 855 °C and single tempered at:					
	205 °C	2000	1790	5	15	54
	315 °C	1790	1655	10	30	52
	425 °C	1550	1380	12	35	47
	540 °C	1275	1170	15	45	41
	650 °C	930	760	25	55	30
L6	Annealed	655	380	25	55	93 HRB
	Oil quenched from 845 °C and single tempered at:					
	315 °C	2000	1790	4	9	54
	425 °C	1585	1380	8	20	46
	540 °C	1345	1100	12	30	42
	650 °C	965	830	20	48	32
S1	Annealed	690	415	24	52	96 HRB
	Oil quenched from 925 °C and single tempered at:					
	205 °C	2070	1895	–	–	57.5
	315 °C	2025	1860	4	12	54
	425 °C	1790	1690	5	17	50.5
	540 °C	1680	11 525	9	23	47.5
	650 °C	1345	1240	12	37	42
S5	Annealed	725	440	25	50	96 HRB
	Oil quenched from 870 °C and single tempered at:					
	205 °C	2345	1930	5	20	59
	315 °C	2240	1860	7	24	58
	425 °C	1895	1690	9	28	52
	540 °C	1515	1380	10	30	48
	650 °C	1035	1170	15	40	37
S7	Annealed	640	380	25	55	95 HRB
	Oil quenched from 940 °C and single tempered at:					
	205 °C	2170	1450	7	20	58
	315 °C	1965	1585	9	25	55
	425 °C	1895	1410	10	29	53
	540 °C	1820	1380	10	33	51
	650 °C	1240	1035	14	45	39

[a] In 50 mm

3.1.5.7 Cast Irons

The term cast iron pertains to a large family of multi-component Fe−C−Si alloys which solidify according to the eutectic of the Fe−C system (Sect. 3.1.5.1; Fig. 3.1-99). They are treated extensively in [1.89]. Their comparatively high C and Si contents lead to solidification either according to the metastable equilibria involving Fe_3C or according to the stable equilibria involving graphite, depending, also, on the content of further alloying elements, melt treatment, and rate of cooling. Since, in addition, the metallic phases can be alloyed and their microstructures varied by annealing and transformation treatments as in other ferrous alloys, a multitude of microstructural states and associated properties result.

Classification

The C rich phases determine the basic classification of cast irons. According to the color of their facture surfaces, Fe_3C-containing grades are called white, graphite-containing grades are called gray, and alloys which solidify in mixed states are called mottled. In addition, the shape of the graphite phase particles and the microstructure of the metallic matrix phases are taken into account since they are also characterizing the mechanical properties.

Shape of Graphite Phase Particles. Lamellar (flake) graphite (FG) is characteristic of cast irons with near-zero ductility; spheroidal (nodular) graphite (SG) is characteristic of ductile cast iron; compacted (vermicular) graphite (CG) is a transition form between flake and nodule shape; temper graphite (TG) results from a tempering treatment and consists of small clusters of branched graphite lamellae.

Microstructure of Metallic Matrix Phases. Ferritic, pearlitic, austenitic, bainitic (austempered). More details are presented in Fig. 3.1-119 and Table 3.1-79.

Iron−Carbon−Silicon Equilibria and Carbon Equivalent

Since C and Si are the alloying elements which dominate the solidification behavior and the resulting microstructures of cast irons, their phase equilibria need to be taken into account. Figure 3.1-120 shows a section through the metastable ternary Fe−C−Si diagram at 2 wt% Si which approximates the Si content of many cast irons. Compared to the binary Fe−C system, the addition of Si decreases the stability of Fe_3C and increases the stability of ferrite, as indicated by the expansion of the α-phase field. With increasing Si concentration, the C concentrations of the eutectic and the eutectoid equilibria decrease while their temperatures increase.

These relations are the basis for correlating the C and Si concentrations with the ranges of formation of steels and of the main groups of cast irons as shown in Fig. 3.1-121. The relations are expressed in terms of the carbon equivalent CE = (wt% C) + (1/3) (wt% Si). The concentration of the eutectic (upper dashed line) is given by $CE_e = 4.3$. Accordingly, alloys with CE < 4.3 are hypoeutectic and alloys with CE > 4.3 are hypereutectic. In P-containing cast irons the relation is CE = (wt% C) + (1/3) (wt% Si + wt% P). In addition, Fig. 3.1-121 shows the limit of solubility of C in austenite (lower dashed line), which is the upper limit of the range of steels. It is given by $CE_{\gamma_{max}} = 2.1 = $ (wt% C) + (1/6) (wt% Si).

Grades of Cast Irons

Table 3.1-80 lists the composition ranges of typical unalloyed cast irons, indicating that they are classified mainly by the type of carbon-rich phase formed and by the basic mechanical behavior.

Table 3.1-79 Classification of cast irons according to commercial designation, microstructure and color of fracture surface [1.89]

Commercial designation	Carbon-rich phase	Matrix[a]	Fracture	Final structure after
Gray iron	Lamellar graphite	P	Gray	Solidification
Ductile iron	Spheriodal graphite	F, P, A	Silver-gray	Solidification or heat treatment
Compacted graphite iron	Compacted (vermicular) graphite	F, P	Gray	Solidification
White iron	Fe_3C	P, M	White	Solidification and heat treatment[b]
Mortled iron	Lamellar Gr + Fe_3C	P	Mottled	Solidification
Malleable iron	Temper graphite	F, P	Silver-gray	Heat treatment
Austempered ductile iron	Spheroidal graphite	At	Silver-gray	Heat treatment

[a] F, ferrite; P, pearlite; A, austenite; M, martensite; At, austempered (bainite)
[b] White irons are not usually heat treated, exept for stress relief and to continue austenite transformation

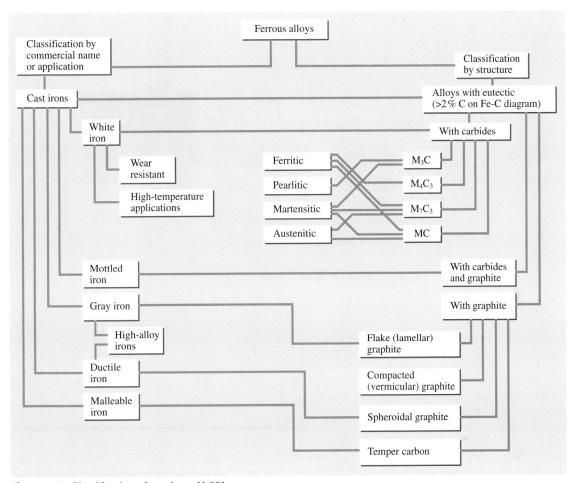

Fig. 3.1-119 Classification of cast irons [1.89]

Table 3.1-80 Range of composition, microstructural and mechanical characteristics of typical unalloyed cast irons [1.89]

Type	Carbon phase	Concentration range (wt%)				
		C	Si	Mn	P	S
Gray	FG	2.5–4.0	1.0–3.0	0.2–1.0	0.002–1.0	0.02–0.25
Compacted graphite	CG	2.5–4.0	1.0–3.0	0.2–1.0	0.01–0.1	0.01–0.03
Ductile	SG	3.0–4.0	1.8–2.8	0.1–1.0	0.01–0.1	0.01–0.03
White	Fe_3C	1.8–3.6	0.5–1.9	0.25–0.8	0.06–0.2	0.06–0.2
Malleable	Fe_3C/TG	2.2–2.9	0.9–1.9	0.15–1.2	0.02–0.2	0.02–0.2

Gray Iron. This most common type of cast iron is characterised by flake graphite and requires a high CE to ensure a sufficient graphitization potential which is also increased by Al addition. Gray irons may be moderately alloyed, e.g., by 0.2–0.6 wt% Cr, 0.2–1 wt% Mo, and 0.1–0.2 wt% V which promote the formation of alloy carbides and pearlite. Upon plastic deformation the flake form of graphite promotes early internal crack formation and, thus, causes the low ductility of gray iron.

Ductile Iron. This cast iron is characterized by the spheroidal graphite phase (SG) in its microstructure. Spheroidal graphite is formed during solidification if the melt has been treated by the addition of a component which promotes the particular nucleation and

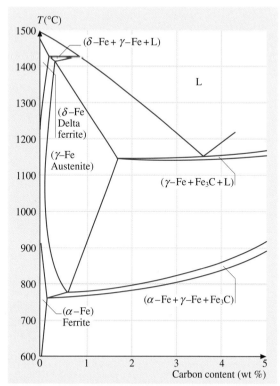

Fig. 3.1-120 Section through the metastable Fe–C–Si phase diagram at 2 wt% Si [1.89]

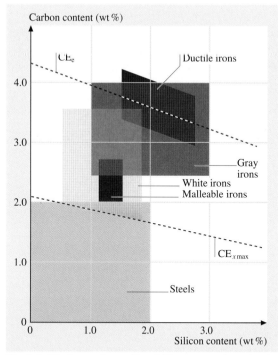

Fig. 3.1-121 Approximate C and Si concentrations for the composition ranges of steels and different grades of cast irons [1.89]

growth behavior of graphite in the form of nodules. The most common alloying component added to nucleate spheroidal graphite is Mg, but Ca, Ce, La, and Y have also been found to favour spheroidal graphite formation. Basically this microstructural modification leads to higher yield strength and higher ductility because the plastic deformation of the metallic matrix phases can be extended to higher strains than in gray iron before fracture sets in.

Malleable Irons. The melt treatment of malleable cast irons involves Mg, Ca, Bi, or Te additions. But malleable irons have an as-cast structure consisting of Fe_3C in a pearlitic matrix. By heat treatment in the range of 800–970 °C the cementite phase is transformed into graphite (TG). The cooling is controlled in such a way as to promote pearlite formation, ferrite formation, or a mixture of the two.

Alloy Cast Irons. Alloying elements beyond the levels mentioned above are added to cast irons almost exclusively to enhance resistance to abrasive wear or chemical corrosion, or to extend their stability for application at elevated temperature. The function of the alloying elements is essentially the same as in steels. Table 3.1-81 lists the groups of grades with typical compositions and the microstructural constituents present in the as-cast state.

Mechanical Properties of Cast Irons

Due to the multitude of as-cast structures as a function of alloy composition, melt treatment, cooling rate (as influenced by the cooling conditions and the cross section of the work piece), and subsequent heat treatment, there is a wide range of mechanical properties which can be achieved according to the requirements of the application. Table 3.1-82 gives a survey in terms of characteristic examples.

Table 3.1-81 Ranges of alloy content of typical alloy cast irons [1.89]

Description	Composition (wt%)[a]									Matrix structure as-cast[c]
	TC[b]	Mn	P	S	Si	Ni	Cr	Mo	Cu	
Abrasion-resistant white irons										
Low-carbon white iron[d]	2.2–2.8	0.2–0.6	0.15	0.15	1.0–1.6	1.5	1.0	0.5	[e]	CP
High-carbon, low-silicon white iron	2.8–3.6	0.3–2.0	0.30	0.15	0.3–1.0	2.5	3.0	1.0	[e]	CP
Martensitic nickel-chromium iron	2.5–3.7	1.3	0.30	0.15	0.8	2.7–5.0	1.1–4.0	1.0	–	M, A
Martensitic nickel, high-chromium iron	2.5–3.6	1.3	0.10	0.15	1.0–2.2	5.0–7.0	7.0–11.0	1.0	–	M, A
Martensitic chromium-molybdenum iron	2.0–3.6	0.5–1.5	0.10	0.06	1.0	1.5	11.0–23.0	0.5–3.5	1.2	M, A
High-chromium iron	2.3–3.0	0.5–1.5	0.10	0.06	1.0	1.5	23.0–28.0	1.5	1.2	M
Corrosion-resistant irons										
High-silicon iron[f]	0.4–1.1	1.5	0.15	0.15	14.0–17.0	–	5.0	1.0	0.5	F
High-chromium iron	1.2–4.0	0.3–1.5	0.15	0.15	0.5–3.0	5.0	12.0–35.0	4.0	3.0	M, A
Nickel-chromium gray iron[g]	3.0	0.5–1.5	0.08	0.12	1.0–2.8	13.5–36.0	1.5–6.0	1.0	7.5	A
Nickel-chromium ductile iron[h]	3.0	0.7–4.5	0.08	0.12	1.0–3.0	18.0–36.0	1.0–5.5	1.0	–	A
High-resistant gray iron										
Medium-silicon iron[i]	1.6–2.5	0.4–0.8	0.30	0.10	4.0–7.0	–	–	–	–	F
Nickel-chromium iron[g]	1.8–3.0	0.4–1.5	0.15	0.15	1.0–2.75	13.5–36.0	1.8–6.0	1.0	7.5	A
Nickel-chromium-silicon iron[j]	1.8–2.6	0.4–1.0	0.10	0.10	5.0–6.0	13.0–43.0	1.8–5.5	1.0	10.0	A
High-aluminium iron	1.3–2.0	0.4–1.0	0.15	0.15	1.3–6.0	–	20.0–25.0 Al	–	–	F
Heat resistant ductile irons										
Medium-silicon ductile iron	2.8–3.8	0.2–0.6	0.08	0.12	2.5–6.0	1.5	–	2.0	–	F
Nickel-chromium ductile iron[h]	3.0	0.7–2.4	0.08	0.12	1.75–5.5	18.0–36.0	1.75–3.5	1.0	–	A
Heat-resistant white irons										
Ferritic grade	1.0–2.5	0.3–1.5	–	–	0.5–2.5	–	30.0–35.0	–	–	F
Austenitic grade	1.0–2.0	0.3–1.5	–	–	0.5–2.5	10.0–15.0	15.0–30.0	–	–	A

[a] Where a single value is given rather than a range, that value is a maximum limit
[b] Total carbon
[c] CP, coarse pearlite; M, martensite; A, austenite; F, ferrite
[d] Can be produced from a malleable-iron base composition
[e] Copper can replace all or part of the nickel
[f] Such as Durion, Durichlor 51, Superchlor
[g] Such as Ni-Resist austenitic iron (ASTM A 436)
[h] Such as Ni-Resist austenitic ductile iron (ASTM A 439)
[i] Such as Silal
[j] Such asd Nicrosilal

Table 3.1-82 Mechanical properties of cast irons

Material	$R_{p0.2}$ (N mm^{-2}) (min.)	R_m (N mm^{-2}) (min.)	A_5 (%) (min.)	Matrix microstructure	Material number[a]	Short code[a]
				Unspecified		
Gray cast irons (FG) DIN 1691		100[b]		Mainly ferritic	0.6010	GG-10
		200[b]			0.6020	GG-20
		350[b]			0.6035	GG-35
Ductile cast irons (SG) DIN 1693	250[c]	390[c]	15[d]	Pearlitic-ferritic	0.7040	GGG-40
	360[c]	600[c]	2[d]	Mainly pearlitic	0.7060	GGG-60
	400[c]	700[c]	2[d]	Wide variation permissible	0.7070	GGG-70

Table 3.1-82 Mechanical properties of cast irons, cont.

Material	$R_{p0.2}$ (N mm^{-2}) (min.)	R_m (N mm^{-2}) (min.)	A_5 (%) (min.)	Matrix microstructure Unspecified	Material number[a]	Short code[a]
White malleable irons (TG) DIN 1692		350[d]	4	Core: ganular pearlite	0.8035	GTW-35-04
	260[b]	450[d]	7	Core: lamellar to granular pearlite	0.8045	GTW-45-07
	220[b]	400[d]	5	Ferrite	0.8040	GTW-40-05
Black malleable irons (TG) DIN 1692	200[c]	350[d]	10	Pearlite, ferrite	0.8135	GTS-35-10
	270[c]	450[d]	6	Pearlite	0.8145	GTS-45-06
	430[c]	650[d]	2	Hardened	0.8165	GTS-65-02
	530[c]	700[d]	2		0.8170	GTS-70-02
Austenitic alloy cast irons (FG) DIN 1694		140/220	–		0.6652	GGL-NiMn13-7
		190/240	12		0.6661	GGL NiCr20-3
		170/240	–		0.6680	GGL-NiSiCr30-5-5
Austenitic cast iron (SG) DIN 1694	210	390	15[d]		0.7652	GGG-NiMn13-7
	210	390	7[d]		0.7661	GGG NiCr20-3
	240	390	–		0.7680	GGG-NiSiCr30-5-5
	210	370	7[d]		0.7685	GGG-NiCr35-3

[a] According to German Materials Standard

[b] The values refer to a cylindrical test specimen of 30 mm in diameter corresponding to a wall thickness of 15 mm. The tensile strength values to be expected depend on the wall thickness Example: GG-20.

Wall thickness (mm)	2.5–5.5	5.5–10	10–20	20–40	40–80	80–150
R_m (N/mm^2)	230	205	180	155	130	115

[c] Diameter of the test rod: 12 or 15 mm. For cast parts with a thickness < 6 mm tensile test specimens.

[d] For a cylindrical test specimen of 12 mm in diameter. The mechanical properties depend on the diameter of the test specimen Example: GTW-45-07

Diameter of the test rod (mm)	9	12	15
R_m (N/mm^2)	400	450	480
$R_{p0.2}$ (N/mm^2)	230	260	280

3.1.6 Cobalt and Cobalt Alloys

Cobalt is applied as a base metal for a number of alloys, as an alloying element, and as a component of numerous inorganic compounds. Table 3.1-83 lists its major applications. Cobalt and cobalt-based materials are treated extensively in [1.90, 91].

Data on the electronic structure of Co and Co alloys may be found in [1.92]. Phase diagrams, crystal structures, and thermodynamic data of binary Co alloys may be found in [1.93].

3.1.6.1 Co-Based Alloys

Cobalt-based alloys with a carbon content in the range of 1 to 3 wt% C are widely used as wear-resistant solid materials and weld overlays. Depending on the alloy composition and heat treatment, $M_{23}C_6$, M_6C, and MC carbides are formed. Materials with lower carbon content are mostly designed for corrosion resistance and for heat resistance, sometimes combined with wear resistance. The metals W, Mo, and Ta are essentially added for solid solution strengthening. In a few alloys Ti and Al are added. They serve to form a coherent ordered $Co_3(Ti, Al)$ phase which precipitates and leads to strengthening by age hardening. The Cr content is generally rather high to provide oxidation and hot corrosion resistance. Table 3.1-84 presents a survey of Co-based alloys.

Table 3.1-83 Applications of cobalt [1.91]

Application	Co consumption (%)	Form	Section
Co-based alloys, Co-based superalloys, steels	24.3	Base and alloying element	Sects. 3.1.5, 3.1.6, 3.1.7
Hard facing and related materials	6.9	Base and alloying element	Sect. 3.1.6
Soft and hard magnetic materials, and controlled thermal expansion alloys	8.5	Base and alloying element, oxide	Chapt. 4.3
Cemented carbides, diamond tooling	15.2	Metal, binder	Sect. 3.1.6
Catalysts	8.0	Metal, sulfates	–
Colorizer for glass, enamel, ceramics, plastics, fabrics	11.6	Oxides, salts	–
Batteries	9.5	Powder, hydroxide, $LiCoO_2$	–
Tire adhesives, soaps, driers	11.0	Soaps, complexes made from metal powder	–
Feedstuff, anodizing, electrolysis, copper winning	5.0	Sulphate, carbonate, hydroxide	–

Table 3.1-84 Compositions of Co-based alloys

Alloy tradename	UNS No.	Nominal composition (wt%)								
		Co	Cr	W	Mo	C	Fe	Ni	Si	Mn	Others
Cast, P/M and weld overlay wear-resistant alloys											
Stellite 1	R30001	bal	30	13	0.5	2.5	3	1.5	1.3	0.5	–
Stellite 3 (P/M)	R30103	bal	30.5	12.5	–	2.4	5 (max)	3.5 (max)	2 (max)	2 (max)	1 B (max)
Stellite 4	R30404	bal	30	14	1 (max)	0.57	3 (max)	3 (max)	2 (max)	1 (max)	–
Stellite 6	R30006	bal	29	4.5	1.5 (max)	1.2	3 (max)	3 (max)	1.5 (max)	1 (max)	–
Stellite 6 (P/M)	R30106	bal	28.5	4.5	1.5 (max)	1	5 (max)	3 (max)	2 (max)	2 (max)	1 B (max)
Stellite 12	R30012	bal	30	8.3	–	1.4	3 (min)	1.5	0.7	2.5	–
Stellite 21	R30021	bal	27	–	5.5	0.25	3 (max)	2.75	1 (max)	1 (max)	0.007 B (max)
Stellite 98M2 (P/M)	–	bal	30	18.5	0.8 (max)	2	5 (max)	3.5	1 (max)	1 (max)	4.2 V, 1 B (max)
Stellite 703	–	bal	32	–	12	2.4	3 (max)	3 (max)	1.5 (max)	1.5 (max)	–
Stellite 706	–	bal	29	–	5	1.2	3 (max)	3 (max)	1.5 (max)	1.5 (max)	–
Stellite 712	–	bal	29	–	8.5	2	3 (max)	3 (max)	1.5 (max)	1.5 (max)	–
Stellite 720	–	bal	33	–	18	2.5	3 (max)	3 (max)	1.5 (max)	1.5 (max)	0.3 B
Stellite F	R30002	bal	25	12.3	1 (max)	1.75	3 (max)	22	2 (max)	1 (max)	–
Stellite Star J (P/M)	R30102	bal	32.5	17.5	–	2.5	3 (max)	2.5 (max)	2 (max)	2 (max)	1 B (max)
Stellite Star J	R31001	bal	32.5	17.5	–	2.5	3 (max)	2.5 (max)	2 (max)	2 (max)	–
Tantung G	–	bal	29.5	16.5	–	3	3.5	7 (max)	–	2 (max)	4.5 Ta/Nb
Tantung 144	–	bal	27.5	18.5	–	3	3.5	7 (max)	–	–	5.5 Ta/Nb
Laves-phase wear-resistant alloys											
Tribaloy T-400	R30400	bal	9	–	29	–	–	–	2.5	–	–
Tribaloy T-800	–	bal	18	–	29	–	–	–	3.5	–	–
Wrought wear-resistant alloys											
Stellite 6B	R30016	bal	30	4	1.5 (max)	1	3 (max)	2.5	0.7	1.4	–
Stellite 6K	–	bal	30	4.5	1.5 (max)	1.6	3 (max)	3 (max)	2 (max)	2 (max)	–

Table 3.1-84 Compositions of Co-based alloys, cont.

Alloy tradename	UNS No.	Nominal composition (wt%)									
		Co	Cr	W	Mo	C	Fe	Ni	Si	Mn	Others
Wrought heat-resistant alloys (see Table 3.1-86 for cast alloy compositions)											
Haynes 25 (L605)	R30605	bal	20	15	–	0.1	3 (max)	10	0.4 (max)	1.5	–
Haynes 188	R30188	bal	22	14	–	0.1	3 (max)	22	0.35	1.25	0.03 La
Inconel 783	R30783	bal	3	–	–	0.03 (max)	25.5	28	0.5 (max)	0.5 (max)	5.5 Al, 3 Nb, 3.4 Ti (max)
UMCo-50	–	bal	28	–	–	0.02 (max)	21	–	0.75	0.75	–
S-816	R30816	40 (min)	20	4	4	0.37	5 (max)	20	1 (max)	1.5	4 Nb
Corrosion-resistant alloys											
Ultimet (1233)	R31233	bal	26	2	5	0.06	3	9	0.3	0.8	0.08 N
MP 159	R30159	bal	19	–	7	–	9	25.5	–	–	3 Ti, 0.6 Nb, 0.2 Al
MP35N	R30035	35	20	–	10	–	–	35	–	–	–
Duratherm 600	R30600	41.5	12	3.9	4	0.05 (max)	8.7	bal	0.4	0.75	2 Ti, 0.7 Al, 0.05 Be
Elgiloy	R30003	40	20	–	7	0.15 (max)	bal	15.5	–	2	1 Be (max)
Havar	R30004	42.5	20	2.8	2.4	0.2	bal	13	–	1.6	0.06 Be (max)

P/M: powder metallurgy; bal: balance

3.1.6.2 Co-Based Hard-Facing Alloys and Related Materials

The behavior of Co-based wear resistant alloys is based on a coarse dispersion of hard carbide phases embedded in a tough Co-rich metallic matrix. The volume fraction of the hard carbide phase is comparatively high: e.g., at 2.4 wt% C the carbide content is 30 wt%. The carbide phases are M_7C_3 (Cr_7C_3 type) and M_6C (W_6C type). Table 3.1-85 lists characteristic properties of Co-based hard facing alloys the compositions of which are listed Table 3.1-83.

3.1.6.3 Co-Based Heat-Resistant Alloys, Superalloys

Both wrought and cast Co-based heat resistant alloys, listed in Tables 3.1-84 and 3.1-86, respectively, are also referred to as Co superalloys. They are based on the face-centered cubic high temperature phase of Co which is stabilized between room temperature and the solidus temperature by alloying with ≥ 10 wt% Ni. They are solid-solution strengthened by alloying with W, Ta, and Mo. Furthermore, they are dispersion strengthened by carbides.

Table 3.1-85 Properties of selected Co-based hard-facing alloys

Property	Stellite 21	Stellite 6	Stellite 12	Stellite 1	Tribaloy T-800
Density, g cm^{-3} (lb in^{-3})	8.3 (0.30)	8.3 (0.30)	8.6 (0.31)	8.6 (0.31)	8.6 (0.31)
Ultimate compressive strength, MPa (ksi)	1295 (188)	1515 (220)	1765 (256)	1930 (280)	1780 (258)
Ultimate tensile strength, MPa (ksi)	710 (103)	834 (121)	827 (120)	620 (90)	690 (100)
Elongation, %	8	1.2	1	1	<1
Coefficient of thermal expansion, °C^{-1} (°F^{-1})	14.8×10^{-6} (8.2×10^{-6})	15.7×10^{-6} (8.7×10^{-6})	14×10^{-6} (7.8×10^{-6})	13.1×10^{-6} (7.3×10^{-6})	12.3×10^{-6} (6.8×10^{-6})
Hot hardness, HV, at:					
445 °C (800 °F)	150	300	345	510	659
540 °C (1000 °F)	145	275	325	465	622
650 °C (1200 °F)	135	260	285	390	490
760 °C (1400 °F)	115	185	245	230	308

Table 3.1-85 Properties of selected Co-based hard-facing alloys, cont.

Property	Stellite 21	Stellite 6	Stellite 12	Stellite 1	Tribaloy T-800
Unlubricated sliding wear[a], mm^3 (in$^3 \times 10^{-3}$) at:					
670 N (150 lbf)	5.2 (0.32)	2.6 (0.16)	2.4 (0.15)	0.6 (0.04)	1.7 (0.11)
1330 N (300 lbf)	14.5 (0.90)	18.8 (1.17)	18.4 (1.14)	0.8 (0.05)	2.1 (0.13)
Abrasive wear[b], mm^3 (in$^3 \times 10^{-3}$)					
OAW	–	29 (1.80)	12 (0.75)	8 (0.50)	–
GTAW	86 (5.33)	64 (3.97)	57 (3.53)	52 (3.22)	24 (1.49)
Unnotched Charpy impact strength, J (ft × lbf)	37 (27)	23 (17)	5 (4)	5 (4)	1.4 (1)
Corrosion resistance[c]:					
65% nitric acid at 65 °C (150 °F)	U	U	U	U	S
5% sulfuric acid at 65 °C (150 °F)	E	E	E	E	–
50% phosphoric acid at 400 °C (750 °F)	E	E	E	E	E

[a] Wear measured from tests conducted on Dow-Corning LFW-1 against 4620 steel ring at 80 rev/min for 2000 rev varying the applied load
[b] Wear measured from dry sand rubber wheel abrasion tests. Tested for 2000 rev at a load of 135 N (30 lbf) using a 230 mm (9 in) diam rubber wheel and American Foundrymen's Society test sand. OAW, oxyacetylene welding; GTAW, gas-tungsten arc welding
[c] E, less than 0.05 mm/yr (2 mils/year); S, 0.5 to less than 1.25 mm/yr (over 20 to less than 50 mils/year); U, more than 1.25 mm/year (50 mils/year)

Table 3.1-86 Nominal compositions of cast cobalt-based heat-resistant alloys

Alloy designation	Nominal composition (wt%)											
	C	Ni	Cr	Co	Mo	Fe	Al	B	Ti	Ta	W	Zr	Other
AiResist 13	0.45	–	21	62	–	–	3.4	–	–	2	11	–	0.1 Y
AiResist 213	0.20	0.5	20	64	–	0.5	3.5	–	–	6.5	4.5	0.1	0.1 Y
AiResist 215	0.35	0.5	19	63	–	0.5	4.3	–	–	7.5	4.5	0.1	0.1 Y
FSX-414	0.25	10	29	52.5	–	1	–	0.010	–	–	7.5	–	–
Haynes 25 (L-605)	0.1	10	20	54	–	1	–	–	–	–	15	–	–
J-1650	0.20	27	19	36	–	–	–	0.02	3.8	2	12	–	–
MAR-M 302	0.85	–	21.5	58	–	0.5	–	0.005	–	9	10	0.2	–
MAR-M 322	1.0	–	21.5	60.5	–	0.5	–	–	0.75	4.5	9	2	–
MAR-M 509	0.6	10	23.5	54.5	–	–	–	–	0.2	3.5	7.5	0.5	–
NASA (Co–W–Re)	0.40	–	3	67.5	–	–	–	–	1	–	25	1	2Re
S-816	0.4	20	20	42	–	4	–	–	–	–	4	–	4 Mo, 4 Nb, 1.2 Mn, 0.4 Si
V-36	0.27	20	25	42	–	3	–	–	–	–	2	–	4 Mo, 2 Nb, 1 Mn, 0.4 Si
WI-52	0.45	–	21	63.5	–	2	–	–	–	–	11	–	2 Nb + Ta
Stellite 23	0.40	2	24	65.5	–	1	–	–	–	–	5	–	0.3 Mn, 0.6 Si
Stellite 27	0.40	32	25	35	5.5	1	–	–	–	–	–	–	0.3 Mn, 0.6 Si
Stellite 30	0.45	15	26	50.5	6	1	–	–	–	–	–	–	0.6 Mn, 0.6 Si
Stellite 31 (X-40)	0.50	10	22	57.5	–	1.5	–	–	–	–	7.5	–	0.5 Mn, 0.5 Si

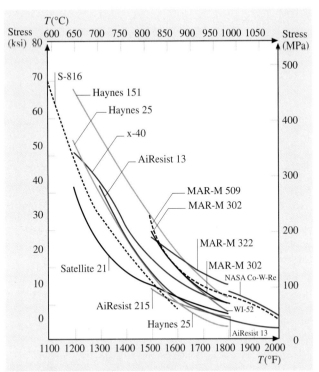

Fig. 3.1-122 Stress–rupture curves for 1000-h life of cast Co-based superalloys

are alloyed with higher Mo contents rather than with W since Mo contributes to their corrosion and oxidation resistance. Table 3.1-87 shows the compositions of various Co-based corrosion-resistant alloys.

The multiphase (MP) alloys MP35N and MP159 combine ultra-high strength, high ductility, and corrosion resistance, including resistance to stress-corrosion cracking in the work-hardened state. The prime strengthening is based on the deformation-induced martensitic transformation of the fcc matrix phase into the hcp phase which has been termed a multiphase reaction. The multiphase microstructure provides an increased density of barriers for slip dislocations. Subsequent annealing leads to a stabilization of the two-phase structure by solute partitioning. Figures 3.1-123 and 3.1-124 show the increase in strength and decrease in ductility for alloys MP35N and Duratherm 600 with work hardening and aging.

Differences of the high-temperature mechanical behavior of these materials are shown in terms of stress-rupture curves in Fig. 3.1-122.

Investment-cast Co alloys are generally used for parts of complex shape such as first- and second-stage vanes and nozzles in gas turbine engines.

3.1.6.4 Co-Based Corrosion-Resistant Alloys

Compared to the heat resistant Co-based alloys, the corrosion-resistant alloys have low C concentrations and

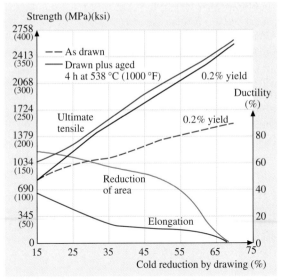

Fig. 3.1-123 Tensile properties of cold-drawn and aged MP35N

Table 3.1-87 Co-based corrosion resistant alloys

Alloy tradename	UNS No.	Nominal Composition (wt%)									
		Co	Cr	W	Mo	C	Fe	Ni	Si	Mn	Others
Ultimet (1233)	R31233	bal	26	2	5	0.06	3	9	0.3	0.8	0.08 N
MP 159	R30159	bal	19	–	7	–	9	25.5	–	–	3 Ti, 0.6 Nb, 0.2 Al
MP35N	R30035	35	20	–	10	–	–	35	–	–	–
Duratherm 600	R30600	41.5	12	3.9	4	0.05 (max)	8.7	bal	0.4	0.75	2 Ti, 0.7 Al, 0.05 Be
Elgiloy	R30003	40	20	–	7	0.15 (max)	bal	15.5	–	2	1 Be (max)
Havar	R30004	42.5	20	2.8	2.4	0.2	bal	13	–	1.6	0.06 Be (max)

bal: balance

3.1.6.5 Co-Based Surgical Implant Alloys

Co-based surgical implant alloys (see Table 3.1-88 for compositions) are used to fabricate a variety of implant parts and devices. These are predominantly implants for hip and knee joint replacements, implants that fix bone fractures such as bone screws, staples, plates, support structures for heart valves, and dental implants. The mechanical properties (shown in Table 3.1-89) depend sensitively on the thermal and thermomechanical treatments of the materials.

3.1.6.6 Cemented Carbides

The term cemented carbides, also called hardmetals, refers to powder-composite materials consisting of carbide particles bonded with metals or alloys. Extensive treatments are given in [1.94, 95]. The most common cemented carbide is WC bonded with Co. Cobalt is used as a binder since it wets the angular WC particles particularly well. Nickel is added to increase corrosion and oxidation resistance of the Co binder phase. The metals Ta, Nb, and Ti may be added to form a (W, Ta, Nb, or Ti) C solid solution carbide phase which is an additional microstructural constituent in the form of rounded particles in the so-called complex grade, multigrade, or steel-cutting grade cemented carbides. Table 3.1-90 lists representative materials.

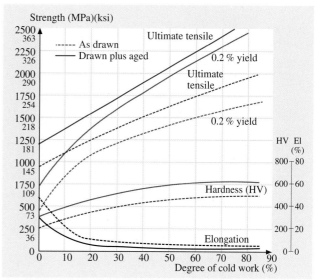

Fig. 3.1-124 Tensile properties of cold-drawn and aged Duratherm 600

Table 3.1-88 Compositions of Co-based surgical implant alloys

ASTM specification	Composition (wt%)								
	Co	Cr	Ni	Mo	Fe	C	Mn	Si	Other
F75	bal	27.0–30.0	1.0	5.0–7.0	0.75	0.35	1.0	1.0	–
F90	bal	19.0–21.0	9.0–11.0	–	3 (max)	0.05–0.15	1.0–2.0	0.4	14.0–16.0 W
F562	bal	19.0–21.0	33.0–37.0	9.0–10.5	1 (max)	0.025 (max)	0.15 (max)	0.15 (max)	1.0 Ti (max)

bal: balance

Table 3.1-89 Mechanical properties of Co-based surgical implant alloys

ASTM specification	Alloy system	Condition	Yield strength (MPa)	(ksi)	Tensile strength (MPa)	(ksi)	Elongation (%)	Elastic modulus (GPa)	(10⁶ ksi)
F75	Co–Cr–Mo	Cast	450	65	655	95	8	248	36
F799	Co–Cr–Mo	Thermomechanically processed	827	120	1172	170	12	–	–
F90	Co–Cr–W–Ni	Wrought	379	55	896	130	–	242	35
F562	Co–Ni–Cr–Mo	Annealed, cold-worked and aged	241–448 1586	35–65 230	793–1000 1793	115–145 260	50 8	228 –	33 –

Table 3.1-90 Compositions, microstructures and properties of representative Co-bonded cemented carbides

Nominal composition	Grain size	Hardness (HRA)	Density (g cm^{-3})	Density (oz in^{-3})	Transverse strength (MPa)	Transverse strength (ksi)	Compressive strength (MPa)	Compressive strength (ksi)	Modulus of elasticit (GPa)	Modulus of elasticit (10^6 psi)	Relative abrasion resistance[a]	Coefficient of thermal expansion (μm/m K) at 200°C (390°F)	at 1000°C (1830°F)	Thermal conductivity (W/m K)
97WC–3Co	Medium	92.5–93.2	15.3	8.85	1590	230	5860	850	641	93	100	4.0	–	121
94WC–6Co	Fine	92.5–93.1	15.0	8.67	1790	260	5930	860	614	89	100	4.3	5.9	–
	Medium	91.7–92.2	15.0	8.67	2000	290	5450	790	648	94	58	4.3	5.4	100
	Coarse	90.5–91.5	15.0	8.67	2210	320	5170	750	641	93	25	4.3	5.6	121
90WC–10Co	Fine	90.7–91.3	14.6	8.44	3100	450	5170	750	620	90	22	–	–	–
	Coarse	87.4–88.2	14.5	8.38	2760	400	4000	580	552	80	7	5.2	–	1,12
84WC–16Co	Fine	89	13.9	8.04	3380	490	4070	590	524	76	5	–	–	–
	Coarse	86.0–87.5	13.9	8.04	2900	420	3860	560	524	76	5	5.8	7.0	88
75WC–25Co	Medium	83–85	13.0	7.52	2550	370	3100	450	483	70	3	6.3	–	71
71WC–12.5TiC–12TaC–4.5Co	Medium	92.1–92.8	12.0	6.94	1380	200	5790	840	565	82	11	5.2	6.5	35
72WC–8TiC–11.5TaC–8.5Co	Medium	90.7–91.5	12.6	7.29	1720	250	5170	750	558	81	13	5.8	6.8	50

[a] Based on a value of 100 for the most abrasion-resistant material

3.1.7 Nickel and Nickel Alloys

Nickel is the base element for a variety of Ni alloys. But it is mainly used as a major alloying element in Fe- and Cu-based alloys, especially in stainless steels. A further major use is Ni plating. A survey of its applications is given in Table 3.1-91. It also indicates where Ni bearing alloys are covered in this handbook. Nickel materials are treated extensively in [1.96] and the special group of heat resistant alloys and superalloys in [1.97]. Alloys of Ni–Fe show ferromagnetism in a wide range of compositions. This, in combination with other intrinsic magnetic properties, is the basis for the Ni–Fe based alloys with soft magnetic and controlled thermal expansion properties, respectively, covered in Chapt. 3.4.3. NiTi shape memory alloys are treated in Sect. 3.1.3.4.

Data on the electronic structure of Ni and Ni alloys may be found in [1.98, 99], phase diagrams, crystal structures and thermodynamic data of binary Ni alloys are contained in [1.100].

3.1.7.1 Commercially Pure and Low-Alloy Nickels

Commercially pure and low-alloy nickels are used as corrosion-resistant materials with very good formability. Some Ni-based alloys are precipitation hardened; Ni–2 wt% Be (Duranickel 301) shows pronounced precipitation hardening. Commonly used, commercially pure and low-alloy nickels are listed in Table 3.1-92. Room-temperature mechanical properties of these and

Table 3.1-91 Uses of nickel

Use	Fraction of total Ni consumption (%)	Section
Stainless steels	63	Sect. 3.1.5.4
Nickel-based alloys	12	Sect. 3.1.7
Nickel plating	10	Sect. 3.1.7
Nickel in alloy steels	9	Sect. 3.1.5
Nickel in foundry products	3.5	Sect. 3.1.5.7
Nickel in copper alloys	1.5	Sect. 3.1.8
Others	1	–

Table 3.1-92 Nominal compositions of corrosion-resistant nickels and nickel-base alloys

Alloy	UNS No.	Composition (wt%)							
		Ni	Cu	Fe	Mn	C	Si	S	Other
Commercially pure nickels									
Nickel 200	N02200	99.0 (min)	0.25	0.40	0.35	0.15	0.35	0.01	–
Nickel 201	N02201	99.0 (min)	0.25	0.40	0.35	0.02	0.35	0.01	–
Nickel 205	N02205	99.0 (min(b))	0.15	0.20	0.35	0.15	0.15	0.008	0.01–0.08 Mg, 0.01–0.05 Ti
Nickel 212	–	97.0 (min)	0.20	0.25	1.5–2.5	0.10	0.20	–	0.20 Mg
Nickel 220	N02220	99.0 (min)	0.10	0.10	0.20	0.15	0.01–0.05	0.008	0.01–0.08 Mg
Nickel 225	N02225	99.0 (min)	0.10	0.10	0.20	0.15	0.15–0.25	0.008	0.01–0.08 Mg, 0.01–0.05 Ti
Nickel 230	N02230	99.0 (min)	0.10	0.10	0.15	0.15	0.01–0.035	0.008	0.04–0.08 Mg, 0.005 Ti
Nickel 270	N02270	99.97 (min)	0.001	0.005	0.02	0.02	0.001	0.001	0.001 Co, 0.001 Cr, 0.001 Ti
Low-alloy nickels									
Nickel 211	N02211	93.7 (min)	0.25	0.75	4.25–5.25	0.20	0.15	0.015	–
Duranickel 301	N03301	93.00 (min)	0.25	0.60	0.50	0.30	1.00	0.01	4.00–4.75 Al, 0.25–1.00 Ti
Alloy 360	N03360	bal	–	–	–	–	–	–	1.85–2.05 Be, 0.4–0.6 Ti

bal: balance

Table 3.1-93 Mechanical properties of nickel-based alloys at room temperature

Alloy	Ultimate tensile strength (MPa)	(ksi)	Yield strength (0.2% offset) (MPa)	(ksi)	Elongation in 50 mm (2 in) (%)	Elastic modulus (tension) (GPa)	(10^6 psi)	Hardness
Nickel 200	462	67	148	21.5	47	204	29.6	109 HB
Nickel 201	403	58.5	103	15	50	207	30	129 HB
Nickel 205	345	50	90	13	45	–	–	–
Nickel 211	530	77	240	35	40	–	–	–
Nickel 212	483	70	–	–	–	–	–	–
Nickel 222	380	55	–	–	–	–	–	–
Nickel 270	345	50	110	16	50	–	–	30 HRB
Duranickel 301 (precipitation hardened)	1170	170	862	125	25	207	30	30–40 HRC
Alloy 400	550	80	240	35	40	180	26	110–150 HB
Alloy 401	440	64	134	19.5	51	–	–	–
Alloy R-405	550	80	240	35	40	180	26	110–140 HB
Alloy K-500 (precipitation hardened)	1100	160	790	115	20	180	26	300 HB
Alloy 600	655	95	310	45	40	207	30	75 HRB
Alloy 601	620	90	275	40	45	207	30	65–80 HRB
Alloy 617 (solution annealed)	755	110	350	51	58	211	30.6	173 HB
Alloy 625	930	135	517	75	42.5	207	30	190 HB
Alloy 690	725	105	348	50.5	41	211	30.6	88 HRB
Alloy 718 (precipitation hardened)	1240	180	1036	150	12	211	30.6	36 HRC
Alloy C-22	785	114	372	54	62	–	–	209 HB
Alloy C-276	790	115	355	52	61	205	29.8	90 HRB
Alloy G3	690	100	320	47	50	199	28.9	79 HRB
Alloy 800	600	87	295	43	44	193	28	138 HB
Alloy 825	690	100	310	45	45	206	29.8	–
Alloy 925[a]	1210	176	815	118	24	–	–	36.5 HRC

Properties are for annealed sheet unless otherwise indicated.
[a] Annealed at 980 °C (1800 °F) for 30 min, air cooled, and aged at 760 °C (1400 °F) for 8 h, furnace at a rate of 55 °C (1150 °F) for 8 h, air cooled

some further Ni alloys which are dealt with in the next section are listed in Table 3.1-93.

3.1.7.2 Highly Alloyed Ni-Based Materials

Nickel forms extensive solid solutions with many alloying elements: complete solid solutions with Fe and Cu, and limited solid solutions with ≤ 35 wt% Cr, ≤ 20 wt% Mo, $\leq 5-10$ wt% Al, Ti, and Mn; and V. Nickel and its alloys are providing favorable properties for uses in corrosive environments and at elevated temperatures.

The extensive solubility of several of the alloying elements is the basis of solid solution hardening which scales roughly with the atomic-size difference of the solute and is, therefore, pronounced with W, Mo, Nb, Ta, and Al. Based on the face-centered cubic structure, Ni-based solid solutions show high ductility, fracture toughness, and formability. The basic corrosion resistance of Ni is strongly increased by alloying additions of Cr, Mo, and W.

On the basis of these possibilities of materials design, a number of corrosion-resistant materials have been developed according to the criteria shown in

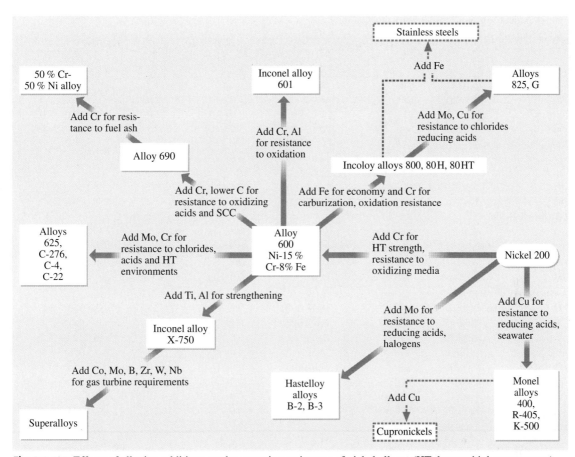

Fig. 3.1-125 Effects of alloying additions on the corrosion resistance of nickel alloys. (HT denotes high temperature)

Fig. 3.1-125. The materials listed in Table 3.1-94 are the main representatives of high-Ni alloys.

Mechanical properties are listed in Table 3.1-93. Characteristic creep data are shown in Fig. 3.1-126.

Table 3.1-94 Nominal compositions of highly-alloyed Ni

Designation	UNS Nr	Ni	Cr	Fe	Co	Mo
Nickel-molybdenum alloys						
Hastelloy B	N10001	bal	1.0	6.0	2.5	26.0–33.0
Hastelloy B-2	N10665	bal	1.0	2.0	1.0	26.0–30.0
Hastelloy B-3	N10675	65.0	1.0–3.0	1.0–3.0	3.0	27.0–32.0
Nicrofer 6629 (B-4)	N10629	bal	0.5–1.5	1.0–6.0	2.5	26.0–30.0
Nickel-chromium-iron alloys						
Inconel 600	N06600	72.0 (min(b))	14.0–17.0	6.0–10.0	–	–
Inconel 601	N06601	58.0–63.0	21.0–25.0	bal	–	–
Inconel 690	N06690	58.0 (min)	27.0–31.0	7.0–11.0	–	–
Haynes 214	N07214	bal	15.0–17.0	2.0–4.0	2.0	0.5
Iron-nickel-chromium alloys						
Incoloy 800	N08800	30.0–35.0	19.0–23.0	39.5 (min)	–	–
Incoloy 800HT	N08811	30.0–35.0	19.0–23.0	39.5 (min)	–	–
Incoloy 801	N08801	30.0–34.0	19.0–22.0	bal	–	–
Incoloy 803	S35045	32.0–37.0	25.0–29.0	37.0 (min)	–	–
Nickel-chromium-molybdenum alloys						
Hastelloy C	N10002	bal	14.5–16.50	4.0–7.0	2.5	15.0–17.0
Hastelloy C-4	N06455	bal	14.0–18.0	3.0	2.0	14.0–17.0
Hastelloy C-22	N06022	bal	20.0–22.5	2.0–6.0	2.5	12.5–14.5
Hastelloy C-276	N10276	bal	14.5–16.50	4.0–7.0	2.5	15.0–17.0
Hastelloy C-2000	N06200	bal	22.0–24.0	3.0	2.0	15.0–17.0
Nicrofer 5923 (Alloy 59)	N06059	bal	22.0–24.0	1.5	0.3	15.0–16.5
Inconel 617	N06617	44.5 (min)	20.0–24.0	3.0	10.0–15.0	8.0–10.0
Inconel 625	N06625	58.0 (min)	20.0–23.0	5.0	1.0	8.0–10.0
Inconel 686	N06686	bal	19.0–23.0	5.0	–	15.0–17.0
Hastelloy S	N06635	bal	14.5–17.0	3.0	2.0	14.0–16.5
Allcorr	N06110	bal	27.0–33.0	–	12.0	8.0–12.0
Nickel-chromium-iron-molybdenum alloys						
Incoloy 825	N08825	38.0–46.0	19.5–23.5	bal	–	2.5–3.5
Hastelloy G	N06007	bal	21.0–23.5	18.0–21.0	2.5	5.5–7.5
Hastelloy G-2	N06975	47.0–52.0	23.0–26.0	bal	–	5.0–7.0
Hastelloy G-3	N06985	bal	21.0–23.5	18.0–21.0	5.0	6.0–8.0
Hastelloy G-30	N06030	bal	28.0–31.5	13.0–17.0	5.0	4.0–6.0
Hastelloy G-50	N06950	50.0 (min)	19.0–21.0	15.0–20.0	2.5	8.0–10.0
Hastelloy D-205	–	65.0	20.0	6.0	–	2.5
Hastelloy N	N10003	bal	6.0–8.0	5.0	0.2	15.0–18.0
Hastelloy X	N06002	bal	20.5–23.0	17.0–20.0	0.5–2.5	8.0–10.0
Nicrofer 3033 (Alloy 33)	R20033	30.0–33.0	31.0–35.0	bal	–	0.5–2.0
Nickel-chromium-tungsten, nickel-iron-chromium, and nickel-cobalt-chromium-silicon alloys						
Haynes 230	N06230	bal	20.0–24.0	3.0	–	1.0–3.0
Haynes HR-120	N08120	35.0–39.0	23.0–27.0	bal	3.0	2.5
Haynes HR-160	N12160	bal	26.0–30.0	3.5	27.0–33.0	1.0
Precipitation-hardening alloys						
Alloy 625 Plus	N07716	57.0–63.0	19.0–22.0	bal	–	7.0–9.50
Inconel 718	N07718	50.0–55.0	17.0–21.0	bal	1.0	2.80–3.30
Inconel 725	N07725	55.0–59.0	19.0–22.5	bal	–	7.0–9.50
Inconel 925	N09925	38.0–46.0	19.5–23.5	22.0 (min)	–	2.50–3.50

(a) Single values are maximum unless otherwise indicated; (b) Nickel plus cobalt content; bal: balance

Table 3.1-94 Nominal compositions of highly alloyed Ni alloys, cont.

W	Nb	Ti	Al	C	Mn	Si	B	Others
(wt%)								
Nickel-molybdenum alloys								
–	–	–	–	0.12	1.0	1.0	–	0.60 V
–	–	–	–	0.02	1.0	0.10	–	–
3.0	0.20	0.20	0.50	0.01	3.0	0.10	–	0.20 Ta
–	–	–	0.1–0.5	0.01	1.5	0.05	–	–
Nickel-chromium-iron alloys								
–	–	–	–	0.15	1.0	0.5	–	0.5 Cu
–	–	–	1.0–1.7	0.10	1.0	0.50	–	1.0 Cu
–	–	–	–	0.05	0.50	0.50	–	0.50 Cu
0.5	–	0.5	4.0–5.0	0.05	0.5	0.2	0.006	0.05 Zr, 0.002–0.040 Y
Iron-nickel-chromium alloys								
–	–	0.15–0.60	0.15–0.60	0.10	1.5	1.0	–	–
–	–	0.15–0.60	0.15–0.60	0.06–0.10	1.5	1.0	–	0.895–1.20 Al+Ti
–	–	0.75–1.5	–	0.10	1.5	1.0	–	0.5 Cu
–	–	0.15–0.60	0.15–0.60	0.06–0.10	1.5	1.0	–	0.75 Cu
Nickel-chromium-molybdenum alloys								
3.0–4.5	–	–	–	0.08	1.0	1.0	–	0.35 V
–	–	0.70	–	0.015	1.0	0.08	–	–
2.5–3.5	–	–	–	0.015	0.5	0.08	–	0.35 V
3.0–4.5	–	–	–	0.02	1.0	0.08	–	0.35 V
–	–	–	0.5	0.010	0.5	0.08	–	1.3–1.9 Cu
–	–	–	0.1–0.4	0.010	0.5	0.10	–	–
–	–	0.6	0.8–1.5	0.05–0.15	1.0	1.0	0.006	0.5 Cu
–	3.15–4.15	0.40	0.40	0.10	0.50	0.50	–	–
3.0–4.4	–	0.02–0.25	–	0.010	0.75	0.08	–	–
1.0	–	–	0.1–0.50	0.02	0.3–1.0	0.20–0.75	0.015	0.35 Cu, 0.01–0.10 La
4.0	2.0	1.50	1.50	0.15	–	–	–	–
Nickel-chromium-iron-molybdenum alloys								
–	–	0.6–1.2	0.2	0.05	1.0	0.5	–	–
1.0	1.75–2.5	–	–	0.05	1.0–2.0	1.0	–	–
–	–	0.70–1.5	–	0.03	1.0	1.0	–	0.7–1.20 Cu
1.5	–	–	–	0.015	1.0	1.0	–	1.5–2.5 Cu, 0.50 Nb+Ta
1.5–4.0	0.3–1.5	–	–	0.03	1.5	0.8	–	1.0–2.4 Cu
1.0	0.5	–	–	0.015	1.0	1.0	–	0.5 Cu
–	–	–	–	0.03	–	5.0	–	2.0 Cu
0.5	–	–	–	0.010	1.0	1.0	0.010	0.50 V, 0.35 Cu
0.20–1.0	–	–	–	0.05–0.15	1.0	1.0	–	–
–	–	–	–	0.015	2.0	0.5	–	0.3–1.2 Cu
Nickel-chromium-tungsten, nickel-iron-chromium, and nickel-cobalt-chromium-silicon alloys								
13.0–15.0	–	–	0.20–0.50	0.05–0.15	0.3–1.0	0.25–0.75	0.015	0.005–0.05 La
2.5	0.4–0.9	0.20	0.40	0.02–0.1	1.5	1.0	0.010	0.5 Cu
1.0	–	0.20–0.80	–	0.15	1.5	2.4–3.0	–	–
Precipitation-hardening alloys								
–	2.75–4.0	1.0–1.60	0.35	0.03	0.20	0.20	–	–
–	4.75–5.50	0.35	0.20–0.80	0.08	0.35	0.35	0.06	0.3 Cu
–	2.75–4.0	1.0–1.70	0.35	0.03	0.35	0.50	–	–
–	0.50	1.9–2.40	0.10–0.50	0.03	1.0	0.50	–	1.5–3.0 Cu

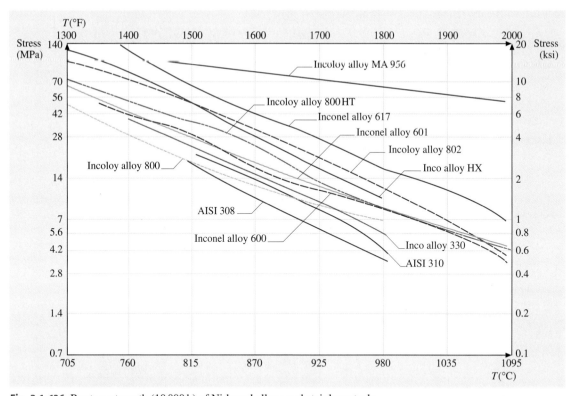

Fig. 3.1-126 Rupture strength (10 000 h) of Ni-based alloys and stainless steels

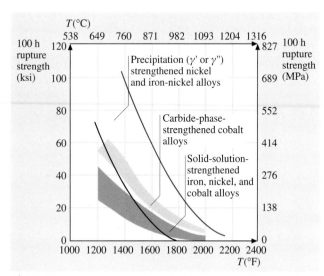

Fig. 3.1-127 Stress–rupture characteristics of wrought superalloys

3.1.7.3 Ni-Based Superalloys

The term superalloy is used for a group of nickel-, iron–nickel-, and cobalt-based high-temperature materials for applications at temperatures $\geq 540\,°C$. It is useful to compare the main subgroups in terms of the strengthening mechanisms applied and stress–rupture characteristics achieved, as shown in Fig. 3.1-127. In this section iron–nickel- and nickel-based superalloys are covered whereas cobalt-based superalloys are dealt with in Sect. 3.1.6.3. Nickel-based superalloys are among the most complex metallic materials with numerous alloying elements serving particular functions, as briefly outlined here.

All Ni and Fe–Ni-based superalloys are alloyed with Al. This leads to a two-phase matrix consisting of the γ-(Ni, Fe, Al) solid solution phase (fcc, A1 structure) and the intermetallic γ'-Ni$_3$Al phase (L1$_2$ structure), which has a superlattice structure relative to the fcc structure of the γ phase. The binary Al–Ni phase diagram in Fig. 3.1-128 shows clearly that the γ' phase is stable up to the melting range. The matrix phase γ is solid solution strengthened by alloying additions of Cr, Mo, W, and

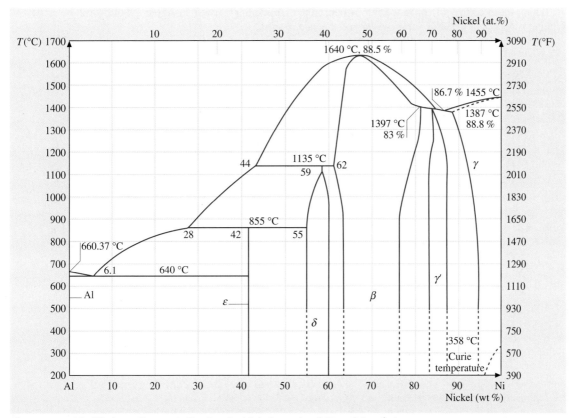

Fig. 3.1-128 Al–Ni phase diagram indicating the high thermal stability of the γ' phase

Ta. The γ'-Ni$_3$Al phase is precipitated coherently from the γ phase by appropriate heat treatment in amounts ranging from 20 to 45 vol% γ' in wrought superalloys and about 60 vol% γ' in cast superalloys. The hardening effect of γ' in these alloys is due to its high yield stress which increases with increasing temperature to a maximum at about 650 °C, as shown in Fig. 3.1-129. The anomalous temperature dependence is due to the fact that the plastic deformation is associated with the formation and motion of partial dislocations, giving rise to local disordering of the ordered structure. Figure 3.1-129 shows also that alloying additions such as Cr, Ti, and Nb increase the yield stress and tend to shift its maximum to higher temperatures. This is because several alloying elements substitute Ni (Fe, Co, Cr) or Al (Ti, Nb) in the Ni$_3$Al structure and raise its critical resolved shear stress. Another major alloying addition is carbon in combination with carbide forming alloy components such as Ti, Cr, Mo, W, and Nb (which are partly added to achieve hardening of the metallic phases as well, as described above). The carbides (MC, M$_7$C$_3$, M$_{23}$C$_6$, and M$_6$C) are

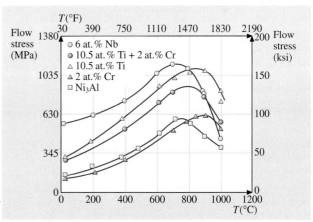

Fig. 3.1-129 Temperature dependence of the flow stress of the γ'-Ni$_3$Al phase and effects of Nb, Ti, and Cr alloying additions

precipitated or formed by secondary reactions as intragrain or grain boundary precipitates further stabilizing the microstructure against creep deformation.

Taking these factors of influence into account the wrought superalloys listed in Table 3.1-95 and the cast superalloys listed in Table 3.1-96 have been developed.

Table 3.1-95 Nominal compositions of Fe–Ni-based and Ni-based wrought superalloys

Alloy	UNS No.	Composition (wt%)										
		Cr	Ni	Co	Mo	W	Nb	Ti	Al	Fe	C	Other
Solid-solution alloys												
Iron-nickel-based												
Alloy N-155 (Multimet)	R30155	21.0	20.0	20.0	3.00	2.5	1.0	–	–	32.2	0.15	0.15 N, 0.2 La, 0.02 Zr
Haynes 556	R30556	22.0	21.0	20.0	3.0	2.5	0.1	–	0.3	29.0	0.10	0.50 Ta, 0.02 La, 0.002 Zr
19-9 DL	S63198	19.0	9.0	–	1.25	1.25	0.4	0.3	–	66.8	0.30	1.10 Mn, 0.60 Si
Incoloy 800	N08800	21.0	32.5	–	–	–	–	0.38	0.38	45.7	0.05	–
Incoloy 800H	N08810	21.0	33.0	–	–	–	–	–	–	45.8	0.08	–
Incoloy 800HT	N08811	21.0	32.5	–	–	–	–	0.4	0.4	46.0	0.08	0.8 Mn, 0.5 Si, 0.4 Cu
Incoloy 801	N08801	20.5	32.0	–	–	–	–	1.13	–	46.3	0.05	–
Incoloy 802	–	21.0	32.5	–	–	–	–	0.75	0.58	44.8	0.35	–
Nickel-based												
Haynes 214	–	16.0	76.5	–	–	–	–	–	4.5	3.0	0.03	–
Haynes 230	N06230	22.0	55.0	5.0 (max)	2.0	14.0	–	–	0.35	3.0 (max)	0.10	0.015 (max) B, 0.02 La
Inconel 600	N06600	15.5	76.0	–	–	–	–	–	–	8.0	0.08	0.25 Cu
Inconel 601	N06601	23.0	60.5	–	–	–	–	–	1.35	14.1	0.05	0.5 Cu
Inconel 617	N06617	22.0	55.0	12.5	9.0	–	–	–	1.0	–	0.07	–
Inconel 625	N06625	21.5	61.0	–	9.0	–	3.6	0.2	0.2	2.5	0.05	–
RA 333	N06333	25.0	45.0	3.0	3.0	3.0	–	–	–	18.0	0.05	–
Hastelloy B	N01001	1.0 (max)	63.0	2.5 (max)	28.0	–	–	–	–	5.0	0.05 (max)	0.03 V
Hastelloy N	N01003	7.0	72.0	–	16.0	–	–	0.5 (max)	–	5.0 (max)	0.06	–
Hastelloy S	N06635	15.5	67.0	–	15.5	–	–	–	0.2	1.0	0.02 (max)	0.02 La
Hastelloy W	N10004	5.0	61.0	2.5 (max)	24.5	–	–	–	–	5.5	0.12 (max)	0.6 V
Hastelloy X	N06002	22.0	49.0	1.5 (max)	9.0	0.6	–	–	2.0	15.8	0.15	–
Hastelloy C-276	N10276	15.5	59.0	–	16.0	3.7	–	–	–	5.0	0.02 (max)	–
Haynes HR-120	N08120	25.0	37.0	3.0	2.5	2.5	0.7	–	0.1	33.0	0.05	0.7 Mn, 0.6 Si, 0.2 N, 0.004 B
Haynes HR-160	N12160	28.0	37.0	29.0	–	–	–	–	–	2.0	0.05	2.75 Si, 0.5 Mn
Nimonic 75	N06075	19.5	75.0	–	–	–	–	0.4	0.15	2.5	0.12	0.25 (max) Cu
Nimonic 86	–	25.0	65.0	–	10.0	–	–	–	–	–	0.05	0.03 Ce, 0.015 Mg
Precipitation-hardening alloys												
Iron-nickel-based												
A-286	S66286	15.0	26.0	–	1.25	–	–	2.0	0.2	55.2	0.04	0.005 B, 0.3 V
Discaloy	S66220	14.0	26.0	–	3.0	–	–	1.7	0.25	55.0	0.06	–
Incoloy 903	N19903	0.1 (max)	38.0	15.0	0.1	–	3.0	1.4	0.7	41.0	0.04	–
Pyromet CTX-1	–	0.1 (max)	37.7	16.0	0.1	–	3.0	1.7	1.0	39.0	0.03	–
Incoloy 907	N19907	–	38.4	13.0	–	–	4.7	1.5	0.03	42.0	0.01	0.15 Si
Incoloy 909	N19909	–	38.0	13.0	–	–	4.7	1.5	0.03	42.0	0.01	0.4 Si
Incoloy 925	N09925	20.5	44.0	–	2.8	–	–	2.1	0.2	29	0.01	1.8 Cu
V-57	–	14.8	27.0	–	1.25	–	–	3.0	0.25	48.6	0.08 (max)	0.01 B, 0.5 (max) V
W-545	S66545	13.5	26.0	–	1.5	–	–	2.85	0.2	55.8	0.08 (max)	0.05 B

Table 3.1-95 Nominal compositions of Fe−Ni-based and Ni-based wrought superalloys, cont.

Alloy	UNS No.	Cr	Ni	Co	Mo	W	Nb	Ti	Al	Fe	C	Other
Nickel-based												
Astroloy	N13017	15.0	56.5	15.0	5.25	−	−	3.5	4.4	<0.3	0.06	0.03 B, 0.06 Zr
Custom Age 625 PLUS	N07716	21.0	61.0	−	8.0	−	3.4	1.3	0.2	5.0	0.01	−
Haynes 242	−	8.0	62.5	2.5 (max)	25.0	−	−	−	0.5 (max)	2.0 (max)	0.10 (max)	0.006 (max) B
Haynes 263	N07263	20.0	52.0	−	6.0	−	−	2.4	0.6	0.7	0.06	0.6 Mn, 0.4 Si, 0.2 Cu
Haynes R-41	N07041	19.0	52.0	11.0	10.0	−	−	3.1	1.5	5.0	0.09	0.5 Si, 0.1 Mn, 0.006 B
Inconel 100	N13100	10.0	60.0	15.0	3.0	−	−	4.7	5.5	<0.6	0.15	1.0 V, 0.06 Zr, 0.015 B
Inconel 102	N06102	15.0	67.0	−	2.9	3.0	2.9	0.5	0.5	7.0	0.06	0.005 B, 0.02 Mg, 0.03 Zr
Incoloy 901	N09901	12.5	42.5	−	6.0	−	−	2.7	−	36.2	0.10 (max)	−
Inconel 702	N07702	15.5	79.5	−	−	−	−	0.6	3.2	1.0	0.05	0.5 Mn, 0.2 Cu, 0.4 Si
Inconel 706	N09706	16.0	41.5	−	−	−	−	1.75	0.2	37.5	0.03	2.9 (Nb+Ta), 0.15 (max) Cu
Inconel 718	N07718	19.0	52.5	−	3.0	−	5.1	0.9	0.5	18.5	0.08 (max)	0.15 (max) Cu
Inconel 721	N07721	16.0	71.0	−	−	−	−	3.0	−	6.5	0.04	2.2 Mn, 0.1 Cu
Inconel 722	N07722	15.5	75.0	−	−	−	−	2.4	0.7	7.0	0.04	0.5 Mn, 0.2 Cu, 0.4 Si
Inconel 725	N07725	21.0	57.0	−	8.0	−	3.5	1.5	0.35 (max)	9.0	0.03 (max)	
Inconel 751	N07751	15.5	72.5	−	−	−	1.0	2.3	1.2	7.0	0.05	0.25 (max) Cu
Inconel X-750	N07750	15.5	73.0	−	−	−	1.0	2.5	0.7	7.0	0.04	0.25 (max) Cu
M-252	N07252	19.0	56.5	10.0	10.0	−	−	2.6	1.0	<0.75	0.15	0.005 B
Nimonic 80A	N07080	19.5	73.0	1.0	−	−	−	2.25	1.4	1.5	0.05	0.10 (max) Cu
Nimonic 90	N07090	19.5	55.5	18.0	−	−	−	2.4	1.4	1.5	0.06	−
Nimonic 95	−	19.5	53.5	18.0	−	−	−	2.9	2.0	5.0 (max)	0.15 (max)	+B, +Zr
Nimonic 100	−	11.0	56.0	20.0	5.0	−	−	1.5	5.0	2.0 (max)	0.30 (max)	+B, +Zr
Nimonic 105	−	15.0	54.0	20.0	5.0	−	−	1.2	4.7	−	0.08	0.005 B
Nimonic 115	−	15.0	55.0	15.0	4.0	−	−	4.0	5.0	1.0	0.20	0.04 Zr
C-263	N07263	20.0	51.0	20.0	5.9	−	−	2.1	0.45	0.7 (max)	0.06	−
Pyromet 860	−	13.0	44.0	4.0	6.0	−	−	3.0	1.0	28.9	0.05	0.01 B
Pyromet 31	N07031	22.7	55.5	−	2.0	−	1.1	2.5	1.5	14.5	0.04	0.005 B
Refractaloy 26	−	18.0	38.0	20.0	3.2	−	−	2.6	0.2	16.0	0.03	0.015 B
René 41	N07041	19.0	55.0	11.0	10.0	−	−	3.1	1.5	<0.3	0.09	0.01 B
René 95	−	14.0	61.0	8.0	3.5	3.5	3.5	2.5	3.5	<0.3	0.16	0.01 B, 0.05 Zr
René 100	−	9.5	61.0	15.0	3.0	−	−	4.2	5.5	1.0 (max)	0.16	0.015 B, 0.06 Zr, 1.0 V
Udimet 500	N07500	19.0	48.0	19.0	4.0	−	−	3.0	3.0	4.0 (max)	0.08	0.005 B
Udimet 520	−	19.0	57.0	12.0	6.0	1.0	−	3.0	2.0	−	0.08	0.005 B
Udimet 630	−	17.0	50.0	−	3.0	3.0	6.5	1.0	0.7	18.0	0.04	0.004 B
Udimet 700	−	15.0	53.0	18.5	5.0	−	−	3.4	4.3	<1.0	0.07	0.03 B
Udimet 710	−	18.0	55.0	14.8	3.0	1.5	−	5.0	2.5	−	0.07	0.01 B
Unitemp AF2-1DA	N07012	12.0	59.0	10.0	3.0	6.0	−	3.0	4.6	<0.5	0.35	1.5 Ta, 0.015 B, 0.1 Zr
Waspaloy	N07001	19.5	57.0	13.5	4.3	−	−	3.0	1.4	2.0 (max)	0.07	0.006 B, 0.09 Zr

Table 3.1-96 Nominal compositions of Ni-based cast superalloys

Alloy designation	Nominal composition (wt%)												
	C	Ni	Cr	Co	Mo	Fe	Al	B	Ti	Ta	W	Zr	Other
Nickel-based													
B-1900	0.1	64	8	10	6	–	6	0.015	1	4[a]	–	0.10	–
CMSX-2	–	66.2	8	4.6	0.6	–	56	–	1	6	8	6	–
Hastelloy X	0.1	50	21	1	9	18	–	–	–	–	1	–	–
Inconel 100	0.18	60.5	10	15	3	–	5.5	0.01	5	–	–	0.06	1 V
Inconel 713C	0.12	74	12.5	–	4.2	–	6	0.012	0.8	1.75	–	0.1	0.9 Nb
Inconel 713LC	0.05	75	12	–	4.5	–	6	0.01	0.6	4	–	0.1	–
Inconel 738	0.17	61.5	16	8.5	1.75	–	3.4	0.01	3.4	–	2.6	0.1	2 Nb
Inconel 792	0.2	60	13	9	2.0	–	3.2	0.02	4.2	–	4	0.1	2 Nb
Inconel 718	0.04	53	19	–	3	18	0.5	–	0.9	–	–	–	0.1 Cu, 5 Nb
X-750	0.04	73	15	–	–	7	0.7	–	2.5	–	–	–	0.25 Cu, 0.9 Nb
M-252	0.15	56	20	10	10	–	1	0.005	2.6	–	–	–	–
MAR-M 200	0.15	59	9	10	–	1	5	0.015	2	–	12.5	0.05	1 Nb[b]
MAR-M 246	0.15	60	9	10	2.5	–	5.5	0.015	1.5	1.5	10	0.05	–
MAR-M 247	0.15	59	8.25	10	0.7	0.5	5.5	0.015	1	3	10	0.05	1.5 Hf
PWA 1480	–	bal	10	5.0	–	–	5.0	–	1.5	12	4.0	–	–
René 41	0.09	55	19	11.0	10.0	–	1.5	0.01	3.1	–	–	–	–
René 77	0.07	58	15	15	4.2	–	4.3	0.015	3.3	–	–	0.04	–
René 80	0.17	60	14	9.5	4	–	3	0.015	5	–	4	0.03	–
René 80 Hf	0.08	60	14	9.5	4	–	3	0.015	4.8	–	4	0.02	0.75 Hf
René 100	0.18	61	9.5	15	3	–	5.5	0.015	4.2	–	–	0.06	1 V
René N4	0.06	62	9.8	7.5	1.5	–	4.2	0.004	3.5	4.8	6	–	0.5 Nb, 0.15 Hf
Udimet 500	0.1	53	18	17	4	2	3	–	3	–	–	–	–
Udimet 700	0.1	53.5	15	18.5	5.25	–	4.25	0.03	3.5	–	–	–	–
Udimet 710	0.13	55	18	15	3	–	2.5	–	5	–	1.5	0.08	–
Waspaloy	0.07	57.5	19.5	13.5	4.2	1	1.2	0.005	3	–	–	0.09	–
WAX-20 (DS)	0.20	72	–	–	–	–	6.5	–	–	–	20	1.5	–

[a] B-1900 + Hf also contains 1.5% Hf
[b] MAR-M 200 + Hf also contains 1.5% Hf
bal: balance

The mechanical properties as a function of temperature of the wrought superalloys are listed in Tables 3.1-97 and 3.1-98, those for the cast alloys in Table 3.1-99. The temperature dependence of the mechanical properties is essentially dependent on the volume fraction, composition, and resulting flow stress of the γ' phase.

Figure 3.1-126 shows some typical results of stress rupture tests characterising the time-dependent high temperature behavior dominated by creep deformation. A compilation of creep data may be found in [1.101].

3.1.7.4 Ni Plating

Nickel plating is used extensively for decorative applications (80%) and for engineering and electroforming purposes (20%).

Electroless Ni coatings are produced by autocatalytical reduction of Ni ions from aqueous solution. Three electroless coatings are applied most frequently: nickel–phosphorus (6–12 wt% P), nickel–boron ($\approx$ 5 wt% B), and composite coatings (Ni–P with SiC, fluorocarbons, and diamond). A more extensive account is given in [1.96].

Table 3.1-97 Effect of temperature on the ultimate tensile strength of wrought Ni-based superalloys

Alloy	Form	Ultimate tensile strength at:										Condition of test material [a]
		21 °C (70 °F)		540 °C (1000 °F)		650 °C (1200 °F)		760 °C (1400 °F)		870 °C (1600 °F)		
		(MPa)	(ksi)	(MPa)	(ksi)	(MPa)	(ksi)	(MPa)	(ksi)	(MPa)	(ksi)	
Nickel-based												
Astroloy	Bar	1415	205	1240	180	1310	190	1160	168	775	112	1095 °C (2000 °F)/4 h/OQ + 870 °C (1600 °F)/8 h/AC + 980 °C (1800 °F)/4 h/AC + 650 °C (1200 °F)/24 h/AC + 760 °C (1400 °F)/8 h/AC
Cabot 214	–	915	133	715	104	675	98	560	84	440	64	1120 °C (2050 °F)
D-979	Bar	1410	204	1295	188	1105	160	720	104	345	50	1040 °C (1900 °F)/1 h/OQ + 845 °C (1550 °F)/6 h/AC + 705 °C (1300 °F)/16 h/AC
Hastelloy C-22	Sheet	800	116	625	91	585	85	525	76	–	–	1120 °C (2050 °F)/RQ
Hastelloy G-30	Sheet	690	100	490	71	–	–	–	–	–	–	1175 °C (2150 °F)/RAC-WQ
Hastelloy S	Bar	845	130	775	112	720	105	575	84	340	50	1065 °C (1950 °F)/AC
Hastelloy X	Sheet	785	114	650	94	570	83	435	63	255	37	1175 °C (2150 °F)/RAC
Haynes 230	–	870	126	720	105	675	98	575	84	385	56	1230 °C (2250 °F)/AC
Inconel 587[b]	Bar	1180	171	1035	150	1005	146	830	120	525	76	–
Inconel 597[b]	Bar	1220	177	1140	165	1060	154	930	135	–	–	–
Inconel 600	Bar	660	96	560	81	450	65	260	38	140	20	1120 °C (2050 °F)/2 h/AC
Inconel 601	Sheet	740	107	725	105	525	76	290	42	160	23	1150 °C (2100 °F)/2 h/AC
Inconel 617	Bar	740	107	580	84	565	82	440	64	275	40	1175 °C (2150 °F)/AC
Inconel 617	Sheet	770	112	590	86	590	86	470	68	310	45	1175 °C (2150 °F)/0.2 h/AC
Inconel 625	Bar	965	140	910	132	835	121	550	80	275	40	1150 °C (2100 °F)/1 h/WQ
Inconel 706	Bar	1310	190	1145	166	1035	150	725	105	–	–	980 °C (1800 °F)/1 h/AC + 845 °C (1550 °F)/3 h/AC + 720 °C (1325 °F)/8 h/FC + 620 °C (1150 °F)/8 h/AC
Inconel 718	Bar	1435	208	1275	185	1228	178	950	138	340	49	980 °C (1800 °F)/1 h/AC + 720 °C (1325 °F)/8 h/FC + 620 °C (1150 °F)/8 h/AC
Inconel 718 Direct Age	Bar	1530	222	1350	196	1235	179	–	–	–	–	735 °C (1325 °F)/8 h/SC + 620 °C (1150 °F)/8 h/AC
Inconel 718 Super	Bar	1530	196	1200	174	1130	164	–	–	–	–	925 °C (1700 °F)/1 h/AC + 735 °C (1325 °F)/8 h/SC + 620 °C (1150 °F)/8 h/AC

Table 3.1-97 Effect of temperature on the ultimate tensile strength of wrought Ni-based superalloys, cont.

Alloy	Form	Ultimate tensile strength at:								Condition of test material[a]
		21 °C (70 °F)		540 °C (1000 °F)		650 °C (1200 °F)		760 °C (1400 °F)		
		(MPa)	(ksi)	(MPa)	(ksi)	(MPa)	(ksi)	(MPa)	(ksi)	

Continued:

Alloy	Form	21 °C (MPa)	(ksi)	540 °C (MPa)	(ksi)	650 °C (MPa)	(ksi)	760 °C (MPa)	(ksi)	870 °C (MPa)	(ksi)	Condition of test material[a]
Nickel-based												
Inconel X750	Bar	1200	174	1050	152	940	136	–	–	–	–	1150 °C (2100 °F)/2 h/AC + 845 °C (1550 °F)/24 h/AC + 705 °C (1300 °F)/20 h/AC
M-252	Bar	1240	180	1230	178	1160	168	945	137	510	74	1040 °C (1900 °F)/4 h/AC + 760 °C (1400 °F)/16 h/AC
Nimonic 75	Bar	745	108	675	98	540	78	310	45	150	22	1050 °C (1925 °F)/1 h/AC
Nimonic 80A	Bar	1000	145	875	127	795	115	600	87	310	45	1080 °C (1975 °F)/8 h/AC + 705 °C (1300 °F)/16 h/AC
Nimonic 90	Bar	1235	179	1075	156	940	136	655	95	330	48	1080 °C (1975 °F)/8 h/AC + 705 °C (1300 °F)/16 h/AC
Nimonic 105	Bar	1180	171	1130	164	1095	159	930	135	660	96	1150 °C (2100 °F)/4 h/AC + 1060 °C (1940 °F)/16 h/AC + 850 °C (1560 °F)/16 h/AC
Nimonic 115	Bar	1240	180	1090	158	1125	163	1085	157	830	120	1190 °C (2175 °F)/1.5 h/AC + 1100 °C (2010 °F)/6 h/AC
Nimonic 263	Sheet	970	141	800	116	770	112	650	94	280	40	1150 °C (2100 °F)/0.2 h/WQ + 800 °C (1470 °F)/8 h/AC
Nimonic 942	Bar	1405	204	1300	189	1240	180	900	131	–	–	–
Nimonic PE.11	Bar	1080	157	1000	145	940	136	760	110	–	–	–
Nimonic PE.16	Bar	885	128	740	107	660	96	510	74	215	31	1040 °C (1900 °F)/4 h/AC + 800 °C (1470 °F)/2 h/AC + 1100–1115 °C (2010–2040 °F)/0.25 h/AC + 850 °C (1500 °F)/4 h/AC
Nimonic PK.33	Sheet	1180	171	1000	145	1000	145	885	128	510	74	1100–1115 °C (2010–2040 °F)/0.25 h/AC + 850 °C (1500 °F)/4 h/AC
Pyromet 860	Bar	1295	188	1255	182	1110	161	910	132	–	–	1095 °C (2000 °F)/2 h/WQ + 830 °C (1525 °F)/2 h/AC + 760 °C (1400 °F)/24 h/AC
René 41	Bar	1420	206	1400	203	1340	194	1105	160	620	90	1065 °C (1950 °F)/4 h/AC + 760 °C (1400 °F)/16 h/AC
René 95	Bar	1620	235	1550	224	1460	212	1170	170	–	–	900 °C (1650 °F)/24 h + 1105 °C (2025 °F)/1 h/OQ + 730 °C (1350 °F)/64 h/AC
Udimet 400	Bar	1310	190	1185	172	–	–	–	–	–	–	–
Udimet 500	Bar	1310	190	1240	180	1215	176	1040	151	640	93	1080 °C (1975 °F)/4 h/AC + 845 °C (1550 °F)/24 h/AC + 760 °C (1400 °F)/16 h/AC
Udimet 520	Bar	1310	190	1240	180	1175	170	725	105	515	75	1105 °C (2025 °F)/4 h/AC + 845 °C (1550 °F)/24 h/AC + 760 °C (1400 °F)/16 h/AC
Udimet 630	Bar	1520	220	1380	200	1275	185	965	140	–	–	–
Udimet 700	Bar	1410	204	1275	185	1240	180	1035	150	690	100	1175 °C (2150 °F)/4 h/AC + 1080 °C (1975 °F)/4 h/AC + 845 °C (1550 °F)/24 h/AC + 760 °C (1400 °F)/16 h/AC

Table 3.1-97 Effect of temperature on the ultimate tensile strength of wrought Ni-based superalloys, cont.

Alloy	Form	Ultimate tensile strength at:									Condition of test material[a]
		21 °C (70 °F)		540 °C (1000 °F)		650 °C (1200 °F)		760 °C (1400 °F)		870 °C (1600 °F)	
		(MPa)	(ksi)	(MPa)	(ksi)	(MPa)	(ksi)	(MPa)	(ksi)	(MPa)	(ksi)	
Nickel-based												
Udimet 710	Bar	1185	172	1150	167	1290	187	1020	148	705	102	1175 °C (2150 °F)/4 h/AC + 1080 °C (1975 °F)/4 h/AC + 845 °C (1550 °F)/24 h/AC + 760 °C (1400 °F)/16 h/AC
Udimet 720	Bar	1570	228	–	–	1455	211	1455	211	1150	167	1115 °C (2035 °F)/2 h/AC + 1080 °C (1975 °F)/4 h/OQ + 650 °C (1200 °F)/24 h/AC + 760 °C (1400 °F)/8 h/AC
Unitemp AF2-1DA6	Bar	1560	226	1480	215	1400	203	1290	187	–	–	1150 °C (2100 °F)/4 h/AC + 760 °C (1400 °F)/16 h/AC
Waspaloy	Bar	1275	185	1170	170	1115	162	650	94	275	40	1080 °C (1975 °F)/4 h/AC + 845 °C (1550 °F)/24 h/AC + 760 °C (1400 °F)/16 h/AC
Iron base												
A-286	Bar	1005	146	905	131	720	104	440	64	–	–	980 °C (1800 °F)/1 h/OQ + 720 °C (1325 °F)/16 h/AC
Alloy 901	Bar	1205	175	1030	149	960	139	725	105	–	–	1095 °C (2000 °F)/2 h/WQ + 790 °C (1450 °F)/2 h/AC + 720 °C (1325 °F)/24 h/AC
Discaloy	Bar	1000	145	865	125	720	104	485	70	–	–	1010 °C (1850 °F)/2 h/OQ + 730 °C (1350 °F)/20 h/AC + 650 °C (1200 °F)/20 h/AC
Haynes 556	Sheet	815	118	645	93	590	85	470	69	330	48	1175 °C (2150 °F)/AC
Incoloy 800	Bar	595	86	510	74	405	59	235	34	–	–	–

[a] OQ, oil quench; AC, air cool; RQ, rapid quench; RAC-WQ, rapid air cool-water quench; FC, furnace cool; SC, slow cool; CW, cold worked
[d] Annealed
[e] Precipitation hardened
[f] Ref 15.
[g] Ref 1.
[h] Work strengthened and aged
[i] At 700 °C (1290 °F)
[j] At 900 °C (1650 °F). Source: Ref 12, except as noted

Table 3.1-98 Effect of temperature on the mechanical properties of wrought Ni-based superalloys

Alloy	Form	Yield strength at 0.2% offset at								Tensile elongation (%) at				
		21 °C (70 °F)		540 °C (1000 °F)		650 °C (1200 °F)		760 °C (1400 °F)		870 °C (1600 °F)		21 °C (70 °F)	540 °C (1000 °F)	650 °C (1200 °F)	760 °C (1400 °F)	870 °C (1600 °F)
		(MPa)	(ksi)	(MPa)	(ksi)	(MPa)	(ksi)	(MPa)	(ksi)	(MPa)	(ksi)					
Nickel-based																
Astroloy	Bar	1050	152	965	140	965	140	910	132	690	100	16	16	18	21	25
Cabot 214	—	560	81	510	74	505	73	495	72	310	45	38	19	14	9	11
D-979	Bar	1005	146	925	134	980	129	655	95	305	44	15	15	21	17	18
Hastelloy C-22	Sheet	405	59	275	40	250	36	240	35	—	—	57	61	65	63	—
Hastelloy G-30	Sheet	315	46	170	25	—	—	—	—	—	—	64	75	—	—	—
Hastelloy S	Bar	455	65	340	49	320	47	310	45	220	32	49	50	56	70	47
Hastelloy X	Sheet	360	52	290	42	275	40	260	38	180	26	43	45	37	37	50
Haynes 230	a	390	57	275	40	270	39	285	41	225	32	48	56	55	46	59
Inconel 587	Bar	705	102	620	90	615	89	605	88	400	58	28	22	21	20	16
Inconel 597	Bar	760	110	720	104	675	98	665	96	—	—	15	15	15	16	—
Inconel 600	Bar	285	41	220	32	205	30	180	26	40	6	45	41	49	70	80
Inconel 601	Sheet	455	66	350	51	310	45	220	32	55	8	40	34	33	78	128
Inconel 617	Bar	295	43	200	29	170	25	180	26	195	28	70	68	75	84	118
Inconel 617	Sheet	345	50	230	33	220	32	230	33	205	30	55	62	61	59	73
Inconel 625	Bar	490	71	415	60	420	61	415	60	275	40	50	50	34	45	125
Inconel 706	Bar	1005	146	910	132	860	125	660	96	—	—	20	19	24	32	—
Inconel 718	Bar	1185	172	1065	154	1020	148	740	107	330	48	21	18	19	25	88
Inconel 718	Bar	1365	198	1180	171	1090	158	—	—	—	—	16	15	23	—	—
Direct Age																
Inconel 718 Super	Bar	1105	160	1020	148	960	139	—	—	—	—	16	18	14	—	—
Inconel X750	Bar	815	118	725	105	710	103	—	—	—	—	27	26	10	—	18
M-252	Bar	840	122	765	111	745	108	720	104	485	70	16	15	11	10	18
Nimonic 75	Bar	285	41	200	29	200	29	160	23	90	13	40	40	46	67	68
Nimonic 80A	Bar	620	90	530	77	550	80	505	73	260	38	39	37	21	17	30
Nimonic 90	Bar	810	117	725	105	685	99	540	78	260	38	33	28	14	12	23
Nimonic 105	Bar	830	120	775	112	765	111	740	107	490	71	16	22	24	25	27
Nimonic 115	Bar	865	125	795	115	815	118	800	116	550	80	27	18	23	24	16
Nimonic 263	Sheet	580	84	485	70	485	70	460	67	180	26	39	42	27	21	25
Nimonic 942	Bar	1060	154	970	141	1000	145	860	125	—	—	—	—	—	—	—
Nimonic PE.11	Bar	720	105	690	100	670	97	560	81	—	—	—	—	—	—	—
Nimonic PE.16	Bar	530	77	485	70	485	70	370	54	140	20	37	26	46	42	80
Nimonic 860	Sheet	780	113	725	105	725	105	670	97	420	61	30	30	26	18	24
Pyromet 860	Bar	835	121	840	122	850	123	835	121	—	—	22	15	17	18	—
René 41	Bar	1060	154	1020	147	1000	145	940	136	550	80	14	14	14	11	19
René 95	Bar	1310	190	1255	182	1220	177	1100	160	—	—	15	12	14	15	—

Table 3.1-98 Effect of temperature on the mechanical properties of wrought Ni-based superalloys, cont.

Alloy	Form	Yield strength at 0.2% offset at									Tensile elongation (%) at				
		21 °C (70 °F)		540 °C (1000 °F)		650 °C (1200 °F)		760 °C (1400 °F)		870 °C (1600 °F)		21 °C (70 °F)	540 °C (1000 °F)	650 °C (1200 °F)	760 °C (1400 °F)	870 °C (1600 °F)
		(MPa)	(ksi)	(MPa)	(ksi)	(MPa)	(ksi)	(MPa)	(ksi)	(MPa)	(ksi)					
Udimet 400	Bar	930	135	830	120	–	–	–	–	–	–	30	26	–	–	–
Udimet 500	Bar	840	122	795	115	760	110	730	106	495	72	32	28	28	39	20
Udimet 520	Bar	860	125	825	120	795	115	725	105	520	75	21	20	17	15	20
Udimet 630	Bar	1310	190	1170	170	1105	160	860	125	–	–	15	15	7	5	–
Udimet 700	Bar	965	140	895	130	855	124	825	120	635	92	17	16	16	20	27
Udimet 710	Bar	910	132	850	123	860	125	815	118	635	92	7	10	15	25	29
Udimet 720	Bar	1195	173	–	–	1130	164	1050	152	–	–	13	–	17	9	–
Unitemp AF2-1DA6	Bar	1015	147	1040	151	1020	148	995	144	–	–	20	19	18	16	–
Waspaloy	Bar	795	115	725	105	690	100	675	98	520	75	25	23	34	28	35
Iron-based																
A-286	Bar	725	105	605	88	605	88	430	62	–	–	25	19	13	19	–
Alloy 901	Bar	895	130	780	113	760	110	635	92	–	–	14	14	13	19	–
Discaloy	Bar	730	106	650	94	630	91	430	62	–	–	19	16	19	–	–
Haynes 556	Sheet	410	60	240	35	225	33	220	32	195	29	48	54	52	49	53
Incoloy 800	Bar	250	36	180	26	180	26	150	22	–	–	44	38	51	83	–
Incoloy 801	Bar	385	56	310	45	305	44	290	42	–	–	30	28	26	55	–
Incoloy 802	Bar	290	42	195	28	200	29	200	29	150	22	44	39	25	15	38
Incoloy 807	Bar	380	55	255	37	240	35	225	32.5	185	26.5	48	40	35	34	71
Incoloy 825	–	310	45	≈234	≈34	≈220	≈32	180	≈26	≈105	≈15	45	≈44	≈35	≈86	≈100
Incoloy 903	Bar	1105	160	–	–	895	130	–	–	–	–	14	–	18	–	–
Incoloy 907	–	≈1110	≈161	≈960	≈139	≈895	≈130	≈565	≈82	–	–	≈12	≈11	≈10	≈20	–
Incoloy 909	Bar	1020	148	945	137	870	126	540	78	–	–	16	14	24	34	–
N-155	Bar	400	58	340	49	295	43	250	36	175	25	40	33	32	32	33
V-57	Bar	830	120	760	110	745	108	485	70	–	–	26	19	22	34	–
19-9 DL	–	570	83	395	57	360	52	–	–	–	–	43	30	30	–	–
16-25-6	–	770	112	–	–	517	75	345	50	255	37	23	–	12	11	9

[a] Cold-rolled and solution-annealed sheet, 1.2 to 1.6 mm thick

Table 3.1-99 Effect of temperature on the mechanical properties of cast Ni-based superalloys

Alloy	Ultimate tensile strength at:						0.2% yield strength at:					
	21 °C (70 °F)		538 °C (1000 °F)		1093 °C (2000 °F)		21 °C (70 °F)		538 °C (1000 °F)		1093 °C (2000 °F)	
	MPa	ksi	MPa	ksi	MPa	ksi	MPa	ksi	MPa	ksi	MPa	ksi
Nickel-base												
IN-713 C	850	123	860	125	–	–	740	107	705	102	–	–
IN-713 LC	895	130	895	130	–	–	750	109	760	110	–	–
B-1900	970	141	1005	146	270	38	825	120	870	126	195	28
IN-625	710	103	510	74	–	–	350	51	235	34	–	–
IN-718	1090	158	–	–	–	–	915	133	–	–	–	–
IN-100	1018	147	1090	150	(380)	(55)	850	123	885	128	(240)	(35)
IN-162	1005	146	1020	148	–	–	815	118	795	115	–	–
IN-731	835	121	–	–	275	40	725	105	–	–	170	25
IN-738	1095	159	–	–	–	–	950	138	–	–	–	–
IN-792	1070	170	–	–	–	–	1060	154	–	–	–	–
IN-22	730	106	780	113	–	–	685	99	730	106	–	–
MAR-M 200	930	135	945	137	325	47	840	122	880	123	–	–
MAR-M 246	965	140	1000	145	345	50	860	125	860	125	–	–
MAR-M 247	965	140	1035	150	–	–	815	118	825	120	–	–
MAR-M 421	1085	157	995	147	–	–	930	135	815	118	–	–
MAR-M 432	1240	180	1105	160	–	–	1070	155	910	132	–	–
MC-102	675	98	655	95	–	–	605	88	540	78	–	–
Nimocast 75	500	72	–	–	–	–	179	26	–	–	–	–
Nimocast 80	730	106	–	–	–	–	520	75	–	–	–	–
Nimocast 90	700	102	595	86	–	–	520	75	420	61	–	–
Nimocast 242	460	67	–	–	–	–	300	44	–	–	–	–
Nimocast 263	730	106	–	–	–	–	510	74	–	–	–	–
René 77	–	–	–	–	–	–	–	–	–	–	–	–
René 80	–	–	–	–	–	–	–	–	–	–	–	–
Udimet 500	930	135	895	130	–	–	815	118	725	105	–	–
Udimet 710	1075	156	–	–	240	35	895	130	–	–	170	25
CMSX-2[a]	1185	172	1295[b]	188[b]	–	–	1135	165	1245[b]	181[b]	–	–
GMR-235	710	103	–	–	–	–	640	93	–	–	–	–
IN-939	1050	152	915[b]	133[b]	325[c]	47[c]	800	116	635[b]	92[b]	205[c]	30[c]
MM 002[d]	1035	150	1035[b]	150[b]	550[c]	80[c]	825	120	860[b]	125[b]	345[c]	50[c]
IN-713 Hf[e]	1000	145	895[b]	130[b]	380[c]	55[c]	760	110	620[b]	90[b]	240[c]	35[c]
René 125 Hf[f]	1070	155	1070[b]	155[b]	550[c]	80[c]	825	120	860[b]	125[b]	345[c]	50[c]
MAR-M 246 Hf[g]	1105	160	1070[b]	155[b]	565[c]	82[c]	860	125	860[b]	125[b]	345[c]	50[c]
MAR-M 200 Hf[h]	1035	150	1035[b]	150[b]	540[c]	78[c]	825	120	860[b]	125[b]	345[c]	50[c]
PWA-1480[a]	–	–	1130[b]	164[b]	685[c]	99[c]	895	130	905[b]	131[b]	495[c]	72[c]
SEL	1020	148	875[b]	127[b]	–	–	905	131	795[b]	115[b]	–	–
UDM 56	945	137	945[b]	137[b]	–	–	850	123	725[b]	105[b]	–	–
SEL 15	1060	154	1090[b]	158[b]	–	–	895	130	815[b]	118[b]	–	–

(a) Single-crystal [001]
[b] At 760 °C (1400 °F)
[c] At 980 °C (1800 °F)
[d] RR-7080
[e] MM 004

Table 3.1-99 Effect of temperature on the mechanical properties of cast Ni-base superalloys, cont.

Tensile elongation % at:			Dynamic modulus of elasticity at:					
21 °C (70 °F)	538 °C (1000 °F)	1093 °C (2000 °F)	21 °C (70 °F)		538 °C (1000 °F)		1093 °C (2000 °F)	
			GPa	10⁶ psi	GPa	10⁶ psi	GPa	10⁶ psi
Nickel-base								
8	10	–	206	29.9	179	26.2	–	–
15	11	–	197	28.6	172	25.0	–	–
8	7	11	214	31.0	183	27.0	–	–
48	50	–	–	–	–	–	–	–
11	–	–	–	–	–	–	–	–
9	9	–	215	31.2	187	27.1	–	–
7	6.5	–	197	28.5	172	24.9	–	–
6.5	–	–	–	–	–	–	–	–
–	–	–	201	29.2	175	25.4	–	–
4	–	–	–	–	–	–	–	–
5.5	4.5	–	–	–	–	–	–	–
7	5	–	218	31.6	184	26.7	–	–
5	5	–	205	29.8	178	25.8	145	21.1
7	–	–	–	–	–	–	–	–
4.5	3	–	203	29.4	–	–	141	20.4
6	–	–	–	–	–	–	–	–
5	9	–	–	–	–	–	–	–
39	–	–	–	–	–	–	–	–
15	–	–	–	–	–	–	–	–
14	15	–	–	–	–	–	–	–
8	–	–	–	–	–	–	–	–
18	–	–	–	–	–	–	–	–
–	–	–	–	–	–	–	–	–
–	–	–	208	30.2	–	–	–	–
13	13	–	–	–	–	–	–	–
8	–	–	–	–	–	–	–	–
10	17[b]	–	–	–	–	–	–	–
3	–	18[c]	–	–	–	–	–	–
5	7[b]	25[c]	–	–	–	–	–	–
7	5[b]	12[c]	–	–	–	–	–	–
11	6[b]	20[c]	–	–	–	–	–	–
5	5[b]	12[c]	–	–	–	–	–	–
6	7[b]	14[c]	–	–	–	–	–	–
5	5[b]	10[c]	–	–	–	–	–	–
4	8[b]	20[c]	–	–	–	–	–	–
6	7[b]	–	–	–	–	–	–	–
3	5[b]	–	–	–	–	–	–	–
9	5[b]	–	–	–	–	–	–	–

[f] M 005
[g] MM 006
[h] MM 009
[i] Data from Metals Handbook, 9th edn., Vol. 3, 1980
[j] At 650 °C (1200 °F)
Source: Nickel Development institute, except as noted

3.1.8 Copper and Copper Alloys

Copper and copper alloys are used as materials mainly because of the high electrical and thermal conductivity of Cu and because of the variability and favorable combination of electrical, mechanical, and corrosion properties and of color (from copper red to silver white) by alloying and heat treatment. The basic properties of Cu are covered in Sect. 2.2.3. The electrical resistivity of Cu, $\sigma = 16.78\,\text{n}\Omega\,\text{m}$ (293 K), is the highest among the metals of moderate cost. The corresponding high thermal conductivity $\lambda = 397\,\text{W}\,\text{m}^{-1}\text{K}^{-1}$ of Cu gives rise to numerous applications in heat exchangers. Its high corrosion resistance and good formability are the basis of its use as gas and water pipes, in the chemical and food industries, and – combined with its color – in architecture.

The alloying of Cu with group IIB to IVB metals results in a series of alloy systems with a characteristic sequence of intermetallic phases characterised by their outer electron to atom ratio (e/a), as first recognised by and named after Hume-Rothery. This behavior is attributed to the fact that the electronic structure rather than ionic radius, directed bonds, or other factors of influence for alloy formation is dominating the and crystal structure formation of the intermetallic phases. The same is true for Ag- and Au-based alloy systems. A survey is given in Table 3.1-100.

The solid solution phase α of Cu, dominating in most Cu materials, is hardened by the solute elements s through solid solution hardening which is proportional to the misfit parameter η_s, given by the relative difference in atomic radius r_s to r_{Cu}: $\eta_s = 2(r_s - r_{Cu})/(r_s + r_{Cu})$. Since $r_{Cu} = 1.28$, $r_{Ni} = 1.25$, $r_{Zn} = 1.33$, and $r_{Sn} = 1.51$, it follows that the increase in yield stress per at.% of solute increases in the order Ni < Zn < Sn.

The materials on copper basis may be subdivided into the groups shown in Table 3.1-101. Standards for copper and copper materials have been issued by ASTM, DIN, DKE, CEN, CENELEC, ISO, and IEC. In the USA, the Unified Numbering System for Metals and Alloys (UNS) applied to Cu materials consists of the letter C and a 5 digit number. More extensive accounts of Cu and Cu alloys as materials are given in [1.102–104].

3.1.8.1 Unalloyed Coppers

Cu is used as the technical standard for the conductivity of metals and alloys. The International Annealed Copper Standard (IACS) is defined by a Cu wire, 1 m long and weighing 0.1 kg, the resistance of which is $0.15327\,\Omega$ at $20\,°\text{C}$. Accordingly, a conductivity of 100% IACS $= 58.00\,\text{MS}\,\text{m}^{-1}$ (or $\text{m}\,\Omega^{-1}\,\text{mm}^{-2}$), which corresponds to a resistivity of $1.7243793\,(\mu\Omega\,\text{cm})$. The electrical conductivity is specifically and strongly dependent on the kind and concentration of impurities (Fig. 3.1-130).

Table 3.1-100 Characteristics of Hume-Rothery phases in Cu-, Ag- and Au-based alloy systems

e/a	Phase designation	Crystal structure	Example
3/2	β	bcc	CuZn
3/2	ζ	hcp	CuGa, CuGe
21/13	γ	γ brass structure	Cu_5Zn_8
7/4	ε	hcp	$CuZn_3$

Table 3.1-101 Groups of copper materials according to ASTM

Designation	Maximum alloying range, wt%, further alloying elements	Wrought alloys	Cast alloys	Particular alloying range for castings
Coppers	–	+	+	–
High copper alloys	1.2Cd, 2Be, 3Fe, 2.7Co, 3Ni, 2.5Sn, 1.5Cr, 0.7Zn, 0.25Si, 3.4Ti, 1Zr, 3.5Pb	+	+	
Brasses	43Zn, 3Pb, 2Al, Co, Si, Mn, Fe	+	+	
Bronzes	8Sn, Zn	+	+	12Sn, 2Pb, 2Ni
Cu–Pb–Sn	–	–	+	10Sn, 20Pb
Copper–Nickels	30Ni, 1Fe, 1Mn, 2Sn	+	+	–
Nickel–Silvers	18Ni, 39Zn, 3Pb, 2Mn	+	–	–
Cu–Sn–Zn	–	–	+	7Sn, 8Zn, 7Pb
Cu–Al	8Al	+	+	

Small oxygen additions to Cu in the range of 0.005 to 0.04 wt% O are used to oxidize the metallic impurities which may be dissolved in small Cu_2O particles or form their own oxides. Thus the conductivity of the Cu is increased. Alternatively ultra-high purity cathodes from the electrolytic copper refining process are used as raw material. Oxygen-bearing Cu is unsuitable for applications requiring exposure to hydrogen, or requiring bonding by soldering or by brazing, because reactions between oxygen and hydrogen or between oxygen and the braze or solder components, respectively, lead to embrittlement. Gas evolution may occur in ultra-high-vacuum equipment if the Cu applied contains oxygen additions or higher contents of other impurities with a higher vapor pressure. Therefore, use of oxygen-free high purity copper is mandatory in these cases. In less critical cases, deoxidation by phosphorus is applied to the copper with an ensuing loss in conductivity. These measures lead to the different technical coppers ranging from about 40 to 58 MS m^{-1} in conductivity and from 99.90 to 99.99 wt% in purity. Table 3.1-102 lists major unalloyed coppers.

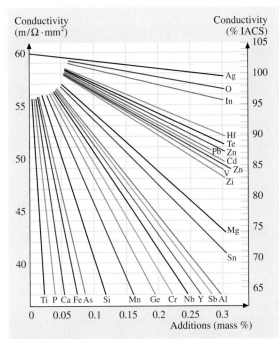

Fig. 3.1-130 Effect of solute elements on the conductivity of Cu

Unalloyed coppers contain ≥ 99.3 wt% Cu. The elements O, P, and/or As are deliberate additions.

3.1.8.2 High Copper Alloys

This group of materials comprises essentially a number of age-hardenable alloys which are listed in Table 3.1-103.

Table 3.1-102 Composition and properties of characteristic unalloyed coppers

Material	UNS No.	Purity; other elements (wt%)	Yield stress $R_{p0.2}$ (MPa)	Ultimate tensile strength R_m (MPa)	Fracture strain A_f (%)	Thermal conductivity κ (W m^{-1} K^{-1})	Electrical resistivity ρ ($\mu\Omega$ cm)
Pure Cu (oxygen-free electronic)	C10100	99.99 Cu	69–365	221–455	4–55	392	1.741
Pure Cu (oxygen-free)	C10200	99.95 Cu	69–365	221–455	4–55	397	1.741
Electrolytic tough pitch Cu	C11000	99.90 Cu–0.04 O	69–365	224–455	4–55	397	1.707
Oxygen-free low phosphorus Cu	C10800	99.95 Cu–0.009 P	69–345	221–379	4–50	397	2.028
Phosphorus deoxidized arsenical Cu	C14200	99.68 Cu–0.35 As –0.02 P	69–345	221–379	8–45	397	3.831

Table 3.1-103 Composition of high copper alloys

Basic alloy system	UNS No.	Major alloying elements (mass%)	Minor elements
Cu–Cd	C16200	0.7–1.2 Cd	
Cu–Be	C17000–C17300	1.6–2.0 Be	Al, Pb
Cu–Ni–Be	C17450–C17460	0.5–1.4 Ni, 0.15–0.50 Be	Al, Zr
Cu–Co–Be	C17500	2.4–2.7 Co, 0.4–0.7 Be	
Cu–Ni–Cr–Si	C18000	1.8–3.0 Ni, 0.1–0.8 Cr, 0.4–0.8 Si	
Cu–Sn–Cr	C18030–C18040	0.08–0.3 Sn, 0.1–0.35 Cr	P

Table 3.1-104 Properties of high copper alloys

Material	UNS No.	Yield stress $R_{p0.2}$ (MPa)	Ultimate tensile strength R_m (MPa)	Fracture strain A_f (%)	Brinell Hardness (HB)	Thermal conductivity κ (W m^{-1} K^{-1})	Electrical resistivity ρ ($\mu\Omega$ cm)
Beryllium copper	C17200	172–1344	469–1462	1–48	100–363	≤95	4.009
Beryllium copper	C17000	221–1172	483–1310	3–45	100–363	≤95	2.053
Cadmium copper	C16200	600	649	n.a.	n.a.	376	2.028
Chromium copper	C18200	479–531	232–593	14–60	58–140	188	3.831
Cobalt beryllium copper	C17500	172–758	310–793	5–28	67–215	84	7.496
Lead copper	C18700	69–345	221–379	8–45	n.a.	n.a.	n.a.
Silver-bearing copper	C11300	69–365	221–455	4–55	55–90	397	1.741
Sulfur copper	C14700	69–379	221–393	8–52	55–85	373	1.815
Tellurium copper	C14500	69–352	221–386	8–45	49–80	382	1.759
Zirconium copper	C15000	41–496	200–524	2–54	n.a.	n.a.	n.a.

n.a. not available

The age-hardening behavior is based on the fact that the solubility of Cd, Be, and Cr in Cu decreases with decreasing temperature. On this basis a heat treatment is applied, that leads to age-hardening. A high-temperature solution treatment is followed by a low-temperature annealing, or aging, treatment. This serves to precipitate a dispersion of small metastable or stable particles which give rise to hardening by several hardening mechanisms that depend in detail on the microstructural, chemical, crystallographic and mechanical properties of the precipitated particles (see Table 3.1-104).

3.1.8.3 Brasses

The term brass was originally applied to binary Cu−Zn alloys (Fig. 3.1-131) but was extended to Cu−Zn-based multi-component alloys, mainly containing Fe, Al, Ni, and Si.

The phases of technical interest are the fcc α phase and the bcc β phase. The high-temperature β phase transforms to its ordered (CsCl type) variant β' at temperatures below 727 to 741 K depending on composition, as shown in Fig. 3.1-131. The intermetallic phases with higher Zn content are brittle and have no technical relevance.

Brasses and other Cu-based alloys are mainly supplied as wrought alloys which designate the state of the final material which is obtained by a final shaping operation (rolling, rod, and wire drawing etc.) with a controlled degree of cold work. The work-hardening behavior of the alloys is, thus, exploited to vary the final mechanical properties in comparatively wide limits (see Table 3.1-105). By way of an example, Fig. 3.1-132 shows the mechanical properties of CuZn15 as a function of the degree of cold work.

The dependence of the mechanical properties on the effect of annealing is similarly important. Figure 3.1-133 shows the changes upon isothermal annealing for the same alloy as in Fig. 3.1-132. Table 3.1-105 lists the nominal compositions and property ranges of brasses.

3.1.8.4 Bronzes

Bronze is the original term for Cu−Sn alloys, Fig. 3.1-134. The term bronze has, also, been extended to Cu−Al and Cu−Si based alloys. Table 3.1-106 lists the composition and properties of charactristic bronzes. As referred to in the introduction the α solid solution of Cu−Sn is increasing in yield stress as a function of solute content most strongly of all common Cu alloys because of the high misfit of Sn in Cu.

Aluminium bronzes, especially ternary alloys containing Ni and Mn, have an increased high temperature oxidation resistance due to the formation of a protective Al_2O_3 layer upon exposure to higher temperature. The Cu−Al alloys and the related systems have been studied extensively as prototype systems which show martensitic transformations upon cooling and upon plastic deformation [1.105]. This transformation behavior is, also, the basis of shape memory and superplastic properties. Table 3.1-106 lists the nominal compositions and property ranges of bronzes. Figure 3.1-135 shows a phase diagram of Cu−Al.

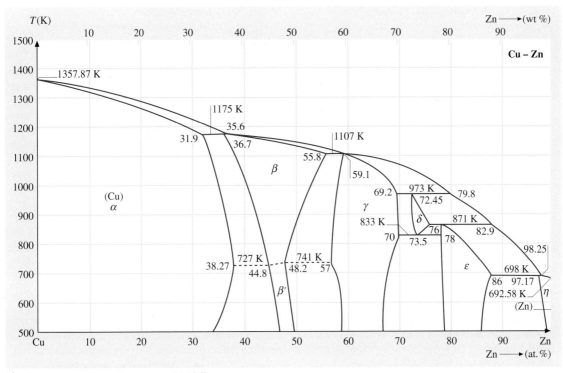

Fig. 3.1-131 Cu–Zn phase diagram [1.106]

Table 3.1-105 Composition and properties of characteristic brasses

Material	UNS No.	Composition (wt%)	Yield stress $R_{p0.2}$ (MPa)	Ultimate tensile strength R_m (MPa)	Fracture strain A_f (%)	Thermal conductivity κ (W m^{-1} K^{-1})	Electrical resistivity ρ ($\mu\Omega$ cm)
Admiralty brass	C44300	71 Cu–28 Zn –1 Sn	124–152	331–379	60–65		
Aluminium brass	C68700	77.5 Cu–20.5 Zn –2 Al	186	414	55	101	7.496
Cartridge brass	C68700	70 Cu–30 Zn	76–448	303–896	3–66	121	6.152
Free-cutting brass	C36000	61.5 Cu–35.5 Zn –3 Pb	124–310	338–469	18–53	109	6.631
Gilding metal (cap copper)	C21000	95.0 Cu–5.0 Zn	69–400	234–441	8–45	234	3.079
High tensile brass (architectural bronze)	C38500	57 Cu–40 Zn –3 Pb	138	414	30	88–109	8.620
Hot stamping brass (forging)	C37700	59 Cu–39 Zn –2 Pb	138	359	45	109	6.631
Low brass	C24000	80 Cu–20 Zn	83–448	290–862	3–55	138	4.660
Muntz metal	C28000	60 Cu–40 Zn	145–379	372–510	10–52	126	6.157
Naval brass	C46400	60 Cu–39.25 Zn –0.75 Sn	172–455	379–607	17–50	117	6.631
Red brass	C23000	85 Cu–15 Zn	69–434	269–724	3–55	159	3.918
Yellow brass	C26800	65 Cu–35 Zn	97–427	317–883	3–65	121	6.631

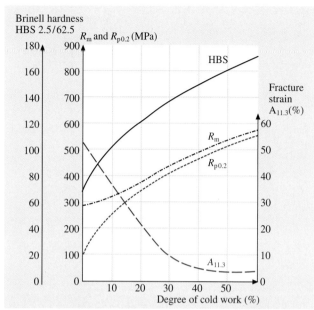

Fig. 3.1-132 Effect of work hardening, expressed as % reduction in thickness by rolling, on the mechanical properties of CuZn15 brass. This relation is a typical example for the control of the mechanical properties of wrought copper alloys by work hardening

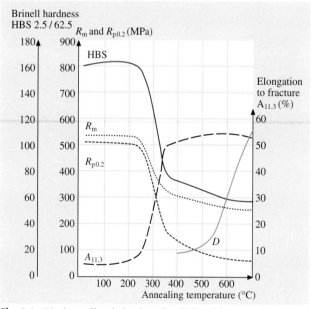

Fig. 3.1-133 Annealing behavior of a CuZn15 brass (cold worked 50%; Annealing time 3 h)

3.1.8.5 Copper–Nickel and Copper–Nickel–Zinc Alloys

Copper–Nickel Alloys

Figure 3.1-136 shows the Cu–Ni phase diagram which is characterized by a continuous solid solution which extends to room temperature. The metal Cu is solution-hardened only weakly by alloying with Ni because of the low misfit between the two components. But Cu–Ni alloys show an excellent corrosion resistance against seawater and are used, e.g., in desalination plants, accordingly. Another outstanding effect of Ni in Cu is the drastic decrease in electrical and thermal conductivity with increasing Ni content, as shown in Figs. 3.1-137 and 3.1-138. The interrelation through the Wiedemann–Franz law is obvious. A prominent application of the low thermal conductivity of Cu–Ni material is its use in equipment for operation at cryogenic temperature where heat loss by conduction is a major concern.

Copper–Nickel–Zinc Alloys. Nickel–Silvers

The designation of these Cu–Ni–Zn alloys is based on the silver-like color. Their composition ranges from 45 to 49 wt% Cu, 10 to 12 wt% Ni, with the balance composing Zn. Up to 2 wt% Pb is added for better drilling and turning behavior. Ni increases the yield stress and decreases the electrical and thermal conductivity (see Table 3.1-107).

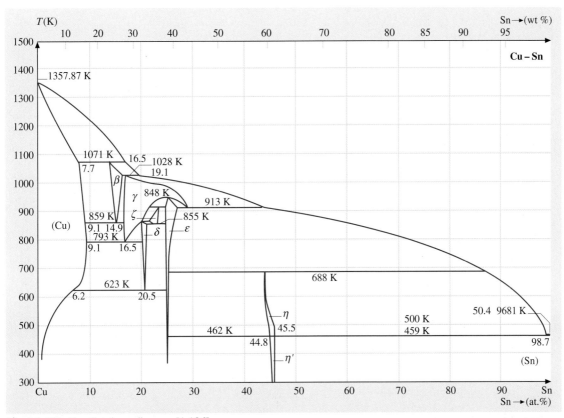

Fig. 3.1-134 Cu–Sn phase diagram [1.106]

Table 3.1-106 Composition and properties of characteristic bronzes

Material	UNS No.	Purity; other elements (wt%)	Yield stress $R_{p0.2}$ (MPa)	Ultimate tensile strength R_m (MPa)	Fracture strain A_f (%)	Thermal conductivity κ (W m^{-1} K^{-1})	Electrical resistivity ρ ($\mu\Omega$ cm)
Phosphor bronze	C51100	96 Cu–3.5 Sn –0.12 P	345–552	317–710	2–48	85	9.171
Phosphor bronze A	C51000	95 Cu–5 Sn –0.09 P	131–552	324–965	2–64	75	10.26
Phosphor bronze C	C52100	92 Cu–7 Sn –0.12 P	165–552	379–965	2–70	67	12.32
Phosphor bronze D	C52400	90 Cu–10 Sn	193	455–1014	3–70	63	12.32
Silicon bronze A	C65500	97 Cu–3.0 Si	145–483	386–1000	3–63	50	21.29
Silicon bronze B	C65100	98.5 Cu–1.5 Si	103–476	276–655	11–55	not available	not available
Aluminium bronze D	C61400	91 Cu–7 Al–2 Fe	228–414	524–614	32–45	not available	12.32
Aluminium bronze	C60800	95 Cu–5 Al	186	414	55	85	9.741
Aluminium bronze	C63000	Cu–9.5 Al–4 Fe –5 Ni–1 Mn	345–517	621–814	15–20	62	13.26

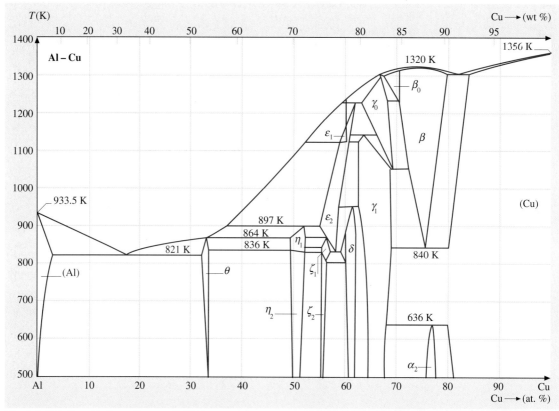

Fig. 3.1-135 Cu–Al phase diagram [1.106]

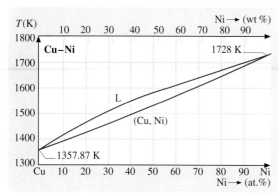

Fig. 3.1-136 Cu–Ni phase diagram [1.106]

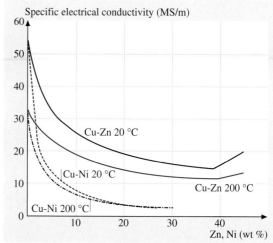

Fig. 3.1-137 Electrical conductivity of Cu–Ni and Cu–Zn alloys [1.103]

Table 3.1-107 Composition and properties of characteristic copper–nickel and copper–nickel–zinc alloys

Material	UNS No.	Purity; other elements (wt%)	Yield stress $R_{p0.2}$ (MPa)	Ultimate tensile strength R_m (MPa)	Fracture strain A_f (%)	Thermal conductivity κ (W m^{-1} K^{-1})	Electrical resistivity ρ ($\mu\Omega$ cm)
Copper nickel	C70400	92.4 Cu–5.5 Ni –1.5 Fe–0.6 Mn	276–524	262–531	2–46	67	13.79
Copper nickel	C70600	87.7 Cu–10 Ni –1.5 Fe–0.6 Mn	110–393	303–414	10–42	42	21.55
Copper nickel	C71500	67 Cu–31 Ni –0.7 Fe–0.5 Be	138–483	372–517	15–45	21	38.31
Nickel silver 10%	C74500	65 Cu–25 Zn–10 Ni	124–524	338–896	1–50	37	20.75
Nickel silver 12%	C74500	65 Cu–23 Zn–12 Ni	124–545	359–641	2–48	30	22.36
Nickel silver 15%	C75200	65 Cu–20 Zn–15 Ni	124–545	365–634	2–43	27	24.59
Nickel silver 18%	C74500	65 Cu–17 Zn–18 Ni	172–621	386–710	3–45	28	27.37

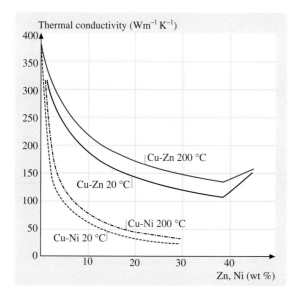

Fig. 3.1-138 Thermal conductivity of Cu–Ni and Cu–Zn alloys [1.103]

3.1.9 Refractory Metals and Alloys

Several books and reviews have been published on the technology and properties of refractory metals and their alloys [1.107–116]. According to the most common definition, refractory metals comprise elements of the Group Va and VIa with a melting point higher than 2000 °C; these are niobium, tantalum, molybdenum, and tungsten. In some publications the VIIa metal rhenium is also included, as it does not fit in any other classification. Less common definitions describe a refractory metal as metal with a melting point equal to or greater than that of chromium, thus additionally including vanadium, technetium, the reactive metal hafnium, and the noble metals ruthenium, osmium, ruthenium, and iridium. This chapter will give data on molybdenum, tungsten, tantalum, niobium, and their alloys. Powder metallurgy (P/M) is the only production route for commercial W and W alloys and for 97% of Mo, the remainder is processed by electron-beam melting (EB) and vacuum-arc casting (VAC) [1.117]. The finer grain structure of P/M-material is advantageous for both the further processing and the mechanical properties of the finished product. For some alloys such as those doped with potassium silicate, La_2O_3, and Y_2O_3, P/M is the only possible production technique.

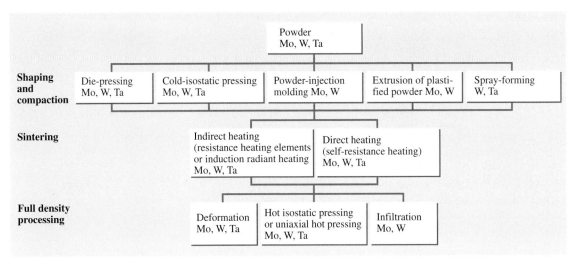

Fig. 3.1-139 Large-scale P/M production routes for Mo, W and Ta [1.118]

The mechanical properties and homogenous microstructure required are again the reasons to apply P/M in producing Ta wire as employed extensively in the manufacture of capacitors. The larger fraction of the Ta sheet production is based on the use of EB sheet bars, as this is more economical. The techniques of VAC and EB dominate in the production of Nb and Nb alloys. The industrial P/M production routes of Mo, W, Ta, and their alloys are given in Fig. 3.1-139.

Net shape techniques such as powder injection molding and spraying on a lost core (vacuum-plasma-spraying, chemical vapor deposition) are limited to some specific applications in the field of electronic devices and aerospace products. Less than 10% of the production quantity is delivered in the as-sintered state. The most common fully-dense-processing techniques are deformation by rolling, forging, swaging, and drawing.

Initially refractory metals were applied mainly in the pure state. Extensive developments mainly driven by US aerospace programs led to a wide variety of alloys now commercially available. The compositions of solid-solution, precipitation- and dispersion-strengthened alloys are given in Table 3.1-108. Tungsten heavy metals and refractory-metal-based composite materials are not included here.

Carbide precipitation hardening (in TZM, MHC) is effective up to $1400\,°C$. The addition of the deformable oxides La_2O_3 (in ML) and $xK_2O \cdot y SiO_2$ (in K−Si−Mo), respectively, results in oxide refinement by deformation and in the possibility of tailoring the mechanical properties. This is the main alloying mechanism for Mo applied in lamps [1.119].

The tungsten alloy which is commercially most important is Al−K-silicate doped W (AKS−W). It is a dispersion-strengthened, micro-alloyed metal with a directionally recrystallized microstructure. Spherical bubbles stabilized by containing potassium gas are interacting with dislocations, sub-boundaries and high-angle boundaries [1.120, 121]. The high stability of these bubbles can be explained by the low solubility of K in W, even at operating temperatures up to $3200\,°C$.

A current task is to reduce the low-temperature brittleness of Mo and W, which is essentially due to a rigid covalent component of the interatomic bond along the edges of the bcc unit cell. This causes a low solubility for interstitial elements which occupy the octahedral sites of the lattice and give rise to its tetragonal distortion and a strong interaction of dislocations with the elastic strain field surrounding the interstitial solutes, thus impeding the dislocation movement [1.122]. One possibility to increase the low-temperature ductility is alloying with rhenium which lowers the brittle-to-ductile-transition temperature of both W and Mo [1.123–125]. But the insufficient supply and the high price of rhenium limit the application of these alloys.

The addition of oxides such as ThO_2, BaO, SrO, Y_2O_3, and Sc_2O_3 lowers the electron work function of W, which is important for its application in electrodes. The production quantities of $W-La_2O_3$ and $W-Ce_2O_3$ as electron emitting materials are increasing at the expense of the slightly radioactive $W-ThO_2$ material.

The production of capacitors, the dominating application of Ta, requires material in its purest state.

Table 3.1-108 Typical compositions of commercial refractory metal alloys (in weight percent, analyses of base-metal correspond to metallic purity) [1.118]

Alloy designation	Mo	W	Ta	Nb	Re	C	O	Si	K	Y	La	Ce	Th	Ti	Zr	Hf
Molybdenum alloys																
Pure Mo	99.97															
UHP-Mo	99.9995															
TZM	99.3					0.025	0.02							0.5	0.08	
MHC	98.6					0.08	0.035									1.2
Mo–La$_2$O$_3$ (ML)	99.2–99.6						0.048–0.1				0.27–0.6					
Mo–Y$_2$O$_3$ (MY)	99.42						0.10–0.12			0.37–0.43		0–0.06				
K–Si–Mo	99.8–99.9					0.01–0.07	0.013–0.07		0.005–0.03							
Mo50Re	52.4				47.5											
Mo30W	69.7	30														
Tungsten alloys																
Pure W		99.99														
UHP-W		99.9995														
AKS-W		99.98							0.004–0.01							
W–La$_2$O$_3$ (WL)		97.9–98.9					0.015–0.03				0.85–1.7					
W–Ce$_2$O$_3$ (WC)		98.0					0.28					1.62				
W–ThO$_2$ (WT)		98.0					0.24						1.71			
AKS-W–ThO$_2$		98.0–99.0					0.12–0.24		0.001–0.005				0.86–1.71			
W5Re		94.9			5.0											
W26Re		73.9			26											
AKS-W3Re		96.9			3.0				0.004							
Tantalum alloys																
Pure Ta			99.95													
Ta2.5W		2.5	97.4													
Ta10W		10.0	89.9													
Niobium alloys																
Pure Nb				99.9												
Nb1Zr				98.9											1.0	
C-103				88.8										1	1	10
FS-85		10	28	60.8												
WC-3009		9	60.8				0.1									30
Nb – 46.5Ti														46.5		

Solid-solution-strengthened Ta2.5W is used for components in chemical apparatus. Superconducting Nb46.5Ti accounts for more than half of all Nb alloys produced. Hafnium is the main addition for niobium-based alloys used by the aerospace industry.

Refractory metals and their alloys are used in a wide variety of fields of application and products such as electrical and electronic devices; light sources; medical equipment; automotive, aerospace, and defense industry; chemical and pharmaceutical industry; or premium and sporting goods.

The producers of electrical and electronic devices, including the lighting industry, are the largest consumers of refractory metal products. In 1998, 1850 t of W products was used for filaments and electrodes in lamps only. Significant quantities of Mo are used for semiconductor base-plates for power rectifiers and in various products for lamps, such as dipped beam shields or support wires. Rapid growth in multimedia and wireless communication networks systems has boosted the need for W–Cu and Mo–Cu heat sink materials. These materials possess a high thermal conductivity combined with a low thermal expansion, close to those of Si and GaAs semiconductors or certain packaging materials. Also, the amount of Mo sputtering targets applied in the production of wiring for large format thin-film transistor LCDs and PDPs has risen significantly owing to the unique combination of low resistivity and high resistance against hillock formation. The electronic industry is the largest market (around 70%) for Ta products, employing the metal mainly in the manufacture of capacitors.

Refractory metals are also widely used by the materials processing industry. Molybdenum glass melting electrodes, TZM and MHC isothermal forging tools, weighing several tons per part, TZM piercing plugs for the production of stainless steel tubes, Mo and Ta crucibles for synthesizing artificial diamond, or TIG-welding electrodes are examples of products in this field. In order to improve the tribological properties of transmission and engine components for automobiles they are coated with Mo.

Recent products in the field of aerospace and defense industry are shaped charge liners made of Mo and Ta penetrators formed explosively. The X-ray anode, a composite product made of W5Re or W10Re, TZM and optionally graphite, is the essential item of computer tomography equipment, a very demanding application that critically depends on the users' expertise.

3.1.9.1 Physical Properties

The atomic and structural properties of the pure refractory metals are listed in Chapt. 2.1, Tables 2.1-12 and 2.1-14. Special features of refractory metals are their low vapor pressure, low coefficient of thermal expansion, and the high thermal and electrical conductivity of Mo and W. This combination of physical properties has opened up a wide range of new applications during the last decade, especially in the field of electronics.

The coefficient of linear thermal expansion, the thermal conductivity, the specific heat, and the electrical resistivity as function of temperature are shown in Figs. 3.1-140 – 3.1-143. The vapor pressure and rate of evaporation are shown in Fig. 3.1-144. In the case of precipitation- and dispersion-strengthened molybdenum alloys, such as TZM, MHC, ML, MY, and K–Si–Mo,

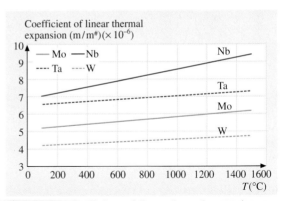

Fig. 3.1-140 Coefficient of linear thermal expansion versus temperature of molybdenum [1.126], tungsten [1.126], niobium [1.127], and tantalum [1.128]

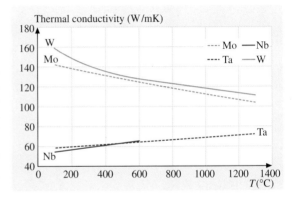

Fig. 3.1-141 Thermal conductivity versus temperature of molybdenum [1.126], tungsten [1.129], niobium [1.127], and tantalum [1.127]

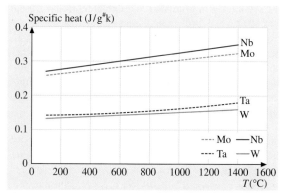

Fig. 3.1-142 Specific heat versus temperature of molybdenum, tungsten, niobium, and tantalum [1.126]

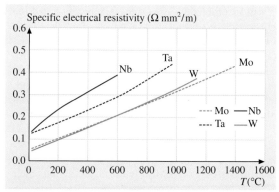

Fig. 3.1-143 Specific electrical resistivity versus temperature of molybdenum [1.126], tungsten [1.126], niobium [1.127], and tantalum [1.127]

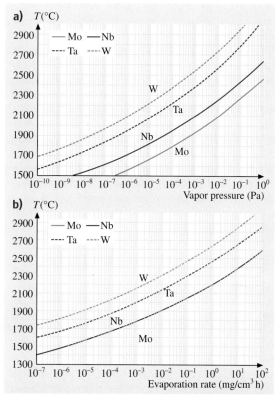

Fig. 3.1-144a,b Vapor pressure (**a**) and evaporation rate (**b**) versus temperature of molybdenum, tungsten, niobium, and tantalum [1.129]

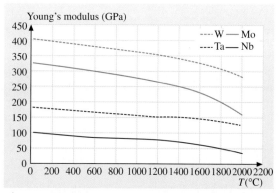

Fig. 3.1-145 Young's moduli versus temperature of molybdenum, tungsten, niobium, and tantalum [1.128]

as well as the tungsten based alloys AKS–W, WL, WC, and WT, the physical properties do not differ significantly from those of the pure metals. Values for the Young's modulus and its temperature dependence are plotted in Fig. 3.1-145. The Young's moduli of the Group Va metals are considerably lower than those of the Group VIa metals due to the differences in electronic structure.

The strong influence of the surface conditions on the emissivity and lack of information in the literature concerning pre-treatment make it difficult to interpret emissivity data. An overview of emissivity measurements for W, Nb, and Ta at 684.5 nm from 1500 °C up to the liquid phase using laser polarimetry is given in [1.130].

3.1.9.2 Chemical Properties

Refractory metals are highly resistant to many chemical agents. Tantalum is outstanding in its performance as it is inert to all concentrations of hydrochloric and nitric acid, 98% sulphuric acid, 85% phosphoric acid, and aqua regia below 150 °C. Tantalum can be attacked by hydrofluoric acid and strong alkalis, however. The excellent corrosion resistance of Ta is attributed to dense natural oxide layers which prevent the chemical attack of the metal. Niobium is less resistant than Ta and embrittles more easily. Nevertheless the more economical Nb has replaced Ta in some applications. Molybdenum and W are highly resistant to many molten glasses and metals as long as free oxygen is absent.

The interaction with H, N, and O is widely different for Mo and W on the one hand, and Nb and Ta on the other. Molybdenum and W have almost no solubility, whereas Nb and Ta can dissolve a considerable amount of these elements. Hydrogen can be removed from Nb at 300 °C to 1600 °C and from Ta at 800 °C to 1800 °C without metal loss by degassing in high vacuum. For the removal of N, temperatures higher than 1600 °C are recommended. The evaporation of volatile oxides at temperatures above 1600 °C in high vacuum leads to a reduction of the oxygen content in Nb and Ta. But during such heat treatments, metal is evaporated simultaneously.

An overview on the resistance of pure Mo, W, Nb, and Ta against different media is given in Tables 3.1-109 to 3.1-111. In accordance with the definition given in [1.131], a material is considered "stable" against a corrosive medium if the metal loss is < 0.1 mm per year. If the loss of material is between 0.1 and 1.0 mm per year, the material is "considerably stable;" and it is "fairly unstable" if the loss of material is between 1.0 and 3.0 mm. The material is "unsuitable" in an environment if the metal loss is > 3.0 mm per year.

Oxidation Behavior

Refractory metals require protection from oxidizing environment as they do not form protective oxide layers. Oxidation of Mo and W leads to a loss of material by the formation of volatile oxides above 600 °C, but without any significant impact on the mechanical properties.

The low temperature oxidation of Mo and W often causes problems in practical use as thin corrosion films are formed during storage in moist air. The surface topography of Mo has a strong impact on the reaction rate. The corrosion film contains oxygen but also chemically-bonded nitrogen, which is incorporated in the film during film growth. C_xH_y residues are the nuclei for the oxidative attack in the early stage of film growth [1.132].

There is extensive evidence that between 300 and 500 °C, oxide-dispersion-strengthened (ODS) refractory metal alloys (e.g., Mo−La_2O_3 and Mo−Y_2O_3 grades) possess markedly reduced oxidation rates compared to the pure metals [1.133]. In-situ oxidation and evaluation of the binding state reveal that MoO_2 dominates over MoO_3 both for Mo and Mo−

Table 3.1-109 Metal loss of Mo, W, Nb and Ta in millimeter per year (mm/yr) in acids, alkalis and salt solutions

Corroding agent (aqueous solution)	Temperature	Metal loss (mm/yr) Mo	W	Nb	Ta
10% HCl	20 °C	< 0.003	0.002	0	< 0.001
10% H_2SO_4	20 °C	< 0.005	< 0.1	< 0.1	< 0.1
10% HNO_3	20 °C	18.6	< 0.25	< 0.013	< 0.013
3% HF	20 °C	< 0.001	< 0.1	> 3.0	> 3.0
10% CH_3COOH	20 °C	0.07	< 0.05	< 0.1	< 0.013
10% KOH	20 °C	< 0.1	< 0.1	< 0.2	< 0.1
3% NaCl	20 °C	< 0.1	< 0.1	< 0.1	< 0.1
10% HCl	100 °C	< 0.025	0.005	0.005 (embrittlement)	< 0.025
10% H_2SO_4	100 °C	0.17	< 0.25	< 0.001	0
10% HNO_3	100 °C	150	< 0.25	< 0.076	< 0.025
3% HF	100 °C	0.18	0.15	> 3.0	> 3.0
10% CH_3COOH	100 °C	0.033	< 0.05	< 0.1	< 0.013
10% KOH	100 °C	0.054	0.01	1.2 (embrittlement)	< 0.003 (embrittlement)
3% NaCl	100 °C	< 0.1	< 0.1	< 0.1	< 0.1

Table 3.1-110 Maximum temperatures for the resistance of Mo, W, Nb, and Ta against metal melts (up to the stated temperatures the solubility of the refractory metal in the metal melts and vice versa is negligible)

Metal melt	Maximum resistance temperature (°C)			
	Mo	W	Nb	Ta
Al	Not resistant	680	Not resistant	Not resistant
Pb	1100	1100	850	1000
Fe	Not resistant	Not resistant	Not resistant	Not resistant
Ga	300	800	400	450
K	1200	900	1000	1000
Cu	1300	Resistant	Resistant	Resistant
Mg	1000	600	950	1150
Na	1030	900	1000	1000
Hg	600	600	600	600
Zn	500	750	Not resistant	500
Sn	550	980	Not resistant	260
Ag	Resistant	Resistant	Resistant	1200

Table 3.1-111 Maximum temperatures for the resistance of Mo, W, Nb, and Ta against gaseous media at atmospheric pressure

Gaseous media	Maximum resistance temperature			
	Mo	W	Nb	Ta
Air and oxygen	see Sect. 3.1.9.2			
Hydrogen	Resistant	Resistant	250 °C ($T > 250$ °C embrittlement owing to dissolved hydrogen)	300 °C ($T > 300$ °C embrittlement owing to dissolved hydrogen)
Nitrogen	Resistant	Resistant	300 °C ($T > 300$ °C embrittlement owing to dissolved nitrogen)	400 °C ($T > 400$ °C embrittlement owing to dissolved nitrogen)
Ammonia	Resistant (except 1000–1100 °C: nitration)	Resistant (except 700–1150 °C: nitration)	300 °C ($T > 300$ °C embrittlement owing to dissolved nitrogen)	700 °C ($T > 700$ °C embrittlement owing to dissolved nitrogen)
Water vapor	700 °C	700 °C	200 °C	200 °C
Carbon monoxide	800 °C ($T > 800$ °C carburization)	1000 °C ($T > 800$ °C carburization)	800 °C	1100 °C
Carbon dioxide	1200 °C ($T > 1200$ °C oxidation)	1200 °C ($T > 1200$ °C oxidation)	400 °C	500 °C

0.47 wt%Y_2O_3–0.08 wt%Ce_2O_3 (Fig. 3.1-146). These results are confirmed when comparing the thickness of the oxide layer formed at 500 °C in air (Fig. 3.1-147). The enhanced oxidation resistance of doped samples cannot be attributed to a chemical but rather to a morphological-mechanical effect. Investigations of the surface of pure Mo samples subjected to long-term oxidation reveal a network of cracks in the MoO_2 layer which drastically diminishes the passivating effect of this layer by facilitating a further attack by O. Such cracks cannot be found on the surface of oxidized Mo–La_2O_3 and Mo–Y_2O_3 samples [1.133].

Above 700 °C for Mo and approximately 900 °C for W, evaporation of the volatile oxides MoO_3 and WO_3 is the rate-controlling process and the oxidation follows a linear time dependence. Above 2000 °C for Mo and 2400 °C for W, the metal loss increases because of the increasing vapor pressure of the pure metals. The time

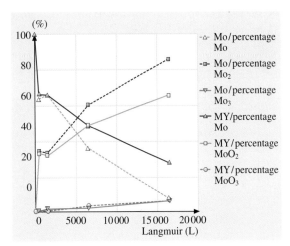

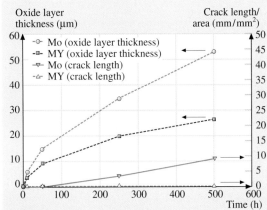

Fig. 3.1-146 ESCA measurements and evaluation of Mo/MoO$_2$/MoO$_3$ fraction versus in-situ oxidation conditions in Langmuir (1 L = 10^{-6} Torr s). Mo: 4N5 Mo ribbon, electro-polished surface; MY: Mo–0.47 wt%Y$_2$O$_3$–0.08 wt%Ce$_2$O$_3$ ribbon, electro-polished surface, test conditions: 500 °C/air [1.133]

Fig. 3.1-147 Thickness of oxide layer and crack length per area versus testing time. Mo: 4N5 Mo ribbon, electropolished surface; MY: Mo–0.47 wt%Y$_2$O$_3$–0.08 wt%Ce$_2$O$_3$ ribbon, electropolished surface, test conditions: 500 °C/air [1.133]

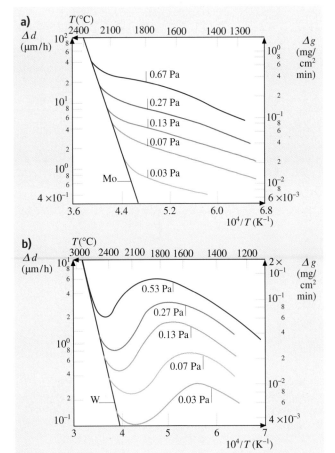

Fig. 3.1-148a,b Metal loss in air for Mo (**a**) and W (**b**) at $T \geq$ 1200 °C [1.134]

dependence of metal loss for Mo and W at $T \geq 1200$ °C is shown in Fig. 3.1-148 [1.134].

The oxidation rates of Nb and Ta strongly depend on temperature, pressure, and time. Different mechanisms causing different oxidation rates can be observed and metastable sub-oxides are formed during use in O-containing atmospheres [1.131, 134]. Oxygen is dissolved in the metal matrix which leads to significant changes of the mechanical properties. A poorly adherent, porous pentoxide is formed on the metal surface which does not protect the metal from further attack. Oxygen which has penetrated the porous oxide layer diffuses along the grain boundaries of the metal, leading to drastic embrittlement.

The metal loss at 1100 °C due to oxidation in air is shown for Mo, W, Nb, and Ta in Fig. 3.1-149 [1.135]. The embrittled zone caused by oxygen diffusion into the substrate is not considered in this diagram.

Only a few coating systems have been found to prevent the refractory metals from oxidation. In the case of Mo, Pt-cladded components with a diffusion barrier interlayer on the basis of alumina, which prevents the formation of brittle intermetallic molybdenum–platinum

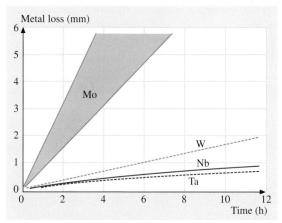

Fig. 3.1-149 Oxidation behavior of Mo, W, Nb, and Ta at 1100 °C [1.135]

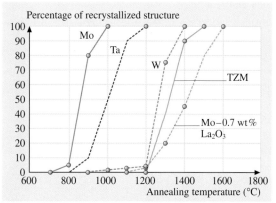

Fig. 3.1-150 Fraction of recrystallized structure versus annealing temperature (annealing time $t_a = 1$ h) for Mo, P/M Ta, W, TZM, and Mo–0.7 wt% La_2O_3 sheets with a thickness of 1 mm, degree of deformation, $\varphi = \frac{th_p - th_s}{th_p} 100[\%]$, where th_p = thickness of sintered plate, and th_s = thickness of rolled sheet. Mo: $\varphi = 94\%$, P/M Ta: $\varphi = 98\%$, W: $\varphi = 94\%$, TZM: $\varphi = 98\%$, Mo–0.7 wt% La_2O_3: $\varphi = 99\%$ [1.126]

phases, are used; e.g., for stirrers used to homogenize special glasses. Mo components for glass tanks and glass melting electrodes are protected from oxidation by coatings based on silicides, such as silicon–boron [1.136]. Silicon–boron and other silicide based coatings, such as silicon–chromium–iron or silicon–chromium–titanium, are also used to protect Mo and Nb based alloys from oxidation in aerospace applications [1.135, 137–139]

3.1.9.3 Recrystallization Behavior

The mechanisms and kinetics of recovery and recrystallization processes significantly affect the processing and application of refractory metals. The homologous temperatures for obtaining 50% recrystallized structure during annealing for one hour, range from 0.39 for Mo, 0.41 for W, 0.42 for Nb, to 0.43 for P/M Ta. The fraction of recrystallized structure as a function of the annealing temperature of Mo, W, P/M Ta, TZM, and Mo–0.7 wt% La_2O_3 sheets of 1-mm thickness is shown in Fig. 3.1-150. Experimental data on the evolution of grain size with the annealing temperature for pure Mo and Ta sheets of 1-mm thickness are shown in Fig. 3.1-151 [1.126, 140]. A recrystallization diagram for Mo is published in [1.141], and for P/M Ta in [1.142].

Compared to pure Mo, the recrystallization temperature of the carbide-precipitation-hardened alloy TZM is increased by 450 °C and that of Mo–0.7 wt%La_2O_3 by 550 °C. The data listed in Table 3.1-112 show that all those alloys containing particles, which deform together with the matrix metal (ML, K−Si−Mo, WL10, WL15, WT20, and AKS−W−ThO$_2$), reveal a significantly increased recrystallization temperature in the highly de-

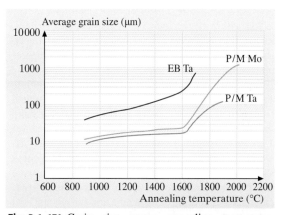

Fig. 3.1-151 Grain size versus annealing temperature (annealing time $t_a = 1$ h) for Mo, P/M Ta, and EB (electron-beam melted) Ta sheets with a thickness of 1 mm. Mo: $\varphi = 94\%$, P/M Ta: $\varphi = 98\%$, EB Ta: $\varphi = 98\%$ [1.126, 140]

formed state. Increasing the degree of deformation from $\varphi = 90\%$ to $\varphi = 99.99\%$ leads to a further increase of the recrystallization temperature (100% recrystallized structure, 1 h annealing time) ranging from 600 °C for K−Si−Mo, 700 °C for ML, 950 °C for WT20, 1000 °C for WL10, to 1050 °C for WL15 and WC20. This can be explained by particle refinement. During the deformation process the particles are elongated. During annealing they break up and rows of smaller particles

Table 3.1-112 Typical recrystallization temperature and ultimate tensile strength of commercial Mo and W based rod and wire materials with a defined degree of total deformation φ [1.118]

Alloy designation	Composition (wt%)	Temperature for 100% recrystallized structure ($t = 1$ h) (°C)	Typical ultimate tensile strength at 1000 °C (MPa)
Pure Mo		1100 ($\varphi = 90\%$)	250 ($\varphi = 90\%$)
TZM	Mo, 0.5% Ti, 0.08% Zr, 0.025% C	1400 ($\varphi = 90\%$)	600 ($\varphi = 90\%$)
MHC	Mo, 1.2% Hf, 0.08% C	1550 ($\varphi = 90\%$)	800 ($\varphi = 90\%$)
ML	Mo, 0.3% La$_2$O$_3$	1300 ($\varphi = 90\%$), 2000 ($\varphi = 99.99\%$)	300 ($\varphi = 90\%$)
MY	Mo, 0.48% La$_2$O$_3$, 0.07% Ce$_2$O$_3$	1100 ($\varphi = 90\%$), 1350 ($\varphi = 99.99\%$)	300 ($\varphi = 90\%$)
K–Si–Mo	Mo, 0.05% Si, 0.025% K	1200 ($\varphi = 90\%$), 1800 ($\varphi = 99.99\%$)	300 ($\varphi = 90\%$)
Mo50Re	Mo, 47.5% Re	1300 ($\varphi = 90\%$)	600 ($\varphi = 90\%$)
Mo30W	Mo, 30% W	1200 ($\varphi = 90\%$)	350 ($\varphi = 90\%$)
Pure W		1350 ($\varphi = 90\%$)	350 ($\varphi = 90\%$)
AKS-W	W, 0.005% K	2000 ($\varphi = 99.9\%$)	800 ($\varphi = 99.9\%$)
WL10	W, 1.0% La$_2$O$_3$	1500 ($\varphi = 90\%$), 2500 ($\varphi = 99.99\%$)	400 ($\varphi = 90\%$)
WL15	W, 1.5% La$_2$O$_3$	1550 ($\varphi = 90\%$), 2600 ($\varphi = 99.99\%$)	420 ($\varphi = 90\%$)
WC20	W, 1.9% Ce$_2$O$_3$	1550 ($\varphi = 90\%$), 2600 ($\varphi = 99.99\%$)	420 ($\varphi = 90\%$)
WT20	W, 2% ThO$_2$	1450 ($\varphi = 90\%$), 2400 ($\varphi = 99.99\%$)	400 ($\varphi = 90\%$)
AKS-W–ThO$_2$	W, 1% ThO$_2$, 0.004% K	2400 ($\varphi = 99.9\%$)	1000 ($\varphi = 99.9\%$)
W5Re	W, 5 Re	1700 ($\varphi = 90\%$)	500 ($\varphi = 90\%$)
W26Re	W, 26 Re	1750 ($\varphi = 90\%$)	900 ($\varphi = 90\%$)

are formed. With the increase in number of particles, the subgrain boundaries are pinned more effectively, resulting in an increase of the recrystallization temperature [1.119].

Experiments with various oxide-dispersion-strengthened (ODS) Mo materials with 2 vol.% of oxide, mean oxide particle sizes in the as-sintered state of around 0.8 μm, and a degree of deformation $\ln(A_0/A) = 8.5$ ($A_0 =$ cross section as-sintered, $A =$ cross section as-deformed), revealed differences in the recrystallization temperature of up to 750 °C depending on the oxide used. It could be shown that this effect is caused by particle refinement during deformation and subsequent heat treatment. Particles which increase the recrystallization temperature very effectively, as is the case with La$_2$O$_3$, show a high particle deformability [1.143].

Whether oxide particles deform in a pseudo-plastic manner or not depends on a multitude of parameters, such as the yield stress of the particles, the yield stress of the matrix, the particle/matrix bonding strength, the crystallite size, the defect density, or the state of stresses. Most of these parameters are unknown or difficult to measure. Good correlation could be found between the particle deformability, with its effect on the increase of the recrystallization temperature, and the fraction of ionic bonding character of the oxide, following the definition of Pauling [1.143]. Figure 3.1-152 shows that compounds with a high fraction of ionic bonding character, such as La$_2$O$_3$ or SrO, raise the recrystallization temperature very effectively. Slight particle multiplication could also be found for Al$_2$O$_3$, ZrO$_2$, and HfO$_2$ compounds with a marked covalent bonding character, but because of breakage of the particles during the deformation process.

The recrystallization temperature can be tailored by varying both the type and the content of the oxide, the lat-

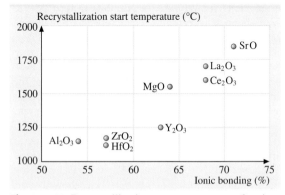

Fig. 3.1-152 Recrystallization start temperature of various ODS Mo materials versus fraction of ionic bonding (according to Pauling) of the respective oxides. Oxide content $= 2$ vol.%, wire diameter $= 0.6$ mm [1.143]

ter shown in Fig. 3.1-153 for Mo–0.03 wt%La$_2$O$_3$ (ILQ) and Mo–0.3 wt%La$_2$O$_3$ (ML).

AKS–W shows a similar effect where the K-containing bubbles are effective pinning centers. The recovery and recrystallization mechanisms of AKS–W as a function the annealing temperature are summarized in Table 3.1-113.

The recrystallization temperature of AKS–W is determined by the relation of the driving to dragging forces. The driving forces are determined by the thermomechanical treatment (TMT), the dragging forces depend on the number, size, and distribution of the K bubbles. With increasing degree of deformation, both the driving forces (increasing dislocation density and low-angle/high-angle boundary volume) and dragging forces (increasing length of stringers of K-filled pores and, as a consequence, increasing number of K bubbles formed) become larger. During the first deformation steps the

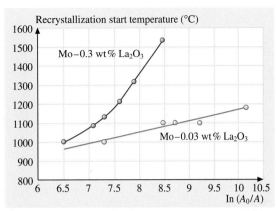

Fig. 3.1-153 Recrystallization start temperature of Mo–0.03 wt% La$_2$O$_3$ and Mo–0.3 wt% La$_2$O$_3$ wires versus degree of deformation $\ln(A_0/A)$ [1.133]

Table 3.1-113 Recrystallization mechanisms for AKS-W wires [1.120, 144, 145]

Microstructural state	Processes
Evolution of the microstructure during wire drawing	• Formation of a dislocation cell structure by static and dynamic recovery processes
Coarsening of the microstructure during annealing	• Annealing temperature 800 to 1400 °C: – Reduction of the dislocation density within the cell walls; – Break-up of the K-containing stringers into pearl rows of bubbles; – Migration of longitudinal grain boundaries is strongly reduced by pearl rows of bubbles; – Similar $\langle 110 \rangle$ texture as in the as-worked state. • For low heating rates, partial or even entire bubble rows can be dragged by the boundaries moving in the transverse direction. As a result, row/row collision and bubble coalescence can occur • The temperature up to which the coarsened substructure remains stable depends on the degree of deformation (e.g., diameter 0.18 mm: 2100 °C/15 min). The coarsened substructure has a significant portion of high-angle grain boundaries (misorientation angles higher than 15°)
Exaggerated grain growth	Nucleation and growth of large interlocking grains occur by primary recrystallization, whereby a subgrain begins to grow laterally at the expense of neighboring polygonized subgrains. • Because of the pearl row of bubbles, the rate of grain boundary movement is higher in the axial, than in the transverse direction • Because of the interaction between the bubbles and the growing grains, the grain boundaries possess a wave-like, interlocking structure • The grain aspect ratio of the recrystallized structure increases with increasing number of K bubbles. The number of K bubbles is a function of the K-content, the degree of deformation and the TMT. • Highly deformed, recrystallized wires reveal a $\langle 531 \rangle$ texture.

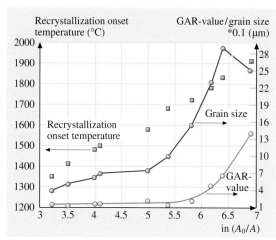

Fig. 3.1-154 Recrystallization onset temperature, grain size, and grain aspect ratio (GAR) versus $\ln(A_0/A)$. AKS-W grade with K content of 42 μg/g; deformation temperature below the onset temperature of recrystallization; no intermediate heat treatments above recrystallization temperature; annealing time = 15 min; heating rate = 3 K/s; annealing atmosphere = hydrogen; annealing temperature for GAR evaluation = 2200 °C (t_a = 5 min); grain size measured in transverse direction [1.133]

increase of the driving forces outweighs that of the dragging forces. Then the rise of the dragging forces starts to dominate, resulting in an increase of the recrystallization temperature (see Fig. 3.1-154). With increasing length of the potassium stringers/number of potassium bubbles, the significance of the dragging effect in the transverse direction increases, resulting in an increase of the grain aspect ratio (GAR) in the recrystallized state. A marked increase of the GAR starts at a degree of deformation that coincides with a pronounced increase of the transverse grain size, as illustrated in Fig. 3.1-154.

3.1.9.4 Mechanical Properties

Influence of Thermomechanical Treatment (TMT) and Impurities

Molybdenum and Tungsten Alloys [1.133]. The formation of a cellular dislocation structure increases both strength and fracture toughness [1.146, 147]. Additionally, the mechanical properties depend on the type of deformation process, purity, and heat treatment. The thermomechanical treatment (TMT) serves to obtain the specified shape, to eliminate the porosity, and to adjust the mechanical and structural properties. In particular, the evolution of the density distribution, pore size and shape, and its interaction with the mechanical properties are of high importance [1.120, 148–150].

During hot deformation which is usually the primary working step for both Mo and W, high-angle grain boundaries are formed and migrate, grains are subdivided by low-angle boundaries, and new large-angle boundaries are formed by coarsening of the substructure [1.151]. Grains with an aspect ratio close to one and a low dislocation density are formed.

Following this hot deformation, the material is processed at temperatures below the recrystallization onset temperature, but above the onset temperature of polygonization, leading to a cellular dislocation structure. The cell boundaries become impenetrable for slip dislocations and behave like grain boundaries, when the misorientation between neighboring cells is higher than a critical value of about 4° for Mo and W [1.146]. The formation of a misoriented cellular dislocation structure results both in an increase of strength and a decrease of the ductile-brittle-transition temperature (DBTT). Both effects become more and more significant with increasing degree of deformation which results in a smaller effective grain size [1.146]. The size of the misoriented cellular dislocation structure depends sensitively on the deformation temperature. A high deformation temperature implies large grains [1.152]. It is essential to control the deformation temperature carefully. The control of the microstructure at intermediate steps has also been recommended [1.153].

The production yield is mainly decreased by the formation of grain boundary cracks. The cohesion of the grain boundaries is believed to be the controlling factor limiting the ductility of Mo and W [1.154]. Impurities segregated at the grain boundaries can lead to a strong decrease in ductility. Based on both semi-empirical and first principle modeling of the energetics and the electronic structures of impurities on a $\Sigma 3$ (111) grain boundary in W, it was concluded that the impurities N, O, P, S, and Si weaken the intergranular cohesion, while B and C enhance the interatomic interaction across the grain boundary [1.154].

The amount of O in Mo ($\approx 10\,\mu g/g$) and W ($\approx 5\,\mu g/g$) is sufficient to form a monolayer of O on the grain boundaries as long as the grain size is not smaller than 10 μm. During recrystallization a migrating grain boundary can be saturated by collecting impurities while sweeping the volume. Based on the investigation of arc-cast Mo samples, a beneficial effect of C was found and attributed to the following mechanisms [1.155]:

- Suppression of oxygen segregation
- Precipitation of carbides, acting as dislocation sources
- Formation of an epitaxial relationship between precipitates and bulk crystals at grain boundaries

From these findings it was proposed that a C/O atomic ratio of >2 should improve the mechanical properties of Mo. The results obtained with arc-cast samples published in [1.155] could not be reproduced for samples produced by a P/M route [1.157].

As a consequence of the above mentioned effects, but in contrast to many other metallic materials, the fracture toughness of molybdenum and tungsten is strongly reduced with increasing degree of recrystallization. With increasing plastic deformation, the fracture toughness increases (see Sect. 3.1.9.4), combined with a transition from intercrystalline to transcrystalline cleavage and to a transcrystalline ductile fracture [1.147, 158, 159].

With increasing degree of deformation the working temperature can be progressively reduced. In particular, products with a high degree of deformation such as wires, thin sheets, and foils can be subjected to a high degree of deformation at a temperature below the polygonization temperature. The reduction of grain boundaries oriented transversely to the drawing direction increases the bending ductility of W [1.160]. Tungsten wire with an optimum ductility can only be obtained, when deformed in such a way that dynamic recovery processes occur without polygonization [1.161]. Other recovery phenomena, besides polygonization, are described in [1.162]. However a high degree of deformation below the onset temperature for dynamic polygonization favors the formation of longitudinal cracks.

Thin sheets and foils, annealed under conditions resulting in a small fraction of primarily recrystallized grains, can show a very specific fracture behavior, i.e., cracks running at an angle of 45° to the rolling direction [1.163, 164]. Such 45° embrittlement is caused by the nucleation of critical cracks at isolated grains formed by recrystallization of weak secondary components of the texture. Cracks propagate under 45° to the rolling direction owing to the alignment of the cleavage planes in the rolling texture.

Niobium and Tantalum Alloys. Contrary to Mo and W, pure Nb and Ta are deformed at room temperature. Only highly-alloyed materials require breaking down the ingot microstructure either by forging or extrusion at elevated temperatures. In these cases the ingot has to be protected to prevent an interaction with the atmosphere. The mechanical behavior of pure annealed niobium and tantalum is characterized by a high ductility and low work-hardening rate. The influence of deformation on the yield strength and fracture elongation of pure tantalum is shown in Fig. 3.1-155.

Mechanical properties of niobium- and tantalum-based alloys are strongly influenced by interstitial impurities, e.g., oxygen, nitrogen, carbon, and hydrogen. The generally lower content of impurities and the larger grain size are the reasons why melt-processed niobium and tantalum have a lower room-temperature tensile strength compared to sintered material. As an example, the influence of oxygen on the mechanical properties of tantalum is presented in Fig. 3.1-156.

After deformation both niobium- and tantalum-based alloys are usually heat treated in high vacuum before delivering in order to achieve a fine grained primarily recrystallized microstructure.

Static Mechanical Properties

Properties at Low Temperatures and Low Strain Rates. The flow stress of the transition metals Mo, W, Nb, and Ta is strongly dependent on temperature and a strain rate below a characteristic transition temperature T_K (corresponding to 0.1 to 0.2 of the absolute melting temperature) and plastic strain rates below $1 \times 10^{-5}\,\text{s}^{-1}$ [1.165]. As an example, experimental data on the temperature dependence of the flow stress of recrystallized Ta are shown in Fig. 3.1-157. This dependence has

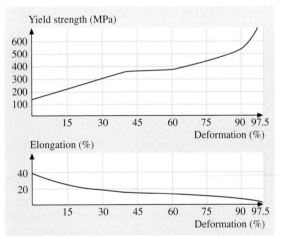

Fig. 3.1-155 Yield strength and fracture elongation of as-worked tantalum versus degree of deformation [1.156]

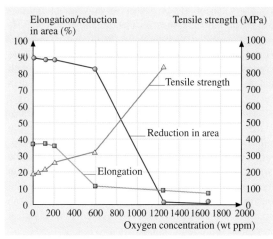

Fig. 3.1-156 Tensile strength, fracture elongation, and reduction in area versus oxygen concentration of tantalum specimens tested at room temperature [1.168]

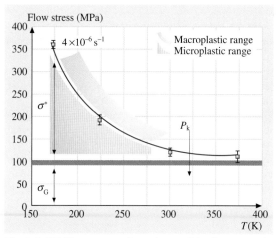

Fig. 3.1-157 Temperature dependence of the flow stress (at $\varepsilon_{pl} = 1 \times 10^{-5}$) of recrystallized Ta under monotonic loading and $d\varepsilon/dt = 2 \times 10^{-6}\,\text{s}^{-1}$, $T_K =$ "knee" temperature, $\sigma_G =$ athermal range, $\sigma^* =$ thermal range [1.140]

been attributed to the characteristic behavior of screw dislocations [1.166]. The transition temperature T_K was shown to depend on the strain rate (Fig. 3.1-158) [1.140, 167].

For test temperatures $T < T_K$ the flow stress increases markedly. Borderlines between elastic/anelastic strain (σ_A) and microstrain/macrostrain deformation ranges can be deduced, subdivided into athermal (σ_G) and thermal (σ^*) ranges. The lower borderline of the microplastic region is almost independent of tem-

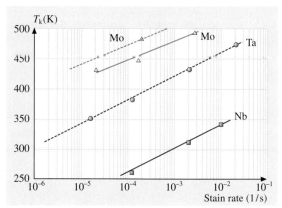

Fig. 3.1-158 Dependence of T_K on strain rate [1.140, 167]

perature and may be called the intrinsic flow stress which is much lower than the conventionally determined flow stress (technical flow stress). In the stress range between the intrinsic and the technical flow stress below T_K, strains of up to several percent were observed after extended loading times for Mo and Ta [1.169, 170]. The effect of temperature and strain rate on the monotonic microflow behavior of bcc metals as functions of temperature and strain rate was presented schematically [1.171]. Experimental data for Mo and Ta in constant load tests (low temperature creep tests) for stresses considerably below the technical flow stress are shown in Figs. 3.1-159 and 3.1-160. After a considerable incubation period, depending on the test temperature and stress, a rapid increase in strain can be noticed approaching a saturation strain which depends on the stress level. The effect of the loading rate on the instantaneous plas-

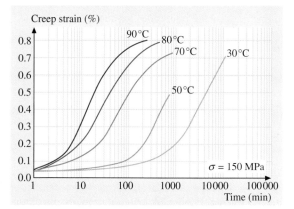

Fig. 3.1-159 Creep elongation of recrystallized Mo at $\sigma = 150\,\text{MPa}$ for $30\,°\text{C} \leq T \leq 90\,°\text{C}$ [1.126]

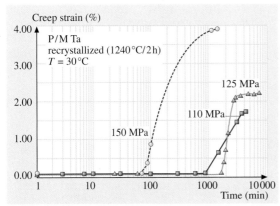

Fig. 3.1-160 Creep elongation of recrystallized Ta sheets (thickness = 2 mm) at 30 °C for 110 MPa ≤ σ ≤ 150 MPa [1.126]

tic strain can be revealed with high resolution in loading–unloading tests under various constant loading rates [1.140, 170]. This microflow behavior may be significant for components at low temperatures and low loads, e.g., under storage conditions. Internal stresses in semi-finished products may be reduced even at room temperature to levels corresponding to the intrinsic flow stress [1.170].

A significant influence of the strain rate on the tensile properties at room temperature was determined for recrystallized Mo and Ta (Fig. 3.1-161) [1.140], in close agreement with literature data [1.172]. This strain rate effect makes it imperative for comparison of test data to list the test conditions.

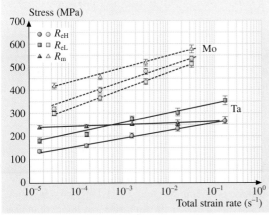

Fig. 3.1-161 Effect of strain rate on tensile properties at room temperature of recrystallized Mo and recrystallized Ta [1.140]

Properties at Elevated Temperatures. A rough ranking of the high temperature strength of Mo and W alloys can be obtained from the comparison presented in Table 3.1-112 (Sect. 3.1.9.3). Carbide-precipitation-strengthened Mo-based alloys (MHC, TZM) and alloys high in Re (Mo–50 wt%Re, W–26 wt%Re) have the highest tensile strength. Alloys containing potassium (AKS–W, AKS–W–ThO$_2$) exhibit high strength only in the case of a high preceding plastic deformation.

A comparison of the high-temperature strength of rods made of Mo, W, Nb, and Ta in their usual state of delivery is given in Fig. 3.1-162. The typical microstructure of stress-relieved Mo is a highly polygonized structure with up to 5% recrystallized grains. Depending on the product shape W is delivered in the as-worked state, especially in case of sheet material and wires, or stress-relieved with a polygonized microstructure.

The decrease of tensile strength and the increase of reduction in area with increasing test temperature can be related to changes in the fracture mode (Fig. 3.1-163), i. e., cleavage fracture, brittle grain boundary failure, and ductile transcrystalline failure [1.158].

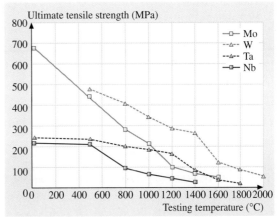

Fig. 3.1-162 Ultimate tensile strength versus test temperature for Mo, W, Ta, and Nb rods in their usual delivering condition. Mo, W: diameter = 25 mm (stress relieved); Ta, Nb: diameter = 12 mm (recrystallized); technical strain rates = 1.0×10^{-4} s^{-1} up to the 0.2% yield strength followed by 3.3×10^{-3} s^{-1} (Mo, room temperature), 1.7×10^{-3} s^{-1} (Mo, elevated temperatures), 8.3×10^{-4} s^{-1} (W, elevated temperatures), 6.7×10^{-4} s^{-1} up to the 0.2% yield strength followed by 3.3×10^{-3} s^{-1} (Nb, Ta for all test temperatures) [1.118]

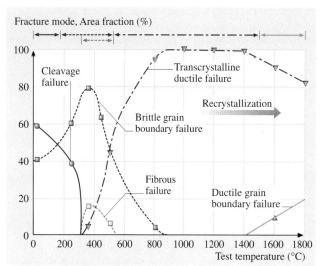

Fig. 3.1-163 Effect of test temperature on fracture modes of pure W, stress relieved at 1000 °C/6 h [1.158]

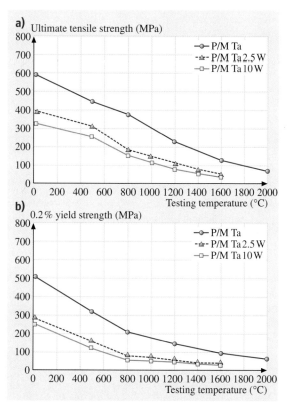

Fig. 3.1-164a,b The ultimate tensile strength R_m (**a**) and the yield strength $R_{p0.2}$ (**b**) versus test temperature of P/M Ta, P/M Ta2.5W, and P/M Ta10W sheets that are 1 mm thick, with impurity content (μg/g) Ta: O = 60, N = 10, H = 1.8, C < 5; Ta2.5W: O = 70, N = 12, H = 2.4, C = 5; Ta10W: O = 31, N < 5, H < 1, C = 21; material condition = recrystallized, technical strain rates = $6.7 \times 10^{-4}\,\mathrm{s}^{-1}$ up to $R_{p0.2}$, followed by $3.3 \times 10^{-3}\,\mathrm{s}^{-1}$ [1.126]

The influence of alloying Ta with W is illustrated in Fig. 3.1-164 for Ta2.5W and Ta10W, which are the main commercial Ta-based alloys.

Figure 3.1-165 summarizes values of $R_{p0.2}$ of the most common Nb-based alloys at elevated temperatures. All of these alloys are hardened primarily by solid solution strengthening; however, small amounts of precipitates are present.

For comparison, the high-temperature strength of stress-relieved 1-mm sheets made of Mo- and W-based materials is shown in Fig. 3.1-166. For short-term application under high stresses, the precipitation-strengthened Mo alloys TZM and MHC offer the best performance up to a service temperature of 1500 °C. For higher temperatures, W-based materials should be applied. Tantalum-based alloys are used only if additional high ductility is required after cooling to room temperature.

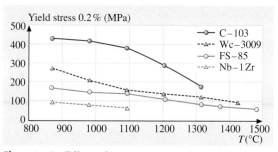

Fig. 3.1-165 Effect of temperature on $R_{p0.2}$ for common Nb-based alloys [1.173]

Dynamic Properties

Microplasticity Effects under Cyclic Loading at Low Temperatures. Microplasticity effects under monotonic loading have been reported in the literature for single- and polycrystalline Mo and Ta [1.171, 174]; information regarding effects of strain rate and temperature on the cyclic stress–strain response is given in [1.140, 170]. Most experiments on cyclic stress–strain behavior have

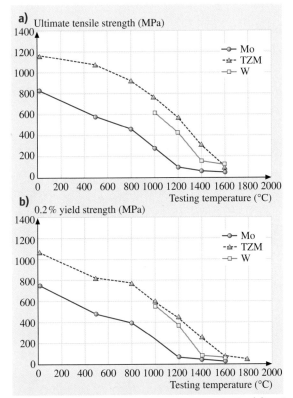

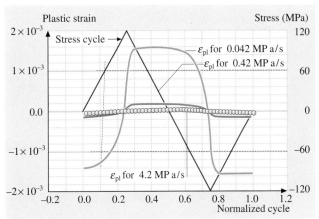

Fig. 3.1-167 Effect of loading rate on cyclic plastic strain of recrystallized Ta during a tension-compression cycle ($R = -1$). Stress amplitude = 120 MPa; test temperature = 25 °C; duration of a cycle at 0.042 MPa/s = 190 min, at 4.2 MPa/s = 1.9 min; maximum plastic strain rate at all loading rates = approx. 3×10^{-6} s^{-1} [1.140]

Fig. 3.1-166a,b The ultimate tensile strength R_m (**a**) and the yield strength $R_{p0.2}$ (**b**) versus test temperature for Mo, TZM, and W sheets that are 1 mm thick. Material condition = stress relieved, technical strain rates = 2.0×10^{-3} s^{-1} (Mo, TZM, room temperature), 1.3×10^{-3} s^{-1} (Mo, TZM, elevated temperatures), 3.3×10^{-4} s^{-1} up to $R_{p0.2}$, followed by 6.7×10^{-4} s^{-1} (W, elevated temperatures) [1.126]

been carried out at room temperature. When bcc metals are deformed at $T < 0.2\,T_m$, microstrain ($\varepsilon \leq 10^{-3}$) is characterized as the plastic strain, accommodated by the motion of non-screw dislocations [1.166, 169]. These differences are manifested in the temperature and strain-rate dependence of the flow behavior [1.171, 175]. Investigations of Mo and Ta showed that a true microplastic deformation can only be considered at plastic strains of less than 5×10^{-4} [1.140]. The critical temperature below which the marked increase in the cyclic flow stress occurs is in the range of 25 °C to 80 °C for Ta and between 200 °C and 280 °C for Mo, depending on the strain rate. The experimental results for Ta showed that the highest cyclic plastic strains are obtained under strain rates between 1×10^{-8} s^{-1} and 2×10^{-6} s^{-1}. As an example, the effect of the loading rate on the cyclic plastic strain of recrystallized tantalum during tension–compression cycles at loading rates between 0.042 MPa/s (duration of one cycle = 190 min) and 4.2 MPa/s (duration of one cycle = 1.9 min) is shown in Fig. 3.1-167 [1.140].

High-Cycle Fatigue Properties. Most of the fatigue data are reported in form of stress vs. number of cycles (S–N) curves. For Mo a fatigue limit may be approached for $N > 10^7$ under stress-controlled conditions [1.176]. Experiments were conducted at test frequencies up to 20 kHz. The results of such tests should be considered with caution, taking into account the temperature and strain-rate sensitivity of bcc metals. Representative S–N curves for as-worked and recrystallized Mo, and a comparison of push–pull- and bending-fatigue-tested Mo sheet specimens are shown in Figs. 3.1-168 [1.126] and 3.1-169 [1.126].

The reported fatigue test data for various test methods are summarized in Table 3.1-114 with the fatigue limit (Se) and the ratio of fatigue limit to tensile-stress (Se/Rm) as characteristic parameters. Methods of statistical evaluation of test data were published [1.176, 177]. A decrease in fatigue limits with decreasing cyclic frequency was found.

Cyclic hardening/softening was deduced and cyclic stress–strain curves over wide ranges of plastic strain amplitudes were published in [1.178, 179] for Mo, in [1.172, 180, 181] for Ta, and in [1.182] for Nb and

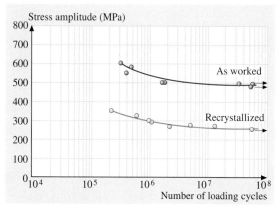

Fig. 3.1-168 Rotating–bending fatigue test results for as-received and recrystallized Mo rods (diameter = 25 mm) at room temperature [1.126]

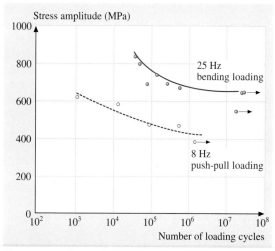

Fig. 3.1-169 Comparison of test results of bending (at 25 Hz) and push–pull fatigue (at 8 Hz) tests of stress-relieved Mo sheet specimens (800 °C/6 h) [1.126] at room temperature

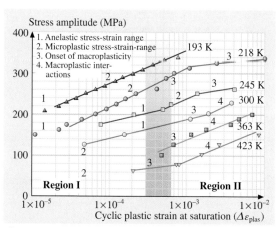

Fig. 3.1-170 Cyclic-stress–plastic-strain curves of recrystallized Mo at different temperatures at a loading rate of 60 MPa/s [1.140]

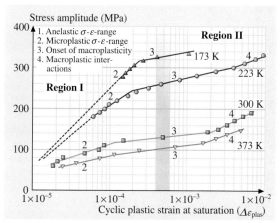

Fig. 3.1-171 Cyclic-stress–plastic-strain curves of recrystallized Ta at different temperatures at a loading rate of 0.42 MPa/s [1.140]

Nb1Zr. Cyclic stress–plastic-strain curves are given in Figs. 3.1-170 and 3.1-171 for Mo and Ta at various test temperatures. The ranges of microplastic and macroplastic strain can be differentiated, based on the different slopes.

The elevated temperature fatigue behavior of TZM was investigated for test temperatures between 300 °C and 500 °C [1.183]. Brittle failure under high cycle fatigue conditions was found over the entire temperature range, with a significant decrease in fatigue strength with increasing temperature.

Low-Cycle Fatigue Properties. Results of low-cycle fatigue experiments under strain control on as-worked W plate material at 815 °C are shown in Fig. 3.1-172. Low-cycle fatigue tests of pure W were performed in the temperature range between 1650 °C and 3300 °C [1.184]. A relationship $N_{\text{failure}} = \exp(-\alpha T)$ was found to be valid up to test temperatures of 2700 °C [1.185]. In all cases the failure mode was intercrystalline. Similar results were also obtained at a test temperature of 1232 °C [1.186]. The deformation behavior of Nb and Nb1Zr under plastic-strain control at room temperature was investigated and cyclic stress–strain curves published [1.182].

Table 3.1-114 Summary of fatigue data of refractory metals, pretreatment: Aw = as worked, Sr = stress relieved, Rxx = recrystallized. RT = room temperature

Test mode	Material	Production process	Dimension (mm) (plate: thickness, bar: diameter)	Pretreatment	$T(°C)/t_a$ (h)	Test conditions, temperature, frequency	S_e (MPa)	S_e/R_m	Ref.
Rotating bending fatigue, $R = -1$, fatigue limit S_e for 50% probability at $N = 5 \times 10^7$	Mo	P/M	bar 12	Aw		RT, 100 Hz	450	0.68	[1.126]
				Rxx	1200/1	RT, 100 Hz	220	0.44	
	Mo5Re	P/M	bar 12	Aw		RT, 100 Hz	450	0.68	[1.126]
				Sr	900/6	RT, 100 Hz	400	0.68	
				Rxx	1300/1	RT, 100 Hz	230	0.48	
	Mo41Re	P/M	bar 12	Aw		RT, 100 Hz	690	0.62	[1.126]
				Sr	1050/6	RT, 100 Hz	620		
				Rxx	1400/1	RT, 100 Hz	320	0.36	
	TZM	P/M	bar 12	Aw		RT, 100 Hz	550	0.62	[1.126]
				Sr		RT, 100 Hz	560		
				Rxx	1500/1	RT, 100 Hz	340	0.57	
	W	P/M	bar 12	Aw		RT, 100 Hz	760	0.48	[1.126]
				Rxx	1600/1	RT, 100 Hz	310	0.70	
	W5Re	P/M	bar 12	Aw		RT, 100 Hz	770	0.71	[1.126]
				Rxx	1800/1	RT, 100 Hz	440	0.65	
	W26Re	P/M	bar 12	Aw		RT, 100 Hz	820	0.54	[1.126]
				Rxx	1700/1	RT, 100 Hz	450	0.39	
	W2ThO$_2$	P/M	bar 12	Rxx	1900/1	RT, 100 Hz	365	0.64	[1.126]
	Nb	P/M	bar 5.9	Aw		RT, 100 Hz	225	0.42	[1.190]
				Rxx		RT, 100 Hz	220	0.51	
	Ta	P/M	bar 5.9	Aw		RT, 100 Hz	290	0.92	[1.190]
				Rxx		RT, 100 Hz	270	0.95	
Push-pull fatigue, $R = -1$, fatigue limit S_e for $N = 1 \times 10^7$	Mo	P/M	plate 1	Rxx	1200/1	RT, 0.05 Hz	195		[1.179]
	TZM	P/M	bar 50	Aw		RT, 25 Hz	440		[1.191]
				Aw		RT, 25 Hz	500		
				Aw		850 °C, 25 Hz	250		
	Ta	P/M	plate 2	Aw		RT, 0.05 Hz	205		[1.179]
				Aw		RT, 10 Hz	225		
				Rxx	1200/2	RT, 0.05 Hz	180		
				Rxx	1200/2	RT, 10 Hz	210		
Bending fatigue, $R = -1$, fatigue limit S_e for 50% fracture probability at $N = 1 \times 10^7$	Mo	P/M	plate 2	Aw		RT, 25 Hz	520		[1.126]
				Sr	780/6	RT, 25 Hz	540		
				Rxx	1200/1	RT, 25 Hz	280		
	Mo5Re	P/M	plate 1.6	Sr		RT, 25 Hz	460	0.56	[1.126]
	Mo41Re	P/M	plate 1.6	Sr		RT, 25 Hz	680	0.61	[1.126]
	TZM	P/M	plate 2	Aw		RT, 25 Hz	650		[1.126]
				Sr	1150/6	RT, 25 Hz	750		
				Rxx	1500/1	RT, 25 Hz	460		
	W	P/M	plate 2	Aw		RT, 25 Hz	520		[1.126]
				Rxx	1600/1	RT, 25 Hz	225		
	Ta	P/M	plate 1	Aw		RT, 25 Hz	335	0.56	[1.126]
				Rxx	1200/1	RT, 25 Hz	240	0.75	
	Ta	EB	plate 1	Aw		RT, 25 Hz	270	0.61	[1.126]
				Rxx	1200/1	RT, 25 Hz	220	0.96	
	Ta2.5W	P/M	plate 1	Rxx	1300/1	RT, 25 Hz	310	0.79	[1.126]
	Ta2.5W	EB	plate 1	Rxx	1300/1	RT, 25 Hz	270	0.69	[1.126]
	Ta10W	EB	plate 0.64	Aw		RT, 25 Hz	480		[1.168]

Table 3.1-114 Summary of fatigue data of refractory metals, pretreatment: Aw = as worked, Sr = stress relieved, Rxx = recrystallized, cont.

Test mode	Material	Production process	Dimension (mm) plate: thickness, bar: diameter	Pretreatment T(°C)/t_a (h)		Test conditions, temperature, frequency	S_e (MPa)	S_e/R_m	Ref.
High-frequency push-pull fatigue, $R = -1$, fatigue limit S_e for 50% fracture probability at $N = 10^8$	Mo	P/M	bar 11	Rxx	1300/4	RT, 20 kHz	278	0.56	[1.176]
	TZM	P/M	bar 11	Rxx	1600/2	RT, 20 kHz	383	0.69	[1.176]
	Ta	EB	bar 3	Aw		RT, 20 kHz	383	0.72	[1.126]
				Sr		RT, 20 kHz	286	0.92	
	Ta2.5W	P/M	plate 2	Rxx		RT, 20 kHz	300		[1.126]
	Ta10W	P/M	bar 12	Rxx		RT, 20 kHz	580		[1.126]
	Nb		bar 3	Aw		RT, 20 kHz	230	0.80	[1.177]
				Rxx		RT, 20 kHz	220	0.65	

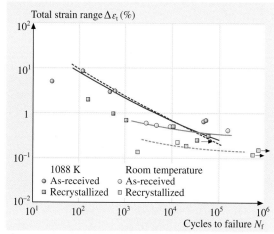

Fig. 3.1-172 Low-cycle fatigue data of as-received and recrystallized W at room temperature and 815 °C [1.186]

Low-cycle fatigue test data of Mo at high test temperatures were reported in [1.187]. The influence of microstructural changes in cold-worked molybdenum on the low-cycle fatigue behavior was reported for test temperatures between 300 °C and 950 °C [1.188]. Deformation experiments under low-cycle fatigue conditions between room temperature and 100 °C showed that recrystallized Mo, in spite of the low temperature and stress level, exhibits considerable plastic strains which depend sensitively on the loading frequency [1.189]. Data on the high temperature (350 °C and 500 °C) isothermal mechanical fatigue behavior of TZM were reported and a model for lifetime prediction was proposed [1.183].

Fracture Mechanics Properties

Fracture Toughness. Fracture toughness properties are affected by many parameters (thermomechanical pretreatments, microstructure, specimen and crack plane orientation; testing procedures as well as the preparation of the starting notch and of the fatigue precrack). Due to the peculiarities of the P/M production process, it is frequently not possible to introduce sufficient deformation into products of larger dimension in order to completely eliminate sinter pores which may affect the dynamic properties. The increase of fracture toughness with increasing degree of hot working of disc-shaped compact tension specimens, cut from a hot forged Mo bar is shown in Fig. 3.1-173 and the decrease of fracture toughness with increasing fraction of recrystallized microstructure in Fig. 3.1-174 [1.147].

Fracture toughness data for Mo, TZM, and W materials are summarised in Table 3.1-115. Data taken at elevated temperature for Mo and W are shown in Fig. 3.1-175 [1.192], and for TZM in Fig. 3.1-176 [1.193]. Data for Nb could only be

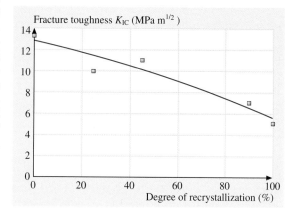

Fig. 3.1-173 K_{IC} of forged Mo rods ($\varphi = 74\%$) versus degree of recrystallization [1.147]

Table 3.1-115 Fracture mechanical data for various refractory metals, pretreatment: Aw = as worked, Sr = stress relieved, Rxx = recrystallized. RT = room temperature

Fracture toughness data				Specimen type: CT compact tension, SCT center surface cracked tension, DCT disk-shaped compact tension, SNB side notched bend CNT center through-thickness notched tension				
Material	Production process	Dimension (mm) plate: thickness, bar: diameter	Pretreatment T (°C)/t_a (h)	Specimen type	Crack plane ASTM 399	Test temperature	Fracture toughness (MPa m$^{1/2}$)	Ref.
Mo	P/M	plate 6.5	Rxx 1300/2	CNT	L-T	RT	9.5	[1.126]
Mo	P/M	bar 50	Sintered	DCT	R-C	RT	6	[1.147]
			Hot forged	DCT	R-C	RT	12	
			Rxx 1400/2	DCT	R-C	RT	5	
Mo	VAC	plate 6.4	Sr	CT	L-T/T-L	RT	20/21	[1.195]
			Rxx	CT	L-T/T-L	RT	18/19	
			Sr	CT	L-T/T-L	300 °C	74/70	
			Rxx	CT	L-T/T-L	300 °C	60/67	
Mo5Re		bar 54	Rxx 1300/1	DCT	R-C	RT/450 °C	18/27	[1.193]
Mo–0.3 wt% La$_2$O$_3$	P/M	plate	Sintered	CT	L-T	RT	25	[1.196]
TZM	VAC	plate 6.4	Sr	CT	T-L/L-T	RT	19/15	[1.195]
			Rxx	CT	T-L/L-T	RT	15/18	
			Sr	CT	T-L/L-T	300 °C	85/89	
			Rxx	CT	T-L/L-T	300 °C	59/68	
TZM	P/M	bar 54	Aw	DCT	R-C	RT	19	[1.159, 197]
		Machined to 12	Aw	SCT	L-C	RT	37	
		Swaged to 12	Rxx 1400/2	SCT	L-C	RT	19	
W		bar 10		SNB	C-R/L-R	−196 °C	11/6	[1.198]
				SNB	C-R/L-R	RT	13/8	
W		bar 25	Aw	SNB	R-L	RT/150 °C/ 300 °C/500 °C	9/10/13/14	[1.199]
W	Single crystal	bar 16 (111)[011]		SNB		−196 °C/RT	9/31	[1.198]
		bar 16 (011)[100]		SNB		−196 °C/RT	4/20	
		bar 16 (100)[001]		SNB		−196 °C/RT	3/9	
W			Aw	SNB		RT/400/600/ 800	15/21/26/39	[1.200]
W5Re			Aw	SNB		RT/400 °C/ 600 °C	11/26/58	[1.200]
W-1 wt% La$_2$O$_3$	P/M	bar 25	Aw	SNB	R-L	RT/150 °C/ 300 °C/500 °C	9/10/11/13	[1.199]
Nb		plate 1.25	Sr	CNT	L-T	−253 °C	40	[1.192]
			Rxx	CNT	L-T	−253 °C	72	
Nb			Rxx	SNB		−196 °C	37	[1.194]
			Rxx	SNB		RT	37	

determined below −200 °C [1.192]. Data on the impact and dynamic toughness of Nb between −196 °C and 25 °C were reported in [1.194]. The dynamic cleavage fracture toughness was shown to be 37 MPa m$^{1/2}$, relatively independent on grain size and test temperature.

Table 3.1-115 Fracture mechanical data for various refractory metals, pretreatment: Aw = as worked, Sr = stress relieved, Rxx = recrystallized. RT = room temperature, cont.

Fatigue crack growth and threshold data				Specimen type: CNT center through-thickness notched tension, CCT corner cracked tension					SCT center surface cracked tension, SNT side notched tension			
Material	**Production process**	**Dimension (mm) plate: thickness, bar: diameter**	**Pretreatment T (°C)/t_a (h)**	**Specimen type**	**Crack plane ASTM 399**	**Test conditions, temperature, frequency**	**Stress ratio R**	**da/dN-range, (m/cycle)**	**ΔK_{th}**	**$\Delta K_{th,eff}$**	**ASTM**	**Ref.**
Mo	P/M	plate 6.5	Rxx 1300/2	CNT	L-T	RT, 20 kHz	−1	10^{-13}–10^{-9}	5.8		E647	[1.126]
Mo	P/M	bar 12	Aw	SCT	L-C	RT, 20 kHz	−1	10^{-13}–10^{-9}	11.3	11	E647	[1.159]
			Rxx 1400/2						10.7	11		
Mo	P/M	bar 12	Sr 850/1	SNT	C-R	RT, 20 kHz	−1	10^{-13}–10^{-9}	10.3	7.0	E647	[1.201]
			Rxx 1400/2						8.5	6.6		
TZM	P/M	bar 12	Rxx 1700/2	SNT	C-R	RT, 20 kHz	−1	10^{-13}–10^{-9}	10.1	6.3	E647	[1.201]
Mo5Re	P/M	bar 12	Rxx 1300/1	SCT	L-C	RT, 20 kHz	−1	10^{-13}–10^{-9}	11.4	9.9	E647	[1.126]
Ta10W			Rxx 1450/1	CCT		RT	0.4	10^{-9}–10^{-5}				[1.202]

20 kHz resonance test method, crack growth monitored by traveling light microscope,
ΔK_{th} corresponding to a crack growth rate of $da/dN < 10^{-13}$ m/cycle,
$\Delta K_{th,eff}$ calculated from crack closure measurements based on strain gauge method

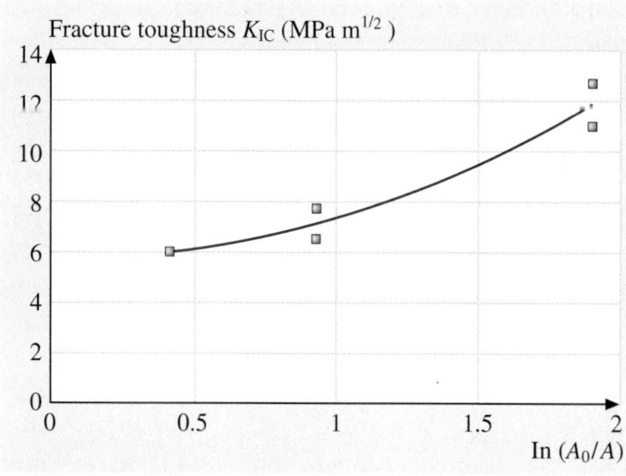

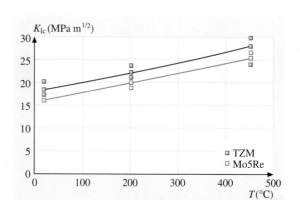

Fig. 3.1-174 K_{IC} of forged Mo rods versus $\ln(A_0/A)$ [1.147]

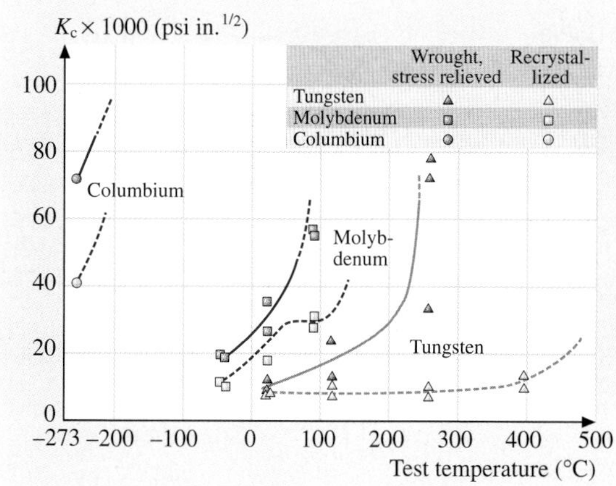

Fig. 3.1-175 Effect of test temperature on fracture toughness of unalloyed W, Mo, and Nb sheet specimens [1.192]

Fig. 3.1-176 Effect of test temperature on static fracture toughness of TZM and Mo5Re specimens [1.193]

Fatigue Crack Growth. Few data on the linear region of crack growth were published for Ta10W (Fig. 3.1-177), and a Nb−W−Zr alloy [1.202, 203].

Threshold Stress Intensity for Fatigue Crack Growth. The fatigue crack growth behavior of Mo, TZM, and W in the region near the threshold stress intensity, ΔK_{th}, considered to correspond to fatigue crack growth rates of $da/dN < 10^{-13}$ m/cycle, is shown in Fig. 3.1-178 [1.126]. The available crack growth and threshold data are included in Table 3.1-115.

An effective threshold value for fatigue crack growth, $\Delta K_{th,eff}$, can be computed. Methods for the determination of this effective threshold stress intensity range are described in [1.197]. The available data on $\Delta K_{th,eff}$ of Mo and TZM are listed in Table 3.1-115.

Short Fatigue Crack Growth Behavior. The nucleation and growth behavior of short fatigue cracks is of considerable practical and theoretical significance. Differences in the growth behavior exist between initial short cracks (length comparable with microstructural features) and long cracks of macroscopic dimensions. The irregular growth rate of such short cracks

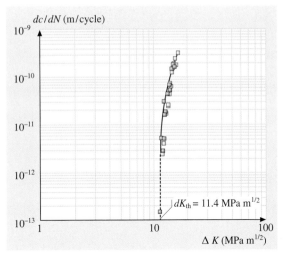

Fig. 3.1-178 Crack growth curve near threshold stress intensity of a center surface cracked specimen machined from a recrystallized Mo5Re rod, tested at a stress ratio of $R = -1$, test temperature = 50 °C, and 20 kHz cyclic frequency; *open symbols* test under increasing load, *solid symbols* tests under decreasing load [1.126]

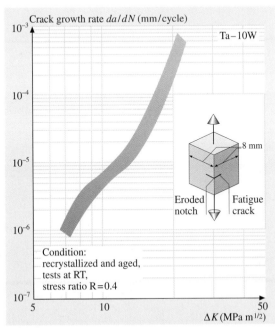

Fig. 3.1-177 Crack propagation behavior of recrystallized P/M Ta10W specimens at room temperature at a stress ratio of $R = 0.4$ [1.202]

(Fig. 3.1-179) may pose problems for a conservative prediction of fatigue life [1.201]. Microscopic observations during fatigue exposure of specimens, loaded at stress amplitudes slightly above the fatigue limit, show initially a surface deformation very early in fatigue life, followed by short crack initiation and growth up to the final long crack growth. The number of fatigue cycles of the various stages depends on the microstructure and the presence of second phase particles (Fig. 3.1-180) [1.204].

It is known that fatigue failures occur in defect-containing materials after a high number of loading cycles ($N > 10^8$) at stresses considerably below the fatigue limit determined by conventional test procedures ($N < 10^7$). A fracture mechanics approach to this problem was proposed by *Kitagawa* et al. [1.205]. Based on a diagram relating a cyclic stress amplitude with crack length, a critical defect size (c_t) can be deduced which, when exceeded, causes a reduction of the fatigue strength. A modification of this diagram by introducing the value of the effective stress intensity range is shown in Fig. 3.1-181 [1.201]. Good agreement between predicted values and experimental results are obtained for hemispherical surface notches of various sizes [1.206, 207].

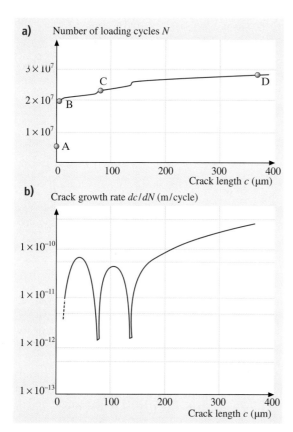

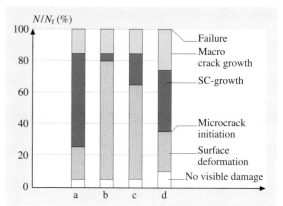

Fig. 3.1-179a,b Growth behavior of short surface cracks in a TZM specimen, tested at a stress amplitude of 375 MPa [1.201, 204]. (**a**) Crack length as function of number of loading cycles. (**b**) Crack growth rate as function of crack length

Fig. 3.1-180 Fraction of total fatigue life spent on damage accumulation and crack growth in specimens of stress relieved Mo, recrystallized Mo, TZM with fine particles, and TZM with coarse particles [1.204]

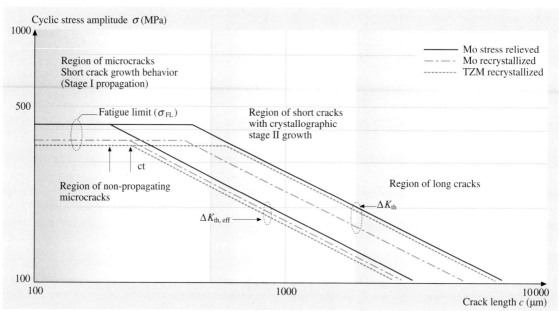

Fig. 3.1-181 Effect of crack length on stress amplitude for crack growth in specimens of stress relieved Mo, recrystallized Mo, and recrystallized TZM (modified Kitagawa diagram) [1.201]

Creep Properties

Creep-rupture data available up to 1970 were collected in [1.107]. The fairly wide scatter results from minor differences in microstructure, thermomechanical pretreatment, and impurity levels, but possibly also from impurities picked up from the environment in the high-temperature test systems. A summary of 10 000 h/1% creep data for Mo, W, and TZM is given in Fig. 3.1-182; the 100 h creep-rupture data for Mo, W, Nb, Ta, W25Re and TZM are given in Fig. 3.1-183, based on [1.107]. Up to 1100 °C the carbide-precipitation-hardened material TZM reveals the highest creep strength, only outperformed by Mo and W alloys precipitation hardened with hafnium carbide, which are not considered in these figures. Comparing the stress causing a steady state creep rate of 1×10^{-4} /h, as illustrated in Fig. 3.1-184, it can be concluded that precipitation hardening is effective up to 1600 °C. At higher temperatures, the creep strength of TZM deteriorates to the level of Mo or even below. Above 1600 °C, W-based materials offer the best performance.

The influence of the microstructure on the creep mechanisms of Mo is illustrated in Fig. 3.1-185 for 1450 °C [1.208]. For a test stress of 35 MPa the steady state creep rate is almost independent of the grain size. In this stress regime a stress exponent of 5.4 was obtained, indicating dislocation creep as the rate-controlling mechanism. Lowering the test stress to 14 MPa and 7 MPa, grain-size-dependent creep mechanisms, such as diffusion creep and grain boundary sliding, become active, and as a consequence, the steady state creep rate increases with decreasing grain size [1.208].

The steady state creep rates of Mo rods made of VAC ingots, VAC and P/M Mo sheets, and P/M Mo–0.7 wt%La_2O_3 sheets are summarized in Fig. 3.1-186. Stress exponents of 4.3 (P/M Mo sheet/1800 °C), 4.5–4.7 (VAC Mo sheet/1600 °C), 4.6 (VAC Mo rod/1600 °C), 5.0 (Mo–0.7 wt% La_2O_3/1800 °C), and 5.5 (VAC Mo sheet/2200 °C) indicate that dislocation-controlled creep is rate-controlling in the stress regime investigated.

For AKS–W wires, used as lamp filaments, creep resistance is one of the most important requirements. The fine, potassium filled bubbles act as an effective barrier against dislocation movement, thereby reducing the deformation rate in the power-

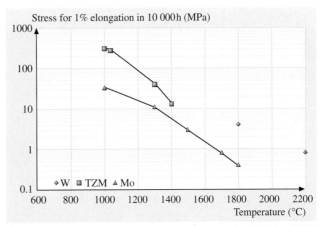

Fig. 3.1-182 Comparison of 10 000 h/1% creep data for Mo, TZM, and W based on [1.107]

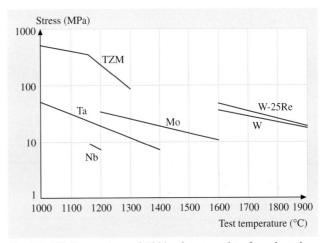

Fig. 3.1-183 Comparison of 100 h of rupture data for selected refractory metals based on [1.107]

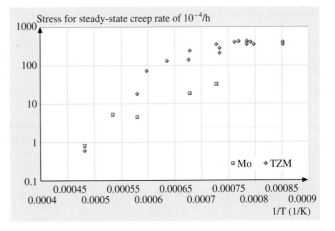

Fig. 3.1-184 Stress for steady state creep rate of 1×10^{-4} /h versus $1/T$ for Mo and TZM, deformed samples, various shapes [1.208]

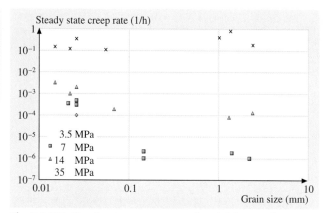

Fig. 3.1-185 Steady state creep rate of molybdenum sheets versus grain size. Sheet thickness = 2 mm/6 mm; test temperature = 1450 °C; test atmosphere = hydrogen [1.208]

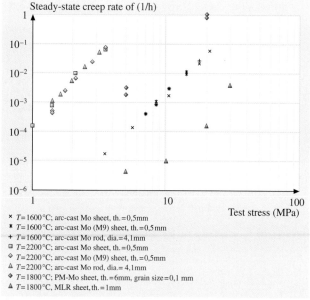

Fig. 3.1-186 Steady state creep rate at various test temperatures of vacuum-arc-cast (VAC) Mo sheet (thickness = 0.5 mm), VAC Mo rod (diameter = 4.1 mm), P/M Mo sheet (thickness = 6 mm), and Mo–0.7 wt% La_2O_3 sheet (thickness = 1 mm) versus test stress [1.107, 208]

with different grain aspect ratios (GAR) is given in Fig. 3.1-187.

In the high stress regime (> 60 MPa) and at temperatures between 2500 °C and 3000 °C, stress exponents between 8 and 25 were found [1.162, 209, 210]. This high stress dependence led to the introduction of a threshold stress (σ_{th}) below which a component does not reveal any measurable creep deformation under usual service conditions. For this threshold stress, which is lower than the Orowan stress, the detachment of the dislocations from the second phase particles or bubbles is the controlling factor [1.211, 212].

In the second phase particle/metal matrix interface, the dislocation line energy is lower compared to the dislocation line energy in the metal matrix. Of all dispersion strengthened materials investigated, potassium bubbles in AKS–W exert the most attractive interaction on dislocations [1.213].

For material produced in the 1970s with a mean, but strongly scattered, grain aspect ratio of around 35, dislocation creep dominated at stresses > 60 MPa (stress exponent = 25), as can be seen in Fig. 3.1-188. For material produced 20 years later with a similar grain aspect ratio of 31, a stress exponent of 1.2 was found in the stress regime from 30 to 80 MPa, indicating a diffusion-controlled creep process [1.214]. The evaluation of the strain rate/stress dependence of the values

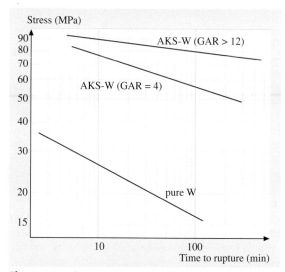

Fig. 3.1-187 Creep rupture data for AKS–W wires in comparison with pure tungsten. Wire diameter = 0.183 mm; 4 < GAR < 12; test temperature = 2527 °C; atmosphere = vacuum better than 7×10^{-5} Pa; heating rate = approximately 2000 °C/s [1.162]

law creep regime. Nabarro–Herring and/or Coble creep is suppressed because of the large diffusion distances in a structure with large, highly elongated grains. Grain boundaries resist sliding because of the interlocking structure. A comparison of creep rupture data of pure tungsten and two AKS–W grades

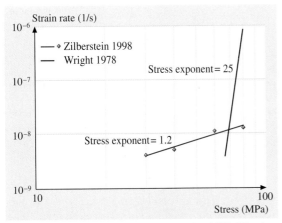

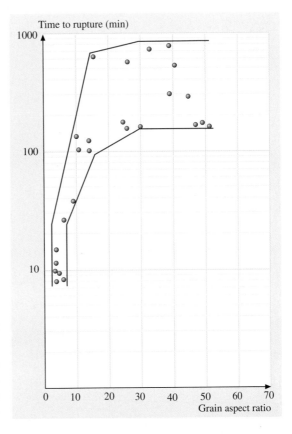

Fig. 3.1-188 Strain rate versus stress for AKS–W wires tested at 2527 °C. *Wright* 1978 [1.162]: AKS–W wire with a diameter of 0.183 mm; atmosphere = vacuum better than 7×10^{-5} Pa; heating rate = approximately 2000 °C/s; GAR = 35 ± 10; pre-recrystallized at 2527 °C/10 min. *Zilberstein* 1998 [1.215]: AKS–W wire with a diameter of 0.178 mm; atmosphere = vacuum; GAR = 31 ± 1; pre-recrystallized at 2527 °C/15 min

Fig. 3.1-189 Time-to-rupture versus grain aspect ratio for AKS–W wires with a diameter of 0.183 mm. Test temperature = 2527 °C; test stress = 73.6 MPa; atmosphere = vacuum better than 7×10^{-5} Pa; heating rate = approximately 2000 °C/s [1.162]

generated under vacuum also reveals a stress exponent close to 1 [1.215]. Data for the stress exponents are summarized in [1.113].

By lowering the grain aspect ratio, the transition temperature between dislocation and diffusion creep is shifted towards lower stresses. In the low GAR regime a strong dependence of the creep resistance on microstructural features can be observed, as grain boundary related phenomena, such as grain boundary sliding and diffusion creep resulting in cavitations become rate-controlling. The influence of the GAR on time to creep rupture is demonstrated in Fig. 3.1-189.

3.1.10 Noble Metals and Noble Metal Alloys

The noble metals Ag, Au, Pd, Pt, Rh, Ir, Ru, and Os are characterized by their positive reduction potentials against hydrogen, high densities, high melting temperatures, high vapor pressures (Fig. 3.1-190), high electrical and thermal conductivities, optical reflectivity (Fig. 3.1-191), and catalytic properties. The electronic density of states (DOS) near the Fermi surface is nearly the same for all noble metals. Individual differences of electrical conductivity, magnetic, and optical behavior are related to different positions of the Fermi level relative to the DOS function (Fig. 3.1-192). Small energy differences between their outer s and d electronic states result in multiple oxidation states.

Silver, Au, Pd, and Pt are comparatively soft and ductile. Their hardness increases in the order Rh < Ir < Ru < Os. Strengthening of the alloys is affected by solid solution and dispersion hardening. The corrosion resistance against different agents decreases in the order Ir > Ru > Rh > Os > Au > Pt > Pd > Ag.

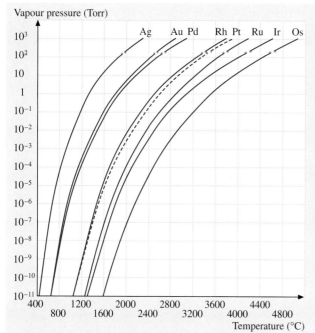

Fig. 3.1-190 Vapour pressures of the noble metals [1.216, p. 43]

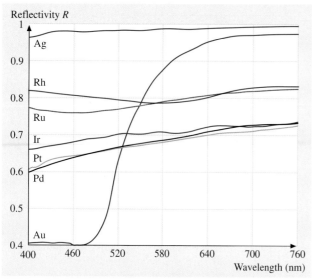

Fig. 3.1-191 Optical reflectivity in the visible spectral range [1.217, p. 173]

The purity grades of the elements are standardized according to ASTM (American Society for Testing and Materials) standards from 99.8 to 99.999 wt%: Ag (B 413-69), Au (B 562-86), Pd (B 589-82), Pt (B 561-86), Rh (B 616-78), Ir (B 671-91), Ru (B 717).

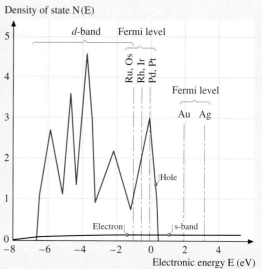

Fig. 3.1-192 Schematic density of states (DOS) curve of the noble metals [1.218, p. 30]

The pure elements and their alloys are key materials in electronics and electrical engineering (Ag, Au, Pd, Pt, and Ru) and serve to manufacture high strength, corrosion-resistant, high temperature, and highly oxidation-resistant structural parts (Pt, Au, Rh, and Ir). The platinum group metals, Ag, and Au, in both the metallic state and in the form of chemical compounds are effective heterogeneous or homogeneous catalysts for a wide variety of chemical reactions. Traditional applications of noble metals and their alloys are in dentistry (Au, Pt, Ag, Pd, and Ir), jewellery (Au, Ag, Pt, Pd, Rh, and Ir), and in coins and medals (Au and Ag).

3.1.10.1 Silver and Silver Alloys

Application

Silver and silver alloys are used for electrical contacts, connecting leads in semiconductor devices, solders and brazes, corrosion-resistant structural parts, batteries, oxidation catalysts, optical and heat reflecting mirrors, table ware, jewellery, dentistry, and coins. Silver halides are base components in photographic emulsions.

Production

Silver is extracted from ores through lead melts and precipitation with zinc by the Parkes process. Zinc is removed by distillation, while the remaining lead and base metals are removed by oxidation (cupellation) up to ≈ 99% Ag. True silver ores are extracted by cyanide leaching. High purity grades are produced by electrolysis. Bars, sheets, and wires are produced by classical metallurgical processing, powder by chemical and by electrolytic precipitation from solutions, and nano-crystalline powder grades by dispersion in organic solutions. Coatings and laminate structures are produced by cladding, by electroplating, in thick film layers by applying pastes of silver or in silver alloy powder with organic binder and glass frits onto ceramic surfaces and firing, in thin film coatings by evaporation, and by sputtering Composite materials are made by powder technology, or by infiltration of liquid Ag into sintered refractory metals skeletons. Commercial grades of Ag are listed in Table 3.1-116. Standard purities of crystal powder and bars range from 99.9% to 99.999 wt% (ASTM B 413-69) [1.217].

Phases and Phase Equilibria

Selected phase diagrams are shown in Figs. 3.1-193–3.1-198 [1.219, 220]. Silver forms continuous solid solutions with Au and Pd, with miscibility gaps occurring in alloy systems with Mn, Ni, Os, P, Rh. Data for the solubility of oxygen are given in Table 3.1-117 [1.216]. Thermodynamic data are given in Tables 3.1-118–3.1-121. The entropy of fusion (L/T) of completely disordered intermetallic phases can generally be calculated by fractional addition from those of the components. For the completely ordered state the term $-19.146\,(N_1 \log N_1 + N_2 \log N_2)$ is to be added to the calculated entropy of fusion [1.216–218, 221, 222]. The molar heat capacity of the homogeneous alloy phases and intermetallic compounds, as calculated approximately from the atomic heat capacities of the components using Neumann–Kopps' rule, is obeyed to within ±3% in the temperature range 0–500 °C in the Ag–Au, Ag–Al, Ag–Al, and Ag–Mg alloy systems. The heat capacities of heterogeneous alloys may be calculated by fractional addition from those of the components by the empirical relation $c_p = 4.1816(a + 10^{-3}bT + 10^5 cT^{-2})\,\text{J/(K mol)}$ to satisfactory accuracy.

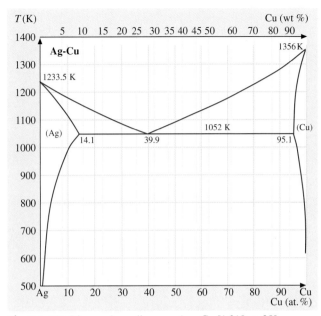

Fig. 3.1-193 Binary phase diagram: Ag–Cu [1.219, p. 38]

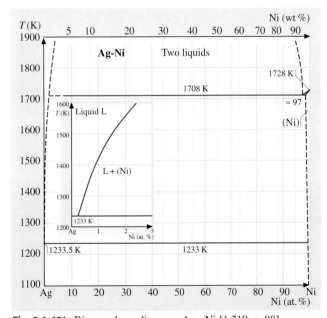

Fig. 3.1-194 Binary phase diagram: Ag–Ni [1.219, p. 88]

Table 3.1-116 Specifications of fine silver grades [1.217, p. 51]

Designation	Grade (wt%)	Impurity	Maximum content (ppm)
"Good delivery"	> 99.9	any, Cu	1000
Fine silver	> 99.97	Cu/Pb/Bi/Se/Te	300/10/10/5/5
Fine silver 999.9	> 99.99	Cu/Pb/Bi/Se/Te	100/10/10/5/5
F. silver high pure	< 99.999	Fe/Pb/Au/Cu/Cd/Bi/Se/Te	2/1/1/1/0.5/0.5/0.5/0.5

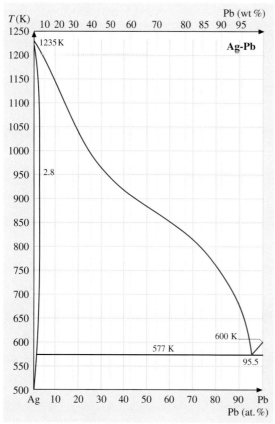

Fig. 3.1-195 Binary phase diagram: Ag−Pb [1.219, p. 92]

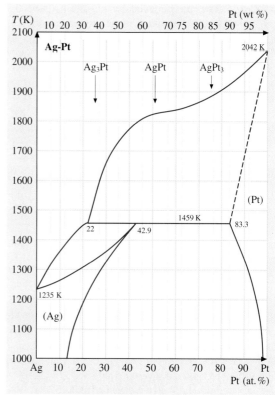

Fig. 3.1-196 Binary phase diagram: Ag−Pt [1.219, p. 101]

Table 3.1-117 Solubility L (ppm) of oxygen in solid and liquid Ag (O_2 pressure 1 bar) [1.216, p. 57]

T (°C)	200	400	600	800	973	1000	1200
L_{ppm}	0.03	1.4	10.6	38.1	3050	3000	2500

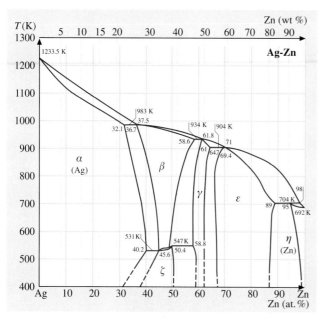

Fig. 3.1-197 Binary phase diagram: Ag−Zn [1.219, p. 144]

Table 3.1-119 Molar heat capacities of solid Ag and Au, $c_p = 4.1868\,(a + 10^{-3}bT + 10^{-5}T^{-2})$ J/K [1.222, p. 219]

Element	a	b	c	Temperature range (K)
Ag	5.09	2.04	0.36	298–mp*
Au	5.66	1.24	–	298–mp*

* = melting point

Table 3.1-120 Latent heat and temperatures of transition of Ag and Au intermediate compounds [1.222, p. 189]

Phase	N_2	Transition	T_t	L_t
β-AgCd	50	β'–β	211	712
AgZn	50	order–disorder	258	2449
AuCu	50	order–disorder	408	1779
AuCu$_3$	75	order–disorder	390	1214
AuSb$_2$	66.7	β–γ	355	335

T_t = transition temperature, L_t = latent heat of transition, N_2 = mole fraction of the second component

Table 3.1-121 Latent heats and temperatures of fusion of Ag and Au intermediate compounds [1.222, p. 188]

Phase	N_2	T_m (°C)	L_m (hJ/g-at.)	
β-AgCd	67.5	592	8.46	0.42
γ-AgZn	61.8	664	7.79	0.33
AgZn	72.1	632	8.75	0.42
δ-AuCd	50.0	627	8.96	0.50
AuSn	50.0	418	12.81	0.33
β-AuZn	50.0	760	12.31	0.54

N_2 = mole fraction of the second component, T_m = melting point, L_m = Latent heat of fusion.

Table 3.1-118 Thermodynamic data of Ag [1.217, p. 107]

T (K)	c_p (J/K mol)	H (J/K mol)	S (J/mol)	G (J/mol)	p (at)
298.15	25.397	42.551	0	−12.687	1.09×10^{-43}
400	25.812	50.069	2.606	−17.421	5.02×10^{-31}
800	28.279	68.661	13.392	−41.537	1.45×10^{-12}

T = Temperature, c_p = specific heat capacity, S = Entropy, H = Enthalpy, G = free Enthalpy, p = partial pressure of the pure elements

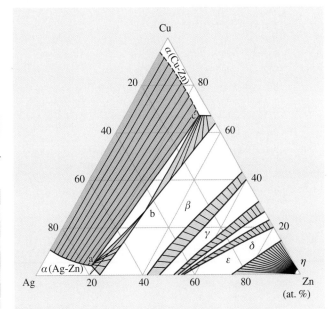

Fig. 3.1-198 Ternary phase diagram: Ag–Cu–Zn [1.220, p. 153]

For compositions and crystal structures, see Tables 3.1-122–3.1-124 [1.217, 218, 223, 224]. Primary solid solutions have the fcc structure of Ag and the lattice parameters correspond roughly to Vegard's rule with a few exceptions. Alloys with Pt, In, Mg, Cd, and Zn form superlattice phases with tetrahedral and rhombohedral symmetry. A characteristic series of structures of intermetallic phases are formed with B-metals at compositions corresponding to e/a values (valence electrons per atom) of 3/2, 21/13, and 7/4 (Hume-Rothery phases) [1.225].

Table 3.1-122 Structure and lattice parameter of intermediate Ag compounds [1.217, p. 112]

Phase	Pearson symbol	a (nm)	b (nm)	c (nm)	c/a	Remarks	Concentration x: $A(1-x)B(x)$
Ag–Al	cF4	0.4064					0.18
Ag_2Al	hP2	0.28777		0.46223	1.6062		
Ag–Cd	hP2	0.2987		0.553	1.8514		0.97
AgCd	cP2	0.3332				293 K	0.5
AgCd	hP2	0.3016		0.4863	1.6124	673 K	0.5
Ag–Cu	cF4	0.3603					0.95
Ag–Mg	cF4	0.4116					0.26
Ag–Mg	hP2	0.3197		0.51838	1.6215		0.98
AgO	cF8	0.4816					
AgO	mP8	0.5852	0.3478	0.5495			
Ag–O	cP6	0.4728					
Ag–O	hP3	0.3072		0.4941	1.6084	HP/HT	
Ag–Zn	hP2	0.28227		0.44274	1.5685		0.775
Ag–Zn	hP9	0.7636		0.28179	0.369		0.38–0.52

HT = High-temperature modification, HP = High-pressure modification

Table 3.1-123 Composition and structures of superlattices in NM-alloy systems [1.218, p. 105]

Atomic ratio	Composition				Superlattice structure	Fundamental structure
3:1 or 1:3	Pd_3Fe	$PdCu_3$	Au_3Cu	$AuCu_3$		
	Pt_3Fe	$PtCu_3$	Au_3Pd			
	Pt_3Ti	$PtNi_3$	Ag_3Pt		$L1_2$	fcc
		$PtMn_3$				
	Rh_3Mo		Ag_3In			
	Rh_3W				DO_{14}	
	Ir_3Mo					
	Ir_2W		(Au_2Mn)		DO_{22}	fcc
	Pd_3V					
	Pd_3Nb					
	Pt_3V					
			Au_3Cd		DO_{23}	fcc
			Au_3Zn			
			Ag_3Mg			
2:1	Pd_2V				Ni_2Cr	fcc
	Pt_2V					
	Pt_2Mo					
1:1	PdFe		AuCu			
	PtFe				$L1_0$	fcc
	PtV					
	PtNi					
	PtCo					
	PtCu					fcc
	PdCu		AuMn		$L1_1$	bcc
			AgCd		$L2_0$ or B2	
			AgZn			
	RhMo					hcp
	IrMo				B19	
	PtMo					
	IrW					
	PtNb					

1.10 Noble Metals and Noble Metal Alloys

Table 3.1-124 Compositions and structures of e/a (Hume-Rothery) compounds [1.225, p. 197]

Electron: atom ratio = 3:2			Electron: atom ratio = 21:13	Electron: atom ratio = 7:4
Body-centered cubic structure	Complex cubic (β-manganese) structure	Close-packed hexagonal structure	γ-brass structure	Close packed hexagonal structure
AgMg	AgHg	AgZn	Ag_5Zn_8	$AgZn_3$
AgZn	Ag_3Al	AgCd	Ag_5Cd_8	$AgCd_3$
AgCd	Au_3Al	Ag_3Al	Ag_5Hg_8	Ag_3Sn
Ag_3Al		Ag_3Ga	Ag_9In_4	Ag_5Al_3
Ag_3In		Ag_3In	Au_3Zn_8	$AuZn_3$
AuMg		Ag_5Sn	Au_5Cd_8	$AuCd_3$
AuZn		Ag_7Sb	Au_9In_4	Au_3Sn
AuCd		Au_3In	Rh_5Zn_{21}	Au_5Al_3
		Au_5Sn	Pd_5Zn_{21}	
			Pt_5Be_{21}	
			Pt_5Zn_{21}	

Mechanical Properties

In Tables 3.1-125–3.1-135 and Figs. 3.1-199–3.1-204 characteristic data are shown [1.217, 220, 225–229]. References for data of elastic constants of Ag alloys are given in [1.222]. Pure silver is very soft. Strengthening is affected by solid solution and by dispersion hardening [1.216, 230]. Alloying with 0.15 wt% Ni affects grain refinement and stabilizes against recrystallization. The high solubility of oxygen in silver (Table 3.1-117) permits the inducement of dispersion hardening by internal oxidation of Ag alloys containing Al, Cd, Sn, and/or Zr.

Table 3.1-125 Module of elasticity of Ag in crystal directions (GPa) [1.217, p. 204]

$E\langle 100\rangle$	$E\langle 110\rangle$	$E\langle 111\rangle$
44	82	115

Table 3.1-126 Elastic constants of Ag (GPa) [1.217, p. 204]

T (°C)	c11	c12	c14
−273	131.4	97.3	51.1
+20	124.0	93.4	46.1

Table 3.1-127 Mechanical properties of Ag (99.97%) at different temperatures (°C) [1.217, p. 204]

T (°C)	E (GPa)	R_m (MPa)	A (%)	$R_{p0.2}$ (MPa)	HV
20	82	150	50	28	26
200	77	130	–	25	22
400	67	100	30	20	17
800	46	35	–	17	5

A = Elongation, E = Module of elasticity, R_p = Limit of proportionality, HV = Vickers hardness, R_m = Tensile strength

Table 3.1-128 Tensile strength R_m (MPa) of binary Ag alloys [1.217, p. 207]

Content (wt%)	2	5	10	20
Alloying elements	R_m (MPa)			
Au	160	170	180	200
Cd	160	170	180	210
Cu	190	240	280	310
Pd	160	180	21	
Sb	190	240	300	
Sn	190	240	300	

Table 3.1-129 Hardness of Ag−Mn alloys [1.226, p. 473]

wt% Mn	0.5	4.9	6.2	12.0
HV (kg/mm²)	31	39	40	58

Table 3.1-130 Mechanical properties of Ag−Cu−P alloys [1.226, p. 477]

Composition (wt%)			BS[a] 1845	Density (g/cm³)	Melting range (°C)	Tensile strength (kg/mm²)	Elongation (%)
Ag	Cu	P					
15	80	5	CP1	8.40	645−719	25	10
5	89	6	CP4	8.20	645−750	25	5
2	91	7	CP3	8.15	645−770	25	5

[a] = British standard (1966/1971) designation

Table 3.1-131 Mechanical properties of Ag−Cu−Zn alloys [1.220, p. 154]

Composition (wt%)			Specific weight (cast)	Tensile strength (kg/mm²)	Elongation (%)	Electrical conductivity (% of Cu)
Cu	Zn	Ag				
36	24	40	9.11	40.5	6.2	19.7
25	15	60	9.52	45.0	7.7	20.5
22	3.0	75	10.35	29.3	5.3	53.4
16	4.0	80	10.05	35.1	16.0	45.8

Table 3.1-132 Mechanical properties of Ag−Pd alloys in annealed (s) and hard (h) condition [1.217, p. 208]

Alloy Ag−X (wt%)	HV5		$R_{p0.2}$ (MPa)		R_{on} (MPa)		A (%)	
	s	h	s	h	s	h	s	h
Pd27.5	55		80		230		33	
Pd27.4Cu10.5	140	310	320	940	510	950	31	3
Pd39.9Zn4	160	270	285	595	560	790	18	6

s = soft, h = hard

Table 3.1-133 Tensile strength R_m (MPa) and elongation A (%) of Ag alloys at different temperatures [1.217, p. 207]

Alloy AgX (wt%)	Temperature (°C)				
	20	200	400	600	800
	R_m/A				
Cd 4	170/60	150/50	−	−	−
Cd 8	480/5	260/20	200/55	−	−
Cu 3	190/35	170/40	140/40	90/80	30/150
Cu 7.5	250.46	220/48	180/47	120/55	60/78
Ni 0.3	190/60	170/60	130/65	100/65	60/50
Si 3	260/39	220/33	180/35	90/52	20/55

Table 3.1-134 Strengthening of Ag (99.975%) by cold forming as a function of reduction in cross section in % [1.217, p. 204]

V (%)	R_m (MPa)	A (%)	HV
0	150	50	26
10	180	30	54
30	260	5	70
80	2	90	

V = reduction of cross section

Table 3.1-135 Strengthening of Ag alloys by cold forming (HV 10) (reduction in % of thickness) [1.217, p. 207]

Reduction (%) Alloying element	0	40	80
Cu 5	58	108	134
Cu 15	76	126	158
Cu 28	98	136	177
Cu 50	84	130	166
Ni 0.15	40	86	100
Pd 30	70	132	164
Pd 30 Cu 5	92	174	2165

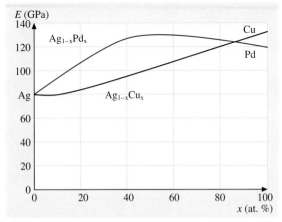

Fig. 3.1-199 Module of elasticity of Ag−Pd and Ag−Cu alloys [1.217, p. 206]

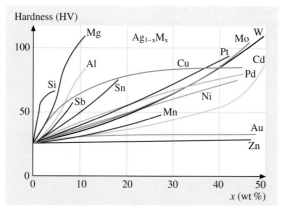

Fig. 3.1-200 Influence of alloying elements on the hardness of binary Ag alloys [1.217, p. 206]

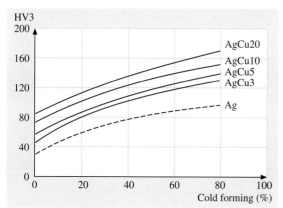

Fig. 3.1-201 Hardening of Ag–Cu alloys by cold forming [1.217, p. 432]

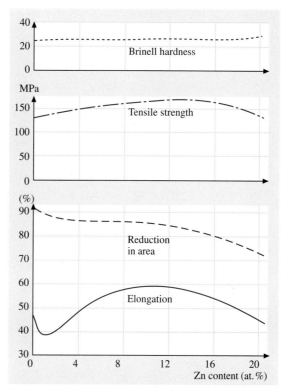

Fig. 3.1-202 Plastic properties of Ag–Zn crystals [1.220, p. 135]

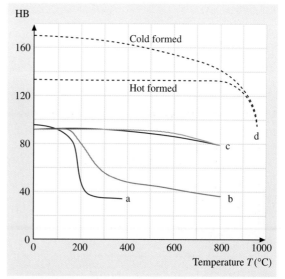

Fig. 3.1-203 Hardness of fine grain and dispersion-hardened Ag [1.216, p. 36]

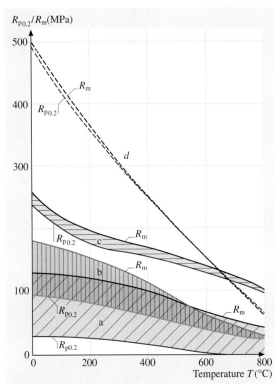

Fig. 3.1-204 Tensile strength and 0.2% proof stress of silver grades at different temperatures [1.216, p. 37]

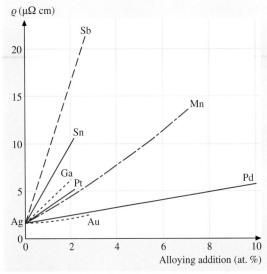

Fig. 3.1-205 Influence of alloying elements on the electrical conductivity of binary Ag alloys [1.231, p. 20]

Electrical Properties

Tables 3.1-136–3.1-139 and Fig. 3.1-205 [1.217, 231–233] show characteristic data. The residual resistivity ratio (RRR) of pure Ag ranges up to 2100. Ag alloys with Pb and Sn show superconductivity in the composition ranges: $Ag_{0.95-0.66}Pb_{0.05-0.34}$ with $T_c = 6.6-7.3$ K and $Ag_{0.72-0.52}Sn_{0.28-0.48}$ with $T_c = 3.5-3.65$ K [1.234].

Table 3.1-136 Specific electrical resistivity $\rho(T) = \rho_0 + \rho_i(T)$ of Ag ($\rho_0 = 0.0008\ \mu\Omega\,cm$) at different temperatures [1.217, p. 156]

T (K)	$\rho(T)$ ($\mu\Omega$ cm)
10	0.0001
50	0.1032
120	0.5448
273	1.470
500	2.860
700	4.172
900	5.562
1100	7.031

Table 3.1-137 Increase of atomic electrical resistivity of Ag by alloying elements $\Delta\rho/C$ ($\mu\Omega$ cm/at.%) [1.217, p. 157]

Base element	Alloying elements
Ag	Al 1.87, As 8, Au 0.36, Cd 0.35, Cr 6.5, Cu 0.07, Fe 3.2, Ga 2.3, Ge 5, Hg 0.8, In 1.6, Pb 4.6, Pd 0.44, Pt 1.5, Sn 4.5, Zn 0.63

Table 3.1-138 Specific electrical resistivity ρ_i and coefficient of electrical resistivity (TCR) of noble metal solid solution alloy phases [1.217, p. 158]

Base/solute-metal		Solute content			
		20	40	60	80
Rh/Ni	ρ_i	21	37	51	50
	TCR	3.8	1.8	< 0.1	3
Ag/Au	ρ_i	8.3	11.0	11.0	8.1
	TCR	0.93	0.83	0.84	1.1
Ag/Pd	ρ_i	11	22	41	34
	TCR	0.58	0.40	–	0.75
Ag/Pt	ρ_i	33	60	46	35
Au/Pd	ρ_i	9.8	17	30	26
	TCR	0.88	0.61	0.45	1.2
Au/Pt	ρ_i	28	44	0.82	0.8
	TCR	0.28	0.26	0.82	0.8

Table 3.1-139 Specific electrical resistivity ρ_{25} of annealed (8 h at 550 °C) Ag–Cu wire at 25 °C and 100 °C (ρ_{100}) and temperature coefficient of resistivity (TCR) α for 25–100 °C [1.226, p. 380]

at.% Ag	5	15	45	75	96
$\rho_{25} \times 10^{-6}$ (μΩ cm)	1.832	1.895	1.913	1.645	1.822
$\rho_{100} \times 10^{-6}$ (μΩ cm)	2.369	2.320	2.411	2.308	2.297
TCR $\times 10^5$	389	387	380	365	381

Thermoelectric Properties

In Tables 3.1-140–3.1-142 and Fig. 3.1-206, characteristic data are shown: absolute thermoelectric power, thermo-electromotive force of pure Ag as well as Ag–Au, Ag–Pd, and Ag–Pt alloys at different temperatures against a reference junction at 0 °C [1.217, 235, 236].

Table 3.1-140 Thermal electromotive force $E_{Ag,Pt}$ (mV) force of Ag at different temperatures; reference junction at 0 °C [1.217, p. 159]

T (°C)	−200	−100	−50	+100	+200	+400	+800
$E_{Ag,Pt}$ (mV)	−0.39	−0.21	−0.10	0.74	1.77	4.57	13.36

Table 3.1-142 Thermal electromotive force of Ag–Au alloys in mV at 100 °C and 700 °C reference junction at 0 °C [1.217, p. 160]

T (°C)	Composition (wt%)					
	0	20	40	60	80	100
100	0.74	0.47	0.42	0.42	0.49	0.78
700	10.75	7.7	6.7	6.8	7.3	10.15

Magnetic Properties

Silver is diamagnetic (Table 3.1-143). The magnetic susceptibility remains constant from 0 K to the melting point. Alloying with B metals causes only minor variations compared to pure Ag. In the continuous solid solution range the molar susceptibilities remain negative and the alloys are diamagnetic. Ni, Pd, and Pt dissolve up to 25 at.% diamagnetically. Cr, Fe, and Mn give rise to paramagnetism, while Co causes ferromagnetism [1.216, 217].

Table 3.1-143 Atom susceptibility of Ag and Au alloys at room temperature [1.216, p. 90]

Base metal	Alloying element	Base metal content (at.%)			
		100	99	95	90	80
Ag	Au	−19		−20.2	−20.8	−22.2
	Pd	−20			−21	−22
Au	Cu	−26		−25.2	−24.2	−22.4
	Ni	−28	−22	±0	+16	–
	Pd	−28			−20	−15
	Pt	−28			−6	±0

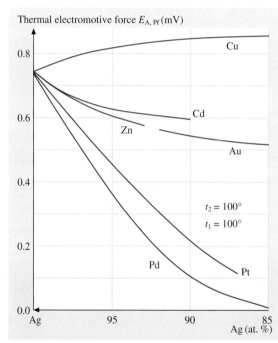

Fig. 3.1-206 Thermal electromotive force of binary Ag alloys [1.216, p. 98]

Table 3.1-141 Absolute thermoelectric power of Ag at different temperatures [1.216, p. 94]

Temperature (°C)	−255	−200	−100	−20	0	100	300	500	800
Thermoelectric power (μV/°C)	+0.62	+0.82	+1.00	–	+1.4	+1.9	+3.0	+4.6	+8.3

Thermal Properties

Selected data of thermal expansion, thermal conductivity, and melting temperatures of Ag alloys are given in Tables 3.1-144–3.1-150 and in Fig. 3.1-207 [1.216, 220].

Table 3.1-144 Recrystallization temperatures (°C) of Ag 99.95 and 99.995% purity after different degrees of deformation V; annealing time 1 h [1.217, p. 205]

V (%)	$T_{recryst.}$ Purity	
	99.95%	99.995%
40	190	125
60	160	100
99	127	70

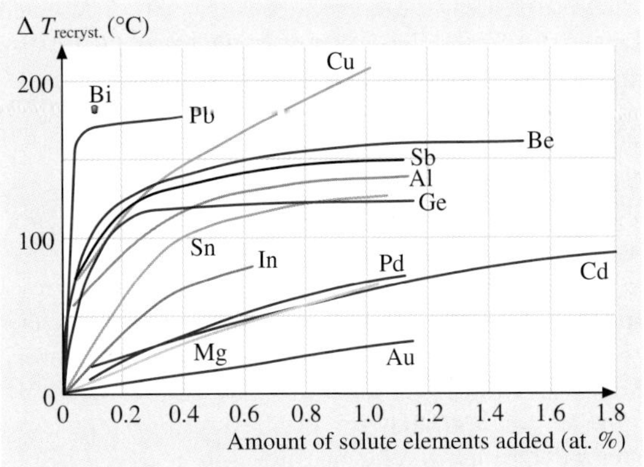

Fig. 3.1-207 Increase of the recrystallization temperature of Ag by solute elements [1.218, p. 452]

Table 3.1-145 Mean coefficients of thermal expansion α (10^{-6} K^{-1}) of Ag and Au [1.217, p. 154]

T (K)	α (10^{-6} K^{-1})	
	Ag	Au
373	17.9	13.9
473	19.1	14.8
673	19.9	15.3
873	21.1	15.7
1073	23.6	16.2

Table 3.1-146 Mean coefficients of thermal expansion α (10^{-6} K^{-1}) of Ag and Au alloys [1.217, p. 154]

Base metal	Au	Au	Au	Au	Ag
2nd metal	Ag	Ag	Cu	Ni	Pd
Temp. range (K)	273 373	293 1073	273 593	273 373	373 473
wt% of 2nd metal	α (10^{-6} K^{-1})				
20	15.5	17.0	14.9	14.4	16.2
40	16.5	18.7	–	13.9	–
50	17.0	–	15.7	14.0	14.7
60	17.5	20.1	15.8	13.9	–
80	18.2	20.8	15.9	13.4	12.4

Table 3.1-149 Thermal conductivity λ (W/m K) of Ag and Au at different temperatures [1.217, p. 153]

T (K)	λ (W/m K)	
	Ag	Au
40	1050	420
100	475	360
273	435	318
600	411	296
800	397	284

Table 3.1-150 Melting range of Ag–Cu–Sn and Ag–Cu–In solder alloys

Composition (wt%)				Melting range (°C)
Ag	Cu	Sn	In	
60	23	17		557–592
50	30	20		555–578
47	20	33		472–515
50	10	40		435–456
49	19		32	549–556
50	25		25	594–601
45	17		38	534–548
25	45		30	581–610

Table 3.1-147 Specific heat of Ag at different temperatures [1.220, p. 9]

T (°C)	−259.46	−240.86	−141.93	−67.90	+98.7	+249.0	+399.5	+652.2
c_p (cal/g)	0.001177	0.01259	0.04910	0.05334	0.0569	0.0583	0.0600	0.0635

Table 3.1-148 Vapor pressure of liquid Ag [1.220, p. 8]

T (°C)	1550	1611	1742	1838	1944	2152
Vapour pressure (mm Hg)	8.5	15.7	54	100	190	760 (extrapol.)

Optical Properties

Table 3.1-151 and Figs. 3.1-208, 3.1-209 [1.237] show characteristic data of optical properties. Ag has the highest reflectivity of all noble metals. An interband transition takes place in the ultraviolet range at 3.9 eV. Ag–Al alloys between 10 at.% and 28 at.% Ag show higher reflectance in the low wavelength range than the pure elements. In Ag–Pd alloys, the threshold energy at 3.9 eV for the interband transition remains constant up to ≈ 34 at.% Pd. Examples of colored Ag alloys are given in Table 3.1-152 [1.237].

Table 3.1-151 Spectral emissivity of Ag and Au at different temperatures [1.217, p. 171]

	Surface	Temperature (°C)	Spectral degree of emission
Ag	Solid	940	0.044
	Liquid	1060	0.0722
Au	Solid	1000	0.154
	Liquid	1067	0.222

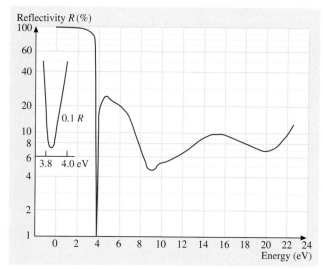

Fig. 3.1-208 Reflectance versus radiation energy of Ag [1.237, p. 158]

Table 3.1-152 Colored noble metal alloys [1.237, p. 178]

Alloy	Color	Remarks
Ag–Zn (β-phase)	Rose	
Ag–Au(70)	Green-yellow	
Al_2Au	Violet	
KAu_2	Violet	
Au–Zn–Cu–Ag	Green	
$AuIn_2$	Blue	
Zintl Phases		
Li_2AgAl	Yellow-rose	VEC 1.5
Li_2AgGa	Yellowish	VEC 1.5
Li_2AgIn	Gold-yellow	VEC 1.5
Li_2AgTl	Violet-rose	VEC 1.5
Li_2AuTl	Green-yellow	VEC 1.5
Li_2AgSi	Rose-violet	VEC 1.75
Li_2AgGe	Rose-violet	VEC 1.75
Li_2AgSn	Violet	VEC 1.75
Li_2AgPb	Blue-violet	VEC 1.75
Li_2AuPb	Violet	VEC 1.75

VEC = Valence electron concentration

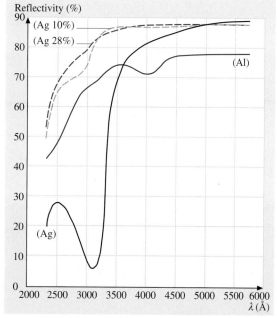

Fig. 3.1-209 Reflectivity of Ag–Al alloys [1.220, p. 142]

Diffusion

Data for self-diffusion of Ag in Ag alloys and diffusion of tracer impurity elements are shown in Tables 3.1-153–3.1-158 and Figs. 3.1-210–3.1-212. Diffusion of H and O is of importance for annealing treatments and dispersion hardening [1.217, 220, 226, 235].

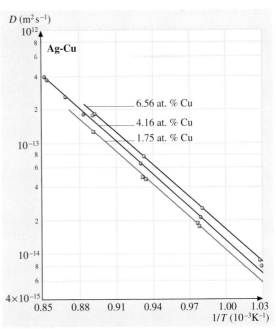

Fig. 3.1-210 Self-diffusion of Ag(110w) in Ag–Cu (1.75–6.56 at.% Cu) alloys [1.238, p. 189]

Table 3.1-153 Self-diffusion in binary homogeneous Ag–Au alloys [1.217, p. 150]

Au (at.%)	ΔT (K)	D^0 (10^4 m^2/s)	Q (kJ/mol)
Ag–Au (^{110m}Ag diffusion)			
8	927–1218	0.52	187.5
17	908–1225	0.32	184.4
83	923–1284	0.09	171.7
94	936–1234	0.072	168.5
Ag–Au (^{198}Au diffusion)			
8	991–1213	0.82	202.2
17	991–1220	0.48	198.0
83	985–1274	0.12	180.2
94	991–1283	0.09	176.1

Table 3.1-154 Self-diffusion in pure Ag and Au [frequency factor D^0 (10^{-4} m^2/s)], activation energy Q (kJ/mol) [1.217, p. 149]

	Element	D^0 (m^2/s)	Q (kJ/mol)	T (K)
Lattice	Ag	0.278	181.7	1038–1218
	Ag(s)^a	0.67	190.1	913–1221
	Au	0.107	176.9	623–733
	Au(s)^a	0.027	165	603–866
Surface	Ag	1×10^4	264	873–1173 (H$_2$)
	Ag	3×10^{-5}	49	580–730 (Vac.)
	Au(110)	1×10^2	227	1138–1329 (H$_2$)
	Au	8×10^2	272	1200–1300 (Vac.)
Grain-boundary	Ag	7.24×10^{-5}	190.4	~790–680
	Au	9.10×10^{-6}	174.6	625–521

^{a} (s) = single crystal

Table 3.1-155 Diffusion of Ag in Cu and Cu in Ag [1.226, p. 365f.]

Ag in Cu						
T (°C)	485	574	625	731	794	
D_{Ag}	4.9×10^{-14}	8.2×10^{-13}	2.91×10^{-12}	7.7×10^{-11}	1.65×10^{-10}	
	±0.1	±0.6	±0.08			

$D = D_0 \exp(-E_A/RT)$ with $D_0 = 0.61$ cm^2/s, $E_A = 46.5$ kcal/g-at.

Cu in Ag						
T (°C)	498	597	760	800	895	
D_{Cu}	2.87×10^{-13}	5.08×10^{-12}	3.55×10^{-10}	5.9×10^{-10}	9.4×10^{-10}	
	±0.45	±0.54				

$D_0 = 1.23$ cm^2/s, $E_A = (46.1 \pm 0.9)$ kcal/g-at.

Table 3.1-156 Diffusion of impurities in Ag, Au, Pt and Pd [1.217, p. 151]

Tracer	D^0 (10^{-4} m²/s)	Q (kJ/mol)	ΔT (K)	Tracer	D^0 (10^{-4} m²/s)	Q (kJ/mol)	ΔT (K)
Matrix: Silver (Ag)				Matrix: Gold (Au)			
Cu	1.23	193.0	990–1218	Pd	0.076	195.1	973–1273
	0.029	164.1	699–897	Pt	0.095	201.4	973–1273
Au	0.262	190.5	923–1223	Ag	0.072	168.3	943–1281
	0.41	194.3	929–1178		0.08	169.1	1046–1312
	0.85	202.1	991–1198		0.086	169.3	1004–1323
	0.62	199.0		Hg	0.116	156.5	877–1300
Pd	9.57	237.6	1009–1212	Matrix: Platin (Pt)			
Pt	6.0	238.2	923–1223	Ag	0.13	258.1	1473–1873
	1.9	235.7	1094–1232	Au	0.13	252.0	850–1265
Ru	180	275.5	1066–1219	Matrix: Palladium (Pd)			
				Fe	0.18	260	1373–1523

Table 3.1-157 Grain boundary tracer diffusion in pure Ag [1.217, p. 150]

Matrix metal	Tracer	D^0 (m²/s)	Q (kJ/mol)	T (K)
Ag	Cd	5.04×10^4	176.6	772–557
	In	5.50×10^4	174.8	764–469
	Sb	2.34×10^4	163.5	771–471
	Sn	4.72×10^4	170.9	776–527
	Te	2.10×10^4	154.7	970–650

Table 3.1-158 Diffusion of hydrogen and oxygen in Ag, Pd, Pt, and Au [1.217, p. 151]

Matrix	Gas	D^0 (10^{-4} m²/s)	Q (kJ/mol)	ΔT (K)
Ag	H	8.55×10^{-3}	30.11	947–1123
Ag	O	3.66×10^{-3}	46.1	680–1140
Pd	H	2.9×10^{-3}	22.19	473–1548
	D	1.7×10^{-3}	19.88	218–233
	T	7.2×10^{-3}	23.8	273–323
Pt	H	6×10^{-3}	24.70	600–900
Au	H	5.6×10^{-4}	23.61	773–1213

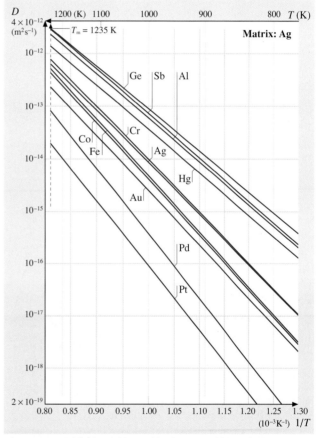

Fig. 3.1-211 Diffusion of impurities in Ag [1.238, p. 244]

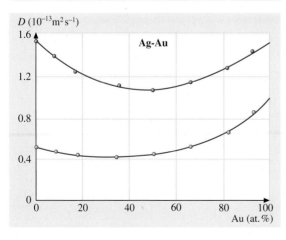

Fig. 3.1-212 Self-diffusion of Ag(110w) (*brown circles*) and Au(198) (*gray circles*) in Ag–Au (8–94 at.%) Ag–Au alloys [1.238, p. 190]

Chemical Properties

Silver has the reduction potential of $E_0 = +0.8\,\text{V}$ for Ag/Ag$^+$. It is resistant against dry oxygen, air, non-oxidizing acids, organic acids, and alkali. Water and water vapor do not attack Ag up to 600 °C. Ag is dissolved in alkaline cyanidic solutions in the presence of oxidizing agents, air, and oxygen. H$_2$S attacks Ag readily at room temperature, forming black Ag$_2$S layers (tarnish) [1.217].

Metallic Ag and Ag–Au alloys are heterogeneous catalysts for oxidation processes, e.g., in the production of ethylene oxide and formaldehyde applied as grids or as powder preparations on Al$_2$O$_3$ or carbon substrates [1.217].

Ag-Based Materials

Binary alloys (Tables 3.1-159–3.1-161) [1.218]: Ag–Ni alloys are grain-stabilized materials usually containing 0.15 wt% Ni. Ag–Cu alloys have manifold applications in jewelry, silverware, brazes, and solders. Jewelry, sil-

Table 3.1-161 Physical properties of noble-metal-containing vacuum braze alloys [1.217, p. 560]

	Composition	Solidus (°C)	Liquidus (°C)
Ag–Cu	Ag40Cu19Zn21Cd20	595	630
	Ag60Cu27In13	605	710
	Ag44Cu30Zn26	675	735
	Ag72Cu28	779 E [a]	
Ag–Mn	Ag85Mn15	950	950
Ag–Pd	Ag68.4Pd5Cu26.6	807	810
	Ag54Pd25Cu21	901	950
	Ag75Pd20Mn5	100	1120
Pd–Cu	Pd18Cu72	1080	1090
Pd–Ni	Pd60Ni40	1237	1237
Au–Cu	Au80Cu20	890	890
	Au33Cu65	990	1010
Au–Ni	Au82Ni18	950	950

[a] E = Eutectic composition

Table 3.1-159 Noble metal containing soft solders [1.217]

Alloy (wt%)	Melting range (°C)	Density (g/cm^3)	Tensile strength (N/mm^2)	Elongation (%)	Eleasticity module (N/mm^2)	Thermal expansion (10^{-6} K^{-1})	Electrical conductivity (m/Ω mm^2)	Thermal conductivity (W/m K)
AuSn20	280	14.57	275	< 1	59 200	16	< 5	57.3
AuGe12	356	14.70	150–200	< 1	69 300	13.4	7	44.4
AuSi2	363–740	14.50	500–600 [a]	0.5–3 [a]		12.6	33	50
SnAg25Sb10	228–395	7.86	80–120	1–4	23 000	19	6.5	55
SnAg3.5	221	7.38	25–35	20–30	41 100	27.9	7.5	57
PbSn5Ag2.5	280	11.70	25–35	20–30	21 300	29	< 5	44
PbIn5Ag2.5	307	11.60	35–40	28–34	19 900	29	< 5	42

[a] Hard rollded condition

Table 3.1-160 Noble-metal-containing brazing alloys [1.217, p. 560]

Alloy (wt%)	Melting range (°C)	Density (g/cm^3)	Tensile strength (N/mm^2)	Elongation (%)	E-Modul. (N/mm^2)	Thermal expansn. (10^{-6} K^{-1})	Electrical condct. (m/Ω mm^2)	Thermal condct. (W/m K)	ISO Type 3677
AgCu28	779	10.00	250–350	20–28	100 000	19.8	10		B Ag72 Cu780
AgCu27In13	605–710	9.70	400–500	20–30	85 000	17.8	46.1	352	B Ag60CuIn 695-710
AgCu26.6Pd5	807–810	10.10	370–410	12–20	120 000	22	26	215	B Ag68CuPd 807-810
AgCu31.5Pd10	824–852	10.10	500–540	2–5	140 000	17.5	19	150	B Ag58CuPd 824-852
AgCu20Pd15	850–900	10.40	510–550	5–9	140 000	22	15	100	B Ag65CuPd 850-900
AgCu21Pd25	910–950	10.50	540–580	13–21	140 000	17.5	8	80	B Ag54PdCu 901-950
AgPd5	970–1010	10.50	180–220	26–34	40 000	22	25	220	B Ag95Pd 970-1010
CuPd18	1080–1090	9.40	380–420	31–39	135 000	18.9	9.1	100	B Cu82Pd 1080-1090
AgCu28Pd20	879–898	10.30	580–620	6–10	100 000	18.6	9.5	95	B Ag52CuPd 879-898
AuNi18	950	15.96	550–650	8		14.6	5.9		B Au82Ni 950

Table 3.1-162 Typical powder grades of Ag, Pd, and Ag−Pd preparations for capacitors

Metal	Manufacturing method	Grain shape	Grain size (μm)	Tap density (g/cm^3)	Specific surface[a] (m^2/g)
Ag	Chemical reduction	Microcrystalline	0.5–2.0	0.8–5.0	0.1–2.0
	Electrolytic deposition	Dendritic ↓	1–200	4.5–4	0.1–0.5
	Ball milling	Flakes	2–40	2.5–5	0.2–1.8
Pd	Chemical reduction	Crystalline spheres	< 20	0.8	2.5
			< 1.2	4.0	2.3
Ag−Pd30	Coprecipitation	Spheres ↓	< 1.2	3.3	2.2
		Flakes	< 9	3.4	3.3

[a] BET in N_2 adsorption

ver ware alloys, and coins usually contain between 7.5 wt% Cu ("sterling silver") and 20 wt% Cu. The material Ag–28 wt% Cu is the most common silver brazing alloy. Alloys of Ag−Mn are special solders for hard metal and refractory metals (Mo, W). The alloy Ag–1 wt% Pt is applied in thick film layers for conductor paths in passive electronic devices. Ag−Pd powder preparations containing 10 to 30 wt% Pd form the conductor layers in multilayer capacitors (Table 3.1-162) [1.217].

Ternary and Higher Alloys

Alloys of the systems Ag−Cu−Sn, Ag−Cu−Zn, and Ag−Cu−Cu$_3$P are used as solders and brazes. Ag−Cu−P solder alloys can be applied without flux. Ti-containing solder alloys (active solders) allow direct bonding to ceramics (Table 3.1-162) [1.239]. Alloys of the systems Ag−Au−Cu, Ag−Au−Ni, and Ag−Cu−Pd are applied in jewelry and dentistry (Tables 3.1-186, 3.1-187).

The oxide AgO forms the cathode of AgO/Zn button type batteries with a cell voltage of 1.55 V and with energy densities in the range of 80–250 W h/dm^2 [1.218, 240–242].

Composite materials with SnO_2, CdO, carbon, Ni, and refractory carbides as dispersoids are base materials of electrical (Tables 3.1-163, 3.1-164, and Fig. 3.1-213) [1.226, 231, 243]. Extruded powder composites show preferred alignment of the dispersoid particles along the rod axes. Silver–nickel fiber composites are magnetic. Their coercivity increases with decreasing diameter of the Ni fibers [1.244].

Table 3.1-163 Hardness and electrical conductivity of Ag−Ni−C (contact) alloys [1.226, p. 436]

Alloy content wt% C	3	2	2.5	2.5	2.5
Alloy content wt% Ni	0.5	0.5	1	5	10
HV (kg/mm^2)$^{0.5}$	43	53	50	53	56–57
Electrical conductivity in % of standard Cu	63.7	74.8	69.8	67.3	64.5

Table 3.1-164 Silver bearing composite contact materials [1.243, p. 156]

Type	Composition (at.%)	Hardness HV	Electrical conductivity (m/Ω mm²)
Alloys	AgNi(0.15)	100	58
	AgCu(3)	120	52
	↓		
	AgCu(20)	150	49
Composites with:			
Metals	Ag–Ni(10)	90	54
	↓		
	Ag–Ni(40)	115	37
Oxides	Ag–CdO(10)	80	48
	↓		
	Ag–CdO(15)	115	45.5
	Ag–ZnO(88)	95	49
	Ag–SnO$_2$(8)	92	51
	Ag–SnO$_2$(12)	100	42
Carbon	Ag–C(2)	40	48
	Ag–C(5)	40	43.5
Refractory metal compounds	Ag–W(20)	240	26–28
	↓		
	Ag–W(80)	80	42
	Ag–WC(40)	130	24–30
	↓		
	Ag–WC(80)	470	

[a] Composite made by infiltration of liquid silver into a tungsten skeleton

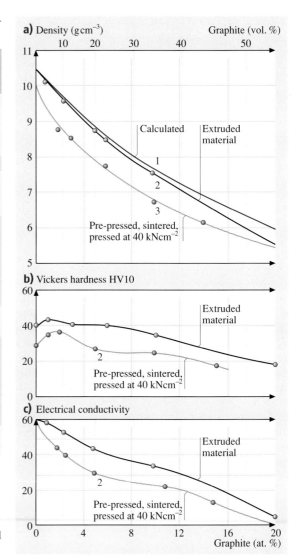

Fig. 3.1-213 (a) Density, (b) hardness, and (c) electrical conductivity of Ag–C alloys [1.231, p. 44f.]

3.1.10.2 Gold and Gold Alloys

Application

Gold and gold alloys are used for electrical contacts, bonding wires and conductor paths in semiconductor devices, chemical and corrosion resistant materials, thin surface coatings for optical and heat reflecting mirrors, special thermocouples, and catalysts for organic chemical reactions. Classical applications are jewelry, dentistry, monetary bars, and coins. Commercial grades: Table 3.1-165. The purity grades of gold bars are standardized in the range of 99.9 to 99.999 wt% (ASTM B 562-86), Tables 3.1-165, and 3.1-166 [1.217].

Production

Elementary gold is extracted from ores by cyanide leaching and precipitated with zinc, and by electrolysis. Refining is achieved by application of chlorine gas up to 99.5%, and to 99.9% and higher by electrolysis. Bars, sheets and wires are made by casting, rolling and drawing; powder is formed by chemical and by electrolytic precipitation from solutions; and nanocrystalline powders are formed by dispersion in organic solutions. Coatings are produced by cladding; electroplating; and applying powder preparations followed by firing. Thin films are produced by evaporation and cathode sputtering. Very fine gold leaves are made by traditional hammering to a thickness of $\sim 0.2\,\mu\text{m}$, or by cathode sputtering.

Phases and Phase Equilibria

Selected phase diagrams are shown in Figs. 3.1-214–3.1-223 [1.245, 246]. Continuous solid solutions are formed with Ag, Co, Cu, Fe, Ni, Pd, and Pt. Miscibility gaps occur with Be, Ni, Pt, Rh, and Ru. Thermochemical data are listed in Tables 3.1-167,

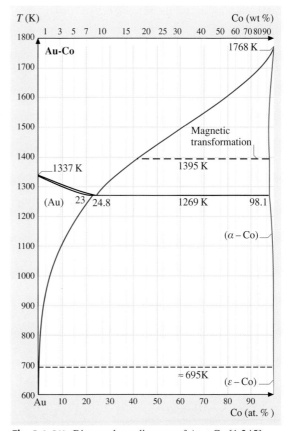

Fig. 3.1-214 Binary phase diagram of Au–Co [1.245]

Table 3.1-165 Specifications of fine gold [1.217, p. 52]

Designation	Grade (wt%)	Impurity	Maximum content (ppm)
"Good delivery" gold	99.5	any, total	5000
Fine gold	99.99	Ag/Cu/others/total	100/20/30/100
Fine gold, chemically pure	99.995	Ag/others/total	25/25/50
Fine gold, high purity	99.999	Ag/Fe/Bi/Al/Cu/Ni/Pd + Pt/total	3/3/2/0.5/0.5/0.5/5/10

Table 3.1-166 Standard fineness of noble metal alloys and corresponding carat of jewelry [1.217, p. 52]

	Fineness (wt‰)					
Au		375	585	750	916,999	
Ag					800,925,999	
Pd			500		850,900,950,999	
Pt			500		950,999	
	333	375	585	750		1000
Carat	8	9	14	18		24

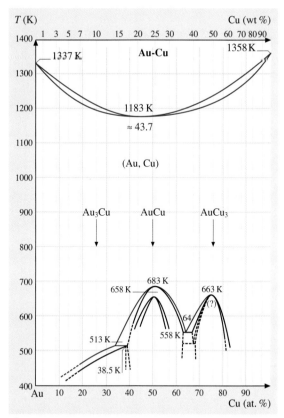

Fig. 3.1-215 Binary phase diagram of Au–Cu [1.245]

and 3.1-168 [1.217, 222]. Compositions, crystal structures and lattice parameters of selected intermetallic compounds are given in Table 3.1-169 [1.217] and in Figs. 3.1-224 and 3.1-225 [1.245]. Primary solid solutions have the fcc structure of Au. The lattice parameters of the substitutional solid solutions correspond roughly to Vegard's law with a few exceptions [1.247].

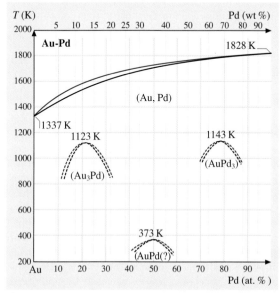

Fig. 3.1-217 Binary phase diagram of Au–Pd [1.245]

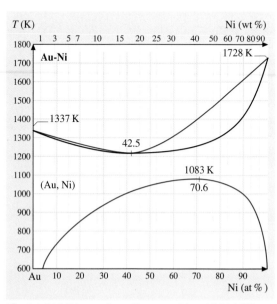

Fig. 3.1-216 Binary phase diagram of Au–Ni [1.245]

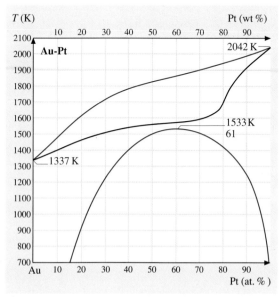

Fig. 3.1-218 Binary phase diagram of Au–Pt [1.245]

Table 3.1-167 Thermodynamic data of Au [1.217, p. 107]

T (K)	c_p (J/K mol)	S (J/K mol)	H (J/mol)	G (J/mol)	p (at)
300	25.303	47.645	0.047	−14.247	6.87×10^{-58}
500	26.158	60.757	5.188	−25.191	2.82×10^{-32}
700	27.028	69.701	10.509	−38.282	2.49×10^{-21}

T = Temperature, c_p = specific heat capacity, S = Entropy, H = Enthalpy, p = partial pressure of the pure elements

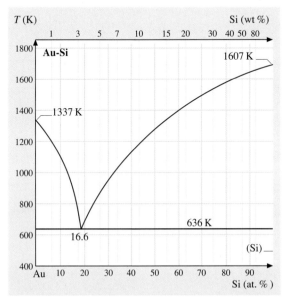

Fig. 3.1-219 Binary phase diagram of Au−Si [1.245]

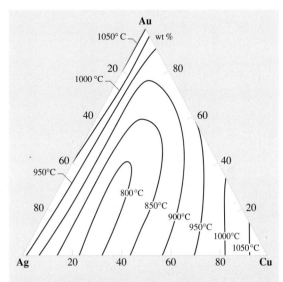

Fig. 3.1-220 Liquidus sections through the miscibility gap of the Au−Ag−Cu system [1.246]

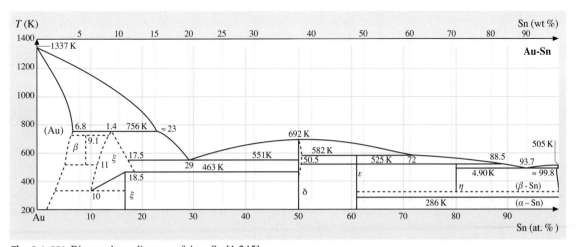

Fig. 3.1-221 Binary phase diagram of Au−Sn [1.245]

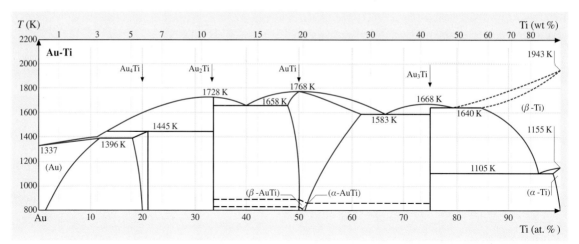

Fig. 3.1-222 Binary phase diagram of Au–Ti [1.245]

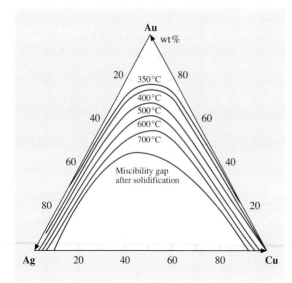

Fig. 3.1-223 Isothermal sections through the miscibility gap of the Au–Ag–Cu system [1.246]

Superlattice phases occur in alloys with Cd, Cu, Mn, Pd, Pt, Rh, Ru, and Zn. The superlattice structures have tetrahedral or rhombohedral symmetry. Typical compositions are AB_3 and A_3B [1.218]. If they are not precipitated as second phases, superlattice phases form antiphase domains on different sublattices separated by antiphase domain boundaries. Intermetallic compounds are formed with numerous elements with different and complex crystal structures [1.225]. Metastable phases exist with Ni and Pt. Alloys with B-metals form intermetallic phases at compositions corresponding to e/a values of 3/2, 21/13, and 7/4 (Hume-Rothery phases). Structural types of intermetallic compounds of gold with rare earth metals are listed in [1.248].

Table 3.1-168 Heats, entropies and free energies of formation of Au compounds [1.222, p. 196]

Phase	N_2	T (°C)	H (J/mol)	G (J/mol)	S (J/K mol)
AuCu s.s.	0.58	500	5.32	9.67	5.65
Au_3Cu	0.26	25	4.03	4.90	2.97
AuCu I	0.50	25	8.96	8.96	–
AuCu II	0.50	400	6.03	8.79	4.10
$AuCu_3$	0.75	25	6.87	7.24	1.26
AuNi s.s.	0.53	877	–7.5	–	8.71
AuSn	0.50	25	14.24	–	–0.4

N_2 = mole fraction of second compound,
ΔS = entropy of formation, s.s. = solid solution
s.s. = solid solution

Table 3.1-169 Structure and lattice parameters of selected intermediate Au compounds [1.217, p. 113]

Phase	Pearson Symbol	a (nm)	b (nm)	c (nm)	c/a	Remarks	Concentration $(A/1-x)B(x)$
AuAl	mP8	0.6415	0.3331	0.6339			
Au$_2$Al	oP30	0.8801	1.6772	0.3219		LT	0.664–0.667
Au$_5$Al	cP20	0.69208				LT	
Au–Al$_2$	cF12	0.59973					0.334
AuCu	oI40	0.3676	0.3972				
AuCu	oP8	0.456	0.892	0.283			
AuCu	tP4	0.3966		0.3673	0.9261	773 K	0.46–0.54
Au$_3$Cu	cP4	0.39853					
Au–Pd	cP4	0.3991					0.552
Au–Pt	cF4	0.3996					0.5
Au–Sn	hP2	0.29228		0.47823	1.6329		0.14
AuSn	hP4	0.43218		0.5523	1.2779		
AuSn$_4$	oC20	0.6502	0.6543	1.1705		475 K	
Au$_4$Ti	tI10	0.6485		0.4002	0.6171		

LT = low-temperature modification

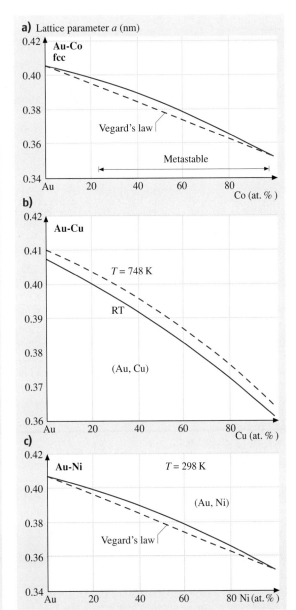

Fig. 3.1-224a–c Lattice parameter versus composition in the systems (**a**) Au–Co, (**b**) Au–Cu, (**c**) Au–Ni [1.245]

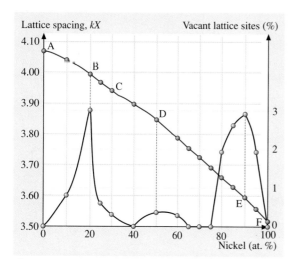

Fig. 3.1-225 Vacant lattice sites in Au–Ni alloys [1.225, p. 141]

Mechanical Properties

The mechanical properties of gold are given in Tables 3.1-170–3.1-176 and Figs. 3.1-226–3.1-237 [1.216, 217]. References for data of elastic constants of Au alloys are given in [1.217]. Pure gold is very soft. It can be cold-worked to more than 90% by

Table 3.1-170 Modulus of elasticity of Au in crystal directions (GPa) [1.217, p. 208]

$E \langle 100 \rangle$	$E \langle 110 \rangle$	$E \langle 111 \rangle$
42	81	114

Table 3.1-171 Elastic constants of Au (GPa) [1.217, p. 209]

T (°C)	c_{11}	c_{12}	c_{14}
−273	131.4	97.3	51.1
20	124.0	93.4	46.1

Table 3.1-172 Mechanical properties of Au (99.99%) at different temperatures [1.217, p. 209]

T (°C)	E (GPa)	R_m (MPa)	$R_{p\,0.2}$ (MPa)	HV
20	79	125	30	28
200	75	110	20	19
400	70	92	–	16
700	58	40	–	5

E = Modul of elasticity, R_m = Tensile strength, R_p = Limit of proportionality, HV = Vickers hardness

Table 3.1-176 Mechanical properties of AuPt alloys in annealed and aged condition, Pt in wt% [1.217, p. 211] ▶

Table 3.1-173 Tensile strength R_m (MPa) of binary Au alloys [1.217, p. 210]

Content (wt%) Alloying element	2	5	10	20
Ag	140	150	170	190
Co	240	–	–	–
Cr	200	–	–	–
Cu	190	290	400	500
Fe	190	–	–	–
Ni	220	350	470	680
Pd	150	170	220	290
Pt	150	189	240	370

Table 3.1-174 Mechanical properties of Au (99.99%) as a function of the reduction V (%) in thickness by cold forming [1.217, p. 209]

V (%)	R_m (MPa)	A (%)	HV
0	120	45	28
10	140	22	55
30	180	75	63
50	220	4	65

R_m = Tensile strength, A = Elongation of rupture, HV = Vickers hardness

Table 3.1-175 Change of hardness (HV 10) of Au alloys by cold forming (Degree of reduction in thickness in %) [1.217, p. 210]

Degree of reduction in thickness (%) Alloying element	0	40	80
Ag20		40	95	1141
Ag25Cu5		92	160	188
Ag20Cu10		120	190	240
Co5		92	126	154
Ni5		120	162	188
Pt10		78	102	118
Pd30Cu5		92	174	216

Alloy	R_m (MPa)	A (%)	HV
Pt10[a]	250	38	42
Pt30[a]	430	12	120
Pt30[b]	740	–	300
Pt50[a]	900	3	240
Pt50[c]	1460	2	420

[a] annealed at 1000–11 150 °C and quenched
[b] stored 70 h at 500 °C,
[c] stored 25 h at 500 °C

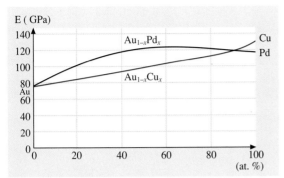

Fig. 3.1-226 Modulus of elasticity of Au–Cu and Au–Pd alloys [1.217, p. 219]

Fig. 3.1-227 Influence of alloying elements on the hardness of binary Au alloys [1.217, p. 211] ▶

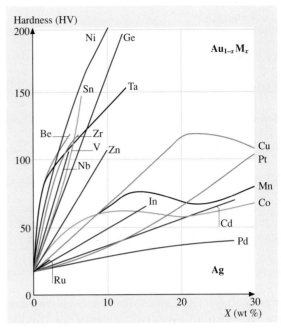

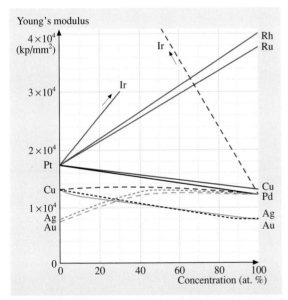

Fig. 3.1-228 Modulus of elasticity versus composition of binary noble-metal alloys [1.216, p. 77]

rolling or drawing. Cold hard drawn wires (about 90% deformation) have predominantly ⟨111⟩ fiber texture, which is converted by annealing into ⟨100⟩ orientation [1.249]. Strengthening of pure gold is affected by alloying (solid solution hardening, precipitation hardening) or by dispersion hardening. Ternary Au–Ag–Cu

Fig. 3.1-229a,b Hardness of Au–Co alloys by annealing; (**a**) influence of time, (**b**) influence of temperature T [1.231, p. 62]

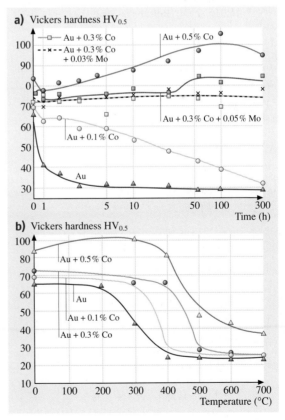

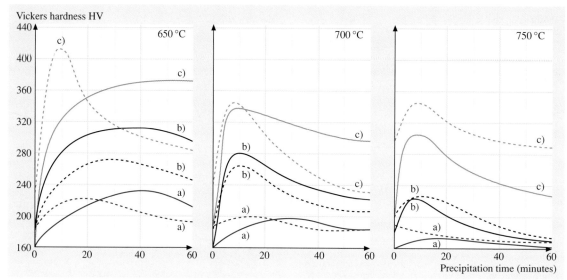

Fig. 3.1-230 Precipitation hardening of Au/Pt-40 (*solid curve*) and Au/Pt-50 (*broken curve*): solution treatment 15 min at a) 950 °C, b) 1050 °C and c) 1150 °C. Precipitation hardening performed at 650 °C, 700 °C, and 750 °C [1.218, p. 610]

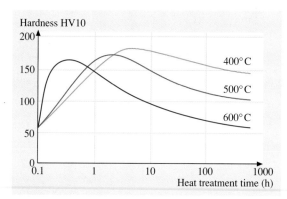

Fig. 3.1-231 Precipitation-hardening characteristic of Au–1% Ti alloy by annealing [1.252, p. 139]

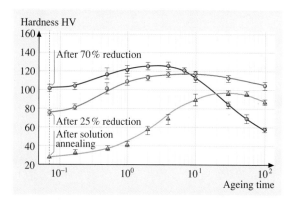

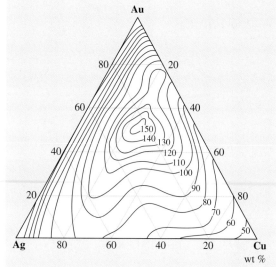

Fig. 3.1-232 Hardness of annealed and quenched Au–Ag–Cu alloys [1.253, p. 517]

alloys can be hardened by decomposition into Cu-rich Cu–Au and Ag-rich Ag–Au phases during annealing below the critical temperature of the miscibility gap and

Fig. 3.1-233 Hardness of the alloy AuSb0.3Co0.2 as a function of the reduction in thickness and of annealing time [1.254, p. 49]

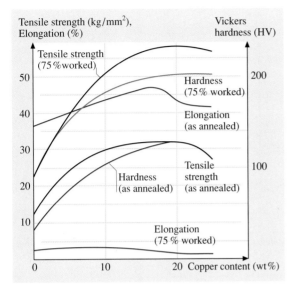

Fig. 3.1-235 Tensile strength and hardness of 18 carat Au–Ag–Cu alloys as a function of Cu content [1.218, p. 437]

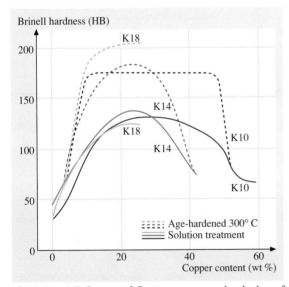

Fig. 3.1-236 Influence of Cu content on age hardening of 14 carat and 18 carat Au–Ag–Cu alloys [1.218, p. 437]

by formation of the ordered Au–Cu-phase at more than 75 wt% Au. Hardening of Au by alloying with rare-earth metals is described in detail in [1.248]. Grain refinement, applied especially to jewelry and dentistry alloys, is affected by the addition of 0.05–1 at.% of Ir, Ru, or Co [1.250, 251].

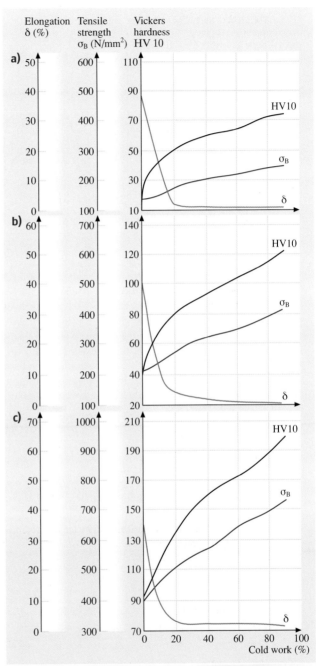

Fig. 3.1-234a–c Mechanical properties of (**a**) Au, (**b**) AuAg30, and (**c**) AuAg25Cu5 as a function of the reduction in thickness (%) [1.231, p. 58f.]

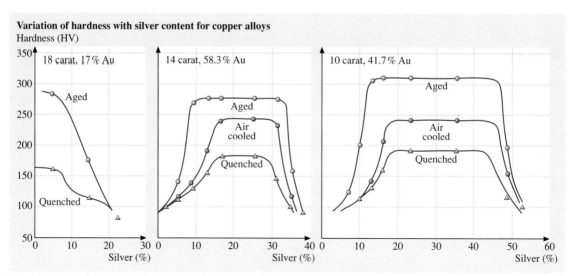

Fig. 3.1-237 Variation of hardness with silver content for Au−Ag−Cu alloys [1.217]

Electrical Properties

Tables 3.1-177–3.1-179 and Figs. 3.1-238, 3.1-239 [1.217, 231, 255] summarize the electrical properties of gold and gold alloys. The residual resistivity ratio for high purity gold amounts to 300. The electrical conductivity of gold alloys decreases in the low concentration range roughly linearly with the atomic concentration of the solute. Au alloys with 1.15 at.% Mn show increasing temperature coefficients of the electrical resistivity (positive TCR) due to the Kondo effect. This behavior is applied in resistance thermometers for temperature measurements below 20 K. Superconductivity occurs in intermetallic phases of Au−Ge with $2.99 < T_c < 3.16$ K and Au−Sn with $T_c = 1.25$ K [1.256, 257].

Table 3.1-177 Specific electrical resistivity $\rho = \rho_0 + \rho_i(T)$ of Au at different temperatures ($\rho_0 = 0.0222\,\mu\Omega$ cm) [1.217, p. 157]

T (K)	20	60	120	273.2	400	800	1000	1200
$\rho(\mu\Omega$ cm)	0.0138	0.287	0.796	2.031	3.094	6.742	8.871	11.299

at $T < 400$ K, $\rho_0 = 0.014\,\mu\Omega$ cm at $T > 400$ K

Table 3.1-178 Specific electrical resistivity (ρ_{25}) and temperature coefficient of resistivity (TCR) of Au−Pd and Au−Pt alloys [1.217, p. 158]

Solute		Content			
Content		80	60	40	30
Pd	ρ_{25}	9.8	17	30	26
	TCR	0.88	0.61	0.45	1.2
Pt	ρ_{25}	28	44	37	34
	TCR	0.28	0.26	0.82	0.8

$$\text{TCR}_{T_1\,T_2} = \frac{1}{\rho_1}\frac{\rho_{T_2} - \rho_{T_1}}{T_2 - T_1}$$

Table 3.1-179 Increase of atomic electrical resistivity of Au by alloying elements $\Delta\rho/C$ (μΩ cm/at.%) [1.217, p. 157]

Base element	Alloying elements	$\Delta\rho/C$ (μΩ cm/at.%)
Au	Ag	0.35
	Al	1.9
	Cd	0.60
	Co	6.2
	Cr	4.5
	Cu	0.4
	Fe	8
	Ga	2.2
	Ge	5.5
	Hg	0.4
	In	1.4
	Mn	2.4
	Mo	4
	Ni	0.8
	Pb	3.9
	Pt	1
	Rh	3.3
	Ru	1.6
	Sn	3.5
	Ti	13
	V	13
	Zn	0.94

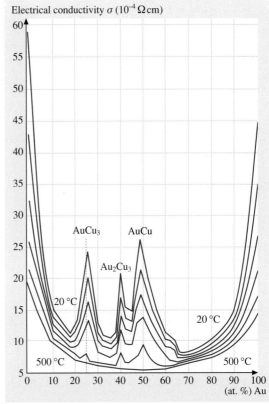

Fig. 3.1-238 Specific electrical conductivity of Au–Cu alloy phases [1.255, p. 619]

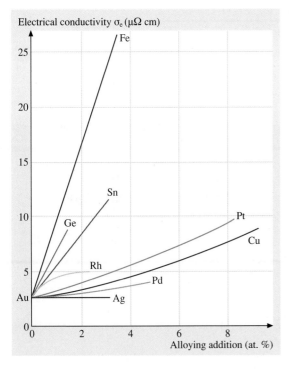

Fig. 3.1-239 Influence of alloying elements on the electrical conductivity of binary Au alloys [1.231, p. 50]

Thermoelectric Properties

Tables 3.1-180–3.1-183 [1.216] and Figs. 3.1-240–3.1-242 [1.216, 218] list the thermoelectric properties of gold and its alloys. Au–Fe and Au–Co alloys are used in thermocouples for measuring very low temperatures [1.258], Au–Pd and Au–Pd–Pt alloys in thermocouples working under highly corrosive conditions.

Magnetic Properties

Figure 3.1-243 [1.217] illustrates the metal's magnetic properties. Gold is diamagnetic. The magnetic susceptibility remains constant from 0 K to the melting point. Alloying of gold with B metals causes only weak variations compared to pure gold. In the range of continuous solid solutions, the molar susceptibilities remain negative, the alloys are diamagnetic. Ni, Pd, and Pt dissolve diamagnetically up to 25 at.%. Cr, Fe, and Mn give rise to paramagnetism. Magnetic transformations are reported

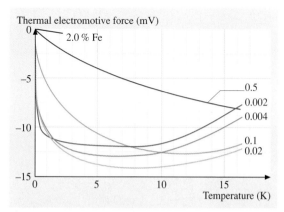

Fig. 3.1-240 Thermal electromotive force of Au–Fe alloys [1.218, p. 58]

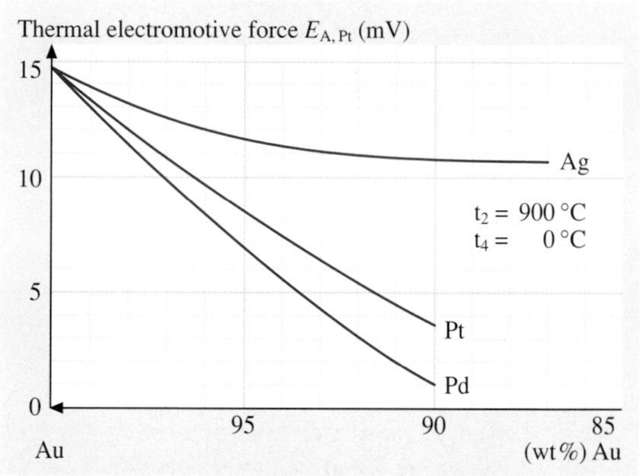

Fig. 3.1-241 Thermal electromotive force of Au alloys [1.216, p. 97]

Table 3.1-180 Absolute thermoelectric power of gold [1.216, p. 94]

Temperature (°C)	−255	−200	−160	−100	0	100	300	500	700
Thermoelectr. power (μV/K)	−0.93	−0.78	+0.80	+1.00	+1.1	+1.8	+3.1	+3.3	+3.7

Table 3.1-181 Thermal electromotive force E of Au and Pt (mV) at different temperatures; reference junction at 0 °C [1.216, p. 159]

T (°C)	−200	−100	−50	+100	+200	+400	+800
$E_{Au,Pt}$ (mV)	−0.39	−0.21	−0.10	0.77	1.834	4.623	12.288

Table 3.1-182 Thermal electromotive force of Au–Fe and Au–Co-alloys [1.216, p. 100]

T_1 (K)	T_2 (K)	Au–Co$_{2.1}$Cu (at.%)	Au–Fe$_{0.02}$Cu (at.%)
4.2	10	0.044	0.093
	20	0.173	0.208
4.2	40	0.590	0.423

Table 3.1-183 Thermocouples for very low temperatures [1.216, p. 97, 99]

- AuFe(0.03 at.% Fe)–chromel from 4.2 to 273 K
- AuCo(2.11 at.% Co)–AuAg (0.37 at.% Ag) or Cu from −240 to 0 °C
- AuFe(0.02 at.% Fe)–Cu from −270 to −230 °C

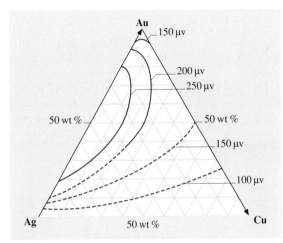

Fig. 3.1-242 Thermal electromotive force of Au–Ag–Cu alloys [1.216, p. 99]

for Au–Co alloys between ≈ 18 and 92 wt% Co at 1122 °C and for Au–Ni alloys between ≈ 3 and 95 wt% at ≈ 340 °C [1.259].

Thermal Properties

Data for thermal expansion and thermal conductivity of Au and Au alloys are listed in Tables 3.1-145–3.1-149. Table 3.1-184 [1.217] shows the recrystallization temperatures of gold of different purity. After 90% cold work, the hardness decreases by about 50%.

Optical Properties

For the optical properties of colored Au alloys, see Tables 3.1-151, 3.1-152 and Figs. 3.1-244–3.1-247 [1.260–263]. The reflectivity of gold shows a marked decrease at ≈ 550 nm in the visible range with a minimum of $R \approx 0.25$ in the near ultraviolet. Interband transitions occur at ≈ 2.17 eV. The reflected light contains all wavelengths above 550 nm, which accounts for the typical gold color.

Table 3.1-184 Recrystallization temperatures of Au 3N, 4N and 5N purity [1.217, p. 210]

Purity (%)	Decrease of hardness	
	50% $T_{recryst.}$	100% $T_{recryst.}$
99.9 (3N)	200	
99.99 (4N)	160	200
99.999 (5N)	112	149

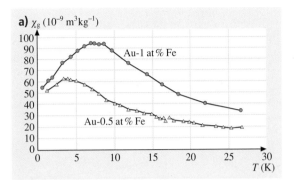

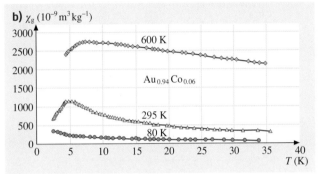

Fig. 3.1-243a,b Magnetic susceptibility of (a) Au–Fe and (b) Au–Co alloys [1.217, p. 169]

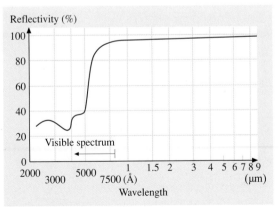

Fig. 3.1-244 Reflectance as a function of wavelength of pure Au [1.260, p. 53]

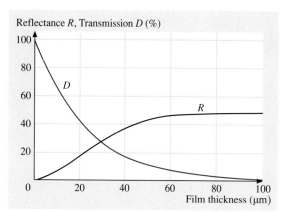

Fig. 3.1-245 Reflectance and transmission of thin Au films at $\lambda = 492\,\mu m$ [1.260]

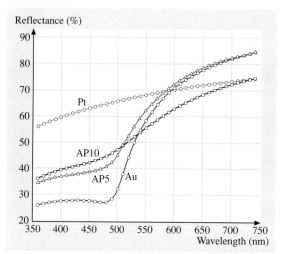

Fig. 3.1-246 Reflectance-wavelength curves for Au–Pt and binary Au–Pt alloys [1.261, p. 130]

Diffusion

Characteristic data are shown on Tables 3.1-153–3.1-156, 3.1-158 and Figs. 3.1-212, and 3.1-248 [1.217, 238].

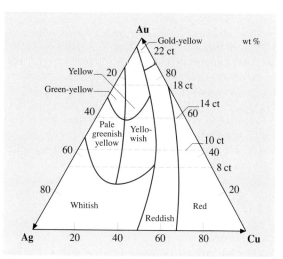

Fig. 3.1-247 Color ranges of Au–Ag–Cu alloys [1.262, p. 37]

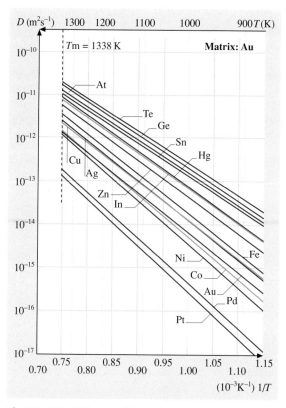

Fig. 3.1-248 Diffusion of impurities in Au [1.238, p. 191]

Chemical Properties

Figures 3.1-249 and 3.1-250 show that gold has the reduction potential of $E_0 = +1.42$ V for Au/Au^{3+}. At room temperature it is resistant against dry and wet atmospheres, H_2O, O_2, F, I, S, alkali, non-oxidizing acids, and ozone below 100 °C. It is dissolved in 3 HCl + 1 HNO$_3$, HCl + Cl$_2$ in acid concentration above 6 mol/l, in NaCN/H_2O/O_2, and other oxidizing solutions. Halogens generally attack gold, except for dry fluorine below 300 °C. Gold alloys are corrosion-resistant against acids if the base metal content is lower than 50% and also if each base metal present contains more than 50% of noble metal. Detailed information of chemical properties of Au and Au alloys are given in [1.217].

Gold and gold alloys (with Ag, Ir, Pt) and cationic gold (I) phosphines act as selective catalysts in hydrogenation, oxidation, and reduction reactions [1.264–266]. Nanometer-sized Au particles (≈ 5 nm) in the presence of ceria or a transition-metal oxide have superior catalytic activities [1.267–269, 269].

Special Alloys

Binary Alloys. The material Au–20 wt% Ag is used for low-voltage electrical contacts. Gold–copper alloys form the ordered phases Au$_3$Cu [60748-60-9], AuCu [12006-51-8], and AuCu$_3$ [12044-96-1]. Gold–nickel alloys decompose into gold-rich and nickel-rich solid

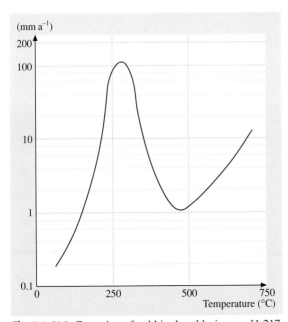

Fig. 3.1-249 Corrosion of gold in dry chlorine gas [1.217, p. 183]

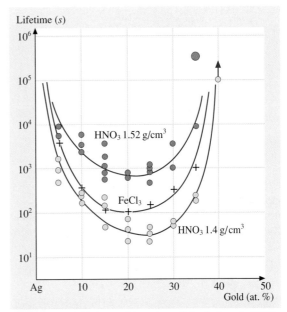

Fig. 3.1-250 Lifetime of Ag–Au solid solutions in HNO$_3$ and FeCl$_3$ solution [1.217, p. 197]

solution phases in a miscibility gap below 800 °C. The alloy Au–18 wt% Ni is a structural material for turbine blades in jet engines and nuclear and space technology materials.

Alloys of Au–Co(Fe, Ni) with 1–3 wt% Co, Fe, or Ni serve as hard and wear-resistant surface coatings on electrical contacts. The gold-cobalt alloy of Au–5 wt% Co is resistant against silver migration. The gold-platinum alloy of Au–10 wt% Pt is used for electrical contacts working under highly corrosive conditions. The high Pt content alloy Au–30 wt% Pt serves as a material for spinnerets for rayon and as a high-melting platinum solder ($T_{\text{liquidus}} = 1450$ °C, $T_{\text{solidus}} = 1228$ °C), additions $\approx 0.5\%$ of Rh, Ru, or Ir suppress segregation. Gold-Platinum alloys containing 40 to 65 wt% Au harden by quenching from 1100 °C and annealing at 500 °C to yield strengths up to ≈ 1400 N/mm^2. Au–1 wt% Ti (Figs. 3.1-222, 3.1-231 [1.246, 252, 270, 271]) is of importance for bonding wires, electrical conductors, and as hard high-carat gold alloy for jewelry. Strengthening can be induced by precipitation of the intermetallic compound Au$_4$Ti and by formation of highly-dispersed Ti oxide on annealing in an oxidizing atmosphere. The alloys Au–12 wt% Ge, Au–3.1 wt% Si, and Au–20 wt% Sn are low melting eutectic solders of high strength, corrosion resistance and stability against temperature cycling,

Table 3.1-185 Physical properties of the eutectic alloys Au−20Sn, Au−12Ge, and Au−3Si [1.272, p. 194]

Composition (wt%)	Melting point (°C)	Young's modulus versus temperature (GPa)				Thermal conductivity (W/m K)	Coefficient of thermal expansion (ppm/°C)
		−60 °C	23 °C	100 °C	150 °C		
Au−20Sn	280	59.5	59.2	48.5	35.8	57.3	15.93 ± 0.88 (−50 to 170)
Au−12Ge	356	69.8	69.3	68.2	62.7	44.4	13.35 ± 3.13 (10 to 250)
Au−3Si	363	77.0	83.0	82.8	83.0	27.2	12.33 ± 0.86 (10 to 250)

used for the hermetic sealing of electronic devices (Table 3.1-185 [1.272]).

Ternary and Higher Alloys. Au−Ag−Cu, Au−Ag−Ni and Au−Ag−Pd alloys are of major importance for jewelry and dentistry (Tables 3.1-186, 3.1-187) [1.250, 251]. The microstructures and thus the mechanical properties are determined by wide miscibility gaps. Additions of Zn and In serve to adjust the melting ranges. The high-carat Au alloy AuSb0.3Co0.2 can be hardened by cold working and precipitation annealing to 142 HV5 (Fig. 3.1-233) [1.254].

AuAg25Pt5, AuAg26Ni3, and AuCu14Pt9Ag4 are used for electrical contacts working under highly corrosive conditions. AuNi22Cr6 is a hard solder of high mechanical stability [1.231]. Au−Ag−Ge alloys of various compositions are solders applicable under H_2, Ar, or vacuum in melting ranges between 400 and 600 °C. Additions of 0.5–2 wt% Pd, Cd, or Zn improve their ductility [1.273–275].

Table 3.1-186 Basic compositions (wt%) of gold-based jewellery alloys [1.250, p. 271]

Jewelry alloys			
Colored gold fineness	Ag (wt%)	Cu (wt%)	
750	0–20	5–25	
585	5–35	5–35	
375	5–15	45–50	
(333)	5–40	25–60	
White gold fineness	Ag (wt%)	Cu (wt%)	Pd(Ni) (wt%)
750	0–10	0–10	10–20
585	0–25	5–30	5–20
375	0–35	5–50	5–20
(333)	0–35	10–50	5–25

Zinc max. 20%. Tin, indium, gallium each max. 4%

Table 3.1-187 Basic compositions of noble-metal-based dental alloys [1.251, p. 251]

Noble metal base	Most common alloying elements
Crown and bridge alloys	
Au	Ag, Cu, Pt, Zn
Au−Ag	Pd, Cu, Zn, In
Ag−Pd	Cu, In, Au, Zn
Porcelain fused to metal alloys	
Au	Pt, Pd, In, Sn
Au−Pd	Sn, In, Ga, Ag
Pd	Cu, Ga, Sn, In
Pd−Ag	Sn, In, Zn

3.1.10.3 Platinum Group Metals and Alloys

Characteristic properties of the platinum-group metals (PGM) Pd, Pt, Rh, Ir, Ru, and Os are their high chemical stability; mechanical strength; thermoelectric and magnetic behavior; and their catalytic activities in heterogeneous and homogeneous chemical reactions, automobile exhaust gas purification, and the stereospecific synthesis of enantiomeric compounds. Their melting temperatures, $T_m(\text{Os}) = 3045\,°\text{C}$, $T_m(\text{Pd}) = 1554\,°\text{C}$, hardness, brittleness, and the recrystallization temperatures decrease with increasing nuclear charge, while their thermal expansion and ductility increase.

The catalytic properties of the PGM in the heterogeneous catalysis are based on the moderate values of the heats of adsorption which correspond to the dissociation energies of the reactant molecules. Figure 3.1-251 [1.276] and Table 3.1-188 [1.218] give some values of the heat of adsorption and binding energies between adsorbates and surface atoms on various noble metal single crystals. The heat of adsorption increases for different orientations of the crystal surface planes of the fcc crystals in the order [111] < [100] < [110] (Table 3.1-189 [1.218]). The catalytic activities are element-specific for different reactions. Reactivity and selectivity of the reactions are presumably controlled by the dimensional fit between adsorbed molecules and catalyst surface, and the alloy composition. A survey of PGM catalyst activities is given in [1.217, 218, 243].

All platinum metals are paramagnetic ($\chi > 0$). The magnetic susceptibilities of palladium and platinum decrease with increasing temperature, the magnetic susceptibilities of rhodium, iridium, ruthenium, and osmium increase with increasing temperature (Fig. 3.1-272 [1.218]).

The platinum group metals occur jointly as alloys and as mineral compounds in placer deposits of varying compositions. Ru and Os are separated from the PGM mix by distillation of their volatile oxides, whereas platinum, iridium, palladium, and rhodium are separated by repeated solution and precipitation as complex PGM chlorides, or by solvent extraction and thermal decomposition to sponge or powder. PGM scrap is recycled by melting with collector metals (lead, iron, or copper) followed by element-specific extraction.

Table 3.1-188 Binding energies (kcal/mol) between adsorbates and surface atoms on noble metal single crystals [1.218, p. 267]

Precious metals	N	O	H	CO	NO
		Binding energy (kcal/mol)			
Ru (0001)			61	29	
Ir (111)	127	93	63	34	20
Pd (111)	130	87	62	34	31
Pt (111)	127		57	30	27
Ag (111)		80		6.5	25

Table 3.1-189 Heat of adsorption of diatomic molecules on different single crystals planes of various transition metals (kcal/mol) [1.218, p. 267]

Adsorption system	(111)	(100)	(110)
	Heat of adsorption (kcal/mol)		
O_2Pd	50	55	80
Co/Ni	27	30	30
Co/Pd	34	37	40
Co/Pt	30	32	32
H_2/Pd	21		24
H_2/W	37	33	35
N_2/Fe	51	53	49

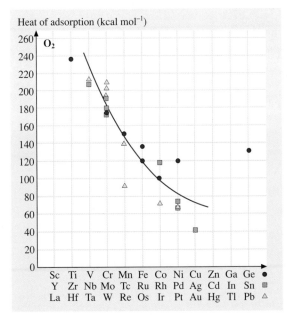

Fig. 3.1-251 Heat of adsorption of molecular oxygen on polycrystaline transition metal surfaces [1.218, p. 265]

Palladium and Palladium Alloys

Applications. Palladium and palladium alloys are important constituents of catalysts of chemical reactions and automobile exhaust gas cleaning, of electrical contacts, capacitors, permanent magnetic alloys, thermocouples, and for the production of high purity hydrogen. The low thermal neutron cross section permits their use in solders and brazes of nuclear structural parts. Classical applications are jewelry and dentistry alloys.

Commercial grades of palladium are sponge and powder in purities of 99.9 wt% to 99.95–99.98 wt% (ASTM (B 589-82)). High purity electronic grade is 99.99 wt%.

Production. Palladium sponge or powder are compacted by pressing and sintering. Melting and alloying is performed in electrical heated furnaces, vacuum arc, or by electron beam melting. Crucible materials are Al_2O_3 and MgO.

Phases and Phase Equilibria. Selected phase diagrams are shown in Figs. 3.1-252 – 3.1-257 [1.219]. Pd forms continuous solid solutions with all other noble metals and with Co, Cu, Fe, and Ni. Miscibility gaps exist in alloys with C, Co, Ir, Pt, Rh, and ternary Pd–Ag–Cu alloys (Fig. 3.1-257) [1.220]. All platinum-group metals (PGM) lower the γ–α transition temperature in Fe-alloys considerably (Fig. 3.1-343). Thermodynamic data are given in Tables 3.1-190 – 3.1-194. Numerous intermediate phases exist also in alloys with rare earth metals [1.216, 217, 217, 222]. The solubility of

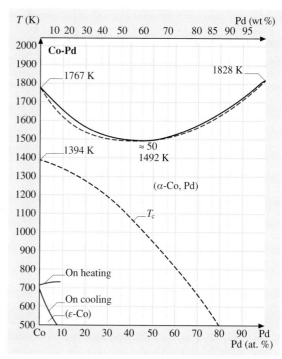

Fig. 3.1-253 Binary phase diagram Pd–Co [1.219]

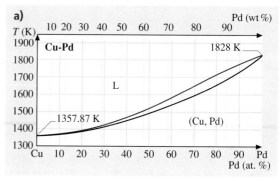

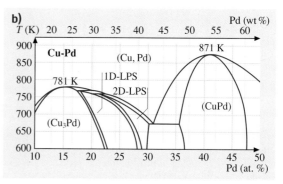

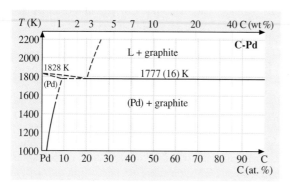

Fig. 3.1-252 Binary phase diagram Pd–C [1.219]

Fig. 3.1-254a,b Binary phase diagrams: Pd–Cu. (**a**) liquid–solid equilibrium; (**b**) low temperature (600–900 °C). 1-D LPS = one-dimensional long-period superstructure; 2-D LPS = two-dimensional long-period superstructure [1.219]

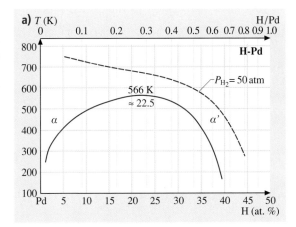

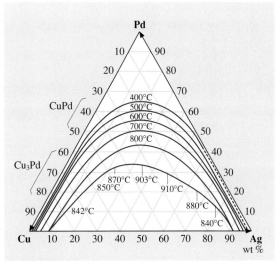

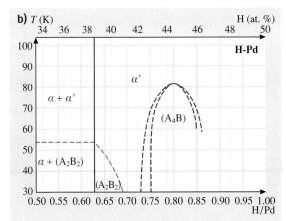

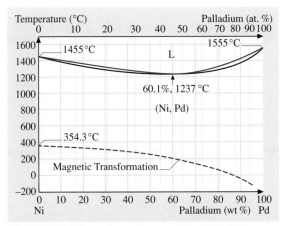

Fig. 3.1-255a,b Binary phase diagrams: Pd–H. (a) Phase diagram; (b) low temperature phase [1.219]

Fig. 3.1-256 Binary phase diagram: Pd–Ni [1.219]

Fig. 3.1-257 Miscibility gap in the Pd–Ag–Cu alloy system [1.220]

Table 3.1-190 Molar heat capacities of solid PGMs [1.222, p. 219]

Element	$c_p = 4.1868(a + 10^{-3}bT + 10^{-5}T^2)$ J/K			Temperature range (K)
	a	b	c	
Ir	5.56	1.42	–	298–1800
Os	5.69	0.88	–	298–1900
Pd	5.80	1.38	–	298–1828
Pt	5.80	1.28	–	298–2043
Rh	5.49	2.06	–	298–1900
Ru	5.20	1.50	–	298–1308
Ru	7.20	–	–	1308–1773
Ru	7.50	–	–	1773–1900

Table 3.1-191 Latent heat and temperatures of transition of Pd and Pt intermediate compounds [1.222, p. 189]

Phase	N_2	Transition	T_t	L_t
CuPt	50	order-disorder	800	3810
Cu_3Pt	20	order-disorder	610	1968
Pd_3Sb	25	order-disorder	950	10300

T_t = transition temperature, L_t = latent heat of transition

carbon rises from 0.04 wt% at 800 °C to 0.45 wt% at 1400 °C, with the hardness increasing from 80 to 180 HV_{25g} [1.277]. The continuous series of solid solutions of Pd–H-alloys (Fig. 3.1-258) [1.277] splits up below 295 °C into a fcc palladium-rich β phase and an fcc hydrogen-rich phase, forming a miscibil-

Table 3.1-192 Thermodynamic data of Pd [1.217, p. 108]

T (K)	c_p (J/K mol)	S (J/K mol)	H (J/mol)	G (J/mol)	p (at)
298.15	25.99	37.823	0	−11.277	5.97×10^{-60}
400	26.706	45.568	2.686	−13.541	3.65×10^{-43}
600	27.768	56.602	8.136	−25.825	8.92×10^{-27}
800	28.827	64.733	13.704	−37.903	1.11×10^{-18}

T = Temperature, c_p = specific heat capacity, S = Entropy, H = Enthalpy, G = free Enthalpy, p = partial pressure of the pure elements

Table 3.1-193 Enthalpy of formation H_T of Pd and Pt alloys at temperatures of reaction [1.216, p. 89]

Base metal	Alloy comp.	Temp. (K)	20 at.%	40 at.%	60 at.%	80 at.%
Enthalpy of formation H_T						
Pd	Ag	915				1050
		1200	283	897	1290	887
Pt	Co	914	1680	2580		
	Cu	1625	(30%):3110		3375	(90%):1950
	Fe	1123	−680	0	1600	1800

Table 3.1-194 Maximum hydrogen inclusion by platinum-group metals in ml/g (elements) [1.217, p. 616]

Element	Max H_2 (ml/g)	Composition
Ru	123	$RuH_{1.1}$
Rh	24	$RhH_{0.2}$
Pd	75	$PdH_{0.7}$
Os	75	$OsH_{1.2}$
Ir	35	$IrH_{0.6}$
Pt	2.4	$PtH_{0.04}$

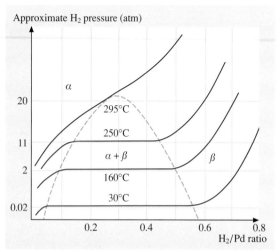

Fig. 3.1-258 Hydrogen pressure in the Pd−H system [1.278, p. 130]

ity gap which broadens with decreasing temperature. The equilibrium hydrogen-pressure at 295 °C amounts to 19.87 atm with 21 at.% hydrogen. The α-phase takes hydrogen up to 1300 times of the volume of palladium, corresponding 50 at.% hydrogen. Further quantities up to 2800 times of the Pd-volume can be loaded by cathodic deposition. The lattice parameters increase with increasing hydrogen content from 3.891 A to 4.06 A at 75 at.% hydrogen. The dissolved hydrogen moves easily and diffuses quickly through thin Pd-membranes. This effect is used for the production of high-purity Pd and for the separation of H isotopes.

Thermal cycling of Pd−H-alloys in the duplex phase causes brittleness due to stresses generated by changes of the lattice dimensions for different quantities of dissolved hydrogen. Palladium-silver alloys with 20−25 wt% silver dissolve higher amounts of hydrogen than pure palladium (Fig. 3.1-259).

For composition and crystal structures, see Tables 3.1-195 and 3.1-196 [1.217, 219, 277]. Primary solid solutions have the fcc structure of Pd. The lattice parameters correspond with few exceptions roughly to Vegard's law. Superlattices occur in alloys with Cu, Fe, Nb, V, in atomic ratios from 1:1, 2:1, and 3:1 (Tables 3.1-123, 3.1-197).

Ordered A_3B-phases of Pd−Mn and Pd−Fe alloys show higher solubility for hydrogen than the disordered phases. In Pd−Mn alloys, hydrogen uptake lowers the temperature of the ordering process.

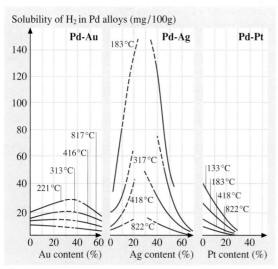

Fig. 3.1-259 Solubility of hydrogen at 1 atm in Pd–Au, Pd–Ag, and Pd–Pt alloys [1.279, p. 44]

Table 3.1-196 Structures of platinum-group metal oxides [1.219, 277]

Oxide	Structure type	Unit cell dimensions (Å)		
		a	b	c
α-PtO_2	primitive hexagonal	3.08		4.19
β-PtO_2	primitive othorhombic	4.486	4.537	3.138
Pt_3O_4	primitive cubic	5.585		
PdO	tetragonal	3.043		
PdRhO	hexagonal	5.22		6.0
Rh_2O_3 (LT)	hexagonal (corundum)	5.108		13.87
Rh_2O_2	tetragonal (rutile)	4.4862	3.0884	
PdO	tetragonal PdO	3.03		5.33
PdO_2	tetragonal TiO_2 (rutile)	4.483		3.101

LT = low temperature modification

Table 3.1-195 Structure and lattice parameters of selected intermediate Pd compounds [1.217, p. 118]

Phase	Pearson Symbol	a (nm)	b (nm)	c (nm)	c/a	Remarks	Concentration x $A(1-x)B(x)$
C–Pd	cF4	0.3735					0.97
Co–Pd	cF4	0.3735					
Cu–Pd	cF4	0.377				HT	0.5
Cu_3Pd	tP4	0.3701	0.3666	0.9905			
Cu_3Pd	tP28	0.371	2.5655	6.9151			0.19
Fe–Pd	cF4	0.38873					0.936
FePd	tP4	0.386		0.3731	0.9666		
$FePd_3$	cP4	0.3851					
H_3Pd_5	cF*	0.4018					
H_4Pd_3	cP*	0.2995				HT > 923 K	
H_4Pd_3	tP4	0.2896		0.333	1.1499		
Ni–Pd	cF4	0.373				298 K	0.474
Pd_3Zr	hP16	0.5612		0.9235	1.6456		

HT = high temperature modification

Table 3.1-197 Superlattice structures of the platinum-group metals [1.280, p. 22]

Ordered structure type	Examples
Tetragonal ($L1_0$-type)	CdPt, CoPt, Cu_4Pd, FePd, FePt, MnPt, NiPt
Face-centered cubic ($L1_2$-type)	$CoPt_3$, Cu_3Pt, $FePd_3$, $FePt_3$, Fe_3Pt, $MnPt_3$, Ni_3Pt
Body-centered cubic (B2-type)	BePd, CuPd, FeRh, RhSc, RhTi
Rhombohedral ($L1_1$-type)	CuPt (unique)
Close-packed hexagonal (DO_{19}-type)	Pt_3U

Mechanical Properties. Characteristic data are shown in Tables 3.1-198–3.1-202 and Figs. 3.1-260–3.1-266 [1.217, 220, 231, 277]. At room temperature Pd is very ductile and can be easily rolled or drawn to form a sheet, foil, and wire. The recrystallization temperatures (Table 3.1-212) depend on purity grade, degree of cold forming and annealing time. Strengthening is affected by solid solution and by order hardening in alloys, forming superlattice structures. Solid solution hardening is also effected by alloying with rare earth metals in concentrations of 0.1–0.6 at.%.

Table 3.1-198 Modulus of elasticity in crystal directions (GPa) [1.217, p. 214]

$E\langle 100\rangle$	$E\langle 110\rangle$	$E\langle 111\rangle$
65	129	186

Table 3.1-199 Elastic constants of Pd [1.217, p. 216]

T (°C)	$c11$	$c12$	$c14$
−273	234.1	176.1	71.2
7	226.2	175.2	71.5

Table 3.1-200 Mechanical properties of Pd (99.9%) at different temperatures (°C) [1.217, p. 215]

T (°C)	E (GPa)	R_m (MPa)	A (%)	$R_{p0.2}$ (MPa)	HV
20	124	190	25	50	50
250	121	180	16	90	47
500	117	68	94	50	39
750	98	28	42	20	17

A = Elongation, E = Modul of elasticity,
R_p = Limit of proportionality, HV = Vickers hardness,
R_m = Tensile strengh

Table 3.1-201 Tensile strength (MPa) of binary Pd and Pt alloys [1.217, p. 223]

Alloying element	Weight % of alloying element					
	2		5		10	
	Pd	Pt	Pd	Pt	Pd	Pt
Ag	230	370	270	550	310	830
Au	200	200	220	320	230	540
Co	190	360	210	–	270	–
Cu	240	290	280	400	300	–
Fe	200	360	230	–	340	–
Ni	190	260	219	450	270	640
Pd	–	170	–	190	–	200
Pt	200	–	220	–	240	–
Rh	230	170	290	230	380	330
Ru	230	250	350	380	–	550

Table 3.1-202 Mechanical properties of Pd by cold forming as a function of reduction in thickness V in % [1.217, p. 215]

V (%)	R_m (MPa)	A (%)	HV
0	220	60	50
20	250	170	80

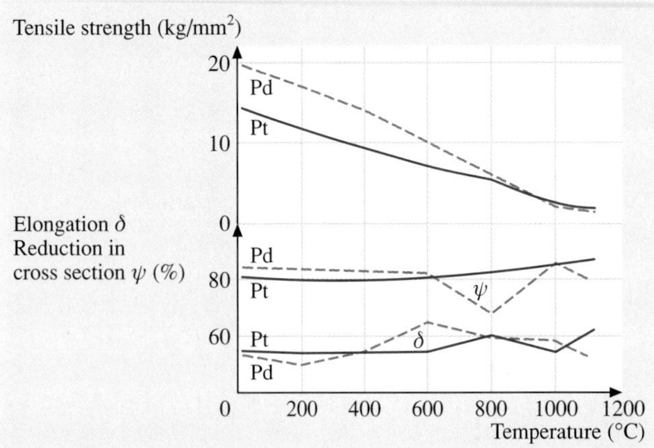

Fig. 3.1-261 Tensile strength of Pd and Pt at different temperatures [1.220, p. 82]

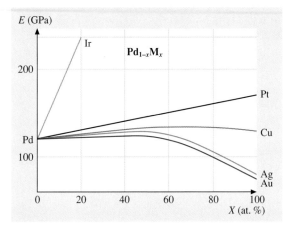

Fig. 3.1-260 Modulus of elasticity of Pd alloys [1.217, p. 216]

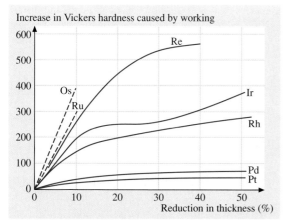

Fig. 3.1-262 Work hardening of the platinum group metals [1.277, p. 93]

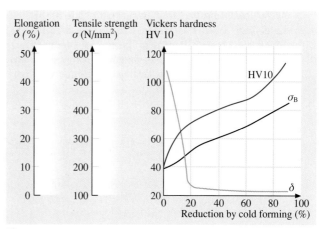

Fig. 3.1-263 Work hardening of Pd (99.99%) (% reduction by cold forming) [1.231, p. 71]

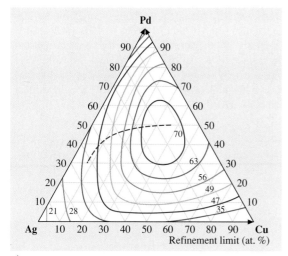

Fig. 3.1-265 Tensile strength of Pd–Ag–Cu alloys. *Dashed line*: refinement limit [1.220, p. 268]

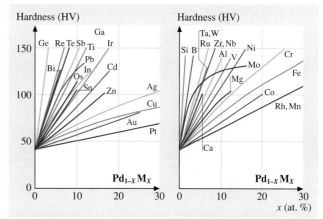

Fig. 3.1-264 Solid solution hardening of Pd by various elements [1.217, p. 217]

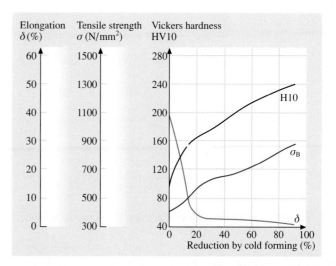

Fig. 3.1-266 Work hardening of PdCu15 alloys (% reduction by cold forming) [1.231, p. 72]

Electrical Properties. In Tables 3.1-203 – 3.1-205 [1.217] and Figs. 3.1-267, 3.1-268 [1.228, 231] characteristic data are shown. Pure Pd shows no superconductivity, PdH and some intermetallic compounds are superconducting at low critical temperatures, e.g., $T_c(\text{Bi}_2\text{Pd}) = 3.7$ K.

Table 3.1-203 Residual electrical resistivity ratio (RRR) of pure noble metals [1.217, p. 156]

273.2 K/4.2 K							
Ru	Rh	Pd	Ag	Os	Ir	Pt	Au
25 000	570	570	2100	400	85	5000	300

Table 3.1-204 Increase of atomic resistivity of Pd and Pt [1.217, p. 158]

Basic element	$\Delta\rho/C$ ($\mu\Omega$ cm/at.%)
Pd	Ag 1.17 Al 2.17 Au 0.65 B 1.43 Bi 5.45 Cd 1.36 Co 2.04 Cr 2.98 Cu 1.35 Fe 2.06 Ga 2.25 Ge 4.13 In 1.96 Ir 7.0 Mn 167 Mo 4.49 Ni 0.72 Pb 3.5 Pt 0.88 Rh 1.67 Ru 3.3 Sn 2.89 V 3.2 Zn 1.73 Zr 2.49
Pt	Ag 2.2 Au 1.3 Be 3 Co 1.7 Cr 6.8 Cu 3 Fe 3.9 In 3.4 Mn 2.95 Mo 6.2 Nb 5.4 Ni 0.9 Os 2.4 Pd 0.6 Rh 1.0 Ru 2.4 Sn 3.9 W 5.7 Zr 4.7

Table 3.1-205 Specific electrical resistivity ($\mu\Omega$ cm) of Pd at different temperatures (K) [1.217, p. 156]

T (K)	90	175	273	500	800	1300
	2.147	5.821	9.725	17.848	26.856	38.061

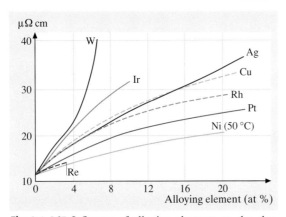

Fig. 3.1-267 Influence of alloying elements on the electrical conductivity of Pd [1.231, p. 67]

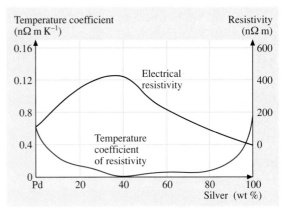

Fig. 3.1-268 Electrical resistivity and temperature coefficient of resistivity of Pd–Ag alloys as a function of Ag content [1.228, p. 702]

Thermoelectric Properties. Tables 3.1-206 – 3.1-209 [1.216, 217] and Figs. 3.1-269, 3.1-270 [1.216, 218] give data of absolute thermoelectric power, thermal electromotive force of pure Pd and Pd alloys at different temperatures. Special alloys for thermocouples with high corrosion resistance are shown in Table 3.1-210 [1.217].

Table 3.1-206 Absolute thermoelectric power (μV/grd) of the platinum-group metals at different temperatures [1.216, p. 89]

Metal	Absolute thermoelectric power (μV/grd) T (°C)								
	−255	−200	−100	−20	0	100	300	500	800
Pd	+1.02	+3.96	−3.16	−7.94	−9.6	−13.4	−18.8	–	−35
Pt	+1.8	+5.9	+0.1	−3.6	−4.4	−7.3	−10.9	−14.0	−18.6
Rh	+1.6	+1.8	–	+1.7	–	+1.2	+0.3	−0.3	–
Ir	–	–	+1.8	–	+1.5	+0.9	−0.3	−1.3	–
Ru	–	–	–	–	–	–	−32.5	−42	−43

Table 3.1-207 Thermal electromotive force $E_{A,Pt}$ (mV) of the thermocouples of noble metals and pure Pt at different temperatures, reference junction at 0 °C [1.217, p. 159]

T (°C)	Ru	Rh	Ir	Pd	Ag	Au
−100		−0.32	−0.35	0.48	−0.21	−0.21
0	0	0	0	0	0	0
100	0.684	0.70	0.660	0.570	0.740	0.770
300	2.673	2.68	2.522	−1.990	3.050	3.127
600	6.485	6.77	6.201	−5.030	8.410	8.115
900	11.229	12.04	10.943	−9.720	10.943	14.615
1200	16.864	18.42	16.665	–	–	–
1400	–	22.56	20.819	−20.41	–	–

Table 3.1-208 Thermal electromotive force $E_{A,Pt}$ (mV) of Pd alloys at different temperatures, reference junction at 0 °C [1.217, p. 160]

Alloying element	T (°C)	Composition (wt% Ir)				
		10	30	50	70	90
Ag	100	−1.1	−2.4	−3.3	−0.5	0.1
	1000	−23.5	−44.4	−45.8	−11.5	–
Au	100	−1.0	−1.7	−2.7	−2.7	0
	1000	−14.5	−24.1	−38.5	−33.5	3.0
	1300	−22.0	−34.0	−52.0	−48.0	–
Cu	100	−1.05	−1.49			
Ir	100	2.01	2.02			
	1000	22.1	26.1			
Ni	100	−0.80	−1.47	−1.75	−1.75	−1.65
	1000	−9.4	−9.4	−10.2	−11.6	−11.5
Pt	100	0.32	0.83	0.75	0.53	0.22
	1000	−2.1	7.8	9.4	7.8	4.6
	1300	−5.2	8.0	11.7	10.7	5.3

Table 3.1-209 Basic data of thermal electromotive force of thermocouples according to Table 3.1-210 [1.217, p. 474]

T (°C)	Th.-C.1 (mV)	Th.-C.2 (mV)	Th.-C.3 (mV)
100	3.31	4.6	3.3
400	15.70	21.5	15.4
600	24.70	34.2	24.6
800	33.50	46.9	33.4
1000	41.65	59.6	41.3

Table 3.1-210 Palladium alloys for thermocouples (Th.-C.) of high corrosion resistance [1.217, p. 472]

Th.-C.	1.:	AuPd40	Pd38Pt14Au3
Th.-C.	2.:	AuPd46	PtIr10
Th.-C.	3.:	AuPd35	PtPd12.5

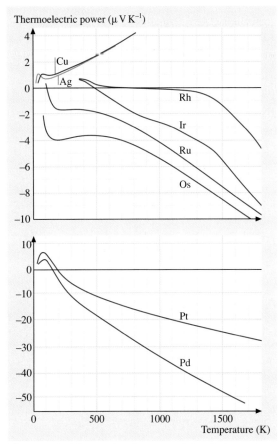

Fig. 3.1-269 Thermoelectric power of the platinum group metals [1.218, p. 58]

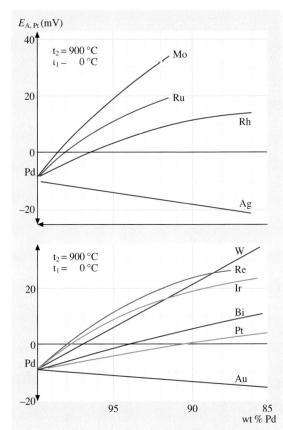

Fig. 3.1-270 Thermal electromotive force of Pd alloys at 900 °C (reference junction at 0 °C) [1.216, p. 97]

Magnetic Properties. All PGMs show magnetostriction in a magnetic field. The reversible change of length is proportional to the square of the applied magnetic field (Table 3.1-211) [1.217, 218]. The paramagnetic susceptibilities of Pd and Pd alloys decrease with increasing temperature (Figs. 3.1-271, 3.1-272) [1.217, 218]. Alloying with 0.05 wt% Rh raises the susceptibility from 88×10^{-10} m^3/mol to 160×10^{-10} m^3/mol. Pd–Cu alloys are diamagnetic up to 50 at.% Pd. The susceptibilities of the ordered phases in this system are higher than those of the disordered solid solution phase. The paramagnetism of Pd decreases by dissolution of H$_2$ to reach zero at PdH$_{0.66}$ and above. Partial ordering within FePd raises its coercive field from the disordered value of 2 to 260 Oe [1.281, 282].

Metal	S_l
Ru	−1.4
Rh	11
Pd	−39.4
Ir	3.8
Pt	−32
Rh$_{0.50}$Ir$_{0.50}$	9.5
Rh$_{0.50}$Pd$_{0.50}$	27
Ir$_{0.60}$Pd$_{0.40}$	13.4
Pd$_{0.67}$Pt$_{0.33}$	−17.4
Pd$_{0.33}$Pt$_{0.67}$	−79

Table 3.1-211 Magnetostriction of platinum-group metal and platinum-group metal alloys, expressed by the factor S_l of proportionality according to $\Delta l/l = S_l H^2$ [1.217, p. 161]

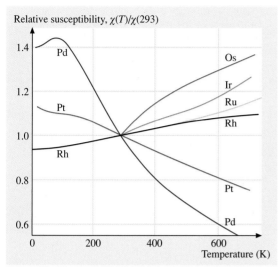

Fig. 3.1-271 Temperature dependence of the magnetic susceptibility of the platinum group metals [1.218, p. 98]

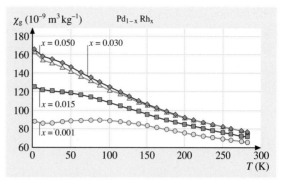

Fig. 3.1-272 Temperature dependence of the magnetic mass susceptibility of Pd–Rh alloys [1.217, p. 169]

Thermal Properties. Selected data of thermal conductivity and thermal expansion of PGM and PgAg alloys are given in Tables 3.1-212–3.1-215, Fig. 3.1-273. FePd-alloys exhibit around the Fe_3Pd stoechiometry in the disordered state zero coefficient of thermal expansion (Invar effect) [1.281, 282].

Table 3.1-212 Recrystallization temperatures of platinum-group metal (0 °C) (Depending on purity, degree of cold forming an annealing time) [1.217, p. 216]

Metal	Recryst. temperature (°C)
Ir	1200–1400
Pd	485–600
Pt	350–600
Rh	700–800
Ru	1200–1300

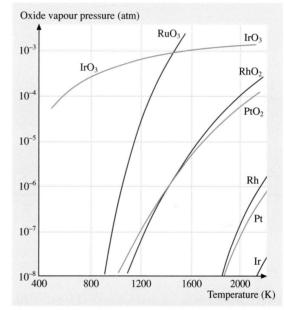

Fig. 3.1-273 Vapor pressures of platinum group metal oxides [1.277, p. 178]

Table 3.1-213 Thermal conductivity of platinum-group metal at different temperatures [1.217, p. 153]

Temperature (°C)	Thermal conductivity (W/m K)							
	Pd	Pt	Rh	Ir	Ru[a]	Ru[b]	Ru[p]	Os
100	76	85.6	185	–	140	180	150	–
273	75.6	75.0	153	149	110	134	119	88
600	79.0	73.0	135	130	95	129	105	85
800	83.0	74.8	126	125	87	112	96	–
1200	88.2	83.2	118	117	77	101	83	–

[a] vertical to the crystal c axis,
[b] parallel to the crystal c axis
[p] polycrystalline

Table 3.1-214 Thermal expansion coefficient of the platinum-group metals [1.217, p. 154]

Temperature (°C) Metal	Thermal expansion coefficients (10^{-6} K^{-1})									
	Pd	Pt	Rh	Ir	Ru^a	Ru^b	Ru^p	Os^a	Os^b	Os^p
323	–	–	–	–	5.9	8.8	6.9	4.0	5.8	4.8
373	11.9	9.1	8.5	6.7	–	–	–	–	–	–
423	–	–	–	–	6.1	9.3	7.2	4.3	6.2	5.0
473	12.1	9.2	9.0	–	–	–	–	–	–	–
623	–	–	–	–	6.8	10.5	8.0	4.0	7.1	5.7
673	12.6	9.5	9.6	–	–	–	–	–	–	–
723	–	–	–	–	7.2	11.0	8.4	5.3	7.6	62
823	–	–	9.6	–	–	11.7	8.8	5.8	8.3	6.9
1073	13.4	10.0	10.3	–	–	–	–	–	–	–

^{a} vertical to the crystal c axis,
^{b} parallel to the crystal c axis
^{p} polycrystalline

Table 3.1-215 Thermal expansion coefficient of Pd–Ag alloys [1.217, p. 154]

Temp. range (K) 373–473	
Pd-content (%)	Thermal expansion coefficient (10^{-6} K^{-1})
20	16.2
50	14.7
80	12.4

Optical Properties. In Table 3.1-216 and Fig. 3.1-274 characteristic data are given. The optical reflectance of Pd is increased by alloying with Ru (Fig. 3.1-275).

Table 3.1-216 Spectral degree of emission ε of the platinum-group metals at different temperatures [1.217, p. 171]

	Surface	Temperature (°C)	Spectral emission ϵ
Ru^b	solid	1000	0.421
	solid	2000	0.314
Os^b	solid	1000	0.526
	solid	2000	0.383
Rh^a	solid	< 1966	0.29
	liquid	> 1966	0.3
Ir^a	solid	927–2027	0.3
	solid	2000	0.383
Pd^a	solid	900–1530	0.33
	liquid	1555	0.37
Pt^a	solid	1000	0.371
	solid	1400	0.421
	liquid	1800	0.38

^{a} 650 nm; ^{b} 655 nm

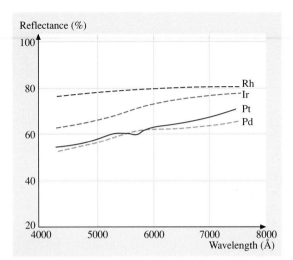

Fig. 3.1-274 Optical reflectance of the platinum group metals [1.220, p. 78]

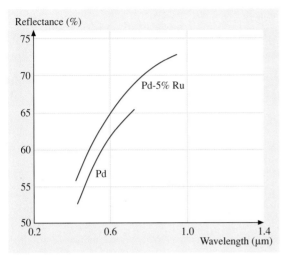

Fig. 3.1-275 Optical reflectance (%) of PdRu5 alloy [1.228, p. 700]

Diffusion. Data for selfdiffusion, diffusion of tracer elements and of hydrogen and oxygen are shown in Tables 3.1-156, 3.1-158, 3.1-217.

Carbon diffuses very rapidly through Pd at elevated temperatures in presence of a concentration gradient on the surface.

Table 3.1-217 Self-diffusion in pure platinum-group metals [1.217, p. 149]

	Element	D^0	O	T (K)
Lattice diffusion	Ir(S)	0.36	438.8	2092–2664
	Pd(S)	0.205	266.3	1323–1773
	Pt	0.33	285.6	1598–1837
	Pt(S)	0.05	257.6	850–1265
Surface diffusion	Rh(111)	4×10^{-6}	174	1200–1500 (Vac.)
	Pt	4×10^{-7}	108	1160–1580 (Vac.)

Chemical Properties. Pd has the reduction potential of $E_0 = 0.951$ for Pd/Pd^{2+}. It is resistant against reducing acids and in oxydizing media above pH 2. Alkali melts attack above $\sim 400\,°$C. In oxygen atmosphere between 400 and 800 °C are thin PdO-surface layers formed, which dissociate above 800 °C. Above 1100 °C occur increasing weight losses by evaporation (Fig. 3.1-276).

Catalysis: Pd and Pd alloys are effective catalysts in numerous chemical reactions. In heterogenous catalysis,

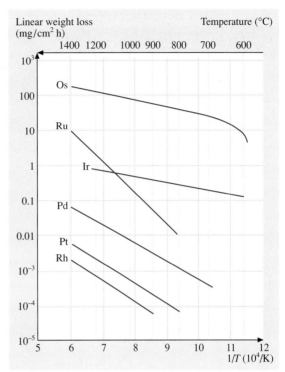

Fig. 3.1-276 Weight losses of the platinum group metals at annealing on air [1.217, p. 183]

PGM are applied in form of wire nets and of powders with high specific surfaces (20 to 1000 m^2/g, "platinum black", "palladium black") on carbon or Al$_2$O$_3$ supports. Automotive gas cleaning catalysts use of Pd–Pt–Rh alloys in different compositions.

Special Alloys. Tables 3.1-159, 3.1-160, 3.1-218 show typical compositions of Pd containing brazing alloys, Table 3.1-219 Pd containing jewelry alloys. PdAg40 has a very low temperature coefficient of resistivity (0.00003/°C between 0 and 100 °C, electrical resistivity 42 μcm). It is used for precision resistance wires (Fig. 3.1-268). Pd60Ni35Cr5 is corrosion resistant against molten salt mixtures up to 700 °C, suited for brazing graphit, Mo and W.

Ti–Pd–Ni and Fe–Pd alloys show shape memory effects. Partial replacement of Pd in the alloy Fe30 at.% Pd by > 4 at.% Pt decreases the temperature of the f.c.c./f.c.t. martensite transformation and effects strengthening.

Table 3.1-218 Physical properties of some technical Pd and Pt alloys [1.231, p. 67]

Material	Density (g/cm^3)	Melting point (interval) (°C)	Electrical conductivity (m/Ω mm^2)	Temperature coefficient of electrical resistance (10^3 K^{-1})	Modulus of elasticity (kN/mm^2)
Pt(99.9)	21.45	1773	9.4	3.92	16–17
PtIr5	21.5	1774–1776	4.5	–	18.5–19.5
PtIr10	21.6	1780–1785	5.6	2.0	ca. 22
PtRu10	20.6	ca. 1800	3.0	0.83	ca. 23.5
PtNi8	19.2	1670–1710	3.3	1.5	ca. 18
PtW5	21.3	1830–1850	2.3	0.7	ca. 18.5
Pd(99.99)	12.0	1552	9.3	3.77	ca. 12.5
PdCu15	11.3	1370–1410	2.6	0.49	ca. 17.5
FdCu40	10.4	1200–1230	3.0	0.28	ca. 17.5
PdNi5	11.8	1455–1485	5.9	2.47	ca. 17.5

Table 3.1-219 Composition and melting temperature range of selected Pd-jewellery alloys [1.217, p. 511]

Alloy	Melting temperature range (°C)
Pd95Cu3Ga2	1340–1400
Pd95Cu5	1400–1460
Pd95Ni5	1450–1490
Pd50Ag47.5Cu2.5	1200–1280

Platinum and Platinum Alloys

Applications. Platinum and platinum alloys are important constituents of catalysts (chemistry, automotive exhaust gas cleaning, fuel cells), sensor materials (thermocouples, resistance thermometers), strong permanent magnet alloys, magnetic and magnetooptical (memory) devices, high temperature and corrosion resistant structural parts, and electrical contacts and connecting elements. Classical applications are jewelry and dentistry alloys.

Commercial grades are sponge and powder in purities varying from minimum 99.9% to 99.95% (ASTM B 561-86). High purity electronic grade is 99.99%.

Production. Platinum sponge or powder are compacted by pressing and sintering. Melting and alloying is done in electrical heated furnaces in Al$_2$O$_3$ or MgO crucibles, by vacuum arc and by electron beam melting 99.98%.

Phases and Phase Equilibria. Selected phase diagrams are shown in Figs. 3.1-277–3.1-281 [1.219]. Thermodynamic data are given in Tables 3.1-190 – 3.1-193 and 3.1-220 [1.216, 217, 222]. For compositions and crystal structures, see Tables 3.1-196, 3.1-219, 3.1-222, 3.1-221 [1.217, 219]. Platinum forms continuous solid

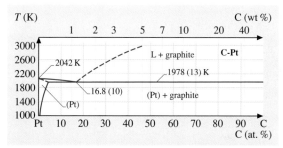

Fig. 3.1-277 Binary phase diagram Pt–C [1.219]

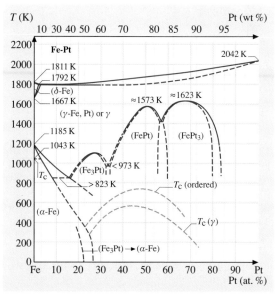

Fig. 3.1-278 Binary phase diagram Pt–Fe (*dash-dotted line*: Curie temperature) [1.219]

Table 3.1-220 Thermodynamic data of Pt [1.217, p. 109]

T (K)	c_p (J/K mol)	S (J/K mol)	H (J/mol)	G (J/mol)	p (at)
298.15	25.857	41.631	0	12.412	8.26×10^{-92}
400	26.451	49.314	2.664	-17.961	1.30×10^{-66}
800	28.593	68.313	13.677	-40.973	9.75×10^{-30}
1400	31.731	85.111	31.776	-87.38	5.84×10^{-14}

T = Temperature, c_p = specific heat capacity, S = Entropy, H = Enthalpy, G = free Enthalpy, p = partial pressure of the pure elements

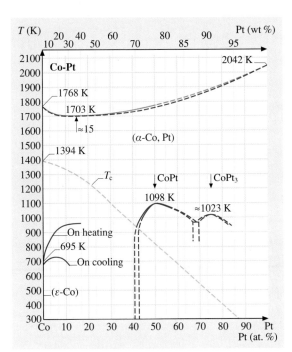

Fig. 3.1-279 Binary phase diagram Pt–Co [1.219]

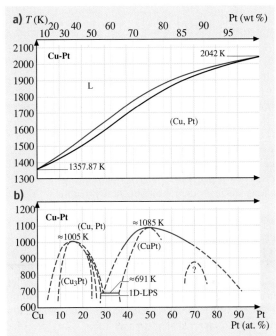

Fig. 3.1-280a,b Binary phase diagrams Pt–Cu. (**a**) Liquid–solid. (**b**) Solid–solid (1-D LPS = one-dimensional long-period superstructure) [1.219]

Table 3.1-221 Crystal structure and lattice parameters of intermediate phases of Pt oxides [1.219]

Phase	Structure	Type	a (nm)	b (nm)	c (nm)
PtO	tetragonal	PtO	0.304		0.534
Pt$_3$O$_4$	cubic	Pt$_3$O$_4$	0.6226		
PtO$_2$	orthorhombic	Fe$_2$O$_3$	0.4533	0.4488	0.3138
PtO$_2$	hexagonal		0.310		0.435

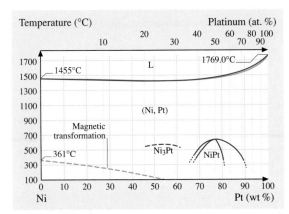

Fig. 3.1-281 Binary phase diagram Pt–Ni [1.219]

solutions with all other noble metals and with Co, Cu, Fe, and Ni. Miscibility gaps exist with C, Co, Ir, Pt, and Rh. Primary solid solutions have fcc structure and the lattice parameters correspond with few exceptions roughly to Vegard's law. Numerous intermediate phases exist in al-

Table 3.1-222 Structure and lattice parameter of selected intermediate Pt compounds [1.217, p. 119]

Phase	Pearson Symbol	a (nm)	b (nm)	c (nm)	c/a	Remarks	Concentration x A$(1-x)$B(x)
CoPt	tP4	0.3806	0.3684	0.9679			
CoPt$_3$	cP4	0.3831					
CuPt	hR32	0.7589					0.5
CuPt$_3$	cF4	0.3849					
Cu$_3$Pt	o**	0.7596	0.2745	0.777			
Cu$_3$Pt	cP4	0.3682					
Fe−Pt	cF4	0.376					0.245
FePt	tP4	0.3861		0.3788	0.9811		
Fe$_3$Pt	cP4	0.3727					
NiPt	tP4	0.3823		0.3589	0.9388		
PtZr	oC8	0.3409	1.0315	0.4277			
Pt$_3$Zr	hP16	0.5624		0.9213	1.6328		

loy systems with rare earth metals. The formation and crystal structures of the intermediate phases have been related to the electron configuration of the alloy components (Engel–Brewer correlation) [1.283, 284]. Phases with superlattice structures are formed with Co, Cu, Fe, Nb, and V in atomic ratios of 1:1, 2:1, and 3:1 (Tables 3.1-123, 3.1-197). The ordered CuPt phase has a long-range ordered rhombohedral structure.

Mechanical Properties. Characteristic data are shown in Tables 3.1-223–3.1-226 [1.217], Figs. 3.1-282–3.1-290 [1.217, 228, 228, 231]. For elastic properties of PGMs at different temperatures, see [1.217]. Strengthening is affected by solid solution hardening, order hardening (Pt−Co, Pt−Cu), and dispersion hardening. Dispersion-strengthened Pt and Pt alloys are remarkably resistant to creep at high temperatures. They are pro-

Table 3.1-223 Elastic constants of Pt [1.217, p. 219]

c_{11}	c_{12}	c_{44}
347	173	76.5

Table 3.1-224 Mechanical properties of Pt (99.9%) at different temperatures (°C) [1.217, p. 220]

T (°C)	E (GPa)	R_m (MPa)	A (%)	$R_{p0.2}$ (MPa)	HV
20	173	135	41	50	55
250	169	110	40	40	53
500	159	78	42	30	50
750	140	44	46	20	35
900	126	34	44	17[a]	23

[a] interpolation

A = Elongation, E = Modulus of elasticity, R_p = Limit of proportionality, HV = Vickers hardness, R_m = Tensile strength

Table 3.1-225 Mechanical properties of Pt as function of reduction in thickness (%) by cold rolling [1.217, p. 220]

Reduction (%)	R_m (MPa)	R_p (MPa)	HV	
			a	b
0	250	140	50	40
20	350	310	70	63
59	400	380	84	73

a = Pt > 99.5%, b = Pt > 99.99%

Table 3.1-226 Tensile strength R_m (MPa) and elongation A (%) of binary Pt alloys at different temperatures [1.217, p. 217, 220]

Alloy compound	Temperature (°C)		
	20	400	600
(wt%)	R_m/A		
Au5	340/18	290/10	250/10
Ir10	260/33	240/27	180/33
Ni5	470/26	420/26	320/25
Pd20Rh5	370/30	290/18	240/23
Rh5	225/44	150/40	120/43
Rh10	287/39	200/33	170/38

duced either by co-precipitation with refractory oxides (e.g., 0.16 vol% ZrO$_2$) or by internal oxidation of alloys with 0.2 wt% Cr or 0.8 wt% Zr. Rh additions improve the solubility for oxygen. TiC powder affects dispersion strengthening in concentrations of 0.04–0.08 wt% (Fig. 3.1-291) [1.277].

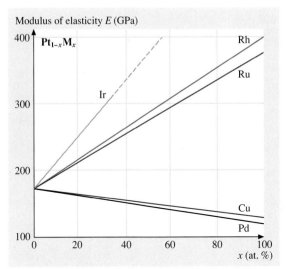

Fig. 3.1-282 Modulus of elasticity of binary Pt alloys [1.217, p. 222]

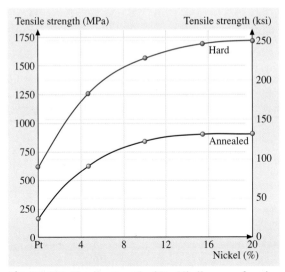

Fig. 3.1-284 Tensile strength of Pt–Ni alloys as a function of Ni content [1.228, p. 695]

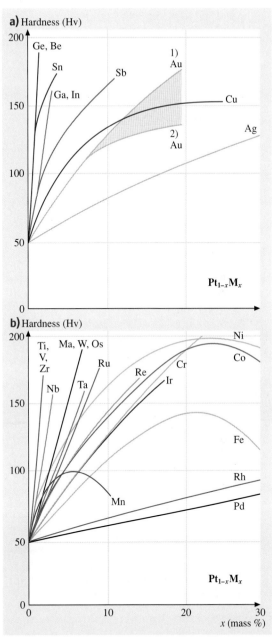

Fig. 3.1-283a,b Solid solution hardening of binary Pt alloys (**a**) de-alloyed at 900 °C; (**b**) solution annealed at 1200 °C [1.217, p. 223]

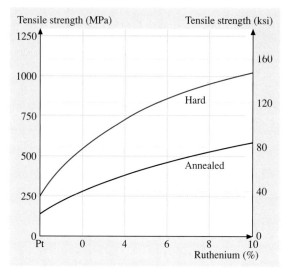

Fig. 3.1-285 Tensile strength of Pt–Ru alloys as a function of Ru content [1.228, p. 694]

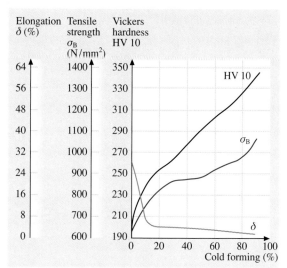

Fig. 3.1-287 Mechanical properties of PtNi8 by cold forming as a function of reduction of cross section [1.231, p. 69]

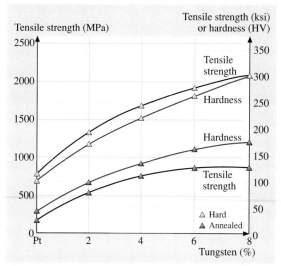

Fig. 3.1-286 Tensile strength of Pt–W alloys as a function of W content [1.228, p. 697]

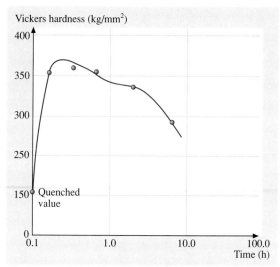

Fig. 3.1-288 Order hardening of stoichiometrc CuPt alloy [1.285, p. 49]

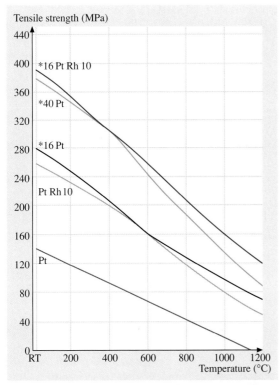

Fig. 3.1-289 Tensile strength of dispersion hardened Pt and PtRh10 (∗ grain stabilized with 0.16 and 0.40 vol.% ZrO$_2$, respectively) [1.217, p. 222]

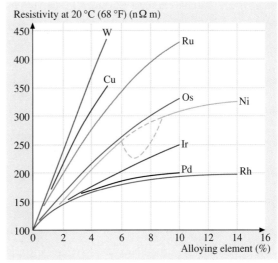

Fig. 3.1-291 Effect of various alloying additions on the electrical resistivity of binary Pt alloys [1.231, p. 67]

Electrical Properties. Characteristic data are shown in Tables 3.1-203, 3.1-205, 3.1-227 [1.217], and Figs. 3.1-292 – 3.1-294 [1.217, 228, 231]. Mo–28 at.% Pt (A15 structure) shows superconductivity at $T_c \approx$ 4.2–5.6 K [1.286].

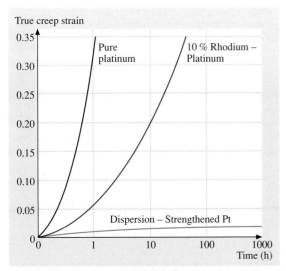

Fig. 3.1-290 Creep curves of TiC-dispersion-strengthened Pt and PtRh wire at 1400 °C in air [1.277]

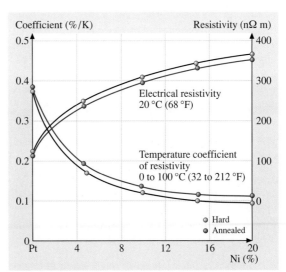

Fig. 3.1-292 Electrical resistivity and temperature coefficient of resistivity (TCR) of Pt–Ni alloys as a function of composition [1.228, p. 696]

Table 3.1-227 Specific electrical resistivity (μΩ cm) of Pt at temperature (K) [1.217, p. 157]

T (K)	10	50	120	273	673	1273	1673
ρ (μΩ cm)	0.0029	0.719	3.56_5	9.83	24.57	37.45	53.35

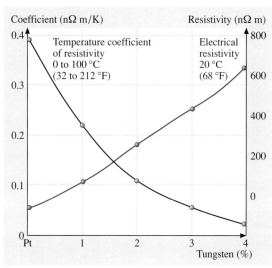

Fig. 3.1-293 Electrical resistivity and temperature coefficient of resistivity (TCR) of Pt−W alloys as a function of composition [1.228, p. 697]

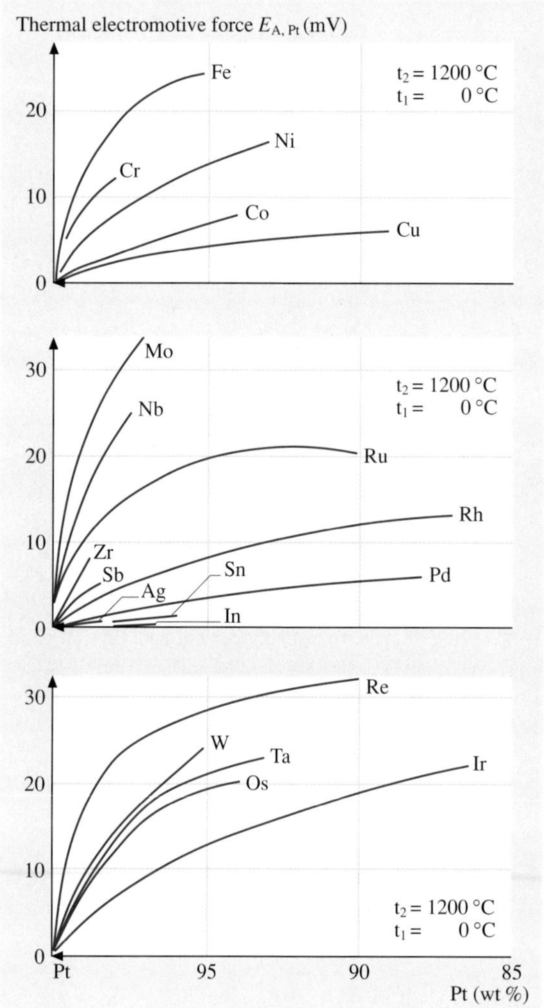

Thermoelectric Properties. Selected values of thermal electromotive force of Pt and Pt alloys are given in Tables 3.1-206, 3.1-207, 3.1-228 – 3.1-231 [1.216, 217, 222], and Fig. 3.1-295 [1.216]. Thermocouples that are Pt−Rh-based are especially suited for high temperatures (see Fig. 3.1-296).

Table 3.1-228 Pt−Rh thermocouples according IEC 5845 (see Fig. 3.1-296) [1.217, p. 472]

Class	Alloy	Maximum applicable temperature (°C)
Type R:	PtRh(87/13)−Pt	1500−1600
Type S:	PtRh(90/10)−Pt	1500−1600
Type B:	PtRh(70/30)−Pt	1750−1800

Fig. 3.1-294 Thermal electromotive force of binary Pt alloys [1.216, p. 97]

Table 3.1-229 Absolute thermoelectric power of Pt [1.222, p. 1009]

Temperature (K)	300	400	500	600	700	800	900	1000	1100	1200
Thermoelectric power (μV/K)	−5.05	−7.66	−9.69	−11.33	−12.87	−14.38	−15.97	−17.58	−19.03	−20.56

Table 3.1-230 Thermal electromotive force of Pt alloys (mV) at different temperatures, reference junction at 0 °C [1.217, p. 160]

Alloy const.	T (°C)	Composition (wt%Ir)					
		10	30	40	70	80	90
Ag	100				−0.4	−0.1	0.2
	900				6.8	4.0	4.5
Au	100					0.8	0.4
	900					11.9	13.5
Ir	100	1.3	1.2				
	1000	15.7	19.1	19.4			
Rh	100	0.64	0.62	0.60			
	1000	9.57	12.3	13.3			
	1300	13.1	17.9	19.0			

Table 3.1-231 Basic values of thermal electromotive force (mV) of common PGM-based thermocouples [1.216, p. 100]

T_1 (°C)	T_2 (°C)	Pt – Rh10/Pt	Pt – Rh20/Pt – Rh5	Rh – Ir60	Pt-el[b]	Pd-or[c]
0	100	0.643	0.074	0.371	3.31	4.6
	500	4.221	1.447	2.562	20.20	27.9
	1000	9.570	4.921	5.495	41.65	59.6
T_1 (K)	T_2 (K)	Au – Co2.1/Cu[d]	Au – Fe0.02/Cu[d]			
4.2	10	0.044	0.093			
	20	0.173	0.208			
	40	0.590	0.423			

[b] Pt-el = Platinel, Pd83Pt14Au3/AuPd35
[c] Pd-or = Pallador, PtIr10/AuPd40
[d] at.%

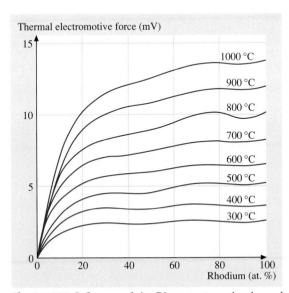

Fig. 3.1-295 Influence of the Rh content on the thermal emf of Pt–Rh alloys against Pt [1.217, p. 473]

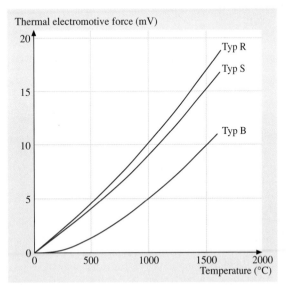

Fig. 3.1-296 Thermal electromotive force of Pt–Rh thermocouples according to IEC 5845 (Type R: PtRh(87/13)–Pt; type S: PtRh(90/10)–Pt; type B: PtRh(70/30)–PtRh(94/6)) [1.217, p. 473]

Magnetic Properties. Selected data are shown in Tables 3.1-211, 3.1-232 [1.217, 222], and Figs. 3.1-297 [1.220] and 3.1-298. The paramagnetic susceptibility of Pt (25.2×10^{-10} m^3 mol^{-1} at 0 K) rises by alloying with 0.1 at.% Rh to 42.5×10^{-10} m^3 mol^{-1}. CoPt is a hard magnetic material ($H_c = 3500$–4700 Oe) but has been replaced by rare-earth transition metal magnetic materials in recent years. Superlattice phases in PtCr-alloys in the composition ranges of 17–65 wt% Cr are ferromagnetic, with the maximum of T_c at ~ 30 at.% Cr. The superlattice structure in FePt and CoPt with tetragonal crystal symmetry gives rise to high values of magnetic anisotropy. The coercivity of sputtered Pt–Co multilayers is increased by annealing in air, caused by the formation of cobalt oxide at the grain boundaries. The

Table 3.1-232 Characteristic properties of technical permanent magnet alloys [1.222, p. 1053]

Alloy (wt%)	H_{max} (kJ/m^3)	H_c (kA/m)	B_r (T)
Pt77Co23	75.0	380	0.60
Sm34Co66	110–160	560	0.80
Co52Fe35V13	22.4	36	1.00

B = magnetic flux density, H = magnetic field, B_r = Remanence

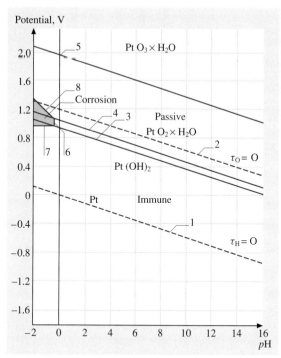

Fig. 3.1-298 Potential pH-diagram of the system Pt/H$_2$O at 25 °C (see Table 3.1-232) [1.217, p. 201]

oxide layer gives rise to domain pinning and to magnetic isolation of the grains, thus leading to a high perpendicular anisotropy [1.217].

Table 3.1-233 Thermal conductivity λ of Pt–(Au, Rh, Ir) alloys (W/m K) [1.217, p. 153]

PtAu5	PtRh5	PtRh10	PtRh20	PtIr5	PtIr10
43	33[a]	30	28[a]	42	31

[a] = calculated with Wiedemann-Franz' law $\lambda = L\sigma T$, Lorenz number $L = 2.45 \times 10^{-6}$ W/K from λ(PtRh10) and the specific electrical conductivity of PtRh-alloys

Table 3.1-234 Thermal expansion coefficient α (10^{-6} K^{-1}) of Pt–Rh alloys at different temperature ranges [1.217, p. 155]

	α (10^{-6}/K)	
Temperature range (K)	273–983	293–1473
Rh-content (wt%)	(10^{-6}/K)	(10^{-6}/K)
6	10.7	11.3
10	10.7	11.2
20	10.9	11.5
30	10.8	11.4

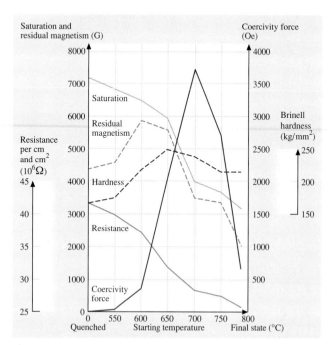

Fig. 3.1-297 Change of magnetic properties of PtCo50 alloy by annealing [1.220, p. 263]

Thermal Properties. Tables 3.1-212–3.1-215, 3.1-233, and 3.1-234 [1.217, 217] provide selected data of thermal conductivity and thermal expansion. In the disordered state the Fe–Pt alloy system exhibits a negative thermal expansion coefficient at room temperature near Fe_3Pt (Invar effect) [1.281, 282].

Optical Properties. Values of the spectral degree of emission and the optical reflectivity are given in Table 3.1-216 [1.217] and Fig. 3.1-275 [1.220].

Diffusion. Data for self-diffusion, diffusion of tracer elements and of hydrogen and oxygen are shown in Tables 3.1-158, 3.1-216, 3.1-217 [1.217].

Chemical Properties. Platinum has the reduction potential of $E_0 = +1.118$ for Pt/Pt^{2+}. It is resistant against reducing acids in all pH ranges, but is attacked by alkali and oxidizing media. Alloying with 30 at.% Rh improves the corrosion resistance against alkali hydroxides. Figure 3.1-298 and Table 3.1-235 [1.217] give the potential pH diagram of the system Pt/H_2O at 25 °C. Dry Chlorine attacks with rising temperature (Fig. 3.1-299 [1.217]). Detailed information about chemical behavior is given in [1.217].

Platinum reacts with ZrC to form Pt_3Zr. It also reacts in the presence of hydrogen with ZrO_2, Al_2O_3, and rare earth oxides at temperatures between 1200 and 1500 °C [1.283, 284]. The solubility of oxygen in platinum is very low. Thin coatings of Pt on reactive materials are an effective protection against oxidation. Alloying of Pt with 2 wt% or higher Al improves the oxidation resistance up to 1400 °C by forming protective dense oxide coatings [1.287]. Superalloys that are Pt–Al-based have high compression strength at high temperatures. Third alloying elements (e.g., Ru) stabilize the high-temperature phase down to room temperature and affects solid-solution strengthening [1.288].

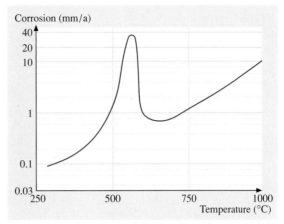

Fig. 3.1-299 Corrosion of Pt in dry Cl_2 gas [1.217, p. 186]

Catalysis. Platinum and Pt alloys are preferably applied in heterogeneous catalysis as wire nets or powders with a high specific surface area ranging from 20 to $1000 \, m^2/g$ ("platinum black," "palladium black") on carbon or Al_2O_3 supports. The catalytic effectivity is structure-sensitive. Figure 3.1-300 show an example of the catalytic action of Pt for the reaction rate and the product selectivity on different crystal planes [1.218]. Pt–Pd–Rh alloys are the main active constituents of catalytic converters for automobile exhaust gas cleaning.

Special Alloys. Molybdenum clad with Pt serves as glass handling equipment up to 1200 °C. Binary Pt alloys with Cu(4), Co(5), W(5), and Ir(10) at.%; and ternary alloys of Pt–Pd–Cu and Pt–Pd–Co are standard jewelry alloys. Alloys of Pt–Au and Pt–Au–Rh surpass the strength of pure Pt at 1000 °C and resist wetting of molten glass. The materials $PtIr_3$, $PtAu_5$ are suitable for laboratory crucibles and electrodes with high mechanical stability.

Table 3.1-235 Reaction and potentials corresponding to graphs of Fig. 3.1-298 [1.217, p. 200]

Number	Reaction equation	Potential E_0 (V)
1	$2H^+ + 2e^- \rightarrow H_2$	$0.000 - 0.0591 \, pH$
2	$2H_2O \rightarrow O_2 + 4H^+ + 4e^-$	$1.228 - 0.0591 \, pH$
3	$Pt + 2H_2O \rightarrow Pt(OH)_2 + 2H^+ + 2e^-$	$0.980 - 0.0591 \, pH$
4	$Pt(OH)_2 \rightarrow PtO_2 + 2H^+ + 2e^-$	$1.045 - 0.0591 \, pH$
5	$PtO_2 + H_2O \rightarrow PtO_3 + 2H^+ + 2e^-$	$2.000 - 0.0591 \, pH$
6	$Pt + H_2O \rightarrow PtO + 2H^+$	$\log[Pt^{++}] = -7.06 - 2 \, pH$
7	$Pt \rightarrow Pt^{++} + 2e^-$	$1.188 + 0.0259 \log[Pt^{++}]$
8	$Pt^{++} + 2H_2O \rightarrow PtO_2 + 4H^+ + 2e^-$	$0.837 - 0.1182 \, pH - 0.0259 \log[Pt^{++}]$

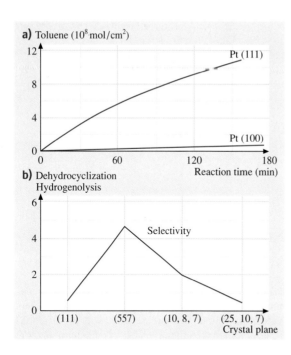

Fig. 3.1-300 (**a**) Rate of reaction of n-heptane dehydrocyclization to toluol on Pt(111) and Pt(100). (**b**) Variation of selectivity at different crystal planes [1.218, p. 279]

3.1.10.4 Rhodium, Iridium, Rhutenium, Osmium, and their Alloys

Rhodium and Rhodium Alloys

Applications. Rhodium is an essential component of catalysts in numerous chemical reactions and automobile exhaust-gas cleaning. In heterogeneous catalysis it is applied in alloyed form, in homogeneous catalysis as complex organic compounds. Rhodium is an alloy component of corrosion- and wear-resistant tools in the glass industry and a constituent in platinum-group-metal-based thermocouples. Rhodium coatings on silverware and mirrors protect them against corrosion. Commercial grades available are powder, shot, foil, rod, plate, and wires with purity from 98–99.5% (ASTM B 616-78; reappraised 1983).

Production. Rhodium is produced as powder and sponge by chemical reduction or thermal decomposition of the chloro–ammonia complex $(NH_4)_3[RHCl_6]$. Bars, rods, and wires are produced by powder compacting and extrusion, while coatings are produced galvanically, by evaporation or by sputtering.

Phases and Phase Equilibria. Selected phase diagrams of Rh are shown in Fig. 3.1-301a–c. Rhodium forms continuous solid solutions with Fe, Co, Ni, Ir, Pd, and Pt. Miscibility gaps exist in alloys with Fe, Co, Ni, Cu, Ag, Au, Pd, Pt, Ru, and Os. Thermodynamic data are given in Table 3.1-236 (see also

Table 3.1-236 Thermodynamic data of Rh [1.217, p. 110]

T (K)	c_p (J/K mol)	S (J/K mol)	H (J/mol)	G (J/mol)	p (atm.)
298.15	24.978	31.506	0	−9.393	1.43×10^{-89}
400	26.044	38.993	2.598	−13	6.59×10^{-65}
800	39.155	58.333	13.853	−32.813	7.21×10^{-29}
1400	35.195	76.556	33.532	−73.646	1.71×10^{-13}

T = Temperature, c_p = specific heat capacity, S = Entropy, H = Enthalpy, G = free Enthalpy, p = partial pressure of the pure elements

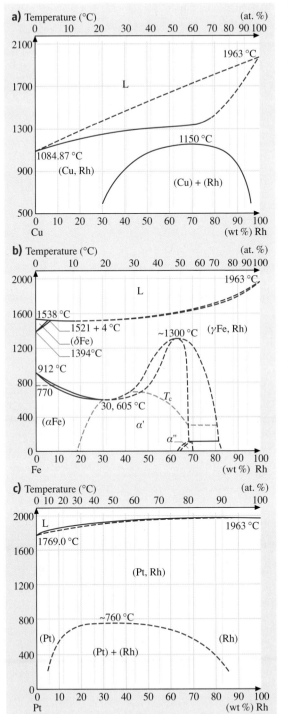

Table 3.1-190) and the maximum hydrogen inclusion is listed in Table 3.1-194.

The compositions and crystal structures of intermediate compounds are shown in Table 3.1-237 (see Table 3.1-197 for superlattice structures).

Mechanical Properties. Characteristic mechanical data of Rh are given in Tables 3.1-238–3.1-241 and Figs. 3.1-302–3.1-307. The modulus of rigidity $G = 153$ GPa; Poisson's ratio is 0.26; the elastic constants are $c_0 = 413$, $c_{12} = 194$, and $c_{44} = 184$.

Rhodium is very hard but can be deformed at temperatures above 200 °C. For strong hardening by deformations, repeated annealing is needed at temperatures higher than 1000 °C. Rh is an effective hardener in Pd and Pt alloys.

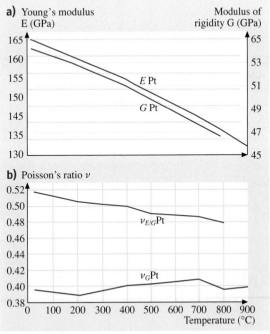

Fig. 3.1-301a–c Phase diagrams of Rh alloys with (**a**) Cu, (**b**) Fe, and (**c**) Pt [1.216, p. 101, 104]

Fig. 3.1-302 (**a**) Young's modulus E (GPa) and modulus of rigidity G (GPa) of Pt at different temperatures. (**b**) Poisson's ratio for Pt at different temperatures [1.289, p. 76]

Table 3.1-237 Structure and lattice parameter of selected Rh compounds [1.217, p. 117ff.]

Phase	Pearson symbol	a (nm)	b (nm)	c (nm)	Concentration x A$(1-x)$B(x)
Cu–Rh	cF4	0.3727			0.5
Fe–Rh	cI2	0.288885			
Fe–Rh	cF4	0.374			0.5
FeRh	cP2	0.2998			
Fe–Ru	hP2	0.258		0.414	0.2
Ni–Rh	cF4	0.36845			0.494

Table 3.1-238 Mechanical properties of Rh (99%) at different temperatures [1.217, p. 227]

T (°C)	E (GPa)	R_m (MPa)	A (%)	$R_{p0.2}$ (MPa)	HV
20	386	420	9	70	130
500	336	370	19	80	91
700	315	340	16	40	73
1000	–	120	10	30	52

A = elongation of rupture, E = modul of elasticity, R_p = limit of proportionality, HV = Vickers hardness, R_m = tensile strength

Table 3.1-239 Increase of Rh hardness by cold forming [1.217, p. 227]

V (%)	HV
0	130
10	275
30	50
50	400

Table 3.1-241 Hardness of Rh–Ni alloys at 300 K [1.217, p. 228]

Alloy	HV 5
RhNi27	275
RhNi40	220
RhPd20	178
RhPt20	165

Table 3.1-240 Hardness of Pd/Rh and Pt/Rh alloys as a funtion of composition [1.217, p. 218]

2nd metal	HV 5 Content (mass%)				
	0	20	40	60	80
Pd	130	178	235	229	138
Pt	130	165	164	145	123

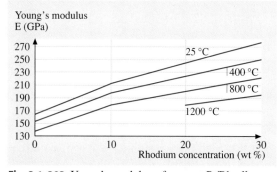

Fig. 3.1-303 Young's modulus of as cast Pt/Rh-alloys at various temperatures [1.289, p. 80]

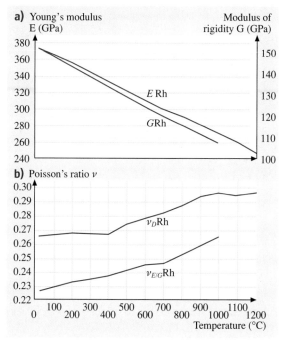

Fig. 3.1-305 Vickers hardness of Pt/Rh-10 as a function of reduction (%) and various annealing temperatures [1.218, p. 608]

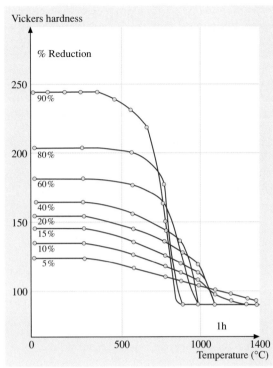

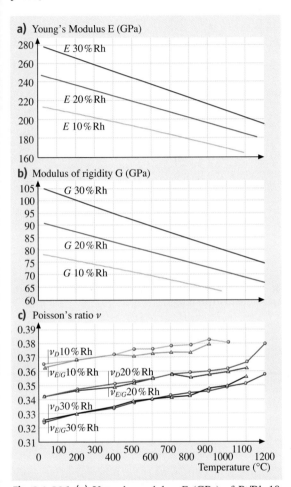

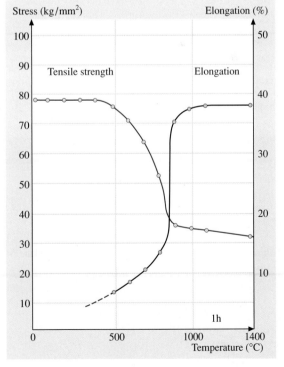

Fig. 3.1-306 (a) Young's modulus E (GPa) of Pt/Rh-10, Pt/Rh-20 and Pt/Rh-30 alloys at different temperatures. (b) Modulus of rigidity G (GPa) of Pt/Rh-10, Pt/Rh-20 and Pt/Rh-30 alloys at different temperatures. (c) Poisson's ratio of Pt/Rh 10, Pt/Rh 20 and Pt/Rh 30 alloys at different temperatures [1.289, p. 79]

◂ **Fig. 3.1-304** (a) Young's modulus E (GPa) and the modulus of rigidity G (GPa) of forged Rh at different temperatures. (b) Poissons ratio for forged Rh at different temperatures [1.289, p. 77]

Fig. 3.1-307 Mechanical properties of Pt/Rh-10 alloy [1.218, p. 609]

Electrical Properties. Characteristic electrical properties are given in Tables 3.1-242 and 3.1-243 (see also Table 3.1-203). Rhodium shows superconductivity below 0.9 K [1.216]. Superconducting Rh alloys are shown in Table 3.1-244. Among the three-element alloys containing precious metals there exists a special group known as magnetic superconductors [1.218]. Figure 3.1-308 shows as example the alloy $ErRh_4B_4$ with the coexistence of superconductivity and magnetic order changing in the region below the critical temperature of beginning superconductivity [1.218]. Data for light and thermoelectric emission are given in Table 3.1-245 [1.290].

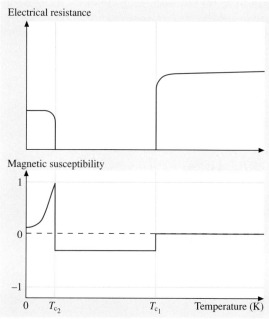

Fig. 3.1-308 Coexistence of superconductivity and magnetic order in ErRh4B4 [1.218, p. 637]

Table 3.1-242 Increase of atomic resistivity $\Delta\rho/C$ of Rh [1.217, p. 158]

Base element	$\Delta\rho/C$ ($\mu\Omega$ cm/at.%)
Rh	Co 0.34, Cr 4, Fe 0.6, Mn 1.4

Table 3.1-243 Specific electric resistivity $\rho_i(T)$ ($\mu\Omega$ cm) of Rh at temperature T ($\rho_0 = 0.0084$) [1.217, p. 156]

T (K)	25	70	160	273	500	1250	1750
$\rho_i(T)$	0.0049	0.34	2.12	4.3	9.20	26.7	40.9

Table 3.1-244 Superconducting Pd, Pt, and Rh alloys [1.218, p. 636]

Pd, Pt Compound	T_c (K)	Rh Compound	T_c (K)
PtNb3	10.6	Rh2P	1.3
Pd1.1Te	4.07	Rh4P3	2.5
PdTe1.02	2.56	ErRhB4	8.5
PdTe2	1.69		

Table 3.1-245 Light and thermoelectric emission of Rh, Pd, and Pt [1.220, p. 79]

Metal	Light-electric constants			Thermoelectric work function (V)
	Wavelength limit (Å)	Work function (V)	Temperature (°C)	
Rh	2500	4.57	20	4.58 (at 1550° abs.)
	2700	–	240	
Pd	2490	4.96	RT	4.99 (at 1550° abs.)
Pt	1962	6.30	RT	6.37 (at 1550° abs.)

Thermoelectrical Properties. Tables 3.1-206 and 3.1-207 give data for the absolute thermoelectric power; Tables 3.1-246 and 3.1-247 give the thermal electromotive force of Rh and of Rh/Ni alloys at different temperatures. Rh is also used as a component in Pt-based thermocouples (Tables 3.1-230–3.1-228, Fig. 3.1-295).

Magnetical Properties. Data of the magnetic susceptibility of Rh and Rh alloys are given in Figs. 3.1-271, 3.1-272, and 3.1-309–3.1-312. For magnetostriction data see Table 3.1-211. The superlattice alloy FeRh shows a transition from the antiferro- to the ferromagnetic state near room temperature Fig. 3.1-312 [1.218] where small additions of Pd, Pt, Ir, Ru, or Os enhance this effect.

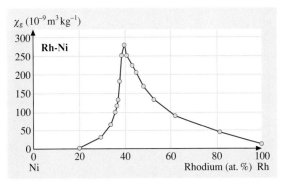

Fig. 3.1-310 Mass susceptibility of Rh–Ni alloys as a function of alloy composition at 4.2 K [1.217, p. 166]

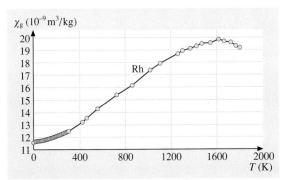

Fig. 3.1-309 Temperature dependence of the mass susceptibility of Rh [1.217, p. 163]

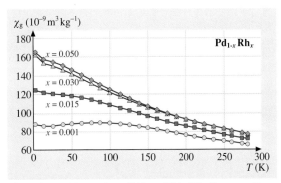

Fig. 3.1-311 Mass susceptibility of Pd–Rh alloys as a function of alloy composition [1.217, p. 169]

Table 3.1-246 Thermal electromotive force of Rh at different temperatures [1.217, p. 159]

T (°C)	−200	−100	+100	+400	+800	+1000	+1300
$E_{Rh/Pt}$	−0.23	−0.32	+0.70	+3.92	+10.16	+14.05	+20.34

Table 3.1-247 Thermal electromotive force of Rh/Ir alloys at different temperatures [1.217, p. 160]

| T (°C) | Composition (wt% Ir) | | | | | |
	10	30	50	70	90	
1000		15.15	17.40	18.40	17.75	15.45

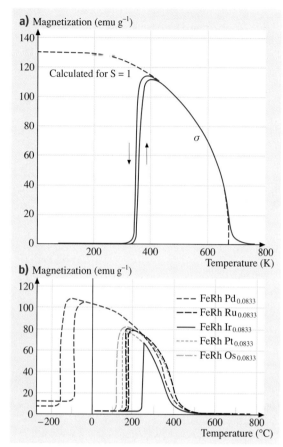

Fig. 3.1-312a,b Metamagnetic behavior of (**a**) FeRh superlattice alloy [1.218, p. 102]. (**b**) Variation by addition of small amounts of Pd, Ru, Ir, Pt, Os [1.218, p. 102]

Thermal Properties. Tables 3.1-212–3.1-214 show the recrystalization temperature, thermal conductivity and thermal expansion at different temperatures. Vapor pressure at different temperatures is shown in Fig. 3.1-273.

Optical Properties. Rhodium has the highest optical reflectivity of all platinum-group metals (Fig. 3.1-274), ranging about 20% below the reflectivity of Ag. It is used as hard and corrosion-resistant coating on silver jewelry and for optical reflectors. Data of the spectral emissivity are given in Table 3.1-216.

Diffusion. Data for self-diffusion are given in Table 3.1-217 (see [1.216] for further data).

Chemical Properties. Rhodium is not attacked by acids or alkali even under oxidizing conditions (aqua regia) (Fig. 3.1-313). Sodium hypochlorite attacks in the order of increasing strength: Pt − Rh − Ir − Ru < Pd < Os. Heating in air causes the formation of thin oxide layers above 600 °C which decompose above 1100 °C (Fig. 3.1-314). Pt alloys with 5–40 wt% Rh are corrosion-resistant against H_2F_2. A detailed survey about these chemical properties is given in [1.216].

Rhodium is the effective component of the three-way Pt/Pd/Rh alloy autocatalyst for the reduction of

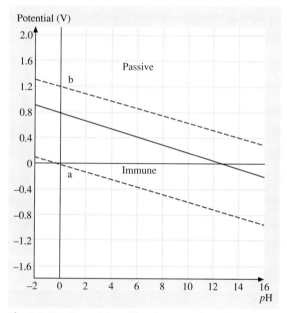

Fig. 3.1-313 Potential pH diagram for the system Rh–H_2O [1.217, p. 202]

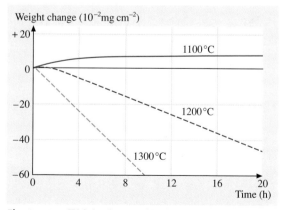

Fig. 3.1-314 Weight change of Rh in oxygen [1.217, p. 185]

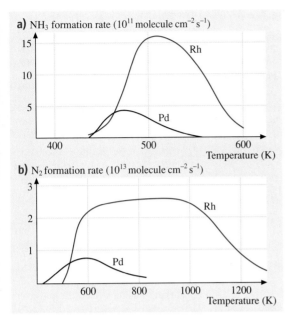

Fig. 3.1-315a,b Product formation rates in $N = -H_2$ reactions on Pd- and Rh-catalyst foils. (**a**) NH_3 formation rates $p(NO) = 9.4 \times 10^{-5}$ Pa, $p(H_2) = 4.9 \times 10^{-5}$ Pa. (**b**) N_2 formation rates $p(NO) = 1.1 \times 10^{-5}$ Pa, $p(H_2) = 1.9 \times 10^{-5}$ Pa [1.218, p. 287]

NO_x of exhaustion gases (Figs. 3.1-315a,b). Rh-catalysts surpass the group homolog, Co-based catalysts, with lower reaction pressures and temperatures and higher yields [1.216]. Complex organic rhodium compounds on the basis of RhCl (PPH$_3$) with different substitute ligands are important homogeneous catalysts in the technical production processes for hydrogenation and hydroformulation ("oxo"-processes, e.g., synthesis of aldehydes and acetic acid). Replacement of PPH$_3$ by complex chiral phosphan ligands enables the synthesis of asymmetric compounds, e.g., L-DOPA and L-menthol (Fig. 3.1-316a,b) [1.291].

Iridium and Iridium Alloys

Applications. Iridium is used for crucibles to grow high-purity crystals for lasers, medical scanners etc., anodes to prevent corrosion of shipping vessels and under-water structures, coatings of electrodes for the manufacturing of chlorine and caustic soda, as an alloy component of automotive exhaust catalysts, and as alloy component and compounds of chemical process catalysts for the production of acetic acid and complex organic compounds. Iridium is an effective hardener for materials used at high temperature, high wear, and high corrosion conditions (e.g., spark plugs). It is also used as fine-grain forming addition in jewelry and dental gold alloys. Commercial grades available are powder, shot, ingot, and wire in a purity ranging from 98–99.9% (ASTM 671-81, reappraised 1987).

Production. Iridium is produced as powder and sponge by chemical reduction or thermal decomposition of the chloro–ammonia compound $(NH_4)_2[IrCl_6]$. Bars, rods, ingot, and wires are produced by compacting of powder followed by extrusion. Coatings are produced galvanically, by evaporation, or by sputtering.

Phases and Phase Equilibria. Figures 3.1-317–3.1-319 [1.216] show the binary phase diagrams of Ir alloys with Pt, Rh, and Ru. Miscibility gaps exist in the solid state also in the alloy systems with Cu, Os, Re, and Ru. Iridium alloyed in Fe lowers the α–λ transition temperature consider-

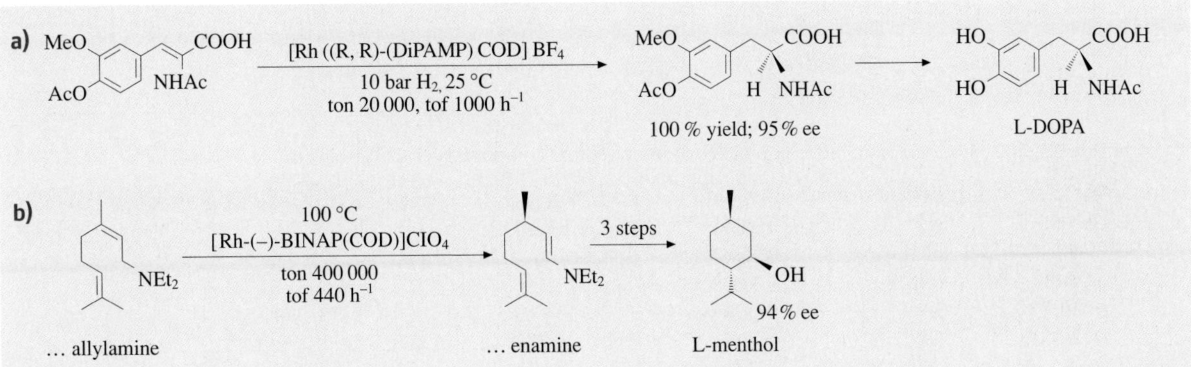

Fig. 3.1-316a,b Examples of organic synthesis of chiral compounds catalysed by complex Rh compounds. (**a**) L-DOPA. (**b**) L-menthol [1.291, p. 83]

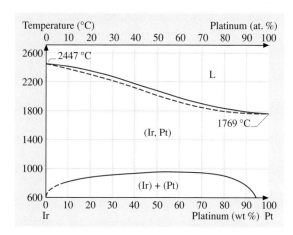

Fig. 3.1-317 Phase diagram of Ir−Pt [1.217, p. 88]

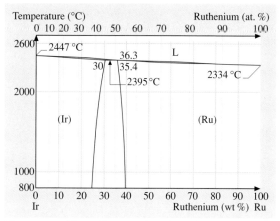

Fig. 3.1-319 Phase diagram of Ir−Ru [1.217, p. 89]

ably (Fig. 3.1-343). Thermodynamic data are given in Table 3.1-248 [1.216].

Structure and lattice parameters of selected intermediate compounds are given in Table 3.1-249 [1.216].

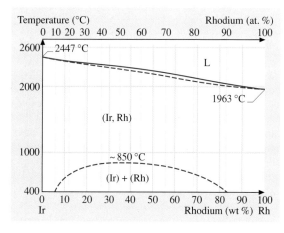

Fig. 3.1-318 Phase diagram of Ir−Rh [1.217, p. 89]

Table 3.1-248 Thermodynamic data of Ir [1.217, p. 109]

T (K)	c_p (J/K mol)	S (J/K mol)	H (J/mol)	G (J/mol)	p (atm)
298.15	24.979	35.505	0	−10.586	
400	25.695	42.946	2.581	−14.598	6.70×10^{-80}
800	28.51	61.62	13.442	−35.875	2.81×10^{-36}
1400	32.733	78.647	31.795	−78.311	1.15×10^{-17}

T = Temperature, c_p = specific heat capacity, S = Entropy, H = Enthalpy, G = free Enthalpy, p = partial pressure of the pure elements

Table 3.1-249 Structure and lattice parameter of intermediate compounds [1.217, p. 117ff,]

Phase	Pearson symbol	a (nm)	b (nm)	c (nm)	Concentration x A$(1-x)$B(x)
Cu−Ir	cF4	0.3629			
Ir−Os	hP2	0.27361		0.43417	0.65
Ir−Os	cF4	0.38358			0.2
Ir−Rh	cF4	0.3824			0.5
Ir−Ru	hP2	0.2718		0.4331	0.56
Ir−Ru	cF4	0.3818			0.47

HT = high temperature modification, LT = low temperature modification

Mechanical Properties. Iridium is extremely hard and can only be deformed at temperatures above 600 °C, with repeated annealing steps at temperatures higher than 1200 °C. The Young's modulus is different for different crystal directions (Table 3.1-250) [1.216], the modulus of rigidity is 214 GPa, and the Poisson's ratio amounts to 0.26.

Characteristic data of mechanical properties of Ir and Pt/Ir alloys are given in Tables 3.1-251–3.1-253 [1.216] and Figs. 3.1-320–3.1-322 [1.289]. Two-phase Ir-based refractory superalloys with fcc and $L1_2$ structure of the components (Ir−12Zr and Ir−17Nb, Ir−15Nb−5Ni) have resist temperatures up to 1200 °C and exhibit marked creep resistance (Figs. 3.1-323–3.1-325) [1.292, 293].

Table 3.1-250 Modulus of elasticity in crystal direction [1.217, p. 212]

$E \langle 110 \rangle$	$E \langle 111 \rangle$
47.4 GPa	662 GPa

Table 3.1-251 Elastic constants of Ir [1.217, p. 212]

$T = 300$ K	$c_{11} = 580$	$c_{12} = 242$	$c_{44} = 256$

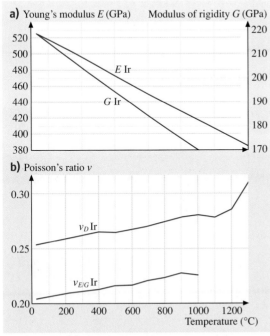

Fig. 3.1-320 (a) Young's modulus of Ir at different temperatures. (b) Poisson's ratio for as cast Ir at different temperatures [1.289]

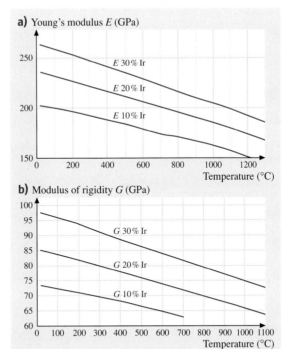

Fig. 3.1-321 (a) Young's modulus of Pt−Ir alloys at different temperatures. (b) Modulus of rigidity of as cast Pt−Ir alloys at different temperatures [1.289]

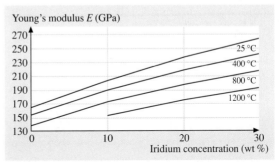

Fig. 3.1-322 Young's modulus of as cast Pt−Ir alloys at different temperatures [1.289]

Table 3.1-252 Mechanical properties of Ir at different temperatures [1.217, p. 213]

T (°C)	E (GPa)	R_m (MPa)	A (%)	$R_{p0.2}$ (MPa)		HV
20	538	623	6.8	234		200
500	488	530	9	12.7	234	138
800	456	450	18	51	142	112
1000	434	331	–	80.6	43.4	97

A = elongation of rupture, E = modul of elasticity, R_p = limit of proportionality, HV = Vickers hardness, R_m = tensile strength

Table 3.1-253 Change of hardness of Ir by degree cold forming V (%) [1.217, p. 312]

V (%)	HV
0	240
10	425
20	485
30	475
59	590

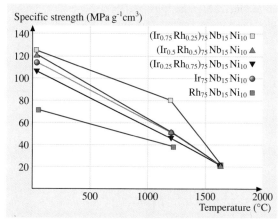

Fig. 3.1-324 Specific strength of Ir–Rh–Nb alloys [1.293, p. 78]

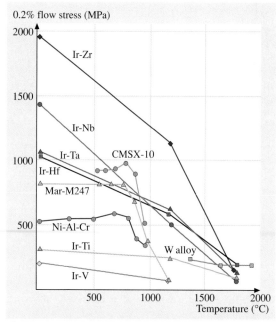

Fig. 3.1-323 High-temperature compression strength of selected Ir-based alloys [1.292, p. 159]

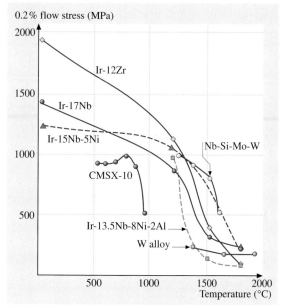

Fig. 3.1-325 Comparison of compressive strength of Ir alloys versus W and Nb/Mo alloys at various temperatures [1.293, p. 77]

Table 3.1-254 Specific electrical resistivity $[\rho_i(T) = \rho_0 + \rho_i(T)]$ of Ir at temperatures T ($\rho_0 = 0.10\,\mu\Omega$ cm) [1.217, p. 157]

T (K)	15	40	140	273	700	1100	1500
$\rho_i(T)$	0.0013	0.10	1.96	4.65	13.90	23.20	34.02

Electrical Properties. The residual resistance ratio (RRR) amounts to 85 (Table 3.1-203). The specific electrical resistivity at different temperatures and the dependence of the atomic resistivity are given in Tables 3.1-254, and 3.1-255 [1.216]. The RRR is listed in Table 3.1-203.

Iridium becomes superconducting below 0.11 K. Some ternary alloys show critical transition temperatures between 3 K and above 8 K (Table 3.1-256) [1.218].

Thermoelectrical Properties. Data for the absolute thermoelectric power and the thermoelectric voltage of Ir, and the thermoelectric voltage of Ir/Rh alloys are shown in Tables 3.1-206–3.1-208, 3.1-257 [1.216] and Fig. 3.1-326 [1.217].

Magnetic Properties. Iridium is paramagnetic. Figures 3.1-271, 3.1-327, 3.1-328 [1.216] show the mass susceptibility of Ir and Pt/Ir alloys at different temperatures. Iridium exhibits magnetostriction according the equation $\Delta l/l = S_l H^2$, with $S_l = +3.8$ (Table 3.1-211).

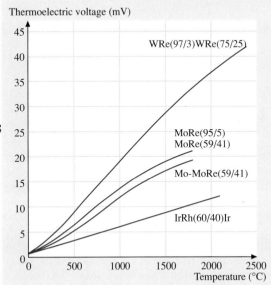

Fig. 3.1-326 Thermoelectric voltage of Ir–Rh alloys compared to Mo–Re and W–Re alloys [1.217, p. 474]

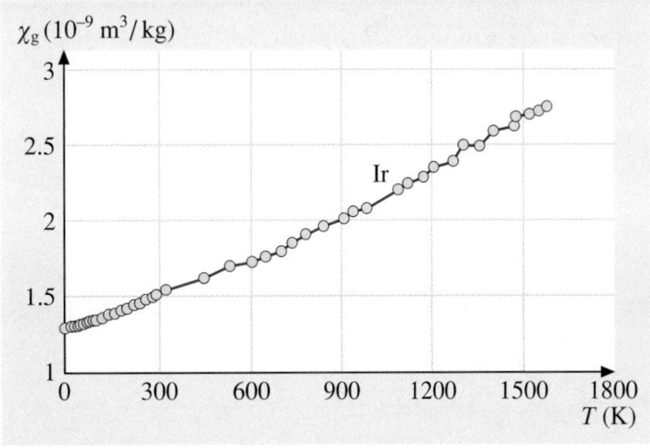

Table 3.1-255 Increase of atomic resistivity [1.217, p. 158]

Base element	$\Delta\rho/C$ ($\mu\Omega$ cm/at.%)
Ir	Pt 1.33, Re 2.7, Cr 2, Mo 3.65, W 3, Fe 0.6

Table 3.1-256 Superconducting Ir alloys [1.218, p. 636]

Ir Compound	T_c (K)
IrTe3	1.18
Sc5Ir4Si10	8.46–8.38
Y5Ir4Si10	3.0–2.3
Lu5Ir4Si10	3.76–3.72

Fig. 3.1-327 Mass susceptibility of Ir at different temperatures [1.217, p. 164]

Table 3.1-257 Thermoelectric voltage of Ir at different temperatures [1.216, p. 95]

T (°C)	−200	−100	−50	+100	+200	+400	+800
$E_{Ir,Pt}$ (mV)	−0.20	−0.35	−0.20	0.66	1.525	3.636	9.246

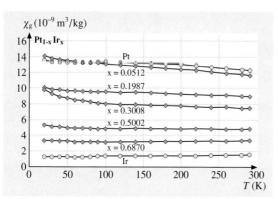

Fig. 3.1-328 Mass susceptibility of Pt–Ir alloys at different temperatures [1.217, p. 68]

Thermal Properties. Tables 3.1-212–3.1-214 give selected data for the recrystallization temperature (varying by purity, degree of cold forming, and annealing time), thermal conductivity, and thermal expansion coefficient.

Optical Properties. The optical reflectivity of Ir is markedly lower than that of Rh increasing in the wavelength range from 0.4 to 0.8 μm (Figs. 3.1-191 and 3.1-274). Data of the spectral emissivity are given in Table 3.1-216.

Diffusion. Table 3.1-217 gives only one value for self diffusion of iridium but further information may be obtained from Landolt–Börnstein [1.238].

Chemical Properties. Iridium is not attacked by acids or alkali even under oxidizing conditions (aqua re-

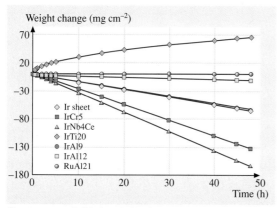

Fig. 3.1-330 Oxidation behavior of various Ir alloys at 1000 °C in air [1.294, p. 100]

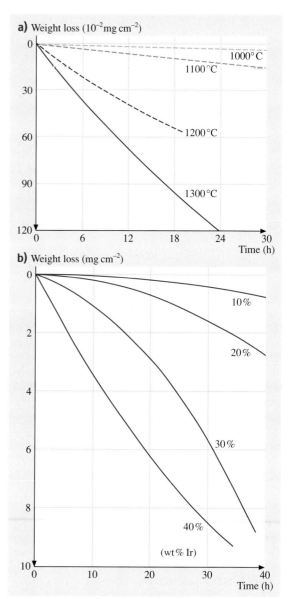

Fig. 3.1-329a,b Evaporation losses. (**a**) Pt loss in oxygen. (**b**) Pt–Ir clad loss in oxygen at 900 °C [1.217, p. 184]

gia). It forms volatile oxides in air above 1000 °C but it can be heated up to 2300 °C without danger of catastrophic oxidation. Pt alloys with 1–30 wt% Ir are corrosion-resistant against H_2F_2. Figures 3.1-329 and 3.1-330 [1.216, 294] show data of the evaporation and oxidation behavior of Ir alloys. A detailed survey on the chemical properties is given in [1.216, 294].

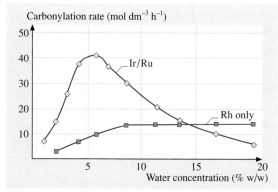

Fig. 3.1-331 Carbonylation rates for Ir–Ru and Rh catalysts in methylacetat reactions [1.295, p. 100]

Metal-organic Ir compounds are effective homogeneous catalysts for organo-chemical reactions such as hydrogenation and carbonylation. The technical production of acetic acid ("Cativa" process). Figure 3.1-331 shows an example for different carbonylation rates of Rh- and Ir/Ru-based catalysts [1.295]. Complex organic Ir catalysts have high stereoselectivity in hydrating cyclic alcohols [1.216].

Ruthenium and Ruthenium Alloys

Applications. Ruthenium is a component of alloys and compounds of chemical process catalysts, and Pt-based catalysts for proton-exchange fuel cells (PEFC). Because of its corrosion resistance, it is used for corrosion-preventing anodes in shipping vessels and under-water structures, pipelines, in geothermal industries, and as coating of electrodes in chlorine and caustic soda production. Ruthenium oxide (RuO_2) and complex Bi/Ba/Pt oxides are materials for electrical resistors. Ruthenium layers on computer hard discs are used for high density data storage improvement of data-storage densities. Ruthenium is an effective hardener of Pd and Pt. Commercial grades available are sponge, powder, grains, and pellets in purity ranging from 99–99.95% (ASTM B 717).

Production. Production of ruthenium starts with chemical reduction of chloro compounds to powder, followed by compacting to pellets. Coatings are produced by galvanic processing, evaporation or sputtering.

Phases and Phase Equilibria. Selected phase diagrams are shown in Figs. 3.1-332–3.1-334, thermodynamic data are listed in Table 3.1-258, and molar heat capacities can be found in Table 3.1-190. Ruthenium alloyed

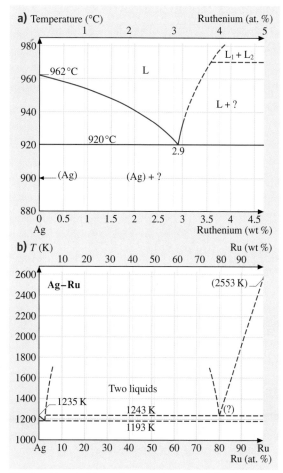

Fig. 3.1-332a,b Phase diagram of Ag–Ru (**a**) and phase diagram of Ag–Ru in the high-temperature range (**b**) [1.217, 245]

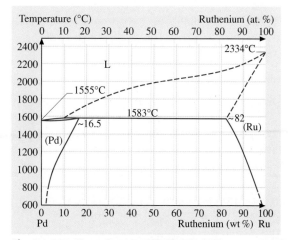

Fig. 3.1-333 Phase diagram of Pd/Ru [1.217]

Table 3.1-258 Thermodynamic data of Ru [1.217, p. 110]

T (K)	c_p (J/K mol)	S (J/K mol)	H (J/mol)	G (J/mol)	p (at)
298.15	23.705	28.535	0	−8.508	
400	24.345	35.595	2.449	−11.79	1.50×10^{-77}
800	26.516	53.121	12.611	−29.885	4.83×10^{-35}
1400	30.97	69.01	29.77	−66.844	7.73×10^{-17}

T = Temperature, c_p = specific heat capacity, S = Entropy, H = Enthalpy, G = free Enthalpy, p = partial pressure of the pure elements

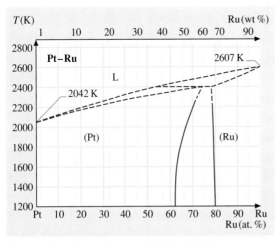

Fig. 3.1-334 Phase diagram of Pt/Ru [1.245]

Table 3.1-261 Hardness (HV 5) of Pd/Ru and Pt/Ru alloys at 300 K [1.217, p. 230]

Alloying metal	HV 5 alloy conc. (wt%)				
	0	20	40	60	80
Pd	350	412	425	284	243
Pt	350	330	446	293	253

pounds. The superlattice structures can be found in Table 3.1-197.

Mechanical Properties. Ruthenium has a Young's modulus of 485 GPa, the Poisson's ratio amounts to 0.29, and the modulus of rigidity is 172 GPa. Characterisitic properties of Ru are given in Tables 3.1-260 and 3.1-261. The mechanical properties are marked anisotropic. The hardness of different single crystal faces varies between 100 HV and 250 HV [1.216]. High compression-strength alloys are formed by two-phase Ru−Al intermetallic structures. Figure 3.1-335 gives

to Fe lowers the γ–α transition temperature considerably (Fig. 3.1-343). Table 3.1-259 gives the structure and lattice parameters of intermediate Co and Fe com-

Table 3.1-259 Structure and lattice parameter of intermediate compounds [1.217, p. 117ff.]

Phase	Pearson symbol	a (nm)	b (nm)	c (nm)	Concentration x A(1 − x)B(x)
Co−Ru	hP2	0.261		0.4181	0.5
Co−Ru	cF4	0.3592			0.2
Fe−Ru	cI2	0.2883			0.06
Fe−Ru	hP2	0.258		0.414	0.2
Os−Ru	hP2	0.27193		0.4394	0.5

Table 3.1-260 Mechanical properties of Ru at different temperatures [1.217, p. 229], [1.216, p. 31]

T (°C)	R_m (MPa)	A (%)	$R_{p0.2}$ (MPa)	HV
20	500	3	380	250–500 [a]
750	300	15	230	160–280 [b]
1000	220 (430)	14	190	100–200 [a]

[a] at different crystal planes,
[b] at 600 °C

A = Elongation of rupture, R_p = Limit of proportionality, HV = Vickers hardness, R_m = Tensile strength

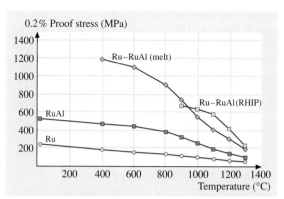

Fig. 3.1-335 High-temperature compression strength of eutectic Ru-70/Al-30 in relation to its constituent phases [1.288, p. 164]

an example of molten and hot isostatic-pressed eutectic Ru (Ru-70/Al-30) in relation to the constituent phases [1.296].

Electrical Properties. The residual resistance ratio (RRR) amounts to 25 000 (Table 3.1-203). Characteristic electrical properties of Ru are given in Tables 3.1-203, 3.1-262, and 3.1-263. The specific electrical resistivity of RuO_2 is 3.5×10^{-5} Ω cm (1 Ω cm for PdO for comparison). Together with its low temperature dependence of the coefficient of resistance, Ru is suited for the production of resistors in sintered form or as thick-film layers covering resistors ranging from ≈ 1.5 to 10 MΩ. Conductive components are either RuO_2, $Pb_2Ru_2O_6$, or $Bi_2Ru_2O_7$ together with additions of doping oxides [1.218].

Ruthenium shows superconductivity below 0.47 K [1.218]. Ternary alloys have critical transition temperatures up to 12.7 K (Table 3.1-264).

Thermoelectric Properties. Data of thermoelectric properties of Ru are given in Tables 3.1-206 and 3.1-265.

Magnetic Properties. Figures 3.1-271, 3.1-337, and 3.1-336 present data of the magnetic mass susceptibility of Ru and of Ru/Cr alloy at different temperatures.

Table 3.1-263 Increase of atomic resistivity by alloying ($\Delta \rho / C$ ($\mu\Omega$ cm/at.%)) [1.217, p. 158]

Base element	$\Delta \rho / C$ ($\mu\Omega$ cm/at.%)
Ru	Fe 0.21, Re 2, Y 1.5
Os	Y 10

Table 3.1-264 Critical transition temperature of superconducting Ru alloys [1.218, p. 636]

Ru	
Composit	T_c (K)
TiRuP	1.33
ZrRuP	12.34–10.56
HrRuP	12.70–11.08
TiRuAs	>0.35
ZrRuAs	11.90–1.03
HfRuAs	4.93–4.37
Y3Ru4Ge13	1.7–1.4
Lu3Ru4Ge13	2.3–2.2

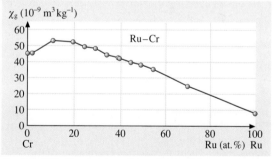

Fig. 3.1-336 Temperature dependence of the mass susceptibility of Ru/Cr alloy [1.217, p. 166]

Table 3.1-262 Specific electrical resistivity $\rho_i(T)$ ($\mu\Omega$ cm) of Ru at different temperatures ($\rho_0 = 0.016$ $\mu\Omega$ cm) $\rho(T) = \rho_0 + \rho_i(T)$ [1.217, p. 156]

T (K)	25	50	100	200	273	300	500
$\rho_i(T)$	0.005	0.105	1.25	4.38	6.69	7.43	13.2

Table 3.1-265 Thermal electromotive force of Ru at different temperatures [1.217, p. 159]

T (°C)	+100	+200	+300	+500	+800	+1000	+1200
$E_{Ru/Pt}$	0.684	1.600	2.673	5.119	9.519	13.003	16.864

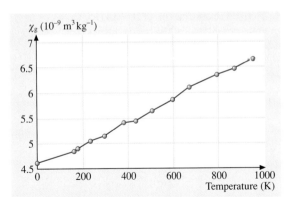

Fig. 3.1-337 Temperature dependence of the mass susceptibility of Ru [1.217, p. 163]

Chemical Properties. Ruthenium is not attacked by acids or alkali even under oxidizing conditions (aqua regia). By heating in air above 800 °C Ru forms the oxides RuO and RuO_2; above 1100 °C Ru forms RuO_3 which vaporizes. Detailed survey about the chemical properties is given in [1.218, 218].

Complex organic Ru compounds are catalysts for the enantioselective hydrogenation of unsaturated carboxylic acids, used in pharmaceutical, agrochemical, flavors and fine chemicals (Fig. 3.1-338) [1.291].

Thermal Properties. Characteristic data of thermal expansion and thermal conductivity are given in Tables 3.1-213 and 3.1-266. Figure 3.1-273 shows the vapor pressure data for Ru.

Optical Properties. The optical reflectivity of Ru is near that of Rh (Fig. 3.1-333). Ruthenium alloyed to Pd enhances the optical reflectivity by 4–5% (Fig. 3.1-275).

Diffusion. Table 3.1-217 gives some values for self diffusion of Ru.

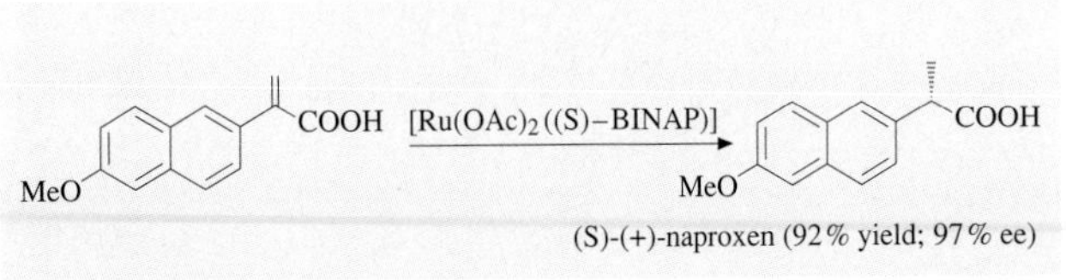

Osmium and Osmium Alloys

Applications. Osmium is used as a component in hard, wear- and corrosion-resistant alloys, as surface coatings of W-based filaments of electric bulbs, cathodes of electron tubes, and thermo-ionic sources. Osmium itself, Os alloys, and Os compounds are strong and selective oxidation catalysts. Commercial grades available are powder in 99.6% and 99.95% purity, OsO_4, and chemical compounds.

Production. The production of Os starts from the mineral osmiridium via soluble compounds and the reduction to metal powder followed by powder-metallurgical compacting.

Phases and Phase Equilibria. Selected phase diagrams are shown in Figs. 3.1-339–3.1-342 [1.216]. Continuous series of solid solution are formed with Re and Ru. Miscibility gaps exist with Ir, Pd and Pt. The solid solubility in the Os−W system are 48.5 at.% for W and ≈ 5 at.% for Os. Osmium alloyed to Fe lowers the $\gamma-\alpha$ transition temperature considerably (Fig. 3.1-343 [1.297]). Thermodynamic data are given in Table 3.1-267 [1.216] and molar heat capacities in Table 3.1-190. Table 3.1-268 gives structures and lattice parameters of intermediate compounds with Ir, Ru, Pt, and W [1.216].

Fig. 3.1-338 Synthesis of (S)-(+)-Naproxen catalysed by Ru-cplx compound [1.291, p. 83]

Table 3.1-266 Thermal expansion coefficients of Ru and Os at different temperatures [1.217, p. 154]

Temperature (°C)	(10^{-6} K^{-1})					
	Ru[a]	Ru[b]	Ru[p]	Os[a]	Os[b]	Os[p]
323	5.9	8.8	6.9	4.0	5.8	4.8
423	6.1	9.3	7.2	4.3	6.2	5.0
623	6.8	10.5	8.0	4.0	7.1	5.7
723	7.2	11.0	8.4	5.3	7.6	6.2
823	7.6	11.7	8.8	5.8	8.3	6.9

[a] vertical to the crystal *c* axis
[b] parallel to the crystal *c* axis
[p] polycrystalline

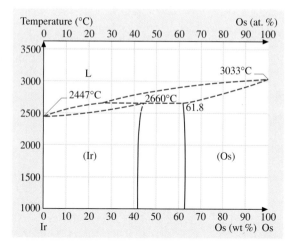

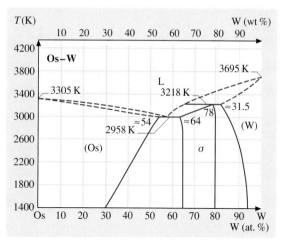

Fig. 3.1-339 Phase diagram of Os−Ir [1.217, p. 87]

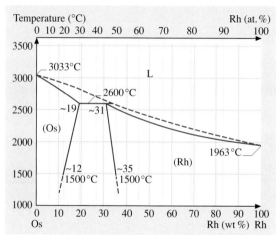

Fig. 3.1-340 Phase diagram of Os−Rh [1.217, p. 90]

Fig. 3.1-342 Phase diagram of Os−W [1.245]

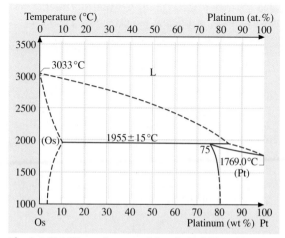

Fig. 3.1-341 Phase diagram of Os−Pt [1.217, p. 90]

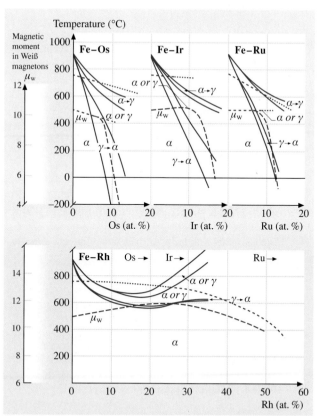

Fig. 3.1-343 Temperature dependence of atomic moments, γ–α transition and magnetic transition of iron alloys [1.220, p. 259]

Table 3.1-267 Thermodynamic data of Os [1.217, p. 110]

T (K)	c_p (J/K mol)	S (J/K mol)	H (J/mol)	G (J/mol)	p (at)
298.15	24.707	32.635	0	−9.73	
400	25.094	39.95	2.536	−13.444	2.95×10^{-95}
800	26.618	57.811	12.879	−33.371	6.31×10^{-44}
1400	28.903	73.287	29.535	−73.067	5.81×10^{-22}

T = Temperature, c_p = specific heat capacity, S = Entropy, H = Enthalpy, G = free Enthalpy, p = partial pressure of the pure elements

Table 3.1-268 Structure and lattice parameter of selected Os alloy phases [1.217, p. 118]

Phase	Pearson symbol	a (nm)	c (nm)	Remarks	Concentration x A(1−x) B(x)
Os	hP2	0.27353	0.43191	293 K	
Os−Ir	hP2	0.27361	0.43417		0.35
Os−Ir	cF4	0.38358			0.8
Os−Pt	hP2	0.27361	0.43247		0.1
Os−Pt	cF4	0.39094			0.8
Os−Ru	hP2	0.27193	0.4394		0.5
Os$_3$W$_7$	tP30	0.9650	0.4990		0.78

Mechanical Properties. Osmium is very hard and brittle. The hardness is, as in the case of Ru, strongly anisotropic. Characterisitic properties for hardness of the element at different temperatures, as well as work hardening and hardness of Os−Pt alloys are given in Tables 3.1-269 and 3.1-270 [1.216, 217] and Fig. 3.1-262. Osmium exhibits a Young's modulus of 570 GPa, a modulus of rigidity of 220 GPa, and the Poisson's ratio is 0.25.

Electrical Properties. The residual electrical resistivity ratio (273.2 K/4.2 K) is 400 [1.216] (Table 3.1-203).

Table 3.1-269 Hardness of Os at different temperatures [1.216]

T (°C)	HV
20	300–680[a]
200	260–580[a]
600	200–410[a]
1200	130–400[a]

[a] all values depending on crystal orientation

Table 3.1-270 Hardness of Os−Pt alloys [1.217]

Pt content (wt%)	HV
0	560
20	578
40	555

Table 3.1-271 Specific electrical resistivity $\rho_i(T)$ (µΩ cm) of Os at temperature T [$\rho(T) = \rho_0 + \rho_i(T)$]; ($\rho_0$ = 0.09 µΩ cm[a]) [1.216]

T (K)	ρ_i (µΩ cm)
25	0.012
100	1.90
273	8.30
900	26.0
1300	38.0

At $T < 273$ K; $\rho_0 = 0.8$ µΩ cm at $T > 273$ K

Table 3.1-271 [1.216] gives the specific electric resistivity of Os at different temperatures. The increase of atomic resistivity is shown in Table 3.1-263. Osmium coatings on W-based dispenser cathodes lower its work function (source Ba−Ca aluminate). It enhances the secondary electron emission (Fig. 3.1-344) and enables the operation at higher current densities in high power klystron and magnetron valves. Osmium shows superconductivity below 0.71 K and Table 3.1-272 gives some examples of superconducting Os alloys [1.218].

Thermoelectric Properties. Figure 3.1-269 shows a comparison of the thermoelectric power of the different noble metals of the platinum group as a function of temperature.

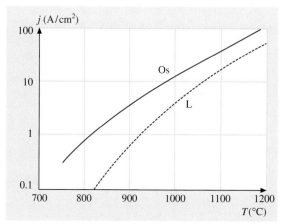

Fig. 3.1-344 Current density as a function of cathodic temperature for a normal cathode (*dashed curve*) and a cathode with a 5 μm thick Os coating [1.298]

Magnetic Properties. Figures 3.1-271 and 3.1-345–3.1-347 [1.216] give a survey and present selected data of the magnetic mass susceptibility for the element and for Os−Cr alloys. This alloy system exhibits antiferromagnetism in compositions from 0.3 to 2.2 at.% Os in the temperature range on the left-hand side of the bold vertical bars in Fig. 3.1-347. In Fe−Os alloys the temperature of the magnetic transition and the atomic magnetic moment decrease with increasing Os content (see Fig. 3.1-343).

Thermal Properties. Data for the thermal expansion coefficient at different temperatures are given in Table 3.1-266.

Table 3.1-272 Superconducting Os-alloys [1.218]

Os	
Compound	T_c (K)
Ce3Os4Ge13	6.1
Pr3Os4Ge13	16
Nd3Os4Ge13	1.9
Eu3Os4Ge13	10.1
Tb3Os4Ge13	14.1
Dy3Os4Ge13	2.1
Er3Os4Ge13	1.9
ZrOsAs	8
HfOsAs	3.2
Y3Os4Ge13	3.9–3.7
Lu3Os4Ge13	3.6–3.1
Y5Os4Ge10	8.68–8.41
TiOsP	<1.2
ZrOsP	7.44–7.1
HfOsP	6.10–4.96

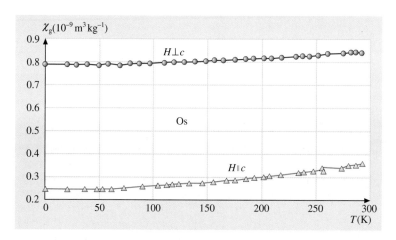

Fig. 3.1-345 Temperature dependence of the mass susceptibility of Os (single crystal) at applied magnet field of 795–700 A/m [1.217, p. 164]

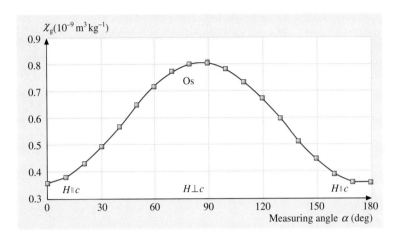

Fig. 3.1-346 Mass susceptibility of an Os single crystal at room temperature as a function of measuring angle [1.217, p. 164]

Fig. 3.1-347 Temperature dependence of the mass susceptibilty of Os−Cr alloys. Small marks indicate the Neel temperature T_N [1.217, p. 167]

Chemical Properties. Osmium is resistant against HCl but is attacked by HNO_3 and aqua regia. The element oxidizes in powder form readily at room temperature, forming OsO_4 which vaporizes above 130 °C. A detailed survey about chemical properties is given in [1.216]. The oxide OsO_4 serves as a catalyst for the synthesis of asymmetric organic compounds. Figure 3.1-348 shows an example for a ligand-supported chiral dihydroxylation [1.291].

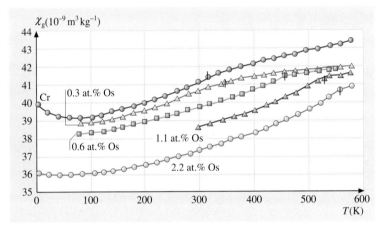

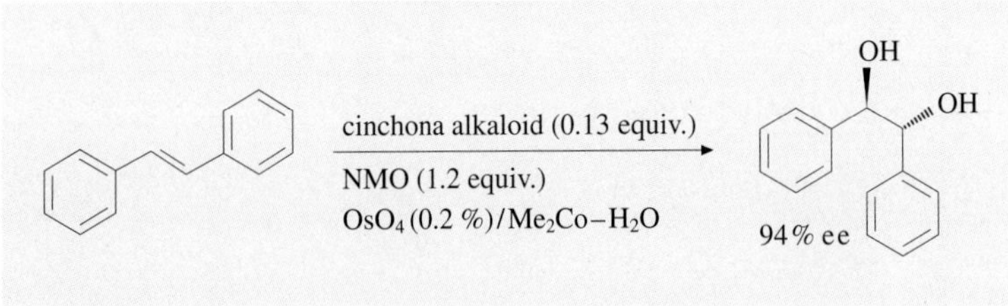

Fig. 3.1-348 Chiral dihydroxylation using OsO_4 as catalyst component [1.291, p. 83]

3.1.11 Lead and Lead Alloys

Lead constitutes only about 12.5 wtppm (weight part per million) of the earth's crust, but concentrated lead ore deposits make it easy to mine. Lead and its alloys are used in a wide range of technical applications because of their low melting point, ease of casting, high density, softness and high formability at room temperature, excellent resistance to corrosion in acidic environments, attractive electrochemical behavior in many chemical environments, chemical stability in air, water and soil, and the high atomic number and stable nuclear structure. Despite their known toxicity, lead and its alloys can be handled safely and it ranks fifth in tonnage consumed (6 Mt/yr), after Fe, Cu, Al, and Zn. The type of data available on different alloys depends to a great extent on the areas of application [1.299].

The most important Pb ore mineral is galena, (87 wt% Pb). The lead ore concentrate is roasted to form Pb oxide. Smelting to reduce the oxide by CO produces Pb. The lead bullion thus obtained contains Sb, As, Te, Sn, Cu, Ni, Co, and Bi besides noble metals, and is further refined to produce various grades of lead. Commercial grade pure lead is produced by the removal of impurities through selective gas phase oxidation, precipitation from molten lead phase as pure elements, and through the formation of intermetallic compounds with low solubility (removal of Fe, Ni, Co, As, Te, and Sb as oxides; precipitation of Cu as elemental Cu, CuS, and Cu arsenates and antimonides; precipitation of Fe as Fe arsenates and antimonides; precipitation of Ag and Au as intermetallic compounds of Zn with Au and Ag; Bi precipitation through the formation of a compound $CaMg_2Bi_2$). Electrolytic refining of commercial purity lead is used to obtain lead with purity to levels down to 99.99 to 99.9995 wt%. Zone melting is used to produce ultrapure grades of Pb.

3.1.11.1 Pure Grades of Lead

The commercial grades of pure lead (Table 3.1-273) are used in chemical plants, sound attenuation, roof-

Table 3.1-273 Impurity levels of commercial lead grades [1.299]

Impurities, additions	Low Bi, low Ag, pure Pb [a,d] L50006	Refined pure Pb [b,d] L50021	Corroding lead [e] L50042	Pure lead (common lead) [e] L50049	Chemical lead [c,e] L51120	Copper bearing lead [e] L51121	Tellurium lead [e] L51123
(wt%)	max.	max.					
Ag, max.	0.0010	0.0025	0.0015	0.005	0.020	0.020	0.020
Ag, min.	–	–	–	–	0.002	–	–
Cu, max.	0.0010	0.0010	0.0015	0.0015	0.080	0.080	0.080
Cu, min.	–	–	–	–	0.040	0.040	0.040
Ag + Cu, max.	–	–	0.0025	–	–	–	–
SbAs, Sn each	0.0005	0.0005	–	–	–	–	–
As + Sb + Sn, max.	–	–	0.002	0.002	0.002	0.002	0.002
Zn, max.	0.0005	0.0005	0.001	0.001	0.001	0.001	0.001
Fe, max.	0.0002	0.001	0.002	0.002	0.002	0.002	0.002
Bi, max.	0.0015	0.025	0.050	0.050 [c]	0.005	0.025	0.025
Te	0.0001	0.0001	–	–	–	–	0.035–0.060
Ni, max.	0.0002	0.0002	–	–	–	–	–
Pb (by difference) min.	99.995	99.97	99.94	99.94	99.90	99.90	99.85

[a] For chemical applications where low Ag and low Bi contents are required
[b] For lead acid battery applications
[c] For applications requiring corrosion protection and formability, as per ASTM B29-92
[d] As per ASTM B29-92
[e] As per ASTM 749-85 (re-approved 1991)

Table 3.1-274 Mechanical properties of pure grades of lead [1.299–301]

Lead grade	Hardness HB	Yield strength (0.125) (MPa)	Tensile strength (MPa)	Comp. strength (25%) (MPa)	Elongation (%)	Fatigue strength at 10^7 cycles (MPa)	Creep strength (0.2%/yr) (MPa)
Pure lead (c)	4.0	5.9	13.1	15.2	45	2.7	
Corroding lead, Pb > 99.94	3.2–4.5	5.5	12–13		30	3.2	
Refined pure (r)	3.8		12.1		53	3.2	1.2
Chemical (c)	5.2	11.3	17.9	20.0	45		
Chemical (r)	5.5		19.3		47	6.9	
Undesilverized (e)			17.2		50		
Undesilverized (r)	4.7	8.6	16.5	17.9	51	5.0	15.8

(r) - rolled; (c) - cast; (e) - extruded

ing, flashings and weather stripping, water-proofing, and radiation shielding.

The mechanical properties of pure grades of lead are listed in Table 3.1-274 [1.299–301]. The near ambient temperatures at which lead and its alloys are used correspond to high homologous temperatures ($T/T_M \sim 0.5$ or higher) for lead and therefore significant diffusion can occur. Consequently, the mechanical properties are affected by dynamic and static recovery, recrystallization effects, and creep deformation. Therefore, caution is advised in the use of short-term mechanical properties. The recrystallization temperatures of different lead grades are shown as a function cold work and grain size in Fig. 3.1-349 [1.303]. The lowest reported value of recrystallization temperature for 99.9999 wt% purity lead is $\sim -59\,°C$; for lead of not very high purity it is $\sim -33\,°C$. The fatigue behavior of 99.99 wt% pure lead in a Haigh push–pull test at a test cycle frequency of 33.67 Hz is presented in the form of S–N (stress to failure versus number of cycles) curves in Fig. 3.1-350 [1.303].

Coefficients of internal friction of relevance to acoustic damping are given in Table 3.1-275 [1.299, 302]. As lead is used in sound attenuation applications, acoustic transmission data of selected single-skin and double-skin partitions with and without lead are given in Table 3.1-276 [1.299]. The sound reduction versus frequency is given in [1.299, 304].

Corrosion rates of lead in H_2SO_4 and HF acids are presented in Figs. 3.1-351 and 3.1-352 [1.301]. Corrosion behavior of chemical lead in some common environments is presented in Table 3.1-277 [1.301]. Corrosion rates of the different lead grades normally fall in the same category.

As lead is extensively used in radiation shielding, the gamma-ray mass-absorption data for lead are presented in Fig. 3.1-353 [1.299, 305].

Table 3.1-275 Experimental values of coefficient of internal friction Q^{-1}. Values in single crystal and polycrystalline lead [1.299, 302] (RT = room temperature)

Material	Frequency (kHz)	Q^{-1}
Pure polycrystalline lead	0.016–2	$0.35 \times 10^{-2} - 4 \times 10^{-2}$
	17–28	$0.2 \times 10^{-2} - 0.8 \times 10^{-2}$
Single-crystal line lead	4–64	$0.2 \times 10^{-2} - 0.7 \times 10^{-2}$
Single-crystal Pb–0.033 wt% Sn	30	0.11×10^{-2} (max. deformation of 10^{-7}, RT)
Single-crystal Pb–0.035 wt% Bi	30	0.22×10^{-3} (max. deformation of 10^{-7}, RT)
Single-crystal Pb–0.0092 wt% Cd	30	0.9×10^{-3} (max. deformation of 10^{-7}, RT)
Single-crystal Pb–0.0022 wt% In	4	2×10^{-3} (max. deformation of 10^{-7})

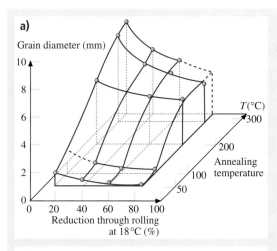

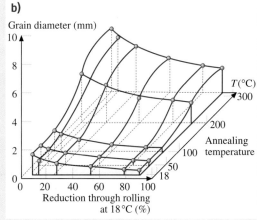

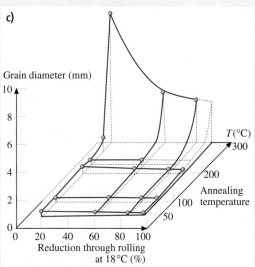

Fig. 3.1-349a–c Recrystallization diagrams: (**a**) Electrolytic Pb. (**b**) Parkes Pb. (**c**) Pattinson Pb [1.303]

Table 3.1-276 Acoustic transmission data of selected single skin and double skin partitions with and without lead [1.299, 304]

Description of test partition	Thickness (mm)	Surface weight (kg/m²)	Average SRI (dB)	RW (dB)	STC (dB)
Single skin – code 1 lead sheet	0.5	5.65	22.7	25	25
Single skin – code 3 lead sheet	1.52	17.16	31.8	35	35
Single skin – 0.5 mm lead equivalent lead impregnated PVC sheet	2.17	7.94	27.4	30	30
Double skin – 12.4 mm Gyproc plasterboard – no infill	117.76	19.04	40.2	42	41
Double skin – code 3 lead sheet bonded to 12.4 mm Gyproc plasterboard – no infill	121.88	52.20	51.8	52	52

SRI: sound reduction index; RW: weighted sound reduction; STC: sound transmission classification

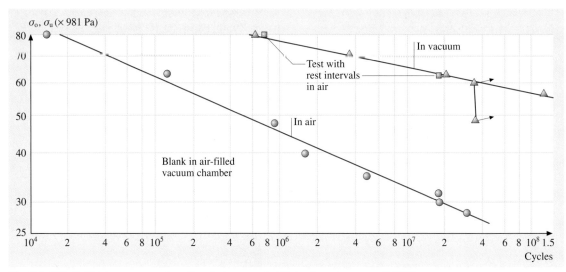

Fig. 3.1-350 S–N (stress vs. number of cycles) curves for Pb in air and vacuum [1.303]

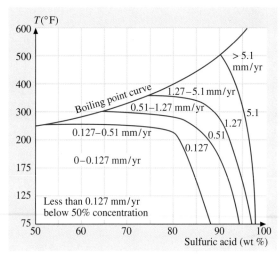

Fig. 3.1-351 Corrosion rates of lead in H_2SO_4 [1.301]

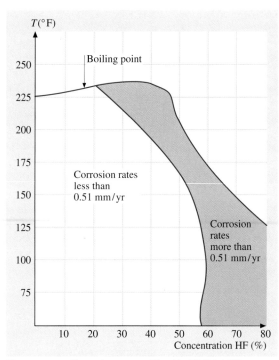

Fig. 3.1-352 Corrosion rates of lead in HF [1.301]

Table 3.1-277 Classifying corrosion behavior of Pb in selected environments [1.301]

Chemical	Temperature (°C)	Concentration (wt%)	Corrosion class
Acetic acid	24	Glacial	B
Acetone	24–100	10–90	A
Acetylene, dry	24	–	A
Ammonia	24–100	10–30	B
Ammonium azide	24	–	B
Ammonium carbonate	24–100	10	B
Ammonium chloride	24	0–10	B
Ammonium hydroxide	27	3.5–40	A
Ammonium nitrate	20–52	10–30	D
Ammonium phosphate	66	–	A
Ammonium sulfate	24	–	B
Arsenic acid	24	10	B
Benzene	24	–	B
Boric acid	24–149	10–100	B
Bromine	24	–	B
Butane	24	–	A
Carbon tetrachloride (dry)	BP	100	A
Chlorine	38	–	B
Citric acid	24–79	10–30	B
DDT	24	–	B
Fluorine	24–100	–	A
Hydrochloric acid	24	0–10	C
Hydrogen chloride (anh HCl)	24	100	A
Mercury	24	100	D
Methanol	30	–	B
Methyl ethyl ketone	24–100	10–100	B
Phosphoric acid	24–93	–	B
Sodium carbonate	24	10	B
Sodium chloride	25	0.5–24	A
Sodium hydroxide	26	0–30	B
Sodium nitrate	24	10	D
Sodium sulfate	24	2–20	A
Sulfur dioxide	24–204	90	B
Natural outdoor atmospheres	24		–A
Industrial, natural and domestic waters	24		–A
Soils	24		–A

Data mostly correspond to chemical lead. The four corrosion performance categories:
A < 0.051 mm/yr: Negligible corrosion – lead recommended for use;
B < 0.51 mm/yr: Practically resistant – lead recommended for use;
C = 0.51–1.27 mm/yr: Lead may be used where this effect on service life can be tolerated;
D > 1.27 mm/yr: Corrosion rate too high to merit any consideration of lead

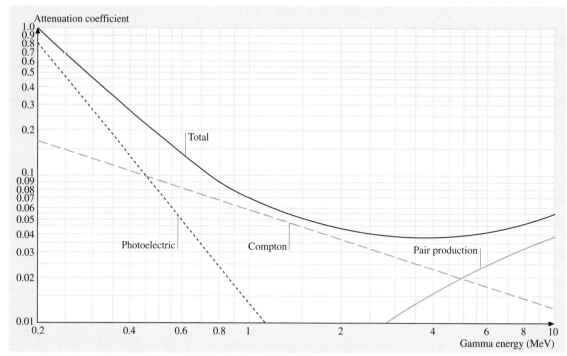

Fig. 3.1-353 Gamma-ray mass-absorption coefficients for lead [1.299, 305]

3.1.11.2 Pb–Sb Alloys

Pb–Sb Binary Alloys

Lead–antimony alloys are widely used for pipe, cable sheathing, collapsible tubes, storage battery grids, anodes, sulfuric acid fittings, and X-ray and gamma ray shielding (in the absence of neutron irradiation). The addition of 1 to 13 wt% Sb to Pb increases tensile strength, fatigue strength, and hardness compared to pure lead (99.99). Lead and Sb form a eutectic system as shown in Fig. 3.1-354 [1.303]. The maximum solubility of Sb in Pb is 3.45 wt% at the eutectic temperature and decreases to 0.3 wt% at 50 °C. Thus considerable age-hardening can be obtained in these alloys. Small additions of As (0.05–0.1 wt%) dramatically increase the rate of aging and final strength. The microstructure at higher Sb contents consists of proeutectic lead-rich phase surrounded by a network of eutectic phase that contributes to enhanced as-cast and high temperature strength. Shrinkage on solidification varies from 3.85% for Pb to 2.06% for a Pb–16 wt% Sb alloy. These alloys have high corrosion resistance in most environments. They form a protective and impermeable film faster than pure lead and, in some cases, even faster than chemical lead. Table 3.1-278 lists the physical properties of selected Pb–Sb alloys [1.300]. Table 3.1-279 lists the mechanical properties of cast

Table 3.1-278 Physical properties of Pb–Sb alloys [1.300]

Alloy composition (wt%)	Solidification range (°C)	Coefficient of thermal expansion (10^{-6} K^{-1})	Specific heat (J kg^{-1} K^{-1})	Thermal conductivity (W m^{-1} K^{-1})	Resistivity (nΩ m)	Density (g cm^{-3})	Volume change on freezing[a] (%)
Pb–1 Sb	322–317	28.8	131	33		11.27	−3.75
Pb–3 Sb	310–269	28.1				11.19	
Pb–6 Sb	285–252	27.2	135 (solid)	29	253	10.88	−3.11
Pb–9 Sb	265–252	26.4	137 (solid)	27	271	10.60	−2.76

[a] Negative values show contraction

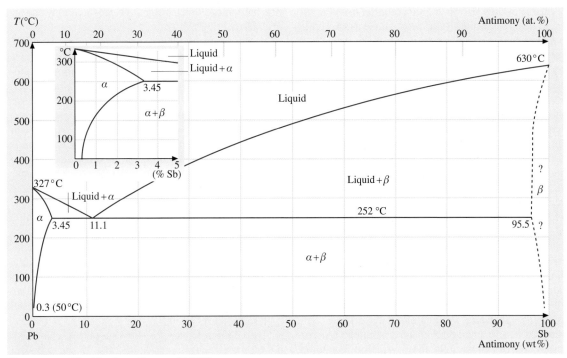

Fig. 3.1-354 Pb–Sb phase diagram [1.303]

Table 3.1-279 Mechanical properties of Pb–Sb alloys [1.299–301, 306]

Sb content (wt%)		Tensile strength (MPa)	Elongation at fracture (%)	Hardness HB	Yield strength $R_{p0.125}$ (MPa)	Fatigue strength for 2×10^7 cycles (MPa)	Creep strength	Young's modulus (GPa)
1	(a)	37.9	20		19.3	7.6	190 h at 20.7 MPa	–
1	(c)	20	50	7.0	–	7.6		–
3	(a)	65.5	10		55.2	–	630 h at 27.6 MPa	–
3	(c)	32.43	15	9.1				–
6	(a)	73.8	8		71.0		1000 h at 27.6 MPa	
6	(c)	47	24	13		17.2		24.15
6	(r)	29.6	42	8.7	15.2	10.3	–	–

(c) cast; (a) cast and stored 30 days at room temperature; (r) rolled

alloys [1.299–301, 306]. Data for one of the rolled alloys are also given to indicate that they exhibit poorer properties.

Pb–Sb-Based Lead Acid Battery Grid Alloys

Lead–acid batteries are the most widely used secondary battery type in current automotive and industrial applications due to the relatively low cost and high availability of the raw materials, room temperature operation, ease of manufacture, long life cycle, versatility, and the excellent reversibility of the electrochemical system. Lead alloys are used as electrode grids, connectors, and grid posts. The two classes of alloys that are in extensive use are (i) Pb–Sb based ternary and multi-component alloys with As, Sn, Ag, Se, Cu, S, and Cd and (ii) Pb–Ca based ternary alloys and multi-component alloys with Sn, Ag, and Al. In the Pb–Sb based alloys, Sb addition to Pb enhances castability, tensile strength, creep strength, corrosion resistance under battery operating conditions, and resistance to structural changes during deep charge–

discharge cycling. However, Sb migration from Pb–Sb based positive grid alloys to negative electrode results in the reduction of hydrogen over-voltage and consequent decrease in cell voltage. This led to increased degassing and water loss. To minimize this poisoning of negative plates, lower Sb contents (1–3 wt% Sb) are now used in battery grids. The posts and straps use about 3 wt% Sb. Low Sb content promotes the formation of solidification shrinkage porosity and cracking but the cracking tendency is overcome by the use of grain-refining additions of S, Cu, and Se. Arsenic additions to Pb–Sb alloys increase the rate of age-hardening and reduce the time of grid storage required after casting. Arsenic addition also increases the creep resistance which is very beneficial in deep cycling conditions. The addition of tin is used to act synergistically with As and Sb to improve fluidity and castability. It also increases cycle life of deep cycling batteries containing thin plates. Silver additions increases both the corrosion and creep resistance in Pb–Sb alloy grids. Cast Pb–Sb based alloys are typically used in grid alloys as the Pb–Sb based wrought alloys have lower yield strength, tensile strength, and creep strength. The corrosion behavior of wrought alloys is inferior due to the nature of distribution of the PbSb eutectic phase and lower creep resistance. Corrosion of cast Pb–Sb occurs by the attack of Pb–Sb eutectic. It solubilizes some Sb and stresses of corrosion product are accommodated. In rolled alloys the eutectic phase is isolated, which leads to stresses in the grid. The current choice of alloy composition is Pb–1.6Sb–0.2As–0.2Sn–0.2Se. Table 3.1-280 lists the compositions of common lead–antimony battery grid alloys [1.300, 307].

Lead–antimony alloys containing 0.2 to 1 wt% Sb are used to form barrier sheaths in high voltage cables. Properties of Pb–0.85Sb cable sheathing alloy are presented in the section on cable sheathing alloys. Lead alloys with 6–8 wt% Sb are used to fabricate a wide variety of equipment such as tank linings, pipe and one type of anode used in chromium plating. Alloys with 13 wt% Sb are used to make castings when hardness is of key importance. About 6% of the Pb produced in the world was used in the production of sports and military ammunition due to its high density and low cost. Lead containing up to 8 wt% Sb and 2 wt% As is used.

Pb–Sb–Sn Alloys

These alloys have low melting points, high hardness, and excellent high temperature strength and fluidity. These characteristics together with its applicability for the replication of detail make them suitable as printing types. Table 3.1-281 lists the characteristics of selected type-metal alloys [1.300].

The Pb-rich ternary Pb–Sb–Sn white metal alloys are also used in journal bearings due to the excellent anti-friction (and anti-seizure) characteristics and hardness. These Pb-rich white metal alloys, also referred to as Babbit alloys, contain 9–15 wt% Sb, 1–20 wt% Sn, and small amounts of Cu and As. Table 3.1-282 lists the physical properties of different bearing alloys [1.299, 300]. The mechanical property data for some of these alloys are presented in Table 3.1-283 [1.299, 308]. Most of the alloys lie in the primary crystallization field of Sb or SbSn of the ternary system. They contain primary crystals of Sb (or SbSn) in a binary (or ternary) eutectic matrix apart from the high-melting Cu-rich phases. Copper contents above 1.5 wt% also increase the hardness. Arsenic addition leads to a fine and uniform structure, improves fatigue strength, and minimizes softening. Arsenic is present in solid solution in Pb, Sb, and Sb-containing phases such as SbSn. The lead–alkali alloys, e.g., Bahnmetall, or the Pb–Sn–alkali alloys, also have a limited significance as bearing metals.

Table 3.1-280 Alloying components of common lead–antimony battery grid alloys [1.300, 307]

Alloy concentration (wt%)				
Sb	Sn	As	Cu	Se
2.75	0.2	0.18	0.075	–
2.75	0.3	0.3	0.075	–
2.9	0.3	0.15	0.04	–
2.9	0.3	0.15	0.05	–
1.6	0.2	0.2	–	0.2

Table 3.1-281 Typical compositions and properties of selected type metals [1.299, 300]

Alloy	Composition (wt%)			Hardness HB	Liquidus temperature (°C)	Solidus temperature (°C)
	Pb	Sn	Sb			
Electrotype – General	94	3	3	14	299	245
Linotype – Special	84	5	11	22	246	239
Stereotype – Flat	80	6	14	23	256	239
Monotype – Ordinary	78	7	15	24	262	239

Table 3.1-282 Composition and physical properties of selected lead-based white metal bearing alloys [1.299, 300]

Alloy composition (wt%)		Freezing range	Density	Volume change on freezing[a]	Coefficient of thermal expansion	Specific heat	Latent heat of fusion	Electrical resistivity	Thermal conductivity
Sb	Sn	(°C)	(g cm^{-3})	(%)	(10^{-6} K^{-1})	(J kg^{-1} K^{-1})	(kJ kg^{-1})	(nΩ m)	(W m^{-1} K^{-1})
9.5–10.5	5.5–6.5	256–240	10.50	−2		150	0.9	287	
14–16	4.5–5.5	272–240	9.96	−2	24	150	0.1	282	
14–16	9.3–9.7	268–240	9.70	−2.3	19.6	160		286	24
14.5–17.5	0.8–1.2	353–247	10.10	−2.5					

[a] Negative values show contraction

Table 3.1-283 Composition and mechanical properties of ASTM B23 Pb-based white metal bearing alloys [1.299, 308]

Alloy no.	Nominal alloy content (wt%)				Yield stress (MPa)		Elongation at fracture (%)	Ultimate strength in compression (MPa)		Hardness HB 500/30		Melting range	Fatigue strength 2×10^7 cycles
	Sb	Sn	Cu	As	20 °C	100 °C	20 °C	20 °C	100 °C	20 °C	100 °C	(°C)	(MPa)
7	15	10.0	<0.5	0.45	24.5	11.0	4	107.9	42.4	22.5	10.5	240–268	28
8	15	5.0	<0.5	0.45	23.4	12.1	5	107.6	42.4	20.0	9.5	237–272	27
15	16	1.0	<0.5	0.8–1.4			2			21.0	13.0	248–281	30

3.1.11.3 Pb–Sn Alloys

Lead-tin alloys serve as materials for a number of applications as summarized in Table 3.1-284.

The Pb–Sn system (Fig. 3.1-355) shows an extended and strongly temperature-dependent solid solubility of Sn in Pb decreasing from 19.2 wt% at the eutectic temperature to about 1.3 wt% Sn at room temperature. This can lead to significant age-hardening on rapid cooling from the range of the homogeneous α phase. Streaky and granular Sn precipitates lead to hardness increases from about HB = 4 in pure lead to around HB = 12 at the solid solubility limit. In the stable $\alpha + \beta$ range hardness increases less rapidly to about HB = 18 at the eutectic composition.

Pb–Sn-Based Solder Alloys

Lead–tin alloys in the Pb-rich hypoeutectic region are the most widely used of all solders. In the melt, the surface tension increases with Sn content. Table 3.1-285 gives the melting characteristics of some Pb–Sn solders and lists their typical applications. When referring to Pb–Sn solders, the Sn content is customarily given first, for example 40/60 refers to 40 wt% Sn and 60 wt% Pb. Table 3.1-286 summarizes the mechanical and physical properties of different soft solders [1.309]. Further alloying additions are Cd, Bi, Sb, and Ag. Silver is added to increase tensile, creep, fatigue, and shear strengths, and to reduce the dissolution of Ag from Ag alloy coatings. Addition of 5 to 6% in Sn content increases the tensile and creep strengths. Addition

Table 3.1-284 Applications and typical compositions of Pb–Sn materials

Application	Alloying elements (wt%)		Remarks
	Sn	Others	
Cable sheathing	<0.5	Sb (0.2), Ca (0.33) or Cd (0.15)	
Solders	2–63		See below
Pressure die castings	∼62 (near eutectic)		
Organ pipes	(45) 55–74		
Sliding layer on bearings	10–20	5 Cu	Electroplated
Type metals	3–12	Sb (3–25)	See Table 3.1-281
Terne steel coatings	12–20		Corrosion protection
Anodes	7		Cr plating

Table 3.1-285 Melting characteristics and applications of Sn−Pb solders [1.309]

Solder alloy Sn/Pb	Composition (wt%) Sn	Pb	Temperature (°C) Solidus	Liquidus	Pasty range	Uses
2/98	2	98	316	322	10	Side seams for can manufacture
5/95	5	95	305	312	13	For coating and joining metals
10/90	10	90	268	302	62	For coating and joining metals
15/85	15	85	227	288	110	For coating and joining metals
20/80	20	80	183	277	170	For coating and joining metals. For filling dents or seams in automobile bodies
25/75	25	75	183	266	150	For machine and torch soldering
30/70	30	70	183	255	130	For machine and torch soldering
35/65	35	65	183	247	116	General purpose and wiping solder
40/60	40	60	183	238	99	Wiping solder for joining lead pipes and cable sheaths. For automobile radiator cores and heating units
45/55	45	55	183	227	80	For automobile radiator cores and roofing seams
50/50	50	50	183	216	60	For general purpose. Most popular of all
60/40	60	40	183	190	13	Primarily used in electronic soldering applications where low soldering temperatures are required
3/37	63	37	183	183	0	Lowest melting (Eutectic) solder for electronic applications

Table 3.1-286 Physical and mechanical property data on Pb–Sn solders [1.309]

Material (wt%)	Tensile strength (MPa)	Shear strength (MPa)	Density (g cm^{-3})	Brinell hardness HB	Electrical conductivity (% of σ_{Cu})	Young's modulus (GPa)	Surface tension (290 °C) (10^{11} N m^{-1})
Pb	12.30	12.44	11.34	4.0	7.9	18.04	
5/95 Sn/Pb	28.96	20.73	11.06	9.0	8.0		
10/90 Sn/Pb	32.48	26.96	10.44	11.0	8.2	19.08	
15/85 Sn/Pb	34.56	30.89	10.50	11.3	8.4		
20/80 Sn/Pb	35.94	32.76	10.23	11.5	8.7	20.04	467
25/75 Sn/Pb	37.32	36.70	10.00	11.5	8.9		
30/70 Sn/Pb	39.74	38.01	9.74	11.3	9.3	21.08	470
35/65 Sn/Pb	41.82	38.64	9.50	11.0	9.8		
40/60 Sn/Pb	42.85	39.26	9.29	10.5	10.1	23.08	474
45/55 Sn/Pb	42.85	39.53	9.08	10.6	10.4		
50/50 Sn/Pb	44.58	40.57	8.88	11.0	10.9		476
60/40 Sn/Pb	44.23	39.40	8.51	12.0	11.3	30.07	
63/37 Sn/Pb	46.31	41.88	8.41	12.0	11.5		490
62Sn/36Pb/2Ag	46.31	43.20	8.42	16.0	11.6	23.57	
10Sn/88Pb/2Ag	33.87	29.72	10.75	12.0	8.4	19.35	
1Sn/97.5Pb/1.5Ag	24.88	24.88	11.28	11.0	8.8		

of 0.18 wt% Cu causes a further increase. Bismuth-containing solders, the so-called fusible alloys, are used for low temperature soldering and are discussed in the section on fusible alloys. Alloys of Pb−In are primarily used for soldering at low temperatures and where reduction in gold-scavenging is desired. They are also extremely ductile, making them suitable for use in areas where there is a thermal mismatch. Compositions of other commonly used Pb solders are given in Table 3.1-287 [1.309].

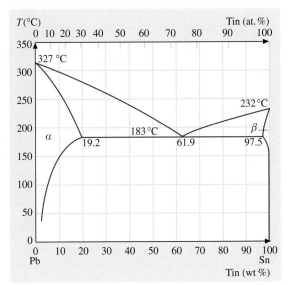

Fig. 3.1-355 The Pb–Sn binary phase diagram [1.303]

3.1.11.4 Pb–Ca Alloys

The Pb–Ca and Pb–Ca–Sn alloys are used in storage battery grid, pipe, wire, cable sheathing, anodes, chemical handling equipment, radiation shielding, and other applications [1.299]. Calcium is also used as a secondary additive in hardened lead-bearing metals. Its solubility in Pb decreases from 0.1 wt% Ca at 328.3 °C to ~ 0.01 wt% Ca at room temperature (Fig. 3.1-356) and pronounced age-hardening can be obtained. A peritectic

Table 3.1-287 Other common Pb-alloy solders [1.309]

Composition (wt%)				Temperature (°C)		Pasty range
Pb	Sn	Ag	In	Solidus	Liquidus	
97.5	1	1.5	–	309	309	Eutectic
36.0	62	2.0	–	179	189	18
97.5	–	2.5	–	304	304	62
50	–	–	50	124	209	52
92.5	–	2.5	5	285	305	36

reaction involving liquid Pb–Ca and Pb_3Ca to form the α-Pb–0.1 wt% Ca phase occurs at 328.3 °C [1.303]. At > 0.07 wt% Ca, Pb_3Ca crystallizes directly on solidification. At > 0.1 wt% Ca, the microstructure consists of primary crystals of Pb_3Ca and a Pb matrix with finer Pb_3Ca precipitates. The two-phase structure present at these higher Ca contents leads to grain refinement. Supersaturated solid solutions of Ca in Pb at room temperature can be obtained at high cooling rates upto about 0.13 wt% Pb. The hardness increase observed on aging increases with Ca content. Upon aging, the hardness of a 0.07 wt% Ca alloy at room temperature increases from HB = 4 to HB = 8.25 (10 mm-31.2 kg-120 s) in 6 h. Electrical resistivity drops from 22.63 to 22.25×10^{-6} Ω cm. Maximum in hardness for quenched alloys occurs at 0.13 wt% Ca and in air-cooled alloys at 0.085 wt% Ca.

Further additions of Li, Ba, and Na increase the hardness. The addition of Sn to Pb–Ca alloys increases the hardness, tensile strength, and stress rupture properties. The hardness decreases as Sb and Bi form

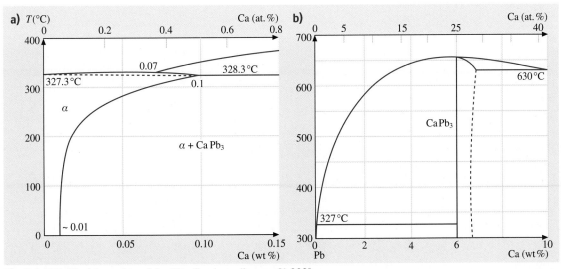

Fig. 3.1-356 Pb-rich portion of the Pb–Ca phase diagram [1.303]

intermetallic compounds with Ca and segregate from the melt. Due to the presence of finely distributed precipitate phase, the age-hardened alloys show a high resistance to recrystallization after room temperature working. The creep rate increases with Ca content, mainly due to the smaller grain size. The mechanical properties of binary Pb–Ca and Pb–Ca–Sn alloys are presented in Table 3.1-288 [1.299, 306]. The fine-grained wrought Pb–Ca and Pb–Ca–Sn alloys possess improved material integrity and also exhibit improved corrosion performance, as they tend to undergo uniform corrosion. The corrosion resistance of these alloys is higher than that of Pb–Sb alloys in many applications.

Pb–Ca–Sn Battery Grid Alloys

Most batteries produced currently use Pb–Ca alloys for grids and connectors and the use of Pb–Sb based alloys is declining [1.307]. The Ca content varies from 0.03 to 0.13 wt%. The narrow freezing range of a few degrees allows continuous casting and grid production. The mechanical properties in Pb–Ca binary alloys peak at 0.07 wt%. Above 0.06 wt% Ca, cellular precipitation of Pb_3Ca leads to fine grain size. Increasing Ca contents above 0.07 wt% level accelerate corrosion and this is believed to be due to fine grains and primary Pb_3Ca. Addition of Sn dramatically improves the properties by promoting the formation of the more effective and stable Sn_3Ca. The phase relations of Pb rich Pb–Ca–Sn alloys according to [1.310] are shown in Fig. 3.1-357. Tin additions also improve the corrosion resistance and the charge–discharge process by Sn enrichment of the corrosion product layer. Aluminium acts as a deoxidant and prevents drossing loss of Ca. The addition of Sn aids electrochemical properties by preventing passivation of the grid and permitting recharge of batteries from a deeply discharged condition. Additions of Ag to Pb–Ca and Pb–Ca–Sn alloys increase the creep and corrosion resistance. However, Sn and Ag additions are not re-

Table 3.1-288 Mechanical properties of Pb–Ca–Sn alloys and corrosion rates in battery environments [1.299, 306]

Composition (wt%)		Tensile strength (MPa)	Yield stress ($R_{p0.125}$) (MPa)	Elongation (%)	Creep to failure, at 20.7 MPa (h)	Corrosion rate in battery service (mm yr^{-1})
Ca	Sn					
Cast						
0.025		25.1	17.7		1	0.279
0.050		37.2	29.0		30	0.345
0.075		46.4	35.3		40	0.358
0.100		47.8	32.5		10	0.411
0.025	0.5	25.5	19.3		10	0.256
0.050	0.5	48.2	38.5		70	0.310
0.065	0.5	48.9	40.0		200	0.325
0.075	0.5	50.3	40.2		300	0.325
0.100	0.5	51.7	38.7		70	0.343
0.025	1.5	45.8	34.3		30	0.246
0.050	1.5	55.1	46.8		100	0.271
0.075	1.5	60.1	49.2		1000	0.297
0.100	1.5	57.9	43.7		250	0.345
Wrought						
0.050	0.5	55.2	45.3	30	10[a]	
0.07	0.5	62.1	45.0	30	20[a]	
0.05	1.0	61.4	52.8	25	150[a]	
0.07	1.0	68.9	64.0	15	400[a]	
0.050	1.5	63.8	57.4	15	300[a]	
0.07	1.5	71.0	65.3	14	1000[a]	

[a] At 27.6 MPa

Table 3.1-289 Currently preferred compositions of automotive battery-grid alloys [1.307]

Manufacturing process; grid type	Composition (wt%)			
	Ca	Sn	Ag	Al
Book mold cast; positive	0.03–0.06	0.5–0.9	0.010–0.045	0.01–0.02
Book mold cast; negative	0.09–0.13	0–0.3	Trace	0.015–0.03
Rolled, expanded; positive	0.06–0.08	1.2–1.6	Trace	0.003–0.008
Rolled, expanded; negative	0.06–0.08	0–0.5	Trace	0.003–0.008
Concast strip, expanded; positive	0.03–0.06	0.4–0.6	0.03–0.045	0.005–0.010
Concast strip, expanded; negative	0.07–0.1	0–0.2	Trace	0.005–0.010
Continuously cast; negative	0.07–0.1	0–0.3	Trace	0.005–0.010

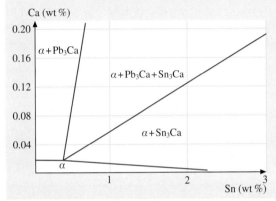

Fig. 3.1-357 Suggested phase relations of Pb-rich Pb–Ca–Sn alloys [1.310]

quired in negative grids to provide the corrosion and creep resistance. The Ca content in positive grids is lower than in negative grids to reduce corrosion. Table 3.1-289 presents the compositions of Pb–Ca–Sn alloys currently preferred in lead acid automotive batteries [1.307]. Table 3.1-290 presents mechanical and properties of commonly used Pb–Ca–Sn battery alloys [1.299, 306].

3.1.11.5 Pb–Bi Alloys

Pb–Bi Binary Alloys

Even though the relative difference of atomic radii of Bi (1.56) and Pb (1.75) amounts to about 12%, the solubility of Bi reaches 23.5 wt% at 184 °C and 7.5 wt% at room temperature. While Bi addition has very little influence on mechanical properties, Pb–Bi alloys' excellent wetting properties make them valuable as solders for glass-to-metal joints. Their desirable solidification shrinkage characteristics and casting properties (that provide an ability to reproduce surface details) make them useful in printing and prototyping applications.

The Pb–Bi alloy system exhibits behavior of an eutectic between an hcp intermetallic β phase and the Bi terminal phase at 56.5 wt% Bi and 125 °C. Both Pb and Bi have low cross sections for neutron absorption such that these alloys are attractive in heat transfer applications in nuclear reactor systems [1.299].

Fusible Alloys

Lead forms a number of extremely useful low-melting alloys when combined with Bi, Sn, Cd, or a combination of these metals. The metals In, Sb, and Ag are also

Table 3.1-290 Properties of some of selected Pb–Ca–Sn battery-grid alloys [1.299, 300]

Alloy composition (wt%)	Liquidus (°C)	Solidus (°C)	Ultimate tensile strength (MPa)	Elongation (%)	Hardness (HR)	Coefficient of thermal expansion (10^{-6} K^{-1})	Resistivity (nΩ m)
Pb–0.065Ca–0.7Sn		327					219
Pb–0.065Ca–1.3Sn		323				26.6	220
Pb–0.07Ca		328	36–39	35–40	70–80	30.2	218
Pb–0.1Ca–0.3Sn	338	328	41–45	20–35	90–95		219
Pb–0.1Ca–0.5Sn	336	327	44.8–51.7	25–35	85–90		219
Pb–0.1Ca–1Sn	332	325	52–55	20–35	90–95		212

added in some of the alloys. Some of these alloys melt at a temperature lower than the boiling point of water, and those containing appreciable amounts of Bi (> 55 wt%) expand slightly upon solidification.

The melting point of the Pb–Sn–Cd ternary eutectic is only 145 °C. By the addition of Zn, a quaternary eutectic can be obtained with a melting point of 138 °C at a composition of Pb–16.7 wt% Cd–52.45 wt% Sn–2.25 wt% Zn. A further effective decrease in the melting point is obtained by additions of Bi to Pb–Cd–Sn alloys. The quaternary Bi–Pb–Cd–Sn eutectic has a melting point of about 70 °C and a composition of Pb–50 wt% Bi–12.5 wt% Cd–12.5 wt% Sn (Wood's metal). The phases of the quaternary eutectic are solid solutions corresponding to the Pb, Sn, and Cd phases, as well as and the β-phase of the Pb–Bi system. The quaternary Pb–Bi–Sn–Cd eutectic alloy is brittle when cast, and becomes ductile on storage for two to three hours. The melting point of the quaternary eutectic alloy can be further lowered to 47 °C by additions of In. Addition of Hg instead of In could also lower the melting point of Pb–Bi–Sn–Ca eutectic but is not used due to its high vapor pressure and toxicity. In addition to the above quaternary eutectic alloys, alloys of the ternary-Pb–Bi–Sn system are also of great technical importance. One of these alloys is the Newton metal (Pb 50 wt% Bi 20 wt% Sn) that has approximately the ternary eutectic composition. The melting temperature of this eutectic alloy is 90 °C. Another important alloy is Rose's metal, with a composition Pb–50 wt% Bi–25 wt% Sn and a melting temperature of 100 °C. Table 3.1-291 presents the compositions and properties of selected fusible alloys [1.299, 300].

Low melting alloys are employed in safety devices such as sprinkler systems and boiler plugs, as special solders where high temperatures cannot be used, hermetic seals for molds, patterns, punches and dies, for anchoring punches in punch plates, and for bending tubing (Table 3.1-292).

3.1.11.6 Pb–Ag Alloys

The addition of Ag to Pb in the range of 0.01–0.1 wt% provides high resistance to recrystallization, grain refinement, and high creep strength. Eutectic Pb–Ag

Table 3.1-291 Compositions and properties of selected fusible alloys [1.299, 300]

Alloy composition[a] (wt%)	Liquidus (°C)	Solidus (°C)	Volume change on freezing (vol.%)[b]	Density (g cm^{-3})	Conductivity (% of IACS)	Coefficient of thermal expansion (10^{-6} K^{-1})	Specific heat (J kg^{-1} K^{-1})	Latent heat of fusion (kJ kg^{-1})	Thermal conductivity (W m^{-1} K^{-1})
Pb–42Bi–11Sn–9Cd	88	70	2	9.45	4	24	168	23	21
Pb–42.9Bi–5.1Cd–7.9Sn–4Hg–18.3In	43	38							
Pb–44.7Bi–5.3Cd–8.3Sn–9.1In ASTM Alloy 117	47	47	1.4	8.85	4.5		147	14	
Pb–48Bi–14.5Sn–9Sb	227	103	1.5	9.50	3	22	189		
Pb–49Bi–21In–12Sn ASTM Alloy 136	58	58	1.5	8.60	3		134.4		21
Pb–50Bi–9.3Sn–6.2Cd	70	78							
Pb–50Bi–10Cd–13.3Sn ASTM Alloy 158	70	70	1.7	9.40	4	22	168	32	
Pb–51.7Bi–8.1Cd	92	92		10.25					
Pb–52.5Bi–15.5Sn ASTM Alloy 203	96	96		9.71					
Pb–55Bi ASTM Alloy 255	124	124	1.5	10.30	3		126	16	16.8

[a] Alloys may contain small amounts of Ag, Cu, Sb and Zn
[b] Positive values indicate expansion on freezing

Table 3.1-292 Typical applications of some common fusible alloys [1.299, 300]

Alloy	Melting temperature (°C)	Typical applications
ASTM 117	47	Dental models, part anchoring, and lens chucking
ASTM 158 (Woods metal)	70	Bushings and locators in jigs and fixtures, lens chucking, reentrant tooling, founding cores and patterns, light sheet-metal embossing dies, tube bending, and Wood's metal-sprinkler heads
ASTM 255	124	Inserts in wood, plastics, bolt anchors, founding cores and patterns, embossing dies, press-form blocks, duplicating plaster patterns, tube bending, and hobbyist pans
ASTM 281	138	Locator members in tools and fixtures, electroforming cores, dies for lost-wax patterns, plastic casting molds, prosthetic development work, encapsulating avionic components, spray metalizing, and pantograph tracer molds
Pb–48 wt% Bi–14.5 wt% Sn–9 wt% Sb	103–217	Punch and die assemblies, small bearings, anchoring for machinery, tooling, forming blocks, and stripper plates in stamping dies

alloys containing about 2.5 wt% Ag are used as soft solders of high melting point. The alloy Pb–0.1 wt% Ag is used as a precoat to metallurgically bond lead to steel. Alloys of Pb–0.8–1 wt% Ag are used as insoluble anodes for electrowinning of metals from leach solutions and for electrogalvanizing. The alloys Pb–6 wt% Sb–1 wt% Ag, Pb–1 wt% Ag, Pb–2 wt% Ag, and Pb–1 wt% Ag–1 wt% Sn are used as anodes for cathodic protection.

3.1.11.7 Pb–Cu, Pb–Te, and Pb–Cu–Te Alloys

Maximum solubility in Pb is very low for both Cu (< 0.007%) and Te (< 0.005%). Copper contents of less than 0.1% provide considerable grain refinement and structural stability at high temperature. Pb–Cu alloys are used in cable sheathing and chemical applications. A Te content of 0.01% refines and stabilizes the grain size and increases the work-hardening in Pb–Te alloys. Significant age-hardening is also obtained from supersaturated solid solutions containing up to 0.1 wt% Te. The optimal Te content is 0.04–0.05% with Cu addition of 0.06%. Alloys of Pb–Te have very high fatigue strengths and are used in cable sheathing, radiation shielding, and steam-heating coil applications.

3.1.11.8 Pb–As Alloys

The alloy Pb–0.85 wt% As has very low volume shrinkage on solidification and pore free castings obtained with this alloy are used in radiation protection applications. Arsenic is added in Pb–Sb alloys to accelerate age hardening. It is also used in Pb cable sheathing alloys to enhance the bending and creep resistance.

3.1.11.9 Lead Cable Sheathing Alloys

Lead alloys are used as cable sheath in the construction of electrical cables for communication and high voltage transmission. The cable sheath serves as an impermeable barrier to prevent access of moisture to the insulated core, a containment of the oil or gas in oil- or gasfilled cables, and a ground to the power cables under short circuit conditions. The sheaths need to be easily applied in long lengths and should have good creep resistance, fatigue resistance, corrosion resistance, and microstructural stability. Table 3.1-293 lists some of cable sheathing alloys and their typical applications.

3.1.11.10 Other Lead Alloys

Alloys of Pb–Li are attractive in some nuclear shielding applications due to their ability to thermalize neutrons. Lead containing more than 0.5 wt% In wets glass, and Pb–In alloys with up to 5 wt% In can be used for soldering glass over a narrow temperature range. Additions over 25 wt% of In are made to Pb–Sn solders to increase their alkali resistance. An addition of 1–2 wt% In in Pb–Ag solders increases their strength. Indium is also used in multi-component fusible alloy systems.

Table 3.1-293 Compositions of commonly-used cable sheathing alloys [1.299]

Alloy name	Nominal composition (wt%)	Fatigue strength (MPa) at 10^7 cycles 37–50 Hz	Applications
Alloy B	Pb-0.8-0.95 Sb	9.6	Solid type cables and telecommunication cables subjected to severe vibrations
Alloy 1/2B	Pb-0.5 Sb	7.5	Solid type cables and telecommunication cables subjected to severe vibrations
Alloy C	Pb-0.35–0.45 Sn-0.12–0.18 Cd	5.4	Power cables in ships. Acceptable for most types of cables
Alloy 1/2C	Pb-0.18–0.22% Sb-0.06–0.09 Cd	4.2 0.07%[a]	Oil-filled and submarine power cables. Power cables subjected to severe vibrations in service. Acceptable for most types of reinforced cables
Alloy E	Pb-0.35–0.45 Sn-0.15–0.25 Sb	6.3	Solid type power cables, telecommunication cables, and reinforced gas or oil pressurized power cables under moderate vibrations
Pb–Cu	Pb-0.06 Cu	3.34	Gas or oil pressurized power cables
Cu–Te	Pb-0.06 Cu-0.045 Te	6.9	Cables subjected to high vibrations and oil-filled cables
F3	Pb-0.15 As-0.15 Sn-0.1% Bi	6.1	PILC cables, submarine cables, power cables subjected to severe bending or vibration conditions
Pb–Ca–Sn	Pb-0.033 Ca-0.38 Sn	0.11%[a]	High-voltage DC and AC submarine cables

[a] Reversible strain for 10^7 cycle at 30 Hz in reversible cantilever bending

References

1.1 E. G. Emley: *Principles of Magnesium Technology* (Pergamon, New York 1966)

1.2 IMA: *Annual Report* (Int. Magnesium Association, Washington, DC 2001)

1.3 M. M. Avedesian, H. Baker: *Magnesium and Magnesium Alloys*, ASM Specialty Handbook (ASM, Metals Park 1999)

1.4 I. J. Polmear: *Light Metals, Metallurgy of the Light Metals* (Wiley, New York 1995)

1.5 G. Neite: Structure and properties of nonferrous alloys. In: *Materials Science and Technologie*, Vol. 8, ed. by K. H. Matucha (Verlag Chemie, Weinheim 1996)

1.6 C. S. Roberts: *Magnesium and Its Alloys* (Wiley, Chichester 1960)

1.7 C. Kammer: *Magnesium Taschenbuch* (Aluminium Verlag, Düsseldorf 2000)

1.8 G. L. Song, A. Atrens: Corrosion mechanisms of magnesium alloys, Adv. Eng. Mater. **1** (1999)

1.9 C. Kammer: *Aluminium Handbook 1, Fundamentals and Materials* (Aluminium Verlag, Düsseldorf 2002)

1.10 J. R. Davis (Ed.): *Aluminium and Aluminium Alloys*, ASM Specialty Handbook (ASM, Metals Park 1993)

1.11 H. W. L. Phillips: *Equilibrium Diagrams of Aluminium Alloy System* (The Aluminium Development Association, London 1961)

1.12 J. Thonstad, P. Fellner, G. M. Haarberg, J. Hives, H. Kvande, A. Sterten: *Aluminium Electrolysis - Fundamentals of the Hall-H'eroult Process*, 3rd edn. (Aluminium Verlag, Düsseldorf 2001)

1.13 International Aluminium Institute (AIA, London 2003), www.world-aluminium.org and www.world-aluminum.org

1.14 E. Nachtigall, H. Landerl: The treatment of the conductor alloy E-AlMgSi, Aluminium Ranshofen Mitteilungen **2**, 40–43 (in German) (1955)

1.15 E. Nachtigall, G. Lang: Electrical conductivity of aluminium castings, Mitt. Verein. Metallwerke Ranshofen-Berndorf, 16–19 (in German) (1965)

1.16 L. F. Mondolfo: *Aluminium Alloys - Structure and Properties* (Butterworths, London 1976)

1.17 C. Kammer: Thermomechanical treatment of Al strip casting. Ph.D. Thesis (TU Bergakademie Freiberg, Freiberg 1989) (in German)

1.18 E. Schürmann, I. K. Geissler: Solid-state phase equilibria in the Al and Mg rich areas of the Al-Mg-Li system, Giessereiforschung **32**, 163–174 (in German) (1981)

1.19 A. Dons: AlFeSi-Particles in industrially cast aluminium alloys, Z. Metallkunde **76**, 609–612 (1985)

1.20 A. Cziraki, B. Fogarassy, I. Szábo: Structure of high purity Al-Fe-Si-cast with different cooling rates, Cryst. Res. Technol. Berlin **20**, 279–281 (1985)

1.21 H. Westengen: Structure inhomogenities in direct chill cast sheet ingots of commercial pure aluminium, Z. Metallkunde **73**, 360–368 (1982)

1.22 H. P. Godard, W. B. Jepson, M. R. Bothwell, R. L. Kane: *The Corrosion of Light Metals* (Wiley, New York 1967)

1.23 K. R. Van Horn: *Aluminium. Bd. 1, Properties, Physical Metallurgy and Phase Diagrams* (ASM, Metals Park 1967)

1.24 D. Altenpohl: *Aluminium and Aluminium-Alloys* (Springer, Berlin, Heidelberg 1965 (in German))

1.25 U. Hielscher, H. Arbenz, H. Diekmann: Properties of AlSi-casting alloys with low iron content, Giesserei **53**, 125–133 (in German) (1966)

1.26 U. Hielscher: Ductile aluminium-silicon casting alloys for safety components in cars, Schweiz. Alum. Rundsch. **29**, 13–15 (in German) (1979)

1.27 U. Hielscher, R. Klos: A new low iron diecasting alloy, Aluminium **71**, 676–685 (in German) (1995)

1.28 W. Jung-König, U. Zwicker: Behaviour of light metal alloys on heating, Aluminium **34**, 337–345 (in German) (1958)

1.29 H. Vosskühler: Aluminium-Gusslegierungen hoher Dauerstandfestigkeit mit Magnesium und Silicium, Aluminium **31**, 219–222 (1955)

1.30 K. Wellinger, E. Keil, G. Maier: Strength of aluminium and its alloys up to 300 °C, Aluminium **34**, 458–463 (in German) (1958)

1.31 K. Wellinger, E. Keil et al.: On the mechanical properties of aluminium and aluminium alloys at elevated temperatures, Aluminium **39**, 372–377 (in German) (1963)

1.32 E. Richter, E. Hanitzsch: Elastic modulus and other physical properties of aluminium-base materials. Part 1., Aluminium **70**, 570–574 (in German) (1994)

1.33 D. Lenz, G. M. Renouard: Definition of cold rolled tempers by means of flow curves and energy of deformation, Aluminium **46**, 694–699 (in German) (1970)

1.34 A. Odok, G. Thym: The technical and economic advantages of continuous strip casting, Aluminium Engl. Suppl. **50**, E9–E11 (English transl.) (1974)

1.35 C. Kammer, M. Krumnacker et al.: Thermomechanical treatment of continuously cast and rolled Al99.5 alloy, Neue Hütte **35**, 418–421 (in German) (1990)

1.36 C. Kammer, M. Krumnacker et al.: Comparison of the strengthening effects in strip cast Al99.5, AlMn1 and AlMn1Fe1, Metall **45**, 135–138 (in German) (1991)

1.37 J. Althoff: Properties and uses of a new heat-resistant Al-Mn alloy, Metall **31**, 263–267 (in German) (1977)

1.38 J. Althoff: Examples of the application-orientated development of high-strength aluminium manganese alloys, Aluminium Engl. Suppl. **56**, E37–E39 (English transl.) (1982)

1.39 P. Brenner, H. Kostron: Treatment of AlMgSi-alloys, Z. Metallkunde **31**, 89–97 (1939)

1.40 I. Novikov: *Theory of Heat Treatment of Metals*, 1st edn. (Metalurgija, Moscow 1978)

1.41 Y. Takeuchi: Effect of plastic deformation on the natural ageing of AlCuMg1, AlCuMg2 and AlZnMgCu0.5, Aluminium **47**, 665–670 (in German) (1971)

1.42 K. Lücke, P. Stüwe: On the theory of impurity cotrolled grain boundary motion, Acta Met. **19**, 1067–1099 (1971)

1.43 H. Warlimont: Effect of segregation and precipitation on the recrystallisation and grain size of non-ferrous metals, Freiberg. Forschungsh. B **200**, 31–57 (in German) (1979)

1.44 D. B. Goel, P. Furrer, H. Warlimont: Precipitation behaviour of AlMnCuFe-alloys, Aluminium **50**, 511–516 (in German) (1974)

1.45 E. Nes, S. Slevolden: Casting and annealing structures in strip cast alloy, Aluminium **55**, 319–324 (1979)

1.46 E. Nes, J. D. Embury: The influence of a fine particle dispersion on the recrystallisation behaviour of a two phase aluminium alloy, Z. Metallkunde **66**, 589–593 (1975)

1.47 E. Nes, S. Slevolden: The concept of a grain size diagramm in the analysis of the recrystallisation behaviour of AlMn-alloys, Aluminium **52**, 560–563 (1976)

1.48 E. Nes: The effect of a fine particle dispersion on heterogeneous recrystallisation, Acta Met. **24**, 391–398 (1976)

1.49 C. Kammer, M. Krumnacker et al.: Thermomechanical treatment of strip cast AlMn1Fe1 alloy, Metall **43**, 1162–1168 (in German) (1993)

1.50 W. Huppatz: The fundamentals of corrosion protection of aluminium alloys used as structural materials. Part 1, Metall **49**, 505–509 (in German) (1995)

1.51 K. H. Matucha: Structure and properties of nonferrous alloys. In: *Materials Science and Technology*, Vol. 8, ed. by R. W. Cahn, P. Haasen, E. J. Kramer (VCH, Weinheim 1996)

1.52 J. R. Davis (Ed.): *Heat-Resistant Materials*, ASM Specialty Handbook (ASM, Metals Park 1997)

1.53 H. Sibum, G. Volker, O. Roidl, H. U. Wolf: Titanium and titanium alloys. In: *Ullmann's Encyclopedia of Industrial Chemistry*, Vol. A27 (VCH, Weinheim 1996) pp. 95–122

1.54 J. L. Murray: *Phase Diagrams of Binary Titanium Alloys* (ASM, Metals Park 1990)

1.55 D. F. Williams: Medical and dental materials. In: *Materials Science and Technology*, Vol. 14, ed. by R. W. Cahn, P. Haasen, E. J. Kramer (VCH, Weinheim 1992)

1.56 S. Steiner: *Properties and Selection: Irons, Steels and High Performance Alloys*, Metals Handbook, Vol. 1, 8th edn. (ASM, Metals Park 1961)

1.57 G. Sauthoff: *Intermetallic Materials*, Landolt-Börnstein, New Series VIII/2, ed. by P. Beiss, R. Ruthardt, H. Warlimont (Springer, Berlin, Heidelberg 2002)
1.58 T. K. Roy, R. Balasubramanian, A. Ghosh: Metall. Mater. Trans. Trans. A **27**, 3993–4003 (1996)
1.59 Y. W. Kim: JOM-J. Min. Met. Mater. Soc. **46** (1994)
1.60 N. S. Stoloff, V. K. Sikka: *Pysical Metallurgy and Processing of Intermetallic Compounds*, 1st edn. (Chapman & Hall, London 1996)
1.61 Y. W. Kim, R. Wagner, M. Yamaguchi: Gamma titanium aluminides, Proc. ISGTA'95 (TMS, Warrendale 1995)
1.62 R. Darolia, J. J. Lewandowski, C. T. Liu, P. L. Martin, D. B. Miracle, M. V. Nanthal: Structural intermetallics, Proc. First Intl. Symp. (TMS, Warrendale 1993)
1.63 K. Otsuka, C. M. Wayman (Eds.): *Shape Memory Materials* (Cambridge University Press, Cambridge 1998)
1.64 K. Otsuka, T. Kakeshita (Guest Eds.): Science and technology of shape-memory alloys: New developments, Mater. Res. Soc. Bull. **27**, 91–129 (2002)
1.65 H. Horikawa: , Proc. 1st Europ. Conf. on Shape Memory and Superelastic Technologies, Antwerp 1999) 256
1.66 A. E. Pelton, S. M. Russell, J. DiCello: The physical metallurgy of nitinol for medical applications, JOM-J. Min. Met. Mater. Soc. **55**, 33–37 (2003)
1.67 W. Chang: Zirconium products: Technical data sheet (Allegheny Technologies Inc. Pittsburgh, PA), www.alleghenytechnologies.com/wahchang
1.68 C. Lemaignan, A. T. Motta: Zirconium alloys in nuclear applications. In: *Materials Science and Technology*, Vol. 10B/II, ed. by B. R. T. Frost (VCH, Weinheim 1994)
1.69 A. Inoue, T. Zhang, T. Masumoto: Production of amorphous cylinder and sheet of $La_{55}Al_{25}Ni_{20}$ alloy by a metallic mold casting method, JIM **31**, 425 (1990)
1.70 A. Peker, W. L. Johnson: Appl. Phys. Lett. **63**, 2342 (1993)
1.71 W. L. Johnson: Fundamental aspects of bulk metallic glass formation in multicomponent alloys, Mater. Sci. Forum **225–227**, 35 (1996)
1.72 A. Inoue: *Bulk Amorphous Alloys* (Trans Tech, Uetikon-Zurich 1998)
1.73 A. Takeuchi, A. Inoue: Mater. Sci. Eng. A **304–306**, 446 (2001)
1.74 T. Wada, T. Zhang, A. Inoue: Mater. Trans. **43**, 2843 (2002)
1.75 C. C. Hays, J. Schroers, U. Geyer, S. Bossuyt, N. Stein, W. L. Johnson: Glass forming ability in the Zr-Nb-Ni-Cu-Al bulk metallic glasses. In: *Metastable, Mechanically Alloyed and Nanocrystalline Materials*, ed. by H. Eckert, H. Schlörb, L. Schultz (Trans Tech, Uetikon-Zurich 1995)
1.76 G. He, W. Löser, J. Eckert, L. Schultz: Mater. Sci. Eng. A **352**, 179 (2003)
1.77 J. E. Truman: Stainless steels. In: *Constitution and Properties of Steels*, Materials Science and Technology, Vol. 7, ed. by F. B. Pickerling (VCH, Weinheim 1992) p. 527
1.78 W. Dahl (Ed.): *Eigenschaften und Anwendungen von Stählen* (Institut für Eisenhüttenkunde IEHK, RWTH Aachen, Aachen 1993) p. 727
1.79 J. R. Davis (Ed.): *Stainless Steels*, ASM Speciality Handbook (ASM, Metals Park 1994)
1.80 J. R. Davis (Ed.): *Carbon and Alloy Steels*, ASM Speciality Handbook (ASM, Metals Park 1996)
1.81 J. E. Bringas (Ed.): *Handbook of Comparative World Steel Standards* (ASTM, West Conshohocken 2001)
1.82 B. Predel (Ed.): *Phase Equilibria, Crystallographic and Thermodynamic Data of Binary Alloys*, Landolt-Börnstein, New Series IV/5 (Springer, Berlin, Heidelberg 1991–1998)
1.83 J. L. Lee, H. K. D. H. Bhadeshia: Mater. Sci. Eng. A **171**, 223–230 (1993)
1.84 G. Vander Voort: *Atlas of Time Temperature Diagrams*, Vol. 1, 2 (ASM, Materials Park 1991)
1.85 H. Schneider: Investment casting of high-hot strength 12% chrome steel, Foundry Trade J. **108**, 562 (1960)
1.86 K. Yagi, G. Merkling, H. Irie, H. Warlimont (Eds.): *Creep Properties of Heat Resistant Steels and Superalloys*, Landolt-Börnstein, New Series VIII (Springer, Berlin, Heidelberg 2004)
1.87 VDEh (Ed.): Stahl-Eisen-Werkstoffblatt (SEW) no. 470, Feb. 1976 (Verlag Stahleisen GmbH, Düsseldorf 1976)
1.88 J. R. Davis (Ed.): *Tool Materials*, ASM Speciality Handbook (ASM, Metals Park 1995)
1.89 J. R. Davis (Ed.): *Cast Irons*, ASM Speciality Handbook (ASM, Metals Park 1996)
1.90 W. Betteridge: *Cobalt and Its Alloys* (Ellis Horwood, New York 1982)
1.91 J. R. Davis (Ed.): *Nickel, Cobalt and Their Alloys*, ASM Specialty Handbook (The Materials Information Society, Materials Park 2000)
1.92 W. Gudat, O. Rader (Eds.): *Electronic Structure of Solids. Photoemission Spectra and Related Data. Magnetic Transition Metals*, Landolt-Börnstein, New Series III/23 (Springer, Berlin, Heidelberg 1999)
1.93 B. Predel: *Phase Equilibria*, Landolt-Börnstein, New Series IV/5 (Springer, Berlin, Heidelberg 1991–1998)
1.94 J. R. Davis (Ed.): *Tool Materials*, ASM Specialty Handbook (The Materials Information Society, Materials Park 1995)
1.95 P. Beiss, R. Ruthardt, H. Warlimont (Eds.): *Powder Metallurgy Data*, Landolt-Börnstein, New Series VIII/2 (Springer, Berlin, Heidelberg 2002)
1.96 J. R. Davis (Ed.): *Nickel, Cobalt and Their Alloys*, ASM Specialty Handbook (ASM International, Materials Park 2000)
1.97 J. R. Davis (Ed.): *Heat-Resistant Materials*, ASM Specialty Handbook (ASM International, Materials Park 1997)

1.98 J. C. Fuggle, U. Hillebrecht, R. Zeller, Z. Zolonierek, P. Bennet, C. Freiburg: Phys. Rev. B **27**, 719 (1982)

1.99 W. Gudat, O. Rader (Eds.): *Electronic Structure of Solids. Photoemission Spectra and Related Data. Magnetic Transition Metals*, Landolt–Börnstein, New Series III/23 (Springer, Berlin, Heidelberg 1999)

1.100 B. Predel: *Phase Equilibria, Crystallographic and Thermodynamic Data of Binary Alloys*, Landolt–Börnstein, New Series IV/5 (Springer, Berlin, Heidelberg 1991–1998)

1.101 K. Yagi, G. Merckling, H. Irie, H. Warlimont (Eds.): *Creep Properties of Heat Resistant Steels and Superalloys*, Landolt–Börnstein, New Series VIII (Springer, Berlin, Heidelberg 2004)

1.102 G. Joseph, K. J. A. Kundig: *Copper, Its Trade, Manufacture, Use, and Environmental Status* (ASM International, Materials Park 1998)

1.103 Wieland-Werke AG: *Wieland-Kupferwerkstoffe* (Ulm, Germany 1999)

1.104 J. R. Davis (Ed.): *Copper and Copper Alloys*, ASM Specialty Handbook (ASM, Metals Park 2001)

1.105 H. Warlimont, L. Delay: *Martensitic Transformations in Copper-, Silver-, and Gold-Based Alloys* (Pergamon, Oxford 1974)

1.106 B. Predel: *Cr–Cs–Cu–Zr*, Landolt–Börnstein, New Series IV/5 (Springer, Berlin, Heidelberg 1991–1998)

1.107 J. B. Conway, B. N. Flagella: *Creep Rupture Data for the Refractory Metals to High Temperatures* (Gordon Breach, New York 1971)

1.108 R. Kieffer, G. Jangg, P. Ettmayer: *Sondermetalle* (Springer, Vienna 1971) in German

1.109 American Society for Metals: *Properties and Selection: Nonferrous Alloys and Pure Metals*, Metals Handbook, Vol. 2, 9th edn. (American Society for Metals, Metals Park 1979)

1.110 W. C. Hagel, J. A. Shields, S. M. Tuominen: Processing and production of molybdenum and tungsten alloys, Proc. Symp. on Refractory Technology for Space Nuclear Power Applications, CONF-8308130 (Oak Ridge National Laboratory 1983) 98

1.111 K. H. Miska, M. Semchyshen, E. P. Whelan (Eds.): *Physical Metallurgy and Technology of Molybdenum and its Alloys* (AMAX, Michigan 1985)

1.112 J. Wadsworth, T. G. Nieh, J. J. Stephens: Recent advances in aerospace refractory metal alloys, Inter. Mater. Rev. **33**(3), 131 (1988)

1.113 E. Pink, I. Gaal: Mechanical properties and deformation mechanisms of non-sag tungsten wires. In: *The Metallury of Doped, Non-Sag Tungsten*, ed. by E. Pink, L. Bartha (Elsevier, New York 1989) p. 209

1.114 T. G. Nieh, J. Wadsworth: Recent advances and developments in refractory alloys, Mat. Res. Soc. Symp. Proc. **322**, 315 (1994), ISBN: 1-55899-221-9

1.115 E. Pink, R. Eck: Refractory metals and their alloys. In: *Materials Science and Technology – A Comprehensive Treatment*, Vol. 8, ed. by R. W. Cahn, P. Haasen, E. J. Kramer (VCH Verlag, Weinheim 1997) p. 589

1.116 E. Lassner, W. D. Schubert: *Tungsten: Properties, chemistry, technology of the element, alloys, and chemical compounds* (Kluwer/Plenum, New York 1999)

1.117 G. Leichtfried: *Handbook of Extractive Metallurgy* (Wiley-VCH, Weinheim 1997) p. 1371

1.118 G. Leichtfried: *Powder Metallurgy Data*, Landolt–Börnstein, New Series /2 (Springer, Berlin, Heidelberg, New York 2002)

1.119 G. Leichtfried: Molybdenum lanthanum oxide: Special material properties by dispersoid refining during deformation. In: *Advances in Powder Metallurgy and Particulate Materials*, Vol. 9 (MPIF, Princeton 1992) p. 123

1.120 D. M. Moon, R. C. Koo: Mechanism and kinetics of bubble formation in doped W, Metall. Trans. **2**, 2125 (1971)

1.121 H. G. Sell, D. F. Stein, R. Stickler, A. Joshi, E. Berkey: The identification of bubble forming impurities in doped tungsten, J. Inst. Met. **100**, 275 (1972)

1.122 P. Makarov, K. Povarova: Principles of the alloying of tungsten and development of the manufacturing technology for the tungsten alloys, Proc. 15th Plansee Seminar, Vol. 3 (Plansee AG, Reutte 2001) p. 464

1.123 G. A. Geach, J. E. Hughes: The alloy of rhenium with molybdenum or with tungsten and having good high temperature properties, Proc. 2nd Plansee Seminar (Plansee AG, Reutte 1955) p. 245

1.124 R. I. Jaffee, C. T. Sims, J. J. Harwood: The effect of rhenium on the fabricability and ductility of molybdenum and tungsten, Proc. 3rd Plansee Seminar (Plansee AG, Reutte 1958) p. 380

1.125 J. G. Booth, R. I. Jaffee, E. I. Salkovitz: The mechanisms of the rhenium-alloying effect in group VI-A metals, Proc. 5th Plansee Seminar (Plansee AG, Reutte 1964) p. 547

1.126 Plansee Aktiengesellschaft: Material Data Base, Reutte (2000)

1.127 H. Borchers, E. Schmidt (Eds.): *Stoffwerte und Verhalten von metallischen Werkstoffen*, Landolt–Börnstein, New Series IV/2b, 6th edn. (Springer, Berlin, Heidelberg 1964)

1.128 T. E. Tietz, J. W. Wilson: *Behavior and Properties of Refractory Metals* (Stanford Univ. Press, Stanford 1965) p. 325

1.129 Plansee Aktiengesellschaft: Tungsten Brochure, Reutte (1997)

1.130 C. Cagran, C. Brunner, A. Seifter, G. Pottlacher: Liquid-phase behaviour of normal spectral emissivity at 684.5 nm of some selected metals, High Temp.-High Press. **34**, 669 (2002)

1.131 Dechema-Werkstoff-Tabelle: *Oxidierende Heißgase* (Dechema, Frankfurt 1981)

1.132 A. Schintlmeister, H.-P. Martinz, P. Wilhartitz, F. P. Netzer: Low-temperature oxidation of industrial molybdenum surfaces, Powder Metallurgy

1.133 G. Leichtfried: Powder metallurgical components for light sources, Habilitation Thesis (Montanuniversität, Leoben 2003)
1.134 E. Fromm, E. Gebhardt: *Gase und Kohlenstoff in Metallen* (Springer, Berlin, Heidelberg 1976) p. 747 in German
1.135 R. Speiser, G. R. St. Pierre: , Proc. AGARD (Advisory Group for Aerospace Research and Development) Conf. on refractory metals, Oslo 1963 (AGARD
1.136 J. Disam, H.-P. Martinz, M. Sulik: European Patent Specification EP798402
1.137 C. A. Krier: *Coatings for the Protection of Refractory Metals from Oxidation*, Defense Metals Information Center Report 162 (Battelle Memorial Institute, Columbus 1961)
1.138 W. Knabl: Oxidationsschutz von Refraktärmetallen auf der Basis von Silizid- und Aluminidschichten. Ph.D. Thesis (Montanuniversität, Leoben 1995)
1.139 H.-P. Martinz, M. Sulik: Oxidation protection of refractory metals in the glass industry, Glastechnische Berichte, Glas Sci. Technol. **73**(C2) (2000)
1.140 C. Stickler: Mikroplastizität und zyklisches Spannungs-Dehnungsverhalten von Ta und Mo bei Temperaturen unter $0.2\,T_m$. Ph.D. Thesis (University of Vienna, Vienna 1998)
1.141 F. Benesovsky: *Pulvermetallurgie und Sinterwerkstoffe* (Plansee AG, Reutte 1982) p. 95
1.142 E. Pink, H. Kärle: Zum Rekristallisationsverhalten von Sintertantal, Planseeberichte für Pulvermetallurgie **16**, 105 (1968)
1.143 G. Leichtfried, G. Thurner, R. Weirather: Molybdenum alloys for glass-to-metal seals, Proc. 14th Plansee Seminar, Vol. 4 (Plansee AG, Reutte 1997) p. 26
1.144 H. H. R. Jansen: The recrystallization texture of non-sag wire. In: *The Metallurgy of Doped, Non-Sag Tungsten*, ed. by E. Pink, L. Bartha (Elsevier, New York 1989) p. 203
1.145 D. B. Snow: The recrystallization of non-sag tungsten wire. In: *The Metallurgy of Doped, Non-Sag Tungsten* (Elsevier, New York 1989) p. 189
1.146 V. I. Trefilov, Y. V. Milman: Physical basis of thermomechanical treatment of refractory metals, Proc. 12th Plansee Seminar, Vol. 1 (Plansee AG, Reutte 1989) p. 107
1.147 E. Parteder, W. Knabl, R. Stickler, G. Leichtfried: Bruchzähigkeit und Porenverteilung von Molybdän Stabmaterial in Abhängigkeit des Reckgrades und des Rekristallisationsgrades, Proc. 14th Plansee Seminar, Vol. 1 (Plansee AG, Reutte 1997) p. 984
1.148 E. Parteder, H. Riedel, R. Kopp: Densification of sintered molybdenum during hot upsetting: Experiments and modeling, Mat. Sci. Eng. A **264**, 17 (1999)
1.149 E. Parteder: Ein Modell zur Simulation von Umformprozessen pulvermetallurgisch hergestellter hochschmelzender Metalle. Ph.D. Thesis (RWTH, Aachen 2000)
1.150 E. Parteder, H. Riedel: Simulating of hot forming processes of refractory metals using porous metal plasticicty models, Proc. 15th Plansee Seminar, Vol. 3 (Plansee AG, Reutte 2001) p. 60
1.151 B. P. Bewlay, C. L. Briant: Discussion of "Evidence for the Existence of Potassium Bubbles in AKS-Doped Tungsten Wire" and Reply, Met. Trans. A **22A**, 2153 (1991)
1.152 C. L. Briant: The effect of thermomechanical processing on the microstructure of tungsten rod, Proc. 13th Plansee Seminar, Vol. 1 (Plansee AG, Reutte 1993) p. 321
1.153 J. L. Walter, C. L. Briant: Tungsten wire for incandescent lamps, J. Mat. Res. **5**, 2004 (1990)
1.154 G. L. Krasko: Effect of impurities on the electronic structure of grain boundaries and intergranular cohesion in tungsten, Proc. 13th Plansee Seminar, Vol. 1 (Plansee AG, Reutte 1993) p. 27
1.155 A. Kumar, B. L. Eyre: Grain boundary segregation and intergranular fracture in molybdenum, Proc. R. Soc. London A **370**, 431 (1980)
1.156 St. M. Cardonne: Tantalum and its alloys, Advanced Mat. & Processes **9**, 16 (1992)
1.157 P. Wilhartitz, G. Leichtfried, H. P. Martinz, H. Hutter, A. Virag, M. Grasserbauer: Applications of 3D-SIMS for the development of refractory metal products, Proc. 2nd Europ. Conf. on Advanced Materials and Processes, ed. by T. W. Clyne, P. J. Withers, London 1992) 323
1.158 J. Fembök, R. Stickler, A. Vinckier: The effect of strain rate and heating rate on the tensile behavior of W and W–ThO$_2$ between room temperature and 1400 °C, Proc. 11th Plansee Seminar, Vol. 1 (Plansee AG, Reutte 1985) p. 361
1.159 D. L. Chen, B. Weiss, R. Stickler, M. Witwer, G. Leichtfried, H. Hödl: Fracture toughness of high melting point materials, Proc. 13th Plansee Seminar, Vol. 1 (Plansee AG, Reutte 1993) p. 621
1.160 E. S. Meiren, D. A. Thomas: Effect of grain boundaries on the bending ductility of tungsten, Metall. Trans. **233**, 937 (1965)
1.161 P. F. Browning, C. L. Briant, B. A. Knudsen: Dependence of material properties on processing history during wire drawing of commercially doped tungsten lamp wire, Proc. 13th Plansee Seminar, Vol. 1 (Plansee AG, Reutte 1993) p. 336
1.162 P. K. Wright: High temperature creep behavior of doped tungsten wire, Metall. Trans. **9**, 955 (July 1978)
1.163 J. Neges, B. Ortner, G. Leichtfried, H. P. Stüwe: On the 45° embrittlement of tungsten sheets, Mat. Sci. Eng. A **196**, 129 (1995)
1.164 Y. V. Milman: unpublished results
1.165 A. Seeger: The temperature dependence of the critical shear stress and of work hardening of metal crystals, Philos. Mag. **7**, 771 (1954)

1.166 J. W. Christian: Plastic deformation of bcc metals, Proc. International Conference on the Strength of Materials (ICSMA-2), Asilomar (ASTM, Philadelphia 1970) 31

1.167 H. Mughrabi: unpublished results

1.168 H. Ullmaier: Design properties of tantalum or everything you always wanted to know about tantalum but were afraid to ask, ESS (European Spallation Source) report ISSN 1433-559X, 03-131-T (2003)

1.169 W. Rinnerthaler, F. Benesovsky: Untersuchungen über das Mikrodehnungsverhalten von Molybdän, Planseeberichte für Pulvermetallurgie **21**, 253 (1973)

1.170 C. Stickler, D. L. Chen, B. Weiss, R. Stickler: Time dependent microplastic deformation of Mo and Ta at low temperatures, Proc. 14th Plansee Seminar, Vol. 1 (Plansee AG, Reutte 1997) p. 1004

1.171 K. J. Bowman, R. Gibala: Cyclic deformation of W single crystals, Scripta Met. **20**, 1451 (1986)

1.172 M. A. Meyers, Y.-J. Chen, F. D. S. Marquis, J. B. Isaacs: High strain rate behavior of Ta, The Univ. of Cal., Inst. for Mechanics and Materials, Report **94-25** (1994)

1.173 C. C. Wojcik: Thermomechanical processing and properties of niobium alloys, Proc. of the Internat. Symposium Niobium 2001 (Niobium 2001 Limited, Orlando 2001) 163

1.174 H. Mughrabi, K. Herz, X. Stark: Cyclic deformation and fatigue behavior of α-Fe mono- and polycrystals, Int. J. Fracture **17**, 193 (1981)

1.175 M. Werner: Temperature and strain rate dependence of the flow stress of Ta single crystals in cyclic deformation, Revue de Physique Appliquee **23**, 672 (1988)

1.176 J. Femböck, K. Pfaffinger, B. Weiss, R. Stickler: Verhalten von Mo-Werkstoffen unter zyklischer Beanspruchung, Proc. 10th Plansee Seminar, Vol. 2 (Plansee AG, Reutte 1981) p. 27

1.177 K. Pfaffinger, J. Femböck: Versuchsplanung und statistische Auswertung von Schwingfestigkeitsdaten von Mo-Werkstoffen, Proc. 10th Plansee Seminar, Vol. 2 (Plansee AG, Reutte 1981) p. 233

1.178 K. Mecke, C. Holste, W. F. Terentjev: Dislocation arrangement in cyclically deformed Mo, Krist. Tech. **15**, 83 (1980)

1.179 S. Kong, B. Weiss, R. Stickler, M. Witwer, H. Hödl: Cyclic stress strain behavior of high melting point metals, Proc. 13th Plansee Seminar, Vol. 1 (Plansee AG, Reutte 1993) p. 720

1.180 D. R. Helebrand, R. I. Stephens: Cyclic yield behavior of Ta, J. Mater. Sci. **7**, 530 (1972)

1.181 C. Stickler, W. Knabl, R. Stickler, B. Weiss: Cyclic behavior of Ta at low temperatures under low stresses and strain rates, Proc. 15th Plansee Seminar, Vol. 3 (Plansee AG, Reutte 2001) p. 34

1.182 J. M. Meiniger, J. C. Gibeling: LCF of Nb and Nb-1Zr alloys, Met. Trans. **23A**, 3077 (1992)

1.183 H. J. Shi, L. S. Niu, C. Korn, G. Pluvinage: High temperature fatigue behavior of Mo-TZM alloy under mechanical and thermomechanical cyclic load, J. Nuclear Mat. **278**, 328 (2000)

1.184 R. F. Brodrick: LCF-data of P/M-W between 1650 and 3300 °C, Proc. ASTM **64**, 505 (1965)

1.185 S. S. Manson: *Thermal Stress and Low Cycle Fatigue* (McGraw-Hill, New York 1981) p. 187

1.186 R. E. Schmunk, G. E. Korth, M. Ulrickson: Tensile and LCF measurements on cross rolled tungsten, J. Nuclear Mat. **103**, 943 (1981)

1.187 T. Kimishima, M. Sukekawa, K. Owada, M. Shimizu: Fatigue data of Mo, 9th Symp. on Engineering Problems of Fusion Research 1981 (IEEE, New York 1981) 255

1.188 H. Nishi, T. Oku, T. Kodeira: Influence of microstructural change caused by cyclic strain on the LCF strength of sintered Mo, Fusion-Engineering-Design **9**, 123 (1989)

1.189 Z. M. Sun, Z. G. Wang, H. Hödl, R. Stickler, B. Weiss: Low cycle fatigue and creep behavior of recrystallized Mo near room temperature, Materialwissenschaft und Werkstofftechnik **26**, 483 (1995)

1.190 M. Papakyriacou, H. Mayer, C. Pypen, H. Plenk, S. Stanzl-Tschegg: Influence of loading frequency on high cycle fatigue properties of bcc and hcp metals, Mat. Sci. Eng. A **A308**, 143 (2001)

1.191 H. A. Calderon, G. Kostorz: Microstructure and plasticity of two molybdenum-base alloys (TZM), Mat. Sci. Eng. A **A160**, 189 (1993)

1.192 C. W. Marschall, F. C. Holden: Fracture toughness of refractory metals and alloys. In: *High Temperature Refractory Metals*, ed. by L. Richardson (Gordon Breach, New York 1964) p. 129

1.193 M. Rödig, H. Derz, G. Pott, B. Werner: Fracture mechanics investigations of TZM and Mo5Re, Proc. 14th Plansee Seminar, Vol. 1 (Plansee AG, Reutte 1997) p. 781

1.194 D. Padhi, J. J. Lewandowski: Effects of test temperature and grain size on the charpy impact toughness and dynamic toughness (KID) of polycrystalline niobium, Met. Mat. Trans. A **34**, 967 (2003)

1.195 J. A. Shields, P. Lipetzly, A. J. Mueller: Fracture toughness of 6.4 mm arc cast Mo and TZM Plate at RT and 300 °C, Proc. 15th Plansee Seminar, Vol. 4 (Plansee AG, Reutte 2001) p. 187

1.196 J. X. Zhang, L. Liu, M. L. Zhou, Y. C. Hu, T. Y. Zuo: Fracture toughness of sintered Mo–La_2O_3, Internat. J. Refract. Met. Hard Mat. **17**, 405 (1999)

1.197 D. L. Chen, B. Weiss, R. Stickler: The effective fatigue threshold: Significance of the loading cycle below the crack opening load, Internat. J. Fatigue **16**, 485 (1994)

1.198 J. Riedle: Bruchwiderstand in Wolfram-Einkristallen: Einfluß der kristallographischen Orientierung, der Temperatur und der Lastrate. In: *Fortschrittsberichte VDI*, Reihe 18, Mechanik/Bruchmechanik, Vol. 184 (VDI, Düsseldorf 1995)

1.199 R. Pippan: *Bruchzähigkeitsuntersuchungen an W Proben* (Erich Schmid Institut, Leoben 1999) Report

1.200 Y. Mutoh, K. Ichikawa, K. Nagata, M. Takeuchi: Effect of Re addition on fracture toughness of W at elevated temperatures, J. Mat. Sci. **30**, 770 (1995)

1.201 A. Fathulla, B. Weiss, R. Stickler: Short fatigue cracks in technical P/M-Mo alloys. In: *The Behavior of Short Fatigue Cracks, Mechanical Engineering Publications*, Vol. 1 (EGF Pub., Suffolk 1986) p. 115

1.202 R. Grill, H. Clemens, P. Rödhammer, A. Voiticek: P/M processing, characterization and application of Ta-10W, Proc. 14th Plansee Seminar, Vol. 4 (Plansee AG, Reutte 1997) p. 211

1.203 R. Heidenreich, R. Schäfer, H. Clemens, M. Witwer: Mechanical properties of high-temperature fasteners from refractory alloys, Proc. 13th Plansee Seminar, Vol. 1 (Plansee AG, Reutte 1993) p. 664

1.204 A. Fathulla, B. Weiss, R. Stickler, J. Femböck: The initiation and growth of short cracks in pm-Mo, Proc. 11th Plansee Seminar, Vol. 1 (Plansee AG, Reutte 1985) p. 45

1.205 H. Kitagawa, S. Takahashi: Applicability of fracture mechanics to very small cracks or the cracks in the early stage, Proc. second international conference on mechanical behavior of materials (ASM, Metals Park 1976) 627

1.206 B. Weiss, R. Stickler: Methods for predicting the fatigue strength of P/M-materials, Proc. International Powder Metallurgy Conf. PM'88, Orlando 1988, 3

1.207 B. Weiss, R. Stickler, A. F. Blom: A model for the description of the influence of small 3-dimensional defects on the HCF limit, Proc. Conf.: Short Fatigue Cracks, Sheffield 1990 (Mechanical Engineering Publications Limited, Suffolk 1992) 423

1.208 G. Leichtfried: Die Entwicklung von kriechfesten Molybdän - Seltenerdoxid - Werkstoffen für Hochtemperaturanwendungen. Ph.D. Thesis (Montanuniversität, Leoben 1997)

1.209 D. M. Moon, R. Stickler: Creep behavior of fine wires of P/M pure, doped and thoriated tungsten, High Temp. High Press. **3**, 503 (1971)

1.210 J. W. Pugh: On the short time creep rupture properties of lamp wire, Metall. Trans. **4**, 533 (1973)

1.211 J. H. Schröder, E. Arzt: Weak beam studies of dislocation/dispersoid interaction in an ODS superalloy, Scripta Met. **19**, 1129 (1985)

1.212 J. Rössler, E. Arzt: Kinetics of dislocation climb over hard particles – Climb without attractive particle-dislocation interaction, Acta Met. **36**, 1043 (1988)

1.213 J. Rössler, E. Arzt: A new model-based creep equation for dispersion strengthened materials, Acta Met. Mat. **38**(4), 671 (1990)

1.214 G. Zilberstein, J. Selverian: Creep deformation of non-sag tungsten in argon doped with low oxygen concentrations, Proc. 13th Plansee Seminar, Vol. 1 (Plansee AG, Reutte 1993) p. 132

1.215 G. Zilberstein: Creep properties of non-sag tungsten recrystallized in stagnant oxygen-doped argon, Int. J. Refract. Met. Hard Mat. **16**, 71 (1998)

1.216 Degussa AG (Ed.): *Edelmetall-Taschenbuch* (Degussa, Frankfurt 1967)

1.217 Degussa AG (Ed.): *Edelmetall-Taschenbuch*, 2nd edn. (Hüthig, Heidelberg 1995)

1.218 L. S. Benner, I. Suzuki, K. Meguro, S. Tanaka (Eds.): *Precious Metals, Science, Technology* (Int. Precious Metals Institute, Allentown 1991)

1.219 B. Predel (Ed.): *Phase Equilibria, Crystallographic, Thermodynamic Data of Binary Alloys*, Landolt–Börnstein, New Series IV/5 (Springer, Berlin, Heidelberg 1991–1998)

1.220 E. Raub: *Die Edelmetalle und ihre Legierungen* (Springer, Berlin, Heidelberg 1940)

1.221 P. J. Spencer, K. Hack: Swiss Materials **2**, 69–73 (1990)

1.222 C. J. Smithells, E. A. Brandes: *Metals Reference Book*, 5th edn. (Butterworth, London 1977)

1.223 P. Villars, L. D. Calvert: *Pearson's Handbook of Crystallographic Data for Intermetallic Phases*, Vol. 2, 3 (American Society for Metals, Metals Park 1985)

1.224 Landolt–Börnstein: *Technik*, Landolt–Börnstein, New Series IV/4, 6th edn. (Springer, Berlin, Heidelberg 1967)

1.225 W. Hume-Rothery, G. V. Raynor: *The Structure of Metals and Alloys* (Institute of Metals, London 1956)

1.226 Gmelin: *Handbuch der anorganischen Chemie*, Syst. Nr. 61 (Springer, Berlin, Heidelberg 1970–1975)

1.227 V. Behrens, K. H. Schröder: *Werkstoffe für elektrische Kontakte und ihre Anwendungen*, Kontakt u. Studium, Vol. 366 (Expert, Ehningen 1992)

1.228 W. H. Cubberly, H. Baker, D. Benjamin (Eds.): *Metals Handbook*, Vol. 2, 9th edn. (American Society for Metals, Metals Park 1979) pp. 671–678

1.229 E. M. Savitskii, A. Prince: *Handbook of Precious Metals* (Hemisphere, New York 1989) pp. 117–128

1.230 K.-H. Hellwege, A. M. Hellwege (Eds.): *Elastics, Piezolelectric, Pyroelectric, Electrooptic Constants, Nonlinear Dielectric Susceptibilities of Crystals*, Landolt–Börnstein, New Series III/18 (Springer, Berlin, Heidelberg 1984) p. 66

1.231 Doduco: *Datenbuch, Handbuch für Techniker*, 2. Aufl. (Doduco, Pforzheim 1977)

1.232 H. Spengler: Metall **18**, 36 (1964)

1.233 K.-H. Hellwege, J. L. Olsen: *Metals – Electronic Transport Phenomena*, Landolt–Börnstein, New Series III/15 (Springer, Berlin, Heidelberg 1982) p. 167

1.234 D. D. Pollok: Trans. Metall. Soc. AIME **230**, 753 (1964)

1.235 H. Flükiger, W. Klose (Eds.): *Superconductors Ac-Na*, Landolt–Börnstein, New Series III/21 (Springer, Berlin, Heidelberg 1990)

1.236 Chr. Raub: Z. Metallkde. **55**, 195 (1964)

1.237 R. E. Hummel: *Optische Eigenschaften von Metallen und Legierungen* (Springer, Berlin, Heidelberg pp. 158,178

1.238 H. Mehrer (Ed.): *Diffusion in Solid Metals and Alloys*, Landolt–Börnstein, New Series III/26 (Springer, Berlin, Heidelberg 1990)

1.239 G. Schlamp: Mater. Sci. Technol. **8**, 471–587 (1996)

1.240 D. Lee: *Modern Chlor-Alkali Technology*, Vol. 2, ed. by C. Jackson (Horwood, Chichester 1983)

1.241 S. U. Falk, A. J. Salkind: *Alkaline Storage Batteries* (Wiley, New York 1971)

1.242 A. Fleischer, J. J. Lander: *Zinc-Silver-Oxide Batteries* (Wiley, New York 1971)

1.243 H. Renner: *Ullmann's Encyclopedia of Industrial Chemistry* (Wiley-VCH, Weinheim 2002)

1.244 D. Z. Stöckel: Z. Werkstofftechnik **10**, 238 (1979)

1.245 B. Bredel: *Phase Equilibria, Crystallophic and Thermodynamik Data of Binary Alloys*, Landolt-Börnstein, New Series IV/5 (Springer, Berlin, Heidelberg 1991–1998)

1.246 H. Renner: *Ullmann's Encyclopedia of Industrial Chemistry*, 6th edn. (Wiley-VCH, Weinheim 2001)

1.247 R. Forro: Gold Bull. **36**(2), 39–58 (2003)

1.248 N. Yuantao: Gold Bull. **34**(3), 77–87 (2001)

1.249 H. Ahlborn, G. Wassermann: Z. Metallkde. **55**, 685 (1964)

1.250 E. Drost, J. H. Hausselt: Interdisc. Sci. Rev. **17**, 271–280 (1992)

1.251 B. Kempf, J. Hausselt: Interdisc. Sci. Rev. **17**, 251–260 (1992)

1.252 G. Humpston: Gold Bull. **26**, 139 (1993)

1.253 C. J. Raub D. Ott: Z. Metallkde. **25**(4), 629 (1992)

1.254 M. Du Toit, E. van der Lingen, L. Glaner, R. Süss: *Gold Bulletin*, quarterly reviews, Vol. 35, ed. by C. Corti, D. Thompson (World gold council, London 2002) p. 49

1.255 Gmelin: *Gmelin Handbuch der anorganischen Chemie*, Vol. 62 (Springer, Berlin, Heidelberg p. 619

1.256 H. R. Khan, Ch. R. Raub: *Gold Bulletin*, quarterly reviews, Vol. 8, ed. by C. Corti, D. Thompson (World gold council, London 1975) pp. 114–118

1.257 H. R. Khan: *Gold Bulletin*, quarterly reviews, Vol. 17, ed. by C. Corti, D. Thompson (World gold council, London 1984) pp. 94–100

1.258 J. Kopp: *Gold Bulletin*, quarterly reviews, Vol. 9, ed. by C. Corti, D. Thompson (World gold council, London 1976) p. 55

1.259 M. Hansen, K. Anderko: *Constitution of Binary Alloys*, 2nd edn. (McGraw-Hill, New York 1958) pp. 195, 221

1.260 E. Raub: *Die Edelmetalle und ihre Legierungen*, Reine angewandte Metallkunde, Vol. 5, ed. by W. Köster (Springer, Berlin, Heidelberg 1955) Chap. II, A5, pp. 53, 54

1.261 T. Shiraishi, K. Hisatsune, Y. Tanaka, E. Miura, Y. Takuma: Gold Bull. **34**, 130 (2001)

1.262 W. S. Rapson, T. Groenewald: *Gold Usage* (Academic Press, New York 1978) p. 37

1.263 D. Compton et al.: *Gold Bulletin*, quarterly reviews, Vol. 10, ed. by C. Corti, D. Thompson (World gold council, London 1977) p. 51

1.264 G. C. Bond: *Gold Bulletin*, quarterly reviews, Vol. 34, ed. by C. Corti, D. Thompson (World gold council, London 2001) p. 117

1.265 A. St. K. Hashmi: *Gold Bulletin*, quarterly reviews, Vol. 36, ed. by C. Corti, D. Thompson (World gold council, London 2003) p. 3

1.266 M. Gupta, A. K. Tripathi: *Gold Bulletin*, quarterly reviews, Vol. 34, ed. by C. Corti, D. Thompson (World gold council, London 2001) p. 120

1.267 C. Corti, R. J. Holliday, D. T. Thompson: *Gold Bulletin*, quarterly reviews, Vol. 35 (World gold council, London 2002) p. 111

1.268 R. Grisel, K. J. Weststrate, A. Gluhoi, B. E. Nieuwenhuys: Gold Bull. **35**, 39–45 (2002)

1.269 H. Knosp, R. J. Holliday, Ch. W. Corti: *Gold in dentistry*, Vol. 36 2003) pp. 93–102

1.270 G. Gafner: *Gold Bulletin*, quarterly reviews, Vol. 22, ed. by C. Corti, D. Thompson (World gold council, London 1989) p. 112

1.271 G. Humston, D. M. Jacobson: *Gold Bulletin*, quarterly reviews, Vol. 25, ed. by C. Corti, D. Thompson (World gold council, London 1992) p. 139

1.272 D. R. Olsen, H. M. Berg: , Proc. 27th Electronic Components Conf. (IEEE, Piscataway 1977)

1.273 G. Zwingmann: Z. Metall. **34**(18), 726 (1964)

1.274 G. Petzow, G. Effenberg: *Ternary Alloys* (Verlag Chemie, Weinheim 1988)

1.275 W. Müller: *Metallische Lotwerkstoffe* (DVS, Düsseldorf 1990)

1.276 K. Toyoshima, Somorjai: Cat. Rev. Sci. Eng. **19** (1979)

1.277 A. S. Darling: , Review 175 (Institute of Metals, London 1973)

1.278 J. B. Hunter: Platinum Met. Rev. **4**(4), 130 (1960)

1.279 A. G. Knapton: Platinum Met. Rev. **21**(2), 44 (1977)

1.280 R. S. Irani: Metals Rev. **17**, 22 (1973)

1.281 A. Kussmann, G. Von Rittberg: Ann. Phys. **7**, 173 (1950)

1.282 A. Kussmann, K. Jesson: J. Phys. Soc. Jpn. **17**, 272 (1962)

1.283 A. S. Darling: Platinum Met. Rev. **11**, 138 (1967)

1.284 A. S. Darling: Platinum Met. Rev. **13**, 53 (1969)

1.285 R. S. Irani, R. W. Cahn: Met. Rev. **16**, 49 (1972)

1.286 H. Ocken, J. H. N. Van Vucht: J. Less-Comm. Met. **15**, 196–199 (1968)

1.287 I. Gurappa: Platinum Met. Rev. **45**(3), 124 (2001)

1.288 I. Wolff, P. J. Hill: Platinum Met. Rev. **44**(4), 161 (2000)

1.289 J. Merker, D. Lupton, M. Töpfer, H. Knake: Platinum Met. Rev. **45**, 76–80 (2001)

1.290 G. Borelius et al.: Proc. K. Acad. Wet. **33**, 17 (1930)

1.291 Th. J. Colacot: Platinum Met. Rev. **46**(2), 83 (2002)

1.292 I. M. Wolff, P. J. Hill: Platinum Met. Rev. **44**(4), 159 (2000)

1.293 Y. Yamabe-Mitarai, Y. F. Gu, H. Harada: Platinum Met. Rev. **46**(2), 77, 78 (2002)

1.294 M. Graff, B. Kempf, J. Breme: Metall **53**, 616–621 (1999)

1.295 J. H. Jones: Platinum Met. Rev. **44**(3), 100 (2000)

1.296 I. M. Wolff, P. J. Hill: Platinum Met. Rev. **44**(4), 164 (2000)

1.297 M. Fallot: Ann. Phys. (Paris) **10**, 29 (1938)

1.298 Philips AG: Philips Techn. Rev. , 73 (1966)

1.299 S. Guruswamy: *Engineering Properties and Applications of Lead Alloys* (Dekker, New York 2000) pp. 1–635

1.300 Lead Industries Association: *Properties of Lead and Lead Alloys*, Booklet No. 5M (Lead Industries Association, New York 1983) pp. 12–83

1.301 Lead Industries Association: *Lead for Corrosion Resistant Applications – A Guide* (Lead Industries Association, New York 1974)

1.302 G. Baralis, I. Tangerini: A study of some dynamic properties of lead and its alloys, Proc. 3rd Intl. Lead Conference, Venice, ed. by European Lead Development Committee and Lead Development Association (Pergamon, London 1968) 309–319

1.303 W. Hofmann: *Lead and Lead Alloys* (Springer, Berlin, Heidelberg 1970)

1.304 G. Kerry, P. Lord: *Project Report LM-375*, International Lead Zinc Research Organization (ILZRO) (University Salford, Salford 1990)

1.305 Lead Industries Association: *Lead Shielding for the Nuclear Industry* (Lead Industries Association, New York 1985)

1.306 R. D. Prengaman: Wrought lead-calcium-tin alloys for tubular lead/acid battery grids, J. Power Sources **53**, 207–214 (1995)

1.307 A. Siegmund, R. D. Prengaman: Grid alloys for automobile batteries in the new millenium, J. Metals **53**, 38 (2001)

1.308 ASTM: *Standard Specification for White Metal Bearing Alloys*, ASTM B23 (American Society for Testing Materials, West Conshohocken 1996)

1.309 Lead Industries Association: *Solders and Soldering – A Primer* (Lead Industries Association, Sparta 1996)

1.310 P. Adeva, G. Caruana, M. Aballe, M. Torralba: The lead-rich corner of the Pb–Ca–Sn phase diagram, Mater. Sci. Eng. **54**, 229–236 (1982)

3.2. Ceramics

Ceramics have various definitions, because of their long history of development as one of the oldest and most versatile groups of materials and because of the different ways in which materials can be classified, such as by chemical composition (silicates, oxides and non-oxides), properties (mechanical and physical), or applications (building materials, high-temperature materials and functional materials). The most widely used, minimal definition of ceramics is that they are inorganic nonmetallic materials. The broadest subdivision is into traditional, and technical or engineering ceramics. Detailed groupings and definitions of technical ceramics are given in DIN EN 60672. In the present section we differentiate between traditional ceramics and cements, silicate ceramics, refractory ceramics, oxide ceramics, and non-oxide ceramics, being aware that there are overlaps. It should also be noted that other sections of this Handbook cover particular groups of ceramics: glasses (Chapt. 3.4), semiconductors (Chapt. 4.1), nonmetallic superconductors (Sect. 4.2.2), magnetic oxides (Sect. 4.3.4), dielectrics and electrooptics (Chapt. 4.4) and ferroelectrics and related materials, (Chapt. 4.5).

3.2.1	**Traditional Ceramics and Cements**		432
	3.2.1.1	Traditional Ceramics	432
	3.2.1.2	Cements	432
3.2.2	**Silicate Ceramics**		433
3.2.3	**Refractory Ceramics**		437
3.2.4	**Oxide Ceramics**		437
	3.2.4.1	Magnesium Oxide	444
	3.2.4.2	Alumina	445
	3.2.4.3	Al–O–N Ceramics	447
	3.2.4.4	Beryllium Oxide	447
	3.2.4.5	Zirconium Dioxide	447
	3.2.4.6	Titanium Dioxide, Titanates, etc.	450
3.2.5	**Non-Oxide Ceramics**		451
	3.2.5.1	Non-Oxide High-Temperature Ceramics	451
	3.2.5.2	Borides	451
	3.2.5.3	Carbides	458
	3.2.5.4	Nitrides	467
	3.2.5.5	Silicides	473
References			476

Detailed treatments of ceramics are given in [2.1–3]. Reference [2.4] is a comprehensive handbook on materials, emphasizing ceramics and minerals. Structural ceramics are treated in [2.5]. Reference [2.6] is a hands-on practical reference book for technical ceramics, and recent data on technical ceramics can be found in conference proceedings such as [2.7].

3.2.1 Traditional Ceramics and Cements

3.2.1.1 Traditional Ceramics

Traditional ceramics are obtained by the firing of clay-based materials. They are commonly composed of a clay mineral (kaolinite, montmorillonite, or illite), fluxing agents (orthoclase and plagioclase), and filler materials (SiO_2, Al_2O_3, and MgO). The processing steps are mixing, forming, drying, firing (high-temperature treatment in air), and finishing (enameling, cleaning, and machining). The main classes of traditional ceramics are fired bricks, whiteware (china, stoneware, and porcelain), glazes, porcelain enamels, high-temperature refractories, cements, mortars, and concretes. Table 3.2-1 lists some of main groups of traditional ceramics, with some of their properties and applications.

3.2.1.2 Cements

Cement is the common binder of traditional building ceramics. It is produced by mixing about 80 wt% of low-magnesium (< 3 wt% MgO) calcium carbonate ($CaCO_3$) (limestone, marl, or chalk) with about 20 wt% clay (which may be obtained from clays, shale, or slag). In terms of oxide content, this corresponds to a ratio of CaO to SiO_2 of 3:1 by weight. The common term "Portland cement" is based on the early use of a particular limestone called Portland stone. The processing steps are milling and mixing, heating to 260 °C, pre-calcining at 900 °C, and calcining in a rotary kiln at temperatures ≤ 1450 °C. The resulting product is termed "clinker" and consists of a vitreous nodular material composed of calcium silicates and aluminates. This is mixed with 2 to 4 wt% gypsum ($CaSO_4 \cdot 2H_2O$) to adjust the setting time, and ground to the final product. The ranges of the percentages of the constituents are given in Table 3.2-2.

Some standardized grades of Portland cement and their uses are listed in Table 3.2-3. Mortars and concretes are mixtures of cement with specified amounts of sand, gravel, or crushed stones with specified particle sizes.

Table 3.2-2 Chemical composition of Portland cement [2.4]

Component	Average mass fraction (wt%)
SiO_2	21.8–21.9
Al_2O_3	4.9–6.9
Fe_2O_3	2.4–2.9
CaO	63.0–65.0
MgO	1.1–2.5 (max. 3.0)
SO_3	1.7–2.6
Na_2O	0.2
K_2O	0.4
H_2O	1.4–1.5

Table 3.2-1 Examples of traditional ceramics [2.4]

Type	Properties	Applications
Fired brick	Porosity 15–30% Firing temperature 950–1050 °C May be enameled or not	Bricks, pipes, ducts, walls, floor tiles
China	Porosity 10–15% Firing temperature 950–1200 °C Enameled, opaque	Sanitary, tiles
Stoneware	Porosity 0.5–3% Firing temperature 1100–1300 °C Glassy surface	Crucibles, labware, pipes
Porcelain	Porosity 0–2% Firing temperature 1100–1400 °C Glassy, translucent	Insulators, labware, cookware

Table 3.2-3 ASTM Portland cement types [2.4]

ASTM type	Name	Compressive strength after 28 days (MPa)	Applications
Type I	Normal or ordinary Portland cement (NPC)	42	General uses, and hence used when no special properties are required.
Type II	Modified Portland cement (MPC)	47	Low heat generation during the hydration process. Most useful in structures with large cross sections and for drainage pipes where sulfate levels are low.
Type III	Rapid-hardening Portland cement (RHPC)	52	Used when high strength is required after a short period of curing.
Type IV	Low-heat Portland cement (LHPC)	34	Less heat generation during hydration than for Type II. Used for mass concrete construction where large heat generation could create problems. The tricalcium aluminate content must be maintained below 7 wt%.
Type V	Sulfate-resisting Portland cement (SRPC)	41	Has high sulfate resistance. It is a special cement used when severe attack is possible.

3.2.2 Silicate Ceramics

Silicates are the salts or esters of orthosilicic acid (H_4SiO_4) and of its condensation products. The silicates are categorized according to the arrangement of the [SiO_4] tetrahedra in their crystal structure. Figure 3.2-1 shows some simple, planar arrangements. In the tectosilicates, the array of the SiO_4 tetrahedra is three-dimensional. Typical examples are talc, feldspar, and the zeolites.

As silicates are the most important constituents of the earth's crust, they became the basis of the traditional

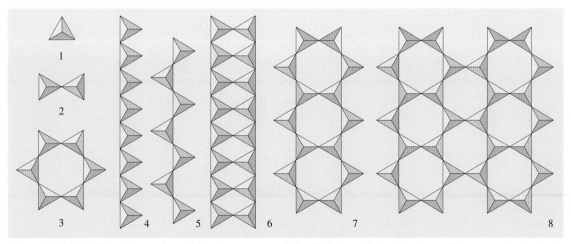

Fig. 3.2-1 Schematic representation of the arrangement of the [SiO_4] tetrahedra in planar silicate crystal structures: 1, nesosilicate; 2, sorosilicate; 3, cyclosilicate; 4 and 5, inosilicates; 6 and 7, ribbon silicates; 8, layered silicate or phyllosilicate

ceramics, from which a variety of technical ceramic materials have been developed, particularly for electrotechnical and electronic applications. Some silicate ceramics are also used for high-temperature applications in the processing of materials. Silicate ceramics are composed essentially of porcelain, steatite, cordierite, and mullite ($3Al_2O_3 \cdot 2SiO_2$).

Recent standards that classify and characterize technical silicate ceramics are listed in Tables 3.2-4 – 3.2-7.

Table 3.2-4 Properties of Alkali aluminium silicates according to DIN EN 60672 [2.6]

			Designation					
			C 110	C 111	C 112	C 120	C 130	C 140
Mechanical properties	Symbol	Units	Quartz porcelain, plastically formed	Quartz porcelain, pressed	Cristobalite porcelain	Alumina porcelain	Alumina porcelain, high-strength	Lithium porcelain
Density, minimum	ϱ	t/m^3	2	2.2	2.3	2.3	2.5	2.0
Bending strength, unglazed	σ_{ft}	MPa	50	40	80	90	140	50
Bending strength, glazed	σ_{fg}	MPa	60	–	100	110	160	60
Young's modulus	E	GPa	60	–	70	–	100	–
Electrical properties								
Breakdown strength	E_d	kV/mm	20	–	20	20	20	15
Withstand voltage	U	kV	30	–	30	30	30	20
Permittivity at 48–62 Hz	ε_r		6–7	–	5–6	6–7	6–7.5	5–7
Loss factor at 20 °C, 1 kHz	$\tan\delta_{pf}$	10^{-3}	25	–	25	25	30	10
Loss factor at 20 °C, 1 MHz	$\tan\delta_{1M}$	10^{-3}	12	–	12	12	15	10
Resistivity at 20 °C	ρ_{20}	Ω m	10^{11}	10^{10}	10^{11}	10^{11}	10^{11}	10^{11}
Resistivity at 600 °C	ρ_{600}	Ω m	10^2	10^2	10^2	10^2	10^2	10^2
Thermal properties								
Average coefficient of thermal expansion at 30–600 °C	α_{30-600}	10^{-6} K^{-1}	4–7	4–7	6–8	4–7	5–7	1–3
Specific heat capacity at 30–600 °C	$c_{p,30-600}$	J kg^{-1}K^{-1}	750–900	800–900	800–900	750–900	800–900	750–900
Thermal conductivity	λ_{30-100}	W m^{-1}K^{-1}	1–2.5	1.0–2.5	1.4–2.5	1.2–2.6	1.5–4.0	1.0–2.5
Thermal fatigue resistance			(Rated) Good	Good	Good	Good	Good	Good

Table 3.2-5 Properties of magnesium silicate according to DIN EN 60672 [2.6]

Mechanical properties	Symbol	Units	Designation					
			C 210	C 220	C 221	C 230	C 240	C 250
			Steatite for low voltage	Steatite, standard	Steatite, low-loss	Steatite, porous	Forsterite, porous	Forsterite, dense
Open porosity		vol. %	0.5	0.0	0.0	35.0	30.0	0.0
Density, minimum	ϱ	t/m^3	2.3	2.6	2.7	1.8	1.9	2.8
Bending strength	σ_B	MPa	80	120	140	30	35	140
Young's modulus	E	GPa	60	80	110	–	–	–
Electrical properties								
Breakdown strength	E_d	kV/mm	–	15	20	–	–	20
Withstand voltage	U	kV	–	20	30	–	–	20
Permittivity at 48–62 Hz	ε_r		6	6	6	–	–	7
Loss factor at 20 °C, 1 kHz	$\tan \delta_{pf}$	10^{-3}	25	5	1.5	–	–	1.5
Loss factor at 20 °C, 1 MHz	$\tan \delta_{1M}$	10^{-3}	7	3	1.2	–	–	0.5
Resistivity at 20 °C	ρ_{20}	Ω m	10^{10}	10^{11}	10^{11}	–	–	10^{11}
Resistivity at 600 °C	ρ_{600}	Ω m	10^3	10^3	10^5	10^5	10^5	10^5
Thermal properties								
Average coefficient of thermal expansion at 30–600 °C	α_{30-600}	10^{-6} K^{-1}	6–8	7–9	7–9	8–10	8–10	9–11
Specific heat capacity at 30–100 °C	$c_{p,30-600}$	J kg^{-1} K^{-1}	800–900	800–900	800–900	800–900	800–900	800–900
Thermal conductivity	λ_{30-100}	W m^{-1} K^{-1}	1–2.5	2–3	2–3	1.5–2	1.4–2	3–4
Thermal fatigue resistance		(Rated)	Good	Good	Good	Good	Good	Good

Table 3.2-6 Properties of alkaline-earth aluminium silicates according to DIN EN 60672 [2.6]

Mechanical properties	Symbol	Units	Designation			
			C 410	C 420	C 430	C 440
			Cordierite, dense	Celsians, dense	Calcium-based, dense	Zirconium-based, dense
Open porosity		vol. %	0.5	0.5	0.5	0.5
Density, minimum	ϱ	t/m^3	2.1	2.7	2.3	2.5
Bending strength	σ_B	MPa	60	80	80	100
Young's modulus	E	GPa	–	–	80	130

Table 3.2-6 Properties of alkaline-earth aluminium silicates according to DIN EN 60672 [2.6], cont.

Electrical properties	Symbol	Units	Designation			
			C 410	C 420	C 430	C 440
			Cordierite, dense	Celsians, dense	Calcium-based, dense	Zirconium-based, dense
Breakdown strength	E_d	kV/mm	10	20	15	15
Withstand voltage	U	kV	15	30	20	20
Permittivity at 48–62 Hz	ε_r		5	7	6–7	8–12
Loss factor at 20 °C, 1 kHz	$\tan\delta_{pf}$	10^{-3}	25	10	5	5
Loss factor at 20 °C, 1 MHz	$\tan\delta_{1M}$	10^{-3}	7	0.5	5.0	5
Resistivity at 20 °C	ρ_{20}	$\Omega\,m$	10^{10}	10^{12}	10^{11}	10^{11}
Resistivity at 600 °C	ρ_{600}	$\Omega\,m$	10^{3}	10^{7}	10^{2}	10^{2}
Thermal properties						
Average coefficient of thermal expansion at 30–600 °C	α_{30-600}	$10^{-6}\,K^{-1}$	2–4	3.5–6	–	–
Specific heat capacity at 30–100 °C	$c_{p,30-600}$	$J\,kg^{-1}K^{-1}$	800–1200	800–1000	700–850	550–650
Thermal conductivity	λ_{30-100}	$W\,m^{-1}K^{-1}$	1.2–2.5	1.5–2.5	1–2.5	5–8
Thermal fatigue resistance			(Rated)	Good	Good	Good

Table 3.2-7 Properties of porous aluminium silicates and magnesium silicates according to DIN EN 60672 [2.6]

Mechanical properties	Symbol	Units	Designation				
			C 510	C 511	C 512	C 520	C 530
			Aluminosilicate-based	Magnesia–aluminosilicate-based	Magnesia–aluminosilicate-based	Cordierite-based	Aluminosilicate-based
Open porosity		vol. %	30	20	40	20	30
Density, minimun	ϱ	t/m^3	1.9	1.9	1.8	1.9	2.1
Bending strength	σ_B	MPa	25	25	15	30	30
Young's modulus	E	GPa	–	–	–	40	–
Electrical properties							
Resistivity at 600 °C	ρ_{600}	$\Omega\,m$	10^{3}	10^{3}	10^{3}	10^{3}	10^{4}
Thermal properties							
Average coefficient of thermal expansion at 30–600 °C	α_{30-600}	$10^{-6}K^{-1}$	3–6	4–6	3–6	2–4	4–6
Specific heat capacity at 30–100 °C	$c_{p,30-600}$	$J\,kg^{-1}K^{-1}$	750–850	750–850	750–900	750–900	800–900
Thermal conductivity	λ_{30-100}	$W\,m^{-1}K^{-1}$	1.2–1.7	1.3–1.8	1–1.5	1.3–1.8	1.4–2.0
Thermal fatigue resistance			Rated	Good	Good	Good	Good	Good

3.2.3 Refractory Ceramics

Traditional refractory ceramics are produced as bricks from a broad variety of materials for various applications, ranging from building bricks used at ambient and moderately elevated temperatures to high-temperature refractory grades with particularly high melting temperatures and refractory stability, such as magnesite, silicon carbide, stabilized zirconia, and chrome-magnesite. Table 3.2-8 gives a survey of fired refractory brick materials.

It should be noted that these refractory ceramics in brick form differ from materials classified as technical refractory ceramics in the amounts of material used and in the composition, including the impurity content, but they rely basically on the same oxide or non-oxide ceramic compounds. These compounds are used in technical ceramics and are treated in more detail below.

Table 3.2-8 Properties of fired refractory brick materials [2.4]

Brick (major chemical components)	Density ϱ (kg/m^3)	Melting temperature (°C)	Thermal conductivity κ (W/m K)
Alumina brick (6–65 wt% Al_2O_3)	1842	1650–2030	4.67
Building brick	1842	1600	0.72
Carbonbrick (99 wt% graphite)	1682	3500	3.6
Chromebrick (100 wt% Cr_2O_3)	2900–3100	1900	2.3
Crome–magnesite brick (52 wt% MgO, 23 wt% Cr_2O_3)	3100	3045	3.5
Fireclay brick (54 wt% SiO_2, 40 wt% Al_2O_3)	2146–2243	1740	0.3–1.0
Fired dolomite (55 wt% CaO, 37 wt% MgO)	2700	2000	–
High-alumina brick (90–99 wt% Al_2O_3)	2810–2970	1760–2030	3.12
Magnesite brick (95.5 wt% MgO)	2531–2900	2150	3.7–4.4
Mullite brick (71 wt% Al_2O_3)	2450	1810	7.1
Silica brick (95–99 wt% SiO_2)	1842	1765	1.5
Silicon carbide brick (80–90 wt% SiC)	2595	2305	20.5
Zircon (99 wt% $ZrSiO_4$)	3204	1700	2.6
Zirconia (stabilized) brick	3925	2650	2.0

3.2.4 Oxide Ceramics

Oxides are the most common constituents of all ceramics, traditional and technical. An extensive account is given in [2.8]. They are used for their physical as well as refractory properties. Table 3.2-9 presents a systematic listing of their composition, structure and properties. Some of the more important oxides are dealt with in more detail further below.

Table 3.2-9 Physical properties of oxides and oxide-based high-temperature refractories [2.4]

IUPAC name (synonyms and common trade names)	Theoretical chemical formula, [CASRN], relative molecular mass ($^{12}C = 12.000$)	Crystal system, lattice parameters, *Strukturbericht* symbol, Pearson symbol, space group, structure type, Z	Density (ϱ, kg m^{-3})	Electrical resistivity (ρ, $\mu\Omega$ cm)	Melting point (°C)	Thermal conductivity (κ, W m^{-1} K^{-1})	Specific heat capacity (c_p, J kg^{-1} K^{-1})	Coefficient linear thermal expansion (α, 10^{-6} K^{-1})
Aluminium sesquioxide (alumina, corundum, sapphire)	α-Al$_2$O$_3$ [1344-28-1] [1302-74-5] 101.961	Trigonal (rhombohedral) $a = 475.91$ pm $c = 1298.4$ pm D5$_1$, hR10, $R\bar{3}c$, corundum type ($Z = 2$)	3987	2×10^{23}	2054	35.6–39	795.5–880	7.1–8.3
Beryllium monoxide (beryllia)	BeO [1304-56-9] 25.011	Trigonal (hexagonal) $a = 270$ pm $c = 439$ pm B4, hP4, $P6_3mc$, wurtzite type ($Z = 2$)	3008–3030	10^{22}	2550–2565	245–250	996.5	7.5–9.7
Calcium monoxide (calcia, lime)	CaO [1305-78-8] 56.077	Cubic $a = 481.08$ pm B2, cP2, $Pm3m$, CsCl type ($Z = 1$)	3320	10^{14}	2927	8–16	753.1	3.88
Cerium dioxide (ceria, cerianite)	CeO$_2$ [1306-38-3] 172.114	Cubic $a = 541.1$ pm C1, cF12, $Fm3m$, fluorite type ($Z = 4$)	7650	10^{10}	2340	–	389	10.6
Chromium oxide (eskolaite)	Cr$_2$O$_3$ [1308-38-9] 151.990	Trigonal (rhombohedral) $a = 538$ pm $\alpha = 54°50'$ D5$_1$, hR10, $R\bar{3}c$, corundum type ($Z = 2$)	5220	1.3×10^9 (346 °C)	2330	–	921.1	10.9
Dysprosium oxide (dysprosia)	Dy$_2$O$_3$ [1308-87-8] 373.00	Cubic D5$_3$, cI80, $Ia\bar{3}$, Mn$_2$O$_3$ type ($Z=16$)	8300	–	2408	–	–	7.74
Europium oxide (europia)	Eu$_2$O$_3$ [1308-96-9] 351.928	Cubic D5$_3$, cI80, $Ia\bar{3}$, Mn$_2$O$_3$ type ($Z=16$)	7422	–	2350	–	–	7.02
Hafnium dioxide (hafnia)	HfO$_2$ [12055-23-1] 210.489	Monoclinic (1790°C) $a = 511.56$ pm $b = 517.22$ pm $c = 529.48$ pm C43, mP12, $P2_1c$, baddeleyite type ($Z = 4$)	9680	5×10^{15}	2900	1.14	121	5.85
Gadolinium oxide (gadolinia)	Gd$_2$O$_3$ [12064-62-9] 362.50	Cubic D5$_3$, cI80, $Ia\bar{3}$, Mn$_2$O$_3$ type ($Z = 16$)	7630	–	2420	–	276	10.44
Lanthanum dioxide (lanthana)	La$_2$O$_3$ [1312-81-8] 325.809	Trigonal (hexagonal) D5$_2$, hP5, $P\bar{3}m1$, lanthana type ($Z=1$)	6510	10^{14} (550 °C)	2315	–	288.89	11.9

Table 3.2-9 Physical properties of oxides and oxide-based high-temperature refractories [2.4], cont.

Young's modulus (E, GPa)	Flexural strength (τ, MPa)	Compressive strength (α, MPa)	Vickers hardness HV (Mohs hardness HM)	Other physicochemical properties, corrosion resistance,[a] and uses	IUPAC name (synonyms and common trade names)
365–393	282	2549–3103	2100–3000 (HM 9)	White and translucent; hard material used as abrasive for grinding. Excellent electrical insulator and also wear-resistent. Insoluble in water, insoluble in strong mineral acids, readily soluble in strong solutions of alkali metal hydroxides, attacked by HF and NH_4HF_2. Owing to its corrosion resistance under an inert atmosphere in molten metals such as Mg, Ca, Sr, Ba, Mn, Sn, Pb, Ga, Bi, As, Sb, Hg, Mo, W, Co, Ni, Pd, Pt, and U, it is used for crucibles for these liquid metals. Alumina is readily attacked under an inert atmosphere by molten metals such as Li, Na, Be, Al, Si, Ti, Zr, Nb, Ta, and Cu. Maximum service temperature 1950 °C.	Aluminium sesquioxide (alumina, corundum, sapphire)
296.5–345	241–250	1551	1500 (HM 9)		Beryllium monoxide (beryllia)
–	–	–	560 (HM 4.5)	Forms white or grayish ceramics. It is readily absorbs CO_2 and water from air to form calcium carbonate and slaked lime. It reacts readily with water to give $Ca(OH)_2$. Volumetric expansion coefficient 0.225×10^{-9} K^{-1}. It exhibits outstanding corrosion resistance in the following liquid metals: Li and Na.	Calcium monoxide (calcia, lime)
181	–	589	(HM 6)	Pale yellow cubic crystals. Abrasive for polishing glass; used in interference filters and antireflection coatings. Insoluble in water, soluble in H_2SO_4 and HNO_3, but insoluble in dilute acids.	Cerium dioxide (ceria, Cerianite)
–	–	–	(HM > 8)		Chromium oxide (eskolaite)
–	–	–	–		Dysprosium oxide (dysprosia)
–	–	–	–		Europium oxide (europia)
57	–	–	780–1050		Hafnium dioxide (hafnia)
124	–	–	480		Gadolinium oxide (gadolinia)
–	–	–	–	Insoluble in water, soluble in dilute strong mineral acids.	Lanthanum dioxide (lanthana)

Table 3.2-9 Physical properties of oxides and oxide-based high-temperature refractories [2.4], cont.

IUPAC name (synonyms and common trade names)	Theoretical chemical formula, [CASRN], relative molecular mass ($^{12}C = 12.000$)	Crystal system, lattice parameters, Strukturbericht symbol, Pearson symbol, space group, structure type, Z	Density (ϱ, kg m^{-3})	Electrical resistivity (ρ, $\mu\Omega$ cm)	Melting point (°C)	Thermal conductivity (κ, W m^{-1} K^{-1})	Specific heat capacity (c_p, J kg^{-1} K^{-1})	Coefficient of linear thermal expansion (α, 10^{-6} K^{-1})
Magnesium monoxide (magnesia, periclase)	MgO [1309-48-4] 40.304	Cubic $a = 420$ pm B1, $cF8$, $Fm3m$, rock salt type ($Z = 4$)	3581	1.3×10^{15}	2852	50–75	962.3	11.52
Niobium pentoxide (columbite, niobia)	Nb$_2$O$_5$ [1313-96-8] 265.810	Numerous polytypes	4470	5.5×10^{12}	1520	–	502.41	–
Samarium oxide (samaria)	Sm$_2$O$_3$ [12060-58-1] 348.72	Cubic D5$_3$, $cI80$, $Ia\bar{3}$, Mn$_2$O$_3$ type ($Z=16$)	7620	–	2350	2.07	331	10.3
Silicon dioxide (silica, α-quartz)	α-SiO$_2$ [7631-86-9] [14808-60-7] 60.085	Trigonal (rhombohedral) $a = 491.27$ pm $c = 540.46$ pm C8, $hP9$, $R\bar{3}c$, α-quartz type ($Z = 3$)	2202–2650	10^{20}	1710	1.38	787	0.55
Tantalum pentoxide (tantalite, tantala)	Ta$_2$O$_5$ [1314-61-0] 441.893	Trigonal (rhombohedral) columbite type	8200	10^{12}	1882	–	301.5	–
Thorium dioxide (thoria, thorianite)	ThO$_2$ [1314-20-1] 264.037	Cubic $a = 559$ pm C1, $cF12$, $Fm3m$, fluorite type ($Z = 4$)	9860	4×10^{19}	3390	14.19	272.14	9.54
Titanium dioxide (anatase)	TiO$_2$ [13463-67-7] [1317-70-0] 79.866	Tetragonal $a = 378.5$ pm $c = 951.4$ pm C5, $tI12$, $I4_1amd$, anatase type ($Z = 4$)	3900	–	700 °C (rutile)	–	–	–
Titanium dioxide (brookite)	TiO$_2$ [13463-67-7] 79.866	Orthorhombic $a = 916.6$ pm $b = 543.6$ pm $c = 513.5$ pm C21, $oP24$, $Pbca$, brookite type ($Z = 8$)	4140	–	1750	–	–	–

Table 3.2-9 Physical properties of oxides and oxide-based high-temperature refractories [2.4], cont.

Young's modulus (E, GPa)	Flexural strength (τ, MPa)	Compressive strength (α, MPa)	Vickers hardness HV (Mohs hardness HM)	Other physicochemical properties, corrosion resistance,[a] and uses	IUPAC name (synonyms and common trade names)
303.4	441	1300–1379	750 (HM 5.5–6)	Forms ceramics with a high reflection coefficient in the visible and near-UV region. Used in linings for steelmaking furnaces and in crucibles for fluoride melts. Very slowly soluble in pure water; but soluble in dilute strong mineral acids. It exhibits outstanding corrosion resistance in the following liquid metals: Mg, Li, and Na. It is readily attacked by molten metals such as Be, Si, Ti, Zr, Nb, and Ta. MgO reacts with water, CO_2, and dilute acids. Maximum service temperature 2400 °C. Transmittance of 80% and refractive index of 1.75 in the IR region from 7 to 300 μm.	Magnesium monoxide (magnesia, periclase)
–	–	–	1500	Dielectric used in film supercapacitors. Insoluble in water; soluble in HF and in hot concentrated H_2SO_4.	Niobium pentoxide (columbite, niobia)
183	–	–	438		Samarium oxide (samaria)
72.95	310	680–1380	550–1000 (HM 7)	Colorless amorphous (fused silica) or crystalline (quartz) material having a low thermal expansion coefficient and excellent optical transmittance in the far UV. Silica is insoluble in strong mineral acids and alkalis except HF, concentrated H_3PO_4, NH_4HF_2 and concentrated alkali metal hydroxides. Owing to its good corrosion resistance to liquid metals such as Si, Ge, Sn, Pb, Ga, In, Tl, Rb, Bi, and Cd, it is used in crucibles for melting these metals. Silica is readily attacked under an inert atmosphere by molten metals such as Li, Na, K, Mg, and Al. Quartz crystals are piezoelectric and pyroelectric. Maximum service temperature 1090 °C.	Silicon dioxide (silica, α-quartz)
–	–	–	–	Dielectric used in film supercapacitors. Tantalum oxide is a high-refractive index, low-absorption material usable for making optical coatings from the near-UV (350 nm) to the IR (8 μm). Insoluble in most chemicals except HF, HF–HNO_3 mixtures, oleum, fused alkali metal hydroxides (e.g. NaOH and KOH) and molten pyrosulfates.	Tantalum pentoxide (tantalite, tantala)
144.8	–	1475	945 (HM 6.5)	Corrosion-resistant container material for the following molten metals: Na, Hf, Ir, Ni, Mo, Mn, Th, U. Corroded by the following liquid metals: Be, Si, Ti, Zr, Nb, Bi. Radioactive.	Thorium dioxide (thoria, thorianite)
–	–	–	(HM 5.5–6)		Titanium dioxide (anatase)
–	–	–	(HM 5.5–6)		Titanium dioxide (brookite)

Table 3.2-9 Physical properties of oxides and oxide-based high-temperature refractories [2.4], cont.

IUPAC name (synonyms and common trade names)	Theoretical chemical formula, [CASRN], relative molecular mass ($^{12}C = 12.000$)	Crystal system, lattice parameters, Strukturbericht, symbol, Pearson symbol, space group, structure type, Z	Density (ϱ, kg m^{-3})	Electrical resistivity (ρ, $\mu\Omega$ cm)	Melting point (°C)	Thermal conductivity (κ, W m^{-1} K^{-1})	Specific heat capacity (c_p, J kg^{-1} K^{-1})	Coefficient of linear thermal expansion (α, 10^{-6} K^{-1})
Titanium dioxide (rutile, titania)	TiO_2 [13463-67-7] [1317-80-2] 79.866	Tetragonal $a = 459.37$ pm $c = 296.18$ pm C4, $tP6$, $P4/mnm$, rutile type ($Z = 2$)	4240	10^{19}	1855	10.4 ($\parallel$ c) 7.4 ($\perp$ c)	711	7.14
Uranium dioxide (uraninite)	UO_2 [1344-57-6] 270.028	Cubic $a = 546.82$ pm C1, $cF12$, $Fm3m$, fluorite type ($Z = 4$)	10 960	3.8×10^{10}	2880	10.04	234.31	11.2
Yttrium oxide (yttria)	Y_2O_3 [1314-36-9] 225.81	Trigonal (hexagonal) D5$_2$, $hP5$, $P\bar{3}m1$, lanthana type ($Z=1$)	5030	–	2439	–	439.62	8.10
Zirconium dioxide (baddeleyite)	ZrO_2 [1314-23-4] [12036-23-6] 123.223	Monoclinic $a = 514.54$ pm $b = 520.75$ pm $c = 531.07$ pm $\alpha = 99.23°$ C43, $mP12$, $P2_1c$, baddeleyite type ($Z = 4$)	5850	–	2710	–	711	7.56
Zirconium dioxide PSZ (stabilized with MgO) (zirconia > 2300 °C)	ZrO_2 [1314-23-4] [64417-98-7] 123.223	Cubic C1, $cF12$, $Fm3m$, fluorite type ($Z = 4$)	5800–6045	–	2710	1.8	400	10.1
Zirconium dioxide TTZ (stabilized with Y_2O_3) (zirconia > 2300 °C)	ZrO_2 [1314-23-4] [64417-98-7] 123.223	Cubic C1, $cF12$, $Fm3m$, fluorite type ($Z = 4$)	6045	–	2710	–	–	–
Zirconium dioxide TZP (zirconia > 1170 °C)	ZrO_2 [1314-23-4] 123.223	Tetragonal C4, $tP6$, $P4_2/mnm$, rutile type ($Z = 2$)	5680–6050	7.7×10^7	2710	–	–	10–11
Zirconium dioxide (stabilized with 10–15% Y_2O_3) (zirconia > 2300 °C)	ZrO_2 [1314-23-4] [64417-98-7] 123.223	Cubic C1, $cF12$, $Fm3m$, fluorite type ($Z = 4$)	6045	–	2710	–	–	–

a Corrosion data in molten salts from [2.9].

Table 3.2-9 Physical properties of oxides and oxide-based high-temperature refractories [2.4], cont.

Young's modulus (E, GPa)	Flexural strength (τ, MPa)	Compressive strength (α, MPa)	Vickers hardness HV (Mohs hardness HM)	Other physicochemical properties, corrosion resistance,[a] and uses	IUPAC name (synonyms and common trade names)
248–282	340	800–940	(HM 7–7.5)	White, translucent, hard ceramic material. Readily soluble in HF and in concentrated H_2SO_4, and reacts rapidly with molten alkali hydroxides and fused alkali carbonates. Owing to its good corrosion resistance to liquid metals such as Ni and Mo, it is used in crucibles for melting these metals. Titania is readily attacked under an inert atmosphere by molten metals such as Be, Si, Ti, Zr, Nb, and Ta.	Titanium dioxide (rutile, titania)
145	–	–	600 (HM 6–7)	Used in nuclear power reactors in sintered nuclear-fuel elements containing either natural or enriched uranium.	Uranium dioxide (uraninite)
114.5	–	393	700	Yttria is a medium-refractive-index, low-absorption material usable for optical coating, in the near-UV (300 nm) to the IR (12 µm) regions. Hence used to protect Al and Ag mirrors. Used for crucibles containing molten lithium.	Yttrium oxide (yttria)
241	–	2068	(HM 6.5) 1200	Zirconia is highly corrosion-resistant to molten metals such as Bi, Hf, Ir, Pt, Fe, Ni, Mo, Pu, and V, while is strongly attacked by the following liquid metals: Be, Li, Na, K, Si, Ti, Zr, and Nb. Insoluble in water, but slowly soluble in HCl and HNO_3; soluble in boiling concentrated H_2SO_4 and alkali hydroxides and readily attacked by HF. Monoclinic (baddeleyite) below 1100 °C, tetragonal between 1100 and 2300 °C, cubic (fluorite type) above 2300 °C. Maximum service temperature 2400 °C.	Zirconium dioxide (baddeleyite)
200	690	1850	1600		Zirconium dioxide PSZ (stabilized with MgO) (zirconia > 2300 °C)
–	–	–	–		Zirconium dioxide TTZ (stabilized with Y_2O_3) (zirconia > 2300 °C)
200–210	> 800	> 2900	–		Zirconium dioxide TZP (zirconia > 1170 °C)
–	–	–	–		Zirconium dioxide (stabilized 10–15% Y_2O_3) (zirconia > 2300 °C)

3.2.4.1 Magnesium Oxide

Table 3.2-10 lists the properties of the magnesium oxide rich raw material used in the production of oxide ceramics. MgO occurs rarely, as the mineral periclase, but is abundant as magnesite ($MgCO_3$) and dolomite (($Mg,Ca)CO_3$).

Magnesium oxide is mainly processed into a high-purity single-phase ceramic material, in either porous or gas-tight form. Some properties of porous MgO are listed in Table 3.2-11. The properties of MgO ceramics that are of particular benefit in applications are their high electrical insulation capability and their high thermal conductivity. Typical examples of their application are for insulation in sheathed thermocouples and in resistive heating elements.

Table 3.2-10 Physical properties of MgO-rich raw materials used in the production of refractory ceramics [2.3]

Composition (wt%)	Sintered MgO clinker	Sintered doloma clinker	Fused MgO
MgO	96–99	39–40	97–98
Al_2O_3	0.05–0.25	0.3–0.8	0.1–0.2
Fe_2O_3	0.05–0.2	0.6–1.0	0.1–0.5
CaO	0.6–2.4	57.5–58.5	0.9–2.5
SiO_2	0.1–0.5	0.6–1.1	0.3–0.9
B_2O_3	0.005–0.6		0.004–0.01
Bulk density (g/cm^3)	3.4–3.45	3.15–3.25	3.5
Grain porosity (vol.%)	2–3	5.5–7	0.5–2

Table 3.2-11 Properties of MgO according to DIN EN 60672 [2.6]

	Symbol	Units	Designation C 820 Porous
Mechanical properties			
Open porosity		vol.%	30
Density, minimum	ϱ	Mg/m^3	2.5
Bending strength	σ_B	MPa	50
Young's modulus	E	GPa	90
Electrical properties			
Permittivity at 48–62 Hz	ε_r		10
Thermal properties			
Average coefficient of thermal expansion at 30–600 °C	α_{30-600}	10^{-6} K^{-1}	11–13
Specific heat capacity at 30–100 °C	$c_{p,30-600}$	J kg^{-1} K^{-1}	850–1050
Thermal conductivity	λ_{30-100}	W m^{-1} K^{-1}	6–10
Thermal fatigue resistance		(Rated)	Good

3.2.4.2 Alumina

The properties of commercial grades of alumina (corundum) are closely related to the microstructure. Pure α-Al_2O_3 is denser, harder, stiffer, and more refractory than most silicate ceramics so that increasing the proportion of the second phase in an alumina ceramic tends to decrease the density, Young's modulus, strength, hardness, and refractoriness. Sintered alumina is produced from high-purity powders, which densify to give single-phase ceramics with a uniform grain size. Table 3.2-12 gives typical examples of properties of high-density alumina, and Table 3.2-13 lists further properties of various alumina ceramics.

Alumina is widely applied both as an electronic ceramic material and in cases where its nonelectronic properties, such as fracture toughness, wear and high-temperature resistance, are required. Some typical applications are in insulators in electrotechnical equipment; substrates for electronic components; wear-resistant machine parts; refractory materials in the chemical industry, where resistance against vapors, melts, and slags is important; insulating materials in sheathed thermocouples and heating elements; medical implants; and high-temperature parts such as burner nozzles.

Table 3.2-12 Typical properties of high-density alumina [2.3]

	Al_2O_3 content (wt%)			
	> 99.9	> 99.7[a]	> 99.7[b]	99–99.7
Density (g/cm^3)	3.97–3.99	3.6–3.85	3.65–3.85	3.89–3.96
Hardness (GPa), HV 500 g	19.3	16.3	15–16	15–16
Fracture toughness K_{IC} at room temperature (MPa m$^{1/2}$)	2.8–4.5	–	–	5.6–6
Young's modulus (GPa)	366–410	300–380	300–380	330–400
Bending strength (MPa) at room temperature	550–600	160–300	245–412	550
Thermal expansion coefficient (10^{-6}/K) at 200–1200 °C	6.5–8.9	5.4–8.4	5.4–8.4	6.4–8.2
Thermal conductivity at room temperature (W/m K)	38.9	28–30	30	30.4
Firing temperature range (°C)	1600–2000	1750–1900	1750–1900	1700–1750

[a] "Recrystallized" without MgO.
[b] With MgO.

Table 3.2-13 Properties of Al_2O_3 according to DIN EN 60672 [2.6]

Mechanical properties	Symbol	Units	Designation							
			C 780	C 786	C 795	C 799	Al_2O_3	Al_2O_3	Al_2O_3	Al_2O_3
			Alumina 80–86%	Alumina 86–95%	Alumina 95–99%	Alumina >99%	Alumina <90%	Alumina 92–96%	Alumina 99%	Alumina >99%
Open porosity		vol.%	0.0	0.0	0.0	0.0	0	0	0	0
Density, min.	ϱ	Mg/m^3	3.2	3.4	3.5	3.7	> 3.2	3.4–3.8	3.5–3.9	3.75–3.98
Bending strength	σ_B	MPa	200	250	280	300	> 200	230–400	280–400	300–580
Young's modulus	E	GPa	200	220	280	300	> 200	220–340	220–350	300–380
Vickers hardness	HV	100	–	–	–	–	12–15	12–15	12–20	17–23
Fracture toughness	K_{IC}	MPa $\sqrt{m}$	–	–	–	–	3.5–4.5	4–4.2	4–4.2	4–5.5

Table 3.2-13 Properties of Al_2O_3 according to DIN EN 60672 [2.6], cont.

Thermal properties / Electrical properties	Symbol	Units	Designation							
			C 780	C 786	C 795	C 799	Al_2O_3	Al_2O_3	Al_2O_3	Al_2O_3
			Alumina 80–86%	Alumina 86–95%	Alumina 95–99%	Alumina >99%	Alumina <90%	Alumina 92–96%	Alumina 99%	Alumina >99%
Breakdown strength	E_d	kV/mm	10	15	15	17	10	15–25	15	17
Withstand voltage	U	kV	15	18	18	20	15	18	18	20
Permittivity at 48–62 Hz	ε_r	–	8	9	9	9	9	9–10	9	9
Loss factor at 20 °C, 1 kHz	$\tan \delta_{pf}$	10^{-3}	1.0	0.5	0.5	0.2	0.5–1.0	0.3–0.5	0.2–0.5	0.2–0.5
Loss factor at 20 °C, 1 MHz	$\tan \delta_{1M}$	10^{-3}	1.5	1	1	1	1	1	1	1
Resistivity at 20 °C	ρ_{20}	Ω m	10^{12}	10^{12}	10^{12}	10^{12}	10^{12}–10^{13}	10^{12}–10^{14}	10^{12}–10^{15}	10^{12}–10^{15}
Resistivity at 600 °C	ρ_{600}	Ω m	10^5	10^6	10^6	10^6	10^6	10^6	10^6	10^6
Ave. coeff. of thermal expansion at 30–600 °C	α_{30-600}	10^{-6}K^{-1}	6–8	6–8	6–8	7–8	6–8	6–8	6–8	7–8
Specific heat at 30–100 °C	$c_{p,30-600}$	J kg^{-1}K^{-1}	850–1050	850–1050	850–1050	850–1050	850–1050	850–1050	850–1050	850–1050
Thermal conductivity	λ_{30-100}	W m^{-1}K^{-1}	10–16	14–24	16–28	19–30	10–16	14–25	16–28	19–30
Thermal fatigue resistance		(Rated)	Good	Good	Good	Good	Good	Good	Good	Good
Typ. max. application temperature	T	°C	–	–	–	–	1400–1500	1400–1500	1400–1500	1400–1700

3.2.4.3 Al–O–N Ceramics

A class of variants of the ceramics related to alumina is based on the ternary Al–O–N system, which forms 13 different compounds, some of which are polytypes of aluminium nitride based on the wurtzite structure. Table 3.2-14 lists some typical properties compiled from various publications. Al–O–N ceramics have optical and dielectric properties which make them suitable, for use, for example, in windows for elctromagnetic radiation.

Table 3.2-14 Selected properties of Al–O–N ceramics [2.3]

Property	Value
Refractive index at $\lambda = 0.55\,\mu m$	1.77–1.88
IR cutoff	$5.12\,\mu m$
UV cutoff	$0.27\,\mu m$
Dielectric constant at 20 °C, 100 Hz	8.5
Dielectric constant at 500 °C, 100 Hz	14.0
Loss tangent at 20 °C, 100 Hz	0.002
Loss tangent at 500 °C, 100 Hz	1.0
Thermal conductivity at 20 °C	$10.89\,\mathrm{W\,m^{-1}\,K^{-1}}$
Thermal expansion coefficient at 25–1000 °C	$7.6\times 10^6\,\mathrm{K^{-1}}$
Flexural strength at 20 °C	306 MPa
Flexural strength at 1000 °C	267 MPa
Fracture toughness	$2.0\text{–}2.9\,\mathrm{MPa\,m^{1/2}}$

3.2.4.4 Beryllium Oxide

Beryllium oxide (BeO, beryllia) is the only material apart from diamond which combines high thermal-shock resistance, high electrical resistivity, and high thermal conductivity at a similar level. Hence its major application is in heat sinks for electronic components. BeO is highly soluble in water, but dissolves slowly in concentrated acids and alkalis. It is highly toxic. It is highly corrosion-resistant in several liquid metals, such as Li, Na, Al, Ga, Pb, Ni, and Ir. Its maximum service temperature is 2400 °C. Some properties are listed in Table 3.2-15.

Table 3.2-15 Properties of BeO according to DIN EN 60672 [2.6]

			Designation
			C 810
Mechanical properties	**Symbol**	**Units**	**Beryllium oxide, dense**
Open porosity		vol. %	0.0
Density, minimum	ϱ	Mg/m^3	2.8
Bending strength	σ_B	MPa	150
Young's modulus	E	GPa	300

Table 3.2-15 Properties of BeO according to DIN EN 60672 [2.6], cont.

			Designation C 810
Electrical properties	**Symbol**	**Units**	**Beryllium oxide, dense**
Breakdown strength	E_d	kV/mm	13
Withstand voltage	U	kV	20
Permittivity at 48–62 Hz	ε_r		7
Loss factor at 20 °C, 1 kHz	$\tan \delta_{pf}$	10^{-3}	1
Loss factor at 20 °C, 1 MHz	$\tan \delta_{1M}$	10^{-3}	1
Resistivity at 20 °C	ρ_{20}	$\Omega\,m$	10^{12}
Resistivity at 600 °C	ρ_{600}	$\Omega\,m$	10^{7}
Thermal properties			
Average coefficient of thermal expansion at 30–600 °C	α_{30-600}	$10^{-6}\,K^{-1}$	7–8.5
Specific heat capacity at 30–100 °C	$c_{p,30-100}$	$J\,kg^{-1}K^{-1}$	1000–1250
Thermal conductivity	λ_{30-100}	$W\,m^{-1}K^{-1}$	150–220
Thermal fatigue resistance		(Rated)	Good

3.2.4.5 Zirconium Dioxide

Zirconium dioxide (ZrO_2), commonly called zirconium oxide, forms three phases, with a monoclinic, a tetragonal, and a cubic crystal structure. Dense parts may be obtained by sintering of the cubic or tetragonal phase only. In order to stabilize the cubic phase, "stabilizers" such as MgO, CaO, Y_2O_3, and CeO_2 are added.

An important class of zirconium dioxide ceramics is known as partially stabilized zirconia (PSZ), and Table 3.2-16 gives typical property data for several different grades. Further properties of PSZ are listed in Table 3.2-17. It should be noted that, as with most ceramic materials, the K_{1C} data depend on the test method applied, and all the other mechanical properties are strongly affected by the microstructure (grain size, stabilizer content, etc.) and further variables such as temperature and atmospheric conditions.

Table 3.2-16 Physical properties reported for commercial grades of partially stabilized zirconia (PSZ) [2.3]

	Mg-PSZ	Ca-PSZ	Y-PSZ	Ca/Mg-PSZ
wt% stabilizer	2.5–3.5	3–4.5	5–12.5	3
Hardness (GPa)	14.4[a]	17.1[b]	13.6[c]	15
Fracture toughness K_{IC} at room temperature (MPa m$^{1/2}$)	7–15	6–9	6	4.6
Young's modulus (GPa)	200[a]	200–217	210–238	–
Bending strength (MPa) at room temperature	430–720	400–690	650–1400	350
Thermal expansion coefficient (10^{-6}/K) at 1000 °C	9.2[a]	9.2[b]	10.2[c]	–
Thermal conductivity at room temperature (W m^{-1}K^{-1})	1–2	1–2	1–2	1–2

[a] 2.8% MgO.
[b] 4% CaO.
[c] 5% Y_2O_3.

Table 3.2-17 Properties of PSZ according to DIN EN 60672 [2.6]

			Designation
			PSZ
Mechanical properties	Symbol	Units	**Partially stabilized ZrO$_2$**
Open porosity		vol. %	0
Density, minimum	ϱ	Mg/m^3	5–6
Bending strength	σ_B	MPa	500–1000
Young's modulus	E	GPa	200–210
Vickers hardness	HV	100	11–12.5
Fracture toughness	K_{IC}	MPa $\sqrt{m}$	5.8–10.5
Electrical properties			
Permittivity at 48–62 Hz	ε_r		22
Resistivity at 20 °C	ρ_{20}	Ω m	10^8–10^{13}
Resistivity at 600 °C	ρ_{600}	Ω m	10^3–10^6
Thermal properties	Symbol	Units	**Partially stabilized ZrO$_2$**
Average coefficient of thermal expansion at 30–600 °C	$\alpha_{30-1000}$	10^{-6} K^{-1}	10–12.5
Specific heat capacity at 30–100 °C	$c_{p,30-1000}$	J kg^{-1} K^{-1}	400–550
Thermal conductivity	λ_{30-100}	W m^{-1} K^{-1}	1.5–3
Thermal fatigue resistance		(Rated)	Good
Typical maximum application temperature	T	°C	900–1600

3.2.4.6 Titanium Dioxide, Titanates, etc.

Titanium dioxide (TiO_2) is widely used in powder form as a pigment and filler material and in optical and catalytic applications. Technical ceramics made from titanium dioxide or titanates have the characteristic feature that their permittivity and its temperature coefficient can be adjusted over a wide range. Furthermore, they have a low loss factor. Some properties are listed in Table 3.2-18.

Table 3.2-18 Properties of titanium dioxide, titanates etc. according to DIN EN 60672 [2.6]

			Designation					
			C 310	C 320	C 330	C 331	C 340	C 350
Mechanical properties	Symbol	Units	Titanium dioxide chief constituent	Magnesium titanate	Titanium dioxide with other oxides	Titanium dioxide with other oxides	Bismuth titanate, basic	Perovskites, middle ε_r
Open porosity		vol. %	0.0	0.0	0.0	0.0	0.0	0.0
Density, minimum	ϱ	Mg/m³	3.5	3.1	4.0	4.5	3.0	4.0
Bending strength	σ_B	MPa	70	70	80	80	70	50
Breakdown strength	E_d	kV/mm	8	8	10	10	6	2
Withstand voltage	U	kV	15	15	15	15	8	2
Permittivity at 48–62 Hz	ε_r		40–100	12–40	25–50	30–70	100–700	350–3000
Loss factor at 20 °C, 1 kHz	$\tan \delta_{1k}$	10^{-3}	6.5	2	20.0	7	–	–
Loss factor at 20 °C, 1 MHz	$\tan \delta_{1M}$	10^{-3}	2	1.5	0.8	1	5	35.0
Resistivity at 20 °C	ρ_{20}	Ω m	10^{10}	10^9	10^9	10^9	10^9	10^8
Thermal properties								
Average coefficient of thermal expansion at 30–600 °C	α_{30-600}	10^{-6} K^{-1}	6–8	6–10	–	–	–	–
Specific heat capacity at 30–100 °C	$c_{p,30-100}$	J kg^{-1} K^{-1}	700–800	900–1000	–	–	–	–
Thermal conductivity	λ_{30-100}	W m^{-1} K^{-1}	3–4	3.5–4	–	–	–	–

3.2.5 Non-Oxide Ceramics

The non-oxide ceramics comprise essentially borides, carbides, nitrides, and silicides. Like oxide ceramics they have two kinds of uses, which frequently overlap: application of their physical properties and of their refractory high-temperature properties. Extensive accounts can be found in [2.1–3, 8, 10].

3.2.5.1 Non-Oxide High-Temperature Ceramics

The two dominating design variables to be considered for high-temperature materials are hardness and thermal conductivity as a function of temperature. Figure 3.2-2 gives a survey. It is obvious that borides and carbides are superior to most oxide ceramics. Depending upon the microstructure, SiC has a higher hardness than that of β-Si_3N_4, but it decreases somewhat more rapidly, with increasing temperature. Both Si_3N_4 and SiC ceramics possess a high thermal conductivity and thus excellent thermal-shock resistance.

3.2.5.2 Borides

Borides and boride-based high-temperature refractories are treated extensively in [2.2, 3, 10]. Some properties are listed in Table 3.2-19.

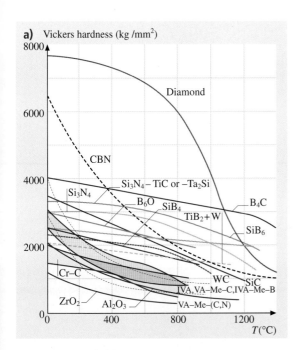
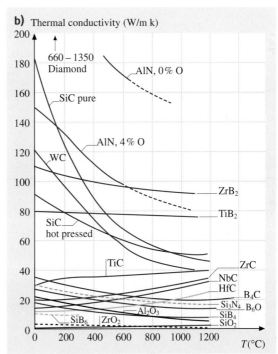

Fig. 3.2-2a,b Temperature dependence of **(a)** the hardness and **(b)** the thermal conductivity of ceramic materials in comparison with diamond and cubic boron nitride (CBN) [2.3]

Table 3.2-19 Physical properties of borides and boride-based high-temperature refractories [2.4]

IUPAC name	Theoretical chemical formula, [CASRN], relative molecular mass ($^{12}C = 12.000$)	Crystal system, lattice parameters, *Strukturbericht* symbol, Pearson symbol, space group, structure type, Z	Density (ϱ, kg m^{-3})	Electrical resistivity (ρ, $\mu\Omega$ cm)	Melting point (°C)	Thermal conductivity (κ, W m^{-1} K^{-1})	Specific heat capacity (c_p, J kg^{-1} K^{-1})	Coefficient of linear thermal expansion (α, 10^{-6}K^{-1})
Aluminium diboride	AlB$_2$ [12041-50-8] 48.604	Hexagonal $a = 300.50$ pm $c = 325.30$ pm C32, $hP3$, $P/6mmm$, AlB$_2$ type ($Z = 1$)	3190	–	1654	–	897.87	–
Aluminium dodecaboride	AlB$_{12}$ [12041-54-2] 156.714	Tetragonal $a = 1016$ pm $c = 1428$ pm	2580	–	2421	–	954.48	–
Beryllium boride	Be$_4$B [12536-52-6] 46.589	–	–	–	1160	–	–	–
Beryllium diboride	BeB$_2$ [12228-40-9] 30.634	Hexagonal $a = 979$ pm $c = 955$ pm	2420	10 000	1970	–	–	–
Beryllium hemiboride	Be$_2$B [12536-51-5]	Cubic $a = 467.00$ pm C1, $cF12$, $Fm3m$, CaF$_2$ type ($Z = 4$)	1890	1000	1520	–	–	–
Beryllium hexaboride	BeB$_6$ [12429-94-6]	Tetragonal $a = 1016$ pm $c = 1428$ pm	2330	10^{13}	2070	–	–	–
Beryllium monoboride	BeB [12228-40-9]	–	–	–	1970	–	–	–
Boron	β-B [7440-42-8] 10.811	Trigonal (rhombohedral) $a = 1017$ pm $\alpha = 65°$ 12′ $hR105$, $R3m$, β-B type	2460	18 000	2190	–	–	–
Chromium boride	Cr$_5$B$_3$ [12007-38-4] 292.414	Orthorhombic $a = 302.6$ pm $b = 1811.5$ pm $c = 295.4$ pm D8$_1$, $ti32$, $I4/mcm$, Cr$_5$B$_3$ type ($Z = 4$)	6100	–	1900	15.8	–	13.7
Chromium diboride	CrB$_2$ [12007-16-8] 73.618	Hexagonal $a = 292.9$ pm $c = 306.6$ pm C32, $hP3$, $P6/mmm$, AlB$_2$ type ($Z = 1$)	5160–5200	21	1850–2100	20–32	712	6.2–7.5
Chromium monoboride	CrB [12006-79-0] 62.807	Tetragonal $a = 294.00$ pm $c = 1572.00$ pm B$_f$, $oC8$, $Cmcm$, CrB type ($Z = 4$)	6200	64.0	2000	20.1	–	12.3

Table 3.2-19 Physical properties of borides and boride-based high-temperature refractories [2.4], cont.

Young's modulus (E, GPa)	Flexural strength (τ, MPa)	Compressive strength (α, MPa)	Vickers hardness HV (Mohs hardness HM)	Other physicochemical properties, corrosion resistance,[a] and uses	IUPAC name
–	–	–	2500	Phase transition to AlB_{12} at 920 °C. Soluble in dilute HCl. Nuclear shielding material.	Aluminium diboride
–	–	–	–	Soluble in hot HNO_3, insoluble in other acids and alkalis. Neutron-shielding material.	Aluminium dodecaboride
–	–	–	–		Beryllium boride
–	–	–	–		Beryllium diboride
–	–	–	870		Beryllium hemiboride
–	–	–	–		Beryllium hexaboride
–	–	–	–		Beryllium monoboride
320	–	–	2055 (11)	Brown or dark powder, unreactive to oxygen, water, acids and alkalis. $\Delta H_{vap} = 480\,\text{kJ}\,\text{mol}^{-1}$.	Boron
–	–	–	–		Chromium boride
211	607	1300	1800	Strongly corroded by molten metals such as Mg, Al, Na, Si, V, Cr, Mn, Fe, and Ni. It is corrosion-resistant to the following liquid metals: Cu, Zn, Sn, Rb, and Bi.	Chromium diboride
–	–	–	–		Chromium monoboride

Table 3.2-19 Physical properties of borides and boride-based high-temperature refractories [2.4], cont.

IUPAC name	Theoretical chemical formula, [CASRN], relative molecular mass ($^{12}C = 12.000$)	Crystal system, lattice parameters, Strukturbericht symbol, Pearson symbol, space group, structure type, Z	Density (ϱ, kg m^{-3})	Electrical resistivity (ρ, $\mu\Omega$ cm)	Melting point (°C)	Thermal conductivity (κ, W m^{-1} K^{-1})	Specific heat capacity (c_p, J kg^{-1} K^{-1})	Coefficient of linear thermal expansion (α, 10^{-6}K^{-1})
Hafnium diboride	HfB$_2$ [12007-23-7] 200.112	Hexagonal $a = 314.20$ pm $c = 347.60$ pm C32, $hP3$, $P6/mmm$, AlB$_2$ type ($Z = 1$)	11 190	8.8–11	3250–3380	51.6	247.11	6.3–7.6
Lanthanum hexaboride	LaB$_6$ [12008-21-8] 203.772	Cubic $a = 415.7$ pm D2$_1$, $cP7$, $Pm3m$, CaB$_6$ type ($Z = 1$)	4760	17.4	2715	47.7	–	6.4
Molybdenum boride	Mo$_2$B$_5$ [12007-97-5] 245.935	Trigonal $a = 301.2$ pm $c = 2093.7$ pm D8$_1$, $hR7$, $R\bar{3}m$, Mo$_2$B$_5$ type ($Z = 1$)	7480	22–55	1600	50	–	8.6
Molybdenum diboride	MoB$_2$ 117.59	Hexagonal $a = 305.00$ pm $c = 311.30$ pm C32, $hP3$, $P6/mmm$, AlB$_2$ type ($Z = 1$)	7780	45	2100	–	527	7.7
Molybdenum hemiboride	Mo$_2$B [12006-99-4] 202.691	Tetragonal $a = 554.3$ pm $c = 473.5$ pm C16, $tI2$, $I4/mcm$, CuAl$_2$ type ($Z = 4$)	9260	40	2280	–	377	5
Molybdenum monoboride	MoB 106.77	Tetragonal $a = 311.0$ $c = 169.5$ B$_g$, $tI4$, $I4_1/amd$, MoB type ($Z = 2$)	8770	α-MoB 45, β-MoB 24	2180	–	368	–
Niobium diboride	ϵ-NbB$_2$ [12007-29-3] 114.528	Hexagonal $a = 308.90$ pm $c = 330.03$ pm C32, $hP3$, $P6/mmm$, AlB$_2$ type ($Z = 1$)	6970	26–65	2900	17–23.5	418	8.0–8.6
Niobium monoboride	δ-NbB [12045-19-1] 103.717	Orthorhombic $a = 329.8$ pm $b = 316.6$ pm $c = 87.23$ pm B$_f$, $oC8$, $Cmcm$, CrB type ($Z = 4$)	7570	40–64.5	2270–2917	15.6	–	12.9
Silicon hexaboride	SiB$_6$ [12008-29-6]	Trigonal (rhombohedral)	2430	200 000	1950	–	–	–
Silicon tetraboride	SiB$_4$ [12007-81-7] 71.330	–	2400	–	– (dec.)	–	–	–
Tantalum diboride	TaB$_2$ [12077-35-1] 202.570	Hexagonal $a = 309.80$ pm $c = 324.10$ pm C$_{32}$, $hP3$, $P6/mmm$, AlB$_2$ type ($Z = 1$)	12 540	33	3037–3200	10.9–16.0	237.55	8.2–8.8

Table 3.2-19 Physical properties of borides and boride-based high-temperature refractories [2.4], cont.

Young's modulus (E, GPa)	Flexural strength (τ, MPa)	Compressive strength (α, MPa)	Vickers hardness HV (Mohs hardness HM)	Other physicochemical properties, corrosion resistance,[a] and uses	IUPAC name
500	350	–	2900	Gray crystals, attacked by HF, otherwise highly resistant.	Hafnium diboride
479	126	–	–	Wear-resistant, semiconducting, thermoionic-conductor films.	Lanthanum hexaboride
672	345	–	–	Corroded by the following molten metals: Al, Mg, V, Cr, Mn, Fe, Ni, Cu, Nb, Mo, and Ta. It is corrosion resistant to molten Cd, Sn, Bi, and Rb.	Molybdenum boride
–	–	–	1280		Molybdenum diboride
–	–	–	– (HM 8–9)	Corrosion-resistant films.	Molybdenum hemiboride
–	–	–	1570		Molybdenum monoboride
637	–	–	3130 (HM > 8)	Corrosion-resistant to molten Ta; corroded by molten Re.	Niobium diboride
–	–	–	–	Wear-resistant, semiconducting films; neutron-absorbing layers on nuclear fuel pellets.	Niobium monoboride
–	–	–	–		Silicon hexaboride
–	–	–	–		Silicon tetraboride
257	–	–	(HM > 8)	Gray metallic powder. Severe oxidation in air above 800 °C. Corroded by the following molten metals: Nb, Mo, Ta, and Re.	Tantalum diboride

Table 3.2-19 Physical properties of borides and boride-based high-temperature refractories [2.4], cont.

IUPAC name	Theoretical chemical formula, [CASRN], relative molecular mass ($^{12}C = 12.000$)	Crystal system, lattice parameters, *Strukturbericht* symbol, Pearson symbol, space group, structure type, Z	Density (ϱ, kg m^{-3})	Electrical resistivity (ρ, μΩ cm)	Melting point (°C)	Thermal conductivity (κ, W m^{-1} K^{-1})	Specific heat capacity (c_p, J kg^{-1} K^{-1})	Coefficient of linear thermal expansion (α, 10^{-6}K^{-1})
Tantalum monoboride	TaB [12007-07-7] 191.759	Orthorhombic $a = 327.6$ pm $b = 866.9$ pm $c = 315.7$ pm B_f, $oC8$, $Cmcm$, CrB type ($Z = 4$)	14 190	100	2340–3090	–	246.85	–
Thorium hexaboride	ThB$_6$ [12229-63-9] 296.904	Cubic $a = 411.2$ pm $D2_1$, $cP7$, $Pm3m$, CaB$_6$ type ($Z = 1$)	6800	–	2149	44.8	–	7.8
Thorium tetraboride	ThB$_4$ [12007-83-9] 275.53	Tetragonal $a = 725.6$ pm $c = 411.3$ pm $D1_e$, $tP20$, $P4/mbm$, ThB$_4$ type ($Z = 4$)	8450	–	2500	25	510	7.9
Titanium diboride	TiB$_2$ [12045-63-5] 69.489	Hexagonal $a = 302.8$ pm $c = 322.8$ pm $C32$, $hP3$, $P6/mmm$, AlB$_2$ type ($Z = 1$)	4520	16–28.4	2980–3225	64.4	637.22	7.6–8.6
Tungsten hemiboride	W$_2$B [12007-10-2] 378.491	Tetragonal $a = 556.4$ pm $c = 474.0$ pm $C16$, $tI12$, $I4/mcm$, CuAl$_2$ type ($Z = 4$)	16 720	–	2670	–	168	6.7
Tungsten monoboride	WB [12007-09-9] 194.651	Tetragonal $a = 311.5$ pm $c = 1692$ pm	15 200 16 000	4.1	2660	–	–	6.9
Uranium diboride	UB$_2$ [12007-36-2] 259.651	Hexagonal $a = 313.10$ pm $c = 398.70$ pm $C32$, $hP3$, $P6/mmm$, AlB$_2$ type	12 710	–	2385	51.9	–	9
Uranium dodecaboride	UB$_{12}$ 367.91	Cubic $a = 747.3$ pm $D2_f$, $cF52$, $Fm3m$, UB$_{12}$ type ($Z = 4$)	5820	–	1500	–	–	4.6
Uranium tetraboride	UB$_4$ [12007-84-0] 281.273	Tetragonal $a = 707.5$ pm $c = 397.9$ pm $D1_e$, $tP20$, $P4/mbm$, ThB$_4$ type ($Z = 4$)	5350	–	2495	4.0	–	7.0
Vanadium diboride	VB$_2$ [12007-37-3] 72.564	Hexagonal $a = 299.8$ pm $c = 305.7$ pm $C32$, $hP3$, $P6/mmm$, AlB$_2$ type ($Z = 1$)	5070	23	2450–2747	42.3	647.43	7.6–8.3

Table 3.2-19 Physical properties of borides and boride-based high-temperature refractories [2.4], cont.

Young's modulus (E, GPa)	Flexural strength (τ, MPa)	Compressive strength (α, MPa)	Vickers hardness HV (Mohs hardness HM)	Other physicochemical properties, corrosion resistance,[a] and uses	IUPAC name
–	–	–	2200 (HM > 8)	Severe oxidation above 1100–1400 °C in air.	Tantalum monoboride
–	–	–	–		Thorium hexaboride
148	137	–	–		Thorium tetraboride
372–551	240	669	3370 (HM > 9)	Gray crystals, superconducting at 1.26 K. High-temperature electrical conductor, used in the form of a cermet as a crucible material for handling molten metals such as Al, Zn, Cd, Bi, Sn, and Rb. It is strongly corroded by liquid metals such as Tl, Zr, V, Nb, Ta, Cr, Mn, Fe, Co, Ni, and Cu. Begins to be oxidized in air above 1100–1400 °C. Corrosion-resistant in hot concentrated brines. Maximum operating temperature 1000 °C (reducing environment) and 800 °C (oxidizing environment).	Titanium diboride
–	–	–	2420 (HM 9)	Black powder.	Tungsten hemiboride
–	–	–	(HM 9)	Black powder.	Tungsten monoboride
–	–	–	1390		Uranium diboride
–	–	–	–		Uranium dodecaboride
440	413	–	2500		Uranium tetraboride
268	–	–	(HM 8–9)	Wear-resistant, semiconducting films.	Vanadium diboride

Table 3.2-19 Physical properties of borides and boride-based high-temperature refractories [2.4], cont.

IUPAC name	Theoretical chemical formula, [CASRN], relative molecular mass ($^{12}C = 12.000$)	Crystal system, lattice parameters, *Strukturbericht* symbol, Pearson symbol, space group, structure type, Z	Density (ϱ, kg m^{-3})	Electrical resistivity (ρ, $\mu\Omega$ cm)	Melting point (°C)	Thermal conductivity (κ, W m^{-1} K^{-1})	Specific heat capacity (c_p, J kg^{-1} K^{-1})	Coefficient of linear thermal expansion (α, 10^{-6} K^{-1})
Zirconium diboride	ZrB$_2$ [12045-64-6] 112.846	Hexagonal $a = 316.9$ pm $c = 353.0$ pm C32, $hP3$, $P6/mmm$, AlB$_2$ type ($Z = 1$)	6085	9.2	3060–3245	57.9	392.54	5.5–8.3
Zirconium dodecaboride	ZrB$_{12}$ 283.217	Cubic $a = 740.8$ pm D2$_f$, $cF52$, $Fm3m$, UB$_{12}$ type ($Z = 4$)	3630	60–80	2680	–	523	–

3.2.5.3 Carbides

Carbides are treated extensively in [2.1–3]. Some properties of carbides and carbide-based high-temperature refractories are listed in Table 3.2-20 and Table 3.2-21. It should be noted that carbides also play a major role as hardening constituents in all carbon-containing steels (Sect. 3.1.5).

Table 3.2-20 Physical properties of carbides and carbide-based high-temperature refractories [2.4]

IUPAC name (synonyms and common trade names)	Theoretical chemical formula, [CASRN], relative molecular mass ($^{12}C = 12.000$)	Crystal system, lattice parameters, *Strukturbericht* symbol, Pearson symbol, space group, structure type, Z	Density (ϱ, kg m^{-3})	Electrical resistivity (ρ, $\mu\Omega$ cm)	Melting point (°C)	Thermal conductivity (κ, W m^{-1} K^{-1})	Specific heat capacity (c_p, J kg^{-1} K^{-1})	Coefficient of linear thermal expansion (α, 10^{-6} K^{-1})
Aluminium carbides	Al$_4$C$_3$ [1299-86-1] 143.959	Numerous polytypes	2360	–	2798	n.a	n.a	n.a
Beryllium carbide	Be$_2$C [506-66-1] 30.035	Cubic $a = 433$ pm C1, $cF12$, $Fm3m$, CaF$_2$ type ($Z = 4$)	1900	–	2100	21.0	1397	10.5
Boron carbide (Norbide®)	B$_4$C [12069-32-8] 55.255	Hexagonal $a = 560$ pm $c = 1212$ pm D1$_g$, $hR15$, $R\bar{3}m$, B$_4$C type	2512	4500	2350–2427	27	1854	2.63–5.6

Table 3.2-19 Physical properties of borides and boride-based high-temperature refractories [2.4], cont.

Young's modulus (E, GPa)	Flexural strength (τ, MPa)	Compressive strength (α, MPa)	Vickers hardness HV (Mohs hardness HM)	Other physicochemical properties, corrosion resistance,[a] and uses	IUPAC name
343–506	305	–	1900–3400 (HM 8)	Gray metallic crystals, excellent thermal-shock resistance, greatest oxidation inertness of all refractory hard metals. Hot-pressed material is used in crucibles for handling molten metals such as Zn, Mg, Fe, Cu, Zn, Cd, Sn, Pb, Rb, Bi, Cr, brass, carbon steel, and cast iron, and also molten cryolite, yttria, zirconia, and alumina. It is readily corroded by liquid metals such as Si, Cr, Mn, Co, Ni, Nb, Mo, and Ta, and attacked by molten salts such as Na_2O, alkali carbonates, and NaOH. Severe oxidation in air occurs above 1100–1400 °C. Stable above 2000 °C under inert or reducing atmosphere.	Zirconium diboride
–	–	–	–		Zirconium dodecaboride

on data in molten salts from [2.9].

Table 3.2-20 Physical properties of carbides and carbide-based high-temperature refractories [2.4], cont.

Young's modulus (E, GPa)	Flexural strength (τ, MPa)	Compressive strength (α, MPa)	Vickers hardness HV (Mohs hardness HM)	Other physicochemical properties, corrosion resistance,[a] and uses	IUPAC name (synonyms and common trade names)
n.a	n.a	n.a	n.a	Decomposed in water with evolution of CH_4.	Aluminium carbides
314.4	–	723	2410 HK	Brick-red or yellowish-red octahedra. Used in nuclear-reactor cores.	Beryllium carbide
440–470	–	2900	3200–3500 HK (HM 9)	Hard, black, shiny crystals, the fourth hardest material known after diamond, cubic boron nitride, and boron oxide. It does not burn in an O_2 flame if the temperature is maintained below 983 °C. Maximum operating temperature 2000 °C (inert or reducing environment) or 600 °C (oxidizing environment). It is not attacked by hot HF or chromic acid. Used as abrasive, and in crucibles for molten salts, except molten alkalimetal hydroxides. In the form of molded shapes, it is used for pressure blast nozzles, wire-drawing dies, and bearing surfaces for gauges. For grinding and lapping applications, the available mesh sizes cover the range 240 to 800.	Boron carbide (Norbide®)

Table 3.2-20 Physical properties of carbides and carbide-based high-temperature refractories [2.4], cont.

IUPAC name (synonyms and common trade names)	Theoretical chemical formula, [CASRN], relative molecular mass ($^{12}C = 12.000$)	Crystal system, lattice parameters, *Strukturbericht* symbol, Pearson symbol, space group, structure type, Z	Density (ϱ, kg m^{-3})	Electrical resistivity (ρ, $\mu\Omega$ cm)	Melting point (°C)	Thermal conductivity (κ, W m^{-1} K^{-1})	Specific heat capacity (c_p, J kg^{-1} K^{-1})	Coefficient of linear thermal expansion (α, 10^{-6} K^{-1})
Chromium carbide	Cr_7C_3 400.005	Hexagonal $a = 1389.02$ pm $c = 453.20$ pm	6992	109.0	1665	–	–	11.7
Chromium carbide	Cr_3C_2 [12012-35-0] 180.010	Orthorhombic $a = 282$ pm $b = 553$ pm $c = 1147$ pm $D5_{10}$, $oP20$, $Pbnm$, Cr_3C_2 type ($Z = 4$)	6680	75.0	1895	19.2	–	10.3
Diamond	C [7782-40-3] 12.011	Cubic $a = 356.683$ pm A4, $cF8$, $Fd3m$, diamond type ($Z = 8$)	3515.24	>10^{16} (types I and IIa) >10^3 (type IIb)	3550	900 (type I) 2400 (type IIa)	–	2.16
Graphite	C [7782-42-5] 12.011	Hexagonal $a = 246$ pm $b = 428$ pm $c = 671$ pm A9, $hP4$, $P6_3/mmc$, graphite type ($Z = 4$)	2250	1385	3650	–	–	0.6–4.3
Hafnium monocarbide	HfC [12069-85-1] 190.501	Cubic $a = 446.0$ pm B1, $cF8$, $Fm3m$, rock salt type ($Z = 4$)	12 670	45.0	3890–3950	22.15	–	6.3
Lanthanum dicarbide	LaC_2 [12071-15-7] 162.928	Tetragonal $a = 394.00$ pm $c = 657.20$ pm C11a, $tI6$, $I4/mmm$, CaC_2 type ($Z = 2$)	5290	68.0	2360–2438	–	–	12.1
Molybdenum hemicarbide	β-Mo_2C [12069-89-5] 203.891	Hexagonal $a = 300.20$ pm $c = 427.40$ pm $L'3$, $hP3$, $P6_3/mmc$, Fe_2N type ($Z = 1$)	9180	71.0	2687	–	29.4	7.8
Molybdenum monocarbide	MoC [12011-97-1] 107.951	Hexagonal $a = 290$ pm $c = 281$ pm B_k, $P6_3/mmc$, BN type ($Z = 4$)	9159	50.0	2577	–	–	5.76
Niobium hemicarbide	Nb_2C [12011-99-3] 197.824	Hexagonal $a = 312.70$ pm $c = 497.20$ pm $L'3$, $hP3$, $P6_3/mmc$, Fe_2N type ($Z = 1$)	7800	–	3090	–	–	–
Niobium monocarbide	NbC [12069-94-2] 104.917	Cubic $a = 447.71$ pm B1, $cF8$, $Fm3m$, rock salt type ($Z = 4$)	7820	51.1–74.0	3760	14.2	–	6.84

Table 3.2-20 Physical properties of carbides and carbide-based high-temperature refractories [2.4], cont.

Young's modulus (E, GPa)	Flexural strength (τ, MPa)	Compressive strength (α, MPa)	Vickers hardness HV (Mohs hardness HM)	Other physicochemical properties, corrosion resistance,[a] and uses	IUAPC name (synonyms and common trade names)
–	–	–	1336	Resists oxidation in the range 800–1000 °C. Corroded by the following molten metals: Ni and Zn.	Chromium carbide
386	–	1041	2650	Corroded by the following molten metals: Ni, Zn, Cu, Cd, Al, Mn, and Fe. Corrosion-resistant in molten Sn and Bi.	Chromium carbide
930	–	7000	8000 HK (HM 10)	Type I contains 0.1–0.2% N, type IIa is N-free, and type IIb is very pure, generally blue in color. Electrical insulator ($E_g = 7$ eV). Burns in oxygen.	Diamond
6.9	–	–	(HM 2)	High-temperature lubricant, used in crucibles for handling molten metals such as Mg, Al, Zn, Ga, Sb, and Bi.	Graphite
424	–	–	1870–2900	Dark, gray, brittle solid, the most refractory binary material known. Used in control rods in nuclear reactors and in crucibles for melting HfO_2 and other oxides. Corrosion-resistant to liquid metals such as Nb, Ta, Mo, and W. Severe oxidation in air above 1100–1400 °C, but stable up to 2000 °C in helium.	Hafnium monocarbide
–	–	–	–	Decomposed by H_2O.	Lanthanum dicarbide
221	–	n.a	1499 (HM > 7)	Gray powder. Wear-resistant film. Oxidized in air at 700–800 °C. Corroded in the following molten metals: Al, Mg, V, Cr, Mn, Fe, Ni, Cu, Zn, and Nb. Corrosion-resistant in molten Cd, Sn, and Ta.	Molybdenum hemicarbide
197	n.a.	n.a.	1800 (HM > 9)	Oxidized in air at 700–800 °C.	Molybdenum monocarbide
n.a.	n.a.	n.a.	2123		Niobium hemicarbide
340	n.a.	n.a.	2470 (HM > 9)	Lavender-gray powder, soluble in HF-HNO_3 mixture. Wear-resistant film, used for coating graphite in nuclear reactors. Oxidation in air becomes severe only above 1000 °C.	Niobium monocarbide

Table 3.2-20 Physical properties of carbides and carbide-based high-temperature refractories [2.4], cont.

IUPAC name (synonyms, and common trade names)	Theoretical chemical formula, [CASRN], relative molecular mass ($^{12}C = 12.000$)	Crystal system, lattice parameters, *Strukturbericht* symbol, Pearson symbol, space group, structure type (Z)	Density (ϱ, kg m^{-3})	Electrical resistivity (ρ, $\mu\Omega$ cm)	Melting point (°C)	Thermal conductivity (κ, W m^{-1} K^{-1})	Specific heat capacity (c_p, J kg^{-1} K^{-1})	Coefficient of linear thermal expansion (α, 10^{-6} K^{-1})
Silicon monocarbide (moissanite, Carbolon®, Crystolon®, Carborundum®)	α-SiC [409-21-2] 40.097	Hexagonal $a = 308.10$ pm $c = 503.94$ pm B4, hP4, P6$_3$/mmc, wurtzite type (Z = 2)	3160	4.1×10^5	2093 transformation temperature	42.5	690	4.3–4.6
Silicon monocarbide (Carbolon®, Crystolon®, Carborundum®)	β-SiC [409-21-2] 40.097	Cubic $a = 435.90$ pm B3, cF8, F43m, ZnS type (Z = 4)	3160	107–200	∼ 2700, subl.	135	1205	4.5
Tantalum hemicarbide	Ta$_2$C [12070-07-4] 373.907	Hexagonal $a = 310.60$ pm $c = 493.00$ pm L'3, hP3, P6$_3$/mmc, Fe$_2$N type (Z = 2)	15 100	80.0	3327	–	–	–
Tantalum monocarbide	TaC [12070-06-3] 194.955	Cubic $a = 445.55$ pm B1, cF8, Fm3m, rock salt type (Z = 4)	14 800	30–42.1	3880	22.2	190	6.64–8.4
Thorium dicarbide	α-ThC$_2$ [12071-31-7] 256.060	Tetragonal $a = 585$ pm $c = 528$ pm C11a, tI6, I4mmm, CaC$_2$ type (Z = 2)	8960–9600	30.0	2655	23.9	–	8.46
Thorium monocarbide	ThC [12012-16-7] 244.089	Cubic $a = 534.60$ pm B1, cF8, Fm3m, rock salt type (Z=4)	10 670	25.0	2621	28.9	–	6.48
Titanium monocarbide	TiC [12070-08-5] 59.878	Cubic $a = 432.8$ pm B1, cF8, Fm3m, rock salt type (Z = 4)	4938	52.5	3140 ± 90	17–21	–	7.5–7.7
Tungsten hemicarbide	W$_2$C [12070-13-2] 379.691	Hexagonal $a = 299.82$ pm $c = 472.20$ pm L'3, hP3, P6$_3$/mmc, Fe$_2$N type (Z = 1)	17 340	81.0	2730	–	–	3.84
Tungsten monocarbide (Widia®)	WC [12070-12-1] 195.851	Hexagonal $a = 290.63$ pm $c = 283.86$ pm L'3, hP3, P6$_3$/mmc, Fe$_2$N type (Z = 1)	15 630	19.2	2870	121	–	6.9

Table 3.2-20 Physical properties of carbides and carbide-based high-temperature refractories [2.4], cont.

Young's modulus (E, GPa)	Flexural strength (τ, MPa)	Compressive strength (α, MPa)	Vickers hardness HV (Mohs hardness HM)	Other physicochemical properties, corrosion resistance,[a] and uses	IUPAC name (synonyms and common trade names)
386–414	–	500	2400–2500 (HM 9.2)	Semiconductor ($E_g = 3.03$ eV). Soluble in fused alkalimetal hydroxides.	Silicon monocarbide (moissanite, Carbolon®, Crystolon®, Carborundum®)
262–468	–	1000	2700–3350 (HM 9.5)	Green to bluish-black, iridescent crystals. Soluble in fused alkalimetal hydroxides. Abrasives made from this material are best suited to the grinding of low-tensile-strength materials such as cast iron, brass, bronze, marble, concrete, stone, glass, optical, structural, and wear-resistant components. Corroded by molten metals such as Na, Mg, Al, Zn, Fe, Sn, Rb, and Bi. Resistant to oxidation in air up to 1650 °C. Maximum operating temperature of 2000 °C in reducing or inert atmosphere.	Silicon monocarbide (Carbolon®, Crystolon®, Carborundum®)
–	–	–	1714–2000		Tantalum hemicarbide
364	–	–	1599–1800 (HM 9–10)	Golden-brown crystals, soluble in HF-HNO$_3$ mixture. Used in crucibles for melting ZrO$_2$ and similar oxides with high melting points. Corrosion-resistant to molten metals such as Ta, and Re. Readily corroded by liquid metals such as Nb, Mo, and Sn. Burning occurs in pure oxygen above 800 °C. Severe oxidation in air above 1100–1400 °C. Maximum operating temperature of 3760 °C under helium.	Tantalum monocarbide
–	–	–	600	$\alpha-\beta$ transition at 1427 °C and $\beta-\gamma$ transition at 1497 °C. Decomposed by H$_2$O with evolution of C$_2$H$_6$.	Thorium dicarbide
–	–	–	1000	Readily hydrolyzed in water, evolving C$_2$H$_6$.	Thorium monocarbide
310–462	–	1310	2620–3200 (HM 9–10)	Gray crystals. Superconducting at 1.1 K. Soluble in HNO$_3$ and aqua regia. Resistant to oxidation in air up to 450 °C. Maximum operating temperature 3000 °C under helium. Used in crucibles for handling molten metals such as Na, Bi, Zn, Pb, Sn, Rb, and Cd. Corroded by the following liquid metals: Mg, Al, Si, Ti, Zr, V, Nb, Ta, Cr, Mo, Mn, Fe, Co, and Ni. Attacked by molten NaOH.	Titanium monocarbide
421	–	–	3000	Black. Resistant to oxidation in air up to 700 °C. Corrosion-resistant to Mo.	Tungsten hemicarbide
710	–	530	2700 (HM > 9)	Gray powder, dissolved by HF-HNO$_3$ mixture. Cutting tools, wear-resistant semiconductor films. Corroded by the following molten metals: Mg, Al, V, Cr, Mn, Ni, Cu, Zn, Nb, and Mo. Corrosion-resistant to molten Sn.	Tungsten monocarbide (Widia®)

Table 3.2-20 Physical properties of carbides and carbide-based high-temperature refractories [2.4], cont.

IUPAC name (synonyms, and common trade names)	Theoretical chemical formula, [CASRN], relative molecular mass ($^{12}C = 12.000$)	Crystal system, lattice parameters, *Strukturbericht* symbol, Pearson symbol, space group, structure type (Z)	Density (ϱ, kg m^{-3})	Electrical resistivity (ρ, μΩ cm)	Melting point (°C)	Thermal conductivity (κ, W m^{-1} K^{-1})	Specific heat capacity (c_p, J kg^{-1} K^{-1})	Coefficient of linear thermal expansion (α, 10^{-6} K^{-1})
Uranium carbide	U_2C_3 [12076-62-9] 512.091	Cubic $a = 808.89$ pm D5c, *cI*40, $I\bar{4}3d$, Pu_2C_3 type ($Z = 8$)	12 880	–	1777	–	–	11.4
Uranium dicarbide	UC_2 [12071-33-9] 262.051	Tetragonal $a = 352.24$ pm $c = 599.62$ pm C11a, *tI*6, $I4/mmm$, CaC_2 type ($Z = 2$)	11 280	–	2350–2398	32.7	147	14.6
Uranium monocarbide	UC [12070-09-6] 250.040	Cubic $a = 496.05$ pm B1, *cF*8, $Fm\bar{3}m$, rock salt type ($Z = 4$)	13 630	50.0	2370–2790	23.0	–	11.4
Vanadium hemicarbide	V_2C [2012-17-8] 113.89	Hexagonal $a = 286$ pm $c = 454$ pm L'3, *hP*3, $P6_3mmc$, Fe_2N type ($Z = 2$)	5750	–	2166	–	–	–
Vanadium monocarbide	VC [12070-10-9] 62.953	Cubic $a = 413.55$ pm B1, *cF*8, $Fm\bar{3}m$, rock salt type ($Z = 4$)	5770	65.0–98.0	2810	24.8	–	4.9
Zirconium monocarbide	ZrC [12020-14-3] 103.235	Cubic $a = 469.83$ pm B1, *cF*8, $Fm\bar{3}m$, rock salt type ($Z = 4$)	6730	68.0	3540–3560	20.61	205	6.82

[a] Corrosion data in molten salts from [2.9].

Table 3.2-20 Physical properties of carbides and carbide-based high-temperature refractories [2.4], cont.

Young's modulus (E, GPa)	Flexural strength (τ, MPa)	Compressive strength (α, MPa)	Vickers hardness HV (Mohs hardness HM)	Other physicochemical properties, corrosion resistance,[a] and uses	IUPAC name (synonyms and common trade names)
179–221	–	434	–		Uranium carbide
–	–	–	600	Transition from tetragonal to cubic at 1765 °C. Decomposed in H_2O, slightly soluble in alcohol. Used in microsphere pellets to fuel nuclear reactors.	Uranium dicarbide
172.4	–	351.6	750–935 (HM > 7)	Gray crystals with metallic appearance, reacts with oxygen. Corroded by the following molten metals: Be, Si, Ni, and Zr.	Uranium monocarbide
–	–	–	3000	Corroded by molten Nb, Mo, and Ta.	Vanadium hemicarbide
614	–	613	2090	Black crystals, soluble in HNO_3 with decomposition. Used in wear-resistant films and cutting tools. Resistant to oxidation in air up to 300 °C.	Vanadium monocarbide
345	–	1641	1830–2930 (HM > 8)	Dark gray, brittle solid, soluble in HF solutions containing nitrates or peroxide ions. Used in nuclear power reactors and in crucibles for handling molten metals such as Bi, Cd, Pb, Sn, and Rb, and molten zirconia (ZrO_2). Corroded by the following liquid metals: Mg, Al, Si, V, Nb, Ta, Cr, Mo, Mn, Fe, Co, Ni, and Zn. In air, oxidized rapidly above 500 °C. Maximum operating temperature of 2350 °C under helium.	Zirconium monocarbide

Table 3.2-21 Properties of carbides according to DIN EN 60672 [2.6]

Mechanical properties	Symbol	Units	Designation SSIC Silicon carbide, sintered	SISIC Silicon carbide, silicon-infiltrated	RSIC Silicon carbide, recrystallized	NSIC Silicon carbide, nitride-bonded	BC Boron carbide
Open porosity		vol. %	0	0	0–15	–	0
Density, minimum	ϱ	Mg/m^3	3.08–3.15	3.08–3.12	2.6–2.8	2.82	2.50
Bending strength	σ_B	MPa	300–600	180–450	80–120	200	400
Young's modulus	E	GPa	370–450	270–350	230–280	150–240	390–440
Hardness	HV	100	25–26	14–25	25	–	30–40
Fracture toughness	K_{IC}	MPa $\sqrt{m}$	3.0–4.8	3.0–5.0	3.0	–	3.2–3.6
Electrical properties							
Resistivity at 20 °C	ρ_{20}	Ω m	10^3–10^4	2×10^1–10^3	–	–	–
Resistivity at 600 °C	ρ_{600}	Ω m	10	5	–	–	–
Thermal properties							
Average coefficient of thermal expansion at 30–600 °C	$\alpha_{30-1000}$	10^{-6} K^{-1}	4–4.8	4.3–4.8	4.8	4.5	6
Specific heat capacity at 30–100 °C	$c_{p,30-1000}$	J kg^{-1} K^{-1}	600–1000	650–1000	600–900	800–900	–
Thermal conductivity	λ_{30-100}	W m^{-1} K^{-1}	40–120	110–160	20	14–15	28
Thermal fatigue resistance		(Rated)	Very good	Very good	Very good	Very good	–
Typical maximum application temperature	T	°C	1400–1750	1380	1600	1450	700–1000

3.2.5.4 Nitrides

Nitrides are treated extensively in [2.1–3]. Some properties are listed in Tables 3.2-22 and 3.2-23.

Silicon Nitride

Si_3N_4 is the dominant nitride ceramic material because of its favorable combination of properties. The preparation of powders for the formation of dense silicon nitride materials requires the use of precursors. Four routes for the production of Si_3N_4 powders are used in practice: nitridation of silicon, chemical vapor deposition from $SiCl_4 + NH_3$, carbothermal reaction of SiO_2, and precipitation of silicon diimide $Si(NH)_2$ followed by decomposition. Table 3.2-24 gives examples of the properties of the resulting powders.

Table 3.2-22 Properties of nitrides according to DIN EN 60672 [2.6]

			Designation				
			SSN	RBSN	HPSN	SRBSN	AlN
Mechanical properties	Symbol	Units	Silicon nitride, sintered	Silicon nitride, reaction-bonded	Silicon nitride, hot-pressed	Silicon nitride, reaction-bonded	Aluminium nitride
Open porosity		vol.%	–	–	0	–	0
Density, minimum	ϱ	Mg/m³	3–3.3	1.9–2.5	3.2–3.4	3.1–3.3	3.0
Bending strength	σ_B	MPa	700–1000	200–330	600–800	700–1200	200
Young's modulus	E	GPa	250–330	80–180	600–800	150–240	320
Hardness	HV	100	4–18	8–10	15–16	–	11
Fracture toughness	K_{IC}	MPa$\sqrt{m}$	5–8.5	1.8–4.0	6.0–8.5	3.0–6?	3.0
Electrical properties							
Resistivity at 20 °C	ρ_{20}	Ω m	10^{11}	10^{13}	10^{13}	–	10^{13}
Resistivity at 600 °C	ρ_{600}	Ω m	10^2	10^{10}	10^9	–	10^{12}
Thermal properties							
Average coefficient of thermal expansion at 30–600 °C	$\alpha_{30-1000}$	10^{-6} K^{-1}	2.5–3.5	2.1–3	3.1–3.3	3.0–3.4	4.5–5
Specific heat capacity at 30–100 °C	$c_{p,30-1000}$	J kg^{-1}K^{-1}	700–850	700–850	700–850	700–850	–
Thermal conductivity	λ_{30-100}	W m^{-1}K^{-1}	15–45	4–15	15–40	14–15	> 100
Thermal fatigue resistance		(Rated)	Very good	Very good	Very good	Very good	Very good
Typical maximum application temperature	T	°C	1250	1100	1400	1250	–

Table 3.2-23 Physical properties of nitrides and nitride-based high-temperature refractories [2.4]

IUPAC name (synonyms and common trade names)	Theoretical chemical formula, [CASRN], relative molecular mass ($^{12}C = 12.000$)	Crystal system, lattice parameters, *Strukturbericht* symbol, Pearson symbol, space group, structure type, Z	Density (ϱ, kg m^{-3})	Electrical resistivity (ρ, $\mu\Omega$ cm)	Melting point (°C)	Thermal conductivity (κ, W m^{-1} K^{-1})	Specific heat capacity (c_p, J kg^{-1} K^{-1})	Coefficient of linear thermal expansion (α, 10^{-6} K^{-1})
Aluminium mononitride	AlN [24304-00-5] 40.989	Hexagonal $a = 311.0$ pm $c = 497.5$ pm B4, hP4, P6$_3$mc, wurtzite type (Z = 2)	3050	10^{17}	2230	29.96	820	5.3
Beryllium nitride	α-Be$_3$N$_2$ [1304-54-7] 55.050	Cubic $a = 814$ pm D5$_3$, cI80, Ia3, Mn$_2$O$_3$ type (Z = 16)	2710	–	2200	–	1221	–
Boron mononitride	BN [10043-11-5] 24.818	Hexagonal $a = 250.4$ pm $c = 666.1$ pm B$_k$, hP8, P6$_3$/mmc, BN type (Z = 4)	2250	10^{19}	2730 (dec.)	15.41	711	7.54
Boron mononitride (Borazon®, CBN)	BN 24.818	Cubic $a = 361.5$ pm	3430	1900 (200°C)	1540	–	–	–
Chromium heminitride	Cr$_2$N [12053-27-9] 117.999	Hexagonal $a = 274$ pm $c = 445$ pm L'3, hP3, P6$_3$/mmc, Fe$_2$N type (Z = 1)	6800	76	1661	22.5	630	9.36
Chromium mononitride	CrN [24094-93-7] 66.003	Cubic $a = 415.0$ pm B1, cF8, Fm3m, rock salt type (Z = 4)	6140	640	1499 (dec.)	12.1	795	2.34
Hafnium mononitride	HfN [25817-87-2] 192.497	Cubic $a = 451.8$ pm B1, cF8, Fm3m, rock salt type (Z = 4)	13 840	33	3310	21.6	210	6.5
Molybdenum heminitride	Mo$_2$N [12033-31-7] 205.887	Cubic $a = 416$ pm L'1, cP5, Pm3m, Fe$_4$N type (Z = 2)	9460	19.8	760–899	17.9	293	6.12
Molybdenum mononitride	MoN [12033-19-1] 109.947	Hexagonal $a = 572.5$ pm $c = 560.8$ pm B$_h$, hP2, P6/mmm, WC type (Z = 1)	9180	–	1749	–	–	–
Niobium mononitride	NbN [24621-21-4] 106.913	Cubic $a = 438.8$ pm B1, cF8, Fm3m, rock salt type (Z = 4)	8470	78	2575	3.63	–	10.1
Silicon nitride	β-Si$_3$N$_4$ [12033-89-5] 140.284	Hexagonal $a = 760.8$ pm $c = 291.1$ pm P6/3m	3170	10^6	1850	28	713	2.25

Table 3.2-23 Physical properties of nitrides and nitride-based high-temperature refractories [2.4], cont.

Young's modulus (E, GPa)	Flexural strength (τ, MPa)	Compressive strength (α, MPa)	Vickers hardness HV (Mohs hardness HM)	Other physicochemical properties, corrosion resistance,[a] and uses	IUPAC name (synonyms and common trade names)
346	–	2068	1200 (HM 9–10)	Insulator ($E_g = 4.26$ eV). Decomposed by water, acids, and alkalis to Al(OH)$_3$ and NH$_3$. Used in crucibles for GaAs crystal growth.	Aluminium mononitride
–	–	–	–	Hard white or grayish crystals. Oxidized in air above 600 °C. Slowly decomposed in water, quickly in acids and alkalis, with evolution of NH$_3$.	Beryllium nitride
85.5	–	310	230 (HM 2.0)	Insulator ($E_g = 7.5$ eV). Used in crucibles for molten metals such as Na, B, Fe, Ni, Al, Si, Cu, Mg, Zn, In, Bi, Rb, Cd, Ge, and Sn. Corroded by these molten metals: U, Pt, V, Ce, Be, Mo, Mn, Cr, V, and Al. Attacked by the following molten salts: PbO$_2$, Sb$_2$O$_3$, Bi$_2$O$_3$, KOH, and K$_2$CO$_3$. Used in furnace insulation, diffusion masks, and passivation layers.	Boron mononitride
–	–	7000	4700–5000 (HM 10)	Tiny reddish to black grains. Used as an abrasive for grinding tool and die steels and high-alloy steels when chemical reactivity of diamond is a problem.	Boron mononitride (Borazon®, CBN)
–	–	–	1200–1571		Chromium heminitride
–	–	–	1090		Chromium mononitride
–	–	–	1640 (HM > 8–9)	Most refractory of all nitrides.	Hafnium mononitride
–	–	–	1700	Phase transition at 5.0 K.	Molybdenum heminitride
–	–	–	650		Molybdenum mononitride
–	–	–	1400 (HM > 8)	Dark gray crystals. Transition temperature 15.2 K. Insoluble in HCl, HNO$_3$, and H$_2$SO$_4$, but attacked by hot caustic solutions, lime, or strong alkalis, evolving NH$_3$.	Niobium mononitride
55	–	–	(HM > 9)		Silicon nitride

Table 3.2-23 Physical properties of nitrides and nitride-based high-temperature refractories [2.4], cont.

IUPAC name (synonyms and common trade names)	Theoretical chemical formula, [CASRN], relative molecular mass ($^{12}C = 12.000$)	Crystal system, lattice parameters, *Strukturbericht* symbol, Pearson symbol, space group, structure type, Z	Density (ϱ, kg m^{-3})	Electrical resistivity (ρ, $\mu\Omega$ cm)	Melting point (°C)	Thermal conductivity (κ, W m^{-1} K^{-1})	Specific heat capacity (c_p, J kg^{-1} K^{-1})	Coefficient of linear thermal expansion (α, 10^{-6} K^{-1})
Silicon nitride (Nitrasil®)	α-Si$_3$N$_4$ [12033-89-5] 140.284	Hexagonal $a = 775.88$ pm $c = 561.30$ pm $P31c$	3184	10^{19}	1900 (sub.)	17	700	2.5–3.3
Tantalum heminitride	Ta$_2$N 375.901	Hexagonal $a = 306$ pm $c = 496$ pm $L'3, hP3, P6_3/mmc$, Fe$_2$N type ($Z = 1$)	15 600	263	2980	10.04	126	5.2
Tantalum mononitride (ϵ)	TaN [12033-62-4] 194.955	Hexagonal $a = 519.1$ pm $c = 290.6$ pm	13 800	128–135	3093	8.31	210	3.2
Thorium mononitride	ThN [12033-65-7] 246.045	Cubic $a = 515.9$ $B1, cF8, Fm3m$, rock salt type ($Z = 4$)	11 560	20	2820	–	–	7.38
Thorium nitride	Th$_2$N$_3$ [12033-90-8]	Hexagonal $a = 388$ pm $c = 618$ pm $D5_2, hP5, \bar{3}ml$, La$_2$O$_3$ type ($Z = 1$)	10 400	–	1750	–	–	–
Titanium mononitride	TiN [25583-20-4] 61.874	Cubic $a = 424.6$ pm $B1, cF8, Fm3m$, rock salt type ($Z = 4$)	5430	21.7	2930 (dec.)	29.1	586	9.35
Tungsten dinitride	WN$_2$ [60922-26-1] 211.853	Hexagonal $a = 289.3$ pm $c = 282.6$ pm	7700	–	600 (dec.)	–	–	–
Tungsten heminitride	W$_2$N [12033-72-6] 381.687	Cubic $a = 412$ pm $L'1, cP5, Pm3m$, Fe$_4$N type ($Z = 2$)	17 700	–	982	–	–	–
Tungsten mononitride	WN [12058-38-7]	Hexagonal	15 940	–	593	–	–	–
Uranium nitride	U$_2$N$_3$ [12033-83-9] 518.259	Cubic $a = 1070$ pm $D5_3, cI80, Ia3$, Mn$_2$O$_3$ type ($Z = 16$)	11 240	–	–	–	–	–
Uranium mononitride	UN [25658-43-9] 252.096	Cubic $a = 489.0$ pm $B1, cF8, Fm3m$, rock salt type ($Z = 4$)	14 320	208	2900	12.5	188	9.72
Vanadium mononitride	VN [24646-85-3] 64.949	Cubic $a = 414.0$ pm $B1, cF8, Fm3m$, rock salt type ($Z = 4$)	6102	86	2360	11.25	586	8.1
Zirconium mononitride	ZrN [25658-42-8] 105.231	Cubic $a = 457.7$ pm $B1, cF8, Fm3m$, rock salt type ($Z = 4$)	7349	13.6	2980	20.90	377	7.24

Table 3.2-23 Physical properties of nitrides and nitride-based high-temperature refractories [2.4], cont.

Young's modulus (E, GPa)	Flexural strength (τ, MPa)	Compressive strength (α, MPa)	Vickers hardness HV (Mohs hardness HM)	Other physicochemical properties, corrosion resistance,[a] and uses	IUAPC name (synonyms and common trade names)
304	–	–	(HM > 9)	Gray amorphous powder or crystals. Corrosion-resistant to molten metals such as Al, Pb, Zn, Cd, Bi, Rb, and Sn, and molten salts NaCl-KCl, NaF, and silicate glasses. Corroded by molten Mg, Ti, V, Cr, Fe, Co, cryolite, KOH, and Na_2O.	Silicon nitride (Nitrasil®)
–	–	–	3200	Decomposed by KOH with evolution of NH_3.	Tantalum heminitride
–	–	–	1110 (HM > 8)	Bronze-colored or black crystals. Transition temperature 1.8 K. Insoluble in water, slowly attacked by aqua regia, HF, and HNO_3.	Tantalum mononitride (ϵ)
–	–	–	600	Gray solid. Slowly hydrolyzed by water.	Thorium mononitride
–	–	–	–		Thorium nitride
248	–	972	1900 (HM 8–9)	Bronze-colored powder. Transition temperature 4.2 K. Corrosion-resistant to molten metals such as Al, Pb, Mg, Zn, Cd, and Bi. Corroded by molten Na, Rb, Ti, V, Cr, Mn, Sn, Ni, Cu, Fe, and Co. Dissolved by boiling aqua regia; decomposed by boiling alkalis, evolving NH_3.	Titanium mononitride
–	–	–	–	Brown crystals.	Tungsten dinitride
–	–	–	–	Gray crystals.	Tungsten heminitride
–	–	–	–	Gray solid. Slowly hydrolyzed by water.	Tungsten mononitride
–	–	–	–		Uranium nitride
149	–	–	455		Uranium mononitride
–	–	–	1520 (HM 9–10)	Black powder. Transition temperature 7.5 K. Soluble in aqua regia.	Vanadium mononitride
–	–	979	1480 (HM > 8)	Yellow solid. Transition temperature 9 K. Corrosion-resistant to steel, basic slag, and cryolite, and molten metals such as Al, Pb, Mg, Zn, Cd, and Bi. Corroded by molten Be, Na, Rb, Ti, V, Cr, Mn, Sn, Ni, Cu, Fe, and Co. Soluble in concentrated HF, slowly soluble in hot H_2SO_4.	Zirconium mononitride

[a] Corrosion data in molten salts from [2.9].

Table 3.2-24 Characteristics of Si_3N_4 powders processed by different preparation techniques [2.3]

	Technique						
	Nitridation of Si		Chemical vapor deposition		Carbothermal reduction	Diimide precipitation	
Sample no.	1	2	1	2		1	2
Specific surface area (m^2/g)	23	11	4	10	10	11	13
O (wt%)	1.4	1.0	1.0	3.0	2.0	1.4	1.5
C (wt%)	0.2	0.25	–	–	0.9	0.1	0.1
Fe, Al, Ca (wt%)	0.07	0.4	0.005	0.005	0.22	0.01	0.015
Other impurities (wt%)			Cl 0.04 Mo + Ti 0.02			Cl 0.1	Cl 0.005
Crystallinity (%)	100	100	60	0	100	98	–
$\alpha/(\alpha+\beta)$ (%)	95	92	95	–	98	86	95
Morphology[a]	E	E	E + R	E + R	E + R	E	E

[a] E, equiaxed; R, rod-like.

Table 3.2-25 Physical properties of silicides and silicide-based high-temperature refractories [2.4]

IUPAC name	Theoretical chemical formula, [CASRN], relative molecular mass (^{12}C = 12.000)	Crystal system, lattice parameters, *Strukturbericht* symbol, Pearson symbol, space group, structure type, Z	Density (ϱ, kg m^{-3})	Electrical resistivity (ρ, $\mu\Omega$ cm)	Melting point (°C)	Thermal conductivity (κ, W m^{-1} K^{-1})	Specific heat capacity (c_p, J kg^{-1} K^{-1})	Coefficient of linear thermal expansion (α, 10^{-6} K^{-1})
Chromium disilicide	$CrSi_2$ [12018-09-6] 108.167	Hexagonal a = 442 pm c = 635 pm C40, $hP9$, $P6_222$, $CrSi_2$ type (Z = 3)	4910	1400	1490	106	–	13.0
Chromium silicide	Cr_3Si [12018-36-9] 184.074	Cubic a = 456 pm A15, $cP8$, $Pm3n$, Cr_3Si type (Z = 2)	6430	45.5	1770	–	–	10.5
Hafnium disilicide	$HfSi_2$ [12401-56-8] 234.66	Orthorhombic a = 369 pm b = 1446 pm c = 346 pm C49, $oC12$, $Cmcm$, $ZrSi_2$ type (Z = 4)	8030	–	1699	–	–	–
Molybdenum disilicide	$MoSi_2$ [12136-78-6] 152.11	Tetragonal a = 319 pm c = 783 pm C11b, $tI6$, $I4/mmm$, $MoSi_2$ type (Z = 2)	6260	21.5	1870	58.9	–	8.12
Niobium disilicide	$NbSi_2$ [12034-80-9] 149.77	Hexagonal a = 479 pm c = 658 pm C40, $hP9$, $P6_222$, $CrSi_2$ type (Z = 3)	5290	50.4	2160	–	–	–
Tantalum disilicide	$TaSi_2$ [12039-79-1] 237.119	Hexagonal a = 477 pm c = 655 pm C40, $hP9$, $P6_222$, $CrSi_2$ type (Z = 3)	9140	8.5	2299	–	–	8.8–9.54

3.2.5.5 Silicides

Silicides, being compounds of silicon with metals, mostly show a metallic luster. Like intermetallic phases the silicides of a metal may occur in different stoichiometric variants, e.g. Ca_2Si, Ca_5Si_3, and $CaSi$. Silicides of the non-noble metals are unstable in contact with water and oxidizing media. Silicides of transition metals are highly oxidation-resistant. Silicides are used as ceramic materials mainly in high-temperature applications. $MoSi_2$ is used in resistive heating elements. Some properties are listed in Table 3.2-25.

Table 3.2-25 Physical properties of silicides and silicide-based high-temperature refractories [2.4], cont.

Young's modulus (E, GPa)	Flexural strength (τ, MPa)	Compressive strength (α, MPa)	Vickers hardness HV (Mohs hardness HM)	Other physicochemical properties, corrosion resistance and uses	IUPAC name
–	–	–	1000–1130		Chromium disilicide
–	–	–	1005		Chromium silicide
–	–	–	865–930		Hafnium disilicide
407	–	2068–2415	1260	The compound is thermally stable in air up to 1000 °C. Corrosion-resistant to molten metals such as Zn, Pd, Ag, Bi, and Rb. It is corroded by the following liquid metals: Mg, Al, Si, V, Cr, Mn, Fe, Ni, Cu, Mo, and Ce.	Molybdenum disilicide
–	–	–	1050		Niobium disilicide
–	–	–	1200–1600	Corroded by molten Ni.	Tantalum disilicide

Table 3.2-25 Physical properties of silicides and silicide-based high-temperature refractories [2.4], cont.

IUPAC name	Theoretical chemical formula, [CASRN], relative molecular mass ($^{12}C = 12.000$)	Crystal system, lattice parameters, *Strukturbericht* symbol, Pearson symbol, space group, structure type, Z	Density (ϱ, kg m^{-3})	Electrical resistivity (ρ, $\mu\Omega$ cm)	Melting point (°C)	Thermal conductivity (κ, W m^{-1} K^{-1})	Specific heat capacity (c_p, J kg^{-1} K^{-1})	Coefficient of linear thermal expansion (α, 10^{-6} K^{-1})
Tantalum silicide	Ta$_5$Si$_3$ [12067-56-0] 988.992	Hexagonal	13060	–	2499	–	–	–
Thorium disilicide	ThSi$_2$ [12067-54-8] 288.209	Tetragonal $a = 413$ pm $c = 1435$ pm Cc, tI12, $I4amd$, ThSi$_2$ type ($Z = 4$)	7790	–	1850	–	–	–
Titanium disilicide	TiSi$_2$ [12039-83-7] 104.051	Orthorhombic $a = 360$ pm $b = 1376$ pm $c = 360$ pm C49, oC12, $Cmcm$, ZrSi$_2$ type ($Z = 4$)	4150	123	1499	–	–	10.4
Titanium trisilicide	Ti$_5$Si$_3$ [12067-57-1] 323.657	Hexagonal $a = 747$ pm $c = 516$ pm D8$_8$, hP16, $P6_3mcm$, Mn$_5$Si$_3$ type ($Z = 2$)	4320	55	2120	–	–	110
Tungsten disilicide	WSi$_2$ [12039-88-2] 240.01	Tetragonal $a = 320$ pm $c = 781$ pm C11b, tI6, $I4/mmm$, MoSi$_2$ type ($Z = 2$)	9870	33.4	2165	–	–	8.28
Tungsten silicide	W$_5$Si$_3$ [12039-95-1] 1003.46		12210	–	2320	–	–	–
Uranium disilicide	USi$_2$ 294.200	Tetragonal $a = 397$ pm $c = 1371$ pm Cc, tI12, $I4/amd$, ThSi$_2$ type ($Z = 4$)	9250	–	1700	–	–	–
Uranium silicide	β-U$_3$Si$_2$ 770.258	Tetragonal $a = 733$ pm $c = 390$ pm D5a, tP10, $P4/mbm$, U$_3$Si$_2$ type ($Z = 2$)	12200	150	1666	14.7	–	14.8
Vanadium disilicide	VSi$_2$ [12039-87-1] 107.112	Hexagonal $a = 456$ pm $c = 636$ pm C40, hP9, $P6_222$, CrSi$_2$ type ($Z = 3$)	5100	9.5	1699	–	–	11.2
Vanadium silicide	V$_3$Si [12039-76-8] 147.9085	Cubic $a = 471$ pm A15, cP8, $Pm3n$, Cr$_3$Si type ($Z = 2$)	5740	203	1732	–	–	8.0
Zirconium disilicide	ZrSi$_2$ [12039-90-6] 147.395	Orthorhombic $a = 372$ pm $b = 1469$ pm $c = 366$ pm C49, oC12, $Cmcm$, ZrSi$_2$ type ($Z = 4$)	4880	161	1604	–	–	8.6

Table 3.2-25 Physical properties of silicides and silicide-based high-temperature refractories [2.4], cont.

Young's modulus (E, GPa)	Flexural strength (τ, MPa)	Compressive strength (α, MPa)	Vickers hardness HV (Mohs hardness HM)	Other physicochemical properties, corrosion resistance, and uses	IUPAC name
–	–	–	1200–1500	The compound is thermally stable in air up to 400 °C.	Tantalum silicide
–	–	–	1120	Corrosion-resistant to molten Cu, while corroded by molten Ni.	Thorium disilicide
–	–	–	890–1039		Titanium disilicide
–	–	–	986		Titanium trisilicide
–	–	–	1090		Tungsten disilicide
–	–	–	770	Corroded by molten Ni.	Tungsten silicide
–	–	–	700		Uranium disilicide
77.9	–	–	796		Uranium silicide
–	–	–	1400		Vanadium disilicide
–	–	–	1500		Vanadium silicide
–	–	–	1030–1060		Zirconium disilicide

References

2.1 L. E. Toth: *Transition Metals, Carbides and Nitrides* (Academic Press, New York 1971)

2.2 R. Freer: *The Physics and Chemistry of Carbides, Nitrides and Borides* (Kluwer, Boston 1989)

2.3 M. V. Swain (Ed.): Structure and properties of ceramics. In: *Materials Science and Technology*, Vol. 11 (Verlag Chemie, Weinheim 1994)

2.4 F. Cardarelli: *Materials Handbook* (Springer, London 2000)

2.5 J. R. Davis (Ed.): *Heat-Resistant Materials*, ASM Specialty Handbook (ASM International, Materials Park 1997)

2.6 Verband der keramischen Industrie: *Brevier Technische Keramik* (Fahner Verlag, Lauf 1999) (in German)

2.7 P. Otschick (Ed.): *Langzeitverhalten von Funktionskeramiken* (Werkstoff-Informationsgesellschaft, Frankfurt 1997) (in German)

2.8 G. V. Samsonov: *The Oxides Handbook* (Plenum, New York 1974)

2.9 G. Geirnaert: Céramiques et mètaux liquides: Compatibilitiés et angles de mouillages, Bull. Soc. Fr. Ceram **106**, 7 (1970)

2.10 V. I. Matkovich (Ed.): *Boron and Refractory Borides* (Springer, Berlin, Heidelberg 1977)

3.3. Polymers

The physical properties of polymers depend not only on the kind of material but also on the molar mass, the molar-mass distribution, the kind of branching, the degree of branching, the crystallinity (amorphous or crystalline), the tacticity, the end groups, any superstructure, and any other kind of molecular architecture. In the case of copolymers, the physical properties are additionally influenced by the type of arrangement of the monomers (statistical, random, alternating, periodic, block, or graft). Furthermore, the properties of polymers are influenced if they are mixed with other polymers (polymer blends), with fibers (glass fibers, carbon fibers, or metal fibers), or with other fillers (cellulose, inorganic materials, or organic materials).

The tables and figures include the physical and physicochemical properties of those polymers, copolymers, and polymer blends which are widely used for scientific applications and in industry. The figures include mainly the following physical properties: stress versus strain, viscosity versus shear rate, and creep modulus versus time. However, other physical properties are also included. Additionally, the most relevant applications of the materials are given.

3.3.1	**Structural Units of Polymers**	480
3.3.2	**Abbreviations**	482
3.3.3	**Tables and Figures**	483
	3.3.3.1 Polyolefines	483
	3.3.3.2 Vinyl Polymers	489
	3.3.3.3 Fluoropolymers	492
	3.3.3.4 Polyacrylics and Polyacetals	497
	3.3.3.5 Polyamides	501
	3.3.3.6 Polyesters	503
	3.3.3.7 Polysulfones and Polysulfides	506
	3.3.3.8 Polyimides and Polyether Ketones	508
	3.3.3.9 Cellulose	509
	3.3.3.10 Polyurethanes	511
	3.3.3.11 Thermosets	512
	3.3.3.12 Polymer Blends	515
References		522

The tables and figures include the physical and physicochemical properties of the most important polymers, copolymers, and polymer blends. "Most important" here means that these materials are widely used for scientific applications and in industry. The values in the main tables are given for room temperature, that is, $\approx 25\,°C$; otherwise, the temperature is given in parentheses. The tables and figures include the following physical properties:

Melting temperature T_m: heating rate $10\,K/\min$ (ISO 11357).

Enthalpy of fusion ΔH_u: the amount of enthalpy (given per monomer unit of the polymer) needed for the transition of the polymer from the solid state to the molten state.

Entropy of fusion ΔS_u: amount of entropy (given per monomer unit of the polymer) which is needed for the transition of a polymer from the solid state to the molten state.

Heat capacity $c_p = (\partial H/\partial T)_p \approx \Delta H/\Delta T$; ΔH = quantity of heat per mass unit, ΔT = temperature increase.

Enthalpy of combustion ΔH_c: amount of enthalpy released in flaming combustion per unit mass of the polymer.

Glass transition temperature T_g: heating rate $10\,K/\min$ (ISO 11357).

Vicat softening temperature : $T_V 10/50$, force $10\,N$, heating rate $50\,K/h$; $T_V 50/50$, force $50\,N$, heating rate $50\,K/h$ (ISO 306).

Thermal conductivity λ: $dq/dt = A\lambda\, dT/dx$; dq/dt = heat flux, A = area, dT/dx = temperature gradient.
Density $\varrho = m/V$ (ISO 1183).
Coefficient of expansion $\alpha = (1/V_0)(\partial V/\partial T)_p$: $T = 23\text{–}55\,°\text{C}$ (ISO 11359).
Compressibility $\kappa = -(1/V)(\partial V/\partial p)_T$.
Elastic modulus $E = \sigma/\varepsilon$ (σ = stress, ε = strain (elongation)); elongation rate 1 mm/min (ISO 527).
Shear modulus $G = \tau/\gamma$ (τ = shear stress, γ = shear angle).
Poisson's ratio $\mu = 0.5[1-(E/\sigma)(\Delta V/V)]$; $\Delta V/V$ = relative volume change.
Stress at yield σ_y, strain (elongation) at yield ε_y: see Fig. 3.3-1; elongation rate 50 mm/min (ISO 527).
Stress at 50% strain (elongation) σ_{50}: see Fig. 3.3-1; elongation rate 50 mm/min (ISO 527).
Stress at fracture σ_b, strain (elongation) at fracture ε_b: see Fig. 3.3-1; elongation rate 5 mm/min (ISO 527).
Impact strength, and notched impact strength (Charpy) (ISO 179).
Sound velocity v_s, longitudinal (long) and transverse (trans).
Shore hardness A, D (ISO 868).
Volume resistivity ρ_e, surface resistivity σ_e: contact electrodes, voltage 500 V (DIN 0303 T30, ISO 93, IEC 60093).
Electric strength E_B: specimen of thickness 1.0 ± 0.1 mm (ISO 10350, IEC 60243).
Relative permittivity ε_r, dielectric loss (dissipation factor) $\tan\delta$ (IEC 60250).
Refractive index n_D, temperature coefficient of refractive index dn_D/dT.
Steam permeation: 20–25 °C, 85% relative humidity gradient (DIN 53122, ISO 15106).
Gas permeation: 20–25 °C, reduced to 23 °C, 1 bar (ISO 2556, DIN 53380, ISO 15105).
Melt-viscosity–molar-mass relation.
Viscosity–molar-mass relation: $[\eta] = KM^a$ means $[\eta]/[\eta_0] = K(M/M_0)^a$, where $[\eta_0] = 1\,\text{cm}^3/\text{g}$, $M_0 = 1\,\text{g/mol}$, and $[\eta]$ = intrinsic viscosity number at concentration $C = 0\,\text{g/cm}^3$ [3.1] (DIN 53726, DIN 53727).
Stress $\sigma(\varepsilon, T)$; ε = strain (elongation), T = temperature (ISO 527).
Viscosity $\eta(d\gamma/dt, T)$; $d\gamma/dt$ = shear rate, T = temperature (ISO 11443).
Creep modulus $E_{tc}(t, p, T)$; t = time, p = pressure, T = temperature; $E_{tc} = \sigma_{tc}/\varepsilon(t)$ (σ_{tc} = creep stress, $\varepsilon(t)$ = creep strain (creep elongation)); strain $\leq 0.5\%$ (ISO 899).

For selected polymers, the temperature dependence of some physical properties is given. Additionally, the most relevant applications of the materials are given. The tables and figures include the physical properties given in the table below (see [3.1–3]).

As the physical and physicochemical properties of each polymer vary with its molecular architecture, the tables show the ranges of the physical and physicochemical properties, whereas the diagrams show the functional relationships for a typical species of the polymer, copolymer, or polymer blend. The table on page 479 shows the selected 77 polymers, copolymers and polymer blends.

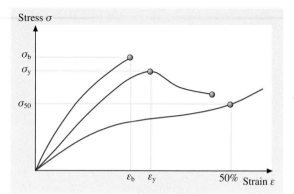

Fig. 3.3-1 Stress σ as a function of the strain ε for different kinds of polymers (see page 478)

3.3.3.1 Polyolefines
Polyolefines I
Polyethylene: high density HDPE, medium density MDPE, low density LDPE, linear low density LLDPE, ultra high molecular weight UHMWPE
Polyolefines II
Poly(ethylene-co-vinylacetate) EVA, Polyethylene ionomer EIM, Cycloolefine copolymer COC [Poly(ethylene-co-norbornene)], Poly(ethylene-co-acrylic acid) EAA
Polyolefines III
Polypropylene PP, Polybutene-1 PB, Polyisobutylene PIB, Poly(4-methylpentene-1) PMP

3.3.3.2 Vinylpolymers
Vinylpolymers I
Polystyrene PS, Poly(styrene-co-butadiene) SB, Poly(styrene-co-acrylonitrile) SAN
Vinylpolymers II
Poly(vinyl carbazole) PVK, Poly(acrylonitrile-co-butadiene-co-styrene) ABS, Poly(acrylonitrile-co-styrene-co-acrylester) ASA
Vinylpolymers III
Poly(vinyl chloride): unplastisized PVC-U, plastisized (75/25) PVC-P1, plastisized (60/40) PVC-P2

3.3.3.3 Fluoropolymers
Polytetrafluoroethylene PTFE, Polychlorotrifluoroethylene PCTFE, Poly(tetrafluoroethylene-co-hexafluoropro-pylene) FEP, Poly(ethylene-co-tetrafluoroethylene) ETFE, Poly(ethylene-co-chlorotrifluoroethylene), ECTFE

3.3.3.4 Polyacrylics, Polyacetals
Poly(methyl methacrylate) PMMA; Poly(oxymethylene) POM-H, Poly(oxymethylene-co-ethylene) POM-R

3.3.3.5 Polyamides
Polyamide 6 PA6, Polyamide 66 PA66, Polyamide 11 PA11, Polyamide 12 PA12, Polyamide 610 PA610

3.3.3.6 Polyesters
Polycarbonate PC, Poly(ethylene terephthalate) PET, Poly(butylene terephthalate) PBT, Poly(phenylene ether) PPE

3.3.3.7 Polysulfones, Polysulfides
Polysulfon PSU, Poly(phenylene sulfide) PPS, Poly(ether sulfone) PES

3.3.3.8 Polyimides, Polyether ketones
Poly(amide imide), PAI; Poly(ether imide), PEI; Polyimide, PI; Poly(ether ether ketone), PEEK

3.3.3.9 Cellulose
Cellulose acetate CA, Cellulose propionate CP, Cellulose acetobutyrate CAB, Ethyl cellulose EC, Vulcanized fiber VF

3.3.3.10 Polyurethanes
Polyurethane PUR, Thermoplastic polyurethane elastomer TPU

3.3.3.11 Thermosets
Thermosets I
Phenol formaldehyde PF, Urea formaldehyde UF, Melamine formaldehyde MF
Thermosets II
Unsaturated polyester UP, Diallylphthalat DAP, Silicone resin SI, Epoxy resin EP

3.3.3.12 Polymer Blends
Polymer Blends I
Polypropylene + Ethylene/propylene/diene-rubber PP + EPDM, Poly(acrylonitrile-co-butadiene-co-styrene) + Polycarbonate ABS + PC, Poly(acrylonitrile-co-butadiene-co-styrene) + Polyamide ABS + PA, Poly(acrylonitrile-co-butadiene-co-acrylester) + Polycarbonate ASA + PC

Polymer Blends II

Poly(vinyl chloride) + Poly(vinylchloride-co-acrylate) PVC + VC/A, Poly(vinyl chloride) + chlorinated Polyethylene PVC + PE-C, Poly(vinyl chloride) + Poly(acrylonitrile-co-butadiene-co-acrylester) PVC + ASA

Polymer Blends III

Polycarbonate + Poly(ethylene terephthalate) PC + PET, Polycarbonate + Liquid crystal polymer PC + LCP, Polycarbonate + Poly(butylene terephthalate) PC + PBT, Poly(ethylene terephthalate) + Polystyrene PET + PS, Poly(butylene terephthalate) + Polystyrene PBT + PS

Polymer Blends IV

Poly(butylene terephthalate) + Poly(acrylonitrile-co-butadiene-co-acrylester) PBT + ASA, Polysulfon + Poly(acrylonitrile-co-butadiene-co-styrene) PSU + ABS, Poly(phenylene ether) + Poly(styrene-co-butadiene) PPE + SB, Poly(phenylene ether) + Polyamide 66 PPE + PA66, Poly(phenylene ether) + Polystyrene PPE + PS

3.3.1 Structural Units of Polymers

The polymers given in this chapter are divided into polyolefines, vinyl polymers, fluoropolymers, polyacrylics, polyacetals, polyamides, polyesters, polysulfones, polysulfides, polyimides, polyether ketones, cellulose, polyurethanes, and thermosets. The structural units of the polymers are as follows:

Polyolefines

Polyethylene, PE: $-CH_2-CH_2-$

Polypropylene, PP: $-CH_2-CH(CH_3)-$

Poly(butene-1), PB: $-CH_2-CH(CH_2CH_3)-$

Poly(isobutylene), PIB: $-CH_2-C(CH_3)_2-$

Poly(4-methylpentene-1), PMP: $-CH_2-CH(CH_2-CH(CH_3)_2)-$

Polynorbornene: cyclopentane–CH=CH–

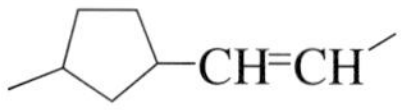

Poly(1,4-butadiene), BR: $-CH_2-CH=CH-CH_2-$

Vinyl Polymers

Polystyrene, PS: $-CH_2-CH(C_6H_5)-$

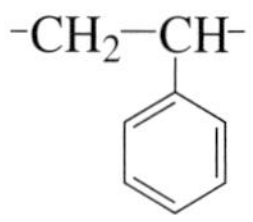

Poly(acrylonitrile), PAN: $-CH_2-CH(CN)-$

Poly(vinyl acetate), PVAC: $-CH_2-CH(O-CO-CH_3)-$

Poly(vinyl chloride), PVC: $-CH_2-CH(Cl)-$

Poly(vinyl carbazole), PVK: $-CH_2-CH(N(C_6H_5)_2)-$

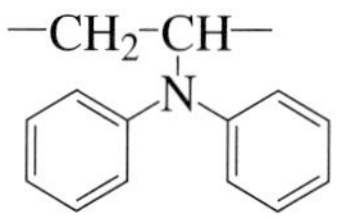

Fluoropolymers

Poly(tetrafluoroethylene), PTFE: $-CF_2-CF_2-$

Poly(chlorotrifluoroethylene), PCTFE: $-CFCl-CF_2-$

Poly(hexafluoropropylene): $-CF_2-CF(CF_3)-$

Polyacrylics and Polyacetals

Poly(methyl methacrylate), PMMA: $-CH_2-C(CH_3)(COOCH_3)-$

Poly(acrylic acid), PAA: $-CH_2-CH(H)(COO^-)-$

Poly(oxymethylene), POM: $-CH_2-O-$

Polyamides

Polyamide 6, PA6: $-CO-(CH_2)_5-NH-$

Polyamide 66, PA66:
   $-NH-(CH_2)_6-NH-CO-(CH_2)_4-CO-$

Polyamide 11, PA11: $-CO-(CH_2)_{10}-NH-$

Polyamide 12, PA12: $-CO-(CH_2)_{11}-NH-$

Polyamide 610, PA610:
   $-NH-(CH_2)_6-NH-CO-(CH_2)_8-CO-$

Polyamide 612, PA612:
   $-NH-(CH_2)_6-NH-CO-(CH_2)_{10}-CO-$

Polyesters

Polycarbonate, PC:

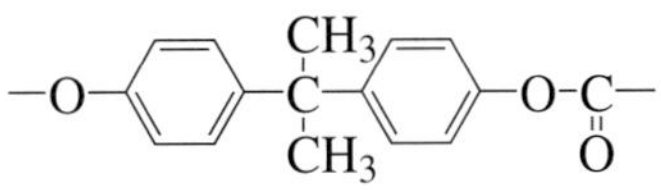

Poly(ethylene terephthalate), PET:

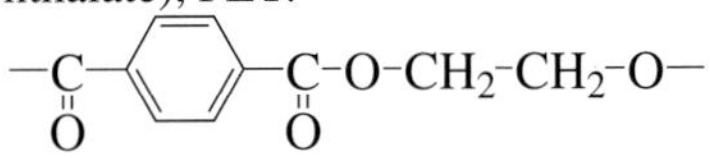

Poly(butylene terephthalate), PBT:

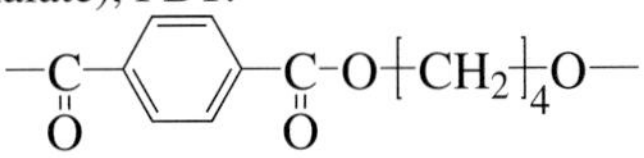

Poly(phenylene ether), PPE:

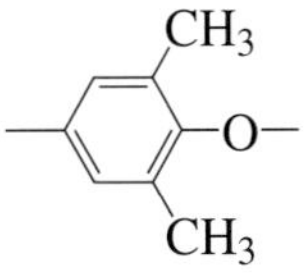

Polysulfones and Polysulfides

Polysulfone, PSU:

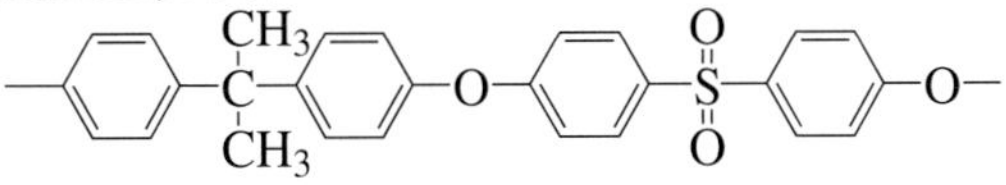

Poly(phenylene sulfide), PPS:

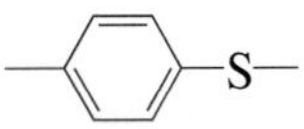

Poly(ether sulfone), PES:

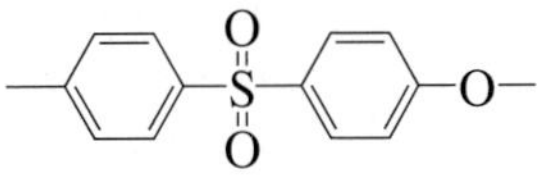

Polyimides and Polyether Ketones

Poly(amide imide), PAI:

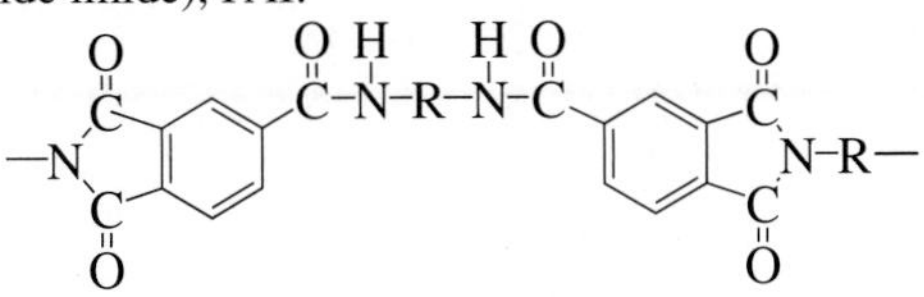

Poly(ether imide), PEI:

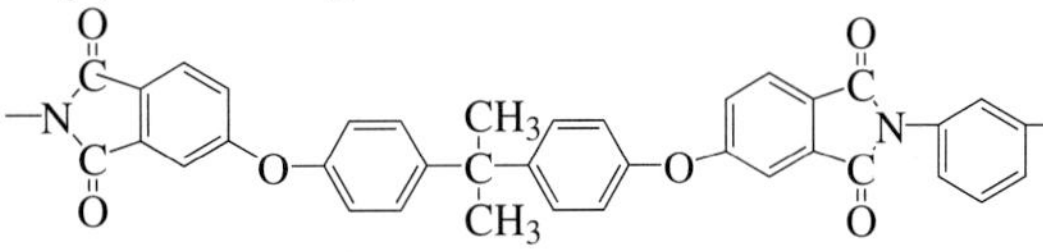

Polyimide, PI:

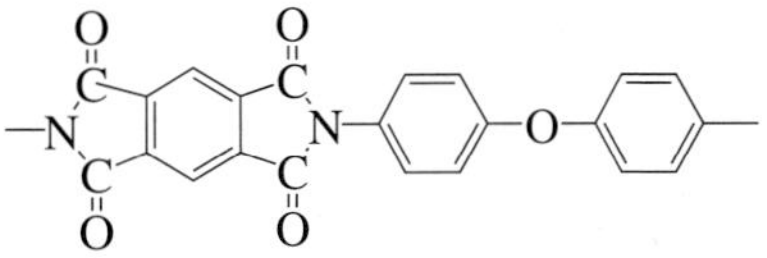

Poly(ether ether ketone), PEEK:

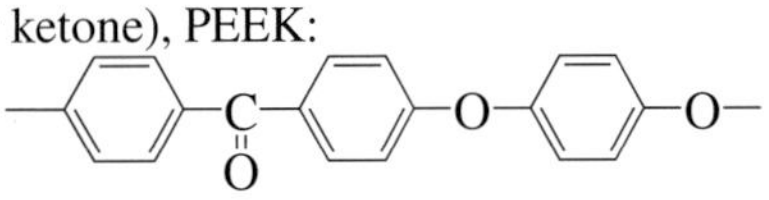

Cellulose

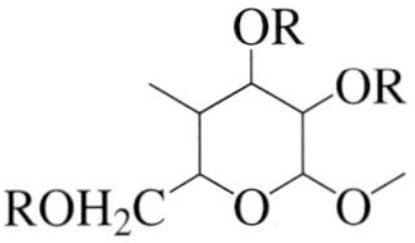

Cellulose acetate, CA: $R = -COCH_3$

Cellulose propionate, CP: $R = -COCH_2CH_3$

Cellulose acetobutyrate, CAB: $R = -COCH_3$ and
                              $R = -COCH_2CH_2CH_3$

Ethyl cellulose, EC: $R = -CH_2CH_3$

Polyurethanes

Polyurethane, PUR, TPU:
   $-CO-NH-(CH_2)_6-NH-CO-O-(CH_2)_4-O-$

Thermosets

Phenol formaldehyde, PF:

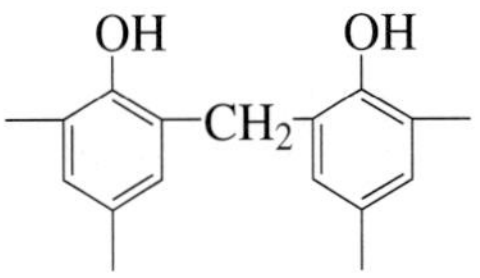

Urea formaldehyde, UF:

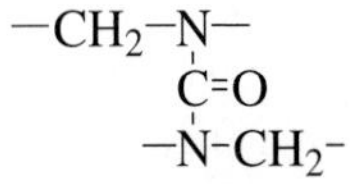

Melamine formaldehyde, MF:

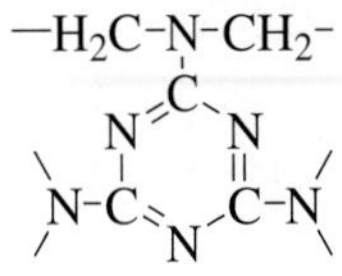

3.3.2 Abbreviations

The following abbreviations are used in this chapter. The abbreviations are in accordance with international rules.

am	amorphous
C	chlorinated
co	copolymer
cr	crystalline
DOP	dioctyl phthalate
HI	high impact (modifier)
iso	isotactic
long	longitudinal
LCP	liquid crystal polymer
mu	monomer unit
pcr	partially crystalline
syn	syndiotactic
str	strained

THF	tetrahydrofuran
trans	transverse
DSC	differential scanning calorimetry
DTA	differential thermal analysis
CFa	carbon fiber content = a mass%, e.g. PS-CF20 means polystyrene with 20% carbon fiber
GBa	content of glass beads, spheres, or balls = a mass%, e.g. PS-GB20 means polystyrene with 20% glass beads
GFa	glass fiber content = a mass%, e.g. PS-GF20 means polystyrene with 20% glass fiber
MeFa	metal fiber content = a mass%

Abbreviated notations for polymer names

ABS	Poly(acrylonitrile-co-butadiene-co-styrene)		PEI	Poly(ether imide)
ASA	Poly(acrylonitrile-co-styrene-co-acrylester)		PES	Poly(ether sulfone)
CA	Cellulose acetate		PET	Poly(ethylene terephthalate)
CAB	Cellulose acetobutyrate		PF	Phenol formaldehyde
COC	Cycloolefine copolymer		PI	Polyimide
CP	Cellulosepropionate		PIB	Poly(isobutylene)
DAP	Diallylphthalat		PMMA	Poly(methyl methacrylate)
EAA	Poly(ethylene-co-acrylic acid)		PMP	Poly(4-methylpenten-1)
EC	Ethylcellulose		POM	Poly(oxymethylene)
ECTFE	Poly(ethylene-co-chlorotrifluoroethylene)		PP	Polypropylene
EPDM	Ethylene/propylene/diene-rubber		PPE	Poly(phenylene ether)
EIM	Polyethylene ionomer		PPS	Poly(phenylene sulfide)
EP	Epoxide; epoxy		PTFE	Poly(tetrafluoro ethylene)
ETFE	Poly(ethylene-co-tetrafluoroethylene)		PVC-P	Poly(vinyl chloride), plasticized with DOP
EVA	Poly(ethylene-co-vinylacetate)		PVC-U	Poly(vinyl chloride), unplasticized
FEP	Poly(tetrafluoroethylene-co-hexafluoropropylene)		PVK	Poly(vinyl carbazole)
HDPE	High density polyethylene		PS	Polystyrene
LDPE	Low density polyethylene		PSU	Polysulfone
LLDPE	Linear low density polyethylene		PUR	Polyurethane
MDPE	Medium density polyethylene		SAN	Poly(styrene-co-acrylonitrile)
MF	Melamine formaldehyde		SB	Poly(styrene-co-butadiene)

PA	Polyamide	SI	Silicone resin
PAI	Poly(amide imide)	TPU	Thermoplastic polyurethane elastomer
PB	Polybutene-1	UF	Urea formaldehyde
PBT	Poly(butylene terephthalate)	UHMWPE	Ultra high molecular weight polyethylene
PC	Polycarbonate	UP	Unsaturated polyester
PCTFE	Poly(trifluorochloroethylene)	VC/A	Poly(vinyl chloride-co-acrylate)
PE	Polyethylene	VF	Vulcanized fiber
PEEK	Poly(ether ether ketone)		

3.3.3 Tables and Figures

The following tables and diagrams contain physical and physicochemical properties of common polymers, copolymers, and polymer blends. The materials are arranged according to increasing number of functional groups, i.e. polyolefines, vinyl polymers, fluoropolymers, polyacrylics, polyacetals, polyamides, polyesters, and polymers with special functional groups [3.2–16].

3.3.3.1 Polyolefines

Polyethylene, HDPE, MDPE. Applications: injection molding for domestic parts and industrial parts; blow molding for containers and sports goods; extrusion for pressure pipes, pipes, electrical insulating material, bags, envelopes, and tissue.

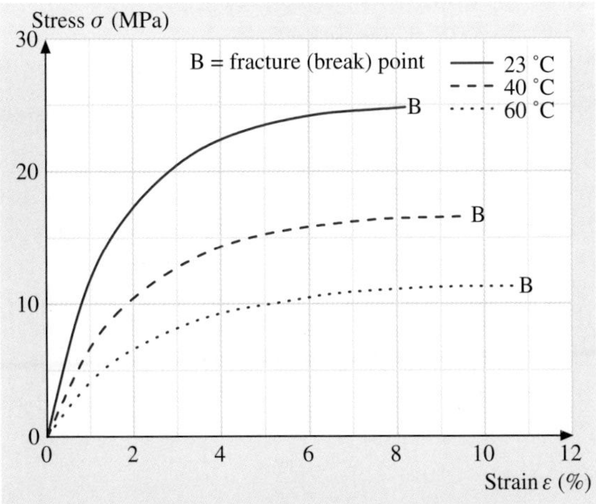

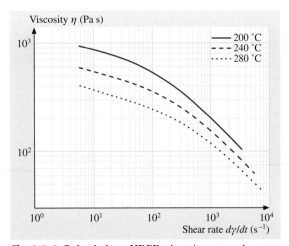

Fig. 3.3-2 Polyethylene, HDPE: stress versus strain

Fig. 3.3-3 Polyethylene, HDPE: viscosity versus shear rate

Table 3.3-1 Polyethylene: high-density, HDPE; medium-density, MDPE; low-density, LDPE; linear low-density, LLDPE; ultrahigh-molecular-weight, UHMWPE

	HDPE	MDPE	LDPE	LLDPE	UHMWPE
Melting temperature T_m (°C)	126–135	120–125	105–118	126	130–135
Enthalpy of fusion ΔH_u (kJ/mol) (mu)	3.9–4.1		3.9–4.1		
Entropy of fusion ΔS_u (J/(K mol)) (mu)	9.6–9.9		9.6–9.9		
Heat capacity c_p (kJ/(kg K))	2.1–2.7		2.1–2.5		1.7–1.8
Temperature coefficient dc_p/dT (kJ/(kg K^2))					
Enthalpy of combustion ΔH_c (kJ/g)	−46.4	−46.5	−46.5		
Glass transition temperature T_g (°C)	−110	−110		−110	−110
Vicat softening temperature T_V 50/50 (°C)	60–80		45–60		74
Thermal conductivity λ (W/(m K))	0.38–0.51		0.32–0.40		0.41
Density ϱ (g/cm^3)	0.94–0.96	0.925–0.935	0.915–0.92	≈ 0.935	0.93–0.94
Coefficient of thermal expansion α (10^{-5}/K) (linear) (296–328 K)	14–18	18–23	23–25	18–20	15–20
Compressibility κ (10^{-4}/MPa) (cubic)			2.2		
Elastic modulus E (GPa)	0.6–1.4	0.4–0.8	0.2–0.4	0.3–0.7	0.7–0.8
Shear modulus G (GPa)	0.85	0.66	0.16–0.25		
Poisson's ratio μ					
Stress at yield σ_y (MPa)	18–30	11–18	8–10	20–30	≈ 22
Stress at 50% strain σ_{50} (MPa)					
Strain at yield ε_y (%)	8–12	10–15	≈ 20	≈ 15	≈ 15
Stress at fracture σ_b (MPa)	18–35		8–23		
Strain at fracture ε_b (%)	100–1000		300–1000		
Impact strength (Charpy) (kJ/m^2)			13–25		
Notched impact strength (Charpy) (kJ/m^2)			3–5		
Sound velocity v_s (m/s) (longitudinal)	2430		2400		
Sound velocity v_s (m/s) (transverse)	950		1150		
Shore hardness D	58–63	45–60	45–51	38–60	62
Volume resistivity ρ_e (Ω m) > 10^{15}	> 10^{15}	> 10^{15}	> 10^{15}	> 10^{15}	
Surface resistivity σ_e (Ω)	> 10^{14}	> 10^{14}	> 10^{14}	> 10^{14}	> 10^{14}
Electric strength E_B (kV/mm)	30–40	30–40	30–40	30–40	30–40
Relative permittivity ε_r (100 Hz)	2.4	2.3	2.3	2.3	2–2.4
Dielectric loss $\tan \delta$ (10^{-4}) (100 Hz)	1–2	2	2–2.4	2	2
Refractive index n_D (589 nm)	1.53		1.51–1.42		
Temperature coefficient dn_D/dT (10^{-4}/K)					
Steam permeation (g/(m^2 d))	0.9 (40 µm)		1 (100 µm)		
Gas permeation (cm^3/(m^2 d bar)), 23 °C, 100 µm	700 (N$_2$) 1800 (O$_2$) 10 000 (CO$_2$) 1100 (air)		700 (N$_2$) 2000 (O$_2$) 10 000 (CO$_2$) 1100 (air)		
Melt viscosity–molar-mass relation					
Viscosity–molar-mass relation	$[\eta] = 62 \times 10^{-3} M^{0.70}$ (decalin, 135 °C) $[\eta] = 51 \times 10^{-3} M^{0.725}$ (tetralin, 130 °C)				

Table 3.3-2 Polyethylene, HDPE: heat capacity, thermal conductivity, and coefficient of thermal expansion

Temperature T (°C)	−200	−150	−100	−50	0	20	50	100	150
Heat capacity c_p (kJ/(kg K))	0.55	0.84	1.10	1.34	1.64		2.05	2.86	
Thermal conductivity λ (W/(m K))		0.62	0.56	0.50	0.44		0.38	0.32	0.25
Coefficient of thermal expansion α (10^{-5}/K) (linear)	4.5	6.8	9.5	12.4		16.9	33.0	69.0	

Table 3.3-3 Polyethylene, LDPE: heat capacity and thermal conductivity

Temperature T (°C)	−200	−150	−100	−50	0	20	50	100	150
Heat capacity c_p (kJ/(kg K))	0.55	0.84	1.10	1.43	1.90		2.73		
Thermal conductivity λ (W/(m K))		0.36	0.38	0.38	0.35		0.31	0.24	0.25

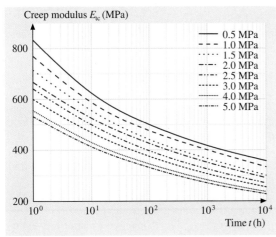

Fig. 3.3-4 Polyethylene, HDPE: creep modulus versus time, at 23 °C

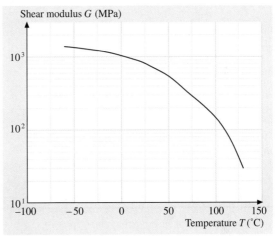

Fig. 3.3-5 Polyethylene, HDPE: shear modulus versus temperature

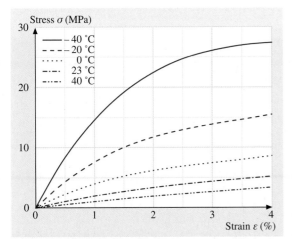

Fig. 3.3-6 Polyethylene, LDPE: stress versus strain

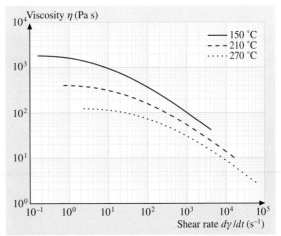

Fig. 3.3-7 Polyethylene, LDPE: viscosity versus shear rate

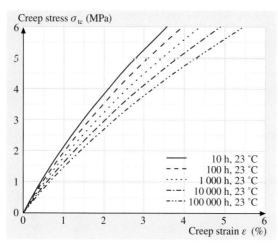

Fig. 3.3-8 Polyethylene, UHMWPE: isochronous stress versus strain

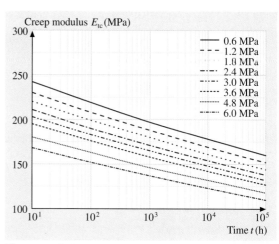

Fig. 3.3-9 Polyethylene, UHMWPE: creep modulus versus time

Polyethylene, LDPE. Applications: all kinds of sheeting, bags, insulating material, hollow bodies, bottles, injection-molded parts.

Polyethylene, LLDPE. Applications: foils, bags, waste bags, injection-molded parts, rotatory-molded parts.

Polyethylene UHMWPE, Applications: mechanical engineering parts, food wrapping, commercial packaging, textile industry parts, electrical engineering parts, paper industry parts, low temperature materials, medical parts, chemical engineering parts, electroplating.

Poly(ethylene-co-vinyl acetate), EVA. Applications: films, deep-freeze packaging, laminates, fancy leather, food packaging, closures, ice trays, bags, gloves, fittings, pads, gaskets, plugs, toys.

Polyethylene ionomer, EIM. Applications: transparent tubes for water and liquid foods, transparent films, bottles, transparent coatings, adhesion promoters.

Cycloolefine copolymer, COC. Applications: precision optics, optical storage media, lenses, medical and labware applications.

Poly(ethylene-co-acrylic acid), EAA. Applications: packaging foils, sealing layers, tubing.

Polypropylene, PP. Applications: injection molding for domestic parts, car parts, electric appliances, packing materials, and pharmaceutical parts; blow

Table 3.3-4 Poly(ethylene-co-vinyl acetate), EVA; polyethylene ionomer, EIM; cycloolefine copolymer, COC (poly-(ethylene-co-norbornene)); poly(ethylene-co-acrylic acid), EAA

	EVA	EIM	COC	EAA
Melting temperature T_m (°C)	90–110	95–110		92–103
Enthalpy of fusion ΔH_u (kJ/mol) (mu)				
Entropy of fusion ΔS_u (J/(K mol)) (mu)				
Heat capacity c_p (kJ/(kg K))		2.2		
Temperature coefficient dc_p/dT (kJ/(kg K^2))				
Enthalpy of combustion ΔH_c (kJ/g)				
Glass transition temperature T_g (°C)	66		80–180	
Vicat softening temperature T_V 50/50 (°C)	63–96 VST/A50			
Thermal conductivity λ (W/(m K))	0.28–0.35	0.24	0.16	
Density ϱ (g/cm^3)	0.93–0.94	0.94–0.95	1.02	0.925–0.935
Coefficient of thermal expansion α (10^{-5}/K) (linear)	≈ 25	10–15	6–7	≈ 20

Table 3.3-4 Poly(ethylene-co-vinyl acetate), EVA; polyethylene ionomer, EIM; ..., cont.

	EVA	EIM	COC	EAA
Compressibility κ (10^{-4}/MPa) (cubic)				
Elastic modulus E (GPa)	0.03–0.1	0.15–0.2	2.6–3.2	0.04–0.13
Shear modulus G (GPa)	0.04–0.14			
Poisson's ratio μ				
Stress at yield σ_y (MPa)	5–8	7–8	66	4–7
Strain at yield ε_y (%)		> 20	3.5–10	> 20
Stress at 50% strain σ_{50} (MPa)	4–9			
Stress at fracture σ_b (MPa)	16–23	21–35	66	
Strain at fracture ε_b (%)	700	250–500	3–10	
Impact strength (Charpy) (kJ/m^2)			13–20	
Notched impact strength (Charpy) (kJ/m^2)			1.7–2.6	
Sound velocity v_s (m/s) (longitudinal)				
Sound velocity v_s (m/s) (transverse)				
Shore hardness D	34–44			
Volume resistivity ρ_e (Ω m)	> 10^{14}	> 10^{15}	> 10^{13}	> 10^{14}
Surface resistivity σ_e (Ω)	> 10^{13}	> 10^{13}	10^{14}	> 10^{13}
Electric strength E_B (kV/mm)	30–35	40		30–40
Relative permittivity ε_r (100 Hz)	2.5–3	$\approx$ 2.4	2.35	2.5–3
Dielectric loss tan δ (10^{-4}) (100 Hz)	20–40	$\approx$ 30	0.2	30–130
Refractive index n_D (589 nm)		1.51	1.53	
Temperature coefficient dn_D/dT (10^{-4}/K)				
Steam permeation (g/(m^2 d)) (25 µm)		25	1–1.8	
Gas permeation (cm^3/(m^2 d bar)) (25 µm)		9300 (O$_2$)		
Melt viscosity–molar-mass relation				
Viscosity–molar-mass relation				

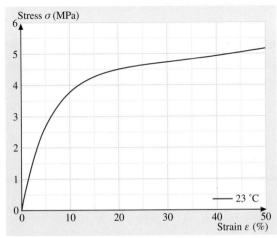

Fig. 3.3-10 Poly(ethylene-co-vinylacetate), EVA: stress versus strain

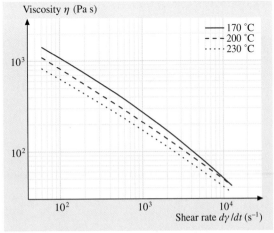

Fig. 3.3-11 Poly(ethylene-co-vinylacetate), EVA: viscosity versus shear rate

Table 3.3-5 Polypropylene, PP; polybutene-1, PB; polyisobutylene, PIB; poly(4-methylpentene-1), PMP

	PP	PB	PIB	PMP
Melting temperature T_m (°C)	160–170	126		230–240
Enthalpy of fusion ΔH_u (kJ/mol) (mu)	7–10			
Entropy of fusion ΔS_u (J/(K mol)) (mu)	15–20			
Heat capacity c_p (kJ/(kg K))	1.68			
Temperature coefficient dc_p/dT (kJ/(kg K^2))				
Enthalpy of combustion ΔH_c (kJ/g)	−44		−45	
Glass transition temperature T_g (°C)	0 to −10	78	−70	
Vicat softening temperature $T_V 50/50$ (°C)	60–102	108–113		
Thermal conductivity λ (W/(m K))	0.22	0.22	0.12–0.20	0.17
Density ϱ (g/cm^3)	0.90–0.915	0.905–0.920	0.91–0.93	0.83
Coefficient of thermal expansion α (10^{-5}/K) (linear)	12–15	13	8–12	12
Compressibility κ (10^{-4}/MPa) (cubic)	2.2			
Elastic modulus E (GPa)	1.3–1.8	0.21–0.26		1.1–2.0
Shear modulus G (GPa)				
Poisson's ratio μ				
Stress at yield σ_y (MPa)	25–40	16–24		
Strain at yield ε_y (%)	8–18	24		
Stress at 50% elongation σ_{50} (MPa)				
Stress at fracture σ_b (MPa)		30–38	2–6	25–28
Strain at fracture ε_b (%)	> 50	300–380	> 1000	10–50
Impact strength (Charpy) (kJ/m^2)	60–140			
Notched impact strength (Charpy) (kJ/m^2)	3–10			
Sound velocity v_s (m/s) (longitudinal)	2650		1950	2180
Sound velocity v_s (m/s) (transverse)	1300			1080
Shore hardness D	69–77			
Volume resistivity ρ_e (Ω m)	> 10^{14}	> 10^{14}	> 10^{13}	> 10^{14}
Surface resistivity σ_e (Ω)	> 10^{13}			
Electric strength E_B (kV/mm)	30–70	18–40	23–25	
Relative permittivity ε_r (100 Hz)	2.3	2.5	2.3	2.1
Dielectric loss $\tan\delta$ (10^{-4}) (100 Hz)	2.5	20–50	< 4	15
Refractive index n_D (589 nm)	1.50			1.46
Temperature coefficient dn_D/dT (10^{-4}/K)				
Steam permeation (g/(m^2 d)) (40 μm)	2.1			
Gas permeation (cm^3/(m^2 d bar)) (40 μm)	430 (N$_2$) 1900 (O$_2$) 6100 (CO$_2$) 700 (air)			
Melt viscosity–molar-mass relation				
Viscosity–molar-mass relation	$[\eta] = 20 \times 10^{-3} M^{0.67}$ (PIB, toluene, 30 °C)			

Table 3.3-6 Polypropylene, PP: heat capacity, thermal conductivity, and coefficient of thermal expansion

Temperature T (°C)	−200	−150	−100	−50	0	20	50	100	150
Heat capacity c_p (kJ/(kg K))	0.46	0.77	1.03	1.28	1.57	1.68	1.92	2.35	
Thermal conductivity λ (W/(m K))	2.3	0.17	0.19	0.21	0.22	0.22	0.22	0.20	
Coefficient of thermal expansion α (10^{-5}/K) (linear)		5.8	6.9	7.6	19.1	19.4	14.3	22.6	29.4

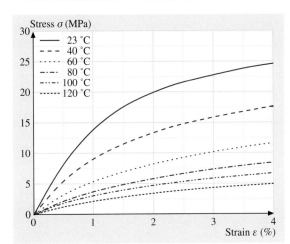

Fig. 3.3-12 Polypropylene, PP: stress versus strain

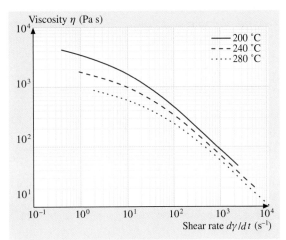

Fig. 3.3-13 Polypropylene, PP: viscosity versus shear rate

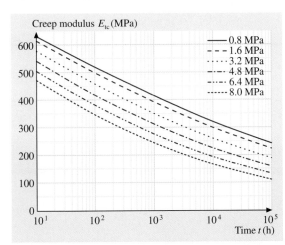

Fig. 3.3-14 Polypropylene, PP: creep modulus versus time, at 23 °C

molding for cases, boxes, and bags; pressure molding for blocks and boards; tissue coating; extrusion for pressure pipes, pipes, semiproducts, flexible tubing, textile fabrics, fibers, sheets; foamed polymers for sports goods, boxes, cases, handles, and domestic parts.

Polybutene-1, PB. Applications: tubing, pipes, fittings, hollow bodies, cables, foils, packing.

Polyisobutylene, PIB. Applications: adhesives, sealing compounds, electrical insulating oils, viscosity improvers.

Poly(4-methylpentene-1), PMP. Applications: glasses, light fittings, molded parts, foils, packing, rings, tubes, fittings, cable insulation.

3.3.3.2 Vinyl Polymers

Polystyrene, PS. Applications: packaging, expanded films, toys, artificial wood, food holders, housings, plates, cases, boxes, glossy sheets, containers, medical articles, panels.

Poly(styrene-co-butadiene), SB. Applications: flame-retardant articles, housings.

Poly(styrene-co-acrylnitrile), SAN. Applications: food holders, cases, shelves, covers, automotive and electrical parts, sheets, profiles, containers, plastic windows, doors, cosmetic items, medical items, pharmaceutical items, fittings, components, musical items, toys, displays, lighting, food compartments, cassettes, tableware, engineering parts, housings.

Table 3.3-7 Polystyrene, PS; poly(styrene-co-butadiene), SB; poly(styrene-co-acrylonitrile), SAN

	PS	SB	SAN
Melting temperature T_m (°C)	> 243.2 (iso), > 287.5 (syn)		
Enthalpy of fusion ΔH_u (KJ/mol) (mu)	8.682 (iso), 8.577 (syn)		
Entropy of fusion ΔS_u (J/(K mol)) (mu)	16.8 (iso), 15.3 (syn)		
Heat capacity c_p (kJ/(kg K))	1.2–1.4	1.3	1.3
Temperature coefficient dc_p/dT (kJ/(kg K^2))	4.04×10^{-3} (323 K)		
Enthalpy of combustion ΔH_c (kJ/g)	−41.6	−36 to −38	−36 to −38
Glass transition temperature T_g (°C)	95–100	90–95	110
Vicat softening temperature T_V 50/50 (°C)	80–100	80–110	105–120
Thermal conductivity λ (W/(m K))	0.16	0.18	0.18
Density ϱ (g/cm^3)	1.05	1.05–1.06	1.07–1.08
Coefficient of thermal expansion α (10^{-5}/K) (linear)	6–8	8–10	7–8
Compressibility κ (10^{-4}/MPa) (cubic)	2.2		
Elastic modulus E (GPa)	3.1–3.3	2.0–2.8	3.5–3.9
Shear modulus G (GPa)	1.2	0.6	1.5
Poisson's ratio μ	0.38		
Stress at yield σ_y (MPa)	50	25–45	
Strain at yield ε_y (%)		1.1–2.5	
Stress at 50% elongation σ_{50} (MPa)			
Stress at fracture σ_b (MPa)	30–55	26–38	65–85
Strain at fracture ε_b (%)	1.5–3	25–60	2.5–5
Impact strength (Charpy) (kJ/m^2)	13–25	50–105	
Notched impact strength (Charpy) (kJ/m^2)	3–5	5–10	2–4
Sound velocity v_s (m/s) (longitudinal)	2400		
Sound velocity v_s (m/s) (transverse)	1150		
Shore hardness D	78		
Volume resistivity ρ_e (Ω m)	> 10^{14}	> 10^{14}	10^{14}
Surface resistivity σ_e (Ω)	> 10^{14}	> 10^{13}	10^{14}
Electric strength E_B (kV/mm)	30–70	45–65	30–60
Relative permittivity ε_r (100 Hz)	2.4–2.5 (am); 2.61 (cr)	2.4–2.6	2.8–3
Dielectric loss tan δ (10^{-4}) (100 Hz)	1–2	1–3	40–50
Refractive index n_D (589 nm)	1.58–1.59		1.57
Temperature coefficient dn_D/dT (10^{-4}/K)	−1.42		
Steam permeation (g/(m^2 d)) (100 µm)	12		
Gas permeation (cm^3/(m^2 d bar)) (100 µm)	2500 (N$_2$) 1000 (O$_2$) 5200 (CO$_2$)		
Melt viscosity–molar-mass relation	$\eta = 13.04 M^{3.4}$ (217 °C)		
Viscosity–molar-mass relation	$[\eta] = 11.3 \times 10^{-3} M^{0.73}$ (PS, toluene, 25 °C)		

Table 3.3-8 Polystyrene, PS: heat capacity, thermal conductivity, and coefficient of thermal expansion

Temperature T (°C)	−200	−150	−100	−50	0	20	50	100	150	200
Heat capacity c_p (kJ/(kg K))				0.89	1.09		1.34	1.68	2.03	2.19
Thermal conductivity λ (W/(m K))		0.13	0.14	0.14	0.16	0.16	0.16	0.16		
Coefficient of thermal expansion α (10^{-5}/K) (linear)	3.9	5.1	6.1	6.7		7.1	10.0	17.6	18.0	17.4

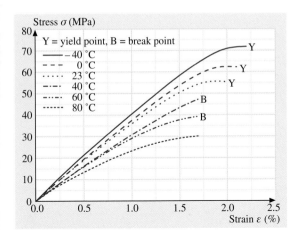

Fig. 3.3-15 Polystyrene, PS: stress versus strain

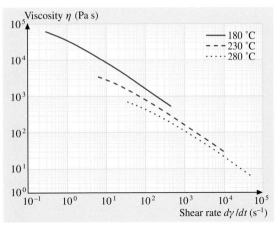

Fig. 3.3-16 Polystyrene, PS: viscosity versus shear rate

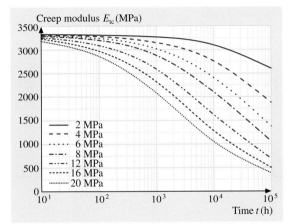

Fig. 3.3-17 Polystyrene, PS: creep modulus versus time, at 23 °C

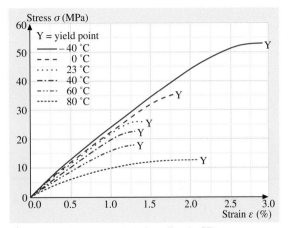

Fig. 3.3-18 Poly(styrene-co-butadiene), SB: stress versus strain

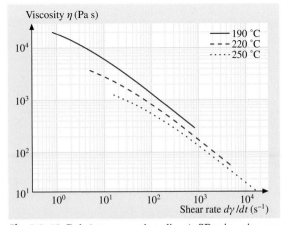

Fig. 3.3-19 Poly(styrene-co-butadiene), SB: viscosity versus shear rate

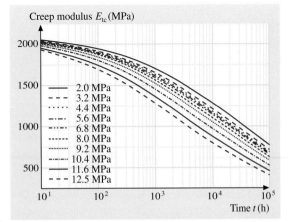

Fig. 3.3-20 Poly(styrene-co-butadiene), SB: creep modulus versus time

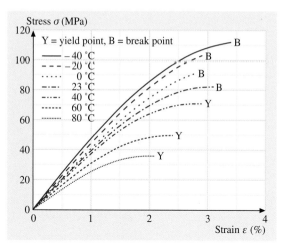

Fig. 3.3-21 Poly(styrene-co-acrylonitrile), SAN: stress versus strain

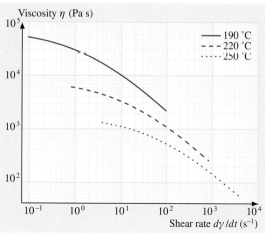

Fig. 3.3-22 Poly(styrene-co-acrylonitrile), SAN: viscosity versus shear rate

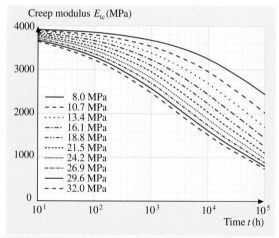

Fig. 3.3-23 Poly(styrene-co-acrylonitrile), SAN: creep modulus versus time

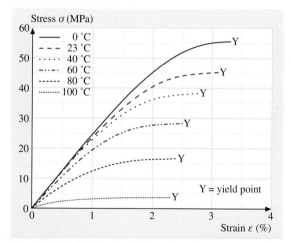

Fig. 3.3-24 Poly(acrylonitrile-co-butadiene-co-styrene), ABS: stress versus strain

Poly(vinyl carbazole), PVK. Applications: electrical insulating materials.

Poly(acrylonitrile-co-butadiene-co-styrene), ABS. Applications: housings, boxes, cases, tools, office equipment, helmets, pipes, fittings, frames, toys, sports equipment, housewares, packaging, panels, covers, automotive interior parts, food packaging, furniture, automobile parts.

Poly(acrylonitrile-co-styrene-co-acrylester), ASA. Applications: outdoor applications, housings, covers, sports equipment, fittings, garden equipment, antennas.

Poly(vinyl chloride), PVC-U. Applications: extruded profiles, sealing joints, jackets, furniture, claddings, roller shutters, fences, barriers, fittings, injection-molded articles, sheets, films, plates, bottles, coated fabrics, layers, toys, bumpers, buoys, car parts, cards, holders, sleeves, pipes, boxes, inks, lacquers, adhesives, foams.

Poly(vinyl chloride), PVC-P1, PVC-P2. Applications: flexible tubes, technical items, compounds for insulation and jacketing, joints, seals, pipes, profiles, soling, tubes, floor coverings, wall coverings, car undersealing, mastics, foams, coatings, cap closures,

Table 3.3-9 Poly(vinyl carbazole), PVK; poly(acrylonitrile-co-butadiene-co-styrene), ABS; poly(acrylonitrile-co-styrene-co-acrylester), ASA

	PVK	ABS	ASA
Melting temperature T_m (°C)			
Enthalpy of fusion ΔH_u (kJ/mol) (mu)			
Entropy of fusion ΔS_u (J/(K mol)) (mu)			
Heat capacity c_p (kJ/(kg K))		1.3	1.3
Temperature coefficient dc_p/dT (kJ/(kg K^2))			
Enthalpy of combustion ΔH_c (kJ/g)		−35	
Glass transition temperature T_g (°C)	173	80–110	100
Vicat softening temperature T_V 50/50 (°C)		95–105	90–102
Thermal conductivity λ (W/(m K))	0.29	0.18	0.18
Density ϱ (g/cm^3)		1.03–1.07	1.07
Coefficient of thermal expansion α (10^{-5}/K) (linear)	1.19	8.5–10	9.5
Compressibility κ (1/MPa) (cubic)			
Elastic modulus E (GPa)	3.5	2.2–3.0	2.3–2.9
Shear modulus G (GPa)		0.7–0.9	0.7–0.9
Poisson's ratio μ			
Stress at yield σ_y (MPa)		45–65	40–55
Strain at yield ε_y (%)		2.5–3	3.1–4.3
Stress at 50% elongation σ_{50} (MPa)			
Stress at fracture σ_b (MPa)	20–30	15–30	
Strain at fracture ε_b (%)		55–80	
Impact strength (Charpy) (kJ/m^2)		40–1000	105–118
Notched impact strength (Charpy) (kJ/m^2)			
Sound velocity v_s (m/s) (longitudinal)		2040–2160	
Sound velocity v_s (m/s) (transverse)		830–930	
Shore hardness D			
Volume resistivity ρ_e (Ω m)	> 10^{14}	10^{12}–10^{13}	10^{12}–10^{14}
Surface resistivity σ_e (Ω)	10^{14}	> 10^{13}	10^{13}
Electric strength E_B (kV/mm)		30–40	
Relative permittivity ε_r (100 Hz)		2.8–3.1	3.4–4
Dielectric loss tan δ (10^{-4}) (100 Hz)	6–10	90–160	90–100
Refractive index n_D (589 nm)		1.52	
Temperature coefficient dn_D/dT (1/K)			
Steam permeation (g/(m^2 d)) (100 μm)		27–33	30–35
Gas permeation (cm^3/(m^2 d bar)) (100 μm)		100–200 (N$_2$)	60–70 (N$_2$)
		400–900 (O$_2$)	150–180 (O$_2$)
			6000–8000 (CO$_2$)
Melt viscosity–molar-mass relation			
Viscosity–molar-mass relation			

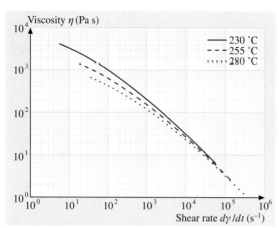

Fig. 3.3-25 Poly(acrylonitrile-co-butadiene-co-styrene), ABS: viscosity versus shear rate

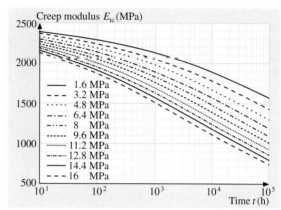

Fig. 3.3-26 Poly(acrylonitrile-co-butadiene-co-styrene), ABS: creep modulus versus time

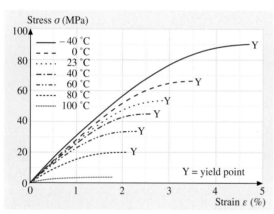

Fig. 3.3-27 Poly(acrylonitrile-co-styrene-co-acrylester), ASA: stress versus strain

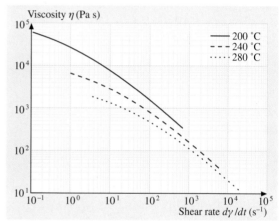

Fig. 3.3-28 Poly(acrylonitrile-co-styrene-co-acrylester), ASA: viscosity versus shear rate

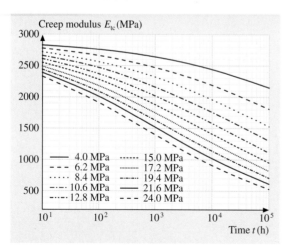

Fig. 3.3-29 Poly(acrylonitrile-co-styrene-co-acrylester), ASA: creep modulus versus time

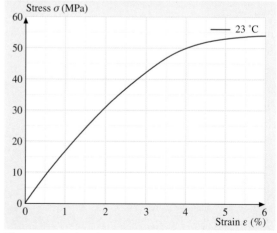

Fig. 3.3-30 Poly(vinyl chloride), PVC-U: stress versus strain

Table 3.3-10 Poly(vinyl chloride): unplasticized, PVC-U; plasticized (75/25), PVC-P1; plasticized (60/40), PVC-P2

	PVC-U	PVC-P1	PVC-P2
Melting temperature T_m (°C)	175		
Enthalpy of fusion ΔH_u (kJ/mol) (mu)	3.28		
Entropy of fusion ΔS_u (J/(K mol)) (mu)			
Heat capacity c_p (kJ/(kg K))	0.85–0.9	0.9–1.8	0.9–1.8
Temperature coefficient dc_p/dT (kJ/(kg K^2))			
Enthalpy of combustion ΔH_c (kJ/g)	−18 to −19		
Glass transition temperature T_g (°C)	85	≈ 80	≈ 80
Vicat softening temperature T_V 50/50 (°C)	63–82	≈ 42	
Thermal conductivity λ (W/(m K))	0.16	0.15	0.15
Density ϱ (g/cm^3)	1.38–1.4	1.24–1.28	1.15–1.20
Coefficient of thermal expansion α (10^{-5}/K) (linear)	7–8	18–22	23–25
Compressibility κ (1/MPa) (cubic)			
Elastic modulus E (GPa)	2.7–3.0	2.9–37	
Shear modulus G (GPa)	0.12	0.07	
Poisson's ratio μ			
Stress at yield σ_y (MPa)	50–60		
Strain at yield ε_y (%)	4–6		
Stress at 50% elongation σ_{50} (MPa)			
Stress at fracture σ_b (MPa)	40–80	10–25	10–25
Strain at fracture ε_b (%)	10–50	170–400	170–400
Impact strength (Charpy) (kJ/m^2)	13–25		
Notched impact strength (Charpy) (kJ/m^2)	3–5		
Sound velocity v_s (m/s) (longitudinal)	2330	2126	
Sound velocity v_s (m/s) (transverse)	1070		
Shore hardness D	74–94		
Volume resistivity ρ_e (Ω m)	> 10^{13}	10^{12}	10^{11}
Surface resistivity σ_e (Ω)	10^{14}	10^{11}	10^{10}
Electric strength E_B (kV/mm)	20–40	30–35	≈ 25
Relative permittivity ε_r (100 Hz)	3.5	4–5	6–7
Dielectric loss tan δ (10^{-4}) (100 Hz)	110–140	0.05–0.07	0.08–0.1
Refractive index n_D (589 nm)	1.52–1.54		
Temperature coefficient dn_D/dT (1/K)			
Steam permeation (g/(m^2 d)) (100 μm)	2.5	20	20
Gas permeation (cm^3/(m^2 d bar)) (100 μm)	2.7–3.8 (N$_2$)	350 (N$_2$)	350 (N$_2$)
	33–45 (O$_2$)	1500 (O$_2$)	1500 (O$_2$)
	120–160 (CO$_2$)	8500 (CO$_2$)	8500 (CO$_2$)
	28 (air)	550 (air)	550 (air)
Melt viscosity–molar-mass relation			
Viscosity–molar-mass relation	$[\eta] = 14.5 \times 10^{-3} M^{0.851}$ (PVC-U, THF, 25 °C)		

Table 3.3-11 Poly(vinyl chloride), PVC-U: heat capacity and thermal conductivity

Temperature T (°C)	−150	−100	−50	0	20	50	100
Heat capacity c_p (kJ/(kg K))	0.47	0.61	0.75	0.92		1.04	1.53
Thermal conductivity λ (W/(m K))	0.13	0.15	0.15	0.16	0.16	0.17	0.17

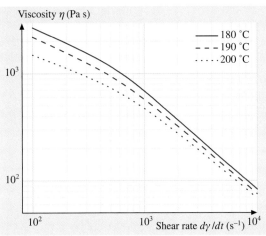

Fig. 3.3-31 Poly(vinyl chloride), PVC-U: viscosity versus shear rate

protective clothes, cable insulation, fittings, leads, films, artificial leathers.

3.3.3.3 Fluoropolymers

Polytetrafluoroethylene, PTFE. Applications: molded articles, foils, flexible tubing, coatings, jackets, sealings, bellows, flasks, machine parts, semifinished goods, printed networks.

Polychlorotrifluoroethylene, PCTFE. Applications: fittings, flexible tubing, membranes, printed networks, bobbins, insulating foils, packings.

Poly(tetrafluoroethylene-co-hexafluoropropylene), FEP. Applications: cable insulation, coatings, coverings, printed networks, injection-molded parts, packaging foils, impregnations, heat-sealing adhesives.

Table 3.3-12 Polytetrafluoroethylene, PTFE; polychlorotrifluoroethylene, PCTFE; poly(tetrafluoroethylene-co-hexafluoro-propylene), FEP; poly(ethylene-co-tetrafluoroethylene), ETFE; poly(ethylene-co-chlorotrifluoroethylene), ECTFE

	PTFE	PCTFE	FEP	ETFE	ECTFE
Melting temperature T_m (°C)	325–335	210–215	255–285	265–270	240
Enthalpy of fusion ΔH_u (kJ/kg)		1.2	24.3		
Entropy of fusion ΔS_u (J/(kg K))					
Heat capacity c_p (kJ/(kg K))	1.0	0.9	1.12	0.9	
Temperature coefficient dc_p/dT (kJ/(kg K^2))					
Enthalpy of combustion ΔH_c (kJ/g)	−5.1				
Glass transition temperature T_g (°C)	127				
Vicat softening temperature T_V 50/50 (°C)	110			134	
Thermal conductivity λ (W/(m K))	0.25	0.22	0.25	0.23	
Density ϱ (g/cm^3)	2.13–2.23	2.07–2.12	2.12–2.18	1.67–1.75	1.68–1.70
Coefficient of thermal expansion α (10^{-5}/K) (linear)	11–18	6–7	8–12	7–10	7–8
Compressibility κ (1/MPa) (cubic)					
Elastic modulus E (GPa)	0.40–0.75	1.30–1.50	0.40–0.70	0.8–1.1	1.4–1.7
Shear modulus G (GPa)					
Poisson's ratio μ					
Stress at yield σ_y (MPa)	11.7			25–35	
Strain at yield ε_y (%)				15–20	

Table 3.3-12 Polytetrafluoroethylene, PTFE; polychlorotrifluoroethylene, PCTFE; ... , cont.

	PTFE	PCTFE	FEP	ETFE	ECTFE
Stress at 50% elongation σ_{50} (MPa)					
Stress at fracture σ_b (MPa)	25–36	32–40	22–28	35–54	
Strain at fracture ε_b (%)	350–550	120–175	250–330	400–500	
Impact strength (Charpy) (kJ/m^2)					
Notched impact strength (Charpy) (kJ/m^2)					
Sound velocity v_s (m/s) (longitudinal)	1410				
Sound velocity v_s (m/s) (transverse)	730				
Shore hardness D	50–60	78	55–58	67–75	
Volume resistivity ρ_e (Ω m)	$> 10^{16}$	$> 10^{16}$	$> 10^{16}$	$> 10^{14}$	$> 10^{13}$
Surface resistivity σ_e (Ω)	$> 10^{16}$	$> 10^{16}$	$> 10^{16}$	$> 10^{14}$	10^{12}
Electric strength E_B (kV/mm)	48	55	55	40	
Relative permittivity ε_r (100 Hz)	2.1	2.5–2.7	2.1	2.6	2.3–2.6
Dielectric loss tan δ (10^{-4}) (100 Hz)	0.5–0.7	90–140	0.5–0.7	5–6	10–15
Refractive index n_D (589 nm)	1.35	1.43	1.344	1.403	
Temperature coefficient dn_D/dT (10^{-5}/K)					
Steam permeation (g/(m^2 d))	0.03 (300 µm)	0.4–0.9 (25 µm)		0.6 (25 µm)	9 (25 µm)
Gas permeation (cm^3/(m^2 d bar))	60–80 (N$_2$)	39 (N$_2$)		470 (N$_2$)	150 (N$_2$)
	160–250 (O$_2$)	110–230 (O$_2$)		1560 (O$_2$)	39 (O$_2$)
	450–700 (CO$_2$)	250–620 (CO$_2$)		3800 (CO$_2$)	1700 (CO$_2$)
	80–100 (air)				
Melt viscosity–molar-mass relation					
Viscosity–molar-mass relation					

Table 3.3-13 Polytetrafluoroethylene, PTFE: heat capacity, thermal conductivity, and coefficient of thermal expansion

Temperature T (°C)	−200	−150	−100	−50	0	20	50	100	150	200	250
Heat capacity c_p (kJ/(kg K))				0.8	0.96	1.0	1.06	1.10	1.15	1.23	
Thermal conductivity λ (W/(m K))		0.23	0.24	0.25	0.25	0.25	0.26	0.26	0.26		
Coefficient of thermal expansion α (10^{-5}/K) (linear)	3.4	4.5	7.0	9.5	11.6		11.9	13.1	16.7	22.2	30.5

Poly(ethylene-co-tetrafluoroethylene), ETFE. Applications: gear wheels, pump parts, packaging, laboratory articles, coverings, cable insulation, blown films.

Poly(ethylene-co-chlorotrifluoroethylene), ECTFE. Applications: cable coverings, coverings, molded articles, packaging, printed networks, foils, fibers.

3.3.3.4 Polyacrylics and Polyacetals

Poly(methyl methacrylate), PMMA. Applications: extruded articles, injection-molded parts, injection blow-molded parts, houseware, medical devices, sanitary ware, automotive components, lighting, tubes, sheets, profiles, covers, panels, fiber optics, optical lenses, displays, disks, cards.

Poly(oxymethylene), POM-H. Applications: molded parts, extruded parts, gears, bearings, snap-fits, fuel system components, cable ties, automotive components, seatbelt parts, pillar loops, tubes, panels.

Poly(oxymethylene-co-ethylene), POM-R. Applications: extruded articles, injection-molded articles, gear wheels, bushes, bearings, rollers, guide rails, tubing, films, clips, zippers, boards, pipes.

Table 3.3-14 Poly(methyl methacrylate), PMMA; poly(oxymethylene), POM-H; poly(oxymethylene-co-ethylene), POM-R

	PMMA	POM-H	POM-R
Melting temperature T_m (K)	175	175	164–172
Enthalpy of fusion ΔH_u (kJ/kg)		222	211
Entropy of fusion ΔS_u (J/(kg K))		1.49	1.44
Heat capacity c_p (kJ/(kg K))	1.255	1.46	1.47
Temperature coefficient dc_p/dT (kJ/(kg K^2))			
Enthalpy of combustion ΔH_c (kJ/g)	−26.2	−16.7	−17
Glass transition temperature T_g (°C)	104–105	25	
Vicat softening temperature T_V 50/50 (°C)	85–110	160–170	151–162
Thermal conductivity λ (W/(m K))	0.193	0.25–0.30	0.31
Density ϱ (g/cm^3)	1.17–1.19	1.40–1.42	1.39–1.41
Coefficient of thermal expansion α (10^{-5}/K) (linear)	7–8	11–12	10–11
Compressibility κ (10^{-4}/MPa) (cubic)	2.45	1.5	
Elastic modulus E (GPa)	3.1–3.3	3.0–3.2	2.8–3.2
Shear modulus G (GPa)	1.7	0.7–1.0	
Poisson's ratio μ			
Stress at yield σ_y (MPa)	50–77	60–75	65–73
Strain at yield ε_y (%)		8–25	8–12
Stress at 50% elongation σ_{50} (MPa)			
Stress at fracture σ_b (MPa)	60–75	62–70	59
Strain at fracture ε_b (%)	2–6	25–70	
Impact strength (Charpy) (kJ/m^2)	18–23	180	
Notched impact strength (Charpy) (kJ/m^2)	2	9	6.5
Sound velocity v_s (m/s) (longitudinal)	2690	2440	
Sound velocity v_s (m/s) (transverse)	1340	1000	
Shore hardness D	85	80	
Volume resistivity ρ_e (Ω m)	$> 10^{13}$	$> 10^{13}$	$> 10^{13}$
Surface resistivity σ_e (Ω)	$> 10^{13}$	$> 10^{14}$	$> 10^{13}$
Electric strength E_B (kV/mm)	30	25–35	35
Relative permittivity ε_r (100 Hz)	3.5–3.8	3.5–3.8	3.6–4
Dielectric loss $\tan\delta$ (10^{-4}) (100 Hz)	500–600	30–50	30–50
Refractive index n_D (589 nm)	1.492	1.49	
Temperature coefficient dn_D/dT (10^{-4}/K)	−1.2		
Steam permeation (g/(m^2 d)) (80 μm)	12	12	
Gas permeation (cm^3/(m^2 d bar))	2500 (N$_2$)	5 (N$_2$)	
	1000 (O$_2$)	24 (O$_2$)	
	5200 (CO$_2$)	470 (CO$_2$)	
		8 (air)	
Melt viscosity–molar-mass relation			
Viscosity–molar-mass relation	$[\eta] = 10.4 \times 10^{-3} M^{0.70}$ (PMMA, THF, 25 °C)		
	$[\eta] = 11.3 \times 10^{-3} M^{0.76}$ (POM-H, Phenol, 90 °C)		

Table 3.3-15 Poly(methyl methacrylate), PMMA: heat capacity, thermal conductivity, and coefficient of thermal expansion

Temperature T (°C)	−200	−150	−100	−50	0	20	50	100	150	200
Heat capacity c_p (kJ/(kg K))		0.67	0.90	1.06	1.26	1.26	1.42	1.85		
Thermal conductivity λ (W/(m K))		0.16	0.18	0.19	0.19	0.19	0.19	0.20	0.19	0.18
Coefficient of thermal expansion α (10^{-5}/K) (linear)		3.0	3.7	4.5	5.7	6.9	7.5	12.0	18.4	

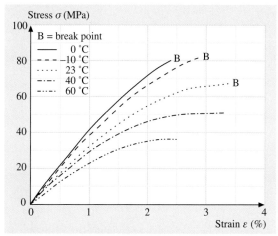

Fig. 3.3-32 Poly(methyl methacrylate), PMMA: stress versus strain

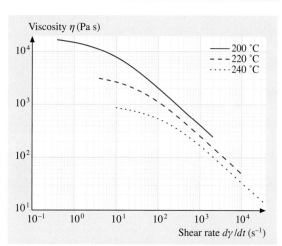

Fig. 3.3-33 Poly(methyl methacrylate), PMMA: viscosity versus shear rate

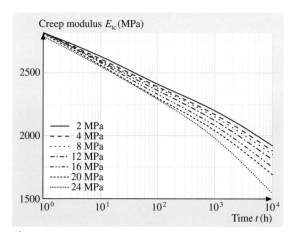

Fig. 3.3-34 Poly(methyl methacrylate), PMMA: creep modulus versus time, at 23 °C

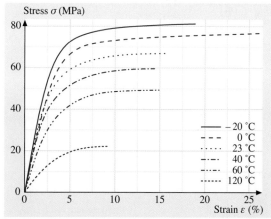

Fig. 3.3-35 Poly(oxymethylene), POM-H: stress versus strain

Table 3.3-16 Poly(oxymethylene), POM-H: heat capacity, thermal conductivity, and coefficient of thermal expansion

Temperature T (°C)	−200	−150	−100	−50	0	20	50	100	150	200
Heat capacity c_p (kJ/(kg K))	0.47	0.65	0.82	1.08	1.27		1.46	1.85		
Thermal conductivity λ (W/(m K))		0.47	0.45	0.43	0.42		0.41			
Coefficient of thermal expansion α (10^{-5}/K) (linear)					9.0	9.5	10.0	16.5	41.0	23.0

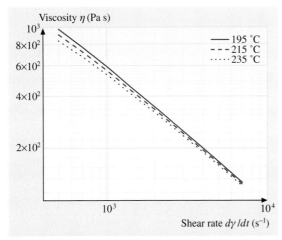

Fig. 3.3-36 Poly(oxymethylene), POM-H: viscosity versus shear rate

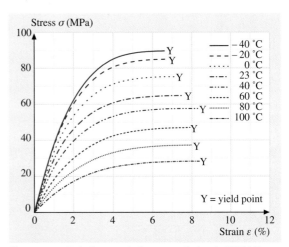

Fig. 3.3-37 Poly(oxymethylene-co-ethylene), POM-R: stress versus strain

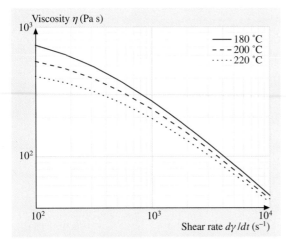

Fig. 3.3-38 Poly(oxymethylene-co-ethylene), POM-R: viscosity versus shear rate

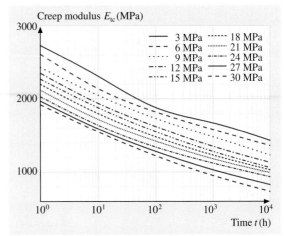

Fig. 3.3-39 Poly(oxymethylene-co-ethylene), POM-R: creep modulus versus time at 23 °C

3.3.3.5 Polyamides

Polyamide 6, PA6; polyamide 66, PA66; polyamide 11, PA11; polyamide 12, PA12; polyamide 610, PA610. Applications: technical parts, bearings, gear wheels, rollers, screws, gaskets, fittings, coverings, housings, automotive parts, houseware, semifinished goods, sports goods, membranes, foils, packings, blow-molded parts, fibers, tanks.

Table 3.3-17 Polyamide 6, PA6; polyamide 66, PA66; polyamide 11, PA11; polyamide 12, PA12; polyamide 610, PA610

	PA6	PA66	PA11	PA12	PA610
Melting temperature T_m (°C)	220–225	255–260	185	175–180	210–220
Enthalpy of fusion ΔH_u (J/mol) (mu)					
Entropy of fusion ΔS_u (J/(K mol)) (mu)					
Heat capacity c_p (kJ/(kg K))	1.7	1.7	1.26	1.26	1.7
Temperature coefficient dc_p/dT (kJ/(kg K^2))					
Enthalpy of combustion ΔH_c (kJ/g)	−31.4	−31.4			
Glass transition temperature T_g (°C)	55	80	50	50	55–60
Vicat softening temperature T_V 50/50 (°C)	180–220	195–220	180–190	140–160	205–215
Thermal conductivity λ (W/(m K))	0.29	0.23	0.23	0.23	0.23
Density ϱ (g/cm^3)	1.12–1.14	1.13–1.15	1.04	1.01–1.03	1.06–1.09
Coefficient of thermal expansion α (10^{-5}/K) (linear)	7–10	7–10	13	10–12	8–10
Compressibility κ (1/MPa) (cubic)					
Elastic modulus E (GPa)	2.6–3.2	2.7–3.3	1.0	1.3–1.6	2.0–2.4
Shear modulus G (GPa)	1.1–1.5	1.3–1.7	0.4–0.5	0.5	0.8
Poisson's ratio μ					
Stress at yield σ_y (MPa)	70–90	75–100		45–60	60–70
Strain at yield ε_y (%)	4–5	4.5–5		4–5	4
Stress at 50% elongation σ_{50} (MPa)					
Stress at fracture σ_b (MPa)	70–85	77–84	56	56–65	40
Strain at fracture ε_b (%)	200–300	150–300	500	300	500
Impact strength (Charpy) (kJ/m^2)					
Notched impact strength (Charpy) (kJ/m^2)	3–6	2–3			4–10
Sound velocity v_s (m/s) (longitudinal)	2700	2710			
Sound velocity v_s (m/s) (transverse)	1120	1120			
Shore hardness D	72	75			
Volume resistivity ϱ_e (Ω m)	$> 10^{13}$	$> 10^{12}$	10^{11}	$> 10^{13}$	$> 10^{13}$
Surface resistivity σ_e (Ω)	$> 10^{12}$	$> 10^{10}$	10^{11}	$> 10^{13}$	$> 10^{12}$
Electric strength E_B (kV/mm)	30	25–35	42.5	27–29	
Relative permittivity ε_r (100 Hz)	3.5–4.2	3.2–4	3.7	3.7–4	3.5
Dielectric loss $\tan\delta$ $\tan\delta$ (10^{-4}) (100 Hz)	60–150	50–150	600	300–700	70–150
Refractive index n_D (589 nm)	1.52–1.53	1.52–1.53	1.52–1.53	1.52–1.53	1.53
Temperature coefficient dn_D/dT (1/K)					
Steam permeation (g/(m^2 d)) (100 µm)	10–20	10–20	2.4–4	2.4–4	
Gas permeation (cm^3/(m^2 d bar)) (100 µm)	1–2 (N$_2$)	1–2 (N$_2$)	0.5–0.7 (N$_2$)	0.5–0.7 (N$_2$)	
	2–8 (O$_2$)	2–8 (O$_2$)	2–3.5 (O$_2$)	2–3.5 (O$_2$)	
	80–120 (CO$_2$)	80–120 (CO$_2$)	6–13 (CO)	6–13 (CO)	
Melt viscosity–molar-mass relation					
Viscosity–molar-mass relation					

Table 3.3-18 Polyamide 6, PA6; polyamide 66, PA66; polyamide 610, PA610: heat capacity, thermal conductivity, and coefficient of thermal expansion

Temperature T (°C)	−200	−150	−100	−50	0	20	50	100	150	200
Heat capacity c_p (kJ/(kg K)), PA	0.47	0.73	0.93	1.15	1.38		1.68	2.15	2.60	
Thermal conductivity λ (W/(m K)), PA6		0.29	0.31	0.32	0.32		0.29	0.27	0.25	
Thermal conductivity λ (W/(m K)), PA66			0.32	0.33	0.33	0.33		0.33		
Thermal conductivity λ (W/(m K)), PA610			0.31	0.32	0.33	0.33		0.32	0.31	
Coefficient of thermal expansion α (10^{-5}/K) (linear), PA6				5.0	6.6	8.0	9.1	40.1	15.1	14.0	34.6

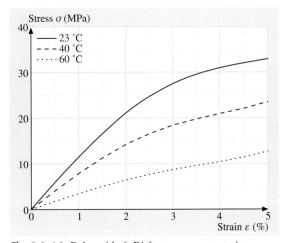

Fig. 3.3-40 Polyamide 6, PA6: stress versus strain

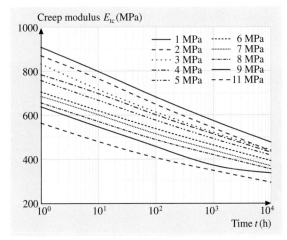

Fig. 3.3-41 Polyamide 6, PA6: creep modulus versus time

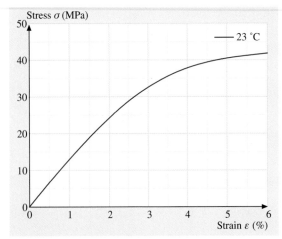

Fig. 3.3-42 Polyamide 66, PA66: stress versus strain

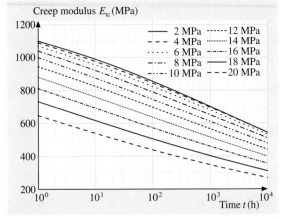

Fig. 3.3-43 Polyamide 66, PA66: creep modulus versus time

3.3.3.6 Polyesters

Polycarbonate, PC. Applications: injection-molded parts, extruded parts, disks, lamp housings, electronic articles, optical articles, houseware, foils.

Poly(ethylene terephthalate), PET. Applications: bearings, gear wheels, shafts, couplings, foils, ribbons, flexible tubing, fibers.

Poly(butylene terephthalate), PBT. Applications: bearings, valve parts, screws, plugs, housings, wheels, houseware.

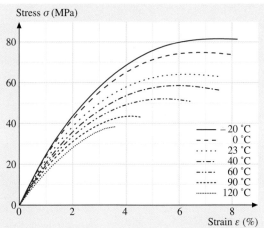

Fig. 3.3-44 Polycarbonate, PC: stress versus strain

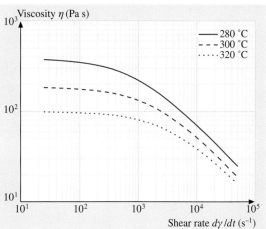

Fig. 3.3-45 Polycarbonate, PC: viscosity versus shear rate

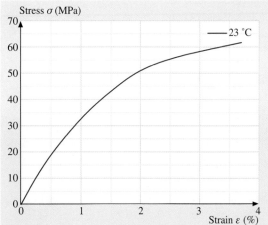

Fig. 3.3-46 Polycarbonate, PC: creep modulus versus time

Fig. 3.3-47 Poly(ethylene terephthalate), PET: stress versus strain

Table 3.3-19 Polycarbonate, PC: heat capacity and thermal conductivity

Temperature T (°C)	−200	−150	−100	−50	0	20	50	100	150
Heat capacity c_p (kJ/(kg K))	0.34	0.50	0.70	0.90	1.10	1.17	1.30	1.50	1.70
Thermal conductivity λ (W/(m K))		0.17	0.19	0.21	0.23		0.24	0.24	0.24

Table 3.3-20 Polycarbonate, PC; poly(ethylene terephthalate), PET; poly(butylene terephthalate), PBT; poly(phenylene ether), PPE

	PC	PET	PBT	PPE
Melting temperature T_m (°C)	220–260	250–260 (pcr)	220–225	
Enthalpy of fusion ΔH_u (kJ/mol) (mu)	8.682 (iso), 8.577 (syn)	2.69	54 kJ/kg	
Entropy of fusion ΔS_u (J/(K mol)) (mu)	16.8 (iso), 15.3 (syn)	48.6		
Heat capacity c_p (kJ/(kg K))	1.17	1.223	1.35	
Temperature coefficient dc_p/dT (kJ/(kg K^2))	4.04×10^{-3}			
Enthalpy of combustion ΔH_c (kJ/g)	−30.7	−21.6		
Glass transition temperature T_g (°C)	150	98	60	110–150
Vicat softening temperature T_V 50/50 (°C)	83	79	165–180	105–132
Thermal conductivity λ (W/(m K))	0.21	0.11	0.21	0.17–0.22
Density ϱ (g/cm^3)	1.2	1.33–1.35 (am)	1.30–1.32	1.04–1.10
	1.38–1.40 (pcr)			
Coefficient of thermal expansion α (10^{-5}/K) (linear)	6.5–7	8 (am), 7(pcr)	8–10	8–9
Compressibility κ (1/MPa) (cubic)	2.2×10^{-4}	6.99×10^{-6} (melt)		
Elastic modulus E (GPa)	2.3–2.4	2.1–2.4 (am)	2.5–2.8	2.0–3.1
		2.8–3.1 (pcr)		
Shear modulus G (GPa)	0.72	1.2	1.0	
Poisson's ratio μ		0.38		
Stress at yield σ_y (MPa)	55–65	55 (am), 60–80 (pcr)	50–60	45–70
Strain at yield ε_y (%)	6–7	4 (am), 5–7 (pcr)	3.5–7	3–6
Stress at 50% elongation σ_{50} (MPa)				
Stress at fracture σ_b (MPa)	69–72	30–55	52	35–55
Strain at fracture ε_b (%)	120–125	1.5–3		15–40
Impact strength (Charpy) (kJ/m^2)		13–25		
Notched impact strength (Charpy) (kJ/m^2)		3–5		6–16
Sound velocity v_s (m/s) (longitudinal)	2220	2400		2220
Sound velocity v_s (m/s) (transverse)	909	1150		1000
Shore hardness D			80	
Volume resistivity ρ_e (Ω m)	$> 10^{14}$	$> 10^{13}$	$> 10^{13}$	10^{14}–10^{15}
Surface resistivity σ_e (Ω)	$> 10^{14}$	$> 10^{14}$	$> 10^{14}$	10^{16}–10^{17}
Electric strength E_B (kV/mm)	30–75	42	42	25–35
Relative permittivity ε_r (100 Hz)	2.8–3.2	3.4–3.6	3.3–4.0	2.6–2.9
Dielectric loss $\tan \delta$ (10^{-4}) (100 Hz)	7–20	20	15–20	10–40
Refractive index n_D (589 nm)	1.58–1.59	1.57–1.64	1.58	
Temperature coefficient dn_D/dT (1/K)	-1.42×10^{-4}			
Steam permeation (g/(m^2 d))	4	4.5–5.5, 0.6 (str)		
Gas permeation (cm^3/(m^2 d bar))	680 (N$_2$), 4000 (O$_2$)	6.6 (N$_2$), 30 (O$_2$)		
	14 500 (CO$_2$)	140 (CO$_2$), 12 (air)		
		9–15 (str, N$_2$)		
		80–110 (str, O$_2$)		
		200–340 (str, CO$_2$)		
Melt viscosity–molar-mass relation				
Viscosity–molar-mass relation	$[\eta] = 4.25 \times 10^{-3} M^{0.69}$ (PET)			

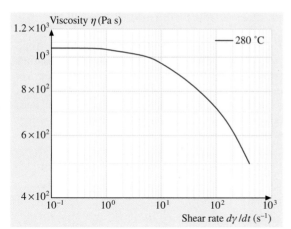

Fig. 3.3-48 Poly(ethylene terephthalate), PET: viscosity versus shear rate

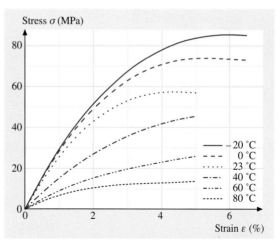

Fig. 3.3-49 Poly(butylene terephthalate), PBT: stress versus strain

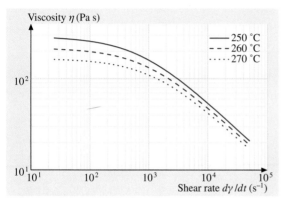

Fig. 3.3-50 Poly(butylene terephthalate), PBT: viscosity versus shear rate

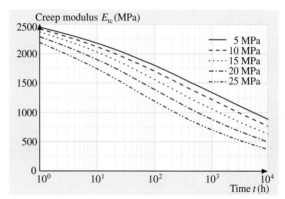

Fig. 3.3-51 Poly(butylene terephthalate), PBT: creep modulus versus time

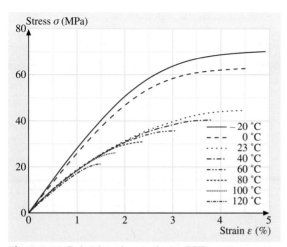

Fig. 3.3-52 Poly(phenylene ether), PPE: stress versus strain

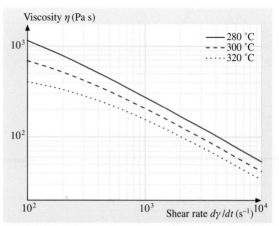

Fig. 3.3-53 Poly(phenylene ether), PPE: viscosity versus shear rate

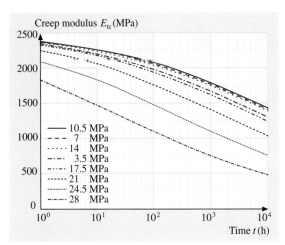

Fig. 3.3-54 Poly(phenylene ether), PPE: creep modulus versus time

3.3.3.7 Polysulfones and Polysulfides

Polysulfone, PSU; Poly(ether sulfone), PES. Applications: injection molded parts, coatings, electronic articles, printed circuits, houseware, medical devices, membranes, lenses, optical devices.

Poly(phenylene sulfide), PPS. Applications: injection-molded parts, sintered parts, foils, fibers, housings, sockets, foams.

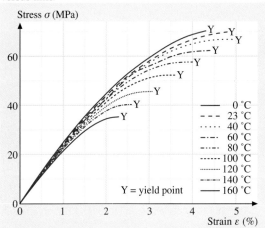

Fig. 3.3-55 Polysulfone, PSU: stress versus strain

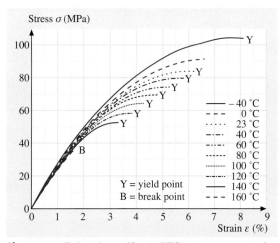

Fig. 3.3-57 Poly(ether sulfone), PES: stress versus strain

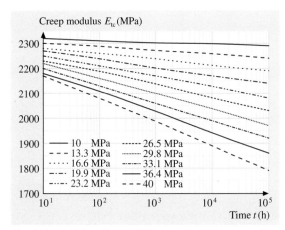

Fig. 3.3-56 Polysulfone, PSU: creep modulus versus time

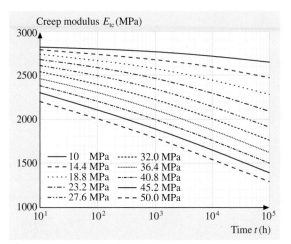

Fig. 3.3-58 Poly(ether sulfone), PES: creep modulus versus time

Table 3.3-21 Polysulfone, PSU; poly(phenylene sulfide), PPS; poly(ether sulfone), PES

	PSU	PPS	PES
Melting temperature T_m (°C)		285	
Enthalpy of fusion ΔH_u (J/mol) (mu)			
Entropy of fusion ΔS_u (J/(K mol)) (mu)			
Heat capacity c_p (kJ/(kg K))	1.30		1.10
Temperature coefficient dc_p/dT (kJ/(kg K²))			
Enthalpy of combustion ΔH_c (kJ/g)	17		36
Glass transition temperature T_g (°C)	190	85	225
Vicat softening temperature T_V 50/50 (°C)	175–210	200	215–225
Thermal conductivity λ (W/(m K))	0.28	0.25	0.18
Density ϱ (g/cm³)	1.24–1.25	1.34	1.36–1.37
Coefficient of thermal expansion α (10^{-5}/K) (linear)	5.5–6	5.5	5–5.5
Compressibility κ (1/MPa) (cubic)			
Elastic modulus E (GPa)	2.5–2.7	3.4	2.6–2.8
Shear modulus G (GPa)			
Poisson's ratio μ			
Stress at yield σ_y (MPa)	70–80		80–90
Strain at yield ε_y (%)	5.5–6		5.5–6.5
Stress at 50% elongation σ_{50} (MPa)			
Stress at fracture σ_b (MPa)	50–100	75	85
Strain at fracture ε_b (%)	25–30	3	30–80
Impact strength (Charpy) (kJ/m²)			
Notched impact strength (Charpy) (kJ/m²)			
Sound velocity v_s (m/s) (longitudinal)	2260		2260
Sound velocity v_s (m/s) (transverse)	920		
Shore hardness D			
Volume resistivity ρ_e (Ω m)	$> 10^{13}$	$> 10^{13}$	10^{13}
Surface resistivity σ_e (Ω)	$> 10^{15}$		$> 10^{13}$
Electric strength E_B (kV/mm)	20–30	59.5	20–30
Relative permittivity ε_r (100 Hz)	3.2	3.1	3.5–3.7
Dielectric loss tan δ (10^{-4}) (100 Hz)	8–10	4	10–20
Refractive index n_D (589 nm)	1.63		1.65
Temperature coefficient dn_D/dT (1/K)			
Steam permeation (g/(m² d)) (25 µm)	6		
Gas permeation (cm³/(m² d bar)) (25 µm)	630 (N_2) 3600 (O_2) 15 000 (CO_2)		
Melt viscosity–molar-mass relation			
Viscosity–molar-mass relation			

3.3.3.8 Polyimides and Polyether Ketones

Poly(amide imide), PAI. Applications: constructional parts, wheels, rotors, bearings, housings, slip fittings, lacquers.

Poly(ether imide), PEI. Applications: electrical housings, sockets, microwave parts, bearings, gear wheels, automotive parts.

Polyimide, PI. Applications: foils, semifinished goods, sintered parts.

Table 3.3-22 Poly(amide imide), PAI; poly(ether imide), PEI; polyimide, PI; poly(ether ether ketone), PEEK

	PAI	PEI	PI	PEEK
Melting temperature T_m (°C)			335–345	
Enthalpy of fusion ΔH_u (J/mol) (mu)				
Entropy of fusion ΔS_u (J/(K mol)) (mu)				
Heat capacity c_p (kJ/(kg K))		1.1		
Temperature coefficient dc_p/dT (kJ/(kg K^2))				
Enthalpy of combustion ΔH_c (kJ/g)				
Glass transition temperature T_g (°C)	240–275	215	250–270	145
Vicat softening temperature T_V 50/50 (°C)		220	260	
Thermal conductivity λ (W/(m K))	0.26	0.22	0.29–0.35	0.25
Density ϱ (g/cm^3)	1.38–1.40	1.27	1.43	1.32
Coefficient of thermal expansion α (10^{-5}/K) (linear)	3.0–3.5	5.5–6.0	5–6	4.7
Compressibility κ (1/MPa) (cubic)				
Elastic modulus E (GPa)	4.5–4.7	2.9–3.0	3.0–3.2	4.7
Shear modulus G (GPa)				
Poisson's ratio μ			0.41	
Stress at yield σ_y (MPa)				
Strain at yield ε_y (%)				5
Stress at 50% elongation σ_{50} (MPa)				
Stress at fracture σ_b (MPa)	150–160	105	75–100	100
Strain at fracture ε_b (%)	7–8	60	88	50
Impact strength (Charpy) (kJ/m^2)				
Notched impact strength (Charpy) (kJ/m^2)				
Sound velocity v_s (m/s) (longitudinal)				
Sound velocity v_s (m/s) (transverse)				
Shore hardness D				
Volume resistivity ρ_e (Ω m)	10^{15}	10^{16}	> 10^{14}	10^{14}
Surface resistivity σ_e (Ω)	> 10^{16}	> 10^{16}	> 10^{15}	
Electric strength E_B (kV/mm)	25	25	200	
Relative permittivity ε_r (100 Hz)	3.5–4.2	3.2–3.5	3.4	3.2
Dielectric loss tan δ (10^{-4}) (100 Hz)	10	10–15	52	30
Refractive index n_D (589 nm)		1.66		
Temperature coefficient dn_D/dT (1/K)				
Steam permeation (g/(m^2 d)) (25 µm)			25	
Gas permeation (cm^3/(m^2 d bar)) (25 µm)			94 (N$_2$) 390 (O$_2$) 700 (CO$_2$)	
Melt viscosity–molar-mass relation				
Viscosity–molar-mass relation				

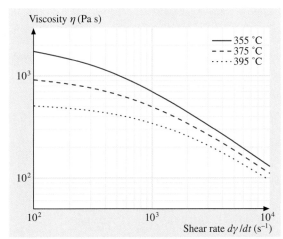

Fig. 3.3-59 Poly(ether imide), PEI: viscosity versus shear rate

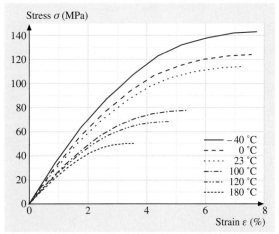

Fig. 3.3-60 Poly(ether imide), PEI: stress versus strain

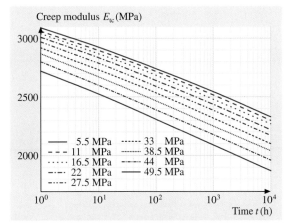

Fig. 3.3-61 Poly(ether imide), PEI: creep modulus versus time

Poly(ether ether ketone), PEEK. Applications: injection-molded parts, automotive parts, aircraft parts, electronic parts, cable insulation, foils, fibers, tapes, plates.

3.3.3.9 Cellulose

Cellulose acetate, CA. Applications: tool handles, pens, combs, buckles, buttons, electronic parts, spectacles, hollow fibers, fibers, foils, lacquers, resin adhesives, sheet molding compounds (SMCs), bulk molding compounds (BMCs).

Cellulose propionate, CP. Applications: fibers, foils, lacquers, resin adhesives, sheet molding compounds, bulk molding compounds (BMC), spectacles, houseware goods, boxes.

Table 3.3-23 Cellulose acetate, CA; cellulose propionate, CP; cellulose acetobutyrate, CAB; ethylcellulose, EC; vulcanized fiber, VF

	CA	CP	CAB	EC	VF
Melting temperature T_m (°C)					
Enthalpy of fusion ΔH_u (J/mol) (mu)					
Entropy of fusion ΔS_u (J/(K mol)) (mu)					
Heat capacity c_p (kJ/(kg K))	1.6	1.7	1.6		
Temperature coefficient dc_p/dT (kJ/(kg K^2))					
Enthalpy of combustion ΔH_c (kJ/g)					
Glass transition temperature T_g (°C)					
Vicat softening temperature $T_V 50/50$ (°C)	77–110	70–100	65–100		

Table 3.3-23 Cellulose acetate, CA; cellulose propionate, CP; cellulose acetobutyrate, CAB; ..., cont.

	CA	CP	CAB	EC	VF
Thermal conductivity λ (W/(m K))	0.22	0.21	0.21		
Density ϱ (g/cm^3)	1.26–1.32	1.17–1.24	1.16–1.22	1.12–1.15	1.1–1.45
Coefficient of thermal expansion α (10^{-5}/K) (linear)	10–12	11–15	10–15	10	
Compressibility κ (1/MPa) (cubic)					
Elastic modulus E (GPa)	1.0–3.0	1.0–2.4	0.8–2.3	1.2–1.3	
Shear modulus G (GPa)		0.75	0.85		
Poisson's ratio μ					
Stress at yield σ_y (MPa)	25–55	20–50	20–55	35–40	
Strain at yield ε_y (%)	2.5–4	3.5–4.5	3.5–5		
Stress at 50% elongation σ_{50} (MPa)					
Stress at fracture σ_b (MPa)	38	14–55	26		85–100
Strain at fracture ε_b (%)	3	30–100	4		
Impact strength (Charpy) (kJ/m^2)					
Notched impact strength (Charpy) (kJ/m^2)					
Sound velocity v_s (m/s) (longitudinal)					
Sound velocity v_s (m/s) (transverse)					
Shore hardness D					
Volume resistivity ρ_e (Ω m)	10^{10}–10^{14}	10^{10}–10^{14}	10^{10}–10^{14}	10^{11}–10^{13}	
Surface resistivity σ_e (Ω)	10^{10}–10^{14}	10^{12}–10^{14}	10^{12}–10^{14}	10^{11}–10^{13}	
Electric strength E_B (kV/mm)	25–35	30–35	32–35	≈ 30	
Relative permittivity ε_r (100 Hz)	5–6	4.0–4.2	3.7–4.2	≈ 4	
Dielectric loss tan δ (10^{-4}) (100 Hz)	70–100	50	50–70	100	
Refractive index n_D (589 nm)	1.47–1.50	1.47–1.48	1.48		
Temperature coefficient dn_D/dT (1/K)					
Steam permeation (g/(m^2 d)) (25 µm)	150–600		460–600		
Gas permeation (cm^3/(m^2 d bar)) (25 µm)	470–630 (N$_2$) 13 000–15 000 (O$_2$) 14 000 (CO$_2$) 1800–2300 (air)		3800 (N$_2$) 15 000 (O$_2$) 94 000 (CO$_2$)		
Melt viscosity–molar-mass relation					
Viscosity–molar-mass relation					

Cellulose acetobutyrate, CAB. Applications: fibers, foils, lacquers, resin adhesives, sheet molding compounds, bulk molding compounds, automotive parts, switches, light housings, spectacles.

Ethylcellulose, EC. Applications: foils, injection-molded parts, lacquers, adhesives.

Vulcanized fiber, VF. Applications: gear wheels, abrasive wheels, case plates.

3.3.3.10 Polyurethanes

Polyurethane, PUR; thermoplastic polyurethane elastomer, TPU. Applications: blocks, plates, formed pieces, cavity forming, composites, coatings, impregnations, films, foils, fibers, molded foam plastics, insulation boards, sandwich boards, installation boards, reaction injection molding (RIM), automotive parts, houseware goods, sports goods.

Table 3.3-24 Polyurethane, PUR; thermoplastic polyurethane elastomer, TPU

	PUR	TPU
Melting temperature T_m (°C)		
Enthalpy of fusion ΔH_u (J/mol) (mu)		
Entropy of fusion ΔS_u (J/(K mol)) (mu)		
Heat capacity c_p (kJ/(kg K))	1.76	0.5
Temperature coefficient dc_p/dT (kJ/(kg K^2))		
Enthalpy of combustion ΔH_c (kJ/mol) (mu)		
Glass transition temperature T_g (K)	15–90	−40
Vicat softening temperature T_V 50/50 (°C)	100–180	
Thermal conductivity λ (W/(m K))	0.58	1.70
Density ϱ (g/cm^3)	1.05	1.1–1.25
Coefficient of thermal expansion α (10^{-5}/K) (linear)	1.0–2.0	15
Compressibility κ (1/MPa) (cubic)		
Elastic modulus E (GPa)	4.0	0.015–0.7
Shear modulus G (GPa)		0.006–0.23
Poisson's ratio μ		
Stress at yield σ_y (MPa)		
Strain at yield ε_y (%)		
Stress at 50% elongation σ_{50} (MPa)		
Stress at fracture σ_b (MPa)	70–80	30–50
Strain at fracture ε_b (%)	3–6	300–500
Impact strength (Charpy) (kJ/m^2)		
Notched impact strength (Charpy) (kJ/m^2)		
Sound velocity v_s (m/s) (longitudinal)	1550–1750	
Sound velocity v_s (m/s) (transverse)		
Shore hardness D	60–90	30–70
Volume resistivity ρ_e (Ω m)	10^{14}	10^{10}
Surface resistivity σ_e (Ω)	10^{14}	10^{11}
Electric strength E_B (kV/mm)	24	30–60
Relative permittivity ε_r (100 Hz)	3.6	6.5
Dielectric loss $\tan\delta$ (10^{-4}) (100 Hz)	500	300
Refractive index n_D (589 nm)		
Temperature coefficient dn_D/dT (1/K)		
Steam permeation (g/(m^2 d)) (25 μm)		13–25
Gas permeation (cm^3/(m^2 d bar)) (25 μm)		550–1600 (N$_2$) 1000–4500 (O$_2$) 6000–22 000 (CO$_2$)
Melt viscosity–molar-mass relation		
Viscosity–molar-mass relation		

Table 3.3-25 Polyurethane, PUR: thermal conductivity and coefficient of thermal expansion

Temperature T (°C)	−200	−150	−100	−50	0	20	50	100
Thermal conductivity λ (W/(m K))		0.20	0.21	0.22	0.21		0.20	0.20
Average coefficient of thermal expansion α^* (10^{-5}/K) (linear)[a]	9.9	12.7	16.0	26.0	20.0			

[a] $\alpha^*(T) = (1/(T-20)) \int_{T=20}^{T'} \alpha(T)\, dT,\quad T$ in °C

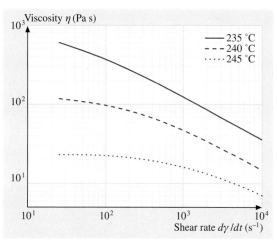

Fig. 3.3-62 Thermoplastic polyurethane elastomer, TPU: viscosity versus shear rate

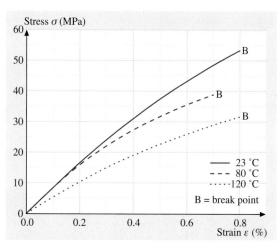

Fig. 3.3-63 Phenol formaldehyde, PF: stress versus strain

3.3.3.11 Thermosets

Phenol formaldehyde, PF; urea formaldehyde, UF; melamine formaldehyde, MF. Applications: molded pieces, granulated molding compounds (GMCs), bulk molding compounds, dough molding compounds (DMCs), sheet molding compounds, plates, gear wheels, abrasive wheels, lacquers, coatings, chipboards, adhesives, foams, fibers.

Unsaturated polyester, UP; diallyl phthalate, DAP. Applications: molding compounds, cast resins, electronic parts, automotive parts, containers, tools, houseware goods, sports goods.

Table 3.3-26 Phenol formaldehyde, PF; urea formaldehyde, UF; melamine formaldehyde, MF

	PF	UF	MF
Melting temperature T_m (°C)			
Enthalpy of fusion ΔH_u (J/mol) (mu)			
Entropy of fusion ΔS_u (J/(K mol)) (mu)			
Heat capacity c_p (kJ/(kg K))	1.30	1.20	1.20
Temperature coefficient dc_p/dT (kJ/(kg K^2))			
Enthalpy of combustion ΔH_c (kJ/g)			
Glass transition temperature T_g (°C)			
Vicat softening temperature $T_V 50/50$ (°C)			
Thermal conductivity λ (W/(m K))	0.35	0.40	0.50
Density ϱ (g/cm^3)	1.4	1.5	1.5
Coefficient of thermal expansion α (10^{-5}/K) (linear)	3–5	5–6	5–6

Table 3.3-26 Phenol formaldehyde, PF; urea formaldehyde, UF; melamine formaldehyde, MF, cont.

	PF	UF	MF
Compressibility κ (1/MPa) (cubic)			
Elastic modulus E (GPa)	5.6–12	7.0–10.5	4.9–9.1
Shear modulus G (GPa)			
Poisson's ratio μ			
Stress at yield σ_y (MPa)			
Strain at yield ε_y (%)			
Stress at 50% elongation σ_{50} (MPa)			
Stress at fracture σ_b (MPa)	25	30	30
Strain at fracture ε_b (%)	0.4–0.8	0.5–1.0	0.6–0.9
Impact strength (Charpy) (kJ/m^2)	4.5–10		
Notched impact strength (Charpy) (kJ/m^2)	1.5–3		
Sound velocity v_s (m/s) (longitudinal)			
Sound velocity v_s (m/s) (transverse)			
Shore hardness D	82		90
Volume resistivity ρ_e (Ω m)	10^9	10^9	10^9
Surface resistivity σ_e (Ω)	$> 10^8$	$> 10^{10}$	$> 10^8$
Electric strength E_B (kV/mm)	30–40	30–40	29–30
Relative permittivity ε_r (100 Hz)	6	8	9
Dielectric loss tan δ (10^{-4}) (100 Hz)	1000	400	600
Refractive index n_D (589 nm)	1.63		
Temperature coefficient dn_D/dT (1/K)			
Steam permeation (g/(m^2 d)) (40 µm)	43		400
Gas permeation (cm^3/(m^2 d bar))			
Melt viscosity–molar-mass relation			
Viscosity–molar-mass relation			

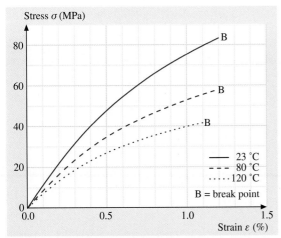

Fig. 3.3-64 Melamine formaldehyde, MF: stress versus strain

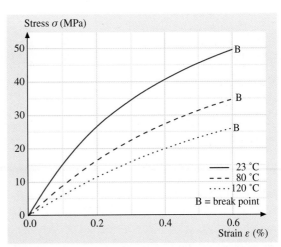

Fig. 3.3-65 Unsaturated polyester, UP: stress versus strain

Table 3.3-27 Unsaturated polyester, UP; diallyl phthalate, DAP; silicone resin, SI; epoxy resin, EP

	UP	DAP	SI	EP
Melting temperature T_m (°C)				
Enthalpy of fusion ΔH_u (J/mol) (mu)				
Entropy of fusion ΔS_u (J/(K mol)) (mu)				
Heat capacity c_p (kJ/(kg K))	1.20		0.8–0.9	0.8–1.0
Temperature coefficient dc_p/dT (kJ/(kg K^2))				
Enthalpy of combustion ΔH_c (kJ/g)				
Glass transition temperature T_g (°C)	70–120			
Vicat softening temperature T_V 50/50 (°C)				
Thermal conductivity λ (W/(m K))	0.7	0.6	0.3–0.4	0.88
Density ϱ (g/cm^3)	2.0	1.51–1.78	1.8–1.9	1.9
Coefficient of thermal expansion α (10^{-5}/K) (linear)	2–4	1–3.5	2–5	1.1–3.5
Compressibility κ (1/MPa) (cubic)				
Elastic modulus E (GPa)	14–20	9.8–15.5	6–12	21.5
Shear modulus G (GPa)				
Poisson's ratio μ				
Stress at yield σ_y (MPa)				
Strain at yield ε_y (%)				
Stress at 50% elongation σ_{50} (MPa)				
Stress at fracture σ_b (MPa)	30	40–75	28–46	30–40
Strain at fracture ε_b (%)	0.6–1.2			4
Impact strength (Charpy) (kJ/m^2)				
Notched impact strength (Charpy) (kJ/m^2)				
Sound velocity v_s (m/s) (longitudinal)				
Sound velocity v_s (m/s) (transverse)				
Shore hardness D	82			
Volume resistivity ρ_e (Ω m)	$> 10^{10}$	10^{11}–10^{14}	10^{12}	$> 10^{12}$
Surface resistivity σ_e (Ω)	$> 10^{10}$	10^{13}	10^{12}	$> 10^{12}$
Electric strength E_B (kV/mm)	25–53	40	20–40	30–40
Relative permittivity ε_r (100 Hz)	6	5.2	4	3.5–5
Dielectric loss $\tan \delta$ (10^{-4}) (100 Hz)	400	400	300	10
Refractive index n_D (589 nm)	1.54–1.58			1.47
Temperature coefficient dn_D/dT (1/K)				
Steam permeation (g/(m^2 d))				
Gas permeation (cm^3/(m^2 d bar))				
Melt viscosity–molar-mass relation				
Viscosity–molar-mass relation				

Table 3.3-28 Unsaturated polyester, UP: coefficient of thermal expansion

Temperature T (°C)	−200	−150	−100	−50	0	20	50	100
Coefficient of thermal expansion α (10^{-5}/K) (linear)	3.0	4.1	4.9	5.8	7.3	8.4	10.7	15.0

Table 3.3-29 Epoxy resin, EP: thermal conductivity and coefficient of thermal expansion

Temperature T (°C)	−200	−150	−100	−50	0	20	50	100	150
Thermal conductivity λ (W/(m K))					0.20	0.20	0.20	0.20	
Coefficient of thermal expansion α (10^{-5}/K) (linear)	1.8	2.8	3.8	4.9	6.1	6.2	6.3	7.5	13.0

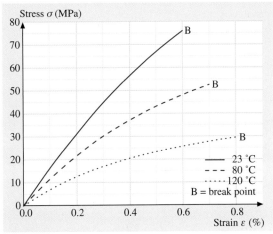

Fig. 3.3-66 Epoxy resin, EP: stress versus strain

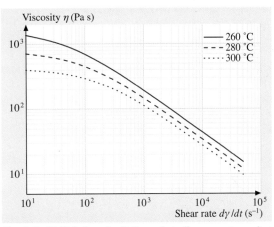

Fig. 3.3-67 Poly(acrylonitrile-co-butadiene-co-styrene) + polycarbonate, ABS + PC: viscosity versus shear rate

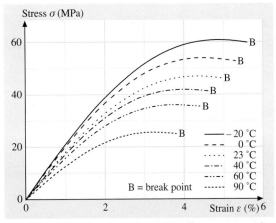

Fig. 3.3-68 Poly(acrylonitrile-co-butadiene-co-styrene) + polycarbonate, ABS + PC: stress versus strain

Silicone resin, SI. Applications: molding compounds, laminates.

Epoxy resin EP, Applications: moulding compounds.

3.3.3.12 Polymer Blends

Polypropylene + ethylene/propylene/diene rubber, PP + EPDM. Applications: automotive parts, flexible tubes, sports goods, toys.

Poly(acrylonitrile-co-butadiene-co-styrene) + polycarbonate, ABS + PC; Poly(acrylonitrile-co-butadiene-co-acrylester) + polycarbonate ASA + PC; poly(acrylonitrile-co-butadiene-co-styrene) + polyamide, ABS + PA. Applications: semifinished goods, automotive parts, electronic parts, optical parts, houseware goods.

Poly(vinyl chloride) + poly(vinyl chloride-co-acrylate), PVC + VC/A; poly(vinyl chloride) + chlorinated polyethylene, PVC + PE-C, poly(vinyl chloride) + poly(acrylonitrile-co-butadiene-co-acrylester), PVC + ASA. Applications: semifinished goods, foils, plates, profiles, pipes, fittings, gutters, window frames, door frames, panels, housings, bottles, blow molding, disks, blocking layers, fibers, fleeces, nets.

Polycarbonate + poly(ethylene terephthalate), PC + PET; polycarbonate + liquid crystal polymer, PC + LCP; polycarbonate + poly(butylene terephthalate), PC + PBT. Applications: optical devices, covering devices, panes, safety glasses, semifinished goods, houseware goods, compact discs, bottles.

Poly(ethylene terephthalate) + polystyrene, PET + PS; poly(butylene terephthalate) + polystyrene, PBT + PS. Applications: same as for PET, PBT, and PS.

Table 3.3-30 Polypropylene + ethylene/propylene/diene rubber, PP + EPDM; poly(acrylonitrile-co-butadiene-co-styrene) + polycarbonate, ABS + PC; poly(acrylonitrile-co-butadiene-co-styrene) + polyamide, ABS + PA; poly(acrylonitrile-co-butadiene-co-acrylester) + polycarbonate, ASA + PC

	PP + EPDM	ABS + PC	ABS + PA	ASA + PC
Melting temperature T_m (°C)	160–168			
Enthalpy of fusion ΔH_u (J/mol) (mu)				
Entropy of fusion ΔS_u (J/(K mol)) (mu)				
Heat capacity c_p (kJ/(kg K))				
Temperature coefficient dc_p/dT (kJ/(kg K^2))				
Enthalpy of combustion ΔH_c (kJ/g)				
Glass transition temperature T_g (°C)				
Vicat softening temperature $T_V 50/50$ (°C)				
Thermal conductivity λ (W/(m K))				
Density ϱ (g/cm^3)	0.89–0.92	1.08–1.17	1.07–1.09	1.15
Coefficient of thermal expansion α (10^{-5}/K) (linear)	15–18	7–8.5	9	7–9
Compressibility κ (1/MPa) (cubic)				
Elastic modulus E (GPa)	0.5–1.2	2.0–2.6	1.2–1.3	2.3–2.6
Shear modulus G (GPa)				
Poisson's ratio μ				
Stress at yield σ_y (MPa)	10–25	40–60	30–32	53–63
Strain at yield ε_y (%)	10–35	3–3.5		4.6–5
Stress at 50% elongation σ_{50} (MPa)				
Stress at fracture σ_b (MPa)				
Strain at fracture ε_b (%)	> 50	> 50	> 50	> 50
Impact strength (Charpy) (kJ/m^2)				
Notched impact strength (Charpy) (kJ/m^2)				
Sound velocity v_s (m/s) (longitudinal)				
Sound velocity v_s (m/s) (transverse)				
Shore hardness D				
Volume resistivity ρ_e (Ω m)	> 10^{14}	> 10^{14}	2 × 10^{12}	10^{11}–10^{13}
Surface resistivity σ_e (Ω)	> 10^{13}	10^{14}	3 × 10^{14}	10^{13}–10^{14}
Electric strength E_B (kV/mm)	35–40	24	30	
Relative permittivity ε_r (100 Hz)	2.3	3		3–3.5
Dielectric loss tan δ (10^{-4}) (100 Hz)	2.5	30–60		20–160
Refractive index n_D (589 nm)				
Temperature coefficient dn_D/dT (1/K)				
Steam permeation (g/(m^2 d))				
Gas permeation (cm^3/(m^2 d bar))				
Melt viscosity–molar-mass relation				
Viscosity–molar-mass relation				

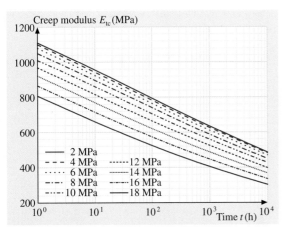

Fig. 3.3-69 Poly(acrylonitrile-co-butadiene-co-styrene) + polycarbonate, ABS + PC: creep modulus versus time

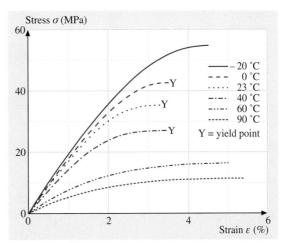

Fig. 3.3-70 Poly(acrylonitrile-co-butadiene-co-styrene) + polyamide, ABS + PA: stress versus strain

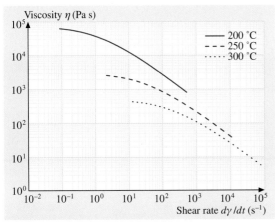

Fig. 3.3-71 Poly(acrylonitrile-co-butadiene-co-acrylester) + polycarbonate, ASA + PC: viscosity versus shear rate

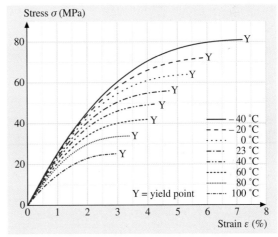

Fig. 3.3-72 Poly(acrylonitrile-co-butadiene-co-acrylester) + polycarbonate, ASA + PC: stress versus strain

Poly(butylene terephthalate) + poly(acrylonitrile-co-butadiene-co-acrylester), PBT + ASA. Applications: same as for PBT and ASA.

Polysulfone + poly(acrylonitrile-co-butadiene-co-styrene), PSU + ABS. Applications: same as for PSU.

Poly(styrene-co-butadiene), PPE + SB. Applications: same as for PPE.

Poly(phenylene ether) + polyamide 66, PPE + PA66. Applications: automotive parts, semifinished goods.

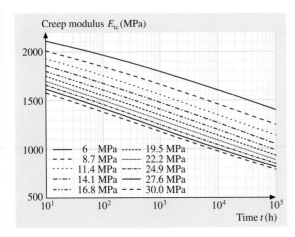

Fig. 3.3-73 Poly(acrylonitrile-co-butadiene-co-acrylester) + polycarbonate, ASA + PC: creep modulus versus time

Table 3.3-31 Poly(vinyl chloride) + poly(vinyl chloride-co-acrylate), PVC + VC/A; poly(vinyl chloride) + chlorinated polyethylene, PVC + PE-C; poly(vinyl chloride) + poly(acrylonitrile-co-butadiene-co-acrylester), PVC + ASA

	PVC + VC/A	PVC + PE-C	PVC + ASA
Melting temperature T_m (°C)			
Enthalpy of fusion ΔH_u (J/mol) (mu)			
Entropy of fusion ΔS_u (J/(K mol)) (mu)			
Heat capacity c_p (kJ/(kg K))			
Temperature coefficient dc_p/dT (kJ/(kg K^2))			
Enthalpy of combustion ΔH_c (kJ/g)			
Glass transition temperature T_g (°C)			
Vicat softening temperature $T_V 50/50$ (°C)			
Thermal conductivity λ (W/(m K))			
Density ϱ (g/cm^3)	1.42–1.44	1.36–1.43	1.28–1.33
Coefficient of thermal expansion α (10^{-5}/K) (linear)	7–7.5	8	7.5–10
Compressibility κ (1/MPa) (cubic)			
Elastic modulus E (GPa)	2.5–2.7	2.6	2.6–2.8
Shear modulus G (GPa)			
Poisson's ratio μ			
Stress at yield σ_y (MPa)	45	40–50	45–55
Strain at yield ε_y (%)	4–5	3	3–3.5
Stress at 50% elongation σ_{50} (MPa)			
Stress at fracture σ_b (MPa)			
Strain at fracture ε_b (%)	> 50	10 to >50	≈ 8
Impact strength (Charpy) (kJ/m^2)			
Notched impact strength (Charpy) (kJ/m^2)			
Sound velocity v_s (m/s) (longitudinal)			
Sound velocity v_s (m/s) (transverse)			
Shore hardness D			
Volume resistivity ρ_e (Ω m)	10^{13}	10^{12} to >10^{13}	10^{12} to >10^{14}
Surface resistivity σ_e (Ω)	> 10^{13}	> 10^{13}	10^{12}–10^{14}
Electric strength E_B (kV/mm)			
Relative permittivity ε_r (100 Hz)	3.5	3.1	3.7–4.3
Dielectric loss tan δ (10^{-4}) (100 Hz)	120	140	100–120
Refractive index n_D (589 nm)			
Temperature coefficient dn_D/dT (1/K)			
Steam permeation (g/(m^2 d))			
Gas permeation (cm^3/(m^2 d bar))			
Melt viscosity–molar-mass relation			
Viscosity–molar-mass relation			

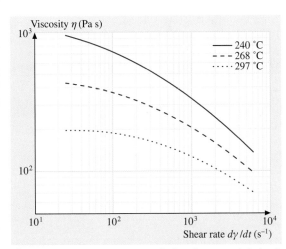

Fig. 3.3-74 Polycarbonate + poly(butylene terephthalate), PC + PBT: viscosity versus shear rate

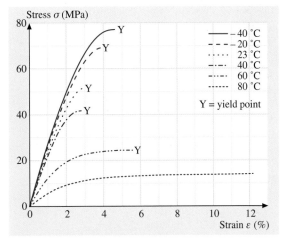

Fig. 3.3-75 Polycarbonate + poly(butylene terephthalate), PC + PBT: stress versus strain

Table 3.3-32 Polycarbonate + poly(ethylene terephthalate), PC + PET; polycarbonate + liquid crystal polymer, PC + LCP; polycarbonate + poly(butylene terephthalate), PC + PBT; poly(ethylene terephthalate) + polystyrene, PET + PS; poly(butylene terephthalate) + polystyrene, PBT + PS

	PC + PET	PC + LCP	PC + PBT	PET + PS	PBT + PS
Melting temperature T_m (°C)					
Enthalpy of fusion ΔH_u (J/mol) (mu)					
Entropy of fusion ΔS_u (J/(K mol)) (mu)					
Heat capacity c_p (kJ/(kg K))				1.05	1.30
Temperature coefficient dc_p/dT (kJ/(kg K^2))					
Enthalpy of combustion ΔH_c (kJ/g)					
Glass transition temperature T_g (°C)					
Vicat softening temperature T_V 50/50 (°C)					
Thermal conductivity λ (W/(m K))				0.24	0.21
Density ϱ (g/cm^3)	1.22		1.2–1.26	1.37	1.31
Coefficient of thermal expansion α (10^{-5}/K) (linear)	9–10		8–9	7	6.0
Compressibility κ (1/MPa) (cubic)					
Elastic modulus E (GPa)	2.1–2.3	2.6–4.0	2.3	3.1	2.0
Shear modulus G (GPa)					
Poisson's ratio μ					
Stress at yield σ_y (MPa)	50–55	66	50–60	47	40
Strain at yield ε_y (%)	5	5.6–6.9	4–5		
Stress at 50% elongation σ_{50} (MPa)					
Stress at fracture σ_b (MPa)		74–82			15
Strain at fracture ε_b (%)	> 50		25 to > 50	50	
Impact strength (Charpy) (kJ/m^2)					
Notched impact strength (Charpy) (kJ/m^2)					
Sound velocity v_s (m/s) (longitudinal)					

Table 3.3-32 Polycarbonate + poly(ethylene terephthalate), PC + PET; polycarbonate + liquid crystal polymer, PC + LCP; polycarbonate + poly(butylene terephthalate), PC + PBT; poly(ethylene terephthalate) + polystyrene, PET + PS; poly(butylene terephthalate) + polystyrene, PBT + PS, cont.

	PC + PET	PC + LCP	PC + PBT	PET + PS	PBT + PS
Sound velocity v_s (m/s) (transverse)					
Shore hardness D					
Volume resistivity ρ_e (Ω m)	$> 10^{13}$		10^{14}	$> 10^{12}$	
Surface resistivity σ_e (Ω)	$> 10^{15}$		10^{14}	$> 10^{12}$	
Electric strength E_B (kV/mm)	30		35		
Relative permittivity ε_r (100 Hz)	3.3		3.3		
Dielectric loss $\tan\delta$ (10^{-4}) (100 Hz)	200		20–40		
Refractive index n_D (589 nm)					
Temperature coefficient dn_D/dT (1/K)					
Steam permeation (g/(m² d))					
Gas permeation (cm³/(m² d bar))					
Melt viscosity–molar-mass relation					
Viscosity–molar-mass relation					

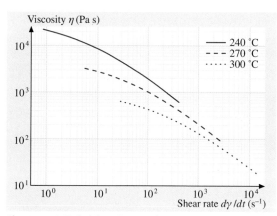

Fig. 3.3-76 Poly(phenylene ether) + poly(styrene-co-butadiene), PPE + SB: viscosity versus shear rate

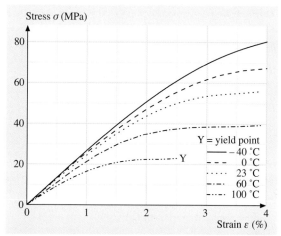

Fig. 3.3-77 Poly(phenylene ether) + poly(styrene-co-butadiene), PPE + SB: stress versus strain

Poly(phenylene ether) + polystyrene, PPE + PS. Applications: automotive parts, foams, office goods, electrical parts, houseware goods.

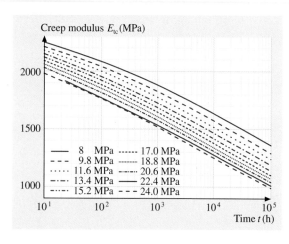

Fig. 3.3-78 Poly(phenylene ether) + poly(styrene-co-butadiene), PPE + SB: creep modulus versus time

Table 3.3-33 Poly(butylene terephthalate) + poly(acrylonitrile-co-butadiene-co-acrylester), PBT + ASA; polysulfon + poly(acrylonitrile-co-butadiene-co-styrene), PSU + ABS; poly(phenylene ether) + poly(styrene-co-butadiene), PPE + SB; poly(phenylene ether) + polyamide 66, PPE + PA66; poly(phenylene ether) + polystyrene PPE + PS

	PBT + ASA	PSU + ABS	PPE + SB	PPE + PA66	PPE + PS
Melting temperature T_m (°C)	225				
Enthalpy of fusion ΔH_u (J/mol) (mu)					
Entropy of fusion ΔS_u (J/(K mol)) (mu)					
Heat capacity c_p (kJ/(kg K))					1.40
Temperature coefficient dc_p/dT (kJ/(kg K^2))					
Enthalpy of combustion ΔH_c (kJ/g)					
Glass transition temperature T_g (°C)					
Vicat softening temperature $T_V 50/50$ (°C)					
Thermal conductivity λ (W/(m K))					0.23
Density ϱ (g/cm^3)	1.21–1.22	1.13	1.04–1.06	1.09–1.10	1.06
Coefficient of thermal expansion α (10^{-5}/K) (linear)	10	6.5	6.0–7.5	8–11	6.0
Compressibility κ (1/MPa) (cubic)					
Elastic modulus E (GPa)	2.5	2.1	1.9–2.7	2.0–2.2	2.5
Shear modulus G (GPa)					
Poisson's ratio μ					
Stress at yield σ_y (MPa)	53	50	45–65	50–60	55
Strain at yield ε_y (%)	3.6	4	3–7	5	
Stress at 50% elongation σ_{50} (MPa)					
Stress at fracture σ_b (MPa)			20 to >50	> 50	
Strain at fracture ε_b (%)	> 50	> 50			50
Impact strength (Charpy) (kJ/m^2)					
Notched impact strength (Charpy) (kJ/m^2)					
Sound velocity v_s (m/s) (longitudinal)					
Sound velocity v_s (m/s) (transverse)					
Shore hardness D					
Volume resistivity ρ_e (Ω m)	> 10^{14}	> 10^{13}	> 10^{14}	> 10^{11}	10^{12}
Surface resistivity σ_e (Ω)	> 10^{15}	> 10^{14}	> 10^{14}	> 10^{12}	10^{12}
Electric strength E_B (kV/mm)	30	20–30	35–40	95	
Relative permittivity ε_r (100 Hz)	3.3	3.1–3.3	2.6–2.8	3.1–3.4	
Dielectric loss tan δ (10^{-4}) (100 Hz)	200	40–50	5–15	450	
Refractive index n_D (589 nm)					
Temperature coefficient dn_D/dT (1/K)					
Steam permeation (g/(m^2 d))					
Gas permeation (cm^3/(m^2 d bar))					
Melt viscosity–molar-mass relation					
Viscosity–molar-mass relation					

References

3.1 M. D. Lechner, K. Gehrke, E. H. Nordmeier: *Makromolekulare Chemie* (Birkhäuser, Basel 2003)

3.2 G. W. Becker, D. Braun (Eds.): *Kunststoff Handbuch* (Hanser, Munich 1969–1990)

3.3 Kunststoffdatenbank CAMPUS, M-Base, Aachen (www.m-base.de)

3.4 J. Brandrup, E. H. Immergut, E. A. Grulke (Eds.): *Polymer Handbook* (Wiley, New York 1999)

3.5 J. E. Mark (Ed.): *Physical Properties of Polymers Handbook* (AIP, Woodbury 1996)

3.6 C. C. Ku, R. Liepins: *Electrical Properties of Polymers* (Hanser, Munich 1987)

3.7 H. Saechtling: *Kunststoff Taschenbuch* (Hanser, Munich 1998)

3.8 W. Hellerich, G. Harsch, S. Haenle: *Werkstoff-Führer Kunststoffe* (Hanser, Munich 1996)

3.9 B. Carlowitz: *Kunststoff Tabellen* (Hanser, Munich 1995)

3.10 G. Allen, J. C. Bevington (Eds.): *Comprehensive Polymer Science* (Pergamon, Oxford 1989)

3.11 H. Domininghaus: *Die Kunststoffe und ihre Eigenschaften* (VDI, Düsseldorf 1992)

3.12 H. J. Arpe (Ed.): *Ullmann's Encyclopedia of Industrial Chemistry*, 5th edn. (VCH, Weinheim 1985–1996)

3.13 R. E. Kirk, D. F. Othmer (Eds.): *Encyclopedia of Chemical Technology*, 4th edn. (Wiley, New York 1978–1984)

3.14 H. F. Mark, N. Bikales, C. G. Overberger, G. Menges, J. I. Kroschwitz: *Encyclopedia of Polymer Science and Engineering* (Wiley, New York 1985–1990)

3.15 Werkstoff-Datenbank POLYMAT, Deutsches Kunststoff-Institut, Darmstadt

3.16 M. Neubronner: Stoffwerte von Kunststoffen. In: *VDI-Wärmeatlas*, 8th edn. (VDI, Düsseldorf 1997)

3.4. Glasses

This chapter has been conceived as a source of information for scientists, engineers, and technicians who need data and commercial-product information to solve their technical task by using glasses as engineering materials. It is not intended to replace the comprehensive scientific literature. The fundamentals are merely sketched, to provide a feeling for the unique behavior of this widely used class of materials.

The properties of glasses are as versatile as their composition. Therefore only a selection of data can be listed, but their are intended to cover the preferred glass types of practical importance. Wherever possible, formulas, for example for the optical and thermal properties, are given with their correct constants, which should enable the reader to calculate the data needed for a specific situation by him/herself.

For selected applications, the suitable glass types and the main instructions for their processing are presented. Owing to the availability of the information, the products of Schott AG have a certain preponderance here. The properties of glass types from other manufacturers have been included whenever available.

3.4.1 **Properties of Glasses – General Comments** 526
3.4.2 **Composition and Properties of Glasses** . 527
3.4.3 **Flat Glass and Hollowware** 528
    3.4.3.1 Flat Glass 528
    3.4.3.2 Container Glass 529
3.4.4 **Technical Specialty Glasses** 530
    3.4.4.1 Chemical Stability of Glasses 530
    3.4.4.2 Mechanical and Thermal Properties 533
    3.4.4.3 Electrical Properties 537
    3.4.4.4 Optical Properties 539
3.4.5 **Optical Glasses** 543
    3.4.5.1 Optical Properties 543
    3.4.5.2 Chemical Properties 549
    3.4.5.3 Mechanical Properties 550
    3.4.5.4 Thermal Properties 556
3.4.6 **Vitreous Silica** 556
    3.4.6.1 Properties of Synthetic Silica 556
    3.4.6.2 Gas Solubility and Molecular Diffusion 557
3.4.7 **Glass-Ceramics** 558
3.4.8 **Glasses for Miscellaneous Applications** . 559
    3.4.8.1 Sealing Glasses 559
    3.4.8.2 Solder and Passivation Glasses .. 562
    3.4.8.3 Colored Glasses 565
    3.4.8.4 Infrared-Transmitting Glasses ... 568
References 572

Glasses are very special materials that are formed only under suitable thermodynamic conditions; these conditions may be natural or man-made. Most glass products manufactured on a commercial scale are made by quenching a mixture of oxides from the melt (Fig. 3.4-1).

For some particular applications, glasses are also made by other technologies, for example by chemical vapor deposition to achieve extreme purity, as required in optical fibers for communication, or by roller chilling in the case of amorphous metals, which need extremely high quenching rates. The term "amorphous" is a more general, generic expression in comparison with the term "glass". Many different technological routes are described in [4.1].

Glasses are also very universal engineering materials. Variation of the composition results in a huge variety of glass types, families, or groups, and a corresponding variety of properties. In large compositional areas, the properties depend continuously on composition, thus allowing one to design a set of properties to fit a specific application. In narrow ranges, the properties depend linearly on composition; in wide ranges, nonlinearity and step-function behavior have to be con-

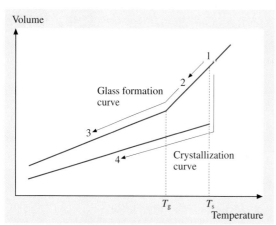

Fig. 3.4-1 Schematic volume–temperature curves for glass formation along path 1–2–3 and crystallization along path 1–4. T_s, melting temperature; T_g, transformation temperature. 1, liquid; 2, supercooled liquid; 3, glass; 4, crystal

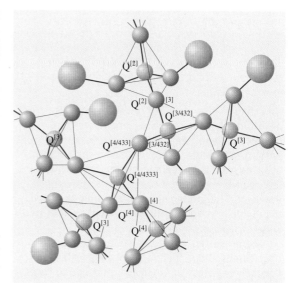

Fig. 3.4-2 Fragment of a sodium silicate glass structure. The SiO_2 tetrahedra are interconnected by the bridging oxygen atoms, thus forming a three-dimensional network. Na_2O units form nonbridging oxygen atoms and act as network modifiers which break the network. The $Q^{[i/klmn]}$ nomenclature describes the connectivity of different atom shells around a selected central atom (after [4.2])

sidered. The most important engineering glasses are mixtures of oxide compounds. For some special requirements, for example a particular optical transmission window or coloration, fluorides, chalcogenides, and colloidal (metal or semiconductor) components are also used.

A very special glass is single-component silica, SiO_2, which is a technological material with extraordinary properties and many important applications.

On a quasi-macroscopic scale (> 100 nm), glasses seem to be homogeneous and isotropic; this means that all structural effects are, by definition, seen only as average properties. This is a consequence of the manufacturing process. If a melt is rapidly cooled down, there is not sufficient time for it to solidify into an ideal, crystalline structure. A structure with a well-defined short-range order (on a scale of less than 0.5 nm, to fulfill the energy-driven bonding requirements of structural elements made up of specific atoms) [4.2] and a highly disturbed long-range order (on a scale of more than 2 nm, disturbed by misconnecting lattice defects and a mixture of different structural elements) (Fig. 3.4-2).

Crystallization is bypassed. We speak of a frozen-in, supercooled, liquid-like structure. This type of quasi-static solid structure is thermodynamically controlled but not in thermal equilibrium and thus is not absolutely stable; it tends to relax and slowly approach an "equilibrium" structure (whatever this may be in a complex multicomponent composition, it represents a minimum of the Gibbs free enthalpy). This also means that all properties change with time and temperature, but in most cases at an extremely low rate which cannot be observed under the conditions of classical applications (in the range of ppm, ppb, or ppt per year at room temperature). However, if the material is exposed to a higher temperature during processing or in the final application, the resulting relaxation may result in unacceptable deformation or internal stresses that then limit its use.

If a glass is reheated, the quasi-solid material softens and transforms into a liquid of medium viscosity in a continuous way. No well-defined melting point exists. The temperature range of softening is called the "transition range". By use of standardized measurement techniques, this imprecise characterization can be replaced by a quasi-materials constant, the "transformation temperature" or "glass temperature" T_g, which depends on the specification of the procedure (Fig. 3.4-3).

As illustrated in Fig. 3.4-4, the continuous variation of viscosity with temperature allows various technologies for hot forming to be applied, for example casting, floating, rolling, blowing, drawing, and pressing, all of which have a specific working point.

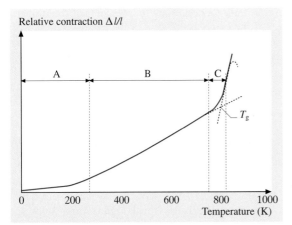

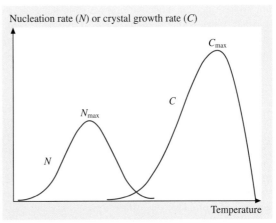

Fig. 3.4-3 Definition of the glass transition temperature T_g by a diagram of length versus temperature during a measurement, as the intersection of the tangents to the elastic region B and to the liquid region to the right of C. In the transformation range C, the glass softens

Fig. 3.4-5 Nucleation rate (N) and crystal growth rate (C) of glass as a function of temperature

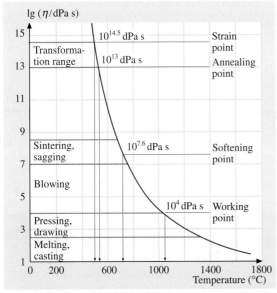

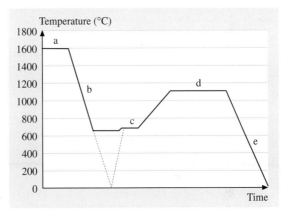

Fig. 3.4-4 Typical viscosity–temperature curve: viscosity ranges for the important processing technologies, and definitions of fixed viscosity points

Fig. 3.4-6 Temperature–time schedule for glass-ceramic production: a, melting; b, working; c, nucleation; d, crystallization; e, cooling to room temperature

The liquid-like situation results in thermodynamically induced density and concentration fluctuations, with a tendency toward phase separation and crystallization, which starts via a nucleation step. This can also be used as a technological route to produce unconventional, i. e. inhomogeneous, glasses. Some examples are colored glasses with colloidal inclusions, porous glasses, and glass-ceramics, to name just a few.

The commercially important glass-ceramics consist of a mixture of 30–90% crystallites (< 50 nm in diameter) and a residual glass phase, without any voids. The material is melted and hot-formed as a glass, cooled down, then annealed for nucleation, and finally tempered to allow crystal growth to a percentage that depends on the desired properties [4.3, 4]. Some characteristic features of the manufacturing process are shown in Figs. 3.4-5 and 3.4-6. As the melt is cooled, a region of high velocity of crystal growth has to be passed through, but no nuclei exist in that region. The nuclei are formed at a much lower temperature, where the crystallization is slow again. Thus the material solidifies as a glass in the first process step. In a second process step, the glass is reheated, annealed in the range of maximum nucleation rate N_{max}, and then annealed at a higher temperature in the range of maximum crystal growth rate C_{max}. The overall bulk properties are the averages of the properties of the components. One of the major properties of commercially important glass-ceramics is a near-zero coefficient of thermal expansion.

3.4.1 Properties of Glasses – General Comments

As for all materials, there are many properties for every type of glass described in the literature. In this section only a limited selection can be given. We also restrict the presentation to commercially important glasses and glass-ceramics. For the huge variety of glasses that have been manufactured for scientific purposes only, the original literature must be consulted. Extensive compilations of data of all kind may be found in books, e.g. [4.5–7], and in software packages [4.8, 9], which are based upon these books and/or additional original data from literature, patents, and information from manufacturers worldwide.

Almost all commercially important glasses are silicate-based. For practical reasons, these glasses are subdivided into five major groups which focus on special properties. These groups are:

1. Mass-production glasses, such as window and container glasses, which are soda–lime–silicate glasses. Besides the application-dependant properties such as transparency, chemical resistance, and mechanical strength, the main concern is cost.
2. Technical specialty glasses, such as display or television glasses, glasses for tubes for pharmaceutical packaging, glasses for industrial ware and labware, glasses for metal-to-glass sealing and soldering, and glasses for glassware for consumers. The dominant properties are chemical inertness and corrosion resistance, electrical insulation, mechanical strength, shielding of X-ray or UV radiation, and others. To fulfill these specifications, the glass composition may be complex and even contain rare components.
3. Optical glasses have the greatest variety of chemical components and do not exclude even the most exotic materials, such as rare earth and non-oxide compounds. The dominant properties are of optical origin: refraction and dispersion with extremely high homogeneity, combined with low absorption and light scattering within an extended transmission range, including the infrared and ultraviolet parts of the electromagnetic spectrum. They are produced in comparably small volumes but with raw materials of high purity, and are thus quite expensive.
4. Vitreous silica, as a single-component material, has some extraordinary properties: high transparency from 160 nm to 1800 nm wavelength, high electrical insulation if it is of low hydrogen content, high chemical and corrosion resistance, and quite low thermal expansion with high thermal-shock resistance. A technological handicap is the high glass temperature $T_g \approx 1250\,°C$, depending on the water content.
5. Glass-ceramics are glassy in the first production step. An important property is the final (after ceramization) coefficient of thermal expansion, often in combination with a high elastic modulus and low specific weight. High thermal stability and thermal-shock resistance are a prerequisite for the major applications. This group has the potential for many other applications which require different property combinations [4.3].

This separation into groups seems to be somewhat artificial, in view of the material properties alone, and is justified only by the very different technological conditions used for manufacturing and processing. It corresponds to the specialization of the industry and to a traditional structuring in the literature.

For all these groups, various classes of product defects exist, and these defects may occur in varying concentration. Among these defects may be bubbles, striations, crystalline inclusions, and metal particles, which are relics of a nonideal manufacturing process. There may also be inclusions of foreign components which were introduced with the raw materials as impurities or contamination. If a three-dimensional volume of glass is cooled down, the finite heat conductivity causes the volume elements to have nonidentical histories in time and temperature. As a consequence, the time- and temperature-dependent relaxation processes produce internal mechanical stress, which results in optical birefringence. These "technical" properties are often very important for the suitability of a piece of glass for a specific application, but they are not "intrinsic" properties of the material, and are not considered in this section.

3.4.2 Composition and Properties of Glasses

A very common method for describing the composition of a glass is to quote the content of each component (oxide) by weight fraction (w_i in wt%) or mole fraction (m_i in mol%). If M_i is the molar mass of the component i, the relationship between the two quantities is given by

$$m_i = \frac{100 w_i}{M_i \sum_{j=1}^{n} w_j / M_j}, \quad 1 \leq i \leq n, \quad (4.1)$$

and

$$w_i = \frac{100 m_i M_i}{\sum_{j=1}^{n} m_j M_j}, \quad 1 \leq i \leq n, \quad (4.2)$$

with the side conditions

$$\sum_{i=1}^{n} m_i = 1 \quad \text{and} \quad \sum_{i=1}^{n} w_i = 1. \quad (4.3)$$

If components other than oxides are used, it is advantageous to define fictitious components, for example the component F_2-O (= fluorine–oxygen, with $M = 2 \times 19 - 16 = 22$); this can be used to replace a fluoride with an oxide and this hypothetical oxide component, for example $CaF_2 = CaO + F_2-O$. This reduces the amount of data required

Table 3.4-1 Factors ϱ_{w_i} for the calculation of glass densities [4.10]

Component i	ϱ_{w_i} (g/cm³)
Na₂O	3.47
MgO	3.38
CaO	5.0
Al₂O₃	2.75
SiO₂	2.20

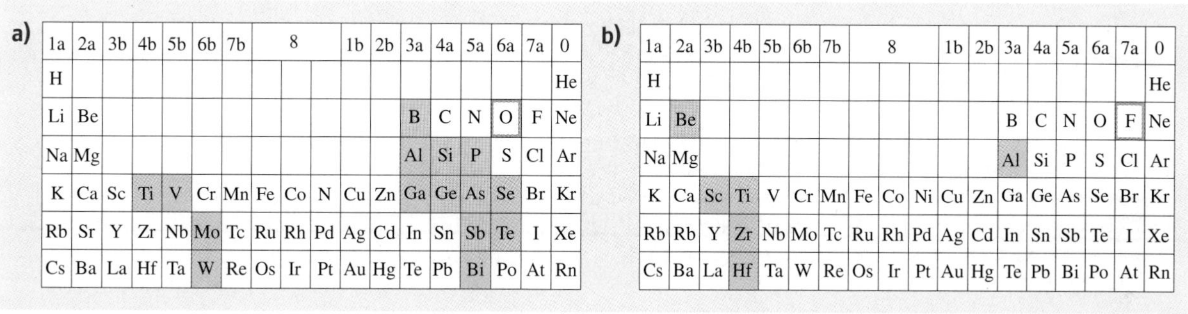

Fig. 3.4-7 (a) Elements whose oxides act as glass-formers (*gray*) and conditional glass-formers (*brown*) in oxide systems (after [4.11]). (b) Elements whose fluorides act as glass-formers (*gray*) and conditional glass-formers (*brown*) in fluoride glass systems (after [4.12])

Table 3.4-2 Composition of selected technical glasses

Glass type	Code[a]	Main components (wt%)					Minor components
		SiO₂	B₂O₃	Al₂O₃	Na₂O	PbO	< 10%
Lead glass	S 8095	57				28	Al₂O₃, Na₂O, K₂O
Low dielectric loss glass	S 8248	70	27				Al₂O₃, Li₂O, Na₂O, K₂O, BaO
Sealing glass	S 8250	69	19				Al₂O₃, Li₂O, Na₂O, K₂O, ZnO
Duran®	S 8330	80	13				Al₂O₃, Na₂O, K₂O
Supremax®	S 8409	52		22			B₂O₃, CaO, MgO, BaO, P₂O₅
Sealing glass	S 8465		11	11	75		SiO₂
Sealing glass	S 8487	75	16.5				Al₂O₃, Na₂O, K₂O
Vycor™	C 7900	96					Al₂O₃, B₂O₃, Na₂O

[a] Code prefix: S = Schott AG, Mainz; C = Corning, New York.

to handle the different cation–anion combinations (Fig. 3.4-7).

With given weight or mole fractions, a property P can be calculated (or approximated) via a linear relation of the type

$$P = \frac{1}{100}\sum_{i=1}^{n} p_{w_i} w_i \quad \text{or} \quad P = \frac{1}{100}\sum_{i=1}^{n} p_{m_i} m_i \,. \tag{4.4}$$

The component-specific factors p_i were first systematically determined by *Winckelmann* and *Schott* [4.13], and are listed in detail in the standard literature, e.g. [4.14].

As an example, the density ϱ can be calculated via (4.4) from: $P = 1/\varrho$, $p_{w_i} = 1/\varrho_{w_i}$, and the data in Table 3.4-1.

The composition of selected technical glasses is given in Table 3.4-2.

3.4.3 Flat Glass and Hollowware

By far the highest glass volume produced is for windows and containers. The base glass is a soda–lime composition, which may be modified for special applications. Owing to the Fe content of the natural raw materials, a green tint is observed for thicknesses that are not too small.

If the product is designed to have a brown or green color (obtained by the use of reducing or oxidizing melting conditions), up to 80% waste glass can be used as a raw material.

3.4.3.1 Flat Glass

Plate glass for windows is mainly produced by floating the melt on a bath of molten tin. All manufacturers worldwide use a soda–lime-type glass, with the average composition given in Table 3.4-3. For some other applications, this base glass may be modified by added coloring agents, for example Cr and Fe oxides.

For architectural applications, the surface(s) may be modified by coatings or grinding to achieve specific

Table 3.4-3 Average composition in wt% and viscosity data, for soda–lime glass, container glass, and Borofloat glass

	Soda–lime glass	Container glass	Borofloat® 33
Composition (wt%)			
SiO_2	71–73	71–75	81
$CaO + MgO$	9.5–13.5	10–15	
$Na_2O + K_2O$	13–16	12–16	4
B_2O_3	0–1.5		13
Al_2O_3	0.5–3.5		2
Other	0–3	0–3	
Temperature (°C)			
at viscosity of $10^{14.5}$ dPa s			518
at viscosity of 10^{13} dPa s	525–545		560
at viscosity of $10^{7.6}$ dPa s	717–735		820
at viscosity of 10^{4} dPa s	1015–1045		1270
Density (g/cm^3)	2.5		2.2
Young's modulus (GPa)	72		64
Poisson's ratio			0.2
Knoop hardness			480
Bending strength (MPa)	30		25
Thermal expansion coefficient α, 20–300 °C (10^{-6}/K)	9.0		3.25
Specific heat capacity c_p, 20–100 °C (kJ/kg K)			0.83

Table 3.4-3 Average composition in wt% and viscosity data for soda–lime glass, cont.

	Soda–lime glass	Container glass	Borofloat®33
Heat conductivity λ at 90 °C (W/m kg)			1.2
Chemical resistance			
Water			HGB1/HGA1
Acid			1
Alkali			A2
Refractive index n_d			1.47140
Abbe value ν_e			65.41
Stress-optical constant K (10^{-6} mm^2/N)			4.0
Dielectric constant ε_r			4.6
Dielectric loss $\tan\delta$ at 25 °C, 1 MHz			37×10^{-4}
Volume resistivity ρ			
at 250 °C (Ω cm)			10^8
at 350 °C (Ω cm)			$10^{6.5}$

optical effects, such as decorative effects, a higher reflectivity in certain spectral regions (e.g. the IR), or an opaque but translucent appearance. Some other applications require antireflection coatings.

If steel wires are embedded as a web into rolled glass, the resulting product can be used as a window glass for areas where improved break-in protection is required.

Fire-protecting glasses are designed to withstand open fire and smoke for well-defined temperature–time programs, for example from 30 min to 180 min with a maximum temperature of more than 1000 °C (test according to DIN 4102, Part 13). The main goal is to avoid the glass breaking under thermal load. There are various ways to achieve this, such as mechanical wire reinforcement, prestressing the plate by thermal or surface-chemical means, or reducing the thermal expansion by using other glass compositions, for example the floated borosilicate glass Pyran® (Schott).

In automotive applications, flat glass is used for windows and mirrors. Single-pane safety glass is produced by generating a compressive stress near to the surface by a special tempering program: a stress of 80–120 MPa is usual. In the case of a breakage the window breaks into small pieces, which greatly reduces the danger of injuries.

Windshields are preferably made of laminated compound glass. Here, two (or more) layers are joined under pressure with a tough plastic foil.

The compound technique is also used for armor-plate glass: here, at least four laminated glass layers with a total thickness of at least 60 mm are used.

For plasma and LCD displays and photovoltaic substrates, extremely smooth, planar, and thin sheets of glass are needed which fit the thermal expansion of the electronic materials in direct contact with them. These panes must remain stable in geometry (no shrinkage) when they are processed to obtain the final product.

3.4.3.2 Container Glass

For packaging, transportation, storage of liquids, chemicals, pharmaceuticals, etc., a high variety of hollowware is produced in the shape of bottles and tubes. Most of the products are made from soda–lime glass (Table 3.4-3) directly from the glass melt. Vials and ampoules are manufactured from tubes; for pharmaceuticals these are made from chemically resistant borosilicate glass.

Clear glass products have to be made from relatively pure raw materials. To protect the contents from light, especially UV radiation, the glass may be colored brown or green by the addition of Fe or Cr compounds. For other colors, see Sect. 3.4.8.3.

3.4.4 Technical Specialty Glasses

Technical glasses are special glasses manufactured in the form of tubes, rods, hollow vessels, and a variety of special shapes, as well as flat glass and granular form. Their main uses are in chemistry, pharmaceutical packaging, electronics, and household appliances. A list of typical applications of different glass types is given in Table 3.4-10, and a list of their properties is given in Table 3.4-11.

The multitude of technical glasses can be roughly arranged in the following four groups, according to their oxide composition (in weight percent). However, certain glass types fall between these groups or even completely outside.

Borosilicate Glasses. The network formers are substantial amounts of silica (SiO_2) and boric oxide (B_2O_3 > 8%). The following subtypes can be differentiated:

- *Non-alkaline-earth borosilicate glass*: with SiO_2 > 80% and B_2O_3 at 12–13%, these glasses show high chemical durability and low thermal expansion (about $3.3 \times 10^{-6}\,\text{K}^{-1}$).
- *Alkaline-earth-containing glasses*: with $SiO_2 \approx$ 75%, B_2O_3 at 8–12%, and alumina (Al_2O_3) and alkaline earths up to 5%, these glasses are softer in terms of viscosity, and have high chemical durability and a thermal expansion in the range of $4.0\text{–}5.0 \times 10^{-6}\,\text{K}^{-1}$.
- *High-borate glasses*: with 65–70% SiO_2, 15–25% B_2O_3, and smaller amounts of Al_2O_3 and alkalis, these glasses have a low softening point and a thermal expansion which is suitable for glass-to-metal seals.

Alkaline-Earth Aluminosilicate Glasses. These glasses are free from alkali oxides, and contain 52–60% SiO_2, 15–25% Al_2O_3, and about 15% alkaline earths. A typical feature is a very high transformation temperature.

Alkali–Lead Silicate Glasses. These glasses contain over 10% lead oxide (PbO). Glasses with 20–30% PbO, 54–58% SiO_2, and about 14% alkalis are highly insulating and provide good X-ray shielding. They are used in cathode-ray tube components.

Alkali–Alkaline-Earth Silicate Glasses (Soda–Lime Glasses). This is the oldest glass type, which is produced in large batches for windows and containers (see Sect. 3.4.3). Such glasses contain about 71% SiO_2, 13–16% alkaline earths $CaO + MgO$, about 15% alkali (usually Na_2O), and 0–2% Al_2O_3.

Variants of the basic composition can contain significant amounts of BaO with reduced alkali and alkaline-earth oxides: these are used for X-ray shielding of cathode-ray tube screens.

3.4.4.1 Chemical Stability of Glasses

Characteristically, glasses are highly resistant to water, salt solutions, acids, and organic substances. In this respect, they are superior to most metals and plastics. Glasses are attacked to a significant degree only by hydrofluoric acid, strongly alkaline solutions, and concentrated phosphoric acid; this occurs particularly at higher temperatures.

Chemical reactions with glass surfaces, induced by exchange, erosion, or adsorption processes, can cause the most diverse effects, ranging from virtually invisible surface modifications to opacity, staining, thin films with interference colors, crystallization, holes, and rough or smooth ablation, to name but a few such effects. These changes are often limited to the glass surface, but in extreme cases they can completely destroy or dissolve the glass. The glass composition, stress medium, and operating conditions will decide to what extent such chemical attacks are technically significant.

Chemical Reaction Mechanisms with Water, Acid, and Alkaline Solutions

Chemical stability is understood as the resistance of a glass surface to chemical attack by defined agents; here, temperature, exposure time, and the condition of the glass surface play important roles.

Every chemical attack on glass involves water or one of its dissociation products, i.e. H^+ or OH^- ions. For this reason, we differentiate between hydrolytic (water), acid, and alkali resistance. In water or acid attack, small amounts of (mostly monovalent or divalent) cations are leached out. In resistant glasses, a very thin layer of silica gel then forms on the surface, which normally inhibits further attack (Fig. 3.4-8a,b). Hydrofluoric acid, alkaline solutions, and in some cases phosphoric acid, however, gradually destroy the silica framework and thus ablate the glass surface in total (Fig. 3.4-8c). In contrast, water-free (i.e. organic) solutions do not react with glass.

Chemical reactions are often increased or decreased by the presence of other components. Alkali attack on glass is thus hindered by certain ions, particularly those

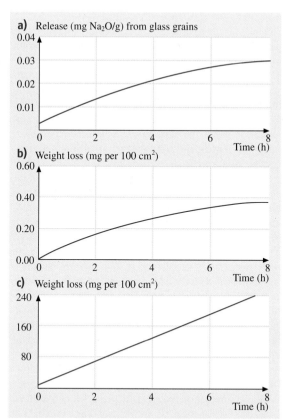

Fig. 3.4-8a–c Attack by (**a**) water, (**b**) acid, and (**c**) alkaline solution on chemically resistant glass as a function of time

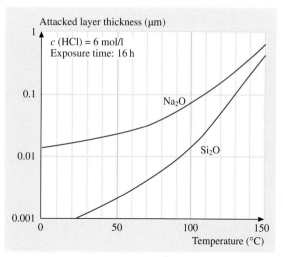

Fig. 3.4-9 Acid attack on Duran® 8330 as a function of temperature, determined from leached amounts of Na_2O and SiO_2

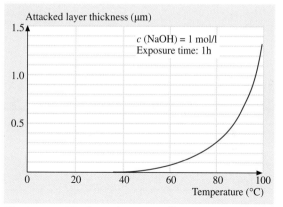

Fig. 3.4-10 Alkali attack on Duran® 8330 as a function of temperature, determined from weight loss

of aluminium. On the other hand, complex-forming compounds such as EDTA, tartaric acid, and citric acid can increase the solubility. In general terms, the glass surface reacts with solutions which induce small-scale exchange reactions and/or adsorption. Such phenomena are observed, for example, in high-vacuum technology when residual gases are removed, and in certain inorganic-chemical operations, when small amounts of adsorbed chromium, resulting from treatment with chromic acid, are removed.

Because acid and alkali attacks on glass are fundamentally different, silica-gel layers produced by acid attack obviously are not necessarily effective against alkali solutions and may be destroyed. Conversely, the presence of ions that inhibit alkali attack does not necessarily represent protection against acids and water. The most severe chemical exposure is therefore an alternating treatment with acids and alkaline solutions. As in all chemical reactions, the intensity of the interaction increases rapidly with increasing temperature (Figs. 3.4-9 and 3.4-10).

In the case of truly ablative solutions such as hydrofluoric acid, alkaline solutions, and hot concentrated phosphoric acid, the rate of attack increases rapidly with increasing concentration (Fig. 3.4-11). As can be seen in Fig. 3.4-12, this is not true for the other frequently used acids.

Determination of the Chemical Stability
In most cases, either is the glass surface is analyzed in its "as delivered" condition (with the original fire-polished surface), or the basic material is analyzed with its fire-

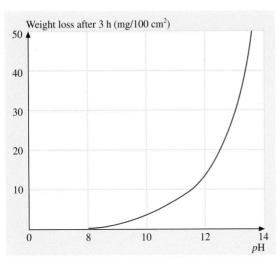

Fig. 3.4-11 Alkali attack on Duran® 8330 at 100 °C as a function of pH value

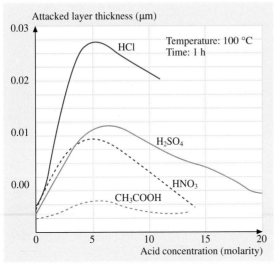

Fig. 3.4-12 Acid attack on Duran® 8330 as a function of concentration

in [4.2, 15] and summarized in Tables 3.4-4–3.4-6. The DIN classes of hydrolytic, acid, and alkali resistance of technical glasses are also listed Table 3.4-11, second page, last three columns.

Significance of the Chemical Stability

Release of Glass Constituents. In various processes in chemical technology, pharmaceutical manufacture, and laboratory work, the glass material used is expected to release no constituents (or a very minimal amnount) into reacting solutions or stored specimens.

Because even highly resistant materials such as non-alkaline-earth and alkaline-earth borosilicate glasses do react to a very small degree with the surrounding media, the fulfillment of this requirement is a question of quantity and of detection limits. Concentrations of 10^{-6}–10^{-9} (i.e. trace amounts), which are measurable today with highly sophisticated analytical instruments, can be released even from borosilicate glasses in the form of SiO_2, B_2O_3, and Na_2O, depending on the conditions. However, solutions in contact with high-grade colorless Duran® laboratory glass will not be contaminated by Fe, Cr, Mn, Zn, Pb, or other heavy-metal ions.

Undesirable Glass Surface Modifications. When an appreciable interaction between a glass surface and an aqueous solution occurs, there is an ion exchange in which the easily soluble glass components are replaced by H^+ or OH^- ions. This depletion of certain glass components in the surface leads to a corresponding enrichment in silica, which is poorly soluble, and thus to the formation of a silica-gel layer. This layer proves, in most cases, to be more resistant than the base glass. When its thickness exceeds about 0.1–0.2 μm, interference colors caused by the different refractive indices of the layer and the base glass make this silica-gel layer visible to the unaided eye. With increasing layer thickness it becomes opaque and finally peels off, destroying the glass. Between these stages there is a wide range of possible surface modifications, some of which, although optically visible, are of no practical significance, whereas others must be considered.

In the case of less resistant glasses, small amounts of water (from air humidity and condensation) in the presence of other agents such as carbon dioxide or sulfur oxides can lead to surface damage. In the case of sensitive glasses, hand perspiration or impurities left by detergents can sometimes induce strongly adhering surface defects, mostly recognizable as stains. If a contaminated glass surface is reheated (> 350–400 °C), the

polished surface removed by mechanical or chemical ablation, or after crushing.

The standardized DIN (Deutsches Institut für Normung, German Institute for Standardization) test methods, which are universally and easily applicable, are the most reliable analysis methods. They include the determination of hydrolytic resistance (by two grain-titration methods and one surface method), of acid resistance to hydrochloric acid, and of alkali resistance to a mixture of alkaline solutions. Details are described

Table 3.4-4 Hydrolytic classes according to DIN 12111 (ISO 719)

Hydrolytic class	Acid consumption of 0.01 mol/l hydrolytic acid per g glass grains (ml/g)	Base equivalent as Na_2O per g glass grains (μg/g)	Possible designation
1	Up to 0.10	Up to 31	Very high resistance
2	Above 0.10, up to 0.20	Above 31, up to 62	High resistance
3	Above 0.20, up to 0.85	Above 62, up to 264	Medium resistance
4	Above 0.85, up to 2.0	Above 264, up to 620	Low resistance
5	Above 2.0, up to 3.5	Above 620, up to 1085	Very low resistance

Table 3.4-5 Acid classes according to DIN 12116

Acid class	Designation	Half loss in weight after 6 h (mg/100 cm^2)
1	High acid resistance	Up to 0.7
2	Good acid resistance	Above 0.7, up to 1.5
3	Medium acid attack	Above 1.5, up to 15
4	High acid attack	Above 15

Table 3.4-6 Alkali classes according to DIN ISO 695

Alkali class	Designation	Loss in weight after 3 h (mg/100 cm^2)
1	Low alkali attack	Up to 75
2	Medium alkali attack	Above 75, up to 175
3	High alkali attack	Above 175

contaminants or some of their components may burn in. Normal cleaning processes will then be ineffective and the whole surface layer has to be removed (e.g. by etching).

Desirable Chemical Reactions with the Glass Surface (Cleaning and Etching). Very strong reactions between aqueous agents and glass can be used for the thorough cleaning of glass. The complete ablation of a glass layer leads to the formation of a new surface.

Hydrofluoric acid reacts most strongly with glass. Because it forms poorly soluble fluorides with a great number of glass constituents, it is mostly used only in diluted form. The best etching effect is usually achieved when another acid (e.g. hydrochloric or nitric acid) is added. A mixture of seven parts by volume of water, two parts of concentrated hydrochloric acid ($c = 38\%$) and one part of hydrofluoric acid ($c = 40\%$) is recommended for a moderate surface ablation of highly resistant borosilicate glasses. When chemically less resistant glasses (e.g. Schott 8245 and 8250) are exposed for 5 min to a stirred solution at room temperature, a surface layer with a thickness of $1-10\,\mu$m is ablated, and a transparent, smooth, completely new surface is produced.

Glasses can also be ablated with alkaline solutions, but the alkaline etching process is much less effective.

3.4.4.2 Mechanical and Thermal Properties

Viscosity

As described earlier, the viscosity of glasses increases by 15–20 orders of magnitude during cooling. Within this viscosity range, glasses are subject to three different thermodynamic states:

- the melting range, above the liquidus temperature T_s;
- the range of the supercooled melt, between the liquidus temperature T_s and the transformation temperature T_g, which is defined by ISO 7884-8;
- the frozen-in, quasi-solid melt range ("glass range"), below the transformation temperature T_g.

The absence of any significant crystallization in the range of the supercooled melt (see Fig. 3.4-1, line segment 2) is of the utmost importance for glass formation. Hence a basically steady, smooth variation in the viscosity in all temperature regions is a fundamental characteristic of glasses (Fig. 3.4-4) and a crucial property for glass production. Figure 3.4-13 shows the strongly differing temperature dependences of the viscosity for some glasses. The best mathematical expression for practical purposes is the VFT (Vogel, Fulcher, and Tammann) equation,

$$\log \eta(T) = A + B/(T - T_0), \qquad (4.5)$$

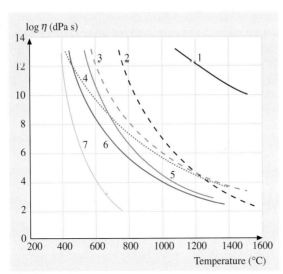

Fig. 3.4-13 Viscosity–temperature curves for some important technical glasses. 1, fused silica; 2, 8405; 3, 8330; 4, 8248; 5, 8350; 6, 8095; 7, 8465. Glasses with steep gradients (such as 7) are called "short" glasses, and those with relatively shallow gradients (such as 4) are called "long" glasses

where A, B, and T_0 are glass-specific constants (Table 3.4-7).

Somewhat above 10^{10} dPa s, the viscosity becomes increasingly time-dependent. With increasing viscosity (i.e. with decreasing temperature), the delay in establishing structural equilibrium finally becomes so large that, under normal cooling conditions, the glass structure at 10^{13} dPa s can be described as solidified or "frozen-in". This temperature (for which a method of measurement is specified by ISO 7884-4) is called the "annealing point". At this viscosity, internal stresses in the glass are released after ≈ 15 min annealing time, while the dimensional stability of the glass is sufficient for many purposes, and its brittleness (susceptibility to cracking) is almost fully developed.

The lower limit of the annealing range is indicated by the "strain point", at which the glass has a viscosity of $10^{14.5}$ dPa s (determined by extrapolation from the viscosity–temperature curve). For most glasses, the strain point lies about 30–40 K below the annealing point. Relaxation of internal stresses here takes 3–5 h. The strain point marks the maximum value for short-term heat load. Thermally prestressed glasses, in contrast, show significant stress relaxation even at 200–300 K below T_g. For glass objects with precisely defined dimensions (e.g. etalons and gauge blocks) and in the case of extreme demands on the stability of certain properties of the glass, application temperatures of 100–200 °C can be the upper limit.

Strength

The high structural (theoretical) strength of glasses and glass-ceramics ($> 10^4$ N/mm² = 10 GPa) is without practical significance, because the strength of glass articles is determined by surface defects induced by wear, such as tiny chips and cracks (Griffith flaws), at whose tips critical stress concentrations may be induced by a mechanical load, especially if the load is applied perpendicular to the plane of the flaw (fracture mode I). Glasses and glass-ceramics, in contrast to ductile materials such as metals, show no plastic flow and behave under a tensile stress σ in as brittle a manner as ceramics. A flaw will result in a fracture if the "stress intensity factor"

$$K_I = 2\sigma\sqrt{a} > K_{Ic}, \qquad (4.6)$$

where a is the depth of the flaw and K_{Ic} is the "critical stress intensity factor", a material constant which is temperature- and humidity-dependent: see Table 3.4-8.

For $K_{Ic} = 1$ MPa $\sqrt{m}$ and a stress $\sigma = 50$ MPa, the critical flaw depth a_c is 100 μm. Thus very small flaws

Table 3.4-7 Parameters of the VFT equation (4.5) for the glasses in Fig. 3.4-13

Glass	A	B (°C)	T_0 (°C)
8095	−1.5384	4920.68	96.54
8248	−0.2453	4810.78	126.79
8330	−1.8500	6756.75	105.00
8350	−1.5401	4523.85	218.10
8405	−2.3000	5890.50	65.00
8465	−1.6250	1873.73	256.88
Fused silica	−7.9250	31 282.9	−415.00
Soda–lime glass	−1.97	4912.5	475.4

can cause cracking at a comparatively low stress level, and the practical strength of a glass is not a materials constant!

Surface Condition. As a result of wear-induced surface defects, glass and glass-ceramic articles have practical tensile strengths of $20\text{--}200\,\text{N/mm}^2 = 20\text{--}200\,\text{MPa}$, depending on the surface condition and the atmospheric-exposure condition. To characterize the strength, a Weibull distribution for the cumulative failure probability F is assumed:

$$F(\sigma) = 1 - \exp\left[-(\sigma/\sigma_c)^m\right], \qquad (4.7)$$

where $0 \leq F(\sigma) \leq 1$ is the probability of a fracture if the applied stress is less than σ; σ_c denotes the characteristic value (approximately the mean value of the distribution), and m is the Weibull modulus of the distribution (which determines the standard deviation). To obtain reproducible measurements, the surface is predamaged by grinding with a narrow grain-size distribution (Fig. 3.4-14).

Only a slight – as a rule with negligible – dependence on the chemical composition is found for silicate glasses (Table 3.4-8).

Stress Rate. The rate of increase of the stress and the size of the glass area exposed to the maximum stress have to be considered for the specification of a strength value. In contrast to the rapid stress increase occurring in an impact, for example, a slowly increasing tensile stress or continuous stress above a certain critical limit may – as a result of stress corrosion cracking – cause the propagation of critical surface flaws and cracks and thus enhance their effect. Hence the tensile strength is time- and stress-rate-dependent (this is mainly important for test loads), as shown in Fig. 3.4-15. Independent of surface damage or the initial tensile strength, increasing

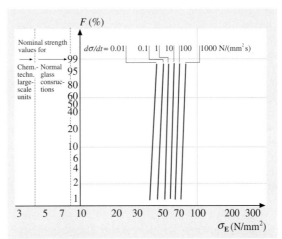

Fig. 3.4-15 Failure probability F of a predamaged surface ($100\,\text{mm}^2$; grain size 600) for various rates of increase of stress $d\sigma/dt$. A, range of nominal strength for large-scale units in chemical technology; B, range of nominal strength for normal glass structures

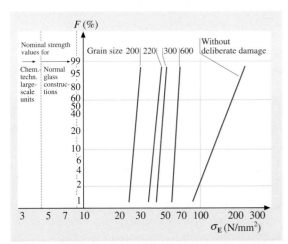

Fig. 3.4-14 Failure probability F for samples abraded by variously sized grains. Predamaged surface area $100\,\text{mm}^2$, rate of stress increase $d\sigma/dt = 10\,\text{MPa}\,\text{s}^{-1}$. A, range of nominal strength for large-scale units in chemical technology; B, range of nominal strength for normal glass structures

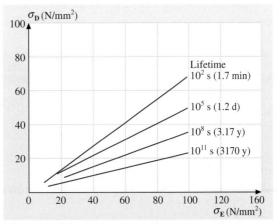

Fig. 3.4-16 Time-related strength σ_D (strength under constant loading) of soda–lime glass compared with the experimental strength σ_E at $d\sigma/dt = 10\,\text{N/mm}^2\,\text{s}$ for various lifetimes, in a normal humid atmosphere

the rate of increase of the stress by a factor of 10 results in an increase in the strength level of about 15%.

Constant Loading. Fracture analysis of the effect and behavior of cracks in glasses and glass-ceramics yield further information about the relationship between the experimentally determined tensile strength σ_E (usually measured for a rapidly increasing load) and the tensile strength σ_D expected under constant loading (= fatigue strength), as shown in Fig. 3.4-16. Such analyses show that, depending on the glass type, the tensile strength under constant loading σ_D (for years of loading) may amount to only about 1/2 to 1/3 of the experimental tensile strength σ_E.

Area Dependence. The larger the stressed area, the higher is the probability of large defects (large crack depths) within this area. This relationship is important for the transfer of experimental tensile strengths, which are mostly determined with relatively small test samples, to practical glass applications such as pipelines, where many square meters of glass can be uniformly stressed (Fig. 3.4-17).

Elasticity

The ideal brittleness of glasses and glass-ceramics is matched by an equally ideal elastic behavior up to breaking point. The elastic moduli for most technical glasses lie within a range of 50–90 kN/mm². The mean value of 70 kN/mm² is about equal to the Young's modulus of aluminium (see Table 3.4-11, first page, column 7).

Coefficient of Linear Thermal Expansion

With few exceptions, the length and the volume of glasses increase with increasing temperature (positive coefficient).

The typical curve begins with a zero gradient at absolute zero (Fig. 3.4-3) and increases slowly. At about room temperature (section A in Fig. 3.4-3), the curve shows a distinct bend and then gradually increases (section B, the quasi-linear region) up to the beginning of the experimentally detectable plastic behavior. Another distinct bend in the expansion curve characterizes the transition from a predominantly elastic to a more plastic behavior of the glass (section C, the transformation range). As a result of increasing structural mobility, the temperature dependence of almost all glass properties changes distinctly in this range. Figure 3.4-18 shows the linear thermal expansion curves of five glasses; 8330 and 4210 roughly define the normal range of technical glasses, with expansion coefficients

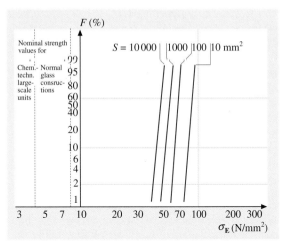

Fig. 3.4-17 Failure probability F for differently sized stressed areas S. All samples abraded with 600 mesh grit, stress rate $d\sigma/dt = 10\,\mathrm{N/mm^2\,s}$. A, range of nominal strength for large-scale units in chemical technology; B, range of nominal strength for normal glass structures

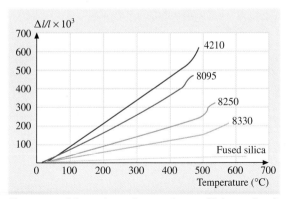

Fig. 3.4-18 Linear thermal expansion coefficients of various technical glasses and of fused silica

$\alpha_{(20\,°C/300\,°C)} = 3.3\text{--}12.0 \times 10^{-6}/\mathrm{K}$ (see Table 3.4-11, first page, third column).

The linear thermal expansion is an essential variable in determining the sealability of glasses to other materials and in determining thermally induced stress formation, and is therefore of prime importance for applications of glasses.

Thermal Stresses. Owing to the low thermal conductivity of glasses (typically 0.9–1.2 W/m K at 90 °C, and a minimum of 0.6 W/m K for high-lead-content glasses), temperature changes produce relatively high temperature differences ΔT between the surface and the interior, which, depending on the elastic properties E (Young's modulus) and μ (Poisson's ratio) and on the coefficient of linear thermal expansion α, can result in stresses

$$\sigma = \frac{\Delta T \alpha E}{(1-\mu)} \quad \text{N/mm}^2 \, . \tag{4.8}$$

In addition to the geometric factors (shape and wall thickness), the material properties α, E, and μ decisively influence the thermal strength of glasses subjected to temperature variations and/or thermal shock. Thermal loads in similar articles made from different glasses are easily compared by means of the characteristic material value

$$\varphi = \frac{\sigma}{\Delta T} = \frac{\alpha E}{(1-\mu)} \quad \text{N/(mm}^2\,\text{K)} \, , \tag{4.9}$$

which indicates the maximum thermally induced stress to be expected in a flexure-resistant piece of glass for a local temperature difference of 1 K. Because cracking originates almost exclusively from the glass surface and is caused there by tensile stress alone, cooling processes are usually much more critical than the continuous rapid heating of glass articles.

3.4.4.3 Electrical Properties

Glasses are used as electrically highly insulating materials in electrical engineering and electronics, in the production of high-vacuum tubes, lamps, electrode seals, hermetically encapsulated components, high-voltage insulators, etc. Moreover, glasses may be used as insulating substrates for electrically conducting surface layers (in surface heating elements and data displays).

Volume Resistivity

Electrical conductivity in technical silicate glasses is, in general, a result of the migration of ions – mostly alkali ions. At room temperature, the mobility of these ions is usually so small that the volume resistivity, with values above $10^{15}\,\Omega\,\text{cm}$, is beyond the range of measurement. The ion mobility increases with increasing temperature. Besides the number and nature of the charge carriers, the structural effects of other components also influence the volume resistivity and its relationship to temperature. The Rasch and Hinrichsen law applies to this relationship at temperatures below the transformation range:

$$\log \rho = A - B/T \, , \tag{4.10}$$

where ρ is the electrical volume resistivity in $\Omega\,\text{cm}$, A, B are constants specific to the particular glass, and T is the absolute temperature in K.

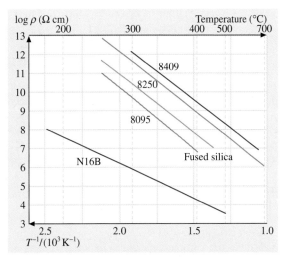

Fig. 3.4-19 Electrical volume resistivity of various technical glasses and fused silica as a function of reciprocal absolute temperature

A plot of $\log \rho = f(1/T)$ thus yields straight lines (Fig. 3.4-19). Because of the relatively small differences in slope for most glasses, the electrical insulation of glasses is often defined only by the temperature at which the resistivity is $10^8\,\Omega\,\text{cm}$. According to DIN 52326, this temperature is denoted by T_{k100}. The international convention is to quote volume resistivities at 250 °C and 350 °C (Table 3.4-11, second page, second column), from which the constants A and B

Table 3.4-8 Fracture toughness of some glasses

Glass	K_{Ic} (MPa $\sqrt{\text{m}}$)
BK7	1.08
F5	0.86
SF6	0.74
K50	0.77
Duran®	0.85

Table 3.4-9 Parameters of the volume resistivity (4.10) of the glasses in Fig. 3.4-19

Glass	A	B (K)
8095	− 2.863	− 6520.0
8250	− 0.594	− 5542.0
8409	− 0.463	− 6520.0
N16B	− 1.457	− 3832.4
Fused silica	− 0.394	− 6222.4

(Table 3.4-9) and various other values below T_g can be calculated.

Surface Resistivity

The generally very high volume resistivity of glasses at room temperature has superimposed on it in a normal atmosphere a surface resistivity which is several orders of magnitude lower. The all-important factor is the adsorption of water on the glass surface. Depending on the glass composition, surface resistivities of $10^{13}–10^{15}\,\Omega$ occur at low relative humidities, and $10^{8}–10^{10}\,\Omega$ at high relative humidities. Above $100\,°C$, the effect of this hydrated layer disappears almost completely. (Treatment with silicones also considerably reduces this effect.)

Dielectric Properties

With dielectric constants generally between 4.5 and 8, technical glasses behave like other electrically insulating materials. The highest values are obtained for lead glasses such as 8531 ($\varepsilon_r = 9.5$) and for ultra-high-lead-content solder glasses ($\varepsilon_r \sim 20$). The dependence of the dielectric constants ε_r on frequency and temperature is relatively small (Fig. 3.4-20). For a frequency range of $50–10^9$ Hz, ε_r values generally do not vary by more than 10%.

The dielectric dissipation factor $\tan\delta$ is frequency- and temperature-dependent. Owing to the diverse mechanisms which cause dielectric losses in glasses, there is a minimum of $\tan\delta$ in the region of $10^6–10^8$ Hz, and increasing values at lower and higher frequencies (Fig. 3.4-21).

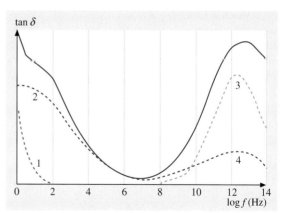

Fig. 3.4-21 Schematic representation of the frequency spectrum of dielectric losses in glasses at room temperature. The *solid curve* gives the total losses, made up of (1) conduction loss, (2) relaxation loss, (3) vibration loss, and (4) deformation loss

At 10^6 Hz, the dissipation factors $\tan\delta$ for most glasses lie between 10^{-2} and 10^{-3}; fused silica, with a value of 10^{-5}, has the lowest dissipation factor of all glasses. The special glass 8248 has relatively low losses, and in this cases $\tan\delta$ increases only slightly up to 5.5 GHz (where $\tan\delta = 3\times 10^{-3}$).

The steep increase in dielectric losses with increasing temperature (Fig. 3.4-22) can lead to instability, i.e. overheating of the glass due to dielectric loss energy in the case of restricted heat dissipation and corresponding electrical power.

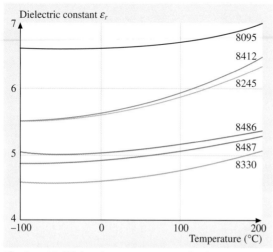

Fig. 3.4-20 Dielectric constant ε_r of electrotechnical glasses as a function of temperature, measured at 1 MHz

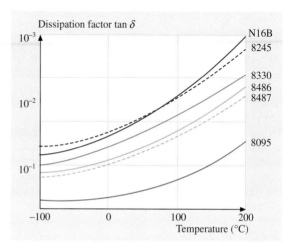

Fig. 3.4-22 Dissipation factor $\tan\delta$ as a function of temperature in the range $-100\,°C < T < +200\,°C$, measured at 1 MHz

Dielectric Strength

Some approximate values for the dielectric strength of glasses are a field strength of 20–40 kV/mm for a glass thickness of 1 mm at 50 Hz at 20 °C, and 10–20 kV/mm for greater thicknesses. At higher temperatures and frequencies, decreasing values can be expected.

3.4.4.4 Optical Properties

Refraction of Light

The refractive indices n_d of technical glasses at a wavelength of $\lambda_d = 587.6$ nm generally lie within the range 1.47–1.57. The exceptions to this rule are lead glasses with PbO contents of over 35% (e.g. glass 8531, which has $n_d = 1.7$; see Table 3.4-11, second page, fourth column). The principal dispersion $n_F - n_C$ ($\lambda_F = 486.1$ nm, $\lambda_C = 656.3$ nm) of technical glasses lies between 0.007 and 0.013.

At perpendicular incidence, the reflectance R_d of a glass–air interface is 3.6% to 4.9%.

The transmittance τ_d and the reflectance ρ_d of a nonabsorbing, planar, parallel-sided glass plate with two glass–air interfaces, with multiple reflections taken into account, can be calculated from the refractive index as

$$\tau_d = \frac{2n_d}{n_d^2 + 1} \tag{4.11}$$

and

$$\rho_d = \frac{(n_d - 1)^2}{n_d + 1} \,. \tag{4.12}$$

The transmittance τ_d at perpendicular incidence has values between 90.6% and 93.1%.

Stress Birefringence

Owing to its structure, glass is an isotropic material. Mechanical stress causes anisotropy, which manifests itself as stress-induced birefringence. A light beam, after passing through a plate of thickness d which is subjected to a principal-stress difference $\Delta\sigma$, shows an optical path difference Δs between the two relevant polarization directions. This path difference can either be estimated by means of the birefringence colors or be measured with a compensator, and is given by

$$\Delta s = K d \Delta\sigma \text{ nm} , \tag{4.13}$$

where K is the stress-optical coefficient of the glass (determined according to DIN 52314),

$$K = \frac{\Delta s}{d} \frac{1}{\Delta\sigma} \text{ mm}^2/\text{N} \,. \tag{4.14}$$

Many glasses have stress-optical coefficients of about 3×10^{-6} mm^2/N, and borosilicate glasses have values of up to 4×10^{-6} mm^2/N. High-lead-content glasses can have values down to nil or even negative (Table 3.4-11, second page, fifth column).

Light Transmittance

The transmittance due to the refractive index can be further reduced by coloring agents (oxides of transition elements or colloids) or by fine particles in the glass which have a different refractive index (in this case light scattering occurs, giving opal glasses).

Absorption caused by impurities such as Fe_2O_3 and by some major glass components such as PbO strongly reduces transparency in the UV range. Particularly good UV-transmitting multicomponent glasses have a cutoff

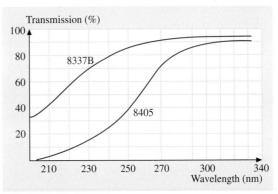

Fig. 3.4-23 UV transmission of highly UV-transparent glasses 8337B and 8405 for 1 mm glass thickness

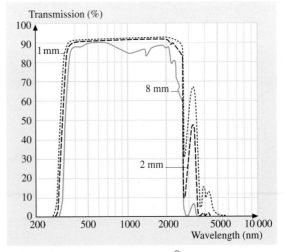

Fig. 3.4-24 Transmission of Duran® 8330 for thicknesses of 1, 2, and 8 mm

(50% value) at a wavelength of 220 nm (Fig. 3.4-23); normal technical glasses already absorb considerably at 300 nm.

In the IR range, absorption caused by impurities such as H_2O and by lattice vibrations limits the transmittance (Fig. 3.4-24).

Table 3.4-10 Schott technical specialty glasses and their typical applications

8095	Lead glass (28% PbO), electrically highly insulating, for general electrotechnical applications
8245	Sealing glass for Fe–Ni–Co alloys and molybdenum, minimum X-ray absorption, chemically highly resistant
8248	Borosilicate glass (of high B_2O_3 content), minimum dielectric losses up to the GHz range, electrically highly insulating
8250	Sealing glass for Ni–Fe–Co alloys and molybdenum, electrically highly insulating
8252	Alkaline-earth aluminosilicate glass for high-temperature applications, for sealing to molybdenum
8253	Alkaline-earth aluminosilicate glass for high-temperature applications, for sealing to molybdenum
8321	Alumino-borosilicate glass for TFT displays
8326	SBW glass, chemically highly resistant
8330	Duran®, borosilicate glass, general-purpose glass for apparatus for the chemical industry, pipelines, and laboratory glassware
8337B	Borosilicate glass, highly UV-transmitting, for sealing to glasses and to metals of the Kovar and Vacon-10 ranges and tungsten
8350	AR glass®, soda–lime silicate glass tubing
8405	Highly UV-transmitting soft glass
8409	Supremax® (black identification line), alkali-free, for high application temperatures in thermometry, apparatus construction, and electrical engineering
8412 [a]	Fiolax®, clear (blue identification line), neutral, glass tubing (chemically highly resistant) for pharmaceutical packaging
8414	Fiolax®, amber (blue identification line), neutral, glass tubing (chemically highly resistant) for pharmaceutical packaging
8415	Illax®, amber tubing glass for pharmaceutical packaging
8421	Sealing glass for seals to NiFe45 (DIN 17745) and compression seals
8422	Sealing glass for seals to NiFe47 or 49 (DIN 17745) and compression seals
8436	Particularly resistant to sodium vapor and alkaline solutions, suitable for sealing to sapphire
8486	Suprax®, borosilicate glass, chemically and thermally resistant, suitable for sealing to tungsten
8487	Sealing glass for tungsten, softer than 8486
8488	Borosilicate glass, chemically and thermally resistant
8490	Black glass, light-transmitting in the UV region, highly absorbing in the visible region
8512	IR-absorbing sealing glass for Fe–Ni, lead-free (reed switches)
8516	IR-absorbing sealing glass for NiFe, lead-free, slow-evaporating (reed switches)
8531	Soft glass, Na-free, high lead-content, for low temperature encapsulation of semiconductor components (diodes)
8532	Soft glass, Na-free, high lead-content, for low-temperature encapsulation of semiconductor components (diodes)
8533	IR-absorbing sealing glass for Ni–Fe, lead- and potassium-free, slow-evaporating (reed switches)
8625	IR-absorbing biocompatible glass for (implantable) transponders
8650	Alkali-free sealing glass for molybdenum, especially for implosion diodes; high lead content
8651	Tungsten sealing glass for power diodes
8652	Tungsten sealing glass, low-melting, for power diodes
8656	Borofloat® 40, borosilicate float glass adapted for prestressing

[a] Also known as 8258, Estax®, low-potassium glass tubing for the manufacture of counting vials.

Table 3.4-11 Characteristic data of technical specialty glasses

Glass No.	Shapes produced[a]	Thermal expansion coefficient $\alpha_{(20/300)}$ (10^{-6}/K)	Transformation temperature T_g (°C)	Temperature at viscosity 10^{13} dPa s (°C)	$10^{7.6}$ dPa s (°C)	10^4 dPa s (°C)	Density at 25 °C (g/cm³)	Young's modulus (10^3 N/mm²)	Poisson's ratio μ	Heat conductivity λ at 90 °C (W/m K)
8095	TP	9.2	435	435	635	985	3.01	60	0.22	0.9
8245	MTRP	5.1	505	515	720	1040	2.31	68	0.22	1.2
8248	BP	3.1	445	490	740	1260	2.12	44	0.22	1.0
8250	MTBPC	5.0	490	500	720	1055	2.28	64	0.21	1.2
8252	TP	4.6	725	725	935	1250	2.63	81	0.24	1.1
8253	TP	4.7	785	790	1000	1315	2.65	83	0.23	1.1
8261	SP	3.7	720	725	950	1255	2.57	79	0.24	1.1
8326	MTP	6.6	560	565	770	1135	2.46	75	0.20	1.2
8330	MSTRPC	3.3	525	560	820	1260	2.23	63	0.20	1.12
8337B	TP	4.1	430	465	715	1090	2.21	51	0.22	1.0
8350	TRP	9.1	525	530	715	1040	2.50	73	0.22	1.1
8405	MTP	9.8	460	450	660	1000	2.51	65	0.21	1.0
8409	MTRP	4.1	745	740	950	1230	2.57	85	0.24	1.2
8412	TP	4.9	565	565	780	1165	2.34	73	0.20	1.2
8414	TP	5.4	560	560	770	1155	2.42	71	0.19	1.2
8415	TP	7.8	535	530	720	1050	2.50	74	0.21	1.1
8421	P	9.7	525	535	705	1000	2.59	74	0.22	1.0
8422	P	8.7	540	535	715	1010	2.46	76	0.21	1.1
8436	TRP	6.7	630	630	830	1110	2.76	85	0.22	1.1
8486	MP	4.1	555	580	820	1220	2.32	66	0.20	1.1
8487	TRP	3.9	525	560	775	1135	2.25	66	0.20	1.2
8488	M	4.3	545	560	800	1250	2.30	67	0.20	1.2
8490	MP	9.6	475	480	660	1000	2.61	70	0.22	1.0
8512	TP	9.0	445	460	665	980	2.53	68	0.22	1.0
8516	TP	8.9	440	445	650	990	2.56	72	0.21	1.1
8531	TP	9.0	440	430	590	830	4.34	52	0.24	0.57
8532	TP	8.8	430	425	565	760	4.47	56	0.24	0.7
8533	TP	8.7	475	480	645	915	2.57	79	0.21	1.1
8625	TP	9.0	510	520	710	1030	2.53	73	0.22	1.1
8650	TP	5.2	475	475	620	880	3.57	62	0.23	0.5
8651	TP	4.5	540	540	735	1040	2.87	59	0.24	0.9
8652	TP	4.5	495	490	640	915	3.18	58	0.25	0.9
8656	SP	4.1	590	600	850	1270	2.35	–	–	–

[a] Shapes produced: B = block glass; C = capillaries; M = molded glass (blown or pressed); P = powder, spray granulates, or sintered parts; R = rods; S = sheet glass; T = tubing

Table 3.4-11 Characteristic data of technical specialty glasses, cont.

T_{k100} (°C)	Logarithm of electrical volume resistivity (Ω cm) at		Dielectric properties at 1 MHz and 25 °C		Refractive index n_d ($\lambda_d = 587.6$ nm)	Stress-optical coefficient K (10^{-6} mm^2/N)	Classes of chemical stability against		
	250 °C	350 °C	ε_r	tan δ (10^{-4})			Water	Acid	Alkaline solution
330	9.6	7.6	6.6	11	1.556	3.1	3	2	3
215	7.4	5.9	5.7	80	1.488	3.8	3	4	3
–	12	10	4.3	10	1.466	5.2	3	3	3
375	10	8.3	4.9	22	1.487	3.6	3	4	3
660	–	12	6.1	11	1.538	3.3	1	3	2
630	–	11	6.6	15	1.547	2.7	1	2	2
585	–	–	5.8	14	1.534	3.1	1	4	2
210	7.3	6.0	6.4	65	1.506	2.8	1	1	2
250	8.0	6.5	4.6	37	1.473	4.0	1	1	2
315	9.2	7.5	4.7	22	1.476	4.1	3	4	3
200	7.1	5.7	7.2	70	1.514	2.7	3	1	2
280	8.5	6.9	6.5	45	1.505	2.8	5	3	2
530	12	10	6.1	23	1.543	2.9	1	4	3
215	7.4	6.0	5.7	80	1.492	3.4	1	1	2
200	7.1	5.6	6.3	107	1.523	2.2	1	2	2
180	6.7	5.3	7.1	113	1.521	3.2	2	2	2
255	8.1	6.4	7.4	43	1.526	2.7	3	3	2
205	7.3	5.8	7.3	60	1.509	2.9	2	3	3
245	7.9	6.5	7.9	75	1.564	2.9	1–2	1–2	1
230	7.5	6.1	5.1	40	1.487	3.8	1	1	2
300	8.3	6.9	4.9	36	1.479	3.6	4	3	3
200	7.1	5.8	5.4	96	1.484	3.2	1	1	2
235	7.7	6.1	6.7	32	1.52	–	3	2	2
320	9.5	7.5	6.5	21	1.510	3.0	3	1–2	2
250	8.1	6.4	6.5	25	1.516	3.0	3	1	2
450	11	9.8	9.5	9	1.700	2.2	1	4	3
440	11	9.4	10.2	9	1.724	1.7	1	4	3
200	7.0	5.5	6.9	55	1.527	3.0	1	2	2
210	7.2	5.8	7.1	68	1.525	–	3	1	2
–	–	–	7.6	33	1.618	2.8	1	4	3
–	11.2	10.0	6.0	31	1.552	3.6	1	4	3
–	–	–	6.9	35	1.589	3.4	1	4	3
265	8.3	6.8	5.5	51	1.493	3.6	1	1	1

3.4.5 Optical Glasses

Historically, optical glasses were developed to optimize imaging refractive optics in the visible part of the spectrum. With new mathematical tools, new light sources, and new detectors, the resulting lens systems, which used a combination of different glasses (or even sometimes crystalline materials), changed the requirement from high versatility in terms of material properties to homogeneity, precision, and reproducibility, and an extended spectral range in the ultraviolet and infrared regions; environmental aspects also became important. Consequently, the high number of glass types has been reduced.

3.4.5.1 Optical Properties

Refractive Index, Abbe Value, Dispersion, and Glass Designation

The most common identifying features used for characterizing an optical glass are the refractive index n_d in the middle range of the visible spectrum, and the Abbe value $v_d = (n_d - 1)/(n_F - n_C)$ as a measure of the dispersion. The difference $n_F - n_C$ is called the principal dispersion. The symbols have subscripts that identify spectral lines that are generally used to determine refractive indices; these spectral lines are listed in Table 3.4-12.

The quantities n_e and $v_e = (n_e - 1)/(n_{F'} - n_{C'})$ based on the e-line are usually used for specifying optical components.

For the comparison of glass types from different manufacturers, an abbreviated glass code is defined in the following way: the first three digits correspond to $n_d - 1$, the second three digits represent the Abbe value v_d, and, after a dot, three more digits characterize the density (see Table 3.4-13).

Glasses can be grouped into families in an n_d/v_d Abbe diagram (Fig. 3.4-25). These glass families differ in chemical composition as shown in Fig. 3.4-26.

Table 3.4-12 Wavelengths of a selection for frequently used spectral lines

Wavelength (nm)	Designation	Spectral line used	Element
2325.42		Infrared mercury line	Hg
1970.09		Infrared mercury line	Hg
1529.582		Infrared mercury line	Hg
1060.0		Neodymium glass laser	Nd
1013.98	t	Infrared mercury line	Hg
852.11	s	Infrared cesium line	Cs
706.5188	r	Red helium line	He
656.2725	C	Red hydrogen line	H
643.8469	C'	Red cadmium line	Cd
632.8		Helium–neon gas laser	He–Ne
589.2938	D	Yellow sodium line (center of the double line)	Na D
587.5618	d	Yellow helium line	He
546.0740	e	Green mercury line	Hg
486.1327	F	Blue hydrogen line	H
479.9914	F'	Blue cadmium line	Cd
435.8343	g	Blue mercury line	Hg
404.6561	h	Violet mercury line	Hg
365.0146	i	Ultraviolet mercury line	Hg
334.1478		Ultraviolet mercury line	Hg
312.5663		Ultraviolet mercury line	Hg
296.7278		Ultraviolet mercury line	Hg
280.43		Ultraviolet mercury line	Hg
248.00		Excimer laser	KrF
248.35		Ultraviolet mercury line	Hg
194.23			
193.00		Excimer laser	ArF

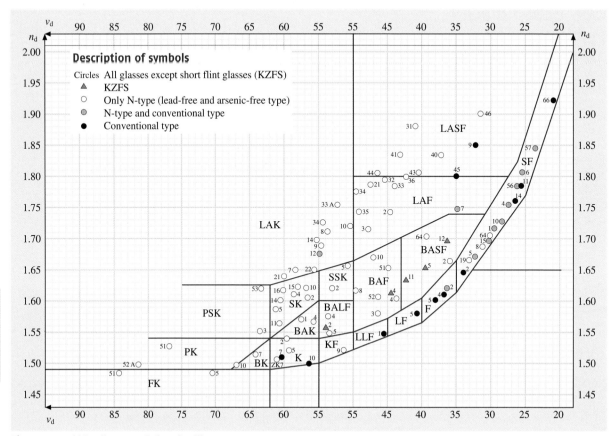

Fig. 3.4-25 Abbe diagram of glass families

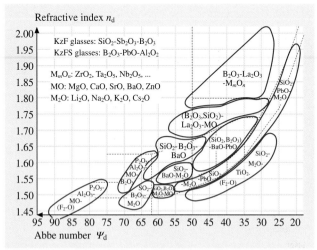

Fig. 3.4-26 Abbe diagram showing the chemical composition of the glass families

Table 3.4-13 Examples of glass codes

Glass type	n_d	v_d	Density ϱ (g cm^{-3})	Glass code	Remarks
N-SF6	1.80518	25.36	3.37	805 254.337	Lead- and arsenic-free glass
SF6	1.80518	25.43	5.18	805 254.518	Classical lead silicate glass

The designation of each glass type here is composed of an abbreviated family designation and a number. The glass families are arranged by decreasing Abbe value in the data tables (Table 3.4-14).

Table 3.4-14 gives an overview of the preferred optical glasses from Schott, Hoya, and Ohara. The glass types are listed in order of increasing refractive index n_d.

Table 3.4-14 Comparison of the preferred optical glasses from different manufacturers

Schott Code	Schott Glass type	Hoya Code	Hoya Glass type	Ohara Code	Ohara Glass type
434950	N-FK56				
				439950	S-FPL53
				456903	S-FPL52
487704	N-FK5	487704	FC5	487702	S-FSL5
487845	N-FK51				
497816	N-PK52	497816	FCD1	497816	S-FPL51
498670	N-BK10				
501564	K10				
508612	N-ZK7				
511604	K7				
		517522	CF6		
		517524	E-CF6	517524	S-NSL36
517642	N-BK7	517642	BSC7	516641	S-BSL7
				517696	S-APL1
		518590	E-C3	518590	S-NSL3
				521526	SSL5
522595	N-K5				
				522598	S-NSL5
523515	N-KF9				
				529517	SSL2
529770	N-PK51				
532488	N-LLF6	532488	FEL6		
		532489	E-FEL6	532489	S-TIL6
				540595	S-BAL12
540597	N-BAK2				
		541472	E-FEL2	541472	S-TIL2
		541472	FEL2		
547536	N-BALF5				
548458	LLF1	548458	FEL1		
548458	N-LLF1	548458	E-FEL1	548458	S-TIL1
		551496	SbF1		
552635	N-PSK3				
558542	N-KZFS2				
				560612	S-BAL50
564608	N-SK11	564607	EBaCD11	564607	S-BAL41
		567428	FL6	567428	PBL26
569561	N-BAK4	569563	BaC4	569563	S-BAL14
569713	N-PSK58				
				571508	S-BAL2
				571530	S-BAL3
573576	N-BAK1			573578	S-BAL11
				575415	S-TIL27
580537	N-BALF4				
581409	N-LF5	581407	E-FL5	581407	S-TIL25
581409	LF5	581409	FL5		
583465	N-BAF3			583464	BAM3
		583594	BaCD12	583594	S-BAL42
589613	N-SK5	589613	BaCD5	589612	S-BAL35
592683	N-PSK57				
				593353	S-FTM16
		594355	FF5		
		596392	E-F8	596392	S-TIM8
		596392	F8		
		603380	E-F5	603380	S-TIM5
603380	F5			603380	F5
603606	N-SK14	603606	BaCD14	603607	S-BSM14
				603655	S-PHM53
606439	N-BAF4			606437	S-BAM4
607567	N-SK2	607568	BaCD2	607568	S-BSM2
609464	N-BAF52				
		613370	F3	613370	PBM3
613443	KZFSN4	613443	ADF10	613443	BPM51
613445	N-KZFS4				
613586	N-SK4	613587	BaCD4	613587	S-BSM4
				614550	BSM9
617366	F4				
				617628	S-PHM51
618498	N-SSK8			618498	S-BSM28
		618634	PCD4	618634	S-PHM52
620364	N-F2	620363	E-F2	620363	S-TIM2
620364	F2	620364	F2		
620603	N-SK16	620603	BaCD16	620603	S-BSM16
		620622	ADC1		
620635	N-PSK53				
				621359	TIM11
621603	SK51				
622532	N-SSK2			622532	BSM22
623569	N-SK10	623570	EBaCD10	623570	S-BSM10
623580	N-SK15	623582	BaCD15	623582	S-BSM15
		624470	BaF8		

Table 3.4-14 Comparison of the preferred optical glasses from different manufacturers, cont.

Schott Code	Glass type	Hoya Code	Glass type	Ohara Code	Glass type	Schott Code	Glass type	Hoya Code	Glass type	Ohara Code	Glass type
		626357	F1			699301	N-SF15	699301	E-FD15	699301	S-TIM35
		626357	E-F1	626357	S-TIM1	699301	SF15	699301	FD15		
639421	N-KZFS11									700481	S-LAM51
				639449	S-BAM12			702412	BaFD7	702412	S-BAH27
639554	N-SK18	639554	BaCD18	639554	S-BSM18	704394	NBASF64				
		640345	E-FD7	640345	S-TIM27	706303	N-SF64				
640601	N-LAK21	640601	LaCL60	640601	S-BSM81	713538	N-LAK8	713539	LaC8	713539	S-LAL8
				641569	S-BSM93	717295	N-SF1	717295	E-FD1		
				643584	S-BSM36	717295	SF1	717295	FD1	717295	PBH1
648339	SF2	648339	FD2			717480	N-LAF3	717480	LaF3	717479	S-LAM3
		648338	E-FD2	648338	S-TIM22					720347	BPH8
		649530	E-BaCED20	649530	S-BSM71					720420	LAM58
										720437	S-LAM52
651559	N-LAK22									720460	LAM61
		651562	LaCL2	651562	S-LAL54						
652449	N-BAF51					720506	N-LAK10	720504	LaC10	720502	S-LAL10
652585	N-LAK7	652585	LaC7	652585	S-LAL7					722292	S-TIH18
654396	KZFSN5	654396	ADF50	654397	BPH5	724381	NBASF51			723380	S-BAH28
658509	N-SSK5	658509	BaCED5	658509	S-BSM25	724381	BASF51	724381	BaFD8		
				658573	S-LAL11					726536	S-LAL60
664360	N-BASF2					728284	SF10	728284	FD10		
				667330	S-TIM39	728285	N-SF10	728285	E-FD10	728285	S-TIH10
		667484	BaF11	667483	S-BAH11	729547	N-LAK34	729547	TaC8	729547	S-LAL18
				670393	BAH32			734515	TaC4	734515	S-LAL59
670471	N-BAF10	670473	BaF10	670473	S-BAH10					740283	PBH3W
								741276	FD13	740283	PBH3
				670573	S-LAL52			741278	E-FD13	741278	S-TIH13
673322	N-SF5	673321	E-FD5	673321	S-TIM25			741527	TaC2	741527	S-LAL61
673322	SF5	673322	FD5			743492	N-LAF35	743493	NbF1	743493	S-LAM60
		678507	LaCL9	678507	S-LAL56	744447	N-LAF2	744447	LaF2	744448	S-LAM2
678549	LAKL12					750350	LaFN7	750353	LaF7	750353	LAM7
678552	N-LAK12	678553	LaC12	678553	S-LAL12	750350	N-LAF7				
689312	N-SF8	689311	E-FD8	689311	S-TIM28	754524	N-LAK33	755523	TaC6	755523	S-YGH51
		689312	FD8			755276	N-SF4	755275	E-FD4	755275	S-TIH4
691547	N-LAK9	691548	LaC9	691548	S-LAL9	755276	SF4	755276	FD4		
		694508	LaCL5	694508	LAL58					756251	TPH55
694533	LAKN13	694532	LaC13	694532	S-LAL13			757478	NbF2	757478	S-LAM54
				695422	S-BAH54	762265	N-SF14	762265	FD140	762265	S-TIH14
				697485	LAM59	762265	SF14	762266	FD14		
		697485	LaFL2							762401	S-LAM55
697554	N-LAK14	697555	LaC14	697555	S-LAL14						
				697565	S-LAL64	772496	N-LAF34	772496	TaF1	772496	S-LAH66

Table 3.4-14 Comparison of preferred optical glasses, cont.

Schott		Hoya		Ohara	
Code	Glass type	Code	Glass type	Code	Glass type
785258	SF11	785258	FD11		
		785258	FD110	785257	S-TIH11
785261	SF56A				
785261	N-SF56	785261	FDS30	785263	S-TIH23
786441	N-LAF33	786439	NBFD11	786442	S-LAH51
				787500	S-YGH52
788475	N-LAF21	788475	TAF4	788474	S-LAH64
794454	N-LAF32			795453	S-LAH67
800423	N-LAF36	800423	NBFD12	800422	S-LAH52
801350	N-LASF45			801350	S-LAM66
				804396	S-LAH63
804466	N-LASF44	804465	TAF3	804466	S-LAH65
805254	N-SF6	805254	FD60	805254	S-TIH6
805254	SF6	805254	FD6		
		805396	NBFD3		
		806333	NBFD15		
806407	N-LASF43	806407	NBFD13	806409	S-LAH53
				808228	S-NPH1
		816445	TAFD10	816444	S-LAH54
		816466	TAF5	816466	S-LAH59
834374	N-LASF40	834373	NBFD10	834372	S-LAH60
835430	N-LASF41	835430	TAFD5	835427	S-LAH55
847238	N-SF57	847238	FDS90	847238	S-TIH53
847236	SFL57				
847238	SF57	847238	FDS9	847238	TIH53
850322	LASFN9				
				874353	S-LAH75
881410	N-LASF31				
		883408	TAFD30	883408	S-LAH58
901315	N-LASF46			901315	LAH78
923209	SF66	923209	E-FDS1		
				923213	PBH71
				1003283	S-LAH79
1022291	N-LASF35				

Formulas for Optical Characterization

The characterization of optical glasses through the refractive index and Abbe value alone is insufficient for high-quality optical systems. A more accurate description of the properties of a glass can be achieved with the aid of the relative partial dispersion.

Relative Partial Dispersion. The relative partial dispersion $P_{x,y}$ for the wavelengths x and y based on the blue F hydrogen line and red C hydrogen line is given by

$$P_{x,y} = (n_x - n_y)/(n_F - n_C) \,. \quad (4.15)$$

The corresponding value based on the blue F′ cadmium line and red C′ cadmium line is given by

$$P'_{x,y} = (n_x - n_y)/(n_{F'} - n_{C'}) \,. \quad (4.16)$$

Relationship Between the Abbe Value and the Relative Partial Dispersion. A linear relationship exists between the Abbe value and the relative partial dispersion for what are known as "normal glasses":

$$P_{x,y} \approx a_{xy} + b_{xy}\nu_d \,. \quad (4.17)$$

Deviation from the "Normal Line". All other glasses deviate from the "normal line" defined by $\Delta P_{x,y}$. For the selected wavelength pairs the ΔP-value are calculated from the following equations:

$$P_{x,y} = a_{xy} + b_{xy}\nu_d + \Delta P_{x,y} \,, \quad (4.18)$$

$$\Delta P_{C,t} = (n_C - n_t)/(n_F - n_C)$$
$$- (0.5450 + 0.004743\nu_d) \,, \quad (4.19)$$

$$\Delta P_{C,s} = (n_C - n_s)/(n_F - n_C)$$
$$- (0.4029 + 0.002331\nu_d) \,, \quad (4.20)$$

$$\Delta P_{F,e} = (n_F - n_e)/(n_F - n_C)$$
$$- (0.4884 - 0.000526\nu_d) \,, \quad (4.21)$$

$$\Delta P_{g,F} = (n_g - n_F)/(n_F - n_C)$$
$$- (0.6438 - 0.001682\nu_d) \,, \quad (4.22)$$

$$\Delta P_{i,g} = (n_i - n_g)/(n_F - n_C)$$
$$- (1.7241 - 0.008382\nu_d) \,. \quad (4.23)$$

The "normal line" has been determined based on value pairs of glasses types K7 and F2. The term $\Delta P_{x,y}$ quantitatively describes the deviation of the behavior of the dispersion from that of "normal glasses".

The Sellmeier dispersion formula for the refractive index,

$$n^2(\lambda) - 1 = B_1\lambda^2/(\lambda^2 - C_1) + B_2\lambda^2/(\lambda^2 - C_2)$$
$$+ B_3\lambda^2/(\lambda^2 - C_3) \,, \quad (4.24)$$

can be derived from classical dispersion theory with the assumption of three resonance wavelengths. It is valid only for interpolation within a spectral region in which the refractive index has been measured. The vacuum wavelength λ in μm has to be used. The precision of the calculation achievable is generally better than 1×10^{-5}.

Temperature Dependence of the Refractive Index. The refractive index is also dependent on temperature. This temperature dependence is represented by $\Delta n_{\text{rel}}/\Delta T$ for an air pressure of 1013.3 hPa and $\Delta n_{\text{abs}}/\Delta T$ in vacuum. The following equation is derived from the Sellmeier formula and is valid with the given coefficients in the temperature range $-40\,°\text{C} < T < +80\,°\text{C}$ and within the wavelength range $435.8\,\text{nm} < \lambda < 643.8\,\text{nm}$:

$$\frac{dn_{\text{abs}}(\lambda, T)}{dT} = \frac{n^2(\lambda, T_0) - 1}{2n(\lambda, T_0)}$$
$$\times \left(D_0 + 2D_1 \Delta T + 3D_2 \Delta T^2 + \frac{E_0 + 2E_1 \Delta T}{\lambda^2 - \lambda_{\text{TK}}^2} \right), \tag{4.25}$$

where ΔT is the temperature difference in °C from 20 °C, and λ_{TK} is an effective resonance wavelength.

The changes in the refractive index and Abbe value caused by a change in the annealing rate are given by

$$n_d(h_x) = n_d(h_0) + m_{nd} \log(h_x/h_0), \tag{4.26}$$

$$v_d(h_x) = v_d(h_0) + m_{vd} \log(h_x/h_0), \tag{4.27}$$

$$m_{vd} = \frac{m_{nd} - v_d(h_0)m_{nF-nC}}{(n_F - n_C) + 2m_{nF-nC}\log(h_x/h_0)}, \tag{4.28}$$

where h_0 is the original annealing rate in °C/h, h_x is the new annealing rate in °C/h, m_{nd} is the annealing coefficient for the refractive index (Table 3.4-15), m_{vd} is the annealing coefficient for the Abbe value and m_{nF-nC} is the annealing coefficient for the principal dispersion. The last three quantities depend on the glass type.

The measurement accuracy of the Abbe value can be calculated using

$$\sigma(v_d) \approx \sigma(n_F - n_C)v_d/(n_F - n_C). \tag{4.29}$$

The accuracy of precision measurements of the refractive indices is better than $\pm 1 \times 10^{-5}$, and the accuracy of the dispersion is $\pm 3 \times 10^{-6}$. In the infrared wavelength range above 2 μm, the corresponding accuracies are $\pm 2 \times 10^{-5}$ and $\pm 5 \times 10^{-6}$.

Table 3.4-15 Annealing coefficients for selected glass types

Glass type	m_{nd}	m_{nF-nC}	m_{vd}
N-BK7	−0.00087	−0.000005	−0.0682
N-FK51	−0.00054	−0.000002	−0.0644
SF6	−0.00058	+0.000035	−0.0464
N-SF6	−0.0025	−0.000212	+0.0904

Transmission. The transmittance of glasses is limited by electronic excitations and light scattering in the UV, by vibronic excitations in the IR, and by reflections and impurity absorptions within the transmission window (in the visible part of the spectrum): Fig. 3.4-27. The UV absorption edge is temperature dependent. An example is shown in Fig. 3.4-28.

Spectral Internal Transmittance. The spectral internal transmittance is given by

$$\tau_{i\lambda} = \Phi_{e\lambda}/\Phi_{i\lambda}, \tag{4.30}$$

where $\Phi_{i\lambda}$ is the incident light intensity and $\Phi_{e\lambda}$ is the intensity at the exit.

Spectral Transmission. The spectral transmission is given by

$$\tau_\lambda = \tau_{i\lambda} P_\lambda, \tag{4.31}$$

where P_λ is the reflection factor.

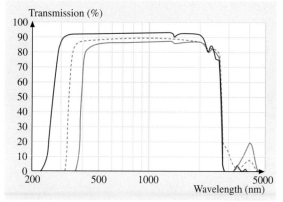

Fig. 3.4-27 Transmission of three glasses for a thickness of 5 mm: *Brown line* FK5; *dashed line* SF2; *gray line* SF11

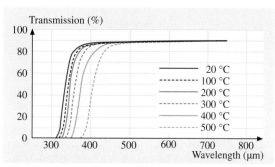

Fig. 3.4-28 Influence of temperature on the UV transmission of glass F2 for a thickness of 10 mm

Fresnel Reflectivity. For a light beam striking the surface perpendicularly, the Fresnel reflectivity is, independent of polarization,

$$R = (n-1)^2/(n+1)^2 \,. \tag{4.32}$$

Reflection Factor. The reflection factor, taking account of multiple reflection, is given by

$$P = (1-R)^2/(1-R^2) = 2n/(n^2+1) \,, \tag{4.33}$$

where n is the refractive index for the wavelength λ.

Conversion of Internal Transmittance to Another Layer Thickness. The conversion of data for internal transmittance to another sample thickness is accomplished by the use of the equation

$$\log \tau_{i1} / \log \tau_{i2} = d_1/d_2 \tag{4.34}$$

$$\text{or } \tau_{i2} = \tau_{i1}^{(d_2/d_1)} \,, \tag{4.35}$$

where τ_{i1} and τ_{i2} are the internal transmittances for the thicknesses d_1 and d_2, respectively.

Stress Birefringence. The change in optical path length for existing stress birefringence can be calculated from

$$\Delta s = (n_\parallel - n_\perp)d = (K_\parallel - K_\perp)d\sigma = Kd\sigma \,, \tag{4.36}$$

where K is the stress optical coefficient, dependent on the glass type, d is the length of the light path in the sample, and σ is the mechanical stress (positive for tensile stress). If K is given in $10^{-6}\,\text{mm}^2/\text{N}$, d is given in mm, and σ is measured in $\text{MPa} = \text{N/mm}^2$, Δs comes out in mm.

For the Pockels glass SF57, the stress optical coefficient K is close to 0 in the visible wavelength range.

Homogeneity. The homogeneity of the refractive index of a sample can be measured from the interferometrically measured wavefront deformation using the equation

$$\Delta n = \Delta W/2d$$
$$= \Delta W(\lambda) \times 633 \times 10^{-6}/(2d[\text{mm}]) \,, \tag{4.37}$$

where the wavefront deformation is in units of the wavelength and is measured using a test wavelength of 633 nm (He–Ne laser); ΔW is the wavefront deformation for double beam passage; and d is the thickness of the test piece. With special effort during melting and careful annealing, it is possible to produce pieces of glass having high homogeneity. The refractive-index homogeneity achievable for a given glass type depends on the volume and the form of the individual glass piece. Values of $\pm 5 \times 10^{-7}$ (class H5) cannot be achieved for all dimensions and glass types.

The properties of a selection of optical glasses are collected together in Table 3.4-16.

Internal Transmittance and Color Code
The internal transmittance, i.e. the light transmission excluding reflection losses, is closely related to the optical position of the glass type, according to general dispersion theory. This can be achieved, however, only by using purest raw materials and costly melting technology.

The internal transmittance of lead- and arsenic-free glasses, in which lead has been replaced by other elements, is markedly less than in the lead-containing predecessor glasses.

The limit of the transmission range of optical glasses towards the UV area is of special interest and is characterized by the position and slope of the UV absorption curve, which is described by a color code. The color code gives the wavelengths λ_{80} and λ_5, at which the transmission (including reflection losses) is 0.80 and 0.05, respectively, at 10 mm thickness. The color code 33/30 means, for example, $\lambda_{80} = 330\,\text{nm}$ and $\lambda_5 = 300\,\text{nm}$.

3.4.5.2 Chemical Properties

The composition of optical glasses includes elements that reduce chemical resistance. For these glasses, five test methods are used to assess the chemical behavior of polished glass surfaces in typical applications. The test methods and classification numbers take the place of those described for technical glasses in Sect. 3.4.4. Data for optical properties are found in Table 3.4-16c.

Climatic Resistance (ISO/WD 13384): Division into Climatic Resistance Classes CR 1–4
Climatic resistance describes the behavior of optical glasses at high relative humidity and high temperatures. In the case of sensitive glasses, a cloudy film can appear that generally cannot be wiped off.

The classifications are based on the increase in transmission haze ΔH after a 30 h test period. The glasses in class CR 1 display no visible attack after being subjected to 30 h of climatic change.

Under normal humidity conditions, no surface attack should be expected during the fabrication and storage of optical glasses in class CR 1. On the other hand, the fabrication and storage of optical glasses in class CR 4 should be done with caution because these glasses are very sensitive to climatic influences.

Stain Resistance: Division into Stain Resistance Classes FR 0–5

The test procedure gives information about possible changes in the glass surface (stain formation) under the influence of lightly acidic water (for example perspiration and acidic condensates) without vaporization. Two test solutions are used. Test solution I is a standard acetate solution with pH = 4.6, for classes FR 0 to 3. Test solution II is a sodium acetate buffer solution with pH = 5.6, for classes FR 4 and FR 5.

Interference color stains develop as a result of decomposition of the surface of the glass by the test solution. The measure used for classifying the glasses is the time that elapses before the first brown–blue stain occurs at a temperature of 25 °C.

Stain resistance class FR 0 contains all glasses that exhibit virtually no interference colors even after 100 h of exposure to test solution I.

Glasses in classification FR 5 must be handled with particular care during processing.

Acid Resistance (ISO 8424: 1987): Division into Acid Resistance Classes SR 1–4, 5, and 51–53

Acid resistance classifies the behavior of optical glasses that come into contact with large quantities of acidic solutions (from a practical standpoint, these may be perspiration, laminating substances, carbonated water, etc.).

The time t required to dissolve a layer with a thickness of 0.1 μm serves as a measure of acid resistance. Two aggressive solutions are used in determining acid resistance. A strong acid (nitric acid, $c = 0.5$ mol/l, pH = 0.3) at 25 °C is used for the more resistant glass types. For glasses with less acid resistance, a weakly acidic solution with a pH value of 4.6 (standard acetate) is used, also at 25 °C.

Alkali Resistance (ISO 10629) and Phosphate Resistance (ISO 9689): Division into Alkali Resistance Classes AR 1–4 and Phosphate Resistance Classes PR 1–4

These two test methods indicate the resistance to aqueous alkaline solutions in excess and use the same classification scheme. The alkali resistance indicates the sensitivity of optical glasses when they are in contact with warm, alkaline liquids, such as cooling liquids used in grinding and polishing processes. The phosphate resistance describes the behavior of optical glasses during cleaning with phosphate-containing washing solutions (detergents).

The alkali resistance class AR is based on the time required to remove a layer of glass of thickness 0.1 μm in an alkaline solution (sodium hydroxide, $c = 0.01$ mol/l, pH = 12) at a temperature of 50 °C.

The phosphate resistance class PR is based on the time required to remove a layer of glass of thickness 0.1 mm in an alkaline phosphate-containing solution (pentasodium triphosphate, $Na_5P_3O_{10}$, $c = 0.01$ mol/l, pH = 10) at a temperature of 50 °C. The thickness is calculated from the weight loss per unit surface area and the density of the glass.

3.4.5.3 Mechanical Properties

Young's Modulus and Poisson's Ratio

The adiabatic Young's modulus E (in units of 10^3 N/mm^2) and Poisson's ratio μ have been determined at room temperature and at a frequency of 1 kHz using carefully annealed test samples. Data are listed in Table 3.4-16c. In most cases, the values decrease slightly with temperature.

The torsional modulus can be calculated from

$$G = E/[2(1+\mu)] \,. \tag{4.38}$$

The longitudinal sound velocity is

$$v_{\text{long}} = \sqrt{\frac{E(1-\mu)}{\varrho(1+\mu)(1-2\mu)}} \,, \tag{4.39}$$

where ϱ is the density.

Knoop Hardness

The Knoop hardness (HK) of a material is a measure of the residual surface changes after the application of pressure with a test diamond. The standard ISO 9385 describes the measurement procedure for glasses. In accordance with this standard, values for Knoop hardness HK are listed in the data sheets for a test force of 0.9807 N (corresponds to 0.1 kp) and an effective test period of 20 s. The test was performed on polished glass surfaces at room temperature. The data for hardness values are rounded to 10 HK 0.1/20. The microhardness is a function of the magnitude of the test force and decreases with increasing test force.

Table 3.4-16a Properties of optical glasses. Refractive index and Sellmeier constants

Glass type	Refractive index		Abbe value		Constants of the Sellmeier dispersion formula					
	n_d	n_e	v_d	v_e	B_1	B_2	B_3	C_1	C_2	C_3
F2	1.62004	1.62408	36.37	36.11	$1.34533359 \times 10^{+00}$	$2.09073176 \times 10^{-01}$	$9.37357162 \times 10^{-01}$	$9.97743871 \times 10^{-03}$	$4.70450767 \times 10^{-02}$	$1.11886764 \times 10^{+02}$
K10	1.50137	1.50349	56.41	56.15	$1.15687082 \times 10^{+00}$	$6.42625444 \times 10^{-02}$	$8.72376139 \times 10^{-01}$	$8.09424251 \times 10^{-03}$	$3.86051284 \times 10^{-02}$	$1.04747730 \times 10^{+02}$
LASF35	2.02204	2.03035	29.06	28.84	$2.45505861 \times 10^{+00}$	$4.53006077 \times 10^{-01}$	$2.38513080 \times 10^{+00}$	$1.35670404 \times 10^{-02}$	$5.45803020 \times 10^{-02}$	$1.67904715 \times 10^{+02}$
LF5	1.58144	1.58482	40.85	40.57	$1.28035628 \times 10^{+00}$	$1.63505973 \times 10^{-01}$	$8.93930112 \times 10^{-01}$	$9.29854416 \times 10^{-03}$	$4.49135769 \times 10^{-02}$	$1.10493685 \times 10^{+02}$
LLF1	1.54814	1.55098	45.89	45.60	$1.23326922 \times 10^{+00}$	$1.16923839 \times 10^{-01}$	$8.62645379 \times 10^{-01}$	$8.85396812 \times 10^{-03}$	$4.36875155 \times 10^{-02}$	$1.04992168 \times 10^{+02}$
N-BAF10	1.67003	1.67341	47.11	46.83	$1.58514950 \times 10^{+00}$	$1.43559385 \times 10^{-01}$	$1.08521269 \times 10^{+00}$	$9.26681282 \times 10^{-03}$	$4.24489805 \times 10^{-02}$	$1.05613573 \times 10^{+02}$
N-BAF52	1.60863	1.61173	46.60	46.30	$1.43903433 \times 10^{+00}$	$9.67046052 \times 10^{-02}$	$1.09875818 \times 10^{+00}$	$9.07800128 \times 10^{-03}$	$5.08212080 \times 10^{-02}$	$1.05691856 \times 10^{+02}$
N-BAK4	1.56883	1.57125	55.98	55.70	$1.28834642 \times 10^{+00}$	$1.32817724 \times 10^{-01}$	$9.45395373 \times 10^{-01}$	$7.79980626 \times 10^{-03}$	$3.15631177 \times 10^{-02}$	$1.05965875 \times 10^{+02}$
N-BALF4	1.57956	1.58212	53.87	53.59	$1.31004128 \times 10^{+00}$	$1.42038259 \times 10^{-01}$	$9.64929351 \times 10^{-01}$	$7.96596450 \times 10^{-03}$	$3.30672072 \times 10^{-02}$	$1.09197320 \times 10^{+02}$
N-BASF64	1.70400	1.70824	39.38	39.12	$1.65554268 \times 10^{+00}$	$1.71319770 \times 10^{-01}$	$1.33664448 \times 10^{+00}$	$1.04485644 \times 10^{-02}$	$4.99394756 \times 10^{-02}$	$1.18961472 \times 10^{+02}$
N-BK7	1.51680	1.51872	64.17	63.96	$1.03961212 \times 10^{+00}$	$2.31792344 \times 10^{-01}$	$1.01046945 \times 10^{+00}$	$6.00069867 \times 10^{-03}$	$2.00179144 \times 10^{-02}$	$1.03560653 \times 10^{+02}$
N-FK56	1.43425	1.43534	94.95	94.53	$9.11957171 \times 10^{-01}$	$1.28580417 \times 10^{-01}$	$9.83146162 \times 10^{-01}$	$4.50933489 \times 10^{-03}$	$1.53515963 \times 10^{-02}$	$2.23961126 \times 10^{+02}$
N-KF9	1.52346	1.52588	51.54	51.26	$1.19286778 \times 10^{+00}$	$8.93346571 \times 10^{-02}$	$9.20819805 \times 10^{-01}$	$8.39154696 \times 10^{-03}$	$4.04010786 \times 10^{-02}$	$1.12572446 \times 10^{+02}$
N-KZFS2	1.55836	1.56082	54.01	53.83	$1.23697554 \times 10^{+00}$	$1.53569376 \times 10^{-01}$	$9.03976272 \times 10^{-01}$	$7.47170505 \times 10^{-03}$	$3.08053556 \times 10^{-02}$	$7.01731084 \times 10^{+01}$
N-LAF2	1.74397	1.74791	44.85	44.57	$1.80984227 \times 10^{+00}$	$1.57295550 \times 10^{-01}$	$1.09300370 \times 10^{+00}$	$1.01711622 \times 10^{-02}$	$4.42431765 \times 10^{-02}$	$1.00687748 \times 10^{+02}$
N-LAK33	1.75398	1.75740	52.43	52.20	$1.45796869 \times 10^{+00}$	$5.55403936 \times 10^{-01}$	$1.19938794 \times 10^{+00}$	$6.80545280 \times 10^{-03}$	$2.25253283 \times 10^{-02}$	$8.27543327 \times 10^{+01}$
N-LASF31	1.88067	1.88577	41.01	40.76	$1.71317198 \times 10^{+00}$	$7.18575109 \times 10^{-01}$	$1.72332470 \times 10^{+00}$	$8.19172228 \times 10^{-03}$	$2.97801704 \times 10^{-02}$	$1.38461313 \times 10^{+02}$
N-PK51	1.52855	1.53019	76.98	76.58	$1.15610775 \times 10^{+00}$	$1.53229344 \times 10^{-01}$	$7.85618966 \times 10^{-01}$	$5.85597402 \times 10^{-03}$	$1.94072416 \times 10^{-02}$	$1.40537046 \times 10^{+02}$
N-PSK57	1.59240	1.59447	68.40	68.01	$9.88511414 \times 10^{-01}$	$5.10855261 \times 10^{-01}$	$7.58837122 \times 10^{-01}$	$4.78397680 \times 10^{-03}$	$1.58020289 \times 10^{-02}$	$1.29709222 \times 10^{+02}$
N-SF1	1.71736	1.72308	29.62	29.39	$1.60865158 \times 10^{+00}$	$2.37725916 \times 10^{-01}$	$1.51530653 \times 10^{+00}$	$1.19654879 \times 10^{-02}$	$5.90589722 \times 10^{-02}$	$1.35521676 \times 10^{+02}$
N-SF56	1.78470	1.79179	26.10	25.89	$1.73562085 \times 10^{+00}$	$3.17487012 \times 10^{-01}$	$1.95398203 \times 10^{+00}$	$1.29624742 \times 10^{-02}$	$6.12884288 \times 10^{-02}$	$1.61559441 \times 10^{+02}$
N-SK16	1.62041	1.62286	60.32	60.08	$1.34317774 \times 10^{+00}$	$2.41144399 \times 10^{-01}$	$9.94317969 \times 10^{-01}$	$7.04687339 \times 10^{-03}$	$2.29005000 \times 10^{-02}$	$9.27508526 \times 10^{+01}$
N-SSK2	1.62229	1.62508	53.27	52.99	$1.43060270 \times 10^{+00}$	$1.53150554 \times 10^{-01}$	$1.01390904 \times 10^{+00}$	$8.23982975 \times 10^{-03}$	$3.33736841 \times 10^{-02}$	$1.06870822 \times 10^{+02}$
SF1	1.71736	1.72310	29.51	29.29	$1.55912923 \times 10^{+00}$	$2.84246288 \times 10^{-01}$	$9.68842926 \times 10^{-01}$	$1.21481001 \times 10^{-02}$	$5.34549042 \times 10^{-02}$	$1.12174809 \times 10^{+02}$
SF11	1.78472	1.79190	25.76	25.55	$1.73848403 \times 10^{+00}$	$3.11168974 \times 10^{-01}$	$1.17490871 \times 10^{+00}$	$1.36068604 \times 10^{-02}$	$6.15960463 \times 10^{-02}$	$1.21922711 \times 10^{+02}$
SF2	1.64769	1.65222	33.85	33.60	$1.40301821 \times 10^{+00}$	$2.31767504 \times 10^{-01}$	$9.39056586 \times 10^{-01}$	$1.05795466 \times 10^{-02}$	$4.93226978 \times 10^{-02}$	$1.12405955 \times 10^{+02}$
SF66	1.92286	1.93325	20.88	20.73	$2.07842233 \times 10^{+00}$	$4.07120032 \times 10^{-01}$	$1.76711292 \times 10^{+00}$	$1.80875134 \times 10^{-02}$	$6.79493572 \times 10^{-02}$	$2.15266127 \times 10^{+02}$
SK51	1.62090	1.62335	60.31	60.02	$1.44112715 \times 10^{+00}$	$1.43968387 \times 10^{-01}$	$8.81989862 \times 10^{-01}$	$7.58546975 \times 10^{-03}$	$2.87396017 \times 10^{-02}$	$9.46838154 \times 10^{+01}$
K7	1.51112	1.51314	60.41	60.15	$1.12735550 \times 10^{+00}$	$1.24412303 \times 10^{-01}$	$8.27100531 \times 10^{-01}$	$7.20341707 \times 10^{-03}$	$2.69835916 \times 10^{-02}$	$1.00384588 \times 10^{+02}$
N-SF6	1.80518	1.81266	25.36	25.16	$1.77931763 \times 10^{+00}$	$3.38149866 \times 10^{-01}$	$2.08734474 \times 10^{+00}$	$1.33714182 \times 10^{-02}$	$6.17533621 \times 10^{-02}$	$1.74017590 \times 10^{+02}$
SF6	1.80518	1.81265	25.43	25.24	$1.72448482 \times 10^{+00}$	$3.90104889 \times 10^{-01}$	$1.04572858 \times 10^{+00}$	$1.34871947 \times 10^{-02}$	$5.69318095 \times 10^{-02}$	$1.18557185 \times 10^{+02}$
N-FK51	1.48656	1.48794	84.47	84.07	$9.71247817 \times 10^{-01}$	$2.16901417 \times 10^{-01}$	$9.04651666 \times 10^{-01}$	$4.72301995 \times 10^{-03}$	$1.53575612 \times 10^{-02}$	$1.68681330 \times 10^{+02}$
Lithosil™ Q	1.45843	1.46004	67.87	67.67	$6.69422575 \times 10^{-01}$	$4.34583937 \times 10^{-01}$	$8.71694723 \times 10^{-01}$	$4.48011239 \times 10^{-03}$	$1.32847049 \times 10^{-02}$	$9.53414824 \times 10^{+01}$

Table 3.4-16b Data for dn/dT

Glass type	Data for dn/dT					
	$10^6 D_0$	$10^8 D_1$	$10^{11} D_2$	$10^7 E_0$	$10^{10} E_1$	λ_{TK} (μm)
F2	1.51	1.56	−2.78	9.34	10.4	0.250
K10	4.86	1.72	−3.02	3.82	4.53	0.260
LASF35	0.143	0.871	−2.71	10.2	15.0	0.263
LF5	−2.27	0.971	−2.83	8.36	9.95	0.228
LLF1	0.325	1.74	−6.12	6.53	2.58	0.233
N-BAF10	3.79	1.28	−1.42	5.84	7.60	0.220
N-BAF52	1.15	1.27	−0.508	5.64	6.38	0.238
N-BAK4	3.06	1.44	−2.23	5.46	6.05	0.189
N-BALF4	5.33	1.47	−1.58	5.75	6.58	0.195
N-BASF64	1.60	1.02	−2.68	7.87	9.65	0.229
N-BK7	1.86	1.31	−1.37	4.34	6.27	0.170
N-FK56	−20.4	−1.03	0.243	3.41	4.37	0.138
N-KF9	−1.66	0.844	−1.01	6.10	6.96	0.217
N-KZFS2	6.77	1.31	−1.23	3.84	5.51	0.196
N-LAF2	−3.64	0.920	−0.600	6.43	6.11	0.220
N-LAK33	2.57	1.16	−7.29	6.01	1.59	0.114
N-LASF31	2.29	0.893	−1.59	6.52	8.09	0.236
N-PK51	−19.8	−0.606	1.60	4.16	5.01	0.134
N-PSK57	−22.3	−0.560	0.997	4.47	5.63	−
N-SF1	−3.72	0.805	−1.71	8.98	13.4	0.276
N-SF56	−4.13	0.765	−1.12	9.90	15.7	0.287
N-SK16	−0.0237	1.32	−1.29	4.09	5.17	0.170
N-SSK2	5.21	1.34	−1.01	5.21	5.87	0.199
SF1	4.84	1.70	−4.52	13.8	12.6	0.259
SF11	11.2	1.81	−5.03	14.6	15.8	0.282
SF2	1.10	1.75	−1.29	10.8	10.3	0.249
SF66	−	−	−	−	−	−
SK51	−5.63	0.738	−6.20	3.91	2.64	0.230
K7	−1.67	0.880	−2.86	5.42	7.81	0.172
N-SF6	−4.93	0.702	−2.40	9.84	15.4	0.290
SF6	6.69	1.78	−3.36	17.7	17.0	0.272
N-FK51	−18.3	−0.789	−0.163	3.74	3.46	0.150
Lithosil™Q	20.6	2.51	−2.47	3.12	4.22	0.160

Table 3.4-16c Chemical and physical data

Glass type	Stress-optical coefficient K ($10^{-6}\,\text{mm}^2/\text{N}$)	Chemical properties					Density (g/cm^3)	Viscosity (dPa s) at temperature (°C)			Thermal properties		Thermal expansion		Mechanical properties		Knoop hardness HK
		CR	FR	SR	AR	PR		$10^{14.5}$	10^{13}	$10^{7.6}$	Heat capacity c_p (J/g K)	Heat conductivity λ (W/m K)	$\alpha_{(30/70)}$ ($10^{-6}/\text{K}$)	$\alpha_{(20/300)}$ ($10^{-6}/\text{K}$)	Young's modulus E ($10^3\,\text{N/mm}^2$)	Poisson's ratio μ	
F2	2.81	1	0	1	2.3	1.3	3.61	432	421	593	0.557	0.780	8.20	9.20	57	0.220	420
K10	3.12	1	0	1	1	1.2	2.52	459	453	691	0.770	1.120	6.50	7.40	65	0.190	470
LASF35	0.73	1	0	1.3	1	1.3	5.41	774	–	–	0.445	0.920	7.40	8.50	132	0.303	810
LF5	2.83	2	0	1	2.3	2	3.22	419	411	585	0.657	0.866	9.10	10.60	59	0.223	450
LLF1	3.05	1	0	1	2	1	2.94	448	426	628	0.650	–	8.10	9.20	60	0.208	450
N-BAF10	2.37	1	0	4.3	1.3	1	3.75	660	652	790	0.560	0.780	6.18	7.04	89	0.271	620
N-BAF52	2.42	1	0	1	1.3	1	3.05	594	596	723	0.680	0.960	6.86	7.83	86	0.237	600
N-BAK4	2.90	1	0	1.2	1	1	3.05	581	569	725	0.680	0.880	6.99	7.93	77	0.240	550
N-BALF4	3.01	1	0	1	1	1	3.11	578	584	661	0.690	0.850	6.52	7.41	77	0.245	540
N-BASF64	2.38	1	0	3.2	1.2	1	3.20	582	585	712	–	–	7.30	8.70	105	0.264	650
N-BK7	2.77	2	0	1	2	2.3	2.51	557	557	719	0.858	1.114	7.10	8.30	82	0.206	610
N-FK56	0.68	1	0	52.3	4.3	4.3	3.54	422	416	–	0.750	0.840	–	16.16	70	0.293	350
N-KF9	2.74	1	0	1	1	1	2.50	476	476	640	0.860	1.040	9.61	10.95	66	0.225	480
N-KZFS2	4.02	1	4	52.3	4.3	4.2	2.55	491	488	600	0.830	0.810	4.43	5.43	66	0.266	490
N-LAF2	1.42	2	3	52.2	1	2.2	4.30	653	645	742	0.510	0.670	8.06	9.10	94	0.288	530
N-LAK33	1.49	1	1	51.3	1	2.3	4.26	652	648	–	0.554	0.900	6.00	7.00	124	0.291	780
N-LASF31	1.10	1	0	2	1	1	5.41	758	756	–	–	0.910	6.80	7.70	124	0.299	770
N-PK51	0.54	2	0	51.2	3.3	4.3	3.96	496	486	–	–	0.560	12.70	14.40	74	0.295	400
N-PSK57	0.13	1	0	51.3	1.2	4.3	4.48	497	499	–	0.490	–	13.17	14.75	69	0.298	370
N-SF1	2.72	1	0	1	1	1	3.03	553	554	660	0.750	1.000	9.13	10.54	90	0.250	540
N-SF56	2.87	1	0	1	1.3	1	3.28	592	585	691	0.700	0.940	8.70	10.00	91	0.255	560
N-SK16	1.90	4	4	53.3	3.3	3.2	3.58	636	633	750	0.578	0.818	6.30	7.30	89	0.264	600
N-SSK2	2.51	1	0	1.2	1	1	3.53	653	655	801	0.580	0.810	5.81	6.65	82	0.261	570
SF1	1.80	2	1	3.2	2.3	3	4.46	417	415	566	–	0.737	8.10	8.80	56	0.232	390
SF11	1.33	1	0	1	1.2	1	4.74	503	500	635	0.431	–	6.10	6.80	66	0.235	450
SF2	2.62	1	0	2	2.3	2	3.86	441	428	600	0.498	0.735	8.40	9.20	55	0.227	410
SF66	–1.20	2	5	53.4	2.3	4.2	6.03	384	385	482	0.340	0.530	9.01	11.48	51	0.258	310
SK51	1.47	2	3	52.3	1.3	4.3	3.52	597	579	684	–	–	8.90	10.10	75	0.291	450
K7	2.95	3	0	2	1	2.3	2.53	513	–	712	–	–	8.4	9.7	69	0.214	520
N-SF6	2.82	1	0	2	1	1	3.37	594	591	694	0.69	0.96	9.03	10.39	93	0.262	550
SF6	0.65	2	3	51.3	2.3	3.3	5.18	423	410	538	0.389	0.673	8.1	9	55	0.244	370
N-FK51	0.70	2	0	52.3	2.2	4.3	3.73	420	403	–	0.636	0.911	13.3	15.3	81	0.293	430
Lithosil™ Q	3.40	–	–	1	1	–	2.20	980	1080	1600	0.790	1.310	0.50	–	72	0.170	580

Table 3.4-16d Internal transmission and color code

Glass type	Color code	Internal transmission measured for 25 mm sample thickness at wavelength λ (nm)											
		2500	2325	1970	1530	1060	700	660	620	580	546	500	460	436
F2	35/32	0.610	0.700	0.890	0.990	0.998	0.998	0.996	0.997	0.997	0.997	0.996	0.993	0.991
K10	33/30	0.520	0.630	0.850	0.983	0.996	0.997	0.994	0.993	0.993	0.992	0.991	0.990	0.988
LASF35	–/37	0.690	0.880	0.972	0.992	0.990	0.978	0.970	0.962	0.950	0.920	0.810	0.630	0.470
LF5	34/31	–	0.660	0.870	0.992	0.998	0.998	0.998	0.998	0.997	0.997	0.996	0.995	0.994
LLF1	33/31	0.500	0.610	0.840	0.990	0.996	0.997	0.996	0.996	0.997	0.997	0.996	0.996	0.996
N-BAF10	39/35	0.450	0.680	0.920	0.980	0.994	0.994	0.990	0.991	0.990	0.990	0.981	0.967	0.954
N-BAF52	39/35	0.390	0.630	0.890	0.975	0.994	0.993	0.990	0.989	0.990	0.989	0.980	0.967	0.954
N-BAK4	36/33	0.540	0.710	0.900	0.982	0.995	0.997	0.995	0.995	0.996	0.996	0.994	0.989	0.988
N-BALF4	37/33	0.580	0.740	0.920	0.984	0.993	0.997	0.995	0.995	0.996	0.995	0.993	0.986	0.983
N-BASF64	40/35	0.450	0.670	0.900	0.970	0.985	0.970	0.955	0.949	0.949	0.950	0.940	0.920	0.900
N-BK7	33/29	0.360	0.560	0.840	0.980	0.997	0.996	0.994	0.994	0.995	0.996	0.994	0.993	0.992
N-FK56	33/28	–	–	0.979	0.991	0.996	0.996	0.996	0.996	0.996	0.996	0.996	0.996	0.995
N-KF9	37/34	0.300	0.430	0.740	0.981	0.995	0.997	0.995	0.994	0.996	0.996	0.994	0.990	0.988
N-KZFS2	34/30	0.040	0.260	0.800	0.940	0.991	0.996	0.994	0.994	0.994	0.994	0.992	0.987	0.981
N-LAF2	40/34	0.400	0.690	0.930	0.990	0.997	0.996	0.993	0.992	0.993	0.994	0.983	0.962	0.940
N-LAK33	39/32	0.090	0.400	0.850	0.975	0.995	0.991	0.990	0.990	0.990	0.990	0.987	0.977	0.967
N-LASF31	45/32	0.540	0.810	0.960	0.992	0.993	0.994	0.994	0.993	0.993	0.990	0.973	0.940	0.910
N-PK51	35/29	0.890	0.920	0.965	0.985	0.992	0.991	0.991	0.992	0.994	0.995	0.993	0.989	0.987
N-PSK57	34/29	–	–	0.950	0.970	0.982	0.996	0.996	0.996	0.996	0.996	0.992	0.991	0.991
N-SF1	41/36	0.460	0.580	0.850	0.973	0.995	0.990	0.986	0.987	0.990	0.986	0.968	0.940	0.910
N-SF56	44/37	0.590	0.680	0.900	0.981	0.996	0.986	0.981	0.981	0.983	0.976	0.950	0.910	0.860
N-SK16	36/30	0.260	0.540	0.880	0.973	0.995	0.996	0.994	0.993	0.994	0.994	0.991	0.984	0.981
N-SSK2	37/33	0.500	0.720	0.930	0.981	0.992	0.996	0.994	0.993	0.995	0.995	0.992	0.985	0.980
SF1	39/34	0.650	0.730	0.900	0.985	0.996	0.996	0.995	0.995	0.996	0.996	0.993	0.984	0.976
SF11	44/39	0.610	0.700	0.930	0.982	0.997	0.993	0.991	0.991	0.991	0.989	0.976	0.940	0.860
SF2	37/33	0.620	0.710	0.880	0.985	0.996	0.996	0.994	0.995	0.995	0.995	0.993	0.988	0.982
SF66	48/38	0.700	0.740	0.920	0.990	0.995	0.990	0.989	0.989	0.988	0.985	0.965	0.890	0.770
SK51	36/31	0.270	0.520	0.830	0.959	0.993	0.993	0.993	0.993	0.993	0.993	0.990	0.981	0.975
K7	33/30	0.340	0.500	0.790	0.980	0.994	0.996	0.995	0.995	0.994	0.994	0.993	0.990	0.990
N-SF6	45/37	0.850	0.880	0.962	0.994	0.994	0.987	0.980	0.979	0.980	0.970	0.940	0.899	0.850
SF6	42/36	0.730	0.780	0.930	0.990	0.996	0.996	0.995	0.995	0.995	0.994	0.989	0.972	0.940
N-FK51	34/28	0.750	0.840	0.940	0.980	0.994	0.995	0.995	0.996	0.997	0.997	0.996	0.993	0.992
Lithosil™Q	17/16	0.780	–	–	–	–	–	–	–	–	–	–	–	–

Internal transmission measured for 25 mm sample thickness at wavelength λ (nm)

420	405	400	390	380	370	365	350	334	320	310	300	290	248	200	193
0.990	0.986	0.984	0.977	0.963	0.940	0.920	0.780	0.210	–	–	–	–	–	–	–
0.988	0.987	0.986	0.982	0.973	0.966	0.958	0.910	0.720	0.310	0.130	0.020	–	–	–	–
0.320	0.170	0.120	0.050	0.010	–	–	–	–	–	–	–	–	–	–	–
0.993	0.992	0.992	0.984	0.973	0.961	0.954	0.880	0.570	0.040	–	–	–	–	–	–
0.995	0.994	0.993	0.992	0.988	0.984	0.981	0.955	0.810	0.300	0.010	–	–	–	–	–
0.940	0.900	0.880	0.800	0.660	0.440	0.310	0.010	–	–	–	–	–	–	–	–
0.938	0.900	0.880	0.800	0.650	0.370	0.210	–	–	–	–	–	–	–	–	–
0.987	0.983	0.980	0.967	0.940	0.890	0.840	0.550	0.070	–	–	–	–	–	–	–
0.981	0.970	0.964	0.940	0.900	0.820	0.750	0.380	–	–	–	–	–	–	–	–
0.880	0.840	0.820	0.750	0.610	0.370	0.220	–	–	–	–	–	–	–	–	–
0.993	0.993	0.992	0.989	0.983	0.977	0.971	0.920	0.780	0.520	0.250	0.050	–	–	–	–
0.994	0.996	0.996	0.995	0.992	0.985	0.975	0.920	0.760	0.460	0.210	0.060	0.010	–	–	–
0.985	0.975	0.965	0.940	0.880	0.770	0.680	0.210	–	–	–	–	–	–	–	–
0.975	0.967	0.963	0.950	0.930	0.910	0.890	0.800	0.590	0.240	0.030	–	–	–	–	–
0.915	0.865	0.840	0.760	0.630	0.430	0.310	0.025	–	–	–	–	–	–	–	–
0.954	0.928	0.910	0.860	0.790	0.690	0.630	0.400	0.140	0.020	–	–	–	–	–	–
0.880	0.840	0.820	0.750	0.650	0.530	0.460	0.210	0.040	0.020	–	–	–	–	–	–
0.986	0.985	0.984	0.977	0.965	0.940	0.910	0.750	0.430	0.120	0.030	–	–	–	–	–
0.991	0.991	0.992	0.992	0.989	0.975	0.965	0.880	0.680	0.380	0.130	0.020	–	–	–	–
0.870	0.760	0.700	0.520	0.250	0.030	–	–	–	–	–	–	–	–	–	–
0.780	0.640	0.570	0.370	0.130	–	–	–	–	–	–	–	–	–	–	–
0.979	0.974	0.970	0.956	0.930	0.890	0.860	0.700	0.400	0.110	0.020	–	–	–	–	–
0.975	0.963	0.954	0.920	0.860	0.750	0.670	0.250	–	–	–	–	–	–	–	–
0.961	0.930	0.920	0.870	0.790	0.640	0.500	0.030	–	–	–	–	–	–	–	–
0.700	0.340	0.200	0.010	–	–	–	–	–	–	–	–	–	–	–	–
0.975	0.962	0.954	0.920	0.870	0.790	0.720	0.370	–	–	–	–	–	–	–	–
0.610	0.340	0.240	0.050	–	–	–	–	–	–	–	–	–	–	–	–
0.971	0.963	0.958	0.940	0.910	0.850	0.800	0.600	0.300	0.100	0.030	–	–	–	–	–
0.990	0.990	0.990	0.988	0.983	0.976	0.971	0.940	0.780	0.420	0.100	–	–	–	–	–
0.780	0.640	0.570	0.370	0.140	–	–	–	–	–	–	–	–	–	–	–
0.900	0.810	0.760	0.620	0.370	0.100	0.020	–	–	–	–	–	–	–	–	–
0.992	0.993	0.993	0.992	0.988	0.976	0.963	0.875	0.630	0.300	0.120	0.035	0.010	–	–	–
–	–	–	–	–	–	–	–	–	–	–	–	–	0.995	0.990	0.980

Viscosity

As explained in the introduction, glasses pass through three viscosity ranges between the melting temperature and room temperature: the melting range, the supercooled melt range, and the solidification range. The viscosity increases during the cooling of the melt, starting from 10^0–10^4 dPa s. A transition from a liquid to a plastic state is observed between 10^4 and 10^{13} dPa s.

The softening point, i.e. the temperature where the viscosity is $10^{7.6}$ dPa s, identifies the plastic range in which glass parts rapidly deform under their own weight. The glass structure can be described as solidified or "frozen" above 10^{13} dPa s. At this viscosity, the internal stresses in glass anneal out equalize in approximately 15 min. The temperature at which the viscosity is 10^{13} dPa s is called the upper annealing point, and is important for the annealing of glasses.

In accordance with ISO 7884-8, the rate of change of the relative linear thermal expansion can be used to determine the transformation temperature T_g, which is close to the temperature at which the viscosity is 10^{13} dPa s.

Precision optical surfaces may deform and refractive indices may change if a temperature of $T_g - 200$ K is exceeded during any thermal treatment.

Coefficient of Linear Thermal Expansion

The typical curve of the linear thermal expansion of a glass begins with an increase in slope from absolute zero to approximately room temperature. Then a nearly linear increase to the beginning of the plastic behavior follows. The transformation range is distinguished by a distinct bending of the expansion curve, which results from the increasing structural rearrangement in the glass. Above this range, the expansion again exhibits a nearly linear increase, but with a noticeably greater slope.

Two averaged coefficients of linear thermal expansion α are usually given: $\alpha_{30/70}$, averaged from $-30\,°\mathrm{C}$ to $+70\,°\mathrm{C}$, which is the relevant value for room temperature; and $\alpha_{20/300}$, averaged from $+20\,°\mathrm{C}$ to $+300\,°\mathrm{C}$, which is the standard international value. These values are listed in Table 3.4-16.

3.4.5.4 Thermal Properties

Thermal Conductivity

The range of values for the thermal conductivity of glasses at room temperature extends from 1.38 W/m K (pure vitreous silica) to about 0.5 W/m K (high-lead-content glasses). The most commonly used silicate glasses have values between 0.9 and 1.2 W/m K. All data in Table 3.4-16c are given for a temperature of 90 °C, with an accuracy of $\pm 5\%$.

Specific Thermal Capacity

The mean isobaric specific heat capacities c_p (20 °C; 100 °C) listed in Table 3.4-16c were measured from the heat transfer from a hot glass sample at 100 °C into a liquid calorimeter at 20 °C. The values of $c_p(20\,°\mathrm{C};\ 100\,°\mathrm{C})$ and also of the true thermal capacity $c_p(20\,°\mathrm{C})$ for silicate glasses range from 0.42 to 0.84 J/g K.

3.4.6 Vitreous Silica

Vitreous silica has a unique set of properties. It is produced either from natural quartz by fusion or, if extreme purity is required, by chemical vapor deposition or via a sol–gel routes. Depending on the manufacturing process, variable quantities impurities are incorporated in the ppm or ppb range, such as Fe, Mg, Al, Mn, Ti, Ce, OH, Cl, and F. These impurities and radiation-induced defects, as well as complexes of impurities and defects, and also overtones, control the UV and IR transmittance. In the visible part of the spectrum, Rayleigh scattering from thermodynamically caused density fluctuations dominates. Defects are also responsible for the damage threshold under radiation load, and for fluorescence. The refractive index n and the absorption constant K as a function of wavelength are found in Fig. 3.4-29.

The highest transmittance is required for applications in optical communication networks, in optics for lithography, and in high-power laser physics. For certain applications, for example to increase the refractive index in the IR in fiber optics, the silica is "doped" with GeO_2, P_2O_5, B_2O_3, etc. in the range of 5–10%. In such cases the scattering loss increases owing to concentration fluctuations.

There are also many technical applications which make use of the chemical inertness, light weight, high temperature stability, thermal-shock resistance, and low thermal expansion of vitreous silica. A very low thermal

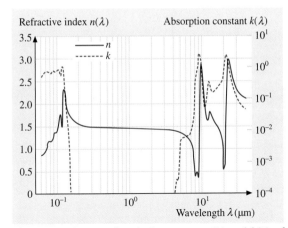

Fig. 3.4-29 Measured optical constants $n(\lambda)$ and $k(\lambda)$ of vitreous silica according to [4.16]

expansion is obtained in ULE glass (Corning "ultralow expansion" glass) by doping with $\approx 9\%$ TiO$_2$.

3.4.6.1 Properties of Synthetic Silica

The precise data for materials from various suppliers differ slightly, depending on the thermal history and impurity concentration. The data listed in Table 3.4-16a–d and in Table 3.4-17, are for Lithosil^{TM}Q0 (Schott Lithotec). The various quantities are defined in the same way as for optical glasses, as described in Sect. 3.4.5.

Table 3.4-17 Electrical properties of vitreous silica (Lithosil^{TM})

Dielectric constant ε_r	3.8 ± 0.2
Dielectric loss angle φ	$89.92° \pm 0.03°$ at 25 °C and 1 MHz
$\tan \delta$ ($\delta = 90° - \varphi$)	$14 \pm 5 \times 10^{-4}$
Electrical resistivity	1.15×10^{18} (Ω cm) at 20 °C

3.4.6.2 Gas Solubility and Molecular Diffusion

The relatively open structure of vitreous silica provides space for the incorporation and diffusion of molecular species. The data in the literature are not very consistent; Table 3.4-18 should serve as an orientation.

The pressure dependence of the solubility is small up to about 100 atm.

The diffusion coefficient depends on temperature as

$$D = D_0 T \exp(-Q/RT). \quad (4.40)$$

Water can react with the silica network:

$$H_2O + Si-O-Si = 2\,Si-OH. \quad (4.41)$$

The reaction has a strong influence on the concentration and apparent diffusion of dissolved molecular water.

Table 3.4-18 Solubility and diffusion of molecular gases in vitreous silica (Lithosil^{TM})

Gas	Molecular diameter (nm)	c_{glass}/c_{gas} at 200–1000 °C	Dissolved molecules S (cm^{-3} atm^{-1}) at 200 °C	Diffusion coefficient D_0 (cm^2/s) 25 °C	Diffusion coefficient D_0 (cm^2/s) 1000 °C	Activation energy Q (kJ/mole)
Helium	0.20	0.025	3.9×10^{-17}	2.4×10^{-8}	5.5×10^{-5}	20
Neon	0.24	0.019	3.1×10^{-17}	5.0×10^{-12}	2.5×10^{-6}	37
Hydrogen	0.25	0.03	4.7×10^{-17}	2.2×10^{-11}	7.3×10^{-6}	36
Argon	0.32	0.01	1.5×10^{-17}	–	1.4×10^{-9}	111
Oxygen	0.32	0.01	1.5×10^{-17}	–	6.6×10^{-9}	105
Water	0.33	–	–	–	$\approx 3.0 \times 10^{-7}$	71
Nitrogen	0.34	–	–	–	–	110
Krypton	0.42	–	–	–	–	≈ 190
Xenon	0.49	–	–	–	–	≈ 300

3.4.7 Glass-Ceramics

Glass-ceramics are distinguished from glasses and from ceramics by the characteristics of their manufacturing processes (see introduction to this chapter 3.4) as well as by their physico-chemical properties.

They are manufactured in two principal production steps. In the first step, a batch of exactly defined composition is melted (as for a conventional glass). The composition is determined by the desired properties of the endproduct and by the necessary working properties of the glass. After melting, the product is shaped by pressing, blowing, rolling, or casting, and then annealed. In this second step, "glassy" articles are partly crystallized by use of a specific temperature–time program between 800 and 1200 °C (this program must be defined for each composition). Apart from the crystalline phase, with crystals 0.05–5 μm in size, this material contains a residual glass phase that amounts to 5–50% of the volume.

In the temperature range between 600 and 700 °C, small amounts of nucleating agents (e.g. TiO_2, ZrO_2, or F) induce precipitation of crystal nuclei. When the temperature is increased, crystals grow on these nuclei. Their type and properties, as well as their number and size, are predetermined by the glass composition and the annealing program. By selection of an appropriate program, either transparent, slightly opaque, or highly opaque, nontransparent glass-ceramics can be produced. Unlike conventional ceramics, these glass ceramics are fully dense and pore-free.

Like the composition of glasses, the composition of glass-ceramics is highly variable. Some well-known compositions lie within the following systems: Li_2O–Al_2O_3–SiO_2, MgO–Al_2O_3–SiO_2, and CaO–P_2O_5–Al_2O_3–SiO_2.

Glass-ceramics of the Li_2O–Al_2O_3–SiO_2 system, which contain small amounts of alkali and alkaline-earth oxides, as well as TiO_2 and ZrO_2 as nucleating agents, have achieved the greatest commercial importance. On the basis of this system, glass-ceramics with coefficients of linear thermal expansion near to zero can be produced (Fig. 3.4-30 and Table 3.4-19). This exceptional property results from the bonding of crystalline constituents (such as solid solutions of h-quartz, h-eucryptite, or h-spodumene) which have negative coefficients of thermal expansion with the residual glass phase of the system, which has a positive coefficient of thermal expansion.

Such "$\alpha = 0$ glass-ceramics" can be subjected to virtually any thermal shock or temperature variation below 700 °C. Wall thickness, wall thickness differences, and complicated shapes are of no significance.

Another technical advantage is the exceptionally high dimensional and shape stability of objects made from these materials, even when the objects are subjected to considerable temperature variations.

The Zerodur® glass-ceramic, whose coefficient of linear thermal expansion at room temperature can be kept at $\leq 0.05 \times 10^{-6}$ /K (Table 3.4-19), was especially developed for the production of large mirror blanks for astronomical telescopes. Zerodur® has further applications in optomechanical precision components such as length standards, and mirror spacers in lasers. With

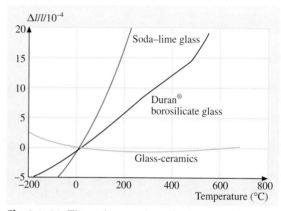

Fig. 3.4-30 Thermal expansion of glass-ceramics compared with borosilicate glass 3.3 and soda–lime glass

Table 3.4-19 Coefficient of linear thermal expansion α, density, and elastic properties of Zerodur® and Ceran® glass-ceramics

	Zerodur®	Ceran®	Units	Product class
$\alpha_{0/50}$	0 ± 0.05	–	10^{-6}/K	1
	0 ± 0.1	–	10^{-6}/K	2
	0 ± 0.15	–	10^{-6}/K	3
$\alpha_{20/300}$	$+0.1$	-0.2	10^{-6}/K	
$\alpha_{20/500}$	–	-0.01	10^{-6}/K	
$\alpha_{20/600}$	$+0.2$	–	10^{-6}/K	
$\alpha_{20/700}$	–	$+0.15$	10^{-6}/K	
Density	2.53	2.56	g/cm³	
Young's modulus E	91×10^3	92×10^3	N/mm²	
Poisson's ratio μ	0.24	0.24		

a length aging coefficient A (where $L = L_0(1 + A\Delta t)$, Δt = time span) below 1×10^{-7}/y, Zerodur® has excellent longitudinal stability.

The Ceran® glass-ceramic is colored and is designed for applications in cooker surface panels.

As in glasses, the variability of the composition can be used to design very different sets of properties of glass-ceramics. Some examples are:

- photosensitive, etchable glass-ceramics based on Ag doping: Foturan (Schott), and Fotoform and Fotoceram (Corning);
- machinable glass-ceramics based on mica crystals, for example for electronic packaging: Macor and Dicor (Corning), and Vitronit (Vitron and Jena);
- glass-ceramics used as substrates for magnetic disks, based on spinel or gahnite crystals, resulting in a very high elastic modulus and thus stiffness: Neoceram (NEG), and products from Corning and Ohara;
- glass-ceramics with extremely good weathering properties for architectural applications: Neoparies (NEG) and Cryston (Asahi Glass);
- biocompatible, bioactive glass-ceramics based on apatite and orthophosphate crystals for dental restoration or bone replacement in medicine: Cerabone (NEG), Bioverit (Vitron), Ceravital, IPS Empress, etc.;
- highly transparent glass-ceramics and glass-ceramics with specific dopings for temperature-resistant fiber optic components, high-temperature loaded color filters, and luminescent solar collectors.

An excellent overview and many details can be found in [4.4].

3.4.8 Glasses for Miscellaneous Applications

3.4.8.1 Sealing Glasses

Glasses are very well suited for the production of mechanically reliable, vacuum-tight fusion seals with metals, ceramics, and mica. Some particularly favorable properties are the viscosity behavior of glass and the direct wettability of many crystalline materials by glasses. As a result, the production technology for such seals is characterized by uncomplicated procedures with few, easily manageable, well-controllable process steps.

A necessary condition for the stability and mechanical strength of glass seals is the limitation of the mechanical stress in the glass component at temperatures encountered during production and use. To ensure "sealability" (which means that the thermal contractions of the two sealing components match each other below the transformation temperature of the glass), glasses of special compositions, called sealing glasses, have been developed. Apart from sealability, such glasses must very often fulfill other requirements such as high electrical insulation or special optical properties. The sealability can be tested and evaluated with sufficient accuracy and certainty by stress-optical measurements in the glass portion of a test seal (ISO 4790).

Apart from characteristic material values such as the coefficient of linear thermal expansion, transformation temperature, and elastic properties, the cooling rate (Fig. 3.4-31) and the shape can also have a considerable influence on the degree and distribution of seal stresses. The material combinations for sealing between metals and ceramics recommended for Schott glasses are shown in Fig. 3.4-32.

Types of Sealing Glasses

Sealing glasses may be classified by reference to the expansion coefficients of metals (e.g. tungsten and molybdenum) and alloys (Ni–Fe–Co, Ni–Fe–Cr, and other alloys) with which they are used. Hence sealing

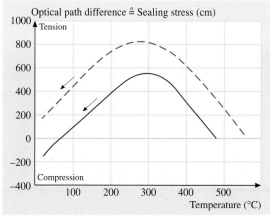

Fig. 3.4-31 Influence of the cooling rate on the sealing stress in an 8516–Ni/Fe combination. The *lower curve* corresponds to a low cooling rate; the *upper curve* corresponds to a high cooling rate

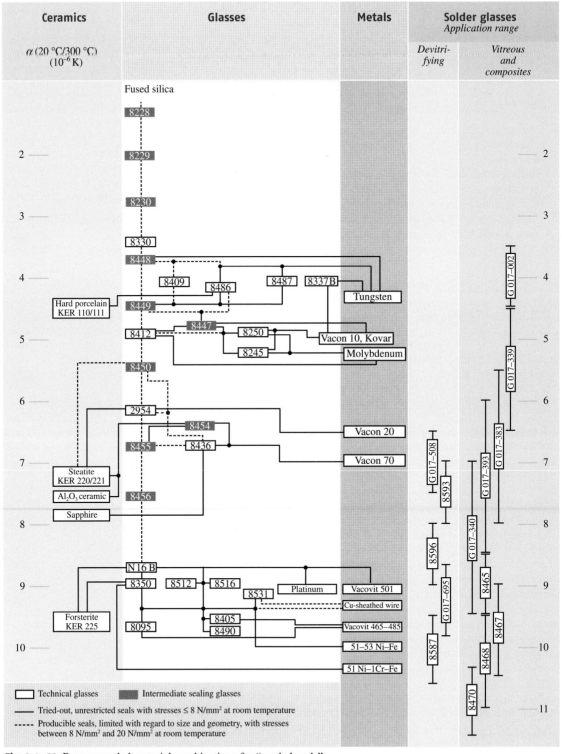

Fig. 3.4-32 Recommended material combinations for "graded seals"

Table 3.4-20 Special properties and principal applications of technically important sealing glasses, arranged according to their respective sealing partners

Metal $\alpha_{20/300}$ (10^{-6}/K)	Glass number	Glass characteristics	Principal applications as sealing glass
Tungsten (4.4)	8486	Alkaline-earth borosilicate, high chemical resistance, high working temperature. Suprax®	Lamp bulbs
	8487	High boron content, low melting temperature	Discharge lamps, surge diverters
Molybdenum (5.2)	8412	Alkaline-earth borosilicate, high chemical resistance. Fiolax® clear	Lamp bulbs
	8253	Alkaline-earth aluminosilicate glass	Lamp interior structures, lamp bulbs
Molybdenum and 28Ni/18Co/Fe (5.1)	8250	High boron content, low melting temperature, high electrical insulation, low dielectric losses	Transmitter tubes, image converters, TV receiver tubes
	8245	High boron content, low melting temperature, low X-ray absorption	X-ray tubes
28Ni/23Co/Fe (7.7)	8454	Alkali–alkaline-earth silicate, sealable with steatite and Al_2O_3 ceramics	Intermediate sealing glass
	8436	Alkali–alkaline-earth silicate, sealable with sapphire, resistant to Na vapor and alkalis	Special applications
51Ni/1Cr/Fe (10.2)	8350	Soda–lime silicate glass. AR glass	Tubes
Cu-sheathed wire ($\alpha_{20/400}$ radial 99, $\alpha_{20/400}$ axial 72)	8095	Alkali–lead silicate, high electrical insulation	Lead glass, stem glass for electric lamps and tubes
	8531	Dense lead silicate, Na- and Li-free, low melting temperature	Low-temperature encapsulation of diodes
	8532	High electrical insulation	
52–53Ni/Fe (10.2–10.5)	8512	Contains FeO for hot forming by IR, lead-free	Reed switches
	8516	Contains FeO for hot forming by IR, low volatilization, lead-free	Reed switches

glasses may be referred to as "tungsten sealing glasses", "Kovar glasses", etc. (see Table 3.4-20).

Alkaline-earth borosilicate glasses (8486 and 8412) and aluminosilicate glasses (8252 and 8253) have the necessary sealability and thermal resistance to be particularly suitable for the tungsten and molybdenum seals frequently used in heavy-duty lamps.

Ni–Fe–Co alloys, which often substitute for molybdenum, require that the transformation temperature be limited to 500 °C maximum. Suitable glasses (8250 and 8245) characteristically contain relatively high amounts of B_2O_3. These glasses have additional special properties, such as high electrical insulation, low dielectric loss, and low X-ray absorption, and meet the most stringent requirements for vacuum-tube technology and electronic applications.

For Ni–Fe–(Cr) alloys, which are frequently used in technological applications, as well as for copper-sheathed wire, glass groups belonging to the soft-glass category are recommended. Such glasses usually meet certain special requirements, such as high electrical insulation (e.g. alkali–lead silicate, 8095) or an exceptionally low working temperature (e.g. the dense-lead glasses 8531 and 8532).

FeO-containing glasses (8512 and 8516) are frequently used for hermetic encapsulation of electrical switches and electronic components in an inert gas. Hot forming and sealing are easily achieved by the absorption of IR radiation with an intensity maximum at 1.1 μm wavelength (Fig. 3.4-33). The presence of a proportion of Fe_2O_3 makes these glasses appear green. At appropriately high IR intensities, they require considerably shorter processing times than do flame-heated clear glasses.

Compression Seals

A common feature of all compression seals is that the coefficient of thermal expansion of the external metal part is considerably higher than the thermal expansion coefficients of the sealing glass and the metallic inner partner (conductor). As a result, the glass body is under overall radial pressure after sealing. This prestressing protects the glass body against dangerous mechanical loads. Because the compressive stress of the glass is compensated by a tensile stress in the jacket, the jacket wall must be sufficiently thick (at least 0.5 mm, even for small seals) to be able to permanently withstand such tension. If the thermal expansion of the metallic inner partner is lower than that of the sealing glass, an additional prestressing of the glass body results.

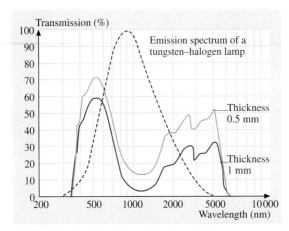

Fig. 3.4-33 IR absorption of Fe-doped glasses compared with the emission of a tungsten–halogen lamp at 3000 K (in relative units). The transmission of reed glass 8516 with thicknesses 0.5 mm and 1 mm is shown

Glasses for Sealing to Ceramics

Dielectrically superior, highly insulating ceramics such as hard porcelain, steatite, Al_2O_3 ceramics, and forsterite exhaust almost the complete expansion range offered by technical glasses. Hard porcelain can generally be sealed with alkaline-earth borosilicate glasses (for example 8486), which are also compatible with tungsten. Glass seals to Al_2O_3 ceramics and steatite are possible with special glasses such as 8454 and 8436, which will also seal to a 28Ni/18Co/Fe alloy. Soft glasses with thermal expansions around 9×10^{-6}/K are suitable for sealing to forsterite.

Intermediate Sealing Glasses

Glasses whose thermal expansion differs so widely from that of the partner component that direct sealing is impossible for reasons of stress must be sealed with intermediate sealing glasses. These glasses are designed in such a way that for the recommended combinations of glasses, the sealing stress does not exceed $20\,\text{N/mm}^2$ at room temperature (Table 3.4-21).

3.4.8.2 Solder and Passivation Glasses

Solder glasses are special glasses with a particularly low softening point. They are used to join glasses to other glasses, ceramics, or metals without thermally damaging the materials to be joined. Soldering is carried out in the viscosity range $10^4–10^6$ dPa s of the solder glass; this corresponds to a temperature range $T_{\text{solder}} = 350–700\,°\text{C}$.

One must distinguish between vitreous solder glasses and devitrifying solder glasses, according to their behavior during the soldering process.

Vitreous solder glasses behave like traditional glasses. Their properties do not change during soldering; upon reheating of the solder joint, the temperature dependence of the softening is the same as in the preceding soldering process.

Unlike vitreous solder glasses, devitrifying solder glasses have an increased tendency to crystallize. They change into a ceramic-like polycrystalline state during soldering. Their viscosity increases by several orders of magnitude during crystallization so that further flow is suppressed. An example of this time-dependent viscosity behavior is shown in Fig. 3.4-34 for a devitrifying solder glass processed by a specific temperature–time program. Crystallization allows a stronger thermal reloading of the solder joint, up to the temperature range of the soldering process itself (e.g. glass 8596 has a soldering temperature of approx-

4.8 Glasses for Miscellaneous Applications

Table 3.4-21 Intermediate sealing glasses and the combinations of sealing partners in which they are used

Glass no.	Sealing partners [a]	$\alpha_{20/300}$ (10^{-6}/K)	Transformation temperature T_g (°C)	Temperature at viscosity 10^{13} dPa s (°C)	$10^{7.6}$ dPa s (°C)	10^4 dPa s (°C)	Density ϱ (g/cm³)	T_{k100} (°C)
N16B	KER 250, Vacovit 501, Platinum — N16B, 8456 (Red Line®)	8.8	540	540	720	1045	2.48	128
2954	KER 220, KER 221, Vacon 20 — 2954	6.3	600	604	790	1130	2.42	145
4210	Iron–4210	12.7	450	455	615	880	2.66	–
8228	Fused silica–8228–8229	1.3	~700	726	1200	1705	2.15	355
8229	8228–8229–8230	2.0	630	637	930	1480	2.17	350
8230	8229–8230–8330	2.7	570	592	915	1520	2.19	257
8447	8412–8447–Vacon 10	4.8	480	505	720	1035	2.27	271
8448	8330–8448–8449, 8486, 8487	3.7	510	560	800	1205	2.25	263
8449	8486, 8487 — 8449 — 8447, 8412	4.5	535	550	785	1150	2.29	348
8450	8412–8450 – KER 220, 2954, 8436	5.4	570	575	778	1130	2.44	200
8454	KER 221, Al₂O₃ — 8454 — Vacon 70	6.4	565	575	750	1070	2.49	210
8455	2954, 8436, 8454 — 8455–8456	6.7	565	–	740	1030	2.44	–
8456	8455–8456 – N16B, 8350	7.4	445	–	685	1145	2.49	–

[a] Type designation of ceramics according to DIN 40685; manufacturer of Vacon alloys Vacuumschmelze Hanau (VAC).

imately 450 °C and a maximum reload temperature of approximately 435 °C).

The development of solder glasses (Table 3.4-22) with very low soldering temperatures is limited by the fact that reducing the temperature generally means increasing the coefficient of thermal expansion. This effect is less pronounced in devitrifying solder glasses. It can be avoided even more effectively by adding inert (nonreacting) fillers with low or negative coefficients of thermal expansion (for example ZrSiO₄ or β-eucryptite). The resulting glasses are called *composite solder glasses*. As a rule, the coefficient of thermal expansion of a solder glass should be smaller than the expansion coefficients of the sealing partners by $\Delta \alpha = 0.5 – 1.0 \times 10^{-6}$/K.

Up to their maximum service temperature, solder glasses are moisture- and gas-proof. Their good elec-

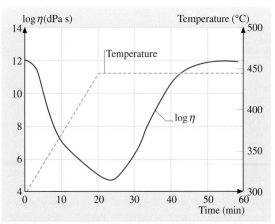

Fig. 3.4-34 Variation of the viscosity of a crystallizing solder glass during processing

Table 3.4-22 Schott solder glasses

Glass number	$\alpha_{(20/300)}$ (10^{-6}/K)	T_g (°C)	T at viscosity $10^{7.6}$ dPa s (°C)	Firing conditions T (°C)	Firing conditions t_{hold} (min)	Density ϱ (g/cm³)	T_{k100} (°C)	ε_r	tan δ (10^{-4})
Vitreous and composite solder glasses									
G017-002[a]	3.6	540	650	700	15	3.4	–	6.8	37
G017-339[a]	4.7[b]	325	370	450	30	4.0	320	11.5	19
G017-383[a]	5.7[b]	325	370	430	15	4.7	325	13.0	15
G017-393[a]	6.5[b]	320	370	425	15	4.8	305	11.6	15
G017-340[a]	7.0[b]	315	360	420	15	4.8	320	13.4	14
8465	8.2	385	460	520	60	5.4	375	14.9	27
8467	9.1	355	420	490	60	5.7	360	15.4	29
8468	9.6	340	405	450	60	6.0	335	16.3	31
8470	10.0	440	570	680	60	2.8	295	7.7	15.5
8471	10.6[b]	330	395	440	30	6.2	–	17.1	52
8472	12.0[b]	310	360	410	30	6.7	–	18.2	47
8474	19.0[b]	325	410	480	30	2.6	170	7.2	5
Devitrifying solder glasses									
G017-508	6.5	365	–	530[c]	60	5.7	340	15.6	206
8593	7.7	300	–	520[c]	30	5.8	230	21.3	260
8596	8.7	320	–	450[c]	60	6.4	280	17.4	58
G017-695	8.9	310	–	425[c]	45	5.7	275	15.4	54
8587	10.0	315	–	435[c]	40	6.6	265	22.1	33

[a] Composite. [b] $\alpha_{20/250}$. [c] Heating rate 7–10 °C/min.

trical insulating properties are superior to those of many standard technical glasses. They are therefore also suitable as temperature-resistant insulators. The chemical resistance of solder glasses is generally lower than that of standard technical glasses. Therefore, solder glass seals can be exposed to chemically aggressive environments (e.g. acids or alkaline solutions) only for a limited time.

Passivation glasses are used for chemical and mechanical protection of semiconductor surfaces. They are generally zinc–borosilicate or lead–alumina–silicate glasses.

To avoid distortion and crack formation, the different coefficients of thermal expansion of the passivation glass and the semiconductor component must be taken into account. If the mismatch is too large, a network of cracks will originate in the glass layer during cooling or subsequent processing and destroy the hermetic protection of the semiconductor surface. There are three ways to overcome this problem:

- A thinner passivation glass layer. Schott recommends a maximum thickness for this layer.
- Slow cooling in the transformation range. As a rough rule, a cooling rate of 5 K/min is suitable for passivation layers in the temperature range $T_g \pm 50$ K.
- Use of composite glasses. Composites can be made with an inert filler such as a powdered ceramic with a very low or negative thermal expansion.

Properties of Passivation Glasses

The electrical insulation, including the dielectric breakdown resistance, generally depends on the alkali content, particularly the Na^+ content. Typical contents are below 100 ppm for Na_2O and K_2O, and below 20 ppm for Li_2O. Heavy metals which are incompatible with semiconductors are controlled as well. The CuO content, for example, is below 10 ppm.

Because the mobility of charge carriers increases drastically with increasing temperature, a temperature limit, called the junction temperature T_j, is defined up to which glass-passivated components can be used in blocking operations.

Various types of passivation glasses are listed in Table 3.4-23.

3.4.8.3 Colored Glasses

In physics, color is a phenomenon of addition or subtraction of parts of the visible spectrum, due to selective absorption or scattering in a material. The light transmission through a sample of thickness d at a wavelength λ is described by Lambert's law,

$$\tau_i(\lambda) = \exp\left[-\sum\sum \varepsilon_n(\lambda, c_n, c_m) d\right], \quad (4.42)$$

where ε is the extinction coefficient, which depends on the wavelength and the concentration of the active agents. For low concentrations, ε is additive and proportional to the concentration, and we obtain Beer's law,

$$\tau_i(\lambda) = \exp\left[-\sum \varepsilon_n(\lambda) c_n d\right], \quad (4.43)$$

where ε now depends only on the wavelength and the specific species or process n. In glasses, the extinction is caused by electronic and phononic processes in the UV and IR regions, respectively, and by absorption and scattering by ions, lattice defects, and colloids and microcrystals in the visible region. Different oxidation states of one atom, for example Fe^{2+} and Fe^{3+}, must be treated as different species. Charge transfer and ligand fields are examples of multiatom mechanisms that modify the absorption characteristics. The position of the maximum-extinction peak depends on the refractive index of the base glass; for example, for Ag metal colloids the position of the peak shifts from $\lambda_{max} = 403$ nm in Duran®, with $n_d = 1.47$, to $\lambda_{max} = 475$ nm in SF 56, with $n_d = 1.79$.

Table 3.4-23 Schott passivation glasses

Glass number	Type	Typical applications	$\alpha_{(20/300)}$ (10^{-6}/K)	T_g (°C)	Pb content (wt%)	Sealing temp. (°C)	Sealing time (min)	T_j (°C)	Layer thickness (µm)
G017-057	Zn–B–Si	Sintered glass diodes	4.5	546	1–5	690	10	180	–
G017-388	Zn–B–Si composite	Thyristors, high-blocking rectifiers	3.6	550	1–5	700	5	180	≤ 30
G017-953	Zn–B–Si composite		2.81	a	1–5	770	30	180	–
G017-058	Zn–B–Si	Sintered glass diodes	4.5	543	1–5	690	10	180	–
G017-002	Zn–B–Si composite	Sintered glass diodes	3.7	545	1–5	700	10	180	–
G017-984	Zn–B–Si	Stack diodes	4.6	538	5–10	720	10	180	–
G017-096R	Pb–B–Si	Sintered glass diodes, planar and mesa diodes	4.8	456	10–50	680	5	160	–
G017-004	Pb–B–Si	Mesa diodes	4.1	440	10–50	740	5	160	≤ 30
G017-230	Pb–B–Si composite	Power transistors	4.2	440	10–50	700	5	160	≤ 25
G017-725	Pb–B–Si	Sintered glass diodes	4.9	468	10–50	670	10	180	–
G017-997	Pb–B–Si composite	Wafers	4.4	485	10–50	760	20	180	–
G017-209	Pb–Zn–B	ICs, transistors	6.6	416	10–50	510	10	180	≤ 5
G017-980	Pb–Zn–B	Varistors			10–50			–	–
Vitreous			6.5	393		520	30	–	–
Devitrified			5.8	a		620	30	–	–
G018-088	Pb–Zn–B composite	Varistors	4.88	425	10–50	560	30	–	–

[a] Cannot be determined.

Colored glasses are thus technical or optical colorless glasses with the addition of coloring agents. A collection of data can be found in [4.17]. They are widely used as optical filters for various purposes, such as short-pass or long-pass edge filters, or in combinations of two or more elements as band-pass or blocking filters.

Owing to the absorbed energy, inhomogeneous heating occurs (between the front and rear sides, and, especially, in the radial direction), which results in internal stress via the thermal expansion under intense illumination. Precautions have to be taken in the mechanical mounting to avoid breakage. The application temperature T should satisfy the conditions $T \leq T_g - 300\,°C$ in the long term and $T \leq T_g - 250\,°C$ for short periods. Prestressing may be necessary to improve the breaking strength in heavy-load applications.

The Schott filter glasses are classified into groups listed in Table 3.4-24:

Table 3.4-24 Groups of Schott Filter glasses

UG	Black and blue glasses, ultraviolet-transmitting
BG	Blue, blue–green, and multiband glasses
VG	Green glasses
OG	Orange glasses, IR transmitting
RG	Red and black glasses, IR-transmitting
NG	Neutral glasses with uniform attenuation in the visible
WG	Colorless glasses with different cutoffs in the UV, which transmit in the visible and IR
KG	Virtually colorless glasses with high transmission in the visible and absorption in the IR (heat protection filters)
FG	Bluish and brownish color temperature conversion glasses

DIN has defined a nomenclature to allow one to see the main optical properties for a reference thickness d from the identification symbol (see Table 3.4-25):
Multiband filters and color conversion filters are not specified by DIN.

Ionically Colored Glasses

Ions of heavy metals or rare earths influence the color when in true solution. The nature, oxidation state, and quantity of the coloration substance, as well as the type of the base glass, determine the color (Figs. 3.4-35 – 3.4-38, and Table 3.4-26).

Table 3.4-25 DIN nomenclature for optical filter glasses

Band-pass filters	BP $\lambda_{max}/\Delta\lambda_{HW}$, where λ_{max} = wavelength of maximum internal transmission, and $\Delta\lambda_{HW}$ = bandwidth at 50% internal transmission.
Short-pass filters	KP $\lambda_{50\%}$, where $\lambda_{50\%}$ = cutoff wavelength at 50% internal transmission.
Long-pass filters	LP $\lambda_{50\%}$, where $\lambda_{50\%}$ = cutoff wavelength at 50% internal transmission.
Neutral-density filters	N τ, where τ = internal transmission at 546 nm.

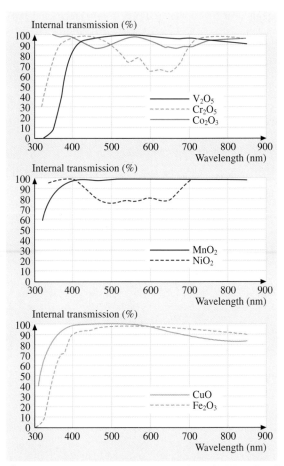

Fig. 3.4-35 Spectral internal transmission of BK7, colored with various oxides; sample thickness 100 mm

4.8 Glasses for Miscellaneous Applications

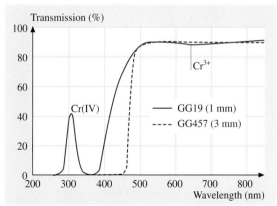

Fig. 3.4-36 Transmission spectra of yellow glasses

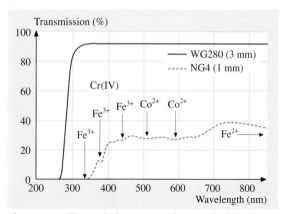

Fig. 3.4-38 Transmission spectra of gray and white glasses

Colloidally Colored Glasses

The colorants of these glasses are, in most cases, rendered effective by a secondary heat treatment (striking) of the initially nearly colorless glass. Particularly important glasses of this type are the yellow, orange, red, and black filter glasses, with their steep absorption edges. As with ionic coloration, the color depends on the type and concentration of the additives, on the type of the base glass, and on the thermal history, which determines the number and diameter of the precipitates (Fig. 3.4-39 and Table 3.4-27).

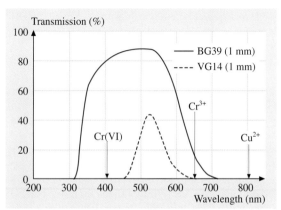

Fig. 3.4-37 Transmission spectra of blue and green glasses

Table 3.4-27 Colors of some metal colloids

Element	Peak position (nm)	n_d	Color
Ag	410	1.5	Yellow
Cu	530–560	?	Red
Au	550	1.55	Red
Se	500	?	Pink

Table 3.4-26 Colors of some ions in glasses

Element	Valency	Color
Fe	2+	Green, sometimes blue
Fe	3+	Yellowish brown
Cu	2+	Light blue, turquoise
Cr	3+	Green
Cr	6+	Yellow
Ni	2+	Violet (tetrahedral coordination)
Ni	2+	Yellow (octahedral coord.)
Co	2+	Intense blue
Co	3+	Green
Mn	2+	Pale yellow
Mn	3+	Violet
V	3+	Green (silicate), brown (borate)
Ti	3+	Violet (reducing melt)
Pr	3+	Light green
Nd	3+	Reddish violet
Er	3+	Pale red

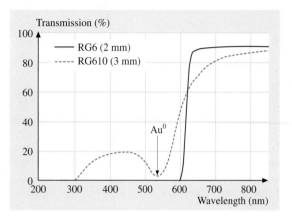

Fig. 3.4-39 Transmission spectra of red glasses

Doping with semiconductors results in microcrystalline precipitates which have band gap energies in the range 1.5–3.7 eV, corresponding to wavelengths in the range 827–335 nm. The preferred materials are ZnS, ZnSe, ZnTe, CdS, CdSe, and CdTe, which also form solid solutions. By mixing these dopants, any cutoff wavelength between 350 nm and 850 nm can be achieved (Table 3.4-30).

3.4.8.4 Infrared-Transmitting Glasses

The transmission in the infrared spectral region is limited by phonons and by local molecular vibrations and their overtones. The vibration frequencies decrease with increasing atomic mass. This intrinsic absorption has extrinsic absorption caused by impurities and lattice defects, such as hydroxyl ions, dissolved water, and microcrystals, superimposed on it. An overview can be found in [4.12].

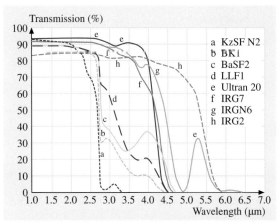

Fig. 3.4-41 Infrared transmission spectra of some Schott optical and special IR glasses; thickness 2.5 mm. The OH absorption of the glasses may vary owing to differences in the raw materials and the melting process

Oxide Glasses

The transmission is determined by the vibrations of the common network formers [4.19] (Fig. 3.4-40). The vibrations of the network modifiers are found at longer wavelengths.

The heavy-metal oxide (HMO) glasses are transparent up to approximately 7 μm for a 1 mm thickness. But often they show a strong tendency toward devitrification, which very much limits the glass-forming compositions. The transmission of some commercial glasses is shown in Fig. 3.4-41.

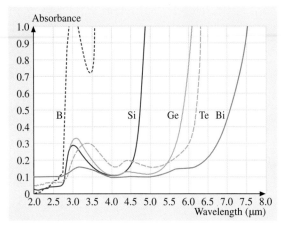

Fig. 3.4-40 Infrared absorbance of analogous oxide glasses with different network-forming cations [4.18]; sample thickness 1.85 mm, except for Bi, where the thickness was 2.0 mm

Halide Glasses

These glasses use F, Cl, Br, and I (halogens) as anions instead of oxygen. The transmission range is extended up to approximately 15 μm [4.20]. The oldest halide glasses are BeF_2, $ZnCl_2$, and AlF_3, which, however, have limited application owing to their toxicity, tendency toward crystallization, and hygroscopic behavior.

Some new glasses use ZrF_4, HfF_4, and ThF_4 as glass formers, and BaF_2, LaF_3 (heavy-metal flourides, HMFs) as modifiers. They are often named after their cation composition; for example, a glass with a cation composition $Zr_{55}Ba_{18}La_6Al_4Na_{17}$ would be called a ZBLAN glass.

Chalcogenide Glasses

These glasses use S, Se, and Te (chalcogens) as anions instead of oxygen. The transmission range is extended up to approximately 30 μm. Stable glass-forming regions are found in the Ge–As–S, Ge–As–Se, and Ge–Sb–Se systems; an example of a commercial glass is $Ge_{30}As_{15}Se_{55}$.

A combination of chalcogenides with halides is found in the TeX glasses, for example Te_3Cl_2S and Te_2BrSe.

The main properties of infrared-transmitting glasses are compiled in Tables 3.4-29 and 3.4-30.

Table 3.4-28 Properties of Schott filter glasses

Glass type	DIN identification	Reference thickness d_r (mm)	Density ϱ (g/cm³)	Refractive index n_d	Stain FR	Acid SR	Alkali AR	Transformation temperature T_g (°C)	$\alpha_{-30/+70\,°C}$ (10^{-6}/K)	$\alpha_{20/300\,°C}$ (10^{-6}/K)	Temp. coeff. T_K (nm/K)
UG1	BP 351/78	1	2.77	1.54	0	1.0	1.0	603	7.9	8.9	–
UG5	BP 318/173	1	2.85	1.54	0	3.0	2.0	462	8.1	9.4	–
UG11	BP 324/112 + BP 720/57	1	2.92	1.56	0	3.0	2.0	545	7.8	9	–
BG3	BP 378/185	1	2.56	1.51	0	1.0	1.0	478	8.8	10.2	–
BG4	BP 378/165	1	2.66	1.53	0	1.0	1.0	536	7.7	9	–
BG7	BP 466/182	1	2.61	1.52	0	1.0	1.0	468	8.5	19.9	–
BG12	BP 409/140	1	2.58	1.52	0	1.0	1.0	480	8.6	10.1	–
BG18	BP 480/250 + KP 605	1	2.68	1.54	0	2.0	2.0	459	7.4	8.8	–
BG20	Multiband	–	2.86	1.55	0	1.0	1.0	561	8.3	9.3	–
BG23	BP 459/232 + KP 575	1	2.37	1.52	0	1.0	1.0	483	8.9	10.2	–
BG24A	BP 342/253	1	2.72	1.53	0	3.0	1.0	460	8.5	9.7	–
BG25	BP 401/156	1	2.56	1.51	0	1.0	1.0	487	8.7	10.1	–
BG26	–	1	2.56	1.51	0	1.0	1.0	494	8.8	10.3	–
BG28	BP 436/156 + KP 514	1	2.60	1.52	0	1.0	1.0	474	8.7	10	–
BG34	Color conversion	2	3.23	1.59	0	1.0	1.0	441	9.9	10.7	–
BG36	Multiband	–	3.62	1.69	1	52.2	1.2	660	6.1	7.2	–
BG38	BP 487/334 + KP 654	1	2.62	1.53	0	2.0	2.0	466	7.5	8.9	–
BG39	BP 475/269 + KP 609	1	2.73	1.54	0	5.1	3.0	321	11.6	13.1	–
BG40	BP 482/318 + KP 641	1	2.67	1.53	0	5.1	3.0	305	11.9	13.7	–
BG42	BP 478/253 + KP 604	1	2.69	1.54	0	2.0	2.0	477	7.3	8.7	–
VG6	BP 523/160	1	2.90	1.55	0	1.0	1.0	470	9.1	10.6	–
VG9	BP 530/114	1	2.87	1.55	0	1.0	1.0	470	9.2	10.6	–
VG14	BP 524/87	1	2.89	1.56	0	1.0	1.0	470	9.2	10.6	–
GG385	LP 385	3	3.22	1.58	0	2	2.3	459	7.7	8.8	0.07
GG395	LP 395	3	3.61	1.62	0	1	2.3	438	7.7	8.6	0.08
GG400	LP 400	3	2.75	1.54	3	4.4	1.0	595	9.6	10.5	0.07
GG420	LP 420	3	2.76	1.54	3	4.4	1.0	586	9.6	10.5	0.07

Table 3.4-28 Properties of Schott filter glasses, cont.

GG435	LP 435	3	2.75	1.54	3	4.4	1.0	605	9.5	10.5	0.07
GG455	LP 455	3	2.75	1.54	3	4.4	1.0	600	9.7	10.5	0.08
GG475	LP 475	3	2.75	1.54	3	4.4	1.0	594	9.8	10.6	0.09
GG495	LP 495	3	2.75	1.54	3	4.4	1.0	600	9.6	10.6	0.10
OG515	LP 515	3	2.76	1.54	3	4.4	1.0	597	9.7	10.6	0.11
OG530	LP 530	3	2.75	1.54	3	4.4	1.0	595	9.7	10.6	0.12
OG550	LP 550	3	2.75	1.54	3	4.4	1.0	597	9.6	10.7	0.13
OG570	LP 570	3	2.75	1.54	3	4.4	1.0	596	9.7	10.7	0.14
OG590	LP 590	3	2.75	1.54	3	4.4	1.0	599	9.8	10.6	0.15
RG9	BP 885307 + LP 731	3	2.76	1.54	3	4.4	1.0	581	9.8	10.7	0.07
RG610	LP 610	3	2.75	1.54	3	4.4	1.0	595	9.8	10.7	0.16
RG630	LP 630	3	2.76	1.54	3	4.4	1.0	597	9.6	10.7	0.17
RG645	LP 645	3	2.76	1.54	3	4.4	1.0	597	9.6	10.7	0.17
RG665	LP 665	3	2.75	1.54	3	4.4	1.0	592	9.8	10.8	0.17
RG695	LP 695	3	2.76	1.54	3	4.4	1.0	599	9.6	10.6	0.18
RG715	LP 715	3	2.75	1.54	3	3	1.0	589	9.8	10.7	0.18
RG780	LP 780	3	2.9	1.56	3	52.4	1.0	571	9.7	10.7	0.22
RG830	LP 830	3	2.94	1.56	5	53.4	1.0	569	9.5	10.5	0.23
RG850	LP 850	3	2.93	1.56	5	53.4	1.0	571	9.5	10.5	0.24
RG1000	LP 1000	3	2.75	1.55	0	1	1.2	478	9.2	9.9	0.38
NG1	N 10-4	1	2.49	1.52	1	2.2	1.0	466	6.6	7.2	–
NG3	N 0.09	1	2.44	1.51	1	2.2	1.0	462	6.5	7.3	–
NG4	N 0.27	1	2.43	1.51	1	3.2	2.0	483	6.7	7.2	–
NG5	N 0.54	1	2.43	1.5	1	3.2	2.0	474	6.6	7.3	–
NG9	N 0.04	1	2.44	1.51	1	3.2	2.0	487	6.4	7.3	–
NG10	N 0.004	1	2.47	1.52	1	3.2	2.0	468	6.4	7.2	–
NG11	N 0.72	1	2.42	1.5	1	3.4	2.0	473	6.9	7.5	–
NG12	N 0.89	1	2.34	1.49	4	51.4	2.0	460	5.9	6.4	–
WG225	LP 225	2	2.17	1.47	3	51.3	3.3	437	3.8	4.1	0.02
WG280	LP 280	2	2.51	1.52	0	1	2.0	563	7.0	8.3	0.04
WG295	LP 295	2	2.51	1.52	0	1	2.0	557	7.1	8.3	0.06
WG305	LP 305	2	2.59	1.52	0	1	1.0	546	8.2	9.6	0.06
WG320	LP 320	2	3.22	1.58	0	1	2.3	413	9.1	10.6	0.06
KG1	KP 751	2	2.53	1.52	0	2.0	3.0	599	5.3	6.1	–
KG2	KP 814	2	2.52	1.51	0	2.0	3.0	605	5.4	6.3	–
KG3	KP 708	2	2.52	1.51	0	2.0	4.0	581	5.3	6.1	–
KG4	KP 868	2	2.53	1.51	0	2.0	3.0	613	5.4	6.2	–
KG5	KP 689	2	2.53	1.51	0	3.0	4.0	565	5.4	6.2	–
FG3	Color conversion	2	2.37	1.5	0	1.0	1.0	564	5.4	6	–
FG13	Color conversion	2	2.78	1.56	1	3.4	1.0	556	9.6	10.5	–

Table 3.4-29 Property ranges of various infrared-transmitting glass types

Glass type	Transparency range[a] (μm)	Refractive index n_d	Abbe value v_d	Thermal expansion α (10^{-6}/K)	Microhardness HV	Transformation temperature T_g (°C)	Density ϱ (g/cm³)
Fused silica	0.17–3.7	1.46	67	0.51	800	1075	2.2
Oxide glasses	0.25–6.0	1.45–2.4	20–100	3–15	330–700	300–700	2–8
Fluorophosphate glasses	0.2–4.1	1.44–1.54	75–90	13–17	–	400–500	3–4
Fluoride glasses	0.3–8.0	1.44–1.60	30–105	7–21	225–360	200–490	2.5–6.5
Chalcogenide glasses	0.7–25	2.3–3.1[b]	105–185[c]	8–30	100–270	115–370	3.0–5.5

[a] 50% internal transmission for 5 mm path length.
[b] Infrared refractive index at 10 μm.
[c] Infrared Abbe value $v_{8-12} = (n_{10} - 1)/(n_8 - n_{12})$.

Table 3.4-30 Commercial infrared-transmitting glasses

Glass type	Glass name	Transparency range[a] (μm)	Refractive index n	Density ϱ (g/cm³)	Thermal expansion α (10^{-6}/K)	Transformation temperature T_g (°C)	Manufacturer
Fused silica	SiO$_2$	0.17–3.7	1.4585[b]	2.20	0.51	1075	Corning, Heraeus, General Electric, Quartz et Silice, Schott Lithotec
Silicates	IRG7	0.32–3.8	1.5644[b]	3.06	9.6	413	Schott
	IRG15	0.28–4.0	1.5343[b]	2.80	9.3	522	Schott
	IRG3	0.40–4.3	1.8449[b]	4.47	8.1	787	Schott
Fluorophosphate	IRG9	0.36–4.2	1.4861[b]	3.63	16.1	421	Schott
Ca aluminate	9753[c]	0.40–4.3	1.597[d]	2.80	6.0	830	Corning
	IRG N6[c]	0.35–4.4	1.5892[b]	2.81	6.3	713	Schott
	WB37A[c]	0.38–4.7	1.669[b]	2.9	8.3	800	Sasson
	VIR6	0.35–5.0	1.601[b]	3.18	8.5	736	Corning France
	IRG11	0.38–5.1	1.6809[b]	3.12	8.2	800	Schott
	BS39B	0.40–5.1	1.676[b]	3.1	8.4	–	Sasson
Germanate	9754	0.36–5.0	1.664[b]	3.58	6.2	735	Corning
	VIR3	0.49–5.0	1.869[b]	5.5	7.7	490	Corning France
	IRG2	0.38–5.2	1.8918[b]	5.00	8.8	700	Schott
Heavy-metal oxide	EO	0.5–5.8	2.31[d]	8.2	11.1	320	Corning
Heavy-metal fluoride	ZBLA	0.30–7.0	1.5195[b]	4.54	16.8	320	Verre Fluoré
	Zirtrex	0.25–7.1	1.50[b]	4.3	17.2	260	Galileo
	HTF-1	0.22–8.1	1.517[d]	3.88	16.1	385	Ohara

Table 3.4-30 Commercial infrared-transmitting glasses, cont.

Glass type	Glass name	Transparency range[a] (μm)	Refractive index n	Density ϱ (g/cm^3)	Thermal expansion α (10^{-6}/K)	Transformation temperature T_g (°C)	Manufacturer
Chalcogenide	AMTIR1	0.8–12	2.5109[e]	4.4	12	362	Amorph. Materials
	IG1.1	0.8–12	2.4086[e]	3.32	24.6	150	Vitron
	AMTIR3	1.0–13	2.6173[e]	4.7	13.1	278	Amorph. Materials
	IG6	0.9–14	2.7907[e]	4.63	20.7	185	Vitron
	IG2	0.9–15	2.5098[e]	4.41	12.1	368	Vitron
	IG3	1.4–16	2.7993[e]	4.84	13.4	275	Vitron
	As$_2$S$_3$	1.0–16	2.653[e]	4.53	30	98	Corning France
	IG5	1.0–16	2.6187[e]	4.66	14.0	285	Vitron
	IG4	0.9–16	2.6183[e]	4.47	20.4	225	Vitron
	1173	0.9–16	2.616[e]	4.67	15	300	Texas Instruments

[a] 50% internal transmission for 5 mm path length.
[b] Refractive index at 0.587 μm.
[c] This glass contains SiO$_2$.
[d] Refractive index at 0.75 μm.
[e] Refractive index at 5 μm.

References

4.1 S. R. Elliott: *Physics of Amorphous Materials* (Longman, Harlow 1990)

4.2 H. Bach, D. Krause (Eds.): *Analysis of the Composition and Structure of Glass and Glass Ceramics* (Springer, Berlin, Heidelberg 1999) 2nd printing

4.3 H. Bach (Ed.): *Low Thermal Expansion Glass Ceramics* (Springer, Berlin, Heidelberg 1995)

4.4 W. Höland, G. Beall: *Glass Ceramic Technology* (American Ceramic Society, Westerville 2002)

4.5 N. P. Bansal, R. H. Doremus: *Handbook of Glass Properties* (Academic Press, Orlando 1986)

4.6 O. V. Mazurin, M. V. Streltsina, T. P. Shvaiko-Shvaikovskaya: *Handbook of Glass Data*, Part A, *Silica Glass and Binary Silicate Glasses*; Part B, *Single Component and Binary Non-silicate Oxide Glasses*; Part C, *Ternary Silicate Glasses*, Physical Sciences Data Series, Vol. 15, (Elsevier, Amsterdam 1983–1987)

4.7 R. Blachnik (Ed.): *Taschenbuch für Chemiker und Physiker*, D'Ans-Lax, Vol. 3, 4th edn. (Springer, Berlin, Heidelberg 1998)

4.8 *MDL SciGlass*, Version 4.0 (Elsevier, Amsterdam)

4.9 *INTERGLAD* (International Glass Database), Version 5, New Glass Forum

4.10 S. English, W. E. S. Turner: Relationship between chemical composition and the thermal expansion of glasses, J. Am. Ceram. Soc. **10**, 551 (1927); J. Am. Ceram. Soc. **12** (1929) 760

4.11 A. Paul: *Chemistry of Glasses* (Chapman and Hall, New York 1982)

4.12 H. Bach, N. Neuroth (Eds.): *The Properties of Optical Glass*, Schott Series on Glass and Glass Ceramics, 2nd printing (Springer, Berlin, Heidelberg 1998)

4.13 A. Winckelmann, F. O. Schott: Über thermische Widerstandskoeffizienten verschiedener Gläser in ihrer Abhängigkeit von der chemischen Zusammensetzung, Ann. Phys. (Leipzig) **51**, 730–746 (1894)

4.14 H. Scholze: *Glass* (Vieweg, Braunschweig 1965)

4.15 Schott: *Schott Technical Glasses* (Schott Glas, Mainz 2000)

4.16 H. R. Philipp: Silicon dioxide (SiO$_2$) (glass). In: *Handbook of Optical Constants of Solids*, ed. by E. D. Palik (Academic Press, New York 1985) pp. 749–763

4.17 C. R. Bamford: *Colour Generation and Control in Glass* (Elsevier, Amsterdam 1977)

4.18 W. H. Dumbaugh: Infrared-transmitting oxide glasses, Proc. SPIE **618**, 160–164 (1986)

4.19 N. Neuroth: Zusammenstellung der Infrarotspektren von Glasbildnern und Gläsern, Glastechn. Ber. **41**, 243–253 (1968)

4.20 J. Lucas, J.-J. Adam: Halide glasses and their optical properties, Glastechn. Ber. **62**, 422–440 (1989) W. Vogel: *Glass Chemistry* (Springer, Berlin, Heidelberg 1994)

Part 4 Functional Materials

1 Semiconductors
  Werner Martienssen, Frankfurt/Main, Germany

2 Superconductors
  Claus Fischer, Dresden, Germany
  Günter Fuchs, Dresden, Germany
  Bernhard Holzapfel, Dresden, Germany
  Barbara Schüpp-Niewa, Dresden, Germany
  Hans Warlimont, Dresden, Germany

3 Magnetic Materials
  Hideki Harada, Higashikaya, Fukaya, Saitama, Japan
  Manfred Müller, Dresden, Germany
  Hans Warlimont, Dresden, Germany

4 Dielectrics and Electrooptics
  Gagik G. Gurzadyan, Garching, Germany
  Pancho Tzankov, Berlin, Germany

5 Ferroelectrics and Antiferroelectrics
  Toshio Mitsui, Takarazuka, Japan

4.1. Semiconductors

The organization of this section follows a two-step approach. The first step corresponds to searching for the substance of interest, that is, the relevant group of substances. The second step corresponds to the physical property of interest.

This section has three subsections, characterized by the groups of the Periodic Table that the constituent elements belong to. The first subsection, Sect. 4.1.1, deals with the elements of Group IV of the Periodic Table and semiconducting binary compounds between elements of this group (IV–IV compounds). The second subsection, Sect. 4.1.2, treats the semiconducting binary compounds between the elements of Groups III and V (III–V compounds); Sect. 4.1.3 treats compounds between the elements of Groups II and VI (II–VI compounds). These two subsections are subdivided further according to the first element in the formula of the compound.

The elements and compounds treated in Sect. 4.1.1 (Group IV and IV–IV compounds) are treated as one group; the data in the tables are given for the whole group in all cases. In Sects. 4.1.2 (III–V compounds) and 4.1.3 (II–VI compounds), data are given separately for each subdivision of those subsections.

For each group of substances, the physical properties are organized into four classes. These are

A. Crystal structure, mechanical and thermal properties.
B. Electronic properties.
C. Transport properties.
D. Electromagnetic and optical properties.

These property classes, finally, are subdivided into individual properties, which are described in the text, tables, and figures.

4.1.1	**Group IV Semiconductors and IV–IV Compounds**	578
4.1.2	**III–V Compounds**	604
	4.1.2.1 Boron Compounds	604
	4.1.2.2 Aluminium Compounds	610
	4.1.2.3 Gallium Compounds	621
	4.1.2.4 Indium Compounds	638
4.1.3	**II–VI Compounds**	652
	4.1.3.1 Beryllium Compounds	652
	4.1.3.2 Magnesium Compounds	655
	4.1.3.3 Oxides of Ca, Sr, and Ba	660
	4.1.3.4 Zinc Compounds	665
	4.1.3.5 Cadmium Compounds	676
	4.1.3.6 Mercury Compounds	686
References		691

This section deals with the physical properties of semiconductors. Semiconductors are substances which, like metals, are electronic conductors. In contrast to metals, however, the density of freely mobile charge carriers in semiconductors is, under normal conditions, smaller by orders of magnitude than it is in metals. Therefore, in semiconductors, a *small change* in the absolute value of the charge carrier density can induce a *large relative change* in this carrier density and in the electrical conductivity. In metals, on the other hand, the carrier density is so high from the beginning that it is practically impossible to produce a reasonable relative change by small changes of the absolute value of the carrier density. In conclusion, we can say that in semiconductors, and only in semiconductors, is it possible to *manipulate* the electronic conduction by small changes of the carrier density.

Such changes can be effected by a number of techniques, for instance by chemical doping, by temperature changes, by the application of an electric field, or by light. The electronic conductivity of a semiconductor can be changed *intentionally* by these techniques by orders of magnitude; some techniques allow stationary changes, and some techniques also allow time-dependent changes on a very short timescale.

Semiconducting materials are functional materials thanks to the above properties: they can be used as very small, robust, energy-efficient devices to control the current in electrical networks, either on the basis of external driving or on the basis of their internal, tailor-made device characteristics.

During the last 50 years, a tremendous amount of knowledge and experience has been collected worldwide in research and development laboratories in the field of semiconductor physics, semiconductor engineering, and semiconductor chemistry. During this time, semiconductor technology laid the foundations for the development of data processing and of communication technology and, more generally, for the establishment of the information society. Today, semiconductor technology is a basic technology of our economy, business practice, and daily life with its modern comforts.

We can give an account of only a very small part of the empirical knowledge about semiconductors in this Handbook. Thus, a very limited selection of semiconducting substances and of physical properties is treated here. Only three major groups of substances are considered. These are, first, the elements of the fourth group of the Periodic Table, C, Si, Ge, and Sn, and binary compounds between these (IV–IV compounds); second, binary compounds between one element from the third group of the Periodic Table, namely B, Al, Ga, or In, and one element from the fifth group of the Periodic Table, N, P, As, or Sb, the so-called III–V semiconductors; and third, binary compounds between one element from the second group of the Periodic Table, namely Mg, Ca, Sr, Ba, Zn, Cd, or Hg, and one element from the sixth group of the Periodic Table O, S, Se, or Te, the so-called II–VI semiconductors.

In fact, these three major groups form a kind of basic set for the physical understanding of semiconductor phenomena and for the wide field of semiconductor applications. It has to be kept in mind, however, that ternary and higher compounds play a very important role in many semiconductor applications and, furthermore, that a large number of semiconducting substances containing elements other than those mentioned above have been developed for special applications.

The limited number of pages of this Handbook has forced us, furthermore, to be even more restrictive with regard to the properties of semiconductors, which are considered as being of first-order importance for this data collection. So, semiconductor chemistry is beyond the scope of this Handbook. The very broad and important field of semiconductor technology could not be included at all. Also, the wide field of the influence of chemical doping, impurities, and defects on the properties of semiconductors had to be left out. The Handbook's emphasis is on the physical properties of a restricted number of most important pure semiconducting materials.

All the data have been compiled from Landolt–Börnstein, *Numerical Data and Functional Relationships in Science and Technology*, New Series, Group III, Vol. 41, *Semiconductors*, published in eight subvolumes between 1998 and 2003 [1.1].

The CD-ROM and the online version of the Landolt–Börnstein volume present a completely revised and supplemented version of the older Landolt–Börnstein volumes III/17 and III/22 on semiconductors. The printed edition of Vol. III/41 contains only a supplement to the earlier Volumes III/17 and III/22, but nevertheless the printed version of Vol. III/41 extends to more than 4000 pages in the eight subvolumes. If, in addition to the data in this Handbook, further or more detailed information is needed, the reader is recommended to consult [1.1].

Only a very restricted number of references is given in this chapter to the original publications in journals.

A very large number of references to all the original papers are given, however, in the Landolt–Börnstein volume mentioned above [1.1] (see also [1.2]).

The data are presented mainly in tables, which always cover data concerning one or a few related properties for a full group of substances. It is felt that this synoptic presentation will provide an immediate overview of the data for the group of substances. This might be especially helpful if one is interested in development or modeling of materials.

In many cases, functional relationships, in particular temperature dependences, are of the utmost importance in condensed-matter physics and materials science. Therefore a large number of figures have been included, in addition to short sections of text and the numerical data in the tables. Figures in the form of graphs allow one to see the general features of a functional relationship at a glance; they also, of course, present information in a very condensed form.

The physical properties of semiconductors treated in this section are the following:

A. Crystal structure, mechanical and thermal properties:

- Crystal structure (space group, crystal system, and structure type) (tables and figures).
- Lattice parameters a and c, given in nm.
- Density ϱ, given in g/cm^3.
- Elastic constants c_{ik}, given in GPa.
- Melting point T_m, given in K.
- Temperature dependence of the lattice parameters, i.e. the functional relationships $a(T)$ and $c(T)$.
- Linear thermal expansion coefficient α, given in 10^{-6} K^{-1}.
- Heat capacity c_p, given in J/mol K.
- Debye temperature Θ_D, given in K.
- Phonon wavenumbers $\tilde{\nu}$, frequencies ν, energies E_phon at symmetry points, given in cm^{-1}, THz, and eV, respectively.
- Phonon dispersion relations, i.e. wavenumber $\tilde{\nu}$ versus wave vector $\boldsymbol{q}$.

B. Electronic properties:

- First Brillouin zone (figures).
- Band structure (text and figures).
- Energy gap, given in eV.
- Exciton binding energies, given in eV.
- Spin–orbit splitting energies, given in eV.
- Effective masses, in units of the electron mass m_0.
- g-factor g_c of conduction electrons.

C. Transport properties:

- Electronic transport, general description.
- Intrinsic carrier concentrations, n and p, given in cm^{-3}.
- Electrical conductivity σ, given in Ω^{-1} cm^{-1}.
- Electron and hole mobilities μ_n and μ_p, given in cm^2/V s.
- Thermal conductivity, given in W/cm K.

D. Electromagnetic and optical properties:

- Dielectric constant ε.
- Refractive index n.
- Spectral dependence of the optical constants (tables and figures).

The data given in this Handbook, if not stated otherwise, are numerical data valid at room temperature (RT) and at normal pressure (100 kPa). If the material crystallizes in different modifications, the data given, if not stated otherwise, concern the phase which is stable under normal conditions (i.e. RT and 100 kPa).

4.1.1 Group IV Semiconductors and IV–IV Compounds

A. Crystal Structure, Mechanical and Thermal Properties
Tables 4.1-1 – 4.1-10.

Table 4.1-1 Crystal structures of Group IV semiconductors and IV–IV compounds

Crystal		Space group		Crystal system	Structure type	Figure
Diamond	C	O_h^7	$Fd3m$	fcc	Diamond	Fig. 4.1-1
Silicon	Si	O_h^7	$Fd3m$	fcc	Diamond	Fig. 4.1-1
Germanium	Ge	O_h^7	$Fd3m$	fcc	Diamond	Fig. 4.1-1
Gray tin[a]	α-Sn	O_h^7	$Fd3m$	fcc	Diamond	Fig. 4.1-1
Silicon carbide[b]	3C-SiC [c]	T_d^2	$F\bar{4}3m$	fcc	Zinc blende	Fig. 4.1-2
	2H-SiC [c]	C_{6v}^4	$P6_3mc$	Hexagonal	Wurtzite	Fig. 4.1-3
Silicon–germanium alloys [d]		O_h^7	$Fd3m$	fcc	Diamond	Fig. 4.1-1

[a] Below 290 K (17 °C) and at ambient pressure.
[b] Silicon carbide is said to be polytypic because it crystallizes in more than one hundred different modifications. This is due to the small energy differences between the different structures.
[c] The labels "3C", "2H", etc. indicate the stacking order along the c axis in a hexagonal system; the cubic form can be considered as fitting into this system by taking the [111] direction as the "c axis".
[d] Silicon and germanium form a continuous series of solid solutions with gradually varying properties.

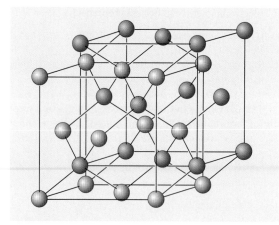

Fig. 4.1-1 The diamond lattice. The elementary cubes of the two face-centered cubic lattices are shown. In the diamond lattice all the atoms in the two elementary cubes are identical, they are atoms from the same chemical element. Each atom in this lattice is surrounded tetrahedrally by four nearest neighbour atoms

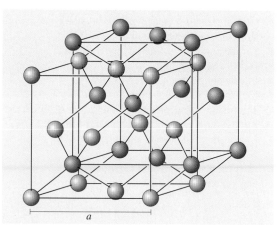

Fig. 4.1-2 The zinc blende lattice (a is the lattice parameter). This is a diamond lattice in which the two elementary cubes are occupied by different atomic species, for example by Zn-ions and by S-ions of opposite charge. Each Zn-ion is surrounded tetrahedrally by four nearest neighbour S-atoms and vice versa

Table 4.1-2 Lattice parameters of Group IV semiconductors and IV–IV compounds

Crystal		Lattice parameters (nm)	Temperature (K)	Remarks
Diamond	C	$a = 0.356685$	298	X-ray diffraction
Silicon	Si	$a = 0.543102018\,(34)$	295.7	High-purity single crystal, vacuum
Germanium	Ge	$a = 0.56579060$	298.15	Single crystal
Gray tin	α-Sn	$a = 0.64892\,(1)$	293.15	X-ray diffraction
Silicon carbide	3C-SiC	$a = 0.43596$	297	Debye–Scherrer method
	2H-SiC	$a = 0.30763\,(10)$ $c = 0.50480\,(10)$	300	Laue and Weissenberg patterns
Silicon–germanium alloys		There is a small deviation from Vegard's law (Fig. 4.1-4)		

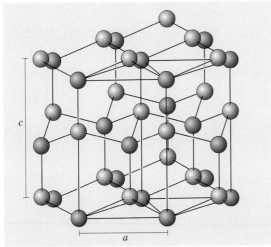

Fig. 4.1-3 The wurtzite lattice (a and c are the lattice parameters)

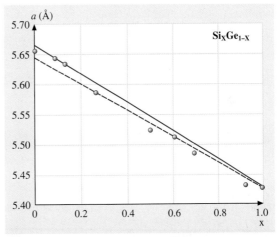

Fig. 4.1-4 Si_xGe_{1-x}. Compositional dependence of the lattice parameter. *Circles*, experimental results from X-ray diffraction; *dashed curve*, calculated by Vegard's law from the values for pure Ge and Si; *solid curve*, experimental results from [1.3]

Table 4.1-3 Densities of Group IV semiconductors and IV–IV compounds

Crystal		Density (g/cm^3)	Temperature (K)	Remarks
Diamond	C	3.51537 (5)	298	Flotation, type Ia
		3.51506 (5)	298	Flotation, type IIb
Silicon	Si	2.329002	298.15	High-purity crystal, hydrostatic weighing
Germanium	Ge	5.3234	298	Hydrostatic weighing
Gray tin	α-Sn	7.285	291	
Silicon carbide	3C-SiC	3.166	293	Polycrystalline sample
	6H-SiC	3.211	300	

Table 4.1-4 Elastic constants c_{ik} of Group IV semiconductors and IV–IV compounds

Crystal		c_{11} (GPa)	c_{12} (GPa)	c_{33} (GPa)	c_{44} (GPa)	Temperature (K)	Remarks
Diamond	C	1076.4 (2)	125.2 (2.3)		577.4 (1.4)	296	Brillouin scattering
Silicon	Si	165.77	63.93		79.62	298	Ultrasound measurement [a]
Germanium	Ge	124.0	41.3		68.3	298	Ultrasound measurement
Gray tin	α-Sn	69.0	29.3		36.2		[b]
Silicon carbide	3C-SiC	289	234		55.4	300	
	4H-SiC	507 (6)	108 (8)	547 (6)	159 (7)	RT	Brillouin scattering

[a] p-type sample, $\rho = 410\,\Omega\,\text{cm}$.
[b] Calculated from an 11-parameter shell model fitted to experimental data.

Table 4.1-5 Melting point T_m of Group IV semiconductors and IV–IV compounds

Crystal		Melting point (K)	Remarks
Diamond	C	4100	Diamond–graphite–liquid eutectic at $p = 12.5\,\text{GPa}$
Silicon	Si	1687	The coordination number changes from 4 in the diamond phase to 6.4 in the liquid phase with an increase in bond length from 0.235 to 0.25 nm. The liquid phase shows metallic conduction
Germanium	Ge	1210.4	
Gray tin	α-Sn		Transforms into metallic β-Sn slightly below RT
Silicon carbide	3C-SiC	3103 (40)	Peritectic decomposition temperature at $p = 35\,\text{bar}$

Table 4.1-6 Temperature dependence of the lattice parameters of Group IV semiconductors and IV–IV compounds

Crystal		Temperature dependence of lattice parameter
Diamond	C	Fig. 4.1-5
Silicon	Si	Fig. 4.1-6 [a]
Germanium	Ge	Fig. 4.1-7
Gray tin	α-Sn	$da/dT = 3.1 \times 10^{-6}$ nm/K; temperature range -130 to $25\,°\text{C}$
Silicon carbide	3C-SiC	Fig. 4.1-8
	6H-SiC	Fig. 4.1-9

[a] The lattice parameter a of high-purity Si can be approximated in the range 20–800 °C by $a(T) = 5.4304 + 1.8138 \times 10^{-5}(T - 298.15\,\text{K}) + 1.542 \times 10^{-9}(T - 298.15\,\text{K})^2$.

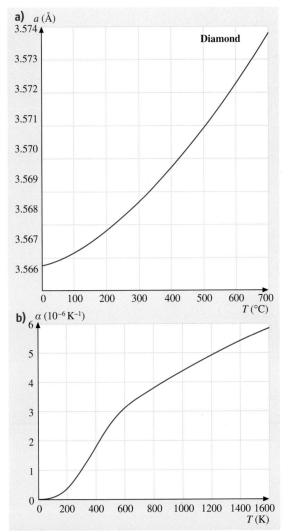

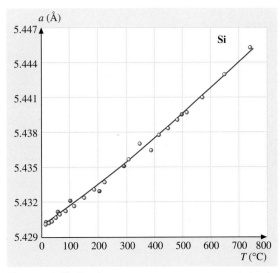

Fig. 4.1-6 Si. Lattice parameter vs. temperature in the range 20–740 °C, measurements on various samples [1.6]

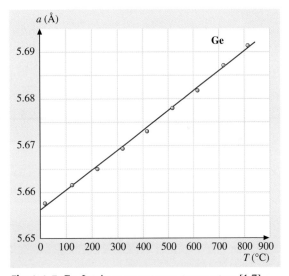

Fig. 4.1-5a,b Diamond. (a) Lattice parameter vs. temperature [1.4]; (b) linear thermal expansion coefficient vs. temperature [1.5]

Fig. 4.1-7 Ge. Lattice parameter vs. temperature [1.7]

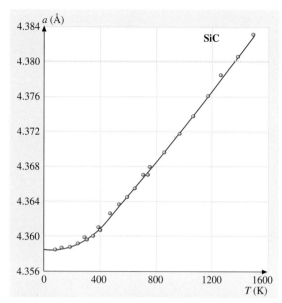

Fig. 4.1-8 SiC (3C). Lattice parameter vs. temperature [1.8]

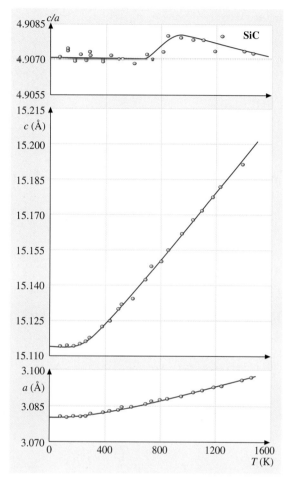

Fig. 4.1-9 SiC (6H). Lattice parameters vs. temperature [1.8]

Table 4.1-7 Linear thermal expansion coefficient α of Group IV semiconductors and IV–IV compounds and its temperature dependence

Crystal		Expansion coefficient α (K^{-1})	Temperature (K)	Remarks
Diamond	C	1.0×10^{-6}	300	See Fig. 4.1-5 for temperature dependence
Silicon	Si	$2.92\,(6) \times 10^{-6}$	293	See Fig. 4.1-10 for temperature dependence[a]
Germanium	Ge	5.90×10^{-6}	300	See Fig. 4.1-11 for temperature dependence
Gray tin	α-Sn	4.7×10^{-6}	293	See Fig. 4.1-12 for temperature dependence
Silicon carbide	3C-SiC	$2.77\,(42) \times 10^{-6}$	300	See Fig. 4.1-13 for temperature dependence

[a] The linear thermal expansion coefficient α of Si can be approximated in the temperature range 120–1500 K by
$\alpha(T) = \left(3.725\left\{1 - \exp[-5.88 \times 10^{-3}(T - 124)]\right\} + 5.548 \times 10^{-4}T\right) \times 10^{-6}$ (K^{-1}) (T measured in K)

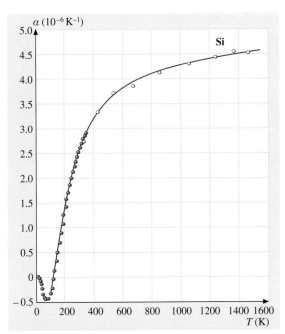

Fig. 4.1-10 Si. Linear thermal expansion coefficient vs. temperature. Experimental data from [1.9] (*filled circles*) and [1.10] (*open circles*)

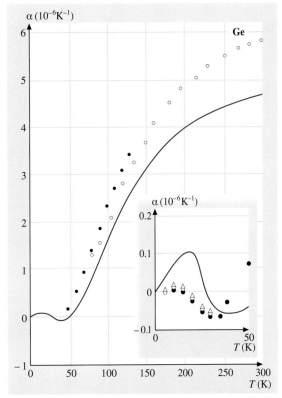

Fig. 4.1-11 Ge. Linear thermal expansion coefficient vs. temperature. Experimental data from various authors (*symbols*) and theoretical results (*solid line*) [1.11]

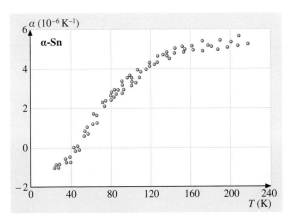

Fig. 4.1-12 α-Sn. Coefficient of linear thermal expansion vs. temperature [1.12, 13]

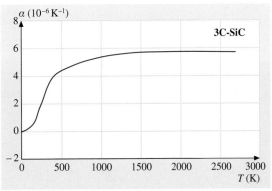

Fig. 4.1-13 3C-SiC. Recommended linear thermal expansion coefficient vs. temperature according to an analysis of data from nine references [1.5]

Table 4.1-8 Heat capacities and Debye temperatures of Group IV semiconductors and IV–IV compounds

Crystal		Heat capacity c_p (J/mol K)	Debye temperature Θ_D	Temperature (K)	Remarks
Diamond	C	6.195	1860 (10)	300	See Fig. 4.1-14 for temperature dependence of c_p, c_V
Silicon	Si	45.358 45.882	636	77 298	See Fig. 4.1-15 for temperature dependence of c_V
Germanium	Ge	51.88	374	298	See Fig. 4.1-16 for temperature dependence of c_V
Gray tin	α-Sn	19.09	220	100	See Fig. 4.1-17 for temperature dependence of c_V
Silicon carbide	3C-SiC	28.5	1270 (20)	293	Polycrystalline sample

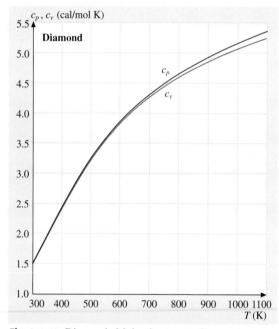

Fig. 4.1-14 Diamond. Molar heat capacity vs. temperature [1.14]

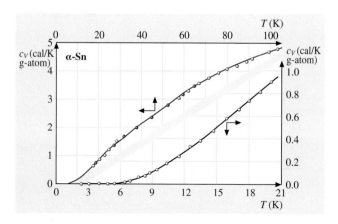

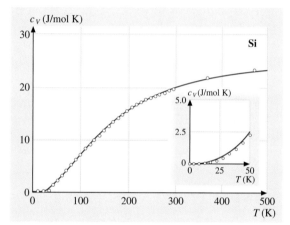

Fig. 4.1-15 Si. Heat capacity at constant volume vs. temperature. Figure from [1.11]

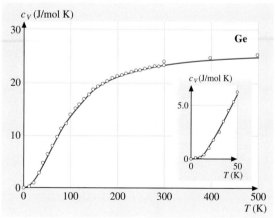

Fig. 4.1-16 Ge. Heat capacity at constant volume vs. temperature

Fig. 4.1-17 α-Sn. Heat capacity c_v vs. temperature. *Lower curve*, temperature range 0–20 K; *upper curve*, temperature range 0–105 K [1.15]

1.1 Group IV Semiconductors and IV–IV Compounds

Table 4.1-9 Phonon frequencies at symmetry points. Diamond (C) (ν in THz, 300 K, from Raman spectroscopy); silicon (Si) (ν in THz, 296 K, from inelastic neutron scattering); germanium (Ge) (ν in THz, 300 K, from coherent inelastic neutron scattering); gray tin (α-Sn) (ν in THz, 90 K, from inelastic thermal neutron scattering); silicon carbide (3C-SiC) (phonon wavenumbers $\tilde{\nu}$ in cm^{-1}, RT, from Raman spectroscopy); silicon carbide (6H-SiC) (ν in THz, derived from photoluminescence data)

Diamond (C)		Silicon (Si)		Germanium (Ge)		Gray tin (Sn)		Silicon carbide (3C-SiC)	
$\nu_{TO/LO}(\Gamma_{25'})$	39.9	$\nu_{LTO}(\Gamma_{25'})$	15.53 (23)	$\nu_{LTO}(\Gamma_{25'})$	9.02 (2)	$\nu_{TO/LO}(\Gamma_{25'})$	6.00 (6)	$\tilde{\nu}_{TA}(L)$	266
$\nu_{TA}(L_3)$	16.9	$\nu_{TA}(X_3)$	4.49 (6)	$\nu_{TA}(X_3)$	2.38 (2)	$\nu_{TA}(L_3)$	1.00 (4)	$\tilde{\nu}_{LA}(L)$	610
$\nu_{LA}(L_1)$	30.2	$\nu_{LAO}(X_1)$	12.32 (20)	$\nu_{LAO}(X_1)$	7.14 (2)	$\nu_{LA}(L_1)$	4.15 (4)	$\tilde{\nu}_{TO}(L)$	766
$\nu_{LO}(L_{2'})$	37.5	$\nu_{TO}(X_4)$	13.90 (30)	$\nu_{TO}(X_4)$	8.17 (3)	$\nu_{LO}(L_{2'})$	4.89 (8)	$\tilde{\nu}_{LO}(L)$	838
$\nu_{TO}(L_{3'})$	36.2	$\nu_{TA}(L_3)$	3.43 (5)	$\nu_{TA}(L_{3'})$	1.87 (2)	$\nu_{TO}(L_{3'})$	5.74 (12)	$\tilde{\nu}_{TA}(X)$	373
$\nu_{TA}(X_3)$	24.2	$\nu_{LA}(L_2)$	11.35 (30)	$\nu_{LA}(L_2)$	6.63 (4)	$\nu_{TA}(X_3)$	1.25 (6)	$\tilde{\nu}_{LA}(X)$	640
$\nu_{LA/LO}(X_1)$	35.5	$\nu_{LO}(L_1)$	12.60 (32)	$\nu_{TO}(L_3)$	8.55 (3)	$\nu_{LA/LO}(X_1)$	4.67 (6)	$\tilde{\nu}_{TO}(X)$	761
$\nu_{TO}(X_4)$	32.0	$\nu_{TO}(L_3)$	14.68 (30)	$\nu_{TO}(L_1)$	7.27 (2)	$\nu_{TO}(X_4)$	5.51 (8)	$\tilde{\nu}_{LO}(X)$	829

Table 4.1-9 Phonon frequencies at symmetry points, cont.

Silicon carbide (6H-SiC)						
ν_{TA}	8.1	8.8	9.5	9.8	10.6	11.2	12.9
ν_{LA}	12.2	12.9	16.2	16.7	18.6		
ν_{TO}	22.9	23.1	23.7				
ν_{LO}	25.2	25.5	25.9				

Table 4.1-10 Phonon dispersion relations of Group IV semiconductors and IV–IV compounds

Crystal		Phonon dispersion curves
Diamond	C	Fig. 4.1-18
Silicon	Si	Fig. 4.1-19
Germanium	Ge	Fig. 4.1-20
Gray tin	α-Sn	Fig. 4.1-21
Silicon carbide	3C-SiC	Fig. 4.1-22
	6H-SiC	

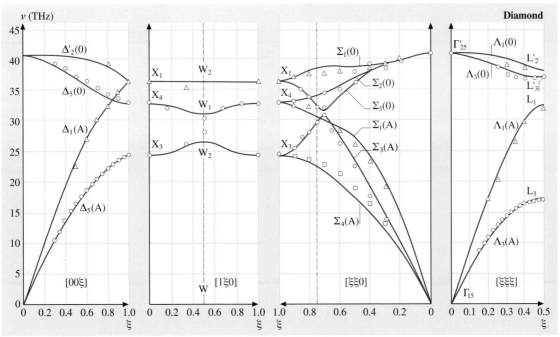

Fig. 4.1-18 Diamond. Phonon dispersion curves. *Symbols*, experimental data from neutron scattering; *solid curves*, shell model calculation [1.16]

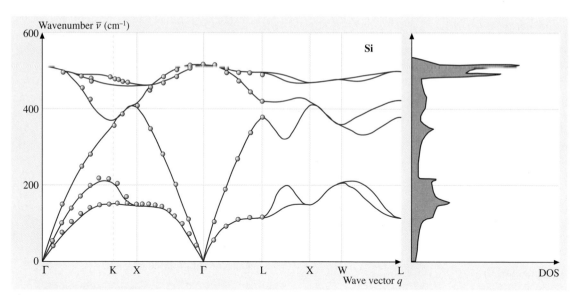

Fig. 4.1-19 Si. Phonon dispersion curves (*left panel*) and phonon density of states (*right panel*) [1.16]. Experimental data points [1.17, 18] and ab initio calculations [1.16]

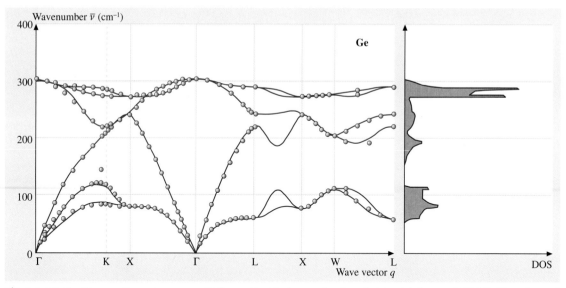

Fig. 4.1-20 Ge. Phonon dispersion curves (*left panel*) and phonon density of states (*right panel*) [1.16]. Experimental data points [1.19] and ab initio calculations [1.16]

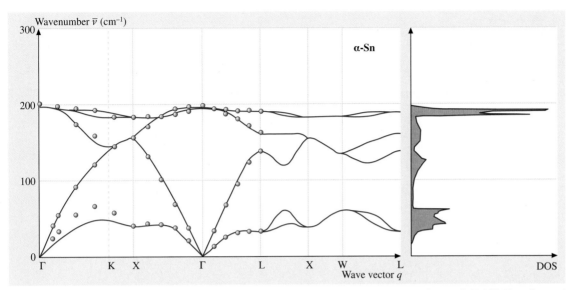

Fig. 4.1-21 α-Sn. Phonon dispersion curves (*left panel*) and phonon density of states (*right panel*) [1.20]. Experimental data points from inelastic neutron scattering at 90 K [1.21]. *Solid lines*, ab initio pseudopotential calculation [1.20]

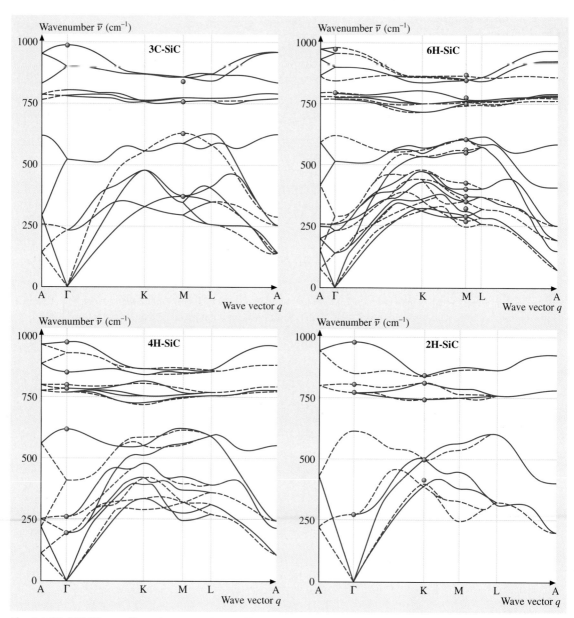

Fig. 4.1-22 SiC. Phonon dispersion curves for the 3C, 2H, 4H, and 6H modifications obtained from a bond-charge model calculation [1.22]. *Symbols* represent experimental data. The different types of *lines* indicate different polarization states

B. Electronic Properties
Tables 4.1-11 – 4.1-18.

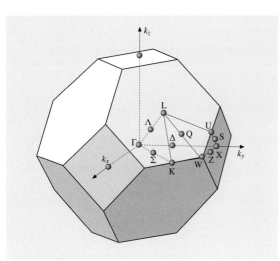

Fig. 4.1-23 Brillouin zone of the diamond and of the zinc blende lattice

Table 4.1-11 First Brillouin zones of Group IV semiconductors and IV–IV compounds

Crystal		Figures
Diamond	C	Fig. 4.1-23
Silicon	Si	Fig. 4.1-23
Germanium	Ge	Fig. 4.1-23
Gray tin	α-Sn	Fig. 4.1-23
Silicon carbide	3C-SiC	Fig. 4.1-23
	2H-SiC	Fig. 4.1-24

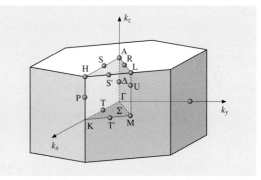

Fig. 4.1-24 Brillouin zone of the Wurtzite lattice

Table 4.1-12 Band structures of Group IV semiconductors and IV–IV compounds

Diamond (C) (Fig. 4.1-25)
Diamond is an indirect-gap semiconductor, the lowest minima of the conduction band being located along the Δ axes. The valence band has the structure common to all group IV semiconductors, namely three bands degenerate at Γ. The spin–orbit splitting of these bands is negligible.

Silicon (Si) (Fig. 4.1-26)
The conduction band is characterized by six equivalent minima along the [100] axes of the Brillouin zone located at about $k_0 = 0.85(2\pi/a)$ (symmetry Δ_1). The surfaces of constant energy are ellipsoids of revolution with their major axes along [100]. Higher minima are located at Γ and along the [111] axes about 1 eV above the [100] minima.

The valence band has its maximum at the Γ point (symmetry Γ_8), the light- and heavy-hole bands being degenerate at this point. Both bands show warping. The third, spin–orbit split-off band has Γ_7 symmetry. The spin–orbit splitting energy at Γ is small compared with most interband energy differences. Thus, spin–orbit interaction is mostly neglected in band structure calculations and the notation of the single group is used to denote the symmetry of a band state.

Table 4.1-12 Band structures of Group IV semiconductors and IV–IV compounds, cont.

Germanium (Ge) (Fig. 4.1-27)

The conduction band is characterized by eight equivalent minima at the end points L of the [111] axes of the Brillouin zone (symmetry L_6). The surfaces of constant energy are ellipsoids of revolution with their major axes along [111]. Higher energy minima of the conduction band are located at the Γ point and (above these) on the [100] axes.

The valence band has its maximum at the Γ point (symmetry Γ_8), the light- and heavy-hole bands being degenerate at this point. Both bands are warped. The third, spin–orbit split-off band has Γ_7 symmetry. In contrast to silicon the spin–orbit splitting energies are considerable. Thus, the symmetry notation of the double group of the diamond lattice is mostly used for Ge.

Gray tin (α-Sn) (Fig. 4.1-28)

The band structure of gray tin (α-Sn) is qualitatively different from those of the other Group IV elements. The s-like Γ_7 conduction band edge, which decreases drastically with atomic number in the sequence C → Si → Ge, is situated below the p-like Γ_8 valence band edge in α-Sn. This causes an inversion of the curvature of the Γ_8 light-hole band. Consequently, α-Sn is a zero-gap semiconductor, with its lowest conduction band and its highest valence band being degenerate at Γ (symmetry Γ_8). A second conduction band, with minima at L_6, follows at a slightly higher energy. This second band determines the properties of n-type samples for $n > 10^{17}$ cm^{-3} ($T > 77$ K for intrinsic samples).

Silicon carbide

Band structure calculations show that the conduction band minima are situated along the cubic axes at the border of the Brillouin zone. The band structure of 3C-SiC is shown in Fig. 4.1-29, and that of 2H-SiC in Fig. 4.1-30.

Silicon–germanium alloys

The band structure is characterized by a crossover in the lowest conduction band edge from Ge-like [111] symmetry to Si-like [100] symmetry at the composition $x = 0.15$. The shape of the band edge (and hence the effective masses) varies only slightly as a function of composition.

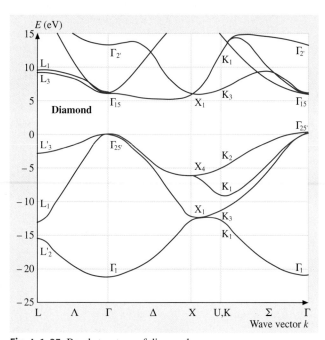

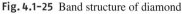

Fig. 4.1-25 Band structure of diamond

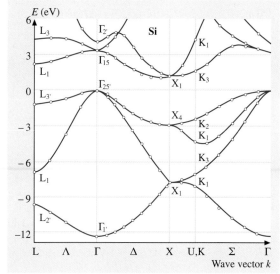

Fig. 4.1-26 Band structure of silicon

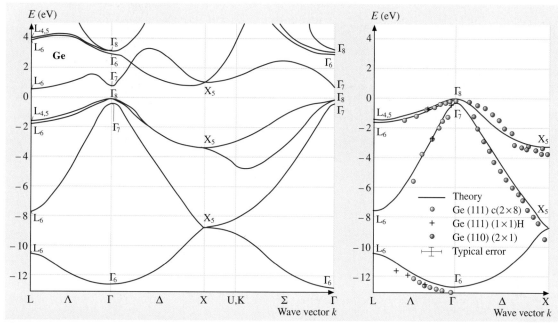

Fig. 4.1-27 Band structure of germanium

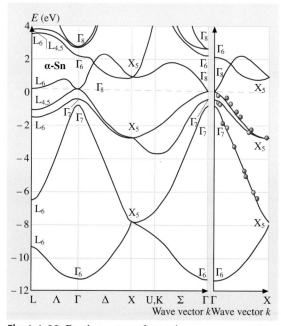

Fig. 4.1-28 Band structure of gray tin

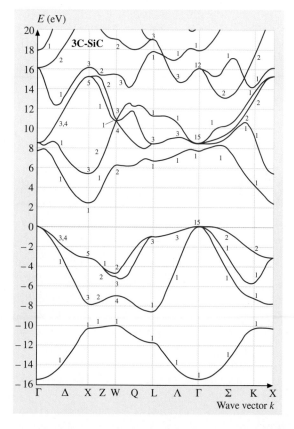

Fig. 4.1-29 Band structure of 3C silicon carbide

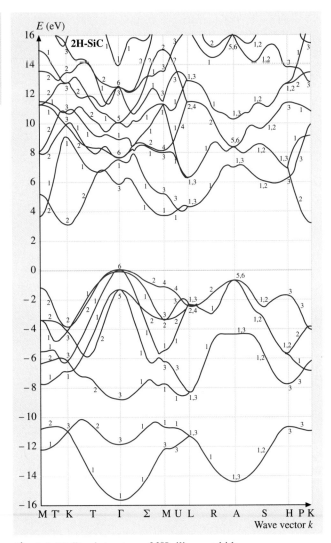

Fig. 4.1-30 Band structure of 2H silicon carbide

Table 4.1-13 Energy gaps of Group IV semiconductors and IV–IV compounds

Crystal		Between bands	Gap	Energy (eV)	Temperature (K)	Remarks
Diamond	C	$\Gamma_{25'v}$ and Δ_{1c}	$E_{g,\,indir}$	5.50 (5)	RT	Quantum photoyield
			$E_{gx,\,indir}$	5.416 (2)	100	Indirect exciton gap with lower valence band
				5.409 (2)		Indirect exciton gap with upper valence band
Silicon	Si	$\Gamma_{25'v}$ and Δ_{1c}	$E_{g,\,ind}$	1.1700	0	Extrapolated, wavelength-modulated transmission
				1.1242	300	Wavelength-modulated transmission
			$E_{g,\,th}$	1.205	0	Extrapolated from temperature dependence of conductivity above 200 K
		$\Gamma_{25'v}$ and $\Gamma_{2'c}$	$E_{g,\,dir}$	4.185 (10)	4.2	Electroreflection
				4.135	190	

Table 4.1-13 Energy gaps of Group IV semiconductors and IV–IV compounds, cont.

Crystal		Between bands	Gap	Energy (eV)	Temperature (K)	Remarks
Germanium	Ge	Γ_{8v} and L_{6c}	$E_{g,\,ind}$	0.744 (1)	1.5	Magnetotransmission
				0.664	291	Optical absorption
			$E_{g,\,th}$	0.785	0	Extrapolated from temperature dependence of intrinsic conductivity
		Γ_{8v} and Γ_{7c}	$E_{g,\,dir}$	0.898 (1)	1.5	Magnetoabsorption
				0.805 (1)	293	
Silicon carbide	3C-SiC	Γ_{15v} and X_{1c}	$E_{g,\,ind}$	2.417 (1)	2	Wavelength-modulated absorption
				2.2	300	Optical absorption
			$E_{g,\,dir}$	6.0	300	Optical absorption
			E_{gx}	2.390	2	Excitonic energy gap, wavelength-modulated absorption
Silicon carbide	2H-SiC		E_{gx}	3.330	4	Excitonic energy gap; optical absorption
Silicon carbide	6H-SiC		$E_{g,\,ind}$	2.86	300	Optical absorption
			E_{gx}	3.023	4	

Table 4.1-14 Exciton energies of Group IV semiconductors and IV–IV compounds (E_b = exciton binding energy)

Crystal	Quantity	State	Exciton energy (eV)	Temperature (K)	Remarks
Diamond	$E_b(1S)$		0.19 (15)		1S core exciton
	E_b		0.07		Indirect exciton binding energy
Silicon	$E(1S)$		1.1552 (3)	1.8	Wavelength-modulated absorption
	E_b		0.01412	4.2	Calculated binding energy
Germanium	$E(1S)$	$1S_{3/2}^{3/2}\,(L_4+L_5)$	0.74046 (3)	2.1, 5.1	Optical absorption at 2.1 K and luminescence at 5.1 K
		$1S_{1/2}^{3/2}\,(L_6)$	0.74158 (3)		
	E_b	$1S_{3/2}^{3/2}\,(L_4+L_5)$	0.00418		
		$1S_{1/2}^{3/2}\,(L_6)$	0.00317		
Silicon carbide (3C-SiC)	E_b		0.027	2	Wavelength-modulated absorption

Table 4.1-15 Spin–orbit splitting energies of Group IV semiconductors and IV–IV compounds

Crystal	Quantity	Bands	Splitting energy (eV)	Temperature (K)	Remarks
Diamond	$\Delta(\Gamma_{25'v})$		0.006 (1)		Cyclotron resonance
Silicon	$\Delta_0(\Gamma_{25'v})$		0.0441 (3)	1.8	Wavelength-modulated absorption
Germanium	Δ_0	Γ_{7v} to Γ_{8v}	0.297	10	Electroreflectance
		Γ_{6v} to Γ_{8c}	0.200		
Silicon carbide (3C-SiC)	Δ_0		0.010	2	Wavelength-modulated absorption

Table 4.1-16 Effective masses of electrons (in units of the electron mass m_0) for Group IV semiconductors and IV–IV compounds

Crystal	Quantity	Value	Temperature (K)	Remarks
Diamond	$m_{n,\,parall}$	1.4		Field dependence of electron drift velocity
	$m_{n,\,perpend}$	0.36		
Silicon	$m_{n,\,parall}$	0.9163 (4)	1.26	Cyclotron resonance with uniaxial stress
	$m_{n,\,perpend}$	0.1905 (1)		

Table 4.1-16 Effective masses of electrons for Group IV semiconductors and IV–IV compounds, cont.

Crystal	Quantity	Value	Temperature (K)	Remarks
Germanium	$m_{n, perpend}(L_6)$	0.0807 (8)	30–100	Cyclotron resonance, magnetophonon resonance
		0.0823	120	
	$m_{n, parall}(L_6)$	1.57 (3)	30–100	Cyclotron resonance magnetophonon resonance
		1.59	120	
	$m_n(\Gamma_7)$	0.0380 (5)	30	Piezomagnetoreflectance
Gray tin	$m_{n, light}(\Gamma_8)$	0.0236 (2)	1.3	Density-of-states mass
	$m_{n, heavy}(L_6)$	0.21	4.2	
Silicon carbide (3C-SiC)	$m_{n, parall}$	0.677 (15)	45	Cyclotron resonance
	$m_{n, perpend}$	0.247 (11)		
Silicon carbide (6H-SiC)	$m_{n, parall}$	3–6		Cyclotron resonance
	$m_{n, perpend}$	0.48 (2)		

Table 4.1-17 Effective masses of holes (in units of the electron mass m_0) for Group IV semiconductors and IV–IV compounds

Crystal	Quantity	Value	Temperature (K)	Field direction	Remarks
Diamond	$m_{p, heavy}$	1.08			Calculated from valence band parameters
	$m_{p, light}$	0.36			
	$m_{spin-orbit\, split}$	0.15			
	$m_{p, dens\, stat}$	0.75	300		Hall effect
Silicon	$m_{p, heavy}$	0.537	4.2		Cyclotron resonance
	$m_{p, light}$	0.153			
	$m_{spin-orbit\, split}$	0.234			
	$m_{p, dens\, stat}$	0.591			Density-of-states mass
Germanium	$m_{p, light}$	0.0438 (3)	4	B parallel to [100]	Cyclotron resonance
		0.0426 (2)		B parallel to [111]	
		0.0430 (3)		B parallel to [110]	
	$m_{p, heavy}$	0.284 (1)		B parallel to [100]	
		0.376 (1)		B parallel to [111]	
		0.352 (4)		B parallel to [110]	
	$m_{spin-orbit\, split}$	0.095 (7)	30		Piezomagnetoabsorption
Gray tin	$m_p(\Gamma_8)$	0.195			Interband magnetoreflection
	$m_p(\Gamma_7)$	0.058			
Silicon carbide (3C-SiC)	m_p	0.45	45		High-field cyclotron resonance
Silicon carbide (6H-SiC)	$m_{p, parall}$	1.85 (3)			Cyclotron resonance
	$m_{p, perpend}$	0.66 (2)			

Table 4.1-18 g-factor g_c of conduction electrons for Group IV semiconductors

Crystal	g_c	Temperature (K)	Remarks
Diamond	2.0030 (3)	140–300	Electron spin resonance
Silicon	1.99893 (28)		Electron spin resonance
Germanium	−3.0 (2)	30	Piezomagnetoabsorption

C. Transport Properties
Tables 4.1-19 – 4.1-21.

Electronic Transport, General Description.
Diamond (C). Owing to the large band gap, most diamonds are insulators at room temperature, and so electronic transport is extrinsically determined and therefore strongly dependent on the impurity content. Natural (type IIb) and synthetic semiconducting diamonds are always p-type. The electron mobility can be derived only from photoconductivity experiments.

Table 4.1-19 Intrinsic carrier concentration n_i and electrical conductivity σ_i of Group IV semiconductors and IV–IV compounds

Crystal	n_i (cm^{-3})	σ_i (Ω^{-1}cm^{-1})	T (K)	Temperature dependence
Silicon	1.02×10^{10}	3.16×10^{-2}	300	Fig. 4.1-31 [a]
Germanium	2.33×10^{13}	2.1×10^{-2}	300	Figs. 4.1-32 and 4.1-33
3C-SiC				Fig. 4.1-34
Si–Ge alloys	Composition dependence of intrinsic conductivity: Fig. 4.1-36			
	Composition dependence of the mobilities: Figs. 4.1-35 and 4.1-37			

[a] The intrinsic conductivity of Si up to 1273 K is given by the phenomenological expression $\log_{10} \sigma_i = 4.247 - 2.924 \times 10^3 T^{-1}$ (σ_i in Ω^{-1}cm^{-1}, T in K).

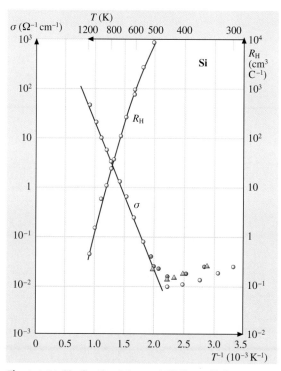

Fig. 4.1-31 Si. Conductivity and Hall coefficient vs. reciprocal temperature in the range of intrinsic conduction; n- and p-type samples with doping concentrations of 1.7×10^{14} cm^{-3} and above [1.23]. The different *symbols* show results from different samples

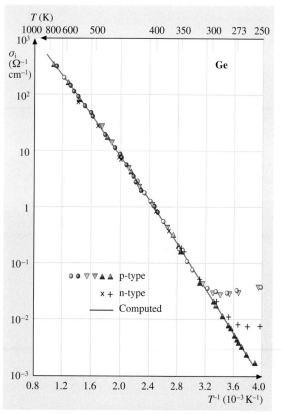

Fig. 4.1-32 Ge. Conductivity vs. reciprocal temperature in the range of intrinsic conduction [1.24]

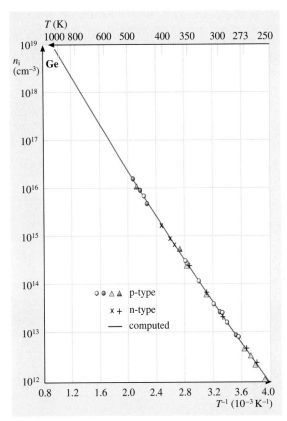

Fig. 4.1-33 Ge. Intrinsic carrier concentration vs. reciprocal temperature [1.24]

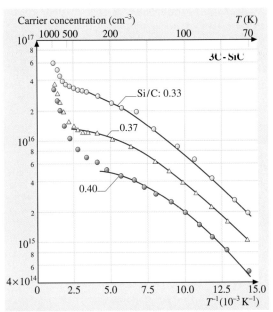

Fig. 4.1-34 3C-SiC. Temperature dependence of carrier concentration for n-type films grown by CVD on Si(100) substrates [1.25]. *Solid lines*, calculated values. Si/C = ratio of source gases

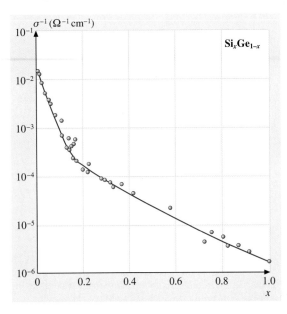

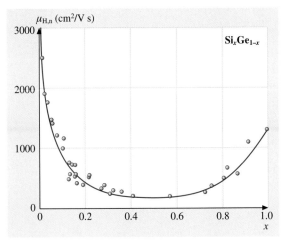

Fig. 4.1-35 Si_xGe_{1-x}. Composition dependence of the electron Hall mobility at room temperature [1.26]

Fig. 4.1-36 Si_xGe_{1-x}. Composition dependence of the intrinsic conductivity at room temperature [1.26]

Table 4.1-20 Electron mobilities μ_n and hole mobilities μ_p of Group IV semiconductors and IV–IV compounds

Crystal	μ_n (cm^2/V s)	μ_p (cm^2/V s)	T (K)	Remarks
Diamond	2000	2100	293	See Figs. 4.1-38 and 4.1-39 for temperature dependence
Silicon	1450	370	300	See Figs. 4.1-40 and 4.1-41 for temperature dependence
		2×10^5	20	
Germanium	3800	1800	300	See Figs. 4.1-33 and 4.1-42 for temperature dependence
Gray tin	$\mu_n(\Gamma_8)$ 0.12×10^6	10×10^3	100	From Hall and magnetoresistance data, via conductivity and Hall coefficient
	$\mu_n(L_6)$ 1.6×10^3			
Silicon carbide (3C-SiC)	510	15–21	296	Epitaxial film grown on Si(100)

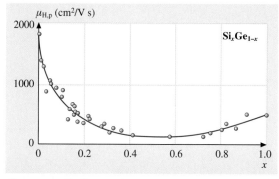

Fig. 4.1-37 Si_xGe_{1-x}. Composition dependence of the hole Hall mobility at room temperature [1.26]

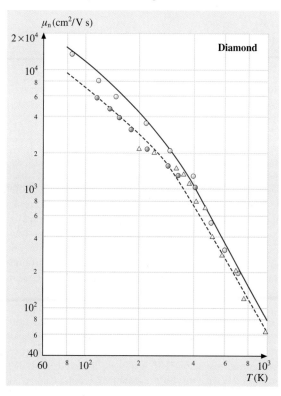

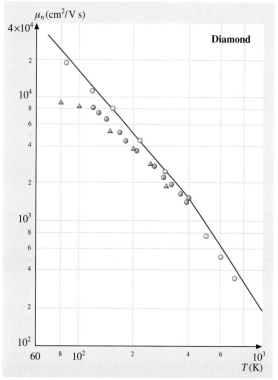

Fig. 4.1-38 Diamond. Electron mobility vs. temperature. *Open circles*, drift mobility data from [1.27]; *filled triangles* and *filled circles*, Hall mobility data from [1.28] and [1.29, 30], respectively. *Continuous curve*, theoretical drift mobility [1.27]

Fig. 4.1-39 Diamond. Hole mobility vs. temperature. *Open circles*, drift mobility data from [1.31]; *filled circles* and *filled triangles*, Hall mobility data from [1.29, 30] and [1.32], respectively. *Solid* and *dashed curves*: calculated drift and Hall mobilities, respectively [1.33]

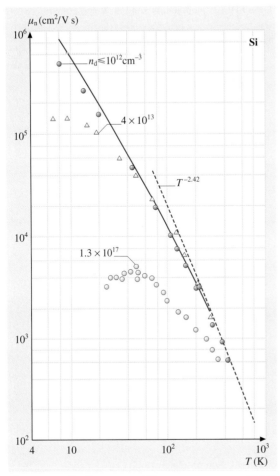

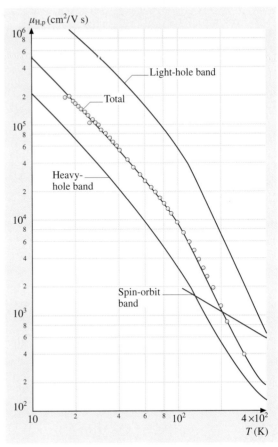

Fig. 4.1-40 Si. Electron mobility vs. temperature [1.34]. Data points for $n_d \leq 10^{12}$ cm^{-3} obtained by time-of-flight technique [1.35]. Other experimental values taken from [1.36] (4×10^{13} cm^{-3}) and [1.23] (1.3×10^{17} cm^{-3}). *Continuous line*, theoretical lattice mobility, after [1.35]. *Dash-dotted line*, $T^{-2.42}$ dependence of μ_n around room temperature

Fig. 4.1-41 Si. (b) Hall hole mobility vs. temperature; *circles* from [1.37], *solid lines* calculated contributions from the three valence bands

Silicon (Si). The electronic transport is due exclusively to electrons in the [100] conduction band minima and holes in the two uppermost (heavy and light) valence bands. In samples with impurity concentrations below 10^{12} cm^{-3}, the mobilities are determined by pure lattice scattering down to temperatures of about 10 K (n-type) or 50 K (p-type), for electrons and holes, respectively. Higher impurity concentrations lead to deviations from the lattice mobility at corresponding higher temperatures. For electrons, the lattice mobility below 50 K is dominated by deformation-potential coupling to accoustic phonons. At higher temperatures, intervalley scattering between the six equivalent minima of the conduction band is added to the intravalley process, modifying the familiar $T^{-1.5}$ dependence of the acoustic-mode-dominated mobility to $T^{-2.42}$. At temperatures below 100 K, the lattice mobility of holes is dominated by acoustic scattering, but does not follow a $T^{-1.5}$ law, owing to the nonparabolicity of the valence bands. The proportionality of μ_p to $T^{-2.2}$ around room temperature is a consequence of optical-phonon scattering.

Germanium (Ge). Low-field electronic transport is provided by electrons in the L_6 minima of the conduction band and holes near the Γ_8 point in the valence bands. At room temperature, the mobility of samples with impurity concentrations below 10^{15} cm^{-3} is limited essentially by lattice scattering; higher donor or acceptor concen-

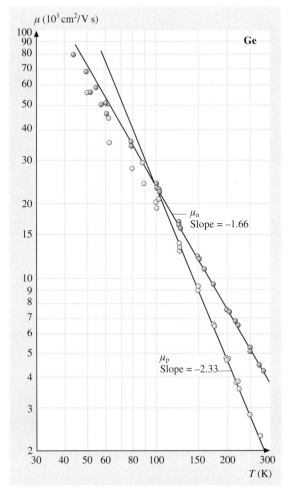

Fig. 4.1-42 Ge. Electron and hole mobilities vs. temperature for a constant carrier concentration (high-purity samples) [1.38]

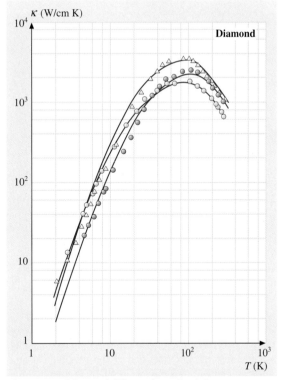

Fig. 4.1-43 Diamond. Thermal conductivity vs. temperature for three type Ia diamonds [1.39]

trations result in an increasing influence of impurity scattering. At 77 K, even for doping concentrations below 10^{13} cm^{-3}, the mobilities depend on the impurity concentration. Low temperatures and high concentrations lead to the replacement of free-carrier conduction by impurity conduction.

Gray Tin (α-Sn). In intrinsic samples, light electrons and holes in the Γ_8 bands determine the transport properties. When the Fermi level crosses the energy of the L_6 minima, heavy electrons have an influence. This leads, for example, to a screening enhancement of the light-electron mobility and, in heavily doped n-type samples, to a dominant role of the L electrons. Consequently,

Table 4.1-21 Thermal conductivity κ of Group IV semiconductors and IV–IV compounds

Crystal		κ (W/cm K)	Temperature (K)	Remarks
Diamond	C	6–10	293	See Fig. 4.1-43 for temperature dependence
Silicon	Si	1.56	300	See Fig. 4.1-44 for temperature dependence
Germanium	Ge			See Fig. 4.1-46 for temperature dependence
Silicon carbide	3C-SiC			
	6H-SiC	4.9	300	Perpendicular to c axis; See Fig. 4.1-45 for temperature dependence

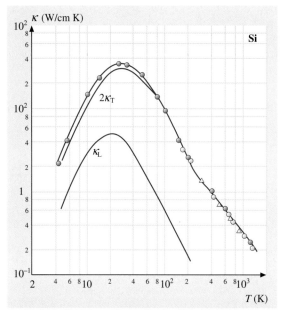

Fig. 4.1-44 Si. Lattice thermal conductivity vs. temperature, according to [1.40] (*filled circles*), [1.41] (*open circles*), and [1.42] (*triangles*). The *curves* marked κ_L and $2\kappa_T$ represent the contributions of the longitudinal and transverse phonons, respectively [1.43]

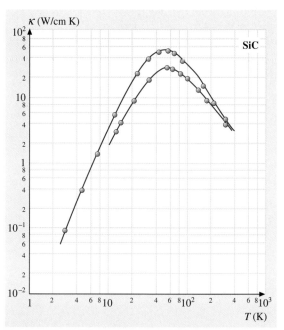

Fig. 4.1-45 6H-SiC. Thermal conductivity ($\perp c$ axis) vs. temperature for two different samples [1.44]

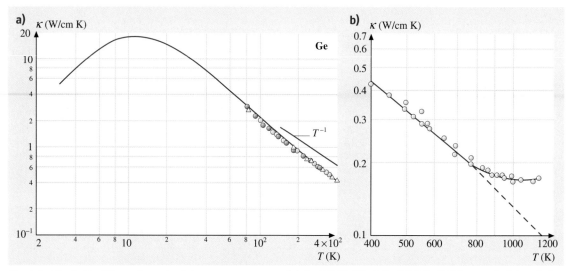

Fig. 4.1-46a,b Ge. Thermal conductivity vs. temperature. (**a**) 3–400 K, (**b**) 400–1200 K. *Solid curve* in (**a**) and data in (**b**) from [1.40]; experimental data in (**a**) from [1.45]. *Dashed line* in (**b**), extrapolated lattice component

heavily doped n-type samples behave as indirect-gap semiconductors with an E_g of about 0.09 eV.
Silicon–Germanium Alloys (Si_xGe_{1-x}). The transport properties have been investigated mostly on single crystals. The mobility is influenced by alloy scattering, which makes a contribution $\mu_{alloy} \propto T^{0.8} x^{-1}(1-x)^{-1}$. Near the band crossover ($x = 0.15$), intervalley scattering has to be taken into account.

D. Electromagnetic and Optical Properties
Tables 4.1-22 – 4.1-25.

Table 4.1-22 High-frequency dielectric constant ε (real part of the complex dielectric constant) of Group IV semiconductors and IV–IV compounds

Crystal		ε	T (K)	Frequency	Method
Diamond	C	5.70 (5)	300	1–10 kHz	Capacitance bridge
Silicon	Si	12.1	4.2	750 MHz	Capacitance bridge
		11.97	300		
Germanium	Ge	16.5	4.2	750 MHz	Capacitance bridge
		16.0	4.2	9200 MHz	Microwave measurement
		16.2	300		
		15.8	77	1 MHz	Capacitance bridge
Gray tin	α-Sn	24	300	Infrared	Infrared reflectance measurement
Silicon carbide	3C-SiC	9.52	300	Low frequency	Infrared transmission
		6.38		High frequency	
Silicon carbide	6H-SiC	9.66	300	Low frequency	ε_{perp}
		10.03			ε_{parall}
		6.52	300	High frequency	ε_{perp}
		6.70			ε_{parall}

Optical Constants. The real part (ε_1) and imaginary part (ε_2) of the dielectric constant at optical frequencies were measured by spectroscopic ellipsometry. The refractive index n, the extinction coefficient k, the absorption coefficient K, and the reflectivity R have been calculated from ε_1 and ε_2. The frequency dependences of ε_1, ε_2, n, and K are shown in Figs. 4.1-48 and 4.1-47.

Diamond. The refractive index n is equal to 3.5 at $\lambda = 177.0$ nm. The spectral dependence of the refractive index n can be approximated by the empirical formula

$$n^2 - 1 = a\lambda^2 / \left(\lambda^2 - \lambda_1^2\right) + b\lambda^2 / \left(\lambda^2 - \lambda_2^2\right),$$

with the parameters $a = 0.3306$, $b = 4.3356$, $\lambda_1 = 175.0$ nm, and $\lambda_2 = 106.0$ nm.

Silicon. (See Table 4.1-23).

Germanium. (See Fig. 4.1-48).

Gray Tin (α-Sn). The index of refraction n, extinction coefficient k, and absorption coefficient K versus photon energy are shown in Fig. 4.1-47.

Silicon Carbide (SiC).

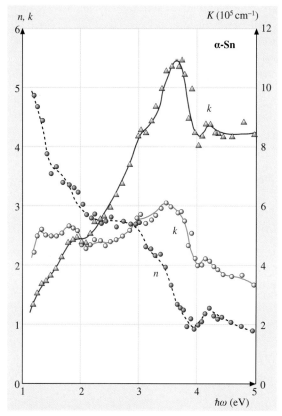

Fig. 4.1-47 α-Sn. Index of refraction n, extinction coefficient k, and absorption coefficient K vs. photon energy [1.46]

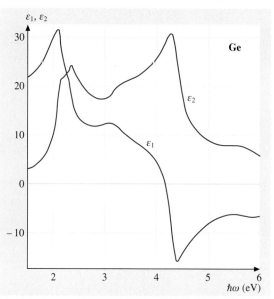

Fig. 4.1-48 Ge. Real and imaginary parts of the dielectric constant vs. photon energy [1.47]

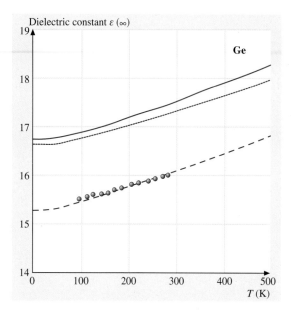

Fig. 4.1-49 Ge. Temperature dependence of the high-frequency dielectric constant [1.48]. Experimental data points [1.49] and ab initio calculations (*solid line*); the *dotted line* is the theoretical result without the effect of thermal expansion, and the *dashed line* is the *solid line* shifted to match the experimental data

Table 4.1-23 Optical constants of silicon

$h\nu$ (eV)	ε_1	ε_2	n	k	K (10^3/cm)	R
1.5	13.488	0.038	3.673	0.005	0.78	0.327
2.0	15.254	0.172	3.906	0.022	4.47	0.351
2.5	18.661	0.630	4.320	0.073	18.48	0.390
3.0	27.197	2.807	5.222	0.269	81.73	0.461
3.5	22.394	33.818	5.610	3.014	1069.19	0.575
4.0	12.240	35.939	5.010	3.586	1454.11	0.591
4.5	−19.815	24.919	2.452	5.082	2317.99	0.740
5.0	−10.242	11.195	1.570	3.565	1806.67	0.675
5.5	−9.106	8.846	1.340	3.302	1840.59	0.673
6.0	−7.443	5.877	1.010	2.909	1769.27	0.677

Table 4.1-24 Optical constants of germanium

$h\nu$ (eV)	ε_1	ε_2	n	k	K (10^3/cm)	R
1.5	21.560	2.772	4.653	0.298	45.30	0.419
2.0	30.361	10.427	5.588	0.933	189.12	0.495
2.5	13.153	20.695	4.340	2.384	604.15	0.492
3.0	12.065	17.514	4.082	2.145	652.25	0.463
3.5	9.052	21.442	4.020	2.667	946.01	0.502
4.0	4.123	26.056	3.905	3.336	1352.55	0.556
4.5	−14.655	16.782	1.953	4.297	1960.14	0.713
5.0	−8.277	8.911	1.394	3.197	1620.15	0.650
5.5	−6.176	7.842	1.380	2.842	1584.57	0.598
6.0	−6.648	5.672	1.023	2.774	1686.84	0.653

Table 4.1-25 Optical constants of silicon carbide

Crystal	Refractive index	Wavelength range (nm)
3C-SiC	$n(\lambda) = 2.55378 + 3.417 \times 10^4/\lambda^2$	467–691
6H-SiC	$n_o(\lambda) = 2.5531 + 3.34 \times 10^4/\lambda^2$ $n_e(\lambda) = 2.5852 + 3.68 \times 10^4/\lambda^2$	

4.1.2 III–V Compounds

4.1.2.1 Boron Compounds

A. Crystal Structure, Mechanical and Thermal Properties
Tables 4.1-26 – 4.1-32.

Table 4.1-26 Crystal structures of boron compounds

Crystal		Space group		Crystal system	Structure type	Remarks	Figure
Boron nitride, hexagonal	BN_{hex}		$P6_3/mmc$	hex	h-BN	Stable under normal conditions	Fig. 4.1-50
Boron nitride, cubic	BN_{cub}	T_d^2	$F\bar{4}3m$	fcc	Zinc blende	Metastable under normal conditions	Fig. 4.1-2
Boron nitride, wurtzite	BN_w	C_{6v}^4	$P6_3mc$	hex	Wurtzite	Metastable under all conditions	Fig. 4.1-3
Boron phosphide	BP	T_d^2	$F\bar{4}3m$	fcc	Zinc blende	Stable under normal conditions	Fig. 4.1-2
Boron arsenide	BAs	T_d^2	$F\bar{4}3m$	fcc	Zinc blende	Stable in presence of As vapor up to 920 °C	Fig. 4.1-2
Boron antimonide	BSb	T_d^2	$F\bar{4}3m$	fcc	Zinc blende	Stable under normal conditions	Fig. 4.1-2

Table 4.1-27 Lattice parameters of boron compounds

Crystal		Lattice parameters (nm)	Temperature (K)	Remarks
Boron nitride, cubic	BN_{cub}	$a = 0.36160\,(3)$	RT	X-ray diffraction
Boron nitride, hexagonal	BN_{hex}	$a = 0.25072\,(1)$	297	Prepared from amorphous BN in
		$c = 0.687\,(1)$	297	N_2 atmosphere at room temperature
Boron phosphide	BP	$a = 0.45383\,(4)$	297	
Boron arsenide	BAs	$a = 0.4777$		
Boron antimonide	BSb	$a = 0.512$		Ab initio pseudopotential calculation

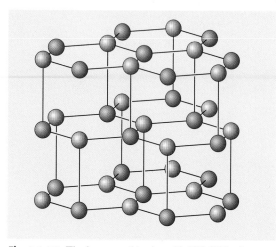

Fig. 4.1-50 The hexagonal lattice of h-BN (BN_{hex})

Table 4.1-28 Densities of boron compounds

Crystal		Density (g/cm³)	Remarks
Boron nitride, cubic	BN_{cub}	3.4863	X-ray diffraction
Boron nitride, hexagonal	BN_{hex}	2.279	X-ray density
Boron phosphide	BP		
Boron arsenide	BAs	5.22	

Table 4.1-29 Melting point T_m or decomposition temperature of boron compounds

Crystal		Melting point T_m (K)	Remarks
Boron nitride, cubic	BN_{cub}	> 3246	
Boron nitride, hexagonal	BN_{hex}	3200–3400	N_2 ambient; pyrometric measurement
Boron phosphide	BP	1400	Decomposition temperature

Table 4.1-30 Linear thermal expansion coefficient α of boron compounds

Crystal		Expansion coefficient α (K^{-1})	Remarks
Boron nitride, cubic	BN_{cub}	1.15×10^{-6}	X-ray measurement
Boron nitride, hexagonal	BN_{hex}	$40.6(4) \times 10^{-6}$	
Boron phosphide	BP	3.65×10^{-6}	At 400 K
Boron arsenide	BAs	4.1×10^{-6}	Calculated

Table 4.1-31 Heat capacities and Debye temperatures of boron compounds

Crystal		Debye temperature Θ_D (K)	Remarks
Boron nitride, cubic	BN_{cub}	1730 (70)	See Fig. 4.1-51 for temp. dependence of c_p
Boron nitride, hexagonal	BN_{hex}	598 (7)	Calorimetry
Boron phosphide	BP	985	See Fig. 4.1-52 for temp. dependence of c_p
Boron arsenide	BAs	800	

Table 4.1-32 Phonon wavenumbers/frequencies of boron compounds. Values for hexagonal boron nitride (BN_{hex}) obtained from infrared reflectivity; values for boron phosphide (BP) from ab initio pseudopotential calculations

Crystal		Phonon wavenumbers (cm^{-1})				Remarks
Boron nitride, cubic	BN_{cub}	$\tilde{\nu}_{LO}$	1305 (1)	$\tilde{\nu}_{TO}$	1054.7 (6)	RT; Raman scattering
Boron arsenide	BAs	$\tilde{\nu}_{LO}(\Gamma)$	714	$\tilde{\nu}_{TO}(\Gamma)$	695	Raman scattering

Table 4.1-32 Phonon wavenumbers/frequencies, cont.

BN_{hex} Phonon branch	Wavenumber (cm^{-1})
$\tilde{\nu}_{TO1}$	783
$\tilde{\nu}_{LO1}$	828
$\tilde{\nu}_{TO2}$	1510
$\tilde{\nu}_{LO2}$	1595
$\tilde{\nu}_{TO1}$	767
$\tilde{\nu}_{LO1}$	778
$\tilde{\nu}_{TO2}$	1367
$\tilde{\nu}_{LO2}$	1610

Table 4.1-32 Phonon wavenumbers/frequencies, cont.

BP Phonon branch	Frequency (THz)
$\nu_{TO}(\Gamma)$	24.25
$\nu_{LO}(X)$	24.00
$\nu_{TO}(X)$	15.81
$\nu_{LO}(X)$	21.04
$\nu_{TO}(X)$	9.20
$\nu_{LO}(L)$	22.91
$\nu_{TO}(L)$	15.18

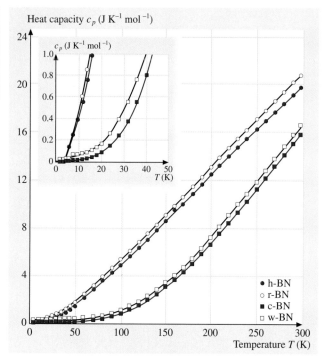

Fig. 4.1-51 BN. Heat capacities for four polymorphic boron nitride modifications. Data taken from [1.50]

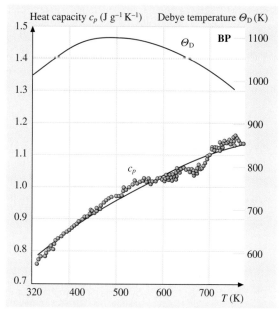

Fig. 4.1-52 BP. Temperature dependence of specific heat capacity and Debye temperature [1.51]

B. Electronic Properties
Tables 4.1-33 – 4.1-35.

Band Structures of Boron Compounds.
Boron Nitride, Cubic (BN_{cub}). All band structure calculations lead to an indirect-gap structure (Fig. 4.1-53).
Boron Nitride, Hexagonal (BN_{hex}). All recent band structure calculations yield an indirect gap (Fig. 4.1-54). The conduction band minimum is located at P.
Boron Phosphide (BP). BP is an indirect-gap semiconductor (Fig. 4.1-55).
Boron Arsenide (BAs). The calculated band structure shows an indirect (Γ–X) gap of a few eV. The conduction

Table 4.1-33 First Brillouin zones of boron compounds

Crystal		Figure
Boron nitride, cubic	BN_{cub}	Fig. 4.1-23
Boron nitride, hexagonal	BN_{hex}	Fig. 4.1-24
Boron phosphide	BP	Fig. 4.1-23
Boron arsenide	BAs	Fig. 4.1-23

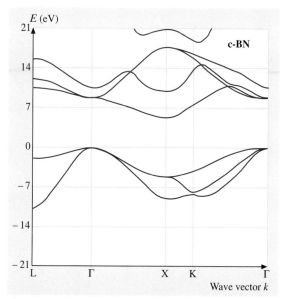

Fig. 4.1-53 Band structure of cubic boron nitride

Table 4.1-34 Energy gaps of boron compounds

Crystal		Quantity	Bands	Energy (eV)	Remarks
Boron nitride, cubic	BN_{cub}	$E_{g,\,ind}$	Γ_{15v} to X_{1c}	6.2	Soft-X-ray emission spectroscopy
		$E_{g,\,dir}$	Γ_{15v} to Γ_{1c}	14.5	Reflectivity
Boron nitride, hexagonal	BN_{hex}	E_g		5.9 (1)	Inelastic electron scattering
		$E_{g,\,th}$		7.1 (1)	Temperature dependence of electrical resistivity
Boron phosphide	BP	$E_{g,\,ind}$		2.1	X-ray emission and absorption
		$E_{g,\,dir}$	Γ	4.4	
			X	6.5	
			L	6.5	
Boron arsenide	BAs	$E_{g,\,ind}$		0.67	Absorption
		$E_{g,\,dir}$		1.46	
Boron antimonide	BSb	E_g	Γ to Δ	0.527	Calculated

Table 4.1-35 Effective masses of electrons (m_n) and holes (m_p) for cubic boron nitride (in units of the electron mass m_0)

Crystal		m_n	$m_{p,\,heavy}$	$m_{p,\,light}$	Remarks
Boron nitride, cubic	BN_{cub}	0.752			Calculated from band structure data
			0.375	0.150	Parallel to [100]
			0.926	0.108	Parallel to [111]

band minimum is on the Δ axis close to the X point. As in silicon, the Γ_{15c} level is below the Γ_{1c} level (Fig. 4.1-56).

Boron Antimonide (BSb). Boron antimonide is an indirect-gap semiconductor (Fig. 4.1-57).

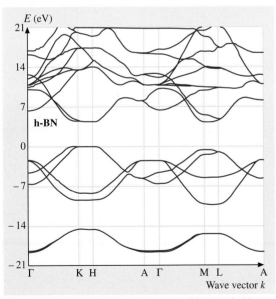

Fig. 4.1-54 Band structure of hexagonal boron nitride

Fig. 4.1-55 Band structure of boron phosphide ▶

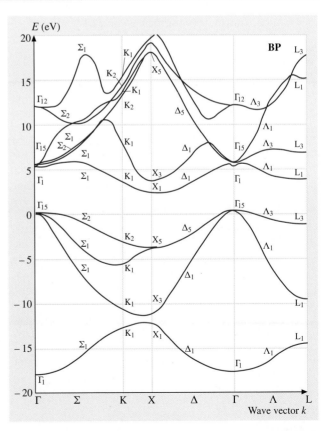

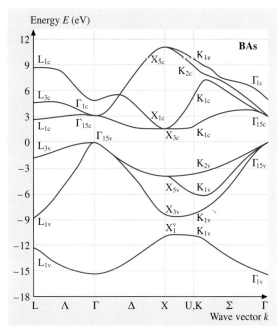

Fig. 4.1-56 Band structure of boron arsenide

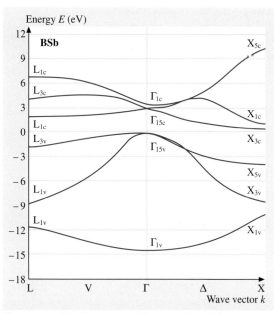

Fig. 4.1-57 Band structure of boron antimonide

C. Transport Properties

Electronic Transport, General Description. Boron Nitride, Cubic (BN_{cub}). Owing to the large gap, all literature data refer to extrinsic conduction. For the temperature dependence of the conductivity, see Fig. 4.1-58. Hall measurements on nominally undoped thin films yield values of the carrier mobility of around $500\,\text{cm}^2/\text{V s}$.

Boron Nitride, hexagonal (BN_{hex}). Figure 4.1-59 shows the temperature dependence of the electrical resistivity at high temperatures. In an undoped nanocrystalline film, resistivity values of the order of $2 \times 10^{11}\,\Omega\,\text{cm}$ have been found.

Boron Phosphide (BP). Boron phosphide is extrinsic at room temperature, and the transport is limited by impurity scattering. Figure 4.1-60 shows the temperature dependence of the conductivity, carrier concentration, and Hall mobility of several samples. In n-doped single crystals, electron mobilities of the order of $30\text{--}40\,\text{cm}^2/\text{V s}$ have been found at 300 K; in p-doped single crystals, mobilities around $500\,\text{cm}^2/\text{V s}$ have been measured at 300 K.

Boron Arsenide (BAs). There is almost no information about the semiconducting properties of BAs.

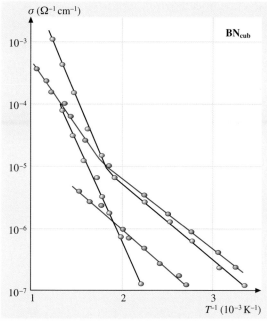

Fig. 4.1-58 c-BN (BN_{cub}). Electrical conductivity vs. reciprocal temperature for several different polycrystalline samples above RT [1.52, 53]

Semiconductors | 1.2 III–V Compounds | 609

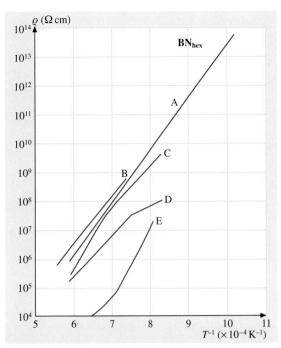

Fig. 4.1-59 h-BN (BN$_{hex}$). Resistivity vs. temperature above 700 °C. A, from bulk resistance of a disk measured in the c direction; B–E, earlier literature data for comparison [1.54]

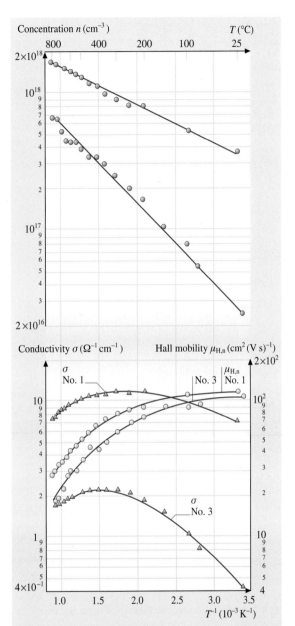

Fig. 4.1-60 Temperature dependence of the conductivity σ_i, the carrier concentration n, and the mobility $\mu_{H,n}$ of n-type BP(100) wafers [1.55]. Sample No. 1 contains autodoped Si with a concentration of 5×10^{18} atoms/cm^3 and sample No. 3 contains 5×10^{19} atoms/cm^3

D. Electromagnetic and Optical Properties

Table 4.1-36.

Table 4.1-36 Dielectric constant ε and refractive index n of boron compounds

Crystal		Temperature (K)	$\varepsilon(0)$	$\varepsilon(\infty)$	n	at λ (nm)	Remarks
Boron nitride, cubic	BN$_{cub}$	300	7.1		2.097	712.5	IR reflectivity
Boron nitride, hexagonal	BN$_{hex}$	300	5.09	4.10	2.13		Parallel to c axis, IR reflectivity
			7.04	4.95	1.65		Perpendicular to c axis, IR reflectivity
Boron phosphide	BP	300	11	7.8			Schottky barrier reflectance
		RT			3.34 (5)	454.5	Brewster angle method
					3.34 (5)	458	
					3.32 (5)	488	
					3.30 (5)	496	
					3.26 (5)	514.5	
					3.00 (5)	632.8	

4.1.2.2 Aluminium Compounds

A. Crystal Structure, Mechanical and Thermal Properties
Tables 4.1-37 – 4.1-44.

Table 4.1-37 Crystal structures of aluminium compounds

Crystal		Space group		Crystal system	Structure type	Remarks	Figure
Aluminium nitride	AlN I	C_{6v}^4	$P6_3mc$	hex	Wurtzite	At normal pressure	Fig. 4.1-3
	AlN II	O_h^5	$Fm3m$	fcc	NaCl structure	At 21 GPa (shock compression)	
Aluminium phosphide	AlP I	T_d^2	$F\bar{4}3m$	fcc	Zinc blende	At normal pressure	Fig. 4.1-2
	AlP II	O_h^5	$Fm3m$	fcc	NaCl structure	High-pressure phase	
Aluminium arsenide	AlAs	T_d^2	$F\bar{4}3m$	fcc	Zinc blende	Transition to NiAs structure at high pressure	Fig. 4.1-2
Aluminium antimonide	AlSb I	T_d^2	$F\bar{4}3m$	fcc	Zinc blende	At normal pressure	Fig. 4.1-2
	AlSb II		$Cmcm$		Orthorhombic	At high pressure	

Table 4.1-38 Lattice parameters of aluminium compounds

Crystal		Lattice parameters (nm)	Temperature	Remarks
Aluminium nitride	AlN	$a = 0.31111$	300 K	X-ray diffraction on ultrafine powder
		$c = 0.49788$		
Aluminium phosphide	AlP	$a = 0.54635$ (4)	25 °C	Epitaxial film on GaP
Aluminium arsenide	AlAs	$a = 0.566139$	291 K	High-resolution X-ray diffraction
Aluminium antimonide	AlSb	$a = 0.61355$ (1)	291.15 K	Powder, X-ray measurement

Table 4.1-39 Densities of aluminium compounds

Crystal		Density (g/cm³)	Temperature (K)	Remarks
Aluminium nitride	AlN	3.255		X-ray diffraction
Aluminium phosphide	AlP	2.40 (1)		
Aluminium arsenide	AlAs	3.760		Calculated from lattice parameter
Aluminium antimonide	AlSb	4.26	293	

Table 4.1-40 Elastic constants c_{ik} of aluminium compounds

Crystal		c_{11} (GPa)	c_{12} (GPa)	c_{13} (GPa)	c_{33} (GPa)	c_{44} (GPa)	Remarks
Aluminium nitride	AlN	296 (18)	130 (11)	158 (06)	267 (18)	241 (20)	a
Aluminium phosphide	AlP	140.50 (28)	62.03 (24)			70.33 (7)	Ultrasound ($f = 10$ and 30 MHz)
Aluminium arsenide	AlAs	190 (1)	53.8 (1)			59.5 (1)	$T = 300$ K
		122.1 (3)	56.6 (3)			59.9 (1)	$T = 77$ K
		112.6	57.1			60.0	$T = 0$ K; extrapolated data
Aluminium antimonide	AlSb	88.34	40.23			43.22	$T = 296$ K; ultrasound

[a] Calculated from the mean square displacement of the lattice atoms measured by X-ray diffraction.

Table 4.1-41 Melting point T_m of aluminium compounds

Crystal		Melting point T_m (K)
Aluminium nitride	AlN	3025
Aluminium phosphide	AlP	2823
Aluminium arsenide	AlAs	2013 (20)
Aluminium antimonide	AlSb	1327 (1)

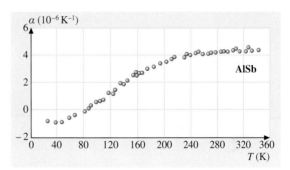

Fig. 4.1-61 AlSb. Coefficient of linear thermal expansion vs. temperature [1.56, 57]

Table 4.1-42 Linear thermal expansion coefficient α of aluminium compounds

Crystal		Expansion coefficient α (K⁻¹)	Temperature	Remarks
Aluminium nitride	AlN	α_{perpend} 4.35×10^{-6} α_{parallel} 3.48×10^{-6}	300 K	Powder; X-ray diffraction
Aluminium phosphide	AlP			
Aluminium arsenide	AlAs	$5.20 (5) \times 10^{-6}$	15–840 °C	X-ray diffraction
Aluminium antimonide	AlSb			Fig. 4.1-61

Table 4.1-43 Phonon wavenumbers of aluminium compounds: aluminium nitride (300 K, from Raman scattering); aluminium Phosphide (from Raman spectroscopy); aluminium arsenide (from Raman spectroscopy; 0.5 μm layer of AlAs on GaAs; $T = 37$ K); aluminium antimonide (from Raman spectroscopy)

AlN		AlP			AlAs		AlSb	
Symmetry point, polarization	Wave-number (cm^{-1})	Symmetry point, polarization	Wave-number (cm^{-1})	Temperature (K)	Symmetry point, polarization	Wave-number (cm^{-1})	Symmetry point, polarization	Wave-number (cm^{-1})
$\tilde{\nu}\,(E_2^{(1)})$	248.6	$\tilde{\nu}_{LO}\,(\Gamma)$	504.5 (4)	5	$\tilde{\nu}_{LO}\,(\Gamma)$	404 (1)	$\tilde{\nu}_{LO}\,(\Gamma)$	334 [a]
$\tilde{\nu}_{TO}\,(A_1)$	611.0	$\tilde{\nu}_{LO}\,(\Gamma)$	501.0 (2)	300	$\tilde{\nu}_{TO}\,(\Gamma)$	361 (1)	$\tilde{\nu}_{TO}\,(\Gamma)$	316 [a]
$\tilde{\nu}\,(E_2^{(2)})$	657.4	$\tilde{\nu}_{TO}\,(\Gamma)$	442.5 (2)	5	$\tilde{\nu}_{LO}\,(X)$	403 (8)	$\tilde{\nu}_{TA}\,(X)$	64 [a]
$\tilde{\nu}_{TO}\,(E_1)$	670.8	$\tilde{\nu}_{TO}\,(\Gamma)$	439.4 (2)	300	$\tilde{\nu}_{TO}\,(X)$	335 (8)	$\tilde{\nu}_{LA}\,(X)$	153 [a]
$\tilde{\nu}_{LO}\,(A_1)$	890.0				$\tilde{\nu}_{TA}\,(X)$	109 (8)	$\tilde{\nu}_{TO}\,(X)$	290 [a]
$\tilde{\nu}_{LO}\,(E_1)$	912.0				$\tilde{\nu}_{LA}\,(X)$	222 (12)	$\tilde{\nu}_{LO}\,(X)$	343 [a]

Table 4.1-43 Wavenumbers of aluminium compounds, cont.

AlSb	
Symmetry point, polarization	Wave-number (cm^{-1})
$\tilde{\nu}_{TA}\,(L)$	49 [a]
$\tilde{\nu}_{LA}\,(K)$	149 [a]
$\tilde{\nu}_{TO}\,(L)$	306 [a]
$\tilde{\nu}_{LO}\,(L)$	327 [a]
$\tilde{\nu}_{LO}\,(\Gamma)$	340.0 (7) [b]
$\tilde{\nu}_{TO}\,(\Gamma)$	318.7 (7) [b]

[a] Ab initio pseudopotential calculation.
[b] First-order Raman scattering.

Table 4.1-44 Phonon dispersion curves of aluminium compounds

Crystal		Figure
Aluminium nitride	AlN	Fig. 4.1-62
Aluminium phosphide	AlP	Fig. 4.1-63
Aluminium arsenide	AlAs	Fig. 4.1-64
Aluminium antimonide	AlSb	Fig. 4.1-65

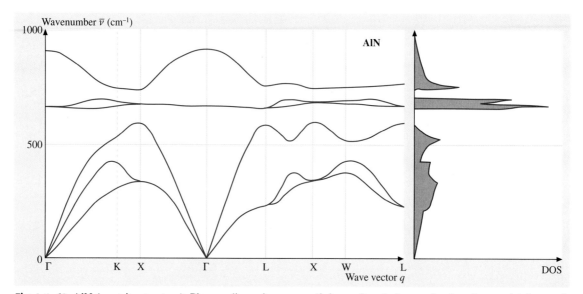

Fig. 4.1-62 AlN (wurtzite structure). Phonon dispersion curves (*left panel*) and phonon density of states (*right panel*), from a rigid-ion model calculation [1.58, 59]

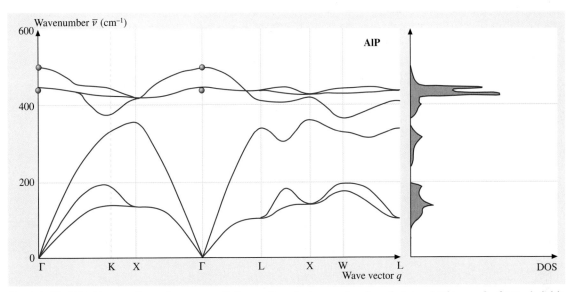

Fig. 4.1-63 AlP. Phonon dispersion curves (*left panel*) and phonon density of states (*right panel*), from ab initio calculations [1.60]. The data points were obtained from Raman scattering [1.61]

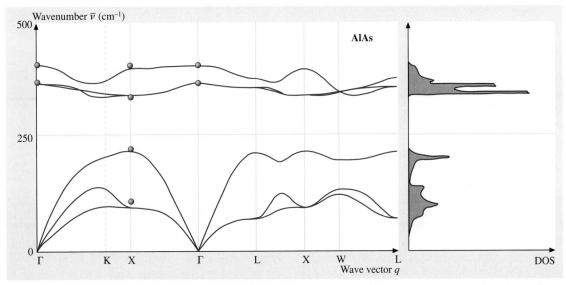

Fig. 4.1-64 AlAs. Phonon dispersion curves (*left panel*) and phonon density of states (*right panel*). Experimental data points [1.61, 62] and ab initio calculations [1.16]. From [1.16]

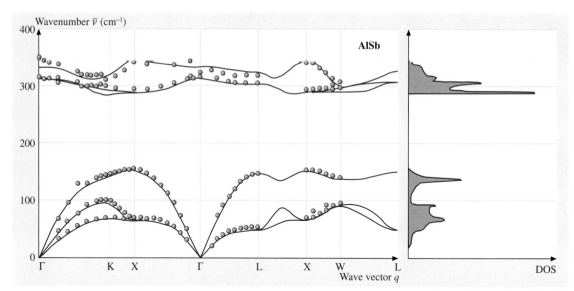

Fig. 4.1-65 AlSb. Phonon dispersion curves (*left panel*) and phonon density of states (*right panel*). Experimental data points [1.63] and ab initio calculations [1.16]. From [1.16]

B. Electronic Properties
Tables 4.1-45 – 4.1-48.

Band Structures of Aluminium Compounds. Aluminium Nitride (AlN). Aluminium nitride is a direct-gap semiconductor. Since it crystallizes in the wurtzite structure, the band structure (Fig. 4.1-66) differs from that of most of the other III–V compounds.

Aluminium Phosphide (AlP). Aluminium phosphide is an indirect, wide-gap semiconductor. The minima of the conduction bands are located at the X point of the Brillouin zone. The top of the valence band has the structure common to all zinc blende semiconductors (Fig. 4.1-67).

Aluminium Arsenide (AlAs). Aluminium arsenide is a wide-gap semiconductor with a band structure (Fig. 4.1-68) similar to that of AlP. Measurements have revealed a camel's back structure near X.

Aluminium Antimonide (AlSb). Aluminium antimonide (like AlAs) is an indirect-gap semiconductor with camel's back conduction band minima near X (Fig. 4.1-69).

Table 4.1-45 First Brillouin zones of aluminium compounds

Crystal		Figure
Aluminium nitride	AlN	Fig. 4.1-24
Aluminium phosphide	AlP	Fig. 4.1-23
Aluminium arsenide	AlAs	Fig. 4.1-23
Aluminium antimonide	AlSb	Fig. 4.1-23

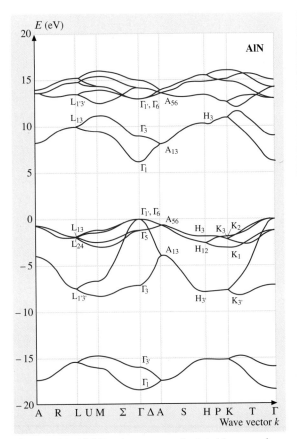

Fig. 4.1-66 AlN. Band structure calculated by a semiempirical tight-binding method [1.64]

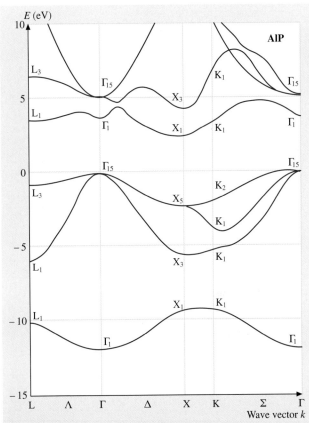

Fig. 4.1-67 AlP. Band structure calculated by an orthogonalized LCAO method [1.65]

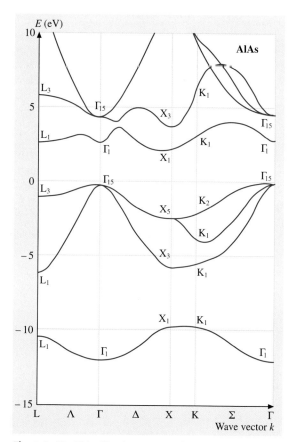

Fig. 4.1-68 AlAs. Band structure calculated by an orthogonalized LCAO method [1.65]

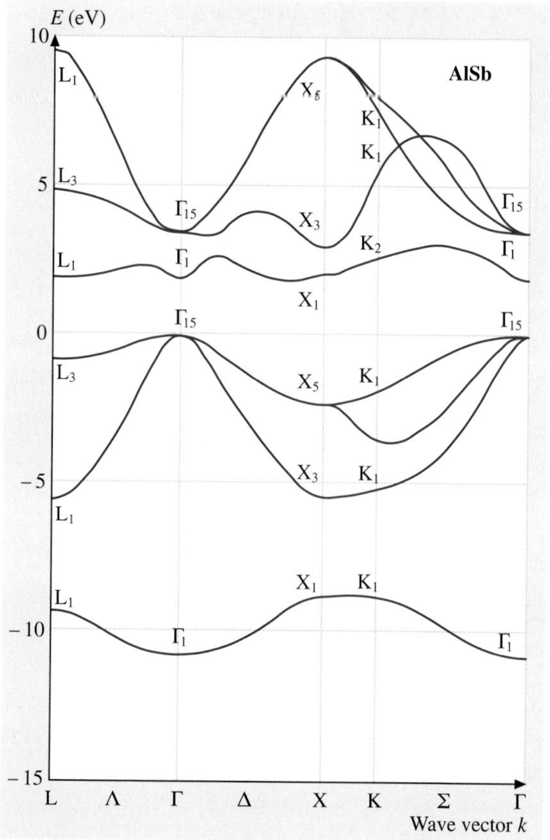

Fig. 4.1-69 Band structure of aluminium antimonide

Table 4.1-46 Energy gaps of aluminium compounds

Crystal		Quantity	Bands	Energy (eV)	Temperature (K)	Remarks
Aluminium nitride	AlN	$E_{g,\,dir}$		6.19	7	Optical absorption
				6.13	300	
Aluminium phosphide	AlP	$E_{g,\,ind}$	Γ_{15v} to X_{1c}	2.5 (1)	2	Excitonic gap;
				2.45	300	Absorption
		$E_{g,\,dir}$	Γ_{15v} to Γ_{1c}	3.63 (2)	4	Excitonic gap;
				3.62 (2)	77	Photoluminescence
Aluminium arsenide	AlAs	$E_{g,\,ind}$	Γ_{15v} to X_{1c}	2.229 (1)	4	Excitonic gap;
				2.153 (2)	300	Photoluminescence
		$E_{g,\,ind}$	Γ_{15v} to L_{1c}	2.363	295	Transport
		$E_{g,\,dir}$	Γ_{15v} to Γ_{1c}	3.13 (1)	4	Excitonic gap;
				3.03 (1)	300	Photoluminescence
				3.14 (1)	300	Transmission
Aluminium antimonide	AlSb	$E_{g,\,ind}$	Γ_{15v} to Δ_{1c}	1.615 (3)	300	Excitonic gap
				1.686 (1)	27	
		$E_{g,\,ind}$	Γ_{15v} to L_{1c}	2.211	295	Electroreflectance
				2.327	35	
		$E_{g,\,dir}$	Γ_{15v} to Γ_{1c}	2.300	295	Electroreflectance
				2.384	25	

Table 4.1-47 Spin–orbit splitting energies of aluminium compounds

Crystal	Spin–orbit splitting	Symmetry points	Energy (eV)	Temperature (K)	Remarks
AlAs	Δ_0	Γ_{15v}	0.275	300	Electroreflectance
	Δ_1	L_{15v}	0.20		
	Δ_0'	Γ_{15c}	0.15		
AlSb	Δ_0	Γ_{8v} to Γ_{7v}	0.673	295	Splitting of Γ_{15v}
	Δ_1	$L_{4,5v}$ to L_{6v}	0.426		Splitting of L_{3v}

Table 4.1-48 Effective masses of electrons (m_n) and holes (m_p) for aluminium compounds (in units of the electron mass m_0). Aluminium nitride (AlN), calculated values; aluminium phosphide (AlP), calculated from band structure; aluminium arsenide (AlAs), calculated from band structure data; Aluminium antimonide (AlSb), theoretical estimates

AlN		AlP		AlAs		AlSb	
$m_{n,\,parall}$	0.33	$m_{n,\,parall}$	3.67	$m_{n,\,parall}$ (X)	1.1[a]	$m_{n,\,perpend}$ (Δ_1)	0.26 estimated
$m_{n,\,perpend}$	0.25	$m_{n,\,perpend}$	0.212	$m_{n,\,perpend}$ (X)	0.19[a]	$m_{n,\,parall}$ (Δ_1)	1.0
$m_{p,\,A,\,parall}$	3.53	$m_{p,\,heavy}$	0.513 parallel to [100]	$m_{n,\,parall}$ (L)	1.32	$m_{p,\,heavy}$	0.336 parallel to [100]
$m_{p,\,A,\,perpend}$	11.14	$m_{p,\,heavy}$	1.372 parallel to [111]	$m_{n,\,perpend}$ (L)	0.15[b]	$m_{p,\,heavy}$	0.872 parallel to [111]
$m_{p,\,B,\,parall}$	3.53	$m_{p,\,light}$	0.211 parallel to [100]	$m_{p,\,heavy}$	0.409 parallel to [100]	$m_{p,\,light}$	0.123 parallel to [100]
$m_{p,\,B,\,perpend}$	0.33	$m_{p,\,light}$	0.145 parallel to [111]	$m_{p,\,heavy}$	1.022 parallel to [111]	$m_{p,\,light}$	0.091 parallel to [111]
$m_{p,\,C,\,parall}$	0.26			$m_{p,\,light}$	0.153 parallel to [100]		
$m_{p,\,C,\,perpend}$	4.05			$m_{p,\,light}$	0.109 parallel to [111]		

[a] Effective masses at X, neglecting camel's back structure.
[b] Effective masses at L.

C. Transport Properties
Tables 4.1-49 and 4.1-50.

Electronic Transport, General Description. Aluminium Nitride (AlN). Owing to the large energy gap, transport is always extrinsic. Typical numerical values for the electrical conductivity of undoped single crystals lie in the range $10^{-13}\,\Omega^{-1}\,\text{cm}^{-1} < \sigma < 10^{-11}\,\Omega^{-1}\,\text{cm}^{-1}$.

Aluminium Phosphide (AlP). At room temperature, electrical conductivities σ of bulk crystals and layers between 1 and $10^5\,\Omega^{-1}\,\text{cm}^{-1}$ have been measured. For the temperature dependence of the electrical conductivity, see Fig. 4.1-70.

Aluminium Arsenide (AlAs). Only a small amount of reliable data exists on the transport properties of AlAs. Most results are extrapolations from data obtained from the technologically more important solid solutions of type $Al_xGa_{1-x}As$. At roomtemperature, values of the electrical conductivity of around $10\,\Omega^{-1}\,\text{cm}^{-1}$ have been found.

Aluminium Antimonide (AlSb). Aluminium antimonide grown without intentional doping is p-type and becomes intrinsic at about 1000 K. For the temperature dependence of the electrical conductivity, see Fig. 4.1-71.

Table 4.1-49 Electron and hole mobilities μ_n and μ_p of aluminium compounds

Crystal		Temperature (K)	μ_n (cm^2/V s)	μ_p (cm^2/V s)	Remarks
Aluminium nitride	AlN	290		14	Doped single crystal
Aluminium phosphide	AlP	298	10–80		Single crystals
Aluminium arsenide	AlAs	300	293–75	105	Single crystal layers, MBE-grown, Be-doped
Aluminium antimonide	AlSb	295	200		Single crystal, Te-doped,
		77	700		$n = 4.7 \times 10^{16}$ cm^{-3}
		300		400	Single crystal, $p = 10^{16}$ cm^{-3}
		77		2000	

Table 4.1-50 Thermal conductivity κ of aluminium compounds

Crystal		Temperature (K)	κ (W/cm K)	Remarks
Aluminium nitride	AlN	300	3.19	See Fig. 4.1-72
Aluminium phosphide	AlP	300	0.9	
Aluminium arsenide	AlAs	300	0.91	
Aluminium antimonide	AlSb			See Fig. 4.1-73

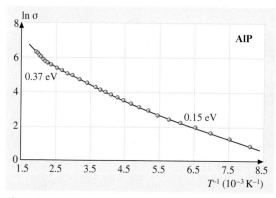

Fig. 4.1-70 AlP. Natural logarithm of electrical conductivity vs. reciprocal temperature for an undoped sample [1.66]. σ in Ω^{-1} cm^{-1}. Activation energies are also shown

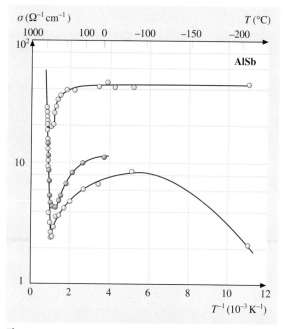

Fig. 4.1-71 AlSb. Electrical conductivity vs. reciprocal temperature for three different polycrystalline p-type samples [1.67]

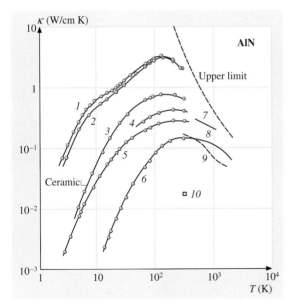

Fig. 4.1-72 AlN. Temperature dependence of the thermal conductivity for two single crystals (*curves 1* and *2*) and several ceramic samples (*curves 3–6*). Other published results are included (*curves 7–10*), as well as a theoretical estimate of the upper limit [1.68]

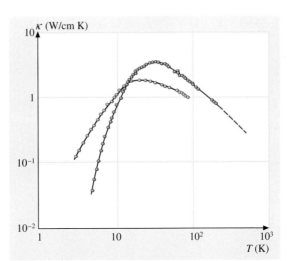

Fig. 4.1-73 AlSb. Thermal conductivity vs. temperature for an n-type and a p-type sample, low-temperature range [1.69, 70]

D. Electromagnetic and Optical Properties
Table 4.1-51.

Table 4.1-51 Dielectric constant ε and refractive index n of aluminium compounds

Crystal		Temperature (K)	$\varepsilon(0)$	$\varepsilon(\infty)$		n	at λ (μm)	Remarks
Aluminium nitride	AlN	300	9.14	$\varepsilon_{\text{perpend}}$	4.71 (22)	2.17–2.34	0.25	Reflectivity and optical inter-
				$\varepsilon_{\text{parall}}$	4.93 (22)	2.04–2.18	0.30	ferometry, $\lambda = 589.0$ nm, CVD films [a]
Aluminium phosphide	AlP		9.8	7.5				From refractive index [b]
Aluminium arsenide	AlAs		10.1	8.2				From infrared reflectivity [c]
Aluminium antimonide	AlSb	300	121.04	10.24				IR reflectance,
		300				3.652	40	transmission and reflectance [d]
						2.995	20	
						2.080	15	
						3.100	10	
						3.182	4	
						3.300	2	
						3.445	1.1	

[a] See Fig. 4.1-74 for the spectral dependence of the refractive index.
[b] See Fig. 4.1-75 for the spectral dependence of the refractive index.
[c] See Fig. 4.1-76 for the spectral dependence of the refractive index.
[d] See Fig. 4.1-77 for the spectral dependence of the refractive index.

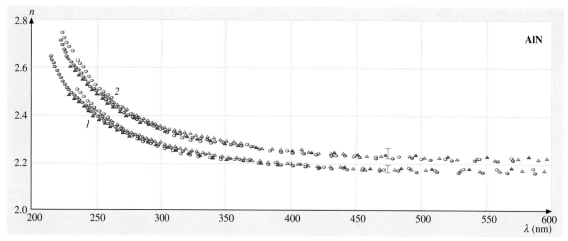

Fig. 4.1-74 AlN. Spectral dependence of the refractive index (*1*, ordinary ray; *2*, extraordinary ray) [1.71]

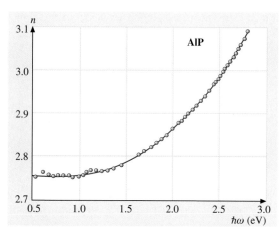

Fig. 4.1-75 AlP. Refractive index vs. photon energy at 300 K [1.72]

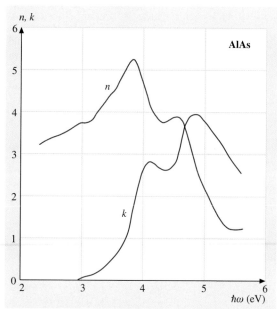

Fig. 4.1-76 AlAs. Spectral variation of the complex refractive index calculated from the dielectric function [1.73]

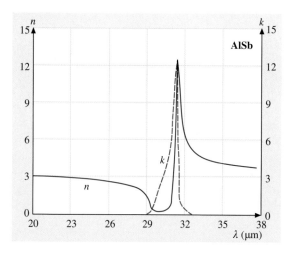

Fig. 4.1-77 AlSb. Refractive index n and extinction coefficient k vs. wavelength in the region of lattice absorption [1.74]

4.1.2.3 Gallium Compounds

A. Crystal Structure, Mechanical and Thermal Properties
Tables 4.1-52 – 4.1-60.

Table 4.1-52 Crystal structures of gallium compounds

Crystal		Space group		Crystal system	Structure type	Remarks	Figure
Gallium nitride	GaN	C_{6v}^4	$P6_3mc$	hex	Wurtzite	Stable at ambient pressure	Fig. 4.1-3
		T_d^2	$F\bar{4}3m$	fcc	Zinc blende	Epitaxial growth on cubic substrate	Fig. 4.1-2
Gallium phosphide	GaP	T_d^2	$F\bar{4}3m$	fcc	Zinc blende	Stable at normal pressure	Fig. 4.1-2
		D_{4h}^{19}	$I4_1/amd$		β-tin structure	Pressure > 21–30 GPa	
Gallium arsenide	GaAs I	T_d^2	$F\bar{4}3m$	fcc	Zinc blende	Stable at normal pressure	Fig. 4.1-2
	GaAs II				Orthorhombic	Pressure > 17 GPa	
Gallium antimonide	GaSb I	T_d^2	$F\bar{4}3m$	fcc	Zinc blende	Stable at normal pressure	Fig. 4.1-2
	GaSb II	D_{4h}^{19}	$I4_1/amd$		β-tin structure	Pressure > 8–10 GPa	

Table 4.1-53 Lattice parameters of gallium compounds

Crystal		Lattice parameters (nm)	Temperature (K)	Remarks
Gallium nitride	GaN	$a = 0.3190\,(1)$		X-ray diffraction, monocrystal
		$c = 0.5189\,(1)$		
Gallium phosphide	GaP	$a = 0.54506\,(4)$	RT	Single crystal
Gallium arsenide	GaAs	$a = 0.565359$	RT	
Gallium antimonide	GaSb	$a = 0.609593\,(4)$	298.15	X-ray powder diffraction

Table 4.1-54 Densities of gallium compounds

Crystal		Density (g/cm³)	Temperature (K)
Gallium nitride	GaN	6.07	
Gallium phosphide	GaP	4.138	300
Gallium arsenide	GaAs	5.3161 (2)	298.15
Gallium antimonide	GaSb	5.6137 (4)	300

Table 4.1-55 Elastic constants c_{ik} of gallium compounds

Crystal		c_{11} (GPa)	c_{12} (GPa)	c_{13} (GPa)	c_{33} (GPa)	c_{44} (GPa)	Temperature (K)	Remarks
Gallium nitride	GaN	377	160	114	209	81.4	300	Ultrasound resonance technique
Gallium phosphide	GaP	140.50 (28)	62.03 (24)	–	–	70.33 (7)		Ultrasound ($f = 10$ and 30 MHz)
Gallium arsenide	GaAs	119 (1)	53.8 (1)	–	–	59.5 (1)	300	
		112.6	57.1	–	–	60.0	0	Extrapolated data
Gallium antimonide	GaSb	88.34	40.23	–	–	43.22	296	Ultrasound

Table 4.1-56 Melting point T_m of gallium compounds

Crystal		Melting point T_m (K)	Remarks
Gallium nitride	GaN	2791	GaN sublimes without decomposition around 800 °C
Gallium phosphide	GaP	1749	
Gallium arsenide	GaAs	1511	
Gallium antimonide	GaSb	991 (1)	

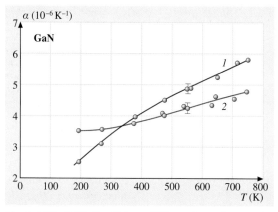

Fig. 4.1-78 GaN. Coefficient of linear thermal expansion vs. temperature (*curve 1*, $\alpha_\perp$; *curve 2*, $\alpha_\parallel$) [1.75]

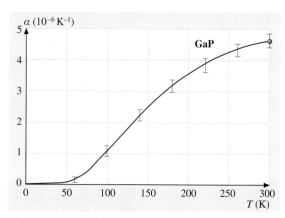

Fig. 4.1-79 GaP. Linear thermal expansion coefficient vs. temperature between 0 K and 300 K [1.76]

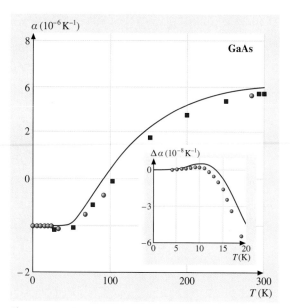

Fig. 4.1-80 GaAs. Coefficient of linear thermal expansion. Experimental data points, and curves from an ab initio pseudopotential calculation [1.77]

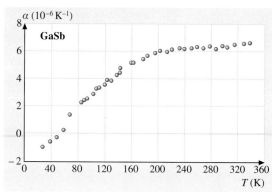

Fig. 4.1-81 GaSb. Linear thermal expansion coefficient vs. temperature measured with a quartz dilatometer. High-temperature range [1.56, 57]

Table 4.1-57 Linear thermal expansion coefficient α of gallium compounds

Crystal		Expansion coefficient α (10^{-6} K^{-1})		Remarks	Temperature dependence
Gallium nitride	GaN	α_{parall}	5.59	X-ray diffraction	Fig. 4.1-78
		$\alpha_{perpend}$	3.17		
Gallium phosphide	GaP		4.65	300 K	Fig. 4.1-79
Gallium arsenide	GaAs		5.87	300 K	Fig. 4.1-80
Gallium antimonide	GaSb		7.75 (50)	Around 300 K; X-ray diffraction	Fig. 4.1-81

Table 4.1-58 Heat capacity c_p and Debye temperature Θ_D of gallium compounds

Crystal		Heat capacity (J/mol K)	Temperature (K)	Debye temperature (K)	Remarks
Gallium nitride	GaN			600	Estimated
Gallium phosphide	GaP	49.65	800	445	
Gallium arsenide	GaAs	46.90	300	344	
Gallium antimonide	GaSb			266	

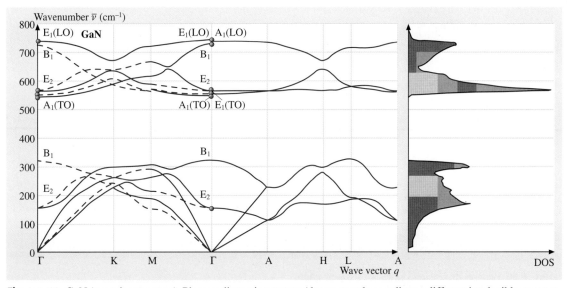

Fig. 4.1-82 GaN (wurtzite structure). Phonon dispersion curves (decomposed according to different irreducible representations along the main symmetry directions) and density of states, from a model-potential calculation. The *circles* show Raman data from [1.78]. From [1.79]

Table 4.1-59 Phonon wavenumbers of gallium compounds. Gallium nitride (GaN), $T = 300$ K, from Raman spectroscopy; gallium phosphide (GaP), RT, from an analysis of Raman, neutron, luminescence, and absorption data; gallium arsenide (GaAs), $T = 296$ K, from coherent inelastic neutron scattering; gallium antimonide (GaSb), $T = 300$ K, from second-order Raman effect

GaN Symmetry point	Branch	Wavenumber (cm^{-1})	GaP Symmetry point	Branch	Frequency (THz)
$\tilde{\nu}_{low}$ (E$_2$)		145	ν (Γ)	TO	10.95 (1)
$\tilde{\nu}$ (A$_1$)	TO	533		LO	12.06 (1)
$\tilde{\nu}$ (E$_1$)	TO	560	ν (X)	TA	3.13 (3)
$\tilde{\nu}_{high}$ (E$_2$)		567		LA	7.46 (6)
$\tilde{\nu}$ (A$_1$)	LO	735		TO	10.58 (3)
$\tilde{\nu}$ (E$_1$)	LO	742		LO	11.09 (6)
			ν (L)	TA	2.58 (3)
				LA	6.45 (3)
				TO	10.64 (3)
				LO	11.24 (3)

GaAs Symmetry point	Branch	Frequency (THz)	GaSb Symmetry point	Branch	Wavenumber (cm^{-1})
ν (Γ_{15})	LO	8.55 (20)	$\tilde{\nu}$ (L)	TA	46
	TO	8.02 (8)	$\tilde{\nu}$ (X)	TA	56
ν (X$_1$)	TA	2.36 (2)	$\tilde{\nu}$ (W)	TA	75
	LO	7.22 (15)	$\tilde{\nu}$ (L)	LA	155
ν (X$_3$)	LA	6.80 (6)		LO	204
ν (X$_5$)	TO	7.56 (8)	$\tilde{\nu}$ (X)	LO	210
ν (L$_1$)	LA	6.26 (10)	$\tilde{\nu}$ (L, X, Σ)	TO	218
	LO	7.15 (7)			
ν (L$_3$)	TA	1.86 (2)			
	TO	7.84 (12)			

Table 4.1-60 Phonon dispersion curves of gallium compounds

Crystal		Figure
Gallium nitride	GaN	Fig. 4.1-82
Gallium phosphide	GaP	Fig. 4.1-83
Gallium arsenide	GaAs	Fig. 4.1-84
Gallium antimonide	GaSb	Fig. 4.1-85

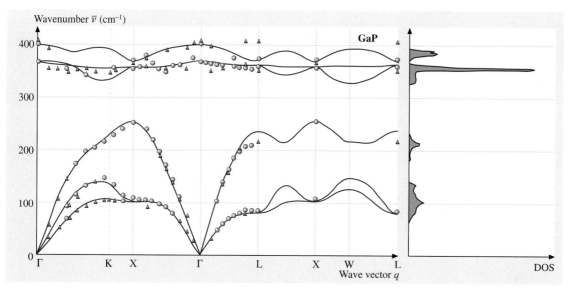

Fig. 4.1-83 GaP. Phonon dispersion curves (*left panel*) and density of states (*right panel*). Experimental data points from [1.80] (*open symbols*) and from [1.81] (*filled symbols*) and theoretical curves from ab initio calculations [1.82]. From [1.82]

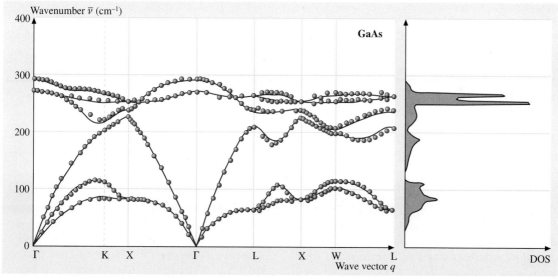

Fig. 4.1-84 GaAs. Phonon dispersion curves (*left panel*) and phonon density of states (*right panel*) [1.16]. Experimental data points [1.83] and ab initio calculations [1.16].

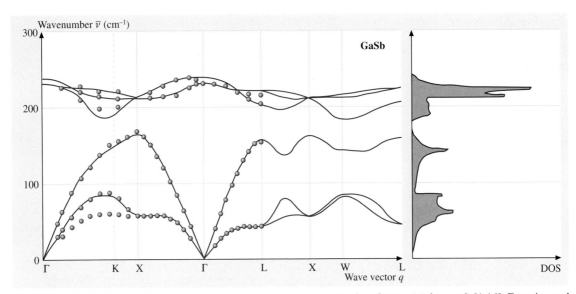

Fig. 4.1-85 GaSb. Phonon dispersion curves (*left panel*) and phonon density of states (*right panel*) [1.16]. Experimental data points [1.84] and ab initio calculations [1.16]

B. Electronic Properties
Tables 4.1-61 – 4.1-66.

Band Structures of Gallium Compounds. Gallium Nitride (GaN). The band structure is shown in Fig. 4.1-87. Owing to spin–orbit interaction, the two top valence bands (Γ_6 and Γ_1) are split into three spin-degenerate bands, one with quantum number $J_z = 3/2(\Gamma_9)$ and two with $J_z = 1/2(\Gamma_7)$'. For convenience, the notation A, B, C is used for these bands.

Gallium Phosphide (GaP). Gallium phosphide is an indirect-gap semiconductor. The lowest set of conduction bands shows a camel's back structure; the band minima are located on the Δ axes near the zone boundary. The valence bands show the usual structure characteristic of zinc blende semiconductors.

The spin–orbit splitting of the top of the valence band is negligible compared with most other energy separations in the band structure. Therefore, Fig. 4.1-86 shows the band structure calculated without inclusion of spin–orbit splitting; the symmetry symbols of the high-symmetry band states are symbols of the single group of the zinc blende structure.

Gallium Arsenide (GaAs). Gallium arsenide is a direct-gap semiconductor. The minimum of the lowest conduction band is located at Γ; higher sets of minima at L and near X (about 10% away from the zone boundary) are also important for the optical and transport properties. The minimum near X most probably has a camel's back-like structure similar to that of GaP, AlAs, and AlSb. The valence bands have the usual structure characteristic of zinc blende crystals (Fig. 4.1-88).

Gallium Antimonide (GaSb). The conduction band of GaSb is characterized by two types of minima, the lowest minimum at Γ and slightly higher minima at the L points at the surface of the Brillouin zone. A third set of minima at the X points has been detected in optical experiments. The valence bands show the usual structure common to zinc blende semiconductors (Fig. 4.1-89).

Table 4.1-61 First Brillouin zones of gallium compounds

Crystal		Figure
Gallium nitride	GaN	Fig. 4.1-24
Gallium phosphide	GaP	Fig. 4.1-23
Gallium arsenide	GaAs	Fig. 4.1-23
Gallium antimonide	GaSb	Fig. 4.1-23

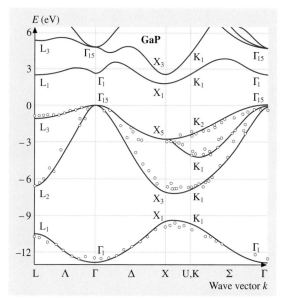

Fig. 4.1-86 Band structure of gallium phosphide

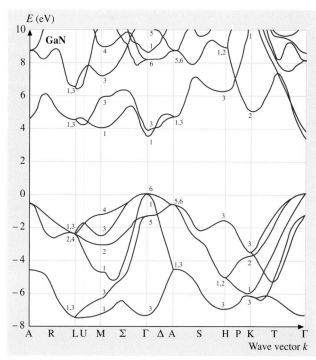

Fig. 4.1-87 Band structure of gallium nitride

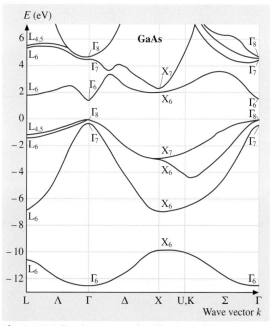

Fig. 4.1-88 Band structure of gallium arsenide

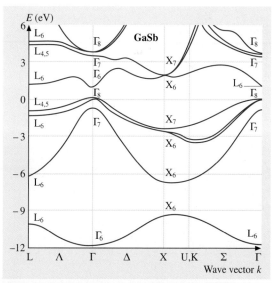

Fig. 4.1-89 Band structure of gallium antimonide

Table 4.1-62 Energy gaps of gallium compounds

Crystal		Quantity	Bands	Energy (eV)	Temperature (K)	Remarks
Gallium nitride	GaN	$E_{g,\,dir}$		3.503 (2)	1.6	Photoluminescence
			A exciton	3.4751 (5)		Transition from Γ_{9v}
			B exciton	3.4815 (10)		Transition from upper Γ_{7v}
			C exciton	3.493 (5)		Transition from lower Γ_{7v}
Gallium phosphide	GaP	$E_{g,\,ind}$	Γ_{8v} to Δ_{5c}	2.350 (1)	0	Extrapolated, from exciton data
				2.272	300	
		$E_{g,\,ind}$	Γ_{15v} to L_{1c}	2.637 (10)	78	Electroabsorption
Gallium arsenide	GaAs	$E_{g,\,dir}$	Γ_{8v} to Γ_{6c}	1.51914	0	Extrapolated, photoluminescence
				1.424 (1)	300	Differentiated reflectivity
		$E_{g,\,th}$		1.604	0	Extrapolated, intrinsic carrier concentration
Gallium antimonide	GaSb	$E_{g,\,dir}$	Γ_{8v} to Γ_{6c}	0.822	0	Extrapolated, electroreflectance
				0.75	300	

Table 4.1-63 Exciton binding energies of gallium compounds

Crystal		Quantity	Energy (meV)	Remarks
Gallium nitride	GaN	E_b^A	20	A exciton, photoreflectance
		E_b^B	18.5	B exciton, photoreflectance
		E_b^C	22.4	C exciton, photoluminescence
Gallium antimonide	GaSb	E_b	1.6	Calculated ground-state binding energy

Table 4.1-64 Spin–orbit splitting energies of gallium compounds

Crystal		Quantity	Bands	Energy (meV)	Temperature (K)	Remarks
Gallium nitride	GaN	Δ_1		9.8		Epitaxial films, reflectance measurements
		Δ_2		5.6		
		Δ_3		5.6		
Gallium phosphide	GaP	Δ_0	Γ_{8v} to Γ_{7v}	80 (3)	100–200	Splitting of Γ_{15v} into Γ_{8v} (upper level) and Γ_{7v} (lower level)
Gallium arsenide	GaAs	Δ_0	Γ_{15v}	346.4 (5)	1.7	Splitting of Γ_{15v} valence band state into Γ_{7v} and Γ_{8v}, magnetoabsorption splitting of Γ_{15} conduction band state into Γ_{7c} and Γ_{8c}, electroreflectance
		Δ_0'	Γ_{15c}	171 (15)	4.2	
Gallium antimonide	GaSb	Δ_0	Γ_{8v} to Γ_{7v}	756 (15)	10	Electroreflectance
		Δ_0'	Γ_{7c} to Γ_{8c}	213 (10)		
		Δ_1	$L_{4,5v}$ to L_{6v}	430 (10)		
		Δ_1'	$L_{4,5c}$ to L_{6c}	130		

Table 4.1-65 Effective masses of electrons (m_n) and holes (m_p) for gallium compounds (in units of the electron mass m_0)

GaN				GaP				
m_n	0.22	6 K	Cyclotron resonance, polaron mass	$m_{n,\,parall}$	Near Δ_{min}	7.25	1–20 K	At bottom of camel's back
$m_{n,\,parall}$	0.20 (6)		From reflectance spectrum	$m_{n,\,perp}$	Near Δ_{min}	0.21		Luminescence
$m_{n,\,perp}$	0.20 (2)			$m_{n,\,parall}$	Away from Δ_{min}	2.2		High above bottom of camel's back
$m_p^{(A)}$	1.01	7 K	Estimated from binding energies	$m_{p,\,heavy}$	Parallel to [111]	0.67 (4)	1.6	Cyclotron resonance at 1.6 K
$m_p^{(B)}$	1.1			$m_{p,\,light}$	Parallel to [111]	0.17 (1)	Parallel to [111]	
$m_p^{(C)}$	1.6			$m_{p,\,so}$		0.4649	0	Spin–orbit mass, calculated from $\mathbf{k} \cdot \mathbf{p}$ model

GaAs					
$m_n\,(\Gamma)$			0.0662 (2)	1.6	Cyclotron resonance
$m_{p,\,heavy}$	Parallel to [100]		0.34	10 K	Electronic Raman scattering
	Parallel to [111]		0.75		
$m_{p,\,light}$	Parallel to [100]		0.094		
	Parallel to [111]		0.082		

GaSb				
$m_n\,(\Gamma)$	0.039 (5)	2 K		Optically detected cyclotron resonance
$m_n\,(L_{6c})$	0.11			Transverse mass
	0.95			Longitudinal mass
	0.226			Density-of-states mass
$m_{p,\,heavy}$	0.29 (9)	30 K	B parallel to [100]	Stress-modulated magnetoreflectance
	0.36 (13)		B parallel to [110]	
	0.40 (16)		B parallel to [111]	
$m_{p,\,light}$	0.042 (2)			

Table 4.1-66 Electron g-factor g_c of gallium antimonide

Crystal		g_c	Temperature (K)	Remarks
Gallium antimonide	GaSb	−9.1 (2)	4.2–50	Magnetoluminescence

C. Transport Properties
Tables 4.1-67 and 4.1-68.

Electronic Transport, General Description. Gallium Nitride (GaN). Undoped GaN is normally an n-type conductor. Carrier concentrations in undoped films can vary from 5×10^{19} cm^{-3} to 5×10^{16} cm^{-3} because of unintentional incorporation of extrinsic impurities, mainly silicon and oxygen. In the purest α-GaN material ($n = 10^{17}$ cm^{-3}), conductivities of the order of 10 Ω cm have been found at 300 K. For the temperature dependence of the electrical resistivity, see Fig. 4.1-90.

Gallium Phosphide (GaP). Even in high-purity samples, intrinsic conduction occurs in gallium phosphide only above 500 °C. Thus, the transport properties are in general determined by the properties of impurities and lattice defects. The data for the electrical conductivity σ of pure n-type material are found to lie around 0.5 Ω cm at 300 K.

As in the case of GaAs, semi-insulating GaP can be obtained by doping with shallow donors and acceptors and with deep centers. Resistivities at room temperature are of the order of 10^8–10^{11} Ω cm. Typical data at high temperatures are shown in Fig. 4.1-91.

Gallium Arsenide (GaAs). At low fields, the electrons are in the Γ_6 minima at the zone center. The dominating scattering process at room temperature is polar optical scattering, while below 60 K the most important contribution to the lattice mobility is from piezoelectric potential scattering. At room temperature, the hole mobility of samples with $p < 5 \times 10^{15}$ cm^{-3} is governed by lattice scattering alone. An intrinsic carrier concentration of $n_i = 2.1 \times 10^6$ cm^{-3} has been found at $T = 300$ K (Fig. 4.1-92).

Owing to the very low intrinsic carrier concentration, the electrical conductivity of GaAs at 300 K is determined by charge carriers provided by impurities. Only under certain doping conditions (e.g. compensation of shallow acceptor levels by donor impurities and overcompensation by deep Cr levels) can the Fermi level be pinned in the middle of the energy gap and nearly intrinsic conduction achieved (semi-insulating GaAs, $\rho = 10^5$–10^9 Ω cm).

Gallium Antimonide (GaSb). Transport in n-type GaSb is complicated by the contribution of three sets of conduction bands with minima located at Γ, L, and X. The data on transport coefficients can be consistently explained by a three-band model, the X bands contributing to transport above 180 °C. Intrinsic carrier concentrations n_i of the order of 10^{14} cm^{-3} have been estimated for a temperature $T = 365$ K.

In p-type GaSb, a multiellipsoidal model has to be used at low temperatures, taking into account the shift of the heavy- and light-hole bands away from $k = 0$. At high temperatures, a warped-sphere model (as in the case of Si and Ge) is adequate.

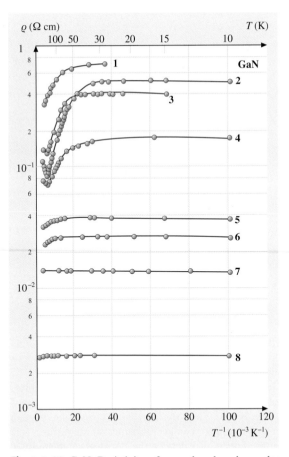

Fig. 4.1-90 GaN. Resistivity of several undoped samples (1–7) and a Zn-doped sample (8) vs. reciprocal temperature [1.85]

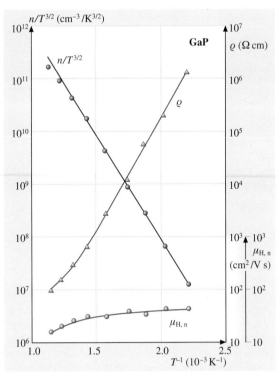

Fig. 4.1-91 GaP. Resistivity, Hall mobility, and electron concentration (divided by $T^{-3/2}$) vs. temperature for a semi-insulating sample [1.86]

Table 4.1-67 Electron and hole mobilities μ_n and μ_p of gallium compounds

Crystal	Temperature (K)	Carrier concentration (cm^{-3})	μ_n (cm^2/V s)	μ_p (cm^2/V s)	Remarks
Gallium nitride	300	$n = 3 \times 10^{16}$	900		Hall mobility [a]
		$p = 5 \times 10^{13}$		350	β-GaN, Hall mobility
Gallium phosphide	RT		160	135	Thin films, Hall mobility, maximum mobility [b]
Gallium arsenide	300		2400–3300		Bulk GaAs, Hall mobility [c]
	300			400	Hall mobility
Gallium antimonide	300	$n = 1.2 \times 10^{16}$	7620		MBE-grown on GaSb substrates
	300			680	Hall mobility

[a] See Fig. 4.1-93 for the temperature dependence of the electron mobility in GaN.
[b] See Figs. 4.1-94a,b for the temperature dependence of the electron and hole mobilities in GaP.
[c] See Figs. 4.1-95 and 4.1-96 for the temperature dependences of the electron and hole mobilities in GaAs.

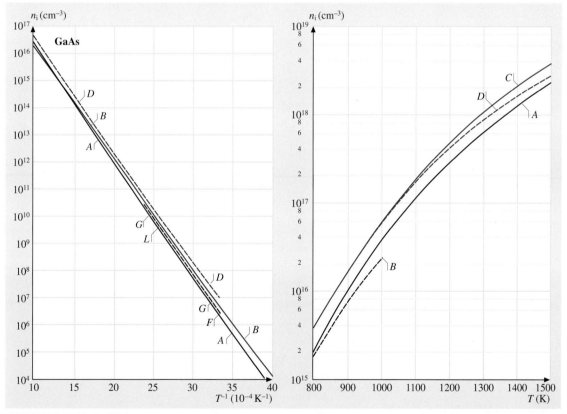

Fig. 4.1-92 GaAs. Intrinsic carrier concentration vs. reciprocal temperature for the range 250–1000 K and vs. temperature for the range 800–1500 K. The *solid curves* (A) are the result of a critical discussion of the literature data. The other curves (B–G) represent data from other sources [1.87]

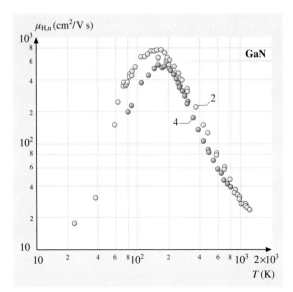

Fig. 4.1-93 GaN. Electron Hall mobility vs. temperature for two samples [1.85]

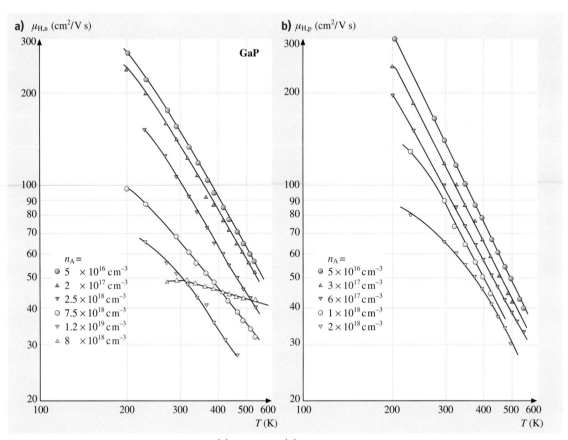

Fig. 4.1-94a,b GaP. Hall mobility in n-type (**a**) and p-type (**b**) LPE-grown layers vs. temperature [1.86]

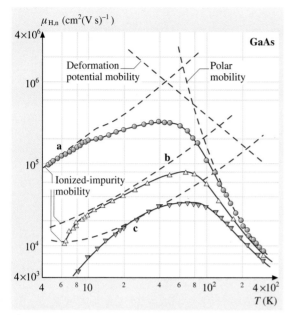

Fig. 4.1-95 GaAs. Temperature variation of Hall mobility at 5 kG for three n-GaAs samples. In the temperature range from 300 K to 77 K, the electron mobility of sample a is dominated by polar optical scattering. Samples b and c show increased effects of ionized-impurity scattering [1.88]

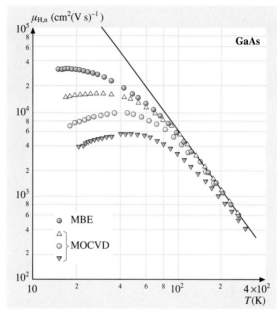

Fig. 4.1-96 GaAs. Hall mobility of holes vs. temperature for MBE- and MOCVD-grown samples. The *solid line* represents the empirical relation $\mu_{H,n} = 450 \times (300/T)^{2.3}$ [1.89]

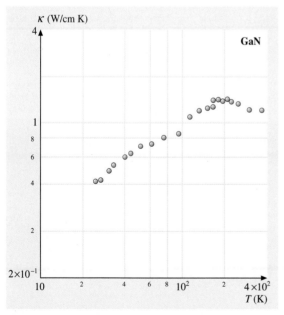

Fig. 4.1-97 GaN. Thermal conductivity along the c axis vs. temperature [1.90]

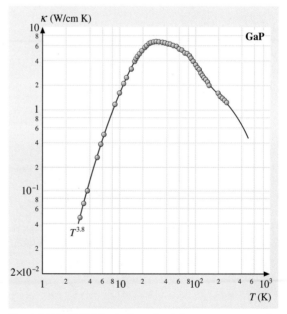

Fig. 4.1-98 GaP. Thermal conductivity vs. temperature for a p-type sample [1.69, 70]

Table 4.1-68 Thermal conductivity κ of gallium compounds

Crystal	κ (W/cm K)	Temperature (K)	Remarks
Gallium nitride	1.3	300	See Fig. 4.1-97 for temperature dependence.
Gallium phosphide	0.77	300	See Fig. 4.1-98 for temperature dependence.
Gallium arsenide	0.455	300	See Fig. 4.1-99 for temperature dependence.
Gallium antimonide			See Fig. 4.1-100 for temperature dependence.

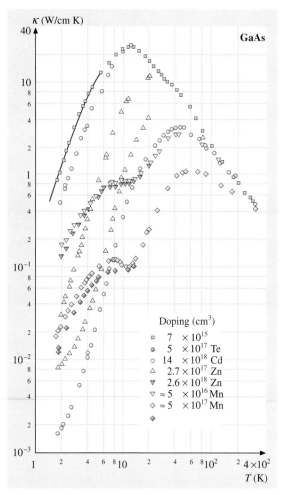

Fig. 4.1-99 GaAs. Thermal conductivity vs. temperature for seven samples with impurity concentrations between 7×10^{15} cm^{-3} and 2.6×10^{18} cm^{-3} [1.91]

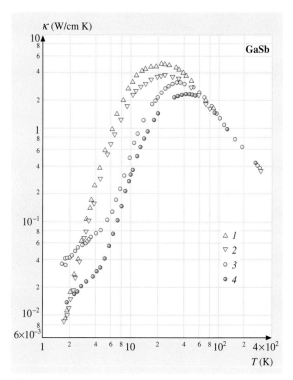

Fig. 4.1-100 GaSb. Thermal conductivity of two n-type samples with impurity contents of 4×10^{18} cm^{-3} (1) and 1.4×10^{18} cm^{-3} (2) and of two p-type samples with impurity contents of 1×10^{17} cm^{-3} (3) and 2×10^{17} cm^{-3} (4) [1.92]

D. Electromagnetic and Optical Properties
Tables 4.1-69 – 4.1-72 and Figs. 4.1-101 – 4.1-105.

Table 4.1-69 Dielectric constant ε and refractive index n of gallium compounds

Crystal		Temperature (K)	$\varepsilon(0)$	$\varepsilon(\infty)$	n at or	λ (nm) $h\nu$ (eV)	Remarks	Figure
Gallium nitride	GaN	300	9.5 (3)				E perpendicular to c	Fig. 4.1-101
			10.4 (3)				E parallel to c	
				5.8 (4)			E parallel to c	
				5.35 (20)			E perpendicular to c	
					2.29 (5)	500 nm	Optical interference	
Gallium phosphide	GaP	300	11.11				Low-frequency capacitance	Fig. 4.1-103
		75.7	10.86				measurements	
		300		9.11			Derived from refractive index	
					3.334	2.0 eV	Spectroscopic ellipsometry	
Gallium arsenide	GaAs	300	12.80 (5)				Microwave technique	Fig. 4.1-102
				10.86			Optical-transmission interference	
					3.878	2.0 eV	Spectroscopic ellipsometry	
Gallium antimonide	GaSb	300	15.7				Reflectance and oscillator fit	Fig. 4.1-105
				14.44				
					5.239	2.0 eV	Spectroscopic ellipsometry	

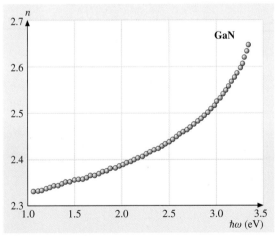

Fig. 4.1-101 GaN. Refractive index vs. photon energy at 300 K; $E \perp c$ [1.93]

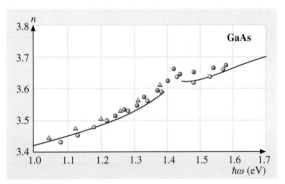

Fig. 4.1-102 GaAs. Refractive index vs. photon energy in the range 1.0–1.7 eV [1.94]. *Solid line*, calculated; *symbols*, experimental data from three sources

Table 4.1-70a Optical constants of gallium phosphide

$h\nu$ (eV)	ε_1	ε_2	n	k	R	K (10^3 cm^{-1})
1.5	10.102	0.000	3.178	0.000	0.272	0.00
2.0	11.114	0.000	3.334	0.000	0.290	0.00
2.5	12.996	0.046	3.605	0.006	0.320	1.63
3.0	16.601	1.832	4.081	0.224	0.369	68.26
3.5	24.833	8.268	5.050	0.819	0.458	290.4
4.0	9.652	16.454	3.790	2.171	0.452	880.1
4.5	11.073	17.343	3.978	2.180	0.461	994.3
5.0	0.218	26.580	3.661	3.631	0.580	1840
5.5	−10.266	10.974	1.543	3.556	0.677	1983
6.0	−5.521	7.041	1.309	2.690	0.583	1636

Table 4.1-71 Optical constants of gallium arsenide

$h\nu$ (eV)	ε_1	ε_2	n	k	R	K (10^3 cm^{-1})
1.5	13.435	0.589	3.66	0.080	0.327	12.21
2.0	14.991	1.637	3.878	0.211	0.349	42.79
2.5	18.579	3.821	4.333	0.441	0.395	111.7
3.0	16.536	17.571	4.509	1.948	0.472	592.5
3.5	8.413	14.216	3.531	2.013	0.425	714.2
4.0	9.279	13.832	3.601	1.920	0.421	778.7
4.5	6.797	22.845	3.913	2.919	0.521	1331
5.0	−11.515	18.563	2.273	4.084	0.688	2070
5.5	−6.705	8.123	1.383	2.936	0.613	1637
6.0	−4.511	6.250	1.264	2.472	0.550	1503

Table 4.1-72a Optical constants of gallium antimonide

$h\nu$ (eV)	ε_1	ε_2	n	k	R	K (10^3 cm^{-1})
1.5	19.135	3.023	4.388	0.344	0.398	52.37
2.0	25.545	14.442	5.239	1.378	0.487	279.4
2.5	13.367	19.705	4.312	2.285	0.484	579.1
3.0	9.479	15.838	3.832	2.109	0.444	641.2
3.5	7.852	19.267	3.785	2.545	0.485	902.8
4.0	−1.374	25.138	3.450	3.643	0.583	1477
4.5	−8.989	10.763	1.586	3.392	0.561	1547
5.0	−5.693	7.529	1.369	2.751	0.585	1394
5.5	−5.527	5.410	1.212	2.645	0.592	1475
6.0	−4.962	4.520	0.935	2.416	0.610	1469

Table 4.1-70b GaP. Refractive index in the infrared

Wavelength (μm)	Refractive index n	Remarks
20	2.529	Reflectance and transmission at 300 K
10	2.90	
5	2.94	
2	3.02	
1	3.17	

Table 4.1-72b GaSb. Refractive index in the infrared

Wavelength (μm)	Refractive index n	Temperature (K)	Remarks
14.9	3.880	300	Reflectance and transmission
10	3.843		
4	3.833		
2	3.789		Prism method
1.9	3.802		
1.8	3.820		

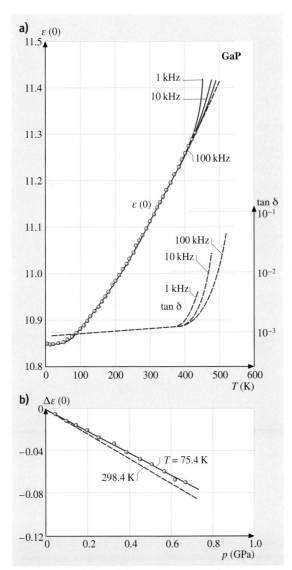

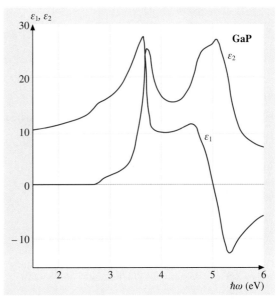

Fig. 4.1-104 GaP. Real and imaginary parts of the dielectric constant vs. photon energy [1.47]

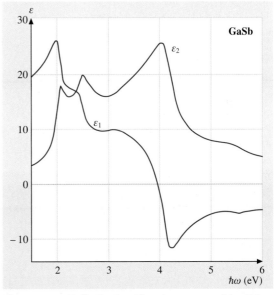

Fig. 4.1-103a,b GaP. (a) Static dielectric constant $\varepsilon(0)$ and dielectric loss $\tan\delta$ vs. temperature, obtained from low-frequency capacitance measurements. For clarity, the actual data points at only one frequency (100 kHz) are shown. The data have been corrected for dimensional changes due to thermal expansion. (b) Change in static dielectric constant vs. pressure at two temperatures [1.95]

Fig. 4.1-105 GaSb. Real and imaginary parts of the dielectric constant vs. photon energy [1.47]

4.1.2.4 Indium Compounds

A. Crystal Structure, Mechanical and Thermal Properties
Tables 4.1-73 – 4.1-81

Table 4.1-73 Crystal structures of indium compounds

Crystal		Space group		Crystal system	Structure type	Remarks	Figure
Indium nitride	InN	C_{6v}^4	$P6_3mc$	hex	Wurtzite	Phase transition to NaCl structure under pressure	Fig. 4.1-3
Indium phosphide	InP I	T_d^2	$F\bar{4}3m$	fcc	Zinc blende	Stable at normal pressure	Fig. 4.1-2
	InP II	O_h^5	$Fm3m$	fcc	NaCl	Pressure > 34 GPa	
Indium arsenide	InAs I	T_d^2	$F\bar{4}3m$	fcc	Zinc blende	Stable at normal pressure	Fig. 4.1-2
	InAs II	O_h^5	$Fm3m$	fcc	NaCl	Pressure > 4 GPa, RT	
Indium antimonide	InSb I	T_d^2	$F\bar{4}3m$	fcc	Zinc blende	Stable at normal pressure [a]	Fig. 4.1-2

[a] Quite a number of phase transitions have been observed at high pressures (> 2.1 GPa).

Table 4.1-74 Lattice parameters of indium compounds

Crystal		Lattice parameters (nm)	Temperature (K)	Remarks
Indium nitride	InN	$a = 0.35446$		Epitaxial layers, X-ray diffraction
		$c = 0.57034$		
Indium phosphide	InP	$a = 0.58687\,(10)$	291.15	Powder, X-ray diffraction
Indium arsenide	InAs	$a = 0.60583$	298.15	
Indium antimonide	InSb	$a = 0.647937$	298.15	Powder, X-ray diffraction

Table 4.1-75 Densities of indium compounds

Crystal		Density (g/cm^3)	Temperature (K)	Remarks
Indium nitride	InN	6.78 (5)		Pycnometric
Indium phosphide	InP	4.81		
Indium arsenide	InAs	5.667	300	X-ray diffraction
Indium antimonide	InSb	5.7747 (4)	300	X-ray diffraction

Table 4.1-76 Elastic constants c_{ik} (in GPa) of indium compounds

Crystal		c_{11}	c_{12}	c_{13}	c_{33}	c_{44}	Remarks
Indium nitride	InN	190 (7)	104 (3)	121 (7)	182 (6)	9.9 (11)	Calculated from the mean square displacements of the lattice atoms measured by X-ray diffraction
Indium phosphide	InP	101.1	56.1	–	–	45.6	Ultrasonic-wave transit times, RT
Indium arsenide	InAs	83.29	45.26	–	–	39.59	Ultrasound, $f = 15$ MHz, n-type, RT
Indium antimonide	InSb	69.18	37.88	–	–	31.32	$T = 0$ K, extrapolated from 4.2 K, ultrasound, $f = 10$ MHz

Table 4.1-77 Melting point T_m of indium compounds

Crystal		Melting point T_m (K)	Remarks
Indium nitride	InN	1900	Estimated; at 80 kbar N_2
Indium phosphide	InP	1327	Optimized
Indium arsenide	InAs	1221 (1)	
Indium antimonide	InSb	800 (1)	Drop calorimetry

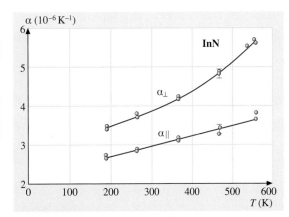

Fig. 4.1-106 InN. Coefficient of linear thermal expansion vs. temperature [1.75]

Table 4.1-78 Linear thermal expansion coefficient α of indium compounds

Crystal		Expansion coefficient α (10^{-6} K^{-1})		Temperature (K)	Temperature dependence	Remarks
Indium nitride	InN	α_a	3.6 (2)	100–673	Fig. 4.1-106	X-ray diffraction
		α_c	2.6 (3)			
Indium phosphide	InP		4.75 (10)	298	Fig. 4.1-107	
Indium arsenide	InAs		4.52	20–250	Fig. 4.1-108	Average value, X-ray diffraction
Indium antimonide	InSb		5.37 (50)	283–343	Fig. 4.1-109	X-ray diffraction

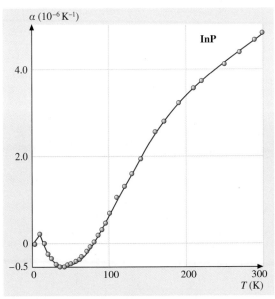

Fig. 4.1-107 InP. Temperature dependence of the coefficient of linear thermal expansion, obtained by the Bond method [1.96]

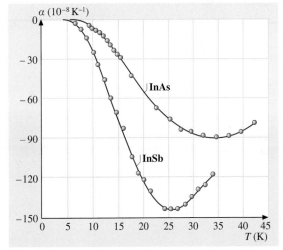

Fig. 4.1-108 InAs, InSb. Linear thermal expansion coefficient vs. temperature, measured with a variable-transformer dilatometer [1.97]

Table 4.1-79 Heat capacity c_p and Debye temperature Θ_D of indium compounds

Crystal		Heat capacity (J/mol K)	Temperature (K)	Debye temperature (K)	Remarks
Indium nitride	InN	41.74	300	660	Calorimetry
Indium phosphide	InP			321	
Indium arsenide	InAs			247	
Indium antimonide	InSb			206	Average value

Table 4.1-80 Phonon wavenumbers of indium compounds. Indium nitride (InN), $T = 300$ K; indium phosphide (InP), RT, from coherent inelastic neutron scattering, carrier concentration $n = 10^{12}$ cm^{-3}; indium arsenide (InAs), from Raman scattering; indium antimonide (InSb), $T = 300$ K, from inelastic neutron scattering, carrier concentration $n = 8 \times 10^{13}$ cm^{-3}

InN					InP		
Symmetry point	Branch	Wavenumber (cm^{-1})	Temperature (K)	Remarks	Symmetry point	Branch	Frequency (THz)
$\tilde{\nu}(\Gamma)$	TO	478	300	Reflectivity, Kramers–Kronig analysis	$\nu(111)(0.05)$	LO	10.3 (3)
	LO	694		Reflectivity, Kramers–Kronig analysis	$\nu(\Gamma_{15})$	TO	9.2 (2)
$\tilde{\nu}(A_1)$	TO	400		Experimental data	$\nu(X_5)$	TA	2.05 (10)
$\tilde{\nu}(E_1)$	TO	484			$\nu(X_3)$	LA	5.8 (3)
	LO	570			$\nu(X_5)$	TO	9.70 (10)
$\tilde{\nu}(E_2^{(1)})$	TO	190			$\nu(X_1)$	LO	9.95 (20)
$\tilde{\nu}(E_2^{(2)})$	TO	590			$\nu(L_3)$	TA	1.65 (2)
					$\nu(L_1)$	LA	5.00 (10)
					$\nu(L_3)$	TO	9.50 (15)
					$\nu(L_1)$	LO	10.2 (3)

InAs				InSb		
Symmetry point	Branch	Wavenumber (cm^{-1})	Temperature (K)	Symmetry point	Branch	Frequency (THz)
$\tilde{\nu}(\Gamma)$	TO	217.3	300	$\nu(\Gamma_{15})$	LO	5.90 (25)
	LO	238.6			TO	5.54 (5)
$\tilde{\nu}(X)$	TA	53		$\nu(X_5)$	TA	1.12 (5)
	LA	160		$\nu(X_3)$	LA	4.30 (10)
	TO	216		$\nu(X_1)$	LO	4.75 (20)
	LO	203	100	$\nu(X_5)$	TO	5.38 (17)
$\tilde{\nu}(L)$	TA	44	300	$\nu(L_3)$	TA	0.98 (5)
	LA	139.5		$\nu(L_1)$	LA	3.81 (6)
	TO	216			LO	4.82 (10)
	LO	203	100	$\nu(L_3)$	TO	5.31 (6)

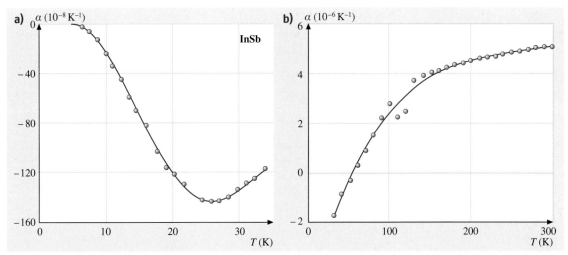

Fig. 4.1-109a,b InSb. Linear thermal expansion coefficient vs. temperature: (**a**) low-temperature region [1.97], (**b**) high-temperature region [1.98]

Table 4.1-81 Phonon dispersion curves of indium compounds

Crystal		Figure
Indium phosphide	InP	Fig. 4.1-110
Indium arsenide	InAs	Fig. 4.1-111
Indium antimonide	InSb	Fig. 4.1-112

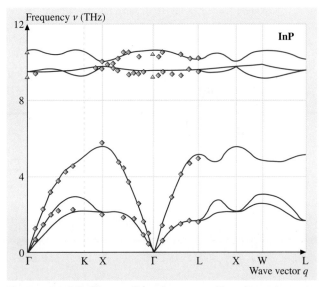

Fig. 4.1-110 InP. Phonon dispersion curves. Experimental neutron data (*diamonds*) [1.99] and Raman data (*triangles*) [1.100] and ab initio pseudopotential calculations (*solid curves*) [1.101]. From [1.101]

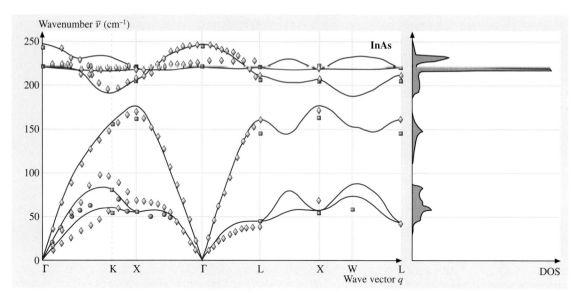

Fig. 4.1-111 InAs. Phonon dispersion curves (*left panel*) and density of states (*right panel*) [1.82]. Experimental neutron data (*circles*, $T = 300$ K) [1.102], thermal-diffuse-scattering X-ray data (*diamonds*, $T = 80$ K) [1.102], Raman data (*squares*, $T = 330$ K) [1.103], and ab initio calculations (*solid curves* [1.82]). From [1.82]

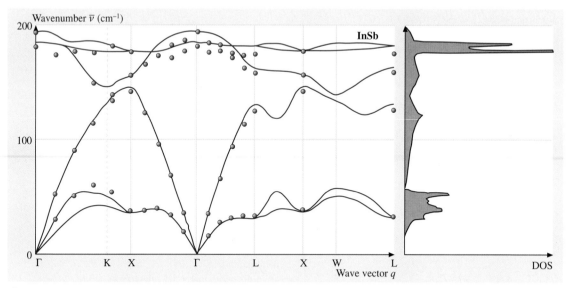

Fig. 4.1-112 InSb. Phonon dispersion curves (*left panel*) and density of states (*right panel*). Experimental neutron data (*circles*) [1.21] and ab initio pseudopotential calculations (*solid curves*) [1.60]. From [1.60]

B. Electronic Properties
Tables 4.1-82 – 4.1-87.

Band Structures of Indium Compounds. Indium Nitride (InN). The band structure (Fig. 4.1-114) shows a direct gap at Γ very similar to that of GaN. A low-lying many-valley conduction band leading to an indirect gap has been suggested.

Indium Phosphide (InP). Indium phosphide is a direct-gap semiconductor. The conduction band minimum is situated at Γ. Higher conduction band minima at L and X have been detected in optical experiments. The X-band minima show no camel's back structure, in contrast to most other III–V compounds with the zinc blende structure. The valence band has the structure common to all zinc blende-type semiconductors (Fig. 4.1-113).

Indium Arsenide (InAs). Indium arsenide resembles InSb in its band structure, having only a slightly larger energy gap and a smaller spin–orbit splitting of the top of the valence band. The conduction band minimum (Γ_6) is situated in the center of the Brillouin zone. Near the minimum, $E(k)$ is isotropic but nonparabolic. The valence band shows the usual structure common to all zinc blende-type III–V compounds (Fig. 4.1-116).

Table 4.1-82 First Brillouin zones of indium compounds

Crystal		Figure
Indium nitride	InN	Fig. 4.1-24
Indium phosphide	InP	Fig. 4.1-23
Indium arsenide	InAs	Fig. 4.1-23
Indium antimonide	InSb	Fig. 4.1-23

Indium Antimonide (InSb). Indium antimonide is a direct-gap semiconductor. The minimum of the conduction band (Γ_6) is located in the center of the Brillouin zone. Near the minimum, $E(k)$ is isotropic but nonparabolic. Thus the effective mass of the electrons is scalar and depends strongly on the electron concentration. Higher band minima (about 0.63 eV above the lowest minimum) seem to have been established by transport measurements in heavily doped n-InSb. The valence band shows the structure common to all zinc blende semiconductors, i.e. two subbands degenerate at Γ_8 and one spin-split band at Γ_7 (Fig. 4.1-115).

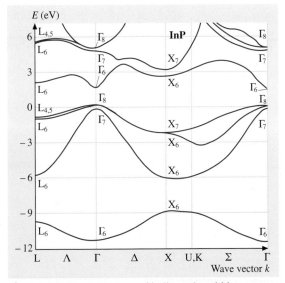

Fig. 4.1-113 Band structure of indium phosphide

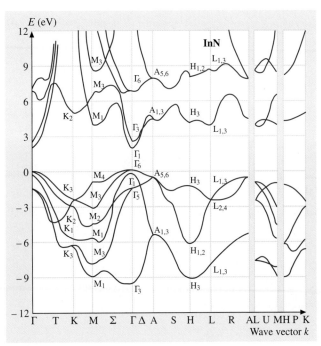

Fig. 4.1-114 Band structure of indium nitride

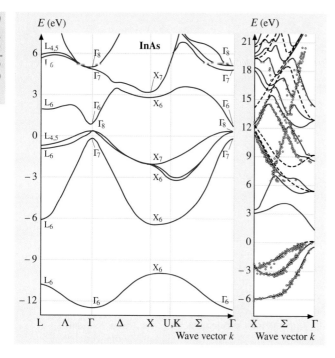

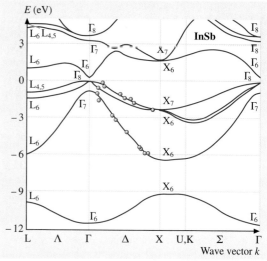

Fig. 4.1-115 Band structure of indium antimonide

Fig. 4.1-116 Band structure of indium arsenide

Table 4.1-83 Energy gaps of indium compounds

Crystal		Quantity	Bands	Energy (eV)	Temperature (K)	Remarks
Indium nitride	InN	$E_{g,dir}$		1.95	300	Polycrystalline film, absorption edge
Indium phosphide	InP	$E_{g,dir}$	Γ_{8v} to Γ_{6c}	1.4236 (1)	1.6	From excitonic gap
				1.344	300	Absorption, photoluminescence
Indium arsenide	InAs	$E_{g,dir}$	Γ_{8v} to Γ_{6c}	0.4180 (5)	4.2	Magnetotransmission
				0.354 (3)	295	Electroreflectance
		$E_{g,x}$		0.41565 (1)	1.4	Excitonic band gap, photoluminescence
Indium antimonide	InSb	$E_{g,dir}$	Γ_{8v} to Γ_{6c}	0.2352	1.7	Emission spectra, from exciton data
				0.180	300	Reflectance, magnetoabsorption

Table 4.1-84 Spin–orbit splitting energies of indium compounds

Crystal		Spin–orbit splitting	Bands	Numerical value (eV)	Temperature (K)	Remarks
Indium phosphide	InP	Δ_0	Γ_{8v} to Γ_{7v}	0.108 (8)	4.2	Wavelength-modulated photovoltaic effect
		Δ_1	$\Lambda_{4,5v}$ to Λ_{6v}	0.13	77	Thermoreflectance
				0.15 (5)	300	Electroreflectance
		Δ_0'	Γ_{8c} to Γ_{7c}	0.07	300	Electroreflectance
Indium arsenide	InAs	Δ_0	Γ_{8v} to Γ_{7v}	0.38 (1)	1.5	Magnetoelectroreflectance
		Δ_1	$L_{4,5v}$ to L_{6v}	0.267	5	Wavelength-modulated reflectance
Indium antimonide	InSb	Δ_0	Γ_{8v} to Γ_{7v}	0.81 (1)	1.5	Magnetoelectroreflectance
		Δ_0'	Γ_{7v} to Γ_{8c}	0.39	5	Wavelength-modulated reflectance
		Δ_1	$\Lambda_{4,5v}$ to Λ_{6v}	0.495		

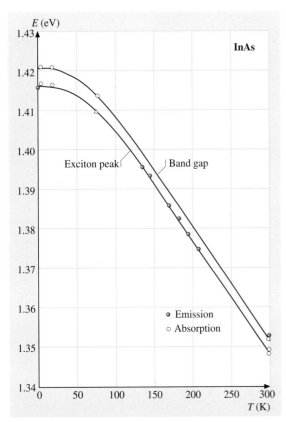

Fig. 4.1-117 InP. Energy gap and exciton peak energy vs. temperature from absorption and emission data [1.104]

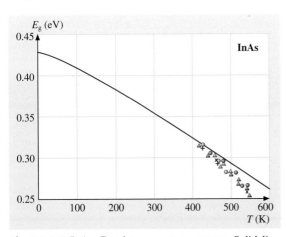

Fig. 4.1-118 InAs. Band gap vs. temperature. *Solid line*, optical band gap calculated from parameters used for fitting various transport data; *symbols*, thermal band gap of six samples, obtained from conductivity and Hall coefficient [1.105]

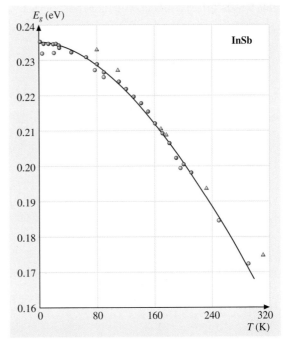

Fig. 4.1-119 InSb. Energy gap vs. temperature below RT, measured by resonant two-photon photo-Hall effect (*filled circles*); the *open circles* and *triangles* show earlier literature data for comparison. *Solid curve*: fit by Varshni's formula [1.106]

Table 4.1-85 Temperature dependence of the energy gaps of indium compounds

Crystal		Figure
Indium phosphide	InP	Fig. 4.1-117
Indium arsenide	InAs	Fig. 4.1-118
Indium antimonide	InSb	Fig. 4.1-119

Table 4.1-86 Effective masses of electrons (m_n) and holes (m_p) for indium compounds (in units of the electron mass m_0)

InN			
m_n	0.11	300 K	Optical mass, plasma edge
$m_{n, calc}$	0.12		Calculated effective mass
$m_{p, heavy}$	0.5		
$m_{p, light}$	0.17		
InP			
m_n	0.0808(10)	0 K	Magnetophonon resonance
	0.073	300 K	Faraday rotation
$m_{p, heavy}$	0.45 (5)	4.2 K	Piezomodulated photovoltaic effect
$m_{p, light}$	0.12 (1)	4.2 K	
$m_{p, so}$	0.121 (1)	110 K	B parallel to [111] and [100]
InAs			
m_n	0.0265	4.2 K	
	0.023	300 K	
$m_{p, heavy}$	0.57	300 K	
$m_{p, [100]}$	0.35		
$m_{p, [111]}$	0.85		
InSb			
m_n (Γ_{6c})	0.01359 (3)	4.2 K	Faraday effect, $n = 4.6 \times 10^{13}$ cm^{-3}
	0.0118	300 K	Cyclotron resonance
$m_{p, heavy}$	0.44	20 K	Parallel to [111], magnetoabsorption, Faraday rotation
	0.42		Parallel to [110]
	0.32		Parallel [100]
$m_{p, light}$	0.016	20 K	Magnetoabsorption, Faraday rotation

Table 4.1-87 Electron g-factor g_c of indium compounds

Crystal	g_c	Temperature (K)	Remarks
Indium phosphide	1.48 (5)	4.2	Modulated photovoltaic effect
Indium arsenide	−15.3 (2)	1.4	Magnetoluminescence
Indium antimonide	−51.31 (1)	1.4	Electron spin resonance

C. Transport Properties
Tables 4.1-88 – 4.1-90.

Electronic Transport, General Description. Indium Phosphide (InP). The transport properties are determined mainly by the electrons in the Γ_{6c} minimum. Above 800 K, multivalley conduction, where the L_{6c} minimum is involved, becomes important.

Indium Arsenide (InAs). The transport properties are determined mainly by the electrons in the Γ_6 minimum. Pure material with intrinsic conduction down to 450 K is available.

Indium Antimonide (InSb). The transport properties are mainly determined by the extremely high mobility of the electrons in the Γ_6 minimum of the conduction band. Pure material with intrinsic conduction down to 200 K is available. At low temperatures and in doped material, impurity scattering limits the electron mobility.

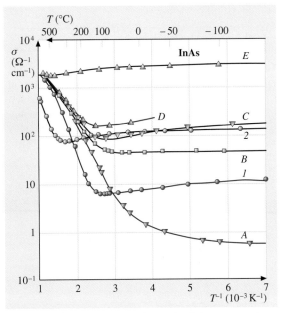

Fig. 4.1-120 InAs. Electrical conductivity vs. reciprocal temperature [1.107]. A, $n = 1.2 \times 10^{15}$ cm^{-3}; B, $n = 1.7 \times 10^{16}$ cm^{-3}; C, $n = 4 \times 10^{16}$ cm^{-3}; D, $n = 7 \times 10^{16}$ cm^{-3}; E, $n = 2 \times 10^{18}$ cm^{-3}; 1, $p = 2 \times 10^{17}$ cm^{-3}; 2, $p = 7 \times 10^{18}$ cm^{-3}

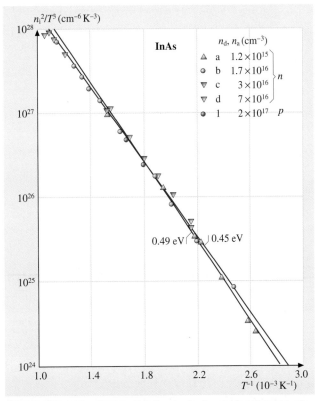

Fig. 4.1-121 InAs. Square of the intrinsic carrier concentration divided by T^3 vs. reciprocal temperature for several different samples [1.107]

Table 4.1-88 Carrier concentrations n_i and electrical conductivity σ of indium compounds

Crystal		Temperature (K)	Carrier concentration n_i (cm^{-3})	Conductivity σ (Ω^{-1} cm^{-1})	Remarks
Indium nitride	InN	300		$2-3 \times 10^2$	Temperature coefficient of resistivity 3.7×10^{-3} K^{-1} at 200–300 K for pressed powder
Indium phosphide	InP	300	6.3×10^{15}	4–5	n-type pure material
Indium arsenide	InAs	300		50	See Fig. 4.1-120 for temperature dependence
Indium antimonide	InSb	300	1.9×10^{16}	220	See Fig. 4.1-122 for temperature dependence

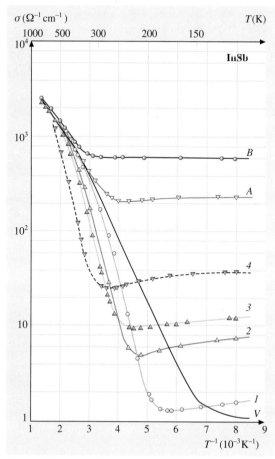

Fig. 4.1-122 InSb. Electrical conductivity vs. reciprocal temperature [1.108]. *Curve V*, $n \approx 10^{13}$ cm^{-3}; *A*, $n = 1.3 \times 10^{16}$ cm^{-3}; *B*, $n = 1 \times 10^{16}$ cm^{-3}; *1*, $p = 4 \times 10^{15}$ cm^{-3}; *2*, $p = 2.2 \times 10^{16}$ cm^{-3}; *3*, $p = 6 \times 10^{16}$ cm^{-3}; *4*, $p = 2 \times 10^{17}$ cm^{-3}

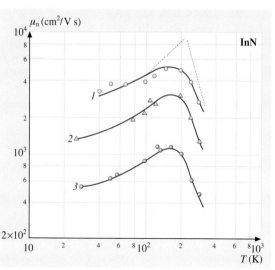

Fig. 4.1-123 InN. Electron mobility vs. temperature for three samples with RT carrier concentrations of 5.3×10^{16} (1), 7.5×10^{16} (2), and 1.8×10^{17} cm^{-3} (3). *Left broken line*, calculated ionized-impurity-scattering mobility; *right broken line*, empirical high-temperature mobility ($\mu \propto T^{-3}$) for sample 1. *Solid lines*, total mobility calculated for each sample [1.109]

Table 4.1-89 Electron and hole mobilities μ_n and μ_p of indium compounds

Crystal	Temperature (K)	Carrier concentration (cm^{-3})	μ_n (cm^2/V s)	μ_p (cm^2/V s)	Remarks	Temperature dependence of mobilities
Indium nitride	300	$5-18 \times 10^{16}$	250 (50)			Fig. 4.1-123
	300		20			
Indium phosphide	300	$n = 0.5-1 \times 10^{16}$	$4-5.4 \times 10^3$		Pure material, Hall mobilities	Fig. 4.1-124
	77	$n = 6 \times 10^{13}$	130×10^3			
	300			190		Fig. 4.1-125
Indium arsenide	300	$n \sim 10^{16}$	$2-3.3 \times 10^4$		Pure material	Fig. 4.1-121
	77		$0.8-1 \times 10^5$			
	300			100–450		
Indium antimonide	300		70×10^3		Hall mobilities	Fig. 4.1-126
	300			850		Fig. 4.1-127

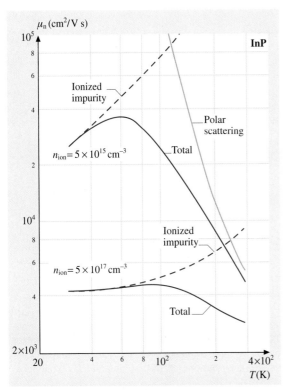

Fig. 4.1-124 InP. Electron mobility vs. temperature, calculated for two concentrations of ionized impurities [1.110]

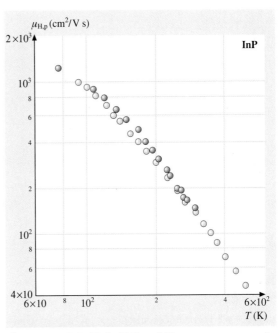

Fig. 4.1-125 InP. Hole Hall mobility vs. temperature for pure p-type samples, after [1.111]

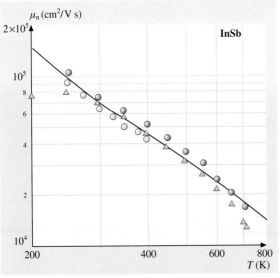

Fig. 4.1-126 InSb. Electron mobility vs. temperature. The experimental data are Hall mobilities. *Triangles*, [1.112]; *filled circles* [1.113]; *open circles*, [1.114, 115]; *solid line*, calculated drift mobility [1.116]

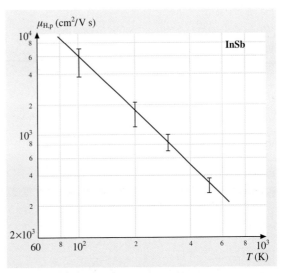

Fig. 4.1-127 InSb. Hole Hall mobility vs. temperature [1.111]. The *vertical bars* indicate the ranges of the experimental values. The *solid line* has a slope -1.8

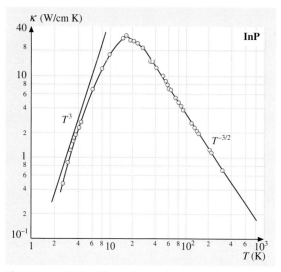

Fig. 4.1-128 InP. Thermal conductivity vs. temperature [1.117, 118]

Table 4.1-90 Thermal conductivity κ of indium compounds

Crystal		κ (W/m K)	Remarks
Indium nitride	InN	38.4	
Indium phosphide	InP		See Fig. 4.1-128
Indium arsenide	InAs		
Indium antimonide	InSb		See Fig. 4.1-129

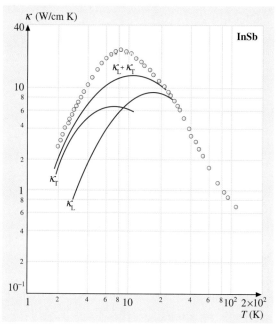

Fig. 4.1-129 InSb. Thermal conductivity vs. temperature for an n-type sample, and theoretical curve showing the contributions of the longitudinal and transverse phonons [1.119]

D. Electromagnetic and Optical Properties
Tables 4.1-91 – 4.1-93.

Table 4.1-91 Dielectric constant ε and refractive index n of indium compounds

Crystal		Temperature (K)	$\varepsilon(0)$	$\varepsilon(\infty)$	n	at λ (μm)	Remarks
Indium nitride	InN		9.3				Infrared reflectivity, heavily doped film
				8.4			
					2.56	1.0 μm	Interference method, $n = 3 \times 10^{20}$ cm^{-3}
					2.93	0.82 μm	
					3.12	0.66 μm	
Indium phosphide	InP	300	12.56 (20)				Table 4.1-92	Capacitance measurements
		77	11.93 (20)					
		297		10.9 (10)			Film interference
Indium arsenide	InAs	300	15.15	12.37		Table 4.1-92	Infrared reflectance and oscillator fit
Indium antimonide	InSb		16.8 (2)			Table 4.1-92	Gyroscopic sphere resonance, infrared reflectance, and oscillator fit
				15.68			

Table 4.1-92 Wavelength dependence of refractive index n of indium compounds at $T = 300$ K

InP			InAs			InSb		
λ (μm)	n	Remarks	λ (μm)	n	Remarks	λ (μm)	n	Remarks
14.85	3.03	Transmission and reflectance	25	3.26	Interference	45	2.57	Reflectance and transmission, $n = 6 \times 10^{15}$ cm^{-3}
5	3.08		10	3.42		21.15	3.814	
2	3.134	Prism method	3.74	3.52		19.98	3.826	Interference technique, $n = 2 \times 10^{16}$ cm^{-3}
1	3.327		1.38	3.516	Reflectance and dispersion relations	10.06	3.953	
						7.87	4.001	
0.652	3.410	Reflectance and dispersion relations	0.517	4.558		2.07	4.03	Reflectance and dispersion relations
0.399	4.100		0.282	3.800		0.689	5.13	
0.200	1.525		0.049	1.139		0.062	1.17	
0.062	0.793							

Table 4.1-93 Energy dependence of optical constants of indium compounds. Real and imaginary parts of the complex dielectric constant measured by spectroscopic ellipsometry; n, k, R, K calculated from these data

InP

$h\nu$ (eV)	ε_1	ε_2	n	k	R	K (10^3 cm^{-1})
1.5	11.904	1.400	3.456	0.203	0.395	30.79
2.0	12.493	2.252	3.549	0.317	0.317	64.32
2.5	14.313	3.062	3.818	0.511	0.349	129.56
3.0	17.759	10.962	4.395	1.247	0.427	379.23
3.5	5.400	12.443	3.193	1.948	0.403	691.21
4.0	6.874	10.871	3.141	1.730	0.376	701.54
4.5	8.891	16.161	3.697	2.186	0.449	996.95
5.0	−7.678	14.896	2.131	3.495	0.613	1771.5
5.5	−4.528	7.308	1.426	2.562	0.542	1428.1
6.0	−2.681	5.644	1.336	2.113	0.461	1285.1

InAs

$h\nu$ (eV)	ε_1	ε_2	n	k	R	K (10^3 cm^{-1})
1.5	13.605	3.209	3.714	0.432	0.337	65.69
2.0	15.558	5.062	3.995	0.634	0.370	128.43
2.5	15.856	15.592	4.364	1.786	0.454	452.64
3.0	6.083	13.003	3.197	2.034	0.412	618.46
3.5	5.973	10.550	3.008	1.754	0.371	622.13
4.0	7.744	11.919	3.313	1.799	0.393	729.23
4.5	−1.663	22.006	3.194	3.445	0.566	1571.2
5.0	−5.923	8.752	1.524	2.871	0.583	1455.3
5.5	−3.851	6.008	1.282	2.344	0.521	1306.6
6.0	−2.403	6.005	1.434	2.112	0.448	1284.2

InSb

$h\nu$ (eV)	ε_1	ε_2	n	k	R	K (10^3 cm^{-1})
1.5	19.105	5.683	4.418	0.643	0.406	97.79
2.0	14.448	14.875	4.194	1.773	0.443	359.46
2.5	7.811	15.856	3.570	2.221	0.447	562.77
3.0	7.354	13.421	3.366	1.994	0.416	606.27
3.5	5.995	17.673	3.511	2.517	0.474	892.82
4.0	−6.722	19.443	2.632	3.694	0.608	1497.8
4.5	−6.297	8.351	1.443	2.894	0.598	1320.2
5.0	−4.250	6.378	1.307	2.441	0.537	1237.0
5.5	−4.325	4.931	1.057	2.333	0.563	1300.6
6.0	−3.835	3.681	0.861	2.139	0.572	1300.9

4.1.3 II–VI Compounds

4.1.3.1 Beryllium Compounds

A. Crystal Structure, Mechanical and Thermal Properties
Tables 4.1-94 – 4.1-99.

Table 4.1-94 Crystal structures of beryllium compounds

Crystal		Space group		Crystal system	Structure type	Figure
Beryllium oxide	BeO	C_{6v}^4	$P6_3mc$	hex	Wurtzite	Fig. 4.1-130
Beryllium sulfide	BeS	T_d^2	$F\bar{4}3m$	fcc	Zinc blende	Fig. 4.1-131
Beryllium selenide	BeSe	T_d^2	$F\bar{4}3m$	fcc	Zinc blende	Fig. 4.1-131
Beryllium telluride	BeTe	T_d^2	$F\bar{4}3m$	fcc	Zinc blende	Fig. 4.1-131

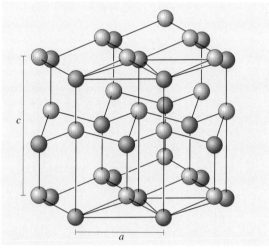

Fig. 4.1-130 The wurtzite lattice (a and c are the lattice parameters)

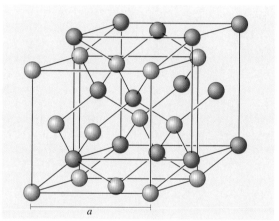

Fig. 4.1-131 The zinc blende lattice (a is the lattice parameter)

Table 4.1-95 Lattice parameters of beryllium compounds

Crystal		Lattice parameters (nm)	Temperature (K)	Remarks
Beryllium oxide	BeO	$a = 0.26979$ $c = 0.4380$		X-ray diffraction
Beryllium sulfide	BeS	$a = 0.48630\,(5)$	RT	X-ray diffraction
Beryllium selenide	BeSe	$a = 0.51520\,(23)$	RT	Epitaxial film, X-ray diffraction
Beryllium telluride	BeTe	$a = 0.56270\,(15)$	300	Epitaxial film, X-ray diffraction

Table 4.1-96 Elastic constants c_{ik} (in GPa) of beryllium oxide

Crystal		c_{11}	c_{12}	c_{13}	c_{33}	c_{44}	Remarks
Beryllium oxide	BeO	460.6	126.5	88.5	491.6	147.7	Ultrasonic data, RT

Table 4.1-97 Densities of beryllium compounds

Crystal		Density (g/cm³)
Beryllium oxide	BeO	3.01
Beryllium sulfide	BeS	2.36

Table 4.1-98 Melting point T_m of beryllium oxide

Crystal		Melting point T_m
Beryllium oxide	BeO	2507 °C

Table 4.1-99 Linear thermal expansion coefficient α of beryllium compounds

Crystal		Expansion coefficient α (K^{-1})	Temperature (K)
Beryllium oxide	BeO	9.7×10^{-6}	373
Beryllium telluride	BeTe	$7.66(21) \times 10^{-6}$	300

B. Electronic Properties

Tables 4.1-100 – 4.1-102.

Table 4.1-100 First Brillouin zones of beryllium compounds

Crystal		Figures
Beryllium oxide	BeO	Fig. 4.1-132
Beryllium sulfide	BeS	Fig. 4.1-133
Beryllium selenide	BeSe	Fig. 4.1-133
Beryllium telluride	BeTe	Fig. 4.1-133

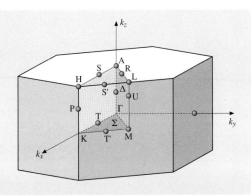

Fig. 4.1-132 The Brillouin zone of the wurtzite lattice

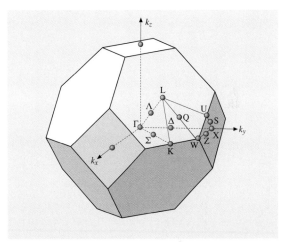

Fig. 4.1-133 The Brillouin zone of the zinc blende and rock salt lattices

Table 4.1-101 Band structures of beryllium compounds

Crystal		Figure	Remarks
Beryllium oxide	BeO	Fig. 4.1-135	
Beryllium sulfide	BeS	Fig. 4.1-134	a
Beryllium selenide	BeSe	Fig. 4.1-136	b
Beryllium telluride	BeTe	Fig. 4.1-137	

[a] According to theory, BeS is an indirect-gap material with the highest valence band edge at Γ and the lowest conduction band edge at X.
[b] According to theory, BeSe is an indirect-gap material with the highest valence band edge at Γ and the lowest conduction band edge at X.

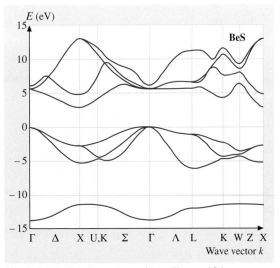

Fig. 4.1-134 Band structure of beryllium sulfide

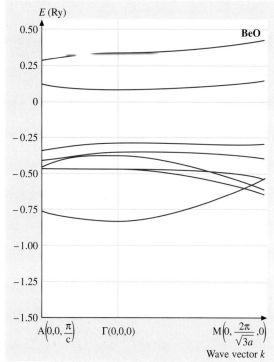

Fig. 4.1-135 Band structure of beryllium oxide

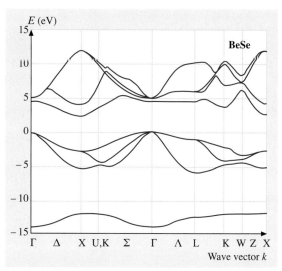

Fig. 4.1-136 Band structure of beryllium selenide

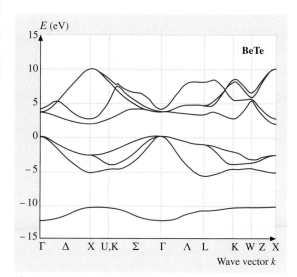

Fig. 4.1-137 Band structure of beryllium telluride

Table 4.1-102 Energy gaps of beryllium compounds

Crystal		Band gap		Remarks
Beryllium oxide	BeO	10.585 eV		
Beryllium sulfide	BeS	>5.5 eV		Absorption
Beryllium selenide	BeSe	5.6 eV		RT
Beryllium telluride	BeTe	$E_{g,\,ind}$ 2.8 eV	$E_{g,\,dir}$ 4.1 eV	300 K optical reflection, ellipsometry

C. Transport Properties
Table 4.1-103.

Table 4.1-103 Electronic transport in beryllium compounds, general description

Crystal		Transport properties
Beryllium oxide	BeO	BeO, synonym beryllia, is an electrical insulator like a ceramic, but conducts heat like a metal. The electrical resistivity is larger than $10^{16}\,\Omega\,\text{cm}$
Beryllium sulfide	BeS	The electrical conductivity of polycrystalline BeS disks was found to be p-type. This is attributed to Be vacancies or S interstitials
Beryllium telluride	BeTe	BeTe, like ZnTe, tends towards p-type conductivity. Using plasma-assisted nitrogen doping during molecular-beam epitaxy, very high p-type concentrations above $10^{20}\,\text{cm}^{-3}$ could be achieved; n-type doping of BeTe has not yet been reported

D. Electromagnetic and Optical Properties
Table 4.1-104.

Table 4.1-104 Dielectric constants ε of beryllium compounds

Crystal		Temperature (K)	$\varepsilon(0)$		$\varepsilon(\infty)$	Method	Remarks
Beryllium oxide	BeO	300	$\varepsilon_{\text{parall}}$ 7.65	$\varepsilon_{\text{perpend}}$ 6.94		Reflectance	Crystalline BeO[a]
Beryllium selenide	BeSe	300					[b]
Beryllium telluride	BeTe	RT			7	Infrared reflection	[c]

[a] BeO is transparent up to 9.4 eV (vacuum UV) and quite resistant to UV damage.
[b] The interest in BeSe originates from the fact that ternary BeZnSe and quaternary BeMgZnSe alloys can be lattice-matched to silicon substrates and can also be used as base materials for blue–green-emitting ZnSe-based laser diodes grown on GaAs. The ternary alloy has a band gap of approximately 4 eV if lattice-matched to silicon.
[c] BeTe is used in efficient pseudomorphic p-contacts for ZnSe laser diodes.

4.1.3.2 Magnesium Compounds

A. Crystal Structure, Mechanical and Thermal Properties
Tables 4.1-105 – 4.1-111.

Table 4.1-105 Crystal structures of magnesium compounds

Crystal		Space group		Crystal system	Structure type	Figure
Magnesium oxide	MgO	O_h^5	$Fm3m$	fcc	NaCl	Fig. 4.1-138
Magnesium sulfide	MgS	O_h^5	$Fm3m$	fcc	NaCl	Fig. 4.1-138
Magnesium selenide	MgSe	O_h^5	$Fm3m$	fcc	NaCl	Fig. 4.1-138
Magnesium telluride	MgTe	C_{6v}^4	$P6_3mc$	hex	Wurtzite	Fig. 4.1-130

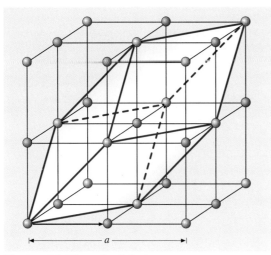

Fig. 4.1-138 The rock salt lattice

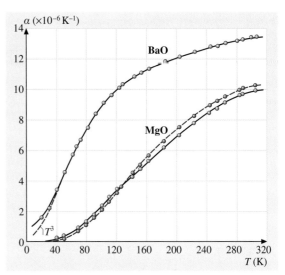

Fig. 4.1-139 MgO, BaO. Linear thermal expansion coefficient α vs. temperature. *Open circles*, [1.120]; *filled circles*, [1.121]

Table 4.1-106 Lattice parameters of magnesium compounds

Crystal		Lattice parameters (nm)	Temperature (K)	Remarks
Magnesium oxide	MgO	$a = 0.4216$	298	X-ray diffraction, powder samples
Magnesium sulfide	MgS	$a = 0.5203$	300	X-ray diffraction
Magnesium selenide	MgSe	$a = 0.546$	300	X-ray diffraction
Magnesium telluride	MgTe	$a = 0.45303$	300	
		$c = 0.74056$		

Table 4.1-107 Densities of magnesium compounds

Crystal		Density (g/cm³)	Temperature (K)
Magnesium oxide	MgO	3.576	298
Magnesium sulfide	MgS	2.86	
Magnesium selenide	MgSe	4.21	
Magnesium telluride	MgTe		

Table 4.1-108 Melting point T_m of magnesium oxide

Crystal		Melting point T_m
Magnesium oxide	MgO	2827 °C

Table 4.1-109 Elastic constants c_{ik} (in 10^{12} dyn/cm²) of magnesium oxide

Crystal		c_{11}	c_{12}	c_{44}	Temperature (K)	Remarks
Magnesium oxide	MgO	2.971 (6)	0.965 (7)	1.557 (2)	293	Single crystals

Table 4.1-110 Linear thermal expansion coefficient α of magnesium oxide

Crystal		Expansion coefficient α (K⁻¹)	Temperature (K)	Remarks
Magnesium oxide	MgO	9.84×10^{-6}	283	Capacitance cell [a]

[a] See Fig. 4.1-139 for the temperature dependence.

Table 4.1-111 Phonon frequencies/wavenumbers of magnesium compounds

Crystal		ν_{TO} (THz)	ν_{LO} (THz)	$\tilde{\nu}_{TO}$ (cm^{-1})	$\tilde{\nu}_{LO}$ (cm^{-1})	Temperature (K)	Symmetry point	Remarks
Magnesium oxide	MgO	12.05	21.52			300	Γ	Thin films, polarized reflectance
Magnesium telluride	MgTe			235	292	300		Raman spectroscopy

B. Electronic Properties
Tables 4.1-112 – 4.1-114.

Table 4.1-112 First Brillouin zones of magnesium compounds

Crystal		Figure
Magnesium oxide	MgO	Fig. 4.1-133
Magnesium sulfide	MgS	Fig. 4.1-133
Magnesium selenide	MgSe	Fig. 4.1-133
Magnesium telluride	MgTe	Fig. 4.1-132

Table 4.1-113 Band structures of magnesium compounds

Crystal		Figure
Magnesium oxide	MgO	Fig. 4.1-140 [a]
Magnesium sulfide	MgS	Fig. 4.1-141
Magnesium selenide	MgSe	Fig. 4.1-142
Magnesium telluride	MgTe	Fig. 4.1-143

[a] Disagreement exists about the position of the minimum of the lowest conduction band. Some authors place the minimum at Γ_1 and anticipate a direct gap; others claim that the minimum is at X_3, indicating an indirect gap.

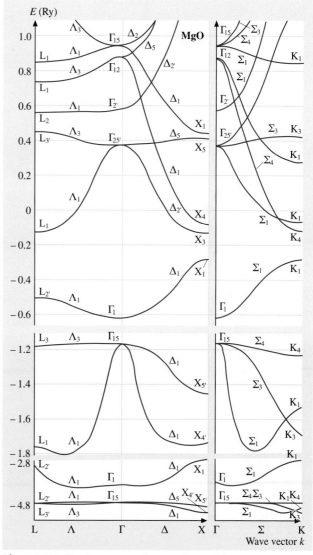

Fig. 4.1-140 Band structure of magnesium oxide

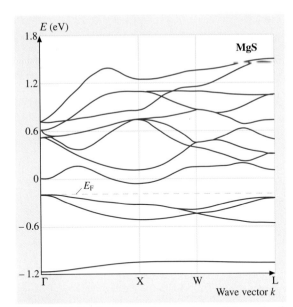

Fig. 4.1-141 Band structure of magnesium sulfide, rock salt structure

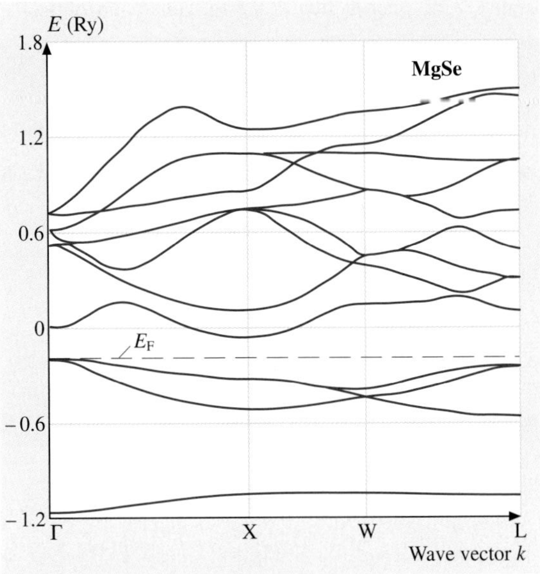

Fig. 4.1-142 Band structure of magnesium selenide, rock salt structure

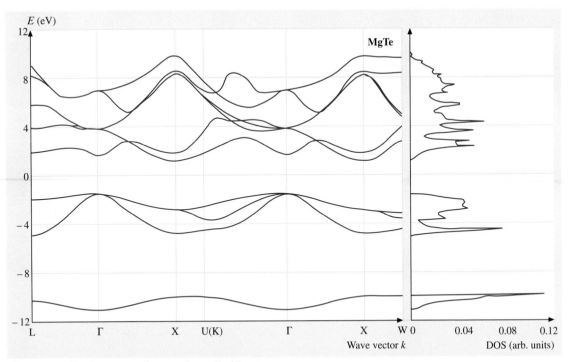

Fig. 4.1-143 Band structure of magnesium telluride

Table 4.1-114 Energy gaps of magnesium compounds

Crystal		Band gap (eV)		Temperature (K)	Remarks
Magnesium oxide	MgO	$E_{gx, dir}$	7.672 (10)	85	Exciton states
		$E_{g, theor}$	7.9 (7)	0	Temperature dependence of microwave conductivity
Magnesium sulfide	MgS	4.5		77	Zinc blende structure
Magnesium selenide	MgSe	4.05		300	Zinc blende structure
Magnesium telluride	MgTe	3.49		RT	Epitaxial films, photoluminescence, zinc blende structure

C. Transport Properties

Electronic Transport, General Description. Magnesium Oxide (MgO). Electrical transport measurements on alkaline earth oxides encounter several difficulties, such as high resistance at low temperatures, a strong influence of surface layers, and high-temperature thermionic emission. The partly contradictory results depend considerably on the purity and nature of the samples (pressed porous powders, sintered samples, polycrystals, and single crystals) and on the experimental conditions.

As in several semiconducting oxides, the conductivity and the type of conduction are determined by the oxygen partial pressure of the ambient atmosphere. In high-purity MgO crystals, temperatures below 1600 K favor ionic conduction, whereas at temperatures above 2000 K, electronic conduction is increasingly predominant. At low oxygen pressures ($< 10^{-3}$ mbar) the conduction is n-type; at high pressures ($> 10^{-1}$ mbar) ionic and electronic hole conduction seem to be superimposed.

See Fig. 4.1-144 for the temperature dependence of the electrical conductivity of MgO and Fig. 4.1-145 for the thermal conductivity of MgO.

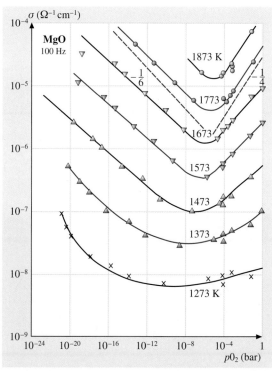

Fig. 4.1-144 MgO. Electrical conductivity σ of a high-purity single crystal at various temperatures vs. oxygen partial pressure p_{O_2} [1.122]. The numbers between the curves for 1673 K and 1773 K indicate the slopes of the *dashed* straight lines

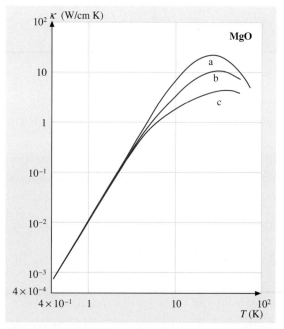

Fig. 4.1-145 MgO. Thermal conductivity κ vs. temperature [1.123]. *Curve* a, high-purity crystal; b, after additive coloration in Mg vapor; c, after subsequent bleaching with UV light

D. Electromagnetic and Optical Properties

Table 4.1-115.

Table 4.1-115 Dielectric constant ε of magnesium compounds

Crystal		$\varepsilon(0)$	$\varepsilon(\infty)$	Temperature (K)	Remarks
Magnesium oxide	MgO	9.830	2.944	300	Single crystals
Magnesium sulfide	MgS				
Magnesium selenide	MgSe		3.8	RT	Zinc blende structure, Raman scattering
Magnesium telluride	MgTe				

4.1.3.3 Oxides of Ca, Sr, and Ba

A. Crystal Structure, Mechanical and Thermal Properties

Tables 4.1-116 – 4.1-123.

Table 4.1-116 Crystal structures of oxides of Ca, Sr, and Ba

Crystal		Space group		Crystal system	Structure type	Figure
Calcium oxide	CaO	O_h^5	$Fm3m$	fcc	NaCl	Fig. 4.1-138
Strontium oxide	SrO	O_h^5	$Fm3m$	fcc	NaCl	Fig. 4.1-138
Barium oxide	BaO	O_h^5	$Fm3m$	fcc	NaCl	Fig. 4.1-138

Table 4.1-117 Lattice parameters of oxides of Ca, Sr, and Ba

Crystal		Lattice parameters (nm)	Temperature (K)	Remarks
Calcium oxide	CaO	$a = 0.48110\,(2)$	297	Single crystal, X-ray diffraction
Strontium oxide	SrO	$a = 0.5159$		Semiempirical calculation
Barium oxide	BaO	$a = 0.55363\,(8)$	300	Powder samples, negligible oxygen pressure

Table 4.1-118 Densities of oxides of Ca, Sr, and Ba

Crystal		Density (g/cm³)
Calcium oxide	CaO	3.335
Strontium oxide	SrO	4.75
Barium oxide	BaO	5.68

Table 4.1-119 Melting point T_m of oxides of Ca, Sr, and Ba

Crystal		Melting point T_m (K)
Calcium oxide	CaO	2900
Strontium oxide	SrO	2930 (30)
Barium oxide	BaO	2290 (30)

Table 4.1-120 Elastic constants c_{ik} of oxides of Ca, Sr, and Ba

Crystal		c_{11} (GPa)	c_{12} (GPa)	c_{44} (GPa)	Temperature (K)	Remarks
Calcium oxide	CaO	221.89 (60)	57.81 (66)	80.32 (10)	298	Calculated from sound velocities
Strontium oxide	SrO	175.47 (37)	49.08 (35)	55.87 (13)	298	
Barium oxide	BaO	126.14 (66)	50.03 (60)	33.68 (30)	298	

Table 4.1-121 Linear thermal expansion coefficient α of oxides of Ca, Sr, and Ba

Crystal		Expansion coefficient α (K^{-1})	Temperature (K)	Remarks
Calcium oxide	CaO	15.2×10^{-6}		Single crystal, X-ray diffraction
Strontium oxide	SrO	13.72×10^{-6}	293–593	Mean value, X-ray diffraction
Barium oxide	BaO	$12.8\,(9) \times 10^{-6}$	200–400	Powder samples, X-ray diffraction

Table 4.1-122 Heat capacity and Debye temperature of oxides of Ca, Sr, and Ba

Crystal		Heat capacity c_p (J/mol K)	Debye temperature Θ_D (K)	Remarks
Calcium oxide	CaO	14.70 at 100 K 42.21 at 300 K	605 (2)	$\Theta_D(0)$ from heat capacity at low temperatures
Strontium oxide	SrO		446 (5)	Θ_D from heat capacity at very high temperatures
Barium oxide	BaO	47.28 at 298 K	370 (4)	Θ_D from heat capacity at very high temperatures

Table 4.1-123 Phonon frequencies at symmetry points for oxides of Ca, Sr, and Ba. Calcium oxide (CaO), 293 K; strontium oxide (SrO), 300 K; barium oxide (BaO), 300 K. All data from inelastic thermal-neutron scattering

CaO			SrO			BaO		
Mode	Symmetry point	Frequency (THz)	Mode	Symmetry point	Frequency (THz)	Mode	Symmetry point	Frequency (THz)
$\tilde{\nu}_{LO}$	$q/q_{max} = 0.2$	16.40 (25)	$\tilde{\nu}_{TO}$	Γ	7.02 (10)	$\tilde{\nu}_{TO}$	Γ	4.32 (6)
$\tilde{\nu}_{TO}$	Γ	8.8 (4)	$\tilde{\nu}_{LO}$	Γ	14.47 (15)	$\tilde{\nu}_{LO}$	Γ	13.02 (3)
$\tilde{\nu}_{LO}$	X	12.00 (22)	$\tilde{\nu}_{TA}$	L	3.33 (5)	$\tilde{\nu}_{TA}$	L	1.99 (7)
$\tilde{\nu}_{TO}$	X	9.4 (5)	$\tilde{\nu}_{LA}$	L	7.14 (10)	$\tilde{\nu}_{LA}$	L	4.88 (27)
$\tilde{\nu}_{TA}$	X	6.48 (11)	$\tilde{\nu}_{TO}$	L	8.35 (12)	$\tilde{\nu}_{TO}$	L	6.58 (9)
			$\tilde{\nu}_{LO}$	L	12.94 (24)			

B. Electronic Properties
Tables 4.1-124 – 4.1-127.

Table 4.1-124 First Brillouin zones of oxides of Ca, Sr, and Ba

Crystal		Figure
Calcium oxide	CaO	Fig. 4.1-133
Strontium oxide	SrO	Fig. 4.1-133
Barium oxide	BaO	Fig. 4.1-133

Table 4.1-125 Energy gaps of oxides of Ca, Sr, and Ba

Crystal		Band gap (eV)		Temperature (K)	Remarks
Calcium oxide	CaO	$E_{g, dir}$	6.931 (6)	85	Exciton states
		$E_{g, theor}$	7.8	0	
Strontium oxide	SrO	$E_{g, opt}$	5.22	300	
		$E_{g, theor}$	6.4	0	Estimated from temperature dependence of conductivity
Barium oxide	BaO	$E_{g, theor}$	4.4	0	Estimated from temperature dependence of conductivity

Table 4.1-126 Effective masses of electrons for oxides of Ca, Sr, and Ba (in units of the electron mass m_0)

Crystal		Quantity	Mass	Temperature (K)	Remarks
Calcium oxide	CaO	m_n	0.50	700	From Hall mobility
		$m_{n, polaron}$	0.68	700–800	Polaron mass, Hall mobility
Strontium oxide	SrO	m_n	0.54	600–700	From Hall mobility
		$m_{n, polaron}$	0.77		Polaron mass, Hall mobility
Barium oxide	BaO	m_n	0.59	600	From Hall mobility
		$m_{n, polaron}$	0.86	500–700	Polaron mass, Hall mobility

Table 4.1-127 Band structure of calcium oxide

Crystal	Figure	Remarks
Calcium oxide CaO	Fig. 4.1-146	a

[a] Some controversy exists in the literature about the positions of the lowest conduction band minima. One result shows the lowest minimum at Γ_1 and the minimum of the d states at X_3, below some s states but above Γ_1. Other results claim the symmetry point of the conduction band bottom is at X_3 instead of Γ_1, whereas according to other authors Γ_1 and X_3 are nearly degenerate.

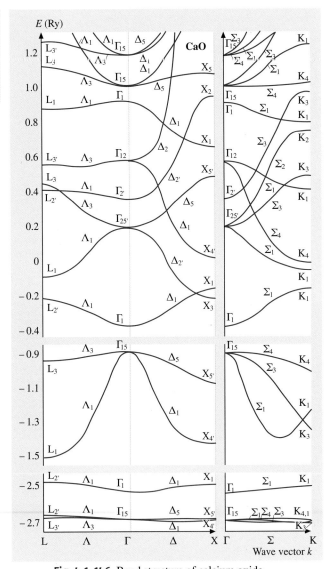

Fig. 4.1-146 Band structure of calcium oxide

C. Transport Properties
Tables 4.1-128 – 4.1-130.

Table 4.1-128 Electronic transport in oxides of Ca, Sr, and Ba, general description

Crystal		Transport properties
Calcium oxide	CaO	For general remarks, see the comments on the transport properties of MgO on p. 659. The transport of charge carriers in CaO has been discussed in various models, e.g. polaron band conduction and hopping conduction through compensated semiconducting samples, and also a considerable contribution from ionic conductivity has been taken into account. From measurements of the oxygen pressure dependence of the conductivity of undoped CaO, it is concluded that n-type conduction occurs at low pressures ($< 10^{-2} - 10^{-4}$ mbar) and that p-type conduction occurs at higher pressures ($> 10^{-2} - 1$ mbar). In some cases this conclusion is confirmed by Hall and thermoelectric measurements. For intermediate pressures, it is supposed that the lattice disorder (presumably of Schottky type) gives rise to predominantly ionic conduction
Strontium oxide	SrO	The transport of charge carriers has been discussed in various models as in the case of CaO (see above). It is supposed that the dominating scattering mechanism is optical-mode scattering
Barium oxide	BaO	The electrical transport has been discussed mainly on the assumption that it is due to electronic conduction. The ionic contribution is small. Polaron conduction and predominant optical-mode scattering are assumed

Table 4.1-129 Electrical conductivity σ and electron mobility μ_n of oxides of Ca, Sr, and Ba

Crystal		Conductivity	Electron mobility			Remarks
			μ_n (cm^2/V s)	at temperature (K)	at O$_2$ pressure (mbar)	
Calcium oxide	CaO		8	700	$< 10^{-6}$	Single crystals, Hall mobility
Strontium oxide	SrO		5	700	$< 10^{-6}$	Single crystals, Hall mobility
Barium oxide	BaO	Fig. 4.1-148	5	600	$< 10^{-6}$	Polycrystals, Hall mobility

Table 4.1-130 Thermal conductivity κ of oxides of Ca, Sr, and Ba

Crystal		Temperature (K)	κ (W/cm K)	Remarks
Calcium oxide	CaO	300	0.3	See Fig. 4.1-147 for temperature dependence
Strontium oxide	SrO	300	0.1	See Fig. 4.1-147 for temperature dependence
Barium oxide	BaO	300	0.03	See Fig. 4.1-147 for temperature dependence

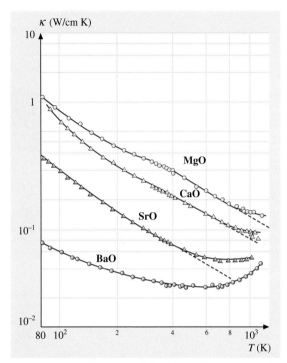

Fig. 4.1-147 MgO, CaO, SrO, BaO. Temperature dependence of the thermal conductivity κ of single crystals [1.124]. *Dashed lines*, $\kappa \propto T^{-1}$

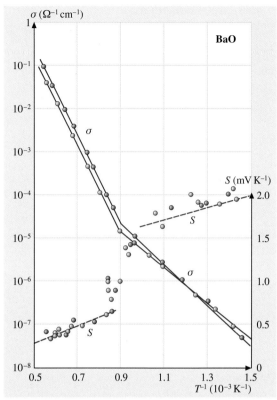

Fig. 4.1-148 BaO. Electrical conductivity σ (*solid lines*) and thermoelectric power S (*dashed lines*) of two crystals vs. $1/T$ [1.125]. $p_{\mathrm{O}_2} = 3 \times 10^{-5}$ mbar. n-type conduction at temperatures below the discontinuity

D. Electromagnetic and Optical Properties
Table 4.1-131.

Table 4.1-131 Dielectric constant ε of oxides of Ca, Sr, and Ba

Crystal		Temperature (K)	$\varepsilon(0)$	$\varepsilon(\infty)$	Remarks
Calcium oxide	CaO	273	12.01 (10)		Capacitance measurement at 1 kHz and 10 kHz
				3.27	Compilation
Strontium oxide	SrO				
Barium oxide	BaO	296	3.903–4.197		Calculated from refractive-index values in the spectral range 624–436 nm

4.1.3.4 Zinc Compounds

A. Crystal Structure, Mechanical and Thermal Properties
Tables 4.1-132 – 4.1-139.

Table 4.1-132 Crystal structures of zinc compounds

Crystal		Space group		Crystal system	Structure type	Figure
Zinc oxide	ZnO	C_{6v}^4	$P6_3mc$	hex	Wurtzite	Fig. 4.1-130
Zinc sulfide	ZnS	T_d^2	$F\bar{4}3m$	fcc	Zinc blende	Fig. 4.1-131[a]
Zinc selenide	ZnSe	T_d^2	$F\bar{4}3m$	fcc	Zinc blende	Fig. 4.1-131
Zinc telluride	ZnTe	T_d^2	$F\bar{4}3m$	fcc	Zinc blende	Fig. 4.1-131

[a] At room temperature, ZnS bulk material consists predominantly of the cubic phase, often with hexagonal inclusions, leading to polytypic material. Epitaxial ZnS has mostly been grown on GaAs and thus has pseudomorphically assumed the cubic structure of the substrate material. In this section, all data refer to the cubic modification unless explicitly stated otherwise.

Table 4.1-133 Lattice parameters of zinc compounds

Crystal		Lattice parameters (nm)	Temperature	Remarks
Zinc oxide	ZnO	$a = 0.3249\,(6)$		X-ray diffraction
		$c = 0.52042\,(20)$		
Zinc sulfide	ZnS	$a = 0.54053$	RT	X-ray diffraction
Zinc selenide	ZnSe	$a = 0.5667\,(4)$	300 K	X-ray diffraction
Zinc telluride	ZnTe	$a = 0.60882$	RT	X-ray diffraction

Table 4.1-134 Densities of zinc compounds

Crystal		Density (g/cm^3)	Temperature	Remarks
Zinc oxide	ZnO	5.67526 (19)	293 K	Hydrostatic weighing
Zinc sulfide	ZnS	Cubic: 4.088	RT	
		Hexagonal: 4.087	RT	
Zinc selenide	ZnSe	5.266		Shock wave experiment
Zinc telluride	ZnTe	5.636	298 K	

Table 4.1-135 Elastic constants c_{ik} (in GPa) of zinc compounds

Crystal		c_{11} (GPa)	c_{13} (GPa)	c_{33} (GPa)	c_{44} (GPa)	c_{66} (GPa)	Remarks
Zinc oxide	ZnO	206 (4)	118 (10)	211 (4)	44.3 (10)	44.0 (10)	ZnO film on Si substrate, Brillouin scattering

Table 4.1-135 Elastic constants c_{ik} (in GPa) of zinc compounds, cont.

Crystal		c_{11} (GPa)	c_{12} (GPa)	c_{44} (GPa)	Temperature	Remarks
Zinc sulfide	ZnS	103.2 (5)	64.6 (5)	46.2 (4)	293 K	Resonance method
Zinc selenide	ZnSe	90.3 (19)	53.6 (23)	39.4 (12)		Brillouin scattering
Zinc telluride	ZnTe	72.2 (2)	40.9 (6)	30.8 (3)	RT	Brillouin scattering

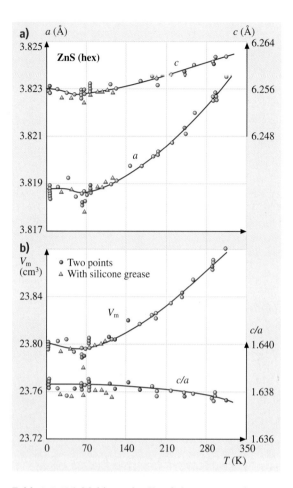

Fig. 4.1-149a,b ZnS, hexagonal. (**a**) Lattice parameters a and c vs. temperature; (**b**) molar volume and c/a ratio vs. temperature [1.126]

Table 4.1-136 Melting point T_m of zinc compounds

Crystal		Melting point T_m (K)	Remarks
Zinc oxide	ZnO	2242 (5)	
Zinc sulfide	ZnS	1991	Under normal pressure, ZnS sublimes before melting
Zinc selenide	ZnSe	1799	
Zinc telluride	ZnTe	1563 (8)	

Table 4.1-137 Linear thermal expansion coefficient α of zinc compounds

Crystal		Expansion coefficient α (K^{-1})	Temperature (K)	Remarks
Zinc oxide	ZnO	$\alpha_{\text{parall } c}$: 2.92×10^{-6}	399	Interferometric and capacitance methods
		$\alpha_{\text{perpend } c}$: 4.75×10^{-6}	300	
Zinc sulfide	ZnS	See Fig. 4.1-149		
Zinc selenide	ZnSe	7.4×10^{-6}	300	Polycrystal, capacitance dilatometer
Zinc telluride	ZnTe	0.083×10^{-6}	300	

Table 4.1-138 Heat capacities and Debye temperatures of zinc compounds

Crystal		Heat capacity c_p (J/mol K)	Debye temperature Θ_D (K)	Temperature (K)	Remarks
Zinc oxide	ZnO		440 (25)	300	Calorimetric data
Zinc sulfide	ZnS	45.358	352	77	Zinc blende lattice, Θ_D calculated from c_p
		45.882	351	298	Wurtzite lattice; Θ_D calculated from c_p
Zinc selenide	ZnSe	51.88	339 (2)	298	
Zinc telluride	ZnTe		180 (6)	93–298	Θ_D from X-ray intensities from powders

Table 4.1-139 Phonon frequencies/wavenumbers at symmetry points for zinc compounds. Zinc oxide (ZnO), fundamental optical modes, $T = 300$ K, from Raman spectroscopy; zinc sulfide (ZnS), $T = 300$ K; zinc selenide (ZnSe), from Raman spectroscopy and luminescence; zinc telluride (ZnTe), RT, from neutron scattering

ZnO Mode	Symmetry point	Wavenumber (cm^{-1})	ZnS Mode	Symmetry point	Wavenumber (cm^{-1})
$\tilde{\nu}$	E_2	101	$\tilde{\nu}_{LO}$	Γ	350
$\tilde{\nu}$	E_2	437	$\tilde{\nu}_{TO}$	Γ	274
$\tilde{\nu}_{TO}$	$E_{1, \text{perp } c}$	407	$\tilde{\nu}_{LO}$	X	332 (1)
$\tilde{\nu}_{TO}$	$A_{1, \text{parall } c}$	380	$\tilde{\nu}_{TO}$	X	318 (1)
$\tilde{\nu}_{LO}$	$E_{1, \text{perp } c}$	583	$\tilde{\nu}_{LA}$	X	212 (1)
$\tilde{\nu}_{LO}$	$A_{1, \text{parall } c}$	574	$\tilde{\nu}_{TA}$	X	88 (1)
			$\tilde{\nu}_{LO}$	L	334 (1)
			$\tilde{\nu}_{TO}$	L	298 (1)
			$\tilde{\nu}_{LA}$	L	192 (1)
			$\tilde{\nu}_{TA}$	L	72 (1)

ZnSe Mode	Symmetry point	Phonon energy (meV)	ZnTe Mode	Symmetry point	Frequency (THz)
$h\nu_{LO}$	Γ_1	30.99	ν_{LO}	Γ	6.20 (5)
$h\nu_{TO}$	Γ_{15}	25.17	ν_{TO}	Γ	5.30 (7)
$h\nu_{LA}$	Γ	19.8	ν_{LO}	X	5.51 (10)
$h\nu_{TA}$	Γ	8.0	ν_{TO}	X	5.21 (10)
$h\nu_{LO}$	X	27.64	ν_{LA}	X	4.29 (5)
$h\nu_{TO}$		25.54	ν_{TA}	X	1.62 (2)
$h\nu_{LA}$		23.55	ν_{LO}	L	5.39 (5)
$h\nu_{LO}$	L	27.77	ν_{TO}	L	5.20 (9)
$h\nu_{TO}$		25.54	ν_{LA}	L	4.06 (5)
$h\nu$	W_3	24.9	ν_{TA}	L	1.25 (2)
$h\nu$	W_1	18.59			
$h\nu$	W_2'	11.53			
$h\nu$	W_2''	26.53			
$h\nu$	W_4'	14.26			
$h\nu$	W_4''	26.41			

B. Electronic Properties
Tables 4.1-140 – 4.1-145.

Band Structures of Zinc Compounds

Zinc oxide (ZnO) (Fig. 4.1-150)
The topmost valence band ($\Gamma_5 + \Gamma_1$) is split owing to crystal-field splitting and spin–orbit coupling into three spin-degenerate states (Γ_7, Γ_9, Γ_7). The exciton states formed with holes in these valence band states are referred to as the A, B, and C excitons, respectively. Owing to the negative spin–orbit coupling (there is a contribution from Zn 3d states), the positions of the edges of the two highest valence band states are reversed as compared with other II–VI compounds with the wurtzite structure.

The conduction band edge, originating from the 4s states of Zn, possesses Γ_7 symmetry.

Zinc sulfide ZnS (Fig. 4.1-150)
Cubic ZnS is a direct-gap semiconductor with the smallest energy gap at the center of the Brillouin zone (Γ). When spin–orbit splitting is taken into account, the topmost valence band state Γ_{15v} splits into Γ_{8v} and Γ_{7v}; further splitting into A, B, and C levels is caused by the crystal field.

Table 4.1-140 First Brillouin zones of zinc compounds

Crystal		Figure
Zinc oxide	ZnO	Fig. 4.1-132
Zinc sulfide	ZnS	Fig. 4.1-133
Zinc selenide	ZnSe	Fig. 4.1-133
Zinc telluride	ZnTe	Fig. 4.1-133

Zinc selenide ZnSe (Fig. 4.1-151)
Zinc selenide is a direct-gap semiconductor with the smallest energy gap at the center of the Brillouin zone (Γ). The topmost valence band (Γ_{15}) is split owing to spin–orbit coupling into a fourfold (Γ_8) and a twofold (Γ_7) state.

Zinc telluride ZnTe (Fig. 4.1-152)
Zinc telluride is a direct-gap semiconductor with the smallest energy gap at the center of the Brillouin zone (Γ). The topmost valence band (Γ_{15}) is split owing to spin–orbit coupling into a fourfold (Γ_8) and a twofold (Γ_7) state.

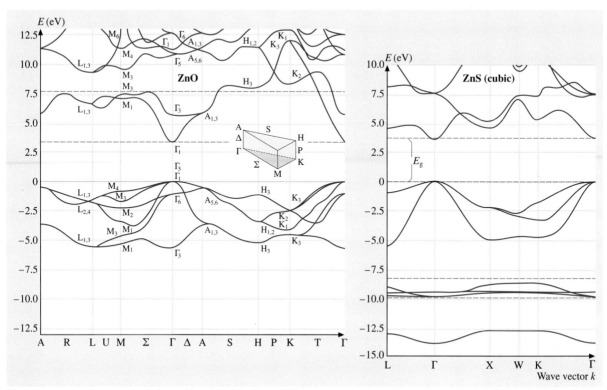

Fig. 4.1-150 Band structure of zinc oxide and cubic zinc sulfide

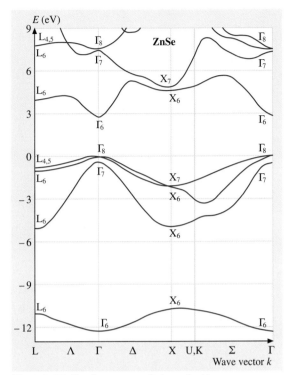

Fig. 4.1-151 Band structure of zinc selenide

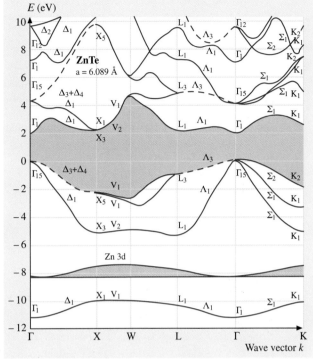

Fig. 4.1-152 Band structure of zinc telluride

Table 4.1-141 Energy gaps of zinc compounds

Crystal		Band gap (eV)		Between bands	Temperature (K)	Remarks
Zinc oxide	ZnO	E_g^A	3.4410 (1)		6	Two-photon absorption,
		E_g^B	3.4434 (1)			distance from A, B, and C valence bands
		E_g^C	3.4817 (2)			to conduction band
Zinc sulfide	ZnS	$E_{g,\,dir}$	3.723 (1)	Γ_{15v} and Γ_{1c}	300	Luminescence;
		$E_{g,\,dir}$	3.78	Γ_{8v} and Γ_{6c}	295	For temperature dependence
		$E_{g,\,dir}$	3.76	Γ_{7v} and Γ_{6c}	298	see Fig. 4.1-162
Zinc selenide	ZnSe	$E_{g,\,dir}$	2.8222 (1)	Γ_{8v} and Γ_{6c}	6	Two-photon spectroscopy
Zinc telluride	ZnTe	$E_{g,\,dir}$	2.3945	Γ_{8v} and Γ_{6c}	< 2	Transmission spectroscopy
			2.35	Γ_{8v} and Γ_{6c}	300	Reflectivity

Table 4.1-142 Exciton binding energies of zinc compounds

Crystal		Binding energy (meV)		Temperature (K)	Remarks
Zinc oxide	ZnO	$E_b(A)$	63.1	6	Two-photon absorption,
		$E_b(B)$	50.4		
		$E_b(C)$	48.9		
Zinc sulfide	ZnS	$E_b(1S)$	38 (1)	10	Absorption,
		$E_b(2S)$	10 (2)		Heavy-hole exciton
Zinc selenide	ZnSe	$E_b(1S)$	20.8	1.8	Absorption, strained layers
		$E_b(2P_{1/2})$	4.19	1.6	Two-photon absorption
		$E_b(2P_{5/2},\Gamma_7)$	4.80		
		$E_b(2P_{5/2},\Gamma_8)$	5.15		

Table 4.1-143 Spin–orbit splitting energy Δ_{so} of zinc compounds

Crystal		Splitting energy	Bands	Temperature (K)	Remarks
Zinc oxide	ZnO	−3.5 (2) meV		6	Two-photon absorption
Zinc sulfide	ZnS	64 meV	Γ_{8v} to Γ_{7v}	293	Reflectivity
Zinc selenide	ZnSe	0.42 eV	Γ_{8v} to Γ_{7v}	295	Reflectivity
		0.20 eV	Γ_{4-5v} to Γ_{6v}	300	
Zinc telluride	ZnTe	0.97 eV	Γ_{8v} to Γ_{7v}	80	Reflectivity

Table 4.1-144 Effective masses of electrons (in units of the electron mass m_0) for zinc compounds

Crystal		Quantity	Mass	Temperature (K)	Remarks
Zinc oxide	ZnO	m_n	0.275	6	Cyclotron resonance
		$m_{n, polaron}$	0.3	80	Polaron mass
Zinc sulfide	ZnS	$m_{n, polaron}$	0.22		Polaron mass, cyclotron resonance
Zinc selenide	ZnSe	m_n	0.160 (2)	4.2	Photoluminescence
Zinc telluride	ZnTe	m_n	0.122 (2)	3.5	Cyclotron resonance
		$m_{n, polaron}$	0.124 (2)	1.5	Polaron mass

Table 4.1-145 Effective masses of holes (in units of the electron mass m_0) for zinc compounds

Crystal		Quantity	Mass	Temperature (K)	Remarks
Zinc oxide	ZnO	$m_{p, parall}$	0.59	1.6	Magnetoreflection
		$m_{p, perpend}$	0.59		Polaron mass
Zinc sulfide	ZnS	$m_{p, light}$	0.23		Calculated
		$m_{p, heavy}$	1.76		
Zinc selenide	ZnSe	m_p	0.75	2.1–200	Phonon-assisted exciton absorption
Zinc telluride	ZnTe	m_p	0.6		Estimated from hole mobility

C. Transport Properties
Table 4.1-146.

Electronic Transport, General Description.
Zinc oxide (ZnO)
The electronic conductivity of pure, stoichiometric ZnO is still unknown. The concentration of foreign admixtures in undoped crystals is of the order of 10^{15}–10^{16} cm^{-3}. Since $E_{g, opt} = 3.2$ eV and impurity ionization energies are about 0.01–0.1 eV at temperatures below 900 K, impurity conduction is always observed. At temperatures above 900 K, dissociation of the intrinsic material occurs.

The conductivity depends on the surrounding atmosphere (which my be O$_2$, Zn, or Ar). Nevertheless, it is possible to investigate the influence of intentional admixtures on the conductivity, the charge-carrier concentration, and the mobility as a function of temperature and current direction.

Data for the electrical resistivity vary between values of the order of 10^8–10^9 Ω cm for ultrapure bulk single crystals to values of the order of 10^{-2}–10^{-4} Ω cm for doping concentrations of up to 10^{20} cm^{-3}. Data for the

Table 4.1-146 Thermal conductivity κ of zinc compounds

Crystal		κ (W/cm K)	Temperature (K)	Temperature dependence	Remarks
Zinc oxide	ZnO	0.54	300	Fig. 4.1-154	Steady-state heat flow
Zinc sulfide	ZnS	0.27	300		Pulse method
		3.6	30		
Zinc selenide	ZnSe	0.19	300	Fig. 4.1-155	
Zinc telluride	ZnTe	0.18	300		

electron mobility μ_n at room temperature range between 0.5 and 200 cm^2/V s. Figure 4.1-153 shows the electrical conductivity of a crystal without intentional admixtures.

Zinc sulfide (ZnS)
The (photo)conductivity of ZnS depends strongly on the growth conditions, dopants, doping characteristics, etc. Data for the electrical resistivity vary between 10^{-3} and 10^2 Ω cm depending on sample preparation. Values of the electron mobility around 100 cm^2/V s and 600 cm^2/V s have been found at room temperature, and values around 3000 cm^2/V s have been found at 77 K. A value of 40 cm^2/V s has been given for the hole mobility.

Zinc selenide (ZnSe)
The (photo)conductivity of ZnSe depends strongly on the growth conditions, dopants, doping characteristics, etc. In undoped material, values of the electrical resistivity between 10^5 and a few Ω cm have been found. Values of the electron mobility at room temperature reach up to 400 cm^2/V s; for the hole mobility, $\mu_p = 110$ cm^2/V s has been calculated as an upper limit.

Zinc telluride (ZnTe)
ZnTe is a p-type conductor owing to deviations from stoichiometry when it is not intentionally doped; n-type doping is extremely difficult. The transport properties of ZnTe depend strongly on the growth conditions, dopants, doping characteristics, etc.

Typical data for the electrical resistivity at room temperature are of the order of a few times 10^4 Ω cm; at temperatures around 77 K, values of a few times 10^6 Ω cm have been found. The experimental values of the electron and hole mobilities at room temperature range from 300 to 350 cm^2/V s.

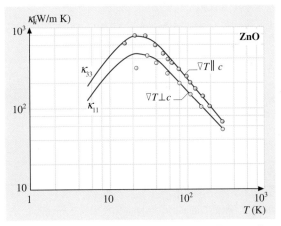

Fig. 4.1-154 ZnO. Thermal conductivity (lattice contribution) vs. temperature. *Solid lines*, calculated values [1.128]

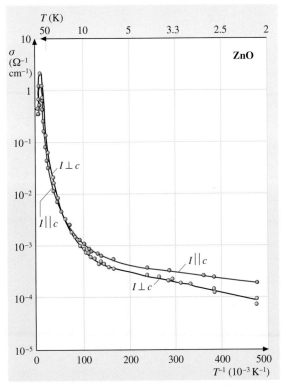

Fig. 4.1-153 ZnO. Electronic conductivity parallel and perpendicular to the c axis vs. temperature. Crystal grown in Al$_2$O$_3$ ceramic without intentional admixtures [1.127]

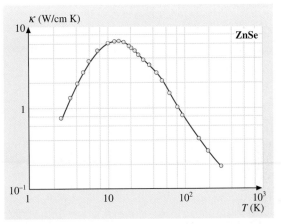

Fig. 4.1-155 ZnSe. Thermal conductivity vs. temperature [1.129]

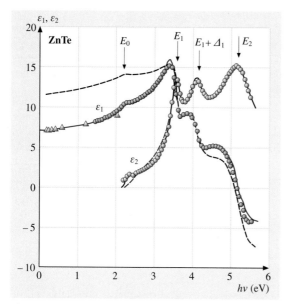

Fig. 4.1-156 ZnTe. Dielectric-function spectrum $\varepsilon(\omega)$, measured by spectroscopic ellipsometry at RT (*filled* and *open circles*); the *solid* and *dashed lines* show calculated values [1.130]. The *triangles* are experimental data from [1.131]

D. Electromagnetic and Optical Properties
Tables 4.1-147 – 4.1-149.

Table 4.1-147 Dielectric constant ε of zinc compounds

Crystal		$\varepsilon(0)$		$\varepsilon(\infty)$		Temperature (K)	Remarks
		$\varepsilon_{\text{parall } c}$	$\varepsilon_{\text{perpend } c}$	$\varepsilon_{\text{parall } c}$	$\varepsilon_{\text{perpend } c}$		
Zinc oxide	ZnO	8.75	7.8	3.75	3.70	300	
Zinc sulfide	ZnS					300	See Fig. 4.1-157 for spectral dependence and Fig. 4.1-158 for the temperature dependence
Zinc selenide	ZnSe	8.6		5.7		300	Ellipsometry
Zinc telluride	ZnTe	10.3				RT	Reflectivity derived from refraction data
				7.28		300	See Fig. 4.1-160 for spectral dependence

Table 4.1-148 Refractive index n of zinc compounds

Crystal		Refractive index n		λ (nm)	Temperature (K)	Spectral dependence
Zinc oxide	ZnO	$n_{\text{parall } c}$	2.151	436	293	Fig. 4.1-161
		$n_{\text{perpend } c}$	2.137			
Zinc sulfide	ZnS				300	
Zinc selenide	ZnSe				300	Fig. 4.1-163
Zinc telluride	ZnTe				300	Fig. 4.1-164

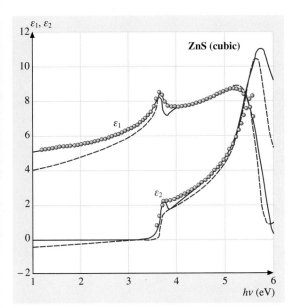

Fig. 4.1-157 ZnS, cubic. Dielectric-function spectrum $\varepsilon(E)$, measured by spectroscopic ellipsometry at 300 K after chemomechanical polishing of the sample (*circles*). The *solid lines* show values calculated from a model dielectric function for interband critical points. The *dashed lines* represent the best-fit standard critical-point lineshapes [1.132]

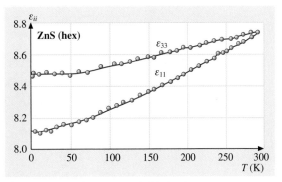

Fig. 4.1-158 ZnS, hexagonal. Dielectric constant vs. temperature [1.133]. At 300 K, the dielectric constant of hexagonal ZnS is isotropic

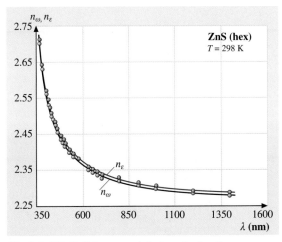

Fig. 4.1-159 ZnS, hexagonal. Index of refraction vs. wavelength. n_ω, ordinary ray; n_ε, extraordinary ray [1.134]

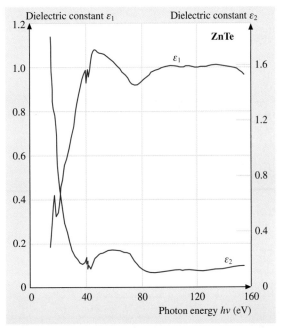

Fig. 4.1-160 ZnTe. Spectral dependence of the real and imaginary parts of the complex dielectric constant, ε_1 and ε_2 [1.135]

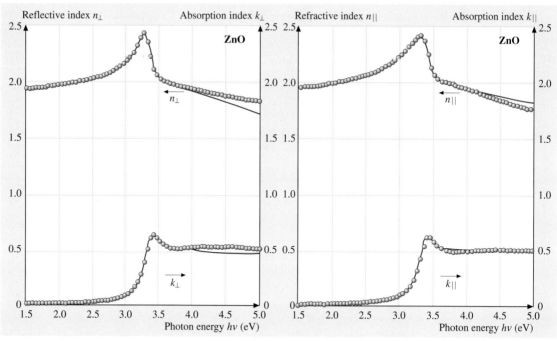

Fig. 4.1-161 ZnO. Numerically calculated spectral dependence of the complex refractive index $n^* = n(E) + ik(E)$ for ZnO (*solid lines*). The *circles* represent experimental data at 300 K [1.136]

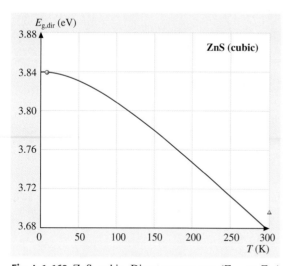

Fig. 4.1-162 ZnS, cubic. Direct energy gap (Γ_{15v} to Γ_{1c}) vs. temperature [1.137]. *Solid line*, theoretical result. *Circle*, experimental data from [1.138]. *Triangle*, experimental data from [1.139]

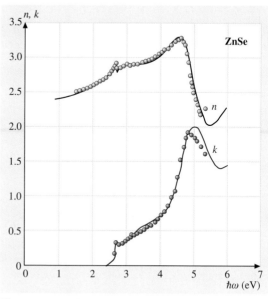

Fig. 4.1-163 ZnSe. Numerically calculated spectral dependence of the real refractive index n and the extinction coefficient k (*solid lines*). The *filled circles* (n) and *open circles* (k) are experimental data [1.140]

Table 4.1-149 Electro-optical constants of zinc compounds. Under the influence of an electric field, the refractive index changes in accordance to the nonlinearity of the dielectric polarization (Pockels effect). Crystals with hexagonal symmetry have three electro-optical constants r_{31}, r_{33}, r_{51}; crystals with cubic symmetry have only one electro-optical constant r_{41}

Crystal		Electro-optical constant (10^{-9} cm/V)		λ (nm)	Temperature (K)	Remarks
Zinc oxide	ZnO	r_{13}, r_{33}	−1.4	400	295	Electrotransmission
		r_{51}	−0.31			
Zinc sulfide	ZnS	r_{41}	0.21	650		Retardation angle
Zinc telluride	ZnTe	r_{41}	−4.71			Calculated

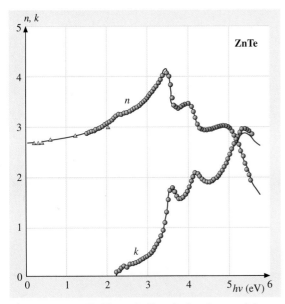

Fig. 4.1-164 ZnTe. Numerically calculated spectral dependence of the complex refractive index (n and k) (*solid lines*). The *circles* are measured ellipsometry data [1.130]. The *triangles* are experimental data taken from [1.131]

4.1.3.5 Cadmium Compounds

A. Crystal Structure, Mechanical and Thermal Properties
Tables 4.1-150 – 4.1-157.

Table 4.1-150 Crystal structures of cadmium compounds

Crystal		Space group		Crystal system	Structure type	Figure
Cadmium oxide	CdO	O_h^5	$Fm3m$	fcc	NaCl	Fig. 4.1-138[a]
Cadmium sulfide	CdS	C_{6v}^4	$P6_3mc$	hex	Wurtzite	Fig. 4.1-130[b]
Cadmium selenide	CdSe	T_d^2	$F\bar{4}3m$	fcc	Zinc blende	Fig. 4.1-131[c]
Cadmium telluride	CdTe	T_d^2	$F\bar{4}3m$	fcc	Zinc blende	Fig. 4.1-131

[a] Cadmium oxide is the only II–VI semiconductor which crystallizes in the NaCl structure at standard pressure.
[b] CdS crystallizes in several different structures. Besides the most common wurtzite structure, it is found also in the zinc blende structure and in the high-pressure NaCl structure. Epitaxial layers of CdS are frequently found in one of the cubic modifications.
[c] Under ambient conditions, CdSe crystallizes in the zinc blende structure. Ab initio calculations show that the energy of the zinc blende structure is lower than that of the wurtzite structure by 1.4 meV/atom. Under increasing pressure, the wurtzite structure transforms to the NaCl structure, the β-tin structure, and as yet unidentified structures. If not stated otherwise, the data in this section concern CdSe in the zinc blende structure.

Table 4.1-151 Lattice parameters of cadmium compounds

Crystal		Lattice parameters (nm)	Remarks
Cadmium oxide	CdO	$a = 0.4689$	
Cadmium sulfide	CdS	$a = 0.41348$	X-ray diffraction
		$c = 0.67490$	
Cadmium selenide	CdSe	$a = 0.6077\,(5)$	Perpendicular to surface, X-ray diffraction
		$a = 0.6078\,(1)$	Parallel to surface
Cadmium telluride	CdTe	$a = 0.646$	X-ray diffraction

Table 4.1-152 Densities of cadmium compounds

Crystal		Density (g/cm³)	Temperature (K)	Remarks
Cadmium oxide	CdO	8.15		X-ray crystal density method
Cadmium sulfide	CdS	4.82	300	
Cadmium selenide	CdSe	5.81	300	
Cadmium telluride	CdTe	5.87	4	

Table 4.1-153 Elastic constants c_{ik} of cadmium compounds

Crystal		c_{11} (GPa)	c_{13} (GPa)	c_{33} (GPa)	c_{44} (GPa)	c_{66} (GPa)	Remarks
Cadmium sulfide	CdS	90.7	51.0	93.8	15.04		
Cadmium selenide	CdSe	74.6	46.1	81.7	13.0	14.3	Ultrasound measurements, RT

Table 4.1-153 Elastic constants c_{ik} of cadmium compounds, cont.

Crystal		c_{11} (GPa)	c_{12} (GPa)	c_{44} (GPa)	Temperature (K)
Cadmium telluride	CdTe	53.8	37.4	20.18	298
		56.2	39.4	20.61	77

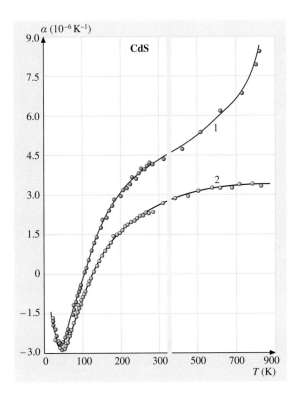

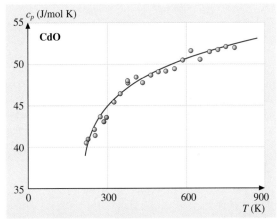

Fig. 4.1-165 CdS, hexagonal. Temperature dependence of the linear thermal expansion coefficients $\alpha_\perp$ (perpendicular to the c axis, *curve 1*) and $\alpha_\parallel$ (parallel to the c axis, *curve 2*), measured by dilatometer experiments [1.141, 142]

Fig. 4.1-166 CdO. Heat capacity c_p vs. temperature. Measurements from two authors [1.143]

Table 4.1-154 Melting point T_m of cadmium compounds

Crystal		Melting point T_m (K)	Remarks
Cadmium oxide	CdO	>1773	
Cadmium sulfide	CdS	1750	Under pressure $p > 3.85$ bar
Cadmium selenide	CdSe	1537	Assessed
Cadmium telluride	CdTe	1366.6 (70)	Mean of experimental values

Table 4.1-155 Linear thermal expansion coefficient α of cadmium compounds

Crystal		Expansion coefficient α (K^{-1})	Temperature (K)	Remarks
Cadmium oxide	CdO	14×10^{-6}	300–770	Dilatometer measurements
Cadmium sulfide	CdS			See Fig. 4.1-165 for temperature dependence
Cadmium selenide	CdSe			
Cadmium telluride	CdTe	4.96×10^{-6}	20–420	

Table 4.1-156 Heat capacity c_p and Debye temperature Θ_D of cadmium compounds

Crystal		Heat capacity c_p (J/mol K)	Temperature (K)	Debye temperature Θ_D (K)	Remarks
Cadmium oxide	CdO	43.639	298.15		See Fig. 4.1-166 for temperature dependence
				255 (6)	Calculated from X-ray diffraction measurements
Cadmium sulfide	CdS	55.10	300		$T = 298$–1678 K
				219.32	From elastic constants
Cadmium selenide	CdSe	49.58	300		Monocrystalline sample, $T = 400$–700 K
				181.7	$T = 0$ K
Cadmium telluride	CdTe	25.63	300		Polycrystalline sample, $T = 300$–530 K
				140 (3)	Single crystal, X-ray diffraction

Table 4.1-157 Phonon frequencies/wavenumbers at symmetry points for cadmium compounds. Cadmium oxide (CdO), Fundamental optical-mode frequencies; cadmium sulfide (CdS), 25 K; cadmium selenide (CdSe), 300 K, from infrared and Raman spectroscopy; cadmium telluride (CdTe), 300 K, from inelastic neutron scattering

CdO					CdS		
Mode	Symmetry point	Frequency (THz)	Temperature (K)		Mode	Symmetry point	Wavenumber (cm^{-1})
$\tilde{\nu}_O$	Γ	7.85	300		$\tilde{\nu}_{LO}$	Γ_5, perpendicular to c	307
$\tilde{\nu}_{AO}$	Γ	15.83	300		$\tilde{\nu}_{LO,1}$	Γ_1, parallel to c	305
$\tilde{\nu}_{LO}$	L	14.4	2		$\tilde{\nu}_{TO,1}$	Γ_5, perpendicular to c	242
$\tilde{\nu}_{LO}$	X	15.6	2		$\tilde{\nu}_{TO,1}$	Γ_1, parallel to c	234
$\tilde{\nu}_{TO}$	X	7.2	2		$\tilde{\nu}_{LO,2}$	Γ_6, perpendicular to c	256
					$\tilde{\nu}_{TO,3}$	Γ_6, perpendicular to c	43

CdSe				CdTe		
Mode	Symmetry point	Wavenumber (cm^{-1})		Mode	Symmetry point	Frequency (THz)
$\tilde{\nu}_{LO,1}$	Γ_1, parallel to c	210		ν_{LO}	Γ	5.08 (10)
$\tilde{\nu}_{TO,1}$	Γ_5, perpendicular to c	169		ν_{TO}	Γ	4.20 (10)
$\tilde{\nu}_{LO,2}$	Γ_5, perpendicular to c	212		ν_{LO}	X	4.44 (6)
$\tilde{\nu}_{TO,2}$	Γ_1, parallel to c	166		ν_{TA}	X	1.05 (3)
$\tilde{\nu}_{TA}$	A_3	43.5		ν_{LO}	L	4.33 (6)
				ν_{TAO}	L	4.33 (7)
				ν_{LA}	L	3.25 (8)
				ν_{TA}	L	0.88 (3)

B. Electronic Properties
Tables 4.1-158 – 4.1-164.

Table 4.1-158 First Brillouin zones of cadmium compounds

Crystal		Figure
Cadmium oxide	CdO	Fig. 4.1-133
Cadmium sulfide	CdS	Fig. 4.1-132
Cadmium selenide	CdSe	Fig. 4.1-133
Cadmium telluride	CdTe	Fig. 4.1-133

Table 4.1-159 Band structures of cadmium compounds

Cadmium oxide (CdO) (Fig. 4.1-167)
The lowest conduction band has its minimum at the Γ_1 point in the center of the Brillouin zone. Within an energy range of a few tenths of an electron volt above its bottom, the band is isotropic but highly nonparabolic. The nonparabolicity is revealed by the strong dependence of the optical electron mass on the electron concentration. The lowest conduction band is separated from the higher bands by a small gap. The valence band consists of several branches, with maxima at L_3 and Σ.

Cadmium sulfide (CdS) (Fig. 4.1-168)
Hexagonal CdS is a direct-gap semiconductor with the smallest energy gap at the center of the Brillouin zone (Γ). The topmost valence band ($\Gamma_5 + \Gamma_1$) is split owing to the crystal field and spin–orbit coupling into three spin-degenerate states. The exciton states formed with holes in these valence band states are referred to as the A, B, and C excitons, respectively.

Cadmium selenide (CdSe)
See Fig. 4.1-170.

Cadmium telluride (CdTe) (Fig. 4.1-169)
Cadmium telluride is a direct-gap semiconductor with the smallest energy gap at the center of the Brillouin zone (Γ). The topmost valence band is split owing to spin–orbit coupling into a fourfold (Γ_8) and a twofold (Γ_7) state.

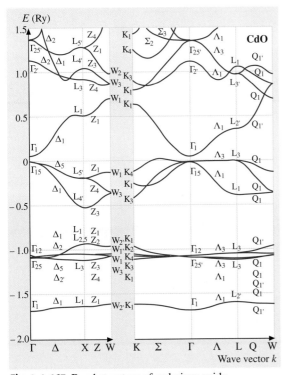

Fig. 4.1-167 Band structure of cadmium oxide

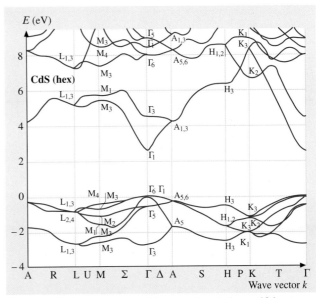

Fig. 4.1-168 Band structure of hexagonal cadmium sulfide

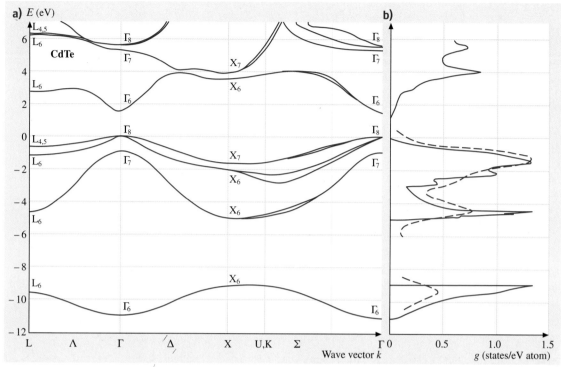

Fig. 4.1-169a,b CdTe. (**a**) Band structure and (**b**) electron density of states from a pseudopotential calculation [1.144]. *Dashed curve*: Experimental data

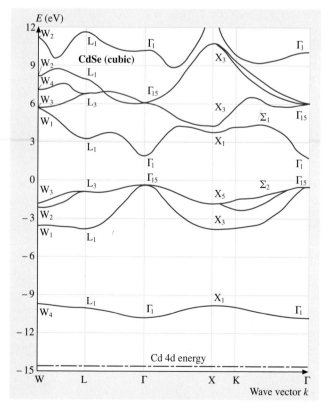

Fig. 4.1-170 Band structure of cubic cadmium selenide

Table 4.1-160 Energy gaps of cadmium compounds (for the letters A, B, C see CdS band structure, Fig. 4.1-168)

Crystal		Band gap (eV)		Between bands	Temperature (K)	Remarks
Cadmium oxide	CdO	$E_{g,\,ind}^{(1)}$	1.09 (5)	Σ_{3v} and Γ_{1c}	100	Thermoreflectance
		$E_{g,\,ind}^{(2)}$	0.84	L_{3v} and Γ_{1c}		
		$E_{g,\,dir}$	2.28	Γ_{15v} and Γ_{1c}		
Cadmium sulfide	CdS	$E_{g,\,dir}^{A}$	2.482	Γ_{9v} and Γ_{7c}	300	Band gaps of the three subbands;
		E_{g}^{B}	2.496	Γ_{7v} and Γ_{7c}		Ellipsometry
		E_{g}^{C}	2.555	Γ_{7v} and Γ_{7c}		
Cadmium selenide	CdSe	$E_{g,\,dir}$	1.74	E_0	300	Ellipsometry
Cadmium telluride	CdTe	$E_{g,\,dir}$	1.475	Γ_{8v} and Γ_{6c}	RT	Photoelectrochemistry

Table 4.1-161 Spin–orbit splitting energy Δ_{so} of cadmium compounds

Crystal		Splitting energy (meV)	Bands	Temperature (K)	Remarks
Cadmium selenide	CdSe	470			Calculation
Cadmium telluride	CdTe	950 (20)	Γ_{7v} to Γ_{78}	300	Photoemission

Table 4.1-162 Effective masses of electrons (in units of the electron mass m_0) for cadmium compounds

Crystal		Quantity	Mass	Temperature (K)	Remarks
Cadmium oxide	CdO	m_n	0.33–0.46	81	Comparison of low-temperature mobilities, $n = 0.4$–6.3×10^{19} cm^{-3}
Cadmium sulfide	CdS	m_n	0.25	300	Thermoelectric power
Cadmium selenide	CdSe	m_n	0.12		Calculation
Cadmium telluride	CdTe	m_n [110]	0.096 (3)	1.5	Cyclotron resonance
		m_n [100]	0.094 (4)	1.8	
		m_n [111]	0.095 (4)	1.8	

Table 4.1-163 Effective masses of holes (in units of the electron mass m_0) for cadmium compounds

Crystal		Quantity	Mass	Temperature (K)	Remarks
Cadmium sulfide	CdS	$m_{p,\,perpend}^{A}$	0.7 (1)	1.6	Exciton magnetoabsorption
		$m_{p,\,parall}^{A}$	5		
Cadmium selenide	CdSe	$m_{p,\,heavy}$ [111]	2.14		Calculation
		$m_{p,\,heavy}$ [100]	0.9		
		$m_{p,\,heavy}$ [110]	1.7		
		$m_{p,\,light}$ [111]	0.16		
		$m_{p,\,light}$ [100]	0.18		
		$m_{p,\,light}$ [110]	0.16		
Cadmium telluride	CdTe	$m_{p,\,heavy}$ [110]	0.81 (5)	1.5	Cyclotron resonance
		$m_{p,\,light}$ [110]	0.12 (2)		

Table 4.1-164 g-factors in the conduction band (g_c) and valence band (g_v) for cadmium compounds

Crystal		Quantity	g-factor	Temperature (K)	Remarks
Cadmium oxide	CdO	g_c	1.806 (5)	77	Electron spin resonance measurements
Cadmium sulfide	CdS	$g_{c,\,perpend}$	1.78 (5)	1.6	Exciton absorption
		$g_{c,\,parall}$	1.72 (10)		(for the letters A, B, C see
		$g_{v,\,parall}^{A}$	1.15 (5)	16	CdS band structure, Fig. 4.1-168)
		$g_{v,\,perpend}^{B}$	0.8	1.8	Lineshape analysis of excitonic polaron reflectivity
Cadmium selenide	CdSe	g	0.23		Calculation
Cadmium telluride	CdTe	g_c	1.652	1.5	Spin-flip Raman measurements

C. Transport Properties

Tables 4.1-165 – 4.1-167.

Table 4.1-165 Electronic transport, general description

Cadmium oxide (CdO)
Owing to a high degree of nonstoichiometry, CdO is generally a highly degenerate n-type semiconductor. The conductivity is related to an excess of Cd in the lattice. Usually the electron concentration is several times 10^{19} cm^{-3}. Values of the electrical resistivity at 300 K range from 1.2 to 1.5×10^{-3} Ω cm.

Cadmium sulfide (CdS)
Electrical transport in CdS is performed by electrons in the Γ_7 conduction band and holes in the Γ_9 valence band. Because of compensation effects during the preparation of the samples and the large ratio $\mu_n/\mu_p > 10$, CdS is normally n-type conducting; p-type conductivity is observed only in a few special cases. The carrier concentration depends on the impurity or defect content and on temperature. Furthermore, the carrier concentration can be changed markedly by illumination: CdS is known to be a model substance for photoconductivity.

Cadmium selenide (CdSe)
Electrical transport in CdSe is governed by electrons in the Γ_7 conduction band and holes in the Γ_9 valence band. CdSe is mainly n-type conducting, owing to compensation effects during the preparation of the samples and to a ratio of the mobilities of electrons and holes of order 10. The carrier concentration depends on the impurity or defect content and on temperature. Furthermore, it depends strongly on irradiation. Photoconductivity in CdSe can be produced by electromagnetic illumination (photon energy > band gap) and by high-energy particles.

Cadmium telluride (CdTe)
CdTe bulk crystals can be doped to become n- or p-type conductive. High-resistivity crystals ($\rho > 10^9$ Ω cm) can be obtained under various growth and doping conditions.

Table 4.1-166 Electron and hole mobilities μ_n and μ_p of cadmium compounds

Crystal		μ_n (cm^2/V s)	μ_p (cm^2/V s)	Temperature (K)	Remarks
Cadmium oxide	CdO	180		300	Hall mobility, cold-pressed CdO, $n = 1.1 \times 10^{19}$ cm^{-3}
Cadmium sulfide	CdS	160		300	In-doped, $n = 5 \times 10^{19}$ cm^{-3}
		>10 000		30–40	Peak mobility, ultrapure crystals
			15	300	Highly resistive crystals
Cadmium selenide	CdSe	660		300	Polycrystal and single-crystal material
		5000		80	Dc Hall effect
		200		800	
			40	300	N-doped epitaxial films
Cadmium telluride	CdTe	<110 000		32	Hall mobilities
			60	300	
			1200	170	

Table 4.1-167 Thermal conductivity κ of cadmium compounds

Crystal		κ (W/cm K)	Temperature (K) dependence	Temperature	Remarks
Cadmium sulfide	CdS	0.20	300	Fig. 4.1-171	
Cadmium selenide	CdSe	0.09	300		
Cadmium telluride	CdTe	0.071	298	Fig. 4.1-172	Single crystal
		10	8		Theoretical maximum for a pure crystal

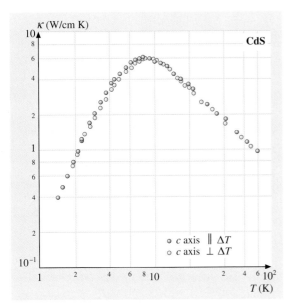

Fig. 4.1-171 CdS. Thermal conductivity vs. temperature for c axis parallel and perpendicular to the heat flow ΔT [1.145, 146]

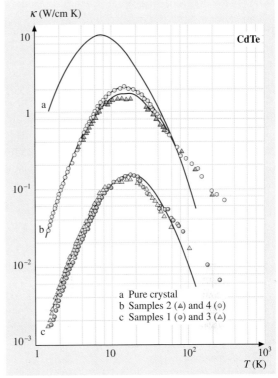

Fig. 4.1-172 CdTe. Thermal conductivity of several different specimens vs. temperature. The results for specimens 1 and 3 have been plotted reduced by a factor of 10. The point defect concentrations of the samples are (1) 3.42×10^{18} cm^{-3}; (2) 3.99×10^{18} cm^{-3}; (3) 3.42×10^{18} cm^{-3}; (4) 2.66×10^{18} cm^{-3} [1.147]

D. Electromagnetic and Optical Properties
Tables 4.1-168 and 4.1-169.

Table 4.1-168 Dielectric constant ε of cadmium compounds

Crystal		$\varepsilon(0)$		$\varepsilon(\infty)$		Temperature (K)	Remarks
Cadmium oxide	CdO		21.9			300	Infrared reflectivity
					2.1	300	Transmittance, absorption
Cadmium sulfide	CdS	$\varepsilon(0)_{\text{perpend}}$	8.28	$\varepsilon(\infty)_{\text{perpend}}$	5.23	300	Spectroscopic ellipsometry
		$\varepsilon(0)_{\text{parall}}$	8.73	$\varepsilon(\infty)_{\text{parall}}$	5.29		
Cadmium selenide	CdSe	$\varepsilon(0)_{\text{perpend}}$	9.29			300	Infrared spectroscopy
			9.15			100	
		$\varepsilon(0)_{\text{parall}}$	10.16			300	
			9.29			100	
Cadmium telluride	CdTe		10.4		7.1 (1)	300	

Table 4.1-169 Refractive index n of cadmium compounds

Crystal		Refractive index n	λ (nm)	Temperature (K)	Spectral dependence	Remarks
Cadmium oxide	CdO	1.95–2.65	600		Fig. 4.1-173	
Cadmium sulfide	CdS	$n_{\text{parall } c}$			Fig. 4.1-174	
		$n_{\text{perpend } c}$				
Cadmium selenide	CdSe				Fig. 4.1-175	
Cadmium telluride	CdTe	2.70	2500	300		Prism refraction

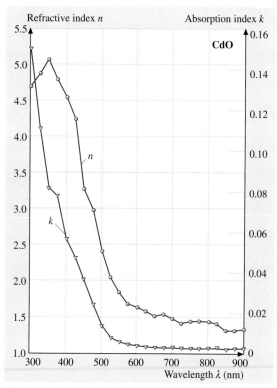

Fig. 4.1-173 CdO. Dispersion of the optical constants n and k of a film [1.148]

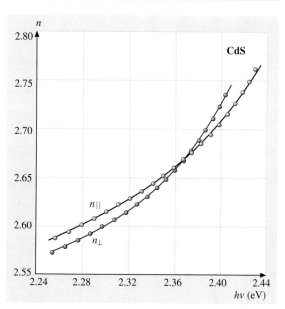

Fig. 4.1-174 CdS, hexagonal. Dispersion of the refractive index. The *symbols* are experimental values and the *curves* correspond to polynomial fits [1.149]

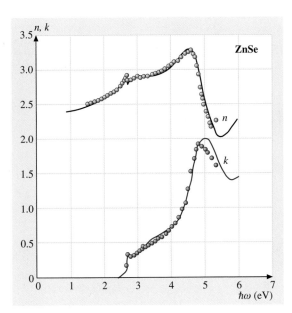

Fig. 4.1-175 CdSe, hex, cub. The real refractive index $n(E)$ and the extinction coefficient $k(E)$, forming the complex refractive index $n*(E) = n(E) + \mathrm{i}k(E)$, for both hexagonal and cubic CdSe. The *circles* represent experimental data, the *solid lines* numerically calculated energy dependences [1.150]

4.1.3.6 Mercury Compounds

A. Crystal Structure, Mechanical and Thermal Properties
Tables 4.1-170 – 4.1-177.

Table 4.1-170 Crystal structures of mercury compounds

Crystal		Space group		Crystal system	Structure type	Figure
Mercury oxide	HgO	D_{2h}^{16}		Trigonal		
Mercury sulfide	HgS				Red α-HgS	
Mercury selenide	HgSe	T_d^2	$F\bar{4}3m$	fcc	Zinc blende	Fig. 4.1-131
Mercury telluride	HgTe	T_d^2	$F\bar{4}3m$	fcc	Zinc blende	Fig. 4.1-131

Table 4.1-171 Lattice parameters of mercury compounds

Crystal		Lattice parameters (nm)	Remarks
Mercury oxide	HgO	$a = 0.3577$ $b = 0.8681$ $c = 0.2427$	X-ray diffraction
Mercury sulfide	HgS	$a = 0.414$ $b = 0.949$ $c = 0.2292$	X-ray diffraction
Mercury selenide	HgSe	$a = 0.5997$ (5)	Angle-dispersion X-ray diffraction, pressure $p = 2.25$ GPa
Mercury telluride	HgTe	$a = 0.6453$	Energy-dispersive X-ray diffraction

Table 4.1-172 Densities of mercury compounds

Crystal		Density (g/cm³)
Mercury oxide	HgO	11.080
Mercury selenide	HgSe	8.11
Mercury telluride	HgTe	8.21

Table 4.1-173 Melting point T_m of mercury compounds

Crystal		Melting point T_m (K)	Remarks
Mercury sulfide	HgS	1093	Zinc blende structure
Mercury selenide	HgSe	1072	Assessed
Mercury telluride	HgTe	943	

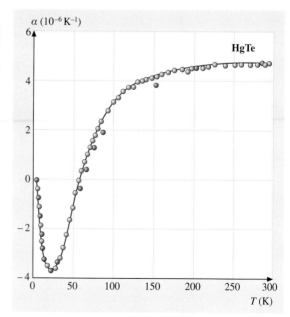

Fig. 4.1-176 HgTe. Linear thermal expansion coefficient α vs. T. *Symbols*, experimental points; *solid line*, calculated values [1.151]

Table 4.1-174 Elastic constants c_{ik} of mercury compounds

Crystal		c_{11} (GPa)	c_{12} (GPa)	c_{33} (GPa)	c_{44} (GPa)	c_{66} (GPa)	Remarks
Mercury sulfide	HgS	35.0 (14)		48.6 (5)		13.0 (5)	300 K
Mercury selenide	HgSe	62.2 (1)	46.4		22.7 (1)		292 K, ultrasound measurements
Mercury telluride	HgTe	53.6	36.6		21.10		300 K
		58.7	40.5		22.3		77 K

Table 4.1-175 Linear thermal expansion coefficient α of mercury compounds

Crystal		Linear thermal expansion coefficient α (10^{-6} K^{-1})		Temperature	Remarks
Mercury sulfide	HgS	α_{parall}	18.8 (8)	20–200 °C	Parallel to c axis
		α_{perpend}	18.1 (5)		Perpendicular to c axis, trigonal HgS
Mercury selenide	HgSe		28.61	30–380 °C	From temperatur dependence of lattice constant a
Mercury telluride	HgTe		4.0	77–300 K	See Fig. 4.1-176 for temperature dependence

Table 4.1-176 Heat capacity c_p and Debye temperature Θ_D of mercury compounds

Crystal		Heat capacity c_p (J/mol K)	Temperature (K)	Debye temperature Θ_D (K)	Remarks
Mercury sulfide	HgS	50.25	300	152	$T = 0$ K, calorimetry, extrapolated
Mercury selenide	HgSe	49.79	300	142	$T = 0$ K, calorimetry, extrapolated
Mercury telluride	HgTe	54.81	300	141.5	

Table 4.1-177 Phonon frequencies/wavenumbers at symmetry points for mercury compounds. Mercury sulfide (HgS), 80 K, from infrared reflectance; mercury selenide (HgSe), RT, from resonance Raman scattering; mercury telluride (HgTe), 290 K, from neutron scattering

HgS			HgSe		HgTe		
Mode	Symmetry point	Wavenumber (cm^{-1})	Mode	Wavenumber (cm^{-1})	Mode	Symmetry point	Phonon energy (meV)
$\bar{\nu}_{LO}$	E_1	356	$\bar{\nu}_{TO}$	130	$h\nu_{LO}$	Γ	14.86 (16)
$\bar{\nu}_{TO}$	E_1	347.6	$\bar{\nu}_{LO}$	174	$h\nu_{TO}$	Γ	14.63 (17)
$\bar{\nu}_{LO}$	E_2	296			$h\nu_{LO}$	X	16.86 (12)
$\bar{\nu}_{TO}$	E_2	284.2			$h\nu_{TO}$	X	16.64 (10)
$\bar{\nu}_{LO}$	E_3	156			$h\nu_{LA}$	X	10.56 (7)
$\bar{\nu}_{TO}$	E_3	115.5			$h\nu_{TA}$	X	1.97 (2)
$\bar{\nu}_{LO}$	E_4	98			$h\nu_{LO}$	L	18.07 (24)
$\bar{\nu}_{TO}$	E_4	92			$h\nu_{TO}$	L	15.86 (18)
$\bar{\nu}_{LO}$	E_5	50			$h\nu_{LA}$	L	9.97 (6)
$\bar{\nu}_{TO}$	E_5	42			$h\nu_{TA}$	L	2.28 (2)

B. Electronic Properties
Tables 4.1-179 – 4.1-183.

Table 4.1-178 Band structures of mercury compounds

Mercury selenide (HgSe)
Mercury selenide is a zero-gap material (semimetal). The lowest conduction band minimum and the top of the valence band are degenerate at the center of the Brillouin zone (Γ_8). The Γ_6 level, which for most cubic semiconductors is the conduction band minimum and has an energy larger than that of the Γ_8 state, is found to be below the Γ_8 state in HgSe ("negativ energy gap", inverted band structure).

Mercury telluride (HgTe)
Mercury telluride, like mercury selenide, is a zero-gap material (semimetal). The lowest conduction band minimum and the top of the valence band are degenerate at the center of the Brillouin zone (Γ_8). The Γ_6 level, which for most cubic semiconductors is the conduction band minimum and has an energy larger than that of the Γ_8 state, is found to be below the Γ_8 state in HgTe ("negativ energy gap", inverted band structure) (see Fig. 4.1-177).

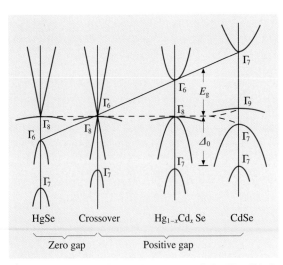

Fig. 4.1-177 Schematic band structure of the Hg–Cd–Se system near the Γ point [1.152]

Table 4.1-179 First Brillouin zones of mercury compounds

Crystal		Figure
Mercury selenide	HgSe	Fig. 4.1-133
Mercury telluride	HgTe	Fig. 4.1-133

Table 4.1-180 Energy gaps of mercury compounds

Crystal		Band gap (eV)	Between bands	Temperature (K)	Remarks
Mercury oxide	HgO	2.19			Photoconductivity
		2.80			
Mercury sulfide	HgS	2.03		300	Absorption
Mercury selenide	HgSe	−0.274	Γ_{8v} and Γ_{6c}	4.2	Shubnikov–de Haas effect
		−0.061	Γ_{8v} and Γ_{6c}	300	
Mercury telluride	HgTe	−0.304	Γ_{8v} and Γ_{6c}	0	Extrapolated optical and
		−0.141 (13)	Γ_{8v} and Γ_{6c}	300	magneto-optical data

Table 4.1-181 Spin–orbit splitting energies Δ_{so} of mercury compounds

Crystal		Splitting energy (eV)	Bands	Temperature (K)	Remarks
Mercury selenide	HgSe	0.383 (2)	Γ_{8v} to Γ_{7v}	4.2	Interband magnetoabsorption
		0.45	Γ_{15v}	12	Reflectivity
Mercury telluride	HgTe	1.08 (2)	Γ_{7v} to Γ_{8c}	300	Electroreflectance

Table 4.1-182 Effective masses of electrons (in units of the electron mass m_0) for mercury compounds

Crystal		Quantity	Band	Mass	Temperature (K)	Remarks
Mercury selenide	HgSe	m_n				Strongly dependent on electron concentration
Mercury telluride	HgTe	m_n	Γ_8	0.031 (1)	4.4	Interband magnetoabsorption

Table 4.1-183 Effective masses of holes (in units of the electron mass m_0) for mercury compounds

Crystal		Quantity	Band	Mass	Temperature (K)	Remarks
Mercury selenide	HgSe	m_p		0.78		
Mercury telluride	HgTe	m_p [100]	Γ_8	0.320	4.2	Interband magnetoreflection
		m_p [110]		0.406		
		m_p [111]		0.445		
		m_p	Γ_6	0.3	4–100	Transport measurements
				0.028 (1)	4.4	Interband magnetoabsorption

C. Transport Properties

Tables 4.1-184 and 4.1-185. See Fig. 4.1-179 for the thermal conductivity of mercury telluride.

Table 4.1-184 Electronic transport, general description

Mercury sulfide (HgS)

In natural red cinnabar crystals (α-HgS), carrier concentrations $n = 10^{10}$–10^{12} cm^{-3} are found; in synthetic crystals, the values range from 10^{11} to 10^{12} cm^{-3}. The observed values of the resistivity ρ are: parallel to the c axis, 6400 Ω cm at 77 K, and perpendicular to the c axis, 3450 Ω cm at 300 K, both values being obtained from natural crystals.

Mercury selenide (HgSe)

All HgSe samples show n-type conductivity. As-grown crystals have carrier concentrations of about 10^{17} electrons/cm^3.

Mercury telluride (HgTe)

See Fig. 4.1-178 for the temperature dependence of the electrical conductivity.

Table 4.1-185 Electron mobilities μ_n of mercury compounds

Crystal		μ_n (cm²/V s)		Temperature (K)	Remarks
Mercury sulfide	HgS	μ_{parall}	157	77	Natural crystal, maximum values
			30	300	
		μ_{perpend}	49	77	
			10	300	
Mercury selenide	HgSe		15×10^3	300	All HgSe samples show n-type conductivity
			55×10^3	95	
Mercury telluride	HgTe		35×10^3	300	Maximum mobility values
			120×10^3	77	
			800×10^3	4.2	

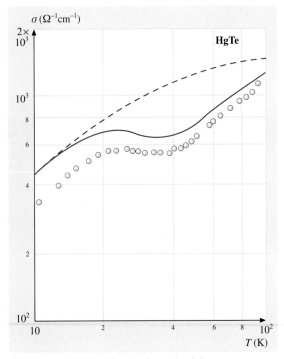

Fig. 4.1-178 HgTe. Electrical conductivity vs. temperature. Experimental data (*circles*), in comparison with calculated curves assuming mixed scattering modes (*solid line*) and neglecting interband optical-phonon scattering (*broken line*) [1.153]

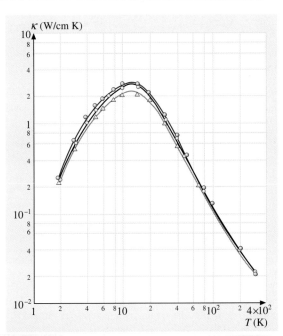

Fig. 4.1-179 HgTe. Thermal conductivity vs. temperature for three different samples; the *solid lines* represent theoretical fits [1.154]

D. Electromagnetic and Optical Properties
Tables 4.1-186 and 4.1-187.

Table 4.1-186 Dielectric constant ε of mercury compounds

Crystal		$\varepsilon(0)$		$\varepsilon(\infty)$		Temperature (K)	Remarks
Mercury sulfide	HgS	$\varepsilon(0)_{\text{perpend}}$	16.85	$\varepsilon(\infty)_{\text{perpend}}$	5.38	80	Synthetic crystal
			18.2		6.25	300	
		$\varepsilon(0)_{\text{parall}}$	21.5	$\varepsilon(\infty)_{\text{parall}}$	7.4	80	
			23.5		7.9	300	
Mercury selenide	HgSe		25.6		12–21		Reflectivity and Kramers–Kronig analysis of fits to data on plasmon–LO-phonon coupled mode
Mercury telluride	HgTe		21.0		15.2	77	Reflectance

Table 4.1-187 Refractive indices n of mercury compounds

Crystal		Refractive index n		λ (nm)	Temperature (K)
Mercury oxide	HgO		2.5	550	
Mercury sulfide	HgS	n_o	2.9028	620	298
		n_e	3.2560		

References

1.1 O. Madelung, U. Rössler, M. Schulz (Eds.): *Semiconductors*, Landolt–Börnstein, New Series III/41 (Springer, Berlin, Heidelberg 1998–2003)
1.2 O. Madelung (Ed.): *Semiconductors: Data Handbook*, 3rd edn. (Springer, Berlin, Heidelberg 2004)
1.3 E. R. Johnson, S. M. Christian: Phys. Rev. **95**, 560 (1954)
1.4 B. J. Skinner: Am. Mineral. **42**, 39 (1957)
1.5 G. A. Slack, S. F. Bartram: J. Appl. Phys. **46**, 89 (1975)
1.6 R. O. A. Hall: Acta Crystallogr. **14**, 1004 (1961)
1.7 H. P. Singh: Acta Crystallogr. **24a**, 469 (1968)
1.8 A. Taylor, R. M. Jones: *Silicon Carbide – A High Temperature Semiconductor*, ed. by J. R. O'Connor, J. Smiltens (Pergamon, Oxford 1960)
1.9 K. G. Lyon, G. L. Salinger, C. A. Swenson, G. K. White: J. Appl. Phys. **48**, 865 (1977)
1.10 Y. Okada, Y. Tokumaru: J. Appl. Phys. **56**, 314 (1984)
1.11 H.-M. Kagaya, T. Soma: Phys. Status Solidi (b) **127**, K5 (1985)
1.12 S. I. Novikova: Sov. Phys. Solid State (English Transl.) **2**, 2087 (1961)
1.13 S. I. Novikova: Fiz. Tverd. Tela **2** (1960)
1.14 A. C. Victor: J. Chem. Phys. **36**, 1903 (1962)
1.15 J. L. Warren, J. L. Yarnell, G. Dolling, R. A. Cowley: Phys. Rev. **158**, 805 (1967)
1.16 P. Giannozzi, S. de Gironcoli, P. Pavone, S. Baroni: Phys. Rev. B **43**, 7231 (1991)
1.17 G. Dolling: *Inelastic Scattering of Neutrons in Solids and Liquids*, Vol. 11 (International Atomic Energy Agency, Vienna 1963) p. 37
1.18 G. Nilsson, G. Nelin: Phys. Rev. B **6**, 3777 (1972)
1.19 G. Nilsson, G. Nelin: Phys. Rev. B **3**, 364 (1971)
1.20 P. Pavone, R. Bauer, K. Karch, O. Schütt, S. Vent, W. Windl, D. Strauch, S. Baroni, S. de Gironcoli: Physica B **219 & 220**, 439 (1996)
1.21 D. L. Price, J. M. Rowe, R. M. Nicklow: Phys. Rev. B **3**, 1268 (1971)
1.22 M. Hofmann, A. Zywietz, K. Karch, F. Bechstedt: Phys. Rev. B **50**, 13401 (1994)
1.23 F. J. Morin, J. P. Maita: Phys. Rev. **96**, 28 (1954)
1.24 F. J. Morin, J. P. Maita: Phys. Rev. **94**, 1525 (1954)
1.25 A. Suzuki, A. Uemoto, M. Shigeta, K. Furukawa, S. Nakajima: Appl. Phys. Lett. **49**, 450 (1986)
1.26 G. Busch, O. Vogt: Helv. Phys. Acta **33**, 437 (1960)
1.27 F. Nava, C. Canali, C. Jacoboni, L. Reggiani: Solid State Commun. **33**, 475 (1980)

1.28 A. G. Redfield: Phys. Rev. **94**, 526 (1954)
1.29 E. A. Konorova, S. A. Shevchenko: Sov. Phys. Semicond. (English Transl.) **1**, 299 (1967)
1.30 E. A. Konorova, S. A. Shevchenko: Fiz. Tekh. Poluprovodn. **1**, 364 (1967)
1.31 L. Reggiani, S. Bosi, C. Canali, F. Nava: Phys. Rev. B **23**, 3050 (1981)
1.32 P. J. Dean, E. C. Lightowlers, D. R. Wright: Phys. Rev. A **140**, 352 (1965)
1.33 L. Reggiani, D. Waechter, S. Zukotynski: Phys. Rev. B **28**, 3550 (1983)
1.34 C. Jacoboni, C. Canali, G. Ottaviani, A. Alberigi Quaranta: Solid State Electron. **20**, 77 (1977)
1.35 C. Canali, C. Jacoboni, F. Nava, G. Ottaviani, A. Alberigi Quaranta: Phys. Rev. B **12**, 2265 (1975)
1.36 P. Norton, T. Braggins, H. Levinstein: Phys. Rev. B **8**, 5632 (1973)
1.37 W. C. Mitchel, P. M. Hemenger: J. Appl. Phys. **53**, 6880 (1982)
1.38 F. J. Morin: Phys. Rev. **93**, 62 (1954)
1.39 R. Berman, M. Martinez: Diamond Res. (Suppl. Ind. Diamond, Rev.) **7** (1976)
1.40 C. J. Glassbrenner, G. A. Slack: Phys. Rev. **134**, A1058 (1964)
1.41 W. Fulkerson, J. P. Moore, R. K. Williams, R. S. Graves, D. L. McElroy: Phys. Rev. **167**, 765 (1968)
1.42 H. R. Shanks, P. D. Maycock, P. H. Sidles, G. C. Danielson: Phys. Rev. **130**, 1743 (1963)
1.43 Y. P. Joshi, G. S. Verma: Phys. Rev. B **1**, 750 (1970)
1.44 G. A. Slack: J. Appl. Phys. **35**, 3460 (1964)
1.45 S. R. Bakhchieva, N. P. Kekelidse, M. G. Kekua: Phys. Status Solidi (a) **83**, 139 (1984)
1.46 T. Hanyu: J. Phys. Soc. Jpn. **31**, 1738 (1971)
1.47 D. E. Aspnes, A. A. Studna: Phys. Rev. B **27**, 985 (1983)
1.48 S. Klotz, J. M. Besson, M. Braden, K. Karch, F. Bechstedt, D. Strauch, P. Pavone: Phys. Status Solidi (b) **198**, 105 (1996)
1.49 H. W. Icenogle, B. C. Platt, W. L. Wolfe: Appl. Opt. **15**, 2348 (1976)
1.50 V. L. Solozhenko: *Properties of Group III Nitrides*, ed. by J. H. Edgar (Inspec, London 1994) Chap. 2.1, p. 43
1.51 Y. Kumashiro: J. Mater. Res. **5**, 2933 (1990)
1.52 I. S. Ham, V. M. Davidenko, V. G. Sidorov, L. I. Fel'dgun, M. D. Skagalov, Y. K. Shalabutov: Sov. Phys. Semicond. (English Transl.) **10**, 331 (1976)
1.53 I. S. Ham, V. M. Davidenko, V. G. Sidorov, L. I. Fel'dgun, M. D. Skagalov, Y. K. Shalabutov: Fiz. Tekh. Poluprov. **10**, 554 (1976)
1.54 L. G. Carpenter, P. J. Kirby: J. Phys. D **15**, 1143 (1982)
1.55 Y. Kumashiro, M. Hirabayashi, T. Kosiro: J. Less-Common Metals **143**, 159 (1988)
1.56 S. I. Novikova, N. Kh. Abrikhosov: Sov. Phys. Solid State (English Transl.) **5**, 1558 (1963)
1.57 S. I. Novikova, N. Kh. Abrikhosov: Fiz. Tverd. Tela **5**, 2138 (1963)
1.58 J. C. Nipko, C. K. Loong: Phys. Rev. B **57**, 10550 (1998)
1.59 C. K. Loong: Gallium nitride and related materials, MRS Symposia Proceedings No. 395, ed. by F. A. Ponce, R. D. Dupuis, S. Nakamura, J. A. Edmond (Materials Research Society, Pittsburgh 1996) 423
1.60 T. Pletl, P. Pavone, U. Engel, D. Strauch: Physics B **263/264**, 392 (1999)
1.61 A. Onton: Proceedings of the 10th International Conference on the Physics of Semiconductors, Cambridge 1970 (USAEC, Oak Ridge 1970) 107
1.62 B. Monemar: Phys. Rev. B **8**, 5711 (1973)
1.63 D. Strauch, B. Dorner, K. Karch: *Phonons 89*, ed. by S. Hunklinger, W. Ludwig, G. Weiss (World Scientific, Singapore 1990) p. 82
1.64 A. Kobayashi, O. F. Sankey, S. M. Volz, J. D. Dow: Phys. Rev. B **28**, 935 (1983)
1.65 M. Huang, W. Y. J. Ching: Phys. Chem. Solids **46**, 977 (1985)
1.66 H. G. Grimmeiss, W. Kischio, A. Rabenau: J. Phys. Chem. Solids **16**, 302 (1960)
1.67 H. Welker: Z. Naturforsch. **5a**, 248 (1953)
1.68 G. A. Slack: J. Phys. Chem. Solids **34**, 321 (1973)
1.69 V. M. Muzhdaba, A. Ya. Nashel'skii, P. V. Tamarin, S. S. Shalyt: Sov. Phys. Solid State (English Transl.) **10**, 2265 (1969)
1.70 V. M. Muzhdaba, A. Ya. Nashel'skii, P. V. Tamarin, S. S. Shalyt: Fiz. Tverd. Tela **10**, 2866 (1968)
1.71 J. Pastrnak, L. Roskovcova: Phys. Status Solidi **14**, K5 (1966)
1.72 B. Monemar: Solid State Commun. **8**, 1295 (1970)
1.73 S. Adachi: *GaAs and Related Materials: Bulk Semiconducting and Superlattice Properties* (World Scientific, Singapore 1994)
1.74 W. J. Turner, W. E. Reese: Phys. Rev. **127**, 126 (1962)
1.75 A. U. Sheleg, V. A. Savastenko: Vesti Akad. Nauk BSSR, Ser. Fiz. Mat. Nauk **3**, 126 (1976)
1.76 T. Soma: Solid State Commun. **34**, 375 (1980)
1.77 A. Debernardi, M. Cardona: Phys. Rev. B **54**, 11305 (1996)
1.78 T. Azuhata, T. Matsunaga, K. Shimada, K. Yoshida, T. Sota, K. Suzuki, S. Nakamura: Physica B **219 & 220**, 493 (1996)
1.79 V. Yu. Davydov, Yu. E. Kitaev, I. N. Goncharuk, A. N. Smirnov, J. Graul, O. Semchinova, D. Uffmann, M. B. Smirnov, A. P. Mirgorodsky, R. A. Evarestov: Phys. Rev. B **58**, 12899 (1998)
1.80 J. L. Yarnell, J. L. Warren, R. G. Wenzel, P. J. Dean: *Neutron Inelastic Scattering* (International Atomic Energy Agency, Vienna 1968) p. 301
1.81 P. H. Borcherds, K. Kunc, G. F. Alfrey, R. L. Hall: J. Phys. C: Solid State Phys. **12**, 4699 (1979)
1.82 C. Eckl, P. Pavone, J. Fritsch, U. Schröder: *The Physics of Semiconductors*, Vol. 1, ed. by M. Scheffler, R. Zimmermann (World Scientific, Singapore 1996) p. 229
1.83 D. Strauch, B. Dorner: J. Phys. Condens. Matter **2**, 1457 (1990)
1.84 M. K. Farr, J. G. Traylor, S. K. Sinha: Phys. Rev. B **11**, 1587 (1975)
1.85 M. Ilegems, H. C. Montgomery: J. Phys. Chem. Solids **34**, 885 (1972)
1.86 Y. C. Kao, O. Eknoyan: J. Appl. Phys. **54**, 2468 (1983)

1.87   J. S. Blakemore: J. Appl. Phys. **53**, R123 (1982)
1.88   R. A. Stradling, R. A. Wood: J. Phys. C **3**, L94 (1970)
1.89   M. H. Kim, S. S. Bose, B. J. Skromme, B. Lee, G. E. Stillman: J. Electron. Mater. **20**, 671 (1991)
1.90   E. K. Sichel, J. I. Pankove: J. Phys. Chem. Solids **38**, 330 (1977)
1.91   M. G. Holland: Proc. 7th Int. Conf. Phys. Semiconductors, Paris 1964 (Dunod, Paris 1964) 1161
1.92   M. G. Holland: Proc. Int. Conf. Phys. Semiconductors, Paris 1964 (Dunod, Paris 1964) 713
1.93   E. Ejder: Phys. Status Solidi (a) **6**, K39 (1971)
1.94   B. Clerjaud, C. Naud, B. Deveaud, B. Lambert, B. Plot, C. Bremond, C. Benjeddou, G. Guillot, A. Nouailhat: J. Appl. Phys. **58**, 4207 (1985)
1.95   G. A. Samara: Phys. Rev. B **27**, 3494 (1983)
1.96   K. Haruna, H. Maeta, K. Ohashi, T. Koike: J. Phys. C: Solid State Phys. **20**, 5275 (1987)
1.97   P. W. Sparks, C. A. Swenson: Phys. Rev. **163**, 779 (1967)
1.98   D. F. Gibbons: Phys. Rev. **112**, 779 (1958)
1.99   P. H. Borcherds, G. F. Alfrey, D. H. Saunderson, A. D. B. Woods: J. Phys. C **8**, 2022 (1975)
1.100 A. Mooradian, G. B. Wright: Solid State Commun. **4**, 431 (1966)
1.101 J. Fritsch, P. Pavone, U. Schröder: Phys. Rev. B **52**, 11326 (1995)
1.102 N. S. Orlova: Phys. Status Solidi (b) **119**, 541 (1983)
1.103 R. Carles, N. Saint-Cricq, J. B. Renucci, M. A. Renucci, A. Zwick: Phys. Rev. B **22**, 4804 (1980)
1.104 W. J. Turner, W. E. Reese, G. D. Pettit: Phys. Rev. **136**, A1467 (1964)
1.105 Y. J. Jung, B. H. Kim, H. J. Lee, J. C. Wolley: Phys. Rev. **26**, 3151 (1982)
1.106 S. Logothetidis, L. Vina, M. Cardona: Phys. Rev. B **31**, 947 (1985)
1.107 O. G. Folberth, O. Madelung, H. Weiss: Z. Naturforsch. **9a**, 954 (1954)
1.108 O. Madelung, H. Weiss: Z. Naturforsch. **9a**, 527 (1954)
1.109 T. L. Tansley, C. P. Foley: Electron. Lett. **20**, 1087 (1984)
1.110 W. Walukiewicz, J. Lagowski, L. Jastrzebski, P. Rava, M. Lichtensteiger, C. H. Gatos, H. C. Gatos: J. Appl. Phys. **51**, 2659 (1980)
1.111 J. D. Wiley: *Semiconductors and Semimetals*, Vol. 10, ed. by R. K. Willardson, A. C. Beer (Academic Press, New York 1975)
1.112 U. Busch, E. Steigmeier: Helv. Phys. Acta **34**, 1 (1961)
1.113 H. J. Hrostowski, F. J. Morin, T. H. Geballe, G. H. Wheatley: Phys. Rev. **100**, 1672 (1955)
1.114 N. I. Volokobinskaya, V. V. Galavanov, D. N. Nasledov: Sov. Phys. Solid State (English Transl.) **1**, 687 (1959)
1.115 N. I. Volokobinskaya, V. V. Galavanov, D. N. Nasledov: Fiz. Tverd. Tela **1**, 756 (1959)
1.116 D. L. Rode: Phys. Rev. **3**, 3287 (1971)
1.117 S. A. Aliev, A. Ya. Nashelskii, S. S. Shalyt: Sov. Phys. Solid State (English Transl.) **7**, 1287 (1965)
1.118 S. A. Aliev, A. Ya. Nashelskii, S. S. Shalyt: Fiz. Tverd. Tela **7**, 1590 (1965)
1.119 H. Katzman, J. Moss, W. F. Libby: J. Phys. Chem. Solids **32**, 2786 (1971)
1.120 K.-Oh Park, J. M. Sivertsen: J. Am. Ceram. Soc. **62**, 218 (1979)
1.121 G. K. White, O. L. Anderson: J. Appl. Phys. **37**, 430 (1966)
1.122 C. M. Osburn, R. W. Vest: J. Am. Ceram. Soc. **54**, 428 (1971)
1.123 D. S. Kupperman, H. Weinstock, Y. Chen: J. Low Temp. Phys. **14**, 277 (1974)
1.124 O. Kamada, T. Takizawa, T. Sakurai: Jpn. J. Appl. Phys. **10**, 485 (1971)
1.125 N. N. Kovalev, M. V. Krasin'kova: Sov. Phys. Solid State **16**, 1960 (1975)
1.126 R. R. Reeber, G. W. Powell: J. Appl. Phys. **38**, 1531 (1967)
1.127 R. Helbig, P. Wagner: J. Phys. Chem. Solids **35**, 327 (1974)
1.128 P. Wagner: Ph.D Thesis (Erlangen-Nürnberg 1978)
1.129 G. A. Slack: *Physics and Chemistry of II–VI Compounds*, ed. by M. Aven, J. S. Prener (1967) p. 557
1.130 K. Sato, S. Adachi: J. Appl. Phys. **73**, 926 (1993)
1.131 L. Ward: *Handbook Optical Constants Solids*, Vol. 2, ed. by E. D. Palik (Academic Press, New York 1991) p. 737
1.132 S. Ozaki, S. Adachi: Jpn. J. Appl. Phys. **32**, 5008 (1993)
1.133 T. B. Kobyakov, G. S. Pado: Sov. Phys. Solid State (English Transl.) **9**, 1707 (1968)
1.134 T. M. Bieniewski, S. J. Czyzak: J. Opt. Soc. Am. **53**, 496 (1963)
1.135 Q. Guo, M. Ikejira, M. Nishio, H. Ogawa: Solid State Commun. **100**, 813 (1996)
1.136 H. Yoshikawa, S. Adachi: Jpn. J. Appl. Phys. **36**, 6237 (1997)
1.137 Y. F. Tsay, S. S. Mitra, J. F. Vetelino: J. Phys. Chem. Solids **34**, 2167 (1973)
1.138 J. L. Birman, H. Samelson, A. Lempicki: G. T. & E. Res. Dev. **1**, 1 (1961)
1.139 M. Cardona, G. Harbeke: Phys. Rev. A **137**, 1467 (1965)
1.140 S. Adachi, T. Taguchi: Phys. Rev B **43**, 9569 (1991)
1.141 V. S. Oskotskii, I. B. Kobyakov, A. V. Solodukhin: Fiz. Tverd. Tela **22**, 1479 (1980)
1.142 V. S. Oskotskii, I. B. Kobyakov, A. V. Solodukhin: Sov. Phys. Solid State (English Transl.) **22**, 861 (1980)
1.143 K. C. Mills: High Temp. High Press. **4**, 371 (1972)
1.144 J. R. Chelikowsky, M. L. Cohen: Phys. Rev. B **14**, 556 (1976)
1.145 G. E. Moore, M. V. Klein: Phys. Rev. **179**, 722 (1969)
1.146 J. L. Beene, G. Contwell: J. Appl. Phys. **57**, 1171 (1985)
1.147 M. S. Kushwaha, S. S. Kushwaha: Can. J. Phys. **58**, 351 (1980)
1.148 A. J. Varkey, A. F. Fort: Thin Solid Films **239**, 211 (1994)
1.149 J. Oberlé, B. Kippelen, A. Daunois, J.-B. Grun: Opt. Commun. **90**, 339 (1992)
1.150 S. Ninomiya, S. Adachi: J. Appl. Phys. **78**, 4681 (1995)
1.151 D. Bagot, R. Granger, S. Rolland: Phys. Status Solidi (b) **177**, 295 (1993)

1.152 C. R. Whitsett, J. G. Broerman, C. J. Summers: *Semiconductors and Semimetals*, Vol. 16, ed. by R. K. Willardson, A. C. Beer (Academic Press, New York 1981) p. 54

1.153 J. J. Dubowski, T. Dietl, W. Szymanska, R. R. Galazka: J. Phys. Chem. Solids **42**, 351 (1981)

1.154 A. Noguera, S. M. Wasim: Phys. Rev. B **32**, 8046 (1985)

4.2. Superconductors

Superconductors are characterized by an anomalous temperature dependence of the electrical resistivity. Below a critical temperature T_c, their resistivity drops by more than a factor of 10^{10}. In superconductors the magnetic flux density $B = \mu_r B_a$ induced by an externally applied field H_a is zero, like in ideal diamagnets with $\mu_r = 0$ (Meißner-Ochsenfeld effect). If H_a exceeds a critical value H_c the superconductor becomes normal conducting. But the magnetic induction B decreases from B_a at the free surface to $B = 0$ in the interior through a layer of finite thickness characterized by the Landau penetration depth λ. The critical field varies with temperature as $H_c(T) = H_c(0)[1 - (T/T_c)^2]$, where $H_c(0) = H_c(T = 0\,\text{K})$. According to the isothermal field dependence of the magnetization $I(H_a) = -\mu_0 H_a$, two types of superconductors may be differentiated, as shown in Fig. 4.2-1:

- Type I superconductors such as Pb with a sudden drop of $-I$, at H_c; all pure metallic elements and their dilute solid solutions belong to this group;
- Type II superconductors such as Pb–In15 which are characterized by a lower critical field H_{c1} at which the drop of $-I$ sets in and an upper critical field H_{c2} at which $-I$ reaches 0.

4.2.1	**Metallic Superconductors**	696
4.2.1.1	Elements	696
4.2.1.2	Practical Metallic Superconductors	704
4.2.2	**Non-Metallic Superconductors**	711
4.2.2.1	Oxide Superconductors	711
4.2.2.2	Superconductors Based on the Y–Ba–Cu–O System	720
4.2.2.3	Superconductors Based on the Bi–Sr–Ca–Cu–O System	736
4.2.2.4	Carbides, Borides, Nitrides	744
References		749

In Type I superconductors, the transition from the fully superconducting to the fully normal conducting state as a function of the applied magnetic field H_a passes through an intermediate, mixed state in which the total volume decomposes into lamellae in the superconducting and the normal conducting state. At their interface, the transition from the superconducting lamellae of the maximum density of electron pairs (Cooper pairs) n_e, which carry the superconducting current, to the normal conducting lamellae where $n_e = 0$ occurs through a transition layer characterized by the coherence length ξ. In terms of the microscopic theory of superconductivity, ξ is the size of the Cooper pairs which is of the order of 1 µm in pure metals. The coherence length ξ as well as the penetration depth λ can be highly anisotropic in compounds with crystal structures of low symmetry, which is the case in low-T_c oxide superconductors (see Sect. 4.2.2.1).

Since superconductivity is caused by electron–phonon interaction, the upper critical field H_{c2} depends on the mean free path of the electrons, and thus on the electrical resistivity ρ in the normal conducting state, and on the electronic specific heat γ as

$$\mu_0 H_{c2}(0) = 3.11 \rho \gamma T_c .$$

As both ρ and T_c are affected by impurities and structural defects, H_{c2} depends not only on composition and structure but also on structural defects accordingly.

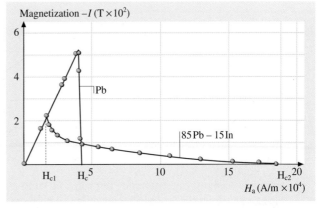

Fig. 4.2-1 $-I(H_a)$ curves for pure Pb as a Type I and the Pb–In15 alloy as a Type II superconductor [2.1]

Type II superconductors in a magnetic field of intermediate strength $H_{c1} < H_a < H_{c2}$ are in a mixed state (Shubnikow phase) consisting of a superconducting matrix through which magnetic flux "lines" of finite thickness, also termed fluxoids or vortices are penetrating, each carrying a quantum of magnetic flux $\Phi = h/2e$. The movement of these flux lines may be pinned by point defects, dislocations, and interfaces due to various modes of interaction. Deliberate introduction of pinning defects is the basis of increasing the critical magnetic field such that the materials – termed non-ideal Type II, Type III or hard superconductors – can bear an increased critical external field and/or critical current at a given temperature. High-field superconducting materials are developed on this basis of microstructural design.

Practical superconductors are mainly applied to generate high magnetic fields in wound magnet coils. Accordingly they are produced as wires or cables. The materials used are based on non-ideal Type II superconductors with a specially optimized microstructure such that strong flux line pinning persists up to a high critical current density $J_c(T)$. A second field of application are superconducting electronic devices, most frequently based on the Josephson effect. The present practical superconductors for high field applications are highly complex composite materials, based essentially on two groups of superconducting alloy phases, the intermetallic phase Nb_3Sn and the solid solution phase Nb–Ti. Recent developments have also led to high field oxide high-T_c superconductors for practical applications in wire, electronic device, and massive form, the latter serving as high field permanent magnets.

Superconducting substances of fundamental as well as practical interest may be divided into 3 main groups which are dealt with in the subsequent sections:

- Metallic superconductors
- Superconducting oxides
- Superconducting carbides, nitrides and borides

An extensive listing of individual publications containing data of both metallic and non-metallic superconductors is given in [2.2]. An encyclopaedic survey of all essential aspects of superconductivity may be found in [2.3].

4.2.1 Metallic Superconductors

The basic superconducting properties of metals and alloys have been compiled comprehensively in [2.1, 2]. While the superconducting properties of all pure metals have been determined, the studies of alloys have been concentrated on those of the elements with the highest critical temperatures, i.e., Pb ($T_c = 7.2$ K), V ($T_c = 5.5$ K), Nb ($T_c = 9.3$ K), and Tc ($T_c = 7.7$ K).

4.2.1.1 Elements

Table 4.2-1 lists T_c and H_c of the superconducting metallic elements along with their relevant thermodynamic properties, the Debye temperature Θ_D and the Sommerfeld constant γ which determines the electronic specific heat in the superconducting state $C_{p,el} = 9.17 \gamma T_c \exp(-1.5 T_c/T)$. Table 4.2-2 lists the pressure dependence of T_c.

Pb Alloys

Ready availability and easy alloy formation of Pb have led to extensive studies of superconductivity of Pb alloys results of which are given in Tables 4.2-3 and 4.2-4, and Figs. 4.2-1 and 4.2-2.

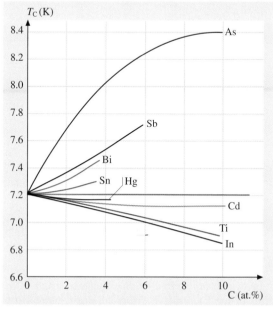

Fig. 4.2-2 Composition dependence of T_c in Pb solid solution alloys [2.1]

Table 4.2-1 Superconducting properties of metallic elements [2.1]

Element	Purity (%)	Residual resistivity ratio $\rho_{293\,K}/\rho_{4.2\,K}$	T_c (K)	$\mu_0 H_c$ (mT)	Θ_D (K)	γ (mJ mol^{-1} K^{-2})
Al		2000–4100	1.175	10.49	420	1.35
Am			0.6			
Be	99.996		0.026		1390	0.21
Cd		>38 000	0.517	2.805	209	0.69
Ga		46 500	1.0833	5.93	325	0.60
Hf		80	0.128	1.27	256	2.21
Hg	99.9999		4.154	41.1	87	1.81
In		9000	3.4087	28.15	109	1.672
Ir		>2000	0.1125	1.6	425	3.19
La	>99.9		4.87	9.8	151	9.8
Lu		15	0.1	35.0	210	10.2
Mo		17 000	0.916	9.686	460	1.83
Nb		500–16 500	9.25	206	276	7.80
Os			0.66	7.0	500	2.35
Pa			1.4			
Pb		15 000	7.196	80.34	105	3.36
Re		1700	1.697	20.1	415	2.35
Ru			0.493	6.9	580	2.8
Sn			3.722	30.55	195	1.78
Ta		29–120	4.47	82.9	258	6.15
Tc		80–100	7.77	141	411	6.28
Th		1200	1.374	16.0	165	4.32
Ti	>99.9		0.40	5.6	415	3.3
Tl		53	2.38	17.65	87.5	1.47
V		24–430	5.46	140	383	9.82
U			0.68	10.0	206	10.60
W		57 000	0.0154	0.115	383	0.90
Zn	99.9999		0.857	5.41	310	0.66
Zr	99.9		0.63	4.7	290	2.77

Table 4.2-2 Pressure dependence of the superconductivity of metals [2.1]

Metal	Pressure range [a] (10^2 MPa)	T_c (K)	$T_{c(max)}/p$ [a] (K 10^{-2} MPa^{-1})	$\delta T_c/\delta p$ (10^{-4} K MPa^{-1})
Al	0–62	2.1–1.7	2.1/0	−4.5
As I	<100	<0.1		
As II	100–140	0.2–0.25	0.25/100	>0
As III	140–220	0.31–0.5	0.5/140	<0
Ba I	0–55	<1		
Ba II	55–85	1–1.8	1.8/83	100
Ba III	85–144	1.8–5	5.0/100	−15
Ba IV	144–192	4.5–5.4	5.4/175	1.3
Bi II	25–27	≦3.9	3.9/25	<0

Table 4.2-2 Pressure dependence of the superconductivity of metals [2.1], cont.

Metal	Pressure range[a] (10^2 MPa)	T_c (K)	$T_{c(max)}/p$ [a] (K 10^{-2} MPa^{-1})	$\delta T_c/\delta p$ (10^{-4} K MPa^{-1})
Bi II′	25–30	≦8.2	8.2/30	−3.2
Bi III	27–37	6.55–7.25	7.25/27	−4.0
Bi VI	80–200	≦8.55	8.55/90	−2.3
Cd	0–28	≦0.52	0.52/0	<0
Ce (α)	20–35	0.020–0.045	0.045/35	>0
Ce (α)′	45–125	1.9–1.3	1.9/45	<0
Cs V	125–150	1.5–1.6	1.6/125	<0
Ga II	30–38	6.24–6.38	6.38/35	−3
Ga II′	≥ 35	7.5		
Ge	115–120	≦5.35	5.35/115	<0
Hf	0–160	0–0.25	0.25/160	0.06
α-Hg	>10	4.2–3.9	4.156/0	−3.47
β-Hg	>10	4.0–3.9	4.017/0	−4.48
α-La	0–23	4.9–8.2	8.2/23	14.1
	0–225	4.8–11.5	11.5/225	
β-La	23–250	5.5–12.9	12–12.9/140–200	11.3
Lu	45–190	0.024–1	1.0/180	1.2
Nb	0–24	7–9.2	9.214/0	−0.25
	25–250	9.7	9.7/250	2.6
P	0–260	5.8–3.6	5.8/170	<0
Pb I	0–110	≈7.2–4.2	7.2/0	−4.9
Pb II	135–200	3.55–2.9	3.55/160	−2.2
Re	0–20	≦1.7	1.7/0	Minimum at
Re II		≦2.3		700 MPa
Sb	0–120	2.6–2.7		
Sb III	85–150	3.40–3.55	3.55/85	<0
Sb II	0–130	6.75–6.95	6.95/130	<0
Si	120–130	6.7–7.1	7.1/120	<0
Sn I	0–113	≦3.7	3.7/0	0
Sn II	125–160	5.2–4.85	5.2/125	−5.04
Sn III	113–270	≦5.3	5.3/113	−4.95
Ta	0–250	≦4.3	4.3/0	−0.26
Tc	0–15	8.0–7.8	8.0/15	
Te II	39–70	2.5–3.9	3.9/70	7.7–8.5
Te III	68–80	4.28–4.15	4.3/70	−0.3–1.7
Bi IV	43–62	6–8.7	8.7/43	−4.6
Bi V	68–81	6.7–8.3	6.7/81	−3.0
Te IV	80–260	4.3–2.8	4.3/80	−4.2
Th	0–170	1.4–0.64	1.4/0	Minimum at 700 MPa
Ti	0–25	–	–	0.06
Tl II	0–35		2.395/2	−1.0
Tl III	35–50		1.45/35 (bcc)	−4.95

Table 4.2-2 Pressure dependence of the superconductivity of metals [2.1], cont.

Metal	Pressure range[a] (10^2 MPa)	T_c (K)	$T_{c(max)}/p$[a] (K 10^{-2} MPa^{-1})	$\delta T_c/\delta p$ (10^{-4} K MPa^{-1})
Tl IV	40–60		2.32/40 (fcc)	−2.0
α-U	10–85	2.4–0.4	2.4/10	
β-U	0–12		1.15/95	>0
U	90–160	<0.35		
V	0–250	5.5–7.15		0.7
Y	110–170	2.7–1.7	2.7/160	3.7
Zn	0–26	≤0.87	0.87/0	<0
α-Zr	0–45		0.69/45	0.35–0.8
ω-Zr	60–130	1.7	1.17/130	0.77

[a] 0 = ambient pressure

Table 4.2-3 Critical temperature of superconductivity of binary Pb compounds [2.1]

Compound	Crystal structure	T_c (K)	Compound	Crystal structure	T_c (K)
PbLi[a]	Cub. B2	7.2	Pb$_3$Th	Cub.	5.5
Pb$_3$Na[a]	Cub.	5.62	PbZr$_3$	Cub. A15	0.76
PbMg$_2$	C1	5.6	PbZr$_5$	Hex. D8$_8$	4.6
Pb$_3$Ca	Cub. L1$_2$	0.84	PbV$_3$[c]	Cub. A15	<4.2
Pb$_3$Sr	Tetr. L1$_2$[b]	1.85	PbNb$_3$[c]	Cub. A15	9.6
Pb$_3$Y	Cub. L1$_2$	4.72	PbTa$_3$[c]	Cub. A15	17
Pb$_3$La$_5$		<4.2[b]	PbRh$_2$		<0.32
Pb$_3$La$_4$		<4.2	Pb$_2$Rh[a]	Tetr. C16	2.66–1.32
Pb$_4$La$_5$		<4.2	PbPd$_3$	Cub. L1$_2$	<0.10
Pb$_{10}$La$_{11}$		<4.2	Pb$_2$Pd	Tetr. C16	3.01–2.95
Pb$_4$La$_3$		<4.2	Pb$_4$Pt	Tetr. C16	2.80
Pb$_2$La		<4.2	PbAu$_2$	C15	1.18
Pb$_3$La	Cub. L1$_2$	4.10–4.05	Pb$_2$Au	Tetr. C16	3.15
PbSm$_3$		<4.2	Pb$_3$Au		4.4
Pb$_3$Sm$_5$		<4.2	PbIn$_3(\alpha)$	Hex.	6.05–5.65
Pb$_4$Sm$_5$		<4.2	Pb$_2$Bi(ϵ)[a]	A3$_3$	8.2–8.5
Pb$_{10}$Sm$_{11}$		<4.2	PbO$_2$		<1.02
Pb$_2$Sm		6.7	PbS	B1	<1.0
Pb$_3$Sm	Cub. L1$_2$	<4.2	PbSe	B1	<1.26
Pb$_3$Yb	Cub. L1$_2$	0.23	PbP		7.8

[a] Compound has a range of homogeneity
[b] No superconductivity is observed above 4.2 K
[c] Compound is formed under high pressure and high temperature

Table 4.2-4 Critical magnetic field $\mu_0 H_{c2}$ of binary Pb alloys

Material[a]	$\mu_0 H_{c2}$ (mT)	Temperature (K)
Pb–1.6–8.6% Na	205–600	4.2
Pb–3–5% Ag	82	0[b]
Pb–~25% Cd	135	0
Pb–5–15% Hg	230–>900	4.2
Pb–2–30% In	98–390	4.2
Pb–14% Sn	100	
Pb–12.9–64% Sn	110–204	1.3
Pb–30% Tl	280	4.2
Pb–0.8–40% Tl	70–290	4.2
Pb–2–50% Bi	73–190	4.2
Pb–30% Bi(ε)	3500	0
Pb–30–45 nm[c]	2200	4.2
200–700 nm[c]	80–110	4.2
Pb pore diameter 3.2–5.8 nm[d]	5500–9600	0
Pb–25% Bi, 126 nm[c]	14 500	4.2
Pb–40% Bi, 2–5 nm[d]	12 500–11 300	4.2

[a] At.%
[b] Extrapolated to 0 K
[c] Layer thickness
[d] Particle diameter in porous glass

V Alloys

Table 4.2-5 and Fig. 4.2-3 contain T_c data of binary V alloys. Even though V_3Ga with its cubic A15 structure has a $T_c \leq 16.8$ K and was, therefore, studied extensively, it has never been considered as a practical superconducting material seriously because of the cost and difficulties of handling of Ga.

Table 4.2-5 Critical temperature T_c of superconductivity of V compounds [2.1]

Compound	Crystal structure	T_c (K)	Compound	Crystal structure	T_c (K)
V_3Al[a]	Cub. A15[b]	9.6–11.65	V_2O_5	Rhomb.	<4.2
V_5Al$_8$[a]	Cub. D8$_2$	<1.2[c]	V_3Os$_2$[a]	Cub. B2	<0.37
VAl$_3$	Tetr. D0$_{22}$	<4.2	VOs[a]	Cub. A15	5.7–3.0
V_4Al$_{23}$	Hex.	<4.2	V_3P	Tetr.	0.07
V_7Al$_{45}$	Monocl.	<4.2	VP[a]	Hex. B8$_1$	<1.02
VAl$_{11}$[a]	Fcc	<4.2	VP$_2$	Monocl.	<0.035
V_3As[a]	Cub. A15	<1.0	V_3Pb[a]	Cub. A15[a]	<4.2
VAs[a]	Rhomb.	<1.0	V_3Pd[a]	Cub. A15	0.082
VAs$_2$	Monocl.	<0.33	VPd	Tetr.	–[e]
V_3Au	Cub. A15	3.22–0.015	VPd$_2$	Tetr.	–
VAu$_2$[a]	Rhomb.	<1.2	VPd$_3$	Tetr.	–
VAu$_4$	Tetr.	<1.2	V_3Pt[a]	Cub. A15	3.62–0.98
V_3B_2	Tetr.	<0.1	VPt[a]	Cub. L1$_0$	–
VB	Rhomb. B$_f$	1.2	VPt$_2$[a]	Tetr.	1.02

Table 4.2-5 Critical temperature T_c of superconductivity of V compounds [2.1], cont.

Compound	Crystal structure	T_c (K)	Compound	Crystal structure	T_c (K)
V_3B	Rhomb.	<4.2	VPt_3	Tetr.[h]	0.07
VB_2[a]	Hex. C32	<1.2		Cub. $L1_2$[i]	0.07
V_2B_5[b]	Hex.	–[e]	V_3Re_7	Cub. A15	8.4–7.6
VBe_2	Hex. C14	<4.2	VRe_3[a]	Tetr. $D8_b$	6.26–4.52
VBe_{12}	Tetr.	–	V_3Rh[a]	Cub. A15	1.075^g–<0.015
V_3Bi	Cub. A15[b]	<4.2			
V_2C[a]	Hex.	<1.2	VRh[a](α_1)	–	–
V_2C_3[a]	Fcc	–	VRh[a](α_2)	Tetr. $L1_0$	–
VC[a]	Cub. B1	<0.03	VRh[a](α_3)	Rhomb.	–
V_3Cd[a]	Cub. A15[b]	<4.2	V_3Rh_5[a]	Rhomb.	–
V_3Co[a]	Cub. A15	0.015	VRh_3[a]	Cub. $L1_2$	1.2
VCo	Tetr. $D8_b$	<4.2	VRu	Cub. B2	5^g–<0.14
VCo_3[a]	Hex.	<4.2	V_3S	Tetr.	<1.15
VFe[a]	Tetr. $D8_b$	<0.3	V_5S_4	Tetr.	<4.2
VFe	Cub. B2[b]	<4.2	VS[a]	Hex. $B8_1$	<4.2
V_3Ga[a]	Cub. A15	16.8–14.2	V_2Sb_3[a]	Hex.	<4.2
V_5Ga_3	Hex. $D8_8$[f]	<4.2	V_3Sb	Cub. A15	0.80
V_6Ga_5	Hex.	<4.2	V_5Sb_4	Tetr.	<4.2
V_6Ga_7[a]	Cub. $D8_2$	4.2	VSb	Hex. $B8_1$	<4.2
V_2Ga_5	Tetr.	2.1–4.07	VSb_2	Tetr. C16	<0.06
VGa_4	Tetr.	<4.2	V_5Se_4	Tetr.	–
V_3Ge[a]	Cub. A15	6.3–5.88 (7.5–11.2)[b]	VSe[a]	Hex. $B8_1$	–
V_5Ge_3[a]	Tetr. T1	<4.2	V_2Se_3	Low Sym.	–
	Hex. $D8_8$[f]	<1.02			
			V_3Si[a]	Cub. A15	17.2–17
$V_{11}Ge_8$	Rhomb.	<1.2	V_5Si_3	Tetr. T1	<0.30
$V_{17}Ge_{31}$	Tetr.	<1.2	V_5Si_3	Hex. $D8_8$[f]	<0.30
V_2H[a]	Tetr. L_{2b}	<1.02	V_5Si_4	Tetr.	<1.2
VH	Cub.	<4.2	VSi_2	Hex. C40	<1.2
V_2Hf[a]	Cub. C15	9.4–9.0	V_3Sn[a]	Cub. A15	7.0–3.8
V_3In[a]	Cub. A15[b]	13.9			$(12.3–17.9)^b$
V_3Ir[a]	Cub. A15	1.71^g–<0.015	V_2Sn_3	Rhomb.	<1.15
			V_2Ta[a]	Hex. C14[i]	10.0
$VIr(\alpha)$[a]	Rhomb.	<1.6	VTc[a]	Cub. C15[h]	3.6
β-$V_{1-x}Ir_{1+x}$	Tetr. $L1_0$	<1.36		Cub. B2	<1.39
VIr_3[a]	Cub. $L1_2$	<4.2	V_2Tc_3[a]	–	–
VMn_4	Tetr. $D8_b$	<4.2	VTc_3[a]	Cub.	7.8–4.0
V_3N[a]	Hex.	<1.2	V_4Te[a]	–	–
VN	Cub. B1	8.5–7.5	VTe[a]	Hex. $B8_1$	–
V_3Ni[a]	Cub. A15	<0.35	VTe_3	Monocl.	<1.13
V_2Ni_3[a]	Tetr. $D8_b$	<4.2	VTe_2	Monocl.	<0.05

Table 4.2-5 Critical temperature T_c of superconductivity of V compounds [2.1], cont.

Compound	Crystal structure	T_c (K)	Compound	Crystal structure	T_c (K)
VNi_2 [a]	Rhomb.	<4.2	V_3Ti [a]	Cub. A15 [b]	<4.2
VNi_3	Tetr. DO_{22}	<4.2	V_4Zn_5	Tetr.	<4.2
V_4O [a]	Tetr.	<4.2	VZn_3	Cub. $L1_2$	<4.2
V_2O [a]	Hex.	<4.2	VZn_{16} [d]	–	–
VO [a]	Cub. B1	<0.07	V_2Zr [d]	Cub. C15	8.8–6.5
V_2O_3	Rhomb.	<1.28		Rhomb. [j]	8.5

[a] Compound with a range of homogeneity
[b] Metastable phase
[c] "<" means that no superconductivity was found above the temperature indicated
[d] Supposed compound
[e] Superconducting properties not investigated
[f] Unstable in the absence of carbon
[g] Non-stoichiometric composition
[h] Low-temperature modification
[i] High-temperature modification
[j] Formed by martensitic transformation

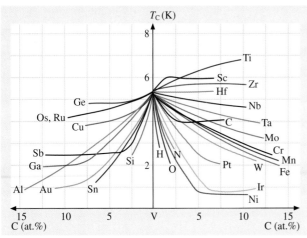

Fig. 4.2-3 Composition dependence of T_c of superconductivity in V solid solution alloys [2.1]

Nb Alloys

Based on its highest T_c among all elements Nb has been considered the most likely base metal for superconducting materials from the beginning and its alloys and intermetallic compounds have been investigated most extensively. Table 4.2-6 shows T_c values for binary Nb compounds, where Nb_3Al, Nb_3Ga, Nb_3Ge, and Nb_3Sn with the cubic A15 type crystal structure stand out for their particularly high T_c. Of these, Nb_3Sn has been developed into a high-field superconducting material (see Sect. 4.2.1.2). Figure 4.2-4 shows the effect of alloying additions on T_c in Nb solid solutions. The Nb–Ti alloys have, finally, been selected to form the most versatile and widely applied high-field superconducting material for present applications (see Sect. 4.2.1.2).

Table 4.2-6 Superconductivity of Nb compounds [2.1]

Compound	Crystal structure	T_c (K)	Compound	Crystal structure	T_c (K)
Nb_3Al	Cub. A15	18.8–15.0	Nb_3Os	Cub. A15	1.5–0.5
NbOAl	Tetr. $D8_b$	0.74	Nb_3Os_2	Tetr. $D8_b$	1.86–1.40
$NbAl_3$	Tetr. $D0_{22}$	0.64	$NbOs_2$	Cub. A12	2.86^d–2.52
Nb_3As	Tetr. Ti_3P	0.31	Nb_3P	Tetr. Ti_3P	2–1.83
					$(7.5–6.0)^g$
$NbAs_2$	Monocl.	$<0.012^a$	Nb_3Pb^e	Cub. A15	9.6<1.5
Nb_3Au	Cub. A15	11.5–8.99	Nb_3Pd_2	Cub. A12	2.47–1.7
Nb_3B_2	Tetr.	<1.0	NbPd	Tetr. $D8_b$	2.0
NbB^b	Rhomb. B_f	8.25	Nb_3Pt	Cub. A15	10.9–6.0
NbB^c	–	<0.05	Nb_2Pt	Tetr. $D8_b$	4.2–3.73
Nb_3B_4	Rhomb. $D7_b$	<1.28	$NbPt(\alpha)$	Rhomb. B19	2.4–<1.39
NbB_2	Hex. C32	6.4^d–<1.0	$NbPt_2$	Rhomb.	<1.46
Nb_3Be_2	Tetr.	2.3	Nb_7Re_9	Tetr. $D8_b$	3.8–2.5
$NbBe_2$	Cub. A15	2.15	$NbRe_4$	Cub. A12	2.45–9.7
$NbBe_3$	Rhomb.	<1.15	Nb_3Rh	Cub. A15	2.64–2.4
Nb_2Be_{17}	Rhomb.	1.47–<1.38	Nb_5Rh_3	Tetr. $D8_b$	4.1–4.04
					10.2^g
$NbBe_{12}$	Tetr.	<1.38	$NbRh(\gamma)$	Tetr. $L1_0$	3.76
Nb_3Bi^e	Cub. A15	3.05 (4.2–2)	$NbRh(\gamma)$	Rhomb.	3.07
Nb_2C	Hex.	9.11–<1.98	$NbRh(\varepsilon)$	Rhomb. B19	3.00
NbC	Cub. B1	11.7–<1.15 (14^f)	$Nb_3Rh_5(\xi)$	Monocl. Sm	2.7
Nb_3Ga	Cub. A15	16.8–14.1 $(20.7)^g$	$NbRh_3$	Cub. $L1_3$	<1.43
Nb_5Ga_3	Tetr.	1.35	Nb_3Ru_2	Tetr.	1.2
$Nb_3Ga_2^h$	Tetr.	<2	Nb_2Ru_3		1.2
Nb_5Ga_4	Hex.	<2	NbS	Hex. $B8_1$	<1.28 $(3.3–3.8)^g$
$NbGa^h$	–	<2	NbS_2^i	Hex. C7	6.15–5.4
Nb_2Ga_3	–	<2	Nb_3Sb	Cub. A15	1.95–0.2
Nb_5Ga_{13}	Rhomb.	<2	Nb_3Sb_2		<1.02
$NbGa_3$	Rhomb.	<2	NbSb	Hex. $B8_1$	<4.2
Nb_3Ge	Cub. A15	6.91 $(23.2)^g$	Nb_4Sb_5	Tetr.	4.2–<1.15
Nb_3Ge	Tetr. Ti_3P	<0.2	$NbSb_2$	Monocl.	4.2–<1.15
Nb_2Ge	Hex. $D8_8$	1.90	Nb_3Se_4	Hex.	1.61
Nb_5Ge_3	Tetr.	<1.02	$NbSe_2$	Hex. C7	7.5–5.4 $(8.5–9)^g$
$NbGe_2$	Hex. C40	2.23–2.09	$NbSe_3$	Monocl.	<1.0
NbH	Cub.	<1.30	Nb_4Si^h	Hex. $\varepsilon – Fe_3N$	<1.5
NbH_2	Cub.	<0.47	Nb_3Si	Tetr.	$0.29(5.45)^e$
Nb_3In	Cub. A15	9.2–4	Nb_3Si^e	Cub. A15	19–13
Nb_3Ir	Cub. A15	1.76–1.63	$Nb_5Si_3(\alpha)$	Tetr.^{b}	<1.02
Nb_5Ir_3	Tetr. $D8_b$	9.8^i–2.4^d	Nb_5Si_3	Hex. $D8_8$	<1.02
$NbIr_{1+x}$	Tetr. $L1_0$	4.75	$NbSi_2$	Hex. C40	<1.20
$Nb_2Ir(\alpha_2)$	Rhomb.	4.6	Nb_3Sn	Cub. A15	18.5–4.0
$NbIr_3$	Cub. $L1_2$	<1.2	Nb_6Sn_5	Rhomb.	2.07

Table 4.2-6 Superconductivity of Nb compounds [2.1], cont.

Compound	Crystal structure	T_c (K)	Compound	Crystal structure	T_c (K)
$Nb_2N(\beta)$	Hex.	9.5–<1.2	$NbSn_2$	Rhomb.	2.68
$Nb_4N_3(\gamma)$	Tetr.	12.2–7.8	$NbTc_3$	Cub. A12	12.9–10.5
$NbN(\delta')$	Tetr.	7.2	Nb_3Te^g	Cub. A15	<2.5
$NbN(\delta)$	Cub. B1	16.5–9.7	Nb_5Te_4	–	<1.1
$NbN(\varepsilon)$	Hex.	<1.20	Nb_3Te_4	Hex.	1.49
Nb_5N_6	Hex.	<1.77	$NbTe_2$	Hex. C7	0.74–0.5
Nb_4N_5	Tetr.	8.0–8.5	$NbTe_4$	–	<0.025
NbO	Cub. B1	1.61–1.38	Nb_3Tl^b	Cub. A15	9.0
NbO_2	Tetr.	<1.2	$NbZn_3$	Cub. $L1_2$	<1.02

[a] "<" means that no superconductivity was found above the temperature indicated
[b] Low-temperature modification
[c] High-temperature modification
[d] Non-stoichiometric composition
[e] Metastable phase
[f] Extrapolated maximum T_c
[g] Under extreme conditions
[h] Stabilized by adding interstitial elements
[i] Phase is stable at >850 °C only

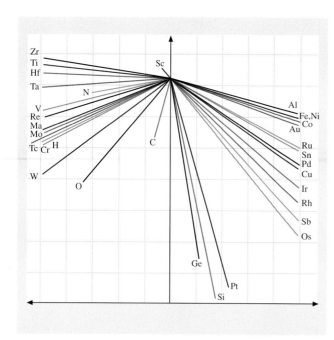

Fig. 4.2-4 Composition dependence of T_c of superconductivity in Nb solid solution alloys [2.1]

Tc Alloys

Since Tc is rare and difficult to prepare in the alloy form, its superconducting properties have been studied to a limited extent only (see Table 4.2-7 and Fig. 4.2-5).

4.2.1.2 Practical Metallic Superconductors

Practical metallic superconductors for DC or AC applications are invariably composite wires with superconducting filaments which are embedded in a normal conducting matrix, usually Cu. They have a high longitudinal conductivity and may contain further components such as low conductivity barriers consisting of a Cu–Ni alloy, or diffusion barriers of Nb or Ta. Accounts of combining the aspects of the physics of superconductivity, materials science and technology, and electrical and mechanical performance criteria which have to be taken into account and are mastered in producing superconducting wires are given in [2.4–8].

Table 4.2-7 T_c of superconductivity of Tc compounds [2.1]

Compound	Crystal structure	T_c (K)	Compound	Crystal structure	T_c (K)
Tc_3As_7	Cub. $D8_f$	<0.3 [a]	Tc_6Ti [b]	Cub. A12	8.10–7.73
TcBe	Cub.	5.21	TcTi	Cub. B2	<1.7
$TcBe_{22}$	Cub. $ZrZn_{22}$	5.25	Tc_2Th	Hex. C14	5.3 [c]
TcC	Cub.	3.85	$Tc_3V(\delta)$	Cub.	7.8–4.0
Tc_2Hf	Hex. C14	5.6	$TcV(\varepsilon)$	Cub. B2	<1.39
Tc_7Mo_3	Tetr. $D8_b$	15.8–14.7	Tc_3W_2	Tetr. $D8_b$	7.88–8.35
Tc_2Mo_3	Cub. A15	14–12	Tc_6Zr	Cub. A12	9.7
Tc_3Nb	Cub. A12	12.9–10.5	Tc_2Zr	Hex. C14	7.6
Tc_3Sn [d]	Hex.	5.92			

[a] "$<$" means that no superconductivity was found above the temperature indicated
[b] Formed as an ordered phase by extended annealing of the solid solution phase
[c] Non-stoichiometric composition
[d] Compound not determinded definitely; T_c may be related to the solid solution

Only two superconducting metallic phases are used routinely in superconducting wires for applications: the Nb–Ti solid solution phase and the Nb_3Sn intermetallic phase. Due to the difference of their intrinsic superconducting properties T_c and B_{c2}, they have different ranges of application as indicated in Table 4.2-8.

The critical current density $J_c(T, B)$ obtained by applying optimal microstructural pinning of the flux lines in conductors of different composition and at different temperatures is shown in Fig. 4.2-6. Two factors of influence may be applied to obtain a higher critical current density $J_c(T, B)$ through effects of the intrinsic properties: either a decrease in temperature of application, e.g., from 4.2 to 1.8 K, or an increase of B_{c2} by alloying as shown for $(Nb, Ta, Ti)_3Sn$ in both Table 4.2-8 and Fig. 4.2-6. It should be noted that the intrinsic properties are affected only marginally by differences in processing of the conductors.

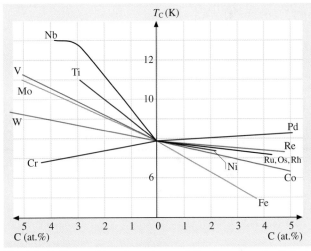

Fig. 4.2-5 Composition dependence of T_c of superconductivity in Tc solid solution alloys [2.1]

Table 4.2-8 Characteristic properties of practical metallic superconductors (After [2.8])

Superconducting phase	Composition	T_c (K)	B_{c2} (T)	Magnetic field in application B (T)
Nb–Ti	46–52 wt% Ti, ≈ 47 wt% Ti optimal	≈ 10	≈ 10.5 (4.2 K)	≤ 9 (4.2 K) > 9 (1.8 K)
Nb_3Sn $(Nb, Ta, Ti)_3Sn$	25 at.% Sn, ≤ 7.5 wt% Ta, ≤ 0.2 wt% Ti	≈ 18	≈ 23 (4.2 K): $\approx 26–29$ (4.2 K)	≤ 20 T (4.2 K) several T (< 4.2 K)

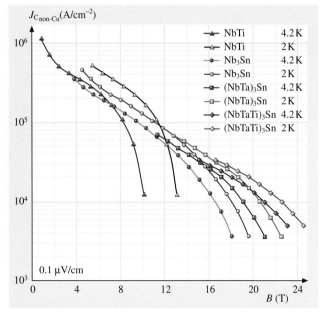

Fig. 4.2-6 Non-Cu J_c versus B characteristics of Nb–Ti, binary Nb$_3$Sn, alloyed (Nb, Ta)$_3$Sn, and alloyed (Nb, Ta, Ti)$_3$Sn multifilamentary superconductors [2.8]

Nb–Ti Superconductors

Extensive accounts are given in [2.4, 6, 7]. The Nb–Ti alloy phase diagram combined with the low-temperature $T_c(c_{Ti})$ and $B_{c2}(c_{Ti})$ relations of the superconducting β phase (Fig. 4.2-7a and b), provide the basic features of both the intrinsic superconducting properties of the β phase and of the potential for flux pinning by precipitation of the α phase induced by suitable heat treatment in the $\alpha + \beta$ two-phase range.

The subsequent characteristic data are taken from systematic treatments and measurements of a Cu stabilized Nb–46.5 wt% Ti superconductor [2.9] which is in the range of optimum composition according to Table 4.2-8 and Fig. 4.2-7b. The processing of the composite consists essentially of a succession of deformation treatments, usually by wire drawing to different strains (true strain $\varepsilon_t = 2\ln(d_i/d)$, where d_i is the initial rod diameter and d is the filament diameter at the drawing strain considered), followed by heat treatments for 20 h to 40 h, mainly at temperatures in the range 360 to 380 °C, but also up to 420 °C. The first stage of deformation of the homogeneous metastable β phase leads to the formation of a high density of dislocations, thus providing a high density of heterogeneous nucleation sites for the precipitation of a fine dispersion of α phase particles during the first heat treatment. During the subsequent cycles of deformation and heat treatment the α particles previously formed are elongated and partially broken up. Additional particles may be newly formed along dislocation cell walls and subgrain boundaries. The particle spacing transverse to the drawing direction is reduced. These variations of the dispersion are strongly affecting the pinning behavior and, thus, the J_c values obtained, referring to a composite resistivity of $10^{-14}\,\Omega\,\mathrm{m}$.

The diagrams of Fig. 4.2-8a–e show J_c as a function of the true drawing strain ε_t. The parameters are different thermal treatments, and different conditions (magnetic field, temperature) of measurement.

From measurements such as those shown in Fig. 4.2-8e, the effective bulk pinning force F_p can be evaluated as shown in Fig. 4.2-9.

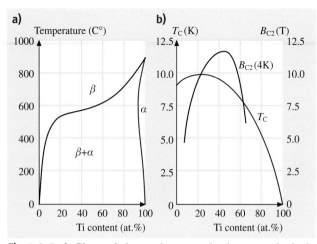

Fig. 4.2-7a,b Phase relations and superconducting properties in the Nb–Ti alloy system. (**a**) Nb–Ti phase diagram in the solid state. (**b**) T_c and B_{c2} (4 K) of the metastable β phase

Fig. 4.2-8a–e ▶ Critical current variations as a function of different thermal treatments, final drawing strain and field of measurement, respectively, of an Nb–46.5 wt%Ti superconductor at 4.2 K [2.9]. (**a**) □ 160 h at 405 °C, ○ 80 h at 405 °C; ×40 h at 375 °C. — $B = 5$ T, --- $B = 8$ T. (**b**) All treatments for 80 h: 375 °C, ○ 405 °C, △ 420 °C, ◇ 435 °C. — $B = 5$ T, --- $B = 8$ T. (**c**) All treatments at 420 °C: ×160 h, 80 h, ○ 40 h, △ 10 h, ◇ 5 h. $B = 5$ T. (**d**) 160 h at 420 °C: ○ two heat treatments, three heat treatments. $B = 5$ T. (**e**) Full field optimum: ×80 h at 420 °C, 40 h at 375 °C, ○ 5 h at 405 °C

Superconductors | 2.1 Metallic Superconductors | 707

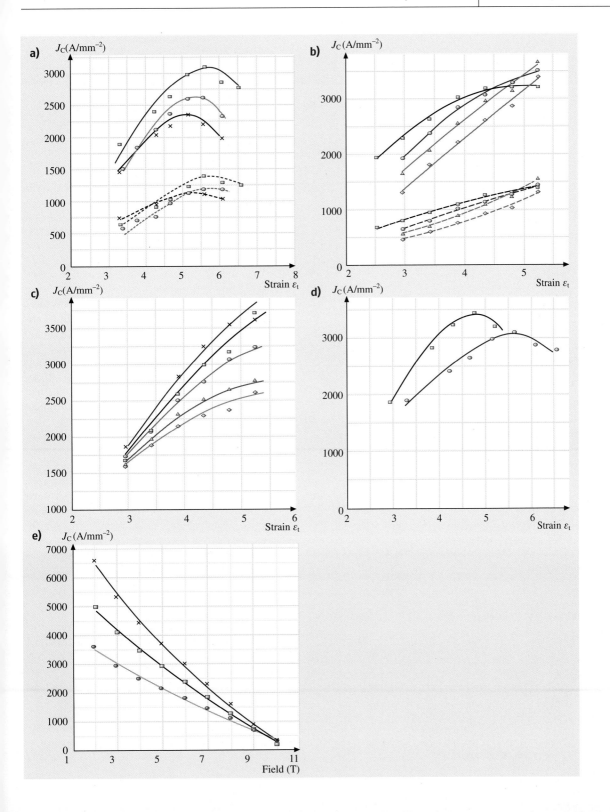

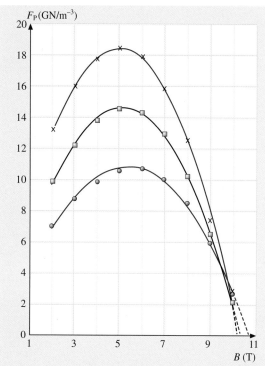

Fig. 4.2-9 Bulk pinning force F_p versus magnetic field, corresponding to the J_c data shown in Fig. 4.2-8e. Typical particle spacings for the optimum J_c is $d \approx 20$ nm and $d \approx 3–5$ nm, respectively [2.9]

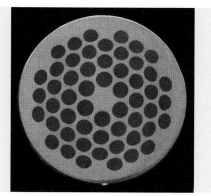

Fig. 4.2-10 Typical cross section of a Cu-stabilized Nb–Ti superconducting wire

Nb$_3$Sn Superconductors

The production of superconductors based on Nb$_3$Sn as the superconducting phase [2.7, 8] is hampered by the brittle behavior of this intermetallic phase which cannot be deformed by wire drawing. Consequently the conductors are produced by composing ductile components first, reducing the cross section of the composite into the final wire form by extrusion and drawing processes, and finally forming the Nb$_3$Sn phase by a diffusion–reaction treatment after the specified wire diameter has been obtained. Five routes of processing have been developed:

1. *Bronze process:* The initial rod composite contains Nb rods surrounded by Cu–Sn alloy rods (13–15 wt% Sn) in a suitable arrangement such that after processing into wire form the diffusion–reaction treatment leads to the formation of Nb$_3$Sn at the interface of Nb and Cu–Sn. The initial composite, if composed in a suitable configuration, has excellent deformation properties such that a geo-

Based on systematic variations and optimizations such as those shown in Fig. 4.2-8, commercial Nb–Ti superconducting composite wires are provided for a wide range of specifications. Figure 4.2-10 shows a typical wire cross section and Table 4.2-9 lists some characteristic data.

Table 4.2-9 Characteristic data at 4.2 K of commercial Nb–Ti superconductors, round wires

Filaments	A_{Cu}/A_{NbTi} [a]	Wire diameter (mm)	Filament diameter (μm)	Critical current J_c, (A) at $\mu_0 H$ (T)			
				3 T	5 T	7 T	9 T
54	1.35	0.3	27	100	70	45	–
54	1.35	0.6	53	380	265	170	–
54	1.35	0.85	75	–	480	310	140
45	1.8	0.4	36	150	105	70	–
45	1.8	0.7	62	420	295	185	–

[a] Ratio of cross sectional area
[b] Data extracted from a commercial brochure of Vacuumschmelze GmbH (9/1990)

Table 4.2-10 Characteristic data at 4.2 K of commercial Nb_3Sn superconductors produced according to the bronze process; filament diameters 4 to 6 μm

Stabilization	Filament material	Wire diameter (mm)	Critical current density I_c (A) at $\mu_0 H =$				Code[a]
			8 T	10 T	12 T	14 T	
None	Nb	0.5	202	136	78	–	NS 4500 (0.5)
None	Nb	0.7	374	252	144	–	NS 6000 (0.7)
None	Nb	1.0	753	504	289	–	NS 10 000 (1.0)
None	Nb–Ta	0.5	171	123	86	58	HNST 4500 (0.5)
None	Nb–Ta	0.7	316	228	159	107	HNST 6000 (0.7)
None	Nb–Ta	1.0	636	456	321	215	HNST 10 000 (1.0)
Cu/Ta core	Nb	0.7	305	204	117	–	NS 6000 (0.7) Ta I
Cu/Ta core	Nb	1.0	623	417	239	–	NS 10 000 (1.0) Ta I
Cu/Ta core	Nb	1.25	973	652	373	–	NS 13 000 (1.25) Ta I
Cu/Ta core	Nb–Ta	0.7	258	185	130	87	HNST 6000 (0.7) Ta I
Cu/Ta core	Nb–Ta	1.0	526	378	266	178	HNST 10 000 (1.0) Ta I
Cu/Ta core	Nb–Ta	1.25	822	591	416	279	HNST 13 000 (1.25) Ta I

[a] According to the VACRYFLUX® designations of Vacuumschmelze GmbH (1990); the figures without parentheses indicate the approximate number of filaments

metrically well controlled wire results, in which decoupling of the filaments can be optimized and no bridging of filaments occurs. Table 4.2-10 shows the characteristic data of commercial Nb_3Sn superconductors produced according to the bronze process.

2. *External Sn diffusion process:* Nb filaments in a Cu matrix are produced from a mechanical composite of rods by processing into wire form; at its final diameter the wire is coated with Sn; during a series of heat treatments with increasing temperature, Sn and Cu react to form a Cu–Sn alloy and this alloy reacts with the Nb filaments to form Nb_3Sn. This process is limited to wire with diameters ≤ 0.2 mm.

3. *Internal Sn diffusion process:* Pure Sn or dilute Sn alloys with Cu or Mg are incorporated in the initial rod composite: after processing into wire form the Nb_3Sn phase is formed by a diffusion–reaction process with the Nb filaments. However, due to the low melting point of Sn and its low yield stress, the processing cannot use hot extrusion and the drawing processes can readily lead to inhomogeneous deformation and mechanical instabilities.

4. *Jelly roll process:* Nb foil or mesh and Cu foil are wound into a roll as a starting rod which yields an Nb filament bundle after processing into wire form. In combination with the bronze or internal-Sn process this results in Nb_3Sn filaments with extremely low diameter. It is difficult to scale the process to large quantities and wire cross-sections.

The four processes mentioned require a final diffusion–reaction treatment at 650 to 700 °C for 50 to 200 h. Since the stabilizing Cu constituent in the composite must not be exposed to Sn interdiffusion in order to maintain its high conductivity, a diffusion barrier is incorporated in the composite which usually consists of Ta in the form of a tube.

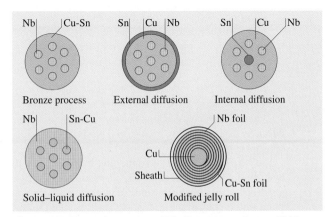

Fig. 4.2-11 Processing routes of Nb_3Sn superconductors [2.7]

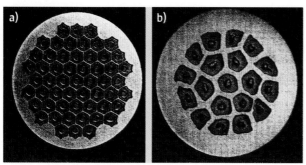

Fig. 4.2-12a,b Examples of cross sections of superconductors produced by the internal-Sn ((**a**) 0.8 mm diameter) and Nb-tube ((**b**) 0.5 mm diameter) processes [2.8]

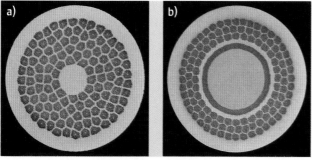

Fig. 4.2-13a,b Superconductors produced according to the bronze process (Vacuumschmelze). (**a**) Non-stabilized, 10 080 Nb filaments in a Cu−Sn matrix. (**b**) Internally stabilized by Cu in a Ta tube, 6156 Nb filaments in a Cu−Sn matrix

only regarding the rate of the diffusion–reaction process but also regarding the resulting grain size.

Moreover, the critical current density depends strongly on the state of strain in Nb_3Sn, in particular close to B_{c2}. Figure 4.2-15 shows the strain sensitiv-

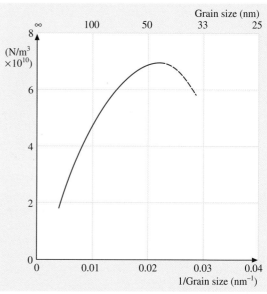

Fig. 4.2-14 Dependence of the volume pinning force F_p in Nb_3Sn on the grain size

5. **Nb-*tube process*:** The need of an extra barrier element can be avoided by using Nb tubes, filled with Sn or with a Sn compound such as $NbSn_2$, embedded in a Cu matrix. During heat treatment, a Nb_3Sn layer is formed inside the tube while its outer part acts as a diffusion barrier. This processing route leads to wires with high critical current densities but is difficult to scale up for industrial production.

Schematic drawings of the composites according to these five processing routes are shown in Fig. 4.2-11.

Examples of cross sections of superconductors produced by the internal-Sn, bronze, and Nb-tube processes are shown in Figs. 4.2-12 and 4.2-13.

In Nb_3Sn superconductors the flux lines are pinned by the grain boundaries in the Nb_3Sn filaments. The dependence of the effective volume pinning force F_p on the grain size is shown in Fig. 4.2-14. This leads to the requirement that the reaction temperature is optimized not

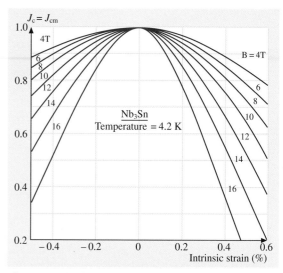

Fig. 4.2-15 Strain sensitivity $J_c(\varepsilon)$ of Nb_3Sn superconductors [2.10]

ity $J_c(\varepsilon)$ for binary Nb_3Sn under uniaxial longitudinal strain. In its state of formation, i.e., in the absence of an externally applied strain, the Nb_3Sn phase as formed has an intrinsic compressive strain of $\varepsilon_i = 0.2-0.3\%$, depending on the conductor design and processing parameters. Thus, applying a tensile strain will first lead to an increase in J_c. However, degradation sets in irreversibly if strains of a few tenths of one per cent are exceeded due to crack formation in the brittle Nb_3Sn phase.

Table 4.2-10 lists characteristic data of commercial Nb_3Sn superconductors.

4.2.2 Non-Metallic Superconductors

4.2.2.1 Oxide Superconductors

Low-T_c Superconducting Oxides

Only a few oxides with critical temperatures above the boiling temperature of liquid helium (4.2 K) were known up to 1986. Their T_c did not exceed 14 K. The stoichiometric composition of nearly all of these compounds ($T_c > 4.2$ K) is related to one of the formulas:

$BaPb_{1-x}Bi_xO_3$, $Li_{1+x}Ti_{2-x}O_4$, or A_yWO_3 ($y < 0.5$; $A = Cs$; Rb).

Relatively high T_c values of the Ba–Pb–Bi–O compounds were also observed after substitution of small amounts of K and Sr for Ba, respectively, and after substitution of Sn for Pb (Table 4.2-11).

Table 4.2-11 Electrical and structural data of selected low-T_c oxide superconductors

Compound	Structure Lattice parameters a, b, c (nm)	T_c (K)	$\mu_0 H_{c2}$ (T)	$\mu_0 H_{c1}$ (mT)
$BaPb_{1-x}Bi_xO_3$ $x = 0.2-0.3$[a]		8–12		
Polycrystalline	Orthorhombic ($x = 0.30$) $a = 0.6075$, $b = 0.610$, $c = 0.857$ Tetragonal ($x = 0.275$) $a = 0.6055$, $c = 0.8633$		For $x = 0.25$: 5.4 (1.8 K) 2.3 (8 K) For $x = 0.2$: 7.0 (1.15 K) 2.3 (6 K)	For $x = 0.25$: 23 (4.2 K)
Single crystal	Orthorhombic ($x \approx 0.25$) $a = 0.6058$, $b = 0.6081$, $c = 0.8552$ Tetragonal ($x \approx 0.25$) $a = 0.6045$, $b = 0.8613$		For $x = 0.27$: 3.8 (1.5 K) For $x = 0.2$: 1.75 (3.5 K)	
$Ba_{0.9}K_{0.1}Pb_{0.75}$–$Bi_{0.25}O_3$		11.5		
$BaSn_{0.01}Pb_{0.74-0.69}$–$Bi_{0.25-0.3}O_3$		10–11		
$Li_{1...1.3}Ti_{2-1.67}O_4$	Cubic $a = 0.841-0.836$	9.6–12	3.3 (0 K)	
$Li_{2.6}Ti_{1.5-2.7}O_4$	Cubic	10.9–11.4	9.6–16.2 (4.2 K)	
$Rb_{0.3}WO_3$	Hexagonal	7.5–0.6		
$Cs_{0.1}WO_{2.9}F_{0.1}$ (Single crystal)	Hexagonal $a = 0.74$; $c = 0.76$	≈ 4.9	0.13–0.36 (dependent on temperature and crystal orientation relative to H)	

[a] The electrical and structural data of $BaPb_{1-x}Bi_xO_3$ are values of different authors published in [2.2]

An extensive compilation of the oxide superconductors of the class called low-T_c superconductors is given in [2.11]. Some data, which are relevant to bulk materials revealing relatively high T_c values are summarized in Table 4.2-11.

For metallic superconductors the electron–phonon interaction accounts for the superconductivity of low-T_c oxide conductors. For all of these materials the electrical parameters indicated in Table 4.2-11 depend on the electron mean free path, and thus on the resistivity ρ, of the material in the normal state. The quantitative relation for H_{c2} is given by

$$\mu_0 H_{c2}(0) = 3.11 \, \rho \gamma \, T_c \,,$$

where γ is the electronic specific heat. As a consequence of the influence of ρ on T_c and H_c, both parameters depend not only on the stoichiometric composition of the multi-component oxides but also on impurities and lattice defects introduced by the preparation. Accordingly, in the literature the electrical parameters are not all identical to those of Table 4.2-11. For $BaPb_{1-x}Bi_xO_3$ compounds the following values of coherence length ξ_0 and London penetration depth λ were published: $\xi_0 = 2400, 250, 50\,\text{nm}$, $\lambda = 0.5, 0.7, 1.5\,\text{nm}$ for $x = 0$, 0.12 and 0.3, respectively [2.11].

High-T_c Superconducting Oxides

Chemical Composition. Bednorz and Müller carried out a series of investigations aiming to enhance the electron–phonon interaction in superconducting oxides. As a result, they found superconductivity at around 30 K in the Ba−La−Cu−O system [2.12]. These compounds were the first ones of the class of cuprates called high-T_c superconductors.

After this discovery many new compounds revealing T_c values up to 138 K under ambient pressure were detected. The stoichiometric compositions of most high-T_c cuprates can be assigned to one of the formulas presented in Table 4.2-12. A similar classification is given by [2.13]. The A-elements as well as the E-elements can be partially substituted by one ore more other metallic elements as shown in Table 4.2-13.

The partial substitution of oxygen by non-metallic elements, such as fluorine or chlorine, is also possible for some compounds. Sometimes this even leads to a slight enhancement of T_c. The introduction of rows of carbonate groups between two CuO_2 layers (see structure principles below) leads to superconducting layered oxy-carbonates with the formulation $(A_2CO_3)_m Ca_{n-m}(CuO_2)_n$. The "$m = 1, n = 2$" and "$m = 1, n = 3$" members with A = Ba are considered as the basic high-T_c compounds of this class [2.14].

The Structure of the Superconducting Compounds. In Fig. 4.2-16, which shows the crystal structure of the $Bi_m Sr_2 Ca_{n-1} Cu_n O_{2n+m+2}$ homologous series, the structural principles of most of the superconducting high-T_c cuprates are indicated. The structure element carrying the mobile charges is a stack of a certain number n ($n = 1, 2, 3$ in Fig. 4.2-16) of CuO_2 monolayers, which are separated by intermediate monolayers of alkaline earth elements E (E = Ca in Fig. 4.2-16) or monolayers of lanthanide elements L (in $LBa_2Cu_3O_{7-x}$, Table 4.2-12). These stacks of CuO_2/E and CuO_2/L, re-

Table 4.2-12 General chemical formulas of high-T_c cuprates

Formula	Examples A-elements	Stoichiometric coefficients m	n	Examplary E-elements
$A_m E_2 Ca_{n-1} Cu_n O_{2n+m+2}$	Tl	1–2	1–4	Ba, Sr
	Bi	1–2	1–4	
	(Bi,Pb)	1–2	1–4	
	Hg	1–2	1–6	
	Cu	1–2	3–5	
	B	1	3–5	
	Pb	1	1–2	
	Au	1	2	
$LBa_2Cu_3O_7$	Y, La, Pr, Nd, Sm, Eu, Gd, Tb, Dy, Ho, Er, Tm, Yb, Lu			
A_2CuO_4	La, Pr, Nd, Sm, Eu			
AE_2CuO_{4-5}	Hg, Tl			Ba, Sr
$A_2Ca_{n-1}Cu_nO_{2n+2}$	Ba, La		2–4	

Table 4.2-13 Electrical data of high-T_c cuprates

Compound	T_c (K)	ΔT_c (K)	$\rho(0); \rho(T_c)$ ($\mu\Omega$ cm)	$H_{c2}(0)$ (T)	$\lambda_{ab}(0)$ (nm)	Ref.
A_2CuO_4 family						
Sr_2CuO_4	25					[2.13]
$La_{1.9}Sr_{0.1}CuO_4$	33	6.5			320 ± 16	[2.15]
$La_{1.875}Sr_{0.125}CuO_4$	36	4.5			270 ± 13.5	[2.15]
$La_{1.85}Sr_{0.15}CuO_4$	39	2	$0; 33 \pm 6$	45 ± 10	218.5 ± 10	[2.15]
$La_{1.775}Sr_{0.225}CuO_4$	29	7.5				[2.15]
$La_{1.725}Sr_{0.275}CuO_4$	22	10				[2.15]
$Eu_{2-x}Ce_xCuO_4$	23					[2.13]
$(Nd_{2-x}Ce_x)CuO_4$	24, 30	2	250; 250			[2.13, 16]
$Pr_{2-x}Ce_xCuO_4$	24					[2.13]
$Sm_{2-x}Ce_xCuO_4$	22					[2.13]
$ABa_2Cu_3O_7$ family						
$YBa_2Cu_3O_{6.67}$	60	1	$x; 63 \pm 8$	87	255 ± 12.5	[2.15]
$YBa_2Cu_3O_7$	92	1	$0; 40 \pm 5$	140 ± 30	141.5 ± 1	[2.15]
$(Y_{0.95}Pr_{0.05})Ba_2Cu_3O_7$	90	1.8	50; 100		150 ± 7.5	[2.15]
$(Y_{0.9}Pr_{0.1})Ba_2Cu_3O_7$	86	2.2	50; 100		173 ± 8.5	[2.15]
$(Y_{0.8}Pr_{0.2})Ba_2Cu_3O_7$	78	3.2	100; 140		198 ± 10	[2.15]
$(Y_{0.7}Pr_{0.3})Ba_2Cu_3O_7$	60	7.5	250; 280		215 ± 10	[2.15]
$(Y_{0.6}Pr_{0.4})Ba_2Cu_3O_7$	40	13.2	300; 335		290 ± 15	[2.15]
$YBa_2(Cu_{0.99}Fe_{0.01})_3O_7$	91	1.3				[2.15]
$YBa_2(Cu_{0.975}Fe_{0.025})_3O_7$	85	3				[2.15]
$YBa_2(Cu_{0.95}Fe_{0.05})_3O_7$	75	5				[2.15]
$YBa_2(Cu_{0.99}Zn_{0.01})_3O_7$	78	1				[2.15]
$YBa_2(Cu_{0.975}Zn_{0.025})_3O_7$	62	1	150; 200			[2.15]
$YBa_2(Cu_{0.95}Zn_{0.05})_3O_7$	36	1.5	240; 300			[2.15]
$HoBa_2Cu_3O_7$	92	1				[2.15]
$NdBa_2Cu_3O_7$	96					[2.13]
$GdBa_2Cu_3O_7$	94					[2.13]
$ErBa_2Cu_3O_7$	92					[2.13]
$YbBa_2Cu_3O_7$	89					[2.13]
$A_mE_2Ca_{n-1}Cu_nO_{2n+m+2}$ family						
$Bi_2Sr_2CaCu_2O_7$	107					[2.13]
$Bi_2Sr_2CuO_6$	34					[2.13]
$Bi_2(Sr_{1.6}La_{0.4})CuO_{6+x}$	23.5		320; 360			[2.15]
$(Bi, Pb)_2(Sr_{1.75}La_{0.25})CuO_6$	24					[2.15]
$Bi_2Sr_2CaCu_2O_8$ (no anneal)	90	1	$0; 38 \pm 2$	107	250 ± 50	[2.15]
$Bi_2Sr_2CaCu_2O_{8+x}$ (O_2-annealed)	75	1				[2.15]
$Bi_2Sr_2Ca_2Cu_3O_{10}$	107, 110	10	$0; <60$			[2.13, 15]
$(Bi_{1.6}Pb_{0.4})Sr_2Ca_2Cu_3O_{10}$	110	1	$0; x$	184	253 ± 30	[2.15]
$Bi_2Sr_2Ca_3Cu_4O_{12}$	110					[2.13]
$PbSr_2CaCu_2O_7$	70					[2.13]
$Pb_2Sr_2Ca_{0.5}Ln_{0.5}Cu_3O_8$	70					[2.16]
$PbSr_2Ca_2Cu_3O_9$	122					[2.13]
$AuBa_2CaCu_2O_7$	82					[2.13]

Table 4.2-13 Electrical data of high-T_c cuprates, cont.

Compound	T_c (K)	ΔT_c (K)	$\rho(0); \rho(T_c)$ ($\mu\Omega$ cm)	$H_{c2}(0)$ (T)	$\lambda_{ab}(0)$ (nm)	Ref.
$A_m E_2 Ca_{n-1} Cu_n O_{2n+m+2}$ family						
$Tl_2Ba_2CuO_6$	85		40; 100		170 ± 10	[2.13, 15]
	90					
$Tl_2Ba_2CaCu_2O_8$	99		0 ± 20;	99	221 ± 10	[2.13, 15, 16]
	110		300 ± 80			
	119					
$Tl_2Ba_2Ca_2Cu_3O_{10}$	125		20; 83	75	196 ± 10	[2.13, 15]
	128					
$Tl_2Ba_2Ca_3Cu_4O_{12}$	119					[2.13]
$TlBa_2CaCu_2O_7$	82					[2.13, 16]
	103					
$TlBa_2Ca_2Cu_3O_9$	135					[2.13, 16]
	110					
$TlBa_2Ca_3Cu_4O_{11}$	127					[2.13]
$(Tl_{0.5}Pb_{0.5})Sr_2CaCu_2O_7$	80				182 ± 9	[2.15]
$(Tl_{0.5}Pb_{0.5})Sr_2(Ca_{0.8}Y_{0.2})Cu_2O_7$	107		0; 250			[2.15]
$(Tl_{0.5}Pb_{0.5})Sr_2Ca_2Cu_3O_9$	122				158 ± 8	[2.15]
$TlSr_2Ca_{0.5}Y_{0.5}Cu_2O_7$	90					[2.16]
$CuBa_2Ca_2Cu_3O_9$	60					[2.13]
$CuBa_2Ca_3Cu_4O_{11}$	117					[2.13]
$Cu_2Ba_2Ca_2Cu_3O_{10}$	67					[2.13]
$Cu_2Ba_2Ca_3Cu_4O_{12}$	113					[2.13]
$BSr_2Ca_2Cu_3O_9$	75					[2.13]
$BSr_2Ca_3Cu_4O_{11}$	110					[2.13]
$BSr_2Ca_4Cu_5O_{13}$	85					[2.13]
$Pb_2(Y_{1-x}Ca_x)Sr_2Cu_3O_8$	80		50; 250			[2.15]
$HgBa_2CaCu_2O_7$	128					[2.13]
$HgBa_2Ca_2Cu_3O_9$	135					[2.13]
$HgBa_2Ca_3Cu_4O_{11}$	127					[2.13]
$HgBa_2Ca_4Cu_5O_{13}$	110					[2.13]
$HgBa_2Ca_5Cu_6O_{15}$	107					[2.13]
$Hg_2Ba_2CaCu_2O_7$	44					[2.13]
$Hg_2Ba_2Ca_2Cu_3O_9$	45					[2.13]
$Hg_2Ba_2Ca_3Cu_4O_{12}$	114					[2.13]
$AE_2CuO_{4-\delta}$ family						
$HgBa_2CuO_5$	97					[2.13]
$TlBa_2CuO_5$	50					[2.13]
$(Hg_{0.8}V_{0.2})Ba_2CuO_{4.3}$	89					[2.14]
$(Hg_{0.8}Mo_{0.2})Ba_2CuO_{4.4}$	66					[2.14]
$(Hg_{0.9}Mo_{0.1})Ba_2CuO_{4.2}$	68					[2.14]
$(Hg_{0.8}W_{0.2})Ba_2CuO_{4.4}$	37					[2.14]
$(Hg_{0.9}W_{0.1})Ba_2CuO_{4.2}$	83					[2.14]
$(Hg_{0.8}Ru_{0.2})Ba_2CuO_{4.2}$	91					[2.14]
$(Hg_{0.7}Cr_{0.3})Sr_2CuO_{4.15}$	60					[2.14]
$(Hg_{0.8}Mn_{0.2})Ba_2CuO_{4.2}$	90					[2.14]
$(Hg_{0.8}Nb_{0.2})Ba_2CuO_{4.3}$	78					[2.14]

Table 4.2-13 Electrical data of high-T_c cuprates, cont.

Compound	T_c (K)	ΔT_c (K)	$\rho(0); \rho(T_c)$ ($\mu\Omega$ cm)	$H_{c2}(0)$ (T)	$\Lambda_{ab}(0)$ (nm)	Ref.
$AE_2CuO_{4-\delta}$ family						
$(Hg_{0.8}Mo_{0.2})Sr_2CuO_x$	78					[2.14]
$(Hg_{0.9}Re_{0.1})Sr_2CuO_x$	70					[2.14]
$A_2Ca_{n-1}Cu_nO_{2n+2}$ family						
$Ba_2CaCu_2O_6$	90					[2.13]
$Ba_2Ca_2Cu_3O_8$	120					[2.13]
$Ba_2Ca_3Cu_4O_{10}$	105					[2.13]
$Ba_2Ca_4Cu_5O_{12}$	90					[2.13]
$La_{1.6}Sr_{0.4}CaCu_2O_6$	60	10				
Other compounds						
$YBa_2Cu_4O_8$	80	2			196 ± 10	[2.15]
$HoBa_2Cu_4O_8$	80	5	100; 400		161 ± 8	[2.15]
$Bi_2Sr_2(Ln_{1-x}Ce_x)_2Cu_2O_{10}$	25					[2.16]
$Ca_{1-x}Sr_xCu_2O_2$	110					[2.16]
$(Pb, Cu)Sr_2(Ln, Ca)Cu_2O_7$	50					[2.16]
$(Pb, Cu)(Sr, Eu)(Eu, Ce)Cu_2O_x$	25					[2.16]
$Sr_{1-x}Nd_xCuO_2$	40					[2.16]
$Tl(Ba, La)CuO_5$	40					[2.16]
$Tl(Sr, La)CuO_5$	40					[2.16]
$(Tl_{0.5}Pb_{0.5})Sr_2CuO_6$	40					[2.16]
$(Tl_{0.7}Cd_{0.3})BaLaCuO_5$	48					[2.15]

spectively, are sandwiched between insulating blocks consisting of m monolayers of A-oxide ($m = 2$ and A = Bi in Fig. 4.2-16) and two monolayers of E-oxide (E = Sr in Fig. 4.2-16). These blocks can act as electronically active charge-reservoirs for hole or electron donation to the copper-oxygen layers.

Superconducting Properties. By means of flux quantization and Josephson tunneling experiments, it has been firmly established that superconductivity in the cuprate compounds is based on electron pairs similar to that of conventional Bardeen–Cooper–Schreiffer (BCS) superconductors. However, there are currently widely differing views as to the pairing mechanism and no consensus exists on a proper microscopic theory. In addition

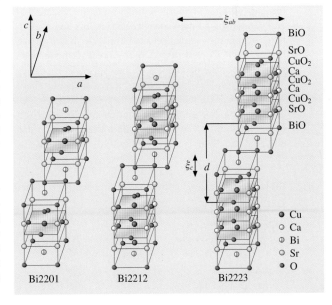

Fig. 4.2-16 Schematic crystal structures of the homologous series of $Bi_2Sr_2Ca_{n-1}Cu_nO_{2n+4}$ superconductors with $n = 1$ ($Bi_2Sr_2CuO_6$, abbreviated as Bi2201), $n = 2$ ($Bi_2Sr_2CaCu_2O_8$ – Bi2212) and $n = 3$ ($Bi_2Sr_2Ca_2Cu_3O_{10}$ – Bi2223). The coherence length in the c-direction, ξ_c, is compared to that perpendicular to the c-direction, ξ_{ab}, and to the distance d between the superconducting CuO_2/Ca stacks

to the conventional electron–phonon interaction other explanations have been proposed [2.17].

Compared to the conventional superconductors, the superconducting cuprate compounds exhibit significantly higher values of T_c, much smaller coherence lengths ξ, and a high anisotropy of the electronic transport properties. The anisotropy is caused by the layered structure of the cuprates (Fig. 4.2-16) and by the fact, concluded from both experiments and theoretical considerations, that superconductivity is essentially confined to the CuO_2-planes. The superconductive coupling between the CuO_2 layers within a CuO_2/E stack and within a CuO_2/L stack, respectively, is much weaker than the coupling within the CuO_2 layers. This is demonstrated qualitatively in Fig. 4.2-16 by the coherence lengths ξ_{ab} and ξ_c within the CuO_2 plane and in the c-direction, respectively, for the compound $Bi_2Sr_2Ca_2Cu_3O_{10}$, ξ_c is of the order of the distance between two CuO_2 layers within the CuO_2/Ca stack. The coupling between neighboring CuO_2/Ca stacks separated by the BiO- and SrO-layers is even weaker. This coupling can be described as Josephson type. For the two compounds $YBa_2Cu_3O_{7-\delta}$ and Pb doped $Bi_2Sr_2Ca_2Cu_3O_{10-x}$, which are the most promising high-T_c conductors for technical applications, the anisotropy of the superconducting properties caused by the quasi 2-dimensional nature of superconductivity in high-T_c conductors is demonstrated and compared to the typical low temperature superconductor Nb_3Sn in Table 4.2-14; γ is the anisotropic parameters defined as $\gamma = \xi_{ab}/\xi_c \equiv \lambda_c/\lambda_{ab}$. The identity is derived from the anisotropic Ginzburg–Landau equations. Here λ_{ab} and λ_c are the London penetration depths for a magnetic field in the c-direction and parallel to the ab-plane, respectively. Considerably different values have been published for the parameters of the high-T_c compounds given in Table 4.2-14. For more details and references see Sects. 4.2.2.2 and 4.2.2.3.

All high-T_c cuprate compounds are Type II superconductors. While a microscopic theory which can describe the high-T_c superconductivity sufficiently is missing, an understanding of this field is reached on the basic assumption that the phenomenological theories [2.18] of London and of Ginsburg and Landau also hold for high-T_c superconductors. Following these theories, the critical fields can be expressed through the corresponding values of the coherence length ξ and the London penetration depth λ. The lower critical field is given by

$$H_{c1\parallel} = \frac{\Phi_0}{4\pi\lambda_{ab}^2}\left(\ln\frac{\lambda_{ab}}{\xi_{ab}} + 0.5\right)$$

and

$$H_{c1\perp} = \frac{\Phi_0}{4\pi\lambda_{ab}\lambda_c}\left[\ln\left(\frac{\lambda_{ab}\lambda_c}{\xi_{ab}\xi_c}\right)^{1/2} + 0.5\right],$$

and the relations for the upper critical fields are

$$H_{c2\parallel} = \frac{\Phi_0}{2\pi\xi_{ab}^2}$$

and

$$H_{c2\perp} = \frac{\Phi_0}{2\pi\xi_{ab}\xi_c}.$$

$H_{c\parallel}$ and $H_{c\perp}$ are the critical fields along the c axis and along the ab-plane, respectively; $\Phi_0 = hc/2e$ is the flux quantum.

Magnetic fields $\geq H_{c1}$ penetrate into bulk superconductors in the form of vortices. The structure of the vortices in layered superconducting cuprates is quite different from that in low-T_{ic} conductors. Within conventional low-T_{ic} superconductors the vortices are always line-like Abrikosov vortices (vortex lines). Each of them carries one flux quantum. In principle, it is a tube with the radius of the London penetration depth λ in which superconducting screening currents circulate around a small non-superconducting core of the radius of the coherence length ξ. But the structure of the vortices in high-T_c superconductors depends on the anisotropy of the compounds and the angle of the applied field

Table 4.2-14 Critical temperature T_c, anisotropy coefficients and coherence lengths of three superconducting compounds

	Nb_3Sn	$YBa_2Cu_3O_{7-\delta}$	$Bi_2Sr_2Ca_2Cu_3O_{10-x}$
T_c (K)	18	92	110
γ	1	≈ 5	≈ 50
ξ	3 (0 K)		
ξ_{ab} (nm)		2.5 (0 K)	2 (0 K)
ξ_c (nm)		0.5 (0 K)	0.05 (0 K)

with regard to the crystal planes. In a field parallel to the c-direction in cuprates of low anisotropy such as YBa$_2$Cu$_3$O$_{7-\delta}$ the structure of the vortices is similar to that of conventional superconductors. However, in compounds of large anisotropy, in which the distance, d, between the Cu$_2$-stacks is larger than two times the coherence length, ξ_c, the vortex lines tend to disintegrate into stacks of vortices in different layers that are shaped like pancakes, due to the low Cooper pair density (Fig. 4.2-17).

Only a weak attractive interaction occurs between the pancake vortices of different layers, the strength of which depends on the magnetic field and the temperature. Therefore the pancake vortices are much more flexible than the continuous, relatively rigid vortex lines in conventional superconductors.

An external field parallel to the superconducting ab-planes penetrates into the sample in the form of so-called Josephson vortices. They differ from Abrikosov ones by the absence of a vortex core. The superconducting shielding currents are strongly confined to the ab-planes. In fields occurring at a small angle to the ab-plane the vortices have a step-like character (Fig. 4.2-18). They consist of a periodic sequence of parts parallel (Josephson vortices) and perpendicular (Abrikosov vortices) to the ab-planes, respectively. The distance between the "Abrikosov-parts" depends on Θ.

For large-scale applications of the superconducting cuprate phases, an effective pinning of the vortices is an indispensable precondition. Pinning centers can be small defects of the crystal structure such as oxygen vacancies, dislocation loops, screw dislocations (e.g., in thin films), stacking faults, precipitates, and substitutional atoms of foreign elements in the superconducting compound. In a magnetic field parallel to the ab-planes, an intrinsic pinning is effective due to the interaction of

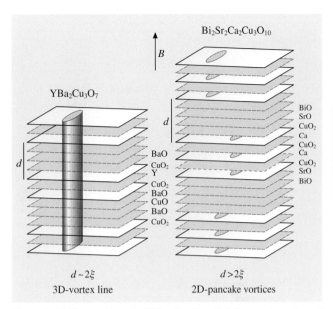

Fig. 4.2-17 Vortex line in YBa$_2$Cu$_3$O$_7$ (*left-hand picture*) and pancake vortices in Bi$_2$Sr$_2$Ca$_2$Cu$_3$O$_{10}$ (*right-hand picture*); the Cu$_2$-planes are drawn with *bold lines* and the isolating planes with *broken lines*

the Josephson vortices with the periodic potential originating from the CuO$_2$ planes. This kind of pinning is much stronger than that caused by crystal defects.

In contrast to conventional superconductors, thermal fluctuations of the vortex positions become very important in high-T_c materials, in particular close to $H_{c2}(T)$. This is a consequence of both the small coherence length ξ_c and the fact that the pinning centers in the cuprates are mainly provided by point defects.

In view of the thermal fluctuation and pinning, the behavior of the vortices determining the resistivity of the system can be demonstrated by the H–T phase diagram proposed by [2.19]. A schematic representation is given in Fig. 4.2-19.

An important new feature compared to conventional superconductors is the existence of new vortex phases: a vortex glass phase and a vortex liquid. In cuprates without vortex pinning, a vortex lattice exists in place of the vortex glass. The presence of pinning centers leads to the destruction of the translational long-range order of the vortices, forming the vortex glass phase in which the resistivity is exponentially small. A truly superconducting state with essentially zero resistivity, $\rho(J \to 0) \to 0$ (J = current density) exists only for this phase. Upon increasing the temperature and magnetic field, respec-

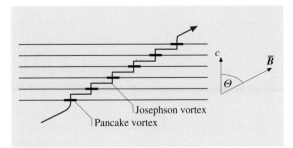

Fig. 4.2-18 Vortex line caused by an external field exhibiting an angle Θ to the ab-planes

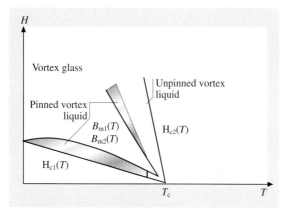

Fig. 4.2-19 Schematic phenomenological phase diagram for the high-T_c superconductors, including the effects of thermal fluctuations and pinning

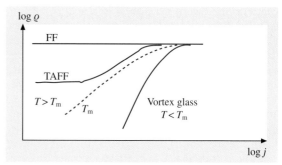

Fig. 4.2-20 Schematic representation of the resistivity ρ and its dependence on the current density j for three different regions in the phenomenological phase diagram (T_m is the melting line) [2.19]

tively, crossing the "melting line" $B_{m1}(T)$, the coupling between the vortices vanishes completely (i.e., the shear modulus c_{66} of the vortex system becomes zero) and the vortex glass "melts", forming the vortex liquid. Close to $B_{m1}(T)$ the vortex liquid is in the pinned regime because the activation energy U of the vortex motion is still large compared to the thermal energy $k_B T$ (k_B = Boltzmann constant). However, in this field-temperature region the resistivity is not zero and its temperature dependence $P = E_0/J_c^0 \, U/k_B T \, \exp[-U/k_B T]$ for small J follows an Arrhenius law, i.e., the system is in a thermally activated flux flow (TAFF) regime where E_0 is the electrical field in the absence of energy barriers and J_c^0 the critical current density at $T = 0$ K [2.20]. When U is considerable larger then $k_B T$ ($U/k_B T \geq 10$) a measurable "flux creep" is observed. Closer to the upper critical field $H_{c2}(T)$ the activation energy U is small, $U < k_B T$, and the vortex liquid cannot be pinned. The system is in the flux flow (FF) regime. With increasing temperature in this H–T region ρ varies smoothly to the resistivity of the normal state.

The existence of a characteristic irreversibility line $B_{irr}(T)$, which is the boundary between the reversible and irreversible magnetization of the superconductor, is closely related to the vortex glass/vortex liquid transition. By crossing this line from lower to higher B values, the critical current density goes to zero. It should be noted that the vortex glass melts not only on crossing $B_{m1}(T)$ but also with decreasing magnetic field near $H_{c1}(T)$ crossing the line $B_{m2}(T)$, as shown in Fig. 4.2-19. As the field decreases, the distance between the vortices increases and eventually exceeds the London penetration depth λ. In this case, the vortex–vortex interaction is exponentially small, and consequently the shear modulus c_{66} decays rapidly, leading to a "melting" of the vortex glass.

For each vortex state a specific resistivity–current relation exists, as schematically shown in Fig. 4.2-20. A linear relation occurs for the "flux flow" regime only. Therefore a voltage criterion for the critical current has to be defined. The one generally used is $1\,\mu\text{V}\,\text{cm}^{-1}$. Extensive reviews of the structure and behavior of vortices in high-T_c superconductors are given in [2.19, 20].

A significant limitation to the critical current density J_c of high-T_c materials is given by the grain boundaries which act as strong resistive barriers to current flow. Figure 4.2-21 shows that the critical current density J_c of $YBa_2Cu_3O_{7-x}$ exhibits an exponential dependence on the misorientation angle θ between the neighboring crystallites [2.21]. Similar angular dependencies of the grain boundary critical current density were observed for all other high-T_c superconductors [2.21].

This weak link (current obstruction) behavior results from the fact that the width of the disturbed crystal lattice region at the grain boundary is on the order of the coherence length. Therefore, for higher angles, a grain boundary becomes a Josephson weak link due to the local suppression of the superconducting order parameter. The reasons which may be responsible for the suppression of the order parameter leading to the strong dependence of J_c on θ are discussed extensively in [2.21].

Applications. The advantage of high-T_c superconductors is that superconductivity is achieved at temperatures above 77 K where liquid nitrogen can be used as a coolant. However, for many potential applications,

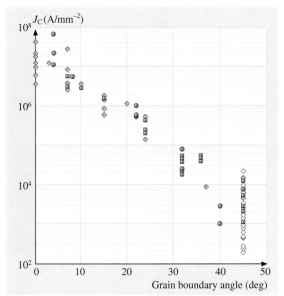

Fig. 4.2-21 Critical current densities of [001] tilt grain boundaries in YBa$_2$Cu$_3$O$_{7-x}$ films as a function of tilt angle. The data published by different authors were compiled by [2.21]

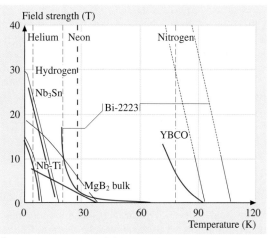

Fig. 4.2-22 Magnetic field strength versus temperature diagram for various superconducting materials. The upper critical field H_{c2} is indicated by *thin lines*, while the irreversibility field *thick lines* indicate H_{irr}

round or flat wires (tapes) with a high critical current density in strong magnetic fields are required. This is true in particular for the use of high-T_c materials in electric power utility devices. Some performance requirements for various electric power applications are summarized in Table 4.2-15 [2.22].

Two main problems discussed above have to be overcome in order to achieve high J_c values. The first one is the current barrier behavior of the grain boundaries, which can be solved by texturing of the crystalline microstructure of the superconducting phase. The second one is the flux flow and flux creep particularly in highly anisotropic materials which have to be avoided and minimized, respectively, in order to obtain a low-loss conductor. Considerable efforts are directed towards flux pinning by the creation of effective pinning centers for the improvement of J_c in high fields at 77 K.

A measurable parameter characterizing the pinning behavior is $H_{irr}(T)$ as mentioned above. In Fig. 4.2-22 this parameter is plotted for the two high-T_c materials Yba$_2$Cu$_3$O$_x$ (YBCO) and (Bi, Pb)$_2$Sr$_2$Ca$_2$Cu$_3$O$_x$ (Bi-2223), which have the most promising H_{irr} characteristics for large scale applications in comparison with H_{irr} of conventional superconductors.

Table 4.2-15 Requirements for industrial wire performance in various utility device applications

Application	J_c (A/cm^2)	Prevailing field strength (T)	Temperature (K)	J_c (A)	Wire length (m)	Strain (%)	Band radius (m)
Fault current limiter	10^4–10^5	0.1–3	20–77	10^3–10^4	1000	0.2	0.1
Large motor	10^5	4–5	20–77	500	1000	0.2–0.3	0.05
Generator	10^5	4–5	20–50	>1000	1000	0.2	0.1
SMES[a]	10^5	5–10	20–77	≈ 10^4	1000	0.2	1
Transmission cable	10^4–10^5	<0.2	65–77	100 per strand	100	0.4	2 (cable)
Transformer	10^5	0.1–0.5	65–77	10^2–10^3	1000	0.2	1

[a] SMES: superconducting magnetic-energy storage

At present, the only high-T_c conductor produced in the form of long wires and tapes is Ag-sheathed Bi-2223. However, a shortcoming of this conductor is the fact, shown in Fig. 4.2-22, that it exhibits an enormous suppression of $H_{irr}(T)$ to the very low value of 0.2 T due to the large anisotropy. Although this does not preclude the use of Bi-2223 for power cables, there is not much scope for use at 77 K in any magnetic field of significant magnitude. This shortcoming of Bi-2223 is an important driving force for the development of a technology based on YBCO, for which $H_{irr}(77\,\text{K}) \approx 7\,\text{T}$.

Specific properties of superconductors at 77 K may be used for electronic applications, e.g., for passive microwave devices such as transmission lines and high-quality resonators. YBCO is widely used in active devices such as SQUIDs and detectors based on Josephson and quasi-particle tunnelling. A further field of application is the use for high-quality reliable and reproducible high-T_c Josephson junctions. A more extensive account of real and potential large-scale applications, respectively, is given in [2.23].

Electrical and Structural Data of High-T_c Superconductors. In Tables 4.2-13 and 4.2-16 electrical and structural data, respectively, of selected high-T_c cuprates are summarized. The data were compiled on the basis of review articles [2.13–16]. In many cases the real stoichiometric coefficients of oxygen are a few tenths higher or lower than the one-digit numbers indicated in the formulas of the tables. The value ΔT_c is the difference between the temperatures at which the resistance reached 90% and 10% of the normal state resistance, respectively, during cooling of the sample; $\rho(0)$ is the residual resistivity expressed by the formula $\rho(T > T_c) = \rho(0) + \beta(T)$ with $\beta = d\rho/dT$.

Table 4.2-16 Structural data of high-T_c cuprates

Compound	Space Group	a (nm)	b (nm)	c (nm)	Ref.
A_2CuO_4 family					
$La_{1.9}Sr_{0.1}CuO_4$	I4/mmm	0.37839	0.37839	1.3211	[2.15]
$La_{1.875}Sr_{0.125}CuO_4$	I4/mmm	0.37784	0.37784	1.3216	[2.15]
$La_{1.85}Sr_{0.15}CuO_4$					
$La_{1.85}Sr_{0.15}CuO_4$	I4/mmm	0.37793	0.37793	1.32260	[2.15]
$La_{1.775}Sr_{0.225}CuO_4$	I4/mmm	0.37708	0.37708	1.3247	[2.15]
$La_{1.725}Sr_{0.275}CuO_4$	I4/mmm	0.37666	0.37666	1.3225	[2.15]
$(Nd_{2-x}Ce_x)CuO_4$	I4/mmm	0.39469	0.39469	1.20776	[2.15, 16]
	I4/mmm	0.395		1.207	
$LBa_2Cu_3O_{7-x}$ family					
$YBa_2Cu_3O_{6.67}$	Pmmm	0.3831	0.3889	1.1736	[2.15]
$YBa_2Cu_3O_7$	Pmmm	0.38198	0.38849	1.16762	[2.15]
$(Y_{0.95}Pr_{0.05})Ba_2Cu_3O_7$	Pmmm				[2.15]
$(Y_{0.9}Pr_{0.1})Ba_2Cu_3O_7$	Pmmm				[2.15]
$(Y_{0.8}Pr_{0.2})Ba_2Cu_3O_7$	Pmmm				[2.15]
$(Y_{0.7}Pr_{0.3})Ba_2Cu_3O_7$	Pmmm				[2.15]
$(Y_{0.6}Pr_{0.4})Ba_2Cu_3O_7$	Pmmm				[2.15]
$YBa_2(Cu_{0.99}Fe_{0.01})_3O_7$	Pmmm				[2.15]
$YBa_2(Cu_{0.975}Fe_{0.025})_3O_7$	Pmmm				[2.15]
$YBa_2(Cu_{0.95}Fe_{0.05})_3O_7$	Pmmm				[2.15]
$YBa_2(Cu_{0.99}Zn_{0.01})_3O_7$	Pmmm				[2.15]
$YBa_2(Cu_{0.975}Zn_{0.025})_3O_7$	Pmmm	0.3820	0.3890	1.1673	[2.15]
$YBa_2(Cu_{0.95}Zn_{0.05})_3O_7$	Pmmm	0.3820	0.38851	1.1671	[2.15]
$HoBa_2Cu_3O_7$	Pmmm	0.38460	0.3881	1.1640	[2.15]

Table 4.2-16 Structural data of high-T_c cuprates, cont.

Compound	Space Group	a (nm)	b (nm)	c (nm)	Ref.
$A_m E_2 Ca_{n-1} Cu_n O_{2n+m+2}$ family					
$Bi_2(Sr_{1.6}La_{0.4})CuO_{6+x}$	Cmmm	0.5370	0.5400	0.2450	[2.15]
$(Bi,Pb)_2(Sr_{1.75}La_{0.25})CuO_6$	Cmmm	0.5282	0.5410	2.462	[2.15]
$Bi_2Sr_2CaCu_2O_8$ (no-anneal)	Fmmm	0.5413	0.5411	3.091	[2.15]
$Bi_2Sr_2CaCu_2O_{8+x}$ (O_2-annealed)	Fmmm	0.5408	0.5413	3.081	[2.15]
$Bi_2Sr_2Ca_2Cu_3O_{10}$	Fmmm	0.539	0.539	3.71	[2.15]
$(Bi_{1.6}Pb_{0.4})Sr_2Ca_2Cu_3O_{10}$	Fmmm	0.5413	0.5413	3.7100	[2.15]
$Pb_2Sr_2Ca_{0.5}Ln_{0.5}Cu_3O_8$	C/mmm	0.5435	0.5463	1.5817	[2.16]
$Tl_2Ba_2CuO_6$	I4/mmm	0.3866	0.38662	2.3239	[2.15]
$Tl_2Ba_2CaCu_2O_8$	I4/mmm	0.38550	0.38550	2.9318	[2.15, 16]
$Tl_2Ba_2Ca_2Cu_3O_{10}$	I4/mmm	0.38503	0.38503	3.588	[2.15]
$TlBa_2Ca_2Cu_3O_9$	P4/mmm	0.3853		1.5913	[2.16]
$(Tl_{0.5}Pb_{0.5})Sr_2CaCu_2O_7$	P4/mmm	0.38023	0.38023	1.2107	[2.15]
$(Tl_{0.5}Pb_{0.5})Sr_2(Ca_{0.8}Y_{0.2})Cu_2O_7$	P4/mmm	0.38075	0.38075	1.2014	[2.15]
$(Tl_{0.5}Pb_{0.5})Sr_2Ca_2Cu_3O_9$	P4/mmm	0.38206	0.38206	1.5294	[2.15]
$TlSr_2Ca_{0.5}Y_{0.5}Cu_2O_7$	P4/mmm	0.380		1.210	[2.16]
$Pb_2(Y_{1-x}Ca_x)Sr_2Cu_3O_8$	Cmmm	0.53933	0.54311	1.57334	[2.15]
$AE_2CuO_{4-\delta}$ family					
$(Hg_{0.8}V_{0.2})Ba_2CuO_{4.3}$		0.38860		0.9338	[2.14]
$(Hg_{0.8}Mo_{0.2})Ba_2CuO_{4.4}$		0.3882		0.9378	[2.14]
$(Hg_{0.9}Mo_{0.1})Ba_2CuO_{4.2}$		0.3875		0.9435	[2.14]
$(Hg_{0.8}W_{0.2})Ba_2CuO_{4.4}$		0.3871		0.9416	[2.14]
$(Hg_{0.9}W_{0.1})Ba_2CuO_{4.2}$		0.3875		0.9450	[2.14]
$AE_2CuO_{4-\delta}$ family					
$(Hg_{0.8}Ru_{0.2})Ba_2CuO_{4.2}$		0.3879		0.9473	[2.14]
$(Hg_{0.7}Cr_{0.3})Sr_2CuO_{4.15}$		0.3845		0.8683	[2.14]
$(Hg_{0.8}Mn_{0.2})Ba_2CuO_{4.2}$		0.3890		0.9343	[2.14]
$(Hg_{0.8}Nb_{0.2})Ba_2CuO_{4.3}$		0.3885		0.9461	[2.14]
$(Hg_{0.8}Mo_{0.2})Sr_2CuO_x$		0.3797		0.8818	[2.14]
$(Hg_{0.9}Re_{0.1})Sr_2CuO_x$		0.3783		0.8883	[2.14]
$A_2Ca_{n-1}Cu_nO_{2n+2}$ family					
$La_{1.6}Sr_{0.4}CaCu_2O_6$	I4/mmm	0.38208	0.38208	1.95993	[2.15]
Other Compounds					
$YBa_2Cu_4O_8$	Ammm	0.386	0.386	2.724	[2.15]
$HoBa_2Cu_4O_8$	Ammm	0.3855	0.3874	2.7295	[2.15]
$Bi_2Sr_2(Ln_{1-x}Ce_x)_2Cu_2O_{10}$	P4/mmm	0.3888		1.728	[2.16]
$YBa_2(Cu_{0.99}Fe_{0.01})_3O_7$	Pmmm				[2.15]
$YBa_2(Cu_{0.975}Fe_{0.025})_3O_7$	Pmmm				[2.15]
$YBa_2(Cu_{0.95}Fe_{0.05})_3O_7$	Pmmm				[2.15]
$YBa_2(Cu_{0.99}Zn_{0.01})_3O_7$	Pmmm				[2.15]
$YBa_2(Cu_{0.975}Zn_{0.025})_3O_7$	Pmmm	0.3820	0.3890	1.1673	[2.15]
$YBa_2(Cu_{0.95}Zn_{0.05})_3O_7$	Pmmm	0.3820	0.38851	1.1671	[2.15]
$HoBa_2Cu_3O_7$	Pmmm	0.38460	0.3881	1.1640	[2.15]

Table 4.2-16 Structural data of high-T_c cuprates, cont.

Compound	Space Group	a (nm)	b (nm)	c (nm)	Ref.
$A_m E_2 Ca_{n-1} Cu_n O_{2n+m+2}$ family					
$Bi_2(Sr_{1.6}La_{0.4})CuO_{6+x}$	Cmmm	0.5370	0.5400	0.2450	[2.15]
$(Bi, Pb)_2(Sr_{1.75}La_{0.25})CuO_6$	Cmmm	0.5282	0.5410	2.462	[2.15]
$Bi_2Sr_2CaCu_2O_8$ (no-anneal)	Pmmm	0.5413	0.5411	3.091	[2.15]
$Bi_2Sr_2CaCu_2O_{8+x}$ (O_2-annealed)	Pmmm	0.5408	0.5413	3.081	[2.15]
$Bi_2Sr_2Ca_2Cu_3O_{10}$	Pmmm	0.539	0.539	3.71	[2.15]
$(Bi_{1.6}Pb_{0.4})Sr_2Ca_2Cu_3O_{10}$	Pmmm	0.5413	0.5413	3.7100	[2.15]
$Pb_2Sr_2Ca_{0.5}Ln_{0.5}Cu_3O_8$	C/mmm	0.5435	0.5463	1.5817	[2.16]
$Tl_2Ba_2CuO_6$	I4/mmm	0.3866	0.38662	2.3239	[2.15]
$Tl_2Ba_2CaCu_2O_8$	I4/mmm	0.38550	0.38550	2.9318	[2.15, 16]
$Tl_2Ba_2Ca_2Cu_3O_{10}$	I4/mmm	0.38503	0.38503	3.588	[2.15]
$TlBa_2Ca_2Cu_3O_9$	P4/mmm	0.3853		1.5913	[2.16]
$(Tl_{0.5}Pb_{0.5})Sr_2CaCu_2O_7$	P4/mmm	0.38023	0.38023	1.2107	[2.15]
$(Tl_{0.5}Pb_{0.5})Sr_2(Ca_{0.8}Y_{0.2})Cu_2O_7$	P4/mmm	0.38075	0.38075	1.2014	[2.15]
$(Tl_{0.5}Pb_{0.5})Sr_2Ca_2Cu_3O_9$	P4/mmm	0.38206	0.38206	1.5294	[2.15]
$TlSr_2Ca_{0.5}Y_{0.5}Cu_2O_7$	P4/mmm	0.380		1.210	[2.16]
$Pb_2(Y_{1-x}Ca_x)Sr_2Cu_3O_8$	Cmmm	0.53933	0.54311	1.57334	[2.15]
AE_2CuO_{4-5} family					
$(Hg_{0.8}V_{0.2})Ba_2CuO_{4.3}$		0.38860		0.9338	[2.14]
$(Hg_{0.8}Mo_{0.2})Ba_2CuO_{4.4}$		0.3882		0.9378	[2.14]
$(Hg_{0.9}Mo_{0.1})Ba_2CuO_{4.2}$		0.3875		0.9435	[2.14]
$(Hg_{0.8}W_{0.2})Ba_2CuO_{4.4}$		0.3871		0.9416	[2.14]
$(Hg_{0.9}W_{0.1})Ba_2CuO_{4.2}$		0.3875		0.9450	[2.14]
$(Hg_{0.8}Ru_{0.2})Ba_2CuO_{4.2}$		0.3879		0.9473	[2.14]
$(Hg_{0.7}Cr_{0.3})Sr_2CuO_{4.15}$		0.3845		0.8683	[2.14]
$(Hg_{0.8}Mn_{0.2})Ba_2CuO_{4.2}$		0.3890		0.9343	[2.14]
$(Hg_{0.8}Nb_{0.2})Ba_2CuO_{4.3}$		0.3885		0.9461	[2.14]
$(Hg_{0.8}Mo_{0.2})Sr_2CuO_x$		0.3797		0.8818	[2.14]
$(Hg_{0.9}Re_{0.1})Sr_2CuO_x$		0.3783		0.8883	[2.14]
$A_2Ca_{n-1}Cu_nO_{2n+2}$ family					
$La_{1.6}Sr_{0.4}CaCu_2O_6$	I4/mmm	0.38208	0.38208	1.95993	[2.15]
Other Compounds					
$YBa_2Cu_4O_8$	Ammm	0.386	0.386	2.724	[2.15]
$HoBa_2Cu_4O_8$	Ammm	0.3855	0.3874	2.7295	[2.15]
$Bi_2Sr_2(Ln_{1-x}Ce_x)_2Cu_2O_{10}$	P4/mmm	0.3888		1.728	[2.16]
$Ca_{1-x}Sr_xCu_2O_2$	P4/mmm	0.3902		0.335	[2.16]
$(Pb, Cu)Sr_2(Ln, Ca)Cu_2O_7$	P4/mmm	0.3820		1.1826	[2.16]
$(Pb, Cu)(Sr, Eu)(Eu, Ce)Cu_2O_x$	I4/mmm	0.3837		2.901	[2.16]
$Sr_{1-x}Nd_xCuO_2$	P4/mmm	0.3942		0.3393	[2.16]
$Tl(Ba, La)CuO_5$	P4/mmm	0.383		0.955	[2.16]
$Tl(Sr, La)CuO_5$	P4/mmm	≈ 0.37		≈ 0.9	[2.16]
$(Tl_{0.5}Pb_{0.5})Sr_2CuO_6$					[2.16]
$(Tl_{0.7}Cd_{0.3})BaLaCuO_5$	P4/mmm	0.3844	0.38440	0.916	[2.15]

4.2.2.2 Superconductors Based on the Y–Ba–Cu–O System

The first superconducting compound with T_c above the boiling point of liquid nitrogen ($T \approx 77$ K) was discovered as a mixed copper oxide representative of the system Y–Ba–Cu–O at ambient pressure by *Wu* et al. in 1987 [2.25], and marked the beginning of the worldwide high-temperature superconductor (HTSC) research. Since then, $YBa_2Cu_3O_{7-\delta}$ is one of the compounds studied most frequently. It is often abbreviated as "Y-123" or "YBCO".

Structural Properties

Crystal Structure. The crystal structure of $YBa_2Cu_3O_{7-\delta}$ has been analysed by diffraction studies and high resolution electron microscopy. The structure is of an orthorhombic perovskite type with oxygen deficiency and an ordered arrangement of the cations Y and Ba (Fig. 4.2-23). Compared with octahedral coordination in the perovskite type structure, 2/9 of the oxygen sites in YBCO are unoccupied, leading to coordination numbers $CN = 5$ and $CN = 4$ for two different Cu sites, respectively. For $\delta = 0$ the unit cell contains one Cu atom in the oxidation state +3 and two Cu atoms in the oxidation state +2. The latter Cu atoms are coordinated by five O atoms forming a square-pyramidal coordination sphere (4 shorter Cu−O bonds, $d(Cu-O) < 2$ Å; one longer bond to the apex of the pyramid, $d \approx 2.29$ Å), while the remaining Cu atom shows a fourfold, nearly square-planar coordination ($d(Cu-O) < 2$ Å). These rectangles are connected via vertices forming chains along the b axis [010]. The pyramids are also connected via vertices forming a two-dimensional arrangement

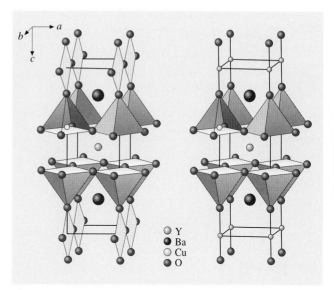

Fig. 4.2-23 Crystal structure of $YBa_2Cu_3O_7$ ($\delta = 0$, *left*) and $YBa_2Cu_3O_6$ ($\delta = 1$, *right*)

within the *ab*-plane. The O−Cu−O angles within the layers are smaller than 180°, thus distorting the planar character of the layers. Ba atoms are coordinated by ten O atoms whereas Y atoms show $CN = 8$.

The high-T_c compound $YBa_2Cu_3O_{7-\delta}$ crystallizes in the orthorhombic space group *Pmmm* (No. 47) with $T_c \approx 91$ K for $0 \leq \delta \leq 0.2$ (Table 4.2-17). The superconducting properties depend on the oxygen content $7 - \delta$. Oxygen loss on the 1e site [O(1) in $YBa_2Cu_3O_7$] leads to a destruction of the Cu−O chains. For $\delta = 1$ only linear coordinated Cu(1) atoms are present.

Table 4.2-17 Crystal structure data of orthorhombic $YBa_2Cu_3O_7$ at $T = 300$ K [2.24]

Formula		$YBa_2Cu_3O_7$		
Space group		*Pmmm* (No. 47)		
Lattice parameters		$a = 3.8206(1)$ Å		
		$b = 3.8851(1)$ Å		
		$c = 11.6757(4)$ Å		
Atom	Site	x	y	z
Y	1h	1/2	1/2	1/2
Ba	2t	1/2	1/2	0.1841(4)
Cu(1)	1a	0	0	0
Cu(2)	2q	0	0	0.3549(3)
O(1)	1e	0	1/2	0
O(2)	2q	0	0	0.1581(4)
O(3)	2r	0	1/2	0.3777(5)
O(4)	2s	1/2	0	0.3779(4)

Table 4.2-18 Crystal structure data of tetragonal YBa$_2$Cu$_3$O$_6$ [2.26]

Formula Space group Lattice parameters		YBa$_2$Cu$_3$O$_6$ $P4/mmm$ (No. 123) $a = 3.865(1)$ Å $c = 11.852(3)$ Å		
Atom	Site	x	y	z
Y	1d	1/2	1/2	1/2
Ba	2h	1/2	1/2	0.1951(2)
Cu(1)	1a	0	0	0
Cu(2)	2g	0	0	0.3609(1)
O(1)	2g	0	0	0.1521(5)
O(2)	4i	0	1/2	0.3793(3)

The non-superconducting YBa$_2$Cu$_3$O$_{7-\delta}$ phase with $0.5 \leq \delta \leq 1$, an antiferromagnetic insulator, crystallizes in a tetragonal unit cell, space group $P4/mmm$ (No. 123) (Table 4.2-18).

The temperature dependence of the lattice parameters of orthorhombic YBa$_2$Cu$_3$O$_{6.91}$ (see Fig. 4.2-24) was determined [2.27] in the range $5\,\text{K} \leq T \leq 320\,\text{K}$. With decreasing temperature the lattice contracts mainly along [001] perpendicular to the Cu–O layers. Small anomalies of the lattice parameters occur at 90 K.

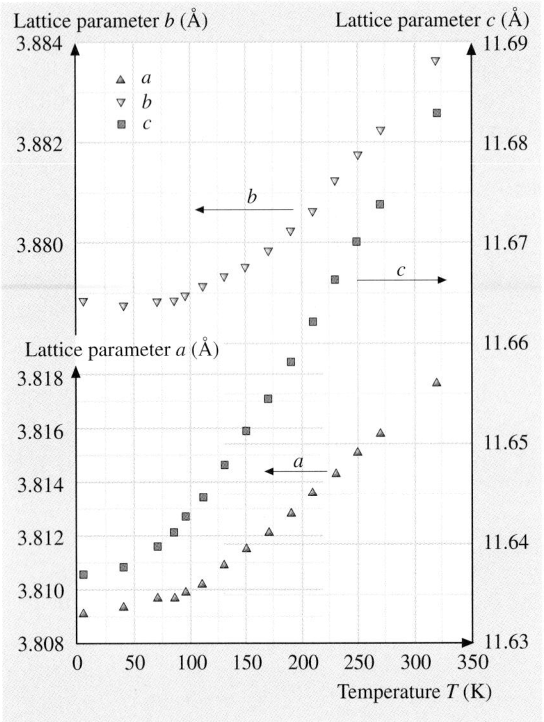

Oxygen Content in YBa$_2$Cu$_3$O$_{7-\delta}$. The oxygen content of YBa$_2$Cu$_3$O$_{7-\delta}$ depends on temperature and oxygen partial pressure. YBa$_2$Cu$_3$O$_7$ can be prepared at $T < 400\,°\text{C}$ in an oxygen atmosphere. The oxygen content can be controlled by annealing in an inert atmosphere or vacuum, or by O-gettering with Zr or Ti. The lower the O partial pressure during annealing of YBa$_2$Cu$_3$O$_7$, the larger is the weight loss (Table 4.2-19, Fig. 4.2-25).

Increasing O loss leads to an increase of the lattice parameters a and c, whereas b decreases. At the phase transformation from the orthorhombic to the tetragonal symmetry ($7 - \delta \approx 6.35$) a discontinuous length change

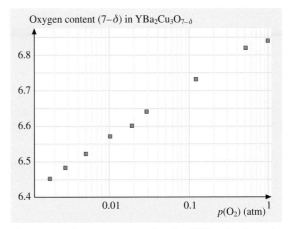

Fig. 4.2-24 Lattice parameters of YBa$_2$Cu$_3$O$_{6.91}$ as function of the temperature [2.27].

Fig. 4.2-25 Oxygen content $(7 - \delta)$ of YBa$_2$Cu$_3$O$_{7-\delta}$ determined by weight loss, annealing conditions: $T = 520\,°\text{C}$, t between 21 h and 27 h at different oxygen partial pressures $p(\text{O}_2)$ [2.28]

Table 4.2-19 Annealing conditions according to [2.28] for oxygen-deficient samples YBa$_2$Cu$_3$O$_{7-\delta}$, $T = 520\,°\text{C}$

Annealing conditions		Oxygen content $(7-\delta)$
$p(\text{O}_2)$(atm)	t(h)	Weight loss
1.00	21	6.84
0.515	25	6.82
0.12	22	6.73
0.29	27	6.64
0.019	23	6.60
0.010	21	6.57
0.005	22	6.52
0.0027	21	6.48
0.00175	22	6.45
0.00107	21	6.40
0.001	22	6.40
0.000415	23	6.34

in the c axis occurs, whereas a and b become equal (Table 4.2-20, Fig. 4.2-26).

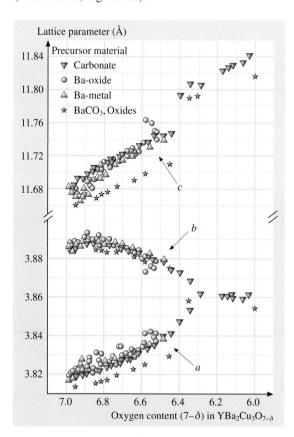

Fig. 4.2-26 Lattice parameters of YBa$_2$Cu$_3$O$_{7-\delta}$ as a function of the oxygen content $7-\delta$; samples prepared from different precursor materials ▼ [2.29]; ○ [2.29]; △ [2.29]; ∗ [2.30]

Chemical Substitution. $R\text{Ba}_2\text{Cu}_3\text{O}_{7-\delta}$: $R = Y$ and Rare-earth Metals; $7-\delta = 6$ and 7. The substitution of Y by trivalent rare earth metals has been studied extensively. The materials Ce and Tb do not form the orthorhombic 123-phase, and PmBa$_2$Cu$_3$O$_7$ has not been synthesized due to the short lifetime of radioactive Pm. The Lu compound is unstable in the polycrystalline ceramic form, but can be prepared as thin films.

For $R\text{Ba}_2\text{Cu}_3\text{O}_{7-\delta}$ ($7-\delta = 6$ and 7), the lattice parameters as well as a large number of interatomic distances show linear dependence on the ionic radius of the trivalent rare earth metal according to the lanthanide contraction, except for PrBa$_2$Cu$_3$O$_7$ which is semiconducting, antiferromagnetic and shows no superconductivity. The superconducting transition temperature reaches a maximum for the Nd phase (Tables 4.2-21–4.2-22, Figs. 4.2-27–4.2-29).

Further Substitutions. Calcium atoms can substitute Y. The higher the Ca content, the lower is T_c. Substitution on the Ba site is reported for La, Sr, Nd, as well as for Na and K. Copper atoms can be substituted by transition metals M (e.g. $M = $ Ti, V, Cr, Mn, Fe, Co, Ni, Zn, Pd, Ag) and main group metals (e.g. Li, Mg, Al, Ga). Concerning the substitution of oxygen, only fluorine has

Table 4.2-20 Dependence of the lattice parameters and T_c on the oxygen content [2.30] (n.d. = not determined)

Oxygen content $(7-\delta)$ in $YBa_2Cu_3O_{7-\delta}$	T_c (K)	a (Å)	b (Å)	c (Å)	V (Å^3)
6.95	90	3.8136	3.8845	11.6603	172.73
6.84	88	3.8153	3.8848	11.6692	172.96
6.81	86	3.8163	3.8845	11.6739	173.06
6.78	n.d.	3.817	3.8836	11.6768	173.09
6.73	69	3.8193	3.8835	11.6832	173.29
6.64	59	3.8224	3.8811	11.6912	173.44
6.58	56	3.8252	3.8786	11.6987	173.57
6.45	56	3.8293	3.875	11.7101	173.76
6.35	–	3.858	3.858	11.7913	175.50
6	–	3.8544	3.8544	11.8175	175.57

Table 4.2-21 Lattice parameters of $RBa_2Cu_3O_{7-\delta}$ (orthorhombic, $Pmmm$) according to *Guillaume* et al. [2.32] at $T = 10$ K, superconducting transition temperature T_c [2.32] and structural phase transformation temperature orthorhombic → tetragonal T_{trans} [2.33]; ionic radii according to *Shannon* [2.34]

Atom R	Ionic radius r of R^{3+} (Å), $CN=8$	a (Å)	b (Å)	c (Å)	V (Å^3)	Oxygen content $(7-\delta)$	T_c (K)	T_{trans} (K)
Y	1.019	3.817(1)	3.883(1)	11.637(1)	172.48	7.00	90(1)	959
La	1.160	3.903(1)	3.920(1)	11.735(1)	179.54	7.06(5)	50(3)	620
Ce	1.143							
Pr	1.126	3.863(1)	3.918(1)	11.650(1)	176.33	6.98(3)	0	
Nd	1.109	3.856(1)	3.912(1)	11.719(1)	176.78	6.98(3)	96(1)	839
Pm	1.093							
Sm	1.079	3.844(1)	3.901(1)	11.692(1)	175.33	6.98(3)	94(1)	888
Eu	1.066	3.836(1)	3.897(1)	11.682(1)	174.63	6.98(3)	95(1)	906
Gd	1.053	3.831(1)	3.893(1)	11.667(1)	174.00	6.99(3)	94(1)	914
Tb	1.040							
Dy	1.027	3.820(1)	3.885(1)	11.646(1)	172.83	6.98(3)	92(1)	945
Ho	1.015	3.815(1)	3.882(1)	11.635(1)	172.31	7.02(3)	90(1)	964
Er	1.004	3.810(1)	3.879(1)	11.625(1)	171.81	7.02(3)	90(1)	972
Tm	0.994	3.805(1)	3.876(1)	11.616(1)	171.31	7.02(3)	90(1)	979
Yb	0.985	3.800(1)	3.872(1)	11.607(1)	170.78	7.02(3)	90(1)	954
Lu	0.977						92	

Table 4.2-22 Lattice parameters of $RBa_2Cu_3O_{7-\delta}$ (tetragonal, $P4/mmm$) according to *Guillaume* et al. [2.32] at $T = 10$ K, ionic radii according to *Shannon* [2.34]

Atom R	Ionic radius r of R^{3+} (Å), $CN=8$	a (Å)	c (Å)	V (Å^3)	Oxygen content $(7-\delta)$
Y	1.019	3.856(1)	11.793(1)	175.35	6.11(3)
La	1.160	3.915(1)	11.850(1)	181.63	6.04(3)
Ce	1.143				
Pr	1.126	3.900(1)	11.832(1)	179.96	6.15(3)

Table 4.2-22 Lattice parameters of $R\text{Ba}_2\text{Cu}_3\text{O}_{7-\delta}$ (tetragonal, $P4/mmm$) according to *Guillaume* et al. [2.32] at $T = 10\,\text{K}$, ionic radii according to *Shannon* [2.34], cont.

Atom R	Ionic radius r of R^{3+} (Å), $CN = 8$	a (Å)	c (Å)	V (Å^3)	Oxygen content $(7-\delta)$
Nd	1.109	3.893(1)	11.830(1)	179.29	6.12(3)
Pm	1.093				
Sm	1.079	3.880(1)	11.815(1)	177.87	6.11(3)
Eu	1.066	3.879(1)	11.811(1)	177.72	6.13(3)
Gd	1.053	3.872(1)	11.807(1)	177.02	6.06(3)
Tb	1.040				
Dy	1.027	3.860(1)	11.796(1)	175.76	6.12(3)
Ho	1.015	3.855(1)	11.792(1)	175.24	6.10(3)
Er	1.004	3.850(1)	11.789(1)	174.74	6.08(3)
Tm	0.994	3.843(1)	11.785(1)	174.05	6.10(3)
Yb	0.985	3.839(1)	11.781(1)	173.63	6.14(3)
Lu	0.977				

been studied intensively. T_c decreases with increasing substitution nearly without exception (see "Superconducting transition temperature"). For additional partial substitution possibilities and advanced literature: [2.31].

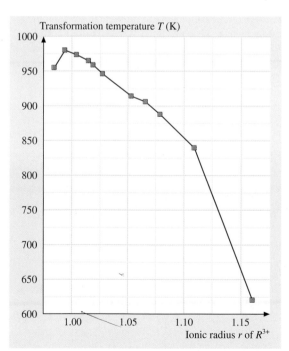

Fig. 4.2-27 Orthorhombic to tetragonal transformation temperature T_{trans} at an oxygen partial pressure of $p = 1\,\text{atm}$ as a function of the ionic radii of the R^{3+} ions ($CN = 8$)

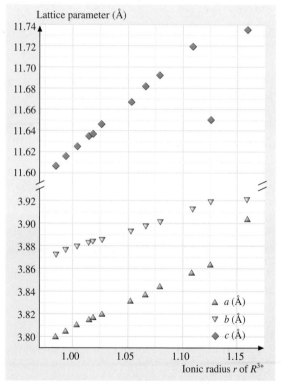

Fig. 4.2-28 Lattice parameters of $R\text{Ba}_2\text{Cu}_3\text{O}_7$ (orthorhombic, $Pmmm$) [2.32] at $T = 10\,\text{K}$, with ionic radii according to *Shannon* [2.34]. Note the large deviation of the lattice parameter c of non-superconducting $\text{PrBa}_2\text{Cu}_3\text{O}_7$

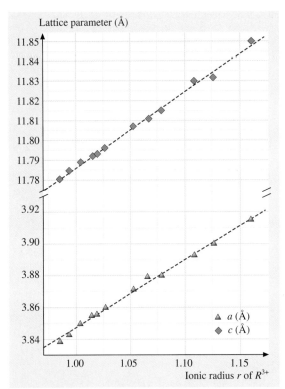

Fig. 4.2-29 Lattice parameters of $RBa_2Cu_3O_6$ (tetragonal, $P4/mmm$) [2.32] at $T = 10$ K, ionic radii according to *Shannon* [2.34]

Microstructural Features. Crystal defects in YBCO grown from the melt may be grouped into crystal defects which are associated with the solidification process and those which are influenced by volume fraction and the size of the secondary phase Y_2BaCuO_5 particles, the so-called 211-phase. *Diko* [2.37] specified the groups of crystal defects as follows:

Crystal defects associated with crystallization from the melt:

- Porosity
- Subgrains
- 211-particles and their inhomogeneity in the sample
- Shape change of the sample
- High angle grain boundaries
- Macrocracks
- Secondary phases

Crystal defects influenced by the volume fraction and size of 211-particles:

- Residual dilation stresses around 211 particles
- Microcracks in *a-b* planes
- Twin structures

211-Particles. Non-superconducting 211-particles may be formed as a dispersion in the 123-phase as the matrix. These particles have a strong pinning effect on the magnetic flux lines. In agreement with theory J_c increases proportionally to V_f/d, where V_f is the volume fraction of the 211-phase particles and d is their average size, as shown in Fig. 4.2-30.

Macrocracks. The stresses arising between the 123-grains on cooling lead to the formation of macrocracks. These stresses stem from the anisotropy of the thermal expansion of the lattice of the 123-phase. The c axis contracts to a higher degree than the a and b axes and the residual stresses puts c-grain boundaries under tension. Data of the temperature dependence of the thermal expansion coefficient $\alpha = d[\ln L(T)]/dT$ at low temperatures are plotted for all three axes in Fig. 4.2-31 according to [2.38]. For further studies on Y-123 up to $T = 500$ K and for $YBa_2Cu_3O_{7-\delta}$, $\delta = 0$, and $\delta = 0.05$, see [2.39].

As an important microstructural feature twinning domains have to be considered. These twinning domains occur at oxygen-rich compositions. The formation of twins arises as a strain relief effect during the structural

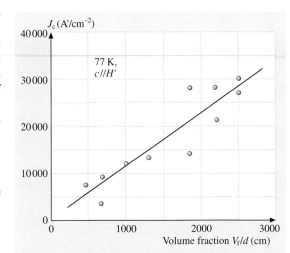

Fig. 4.2-30 Critical current density as a function of V_f/d of the 211-phase particles for melt-processed samples [2.35, 36]

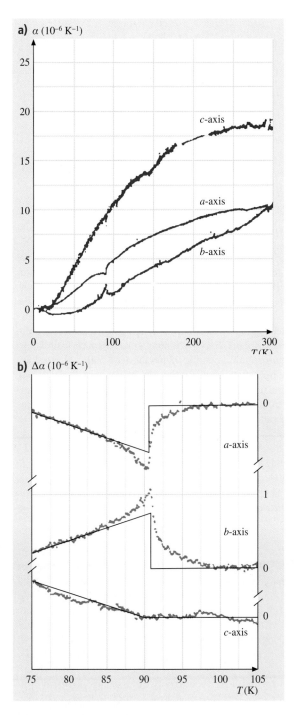

Fig. 4.2-31 Coefficient of thermal expansion α of $YBa_2Cu_3O_{7-\delta}$ at $T \leq 300$ K (**a**) and (**b**) expanded view of the change in expansivity $\Delta\alpha$ near the transition temperature T_c obtained by subtracting a linear fit above T_c from the data

Grain Boundaries. The critical current density as a function of the magnetic field strength for a typical sintered polycrystalline sample and for a single crystal film is shown in Fig. 4.2-32. The critical current densities of polycrystalline samples are more sensitive to applied magnetic fields than those of single crystalline materials. Furthermore, the maximum critical current densities measured on epitaxial films exceed 10^6 A cm^{-2}, which is some 10^3–10^4 times larger than for bulk polycrystalline samples. The critical current density of polycrystalline samples is controlled by the weak link effect of the grain boundaries. This detrimental effect of phase transformation from tetragonal to orthorhombic symmetry.

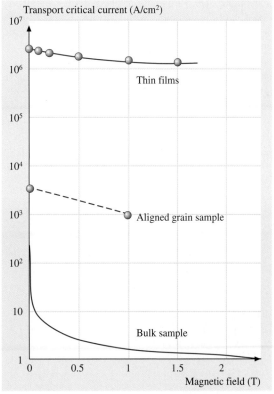

Fig. 4.2-32 Critical current density on dependence on the applied magnetic field for a single crystal film and a sintered sample [2.40]

the grain boundaries is also evidenced by the occurrence of a rapid decrease in J_c with an applied field (a field of few mT will reduce J_c to zero), whereas such fields do not lower the J_c values of single crystals significantly

High-Angle Grain Boundaries. The critical current density across a grain boundary decreases exponentially with increasing grain-boundary misorientation angle. At low misorientation angles between 2° and 5°, a plateau without observable reduction of J_c is suggested [2.44, 45]. Possible mechanisms could be connected with the reduction of the current-carrying cross-section by local suppression of the order parameter at the grain boundaries or by insulating dislocation cores. Figure 4.2-21 shows the critical current densities of [001]-tilt grain boundaries in YBCO films [2.21].

Electrical Resistivity. According to the crystal structure the electrical resistivity of YBCO is anisotropic. Figure 4.2-33 shows the temperature dependence of ρ for a twin-free YBCO single crystal along the orthorhombic axes. Resistivity measurements on sintered YBa$_2$Cu$_3$O$_{7-\delta}$ samples with various oxygen deficiencies show a large increase in resistivity with increasing δ [2.42, 43] (Fig. 4.2-34).

Resistivity as a Function of Temperature and Oxygen Content. The in-plane (ρ_a) as well as the out-of-plane (ρ_c) resistivity of detwinned YBCO crystals

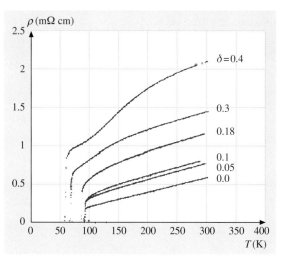

Fig. 4.2-34 Temperature dependence of the resistivity of YBa$_2$Cu$_3$O$_{7-\delta}$ [2.42, 43]

decrease with increasing oxygen content. Moreover, the out-of-plane resistivity shows a crossover from high-temperature metallic behavior ($d\rho_c/dT > 0$) to low-temperature semiconducting behavior ($d\rho_c/dT < 0$). The in-plane resistivity (perpendicular to the CuO chain) deviates in the low-temperature region from linear temperature dependence (Fig. 4.2-35).

Thermal Conductivity. The thermal conductivity $\kappa(T)$ shows a maximum in the temperature range $40\,\text{K} \leq T \leq 90\,\text{K}$ and a change of the slope at T_c, as shown in Fig. 4.2-36. The experimental data depend on the sample microstructure, the oxygen content, crystalline borders and the admixture.

Phase Diagram. Figure 4.2-37 shows a schematic partial phase diagram for YBa$_2$Cu$_3$O$_{7-\delta}$ as a function of the oxygen content $7-\delta$. At $7-\delta = 6$ the compound is an antiferromagnetic insulator with $T_N \approx 500\,\text{K}$. At high δ values two antiferromagnetic phases are marked, with only Cu spins in the plane layers ordered (AFM), or Cu spins in the chain layers also ordered (AFM$_c$). With decreasing δ the hole concentration increases, while T_N decreases. At $7-\delta \approx 6.45$, antiferromagnetism disappears and superconductivity occurs at low temperatures. For $7-\delta \approx 6.75$, a plateau at $T_c \approx 60\,\text{K}$ is reached which corresponds to a partial ordering of oxygen atoms. At high temperatures or at high oxygen content the compounds are paramagnetic.

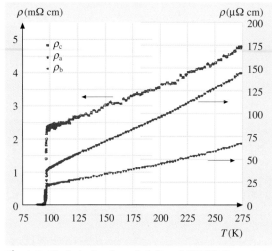

Fig. 4.2-33 Anisotropic resistivity of YBCO for a twin-free single crystal along the orthorhombic axes [2.41]

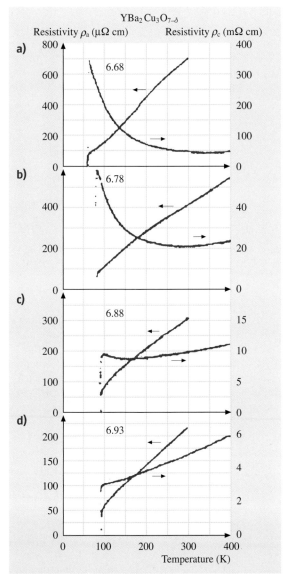

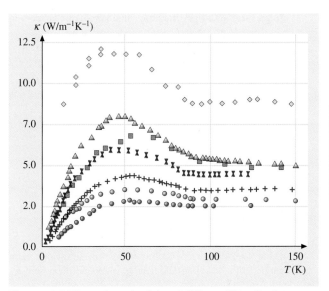

Fig. 4.2-36 Thermal conductivity of different $YBa_2Cu_3O_{7-\delta}$ samples ($\diamond$ single crystals, others: polycrystalline samples) according to [2.47]

Fig. 4.2-35a–d In-plane (ρ_a) and out-of-plane (ρ_c) resistivity of $YBa_2Cu_3O_{7-\delta}$ as a function of the temperature for various oxygen contents [2.46]

Superconducting Properties

The intrinsic superconducting properties which are independent of the microstructure are the lower and the upper critical field, H_{c1} and H_{c2}, respectively and the superconducting transition temperature T_c. These superconducting properties depend on the composition, especially the oxygen content (Table 4.2-23). Further

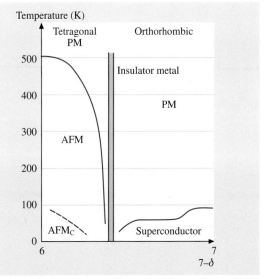

Fig. 4.2-37 Schematic phase diagram for YBCO ($YBa_2Cu_3O_{7-\delta}$) in dependence of the oxygen content $7 - \delta$ [2.48]

parameters such as the effective coherence length ξ, the magnetic penetration depth λ, and the Ginzburg–Landau parameter that is dependent on the microstructure.

Table 4.2-23 Typical properties of sintered YBCO according to [2.49]

Flexural strength (MPa)	216 ± 16
Young's modulus (GPa)	141.8 ($\varrho = 87\%$); 180–200 ($\varrho = 100\%$)
Fracture toughness, K1c (MPa m$^{1/2}$)	1.07 ± 0.18
Critical flow size (μm)	15
Thermal expansion, 50–450 °C ($\times 10^{-6}$ K^{-1})	11.5
Specific heat at 82% theoretical density (J kg^{-1} K^{-1}))	431
Thermal conductivity (W m^{-1} K^{-1})) 300 K	2.67 ($\varrho = 85\%$)
77 K	1 ($\varrho = 85\%$)
Resistivity (μΩ cm) 300 K	670
100 K	220
T_c (K)	92 (width 1 K)
J_c(77 K, $H = 0$) (A cm^{-2})	500–1000 ($\varrho = 85\%$)

ϱ: bulk density

Table 4.2-24 Superconducting properties (transition temperature T_c, coherence length ξ, penetration depth λ, GL coefficient κ, lower critical field H_{c1}, upper critical field H_{c2}) for YBCO [2.54]

T_c (K)		92	[2.25]
Coherence length (nm)	$\xi_0^{\parallel}$	0.6	[2.50]
		0.4	[2.51]
	$\xi_0^{\perp}$	2.7	[2.50]
		3.1	[2.51]
Penetration depth (nm)	$\lambda_L^{\parallel}$	26	[2.52]
		90	[2.51]
		130	[2.53]
	$\lambda_L^{\perp}$	125	[2.52]
		800	[2.51]
		450	[2.53]
GL coefficient	$\kappa^{\parallel}$	7.6	[2.54]
	$\kappa^{\perp}$	37	[2.54]
H_{c1} (T)	$H_{c1}^{\parallel}$	53	[2.55]
	$H_{c1}^{\perp}$	520	[2.55]
H_{c2} (T)	$H_{c2}^{\parallel}$	140	[2.52]
	$H_{c2}^{\perp}$	650	[2.52]

Table 4.2-25 Basic material and critical current density relevant parameters for YBCO according to *Larbalestier* et al. [2.22]

Material	YBCO
Anisotropy	7
T_c	92 K
H_{c2}	> 100 T (4 K)
H^*	5–7 T (77 K)
In-plane coherence length $\xi(0)$	1.5 nm
In-plane penetration depth $\lambda(0)$	150 nm
Depairing current density	3×10^8 A cm^{-2} (4.2 K)
Critical current density	$\approx 10^7$ A cm^{-2}
$\rho(T_c)$	≈ 40–60 μΩ cm

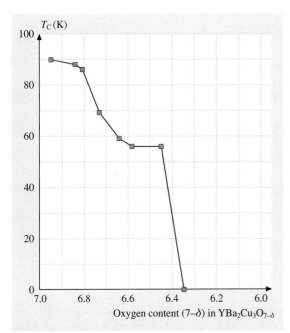

Fig. 4.2-38 Critical temperature T_c versus oxygen content $7-\delta$ [2.30]

actually constant, slight decrease in T_c with decreasing $(7-\delta)$; second plateau: $6.55 < 7-\delta < 6.65$, $T_c \approx 56$ K) (Fig. 4.2-38). Complete substitution of yttrium by rare-earth metals do not show a significant change for the value of T_c, with the exception of the lanthanum ($T_c = 50$ K) and praseodym compounds (not superconducting) (Table 4.2-21). Other dopants on the different atomic sizes as a general rule show decreasing T_c values with increasing substitution rates. Selected examples are shown in Fig. 4.2-39.

Critical Current Density. The critical current density $J_c(B)$ of YBCO thin films at three different temperatures as a function of the magnetic field $B_{\perp c}$ and $B_{\parallel c}$ is given in Fig. 4.2-40. At temperatures $T \leq 60$ K the maximum of $J_c(B)$ occurs when the magnetic field is aligned along the film plane. At $T = 77$ K *Roas* et al. [2.61] observed higher J_c values at $B < 2$ T and a crossover of the curves perpendicular and parallel to the film plane. The crossover seems to depend on microstructural features, e.g., defect concentration, twin-boundary distance etc.

Hole Concentration. The oxygen content of YBCO (YBa$_2$Cu$_3$O$_{7-\delta}$) determines the hole concentration in the CuO$_2$-planes. As shown by *Breit* et al. [2.62] the maximum value of T_c is achieved at an oxygen content of $7-\delta = 6.94$ (see Fig. 4.2-41). This maximum is due

Superconducting Transition Temperature. For the T_c versus oxygen-content behavior, a two plateau dependence is observed (first plateau: $6.8 < 7-\delta < 7$, T_c not

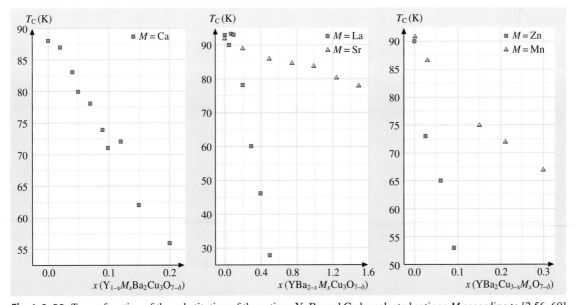

Fig. 4.2-39 T_c as a function of the substitution of the cations Y, Ba and Cu by selected cations M according to [2.56–60]

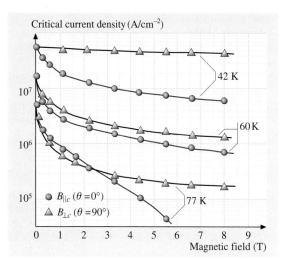

Fig. 4.2-40 Critical current density with $B_{\perp c}$ and $B_{\parallel c}$ at various temperatures in magnetic fields up to 8 T according to [2.61]

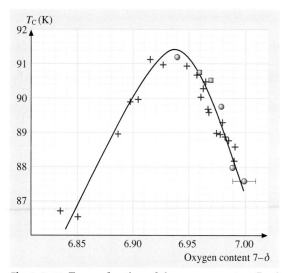

Fig. 4.2-41 T_c as a function of the oxygen content $7 - \delta$ in $YBa_2Cu_3O_{7-\delta}$ according to [2.62] (data from neutron scattering experiments: ○ [2.62], □ [2.63]; data from normal-state resistivity: + [2.64, 65]). The line is a guide to the eyes

to an optimum hole doping of the CuO_2 planes. At an overdoped state $7 - \delta > 6.94$, T_c decreases as the holes in the planes exceed the optimum concentration.

Lower Critical Field. Figure 4.2-42 presents the temperature dependence of the lower critical field H_{c1} parallel

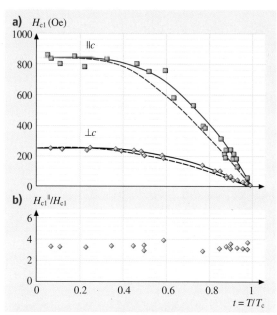

Fig. 4.2-42a,b Temperature dependence of H_{c1} (**a**) and of the anisotropy ratio $H_{c1}^{\parallel}/H_{c1}^{\perp}$ (**b**) according to [2.66]

and perpendicular to c. The anisotropy of the H_{c1} is temperature-independent. The temperature dependence of H_{c1} is in good agreement with BCS calculations (*solid lines* in Fig. 4.2-42a) as shown by *Wu* and *Shridhar* [2.66]. The authors obtained $H_{c1}^{\parallel}(0) = 850 \pm 40$ Oe and $H_{c1}^{\perp}(0) = 250 \pm 20$ Oe.

Upper Critical Field. The upper critical fields H_{c2} in high-T_c superconductors give information about microscopic parameters, e.g., the coherence length and their anisotropies in the superconducting state. The temperature dependence of the upper critical field $H_{c2}(T)$ for $H_{\perp c}$ and $H_{\parallel c}$ measured on a polydomain sample [2.67] is given in Fig. 4.2-44. A linear dependence is observed for both directions with slopes of the critical field of -1.9 and -10.5 T K^{-1} for $H_{\parallel c}$ and $H_{\perp c}$, respectively. The same dependence measured by the resistivity method is shown by the dashed lines. Corresponding to these measurements one can calculate: $H_{c2}^{\parallel}(T = 0\,\text{K}) = 122$ T, $H_{c2}^{\perp}(T = 0\,\text{K}) = 674$ T, $\xi_0^{\parallel} = 3.0$ Å, and $\xi_0^{\perp} = 16.4$ Å.

In low magnetic fields a region of nonlinear dependence of the upper critical field is observed, which could not be reproduced in monodomain crystals.

Pinning Mechanisms. In high J_c value melt-processed YBCO samples, many defects – such as twin planes,

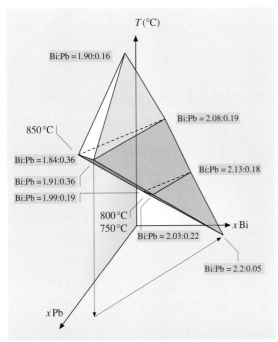

Fig. 4.2-43 Schematical three-dimensional representation of the single-phase region of the Bi2223 phase, the variables being the temperature and the Bi and Pb contents [2.68]

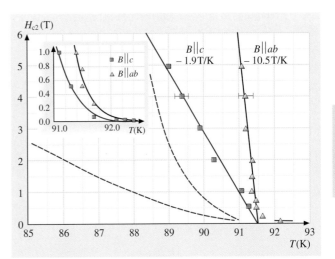

Fig. 4.2-44 Temperature dependence of the upper critical field (*solid line*: magnetization measurements, *dashed line*: resistive measurements) [2.67]

stacking faults, grain boundaries, cracks, oxygen defects, dislocations or non-superconducting particles – can act as pinning centers.

Twin planes are usually introduced into YBCO crystals at the transformation from tetragonal to orthorhombic symmetry. The twin planes only act as effective pinning centers, when the fluxoids are aligned parallel to the twin plane while the Lorentz force acts perpendicular to the plane. Twin planes are not dominant pinning centers in melt-processed YBCO samples.

Stacking faults. If extra CuO layers are inserted adjacent to the chain plane of YBCO, this tends to a local structure of the defects resembling that of the so-called Y124 phase. These may act as important pinning centers for $H_{\perp c}$, but are not important for $H_{\parallel c}$. From a practical point of view these are not suitable pinning centers as it is no possible to prepare samples with or without stacking faults.

Oxygen defects. Däumling et al. observed a peak in the magnetization hysteresis in oxygen-deficient single crystals at an intermediate field, which they ascribed to flux pinning. This kind of pinning center is not the most important one.

Cracks. As mentioned above, the tetragonal to orthorhombic phase transition associated with oxygen addition causes stresses in YBCO samples leading to cracks in the c-direction. These cracks can contribute in flux pinning as in the case of twin planes or stacking faults.

Dislocations. In as-grown thin films, many screw dislocations may occur, which can act as pinning centers.

Nonsuperconducting inclusions in YBCO may be effective in flux pinning. Especially the influence of the 211-phase (or 422-phase in the case of Nd-123) is discussed. Such inclusions are stable and can be dispersed in a controlled pattern.

Application of YBCO Superconductors. Applications of YBCO high-T_c superconductors are superconducting quantum interference devices (SQUIDS), magnetic shielding devices, infrared sensors, analog signal processing devices, microwave devices, and medical imaging systems. Since melt-textured bulk YBCO material shows magnetic pinning properties, a possible application may be high-field permanent magnets. For flexible tape conductors the YBCO material is coated onto metallic carrier tapes. Conductors coated with YBCO are studied intensively, because of the potentially high J_c values in high magnetic fields at $T = 77$ K. Typical architectures of these coated conductors are sequences of polycrystalline metallic substrate or biaxially-textured metallic substrate – biaxially-textured

Table 4.2-26 Present status of Y based coated conductors according to *Yamada* and *Shiohara* [2.69] (IBAD: Ion-beam-assisted deposition, RABiTS: Rolling assisted bi-axially textured substrates, PLD: Pulsed laser deposition, TFA-MOD: Trifluoroacetate–metal organic deposition)

Material	YBCO
Typical Process	IBAD, RABiTS, PLD, TFA-MOD
Structure	Hastelloy (SUS) / oxide buffer / YBCO layer, monofilament
J_c	$1-2\,\text{MA}\,\text{cm}^{-2}$
I_c	$-100\,\text{A}\,(77\,\text{K}, 0\,\text{T})$
Non-SC/SC ratio	100
Length achieved	46 m (reported by Fujikura, Japan AS, Jan. 2003)
Wire configuration	Flat tape with thin YBCO layer
Reinforcement	Hastelloy, SUS substates
Application Tested	–
Remarks	R&D stage for 77 K, high field

buffer layer – biaxially-textured superconducting YBCO layer – Ag stabilizer. The texturing is achieved by rolling and recrystallization.

Wafers of YBCO thin films are prepared for microwave applications. Thermal coevaporation is used as the commercial deposition method. With this method substrates of about $20 \times 20\,\text{cm}^2$ can be coated at rates of 20–30 nm/min. A double-sided deposition can be achieved in two deposition steps. The sol–gel method as well as the chemical solution deposition method are promising with respect to lower production costs. Detailed information on applications is given in [2.81] (Table 4.2-26)

4.2.2.3 Superconductors Based on the Bi–Sr–Ca–Cu–O System

General Properties

Three superconducting phases exist in the Bi–Sr–Ca–Cu–O (BSCCO) system: $Bi_2Sr_2Ca_2Cu_3O_{10}$ (abbreviated using the stoichiometric coefficients as Bi2223), $Bi_2Sr_2CaCu_2O_8$ (Bi2212), and $Bi_2Sr_2CuO_6$ (Bi2201). A review about phase diagram studies in the system Bi–Pb–Sr–Ca–Cu–O, particularly in view of the superconducting compounds, is given in [2.68]. The Bi2212 phase is thermodynamically stable over a relatively wide temperature range. By contrast, the

Table 4.2-27 Structural data of superconducting compounds in the Bi–Sr–Ca–Cu–O system and of Pb substituted Bi2223 [2.15]

Compound	Space group	Lattice parameters		
		a (nm)	b (nm)	c (nm)
$Bi_2Sr_2CuO_6$	Cmmm	0.5361	0.5370	2.4369
$Bi_2Sr_2CaCu_2O_8$	Fmmm	0.5408	0.5413	3.081
$Bi_2Sr_2Ca_2Cu_3O_{10}$	Fmmm	0.539	0.539	3.71
$(Bi_{1.6}Pb_{0.4})Sr_2Ca_2Cu_3O_{10}$	Fmmm	0.5413	0.5413	3.7100

Table 4.2-28 Maximum T_c and a compilation of several values of the coherence lengths, London penetration depths and upper critical fields at $T = 0\,\text{K}$ for superconducting compounds of the Bi–Sr–Ca–Cu–O system

Parameter	Bi2201	Bi2212	Bi2223	Pb-doped Bi2223
T_c (K)	34 [2.70][a]	90 [2.71][a]	110.4 [2.72][d]	109 [2.73][c]
$\xi_{ab}(0)$ (nm)	3.5–4.5 [2.74][b]	≥ 1.09 [2.76][a]–3.8 [2.77][a]	2.9 [2.78][c]	2.9 [2.73][c]
	4.0 [2.75][a]			≈ 1 [2.79, 80][e]

Table 4.2-28 Maximum T_c and a compilation of several values of the coherence lengths, London penetration depths and upper critical fields at $T = 0$ K for superconducting compounds of the Bi−Sr−Ca−Cu−O system, cont.

Parameter	Bi2201	Bi2212	Bi2223	Pb-doped Bi2223
$\xi_c(0)$ (nm)	1.5 [2.74][b]	0.16 [2.77][a]	0.093 [2.78][c]	0.02 [2.80][e]
$\lambda_{ab}(0)$ (nm)	310 [2.82] 438 [2.75][a]	180 [2.83][a]–310 [2.84][a]	165–230 [2.72][d]	≈ 250 [2.73][c], [2.15] 194 [2.79][e]
$\lambda_c(0)$ (nm)		1×10^4–4×10^4 [2.85]		
$B_{c2(ab)}(0)$ (T)	43 [2.74][b]	533 [2.77][a]	1210 [2.78][c]	
$B_{c2(c)}(0)$ (T)	20.2 [2.75][a] 16–27 [2.74][b]	60 [2.86][a]; 70[b]	39 [2.78][c]	297 [2.79][e]

Types of the investigated samples
[a] Single crystal
[b] Overdoped single crystal
[c] Whisker
[d] Thin film
[e] Textured Ag-sheathed tape

Bi2223 phase is stable only over a narrow temperature and concentration range. Considering the preparation of Bi2223 materials, it has to be noted that small variations in concentration and/or temperature can result in quite different phase compositions and in a significant decrease of the volume content of the Bi2223 phase. The latter exhibits an excess of the Bi content of about $Bi_{2.5}$. The formation of the Bi2223 phase is significantly promoted by partial substitution of 20–30% Pb for Bi. Figure 4.2-43 represents the single-phase region of the BiPb2223 phase schematically. The BiPb2223 phase contains an excess of Bi + Pb concentration of about $(Bi, Pb)_{2.2}$. A review of the synthesis of bismuth cuprates is given in [2.16].

As shown in Fig. 4.2-16 of Sect. 4.2.1, the Bi superconductors have a layered perovskite structure which is composed of superconducting CuO_2/Ca stacks (one stack in Bi2201, two stacks in Bi2212, and three in Bi2223) sandwiched between insulating BiO and SrO planes. The structural data of the compounds are presented in Table 4.2-27.

Due to this layered structure, the compounds exhibit a more or less pronounced anisotropy of the electrical parameters (Table 4.2-28).

Here, λ_{ab} and λ_c are the London penetration depths for a magnetic field in the c-direction and parallel to the ab-plane, respectively. $B_{c2(c)}$ and $B_{c2(ab)}$ are approximate upper critical fields aligned in the c-direction and parallel to the ab-plane, respectively; ξ_c and ξ_{ab} are the coherence lengths in c-direction and in the ab plane, respectively. Considerably different values have been reported for the parameters listed in Table 4.2-28. This is true in particular for H_{c2} and also for ξ, which is usually

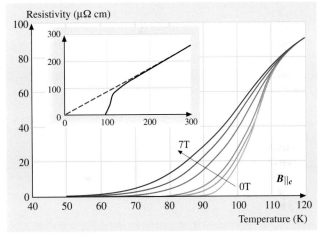

Fig. 4.2-45 Resistivity of a Bi2223 film as a function of the magnetic field aligned parallel to the c axis [2.87]

derived from H_{c2}. The scatter may be partly due to different chemical composition (doping) and type of samples (e.g., thin film, single crystal, polycrystalline textured bulk material). Furthermore, different methods of measuring, e.g., in the case of H_{c2}, magnetization $M(T)$ and resistive measurements $\rho(T)$, respectively, were used and various theoretical models were applied for the determination of the parameters at $T = 0$ K. Different criteria were also used for the resistive determination of H_{c2}: $\rho(T) = 0$ and $\rho(T) = 0.5\rho_n$, respectively. Here ρ_n is the normal state resistivity near T_c. So values of $B_{c2(ab)}(0)$ were determined to 533 T using the criterion $\rho(T) = 0$ and to 2640 T using the criterion $\rho(t) = 0.5\rho_n$ for one and the same sample [2.77]. These high values are questionable in general, as has been shown in [2.88]. A further compilation of data is given in [2.89].

Thin Films, Single and Bi-Crystals

There are essentially four methods of preparation supplying high quality BSCCO films:

- Liquid phase epitaxy (LPE) [2.92]
- Chemical vapor deposition (CVD) [2.93]
- Ion sputtering [2.94]
- Pulsed laser deposition (PLD) [2.90]

Epitaxial deposition on monocrystalline ceramic substrates such as Sr_2TiO_3 or MgO is used to prepare monocrystalline films. Investigation on the influence of grain boundaries on the superconducting properties of

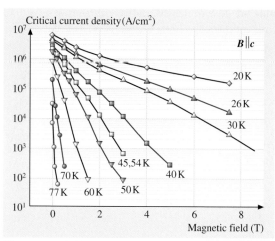

Fig. 4.2-47 Critical current density J_c of an epitaxial Bi2223 film as a function of the magnetic field aligned parallel to the c axis for the temperature range of 20–77 K [2.87]

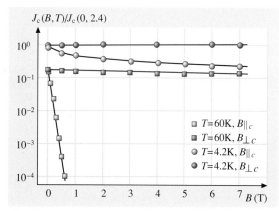

Fig. 4.2-46 Magnetic field dependence of the critical current density of an epitaxial Bi2212 film for the $B_{\|c}$ and $B_{\perp c}$ directions at 4.2 K and 60 K, respectively. $J_c(0, 4.2)$ is the critical current density at 4.2 K in zero field [2.90]

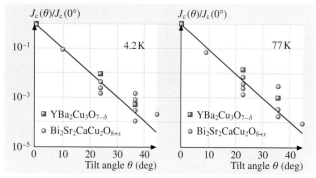

Fig. 4.2-48 Ratio of the intergrain, $J_c(\theta)$ and intragrain critical current densities, $J_c(0)$, as a function of the misorientation angle θ for bicrystalline [001]-tilt Bi2223 (*circles*) and $YBa_2Cu_3O_{7-x}$ films (*squares*) [2.91]

the films' artificial grain boundaries within the films have been carried out using bi-crystalline substrates.

Critical temperatures up to 91 K [2.91] and 110.4 K [2.72] were measured on Bi2212 and Bi2223 films, respectively. Typical thickness values of such films are 200–400 nm. An example for the magnetic-field dependence of T_c is given in Fig. 4.2-45 [2.87]. In contrast to the conventional low-T_c conductors, the onset temperature of the superconducting transition practically does not depend on the field and a characteristic enhancement of the transition width with increasing field is observed.

Critical current densities on the order of 10^5 A cm^{-2} at 77 K in zero field were measured on BSCCO thin films. In Fig. 4.2-46 the field dependence of J_c of a Bi2212 film is shown for two temperatures [2.90] and Fig. 4.2-47 demonstrates the $J_c(B, T)$ dependence of a Bi2223 film [2.87]. In magnetic fields aligned parallel to the plane of the film, that is perpendicular to the c axis of the crystal structure, J_c is practically independent on the field strength, even at higher temperatures such as 60 K. The reasons for this behavior are discussed below in connection with the $J_c(B, T)$ correlation of wires and tapes.

In connection with the manufacturing of real conductors of large length the influence of grain boundaries on the critical current density is of great importance. Investigations of BSCCO bi-crystalline thin films have shown that tilt boundaries act as weak links, causing an exponential dependence of J_c on the misorientation angle, θ, similar to the $J_c(\theta)$ dependence of tilt grain boundaries in YBCO films as shown in Fig. 4.2-48 (compare also with Fig. 4.2-54 of Sect. 4.2.2.1).

Much effort has been expended to grow BSCCO single crystals using several methods, such as crystal growth from the melt, fused-salt reactions in alkali-halide and alkali-carbonate fluxes, respectively, solid-state reaction, and growth in a gaseous phase. Very high quality bulk single Bi2212 crystals with a size of a few mm along the a axis and b axis and about 0.1 mm along the c axis were prepared by the travelling solvent floating zone technique [2.95] or by slow cooling of the melt below the melting temperature and subsequent removing of single crystal fragments from the ingot by crashing [2.96]. Maximum T_c values of 90 K were measured on these crystals. The critical current densities of such crystals were very small, on the order of $100\,\text{A}\,\text{cm}^{-2}$ at 77 K in zero field probably due to the small density of crystal defects. After irradiation of the crystals with 2.2-GeV Au ion beams directed along the c axis at a fluence of $1.0 \times 10^{11}\,\text{cm}^{-2}$, an essential increase of the irreversibility line were observed at temperatures < 70 K. An enhancement of J_c by a few orders of magnitude was also measured at low temperatures of about 30 K in a field of 9 T directed parallel to the c axis [2.97]. These effects can be attributed to columnar tracks within the crystal lattice.

It is much more difficult to grow Bi2223 than Bi2212 single crystals. Very small crystals with a phase purity of 97% Bi2223 and dimensions of 0.1 mm along the a and b axes and 0.001–0.01 mm along the c axis could be prepared by the fused-salt reaction method using a KCl flux [2.98]. The onset temperature of the superconducting transition was determined by dc magnetization measurement to 110 K and the transition width to 5 K. Using the measured magnetization curves and the "Bean" model, the zero field critical shielding-current density in the ab-plane of the crystal has been determined to be $1 \times 10^6\,\text{A}\,\text{cm}^{-2}$ at 5 K.

The preparation of bulk Bi2212 bi-crystals provided the possibility to study the influence of [001] twist boundaries (c axis twist boundaries) on the superconducting properties of Bi2212 conductors [2.95, 96]. Assuming the results can be applied to Bi2223 tapes too, pieces of single crystals were cleaved. One cleave was rotated through an angle about the c axis with respect to the other, and placed atop it. The boundary was formed in a controlled sintering process. Results showed that the transition temperature of a bicrystal agreed with that of the single crystal (the twist angle is not noted [2.96]) and the ratio of critical current densities at $T/T_c \geq 0.9$ across the twist junction to that across the single crystal was unity, independent of the twist angle. However, it has to be noted that in these experiments the critical current densities were very low and it is not clear whether the true J_c values of the grain boundaries could not be measured due to the current limiting J_c of the single crystals. Bicrystals with an enhanced critical current density after irradiation with Au-ion beams revealed, at low temperatures, a more than one order of magnitude higher critical current across the single crystal than that across the 45° twist grain boundary [2.97].

Practical Conductors

Fabrication of Wires and Tapes. The most common approach used to fabricate practical BSCCO conductors is the so called "Powder-in-Tube" (PIT) method. As a first step of PIT precursor, powders of appropriate chemical and phase compositions are produced. In the case of Bi2212 conductors, usually a mixture of Bi_2O_3, $SrCaO_3$, CaO or $CaCO_3$, and CuO with the ratio Bi:Sr:Ca:Cu equal to 2:2:1:2 is calcined at 800–850 °C to remove the residual carbon content. The mixture of oxides and carbonates can be obtained by alternatively applying mechanical mixing (e.g. milling), or a sol-gel process, a co-precipitation method, or spray freeze drying technique. Beside the oxide/carbonate mixture, a powder produced by spray pyrolysis of a metal-nitrate solution with subsequent calcination is often used as an alternate precursor for BiPb2223 conductors. Typical BPSCCO(2223) precursor powders exhibit the concentration ratio of Bi:Pb:Sr:Ca:Cu equal to (1.7–1.8):(0.3–0.4):(1.9–2.0):(2.0–2.1):3.0, and consist of the Bi2212 phase (major phase, about 70%) and other phases such as Ca_2PbO_4, $(Pb, Bi)_3Sr_3Ca_2CuO_x$, alkali earth cuprates, $(Ca, Sr)_{14}Cu_{24}O_x$, or CuO, depending on the preparation conditions (particularly on the final heat treatment of the precursor). The chemical and phase compositions of the precursor have a strong influence on the final properties of the wires and tapes [2.99].

The further steps of PIT excluding the thermal and thermomechanical treatments, respectively, are illustrated in Fig. 4.2-49. The as-produced powders are filled into an Ag tube and then swaged and drawn into wires. Mostly the conductors are produced as multifilamentary (MF) wires due to their more uniform superconducting properties and superior behavior with respect to the mechanical properties compared to monofilamentary conductors. When making MF conductors, pieces of the monofilamentary wire are bundled into a second Ag tube and then this composite is swaged and drawn

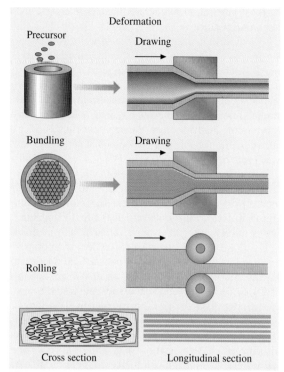

Fig. 4.2-49 Scheme of the "powder-in-tube" method, excluding the thermal and thermomechanical treatments

into wires again. In most cases the wires are finally rolled into tapes since a higher critical current density can be achieved for tape-like conductors than for round wires.

After completion of the deformation process the Bi2212/Ag and BiPb2223/Ag composites have to be subjected to a thermal and thermomechanical process, respectively, in order to transform the Ag-sheathed precursors into superconducting phases and to realize an optimal microstructure of the superconducting phase with respect to high critical current densities. The Bi2212 composites are heated up slightly above the temperature of partial melting (880–900 °C, depending on the oxygen partial pressure of the annealing atmosphere) and then slowly cooled at a rate of 5–10 K h^{-1} to about 830–860 °C, followed by holding the composites at this temperature for a few hours. During this process the Bi2212 phase crystallizes from the melt as platelets with the crystal c axes aligned perpendicular to the wide surface of the platelets. The microstructure reveals a pronounced degree of texture with the c axes perpendicular to the filament/sheath interface.

A more complicated thermomechanical process that usually contains three steps is needed in the case of the BiPb2223/Ag composites. At least the first annealing is performed in a temperature range where a transient melt occurs, accelerating the conversion of the precursor into the superconducting phase. The optimum temperature and dwell time strongly depend on the composition of the precursor and oxygen content of the N_2/O_2 annealing atmosphere. At a lower O_2-content compared to air, lower temperatures and shorter dwell times are suitable and less dependence of J_c on the annealing temperature has been observed. For annealing atmospheres consisting of the inert nitrogen with an O_2 content of 7–21%, the optimal temperature is in the range of 820–835 °C, depending on the composition of the precursor. X-ray diffraction (XRD) investigations of tapes quenched from different states of the annealing process as well as in-situ measurements by means of high-energy synchrotron X-ray diffraction [2.100] and neutron diffraction [2.101], [2.102], respectively, have shown that during heating up and dwelling at the reaction temperature, several reactions such as the incorporation of Pb into the Bi2212 phase, the formation of transient compounds (e.g., $(Ca, Sr)_{14}Cu_{24}O_{41}$ and $Pb_3(Sr, Bi)_3Ca_2CuO_y$) and of a liquid phase mentioned above precede the real conversion of the Bi2212 into Bi2223. These reactions are accompanied by the recrystallization of the Bi2212 phase and improvement of its original texture degree, which was caused by the deformation of the composite into wires and tapes.

During the reaction annealing the density of the BiPb2223 filaments decreases. Therefore the tapes are densified after annealing by a compressive treatment, whereby the electrical connectivity and degree of texture of the grains are enhanced. Uniaxial pressing is the most effective method with respect to J_c. If long tapes are produced, however, the uniaxial pressing has to be replaced by rolling. The lower J_c after rolling, compared to that after pressing, is probably due to additional microcracks induced by rolling and can not be recovered completely by the subsequent annealing which has to be performed after rolling as well as pressing. Mostly, a further increase of J_c can be achieved after a second compressive treatment and subsequent third annealing.

Figure 4.2-50 shows the longitudinal cross-section of a BiPb2223 filament embedded in silver. The microstructure of the superconducting phase in contact with the Ag sheath reveals a higher density and degree of texture compared to areas in the middle of the filament. Obviously, Ag influences the conversion reaction of Bi2212 into BiPb2223. It is known that Ag

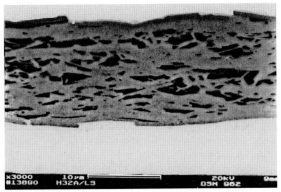

Fig. 4.2-50 Longitudinal cross section of a BiP2223 filament embedded in silver

decreases the solidus temperature of Bi2212 [2.103] and accelerates the BiPb2223 formation process. However, the reaction mechanism has not yet been fully understood. Two different mechanisms have been proposed: the nucleation and growth of the BiPb2223 phase from the melt [2.104] and the formation of the BiPb2223 phase from the Bi2212 grains by intercalation of Ca, Cu, and O [2.105]. In connection with the first-mentioned mechanism beside the homogeneous nucleation, the heterogeneous one is discussed in [2.102]. In the frame of the intercalation process, the occurrence of a transient liquid phase also seems to be an important feature.

In order to improve the tensile strength of the conductors, Ag can be replaced by Ag alloys, such as Ag–Mg and Ag–Mn alloys as sheath material containing a few tenths wt% of the alloying element. Ternary alloys such as Ag–Ni–Mg are also used. Due to the respective internal oxidation of Mg and Mn, during the long-term annealing necessary to form the superconducting phase, the tensile stress which can be applied to the tapes without J_c degradation could be enhanced from 40 MPa for BiPb2223/Ag tapes to 90 MPa [2.107] and 130 MPa [2.108] for AgMn- and AgMg-sheathed tapes, respectively. A high tensile strength protects the conductor against damages by handling and hoop stresses acting on the conductor due to the Lorentz force in magnetic coils. Silver and its alloys are indispensable since no other materials have been found so far which meet the following requirements necessary for the construction of practical conductors: high workability; resistance against oxidation at high temperatures; oxygen permeability since an exchange of oxygen between the precursor and ambient atmosphere during the heat treatment is necessary; no detrimental reaction with elements of the supercon-

ducting phase at high temperatures; and sufficiently high tensile strength.

Superconducting Properties. The largest efforts with respect to the production of practical conductors in long length have been made with BiPb2223 tapes since they are the most promising conductors for electric power applications at present. Figure 4.2-51 shows the critical current density, J_c, of the superconducting phase of BiPb2223 tapes, at 77 K in self-field versus the tape length achieved by various manufacturers [2.106]. With respect to application, the so-called engineering critical current density, J_e, defined by the critical current, I_c, divided by the whole transverse cross-section area (superconductor + sheath) is of great importance. An example of the performance of long tapes is given in Fig. 4.2-52, showing J_e values at 77 K in self-field for 200-m long tapes produced in one 17 000-m manufacturing run [2.109]. Much higher critical current densities were measured on experimental short samples and still higher J_c values were observed locally in tapes, demonstrating the potential of BiPb2223 conductors. In Fig. 4.2-53 the steady improvement of J_c of short multifilamentary samples in the 1990s is shown [2.110]. Values of about 75 kA cm^{-2} at 77 K and 0 T were achieved [2.111]. Magnetooptical investigations on a monofilamentary tape with an average J_c (77 K, 0 T) = 35 kA cm^{-2} revealed some small local regions carrying up to 180 kA cm^{-2} (77 K, 0 T) [2.22].

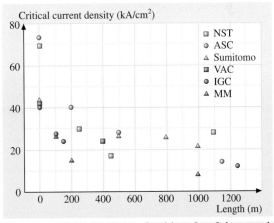

Fig. 4.2-51 Critical current densities J_c of long multifilamentary BiPb2223/Ag tapes produced by various manufacturers as a function of the length (NST = Nordic Superconductor Technologies, ASC = American Superconductor Corporation, VAC = Vacuumschmelze, IGC = Intermagnetics General Corporation [2.106])

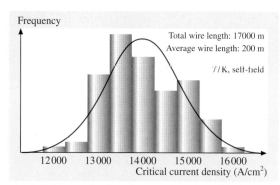

Fig. 4.2-52 Distribution of the engineering critical current density J_e at 77 K in self-field for 200-m long BiPb2223/Ag tapes produced in one 17 000-m manufacturing run [2.109]

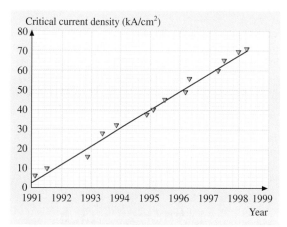

Fig. 4.2-53 Critical current density J_c for short length multifilamentary BiP2223/Ag tapes versus year of production [2.110]

The development of Bi2212 conductors has also reached a level in recent years which allows the manufacture of long-length wires and tapes. For instance, 70-m long wires with an average critical current density, J_c, of 200 kA cm^{-2} at 4.2 K in self-field were produced [2.112]. On tapes of 0.1 to 0.3 mm thickness, J_c values of up to 35 kA cm^{-2} at 77 K and 0 T were measured.

The field dependence of J_c at 77 K is strongly anisotropic for both BiP2222 and Bi2212 conductors. In Fig. 4.2-54a, the $J_c(B, T)$ correlations are demonstrated for BiPb2223 tapes in magnetic fields, $H_{\parallel c}$, aligned perpendicularly to the wide plane of the tape, that is parallel to the crystallographic c axis. At temperatures near 77 K, J_c already drops at $B \approx 0.1$ T to very small values. This tendency, which is also observed in Bi2212 (Fig. 4.2-46) and Bi2223 thin films (Fig. 4.2-47), respectively, is attributed to the thermally activated motion of the pancake vortices near the temperature of liquid nitrogen (see Sect. 4.2.2.1). At 4.2 K, the activation energy of vortex motion is large compared to the thermal energy, $k_B T$, and therefore a much weaker dependence of J_c on B is observed. Figure 4.2-54b represents the $J_c(B, T)$ dependence of BiPb2223 tapes in fields, $H_{\perp c}$, aligned parallel to the plane of the tape (parallel to the crystal ab-plane). A comparison of Fig. 4.2-54b with Fig. 4.2-54a shows that J_c is less dependent on B for $H_{\perp c}$ than for $H_{\parallel c}$ (also at 77 K). This can be attributed to the strong intrinsic pinning of Josephson vortices.

Even the highest J_c values observed for the Ag-sheathed tapes are about one order of magnitude lower than those of monocrystalline thin films. These results pose the question of which microstructural features are responsible for this discrepancy. One cause might be different densities of the flux pinning centers. The flux–pinning interactions in BSCCO superconductors are not well known. In particular, due to the small coherence length of the high-T_c superconductors, it is generally expected that even atomic-sized point defects can pin vortices. It could be shown experimentally that single-atomic impurities of Zn and Ni in Bi2212 [2.113] or line defects such as edge [2.114] and screw dislocations [2.115] in YBa$_2$Cu$_3$O$_{7-x}$ can effectively pin pancake vortices.

The essential reason for the much lower J_c of the BSCCO tapes compared to monocrystalline films stems from the existence of grain boundaries, which act as barriers to current flow as discussed above. Considering the importance of grain boundaries in a frame of mechanism of the supercurrent flow in BiPb2223 tapes, two models were proposed, the "brick-wall" [2.116] and "railway-switch" [2.117] models, which were combined afterwards within the "freeway" model [2.118].

The models were based on microstructural investigations. Figure 4.2-55 shows the cross-sectional detail of a BiPb2223/Ag tape and reveals a platelet-like microstructure of the superconducting phase. Each platelet called "colony", with a typical diameter of 10–20 µm and a thickness of about 1 µm, consists of a group of single crystal grains sharing a common c axis. The grains inside the colony have random ab-orientations, with planar twist basal grain boundaries (BGB's) lying along the ab-planes (Fig. 4.2-56a). Using transmission electron microscopy, the dominant colony-boundaries schematically depicted in Fig. 4.2-56b–c were observed. Considering these microstructural details, the main fea-

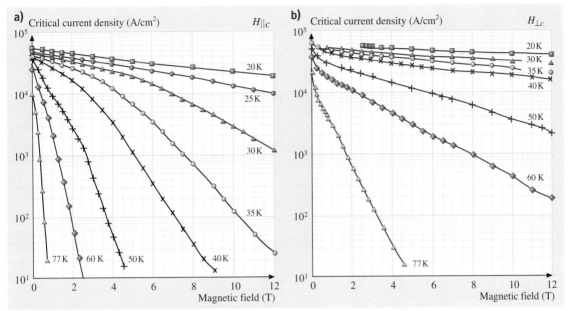

Fig. 4.2-54a,b Critical current density J_c of BiP2223/Ag-alloy tapes as a function of the temperature and magnetic field aligned parallel (**a**) and perpendicular to the to the c axis (**b**) [2.119]

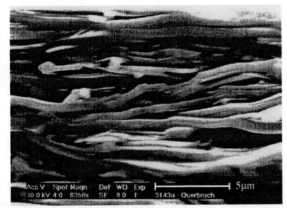

Fig. 4.2-55 REM micrograph of the cross-sectional detail of a BiP2223/Ag tape [2.119]

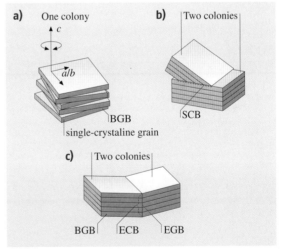

Fig. 4.2-56a–c Scheme of colony and single-crystal grain boundaries proposed for the "railway-switch model" [2.117] and in a modified picture for the "freeway model" [2.118]

tures of the "freeway" model can be summarized as follows:

- The "freeway" represents current flow along ab-planes.
- The current flows primarily across edge colony boundaries (ECB) rather than small-angle colony boundaries (SCB).
- The relative twist orientation of the ab axes across the ECB of adjacent colonies is considered as the dominant factor in the current flow limiting the critical current density.
- Misorientation angles of the ab-axes that are less than 10° are required to achieve strong superconducting links across the edge crystal grain

boundaries (EGB) within an edge colony boundary (ECB) of adjacent colonies.
- The probability of overlap of the *ab*-axes at the boundary of adjacent colonies within the 10° criterion is about 1/9 for any one grain inside the colony. Thus the colony structure can carry a current flow (I) at a fraction $f_{ECB} \approx 1/9$ of what a single crystal with the *ab*-plane parallel to the current direction (I) could carry. The ECB current is expected to be on the scale of the *ab*-current flow.
- An important assumption required for this mechanism is the possibility that high current can shift along the *c* axis between different *ab*-planes, otherwise the strongly coupled *ab*-planes cannot be accessed. This possibility is controversial, since the current must cross many *c* axes boundaries (BGB) inside the colony and the dependence of J_c on the twist angle is not clear yet, as discussed above.

From the model one can derive opportunities to increase J_c of Bi2223 tapes, e.g., through biaxial texturing of the colony grains and increasing the critical current density of the grains. Furthermore, the model provides a coherent explanation of a variety of data observed on Bi2223 tapes such as the $J_c(B, T)$ dependence discussed above.

4.2.2.4 Carbides, Borides, Nitrides

Superconductivity is found in many carbides, nitrides, and borides. Several of these compounds contain transition metals T, forming a close-packed lattice with the small atoms, $X = $ C, N, or B, in the interstices of the lattice. The simplest compound of the form TX has a cubic lattice of the NaCl-type. Compounds of TX with $T = $ Nb, V, Mo, etc., are generally very hard, corrosion-resistant, and of high thermal stability. These desirable properties are understood to arise from strong covalent bonding. According to the BCS theory, superconductivity in these and related compounds can be explained by high-frequency phonon modes of the light X atoms combined with a high density of states of the conduction electrons at the Fermi level. A further precondition for high superconducting transition temperatures (which is not included in the BCS theory) is a strong coupling of the phonon modes to the electrons. Recently, a supercon-

Table 4.2-29 Structure and T_c of carbides

	T_c (K)	Structure type	Ref.
Rb_3C_{60}	28	BiF_3	[2.120]
BEDT-TTF-based salt	12.8 (HP)	organic	[2.120]
MoC_{1-x}	14.3	NaCl	[2.121]
NbC	12	NaCl	[2.121]
TaC	10	NaCl	[2.121]
Mo_2C	12.2	orthorh.	[2.120]
LaBrC	7.1	$Gd_2C_2I_2$	[2.120]
YIC	10	$Gd_2C_2I_2$	[2.120]
Y(Br, I)C	7.1	$Gd_2C_2I_2$	[2.120]
Mo_3Al_2C	10	β-Mn	[2.120]
$MgNi_3C$	8.5	$SrTiO_3$	[2.122]
Nb_2SC	5	Tm_2SC	[2.123]
Nb_2S_2C	7.6	Tm_2S_2C	[2.124]
LaC_2	1.6	CaC_2	[2.121]
LuC_2	3.3	CaC_2	[2.121]
YC_2	4.0	CaC_2	[2.120]
(La, Th)NiC_2	7.9	$CeNiC_2$	[2.120]
U_2PtC_2	1.5	U_2IrC_2	[2.121]
La_2C_3	11	Pu_2C_3	[2.120]
Y_2C_3	8.2	Pu_2C_3	[2.121]
Th_2C_3	4.1	Pu_2C_3	[2.121]
(Y, Th)$_2C_3$	17	Pu_2C_3	[2.120]

Table 4.2-30 Structure and T_c of boron and borides [2.120]

	T_c (K)	Structure type
B	11.2 (at 250 GPa)	
NbB	8.3	CrB
TaB	4.0	CrB
Mo_2B	5.1	$CuAl_2$
Re_3B	4.7	Re_3B
MgB_2	39	AlB_2
ReB_2	6.3	AlB_2
YB_6	7.1	CaB_6
LaB_6	5.7	CaB_6
YB_{12}	4.7	UB_{12}
ZrB_{12}	5.8	UB_{12}
$YRuB_2$	7.8	$LuRuB_2$
$LuRuB_2$	10.0	$LuRuB_2$
YOs_3B_2	6.0	$CeCo_3B_2$
$LuOs_3B_2$	4.7	$CeCo_3B_2$
$ErRh_4B_4$	8–9	$CeCo_4B_4$
YRh_4B_4	10–11	$CeCo_4B_4$
$LuRh_4B_4$	8–12	$CeCo_4B_4$

ducting transition temperature $T_c = 39$ K was found in MgB_2 [2.125], essentially metallic boron bonded by covalent B–B and ionic B–Mg bonds. This is the highest T_c of binary compounds and close to or above the theoretical value predicted by the BCS theory. The material MgB_2 has been known since the early 1950s, but its transport and magnetic properties have not been investigated until recently despite an intensive search for higher values of T_c in the family of binary compounds. The T_c of 23 K obtained for the intermetallic A15 compound Nb_3Ge in the early seventies could not be exceeded until the discovery of the high-T_c superconductors in 1986. Since 1994, there has been a renewed interest in intermetallic superconductors which incorporate light elements such as boron due to the discovery of the new class of quaternary rare-earth transition metal borocarbides RT_2B_2C ($R = $ Y, Lu, or rare-earth metals; $T = $ Ni, Pt, or Pd) [2.126, 127]. These compounds are interesting with regard to the strong interplay between magnetism and superconductivity, including magnetic pair-breaking and the coexistence of superconductivity and magnetism.

Carbides

Several carbon-containing superconductors are listed in Table 4.2-29. Superconductivity in the fullerenes can be achieved by alkali metal intercalation, resulting in electronically-doped bulk material. This was at first demonstrated by *Rosseinsky* et al. [2.128] who found superconductivity in Rb_3C_{60} at 28 K. Among the superconductors in Table 4.2-29 that have not been mentioned so far are: a) organic compounds, e.g., in BEDT-TTF in which superconductivity is induced by high pressure (HP), b) the layered rare-earth carbide halides with $Gd_2C_2I_2$ structure, c) the niobium carbosulfides with Tm_2SC and Tm_2S_2C structure, d) the lanthanoid dicarbides which crystallize in the body-centered tetragonal CaC_2 structure, e) the lanthanoid (and actinide) sesquicarbides with Pu_2C_3 structure, and f) the intermetallic compound $MgNi_3C$ with cubic perovskite structure in which superconductivity was found very recently [2.122]. Not included in Table 4.2-29 are the rare-earth transition metal borocarbides RT2B2C, which will be described in more detail later.

Borides

The superconducting transition temperatures of boron and selected borides are listed in Table 4.2-30. Boron itself, which is an isolator at ambient pressure, was recently found to become superconducting under high pressure [2.129]. The boron atoms in transition metal borides can form chains, nets, and three-dimensional networks. In this respect the borides differ markedly from other interstitial compounds such as carbides

Table 4.2-31 Superconducting parameters (ranges of experimental and derived data) of MgB_2 [2.130]

Parameter		Values reported
Superconducting transition temperature	T_c	39–40 K
Lattic parameters		$a = 0.3086$ nm, $b = 0.3524$ nm
Theoretical density		2.55 g cm^{-3}
Resistivity near T_c	$\rho(40\,\text{K})$	0.4–360 $\mu\Omega$ cm
"Residual" resistance ratio	RRR = $\rho(300\,\text{K})/\rho(40\,\text{K})$	1–27
Coherence lengths	ξ_{ab}	4–12 nm
	ξ_c	1.6–3.6 nm
Penetration depths	$\lambda(0)$	85–180 nm
Upper critical fields	$H_{c2(ab)}(0)$	14–40 T
	$H_{c2(c)}(0)$	2–24 T
Anisotropy parameter	$H_{c2(ab)}(0)/H_{c2(c)}(0)$	2–13
Lower critical fields	$H_{c1}(0)$	27–48 mT
Irreversibility fields	$H_{irr}(0)$	6–35 T

and nitrides. Among the transition metal orthorhombic monoborides, NbN and TaN which crystallize in the CrB structure are superconductors. Some properties of MgB_2, which has the highest T_c among the borides, are presented in the next section.

MgB_2

The boride MgB_2 crystallizes in the simple hexagonal AlB_2-type structure and contains graphite-type boron layers which are separated by hexagonal close-packed layers of magnesium (see Fig. 4.2-57). MgB_2 has not only an attractively high transition temperature, but, in contrast to high-T_c superconductors, the advantage of strongly-linked grains.

Therefore, high currents can flow across the grain boundaries of bulk polycrystalline MgB_2. This is due to the relatively large coherence length ξ of this superconductor (Table 4.2-31). On the other hand, this large coherence length is responsible for the relatively low values of the upper critical field $H_{c2} \propto \xi^{-2}$ observed, especially for clean MgB_2 single crystals in a magnetic field parallel to the c axis. Studies of the electronic structure of MgB_2 revealed two relevant bands which have to be taken into account in order to explain the magnitude and temperature dependence of the upper critical field $H_{c2}(T)$. Considerably higher upper critical fields than for bulk and single-crystalline MgB_2 can be achieved in thin films by the introduction of impurities to decrease the mean free path. In such films, enhanced resistivity, reduced values of T_c, and upper critical fields up to 40 T have been reported. One of the open questions concerns the anisotropy of H_{c2}. Reported values of $H_{c2}(ab)$ and $H_{c2}(c)$ range between 2 and 13.

Borocarbides

Table 4.2-32 gives an overview of superconducting borocarbides. Most of them belong to the family of quaternary rare-earth transition metal borocarbides RT_2B_2C (R = Y, Lu, or other rare-earth metals; T = Ni, Pt, or Pd). In this table the superconducting transition temperature T_c and the magnetic ordering temperature T_N are listed. The $LuNi_2B_2C$-type structure of these compounds can be considered as the $ThCr_2Si_2$-type (space group $I4/mmm$) interstitially modified by carbon. They consist of alternating sheets of T_2B_2 tetrahedra and RC layers as shown in Fig. 4.2-58 for RNi_2B_2C. The

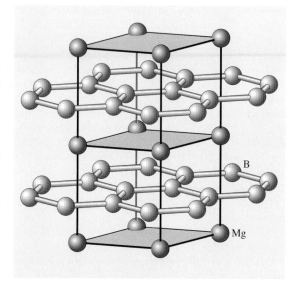

Fig. 4.2-57 Structure of MgB_2

Table 4.2-32 Borocarbide superconductors [2.120]

	T_c (K)	T_N (K)	Structure type
LuB_2C_2	2.4	–	LaB_2C_2
YB_2C_2	3.6	–	LaB_2C_2
Mo_2BC	7.5	–	Mo_2BC
$LuNi_2B_2C$	16	–	$LuNi_2B_2C$
YNi_2B_2C	15.5	–	$LuNi_2B_2C$
$ScNi_2B_2C$	15.5	–	$LuNi_2B_2C$
$ThNi_2B_2C$	15	–	$LuNi_2B_2C$
$CeNi_2B_2C$	0.1	–	$LuNi_2B_2C$
$TmNi_2B_2$	11	1.5	$LuNi_2B_2C$
$ErNi_2B_2C$	10.5	6.8	$LuNi_2B_2C$
$HoNi_2B_2C$	7.5–8	5.8	$LuNi_2B_2C$
$DyNi_2B_2C$	6.2–6.4	11	$LuNi_2B_2C$
YRu_2B_2C	9.7	–	$LuNi_2B_2C$
YPd_2B_2C	23	–	$LuNi_2B_2C$
$ThPd_2B_2C$	14.5	–	$LuNi_2B_2C$
$LaPd_2B_2C$	1.8	–	$LuNi_2B_2C$
YPt_2B_2C	10–11	–	$LuNi_2B_2C$
$LaPd_2B_2C$	10–11	–	$LuNi_2B_2C$
$ThPt_2B_2C$	6.5	–	$LuNi_2B_2C$
$PrPt_2B_2C$	6	–	$LuNi_2B_2C$

structure of RNi_2B_2C is highly anisotropic with a c/a ratio of about 3. Nevertheless, these compounds have a three-dimensional globally isotropic structure, but due to the complicated shape of the Fermi surface there is a pronounced dispersion in the Fermi velocity. This peculiarity of the electronic structure is important for the upper critical field $H_{c2}(T)$ because these superconductors are in the clean limit, i. e., their coherence length is smaller than the mean free path. The magnitude of H_{c2} at low temperatures and an unusual positive curvature of $H_{c2}(T)$ observed near T_c, especially for the nonmagnetic RNi_2B_2C ($R = Y, Lu$) compounds, can be understood by taking their two-band electronic structure into account.

The superconducting transition temperature of nonmagnetic RNi_2B_2C compounds ($R =$ Sc, Lu, Y, Th, La) strongly depends on the lattice parameter a, as shown in Fig. 4.2-59. The highest T_c is obtained for $LuNi_2B_2C$, which has an optimum lattice parameter corresponding to a maximum density of states at the Fermi level. In Fig. 4.2-59, magnetic R elements (Tm, Er, Ho, Dy, Yb, Tb, Gd, Eu, Nd, Pr) are also included. The T_c of the superconducting compounds among the magnetic RNi_2B_2C compounds ($R =$ Tm, Er, Ho, and Dy) scales roughly with the de Gennes factor $dG = (g-1)^2 J(J+1)$ of the R^{3+} Hund's-rule ground state, where g is the Landé factor and J the total angular momentum. Such so-called de Gennes scaling is observed for both T_c and the magnetic or-

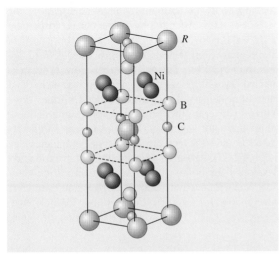

Fig. 4.2-58 Structure of the rare-earth nickel borocarbides

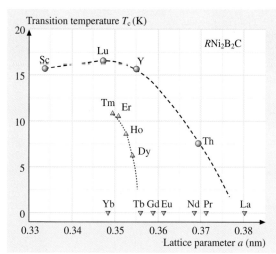

Fig. 4.2-59 Transition temperature T_c and lattice parameter a for $R\mathrm{Ni}_2\mathrm{B}_2\mathrm{C}$ compounds for non-magnetic (*circles*) and magnetic (*triangles*) R elements [2.131]

dering temperature T_N as shown in Fig. 4.2-60. The reason is that both effects, i.e., antiferromagnetism and the suppression of superconductivity, are governed by the exchange interaction of conduction electrons with R 4f electrons.

A striking feature distinguishing the superconducting $R\mathrm{Ni}_2\mathrm{B}_2\mathrm{C}$ compounds from other magnetic superconductors is their high magnetic ordering temperature T_N being comparable with T_c. Therefore, the interplay between superconductivity and magnetism in these compounds is particularly pronounced. In particular, $\mathrm{DyNi}_2\mathrm{B}_2\mathrm{C}$ is the unique member of the $R\mathrm{Ni}_2\mathrm{B}_2\mathrm{C}$ series which becomes superconducting in the antiferromagnetically ordered state.

Nitrides

The refractory transition metal nitrides with NaCl-type crystal structure have been widely studied in the past. The basic features of superconductivity in these compounds indicate a standard BCS mechanism. The highest T_c among these compounds was found for NbN (see Table 4.2-32). Niobium nitride which can easily prepared as a thin film is a promising material for the application in various cryogenic devices such as high resolution X-ray detectors or superconducting quantum mixers. For NbN thin films, very high upper critical fields up to 30 T and high critical currents, e.g.,

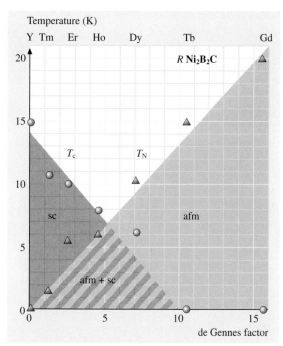

Fig. 4.2-60 Superconducting transition temperature T_c and antiferromagnetic ordering temperature T_N for $R\mathrm{Ni}_2\mathrm{B}_2\mathrm{C}$ compounds, with R = Lu, Tm, Er, Dy, Tb, and Gd as functions of the de Gennes factor (see text). (afm = antiferromagnetic, sc = superconducting, afm + sc = coexistence of superconductivity and antiferromagnetism)

$10^5\,\mathrm{A\,cm^{-2}}$ at 4.2 K in fields of 19 T [2.132] were reported. For MoN with the NaCl structure, $T_c = 29\,\mathrm{K}$ was predicted [2.133], although this structure does not appear in the equilibrium phase diagram of the Mo−N system. Attempts have been made to prepare it by techniques leading to non-equilibrium states. In single-phase films with the NaCl structure, $T_c = 12.5\,\mathrm{K}$ was found [2.134]. This unexpectedly low T_c was explained by the presence of nitrogen vacancies and interstitial defects. The transition metal nitrides crystallize in several other structures, too. Some of these compounds are superconducting and have similar T_c values as the corresponding NaCl-type compounds (see Table 4.2-33). Recently, a new route of preparation was developed for lithium intercalated metal nitride chloride. Superconductivity with $T_c = 14\,\mathrm{K}$ and even 25 K was reported for $\mathrm{Li}_{0.16}\mathrm{ZrNCl}$ and $\mathrm{Li}_{0.48}(\mathrm{THF})_y\mathrm{HfNCl}$, respectively [2.135, 136], where THF = tetrahydrofuran.

Table 4.2-33 Nitrides superconductors

	T_c (K)	Structure type	Reference
TiN	6.5	NaCl	2.121
ZrN	10.6	NaCl	2.121
HfN	8.3	NaCl	2.121
Vn	8.8	NaCl	2.121
NbN	17.3	NaCl	2.121
NbN	9	W	2.132
NbN	15	Fcc	2.121
TaN	10.4	NaCl	2.121
TaN	10.8	Fcc	2.137
MoN	12.5	NaCl	2.121
MoN	13.2	Hexagonal	2.134
ReN	5	NaCl	2.121
Nb_2N	8.6	Hexagonal	2.132
Ta_2N	10.6	Hexagonal	2.137
Mo_2N	6–7	Fcc	2.134
$Li_{0.16}ZrNCL$	14		2.135
$Li_{0.48}(THF)HFNCl$	25.5		2.136

References

2.1 O. Henkel, E. M. Sawitzkij (Eds.): *Supraleitende Werkstoffe* (VEB Deutscher Verlag für Grundstoffindustrie, Leipzig 1982) (in German)

2.2 R. Flükiger, W. Klose (Eds.): *Superconductors*, Landolt–Börnstein, New Series III/21 (Springer, Berlin, Heidelberg 1997)

2.3 J. Evetts (Ed.): *Concise Encyclopedia of Magnetic & Superconducting Materials* (Pergamon Press, Oxford, New York, Seoul, Tokyo 1992)

2.4 H. Hillmann: Superconductor Materials Science. In: *Metallurgy, Fabrication, Applications*, ed. by S. Foner, B. B. Schwartz (Plenum, New York 1981)

2.5 D. C. Larbalastier: Superconductor Materials Science. In: *Metallurgy, Fabrication, Applications*, ed. by S. Foner, B. B. Schwartz (Plenum, New York, London 1981)

2.6 E. W. Collings (Ed.): *A Source Book of Titanium Alloy Superconductivity* (Plenum, New York 1983)

2.7 K. Osamura (Ed.): *Composite Superconductors* (Marcel Dekker, Basel, New York, Hong Kong 1994)

2.8 H. Krauth: *Conductors for d.c. Applications*, Handbook of Applied Superconductivity, ed. by B. Seeber (Institute of Physics Publishing, Bristol, Philadelphia 1998)

2.9 C. Li, D. C. Larbalastier: Development of high critical current densities in niobium 46.5 wt% titanium, Cryogenics **27**, 171 (1987)

2.10 J. Ekin: *Materials at Low Temperature*, ed. by P. R. Reed, F. C. Clark (American Society for Metals, Metals Park OH 1983)

2.11 R. Flükiger, W. Klose (Eds.): *Superconductors*, Landolt–Börnstein, New Series III/21 (Springer, Berlin, Heidelberg 1997) pp. 1–11

2.12 J. G. Bednorz, K. A. Müller: Possible high-T_c superconductivity in the Ba–La–Cu–O system, Z. Phys. **B 64**, 189 (1986)

2.13 R. Hott: *High Temperature Superconductivity, 1: Materials*, ed. by A. V. Narlikar (Springer, Berlin, Heidelberg, New York 2004) p. 1

2.14 B. Raveau: In: *High-T_c superconductivity: Ten Years after the Discovery*, NATO ASI Ser. E: Applied Sciences, Vol. 343, ed. by E. Kaldis, E. Liarokapis, K. A. Müller (Kluwer 1997) p. 109

2.15 D. R. Harshman, A. P. Mills Jr.: Concerning the nature of high-T_c superconductivity: Survey of experimental properties and implications for interlayer coupling, Phys. Rev. **B 45**, 10684 (1992)

2.16 C. N. R. Rao, R. Nagarajan, R. Vijayaraghavan: Synthesis of cuprate superconductors, Supercond. Sci. Technol. **6**, 1 (1993)

2.17 N. M. Plakida: *High-Temperature Superconductivity* (Springer, Berlin, Heidelberg 1995)

2.18 M. Tinkham: *Theory of Superconductivity* (McGraw-Hill, New York 1975)

2.19 G. Blatter, M.V. Feigelmann, V.B. Geschkenbein, A.I. Larkin, V.M. Vinokur: Vortices in high-T_c superconductors, Rev. Mod. Phys. **66**, 1125 (1994)

2.20 K.H. Fischer: Vortices in high-T_c superconductors, Superconductivity Rev. **1**, 153 (1995)

2.21 H. Hilgenkamp, J. Mannhart: Grain boundaries in high-T_c superconductors, Rev. Mod. Phys. **74**, 485 (2002)2

2.22 D. Larbalestier, A. Gurevich, D.M. Feldmann, A. Polyanskii: High-T_c superconducting materials for electric power applications, Nature **414**, 368 (2001)

2.23 B. Seeber (Ed.): *Handbook of Applied Superconductivity*, Vol. 2 (Institut of Physics Publishing, Bristol, Philadelphia 1998)

2.24 J.J. Capponi, C. Chaillout, A.W. Hewat, P. Lejay, M. Marezio, N. Nguyen, B. Raveau, J.L. Soubeyroux, J.L. Tholence, R. Tournier: Structure of the 100 K superconductor $Ba_2YCu_3O_7$ between (5–300) K by neutron powder diffraction, Europhys. Lett. **3**, 1301–1307 (1987)

2.25 M.K. Wu, J.R. Ashburn, C.J. Torng, P.H. Hor, R.L. Meng, L. Gao, Z.J. Huang, Y.Q. Wang, C.W. Chu: Superconductivity at 93 K in a new mixed-phase Y–Ba–Cu–O compound system at ambient pressure, Phys. Rev. Lett. **58**, 908–910 (1987)

2.26 N.L. Ross, R.J. Angel, L.W. Finger, R.M. Hazen, C.T. Prewitt: Oxygen-defect perovskites and the 93-K superconductor, American Chemical Society: Symposium Series **351**, Chemistry of high-temperature superconductors, New Orleans 1987, ed. by E.D. Nelson, M.S. Wittingham, T.F. George (American Chemical Society, New Orleans 1987) 164–172

2.27 M. François, A. Junod, K. Yvon, A.W. Hewat, J.J. Capponi, P. Strobel, M. Marezio, P. Fischer: A study of the Cu–O chains in the high T_c superconductor $YBa_2Cu_3O_7$ by high resolution neutron powder diffraction, Solid State Comm. **66**, 1117–1125 (1988)

2.28 J.D. Jorgensen, B.W. Veal, A.P. Paulikas, L.J. Nowicki, G.W. Crabtree, H. Claus, W.K. Kwok: Structural properties of oxygen-deficient $YBa_2Cu_3O_{7-\delta}$, Phys. Rev. B **41**, 1863–1877 (1990)

2.29 E. Kaldis: Oxygen nonstoichiometry and lattice effects in $YBa_2Cu_3O_x$. In: *Handbook on the Physics and Chemistry of Rare Earths*, High-Temperature Superconductors II, Vol. 31, ed. by K.A. Gschneidner Jr., L. Eyring, M.B. Maple (Elsevier, Amsterdam, London, New York, Oxford, Paris, Shannon, Tokyo 2001) Chap. 195, pp. 1–186

2.30 R.J. Cava, A.W. Hewat, E.A. Hewat, B. Batlogg, M. Marezio, K.M. Rabe, J.J. Krajewski, W.F. Peck Jr., L.W. Rupp Jr.: Structural anomalies, oxygen ordering and superconductivity in oxygen deficient $Ba_2YCu_3O_x$, Physica C **165**, 419–433 (1990)

2.31 J.M.S. Shakle: Crystal chemical substitution and doping of $YBa_2Cu_3O_x$ and related superconductors, Mat. Sci. Engineering R **23**, 1–40 (1998)

2.32 M. Guillaume, P. Allenspach, W. Henggeler, J. Mesot, B. Roessli, U. Staub, P. Fischer, A. Furrer, V. Trounov: A systematic low-temperature neutron diffraction study of the $RBa_2Cu_3O_x$ (R = yttrium and rare earths; $x = 6$ and 7) compounds, J. Phys.: Condens. Matter **6**, 7963–7976 (1994)

2.33 Y. Nakabayashi, Y. Kubo, T. Manato, J. Tabuchi, A. Ochi, K. Utsumi, H. Igarashi, M. Yonezawa: The orthorhombic-tetragonal phase transformation an doxygen deficiency in $LnBa_2Cu_3P_{7-\delta}$, Jpn. J. Appl. Phys. **27**, L64–L66 (1988)

2.34 R.D. Shannon: Revised effective ionic radii and systematic studies of interatomic distances in halides and chalcogenides, Acta Crystallogr. A **32**, 751–767 (1976)

2.35 L.J. Masur, J. Kellers, C.M. Pegrum, D.A. Cardwell: Fundamentals of superconductivity: Applied properties of superconducting materials. In: *Handbook of Superconducting Materials*, Superconductivity, Materials and Processes, Vol. I, ed. by D.A. Cardwell, D.S. Ginley (IOP Publishing, Bristol, Philadelphia 2003) pp. 27–52

2.36 M. Murakami: Processing of bulk YBaCuO, Supercond. Sci. Technol. **5**, 185–203 (1992)

2.37 P. Diko: Characterization techniques: Optical microscopy. In: *Handbook of Superconducting Materials*, Characterization, Applications and Cryogenics, Vol. II, ed. by D.A. Cardwell, D.S. Ginley (IOP Publishing, Bristol 2003) pp. 1147–1175

2.38 C. Meingast, O. Kraut, T. Wolf, H. Wühl, A. Erb, G. Müller-Vogt: Large a-b anisotropy of the expansivity anomaly at T_c in untwinned $YBa_2Cu_3O_{7-\delta}$, Phys. Rev. Lett. **67**, 1634–1637 (1991)

2.39 P. Nagel, V. Pasler, C. Meingast, A.I. Rykov, S. Tajima: Anomalously large oxygen-ordering contribution to the thermal expansion of untwinned $YBa_2Cu_3O_{6.95}$ single crystals: A glasslike transition near room temperature, Phys. Rev. Lett. **85**, 2376–2379 (2000)

2.40 D. Dimos, D.R. Clarke: Surfaces, Inferfaces of Ceramic Materials, NATO ASI series: E173, Proceedings of the NATO Advanced Study Institute on Surfaces and Interfaces of Ceramic Materials, CAES-CNRS, Ile d'Oléron, France 1988, ed. by L.C. Dufour, others (Kluwer Academic, Dordrecht 1989) 301–318

2.41 T.A. Friedmann, M.W. Rabin, J. Giapintzakis, J.P. Rice, D.M. Ginsberg: Direct measurement of the anisotropy of the resistivity in the a-b plane of twin-free single-crystal, superconducting $YBa_2Cu_3O_{7-\delta}$, Phys. Rev. B **42**, 6217–6221 (1990)

2.42 J.R. Cooper, S.D. Obertelli, A. Carrington, J.W. Loram: Effect of oxygen depletion on th etransport properties of $YBa_2Cu_3O_{7-\delta}$, Phys. Rev. B **44**, 12086–10289 (1991)

2.43 A. Carrington, D.J.C. Walker, A.P. Mackenzie, J.R. Cooper: Hall effect and resistivity of oxygen-deficient $YBa_2Cu_3O_{7-\delta}$ thin films, Phys. Rev. B **48**, 13051–13059 (1993)

2.44 D. T. Verebelyi, D. K. Christen, R. Feenstra, C. Cantoni, A. Goyal, D. F. Lee, M. Paranthaman, P. N. Arendt, R. F. DePaula, J. R. Groves, C. Prouteau: Low angle grain boundary transport in $YBa_2Cu_3O_{7-\delta}$ coated conductors, Appl. Phys. Lett. **76**, 1755–1757 (2000)

2.45 B. Holzapfel, D. Verebelyi, C. Cantoni, M. Paranthaman, B. Sales, R. Feenstra, D. Christen, D. P. Norton: Low angle grain boundary transport properties of undoped and doped Y123 thin film bicrystals, Physica C **341**, 1431–1434 (2000)

2.46 K. Takenaka, K. Mizuhashi, H. Takagi, S. Uchida: Interplane charge transport in $YBa_2Cu_3O_{7-y}$: Spingab effect on in-plane and out-of-plane resistivity, Phys. Rev. B **50**, 6534–6537 (1994)

2.47 M. D. Nunez Regueiro, D. Castello: Thermal conductivity of high temperature superconductors, Int. J. Mod. Phys. B **5**, 2003–2035 (1991)

2.48 W.-H. Li, J. W. Lynn, Z. Fisk: Magnetic order of the Cu planes and chains in $RBa_2Cu_3O_{7+x}$, Phys. Rev. B **41**, 4098–4111 (1990)

2.49 J. S. Abell, T. W. Button: Processing: Sintering techniques for YBCO. In: *Handbook of Superconducting Materials*, Superconductivity, Materials and Processes, Vol. I, ed. by D. A. Cardwell, D. S. Ginley (IOP Publishing, Bristol 2003) pp. 251–258

2.50 M. Hikita, Y. Tajima, A. Katsui, Y. Hidaka, S. Iwata, S. Tsurumi: Electrical properties of high-T_c superconducting single-crystal $Eu_1Ba_2Cu_3O-y$, Phys. Rev. B **36**, 7199–7202 (1987)

2.51 W. J. Gallagher: Studies at IBM on anisotropy in single crystals of the high-temperature oxide superconductor $Y_1Ba_2Cu_3O_{7-\delta}$, J. Appl. Phys. **63**, 4216–4219 (1988), invited

2.52 T. K. Worthington, W. J. Gallagher, T. R. Dinger: Anisotropic nature of high-temperature superconductivity in single-crystal $Y_1Ba_2Cu_3O_{7-\delta}$, Phys. Rev. Lett. **59**, 1160–1163 (1987)

2.53 D. Feinberg, C. Villard: Intrinsic pinning and lock-in transition of flux lines in layered type-II superconductors, Phys. Rev. Lett. **65**, 919–922 (1990)

2.54 A. Koblischka-Veneva, N. Sakai, S. Tajima, M. Murakami: *High temperature superconductors: YBCO*, Superconductivity, Materials and Processes, Vol. I, ed. by D. A. Cardwell, D. S. Ginley (IOP Publishing, Bristol, Philadelphia 2003) pp. 893–945

2.55 T. R. Dinger, T. K. Worthington, W. J. Gallagher, R. L. Sandstrom: Direct observation of electronic anisotropy in single-crystal $Y_1Ba_2Cu_3O_{7-x}$, Phys. Rev. Lett. **58**, 2687–2690 (1987)

2.56 G. Böttger, I. Mangelschots, E. Kaldis, P. Fischer, C. Krüger, F. Fauth: The influence of Ca doping on the crystal structure and superconductivity of orthorhombic $YBa_2Cu_3O_{7-\delta}$, J. Phys.: Condens. Matter **8**, 8889–8905 (1996)

2.57 J. R. Grasmeder, M. T. Weller, P. C. Lanchester, C. E. Meats: Superconductivity in the Y–La–Ba–Cu–O system, Solid State Ionics **32/33**, 1115–1124 (1989)

2.58 F. Licci, A. Gauzzi, M. Marezio, G. P. Radaelli, R. Masini, C. Chaillout-Bougerol: Structural and electronic effects of Sr substitution for Ba in $Y(Ba_{1-x}Sr_x)Cu_3O_w$, Phys. Rev. B **58**, 15208–15217 (1998)

2.59 E. Kim, I.-S. Yang: Structural analysis of Y–Zn 123 superconductor using X-ray powder diffraction, New Physics (Korean Physical Society) **32**, 839–846 (1992)

2.60 N. L. Saini, K. B. Garg, H. Rajagopal, A. Sequeira: Neutron diffraction and Hall effect measurements on Y–Ba–Cu(Mn)–O, Solid State Comm. **82**, 895–899 (1992)

2.61 B. Roas, L. Schultz, G. Saemann-Ischenko: Anisotropy of the critical current density in epitaxial $YBa_2Cu_3O_x$ films, Phys. Rev. Lett. **64**, 479–482 (1990)

2.62 V. Breit, P. Schweiss, R. Hauff, H. Wühl, H. Claus, H. Rietschel, A. Erb, G. Müller-Vogt: Evidence for chain superconductivity in near-stoichiometric $YBa_2Cu_3O_x$ single crystals, Phys. Rev. B **52**, R15727–R15730 (1995)

2.63 P. Schweiss, W. Reichardt, M. Braden, G. Collin, G. Heger, H. Claus, A. Erb: Static and dynamic displacement in $RBa_2Cu_3O_{7-\delta}$ ($R = Y$, Ho; $\delta = 0.05, 0.5$): A neutron-diffraction study on single crystals, Phys. Rev. B **49**, 1387–1396 (1994)

2.64 H. Claus, M. Braun, A. Erb, K. Röhberg, B. Runtsch, H. Wühl, G. Bräuchle, P. Schweib, G. Müller-Vogt, H. v. Löhneysen: The "90 K" plateau of oxygen deficient of $YBa_2Cu_3O_{7-\delta}$ single crystals, Physica C **198**, 42–46 (1992)

2.65 H. Claus, U. Gebhard, G. Linker, K. Röhberg, S. Riedling, J. Franz, T. Ishida, A. Erb, G. Müller-Vogt, H. Wühl: Phase separation in $YBa_2Cu_3O_{7-\delta}$ single crystals near $\delta = 0$, Physica C **200**, 271–276 (1992)

2.66 D.-H. Wu, S. Sridhar: Pinning forces and lower critical fields in $YBa_2Cu_3O_y$ crystals: Temperature dependence and anisotropy, Phys. Rev. Lett. **65**, 2074–2077 (1990)

2.67 U. Welp, W. K. Kwok, G. W. Crabtree, K. G. Vandervoort, J. Z. Liu: Magnetic measurements of the upper critical field of $YBa_2Cu_3O_{7-\delta}$ single crystals, Phys. Rev. Lett. **62**, 1908–1911 (1989)

2.68 P. Majewski: Phase diagram studies in the system Bi–Pb–Sr–Ca–Cu–O–Ag, Supercond. Sci. Technol. **10**, 453 (1997)

2.69 Y. Yamada, Y. Shiohara: Progress of high-T_c wires and its application. In: *High Temperature Superconductivity, 1: Materials*, ed. by A. V. Narlikar (Springer, Berlin, Heidelberg, New York 2004) pp. 291–337

2.70 D. L. Feng, A. Damascelli, K. M. Chen, N. Motoyama, D. H. Lu, H. Eisaki, K. Shimizu, J. Shimoyama, K. Kishio, N. Kaneko, M. Greven, G. D. Gu, X. J. Zhou, C. Kim, F. Ronning, N. P. Armitage, Z.-X. Shen: Electronic structure of the trilayer cuprate superconductor $Bi_2Sr_2Ca_2Cu_3O_{10+\delta}$, Phys. Rev. Lett. **88**, 107001-1 (2002)

2.71 A. Maeda, T. Shibauchi, N. Kondo, K. Uchinokura, M. Kobayashi: Magnetic-field penetration depth

2.72 L. Miu, P. Wagner, U. Frey, A. Hadish, D. Miu, H. Adrian: Vortex unbinding and layer decoupling in epitaxial $Bi_2Sr_2Ca_2Cu_3O_{10+\delta}$ films, Phys. Rev. B. **52**, 4553 (1995)

and lower critical field of quasi-two-dimensional superconductor $Bi_2Sr_2Ca_2Cu_2O_y$, Phys. Rev. B **46**, 14234 (1992)

2.73 I. Matsubara, R. Funahashi, K. Ueno, H. Yamashita, T. Kawai: Lower critical field and reversible magnetization of $(Bi,Pb)_2Sr_2Ca_2Cu_3O_x$ superconducting whiskers, Physica C **256**, 33 (1996)

2.74 S. I. Vedeneev, A. G. M. Jansen, E. Haanappel, P. Wyder: Temperature dependence of the upper critical field of $Bi_2Sr_2CuO_x$ single crystals, Phys. Rev. B **60**, 12467 (1999)

2.75 M. Akamatsu, L. X. Chen, H. Ikeda, R. Yoshizaki: Physica C **235-240**, 1619 (1994)

2.76 D. C. Johnston, J. H. Cho: Magnetic-susceptibility anisotropy of single-crystal $Bi_2Sr_2CaCu_2O_8$, Phys. Rev. B **42**, 8710 (1990)

2.77 T. T. M. Palstra, B. Batlogg, L. F. Scheemeyer, R. B. van Dover, J. Waszczak: Angular dependence of the upper critical field of $Bi_{2.2}Sr_2Ca_{0.8}Cu_2O_{8+\delta}$, Phys. Rev. B **38**, 5102 (1988)

2.78 I. Matsubara, H. Tanigawa, T. Ogura, H. Yamashita, M. Kinoshita, T. Kawai: Upper critical field and anisotropy of the high-T_c $Bi_2Sr_2CaCu_2O_x$ phase, Phys. Rev. B **45**, 7414 (1992)

2.79 Q. Li, M. Suenaga, J. Gohng, D. K. Finnemore, T. Hikata, K. Sato: Reversible magnetic properties of c-axis-oriented superconducting $Bi_2Sr_2Ca_2Cu_{10}$, Phys. Rev. B **46**, 3195 (1992)

2.80 Q. Li, M. Suenaga, T. Hikata, K. Sato: Two-dimensional fluctuations in the magnetization of $Bi_2Sr_2Ca_2Cu_3O_{10}$, Phys. Rev. B **46**, 5857 (1992)

2.81 A. V. Narlikar (Ed.): *High Temperature Superconductivity, 1: Materials, 2: Engineering Applications* (Springer, Berlin, Heidelberg, New York 2004)

2.82 S. Martin, A. T. Fiory, R. M. Fleming, L. E. Schneemeyer, J. V. Waszczak: Normal-state transport properties of $Bi_{2+x}Sr_{2-y}CuO_{6\pm\delta}$ crystals, Phys. Rev. B **41**, 846 (1990)

2.83 T. W. Li, A. A. Menovski, J. J. M. Franse, P. H. Kes: Flux pinning in Bi-2212 single crystals with various oxygen contents, Physica **C 257**, 179 (1996)

2.84 D. R. Harshman, R. N. Kleiman, M. Inui, G. P. Espinosa, D. B. Mitzi, A. Kapitulnik, T. Pfiz, D. L. Williams: Magnetic penetration depth and flux dynamics in single crystal $Bi_2Sr_2CaCu_2O_{8-\delta}$, Phys. Rev. Lett. **67**, 3152 (1991)

2.85 H. Enriquez, N. Bontemps, A. A. Zhukov, D. V. Shovkun, M. R. Trunin, A. Buzdin, M. Daumens, T. Tamegai: A compilation of references considering $\lambda_c(0)$, Phys. Rev. B **63**, 144 (2001)

2.86 Y. Ando, G. S. Boebinger, A. Passner, L. F. Schneemeyer, T. Kimura, M. Okuya, S. Watauchi, J. Shimoyama, K. Kishio, K. Tamsaku, N. Ichikawa, S. Uchida: Upper critical fields and irreversibility lines of optimal doped high-T_c cuprates, Phys. Rev. **B 60**, 12475 (1999)

2.87 A. Attenberger: Elektrische Transportmessungen über definierte Korngrenzen-Strukturen in epitaktischen BI2223-Schichten. Ph.D. Thesis (Technische Universität Dresden, Dresden 2000)

2.88 I. L. Landau, H. R. Ott: Temperature dependence of the upper critical field of type-II superconductors from isothermal magnetization data: Application to high-temperature superconductors, Phys. Rev. B **66**, 144506 (2002)

2.89 R. Wesche: *High-Temperature Superconductors: Materials, Properties, and Applications* (Kluwer, Boston, Dordrecht, London 1998)

2.90 P. Schmitt, P. Kummeth, L. Schultz, G. Saemann-Ischenko: Two-dimensional behavior and critical-current anisotropy in epitaxial $Bi_2Sr_2CaCu_2O_{8+x}$ thin films, Phys. Rev. Lett. **67**, 267 (1991)

2.91 T. Amrein, L. Schultz, B. Kabius, K. Urban: Orientation dependence of grain-boundary critical current densities in high-T_c bicrystals, Phys. Rev. B **51**, 6792 (1995)

2.92 G. Balestrino, M. Marinelli, E. Milani, L. Reggiani, R. Vaglio, A. A. Varlamov: Exess conductivity in 2:2:1:2-phase Bi–Sr–Ca–Cu–O epitaxial thin films, Phys. Rev. B **46**, 14919 (1992)

2.93 K. Endo, H. Yamasaki, S. Misawa, S. Yoshida, K. Kajimura: Lett. Nature **355**, 327 (1992)

2.94 P. Wagner, F. Hillmer, U. Frey, H. Adrian, T. Steinborn, L. Ranno, A. Elschner, I. Heyvaert, Y. Bruynseraede: Preparation and structural characterisation of thin epitaxial $Bi_2Sr_2CaCu_2O_{8+\delta}$ films with T_c in the 90 K range, Physica C **215**, 123 (1993)

2.95 Q. Li, Y. N. Tsay, M. Suenaga, R. A. Klemm, G. D. Gu, N. Koshizuka: $Bi_2Sr_2CaCu_2O_{8+\delta}$ bicrystal c-axis twist Josephson junktions: A new phase sensitive test of order parameter symmetry, Phys. Rev. Lett. **83**, 4160 (1999)

2.96 N. Tomita, Y. Takahashi, Y. Ishida: Preparation of bicrystal in a Bi–Sr–Ca–O superconductor, Jpn. J. Appl. Phys. **29**, 1 (1990)L 30

2.97 Q. Li, Y. N. Tsay, M. Suenaga, G. Wirth, G. D. Du, N. Koshizuka: Superconducting transport properties of 2.2-GeV Au-ion irradiated c-axis twist $Bi_2Sr_2CaCu_2O_{8+\delta}$ bicrystals, Appl. Phys. Lett. **74**, 1323 (1999)

2.98 S. Chu, M. E. McHenry: Growth and characterization of $(Bi,Pb)_2Sr_2Ca_2Cu_3O_x$ single crystals, J. Mater. Res. **13**, 589 (1998)3

2.99 B. Sailer, F. Schwaigerrer, K. Gibson, H.-J. Meyer, M. Lehmann, L. Woodall, M. Gerards: Effect of precursor powder properties on magnetic and electrical transport properties of (Bi,Pb)-2223-tapes, IEEE Trans. Appl. Supercond. **11**, 2975 (2001)

2.100 H. F. Poulsen, L. Gottschalck Andersen, T. Frello, S. Prantontep, N. H. Andersen, S. Garbe, J. Madsen, A. Abrahamsen, M. D. Bentzon, M. von Zimmermann: In situ study of equilibrium phenomena and kinetics in a BSCCO/Ag tape, Physica C **315**, 254 (1999)

2.101 E. Giannini, E. Bellingeri, R. Passerini, R. Flükiger: Direct observation of the Bi,Pb (2223) phase formation inside Ag-sheathed tapes and quantitative secondary phase analysis by means of in situ high-temperature neutron diffraction, Physica C **315**, 185 (1999)

2.102 T. Fahr, H. P. Trinks, R. Schneider, C. Fischer: Investigation of the formation of the Bi-2223/Ag tapes by in situ high temperature neutron diffraction, IEEE Trans. Appl. Supercond. **11**, 3399 (2001)

2.103 T. Lang, B. Heeb, D. Buhl, L. J. Gaukler: Proc. of the Forth Intern. Conf. And Exhibition, World Congress on Supercoductivity, June 27–July 1, 1994, Orlando (Florida), Vol. 2, ed. by K. Krishen, C. Burnham, p. 753

2.104 J. C. Grivel, R. Flükiger: Visualization of the formation of the $(Bi,Pb)_2Sr_2Ca_2Cu_3O_{10+\delta}$ phase, Supercond. Sci. Technol. **9**, 555 (1996)

2.105 L. Y. Wang, W. Bian, Y. Zhu, Z. X. Cai, D. O. Welch, R. L. Sabatini, T. R. Thurston, M. Suenaga: A kinetic mechanism for the formation of aligned $(Bi,Pb)_2Sr_2Ca_2Cu_3O_{10}$ in a powder-in-tube processed tape, Appl. Phys. Lett. **69**, 580 (1996)

2.106 G. W. Wang, M. D. Bentzon, P. Vase: High critical current and detect free km long Bi-2223/Ag-alloy multifilament tapes, Proceedings of EUCAS 1999, ed. by X. Obradors, F. Sandiumenge, J. Fontcuberta, Institut of Physics Conference Ser. Number 167, Vol. 1, p. 459

2.107 C. Fischer, T. Fahr, A. Hütten, U. Schläfer, M. Schubert, C. Rodig, H. P. Trinks: Technology of Bi,Pb-2223 tape fabrication with Ag and Ag alloy sheaths for electric power systems, Supercond. Sci. Technol. **11**, 995 (1998)

2.108 P. Vase, R. Flükiger, M. Leghissa, B. Glowcki: Current status of high-T_c wire, Supercond. Sci. Technol. **13**, R.71 (2000)

2.109 L. Masur: Adv. Cryog. Eng. **46**, 871 (2000)

2.110 P. M. Grant: IEEE Trans. Appl. Supercond. **7**, 7 (1997)

2.111 A. P. Malozemoff, W. Carter, S. Fleshler, L. Fritzemeier, Q. Li, L. Masue, P. Miles, D. Parker, R. Parella, E. Podtburg, G. N. Riley Jr., M. Rupich, J. Scudiere, W. Zhang: HTS wire at commercial levels, IEEE Trans. Appl. Supercond. **9**, 2469 (1999)

2.112 T. Hasegawa, N. Ohtani, T. Koizumi, Y. Aoki, S. Nagaya, N. Hirano, L. Motowidlo et al.: Improvement of superconducting properties of B-2212 round wire and primary test results of large capacity Rutherford cable, IEEE Trans. Appl. Supercond. **11**, 3034 (2001)

2.113 H. S. Pan, E. W. Hudson, K. M. Lang, H. Elsaki, S. Uchida, J. C. Davis: Imaging the effects of individual zinc impurity atoms on superconductivity in $Bi_2Sr_2CaCu_2O_{8+\delta}$, Nature **403**, 746 (2000)

2.114 A. Diaz, L. Mechin, P. Berghuis, J. E. Evetts: Evidence for vortex pinning by dislocations in $YBa_2Cu_3O_{7-\delta}$

low-angle grain boundaries, Phys. Rev. Lett. **80**, 3855 (1998)

2.115 B. Dam, J. L. Huijbregtse, F. C. Klaassen, R. C. F. van der Geest, G. Doornbos, J. H. Rector, A. M. Testa, S. Freisem, J. C. Martinez, B. Stäuble-Pümpin, R. Griessen: Origin of high critical currents in $YBa_2Cu_3O_{7-\delta}$ superconducting films, Nature **399**, 439 (1999)

2.116 L. Bulaevskij, L. Daemen, M. Maley, J. Coulter: Limits to the critical current in high-T_c superconducting tapes, Phys. Rev. B **48**, 13798 (1993)

2.117 B. Hensel, G. Grasso, R. Flükiger: Limits to the critical transport current in superconducting $(Bi,Pb)_2Sr_2Ca_2Cu_3O_{10}$ silver sheathed tapes: The railway-switch model, Phys. Rev. B **51**, 15456 (1995)

2.118 G. N. Riley, A. P. Malozemoff, Q. Li, S. Fleshler, T. G. Holesinger: The freeway model: New concepts in understanding supercurrent transport in Bi-2223 tapes, JOM 49 **10**, 24 (1997)

2.119 T. Staiger: Stromtransport in supraleitenden $(Bi,Pb)_2Sr_2Ca_2Cu_3O_x$/Ag-Bändern. Ph.D. Thesis (Technische Universität Dresden, Dresden 1998)

2.120 K-H. Müller, G. Drechsler, S-L. Fuchs, V. N. Narozhnyi: *Magnetic, superconducting properties of rare earth borocarbides of the type* RNi_2B_2C, Handbook of Magnetic Materials, Vol. 14, ed. by K. H. J. Buschow (Elsevier, Amsterdam 2002) pp. 199–306

2.121 R. Flükiger, W. Klose (Eds.): *Superconductors*, Landolt–Börnstein, New Series III/21 (Springer, Berlin, Heidelberg 1990)

2.122 T. Q. Huang He, A. P. Ramirez, Y. Wang, K. A. Regan, N. Rogado, M. A. Hayward, M. K. Haas, J. I. Slusky, K. Inumara, H. W. Zandbergen, N. P. Ong, R. J. Cava: Nature **411**, 54 (2001)

2.123 K. Sakamaki, H. Wada, H. Nozaki, Y. Onuki: Solid State Commun. **112**, 323 (1999)

2.124 K. Sakamaki, H. Wada, H. Nozaki, Y. Onuki: Solid State Commun. **118**, 113 (2001)

2.125 J. Nagamatsu, N. Nakagawa, T. Muranaka, Y. Zenitani, J. Akimitsu: Nature **410**, 63 (2001)

2.126 R. Nagarajan, C. Mazumdar, Z. Hossain, S. K. Dhar, K. V. Gopalakrishnan, L. C. Gupta, C. Godart, B. C. Padalia, R. Vijayaraghavan: Phys. Rev. Lett. **72**, 274 (1994)

2.127 R. J. Cava, H. Takagi, H. W. Zandbergen, J. J. Krajewski, W. F. Peck Jr., T. Siegrist, B. Batlogg, R. B. van Dover, R. J. Felder, K. Mizuhashi, J. O. Lee, S. Uchida: Nature **367**, 252 (1994)

2.128 Rosseinsky et al.: Superconductivity at 28 K in Rb_xC_{60}, Phys. Rev. Lett. **60** (1991)

2.129 M. I. Eremets, V. V. Struzhkin, H.-K. Mao, R. J. Hemley: Science **293**, 272 (2001)

2.130 C. Buzea, T. Yamashita: Supercond. Sci. Technol. **14**, 115 (2001)

2.131 C. C. Lai, M. S. Lin, Y. B. You, H. C. Ku: Phys. Rev. B **51**, 420 (1995)

2.132 R. Flükiger, W. Klose (Eds.): *Superconductors*, Landolt–Börnstein, New Series III/21 (Springer, Berlin, Heidelberg 1994)

2.133 Z. Xou-xiang, H. Shou-an: Solid State Commun. **45**, 281 (1983)

2.134 H. Ihara, Y. Kimura, K. Senzaki, H. Kezuka, M. Hirabayashi: Phys. Rev. B **31**, 3177 (1985)

2.135 S. Yamanaka, H. Kawaji, K. Hotehama, M. Ohashi: Adv. Mater. **8**, 771 (1996)

2.136 S. Yamanaka, K. Hotehama, H. Kawaji: Nature **392**, 580 (1998)

2.137 R. Flükiger, W. Klose (Eds.): *Superconductors*, Landolt–Börnstein, New Series III/21 (Springer, Berlin, Heidelberg 1998)

4.3. Magnetic Materials

Magnetic materials consist of a wide variety of metals and oxides. Their effective properties are given by a combination of two property categories: *intrinsic* properties which are the atomic moment per atom p_{at}, Curie temperature T_c, magnetocrystalline anisotropy coefficients K_i, and magnetostriction coefficients λ_i; and *extrinsic* properties which are essentially their coercivity H_c and their magnetisation M or magnetic Induction J as a function of the applied magnetic field H. Moreover, the effective properties are depending decisively on the microstructural features, texture and, in most cases, on the external geometric dimensions such as thickness or shape of the magnetic part. In some cases non-magnetic inorganic and organic compounds serve as binders or magnetic insulators in multiphase or composite magnetic materials.

- 4.3.1 **Basic Magnetic Properties** ... 755
  - 4.3.1.1 Atomic Moment ... 755
  - 4.3.1.2 Magnetocrystalline Anisotropy ... 756
  - 4.3.1.3 Magnetostriction ... 757
- 4.3.2 **Soft Magnetic Alloys** ... 758
  - 4.3.2.1 Low Carbon Steels ... 758
  - 4.3.2.2 Fe-based Sintered and Composite Soft Magnetic Materials ... 759
  - 4.3.2.3 Iron–Silicon Alloys ... 763
  - 4.3.2.4 Nickel–Iron-Based Alloys ... 769
  - 4.3.2.5 Iron–Cobalt Alloys ... 772
  - 4.3.2.6 Amorphous Metallic Alloys ... 772
  - 4.3.2.7 Nanocrystalline Soft Magnetic Alloys ... 776
  - 4.3.2.8 Invar and Elinvar Alloys ... 780
- 4.3.3 **Hard Magnetic Alloys** ... 794
  - 4.3.3.1 Fe–Co–Cr ... 795
  - 4.3.3.2 Fe–Co–V ... 797
  - 4.3.3.3 Fe–Ni–Al–Co, Alnico ... 798
  - 4.3.3.4 Fe–Nd–B ... 800
  - 4.3.3.5 Co–Sm ... 803
  - 4.3.3.6 Mn–Al–C ... 810
- 4.3.4 **Magnetic Oxides** ... 811
  - 4.3.4.1 Soft Magnetic Ferrites ... 811
  - 4.3.4.2 Hard Magnetic Ferrites ... 813
- **References** ... 814

4.3.1 Basic Magnetic Properties

Basic magnetic properties of metallic systems and materials are treated by *Gignoux* in [3.1]. Extensive data on magnetic properties of metals can be found in [3.2]. Magnetic properties of ferrites are treated by *Guillot* in [3.3]. Extensive data on magnetic and other properties of oxides and related compounds can be found in [3.4] and [3.5].

4.3.1.1 Atomic Moment

The suitability of a metal or oxide to be used as a magnetic material is determined by its mean atomic moment (p_{at}). For metals the Bethe–Slater–Pauling curves, Fig. 4.3-1, indicate how p_{at} depends on the average number (n) of 3d and 4s electrons per atom, and on the crystal structure, i.e., face-centered cubic (fcc) or body-centered cubic (bcc) structure. Alloys based on Fe, Co, and Ni are most suitable from this point of view, corresponding to their actual use. The characteristic temperature dependence of the spontaneous magnetization $I_s(T)$, shown for Fe in Fig. 4.3-2, the Curie temperature T_c and the spontaneous magnetization at room temperature I_s (see Table 4.3-1), are the ensuing properties.

4.3.1.2 Magnetocrystalline Anisotropy

Since the magnetic moment arises from the exchange coupling of neighboring ions, it is also, coupled to their positions in the crystal structures. Basically this is the origin of the magnetocrystalline anisotropy which plays a major role as an intrinsic property for the optimization of both soft and hard magnetic materials because it determines the crystallographic direction and relative magnitude of easy magnetization. As an example, Fig. 4.3-3a shows the magnetization curves for a single crystal of Fe in the three major crystallo-

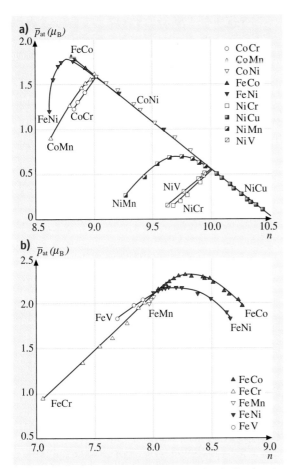

Fig. 4.3-1a,b Bethe–Slater–Pauling relation indicating the dependence of the mean magnetic moment per atom p_{at} on the average number n of 3d and 4s electrons per atom for binary alloys with (**a**) fcc structure and (**b**) bcc structure [3.6]

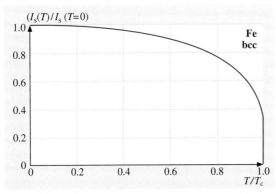

Fig. 4.3-2 Reduced spontaneous magnetization $I_s(T)/I_s(T=0)$ vs. reduced temperature T/T_c for Fe [3.6]

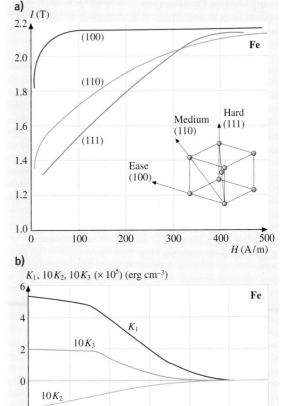

Fig. 4.3-3a,b Magnetization curves of an Fe single crystal indicting the typical characteristics of (**a**) magnetocrystalline anisotropy and (**b**) the temperature dependence of the magnetocrystalline anisotropy constants K_i of Fe [3.6]

Table 4.3-1 Curie temperatures and intrinsic magnetic properties at room temperature

Element	T_c (K)	I_s (T)	K_1 (10^4 J m^{-3})	K_2 (10^4 J m^{-3})	K_3 (10^4 J m^{-3})	λ_{100} (10^{-6})	λ_{111} (10^{-6})	λ_{0001} (10^{-6})	λ_{1010} (10^{-6})	λ_s (10^{-6})
Fe, bcc	1044	2.15	4.81	0.012	−0.012	22.5	−18.8	–	–	−4
Co, hcp	1388	1.62	41	120	–	–	–	−4	−22	−71
Ni, fcc	624	0.55	−0.57	−0.23	0	−46	−24.3	–	–	−34

graphic directions of the body-centered cubic crystal structure.

The anisotropy constants $K_1, K_2, \ldots$ are defined for cubic lattices, such as Fe and Ni, by expressing the free energy of the crystal anisotropy per unit volume as

$$E_a = K_0 + K_1 S + K_2 P + K_3 S^2 + K_4 SP + \ldots ,$$

with

$$S = \alpha_1^2 \alpha_2^2 + \alpha_2^2 \alpha_3^2 + \alpha_3^2 \alpha_1^2 \quad \text{and} \quad P = \alpha_1^2 \alpha_2^2 \alpha_3^2 ,$$

where α_i, α_j, and α_k are the direction cosines of the angle between the magnetization vector and the crystallographic axes. Corresponding definitions pertain to the anisotropy constants for other crystal lattices.

In practice it can be important to know not only the room temperature value of the magnetocrystalline anisotropy constants K_i but also their temperature dependence, which is shown for Fe in Fig. 4.3-3b. For many technical considerations it suffices to take the dominating anisotropy constant K_1 into account.

4.3.1.3 Magnetostriction

Magnetostriction is the intrinsic magnetic property which relates spontaneous lattice strains to magnetization. It is treated extensively by *Cullen* et al. in [3.3]. As a crystal property magnetostriction is purely intrinsic. The phenomenon of macroscopic magnetostrictive strains of a single or polycrystalline sample is due to the presence of magnetic domains whose reorientation occurs under the influence of magnetic fields or applied stresses and, thus, is a secondary property which has to be distinguished.

Magnetostriction plays a role in determining a number of different effects:

- As a major factor of influence on the coercivity of soft magnetic materials because it determines the magnitude of interaction of internal stresses of materials with the movement of magnetic domain walls.

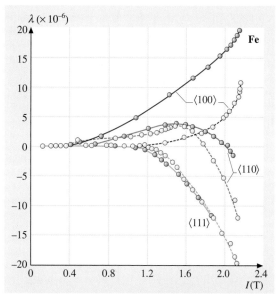

Fig. 4.3-4 Dependence of the magnetostrictive elongation on the magnetization intensity in three different crystallographic directions of Fe single crystals according to two different sources (*open* and *closed circles*) [3.7]

These boundaries are associated with strains themselves unless the material is free of magnetostriction altogether. This is practically impossible because the components of this property vary with the applied magnetic field, as shown for the example of Fe in Fig. 4.3-4. Moreover, they vary with temperature and alloy composition.

- As a decisive variable determining the properties of invar and elinvar alloys (Sect. 4.3.2.8).
- As a property of anomalously high magnitude in cubic Laves phase compounds of rare earth metals such as $Fe_2(Tb_{0.3}Dy_{0.7})$, which are the basis of magnetostrictive materials serving as magnetostrictive transducers and sensors [3.3].

4.3.2 Soft Magnetic Alloys

The basic suitability of a metal or alloy as a soft magnetic material is provided if its Curie temperature T_c and saturation polarization I_s at room temperature are sufficiently high. The main goal in developing soft magnetic materials is to reduce the magnetocrystalline anisotropy constants K_i, in particular K_1, and the magnetostriction constants λ_i, to a minimum. This leads to a minimum energy requirement for magnetization reversal. The appropriate combination of the intrinsic magnetic properties can be achieved by alloying additions and, in some cases, by annealing treatments which induce atomic ordering and, thus, an additional variation of K_1 and/or λ_i. Surveys and data of soft magnetic alloys are given in [3.8].

4.3.2.1 Low Carbon Steels

Low carbon steels are the most widespread magnetic metallic materials for use in electric motors and inductive components such as transformers, chokes, and a variety of other AC applications which require high magnetic induction at moderate to low losses. This is mainly due to the high intrinsic saturation polarization of the base element Fe, $J_s = 2.15\,\text{T}$, and to the low cost of mass produced steels.

Main processing effects on achieving materials with a high relative permeability is the minimization of the formation of particles (mainly carbides and sulfides) which may impede domain boundary motion, and induc-

Table 4.3-2 Standard IEC specification for non-alloyed magnetic steel sheet. The conventional designation of the different grades (first column) comprises the following order: (1) 100 times the maximum specified loss (W kg^{-1}) at 1.5 T; (2) 100 times the nominal sheet thickness in (mm); (3) the characteristic letter "D"; (4) one tenth of the frequency in Hz at which the magnetic properties are specified. The materials are delivered in the semi-processed state. The magnetic properties apply to test specimens heat-treated according to manufacturer's specifications

Grade	Nominal thickness (mm)	Maximum specific total loss for peak induction (W kg^{-1})			Minimum induction in a direct or alternating field at field strength given (T)			Stacking factor	Conventional density (kg dm^{-3})
		1.5 T at at 50 Hz	1.0 T at at 50 Hz	1.5 T at at 60 Hz	at 2500 A/m	at 5000 A/m	at 10 000 A/m		
660-50-D5	0.50	6.60	2.80	8.38	1.60	1.70	1.80	0.97	7.85
890-50-D5	0.50	8.90	3.70	11.30	1.58	1.68	1.79	0.97	7.85
1050-50-D5	0.50	10.50	4.30	13.34	1.55	1.65	1.78	0.97	7.85
800-65-D5	0.65	8.00	3.30	10.16	1.60	1.70	1.80	0.97	7.85
1000-65-D5	0.65	10.00	4.20	12.70	1.58	1.68	1.79	0.97	7.85
1200-65-D5	0.65	12.00	5.00	15.24	1.55	1.65	1.78	0.97	7.85

Table 4.3-3 Standard IEC specifictaion for cold rolled magnetic alloyed steel strip delivered in the semi-processed state. The conventional designation of the different grades comprises the following order (first column): (1) 100 times the maximum specified loss at 1.5 T peak induction in (W kg^{-1}); (2) 100 times the nominal strip thickness, in (mm); (3) the characteristic letter "E"; (4) one tenth of the frequency in Hz, at which the magnetic properties are specified. the magnetic properties apply to test specimens subjected to a reference heat treatment

Grade	Nominal thickness (mm)	Reference treatment temperature (°C)	Maximum specific total loss for peak induction (W kg^{-1})		Minimum induction in a direct or alternating field at field strength given (T)			Conventional density (10^3 kg m^{-3})
			1.5 T	1.0 T	2500 Am^{-1}	5000 Am^{-1}	10 000 Am^{-1}	
340-50-E5	0.50	840	3.40	1.40	1.52	1.62	1.73	7.65
390-50-E5	0.50	840	3.90	1.60	1.54	1.64	1.75	7.70
450-50-E5	0.50	790	4.50	1.90	1.55	1.65	1.76	7.75
560-50-E5	0.50	790	5.60	2.40	1.56	1.66	1.77	7.80

ing a texture – by suitable combinations of deformation and recrystallization – with a preponderance of ⟨100⟩ and ⟨110⟩ components, while suppressing the ⟨111⟩ and ⟨211⟩ components, such that the operating induction can be aligned most closely to the easiest direction of magnetization ⟨100⟩. More details are given by *Rastogi* in [3.9]. Tables 4.3-2 and 4.3-3 show characteristic data of non-alloyed magnetic steel sheet and cold rolled magnetic alloyed steel strip, respectively.

4.3.2.2 Fe-based Sintered and Composite Soft Magnetic Materials

Iron-based sintered and composite soft magnetic materials are treated and listed extensively in [3.10]. Iron and Fe based alloys are used as sintered soft magnetic materials because of the particular advantages of powder metallurgical processing in providing net-shaped parts economically. The sintering process serves to achieve diffusion-bonding of the powder particles with a uniform distribution of the alloying elements. The metallic bond and degree of particle contact determine the magnetic properties in each alloy system.

The operating frequency in an application is limited by the resistivity of the material, which is increased beyond that of pure and dense Fe, primarily by varying the type and concentration of alloying elements. Density and crystal structure have a major impact on the magnetic properties, but only a minor effect on the resistivity. The sintered materials have a resistivity ranging from 10 to 80 μΩ cm and are applicable in DC and very low frequency fields. The sintered materials comprise Fe and Fe−P, Fe−Si, Fe−Si−P, Fe−Sn−P, Fe−Ni, Fe−Cr, and Fe−Co alloys. Characteristic data are listed in Tables 4.3-4 to 4.3-11 [3.10].

It is useful to note that the evaluation of a large number of test data of sintered soft magnetic materials yields consistent empirical relations. For sintered Fe products these are:

$$B_{15}[\text{T}] = 4.47\varrho - 10.38 ,$$

$$B_{\text{r}}[\text{T}] = 3.87\varrho - 17.23 ,$$

$$H_{\text{c}}[\text{A/m}] = 11.47 L^{-0.59} ,$$

$$\rho[\mu\Omega\,\text{cm}] = -4.34\varrho + 44.77 ,$$

and

$$\mu_{\text{max}} = 0.21 L^{0.68} \times 10^3 ,$$

where ϱ = density (g cm^{-3}) and L = grain diameter (average intercept length) (μm).

Dust core materials consist of Fe or Fe alloy particles which are insolated by an inorganic high resistivity barrier. They are applied in the 1 kHz to 1 MHz range. These are to be distinguished from soft magnetic Fe composite materials, which consist of pure Fe particles separated by an insolating organic barrier, providing a medium to high bulk resistance. Compositions and magnetic properties of Fe based composite materials are listed in Tables 4.3-12 and 4.3-13. These materials are applicable in the frequency range from 50 Hz to 1 kHz.

Table 4.3-4 Magnetic properties of sintered Fe products in relation to density [3.10]

Density ϱ (g cm^{-3})	Sintering conditions temp. (°C), atm.	Induction at 1200 A m^{-1} (T)	Coercitivity H_{c} (A m^{-1})	Remanence B_{r} (T)	Max. rel. permeability μ_{max}
6.6	1120, DA	0.90	170[a]	0.78[a]	1700
6.6	1120, H$_2$/Vac.	0.95	140[a]	0.82[a]	1800
6.6	1260, H$_2$/Vac.	1.05	120[a]	0.85[a]	2800
6.9	1120, DA	1.05	170[a]	0.90[a]	2100
6.9	1120, H$_2$/Vac.	1.05	140[a]	0.97[a]	2300
6.9	1260, H$_2$/Vac.	1.20	120[a]	1.0[a]	3300
7.2	1120, DA	1.20	170[a]	1.05[a]	2700
7.2	1120, H$_2$/Vac.	1.20	140[a]	1.10[a]	2900
7.2	1260, H$_2$/Vac.	1.30	120[a]	1.15[a]	3800
7.4		1.25	136	1.20	3500
7.4		1.30	112	1.30	5500
7.6		1.50	80	1.40	6000

[a] Measured from a maximum applied magnetic field strength of 1200 A/m.

Table 4.3-5 Magnetic properties of sintered Fe–0.8 wt% P products in relation to density [3.10]

Density ϱ (g cm^{-3})	Sintering conditions temp. (°C), atm.	Induction at 1200 A m^{-1} (T)	Coercivity H_c (A m^{-1})	Remanence B_r (T)	Max. rel. permeability μ_{max}
6.8	1120, DA	1.05	120[a]	1.00[a]	3500
7.0	1120, H$_2$/Vac.	1.20	100[a]	1.05[a]	4000
7.2	1260, H$_2$/Vac.	1.25	95[a]	1.15[a]	4000
7.0	1120, DA	1.20	120[a]	1.10[a]	4000
7.2	1120, H$_2$/Vac.	1.25	100[a]	1.15[a]	4500
7.4	1260, H$_2$/Vac.	1.30	95[a]	1.20[a]	4500
7.0	1120, DA	1.25	120[a]	1.20[a]	4500
7.2	1120, H$_2$/Vac.	1.30	100[a]	1.30[a]	5000
7.4	1260, H$_2$/Vac.	1.35	95[a]	1.25[a]	5000

[a] Measured from a maximum applied magnetic field strength of 1200 A/m

Table 4.3-6 Magnetic properties of sintered Fe–3 wt% Si products in relation to density[a] [3.10]

Density ϱ (g cm^{-3})	Sintering conditions temp. (°C), time, atm.	Coercivity H_c (A m^{-1})	Remanence B_r (T)	Sat. induction B_s (T)	Max. rel. permeability μ_{max}
7.3		64	1.15	1.90	8000
7.5		48	1.25	2.00	9500
7.2	1250, 30 min, H$_2$[b]	80	1.0		4300
7.01	1120, 60 min, H$_2$[b]	117[d]			2800
7.19	1200, 60 min, H$_2$[b]	88[d]			4200
7.40	1200, 60 min, H$_2$[c]	79[d]	1.25		5600
7.43	1300, 60 min, H$_2$	56[d]			8400
7.55	1371, DA[e]	51	1.21	1.50	
7.55	1371, DA[e]	57	1.07	1.45	

[a] Measured according to ASTM A 596
[b] Conventional compacting at 600 MPa
[c] Warm compacted at 600 MPa
[d] Defined as coercive force i. e. magnetized to a field strength well below saturation
[e] Formed by MIM

Table 4.3-7 Magnetic properties of sintered Fe−S−P in relation to density[a] [3.10]

Density ϱ (g cm^{-3})	Sintering conditions temp. (°C), atm.	Coercivity H_c (A m^{-1})	Remanence B_r (T)	Sat. induction B_s (T)	Max. rel. permeability μ_{max}
7.3[b]		45[b]	1.30[b]	1.90[b]	10 800[b]
7.55[c]		33[c]		2.00[c]	12 500[c]
7.3[c]	1250 30 min, H$_2$	60[c]	1.1[c]		6100[c]
6.8[d]	1120 30 min, H$_2$	100[d]	0.6[d]		2200[d]

[a] Measured according to ASTM A 596
[b] Fe/3 wt% Si/0.45 wt% P
[c] Fe/2 wt% Si/0.45 wt% P
[d] Fe/4 wt% Si/0.45 wt% P

Table 4.3-8 Magnetic properties of sintered Fe–Sn–P in relation to density [3.10]

Density ϱ (g cm^{-3})	Sintering temp. (°C), time, atm.	Coercivity H_c (A m^{-1})	Remanence B_r (T)	Max. rel. permeability μ_{max}
7.2[a]	1120 30 min, H$_2$ [b]	80[a]	1.1[a]	4800[a]
7.4[b]	1250 30 min, H$_2$ [b]	37[b]	1.0[b]	9700[b]

[a] Fe/5 wt% Sn/0.45 wt% P
[b] Fe/5 wt% Sn/0.5 wt% P

Table 4.3-9 Magnetic properties of sintered Fe–Ni in relation to density[a] [3.10]

Density ϱ (g cm^{-3})	Sintering temp. (°C), time atm.	Coercivity H_c (A m^{-1})	Remanence B_r (T)	Sat. induction B_s (T)	Max. rel. permeability μ_{max}
8.0[b]		24[b]	0.25[b]	1.55[b]	6000[b]
8.0[c]		13[c]	0.85[c]	1.60[c]	30 000[c]
8.5[d]		2[d]	0.40[d]	0.80[d]	74 900[d]
6.99[e]	1260, H$_2$/Vac.	20[e]	0.75[e]		
6.99[f]		24[f]	0.76[f]		7800[f]
7.30[e]	1260, H$_2$/Vac.	19[e]	0.90[e]		
7.30[f]		32[f]	0.85[f]		8700[f]
7.50[e]	1260, H$_2$/Vac.	16[e]	0.94[e]		21 000[e]
7.50[f]		32[f]	0.90[f]		9100[f]
7.4[c]	1250, 30 min, H$_2$	25[c]	0.8[c]		13 000[c]
7.66	1371, DA[f]	16	0.42	1.27	

[a] Measured according to ASTM A 596
[b] Fe/35–40 wt% Ni
[c] Fe/45–50 wt% Ni
[d] Fe/72–82 wt% Ni/3–5 wt% Mo
[e] Fe/47–50 wt% Ni
[f] Formed by MIM

Table 4.3-10 Magnetic properties of sintered Fe–Cr in relation to density[a] [3.10]

Density ϱ (g cm^{-3})	Coercivity H_c (A m^{-1})	Remanence B_r (T)	Sat. induction B_s (T)	Max. rel. permeability μ_{max}
7.1[b]	200[b]	0.50[b]		1200[b]
7.45[c]	100[c]		1.70[c]	2500[c]
7.35[d]	100[d]		1.55[d]	1900[d]

[a] Measured according to ASTM A 596
[b] Fe/16–18 wt% Cr/0.5–1.5 wt% Mo
[c] Fe/12 wt% Cr/0.2 wt% Ni/0.7 wt% Si
[d] Fe/17 wt% Cr/0.9 wt% Mo/0.2 wt% Ni/0.8 wt% Si

Table 4.3-11 Magnetic properties of sintered Fe−Co products in relation to density[a] [3.10]

Density ϱ (g cm^{-3})	Coercivity H_c (A m^{-1})	Remanence B_r (T)	Sat. induction B_s (T)	Max. rel. permeability μ_{max}
7.9[b]	136[b]	1.10[b]	2.25[b]	3900[b]

[a] Measured according to ASTM A 596
[b] Fe/47−50 wt% Co/1−3 wt% V

Table 4.3-12 Magnetic properties of composite Fe products in relation to density and insulation [3.10]

Density ϱ (g cm^{-3})	Insulation	Curing (°C) (min)	Coercivity H_c (A m^{-1})	Induction at 3183 A m^{-1} (T)	Max. rel. permeability μ_{max}
5.7−7.26	Polymer		263−374	0.33−0.83	97−245
7.2	Oxide + 0.75 Polymer		374	0.77	210
7.4−7.45	0.75−0.6 % Polymer		381−374	1.09−1.12	400−425
7.54	Oxide				630
7.04	Inorganic (LCM^{TM})				305
7.0	0.5 % Phenolic resin[a]	150, 60	400		270
7.13−6.84[b]	0.4−1.8 % Phenolic[a]	160.30			325−175[c]
7.2[b]	0.8 % Phenolic[a]	150−500.30	445−310		270−390
6.6	3 % resin				200
6.7−7.0	Inorganic & 2 % resin	?−500			200−400
7.4	Inorganic (Somaloy^{TM}500)	500			600

[a] Dry Mixing or wet mixing with phenolic resin
[b] Compacted at 620 MPa and 65 °C
[c] AC permeability at 60 Hz and 1.0 T

Table 4.3-13 Core loss of composite Fe products in relation to lubricant and heat treatment [3.10]

Lubricant	Heat treatment (°C)	(min.)	(Atm.)	Density ϱ (g cm^{-3})	Total loss at 100 Hz, 1.5 T (W kg^{-1})	Total loss at 1000 Hz, 1.5 T (W kg^{-1})
0.1 % Kenolube^{TM} d	500	30	air	7.40[b]	29	
0.5 % Kenolube^{TM}	500	30	air	7.20[a]	32	450
0.5−0.8 % Kenolube^{TM}	500	30	air	7.36−7.26[b]	29−31	330−450
0.5 % Kenolube^{TM}	500	30	nitrogen	7.35[b]	31	350
0.5 % Kenolube^{TM}	250−580	30	steam	7.35[b]	30−70	
0.6 % LBI^{TM}	275	60	air	7.27[b]		700
0.5 % Kenolube^{TM}	500	30	air	7.20[a]	35	480
0.5 % Kenolube^{TM}	500	30	air	7.34[a]	34	420

[a] Conventional compacting at 600 MPa
[b] Conventional compacting at 800 MPa

Water atomized iron powder, particle size > 150 μm < 400 μm, low inorganic insulation thickness, (Somatoy^{TM} 550)

4.3.2.3 Iron–Silicon Alloys

The physical basis of the use of Fe–Si alloys, commonly called silicon steels, as soft magnetic materials is the fact that both the magnetocrystalline anisotropy K_1 and the magnetostriction parameters λ_{100} and λ_{111} of Fe approach zero with increasing Si content (see Fig. 4.3-5a). The lower the magnitude of these two intrinsic magnetic properties is, the lower are the coercivity H_c and the AC magnetic losses p_{Fe}. The total losses p_{Fe} consist of the static hysteresis losses p_h and the dynamic eddy current losses p_w which may be subdivided into a classical p_{wc} and an anomalous p_{wa} eddy current loss term,

$$p_{Fe} = p_h + p_{wc} + p_{wa}$$
$$= c_h(f) H_c B f$$
$$+ c_{wc}(\pi d B f)^2 / 6\rho\gamma + c_{wa}(\alpha_R/\rho)(Bf)^{3/2},$$

where $c_h(f)$ is a form factor of the hysteresis which depends on the frequency f, H_c is the coercivity of the material, B is the peak operating induction, c_{wc} and c_{wa} are terms taking the wave form of the applied field into account, d is the sheet thickness, ρ is the resistivity, γ is the density of the material, and α_R is the Raleigh constant. These are the factors to be controlled to obtain minimal losses. The increase in electrical resistivity with Si content (Fig. 4.3-5b) adds to lowering the eddy current losses as shown by the relation above.

The Fe–Si equilibrium diagram shows a very small stability range of the γ phase, indicating that the ferromagnetic α phase can be heat-treated in a wide temperature range without interference of a phase transformation which would decrease the magnetic softness of the material by the lattice defects induced.

Next to low-carbon steels, Fe–Si steels are the most significant group of soft magnetic materials (30% of the world market). A differentiation is made between non-oriented, isotropic (NO), and grain-oriented (GO) silicon steels. Non-oriented steels are mainly applied in rotating machines where the material is exposed to varying directions of magnetic flux. Grain-oriented steels with GOSS-texture (110) ⟨001⟩ are used predominantly as core material for power transformers.

Since Fe–Si steels are brittle above about 4.0 wt% Si, conventional cold rolling is impossible at higher Si contents.

Non-oriented Silicon Steels (NO)

NO laminations are usually produced with thicknesses between 0.65 mm and 0.35 mm, and Si concentrations up to 3.5 wt%. According to their grade, NO silicon steels are classified in low grade (low Si content) alloys employed in small devices and high-grade (high Si content) alloys for large machines (motors and generators). Suitable microstructural features (optimum grain size) and a low level of impurities are necessary for optimum magnetic properties. Critical factors in processing are the mechanical behavior upon punching of laminations, the application of insulating coatings, and the build-up of stresses in magnetic cores. Table 4.3-14 lists the ranges of typical processing parameters.

In the case of low Si steels (< 1 wt% Si), the last two annealing steps are applied by the user after lamination punching (semi-finished sheet). Table 4.3-15 lists the specifications, including all relevant properties for non-oriented magnetic steel sheet.

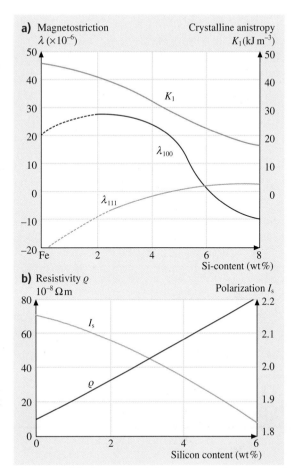

Fig. 4.3-5 (a) Magnetostriction λ_{100} and λ_{111} and magneto-crystalline anisotropy energy K_1. (b) Electrical resistivity ρ and saturation polarization I_s, as a function of the Si content in Fe–Si alloys

Composition (wt%) Si: 0.9–3.4, Al: 0.2–0.6, Mn: 0.1–0.3
Melting, degassing, continous casting
Hot rolling to 1.8–2.3 mm (1000–1250 °C)
Cold rolling to intermediate gauge
Annealing (750–900 °C)
Cold rolling to final gauge (0.65–0.35 mm)
Decarburizing anneal (830–900 °C, wet H_2)
Recrystallization
Grain growth anneal (830–1100 °C)
Coating

Table 4.3-14 Schematic of NO silicon steel processing. Addition of small quantities (50–800 wt ppm) of Sb, Sn, or rare earth metals can be made to improve texture and/or control the morphology of the precipitates. Cold reduction in a single stage repesents a basic variant of the above scheme. The final grain growth annealing aims at an optimum grain size, leading to minimum losses. Coating provides the necessary interlaminar electrial insulation. Phosphate- or chromate-based coatings are applied, which ensure good lamination punchability [3.7]

Table 4.3-15 Standard IEC specification for nonoriented magnetic steel sheet delivered in the final sate. The conventional designation of the different grades comprises the following order (first column): (1) 100 times the maximum specified loss at 1.5 T peak induction in (W kg^{-1}); (2) 100 times the nominal sheet thickness; (3) the characteristic letter "A"; (4) one tenth of the frequency in Hz, at which the magnetic properties are specified. The anisotropy of loss, T, is specified at 1.5 T peak induction according to the formula $T = (P_1 - P_2)/(P_1 + P_2)100$, with P_1 and P_2 the power losses of samples cut perpendicular and parallel to the rolling direction, respectively [3.7]

Quality	Nominal thickness (mm)	Maximum specific total loss (W kg^{-1}) at peak induction		Minimum magnetic flux density (T) in direct or alternating field at field strength			Maximum anisotropy of loss (%)	Minimum stacking factor	Minimum number of bends	Conven-tional density (10^3 kg m^{-3})
		1.5 T	1.0 T	2500 Am^{-1}	5000 Am^{-1}	10 000 Am^{-1}				
250-35-A5	0.35	2.50	1.00	1.49	1.60	1.71			2	7.60
270-35-A5	0.35	2.70	1.10	1.49	1.60	1.71	±18	0.95	2	7.65
300-35-A5	0.35	3.00	1.20	1.49	1.60	1.71			3	7.65
330-35-A5	0.35	3.30	1.30	1.49	1.60	1.71			3	7.65
270-50-A5	0.50	2.70	1.10	1.49	1.60	1.71			2	7.60
290-50-A5	0.50	2.90	1.15	1.49	1.60	1.71	±18		2	7.60
310-50-A5	0.50	3.10	1.25	1.49	1.60	1.71			3	7.65
330-50-A5	0.50	3.30	1.35	1.49	1.60	1.71	±14		3	7.65
350-50-A5	0.50	3.50	1.50	1.50	1.60	1.71			5	7.65
400-50-A5	0.50	4.00	1.70	1.51	1.61	1.72		0.97	5	7.65
470-50-A5	0.50	4.70	2.00	1.52	1.62	1.73				7.70
530-50-A5	0.50	5.30	2.30	1.54	1.64	1.75				7.70
600-50-A5	0.50	6.00	2.60	1.55	1.65	1.76	±12		10	7.75
700-50-A5	0.50	7.00	3.00	1.58	1.68	1.76				7.80
800-50-A5	0.50	8.00	3.60	1.58	1.68	1.78				7.80
350-65-A5	0.65	3.50	1.50	1.49	1.60	1.71			2	7.65
400-65-A5	0.65	4.00	1.70	1.50	1.60	1.71			2	7.65
470-65-A5	0.65	4.70	2.00	1.51	1.61	1.72	±14		5	7.65
530-65-A5	0.65	5.30	2.30	1.52	1.62	1.73		0.97	5	7.70
600-65-A5	0.65	6.00	2.60	1.54	1.64	1.75				7.75
700-65-A5	0.65	7.00	3.00	1.55	1.65	1.76	±12		10	7.75
800-65-A5	0.65	8.00	3.60	1.58	1.68	1.76				7.80
1000-65-A5	0.65	10.00	4.40	1.58	1.68	1.78				7.80

Grain-Oriented Silicon Steels (GO)

Grain-oriented (GO) silicon steel is used mainly as core material for power transformers. Worldwide production is about 1.5 million tons per year. The increasing demand for energy-efficient transformers requiring still lower loss materials has led to continuous improvements of the magnetic properties over the years, where a decrease in the deviation from the ideal Goss texture (110)[001] has played a decisive role. Moreover, the eddy current losses have been reduced by decreasing the lamination thickness from 0.35 mm through 0.30 mm and 0.27 mm to 0.23 mm. Surface treatments of the laminations by mechanical scratching or laser scribing have been introduced. They increase the number of mobile Bloch walls and, thus, decrease the spacing between them, i.e., the domain size. Accordingly, the anomalous eddy current losses were reduced. The reduction of the total losses of grain oriented electrical steel due to these improvements is illustrated in Fig. 4.3-6.

The GO steels are classified in two categories: conventional grain-oriented (CGO) and high permeability (HGO) steels. The latter are characterized by a sharp crystallographic texture, with average misorientation of the [001] axes of the individual crystallites around the rolling direction (RD) on the order of 3°. For CGO the misorientation is about 7°. The relation between the angular deviation of grain orientation and the total loss reduction for HGO material is shown in Fig. 4.3-7.

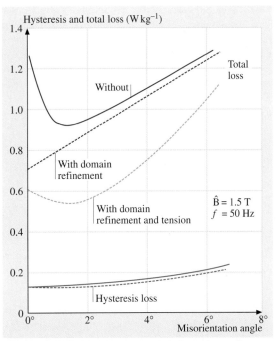

Fig. 4.3-7 Relation between grain orientation and power loss reduction for highly grain-oriented material (single crystal) (*Bölling* and *Hastenrath*) [3.12, 13]

The (GO) manufacturing route is an extraordinarily long sequence of hot and cold processing steps. The final magnetic properties are highly sensitive to even small parameter variations throughout this route. Some of these processes, their microstructure, and the inhibitor element influence are given in Table 4.3-16.

A key factor is the controlled development of the (110)[001] texture during secondary recrystallization. It requires the presence of large Goss textured grains in the surface layers of the annealed hot band, the presence of inhibitors as finely dispersed second-phase particles which strongly impede normal grain growth during primary recrystallization and a primary recrystallized texture having a suitable orientation relationship with respect to the Goss texture. This can be obtained through carefully-controlled chemistry and a precisely-defined sequence of thermomechanical treatments. Abnormal grain growth during secondary recrystallization may increase the size of magnetic domains and consequently greater energy dissipation under dynamic conditions. Refining the domain structure by laser or mechanical scribing core as mentioned above reduces the losses. Standard IEC specifications for grain-oriented magnetic

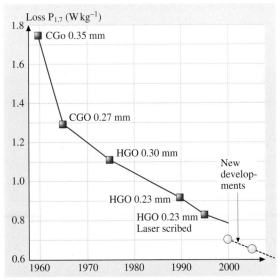

Fig. 4.3-6 Qualitative improvment of GO electrical steel [3.11]

Table 4.3-16 Summary of the processing of grain-oriented silicon steel. The first column relates to the conventional grain-oriented (CGO) laminations. Process for three different types of high permeability (HGO) steels are outlined in columns 2–4. They basically differ for the type of grain growth inhibitors, the cold-rolling sequence, and the annealing temperatures. The processes CGO and HGO-1 adopt a two-stage cold reduction, with intermediate annealing, while HGO-2 and HGO-3 steels are reduced to the final thickness in a single step. Growth inhibition of the primary recrystallized grains is obtained by MnS precipitates in the CGO process. MnSe particles + solute Sb operate in process HGO-1, MnS + AlN particles in HGO-2, and solute B + N + S in HGO-3. Abnormal growth of (110)[001] grains occurs by final box annealing, which also promotes the dissolution of the precipitates [3.7]

Type of Steel							
CGO		HGO-1		HGO-2		HGO-3	
		Composition		(wt%)			
3–3.2	Si	2.9–3.3	Si	2.9–3.3	Si	3.1–3.3	Si
0.04–0.1	Mn	0.05	Mn	0.03	Al	0.02	Mn
0.02	S	0.02	Se	0.015	N	0.02	S
0.03	C	0.04	Sb	0.07	Mn	0.001	B
balance	Fe	0.03–0.07	C	0.03	S	0.005	N
		balance	Fe	0.05–0.07	C	0.03–0.05	C
				balance	Fe	balance	Fe
		Inhibitors					
MnS		MnSe + Sb		MnS + AlN		B + N + S	
		Melting, degassing and continuous casting					
		Reheating - hot rolling					
1320 °C		1320 °C		1360 °C		1250 °C	
		Annealing					
800–1000 °C		900 °C		1100 °C		870–1020 °C	
		Cold rolling					
70 %		60–70 %		87 %		80 %	
Annealing							
800–1000 °C							
Cold rolling							
55 %		65 %					
		Decarburizing anneal					
		800–850 °C (wet H$_2$)					
		MgO coating and coiling					
		Box-annealing					
1200 °C		820–900 °C		1200 °C		1200 °C	
		+ 1200 °C					
		Phospate coating and thermal flattening					

steel sheets are listed in Table 4.3-17. Basic properties of grain-oriented Fe–3.2 wt% Si alloys are given in Table 4.3-18.

A recent technology development in the production of GO electrical steel is the combination of thin slab casting, direct hot rolling, and acquired inhibitor formation. This practice combines the advantages of low temperatures, process-shortening, microstructural homogeneity, improved strip geometry, and better surface condition of the products. The slab thickness is on the order of 50–70 mm. Another future technology with remarkable process shortening is to produce (GO) hot strip in the thickness range of about 2–3 mm by direct casting from the steel melt using a twin-roll casting method. In pilot line tests, good workability and good magnetic properties have been achieved.

Further potential for cost and time saving is expected from replacing box annealing at the end of the cold process by short-time continuous annealing.

Table 4.3-17 Standard IEC specification for grain-oriented magnetic steel sheet: **(a)** normal material; **(b)** material with reduced loss; **(c)** high-permeability material. The conventional designation of the various grades (first column) includes, from left to right, (1) 100 times the maximum power loss, in (W kg^{-1}), at 1.5 T **(a)** or 1.7 T **(b,c)** peak induction; (2) 100 times the nominal sheet thickness, in (mm); (3) the letter "N" for the nominal material **(a)**, or "S" for material with reduced loss **(b)**, or "P" for high-permeability material **(c)**; (4) one tenth of the frequency in Hz, at which the magnetic properties are specified [3.7]

a) Grade	Thickness (mm)	Maximum specific total loss (W kg^{-1}) at peak induction		Minimum magnetic flux density (T) for $H = 800$ Am^{-1}	Minimum stacking factor
		1.5 T	1.7 T		
089-27-N 5	0.27	0.89	1.40	1.75	0.950
097-30-N 5	0.30	0.97	1.50	1.75	0.955
111-35-N 5	0.35	1.11	1.65	1.75	0.960

b) Grade	Thickness (mm)	Maximum specific total loss (W kg^{-1}) at 1.7 T peak induction	Minimum magnetic flux density (T) for $H = 800$ Am^{-1}	Minimum stacking factor
130-27-S 5	0.27	1.30	1.78	0.950
140-30-S 5	0.30	1.40	1.78	0.955
155-35-S 5	0.35	1.55	1.78	0.960

c) Grade	Thickness (mm)	Maximum specific total loss (W kg^{-1}) at 1.7 T peak induction	Minimum magnetic flux density (T) for $H = 800$ Am^{-1}	Minimum stacking factor
111-30-P 5	0.30	1.11	1.85	0.955
117-30-P 5	0.30	1.17	1.85	0.955
125-35-P 5	0.35	1.25	1.85	0.960
135-35-P 5	0.35	1.35	1.85	0.960

Table 4.3-18 Basic properties of grain-oriented Fe–3.2 wt% Si alloys [3.7]

Property	Value
Density	7.65×10^3 kg m^{-3}
Thermal conductivity	16.3 W °C^{-1} kg m^{-3}
Electrical resistivity	48×10^{-8} Ω m
Young's modulus	
Single crystals	
[100] direction	120 GPa
[110] direction	216 GPa
[111] direction	295 GPa
(110)[001] texture	
Rolling direction (RD)	122 GPa
45° to RD	236 GPa
90° to RD	200 GPa
Yield strength	
(110)[001] texture	
Rolling direction	324 MPa
Tensile strength	
(110)[001] texture	
Rolling direction	345 MPa
Saturation induction	2.0 T

Table 4.3-18 Basic properties of grain-oriented Fe–3.2 wt% Si alloys [3.7], cont.

Property	Value
Curie temperature	745 °C
Magnetocrystalline anisotropy	3.6×10^4 J m^{-3}
Magnetostriction constants	
λ_{100}	23×10^{-6}
λ_{111}	-4×10^{-6}

Rapidly Solidified Fe–Si Alloys

The Fe–6.5 wt% Si alloy exhibits good high-frequency soft magnetic properties due to a favorable combination of low values of the saturation magnetostriction λ_s, as well as the low values of the magnetocrystalline anisotropy energy K_1, and a high electrical resistivity. But as mentioned above, Fe–Si alloys which contain more than about 4 wt% Si are brittle and thin sheets cannot be manufactured by rolling. Therefore, Fe–6.5 wt% Si sheets and ribbons are manufactured via two different routes by which the adverse mechanical properties are circumvented: a continuous "siliconizing" process in commercial scale production and a rapid quenching process.

In the siliconizing process Fe–3 wt% Si sheet reacts with a Si-containing gas at 1200 °C. The sheet is held at 1200 °C in order to increase and homogenize the Si content by diffusion. After this treatment the ductility of Fe–6.5 wt% Si sheets amounts to about 5% elongation to fracture.

By rapid quenching from the melt, the formation of the B2 and D0$_3$ type ordered structures, based on conventional cooling after casting of Fe–6.5 wt% Si alloys, may be suppressed. Thus, the ensuing material brittleness can be overcome. The ribbons formed by the rapid quenching process are about 20 to 60 μm thick and are ductile, with a microcrystalline structure. By means of an annealing treatment above 1000 °C fol-

Table 4.3-19 Physical and magnetic properties of rapidly quenched Fe–6.5 wt% Si alloys [3.7]

Property	Value
Density	7.48×10^3 kg m^{-3}
Thermal conductivity (31 °C)	4.5 cal m^{-1}°C^{-1}s^{-1}
Specific heat (31 °C)	128 cal °C^{-1} kg^{-1}
Coefficient of thermal expansion (150 °C)	11.6×10^{-6} °C^{-1}
Electrical resistivity	82×10^{-8} Ω m
Tensile strength (rapidly-quenched ribbons 60 μm thick)	630 MPa
Saturation magnetization	1.8 T
Curie temperature	700 °C
Saturation magnetostriction	0.6×10^{-6}

Table 4.3-20 Magnetic properties of 30–40 μm thick, rapidly-quenched ribbons of Fe–6.5 wt% Si, in the as-quenched state and after 24 h annealing at various temperatures [3.7]

Annealing temperature (°C)	H_c (A m^{-1})	H_{10} (T)	B_r/B_{10}	μ_{max}/μ_0
As-quenched	112	1.25	0.70	3100
500	100	1.27	0.93	4300
700	72	1.32	0.90	5400
800	45	1.31	0.92	9400
850	37.5	1.28	0.94	10 000
900	33	1.30	0.77	12 500
1000	21	1.31	0.83	17 000
1100	18	1.30	0.84	18 000
1200	20	1.32	0.87	22 000

lowed by rapid cooling to restrain D0₃ type ordering, large-grain-sized, recrystalllized (100) ⟨0vw⟩ textured material with good ductility and good soft magnetic properties is obtained, as characterized in Tables 4.3-19 and 4.3-20. Figures 4.3-8 and 4.3-9 show the loss behavior as a function of magnetizing frequency and ribbon thickness.

4.3.2.4 Nickel–Iron-Based Alloys

The fcc phase in the Ni–Fe alloy system and the formation of the ordered Ni₃Fe phase provide a wide range of structural and magnetic properties for developing soft magnetic materials with specific characteristics for different applications. The phase diagram is shown in Sect. 3.1.5. Before amorphous and nanocrystalline soft magnetic alloys were introduced, the Ni–Fe materials

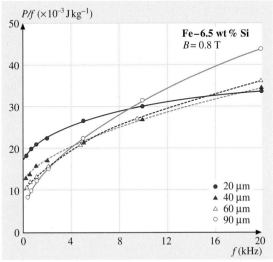

Fig. 4.3-8 Power loss per cycle vs. magnetizing frequency for rapidly quenched Fe–6.5 wt% Si ribbons of various thicknesses [3.7]

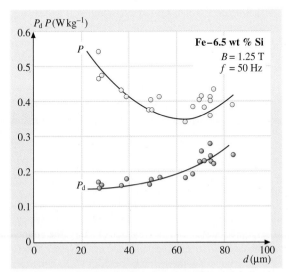

Fig. 4.3-9 Total and dynamic loss at 1.25 T peak induction and 50 Hz vs. ribbon thickness for rapidly quenched Fe–6.5 wt% Si ribbons characterized by a strong (100) [0uv] grain texture induced by vacuum annealing [3.7]

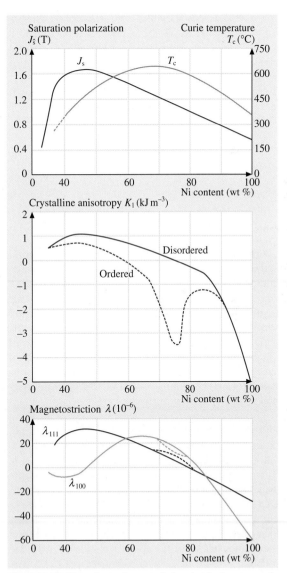

Fig. 4.3-10 The dependence of the intrinsic magnetic parameters I_s, T_c, K_1, λ of Ni–Fe alloys on the Ni concentration [3.12]

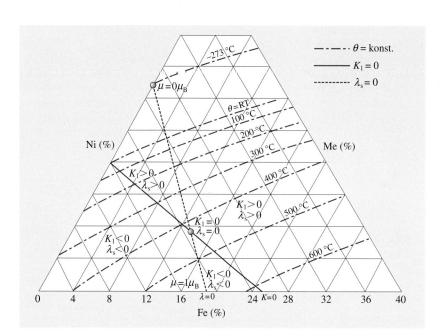

Fig. 4.3-11 Zero lines for $K_1 = 0$ and $\lambda_s = 0$ in the Ni–Fe–Me system [3.14, 15]

containing about 72 to 83 wt% Ni with additions of Mo, Cu, and/or Cr were the magnetically softest materials available.

Based on the low magnetocrystalline anisotropy K_1 and low saturation magnetostriction λ_s, the alloys containing about 80 wt% Ni where K_1 and λ_s pass through zero attain the lowest coercivity $H_c \approx 0.5 \, \text{A m}^{-1}$ and the highest initial permeability $\mu_i \approx 200\,000$. Figure 4.3-10 shows the variation of the decisive intrinsic magnetic parameters I_s, T_c, K_1, λ_{100}, and λ_{111} of binary Ni–Fe alloys with the Ni concentration. The strong effect of structural ordering on the magnitude of K_1 should be noted because it permits control of this intrinsic magnetic property by heat treatment.

In the binary Ni–Fe system, $K_1 = 0$ at about 76 wt% Ni and $\lambda_s = 0$ at about 81 wt% Ni. Small additions of Cu lower the Ni content for which $\lambda_s = 0$ while Mo additions increase the Ni content for $K_1 = 0$. Thus different alloy compositions around 78 wt% Ni are available which have optimal soft magnetic properties. General relations of the effect of alloying elements in Ni–Fe-based alloys on K_1, λ_s, and on the permeability have been developed in [3.14, 15]. Figure 4.3-11 shows the position of the lines for $K_1 = 0$ and $\lambda_s = 0$ in the disordered Ni–Fe–Me system, where Me = Cu, Cr, Mo, W, and V. High permeability regions in the Ni–Fe–Me system with a different valence of Me are delineated in Fig. 4.3-12.

The main fields of application of high permeability Ni–Fe alloys are fault-current circuit breakers, LF and HF transformers, chokes, magnetic shielding, and high sensitivity relays.

It should be noted that annealing treatments in a magnetic field of specified direction induce atomic rearrangements which provide an additional anisotropy termed uniaxial anisotropy K_u. It can be used to modify the field dependence of magnetic induction in such a way that the hysteresis loop takes drastically different forms, as shown in Fig. 4.3-13.

The extremes of a steep loop (Z type) and a skewed loop (F type) are obtained by field annealing with the direction of the field (during annealing) longitudinal and transverse to the operating field of the product, respectively. Materials with Z and F loops produced by magnetic field annealing are used in magnetic amplifiers, switching and storing cores, as well as for pulse and instrument transformers and chokes.

Combined with small alloy variations, primary treatments and field annealing treatments, a wide variety of annealed states can be realized to vary the induction behavior. The field dependence of the permeability of some high-permeability Ni–Fe alloys (designations according to Vacuumschmelze GmbH) are shown in Fig. 4.3-14 [3.16].

Alloys of Ni–Fe in the range of 54–68 wt% Ni combine relatively high permeability with high saturation

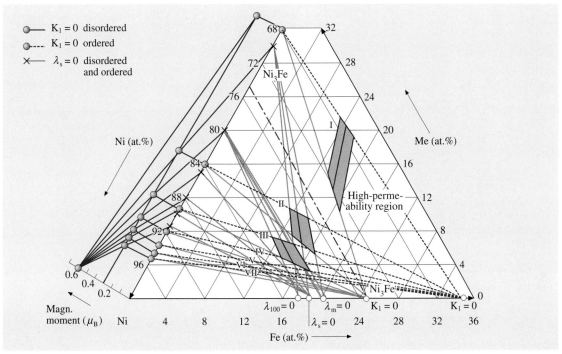

Fig. 4.3-12 Schematic representation of high-permeability regions in the ternary system Ni−Fe−Me by means of the zero curves K_1, $\lambda_{100} = 0$ and $\lambda_{111} = 0$ for additives with valence I–VIII [3.15]

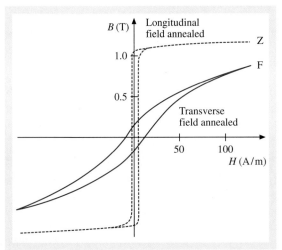

Fig. 4.3-13 Alloy of the 54−68 wt% NiFe-group with Z- and F-loop [3.12]

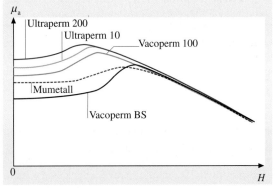

Fig. 4.3-14 Permeability versus field strength for high permeability high Ni-content Ni−Fe-alloys of Vacuumschmelze Hanau

polarization. Magnetic field annealing of these alloys provides a particularly high uniaxial anisotropy with ensuing Z and F type loops [3.12].

Alloys containing 45 to 50 wt% Ni reach maximum saturation polarization of about 1.6 T. Under suitable rolling and annealing conditions a cubic texture with an ensuing rectangular hysteresis loop and further loop variants over a wide range can be realized. The microstructure may vary from fine grained to coarsely grained.

Alloys containing 35 to 40 wt% Ni show a small but constant permeability, $\mu_r = 2000 - 8000$, over a wide range of magnetic field strength. Moreover,

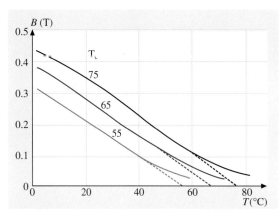

Fig. 4.3-15 Induction vs. temperature of Fe–Ni alloys with approximately 30% Ni as a function of Curie temperature [3.17]

they have the highest resistivity of all Ni–Fe alloys and a saturation polarization between 1.3 and 1.5 T.

The Curie temperature of the alloys with about 30 wt% Ni is near room temperature. Accordingly, the magnetization is strongly temperature-dependent in this vicinity (see Fig. 4.3-15). By slight variation of the Ni content (a composition increase of 0.1 wt% Ni gives rise to an increase of T_c by 10 K) T_c can be varied between 30 °C and 120 °C. These alloys are used mainly for temperature compensation in permanent magnet systems, measuring systems, and temperature sensitive switches [3.16].

4.3.2.5 Iron–Cobalt Alloys

Of all known magnetic materials, Fe–Co alloys with about 35 wt% Co have the highest saturation polarization $I_s = 2.4$ T at room temperature and the highest Curie temperature of nearly 950 °C. The intrinsic magnetic properties I_s, T_c, K_1, and λ_{hkl} as a function of Co content are shown in Fig. 4.3-16.

Since K_1 and λ_s have minima at different Co contents, different compositions for different applications have been developed. A Fe–49 wt% Co–2 wt% V composition is commonly used. The V addition reduces the brittleness by retarding the structural ordering transformation, improves the rolling behavior, and increases the electrical resistivity. The workability of Co–Fe alloys is difficult altogether. Two further alloy variants containing 35 wt% Co and 27 wt% Co, respectively, are of technical interest. They are applied where highest flux density is required, e.g., in magnet yokes, pole shoes, and magnetic lenses. The high T_c makes Fe–Co based alloys applicable as high temperature magnet material.

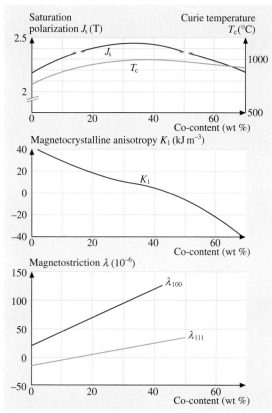

Fig. 4.3-16 The dependence of intrinsic magnetic parameters I_s, T_c, K_1 and λ of Co–Fe alloys on the Co content [3.12]

4.3.2.6 Amorphous Metallic Alloys

By rapid quenching of a suitable alloy from the melt at a cooling rate of about 10^{-5}–10^{-6} K/s, an amorphous metallic state will be produced where crystallization is suppressed. Commonly, casting through a slit nozzle onto a rotating copper wheel is used to form a ribbon-shaped product. The thickness of the ribbons is typically between 20 and 40 μm.

From the magnetic point of view amorphous alloys have several advantages compared to crystalline alloys: they have no magnetocrystalline anisotropy, they combine high magnetic softness with high mechanical hardness and yield strength, their low ribbon thickness and their high electrical resistivity (100–150 μΩ cm)

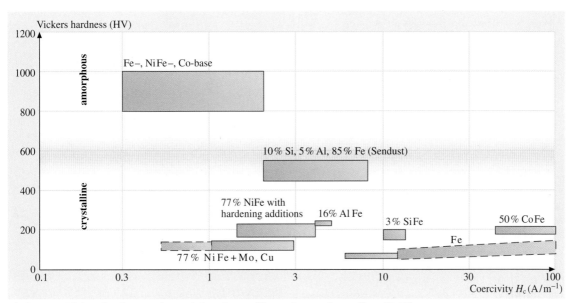

Fig. 4.3-17 Vickers hardness and coercivity of crystalline and amorphous alloys [3.12]

provide excellent soft magnetic material properties for high frequency applications, in particular low losses. A plot of Vickers hardness (HV) vs. coercivity (H_c) for crystalline and amorphous alloys is shown in Fig. 4.3-17 and indicates that a particularly favorable combination for soft magnetic amorphous alloys applies: being magnetically soft and mechanically hard.

The soft magnetic properties of amorphous alloys depend essentially on alloy composition, focusing on a low saturation magnetostriction λ_s, high glass forming ability required for ribbon preparation at technically accessible cooling conditions, and annealing treatments which provide structural stability and field-induced anisotropy K_u.

The soft magnetic amorphous alloys are based on the ferromagnetic elements Fe, Co, and Ni with additions of metalloid elements, the so-called glass forming elements Si, B, C, and P. The most stable alloys contain about 80 at.% transition metal (TM) and 20 at.% metalloid (M) components.

Depending on their base metal they exhibit characteristic differences of technical significance. Accordingly they are classified into three groups: Fe-based alloys, Co-based alloys, and Ni-based alloys. The characteristic variation of their intrinsic magnetic properties saturation polarization I_s, saturation magnetostriction λ_s, and the maximum field induced magnetic

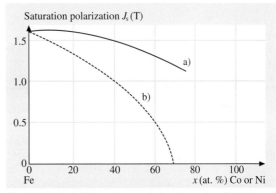

Fig. 4.3-18a,b Saturation polarization J_s of Fe-based amorphous alloys depending on Co and Ni content: **(a)** $Fe_{80-x}Co_xB_{20}$ (*O'Handley*, 1977) [3.18]. **(b)** $Fe_{80-x}Ni_xB_{20}$ (*Hilzinger*, 1980) [3.12, 19]

anisotropy energy K_u, as functions of alloy concentration is shown in Figures 4.3-18 to 4.3-20.

Iron–Based Amorphous Alloys

Of all amorphous magnetic alloys, the iron-rich alloys on the basis $Fe_{\sim 80}(Si, B)_{\sim 20}$ have the highest saturation polarization of 1.5–1.8 T. Because of their relatively high saturation magnetostriction (λ_s) of around 30×10^{-6}, their use as soft magnetic material is limited. The application is focused on transformers at low and

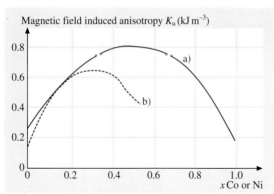

Fig. 4.3-19a,b Field induced anisotropy K_u of FeCo-based and FeNi-based amorphous alloys: **(a)** $(Fe_{1-x}Co_x)_{77}Si_{10}B_{13}$ (*Miyazaki* et al., 1972) [3.20]. **(b)** $(Fe_{1-x}Ni_x)_{80}B_{20}$ (*Fujimori* et al., 1976) [3.12, 21]

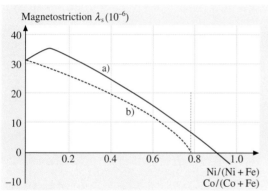

Fig. 4.3-20a,b Magnetostriction λ_s of FeCo-based and FeNi-based amorphous alloys depending on Co and Ni Content: **(a)** $(FeCo)_{80}B_{20}$ [3.18]. **(b)** $(FeNi)_{80}B_{20}$ [3.12, 18]

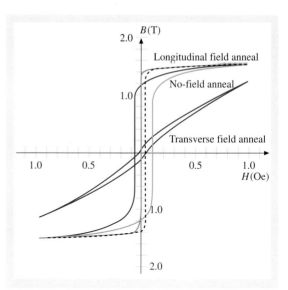

Fig. 4.3-21 Typical dc hysteresis loop of Fe-rich METGLAS alloy SA 1 [3.22]

medium frequencies in electric power distribution systems.

Compared to grain-oriented silicon steels the iron-rich amorphous alloys show appreciably lower coercivity and consequently lower total losses.

The physical and magnetic properties of a characteristic commercial Fe-rich Metglas amorphous alloy are shown in Fig. 4.3-21 and Table 4.3-21 [3.22].

Cobalt–Based Amorphous Alloys

In the $(Fe_xCo_{1-x})_{\sim 80}B_{\sim 20}$ system the saturation magnetostriction λ_s passes through zero. Along with a proper selection of alloy composition this behavior gives rise to a particularly low coercivity, the highest permeability of all amorphous magnetic alloys, low stress sensitivity, and extremely low total losses.

The saturation polarization ranging from 0.55 to 1.0 T is lower than in Fe-rich amorphous alloys but comparable to the ~ 80 wt% Ni crystalline permalloy materials. By applying magnetic field annealing, well-controlled Z-type and F-type loops can be realized.

Nickel–Based Amorphous Alloys

A typical composition of this amorphous alloy group is $Fe_{40}Ni_{40}(Si, B)_{20}$, with a saturation polarization I_s of 0.8 T and a saturation magnetostriction λ_s of 10×10^{-6}. The latter is causing magnetoelastic anisotropy which can be applied to design magnetoelastic sensors where a change in the state of applied stress causes a change in permeability and loop shape, respectively. Upon annealing, Fe–Ni-based alloys will have an R-(round) type

Table 4.3-21 Physical and magnetic properties of Fe-rich METGLAS alloy SA 1 [3.22]

Property	Value
Ribbon Thickness (μm)	25
Density (g/cm^3)	7.19
Thermal Expansion (ppm/°C)	7.6
Crystallization Temperature (°C)	550
Curie Temperature (°C)	415
Countinuous Service Temperature (°C)	155
Tensile Strength (MN/m^2)	1–1.7 k
Elastic Modulus (GN/m^2)	100–110
Vicker's Hardness (50 g load)	860
Saturation Flux Density (Tesla)	1.56
Permeability (depending on gap size)	Variable
Saturation Magnetostriction (ppm)	27
Electrical Resistivity (μΩ cm)	137

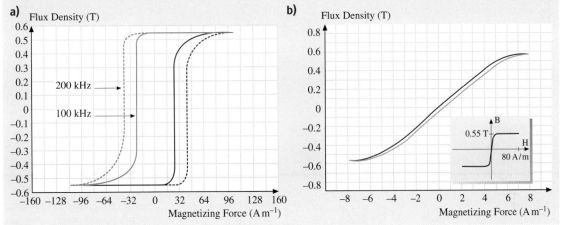

Fig. 4.3-22a,b Typical dc hysteresis loop of Co-based METGLAS alloy: (**a**) alloy 2714 AS (Z-loop) and (**b**) alloy 2714 AF (F-loop) [3.22]

Table 4.3-22 Physical and magnetic properties of the cobalt-based amorphous alloy METGLAS alloy 2714 AF [3.22]

Property	Value
Ribbon Thickness (μm)	18
Density (g/cm^3)	7.59
Thermal Expansion (ppm/°C)	12.7
Crystallization Temperature (°C)	560
Curie Temperature (°C)	225
Countinuous Service Temperature (°C)	90
Tensile Strength (MN/m^2)	1–1.7 k
Elastic Modulus (GN/m^2)	100–110
Vicker's Hardness (50 g load)	960
Saturation Flux Density (Tesla)	0.55
Permeability (μ @ 1 kHz, 2.0 mA/cm)	90 000 ± 20%
Saturation Magnetostriction (ppm)	≪ 1
Electrical Resistivity (μΩ cm)	142

Table 4.3-23 Survey of soft magnetic amorphous and nanocrystalline alloys. Some amorphous alloy of METGLAS (Allied Signal Inc., Morristown/NJ) and VITROVAC (Vacuumschmelze GmbH, Hanau, Germany) have been selected from commercially available alloys [3.12]

Composition	Typical properties						
	Saturation polarization in (T)	Curie temperature in (°C)	Saturation magnetostriction in 10^{-6}	Coercivity (dc) in (A m^{-1})	Permeability[a] at $H = 4$ mA m^{-1} $\times 10^3$	Density in (g cm^{-3})	Specific electrical resistivity[b] in (Ω mm^2 m^{-1})
Amorphorus							
Fe-based:							
$Fe_{78}Si_9B_{13}$	1.55	415	27	3	8	7.18	1.37
$Fe_{67}Co_{18}Si_1B_{14}$	1.80	$\sim 550^c$	35	5	1.5	7.56	1.23
FeNi-based:							
$Fe_{39}Ni_{39}Mo_2Si_{12}B_8$	0.8	260	+8	2	20	7.4	1.35
Co-based:							
$Fe_{67}Fe_4Mo_1Si_{17}B_{11}$	0.55	210	< 0.2	0.3	100	7.7	1.35
$Fe_{74}Fe_2Mn_4Si_{11}B_9$	1.0	480^c	< 0.2	1.0	2	7.85	1.15
Nanocristalline							
$Fe_{73.5}Cu_1Nb_3Si_{13.5}B_9$	1.25	600	+2	1	100	7.35	1.35

[a] Materials with round (R) or flat loops (F), $f = 50$ Hz
[b] 1 Ω mm^2/m = 10^{-4} Ω cm
[c] Extrapolated values ($T_c > T_x$, T_x: crystallization temperature)

hysteresis loop associated with high initial permeability, or an F-type loop with low losses.

Table 4.3-23 gives a survey of the magnetic and physical properties of several soft magnetic amorphous alloys

4.3.2.7 Nanocrystalline Soft Magnetic Alloys

Nanocrystalline soft magnetic alloys are a rather recent class of soft magnetic materials with excellent magnetic properties such as low losses, high permeability, high saturation polarization up to 1.3 T, and near-zero magnetostriction. The decisive structural feature of this alloy type is its ultra-fine microstructure of bcc α-Fe–Si nanocrystals, with grain sizes of 10–15 nm which are embedded in an amorphous residual phase. Originally, this group of materials was discovered in the alloy system Fe–Si–B–Cu–Nb with the composition $Fe_{73.5}Si_{15.5}B_7Cu_1Nb_3$. This material is prepared by rapid quenching like an amorphous Fe–Si–B alloy with a subsequent annealing treatment and comparatively high temperature in the range of 500 to 600 °C which leads to partial crystallization.

The evolution of the nanocrystalline state during annealing occurs by partial crystallization into randomly oriented, ultrafine bcc α-Fe–Si grains that are 10–15 nm in diameter. The residual amorphous matrix phase forms a boundary layer that is 1–2 nm thick. This particular nano-scaled microstructure is the basis for ferromagnetically-coupled exchange interaction of and through these phases, developing excellent soft magnetic properties: $\mu_a \approx 10^5$, $H_c < 1$ A m^{-1}. Annealing above 600 °C gives rise to the precipitation of the borides Fe_2B and/or Fe_3B with grain sizes of 50–100 nm. At higher annealing temperatures, grain coarsening arises. Both of these microstructural changes are leading to a deterioration of the soft magnetic properties.

The influence of the annealing temperature on grain size, H_c, and μ_i of a nanocrystalline type alloy is shown in Fig. 4.3-23 [3.23].

The small additions of Cu and Nb favor the formation of the nanocrystalline structure. Copper is thought to increase the rate of nucleation of α-Fe–Si grains by a preceding cluster formation, and Nb is supposed to lower the growth rate because of its partitioning effect and decrease of diffusivity in the amorphous phase. Figure 4.3-24 illustrates the formation of the nanocrystalline structure schematically.

It is useful to note the influence of the atomic diameter of alloying additions on the grain size of the α-Fe–Si phase starting from the classical alloy composition $Fe_{73.5}Si_{15.5}B_7Cu_1Nb_3$. This effect is shown for

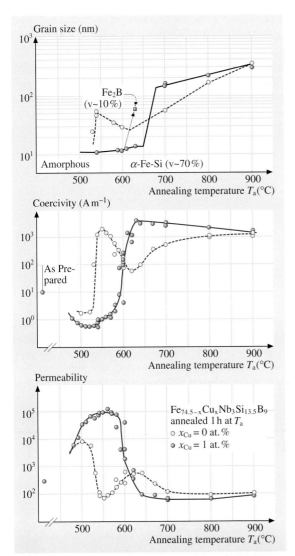

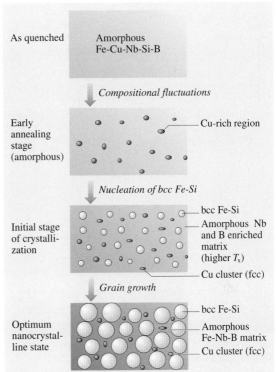

Fig. 4.3-23 Average grain size, coercivity and initial permeability of a nanocrystalline soft magnetic alloy as a function of the annealing temperature [3.23]

partial substitution of Nb by V, Mo, W, and Ta (2 at.% each) in Fig. 4.3-25. The larger the atomic diameter, the smaller the resulting grain size. The elements Nb and Ta have the same atomic diameter. Furthermore, the smaller the atomic diameter of the alloying element, the sooner the crystallization of the α-Fe-Si phase begins.

One of the decisive requirements for excellent soft magnetic properties is the absence of magnetostriction. Amorphous Fe–Si–B–Cu–Nb alloys have a saturation

Fig. 4.3-24 Schematic illustration of the formation of the nanocrystalline structure in Fe–Cu–Nb–Si–B alloys, based on atom probe analysis results and transmission electron microscopy observations by *Hono* et al. [3.23, 24]

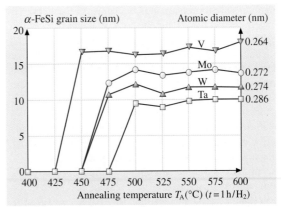

Fig. 4.3-25 Influence of partial substitution of Nb by the refractory elements R = V, Mo, W and Ta on the α-FeSi grain size during annealing of the alloy $Fe_{73.5}Si_{15.5}B_7Cu_1Nb_3R_2$ [3.25]

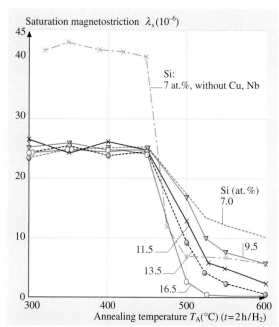

Fig. 4.3-26 Influence of the Si-content on the saturation magnetostriction λ_s during annealing of the nanocrystalline alloys $Fe_{(75-73.5)}Si_{(7-16.5)}B_{(14-6)}Cu_1Nb_3$ [3.26]

magnetostriction $\lambda_s \approx 24 \times 10^{-6}$, with a magnetoelastic anisotropy energy $K_\sigma \approx 50\,\mathrm{J\,m^{-3}}$. With partial crystallization of the α-Fe-Si phase during annealing, λ_s varies significantly. At higher Si contents it decreases strongly and passes through zero at about 16 at.% Si, as shown in Fig. 4.3-26.

This behavior is caused by the compensation of the negative saturation magnetostriction λ_s of the α-Fe-Si phase $\lambda_s \approx -8 \times 10^{-6}$ and the positve values of λ_s of the residual amorphous phase of $\lambda_s \approx +24 \times 10^{-6}$.

The superposition of the local magnetostrictive strains to an effective zero requires a large crystalline volume fraction of about 70 vol.% to compensate the high positive value of the amorphous residual phase of about 30 vol.%. By anealing at about 550 °C this relation can be realized. The second requirement for superior soft magnetic properties is a small or vanishing magnetocrystalline anisotropy energy K_1.

By developing a particular variant of the random anisotropy model, it was shown [3.27] that for grain diameters D smaller than the magnetic exchange length L_0, the averaged anisotropy energy density $\langle K \rangle$ is given by

$$\langle K \rangle \approx v_{cr}^2 K_1 (D L_0^{-1})^6 = v_{cr}^2 D^6 K_1^4 A^{-3},$$

where v_{cr} = crystallized volume, K_1 = intrinsic crystal anisotropy energy of α-Fe-Si, $L_0 = \sqrt{A K_1^{-1}}$, and A = exchange stiffness constant. A schematic representation of this model is given in Fig. 4.3-27. The basic effect of decreasing the grain size consists of local averaging of the magnetocrystalline anisotropy energy K_1 for $D = 10-15$ nm at $L_0 = 30-50$ nm (about equal to

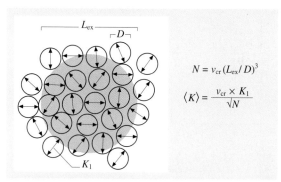

Fig. 4.3-27 Schematic representation of the random anisotropy model for grains embedded in an ideally soft ferromagnetic matrix. The *double arrows* indicate the randomly fluctuating anisotropy axis; the *dark* area represents the ferromagnetic correlation volume determined by the exchange length $L_{ex} = (A/\langle K \rangle)^{1/2}$ [3.23]

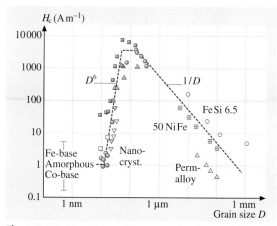

Fig. 4.3-28 Coercivity, H_c, vs. grain size, D, for various soft magnetic metallic alloys [3.27]: Fe–Nb–Si–B (*solid up triangles*) [3.28], Fe–Cu–Nb–Si–B (*solid circles*) [3.29, 30], Fe–Cu–V–Si–B (*solid* and *open down triangles*) [3.31], Fe–Zr–B (*open squares*) [3.32], Fe–Co–Zr (*open diamonds*) [3.33], NiFe-alloys (+ *center squares* and *open up triangles*) [3.34], and Fe–Si(6.5 wt%) (*open circles*) [3.35]

Table 4.3-24 Typical values of grain size D, saturation magnetization J_s, saturation magnetostriction λ_s, coercivity H_c, initial permeability μ_i, electrical resistivity ρ, core losses P_{Fe} at 0.2 T, 100 kHz and ribbon thickness t for nanocrystalline, amorphous, and crystalline soft magnetic ribbons

Alloy	D (nm)	J_s (T)	λ_s (10^{-6})	H_c (A m^{-1})	μ_i (1 kHz)	ρ ($\mu\Omega$ cm)	P_{Fe} (W kg^{-1})	t (μm)	Ref
Fe$_{73.5}$Cu$_1$NB$_3$Si$_{13.5}$B$_9$	13	1.24	2.1	0.5	100 000	118	38	18	a
Fe$_{73.5}$Cu$_1$NB$_3$Si$_{15.5}$B$_7$	14	1.23	~0	0.4	110 000	115	35	21	b
Fe$_{84}$NB$_7$B$_9$	9	1.49	0.1	8	22 000	58	76	22	c
Fe$_{86}$Cu$_1$Zr$_7$B$_6$	10	1.52	~0	3.2	48 000	56	116	20	c
Fe$_{91}$Zr$_7$B$_3$	17	1.63	−1.1	5.6	22 000	44	80	18	c
Co$_{68}$Fe$_4$(MoSiB)$_{28}$	amorphous	0.55	~0	0.3	150 000	135	35	23	b
Co$_{72}$(FeMn)$_5$(MoSiB)$_{23}$	amorphous	0.8	~0	0.5	3000	130	40	23	b
Fe$_{76}$(SiB)$_{24}$	amorphous	1.45	32	3	8000	135	50	23	b
80 % Ni−Fe (permalloys)	~100 000	0.75	<1	0.5	100 000^d	55	>90^e	50	b
50–60 % Ni−Fe	~100 000	1.55	25	5	40 000^d	45	>200^e	70	b

^{a} [3.36]
^{b} Typical commercial grades for low remanence hysteresis loops, Vacuumschmelze GmbH 1990, 1993
^{c} [3.37, 38]
^{d} 50 Hz-values
^{e} Lower bound due to eddy currents

the domain wall thickness), which leads to the extreme variation of $\langle K \rangle$ with the sixth power of grain size. These relations were confirmed experimentally and result in an anomalous variation of the coercivity with grain size, as shown in Fig. 4.3-28.

Another type of nanocrystalline soft magnetic materials is based on Fe−Zr−B−Cu alloys [3.37, 38]. A typical composition is Fe$_{86}$Zr$_7$B$_6$Cu$_1$. As Zr provides high glass-forming ability, the total content of glass-forming elements can be set to ≪ 20 at.%. As a consequence the Fe content is higher, which implies higher saturation polarization. The nanocrystalline microstructure consists of a crystalline α-Fe phase with grain sizes of about 10 nm embedded in an amorphous residual phase. After annealing at 600 °C, an optimum combination of magnetic properties is obtained: $I_s \geq 1.5$ T, $H_c \approx 3$ A m^{-1}, $\lambda_s \approx 0$. Because of the high reactivity of Zr with oxygen the preparation of this type of alloy is difficult. The production of these materials on an industrial scale has not yet succeeded.

Table 4.3-24 shows the magnetic and physical properties of some commercially-available nanocrystalline alloys for comparison to amorphous and Ni−Fe-based crystalline soft magnetic alloys. Magnetic field annealing allows the shape of the hysteresis loops of

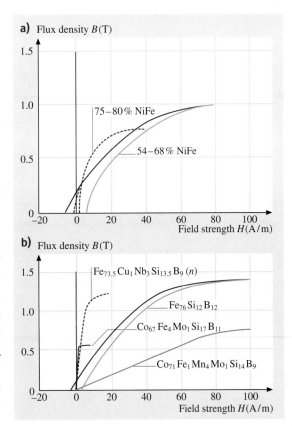

Fig. 4.3-29a,b Soft magnetic alloys with flat hysteresis loops: **(a)** Crystalline. **(b)** Amorphous and nanocrystalline (curve indicated by n) [3.12]

nanocrystalline soft magnetic alloys to be varied according to the demands of the users. Accordingly, different shapes of hystesis loops (Z, F, or R type) may be achieved. For comparison several characteristic hysteresis loops of crystalline, amorphous, and nanocrystalline soft magnetic alloys are shown in Figs. 4.3-29a,b and 4.3-30a,b.

A survey of the field dependence of the amplitude permeability of various crystalline, amorphous, and nanocrystalline soft magnetic alloys is given in Fig. 4.3-31 [3.12]. Figure 4.3-32 [3.23] represents the frequency behavior of the permeability $|\mu|$ of different soft magnetic materials for comparison.

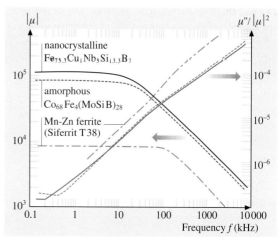

Fig. 4.3-32 Frequency dependence of permeability, $|\mu|$, and the relative loss factor, $\mu''/|\mu|^2$, for nanocrystalline $Fe_{73.5}Cu_1Nb_3Si_{15.5}B_7$ and comparable, low remanence soft magnetic materials used for common mode choke cores [3.23]

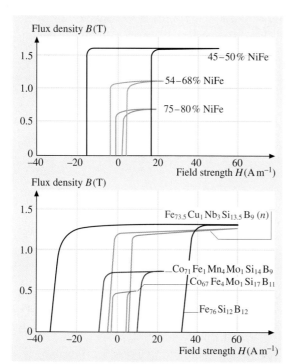

Fig. 4.3-30a,b Soft magnetic alloys with rectangular hysteresis loops: (**a**) Crystalline. (**b**) amorphous and nanocrystalline (curve indicated by n) [3.12]

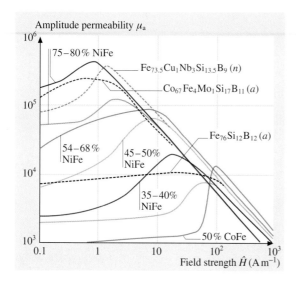

Fig. 4.3-31 Amplitude permeability–field strength curves of soft magnetic alloys ($f = 50$ Hz): amorphous (a); nanocrystalline (n) [3.12]

4.3.2.8 Invar and Elinvar Alloys

The term invar alloys is used for some groups of alloys characterized by having temperature-invariant properties, either temperature-independent volume (invar) or temperature-independent elastic properties (elinvar) in a limited temperature range. A comprehensive survey of the physics and applications of invar alloys is given in [3.17].

Invar Alloys
With the discovery of an Fe–36 wt% Ni alloy with an uncommonly low thermal expansion coefficient (TEC) around room temperature and called "Invar" by Guil-

laume in the 1890s, the history of invar and elinvar type alloys began. Below the magnetic transition temperature, Curie temperature T_c or Néel temperature T_N of ferromagnetic or antiferromagnetic materials, a spontaneous volume magnetostriction ω_s sets in. With some alloy compositions, ω_s is comparable in magnitude to the linear thermal expansion but opposite in sign. As a result the coefficient of linear thermal expansion may become low and even zero.

The linear (α) and volumetric (β) thermal expansion coefficients (TECs) are defined as:

$$\alpha = (1/l)(\Delta l/\Delta T)_P \quad [\text{K}^{-1}]$$

and

$$\beta = (1/V)(\Delta V/\Delta T)_P \quad [\text{K}^{-1}],$$

with l = length, V = volume, and T = temperature. If the alloys are isotropic the volumetric thermal expansion coefficient is equal to three times the linear TEC:

$$\beta = 3\alpha.$$

The spontaneous volume magnetostriction ω_s is in a first approximation:

$$\omega_s = \kappa C M_s^2,$$

with κ = compressibility, C = magnetovolume coupling constant, and M_s = spontaneous magnetization. Figure 4.3-33 [3.42] illustrates schematically the

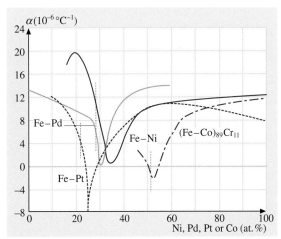

Fig. 4.3-34 Temperature coefficient of linear thermal expansion α at room temperature as a function of the composition in typical invar alloy systems: Fe–Ni [3.39], Fe–Pt [3.40], Fe–Pd [3.40], and Fe–Co–Cr [3.39]. *Vertical dotted lines* show boundary between bcc and fcc phases

temperature-dependent behavior of thermal expansion, ω_s and M_s, which give rise to a small linear expansion coefficient.

Crystalline invar alloys are essentially based on 4 binary alloy systems: Fe–Ni, Fe–Pt, Fe–Pd, and Fe–Co, containing a few percent of Cr. Figure 4.3-34 [3.17] shows the thermal expansion coefficient of these invar type alloy systems. The invar alloy in each system is fcc

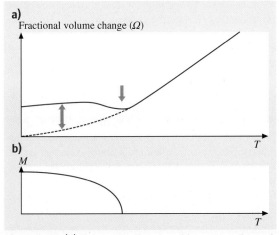

Fig. 4.3-33 (a) Schematic diagram of invar-type thermal-expansion anomaly. The *dashed curve* indicates thermal expansion for hypothetical paramagnetic state. The difference between the two curves corresponds to the spontaneous volume magnetostriction, ω_s. **(b)** Temperature dependence of the spontaneous magnetization

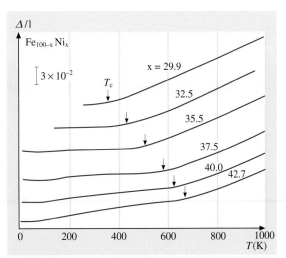

Fig. 4.3-35 Thermal expansion curves of Fe–Ni alloys annealed at 1323 K for 5 days [3.41]

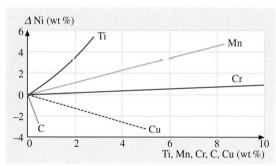

Fig. 4.3-36 Displacement of the composition corresponding to the minimum thermal expansion coefficient of Fe−Ni alloys by the addition of Ti, Mn, Cr, Cu, and C [3.17]

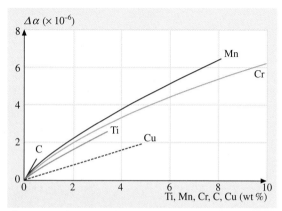

Fig. 4.3-37 Increase of the minimum value of thermal expansion coefficient of Fe−Ni alloys by the addition of Ti, Mn, Cr, Cu, and C [3.17]

and its composition is near the boundary between the bcc and fcc phase fields.

Fe−Ni-Based Invar Alloys. An iron alloy containing 34 to 36.5 wt% Ni is well known as a commercial invar material. The materials C < 0.12 wt%, Mn < 0.50 wt%, and Si < 0.50 wt% are generally added for metallurgical purposes. Figure 4.3-35 [3.41] shows the thermal expansion curves of some Fe−Ni alloys.

The thermal expansion of invar alloys is affected significantly by the addition of third elements,

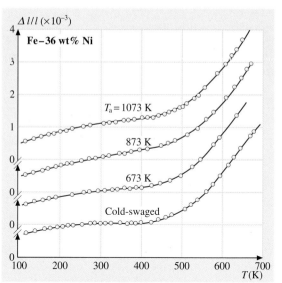

Fig. 4.3-38 Effect of thermal annealing at a temperature T_a and of mechanical treatment on the thermal expansion of Fe–36 at.% Ni invar alloy [3.41]

by cold working and by thermal treatment as shown in Fig. 4.3-36 [3.17], Fig. 4.3-37 [3.17] and Fig. 4.3-38 [3.41].

Invar is an austenitic alloy and cannot be hardened by heat treatment. The effect of heat treatment on the TEC α depends on the method of cooling after annealing. Air cooling or water quenching from the annealing temperature results in a reduction of α but at the same time α becomes unstable. In order to stabilize the material, annealing at low temperature and slow cooling to room temperature are necessary.

Following heat treatments for an optimum and stable magnitude of α are recommended: 830 °C, 1/2 h, water quenching; 315 °C, 1 h, air cooling; 95 °C, 20 h, air cooling. Cold working reduces α. But before use in high precision instruments a stress-relief anneal at 320 to 370 °C for 1 h followed by air cooling is required.

The mechanical properties of some invar-type alloys are listed in Table 4.3-25 [3.42].

Table 4.3-25 Thermal expansion coefficient, α and mechanical properties of invar alloys for practical use [3.42]

	Composition (wt%)	α at RT (10^{-6} K^{-1})	Vickers hardness	Tensile stress (MPa)
Invar	Fe−36Ni	<2	150−200	500−750
Super Invar[a]	Fe−32Ni−4Co	<0.5	150−200	500−800
Stainless Invar[b]	Fe−54Co−9Cr	<0.5		
High strength Invar	Fe−Ni−Mo−C[c]	<4		1250
	Fe−Ni−Co−Ti[d]	<4	300−400	1100−1400

[a] Masumoto, 1931
[b] Masumoto, 1934
[c] Yokota et al., 1982
[d] Yahagi et al., 1980

Table 4.3-26 Thermal expansion coeffcient of invar 36 and free-cut invar [3.17]

Temperature (°C)	α ($\times 10^{-6}$ K^{-1})			
	As annealed		As cold-drawn	
	Invar 36	Free-cut Invar 36	Invar 36	Free-cut Invar 36
25−100	11.18	1.60	0.655	0.89
25−200	1.72	2.91	0.956	1.62
25−300	4.92	5.99	2.73	3.33
25−350	6.60	7.56	3.67	4.20
25−400	7.82	8.88	4.34	4.93
25−450	8.82	9.80	4.90	5.45
25−500	9.72	10.66	5.40	5.92
25−600	11.35	12.00	6.31	6.67
25−700	12.70	12.90	7.06	7.17
25−800	13.45	13.60	7.48	7.56
25−900	13.85	14.60	7.70	8.12

Table 4.3-27 Some physical properties of invar alloys [3.17]

Property	Value	
Melting point	1.425 °C	2600 °F
Density	8 g cm^{-3} (8.0−8.13)	500 lb ft^{-3}
Thermal electromotive force to copper (0−96 °C)	9.8 μV/K	
Specific resistance (Annealed)	82×10^{-6} Ω cm (81−88)	495 Ω circ. mil/ft
Temperature coefficient of electric resistivity	1.21×10^{-3} K^{-1}	0.67×10^{-3} °F^{-1}
Specific heat	0.123 cal/g K (25−100 °C)	0.123 Btu/lb °F (77−212 °F)
Thermal conductivity	0.0262 cal/sec cm K (22−100 °C)	72.6 Btu in./hr ft^2 °F (68−212 °F)
Curie temperature	277 °C (277−280 °C)	530 °F
Inflection temperature	191 °C	375 °F
Modulus of elasticity (in tension)	15.0×10^{10} Pa	21.4×10^6 lb in.$^{-2}$
Temperature coefficient of elastic modulus	$+50 \times 10^{-5}$ K^{-1} (16−50 °C)	$+27 \times 10^{-5}$ °F^{-1} (60−122 °F)
Modulus of rigidity	5.7×10^{10} Pa	8.1×10^6 lb in.$^{-2}$
Temperature coefficient of rigidity modulus	$+58 \times 10^{-5}$ K^{-1}	$+30 \times 10^{-5}$ °F^{-1}
Posson's ratio	0.290	

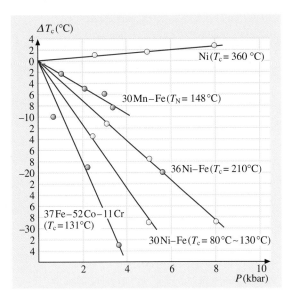

Fig. 4.3-39 The displacement of the Curie point ΔT_c vs. pressure in Fe–Ni invar alloys [3.43] and in Fe–Co–Cr invar alloy [3.44]. The results for Ni [3.43] and for a 30Mn–Fe alloy [3.44] are also shown for comparison. Numerical values in parentheses show the Curie or Néel point at atmospheric pressure

Invar-type alloys show large effects of pressure on magnetization and on the Curie temperature, which suggests a high sensitivity to the interatomic spacing, as shown in Fig. 4.3-39 [3.17].

Thermal expansion coefficients of invar 36 and free-cut invar 36 (containing S and P, or Se) between 25 and 900 °C are listed in Table 4.3-26 [3.17]. Some physical properties of Invar alloys are given in Table 4.3-27 [3.17].

Low thermal expansion coefficients are observed at ternary and quaternary Fe-alloy systems, too. The composition Fe–32 wt% Ni–4 wt% Co was the starting point of superinvar, whose TEC α is in the order of 10^{-7} K^{-1}. The thermal expansion curves of different variants are shown in Fig. 4.3-40 [3.17].

In order to improve the corrosion resistance of invar alloys, "stainless invar" was developed. The basic composition is Fe–54 wt% Co–9.5 wt% Cr. Stainless invar has the bcc structure at room temperature in the equilibrium state. As an invar material, it is used after quenching from a high temperature to retain the fcc structure [3.17].

To improve the mechanical properties two types of high strength invar materials were developed: a work-hardening type based on Fe–Ni–Mo–C, and

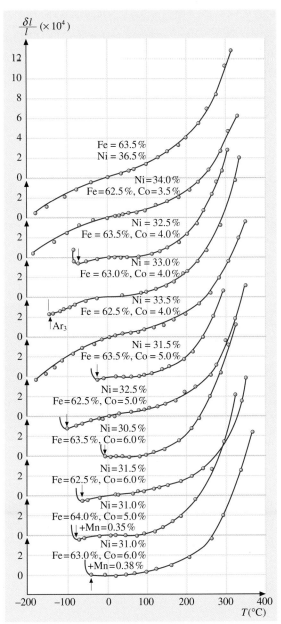

Fig. 4.3-40 Thermal expansion curves of super invar alloys [3.17]

a precipitation-hardening type Fe–Ni–Co–Ti alloy [3.17].

Fe–Pt-Based Invar Alloys. Among the ordered phases of the Fe–Pt system (Fig. 4.3-41 [3.17]), the Fe$_3$Pt phase shows the invar type thermal expansion anomaly. In

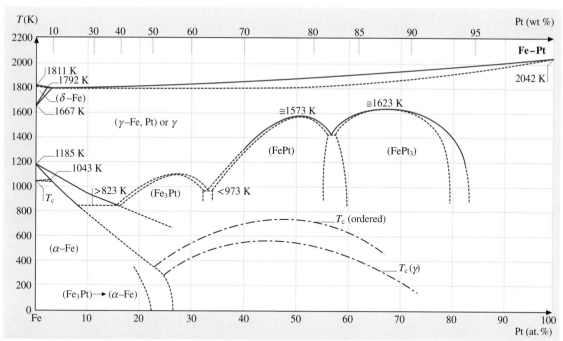

Fig. 4.3-41 Fe–Pt phase diagram. *Dashed-dotted lines*: Curie temperature T_c

order to obtain a well ordered state, a long annealing time is necessary (600 °C, 160 h). Disordered fcc alloys which show invar anomalies, too, may be obtained by rapid quenching from above the order–disorder transformation temperature. The Curie temperature of the disordered state is lower than that of the ordered state. A high negative value of α is observed just below the Curie temperature, particularly for disordered alloys. Annealed alloys containing 52 to 54 wt% Pt have small TEC and those containing 52.5 to 53.5 wt% Pt show negative values of α, Fig. 4.3-42 [3.17].

Fe–Pd-Based Invar Alloys. Alloys of Fe–Pd (see Fig. 4.3-43) containing 28 to 31 at.% Pd show invar characteristics, as seen in Fig. 4.3-44 [3.41]. In order to obtain invar behavior, the alloys are quenched from the high temperature γ phase field such that phase transformations at lower temperatures are suppressed. As shown in Fig. 4.3-45 [3.41], the thermal expansion is strongly decreased by cold deformation, i.e., by disordering, lattice defects, and internal stresses. After cold working, an instability of the invar property is observed.

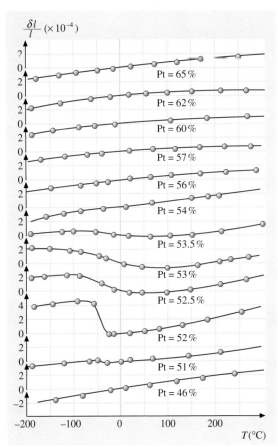

Fig. 4.3-42 Thermal expansivity curves of Fe–Pt alloys [3.17]

Other Alloy Systems with Invar Behavior. Beyond the invar-type alloys mentioned, many other Fe-based alloy systems show invar behavior, for instance, Fe–Ni–Cr, Fe–Ni–V, Fe–Pt–Re, Fe–Ni–Pt, Fe–Ni–Pd, Fe–Pt–Ir, Fe–Cu–Ni, and Fe–Mn–Ni. Some amorphous melt/quenched alloy systems show invar characteristics too, e.g., $Fe_{83}B_{17}$, $Fe_{85}P_{15}$, $Fe_{79}Si_9B_{12}$. Similarly, amorphous alloys prepared by sputtering show invar characteristics: $Fe_{75}Zr_{25}$, $Fe_{72}Hf_{28}$. Some antiferromagnetic Mn- and Cr-based alloys also exhibit a remarkable anomaly of thermal expansion due to magnetic ordering: $Pd_{64.5}Mn_{35.5}$, $Mn_{77}Ge_{23}$, $Cr_{92.5}Fe_{4.3}Mn_{0.5}$, $Cr_{96.5}Si_3Mn_{0.5}$.

Elinvar Alloys

Elinvar or constant-elastic-modulus alloys are based on work by *Guillaume* [3.45] and *Chevenard* [3.46]. They show a nearly temperature independent behavior of the Young's modulus (E). For technical applications, elinvar alloys are designed to show the anomaly in the range of their operating temperature, usually close to room temperature. Since the resonance frequency f_0 of an oscillating body is related to its E modulus by $f_0 \sim \sqrt{E/\varrho}$ (ϱ = density), the thermoelastic properties of elinvar alloys are utilized for components in oscillating systems of the precision instrument industry. In these applications highly constant resonance frequencies are required. Typical examples are: resonators in magnetomechanical filters; balance springs in watches, tuning forks; helical springs in spring balances or seismographs, as well as in pressure or load cells. The condition for temperature compensation of the Young's modulus E is:

$$2\Delta f/f \Delta T = \Delta E/E \Delta T + \alpha \approx 0,$$

with T = temperature and α = linear thermal expansion coefficient. Elinvar characteristics also refer to the temperature independence of the shear modulus G, which is related to the E modulus via

$$3/E = 1/G + 1/3B$$

and

$$E = 2(1+\nu)G = 3(1-\nu)B,$$

with B = bulk modulus and ν = Poisson's ratio.

In ordinary metallic materials, E decreases with increasing temperature according to the variation of the elastic constants with temperature according to the anharmonicity in the phonon energy term. The temperature compensation of the E-modulus in ferromagnetic and antiferromagnetic elinvar alloys is caused by an anomaly in its temperature behavior (ΔE effect). As a consequence in ferromagnetic alloys, the E modulus in the demagnetized state is different from that in the magnetized state. The ΔE effect with ferromagnetic elinvar alloys consists of three parts [3.47]:

$$\Delta E = \Delta E_\lambda + \Delta E_\omega + \Delta E_A.$$

A schematic description of the components of the ΔE effect in ferromagnetic elinvar alloys is given in Fig. 4.3-46.

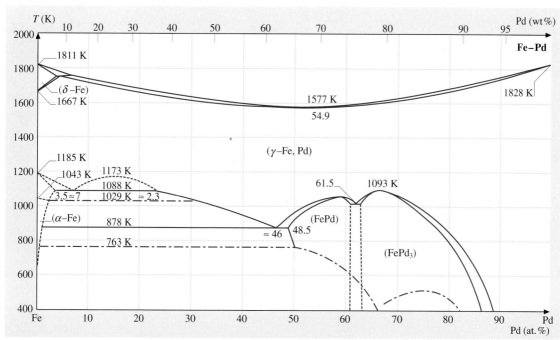

Fig. 4.3-43 Fe–Pd phase diagram. *Dashed-dotted lines*: Curie temperature T_c

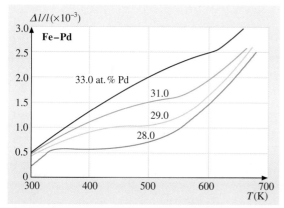

Fig. 4.3-44 Thermal expansion curves of Fe–Pd alloys rapidly cooled from high temperature [3.41]

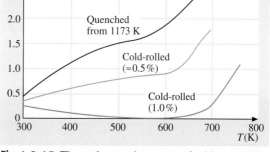

Fig. 4.3-45 Thermal expansion curves of cold-worked Fe–31 at.% Pd alloy. The rolling ratio is given by percentage

These components are attributable to the following relations:

$$\Delta E_\lambda = -2/5 E^2 \lambda_s / \sigma_i ,$$

where ΔE_λ is attributed to the shape magnetostriction (λ_s) that changes the direction of the spontaneous magnetization, owing to domain wall motion and rotation processes as a consequence of the influence of mechanical stresses (σ_i) or magnetic fields on the unsaturated state of material [3.48]. The value ΔE_ω is caused by forced volume magnetostriction (ω) as a consequence of changes of the interatomic distances induced by stress or very high magnetic fields, which lead to a change of the magnetic interaction [3.49]:

$$\Delta E_\omega = -1/9 E^2 [(\partial \omega / \partial H)^2 / \partial J / \partial H] ,$$

where $\partial \omega / \partial H =$ forced volume magnetostriction and $\partial J / \partial H =$ para-susceptibility. The value ΔE_A takes the

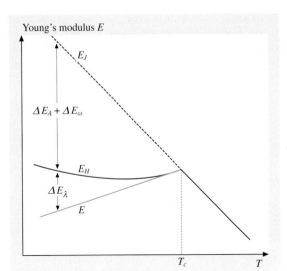

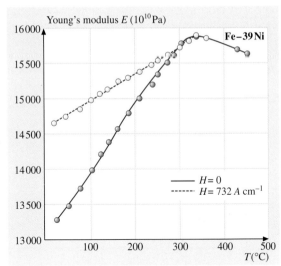

Fig. 4.3-46 Young's modulus E as a function of temperature of ferromagnetic elinvar alloys: $E = E$ modulus in the absence of a magnetic field, $E_H = E$ modulus measured in a magnetic field H, $E_J = E$ modulus at constant polarization J

Fig. 4.3-47 E modulus and its dependence on the temperature of an annealed Fe–39 wt% Ni alloy, with and without influence of a magnetic field H [3.51]

role of the exchange energy into account [3.50]:

$$\Delta E_A \sim -\omega_s \sim J_s^2 \,,$$

where ω_s = volume magnetostriction and J_s = saturation polarization. It originates from the spontaneous volume magnetostriction ω_s as a function of the change of exchange energy with temperature due to the variation of magnetic ordering up to the Curie temperature.

Ferromagnetic Elinvar Alloys. The development of ferromagnetic elinvar alloys is based on affecting the shape magnetostriction λ_s by control of internal stresses resulting from deformation and/or precipitation hardening. But this requires a zero or negative temperature coefficient of the E_H modulus. Based on an Fe–39 wt% Ni alloy, it can be shown how this behavior is achievable. Alloys of Fe–Ni with 36 to 45 wt% Ni have a positive sign of the temperature coefficient of the E_H modulus (see Fig. 4.3-47 [3.51]).

By addition of 7 wt% Cr this coefficient is reduced to zero or to slightly negative values. By a deformation it can be influenced furthermore while the absolute value

Fig. 4.3-48 Young's modulus E of an Fe–39Ni–7Cr–0.8Be–1.0Ti (wt%) alloy as a function of temperature, degree of cold deformation η (%) with and without a magnetic field H [3.51]

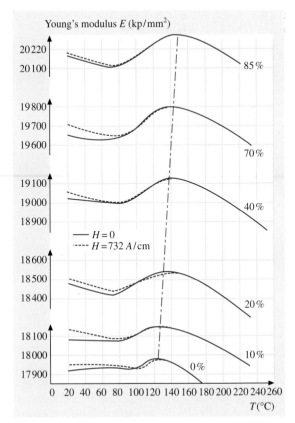

and 1.0 wt% Ti on the $E = f(T)$ characteristic. Cold deformation causes the Curie temperature to rise.

The final processing steps consist of solution annealing at 1150 °C, water quenching, cold deformation, and a final precipitation annealing at about 600 °C. Accordingly commercial elinvar alloys are produced, such as those listed in Tables 4.3-28, 4.3-29, and 4.3-30.

Apart from the well-tried Fe−Ni−Cr−(Be, Ti)-based elinvar-type alloys, Fe−Ni−Mo-based alloys have gained technical application. Addition of Mo improves the elastic properties, lowers the Curie temperature, and increases the resistance to corrosion. While in Europe and the USA, Fe−Ni-based elinvar-type alloys mainly were developed, in Japan Co−Fe-based elinvar alloys were discovered. Ternary Co−Fe−Cr alloys and quaternary alloys containing Ni (Co-elinvar) attained significant technical relevance. Distinguishing marks worth mentioning include: higher Young's modulus than Fe−Ni-based alloys, corrosion resistance, wide range of temperature compensating of E modulus, and easy hardening by cold-working.

Antiferromagnetic Elinvar Alloys. In antiferromagnetic alloys no domains are formed and no ΔE_λ effect occurs. With antiferromagnetic ordering the ΔE_A effect can only be exploited in combination with a pronounced cubic to tetragonal lattice distortion associated with antiferromagnetic ordering. This requires methods to develop temperature compensating elastic behavior which are different from those for ferromagnetic thermoelastic materials. The development of antiferromagnetic alloys with Elinvar properties has been concentrated on Mn−Cu, Mn−Ni, and Fe−Mn base alloys. In [3.52] the distinct anomalies of Young's modulus at the Néel

Table 4.3-28 Compositions in (wt%) of Fe−Ni-based elinvar alloys [3.17]

	Bal. Fe														
	C	Ni	Cr	Ti	Mo	W	Mn	Si	Al	Be	Nb	Cu	V	Co	P	S
Durinal	0.1	42		2.1			2		2							
Elinvar Extra	0.04	43	5	2.75			0.6	0.5	0.3					0.35		
	0.6	42	5.5	2.5			0.5	0.5	0.6							
Elinvar New	1	35	5			2	1									
Elinvar Original		36	12													
	0.5−2	33−35	21−5				1−3	0.5−2	0.5−2							
	0.71	33.5	8.4				2.98	2.4	0.33						0.018	0.01
Iso-elastic		36	8		0.5	(other small constituents)										
	0.1	36	7.5		0.5		0.6	0.5				0.2				
Isoval	0.6	30			2.2	3.2	0.15	0.2			3.8		4.2			
Métélinvar	0.6	40	6		1.5	3	2									
Ni-Span C	0.03	42.2	5.3	2.5			0.4	0.4	0.4			0.05			<0.04	<0.04
Ni-Span C 902	<0.06	42	5.2	2.3			<0.8	<0.1	0.5							
Nivarox CT	0.02	37	8	1			0.8	0.2		0.8						
Nivarox CTC	0.2	38	8	1						1						
Nivarox M	0.03	31			6		0.7	0.1		0.7						
Nivarox M30		30			9					1						
Nivarox M40		40			9					1						
Nivarox W		36		1						1						
Sumi-Span 1		36	9													
Sumi-Span 2	0.4	38	11													
Sumi-Span 3		42.5	5.5	2.4												
Thermelast 4009		40			9						0.5					
Thermelast 5409		40			9						0.5					
Vibralloy		39			9											
		40			10											
YNiC	0.03	41−43	5.1−5.5	2.2−3					0.5−0.6							

Table 4.3-29 Fe–Ni-based commercial elinvar-type alloys and the corresponding values of: density d, melting point T_m, Curie temperature T_c, electrical resistivity ρ, thermal expansion coefficient α, Young's modulus E, its temperature coefficient e, and shear modulus G [3.17]

	d (g cm^{-3})	T_m (°C)	T_c (°C)	ρ (μΩ cm)	α (10^{-6} K^{-1})	E (10^{10} Pa)	e (10^{-5} K^{-1})	G (10^{10} Pa)
Durinal			90				−1.0–1.0	
Elinvar		1420–1450	−100		8	7.8–8.3	−0.3–0.3	
Elinvar Extra[a]	8.15				6.5	18.9	0	6.9
Iso-elastic				88	6.7	18.0	−3.3–2.5	6.4
Métélinvar			260, 295				0	
Ni-Span C	8.15	1450–1480		80	7.1	18.9	−1.7–1.7	
Ni-Span C 902[a]	8.14	1460–1480	160–190	100–120	8.1	17.7–19.6		6.9–7.4
Nivarox CT	8.3		80	97	7.5	18.6	−2.5–2.5	
Sumi-Span 1	8.15		−140	100	−10	18.1	0–2.5	
Sumi-Span 2	8.08		−190	105	−10	18.1	−1.5–0[b]	
Sumi-Span 3[a]	8.05		−190	110	≈8	19.2	−1.0–1.0	7.8
Thermelast 4002[a]	8.3			100	8.5	18.6		6.4
Thermelast 4005[a]	8.3			100	8.3	17.2		6.4
Thermelast 5405[a]	8.3			100	8.0	18.6		6.4
Thermelast 5429[a]	8.3			100	8.0	18.6		6.4
Vibralloy[a]	8.3		300		8	17.4	0	
YNiC	8.15		90–180		8.1	19.6	−1.8–1.5	6.5–6.8

[a] Properties in the fully aged state
[b] Temperature coefficient of the frenquency of proper vibration

Table 4.3-30 Co–Fe-based elinvar-type alloys (annealed state). Composition, thermal expansion coefficient α, Young's modulus E and shear modulus G, and their respective temperature coefficients, e and g [3.17]

	Composition in wt%								α^a (10^{-6} K^{-1})	E^b (10^{10} Pa)	e^c (10^{-5} K^{-1})	G^b (10^{10} Pa)	g^c (10^{-5} K^{-1})
	Co	Fe	Cr	V	Mo	W	Mn	Ni					
Co-elinvar	60.0	30.0	10.0						5.1	17.07		6.91	−0.2
	51.5	38.5	10.0						8.7	18.84	−1.0	7.55	
	47.3	34.5	9.1					9.1	7.8			6.61	0.2
	27.7	39.2	10.0					23.1	8.1			6.48	−0.3
	26.7	50.8	5.8					16.7	7.8			4.99	0.3
	17.9	42.8	10.7					28.6	8.3			6.74	0.9
Elcolloy	40.0	35.0	5.0		5.0		15.0	5.0					−0.2
	35.0	36.0	5.0	4.0	4.0		16.0	9.0					0.5
Mangelinvar	38.0	37.0				15.0	10.0	9.7	18.0	−1.0			
Moelinvar	50.0	32.5	17.5						9.6			7.36	−0.2
	45.0	35.0	10.0				10.0		8.5			6.15	0.7
	20.0	40.0	20.0				20.0		8.4			7.70	0.9
	10.0	45.0	15.0				30.0		9.8			7.85	−0.4
Tungelinvar	50.0	28.5				21.5			7.4			6.45	−0.7
	39.0	32.0				19.0	10.0		7.8			8.13	0.4

Table 4.3-30 Co–Fe-based elinvar-type alloys (annealed state). Composition, thermal expansion coefficient α, cont.

| | Composition in wt% | | | | | | | α^a | E^b | e^c | G^b | g^c |
	Co	Fe	Cr	V	Mo	W	Mn	Ni	(10^{-6} K^{-1})	(10^{10} Pa)	(10^{-5} K^{-1})	(10^{10} Pa)	(10^{-5} K^{-1})
Velinvar	60.0	30.0	10.0						8.1			6.53	0.0
	50.0	31.8	8.2					9.1	11.0			6.70	1.0
	37.5	35.5	7.0					20.0	11.1			6.45	−0.6
	22.5	43.5	4.0					30.0	12.3			5.66	−0.3

[a] For the temperature range 10–50 °C
[b] At 20 °C
[c] For the temperature range 0–50 °C

temperature of Mn–Ni- and Mn–Cu-based alloys were modified by additions of Cr, Fe, Ni, Mo, and W as well as by suitable technological treatment in such a way that useful thermoelastic coefficients could be reached. However, the mechanical workability, the sensitivity of elastic properties to the degree of cold-working, the unsufficient spring properties as well the low corrosion resistance and high mechanical damping could not satisfy the conditions of technical application. Figures 4.3-49 and 4.3-50 [3.17] indicate the temperature dependence of E of some Mn–Ni- and Mn–Cu-based alloys. Table 4.3-31 [3.17] lists the compositions and the thermoelastic and mechanical properties of this group of alloys.

By optimization of the chemical composition and the processing technology, an Fe–24Mn–8Cr–7Ni–0.8Be

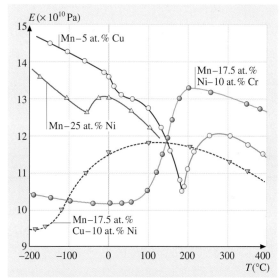

Fig. 4.3-50 Temperature dependence of Young's modulus E [3.17] for various Mn-based alloys annealed at 1223 K for 1 h and then quenched

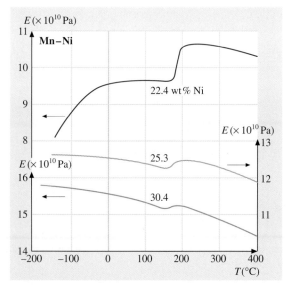

Fig. 4.3-49 Mn–Ni binary alloys. Young's modulus E vs. temperature for alloys annealed at 1223 K for 1 h after cold-working [3.17]

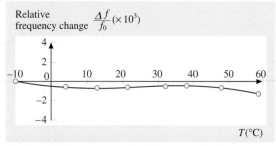

Fig. 4.3-51 Temperature dependence of the frequency of a screw spring [3.53]

Table 4.3-31 Nonferromagnetic Mn-based elinvar-type alloys. Composition, thermal expansion coefficient α, Young's modulus E and shear modulus G, and their respective temperature coefficients, e and g and hardness HV [3.17]

| Composition in wt% | | | | | | | | α^a | E | e^a | G | g^a | HV |
Mn	Cu	Ni	Cr	Fe	Co	Mo	W	Ge	(10^{-6} K^{-1})	(10^{10} Pa)	(10^{-5} K^{-1})	(10^{10} Pa)	(10^{-5} K^{-1})	
87		10	3						23.7	12.3	1.25	5.18	1.05	121
82		15					3		23.0	12.1	0.55	5.07	0.78	149
80		16		4					21.1	12.2	−0.13	4.63	−0.75	150
80		9			11				20.3	11.9	0.05	5.00	1.10	380
80								20	12.3	9.0	1.5			255
79		21								9.8	−2.5	3.60	−2.7	235
67	20	13							22.4	14.4	0.21	4.55	0.29	125
59		16	25						21.6	16.2	0.85	5.21	0.83	250
49	41			10					22.4	13.5	−0.97	5.53	−0.20	250
44	55		1						22.1	13.2	0.11	5.03	0.08	145
43	57								23.6	11.2	0.3	4.2	−0.9	131
43	55				2				22.9	8.50	−1.11	4.02	−2.57	135
42	55					3			23.0	15.2	2.30	6.77	1.88	149
39	56						5		23.2	12.0	−0.25	5.05	−0.56	140

[a] For the temperature range 0–40 °C
[b] At 20 °C

Table 4.3-32 Properties of an antiferromagnetic Fe−24Mn−8Cr−7Ni−0.8Be elinvar alloy [3.53]

Property	Value
Young's modulus E	165–195 GPa
Thermoelastic coefficient TKE	(1–10) MK^{-1}
Compensation range of E	0–50 °C
Shear modulus G	74–82 GPa
Coefficient of thermal expansion α (20–100 °C)	13 MK^{-1}
Tensile strength σ_B	1200–1800 MPa
Yield point σ_s	1100–1650 MPa
Elongation δ	12–2 %
Vicker's Hardness HV 10	420–540 HV
Quality factor Q	20 000–10 000
Specific electrical resistance ρ	80 $\mu\Omega$ cm
Density γ	7.6 g cm^{-3}
Electrochemical breakdown potential ϵ_D	−0.25 V
Melting temperature T_s	1450–1480 °C

antiferromagnetic elinvar alloy was developed which fulfills the complex requirements for an antiferromagnetic, corrosion resistant, temperature compensating thermoelastic spring material for applications near room temperature, Fig. 4.3-51 [3.53]. The thermoelastic, mechanical and physical properties are summarized in Table 4.3-32 [3.53].

Other Nonmagnetic Elinvar-Type Alloys. In general, the elastic constants decrease with increasing temperature. The temperature dependence of the elastic properties of Nb shows highly anisotropic anomalies (see Fig. 4.3-52) [3.54]. In a randomly oriented polycrystal, Nb becomes an elinvar-type material, Fig. 4.3-53 [3.41]. This plot of Young's modulus vs. temperature also shows the influence of alloying elements.

Elinvar behavior is, also, found in concentrated Nb–Zr and Nb–Ti alloys. Furthermore, some amorphous alloys, e.g., in the Fe–B-, Fe–P-, and Fe–Si–B-based systems, show elinvar behavior. Examples for amorphous Fe–B alloys are shown in Fig. 4.3-54.

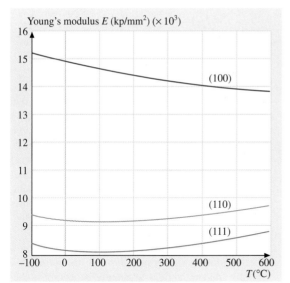

Fig. 4.3-52 Thermoelastic behavior of single crystalline Nb in the major lattice directions [3.54]

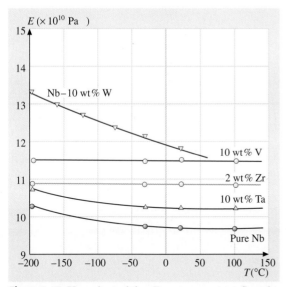

Fig. 4.3-53 Young's modulus E vs. temperature. Samples annealed at 1400 °C for 4 h [3.41] of pure Nb and Nb-based binary alloys

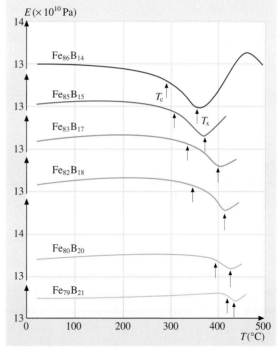

Fig. 4.3-54 Fe–B alloys. Temperature dependence of Young's modulus E for amorphous alloys annealed at 200 °C for 2 h. T_x and T_c show the crystallization and the Curie temperature, respectively [3.41]

4.3.3 Hard Magnetic Alloys

Permanent or hard magnetic materials comprise traditionally some special steels but consist essentially of multiphase alloys and intermetallic and ceramic compounds today. Relatively few magnetic alloys and compounds fulfill the requirement of combining high magnetic efficiency and competitive cost: Fe−Ni−Al−Co (Alnico), Fe−Cr−Co, Mn−Al−C, hard ferrites, and rare-earth transition metal compounds of the Co−Sm, Fe−Nd−B and Fe−Sm−N alloy systems. Surveys may be found in handbooks [3.1, 3, 55] and data collections [3.10, 56]. A survey of permanent magnetic materials is given in Table 4.3-33. The metallic hard magnets are treated in this chapter, the oxidic hard magnetic materials are dealt with in Sect. 4.3.4.

All of the hard magnetic materials are based on choosing a base alloy with a sufficiently high saturation magnetization M_s and a high magnetocrystalline anisotropy constant K_1, and tailoring the microstructure to exploit this crystal anisotropy. In some cases, shape anisotropy is generated in addition. This microstructural control is achieved by

1. Inducing a texture by processing in such a way that a macroscopic direction in the material, e.g., the rolling direction of a sheet or the pressing direction in a sintered material, is an easy direction, and processing at 90°, i.e., the transverse direction, is a hard direction for magnetization. This is the basic magnetic hardening mechanism of hard magnetic steels which have lost their importance in present technology; but it is the basis of producing the more recent high energy magnets made from intermetallic compounds such as Co_5Sm and ferrites.

2. Inducing a two-phase microstructure by coherent precipitation or decomposition and promoting, by suitable magnetic field annealing procedures, the alignment of the elongated precipitates in one direction of easy magnetization, i.e., inducing both

Table 4.3-33 Survey commercially used permanent magnetic materials. Survey

Material	Fe, Co content (wt%)	B_r^a (T)	$_JH_c$ (kA m^{-1})	$(BH)_{max}$ (kJ m^{-3})	appr. T_c (°C)	T_{max}^b (°C)	Process^c
Dense magnets							
3.5Cr steel	94–95	0.95	5	2.3	745		C^d
6W steel	92–93	0.95	6	2.6	760		C
36Co steel	90–91	0.95	19	7.4	890		C
Alnico	67–74	0.52–1.4	40–135	13–69	810–900	450–550	C, P^d
Fe−Cr−Co	65–73	1.1–1.4	40–65	25–55	670	500	C
hard ferrite	58–63	0.37–0.45	160–400	26–40	460	250	P
Pt-Co	23.3	0.64	430	73	480	350	C, P
MnAlC	–	0.55	250	44	500	300	P
Co_5Sm	63–65	0.85–1	>1600	140–200	730	250	P
$TM_{17}Sm_2$	61–68	0.95–1.15	480–2000	190–220	810	330–550	P
Fe−Nd−B	66–72	1.05–1.5	950–2700	240–400	320	60–180	P
Bonded magnets							
hard ferrite	58–63	0.1–0.31	180–300	2–18		140	P
$TM_{17}Sm_2$	61–68			70–120		150	P
$Fe_{14}Nd_2B$	70–72	0.47–0.69	600–1200	35–80		80–110	P
$Fe_{17}Sm_2N_3$		0.77	650	105		100	P

^{a} B_r values are for magnets operated at load lines $B/H \gg 1$
^{b} The maximum operating temperature of bonded magnets is determined by the organic binder used
^{c} Magnets are manufactured either by a casting/heat treatment technique or by a powder metallurgical process. Powder metallurgy is applied for small magnets where small and intricate shapes to precise tolerances are required
^{d} C: magnets produced by cast and heat treatment; P: magnets produced by means of powder metallurgical techniques

magnetocrystalline and shape anisotropy. This is termed magnetic shape anisotropy and is employed in the Alnico and Fe−Co−Cr hard magnetic materials.

3. Inducing a fine grained microstructure with a magnetically insulating phase at the grain boundaries so that the grains are magnetically decoupled and, as a consequence, the nucleation of reverse magnetization is requiring an extremely high nucleation energy. This is applied, for example, to $TM_{17}Sm_2$, $Fe_{14}Nd_2B$ and bonded magnets.

In some of the hard magnetic materials two of these variants of microstructural design are combined.

4.3.3.1 Fe−Co−Cr

Hard magnetic materials made of ternary Fe−Co−Cr alloys are based on the high atomic moment of Fe−Co alloys and the miscibility gap occurring when Cr is added. Intrinsic magnetic properties are compiled in [3.6]. Extensive magnetic materials data are found in [3.56]. Figure 4.3-55 shows the relevant metastable phase relations in the ternary equilibrium diagram. If an alloy is homogenized in the solid solution range above the solvus surface with $T_{max} > 700\,°C$ first and annealed in the miscibility gap subsequently, coherent decomposition occurs, which results in a two-phase microstructure on the nanometer scale. The α_1 phase is rich in Fe and Co and ferromagnetic while the α_2 phase is rich in Cr and antiferromagnetic. This two-phase microstructure on the nm scale has hard magnetic properties which can be varied by adjusting the alloy composition and heat treatment. The term spinodal decomposition is frequently applied to all kinds of coherent decomposition, e.g., in [3.56]. But it is used correctly only if referring to a special mode of compositional evolution associated with particular kinetics in the initial stage of decomposition within the spinodal of a miscibility gap.

Three groups of materials have been developed, differing essentially in the Co content (< 5, 10–15, 23–25 wt% Co), while the Cr content ranges from 22 to 40 wt% Cr. Table 4.3-34 lists data obtained by varying composition, mode of manufacturing, and heat treatment systematically for the group characterized by < 5 wt% Co as an example. The variation of the magnetic properties is determined by the intrinsic properties of the decomposed phases α_1 and α_2 and their microstructural array.

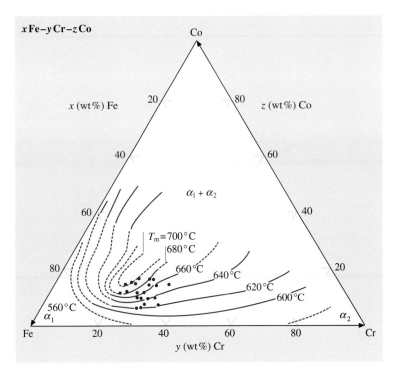

Fig. 4.3-55 The miscibility gap ($\alpha_1 + \alpha_2$) of the bcc α-phase in the Fe−Co−Cr phase diagram [3.56]

Table 4.3-34 Survey of magnetic properties of Fe–Cr–Co (≤ 5 wt% Co) alloys in relation to the composition, mode of manufacturing and heat treatment [3.56]

Alloy components (wt%)			Mode of manufacturing and heat treatment	Magnetic properties					
				B_r		$_BH_c$		$(BH)_{max}$	
Cr	Co	others		(T)	(kG)	(kA m^{-1})	(kOe)	kJ m^{-3}	(MGOe)
33	2	1 Hf	H(700, 15):MCL > T_s, R, 500):FCL	1.25	12.5	16.2	0.203	14.0	1.76
32	3		As above	1.29	12.9	35.9	0.449	32.4	4.08
32	4	0.5 Ti	As above	1.26	12.6	42.7	0.534	40.1	5.06
28	5		As above	1.38	13.8	29.0	0.362	27.9	3.52
30	5		As above	1.34	13.4	42.2	0.528	42.1	5.31
33	5		As above	1.22	12.2	40.8	0.51	36.3	4.58
35	5		As above	1.15	11.5	37.0	0.462	29.3	3.69
30	5	0.1 B	As above	1.31	13.1	42.0	0.525	40.2	5.07
30	5	0.25 B	As above	1.29	12.9	39.8	0.498	34.8	4.39
30	5	0.1 C	As above	1.31	13.1	42.2	0.527	38.8	4.89
30	5	0.8 Ge	As above	1.32	13.2	24.8	0.31	39.1	4.93
30	5	0.25 Ti	As above	1.34	13.4	27.2	0.34	40.2	5.07
30	5	0.5 Ti	As above	1.30	13.0	41.4	0.518	40.1	5.06
30	5	1.5 Ti	As above	1.27	12.7	40.8	0.51	38.1	4.81
30	5	0.25 Hf	As above	1.32	13.2	43.0	0.537	41.2	5.2
30	5	0.5 Hf	As above	1.29	12.9	43.9	0.549	41.4	5.22
30	5	1 Hf	As above	1.30	13.0	43.0	0.537	40.4	5.1
30	5	3 Hf	As above	1.24	12.4	41.5	0.519	34.8	4.39
23	2	1 Hf	MCL(> T_s, 550): CCL (550, 500)	1.24	12.4	36.8	0.46	34.1	4.3
32	3		As above	1.25	12.5	40.0	0.5	34.1	4.3
30	5		As above	1.34	13.4	42.4	0.53	42.0	5.3
32	4	0.5 Ti	As above	1.26	12.6	42.8	0.535	40.4	5.1
28	5	4 Ni	As above	1.27	12.7	29.6	0.37	30.1	3.8
28	7		As above	1.25	12.5	40.8	0.51	41.2	5.2
27	9		MCL: CCL (as above)	1.30	13.0	46.4	0.58	49.2	6.2
33	5		CL(680, 40 K/h):HW(D:67%):H (600): CCL (15–4 K/h, 500)[a]	1.15	11.5	24.8	0.31	19.0	2.4
33	7	2 Cu	As above	1.19	11.9	38.8	0.485	26.2	3.3
33	7		As above	1.18	11.8	42.0	0.525	33.3	4.2
33	9		As above	1.24	12.4	46.4	0.58	32.5	4.1
28	7		CL(> T_s, 60 ($T_s = 645$ °C))	0.97	9.7	26.4	0.33	11.1	1.4
31	5		(Sintered 1400 °C, 4 h; H_2)[b]:WQ: H (700, 30): FCL (700, 640):MCL (640, 0.9 K/h, 500)	1.23	12.3	40.0	0.5	34.9	4.4

[a] For the deformation aging process the initially aged state corresponds to an overaged state
[b] Sintering ST

Commercial materials are characterized by the fact that they can be quenched from temperatures above the miscibility gap first, which results in mechanical properties amenable to forming by conventional processes such as rolling, stamping, drilling. The final annealing treatment in the miscibility gap results in the magnetically hard state. This is associated with a drastic decrease in ductility. If the annealing treatment is carried out in a magnetic field, the final product has an anisotropic behavior. Table 4.3-35 shows the property range of commercial Fe−Cr−Co materials.

4.3.3.2 Fe−Co−V

Magnetic materials based on the Fe−Co−V alloy system were the first ductile magnets. The intrinsic magnetic properties may be found in [3.6] while extensive magnetic materials data are treated in [3.56]. The optimum magnetic behavior is obtained for alloy compositions around Fe−55 wt% Co−10 wt% V. As the isothermal sections of the Fe−Co−V phase diagram Fig. 4.3-56a and Fig. 4.3-56b show, this alloy is mainly in the fcc γ-phase (austenite) state at 900 °C,

Table 4.3-35 Commercial Fe−Cr−Co magnetic materials

Composition nominal wt%	Variant	Remanence (T)	Coercivity (kA m^{-1})	Energy density (kJ m^{-3})	Curie temperature (°C)	Maximum application temperature (°C)	Hardness HV	Commercial designation[a]
Fe−27Cr−11Co−Mo	isotropic	0.85−0.95	36−42	13	640	480	480	12/160
Fe−28Cr−16Co−Mo	isotropic	0.80−0.90	39−45	15	640	480	480	16/160
Fe−27Cr−11Co−Mo	anisotropic	1.15−1.25	47−55	35	640	480	480	12/500
Fe−28Cr−16Co−Mo	anisotropic	1.10−1.20	53−61	37	640	480	480	16/550

[a] Designation of CROVAC® by Vacuumschmelze, Hanau, Germany

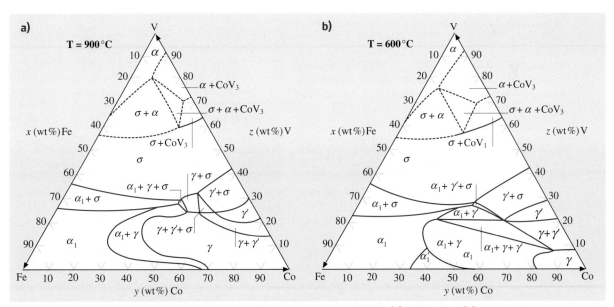

Fig. 4.3-56a,b Isothermal sections of the Fe−Co−V phase diagram at 900 °C (**a**) and 600 °C (**b**). α: bcc disordered; α_1: bcc ordered (CsCl type); γ: fcc disordered (austenite); γ': fcc ordered (Au$_3$Cu type) [3.56]

Table 4.3-36 Commercial Fe−Co−V-based magnetic materials

Composition nominal (wt%)	Remanence (T)	Coercivity (kA m^{-1})	Energy density (kJ m^{-3})	Curie temperature (°C)	Maximum application temperature (°C)	Hardness (HV) As rolled	Hardness (HV) Heat treated	Alloy code[a]
34Fe−52Co−13V	0.80−0.90	25−30	12	700	500	480	900	35U
34Fe−53Co−8.5V−3.5Cr	1.00−1.10	30−35	20	700	500	520	950	93

[a] Designation of MAGNETOFLEX® by Vacuumschmelze, Hanau, Germany

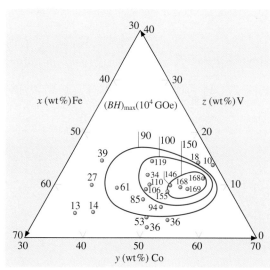

Fig. 4.3-57 Contour map of $(BH)_{max}$ of Fe−Co−V alloys in the optimum annealed state. It is obtained by annealing in the temperature range $T_a = 555-750\,°C$ in the magnetically preferred direction of the anisotropic sample [3.56]

while it decomposes into the bcc ordered $\alpha_1 + \gamma$ state upon annealing at lower temperature such as 600 °C. In combinations of heat treatment with plastic deformation (also serving to form the product, e.g., wire) an optimum anisotropic hard magnetic state can be realized.

If quenched from the γ-phase state, the alloy can be deformed. By a judicious choice of annealing temperatures in the range of 555 to 750 °C the maximum energy product as a function of alloy composition, as shown in Fig. 4.3-57, may be obtained. This annealing treatment is associated with a drastic increase in hardness, as indicated in Table 4.3-36, and a concomitant loss in ductility.

Based on these interrelations of phase equilibria and thermomechanical treatments as well as by optimization through further alloying additions, commercial magnetic materials such as those listed with their properties in Table 4.3-36 have been developed.

4.3.3.3 Fe−Ni−Al−Co, Alnico

The term Alnico refers to two-phase hard magnetic materials based on the Fe−Ni−Al system (Fig. 4.3-58). The intrinsic magnetic properties may be found in [3.6], while extensive magnetic materials data are treated in [3.56]. Table 4.3-37 lists some of the magnetic properties of Alnico type magnets. The magnetically optimized microstructure consists essentially of elongated ferromagnetic Fe-rich precipitates (α_1-phase, bcc disordered) in a non-magnetic matrix of NiAl (α_2-phase, bcc ordered, CsCl type). The remanence B_r is increased significantly by adding Co, which leads to the formation of precipitates rich in Fe−Co. The coercivity H_c is optimized by adding Ti and Cu. The two-phase state is obtained by a homogenization at about 1300 °C, followed by annealing treatments which lead to decomposition into structurally coherent phases on the nanometer scale. The particles are aligned preferentially along the ⟨100⟩ directions of the bcc lattice. This decomposition microstructure is the essential microstructural feature. Higher remanence and coercivity prevails in chill-cast magnets with a columnar microstructure and ⟨100⟩ fiber texture, providing additional magnetocrystalline anisotropy. More extensive treatments and data may be found in [3.10, 56, 57].

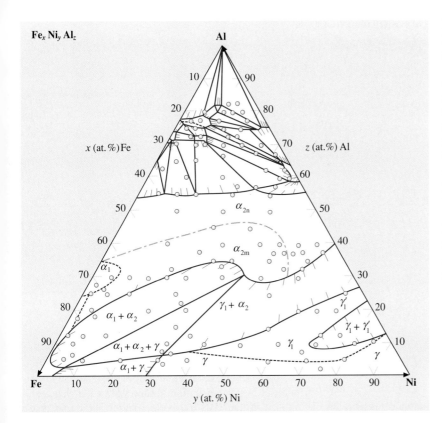

Fig. 4.3-58 Effective $Fe_xNi_yAl_z$ phase diagram after cooling from the melt at 10 K/h. *Broken lines* indicate superlattice phase boundaries; the *point-dash line* the magnetic phase boundary in the Ni(Al,Fe) phase field. α_1: bcc; α'_1: Fe_3Al-type superlattice phase; α_2: (Fe,Ni)Al-type superlattice phase; γ: fcc; γ_1: Ni_3Al-type superlattice phase; γ'_1: as γ_1 but with the larger lattice spacing. The indices m and n indicate magnetic and non-magnetic phases, respectively

Table 4.3-37 Magnetic properties of Alnico type magnets

Designation according to DIN 17410[a]	Remanence B_r		Coercivity $_BH_c$		$_JH_c$		Energy density $(BH)_{max}$		Alloy code[b]
	(mT)	(G)	(kA m^{-1})	(Oe)	(kA m^{-1})	(Oe)	(kJ m^{-3})	(MGOe)	
AlNiCo 9/5i	550	5500	54	679	57	716	10.3	1.3	130
AlNiCo 12/6i	650	6500	57	716	60	754	13.5	1.7	160
AlNiCo 19/11i	640	6400	105	1319	115	1433	22.3	2.8	260
AlNiCo 15/6a	750	7500	60	54	62	57	16.7	2.1	190
AlNiCo 28/6a	1100	11 000	64	804	65	817	31.8	4.0	400
AlNiCo 39/12a	880	8800	115	1445	119	1495	43.8	5.5	450
AlNiCo 37/5a	1240	12 400	51	641	51	641	41.4	5.2	500
AlNiCo 39/15a	740	7400	150	1855	160	2011	43.8	5.5	1800

[a] i = isotropic; a = anisotropic
[b] Commercial designations of Koerzit® by WIDIA Magnettechnik

Table 4.3-38 Intrinsic properties of $Fe_{14}RE_2B$ at $T = 300$ K

RE	K_1 (10^7 erg cm^{-3})	$4\pi M_s$ (T)	$H_a = 2K_1/M_s$ (MA m^{-1})	T_c (K)
Ce	1.7	1.28 (4)	3.0	430 (6)
Pr	4.5	1.59 (8)	6.3	563 (3)
Nd	4.8 (3)	1.68 (9)	5.7	590 (5)
Sm	plane	1.55 (10)		618
Gd	1.0	0.94 (2)	2.3	665 (4)
Tb		0.72 (5)	11.1 [a]	639
Dy	4.1	0.75 (5)	12.6	597 (5)
Ho		0.84 (12)	5.7 [a]	576
Er	plane	1.02 (12)		556 (3)
TM		1.25 (2)		543 (2)
Lu		–	2.1 [a]	538
La		–		543
Y	1.1(1)	1.40(5)	2.2	566 (2)

[a] Data taken from [3.58]. H_a value obtained directly by extrapolation of the magnetization curves for the easy and the hard direction. The average value is given where more than one reference is available; the number in parentheses indicates the standard deviation in the last figure

4.3.3.4 Fe–Nd–B

The most powerful permanent magnets presently available consist essentially of the tetragonal $Fe_{14}Nd_2B$ phase. The intrinsic magnetic properties may be found in [3.6] while extensive magnetic materials treatments and data may be found in [3.1, 10, 56]. Two different production routes are used to prepare dense anisotropic magnets: conventional powder metallurgy and a rapid quenching process to produce flake-shaped powder particles with a nanocrystalline microstructure as a starting material. The flakes are then processed further into dense isotropic or anisotropic magnets by means of a combination of cold pre-forming, hot pressing, and hot deformation steps.

The $Fe_{14}RE_2B$ phase is formed with all rare earth (RE) elements with the exception of Eu. Their intrinsic properties have been investigated extensively. They are listed in Table 4.3-38. Neodymium shows the highest permanent magnet potential based on its combination of high values of K_1 and M_s.

Conventional Powder Metallurgical Processing
Figure 4.3-59 shows the approximate phase relations of Fe–Nd–B at room temperature. According to the phase diagram of the Fe–Nd–B system the $Fe_{14}Nd_2B$ phase forms at 1180 °C. In powder metallurgical processing of the magnets, sintering at 1050 °C leads to the formation of $Fe_{14}Nd_2B$ in equilibrium with a Nd-rich liquid and with the Fe_4NdB_4 boride phase. The liquid phase solidifies below the ternary eutectic at 630 °C. The resulting non-magnetic Nd-rich solid phase spreads along

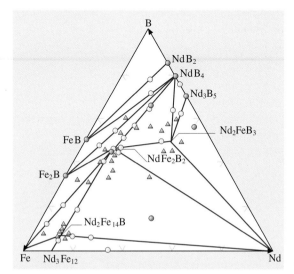

Fig. 4.3-59 Approximate phase relations of Fe–Nd–B at room temperature

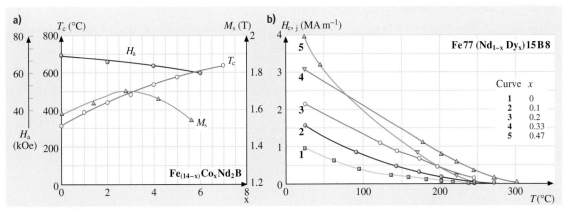

Fig. 4.3-60a,b Influence of substitutional elements on $Fe_{14}Nd_2B$ type magnets. (**a**) Influence of the Co content on the intrinsic properties at room temperature. Data for the magnetization M_s, for the anisotropy field H_a, and for the Curie temperature T_c. (**b**) Influence of the Dy content on the temperature dependence of the coercivity $_JH_c$ of sintered magnets. There is an approximately linear increase of $_JH_c$ at room temperature. The temperature dependence increases with increasing Dy content

the grain boundaries and provides the magnetic decoupling of the $Fe_{14}Nd_2B$ grains, thus providing the basic coercivity of the sintered magnet.

Additions of Dy and Al are increasing the coercivity. Dysprosium enters the RE sites in the $Fe_{14}Nd_2B$ structure, increasing the magnetocrystalline anisotropy but decreasing the magnetic remanence B_r. At compositions of > 2 at.% Al, the anisotropy field H_a decreases linearly at a rate of 0.13 MA m^{-1} per at.% Al. Nevertheless the coercivity increases significantly due to an optimization of the microstructure: Al is enriched in the Nd grain boundary phase which is spreading more uniformly around the magnetic grains, thus leading to better decoupling of exchange interactions. This is a basic condition for the increase in coercivity.

As indicated by Fig. 4.3-60a, Co addition leads to a strong increase of the Curie temperature. However, the anisotropy field H_a is reduced by Co, and the decrease of the coercive field is even larger than expected from this decrease in H_a. On the other hand, there is only a small increase in the magnetic saturation with a maximum at 20 at.% Fe substituted by Co. Accordingly the alloying is limited to 20 at.% Co.

The vulnerability of RE compounds to corrosion is a problem. The corrosion behavior of Fe–Nd–B magnets has been improved by adding elements which influence the electrochemical properties of the Nd-rich grain boundary phase. Additions of small amounts of more noble elements such as Cu, Co, Ga, Nb, and V result in the formation of compounds which replace the highly corrosive Nd-rich phase. Table 4.3-39 lists some of the elements used for manufacturing Fe–Nd–B magnets.

A multitude of grades of Fe–Nd–B magnets is produced by varying chemical composition and processing, such as the press technique applied in order to satisfy the different specifications required for the different fields of application. A maximum remanence is needed, for instance, for disc drive systems in personal computers and for background field magnets in magnetic resonance imaging systems. On the other hand, straight line demagnetization curves up to operating temperatures of 150 °C are specified for application in highly dynamic motors. This requires very high coercive fields at room temperature. Magnetic remanence values of $B_r > 1.4$ T as well as $_JH_c$ values of > 2500 kA/m can be achieved. However, high B_r values are attainable only with lowering the $_JH_c$ value and the operating temperature, and vice versa. Possible combinations of B_r and $_JH_c$ for a given manufacturing process (pressing technique), can be rep-

Table 4.3-39 Elements used for manufacturing Fe–Nd–B magnets

Element	Fe	Nd	B	Dy	Co	Al	Ga	Nb, V
wt%	balance	15–33	0.8–1	0–15	0–15	0.5–2	0–2	0–4

Table 4.3-40 Physical properties of sintered Nd−Fe−B magnets [a]

Density	Curie temperature	Electrical resistivity	Specific heat	Thermal conductivity	Thermal expansion coefficient		Young's modulus	Flexural strength	Compression strength	Vickers Hardness
					∥-c axis	⊥-c axis				
(g cm^{-3})	(K)	(Ω mm^2 m^{-1})	(J kg^{-1} K^{-1})	(W m^{-1} K^{-1})	(10^{-6} K^{-1})	(10^{-6} K^{-1})	(kN mm^{-2})	(N mm^{-2})	(N mm^{-2})	
7.5 (0.05)	580–605	1.50 (0.10)	430 (10)	9	4.4 (0.6)	−1	155 (5)	260 (10)	930 (170)	580 (10)

[a] All values are for 300 K. The values are the averages taken from the companies brochures of: VAC Vacuumschmelze, Hanau, Germany; MS Magnetfabrik Schramberg, Schramberg, Germany; Ugimag Inc, Valparaiso, USA; Neorem Magnets Oy, Ulvila, Finland; TDK Corporation, Tokyo, Japan; Hitachi Metals Ltd, Tokyo, Japan. The numbers in parentheses indicate the maximum deviation

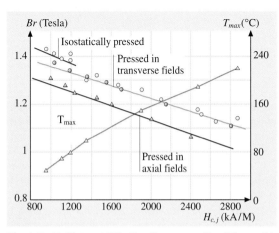

Fig. 4.3-61 Sintered Nd−Fe−B magnets. Possible combinations of B_r and $_JH_c$

resented by a straight line, as shown in Fig. 4.3-61. The physical properties of sintered Nd−Fe−B magnets are given in Table 4.3-40.

Magnets Processed by the Rapid Quenching/Hot Working Technique

This alternative technology uses isotropic or amorphous material processed by a rapid quench technique involving melt spinning the molten alloy through a nozzle on to a rotating wheel. Flakes that are about 30 μm thick are obtained. Their microstructure shows typically an average grain size of 50 to 60 nm. The material is subjected to controlled deformation at elevated temperatures, which gives rise to a texture oriented perpendicular to the direction of mass flow during deformation. Hot working processes applied are the die-upset method and indirect extrusion. The major steps of the process are: alloy preparation, melt spinning, cold forming, hot working, coating, and magnetizing. The first step of the hot working procedure is pressing to 100% density at temperatures between 700 and 800 °C. Isotropic magnets are obtained. The isotropic parts are deformed at about 800 °C by using the die-upset technique. Constant strain rates have to be applied. Typically, the strain rate is about 0.1 sec^{-1}. The strain rate and the degree of deformation determine the alignment factor of the final magnet. A maximum value for the magnetic remanence B_r of about 1.35 T may be achieved under economically reasonable conditions. The relationship between the degree of deformation and B_r is shown in Fig. 4.3-62. Properties of commercially availabe magnets are included in Table 4.3-41. The application of the magnets is limited to special fields where complicated shapes would otherwise require expensive machining

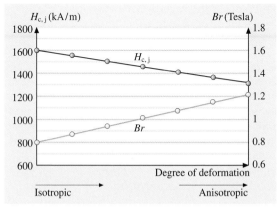

Fig. 4.3-62 Dependence of remanence B_r and coercivity $_JH_r$ of dense magnets, manufactured by the hot pressing/hot working technique, on the degree of deformation. Data provided and authorized for publication by Magnequench Int., Tübingen

Table 4.3-41 Magnetic properties of $Fe_{14}Nd_2B$-based magnetic materials at room temperature. Typical values of commercially available magnets

Remanence	Coercivity		Energy density	Temperature coefficients		Max. operation temperature[a]	Product code[b]
(T)	(kA m^{-1})		(kJ m^{-3})	(% K^{-1})		(°C)	
B_r	$_BH_c$	$_JH_c$	$(BH)_{max}$	$TC(B_r)$	$TC(_JH_c)$	T_{max}	
Sinter Route							
1.47	915	955	415	−0.115	−0.77	50	VD722HR
1.44	1115	1195	400	−0.115	−0.73	70	VD745HR
1.35	1040	1430	350	−0.095	−0.65	110	VD633HR
1.30	980	1035	325	−0.115	−0.80	70	VD335HR
1.18	915	2465	270	−0.085	−0.55	190	VD677HR
1.43	915	9550	395	−0.115	−0.77	50	VD722TP
1.41	1090	1195	385	−0.115	−0.73	70	VD745TP
1.32	1020	1430	335	−0.095	−0.65	110	VD633TP
1.25	965	1195	300	−0.115	−0.75	70	VD335TP
1.14	885	2865	250	−0.080	−0.51	220	VD688TP
1.32	965	1115	335	−0.115	−0.73	80	VD510AP
1.26	965	1510	305	−0.095	−0.64	120	VD633AP
1.22	900	1195	285	−0.115	−0.75	80	VD335AP
1.08	830	2865	225	−0.080	−0.51	230	VD688AP
Hot Working							
0.83	575	1400	120	−0.10	−0.50	180	MQ2-E15
1.28	907	995	302	−0.10	−0.60	125	MQ3-E38
1.25	915	1313	287	−0.09	−0.60	150	MQ3-F36
1.31	979	1274	334	−0.09	−0.60	150	MQ3-F42
1.16	876	1592	255	−0.09	−0.06	200	MQ3-G32SH

[a] Maximum operating temperature is defined by a straight demagnetization line up to an operation point of the magnet of $B/\mu_0 H = -2$
[b] VD = Vacodym, trade name of Vacuumschmelze GmbH for NdFeB based magnets; HR grades = isostatically pressed, TP grades = transverse field pressed, AP grades = axial field pressed. MQ2, MQ3 = trade names of Magnequench International Inc. for hot pressed - hot deformed magnets

work, as for instance screwed arcs for non-cogging motors.

Typical technical alloys and their properties are listed in Table 4.3-41. Characteristic demagnetization curves are shown in Fig. 4.3-63.

4.3.3.5 Co–Sm

Numerous magnetic, binary, rare earth (RE) transition metal (TM) compounds exist, of which the Co_5RE and $Co_{17}RE_2$ phases form the basis for materials with excellent permanent magnetic properties. They combine high saturation magnetization M_s with high crystal anisotropy K_1 and high Curie temperature T_c. The intrinsic magnetic properties of the Co_5RE and $Co_{17}RE_2$ phases are summarized in Tables 4.3-42 and 4.3-43, respectively (Co_5RE and $Co_{17}RE_2$ are, also, referred to as 5/1 and 17/2 phases). It is obvious that Co_5Sm and $Co_{17}Sm_2$ have the best potential for manufacturing permanent magnetic materials.

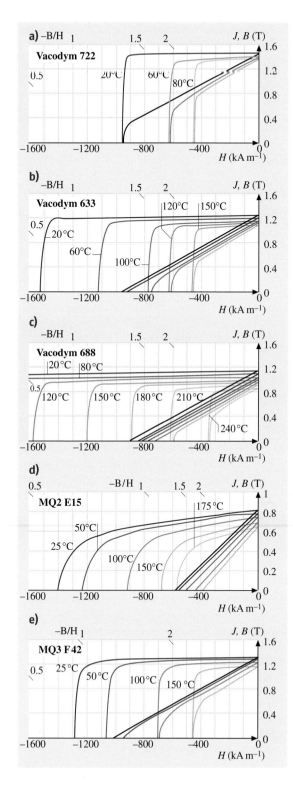

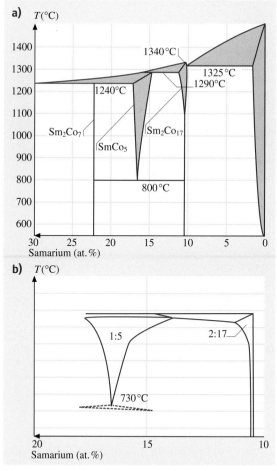

Fig. 4.3-63a–e $Fe_{14}Nd_2B$-based commercial magnets [3.10]. (**a**) Top grade magnet with highest remanence B_r. Isostatically pressed, designed to meet exceptional requirements for maximum energy density at operating temperatures up to 60 °C. (**b**) Magnet, axially pressed, with an optimum combination of high coercivity and energy product (**c**) Magnet, axially pressed, with a very high coercivity; exceptionally well suited for use in highly dynamic servo motor applications. (**d**) Isotropic dense magnet made from rapidly quenched powder by means of the hot pressing technique. (**e**) Magnet made from rapidly quenched powder by means of the hot pressing and hot deforming technique; especially well suited for complicated shaped magnets, such as scewed arcs in special motor applications

Fig. 4.3-64 (**a**) Part of the Co–Sm phase diagram. (**b**) Section of the Co–Sm–Cu phase diagram at 10 at.% Cu [3.10]

Table 4.3-42 Co$_5$RE alloys. Room temperature intrinsic magnetic properties, the crystal anisotropy constant K_1, the magnetic saturation M_s, the anisotropy field $H_a = 2K_1/M_s$, and the Curie temperature, T_c [3.10]

RE[a]	K_1^b (10^7 erg cm^{-3})	$4\pi M_s^b$ (T)	H_a (MA m^{-1})	T_c^b (K)
Ce	5.3 (6)	0.83	13	660 (20)
Pr	8.1 (9)	1.29	12.5	900 (10)
Nd	0.7 (5)	1.34	1.1	910 (2)
Sm	17.2 (9)	1.12 (3)	31	990 (10)
Gd	4.6 (5)	0.35 (3)	26	1035
Tb	NC			973
Dy	NC			961
Ho	3.6 (3)	0.52	13.8	996
Er	4.2 (4)	0.4	14.6	983
La	5.9 (5)	0.91	10.6	834
Y	5.2 (4)	1.11	9.3	980

[a] No data are available for the RE elements Tm, Yb, and Lu
[b] The mean value is given where more than one reference is available; the number in parentheses indicates the standard deviation in the last figure. NC: non-collinear spin structure

Table 4.3-43 Room temperature intrinsic magnetic properties, of Co$_{17}$RE$_2$ alloys [3.10]

RE[a]	K_1^b (10^7 erg cm^{-3})	$4\pi M_s^b$ (T)	H_a (MA m^{-1})	T_c^b (K)
Ce	−0.6	1.15		1068 (10)
Pr	−0.6	1.38		1160 (10)
Nd	−1.1	1.39		1160 (3)
Sm	3.3 (1)	1.25 (3)	5.3 (1)	1196 (1)
Gd	−0.5	0.73		1212 (6)
Tb	−3.3 (6)	0.67		1189 (4)
Dy	−2.6 (5)	0.69		1173 (8)
Ho	−1.0 (2)	0.85		1177 (1)
Er	0.41 (3)	1.05 (5)	0.75	1175 (15)
TM	0.50 (5)	1.21	0.83	1179 (2)
Yb	−0.38	1.35		1180
Lu	−0.20 (5)	1.41		1202 (6)
Y	−0.34 (2)	1.27		1199 (13)

[a] La does not form the 17/2 phase. The 17/2 phases with Sm, Er, and Tm show uniaxial anisotropy while all others have easy plane anisotropy
[b] The mean value is given where more than one reference is available; the number in parentheses indicates the standard deviation in the last figure

Phase Equilibria of Co−RE Systems

The phase diagrams of Co−RE systems are very similar. Figure 4.3-64 shows the magnetically relevant part of the Co−Sm phase diagram. The features to note are: Co and Co$_{17}$Sm$_2$ form a eutectic, while Co$_5$Sm forms as a result of a peritectic reaction between Co$_{17}$Sm$_2$ and liquid. Both Co$_5$Sm and Co$_{17}$Sm$_2$ show a significant range of homogeneity at elevated temperatures. The compound Co$_5$Sm is unstable at room temperature and decomposes via an eutectoid reaction into Co$_7$Sm$_2$ and Co$_{17}$Sm$_2$.

Iron and Cu are two important substitutional elements for Co−RE alloys with respect to manufacturing permanent magnets. In Co$_5$RE compounds, Fe may substitute up to 5 at.% Co while complete solubility occurs in (Co$_{1-x}$Fe$_x$)$_{17}$RE$_2$. Copper is essential in Co$_{17}$Sm$_2$ type alloys. The different solubility of Cu in 5/1 and 17/2 is used to form precipitates in the 17/2 phase, resulting in a microstructure which provides high coercivity.

Co–Sm-Based Permanent Magnets

Powder Metallurgical Processing. Cobalt–Samarium magnets are produced using powder metallurgical techniques. Alloys are prepared by an inductive melting process or a Ca reduction process. The starting alloys are crushed and pulverized to single crystalline particles 3–4 μm in diameter. The powders are compacted in a magnetic field to obtain anisotropic magnets: by uniaxial compaction in magnetic fields, either parallel or transverse to the direction of the applied force; or by isostatic compaction of powders in elastic bags after subjecting the filled bags to a pulsed field. The magnetically aligned green compacts are sintered in an inert atmosphere to achieve an optimum combination of high density and high coercive field.

5/1 Type Magnets. Binary Co$_5$Sm is the basis of 5/1 type magnets. Table 4.3-44 lists some of the magnetic properties at room temperature and Table 4.3-45 lists the physical properties of sintered Co$_5$Sm magnets. Partial substitution of Sm by Pr increases B_s while still yielding sufficiently high $_JH_c$. The microstructure of Co$_5$Sm magnets consists of single domain grains. Magnetization reversal starts by nucleation of domains in a demagnetizing field. The domain wall moves easily through the particle.

The application of permanent magnets in measuring devices or in devices in aircraft or space systems requires a small temperature coefficient (TC) of B_r. The combination of Co$_5$Sm which has a negative TC, with Co$_5$Gd, which has a positive TC, yields magnets with reduced temperature dependence of B_r, reaching about

Table 4.3-44 Co$_5$Sm-based magnetic materials. Magnetic properties at room temperature, typical values [3.10]

B_r (T)	$_BH_c$ (kA m^{-1})	$_JH_c$ (kA m^{-1})	$(BH)_{max}$ (kJ m^{-3})	Press mode[a]	Material	Producer code[b]
1.01	755	1500	200	Iso	Co$_5$Sm	Vacomax 200
0.95	720	1800	180	TR	Co$_5$Sm	Vacomax 170
1.0	775	2400	200	Iso	Co$_5$Sm	Recoma 25
0.94	730	2400	175	TR	Co$_5$Sm	Recoma 22
0.9	700	2400	160	A	Co$_5$Sm	Recoma 20
0.73	570	>2400	105	A	Co$_5$Sm$_{0.8}$Gd$_{0.2}$	EEC 1.5TC-13
0.61	480	>2400	70	A	Co$_5$Sm$_{06}$Gd$_{0.4}$	EEC 1.5TC-9

[a] Iso: isostatically pressed; TR: uniaxially pressed in transverse oriented aligning fields; A: uniaxially pressed in axially oriented aligning fields
[b] Vacomax: Trademark of Vacuumschmelze GmbH, Germany; Recoma: Trademark of Ugimag AG, Switzerland; EEC: Trademark of EEC Electron Energy Corporation, USA

Table 4.3-45 Physical properties of sintered Co$_5$Sm magnets[a] [3.10]

Density	Curie temperature	Electrical resistivity	Specific heat	Thermal conductivity	Thermal expansion coefficient		Young's modulus	Flexural strength	Compression strength	Vickers Hardness
					∥ c axis	⊥ c axis				
(g cm^{-3})	(K)	(Ω mm^2 m^{-1})	(J kg^{-1} K^{-1})	(W m^{-1} K^{-1})	(10^{-6} K^{-1})	(10^{-6} K^{-1})	(kN mm^{-2})	(N mm^{-2})	(N mm^{-2})	
8.40 (0.10)	990 (10)	0.53 (0.03)	372 (3)	11.5 (1.0)	6.0 (1.5)	12.5 (0.5)	150 (40)	125 (35)	900 (300)	580 (50)

[a] All values for 300 K. Average values are taken from the companies brochures of: VAC Vacuumschmelze, Hanau, Germany; MS Magnetfabrik Schramberg, Schramberg, Germany; Ugimag AG, Lupfig, Switzerland; EEC Electron Energy Corporation, Landisville, USA; TDK Corporation, Tokyo Japan; Hitachi Metals Ltd, Tokyo, Japan. The numbers in parentheses indicate the standard deviation

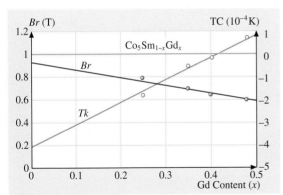

Fig. 4.3-65 Temperature coefficient TC of $Co_5Sm_{1-x}Gd_x$ magnets in the temperature range between 20 and 200 °C. The coercivity of these ternary magnets is comparable to that of binary Co_5Sm magnets due to the high anisotropy field of Co_5Gd $H_a = 24\,\text{MA}\,\text{m}^{-1}$ while for Co_5Sm, $H_a = 31\,\text{MA}\,\text{m}^{-1}$ [3.10]

zero for $Co_5Sm_{0.6}Gd_{0.4}$ between room temperature and 200 °C (Fig. 4.3-65).

Tables 4.3-44 and 4.3-45 list the magnetic properties of typical Co_5Sm based magnets and their physical properties, respectively. Figures 4.3-66a,b show characteristic demagnetisation curves.

17/2 Type Magnets. The permanent magnetic potential of the binary phase $Co_{17}Sm_2$ is increased by partially substituting Co by other transition metals. The general chemical composition of commercial magnets corresponds to $(Co_{bal}Fe_vCu_yZr_x)_zSm$. Copper is the essential addition. Its solubility in the 17/2 phase is strongly temperature-dependent, and this is used for precipitation hardening of 17/2 magnets.

The influence of Fe on the intrinsic magnetic properties is summarized in Fig. 4.3-67. Both Zr and Hf increase coercivity. Figures 4.3-67a,b show the combined effects of Fe, Cu, and Zr additions and the Co/Sm ratio on the temperature dependence of coercivity $_JH_c$. The magnets are sintered between 1200 and 1220 °C. A single phase with $Zn_{17}Th_2$ structure is obtained by homogenization at temperatures between 1160 and 1190 °C. After rapid cooling the magnets are finally annealed between 800 and 850 °C, followed by cooling to 400 °C. The microstructure leading to high coercivity $_JH_c$ consists of 17/2 matrix grains, a 5/1 boundary phase enriched in Cu, and platelet-shaped precipitates enriched in Fe and Zr. The coercivity of these magnets

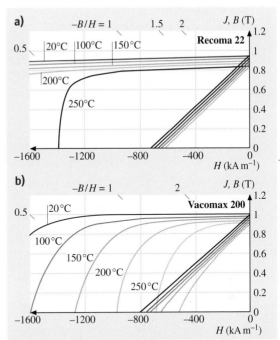

Fig. 4.3-66a,b Co_5Sm magnets. Demagnetization curves of typical commercially available magnets [3.10]. **(a)** Magnet with high intrinsic coercivity $_JH_c$ up to 300 °C, uniaxially pressed in a transverse aligning field; Recoma: trademark of Ugimag AG, Switzerland. **(b)** Magnet with highest energy density obtained by cutting from isostatically pressed block; Vacomax: trademark of Vacuumschmelze GmbH, Germany

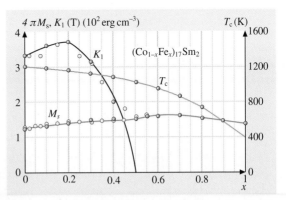

Fig. 4.3-67 $(Co_{1-x}Fe_x)_{17}Sm_2$. Dependence K_1, $4\pi M_s$, and T_c on the Fe content x [3.10]

is based on pinning of the Bloch walls at the 5/1 grain boundary phase.

Table 4.3-46 Chemical composition of commercial high energy 17/2 magnets [3.10]

Element	Sm[a]	Co	Cu	Fe	Zr
wt%	25–27	balance	4.5–8	14–20	1.5–3

[a] The Sm content includes the fraction of Sm which is present as Sm_2O_3, typically 2.5 wt%

Table 4.3-47 Annealing treatments for 17/2 type magnets [3.10]

	Isothermal ageing	Cooling rate to 400 °C	$_JH_c$ (kA m^{-1})	Magnetizing field H_m (kA m^{-1})
high $_JH_c$	850 °C/10 hours	1 K/min	> 2000	5000
low $_JH_c$	800 °C/30 min.	5 K/min	500–800	1500

Table 4.3-48 High energy $TM_{17}Sm_2$ magnets for applications up to 300 °C, typical values [3.10]

B_r (T)	$_BH_c$ (kA m^{-1})	$_JH_c$ (kA m^{-1})	$(BH)_{max}$ (kJ m^{-3})	Press mode[a]	Material	Producer code[b]
1.14	670	800	247	Iso	$TM_{17}Sm_2$	Hicorex -30CH
1.12	730	800	240	Iso	$TM_{17}Sm_2$	Vacomax 240HR
1.05	720	800	210	A	$TM_{17}Sm_2$	Vacomax 240
1.10	820	2070	225	Iso	$TM_{17}Sm_2$	Vacomax 225HR
1.07	800	2000	215	TR	$TM_{17}Sm_2$	Recoma 28
1.04	760	2070	205	A	$TM_{17}Sm_2$	Vacomax 225
0.9	654	> 2000	150	A	$TM_{17}(Sm_{0.8}, Gd_{0.2})_2$	EEC2:17TC-18
0.8	575	> 2000	115	A	$TM_{17}(Sm_{0.6}, Gd_{0.4})_2$	EEC2:17TC-15

[a] Iso: isostatically pressed; TR: uniaxially pressed in transverse oriented aligning fields; A: uniaxially pressed in parallel oriented aligning fields

[b] Hicorex: Trademark of Hitachi Metals Ltd., Japan; Vacomax: Trademark of Vacuumschmelze GmbH, Germany; Recoma: Trademark of Ugimag AG, Switzerland; EEC: Trademark of EEC Electron Energy Corporation, USA

Table 4.3-49 Composition of 17/2 magnets for operating up to 550 °C [3.10]

Element	Sm	Co	Cu	Fe	Zr
wt%	26.5–29	balance	6–12	7	2.5–6

Table 4.3-50 Physical properties of sintered $TM_{17}Sm_2$ magnets[a] [3.10]

Density	Curie temperature	Electrical resistivity	Specific heat	Thermal conductivity	Thermal expansion coefficient		Young's modulus	Flexural strength	Compression strength	Vickers hardness
					∥-c axis	⊥-c axis				
(g cm^{-3})	(K)	(Ω mm^2 m^{-1})	(J kg^{-1} K^{-1})	(W m^{-1} K^{-1})	(10^{-6} K^{-1})	(10^{-6} K^{-1})	(kN mm^{-2})	(kN mm^{-2})	(kN mm^{-2})	
8.38 (0.08)	1079 (8)	0.86 (0.05)	355 (30)	10.5 (1.5)	8.4 (1)	11.5 (0.5)	160 (40)	130 (15)	760 (150)	600 (50)

[a] All values are for 300 K. The values are the average of values taken from company brochures

High Energy Magnets for Applications at Temperatures up to 300 °C. The compositional parameters v, y, x, and z of the 17/2 type magnets are normally optimized for highest $(BH)_{max}$ between room temperature and 300 °C. The range of chemical compositions of these high energy magnets is given in Table 4.3-46.

Tables 4.3-48 and 4.3-50 list the magnetic properties of typical $Co_{17}Sm_2$ based magnets and their physical properties, respectively. Figures 4.3-69a–c show characteristic demagnetization curves.

Two modifications of 17/2 magnets can be obtained from identical chemical compositions according to the annealing treatment applied (see Table 4.3-47). 17/2 magnets with reduced TC of B_r are obtained by substituting part of Sm by Gd. Table 4.3-45 shows some typical values of the magnetic properties of high energy $TM_{17}Sm_2$ magnets.

Magnets for High Temperature Applications. Special applications in aircraft and spacecraft require magnets with linear demagnetization curves up to 550 °C. Suitable high temperature magnets are obtained by increasing Sm and Cu and decreasing Fe. Data are given in Table 4.3-49 and Fig. 4.3-69a.

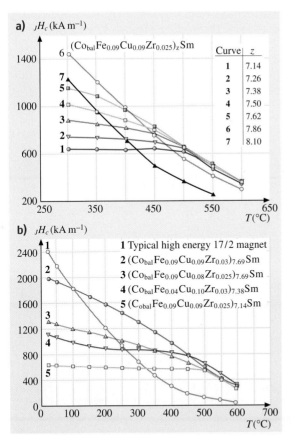

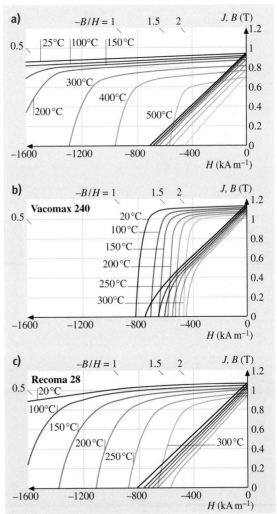

Fig. 4.3-68a,b Temperature dependence of $_JH_c$ for $(Co_{bal}Fe_vCu_yZr_x)_z$ as varied by the concentration of the substitutional elements and of Sm [3.10] (**a**) Dependence on the TM/Sm ratio; (**b**) Dependence on other variations in chemical composition

Fig. 4.3-69a–c $TM_{17}Sm_2$ magnets. Demagnetization curves [3.10] (**a**) Magnet may be used up to 500 °C. Coating is needed for protection against oxidation > 300 °C. (**b**) Low coercivity, easy-to-magnetize magnet. (**c**) High energy, high coercivity magnet

4.3.3.6 Mn–Al–C

In the binary Mn–Al system the ferromagnetic, metastable τ-Mn–Al phase containing 55 wt% Mn is obained by annealing at 923 K, or by controlled cooling of the stable ε phase from ≥ 1073 K. The τ phase has a high magnetocrystalline anisotropy $K_1 \cong 10^6$ J m^{-3} and $I_s = 0.6$ T. The magnetic easy axis is the c axis. The τ phase can be stabilized by adding 0.5 wt% C. Hot deformation at about 973 K permits formation of

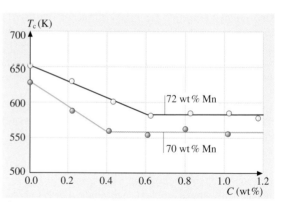

Fig. 4.3-71 Dependence of Curie temperature of τ-Mn–Al–C on the carbon concentration [3.10]

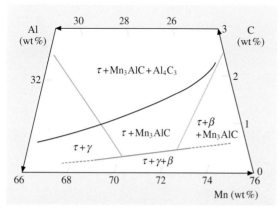

Fig. 4.3-70 Schematic representation of the phase relations of Mn–Al–C at 873 K [3.10]

anisotropic magnets. Figure 4.3-70 shows schematically the 873 K isotherm of the Mn–Al–C phase diagram.

Carbides coexist with the metallic ferromagnetic phases. This is unfavorable for the mechanical properties. The Curie temperature of the τ phase decreases with increasing C content to a constant value (see Fig. 4.3-71) according to the phase diagram.

An alternative process to form τ-Mn–Al–C magnets is to use gas-atomized powder which is canned

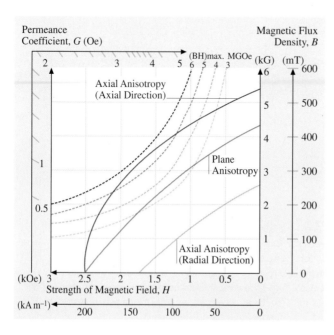

Fig. 4.3-72 Magnetic properties of an extruded Mn–Al–C magnet [3.10]

Table 4.3-51 Properties of a Mn−Al−C magnet [3.10]

		Axial Anisotropy		Plane anisotropy
		Axial direction	Radial direction	
Maximum energy product $(BH)_{max}$	(kJ m^{-3})	44	10	28
Residual magnetic flux density B_r	(mT)	550	270	440
Coercivity H_c	(kA m^{-1})	200	144	200
Optimum permeance coefficient	(µH m^{-1})	1.9	1.9	1.9
Average reversal permeability	(µH m^{-1})	1.4 ∼ 1.6	1.4 ∼ 1.6	1.4 ∼ 1.6
Temperature coefficient of B_r	(% K^{-1})	−0.11	−0.11	−0.11
Curie point T_c	(K)	573	573	573
Maximum operating temperature	(K)	773	773	773
Density	(kg m^{-3})	5100	5100	5100
Hardness	(HR$_c$)	49 ∼ 56	49 ∼ 56	49 ∼ 56
Tensile strength R_m	(N m^{-2})	> 290 × 10^6	> 290 × 10^6	> 290 × 10^6
Compression strength	(N m^{-2})	> 2000 × 10^6	> 2000 × 10^6	> 2000 × 10^6

and extruded at 973 K. By atomizing, the powder is rapidly quenched and consists of the high temperature ε phase. During hot extrusion it transforms to the axially anisotropic τ phase. Repeated die pressing results in plane anisotropy. The magnetic properties are shown in Fig. 4.3-72 and listed in Table 4.3-51.

4.3.4 Magnetic Oxides

The magnetic properties of oxides and related compounds have been tabulated in comprehensive data collections [3.4]. A review of the basic magnetic properties of garnets {A$_3$}[B$_2$](Si$_3$)O$_{12}$ and spinel ferrites MeOFe$_2$O$_3$ or Me^{II}Fe$_2^{III}$O$_4$ is given by *Guillot* in [3.3].

4.3.4.1 Soft Magnetic Ferrites

Both MnZn and NiZn ferrites are the common designations of the two main groups of soft magnetic oxide materials. A more extensive account is given in [3.10]. The chemical formula of MnZn and NiZn ferrites is M^{2+}Fe$_2$O$_4$ and they have spinel structures. The divalent ions (M^{2+}) are elements such as Mn, Fe, Co, Ni, Cu, Mg, Zn, and Cd. They are located at tetrahedral or octahedral sites of the spinel structures. Ceramic processing methods are applied to produce the magnetic parts such as ring-shaped cores for inductive components.

The use of different types of MnZn and NiZn ferrite materials varies with the operating frequency in the application concerned. Figure 4.3-73 and Table 4.3-52 indicate the typical ranges of use and the magnetic properties of characteristic materials. Typically, MnZn ferrites are used in the range of several MHz. Table 4.3-53 lists the major applications with the pertinent operating frequencies. Since a phase shift occurs in an ac-excited magnetic field, the permeability μ is expressed as a complex number, $\mu' - i\mu''$. The imaginary

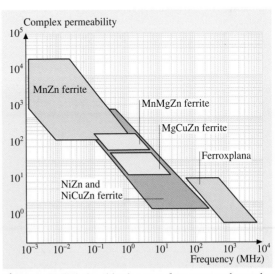

Fig. 4.3-73 Relationships between frequency and complex permeability of groups of characteristic soft magnetic ferrites [3.10]

Table 4.3-52 Typical magnetic properties of characteristic materials [3.10]

Type	MnZn ferrite					NiCuZn ferrite		
Designation	PC40	PC44	PC45	PC50	H5C4	L7H	HF70	L6
μ_i	2300	2400	2500	1400	12 000	800	1300	1500
B_s (mT) at 25 °C	510	510	530	470	380	390	270	280
H_c (A m^{-1}) at 25 °C	14	13	12	37	4	16	16	16
T_c (°C)	>215	>215	>230	>240	>110	>180	>110	>110
ρ_v (m)	6.5					10^6	10^6	10^6
Pcv (kW m^{-3}) At 100 kHz, 200 mT	410 (100 °C)	300 (100 °C)	250 (75 °C)	80^a (100 °C)				

^{a} At 500 kHz, 50 mT

Table 4.3-53 Ferrite applications [3.10]

Application	Frequency	Ferrite material
Communication coils	1 kHz ~ 1 MHz	MnZn
	0.5 ~ 80 MHz	NiZn
Pulse transformers		MnZn, NiZn
Transformers	~ 300 kHz	MnZn
Flyback transformers	15.75 kHz	MnZn
Deflection yoke cores	15.75 kHz	MnZn, MnMgZn, NiZn
Antennas	0.4 ~ 50 MHz	NiZn
Intermediate frequency transformers	0.4 ~ 200 MHz	NiZn
Magnetic heads	1 kHz ~ 10 MHz	MnZn
Isolators		MnMgAl
Circulators	30 MHz ~ 30 GHz	YIG
Splitters		YIG
Temperature responsive switches		MnCuZn

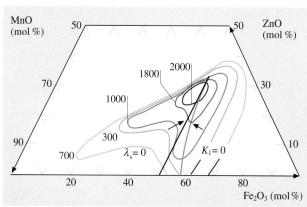

Fig. 4.3-74 Relationship between composition and complex permeability in MnZn ferrites [3.10]

part, μ'', is related to magnetic losses and is high at the resonance frequency. For a specific composition range, the values of magnetic anisotropy and magnetostriction are reduced to near zero leading to a maximum in complex permeability, as shown in Fig. 4.3-74. By suitable control of the microstructure it is possible to favor the formation and mobility of fast-moving domain walls. Consequently, the permeability of MnZn ferrite is the highest among the ferrites with spinel structure. However, the increase of the loss factor is considerably enhanced with increasing frequency because of the lower resistivity of MnZn ferrites compared to other ferrite materials.

The properties of NiZn and NiCuZn ferrites are designed for applications in the radio wave band. Figure 4.3-75 shows the effect of the Zn content on

4.3.4.2 Hard Magnetic Ferrites

A hard magnetic ferrite, also called ferrite magnet, is a magnetic material based on iron oxide. The composition of the typical hard magnetic ferrite compounds is shown in Table 4.3-54. M type material is used most widely and BaO can be replaced by SrO.

Figure 4.3-76 shows the quasi-binary phase diagram $BaO–Fe_2O_3$. The composition of the actual industrial hard magnetic ferrite is selected to deviate slightly from the stoichiometric composition in order to provide easy wetting and permit liquid-phase sintering. The ferromagnetic phase $SrO·6Fe_2O_3$ has a higher magnetocrystalline anisotropy constant K_1 and, thus, a higher intrinsic coercive force $_JH_c$ than the BaO-based compound.

Table 4.3-55 shows the basic magnetic properties of hard magnetic ferrites. The properties of actual products are shown in Fig. 4.3-77. The demagnetization curves of type YBM-9B, which have the best magnetic properties, are shown in Fig. 4.3-78.

Comparing hard ferrites and rare earth magnets, the ratio of remanence B_r is about $1:3$, that of the coercivity $_JH_c$ is also about $1:3$, such that the ratio of the energy product $(BH)_{max}$ is about $1:10$. From the cost/performance point of view, the rare earth magnets are used where weight and size are essential.

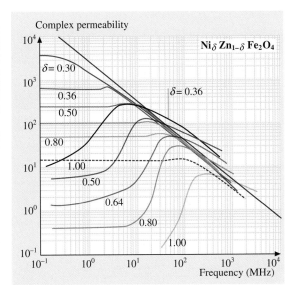

Fig. 4.3-75 Frequency dependence of the complex permeability in NiZn ferrites. *Solid* and *dashed lines* refer to the real and the imaginary part, respectively [3.10]

the complex permeability spectra of NiZn ferrites. In general, the high-permeability materials cause lower resonance frequencies.

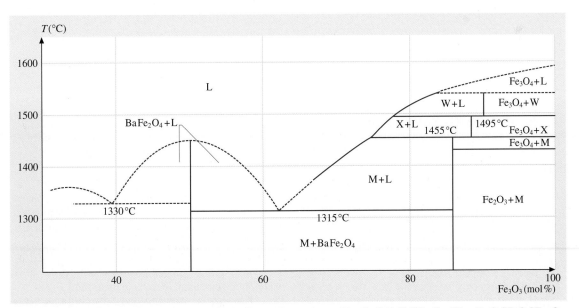

Fig. 4.3-76 Quasi-binary phase diagram $BaO–Fe_2O_3$. $p(O_2) = 1$ atm; $X = BaO·FeO·7Fe_2O_3$; $W = BaO·2FeO·8Fe_2O_3$. van Hook 1964 [3.10, 59]

Table 4.3-54 BaO–MeO–Fe$_2$O$_3$ hexagonal magnetic compounds and their technical designations as hard ferrite materials 3.10

Designation symbol	Molecular formula	Chemical composition (mol%)		
		MeO	BaO	Fe$_2$O$_3$
M	BaO·6Fe$_2$O$_3$	–	14.29	85.71
W	2MeO·BaO·8Fe$_2$O$_3$	18.18	9.09	72.71
Y	2MeO·2BaO·6Fe$_2$O$_3$	20	20	60
Z	2MeO·3BaO·12Fe$_2$O$_3$	11.76	17.65	70.59

Table 4.3-55 Magnetic properties of hard ferrites 3.10

Composition	σ_s (10^{-4} Wb m kg^{-1})	(emu g^{-1})	σ_0 (10^{-4} Wb m kg^{-1})	(emu g^{-1})	J_s (wb m^{-3})	T_c (K)	K_1 (10^3 J m^{-2})	H_a (kA m^{-1})	D_c (μm)
BaFe$_{17}$O$_{19}$	0.89	71	1.257	100	47.8	723	3.2	1.350	0.90
SrFe$_{17}$O$_{19}$	0.905	72	1.357	108	47.8	735	3.5	1.590	0.94

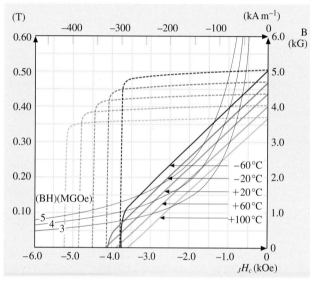

Fig. 4.3-78 Demagnetization behavior and its temperature dependence of a hard ferrite YBM9B (Hitachi Metals) 3.10

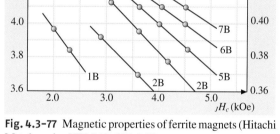

Fig. 4.3-77 Magnetic properties of ferrite magnets (Hitachi Metals series YBM 1B-9B) 3.10

References

3.1 K. H. J. Buschow: *Electronic and Magnetic Properties of Metals and Ceramics, Part I.* In: *Materials Science and Technology*, Vol. 3, ed. by R. W. Cahn, P. Haasen, E. J. Kramer (VCH, Weinheim 1991)

3.2 H. P. J. Wijn (Ed.): *Magnetic Properties of Metals*, Landolt–Börnstein, New Series III/19, 32 (Springer, Berlin, Heidelberg 1986–2001)

3.3 K. H. J. Buschow: *Electronic and Magnetic Properties of Metals and Ceramics, Part II.* In: *Materials Science and Technology*, Vol. 3, ed. by R. W. Cahn, P. Haasen, E. J. Kramer (VCH, Weinheim 1994)

3.4 K.-H. Hellwege, A. M. Hellwege (Eds.): *Magnetic and Other Properties of Oxides and Related Compounds*, Landolt–Börnstein, New Series III/4, 12 (Springer, Berlin, Heidelberg 1970–1982)

3.5 H. P. J. Wijn (Ed.): *Magnetic Properties of Non-Metallic Inorganic Compounds Based on Transition Metals*, Landolt–Börnstein, New Series III/27 (Springer, Berlin, Heidelberg 1988–2001)

3.6 K. Adachi, D. Bonnenberg, J. J. M. Franse, R. Gersdorf, K. A. Hempel, K. Kanematsu, S. Misawa, M. Shiga, M. B. Stearns, H. P. J. Wijn: *Magnetic*

3.7 G. Bertotti, A. R. Ferchmin, E. Fiorello, K. Fukamichi, S. Kobe, S. Roth: *Magnetic Alloys for Technical Applications. Soft Magnetic Alloys, Invar, Elinvar Alloys*, Landolt–Börnstein, New Series III/19, ed. by H. P. J. Wijn (Springer, Berlin, Heidelberg 1994)

Properties of Metals: 3d, 4d, 5d Elements, Alloys, Compounds, Landolt–Börnstein, New Series III/19 (Springer, Berlin, Heidelberg 1986)

3.8 R. Boll: *Soft Magnetic Materials* (Heyden & Son, London 1979)
3.9 J. Evetts: Concise Encyclopedia of Magnetic & Superconducting Materials. In: *Advances in Materials Science, Engineering* (Pergamon Press, Oxford 1992)
3.10 P. Beiss, R. Ruthardt, H. Warlimont (Eds.): *Advanced Materials, Technologies: Materials: Powder Metallurgy Data: Metals, Magnets*, Landolt–Börnstein, New Series VIII/2 (Springer, Berlin, Heidelberg 2003)
3.11 K. Günther: SMM Conference 2003
3.12 R. Boll: Soft Magnetic Metals and Alloys. In: *Materials Science and Technology*, Vol. 3B, ed. by K. H. J. Buschow (VCH, Weinheim 1994) pp. 399–450
3.13 Bölling and Hastenrath
3.14 G. Rassmann, U. Hofmann: Magnetismus, Struktur und Eigenschaften magnetischer Festkörper. In: (VEB Deutscher Verlag für Grundstoffindustrie, Leipzig 1967) pp. 176–198
3.15 G. Rassmann, U. Hofmann: J. Appl. Phys. **39**, 603 (1968)
3.16 R. Boll: *Weichmagnetische Werkstoffe*, 4 edn. (Vacuumschmelze, 1990)
3.17 H. Saito (Ed.): *Physics and Application of Invar Alloys* (Maruzen Company, Ltd., Tokyo 1978)
3.18 O'Handley 1977
3.19 Hilzinger 1980
3.20 Miyazaki 1972
3.21 Fujimori 1976
3.22 Allied Signal: Technical Bulletin 9 (1998)
3.23 H. Herzer: Nanocrystalline Soft Magnetic Alloys. In: *Handbook of Magnetic Materials*, ed. by K. H. J. Buschow (Elsevier, Amsterdam 1997) pp. 416–462
3.24 Hono and Sakurai 1995
3.25 M. Müller, N. Mattern, L. Illgen, H. R. Hilzinger, G. Herzer: Key Eng. Mater. **81-83**, 221–228 (1991)
3.26 M. Müller, N. Mattern, L. Illgen: Z. Metallkunde **82**, 895–901 (1991)
3.27 G. Herzer: IEEE Trans. Magn. Mag. **26**, 1397 (1990)
3.28 Herzer 1990
3.29 Herzer 1990–1995
3.30 Herzer and Warlimont 1992
3.31 Sawa and Takahashi 1990
3.32 Suzuki et al. 1991
3.33 Guo et al. 1991
3.34 Pfeifer and Radeloff 1980
3.35 Arai et al. 1984
3.36 Y. Yoshizawa, S. Oguma, K. Yamauchi: J. Appl. Phys. **64**, 6044 (1988)
3.37 K. Suzuki, A. Makino, N. Kataoka, A. Inoue, T. Masumoto: Mater. Trans. Japn. Inst. Met. (JIM) **32**, 93 (1991)
3.38 K. Suzuki, A. Makino, N. Kataoka, A. Inoue, T. Masumoto: J. Appl. Phys. **74**, 3316 (1993)
3.39 Masumoto
3.40 Kussmann and Jessen
3.41 K. Fukamichi: *Magnetic Properties of Metal*, Landolt–Börnstein, New Series III/19, ed. by H. P. J. Wijn (Springer, Berlin, Heidelberg 1994) pp. 193–238
3.42 M. Shiga: Invar Alloys. In: *Materials Science and Technology*, Vol. 3B, ed. by K. H. J. Buschow (VCH, Weinheim 1994) pp. 161–207
3.43 Patrick
3.44 Fujimori and Kaneko
3.45 C. E. Guillaume: C.R. hebd. Séances Acad. Sci. **1244**, 176–179 (1897)
3.46 P. Chevenard, X. Waché, A. Villechon: Ann. France Chronométrie, 259–294 (1937)
3.47 M. Müller: . Ph.D. Thesis (Akademie der Wissenschaften der DDR, Berlin 1977)
3.48 M. Kersten: Z. Phys. **85**, 708 (1933)
3.49 R. Becker, W. Döring: *Ferromagnetismus* (Springer, Berlin 1939) p. 340
3.50 G. Hausch: Phys. Stat. Sol. (a) **15**, 501 (1973) **16 (1973)**, 371
3.51 M. Müller: . Ph.D. Thesis (Technische Universität Dresden, Dresden 1969)
3.52 H. Masumoto, S. Sawaya, M. Kikuchi: J. Japn. Inst. Met. (JIM) **35**, 1143, 1150 (1971) **36 (1972), 57, 176, 492, 498, 881, 886, 1116**
3.53 M. Müller: J. Magn. Magn. Mater. **78**, 337 (1989)
3.54 P. E. Armstrong, J. M. Dickinson, A. L. Brown: Trans. AIME **236**, 1404 (1966)
3.55 E. P. Wohlfahrt: Hard Magnetic Materials. Advances in Physics, Phil. Mag. **A8** (1959)
3.56 D. Bonnenberg, K. Burzo, K. Fukamichi, H. P. Kirchmayr, T. Nakamichi, H. P. J. Wijn: *Magnetic Properties of Metals: Magnetic Alloys for Technical Applications. Hard Magnetic Alloys*, Landolt–Börnstein, New Series III/19 (Springer, Berlin, Heidelberg 1992)
3.57 WIDIA Magnettechnik, Gesinterte Dauermagnete Koerzit und Koerox, d.2000.D.5
3.58 C. Abache, H. Oesterreicher: J. Appl. Phys. **57**, 4112 (1985)
3.59 van Hook 1964

4.4. Dielectrics and Electrooptics

The present section describes the physical properties of dielectrics and includes the following data:

1. *low-frequency properties*, i.e. density and Mohs hardness, thermal conductivity, static dielectric constant, dissipation factor (loss tangent), elastic stiffness and elastic compliance, and piezoelectric strain;
2. *high-frequency (optical) properties*, i.e. elastooptic and electrooptic coefficients, optical transparency range, two-photon absorption coefficient, refractive indices and their temperature variation, dispersion relations (Sellmeier equations), and second and/or third-order nonlinear dielectric susceptibilities.

4.4.1 **Dielectric Materials:**
   Low-Frequency Properties 822
   4.4.1.1 General Dielectric Properties 822
   4.4.1.2 Static Dielectric Constant
            (Low-Frequency) 823
   4.4.1.3 Dissipation Factor 823
   4.4.1.4 Elasticity 823
   4.4.1.5 Piezoelectricity 824

4.4.2 **Optical Materials:**
   High-Frequency Properties 824
   4.4.2.1 Crystal Optics: General 824
   4.4.2.2 Photoelastic Effect 824
   4.4.2.3 Electrooptic Effect 825
   4.4.2.4 Nonlinear Optical Effects 825

4.4.3 **Guidelines for Use of Tables** 826
4.4.4 **Tables of Numerical Data**
       for Dielectrics and Electrooptics 828
       Isotropic Materials 828
       Cubic, Point Group $m3m$ (O_h) Materials .. 828
       Cubic, Point Group $\bar{4}3m$ (T_d) Materials 834
       Cubic, Point Group 23 (T) Materials 838
       Hexagonal, Point Group $\bar{6}m2$ (D_{3h})
       Materials 838
       Hexagonal, Point Group $6mm$ (C_{6v})
       Materials 840
       Hexagonal, Point Group 6 (C_6) Materials . 842
       Trigonal, Point Group $\bar{3}m$ (D_{3d}) Materials 844
       Trigonal, Point Group 32 (D_3) Materials ... 844
       Trigonal, Point Group $3m$ (C_{3v}) Materials . 848
       Tetragonal, Point Group $4mmm$ (D_{4h})
       Materials 852
       Tetragonal, Point Group $4/m$ (C_{4h})
       Materials 854
       Tetragonal, Point Group 422 (D_4)
       Materials 854
       Tetragonal, Point Group $\bar{4}2m$ (D_{2d})
       Materials 856
       Tetragonal, Point Group $4mm$ (C_{4v})
       Materials 864
       Orthorhombic, Point Group mmm (D_{2h})
       Materials 866
       Orthorhombic, Point Group 222 (D_2)
       Materials 866
       Orthorhombic, Point Group $mm2$ (C_{2v})
       Materials 872
       Monoclinic, Point Group 2 (C_2) Materials . 884

References 890

A large amount of the information in this section is taken from the compilations of low- and high-frequency properties of dielectric crystals in Landolt–Börnstein, Group III, Vols. 29 and 30, especially Vol. 30b. Since 1992–1993, the date of publication of the first of these volumes, a large amount of new data on the physical properties of dielectrics has appeared in the literature. In particular, various linear and nonlinear optical properties of new crystals in the borate family (BBO, LBO, CBO, and CLBO) and of new organic crystals (DLAP, MNMA, etc.) have been (re)measured in recent years. This situation has encouraged us to "refresh" the knowledge about these crystals by adding new data from publications from the last decade. We have also included the most commonly used isotropic materials. The criteria for the selection of 124 dielectrics out of several hundred were their wide range of application and the availability of most of the above-mentioned data.

One of our aims was to produce a "reader-friendly" compilation. For this purpose, all data on dielectrics are presented in a unified form in tables that are similar for both isotropic materials and crystals of various symmetry classes. Moreover, *all* data for each particular material are collected together in one place. Sections 4.4.1–4.4.2 serve as a brief introduction to the various physical phenomena and to the definitions, symbols, and abbreviations used. All numerical data are presented in Sect. 4.4.4. The tables in this section are arranged according to piezoelectric classes in order of decreasing symmetry. Guidelines for searching for (and finding!) the required parameter in these tables can be found in Sect. 4.4.3.

The following table presents a list of the 124 different substances which have been selected to be described in Sect. 4.4.4.

Alphabetical List of Described Crystals and Isotropic Dielectrics

Name of material	Formula	Symbol	Page
α-Aluminium oxide (sapphire)	Al_2O_3		844
Aluminium phosphate (berlinite)	$AlPO_4$		844
Ammonium dideuterium phosphate	$ND_4D_2PO_4$	AD*P or DADP	856
Ammonium dihydrogen arsenate	$NH_4H_2AsO_4$	ADA	856
Ammonium dihydrogen phosphate	$NH_4H_2PO_4$	ADP	856
Ammonium sulfate	$(NH_4)_2SO_4$		872
β-Barium borate	BaB_2O_4	BBO	848
Barium fluoride	BaF_2		828
Barium formate	$Ba(COOH)_2$		866
Barium magnesium fluoride	$BaMgF_4$	BMF	872
Barium nitrite monohydrate	$Ba(NO_2)_2 \cdot H_2O$		842
Barium sodium niobate ("banana")	$Ba_2NaNb_5O_{15}$		872
Barium titanate	$BaTiO_3$		864
Beryllium oxide (bromellite)	BeO		840
Bismuth germanium oxide	$Bi_{12}GeO_{20}$	BGO	838
Bismuth silicon oxide (sillenite)	$Bi_{12}SiO_{20}$	BSO	838
Bismuth triborate	BiB_3O_6	BIBO	884
BK7 Schott glass		BK7	828
Cadmium germanium arsenide	$CdGeAs_2$		856
Cadmium germanium phosphide	$CdGeP_2$		856
Cadmium selenide	$CdSe$		840
Cadmium sulfide (greenockite)	CdS		840
Cadmium telluride (Irtran-6)	$CdTe$		834
Calcite (calcspar, Iceland spar)	$CaCO_3$		844
Calcium fluoride (fluorite, fluorspar, Irtran-3)	CaF_2		828
Calcium tartrate tetrahydrate	$Ca(C_4H_4O_6) \cdot 4H_2O$	L-CTT	874
Cesium dideuterium arsenate	CsD_2AsO_4	CD*A or DCDA	856
Cesium dihydrogen arsenate	CsH_2AsO_4	CsDA or CDA	858

Name of material	Formula	Symbol	Page
Cesium lithium borate	$CsLiB_6O_{10}$	CLBO	858
Cesium triborate	CsB_3O_5	CBO	868
m-Chloronitrobenzene	$ClC_6H_4NO_2$	CNB	875
Copper bromide	CuBr		834
Copper chloride (nantokite)	CuCl		834
Copper gallium selenide	$CuGaSe_2$		858
Copper gallium sulfide	$CuGaS_2$		858
Copper iodide	CuI		834
2-Cyclooctylamino-5-nitropyridine	$C_{13}H_{19}N_3O_2$	COANP	874
Deuterated L-arginine phosphate	$(ND_xH_{2-x})_2^+(CND)(CH_2)_3CH$ $(ND_yH_{3-y})^+COO^- \cdot D_2PO_4^- \cdot D_2O$	DLAP	884
Diamond	C		830
4-($N1N$-Dimethylamino)-3-acetamido-nitrobenzene (N-[2-(dimethylamino)-5-nitrophenyl]-acetamide)	$C_{10}H_{13}N_3O_3$	DAN	884
N, 2-Dimethyl-4-nitrobenzenamine	$C_8H_{10}N_2O_2$	MNMA	874
Dipotassium tartrate hemihydrate	$K_2C_4H_4O_6 \cdot 0.5H_2O$	DKT	886
Gadolinium molybdate	$Gd_2(MoO_4)_3$	GMO	876
Gallium antimonide	GaSb		834
Gallium arsenide	GaAs		834
Gallium nitride	GaN		840
Gallium phosphide	GaP		834
Gallium selenide	GaSe		838
Gallium sulfide	GaS		838
Germanium	Ge		830
Indium antimonide	InSb		836
Indium arsenide	InAs		836
Indium phosphide	InP		836
α-Iodic acid	HIO_3		868
Lead molybdate	$PbMoO_4$		854
Lead titanate	$PbTiO_3$		864
Lithium fluoride	LiF		830
Lithium formate monohydrate	$LiCOOH \cdot H_2O$	LFM	876
Lithium gallium oxide (lithium metagallate)	$LiGaO_2$		876
α-Lithium iodate	$LiIO_3$		842
Lithium niobate	$LiNbO_3$		848

Name of material	Formula	Symbol	Page
Lithium niobate (MgO-doped)	$MgO:LiNbO_3$		848
Lithium sulfate monohydrate	$Li_2SO_4 \cdot H_2O$		886
Lithium tantalate	$LiTaO_3$		848
Lithium tetraborate	$Li_2B_4O_7$		864
Lithium triborate	LiB_3O_5	LBO	878
Magnesium fluoride	MgF_2		852
Magnesium oxide	MgO		830
Magnesium silicate (forsterite)	Mg_2SiO_4		866
α-Mercurie sulfide (cinnabar)	HgS		846
3-Methyl 4-nitropyridine 1-oxide	$C_6N_2O_3H_6$	POM	868
m-Nitroaniline (3-nitrobenzenamine, meta-nitroaniline)	$C_6H_4(NO_2)NH_2$	mNA	878
Poly(methyl methacrylate) (Plexiglas®)	$(C_5H_8O_2)_n$	PMMA	828
Potassium acid phthalate	$KH(C_8H_4O_4)$		878
Potassium bromide	KBr		830
Potassium chloride (sylvine, sylvite)	KCl		830
Potassium dideuterium arsenate	KD_2AsO_4	KD*A or DKDA	858
Potassium dideuterium phosphate	KD_2PO_4	KD*P or DKDP	858
Potassium dihydrogen arsenate	KH_2AsO_4	KDA	860
Potassium dihydrogen phosphate	KH_2PO_4	KDP	860
Potassium fluoroboratoberyllate	$KBe_2BO_3F_2$	KBBF	846
Potassium iodide	KI		830
Potassium lithium niobate	$K_3Li_2Nb_5O_{15}$	KLINBO	864
Potassium niobate	$KNbO_3$		880
Potassium pentaborate tetrahydrate	$KB_5O_8 \cdot 4H_2O$	KB5	880
Potassium sodium tartrate tetrahydrate (Rochelle salt)	$KNa(C_4H_4O_6) \cdot 4H_2O$		870
Potassium titanate (titanyl) phosphate	$KTiOPO_4$	KTP	880
Potassium titanyl arsenate	$KTiOAsO_4$	KTA	880
Rubidium dideuterium arsenate	RbD_2AsO_4	RbD*A, RD*A, or DRDA	860
Rubidium dideuterium phosphate	RbD_2PO_4	RbD*P, RD*P, or DRDP	860
Rubidium dihydrogen arsenate	RbH_2AsO_4	RbDA or RDA	860
Rubidium dihydrogen phosphate	RbH_2PO_4	RbDP or RDP	862
Rubidium titanate (titanyl) phosphate	$RbTiOPO_4$	RTP	882

Name of material	Formula	Symbol	Page
D(+)-Saccharose (sucrose)	$C_{12}H_{22}O_{11}$		888
Silicon	Si		830
α-Silicon carbide	SiC		840
α-Silicon dioxide (quartz)	SiO_2		846
Silicon dioxide (fused silica)	SiO_2		828
Silver antimony sulfide (pyrargyrite)	Ag_3SbS_3		850
Silver arsenic sulfide (proustite)	Ag_3AsS_3		850
Silver gallium selenide	$AgGaSe_2$		862
Silver gallium sulfide (silver thiogallate)	$AgGaS_2$		862
Sodium ammonium tartrate tetrahydrate (ammonium Rochelle salt)	$Na(NH_4)C_4H_4O_6 \cdot 4H_2O$		870
Sodium chlorate	$NaClO_3$		838
Sodium chloride (rock salt, halite)	NaCl		832
Sodium fluoride	NaF		832
Sodium nitrite	$NaNO_2$		882
Strontium fluoride	SrF_2		832
Strontium titanate	$SrTiO_3$		832
Tellurium	Te		846
Tellurium dioxide (paratellurite)	TeO_2		854
Thallium arsenic selenide	Tl_3AsSe_3	TAS	850
Titanium dioxide (rutile)	TiO_2		852
Tourmaline	$(Na, Ca)(Mg, Fe)_3B_3Al_6Si_6(O, OH, F)_{31}$		850
Triglycine sulfate	$(CH_2NH_2COOH)_3 \cdot H_2SO_4$	TGS	888
Urea	$(NH_2)_2CO$		862
Yttrium aluminate	$YAlO_3$	YAP or YALO	866
Yttrium aluminium garnet	$Y_3Al_5O_{12}$	YAG	832
Yttrium lithium fluoride	$YLiF_4$	YLF	854
Yttrium vanadate	YVO_4	YVO	852
Zinc germanium diphosphide	$ZnGeP_2$		862
Zinc oxide (zincite)	ZnO		840
Zinc selenide	ZnSe		836
α-Zinc sulfide (wurtzite)	ZnS		842
Zinc telluride	ZnTe		836

The physical quantities used to describe the properties of the dielectric substances are drawn up as follows.

Used Physical Qunatities, their Symbols and their Units		
B_{ij} (B_m)	relative dielectric impermeability	dimensionless
C	capacitance C/V	
c_{ijkl} (c_{mn})	elastic stiffness tensor	10^9 Pa = GPa
D	electric displacement field (or electric flux density)	C/m^2
d_{ijk} (d_{in})	piezoelectric strain tensor	10^{-12} C/N
E	electric field	V/m
n	refractive index	dimensionless
P	polarization (dipole moment per unit volume of matter)	C/m^2
p_{ijkl} (p_{mn})	elastooptic tensor	dimensionless
q_{ijkl} (q_{mn})	piezooptic tensor	10^{-9} Pa^{-1} = (GPa)$^{-1}$
r_{ijk} (r_{mk})	electrooptic coefficient	10^{-12} m/V = pm/V
r^S_{mk}, r^T_{mk}	electrooptic coefficient at constant strain or stress, respectively	10^{-12} m/V = pm/V
S_{ij} (S_m)	strain tensor	dimensionless
s_{ijkl} (s_{mn})	elastic compliance tensor	10^{-12} Pa^{-1} = (TPa)$^{-1}$
T_{ij} (T_m)	stress tensor	10^9 Pa = GPa
tan δ	dissipation factor (loss tangent)	dimensionless
ϱ	density	g/cm^3
ε_0	dielectric constant (or permittivity) of free space (vacuum),	8.854×10^{-12} C/V m
ε	relative dielectric constant (permittivity)	dimensionless
κ_{ij}	thermal conductivity	W/m K
v^s_{mn}	sound velocity in the direction mn	m/s
χ_{ij} ($\chi^{(1)}_{ij}$)	linear dielectric susceptibility	dimensionless
$\chi^{(2)}_{ijk}$, d_{ijk} (d_{im})	second-order nonlinear dielectric susceptibility	10^{-12} m/V = pm/V
$\chi^{(3)}_{ijkl}$	third-order nonlinear dielectric susceptibility	10^{-22} m^2/V^2

4.4.1 Dielectric Materials: Low-Frequency Properties

4.4.1.1 General Dielectric Properties

Density

The density of a substance is defined as the mass per unit volume of the substance:

$$\varrho = \frac{m}{V}, \qquad (4.1)$$

where V is the volume occupied by a mass m. The density is thus a measure of the volume concentration of mass.

Mohs Hardness Scale

In 1832, Mohs introduced a hardness scale ranging from 1 to 10, based on ten minerals:

1. talc, $Mg_3H_2SiO_{12}$;
2. gypsum, $CaSO_4 \cdot 2H_2O$;
3. iceland spar, $CaCO_3$;
4. fluorite, CaF_2;
5. apatite, $Ca_5F(PO_4)_3$;
6. orthoclase, $KAlSi_3O_8$;

7. quartz, SiO_2;
8. topaz, $Al_2F_2SiO_4$;
9. corundum, Al_2O_3;
10. diamond, C.

Thermal Conductivity

A temperature gradient between different parts of a solid causes a flow of heat. In an isotropic medium, the heat flux of thermal energy h (i.e. the heat transfer rate per unit area normal to the direction of heat flow) is given by

$$h = -\kappa \operatorname{grad} T, \qquad (4.2)$$

where κ is the thermal conductivity. In a crystal, this expression is replaced by

$$h_i = -\kappa_{ij} \frac{\partial T}{\partial x_j}. \qquad (4.3)$$

Here κ_{ij} is the thermal-conductivity tensor. Note that other notations are also used in the literature, e.g. $\kappa \equiv \lambda$ or $\kappa \equiv k$.

4.4.1.2 Static Dielectric Constant (Low-Frequency)

In isotropic and cubic dielectric materials, the electric displacement field D, the electric field E, and the polarization P are connected by the relation

$$D = \varepsilon_0 E + P = \varepsilon_0 (1 + \chi) E, \qquad (4.4)$$

where $\varepsilon_0 = 8.854 \times 10^{-12}$ C/V m is the dielectric constant (or permittivity) of free space (vacuum), and χ is the *dielectric susceptibility*. The relative dielectric constant of the material is defined as

$$\varepsilon = 1 + \chi, \qquad (4.5)$$

and therefore (4.4) becomes

$$D = \varepsilon_0 \varepsilon E. \qquad (4.6)$$

In anisotropic crystals, these equations should be written in tensor form:

$$D_i = \varepsilon_0 \varepsilon_{ij} E_j, \quad \varepsilon_{ij} = 1 + \chi_{ij}. \qquad (4.7)$$

The following relations are valid:

$$\varepsilon_{ij} = \varepsilon_{ji}, \quad \chi_{ij} = \chi_{ji}. \qquad (4.8)$$

Note that other notations are also used in the literature, e.g. $\varepsilon \equiv \varepsilon_r$, or $\varepsilon \equiv \kappa$ and $\varepsilon_0 \equiv \kappa_0$.

4.4.1.3 Dissipation Factor

The capacitance C of a capacitor filled with a dielectric is

$$C = \frac{\varepsilon_0 \varepsilon A}{d}, \qquad (4.9)$$

where A is the area of the two parallel plates and d is the spacing between them. For a lossy dielectric, the relative dielectric constant ε can be represented in a complex form,

$$\varepsilon = \varepsilon' - i\varepsilon''. \qquad (4.10)$$

The imaginary part is the frequency-dependent conductivity

$$\sigma(\omega) = \omega \varepsilon_0 \varepsilon'', \qquad (4.11)$$

where ω is the frequency. The dissipation factor (or loss tangent) is defined as

$$\tan \delta = \varepsilon'' / \varepsilon'. \qquad (4.12)$$

and in anisotropic crystals

$$\tan \delta_1 = \frac{\varepsilon_{11}''}{\varepsilon_{11}'};$$

$$\tan \delta_2 = \frac{\varepsilon_{22}''}{\varepsilon_{22}'}; \quad \tan \delta_3 = \frac{\varepsilon_{33}''}{\varepsilon_3'}; \qquad (4.13)$$

The quality factor Q of the dielectric is the reciprocal of the dissipation factor:

$$Q = 1/\tan \delta. \qquad (4.14)$$

4.4.1.4 Elasticity

Hooke's law states that for sufficiently small deformations the strain is directly proportional to the stress. Thus the strain tensor **S** and the stress tensor **T** obey the relation

$$S_{ij} = s_{ijkl} T_{kl}, \qquad (4.15)$$

where s_{ijkl} is called the *elastic compliance constant* (or compliance, or elastic constant). The *elastic stiffness constant* (or stiffness, or Young's modulus) is the reciprocal tensor

$$c_{ijkl} = s_{ijkl}^{-1}, \qquad (4.16)$$

Table 4.4-1 The relations between ij (tensor notation) and m (matrix notation), jk and n, and kl and n

Tensor notation	11	22	33	23 or 32	31 or 13	12 or 21
Matrix notation	1	2	3	4	5	6

and for the stress tensor we have

$$T_{ij} = c_{ijkl} S_{kl} \, . \qquad (4.17)$$

In the matrix notation for the elastic compliance and stiffness, we have

$$S_m = s_{mn} T_n \text{ and } T_m = c_{mn} S_n \, , \qquad (4.18)$$

where

$$\begin{aligned} s_{ijkl} &= s_{mn} & \text{when both } m \text{ and } n \text{ are } 1, 2, \text{ or } 3 \, , \\ 2 s_{ijkl} &= s_{mn} & \text{when either } m \text{ or } n \text{ is } 4, 5, \text{ or } 6 \\ 4 s_{ijkl} &= s_{mn} & \text{when both } m \text{ and } n \text{ are } 4, 5, \text{ or } 6 \, , \\ c_{ijkl} &= c_{mn} & \text{for all } m \text{ and } n \, ; \end{aligned} \qquad (4.19)$$

and

$$\begin{aligned} S_{ij} &= S_m & \text{when } m \text{ is } 1, 2, \text{ or } 3 \, , \\ S_{ij} &= \tfrac{1}{2} S_m & \text{when } m \text{ is } 4, 5, \text{ or } 6 \, , \\ T_{ij} &= T_m & \text{for all } m \, . \end{aligned} \qquad (4.20)$$

For relations between tensor and matrix notation, see Table 4.4-1. The sound velocity v_{mn}^s in the direction mn in a crystal is given by

$$v_{mn}^s = \sqrt{c_{mn}/\varrho} \, . \qquad (4.21)$$

4.4.1.5 Piezoelectricity

The phenomenon of the development of an electric moment P_i if a stress T_{jk} is applied to a crystal is called the direct piezoelectric effect:

$$P_i = d_{ijk} T_{jk} \, , \qquad (4.22)$$

where d_{ijk} is the piezoelectric strain tensor (or the piezoelectric moduli). The relation $d_{ijk} = d_{ikj}$ reduces the number of independent tensor components to 18. The matrix notation is introduced for the piezoelectric strain as follows:

$$\begin{aligned} d_{ijk} &= d_{in} & \text{when } n = 1, 2, \text{ or } 3 \, , \\ 2 d_{ijk} &= d_{in} & \text{when } n = 4, 5, \text{ or } 6 \, , \end{aligned} \qquad (4.23)$$

and thus

$$P_i = d_{in} T_n \, . \qquad (4.24)$$

The relations between jk and n are presented in Table 4.4-1.

The converse piezoelectric effect is described by

$$S_{jk} = d_{ijk} E_i \qquad (4.25)$$

and, correspondingly,

$$S_n = d_{in} E_i \, . \qquad (4.26)$$

4.4.2 Optical Materials: High-Frequency Properties

4.4.2.1 Crystal Optics: General

The dielectric properties of a medium at optical frequencies are given by

$$D = \varepsilon_0 \varepsilon E \, , \qquad (4.27)$$

where ε_0 is the dielectric constant of free space and ε is the relative dielectric constant of the material. From Maxwell's equations, the velocity of propagation of electromagnetic waves through the medium is given by

$$v = c/\sqrt{\varepsilon} \, , \qquad (4.28)$$

where c is the velocity in vacuum (the relative magnetic permeability is taken as 1). The refractive index $n = c/v$ is therefore $n = \sqrt{\varepsilon}$.

In an anisotropic medium,

$$D_i = \varepsilon_0 \varepsilon_{ij} E_j \, . \qquad (4.29)$$

In this general case, two waves of different velocity may propagate through the crystal. The relative dielectric impermeabilities are defined as the reciprocals of the principal dielectric constants:

$$B_i = 1/\varepsilon_i = 1/n_i^2 \, . \qquad (4.30)$$

4.4.2.2 Photoelastic Effect

The photoelastic effect is the effect in which a change of the refractive index is caused by stress. The changes in the relative dielectric impermeabilities are

$$\Delta B_{ij} = q_{ijkl} T_{kl}, \quad (4.31)$$

where the q_{ijkl} are the piezooptic coefficients. The photoelastic effect can also be expressed in terms of the stress:

$$\Delta B_{ij} = p_{ijrs} S_{rs}, \quad (4.32)$$

where the $p_{ijrs} = q_{ijkl} c_{klrs}$ are the (dimensionless) elastooptic coefficients. In matrix notation,

$$\Delta B_m = q_{mn} T_n \quad \text{and} \quad \Delta B_m = p_{mn} S_n. \quad (4.33)$$

Note that $q_{mn} = q_{ijkl}$ when $n = 1, 2,$ or 3 and $q_{mn} = 2q_{ijkl}$ when $n = 4, 5,$ or 6; $p_{mn} = p_{ijrs}$ (see Table 4.4-1).

4.4.2.3 Electrooptic Effect

The electrooptic effect is the effect in which a change in the refractive index of a crystal is produced by an electric field:

$$n = n_0 + aE_0 + bE_0^2 + \ldots, \quad (4.34)$$

where a and b are constants and n_0 is the refractive index at $E_0 = 0$. The linear electrooptic effect (Pockels effect) is due to the first-order term aE_0. In isotropic dielectrics and in crystals with a center of symmetry, $a = 0$, and only the second-order term bE_0^2 and higher even-order terms exist (Kerr effect).

The changes in the relative dielectric impermeabilities are

$$\Delta B_{ij} = r_{ijk} E_k, \quad (4.35)$$

where the r_{ijk} are the electrooptic coefficients. Since $r_{ijk} = r_{jik}$, the number of independent tensor components is 18, and the above formula can be written in matrix notation (see Table 4.4-1):

$$\Delta B_m = r_{mk} E_k \quad (m = 1, 2, \ldots, 6, \ k = 1, 2, 3). \quad (4.36)$$

4.4.2.4 Nonlinear Optical Effects

The dielectric polarization P is related to the electromagnetic field E at optical frequencies by the material equation of the medium:

$$P(E) = \varepsilon_0 \left(\chi^{(1)} E + \chi^{(2)} E^2 + \chi^{(3)} E^3 + \ldots \right), \quad (4.37)$$

where $\chi^{(1)} = n^2 - 1$ is the linear dielectric susceptibility, and $\chi^{(2)}, \chi^{(3)}$, etc. are the nonlinear dielectric susceptibilities.

The Miller delta formulation is

$$\varepsilon_0 E_i(\omega_3) = \delta_{ijk} P_j(\omega_1) P_k(\omega_2), \quad (4.38)$$

where the Miller coefficient,

$$\delta_{ijk} = \frac{1}{2\varepsilon_0} \frac{\chi_{ijk}^{(2)}(\omega_3)}{\chi_{ii}^{(1)}(\omega_1) \chi_{jj}^{(1)}(\omega_2) \chi_{kk}^{(1)}(\omega_3)}, \quad (4.39)$$

has a small dispersion and is almost constant for a wide range of crystals.

For anisotropic media, the coefficients $\chi^{(1)}$ and $\chi^{(2)}$ are, in general, second- and third-rank tensors, respectively. In practice, the tensor

$$d_{ijk} = (1/2) \chi_{ijk} \quad (4.40)$$

is used instead of χ_{ijk}. Usually the "plane" representation of d_{ijk} in the form d_{il} is used; the relations between jk and l are presented in Table 4.4-1.

The Kleinman symmetry conditions

$$d_{21} = d_{16}, \quad d_{23} = d_{34}, \quad d_{14} = d_{25} = d_{36},$$
$$d_{26} = d_{12}, \quad d_{31} = d_{15}, \quad d_{32} = d_{24}, \quad d_{35} = d_{13} \quad (4.41)$$

are valid in the case of no dispersion of the electronic nonlinear polarizability.

The following three-wave interactions in crystals with a square nonlinearity ($\chi^{(2)} \neq 0$) are possible:

- second-harmonic generation (SHG), $\omega + \omega = 2\omega$;
- sum frequency generation (SFG) or up-conversion, $\omega_1 + \omega_2 = \omega_3$;
- difference frequency generation (DFG) or down-conversion, $\omega_3 - \omega_2 = \omega_1$;
- optical parametric oscillation (OPO), $\omega_3 = \omega_2 + \omega_1$.

For efficient frequency conversion, the phase-matching condition $\mathbf{k}_1 + \mathbf{k}_2 = \mathbf{k}_3$, where the $\mathbf{k}_i$ are the wave vectors for $\omega_1, \omega_2,$ and ω_3, respectively, must be satisfied. Two types of phase matching can be defined:

type I: $o + o \rightarrow e$ or $e + e \rightarrow o$;

and

type II: $o + e \rightarrow e$ or $o + e \rightarrow o$.

These can be represented with a shortened notation as follows:

ooe: $o + o \rightarrow e$ or $e \rightarrow o + o$;

eeo: $e + e \rightarrow o$ or $o \rightarrow e + e$;

eoe: $e + o \rightarrow e$ or $e \rightarrow e + o$;

oeo: $o + e \rightarrow o$ or $o \rightarrow e + o$.

In the shortened notation (ooe, eoe, ...), the frequencies satisfy the condition $\omega_1 < \omega_2 < \omega_3$, i.e. the first symbol refers to the longest-wavelength radiation, and the last symbol refers to the shortest-wavelength radiation. Here the ordinary beam, or o-beam is the beam with its polarization normal to the principal plane of the crystal, i.e. the plane containing the wave vector $\mathbf{k}$ and the crystallophysical axis Z (or the optical axis, for uniaxial crystals). The extraordinary beam, or e-beam is the beam with its polarization in the principal plane. The third-order term $\chi^{(3)}$ is responsible for the optical Kerr effect.

Uniaxial Crystals

For uniaxial crystals, the difference between the refractive indices of the ordinary and extraordinary beams, the birefringence Δn, is zero along the optical axis (the crystallophysical axis Z) and maximum in a direction normal to this axis. The refractive index for the ordinary beam does not depend on the direction of propagation. However, the refractive index for the extraordinary beam $n^e(\theta)$, is a function of the polar angle θ between the Z axis and the vector $\mathbf{k}$:

$$n^e(\theta) = n_o \left(\frac{1 + \tan^2 \theta}{1 + (n_o/n_e)^2 \tan^2 \theta} \right)^{1/2}, \quad (4.42)$$

where n_o and n_e are the refractive indices of the ordinary and extraordinary beams, respectively in the plane normal to the Z axis, and are termed the *principal values*.

If $n_o > n_e$ the crystal is called *negative*, and if $n_o < n_e$ it is called *positive*. For the o-beam, the indicatrix of the refractive indices is a sphere with radius n_o, and for the e-beam it is an ellipsoid of rotation with semiaxes n_o and n_e. In the crystal, in general, the beam is divided into two beams with orthogonal polarizations; the angle between these beams ρ is the *birefringence* (or *walk-off*) angle.

Equations for calculating phase-matching angles in uniaxial crystals are given in [4.1–4].

Biaxial Crystals

For biaxial crystals, the optical indicatrix is a bilayer surface with four points of interlayer contact, which correspond to the directions of the two optical axes. In the simple case of light propagation in the principal planes XY, YZ, and XZ, the dependences of the refractive indices on the direction of light propagation are represented by a combination of an ellipse and a circle. Thus, in the principal planes, a biaxial crystal can be considered as a uniaxial crystal; for example, a biaxial crystal with $n_Z > n_Y > n_X$ in the XY plane is similar to a negative uniaxial crystal with $n_o = n_Z$

$$n^e(\varphi) = n_Y \left(\frac{1 + \tan^2 \varphi}{1 + (n_Y/n_X)^2 \tan^2 \varphi} \right)^{1/2}, \quad (4.43)$$

where φ is the azimutal angle. Equations for calculating phase-matching angles for propagation in the principal planes of biaxial crystals are given in [4.3–6].

4.4.3 Guidelines for Use of Tables

Tables 4.4-3–4.4-21 are arranged according to piezoelectric classes in order of decreasing symmetry (see Table 4.4-2), and alphabetically within each class. They contain a number of columns placed on two pages, even and odd. The following properties are presented for each dielectric material: density ϱ, Mohs hardness, thermal conductivity κ, static dielectric constant ε_{ij}, dissipation factor $\tan \delta$ at various temperatures and frequencies, elastic stiffness c_{mn}, elastic compliance s_{mn} (for isotropic and cubic materials only), piezoelectric strain tensor d_{in}, elastooptic tensor p_{mn}, electrooptic coefficients r_{mk} (the latter two at 633 nm unless otherwise stated), optical transparency range, temperature variation of the refractive indices dn/dT, refractive indices n (the latter two at 1.064 μm unless otherwise stated), dispersion relations (Sellmeier equations), second-order nonlinear dielectric susceptibility d_{ij}, and third-order nonlinear dielectric susceptibility $\chi_{ijk}^{(3)}$ (for isotropic and cubic materials only) (the latter two at 1.064 μm unless otherwise stated). For isotropic materials, the two-photon absorption coefficient β is also included

The numerical values of the elastic and elastooptic constants are often averages of three or more measurements, as presented in [4.7–11]. In such cases, the corresponding Landolt–Börnstein volume is cited together with the most reliable (latest) reference. The standard deviation of the averaged value is given in parentheses. Vertical bars || mean the modulus of the corresponding quantity. The absolute scale for the second-order nonlinear susceptibilities of crystals is based on [4.12–14]. The second-order susceptibilities for all crystals measured relative to a standard crystal have been recalculated accordingly. In particular, all previous measurements relative to KDP and quartz have been normalized to $d_{36}(\text{KDP}) = 0.39\,\text{pm/V}$ and $d_{11}(\text{SiO}_2) = 0.30\,\text{pm/V}$. These data lead to an accurate, self-consistent set of absolute second-order nonlinear coefficients [4.14]. All numerical data are for room temperature (300 K) and in SI units.

Table 4.4-2 Number of independent components of the various property tensors

Symmetry point group	Dielectric tensor[a] ε_{ij}	Elastic tensor c_{mn} or s_{mn}	Piezoelectric tensor d_{imn}	Elasto-optic tensor p_{mn} or q_{mn}	Electro-optic tensor r_{mk}	Nonlinear susceptibility tensors[b] $\chi^{(2)}$	$\chi^{(3)}$	Table number
Isotropic	1(1)	2	0	2	0	0(0)	1(1)	4.4-3
Cubic:								
432 (O)	1(1)	3	0	3	0	0(0)	2(2)	
$m3m$ (O_h)	1(1)	3	0	3	0	0(0)	2(2)	4.4-4
$\bar{4}3m$ (T_d)	1(1)	3	1	3	1	1(1)	2(2)	4.4-5
23 (T)	1(1)	3	1	4	1	1(1)	3(2)	4.4-6
Hexagonal:								
$6/mmm$ (D_{6h})	2(2)	5	0	6	0	0(0)	4(3)	
$6/m$ (C_{6h})	2(2)	5	0	8	0	0(0)	6(3)	
622 (D_6)	2(2)	5	1	6	1	1(1)	4(3)	
$\bar{6}m2$ (D_{3h})	2(2)	5	1	6	1	1(1)	4(3)	4.4-7
$\bar{6}$ (C_6)	2(2)	5	2	8	2	2(2)	6(3)	
$6mm$ (C_{6v})	2(2)	5	3	6	3	3(2)	4(3)	4.4-8
6 (C_6)	2(2)	5	4	8	4	4(3)	6(3)	4.4-9
Trigonal:								
$\bar{3}m$ (D_{3d})	2(2)	6	0	8	0	0(0)	5(4)	4.4-10
$\bar{3}$ (C_{3i})	2(2)	7	0	12	0	0(0)	10(5)	
32 (D_3)	2(2)	6	2	8	2	2(2)	5(4)	4.4-11
$3m$ (C_{3v})	2(2)	6	4	8	4	4(3)	5(4)	4.4-12
3 (C_3)	2(2)	7	6	12	6	6(5)	10(5)	
Tetragonal:								
$4/mmm$ (D_{4h})	2(2)	6	0	7	0	0(0)	5(4)	4.4-13
$4/m$ (C_{4h})	2(2)	7	0	10	0	0(0)	8(6)	4.4-14
422 (D_4)	2(2)	6	1	7	1	1(1)	5(4)	4.4-15
$\bar{4}2m$ (D_{2d})	2(2)	6	2	7	2	2(1)	5(4)	4.4-16
$4mm$ (C_{4v})	2(2)	6	3	7	3	3(2)	5(4)	4.4-17
$\bar{4}$ (S_4)	2(2)	7	4	10	4	4(2)	8(6)	
4 (C_4)	2(2)	7	4	10	4	4(3)	8(6)	
Orthorhombic:								
mmm (D_{2h})	3(3)	9	0	12	0	0(0)	9(6)	4.4-18
222 (D_2)	3(3)	9	3	12	3	3(1)	9(6)	4.4-19
$mm2$ (C_{2v})	3(3)	9	5	12	5	5(3)	9(6)	4.4-20
Monoclinic:								
$2/m$ (C_{2h})	4(3)	13	0	20	0	0(0)	16(9)	
2 (C_2)	4(3)	13	8	20	8	8(4)	16(9)	4.4-21
m (C_s)	4(3)	13	10	20	10	10(6)	16(9)	
Triclinic:								
$\bar{1}$ (C_i)	9(3)	21	0	36	0	0(0)	30(15)	
1 (C_1)	9(3)	21	18	36	18	18(10)	30(15)	

[a] The number of principal refractive indices n_i is given in parentheses.
[b] The number of independent components for the case of Kleinman symmetry conditions is given in parentheses.

4.4.4 Tables of Numerical Data for Dielectrics and Electrooptics

Table 4.4-3 Isotropic materials

Material	General ϱ (g/cm^3) Mohs hardness κ (W/m K)	Static dielectric constant ε_{11}^S ε_{11}^T	Dissipation factor $\tan \delta_1$ (f (Hz))	Elastic stiffness tensor c_{11} c_{12} (10^9 Pa)	Elastic compliance tensor s_{11} s_{12} (10^{-12} Pa^{-1})	Elastooptic tensor p_{11} p_{12}
BK7 Schott glass	2.510 – 1.114 [4.17]			92.325 (145) – [4.15]	12[b] –2[b] [4.16]	– 0.198 (22) [4.15]
Poly(methylmethacrylate), (C$_5$H$_8$O$_2$)$_n$ (PMMA, Plexiglas)	1.190 2–3 0.2 [4.19]	3.65 (19)[a] [4.18]	0.06 (50 Hz) [4.18]	9.282 (30) – [4.15]	300[b] 8.9[b] [4.18]	
Silicon dioxide, SiO$_2$ (Fused silica, fused quartz, vitreous quartz)	2.202 5–6 1.38 [4.21]	3.5[a] [4.20]	14 (5) × 10^{-4} (1 MHz) [4.20]	77.806 (185) – [4.15]	14[b] –2.1[b] [4.20]	– 0.243 (17) [4.15]

[a] This value for ε_{11} is neither at constant stress nor at constant strain.
[b] The elastic compliances are at constant electric field (s^E) and have been calculated from the Young's modulus via $Y_0 = 1/s^E$ and from the shear modulus via $G = \left[2\left(s_{11}^E - s_{12}^E\right)\right]^{-1}$.

Table 4.4-4 Cubic, point group $m3m$ (O_h) materials

Material	General ϱ (g/cm^3) Mohs hardness κ (W/m K)	Static dielectric constant ε_{11}	Dissipation factor $\tan \delta_1$ (f (Hz))	Elastic stiffness tensor c_{11} c_{44} c_{12} (10^9 Pa)	Elastic compliance tensor s_{11} s_{44} s_{12} (10^{-12} Pa^{-1})	Elastooptic tensor p_{11} p_{12} p_{44} $p_{11} - p_{12}$
Barium fluoride, BaF$_2$	4.89 3 11 [4.21, 22]	7.33 [4.21]		91.1(10) 25.3(4) 41.2(15) [4.7]	15.2(1) 39.6(7) –4.7(1) [4.7]	0.11 0.26 0.02 –0.14 (589–633 nm) [4.23–25]
Calcium fluoride, CaF$_2$ (fluorite, fluorspar, Irtran-3)	3.179 4.0 9.71 [4.21, 22]	7.4 [4.26]		165(2) 33.9(3) 47(3) [4.7]	6.93(14) 29.5(3) –1.52(11) [4.7]	0.027 0.198 0.02 –0.17 (550–650 nm) [4.23, 24, 27]

Table 4.4-3 Isotropic materials, cont.

General optical properties Transparency range (μm) dn/dT $(10^{-5}\,\text{K}^{-1})$	Refractive index n	Two-photon absorption coefficient β $(10^{-9}\,\text{cm/W})$	Nonlinear dielectric susceptibility $\chi^{(3)}_{1111}$ $(10^{-22}\,\text{m}^2/\text{V}^2)$
0.35–2.8 [4.16, 28] 0.28 (0.546 μm) [4.16]	**1.5067** [4.16] For dispersion relation, see[c] [4.28]	0.006 (351 nm) [4.29] 0.0029 (532 nm) [4.31]	2.7 ± 0.2 [4.30]
0.27–1.1 –	**1.503** (0.436 μm) **1.493** (0.546 μm) **1.489** (0.633 μm) **1.481** (1.052 μm) [4.32]		
0.17–4.0 [4.22] 1.5 (0.21 μm) 1.0 (0.4 μm) 1.2 (2.0 μm) 1.0 (3.7 μm) [4.39]	**1.5343** (0.2139 μm) **1.4872** (0.3022 μm) **1.4601** (0.5461 μm) **1.4494** (1.083 μm) **1.4372** (2.0581 μm) **1.3994** (3.7067 μm) For dispersion relation, see [d] [4.39]	0.75 (212.8 nm) [4.33] 0.5 (216 nm) [4.35] 0.08 (248 nm) [4.37] 0.045 (266 nm) [4.38]	1.8 [4.34] 1.7 ± 0.3 [4.36]

[cd] Dispersion relations (λ (μm), $T = 20\,°\text{C}$):

[c] $n^2 = 1 + \dfrac{1 + 1.03961212\lambda^2}{\lambda^2 - 0.00600069867} + \dfrac{0.231792344\lambda^2}{\lambda^2 - 0.0200179144} + \dfrac{1.01046945\lambda^2}{\lambda^2 - 103.560653}$;

[d] $n^2 = 1 + \dfrac{1 + 0.6961663\lambda^2}{\lambda^2 - (0.0684043)^2} + \dfrac{0.4079426\lambda^2}{\lambda^2 - (0.1162414)^2} + \dfrac{0.8974794\lambda^2}{\lambda^2 - (9.896161)^2}$.

Table 4.4-4 Cubic, point group $m3m$ (O_h) materials, cont.

General optical properties	Refractive index[a]		Nonlinear dielectric susceptibility
Transparency range (μm) dn/dT $(10^{-5}\,\text{K}^{-1})$	n A B C	D E F G	$\chi^{(3)}_{1111}$ $\chi^{(3)}_{1122}$ $\chi^{(3)}_{1133}$ $(10^{-22}\,\text{m}^2/\text{V}^2)$
0.14–12.2 [4.40] −1.60 (0.633 μm) [4.41] −0.6 (0.150 μm) −1.0 (0.590 μm) −1.1 (15.0 μm) [4.43]	**1.678** (0.150 μm) **1.557** (0.200 μm) **1.5118** (0.266 μm) **1.4744** (0.590 μm) **1.4683** (1.05 μm) **1.4441** (6.00 μm)	**1.4027** (9.00 μm) **1.3865** (11.0 μm) **1.305** (15.0 μm) dispersion relations, see [4.43][a]	1.548 ± 0.17 – 0.636 ± 0.06 (0.575; 0.613 μm) [4.42]
0.135–9.4 [4.40] −1.04 (0.633 μm) [4.44] −0.1 (0.150 μm) −1.5 (0.590 μm) −1.0 (12.0 μm)	**1.577** (0.150 μm) **1.495** (0.200 μm) **1.4621** (0.266 μm) **1.4338** (0.590 μm) **1.4286** (1.05 μm) **1.3856** (6.00 μm)	**1.3268** (9.00 μm) **1.268** (11.0 μm) **1.230** (12.0 μm) dispersion relations, see [4.43][b]	0.8 ± 0.24 0.36 ± 0.1 (0.575; 0.613 μm) [4.42]

Table 4.4-4 Cubic, point group $m3m$ (O_h) materials, cont.

Material	General ϱ (g/cm³) Mohs hardness κ (W/m K)	Static dielectric constant ε_{11}	Dissipation factor $\tan \delta_1$ (f (Hz))	Elastic stiffness tensor c_{11} c_{44} c_{12} (10^9 Pa)	Elastic compliance tensor s_{11} s_{44} s_{12} (10^{-12} Pa^{-1})	Elastooptic tensor p_{11} p_{12} p_{44} $p_{11} - p_{12}$
Diamond, C	3.52 10 660 (0 °C) [4.22, 45]	5.5–10.0 [4.26]		1077(2) 577(1) 124.7(6) [4.7]	0.951(2) 1.732(3) −0.0987(6) [4.7]	−0.25 0.04 0.17 −0.30(1) (514–633 nm) [4.46]
Germanium, Ge	5.33 6 58.61 [4.21]	16.6 [4.21]		129(3) 67.1(5) 48(3) [4.7]	9.73(5) 14.90(12) −2.64(11) [4.7]	
Lithium fluoride, LiF	2.639 4 4.01 [4.21]	9.1 [4.21]		112(2) 63.5(6) 46(3) [4.7]	11.6(1) 15.8(2) −3.35(13) [4.7]	0.02 0.13 −0.04(1) −0.10 (589–633 nm) [4.27, 47]
Magnesium oxide, MgO	3.58 – 42 [4.21]	9.7 [4.26]	0.009 (10 GHz) [4.48]	294(6) 155(3) 93(5) [4.7]	4.01(4) 6.46(12) −0.96(2) [4.7]	−0.25 −0.01 −0.10 −0.24 (589 nm) [4.27, 49]
Potassium bromide, KBr	2.753 1.5 4.816 [4.21]	4.9 [4.21]		34.5(3) 5.1(2) 5.5(4) [4.7]	30.3(6) 196(6) −4.2(3) [4.7]	0.22 0.17 (546–600 nm) [4.50] 0.02 0.04 (488–600 nm) [4.51]
Potassium chloride, KCl (sylvine, sylvite)	1.99 2 6.53 [4.21]	4.6 [4.26]		40.5(4) 6.27(6) 6.9(3) [4.7]	25.9(1) 159(1) −3.8(3) [4.7]	0.23 0.17 0.02 0.05 (546–600 nm) [4.50–53]
Potassium iodide, KI	3.12 – 2.1 [4.21]	5.6 [4.26]		27.4(5) 3.70(4) 4.3(2) [4.7]	38.2(8) 270(3) −5.2(3) [4.7]	1.21 0.15 −0.031 [4.41] –
Silicon, Si	2.329 7 163.3 [4.21]	11.0–12.0 [4.26]		165(2) 79.1(6) 63(1) [4.7]	7.73(8) 12.70(9) −2.15(4) [4.7]	−0.094 0.017 −0.051 −0.111 (3390 nm) [4.54–57]

Table 4.4-4 Cubic, point group $m3m$ (O_h) materials, cont.

General optical properties	Refractive index[a]		Nonlinear dielectric susceptibility
Transparency range (μm)	n	D	$\chi^{(3)}_{1111}$
dn/dT $(10^{-5}\,\text{K}^{-1})$	A	E	$\chi^{(3)}_{1122}$
	B	F	$\chi^{(3)}_{1133}$
	C	G	$(10^{-22}\,\text{m}^2/\text{V}^2)$
0.24–27	**2.7151** (0.2265 μm)	For dispersion relation, see[c] [4.58]	$\chi^{(3)}_{1122} = 4.87(12)$
1.01 (0.546 μm)	**2.4190** (0.578 μm)		$\chi^{(3)}_{1111} + 3\chi^{(3)}_{1122} = 30 \pm 0.8$
0.96 (30 μm)	**2.3914** (1.064 μm)		(both at 0.407 μm) [4.42]
[4.59, 60]			$\chi^{(3)}_{1111} + 3\chi^{(3)}_{1133} = 65 \pm 20$
			(0.53 μm) [4.61]
1.8–15 [4.62]	**4.0038** (10.6 μm)	0.21307	$56\,000 \pm 28\,000$
–	9.28156	3870.1	$34\,000 \pm 1200$
	6.72880	0	–
	0.44105	0 [4.63][d]	(10.6 and 9.5 μm) [4.64]
0.120–6.60	**1.387**	6.96747	1.12(24)
–	1	1075	0.50
	0.9259	0	–
	0.005441	0 [4.65][d]	(0.6943 and 0.7456 μm) [4.66]
0.35–6.8 [4.67]	**1.7217**	0.8460085	+1.4(2)
1.95 (0.365 μm)	1	0.01891186	–
1.65 (0.546 μm)	1.111033	7.808527	+0.77(15)
1.36 (0.768 μm)	0.00507606	723.2345	(0.6943 and 0.7456 μm)
[4.68]		[4.69][d]	[4.66]
0.200–30 [4.40]	**1.5435**		15.7 [4.70]
−3.93 (0.458 μm)	For dispersion relation,		5.8 [4.71]
−4.19 (1.15 μm)	see[e] [4.65]		6.2 (0.6943 and 0.7456 μm)
−4.11 (10.6 μm) [4.41]			[4.66]
0.18–23.3 [4.40]	**1.4792**		6.7 [4.70]
−3.49 (0.458 μm)	For dispersion relation,		1.9 [4.71]
−3.62 (1.15 μm)	see[f] [4.65]		0.8 (0.6943 and 0.7456 μm)
−3.48 (10.6 μm) [4.41]			[4.66]
0.32–42 [4.21]	**1.6393**		1.4 ± 0.3 (0.6943 and 0.7456 μm)
−4.15 (0.458 μm)	For dispersion relation,		[4.66]
−4.47 (1.15 μm)	see[g] [4.65]		
−3.08 (30 μm) [4.41]			–
1.1–6.5 [4.62]	**3.4176** (10.6 μm)	0.003043475	24 300 [4.72]
–	1	1.2876602	–
	10.6684293	1.54133408	33 000 [4.71] at 10.6 and 11.8 μm
	0.09091219	1 218 820 [4.73][d]	$\chi^{(3)}_{1111} = 2800$
		[4.74]	$\chi^{(3)}_{1122} = 1340$ [4.75]

Table 4.4-4 Cubic, point group $m3m$ (O_h) materials, cont.

Material	General ϱ (g/cm^3) Mohs hardness κ (W/m K)	Static dielectric constant ε_{11}	Dissipation factor $\tan \delta_1$ (f (Hz))	Elastic stiffness tensor c_{11} c_{44} c_{12} (10^9 Pa)	Elastic compliance tensor s_{11} s_{44} s_{12} (10^{-12} Pa^{-1})	Elastooptic tensor p_{11} p_{12} p_{44} $p_{11} - p_{12}$
Sodium chloride, NaCl (rock salt, halite)	2.17 2 1.15 [4.21]	5.9 [4.26]		49.1(5) 12.8(1) 12.8(1) [4.7]	22.9(5) 78.3(8) −4.8(1) [4.7]	0.128 0.171 −0.01 −0.04 (589–633 nm) [4.52, 53, 76]
Sodium fluoride, NaF	2.558 – 3.746 [4.21]	6 [4.21]		97.0(4) 28.1(3) 24.2(10) [4.7]	11.50(6) 35.6(4) −2.30(7) [4.7]	0.03 0.14 – −0.10 (589–633 nm) [4.76–78]
Strontium fluoride, SrF$_2$	4.24 – 1.42 [4.21]	7.69 [4.21]		124(1) 31.8(3) 44(1) [4.7]	9.86(4) 31.5(3) −2.57(5) [4.7]	0.080 0.269 0.018 −0.189 [4.23, 24]
Strontium titanate, SrTiO$_3$	5.12 6–6.5 12 [4.79]	300 [4.48]	0.02 (10 GHz) [4.48]	316(2) 123(1) 102(1) [4.7]	3.75(3) 8.15(8) −0.92(1) [4.7]	|0.15| |0.095| |0.072| – [4.80]
Yttrium aluminium garnet, Y$_3$Al$_5$O$_{12}$ (YAG)	4.56 8–8.5 13.4 [4.81]			333(3) 114(1) 111(3) [4.7]	3.61(4) 8.74(4) −0.90(2) [4.7]	−0.029 0.0091 −0.0615 −0.038 [4.82]

[a] The refractive index is given in **bold** type to distinguish it from the constants A, B, etc. in the dispersion relation.

[abcdefghij] Dispersion relations (λ (μm), $T = 20\,°$C):

[a] $n^2 = 1.33973 + \dfrac{0.81070\lambda^2}{\lambda^2 - 0.10065^2} + \dfrac{0.19652\lambda^2}{\lambda^2 - 29.87^2} + \dfrac{4.52469\lambda^2}{\lambda^2 - 53.82^2}$;

[b] $n^2 = 1.33973 + \dfrac{0.69913\lambda^2}{\lambda^2 - 0.09374^2} + \dfrac{0.11994\lambda^2}{\lambda^2 - 21.18^2} + \dfrac{4.35181\lambda^2}{\lambda^2 - 38.46^2}$;

[c] $n^2 = 2.37553 + \dfrac{0.0336440\lambda^2}{\lambda^2 - 0.028} - \dfrac{0.0887524}{(\lambda^2 - 0.028)^2} - 2.40455 \times 10^{-6}\lambda^2 + 2.21390 \times 10^{-9}\lambda^4$;

[d] $n^2 = A + \dfrac{B\lambda^2}{\lambda^2 - C} + \dfrac{D\lambda^2}{\lambda^2 - E} + \dfrac{F\lambda^2}{\lambda^2 - G}$;

[e] $n^2 = 1.39408 + \dfrac{0.79221\lambda^2}{\lambda^2 - 0.0213} + \dfrac{0.01981\lambda^2}{\lambda^2 - 0.0299} + \dfrac{0.15587\lambda^2}{\lambda^2 - 0.0350} + \dfrac{0.17673\lambda^2}{\lambda^2 - 3674} + \dfrac{2.0621\lambda^2}{\lambda^2 - 7695}$.

Table 4.4-4 Cubic, point group $m3m$ (O_h) materials, cont.

General optical properties	Refractive index		Nonlinear dielectric susceptibility
Transparency range (μm) dn/dT $(10^{-5}\,\mathrm{K}^{-1})$	n A B C	D E F G	$\chi^{(3)}_{1111}$ $\chi^{(3)}_{1122}$ $\chi^{(3)}_{1133}$ $(10^{-22}\,\mathrm{m}^2/\mathrm{V}^2)$
0.2–16 [4.40] −3.42 (0.458 μm) −3.54 (0.633 μm) −3.63 (3.39 μm) [4.41]	**1.5313** For dispersion relation, see^h [4.65]		6.7 [4.70] 2.78 [4.71] at 0.6943 and 0.7456 μm – $\chi^{(3)}_{1111}=9.5\pm2$, $\chi^{(3)}_{1133}=4.0\pm1.2$ [4.66]
0.135–11.2 [4.83] −1.19 (0.458 μm) −1.32 (0.633 μm) −1.25 (3.39 μm) [4.41]	**1.32** 1.41572 0.32785 0.0137	3.18248 1646 0 0 [4.65]^d	1.4 [4.70] – –
0.13–11 −0.9 (0.150 μm) −1.2 (0.590 μm) −0.4 (14.0 μm) [4.43]	**1.59** (0.150 μm) **1.50** (0.200 μm) **1.47** (0.266 μm) **1.4380** (0.590 μm) **1.433** (1.05 μm) **1.404** (6.00 μm)	**1.37** (9.00 μm) **1.33** (11.0 μm) **1.25** (14.0 μm) dispersion relations, see [4.43]^i	0.82 ± 0.2 – 0.56 ± 0.8 (0.575; 0.613 μm) [4.42]
0.41–5.1 [4.84, 85] –	**2.3104** 1 3.042143 0.02178287	1.170065 0.08720717 30.83326 1101.3146 [4.86]^d	≈ 2000 – ≈ 1000 (0.6943 and 0.7456 μm) [4.66]
0.21–5.3 [4.87] 0.905 [4.87]	**1.8422** (0.5 μm) **1.8258** (0.7 μm) **1.8186** (0.9 μm) **1.8152** (1.05 μm) **1.8149** (1.064 μm)	For dispersion relation, see^j [4.88]	

^{f} $n^2 = 1.26486 + \dfrac{0.30523\lambda^2}{\lambda^2-0.0100} + \dfrac{0.41620\lambda^2}{\lambda^2-0.0172} + \dfrac{0.18870\lambda^2}{\lambda^2-0.0262} + \dfrac{2.6200\lambda^2}{\lambda^2-4959}$;

^{g} $n^2 = 1.47285 + \dfrac{0.16512\lambda^2}{\lambda^2-0.0166} + \dfrac{0.41222\lambda^2}{\lambda^2-0.0306} + \dfrac{0.44163\lambda^2}{\lambda^2-0.0350} + \dfrac{0.16076\lambda^2}{\lambda^2-0.0480} + \dfrac{0.33571\lambda^2}{\lambda^2-4822} + \dfrac{1.92474\lambda^2}{\lambda^2-9612}$;

^{h} $n^2 = 1.00055 + \dfrac{0.19800\lambda^2}{\lambda^2-0.00250} + \dfrac{0.48398\lambda^2}{\lambda^2-0.0100} + \dfrac{0.38696\lambda^2}{\lambda^2-0.0164} + \dfrac{0.25998\lambda^2}{\lambda^2-0.0250} + \dfrac{0.08796\lambda^2}{\lambda^2-1640} + \dfrac{3.17064\lambda^2}{\lambda^2-3719} + \dfrac{0.30038\lambda^2}{\lambda^2-14482}$;

^{i} $n^2 = 1.33973 + \dfrac{0.7097\lambda^2}{\lambda^2-0.09597^2} + \dfrac{0.1788\lambda^2}{\lambda^2-26.03^2} + \dfrac{3.8796\lambda^2}{\lambda^2-45.60^2}$;

^{j} $n^2 = 3.2968230 - 0.0166197\lambda^2 + 0.0126503\lambda^{-2} + 0.0069986\lambda^{-4} - 0.0013968\lambda^{-6} + 0.0001088\lambda^{-8}$.

Table 4.4-5 Cubic, point group $\bar{4}3m$ (T_d) materials

Material	General ϱ (g/cm³) Mohs hardness κ (W/m K)	Static dielectric constant ε_{11}^S ε_{11}^T	Dissipation factor $\tan \delta_1$ (f (Hz))	Elastic stiffness tensor c_{11} c_{44} c_{12} (10^9 Pa)	Piezoelectric strain tensor d_{14} (10^{-12} C/N)	Elastooptic tensor p_{11} p_{12} p_{44} $p_{11}-p_{12}$
Cadmium telluride, CdTe (Irtran-6)	5.855 3 6.28 [4.79]	9.65 (−196 °C) [4.89] 11.00 (23 °C) [4.90]		53.5(2) 20.2(2) 36.9(3) [4.7]	1.68(8) (−196 °C) [4.89]	−0.152 −0.017 −0.057 −0.135 (10 600 nm) [4.91]
Copper bromide, CuBr		7.9(8) (22 °C) [4.92] 6.6 (−193 °C) [4.93]		43.5 14.7 34.9 [4.95]	16(1) [4.94]	0.072 0.195 −0.083 −0.123 [4.96]
Copper chloride, CuCl (nantokite)	4.136 − −	8.3(5) 9.2(5) [4.97, 98]		45.4 13.6[a] 36.3 [4.99]	27.2(5) [4.94]	0.120 0.250 −0.082 −0.130 [4.96]
Copper iodide, CuI	5.60 − −	− 6.5 (−269 °C) [4.93]		45.1 18.2[a] 30.7 [4.93]	7(1) [4.94]	3.032 0.151 −0.068 −0.119 [4.96]
Gallium antimonide, GaSb	5.614 4.5 32 [4.101]	15 [4.100] 15.7[b] [4.101]		88.4(9) 43.4(9) 40.3(8) [4.7]	−2.9 [4.100]	
Gallium arsenide, GaAs	5.3169 4.5 46.05 [4.16]	12.95(10) [4.102] 13.08 [4.90]	0.001 (2.5–10 GHz) [4.103]	118(1) 59.4(2) 53.5(5) [4.7]	−2.7 [4.100]	−0.16 −0.13 −0.05 −0.03 (1150 nm) [4.96, 104]
Gallium phosphide, GaP	4.138 5 110 [4.101]	11.1[b] [4.105]		141(3) 71.2(21) 62.4(12) [4.7]		−0.151 −0.082 −0.074 −0.069 [4.82]

Table 4.4-5 Cubic, point group $\bar{4}3m$ (T_d) materials, cont.

Electrooptic tensor	General optical properties	Refractive index	Nonlinear dielectric susceptibility Second order	Third order
r^T_{41} (10^{-12} m/V)	Transparency range (μm) dn/dT (10^{-5} K^{-1})	n A B C D E	d_{14} (10^{-12} m/V)	$\chi^{(3)}_{1111}$ $\chi^{(3)}_{1122}$ $\chi^{(3)}_{1133}$ (10^{-22} m^2/V^2)
6.8 (10.6 μm) [4.106]	0.85–29.9 [4.85] For temperature-dependent dispersion relations, see [4.110]	**2.693** (14 μm) 1 6.1977889 0.100533 3.2243821 5193.55 [4.111]^c	170 ± 60 (10.6 μm) 60 ± 24 (28 μm) [4.107–109]	
−2.5(5) [4.112]^f	0.50–20 [4.113] –	**2.3365** (0.4358 μm) **2.152** (0.532 μm) **2.045** (1.064 μm) **2.025** (10.6 μm) [4.92, 113, 114]	−6.5 ± 1.3 (1.064 μm) [4.114] −5.0 ± 1.5 (10.6 μm) [4.113]	
−4.97(50) [4.115]	0.45–15 [4.113] –	**1.9216** 3.580 0.03162 0.1642 0.09288 0 [4.116]^c	−5.7 ± 1.1 (1.064 μm) [4.114] −4.15 ± 1.2 (10.6 μm) [4.113]	
18 [4.117, 118]^g	0.50–20 [4.113] –	**2.5621** (0.4358 μm) **2.378** (0.532 μm) **2.245** (1.064 μm) [4.113, 114]	−4.7 ± 1.0 (1.064 μm) [4.114] −5.0 ± 1.5 (10.6 μm) [4.113]	
	1.8–20 [4.22, 85] –	**3.820** (1.8 μm) **3.898** (3.0 μm) **3.824** (5.0 μm) **3.843** (10 μm) [4.121]	+628 ± 63 (10.6 μm) [4.119, 120]	
1.43(7) (1.15 μm) 1.24(4) (3.39 μm) 1.51(5) (10.6 μm) [4.126]	0.900–17.3 [4.122] 25 (1.15 μm) 20 (3.39 μm) 20 (10.6 μm) [4.127]	**3.5072** 3.5 7.497 0.167 1.935 1382 [4.129]^c	d_{36} = 170 (1.064 μm) [4.13, 14, 123, 124] d_{36} = 83 (10.6 μm) [4.12, 14, 128]	6700 [4.64] 1400 [4.125] 1700 [4.64] (all at 10.6 μm)
−0.76(3) [4.130]	0.54–10.5 20 (0.546 μm) 16 (0.633 μm) [4.131]	**3.1057** For dispersion relation, see [4.132]^d	100 [4.105, 131] 58 (10.6 μm) [4.131]	

Table 4.4-5 Cubic, point group $\bar{4}3m$ (T_d) materials, cont.

Material	General ϱ (g/cm³) Mohs hardness κ (W/m K)	Static dielectric constant ε_{11}^S ε_{11}^T	Dissipation factor $\tan \delta_1$ (f (Hz))	Elastic stiffness tensor c_{11} c_{44} c_{12} (10^9 Pa)	Piezoelectric strain tensor d_{14} (10^{-12} C/N)	Elastooptic tensor p_{11} p_{12} p_{44} $p_{11} - p_{12}$
Indium antimonide, InSb	5.78 – 18 [4.101]	16.8[b] [4.101] 17 [4.100]		66.2(7) 30.2(3) 35.9(22) [4.7]	−2.35 [4.100]	0.46 0.58 0.064 (10.6 μm) – [4.133, 134]
Indium arsenide, InAs	5.70 3.8 27 [4.101]	15.5[b] [4.101] 14.5 [4.100]		84.4(17) 39.6(1) 46.4(19) [4.7]	−1.14 [4.100]	
Indium phosphide, InP	4.78 – 68 [4.101]	12.56(20)[b]	See [4.135]	101.1 45.6 56.1 [4.136]		
Zinc selenide, ZnSe	5.26 3–4 19 [4.137]	9.12 9.12 [4.89]		86.4(39) 40.2(18) 51.5(34) [4.7]	1.1(1) [4.89]	|0.100| |0.065| |0.065| (633 nm) [4.138] −0.10 (10 600 nm) [4.139]
Zinc telluride, ZnTe	6.34 6 18 [4.140]	10.10 10.10 [4.89]		71.5(6) 31.1(3) 40.8(1) [4.7]	0.91(5) [4.89]	−0.144 −0.094 −0.046 [4.141] −0.040 [4.142]

[a] The elastic stiffness was measured at constant electric field (c^E).
[b] This value for ε_{11} is neither at constant stress nor at constant strain.
[cde] Dispersion relations (λ (μm), $T = 20\,°\mathrm{C}$):
[c] $n^2 = A + \dfrac{B\lambda^2}{\lambda^2 - C} + \dfrac{D\lambda^2}{\lambda^2 - E}$
[d] $n^2 = 1 + \dfrac{1.390\lambda^2}{\lambda^2 - 0.0296} + \dfrac{4.131\lambda^2}{\lambda^2 - 0.05476} + \dfrac{2.570\lambda^2}{\lambda^2 - 0.1190} + \dfrac{2.056\lambda^2}{\lambda^2 - 757.35}$.

Table 4.4-5 Cubic, point group $\bar{4}3m$ (T_d) materials, cont.

Electrooptic tensor	General optical properties	Refractive index	Nonlinear dielectric susceptibility Second order	Third order
r_{41}^T (10^{-12} m/V)	Transparency range (μm) dn/dT (10^{-5} K^{-1}) —	n A B C D E	d_{14} (10^{-12} m/V)	$\chi^{(3)}_{1111}$ $\chi^{(3)}_{1122}$ $\chi^{(3)}_{1133}$ (10^{-22} m^2/V^2)
	8–30 [4.85] —	**3.904** (14 μm) **3.745** (28 μm) [4.107]	660 [4.123] 2280 ± 270 (10.6 μm) [4.145] 580 (28 μm) [4.107]	See [4.143, 144]
	3.9–20 [4.85] 50 (4 μm) 40 (6 μm) 30 (10 μm) [4.127]	**3.49** (10.6 μm) 11.1 0.71 6.5076 2.75 2084.8 [4.146]^c	346 (1.058 μm) [4.123] 249 (10.6 μm) [4.119, 120]	≈ 25 000 (10.6 and 9.5 μm) [4.64]
1.45 (1.06 μm) [4.149]	0.98–20 [4.147] 8.3 (5 μm) 8.2 (10.6 μm) 7.7 (20 μm) [4.150]	**3.44** 7.255 2.316 0.39225 2.765 1084.7 [4.151, 152]^c	136 (1.058 μm) [4.148] 105 ± 11 (10.6 μm) [4.145]	
1.8 [4.153, 154] 2.2 (10.6 μm) [4.155]	0.47–19 [4.122] —	**2.48** For dispersion relation, see [4.41]^e	+103 [4.122] +80 (10.6 μm) [4.109, 156]	1.05×10^7 (0.4606 μm) 2650 (0.532 μm) 1680 (1.064 μm) (all for $|\chi^{(3)}|$) [4.157]
4.27 (0.616 μm) [4.153, 154]	0.59–25 [4.14] —	**2.69** (10.6 μm) 4.27 3.01 0.142 0 0 [4.158]^c	3.47 [4.122] +90 (10.6 μm) [4.108, 109]	

^{e} $n^2 = 1 + \dfrac{4.2980149\lambda^2}{\lambda^2 - 0.03688810} + \dfrac{0.62776557\lambda^2}{\lambda^2 - 0.14347626} + \dfrac{2.8955633\lambda^2}{\lambda^2 - 2208.4920}$.

^{f} r_{41}^S.

^{g} $n^3 r_{41}$.

Table 4.4-6 Cubic, point group 23 (T) materials

Material	General ϱ (g/cm³) Mohs hardness κ (W/m K)	Static dielectric constant ε_{11}^S ε_{11}^T	Dissipation factor $\tan \delta_1$ (f (Hz))	Elastic stiffness tensor c_{11} c_{44} c_{12} (10⁹ Pa)	Piezoelectric strain tensor d_{14} (10⁻¹² C/N)	Elastooptic tensor p_{11} p_{12} p_{44} $p_{11} - p_{12}$
Bismuth germanium oxide, Bi₁₂GeO₂₀ (BGO)	9.239 5 [4.159] –	38.0(4) [4.160] 40 [4.161]	0.0035 [4.161]	126.0 26.9 34.2 [4.163]	37.58(4) [4.162]	0.12 0.10 0.09(1) 0.01 [4.164–169]
Bismuth silicon oxide, Bi₁₂SiO₂₀ (BSO, sillenite)	9.2 5 [4.159] –	42(1) [4.170] 64(2) [4.171]	0.0004 (> 1 MHz) [4.171]	129(2) 24.7(2) 29.4(12) [4.7]		0.16 0.13 0.12 0.015 [4.172, 173]
Sodium chlorate, NaClO₃	2.488 – –	4.8 5.85 [4.175]		49.6(5) 11.6(2) 14.7(6) [4.7]	−1.74 [4.174]	0.162 0.24 −0.0198 −0.078 [4.176]

Table 4.4-7 Hexagonal, point group $\bar{6}m2$ (D_{3h}) materials

Material	General ϱ (g/cm³) Mohs hardness $\kappa \perp c$ $\kappa \parallel c$ (W/m K)	Static dielectric tensor ε_{11}^S ε_{33}^S ε_{11}^T ε_{33}^T	Elastic stiffness tensor c_{11} c_{33} c_{44} c_{12} c_{13} (10⁹ Pa)	Elastooptic tensor p_{11} p_{12} p_{13} p_{31} p_{33} p_{44}
Gallium selenide, GaSe	5.0 2 9.0 ($\perp c$) 8.25 ($\parallel c$) [4.178]	7.45[a] 7.1[a] 9.80[a] 8.0(3) [4.179]	106.4(37) 35.8(15) 10.2(5) 30.0(25) 12.1(4) [4.7]	$\|p_{12}/p_{13}\| < 0.05$ [4.177]
Gallium sulfide, GaS	3.86 2 – –		127(19) 42(11) 12.0(73) 35.7(45) 14.3(81) [4.7]	

[a] From IR measurements. "ε^S" is ε at optical frequencies.

Table 4.4-6 Cubic, point group 23 (T) materials, cont.

Electrooptic tensor r_{41}^T (10^{-12} m/V)	General optical properties Transparency range (μm) dn/dT (10^{-5} K^{-1})	Refractive index n A B C D E	Nonlinear dielectric susceptibility d_{14} (10^{-12} m/V)
4.1(1)	0.385–7 [4.159, 180] —	2.55 (0.633 μm) [4.159]	
4.25–5.0 [4.181–183]	0.390–6 [4.159, 180] —	2.54 (0.633 μm) [4.159]	
0.36 (0.4 μm) 0.39 (0.59 μm) [4.185]		1.512 (0.6943 μm) For dispersion relation, see [4.186][a]	0.43 (0.6943 μm) [4.184]

[a] Dispersion relation (λ (μm), $T = 20\,^\circ$C):
$$n^2 = A + \frac{B\lambda^2}{\lambda^2 - C} + \frac{D\lambda^2}{\lambda^2 - E} + \frac{F\lambda^2}{\lambda^2 - G}.$$

Table 4.4-7 Hexagonal, point group $\bar{6}m2$ (D_{3h}) materials, cont.

Electrooptic tensor r_{13}^T r_{33}^T r_{41}^T r_{51}^T (10^{-12} m/V) at $\lambda = 633$ nm	General optical properties Transparency range (μm)	Refractive index n_o A B C D E F	n_e A B C D E F	Nonlinear dielectric susceptibility d_{31} d_{33} (10^{-12} m/V)
22 ($n_o^3 r_{11}$) [4.188, 189]	0.62–20 [4.187]	2.9082 7.443 0.4050 0.0186 0.0061 3.1485 2194 [4.193][b]	2.5676 5.76 0.3879 0.2288 0.1223 1.8550 1780 [4.193][b]	d_{22} = 54 pm/V (10.6 μm) [4.190–192]
	0.420–			d_{16} = 84 pm/V (0.6943 μm) [4.194]

[b] Dispersion relations (λ (mm), $T = 20\,^\circ$C):
$$n^2 = A + \frac{B}{\lambda^2} + \frac{C}{\lambda^4} + \frac{D}{\lambda^6} + \frac{E\lambda^2}{\lambda^2 - F}.$$

Table 4.4-8 Hexagonal, point group $6mm$ (C_{6v}) materials

Material	General ϱ (g/cm³) Mohs hardness $\kappa \perp c$ $\kappa \parallel c$ (W/m K)	Static dielectric tensor ε_{11}^S ε_{33}^S ε_{11}^T ε_{33}^T	Elastic stiffness tensor c_{11} c_{33} c_{44} c_{12} c_{13} (10^9 Pa)	Piezoelectric tensor d_{31} d_{33} d_{15} d_h (10^{-12} C/N)	Elastooptic tensor p_{11} p_{12} p_{13} p_{31} p_{33} p_{44}
Beryllium oxide, BeO (bromellite)	3.010 – 370 – [4.45]	6.82[a] 7.62[a] [4.195]	460.6 491.6 147.7 126.5 88.4 [4.196]	−0.12(3) +0.24(6) – – [4.197]	
Cadmium selenide, CdSe	5.67 3.25 6.2 ($\perp c$) 6.9 ($\parallel c$) [4.200]	9.53 10.20 9.70 10.65 [4.89]	74.1 83.6 13.17 45.2 39.3 [4.89][b]	−3.80(11) 7.81(23) −10.1(4) – [4.201, 202]	See [4.198, 199]
Cadmium sulfide, CdS (greenockite)	4.82 3–3.5 14 ($\perp c$) 16 ($\parallel c$) [4.200]	8.67(7) 9.53(7) 8.92(7) 10.20(7) [4.203, 204]	88.4(47) 95.2(22) 15.0(3) 55.4(39) 48.0(22) [4.7, 205]	−5.09(9) [4.203, 204] +9.71(29) [4.203, 204] −11.91(39) [4.203, 204] <0.05 [4.8]	−0.142 −0.066 −0.057 −0.041 −0.20 ≈|0.054| [4.82, 206]
Gallium nitride, GaN	6.15 – 1300 – [4.208]	9.5(3) 10.4(3) [4.207][a]	296 267 24.1 130 158 [4.209]		
α-Silicon carbide, SiC	3.217 – 490 [4.45] –	$\varepsilon_{33} = 6.65$ [4.213]	502 565 169 95 56 [4.214]		See [4.210–212]
Zinc oxide, ZnO (zincite)	5.605 – 29 [4.45] –	8.33(8) 8.81(10) 8.67(9) 11.26(12) [4.215, 216]	207.0 209.5 44.8 117.7 106.1 [4.215, 216][b]	−5.12(6) [4.215, 216] 12.3(2) [4.215, 216] −8.3(3) [4.215, 216] <0.2 [4.8]	|0.221| |0.099| −0.090 |0.089| −0.263 −0.061 [4.90]

Table 4.4-8 Hexagonal, point group $6mm$ (C_{6v}) materials, cont.

Electrooptic tensor	General optical properties	Refractive index		Nonlinear dielectric susceptibility
r_{13}^T r_{33}^T r_{51}^T (10^{-12} m/V) at $\lambda = 633$ nm	Transparency range (µm) dn_o/dT (10^{-5} K^{-1}) dn_e/dT (10^{-5} K^{-1})	n_o A B C D E	n_e A B C D E	d_{31} d_{33} d_{15} (10^{-12} m/V)
1.33 (r_{51}^T) [4.218, 219]	0.21–7, 15–25 [4.217] 0.818 [4.220] 1.34 [4.220]	**1.7055** 1 1.92274 0.0062536 1.24209 94.344 [4.222]^c	**1.7204** 1 1.96939 0.0073788 1.24209 109.82 [4.222]^c	0.23 0.32 – [4.156, 221]
1.8 (r_{13}^S) 4.3 (r_{33}^S) [4.225]	0.75–20 [4.223, 224] – –	**2.5375** 4.2243 1.768 0.227 3.12 3380 [4.227]^c	**2.5572** 4.2009 1.8875 0.2171 3.6461 3629 [4.227]^c	−29 (10.6 µm) [4.109] +55 (10.6 µm) [4.109] 23 ± 3 (1.054 mm) [4.226]
2.45(8) 2.75(8) 1.7(3) (all at 10.6 µm) [4.126]	0.53–16 [4.122] 5.86 (10.6 µm) 6.24 (10.6 µm) [4.229]	**2.212** (10.6 µm) 5.235 −0.1819 0.1651 0 [4.230]^d –	**2.225** (10.6 µm) 5.239 0.2076 0.1651 0 [4.230]^d –	16 [4.228] 32 [4.228] 18 [4.228] −16 (d_{31}, 10.6 µm) [4.12, 14] −16 (d_{32}, 10.6 µm) [4.12, 14] 32 (d_{33}, 10.6 µm) [4.12, 14]
0.57 ± 0.11 (r_{31}) 1.91 ± 0.35 (r_{33}) [4.231]	0.37– – –	**2.33** 3.60 1.75 0.0655 4.1 319 [4.207]	**2.35** 5.35 5.08 315.4 0 [4.207] –	21 44 25.5 [4.232]
	0.51–4 – –	**2.5830** 1 5.5515 0.026406 0 [4.213]^c –	**2.6225** 1 5.7382 0.028551 0 [4.213]^c –	5 50 [4.233] –
−1.4 (r_{13}^S) +2.6 (r_{33}^S) [4.225, 234, 235] −3.1 (r_{51}^T) at 396 nm [4.236]	0.38–6.0 – –	**1.939** 2.81418 0.87968 0.09254 0.00711 [4.222]^e –	**1.955** 2.80333 0.94470 0.09024 0.00714 [4.222]^e –	+1.7 ± 0.15 −5.6 ± 0.15 1.8 ± 0.15 [4.156, 228]

Table 4.4-8 Hexagonal, point group $6mm$ (C_{6v}) materials, cont.

Material	General ϱ (g/cm³) Mohs hardness $\kappa \perp c$ $\kappa \parallel c$ (W/m K)	Static dielectric tensor ε_{11}^S ε_{33}^S ε_{11}^T ε_{33}^T	Elastic stiffness tensor c_{11} c_{33} c_{44} c_{12} c_{13} (10^9 Pa)	Piezoelectric tensor d_{31} d_{33} d_{15} d_h (10^{-12} C/N)	Elastooptic tensor p_{11} p_{12} p_{13} p_{31} p_{33} p_{44}
α-Zinc sulfide, α-ZnS (wurtzite, zinc blende)	4.09 – 27 [4.45] –	8.58 8.52 8.60(5) 8.57(7) [4.239, 240]	122.0 140.2 28.5 58.0 46.8 [4.239, 240]^b	$-1.1(0)$ $+3.2(1)$ $-2.8(1)$ $+1.0$ [4.237, 238]	-0.115 0.017 0.025 0.0271 -0.13 -0.0627 [4.241]

^{a} These values for ε_{ij} are neither at constant stress nor at constant strain.
^{b} The elastic stiffness constant were measured at constant electric field (c^E).

Table 4.4-9 Hexagonal, point group 6 (C_6) materials

Material	General ϱ (g/cm³) Mohs hardness $\kappa \perp c$ $\kappa \parallel c$ (W/m K)	Static dielectric tensor ε_{11}^S ε_{33}^S ε_{11}^T ε_{33}^T	Elastic stiffness tensor c_{11} c_{33} c_{44} c_{12} c_{13} (10^9 Pa)	Piezoelectric strain tensor d_{31} d_{33} d_{14} d_{15} (10^{-12} C/N)	Elastooptic tensor p_{11} p_{12} p_{13} p_{16} p_{31} p_{33} p_{44} p_{45}
Barium nitrite monohydrate, Ba(NO$_2$)$_2$ · H$_2$O	3.179 – – –	– – 7.56(8) 6.78(7) [4.242]	54.2 29.9 11.2 27.5 17.8 [4.242]^b	$-1.73(17)$ $+3.27(16)$ $+0.47(9)$ $-1.03(10)$ [4.242]	
α-Lithium iodate, LiIO$_3$	4.490 4 1.27 ($\perp c$) 0.65 ($\parallel c$) [4.247, 248]	7.9(4) (-20 °C) [4.245] 5.9(3) (-20 °C) [4.245] 8.25^a (0 °C) [4.246] 6.53^a (0 °C) [4.246]	82.5(8) 55.9(23) 18.0(3) 31.9(8) 20.8(87) [4.7, 249]^b	3.5 [4.8, 243, 244] 48.5 [4.8, 243, 244] 73(7) [4.246] 55.5 [4.8, 243, 244]	|0.32| – |0.24| |0.03| |0.41| |0.23| [4.250] – –

^{a} The values for ε_{ij} are neither at constant stress nor at constant strain.
^{b} The elastic stiffness constants were measured at constant electric field (c^E).

Table 4.4-8 Hexagonal, point group $6mm$ (C_{6v}) materials, cont.

Electrooptic tensor	General optical properties	Refractive index		Nonlinear dielectric susceptibility
r_{13}^T r_{33}^T r_{51}^T (10^{-12} m/V) at $\lambda = 633$ nm	Transparency range (μm) dn_o/dT (10^{-5} K^{-1}) dn_e/dT (10^{-5} K^{-1})	n_o A B C D E	n_e A B C D E	d_{31} d_{33} d_{15} (10^{-12} m/V)
0.92 1.85 [4.234, 235] –	0.35–23 [4.122] – –	**2.213** (10.6 μm) 3.4175 1.7396 0.07166 0 [4.222]^e –	**2.219** (10.6 μm) 3.4264 1.7491 0.07150 0 [4.222]^e –	-18.9 ± 6.3 $+37.3 \pm 12.6$ 21.4 ± 8.4 (all at 10.6 μm) [4.109, 251]

^{cde} Dispersion relations (λ (mm), $T = 20\,°$C):

^{c} $n^2 = A + \dfrac{B\lambda^2}{\lambda^2 - C} + \dfrac{D\lambda^2}{\lambda^2 - E}$.

^{d} $n^2 = A + \dfrac{B}{\lambda^2 - C} - D\lambda^2$.

^{e} $n^2 = A + \dfrac{B\lambda^2}{\lambda^2 - C} - D\lambda^2$.

Table 4.4-9 Hexagonal, point group 6 (C_6) materials, cont.

Electrooptic tensor	General optical properties	Refractive index		Nonlinear dielectric susceptibility
r_{13}^T r_{33}^T r_{41}^T r_{51}^T (10^{-12} m/V) at $\lambda = 633$ nm	Transparency range (μm) dn_o/dT (10^{-5} K^{-1}) dn_e/dT (10^{-5} K^{-1})	n_o A B C D	n_e A B C D	d_{31} d_{33} (10^{-12} m/V)
$+3.47(10)$ $+3.31(13)$ $-0.85(21)$ $-0.41(10)$ [4.242]	0.4–2 [4.252] – –	**1.6266** (0.5 μm) 1.99885 0.542910 0.04128 0.01012 [4.252]^c	**1.5238** (0.5 μm) 1.48610 0.775625 0.01830 0.00090 [4.252]^c	1.14 [4.252] –
6.4 4.2 3.1 7.9 (r_{51}^S) [4.256, 257]	0.3–6.0 [4.253, 254] -9.38 [4.255] -8.25 [4.255]	**1.8571** 3.415716 0.047031 0.035306 0.008801 [4.258]^c	**1.7165** 2.918692 0.035145 0.028224 0.003641 [4.258]^c	4.4 4.5 [4.12, 14]

^{c} Dispersion relation (λ (mm), $T = 20\,°$C):
$n^2 = A + \dfrac{B}{\lambda^2 - C} - D\lambda^2$.

Table 4.4-10 Trigonal, point group $\bar{3}m$ (D_{3d}) materials

Material	General ϱ (g/cm³) Mohs hardness $\kappa \perp c$ $\kappa \parallel c$ (W/m K)	Static dielectric tensor ε_{11}^{S} ε_{33}^{S} ε_{11}^{T} ε_{33}^{T}	Dissipation factor $\tan \delta_1$ $\tan \delta_3$	Elastic stiffness tensor c_{11} c_{33} c_{44} c_{12} c_{13} c_{14} (10^9 Pa)	Elastic compliance tensor s_{11} s_{33} s_{44} s_{12} s_{13} s_{14} (10^{-12} Pa^{-1})	Elastooptic tensor p_{11} p_{12} p_{13} p_{14} p_{31} p_{33} p_{41} p_{44}
α-Aluminium oxide, Al$_2$O$_3$ (sapphire)	3.98 9 33 ($\perp c$) 35 ($\parallel c$) [4.259]	9.4 11.5 [4.48]^a	0.0001 [4.48] –	496(3) 499(4) 146(3) 159(10) 114(3) −23(1) [4.7, 260]	2.35 2.17 6.95 −0.70 −0.38 0.46 [4.260]	−0.23 −0.03 0.02 0.00 0.04 −0.20 0.01 −0.10 (644 nm) [4.261]
Calcite, CaCO$_3$ (calcspar, Iceland spar)	2.71 3 3.00 ($\perp c$) 3.40 ($\parallel c$) (50 °C) [4.262]	8.0^{ab} [4.26]		144(3) 84.3(26) 33.5(7) 54.2(47) 51.2(34) −20.5(2) [4.7, 263]	11.4(2) 17.3(5) 41.4(12) −4.0(4) 4.5(4) 9.5(5) [4.7]	0.062 0.147 0.186 −0.011 0.241 0.139 −0.036 −0.058 (514 nm) [4.264]

^{a} These values for ε_{11} are neither at constant stress nor at constant strain.
^{b} It is not clear which relative dielectric constant ε_{ij} is specified.

Table 4.4-11 Trigonal, point group 32 (D_3) materials

Material	General ϱ (g/cm³) Mohs hardness $\kappa \perp c$ $\kappa \parallel c$ (W/m K)	Static dielectric tensor ε_{11}^{S} ε_{33}^{S} ε_{11}^{T} ε_{33}^{T}	Elastic stiffness tensor c_{11} c_{33} c_{44} c_{12} c_{13} c_{14} (10^9 Pa)	Piezoelectric tensor d_{11} d_{14} (10^{-12} C/N)	Elastooptic tensor p_{11} p_{12} p_{13} p_{14} p_{31} p_{33} p_{41} p_{44}
Aluminium phosphate, AlPO$_4$ (berlinite)	2.620 6.5 $\approx$6 [4.45] –	5.88 [4.8, 265] – 4.73 4.62 [4.268, 269]	67.0(26) 87.2(10) 42.9(4) 9.3(14) 13.1(21) −12.7(5) [4.7, 270]^a	−3.0 1.3 [4.266, 267]	

Table 4.4-10 Trigonal, point group $\bar{3}m$ (D_{3d}) materials, cont.

General optical properties	Refractive index	
Transparency range (μm) dn_o/dT (10^{-5} K^{-1}) dn_e/dT (10^{-5} K^{-1})	n_o A B C D E F	n_e A B C D E F
0.147–5.2 [4.87] – –	**1.7655** (0.7 μm) 1.077 0.0033 1.025 0.0114 5.04 151.2 [4.271]^{cd}	**1.7573** (0.7 μm) 1.041 0.0004 1.030 0.0141 3.55 123.8 [4.271]^{cd}
0.2–2.3 0.21 (0.633 μm) 1.19 (0.633 μm) [4.22]	**1.6428** (1.042 μm) [4.272]	**1.4799** (1.042 μm) [4.272]

^{c} Dispersion relation (λ (μm), $T = 20\,°\text{C}$):
$$n^2 = 1 + \frac{A\lambda^2}{\lambda^2 - B} + \frac{C\lambda^2}{\lambda^2 - D} + \frac{E\lambda^2}{\lambda^2 - F}.$$
^{d} For crystals grown by heat exchanger method (HEM).

Table 4.4-11 Trigonal, point group 32 (D_3) materials, cont.

Electrooptic tensor	General optical properties	Refractive index		Nonlinear dielectric susceptibility
r_{11}^T r_{41}^T (10^{-12} m/V) at $\lambda = 633$ nm	Transparency range (μm) dn_o/dT (10^{-5} K^{-1}) dn_e/dT (10^{-5} K^{-1})	n_o A B C D E	n_e A B C D E	d_{11} (10^{-12} m/V)
		1.5161 (1 μm) [4.273]	**1.5285** (1 μm) [4.273]	0.53 [4.228] $d_{14} = 0.013$ [4.228]

Table 4.4-11 Trigonal, point group 32 (D_3) materials, cont.

Material	General ϱ (g/cm³) Mohs hardness $\kappa \perp c$ $\kappa \parallel c$ (W/m K)	Static dielectric tensor ε_{11}^S ε_{33}^S ε_{11}^T ε_{33}^T	Elastic stiffness tensor c_{11} c_{33} c_{44} c_{12} c_{13} c_{14} (10^9 Pa)	Piezoelectric tensor d_{11} d_{14} (10^{-12} C/N)	Elastooptic tensor p_{11} p_{12} p_{13} p_{14} p_{31} p_{33} p_{41} p_{44}
α-Mercuric sulfide, HgS (cinnabar)	8.05 2–2.5 – –	14.0(3) 25.5(5) 15.0(3) 25.5(5) [4.275]	35.36 50.92 21.40 7.02 8.66 11.3 [4.277]^b	19.1 $\approx \lvert 1.7 \rvert$ [4.8, 274]	– – $\lvert 0.445 \rvert$ – – $\lvert 0.115 \rvert$ [4.276] – –
Potassium fluoroboratoberyllate, KBe$_2$BO$_3$F$_2$ (KBBF)					
α-Silicon dioxide, SiO$_2$ (quartz)	2.6485 7 6.5 ($\perp c$) 11.7 ($\parallel c$) [4.200]	4.435 4.640 4.520 4.640 [4.280]	86.6(3) 106.4(12) 58.0(7) 6.7(9) 12.4(16) 17.8(3) [4.7]^a	+2.3 −0.67 [4.278, 279]	0.16 0.27 0.27 −0.030 0.29 0.10 −0.047 −0.079 (589 nm) [4.281]
Tellurium, Te	6.25 2–2.5 2.0 ($\perp c$) 3.4 ($\parallel c$) [4.285]	33 [4.282] 53 [4.282] – 37 [4.284]	33(1) 71(3) 31.8(25) 9.1(7) 24.5(15) (−)12.4(15) [4.7]	+55 +50 [4.283]	0.164 0.138 0.146 −0.04 −0.086 0.038 0.28 0.14 (10.6 μm) [4.286]

^{a} These values are approximately the constant-field values c^E.
^{b} c^E at constant electric field.
^{cde} Dispersion relations (λ (μm), $T = 20\,°C$):
^{c} $n^2 = A + \dfrac{B\lambda^2}{\lambda^2 - C} + \dfrac{D\lambda^2}{\lambda^2 - E}$.

Table 4.4-11 Trigonal, point group 32 (D_3) materials, cont.

Electrooptic tensor r_{11}^T r_{41}^T (10^{-12} m/V) at $\lambda = 633$ nm	General optical properties Transparency range (µm) dn_o/dT (10^{-5} K^{-1}) dn_e/dT (10^{-5} K^{-1})	Refractive index n_o A B C D E	n_e A B C D E	Nonlinear dielectric susceptibility d_{11} (10^{-12} m/V)
3.1 1.55 [4.289]	0.63–13.5 [4.287] – –	**2.7041** 4.1506 2.7896 0.1328 1.1378 705 [4.227]^c	**2.9909** 4.0101 4.3736 0.1284 1.5604 705 [4.227]^c	50 (10.6 µm) [4.288]
	0.155–3.5 [4.290] – –	**1.487** (0.4 µm) 1 1.169725 0.00624 0.009904 [4.290]^d –	**1.410** (0.4 µm) 1 0.956611 0.0061926 0.027849 [4.290]^d –	0.76 [4.290]
−0.445(10) +0.1904(50) [4.293]	0.15–4.5 [4.262] −0.62 (546 nm) −0.7 (546 nm) [4.68]	**1.5350** (1 µm) For dispersion relation, see [4.68]^e	**1.5438** (1 µm) For dispersion relation, see [4.68]^e	0.30 [4.12–14, 128, 291, 292]
	3.8–32 [4.85] – –	**4.7979** (10.6 µm) 18.5346 4.3289 3.9810 3.78 11.813 [4.227]^c	**6.2483** (10.6 µm) 29.5222 9.3068 2.5766 9.235 13.521 [4.227]^c	598 (10.6 µm) [4.294]

^{d} $n^2 = A + \dfrac{B\lambda^2}{\lambda^2 - C} - D\lambda^2$.

^{e} $n_o^2 = 1 + \dfrac{0.663044\lambda^2}{\lambda^2 - 0.0036} + \dfrac{0.517852\lambda^2}{\lambda^2 - 0.0112} + \dfrac{0.175912\lambda^2}{\lambda^2 - 0.0142} + \dfrac{0.565380\lambda^2}{\lambda^2 - 78.216} + \dfrac{1.675299\lambda^2}{\lambda^2 - 430.23}$;

$n_e^2 = 1 + \dfrac{0.665721\lambda^2}{\lambda^2 - 0.0036} + \dfrac{0.503511\lambda^2}{\lambda^2 - 0.0112} + \dfrac{0.214792\lambda^2}{\lambda^2 - 0.0142} + \dfrac{0.539173\lambda^2}{\lambda^2 - 77.30} + \dfrac{1.807613\lambda^2}{\lambda^2 - 390.85}$.

Table 4.4-12 Trigonal, point group $3m$ (C_{3v}) materials

Material	General ϱ (g/cm^3) Mohs hardness $\kappa \perp c$ $\kappa \parallel c$ (W/m K)	Static dielectric tensor ε_{11}^S ε_{33}^S ε_{11}^T ε_{33}^T	Dissipation factor $\tan \delta_1$ $\tan \delta_3$ (f (kHz))	Elastic stiffness tensor c_{11} c_{33} c_{44} c_{12} c_{13} c_{14} (10^9 Pa)	Piezoelectric tensor d_{22} d_{31} d_{33} d_{15} d_h (10^{-12} C/N)	Elastooptic tensor p_{11} p_{12} p_{13} p_{14} p_{31} p_{33} p_{41} p_{44}
β-Barium borate, BaB$_2$O$_4$ (BBO)	3.849 4 1.2 ($\perp c$) 1.6 ($\parallel c$) [4.200]	– – 6.7 8.1 [4.295]	$< 10^{-3}$ $< 10^{-3}$ (10–50 °C, 1–100 kHz) [4.295]	123.8 53.3 7.8 60.3 49.4 12.3 [4.295]		
Lithium niobate, LiNbO$_3$	4.628 5–5.5 4.6 [4.137]	43.9(22) [4.296] 23.7(12) [4.296] 85.2 [4.298] 28.7 [4.298]	0.0015(4) 0.0011(3) (1 GHz) [4.296]	202(2) 244(5) 60.2(6) 55(2) 72(5) 8.5(7) [4.7]^a	20.7(1) [4.297] −0.86(2) [4.297] +16.2(1) [4.297] 74.0(3) [4.297] 6.310(14) [4.299]	−0.031(9) 0.082(9) 0.135(7) −0.076(15) 0.170(14) 0.069(5) −0.150(10) 0.147(100) [4.11]^b
Lithium niobate (5% MgO-doped) MgO:LiNbO$_3$						
Lithium tantalate, LiTaO$_3$	7.45 5.5	42.6 42.8 53.6 43.4 [4.298]	0.0013(4) 0.0007(3) (1 GHz) [4.296]	230(2) 276(6) 96(1) 42(6) 79(4) −11(1) [4.7, 301, 302]	8.5 [4.300] −3.0 [4.300] 9.2 [4.300] 26 [4.300] 2.000(12) [4.299]	−0.081 0.081 0.093 −0.026 0.089 −0.044 −0.085 0.028 [4.303, 304]

Table 4.4-12 Trigonal, point group $3m$ (C_{3v}) materials, cont.

	Electrooptic tensor	General optical properties	Refractive index		Nonlinear dielectric susceptibility
	r_{13}^T r_{22}^T r_{33}^T r_{51}^T (10^{-12} m/V) at $\lambda = 633$ nm	Transparency range (μm) dn_o/dT (10^{-5} K^{-1}) dn_e/dT (10^{-5} K^{-1}) B C D E F	n_o A B C D E F	n_e A B C D E F	d_{31} d_{33} d_{22} d_{15} (10^{-12} m/V)
β-Barium borate, BaB$_2$O$_4$ (BBO)	+0.27(2) −2.41(3) +0.29(3) +1.7(1) [4.310]	0.189–3.5 [4.295, 305–308] −1.66 [4.295] −0.93 [4.295]	**1.6551** 2.7359 0.01878 0.01822 0.01471 0.0006081 0.0000674 [4.311]^h	**1.5426** 2.3753 0.01224 0.01667 0.01627 0.0005716 0.00006305 [4.311]^h	2.16 ± 0.08 (d_{22}) [4.12, 14, 309]
Lithium niobate, LiNbO$_3$	+9.6 +6.8 +30.9 +32.6 [4.315, 316]	0.33–5.5 [4.312] $dn_o/dT = 0.141$ [4.313]^f $dn_o/dT = 2.0$ [4.314]^g $dn_e/dT = 3.85$ [4.313]^f $dn_e/dT = 7.6$ [4.314]^g	**2.2340** 4.91296 0.116275 0.048398 0.0273 [4.313]^i – –	**2.1554** 4.54528 0.091649 0.046079 0.0303 [4.313]^i – –	−4.6 −25 [4.13, 14] – –
Lithium niobate (5% MgO-doped) MgO:LiNbO$_3$	11.2(5) – 36.0(5) – (for 10.7% MgO) [4.320]	0.4–5 [4.317–319] – –	**2.2272** 4.9017 0.112280 0.049656 0.039636 [4.317, 318]^i – –	**2.1463** 4.5583 0.091806 0.048086 0.032068 [4.317, 318]^i – –	−4.69 [4.309] |25| [4.13] – –
Lithium tantalate, LiTaO$_3$	4.5 0.3 27 15 (all for r^S at 3.39 μm) [4.323]	0.4–5.5 – –	**2.1366** (1.058 μm) [4.321]	**2.1406** (1.058 μm) [4.321]	−1.0 ± 0.2 −16 ± 2 +1.7 ± 0.2 [4.156, 322] –

Table 4.4-12 Trigonal, point group $3m$ (C_{3v}) materials, cont.

Material	General ϱ (g/cm^3) Mohs hardness $\kappa \perp c$ $\kappa \parallel c$ (W/m K)	Static dielectric tensor ε_{11}^S ε_{33}^S ε_{11}^T ε_{33}^T	Dissipation factor $\tan\delta_1$ $\tan\delta_3$ (f (kHz))	Elastic stiffness tensor c_{11} c_{33} c_{44} c_{12} c_{13} c_{14} (10^9 Pa)	Piezoelectric tensor d_{22} d_{31} d_{33} d_{15} d_h (10^{-12} C/N)	Elastooptic tensor p_{11} p_{12} p_{13} p_{14} p_{31} p_{33} p_{41} p_{44}
Silver antimony sulfide, Ag$_3$SbS$_3$ (pyrargyrite)	5.83 2–2.5 – –	– – 21.7(7) 24.7(7) [4.324]		52.7 35.6 11.6 26.3 28.0 −0.4 [4.324, 327]^c	16.2 −19.5 62.2 31.6 – [4.325, 326]	
Silver arsenic sulfide, Ag$_3$AsS$_3$ (proustite)	5.63 2–2.5 – –	20.2(2) 20.2(2) 21.5 22.0 [4.324]		56.76 37.0 9.12 30.4 30.4 −0.16 [4.324]	12.1 −13.8 30 −20 [4.324] –	|0.056| |0.082| |0.068| – |0.103| |0.100| |≈ 0.01| (1150 nm) – [4.328, 329]
Thallium arsenic selenide, Tl$_3$AsSe$_3$ (TAS)	7.83 2–3 – –			See [4.330]		See [4.330]
Tourmaline, (Na, Ca)(Mg, Fe)$_3$B$_3$Al$_6$ Si$_6$(O, OH, F)$_{31}$	3.0–3.2 7–7.5 3.03 ($\perp c$) 2.92 ($\parallel c$) (125 °C) [4.332]^d	6.3^e [4.278] 7.1^e [4.278] 8.2 [4.331] 7.5 [4.331]		277(19) 163(11) 64(3) 65(24) 32(20) −6.9(25) [4.7]	−0.3 [4.331] −0.34 [4.331] −1.8 [4.331] −3.6 [4.331] 4.00(6) [4.333, 334]	

^{a} These values are approximately the constant-field values c^E.
^{b} p^E at constant electric field.
^{c} c^E at constant electric field.
^{d} The thermal conductivities have been interpolated from $\kappa(\perp c) = 0.108 \times T^{0.556}$ and $\kappa(\parallel c) = 0.492 \times T^{0.297}$, valid for temperature ranges of 398.2–723.2 K and 393.2–729.2 K, respectively [4.332].
^{e} These values for ε_{ij} are neither at constant stress nor at constant strain.
^{f} Fabricated by vapor transport equilibration (VTE).
^{g} Stoichiometric melt.

Table 4.4-12 Trigonal, point group $3m$ (C_{3v}) materials, cont.

Electrooptic tensor	General optical properties	Refractive index		Nonlinear dielectric susceptibility
r_{13}^T	Transparency range (µm)	n_o	n_e	d_{31}
r_{22}^T	dn_o/dT (10^{-5} K^{-1})	A	A	d_{33}
r_{33}^T	dn_e/dT (10^{-5} K^{-1})	B	B	d_{22}
r_{51}^T		C	C	d_{15}
(10^{-12} m/V)		D	D	(10^{-12} m/V)
at $\lambda = 633$ nm		E	E	
		F	F	
	0.7–14 [4.335]	**2.9458**	**2.7956**	7.8 (10.6 µm)
	–	1	1	–
	–	6.585	5.845	8.2 (10.6 µm)
		0.16	0.16	–
		0.1133	0.0202	[4.3, 4, 294, 336]
		225 [4.335]^j	225 [4.335]^j	
		–	–	
2.0 [4.337, 338]	0.6–13 [4.339]	**2.8163**	**2.5822**	10.4 (10.6 µm)
1.05(4) [4.340]	–	9.220	7.007	–
0.22 [4.337, 338]	–	0.4454	0.3230	16.6 (10.6 µm)
3.4(5) [4.340]		0.1264	0.1192	10.8 (10.6 µm)
		1733	660	[4.339, 342]
		1000 [4.341]^k	1000 [4.341]^k	
		–	–	
	1.28–17 [4.343]	**3.331** (11 µm)	**3.152** (11 µm)	d_{eff} SHG 10.6 µm:
	–4.52 [4.344]	1	1	67.5 [4.342, 343]
	+3.55 [4.344]	10.210	8.993	36.5 [4.294, 343]
		0.197136	0.197136	29 [4.345]
		0.522	0.308	20 [4.346]
		625 [4.344]^j	625 [4.344]^j	
		–	–	
$r_{13}^S = 1.7$ [4.347]		**1.6274**	**1.6088**	0.13(3)
$r_{22}^T = 0.3$ (589 nm)		1	1	0.47(6)
[4.348]		1.6346	1.57256	0.07(1)
$r_{33}^S = 1.7$ [4.347]		0.010734	0.011346	0.23(4) [4.321]
		0 [4.349]^j	0 [4.349]^j	
		–	–	

^{hijk} Dispersion relations (λ (µm), $T = 20\,°C$):

^{h} $n^2 = A + \dfrac{B}{\lambda^2 - C} - D\lambda^2 + E\lambda^4 - F\lambda^6$.

^{i} $n^2 = A + \dfrac{B}{\lambda^2 - C} - D\lambda^2$.

^{j} $n^2 = A + \dfrac{B\lambda^2}{\lambda^2 - C} + \dfrac{D\lambda^2}{\lambda^2 - E}$.

^{k} $n^2 = A + \dfrac{B}{\lambda^2 - C} + \dfrac{D}{\lambda^2 - E}$.

Table 4.4-13 Tetragonal, point group $4/mmm$ (D_{4h}) materials

Material	General ϱ (g/cm^3) Mohs hardness $\kappa \perp c$ $\kappa \parallel c$ (W/m K)	Static dielectric tensor ε_{11}^S ε_{33}^S ε_{11}^T ε_{33}^T	Dissipation factor $\tan \delta_1$	Elastic stiffness tensor c_{11} c_{33} c_{44} c_{66} c_{12} c_{13} (10^9 Pa)	Elastic compliance tensor s_{11} s_{33} s_{44} s_{66} s_{12} s_{13} (10^{-12} Pa^{-1})	Elastooptic tensor p_{11} p_{12} p_{13} p_{31} p_{33} p_{44} p_{66}
Magnesium fluoride, MgF$_2$	3.177 6 21 ($\perp c$) 30 ($\parallel c$) [4.200]	5.85 4.87 [4.21]^a		138(7) 201(10) 56.5(5) 96(2) 88(7) 62(4) [4.7]	12.6(1) 6.0(3) 17.7(2) 10.5(2) −7.2(2) −1.7(1) [4.7]	0.041 0.119 0.078 −0.099 0.014 − −0.0475 [4.350]
Titanium dioxide, TiO$_2$ (rutile)	4.26 6–6.5 8.8 ($\perp c$) (40 °C) 12.6 ($\parallel c$) [4.262]	85 190 [4.79]^a	0.017 [4.79] −	269(3) 480(5) 124(1) 192(2) 177(4) 146(6) [4.7]	6.8(3) 2.60(1) 8.06(5) 5.21(5) −4.0(3) −0.85(3) [4.7]	0.012(4) 0.162(16) −0.157(12) −0.092(9) −0.059(3) 0.020 −0.060 [4.351]
Yttrium vanadate, YVO$_4$ (YVO)	4.22 5 5.32 ($\perp c$) 5.10 ($\parallel c$) [4.48]					

^{a} These values for ε_{ij} are neither at constant stress nor at constant strain.

Table 4.4-13 Tetragonal, point group 4*mmm* (D_{4h}) materials, cont.

General optical properties	Refractive index		Nonlinear dielectric susceptibility
Transparency range (μm) dn_o/dT (10^{-5} K^{-1}) dn_e/dT (10^{-5} K^{-1})	n_o A B C D E	n_e A B C D E	$\chi^{(3)}$
0.12–7.5 0.17 (0.4 μm) 0.23 (0.4 μm) [4.21]	**1.48** (0.150 μm) **1.423** (0.200 μm) **1.3776** (0.590 μm) **1.373** (1.05 μm) **1.32** (6.00 μm) **1.29** (7.50 μm) dispersion relations, see [4.43]^b	**1.49** (0.150 μm) **1.437** (0.200 μm) **1.3894** (0.590 μm) **1.385** (1.05 μm) **1.33** (6.00 μm) **1.30** (7.50 μm) dispersion relations, see [4.43]^b	[4.21]
0.42–4.0 [4.84] 0.4 (0.4 μm) [4.353] −0.9 (0.4 μm) [4.353]	**2.4851** 5.913 0.2441 0.0803 0 [4.353]^c –	**2.7488** 7.197 0.3322 0.0843 0 [4.353]^c –	See [4.352]
0.4–5 0.85 0.3 [4.354]	**1.9573** 3.77834 0.069736 0.04724 0.0108133 [4.354]^d –	**2.1652** 4.59905 0.110534 0.04813 0.0122676 [4.354]^d –	

^{bcd} Dispersion relations (λ (μm), $T = 20\,°C$):

^{b} $n_o^2 = 1.27620 + \dfrac{0.60967\lambda^2}{\lambda^2 - 0.08636^2} + \dfrac{0.0080\lambda^2}{\lambda^2 - 18.0^2} + \dfrac{2.14973\lambda^2}{\lambda^2 - 25.0^2}$;

$n_e^2 = 1.25385 + \dfrac{0.66405\lambda^2}{\lambda^2 - 0.088504^2} + \dfrac{1.0899\lambda^2}{\lambda^2 - 22.2^2} + \dfrac{0.1816\lambda^2}{\lambda^2 - 24.4^2} + \dfrac{2.1227\lambda^2}{\lambda^2 - 40.6}$;

^{c} $n^2 = A + \dfrac{B\lambda^2}{\lambda^2 - C} + \dfrac{D\lambda^2}{\lambda^2 - E}$;

^{d} $n^2 = A + \dfrac{B}{\lambda^2 - C} - D\lambda^2$.

Table 4.4-14 Tetragonal, point group $4/m$ (C_{4h}) materials

Material	General	Elastic stiffness tensor	Elastic compliance tensor	Elastooptic tensor
	ϱ (g/cm³) Mohs hardness $\kappa \perp c$ $\kappa \parallel c$ (W/m K)	c_{11} c_{33} c_{44} c_{66} c_{12} c_{13} c_{16} (10^9 Pa)	s_{11} s_{33} s_{44} s_{66} s_{12} s_{13} s_{16} (10^{-12} Pa^{-1})	p_{11} p_{12} p_{13} p_{16} p_{31} p_{33} p_{44} p_{45} p_{61} p_{66}
Lead molybdate, PbMoO$_4$	6.95 2.5–3 150 ($\perp c$) 150 ($\parallel c$) [4.355]	107.2(19) 93.2(13) 26.4(3) 34.8(8) 61.9(55) 52.0(8) −15.8(18) [4.7]	20.8(2) 16.3(8) 37.9(4) 42.3(20) −11.8(6) −5.0(5) 14.8(12) [4.356]	0.253(23) 0.253(23) 0.273(43) 0.015(4) 0.163(20) 0.298(12) 0.04(0) −0.01(0) 0.025(21) 0.046(5) (477–633 nm) [4.357]
Yttrium lithium fluoride, YLiF$_4$ (YLF)	3.995 4–5 7.2 ($\perp c$) 5.8 ($\parallel c$) [4.358]	121 156 40.9 17.7 60.9 52.6 −7.7 [4.359]	12.8 7.96 24.4 63.6 −6.0 −2.3 8.16 [4.359]	

[a] Dispersion relation (λ (μm), $\lambda > 0.5$ μm, $T = 20\,°\mathrm{C}$):
$$n^2 = A + \frac{B\lambda^2}{\lambda^2 - C}.$$

Table 4.4-15 Tetragonal, point group 422 (D_4) materials

Material	General	Static dielectric tensor	Dissipation factor	Elastic stiffness tensor	Piezoelectric tensor	Elastooptic tensor
	ϱ (g/cm³) Mohs hardness $\kappa \perp c$ $\kappa \parallel c$ (W/m K)	ε_{11}^S ε_{33}^S ε_{11}^T ε_{33}^T	$\tan \delta_1$ $\tan \delta_3$ (f (kHz))	c_{11} c_{33} c_{44} c_{66} c_{12} c_{13} (10^9 Pa)	d_{14} (10^{-12} C/N)	p_{11} p_{12} p_{13} p_{31} p_{33} p_{44} p_{66}
Tellurium dioxide, TeO$_2$ (paratellurite)	5.99 4 3.0(3) [4.362] –	22.7 [4.360] 24.9[a] [4.361] 22.9(10) [4.360] 24.7(15) [4.360]	0.0011 0.012 (100 kHz) [4.360]	55.9(2) 105.5(4) 26.7(2) 66.3(4) 51.6(3) 23.9(22) [4.7]	8.13(70) [4.360]	0.007 0.187 0.340 0.091 0.240 −0.17 −0.046 [4.363]

[a] This value for ε_{33} is neither at constant stress nor at constant strain.

Table 4.4-14 Tetragonal, point group $4/m$ (C_{4h}) materials, cont.

General optical properties	Refractive index	
Transparency range (μm) dn_o/dT (10^{-5} K^{-1}) dn_e/dT (10^{-5} K^{-1})	n_o A B C	n_e A B C
0.42–5.5 [4.364] $-(3.0\pm 0.2)n_o$ $-(1.8\pm 0.2)n_e$ [4.355]	**2.38** (633 nm) [4.364] 1 4.0650407 0.0536585 [4.355]^a	**2.25** (633 nm) [4.364] 1 3.7037037 0.0400000 [4.355]^a
0.18–6.7 -0.2 -0.43 [4.354]	**1.448** 1.38757 0.70757 0.00931 0.18849 50.99741 [4.354]^c	**1.470** 1.31021 0.84903 0.00876 0.53607 134.9566 [4.354]^c

^{c} $n^2 = A + \dfrac{B\lambda^2}{\lambda^2 - C} + \dfrac{D\lambda^2}{\lambda^2 - E}$.

Table 4.4-15 Tetragonal, point group 422 (D_4) materials, cont.

Electrooptic tensor	General optical properties	Refractive index		Nonlinear dielectric susceptibility
r_{41}^T (10^{-12} m/V) at $\lambda = 633$ nm	Transparency range (μm) dn_o/dT (10^{-5} K^{-1}) dn_e/dT (10^{-5} K^{-1})	n_o A B C D E	n_e A B C D E	d_{14} (10^{-12} m/V)
0.62 [4.365]	0.35–6 [4.366] 0.9 (0.644 μm) 0.8 (0.644 μm) [4.368]	**2.2005** 1 2.584 0.0180 1.157 0.0696 [4.368]^b	**2.3431** 1 2.823 0.0180 1.542 0.0692 [4.368]^b	0.39 [4.367]

^{b} Dispersion relation (λ (μm), $T = 20\,°C$):
$n^2 = A + \dfrac{B\lambda^2}{\lambda^2 - C} + \dfrac{D\lambda^2}{\lambda^2 - E}$

Table 4.4-16 Tetragonal, point group $\bar{4}2m$ (D_{2d}) materials

Material	General ϱ (g/cm^3) Mohs hardness $\kappa \perp c$ $\kappa \parallel c$ (W/m K)	Static dielectric tensor ε_{11}^S ε_{33}^S ε_{11}^T ε_{33}^T	Dissipation factor $\tan \delta_1$ $\tan \delta_3$ (f (kHz))	Elastic stiffness tensor c_{11} c_{33} c_{44} c_{66} c_{12} c_{13} (10^9 Pa)	Piezoelectric tensor d_{14} d_{36} (10^{-12} C/N)	Elastooptic tensor p_{11} p_{12} p_{13} p_{31} p_{33} p_{44} p_{66}
Ammonium dideuterium phosphate, ND$_4$D$_2$PO$_4$ (AD*P, DADP)	1.885 [4.370, 371] – – –	70^a [4.369] 26^a [4.369] 72 [4.372] 22 [4.372]	0.003 0.015 (1 kHz) [4.369]	62.1 29.9 9.1 6.1 –5 14 [4.372]^b	10 75 [4.372]	– – – – – – 0.04 [4.373, 374]
Ammonium dihydrogen arsenate, NH$_4$H$_2$AsO$_4$ (ADA)	2.310 – – –	74 [4.375] 13 [4.375] 75 [4.376] 14 [4.376]		67.5 30.2 6.85 6.39 –10.6 16.5 [4.379]	36.5 27.5 [4.377, 378]	
Ammonium dihydrogen phosphate, NH$_4$H$_2$PO$_4$ (ADP)	1.80 2.0 1.26 ($\perp c$) 0.71 ($\parallel c$) (315 K) [4.262]	55.5(15) 15.0(5) 56.0(15) 15.5(5) [4.97, 98]	0.001 0.004 (1 kHz) [4.369]	67.3(27) 33.7(7) 8.6(2) 6.02(6) 5.0(13) 19.8(6) [4.7]	1.76 48.31 [4.380]	0.302(12) 0.252(17) 0.204(33) 0.191(6) 0.219(49) –0.058 –0.088(17) (589–633 nm) [4.381]
Cadmium germanium arsenide, CdGeAs$_2$	5.60 3.5–4 4.18 [4.140] –			98.0 86.6 43.2 42.3 60.5 59.6 [4.382]		
Cadmium germanium phosphide, CdGeP$_2$						
Cesium dideuterium arsenate, CsD$_2$AsO$_4$ (CD*A, DCDA)		– – 74 61 [4.369]^{cd}	0.57 0.38 (1 kHz) [4.369]^{cd}		10.5 125.4 [4.383]^c	0.223 0.271 0.160 0.206 0.133 –0.061 [4.384, 385] –

Table 4.4-16 Tetragonal, point group $\bar{4}2m$ (D_{2d}) materials, cont.

Electrooptic tensor	General optical properties	Refractive index		Nonlinear dielectric susceptibility
r_{41}^T r_{63}^T (10^{-12} m/V)	Transparency range (μm) dn_o/dT (10^{-5} K^{-1}) dn_e/dT (10^{-5} K^{-1})	n_o A B C D E	n_e A B C D E	d_{36} (10^{-12} m/V)
	0.22–1.7 [4.386–389] – –	**1.5049** 2.279481 1.215879 57.97555433 0.010761 0.013262977 [4.386]^e	**1.4659** 2.151161 1.199009 126.6005279 0.009652 0.009712103 [4.386]^e	0.43 (0.6943 μm) [4.390, 391]
33.5 (550 nm) 9.2 (550 nm) [4.284, 369]	0.22–1.2 [4.386] −4.56 [4.392] 1.25 [4.392]	**1.5550** 2.443449 2.017752 57.83514282 0.016757 0.018272821 [4.386]^e	**1.5081** 2.275962 1.59826 126.8851303 0.014296 0.016560859 [4.386]^e	0.43 [4.386]
26.0 (633 nm) 8.7 (633 nm)	0.18–1.53 [4.386, 393] −4.93 [4.392] 0 [4.392]	**1.5065** 2.302842 15.102464 400 0.011125165 0.013253659 [4.394]^e	**1.4681** 2.163510 5.919896 400 0.009616676 0.012989120 [4.394]^e	0.47 [4.12, 14]
	2.5–15 [4.395] – –	**3.5046** (10.6 μm) 10.1064 2.2988 1.0872 1.6247 1370 [4.227]^f	**3.5911** (10.6 μm) 11.8018 1.2152 2.6971 1.6922 1370 [4.227]^f	282 (10.6 μm) [4.395, 396]
	0.9–12 [4.396] – –	**3.1422** (10.6 μm) 5.9677 4.2286 0.2021 1.6351 671.33 [4.397]^f	**3.1563** (10.6 μm) 6.1573 4.0970 0.2330 1.4925 671.33 [4.397]^f	100 (10.6 μm) [4.396]
– 38.8 (700 nm) [4.399]	0.27–1.66 [4.398] −2.33 [4.392] −1.67 [4.392]	**1.5499** 2.40817 2.2112173 126.871163 0.015598 0.019101728 [4.386]^e	**1.5341** 2.345809 0.651843 127.3304614 0.015141 0.016836101 [4.386]^e	0.402 [4.398]

Table 4.4-16 Tetragonal, point group $\bar{4}2m$ (D_{2d}) materials, cont.

Material	General ϱ (g/cm^3) Mohs hardness $\kappa \perp c$ $\kappa \parallel c$ (W/m K)	Static dielectric tensor ε_{11}^S ε_{33}^S ε_{11}^T ε_{33}^T	Dissipation factor $\tan \delta_1$ $\tan \delta_3$ (f (kHz))	Elastic stiffness tensor c_{11} c_{33} c_{44} c_{66} c_{12} c_{13} (10^9 Pa)	Piezoelectric tensor d_{14} d_{36} (10^{-12} C/N)	Elastooptic tensor p_{11} p_{12} p_{13} p_{31} p_{33} p_{44} p_{66}
Cesium dihydrogen arsenate, CsH$_2$AsO$_4$ (CsDA, CDA)	3.53 – – –	58.0^a [4.400]^d 34.0^a [4.400]^d 61 [4.369]^d 29 [4.369]^d	15 13 (1 kHz) [4.369]	51.6 39.9 6.66 1.7 0.56 1.33 [4.403, 404]	5.6(10) 123(8) [4.401, 402]	0.238(58) 0.225(0) 0.211(19) 0.200(4) 0.220(12) −0.042 |0.065| (633 nm) [4.384, 385]
Cesium lithium borate, CsLiB$_6$O$_{10}$ (CLBO)	– 4 – –					
Copper gallium selenide, CuGaSe$_2$						
Copper gallium sulfide, CuGaS$_2$	4.45 – – –	9.3^a 10^a [4.112]				
Potassium dideuterium arsenate, KD$_2$AsO$_4$ (KD*A, DKDA)	2.890 [4.405]^c – –	71^a 33^{ad} [4.369] ($\approx 90\%$ deuterated)	0.02 0.20 (1 kHz) [4.369]^d ($\approx 90\%$ deuterated)	74.6 69.3 9.88 6.17 12.4 33.9 [4.406]	22.5 24.8 [4.406] ($\approx 90\%$ deuterated)	
Potassium dideuterium phosphate, KD$_2$PO$_4$ (KD*P, DKDP)	2.355 2.5 2.09 ($\perp c$) 1.86 ($\parallel c$) [4.386]	47.1(24)^a [4.177]^c 48 [4.375]^c – 50(2) [4.407]^c	0.0068(17) 0.0072(18) (1 GHz) [4.296]^c	67.4 54.5 12.6 5.94 −5.8 12.2 [4.408]	– 58 [4.407]^c	0.241 0.247 0.245 0.236 0.245 −0.035 −0.072 [4.409, 410]

Table 4.4-16 Tetragonal, point group $\bar{4}2m$ (D_{2d}) materials, cont.

Electrooptic tensor	General optical properties	Refractive index		Nonlinear dielectric susceptibility
r_{41}^T r_{63}^T (10^{-12} m/V)	Transparency range (μm) dn_o/dT (10^{-5} K^{-1}) dn_e/dT (10^{-5} K^{-1})	n_o A B C D E	n_e A B C D E	d_{36} (10^{-12} m/V)
– 19.1 (700 nm) [4.284]	0.26–1.43 [4.398] −2.87 [4.392] −2.21 [4.392]	**1.5514** 2.420405 1.403336 57.82416181 0.016272 0.018005614 [4.386][e]	**1.5356** 2.350262 0.685328 127.2688578 0.015645 0.014820871 [4.386][e]	0.402 [4.398]
	0.18–2.75 [4.411] $-0.104\lambda^2 + 0.035\lambda - 1.291$ [4.412] $0.331\lambda^2 - 0.243\lambda - 0.84$ [4.412]	**1.4854** 2.2145 0.00890 0.02051 0.01413 [4.412][g] –	**1.4352** 2.0588 0.00866 0.01202 0.00607 [4.412][g] –	0.86 [4.411]
	0.73–17 [4.413] – –	**2.8358** (1.0 μm) **2.7430** (2.0 μm) **2.7133** (6.0 μm) [4.413]	**2.8513** (1.0 μm) **2.7510** (2.0 μm) **2.7192** (6.0 μm) [4.413]	27 (10.6 μm) [4.413]
1.76 (633 nm) – [4.112]	0.52–12 [4.414] 5.9 [4.415] 6.0 [4.415]	**2.4360** (10.6 μm) 4.0984 2.1419 0.1225 1.5755 738.43 [4.414][f]	**2.4201** (10.6 μm) 4.4834 1.7316 0.1453 1.7785 738.43 [4.414][f]	9.0 (10.6 μm) [4.414]
– 18.2 (550 nm) [4.284]	0.22–2.3 [4.416] – –			0.39 [4.386]
8.8(4) (546 nm) [4.418] 25.8 (633 nm) [4.419]	0.2–2.1 [4.387, 417] −3.1 [4.392] −2.1 [4.392]	**1.4928** 1.661145 0.586015 0.016017 0.691194 30 [4.397][f]	**1.4555** 1.687499 0.44751 0.017039 0.596212 30 [4.397][f]	0.37 [4.12, 14]

Table 4.4-16 Tetragonal, point group $42m$ (D_{2d}) materials, cont.

Material	General ϱ (g/cm³) Mohs hardness $\kappa \perp c$ $\kappa \parallel c$ (W/m K)	Static dielectric tensor ε_{11}^S ε_{33}^S ε_{11}^T ε_{33}^T	Dissipation factor $\tan\delta_1$ $\tan\delta_3$ (f (kHz))	Elastic stiffness tensor c_{11} c_{33} c_{44} c_{66} c_{12} c_{13} (10^9 Pa)	Piezoelectric tensor d_{14} d_{36} (10^{-12} C/N)	Elastooptic tensor p_{11} p_{12} p_{13} p_{31} p_{33} p_{44} p_{66}
Potassium dihydrogen arsenate, KH_2AsO_4 (KDA)	2.87 – – –	53 [4.420] 18 [4.420] 53.7 [4.421] 21.0 [4.421]	0.007 0.008 (10 GHz) [4.420]	64.8 48.2 10.75 6.63 0.77 13.6 [4.422]	17.1 19.1 [4.406]	– – – – – – 0.020 [4.373, 374]
Potassium dihydrogen phosphate, KH_2PO_4 (KDP)	2.3383 2.5 1.34 ($\perp c$) (42 °C) 1.21 ($\parallel c$) [4.262]	42.5(15) 20.0(5) 43.2(15) 20.8(5) [4.97, 98]	0.004 (9.2 GHz) [4.423] 0.0005 (1 GHz) [4.97, 98]	71.2(14) 57(1) 12.6(4) 6.2(1) −5.0(11) 14.1(14) [4.7]	3.5(15) 22.4(10) [4.401, 402]	0.256(18) 0.254(19) 0.224(34) 0.227(8) 0.200(54) −0.265(10) −0.063(6) (589–633 nm) [4.409, 410]
Rubidium dideuterium arsenate, RbD_2AsO_4 (RbD*A, RD*A, DRDA)	3.333 – – –	72ᵃ 41ᵃᶜ [4.369]	1.6 2.3 (1 kHz) [4.369]ᶜ	49.3 38.6 9.48 4.08 −19.3 4.9 [4.424]	14.6 45.8ᶜ [4.424]	
Rubidium dideuterium phosphate, RbD_2PO_4 (RbD*P, RD*P, DRDP)		– 72ᵃ [4.425, 426]		See [4.427, 428]ᶜ	3.3(3) 52(1)	
Rubidium dihydrogen arsenate, RbH_2AsO_4 (RbDA, RDA)	3.28 – – –	54.5ᵃ 28.5ᵃ [4.429]	2.1 5.2 [4.429]	51.0 39.2 10.4 4.31 −18.9 2.3 [4.424]	9.9 28.5 [4.400, 424]	0.227 0.239 0.200 0.205 0.182 [4.430, 431] – 0.023 [4.373, 374]

Table 4.4-16 Tetragonal, point group $\bar{4}2m$ (D_{2d}) materials, cont.

Electrooptic tensor	General optical properties	Refractive index		Nonlinear dielectric susceptibility
r_{41}^T r_{63}^T (10^{-12} m/V)	Transparency range (μm) dn_o/dT (10^{-5} K^{-1}) dn_e/dT (10^{-5} K^{-1})	n_o A B C D E	n_e A B C D E	d_{36} (10^{-12} m/V)
12.5 (550 nm) 10.9 (550 nm) [4.369, 399]	0.216–1.67 [4.386, 387, 390, 391] −3.95 [4.392] −2.27 [4.392]	**1.5509** 2.424647 3.742954 126.9036045 0.015841 0.018624061 [4.386]^e	**1.5059** 2.262579 0.769288 127.0537007 0.013461 0.016165851 [4.386]^e	0.41 [4.228]
−8.277(7) (589 nm) +10.22 (589 nm) [4.433]	0.174–1.57 [4.393, 417, 432] −3.4 [4.392] −2.87 [4.392]	**1.4938** 2.259276 0.01008956 0.012942625 13.00522 400 [4.434, 435]^f	**1.4599** 2.132668 0.008637494 0.012281043 3.2279924 400 [4.434, 435]^f	0.39 [4.12, 14]
– 21.4 (550 nm) [4.399]	0.26–1.7 [4.386] – –	**1.5392** 2.373255 1.979528 126.9867549 0.01543 0.015836964 [4.386]^e	**1.5091** 2.270806 0.275372 58.08499107 0.013592 0.01596609 [4.386]^e	0.31 [4.386]
	0.22–1.5 [4.386, 436] – –	**1.4913** 2.235596 0.010929 0.001414783 2.355322 126.8547185 [4.386]^e	**1.4681** 2.152727 0.010022 0.001379157 0.691253 127.0144778 [4.386]^e	0.38 [4.386]
13.5 (550 nm) [4.369] 14.8 (550 nm) [4.399]	0.26–1.46 [4.437] −3.37 [4.392] −2.21 [4.392]	**1.5405** 2.390661 3.487176 126.7648558 0.015513 0.018112315 [4.386]^e	**1.5105** 2.27557 0.720099 126.6309092 0.013915 0.01459264 [4.386]^e	0.4 (0.694 μm) [4.438, 439]

Table 4.4-16 Tetragonal, point group $\bar{4}2m$ (D_{2d}) materials, cont.

Material	General ϱ (g/cm³) Mohs hardness $\kappa \perp c$ $\kappa \parallel c$ (W/m K)	Static dielectric tensor ε_{11}^S ε_{33}^S ε_{11}^T ε_{33}^T	Dissipation factor $\tan \delta_1$ $\tan \delta_3$ (f (kHz))	Elastic stiffness tensor c_{11} c_{33} c_{44} c_{66} c_{12} c_{13} (10^9 Pa)	Piezoelectric tensor d_{14} d_{36} (10^{-12} C/N)	Elastooptic tensor p_{11} p_{12} p_{13} p_{31} p_{33} p_{44} p_{66}
Rubidium dihydrogen phosphate, RbH₂PO₄ (RbDP, RDP)	2.805 – – –	42.0 26.5 41.4 27 [4.440, 441]	0.034 0.051 (38.6 GHz) [4.440, 441]	63(7) 50.0(55) 10.6(7) 3.56(7) −6.0(6) 10.6(72) [4.7]	4.0(3) 37(1) [4.427, 428]	0.247 0.265 0.248 0.229 0.248 −0.032 −0.032 [4.409, 410]
Silver gallium selenide, AgGaSe₂	5.71 3–3.5 1.1 ($\perp c$) 1.0 ($\parallel c$) [4.200]	– – 10.5 12.0 [4.442]			9.0 3.7 [4.442]	
Silver gallium sulfide (silver thiogallate), AgGaS₂	4.58 3–3.5 1.5 ($\perp c$) 1.4 ($\parallel c$) [4.200]	10 14 [4.443]ᵃ		87.9 75.8 24.1 30.8 58.4 59.2 [4.444, 445]		
Urea, (NH₂)₂CO	1.318 <2.5 – –	5–8ᵃ [4.26]		23.5 51.0 6.2 0.50 −0.50 7.5 [4.446]		
Zinc germanium diphosphide, ZnGeP₂	4.12 5.5 35 ($\perp c$) 36 ($\parallel c$) [4.200]	15 12 [4.112]ᵃ				See [4.447, 448]

ᵃ These values for ε_{ij} are neither at constant stress nor at constant strain.
ᵇ The elastic stiffness constants were measured at constant electric field (c^E).
ᶜ May be only partially deuterated.
ᵈ *Adhav* and *Vlassopoulous* [4.369] measured arsenates with low resistivity ($\approx 10^8 \, \Omega$ cm), and as a result their $\tan \delta$ values are high and ε may also be high. The data for RDA in [4.369] are incorrectly labeled ADA (remarks of the compiler).

Table 4.4-16 Tetragonal, point group $\bar{4}2m$ (D_{2d}) materials, cont.

Electrooptic tensor r_{41}^T r_{63}^T (10^{-12} m/V)	General optical properties Transparency range (μm) dn_o/dT (10^{-5} K^{-1}) dn_e/dT (10^{-5} K^{-1})	Refractive index n_o A B C D E F	n_e A B C D E F	Nonlinear dielectric susceptibility d_{36} (10^{-12} m/V)
10.3 14.3 [4.449, 450]	0.22–1.5 [4.386, 436] −3.74 [4.392] −2.73 [4.392]	**1.4926** 2.249885 0.01056 0.007780475 3.688005 127.1998253 [4.386]^e —	**1.4700** 2.159913 0.009515 0.00847799 0.988431 127.692938 [4.386]^e —	0.36 [4.438]
4.5 (1.15 μm) 3.9 (1.15 μm) [4.453]	0.71–18 [4.451] 4.5 (3.39 μm) [4.452] 7.6 (3.39 μm) [4.452]	**2.7008** 4.6453 2.2057 0.1879 1.8377 1600 [4.227]^f	**2.6800** 5.2912 1.3970 0.2845 1.9282 1600 [4.227]^f	33 (10.6 μm) [4.12, 14]
4.0(2) 3.0(1) [4.455]	0.47–13 [4.454] For dn/dT, see [4.456, 457]	**2.4540** 3.40684 2.40065 0.09311 2.06248 950 [4.457]^f —	**2.4012** 3.60728 1.94792 0.11066 2.24544 1030.7 [4.457]^f —	17.5 (1.06 μm) [4.12, 14] 11.2 (10.6 μm) [4.12, 14]
+1.03 (633 nm) −0.75 (633 nm) [4.460]	0.2–1.8 [4.458] – –	**1.4811** 2.1823 0.0125 0.03 0 [4.461]^h — —	**1.5825** 2.51527 0.0240 0.03 0.0202 1.52 0.08771 [4.461]^h	1.18 (0.6 μm) [4.458, 459]
1.8 (5 μm) – [4.463]	0.74–12 [4.462] 21.18 [4.415] 23.01 [4.415]	**3.2324** 4.4733 5.26576 0.13381 1.49085 662.55 [4.464, 465]^f —	**3.2786** 4.63318 5.34215 0.14255 1.45785 662.55 [4.464, 465]^f —	69 (10.6 μm) [4.12, 462]

^{efgh} Dispersion relations (λ (μm), $T = 20\,°\text{C}$):

^{e} $n^2 = A + \frac{B\lambda^2}{\lambda^2-C} + \frac{D}{\lambda^2-E}$.

^{f} $n^2 = A + \frac{B\lambda^2}{\lambda^2-C} + \frac{D\lambda^2}{\lambda^2-E}$.

^{g} $n^2 = A + \frac{B}{\lambda^2-C} - D\lambda^2$.

^{h} $n^2 = A + \frac{B}{\lambda^2-C} + \frac{D(\lambda^2-E)}{(\lambda^2-E)+F}$.

Table 4.4-17 Tetragonal, point group $4mm$ (C_{4v}) materials

Material	General ϱ (g/cm^3) Mohs hardness $\kappa \perp c$ $\kappa \parallel c$ (W/m K)	Static dielectric tensor ε_{11}^S ε_{33}^S ε_{11}^T ε_{33}^T	Dissipation factor $\tan \delta_1$ $\tan \delta_3$ (f (kHz))	Elastic stiffness tensor c_{11} c_{33} c_{44} c_{66} c_{12} c_{13} (10^9 Pa)	Piezoelectric tensor d_{31} d_{33} d_{15} d_h $(10^{-12}$ C/N)	Elastooptic tensor p_{11} p_{12} p_{13} p_{31} p_{33} p_{44} p_{66}
Barium titanate, BaTiO$_3$[a]	6.02 – 1.34 [4.45] –	1970 109 2920 168 [4.468]	0.13 – (24 GHz) [4.469]	243 147.9 54.9 120 128 123 [4.467][b]	−33.4 +68.5 647.0 [4.467] –	See [4.443, 466]
Lead titanate, PbTiO$_3$	7.9 – – –	102(3) 33.5(10) 130–140 105–110 [4.472]	0.02–0.05 0.02–0.05 (0.1 GHz) [4.472]	$s_{11}^E = 7.2$ $s_{33}^E = 32.5$ $s_{44}^E = 12.2$ $s_{66}^E = 7.9$ $s_{12}^E = -2.1$ [4.470, 471][c]	−25 +117 61 [4.470, 471] –	
Lithium tetraborate, Li$_2$B$_4$O$_7$	2.44 5 – –	78.80 71.45 82.61 87.92 [4.473]		130.9(42) 55.4(11) 57.3(14) 46.4(7) 1.5(12) 30(4) [4.7]		
Potassium lithium niobate, K$_3$Li$_2$Nb$_5$O$_{15}$ (KLINBO)	4.3 – – –	271 83 306 115 (405 °C) [4.474][d]		220 109 68 70 74 59 [4.474][bd]	−14 57 68 [4.474][d] –	

[a] Top-seeded solution grown (TSSG) BaTiO$_3$.
[b] The elastic stiffness constants were measured at constant electric field (c^E).
[c] Units not specified in original, but probably 10^{-12} Pa^{-1}.
[d] The stoichiometry was K$_{2.89}$Li$_{1.55}$Nb$_{5.11}$O$_{15}$ instead of K$_3$Li$_2$Nb$_5$O$_{15}$.

Table 4.4-17 Tetragonal, point group $4mm$ (C_{4v}) materials, cont.

Electrooptic tensor	General optical properties	Refractive index		Nonlinear dielectric susceptibility
r_{13}^T	Transparency range (μm)	n_o	n_e	d_{31}
r_{33}^T	dn_o/dT (10^{-5} K^{-1})	A	A	d_{33}
r_{51}^T	dn_e/dT (10^{-5} K^{-1})	B	B	(10^{-12} m/V)
(10^{-12} m/V)		C	C	
at $\lambda = 633$ nm		D	D	
		E	E	
		F	F	
8	0.14–10	**2.3218**	**2.2894**	−14.4
105	–	1	1	−5.4 [4.475, 476]
1300 [4.477]	–	4.195	4.073	
		0.04964	0.04456	
		0 [4.478]^e	0 [4.478]^e	
		–	–	
		–	–	
13.8 (r_{13}^S)	0.6–6 [4.479]	**2.5715**	**2.5690**	+35.3
5.9 (r_{33}^S) [4.480]	–	1	1	−7.0 [4.479]
	–	5.359	5.365	
		0.0502	0.0471	
		0 [4.479]^e	0 [4.479]^e	
		–	–	
		–	–	
+3.74	0.17–3.5 [4.481, 482]	**1.5968** (1.1 μm)	**1.5422** (1.1 μm)	0.12
+3.67	≈ 0.1 [4.482]	2.564310	2.386510	0.93 [4.483]
−0.11 [4.484]	≈ 0.3 [4.482]	0.012337	0.010664	
		0.013103	0.012878	
		0.019075 [4.482]^f	0.012813 [4.482]^f	
		–	–	
		–	–	
8.9 [4.485, 486]	0.4–5 [4.485–487]	**2.208**	**2.112**	11.8 (0.8 μm) [4.487]
78 [4.485, 486]	–	1	1	10.5 [4.485, 486]
80 [4.488]	–	3.708	3.349	
		0.04601	0.03564	
		0 [4.321]^e	0 [4.321]^e	
		–	–	
		–	–	

^{ef} Dispersion relations (λ (μm), $T = 20\,°C$):
^{e} $n^2 = A + \dfrac{B}{\lambda^2 - C} - D\lambda^2$.
^{f} $n^2 = A + \dfrac{B}{\lambda^2 - C} + \dfrac{D(\lambda^2 - E)}{(\lambda^2 - E) + F}$.

Table 4.4-18 Orthorhombic, point group mmm (D_{2h}) materials

Material	General ϱ (g/cm^3) Mohs hardness κ (W/m K)	Static dielectric tensor ε_{11} ε_{22} ε_{33}	Dissipation factor $\tan \delta_1$	Elastic stiffness tensor c_{11} c_{22} c_{33} c_{44} c_{55} c_{66} c_{12} c_{13} c_{23} (10^9 Pa)	Elastic compliance tensor s_{11} s_{22} s_{33} s_{44} s_{55} s_{66} s_{12} s_{13} s_{23} (10^{-12} Pa^{-1})	Piezooptic tensor
Magnesium silicate, Mg$_2$SiO$_4$ (forsterite)	3.217 7 5.12 [4.332]	6.2^a [4.26]		328.5(13) 200.0(1) 235.3(7) 66.9(6) 81.3(1) 80.9(2) 68.0(16) 68.7(7) 72.8(6) [4.7]	3.37 5.84 4.93 15.0 12.3 12.4 −0.85 −0.71 −1.56 [4.7]	$q_{11}+q_{12}+q_{13}=0.16$ $q_{21}+q_{22}+q_{23}=0.20$ $q_{31}+q_{32}+q_{33}=0.27$ (589 nm) [4.489]
Yttrium aluminate, YAlO$_3$ (YAP, YALO)	5.35 8.5 11 [4.259]	16–20^a [4.48]	0.001 [4.48]			

^{a} It is not clear which relative dielectric constant ε_{ij} is specified.

Table 4.4-19 Orthorhombic, point group 222 (D_2) materials

Material	General ϱ (g/cm^3)	Static dielectric tensor ε_{11}^S ε_{22}^S ε_{33}^S ε_{11}^T ε_{22}^T ε_{33}^T	Elastic stiffness tensor c_{11} c_{22} c_{33} c_{44} c_{55} c_{66} c_{12} c_{13} c_{23} (10^9 Pa)	Piezoelectric tensor d_{14} d_{25} d_{36} (10^{-12} C/N)	Elastooptic tensor p_{11} p_{12} p_{13} p_{21} p_{22} p_{23} p_{31} p_{32} p_{33} p_{44} p_{55} p_{66}
Barium formate, Ba(COOH)$_2$	3.261	6.9^a [4.490, 491] 6.7^a [4.490, 491] 5.2^a [4.490, 491] 7.9 [4.331] 5.9 [4.331] 7.5 [4.331]	$s_{44}^E = 78.5$ $s_{55}^E = 60$ $s_{66}^E = 82.5$ [4.331]	−5.6(5) −10.2(10) +1.9(1) [4.8, 492]	

Table 4.4-18 Orthorhombic, point group mmm (D_{2h}) materials, cont.

General optical properties	Refractive index		
Transparency range (μm) dn_X/dT (10^{-5} K^{-1}) dn_Y/dT (10^{-5} K^{-1}) dn_Z/dT (10^{-5} K^{-1})	n_X A B C	n_Y A B C	n_Z A B C
0.27–5.9 [4.87] 1.45 [4.493] – 0.98 [4.493]	**1.9111** (1 μm) 1 2.61960 0.012338 [4.494][b]	**1.9251** (1 μm) 1 2.67171 0.012605 [4.494][b]	**1.9337** (1 μm) 1 2.70381 0.012903 [4.494][b]

[b] Dispersion relation (λ (μm), $T = 20\,°C$):
$$n^2 = A + \frac{B\lambda^2}{\lambda^2 - C}.$$

Table 4.4-19 Orthorhombic, point group 222 (D_2) materials, cont.

Electrooptic tensor	General optical properties	Refractive index			Nonlinear dielectric susceptibility
r_{41}^T r_{52}^T r_{63}^T (10^{-12} m/V)	Transparency range (μm) dn_X/dT (10^{-5} K^{-1}) dn_Y/dT (10^{-5} K^{-1}) dn_Z/dT (10^{-5} K^{-1})	n_X A B C D	n_Y A B C D	n_Z A B C D	d_{14} d_{25} d_{36} (10^{-12} m/V)
+1.81(18) −2.03(20) +0.48(10) [4.492]	0.245–2.2, 4.8–5.1 [4.495] – – –	**1.6214** 2.619 0.0177 0.039 0 [4.495][b]	**1.5819** 2.491 0.0184 0.035 0 [4.495][b]	**1.5585** 2.421 0.016 0.042 0 [4.495][b]	0.10 0.11 0.11 [4.495]

Table 4.4-19 Orthorhombic, point group 222 (D_2) materials, cont.

Material	General ϱ (g/cm^3)	Static dielectric tensor ε_{11}^S, ε_{22}^S, ε_{33}^S, ε_{11}^T, ε_{22}^T, ε_{33}^T	Elastic stiffness tensor c_{11}, c_{22}, c_{33}, c_{44}, c_{55}, c_{66}, c_{12}, c_{13}, c_{23} (10^9 Pa)	Piezoelectric tensor d_{14}, d_{25}, d_{36} (10^{-12} C/N)	Elastooptic tensor p_{11}, p_{12}, p_{13}, p_{21}, p_{22}, p_{23}, p_{31}, p_{32}, p_{33}, p_{44}, p_{55}, p_{66}
Cesium triborate, CsB$_3$O$_5$ (CBO)	3.357				
α-Iodic acid, HIO$_3$	4.64	7.5 12.4 8.1 [4.331][a]	57(2) 43(1) 30.0(3) 21(1) 16(1) 17.8(13) 6(3) 15(2) 11.5(5) [4.7]	−26(1) −18(1) +28(1) [4.8, 496]	0.418(16) 0.308(330) 0.313(56) 0.304(36) 0.322(31) 0.307(57) 0.527(48) 0.340(31) 0.377(103) −0.077(66) 0.107(12) 0.092(8) [4.11]
3-Methyl 4-nitropyridine 1-oxide, C$_6$N$_2$O$_3$H$_6$ (POM)	1.55	3.77 5.41 3.77 [4.497][a]	13.29 18.14 12.20 7.8 5.2 5.4 4.9 −2.6 10.6 [4.498]		0.45 0.53 0.49 0.53 0.57 0.42 0.36 0.70 0.64 −0.078 −0.074 −0.046 (514.5 nm) [4.498]

Table 4.4-19 Orthorhombic, point group 222 (D_2) materials, cont.

Electrooptic tensor	General optical properties	Refractive index			Nonlinear dielectric susceptibility
		n_X	n_Y	n_Z	
r_{41}^T	Transparency range (µm)	A	A	A	d_{14}
r_{52}^T	dn_X/dT (10^{-5} K^{-1})	B	B	B	d_{25}
r_{63}^T	dn_Y/dT (10^{-5} K^{-1})	C	C	C	d_{36}
(10^{-12} m/V)	dn_Z/dT (10^{-5} K^{-1})	D	D	D	(10^{-12} m/V)
	0.170–3.0 [4.499]	**1.5194**	**1.5505**	**1.5781**	1.5 [4.499]
	–	2.3035	2.3704	2.4753	–
	–	0.01378	0.01528	0.01806	–
	–	0.01498	0.01581	0.01752	
		0.00612 [4.500]^b	0.00939 [4.500]^b	0.01654 [4.500]^b	
6.6(3)	0.35–1.6 $E \parallel a$	**1.8129**	**1.9273**	**1.9500**	8.3 [4.501]
7.0(5)	0.35–2.2 $E \parallel c$	2.5761	2.4701	2.6615	–
6.0(3) [4.502, 503]	[4.504, 505]	0.6973	1.2054	1.1316	–
	–	0.05550736	0.05044516	0.05202961	
	–	0.0201 [4.505]^c	0.0152 [4.505]^c	0.0398 [4.505]^c	
|3.6(6)|	0.4–3.3 [4.506]	**1.625**	**1.668**	**1.829**	5.3–6.0 [4.506]
|5.1(4)|	–	2.4529	2.4315	2.5521	–
|2.6(3)| [4.507]	–	0.1641	0.3556	0.7962	–
	–	0.128	0.1276	0.1289	
		0 [4.506]^c	0.0579 [4.506]^c	0.0941 [4.506]^c	

Table 4.4-19 Orthorhombic, point group 222 (D_2) materials, cont.

Material	General ϱ (g/cm³)	Static dielectric tensor ε_{11}^S ε_{22}^S ε_{33}^S ε_{11}^T ε_{22}^T ε_{33}^T	Elastic stiffness tensor c_{11} c_{22} c_{33} c_{44} c_{55} c_{66} c_{12} c_{13} c_{23} (10^9 Pa)	Piezoelectric tensor d_{14} d_{25} d_{36} (10^{-12} C/N)	Elastooptic tensor p_{11} p_{12} p_{13} p_{21} p_{22} p_{23} p_{31} p_{32} p_{33} p_{44} p_{55} p_{66}
Potassium sodium tartrate tetrahydrate, KNa(C$_4$H$_4$O$_6$)·4H$_2$O (Rochelle salt)	1.767	245 – – 1100 11.1 9.2 [4.331]	40(16) 55(25) 63(23) 11.9(38) 3.1(2) 10.0(16) 24(29) 32(20) 23.8(438) [4.7]	2300 −56 11.8 [4.331]	0.35 0.41 0.42 0.37 0.28 0.34 0.36 0.35 0.36 −0.030 0.0046 −0.025 (589 nm) [4.508]
Sodium ammonium tartrate tetrahydrate, Na(NH$_4$)C$_4$H$_4$O$_6$·4H$_2$O (Ammonium Rochelle salt)		10(1)[a] 10(1)[a] 10(1)[a] [4.509] 9.0 8.9 10.0 [4.331]	36.8 50.9 55.4 10.6 3.03 8.70 27.2 30.8 34.7 [4.331]	≈ 13 38 ≈ 7 [4.510, 511]	−0.48 −0.61 −0.68 −0.55 −0.76 −0.83 −0.44 −0.57 −0.71 0.0077 0.013 −0.0026 (589 nm) [4.512]

Table 4.4-19 Orthorhombic, point group 222 (D_2) materials, cont.

Electrooptic tensor	General optical properties	Refractive index			Nonlinear dielectric susceptibility
r_{41}^T r_{52}^T r_{63}^T (10^{-12} m/V)	Transparency range (μm) dn_X/dT (10^{-5} K^{-1}) dn_Y/dT (10^{-5} K^{-1}) dn_Z/dT (10^{-5} K^{-1})	n_X A B C D	n_Y A B C D	n_Z A B C D	d_{14} d_{25} d_{36} (10^{-12} m/V)
8.3 [4.513, 514] – –	<250 to >850 nm −6.54 [4.515, 516] −5.57 [4.515, 516] −6.00 [4.515, 516]	**1.49540** (585 nm) [4.517] **1.5622** (260 nm) [4.515, 516]	**1.49183** (585 nm) [4.517] **1.5576** (260 nm) [4.515, 516] $A = 1$ $B = 1.1851057$ $C = 0.0113636$ $D = 0$ [4.515, 516]^c	**1.49001** (585 nm) [4.517] **1.5566** (260 nm) [4.515, 516]	$\chi^{(3)}_{zzzz}$ $= 0.8 \times 10^{-22}$ V^{-2} m^2 $\chi^{(3)}_{xxxx}$ $= 0.4 \times 10^{-22}$ V^{-2} m^2 (at 576.8 and 627 nm) [4.518]
$r_{52} = 2.1$ [4.519, 520]		**1.4984** (589 nm) [4.521, 522]	**1.4996** (589 nm) [4.521, 522]	**1.4953** (589 nm) [4.521, 522]	

^{a} The relative dielectric constant ε_{ij} was measured neither at constant stress nor at constant strain.
^{bc} Dispersion relations (λ (μm), $T = 20\,°\text{C}$):
^{b} $n^2 = A + \dfrac{B}{\lambda^2 - C} - D\lambda^2$.
^{c} $n^2 = A + \dfrac{B\lambda^2}{\lambda^2 - C} - D\lambda^2$.

Table 4.4-20 Orthorhombic, point group $mm2$ (C_{2v}) materials

Material	General	Static dielectric tensor	Dissipation factor	Elastic stiffness tensor	Piezoelectric tensor	Elastooptic tensor
	ϱ (g/cm^3) Mohs hardness κ (W/m K)	ε_{11}^S ε_{22}^S ε_{33}^S ε_{11}^T ε_{22}^T ε_{33}^T	$\tan\delta_1$ $\tan\delta_2$ $\tan\delta_3$ (T (°C), f (kHz))	c_{11} c_{22} c_{33} c_{44} c_{55} c_{66} c_{12} c_{13} c_{23} (10^9 Pa)	d_{31} d_{32} d_{33} d_{15} d_{24} d_h (10^{-12} C/N)	p_{11} p_{12} p_{13} p_{21} p_{22} p_{23} p_{31} p_{32} p_{33} p_{44} p_{55} p_{66}
Ammonium sulfate, (NH$_4$)$_2$SO$_4$		– – 8[a] (−60 °C) [4.526, 527]		35.2 29.7 36.0 9.5 7.0 10.3 14.1 15.7 17.3 [4.525]		0.26 [4.523, 524] |0.27| [4.525] |0.26| [4.525] |0.23| [4.525] ≈ |0.27| [4.525] |0.25| [4.525] |0.23| [4.525] ≈ |0.26| [4.525] ≈ 0.26 [4.525] 0.02 [4.525] ≤ |0.02| [4.525] ≈ 0.01 [4.11, 525]
Barium magnesium fluoride, BaMgF$_4$ (BMF)		14 [4.528] 8 [4.528] 8.5 [4.528] 14.75(74)[a] 8.24(41)[a] 8.40(42)[a] [4.529]		104 81 130 32.1 55.1 24.7 28.7 63.7 35.8 [4.529]	+2.5(3) −4.1(4) +8.0(2) −5.3(5) −1.2(1) [4.529] –	
Barium sodium niobate, Ba$_2$NaNb$_5$O$_{15}$ ("banana")	5.41 – 3.5 [4.532]	215 205 20 238(5) 228(5) 430 [4.534][b]	0.0068(17) – 0.0024(6) (25 °C, 1 GHz) [4.296]	239 247 135 65 66 76 104 50 52 [4.535][c]	−6.8 −6.9 34 32 45 [4.533] –	$p_{66} = 0.0021$ [4.530, 531]

Table 4.4-20 Orthorhombic, point group $mm2$ (C_{2v}) materials, cont.

	Electrooptic tensor	General optical properties	Refractive index			Nonlinear dielectric susceptibility
	r_{13}^T r_{23}^T r_{33}^T r_{42}^T r_{51}^T (10^{-12} m/V)	Transparency range (μm) dn_X/dT (10^{-5} K^{-1}) dn_Y/dT (10^{-5} K^{-1}) dn_Z/dT (10^{-5} K^{-1})	n_X A B C D E F	n_Y A B C D E F	n_Z A B C D E F	d_{31} d_{32} d_{33} d_{15} d_{24} (10^{-12} m/V)
Ammonium sulfate, (NH$_4$)$_2$SO$_4$	$r_c = 0.4$ [4.536]					0.27 0.29 0.50 [4.537] – –
Barium magnesium fluoride, BaMgF$_4$		0.185–10 [4.495]	**1.4436** 2.077 0.0076 0.0079 0 [4.495]^h –	**1.4604** 2.1238 0.0086 0 0 [4.495]^h –	**1.4674** 2.1462 0.00736 0.009 0 [4.495]^h –	0.022 0.033 0.009 – 0.024 [4.495]
Barium sodium niobate, Ba$_2$NaNb$_5$O$_{15}$ ("banana")	15(1) 13(1) 48(2) 92(4) 90(4) [4.534]	0.37–5 [4.534, 538, 539] −2.5 [4.534] – 8 [4.534]	**2.2573** 1 3.9495 0.04038894 0 [4.534]^i –	**2.2571** 1 3.9495 0.04014012 0 [4.534]^i –	**2.1694** 1 3.6008 0.03219871 0 [4.534]^i –	−12 −12 −16.5 12 11.4 [4.534]

Table 4.4-20 Orthorhombic, point group $mm2$ (C_{2v}) materials, cont.

Material	General	Static dielectric tensor	Dissipation factor	Elastic stiffness tensor	Piezoelectric tensor	Elastooptic tensor
	ϱ (g/cm^3) Mohs hardness κ (W/m K)	ε_{11}^S ε_{22}^S ε_{33}^S ε_{11}^T ε_{22}^T ε_{33}^T	$\tan\delta_1$ $\tan\delta_2$ $\tan\delta_3$ (T (°C), f (kHz))	c_{11} c_{22} c_{33} c_{44} c_{55} c_{66} c_{12} c_{13} c_{23} (10^9 Pa)	d_{31} d_{32} d_{33} d_{15} d_{24} d_h (10^{-12} C/N)	p_{11} p_{12} p_{13} p_{21} p_{22} p_{23} p_{31} p_{32} p_{33} p_{44} p_{55} p_{66}
Calcium tartrate tetrahydrate, Ca(C$_4$H$_4$O$_6$)·4H$_2$O (L-CTT)		14.3 10 20 [4.540]^a	0.9 — — (30 °C, 100 Hz) [4.540, 541]			
m-Chloronitrobenzene, ClC$_6$H$_4$NO$_2$ (CNB)						
2-Cyclooctylamino-5-nitropyridine, C$_{13}$H$_{19}$N$_3$O$_2$ (COANP)	1.24 — —					
N, 2-Dimethyl-4-nitrobenzenamine, C$_8$H$_{10}$N$_2$O$_2$ (MNMA)						

Table 4.4-20 Orthorhombic, point group $mm2$ (C_{2v}) materials, cont.

Electrooptic tensor	General optical properties	Refractive index			Nonlinear dielectric susceptibility
r^T_{13}	Transparency range (μm)	n_X	n_Y	n_Z	d_{31}
r^T_{23}	dn_X/dT (10^{-5} K^{-1})	A	A	A	d_{32}
r^T_{33}	dn_Y/dT (10^{-5} K^{-1})	B	B	B	d_{33}
r^T_{42}	dn_Z/dT (10^{-5} K^{-1})	C	C	C	d_{15}
r^T_{51}		D	D	D	d_{24}
(10^{-12} m/V)		E	E	E	(10^{-12} m/V)
		F	F	F	
	0.28–1.4 [4.542]	**1.5125**	**1.5220**	**1.5477**	<0.015
	–	1	1	1	0.20
	–	1.26	1.30	1.38	0.14
	–	0.0127273	0.0121495	0.0094521	1.73
	–	0 [4.542]^i	0 [4.542]^i	0 [4.542]^i	0.90 [4.542]
		–	–	–	
		–	–	–	
		1.6557	**1.6626**	**1.624**	4.6
		2.4882	2.5411	2.2469	4
		0.2384	0.2148	0.3722	7.8 [4.543]
		0.1070	0.1122	0.0810	–
		0.0091 [4.543]^i	0.0135 [4.543]^i	0.0092 [4.543]^i	–
		–	–	–	
		–	–	–	
0.63 [4.544]	0.41–... [4.545]	**1.6100**	**1.6383**	**1.7170**	11.3
–	–	2.3320	2.3994	2.5104	24
–	–	0.2215	0.2469	0.3689	10.8 [4.546, 547]
–	–	0.1686	0.1500	0.1780	–
–	–	0 [4.545]^i	0 [4.545]^i	0 [4.545]^i	–
		–	–	–	
		–	–	–	
8 ± 2	0.5–2 [4.548]	**1.936**		**1.506**	13
–	–	1.6797		2.1798	–
7.5 ± 2 [4.548]	–	1.7842		0.0736	2.6
–	–	0.1571		0.1757	12 [4.548]
–	–	0 [4.548]^i		0 [4.548]^i	–
		–		–	
		–		–	

Table 4.4-20 Orthorhombic, point group $mm2$ (C_{2v}) materials, cont.

Material	General ϱ (g/cm³) Mohs hardness κ (W/m K)	Static dielectric tensor ε^S_{11} ε^S_{22} ε^S_{33} ε^T_{11} ε^T_{22} ε^T_{33}	Dissipation factor $\tan\delta_1$ $\tan\delta_2$ $\tan\delta_3$ (T (°C), f (kHz))	Elastic stiffness tensor c_{11} c_{22} c_{33} c_{44} c_{55} c_{66} c_{12} c_{13} c_{23} (10^9 Pa)	Piezoelectric tensor d_{31} d_{32} d_{33} d_{15} d_{24} d_h (10^{-12} C/N)	Elastooptic tensor p_{11} p_{12} p_{13} p_{21} p_{22} p_{23} p_{31} p_{32} p_{33} p_{44} p_{55} p_{66}
Gadolinium molybdate, $Gd_2(MoO_4)_3$ (GMO)		– – 9.6 – – 10.2 [4.550]		55(4) 71(3) 101(3) 25.2(8) 25.8(2) 33.1(3) 12.8(46) 23.5(45) 27.2(76) [4.7]^c	1.5 [4.549] 0.2 [4.8] 0.5 [4.549] – – –	0.19 0.31 0.175 0.215 0.235 0.175 0.185 0.23 0.115 −0.033 −0.028 0.035 [4.551] (515–633 nm)
Lithium formate monohydrate, $LiCOOH \cdot H_2O$ (LFM)	1.46 – –	3.24 [4.552, 553] – – 4.5(5)^a [4.554, 555] 5.0(5)^a [4.554, 555] 6.0(3)^a [4.554, 555]		19.15 33.55 41.98 15.02 5.31 4.89 8.62 8.40 22.77 [4.552, 553]^c	−2.1(2) −6.7(5) +9.2(7) −15.0(10) +1.2(2) [4.554–557]	
Lithium gallium oxide, $LiGaO_2$ (lithium metagallate)	4.187 7.5 –	7.0 [4.8] 6.0 [4.8] 8.3 [4.558] 7.18 [4.8] 6.18 [4.8] 8.78 [4.8]		140 120 140 57.1 47.4 69.0 14 28 31 [4.558]^c	−2.5 [4.558] −4.7 [4.558] +8.6 [4.558] −6.9 [4.558] −6.0 [4.558] +0.9 [4.8, 274]	

Table 4.4-20 Orthorhombic, point group $mm2$ (C_{2v}) materials, cont.

Electrooptic tensor	General optical properties	Refractive index				Nonlinear dielectric susceptibility
r_{13}^T	Transparency range (μm)	n_X	n_Y		n_Z	d_{31}
r_{23}^T	dn_X/dT (10^{-5} K^{-1})	A	A		A	d_{32}
r_{33}^T	dn_Y/dT (10^{-5} K^{-1})	B	B		B	d_{33}
r_{42}^T	dn_Z/dT (10^{-5} K^{-1})	C	C		C	d_{15}
r_{51}^T		D	D		D	d_{24}
(10^{-12} m/V)		E	E		E	(10^{-12} m/V)
		F	F		F	
+2.15(16)		**1.8141**	**1.8145**		**1.8637**	−2.3
−2.31(16)		1	1		1	+2.3
+0.123(15)		2.2450	2.24654		2.41957	−0.04
−		0.022693	0.0226803		0.0245458	−4.1
−		0 [4.321]^i	0 [4.321]^i		0 [4.321]^i	+4 [4.156, 559]
[4.560, 561]		−	−		−	
		−	−		−	
−1.0(1)	0.23–1.2 [4.562, 563]	**1.3595** (1 μm)	**1.4694** (1 μm)		**1.5055** (1 μm)	0.13
+3.2(2)	−	1.4376	1.6586		1.6714	−0.60
−2.6(2)	−	0.4045	0.5006		0.5928	+0.94 [4.12]
+1.0(2)		0.01692601	0.023409		0.02534464	−
+2.4(2)		0.0005 [4.564]^i	0.0127 [4.564]^i		0.0153 [4.564]^i	−
[4.554–557]		−	−		−	
		−	−		−	
	0.3–5 [4.251, 565]	**1.7477**	**1.7768**		**1.7791**	+0.066
	−	(0.5 μm) [4.566]	(0.5 μm) [4.566]		(0.5 μm) [4.566]	−0.14
	−					+0.57 [4.156, 565]
	−					−
						−

Table 4.4-20 Orthorhombic, point group $mm2$ (C_{2v}) materials, cont.

Material	General ϱ (g/cm^3) Mohs hardness κ (W/m K)	Static dielectric tensor ε_{11}^S ε_{22}^S ε_{33}^S ε_{11}^T ε_{22}^T ε_{33}^T	Dissipation factor $\tan\delta_1$ $\tan\delta_2$ $\tan\delta_3$ (T (°C), f (kHz))	Elastic stiffness tensor c_{11} c_{22} c_{33} c_{44} c_{55} c_{66} c_{12} c_{13} c_{23} (10^9 Pa)	Piezoelectric tensor d_{31} d_{32} d_{33} d_{15} d_{24} d_h (10^{-12} C/N)	Elastooptic tensor p_{11} p_{12} p_{13} p_{21} p_{22} p_{23} p_{31} p_{32} p_{33} p_{44} p_{55} p_{66}
Lithium triborate, LiB$_3$O$_5$ (LBO)	2.47 6 3.5 [4.200]					
m-Nitroaniline; (3-nitrobenzenamine, meta-nitroaniline), C$_6$H$_4$(NO$_2$)NH$_2$ (mNA)		3.9 4.2 4.6 [4.567]^a				
Potassium acid phthalate, KH(C$_8$H$_4$O$_4$)		6.00(2) 3.87(2) 4.34(2) [4.568, 569] – – 4.34^a [4.570]		17.6 13.3 17.0 4.86 7.59 6.23 7.4 10.4 5.0 [4.568, 569]	−15.3(0) +8.8(0) +5.5(0) −7.1(0) +4.3(0) [4.568, 569] –	−0.61 −0.25 −0.52 −0.58 −0.36 −0.60 −0.84 −0.36 −0.90 −0.26 −0.63 −0.13 (589 nm) [4.571]

Table 4.4-20 Orthorhombic, point group $mm2$ (C_{2v}) materials, cont.

Electrooptic tensor	General optical properties	Refractive index			Nonlinear dielectric susceptibility
r_{13}^T	Transparency range (μm)	n_X	n_Y	n_Z	d_{31}
r_{23}^T	dn_X/dT (10^{-5} K^{-1})	A	A	A	d_{32}
r_{33}^T	dn_Y/dT (10^{-5} K^{-1})	B	B	B	d_{33}
r_{42}^T	dn_Z/dT (10^{-5} K^{-1})	C	C	C	d_{15}
r_{51}^T		D	D	D	d_{24}
(10^{-12} m/V)		E	E	E	(10^{-12} m/V)
		F	F	F	
	0.16–2.6 [4.572]	**1.5656**	**1.5905**	**1.6055**	−0.67
	−18 [4.573]	2.4542	2.5390	2.5865	0.85
	−136 [4.573]	0.01125	0.01277	0.01310	0.04
	−(63 + 21λ) [4.573]	0.01135	0.01189	0.01223	[4.12, 14, 572–574]
		0.01388	0.01849	0.01862	–
		0	4.3025×10^{-5}	4.5778×10^{-5}	–
		0 [4.575]^j	2.9131×10^{-5} [4.575]^j	3.2526×10^{-5} [4.575]^j	
7.4(7)	0.5–2 [4.576, 577]	**1.631**	**1.678**	**1.719**	20
0.1(6)	–	2.469	2.6658	2.8102	1.6
16.7(2) [4.578]	–	0.1864	0.1626	0.1524	21 [4.543]
–	–	0.16	0.1719	0.175	–
–		0.0199 [4.579]^i	0.0212 [4.579]^i	0.0294 [4.579]^i	–
		–	–	–	
		–	–	–	
$r_b \approx 1.8$ [4.570]	0.3–1.7	**1.63**	**1.64**	**1.48**	0.21 (d_{31}, 1.15 μm)
	with a narrow absorption band at 1.14 μm [4.580]	(1.1 μm) [4.580]	(1.1 μm) [4.580]	(1.1 μm) [4.580]	0.06 (d_{32}, 1.15 μm)
					0.65 (d_{31}, 0.63 μm)
	–				0.15 (d_{32}, 0.63 μm)
	–				–
	–				[4.580]

Table 4.4-20 Orthorhombic, point group $mm2$ (C_{2v}) materials, cont.

Material	General ϱ (g/cm³) Mohs hardness κ (W/m K)	Static dielectric tensor ε_{11}^S ε_{22}^S ε_{33}^S ε_{11}^T ε_{22}^T ε_{33}^T	Dissipation factor $\tan\delta_1$ $\tan\delta_2$ $\tan\delta_3$ (T (°C), f (kHz))	Elastic stiffness tensor c_{11} c_{22} c_{33} c_{44} c_{55} c_{66} c_{12} c_{13} c_{23} (10^9 Pa)	Piezoelectric tensor d_{31} d_{32} d_{33} d_{15} d_{24} d_h (10^{-12} C/N)	Elastooptic tensor p_{11} p_{12} p_{13} p_{21} p_{22} p_{23} p_{31} p_{32} p_{33} p_{44} p_{55} p_{66}
Potassium niobate, KNbO₃	4.617 – >3.5 [4.581]	37(2) 780(50) 24(2) 160(10) 1000(80) 55(5) (25 °C) [4.583]	0.003 0.002 0.01 (25 °C) [4.583]	226 270 280 74.3 25.0 95.5 96 [4.583] – –	+9.8(7) −19.5(20) +24.5(15) [4.582] 215(5) 159(5) [4.583] –	|0.197|[d] |0.115|[d] |0.109|[e] |0.130|[d] |0.234|[d] |0.005|[e] |0.64|[d] |0.153|[d] |0.075|[e] |0.57|[d] |0.45|[d] [4.584] –
Potassium pentaborate tetrahydrate, KB₅O₈·4H₂O (KB5)	1.74 2.5 –	5.5 4.6 4.5 [4.585] – – –		58.2 35.9 25.5 16.4 4.63 5.7 22.9 17.4 23.1 [4.585]	−0.35 −2.3 +5.5 +4.7 +20.3 [4.586] –	
Potassium titanyl arsenate, KTiOAsO₄ (KTA)	3.45 3 –	– – – 12(1) 12(1) 18(1) [4.587, 588]				
Potassium titanate (titanyl) phosphate, KTiOPO₄ (KTP)	3.02 5 2 (‖ X) 3 (‖ Y) 3.3 (‖ Z) [4.590]	11.6(2) 11.0(2) 15.4(3) 11.9(2) 11.3(2) ≥17.5(4) [4.591]	≲ 0.004 ≲ 0.004 < 0.005 (10–10³ MHz) ≈ 0.017 0.017 ≈ 0.35 (100 kHz) [4.591]	159 154 175 [4.589] – – – – – –		

Table 4.4-20 Orthorhombic, point group $mm2$ (C_{2v}) materials, cont.

Electrooptic tensor	General optical properties	Refractive index			Nonlinear dielectric susceptibility
r_{13}^{T}	Transparency range (μm)	n_X	n_Y	n_Z	d_{31}
r_{23}^{T}	dn_X/dT (10^{-5} K^{-1})	A	A	A	d_{32}
r_{33}^{T}	dn_Y/dT (10^{-5} K^{-1})	B	B	B	d_{33}
r_{42}^{T}	dn_Z/dT (10^{-5} K^{-1})	C	C	C	d_{15}
r_{51}^{T}		D	D	D	d_{24}
(10^{-12} m/V)		E	E	E	(10^{-12} m/V)
		F	F	F	
34(2)	0.4–4.5 [4.386]	**2.2576**	**2.2195**	**2.1194**	+11 [4.12–14]
6(1)	For temperature-dependent	1	1	1	−13 [4.12–14]
63.4(10)	dispersion relations,	1.44121874	1.33660410	1.04824955	−19.5 [4.12–14]
450(30)	see [4.592]	0.07439136	0.06664629	0.06514225	16 [4.593]
120(10) [4.584]		2.54336918	2.49710396	2.37108379	17 [4.593]
		0.01877036	0.01666505	0.01433172	
		0.02845018	0.02517432	0.01943289	
		[4.594]^k	[4.594]^k	[4.594]^k	
	0.165–1.4 [4.595]	**1.4917** (0.5 μm)	**1.4380** (0.5 μm)	**1.4251** (0.5 μm)	0.04
	–	1	1	1	0.003
	–	1.1790826	1.0280852	0.9919090	0.05
	–	0.0087815	0.0090222	0.0093289	(all at 0.5 μm)
		0 [4.596, 597]^j	0 [4.596, 597]^j	0 [4.596, 597]^j	–
		–	–	–	
		–	–	–	[4.12, 598, 599]
15(1)	0.35–5.3	**1.782**	**1.790**	**1.868**	2.9
21(1)	[4.587, 588, 600]	1.90713	2.15912	2.14786	5.2
40(1)	–	1.23552	1.00099	1.29559	12.0
–	–	0.0387775	0.0477160	0.0516153	–
–		0.01025 [4.601]^i	0.01096 [4.601]^i	0.01436 [4.601]^i	–
[4.587, 588]		–	–	–	[4.12, 14, 602–604]^l
		–	–	–	
+9.5(5)	0.35–4.5 [4.605, 606]	**1.7381**	**1.7458**	**1.8302**	2.2
+15.7(8)	0.61 [4.607]^m	3.0065	3.0333	3.3134	3.7
+36.3(18)	0.83 [4.607]^m	0.03901	0.04154	0.05694	14.6
9.3(9)	1.45 [4.607]^m	0.04251	0.04547	0.05658	1.9
7.3(7) [4.591]		0.01327 [4.608]^{hm}	0.01408 [4.608]^{hm}	0.01682 [4.608]^{hm}	3.7 [4.13]^l
		–	–	–	
		–	–	–	

Table 4.4-20 Orthorhombic, point group $mm2$ (C_{2v}) materials, cont.

Material	General ϱ (g/cm^3) Mohs hardness κ (W/m K)	Static dielectric tensor ε_{11}^S ε_{22}^S ε_{33}^S ε_{11}^T ε_{22}^T ε_{33}^T	Dissipation factor $\tan\delta_1$ $\tan\delta_2$ $\tan\delta_3$ (T (°C), f (kHz))	Elastic stiffness tensor c_{11} c_{22} c_{33} c_{44} c_{55} c_{66} c_{12} c_{13} c_{23} (10^9 Pa)	Piezoelectric tensor d_{31} d_{32} d_{33} d_{15} d_{24} d_h (10^{-12} C/N)	Elastooptic tensor p_{11} p_{12} p_{13} p_{21} p_{22} p_{23} p_{31} p_{32} p_{33} p_{44} p_{55} p_{66}
Rubidium titanate (titanyl) phosphate, RbTiOPO$_4$ (RTP)		23^f (25 °C) [4.609–611]		143 142 175 33 40 57 [4.589] – – –		
Sodium nitrite, NaNO$_2$	2.168 – –	8 [4.612] 5.2 [4.612] 4.18 [4.612] 7.4 [4.614] 5.5 [4.614] 5.0 [4.614]	0.004 0.006 0.015 (3.3 GHz) [4.612]	30.6(3) 56(2) 64(2) 12(1) 9.9(1) 5.0(3) 12.5(1) 15.6(53) 14.6(48) [4.7]	−1.1 −2.8 +1.6 +9.3 −20.2 – [4.8, 614]	|0.44| [4.613] |0.37| [4.613] |0.36| [4.613] |0.39| [4.613] |0.33| [4.613] |0.27| [4.613] |0.18| [4.613] |0.19| [4.613] |0.15| [4.613] −0.050^g [4.615] −0.30 [4.615] −0.10^g [4.615]

[a] These values for ε_{ij} are neither at constant stress nor at constant strain.
[b] Stoichiometric crystal.
[c] c^E at constant electric field.
[d] p^E at constant electric field.
[e] $p_{i3}^* = p_{i3}^E - r_{i3}^S e_{333}/\varepsilon_0 \varepsilon_{33}^S$, where r_{i3}^S is the electrooptic-tensor element $i3$ at zero strain, e_{333} is the piezoelectric-tensor element 333, and ε_{33}^S is the relative-dielectric-constant element 33 at constant strain.
[f] Element of ε unspecified, but probably ε_{11} or ε_{33}.

Table 4.4-20 Orthorhombic, point group $mm2$ (C_{2v}) materials, cont.

Electrooptic tensor	General optical properties	Refractive index			Nonlinear dielectric susceptibility
r_{13}^T r_{23}^T r_{33}^T r_{42}^T r_{51}^T (10^{-12} m/V)	Transparency range (μm) dn_X/dT (10^{-5} K^{-1}) dn_Y/dT (10^{-5} K^{-1}) dn_Z/dT (10^{-5} K^{-1})	n_X A B C D E F	n_Y A B C D E F	n_Z A B C D E F	d_{31} d_{32} d_{33} d_{15} d_{24} (10^{-12} m/V)
+9.7(9) +10.8(9) +22.5(9) +14.9(9) −7.6(9) [4.618–620]	0.35–4.5 [4.616, 617] – – –	**1.7569** 2.56666 0.53842 0.06374 0.01666 [4.617]^i –	**1.7730** 2.34868 0.77949 0.05449 0.0211 [4.617]^i –	**1.8540** 2.77339 0.63961 0.08151 0.02237 [4.617]^i –	3.3 4.1 17.3 [4.603] – –
$r_{42} = -3.0(2)$ $r_{51} = -1.9(2)$ (at 546 nm) [4.623]	0.35–3.4 and 5–8 [4.621, 622] – – –	**1.3395** 1 0.727454 0.0118285 0 [4.621]^i –	**1.4036** 1 0.978108 0.0112296 0 [4.621]^i –	**1.6365** 1 1.616683 0.0222073 0 [4.621]^i –	0.11 2.9 0.14 0.11 2.8 [4.624]

^{g} $p_{ijkl}^{\text{eff}} = p_{ijkl}^E - r_{ijm}^S a_m a_n e_{nkl}/\boldsymbol{a}\cdot\boldsymbol{\varepsilon}\cdot\boldsymbol{a}$, where r^S is the electrooptic tensor, $\boldsymbol{a}$ is a unit acoustic-wave propagation vector, and $\boldsymbol{\varepsilon}$ is the dielectric-constant tensor at the frequency of the acoustic wave.

^{hijk} Dispersion relations (λ (μm), $T = 20\,°$C):

^{h} $n^2 = A + \dfrac{B}{\lambda^2 - C} - D\lambda^2$.

^{i} $n^2 = A + \dfrac{B\lambda^2}{\lambda^2 - C} - D\lambda^2$.

^{j} $n^2 = A + \dfrac{B}{\lambda^2 - C} - D\lambda^2 + E\lambda^4 - F\lambda^6$.

^{k} $n^2 = A + \dfrac{B\lambda^2}{\lambda^2 - C} + \dfrac{D\lambda^2}{\lambda^2 - E} - F\lambda^2$.

^{l} Note reversals between d_{31} and d_{32} (also between d_{15} and d_{24}) for KTA and KTP given in [4.12] and [4.13].

^{m} For flux-grown crystals.

Table 4.4-21 Monoclinic, point group 2 (C_2) materials

Material	General	Static dielectric tensor	Dissipation factor	Elastic stiffness tensor	Piezoelectric tensor	Elastooptic tensor
	ϱ (g/cm^3) Mohs hardness κ (W/m K)	ε_{11}^S ε_{22}^S ε_{33}^S ε_{13}^S ε_{11}^T ε_{22}^T ε_{33}^T ε_{13}^T	$\tan\delta_1$ $\tan\delta_2$ $\tan\delta_3$ (f (kHz))	c_{11} c_{22} c_{33} c_{44} c_{55} c_{66} c_{12} c_{13} c_{23} c_{15} c_{25} c_{35} c_{46} (10^9 Pa)	d_{21} d_{22} d_{23} d_{14} d_{16} d_{25} d_{34} d_{36} d_h (10^{-12} C/N)	p_{11} p_{12} p_{13} p_{15} p_{21} p_{22} p_{23} p_{25} p_{31} p_{32} p_{33} p_{35} p_{44} p_{46} p_{51} p_{52} p_{53} p_{55} p_{64} p_{66}
Bismuth triborate, BiB$_3$O$_6$ (BIBO)	4.9 [4.625] – –					
Deuterated L-arginine phosphate, (ND$_x$H$_{2-x}$)$_2^+$(CND)(CH$_2$)$_3$CH(ND$_y$H$_{3-y}$)$^+$COO$^-$ · D$_2$PO$_4^-$ · D$_2$O (DLAP)	≈ 1.5 3 –			17.477(111) 31.996(89) 31.575(72) – – – – – – – [4.15]		
4-($N1N$-Dimethylamino)-3-acetamidonitrobenzene (N-[2-(dimethylamino)-5-nitrophenyl]-acetamide, DAN)						

Table 4.4-21 Monoclinic, point group 2 (C_2) materials, cont.

Electrooptic tensor	General optical properties	Refractive index				Nonlinear dielectric susceptibility
		n_X	n_Y	n_Z		
r_{12}^T	Transparency range (μm)	A	A	A		d_{21}
r_{22}^T	dn_X/dT (10^{-5} K^{-1})	B	B	B		d_{22}
r_{32}^T	dn_Y/dT (10^{-5} K^{-1})	C	C	C		d_{23}
r_{41}^T	dn_Z/dT (10^{-5} K^{-1})	D	D	D		d_{25}
r_{43}^T						(10^{-12} m/V)
r_{52}^T						
r_{61}^T						
r_{63}^T (10^{-12} m/V)						
(at $\lambda = 633$ nm)						
	0.27–6.25 [4.625]	**1.9190**	**1.7585**	**1.7854**		2.3(2)
	–	3.6545	3.0740	3.1685		2.53(8)
	–	0.0511	0.0323	0.0373		1.3(1)
	–	0.0371	0.0316	0.0346		2.3(2) [4.626]
		0.0226 [4.626]^a	0.01337 [4.626]^a	0.01750 [4.626]^a		(see also [4.627])
	0.25–1.3 [4.628]	**1.4960**	**1.5584**	**1.5655**		0.48
	−3.64	2.2352	2.4313	2.4484		0.685
	−5.34	0.0118	0.0151	0.0172		−0.80
	−6.69	0.0146	0.0214	0.0229		−0.22
	(all at 532 nm [4.629])	0.00683 [4.628]^a	0.0143 [4.628]^a	0.0115 [4.628]^a		[4.12, 14, 628]
	0.485–2.27 [4.630]	**1.517**	**1.636**	**1.843**		1.1
	–	2.1390	2.3290	2.5379		3.9
	–	0.147408	0.307173	0.719557		37.5
	–	0.3681 [4.630]^b	0.3933 [4.630]^b	0.4194 [4.630]^b		1.1 [4.630, 631]
		–	–	–		

Table 4.4-21 Monoclinic, point group 2 (C_2) materials, cont.

Material	General	Static dielectric tensor	Dissipation factor	Elastic stiffness tensor	Piezoelectric tensor	Elastooptic tensor
	ϱ (g/cm^3) Mohs hardness κ (W/m K)	ε^S_{11} ε^S_{22} ε^S_{33} ε^S_{13} ε^T_{11} ε^T_{22} ε^T_{33} ε^T_{13}	$\tan \delta_1$ $\tan \delta_2$ $\tan \delta_3$ (f (kHz))	c_{11} c_{22} c_{33} c_{44} c_{55} c_{66} c_{12} c_{13} c_{23} c_{15} c_{25} c_{35} c_{46} (10^9 Pa)	d_{21} d_{22} d_{23} d_{14} d_{16} d_{25} d_{34} d_{36} d_h (10^{-12} C/N)	p_{11} p_{12} p_{13} p_{15} p_{21} p_{22} p_{23} p_{25} p_{31} p_{32} p_{33} p_{35} p_{44} p_{46} p_{51} p_{52} p_{53} p_{55} p_{64} p_{66}
Dipotassium tartrate hemihydrate, $K_2C_4H_4O_6 \cdot 0.5H_2O$ (DKT)	1.987 – –	6.44 5.80 6.49 0.005 [4.331]^d – – – –		35.7(86) 39(17) 62(10) 9.0(5) 11.7(19) 8.3(1) 17.8(9) 22.5(99) 13.5(37) 1.8 1.2(1) 5.9(27) 0.54(18) [4.7]	–0.8 4.5 –5.3 7.9 3.4 –6.4 –12.2 –23.2 1.6 [4.632]	
Lithium sulfate monohydrate, $Li_2SO_4 \cdot H_2O$	– – ≈ 4 [4.635]	5.16^d [4.633, 634] 10.3^d [4.633, 634] 4.95^d [4.633, 634] – 5.6 [4.331] 10.3 [4.331] 6.5 [4.331] 0.07 [4.331]		54.9 70.5 61.8 14.0 24.2 27.0 26.3 11.4 17.1 6.5 15.7 –5.2 –26.5 [4.331]^e	–3.6 +16.3 +1.7 +0.7 –2.0 –5.0 –2.13 –4.2 14.4 [4.632]	

Table 4.4-21 Monoclinic, point group 2 (C_2) materials, cont.

Electrooptic tensor	General optical properties	Refractive index			Nonlinear dielectric susceptibility
r_{12}^T r_{22}^T r_{32}^T r_{41}^T r_{43}^T r_{52}^T r_{61}^T r_{63}^T (10^{-12} m/V) (at $\lambda = 633$ nm)	Transparency range (μm) dn_X/dT (10^{-5} K^{-1}) dn_Y/dT (10^{-5} K^{-1}) dn_Z/dT (10^{-5} K^{-1})	n_X A B C D	n_Y A B C D	n_Z A B C D	d_{21} d_{22} d_{23} d_{25} (10^{-12} m/V)
		1.4832 (1.014 μm) [4.636]	1.5142 (1.014 μm) [4.636]	1.5238 (1.014 μm) [4.636]	$d_{21} = 0.11$ $d_{22} = 3.9$ $d_{14} = 0.17$ (all at 0.6943 μm) [4.637]
+8.5(4) +6.5(4) +4.5(5) 0 −1.2(7) −0.7(2) +0.41(2) +0.8(6) [4.638]		1.4521 [4.321]	1.4657 [4.321]	1.4752 [4.321]	$d_{22} = 0.38 \pm 0.06$ $d_{23} = 0.27 \pm 0.04$ $d_{34} = 0.23 \pm 0.04$ [4.321]

Table 4.4-21 Monoclinic, point group 2 (C_2) materials, cont.

Material	General ϱ (g/cm³) Mohs hardness κ (W/m K)	Static dielectric tensor ε^S_{11} ε^S_{22} ε^S_{33} ε^S_{13} ε^T_{11} ε^T_{22} ε^T_{33} ε^T_{13}	Dissipation factor $\tan \delta_1$ $\tan \delta_2$ $\tan \delta_3$ (f (kHz))	Elastic stiffness tensor c_{11} c_{22} c_{33} c_{44} c_{55} c_{66} c_{12} c_{13} c_{23} c_{15} c_{25} c_{35} c_{46} (10^9 Pa)	Piezoelectric tensor d_{21} d_{22} d_{23} d_{14} d_{16} d_{25} d_{34} d_{36} d_h (10^{-12} C/N)	Elastooptic tensor p_{11} p_{12} p_{13} p_{15} p_{21} p_{22} p_{23} p_{25} p_{31} p_{32} p_{33} p_{35} p_{44} p_{46} p_{51} p_{52} p_{53} p_{55} p_{64} p_{66}
D(+)-Saccharose, $C_{12}H_{22}O_{11}$ (sucrose)	– > 2.5 –				1.47 −3.41 0.73 1.25 −2.41 −0.87 −4.21 0.423 [4.639] —	
Triglycine sulfate, $(CH_2NH_2COOH)_3 \cdot H_2SO_4$ (TGS)		9.38[d] [4.640] 20[d] [4.643] 6.00[d] [4.640] 1.17(2)[d] [4.640] 9 [4.644] 40 [4.644] 6.6 [4.644]	0.043 — — (9.6 GHz) [4.643]	41.7 32.1 33.6 9.40 9.39 6.31 17.3 18.1 20.7 4.1 −0.6 7.7 −0.6 [4.646–648]	23.5(0) 7.9(1) 25.3(0) 2.7(0) −4.5(0) 24.3(1) −3.20(1) 2.8(0) — [4.646, 647]	|0.204| [4.641, 642] |0.162| [4.641, 642] |0.175| [4.641, 642] — |0.172| [4.641, 642] |0.208| [4.641, 642] |0.150| [4.641, 642] |0.083| [4.645] |0.204| [4.641, 642] |0.169| [4.641, 642] |0.151| [4.641, 642] — |0.273| [4.645] |0.276| [4.645] — — |0.075| [4.645] — — |0.075| [4.645] —

Table 4.4-21 Monoclinic, point group 2 (C_2) materials, cont.

Electrooptic tensor	General optical properties	Refractive index			Nonlinear dielectric susceptibility
		n_X	n_Y	n_Z	
r_{12}^T	Transparency range (μm)	A	A	A	d_{21}
r_{22}^T	dn_X/dT (10^{-5} K^{-1})	B	B	B	d_{22}
r_{32}^T	dn_Y/dT (10^{-5} K^{-1})	C	C	C	d_{23}
r_{41}^T	dn_Z/dT (10^{-5} K^{-1})	D	D	D	d_{25}
r_{43}^T					(10^{-12} m/V)
r_{52}^T					
r_{61}^T					
r_{63}^T (10^{-12} m/V)					
(at $\lambda = 633$ nm)					
	0.192–1.35 [4.649]	**1.5278**	**1.5552**	**1.5592**	See [4.649, 650]
	–	1.8719	1.9703	2.0526	
	–	0.466	0.4502	0.3909	
	–	0.0214	0.0238	0.252	
		0.0113 [4.649]^c	0.0101 [4.649]^c	0.0187 [4.649]^c	
70 [4.651, 652]					$d_{23} = 0.3$
–					at 0.6943 μm [4.653]
54 [4.651, 652]					
–					
–					
1 [4.654, 655]					
–					
–					

^{abc} Dispersion relations (λ (μm), $T = 20$ °C):

^{a} $n^2 = A + \dfrac{B}{\lambda^2 - C} - D\lambda^2$.

^{b} $n^2 = A + \dfrac{B\lambda^2}{\lambda^2 - C}$.

^{c} $n^2 = A + \dfrac{B\lambda^2}{\lambda^2 - C} - D\lambda^2$.

^{d} These values for ε_{ij} are neither at constant stress nor at constant strain.

^{e} The elastic stiffness constant were measured at constant electric field (c^E).

References

4.1 D. N. Nikogosyan, G. G. Gurzadyan: Kvantovaya Elektron. **13**, 2519–2520 (1986)
4.2 D. N. Nikogosyan, G. G. Gurzadyan: Sov. J. Quantum Electron. [English Transl.] **16**, 1663–1664 (1986)
4.3 V. G. Dmitriev, G. G. Gurzadyan, D. N. Nikogosyan: *Handbook of Nonlinear Optical Crystals*, 2nd edn. (Springer, Berlin, Heidelberg 1997)
4.4 V. G. Dmitriev, G. G. Gurzadyan, D. N. Nikogosyan: *Handbook of Nonlinear Optical Crystals*, 3rd edn. (Springer, Berlin, Heidelberg 1999)
4.5 D. N. Nikogosyan, G. G. Gurzadyan: Kvantovaya Elektron. **14**, 1529–1541 (1987)
4.6 D. N. Nikogosyan, G. G. Gurzadyan: Sov. J. Quantum Electron. [English Transl.] **17**, 970–977 (1987)
4.7 A. G. Every, A. K. McCurdy: *Second and Higher Order Elastic Constants*, Landolt–Börnstein, New Series III/29, ed. by D. F. Nelson, O. Madelung (Springer, Berlin, Heidelberg 1992)
4.8 W. R. Cook Jr.: *Piezoelectric, Electrostrictive, Dielectric Constants, Electromechanical Coupling Factors*, Landolt–Börnstein, New Series III/29, ed. by D. F. Nelson, O. Madelung (Springer, Berlin, Heidelberg 1993)
4.9 W. R. Cook Jr.: *Electrooptic Coefficients*, Landolt–Börnstein, New Series III/30, ed. by D. F. Nelson (Springer, Berlin, Heidelberg 1996)
4.10 D. F. Nelson: *Piezooptic, Electrooptic Constants of Crystals*, Landolt–Börnstein, New Series III/30, ed. by D. F. Nelson (Springer, Berlin, Heidelberg 1996)
4.11 K. Vedam: *Piezooptic, Elastooptic Coefficients*, Landolt–Börnstein, New Series III/30, ed. by D. F. Nelson (Springer, Berlin, Heidelberg 1996)
4.12 D. A. Roberts: IEEE J. Quantum Electron. **28**, 2057 (1992)
4.13 I. Shoji, T. Kondo, A. Kitamoto, M. Shirane, R. Ito: J. Opt. Soc. Am. B **14**, 2268 (1997)
4.14 F. Charra, G. G. Gurzadyan: *Nonlinear Dielectric Susceptibilities*, Landolt–Börnstein, New Series III/30, ed. by D. F. Nelson (Springer, Berlin, Heidelberg 2000)
4.15 G. W. Faris, L. E. Jusinski, A. P. Hickman: J. Opt. Soc. Am. B **3**, 587–599 (1993)
4.16 Opto-Technological Laboratory: Data sheet – Catalog of optical materials (Opto-Technological Laboratory, St. Petersburg 2003)
4.17 Layertec: Data sheet – Material data (Layertec GmbH, Mellingen 2003)
4.18 Röhm Plexiglas: Data sheet – Technical data (Röhm Plexiglas, Darmstadt 2003)
4.19 D. G. Cahill, R. O. Pohl: Phys. Rev. B **35**, 4067–4073 (1987)
4.20 Schott Lithotec: Data sheet – Products and applications (Schott Lithotec AG, Jena 2003)
4.21 Crystran: Data sheet – Materials (Crystran Ltd., Poole 2002)
4.22 D. E. Gray: *American Institute of Physics Handbook* (McGraw-Hill, New York 1977)
4.23 O. V. Shakin, M. F. Bryzhina, V. V. Lemanov: Fiz. Tverd. Tela **13**, 3714–3716 (1971)
4.24 O. V. Shakin, M. F. Bryzhina, V. V. Lemanov: Sov. Phys. Solid State [English Transl.] **13**, 3141–3142 (1972)
4.25 E. D. D. Schmidt, K. Vedam: J. Phys. Chem. Solids **27**, 1563–1566 (1966)
4.26 ASI Instruments: Data sheet – Dielectric constant reference guide (ASI Instruments, Inc., Houston 2003)
4.27 R. Waxler: IEEE J. Quantum Electron. **7**, 166–167 (1971)
4.28 Melles Griot: Data sheet – Optics: Windows and optical flats (Melles Griot, Rochester 2003)
4.29 W. L. Smith: Optical materials, Part I. In: *CRC Handbook of Laser Science and Technology*, Vol. 3, ed. by M. J. Weber (CRC Press, Boca Raton 1986) pp. 229–258
4.30 D. Milam, M. J. Weber: J. Appl. Phys. **47**, 2497–2501 (1976)
4.31 W. T. White III, M. A. Henesian, M. J. Weber: J. Opt. Soc. Am. B **2**, 1402–1408 (1985)
4.32 I. D. Nikolov, C. D. Ivanov: Appl. Opt. **39**, 2067–2070 (2000)
4.33 Yu. A. Repeyev, E. V. Khoroshilova, D. N. Nikogosyan: J. Photochem. Photobiol. B **12**, 259–274 (1992)
4.34 R. Adair, L. L. Chase, S. A. Payne: Phys. Rev. B **39**, 3337–3349 (1989)
4.35 G. G. Gurzadyan, R. K. Ispiryan: Int. J. Nonlin. Opt. Phys. **1**, 533–540 (1992)
4.36 R. DeSalvo, A. A. Said, D. J. Hagan, E. W. Van Stryland, M. Sheik-Bahae: IEEE J. Quantum Electron. **32**, 1324–1333 (1996)
4.37 T. Tomie, I. Okuda, M. Yano: Appl. Phys. Lett. **55**, 325–327 (1989)
4.38 P. Liu, W. L. Smith, H. Lotem, J. H. Bechtel, N. Bloembergen, R. S. Adhav: Phys. Rev. B **17**, 4620–4632 (1978)
4.39 I. H. Malitson: J. Opt. Soc. Am. **55**, 1205–1209 (1965)
4.40 A. Smakula: *Harshaw Optical Crystals* (Harshaw Chemical Co., Cleveland 1967)
4.41 A. Feldman, D. Horowitz, R. M. Waxier, M. J. Dodge: Natl. Bur. Stand. (USA) Techn. Note **993** (1979) erratum (see also [4.656])
4.42 M. D. Levenson, N. Bloembergen: Phys. Rev. B **10**, 4447–4463 (1974)
4.43 H. H. Li: J. Phys. Chem. Ref. Data **9**, 161–289 (1980)
4.44 I. H. Malitson: Appl. Opt. **2**, 1103–1107 (1963)
4.45 S. S. Ballard, J. S. Browder: Thermal properties. In: *Optical Materials, CRC Handbook of Laser Science and Technology*, Vol. 4, Subvol. 2, ed. by M. J. Weber (CRC, Boca Raton 1987) pp. 49–54
4.46 A. D. Papadopoulos, E. Anastassakis: Phys. Rev. B **43**, 9916–9923 (1991)
4.47 H. Braul, C. A. Plint: Solid State Commun. **38**, 227–230 (1981)

4.48 MarkeTech: Data sheet – Non-metallic crystals (MarkeTech International Inc., Port Townsend 2002)

4.49 K. V. Krishna Rao, V. G. Krishna Murty: Acta Crystallogr. **17**, 788–789 (1964)

4.50 K. Vedam, E. D. D. Schmidt, W. C. Schneider: *Optical Properties of Highly Transparent Solids*, ed. by S. S. Mitra, B. Bendow (Plenum, New York 1975) pp. 169–175

4.51 J. P. Szczesniak, D. Cuddeback, J. C. Corelli: J. Appl. Phys. **47**, 5356–5359 (1976)

4.52 V. M. Maevskii, A. B. Roitsin: Fiz. Tverd. Tela **31**, 294–296 (1989)

4.53 V. M. Maevskii, A. B. Roitsin: Sov. Phys. Solid State [English Transl.] **31**, 1448–1449 (1989)

4.54 D. K. Biegelsen: Phys. Rev. Lett. **32**, 1196–1199 (1974)

4.55 D. K. Biegelsen: Erratum, Phys. Rev. Lett. **33**, 51 (1974)

4.56 A. A. Berezhnoi, V. M. Fedulov, K. P. Skornyakova: Fiz. Tverd. Tela **17**, 2785–2787 (1975)

4.57 A. A. Berezhnoi, V. M. Fedulov, K. P. Skornyakova: Sov. Phys. Solid State [English Transl.] **17**, 1855–1856 (1975)

4.58 D. F. Edwards, E. Ochoa: J. Opt. Soc. Am. **71**, 607 (1981)

4.59 G. N. Ramachandran: Proc. Indian Acad. Sci. A **25**, 266–279 (1947)

4.60 J. Fontanella, R. L. Johnston, J. H. Colwell, C. Andeen: Appl. Opt. **16**, 2949–2951 (1977)

4.61 M. D. Levenson, C. Flytzanis, N. Bloembergen: Phys. Rev. B **6**, 3962–3965 (1972)

4.62 D. E. McCarthy: Appl. Opt. **2**, 591–603 (1963)

4.63 N. P. Barnes, M. S. Piltch: J. Opt. Soc. Am. **69**, 178–180 (1979)

4.64 J. J. Wynne: Phys. Rev. **178**, 1295–1303 (1969)

4.65 H. H. Li: J. Phys. Chem. Ref. Data **5**, 329–528 (1976)

4.66 P. D. Maker, R. W. Terhune: Phys. Rev. **137**, A801–A818 (1965)

4.67 A. J. Moses: *Optical Materials Properties*, Handbook of Electronic Materials, Vol. 1 (IFI/Plenum, New York 1971)

4.68 T. Radhakrishnan: Proc. Indian Acad. Sci. A **33**, 22–34 (1951)

4.69 R. F. Stephens, I. H. Malitson: Natl. Bur. Stand. J. Res. **49**, 249 (1952)

4.70 C. C. Wang, E. L. Baardsen: Phys. Rev. **185**, 1079–1082 (1969)

4.71 W. K. Burns, N. Bloembergen: Phys. Rev. B **4**, 3437–3450 (1971)

4.72 N. Bloembergen, W. K. Burns, M. Matsukoda: Opt. Commun. **1**, 195 (1969)

4.73 B. Tatian: Appl. Opt. **23**, 4477–4485 (1984)

4.74 D. F. Edwards, E. Ochoa: Appl. Opt. **19**, 4130 (1980)

4.75 J. J. Wynne, G. D. Boyd: Appl. Phys. Lett. **12**, 191–192 (1968)

4.76 W. Kucharczyk: Physica B **172**, 473–490 (1991)

4.77 A. V. Pakhnev, M. P. Shaskol'skaya, S. S. Gorbach: Izv. Vyssh. Uchebn. Zaved., Fiz. **12**, 28–34 (1975)

4.78 A. V. Pakhnev, M. P. Shaskol'skaya, S. S. Gorbach: Sov. Phys. J. [English Transl.] **18**, 1662 (1975)

4.79 Princeton Scientific: Data sheet – Crystalline materials (Princeton Scientific Corp., Princeton 2003)

4.80 J. Reintjes, M. B. Schulz: J. Appl. Phys. **39**, 5254–5258 (1968)

4.81 S. Musikant: *Optical Materials: An Introduction to Selection and Application* (Marcel Dekker, New York 1985)

4.82 R. W. Dixon: J. Appl. Phys. **38**, 5149–5153 (1967)

4.83 P. Billard: Acta Electron. **6**, 75–169 (1962)

4.84 M. D. Beals, L. Merker: Mater. Des. Eng. **51**, 12–13 (1960)

4.85 D. E. McCarthy: Appl. Opt. **7**, 1997 (1968)

4.86 M. O. Manasreh, D. O. Pederson: Phys. Rev. B **30**, 3482–3485 (1984)

4.87 L. G. DeShazer, S. C. Rand, B. A. Wechsler: Laser crystals. In: *Optical Materials, CRC Handbook of Laser Science and Technology*, Vol. 5, Subvol. 3, ed. by M. J. Weber (CRC, Boca Raton 1987) pp. 281–338

4.88 K. L. Ovanesyan, A. G. Petrosyan, G. O. Shirinyan, A. A. Avetisyan: Izv. Akad. Nauk SSSR, Ser. Neorg. Materiali [in Russian] **17**, 459–462 (1981)

4.89 D. Berlincourt, H. Jaffe, L. R. Shiozawa: Phys. Rev. **129**, 1009–1017 (1963)

4.90 H. Sasaki, K. Tsoubouki, N. Chubachi, N. Mikoshiba: J. Appl. Phys. **47**, 2046–2049 (1976)

4.91 R. Weil, M. J. Sun: Proc. Int. Symp. Cadmium Telluride Materials Gamma-Ray Detectors, Tech. Dig., Strasbourg 1971, pages XIX-1 to XIX-6

4.92 E. H. Turner, I. P. Kaminow, C. Schwab: Phys. Rev. B **9**, 2524–2529 (1974)

4.93 R. C. Hanson, J. R. Hallberg, C. Schwab: Appl. Phys. Lett. **21**, 490–492 (1972)

4.94 A. Boese, E. Mohler, R. Pitka: J. Mater. Sci. **9**, 1754–1758 (1974)

4.95 B. Prevot, C. Carabatos, C. Schwab, B. Hennion, F. Moussa: Solid State Commun. **13**, 1725–1727 (1973)

4.96 D. K. Biegelsen, J. C. Zesch, C. Schwab: Phys. Rev. B **14**, 3578–3582 (1976)

4.97 L. M. Belyaev, G. S. Belikova, G. F. Dobrzhanskii, G. B. Netesov, Yu. V. Shaldin: Fiz. Tverd. Tela **6**, 2526–2528 (1964)

4.98 L. M. Belyaev, G. S. Belikova, G. F. Dobrzhanskii, G. B. Netesov, Yu. V. Shaldin: Sov. Phys. Solid State [English Transl.] **6**, 2007–2008 (1965)

4.99 R. C. Hanson, K. Helliwell, C. Schwab: Phys. Rev. B **9**, 2649–2654 (1974)

4.100 G. Arlt, P. Quadflieg: Phys. Status Solidi **25**, 323–330 (1968)

4.101 Ioffe Institute: Database (Ioffe Institute, St. Petersburg 2003)

4.102 K. S. Champlin, R. J. Erlandson, G. H. Glover, P. S. Hauge, T. Lu: Appl. Phys. Lett. **11**, 348–349 (1967)

4.103 T. E. Walsh: RCA Rev. **27**, 323–335 (1966)

4.104 A. Feldman, R. M. Waxler: J. Appl. Phys. **53**, 1477–1483 (1982)

4.105 D. R. Nelson, E. H. Turner: J. Appl. Phys. **39**, 3337–3343 (1968)

4.106 J. E. Kiefer, A. Yariv: Appl. Phys. Lett. **15**, 26–27 (1969)

4.107 G. H. Sherman, P. D. Coleman: J. Appl. Phys. **44**, 238 (1973)

4.108 R. C. Miller, W. A. Nordland: Phys. Rev. B **5**, 4931–4934 (1972)

4.109 J. Pastrnak, L. Roskovcova: Phys. Status Solidi **14**, K5–K8 (1966)

4.110 N. P. Barnes, M. S. Piltch: J. Opt. Soc. Am. **67**, 628 (1977)

4.111 A. G. DeBell, E. L. Dereniak, J. Harvey, J. Nissley, J. Palmer, A. Selvarajan, W. L. Wolfe: Appl. Opt. **18**, 3114–3115 (1979)

4.112 E. H. Turner, E. Buehler, H. Kasper: Phys. Rev. B **9**, 558–561 (1974)

4.113 D. S. Chemla, P. Kupecek, C. Schwartz, C. Schwab, A. Goltzene: IEEE J. Quantum Electron. **7**, 126 (1971)

4.114 R. C. Miller, W. A. Nordland, S. C. Abrahams, J. L. Bernstein, C. Schwab: J. Appl. Phys. **44**, 3700 (1973)

4.115 E. Mohler, B. Thomas: Phys. Status Solidi (b) **79**, 509–517 (1977)

4.116 A. Feldman, D. Horowitz: J. Opt. Soc. Am. **59**, 1406–1408 (1969)

4.117 T. G. Okroashvili: Opt. Spektrosk. **47**, 798–800 (1979)

4.118 T. G. Okroashvili: Opt. Spectrosc. [English Transl.] **47**, 442–443 (1979)

4.119 J. J. Wynne, N. Bloembergen: Phys. Rev. **188**, 1211 (1969)

4.120 J. J. Wynne, N. Bloembergen: Erratum, Phys. Rev. B **2**, 4306 (1970)

4.121 B. O. Seraphin, H. E. Bennett: *Semiconductors and Semimetals 3*, ed. by R. K. Willard, R. C. Beer (Academic Press, New York 1967)

4.122 R. A. Soref, H. W. Moos: J. Appl. Phys. **35**, 2152 (1964)

4.123 R. K. Chang, J. Ducuing, N. Bloembergen: Phys. Rev. Lett. **15**, 415–418 (1965)

4.124 W. D. Johnston Jr., I. P. Kaminow: Phys. Rev. **188**, 1209–1211 (1969)

4.125 E. Yablonovitch, C. Flytzanis, N. Bloembergen: Phys. Rev. Lett. **29**, 865–868 (1972)

4.126 M. Sugie, K. Tada: Jpn. J. Appl. Phys. **15**, 421–430 (1976)

4.127 M. Bertolotti, V. Bogdanov, A. Ferrari, A. Jascow, N. Nazorova, A. Pikhtin, L. Schirone: J. Opt. Soc. Am. B **7**, 918–922 (1990)

4.128 B. F. Levine, C. G. Bethea: Appl. Phys. Lett. **20**, 272–274 (1972)

4.129 A. H. Kachare, W. G. Spitzer, J. E. Fredrickson: J. Appl. Phys. **47**, 4209 (1976)

4.130 Yu. Berezashvili, S. Machavariani, A. Natsvlishvili, A. Chirakadze: J. Phys. D **22**, 682–686 (1989)

4.131 M. M. Choy, R. L. Byer: Phys. Rev. B **14**, 1693–1705 (1976)

4.132 D. F. Parsons, P. D. Coleman: Appl. Opt. **10**, 1683–1685 (1971)

4.133 Yu. X. Ilisavskii, L. A. Kutakova: Fiz. Tverd. Tela **23**, 3299–3307 (1981)

4.134 Yu. X. Ilisavskii, L. A. Kutakova: Sov. Phys. Solid State [English Transl.] **23**, 1916–1920 (1981)

4.135 L. G. Meiners: J. Appl. Phys. **59**, 1611–1613 (1986)

4.136 D. N. Nichols, D. S. Rimai, R. J. Sladek: Solid State Commun. **36**, 667–669 (1980)

4.137 A. A. Blistanov, V. S. Bondarenko, N. V. Perelomova, F. N. Strizhevskaya, V. V. Tchakalova, M. P. Shaskolskaya: *Acoustic Crystals* [in Russian] (Nauka, Moscow 1982)

4.138 E. Käräjämäki, R. Laiho, T. Levola: Physica **429**, 3–18 (1982)

4.139 G. R. Mariner, K. Vedam: Appl. Opt. **20**, 2878–2879 (1981)

4.140 A. V. Novoselova (Ed.): *Physical-Chemical Properties of Semiconductors – Handbook* [in Russian] (Nauka, Moscow 1979)

4.141 M. Yamada, K. Yamamoto, K. Abe: J. Phys. D **10**, 1309–1313 (1977)

4.142 S. Adachi, C. Hamaguchi: J. Phys. Soc. Jpn. **43**, 1637–1645 (1977)

4.143 C. K. N. Patel, R. E. Slusher, P. A. Fleury: Phys. Rev. Lett. **17**, 1011–1014 (1966)

4.144 H. McKenzie, D. J. Hagan, H. A. Al-Attar: IEEE J. Quantum Electron. **22**, 1328 (1986)

4.145 C. C. Lee, H. Y. Fan: Phys. Rev. B **9**, 3502 (1974)

4.146 O. G. Lorimor, W. G. Spitzer: J. Appl. Phys. **36**, 1841–1844 (1965)

4.147 T. S. Moss: *Optical Properties of Semiconductors* (Academic Press, New York 1959) p. 224

4.148 R. Braunstein, N. Ockman: *Interactions of Coherent Optical Radiation with Solids*, Final Report, Office of Naval Research (Department of Navy, Washington, DC 1964) ARPA Order No. 306-362

4.149 K. Tada, N. Suzuki: Jpn. J. Appl. Phys. **19**, 2295–2296 (1980)

4.150 Y. Tsay, B. Bendow, S. S. Mitra: Phys. Rev. B **5**, 2688–2696 (1972)

4.151 J. Stone, M. S. Whalen: Appl. Phys. Lett. **41**, 1140–1142 (1982)

4.152 A. N. Pikhtin, A. D. Yas'kov: Sov. Phys. Semicond. **12**, 622–626 (1978)

4.153 I. I. Adrianova, A. A. Berezhnoi, K. K. Dubenskii, V. A. Sokolov: Fiz. Tverd. Tela **12**, 2462–2464 (1970)

4.154 I. I. Adrianova, A. A. Berezhnoi, K. K. Dubenskii, V. A. Sokolov: Sov. Phys. Solid State [English Transl.] **12**, 1972–1973 (1970)

4.155 C. Kojima, T. Shikama, S. Kuninobu, A. Kawabata, T. Tanaka: Jpn. J. Appl. Phys. **8**, 1361–1362 (1969)

4.156 R. C. Miller, W. A. Nordland: Phys. Rev. B **2**, 4896–4902 (1970)

4.157 A. Chergui, J. L. Deiss, J. B. Grun, J. L. Loison, M. Robino, R. Besermann: Appl. Surf. Sci. **96**, 874 (1996)

4.158 D. T. F. Marple: J. Appl. Phys. **35**, 539 (1964)

4.159 Alkor Technologies: Data sheet – Optical crystals (Alkor Technologies, St. Petersburg 2002)

4.160 E. A. Kraut, B. R. Tittman, L. J. Graham, T. C. Lim: Appl. Phys. Lett. **17**, 271–272 (1970)

4.161 R.E. Aldrich, S.L. Hou, M.L. Harvill: J. Appl. Phys. **42**, 493–494 (1971)

4.162 J. Zelenka: Czech. J. Phys. B **28**, 165–169 (1978)

4.163 M. Krzesinska: Ultrasonics **24**, 88–92 (1986)

4.164 V.V. Kucha, V.I. Mirgorodskii, S.V. Peshin, A.T. Sobolev: Pis'ma Zh. Tekh. Fiz. **10**, 124–126 (1984)

4.165 V.V. Kucha, V.I. Mirgorodskii, S.V. Peshin, A.T. Sobolev: Sov. Tech. Phys. Lett. [English Transl.] **10**, 51–52 (1984)

4.166 A. Reza, G. Babonas, D. Senuliene: Liet. Fiz. Rinkinys. **26**, 41–47 (1986)

4.167 A. Reza, G. Babonas, D. Senuliene: Sov. Phys. Collection [English Transl.] **26**, 33–38 (1986)

4.168 P.I. Ropot: Opt. Spektrosk. **70**, 371–375 (1991)

4.169 P.I. Ropot: Opt. Spectrosc. [English Transl.] **70**, 217–220 (1991)

4.170 H. Schweppe, P. Quadflieg: IEEE Trans. Sonics Ultrason. **21**, 56–57 (1974)

4.171 Y.R. Reddy, L. Sirdeshmukh: Phys. Status Solidi (a) **103**, K157–K160 (1987)

4.172 G.A. Babonas, A.A. Reza, E.I. Leonov, V.I. Shandaris: Zh. Tekh. Fiz. **55**, 1203–1205 (1985)

4.173 G.A. Babonas, A.A. Reza, E.I. Leonov, V.I. Shandaris: Sov. Phys. Tech. Phys. [English Transl.] **30**, 689–690 (1985)

4.174 R. Bechmann: Proc. R. Phys. Soc. (London) B **64**, 323–337 (1951)

4.175 A.D. Prasad Rao, P. da R. Andrade, S.P.S. Porto: Phys. Rev. B **9**, 1077–1084 (1974)

4.176 T.S. Narasimhamurty: Proc. Indian Acad. Sci. A **40**, 167–175 (1954)

4.177 M. Tanaka, M. Yamada, C. Hamaguchi: J. Phys. Soc. Jpn. **38**, 1708–1714 (1975)

4.178 Eksma: Data sheet – Optics and optomechanics (Eksma Co., Vilnius 2003)

4.179 P.C. Leung, G. Andermann, W.P. Spitzer, C.A. Mead: J. Phys. Chem. Solids **27**, 849–855 (1966)

4.180 V.M. Skorikov, I.S. Zakharov, V.V. Volkov, E.A. Spirin: Inorgan. Mater. **38**, 172–178 (2002)

4.181 F. Vachss, L. Hesselink: Opt. Commun. **62**, 159–165 (1987)

4.182 P. Bayvel, M. McCall, R.V. Wright: Opt. Lett. **13**, 27–29 (1988)

4.183 P. Bayvel: Sensors Actuators **16**, 247–254 (1989)

4.184 H.J. Simon, N. Bloembergen: Phys. Rev. **171**, 1104 (1968)

4.185 H.-J. Weber: Acta Crystallogr. A **35**, 225–232 (1979)

4.186 S. Chandrasekhar: Proc. R. Soc. A **259**, 531 (1961)

4.187 K.L. Vodopyanov, I.A. Kulevskii, V.G. Voevodin, A.I. Gribenyukov, K.R. Allakhverdiev, T.A. Kerimov: Opt. Commun. **83**, 322–326 (1991)

4.188 V.I. Sokolov, V.K. Subashiev: Fiz. Tverd. Tela **14**, 222–228 (1972)

4.189 V.I. Sokolov, V.K. Subashiev: Sov. Phys. Solid State [English Transl.] **14**, 178–183 (1972)

4.190 G.B. Abdullaev, L.A. Kulevsky, A.M. Prokhorov, A.D. Saleev, E.Yu. Salaev, V.V. Smirnov: Pisma Zh. Eksp. Teor. Fiz. **16**, 130 (1972)

4.191 G.B. Abdullaev, L.A. Kulevsky, A.M. Prokhorov, A.D. Saleev, E.Yu. Salaev, V.V. Smirnov: JETP Lett. [English Transl.] **16**, 90–92 (1972)

4.192 P.J. Kupecek, E. Batifol, A. Kuhn: Opt. Commun. **11**, 291–295 (1974)

4.193 K.L. Vodopyanov, I.A. Kulevskii: Opt. Commun. **118**, 375 (1995)

4.194 G.A. Akundov, A.A. Agaeve, V.M. Salmanov, Yu.P. Sharorov, I.D. Yaroshetskii: Sov. Phys. Semicond. **7**, 826–827 (1973)

4.195 C.A. Arguello, D.L. Rousseau, S.P.S. Porto: Phys. Rev. **181**, 1351–1363 (1969)

4.196 C.F. Cline, H.L. Dunegan, G.W. Henderson: J. Appl. Phys. **38**, 1944 (1967)

4.197 S.B. Austerman, D.A. Berlincourt, H.H.A. Krueger: J. Appl. Phys. **34**, 339–341 (1963)

4.198 A.A. Reza, G.A. Babonas: Fiz. Tverd. Tela **16**, 1414–1418 (1974)

4.199 A.A. Reza, G.A. Babonas: Sov. Phys. Solid State [English Transl.] **16**, 909–911 (1974)

4.200 J.D. Beasley: Appl. Opt. **33**, 1000–1003 (1994)

4.201 E.F. Tokarev, G.S. Pado, L.A. Chernozatonskii, V.V. Drachev: Fiz. Tverd. Tela **15**, 1593–1595 (1973)

4.202 E.F. Tokarev, G.S. Pado, L.A. Chernozatonskii, V.V. Drachev: Sov. Phys. Solid State [English Transl.] **15**, 1064–1065 (1973)

4.203 I.A. Dan'kov, G.S. Pado, I.B. Kobyakov, V.V. Berdnik: Fiz. Tverd. Tela **21**, 2570–2575 (1979)

4.204 I.A. Dan'kov, G.S. Pado, I.B. Kobyakov, V.V. Berdnik: Sov. Phys. Solid State [English Transl.] **21**, 1481–1483 (1979)

4.205 I.B. Kobyakov, V.M. Arutyunova: Zh. Tekh. Fiz. [in Russian] **58**, 983 (1988)

4.206 K. Vedam, T.A. Davis: Phys. Rev. **181**, 1196–1201 (1969)

4.207 A.S. Barker, M. Ilegems: Phys. Rev. B **7**, 743–750 (1973)

4.208 E.K. Sichel, J.I. Pankove: J. Phys. Chem. Solids **38**, 330 (1977)

4.209 V.A. Savastenko, A.U. Sheleg: Phys. Status Solidi (a) **48**, K135 (1978)

4.210 M. Gospodinov, P. Sveshtarov, N. Petkov, T. Milenov, V. Tasev, A. Nikolov: Bulg. J. Phys. **16**, 520–522 (1989)

4.211 S.A. Geidur: Opt. Spektrosk. **49**, 193–195 (1980)

4.212 S.A. Geidur: Opt. Spectrosc. [English Transl.] **49**, 105–106 (1980)

4.213 S. Singh, J.R. Potopowicz, L.G. Van Uitert, S.H. Wemple: Appl. Phys. Lett. **19**, 53–56 (1971)

4.214 G. Arlt, G.R. Schodder: J. Acoust. Soc. Am. **37**, 384 (1965)

4.215 E.E. Tokarev, I.B. Kobyakov, I.P. Kuz'mina, A.N. Lobachev, G.S. Pado: Fiz. Tverd. Tela **17**, 980–986 (1975)

4.216 E.E. Tokarev, I.B. Kobyakov, I.P. Kuz'mina, A.N. Lobachev, G.S. Pado: Sov. Phys. Solid State [English Transl.] **17**, 629–632 (1975)

4.217 R.J. Morrow, H.W. Newkirk: Rev. Int. Hautes Temp. Refract. **6**(2), 99–104 (1969)

4.218 D.A. Belogurov, T.G. Okroashvili, Yu.V. Shaldin, V.A. Maslov: Opt. Spektrosk. **54**, 298–301 (1983)

4.219 D.A. Belogurov, T.G. Okroashvili, Yu.V. Shaldin, V.A. Maslov: Opt. Spectrosc. [English Transl.] **54**, 177–179 (1983)

4.220 H.W. Newkirk, D.K. Smith, J.S. Kahn: Am. Mineral. **51**, 141–151 (1966)

4.221 J. Jerphagnon, H.W. Newkirk: Appl. Phys. Lett. **18**, 245–247 (1971)

4.222 M. Bass (Ed.): *Handbook of Optics*, Vol. 2 (McGraw-Hill, New York 1995)

4.223 A.A. Davydov, L.A. Kulevsky, A.M. Prokhorov, A.D. Savelev, V.V. Smirnov: Pisma Zh. Eksp. Teor. Fiz. **15**, 725 (1972)

4.224 A.A. Davydov, L.A. Kulevsky, A.M. Prokhorov, A.D. Savelev, V.V. Smirnov: JETP Lett. [English Transl.] **15**, 513 (1972)

4.225 S.H. Wemple, M. DiDomenico Jr.: Appl. Solid State Sci. **3**, 263–283 (1972)

4.226 A. Penzkofer, M. Schaffner, X. Bao: Opt. Quantum Electron. **22**, 351 (1990)

4.227 G.C. Bhar: Appl. Opt. **15**, 305 (1976)

4.228 R.C. Miller: Appl. Phys. Lett. **5**, 17–19 (1964)

4.229 R. Weil, D. Neshmit: J. Opt. Soc. Am. **67**, 190–195 (1977)

4.230 S.J. Czyzak, W.M. Baker, R.C. Crane, J.B. Howe: J. Opt. Soc. Am. **47**, 240 (1957)

4.231 X.C. Long, R.A. Myers, S.R.J. Brueck, R. Ramer, K. Zheng, S.D. Hersee: Appl. Phys. Lett. **67**, 1349–1351 (1995)

4.232 I. Ishidate, K. Inone, M. Aoki: Jpn. J. Appl. Phys. **19**, 1641–1645 (1980)

4.233 P.M. Lundquist, W.P. Lin, G.K. Wong, M. Razeghi, J.B. Ketterson: Appl. Phys. Lett. **66**, 1883–1885 (1995)

4.234 I.P. Kaminow, E.H. Turner: Appl. Opt. **5**, 1612–1628 (1966)

4.235 I.P. Kaminow, E.H. Turner: Proc. IEEE **54**, 1374–1390 (1966)

4.236 O.W. Madelung, E. Mollwo: Z. Phys. **249**, 12–30 (1971)

4.237 L.B. Kobyakov: Kristallografiya **11**, 419–421 (1966)

4.238 L.B. Kobyakov: Sov. Phys. Crystallogr. [English Transl.] **11**, 369–371 (1966)

4.239 I.A. Dan'kov, I.B. Kobyakov, S.Yu. Davydov: Fiz. Tverd. Tela **24**, 3613–3620 (1982)

4.240 I.A. Dan'kov, I.B. Kobyakov, S.Yu. Davydov: Sov. Phys. Solid State [English Transl.] **24**, 2058–2063 (1982)

4.241 N. Uchida, S. Saito: J. Appl. Phys. **43**, 971–976 (1972)

4.242 S. Haussühl: Acta Crystallogr. A **34**, 547–550 (1978)

4.243 N.A. Zakharov, A.V. Egorov, N.S. Kozlova, O.G. Portnov: Fiz. Tverd. Tela **30**, 3166–3168 (1988)

4.244 N.A. Zakharov, A.V. Egorov, N.S. Kozlova, O.G. Portnov: Sov. Phys. Solid State [English Transl.] **30**, 1823–1824 (1988)

4.245 S. Haussühl: Acustica **23**, 165–169 (1970)

4.246 S. Haussühl: Phys. Status Solidi **29**, K159–K162 (1968)

4.247 Ya.V. Burak, K.Ya. Borman, I.S. Girnyk: Fiz. Tverd. Tela **26**, 3692–3694 (1984)

4.248 Ya.V. Burak, K.Ya. Borman, I.S. Girnyk: Sov. Phys. Solid State [English Transl.] **26**, 2223–2224 (1984)

4.249 J. Wang, H. Li, L. Zhang, R. Wang: Acta Acust. (China) **11**, 338 (1986)

4.250 A.W. Warner, D.A. Pinnow, J.G. Bergman Jr, G.R. Crane: J. Acoust. Soc. Am. **47**, 791–794 (1970)

4.251 R.C. Miller, S.C. Abrahams, R.L. Barns, J.L. Bernstein, W.A. Nordland, E.H. Turner: Solid State Commun. **9**, 1463–1465 (1971)

4.252 A.A. Abdullaev, A.V. Vasil'eva, G.F. Dobrzhanskii, Yu.N. Polivanov: Sov. J. Quantum Electron. **7**, 56–59 (1977)

4.253 G. Nath, H. Nehmanesch, M. Gsänger: Appl. Phys. Lett. **17**, 286–288 (1970)

4.254 S.V. Bogdanov (Ed.): *Lithium Iodate – Growth, Properties and Applications* [in Russian] (Nauka, Novosibirsk 1980)

4.255 D.J. Gettemy, W.C. Harker, G. Lindholm, N.P. Barnes: IEEE J. Quantum Electron. **24**, 2231 (1988)

4.256 O.G. Vlokh, I.A. Velichko, L.A. Laz'ko: Kristallografiya **20**, 430–432 (1975)

4.257 O.G. Vlokh, I.A. Velichko, L.A. Laz'ko: Sov. Phys. Crystallogr. [English Transl.] **20**, 263–264 (1975)

4.258 K. Kato: IEEE J. Quantum Electron. **21**, 119 (1985)

4.259 M.J. Weber: Insulating crystal lasers. In: *CRC Handbook of Lasers with Selected Data on Optical Technology*, ed. by R.J. Pressley (Chemical Rubber Co., Cleveland 1971) pp. 371–417

4.260 T. Goto, O.L. Anderson, I. Ohno, S. Yamamoto: J. Geophys. Res. **94**, 7588 (1989)

4.261 R.M. Waxler, E.N. Farabaugh: J. Res. Natl. Bur. Stand. A **74**, 215–220 (1970)

4.262 E.M. Voronkova, B.N. Grechushnikov, G.I. Distler, I.P. Petrov: *Optical Materials for the Infrared Technique* [in Russian] (Nauka, Moscow 1965)

4.263 N.N. Khromova: Izv. Akad. Nauk SSSR, Ser. Neorg. Materiali [in Russian] **23**, 1500 (1987)

4.264 D.F. Nelson, P.D. Lazay, M. Lax: Phys. Rev. B **6**, 3109–3120 (1972)

4.265 Z.P. Chang, G.R. Barsch: IEEE Trans. Sonics Ultrason. **23**, 127–135 (1976)

4.266 I.M. Sil'vestrova, Yu.V. Pisarevskii, O.V. Zvereva, A.A. Shternberg: Kristallografiya **32**, 792–794 (1987)

4.267 I.M. Sil'vestrova, Yu.V. Pisarevskii, O.V. Zvereva, A.A. Shternberg: Sov. Phys. Crystallogr. [English Transl.] **32**, 467–468 (1987)

4.268 I.M. Sil'vestrova, O.A. Aleshko-Ozhevskii, Yu.V. Pisarevskii, A.A. Shternberg, G.S. Mironova: Fiz. Tverd. Tela **29**, 3454–3456 (1987)

4.269 I.M. Sil'vestrova, O.A. Aleshko-Ozhevskii, Yu.V. Pisarevskii, A.A. Shternberg, G.S. Mironova: Sov. Phys. Solid State [English Transl.] **29**, 1979–1980 (1987)

4.270 H.A.A. Sidek, G.A. Saunders, H. Wang, B. Xu, J. Han: Phys. Rev. B **36**, 7612 (1987)

4.271 A.C. DeFranzo, B.G. Pazol: Appl. Opt. **32**, 2224–2234 (1993)

4.272 E. Carvallo: Compte Rendus **126**, 950 (1898)

4.273 W.L. Bond: J. Appl. Phys. **36**, 1674–1677 (1965)

4.274 H. Jaffe, D. A. Berlincourt: Proc. IEEE **53**, 1372–1386 (1965)
4.275 J. Sapriel, R. Lancon: Proc. IEEE **61**, 678–679 (1973)
4.276 J. Sapriel: Appl. Phys. Lett. **19**, 533–535 (1971)
4.277 I. M. Silvestrova, V. A. Kuznetsov, N. A. Moiseeva, E. P. Efremova, Yu. V. Pisarevskii: Fiz. Tverd. Tela [in Russian] **28**, 180 (1986)
4.278 W. G. Cady: *Piezoelectricity* (McGraw-Hill, New York 1946)
4.279 P. H. Carr: J. Acoust. Soc. Am. **41**, 75–83 (1967)
4.280 R. Bechmann: Phys. Rev. **110**, 1060–1061 (1958)
4.281 T. S. Narasimhamurty: J. Opt. Soc. Am. **59**, 682–686 (1969)
4.282 H. Wagner: Z. Phys. **193**, 218–234 (1966)
4.283 P. Grosse: *Die Festkörpereigenschaften von Tellur*, Springer Tracts in Modern Physics, No. 48 (Springer, Berlin, Heidelberg 1969) p. 65
4.284 G. Arlt, P. Quadflieg: Phys. Status Solidi **32**, 687–689 (1969)
4.285 I. S. Grigoriev, E. Z. Melikhov (Eds.): *Physical Quantities Handbook* [in Russian] (Emergoatomizdat, Moscow 1991)
4.286 D. Souilhac, D. Billeret, A. Gundjan: Appl. Opt. **28**, 3993–3996 (1989)
4.287 W. L. Bond, G. D. Boyd, H. L. Carter Jr.: J. Appl. Phys. **38**, 4090–4091 (1967)
4.288 G. D. Boyd, T. J. Bridges, E. G. Burkhardt: IEEE J. Quantum Electron. **4**, 515 (1968)
4.289 E. H. Turner: IEEE J. Quantum Electron. **3**, 695 (1967)
4.290 C. Chen, Z. Xu, D. Deng, J. Zhang, G. K. L. Wong, B. Wu, N. Ye, D. Tang: Appl. Phys. Lett. **68**, 2930–2932 (1996)
4.291 J. Jerphagnon, S. K. Kurtz: Phys. Rev. B **1**, 1739 (1970)
4.292 F. Hache, A. Zeboulon, G. Gallot, G. M. Gale: Opt. Lett. **20**, 1556–1558 (1995)
4.293 A. J. Rogers: Proc. R. Soc. London A **353**, 177–192 (1977)
4.294 J. H. McFee, G. D. Boyd, P. H. Schmidt: Appl. Phys. Lett. **17**, 57–59 (1970)
4.295 D. Eimerl, L. Davis, S. Velsko, E. K. Graham, A. Zalkin: J. Appl. Phys. **62**, 1968–1983 (1987)
4.296 J. R. Teague, R. R. Rice, R. Gerson: J. Appl. Phys. **46**, 2864–2866 (1975)
4.297 T. Yamada, N. Niizeki, H. Toyoda: Jpn. J. Appl. Phys. **6**, 151–155 (1967)
4.298 R. T. Smith, F. S. Welsh: J. Appl. Phys. **42**, 2219–2230 (1971)
4.299 R. A. Graham: Ferroelectrics **10**, 65–69 (1976)
4.300 T. Yamada, H. Iwasaki, N. Niizeki: Jpn. J. Appl. Phys. **8**, 1127–1132 (1969)
4.301 I. Tomeno, N. Hirano: J. Phys. Soc. Jpn. **50**, 1809 (1981)
4.302 I. Tomeno, N. Hirano: J. Phys. Soc. Jpn. **51**, 339 (1982)
4.303 L. P. Avakyants, D. F. Kiselev, N. N. Shchitov: Fiz. Tverd. Tela **18**, 2129–2130 (1976)
4.304 L. P. Avakyants, D. F. Kiselev, N. N. Shchitov: Sov. Phys. Solid State [English Transl.] **18**, 1242–1243 (1976)
4.305 C. Chen, B. Wu, A. Jiang, G. You: Sci. Sin. B **28**, 235 (1985)
4.306 L. J. Bromley, A. Guy, D. C. Hanna: Opt. Commun. **67**, 316–320 (1988)
4.307 G. G. Gurzadyan, A. S. Oganesyan, A. V. Petrosyan, R. O. Sharkhatunyan: Zh. Tekh. Fiz. **61**, 152–154 (1991)
4.308 G. G. Gurzadyan, A. S. Oganesyan, A. V. Petrosyan, R. O. Sharkhatunyan: Sov. Phys. Tech. Phys. [English Transl.] **36**, 341–342 (1991)
4.309 R. C. Eckardt, H. Masuda, Y. X. Fan, R. L. Byer: IEEE J. Quantum Electron. **26**, 922–933 (1990)
4.310 L. Bohaty, J. Liebertz: Z. Kristallogr. **192**, 91–95 (1990)
4.311 D. Zhang, Y. Kong, J.-Y. Zhang: Opt. Commun. **184**, 485–491 (2000)
4.312 M. V. Hobden, J. Warner: Phys. Lett. **22**, 243 (1966)
4.313 D. H. Jundt, M. M. Fejer, R. L. Byer: IEEE J. Quantum Electron. **26**, 135–138 (1990)
4.314 J. E. Midwinter: J. Appl. Phys. **39**, 3033–3038 (1968)
4.315 K. F. Hulme, P. H. Davies, V. M. Cound: J. Phys. C **2**, 855–857 (1969)
4.316 J. D. Zook, D. Chen, G. N. Otto: Appl. Phys. Lett. **11**, 159–161 (1967)
4.317 A. L. Aleksandrovskii, G. I. Ershova, G. Kh. Kitaeva, S. P. Kulik, I. I. Naumova, V. V. Tarasenko: Kvantovaya Elektron. **18**, 254–256 (1991)
4.318 A. L. Aleksandrovskii, G. I. Ershova, G. Kh. Kitaeva, S. P. Kulik, I. I. Naumova, V. V. Tarasenko: Sov. J. Quantum Electron. [English Transl.] **21**, 225–227 (1991)
4.319 Y. Chang, J. Wen, H. Wang, B. Li: Chin. Phys. Lett. **9**, 427–430 (1992)
4.320 R. J. Holmes, Y. S. Kim, C. D. Brandle, D. M. Smyth: Ferroelectrics **51**, 41–45 (1983)
4.321 S. Singh: Nonlinear optical materials. In: *CRC Handbook of Lasers with Selected Data on Optical Technology*, ed. by R. J. Pressley (Chemical Rubber Co., Cleveland 1971) pp. 489–525
4.322 R. C. Miller, A. Savage: Appl. Phys. Lett. **9**, 169–171 (1966)
4.323 E. H. Turner: Meeting Opt. Soc. Am., Tech. Dig. Paper A13, San Francisco 1966
4.324 C. O'Hara, N. M. Shorrocks, R. W. Whatmore, O. Jones: J. Phys. D **15**, 1289–1299 (1982)
4.325 Ya. M. Olikh: Fiz. Tverd. Tela **25**, 2222–2225 (1983)
4.326 Ya. M. Olikh: Sov. Phys. Solid State [English Transl.] **25**, 1282–1283 (1983)
4.327 Ya. M. Olikh: Phys. Status Solidi (a) **80**, K81 (1983)
4.328 I. I. Zubrinov, V. I. Semenov, D. V. Sheloput: Fiz. Tverd. Tela **15**, 2871–2873 (1973)
4.329 I. I. Zubrinov, V. I. Semenov, D. V. Sheloput: Sov. Phys. Solid State [English Transl.] **15**, 1921–1922 (1974)
4.330 V. V. Lomanov: *Acoustooptical Radioelectronic System Devices* [in Russian] (Nauka, Leningrad 1988) pp. 48–61
4.331 W. P. Mason: *Piezoelectric Crystals and their Applications to Ultrasonics* (Van Nostrand, New York 1950)
4.332 Y. S. Touloukian, R. W. Powell, C. Y. Ho, P. G. Klemens: *Thermal Conductivity, Non-Metallic Solids*, Thermal Properties of Matter, Vol. 2 (IFI/Plenum, New York 1970)

4.333 I. S. Zheludev, M. M. Tagieva: Kristallografiya **7**, 589–592 (1962)

4.334 I. S. Zheludev, M. M. Tagieva: Sov. Phys. Crystallogr. [English Transl.] **7**, 473–475 (1963)

4.335 J. D. Feichtner, R. Johannes, G. W. Roland: Appl. Opt. **9**, 1716–1717 (1970)

4.336 W. B. Gandrud, G. D. Boyd, J. H. McFee, F. H. Wehmeier: Appl. Phys. Lett. **16**, 59–61 (1970)

4.337 I. V. Kityk: Fiz. Tverd. Tela **33**, 1826–1833 (1991)

4.338 I. V. Kityk: Sov. Phys. Solid State [English Transl.] **33**, 1026–1030 (1991)

4.339 K. F. Hulme, O. Jones, P. H. Davies, M. V. Hobden: Appl. Phys. Lett. **10**, 133–135 (1967)

4.340 J. Warner: Brit. J. Appl. Phys. II **1**, 949–950 (1968)

4.341 R. F. Lucy: Appl. Opt. **11**, 1329 (1972)

4.342 D. S. Chemla, P. J. Kupecek, C. A. Schwartz: Opt. Commun. **7**, 225–228 (1973)

4.343 J. D. Feichtner, G. W. Roland: Appl. Opt. **11**, 993–998 (1972)

4.344 M. D. Ewbank, P. R. Newman, N. L. Mota, S. M. Lee, W. L. Wolfe, A. G. DeBell, W. A. Harrison: J. Appl. Phys. **51**, 3848–3852 (1980)

4.345 R. C. Y. Auyeung, D. M. Zielke, B. J. Feldman: Appl. Phys. B **48**, 293 (1989)

4.346 D. R. Suhre: Appl. Phys. B **52**, 367–370 (1991)

4.347 I. P. Kaminow, E. H. Turner: *CRC Handbook of Lasers with Selected Data on Optical Technology*, ed. by R. J. Pressley (Chemical Rubber Co., Cleveland 1971) pp. 447–459

4.348 F. Pockels: Abhandl. Ges. Wiss. Göttingen **39**, 1–204 (1894)

4.349 M. J. Weber (Ed.): *Handbook of Laser Science and Technology*, Vol. 3 (CRC, Boca Raton 1986)

4.350 S. Chang, H. R. Carleton: *Basic Optical Properties of Materials*, National Bureau of Standards Special Publication 574, ed. by A. Feldman (National Bureau of Standards, Boulder 1980) pp. 213–216

4.351 M. H. Grimsditch, A. K. Ramdas: Phys. Rev. B **22**, 4094–4096 (1980)

4.352 H. J. Eichler, H. Fery, J. Knof, J. Eichler: Z. Phys. B **28**, 297–306 (1977)

4.353 J. R. DeVore: J. Opt. Soc. Am. **41**, 416–419 (1951)

4.354 Bluelightning Optics: Data sheet – Product (Bluelightning Optics Co., Ltd., Beijing 2002)

4.355 G. C. Coquin, D. A. Pinnow, A. W. Warner: J. Appl. Phys. **42**, 2162–2168 (1971)

4.356 K. Wu, W. Hua: Acta. Acust. (China) **12**, 64 (1987)

4.357 I. L. Chistyi, L. Csillag, V. F. Kitaeva, N. Kroo, N. N. Sobolev: Phys. Status Solidi (a) **47**, 609–615 (1978)

4.358 B. W. Woods, S. A. Payne, J. E. Marion, R. S. Hughes, L. E. Davis: J. Opt. Soc. Am. B **8**, 970–977 (1991)

4.359 P. Blanchfield, G. A. Saunders: J. Phys. C **12**, 4673 (1979)

4.360 Y. Ohmachi, N. Uchida: J. Appl. Phys. **41**, 2307–2311 (1970)

4.361 G. Arlt, H. Schweppe: Solid State Commun. **6**, 783–784 (1968)

4.362 A. W. Warner, D. L. White, W. A. Bonner: J. Appl. Phys. **11**, 4489–4495 (1972)

4.363 Y. Ohmachi, N. Uchida: Rev. Electr. Commun. Lab. (Tokyo) **20**, 529–541 (1972)

4.364 Almaz Optics Inc.: Data sheet – Optical components (Almaz Optics Inc., Marlton 2002)

4.365 M. D. Ewbank, F. R. Newman: J. Appl. Phys. **53**, 1150–1153 (1982)

4.366 G. H. Sherman, P. D. Coleman: IEEE J. Quantum Electron. **9**, 403–409 (1973)

4.367 Y. Fujii, S. Yoshida, S. Misawa, S. Mackawa, T. Sakudo: Appl. Phys. Lett. **31**, 815–816 (1977)

4.368 N. Uchida: Phys. Rev. **4**, 3736 (1971)

4.369 R. S. Adhav, A. D. Vlassopoulos: IEEE J. Quantum Electron. **10**, 688 (1974)

4.370 E. N. Volkova, A. N. Izrailenko: Kristallografiya **28**, 1217–1219 (1983)

4.371 E. N. Volkova, A. N. Izrailenko: Sov. Phys. Crystallogr. [English Transl.] **28**, 716–717 (1983)

4.372 W. P. Mason, B. T. Matthias: Phys. Rev. **88**, 477–479 (1952)

4.373 A. S. Vasilevskaya, A. S. Sonin: Kristallografiya **14**, 713–716 (1970)

4.374 A. S. Vasilevskaya, A. S. Sonin: Sov. Phys. Crystallogr. [English Transl.] **14**, 611–613 (1970)

4.375 Clevite Corp.: *Reference Data on Linear Electro-Optic Effects* (Clevite Corp., Cleveland 1967)

4.376 H. Jaffe: Meeting Opt. Soc. Am., Tech. Dig., Cleveland 1950

4.377 G. K. Afanaseva: Kristallografiya **13**, 1024 (1968)

4.378 G. K. Afanaseva: Sov. Phys. Crystallogr. [English Transl.] **13**, 1024 (1968)

4.379 R. S. Adhav: J. Acoust. Soc. Am. **43**, 835 (1968)

4.380 F. Hoff, B. Stadnik: Electron. Lett. **2**, 293 (1966)

4.381 T. S. Narasimhamurty, K. Veerabhadra Rao, H. E. Pettersen: J. Mater. Sci. **8**, 577–580 (1973)

4.382 T. Hailing, G. A. Saunders, W. A. Lambson: Phys. Rev. B **26**, 5786 (1982)

4.383 R. S. Adhav: 76th Annual Meeting Acoust. Soc. Am., Tech. Dig. Paper No. 79, Cleveland 1968

4.384 B. A. Strukov, T. P. Spiridonov, N. N. Tkachev, K. A. Minaeva, M. Yu. Kozhevnikov: Izv. Akad. Nauk SSSR Ser. Fiz. **53**, 1320–1326 (1989)

4.385 B. A. Strukov, T. P. Spiridonov, N. N. Tkachev, K. A. Minaeva, M. Yu. Kozhevnikov: Bull. Acad. Sci. USSR, Phys. Ser. [English Transl.] **53**, 86–92 (1989)

4.386 D. Eimerl: Ferroelectrics **72**, 95–139 (1987)

4.387 A. S. Sonin, A. S. Vasilevskaya: *Electrooptic Crystals* [in Russian] (Atomizdat, Moscow 1971)

4.388 A. S. Vasilevskaya, E. N. Volkova, V. A. Koptsik, L. N. Rashkovich, T. A. Regulskaya, I. S. Rez, A. S. Sonin, V. S. Suvorov: Kristallografiya **12**, 518 (1967)

4.389 A. S. Vasilevskaya, E. N. Volkova, V. A. Koptsik, L. N. Rashkovich, T. A. Regulskaya, I. S. Rez, A. S. Sonin, V. S. Suvorov: Sov. Phys. Crystallogr. [English Transl.] **12**, 446 (1967)

4.390 V. S. Suvorov, A. S. Sonin, I. S. Rez: Zh. Eksp. Teor. Fiz. **53**, 491 (1967)
4.391 V. S. Suvorov, A. S. Sonin, I. S. Rez: Sov. Phys. JETP [English Transl.] **26**, 23 (1967)
4.392 N. P. Barnes, D. J. Gettemy, R. S. Adhav: J. Opt. Soc. Am. **72**, 895 (1982)
4.393 F. M. Johnson, J. A. Duardo: Laser Focus **11**, 31 (1967)
4.394 F. Zernike: J. Opt. Soc. Am. **55**, 91 (1965)
4.395 R. L. Byer, H. Kildal, R. S. Feigelson: Appl. Phys. Lett. **19**, 237–240 (1971)
4.396 G. D. Boyd, E. Buehler, F. G. Storz, J. H. Wernick: IEEE J. Quantum Electron. **8**, 419–426 (1972)
4.397 G. C. Ghosh, G. C. Bhar: IEEE J. Quantum Electron. **18**, 143 (1982)
4.398 K. Kato: IEEE J. Quantum Electron. **10**, 616 (1974)
4.399 R. S. Adhav: J. Opt. Soc. Am. **59**, 414–418 (1969)
4.400 R. S. Adhav: J. Appl. Phys. **46**, 2808 (1975)
4.401 M. P. Zaitseva, Yu. I. Kokorin, A. M. Sysoev, I. S. Rez: Kristallografiya **27**, 146–151 (1982)
4.402 M. P. Zaitseva, Yu. I. Kokorin, A. M. Sysoev, I. S. Rez: Sov. Phys. Crystallogr. [English Transl.] **27**, 86–89 (1982)
4.403 A. T. Anistratov, V. G. Martynov, L. A. Shabanova, I. S. Rez: Fiz. Tverd. Tela **21**, 1354 (1979)
4.404 A. T. Anistratov, V. G. Martynov, L. A. Shabanova, I. S. Rez: Sov. Phys. Solid State [English Transl.] **21** (1979)
4.405 C. W. Fairall, W. Reese: Phys. Rev. B **6**, 193–199 (1972)
4.406 R. S. Adhav: J. Appl. Phys. **39**, 4091 (1968)
4.407 T. R. Sliker, S. R. Burlage: J. Appl. Phys. **34**, 1837–1840 (1963)
4.408 L. A. Shuvalov, A. V. Mnatsakanyan: Kristallografiya **11**, 222 (1966)
4.409 L. P. Avakyants, D. F. Kiselev, N. V. Perelomova, V. I. Sugrei: Fiz. Tverd. Tela **25**, 580–582 (1983)
4.410 L. P. Avakyants, D. F. Kiselev, N. V. Perelomova, V. I. Sugrei: Sov. Phys. Solid State [English Transl.] **25**, 329–330 (1983)
4.411 Y. Mori, I. Kuroda, S. Nakajima, T. Sasaki, S. Nakai: Appl. Phys. Lett. **67**, 1818–1820 (1995)
4.412 N. Umemura, K. Kato: Appl. Opt. **36**, 6794 (1997)
4.413 G. D. Boyd, H. M. Kasper, J. H. McFee, F. G. Storz: IEEE J. Quantum Electron. **8**, 900–908 (1972)
4.414 G. D. Boyd, H. Kasper, J. M. McFee: IEEE J. Quantum Electron. **7**, 563–573 (1971)
4.415 G. C. Bhar, G. Ghosh: J. Opt. Soc. Am. **69**, 730–733 (1979)
4.416 R. S. Adhav: J. Appl. Phys. **39**, 4095–4098 (1968)
4.417 R. Danelyus, A. Piskarskas, V. Sirutkaitis, A. Stabinis, Ya. Yasevichyute: *Parametric Generators of Light and Picosecond Spectroscopy* [in Russian] (Mokslas, Vilnius 1983)
4.418 J. H. Ott, T. R. Sliker: J. Opt. Soc. Am. **54**, 1442–1444 (1964)
4.419 M. I. Zugrav, A. Dumitrica, A. Dumitras, N. Comaniciu: Cryst. Res. Technol. **17**, 475–480 (1982)
4.420 I. P. Kaminow: Phys. Rev. A **138**, 1539–1543 (1965)
4.421 D. A. Berlincourt, D. R. Curran, H. Jaffe: *Physical Acoustics*, Vol. 1A, ed. by W. P. Mason (Academic Press, New York 1964) pp. 1169–1270
4.422 S. Haussühl: Z. Kristallogr. **120**, 401 (1964)
4.423 I. P. Kaminow, G. O. Harding: Phys. Rev. **129**, 1562–1566 (1963)
4.424 R. S. Adhav: J. Phys. D **2**, 171–175 (1969)
4.425 E. N. Volkova, B. M. Berezhnoi, A. N. Izrailenko, A. V. Mishchenko, L. N. Rashkovich: Izv. Akad. Nauk SSSR, Ser. Fiz. **35**, 1858–1861 (1971)
4.426 E. N. Volkova, B. M. Berezhnoi, A. N. Izrailenko, A. V. Mishchenko, L. N. Rashkovich: Bull. Acad. Sci. USSR, Phys. Ser. [English Transl.] **35**, 1690–1693 (1971)
4.427 L. A. Shuvalov, I. S. Zheludev, A. V. Mnatsakanyan, Ts.-Zh. Ludupov, I. Fiala: Izv. Akad. Nauk SSSR, Ser. Fiz. **31**, 1919–1922 (1967)
4.428 L. A. Shuvalov, I. S. Zheludev, A. V. Mnatsakanyan, Ts.-Zh. Ludupov, I. Fiala: Bull. Acad. Sci. USSR, Phys. Ser. [English Transl.] **31**, 1963–1966 (1967)
4.429 R. S. Adhav: J. Phys. D **2**, 177–182 (1969)
4.430 K. S. Aleksandrov, A. T. Anistratov, A. R. Zamkov, I. S. Rez: Fiz. Tverd. Tela **19**, 1863–1866 (1977)
4.431 K. S. Aleksandrov, A. T. Anistratov, A. R. Zamkov, I. S. Rez: Sov. Phys. Solid State [English Transl.] **19**, 1090–1091 (1977)
4.432 W. L. Smith: Appl. Opt. **6**, 1798 (1977)
4.433 K. Veerabhadra Rao, T. S. Narasimhamurty: J. Phys. C **11**, 2343–2347 (1978)
4.434 F. Zernike: J. Opt. Soc. Am. **54**, 1215 (1964)
4.435 F. Zernike: Erratum, J. Opt. Soc. Am. **55**, 210 (1965)
4.436 K. Kato, A. J. Alcock, M. C. Richardson: Opt. Commun. **11**, 5 (1974)
4.437 K. Kato: Opt. Commun. **13**, 93 (1975)
4.438 J. E. Pearson, G. A. Evans, A. Yariv: Opt. Commun. **4**, 366–367 (1972)
4.439 K. Kato: IEEE J. Quantum Electron. **10**, 622 (1974)
4.440 I. S. Zheludev, Ts.-Zh. Ludupov: Kristallografiya **10**, 764–766 (1965)
4.441 I. S. Zheludev, Ts.-Zh. Ludupov: Sov. Phys. Crystallogr. [English Transl.] **10**, 645–646 (1966)
4.442 H. Horinaka, H. Nozuchi, H. Sonomura, T. Miyauchi: Jpn. J. Appl. Phys. **22**, 546 (1983)
4.443 V. M. Cound, P. H. Davies, K. F. Hulme, D. Robertson: J. Phys. C **3**, L83–L84 (1970)
4.444 M. H. Grimsditch, G. D. Holah: Phys. Rev. B **12**, 4377 (1975)
4.445 M. H. Grimsditch, G. D. Holah: J. Phys. (Paris) **36**, Suppl. C, C3–C185 (1975)
4.446 A. Yoshihara, E. R. Bernstein: J. Chem. Phys. **77**, 5319 (1982)
4.447 G. Ambrazevicius, G. Babonas: Liet. Fiz. Sbornik. **18**, 765–774 (1978)
4.448 G. Ambrazevicius, G. Babonas: Sov. Phys. Collection [English Transl.] **18**, 52–59 (1978)
4.449 Yu. K. Shaldin, D. A. Belogurov: Opt. Spektrosk. **35**, 693–701 (1973)
4.450 Yu. K. Shaldin, D. A. Belogurov: Opt. Spectrosc. [English Transl.] **35**, 403–407 (1973)

4.451 R. L. Byer, M. M. Choy, R. L. Herbst, D. S. Chemla, R. S. Feigelson: Appl. Phys. Lett. **24**, 65–68 (1974)

4.452 N. P. Barnes, D. J. Gettemy, J. R. Hietanen, R. A. Iannini: Appl. Opt. **28**, 5162–5168 (1989)

4.453 H. Horinaka, H. Sonomura, T. Migauchi: Jpn. J. Appl. Phys. **21**, 1485–1488 (1982)

4.454 D. S. Chemla, P. J. Kupecek, D. S. Robertson, R. C. Smith: Opt. Commun. **3**, 29 (1971)

4.455 M. G. Cohen, M. DiDomenico Jr., S. H. Wemple: Phys. Rev. B **1**, 4334–4337 (1970)

4.456 G. C. Bhar, D. K. Ghosh, P. S. Ghosh, D. Schmitt: Appl. Opt. **22**, 2492 (1983)

4.457 J. J. Zondy, D. Touahri: J. Opt. Soc. Am. B **14**, 1331 (1997)

4.458 J. M. Halbout, S. Blit, W. Donaldson, C. L. Tang: IEEE J. Quantum Electron. **15**, 1176–1180 (1979)

4.459 K. Betzler, H. Hesse, P. Loose: J. Mol. Struct. **47**, 393–396 (1978)

4.460 L. Bohaty: Z. Kristallogr. **163**, 307–309 (1983)

4.461 M. J. Rosker, K. Cheng, C. L. Tang: IEEE J. Quantum Electron. **21**, 1600 (1985)

4.462 G. D. Boyd, E. Buehler, F. G. Storz: Appl. Phys. Lett. **18**, 301–304 (1971)

4.463 S. R. Sashital, R. R. Stephens, J. F. Lotspeich: J. Appl. Phys. **59**, 757–760 (1986)

4.464 G. C. Bhar, L. K. Samanta, D. K. Ghosh, S. Das: Kvantovaya Elektron. **14**, 1361 (1987)

4.465 G. C. Bhar, L. K. Samanta, D. K. Ghosh, S. Das: Sov. J. Quantum Electron. [English Transl.] **17**, 860 (1987)

4.466 K. Tada, K. Kikuchi: Jpn. J. Appl. Phys. **19**, 1311–1315 (1980)

4.467 A. Schaefer, H. Schmitt, A. Dörr: Ferroelectrics **69**, 253–266 (1986)

4.468 D. Berlincourt, H. Jaffe: Phys. Rev. **111**, 143–148 (1958)

4.469 W. J. Merz: Phys. Rev. **76**, 1221–1225 (1949)

4.470 V. G. Gavrilyachenko, E. G. Fesenko: Kristallografiya **16**, 640–641 (1971)

4.471 V. G. Gavrilyachenko, E. G. Fesenko: Sov. Phys. Crystallogr. [English Transl.] **16**, 549–550 (1971)

4.472 A. V. Turik, N. B. Shevchenko, V. G. Gavrilyachenko, E. G. Fesenko: Phys. Status Solidi (b) **94**, 525–528 (1979)

4.473 Molecular Technology: Data sheet – World of crystals (Molecular Technology GmbH, Berlin 2002)

4.474 M. Adachi, A. Kawabata: Jpn. J. Appl. Phys. **17**, 1969–1973 (1978)

4.475 R. C. Miller, D. A. Kleinman, A. Savage: Phys. Rev. Lett. **11**, 146–149 (1963)

4.476 R. C. Miller, W. A. Nordland: Opt. Commun. **1**, 400–402 (1970)

4.477 M. Zgonik, P. Bernasconi, M. Duelli, R. Schlesser, P. Günter, M. H. Garrett, D. Rytz, Y. Zhu, X. Wu: Phys. Rev. B **50**, 5941–5949 (1994)

4.478 S. H. Wemple, M. DiDomenico Jr., I. Camlibel: J. Phys. Chem. Solids **29**, 1797–1803 (1968)

4.479 S. Singh, J. P. Remeika, J. R. Potopowicz: Appl. Phys. Lett. **20**, 135–137 (1972)

4.480 E. H. Turner: unpublished work (quoted in [4.225, 347])

4.481 R. Komatsu, T. Sugawara, K. Sassa, N. Sarukura, Z. Liu, S. Izumida, Y. Segawa, S. Uda, T. Fukuda, K. Yamanouchi: Appl. Phys. Lett. **70**, 3492–3494 (1997)

4.482 T. Sugawara, R. Komatsu, S. Uda: Solid State Commun. **107**, 233 (1998)

4.483 S. I. Furusawa, O. Chikagawa, S. Tange, T. Ishidate, H. Orihara, Y. Ishibashi, K. Miwa: J. Phys. Soc. Jpn. **60**, 2691–2693 (1991)

4.484 L. Bohaty, S. Haussühl, J. Liebertz: Cryst. Res. Technol. **24**, 1159–1163 (1989)

4.485 L. G. Van Uitert, S. Singh, H. J. Levinstein, J. E. Geusic, W. A. Bonner: Appl. Phys. Lett. **11**, 161–163 (1967)

4.486 L. G. Van Uitert, S. Singh, H. J. Levinstein, J. E. Geusic, W. A. Bonner: Erratum, Appl. Phys. Lett. **12**, 224 (1968)

4.487 J. J. E. Reid: Appl. Phys. Lett. **62**, 19–21 (1993)

4.488 M. Adachi, T. Shiosaki, A. Kawabata: Ferroelectrics **27**, 89–92 (1980)

4.489 J. L. Kirk, K. Vedam: J. Phys. Chem. Solids **33**, 1251–1255 (1972)

4.490 R. Kiriyama: Science (Tokyo) **17**, 239–240 (1947)

4.491 R. Kiriyama: Chem. Abstr. **45**, 2278c (1951)

4.492 M. S. Madhava, S. Haussühl: Z. Kristallogr. **141**, 25–30 (1975)

4.493 M. J. Dodge: Refractive index. In: *Optical Materials*, CRC Handbook of Laser Science and Technology, Vol. 4, Subvol. 2, ed. by M. J. Weber (CRC, Boca Raton 1987) pp. 21–47

4.494 K. W. Martin, L. G. DeShazer: Appl. Opt. **12**, 941–943 (1973)

4.495 P. S. Bechthold, S. Haussühl: Appl. Phys. **14**, 403–410 (1977)

4.496 S. Haussühl: Acta Crystallogr. A **24**, 697–698 (1968)

4.497 P. Günter: Ferroelectrics **75**, 5–23 (1987)

4.498 J. Sapriel, R. Hierle, J. Zyss, M. Bossier: Appl. Phys. Lett. **55**, 2594–2596 (1989)

4.499 Y. Wu, T. Sasaki, S. Nakai, A. Yokotani, H. Tang, C. Chen: Appl. Phys. Lett. **62**, 2614–2615 (1993)

4.500 K. Kato: IEEE J. Quantum Electron. **31**, 169–171 (1995)

4.501 G. R. Crane: J. Chem. Phys. **62**, 3571 (1975)

4.502 E. N. Volkova, V. A. Dianova, A. L. Zuev, A. N. Izrailenko, A. S. Lipatov, V. N. Parygin, L. N. Rashkovich, L. E. Chirkov: Kristallografiya **16**, 346–349 (1971)

4.503 E. N. Volkova, V. A. Dianova, A. L. Zuev, A. N. Izrailenko, A. S. Lipatov, V. N. Parygin, L. N. Rashkovich, L. E. Chirkov: Sov. Phys. Crystallogr. [English Transl.] **16**, 284–286 (1971)

4.504 S. K. Kurtz, T. T. Perry, J. G. Bergman Jr.: Appl. Phys. Lett. **12**, 186–188 (1968)

4.505 H. Naito, H. Inaba: Optoelectronics **4**, 335 (1972)

4.506 J. Zyss, D. S. Chemla, J. F. Nicoud: J. Chem. Phys. **74**, 4800–4811 (1981)

4.507 M. Sigelle, R. Hierle: J. Appl. Phys. **52**, 4199–4204 (1981)

4.508 T. S. Narasimhamurty: Phys. Rev. **186**, 945–948 (1969)

4.509 Y. Takagi, Y. Makita: J. Phys. Soc. Jpn. **13**, 272–277 (1958)

4.510 V. V. Gladkii, V. K. Magataev, V. A. Kirikov: Fiz. Tverd. Tela **19**, 1102–1106 (1977)

4.511 V. V. Gladkii, V. K. Magataev, V. A. Kirikov: Sov. Phys. Solid State [English Transl.] **19**, 641–644 (1977)

4.512 M. S. Khan, T. S. Narasimhamurty: Solid State Commun. **43**, 941–943 (1982)

4.513 A. T. Anistratov, S. X. Mel'nikova: Kristallografiya **17**, 149–152 (1972)

4.514 A. T. Anistratov, S. X. Mel'nikova: Sov. Phys. Crystallogr. [English Transl.] **17**, 119–121 (1972)

4.515 N. A. Romanyuk, A. M. Kostetskii: Fiz. Tverd. Tela **18**, 1489–1491 (1976)

4.516 N. A. Romanyuk, A. M. Kostetskii: Sov. Phys. Solid State [English Transl.] **18**, 867–869 (1976)

4.517 J. Valasek: Phys. Rev. **20**, 639–664 (1922)

4.518 G. B. Hadjichristov, P. P. Kircheva, N. Kirov: J. Mol. Struct. **382**, 33–37 (1996)

4.519 I. V. Berezhnoi, R. O. Vlokh: Fiz. Tverd. Tela **30**, 2223–2225 (1988)

4.520 I. V. Berezhnoi, R. O. Vlokh: Sov. Phys. Solid State [English Transl.] **30**, 1282–1284 (1988)

4.521 N. R. Ivanov, D. Khusravov, L. A. Shuvalov, N. M. Shchagina: Izv. Akad. Nauk SSSR, Ser. Fiz. **43**, 1691–1701 (1979)

4.522 N. R. Ivanov, D. Khusravov, L. A. Shuvalov, N. M. Shchagina: Bull. Acad. Sci. USSR, Phys. Ser. [English Transl.] **43**, 121–129 (1979)

4.523 V. G. Martynov, K. S. Aleksandrov, A. T. Anistratov: Fiz. Tverd. Tela **15**, 2922–2926 (1973)

4.524 V. G. Martynov, K. S. Aleksandrov, A. T. Anistratov: Sov. Phys. Solid State [English Transl.] **15**, 1950–1952 (1974)

4.525 Y. Luspin, G. Hauret: C. R. Hebd. Seances Acad. Sci. (Paris), Ser. B **274**, 995 (1972)

4.526 A. T. Anistratov, V. G. Martynov: Kristallografiya **15**, 308–312 (1970)

4.527 A. T. Anistratov, V. G. Martynov: Sov. Phys. Crystallogr. [English Transl.] **15**, 256–260 (1970)

4.528 M. Eibschütz, H. J. Guggenheim, S. H. Wemple, I. Camlibel, M. DiDomenico Jr.: Phys. Lett. A **29**, 409–410 (1969)

4.529 K. Recker, F. Wallrafen, S. Haussühl: J. Cryst. Growth **26**, 97–100 (1974)

4.530 R. O. Vlokh, I. P. Skab: Fiz. Tverd. Tela **34**, 3250–3255 (1992)

4.531 R. O. Vlokh, I. P. Skab: Sov. Phys. Solid State [English Transl.] **34**, 1739–1741 (1992)

4.532 J. D. Barry, C. J. Kennedy: IEEE J. Quantum Electron. **11**, 575–579 (1975)

4.533 T. Yamada, H. Iwasaki, N. Niizeki: J. Appl. Phys. **41**, 4141–4147 (1970)

4.534 S. Singh, D. A. Draegert, J. E. Geusic: Phys. Rev. B **2**, 2709–2724 (1970)

4.535 A. W. Warner, G. A. Coquin, J. L. Fink: J. Appl. Phys. **40**, 4353 (1969)

4.536 J. Fousek, C. Konak: Ferroelectrics **12**, 185–187 (1976)

4.537 K. Suzuki, K. Inoue, M. Shibuya: J. Phys. Soc. Jpn. **43**, 1457–1458 (1977)

4.538 J. E. Geusic, H. J. Levinstein, J. J. Rubin, S. Singh, L. G. Van Uitert: Appl. Phys. Lett. **11**, 269–271 (1967)

4.539 J. E. Geusic, H. J. Levinstein, J. J. Rubin, S. Singh, L. G. Van Uitert: Erratum, Appl. Phys. Lett. **12**, 224 (1968)

4.540 S. K. Gupta, H. B. Gon, K. V. Rao: Ferroelectr. Lett. **7**, 15–19 (1987)

4.541 S. K. Gupta, H. B. Gon, K. V. Rao: J. Mater. Sci. Lett. **6**, 4–6 (1987)

4.542 C. Medrano, P. Günter, H. Arend: Phys. Status Solidi **143**, 749–754 (1987)

4.543 A. Carenco, J. Jerphagnon, A. Perigaud: J. Chem. Phys. **66**, 3806–3813 (1977)

4.544 M. Y. Antipin, T. V. Timofeeva, R. D. Clark, V. N. Nesterov, F. M. Dolgushin, J. Wu, A. J. Leyderman: Mater. Chem. **11**, 351–358 (2001)

4.545 C. Bosshard, K. Sutter, P. Günter, G. Chapuis, R. J. Twieg, D. Dobrowolsky: Proc. SPIE **1017**, 207–211 (1988)

4.546 P. Günter, C. Bosshard, K. Sutter, H. Arend, G. Chapuis, R. J. Twieg, D. Dobrowolski: Appl. Phys. Lett. **50**, 486–488 (1987)

4.547 C. Bosshard, K. Sutter, P. Günter: Ferroelectrics **92**, 387–393 (1989)

4.548 K. Sutter, C. Bosshard, M. Ehrensperger, P. Günter, R. J. Twieg: IEEE J. Quantum Electron. **24**, 2362–2366 (1988)

4.549 C. Scheiding, G. Schmidt, H. D. Kürsten: Krist. Tech. **8**, 311–321 (1973)

4.550 L. E. Cross, A. Fouskova, S. E. Cummins: Phys. Rev. Lett. **21**, 812–814 (1968)

4.551 J. Sapriel, R. Vacher: J. Appl. Phys. **48**, 1191–1194 (1977)

4.552 M. P. Zaitseva, L. A. Shabanova, B. I. Kidyarov, Yu. I. Kokorin, S. I. Burkov: Kristallografiya **28**, 741–744 (1983)

4.553 M. P. Zaitseva, L. A. Shabanova, B. I. Kidyarov, Yu. I. Kokorin, S. I. Burkov: Sov. Phys. Crystallogr. [English Transl.] **28**, 439–440 (1983)

4.554 A. L. Aleksandrovskii, A. N. Izrailenko, L. N. Razhkovich: Kvantovaya Elektron. **1**, 1261–1264 (1974)

4.555 A. L. Aleksandrovskii, A. N. Izrailenko, L. N. Razhkovich: Sov. J. Quantum Electron. [English Transl.] **4**, 699–700 (1974)

4.556 A. L. Aleksandrovskii: Prib. Tekh. Eksp. **1**, 205–206 (1974)

4.557 A. L. Aleksandrovskii: Instrum. Exp. Tech., USSR [English Transl.] **17**, 234–235 (1974)

4.558 S. Nanamatsu, K. Doi, M. Takahashi: Jpn. J. Appl. Phys. **11**, 816–822 (1972)

4.559 R. C. Miller, W. A. Nordland, K. Nassau: Ferroelectrics **2**, 97–99 (1971)

4.560 Yu. V. Shaldin, D. A. Belogurov, T. M. Prokhortseva: Fiz. Tverd. Tela **15**, 1383–1387 (1973)

4.561 Yu. V. Shaldin, D. A. Belogurov, T. M. Prokhortseva: Sov. Phys. Solid State [English Transl.] **15**, 936–938 (1973)

4.562 S. Singh, W. A. Bonner, J. R. Potopowicz, L. G. Van Uitert: Appl. Phys. Lett. **17**, 292–294 (1970)

4.563 F. B. Dunning, F. K. Tittel, R. F. Stebbings: Opt. Commun. **7**, 181 (1973)

4.564 H. Naito, H. Inaba: Optoelectronics **5**, 256 (1973)

4.565 R. C. Miller, W. A. Nordland, E. D. Kolb, W. L. Bond: J. Appl. Phys. **41**, 3008–3011 (1970)

4.566 P. V. Lenzo, E. G. Spencer, J. P. Remeika: Appl. Opt. **4**, 1036–1037 (1965)

4.567 K. Araki, T. Tanaka: Jpn. J. Appl. Phys. **11**, 472–479 (1972)

4.568 L. M. Belyaev, G. S. Belikova, A. B. Gilvarg, I. M. Silvestrova: Kristallografiya **14**, 645–651 (1969)

4.569 L. M. Belyaev, G. S. Belikova, A. B. Gilvarg, I. M. Silvestrova: Sov. Phys. Crystallogr. [English Transl.] **14**, 544–549 (1970)

4.570 A. Miniewicz, A. Samoc, J. Sworakowski: J. Mol. Electron. **4**, 25–29 (1988)

4.571 M. S. Khan, T. S. Narasimhamurty: J. Mater. Sci. Lett. **1**, 268–270 (1982)

4.572 C. Chen, Y. Wu, A. Jiang, B. Wu, G. You, R. Li, S. Lin: J. Opt. Soc. Am. B **6**, 616–621 (1989)

4.573 S. P. Velsko, M. Webb, L. Davis, C. Huang: IEEE J. Quantum Electron. **27**, 2182–2192 (1991)

4.574 S. Lin, Z. Sun, B. Wu, C. Chen: J. Appl. Phys. **67**, 634 (1990)

4.575 K. Kato: IEEE J. Quantum Electron. **30**, 2950–2952 (1994)

4.576 B. L. Davydov, L. G. Koreneva, E. A. Lavrovsky: Radiotekh. Elektron. **19**, 1313 (1974)

4.577 B. L. Davydov, L. G. Koreneva, E. A. Lavrovsky: Radio Eng. Electron. Phys. [English Transl.] **19**(6), 130 (1974)

4.578 J. L. Stevenson: J. Phys. D **6**, L13–L16 (1973)

4.579 S. K. Kurtz, J. Jerphagnon, M. M. Choy: *Elastic, Piezoelectric, Pyroelectric, Piezooptic, Electrooptic Constant, and Nonlinear Dielectric Susceptibilities of Crystals*, Landolt–Börnstein, New Series III/11, ed. by K. H. Hellwege (Springer, Berlin, Heidelberg 1979) pp. 671–743

4.580 G. S. Belyaev, G. S. Belikova, A. G. Gilvarg, M. P. Golovei, I. N. Kalinkina, G. I. Kosourov: Opt. Spectr. (USSR) [English Transl.] **29**, 522 (1970)

4.581 Y. Uematsu, T. Fukuda: Jpn. J. Appl. Phys. **12**, 841–844 (1973)

4.582 P. Günter: Jpn. J. Appl. Phys. **16**, 1727–1728 (1977)

4.583 E. Wiesendanger: Ferroelectrics **6**, 263–281 (1974)

4.584 M. Zgonik, R. Schlesser, I. Biaggio, I. Voit, J. Tscherry, P. Günter: J. Appl. Phys. **74**, 1287–1297 (1993)

4.585 W. R. Cook Jr., H. Jaffe: Acta Crystallogr. **10**, 705–707 (1957)

4.586 T. Krajewski, Z. Tylczynski, T. Breczewski: Acta Phys. Polonica A **47**, 455–466 (1975)

4.587 J. D. Bierlein, H. Vanherzeele, A. A. Ballman: Appl. Phys. Lett. **54**, 783–785 (1989)

4.588 J. D. Bierlein, H. Vanherzeele, A. A. Ballman: Erratum, Appl. Phys. Lett. **61**, 3193 (1992)

4.589 V. V. Aleksandrov, T. S. Velichkina, V. I. Voronkova, L. V. Koltsova, I. A. Yakovlev, V. K. Yanovskii: Solid State Commun. **69**, 877 (1989)

4.590 J. D. Bierlein, H. Vanherzeele: J. Opt. Soc. Am. B **6**, 622–633 (1989)

4.591 J. D. Bierlein, C. B. Arweiler: Appl. Phys. Lett. **49**, 917–919 (1986)

4.592 D. H. Jundt, P. Günter, B. Zysset: Nonlin. Opt. **4**, 341–345 (1993)

4.593 Y. Uematsu: Jpn. J. Appl. Phys. **13**, 1362–1368 (1974)

4.594 B. Zysset, I. Biaggio, P. Günter: J. Opt. Soc. Am. B **9**, 380–386 (1992)

4.595 J. A. Paisner, M. L. Spaeth, D. C. Gerstenberger, I. W. Ruderman: Appl. Phys. Lett. **32**, 476–478 (1978)

4.596 W. R. Cook, L. H. Hubby Jr.: J. Opt. Soc. Am. **66**, 72 (1976)

4.597 F. B. Dunning, R. G. Stickel Jr.: Appl. Opt. **15**, 3131 (1976)

4.598 H. J. Dewey: IEEE J. Quantum Electron. **12**, 303 (1976)

4.599 Y. Wu, C. Chen: Wuli Xuebao (Acta Phys. Sin.) **35**, 1–6 (1986)

4.600 A. H. Kung: Appl. Phys. Lett. **65**, 1082–1084 (1994)

4.601 D. L. Fenimore, K. L. Schepler, U. B. Ramabadran, S. R. McPherson: J. Opt. Soc. Am. B **12**, 794–796 (1995)

4.602 L. T. Cheng, L. K. Cheng, J. D. Bierlein, F. C. Zumsteg: Appl. Phys. Lett. **63**, 2618–2620 (1993)

4.603 L. K. Cheng, L. T. Cheng, J. Galperin, P. A. M. Hotsenpiller, J. D. Bierlein: J. Cryst. Growth **137**, 107–115 (1994)

4.604 K. Kato: IEEE J. Quantum Electron. **30**, 881–883 (1994)

4.605 A. C. Aleksandrovsky, S. A. Akhmanov, V. A. Dyakov, N. I. Zheludev, V. I. Pryalkyn: Kvantovaya Elektron. **12**, 1333 (1985)

4.606 A. C. Aleksandrovsky, S. A. Akhmanov, V. A. Dyakov, N. I. Zheludev, V. I. Pryalkyn: Sov. J. Quantum Electron. [English Transl.] **15**, 885 (1985)

4.607 W. Wiechmann, S. Kubota, T. Fukui, H. Masuda: Opt. Lett. **18**, 1208–1210 (1993)

4.608 K. Kato: IEEE J. Quantum Electron. **QE-27**, 1137–1140 (1991)

4.609 V. K. Yanovskii, V. I. Voronkova, A. P. Leonov, S. Yu. Stefanovich: Fiz. Tverd. Tela **27**, 2516–2517 (1985)

4.610 V. K. Yanovskii, V. I. Voronkova, A. P. Leonov, S. Yu. Stefanovich: Sov. Phys. Solid State [English Transl.] **27**, 1508–1509 (1985)

4.611 V. K. Yanovskii, V. I. Voronkova: Phys. Status Solidi (a) **93**, 665–668 (1980)

4.612 E. Nakamura: J. Phys. Soc. Jpn. **17**, 961–966 (1962)

4.613 H. Shimizu, M. Tsukamoto, Y. Ishibashi, Y. Takagi: J. Phys. Soc. Jpn. **38**, 195–201 (1975)

4.614 K. Hamano, K. Negishi, M. Marutake, S. Nomura: Jpn. J. Appl. Phys. **2**, 83–90 (1963)

4.615 G. Hauret, J. P. Chapelle, L. Taurel: Phys. Status Solidi (a) **11**, 255–261 (1972)

4.616 F. C. Zumsteg, J. D. D. Bierlein, T. E. Gier: J. Appl. Phys. **47**, 4980 (1976)

4.617 Yu. S. Oseledchik, A. I. Pisarevsky, A. L. Prosvirin, V. N. Lopatko, L. E. Kholodenkov, E. F. Titkov, A. A. Demidovich, A. P. Shkadarevich: Proc. of Laser Optics Conf. [in Russian], (Leningrad Univ. Press, Leningrad 1990)

4.618 J. Y. Wang, Y. G. Liu, J. Q. Wei, L. P. Shi, M. Wang: Z. Kristallogr. **191**, 231–238 (1990)

4.619 J. Y. Wang, Y. G. Liu, J. Q. Wei, L. P. Shi, M. Wang: Guisuanyan Xuebao **18**, 165–170 (1990)

4.620 J. Y. Wang, Y. G. Liu, J. Q. Wei, L. P. Shi, M. Wang: Chem. Abstr. **114** (1991) No. 154040x

4.621 K. Iio: J. Phys. Soc. Jpn. **34**, 138 (1973)

4.622 D. S. Chemla, E. Batifol, R. L. Byer, R. L. Herbst: Opt. Commun. **11**, 57 (1974)

4.623 A. R. Johnston, T. Nakamura: J. Appl. Phys. **40**, 3656–3658 (1969)

4.624 K. Inoue, T. Ishidate: Ferroelectrics **7**, 105–106 (1974)

4.625 B. Teng, J. Wang, Z. Wang, H. Jiang, X. Hu, R. Song, H. Liu, Y. Liu, J. Wei, Z. Shao: J. Cryst. Growth **224**, 280–283 (2001)

4.626 H. Hellwig, J. Liebertz, L. Bohaty: J. Appl. Phys. **88**, 240 (2000)

4.627 Z. Lin, Z. Wang, C. Chen, M.-H. Lee: J. Appl. Phys. **90**, 5585 (2001)

4.628 D. Eimerl, S. Velsko, L. Davis, F. Wang, G. Loiacono, G. Kennedy: IEEE J. Quantum Electron. **25**, 179–193 (1989)

4.629 C. E. Barker, D. Eimerl, S. P. Velsko: J. Opt. Soc. Am. B **8**, 2481–2492 (1991)

4.630 P. Kerkoc, M. Zgonik, K. Sutter, C. Bosshard, P. Günter: J. Opt. Soc. Am. B **7**, 313–319 (1990)

4.631 I. Biaggio, P. Kerkoc, L.-S. Wu, P. Günter, B. Zysset: J. Opt. Soc. Am. B **9**, 507–517 (1992)

4.632 R. Bechmann: *Elastic, Piezoelectric, Piezooptic and Electrooptic Constants of Crystals*, Landolt-Börnstein, New Series III/1 (Springer, Berlin, Heidelberg 1966) pp. 40–123

4.633 C. S. Brown, R. C. Kell, R. Taylor, L. A. Thomas: Proc. Inst. Elec. Eng. (London) B **109**, 99–114 (1962)

4.634 C. S. Brown, R. C. Kell, R. Taylor, L. A. Thomas: IRE Trans. Compon. Parts **9**, 193–211 (1962)

4.635 B. M. Suleiman, A. Lundén, E. Karawacki: Solid State Ionics **136-137**, 325–330 (2000)

4.636 M. V. Hobden: J. Appl. Phys. **38**, 4365 (1967)

4.637 A. Sonin, A. A. Filimonov, S. V. Suvorov: Sov. Phys. Solid State **10**, 1481 (1968)

4.638 L. Bohaty, S. Haussühl, G. Nohl: Z. Kristallogr. **139**, 33–38 (1974)

4.639 W. F. Holman: Ann. Phys. **29**, 160–178 (1909)

4.640 G. Brosowki, G. Luther, H. E. Muser: Phys. Status Solidi (a) **14**, K15–K17 (1972)

4.641 B. A. Strukov, K. A. Minaeva: Prib. Tekh. Eksp. **29**, 157–160 (1986)

4.642 B. A. Strukov, K. A. Minaeva: Instrum. Exp. Tech. (USSR) [English Transl.] **29**, Part II, 1413–1416 (1986)

4.643 P. P. Craig: Phys. Lett. **20**, 140–142 (1966)

4.644 F. Jona, G. Shirane: *Ferroelectric Crystals* (MacMillan, New York 1962)

4.645 J. Berdowski, A. Opilski, J. Szuber: J. Tech. Phys. (Poland) **18**, 211–217 (1977)

4.646 V. P. Konstantinova, I. M. Silvestrova, K. S. Aleksandrov: Kristallografiya **4**, 69–73 (1959)

4.647 V. P. Konstantinova, I. M. Silvestrova, K. S. Aleksandrov: Sov. Phys. Crystallogr. [English Transl.] **4**, 63–67 (1960)

4.648 S. Haussühl, J. Albers: Ferroelectrics **15**, 73 (1977)

4.649 J. M. Halbout, C. L. Tang: IEEE J. Quantum Electron. **18**, 410–415 (1982)

4.650 M. J. Rosker, C. L. Tang: IEEE J. Quantum Electron. **20**, 334 (1984)

4.651 N. R. Ivanov, S. Ya. Benderskii, L. A. Shuvalov: Kristallografiya **22**, 115–125 (1977)

4.652 N. R. Ivanov, S. Ya. Benderskii, L. A. Shuvalov: Sov. Phys. Crystallogr. [English Transl.] **22**, 64–69 (1977)

4.653 A. S. Sonin, V. S. Suvorov: Sov. Phys. Solid State **9**, 1437 (1967)

4.654 N. R. Ivanov, L. A. Shuvalov: Kristallografiya **11**, 760–765 (1966)

4.655 N. R. Ivanov, L. A. Shuvalov: Sov. Phys. Crystallogr. [English Transl.] **11**, 648–651 (1967)

4.656 A. Feldman, R. M. Waxler: Phys. Rev. Lett. **45**, 126–129 (1980)

4.5. Ferroelectrics and Antiferroelectrics

Ferroelectric crystals (especially oxides in the form of ceramics) are important basic materials for technological applications in capacitors and in piezoelectric, pyroelectric, and optical devices. In many cases their nonlinear characteristics turn out to be very useful, for example in optical second-harmonic generators and other nonlinear optical devices. In recent decades, ceramic thin-film ferroelectrics have been utilized intensively as parts of memory devices. Liquid crystal and polymer ferroelectrics are utilized in the broad field of fast displays in electronic equipment.

This chapter surveys the nature of ferroelectrics, making reference to the data presented in the Landolt–Börnstein data collection *Numerical Data and Functional Relationships in Science and Technology*, Vol. III/36, *Ferroelectrics and Related Substances* (LB III/36). The data in the figures in this chapter have been taken mainly from the Landolt–Börnstein collection. The Landolt–Börnstein volume mentioned above consists of three subvolumes: Subvolume A [5.1, 2], covering oxides; Subvolume B [5.3], covering inorganic crystals other than oxides; and Subvolume C [5.4], covering organic crystals, liquid crystals, and polymers.

4.5.1	Definition of Ferroelectrics and Antiferroelectrics	903
4.5.2	Survey of Research on Ferroelectrics	904
4.5.3	Classification of Ferroelectrics	906
	4.5.3.1 The 72 Families of Ferroelectrics	909
4.5.4	Physical Properties of 43 Representative Ferroelectrics	912
	4.5.4.1 Inorganic Crystals Oxides [5.1, 2]	912
	4.5.4.2 Inorganic Crystals Other Than Oxides [5.3]	922
	4.5.4.3 Organic Crystals, Liquid Crystals, and Polymers [5.4]	930
References		936

Matter consists of electrons and nuclei. Most of the electrons generally are tightly bound to the nuclei, but some of the electrons are only weakly bound or are freely mobile in a lattice of ions. The physical properties of matter can be considered as being split into two categories. The properties in the first category are determined directly by the electrons and by the interaction of the electrons with lattice vibrations. Examples are the metallic, magnetic, superconductive, and semiconductive properties. The properties in the second category are only indirectly related to the electrons and can be discussed as being due to interaction between atoms, ions, or molecules. In this category we have, for example, the dielectric, elastic, piezoelectric, and pyroelectric properties; we have the dispersion relations of the lattice vibrations; and we have most of the properties of liquid crystals and polymers. The important properties of ferroelectrics are linked to all the latter properties, and they exhibit diverse types of phase transitions together with anomalies in these properties. These specific modifications convey information about cooperative interactions among ions, atoms, or molecules in the condensed phase of matter.

4.5.1 Definition of Ferroelectrics and Antiferroelectrics

A ferroelectric crystal is defined as a crystal which belongs to the pyroelectric family (i.e. shows a spontaneous electric polarization) and whose direction of spontaneous polarization can be reversed by an electric field. An antiferroelectric crystal is defined as a crystal whose structure can be considered as being composed of two sublattices polarized spontaneously in antiparallel directions and in which a ferroelectric phase can

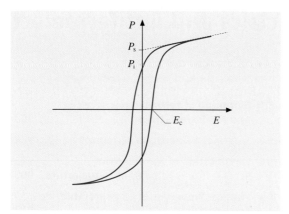

Fig. 4.5-1 Ferroelectric hysteresis loop. P_s, spontaneous polarization; P_r, remanent polarization; E_c, coercive field

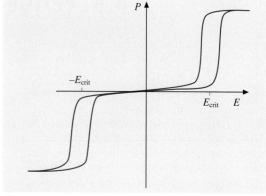

Fig. 4.5-2 Antiferroelectric hysteresis loop. E_{crit}, critical field

be induced by applying an electric field. Experimentally, the reversal of the spontaneous polarization in ferroelectrics is observed as a single hysteresis loop (Fig. 4.5-1), and the induced phase transition in antiferroelectrics as a double hysteresis loop (Fig. 4.5-2), when a low-frequency ac field of a suitable strength is applied.

The spontaneous polarization in ferroelectrics and the sublattice polarizations in antiferroelectrics are analogous to their magnetic counterparts. As described above, however, these polarizations are a necessary but not sufficient condition for ferroelectricity or antiferroelectricity. In other words, ferroelectricity and antiferroelectricity are concepts based not only upon the crystal structure, but also upon the dielectric behavior of the crystal. It is a common dielectric characteristic of ferroelectrics and antiferroelectrics that, in a certain temperature range, the dielectric polarization is observed to be a two-valued function of the electric field.

The definition of ferroelectric liquid crystals needs some comments; see remark *l* in Sect. 4.5.3.1.

4.5.2 Survey of Research on Ferroelectrics

The ferroelectric effect was discovered in 1920 by *Valasek*, who obtained hysteresis curves for Rochelle salt analogous to the *B–H* curves of ferromagnetism [5.5], and studied the electric hysteresis and piezoelectric response of the crystal in some detail [5.6]. For about 15 years thereafter, ferroelectricity was considered as a very specific property of Rochelle salt, until Busch and Scherrer discovered ferroelectricity in KH_2PO_4 and its sister crystals in 1935. During World War II, the anomalous dielectric properties of $BaTiO_3$ were discovered in ceramic specimens independently by Wainer and Solomon in the USA in 1942, by Ogawa in Japan in 1944, and by Wul and Goldman in Russia in 1946. Since then, many ferroelectrics have been discovered and research activity has rapidly increased. In recent decades, active studies have been made on ferroelectric liquid crystals and high polymers, after ferroelectricity had been considered as a characteristic property of solids for more than 50 years.

Figures 4.5-3, 4.5-4, and 4.5-5 demonstrate how ferroelectric research has developed. Figure 4.5-3 indicates the number of ferroelectrics discovered each year for oxide (Fig. 4.5-3a) and nonoxide ferroelectrics (Fig. 4.5-3b). Figure 4.5-4 gives the total number of ferroelectrics known at the end of each year. At present more than 300 ferroelectric substances are known. Figure 4.5-5 indicates the number of research papers on ferroelectrics and related substances published each year.

Advanced experimental methods (e.g. inelastic neutron scattering and hyper-Raman scattering) have been applied effectively to studies of ferroelectrics, and several new concepts (e.g. soft modes of lattice vibrations and the dipole glass) have been introduced to understand the nature of ferroelectrics. Ferroelectric crystals have been widely used in capacitors and piezoelectric devices. Steady developments in crystal growth and in the preparation of ceramics and ceramic thin

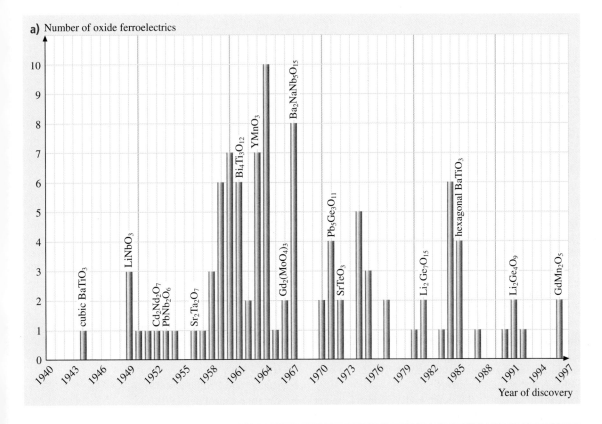

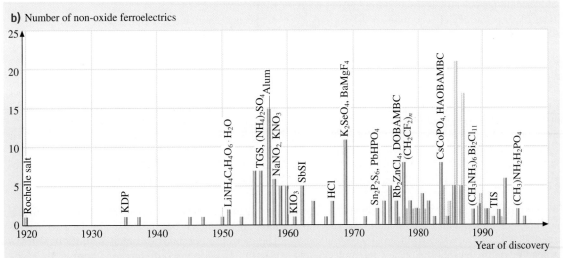

Fig. 4.5-3a,b Number of ferroelectric substances discovered in each year. Representative ferroelectrics are indicated at their year of discovery. (**a**) Oxide ferroelectrics. Only pure compounds are taken into account. (**b**) Nonoxide ferroelectrics. *Gray bars* stand for nonoxide crystals, counting each pure compound as one unit. *Brown bars* stand for liquid crystals and polymers, counting each group of homologues (cf. Chaps. 71 and 72 in [5.4]) as one unit. The figure was prepared by Prof. K. Deguchi using data from LB III/36

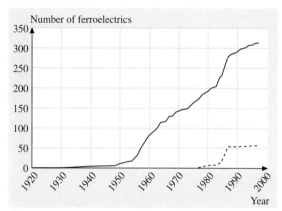

Fig. 4.5-4 Number of ferroelectric substances known at the end of each year. The *solid line* represents all ferroelectrics, including liquid crystals and polymers. For liquid crystals and polymers, each group of homologues is counted as one substance. The *dashed line* represents ferroelectric liquid crystals and polymers alone. Figure prepared by Prof. K. Deguchi

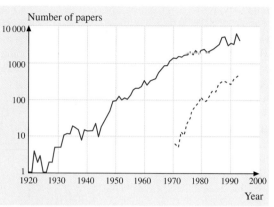

Fig. 4.5-5 Number of research papers on ferroelectrics and related substances published in each year. The *solid line* indicates the number of papers concerning all ferroelectrics (crystals + liquid crystals + polymers). The *dashed line* indicates the number of papers concerning liquid crystals and polymers alone. Prepared by Prof. K. Deguchi

films have opened the way to various other applications (e.g. second-harmonic generation and memory devices). Liquid crystals and high-polymer ferroelectrics are very useful as fast display elements.

Corresponding to this intense development, many textbooks, monographs, and review articles have been published on ferroelectric research during recent years. Some of them, arranged according to the various research fields, are listed here:

- Introduction to ferroelectrics: [5.7–10]
- Applications in general: [5.10, 11]
- Piezoelectricity: [5.12–14]
- Structural phase transitions: [5.15–17]
- Incommensurate phases: [5.18]
- Soft-mode spectroscopy: [5.19]
- Inelastic neutron scattering studies of ferroelectrics: [5.20]
- Raman and Brillouin scattering: [5.21]
- Ceramic capacitors: [5.22]
- Dipole glasses: [5.23]
- Relaxors: [5.24]
- Second-harmonic generation (SHG): [5.25–28]
- Ferroelectric ceramics: [5.29]
- Thin films: [5.30–35]
- Acoustic surface waves (ASWs): [5.36–39]
- Ferroelectric transducers and sensors: [5.40]
- Memory applications: [5.35]
- Ferroelectric liquid crystals: [5.41–47]
- Ferroelectric polymers: [5.48–51]

4.5.3 Classification of Ferroelectrics

Ferroelectricity is caused by a cooperative interaction of molecules or ions in condensed matter. The transition to ferroelectricity is characterized by a phase transition. Depending on the mechanism of how the molecules or ions interact in the material, we can classify the ferroelectric phase transitions and also the ferroelectric materials themselves into three categories: (I) order–disorder type, (II) displacive type, and (III) indirect type. In the order–disorder type (I), the spontaneous polarization is caused by orientational order of dipolar molecules, which is best visualized by the Ising model. The dielectric constants of order–disorder type ferroelectrics increase markedly in the vicinity of the Curie point. In the displacive type (II), the spontaneous polarization results from softening of the transverse optical modes of the lattice vibrations at the origin of the Brillouin zone; again, a marked increase of the dielectric constants is observed near the Curie temperature. The indirect type (III) is further classified into III$_{op}$ and III$_{ac}$. In type III$_{op}$, the phase transition is originally caused by softening of the optical modes at the Brillouin zone boundary; a coupling of the soft modes with the electric polarization (through a complex mechanism) causes the spontaneous polarization (e.g. substance 18, GMO, in Sect. 4.5.4.1); only a very slight anomaly of the dielectric constants is observed above the Curie point (see Fig. 4.5-6). In type III$_{ac}$, softening of the acoustic modes (decrease of the tangent of the dispersion relation) takes place at the origin of the Brillouin zone, and a piezoelectric coupling between the soft modes and the polarization results in the spontaneous polarization (e.g. substance 40, LAT, in Sect. 4.5.4.3). No dielectric-constant anomaly appears above the Curie temperature when the crystal is clamped so that elastic deformation is prohibited (see Fig. 4.5-7). Most of the nonoxide ferroelectrics are of the order–disorder type and most of the oxide ferroelectrics are of the displacive type, while a few ferroelectrics are of the indirect type.

It might be expected that the dielectric dispersion would be dissipative and occur at relatively low frequency (e.g. in the microwave region) in the order–disorder ferroelectrics, while the dispersion would be of the resonance type and occur in the millimeter or infrared region in the displacive ferroelectrics. Actually, however, the situation is more complex owing to phonon–phonon coupling and cluster formation near the Curie temperature. Phonon–phonon coupling is inevitable in displacive ferroelectrics because their ferroelectricity is closely related to the anharmonic term of the ion potential, as first pointed out by *Slater* (see [5.7]). Accordingly, the resonance-type oscillation tends to be overdamped. High dielectric constants

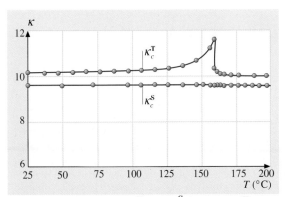

Fig. 4.5-6 Gd$_2$(MoO$_4$)$_3$. κ_c^T, and κ_c^S versus T. κ_c^T is the free dielectric constant κ_c measured at 1 kHz, and κ_c^S is the clamped dielectric constant κ_c measured at 19 MHz

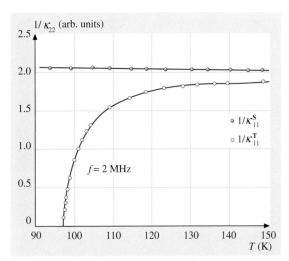

Fig. 4.5-7 LiNH$_4$C$_4$H$_4$O$_6$ · H$_2$O. $1/\kappa_{22}^S$, and $1/\kappa_{22}^T$ versus T. $\kappa_{22} = \kappa_b$. $f = 2$ MHz. κ_{22}^S is κ_{22} of the clamped crystal. κ_{22}^T is κ_{22} of the free crystal

tend to induce the formation of clusters in which unit cell polarizations are aligned, similarly to the domains in the ferroelectric phase, but the boundaries of these clusters fluctuate thermally. Inelastic neutron, hyper-Raman, and hyper-Rayleigh scattering, etc. indicate the existence of such clusters in several displacive ferroelectrics. It is known that a ferromagnetic domain wall can move to follow an ac magnetic field and contribute to the magnetic susceptibility, while a ferroelectric domain wall usually cannot follow an ac electric field and in practice does not contribute to the dielectric constant. The cluster boundaries, however, fluctuate thermally and hence will be able to follow an ac electric field and contribute to the dielectric constant. This contribution will be dissipative in its character. Accordingly, it is possible that dielectric dispersion occurs as a combination of an overdamped-resonance dispersion and a dissipative dispersion in most displacive ferroelectrics.

In second-order or nearly second-order phase transitions, the dielectric dispersion is observed to show a critical slowing-down: a phenomenon in which the response of the polarization to a change of the electric field becomes slower as the temperature approaches the Curie point. Critical slowing-down has been observed in the GHz region in several order–disorder ferroelectrics (e.g. Figs. 4.5-8 and 4.5-9) and displacive ferroelectrics (e.g. Fig. 4.5-10). The dielectric constants at the Curie point in the GHz region are very small in order–disorder

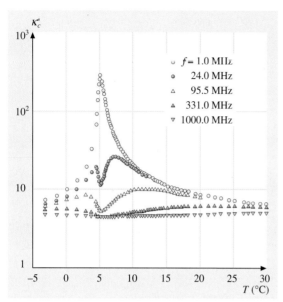

Fig. 4.5-8 $Ca_2Sr(CH_3CH_2COO)_6$. κ'_c versus T. Parameter: f

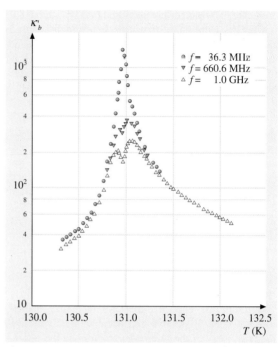

Fig. 4.5-10 $(CH_3NHCH_2COOH)_3 \cdot CaCl_2$. κ'_b versus T. Parameter: f. Critical slowing-down takes place

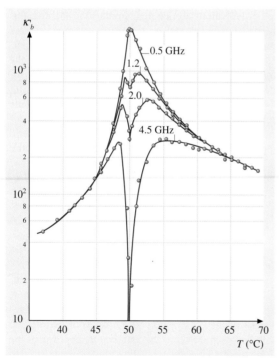

Fig. 4.5-9 $(NH_2CH_2COOH)_3 \cdot H_2SO_4$. κ'_b versus T relation, showing critical slowing-down. Parameter: f

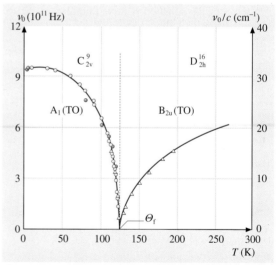

Fig. 4.5-11 $(CH_3NHCH_2COOH)_3 \cdot CaCl_2$. ν_0 versus T. ν_0 is the phonon mode frequency. *Triangles*: measured by millimeter spectroscopy. *Brown circles*: measured from far-infrared spectra. *Gray circles*: measured from electric-field-induced Raman spectra. In the paraelectric phase ($T > \Theta_f$), ν_0 decreases as the temperature decreases toward Θ_f, that is, the phonon mode softens

ferroelectrics, as seen in Figs. 4.5-8 and 4.5-9. The dielectric constant at the Curie point at 1 GHz is not so small in displacive ferroelectrics, as seen in Fig. 4.5-10, where the critical slowing-down seems to be related to cluster boundary motion, and the dielectric constant at the Curie point at 1 GHz contains a contribution from the soft phonon as observed in millimeter spectroscopy (Fig. 4.5-11).

4.5.3.1 The 72 Families of Ferroelectrics

In the Landolt–Börnstein data collection, ferroelectric and antiferroelectric substances are classified into 72 families according to their chemical composition and their crystallographic structure. Some substances which are in fact neither ferroelectric nor antiferroelectric but which are important in relation to ferroelectricity or antiferroelectricity, for instance as an end material of a solid solution, are also included in these families as related substances. This subsection surveys these 72 families of ferroelectrics presented in Landolt–Börnstein Vol. III/36 (LB III/36). Nineteen of these families concern oxides [5.1, 2], 30 of them concern inorganic crystals other than oxides [5.3], and 23 of them concern organic crystals, liquid crystals, and polymers [5.4]. Table 4.5-1 lists these families and gives some information about each family. Substances classified in LB III/36 as miscellaneous crystals (outside the families) are not included.

In the following, remarks are made on 13 of the families, labeled by the letters a–m in Table 4.5-1. The corresponding family numbers are repeated in the headings.

a. Perovskite-Type Family (Family Number 1). The name of this group is derived from the mineral perovskite ($CaTiO_3$). The perovskite-type oxides are cubic (e.g. $CaTiO_3$ above 1260 °C and $BaTiO_3$ above 123 °C) or pseudocubic with various small lattice distortions (e.g. $CaTiO_3$ below 1260 °C and $BaTiO_3$ below 123 °C). Ceramics made from solid solutions of perovskite-type oxides are the most useful ferroelectrics in high-capacitance capacitors, piezoelectric elements, and infrared sensors. Ceramic thin films are useful in memory devices.

The pure compounds are divided into simple perovskite-type oxides and complex perovskite-type oxides. Simple perovskite-type oxides have the chemical formula $A^{1+}B^{5+}O_3$ or $A^{2+}B^{4+}O_3$. Complex perovskite-type oxides have chemical formulas expressed by $(A^{1+}_{1/2}A^{3+}_{1/2})BO_3$, $A^{2+}(B^{2+}_{1/2}B^{6+}_{1/2})O_3$, $A^{2+}(B^{3+}_{1/2}B^{5+}_{1/2})O_3$, $A^{2+}(B^{2+}_{1/3}B^{5+}_{2/3})O_3$, $A^{2+}(B^{3+}_{2/3}B^{6+}_{1/3})O_3$, $A(B, B', B'')O_3$, or $(A, A')(B, B')O_3$. Among the complex perovskite-type oxides, most of the $Pb(B, B')O_3$-type oxides show a diffuse phase transition such that the transition point is smeared out over a relatively wide temperature range and exhibits a characteristic dielectric relaxation; these materials therefore are called "relaxors".

b. $LiNbO_3$ Family (Family Number 2). This family contains $LiNbO_3$ and $LiTaO_3$. Their chemical formulas are similar to those of the simple perovskite oxides, but their structures are trigonal, unlike the perovskite oxides.

c. Stibiotantalite Family (Family Number 5). The members of this family are isomorphous with the mineral stibiotantalite $Sb(Ta, Nb)O_4$. They have the common chemical formula ABO_4, where A stands for Sc, Sb, or Bi and B for Ta, Nb, or Sb.

d. Tungsten Bronze-Type Family (Family Number 6). The tungsten bronzes are a group of compounds having the chemical formula M_xWO_3, where M stands for an alkali metal, an alkaline earth metal, Ag, Tl, etc. (e.g. Na_xWO_3, where $x = 0.1$–0.95). Most of them exhibit a bronze-like luster. The tungsten bronze-type oxides consist of crystals isomorphous with tungsten bronze, including a simple type (e.g. $Pb_{1/2}NbO_3$) and a complex type (e.g. $Ba_2NaNb_5O_{15}$). Single crystals (not ceramics) are used for technological applications.

e. Pyrochlore-Type Family (Family Number 7). The members of this family are isomorphous with the mineral pyrochlore, $CaNaNb_2O_6F$. Most of the members have the general chemical formula $A_2B_2O_7$ or $A_2B_2O_6$ (anion-deficient compounds), where A stands for Cd, Pb, Bi, etc. and B for Nb, Ta, etc.

f. $Sr_2Nb_2O_7$ Family (Family Number 8). This family contains high-temperature ferroelectrics such as $Nb_2Ti_2O_7$ and $La_2Ti_2O_7$. Their Curie points are higher than 1500 °C.

g. Layer-Structure Family (Family Number 9). The common chemical formula of these oxides is $(Bi_2O_2)(A_{n-1}B_nO_{3n-1})$, where A stands for Ca, Sr, Ba, Pb, Bi, etc., B stands for Ti, Nb, Ta, Mo, W, etc., and n varies from 1 to 9. The crystal structure is a repeated stacking of a layer of $(Bi_2O_2)^{2+}$ and a layer of $(A_{n-1}B_nO_{3n-1})^{2-}$, which can be approximately repre-

Inorganic Crystals Oxides [5.1, 2]		Inorganic Crystals other than Oxides [5.3]		Organic Crystals, Liquid Crystals, and Polymers [5.4]	
Family Nr.	Name	Family Nr.	Name	Family Nr.	Name
1	Perovskite-type family (90, 40; 11) a	20	SbSI family (11, 4; 1)	50	$SC(NH_2)_2$ family (1, 1; 1)
2	$LiNbO_3$ family (2, 2; 1) b	21	TlS family (1, 1; 0)	51	CCl_3CONH_2 family (1, 1; 0)
3	$YMnO_3$ family (6, 6; 0)	22	$TlInS_2$ family (5, 2; 0)	52	$Cu(HCOO)_2 \cdot 4H_2O$ family (1, 1; 0)
4	$SrTeO_3$ family (1, 1; 1)	23	Ag_3AsS_3 family (2, 1; 0)	53	$N(CH_3)_4HgCl_3$ family (6, 5; 0)
5	Stibiotantalite family (7, 6; 0) c	24	$Sn_2P_2S_6$ family (2, 2; 0)	54	$(CH_3NH_3)_2AlCl_5 \cdot 6H_2O$ family (3, 2; 0)
6	Tungsten bronze-type family (141, 21; 2) d	25	$KNiCl_3$ family (3, 3; 0)	55	$[(CH_3)_2NH_2]_2CoCl_4$ family (3, 2; 0)
7	Pyrochlore-type family (19, 2; 0) e	26	$BaMnF_4$ family (6, 6; 1)	56	$[(CH_3)_2NH_2]_3Sb_2Cl_9$ family (5, 5; 0)
8	$Sr_2Nb_2O_7$ family (6, 5; 1) f	27	HCl family (2, 2; 1)	57	$(CH_3NH_3)_5Bi_2Cl_{11}$ family (2, 2; 0)
9	Layer-structure family (36, 16; 0) g	28	$NaNO_2$ family (2, 2; 1)	58	DSP $(Ca_2Sr(CH_3CH_2COO)_6)$ family (3, 3; 1)
10	$BaAl_2O_4$-type family (1, 1; 0)	29	$CsCd(NO_2)_3$ family (2, 2; 0)	59	$(CH_2ClCOO)_2H \cdot NH_4$ family (2, 2; 0)
11	$LaBGeO_5$ family (1, 1; 0)	30	KNO_3 family (4, 3; 1)	60	TGS $((NH_2CH_2COOH)_3 \cdot H_2SO_4)$ family (3, 3; 1)
12	$LiNaGe_4O_9$ family (2, 2; 0)	31	$LiH_3(SeO_3)_2$ family (7, 5; 0)	61	$NH_2CH_2COOH \cdot AgNO_3$ family (1, 1; 0)
13	$Li_2Ge_7O_{15}$ family (1, 1; 1)	32	KIO_3 family (3, 3; 0)	62	$(NH_2CH_2COOH)_2 \cdot HNO_3$ family (1, 1; 0)
14	$Pb_5Ge_3O_{11}$ family (1, 1; 0)	33	KDP (KH_2PO_4) family (12, 12; 3)	63	$(NH_2CH_2COOH)_2 \cdot MnCl_2 \cdot 2H_2O$ family (1, 1; 0)
15	$5PbO \cdot 2P_2O_5$ family (1, 1; 0) (exact chemical formula unknown)	34	$PbHPO_4$ family (2, 2; 1)	64	$(CH_3NHCH_2COOH)_3 \cdot CaCl_2$ family (2, 2; 1)
16	$Ca_3(VO_4)_2$ family (2, 2; 0)	35	$KTiOPO_4$ family (23, 15; 0)	65	$(CH_3)_3NCH_2COO \cdot H_3PO_4$ family (3, 3; 0)
17	GMO $(Gd_2(MoO_4)_3)$ family (5, 5; 1)	36	$CsCoPO_4$ family (5, 5; 0)	66	$(CH_3)_3NCH_2COO \cdot CaCl_2 \cdot 2H_2O$ family (1, 1; 1)
18	Boracite-type family (28, 14; 1) h	37	$NaTh_2(PO_4)_3$ family (2, 2; 0)	67	Rochelle salt $(NaKC_4H_4O_6 \cdot 4H_2O)$ family (3, 2; 1)
19	$Rb_3MoO_3F_3$ family (4, 4; 0)	38	$Te(OH)_6 \cdot 2NH_4H_2PO_4 \cdot (NH_4)_2HPO_4$ family (1, 1; 0)	68	$LiNH_4C_4H_4O_6 \cdot H_2O$ family (3, 2; 1) k
		39	$(NH_4)_2SO_4$ family (22, 21; 1)	69	$C_5H_6NBF_4$ family (1, 1; 0)
		40	NH_4HSO_4 family (9, 4; 1)	70	$3C_6H_4(OH)_2 \cdot CH_3OH$ family (1, 1; 0)
		41	NH_4LiSO_4 family (9, 6; 1)	71	Liquid crystal family (97, 90; 2) l
		42	$(NH_4)_3H(SO_4)_2$ family (2, 2; 1)	72	Polymer family (5, 5; 1) m
		43	Langbeinite-type family (16, 5; 1) i		
		44	Lecontite $(NaNH_4SO_4 \cdot 2H_2O)$ family (2, 2; 0)		
		45	Alum family (16, 15; 0) j		
		46	GASH $(C(NH_2)_3Al(SO_4)_2 \cdot 6H_2O)$ family (9, 9; 0)		
		47	Colemanite $(Ca_2B_6O_{11} \cdot 5H_2O)$ family (1, 1; 0)		
		48	$K_4Fe(CN)_6 \cdot 3H_2O$ family (4, 4; 0)		
		49	$K_3BiCl_6 \cdot 2KCl \cdot KH_3F_4$ family (1, 1; 0)		

Table 4.5-1 The 72 families of ferroelectric materials. The number assigned to each family corresponds to the number used in LB III/36. The numbers in parentheses (N_{Sub}, $N_{\text{F+A}}$; n) after the family name serve the purpose of conveying some information about the size and importance of the family. The numbers indicate the following: N_{Sub}, the number of pure substances (ferroelectric, antiferroelectric, and related substances) which are treated as members of this family in LB III/36; $N_{\text{F+A}}$, the number of ferroelectric and antiferroelectric substances which are treated as members of this family in LB III/36; n, the number of representative substances from this family whose properties are surveyed in Sect. 4.5.4. For some of these families, additional remarks are needed: for instance, because the perovskite-type oxide family has many members and consists of several subfamilies; because the liquid crystal and polymer families have very specific properties compared with crystalline ferroelectrics; and because the traditional names of some families are apt to lead to misconceptions about their members. Such families are marked by letters a–m following the parentheses, and remarks on these families are given under the corresponding letter in the text in Sect. 4.5.3.1

sented by a chain of n perovskite-type units of ABO_3 perpendicular to the layer.

h. Boracite-Type Family (Family Number 18). Boracite is a mineral, $Mg_3B_7O_{13}Cl$. The boracite-type family contains crystals isomorphous with the mineral, and has the chemical formula $M_3^{2+}B_7O_{13}X^{1-}$, where M^{2+} stands for a divalent cation of Mg, Cr, Mn, Fe, Co, Ni, Cu, Zn, or Cd, and X^{1-} stands for an anion of Cl, Br, or I.

i. Langbeinite-Type Family (Family Number 43). This family consists of crystals which are basically isomorphous with $K_2Mg_2(SO_4)_3$ (langbeinite), and have the common chemical formula $M_2^{1+}M_2^{2+}(SO_4)_3$, where M^{1+} stands for a monovalent ion of K, Rb, Cs, Tl, or NH_4, and M^{2+} stands for a divalent ion of Mg, Ca, Mn, Fe, Co, Ni, Zn, or Cd.

j. Alum Family (Family Number 45). The alums are compounds with the chemical formula $M^{1+}M^{3+}(SO_4)_2 \cdot 12\,H_2O$, where M^{1+} is a monovalent cation and M^{3+} is a trivalent cation. Ferroelectricity has been found for the monovalent cations of NH_4, CH_3NH_3, etc. and the trivalent cations of Al, V, Cr, Fe, In, and Ga. This family contains a few isomorphous selenates, $M^{1+}M^{3+}(SeO_4)_2 \cdot 12\,H_2O$.

k. $LiNH_4C_4H_4O_6 \cdot H_2O$ Family (Family Number 68). This family contains two ferroelectrics, $LiNH_4C_4H_4O_6 \cdot H_2O$ and $LiTlC_4H_4O_6 \cdot H_2O$. The polar directions of the two ferroelectrics are different from each other.

l. Liquid Crystals (Family Number 71). Ferroelectric and antiferroelectric liquid crystals are very useful as fast display elements.

A. *Ferroelectric liquid crystals (family number 71A).* Ferroelectric liquid crystals are defined as liquid crystals which exhibit a ferroelectric hysteresis loop like that shown in Fig. 4.5-1. Unlike ferroelectric crystals, however, ferroelectric liquid crystals generally have no spontaneous polarization in the bulk state. The chiral smectic phase denoted by Sm C* (e.g. of DOBAMBC) consists of many layers, each of which has a spontaneous polarization parallel to the layer plane, but the spontaneous polarization varies helically in different directions from layer to layer, so that the bulk has no spontaneous polarization as a whole. A sufficiently strong electric field causes a transition from the helical phase to a polar phase. Under an alternating electric field, the helical structure does not have a chance to build up owing to the delay in the transition. Instead, a direct transition occurs between the induced polar phases, resulting in a hysteresis loop of the type shown in Fig. 4.5-1. Accordingly, the hysteresis loop may be regarded as one in which the linear part of the antiferroelectric hysteresis shown in Fig. 4.5-2 is eliminated. It should be noted, however, that the helical structure disappears and two stable states with parallel and antiparallel polarizations appear, similar to the domain structure of a ferroelectric crystal, when a liquid crystal is put in a cell which is thinner than the helical pitch [5.52].

B. *Antiferroelectric liquid crystals (family number 71B).* The phase denoted by Sm C_A^* (e.g. of MHPOBC) exhibits a double hysteresis of the type shown in Fig. 4.5-2, and a liquid crystal showing this phase is called antiferroelectric [5.53].

m. Polymers (Family Number 72). Polyvinylidene fluoride $(CH_2CF_2)_n$ and its copolymers with trifluoroethylene $(CHFCF_2)_n$, etc. are ferroelectric. Ferroelectric polymers are usually prepared as thin films in which crystalline and amorphous regions coexist. The ferroelectric hysteresis loop originates from reversal of the spontaneous polarization in the crystalline regions. The electric-field distribution is expected to be complex in the thin film because of the interposition of the amorphous regions. A ferroelectric hysteresis loop can be observed when the ratio of the volume of the crystalline regions to the total volume is relatively large (e.g. more than 50%). The coercive field is larger (e.g. > 50 MV/m) than that of solid ferroelectrics (usually a few MV/m or smaller). The ferroelectric properties depend sensitively upon the details of sample preparation, for example the use of melt quenching or melt extrusion, the annealing temperature, or the details of the poling procedure. Polymer ferroelectrics are useful for soft transducers.

4.5.4 Physical Properties of 43 Representative Ferroelectrics

This section surveys the characteristic properties of 43 representative ferroelectrics with the aim of demonstrating the wide variety in the behavior of ferroelectrics. The 43 representative ferroelectrics are selected from 29 of the above-mentioned families. The presentation is mainly graphical. Most of the figures are reproduced from LB III/36, where the relevant references can be found. Table 4.5-2 summarizes the meaning of the symbols frequently used in the figure captions and indicates the units in which the data are given.

To facilitate to get more information on each representative substance in LB II/36, the number assigned to the substance in LB III/36 is given in parenthesis following its chemical formula, e.g., $KNbO_3$ (LB number 1A-2).

4.5.4.1 Inorganic Crystals Oxides [5.1, 2]

Perovskite-Type Family

$KNbO_3$ (LB Number 1A-2). This crystal is ferroelectric below about 418 °C. Further phase transitions take place at about 225 °C and about -10 °C, retaining ferroelectric activity. The crystal has large electromechanical coupling constants and is useful in lead-free piezoelectric elements and SAW (surface acoustic wave) filters in communications technology (Fig. 4.5-13, 4.5-14).

Table 4.5-2 Symbols and units frequently used in the figure captions

a, b, c	unit cell vector, units Å
a^*, b^*, c^*	unit cell vector in reciprocal space, units Å^{-1}
E_c	coercive field, units V/m
f	frequency, units Hz = 1/s
P_s	spontaneous polarization, units C/m^2
T	temperature, units K or °C
κ	dielectric constant (or relative permittivity) $= \varepsilon/\varepsilon_0$, where ε is the permittivity of the material and ε_0 is the permittivity of a vacuum. κ is a dimensionless number
κ', κ''	real and imaginary parts of the complex dielectric constant $\kappa^* = \kappa' + i\kappa''$. κ' and κ'' are both dimensionless numbers
κ_{ij}	component of dielectric-constant tensor; dimensionless numbers
$\kappa_a, \kappa_b, \kappa_c$	κ measured along a, b, c axes; dimensionless numbers
$\kappa_{(hkl)}$	κ measured perpendicular to the (hkl) plane; dimensionless number
$\kappa_{[uvw]}$	κ measured parallel to the $[uvw]$ direction; dimensionless number
κ^T	κ of free crystal, i.e. κ measured at constant stress **T**; dimensionless number
κ^S	κ of clamped crystal, i.e. κ measured at constant strain **S**; dimensionless number
Θ_f	ferroelectric transition temperature, units K or °C

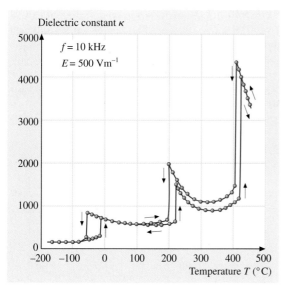

Fig. 4.5-12 KNbO$_3$. Dielectric constant κ versus temperature T

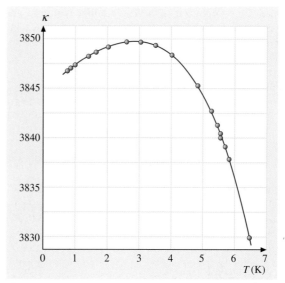

Fig. 4.5-14 KTaO$_3$. Real part of dielectric constant κ' versus T. $f = 1$ kHz

KTaO$_3$ (LB Number 1A-5). KTaO$_3$ is cubic at all temperatures, and its dielectric constant becomes very large at low temperatures without a phase transition (Fig. 4.5-14). It is generally believed that this behavior is related to the zero-point lattice vibrations. Replacement of Nb by Ta generally lowers drastically the ferroelectric Curie temperature, as seen by comparing Fig. 4.5-14 with Fig. 4.5-12. (This effect is well demonstrated later in Figs. 4.5-39 and 4.5-40).

SrTiO$_3$ (LB Number 1A-8). This crystal is cubic at room temperature and slightly tetragonal below 105 K. The phase transition at 105 K is caused by softening of the lattice vibration mode at the (1/2, 1/2, 1/2) Bril-

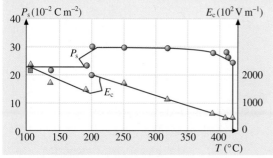

Fig. 4.5-13 KNbO$_3$. P_s and E_c versus T. Measurements were made by applying the electric field parallel to the pseudocubic [100] direction

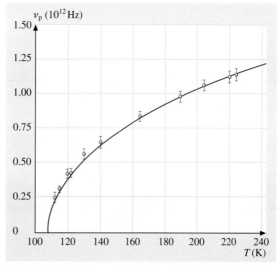

Fig. 4.5-15 SrTiO$_3$. ν_P versus T. ν_P is the frequency of the R_{25} (Γ_{25}) optical phonon

louin zone corner (Fig. 4.5-15) without an appreciable dielectric anomaly. At very low temperatures, the dielectric constants become extraordinarily high without a phase transition (Fig. 4.5-16), owing to softening of the optical phonon at the origin of the Brillouin zone (Fig. 4.5-17). It is generally believed that the absence of a low-temperature transition is related to zero-point lat-

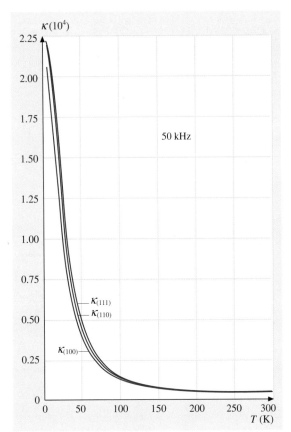

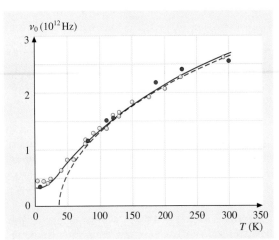

Fig. 4.5-17 SrTiO$_3$. ν_0 versus T. ν_0 is the frequency of the soft phonon at $q=0$. *Open circles*: inelastic neutron scattering. *Filled circles*: Raman scattering. The *solid curve* is $194.4\,\kappa^{1/2}$ and the *dashed curve* is $0.677(T-T_0)^{1/2}$, with $T_0 = 38$ K

Fig. 4.5-16 SrTiO$_3$. $\kappa_{(111)}$, $\kappa_{(110)}$, and $\kappa_{(100)}$ versus T at $f = 50$ kHz. $\kappa_{(111)}$, $\kappa_{(110)}$, and $\kappa_{(100)}$ are slightly different from each other in the tetragonal phase

tice vibrations as in KTaO$_3$. Solid solutions of SrTiO$_3$ are useful as ceramic thin films in memory devices.

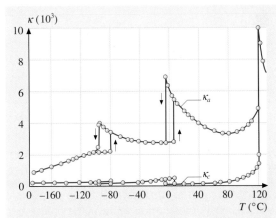

Fig. 4.5-18 BaTiO$_3$. κ_a and κ_b versus T. κ_a and κ_b are the values of κ along the a and b axes, respectively, of the tetragonal phase

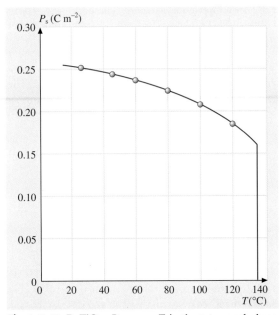

Fig. 4.5-19 BaTiO$_3$. P_s versus T in the tetragonal phase. P_s is parallel to the c axis

BaTiO₃ (LB Number 1A-10). BaTiO$_3$ is the most extensively studied ferroelectric crystal. It is ferroelectric below about 123 °C, where the crystal symmetry changes from cubic to tetragonal. Further phase transitions take place from tetragonal to orthorhombic at about 5 °C and to rhombohedral at about −90 °C (Figs. 4.5-18 to 4.5-20). It is believed that the ferroelectric transition

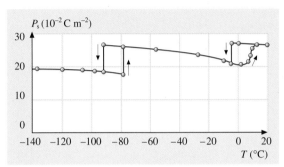

Fig. 4.5-20 BaTiO$_3$. $P_{s[001]}$ versus T below 20 °C, where $P_{s[001]}$ is the component of the spontaneous polarization parallel to the c axis in the tetragonal phase

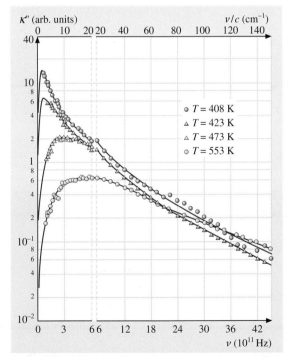

Fig. 4.5-21 BaTiO$_3$. κ'' versus ν. Parameter: T. κ'' is the imaginary part of κ obtained from the hyper-Raman spectrum. *Curves*: calculations based upon the classical dispersion oscillator model. c: light velocity

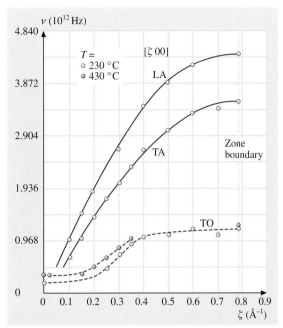

Fig. 4.5-22 BaTiO$_3$. Phonon dispersion relation determined by neutron scattering along the [100] direction in the cubic phase. ν is the phonon frequency. LA, longitudinal acoustic branch; TA, transverse acoustic branch; TO, transverse optical branch. The frequency of the TO branch is lower (softer) at 230 °C than at 430 °C, indicating mode softening

is caused by softening of an optical mode at the center of the Brillouin zone. Hyper-Raman scattering studies support this model (Figs. 4.5-21 and 4.5-22). The results of neutron scattering studies favor this model (Fig. 4.5-23), but the measurement was not easy owing to the intense elastic peaks (Fig. 4.5-24a). These elastic peaks indicate marked cluster formation in the vicinity of the Curie point. As discussed in Sect. 4.5.3, there seem to be two components contributing to the dielectric constant measured in the vicinity of the Curie point: a component due to the soft mode and another one due to the motion of the cluster boundaries. Presumably the two components suggested by infrared and hyper-Raman scattering data correspond to these two components. For theoretical studies of the phase transitions, readers should refer to [5.7]. Solid solutions of BaTiO$_3$ are the most useful ferroelectrics in ceramic condensers and as thin films in memory devices.

When the temperature is raised above 1460 °C, cubic BaTiO$_3$ performs another phase transition to a hexagonal structure. This hexagonal phase can be quenched

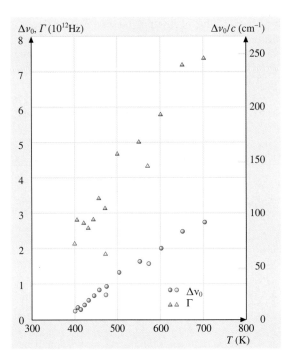

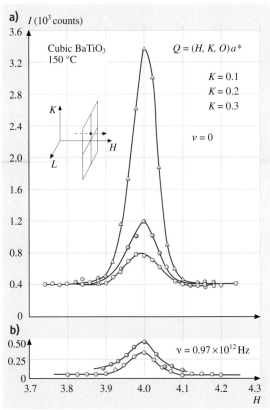

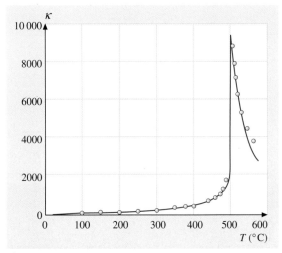

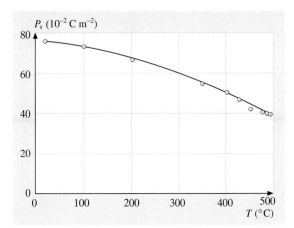

Fig. 4.5-23 BaTiO$_3$. $\Delta\nu_0$ and Γ versus T, obtained from hyper-Raman scattering in the cubic phase. $\Delta\nu_0$ and Γ are the optical mode frequency and damping constant, respectively. The different symbols (*brown* and *gray*) show results from different authors. $\Delta\nu_0$ decreases as the temperature decreases to the Curie point, showing the presence of mode softening. c: ligth velocity

Fig. 4.5-25 PbTiO$_3$. κ versus T

Fig. 4.5-26 PbTiO$_3$. P_s versus T

Fig. 4.5-24a,b BaTiO$_3$. Triple-axis neutron spectrometer scans at constant frequency across the sheet of diffuse scattering at 150 °C. The path of the scans is shown in the *inset*. (**a**) shows the elastic scan ($\nu = 0$), where the high background level is due to nuclear incoherent scattering. (**b**) Inelastic scan ($\nu = 0.97 \times 10^{12}$ Hz)

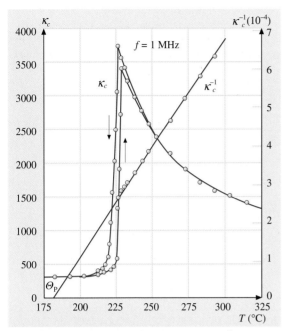

Fig. 4.5-27 PbZrO$_3$ (single crystal). κ_c and κ_c^{-1} versus T

to room temperature by relatively rapid cooling. This hexagonal BaTiO$_3$ also shows ferroelectric activity below $-199\,°$C.

PbTiO$_3$ (LB Number 1A-11). This crystal is ferroelectric below about $500\,°$C (Figs. 4.5-25, 4.5-26). The spontaneous polarization is large (Fig. 4.5-26) and thus the pyroelectric coefficient is large, which makes the crystal useful in infrared sensors. Solid solutions with other perovskite-type oxides provide good dielectric and piezoelectric materials (see PZT, PLZT, and (PbTiO$_3$)$_x$(Pb(Sc$_{1/2}$Nb$_{1/2}$)O$_3$)$_{1-x}$ below).

PbZrO$_3$ (LB Number 1A-15). This crystal is antiferroelectric below about $230\,°$C (Fig. 4.5-27), exhibiting a typical antiferroelectric hysteresis loop (Fig. 4.5-28). Solid solutions of this substance are very important in technological applications (see PZT and PLZT below).

Pb(Mg$_{1/3}$Nb$_{2/3}$)O$_3$ (LB Number 1B-d4). This crystal exhibits a broad ferroelectric phase transition with an average transition temperature of $-8\,°$C determined by the maximum of the low-frequency dielectric constant. A marked frequency dispersion of the

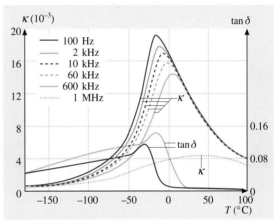

Fig. 4.5-29 Pb(Mg$_{1/3}$Nb$_{2/3}$)O$_3$ (ceramic). κ and $\tan\delta$ versus T. Parameter: f

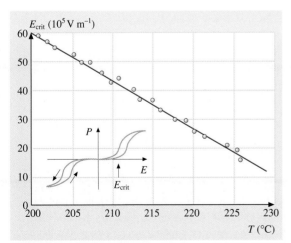

Fig. 4.5-28 PbZrO$_3$. E_{crit} versus T. E_{crit} is the critical field of the antiferroelectric hysteresis loop

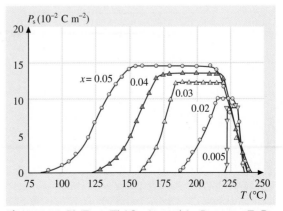

Fig. 4.5-30 Pb(Zr$_{1-x}$Ti$_x$)O$_3$ (ceramic). P_s versus T. Parameter: x

dielectric constant occurs around the transition temperature. These is a typical behavior of a relaxor (Fig. 4.5-29).

Pb(Zr,Ti)O₃ (LB Number 1C-a62). $PbZrO_3$ is antiferroelectric, as described above, while a small addition of $PbTiO_3$ induces a ferroelectric phase (Fig. 4.5-30).

PZT (LB Number 1C-a63). Solid solutions $Pb(Ti_{1-x}Zr_x)O_3$ with $x = 0.5$–0.6 are commonly called PZT. They have very large electromechanical coupling constants and are widely utilized as piezoelectric elements. Thin ceramic films are useful in memory devices (Fig. 4.5-31).

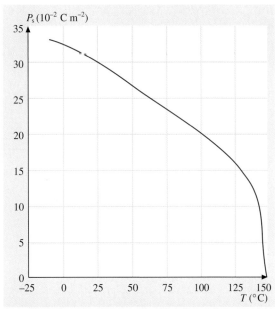

Fig. 4.5-32 $(Pb_{0.92}La_{0.08})(Zr_{0.40}Ti_{0.60})_{0.98}O_3$ (PLZT). P_s versus T

PLZT (LB Number 1C-c66). The acronym PLZT means La-modified PZT. Hot-pressed ceramic PLZT is transparent and useful for optical switches and similar devices (Fig. 4.5-32).

$(PbTiO_3)_x(Pb(Sc_{1/2}Nb_{1/2})O_3)_{1-x}$ (LB Number 1C-b34). These solid solutions have very large electromechan-

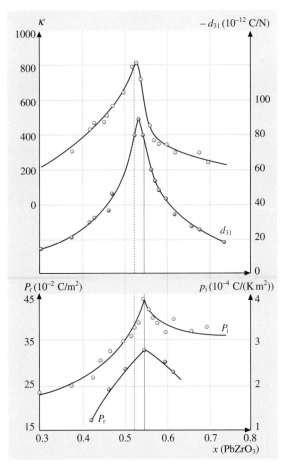

Fig. 4.5-31 $Pb(Ti_{1-x}Zr_x)O_3$ 2% $Zr(Mn_{1/3}Bi_{2/3})O_3$ as an additive (ceramic). κ, p_i, d_{31}, and P_r versus x. p_i is the pyroelectric coefficient, d_{31} is the piezoelectric strain constant, and P_r is the remanent polarization (see Fig. 4.5-1)

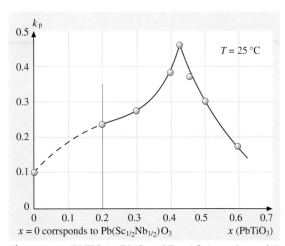

Fig. 4.5-33 $(PbTiO_3)_x(Pb(Sc_{1/2}Nb_{1/2})O_3)_{1-x}$ (ceramic). k_p versus x. k_p is the planar electromechanical coupling factor

ical coupling factors, and are utilized for piezoelectric actuators and similar devices (Fig. 4.5-33).

LiNbO₃ Family

LiTaO₃ (LB Number 2A-2). This crystal is ferroelectric below 620 °C. The coercive field is large. The crystal is useful for piezoelectric elements, for linear and nonlinear optical elements, and for SAW filters in communications technology (Figs. 4.5-34 and 4.5-35).

SrTeO₃ Family

SrTeO₃ (LB Number 4A-1). This crystal is ferroelectric between 312 and 485 °C (Figs. 4.5-36 and 4.5-37).

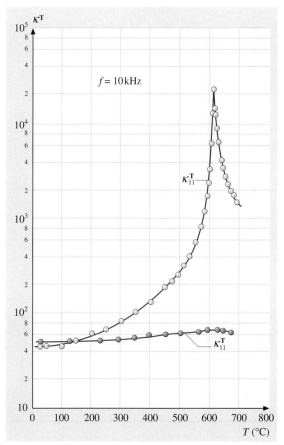

Fig. 4.5-34 LiTaO₃. κ_{11}^T and κ_{33}^T versus T. $f = 10\,\mathrm{kHz}$

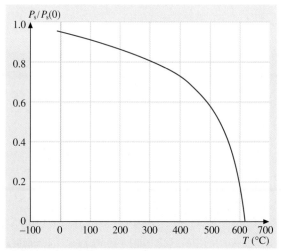

Fig. 4.5-35 LiTaO₃. $P_s/P_s(0)$ versus T. $P_s(0)$ is the value of P_s at 0 K (about $0.5\,\mathrm{C\,m^{-2}}$)

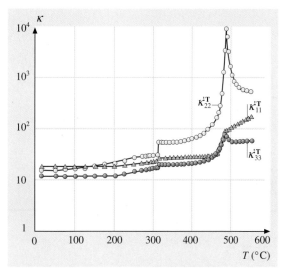

Fig. 4.5-36 SrTeO₃. κ versus T. $f = 10\,\mathrm{kHz}$

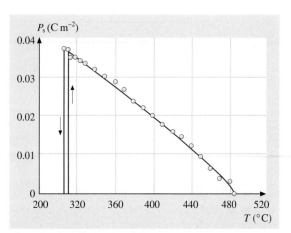

Fig. 4.5-37 SrTeO₃. P_s versus T

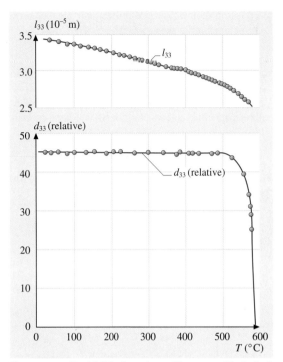

Fig. 4.5-38 $Ba_2NaNb_5O_{15}$. d_{33} and l_{33} versus T. d_{33} is the nonlinear optical susceptibility (relative value), and l_{33} is the coherence length

Tungsten Bronze-Type Family

$Ba_2NaNb_5O_{15}$ (BNN) (LB Number 6B-a7). This crystal is ferroelectric below about 580 °C. The crystal structure is modulated below 300 °C. This material is utilized for optical second-harmonic generation and in optical parametric oscillators (Fig. 4.5-38).

$Ba_2Na(Nb_{1-x}Ta_x)_5O_{15}$ (BNNT) (LB Number 6C-b30). The transition temperature varies over a wide temperature range as x varies from 0 to 1.0 (Fig. 4.5-39).

$Sr_2Nb_2O_7$ Family

$Sr_2(Nb_{1-x}Ta_x)_2O_7$ (LB Number 8B-6). These solid solutions can be made over the whole range of $x = 0-1.0$, and the Curie point varies over a wide temperature range from 1342 °C to −107 °C (Fig. 4.5-40). The solid solutions are very useful as high-temperature dielectric materials, especially because they do not contain Pb, which is volatile at high temperatures.

$Li_2Ge_7O_{15}$ Family

$Li_2Ge_7O_{15}$ (LB Number 13A-1). This crystal is ferroelectric below 283.5 K. The sign of the spontaneous polarization is reversed around 130 K (Figs. 4.5-41 and 4.5-42).

GMO ($Gd_2(MoO_4)_3$) Family

$Gd_2(MoO_4)_3$ (GMO) (LB Number 17A-3). Three crystal structures of GMO are known, α, β, and γ. The β structure is stable above 850 °C but can be obtained at room temperature as a metastable state by

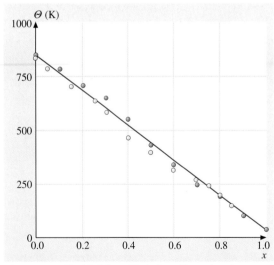

Fig. 4.5-39 $Ba_2Na(Nb_{1-x}Ta_x)_5O_{15}$. Ferroelectric transition temperature Θ versus x. *Brown circles* and *gray circles* represent data measured by different authors

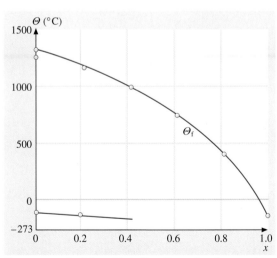

Fig. 4.5-40 $Sr_2(Nb_{1-x}Ta_x)_2O_7$. Ferroelectric transition temperature Θ_f versus x. The *lower curve* shows another phase transition temperature

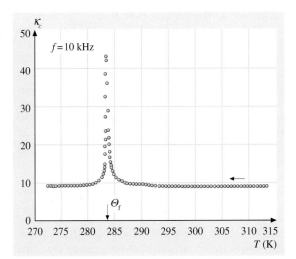

Fig. 4.5-41 Li$_2$Ge$_7$O$_{15}$. κ_c versus T. $f = 10$ kHz

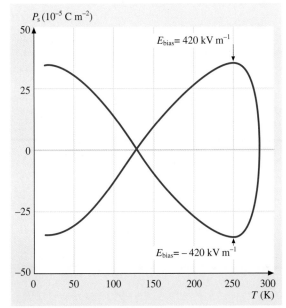

Fig. 4.5-42 Li$_2$Ge$_7$O$_{15}$. P_s versus T. P_s was determined by pyroelectric-charge measurement

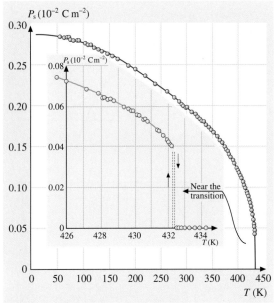

Fig. 4.5-43 Gd$_2$(MoO$_4$)$_3$. P_s versus T

resulting antiparallel displacements produces a spontaneous strain, which in turn causes a spontaneous polarization through the normal piezoelectric coupling (Fig. 4.5-43).

Boracite-Type Family

Ni$_3$B$_7$O$_{13}$I (LB Number 18A-23). This crystal is ferroelectric and ferromagnetic below about 60 K (Figs. 4.5-44, 5-45, 5-46): a rare example where both ferroelectric and ferromagnetic spontaneous polarizations

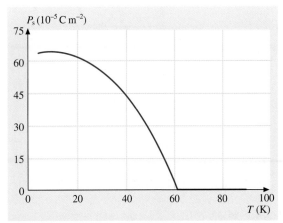

Fig. 4.5-44 Ni$_3$B$_7$O$_{13}$I. P_s versus T. P_s was measured with the specimen parallel to the cubic (001) plane

rapid cooling. Ferroelectric activity takes place in this metastable β-GMO below 159 °C. The phase transition is the indirect type III$_{op}$ discussed in Sect. 4.5.3. The dielectric constant of the clamped crystal (κ_c^S in Fig. 4.5-6) shows no anomaly at the transition point. The ferroelectric phase results from a phonon instability at the (1/2, 1/2, 0) Brillouin zone corner of the parent tetragonal phase. Anharmonic coupling to the

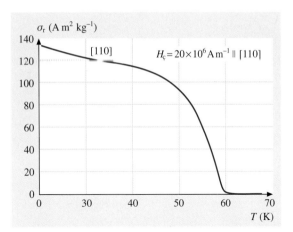

Fig. 4.5-45 $Ni_3B_7O_{13}I$. σ_r versus T. σ_r is the remanent magnetization. The sample was cooled down to 4.2 K in a magnetic field of 1.6×10^6 A/m parallel to [100] prior to the measurement

take place simultaneously. There is a magnetoelectric effect, where the magnetic polarization is reversed by reversal of the electric polarization and vice versa.

4.5.4.2 Inorganic Crystals Other Than Oxides [5.3]

SbSI Family
SbSI (LB Number 20A-7). SbSI is ferroelectric below 20 °C. The phase transition is of the displacive type, a relatively rare characteristic in nonoxide materials. The crystal is photoconductive (Figs. 4.5-47 and 4.5-48).

BaMnF$_4$ Family
BaMnF$_4$ (LB Number 26A-2). This crystal exhibits a dielectric anomaly at about 242 K (Fig. 4.5-49). The coercive field is very large. The crystal is antiferromagnetic below 25 K (Fig. 4.5-50). The dielectric constant varies depending upon the magnetic field at low temperatures (Fig. 4.5-51).

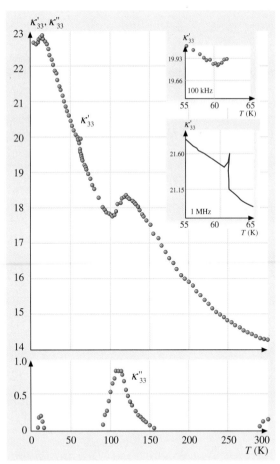

Fig. 4.5-46 $Ni_3B_7O_{13}I$. κ'_{33} and κ''_{33} versus T. $f = 100$ kHz. The *insets* show details of κ'_{33} versus T at 100 kHz and 1 MHz around Θ_f

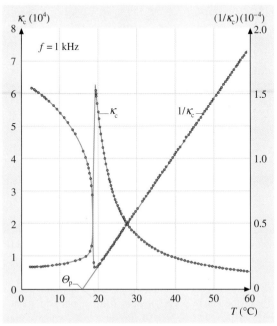

Fig. 4.5-47 SbSI. κ_c and $1/\kappa_c$ versus T

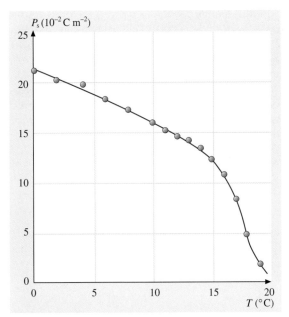

Fig. 4.5-48 SbSI. P_s versus T

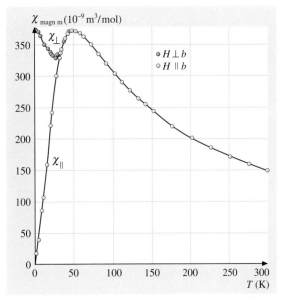

Fig. 4.5-50 BaMnF$_4$. $\chi_{\text{magn m}}$ versus T. $\chi_{\text{magn m}}$ is the magnetic susceptibility, and $\chi_\perp$ and $\chi_\parallel$ are the magnetic susceptibilities measured perpendicular and parallel, respectively, to the b axis

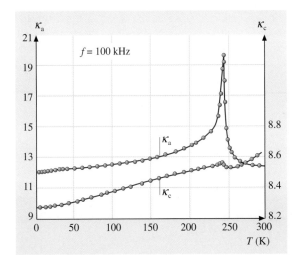

Fig. 4.5-49 BaMnF$_4$. κ_a and κ_c versus T

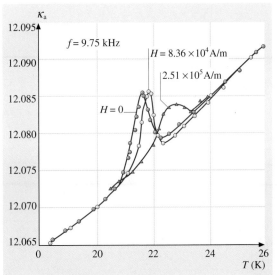

Fig. 4.5-51 BaMnF$_4$. κ_a versus T. Parameter: magnetic field H. $f = 9.75$ kHz

HCl Family

HCl (LB Number 27A-1). This crystal is ferroelectric below 98 K. The chemical formula is the simplest one among all known ferroelectrics. The coercive field is large (Figs. 4.5-52 and 4.5-53).

NaNO₂ Family

NaNO₂ (LB Number 28A-1). This crystal is ferroelectric below 163.9 °C (Figs. 4.5-54 and 4.5-55). The spontaneous polarization results from orientational order of the NO_2^- ions. Between 163.9 and 165.2 °C, the crystal structure is incommensurately modulated with a wave vector δa^*, where δ varies from 0.097 to 0.120 with increasing temperature.

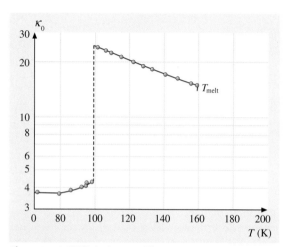

Fig. 4.5-52 HCl (polycrystalline). κ_0 versus T. κ_0 is the static dielectric constant

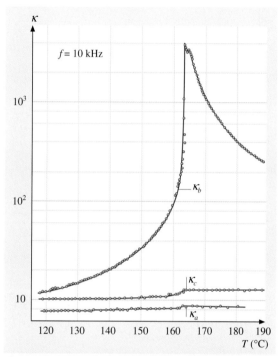

Fig. 4.5-54 NaNO₂. κ_a, κ_b, and κ_c versus T

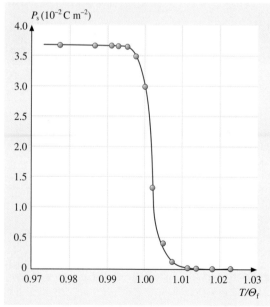

Fig. 4.5-53 HCl. P_s versus T/Θ_f obtained from pyroelectric-current measurement. $\Theta_f = 98$ K

Fig. 4.5-55 NaNO₂. P_s versus T, determined by pyroelectric-charge measurement

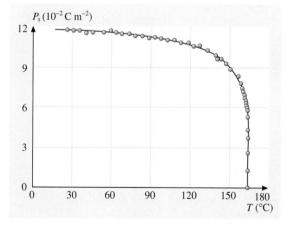

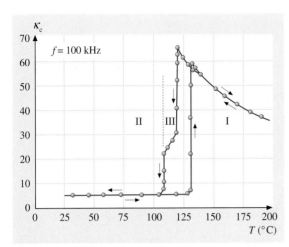

Fig. 4.5-56 KNO$_3$. κ_c versus T

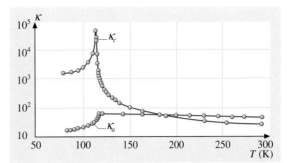

Fig. 4.5-58 KH$_2$PO$_4$. κ_a and κ_c versus T. $f = 800$ Hz

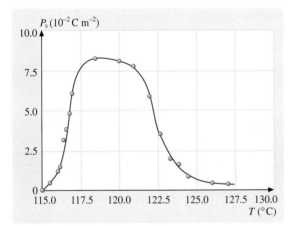

Fig. 4.5-57 KNO$_3$. P_s versus T

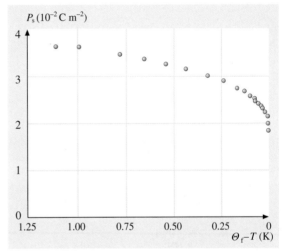

Fig. 4.5-59 KH$_2$PO$_4$. P_s versus $(\Theta_f - T)$. $\Theta_f = 123$ K

KNO$_3$ Family

KNO$_3$ (LB Number 30A-2). This crystal is ferroelectric between about 115 and 125 °C in a metastable phase III which appears on cooling. Hydrostatic pressure stabilizes this phase (Figs. 4.5-56 and 4.5-57).

KDP (KH$_2$PO$_4$) Family

KH$_2$PO$_4$ (KDP) (LB Number 33A-1). KH$_2$PO$_4$ is a classical and extensively studied ferroelectric crystal. It is ferroelectric below 123 K (Figs. 4.5-58 and 4.5-59). The transition is a typical ferroelectric phase transition, related to a configuration change in a three-dimensional hydrogen-bond network. Figures 4.5-61 and 4.5-60 demonstrate changes in the proton configuration associated with the phase transition. The transitions related to hydrogen atom rearrangement in the hydrogen-bond network are characterized by sensitivity to deuteration and hydrostatic pressure, as demonstrated in Figs. 4.5-62 and 4.5-63, respectively. For theoretical studies of the phase transition, readers should refer to [5.7]. The crystal is useful in nonlinear optical devices.

CsH$_2$PO$_4$ (LB Number 33A-3). This crystal is ferroelectric below about 151.5 K. The crystal system of its paraelectric phase (monoclinic) is different from that of KH$_2$PO$_4$ (tetragonal). The temperature dependence of the dielectric constant above the Curie point deviates considerably from the Curie–Weiss law, suggesting that the transition is related to one-dimensional ordering of hydrogen atoms in a hydrogen-bond network. Deuteration changes the transition temperature from 151.5 to 264.7 K (Fig. 4.5-64).

Fig. 4.5-60 KH$_2$PO$_4$. Fourier map of the projection of the proton distribution on (001) in the ferroelectric phase (77 K), determined by neutron diffraction. The proton distribution lies approximately on a line joining two oxygen atoms O(1) and O(2) and closer to O(1)

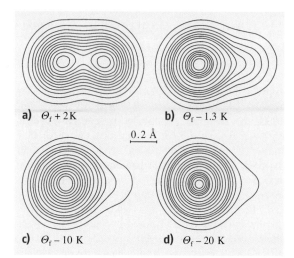

Fig. 4.5-61a–d KH$_2$PO$_4$. Change of proton distribution above and below the Curie point Θ_f, determined by neutron diffraction. Contours are all equally spaced. (**a**) $\Theta_f + 2$ K; (**b**) $\Theta_f - 1.3$ K; (**c**) $\Theta_f - 10$ K; (**d**) $\Theta_f - 20$ K

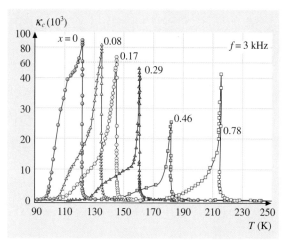

Fig. 4.5-62 KH$_{2(1-x)}$D$_{2x}$PO$_4$. κ_c versus T. Parameter: x

RbH$_2$PO$_4$–NH$_4$H$_2$PO$_4$ (LB Number 33B-5). The mixed crystals Rb$_{1-x}$(NH$_4$)$_x$H$_2$PO$_4$, where $0.2 < x < 0.8$, show a characteristic temperature dependence of the dielectric constants, suggesting that at low temperatures

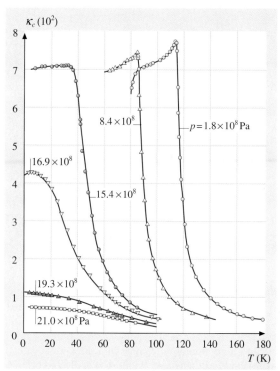

Fig. 4.5-63 KH$_2$PO$_4$. κ_c versus T. Parameter: hydrostatic pressure p

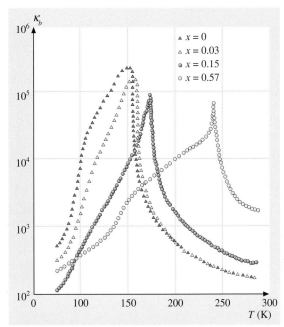

Fig. 4.5-64 CsH_2PO_4 and $CsH_{2(1-x)}D_{2x}PO_4$. κ_b versus T. Parameter: x

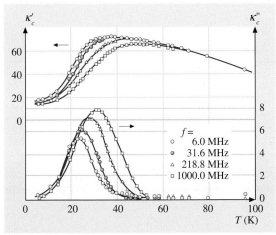

Fig. 4.5-65 $Rb_{0.65}(NH_4)_{0.35}H_2PO_4$. κ'_c and κ''_c versus T. Parameter: f

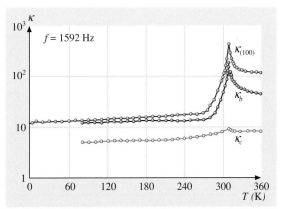

Fig. 4.5-66 $PbHPO_4$. $\kappa_{(100)}$, κ_b, and κ_c versus T. $\kappa_{(100)}$ is the dielectric constant perpendicular to the (100) plane

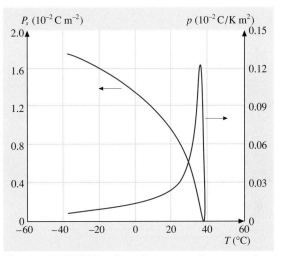

Fig. 4.5-67 $PbHPO_4$. P_s and p versus T, measured on (100) planar specimen. $p = -dP_s/dT$ is the pyroelectric coefficient

a local order becomes predominant in a configuration of hydrogen atoms without definite long-range order, i. e. a dipole glass state develops (Fig. 4.5-65).

PbHPO₄ Family

PbHPO₄ (LB Number 34A-1). This crystal is ferroelectric below $37\,°C$ (Figs. 4.5-66 and 4.5-67). It exhibits characteristic critical phenomena, suggesting that the spontaneous polarization results from an ordered arrangement of hydrogen atoms in a one-dimensional array of hydrogen bonds.

(NH₄)₂SO₄ Family

(NH₄)₂SO₄ (LB Number 39A-1). This crystal is ferroelectric below $-49.5\,°C$. The dielectric constant is practically independent of temperature above the Curie point (Fig. 4.5-68). The spontaneous polarization changes its sign at about $-190\,°C$ (Fig. 4.5-69), suggesting a ferrielectric mechanism for the spontaneous polarization.

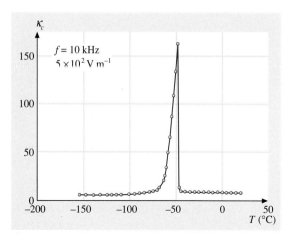

Fig. 4.5-68 $(NH_4)_2SO_4$. κ_c versus T

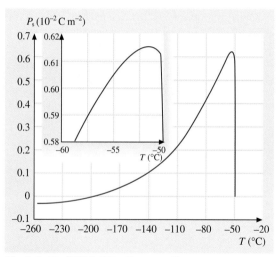

Fig. 4.5-69 $(NH_4)_2SO_4$. P_s versus T

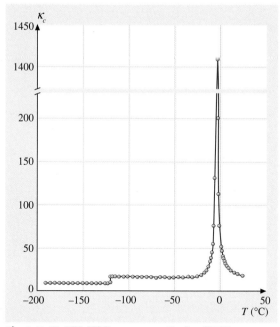

Fig. 4.5-70 NH_4HSO_4. κ_c versus T. $f = 10\,\text{kHz}$

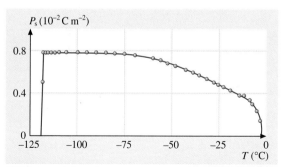

Fig. 4.5-71 NH_4HSO_4. P_s versus T

$(NH_4)HSO_4$ Family

$(NH_4)HSO_4$ *(LB Number 40A-5)*. This crystal is ferroelectric in the temperature range between -119 and $-3\,°C$ (Figs. 4.5-70 and 4.5-71).

$(NH_4)LiSO_4$ Family

$(NH_4)LiSO_4$ *(LB Number 41A-5)*. This crystal is ferroelectric in the temperature range between 10 and $186.5\,°C$ (Figs. 4.5-72 and 4.5-73).

$(NH_4)_3H(SO_4)_2$ Family

$(NH_4)_3H(SO_4)_2$ *(LB Number 42A-1)*. This crystal is ferroelectric in its phase VII below $-211\,°C$. Another

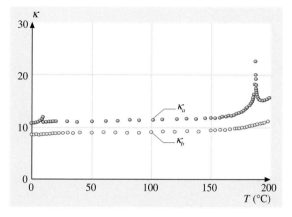

Fig. 4.5-72 NH_4LiSO_4. κ_a and κ_b versus T. $f = 3\,\text{kHz}$

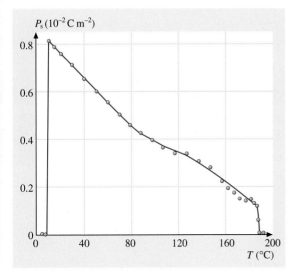

Fig. 4.5-73 NH$_4$LiSO$_4$. P_s versus T

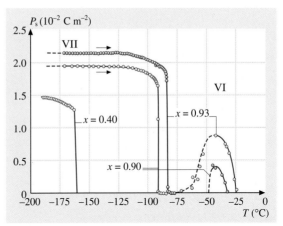

Fig. 4.5-75 ((NH$_4$)$_3$H)$_{1-x}$((ND$_4$)$_3$D)$_x$(SO$_4$)$_2$. P_s versus T. Parameter: x. *Gray circles* (for Phase VI), determined by pyroelectric measurements. *Brown circles* (for phase VII), determined by hysteresis loop measurements

ferroelectric phase, VI, is induced by hydrostatic pressure (Fig. 4.5-74). When H is substituted by D, phase VI appears at atmospheric pressure and the temperature of the transition to phase VII becomes higher. Figure 4.5-75

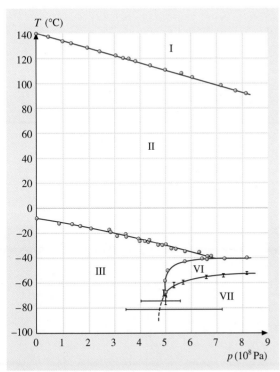

Fig. 4.5-74 (NH$_4$)$_3$H(SO$_4$)$_2$. T versus p phase diagram. p is the hydrostatic pressure. Phases VI and VII are ferroelectric

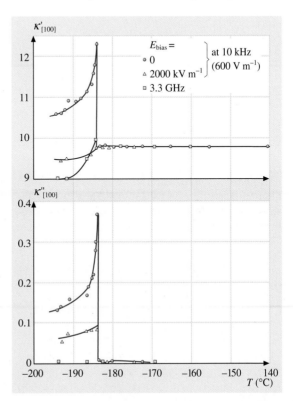

Fig. 4.5-76 (NH$_4$)$_2$Cd$_2$(SO$_4$)$_3$. $\kappa'_{[100]}$ and $\kappa''_{[100]}$ versus T

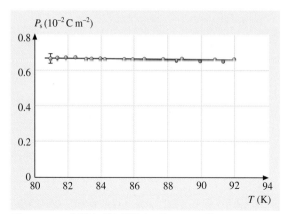

Fig. 4.5-77 $(NH_4)_2Cd_2(SO_4)_3$. P_s versus T

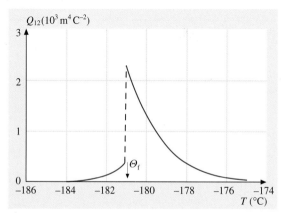

Fig. 4.5-78 $(NH_4)_2Cd_2(SO_4)_3$. Q_{12} versus T. Q_{12} is the electrostrictive constant

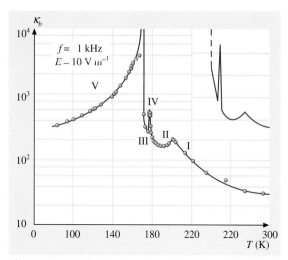

Fig. 4.5-79 $SC(NH_2)_2$. κ_b versus T

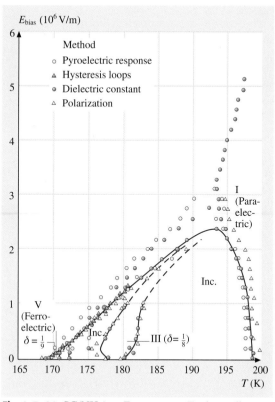

Fig. 4.5-80 $SC(NH_2)_2$. E_{bias} versus T phase diagram. The value of δ means that the phase is commensurately modulated with a vector of wavenumber δc^*. Inc., incommensurate phase

shows temperature dependence of P_s for three values of x in $((NH_4)_3H)_{1-x}((ND_4)_3D)_x(SO_4)_2$.

Langbeinite-Type Family

$(NH_4)_2Cd_2(SO_4)_3$ *(LB Number 43A-13)*. This crystal is ferroelectric below about $-184\,°C$. The dielectric constants are insensitive to temperature above the transition point (Fig. 4.5-76), and the spontaneous polarization does not depend upon temperature (Fig. 4.5-77). The electrostrictive constant Q_{12}, however, exhibits an anomaly at the transition point (Fig. 4.5-78).

4.5.4.3 Organic Crystals, Liquid Crystals, and Polymers [5.4]

$SC(NH_2)_2$ Family

$SC(NH_2)_2$ *(LB Number 50A-1)*. This crystal exhibits at least five phases, I, II, III, IV, and V (Figs. 4.5-79

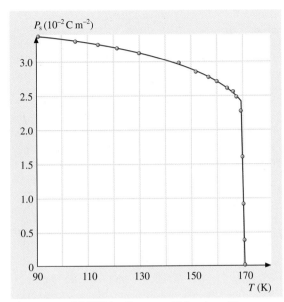

Fig. 4.5-81 $SC(NH_2)_2$. P_s versus T for phase V

and 4.5-80). The crystal is ferroelectric in phase V (Fig. 4.5-81), and slightly ferroelectric with a very small spontaneous polarization in phase III. The crystal structure is modulated commensurately or incommensurately except for phases I and V, as indicated in Fig. 4.5-80.

DSP ($Ca_2Sr(CH_3CH_2COO)_6$) Family

$Ca_2Sr(CH_3CH_2COO)_6$ (DSP) (LB Number 58A-1). This crystal is ferroelectric below about $4\,°C$ (Fig. 4.5-82). The Curie–Weiss constant is small (60 K). Critical slowing-down (see Sect. 4.5.3) takes place (Fig. 4.5-8).

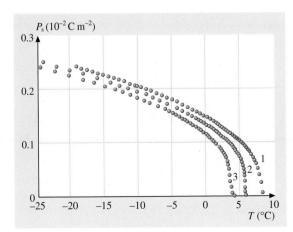

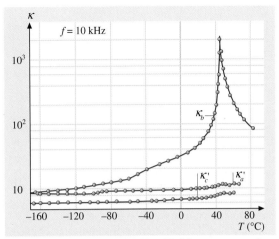

Fig. 4.5-83 $(NH_2CH_2COOH)_3 \cdot H_2SO_4$. $\kappa_{a'}$, $\kappa_{b'}$, and $\kappa_{c'}$ versus T. These quantities are referred to the unit cell vectors: $a' = a+c$, $b' = -b$, and $c' = -c$

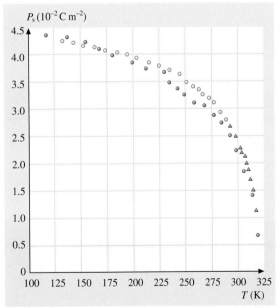

Fig. 4.5-84 $(NH_2CH_2COOH)_3 \cdot H_2SO_4$. P_s versus T. P_s is parallel to the b axis. *Gray circles*: values determined by pyroelectric measurements. *Triangles* and *brown circles*: determined from hysteresis loop by different authors

Fig. 4.5-82 $Ca_2Sr(CH_3CH_2COO)_6$. P_s versus T. The three curves show the effect of annealing: 1, unannealed; 2, annealed at $330\,°C$ for $60\,h$; 3, annealed at $330\,°C$ for $60\,h$ and at $390\,°C$ for $5\,h$

TGS ((NH_2CH_2COOH)$_3 \cdot H_2SO_4$) Family
(NH_2CH_2COOH)$_3 \cdot H_2SO_4$ *(TGS) (LB Number 60A-1)*. This crystal is ferroelectric below 49.4 °C (Figs. 4.5-83 and 4.5-84). Critical slowing down takes place (Fig. 4.5-9).

(CH_3NHCH_2COOH)$_3 \cdot CaCl_2$ Family
(CH_3NHCH_2COOH)$_3 \cdot CaCl_2$ *(LB Number 64A-1)*. This crystal is ferroelectric below 127 K (Figs. 4.5-85 and 4.5-86). The Curie–Weiss constant is small (40 K). As discussed in Sect. 4.5.3, two components are expected in the dielectric constants of displacive-type ferroelectrics: one contribution from the soft optical phonon and one contribution from cluster boundary motion. This crystal provides a good example of this situation. Softening of the optical phonon mode B_{2u} occurs (Fig. 4.5-11), indicating that the transition is of the displacive type, while critical slowing-down takes place in the GHz region (Fig. 4.5-10); this seems to take place in the contribution from cluster boundary motion. The value of the dielectric constant at the Curie point at 1.0 GHz seems to contain a contribution from the optical phonon mode.

(CH_3)$_3NCH_2COO \cdot CaCl_2 \cdot 2H_2O$ Family
(CH_3)$_3NCH_2COO \cdot CaCl_2 \cdot 2H_2O$ *(LB Number 66A-1)*. This crystal is ferroelectric below 46 K. It exhibits at least ten phase transitions (eight can be recognized in Fig. 4.5-87), and the crystal structure is commensurately or incommensurately modulated between 46 and 164 K, with a feature known as a devil's staircase. Multiple hysteresis loops are observed in the modulated phases, as shown in Fig. 4.5-88.

Rochelle Salt ($NaKC_4H_4O_6 \cdot 4H_2O$) Family
$NaKC_4H_4O_6 \cdot 4H_2O$ *(Rochelle Salt, RS) (LB Number 67A-1)*. Rochelle salt was the first ferroelectric crystal to be discovered. It is ferroelectric between −18 and 24 °C (Figs. 4.5-89 and 4.5-90). It is rare that an ordered ferroelectric phase of lower symmetry ($P2_1$ in this case) appears in an intermediate temperature range between disordered phases of the same higher symmetry ($P2_12_12$). Figure 4.5-91 shows X-ray evidence for such space group changes. For theoretical studies, readers should refer to [5.7].

LAT ($LiNH_4C_4H_4O_6 \cdot H_2O$) Family
$LiNH_4C_4H_4O_6 \cdot H_2O$ *(LAT) (LB Number 68A-1)*. Ferroelectric activity appears along the *b* axis below 106 K (Fig. 4.5-92). The dielectric constant of the clamped

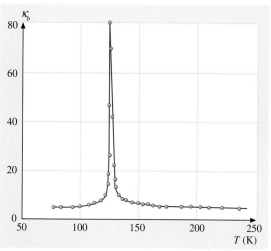

Fig. 4.5-85 (CH_3NHCH_2COOH)$_3 \cdot CaCl_2$. κ_b versus T. $f = 10$ kHz

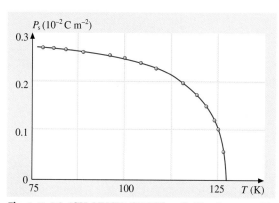

Fig. 4.5-86 (CH_3NHCH_2COOH)$_3 \cdot CaCl_2$. P_s versus T

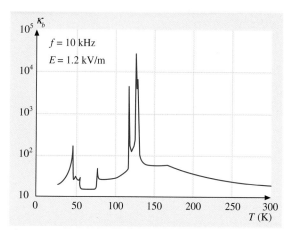

Fig. 4.5-87 (CH_3)$_3NCH_2COO \cdot CaCl_2 \cdot 2H_2O$. κ_b versus T

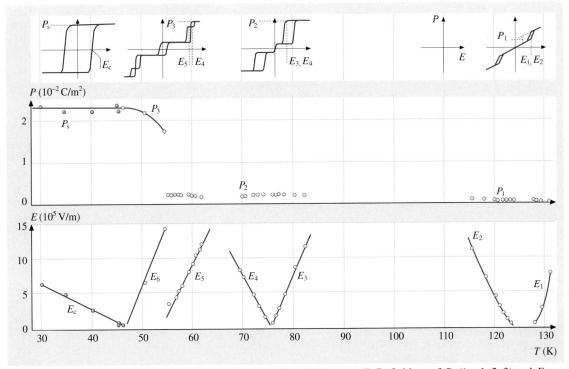

Fig. 4.5-88 $(CH_3)_3NCH_2COO \cdot CaCl_2 \cdot 2H_2O$. P_s, E_c, P_i, and E_i versus T. Definitions of P_i ($i = 1, 2, 3$) and E_i are shown in the *top figure*

crystal (κ_{22}^S) shows no anomaly at the transition temperature (Fig. 4.5-7), while the elastic compliances exhibit a pronounced anomaly (Fig. 4.5-93). These data indicate that the ferroelectric phase transition is of the indirect type III$_{ac}$ triggered by an elastic anomaly (see Sect. 4.5.3). The crystal is piezoelectric in the paraelec-

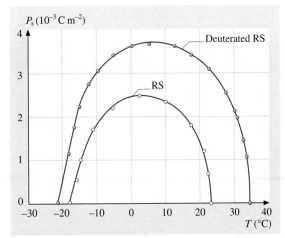

Fig. 4.5-89 $NaKC_4H_4O_6 \cdot 4H_2O$ (RS) and $NaKC_4H_2D_2O_6 \cdot 4D_2O$ (deuterated RS). κ_{11}^T versus T

Fig. 4.5-90 $NaKC_4H_4O_6 \cdot 4H_2O$ (RS) and $NaKC_4H_2D_2O_6 \cdot 4D_2O$ (deuterated RS). P_s versus T

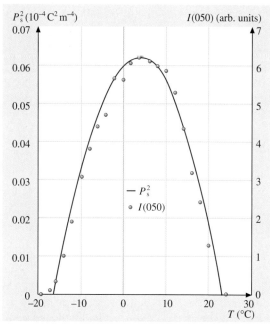

Fig. 4.5-91 NaKC$_4$H$_4$O$_6$ · 4H$_2$O. $I_{(050)}$ versus T. $I_{(050)}$ is the integrated intensity of the X-ray (050) reflection, which should disappear in the space group $P2_12_12$ below $-18\,°$C and above $23\,°$C. The *solid line* is P_s^2 normalized to fit to $I_{(050)}$

tric phase and the elastic anomaly causes a dielectric anomaly in the free crystal (κ_b in Fig. 4.5-92 and κ_{22}^T in Fig. 4.5-7), but no dielectric anomaly in the clamped crystal (κ_{22}^S in Fig. 4.5-7).

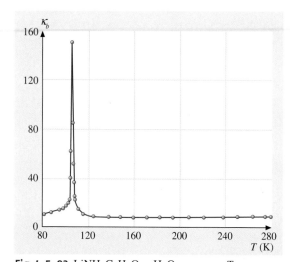

Fig. 4.5-92 LiNH$_4$C$_4$H$_4$O$_6$ · H$_2$O. κ_b versus T

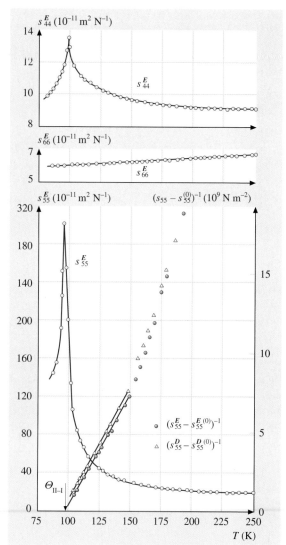

Fig. 4.5-93 LiNH$_4$C$_4$H$_4$O$_6$ · H$_2$O. s_{44}^E, s_{55}^E, and s_{66}^E versus T. s_{44}^E, s_{55}^E, and s_{66}^E are the shear components of the elastic compliance tensor at constant electric field. The two *straight lines* show $\left[s_{55}^E - s_{55}^E(0)\right]^{-1}$ and $\left[s_{55}^D - s_{55}^D(0)\right]^{-1}$, where s_{55}^D is s_{55} at constant electric displacement, $s_{55}^E(0) = 18.0\,\mathrm{m}^2\,\mathrm{N}^{-1}$, and $s_{55}^D(0) = 17.0\,\mathrm{m}^2\,\mathrm{N}^{-1}$

Liquid Crystal Family

DOBAMBC, p-decyloxybenzylidene p′-amino 2-methylbutyl cinnamate (LB Number 71A-1 (A)) (Liquid Crystal). This liquid crystal is ferroelectric in the smectic C* phase between about 76 and 92 °C on heating. On cooling, the C* phase is transformed into a smectic H*

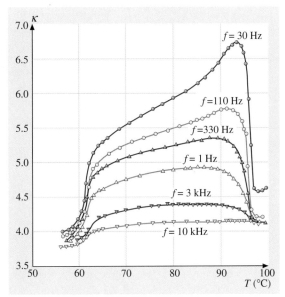

Fig. 4.5-94 DOBAMBC. κ versus T. Parameter: f. Sample cell thickness = 2.3 μm

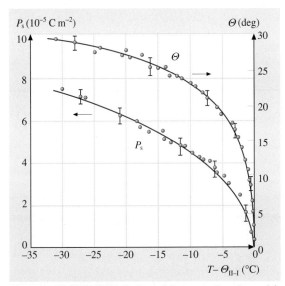

Fig. 4.5-95 DOBAMBC. P_s and θ versus $T - \Theta_{\mathrm{II-I}}$. θ is the apparent tilt angle, and $\Theta_{\mathrm{II-I}}$ is the ferroelectric transition temperature

Fig. 4.5-96a,b MHPOBC. $D-E$ hysteresis loops obtained from switching-current measurements. D is the electric displacement. Sample cell thickness = 3 μm. (a) Double hysteresis loop at 2 Hz. (b) Single hysteresis loop at 100 Hz

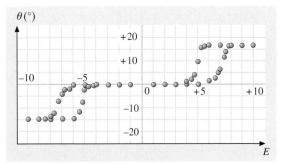

Fig. 4.5-97 MHPOBC. θ versus E. θ is the apparent tilt angle. Sample cell thickness = 3 μm

phase at about 63 °C, retaining the ferroelectric activity (Figs. 4.5-94 and 4.5-95).

MHPOBC, 4-(1-methylheptyl oxycarbonyl)phenyl 4′-octyloxybiphenyl-4-carboxylate (LB Number 71B-1 (A)) (Liquid Crystal). Seven phases are known for this liquid crystal. It exhibits an antiferroelectric hysteresis loop at low frequency, as shown in Fig. 4.5-96a, between

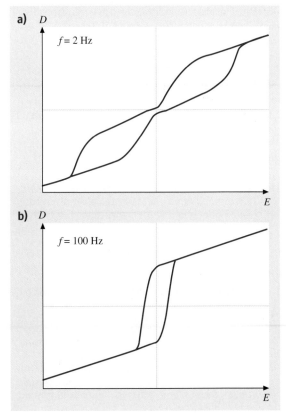

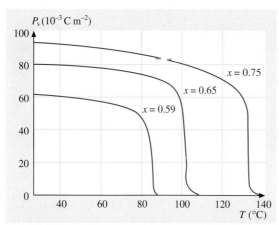

Fig. 4.5-98 $((CH_2CF_2)_x(CF_2CHF)_{1-x})_n$. P_s versus T. Parameter: x. Thickness of specimen film = $50\,\mu m$

65 and 119.7 °C. The antiferroelectric behavior can be best demonstrated by measuring the apparent tilt angle, as shown in Fig. 4.5-97. The hysteresis loop turns into a triple one between 119.7 and 120.6 °C, and into a normal ferroelectric hysteresis loop between 120.6 and 122.1 °C. The phase exhibiting the triple hysteresis loop is called "ferrielectric" by several authors.

Polymer Family
$((CH_2CF_2)_x(CF_2CHF)_{1-x})_n$, Vinylidene fluoride–trifluoroethylene Copolymer, $((VDF)_x(TrFE)_{1-x})_n$ (LB Number 72-2) (Polymer). Random polymers $(CH_2CF_2)_x$ $(CF_2CHF)_{1-x}$ exhibit a ferroelectric hysteresis loop for

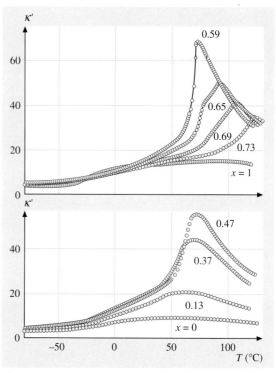

Fig. 4.5-99 $((CH_2CF_2)_x(CF_2CHF)_{1-x})_n$. κ' versus T. Parameter: x. $f = 1\,kHz$. Thickness of specimen film = $50\,\mu m$

$x > 0.5$ (Figs. 4.5-98 and 4.5-99). Ferroelectric activity is preserved up to the melting point for $x > 0.8$.

References

5.1 Y. Shiozaki, E. Nakamura, T. Mitsui (Eds.): *Ferroelectrics and Related Substances: Oxides: Perovskite-Type Oxides and LiNbO₃ Family*, Landolt–Börnstein, New Series III/36 (Springer, Berlin, Heidelberg 2001)

5.2 Y. Shiozaki, E. Nakamura, T. Mitsui (Eds.): *Ferroelectrics and Related Substances: Oxides: Oxides Other Than Perovskite-Type Oxides and LiNbO₃ Family*, Landolt–Börnstein, New Series III/36 (Springer, Berlin, Heidelberg 2002)

5.3 Y. Shiozaki, E. Nakamura, T. Mitsui (Eds.): *Ferroelectrics and Related Substances: Inorganic Crystals Other Than Oxides*, Landolt–Börnstein, New Series III/36 (Springer, Berlin, Heidelberg 2004)

5.4 Y. Shiozaki, E. Nakamura, T. Mitsui (Eds.): *Ferroelectrics and Related Substances: Organic Crystals, Liquid Crystals and Polymers*, Landolt–Börnstein, New Series III/36 (Springer, Berlin, Heidelberg 2005) in preparation

5.5 J. Valasek: Piezoelectric and Allied Phenomena in Rochelle Salt, Phys. Rev. **15**, 537 (1920)

5.6 J. Valasek: Piezo-Electric and Allied Phenomena in Rochelle Salt, Phys. Rev. **17**, 475 (1921)

5.7 T. Mitsui, E. Nakamura, I. Tatsuzaki: *An Introduction to the Physics of Ferroelectrics* (Gordon and Breach, New York 1976)

5.8 F. Jona, G. Shirane: *Ferroelectric Crystals* (Dover Publications, New York 1993)

5.9 B. A. Strukov, A. P. Levanyuk: *Ferroelectric Phenomena in Crystals: Physical Foundations* (Springer, Berlin, Heidelberg 1998)

5.10 M. E. Lines, A. M. Glass: *Principles and Applications of Ferroelectrics and Related Materials* (Clarendon, Oxford 1979)

5.11 Y. Xu: *Ferroelectric Materials and Their Applications* (North-Holland, Amsterdam 1991)

5.12 B. Jaffe, W. R. Cook Jr., H. Jaffe: *Piezoelectric Ceramics* (Academic Press, London 1971)

5.13 J. Zelenka: *Piezoelectric Resonators and Their Applications* (Elsevier, Amsterdam 1986)

5.14 T. Ikeda: *Fundamentals of Piezoelectricity* (Oxford Univ. Press, Oxford 1990)

5.15 R. A. Cowley: *Structural Phase Transitions* (Consultant Bureau, New York 1980)

5.16 K. A. Miller, H. Thomas (Eds.): *Structural Phase Transitions I*, Topics in Current Physics, Vol. 23 (Springer, Berlin, Heidelberg 1981)

5.17 M. Fujimoto: *The Physics of Structural Phase Transitions* (Springer, Berlin, Heidelberg 1997)

5.18 R. Blinc, A. P. Levanyuk (Eds.): *Incommensurate Phases in Dielectrics*, Modern Problems in Condensed Matter Science 14 (Elsevier, Amsterdam 1986)

5.19 F. J. Scott: Soft-mode spectroscopy: Experimental studies of structural phase transitions, Rev. Mod. Phys. **46**, 83 (1974)

5.20 G. Shirane: Neutron scattering studies of structural phase transitions, Rev. Mod. Phys. **46**, 437 (1974)

5.21 C. H. Wang: Raman and Brillouin scattering spectroscopy of phase transitions in solids. In: *Vibrational Spectroscopy of Phase Transitions*, ed. by Z. Iqbal, F. J. Owens (Academic Press, Orlando 1984) pp. 153–207

5.22 J. M. Herbert: *Ceramic Dielectrics and Capacitors*, Electrocomp. Sci. Mon., Vol. 6 (Gordon and Breach, New York 1985)

5.23 B. E. Vugmeister, M. D. Glinchuk: Dipole glass and ferroelectricity in random-site electric dipole systems, Mod. Phys. **62**, 993 (1990)

5.24 L. E. Cross: Relaxor ferroelectrics, Ferroelectrics **76**, 241 (1987)

5.25 S. Singh: Nonlinear optical materials. In: *Handbook of Lasers with Selected Data on Optical Technology*, ed. by R. J. Pressley (The Chemical Rubber Co., Cleveland 1971) pp. 489–525

5.26 V. G. Dmitriev, G. G. Gurzadyan, D. N. Nikogosyan: *Handbook of Nonlinear Optical Crystals*, Springer Series in Optical Science, Vol. 64, ed. by A. E. Siegman (Springer, Berlin, Heidelberg 1991)

5.27 P. Yeh: *Introduction to Photorefractive Nonlinear Optics* (Wiley, New York 1993)

5.28 G. Rosenman, A. Skliar, A. Arie: Ferroelectric domain engineering for quasi-phase-matched nonlinear optical devices, Ferroelectr. Rev. **1**, 263 (1999)

5.29 N. Setter, E. L. Colla: *Ferroelectric Ceramics* (Birkhäuser, Basel 1993)

5.30 B. A. Tuttle, S. B. Desu, R. Ramesh, T. Shiosaki: *Ferroelectric Thin Films IV* (Materials Research Society, Pittsburgh 1995)

5.31 C. P. Araujo, J. F. Scott, G. W. Taylor (Eds.): *Ferroelectric Thin Films: Synthesis and Basic Properties* (Gordon and Breach, New York 1996)

5.32 R. Ramesh (Ed.): *Thin Film Ferroelectric Materials and Devices*, Electronic Materials Science and Technology, Vol. 3 (Kluwer, Dordrecht 1997)

5.33 J. F. Scott: The physics of ferroelectric ceramic thin films for memory applications, Ferroelectr. Rev. **1**, 1 (1998)

5.34 D. Damjanovic: Ferroelectric, dielectric and piezoelectric properties of ferroelectric thin films and ceramics, Rep. Prog. Phys. **62**, 1267 (1998)

5.35 J. F. Scott: *Ferroelectric Memories*, Springer Series in Advanced Microelectronics, Vol. 3 (Springer, Berlin, Heidelberg 2000)

5.36 R. M. White: Surface elastic waves, Proc. IEEE **58**, 1238 (1970)

5.37 G. W. Farnell, E. L. Adler: Acoustic Wave Propagation in Thin Layers. In: *Physical Acoustics*, Vol. 9, ed. by W. P. Mason, R. N. Thurston (Academic Press, New York 1972) pp. 35–127

5.38 A. A. Oliner (Ed.): *Acoustic Surface Waves* (Springer, Berlin, Heidelberg 1978)

5.39 K.-Y. Hashimoto: *Surface Acoustic Wave Devices in Telecommunications, Modeling and Simulation* (Springer, Berlin, Heidelberg 2000)

5.40 J. M. Herbert: *Ferroelectric Transducers and Sensors*, Electrocomp. Sci. Mon., Vol. 3 (Gordon and Breach, New York 1982)

5.41 W. H. de Jeu: *Physical Properties of Liquid Crystalline Materials* (Gordon and Breach, New York 1980)

5.42 J. W. Goodby, R. Blinc, N. A. Clark, S. T. Lagerwall, M. A. Osipov, S. A. Pikin, T. Sakurai, K. Yoshino, B. Zeks: *Ferroelectric Liquid Crystals: Principles, Properties and Applications*, Ferroelectrics and Related Phenomena, Vol. 7 (Gordon and Breach, New York 1991)

5.43 G. W. Taylor (Ed.): *Ferroelectric Liquid Crystals: Principles, Preparations and Applications* (Gordon and Breach, New York 1991)

5.44 L. M. Blinov, V. G. Chigrinov: *Electric Effects in Liquid Crystal Materials* (Springer, Berlin, Heidelberg 1994)

5.45 A. Fukada, Y. Takanishi, T. Isozaki, K. Ishikawa, H. Takezoe: Antiferroelectric Chiral Smectic Liquid Crystals, J. Mater. Chem. **4**, 997 (1994)

5.46 P. J. Collings, J. S. Patel (Eds.): *Handbook of Liquid Crystal Research* (Oxford Univ. Press, Oxford 1997)

5.47 S. T. Lagerwall: *Ferroelectric and Antiferroelectric Liquid Crystals* (Wiley VCH, Weinheim 1999)

5.48 T. Furukawa: Ferroelectric Properties of Vinylidene Fluoride Copolymers, Phase Transitions **18**, 143 (1989)

5.49 D. K. Das-Gupta (Ed.): *Ferroelectric Polymer and Ceramic-Polymer Composites* (Trans Tech Publications, Aedermannsdorf 1994)

5.50 H. S. Nalva (Ed.): *Ferroelectric Polymers: Chemistry, Physics and Applications* (Marcel Dekker, New York 1995)

5.51 H. Kodama, Y. Takahashi, T. Furukawa: Effects of Annealing on the Structure and Switching Characteristics of VDF/TrFE Copolymers, Ferroelectrics **205**, 433 (1997)

5.52 N. A. Clark, S. T. Lagerwall: Submicrosecond Bistable Electro-Optic Switching in Liquid Crystals, Appl. Phys. Lett. **36**, 899 (1980)

5.53 A. D. L. Chandani, E. Gorecka, Y. Ouchi, H. Takezoe, A. Fukuda: Antiferroelectric Chiral Smectic Phases Responsible for the Tristable Switching in MHPOBC, Jpn. J. Appl. Phys. **28**, L1265 (1989)

Part 5 Special Structures

1. **Liquid Crystals**
   Sergei Pestov, Moscow, Russia
   Volkmar Vill, Hamburg, Germany

2. **The Physics of Solid Surfaces**
   Gianfranco Chiarotti, Roma, Italy

3. **Mesoscopic and Nanostructured Materials**
   Fabrice Charra, Gif-sur-Yvette, France
   Susana Gota-Goldmann, Fontenay aux Roses, France

5.1. Liquid Crystals

Liquid crystals (LCs) are widely used in information-processing devices, for optical visualization of physical influences (heat, IR, high-frequency radiation, pressure, etc.), for nondestructive testing, and for thermography.

5.1.1	**Liquid Crystalline State**	941
	5.1.1.1 Chemical Requirements	943
	5.1.1.2 Physical Properties of Liquid Crystals	943
	5.1.1.3 Applications of Liquid Crystals	944
	5.1.1.4 List of Abbreviations	944
	5.1.1.5 Conversion Factors	945
5.1.2	**Physical Properties of the Most Common Liquid Crystalline Substances**	946
	5.1-2 Acids	946
	5.1-3 Two-ring Systems without Bridges	947
	5.1-4 Two-ring Systems with Bridges	955
	5.1-5 Three and Four-ring Systems	964
	5.1-6 Ferroelectric Liquid Crystals	967
	5.1-7 Cholesteryl (cholest-5-ene) Substituted Mesogens	968
	5.1-8 Discotic Liquid Crystals	972
	5.1-9 Liquid Crystal Salts	973
5.1.3	**Physical Properties of Some Liquid Crystalline Mixtures**	975
	5.1-10 Nematic Mixtures	975
	5.1-11 Ferroelectric Mixtures	975
References		977

5.1.1 Liquid Crystalline State

Liquid crystals represent an intermediate state of order (mesophase) between crystals and liquids. Crystals have a three-dimensional long-range order of both position and orientation (Fig. 5.1-1a). Liquids, in contrast, do not show any long-range order (Fig. 5.1-1b). In plastic crystals (disordered crystals, Fig. 5.1-1c), positional order is maintained, but orientational order is lost. In mesophases, imperfect long-range order is observed, and thus they are situated between crystals and liquids. The reasons for the formation of a mesophase can be the molecular shape or a microphase separation of amphiphilic compounds.

More than 100 000 individual liquid crystals have been prepared until now [1.1–4]. About 2000 of them have been tested for physical properties and technical applications [1.5–14]. These materials can be classified by chemical structures and physical characteristics (see Table 5.1-1).

Generally, molecules of liquid crystalline substances have the following shapes:

- rod-like molecules, which form calamitic liquid crystals (nematic and smectic phases);
- disk-like molecules, which form discotic liquid crystals (discoid nematic and discotic phases);
- amphiphilic compounds, which form layered columnar or cubic phases in the pure state and in solution.

The simplest and most widespread liquid crystalline phase is the nematic phase. The molecules are statistically distributed within the medium, but the long axes are orientated in one direction, the director (Fig. 5.1-2a). A special class of nematic phases is the cholesteric phase

Table 5.1-1 Classifications of liquid crystals

Shape	Rod-like molecules	Disk-like molecules
Phase structure	Calamitic liquid crystals	Discotic liquid crystals
Mesophase units	Thermotropic liquid crystals	Lyotropic liquid crystals
Mesophase origin	Amphiphilic liquid crystals	Monophilic liquid crystals

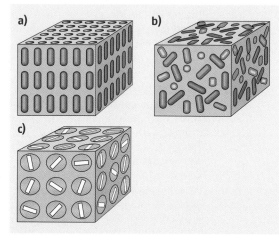

Fig. 5.1-1a–c Types of states: (**a**) crystal, (**b**) isotropic liquid, (**c**) plastic crystal

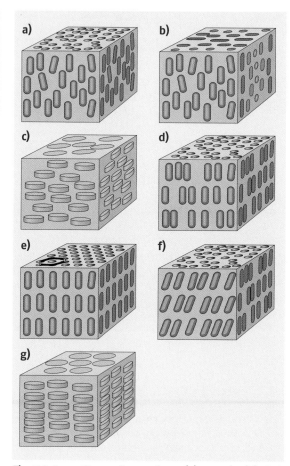

Fig. 5.1-2a–g Types of mesophase: (**a**) nematic, (**b**) cholesteric, (**c**) discoid nematic, (**d**) smectic A, (**e**) smectic B, (**f**) smectic C, (**g**) discotic

(Fig. 5.1-2b). Here the orientation of the director does not apply to the whole medium but rather to a virtual layer. Perpendicular to this layer, the director follows a helix with a certain pitch. In the case of the blue phases, such a helical structure is formed not in one but in all three dimensions. Thus, highly complex arrangements, with chiral cubic symmetry in most cases, are generated. Not only rod-like but also disk-like molecules can form nematic phases. The discoid nematic phase is shown in Fig. 5.1-2c.

Rod-like molecules arranged in layers form smectic phases. They are subdivided into a considerable number of different species. These classifications result from various arrangements of the molecules within the layers and different restrictions on their movement. The smectic A phase, the simplest smectic phase, can be regarded as a two-dimensional liquid. The molecules are arranged normal to the layers (Fig. 5.1-2d). The smectic B phase can be interpreted as the closest packing of rod-like molecules, so that within a layer each molecule has a hexagonal environment (Fig. 5.1-2e). The smectic A phase and the smectic C phase are similar, except that in the latter the molecules are tilted within the layers by a tilt angle (Fig. 5.1-2f). A particular case of smectic C is the chiral smectic C* phase, where the tilt angle varies from layer to layer, forming a helical structure. For discussion of other smectic phases, as well as their further subclassification, the reader should consult the references [1.15, 16].

In discotic phases, the disk-like molecules are arranged in columns. In this group, again various phases are possible, depending of the orientation of the molecules within the columns and the order between the columns. The simplest phase is the hexagonal columnar discotic phase. It can be regarded as a one-dimensional liquid. The columns have a hexagonal order (Fig. 5.1-2g).

Lyotropic liquid crystals are formed by aggregation of micelles. They are multicomponent systems. Normally they consist of an amphiphilic substance and a solvent. In contrast, thermotropic liquid crystals are individual compounds.

Enantiotropic LC phases are formed during both the heating and the cooling process. Monotropic LC phases exist only in the supercooled state below the melting point. Thus, these phases are observed during cooling only.

5.1.1.1 Chemical Requirements

A liquid crystalline compound can be divided into the mesogenic group and the side groups. The mesogenic group is subdivided into fragments of rings and bridges. The side groups are subdivided into links and terminal groups.

Many other types of liquid crystalline compounds exist besides those with rod-like molecules e.g. compounds with disk-shaped, banana-shaped, and bowl-shaped molecules. However, over 80% of all liquid crystals have a rod-like form (e.g. the molecule shown in Fig. 5.1-3).

5.1.1.2 Physical Properties of Liquid Crystals

The order parameter $S = 0.5(3\langle\cos^2\Theta\rangle - 1)$ characterizes the long-range order of molecules in a mesophase, where Θ is the momentary angle between the long axis of the molecule and the director. In an ideal crystal the order parameter S equals 1, and it equals 0 in an isotropic liquid. In a nematic phase the order parameter lies in the range 0.5–0.7.

One of the most useful properties for the application of liquid crystals is the anisotropy of their refractive index $\Delta n = n_e - n_o$, where n_e is the extraordinary and n_o

is the ordinary refractive index. For nematics, n_e corresponds to $n_\parallel$ and $n_o = n_\perp$. For $n_\perp$ the vibration vector of plane-polarized light is perpendicular to the optical axis, i.e. the director, while for $n_\parallel$ the vibration vector of plane-polarized light is parallel to the director. For the majority of LCs the value of Δn is positive, but cholesteryl-substituted compounds are optically negative ($\Delta n < 0$). On increasing the wavelength, Δn usually decreases. For homologues Δn decreases as the length of the alkyl chains increases.

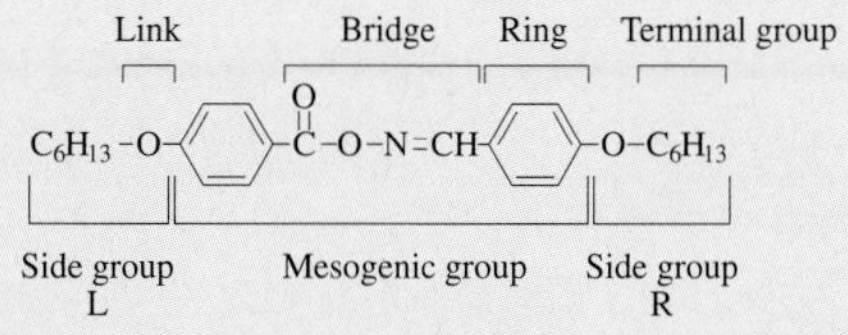

The orientation of molecules in an electric field is determined by the sign of the dielectric anisotropy ($\Delta\varepsilon = \varepsilon_\parallel - \varepsilon_\perp$); $\varepsilon_\parallel$ and $\varepsilon_\perp$ are the dielectric constants measured parallel and perpendicular to the director. Some nematic LCs can change their sign of $\Delta\varepsilon$ depending upon the frequency of the applied field.

The majority of mesogens are diamagnetic. The diamagnetic anisotropy ($\Delta\chi$) characterizes the behavior of a LC under the influence of a magnetic field; $\Delta\chi = \chi_\parallel - \chi_\perp$, where $\chi_\parallel$ and $\chi_\perp$ are the diamagnetic susceptibilities parallel and perpendicular to the director.

The basic methods for the determination of phase transition temperatures (T^{tr}) are DSC (differential scanning calorimetry), DTA (differential thermal analysis), and polarization microscopy. Every method has its advantages and restrictions. DSC allows one to determine the enthalpies of phase transitions (ΔH^{tr}). Microscopy allows one both to determine the phase transition temperatures and to identify the type of mesophase. DTA gives reliable results for melting temperatures of LCs that show solid-state polymorphism, and of LC mixtures. The differences in T^{tr} that can be found in publications by different authors arise from different measurement techniques and the presence of impurities. We have selected the data with the higher T^{tr} values in such cases.

The temperature dependence of the density (ϱ) is practically linear, with the exception of jumps near phase transitions. The volume changes are 3–9%, 0.1–0.4%, and 0.01–0.2% for the crystal–mesophase, nematic–isotropic, and smectic A–nematic transitions, respectively.

The kinematic and dynamic viscosities v and η, respectively, can be determined by a capillary viscometer of the Ubbelohde or Ostwald type ($v = \eta/\varrho$).

The surface tension (γ_{LV}) influences the angle between the surface and the nematic-phase director. It plays an important role in the selection of coatings for the creation of homeotropic alignment (where the molecules are perpendicular to the cell surfaces) or planar alignment (where the molecules are parallel to the surfaces) of LCs. The Friedel–Creagh–Kmetz rule states that if

Fig. 5.1-3 Structure of a calamitic mesogenic compound

γ_L (energy of LC–surface interaction) > γ_S (solid surface energy), then a homeotropic alignment is induced; otherwise, a parallel alignment is induced.

5.1.1.3 Applications of Liquid Crystals

The main requirements on LC materials (LCMs) for electro-optics are a high clearing temperature ($T^{N\text{-}Is}$) and low melting temperature, i. e. a wide temperature range of definite liquid crystalline phases, and a low viscosity for reducing the switching time of electro-optical effects. Wide usage of LCMs in displays became possible after the discovery of mesogenic cyanobiphenyls. At present many of homologous LCs are synthesized for display applications. Phenylcyclohexane- and bicyclohexane-substituted LCs are used as components in LCMs; they have low viscosity and, accordingly, fast switching. Fluorinated three-ring LCs containing two cyclohexane rings and a phenyl ring are used in LCMs with high dielectric anisotropy, which are used at low voltages. For the application of a LCM as a material in LC displays, many properties have to correspond to rigid specifications. For this reason LCMs consist typically of 7 to 15 components. Sometimes LCMs contain nonmesogenic additives, e.g. to reduce the viscosity of the mixture.

In the twist structure, the molecules are parallel to the cell surfaces and the angle between the boundary directors is 90°. In the S effect (Frederics effect), a planar structure is transformed to a homeotropic one, and in the B effect, a homeotropic structure is transformed to a planar one. In the twist effect (or twisted nematic, TN, effect) a twist structure turns into a homeotropic one. The disadvantage of the TN effect is the necessity to use polarizers. The "dynamic scattering mode" of use of LCs depends on the influence of an electric current on the orientation of the molecules. At a high enough voltage, transparent nematic cells becomes turbid. A disadvantage of this effect is that the lower the voltage, the longer the time of switching. One of the modern uses of this effect is use of the dynamic scattering mode for data storage. After the applied voltage is switched off, the mode of the planar cell does not return to its initial state. The cell can be stored in the turbid state for a long time (from some minutes to some months).

The host–guest effect results from a reorientation of dye dopants (1–2% in the LC matrix) in an electric field. In this case the wavelength of maximum absorption of light is shifted and the color of the LC cell changes.

The sign of the dielectric anisotropy of a LCM determines the type of electro-optical effect. LCMs with $\Delta\varepsilon > 0$ are used for the TN (twisted nematic), STN (supertwisted nematic), and TFT (thin-film transistor) effects. LCMs with low negative values of $\Delta\varepsilon$ were formerly used for the dynamic scattering mode. Now LCMs with negative anisotropy $\Delta\varepsilon$ are utilized for MVA-TFT (multidomain vertical-alignment thin-film transistor) displays. The higher the value of $\Delta\varepsilon > 0$, the smaller the working voltage.

Electric current leads to degradation of a LCM and reduces the lifetime of the display. Impurities influence the stability of the material and accelerate electrodegradation. Therefore a multistage purification, consisting for example of recrystallization, and column chromatography, to remove conducting impurities (intermediate products, water, and CO_2), is necessary. Usually the specific conductivity of a LCM is lower than 10^{-11}–10^{-12} C m/cm and corresponds to the intrinsic conductivity.

The elastic constants (K_i) determine the switching time of the electro-optical effects. The elastic constant K_1 corresponds to the S effect, K_3 to the B effect, and K_2 to the TN effect; here K_1 corresponds to splaying K_2 to twisting, and K_3 to bending [1.11, 12].

Cholesteryl compounds were the first materials that found application in thermography.

A huge number of other applications exist, where the chemical and physical requirements are totally different. These applications include reflectors, temperature measurement with thermochromic materials, nonlinear optics, polymer materials, SAMs (self-assembled monolayers) and LB (Langmuir–Blodgett) films, the use of liquid crystals in template synthesis of porous materials, drug delivery, and many more.

5.1.1.4 List of Abbreviations

Cr, Cr'	crystalline phases
S	smectic
A, B, C, E, F, G, H	specific smectic mesophases
C*	chiral smectic C (ferroelectric)
D	discotic

Dh, Dho, Dhd	hexagonal columnar discotic (ordered or disordered)
Dr, Dt	rectangular, tilted columnar discotic
N	nematic
N*	cholesteric
Is	isotropic

Examples	
Cr 95.0 C 99.0 N 107.5 Is	Crystals melt at 95 °C, smectic C transforms to nematic at 99 °C and to isotropic liquid at 107.5 °C.
Cr' 5.0 Cr 67.3 (N 30.3) Is	Crystals melt at 67.3 °C. Compound has a monotropic nematic phase on cooling below 30.3 °C and polymorphism (solid–solid transition) at 5 °C.
(extra)	Data were extrapolated from 10–20 wt% solution in nematic mixture.
dec	Substance decomposes on heating.

Common names are historical names, and trade names of BDH, Licristal® Merck, and Hoffmann-La Roche [1.8, 9, 17].

CAS-RN	Registration number of the Chemical Abstract Service (CAS)
T^{tr}	temperatures of phase transitions
T^{N-Is}	temperature of nematic–isotropic transition (in Kelvin)

Relative temperatures (e.g. $T = 0.977 T^{N-Is}$) are assumed to be in K only. Inverse temperatures are always given in 1/Kelvin (e.g. $1000/T = 2.4$).

Some data are given for supercooled mesophases ($T < T_{melting}$).

ΔH^{tr}	enthalpies of phase transitions (kJ/mol)
C_p	heat capacity at constant pressure (J/(mol K))
v	sound velocity (m/s)
p	helix pitch (for N* phase) (μm)
Δn	anisotropy of the refractive index ($n_e - n_o$)
$\Delta \varepsilon$	dielectric anisotropy
P_s	spontaneous polarization (for ferroelectric LC) (nC/cm²)
D	diffusion coefficient (m²/s)
C_6H_{12}	cyclohexane

5.1.1.5 Conversion Factors

Molar mass	$1 \text{ g/mol} = 1 \times 10^{-3} \text{ kg/mol}$
Dipole moment (μ)	$1 \text{ D (Debye)} = 3.33564 \times 10^{-30}$ Cm (Coulomb meter)
Density ϱ	$1 \text{ g/cm}^3 = 1 \times 10^3 \text{ kg/m}^3$
Temperatures of phase transitions T^{tr}	$T(°C) = T(K) - 273.15$
Dynamic viscosity η	$1 \text{ mPa s} = 1 \times 10^{-3}$ Pa s = 1 cP (centi-Poise)
Kinematic viscosity $v = \eta/\varrho$	$1 \text{ mm}^2/\text{s} = 1 \times 10^{-6} \text{ m}^2/\text{s} = 1$ cSt (centi-Stokes)
Diamagnetic anisotropy $\Delta \chi$	$1 \text{ m}^3/\text{kg} = 1 \times 10^3 \text{ cm}^3/\text{g}$ (CGS unit)
Thermal conductivity	$1 \text{ W/(m K)} = 1 \times 10^{-2}$ W/(cm K)

5.1.2 Physical Properties of the Most Common Liquid Crystalline Substances

The compounds are arranged in the tables on the basis of the number and priority of fragments. The order principles for mesogenic groups are the number of rings and bridges, and the priority of rings, bridges, and side groups.

Priority of rings:

benzene > cyclohexane > heterocycles > halogen-substituted benzenes.

Priority of bridges:

C_nH_m > CH=N > N=N > N=N(O) > COO.

Homologues are arranged in order of increasing number of carbon atoms in the alkyl chains.

We have attempted to include all of the most common liquid crystals, from the traditional "model substances" (e.g. 5CB, MBBA, and PAA) to substances used in modern applications.

Table 5.1-2 Acids

Number/common name	1	2	3
Substance	H_7C_3–O–C$_6$H$_4$–COOH	H_9C_4–O–C$_6$H$_4$–COOH	$H_{11}C_5$–O–C$_6$H$_4$–COOH
Formula	$C_{10}H_{12}O_3$	$C_{11}H_{14}O_3$	$C_{12}H_{16}O_3$
Molar mass (g/mol)	180.205	194.232	208.26
CAS-RN	5438-19-7	1498-96-0	15872-41-0
Temperatures of phase transitions T^{tr} (°C)	Cr 146.0 N 156.0 Is	Cr 147.5 N 161.0 Is	Cr 124.4 N 151.4 Is
Enthalpies of phase transitions ΔH^{tr} (kJ/mol)	16.7 (Cr–N), 2.5 (N–Is)	19.7 (Cr–N), 2.4 (N–Is)	18.0 (Cr–N), 1.8 (N–Is)
Crystallographic space group	$P2_1/c$	$P\bar{1}$	$P2_1$
Order parameter S	0.533 (149 °C)	0.538 (154 °C)	0.587 (137 °C)
Density ϱ (g/cm³)	1.0095 (N, 150 °C), 1.0008 (Is, 159 °C)	1.0079 (N, 150 °C), 0.9883 (Is, 165 °C)	0.9945 (N, 145 °C), 0.9721 (Is, 155 °C)
Refractive index n	n_e 1.632, n_o 1.457 (N, 546 nm, 140.5 °C)	n_e 1.615, n_o 1.449 (N, 589 nm, 149 °C)	n_e 1.602, n_o 1.456 (N, 546 nm, 140.5 °C)
Dielectric anisotropy $\Delta\varepsilon$	–	0.079 ($T = 0.977\, T^{\text{N-Is}}$)	0.042 ($T = 0.977\, T^{\text{N-Is}}$)
Dynamic viscosity η (mPa s)	1.81 (149 °C), 2.63 (159 °C)	1.88 (151 °C), 2.62 (165 °C)	2.37 (145 °C), 3.30 (155 °C)
Diffusion coefficient D (m²/s)	–	$D_\perp = 4\times 10^{-10}$, $D_\parallel = 14\times 10^{-10}$ ($1000/T = 2.38$)	$D_\perp = 4\times 10^{-10}$, $D_\parallel = 12\times 10^{-10}$ ($1000/T = 2.45$)

Number/common name	4	5 (HOBA)	6 (OOBA)
Substance	$H_{13}C_6$–O–C$_6$H$_4$–COOH	$H_{15}C_7$–O–C$_6$H$_4$–COOH	$H_{17}C_8$–O–C$_6$H$_4$–COOH
Formula	$C_{13}H_{18}O_3$	$C_{14}H_{20}O_3$	$C_{15}H_{22}O_3$
Molar mass (g/mol)	222.287	236.314	250.341
CAS-RN	1142-39-8	15872-42-1	2493-84-7
Temperatures of phase transitions T^{tr} (°C)	Cr' 75 Cr 105.4 N 153.2 Is	Cr 94.0 C 102.0 N 147.5 Is	Cr 100.5 C 107.5 N 147.0 Is
Enthalpies of phase transitions ΔH^{tr} (kJ/mol)	13.8 (Cr–N), 3.3 (N–Is), 6.7 (Cr'–Cr)	19.2 (Cr–C), 2.5 (N–Is)	12.7 (Cr–C), 1.1 (C–N), 2.1 (N–Is)

Table 5.1-2 Acids, cont.

Number/common name	4	5 (HOBA)	6 (OOBA)
Crystallographic space group	$P2_1/c$ (Cr), $P1$ (Cr')	$P\bar{1}$	$P\bar{1}$
Order parameter S	0.611 (129 °C)	0.626 (123 °C)	0.615 (128 °C)
Density ϱ (g/cm³)	0.983 (N, 146 °C), 0.969 (Is, 156 °C)	0.970 (N, 140 °C), 0.959 (Is, 150 °C)	0.961 (N, 140 °C), 0.945 (Is, 150 °C)
Refractive index n	n_e 1.595, n_o 1.453 (546 nm, N, 142.5 °C)	n_e 1.580, n_o 1.452 (546 nm, N, 137.5 °C)	n_e 1.578, n_o 1.450 (546 nm, N, 138.5 °C)
Dielectric anisotropy $\Delta\varepsilon$	0.052 ($T = 0.977\, T^{\text{N–Is}}$)	0.038 ($T = 0.977\, T^{\text{N–Is}}$)	0.037 ($T = 0.977\, T^{\text{N–Is}}$)
Dynamic viscosity η (mPa s)	2.61 (146 °C), 3.46 (157 °C)	3.02 (137 °C), 3.95 (150 °C)	3.19 (140.5 °C), 3.97 (151 °C)
Thermal conductivity (W/(m K))	0.137 (N, 115 °C), 0.159 (Is, 157 °C)	0.146 (N, 130 °C), 0.162 (Is, 159 °C)	–
Sound velocity v (m/s)	972 (3.36 GHz, 141 °C)	1021 (3.40 GHz, 135 °C)	1050 (3.63 GHz, 137 °C)
Diffusion coefficient D (m²/s)	$D_\perp = 2\times 10^{-10}$, $D_\parallel = 8\times 10^{-10}$ (1000/T = 2.45)	–	–

Number/common name	7	8	9
Substance	$H_{11}C_5$—⬡—COOH	$H_{13}C_6$—⬡—COOH	$H_{15}C_7$—⬡—COOH
Formula	$C_{12}H_{22}O_2$	$C_{13}H_{24}O_2$	$C_{14}H_{26}O_2$
Molar mass (g/mol)	198.308	212.335	226.362
CAS-RN	32829-29-1	38289-30-4	32829-31-5
Temperatures of phase transitions T^{tr} (°C)	Cr 53.0 (B 45.0) N 105.0 Is	Cr 31.7 B 47.2 N 96.4 Is	Cr' 31.0 Cr 54.0 B 77.0 N 104.0 Is
Enthalpies of phase transitions ΔH^{tr} (kJ/mol)	20.9 (Cr–N), 1.1 (N–Is)	15.7 (Cr–B), 1.0 (B–N), 0.5 (N–Is)	13.4 (Cr'–Cr), 5.2 (Cr–B), 2.3 (B–N), 1.1 (N–Is)
$\Delta\varepsilon$ (N, 90 °C, 500 kHz)	0.05	–	0.05

Table 5.1-3 Two-ring systems without bridges

Number/common name	10 (2CB, K6, RO-CM-5106)	11 (3CB, K9, RO-CM-5109)	12 (4CB, K12, RO-CM-5112)
Substance	H_5C_2—⬡—⬡—CN	H_7C_3—⬡—⬡—CN	H_9C_4—⬡—⬡—CN
Formula	$C_{15}H_{13}N$	$C_{16}H_{15}N$	$C_{17}H_{17}N$
Molar mass (g/mol)	207.278	221.305	235.332
CAS-RN	58743-75-2	58743-76-3	52709-83-8
Temperatures of phase transitions T^{tr} (°C)	Cr 75.0 (N 22.0) Is	Cr 67.3 (N 30.3) Is	Cr 48.0 (N 16.5) Is
Enthalpies of phase transitions ΔH^{tr} (kJ/mol)	17.2 (Cr–Is)	26.8 (Cr–Is), 0.3 (N–Is)	23.0 (Cr–Is)
Crystallographic space group	–	$P2_1/c$	$P2_1/c$
Anisotropy of refractive index Δn	0.206	0.211	0.202
Dielectric constant ε	$\Delta\varepsilon$ +18.9	$\varepsilon_\parallel$ 25.8, $\varepsilon_\perp$ 6.9 (20 °C)	$\Delta\varepsilon$ +17
Kinematic viscosity ν (mm²/s)	5 (70 °C)	23 (extra, 20 °C), 7 (70 °C)	7.5 (70 °C)
Dipole moment μ (D), 25 °C	4.81 (CCl_4), 4.96 (C_6H_{14}), 4.90 (C_6H_{12}), 5.01 (C_6H_6)	4.72 (CCl_4), 5.02 (C_6H_{14}), 4.93 (C_6H_{12}), 4.93 (C_6H_6)	4.73 (CCl_4), 5.04 (C_6H_{14}), 4.94 (C_6H_{12}), 4.99 (C_6H_6)

Table 5.1-3 Two-ring systems without bridges, cont.

Number/common name	13 (5CB, K15, RO-CM-5115)	14 (6CB, K18, RO-CM-5118)	15 (7CB, K21, RO-CM-5121)
Substance	$H_{11}C_5$–⟨⟩–⟨⟩–CN	$H_{13}C_6$–⟨⟩–⟨⟩–CN	$H_{15}C_7$–⟨⟩–⟨⟩–CN
Formula	$C_{18}H_{19}N$	$C_{19}H_{21}N$	$C_{20}H_{23}N$
Molar mass (g/mol)	249.359	263.386	277.413
CAS-RN	40817-08-1	41122-70-7	41122-71-8
Temperatures of phase transitions T^{tr} (°C)	Cr 24.0 N 35.3 Is	Cr 14.3 N 30.1 Is	Cr′ 15.0 Cr 30.0 N 42.8 Is
Enthalpies of phase transitions ΔH^{tr} (kJ/mol)	17.2 (Cr–N), 0.4 (N–Is)	24.3 (Cr–N), 0.4 (N–Is)	25.9 (Cr–N), 0.9 (N–Is)
Crystallographic space group	$P2_1/a$	$P1$	$P1$
Order parameter S	0.64 (24 °C)	–	0.65 (29 °C)
Density ϱ (g/cm³)	1.022 (N, 24 °C), 1.016 (N, 30 °C)	1.017 (N, 20 °C),	1.001 (N, 29 °C), 0.991 (N, 39 °C)
Refractive index n (589 nm)	n_e 1.716, n_o 1.533 (N, 24 °C, 589 nm)	n_e 1.687, n_o 1.531 (N, 20 °C, 633 nm)	n_e 1.679, n_o 1.522 (N, 32 °C, 589 nm)
Dielectric constant ε	$\varepsilon_\parallel$ 20, $\varepsilon_\perp$ 7 (25 °C, 1.5 kHz)	$\varepsilon_\parallel$ 14.45, $\varepsilon_\perp$ 7.2 (28 °C, 1 kHz)	$\varepsilon_\parallel$ 13.55, $\varepsilon_\perp$ 6.8 (41 °C, 1 kHz)
Viscosity	η 28 mPa s (25 °C)	ν 46 mm²/s (20 °C)	ν 26 mm²/s (20 °C)
Surface tension (mN/m)	33 (N, 20 °C), 28 (N, 30 °C), 28.3 (Is, 40 °C)	27.4 (N, 22 °C), 28.0 (Is, 32 °C)	26.3 (N, 37 °C), 27.5 (Is, 47 °C)
Thermal conductivity (W/(m K)) (N, 25 °C)	$k_\perp$ 0.124, $k_\parallel$ 0.242	$k_\perp$ 0.126, $k_\parallel$ 0.223	$k_\perp$ 0.127, $k_\parallel$ 0.269
Magnetic susceptibility $\Delta\chi$ (m³/kg) (×10⁻¹²)	113.5 (25.6 °C)	101.5 (20 °C)	102.0 (33 °C)
Sound velocity v (m/s)	$v_\parallel$ 1740, $v_\perp$ 1680 (14.75 GHz, 30 °C)	$v_\parallel$ 1776, $v_\perp$ 1712 (14.75 GHz, 25 °C)	$v_\parallel$ 1675, $v_\perp$ 1640 (14.75 GHz, 38 °C)
Diffusion coefficient D (m²/s)	$D_\perp$ 6.8×10⁻⁸, $D_\parallel$ 12.7×10⁻⁸ (25 °C)	$D_\perp$ 6.4×10⁻⁸, $D_\parallel$ 10.9×10⁻⁸ (25 °C)	$D_\perp$ 6.8×10⁻⁸, $D_\parallel$ 14.4×10⁻⁸ (25 °C)
Dipole moment μ (D), 25 °C	4.85 (CCl₄), 5.08 (C₆H₁₄), 4.95 (C₆H₁₂), 4.93 (C₆H₆)	4.83 (CCl₄), 5.09 (C₆H₁₄), 5.01 (C₆H₁₂), 4.98 (C₆H₆)	4.74 (CCl₄), 5.11 (C₆H₁₄), 4.94 (C₆H₁₂), 4.97 (C₆H₆)

Number/common name	16 (8CB, K24)	17 (9CB, K27)	18 (10CB, K30)
Substance	$H_{17}C_8$–⟨⟩–⟨⟩–CN	$H_{19}C_9$–⟨⟩–⟨⟩–CN	$H_{21}C_{10}$–⟨⟩–⟨⟩–CN
Formula	$C_{21}H_{25}N$	$C_{22}H_{27}N$	$C_{23}H_{29}N$
Molar mass (g/mol)	291.44	305.467	319.494
CAS-RN	52709-84-9	52709-85-0	59454-35-2
Temperatures of phase transitions T^{tr} (°C)	Cr 21.5 A 33.5 N 40.5 Is	Cr 42.0 A 48.0 N 49.5 Is	Cr 44.0 A 54.5 Is
Enthalpies of phase transitions ΔH^{tr} (kJ/mol)	28.3 (Cr–A), 0.2 (A–N), 0.7 (N–Is)	33.5 (Cr–A), 0.6 (A–N), 1.7 (N–Is)	33.1 (Cr–A), 2.8 (N–Is)
Crystallographic space group	$P2_1/n$	$P\bar{1}$	$P2_1/n$
Density ϱ (g/cm³)	0.991 (A, 32.5 °C), 0.981 (N, 40 °C), 0.978 (Is, 41 °C)	0.973 (A, 47.7 °C), 0.968 (N, 49.5 °C)	–
Refractive index n (589 nm)	n_e 1.657, n_o 1.524 (N, 37.4 °C)	n_e 1.638, n_o 1.519 (N, 49 °C)	–
Dielectric constant ε (1 kHz)	$\varepsilon_\parallel$ 12.8, $\varepsilon_\perp$ 5.3 (N, 34 °C), $\varepsilon_\parallel$ 12.4, $\varepsilon_\perp$ 4.8 (A, 28 °C)	$\varepsilon_\parallel$ 12.5, $\varepsilon_\perp$ 5.1 (44 °C)	ε 8.8 (Is, 64.5 °C)
Dynamic viscosity η (mPa s)	35 (N, 33.5 °C)	–	–

Table 5.1-3 Two-ring systems without bridges, cont.

Number/common name	16 (8CB, K24)	17 (9CB, K27)	18 (10CB, K30)
Surface tension (mN/m)	25.5 (N, 35 °C), 26.2 (Is, 41 °C)	–	–
Thermal conductivity (W/(m K))	$k_\perp$ 0.130, $k_\parallel$ 0.276 (A, 30 °C), $k_\perp$ 0.132, $k_\parallel$ 0.254 (N, 35 °C)	$k_\perp$ 0.130, $k_\parallel$ 0.310 (40 °C)	–
Magnetic susceptibility $\Delta\chi$ (m^3/kg) (×10^{-12})	88 (35 °C)	88.5 (48.9 °C)	–
Sound velocity v (m/s)	$v_\parallel$ 1596, $v_\perp$ 1545 (14.75 GHz, 39 °C)	$v_\parallel$ 1740, $v_\perp$ 1680 (14.75 GHz, 44.5 °C)	–
Diffusion coefficient D (m^2/s)	$D_\perp$ 6×10^{-8}, $D_\parallel$ 12×10^{-8} (A, 30 °C), $D_\perp$ 5.8×10^{-8}, $D_\parallel$ 10×10^{-8} (N, 35 °C)	$D_\perp$ 6.6×10^{-8}, $D_\parallel$ 15.2×10^{-8} (A, 40 °C)	–
Dipole moment μ (D), 25 °C	4.78 (CCl$_4$), 5.12 (C$_6$H$_{14}$), 4.93 (C$_6$H$_{12}$), 5.00 (C$_6$H$_6$)	4.73 (CCl$_4$), 5.11 (C$_6$H$_{14}$), 4.94 (C$_6$H$_{12}$), 4.94 (C$_6$H$_6$)	4.76 (CCl$_4$), 5.10 (C$_6$H$_{14}$), 5.00 (C$_6$H$_{12}$), 5.00 (C$_6$H$_6$)

Number/common name	19 (3OCB, M9, RO-CM-5309)	20 (4OCB, M12)	21 (5OCB, M15, RO-CM-5315)
Substance	H$_7$C$_3$–O–C$_6$H$_4$–C$_6$H$_4$–CN	H$_9$C$_4$–O–C$_6$H$_4$–C$_6$H$_4$–CN	H$_{11}$C$_5$–O–C$_6$H$_4$–C$_6$H$_4$–CN
Formula	C$_{16}$H$_{15}$NO	C$_{17}$H$_{17}$NO	C$_{18}$H$_{19}$NO
Molar mass (g/mol)	237.304	251.331	265.368
CAS-RN	52709-86-1	52709-87-2	52364-71-3
Temperatures of phase transitions T^{tr} (°C)	Cr 71.5 (N 64.0) Is	Cr 78.0 (N 75.5) Is	Cr' 48.0 Cr 53.0 N 68.0 Is
Enthalpies of phase transitions ΔH^{tr} (kJ/mol)	19.2 (Cr–Is)	23.4 (Cr–Is)	28.9 (Cr–N), 0.2 (N–Is)
Crystallographic space group	$P2_1/c$	$Pca2_1$	$P2_1/n$
Order parameter S	–	–	0.63 (55 °C)
Density ϱ (g/cm^3)	1.19 (Cr)	1.14 (Cr)	1.042 (N, 61.5 °C), 1.029 (Is, 70 °C)
Refractive index n (589 nm)	–	–	n_e 1.697, n_o 1.528 (N, 55 °C)
Dielectric constant ε (1 kHz)	–	$\Delta\varepsilon$ +28.7, $\varepsilon_\perp$ +7.5 (20 °C, extra)	$\Delta\varepsilon$ +8.5, $\varepsilon_\parallel$ 16.7 ($T = 0.98 T^{N-Is}$)
Kinemamic viscosity ν (mm^2/s)	10 (Is, 80 °C)	10 (75 °C)	20 (53 °C), 10 (85 °C)
Surface tension (mN/m)	–	–	33.2 (N, 51 °C), 31.9 (Is, 70 °C)
Magnetic susceptibility $\Delta\chi$ (m^3/kg) (×10^{-12})	–	–	87 (58 °C), 98 (50 °C)
Dipole moment μ (D)	–	5.1	–

Number/common name	22 (6OCB, M18)	23 (7OCB, M21)	24 (8OCB, M24, RO-CM-5324)
Substance	H$_{13}$C$_6$–O–C$_6$H$_4$–C$_6$H$_4$–CN	H$_{15}$C$_7$–O–C$_6$H$_4$–C$_6$H$_4$–CN	H$_{17}$C$_8$–O–C$_6$H$_4$–C$_6$H$_4$–CN
Formula	C$_{19}$H$_{21}$NO	C$_{20}$H$_{23}$NO	C$_{21}$H$_{25}$NO
Molar mass (g/mol)	279.385	293.412	307.44
CAS-RN	41424-11-7	52364-72-4	52364-73-5
Temperatures of phase transitions T^{tr} (°C)	Cr' 44.0 Cr 57.0 N 75.5 Is	Cr' 45.5 Cr 53.5 N 75.0 Is	Cr' 46.0 Cr' 51.0 Cr 54.5 A 67.0 N 80.0 Is
Enthalpies of phase transitions ΔH^{tr} (kJ/mol)	29.7 (Cr–N), 0.8 (N–Is)	28.9 (Cr–N), 0.6 (N–Is)	28.5 (Cr–A), 0.1 (A–N), 0.8 (N–Is)

Table 5.1-3 Two-ring systems without bridges, cont.

Number/common name	22 (6OCB, M18)	23 (7OCB, M21)	24 (8OCB, M24, RO-CM-5324)
Crystallographic space group	–	$P2_1/c$, $P\bar{1}$	$P\bar{1}$
Order parameter S	–	0.61 (63 °C)	
Density ϱ (g/cm^3)	1.017 (N, 71 °C), 1.025 (N, 63 °C)	1.0051 (N, 63 °C), 0.9869 (Is, 79.5 °C)	0.990 (N, 75 °C), 0.977 (Is, 85 °C)
Refractive index n (589 nm)	n_e 1.72, n_o 1.50 (60 °C)	n_e 1.6734, n_o 1.5155 (60 °C)	n_e 1.647, n_o 1.509 (71 °C)
Dielectric constant ε (1 kHz)	$\Delta\varepsilon$ +8.0, $\varepsilon_\parallel$ 15.8 ($T = 0.98 T^{\text{N-Is}}$)	$\Delta\varepsilon$ +7.5, $\varepsilon_\parallel$ 15.0 ($T = 0.98 T^{\text{N-Is}}$)	$\Delta\varepsilon$ +8.0 (N, 74 °C)
Kinematic viscosity v (mm^2/s)	17 (60 °C), 12 (85 °C)	10 (70 °C), 30 (50 °C)	16 (70 °C), 12 (85 °C)
Surface tension (mN/m)	–	–	36.5 (A, 60), 35.4 (N, 70), 30.5 (Is, 80)
Magnetic susceptibility $\Delta\chi$ (m^3/kg) (×10^{-12})	88 (67 °C), 80 (71 °C)	71 (70 °C), 87.5 (60 °C)	83 (70.6 °C)
Dipole moment μ (D)	5.2	4.9	5.2 (CCl$_4$)

Number/common name	25 (PCH-3)	26 (PCH-4)	27 (PCH-5)
Substance	NC–⌬–⌬–C$_3$H$_7$	NC–⌬–⌬–C$_4$H$_9$	NC–⌬–⌬–C$_5$H$_{11}$
Formula	C$_{16}$H$_{21}$N	C$_{17}$H$_{23}$N	C$_{18}$H$_{25}$N
Molar mass (g/mol)	227.352	241.38	255.407
CAS-RN	61203-99-4	61204-00-0	61204-01-1
Temperatures of phase transitions T^{tr} (°C)	Cr 42.7 N 45.9 Is	Cr 41.0 N 39.0 Is	Cr 30.0 N 55.0 Is
Enthalpies of phase transitions ΔH^{tr} (kJ/mol)	19.0 (Cr–N), 0.7 (N–Is)	22.2 (Cr–Is)	18.0 (Cr–N), 0.9 (N–Is)
Crystallographic space group	$C2/c$	$P2_1/c$, $P\bar{1}$	–
Order parameter S	0.64 (43 °C)	–	0.54 (43 °C)
Density ϱ (g/cm^3)	0.9672 (N, 43 °C), 0.9571 (Is, 46.5 °C)	0.962 (N, 35 °C), 0.951 (Is, 43 °C)	0.967 (N, 25 °C), 0.932 (Is, 60 °C)
Refractive index n (589 nm)	n_e 1.582, n_o 1.492 (N, 43 °C)	n_e 1.579, n_o 1.498 (N, 37 °C)	n_e 1.600, n_o 1.492 (N, 40 °C)
Dielectric constant ε (1 kHz)	$\varepsilon_\parallel$ 19.3, $\varepsilon_\perp$ 5.5 ($T = 0.95 T^{\text{N-Is}}$)	$\Delta\varepsilon$ +11 (20 °C)	$\varepsilon_\parallel$ 14.4, $\varepsilon_\perp$ 5.2 (39 °C)
Viscosity (20 °C)	η 19.2 mPa s	v 23 mm^2/s	v 23 mm^2/s
Dipole moment μ (D)	3.9	–	4.1

Number/common name	28 (PCH-7)	29 (PCH-8)	30 (PCH-32, RO-CM-4232, ZLI-1484)
Substance	NC–⌬–⌬–C$_7$H$_{15}$	NC–⌬–⌬–C$_8$H$_{17}$	H$_5$C$_2$–⌬–⌬–C$_3$H$_7$
Formula	C$_{20}$H$_{29}$N	C$_{21}$H$_{31}$N	C$_{17}$H$_{26}$
Molar mass (g/mol)	283.461	297.488	230.397
CAS-RN	61204-03-3	83626-40-8	82991-47-7
Temperatures of phase transitions T^{tr} (°C)	Cr 30.0 N 57.8 Is	Cr 37.0 N 55.0 Is	Cr −1 Is

Table 5.1-3 Two-ring systems without bridges, cont.

Number/common name	28 (PCH-7)	29 (PCH-8)	30 (PCH-32, RO-CM-4232, ZLI-1484)
Enthalpies of phase transitions ΔH^{tr} (kJ/mol)	25.5 (Cr–N), 1.0 (N–Is)	–	18.4 (Cr–Is)
Crystallographic space group	–	$P2_1/c$, $P\bar{1}$ (Cr')	–
Order parameter S	0.47 (47 °C)	–	–
Density ϱ (g/cm^3)	0.952 (N, 30 °C), 0.915 (Is, 67 °C)	0.941 (N, 37 °C), 0.915 (Is, 62 °C)	0.909 (20 °C)
Refractive index n (589 nm)	n_e 1.590, n_o 1.482 (47 °C)	–	1.514 (20 °C)
Dielectric constant ε (1 kHz)	$\varepsilon_\parallel$ 12.9, $\varepsilon_\perp$ 4.2 (31 °C)	–	ε 2.293, $\Delta\varepsilon$ +0.5 (20 °C, extra)
Kinematic viscosity ν (mm^2/s)	28 (extra, 20 °C)	–	4 (extra, 20 °C)

Number/common name	31 (PCH-301, ZLI-2446)	32 (PCH-302, ZLI-1476)	33 (PCH-304, ZLI-1477)
Substance	H$_3$C–O–⟨ring⟩–⟨ring⟩–C$_3$H$_7$	H$_5$C$_2$–O–⟨ring⟩–⟨ring⟩–C$_3$H$_7$	H$_9$C$_4$–O–⟨ring⟩–⟨ring⟩–C$_3$H$_7$
Formula	C$_{16}$H$_{24}$O	C$_{17}$H$_{26}$O	C$_{19}$H$_{30}$O
Molar mass (g/mol)	232.369	246.396	274.45
CAS-RN	81936-32-5	80944-44-1	79709-84-5
Temperatures of phase transitions T^{tr} (°C)	Cr 32.0 (N 10.0) Is	Cr 40.8 (N 37.8) Is	Cr 35.5 (N 33.0) Is
Enthalpies of phase transitions ΔH^{tr} (kJ/mol)	17.6 (Cr–Is)	26.4 (Cr–Is), 0.7 (N–Is)	30.6 (Cr–Is), 0.5 (N–Is)
Anisotropy of refractive index Δn (20 °C, 589 nm)	0.09 (extra)	0.09 (extra)	0.09 (extra)
Dielectric anisotropy $\Delta\varepsilon$ (20 °C, 1 kHz)	−0.5 (extra)	−0.5 (extra)	−0.5 (extra)
Kinematic viscosity ν (mm^2/s)	6 (extra, 20 °C)	7 (extra, 20 °C)	10 (extra, 20 °C)

Number/common name	34 (H33, ZLI-1305)	35	36
Substance	H$_7$C$_3$–C(=O)–O–⟨ring⟩–⟨ring⟩–C$_3$H$_7$	H$_5$C$_2$–C(=O)–⟨ring⟩–⟨ring⟩–C$_3$H$_7$	H$_5$C$_2$–C(=O)–⟨ring⟩–⟨ring⟩–C$_5$H$_{11}$
Formula	C$_{19}$H$_{28}$O$_2$	C$_{18}$H$_{26}$O	C$_{20}$H$_{30}$O
Molar mass (g/mol)	288.434	258.407	286.462
Temperatures of phase transitions T^{tr} (°C)	Cr 11.0 B 26.1 N 30.3 Is	Cr 49.2 N 56.5 Is	Cr 56.6 N 68.8 Is
Enthalpies of phase transitions ΔH^{tr} (kJ/mol)	16.4 (Cr–B), 4.9 (B–N), 0.4 (N–Is)	–	–
Refractive index n (589 nm)	Δn 0.09 (20 °C)	n_e 1.589, n_o 1.497 ($T = 0.95T^{\text{N-Is}}$)	n_e 1.574, n_o 1.487 ($T = 0.95T^{\text{N-Is}}$)
Dielectric constant ε (1 kHz)	$\Delta\varepsilon$ −1.5 (20 °C)	$\varepsilon_\parallel$ 9.0, $\varepsilon_\perp$ 7.6 ($T = 0.95T^{\text{N-Is}}$)	$\varepsilon_\parallel$ 8.3 $\varepsilon_\perp$ 6.9 ($T = 0.95T^{\text{N-Is}}$)
Viscosity ν (20 °C)	15 mm^2/s (extra)	–	–
Diamagnetic anisotropy $\Delta\chi$ (m^3/mol) ($\times 10^{-12}$)	–	25.3	26.1
Dipole moment μ (D)	–	3.02	3.1

Table 5.1-3 Two-ring systems without bridges, cont.

Number/common name	37	38	39
Substance	NC–⌬–⌬–C_3H_7	NC–⌬–⌬–C_5H_{11}	NC–⌬–⌬–C_7H_{15}
Formula	$C_{18}H_{23}N$	$C_{20}H_{27}N$	$C_{22}H_{31}N$
Molar mass (g/mol)	253.391	281.445	309.499
CAS-RN	–	74385-67-4	–
Temperatures of phase transitions T^{tr} (°C)	Cr 66.5 N 88.0 Is	Cr 62.0 N 100.0 Is	Cr 61.0 N 95.0 Is
Enthalpies of phase transitions ΔH^{tr} (kJ/mol)	17.4 (Cr–N)	20.7 (Cr–N)	32.7 (Cr–N)
Density ϱ (g/cm^3)	1.007 (N, 66 °C)	0.993 (N, 60 °C)	0.975 (N, 65 °C)
Refractive index n (633 nm)	n_e 1.614, n_o 1.492 (66 °C)	n_e 1.616, n_o 1.490 (60 °C)	n_e 1.598, n_o 1.489 (65 °C)
Dielectric constant ε (10 kHz)	$\varepsilon_\perp$ 5.2, $\varepsilon_\parallel$ 15.4 (66 °C)	$\varepsilon_\perp$ 4.6, $\varepsilon_\parallel$ 13.1 (60 °C)	$\varepsilon_\perp$ 4.4, $\varepsilon_\parallel$ 12.0 (65 °C)
Viscosity ν (mm^2/s)	–	7.8 (70 °C)	–
Dipole moment μ (D)	–	4.1	–

Number/common name	40 (CCH-3)	41 (CCH-5)	42 (CCH-7)
Substance	H_7C_3–⌬–⌬–CN	$H_{11}C_5$–⌬–⌬–CN	$H_{15}C_7$–⌬–⌬–CN
Formula	$C_{16}H_{27}N$	$C_{18}H_{31}N$	$C_{20}H_{35}N$
Molar mass (g/mol)	233.4	261.454	289.509
CAS-RN	65355-35-3	65355-36-4	65355-37-5
Temperatures of phase transitions T^{tr} (°C)	Cr 58.0 (S 18.0 S 44.0 S 48.0 B 57.0) N 80.0 Is	Cr' 59.2 Cr 63.4 (S 40.5 B 49.1) N 86.4 Is	Cr 71.0 N 83.0 Is
Enthalpies of phase transitions ΔH^{tr} (kJ/mol)	26.8 (Cr–N), 1.1 (N–Is)	26.8 (Cr–N), 1.3 (N–Is)	33.9 (Cr–N), 0.9 (N–Is)
Crystallographic space group	$P\bar{1}$	$P2_1/c$, $P2_12_12_1$ (Cr')	$P2_12_12_1$
Density ϱ (g/cm^3)	0.902 (N, 73 °C), 0.885 (Is, 87 °C)	0.902 (N, 72 °C)	0.893 (N, 70 °C)
Order parameter S	0.53 (73 °C)	0.62 (72 °C)	0.52 (71 °C)
Refractive index n (589 nm)	n_e 1.4930, n_o 1.4553 (N, 73 °C)	n_e 1.5061, n_o 1.4568 (N, 72 °C)	n_e 1.502, n_o 1.456 (N, 70 °C)
Dielectric constant ε (1 kHz)	$\varepsilon_\perp$ 5.5, $\varepsilon_\parallel$ 10.0 (73 °C)	$\varepsilon_\perp$ 4.75, $\varepsilon_\parallel$ 9.25 (73 °C)	$\varepsilon_\perp$ 3.89, $\varepsilon_\parallel$ 7.17 (74 °C)
Viscosity ν (mm^2/s) (20 °C)	63 (extra)	66 (extra)	78 (extra)
Dipole moment μ (D)	–	3.8 (xylene)	–

Number/common name	43 (RO-CM-4513)	44 (RO-CM-4535)	45 (CCH-301)
Substance	H_3C–CH=CH–⌬–⌬–CN	H_3C–CH$_2$–CH=CH–⌬–⌬–CN	H_7C_3–⌬–⌬–O–CH_3
Formula	$C_{16}H_{25}N$	$C_{18}H_{29}N$	$C_{16}H_{30}O$
Molar mass (g/mol)	231.384	259.439	238.417
CAS-RN	122705-86-6	–	–

Table 5.1-3 Two-ring systems without bridges, cont.

Number/common name	43 (RO-CM-4513)	44 (RO-CM-4535)	45 (CCH-301)
Temperatures of phase transitions T^{tr} (°C)	Cr 64.9 N 99.7 Is	Cr 79.5 (A 45.0) N 100.0 Is	Cr 10.0 N 17.0 Is
Enthalpies of phase transitions ΔH^{tr} (kJ/mol)	–	–	18.8 (Cr–N)
Refractive index n (589 nm)	Δn 0.066, n_o 1.457 (89.7 °C)	Δn 0.065, n_o 1.456 (89.7 °C)	–
Dielectric constant ε (1 kHz)	$\varepsilon_\perp$ 4.35, $\Delta\varepsilon$ +5.03 (89.7 °C)	$\varepsilon_\perp$ 4.12, $\Delta\varepsilon$ +4.61 (89.7 °C)	$\Delta\varepsilon$ –0.3 (20 °C)
Viscosity v (mm^2/s) (20 °C)	–	–	7 (extra)
Dipole moment μ (D)	3.76	3.83	–

Number/common name	46 (CCN55, ZLI-2395)	47 (C33)	48 (C35)
Substance	H$_{11}$C$_5$–⟨⟩–(NC)–⟨⟩–C$_5$H$_{11}$	H$_7$C$_3$–⟨⟩–⟨⟩–O–C(O)–C$_3$H$_7$	H$_7$C$_3$–⟨⟩–⟨⟩–O–C(O)–C$_5$H$_{11}$
Formula	C$_{23}$H$_{41}$N	C$_{19}$H$_{34}$O$_2$	C$_{21}$H$_{38}$O$_2$
Molar mass (g/mol)	331.59	294.482	322.536
CAS-RN	88510-89-8	–	–
Temperatures of phase transitions T^{tr} (°C)	Cr 25.0 B 30.0 N 66.0 Is	Cr 41.0 B 69.0 N 73.0 Is	Cr 44.0 S 46.0 N 74.0 Is
Enthalpies of phase transitions ΔH^{tr} (kJ/mol)	22.6 (Cr–S)	22.6 (Cr–S)	23.8 (Cr–S)
Anisotropy of refractive index Δn (extra, 20 °C, 589 nm)	0.03	0.04	0.03
Dielectric anisotropy $\Delta\varepsilon$ (20 °C, 1 kHz)	–8.4 (extra)	–1.6 (extra)	–0.8 (extra)
Viscosity v (mm^2/s) (20 °C)	67 (extra)	11 (extra)	13 (extra)

Number/common name	49	50 (RO-CM-7035)	51 (RO-CM-7037, RO-CP-7037)
Substance	H$_{11}$C$_5$–⟨pyridine⟩–⟨⟩–CN	H$_{11}$C$_5$–⟨pyrimidine⟩–⟨⟩–CN	H$_{15}$C$_7$–⟨pyrimidine⟩–⟨⟩–CN
Formula	C$_{17}$H$_{18}$N$_2$	C$_{16}$H$_{17}$N$_3$	C$_{18}$H$_{21}$N$_3$
Molar mass (g/mol)	250.346	251.334	279.388
CAS-RN	77782-82-2	59855-05-9	59854-97-6
Temperatures of phase transitions T^{tr} (°C)	Cr 33.6 N 43.5 Is	Cr 71.0 (N 52.0) Is	Cr 45.0 N 51.0 Is
Enthalpies of phase transitions ΔH^{tr} (kJ/mol)	20.9 (Cr–N)	–	0.4 (N–Is)
Density ϱ (g/cm^3)	1.0483 (N, 30.0 °C)	–	–
Anisotropy of refractive index Δn	0.176 ($T = 0.95 T^{N-Is}$)	0.220 (extra, 20 °C)	0.2098
Dielectric constant ε	$\varepsilon_\parallel$ 28.7, $\Delta\varepsilon$ +17.8 (27.7 °C)	$\Delta\varepsilon$ +21.3, $\varepsilon_\parallel$ 31.3 ($T = 0.98 T^{N-Is}$, 1.592 kHz)	$\Delta\varepsilon$ +16.0, $\varepsilon_\parallel$ 24.6 ($T = 0.98 T^{N-Is}$)
Viscosity	v 50 mm^2/s (extra, 20 °C)	v 55 mm^2/s (extra, 20 °C)	η 25 mPa s (38 °C)
Diamagnetic anisotropy $\Delta\chi$ (m^3/kg) (×10^{-12})	–	–	90.7 ($T = 0.98 T^{N-Is}$)
Dipole moment μ (D)	6.0	6.0	6.7

Table 5.1-3 Two-ring systems without bridges, cont.

Number/common name	52 (PYP605, ZLI-2543)	53 (PYP606, ZLI-2303)	54 (PYP607, ZLI-2304)
Substance	$H_{13}C_6$–[pyrimidine]–[phenyl]–OC_5H_{11}	$H_{13}C_6$–[pyrimidine]–[phenyl]–OC_6H_{13}	$H_{13}C_6$–[pyrimidine]–[phenyl]–OC_7H_{15}
Formula	$C_{21}H_{30}N_2O$	$C_{22}H_{32}N_2O$	$C_{23}H_{34}N_2O$
Molar mass (g/mol)	326.486	340.513	354.54
CAS-RN	57202-28-5	51518-75-3	57202-29-6
Temperatures of phase transitions T^{tr} (°C)	Cr 43.0 N 53.0 Is	Cr 30.5 N 60.8 Is	Cr 35.5 N 58.3 Is
Enthalpies of phase transitions ΔH^{tr} (kJ/mol)	29.4 (Cr–N), 1.2 (N–Is)	19.2 (Cr–N), 1.4 (N–Is)	32.7 (Cr–N), 1.5 (N–Is)
Density ϱ (g/cm^3)	0.9899 (N, 48.0 °C), 0.9673 (Is, 68.0 °C)	0.9857 (N, 48.0 °C), 0.9618 (Is, 68.0 °C)	0.9806 (N, 48.0 °C), 0.9562 (Is, 68.0 °C)
Refractive index n (589 nm)	n_e 1.6217, n_o 1.5042 (48 °C)	n_e 1.6435, n_o 1.4939 (48 °C)	n_e 1.6305, n_o 1.4940 (48 °C)
Dielectric constant (1 kHz)	$\Delta\varepsilon$ +1.2 (extra, 20 °C)	$\varepsilon_\parallel$ 3.92, $\varepsilon_\perp$ 3.14 ($T = 0.97T^{N-Is}$)	$\varepsilon_\parallel$ 3.79, $\varepsilon_\perp$ 3.10 ($T = 0.97T^{N-Is}$)
Viscosity ν (mm^2/s) (20 °C)	50 (extra)	43 (extra)	49 (extra)

Number/common name	55 (PYP609, ZLI-2306)	56 (PYP707, ZLI-2710)	57 (PYP909, ZLI-2713)
Substance	$H_{13}C_6$–[pyrimidine]–[phenyl]–OC_9H_{19}	$H_{15}C_7$–[pyrimidine]–[phenyl]–OC_7H_{15}	$H_{19}C_9$–[pyrimidine]–[phenyl]–OC_9H_{19}
Formula	$C_{25}H_{38}N_2O$	$C_{24}H_{36}N_2O$	$C_{28}H_{44}N_2O$
Molar mass (g/mol)	382.594	368.567	424.676
CAS-RN	51462-26-1	–	99895-85-9
Temperatures of phase transitions T^{tr} (°C)	Cr 35.0 N 61.0 Is	Cr 44.0 C 44.0 A 49.0 N 68.0 Is	Cr 34.0 C 61.0 A 75.0 Is
Enthalpies of phase transitions ΔH^{tr} (kJ/mol)	35.1 (Cr–N), 1.7 (N–Is)	35.5 (Cr–A), 2.0 (N–Is)	35.1 (Cr–C)
Density ϱ (g/cm^3)	0.9707 (N, 48.0 °C), 0.9454 (Is, 68.0 °C)	–	0.9650 (C, 48.0 °C), 0.9494 (N, 68.0 °C)
Refractive index n (589 nm)	n_e 1.6191, n_o 1.4886 (48 °C)	Δn 0.14 (20 °C)	Δn 0.13 (20 °C)
Dielectric anisotropy $\Delta\varepsilon$	+0.49 (56.0 °C)	+1.1 (extra, 20 °C)	+0.9 (extra, 20 °C)
Viscosity ν (mm^2/s) (20 °C)	63 (extra)	53 (extra)	110 (extra)

Number/common name	58 (PDX3, ZLI-1906)	59 (PDX5, ZLI-1908)	60 (PDX7, ZLI-1910)
Substance	H_7C_3–[dioxane]–[phenyl]–CN	$H_{11}C_5$–[dioxane]–[phenyl]–CN	$H_{15}C_7$–[dioxane]–[phenyl]–CN
Formula	$C_{14}H_{17}NO_2$	$C_{16}H_{21}NO_2$	$C_{18}H_{25}NO_2$
Molar mass (g/mol)	231.397	259.351	287.405
CAS-RN	–	–	97128-75-1
Temperatures of phase transitions T^{tr} (°C)	Cr 52.9 (N 39.3) Is	Cr 56.0 (N 49.0) Is	Cr 54.0 (N 53.0) Is
Enthalpies of phase transitions ΔH^{tr} (kJ/mol)	20.9 (Cr–Is)	21.0 (Cr–Is), 0.4 (N–Is)	26.4 (Cr–Is), 0.8 (N–Is)

Table 5.1-3 Two-ring systems without bridges, cont.

Number/common name	58 (PDX3, ZLI-1906)	59 (PDX5, ZLI-1908)	60 (PDX7, ZLI-1910)
Crystallographic space group	–	–	$P2_12_12_1$
Anisotropy of refractive index Δn (589 nm, 20 °C)	0.13	0.14	0.13
Dielectric constant ε (1 kHz)	$\Delta\varepsilon$ +32 (20 °C)	$\Delta\varepsilon$ +29.6, $\varepsilon_\perp$ 8.2 (20 °C)	$\Delta\varepsilon$ +32 (20 °C), $\Delta\varepsilon$ +7.9 ($T = 0.98T^{\text{N-Is}}$)
Kinematic viscosity ν (mm^2/s)	33 (20 °C)	29.4 (25 °C)	34.7 (25 °C)
Dipole moment μ (D)	4.1	6.2 (C_6H_{12})	6.2 (C_6H_{12})

Table 5.1-4 Two-ring systems with bridges

Number/common name	61	62	63 (RO-CM-3952)
Substance	NC–⬡–⬢–C_5H_{11}	NC–⬡–⬢=C_3H_7	H_5C_2–O–⬡–⬢–C_5H_{11}
Formula	$C_{20}H_{29}N$	$C_{20}H_{27}N$	$C_{21}H_{34}O$
Molar mass (g/mol)	283.461	281.445	302.505
Temperatures of phase transitions T^{tr} (°C)	Cr 31.0 N 52.5 Is	Cr 25.1 N 47.5 Is	Cr 27.0 (B 8.0) N 47.0 Is
Enthalpies of phase transitions ΔH^{tr} (kJ/mol)	14.6 (Cr–N), 2.0 (N–Is)	15.2 (Cr–N)	–
Dielectric constant ε (1 kHz)	$\varepsilon_\perp$ 4.98, $\Delta\varepsilon$ +9.77 (42.5 °C)	$\varepsilon_\perp$ 4.61, $\Delta\varepsilon$ +11.13 (22 °C)	$\Delta\varepsilon$ −0.24, $\varepsilon_\perp$ 2.98
Dynamic viscosity η (mPa s)	η 22.4 (22 °C)	η 22.8 (22 °C)	η 12.0 (22 °C)

Number/common name	64	65	66
Substance	$H_{11}C_5$–⬡–≡–⬡–O–CH_3	$H_{11}C_5$–⬡–≡–⬡–O–C_2H_5	$H_{11}C_5$–⬡–≡–⬡–O–C_3H_7
Formula	$C_{20}H_{22}O$	$C_{21}H_{24}O$	$C_{24}H_{30}O$
Molar mass (g/mol)	278.398	292.425	334.506
Temperatures of phase transitions T^{tr} (°C)	Cr 47.0 N 58.0 Is	Cr 60.0 N 80.0 Is	Cr 41.0 N 65.0 Is
Enthalpies of phase transitions ΔH^{tr} (kJ/mol)	18.0 (Cr–N), 0.6 (N–Is)	23.2 (Cr–N), 1.0 (N–Is)	22.6 (Cr–N), 0.9 (N–Is)
Refractive index n (589 nm)	n_e 1.735, n_o 1.526 (44.5 °C)	n_e 1.726, n_o 1.517 (70 °C)	n_e 1.683, n_o 1.510 (60 °C)
Dielectric constant ε (1 kHz)	$\varepsilon_\perp$ 3.58, $\Delta\varepsilon$ −0.11 (50 °C)	$\Delta\varepsilon$ +0.2, (extra, 20 °C)	$\varepsilon_\perp$ 3.67, $\Delta\varepsilon$ −0.18 (55 °C)
Kinematic viscosity (mm^2/s)	–	ν 20 (extra, 20 °C)	–
Surface tension (mN/m)	25 (22 °C)	–	–

Number/common name	67 (MBBA)	68 (EBBA)	69 (CBOOA)
Substance	H_3C–O–⬡–N=⬡–C_4H_9	H_5C_2–O–⬡–N=⬡–C_4H_9	NC–⬡–N=⬡–O–C_8H_{17}
Formula	$C_{18}H_{21}NO$	$C_{19}H_{23}NO$	$C_{22}H_{26}N_2O$
Molar mass (g/mol)	267.374	281.401	334.465
CAS-RN	26227-73-6	29743-08-6	65756-96-9

Table 5.1-4 Two-ring systems with bridges, cont.

Number/common name	67 (MBBA)	68 (EBBA)	69 (CBOOA)
Temperatures of phase transitions T^{tr} (°C)	Cr 22.0 N 48.0 Is	Cr 36.5 N 79.8 Is	Cr 73.1 A 83.3 N 107.9 Is
Enthalpies of phase transitions ΔH^{tr} (kJ/mol)	15.1 (Cr–N), 0.4 (N–Is)	17.2 (Cr–N), 0.6 (N–Is)	27.6 (Cr–N), 0.7 (N–Is)
Crystallographic space group	–	$P2_1/c$	$P2_1/c$
Order parameter S	0.55 (25 °C)	0.39 (55 °C)	–
Density ϱ (g/cm^3)	1.042 (N, 25 °C), 1.015 (Is, 60 °C)	1.020 (N, 40 °C), 0.988 (Is, 80 °C)	1.009 (A, 83 °C), 1.003 (N, 90 °C), 0.981 (Is, 110 °C)
Refractive index n	n_e 1.764, n_o 1.549 (589 nm, 25 °C)	n_e 1.763, n_o 1.524 (578 nm, N, 43.5 °C)	–
Dielectric constant ε	$\varepsilon_\parallel$ 4.72, $\varepsilon_\perp$ 5.31 (30 °C, 1.6 kHz)	$\varepsilon_\parallel$ 4.37, $\varepsilon_\perp$ 4.50 (73.8 °C)	$\varepsilon_\parallel$ 14.4, $\varepsilon_\perp$ 7.5 (95 °C, 1 kHz)
Dynamic viscosity η (mPa s)	23 (N, 30 °C)	11 (N, 69 °C)	13 (N, 84 °C)
Surface tension (mN/m)	34.0 (N, 22 °C), 32.6 (Is, 50 °C)	23.8 (N, 45 °C), 23.3 (Is, 80 °C)	26.1 (A, 80 °C), 26.7 (N, 90 °C), 26.6 (Is, 110 °C)
Thermal conductivity (W/(m K))	0.125 (N, 20 °C), 0.128 (N, 30 °C), 0.145 (N, 50 °C)	0.135 (N, 125 °C), 0.157 (Is, 140 °C)	–
Heat capacity C_p (J/(mol K))	509 (N, 37 °C), 507 (Is, 49 °C)	540 (N, 37 °C), 590 (Is, 82 °C)	401 (A, 80 °C), 474 (N, 100 °C), 514 (Is, 120 °C)
Magnetic susceptibility $\Delta\chi$ (m^3/kg) ($\times 10^{-12}$)	116 (N, 23 °C)	47 ($T = T^{N-Is}$)	–
Sound velocity v (m/s)	1200 (N, 33 °C)	1375 (N, 73 °C, 4.8 GHz)	1340 (N, 90 °C, 2 MHz)
Diffusion coefficient D (m^2/s)	$D_\perp$ 8 × 10^{-10}, $D_\parallel$ 13 × 10^{-10} (1000/T = 3.3)	D 2.5 × 10^{-10} (Is, 89 °C)	$D_\perp$ 6.6 × 10^{-8}, $D_\parallel$ 13.5 × 10^{-8} (N, 88 °C)
Dipole moment μ (D)	3.2 (22 °C)	$\mu_\parallel$ 1.28, $\mu_\perp$ 1.35 (73.8 °C)	5.21 (C$_6$H$_6$, 25 °C)

Number/common name	70 (BBBA)	71 (HBT)	72 (OBT)
Substance	H$_9$C$_4$–O–C$_6$H$_4$–N=CH–C$_6$H$_4$–C$_4$H$_9$	H$_{13}$C$_6$–O–C$_6$H$_4$–N=CH–C$_6$H$_4$–CH$_3$	H$_{17}$C$_8$–O–C$_6$H$_4$–N=CH–C$_6$H$_4$–CH$_3$
Formula	C$_{21}$H$_{27}$NO	C$_{20}$H$_{25}$NO	C$_{22}$H$_{29}$NO
Molar mass (g/mol)	309.455	295.428	323.483
CAS-RN	29743-09-7	25959-51-7	–
Temperatures of phase transitions T^{tr} (°C)	Cr 8.0 G 41.0 B 45.0 A 45.5 N 75.0 Is	Cr 58.0 (G 44.0 B 53.0) N 76.0 Is	Cr 70.0 (B 61.5 A 69.0) N 78.5 Is
Enthalpies of phase transitions ΔH^{tr} (kJ/mol)	3.3 (Cr–G), 0.7 (G–B), 2.9 (B–A), 0.4 (A–N), 0.9 (N–Is)	29.3 (Cr–N), 5.4 (B–N), 1.0 (N–Is)	–
Crystallographic space group	–	–	$P\bar{1}$
Order parameter S	0.36 (52 °C)	–	–
Density ϱ (g/cm^3)	1.007 (B, 40 °C), 0.990 (N, 50 °C), 0.964 (Is, 75 °C)	0.985 (N, 70 °C), 0.976 (Is, 77 °C)	0.9714 (N, 71 °C), 0.9576 (Is, 81 °C)
Refractive index n (589 nm)	n_e 1.670, n_o 1.527 (52 °C)	n_e 1.692, n_o 1.525 (71 °C)	n_e 1.688, n_o 1.500 (72.5 °C)

Table 5.1-4 Two-ring systems with bridges, cont.

Number/common name	70 (BBBA)	71 (HBT)	72 (OBT)
Dielectric constant ε (68 °C)	$\varepsilon_\parallel$ 3.99, $\varepsilon_\perp$ 4.01 (1 MHz)	$\varepsilon_\parallel$ 4.41, $\varepsilon_\perp$ 4.25 (10 kHz)	–
Dynamic viscosity η (mPa s)	18 (N, 57 °C)	6.5 (N, 70 °C), 7.0 (Is, 80 °C)	8.6 (N, 71 °C), 8.3 (Is, 80 °C)
Heat capacity C_p (J/(mol K))	–	689 (N, 67 °C), 624 (Is, 90 °C)	–
Magnetic susceptibility $\Delta\chi$ (m³/kg) (×10⁻¹²)	49 ($T = T^{\text{N-Is}}$)	–	–
Sound velocity v (m/s)	1467 (G, 40 °C), 1405 (N, 60 °C) (2 MHz)	1358 (N, 70 °C), 1328 (Is, 82 °C) (2 MHz)	1360 (N, 70 °C), 1332 (Is, 80 °C) (2 MHz)

Number/common name	73	74	75
Substance	H₃C-O-⌬-CH=N-N=CH-⌬-O-CH₃	H₁₅C₇-⌬-N=N-⌬-C₇H₁₅	H₁₃C₆-O-⌬-N=N-⌬-O-C₆H₁₃
Formula	$C_{16}H_{16}N_2O_2$	$C_{26}H_{38}N_2$	$C_{24}H_{34}N_2O_2$
Molar mass (g/mol)	268.318	378.606	382.551
CAS-RN	2299-73-2	37592-97-5	10225-93-1
Temperatures of phase transitions T^{tr} (°C)	Cr 173.0 N 186.0 Is	Cr 40.0 (A 21.4) N 47.3 Is	Cr 102.6 N 116.2 Is
Enthalpies of phase transitions ΔH^{tr} (kJ/mol)	38.7 (Cr–N), 1.6 (N–Is)	24.4 (Cr–N), 1.1 (N–Is)	39.0 (Cr–N), 1.5 (N–Is)
Crystallographic space group	Cc	–	–
Order parameter S	0.656 (154.2 °C), 0.410 (180 °C)	–	0.676 (105.7 °C)
Density ϱ (g/cm³)	1.044 (N, 178 °C), 1.023 (Is, 195 °C)	0.9430 (40.6 °C)	0.9493 (Is, 115 °C)
Refractive index n (589 nm)	n_e 1.791, n_o 1.549 (178 °C)	n_o 1.5095, Δn 0.191 (40.3 °C)	–
Dielectric constant ε	−0.124 (172.5 °C, 0.8 MHz)	$\varepsilon_\parallel$ 2.8, $\varepsilon_\perp$ 2.5 (23 °C, extra)	$\varepsilon_\parallel$ 3.254, $\varepsilon_\perp$ 3.219 (106 °C, 650 kHz)
Dynamic viscosity η (mPa s)	130 (N, 175 °C), 170 (Is, 197 °C)	–	–
Surface tension (mN/m)	34.6 (N, 180 °C), 32.9 (Is, 190 °C)	–	–
Heat capacity C_p (J/(mol K))	572 (N, 180 °C), 606 (Is, 220 °C)	–	–
Diamagnetic anisotropy $\Delta\chi$ (10⁻¹¹ m³/kg)	–	10.0 (N, 35 °C)	–
Diffusion coefficient D (m²/s)	1.3×10^{-9} (N), 2.0×10^{-9} ($1000/T = 2.15$) (Is)	–	–
Dipole moment μ (D)	–	–	1.87 (C_6H_6)

Table 5.1-4 Two-ring systems with bridges, cont.

Number/common name	76 (PAA)	77 (PAP)	78
Substance	(azoxy structure with H_3C-O and $-O-CH_3$)	(azoxy structure with H_5C_2-O and $-O-C_2H_5$)	(azoxy structure with H_7C_3-O and $-O-C_3H_7$)
Formula	$C_{14}H_{14}N_2O_3$	$C_{16}H_{18}N_2O_3$	$C_{18}H_{22}N_2O_3$
Molar mass (g/mol)	258.279	286.333	314.388
CAS-RN	51437-65-1	51437-64-0	104746-32-9
Temperatures of phase transitions T^{tr} (°C)	Cr 119.5 N 136.5 Is	Cr 136.8 N 168.4 Is	Cr 118.3 N 124.0 Is
Enthalpies of phase transitions ΔH^{tr} (kJ/mol)	29.6 (Cr–N), 0.6 (N–Is)	26.9 (Cr–N), 1.5 (N–Is)	26.9 (Cr–N), 0.7 (N–Is)
Crystallographic space group	$P2_1/c$	Cc	$P2_1/n$
Order parameter S	0.50 (N, 122 °C)	0.57 ($T = 0.90 \times T^{N-Is}$)	0.605 ($T = 0.90 \times T^{N-Is}$)
Density ϱ (g/cm^3)	1.165 (N, 120 °C), 1.140 (Is, 140 °C)	1.096 (N, 142 °C), 1.032 (Is, 198 °C)	1.067 (N, 114 °C), 1.046 (Is, 134 °C)
Refractive index n	n_e 1.804, n_o 1.572 (589 nm, N, 130 °C)	n_e 1.784, n_o 1.522 (589 nm, N, 160 °C)	n_o 1.534 (N, 114 °C, 546 nm)
Dielectric constant ε	$\varepsilon_\parallel$ 5.53, $\varepsilon_\perp$ 5.69 (126 °C)	$\varepsilon_\parallel$ 4.77, $\varepsilon_\perp$ 4.97 (160 °C, 1.2 MHz)	$\Delta\varepsilon$ −0.22 (118.4 °C, 0.8 MHz)
Dynamic viscosity η (mPa s)	2.5 (N, 120 °C), 2.5 (151 °C)	5.2 (N, 164.6 °C)	9.6 (N, 121.2 °C)
Surface tension (mN/m)	38.8 (N, 126 °C), 38.0 (Is, 136 °C)	29.7 (N, 159 °C), 28.8 (Is, 170 °C)	–
Thermal conductivity (W/(m K))	0.135 (N, 125 °C), 0.157 (Is, 140 °C)	–	–
Heat capacity C_p (J/(mol K))	399 (Cr, 100 °C), 507 (N, 120 °C), 508 (Is, 150 °C)	548 (Cr, 126 °C), 682 (N, 160 °C), 643 (Is, 169 °C)	609 (Cr, 100 °C), 696 (N, 117 °C), 665 (Is, 128 °C)
Diamagnetic anisotropy $\Delta\chi$ (10^{-11} m^3/kg)	9.7 (130.8 °C)	13.7 (N, 136 °C)	10.8 (N, 112 °C)
Sound velocity v (m/s)	1240 (N, 130 °C, 3 MHz)	1128 (N, 160 °C, 4.2 MHz)	1247 (N, 121 °C, 3.1 MHz)
Diffusion coefficient D (m^2/s)	$D_\perp$ 5.5×10^{-10}, $D_\parallel$ 9×10^{-10} ($1000/T = 2.55$)	$D_\perp$ 7.5×10^{-10}, $D_\parallel$ 12×10^{-10} ($1000/T = 2.45$)	$D_\perp$ 4×10^{-10}, $D_\parallel$ 7×10^{-10} ($1000/T = 2.60$)
Dipole moment μ (D)	2.3	2.42	2.41

Number/common name	79	80	81
Substance	(azoxy structure with H_9C_4-O and $-O-C_4H_9$)	(azoxy structure with $H_{11}C_5-O$ and $-O-C_5H_{11}$)	(azoxy structure with $H_{13}C_6-O$ and $-O-C_6H_{13}$)
Formula	$C_{20}H_{26}N_2O_3$	$C_{22}H_{30}N_2O_3$	$C_{24}H_{34}N_2O_3$
Molar mass (g/mol)	342.442	370.496	398.55
CAS-RN	113787-54-5	107266-21-7	122055-52-1

Table 5.1-4 Two-ring systems with bridges, cont.

Number/common name	79	80	81
Temperatures of phase transitions T^{tr} (°C)	Cr 105.8 N 136.6 Is	Cr′ 68.0 Cr 76.0 N 124.0 Is	Cr 81.0 (C 74.0) N 129.0 Is
Enthalpies of phase transitions ΔH^{tr} (kJ/mol)	20.9 (Cr–N), 1.0 (N–Is)	14.6 (Cr–N), 0.7 (N–Is)	41.4 (Cr–N), 1.0 (N–Is)
Crystallographic space group	–	$P\bar{1}$	–
Order parameter S	0.590 around ($T = 0.90 \times T^{N-Is}$)	0.590 around ($T = 0.90 \times T^{N-Is}$)	0.610 around ($T = 0.90 \times T^{N-Is}$)
Density ϱ (g/cm³)	1.031 (N, 127 °C), 1.007 (Is, 147 °C)	1.072 (N, 120 °C), 1.063 (Is, 123 °C)	0.990 (N, 125 °C), 0.979 (Is, 132 °C)
Refractive index n (589 nm)	n_o 1.516 (127 °C)	n_e 1.812, n_o 1.518 (103.3 °C)	n_e 1.731, n_o 1.505 (107.5 °C)
Dielectric anisotropy $\Delta\varepsilon$	−0.38 (110 °C, 0.8 MHz)	−0.29 (105 °C)	−0.65 (99.5 °C, 0.8 MHz)
Dynamic viscosity η (mPa s)	6.9 (N, 131.7 °C)	13.5 (N, 120 °C)	10.5 (N, 128.1 °C)
Surface tension (mN/m)	32.0 (N, 110 °C), 29.4 (Is, 140 °C)	30.9 (N, 120 °C), 30.6 (Is, 140 °C)	28.4 (N, 120 °C), 27.4 (Is, 140 °C)
Heat capacity C_p (J/(mol K))	759 (N, 120 °C), 761 (Is, 143 °C)	718 (Cr, 70 °C), 808 (N, 100 °C), 854 (Is, 125 °C)	828 (Cr, 69 °C), 978 (N, 120 °C), 917 (Is, 155 °C)
Diamagnetic anisotropy $\Delta\chi$ (10^{-11} m³/kg)	10.2 (N, 122 °C)	8.9 (N, 112 °C)	9.5 (N, 109 °C)
Sound velocity v (m/s)	1235 (N, 125 °C, 3.1 MHz)	1205 (N, 122 °C, 3.1 MHz)	1160 (N, 125 °C, 3.1 MHz)
Diffusion coefficient D (m²/s)	$D_\perp$ 5.5×10⁻⁸, $D_\parallel$ 12×10⁻⁸ (N, 117.6 °C)	$D_\perp$ 6.0×10⁻⁸, $D_\parallel$ 12×10⁻⁸ (N, 110 °C)	$D_\perp$ 6.0×10⁻⁸, $D_\parallel$ 13×10⁻⁸ (N, 113.6 °C)
Dipole moment μ (D)	2.38	2.35	2.35

Number/common name	82	83 (EPAB)	84 (N-4)
Substance	H₁₅C₇O–C₆H₄–N=N(O)–C₆H₄–OC₇H₁₅	H₅C₂O–C₆H₄–N(O)=N–C₆H₄–COOC₂H₅	H₉C₄–C₆H₄–N=N(O)–C₆H₄–OCH₃
Formula	$C_{26}H_{38}N_2O_3$	$C_{18}H_{18}N_2O_5$	$C_{17}H_{20}N_2O_2$
Molar mass (g/mol)	426.604	342.355	284.361
CAS-RN	70906-50-2	6421-04-1	102135-46-6
Temperatures of phase transitions T^{tr} (°C)	Cr 74.4 C 95.4 N 124.2 Is	Cr′ 102.0 Cr 115.8 A 123.1 Is	Cr 16.0 N 76.0 Is
Enthalpies of phase transitions ΔH^{tr} (kJ/mol)	40.9 (Cr–C), 1.6 (C–N), 1.0 (N–Is)	19.7 (Cr–A), 5.0 (A–Is)	–
Crystallographic space group	$P\bar{1}$	$C2/c$, $P\bar{1}$ (Cr′)	–
Order parameter S	0.615 ($T = 0.90 \times T^{N-Is}$)	–	0.65 (19 °C)
Density ϱ (g/cm³)	0.994 (C, 94 °C), 0.985 (N, 105 °C)	1.146 (Is, 123 °C), 1.138 (Is, 135 °C)	1.1217 (N, 20 °C), 1.1067 (N, 40 °C)
Refractive index n	n_e 1.673, n_o 1.512 (589 nm, N, 117 °C)	–	Δn 0.45 (450 nm), 0.34 (550 nm), 0.30 (650 nm) (20 °C)
Dynamic viscosity η (mPa s)	15.1 (N, 123.5 °C)	–	–
Surface tension (mN/m)	–	26.0 (A, 120 °C), 26.3 (Is, 160 °C)	37.3 (N, 23 °C), 35 (N, 76 °C)

Table 5.1-4 Two-ring systems with bridges, cont.

Number/common name	82	83 (EPAB)	84 (N-4)
Heat capacity C_p (J/(mol K))	887 (C, 77 °C), 1004 (N, 102 °C), 987 (Is, 127 °C)	675 (A, 116.2 °C), 676 (Is, 130 °C)	511 (N, 61 °C)
Diamagnetic anisotropy $\Delta\chi$	7.5×10^{-11} m^3/kg (116 °C)	–	6.9×10^{-10} m^3/mol (75.3 °C)
Sound velocity v (m/s)	1300 (80 °C), 1250 (100 °C) (12 MHz)	1276 (2 MHz, 117.4 °C)	–
Diffusion coefficient D (m^2/s)	$D_\perp$ 3.7×10^{-10}, $D_\parallel$ 6×10^{-10} (1000/T = 2.60)	–	–
Dipole moment μ (D)	2.36	–	–

Number/common name	85 (RO-CM-1500)	86 (RO-CM-1530)	87 (RO-CM-1510)
Substance	H$_9$C$_4$–⟨⟩–COO–⟨⟩–CN	H$_{11}$C$_5$–⟨⟩–COO–⟨⟩–CN	H$_{13}$C$_6$–⟨⟩–COO–⟨⟩–CN
Formula	C$_{18}$H$_{17}$NO$_2$	C$_{19}$H$_{19}$NO$_2$	C$_{20}$H$_{21}$NO$_2$
Molar mass (g/mol)	279.342	293.369	307.396
CAS-RN	38690-77-6	49763-64-6	50793-85-6
Temperatures of phase transitions T^{tr} (°C)	Cr 67.1 (N 42.6) Is	Cr 64.4 (N 55.4) Is	Cr 44.4 N 48.6 Is
Enthalpies of phase transitions ΔH^{tr} (kJ/mol)	28.0 (Cr–Is)	25.1 (Cr–Is)	40.2 (Cr–N)
Crystallographic space group	–	$P2_1/n$	–
Density ϱ (g/cm^3)	1.095 (N, 40 °C), 1.061 (Is, 80 °C)	1.076 (N, 47.4 °C), 1.053 (Is, 67.4 °C)	1.064 (N, 36.6 °C), 1.043 (Is, 56.6 °C)
Refractive index n	n_e 1.639, n_o 1.525 (589 nm, N, 39 °C)	n_e 1.669, n_o 1.512 (546 nm, N, 47.4 °C)	n_e 1.657, n_o 1.514 (546 nm, N, 36.6 °C)
Dielectric constant ε (1.592 kHz, $T = 0.98 T^{N-Is}$)	$\varepsilon_\parallel$ 34.3, $\varepsilon_\perp$ 12.0	$\varepsilon_\parallel$ 30.2, $\varepsilon_\perp$ 10.2	$\varepsilon_\parallel$ 29.3, $\varepsilon_\perp$ 9.7
Surface tension (mN/m)	–	28.2 (N, 38 °C), 30.7 (53 °C)	–
Dipole moment μ (D)	5.77	–	–

Number/common name	88 (RO-CM-1540)	89	90
Substance	H$_{15}$C$_7$–⟨⟩–COO–⟨⟩–CN	H$_{17}$C$_8$–⟨⟩–COO–⟨⟩–CN	H$_{17}$C$_8$–O–⟨⟩–COO–⟨⟩–NO$_2$
Formula	C$_{21}$H$_{23}$NO$_2$	C$_{22}$H$_{25}$NO$_2$	C$_{21}$H$_{25}$NO$_5$
Molar mass (g/mol)	321.423	335.45	371.437
CAS-RN	38690-76-5	50793-86-7	52910-78-8
Temperatures of phase transitions T^{tr} (°C)	Cr 44.0 N 56.5 Is	Cr 47.0 N 55.0 Is	Cr' 47.5 Cr 50.5 A 61.4 N 68.1 Is
Enthalpies of phase transitions ΔH^{tr} (kJ/mol)	31.5 (Cr–N), 1.0 (N–Is)	38.5 (Cr–N)	18.4 (Cr–A), 0.2 (A–N), 0.4 (N–Is)
Crystallographic space group	$P2_1/n$	–	$P2_1/c$
Order parameter S	0.574 (44 °C), 0.42 (56 °C)	–	0.605 (A, 44 °C), 0.31 (N, 66 °C)

Table 5.1-4 Two-ring systems with bridges, cont.

Number/common name	88 (RO-CM-1540)	89	90
Density ϱ (g/cm^3)	1.050 (N, 48 °C), 1.034 (Is, 60 °C)	1.042 (N, 45.8 °C), 1.021 (Is, 65.8 °C)	1.123 (A, 52 °C), 1.110 (N, 66 °C), 1.106 (Is, 70 °C)
Refractive index n	n_e 1.649, n_o 1.505 (546 nm, N, 45 °C)	n_e 1.627, n_o 1.508 (546 nm, N, 50.8 °C)	n_e 1.644, n_o 1.504 (589 nm, N, 55 °C)
Dielectric constant ε	$\varepsilon_\parallel$ 26.4, $\varepsilon_\perp$ 8.7 (1.592 kHz, $T = 0.98 T^{\text{N-Is}}$)	$\varepsilon_\parallel$ 24.3, $\varepsilon_\perp$ 8.5 (1.592 kHz, $T = 0.98 T^{\text{N-Is}}$)	$\Delta\varepsilon$ 11.1 (60 °C), $\Delta\varepsilon$ 11.4 (65 °C) (1 kHz)
Dynamic viscosity η (mPa s)	22 (50 °C), 22 (60 °C)	–	–
Surface tension (mN/m)	27 (22 °C), 25.4 (45 °C)	19.6 (43 °C), 22.5 (58 °C)	–
Sound velocity v (m/s)	1235 (N, 125 °C, 3.1 MHz)	–	1372 (58 °C, 4.74 GHz)
Dipole moment μ (D)	6.1 (CCl$_4$)	–	6.04

Number/common name	91	92	93
Substance	H$_{11}$C$_5$–O–C$_6$H$_4$–C(O)O–C$_6$H$_4$–CN	H$_{15}$C$_7$–O–C$_6$H$_4$–C(O)O–C$_6$H$_4$–CN	H$_{17}$C$_8$–O–C$_6$H$_4$–C(O)O–C$_6$H$_4$–CN
Formula	C$_{19}$H$_{19}$NO$_3$	C$_{21}$H$_{23}$NO$_3$	C$_{22}$H$_{25}$NO$_3$
Molar mass (g/mol)	309.368	337.422	351.449
CAS-RN	50649-73-5	50793-88-9	50793-89-0
Temperatures of phase transitions T^{tr} (°C)	Cr 87.0 (N 78.0) Is	Cr 71.5 N 82.0 Is	Cr 75.6 N 88.0 Is
Enthalpies of phase transitions ΔH^{tr} (kJ/mol)	32.2 (Cr–Is)	34.3 (Cr–N)	40.6 (Cr–N)
Crystallographic space group	$Pnam$	$P2_1/a$	$P\bar{1}$
Density ϱ (g/cm^3)	1.093 (N, 69 °C), 1.075 (Is, 85.5 °C)	1.061 (N, 73 °C), 1.045 (Is, 88 °C)	1.044 (N, 79 °C), 1.028 (Is, 94 °C)
Refractive index n (546 nm)	n_e 1.662, n_o 1.513 (76.7 °C)	n_e 1.633, n_o 1.508 (73.3 °C)	n_e 1.627, n_o 1.502 (79 °C)
Dielectric constant ε	$\varepsilon_\parallel$ 27.8, $\varepsilon_\perp$ 11.7 (71.8 °C)	–	–
Surface tension (mN/m)	29.1 (76 °C), 26.4 (91 °C)	25.0 (73 °C), 22.9 (88 °C)	23.2 (76 °C), 21.1 (91 °C)
Sound velocity v (m/s)	1359 (83 °C, 2 MHz)	–	–
Dipole moment μ (D)	6.6 (CCl$_4$)	–	–

Number/common name	94 (ME105, ZLI-0245)	95 (ME605, ZLI-1004)	96
Substance	H$_3$C–O–C$_6$H$_4$–C(O)O–C$_6$H$_4$–C$_5$H$_{11}$	H$_{13}$C$_6$–O–C$_6$H$_4$–C(O)O–C$_6$H$_4$–C$_5$H$_{11}$	H$_9$C$_4$–C$_6$H$_4$–C(O)O–C$_6$H$_4$–O–C$_6$H$_{13}$
Formula	C$_{19}$H$_{22}$O$_3$	C$_{24}$H$_{32}$O$_3$	C$_{23}$H$_{30}$O$_3$
Molar mass (g/mol)	298.385	368.521	354.494
CAS-RN	–	38444-15-4	38454-28-3
Temperatures of phase transitions T^{tr} (°C)	Cr 29.5 N 43.5 Is	Cr 50.0 N 63.0 Is	Cr 31.3 N 48.6 Is
Enthalpies of phase transitions ΔH^{tr} (kJ/mol)	17.6 (Cr–N), 0.6 (N–Is)	21.9 (Cr–N)	17.6 (Cr–N)

Table 5.1-4 Two-ring systems with bridges, cont.

Number/common name	94 (ME105, ZLI-0245)	95 (ME605, ZLI-1004)	96
Density ϱ (g/cm^3)	1.078 (N, 31 °C), 1.061 (Is, 45 °C)	1.019 (N, 50 °C), 1.001 (Is, 65 °C)	1.020 (N, 41 °C), 1.006 (Is, 53 °C)
Refractive index n (589 nm)	n_e 1.639, n_o 1.515 (35 °C)	n_e 1.613, n_o 1.490 (47.4 °C)	n_e 1.599, n_o 1.498 (41 °C)
Dielectric anisotropy $\Delta\varepsilon$	+0.11 (38 °C)	+0.1 (20 °C, 1 kHz)	–
Kinematic viscosity ν (mm^2/s)	38 (20 °C)	87 (20 °C)	–
Surface tension (mN/m)	24.5 (20 °C)	–	27.5 (41.5 °C)
Dipole moment μ (D)	$\mu_\parallel$ 2.18, $\mu_\perp$ 2.15 (38 °C)	–	2.3 (CCl$_4$)

Number/common name	97	98	99
Substance	H$_{13}$C$_6$–O–C$_6$H$_4$–COO–C$_6$H$_4$–O–C$_4$H$_9$	H$_{17}$C$_8$–O–C$_6$H$_4$–COO–C$_6$H$_4$–O–C$_6$H$_{13}$	H$_{21}$C$_{10}$–O–C$_6$H$_4$–COO–C$_6$H$_4$–O–C$_6$H$_{13}$
Formula	C$_{23}$H$_{30}$O$_4$	C$_{27}$H$_{38}$O$_4$	C$_{29}$H$_{42}$O$_4$
Molar mass (g/mol)	370.493	426.602	454.656
CAS-RN	38454-24-9	54963-63-2	68162-09-4
Temperatures of phase transitions T^{tr} (°C)	Cr 66.0 N 89.5 Is	Cr 55.0 C 66.0 N 89.0 Is	Cr 62.5 (E 38.0 B 44.5) C 77.5 A 83.3 N 88.9 Is
Enthalpies of phase transitions ΔH^{tr} (kJ/mol)	35.0 (Cr–N), 1.3 (N–Is)	38.0 (Cr–C), 0.4 (C–N), 2.1 (N–Is)	45.3 (Cr–C), 4.7 (B–C), 0.7 (A–N), 2.1 (N–Is)
Order parameter S	0.49 (84 °C)	0.49 (84.5 °C)	0.51 (84 °C)
Density ϱ (g/cm^3)	1.015 (N, 88 °C)	0.9821 (N, 83 °C)	0.9735 (N, 84.9 °C)
Refractive index n	n_e 1.518, n_o 1.394 (80 °C, 650 nm)	n_e 1.5811, n_o 1.4828 (83.5 °C, 589 nm)	–
Dielectric constant ε (1 kHz)	$\Delta\varepsilon$ –0.27 (79 °C)	$\Delta\varepsilon$ –0.20 (84 °C)	$\varepsilon_\parallel$ 3.85, $\varepsilon_\perp$ 4.05 (85 °C)
Dynamic viscosity η (mPa s)	–	11.4 (80 °C)	–
Sound velocity v (m/s)	–	1383 (83 °C, 0.36 MHz)	1285 (80 °C, 0.36 MHz)
Dipole moment μ (D), 25 °C	2.67 (CCl$_4$)	2.86 (C$_6$H$_6$)	3.31 (CCl$_4$)

Number/common name	100 (D55, ZLI-1497)	101 (D301, ZLI-2469)	102 (D302, ZLI-1496)
Substance	H$_{11}$C$_5$–cyclohexyl–COO–C$_6$H$_4$–C$_5$H$_{11}$	H$_7$C$_3$–cyclohexyl–COO–C$_6$H$_4$–O–CH$_3$	H$_7$C$_3$–cyclohexyl–COO–C$_6$H$_4$–O–C$_2$H$_5$
Formula	C$_{23}$H$_{36}$O$_2$	C$_{17}$H$_{24}$O$_3$	C$_{18}$H$_{26}$O$_3$
Molar mass (g/mol)	344.542	276.379	290.406
CAS-RN	67589-72-4	–	67589-39-3
Temperatures of phase transitions T^{tr} (°C)	Cr 36.0 (A 29.0) N 48.0 Is	Cr 55.5 N 62.5 Is	Cr 49.0 N 79.8 Is
Enthalpies of phase transitions ΔH^{tr} (kJ/mol)	28.9 (Cr–N), 0.55 (N–Is)	28.0 (Cr–N)	27.6 (Cr–N)
Order parameter S	0.67 (39 °C)	–	0.71 (60 °C)
Density ϱ (g/cm^3)	0.9508 (N, 36 °C)	–	1.0118 (51 °C)
Refractive index n (589 nm)	n_e 1.5304, n_o 1.4730 (36 °C)	Δn 0.09 (20 °C)	n_e 1.5304, n_o 1.4684 (60 °C)
Dielectric constant ε (1 kHz)	$\varepsilon_\parallel$ 2.958, $\varepsilon_\perp$ 3.396 (40 °C)	$\Delta\varepsilon$ –1.8 (20 °C)	$\varepsilon_\parallel$ 3.28, $\varepsilon_\perp$ 4.54 (51 °C)
Kinematic viscosity ν (mm^2/s) 20 °C	13 (extra)	12 (extra)	14 (extra)
Dipole moment μ (D)	1.99	–	2.17

Table 5.1-4 Two-ring systems with bridges, cont.

Number/common name	103 (D401, ZLI-2470)	104 (D402, RO-CM-1942, ZLI-1563)	105 (D501, RO-CM-1951, ZLI-1495)
Substance	H_9C_4—⬡—COO—⌬—O—CH_3	H_9C_4—⬡—COO—⌬—O—C_2H_5	$H_{11}C_5$—⬡—COO—⌬—O—CH_3
Formula	$C_{18}H_{26}O_3$	$C_{19}H_{28}O_3$	$C_{19}H_{28}O_3$
Molar mass (g/mol)	290.406	304.433	304.433
CAS-RN	67589-46-2	67589-47-3	67589-52-0
Temperatures of phase transitions T^{tr} (°C)	Cr 42.0 N 59.0 Is	Cr 36.3 N 74.6 Is	Cr 40.9 N 71.3 Is
Enthalpies of phase transitions ΔH^{tr} (kJ/mol)	23.0 (Cr–N)	21.4 (Cr–N)	21.0 (Cr–N), 0.5 (N–Is)
Order parameter S	–	0.73 (40 °C)	0.71 (50 °C)
Density ϱ (g/cm^3)	–	1.059 (40 °C)	1.050 (48 °C)
Refractive index n (589 nm)	Δn 0.09 (20 °C)	n_e 1.5508, n_o 1.4730 (40 °C)	n_e 1.5389, n_o 1.4742 (50 °C)
Dielectric constant ε (1 kHz)	$\Delta\varepsilon$ −1.8 (20 °C)	$\varepsilon_\perp$ 4.20, $\Delta\varepsilon$ −0.91 (65.4 °C)	$\varepsilon_\parallel$ 3.37, $\varepsilon_\perp$ 4.28 (51 °C)
Viscosity	ν 14 mm^2/s (20 °C)	η 20 mPa s (22 °C)	–
Diamagnetic anisotropy $\Delta\chi$ (10^{-11} m^3/kg)	–	3.0 (52 °C)	3.0 (52.5 °C)
Dipole moment μ (D)	–	2.08 (C_6H_6)	2.23 (C_6H_6)

Number/common name	106 (OS33, RO-CM-1633, ZLI-2155)	107 (OS53, RO-CM-1653, ZLI-2256)	108 (OS55, ZLI-2495)
Substance	H_7C_3—⬡—COO—⬡—C_3H_7	$H_{11}C_5$—⬡—COO—⬡—C_3H_7	$H_{11}C_5$—⬡—COO—⬡—C_5H_{11}
Formula	$C_{19}H_{34}O_2$	$C_{21}H_{38}O_2$	$C_{23}H_{42}O_2$
Molar mass (g/mol)	294.482	322.536	350.59
CAS-RN	73255-62-6	73255-65-9	–
Temperatures of phase transitions T^{tr} (°C)	Cr 22.8 N 36.6 Is	Cr 24.5 B 37.5 N 52.0 Is	Cr 52.0 B 72.0 Is
Enthalpies of phase transitions ΔH^{tr} (kJ/mol)	–	29.3 (Cr–S)	31.0 (Cr–S)
Anisotropy of refractive index Δn (589 nm)	0.036 (25 °C)	0.05 (extra, 20 °C)	–
Dielectric anisotropy $\Delta\varepsilon$ (1 kHz)	−1.1 (25 °C)	−1.5 (20 °C)	−1.6 (20 °C)
Kinematic viscosity ν (mm^2/s) (20 °C)	11	14	15

Number/common name	109	110	111
Substance	$H_{11}C_5$—⬡—COO—⌬—C_5H_{11}	$H_{11}C_5$—⬡—COO—⌬(F)—⬡—C_5H_{11}	NC—⌬(F)—OOC—⬡—C_5H_{11}
Formula	$C_{25}H_{38}O_2$	$C_{23}H_{35}FO_2$	$C_{19}H_{24}FNO_2$

Table 5.1-4 Two-ring systems with bridges, cont.

Number/common name	109	110	111
Molar mass (g/mol)	370.58	362.533	317.407
Temperatures of phase transitions T^{tr} (°C)	Cr 31.0 N 64.5 Is	Cr 19.0 N 37.9 Is	Cr 75.5 N 93.5 Is
Enthalpies of phase transitions ΔH^{tr} (kJ/mol)	29.8 (Cr–N)	33.5 (Cr–N)	–
Refractive index n	n_e 1.5337, n_o 1.4770 ($T = 0.97 \times T^{\text{N-Is}}$, 633 nm)	Δn 0.0608, n_e 1.5273 ($T = 0.97 \times T^{\text{N-Is}}$, 633 nm)	Δn 0.093 ($T = 0.95 \times T^{\text{N-Is}}$, 589 nm)
Dielectric constant ε (1 kHz)	$\varepsilon_\parallel$ 2.79, $\varepsilon_\perp$ 3.18 ($T = 0.97 \times T^{\text{N-Is}}$)	–	$\Delta \varepsilon$ +4.4 ($T = 0.95 \times T^{\text{N-Is}}$)
Viscosity (20 °C)	η 35 mPa s (extra)	ν 16 mm²/s (extra)	–

Table 5.1-5 Three and four-ring systems

Number/common name	112 (RO-CM-5515, T15)	113	114
Substance	$H_{11}C_5$–⬡–⬡–⬡–CN	NC–⬡–⬡(F)–⬡–C_5H_{11}	H_7C_3–⬡–⬡(F,F)–⬡–C_3H_7
Formula	$C_{24}H_{23}N$	$C_{24}H_{22}FN$	$C_{24}H_{24}F_2$
Molar mass (g/mol)	325.458	343.448	350.456
CAS-RN	54211-46-0	–	–
Temperatures of phase transitions T^{tr} (°C)	Cr' 80.0 Cr' 115.0 Cr 131.0 N 240.0 Is	Cr 97.0 N 189.0 Is	Cr 95.0 N 131.0 Is
Enthalpies of phase transitions ΔH^{tr} (kJ/mol)	17.2 (Cr–N), 0.9 (N–Is)	–	24.3 (Cr–N)
Anisotropy of refractive index Δn (extra, 20 °C, 589 nm)	0.30	0.343	0.24
Dielectric anisotropy $\Delta \varepsilon$ (1 kHz)	+13, $\varepsilon_\parallel$ 16.0 ($T = 0.75 \times T^{\text{N-Is}}$)	+10, (20 °C, extra)	−1.7, (20 °C, extra)
Kinematic viscosity ν (mm²/s)	86 (extra, 20 °C)	195 (extra, 20 °C)	31 (extra, 20 °C)

Number/common name	115 (BCH5, ZLI-1131)	116 (BCH32, ZLI-1639)	117 (BCH52, ZLI-1409)
Substance	NC–⬡–⬡–⬢–C_5H_{11}	H_5C_2–⬡–⬡–⬢–C_3H_7	H_5C_2–⬡–⬡–⬢–C_5H_{11}
Formula	$C_{24}H_{29}N$	$C_{23}H_{30}$	$C_{25}H_{34}$
Molar mass (g/mol)	331.505	306.496	334.55
Temperatures of phase transitions T^{tr} (°C)	Cr 96.0 N 222.0 Is	Cr 66.0 S 134.0 N 166.0 Is	Cr 34.0 B 146.0 N 164.0 Is
Enthalpies of phase transitions ΔH^{tr} (kJ/mol)	21.1 (Cr–N), 0.7 (N–Is)	–	18.5 (Cr–B), 6.8 (B–N), 0.5 (N–Is)
Crystallographic space group	$P2_1/c$	–	$C2/c$
Density ϱ (g/cm³)	1.196 (101 °C)	–	–
Refractive index n	n_e 1.734, n_o 1.515 (101 °C, 546 nm)	Δn 0.18 (20 °C, 589 nm)	Δn 0.18 (20 °C, 589 nm)
Dielectric constant ε (1 kHz)	$\varepsilon_\parallel$ 16.8, $\varepsilon_\perp$ 4.9 (20 °C, extra)	$\Delta \varepsilon$ +0.5 (20 °C, extra)	$\Delta \varepsilon$ +0.4 (20 °C, extra)
Kinematic viscosity ν (mm²/s)	90 (extra, 20 °C)	20 (extra, 20 °C)	20 (extra, 20 °C)

Table 5.1-5 Three and four-ring systems, cont.

Number/common name	118 (BCH52F)	119	120
Substance	$H_{11}C_5$—⟨⟩—⟨F⟩—C_2H_5	$H_{11}C_5$—⟨⟩—⟨F⟩—CN	$H_{11}C_5$—⟨⟩—⟨F,F,F⟩
Formula	$C_{25}H_{33}F$	$C_{24}H_{28}FN$	$C_{23}H_{27}F_3$
Molar mass (g/mol)	352.54	349.496	360.467
CAS-RN	–	–	137019-95-5
Temperatures of phase transitions T^{tr} (°C)	Cr 37.0 N 117.0 Is	Cr 84.0 N 175.4 Is	Cr 30.4 N 58.0 Is
Enthalpies of phase transitions ΔH^{tr} (kJ/mol)	23.0 (Cr–N)	–	18.0 (Cr–N), 0.2 (N–Is)
Refractive index n	Δn 0.096 (633 nm) ($T = 0.97 \times T^{N-Is}$)	Δn 0.201 (20 °C, extra, 589 nm)	Δn 0.134 (25 °C, extra, 589 nm)
Dielectric constant ε (1 kHz)	$\varepsilon_\parallel$ 2.97, $\varepsilon_\perp$ 2.85 ($T = 0.97 \times T^{N-Is}$)	$\Delta \varepsilon$ +20.3 (20 °C, extra)	$\Delta \varepsilon$ +11.3 (20 °C, extra)
Viscosity (20 °C, extra)	ν 27 mm²/s	ν 110 mm²/s	η 32.1 mPa s

Number/common name	121	122 (BCN55, ZLI-2769)	123
Substance	$H_{11}C_5$—⟨⟩—⟨⟩—⟨F,F,F⟩	$H_{11}C_5$—⟨⟩—⟨CN⟩—⟨⟩—C_5H_{11}	NC—⟨⟩—⟨⟩—⟨⟩—C_3H_7
Formula	$C_{23}H_{33}F_3$	$C_{29}H_{51}N$	$C_{24}H_{35}N$
Molar mass (g/mol)	366.515	413.737	337.553
Temperatures of phase transitions T^{tr} (°C)	Cr 87.3 N 101.2 Is	Cr 33.0 B 176.0 A 185.0 N 198.7 Is	Cr 69.0 N 196.0 Is
Enthalpies of phase transitions ΔH^{tr} (kJ/mol)	26.4 (Cr–N), 0.6 (N–Is)	17.6 (Cr–B)	–
Anisotropy of refractive index Δn (589 nm)	0.085 (25 °C, extra)	0.06 (20 °C, extra)	0.17 (20 °C, extra)
Dielectric anisotropy $\Delta \varepsilon$ (extra, 20 °C, 1 kHz)	+7.8	–4.7	+12
Kinematic viscosity ν (mm²/s)	20 (extra, 20 °C)	110 (extra, 20 °C)	75 (extra, 20 °C)

Number/common name	124 (I32)	125 (I35)	126 (I52)
Substance	H_5C_2—⟨F⟩—⟨⟩—⟨⟩—C_3H_7	$H_{11}C_5$—⟨F⟩—⟨⟩—⟨⟩—C_3H_7	H_5C_2—⟨F⟩—⟨⟩—⟨⟩—C_5H_{11}
Formula	$C_{25}H_{33}F$	$C_{28}H_{39}F$	$C_{27}H_{37}F$
Molar mass (g/mol)	352.54	394.621	380.594
CAS-RN		100497-33-4	95379-18-3
Temperatures of phase transitions T^{tr} (°C)	Cr 27.0 N 97.0 Is	Cr 30.0 N 106.0 Is	Cr 24.0 (B 14.0) N 103.0 Is
Enthalpies of phase transitions ΔH^{tr} (kJ/mol)	–	–	14.2 (Cr–N), 1.3 (N–Is)
Density ϱ (g/cm³)	1.0156 (N, 25 °C)	0.9975 (N, 25 °C)	1.003 (N, 25 °C)
Refractive index n ($T = 0.95 \times T^{N-Is}$, 633 nm)	n_e 1.6108, n_o 1.4893	n_e 1.6021, n_o 1.4851	n_e 1.6013, n_o 1.4824

Table 5.1-5 Three and four-ring systems, cont.

Number/common name	124 (I32)	125 (I35)	126 (I52)
Dielectric constant ε	$\varepsilon_\parallel$ 2.94, $\varepsilon_\perp$ 2.93 ($T = 0.95 \times T^{N-Is}$, 0.5 kHz)	$\varepsilon_\parallel$ 2.92, $\varepsilon_\perp$ 2.89 ($T = 0.95 \times T^{N-Is}$, 0.5 kHz)	$\Delta\varepsilon$ −0.06, $\varepsilon_\parallel$ (25 °C, 1 kHz) 2.964, $\varepsilon_\perp$ 3.024
Kinematic viscosity ν (mm^2/s)	19 (20 °C)	24.5 (20 °C)	25.1 (25 °C)

Number/common name	127 (HP5N, ZLI-1226)	128 (HP33, ZLI-1222)	129 (HP35, ZLI-2429)
Substance	H$_{11}$C$_{5}$''–⟨⟩–⟨⟩–C(O)O–⟨⟩–CN	H$_{7}$C$_{3}$''–⟨⟩–⟨⟩–C(O)O–⟨⟩–C$_{3}$H$_{7}$	H$_{7}$C$_{3}$''–⟨⟩–⟨⟩–C(O)O–⟨⟩–C$_{5}$H$_{11}$
Formula	C$_{25}$H$_{29}$NO$_{2}$	C$_{25}$H$_{32}$O$_{2}$	C$_{27}$H$_{36}$O$_{2}$
Molar mass (g/mol)	375.515	364.533	392.587
Temperatures of phase transitions T^{tr} (°C)	Cr' 82.0 Cr 111.5 N 226.0 Is	Cr 87.0 N 186.0 Is	Cr 83.0 (S 79.0) N 175.0 Is
Enthalpies of phase transitions ΔH^{tr} (kJ/mol)	5.4 (Cr–Cr'), 19.0 (Cr–N), 1.4 (N–Is)	20.9 (Cr–N)	22.0 (Cr–N)
Density ϱ (g/cm^3)	–	1.141 (N, 99.5 °C)	–
Refractive index n	Δn 0.16 (20 °C, 589 nm)	n_e 1.655, n_o 1.509 (99.5 °C, 546 nm)	Δn 0.14 (20 °C, 589 nm)
Dielectric constant ε (1 kHz)	$\varepsilon_\parallel$ 24.4, $\varepsilon_\perp$ 7.5 (156.0 °C)	$\Delta\varepsilon$ +0.4 (20 °C)	$\Delta\varepsilon$ +0.6 (20 °C)
Kinematic viscosity ν (mm^2/s)	220 (extra, 20 °C)	60 (extra, 20 °C)	41 (extra, 20 °C)

Number/common name	130 (HH33, ZLI-1224)	131 (HH53, ZLI-1223)	132 (HD34, ZLI-1749)
Substance	H$_{7}$C$_{3}$''–⟨⟩–⟨⟩–C(O)O–C$_{3}$H$_{7}$	H$_{11}$C$_{5}$''–⟨⟩–⟨⟩–C(O)O–C$_{3}$H$_{7}$	H$_{9}$C$_{4}$''–⟨⟩–C(O)O–⟨⟩–''C$_{3}$H$_{7}$
Formula	C$_{25}$H$_{38}$O$_{2}$	C$_{27}$H$_{42}$O$_{2}$	C$_{26}$H$_{40}$O$_{2}$
Molar mass (g/mol)	370.58	398.635	384.63
Temperatures of phase transitions T^{tr} (°C)	Cr 94.0 N 158.0 Is	Cr 67.0 (S 43.0 A 55.0) N 155.0 Is	Cr 64.0 S 97.0 S 116.0 N 189.0 Is
Enthalpies of phase transitions ΔH^{tr} (kJ/mol)	18.1 (Cr–N)	21.8 (Cr–N)	13.5 (Cr–S)
Crystallographic space group	$P\bar{1}$	$P\bar{1}$	
Anisotropy of refractive index Δn (extra, 589 nm)	0.19 (20 °C)	0.092 (N, 80 °C)	0.11 (20 °C)
Dielectric anisotropy $\Delta\varepsilon$ (20 °C, 1 kHz)	−1	−1	−1.4
Kinematic viscosity ν (mm^2/s)	130 (extra, 20 °C)	119 (extra, 20 °C)	40 (extra, 20 °C)

Number/common name	133 (CH33)	134 (CH35)	135 (CH45)
Substance	H$_{7}$C$_{3}$''–⟨⟩–⟨⟩–C(O)O''–⟨⟩–C$_{3}$H$_{7}$	H$_{7}$C$_{3}$''–⟨⟩–⟨⟩–C(O)O''–⟨⟩–C$_{5}$H$_{11}$	H$_{9}$C$_{4}$''–⟨⟩–⟨⟩–C(O)O''–⟨⟩–C$_{5}$H$_{11}$
Formula	C$_{25}$H$_{44}$O$_{2}$	C$_{27}$H$_{48}$O$_{2}$	C$_{28}$H$_{50}$O$_{2}$
Molar mass (g/mol)	376.628	404.682	418.71
Temperatures of phase transitions T^{tr} (°C)	Cr 58.0 S 155.0 N 189.0 Is	Cr 53.0 S 169.0 N 188.0 Is	Cr 38.0 S 181.0 N 186.0 Is
Enthalpies of phase transitions ΔH^{tr} (kJ/mol)	24.3 (Cr–S)	30.1 (Cr–S)	22.2 (Cr–S)
Anisotropy of refractive index Δn (extra, 589 nm)	0.06 (20 °C)	0.06 (20 °C)	0.05 (20 °C)

Table 5.1-5 Three and four-ring systems, cont.

Number/common name	133 (CH33)	134 (CH35)	135 (CH45)
Dielectric anisotropy $\Delta\varepsilon$ (20 °C, 1 kHz)	−1.8	−1.6	−1.7
Kinematic viscosity ν (mm^2/s)	31 (extra, 20 °C)	34 (extra, 20 °C)	37 (extra, 20 °C)

Number/common name	136 (CBC53, ZLI-1544)	137 (CBC33, ZLI-1987)	138
Substance	H$_7$C$_{3'}$–◯–◯–◯–C$_5$H$_{11}$	H$_7$C$_{3'}$–◯–◯–◯–C$_3$H$_7$	H$_{11}$C$_{5'}$–◯–◯–◯(F,F,F)
Formula	C$_{32}$H$_{46}$	C$_{30}$H$_{42}$	C$_{29}$H$_{37}$F$_3$
Molar mass (g/mol)	430.723	402.669	442.613
CAS-RN	80955-71-1	85600-56-2	–
Temperatures of phase transitions (°C)	Cr′ 54.0 Cr 58.0 B 232.0 A 251.0 N 311.0 Is	Cr′ 108.0 Cr′ 127.0 Cr 155.0 B 210.0 A 220.0 N 325.0 Is	Cr 87.8 N >250.0 Is
Enthalpies of phase transitions ΔH^{tr} (kJ/mol)	5.1 (Cr–Cr′), 1.4 (Cr–B), 6.1 (B–A), 0.5 (A–N), 1.1 (N–Is)	12.0 (Cr–B), 5.2 (B–A), 1.2 (N–Is)	–
Anisotropy of refractive index Δn (extra, 589 nm)	0.19 (20 °C)	0.19 (20 °C)	0.144 (25 °C)
Dielectric anisotropy $\Delta\varepsilon$ (extra, 20 °C, 1 kHz)	+0.4	+0.5	+11.3
Viscosity (extra, 20 °C)	ν 42 mm^2/s	ν 53 mm^2/s	η 51.1 mPa s

Table 5.1-6 Ferroelectric liquid crystals

Number/common name	139 (DOBAMBC)	140 (C-7)	141 (MHPOBC)
Substance	H$_{21}$C$_{10}$O–◯–N=CH–◯–CH=CH–COO–CH(CH$_3$)C$_2$H$_5$	H$_{15}$C$_7$O–◯–◯–COO–CH(CH$_3$)C$_2$H$_5$, Cl	H$_{17}$C$_8$O–◯–◯–COO–◯–COO–CH(CH$_3$)C$_6$H$_{13}$
Formula	C$_{31}$H$_{43}$NO$_3$	C$_{25}$H$_{33}$ClO$_3$	C$_{36}$H$_{46}$O$_5$
Molar mass (g/mol)	477.693	416.993	558.765
CAS-RN	97335-57-4	100497-43-6	103376-72-3
Temperatures of phase transitions T^{tr} (°C)	Cr 74.6 (I* 62.0) C* 94.0 A 117.0 Is	Cr 55.0 C* 55.0 A 62.0 Is	Cr 84.0 CA* 118.4 Cγ 119.2 C* 120.9 Cα 122.0 A 148.0 Is
Enthalpies of phase transitions ΔH^{tr} (J/mol)	2.58 × 10^4 (Cr–C*), 1.67 × 10^3 (I–C*), 5.15 × 10^3 (A–Is)	–	14.6 (C*–Cα), 16.4 (CA*–Cγ), 18.8 (Cα–C*), 288 (Cα–A), 6420 (A–Is)
Density ϱ (g/cm^3)	0.994 (C*, 92 °C), 0.985 (A, 100 °C), 0.959 (Is, 120 °C)	–	–
Refractive index n	n_e 1.700, n_o 1.490 (A, 110 °C, 633 nm)	–	–
Dynamic viscosity η (mPa s)	6.7 (90 °C)	–	–
Spontaneous polarization P_s (nC/cm^2)	+4.75 (65 °C)	+260 (45 °C)	+70 (139.5 °C)

Table 5.1-7 Cholesteryl (cholest-5-ene) substituted mesogens

Number/common name	142 (CC)	143	144
Substance	(structure)	(structure)	(structure)
Formula	$C_{27}H_{45}Cl_1$	$C_{27}H_{45}Br$	$C_{27}H_{45}I$
Molar mass (g/mol)	405.113	449.569	496.564
CAS-RN	910-31-6	516-91-6	2930-80-5
Temperatures of phase transitions T^{tr} (°C)	Cr 95.7 (N* 67.3) Is	Cr 100 (N* 69) Is	Cr 106.5 (N* 95) Is
Enthalpies of phase transitions ΔH^{tr} (kJ/mol)	54.4 (Cr–Is), 0.38 (N*–Is)	–	–
Crystallographic space group	$P2_1$	$P2_1$	–
Density ϱ (g/cm³)	0.98	–	–
Dielectric constant ε (10 kHz)	3.52 (60 °C), 5.02 (90 °C)	–	–
Dipole moment μ (D)	2.36 (CCl_4), 2.48 (C_6H_6)	2.68 (CCl_4), 2.40 (C_6H_6)	–

Number/common name	145	146	147 (CP)
Substance	(structure)	(structure)	(structure)
Formula	$C_{28}H_{46}O_2$	$C_{29}H_{48}O_2$	$C_{30}H_{50}O_2$
Molar mass (g/mol)	414.678	428.705	442.732
CAS-RN	4351-55-7	604-35-3	633-31-8
Temperatures of phase transitions T^{tr} (°C)	Cr 97.5 (N* 60.5) Is	Cr 116.5 (N* 94.5) Is	Cr 101.6 N* 115.2 Is
Enthalpies of phase transitions ΔH^{tr} (kJ/mol)	22.6 (Cr–Is)	20.1 (Cr–Is), 0.3 (N*–Is)	23.0 (Cr–N*), 0.5 (N*–Is)
Crystallographic space group	–	$P2_1$	–
Density ϱ (g/cm³)	–	0.951 (100 °C), 0.898 (130 °C)	0.919 (110 °C), 0.901 (130 °C)
Refractive index n	–	n_e 1.482, n_o 1.499 (N*, 95 °C)	n_e 1.472, n_o 1.489 (N*, 589 nm, 107.5 °C)
Dielectric constant ε (10 kHz)	2.45 (80 °C), 3.40 (100 °C)	2.35 (90 °C), 2.90 (130 °C)	2.22 (80 °C), 2.79 (120 °C)
Viscosity	η 44 (71 °C), 25 (100 °C) (mPa s)	η 53 (105 °C), 17 (125 °C) (mPa s)	ν 22.1 mm²/s (117 °C)
Surface tension (mN/m)	28.5 (70 °C)	26.6 (110 °C)	24.7 (110 °C)
Sound velocity v (m/s) (2 MHz)	1030 (95 °C), 990 (100 °C)	1107 (100 °C), 1042 (130 °C)	1245 (113 °C), 1228 (120 °C)
Pitch p (µm)	–	–	0.289 (110 °C)
Dipole moment μ (D)	2.50 (CCl_4), 2.55 (C_6H_6)	2.08 (CCl_4), 1.98 (C_6H_6)	2.30 (CCl_4), 2.16 (C_6H_6)

Table 5.1-7 Cholesteryl (cholest-5-ene) substituted mesogens, cont.

Number/common name	148	149	150
Substance	H_7C_3- structure	H_9C_4- structure	$H_{11}C_5$- structure
Formula	$C_{31}H_{52}O_2$	$C_{32}H_{54}O_2$	$C_{33}H_{56}O_2$
Molar mass (g/mol)	456.759	470.786	484.813
CAS-RN	521-13-1	7726-03-6	1062-96-0
Temperatures of phase transitions T^{tr} (°C)	Cr 102.0 N* 113.0 Is	Cr 93.0 N* 101.5 Is	Cr 99.5 N* 101.5 Is
Enthalpies of phase transitions ΔH^{tr} (kJ/mol)	22.0 (Cr–N*), 0.6 (N*–Is)	22.0 (Cr–N*), 0.6 (N*–Is)	30.0 (Cr–N*), 0.7 (N*–Is)
Crystallographic space group	$P2_1$	$P2_12_12_1$	$P2_1$
Density ϱ (g/cm³)	0.992 (110 °C), 0.984 (120 °C)	0.893 (99 °C), 0.853 (119 °C)	–
Refractive index n (589 nm)	n_e 1.4716, n_o 1.4865 (N*, 589 nm, 107.0 °C)	n_e 1.4763, n_o 1.4905 (N*, 589 nm, 92.5 °C)	n_e 1.4735, n_o 1.4880 (N*, 589 nm, 95 °C)
Kinematic viscosity ν (mm²/s)	22.3 (117 °C)	39.8 (97.3 °C)	–
Surface tension (mN/m)	24.6 (110 °C)	24.6 (110 °C)	25.0 (100 °C)
Pitch p (μm)	0.260 (109 °C)	0.2588 (96.2 °C)	0.247 (97 °C)
Dipole moment μ (D)	2.21 (CCl₄), 2.12 (C₆H₆)	2.34 (CCl₄), 2.48 (C₆H₆)	1.86 (45 °C, C₆H₆)

Number/common name	151	152	153 (CN)
Substance	$H_{13}C_6$- structure	$H_{15}C_7$- structure	$H_{17}C_8$- structure
Formula	$C_{34}H_{58}O_2$	$C_{35}H_{60}O_2$	$C_{36}H_{62}O_2$
Molar mass (g/mol)	498.84	512.867	526.894
CAS-RN	1182-07-6	1182-42-9	1182-66-7
Temperatures of phase transitions T^{tr} (°C)	Cr 114.0 (S <92.5 N* 95.5) Is	Cr 110.0 (S 69.5 N* 96.5) Is	Cr 80.5 (S 77.5) N* 92.0 Is
Enthalpies of phase transitions ΔH^{tr} (kJ/mol)	32.2 (Cr–Is), 0.5 (N*–Is)	34.1 (Cr–Is), 0.7 (N*–Is)	23.4 (Cr–N*), 0.7 (N*–Is), 0.5 (S–N*)
Crystallographic space group	$P2_1$	$P2_1$	$P2_1$
Refractive index n (589 nm)	–	–	n_e 1.4728, n_o 1.4895 (N*, 85 °C), n_e 1.5173, n_o 1.4734 (S, 75.5 °C)
Dielectric constant ε	–	–	2.78 (90 °C, 600 kHz)
Dynamic viscosity η (mPa s)	–	31 (90 °C)	98 (N*, 90.2 °C)
Surface tension (mN/m)	20.4 (120 °C)	21.0 (110 °C)	24 (80 °C), 25 (90 °C)
Sound velocity v (m/s) (2 MHz)	1250 (105 °C), 1208 (120 °C)	–	–
Pitch p (μm)	–	–	0.231 (90 °C)
Dipole moment μ (D), 45 °C	1.78 (C₆H₆)	1.79 (C₆H₆)	1.73 (C₆H₆)

Table 5.1-7 Cholesteryl (cholest-5-ene) substituted mesogens, cont.

Number/common name	154	155	156 (CM)
Substance	$H_{19}C_9$–O– (cholesteryl)	$H_{23}C_{11}$–O– (cholesteryl)	$H_{27}C_{13}$–O– (cholesteryl)
Formula	$C_{37}H_{64}O_2$	$C_{39}H_{68}O_2$	$C_{41}H_{72}O_2$
Molar mass (g/mol)	540.921	568.976	597.03
CAS-RN	1183-04-6	1908-11-8	1989-52-2
Temperatures of phase transitions T^{tr} (°C)	Cr 85.5 (S 81.5) N* 92.5 Is	Cr 92.4 (A 80.2 N* 88.9) Is	Cr 73.6 A 80.0 N* 85.6 Is
Enthalpies of phase transitions ΔH^{tr} (kJ/mol)	31.0 (Cr–N*), 0.8 (N*–Is), 0.6 (S–N*)	37.9 (Cr–Is), 0.9 (N*–Is), 0.6 (S–N*)	46.9 (Cr–A), 1.0 (N*–Is), 1.3 (S–N*)
Crystallographic space group	$P2_1$	$P2_1$	$P2_1$
Density ϱ (g/cm^3)	–	0.932 (N*, 86 °C), 0.941 (A, 79 °C)	0.896 (A, 74 °C), 0.890 (N*, 82 °C), 0.885 (Is, 86 °C)
Refractive index n (589 nm)	n_e 1.4724, n_o 1.4863 (N*, 90 °C), n_e 1.5151, n_o 1.4713 (S, 80 °C)	n_e 1.4715, n_o 1.4878 (N*, 85 °C), n_e 1.5148, n_o 1.4705 (S, 80 °C)	n_e 1.4717, n_o 1.4876 (N*, 81 °C), n_e 1.5152, n_o 1.4704 (S, 76 °C)
Dielectric constant ε (10 kHz)	–	2.33 (80 °C)	4.72 (70 °C)
Dynamic viscosity η (mPa s)	–	40 (90.4 °C), 43 (82 °C)	41 (88 °C), 126 (79.6 °C)
Heat capacity C_p (J/(mol K))	1300 (N*, 90 °C)	$C_p = 1008 + 3.67 \times T$ (S, T (°C) = 69–80)	$C_p = 1033 + 4.84 \times T$ (S, T (°C) = 66–71)
Surface tension (mN/m)	24.7 (100 °C), 24.5 (90 °C)	24.7 (100 °C), 24.4 (90 °C)	23.9 (82 °C), 25.0 (94 °C)
Sound velocity v (m/s)	–	1348 (85 °C), 1335 (100 °C) (2 MHz)	1345 (84 °C), 1316 (90 °C)
Diffusion coefficient D (m^2/s)	–	–	1.88×10^{-10} (Is, 90 °C), 1.75×10^{-10} (N*, 80 °C)
Pitch p (µm)	0.225 (90 °C)	0.2125 (82 °C)	0.1994 (83 °C)
Dipole moment μ (D), 45 °C	1.74 (C_6H_6)	1.73 (C_6H_6)	–

Number/common name	157	158	159 (CB)
Substance	$H_{31}C_{15}$–O– (cholesteryl)	$H_{35}C_{17}$–O– (cholesteryl)	benzoate of cholesteryl
Formula	$C_{43}H_{76}O_2$	$C_{45}H_{80}O_2$	$C_{34}H_{50}O_2$
Molar mass (g/mol)	625.084	653.138	490.776
CAS-RN	601-34-3	35602-69-8	604-32-0
Temperatures of phase transitions T^{tr} (°C)	Cr 79.0 (S 78.5) N* 83.0 Is	Cr 83.0 (S 75.5 N* 79.5) Is	Cr 150.5 N* 182.6 Is
Enthalpies of phase transitions ΔH^{tr} (kJ/mol)	59.4 (Cr–N*), 1.3 (N*–Is), 1.7 (S–N*)	67.5 (Cr–Is), 1.7 (N*–Is), 1.8 (S–N*)	22.2 (Cr–N*), 0.6 (N*–Is)
Crystallographic space group	$A2$	$P2_1$	$P2_12_12_1$
Density ϱ (g/cm^3)	0.917 (N*, 80 °C), 0.909 (Is, 100 °C)	0.855 (Is, 90 °C)	0.9574 (N*, 159.2 °C), 0.9354 (Is, 185.4 °C)

Table 5.1-7 Cholesteryl (cholest-5-ene) substituted mesogens, cont.

Number/common name	157	158	159 (CB)
Refractive index n (589 nm)	n_e 1.4752, n_o 1.4881 (N*, 78 °C)	–	n_e 1.4813, n_o 1.5656 (N*, 160 °C)
Dielectric constant ε (10 kHz)	–	2.36 (70 °C)	–
Dynamic viscosity η (mPa s)	47 (85 °C), 68 (76.6 °C)	78.4 (80 °C), 85.2 (78 °C), 117 (74 °C)	35.5 (170 °C), 25.8 (180 °C)
Surface tension (mN/m)	25.0 (90 °C), 24.2 (80 °C)	25.6 (Is, 80 °C), 25.2 (N*, 75 °C), 25.3 (S, 70 °C)	21.8 (Is, 210 °C), 23.8 (N*, 147.4 °C), 22.9 (N*, 177 °C)
Sound velocity v (m/s) (2 MHz)	–	1343 (77 °C), 1332 (85 °C)	1093 (170 °C), 1087 (190 °C)
Pitch p (μm)	0.2215 (77 °C)	–	–
Dipole moment μ (D)	1.69 (45 °C, C_6H_6)	1.62 (45 °C, C_6H_6)	2.08 (CCl_4), 2.11 (C_6H_6)

Number/common name	160 (CO)	161 (COC)	162
Substance			
Formula	$C_{45}H_{78}O_2$	$C_{46}H_{80}O_3$	$C_{30}H_{50}O_3$
Molar mass (g/mol)	651.122	681.149	458.731
CAS-RN	303-43-5	17110-51-9	23836-43-3
Temperatures of phase transitions T^{tr} (°C)	Cr 51 (A 42 N* 47.5) Is	Cr 26.7 (A 20) N* 34 Is	Cr 83.9 N* 105.8 Is
Enthalpies of phase transitions ΔH^{tr} (kJ/mol)	29.0 (Cr–Is), 0.84 (N*–Is), 1.3 (S–N*)	27.3 (Cr–N*), 0.8 (N*–Is), 0.8 (S–N*)	21.1 (Cr–N*), 0.7 (N*–Is)
Crystallographic space group	$P2_1$	–	$P2_1$
Density ϱ (g/cm³)	1.084 (N*, 40 °C), 1.072 (Is, 50 °C)	0.972 (N*, 20 °C), 0.962 (Is, 32 °C)	–
Refractive index n (589 nm)	n_e 1.505, n_o 1.521 (A, 20 °C)	n_e 1.4947, n_o 1.5101 (N*, 25 °C)	n_e 1.472, n_o 1.490 (N*, 90 °C)
Dielectric constant ε (10 kHz)	–	2.44 (A, 3 °C), 2.41 (Is, 40 °C)	–
Sound velocity v (m/s) (10 MHz)	1459 (45 °C), 1406 (58 °C)	–	–
Pitch p (nm)	–	–	232.3 (90 °C)
Dipole moment μ (D)	–	1.14 (45 °C, C_6H_{12})	–

Table 5.1-8 Discotic liquid crystals

Number/common name	163	164	165 (H10TX)
Substance	(structure)	(structure)	(structure)
Formula	$C_{48}H_{78}O_{12}$	$C_{54}H_{90}O_{12}$	$C_{87}H_{126}O_{12}$
Molar mass (g/mol)	847.15	931.312	1363.967
CAS-RN	65201-70-9	65201-71-0	86108-14-7
Temperatures of phase transitions T^{tr} (°C)	Cr 81.2 D 87.0 Is	Cr' 28.8 Cr 82.0 D 84.0 Is	Cr 68 N 85 Drd 138 Dho 280 Is
Enthalpies of phase transitions ΔH^{tr} (kJ/mol)	32.2 (Cr–D), 21.5 (D–Is)	46.1 (Cr–D), 19.2 (D–Is), 49.0 (Cr'–Cr)	21.3 (Cr–N), 1.0 (N–D)
Refractive index n	–	–	n_o 1.531, n_e 1.454 ($T = 0.999 T^{N-D}$)

Number/common name	166 (HAT5)	167 (HAT6)	168 (HAT8)
Substance	(structure)	(structure)	(structure)
Formula	$C_{48}H_{72}O_6$	$C_{54}H_{84}O_6$	$C_{66}H_{108}O_6$
Molar mass (g/mol)	745.105	829.268	997.593
CAS-RN	69079-52-3	70351-86-9	70351-87-0
Temperatures of phase transitions T^{tr} (°C)	Cr 69.0 Dho 122.0 Is	Cr 68.0 Dho 97.0 Is	Cr 67.0 Dho 86.0 Is
Enthalpies of phase transitions ΔH^{tr} (kJ/mol)	32.6 (Cr–D), 8.1 (D–Is)	36.4 (Cr–D), 3.6 (D–Is)	83.3 (Cr–D), 4.2 (D–Is)

Number/common name	169	170 (HAT11)	171
Substance	(structure)	(structure)	(structure)
Formula	$C_{66}H_{96}O_{12}$	$C_{90}H_{144}O_{12}$	$C_{102}H_{120}O_{18}$
Molar mass (g/mol)	1081.494	1418.144	1634.083
CAS-RN	70351-94-9	70187-34-7	75747-38-5

Table 5.1-8 Discotic liquid crystals, cont.

Number/common name	169	170 (HAT11)	171
Temperatures of phase transitions T^{tr} (°C)	Cr 66.0 D 126.0 Is	Cr 80.0 Dh 93.0 Dt 111.0 Dh 122.3 Is	Cr′ 117.8 Cr′ 130.9 Cr 169.1 N 253.1 Is
Enthalpies of phase transitions ΔH^{tr} (kJ/mol)	19.7 (Cr–D), 2.8 (D–Is)	59.4 (Cr–D), 2.4 (D–Is)	9.4 (Cr–D)
Anisotropy of refractive index Δn	–	–	−0.09 (200 °C, 633 nm)
Dielectric constant ε	–	–	$\varepsilon_\parallel$ 3.78, $\varepsilon_\perp$ 3.33 (230 °C)
Dynamic viscosity η (mPa s)	–	–	350 (230 °C)

Table 5.1-9 Liquid crystal salts

Na$^+$	Temperatures of phase transitions (°C)	Tl$^+$	Temperatures of phase transitions (°C)
C$_3$H$_7$–COO–	Cr 251.0 S 327.0 Is	C$_4$H$_9$–COO–	Cr 80.4 S 180.0 A 214.3 Is
C$_4$H$_9$–COO–	Cr 241.0 A 344.0 Is	C$_5$H$_{11}$–COO–	Cr 124.4 S 137.2 S 148.5 A 227.1 Is
C$_5$H$_{11}$–COO–	Cr 210.0 S 235.0 A 361.0 Is	C$_6$H$_{13}$–COO–	Cr 140.5 A 226.0 Is
C$_6$H$_{13}$–COO–	Cr 198.0 S 242.0 A 363.0 Is	C$_7$H$_{15}$–COO–	Cr 135.0 A 221.0 Is
C$_7$H$_{15}$–COO–	Cr 189.0 S 243.0 A 360.0 Is	C$_8$H$_{17}$–COO–	Cr 39.5 S 55.8 S 136.5 A 217.9 Is
C$_8$H$_{17}$–COO–	Cr 185.0 S 243.0 A 355.0 Is	C$_9$H$_{19}$–COO–	Cr 30.2 S 49.8 S 128.4 A 207.6 Is
C$_9$H$_{19}$–COO–	Cr 140.0 S 181.0 S 245.0 S 348.0 Is	C$_{10}$H$_{21}$–COO–	Cr 35.8 S 47.0 S 74.4 S 128.7 A 201.6 Is
C$_{10}$H$_{21}$–COO–	Cr 115.0 S 145.0 S 167.0 S 187.0 S 242.0 A 337.0 Is	C$_{11}$H$_{23}$–COO–	Cr 36.0 S 73.8 S 122.3 A 196.8 Is
C$_{11}$H$_{23}$–COO–	Cr 100.0 S 141.0 S 182.0 S 220.0 S 255.0 A 336.0 Is	C$_{12}$H$_{25}$–COO–	Cr 53.8 S 58.6 S 97.0 S 125.3 A 192.5 Is
C$_{12}$H$_{25}$–COO–	Cr 121.0 S 162.0 S 187.0 S 200.0 S 217.0 S 248.0 A 323.0 Is	C$_{13}$H$_{27}$–COO–	Cr 36.5 S 42.0 S 94.8 S 118.9 A 184.4 Is
C$_{13}$H$_{27}$–COO–	Cr 113.0 S 138.0 S 171.0 S 215.0 S 246.0 A 311.0 Is	C$_{14}$H$_{29}$–COO–	Cr 66.9 S 111.6 S 120.4 A 179.6 Is
C$_{14}$H$_{29}$–COO–	Cr 121.0 S 160.0 S 187.0 S 203.0 S 251.0 S 277.0 A 307.0 Is	C$_{15}$H$_{31}$–COO–	Cr 54.0 S 114.4 S 116.4 A 175.5 Is
C$_{15}$H$_{31}$–COO–	Cr 117.0 S 136.0 S 168.0 S 212.0 S 251.0 A 302.0 Is	C$_{16}$H$_{33}$–COO–	Cr 75.3 S 119.4 A 171.5 Is
C$_{16}$H$_{33}$–COO–	Cr 130.0 S 205.0 S 260.0 A 290.0 Is	C$_{17}$H$_{35}$–COO–	Cr 62.1 S 118.4 A 168.1 Is
C$_{17}$H$_{35}$–COO–	Cr 117.0 S 132.0 S 167.0 S 198.0 S 257.0 A 288.0 Is	C$_{19}$H$_{39}$–COO–	Cr 70.2 S 120.5 A 158.2 Is
C$_{18}$H$_{37}$–COO–	Cr 118.0 S 133.0 S 146.0 S 193.0 S 202.0 S 258.0 A 283.0 Is	C$_{21}$H$_{43}$–COO–	Cr 67.2 S 76.8 S 121.8 A 151.8 Is
C$_{19}$H$_{39}$–COO–	Cr 110.0 S 131.0 S 163.0 S 200.0 A 262.0 Is	C$_{25}$H$_{51}$–COO–	Cr 114.0 S 125.0 Is

Table 5.1-9 Liquid crystal salts, cont.

Li$^+$	Temperatures of phase transitions (°C)	K$^+$	Temperatures of phase transitions (°C)
C$_{11}$H$_{23}$–COO–	Cr 229.0 S 239.0 Is	C$_3$H$_7$–COO–	Cr 353.0 S 404.0 Is
C$_{12}$H$_{25}$–COO–	Cr 224.0 S 232.0 Is	C$_4$H$_9$–COO–	Cr 313.4 S 443.0 Is
C$_{13}$H$_{27}$–COO–	Cr 210.0 S 231.0 S 239.0 Is	C$_5$H$_{11}$–COO–	Cr 308.5 S 452.0 Is
C$_{14}$H$_{29}$–COO–	Cr 206.0 S 229.0 Is	C$_6$H$_{13}$–COO–	Cr 298.1 S 449.0 Is
C$_{15}$H$_{31}$–COO–	Cr 197.0 S 215.0 S 223.0 Is	C$_7$H$_{15}$–COO–	Cr 287.0 S 439.0 Is
C$_{17}$H$_{35}$–COO–	Cr 190.0 S 215.0 S 229.0 Is	C$_8$H$_{17}$–COO–	Cr 276.0 S 434.0 Is
C$_{19}$H$_{39}$–COO–	Cr 189.0 S 226.0 Is	C$_9$H$_{19}$–COO–	Cr 271.0 S 423.0 Is
Cs$^+$	**Temperatures of phase transitions (°C)**	C$_{10}$H$_{21}$–COO–	Cr 268.0 S 418.0 Is
C$_5$H$_{11}$–COO–	Cr 359.0 S 399.0 Is	C$_{11}$H$_{23}$–COO–	Cr 268.0 S 406.0 Is
C$_6$H$_{13}$–COO–	Cr 345.0 S 421.0 Is	C$_{12}$H$_{25}$–COO–	Cr 89.9 S 397.6 Is
C$_7$H$_{15}$–COO–	Cr 334.0 S 425.0 Is	C$_{13}$H$_{27}$–COO–	S 271.0 A 375.0 Is
C$_8$H$_{17}$–COO–	Cr 325.0 S 424.0 Is	C$_{14}$H$_{29}$–COO–	Cr 64.0 S 388.5 Is
C$_9$H$_{19}$–COO–	Cr 314.0 S 415.0 Is	C$_{15}$H$_{31}$–COO–	Cr 195.0 S 269.0 A 362.0 Is
C$_{11}$H$_{23}$–COO–	Cr 278.0 S 355.0 Is	C$_{17}$H$_{35}$–COO–	Cr 170.0 S 238.0 S 264.0 S 356.0 Is
C$_{13}$H$_{27}$–COO–	Cr 287.0 S 386.0 Is	C$_{19}$H$_{39}$–COO–	Cr 74.1 S 339.9 Is
Rb$^+$	**Temperatures of phase transitions (°C)**	**Cu^{2+}**	**Temperatures of phase transitions (°C)**
C$_4$H$_9$–COO–	Cr 367.0 S 430.0 Is	C$_3$H$_7$–COO–	Cr 195.0 D > 200.0 dec
C$_5$H$_{11}$–COO–	Cr 342.2 S 450.0 Is	C$_4$H$_9$–COO–	Cr 111.0 D > 200.0 dec
C$_6$H$_{13}$–COO–	Cr 327.0 S 451.0 Is	C$_5$H$_{11}$–COO–	Cr 95.0 D > 200.0 dec
C$_7$H$_{15}$–COO–	Cr 312.0 A 440.0 Is	C$_6$H$_{13}$–COO–	Cr 93.0 D > 200.0 dec
C$_8$H$_{17}$–COO–	Cr 300.0 A 439.0 Is	C$_7$H$_{15}$–COO–	Cr 88.0 D > 200.0 dec
C$_9$H$_{19}$–COO–	Cr 291.0 A 429.0 Is	C$_8$H$_{17}$–COO–	Cr 102.0 D > 200.0 dec
C$_{11}$H$_{23}$–COO–	Cr 300.0 A 400.0 Is	C$_9$H$_{19}$–COO–	Cr 106.9 D 210.0 dec
C$_{13}$H$_{27}$–COO–	Cr 212.4 S 217.9 S 240.3 S 277.6 S 394.0 Is	C$_{15}$H$_{31}$–COO–	Cr 122.0 D 220.0 dec

Table 5.1-9 Liquid crystal salts, cont.

L–⟨C$_6$H$_4$⟩–R

L	R		L	R	
H–	–C$_3$H$_6$–COO–Tl	Cr 58.4 A 62.4 Is	C$_5$H$_{11}$–O–	–COO–Tl	Cr 160.0 S 294.0 A 344.0 Is
H–	–C$_5$H$_{10}$–COO–Tl	Cr 34.0 A 93.0 Is	C$_6$H$_{13}$–O–	–COO–Tl	Cr 27.0 S 114.0 S 284.0 A 343.0 Is
H–	–C$_7$H$_{14}$–COO–Tl	Cr 39.0 A 107.5 Is	C$_7$H$_{15}$–O–	–COO–Tl	Cr 36.0 S 57.0 S 121.0 S 274.0 A 334.0 Is
H–	–C$_9$H$_{18}$–COO–Tl	Cr 51.0 A 114.0 Is	C$_8$H$_{17}$–O–	–COO–Tl	Cr 30.0 S 43.0 S 131.0 S 269.0 A 330.0 Is
C$_6$H$_{13}$–	–COO–Tl	Cr 118.0 S 264.0 A 334.0 Is	C$_9$H$_{19}$–O–	–COO–Tl	Cr 57.0 S 62.0 S 135.0 S 264.0 A 322.0 Is

5.1.3 Physical Properties of Some Liquid Crystalline Mixtures

Table 5.1-10 Nematic mixtures

Mixture	E49 (Merck)	ZLI-2857 (Merck)	ZLI-3086 (Merck)
Components	Cyanobiphenyls		
Temperatures of phase transitions T^{tr} (°C)	Cr −9 N 100 Is	Cr −19 N 82.3 Is	N 72.0 Is
Anisotropy of refractive index Δn (20 °C)	0.251	0.0776	0.1131
Dielectric anisotropy $\Delta\varepsilon$ (20 °C)		−1.4	0.1
Kinematic viscosity (mm²/s) (20 °C)	46.5	20	

Mixture	E7 (Merck)	ZLI-1132 (Merck)	ZLI-4792 (Merck)
Components	Cyanobiphenyls	Phenylcyclohexanes	Fluorinated LCs
Temperatures of phase transitions T^{tr} (°C)	Cr <−30 N 58 Is	Cr −6 N 71 Is	Cr <−40 N 92 Is
Anisotropy of refractive index Δn (589 nm, 20 °C)	0.2253	0.1396	0.0969
Extraordinary refractive index n_e (589 nm, 20 °C)	1.7464	1.6326	1.5763
Dielectric anisotropy $\Delta\varepsilon$ (1 kHz, 20 °C)	13.8	13.1	5.2
$\varepsilon_\parallel$ (1 kHz, 20 °C)	19.0	17.7	8.3
Kinematic viscosity (mm²/s)	39 (20 °C), 145 (0 °C)	28 (20 °C), 110 (0 °C)	15 (20 °C), 40 (0 °C)
Surface tension (mN/m)	29.3 (22 °C), $\sigma_\perp$ 14.5, $\sigma_\parallel$ 23.7	31.0 (22 °C), $\sigma_\perp$ 13.6, $\sigma_\parallel$ 26.0	−
Elastic-constants ratio K_3/K_1 (20 °C)	1.54	1.95	1.39
V_{10} (V) threshold	1.41	1.77	2.00
V_{50} (V)	1.63	2.05	2.47
V_{90} (V) saturation	1.99	2.49	3.15

Data have been taken from [1.5]. ZLI-4792 is an SFM (superfluorinated material); it is recommended for VIP (viewing-independent panel) twisted nematic displays. Mixtures E7 and ZLI-1132 are recommended for usage in calculators, wristwatches, and measuring instruments. V_{10}, V_{50}, and V_{90} are voltages at 10, 50, and 90%, respectively, of the maximum absorption at 0° viewing angle and 20 °C.

Table 5.1-11 Ferroelectric mixtures

Mixture	CS1024	CS2004	CS4000
CAS-RN	123967-01-1	135976-66-8	150260-44-9
Temperatures of phase transitions T^{tr} (°C)	Cr −12 C* 62 A 82 N* 90 Is	Cr <20 C* 62 N* 71 Is	Cr −10 CA 70.5 Cγ 72.5 C* 74.6 Cα 75.6 A 100 Is
Spontaneous polarization P_s (25 °C) (nC/cm²)	−46.9	−	−
Tilt Θ (25 °C)	25	44	−
Pitch (µm)	>20	−	−

Table 5.1-11 Ferroelectric mixtures, cont.

Mixture	Felix-015-100	SCE8	SCE9
CAS-RN	211365-96-7	145380-16-1	126879-69-4
Temperatures of phase transitions T^{tr} (°C)	Cr −12 C* 72 A 83 N* 86 Is	Cr <20 C 59 A 79 N 100 Is	Cr <20 C* 61 A 91 N* 115 Is
Spontaneous polarization P_s (25 °C) (nC/cm^2)	+33	+630	−
Tilt Θ (25 °C)	25.5	−	−
Switching time τ (25 °C)	−	294 µs	−

Mixture	SCE10	SCE13	TKF 8617
CAS-RN	134499-04-0	133758-42-6	114899-67-1
Temperatures of phase transitions T^{tr} (°C)	Cr <20 C* 61 N* 109 Is	Cr <0 C* 61 A 86 N* 103 Is	Cr 4 C* 54 A 65 Is
Spontaneous polarization P_s (20 °C) (nC/cm^2)	19.5	30.6	−
Tilt Θ (20 °C)	−	29	−
Pitch (µm)	−	10–12	−

Mixture	W22	ZhKS-309C	ZLI-3654
CAS-RN	137988-43-3	205599-65-1	116580-90-6
Temperatures of phase transitions T^{tr} (°C)	Cr 4 C* 51 A 82 N* 92 Is	Cr −1 C* 42 A 91 Is	Cr −30 C* 62 A 76 N 86 Is
Spontaneous polarization P_s (20 °C) (nC/cm^2)	−	+50.8	−29
Tilt Θ (20 °C)	−	26	−
Pitch (µm)	−	0.4	3

Mixture	ZLI-4655-000	ZLI-4851-100, Felix-M-4851-100	ZLI-5014-100
CAS-RN	139352-77-5	158854-82-1	172452-16-3
Temperatures of phase transitions T^{tr} (°C)	Cr <10 C* 60 A 69 N* 72 Is	Cr <−20 C* 67 A 71 N* 76 Is	Cr <−10 C* 65 A 70 N* 72 Is
Spontaneous polarization P_s (20 °C) (nC/cm^2)	+7	+22.8	−20
Tilt Θ (20 °C)	−	30.5	−
Pitch (µm)	−	−	10
Switching time τ (20 °C)	−	38 µs	3 µs

References

1.1 D. Demus, H. Demus, H. Zaschke: *Flüssige Kristalle in Tabellen I* (VEB Deutscher Verlag Grundstoffindustrie, Leipzig 1974)

1.2 D. Demus, H. Zaschke: *Flüssige Kristalle in Tabellen II* (Deutscher Verlag Grundstoffindustrie, Leipzig 1982)

1.3 V. Vill: *Liquid Crystals*, Landolt–Börnstein, New Series IV/7 (Springer, Berlin, Heidelberg 1992–1995)

1.4 V. Vill: Liqcryst 4.3 - Database of liquid crystalline compounds, http://liqcryst.chemie.uni-hamburg.de/ (2003)

1.5 Merck: Product information, liquid crystal mixtures for electro-optic displays (Merck, Darmstadt 1992)

1.6 H. Kelker, R. Hatz: *Handbook of Liquid Crystals* (Verlag Chemie, Weinheim 1980)

1.7 A. Beguin, J. C. Dubois, P. Le Barny, J. Billard, F. Bonamy, J. M. Buisine, P. Cuvelier: Sources of thermodynamic data on mesogens, Mol. Cryst. Liq. Cryst. **115**, 1 (1984)

1.8 BDH Chemical: Product information (BDH, Poole 1986)

1.9 E. Merck: Product information (Merck, Darmstadt 1986)

1.10 S. Chandrasekhar: *Liquid Crystals* (Cambridge Univ. Press, Cambridge 1992)

1.11 D. Demus, J. Goodby, G. W. Gray, H.-W. Spiess, V. Vill (Eds.): *Handbook of Liquid Crystals*, Vol. I – III (Wiley-VCH, Weinheim 1998)

1.12 D. Demus, J. Goodby, G. W. Gray, H.-W. Spiess, V. Vill (Eds.): *Physical Properties of Liquid Crystals* (Wiley-VCH, Weinheim 1999)

1.13 V. G. Chigrinov: *Liquid Crystal Devices: Physics and Applications* (Artech House, Boston 1999)

1.14 S. Pestov: *Physical Properties of Liquid Crystals*, Landolt–Börnstein, New Series VIII/5 (Springer, Berlin, Heidelberg 2003)

1.15 D. Demus, R. Richter: *Textures of Liquid Crystals* (Verlag Chemie, Weinheim 1978)

1.16 G. W. Gray, J. W. G. Goodby: *Smectic Liquid Crystals – Textures and Structures* (Hill, Glasgow 1984)

1.17 Hoffmann-La Roche: Product information (Hoffmann-La Roche, Basel 1988)

5.2. The Physics of Solid Surfaces

The data compiled in this chapter refer to so-called "clean surfaces", i. e. crystalline surfaces that are atomically clean and well characterized. Data on interfaces are dealt with only marginally, in connection with MOS devices.

The values reported in the tables are mainly averages from several different authors. In such cases the errors are given as standard deviations. Reference to the individual measurements and to the original papers is made by referring to larger compilations (mainly the four volumes of Landolt-Börnstein III/24, *Physics of Solid Surfaces*, [2.1] or the single articles therein [2.2–16]). On the other hand, the figures are fully referenced.

5.2.1	**The Structure of Ideal Surfaces**............	979
	5.2.1.1 Diagrams of Surfaces [2.2]	979
	5.2.1.2 Crystallographic Formulas.........	986
5.2.2	**Surface Reconstruction and Relaxation**.	986
	5.2.2.1 Definitions and Notation..........	986
	5.2.2.2 Metals	987
	5.2.2.3 Semiconductors	987
5.2.3	**Electronic Structure of Surfaces**............	996
	5.2.3.1 Metals	997
	5.2.3.2 Semiconductors	1003
	5.2.3.3 Magnetic Surfaces	1007
5.2.4	**Surface Phonons**................................	1012
	5.2.4.1 Metals	1012
	5.2.4.2 Semiconductors and Insulators	1017
	5.2.4.3 Atom–Surface Potential	1019
5.2.5	**The Space Charge Layer at the Surface of a Semiconductor**............................	1020
	5.2.5.1 Definitions and Notation..........	1020
	5.2.5.2 Useful Formulas and Numerical Values	1022
	5.2.5.3 Surface Conductivity	1023
5.2.6	**Most Frequently Used Acronyms**...........	1026
References		1029

5.2.1 The Structure of Ideal Surfaces

An ideal surface is a surface of a half-crystal in which the atoms are held in their original positions. The structure of an ideal surface is identical to that of a parallel crystallographic plane in the bulk. For a 2-D lattice, the elementary Bravais cell can have only one of the five structures shown in Fig. 5.2-1.

5.2.1.1 Diagrams of Surfaces [2.2]

Figure 5.2-2 gives diagrams of ideal surfaces for some common faces of the fcc (face-centered cubic), bcc (body-centered cubic), diamond, and zinc blende systems, as well as the coordinates of the atoms of the first layers of the half-crystal. The atoms are drawn as solid balls with diameters appropriate to close-packed stacking.

Atoms in the surface layer are labeled by O, A, B, C, ... Atoms in the first, second, third, etc. sublayer are labeled by 1, 2, 3, ... When two classes of atoms are present (for example in diamond-like structures), the atoms of the second class are indicated by primed symbols. In such a case, the division of an ideal crystal by a geometrical plane may expose different types of surfaces. A well-known example is the (111) face of NaCl-type crystals, which may be either anion- or cation-terminated.

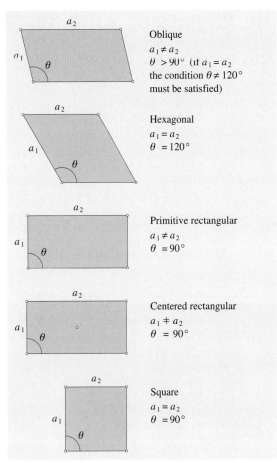

Fig. 5.2-1 Elementary Bravais cells for a 2D lattice

A parameter δ, $0 \leq \delta < 1$, is introduced to specify the distance from the dividing plane to the nearest plane through lattice points.

Coordinates are referred either to crystal axes ($Oxyz$) or to a system $OXYZ$, in which XY are axes in the surface plane; X is parallel to a side of the 2-D Bravais cell and Z is perpendicular to the surface.

Coordinates are always given in terms of $a/2$, a being the lattice parameter.

A more detailed discussion of the structure of ideal surfaces and many other diagrams of surfaces can be found in [2.2].

a)

fcc (100)

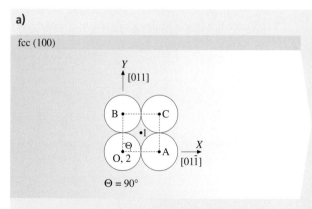

Θ = 90°

Atom	Atomic positions					
	Coordinates					
	Relative to cubic axes			Relative to OXYZ		
A	0	1	−1	$\sqrt{2}$	0	0
B	0	1	1	0	$\sqrt{2}$	0
1	−1	1	0	$\dfrac{\sqrt{2}}{2}$	$\dfrac{\sqrt{2}}{2}$	−1
2	−2	0	0	0	0	−2

fcc (110)

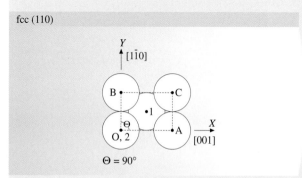

Θ = 90°

Atom	Atomic positions					
	Coordinates					
	Relative to cubic axes			Relative to OXYZ		
A	0	0	2	2	0	0
B	1	−1	0	0	$\sqrt{2}$	0
1	0	−1	1	1	$\dfrac{\sqrt{2}}{2}$	$\dfrac{-\sqrt{2}}{2}$
2	−1	−1	0	0	0	$-\sqrt{2}$

fcc (111)

Θ = 120°

Atom	Atomic positions					
	Coordinates					
	Relative to cubic axes			Relative to OXYZ		
A	0	−1	1	$\sqrt{2}$	0	0
B	1	0	−1	$\dfrac{-\sqrt{2}}{2}$	$\dfrac{3}{\sqrt{6}}$	0
1	0	−1	−1	0	$\dfrac{2}{\sqrt{6}}$	$\dfrac{-2}{\sqrt{3}}$
2	−1	−2	−1	$\dfrac{\sqrt{2}}{2}$	$\dfrac{1}{\sqrt{6}}$	$\dfrac{-4}{\sqrt{3}}$
3	−2	−2	−2	0	0	$\dfrac{-6}{\sqrt{3}}$

Fig. 5.2-2 (**a**) Surface diagrams: face-centered cubic (fcc) positions given in terms of $a/2$ [2.2] (**b**) Surface diagrams: body-centered cubic (bcc) positions given in terms of $a/2$ [2.2] (**c**) Surface diagrams: diamond, GaAs (positions given in terms of $a/2$) Ga atoms are denoted by *unprimed symbols*; As atoms by *primed symbols* and *shaded circles* [2.2]

b)

bcc (100), $\Theta = 90°$

Atom	Atomic positions					
	Coordinates					
	Relative to cubic axes			Relative to $OXYZ$		
A	0	2	0	2	0	0
B	0	0	2	0	2	0
1	−1	1	1	1	1	−1
2	−2	0	0	0	0	−2

bcc (110), $\Theta = 125.26°$

Atom	Atomic positions					
	Coordinates					
	Relative to cubic axes			Relative to $OXYZ$		
A	0	0	2	2	0	0
B	1	−1	−1	−1	$\sqrt{2}$	0
1	−1	−1	1	1	0	$-\sqrt{2}$
2	−2	−2	0	0	0	$-2\sqrt{2}$

bcc (111), $\Theta = 120°$

Atom	Atomic positions					
	Coordinates					
	Relative to cubic axes			Relative to $OXYZ$		
A	−2	2	0	$2\sqrt{2}$	0	0
B	0	−2	2	$-\sqrt{2}$	$\frac{6}{\sqrt{6}}$	0
1	−1	−1	1	0	$\frac{4}{\sqrt{6}}$	$\frac{-1}{\sqrt{3}}$
2	−2	0	0	$\sqrt{2}$	$\frac{2}{\sqrt{6}}$	$\frac{-2}{\sqrt{3}}$
3	−1	−1	−1	0	0	$\frac{-3}{\sqrt{3}}$

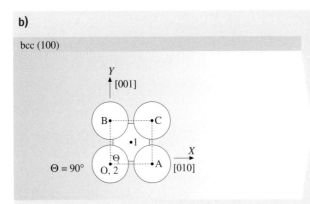

Fig. 5.2-2b Surface diagrams: body-centered cubic (bcc) positions given in terms of $a/2$ [2.2]

c)

Atomic positions					
Atom	Coordinates				
	Relative to cubic axes			Relative to $OXYZ$	
A	0	1	−1	$\sqrt{2}$	0	0
B	0	1	1	0	$\sqrt{2}$	0
1'	$-\frac{1}{2}$	$\frac{1}{2}$	$-\frac{1}{2}$	$\frac{\sqrt{2}}{2}$	0	$-\frac{1}{2}$
2	−1	1	0	$\frac{\sqrt{2}}{2}$	$\frac{\sqrt{2}}{2}$	−1
3'	$-\frac{3}{2}$	$\frac{1}{2}$	$\frac{1}{2}$	0	$\frac{\sqrt{2}}{2}$	$-\frac{3}{2}$
4	−2	0	0	0	0	−2

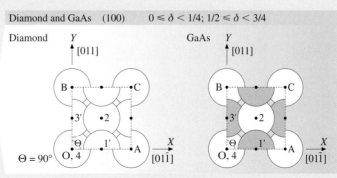

Diamond and GaAs (100) $0 \leq \delta < 1/4;\ 1/2 \leq \delta < 3/4$

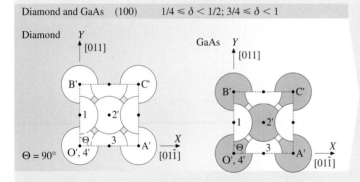

Diamond and GaAs (100) $1/4 \leq \delta < 1/2;\ 3/4 \leq \delta < 1$

Atomic positions					
Atom	Coordinates				
	Relative to cubic axes			Relative to $O'XYZ$	
A'	0	1	−1	$\sqrt{2}$	0	0
B'	0	1	1	0	$\sqrt{2}$	0
1	$-\frac{1}{2}$	$\frac{1}{2}$	$\frac{1}{2}$	0	$\frac{\sqrt{2}}{2}$	$-\frac{1}{2}$
2'	−1	1	0	$\frac{\sqrt{2}}{2}$	$\frac{\sqrt{2}}{2}$	−1
3	$-\frac{3}{2}$	$\frac{1}{2}$	$-\frac{1}{2}$	$\frac{\sqrt{2}}{2}$	0	$-\frac{3}{2}$
4'	−2	0	0	0	0	−2

Atomic positions					
Atom	Coordinates				
	Relative to cubic axes			Relative to $OXYZ$	
A	0	−1	1	$\sqrt{2}$	0	0
B	0	1	1	0	$\sqrt{2}$	0
1'	$\frac{1}{2}$	$\frac{1}{2}$	$\frac{1}{2}$	0	$\frac{\sqrt{2}}{2}$	$-\frac{1}{2}$
2	1	0	1	$\frac{\sqrt{2}}{2}$	$\frac{\sqrt{2}}{2}$	−1
3'	$\frac{3}{2}$	$-\frac{1}{2}$	$\frac{1}{2}$	$\frac{\sqrt{2}}{2}$	0	$-\frac{3}{2}$
4	2	0	0	0	0	−2

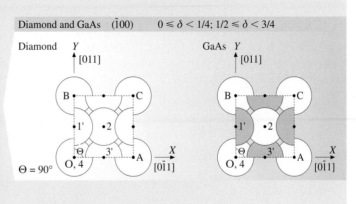

Diamond and GaAs ($\bar{1}$00) $0 \leq \delta < 1/4;\ 1/2 \leq \delta < 3/4$

Fig. 5.2-2c Surface diagrams: diamond, GaAs (positions given in terms of $a/2$) Ga atoms are denoted by *unprimed symbols*; As atoms by *primed symbols* and *shaded circles* [2.2]

c) (cont.)

Atomic positions

Atom	Coordinates					
	Relative to cubic axes			Relative to OXYZ		
A'	0	−1	1	$\sqrt{2}$	0	0
B'	0	1	1	0	$\sqrt{2}$	0
1	$\frac{1}{2}$	$-\frac{1}{2}$	$\frac{1}{2}$	$\frac{\sqrt{2}}{2}$	0	$-\frac{1}{2}$
2'	1	0	1	$\frac{\sqrt{2}}{2}$	$\frac{\sqrt{2}}{2}$	−1
3	$\frac{3}{2}$	$\frac{1}{2}$	$\frac{1}{2}$	0	$\frac{\sqrt{2}}{2}$	$-\frac{3}{2}$
4'	2	0	0	0	0	−2

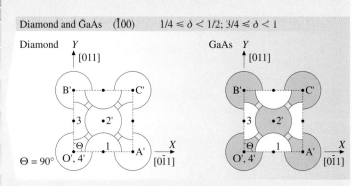

Diamond and GaAs ($\bar{1}00$) $1/4 \leq \delta < 1/2; \ 3/4 \leq \delta < 1$

$\Theta = 90°$

Diamond and GaAs (110)

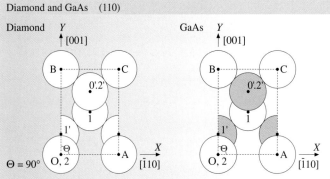

$\Theta = 90°$

Atomic positions

Atom	Coordinates					
	Relative to cubic axes			Relative to OXYZ		
A	−1	1	0	$\sqrt{2}$	0	0
B	0	−1	1	$-\frac{\sqrt{2}}{2}$	$\frac{3}{\sqrt{6}}$	0
1'	$-\frac{1}{2}$	$-\frac{1}{2}$	$\frac{1}{2}$	0	$\frac{2}{\sqrt{6}}$	$-\frac{1}{2\sqrt{3}}$
4	−1	−1	0	0	$\frac{2}{\sqrt{6}}$	$-\frac{2}{\sqrt{3}}$
5'	$-\frac{3}{2}$	$-\frac{1}{2}$	$-\frac{1}{2}$	$\frac{\sqrt{2}}{2}$	$\frac{1}{\sqrt{6}}$	$-\frac{5}{2\sqrt{3}}$
8	−2	−1	−1	$\frac{\sqrt{2}}{2}$	$\frac{1}{\sqrt{6}}$	$-\frac{4}{\sqrt{3}}$
9'	$-\frac{3}{2}$	$-\frac{3}{2}$	$-\frac{3}{2}$	0	0	$-\frac{9}{2\sqrt{3}}$

Atomic positions

Atom	Coordinates					
	Relative to cubic axes			Relative to OXYZ		
A	−1	1	0	$\sqrt{2}$	0	0
B	0	0	2	0	2	0
0'	$-\frac{1}{2}$	$\frac{1}{2}$	$\frac{3}{2}$	$\frac{\sqrt{2}}{2}$	$\frac{3}{2}$	0
1	−1	0	1	$\frac{\sqrt{2}}{2}$	1	$-\frac{\sqrt{2}}{2}$
1'	$-\frac{1}{2}$	$-\frac{1}{2}$	$\frac{1}{2}$	0	$\frac{1}{2}$	$-\frac{\sqrt{2}}{2}$
2	−1	−1	0	0	0	$-\sqrt{2}$
2'	$-\frac{3}{2}$	$-\frac{1}{2}$	$\frac{3}{2}$	$\frac{\sqrt{2}}{2}$	$\frac{3}{2}$	$-\sqrt{2}$

Diamond and GaAs (111) $0 \leq \delta < 3/4$

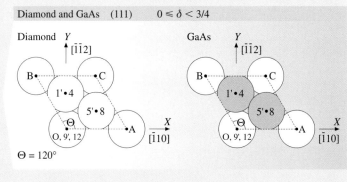

$\Theta = 120°$

Fig. 5.2-2c cont.

c) (cont.)

Atomic positions

Atom	Coordinates					
	Relative to cubic axes			Relative to $O'XYZ$		
A'	-1	1	0	$\sqrt{2}$	0	0
B'	0	-1	1	$-\frac{\sqrt{2}}{2}$	$\frac{3}{\sqrt{6}}$	0
3	$-\frac{1}{2}$	$-\frac{1}{2}$	$-\frac{1}{2}$	0	0	$-\frac{3}{2\sqrt{3}}$
4'	-1	-1	0	0	$\frac{2}{\sqrt{6}}$	$-\frac{2}{\sqrt{3}}$
7	$-\frac{3}{2}$	$-\frac{3}{2}$	$-\frac{1}{2}$	0	$\frac{2}{\sqrt{6}}$	$-\frac{7}{2\sqrt{3}}$
8'	-2	-1	-1	$\frac{\sqrt{2}}{2}$	$\frac{1}{\sqrt{6}}$	$\frac{-4}{\sqrt{3}}$
11	$-\frac{5}{2}$	$-\frac{3}{2}$	$-\frac{3}{2}$	$\frac{\sqrt{2}}{2}$	$\frac{1}{\sqrt{6}}$	$-\frac{11}{2\sqrt{3}}$

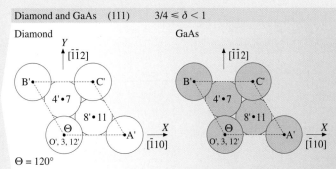

Diamond and GaAs (111) $3/4 \leq \delta < 1$

$\Theta = 120°$

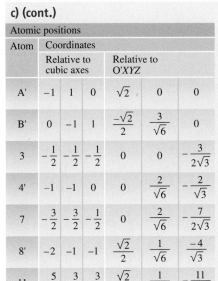

Diamond and GaAs $(\bar{1}\bar{1}\bar{1})$ $0 \leq \delta < 1/4$

$\Theta = 120°$

Atomic positions

Atom	Coordinates					
	Relative to cubic axes			Relative to $O'XYZ$		
A'	1	-1	0	$\sqrt{2}$	0	0
B'	-1	0	1	$-\frac{\sqrt{2}}{2}$	$\frac{3}{\sqrt{6}}$	0
1	$\frac{1}{2}$	$-\frac{1}{2}$	$\frac{1}{2}$	$\frac{\sqrt{2}}{2}$	$\frac{1}{\sqrt{6}}$	$-\frac{1}{2\sqrt{3}}$
4'	1	0	1	$\frac{\sqrt{2}}{2}$	$\frac{1}{\sqrt{6}}$	$-\frac{2}{\sqrt{3}}$
5	$\frac{1}{2}$	$\frac{1}{2}$	$\frac{3}{2}$	0	$\frac{2}{\sqrt{6}}$	$-\frac{5}{2\sqrt{3}}$
8'	1	1	2	0	$\frac{2}{\sqrt{6}}$	$-\frac{4}{\sqrt{3}}$
9	$\frac{3}{2}$	$\frac{3}{2}$	$\frac{3}{2}$	0	0	$-\frac{9}{2\sqrt{3}}$

Atomic positions

Atom	Coordinates					
	Relative to cubic axes			Relative to $OXYZ$		
A	1	-1	0	$\sqrt{2}$	0	0
B	-1	0	1	$-\frac{2}{\sqrt{2}}$	$\frac{3}{\sqrt{6}}$	0
3'	$\frac{1}{2}$	$\frac{1}{2}$	$\frac{1}{2}$	0	0	$-\frac{3}{2\sqrt{3}}$
4	1	0	1	$\frac{\sqrt{2}}{2}$	$\frac{1}{\sqrt{6}}$	$\frac{-2}{\sqrt{3}}$
7'	$\frac{3}{2}$	$\frac{1}{2}$	$\frac{3}{2}$	$\frac{\sqrt{2}}{2}$	$\frac{1}{\sqrt{6}}$	$-\frac{7}{2\sqrt{3}}$
8	1	1	2	0	$\frac{2}{\sqrt{6}}$	$\frac{-4}{\sqrt{3}}$
11'	$\frac{3}{2}$	$\frac{3}{2}$	$\frac{5}{2}$	0	$\frac{2}{\sqrt{6}}$	$-\frac{11}{2\sqrt{3}}$

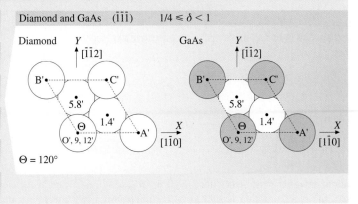

Diamond and GaAs $(\bar{1}\bar{1}\bar{1})$ $1/4 \leq \delta < 1$

$\Theta = 120°$

Fig. 5.2-2c cont.

5.2.1.2 Crystallographic Formulas

Some useful crystallographic formulas are listed in Table 5.2-1.

Table 5.2-1 Some useful crystallographic formulas

	Cubic system		Hexagonal system
Lattice parameter(s)	a		a, c
Plane	(hkl)		$[hkil]$
Plane normal, n	$[hkl]$		$[hkil/\lambda^2]$ $\lambda = \sqrt{2/3}\, c/a$
Cosine of the angle between planes with normals n_1, n_2	$\dfrac{h_1 h_2 + k_1 k_2 + l_1 l_2}{(h_1^2 + k_1^2 + l_1^2)^{1/2}(h_2^2 + k_2^2 + l_2^2)^{1/2}}$		$\dfrac{h_1 h_2 + k_1 k_2 + i_1 i_2 + l_1 l_2/\lambda^2}{(h_1^2 + k_1^2 + i_1^2 + l_1^2/\lambda^2)^{1/2}(h_2^2 + k_2^2 + i_2^2 + l_2^2/\lambda^2)^{1/2}}$
Interlayer distance	fcc $(a/Q)(h^2 + k^2 + l^2)^{-1/2}$ $Q = 1$, except that $Q = 2$ if at least one of hkl is even	bcc $(a/Q)(h^2 + k^2 + l^2)^{-1/2}$ $Q = 1$, except that $Q = 2$ if $h + k + l$ is odd	$(c/\lambda)(h^2 + k^2 + i^2 + l^2/\lambda^2)^{-1/2}$
Volume of the primitive cell	$\tfrac{1}{4}a^3$	$\tfrac{1}{2}a^3$	$\tfrac{1}{2}a^2 c\sqrt{3}$
Area of the surface primitive cell	$\tfrac{1}{4}a^2 Q(h^2 + k^2 + l^2)^{1/2}$	$\tfrac{1}{2}a^2 Q(h^2 + k^2 + l^2)^{1/2}$	$ac[(h^2 + k^2 + i^2 + l^2/\lambda^2)/2]^{1/2}$

5.2.2 Surface Reconstruction and Relaxation

The structure of a real (clean) surface may differ from that of an ideally truncated crystal. Two types of deviation are possible: relaxation and reconstruction.

Relaxation consists of a rigid inward or outward displacement of the uppermost planes of the surface, the symmetry in the surface plane being maintained. In contrast, in the case of reconstruction, the displacements of atoms alter the 2-D symmetry of the surface.

5.2.2.1 Definitions and Notation

If a and b are the basis vectors of an ideally terminated surface and A and B those of the reconstructed surface, then

$$A = m_{11}a + m_{12}b,$$
$$B = m_{21}a + m_{22}b.$$

The matrix

$$\begin{pmatrix} m_{11} & m_{12} \\ m_{21} & m_{22} \end{pmatrix}$$

characterizes the reconstruction. If all the m_{ij} are integers the reconstruction is said to be commensurate. If at least one of the m_{ij} is irrational, the reconstruction is said to be incommensurate. The term "incommensurate", however, is often extended to rational noninteger numbers. For example the reconstruction of Mo(100),

$$\begin{pmatrix} 2 & 0 \\ 0 & 2 \end{pmatrix},$$

is called incommensurate.

If the angle between A and B is the same as that between a and b, a simpler notation introduced by Wood [2.17] can be used, namely:

$$\text{p}(u \times v)\text{R}\Phi \quad \text{or} \quad \text{c}(u \times v)\text{R}\Phi,$$

where $u = A/a$, $v = B/b$; p and c stand for *primitive* and *centered*, respectively, and RΦ means a rotation of Φ of the cell (A, B) with respect to (a, b). When $\Phi = 0$, RΦ is omitted. Also, the prefix p and the rotation symbol R are often omitted.

Figure 5.2-3 shows a few examples of reconstruction for a layer of adatoms on top of a substrate. The matrix and Wood notations are compared. It can be seen from the figure that (i) the elementary cell is not uniquely

5.2.2.2 Metals

Table 5.2-2 summarizes the essential features of reconstruction or relaxation at metal surfaces. Vertical relaxations (in % of the bulk interlayer distances) are given in Table 5.2-3.

Reconstruction Models

In Figs. 5.2-4 and 5.2-6, some accepted models of reconstructed surfaces of metals are shown. Table 5.2-4 gives the parameters for the 2×1 missing-row reconstruction of Au(110), Ir(110), and Pt(110). The missing-row reconstruction stabilizes the surface by relieving the elastic stress.

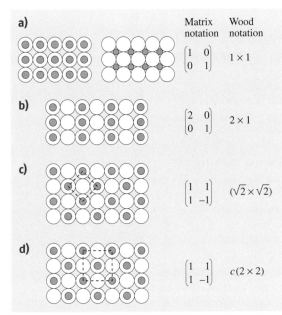

Fig. 5.2-3a–d Examples of reconstruction for a layer of adatoms (*small shaded circles*) on top of a square lattice substrate (*large empty circles*). Matrix and Wood notations are compared. Notice that (**c**) and (**d**) differ only for the choice of the unit cell

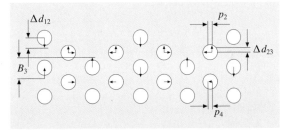

Fig. 5.2-4 Side view of the atomic positions in the 2×1 missing-row reconstruction of the (110) face of Au, Ir, and Pt, and definition of parameters used in Table 5.2-4

defined (as in 3-D), and (ii) the symmetry alone does not define unequivocally the reconstruction. In practice this means that LEED (low-energy electron diffraction) patterns are not enough to define a model for a reconstruction and that other techniques (including dynamical LEED) are necessary.

The energies associated with reconstruction are $\sim 10^{-2}$ eV/atom in metals and $\sim 10^{-1}$ eV/atom in semiconductors. Metals are therefore seldom reconstructed, while semiconductors are very often reconstructed.

A detailed discussion of the various problems connected to relaxation and reconstruction are found in [2.3–8]

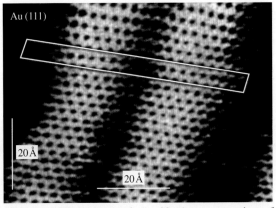

Fig. 5.2-5 STM image of the 23×1 reconstruction of Au(111) [2.18]

Table 5.2-2 Reconstruction and relaxation of metals [1]

Metal	Structure	Face	Reconstruction	Remarks
		(100)	1×1	Unrelaxed
Ag	fcc	(110)	1×1	Relaxed
		(111)	1×1	Unrelaxed
		(100)	1×1	Unrelaxed
Al	fcc	(110)	1×1	Relaxed
		(111)	1×1	Relaxed

Table 5.2-2 Reconstruction and relaxation of metals [1], cont.

Metal	Structure	Face	Reconstruction	Remarks
Au	fcc	(100)	5×20 6×6	$T < 700$ K. Reconstruction model: incommensurate hexagonal layer $700 < T < 820$ K
		(110)	1×2	Reconstruction model: missing-row (Fig. 5.2-4 and Table 5.2-4). Reversible phase transition to 1×1 at $T \cong 690$ K
		(111)	1×23	Reconstruction model: see Fig. 5.2-5 (observed by STM)
Co	hcp cubic phase	(0001) (100) (111)	1×1 1×1 1×1	Unrelaxed Relaxed Unrelaxed
Cu	fcc	(100) (110) (111)	1×1 1×1 1×1	Relaxed Relaxed Essentially unrelaxed
Fe	bcc	(100) (110) (111)	1×1 1×1 1×1	Relaxed Essentially unrelaxed Relaxed
Ir	fcc	(100)	1×5	$T < 2000$ K. Reconstruction model: hexagonal layer with buckling. A metastable 1×1 phase is also observed
		(110)	1×2	Reconstruction model: missing-row (Fig. 5.2-4 and Table 5.2-4). A metastable 1×1 phase is also observed
		(111)	1×1	Essentially unrelaxed
Mo	bcc	(100)	2.2×2.2 1×1	$T < 300$ K, incommensurate $T > 300$ K, relaxed
		(110)	1×1	Relaxed
Ni	fcc	(100) (110) (111)	1×1 1×1 1×1	Relaxed (results controversial) Relaxed Unrelaxed
Pb	fcc	(110)	1×1	Relaxed; annealing at $T > 340$ K gives a c(2×4) reconstruction
Pd	fcc	(100) (110) (111)	1×1 1×1 1×1	Essentially unrelaxed Relaxed Unrelaxed
Pt	fcc	(100)	1×5	77 K $< T < 1900$ K. Reconstruction model: hexagonal layer slightly rotated. Other metastable superstructures, 1×1 and 5×20, are observed
		(110)	1×2	Reconstruction model: missing-row (Fig. 5.2-4 and Table 5.2-4). Reversible transition to 1×1 at $T = 1100$ K
		(111)	1×1	Essentially unrelaxed
Rh	fcc	(100) (110) (111)	1×1 1×1 1×1	Essentially unrelaxed Relaxed Relaxed
Ta	bcc	(100)	1×1	Relaxed
Th	fcc	(111)	1×1	After annealing at $T \approx 1000$ K becomes 9×9
V	bcc	(100) (110)	1×1 1×1	Relaxed Essentially unrelaxed
W	bcc	(100)	c(2×2) 1×1	$T < 400$ K. Reconstruction model: see Fig. 5.2-6 $T > 400$ K. Relaxed
		(110)	1×1	Unrelaxed

[1] Reference to the original papers is given in [2.3]

Table 5.2-3 Vertical relaxation of metals. Data are given as a percentage of the bulk interlayer distance d_{bulk}. The data are average values from various authors.[1] Errors are expressed as standard deviations from the average. The + sign corresponds to expansion. LEIS, low-energy ion scattering; HEIS, high-energy ion scattering; MEIS: medium-energy ion scattering

Crystal	Face	d_{bulk} (Å)	Δd_{12} (%)	Δd_{23} (%)	Δd_{34} (%)	Technique
Ag	(110)	1.445	−7.5 (13)	+1.6 (6)	−2 (2)	LEED
			−8.6 (8)	+5.1 (8)[2]	−	LEIS, MEIS
Al	(100)	2.025	0 (5)	−	−	LEED
			−2.4 (24)	−	−	HEIS
	(110)	1.427	−7 (2)	+5.0 (2)	−1.6	LEED
	(111)	2.329	+1.5 (6)	−	−	LEED
Cu	(100)	1.807	−1.1	+1.7	<1	LEED
			−2.3 (2)	−	−	MEIS
	(110)	1.278	−8.3 (9)	+2.3	−0.9	LEED
			−7 (3)	+2.9 (4)	−	LEIS, MEIS, HEIS
	(111)	2.087	+0.8 (2)	−	−	LEED
Fe	(100)	1.433	−1.4 (30)	−	−	LEED
	(110)	2.027	+0.5 (20)	−	−	LEED
			+1 (2)	−	−	MEIS
	(111)	0.827	−16.9 (30)	−9.8 (30)	+4.2 (36)	LEED
			−29 (7)	+7 (5)[2]	−	MEIS
Ir	(100) 1×1 metastable	1.920	+5.2	−	−	LEED
Mo	(100)	1.574	−9 (2)	+0.25 (175)	+0.5 (30)	LEED
	(110)	2.225	−1.6	−	−	LEED
	(111)	0.909	−18	+4	−	LEED
			−16	+4	−	LEIS
Ni	(110)	1.245	−7 (2)	+3.3 (3)	−0.5 (7)	LEED
			−6.9 (21)	+3.0 (6)	−	MEIS, HEIS
Pb	(110)	1.750	−17.6 (13)	+3.8 (10)	−5.9 (13)	LEED
			−15.8	+7.9	−6.8	MEIS
Pd	(110)	1.375	−5.6 (4)	+1.5 (10)	−	LEED
Rh	(100)	1.902	+0.5 (10)	0 (15)	−	LEED
	(110)	1.345	−3.5 (25)	−	−	LEED
	(111)	2.196	−1	−	−	LEED
Ta	(100)	1.653	−11	+1	−	LEED
			−10.5	−	−	PED
V	(100)	1.514	−6.8 (2)	+1	−	LEED
W	(100)	1.583	−6.2 (14)	+1	−	LEED
	high-T phase		−5.6	−	−	X-ray

[1] References to the original papers and individual errors are given in [2.3–5]
[2] Inconsistency between LEED and ion scattering data

Table 5.2-4 Structural parameters of the 1×2 reconstruction of the (110) face of Au, Ir, and Pt (average values from various authors[1]). Errors are expressed as standard deviations from the average. An illustration of the model and parameters is given in Fig. 5.2-4. The bulk interlayer spacings are 1.445 Å for Au, 1.358 Å for Ir, and 1.387 Å for Pt

Crystal	Δd_{12} (Å)	Δd_{23} (Å)	p_2 (Å)	p_4 (Å)	B_3 (Å)	Technique
Au	−0.25 (3)	0.03	0.07	–	0.24	LEED
	−0.20 (4)	0.06	< 0.1	–	0.20	LEIS, MEIS
	−0.32	–	0.05	0.05	–	X-ray
Ir	−0.17	−0.16	0.04	–	0.23	LEED
	−0.13	–	–	–	–	MEIS
Pt	−0.30 (5)	−0.04 (10)	0.06 (2)	0.08 (3)	0.22 (7)	LEED, RHEED
	−0.22 (4)	0.06 (4)	< 0.04	–	0.10	MEIS
	−0.27	−0.11	0.05	0.04	–	X-ray

[1] References to the original papers and individual errors are given in [2.3]

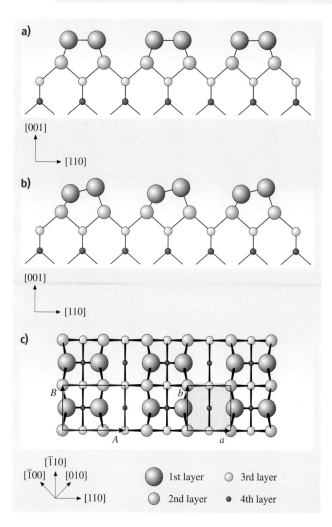

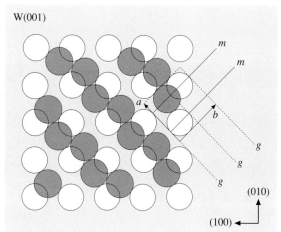

Fig. 5.2-6 Top view of the $c(2\times 2)$ reconstruction of the W(001) face. *g*, glide plane; *m*, mirror plane. *Shaded circles*, surface atoms; *empty circles*, second-layer atoms [2.19]

5.2.2.3 Semiconductors

The essential features of reconstruction and relaxation of semiconductor surfaces are given in Table 5.2-5, together with indications of the methods of preparation and remarks on peculiarities of the specific surfaces.

Fig. 5.2-7a–c Dimer models for the 2×1 reconstruction of Si and Ge(100) faces, with (**a**) symmetric and (**b**) buckled dimers (side views). (**c**) Top view of the symmetric case. Ideal (*shaded*) and reconstructed unit cells are shown

Reconstruction Models

In semiconductors, directional bonds are present that are unsaturated at the surface. The driving force for reconstruction is the tendency to minimize the number of such dangling bonds (DBs).

In sp^3-bonded crystals, the energy associated with the angles between the bonds is rather small. In contrast, the energy associated with the length of the bonds is large. Accordingly, reconstruction of covalent (or moderately ionic) semiconductors occurs through dis-

Table 5.2-5 Reconstruction and relaxation of semiconductors. IBA, ion bombardment and annealing; MBE, molecular-beam epitaxy; DAS, dimer–adatom–stacking fault. References are given in [2.3]

Crystal	Face	Reconstruction symmetry	Preparation	Model	Remarks
Diamond	(100)	2×1	Annealing $T > 1300$ K	Symmetric dimers	
	(110)	1×1	Annealing	Truncated bulk	
	(111)	2×1	Annealing $T > 1200$ K	π-bonded chains	
Si	(100)	2×1	IBA or MBE	Symmetric and/or buckled dimers (Fig. 5.2-7)	
		p(2×2)	Same as above	Ordered arrangement of buckled dimers (Fig. 5.2-8)	Present locally ($\sim 5\%$)
		c(4×2)	Same as above	Ordered arrangement of buckled dimers (Fig. 5.2-10)	Present locally ($\sim 5\%$)
	(110)	16×2	Annealing $T > 1300$ K	Dimers + adatoms + terraces (?)	Other superstructures (probably stabilized by impurities) are observed: 4×5, 5×1, 2×1
		32×2	Same as above	Same as above	
	(111)	2×1	Cleaving	π-bonded chains (Fig. 5.2-11, Table 5.2-6)	Transforms irreversibly into 7×7 at $T \simeq 550$ K
		7×7	Cleaving + annealing at 550 K or Annealing at 900 K or MBE	DAS + vacancies (Fig. 5.2-13)	Equilibrium reconstruction
		5×5	Same as above	Same as above	Present locally near steps
Ge	(100)	2×1	IBA or MBE	Symmetric and/or buckled dimers	
		p(2×2)	Same as above	Ordered arrangement of buckled dimers	Present locally
		c(4×2)	Same as above	Same as above	Same as above
	(111)	2×1	Cleaving at $40 < T < 400$ K	π-bonded chains	Transforms irreversibly into c(2×8) at $T \simeq 400$ K
		c(2×8)	Annealing $T > 400$ K	Adatoms (Fig. 5.2-16)	Equilibrium reconstruction
GaAs	(100)	c(2×8)	IBA $T \simeq 850$ K	Ga (or As) dimers (?)	As coverage 0.22–0.52
		c(4×4)	MBE + annealing $T \simeq 670$ K		As coverage 1–1.25
		4×6	IBA $T \simeq 770$–850 K		As coverage 0.27–0.31
	(110)	1×1	Cleaving, IBA	Rotation/relaxation (Fig. 5.2-17, Table 5.2-8)	
	(111)Ga	2×2	IBA or MBE	Ga vacancies	
	($\bar{1}\bar{1}\bar{1}$)As	2×2	Same as above	As trimers	
GaP	(110)	1×1	Cleaving, IBA	Rotation/relaxation (Fig. 5.2-17, Table 5.2-8)	
GaSb	(110)	1×1	Cleaving	Same as above	

Table 5.2-5 Reconstruction and relaxation of semiconductors, cont.

Crystal	Face	Reconstruction symmetry	Preparation	Model	Remarks
InAs	(110)	1×1	Same as above	Same as above	
InSb	(110)	1×1	Same as above	Same as above	
CdS	(0001)Cd	1×1	IBA		
CdSe	(10$\bar{1}$0)	1×1	Cleaving	Rotation/relaxation (Fig. 5.2-17, Table 5.2-8)	Results controversial
ZnO	(0001)Zn	1×1	Cleaving, IBA		

placements of atoms that conserve the lengths of the bonds but not their relative angles.

Figures 5.2-7 – 5.2-17 give the accepted reconstruction models for a selection of covalent and polar semiconductors, together with STM (scanning tunneling microscopy) images of some of the surfaces. Tables 5.2-6, 5.2-7, and 5.2-8 give the positions of the atoms in reconstructed Si(111) 2×1 and Si(111)7×7 surfaces and the parameters of the rotation/relaxation model of polar semiconductors.

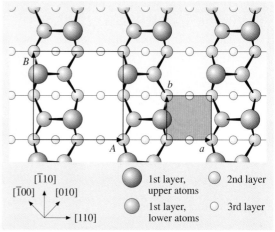

Fig. 5.2-8 Ordered arrangement of buckled dimers that gives rise to p(2×2) reconstruction of the (100) face of Si and Ge (*top views*). Ideal (*shaded*) and reconstructed unit cells are shown

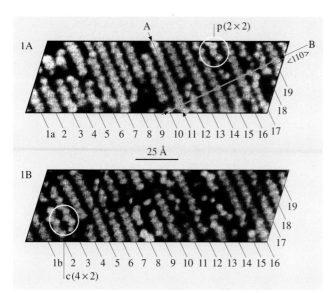

Fig. 5.2-9 STM image of the Si(100) surface showing coexistence of 2×1, p(2×2), and c(4×2) reconstructions [2.20]

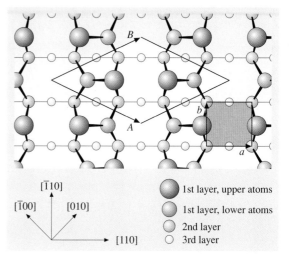

Fig. 5.2-10 Ordered arrangement of buckled dimers that gives rise to c(4×2) reconstruction of the (100) face of Si and Ge (*top views*). Ideal (*shaded*) and reconstructed unit cells are shown

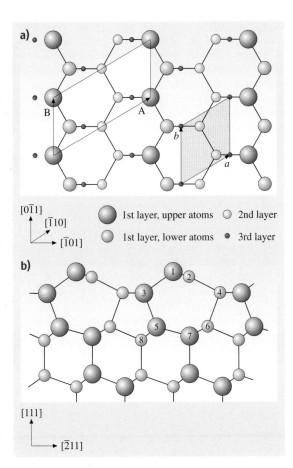

Fig. 5.2-11 (a) top view and (b) side view of the Pandey's π-bonded chain model for the 2×1 reconstruction of the (111) face of diamond, Si and Ge [2.21]. In (a) the ideal (*shaded*) and 2×1 unit cells are shown. The numeration of atoms in (b) refers to the positions given in Table 5.2-6

Table 5.2-6 Coordinates of the atoms of the uppermost layers of Si(111) 2×1 (π-bonded chain geometry). The atoms are labeled as in Fig. 5.2-11b. The origin is at atom 1; the XYZ axes are in the directions $[\bar{1}\bar{1}\bar{1}]$, $[1\bar{1}0]$, and $[\bar{1}\bar{1}2]$, respectively[1]

Atom	X (Å)	Y (Å)	Z (Å)
1	0.0	0.0	0.0
2	0.34	1.92	1.12
3	1.32	0.0	−1.98
4	1.23	1.92	3.28
5	3.51	0.0	−1.14
6	3.40	1.92	2.35
7	4.09	0.0	1.16
8	4.35	1.92	−2.18

[1] Data are average values from various authors. References to the original papers are given in [2.3]

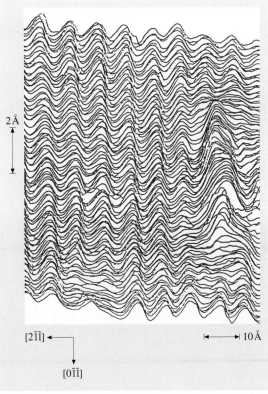

Fig. 5.2-12 STM image of Si(111) 2×1, showing the great anisotropy of the surface, in agreement with the model of Fig. 5.2-11. Optical SDR spectra (see Fig. 5.2-41) show the same anisotropy [2.22]

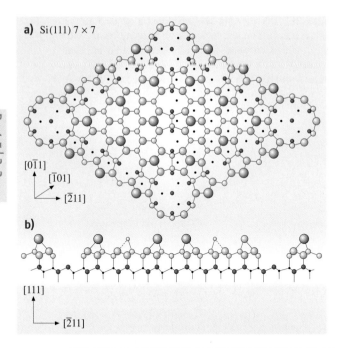

Fig. 5.2-13a,b DAS (dimers, adatoms, and stacking faults) + vacancy model of Takayanagi et al. for Si(111) 7×7. (**a**) Top view; (**b**) side view. In (**a**), the unit cell could be drawn by joining the centers of the four vacancies. The faulted area is the triangular half-cell on the left. The rotation associated with the stacking fault hides atoms in deep unreconstructed layers, which, in contrast, are visible in the unfaulted right-hand side [2.23]

Fig. 5.2-14a–d Symmetry-reduced unit cell for Si(111) 7×7 (DAS model). In (**a**), the unit cell (two triangular halves, one faulted (*hatched*) and one unfaulted) is reduced to smaller subunits and recomposed into a c-hexagonal cell. The atoms are drawn and numbered in (**b**) – (**d**). The numbering is the same as in Table 5.2-7; the origin of the coordinates is at atom 103. (**b**) The four adatoms; (**c**) the first- and second-layer atoms; (**d**) the third- and fourth-layer atoms [2.24, 25]

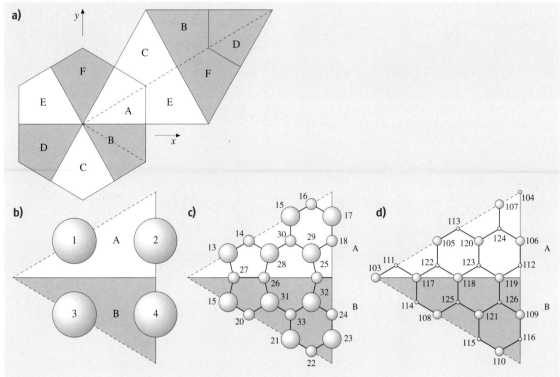

Table 5.2-7 Coordinates of the adatoms and four uppermost layer atoms in the symmetry-reduced unit cell of Si(111) 7×7 (see Fig. 5.2-14 for the numbering of the atoms). The origin is taken at atom 103[1]. X-ray data give a value of 2.11 Å for the adatom–back atom projected bond length (between atoms 1 and 13 and between atoms 1 and 28 in the table) and a value of 2.49 Å for the dimer bond length (betweens atoms 26 and 27 in the table) [2.24–26]

Atom no.	x (Å)	y (Å)	z (Å)	Atom no.	x (Å)	y (Å)	z (Å)
1	5.76	3.33	−4.21	103	0.00	0.00	−0.05
2	13.44	3.33	−4.17	104	13.44	7.76	0.73
3	5.76	−3.33	−4.29	105	5.76	3.33	0.45
4	13.44	−3.33	−4.25	106	13.44	3.33	0.45
13	4.01	2.32	−2.94	107	11.52	6.65	−0.07
14	5.76	3.33	−1.68	108	5.76	−3.33	0.45
15	9.60	5.54	−3.11	109	13.44	−3.33	0.45
16	11.55	6.67	−2.41	110	11.52	−6.65	−0.07
17	13.44	5.32	−2.98	111	1.92	1.11	0.74
18	13.44	3.33	−1.68	112	13.44	1.11	0.87
19	4.01	−2.32	−2.99	113	7.68	4.43	0.80
20	5.76	−3.33	−1.73	114	3.84	−2.22	0.85
21	9.60	−5.54	−3.16	115	9.60	−5.54	0.73
22	11.55	−6.67	−2.46	116	13.44	−5.54	0.80
23	13.44	−5.32	−3.03	117	3.84	0.00	0.00
24	13.44	−3.33	−1.73	118	7.68	0.00	0.00
25	12.21	0.00	−2.24	119	11.52	0.00	0.00
26	6.98	0.00	−2.24	120	9.60	3.33	−0.07
27	4.53	0.00	−2.24	121	9.60	−3.33	−0.07
28	7.49	2.33	−2.98	122	5.76	1.11	0.87
29	11.71	2.33	−2.98	123	9.60	1.11	0.72
30	9.60	3.29	−2.41	124	11.52	4.43	0.80
31	7.49	−2.33	−3.03	125	7.68	−2.22	0.83
32	11.71	−2.33	−3.03	126	11.52	−2.22	0.83
33	9.60	−3.29	−2.46				

[1] A table of the atomic positions for the full 7×7 cell is given in [2.25]. The results were obtained by LEED analysis [2.24]

Table 5.2-8 Surface relaxation and rotation parameters for polar semiconductors as defined in Fig. 5.2-17[1]

Crystal	ω (°)	$\Delta_{1\perp}$ (Å)	$\Delta_{2\perp}$ (Å)	Δ_{1y} (Å)	$d_{12\perp}$ (Å)	d_{12y} (Å)	$d_{23\perp}$ (Å)
GaAs	27.4	0.65	−0.12	4.39	1.43	3.31	2.06
GaP	27.5	0.63	0.0	4.24	1.44	3.20	1.93
GaSb	30	0.77	0.0	4.79	1.61	3.27	2.16
InAs	36.5	0.78	−0.15	4.98	1.50	3.60	2.21
InP	28	0.69	0.0	4.57	1.59	3.45	2.07
InSb	28.8	0.78	−0.18	5.06	1.60	3.82	2.38
CdSe	21.3	1.03	0.0	4.60	0.45	4.10	2.48

[1] The data were obtained by dynamical LEED analysis. References to the original papers are given in [2.3, 4]

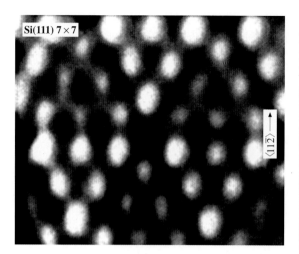

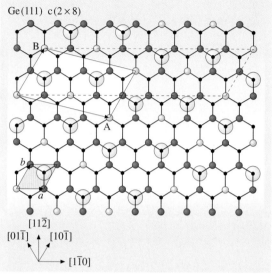

Fig. 5.2-15 STM image of the Si(111)7×7 surface. The picture gives evidence to the 12 adatoms, the 4 vacancies, and the asymmetry between the two triangular half cells of Fig. 5.2-13 [2.7, 27]

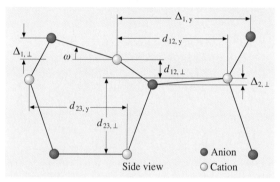

Fig. 5.2-16 Top view of Ge(111)c(2×8) simple adatom model. Large and small empty circles represent adatoms and "rest" atoms respectively. Both ideal (*shaded*) and c(2×8) unit cells are shown [2.28]

Fig. 5.2-17 Side view of the relaxed cell of a GaAs-type crystal, with parameters occurring in the relaxation/rotation model of the (110) face used in Table 5.2-8

5.2.3 Electronic Structure of Surfaces

The truncation of the lattice and/or the reconstruction and relaxation cause the electronic states at the surface or in the uppermost layers to be distinctly different from those of the bulk. Such new states are called surface states. Their wave functions decay exponentially on both sides of the surface. Since their $k_\perp$ is imaginary, the surface band structure is defined in the surface Brillouin zone (SBZ), which is the projection of the 3-D Brillouin zone onto the surface plane. The projection of the bulk bands onto the SBZ is called the projected band structure. When the energy of a surface state is localized in a gap of the projected bulk structure (either an absolute gapp, i.e. one that extends throughout the whole SBZ, or a *partial* gap), one speaks of a true (or bona fide) surface state. When there is degeneracy (both in energy and k) with the bulk bands, one speaks of a surface resonance.

The SBZs, of the most common faces of fcc and bcc crystals are shown in Fig. 5.2-18.

The potential felt by an electron outside the surface (though sufficiently far from it) is the classical image potential: $(-e^2/4z)(\varepsilon-1)/(\varepsilon+1)$, ε being the dielectric function of the crystal. From a microscopic point of view, the image potential arises from the charge density fluctuations induced in the solid by the outside electron. An important consequence of the Coulomb-like image potential is the occurrence of image states in the potential well outside the solid (especially in metals). In surface structure calculations done with the local-density approximation (LDA), such image states are not present,

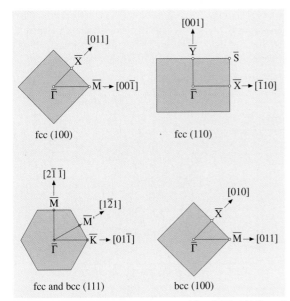

Fig. 5.2-18 Surface Brillouin zones of fcc (100), (110), and (111), bcc (100) and (111) faces

since the potential decays exponentially on the vacuum side. Nonlocal forms of the exchange correlation energies that give the correct long-range potential have been introduced to overcome this difficulty.

Surface states were first detected by optical techniques in semiconductors and are now studied mainly by ARUPS, KRIPES, STM, SDR, and reflectance anisotropy spectroscopy (RAS).

Data, in the form of tables and figures, are presented separately for metals (including jellium), semiconductors, and magnetic surfaces.

5.2.3.1 Metals

The data are ordered alphabetically according to the chemical symbol. Where necessary, introductory remarks are given. A more detailed discussion is given in [2.9, 11].

Work Function Data
Work function data are listed in Table 5.2-9 with the specification of the method of measurement.

Table 5.2-9 Work functions Φ of metals (average values from various authors; references to the original papers are given in [2.9, 16]). P, photoelectric threshold; A, angle-resolved photoemission spectroscopy; T, thermionic emission; F, field emission; th, theory

Metal	Face	Φ (eV)	Technique
Ag	(100)	4.18	P
		4.38	A
		4.2	th
	(111)	4.46	P
		4.50	A
Al	(100)	4.30	P
	(110)	4.17	P
	(111)	4.23	P
		4.5	th
Au	(100)	5.22	A
	(110)	5.20	A
	(111)	5.26	A
Co	(0001)	5.2	A
Cu	(100)	4.59	P
		4.63	A
		5.10	F
		4.5	th
	(110)	4.48	P
		4.5	A
	(111)	4.85	P
		4.5	A
Fe	(100)	4.67	P
	(110)	5.05	A
	(111)	4.81	P
Ir	(110)	5.42	F
	(111)	5.76	F
Mo	(100)	4.53	P
	(110)	4.95	P
	(111)	4.55	P
Ni	(100)	4.89	T
		5.53	F
		5.1	th
	(110)	4.64	T
	(111)	5.22	T
Pb	(111)	4.05	A
Pd	(111)	5.58	A
		5.8	th
Pt	(100)	5.84	P
		5.84	F
	(111)	5.82	A
		5.93	F
Rh	(100)	5.1	th
	(111)	5.35	th
Ta	(100)	4.15	T
	(110)	4.80	T
	(111)	4.00	T
W	(100)	4.66	F
		4.5	th
	(110)	5.57	F
	(111)	4.46	F

Surface Core Level Shifts (SCLS)

A shift in the energy of the core levels is caused by charge transfer between surface and bulk atoms. A positive SCLS means a larger binding energy. It is generally assumed that in metals the change in the width of the band and its position with respect to the Fermi energy are the important factors. Data are shown in Table 5.2-10.

Jellium Model

The jellium model of a metal surface takes into account, in a simplified though significant way, the problem of electron–electron interaction. The atomic potential is smeared out into a uniform positive background extending over the region $z \leq 0$. The electron density is taken into account by a dimensionless parameter r_s, defined by the equation

$$(4/3)\pi(r_s a_0)^3 n = 1 ,$$

where a_0 is the Bohr radius and n the bulk electron density.

Figure 5.2-19 shows the charge density as a function of z for $r_s = 2$ and $r_s = 5$, corresponding roughly to Al and K. Table 5.2-11 gives the values of the work function Φ for various values of r_s and compares the

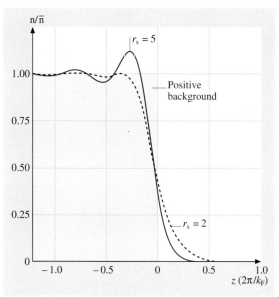

Fig. 5.2-19 Charge density near a jellium surface, for $r_s = 2$ and $r_s = 5$ as a function of z [2.29]

Table 5.2-10 Surface core level shift data. The data are from various authors. References to the original papers are given in [2.9]

Metal	Surface	Core level	SCLS (eV)	Remarks
	(100)		−0.057	Photoemission (partial yield)
Al		2p	−0.120	Theory
	(110) and (111)		0.0	UPS
	(100) 1×1		−0.35	UPS
Au	(100) 5×20	4f	−0.28	UPS
	(111)		−0.35	UPS, no distinct results for (111) 23×1
Cu	(100)	3s	−0.36	Theory
		2p	−0.27	UPS
Ir	(100) 1×5	4f	−0.49	
	(111)		−0.50	
	(110) 1×2		−0.21	UPS, evidence for nonequivalent atoms
Pt		4f	−0.55	
	(111)		−0.40	UPS
	(100)		−0.75 1st layer	Theory
Rh		4d	−0.05 2nd layer	
	(111)		−0.46 1st layer	
			−0.05 2nd layer	
	(110)		−0.30	UPS
W		4f	−0.30 1st layer	Theory
			−0.02 2nd layer	

Table 5.2-11 Work functions for the jellium model. References to the original papers are given in [2.9]

r_s	$\Phi_{jellium}$ (eV)	Element	Φ_{exp} (eV)
2	3.89	Al ($r_s = 2.08$)	4.25
2.5	3.72		
3	3.50	Ag ($r_s = 3.02$)	4.35
3.5	3.26		
4	3.06	Na ($r_s = 3.93$)	2.45
4.5	2.87		
5	2.73	K ($r_s = 4.86$)	2.3

results with experimental values for metallic elements of comparable densities. Figure 5.2-20 gives some useful insight into the problem of the potential near a metallic surface.

Image States

Image states are confined to the region of space between the crystal surface and the image potential which describes the electrostatic force acting on an electron close to the surface. Figure 5.2-21 illustrates the occurrence of image states in Ag(100). The figure shows (i) the electron potential inside and outside the crystal, (ii) the image state levels $n = 1$ and $n = 2$ and the corresponding wave functions, and (iii) the bulk band structure

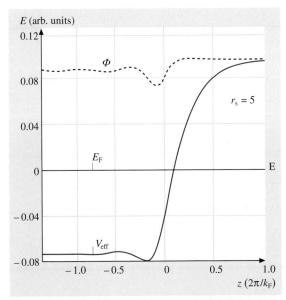

Fig. 5.2-20 Effective one-electron potential V_{eff} (*continuous line*) and its electrostatic part (*dashed line*) near a jellium surface for $r_s = 5$. Notice the characteristic oscillations (Friedel oscillations) [2.29]

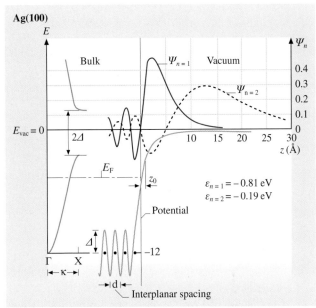

Fig. 5.2-21 Schematic drawing of energy versus distance at the surface of Ag(100), illustrating the origin of the image states. Wave functions for $n = 1$ and $n = 2$ are also displayed in the right part of the figure, corresponding to the levels in the potential well. On the left side of the figure the bulk energy bands as a function of k are also shown. Notice that the image state energies occur in the band gap of the bulk. This allows confinement of the image state electron [2.30, 31]

with an energy gap of width 2Δ. It can be seen that the image state levels have energies corresponding to the bulk gap so that the electron cannot penetrate the crystal [2.32].

Figure 5.2-22 shows the dispersion curves $E(k_\parallel)$ for the (111) faces of Ni, Cu, and Ag. Image states (S_1) and surface and resonance states (S_2, S_3), together with the projected bulk bands (hatched regions), are shown. Table 5.2-12 gives average values of the binding energies E_b (measured downward from the vacuum level) and effective masses m^*/m_0 for some image states.

Surface Dispersion Curves $E(k_\parallel)$

Dispersion curves are given as a function of $k_\parallel$ along symmetry lines of the SBZ in Figs. 5.2-23–5.2-30. Where possible, a comparison with theoretical results and the projected bulk band structure has been presented.

Most of the experimental data were obtained by ARUPS (for filled states) and KRIPES (for empty states). Since the component $k_\parallel$ is conserved across the

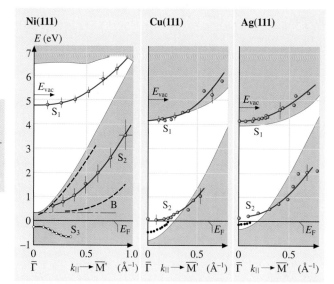

Fig. 5.2-22 Dispersion curves $E(k_\parallel)$ for the empty states of the (111) face of Ni, Cu and Ag, showing image states (S_1), surface states and resonances (S_2, S_3). Experimental data by KRIPES [2.33]

Table 5.2-12 Image state parameters. The data are from various authors. References to the original papers are given in [2.9]

Crystal	Face	E_b (eV)	m^*/m_0
Ag	(100)	0.51	1.32
		0.16 ($n=2$)	
	(111)	0.69	1.35
		0.23 ($n=2$)	
Au	(100)	0.63	1.0
	(111)	0.42	
Cu	(100)	0.65	1.13
		0.18 ($n=2$)	
	(110)	0.48	
	(111)	0.82	1.07
Ni	(100)	0.4	1.2
	(110)	0.6	1.7
	(111)	0.74	1.3
		0.25 ($n=2$)	
		0.10 ($n=3$)	
Pt	(111)	0.63	

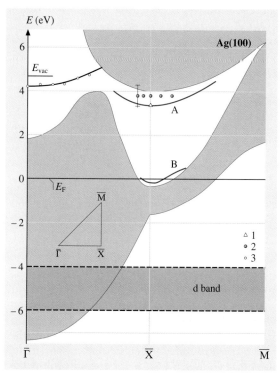

Fig. 5.2-24 Energy dispersion curves $E(k_\parallel)$ around $\overline{X}$ of the SBZ for Ag(100). Projected bands and surface state features A and B are the same as in Fig. 5.2-8a. Experimental points are obtained by: electroreflectance (1) and KRIPES (2 and 3). Curve (3) represents image states [2.34, 35]

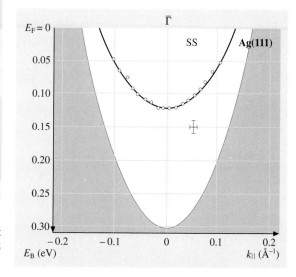

Fig. 5.2-25 Energy dispersion curves $E(k_\parallel)$ around $\overline{\Gamma}$ point of the SBZ for Ag(111). Experimental data from ARUPS are fitted with a parabolic band with $m^*/m_0 = 0.53$ [2.36]

Fig. 5.2-23 Theoretical projected band structures (*shaded areas*) along symmetry lines of the SBZ for Ag(100), Ag(110) and Ag(111). Surface states (*solid lines*) and resonances (*dashed lines*) are shown. SBZ in the insets [2.37–39]

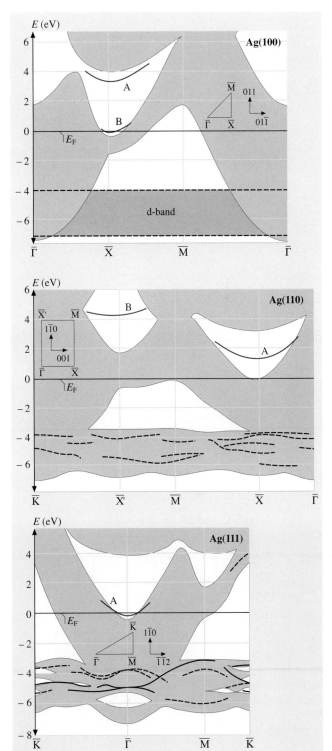

surface, the determination of the direction of the emitted or incident electron in ARUPS or KRIPES, respectively, and its energy allows the determination of the dispersion curve through the relation

$$k_\parallel = (2mE_{\text{kin}}/\hbar^2)^{1/2} \sin\theta ,$$

where E_{kin} is the kinetic energy of the emitted or incident electron and θ the angle of emission or incidence with respect to the surface normal.

Because of the large amount of data available for metals, only a few significant examples have been given. For more detailed information, the reader is referred to [2.9, 11].

Surface Plasmons

Surface plasmons are self-sustained oscillations that decay exponentially on both sides of the surface plane. They occur at the poles of the loss function, i.e. for $\varepsilon_1 = 0$ in the bulk and $\varepsilon_1 = -1$ at the surface, ε_1 being the real part of the dielectric function. The surface plasmon is shifted to lower energies with respect to the bulk.

Table 5.2-13 gives some of the parameters of surface plasmons for various metals. $d_\perp(\omega)$ is, in the jellium model, the distance of the centroid of the induced charge from the surface plane. $d_\perp > 0$ means that the centroid is outside the edge of the jellium. Figure 5.2-31 shows the dispersion curves of surface plasmons for Al(111). For a more detailed discussion see [2.12].

Table 5.2-13 Surface plasmon parameters. The data are from various authors. References to the original papers are given in [2.12]

Metal	Surface plasmon energy (eV)	Bulk plasmon energy (eV)	$d_\perp(\omega_{sp})$ (Å)
Al	10.3	15	
Mg	7.38	10.4	0.82
Li	4.3	7.12	0.48
Na	3.98	5.72	0.73
K	2.73	3.72	0.73
Rb		3.41	
Cs	1.99	2.90	0.88

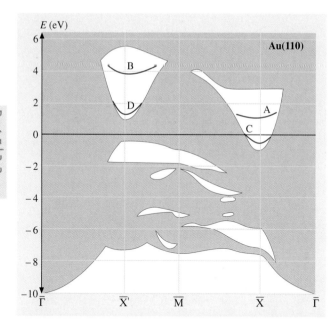

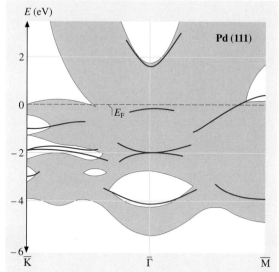

Fig. 5.2-26 Theoretical projected band structure for the unreconstructed Au(110) surface. Note that in the notation of this figure, $\overline{X}'$ and $\overline{M}$ correspond to $\overline{Y}$ and $\overline{S}$, respectively, in the SBZ of Fig. 5.2-18 [2.39]

Fig. 5.2-28 Theoretical projected band structure (*shaded area*) for Pd(111), showing an empty surface state (near $\overline{\Gamma}$) and various occupied surface states and resonances [2.42]

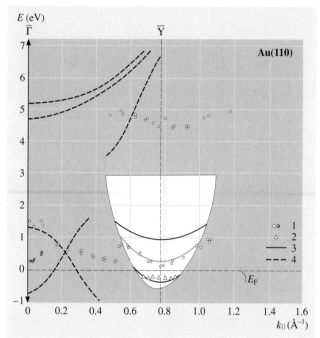

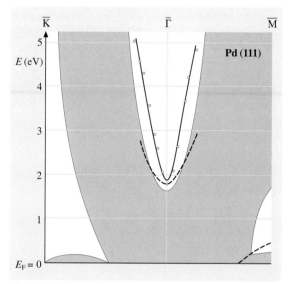

Fig. 5.2-27 Comparison of experimental results [obtained by KRIPES for the empty states (*dots*) and by ARUPS for the initial states (*triangles*)] with the theoretical calculations of Fig. 5.2-26. Notice the splitting of the state near $\overline{Y}(\overline{X}')$ probably due to the 2×1 reconstruction [2.40, 41]

Fig. 5.2-29 Comparison of the dispersion of the empty surface state of Fig. 5.2-28 with experimental results obtained by KRIPES [2.43]

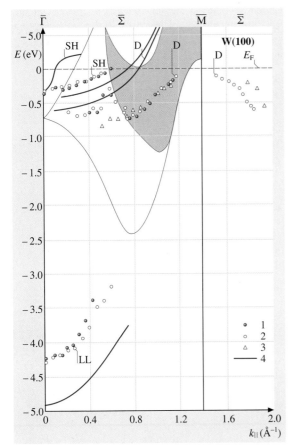

Fig. 5.2-30 Dispersion of surface states for W(100) along the $\overline{\Gamma M}$ line obtained by ARUPS and comparison with theory [2.44, 45]

5.2.3.2 Semiconductors

The data on elemental semiconductors (including diamond) are ordered according to the atomic number. They are followed by data on some of the compound semiconductors. For a more detailed discussion of the various subjects, the reader may consult [2.8, 10, 11] or the books [2.47, 48].

Surface Ionization Energies

The *ionization energy* is defined as the distance of the top of the valence band at the surface from the vacuum level. At low impurity concentrations, the ionization energy is independent of doping and should, preferably, be measured in place of the work function. Table 5.2-14 reports the ionization energies for a number of semiconductors.

Surface Core Level Shifts

In semiconductors, core levels are shifted because of charge transfer between surface atoms. Table 5.2-15

Table 5.2-14 Ionization energies at semiconductor surfaces[1]

Crystal	Face	Reconstruction	Ionization energy (eV)
Diamond	(111)	2×1	6.25 (25)
	(100)	2×1	5.33
Si	(111)	2×1	5.24 (9)
		7×7	5.30
Ge	(111)	2×1	4.77 (2)
		c(2×8)	4.70
GaP	(110)	1×1	5.98 (3)
	(100)	c(2×8)	5.40 (7)
GaAs		c(4×4)	5.34 (5)
		4×6	5.05
	(110)	1×1	5.48 (6)
GaSb	(110)	1×1	4.79 (9)
InP	(110)	1×1	5.77 (8)
InAs	(110)	1×1	5.37 (6)
InSb	(110)	1×1	4.84 (6)
CdS	(110)	1×1	6.7 (6)
CdSe	(110)	1×1	6.62
CdTe	(110)	1×1	5.79 (1)

[1] Data are average values from various authors. Errors are standard deviations. References to the original papers and individual errors are given in [2.47]

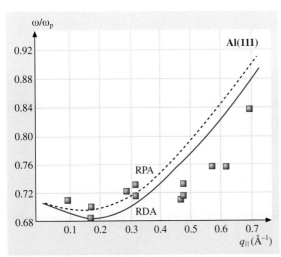

Fig. 5.2-31 Experimental dispersion of the surface plasmon in Al(111) as a function of $q_\parallel$ (*squares*) and comparison with theoretical results obtained with two different approximations [2.46]

Table 5.2-15 Surface core level shifts for elemental semiconductors[1]

Crystal	Face	Reconstruction	Core level	Surface shifts (eV)			
				S_1	S_2	S_3	S_4
Diamond	(111)	2×1	1s	−0.8			
Si	(100)	2×1	2p	−0.49 (3)	−0.230	0.062	0.29 (5)
		c(2×4)		−0.485	−0.205	0.062	0.22
	(111)	2×1		−0.46 (9)	−0.14	0.23 (6)	0.64
		7×7		−0.74 (5)	−0.07 (5)	0.30 (6)	0.52 (3)
Ge	(100)	2×1	3d	−0.49 (7)	−0.21 (3)		
		c(4×2)		−0.60	−0.24		
	(111)	2×1		−0.49 (8)		0.44	
		c(2×8)		−0.73 (3)	−0.24 (3)		

[1] Data are average values from various authors. Errors are standard deviations. References to original papers and individual errors are given in [2.47]

Table 5.2-16 Surface core level shifts at Ga and As sites in GaAs[1]

Face	Reconstruction	Surface shift (eV)	
		Ga(3d)	As(3d)
(100)	c(4×4)	+0.49	−0.285 (5)
		+0.55	
	c(2×8)	+0.3	−0.26 (1)
		+0.55 (0)	
	4×6	+0.4	−0.45 (16)
		−0.21	
(110)	1×1	+0.28 (0)	−0.373 (5)

[1] Data are average values from various authors. Errors are standard deviations. References to the original papers and individual errors are given in [2.47]

Table 5.2-17 Surface core level shifts at (110) surfaces of III–V compounds[1]

Crystal	Surface shifts (eV)	
	Cation	Anion
AlSb	+0.38	−0.38
GaP	+0.31 (2)	−0.41 (0)
GaAs	+0.28 (0)	−0.370 (5)
GaSb	+0.30	−0.36
InP	+0.30 (2)	−0.30 (1)
InAs	+0.27 (1)	−0.30 (0)
InSb	+0.23 (1)	−0.29 (1)

[1] Data are average values from various authors. Errors are standard deviations. References to the original papers and individual errors are given in [2.47]

gives the shifts for the C(1s), Si(2p), and Ge(3d) levels for various faces and reconstructions. The surface shifts of the 3d levels at the Ga and As sites of GaAs (110) and (100) faces are given in Table 5.2-16.

The p and d levels are spin–orbit split and give rise to doublets. Nonequivalent atoms increase further the number of surface shifts. Up to four components are observed in Si and Ge and up to two in GaAs.

Unresolved surface shifts for various III–V compounds are given in Table 5.2-17.

Surface State Bands

The origin and nature of surface states have been discussed in the introduction to Sect. 5.2.3. Here the energy versus $k_\parallel$ along significant directions of the SBZ and/or the density of states are reported for various reconstructed/relaxed surfaces. The experimental data were obtained mainly with ARUPS and KRIPES or by optical, energy loss, and STM techniques.

Because of the large amount of available data, only a selection of dispersion curves and surfaces has been given. For more detailed information, the reader is referred to [2.10, 11].

The rearrangement of unsaturated dangling bonds at the surface is considered to be the driving force for reconstruction of semiconductors. An example of DBs is shown in Fig. 5.2-32a,b that give contours of the charge density for the filled (P) and empty (Ga) surface states on the (110) surface of GaP. In Fig. 5.2-32c, a comparison is made with an STM image taken with a bias voltage that allows the observation of filled and empty states.

Figures 5.2-33 – 5.2-36 and 5.2-38 – 5.2-40 give the $E(k_\parallel)$ curves for selected surfaces of diamond, Si, Ge,

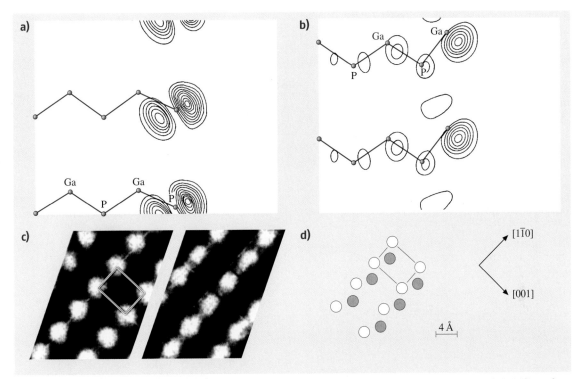

Fig. 5.2-32 (a) Theoretical charge density contours for the filled DB states at P sites on the relaxed GaP(110) surface, (b) the same for the Ga site empty states, (c) comparison with constant current STM images for GaAs(110). The two STM images refer to filled (As) states (*left*) and empty (Ga) states (*right*). The two images are obtained with positive (*left*) and negative (*right*) bias. The rectangular unit cell is in the same (absolute) position. (d) Schematic representation of the unit cell. As atoms are shown as *empty circles*; Ga atoms as *black circles* [2.49, 50]

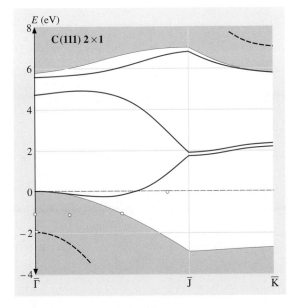

GaAs, and CdSe. The local density of states for Si(111) 7×7 as obtained from ARUPS and STM spectroscopy is shown in Fig. 5.2-37.

Transitions Between Surface States

Transitions between surface state levels give rise to optical absorption and electron energy loss. Such transitions have been observed by SDR, RAS, ellipsometry, and EELS. As an example, the arrows in Figs. 5.2-36 and 5.2-38 show the transitions observed in Si and Ge(111)2×1. Notice the high joint density of states caused by the quasi-parallelism of the bands along $\overline{JK}$.

Fig. 5.2-33 Theoretical surface state bands (*full lines*) and resonances (*dashed lines*) for a relaxed π-bonded chain model of diamond(111)2×1. Comparison with experimental results (*open circles*) obtained by ARUPS along the $\overline{\Gamma J}$ line [2.51, 52]

Fig. 5.2-35 Dispersion curves of Si(100)2×1 observed by ARUPS at 21.2 eV. Prominent peaks are indicated by *open circles*, while weak peaks and shoulders by *dots*. S and B peaks have surface and bulk characters. The experimental bands should be compared with the surface structures of Fig. 5.2-34. No optical transition clearly associated to S_1, S_2 (or D_{down}, D_{up} in Fig. 5.2-34) has been observed in spite of the essential parallelism of the two bands (meaning a high JDOS) [2.54, 55] ▶

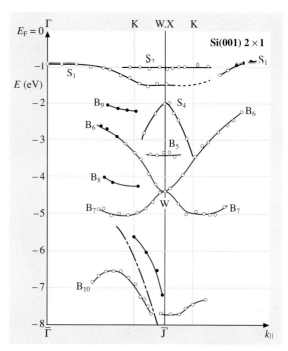

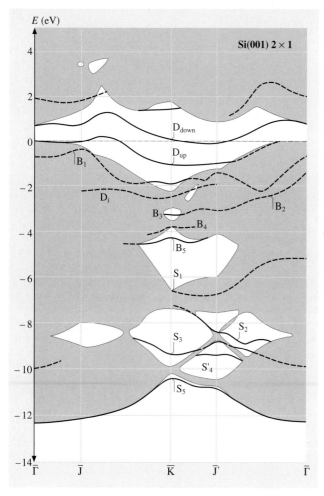

◀ **Fig. 5.2-34** Theoretical surface band structure of Si(100)2×1, obtained with the asymmetric dimer model (see Sect. 5.2.2.3, Fig. 5.2-7). D_{down} and D_{up} refer to the DB bands at the down and up atoms. The bands labeled S_1–S_5 and B_1–B_5 are back-bond states (modified by the surface) [2.53]

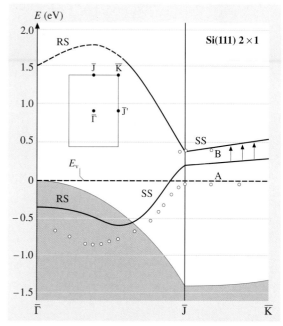

Fig. 5.2-36 Surface band structure of Si(111)2×1. Comparison between theoretical structure based on pseudopotential calculations for the π-bonded chain model (see Sect. 5.2.2.3, Fig. 5.2-11) and experimental results from ARUPS. Photoemission from the excited surface states (around the B point in the figure) is obtained by using highly doped samples. *Arrows* indicate optical transitions observed in SDR and EELS (see Figs. 5.2-41 and 5.2-43) [2.56, 57] ▶

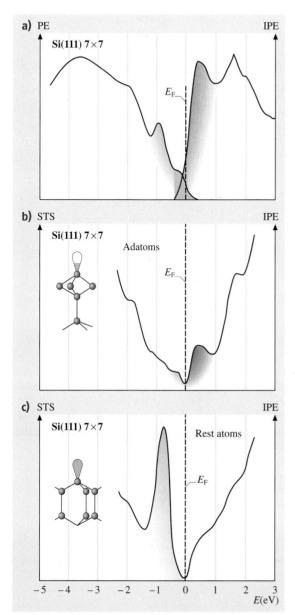

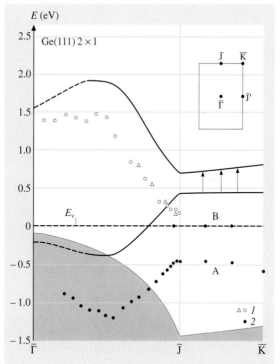

Fig. 5.2-37a–c Density of states associated with the dangling bonds (on the adatoms and "rest atoms") of the Si(111) 7×7 surface (see Fig. 5.2-13), obtained (**a**) by ARUPS (integrated over all atoms), and (**b**) and (**c**) by STM spectroscopy. Notice that the gap (if present) occurs at the Fermi energy [2.57–59]

Fig. 5.2-38 Surface band structure of Ge(111)2×1. Comparison between theoretical structure (in the frame of the chain model) and experimental results obtained with KRIPES (*open dots* and *triangles*) and ARUPS (*black dots*). Photoemission from excited states has been reached by using highly doped Ge. *Arrows* indicate optical transitions observed in SDR and EELS (see Figs. 5.2-42 and 5.2-44) [2.60–62]

In the dielectric theory of the surface response, the transition probability is proportional to $-\mathrm{Im}\,\hat{\varepsilon} = \varepsilon''$ in the optical absorption and to $\mathrm{Im}[1/1+\hat{\varepsilon}] = \varepsilon''/[(1+\varepsilon')^2 + \varepsilon''^2]$ in EELS, $\hat{\varepsilon} = \varepsilon' - \mathrm{i}\varepsilon''$ being the dielectric function of the surface layer. The two techniques give essentially the same results, as shown in Fig. 5.2-45.

The great anisotropy of some reconstructed surfaces, as shown in Figs. 5.2-41 and 5.2-42, is reflected in second-harmonic generation, which is specially sensitive to the symmetry of the surface, being forbidden for centrosymmetric media (see Fig. 5.2-46 for Si(111)2×1).

Surface anisotropy has given rise to the technique of reflectance anisotropy spectroscopy (RAS), in which linearly polarized light is modulated between two principal directions (of the surface tensor) and the difference

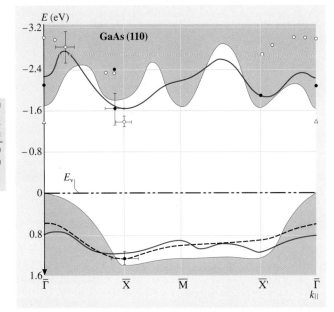

Fig. 5.2-39 Surface band structure of GaAs(110). Comparison between theoretical structure (*continous brown line*) and experimental determinations: KRIPES (*filled* and *open circles*), ARUPS (*dashed line*), and two-step photoemission (*triangles*). It can be noticed that the surface states nearly coincide with the band edges of the projected bulk band structure [2.63–67]

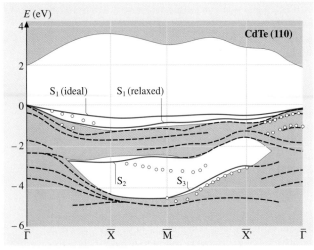

Fig. 5.2-40 Electronic band structure of the CdTe(110) surface. The *open circles* show experimental points obtained by ARUPS. Theoretical surface states and resonances are shown by *solid* and *dashed lines*, respectively [2.68, 69]

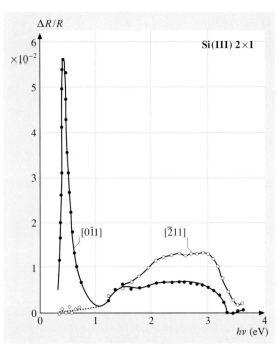

Fig. 5.2-41 Surface differential reflectivity (SDR) spectra for Si(111)2×1, obtained with two polarizations: with the electric vector parallel ([0$\bar{1}$1], *filled circles*) and perpendicular ([$\bar{2}$11], *open circles*) to the chains of the π-bonded chain model. For the transitions shown by *arrows* in Fig. 5.2-36 ($h\nu < 1$ eV) the change in reflectivity can be observed only with the electric vector parallel to the chains. The spectrum for $h\nu > 1$ eV is associated to transitions along $\overline{\Gamma J}$ in the SBZ. The change of anisotropy at approximately 1.2 eV is conform to the "sum rule": $\int \varepsilon_x \omega d\omega = \int \varepsilon_y \omega d\omega$ [2.70]

in the reflected light is recorded. Figure 5.2-47 gives an example of RAS spectra for GaAs(100).

5.2.3.3 Magnetic Surfaces

The magnetic moments of atoms near the surface plane, p_n ($n = 1$ for the surface), differ from those for the bulk, p_b. The total incremental moment at the surface is defined as

$$\Delta p_{sb} = \sum_{n=1}^{\infty} (p_n - p_b) \,.$$

Table 5.2-18 gives the values of p_b, p_n, and Δp_{sb} for various ferromagnetic surfaces.

Temperature is more effective in destroying the magnetic order at the surface than in the bulk. This

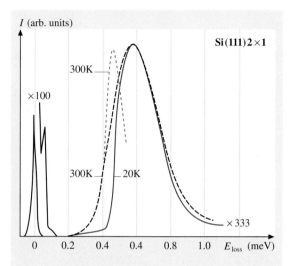

Fig. 5.2-43 EELS spectra at $T = 20\,\text{K}$ and $300\,\text{K}$ for Si(111)2×1 (*full* and *dashed lines*) compared with the results of SDR of Fig. 5.2-41 (*brown dashed line*). The peak of EELS is slightly displaced towards higher energies because of the energy dependence of the factor in front of the loss function $\text{Im}\,(1/\hat{\varepsilon} + 1)$ [2.70, 71]

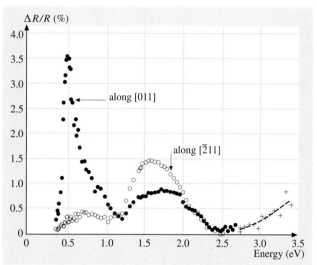

Fig. 5.2-42 Surface differential reflectivity versus energy for Ge(111)2×1 with polarization parallel and perpendicular to the chains of the π-bonded model. The anisotropy for $h\nu < 1\,\text{eV}$ is smaller than for Si(111)2×1. The same considerations done in the caption of Fig. 5.2-41 hold also here [2.72]

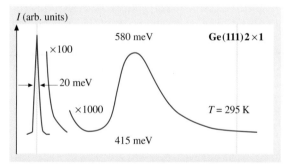

Fig. 5.2-44 EELS spectrum of Ge(111)2×1 at room temperature. The same considerations done in the caption of Fig. 5.2-43 hold also here [2.73]

effect is shown in Fig. 5.2-48, where the ratio of the polarization P at a given temperature to that at absolute zero is plotted versus temperature. The Curie temperature at the surface T_{CS} is approximately equal to that of the bulk (T_{C}) in most cases. Table 5.2-19 reports the ratio $T_{\text{CS}}/T_{\text{C}}$ for a number of materials.

A more detailed discussion of the magnetic properties of surfaces is given in [2.13].

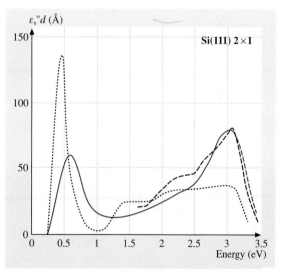

Fig. 5.2-45 Comparison of SDR (*dotted line*) [2.70, 74], EELS (*solid line*) [2.75] and ellipsometry (*dashed line*) [2.76] spectra for Si(111)2×1. The qualitative agreement is remarkable [2.70, 74–76]

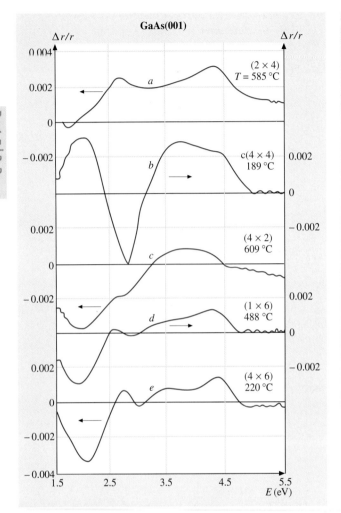

Fig. 5.2-47 Reflectance anisotropy spectra for GaAs(100) grown by MBE at different substrate temperatures in an As$_4$ flux of 2.8×10^{-6} torr. The corresponding reconstructions (as determined by RHEED) are also reported [2.78]

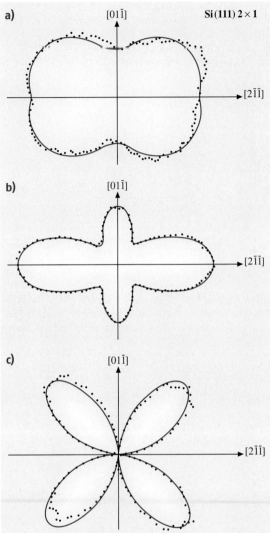

Fig. 5.2-46a–c Intensity of the second harmonic (SH) generation signal as a function of polarization of the incidence pump beam (1.17 eV), for Si(111)2×1. (**a**) total SH signal; (**b**) SH component polarized along [2$\bar{1}\bar{1}$]; (**c**) SH component polarized along [01$\bar{1}$] [2.77]

Table 5.2-18 Surface magnetizations at given temperatures. The theoretical values are averages from various authors (obtained with linearized-augmented-plane-wave or tight-binding methods). References to original papers are given in [2.13]

Crystal	Face	T (K)	Magnetic moment (Bohr magnetons)					Technique
			p_b	p_1	p_2	p_3	Δp_{sb}	
Fe	(100)	0	2.25	2.96	2.34	2.45	1.0	Theory
	(110)	0	2.22	2.60	2.37	2.28	0.66	Theory
		300	2.2	2.9(3)			0.7(3)	SPLEED
Ni	(100)	0	0.57	0.67	0.59	0.60	0.13	Theory
		300					0.03(3)	SPLEED
	(110)	0	0.62	0.63	0.64	0.58	0.11	Theory
	(111)	0	0.58	0.63	0.64	0.58	0.11	Theory
		300					0.06(12)	SPLEED

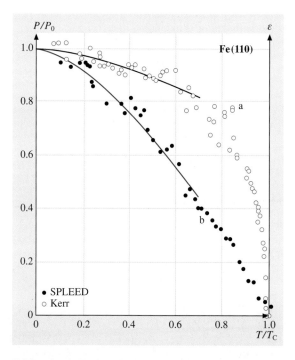

Fig. 5.2-48 Temperature dependence of magnetic ordering. Kerr ellipticity (representing bulk magnetization) and normalized polarization P/P_0 of scattered electrons in SPLEED (representing surface magnetization) as a function of T/T_C. The *solid curves* are fits to a Bloch law: $m_1(0) - m_1(T) = bT^{3/2}$, m_1 being the magnetization of the first layer [2.79]

Table 5.2-19 Surface Curie temperatures (T_{CS}). References to original papers are given in [2.13]

Crystal	Face	T_C (K)	T_{CS}/T_C	Technique
Ni	(100)	630	1	SPLEED
	(110)	630	1	SPLEED
	(111)	630	1	ECS
Fe	(100)	1040	1	SPARPES
	(110)	1040	1	Spin-polarized secondary electrons
EuS	(111)	16.71(5)	1.000(4)	SPLEED
Gd	(0001)	293	1.05–1.07	SPLEED
Tb	(0001)	220	1.14	ECS

5.2.4 Surface Phonons

The lattice dynamics is strongly perturbed by the presence of the surface. The elements of the force constant matrix connecting two atoms at or near the surface differ from those for the bulk. The number of neighbors is also different at the surface. Vibrational frequencies are generally expected to be lower than in the bulk.

Phonon bands occur in the SBZ, similarly to the surface states discussed in Sect. 5.2.3. When the frequency of a surface mode corresponds to a gap in the bulk spectrum, the mode is localized at the surface and is called a surface phonon. If degeneracy with bulk modes exists, one speaks of surface resonances. Surface phonon modes are labeled S_j ($j = 1, 2, 3, \ldots$), and surface resonances by R_j; when strong mixing with bulk modes is present, the phonon is labeled MS_j. The lowest mode that is desired from the (bulk) acoustic band is often called the Rayleigh mode, after Lord Rayleigh, who first predicted (in 1887) the existence of surface modes at lower frequencies than in the bulk.

The experimental results have been obtained mainly through atom and electron scattering, HATOF and EELS.

5.2.4.1 Metals

Surface phonon bands along symmetry lines of the SBZ are given for fcc metals in Figs. 5.2-49 – 5.2-55 and in Table 5.2-20. In all figures the horizontal axis is the reduced wave vector, expressed as the ratio to its value at the zone boundary. Table 5.2-21 gives the surface Debye temperatures for some fcc and bcc metals, as well as the amplitudes of thermal vibrations of atoms in the first layer: ρ_1 as compared with those of the bulk: ρ_b. In the harmonic approximation, the root mean square displacement of the atoms is proportional to the inverse of the Debye temperature.

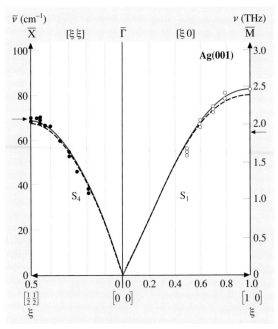

Fig. 5.2-49 Surface phonon dispersion curves for Ag(100) measured by EELS. The *filled* and *open circles* refer to different electron energies. The *full* and *dashed lines* are theoretical fits. Energies are given in cm^{-1}. 1 cm^{-1} = 0.1240 meV [2.80]

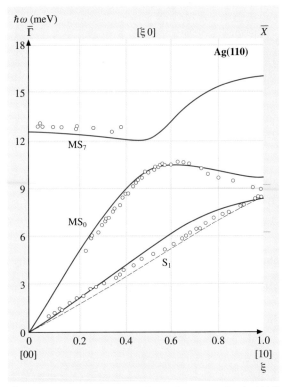

Fig. 5.2-50 Surface phonon dispersion curves for Ag(110) measured by HATOF. The *solid curves* are theoretical fits [2.81]

Table 5.2-20 Surface phonon energies of fcc metals. References to the original articles are given in the figure captions and/or in [2.14, 15]

Metal	Face	Energy (meV) and mode				Figure	Technique
		S_1	S_2	S_3	S_4		
Ag	(100)	10.2 ($\bar{M}$)			8.6 ($\bar{X}$)	Fig. 5.2-49	EELS
	(110)	8.15 ($\bar{X}$)				Fig. 5.2-50	HATOF
		5.0 ($\bar{Y}$)		7.7 ($\bar{Y}$)			
	(111)	8.7 ($\bar{M}$)				Fig. 5.2-51	HATOF
		9.1 ($\bar{K}$)					
Al	(100)				15.1 ($\bar{X}$)		EELS
	(110)	14.6 ($\bar{X}$)				Fig. 5.2-52	HATOF
		8.9 ($\bar{Y}$)		13.5 ($\bar{Y}$)			
	(111)	17 ($\bar{M}$)					HATOF
		18.6 ($\bar{K}$)					
Au	(110) 2×1					Fig. 5.2-53	EELS
	(111)	7.5 ($\bar{M}$)					HATOF
		7.5 ($\bar{K}$)					
Cu	(100)	16.7 ($\bar{M}$)	20.2 ($\bar{M}$)			Fig. 5.2-54	EELS
			14.1 ($\bar{X}$)		13.6 ($\bar{X}$)		
	(110)	≅ 14 ($\bar{X}$)		13 ($\bar{Y}$)			HATOF
		7.5 ($\bar{Y}$)					
	(111)	13.3 ($\bar{M}$)	26.1 ($\bar{M}$)				EELS, HATOF
		14.0 ($\bar{K}$)					
Ni	(100)	19.2 ($\bar{M}$)			16.1 ($\bar{X}$)	Fig. 5.2-55	EELS, HATOF
	(110)	17 ($\bar{X}$)		15.8			EELS
		9.9 ($\bar{Y}$)					
	(111)	17.2 ($\bar{M}$)	32.2 ($\bar{M}$)				EELS
Pd	(110)	11 ($\bar{X}$)					HATOF
		6.7 ($\bar{Y}$)	9 ($\bar{Y}$)				
Pt	(111)	10.8 ($\bar{M}$)					HATOF
		11.1 ($\bar{K}$)					

Table 5.2-21 Surface Debye temperatures of metals. References to the original articles are given in [2.4, 6]

Crystal	Bulk Θ_D (K)	Face	Surface Θ_D (K)	ρ_1/ρ_b	Technique
Ag	225	(100)	104		LEED
		(110)	150–190		LEED
				1.5	LEIS
				1.65	MEIS
				1.44	HEIS
		(111)	⊥ surface 155		LEED
			∥ surface 226		
			247		HAS
Al	428	(100)	200		LEED
				1.25	HEIS
		(110)	140		LEED
		(111)	200		LEED

Table 5.2-21 Surface Debye temperatures of metals, cont.

Crystal	Bulk Θ_D (K)	Face	Surface Θ_D (K)	ρ_1/ρ_b	Technique
Au	165	(100)	8?		LEED
		(110) 2×1	83		LEED
			127		LEIS
				1.55	HEIS
		(111)	83		LEED
Cu	343	(100)	230		LEED
			260		LEIS
				1.5	MEIS
		(110)	260 (150)	1.5	LEIS
				1.55	MEIS
				1.27	HEIS
		(111)	245		LEED
Fe	467	(111)		1.18	MEIS
Ir	420	(110)	150		LEED
			160		LEIS
		(111)	170		LEED
Mo	450	(100)	239		LEED
Ni	450	(100)		1.46	MEIS
				1.35	HEIS
		(110)	⊥ surface 216		LEED
			∥ surface 344		
			395		MEIS
				1.15	HEIS
		(111)	⊥ surface 208		LEED
			∥ surface 348		
				1.20	HEIS
Pt	240	(100)	118		LEED
		(110)	107		LEED
				1.6	MEIS
		(111)	111		LEED
				1.3	MEIS
Rh	480	(111)	197		LEED
V	380	(100)	250		LEED
W	400	(100) 1×1	210		LEED
		(100) c2×2	400 (200)		LEED
				1.5	HEIS
		(110)	190–235		LEED
				2.6, 1	HEIS

Fig. 5.2-51 Surface phonon dispersion curves for Ag(111) measured by HATOF. *Circles* are experimental points; *solid* and *dashed lines* theoretical calculations. The *hatched area* represents the projected band structure [2.82]. Very similar plots hold for Au(111) and Cu(111). For appropriate energy parameters see Table 5.2-20

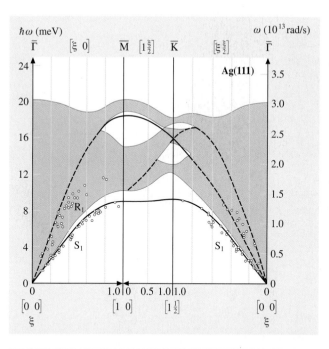

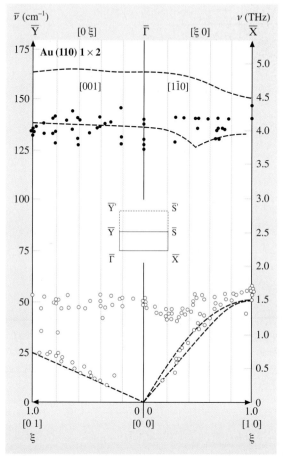

Fig. 5.2-53 Surface phonon dispersion curves along two directions of the SBZ for reconstructed Au(110) 2×1, measured by EELS. The *dashed lines* are theoretical results. In the *inset*, the SBZ shows the folding due to the 2×1 reconstruction that introduces the optical bands. Energies are given in cm^{-1}. 1 cm^{-1} = 0.1240 meV [2.84]

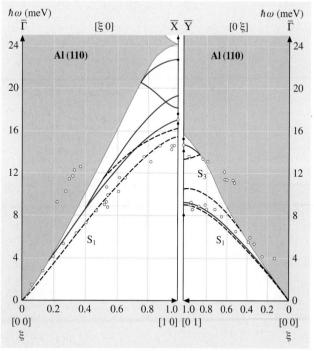

Fig. 5.2-52 Surface phonon dispersion curves for Al(110). The experimental results (*circles*) were obtained by HATOF. The *solid* and *dashed lines* are theoretical fits. The *shaded regions* represent the projected bulk modes [2.83]

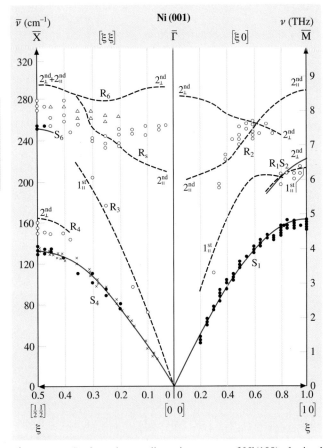

Fig. 5.2-55 Surface phonon dispersion curves of Ni(100) obtained by EELS. S_j, surface phonons; R_j, surface resonances. Energies are given in cm^{-1}. 1 cm^{-1} = 0.1240 meV [2.86]

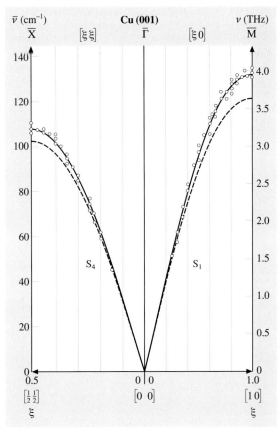

Fig. 5.2-54 Surface phonon dispersion curves for Cu(100) measured by EELS. The *solid* and *dashed lines* are theoretical fits. Energies are given in cm^{-1}. 1 cm^{-1} = 0.1240 meV [2.85]

5.2.4.2 Semiconductors and Insulators

Phonon surface bands of some insulators and semiconductors are given in Figs. 5.2-56 – 5.2-58. Surface phonon energies of alkali halide crytals are summarized in Table 5.2-23. Since insulators and semiconductors have in general more than one atom per unit cell, they display both acoustical and optical branches. Surface Debye temperatures of some semiconductors are given in Table 5.2-22.

Table 5.2-23 Surface phonon energies of alkali halides. References to the original articles are given in [2.14, 15]

Crystal	Face	Energy of mode S_1 (meV)	Technique
LiF	(100)	22.9 ($\overline{M}$)	HATOF
NaF	(100)	16.0 ($\overline{M}$)	HATOF
		18.5 ($\overline{X}$)	
NaCl	(100)	9.9 ($\overline{M}$)	HATOF
		11 ($\overline{X}$)	
NaI	(100)	4.1 ($\overline{M}$)	HATOF
		4.9 ($\overline{X}$)	
KCl	(100)	7.2 ($\overline{M}$)	HATOF
RbCl	(100)	4.7 ($\overline{M}$)	HATOF
		$\cong 5.7$ ($\overline{X}$)	

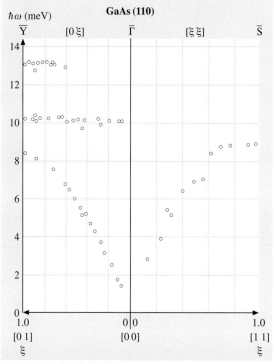

Fig. 5.2-56 Surface phonon dispersion curves for GaAs(110) measured by HATOF (*open circles*). The energies at symmetry points are: Y, 8.3, 10.2, and 13.0 meV; Γ, 10.0 meV; S, 8.9 meV [2.87]

Table 5.2-22 Surface Debye temperatures of semiconductors. References to the original articles are given in [2.4, 6]

Crystal	Bulk Θ_D (K)	Face	Surface Θ_D (K)	ρ_1/ρ_b	Technique
GaSb	265	(110)		1.57	MEIS
InAs	249	(110)		1.65	MEIS
Si	640	(100)	430–475		HEED
				1.4	HEED, LEED
		(111) 2×1		1.45, 2.1	MEIS
			300		HEIS

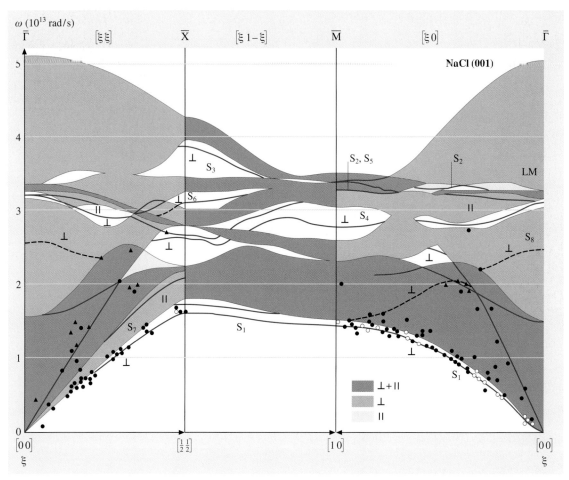

Fig. 5.2-57 Surface phonon dispersion curves for NaCl(100). Theoretical projected band structure (*hatched*) and experimental points (*dots*) measured by HATOF. *Solid* and *dashed lines* represent theoretical surface states and resonances. Ordinates are given in units of 10^{13} rad/s = 6.586 meV [2.88, 89]

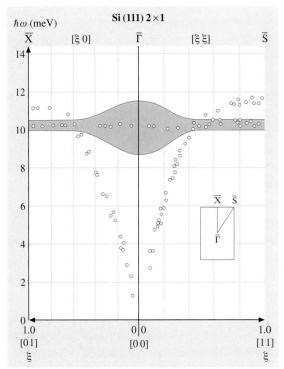
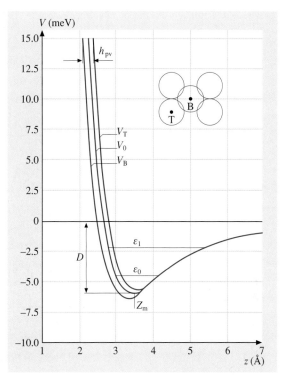

Fig. 5.2-58 Surface phonon dispersion curves for Si(111) 2×1 measured by HATOF. Energies at symmetry points: X, 10.2 and 11.1 meV; S, 10.5 and 11.6 meV. The flat phonon mode at 10.5 meV is associated with the 2×1 reconstruction. The surface mode couples with transverse bulk phonons near the center of the SBZ, giving rise to considerable broadening. The *shaded area* corresponds to the width of the 10.5 meV peak [2.90]. The energy of the optical mode (not shown in the figure) is 56.0 meV [2.91]

Fig. 5.2-59 Typical atom–surface potential, with definitions of the main interaction parameters. The surface unit cell is shown in the *inset*. The numbers on the axes refer to He–Ag(110) [2.92]. V_0: surface averaged potential; V_T: potential at the top position; V_B: potential at the bottom position

5.2.4.3 Atom–Surface Potential

Many results on surface phonon dispersions are obtained through atomic (mainly He) scattering.

The interaction of an incident atom with a surface is described by the atom–surface potential (Fig. 5.2-59), which consists of a hard repulsive part at short distances and a weak van der Waals attraction at larger distances.

The attractive contribution is often assumed in to be the form

$$V_{\text{att}} = -\frac{C_3}{(z - z_{\text{vW}})^3}.$$

In the potential well between the van der Waals potential and the hard repulsive potential, quantized energy levels ε_n exist (see Fig. 5.2-59). Table 5.2-24 summarizes the parameters of the atom–surface potentials for a number of surfaces and impinging atoms (or molecules). A more extended discussion is given in [2.14, 15].

Table 5.2-24 Parameters of the atom–surface potential. Values of ε_n are given in the order $n = 0, 1, 2, 3\ldots$. References to the original papers are given in [2.15]

Crystal	Face	Atom/molecule	D (meV)	z_m (Å)	z_{vW} (Å)	C_3 (meV Å^3)	ε_n (meV)
Ag	(110)	H$_2$	31.7	3.27	0.755	714	25.3, 16.4, 9.7, 5.3, 2.5, 1.1
		He	6.0		0.755	249	4.44, 2.18, 0.92, 0.31, 0.1
	(111)	H$_2$	32.5	2.5	0.2	714	25.5, 16.6, 9.9, 5.4, 2.5, 1.1
Al	(110)	He	5.2	4.47	1.71	202	3.65
Au	(111)	He	8.0	3.98	1.34	274	5.91
Cu	(110)	He	6.27	3.65	1.21	235	4.52, 2.2, 1.0, 0.3
GaAs	(110)	He	(7)	(3.5)	(−0.2)	(473)	(4.2)
NaCl	(100)	He	6.1	3.49	1.03	106	4.1, 1.5, 0.31
LiF	(100)	He	8.5	2.98	0.79	93 (137)	5.9, 2.46, 0.78, 0.21
		Ne	13.5			192	11.7
Ni	(100)	He	4.2 (theory)	3.76 (theory)		218	
Pd	(110)	He	8.05	3.60		211	6.42
Pt	(110) 2×1	He	8.8			251	7.4, 5.1, 3.1, 1.5
Si	(100)	He				174	
W	(110)	He				265	5.6

5.2.5 The Space Charge Layer at the Surface of a Semiconductor

Surface states in the gap of a semiconductor may trap or loose electrons. The localized charge in surface states is neutralized by a macroscopic space charge layer of opposite sign, caused by accumulation or depletion of the bulk carriers. The potential associated with the space charge gives rise to a band bending of the bulk bands at the surface. Band bending alters various properties of semiconductor surfaces and is crucial for many applications, such as chemisorption, catalysis, corrosion, and, most significantly, channel conductivity in MOS field-effect transistors. For this reason, the problem of the surface space charge layer is treated extensively in the following sections.

5.2.5.1 Definitions and Notation

Figure 5.2-60 shows the main parameters that characterize the space charge layer. With reference to the figure, the following definitions can be given:

- $E_i(z)$ is the *intrinsic level*: this level runs parallel to the band edges and coincides, in the bulk, with the intrinsic Fermi level E_{ib}:

$$E_{ib} = \frac{E_c + E_v}{2} + \frac{1}{2}k_B T \ln \frac{N_v}{N_c}.$$

- The following quantities related to band bending:

$$u(z) \equiv \frac{E_F - E_i(z)}{k_B T},$$

$$v(z) = u(z) - u_b \equiv \frac{E_{ib} - E_i(z)}{k_B T}, \quad (2.1)$$

where v_s, u_s, and u_b are the values of $v(z)$, and $u(z)$ at the surface and the value of $u(z)$ in the bulk; $u_b > 0$ for n-type semiconductors and $u_b < 0$ for p-type semiconductors. If $v_s > 0$ the band bends downward; if $v_s < 0$ the band bends upward. $v_s = 0$ is the condition for flat bands, and $u_s = 0$ the condition for an intrinsic surface.

An accumulation layer exists if v_s and u_b have the same sign (as in Fig. 5.2-60).
An depletion layer exists if v_s and u_b have opposite signs and u_s and u_b have the same sign.
An inversion layer exists if both of the pairs v_s, u_b and u_s, u_b have opposite signs.

- Debye length:

$$L_D = \sqrt{\frac{\varepsilon_0 \varepsilon_b k_B T}{q^2(n_b + p_b)}}, \quad (2.2)$$

where n_b and p_b are the bulk carrier concentrations and q is the elementary charge. For small

2.5 The Space Charge Layer at the Surface of a Semiconductor

Table 5.2-25 Intrinsic Debye lengths for various semiconductors

	Si	Ge	GaAs	InP	InAs	InSb
Intrinsic Debye length (cm)	2.9×10^{-3}	7.0×10^{-5}	2.0×10^{-1}	2.7×10^{-2}	8.6×10^{-6}	2.4×10^{-6}

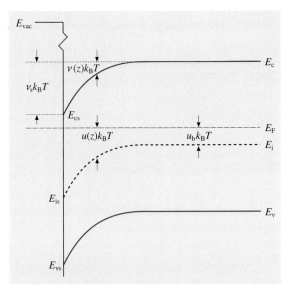

Fig. 5.2-60 Schematic behavior of the bands at the surface of a semiconductor and definition of the main parameters. Notice that band bending $v(z)$ is positive when the band bends downward

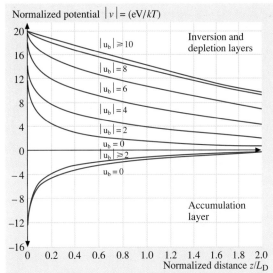

Fig. 5.2-61 Profile of the potential $|v(z)|$ as a function of z/L_D plotted for $|v_s| = 20$ and various values of $|u_b|$. The behavior for other values of $|v_s|$ can be obtained by drawing the parallel to the abscissa through the new value $|v'_s|$ and translating the origin of the abscissa at the value of the interception [2.48, p. 329]

bending ($< k_B T$), the surface barrier decays as $v(z) = v_s e^{-z/L_D}$. For larger bending, the shape of the potential cannot be given in analytic form, though the thickness of the space charge layer is still on the order of the Debye length (see Fig. 5.2-61). Values of the intrinsic Debye length for various semiconductors are given in Table 5.2-25. Figures 5.2-62 and 5.2-63 give L_D as a function of the carrier density and bulk resistivity for various semiconductors.

- Q_{sc} is the space charge density (per unit area), Q_{ss} is the charge density (per unit area) in surface states, and

$$Q_{total} = Q_{sc} + Q_{ss} ; \qquad (2.3)$$

$Q_{total} = 0$ in the absence of an external electric field. In the presence of an external field, Q_{total} is the charge density on the field electrode (the gate in a MOSFET).

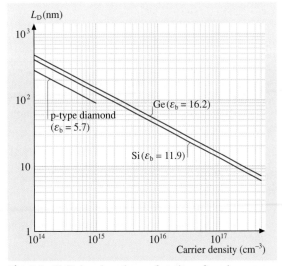

Fig. 5.2-62 Debye length as a function of carrier concentration for Si, Ge and p-type diamond

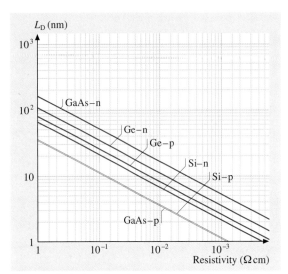

Fig. 5.2-63 Debye length as a function of resistivity for n-type and p-type Si, Ge, GaAs

5.2.5.2 Useful Formulas and Numerical Values

The space charge density and surface potential can be obtained by solving Poisson's equation for the one-dimensional problem at the surface [2.93, 94]. For nondegenerate conditions, one obtains

$$Q_{sc} = \mp q(n_b + p_b) L_D F_s(u_b, v_s) \qquad (2.4)$$

(minus sign if $v_s > 0$), where

$$F(u_b, v) = \sqrt{2} \left(\frac{\cosh(u_b + v)}{\cosh u_b} - v \tanh u_b - 1 \right)^{1/2} . \qquad (2.5)$$

The function F_s is plotted in Fig. 5.2-64 as a function of $k_B T v_s$ for various values of $u_b > 0$.

To obtain values for $u_b < 0$, the following relation can be used:

$$F(u_b, -v) = F(-u_b, v) . \qquad (2.6)$$

The profile of the potential barrier $v(z)$ can be obtained from

$$\frac{z}{L_D} = \int_{v_s}^{v} \frac{dv}{\mp F_s(u_b, v)} , \qquad (2.7)$$

which can be solved numerically.

Figure 5.2-61 shows $v(z)$ as a function of z/L_D for accumulation and depletion/inversion layers and

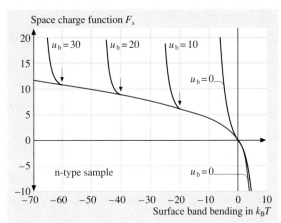

Fig. 5.2-64 The space charge function $F_s(v_s, u_b)$ for n-type semiconductors ($u_b > 0$) plotted versus band bending $v_s k_B T$. The function for p-type semiconductors ($u_b < 0$) may be obtained from the relation: $F(u_b, -v_s) = F(-u_b, v_s)$ [2.47, p. 25]

various values of u_b. The profiles are plotted for $|v_s| = 20$. The behavior, however, has a universal character, since

$$\frac{z'}{L_D} - \frac{z' - z}{L_D} = \int_{v_s}^{v'} \frac{dv}{\mp F_s} + \int_{v'}^{v} \frac{dv}{\mp F_s} , \qquad (2.8)$$

so that curves can be scaled for different values of v_s simply by translating the origin.

The charge Q_{ss} in the surface states can be calculated once their energy in the gap and their donor or acceptor character is known. (A surface state is called a *donor* if it is neutral when filled with electrons, and an *acceptor* if it is neutral when empty.)

For a discrete distribution of surface states of densities N_i^d and N_j^a, where the superscripts "d" and "a" refer to donors and acceptors, respectively,

$$Q_{ss} = -q \sum_j N_j^a(E_j) f(E_j - E_F)$$
$$+ q \sum_i N_i^d(E_i)[1 - f(E_i - E_F)] , \qquad (2.9)$$

where f is the Fermi function. A similar expression holds for a continuous distribution of states. If we define a reduced surface state energy

$$u_j \equiv \frac{E_j - E_{is}}{k_B T} , \qquad (2.10)$$

5.2.5.3 Surface Conductivity

The excess carrier densities, defined as

$$\Delta N = \int_0^\infty (n - n_b)\, dz, \qquad (2.12)$$

excess electron density (per unit area) at the surface, and

$$\Delta P = \int_0^\infty (p - p_b)\, dz, \qquad (2.13)$$

the excess hole density (per unit area) at the surface, can also be obtained from the function F_s:

- for majority carriers in accumulation layers:

$$\Delta N = n_b L_D G^+(u_b, v_s)\ (u_b > 0,\ v_s > 0),$$
$$\Delta P = p_b L_D G^+(u_b, v_s)\ (u_b < 0,\ v_s < 0);$$

- for majority carriers in depletion/inversion layers:

$$\Delta N = n_b L_D G^-(u_b, v_s)\ (u_b > 0,\ v_s < 0),$$
$$\Delta P = p_b L_D G^-(u_b, v_s)\ (u_b < 0,\ v_s > 0);$$

- for minority carriers in accumulation layers:

$$\Delta N = p_b L_D g^-(u_b, v_s)\ (u_b < 0,\ v_s < 0),$$
$$\Delta P = n_b L_D g^-(u_b, v_s)\ (u_b > 0,\ v_s > 0);$$

- for minority carriers in depletion/inversion layers:

$$\Delta N = p_b L_D g^+(u_b, v_s)\ (u_b < 0,\ v_s > 0),$$
$$\Delta P = n_b L_D g^+(u_b, v_s)\ (u_b > 0,\ v_s < 0).$$

The four functions G^+, G^-, g^+, g^- are defined as

$$G^+ = \int_0^{|v_s|} \frac{e^v - 1}{F(|u_b|, v)}\, dv,$$

$$G^- = \int_0^{|v_s|} \frac{e^{-v} - 1}{F(-|u_b|, v)}\, dv, \qquad (2.14)$$

$$g^+ = e^{-2|u_b|} \int_0^{|v_s|} \frac{e^v - 1}{F(-|u_b|, v)}\, dv,$$

$$g^- = e^{-2|u_b|} \int_0^{|v_s|} \frac{e^{-v} - 1}{F(|u_b|, v)}\, dv. \qquad (2.15)$$

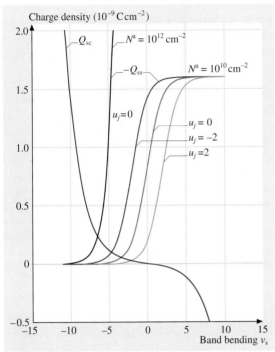

Fig. 5.2-65 Sketch of the graphical solution of the equation $Q_{sc} + Q_{ss} = 0$, for an intrinsic semiconductor ($u_b = 0$) with a single acceptor surface level at reduced energy u_j. Two values of acceptor densities N^a are shown. In case of $u_b \neq 0$, the curves $-Q_{ss}$ should be translated rigidly towards the new origin $v_s = -u_b$. When, on the other hand, $Q_{\text{total}} \neq 0$, the Q_{ss} curves should be translated vertically by Q_{total}. The actual band bending is at the intersection of the two sets of curves

where E_{is} is the intrinsic level at the surface, the Fermi function can be written as

$$f(E_j - E_F) = \frac{1}{e^{u_j - v_s - u_b} + 1}. \qquad (2.11)$$

It is then possible to solve (2.1) numerically to obtain the band bending v_s. A graphical solution (valid in the absence of an external field) is sketched in Fig. 5.2-65 for an intrinsic semiconductor ($u_b = 0$) with a single surface acceptor level and various values of u_j and N_j^a. For extrinsic semiconductors ($u_b \neq 0$) the procedure is similar, provided the origin is translated to $v_s = -u_b$ and the appropriate function for Q_{sc} is chosen. The same procedure can also be used in the presence of an external field by translating the origin of the coordinates.

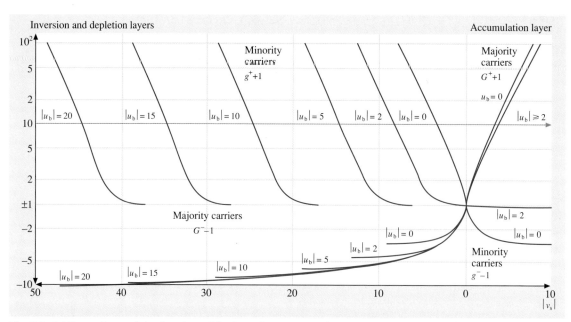

Fig. 5.2-66 Dependence of the functions G^+, G^-, g^+, g^- on $|v_s|$ for various values of $|u_b|$. The functions defined in (2.14), (2.15) can be used to obtain the excess carrier densities for majority and minority carriers in accumulation and depletion/inversion layers [2.93, p. 151]

Table 5.2-26 Field-effect mobilities of semiconductors. References to original papers are given in [2.47]

Semiconductor	Face	Reconstruction	Field-effect mobility (cm^2 V^{-1} s^{-1})
Si	(111)	2×1	−(1±0.5)
		7×7 (?) [1]	−(0.01±0.003)
Ge	(111)	2×1	−(100±50)
		c(2×8)	−(200±50)

[1] Surface prepared in situ by IBA

The superscripts of the functions indicate the signs of the respective integrals. The above functions are plotted in Fig. 5.2-66 as a function of $|v_s|$ and, in greater detail, in Fig. 5.2-67 for an inversion layer (as in a MOSFET). Tabulations are given in [2.94].

The surface excess conductivity is defined as

$$\Delta\sigma = q(\mu_{ns}\,\Delta N + \mu_{ps}\,\Delta P), \quad (2.16)$$

where ΔN and ΔP are given above. The surface mobilities μ_{ns} and μ_{ps} are in general different from the mobilities in the bulk, because of excess scattering at the surface.

The surface field-effect mobility can be defined as

$$\mu_{fe}(u_b, v_s) = \frac{d\Delta\sigma}{dQ_{total}}. \quad (2.17)$$

The presence of surface states, which may capture the carriers, decreases the surface field-effect mobility. Values of μ_{fe} for some surfaces are presented in Table 5.2-26.

The very low field effect mobilities reported in Table 5.2-26, show that the Fermi level is pinned in the gap because of the large concentration of surface states. Pinning positions of the Fermi level with respect to the top of the valence band at the surface are given in Table 5.2-27 for various semiconductors.

In a MOSFET, the surface conductance of the inversion layer channel between the source and the drain (Fig. 5.2-68) is controlled by the gate potential. The above formulas for the surface conductivity are applicable to MOSFETs under equilibrium conditions (low frequency and low voltage). Here the interface states play the role of surface states at the free surface. The density of interface states should be kept to a minimum for optimal performance. Figure 5.2-69 shows the den-

Crystal	Face	Reconstruction	$E_F - E_{VS}$ (eV)
Diamond	(111)	2×1	1.5
Si	(100)	2×1	0.32 (1)
	(111)	2×1	0.38 (2)
		7×7	0.69 (3)
Ge	(111)	2×1	0.01 (2)
		c(2×8)	0.12 (2)
GaAs	(100)	c(2×8)	0.47 (7)
		c(4×4)	0.47 (7)
		4×6	0.49
	(110)	1×1	Flat bands
GaSb	(110)	1×1	Flat bands
GaP	(110)	1×1	Flat bands
InAs	(110)	1×1	Flat bands

Table 5.2-27 Pinning position of the Fermi level at semiconductor surfaces. Data are average values from various authors. Errors are given as standard deviations. References to the original papers and individual errors are given in [2.47]

sity of interface states in the gap of Si for two different orientations.

Channel mobilities are generally lower than bulk mobilities. An example is reported for n- and p-type carriers at the silicon–silicon oxide interface in Fig. 5.2-70.

A more extended discussion of the space charge layer may be found in [2.47, 48, 93, 94].

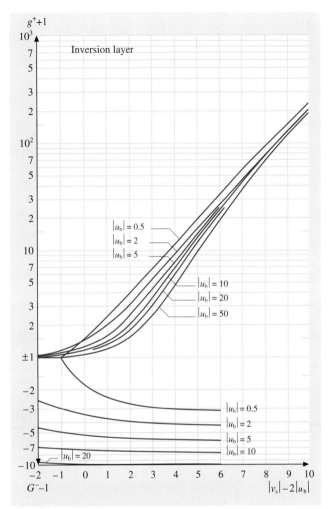

Fig. 5.2-67 More detailed plots of the functions G^-, g^+ for inversion layers. Notice the change of the abscissa to $|v_s| - 2|u_b|$ [2.93, p. 155]

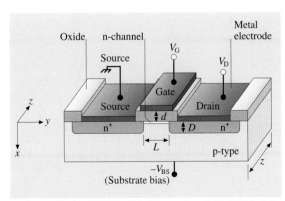

Fig. 5.2-68 Schematic drawing of a MOSFET. The n-inversion layer is obtained with a positive bias of the gate with respect to the source [2.48, p. 345]

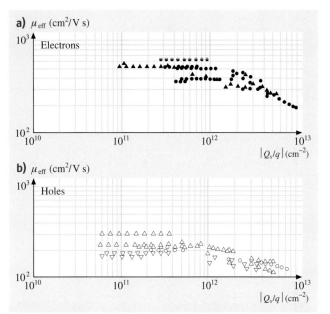

Fig. 5.2-70 Effective mobility of electrons and holes in MOSFET's inversion channel. $Q_s = Q_{total}$ is the charge on the gate electrode. It is seen that mobility decreases at high transverse electric fields. For comparison, bulk mobilities are: $\mu_n = 1600$, $\mu_p = 600\ \text{cm}^2/\text{V s}$ [2.96]

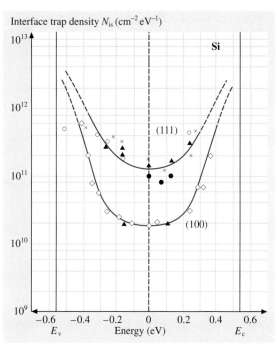

Fig. 5.2-69 Interface state density in thermally oxidized silicon for two different orientations. Notice the low density around mid-gap [2.95, after [2.48] p. 340]

5.2.6 Most Frequently Used Acronyms

Surface physics, like many other fields, is plagued by the use of acronyms. The acronyms used in this chapter, as well as those most frequently found in the literature, are listed below:

AES	Auger electron spectroscopy
AFM	atomic force microscope/microscopy
APS	appearance potential spectroscopy
ARUPS	angle-resolved ultraviolet photoemission spectroscopy
ARXPS	angle-resolved X-ray photoemission spectroscopy
ATR	attenuated total reflection
BB	back bond
BBZ	bulk Brillouin zone
BEP	beam-equivalent pressure
bl	blocking
BZ	Brillouin zone
CB	conduction band
CBM	conduction band minimum
CCV	core–core valence Auger transition
chan	channeling
CISS	collision ion scattering spectroscopy

CITS	current imaging tunneling spectroscopy
CMA	cylindrical mirror analyzer
CVD	chemical vapor deposition
CVV	core valence–valence Auger transition
DAS	dimer–adatom–stacking fault
DB	dangling bond
2D BZ	2-dimensional Brillouin zone
DOS	density of states
DR	differential reflectivity
ECS	electron capture spectroscopy
EDC	energy distribution curve
EELS	electron-energy loss spectroscopy
ELEED	elastic low-energy electron diffraction
ESA	electrostatic analyzer
ESD	electron-stimulated desorption
EXAFS	extended X-ray absorption fine structure
FEED	field emission energy distribution
FEM	field emission microscope/microscopy
FIM	field ion microscope/microscopy
FLAPW	full-potential linearized augmented plane wave
HAS	helium atom scattering
HATOF	helium atom time-of-flight spectroscopy
HEED	high-energy electron diffraction
HEIS	high-energy ion scattering/high-energy ion scattering spectroscopy
HREELS	high-resolution electron energy loss spectroscopy
HR-LEED	high-resolution LEED
HR-RHEED	high-resolution RHEED
IB	ion bombardment
IBA	ion bombardment and annealing
ICISS	impact ion scattering spectroscopy
IPE	inverse photoemission
IPES	inverse photoemission spectroscopy
IS	image state
ISS	ion scattering spectroscopy
JDOS	joint density of states
KRIPES	K-resolved inverse photoelectron spectroscopy
LAPW	linearized augmented-plane-wave method
LDA	local-density approximation
LDOS	local density of states
LEED	low-energy electron diffraction
LEED-I/V	LEED intensity/voltage measurements
LEIS	low-energy ion scattering/low-energy ion scattering spectroscopy
LEPD	low-energy positron diffraction
MBE	molecular-beam epitaxy

MD	molecular dynamics
MEED	medium-energy electron diffraction
MEIS	medium-energy ion scattering/medium-energy ion scattering spectroscopy
ML	monolayer
MOKE	magneto-optical Kerr effect
MR	missing row
NICISS	neutral impact collision ion scattering spectroscopy
PBBS	projected bulk band structure
PDS	photothermal displacement spectroscopy
PED	photoelectron diffraction
PES	photoemission spectroscopy, photoelectron spectroscopy
PLAP	pulsed laser atom probe
2P-PES	2-photon photoemission spectroscopy
RAS	reflectance anisotropy spectroscopy
REM	reflection electron microscope/microscopy
RHEED	reflection high-energy electron diffraction
RPA	random-phase approximation
SAM	scanning Auger microscope/microscopy
SARS	scattering and recoiling ion spectroscopy
SBZ	surface Brillouin zone
SCLS	surface core level shift
SDR	surface differential reflectivity
SEM	scanning electron microscope
sh	shadowing
SHG	second-harmonic generation
SIMS	secondary-ion mass spectroscopy
SPARPES	spin-polarized angle-resolved photoemission spectroscopy
SPIPES	spin-polarized inverse photoemission spectroscopy
SPLEED	spin-polarized LEED
SPV	surface photovoltage spectroscopy
SS	surface state
STM	scanning tunneling microscope/microscopy
STS	scanning tunneling spectroscopy
SXRD	surface X-ray diffraction
TEM	transmission electron microscope/microscopy
TOF	time of flight
TOM	torsion oscillation magnetometry
TRS	truncation rod scattering
UHV	ultra-high vacuum
UPS	ultraviolet photoemission spectroscopy
VB	valence band
VBM	valence band maximum
VLEED	very low-energy electron diffraction
XPS	X-ray photoemission spectroscopy

References

2.1 G. Chiarotti (Ed.): *The Physics of Solid Surfaces*, Landolt–Börnstein, New Series III/24 (Springer, Berlin, Heidelberg 1993–1996)

2.2 J. F. Nicholas: *The Structure of Ideal Surfaces*, Landolt–Börnstein, New Series III/24, ed. by G. Chiarotti (Springer, Berlin, Heidelberg 1993) p. 29

2.3 A. Fasolino, A. Selloni, A. Shkrebtii: *Surface Reconstruction and Relaxation*, Landolt–Börnstein, New Series III/24, ed. by G. Chiarotti (Springer, Berlin, Heidelberg 1993) p. 125

2.4 E. Zanazzi: *Elastic Scattering and Diffraction of Electrons and Positrons*, Landolt–Börnstein, New Series III/24, ed. by G. Chiarotti (Springer, Berlin, Heidelberg 1995) p. 29

2.5 R. Colella: *X-ray Diffraction of Surface Structures*, Landolt–Börnstein, New Series III/24, ed. by G. Chiarotti (Springer, Berlin, Heidelberg 1996) p. 312

2.6 P. Alkemade: *Elastic and Inelastic Scattering of Ions*, Landolt–Börnstein, New Series III/24, ed. by G. Chiarotti (Springer, Berlin, Heidelberg 1995) p. 176

2.7 R. J. Hamers, R. M. Tromp, J. E. Demuth: Phys. Rev. Lett. **56**, 1972 (1986)

2.8 P. Chiaradia: *Optical Properties of Surfaces*, Landolt–Börnstein, New Series III/24, ed. by G. Chiarotti (Springer, Berlin, Heidelberg 1996) p. 29

2.9 K. Jacobi: *Electronic Structure of Surfaces: Metals*, Landolt–Börnstein, New Series III/24, ed. by G. Chiarotti (Springer, Berlin, Heidelberg 1994) p. 29

2.10 C. Calandra, F. Manghi: *Electronic Structure of Surfaces: Semiconductors*, Landolt–Börnstein, New Series III/24, ed. by G. Chiarotti (Springer, Berlin, Heidelberg 1994) p. 352

2.11 A. M. Bradshaw, R. Hemmen, D. E. Ricken, T. Schedel-Niedrig: *Photoemission and Inverse Photoemission*, Landolt–Börnstein, New Series III/24, ed. by G. Chiarotti (Springer, Berlin, Heidelberg 1996) p. 70

2.12 M. Rocca: *Inelastic Scattering of Electrons*, Landolt–Börnstein, New Series III/24, ed. by G. Chiarotti (Springer, Berlin, Heidelberg 1995) p. 113

2.13 U. Gradmann: *Magnetic Properties of Single Crystal Surfaces*, Landolt–Börnstein, New Series III/24, ed. by G. Chiarotti (Springer, Berlin, Heidelberg 1994) p. 506

2.14 R. F. Wallis, S. Y. Tong: *Surface Phonons*, Landolt–Börnstein, New Series III/24, ed. by G. Chiarotti (Springer, Berlin, Heidelberg 1994) p. 433

2.15 V. Celli: *Interaction of Atoms with Surfaces*, Landolt–Börnstein, New Series III/24, ed. by G. Chiarotti (Springer, Berlin, Heidelberg 1995) p. 278

2.16 G. L. Kellog: *Field Emission, Field Ionization and Field Desorption*, Landolt–Börnstein, New Series III/24, ed. by G. Chiarotti (Springer, Berlin, Heidelberg 1996) p. 342

2.17 W. A. Wood: J. Appl. Phys. **35**, 1306 (1963)

2.18 J. V. Barth, H. Brune, G. Ertl, R. J. Behm: Phys. Rev. B **42**, 9307 (1990)

2.19 M. K. Debe, D. A. King: Phys. Rev. Lett. **39**, 708 (1977)

2.20 R. M. Tromp, R. J. Hamers, J. E. Demuth: Phys. Rev. Lett. **55**, 1303 (1985)

2.21 K. C. Pandey: Phys. Rev. Lett. **47**, 1913 (1981)

2.22 R. M. Feenstra, W. A. Thompson, A. P. Fein: Phys. Rev. Lett. **56**, 608 (1986)

2.23 K. Takayanagi, Y. Tanishiro, M. Takahashi, S. Takahashi: J. Vac. Sci. Technol. A **3**, 1502 (1985)

2.24 H. Huang, S. Y. Tong, W. E. Packard, M. B. Webb: Phys. Lett. A **130**, 166 (1988)

2.25 S. Y. Tong, H. Huang, C. M. Wei, W. E. Packard, F. K. Men, G. Glander, M. B. Webb: J. Vac. Sci. Technol. A **6**, 615 (1988)

2.26 I. K. Robinson, W. K. Waskiewics, P. H. Fuoss, L. J. Norton: Phys. Rev. B **37**, 4325 (1988)

2.27 R. J. Hamers: *Scanning Tunneling Microscopy*, Landolt–Börnstein, New Series III/24, ed. by G. Chiarotti (Springer, Berlin, Heidelberg 1996) p. 363

2.28 W. S. Yang, F. Jona: Phys. Rev. B **29**, 899 (1984)

2.29 N. D. Lang, W. Kohn: Phys. Rev. B **1**, 4555 (1970)

2.30 H. Eckardt, L. Fritsche: J. Phys. F **14**, 97 (1984)

2.31 N. Garcia, B. Reihl, K. H. Frank, A. R. Williams: Phys. Rev. Lett. **54**, 591 (1985)

2.32 M. Echenique, J. B. Pendry: Progr. Surf. Sci. **32**, 111 (1990)

2.33 A. Goldmann, V. Dose, G. Borstel: Phys. Rev. B **32**, 1971 (1985)

2.34 D. M. Kolb, W. Boeck, K.-M. Ho, S. H. Liu: Phys. Rev. Lett. **47**, 1921 (1981)

2.35 B. Reihl, K. H. Frank, R. R. Schlittler: Phys. Rev. B **30**, 7328 (1984)

2.36 S. D. Kevan, R. H. Gaylord: Phys. Rev. B **36**, 5809 (1987)

2.37 K. M. Ho, C. L. Liu, D. M. Kolb, G. Piazza: J. Electroanal. Chem. **150**, 235 (1983)

2.38 K. M. Ho, B. N. Harmon, S. H. Liu: Phys. Rev. Lett. **44**, 1531 (1980)

2.39 S. H. Liu, C. Hinnen, C. Nguyen Van Huong, N. R. De Tacconi, K. M. Ho: J. Electroanal. Chem. **176**, 325 (1984)

2.40 R. A. Bartynski, T. Gustafsson: Phys. Rev. B **33**, 6588 (1986)

2.41 P. Heimann, H. Miosga, H. Neddermeyer: Phys. Rev. Lett. **42**, 801 (1979)

2.42 S. G. Louie: Phys. Rev. Lett. **40**, 1525 (1978)

2.43 P. D. Johnson, N. V. Smith: Phys. Rev. Lett. **49**, 290 (1982)

2.44 M. I. Holmes, T. Gustafsson: Phys. Rev. Lett. **47**, 443 (1981)

2.45 M. Posternak, H. Krakauer, A. J. Freeman, D. D. Koelling: Phys. Rev. B **21**, 5601 (1980)

2.46 K. D. Tsuei, E. W. Plummer, A. Liebsch, E. Pehlke, K. Kempa, P. Bakshi: Surf. Sci. **247**, 302 (1991)

2.47 W. Mönch: *Semiconductor Surfaces and Interfaces* (Springer, Berlin, Heidelberg 1995)
2.48 H. Lüth: *Surfaces and Interfaces of Solid Materials* (Springer, Berlin, Heidelberg 1995)
2.49 F. Manghi, C. M. Bertoni, C. Calandra, E. Molinari: Phys. Rev. B **24**, 6029 (1981)
2.50 J. A. Stroscio, R. M. Feenstra, D. M. Newns, A. P. Fein: J. Vac. Sci. Technol. A **6**, 499 (1988)
2.51 V. Vanderbilt, S. G. Louie: Phys. Rev. B **30**, 6118 (1984)
2.52 F. J. Himpsel, D. E. Eastman, P. Heimann, J. F. van der Veen: Phys. Rev. B **24**, 7270 (1981)
2.53 P. Kruger, A. Mazur, G. Wolfgarten: Phys. Rev. Lett. **57**, 1468 (1986)
2.54 P. Koke, A. Goldmann, W. Monch, G. Wolfgarten, J. Polmann: Surf. Sci. **152/153**, 1001 (1985)
2.55 A. Goldmann, P. Koke, W. Monch, G. Wolfgarten, J. Polmann: Surf. Sci. **169**, 438 (1986)
2.56 J. E. Northrup, M. L. Cohen: Phys. Rev. Lett. **49**, 1349 (1982)
2.57 P. Martensson, A. Cricenti, G. V. Hansson: Phys. Rev. B **32**, 6959 (1985)
2.58 F. J. Himpsel, T. Fauster: J. Vac. Sci. Technol. A **2**, 815 (1984)
2.59 R. Wolkov, P. Avouris: Phys. Rev. Lett. **60**, 1049 (1988)
2.60 J. M. Nicholls, B. Reihl: Surf. Sci. **218**, 237 (1989)
2.61 J. M. Nicholls, P. Martensson, G. V. Hansson: Phys. Rev. Lett. **54**, 2363 (1985)
2.62 J. E. Northrup, M. L. Cohen: Phys. Rev. B **27**, 6553 (1983)
2.63 D. Straub, M. Skibowski, F. J. Himpsel: Phys. Rev. B **32**, 5237 (1985)
2.64 B. Reihl, T. Riesterer, M. Tschudy, P. Perfetti: Phys. Rev. B **38**, 13456 (1988)
2.65 J. Henk, W. Shattke, H. P. Hannescheidt, C. Janowitz, R. Manzke, M. Skibowski: Phys. Rev. B **39**, 13286 (1989)
2.66 A. Huijser, J. van Laar, T. L. van Roy: Phys. Lett. A **65**, 337 (1978)
2.67 Zhu, S. B. Zhang, S. G. Louie, M. L. Cohen: Phys. Rev. Lett. **63**, 2112 (1989)
2.68 K. O. Magnusson, S. A. Flodstrom, P. E. S. Persson: Phys. Rev. B **38**, 5384 (1988)
2.69 Y. R. Wang, C. B. Duke, K. O. Magnusson, S. A. Flodström: Surf. Sci., **205**, 2760 (1988)
2.70 S. Selci, P. Chiaradia, F. Ciccacci, A. Cricenti, N. Sparvieri, G. Chiarotti: Phys. Rev. B **31**, 4096 (1987)
2.71 N. J. DiNardo, J. E. Demuth, W. A. Thompson, P. Avouris: Phys. Rev. B **31**, 4077 (1985)

2.72 G. Chiarotti, P. Chiaradia, E. Faiella, C. Goletti: Surf. Sci. **453**, 112 (2000)
2.73 J. E. Demuth, R. Imbihl, W. A. Thompson: Phys. Rev. B **34**, 1330 (1986)
2.74 G. Chiarotti, S. Nannarone, R. Pastore, P. Chiaradia: Phys. Rev. B **4**, 3398 (1971)
2.75 H. Froitzheim, H. Ibach, D. L. Mills: Phys. Rev. B **11**, 4980 (1975)
2.76 M. K. Kelly, S. Zollner, M. Cardona: Surf. Sci. **285**, 282 (1993)
2.77 T. F. Heinz, M. M. T. Loy, W. A. Thompson: Phys. Rev. Lett. **54**, 63 (1985)
2.78 I. Kamiya, D. E. Aspness, L. J. Florez, J. P. Harbison: Phys. Rev. B **46**, 15894 (1992)
2.79 M. Taborelli, O. Paul, O. Zuger, M. Landolt: J. Phys. (Paris) Colloq. **8**, 1659 (1988)
2.80 P. Moretto, M. Rocca, U. Valbusa, J. E. Black: Phys. Rev. B **41**, 12905 (1990)
2.81 G. Bracco, R. Tatarek, F. Tommasini, U. Linke, M. Persson: Phys. Rev. B **36**, 2928 (1987)
2.82 U. Harten, J. P. Toennies, Ch. Woll: Faraday Discuss. Chem. Soc. **80**, 137 (1985)
2.83 J. P. Toennies, C. Wöll: Phys. Rev. B **36**, 4475 (1987)
2.84 B. Voigtlander, S. Lehwald, H. Ibach, K. P. Bohnen, K. M. Ho: Phys. Rev. B **40**, 8068 (1989)
2.85 M. Wuttig, R. Franchy, H. Ibach: Z. Phys. B **65**, 71 (1986)
2.86 M. Rocca, S. Lehwald, H. Ibach, T. S. Rahman: Surf. Sci. **171**, 632 (1986)
2.87 U. Harten, J. P. Toennies: Europhys. Lett. **4**(7), 833 (1987)
2.88 A. Safron, W. P. Brug, G. Chern, J. Duan, J. G. Skofronick: J. Vac. Sci. Technol. A **8**, 2627 (1990)
2.89 G. Benedek, G. Brusdeylins, R. B. Doak, J. G. Skofronick, J. P. Toennies: Phys. Rev. B **28**, 2104 (1983)
2.90 U. Harten, J. P. Toennies, C. Woll: Phys. Rev. Lett. **57**, 2947 (1986)
2.91 H. Ibach: Phys. Rev. Lett. **27**, 253 (1971)
2.92 D. Eichenauer, U. Harten, J. P. Toennies, V. Celli: J. Chem. Phys. **86**, 3693 (1987)
2.93 A. Many, Y. Goldstein, N. B. Grove: *Semiconductor Surfaces* (North-Holland, Amsterdam 1965)
2.94 D. R. Frankl: *Electrical Properties of Semiconductor Surfaces* (Pergamon, Oxford 1967)
2.95 M. H. White, J. R. Cricchi: IEEE Trans. Electron Dev. **19**, 1280 (1972)
2.96 O. Leistiko, A. S. Grove, C. T. Sah: IEEE Trans. Electron Dev. **12**, 248 (1965)

5.3. Mesoscopic and Nanostructured Materials

This chapter addresses the properties of nanostructured materials considered as statistical ensembles of nanostructures. Emphasis is put on size and confinement effects, although enhancements in surface and interface properties are mentioned. After a survey and a summary of basic definitions and concepts in the introductory Sect. 5.3.1, the properties associated with electronic confinement are addressed in Sect. 5.3.2. Electronic confinement affects the spectral properties, i.e. light absorption and luminescence, mainly through quantum size effects, and the electrical conduction properties through the Coulomb blockade. Both two-dimensional systems (quantum wells) and zero-dimensional systems (quantum dots) are reviewed. Particular attention is drawn to semiconductor-doped matrices. The effects associated with confinement of electromagnetic fields are treated in Sect. 5.3.3. Numerical relationships and data for plasmon excitations of various metal nanoparticles can be found in this section. Magnetic nanostructures are addressed in Sect. 5.3.4. The two main applications of nanostructured magnetic materials, namely spin electronics, or spintronics, and ultrahigh-density data storage media, are treated. Finally, we list and briefly describe in Sect. 5.3.5 some generic techniques for the preparation of nanostructured materials, organized into the following groups of methods: molecular-beam epitaxy (MBE), metal-organic chemical vapor deposition (MOCVD), nanolithography, nanocrystal growth in matrices, and ex-situ synthesis of clusters.

- 5.3.1 **Introduction and Survey** 1031
  - 5.3.1.1 Historical Review 1031
  - 5.3.1.2 Definitions 1032
  - 5.3.1.3 Specific Properties 1034
  - 5.3.1.4 Organization of this Chapter 1034
- 5.3.2 **Electronic Structure and Spectroscopy** 1035
  - 5.3.2.1 Electronic Quantum Size Effects 1035
  - 5.3.2.2 Breakdown of the Momentum Conservation Rule 1036
  - 5.3.2.3 Excitons in Quantum-Confined Systems 1036
  - 5.3.2.4 Vibrational Modes and Electron–Phonon Coupling 1040
  - 5.3.2.5 Electron Transport Phenomena 1042
- 5.3.3 **Electromagnetic Confinement** 1044
  - 5.3.3.1 Nanoparticle-Doped Materials 1044
  - 5.3.3.2 Periodic Electromagnetic Lattices 1048
- 5.3.4 **Magnetic Nanostructures** 1048
  - 5.3.4.1 Spin Electronics 1049
  - 5.3.4.2 Ultrahigh-Density Storage Media in Hard Disk Drives 1060
- 5.3.5 **Preparation Techniques** 1063
  - 5.3.5.1 Molecular-Beam Epitaxy 1063
  - 5.3.5.2 Metal-Organic Chemical Vapor Deposition (MOCVD) 1064
  - 5.3.5.3 Lithography 1064
  - 5.3.5.4 Nanocrystals in Matrices 1064
  - 5.3.5.5 Ex Situ Synthesis of Clusters 1065
- **References** 1066

5.3.1 Introduction and Survey

5.3.1.1 Historical Review

Except in the life sciences, there are only a few examples of materials that are naturally structured on scales of the order of a few to a few hundred nanometers. One can cite, however, the natural zeolites, which constitute a group of hydrated crystalline aluminosilicates containing regularly shaped pores with sizes from 1 nm to several nanometers. This provides them the ability to reversibly adsorb and desorb specific molecules.

They were named "zeolite" ("boiling stone") in 1756 by Cronstedt, a Swedish mineralogist, who observed their emission of water vapor when heated. At the other size limit, opals constitute another example of a naturally occurring nanostructured material. These gems are made up mainly of spheres of amorphous silica with sizes ranging from 150 nm to 300 nm. In precious opals, these spheres are of approximately equal size and can thus be arranged in a three-dimensional periodic lattice. The optical interferences produced by this periodic index modulation are the origin of the characteristic iridescent colors (opalescence).

Apart from these few examples, most nanostructured materials are synthetic. Empirical methods for the manufacture of stained glasses have been known for centuries. It is now well established that these methods make use of the diffusion-controlled growth of metal nanoparticles. The geometrical constraints on the electron motion and the electromagnetic field distribution in noble-metal nanoparticles lead to the existence of a particular collective oscillation mode, called the plasmon oscillation, which is responsible for the coloration of the material. It has been noticed recently that the beautiful tone of Maya blue, a paint often used in Mesoamerica, involves simultaneously metal nanoparticles and a superlattice organization [3.1].

The chemistry and color changes of colloidal gold solutions were observed by Faraday during the 19th century [3.2]. These properties were due to highly size-dispersed gold nanoparticles.

Improvements in the diffusion-controlled growth technique opened up the possibility of growing nanocrystallites with better-controlled sizes and densities and permitted its extension to various semiconductors. The fabrication of colored long-wavelength-pass glasses and of photochromic glasses provides well-known examples of commercial technologies based on such methods developed decades ago. Various techniques for the production and assembly of cluster- and nanoparticle-based materials are currently under intense study.

More recently, important technological efforts have been made, driven by the increasing needs of the electronics industry, in order to understand and control the growth of semiconductors at the atomic level. The development of molecular-beam epitaxy (MBE) permitted the control of atomic-layer-by-atomic-layer growth of semiconductors. It has become possible to create structures made up of an alternation of different layers, each of which is only a few atomic layers thick. The first observation of the quantization of energy levels in a quantum well in 1974 [3.3] opened the way to the tailoring of the electronic wave function in one dimension on the nanometer scale, leading to the production of new electronic and also magnetic materials. A new trend in surface science is work aimed at the control of in-plane nanostructuring, such as the formation of wire or dot shapes, through self-organization.

In parallel to developments in the field of electronics, nanostructured materials have been developed by materials scientists and chemists also. The concept of nanocrystalline structures emerged in the field of materials science, and polycrystals with ultrafine grain sizes in the nanometer range have been produced. These "nanophase materials" have been shown to have significant modifications of their mechanical properties compared with the coarse-grain equivalent materials. The huge surface area of nanoporous materials has attracted much attention for applications in chemistry such as molecular sieves, catalysis, and gas sensing. This has motivated intense research aimed at the fabrication of materials with a well-controlled composition and nanoscale structure, such as synthetic zeolites.

The scientific and technological domains of research on nanostructured materials cover a range of disciplines, from biology to physics and chemistry. However, their convergent aspects, as well as, to some extent, a common type of approach, have been recognized recently in the realm of nanoscience and nanotechnology, under the term "nanostructured materials", or simply "nanomaterials".

For an extended review on nanotechnology see the recently published Handbook of Nanotechnology [3.4].

5.3.1.2 Definitions

In their broadest definition, nanostructured materials show structural features with sizes in the range from 1 nm to a few hundred nanometers in at least one dimension. This very general criterion actually includes very diverse physical situations.

First, as is apparent from the previous section, each nanostructured material is associated with a specific novel property or a significant improvement in a specific property resulting from the nanoscale structuring. As a consequence, the type of nanostructuring used must be based on a spatial dependence of some parameter related to the property under consideration. This parameter could be, for example, the material density, transport parameters, or the dielectric constant. Another consequence is that the upper size limit of the structural features varies depending on the property considered, from a molecular size for molecular-sieve properties up to the wavelength of light for the optical properties.

Second, in addition to the nanometer-scale structuring, a larger-scale ordering of the unit patterns may be necessary for the existence of the property sought. For example, the particular optical properties of opals mentioned above require the silica nanospheres to show a long-range order with a coherence length well beyond a micron. The same considerations hold for quantum-well superlattices, for example. In other cases, the nanosized building blocks do not need long-range order to provide a specific property, but still require some degree of short-range organization. For example, an electrical conductivity appears only above a critical percolation density of conducting particles. Finally, some properties of nanostructured materials simply reflect a corresponding intrinsic property of their individual building blocks. This is the case, for instance, in nanoparticles embedded in glass or polymer matrices for optical-filtering applications.

Two main technological approaches may be defined:

- The top-down manufacturing paradigm consists in downscaling the patterning of materials to nanometer sizes. This allows the generation of materials which are coherently and continuously ordered from macroscopic down to nanoscopic sizes.
- The bottom-up paradigm is based on the atomically precise fabrication of entities of increasing size. It is the domain of macromolecular and supramolecular chemistry (dendrimers, engineered DNA, etc.) and of cluster and surface physics (epitaxy, self-assembly, etc.).

Mesoscopic materials form the subset of nanostructured materials for which the nanoscopic scale is large compared with the elementary constituents of the material, i.e. atoms, molecules, or the crystal lattice. For the specific property under consideration, these materials can be described in terms of continuous, homogeneous media on scales less than that of the nanostructure. The term "mesoscopic" is often reserved for electronic transport phenomena in systems structured on scales below the phase-coherence length Λ_Φ of the carriers.

Most of the common nanomaterials can be classified in terms of dimensionality, according to the number of orthogonal directions X, Y, Z in which the structural patterns referred to above have dimensions $L_{X,Y,Z}$ smaller than the nanoscopic limit L_0. This leads to the classical definitions of dimensionality summarized in Table 5.3-1. However, it should be noted that experimental situations

Table 5.3-1 Examples of reduced-dimensional material geometries, and definitions of their dimensionality and of the associated type of confinement

$L_{X,Y,Z} > L_0$	No nanostructures	No confinement	Bulk material
$L_{X,Y} > L_0 > L_Z$	Two-dimensional (2-D) nanostructures	One-dimensional (1-D) confinement	Wells
$L_X > L_0 > L_{Y,Z}$	One-dimensional (1-D) nanostructures	Two-dimensional (2-D) confinement	Wires
$L_0 > L_{X,Y,Z}$	Zero-dimensional (0-D) nanostructures	Three-dimensional (3-D) confinement	Dots

can be encountered in which the dimensionality may not be so obviously defined. For instance, the structural patterns may be formed by nonrectilinear wires which occupy a surface or a volume or may be branched.

The structure of the density of states (DOS) in nanostructures is strongly dependent on the dimensionality. A free 3-D motion yields a band characterized by three wave vectors k_x, k_y, k_z. The corresponding DOS depends smoothly on the energy E, as $(E - E^0)^{1/2}$, where E^0 is the energy of the bottom of the band. Confinement into a 2-D system splits the band into subbands, and leaves only two continuously varying wave vectors k_x, k_y in each subband. The DOS for each subband is then constant above the energy E_N^0 of its bottom state. The overall DOS is discontinuous, with a stepwise structure that is characteristic of quantum wells (Fig. 5.3-1). Confinement in one additional dimension splits each 2-D subband further into a set of 1-D subbands. Each 1-D subband is characterized by only one continuously varying wave vector k_x, and two quantum numbers N, M. The DOS corresponding to the subband N, M has a variation of the form $(E - E_{N,M}^0)^{-1/2}$, with a divergence at the bottom of the subband $E_{N,M}^0$. The DOS of a quantum wire thus has a more pronounced structure than does a 2-D well, with a larger number of subbands, each one starting as a peak (Fig. 5.3-1). Finally, confinement in all three dimensions creates a completely discrete, atom-like set of states. The DOS of a quantum dot thus consists of a series of δ-functions (not represented in Fig. 5.3-1). The sharpening of the DOS at specific energies induced by quantum confinement is the origin of many improvements in the properties of nanostructured materials compared with bulk materials. This spectral concentration enhances all resonant effects and increases the energy selectivity. The preservation of these effects when one is dealing with an ensemble of quantum-confined systems requires a high homogeneity.

5.3.1.3 Specific Properties

The specific properties of nanostructured materials can have two different possible origins:

- Size effects, which result from the spatial confinement of a physical entity inside an element of the nanoscale structural pattern. Such an element is called a low-dimensional system. An example is the confinement of electron wave functions inside a region whose size is smaller than the electron mean free path. This class of effects may give birth to completely new properties.
- Boundary effects, which are a consequence of the significant volume fraction of matter located near surfaces, interfaces, or domain walls. Processes that take place only at such locations may be highly favored, and properties specific to structural boundaries may also be greatly enhanced.

For a recent comprehensive review of the basic principles of the origin of the properties of nanostructured materials, see [3.5].

5.3.1.4 Organization of this Chapter

Many classification schemes for nanostructured materials exist. These may be based on their chemical composition, on the technique for their manufacture, or on their dimensionality. These schemes, however, are often suitable only for a subset of materials. Moreover, they generally address only one particular scientific or technological approach and its associated community of specialists. Since, as noted above, most nanostructured materials are associated with a specific property, we have chosen a presentation based on properties. This scheme allows us to include all nanostructured materials and is accessible to the largest possible readership.

Consistently with the materials approach of this chapter and with its limited size, only the properties of statistical ensembles of nanostructures will be considered. The specific behaviors of individual nano-

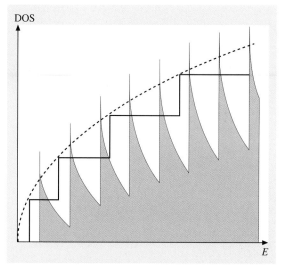

Fig. 5.3-1 Schematic illustration of the density of states for 3-D motion of a free electron (*dashed line*), a 2-D quantum well (*solid line*), and a 1-D quantum wire (*shaded area*) as a function of energy

sized objects or devices are not included. Since surface and interface properties are treated in another section, emphasis will be put here on size effects. Properties related to electronic confinement and its consequences for spectral properties are addressed in Sect. 5.3.2. Effects of the confinement of electromagnetic fields are treated in Sect. 5.3.3. Magnetic size effects are addressed in Sect. 5.3.4. Finally, we list and briefly describe in Sect. 5.3.5 some generic methods for preparation of nanostructured materials.

5.3.2 Electronic Structure and Spectroscopy

5.3.2.1 Electronic Quantum Size Effects

Confined electronic systems are quantum systems in which carriers, either electron or holes, are free to move only in a restricted number of dimensions. In the confined dimension, the sizes of the structural elements are of the order of a few de Broglie wavelengths of the carriers or less. Depending on their dimensionality, these structures can be quantum dots (0-D), quantum wires (1-D), or quantum wells (2-D). Quantum wells are typically produced by the alternate epitaxial growth of two or more different semiconductors. Quantum wires are less commonly encountered, since their fabrication procedures are much more complicated (Sect. 5.3.4).

One of the most dramatic effects, called the quantum size effect, consists in a redistribution of the energy spectrum of the system, the density of states becoming discrete along the confinement direction. In the most simple "particle in a box" model of a quantum well, the energies of the corresponding eigenstates are

$$E_{N,k_x,k_y} = N^2 \frac{\pi^2 \hbar^2}{2md^2} + \left(k_x^2 + k_y^2\right) \frac{\hbar^2}{2m}, \quad (3.1)$$

where m is the effective mass of the carrier, d is the confinement dimension, and N is a quantum number. The first term appears because of the quantized motion in the z direction, whereas the second term represents the energy of the free x, y motion, characterized by wave vectors k_x, k_y. Each value of N defines a semi-infinite subband of energy levels. Although the "particle in a box" model is simplistic compared with real systems, this limiting case is often successful in describing the essential features of quantum size effects [3.6].

The importance of the quantum size effect is mainly determined by the energy differences $E_{N+1} - E_N$. Quantum size effects become observable when this separation exceeds the thermal energy of the carriers, so that adjacent subbands are differently populated. Since the energy difference $E_{N+1} - E_N$ increases with N, it could be anticipated that quantization effects would be more important for processes involving higher subbands. However, the practical observation of such effects in systems with a large Fermi energy ($E_F \gg E_1$), i.e. in metals, is often made difficult because of contributions from many levels and the appearance of collective phenomena such as plasmon oscillations, which wash out quantization effects. Most of the spectral data revealing quantum well states in metal films a few monolayers thick have been obtained by angle-resolved photoemission for Cs on Cu(111) [3.7] and for Ag on Au(111) [3.8]. The Fabry–Perot-like regularly-spaced spectrum of energy levels appears particularly clearly in a study of Ag films on Fe(100) for thicknesses up to more than 100 monolayers (Fig. 5.3-2). The superposition at half-integer coverages (27.5 and 42.5 monolayers) of two sets of peaks corresponding to the two closest integer thicknesses emphasizes the extreme sensitivity to surface homogeneity. Quantitatively, the characteristic roughness of well barriers must be much below the de Broglie wavelength of the quantum states observed. This is another reason why quantum size effects have been observed mainly in semiconductor nanostructures, which have large Fermi wavelengths, rather than in metals, which have Fermi wavelengths comparable to a few crystal lattice periods.

The existence of discrete electronic states of electrons confined in a small metal cluster has been observed to influence the thermodynamic stability of the system, in particular during the production of sodium clusters in supersonic beams composed of the metal vapor and an inert gas. The statistics of the relative abundances of different particle sizes reveal the existence of "magic numbers" for the number of atoms in the cluster, $N = 8, 20, 40, 58, 92, \ldots$ [3.9]. This has been interpreted in terms of the existence of degenerate energy levels in a spherical well with infinite-potential walls. Particularly stable structures are obtained when the number of valence electrons is such that it leads to a closed-shell electronic structure, i.e. a structure with a completely filled energy level and an empty up-

5.3.2.2 Breakdown of the Momentum Conservation Rule

As a result of the Heisenberg uncertainty principle, localization of carriers in a confinement volume spreads their wave functions in reciprocal space. The possible overlap in k space of electron and hole wave functions breaks the rule of conservation of crystal momentum. In other words, the momentum needed for a transition at a different k may be provided by scattering from the boundaries. This consequence of quantum confinement is particularly meaningful for nanostructures based on indirect-band-gap semiconductors, for which luminescence from recombination of the lowest-energy electrons and holes is forbidden in the bulk material, but becomes increasingly possible for systems of decreasing size, with a scaling law of d^{-6} for dots [3.13]. The observation of bright luminescence from nanoporous silicon samples [3.14] was, from the start, interpreted in terms of quantum confinement into silicon nanowires [3.14, 15]. This observation triggered an important research activity motivated by the industrial prospect of being able to incorporate silicon photonic elements into the current silicon technology. Although it has been shown that other phenomena such as surface effects and phonon modes (discussed later) come into the play, and that more complicated geometries including both quantum dots and quantum wires have to be considered, the role of quantum confinement in the luminescence of porous silicon is generally considered to be central, at least for the "red" part of the luminescence. The studies of nanostructure-induced luminescence have been extended to porous GaP [3.16, 17] and SiC [3.18], two other semiconductors with an indirect band gap, and the results parallel those for porous silicon. Quantum-confined luminescence has been observed in indirect-gap semiconductor nanostructures with well-controlled geometries such as silicon and germanium 2-D wells [3.19] and quantum dots (see e.g. [3.20]), which may possibly be integrated into light-emitting devices [3.21]. The observation of optical gain in silicon nanocrystals [3.22] opens up new prospects for the creation of silicon-based lasers [3.23].

5.3.2.3 Excitons in Quantum-Confined Systems

A very general effect of quantum confinement in semiconductors is a widening of their optical band gap. For the model system of an infinite-barrier 2-D quantum well, (3.1) shows that the lowest energy ($N = 1$) of a con-

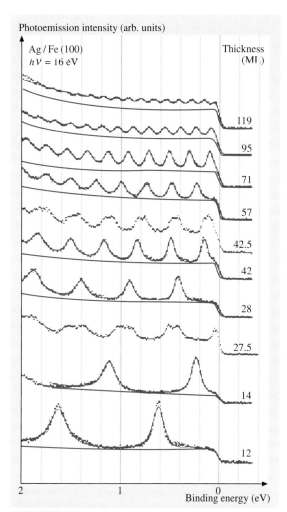

Fig. 5.3-2 *Dots*: experimental normal photoemission spectra for Ag quantum wells deposited on Fe(100), with various thicknesses. *Solid curves*: fits and background functions. (After [3.12])

per level. The electronic shell structure induces similar size dependences in the work function and the electron density, through Friedel-like oscillations [3.10]. The series of "magic numbers" is discernible up to $N = 1430$. For larger particles, the relative abundance of different sizes is controlled instead by the stability and relative sizes of the various facets of the surface of the particle. It is noteworthy that 2-D quantum well states are also predicted to produce oscillatory variations of the energetics of a film as a function of thickness, with similar "magic numbers" for the number of deposited monolayers [3.11].

duction electron is increased compared with the bottom of the conduction band. A similar consideration applies to the lowest energy of a hole in the valence band. Both effects contribute to an increase in the minimum excitation energy of the system compared with the bulk band gap. This blueshift has a comparable magnitude in 2-D and 0-D structures [3.24]. This constitutes the leading contribution to the characteristic blueshifts in the optical spectra of strongly quantum-confined semiconductor systems. However, the confinement of oppositely charged carriers at reduced separations also has dramatic effects on the electron–hole Coulomb energy and thus on exciton formation.

Two regimes of exciton confinement must be distinguished, depending on the confinement size d compared with the Bohr radius a_0^* of the Mott–Wannier exciton in the bulk semiconductor [3.25]. The weak-confinement regime corresponds to sizes such that $d \geq 4a_0^*$. In this regime, the relative electron–hole motion, and in particular its binding energy, is essentially left unchanged. The exciton can still be considered as a quasiparticle, but its center-of-mass translational motion becomes quantized. A simple model [3.26] of a single-point quasiparticle with mass M_{Ex} equal to the total mass of the exciton, $M_{Ex} = m_e^* + m_h^*$, explains the observed frequency shifts. The translational state with the lowest energy, $N = 1$, which is the only optically allowed state, has a nonzero kinetic energy. Compared with an infinite-sized system, the lowest excitation energy in the weak-confinement regime is thus increased by

$$\Delta E_{Weak} = \frac{\pi^2 \hbar^2}{2 M_{Ex} d^2}. \quad (3.2)$$

Like that for electron states, this shift has comparable magnitudes for various dimentionalities. In the weak-confinement regime, the exciton wave function still involves combinations of several conduction-band electron states and valence-band hole states, and not only the lowest-energy states. The quantization of carrier states does not change significantly the average energy of all of the states involved in exciton formation. As a consequence, in the weak-confinement regime, the exciton energy is not affected by the band gap increase which results from the increase in the energy levels of the lowest electron and hole states discussed above. The weak-confinement model is very suitable for describing spectral variations in nanoparticles of semiconductors with low exciton Bohr radii, such as CuCl (Table 5.3-2), as shown in Fig. 5.3-3.

The strong-confinement regime corresponds to the opposite limit, $d \leq 2a_0^*$. In that case the relative electron–hole motion is strongly affected by the barriers. In the direction of restricted motion, the kinetic energies of the lowest-energy electron and hole states determined by the quantum confinement are larger than the Coulomb

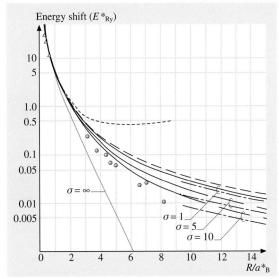

Fig. 5.3-3 Observed values of energy shifts in the luminescence peak of CuCl nanocrystals in NaCl (*circles*) and in the absorption peak of CdS nanocrystals in silicate glass (*triangles*), compared with theoretical models: weak-confinement model (*dashed-dotted lines*), strong confinement model (*short-dashed line*); $\sigma = m_h^*/m_e^*$ is considered as a fixed parameter. (After [3.25])

Table 5.3-2 1s exciton Bohr radius for various semiconductors

Direct-band-gap semiconductors							
Semiconductor	CuCl	Diamond	CdTe	CdS	GaN	CdSe	GaAs
a_B^* (nm)	0.7	0.85	2.8	2.9	3.6	5.6	12
Indirect-band-gap semiconductors							
Semiconductor	GaP	SiC	Si	Ge			
a_B^* (nm)	1.17	2.7	4.9	17.7			

attraction. The exciton states are thus formed from uncorrelated electron and hole states of the "particle in a box" model in the confinement direction, whereas spatially correlated electron hole bound states can still be formed in the unrestricted directions. This yields 2-D, 1-D, and 0-D excitons in quantum wells, quantum wires, and quantum dots, respectively. The binding energy of a 2-D exciton can be deduced from the two-dimensional hydrogen atom problem [3.27]. The exciton Rydberg series changes from $E^*_{3\text{-D}} = -R^*/n^2$ in the 3-D case to $E^*_{3\text{-D}} = -R^*/(n-1/2)^2$ in the 2-D case, where R^* is the Rydberg constant for the exciton. This effect is opposed by the increased energy required for the creation of an electron–hole pair owing to the confinement of the carrier motion. The variation of the excitation energy of the first 1s-like exciton state in the regime of strong 1-D confinement is then

$$\Delta E^{\text{2-D}}_{\text{Strong}} = \frac{\pi^2 \hbar^2}{2\mu d^2} - 3R^* \,, \tag{3.3}$$

where μ is the electron–hole reduced mass, such that $\mu^{-1} = m_\text{e}^{*-1} + m_\text{h}^{*-1}$. The first term is larger than the term due to the weak confinement of exciton motion (3.2) because of the difference between μ and M_{Ex}. The 0-D case is treated in detail in [3.25]. The variation of the minimum excitation energy in the regime of strong 3-D confinement is

$$\Delta E^{\text{0-D}}_{\text{Strong}} = \frac{\pi^2 \hbar^2}{2\mu d^2} - 1.786 \frac{e^2}{4\pi\varepsilon d} - 0.248 R^* \,. \tag{3.4}$$

The divergence as d^{-1} of the second term for small sizes results from the increased Coulomb interaction in this restricted geometry compared with the 2-D case. As could be expected, the case of a 1-D exciton in a quantum wire is intermediate between 0-D and 2-D, with a divergent Coulomb interaction which is sublinear in d^{-1} [3.28]. For all dimensionalities, in the strong-confinement regime, the increased kinetic energy of quantum-confined carriers is dominant, and a blueshift of the optical gap is observed that varies continuously between the weak- and the strong-confinement regimes. However, the considerably increased exciton binding energy in strongly confined systems yields more pronounced exciton effects. For the most widely used semiconductors, excitonic effects are not discernible in room temperature spectra without quantum-confinement. However, in strongly confined geometries of any dimensionality, many semiconductors exhibit well-resolved excitonic peaks since the exciton binding energy can then exceed the thermal excitation energy $k_\text{B}T$. Both the blueshift of the spectra and the increasingly pronounced excitonic absorption are very evident in II–VI semiconductor nanoparticles, as exemplified in Fig. 5.3-4 for CdSe [3.29]. Semiconductor-doped glasses, formed from a dispersion of II–VI semiconductor crystallites in a silicate glass matrix, are the basis for some commercially available yellow-to-red long-wavelength-pass optical filters [3.30]. The glasses with a cutoff wavelength in the visible contain $CdS_{1-x}Se_x$ nanocrystallites, and the glasses used in the infrared filters contain essentially CdTe nanocrystallites. Their abrupt absorption band edge can be tuned by adjusting the composition, the size [3.31], and, as will be discussed later, the concentration of the particles. In other widely used semiconductors, such as Si, Ge, and III–V compounds, the valence bands may have complex features and contain light- and heavy-hole branches. The two corresponding types of excitons will have different characteristics, such as different binding energies.

The second important consequence of the squeezing of excitons in the strong-confinement regime is an increase in the optical matrix element associated with their excitation. A detailed discussion of this effect can be found in [3.32]. Its origin is the stronger overlap of electron and hole wave functions in reduced-dimensional excitons. This effect, associated with the concentration of oscillator strength into a few discrete transitions, has found many applications in optoelectronic devices. Certainly the most important is the fabrication of electrically powered multiple-quantum-well (MQW) lasers in $In_xGa_{1-x}As$ and $Al_xGa_{1-x}As$ (infrared), and $Al_xGa_yIn_{1-x-y}P$ (red). These systems are widely used today in data storage systems, bar code scanners, laser pointers, and printers. The physics and technology of MQW lasers are far beyond the scope of this section. Details may be found in [3.33], for example. 1-D and 0-D systems are also of interests for laser systems [3.34]. Laser applications of quantum-confined systems exploit the enhanced stimulated-emission coefficients that result from the increased transition matrix elements and spectral sharpening in quantum-confined systems. Besides MBE-grown devices, optical gain in chemically synthesized CdSe nanocrystal quantum dots has been achieved [3.35]. The operation of an optically pumped solid-state distributed-feedback (DFB) laser based on a CdSe-doped titania matrix gain medium has been demonstrated [3.36]. The output color of this laser can be selected by choosing appropriately sized nanocrystals and tuning the DFB period accordingly (see Fig. 5.3-5).

Another consequence of the large oscillator strengths associated with confined-exciton transitions is the

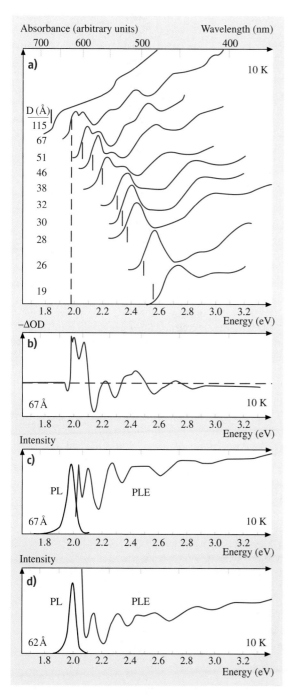

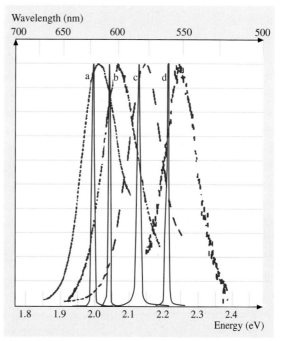

Fig. 5.3-4 (a) Low-temperature (10 K) absorption spectra of CdSe dots dispersed in a polymer matrix, for various mean diameters. (b) Nanosecond saturated-absorption spectra for 67 Å dots. The pump beam had a 7 ns FWHM. The excitation photon energy, 1.984 eV, is prolonged as a *dashed vertical line* from (a) for easier comparison. The *horizontal dashed line* corresponds to $\Delta OD = 0$. (c),(d) Photoluminescence(PL) and photoluminescence-excitation(PLE, observed at 2.022 eV) spectra for 67 Å dots (c) and for 62 Å dots (d). (After [3.29])

Fig. 5.3-5 The photoluminescence emission spectra of CdSe nanocrystals with different radii (*dashed* and *dotted lines*, for radii of 2.7 nm, 2.4 nm, and 2.1 nm) and of 1.7 nm CdSe/ZnS core–shell nanocrystals (the radii are given in order of increasing photon energy) at low pump power, and at a pump power above the laser threshold (*solid lines*). The nanocrystals were dispersed in a titania film used in a distributed-feedback laser configuration at 80 K. (After [3.36])

enhancement of nonlinear optical and electrooptical coefficients. Nonlinear absorption and electroabsorption correspond to changes in absorption coefficients induced by high-intensity optical fields and large DC electric fields, respectively. Exciton–exciton interactions increase the energy of doubly excited states in quantum dots, which results in large saturable absorption effects (see Fig. 5.3-4b). The absorption saturation intensities may be as low as those of saturable dye solutions, but these systems have an increased robustness and the possibility of including them in solid-state media.

PbS-nanocrystal-doped glasses with an absorption saturation intensity of 0.2 MW/cm have been realized [3.37] and used as intracavity passive Q-switches in picosecond [3.37] and femtosecond [3.38] pulsed near-infrared lasers. The Frantz–Keldysh effect is also strongly enhanced in quantum-confined systems [3.39]. Moreover, the well-resolved excitonic transition shows quantum-confined Stark effects (QCSEs) with particularly large magnitudes. In quantum wells, this effect usually consists of a redshift of the absorption and a strong decrease of its oscillator strength, and has been exploited for the realization of fast, low-drive-voltage electrooptic modulators [3.40], which could possibly be integrated with semiconductor lasers [3.41].

5.3.2.4 Vibrational Modes and Electron–Phonon Coupling

The description in terms of phonons of the vibrational properties of solids assumes a translational invariance. It can be used only inside domains of homogeneous mechanical properties with sizes larger than the spatial extension of the phonon considered, typically given by its wavelength λ_{ph}. Similarly to the case of vibrational modes of impurities in crystals [3.43], the confinement of phonons in structures smaller than λ_{ph} leads to the localization and the quantization of the vibrational modes. For typical solids, the phonon frequencies affected by confinement sizes on the 1–100 nm scale have frequencies on the order of 0.1–10 THz, i.e. thermal phonon energies and higher, and both acoustic and optical phonons are affected. The frequencies of the quantized modes, as well as their damping rates, depend on the size, shape, and nature of the environment of the nanostructure and its interfaces. The electron–phonon coupling may be strongly modified by phonon confinement, with manifestations in the transport properties and in the optical absorption and luminescence.

In 2-D systems, two-dimensional phonons appear, with properties and consequences similar to those of surface phonons. The role of in-plane propagating phonons in electron–phonon scattering rates has been demonstrated in metal films [3.44] and in Si/Ge quantum wells [3.45]. Phonon propagation perpendicular to a quantum-well superlattice produces Bragg-like selective transmission [3.46]. The formation of phonon band gaps results from the zone folding of the LA and TA phonons propagating perpendicular to the layers, and is well described by a mesoscopic model of a periodic series of homogeneous layers [3.47].

When we move to 0-D systems, the phonon spectrum changes to a set of size-dependent discrete lines. As in the case of 2-D systems, most data on the various vibrational modes of metal nanoparticles [3.48] and semiconductor nanocrystals (see e.g. [3.49] and references therein) embedded in solid matrices have been obtained by spontaneous-Raman-scattering spectroscopy. The lowest frequencies are well described by the simple Lamb theory [3.50] of the vibration of a homogeneous, elastic sphere. The breathing mode, the mode of lowest energy, is usually the best resolved one. A detailed calculation of the mesoscopic acoustic modes in quantum dots of various compositions embedded in various matrices can be found in [3.51]. The availability of ultrafast pulsed lasers, with a pulse duration shorter than the period of an acoustic vibration mode, offers the possibility to monitor the vibrations in the time domain rather than the frequency domain, as shown in Fig. 5.3-6 for 3 nm PbS quantum dots in a polymer matrix. Similar experiments have been performed on Ag particles of various sizes in a glass matrix [3.52]. These experiments offer a unique opportunity to analyze the damping of these vibrations, which is generally attributed to radiation of acoustic waves into the matrix.

The modes excited by ultrafast laser pulses are usually not the same as those detected by spontaneous-Raman spectroscopy, mainly for symmetry reasons [3.53] (see Fig. 5.3-7). Moreover, the symmetry selection rules depend on the mechanism of excitation by the laser pulse. A thermal mechanism provokes

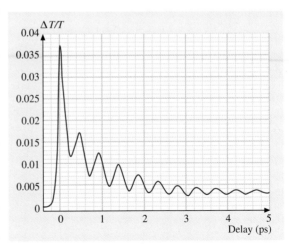

Fig. 5.3-6 Time-domain breathing-mode acoustic oscillations in 3 nm PbS quantum dots embedded in a polymer matrix, observed by transient saturated absorption. (After [3.42])

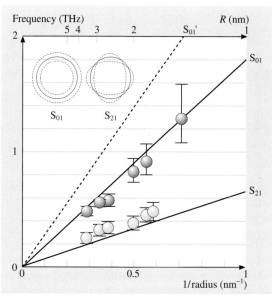

Fig. 5.3-7 Frequency of the coherent-phonon and Raman peaks as function of PbSe dot size. The *solid lines* represent the calculation on the assumption that the stress at the dot boundary is zero, and the *dashed line* represents the calculation with a rigid boundary condition. The matrix was a phosphate glass. The *inset* represents the displacement in the S_{01} and S_{21} spheroidal modes. The coherent-phonon and Raman frequencies are assigned to the S_{01} and S_{21} modes, respectively. (After [3.53])

breathing-mode (S_{01}) oscillations, whereas electrostrictive excitation excites the elongation mode (S_{12}). Figure 5.3-7 also illustrates the parallel dependences of the modes excited by Raman scattering and by ultrafast laser-pulse on particle size. These dependences are consistent with a Lamb model with boundary conditions corresponding to the absence of stress.

The confinement also influences the electron–phonon coupling. For example, in GaAs, each quantized optical-phonon mode produces a Huang–Rhys ladder in the phonon sidebands in the spectrum of the exciton transition [3.55]. A scaling law in d^{-3} is predicted for the effective coupling strength [3.55]. It has been shown that the luminescence spectra of silicon nanocrystals exhibit not only a no-phonon peak but also TO- and TA-phonon-assisted peaks. Hence, besides the breakdown of the momentum conservation rule discussed above, an increased electron–phonon interaction also favors luminescence in this indirect-band-gap semiconductor. This coupling prevails for moderate confinement energies, below 0.7 eV [3.56]. The enhanced electron–phonon coupling results not from phonon quantization but rather from an increased overlap factor [3.57].

In metal particles, the main consequence of changes in the electron–phonon couplings is a variation of the electron relaxation rates. This effect depends strongly on the intensity regime of the excitation, but experiments carried out at weak excitation on Ag and Au nanoparticles in various matrices show a strong increase of the relaxation rate, i.e. a decrease of the electron–lattice energy exchange time, with quantum confinement (see Fig. 5.3-8) [3.54]. In this case also, the increased

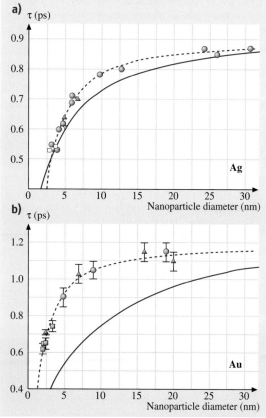

Fig. 5.3-8 (a) Size dependence of the measured electron-lattice energy exchange time τ_{e-ph} for silver clusters in BaO–P$_2$O$_5$ (*circles*), in Al$_2$O$_3$ (*open squares*), in MgF$_2$ (*diamond*), in a polymer (*up triangles*), and deposited on a glass substrate (*down triangle*). The *solid line* is the result of the theoretical model proposed by the authors of [3.54]. The *dashed line* is a guide to the eye. (b) As (a), for gold clusters in Al$_2$O$_3$ (*squares*), in colloidal solutions prepared using a radiolysis technique (*triangles*), and in commercially available colloidal solutions (*circles*). (After [3.54])

electron–phonon coupling is attributed not to phonon confinement but to a reduction of the screening of the electron–ion interaction following electron confinement.

5.3.2.5 Electron Transport Phenomena

The charge transport properties in the direction of free-carrier motion in a restricted-dimensional system have important consequences for the magnetotransport effects. This is treated in Sect. 5.3.4. For disordered systems, when the Mott variable-range hopping mechanism [3.58] dominates the conductivity, the temperature scaling law depends on the dimensionality: the 3-D conductivity $\sigma_{3\text{-D}}$ varies with temperature following the law $\log \sigma_{3\text{-D}} \propto T^{-\frac{1}{4}}$, whereas the 2-D conductivity varies as $\log \sigma_{2\text{-D}} \propto T^{-\frac{1}{3}}$. The transition from 3-D to 2-D behavior has been demonstrated for films of amorphous Ge at a thickness d of 50 nm, as illustrated in Fig. 5.3-9 [3.54].

A vertical conductance, i.e. perpendicular to the free-motion direction, is necessary in quantum wells in many device applications. This requires a finite barrier in order for one to have a continuum of unbound states, in addition to one or a few bound states, as illustrated in Fig. 5.3-10. Practical realizations of such structures are formed from an n-type low-band-gap well, usually GaAs, sandwiched between barrier layers of an intrinsic semiconductor with larger band gap, such as $\text{Al}_x\text{Ga}_{1-x}\text{Al}$. The conduction electron states form a structure of the type shown in Fig. 5.3-10. The choice of the well width and barrier height (through the compos-

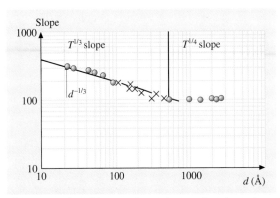

Fig. 5.3-9 Plot of the experimental power-law dependence of the conductivity on temperature for films of amorphous Ge of various thicknesses d. The transition from a 3-D ($T^{-1/4}$) behavior (region marked as $T^{1/4}$ slope) to a 2-D ($T^{-1/3}$) behavior (region marked as $T^{1/3}$ slope), as expected for the variable-range hopping mechanism, is clearly observed $d = 500$ Å (*vertical bar*). (After [3.54])

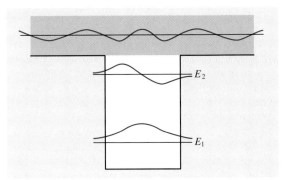

Fig. 5.3-10 Schematic view of the electron states in a finite-barrier quantum well. States with energies below the barrier are bound to the well, and states with energies above the barrier are extended outside the well and form a continuum of conduction states

ition x of the barrier) permits one to adjust the number and energy of bound states, and the n-type doping of the well controls their population.

An important example of a device based on such a structure is the quantum-well infrared photodetector (QWIP) [3.59]. The energy difference $E_1 - E_2$ between the subbands matches the energy of the photons to be detected. E_2 is close to the top of the barrier. Excitation of the intersubband transition by absorption of an IR photon results in the population of the second subband. If an electric field is applied, the excited electron can tunnel through the small remaining barrier towards the continuum and move away from the well under the influence of the electric field. This gives rise to a photocurrent, which can either constitute the signal of the detector or feed a visible-light-emitting junction in order to achieve IR-to-visible conversion. To increase the collection efficiency in practical devices, several quantum wells are associated in a periodic stack. For a large enough bias, once an electron has reached the continuum it is accelerated by the electric field and has a high probability of crossing subsequent wells without being trapped in a bound state.

In applications such as IR detection, the conductivity is obtained by populating the continuum. In that case, the interwell spacing is large enough to prevent direct coupling between neighboring wells, and each well can be considered as isolated. Such systems are usually called multiple quantum wells.

A nonzero conductivity in the ground state can be obtained by reducing the width of the barriers, as originally proposed in [3.60]. Then, the subband states of neighboring wells interact through the evanescent parts of their wave functions in the barrier. The ensemble of bound states consisting of the subband levels from all of the coupled quantum wells forms a band in the confinement direction called a miniband. The periodic set of coupled quantum wells is called a superlattice. At very low applied voltages, a band-type conduction governs the electron transport. However, above a threshold voltage, the conduction mechanism becomes highly complicated. This results in highly nonlinear behaviors, with the appearance of negative differential resistance and oscillations at low temperatures [3.61]. These effects are attributed to the localization of the electron states when the voltage drop between neighboring wells induces an energy difference larger than the coupling energy between them. At even higher fields, the voltage drop spontaneously localizes on one single well [3.61]. A particularly spectacular manifestation of this effect has been observed in weakly coupled wells at low temperature, as shown in Fig. 5.3-11 [3.62].

Electrical conduction is also very important for many devices that exploit the huge area of surface or interface per unit volume in zero-dimensional nanostructured materials such as nanoporous materials, granular materials, nanocomposites, and nanoparticle assemblies. Examples of such devices are chemiresistor-type sensors, solar cells, light-emitting diodes, and energy-storage cells. From the point of view of electron transport, these materials can often be modeled as a dense assembly of metal or semiconductor quantum dots dispersed in an insulating matrix. As in noncrystalline materials [3.63], the elementary mechanism of conduction is phonon-assisted electron tunneling from one dot to another, or hopping. Above a threshold temperature, the limiting factor for the hopping probability is the spatial proximity of two dots. Below this threshold, the weak thermal energy available imposes a strong selection of sites with a low energy difference, tunneling to such sites at larger distances becomes dominant, and the variable-range hopping regime is obtained [3.58]. The temperature dependence of the conductance σ changes from $-\log(\sigma) \propto T^{-\alpha}$ with $1/2 < \alpha < 1$ for nearest-neighbor hopping to $-\log(\sigma) \propto T^{-1/4}$ for variable-range hopping [3.64]. However, in contrast to noncrystalline semiconductors, the sites in such granular materials are spatially expanded to the particle size, and the tunnel gap between neighboring particles is usually much smaller than this size. As a consequence, the threshold temperature is very low and a T^α law with $\alpha \approx 1/2$ is observed over a wide temperature range.

Another important factor comes into play in nanoparticle-based conductors compared with 2-D well systems. The addition of an electron to an initially electrically neutral nanoparticle requires an energy $e^2/2C$, where C is the capacitance between the nanoparticle

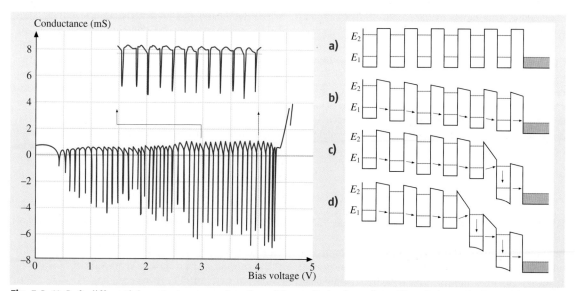

Fig. 5.3-11 *Left*: differential conductance as a function of the applied voltage for a 49-period superlattice at 20 K. *Right*: schematic model to explain the 48 negative peaks in the differential conductance. (**a**) Zero bias; (**b**) ground-state resonant-tunneling conduction; (**c**) first field localization, where resonant tunneling between the ground state and an adjacent excited state takes place; (**d**) expansion of the high-field region by one additional quantum well. (After [3.62])

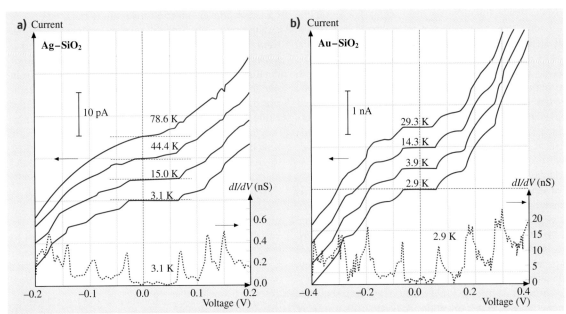

Fig. 5.3-12a,b Temperature dependence of I/V characteristics of films of (**a**) Ag nanoparticles dispersed in SiO_2 (sizes 3.8 ± 1.0 nm, volume fraction 1.5%, film thickness 12 nm) and (**b**) Au nanoparticles dispersed in SiO_2 (sizes 2.9 nm, film thickness 9 nm). In both cases the electrode area was $100 \times 100\ \mu m^2$. (After [3.68])

and its surroundings. Typical values of C can be in the region of 10^{-17} F or less, so that the charging energy approaches or is larger than the thermal energy at room temperature. This produces the Coulomb blockade effect: conduction through a single nanoparticle requires a minimum voltage to overcome the cost in charging energy. The conduction may even show a stepwise dependence on the bias, called a Coulomb staircase, which has been observed for a single nanoparticle at room temperature [3.65]. Extension to three-terminal devices has led to the realization of mesoscopic devices such as single-electron transistors, where the charge state of a confined electron pool is controlled by a gate voltage [3.66]. The influence of the single-electron charging energy on the conductivity of a mesoscopic material made of an assembly of dots has not yet been completely elucidated. It is predicted to reinforce the T^α law with $\alpha \approx 1/2$ for the conductivity [3.67]. However, the most prominent experimental effect is the appearance of remnants of the single-particle blockade effect when the conducting path involves only a few particles, which occurs mainly for the perpendicular conductivity in very thin films with a low particle volume fraction (see Fig. 5.3-12) [3.68, 69]. There are some indications that a regular ordering also favors Coulomb blockade effects [3.70].

5.3.3 Electromagnetic Confinement

5.3.3.1 Nanoparticle-Doped Materials

The polarizability $\alpha(\omega)$ of a spherical inclusion of volume V with a dielectric constant $\varepsilon(\omega)$ at an optical frequency ω, in a matrix with a dielectric constant $\varepsilon_M(\omega)$, is given by

$$\alpha = 9\varepsilon_0 V \frac{\varepsilon - \varepsilon_M}{\varepsilon + 2\varepsilon_M}, \qquad (3.5)$$

where the explicit frequency dependence of α, ε, and ε_M has been dropped in the notation. When the medium is subjected to a propagating optical field, the polarizability of the particle will produce both absorption and elastic scattering of light. The corresponding cross sections, σ_A and σ_S, respectively, are

$$\sigma_A = \frac{18\pi V}{\lambda_M} \frac{\varepsilon_0 \varepsilon''}{(\varepsilon' + 2\varepsilon_M)^2 + \varepsilon''^2} \qquad (3.6)$$

and

$$\sigma_S = \frac{6\varepsilon_0^2 V^2}{\varepsilon_M^2 \lambda_0^3 \lambda_M} \frac{(\varepsilon' - \varepsilon_M)^2 + \varepsilon''^2}{(\varepsilon' + 2\varepsilon_M)^2 + \varepsilon''^2}, \quad (3.7)$$

where λ_0 and λ_M are the wavelengths of the light in vacuum and in the matrix, respectively, and $\varepsilon = \varepsilon' + i\varepsilon''$, with ε' and ε'' real. The dependence of σ_A and σ_S on the volume shows that for nanoparticles with diameters much smaller than the optical wavelength, absorption effects, when present, are preponderant over scattering.

Low-Concentration Systems

For semiconductor particles, the absorption spectrum is changed only slightly compared with the bulk. The scattering losses, present even in the nonabsorbing region, have been studied in detail, in view of their waveguiding applications and account has been taken of the size dependence of the refractive index [3.71]. Some results are reproduced in Fig. 5.3-13.

In contrast, metals often show negative values of ε'. In that case both the absorption and the scattering show resonances when $\varepsilon' + 2\varepsilon_M = 0$, which corresponds to the dipolar plasmon oscillation mode, and a peak appears in the absorption spectrum; this is observed especially clearly with alkali and noble metals (Fig. 5.3-14). For most metals, the plasmon peak shifts towards longer wavelengths and becomes sharper for matrices with a larger refractive index. This is clearly observed for the alkali metals in Fig. 5.3-14; for a computation of

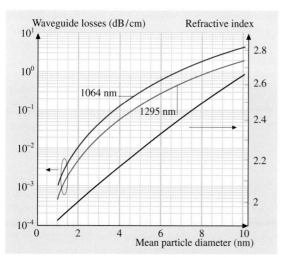

Fig. 5.3-13 Scattering losses of a 5% PbS nanoparticle-doped silica–titania waveguide as a function of the nanoparticle diameter at wavelengths of 1064 nm and 1295 nm (After [3.71])

matrix-dependent spectra for various metals, see [3.72]. Higher-multipolar-order plasmon modes exist [3.73] but cannot be excited optically in the quasi-static regime. However, they have a nonnegligible influence on the optical absorption spectra of larger particles, as shown in Fig. 5.3-15, and can be excited by, for example, electron beams. From the Mie theory, the frequency of the Lth-order plasmon mode of a small particle of a Drude-model metal, with $\varepsilon(\omega) = \varepsilon_0(1 - \omega_p^2/\omega^2)$, where ω_p is the plasma frequency, is given by [3.73]

$$\omega_L = \sqrt{\frac{\omega_p}{1 + (\varepsilon_M/\varepsilon_0)(1 + 1/L)}}. \quad (3.8)$$

A continuous tuning of the positions of the plasmon absorption peak by alloy formation has been demonstrated. The dependence of the absorption spectra of Au–Ag alloy nanoparticles on the composition is reproduced in Fig. 5.3-16 [3.74].

Local-Field Effects

In the quasi-static approximation, the electric field inside a nanoparticle is changed compared with the field outside by a local-field factor f, given by

$$f(\omega) = \frac{3\varepsilon_M(\omega)}{\varepsilon(\omega) + 2\varepsilon_M(\omega)}. \quad (3.9)$$

This factor is taken into account in (3.5)–(3.8) above, but it can have an even greater importance in nonlinear effects, since the second-order and third-order nonlinear optical coefficients, $\chi^{(2)}$ and $\chi^{(3)}$, respectively, are affected by factors f^2 and f^3, respectively, as compared with the bulk material of the nanoparticle. Hence, for large f, a nanostructured material can have a larger optical nonlinearity than its bulk constituents. For typical semiconductor-doped matrices, $\varepsilon > \varepsilon_M$ and $f < 1$. However, particularly strong local-field enhancements are observed for metal nanoparticles in the vicinity of the plasmon resonance [3.75].

Electromagnetic Concentration Effects

For larger particle concentrations, the interactions between particles influence the electromagnetic properties. For interparticle distances much smaller than the wavelength, the Maxwell–Garnett model applies and leads to the Lorentz–Lorenz relation for the effective dielectric constant ε_{eff} of the composite medium, which takes the form

$$\varepsilon_{\text{eff}} = \varepsilon_M \frac{\varepsilon(\omega)(1 + 2p) + 2\varepsilon_0(1 - p)}{\varepsilon(\omega)(1 - p) + \varepsilon_0(2 + p)}, \quad (3.10)$$

where p is the volume fraction. This usually leads to a redshift of absorption peaks. In particular, the frequency of the dipole ($L = 1$) plasmon mode of a particle

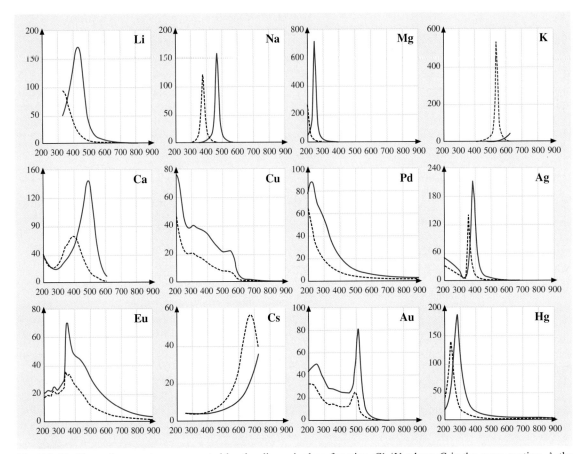

Fig. 5.3-14 Absorption spectra, represented by the dimensionless function $C\lambda/V$ where C is the cross section, λ the wavelength, and V the particle volume, of 10 nm diameter spherical particles of selected metallic elements, as calculated from their frequency-dependent dielectric constant, in vacuum (*dashed lines*) and in a matrix with a refractive index equal to that of water, 1.33 (*solid lines*). Horizontal scale: λ (nm). (After [3.77])

Fig. 5.3-15 Absorption spectra of spherical particles of various materials, as calculated from their frequency-dependent dielectric constant, in vacuum for various particle radii. (After [3.72]) ▶

Fig. 5.3-16 Effect of formation of a gold–silver alloy on the surface plasmon absorption: measured UV–VIS absorption spectra of spherical Au–Ag alloy nanoparticles of various compositions. The gold mole fraction x_{Au} varies between 1 and 0.27. The plasmon absorption maximum is blueshifted with decreasing x_{Au}. (After [3.74])

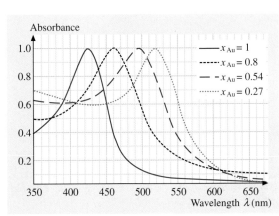

in the Drude model is $\omega_{L=1} = \sqrt{(1-p)/3}\,\omega_p$ instead of $\omega_p/\sqrt{3}$, the value in the low-concentration limit ($p \approx 0$).

This effect has found applications in sensing, in particular the selective colorimetric detection of nucleotides based on aggregation of gold nanoparticles [3.76]. The electromagnetic concentration effect is also used,

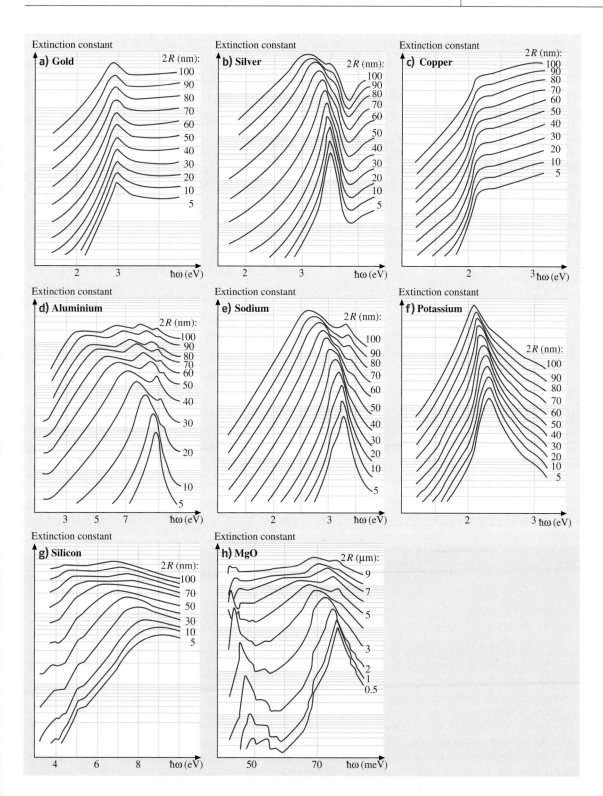

in addition to quantum size effects (Sect. 5.3.2), to tune the absorption band edge of semiconductor-doped glasses.

5.3.3.2 Periodic Electromagnetic Lattices

Coupled Plasmon Modes

The coupling between the plasmon modes of several closely packed nanoparticles yields new electromagnetic eigenmodes and spectroscopic properties. The electromagnetic eigenmodes of some few-particle systems, such as Ag and Au pairs and triplets with various geometries, have been computed, for instance in [3.72, Sect. 2.3.4], and have been observed directly by e-beam excitation [3.78]. For the lowest eigenmodes, a large field enhancement is observed in the small gap between the particles.

Coupled plasmon modes have been observed experimentally in regularly spaced linear chains of gold nanoparticles [3.79] and in 2-D hexagonal arrays of silver nanoparticles [3.80]. The controlled propagation of plasmon excitations in tailored metal nanoparticle structures is a new domain of application, sometimes called "plasmonics" [3.81].

Opals and Photonic-Band-Gap Materials

3-D ordered arrays of nanostructures with periods of the order of a fraction of the optical wavelength may show intense Bragg diffraction for specific wavelengths and diffraction angles. This phenomenon is the origin of the iridescent colors (opalescence) of opals. Opals consist of an fcc-like array of silica nanoparticles with sizes in the range 150–900 nm [3.82], with a size dispersion below 5% [3.83].

For nanoscale objects with large scattering strengths and particular geometries, spectral intervals where no propagating mode is allowed in any direction appear, thus forming "photonic band gaps". The idea of exploiting this phenomenon in optoelectronic devices has been suggested [3.84, 85]. However, up to now, this phenomenon has been realized only in the infrared and microwave regions. Application in the visible, which would require 3-D nanostructured materials, is still under active development.

5.3.4 Magnetic Nanostructures

The driving force for the research on magnetic nanostructures is the possibility of their incorporation into future generations of information storage devices. This could lead to a real technological breakthrough, and a huge economic impact is foreseen. As always, the key question is whether any potential benefit of such technology will be worth the production costs. Overall, the current effort in this research field is considerable. The study of the magnetism of nanoscale objects is characterized by the closeness between fundamental research and applications. What is more, a number of magnetic devices are currently in commercial use without a full understanding of the basic magnetic phenomena underlying them. An instructive example is the following: spin electronic phenomena have been applied rapidly since the first observation of giant magnetoresistance (GMR) was reported in 1988, and magnetic sensors and read heads based on this effect have been available since 1994 and 1997, respectively.

Most digital information is stored in the form of tiny magnetized regions or "bits" within thin magnetic layers on disks. The size of a magnetic bit determines the information storage capacity of a magnetic disk drive. The magnetic storage media used in today's commercial hard disks consist of homogeneous polycrystalline magnetic films. Owing to continuous improvements in both the magnetic properties of the media and the read/write heads, the storage density of hard disk drives has increased considerably during recent years, as shown in Fig. 5.3-17. Since 1990, the storage capacity of disk drives has increased at a 60%

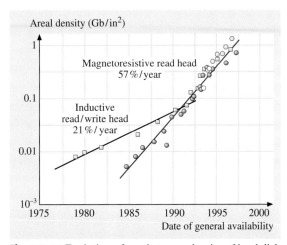

Fig. 5.3-17 Evolution of areal storage density of hard disk drives (data from IBM, Almaden)

annual compound growth rate. The changes of the head technology first from thin-film heads to magnetoresistive heads and then to read heads based on the GMR effect [3.86, 87] have enabled enormous increases in volumetric storage capacity.

The overall gain in areal density has been achieved by scaling the complete disk drive system. The medium has to accommodate smaller bit dimensions and smaller transition widths between written bits. However, the optimization also includes more sensitive read/write heads, improving the signal processing and coding, and improving the head–disk tribology. A good discussion of these aspects can be found in [3.88].

Progress in this field would not be possible without the maturity of growth techniques such as electron- or ion-beam lithography, and molecular-beam epitaxy and the ability to manufacture devices on a nanometric scale by nanoimprinting lithography. Another essential point has been the addition of magnetic contrast to near-field microscopy and electron microscopy (magnetic-force and Lorentz microscopy, respectively). These new tools permit us to establish a link between magnetism and structure on the atomic scale.

We have selected here some examples to illustrate the state of the art in nanometric nanostructures, keeping as the main idea the impact of the underlying technological breakthrough. We focus our attention on two very active fields: magnetoresistive nanometric multilayers for spin electronics, and magnetic dots for ultrahigh-density storage media.

5.3.4.1 Spin Electronics

Conventional electronic devices are based on charge transport, and their performance is limited by the speed of the carriers (electrons) and the dissipation of their energy. Conventional electronics has ignored the spin of the electron. Nevertheless, in every wire or device, approximately 50% of the conducting electrons are generally spin-up and the remainder spin-down (where "up" or "down" relate to some locally induced quantization axis). *Spin electronics*, or *spintronics*, is a new field of electronics which relies on the different transport properties of majority-spin and minority-spin electrons. *Mott* postulated the basis of spin electronics in the mid-1930s: he attributed certain anomalies in the electrical transport behavior of metallic ferromagnets to the fact that the spin-up and spin-down conduction electrons are two *independent* families of charge carriers, each with its own distinct transport properties [3.89].

The Concepts of Spin Accumulation and Spin Diffusion Length

The other necessary ingredient of this model is that the two families contribute very differently to the electrical transport processes. This may be because the number densities of the carrier types are different, or it may because they have different mobilities. In either case, the asymmetry which makes spin-up electrons behave differently from spin-down ones arises because the ferromagnetic exchange field causes a splitting between the spin-up and spin-down conduction bands, leaving different band structures evident at the Fermi surface. In nonmagnetic materials, the two spin channels are equivalent because they have the same density of states (DOS) at the Fermi energy. In a magnetic material, the *spin polarization* is defined as the ratio of the difference between the populations in the up and down spin channels to the total number of carriers at the Fermi level,

$$SP = \frac{N_\uparrow - N_\downarrow}{N_\uparrow + N_\downarrow} \,. \qquad (3.11)$$

Clearly, if there is a nonzero spin polarization, the numbers of electrons participating in the conduction process are different for the two spin channels. More subtly, this implies that the susceptibilities to scattering of the two spin types are different, and this in turn leads to different mobilities [3.90].

Let us consider two spin channels of different mobility. When an electric field is applied to the metal, there is a shift in momentum space Δk of the spin-up and spin-down Fermi surfaces in accordance with the equation

$$F = eE = \hbar \frac{dk}{dt} = \hbar \frac{\Delta k}{\tau} \,, \qquad (3.12)$$

where F is the force on the carrier, E is the electric field, e is the electronic charge, and τ is the electron scattering time. Since the channels have different mobilities, this shift is different for the spin-up and spin-down Fermi surfaces, as illustrated in Fig. 5.3-18.

As a consequence, if a current is passed from such a spin-asymmetric material, for example cobalt, into a paramagnetic material such as silver (which has no spin asymmetry between spin channels), there is a net influx into the silver of up-spins over down-spins. Thus, a surplus of up-spins appears in the silver, and with it a small associated magnetic moment per unit volume. This surplus is known as a *spin accumulation*. Evidently, for a constant current flow, the spin accumulation cannot increase indefinitely. The spin-up electrons injected across

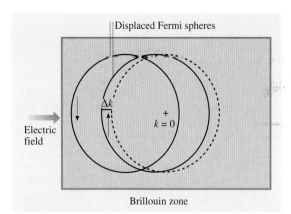

Fig. 5.3-18 Schematic illustration of the shift in the Fermi surface when an electric field is applied to a ferromagnet. The *solid circles* represent the Fermi spheres of up- and down-spin electrons in the field, and the *dashed circle* represents the Fermi sphere in zero external field. (After [3.90])

the cobalt–silver interface are converted into spin-down electrons by spin-flip process, which we have hitherto ignored. So now we have a dynamic equilibrium between an influx of up-spins and their destruction by spin-flipping. This in turn defines a characteristic length scale that describes how far the spin accumulation extends into the silver. In the present description, we have assumed that both the spin-up and spin-down electrons are present in the ferromagnetic material in equal numbers but that their mobilities are different. Nevertheless, we can produce spin accumulation either as a direct consequence of an asymmetric density of states or as an indirect consequence via asymmetry in the electron mobility.

It follows from the above discussion that the spin accumulation decays exponentially away from the interface, on a length scale called the *spin diffusion length* (λ_{sd}). This concept is crucial for realizing a link between spin electronics and magnetic nanostructures. In this section, we shall not go further into this concept, but we shall do a rough calculation to see how large λ_{sd} is and on what parameters it depends. We use the following expression from [3.90]:

$$\lambda_{sd} = \sqrt{\frac{\lambda v_F \tau_{\uparrow\downarrow}}{3}}, \qquad (3.13)$$

where v_F is the Fermi velocity, $\tau_{\uparrow\downarrow}$ is the spin-flip time, and λ is the mean free path. This relationship highlights the critical role of the impurity concentration in determining the spin diffusion length. If the impurity level is increased in the nonmagnetic material, λ_{sd} drops because λ is shortened. This relation also allows us to estimate values for the spin diffusion length. Taking silver as an example, λ_{sd} can vary between several microns for very pure silver to the order of 10 nm for silver with 1% gold impurity. In a realistic material, λ_{sd} is of the order of 10–100 nm. This is the reason why the experimental observation of spin-dependent transport phenomena has been possible when researchers have controlled the growth of nanometric multilayers. It is obvious that the layer thickness has to be small on the scale of the spin diffusion length.

The Discovery of Giant Magnetoresistance

The era of spin electronics began with the discovery of a magnetoresistance effect in a magnetic multilayer consisting of a sequence of thin magnetic layers separated by equally thin nonmagnetic metallic layers: the electric current is strongly influenced by the relative orientation of the magnetizations of the magnetic layers. This experimental observation was made almost simultaneously by two research groups: Albert Fert's group in Orsay [3.86] and Peter Grünberg's group in Jülich [3.87, 91].

In fact, it was found that the resistance of the overall multilayer is low when the magnetizations of all the magnetic layers are parallel (ferromagnetic coupling), but it becomes much higher when the magnetizations of neighboring magnetic layers are antiparallel (antiferromagnetic coupling).

Figure 5.3-19 shows the magnetoresistance curves at 4.2 K of Fe/Cr (magnetic/nonmagnetic) multilayers with thicknesses of the individual layers of the order of 1 nm (from [3.86]). In these multilayers, for certain thicknesses of the Cr interlayer, the magnetizations of adjacent Fe layers are oriented antiparallel by an antiferromagnetic interlayer exchange coupling [3.87]. When a magnetic field is applied, the resistance of the multilayer decreases drastically as the magnetizations of the two layers progressively align in the direction of the field. The GMR ratio is usually defined as

$$\mathrm{GMR} = \frac{R^{\mathrm{AP}} - R^{\mathrm{P}}}{R^{\mathrm{P}}}, \qquad (3.14)$$

where R^{P} and R^{AP} are the resistances in the parallel and the antiparallel state, respectively. In the case of Fe/Cr multilayers, the GMR is 80% at liquid-helium temperature and 20% at room temperature when the thickness of the Cr is 9 Å.

The relative change of the resistance can be larger than 200% [3.92], and that is the reason why the effect is called "giant" magnetoresistance. GMR effects

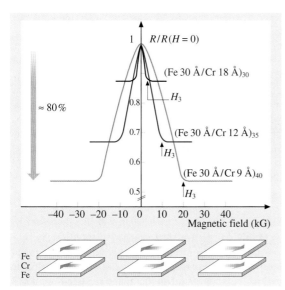

Fig. 5.3-19 Magnetoresistance curves at 4.2 K of Fe/Cr multilayers. (After [3.86])

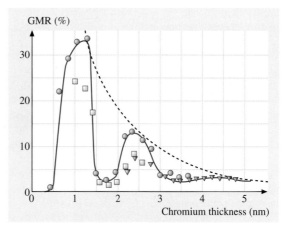

Fig. 5.3-20 Dependence of the GMR ratio of an Fe/Cr multilayer on Cr thickness. (After [3.94])

have been reported in a large number of systems combining ferromagnetic transition metals or alloys with nonmagnetic metals (for a review, see [3.93]). The most commonly used combinations of magnetic and non-magnetic layers are Co/Cu and Fe/Cr, but multilayers based on permalloy as the magnetic component are also frequently used.

The second key ingredient was discovered by *Parkin* and *Mauri* [3.94]. These authors discovered that the orientation of the magnetic moments of two neighboring magnetic layers depends on the thickness of the intervening nonmagnetic layer. In fact, the orientation of the magnetic moments oscillates between parallel and antiparallel as a function of the thickness of the nonmagnetic layer [3.96]. This phenomenon is referred to as an oscillatory exchange coupling. Figure 5.3-20 shows the results of the original experiment of Parkin and Mauri. Oscillations of the GMR as a function of the Cr thickness occur because the magnetoresistive effect is only measurable for those thicknesses of Cr for which the interlayer coupling aligns the magnetic moments of all the Fe layers so that they are antiparallel.

GMR effects have been obtained in two geometries. In the first one, the current is applied in the plane of the layer (hereafter denoted by CIP), as in the experiments for which results are shown in Figs. 5.3-19 and 5.3-20, while in the second one, the current flows perpendicular to the plane of the layers (hereafter denoted by CPP). The first measurements in the CPP configuration were obtained by sandwiching the multilayer between two superconducting Nb layers [3.97]. A CPP-GMR configuration has also been obtained in nanowires, namely multilayers electrodeposited in the pores of a nuclear-track-etched polycarbonate membrane [3.98], and by oblique deposition on a prestructured substrate [3.99]. The CPP-GMR effect is definitely larger than the CIP-GMR effect and occurs at much larger thicknesses.

The mechanism of the GMR effect is illustrated in Fig. 5.3-21 for $\alpha = (\rho_\downarrow / \rho_\uparrow) > 1$, where $\rho_\downarrow$ and $\rho_\uparrow$ are

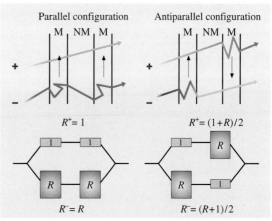

Fig. 5.3-21 Schematic picture of the GMR mechanism. The electron trajectories between two scatterings are represented by *straight lines* and the scatterings by abrupt changes in the direction. The *signs* + and − are for spins $S_z = 1/2$ and $S_z = -1/2$, respectively. The *arrows* represent the majority-spin direction in the magnetic layers. M = magnetic, NM = nonmagnetic. (After [3.95])

the resistivities for spins parallel and antiparallel, respectively, to the magnetization direction; $\alpha > 1$ means that the resistivity is smaller for the majority spins. In the parallel configuration, the spin↑ (up) electrons are always the majority-spin electrons and are always weakly scattered in all layers, resulting in a resistance r for this channel that is smaller than the resistance R for the spin⁻ (down) channel. The shorting of the current by this fast electron channel makes the resistivity low in the parallel state ($R^P = r$). In the antiparallel configuration, each of the spin directions is alternately the majority and the minority one. The resistance is averaged for each channel, and the overall resistance $R^{AP} = (r + R)/4$ is larger than in the parallel state. The GMR ratio is then

$$\text{GMR} = \frac{R^{AP} - R^P}{R^P} = \frac{(r-R)^2}{4rR}. \quad (3.15)$$

This picture holds in both the CIP and the CPP geometries.

An antiparallel configuration can also be obtained in multilayers in which consecutive magnetic layers have different coercivities [3.101], or by combining hard and soft magnetic layers.

Spin Valves

The best known structure for obtaining an antiparallel arrangement is the *spin valve* structure, originally proposed by *Dieny* et al. [3.100] (Fig. 5.3-22). A spin valve has two ferromagnetic layers (alloys of Ni, Fe, and Co) sandwiching a thin nonmagnetic metal (usually Cu), with one of the two magnetic layers being "pinned", i.e. the magnetization in this layer is relatively insensitive to moderate magnetic fields. The other magnetic layer is called the "free layer" and its magnetization can be changed by application of a relatively small magnetic field.

As the alignment of the magnetizations in the two layers changes from parallel to antiparallel, the resistance of the spin valve typically rises by 5–10%. The goal is to obtain the result that the magnetization of the free layer reverses in a very small field (a few oersteds), while the magnetization of the pinned layer remains

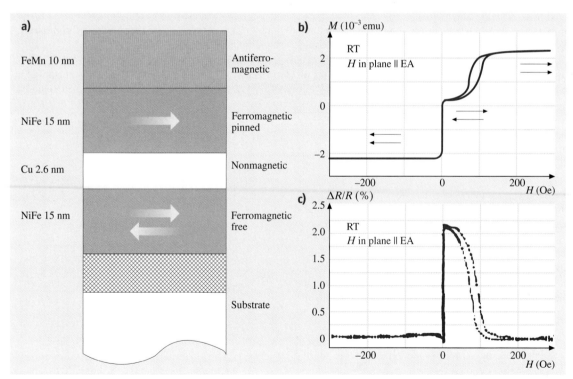

Fig. 5.3-22 (a) Schematic illustration of the spin valve multilayer originally proposed by Dieny et al. (15 nm NiFe/2.6 nm Cu/15 nm NiFe/10 nm FeMn). (b) Magnetization curve, and (c) relative change in resistance. The magnetic field is applied parallel to the exchange anisotropy (EA) field created by the FeMn layer. The current flows perpendicular to this direction. (After [3.100])

fixed. Pinning is usually accomplished by using an antiferromagnetic layer, such as FeMn, that is in close contact with the pinned magnetic layer. The resultant spin valve response is given by $\Delta R \propto \cos(\theta_1 - \theta_2)$, where θ_1 and θ_2 are the angles of the magnetization directions respectively of the free layer and of the pinned layer with respect to the direction parallel to the plane of the magnetization of the medium whose magnetization is being sensed, as shown in Fig. 5.3-23.

The discovery of the GMR effect has created great expectations, since this effect has important applications, particularly in magnetic information storage technology. The use of spin valve multilayers in hard-disk read heads was first proposed by IBM in 1994 [3.102]. The principle of operation of a magnetic read head will be detailed in the section on 'Applications' below. Because the GMR effect is so important for industrial applications, there have been many improvements in recent years. The simplest type of pinned layer has been replaced by a synthetic antiferromagnet (two magnetic layers separated by a very thin (1 nm) nonmagnetic conductor, usually ruthenium) [3.94]. The magnetizations in the two magnetic layers are strongly coupled so as to make them antiparallel, and are thus effectively immune to outside magnetic fields. The second innovation is the nano-oxide layer (NOL), which is formed on the outside surface of the soft magnetic layer. This NOL reduces the resistance due to surface scattering, thus reducing the background resistance and thereby increasing the percentage change in the magnetoresistance of the structure [3.103].

Magnetic Tunnel Junctions

Tunneling belongs to a class of electron transport phenomena known as *quantum transport*, "quantum" because they cannot be explained unless the wave nature of electrons is invoked. In the quantum mechanical picture, an electron is associated with a wave function ψ related to the probability Ψ that the electron can be found in a volume dv: by $\Psi = \psi\psi^* \, dv$. Quantum theory predicts that a particle has a nonzero probability of penetrating through a classically forbidden region separating two classically allowed regions of configuration space, by the process of *tunneling*.

The concept of tunneling has found applications in a vast number of domains in modern physics and chemistry; the particular case of interest here is tuneling in heterostructures consisting of metallic and insulator films, leading to innovative electronic devices. In a metal–insulator–metal (M–I–M) heterostructure, the potential barrier arises from the microscopic interactions between the metals and the insulator layer. By means of tunneling, an electron with an energy lower than the potential barrier can be transferred from one metal to the other, under the condition that the thickness of the potential barrier is not too large compared with the electron's wavelength (Fig. 5.3-24).

For a long period, up to the 1970s, the spin of the electron was neglected in tunnel transport. In 1971, *Tedrow* and *Meservey* [3.104] conducted tunneling measurements on junctions between a very thin superconducting Al film and a ferromagnetic Ni film, through an Al_2O_3 barrier, in a high magnetic field. They showed

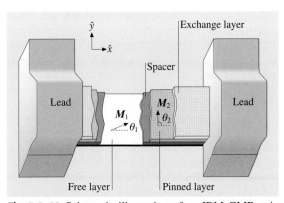

Fig. 5.3-23 Schematic illustration of an IBM GMR spin valve sensor used in read heads. M_1 and M_2 are magnetizations, and θ_1 and θ_2 are angles from the longitudinal direction. The read head moves perpendicular to the xy plane. (After [3.102])

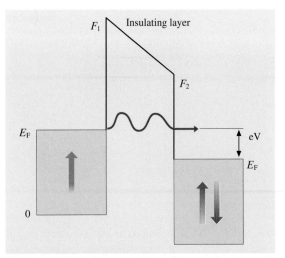

Fig. 5.3-24 One-electron energy diagram for a metal–insulator–metal tunnel junction

that the tunneling current was spin-dependent. The spin splitting of the quasiparticle density of states of the superconductor caused by the application of the magnetic field was used to analyze the spin polarization of the tunneling current for Ni. The same method was used by the same authors to measure the spin polarization of electrons tunneling from a variety of ferromagnetic films: Fe, Co, Ni, and Gd [3.105]. A positive spin polarization was obtained in all cases; +44% was found for Fe, +34% for Co, +11% for Ni, and +4.3% for Gd. These results demonstrated that the tunneling process preserves the character of the electronic spin and inaugurated the exciting research field of spin-dependent tunnel transport. The large amount of progress in recent years in this particular domain has led to the construction of innovative spintronic devices, at the heart of which we find the magnetic tunnel junction (MTJ).

An MTJ behaves as a spin valve. However, it is a hybrid system comprising at least two ferromagnetic metals of different coercivities $H_{C,1}$ and $H_{C,2}$ ($H_{C,1} < H_{C,2}$), separated by a *thin insulator* (Fig. 5.3-25a) instead of a nonmagnetic metallic spacer as in classic spin valves. In Fig. 5.3-25b, the magnetization loop of a magnetic tunnel junction is schematically shown. In such a device, the current flowing perpendicular to the film plane (CPP geometry) depends on the relative orientation of the magnetizations M_1 and M_2 of the two ferromagnetic electrodes of the junction, reaching extreme values for parallel and antiparallel alignment (we assume that the magnetization lies in the plane of the ferromagnetic layers). For an applied magnetic field parallel to the plane of the films with a value $H \geq |H_{C,2}|$, the magnetizations of the two electrodes are aligned parallel to each other, giving rise to a low-resistance state (R^P) for the junction (Fig. 5.3-25c). In the magnetic-field window $|H_{C,1}| < H < |H_{C,2}|$, the electrodes' magnetizations are aligned antiparallel to each other, corresponding to a high-resistance state (R^{AP}) for the junction.

High tunnel magnetoresistance (TMR) at room temperature was discovered in magnetic tunnel junctions in the mid-1990s in Fe/Al$_2$O$_3$/Fe junctions [3.107], Fe/Al$_2$O$_3$/Ni$_{1-x}$Fe$_x$ junctions [3.108], and CoFe/α-Al$_2$O$_3$/Co and NiFe [3.109] junctions, although the first measurements at low temperature were performed in 1975 [3.110]. The tunneling resistance is modulated by the magnetic field in the same way as in a spin valve, but the advantage over a "classical" spin valve is that it has been shown to exhibit a significantly higher GMR ratio (50% or more), while requiring a saturation magnetic field equal to or somewhat less than that required for a spin valve. It was proposed that these MTJs would have potential uses as low-power field sensors and memory elements. Because the tunneling current density is usually small, MTJ devices tend to have high resistances. The resistance of an MTJ depends exponentially on the thickness of the insulating layer. Uniformity over the insulating layer is crucial to device operation. In addition, reproducibility from MTJ element to MTJ element is important for the proper working of arrays of such devices.

An example of such an effect is shown in Fig. 5.3-26, for a Co/Al$_2$O$_3$/NiFe junction. When the magnetizations of the Co and NiFe layers go from an antiparallel

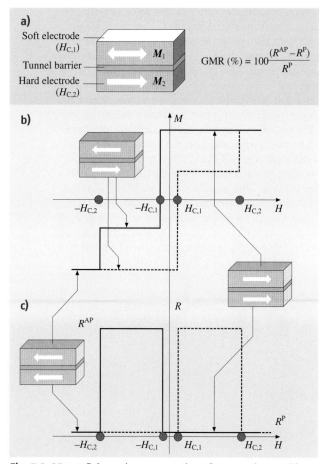

Fig. 5.3-25a–c Schematic representation of a magnetic tunnel junction (**a**) and its magnetization loop (**b**). The resistance of the MTJ is determined by the relative orientation of the magnetizations of the ferromagnetic electrodes of the junction, giving rise to low- and high-resistance states for parallel and antiparallel alignments, respectively (**c**). (After [3.106])

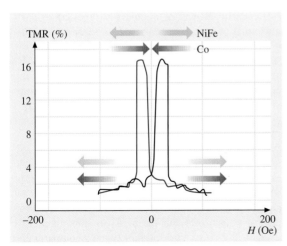

Fig. 5.3-26 Tunnel magnetoresistance curve at room temperature of a Co/AL$_2$O$_3$/NiFe junction. (After [3.111])

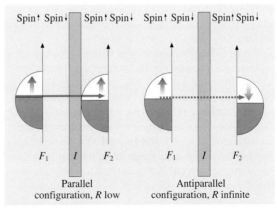

Fig. 5.3-27 Schematic picture of the TMR effect in the case of two half-metallic electrodes. The transmission of the electrons from one layer to the other is only possible in the parallel state, resulting in a 100% TMR effect. (Courtesy of A. Bartélémy)

to a parallel arrangement, the resistance of the junction decreases. In this example, the effect is 16% at room temperature [3.111]. Most of the measurements reported in the literature have been performed using an alumina (oxidized aluminium) barrier.

In Jullière's model [3.110], the TMR ratio is expressed as a function of the spin polarizations SP$_1$ and SP$_2$ of the two magnetic electrodes by

$$\text{GMR} = \frac{R^{\text{AP}} - R^{\text{P}}}{R^{\text{P}}} = \frac{2\text{SP}_1\text{SP}_2}{1 + \text{SP}_1\text{SP}_2},$$

$$\text{where} \quad \text{SP}_i = \frac{N_\uparrow(E_\text{F}) - N_\downarrow(E_\text{F})}{N_\uparrow(E_\text{F}) + N_\downarrow(E_\text{F})}. \quad (3.16)$$

A structure analogous to the spin valve described above may be made by making the two metallic electrodes from a half-metallic ferromagnet (HMF, SP = 100%) and separating them with a thin layer of insulator. Now, if the magnetizations of the electrodes are opposite, no current can flow across the junction, since the electrons which might tunnel have no density of final states on the far side to receive them. However, if the electrode magnetizations are parallel, a tunneling current may flow as usual. This can be explained simply by the schematic picture in Fig. 5.3-27, which represents the case of two half-metals (with only one spin direction at the Fermi level) separated by a thin insulating barrier.

We thus have a spin-electronic switch whose operation mirrors that of a pair of crossed optical polarizers, and which may be switched "on" and "off" by the application of an external magnetic field. If the electrodes are not ideal HMFs, then the on/off conductance ratio

is finite and reflects the majority and minority densities of states of the ferromagnet concerned. In the case of an ideal HMF, in the antiparallel configuration, transmission from one electrode to the other is not allowed, and the resistance of the junction is infinite. In contrast, in the parallel state, transmission occurs, and the resistance is finite. This gives, for half-metallic electrodes, a spin polarization equal to 1 and a maximum TMR effect of 100%. In the case of a transition metal, the effect is more complex to analyze owing to the presence of s and d bands at the Fermi level with majority and minority spin directions. For transition metals, the TMR ratio is limited to 40% at liquid-helium temperature and 29% at room temperature. This is a demonstration of how the properties of metallic systems are controlled exclusively by the "mafia" of electrons at or very near the Fermi surface, whose band-structure properties the metal reflects.

The spin tunnel junctions described above depend for their operation only on the density of states, and not on the carrier mobility. Unlike fully metallic systems, they have a lower conductance per unit area of device, and hence larger signal voltages (of the order of millivolts or more) are realizable for practical values of the operating current. Moreover, the device characteristics, such as the size of the "on" resistance, the current density, the operating voltage, and the total current, may be tuned by varying with device cross section, the barrier height, and the barrier width. This is just one reason why spin tunnel junctions are very promising candidates for the spin injector stages of future spin-electronic devices. They will also be the basis of the next generation of MRAMs, as illustrated later in this section.

Large TMR effects have been obtained using HMFs as electrodes, as can be seen in Fig. 5.3-28 in the case of a junction with $La_{0.7}Sr_{0.3}MnO_3$ (LSMO) electrodes and a $SrTiO_3$ (STO) barrier [3.112, 113]. The spin polarization deduced from this measurement is 83% if one assumes the same polarization for the two magnetic electrodes. With a polarization of 100%, half-metallic compounds would allow true on/off operations, and would be appropriate for gates of nonvolatile logic devices.

Recently, *Bowen* et al. [3.114] reported a TMR ratio of more than 1800% at $T = 4.2$ K in LSMO/$SrTiO_3$/LSMO fully epitaxial MTJs, from which they deduce a spin polarization of least 95% for the LSMO layer. This large TMR value arises both from preserving the quality of the LSMO/STO interfaces during the upgraded patterning process and from the use of junctions of small size ($5.6 \times 5.6\,\mu m^2$) (Fig. 5.3-29). The TMR extends to temperatures of about 280 K, an improvement compared with previous results in the literature [3.115].

The temperature dependence of the TMR is plotted in Fig. 5.3-30a,b for two junctions using $R(H)$ loops measured at $V_{dc} = 10$ mV. The TMR decreases rather quickly with increasing T, but vanishes only at temperatures of about 280 K for the $2 \times 6\,\mu m^2$ junction. TMR ratios of 30% (Fig. 5.3-30c) and 12% are obtained at 250 and 270 K, respectively. This represents a sizable improvement with respect to previous results [3.112, 113] obtained from heterostructures grown under identi-

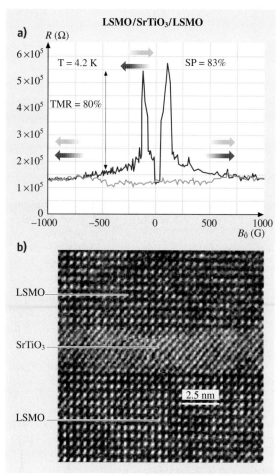

Fig. 5.3-28 (a) TMR curve of an LSMO/STO/LSMO junction at 4.2 K. The spin polarization of the LSMO layer is 83%. After [3.112, 113]. (b) High-resolution transmission electron microscopy (HRTEM) image of an LSMO/STO/LSMO junction (courtesy of J. L. Maurice)

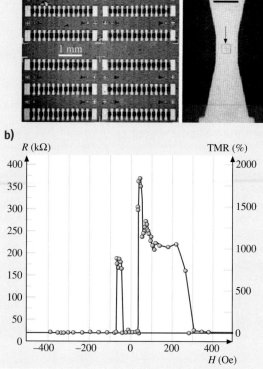

Fig. 5.3-29 (a) Optical images of a processed sample of small-size LSMO/STO/LSMO magnetic tunnel junctions. (b) $R(H)$ loop for a $5.6 \times 5.6\,\mu m^2$ junction measured at 4.2 K. (After [3.114])

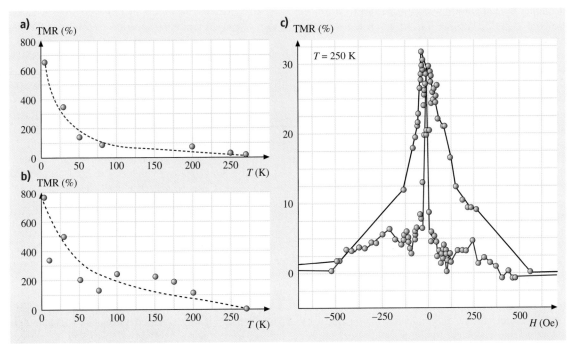

Fig. 5.3-30a–c Temperature dependence of the TMR measured with $V_{dc} = 510$ mV for two junctions: $2 \times 6\,\mu m^2$ (**a**) and $1.4 \times 4.2\,\mu m^2$ (**b**) (the *dashed lines* are guides to the eye). (**c**) $R(H)$ loop at $T = 250$ K and $V_{dc} = 10$ mV, showing 30% TMR. (After [3.114])

cal conditions. Since the results shown in Fig. 5.3-30 were obtained from LSMO/STO/LSMO junctions with a fully strained crystallographic structure, these measurements prove that strain is not a limiting factor in the potential production of manganite-based MTJs with sizable TMR values at relatively high temperatures. High-quality interfaces that limit the disruption of the properties of the manganite are a more important requirement.

This example illustrates fairly well that progress in spin electronics would not be possible without the improvement in the manufacturing of devices on the nanometric scale.

Another promising HMF is Fe_3O_4, because this oxide exhibits the highest Curie temperature ($T_c = 858$ K). Consequently, one can expect that its HMF character will remain significant at room temperature. Figure 5.3-31 shows a cross-sectional high-

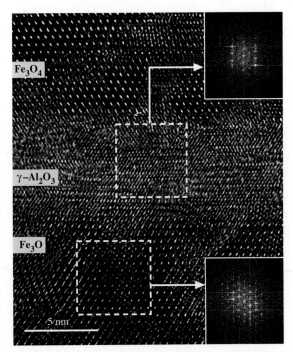

Fig. 5.3-31 HRTEM image of a fully epitaxial Fe_3O_4/Al_2O_3/Fe_3O_4 MTJ grown by MBE. The Fourier transform of the barrier region of the micrograph shows the high crystalline quality of this region. (Courtesy of P. Bayle Guillemaud and P. Warin)

resolution transmission electron micrograph of a fully epitaxial $Fe_3O_4/Al_2O_3/Fe_3O_4$ MJT, grown by MBE assisted by atomic oxygen [3.116]. The interest in this result is that the growth of the interfaces was performed without exposing the surface layers to air, limiting the disruption of the bulk properties. The Fourier transform of the barrier region of the micrograph shows the high crystalline quality of this region. The use of crystalline barriers in MTJs opens the way to investigation of tunnel transport in periodic barriers, which will permit a better understanding of the underlying fundamental physics.

The record for magnetoresistance values has recently been obtained by *Van Dijken* et al. [3.117]. These authors report a large magnetic-field sensitivity of the collector current in a three-terminal magnetic tunnel transistor device with spin-valve metallic base layers. Giant magnetocurrents exceeding 3400% result from strong spin-dependent filtering of electrons traversing perpendicular to the spin-valve layers at energies well above the Fermi energy. With its giant magnetocurrent and reasonable output current, the magnetic tunnel transistor is a promising candidate for future magnetoelectronic devices.

Applications

Magnetic Read Heads and Sensors. Information is stored on a magnetic disk in small magnetic domains arranged in concentric tracks. The conventional read head used to be an induction coil, which sensed the rate of change of the magnetic field as the disk rotated. The signal and hence the density of magnetized bits were thus limited by the speed of rotation of the disk. Magnetoresistive sensors do not suffer from this defect, since they sense the strength of the field rather than its rate of change. They are, therefore, capable of reading disks with a much higher density of magnetic bits. Magnetoresistive read heads are commercially available and will be the leading technology beyond the year 2004.

When two oppositely magnetized domains meet, a domain wall appears, with a characteristic width of 100–1000 Å, depending on the magnetic properties of the material. Although there is no magnetic field emanating from the uniformly magnetized domains themselves, uncompensated magnetic poles at the domain walls emanate stray fields that extend out of the medium. Magnetic read heads contain a GMR or MTJ sensor element, which is sensitive to these domain wall stray fields. A schematic illustration of a GMR read head is shown in Fig. 5.3-32. The element is fabricated in such a way that the magnetization of the free layer is paral-

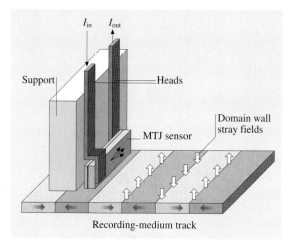

Fig. 5.3-32 Schematic illustration of an MR read head. A spin valve ("all-metal" or MTJ) is used as an ultra-sensitive sensor for the very low magnetic field strength generated by the small magnetic domains (bits). When a weak magnetic field, such as from a bit on a hard disk, passes beneath the read head, the magnetic orientation of the free layer changes its direction relative to the pinned layer, generating a change in electrical resistance due to the GMR effect. (After [3.106])

lel to the plane of the medium. Every time the magnetic head passes over a domain wall, the magnetic field emanating from the wall pushes the magnetization of the free layer upwards or downwards, resulting in a change in the resistance of the MTJ device (the magnetization of the pinned layer is fixed). The design goal is to obtain the maximum rate of change in the resistance for a change in the magnetic field.

The concept used in read heads is also applied in another type of magnetic sensors, with which we can sense the operation of a piece of machinery when a magnetic field is applied, as illustrated in Fig. 5.3-33. A gear made of ferrous metal can be arranged so that it perturbs the magnetic fringing field originating from a permanent magnet close to the rotating gear and the sensor MTJ element. The MTJ element is sensitive to the changing orientation of the magnetic flux as gear teeth pass by the magnet. Thus, the readout resistance is related to the angular position of the rotating gear. Such an arrangement is used to monitor the engine speed in an automobile.

Magnetic Random Access Memories (MRAMs). Anisotropic-magnetoresistance elements were the first to be used for MRAM applications, approximately ten years ago. However, they provided a very small mag-

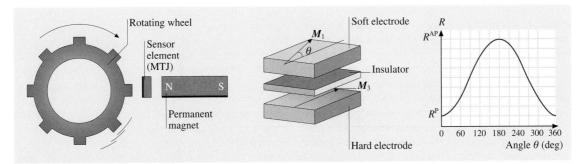

Fig. 5.3-33 Principle of operation of a magnetic sensor using an MTJ element

netoresistive effect of about 2%, making their use disadvantageous in comparison with semiconductor memories in terms of read access time (> 250 ns). The density and cost of this kind of MRAMs were also not competitive with standard semiconductor memories. A major breakthrough in MRAM technology came with the use of GMR multilayers. The magnetoresistive signal is much higher – rising up to 18% – but the resistance of the all-metal GMR element is too small compared with the complementary metal–oxide–semiconductor (CMOS) transistor serving as the diode in an integrated structure. For this reason, a number of GMR cells have to be connected in series with the CMOS transistor, thereby decreasing the effective magnetoresistive signal. This is a serious drawback in the design of a fast-access memory. MTJs are the best candidates for MRAMs, as they provide a high magnetoresistive effect of 20–50% depending on the spin polarization of the magnetic electrodes, combined with a modifiable resistance, which depends on the thickness of the insulator and its degree of oxidation. This permits the utilisation of a single MTJ cell in series with a CMOS transistor (Fig. 5.3-34). This high-density architecture is suitable for fast-memory applications. The idea of the MRAM is to use the magnetization direction of the free magnetic layer for information storage, and the resulting resistance for readout. The resistance of the memory bit is either low or high, depending on the relative orientation (parallel or antiparallel) of the magnetization of the free layer with respect to the pinned layer. An applied magnetic field can switch the free layer between the two states. In an MRAM array, orthogonal lines pass under and over each memory bit, carrying the current which produces the switching magnetic field. The idea is that the current in one line is not sufficient to switch the memory cell; the bit will switch only if current is flowing through both of the lines that cross at the selected

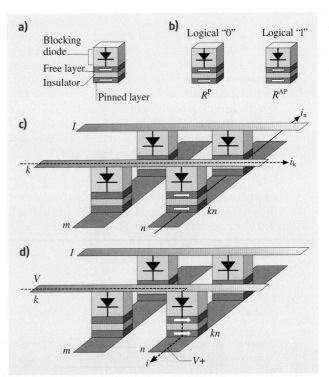

Fig. 5.3-34 (a) Schematic representation of an MRAM cell containing a single MTJ and a blocking diode (CMOS transistor), allowing current flow in only one direction, in series. (b) Representation of a logical "0" by the low-resistance state R^P and of a logical "1" by the high-resistance state R^{AP}. (c) The process of writing information on bit kn situated at the crossing of word lines k and n. The existence of the two currents i_n and i_k is necessary to switch the magnetization of the free layer. For example, in the memory cell km, which is influenced by a single current i_k, no information is stored. (d) The process of reading information from cell kn. A bias is applied between the word lines k and n and the current through the bit cell is measured and compared with a reference value. The determination of the magnetic state of the junction, which can then be related to logical "0" or "1", is thus possible

bit. The readout of the stored information in the selected bit is done by applying a bias between the lines that cross it and measuring the current that flows through it.

For the successful realization of MRAM devices using MTJs, two very critical components – the tunnel barrier and the free layer – have to be optimized. The tunnel barrier needs to show large-scale uniformity over the wafer, in addition to the more or less obvious prerequisites of being pinhole-free and extremely smooth. This is because the tunnel resistance depends exponentially on the barrier thickness and the absolute value of the MTJ resistance is compared with a reference cell during the read operation. As far as the free layer is concerned, its thickness is directly related to the value of the magnetic field or current required for switching a bit. A free layer less than 5 nm thick is desirable, since it results in a low switching field and, as a consequence, low power consumption.

A number of difficulties have to be resolved to create successful devices. These include efficient spin injection into semiconductors and heterostructures, and a search for new spin-polarized materials. Other effects potentially important for spintronic devices include optical and electrical manipulation of ferromagnetism, current-induced switching and precessing of magnetization, and the possibility of a long coherence time for optically excited spins in semiconductors. For a good overview of the issues in spin electronics, including the prospects for spintronic quantum devices, see [3.118].

5.3.4.2 Ultrahigh-Density Storage Media in Hard Disk Drives

The Limits of the Present Technology

In the current storage technology, which uses continuous polycrystalline thin-film media, the signal-to-noise ratio (SNR) of the medium is directly related to the number of grains per bit. At storage densities on the order of $80\,\text{Gb/in}^2$, the grain size of the CoCrPtX alloys (X = Ta, P, etc.) used in the current media is on the order of 8 nm, the number of grains per bit is about 100, and the bit size is on the order of $250\,\text{nm} \times 30\,\text{nm}$. A further reduction in the number of grains per bit below 100 would lead to a strong decrease in the SNR. To continue decreasing the bit cell size in continuous media (while keeping the SNR constant), it is necessary to reduce the grain size further. However, there is a physical limit to how small the grain size can be before thermal variations can affect the stability of the magnetization. This is known as the superparamagnetic limit. In CoCrPt alloys, the minimum grain size for which the criterion for magnetic stability can be satisfied is about 8 nm, which is the grain size used in today's media. Further reducing the grain size requires the use of other materials as FePt, SmCo, and FeNdB alloys. However, this implies that larger write fields will be needed to switch the magnetization of the grains. This would necessitate the development of new materials which are not yet known. In the short term, the use of antiferromagnetically coupled media has allowed the superparamagnetic limit to be extended in combination with GMR read heads [3.119].

The Promise of Perpendicular Media

It is foreseen that within two or three years, magnetic recording will switch to perpendicular discontinuous media for several reasons. In perpendicular media, as opposed to longitudinal granular media, the smaller the bit size, the smaller the effect of the demagnetizing field on each bit is. In addition, the efficiency of the writer can be greatly improved, and higher-anisotropy materials can be used. Both effects contribute to a better thermal stability at small size grain size as compared with longitudinal media. This should help us to reach densities of the order of $200-400\,\text{Gb/in}^2$. However, to progress beyond these densities, a totally different approach will be required, which may consist of using patterned media, i.e. media made of arrays of nanometer-scale individual dots, in which each dot carries a bit of information. The significant advantage of this type of medium is that the shape and edge of each bit are not determined by the grain size as in continuous media, but are determined by the patterning process used to make the dots. Consequently, each dot can be made of a single magnetic domain (1 grain per bit instead of the 100 grains per bit in a continuous medium). The volume of a grain can therefore be much larger in a patterned medium so that the superparamagnetic limit is no longer a problem, at least up to several terabits per square inch. Of course, the use of patterned media raises other questions such as fabrication at low cost, tracking, and reproducibility.

Various techniques have been developed to prepare arrays of submicrometer magnetic dots: optical, electron-beam, or X-ray lithography followed by etching of the magnetic material; patterning using a focused beam [3.120]; and self-organization of nanoparticles [3.121]. So far, however, none of these techniques seems to be able to combine the criteria of low cost, and rapid patterning of large areas (several square inches) down to 25 nm pitch.

Nanoimprinting. Nanoimprinting is a promising method of nanopatterning proposed recently for solving the above-mentioned drawbacks. Nanoimprint lithography is a major breakthrough in nanopatterning because it can produce sub-10 nm feature sizes over a large area with a high throughput and low cost, a feat impossible with current, conventional lithography.

Basically, the process of nanoimprinting starts by first making a mold by electron-beam lithography. This may take a long time, but this mold can be subsequently used a large number of times to imprint layers of resists on almost any kind of flat substrate. The patterned layer of resist is next used as an etching mask to transfer the pattern to the substrate by direct reactive ion etching (RIE) (Fig. 5.3-35a). The original approach to preparing arrays of magnetic dots starts from prepatterned wafers, on top of which a ferromagnetic material is deposited. In that case the nanopatterning of the magnetic material results directly from its deposition on a prepatterned wafer (Fig. 5.3-35b).

Moritz and coworkers have studied the coupling between the dots obtained by deposition of Co/Pt multilayers on prepatterned Si substrates [3.122, 123]. Two patterning process were compared: (a) electron-beam lithography followed by direct RIE [3.124], and (b) nanoimprint lithography followed by lift-off and dry RIE of silicon [3.125, 126]. Subsequent deposition of a magnetic material (Co–Pt) leads to arrays of single-domain dots with a perpendicular magnetization.

By nanoimprinting, a dot size down to 30 nm in diameter with a pitch of 60 nm has been obtained. This represents a density of 200 Gb/in^2. After the nanoimprinting of the Si substrate, a magnetic material was deposited on the prepatterned wafer by sputtering or by some other relatively directional physical deposition process. The magnetic material then covered the top of the dots, the bottom of the trenches separating the dots, and, to a lesser extent, the sidewalls of the dots [3.122].

Corresponding atomic and magnetic images are shown in Fig. 5.3-36.

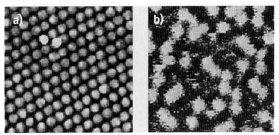

Fig. 5.3-36 Structural (**a**) and magnetic (**b**) images of an array of single-domain dots with a 60 nm pitch prepared by nanoimprinting and coating with a (Pt/Co) multilayer. Images obtained by atomic and magnetic force microscopy

Influence of the Angle of Deposition. In order for one to be able to use these arrays of magnetic dots as recording media, the dots should be weakly coupled so that the magnetization of one dot can be switched without affecting the magnetization of the neighboring dots. In the present example, the sputtered species

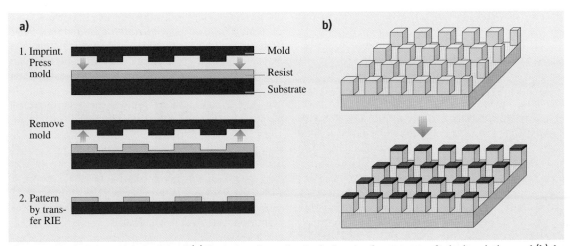

Fig. 5.3-35a,b Schematic principle of (**a**) the successive technological steps of pattern transfer by imprinting, and (**b**) the deposition of a magnetic material on a prepatterned substrate. The magnetic deposit covers the top of the dots, the bottom of the trenches, and, to a lesser extent, the sidewalls of the dots

(Co and Pt) were directed onto the substrate at either normal or oblique (37° from the normal) incidence (Fig. 5.3-37) [3.123].

When the substrate consists of an array of silicon pillars, the structural and magnetic properties obtained with the two deposition modes are quite different, as illustrated in Figs. 5.3-37 and 5.3-38 [3.123]. With normal deposition, a rough deposit made of Co, Pt, and CoPt alloy grains is observed (Fig. 5.3-37a), which induces a ferromagnetic coupling between adjacent dots (Fig. 5.3-38a) via the sidewalls and the magnetic deposit in the bottom of the trenches [3.122].

In contrast, when the Pt is deposited at oblique incidence, one sidewall of the dots is coated with antiferromagnetic CoO, whereas the other side consists of a Pt-rich PtCo alloy, which is nonmagnetic (Fig. 5.3-37b). In this latter case, no direct coupling between the top of the dots and the bottom of the trenches takes place, resulting on no direct ferromagnetic coupling between adjacent dots. A checkerboard-like pattern of single-domain dots with a perpendicular magnetization is observed (Fig. 5.3-38b) [3.122, 123, 126]. This second situation is favorable if one wishes to use this type of array of dots as a discrete recording medium.

Storing the Information on the Sidewalls of the Dots.
So far, the magnetic deposit on the top of the dots carries the recorded information. An alternative method would be to deposit the material on the sidewalls of the dots. This can be achieved by performing a codeposition of Cr at normal incidence and Co at oblique incidence. Thereby, a nonmagnetic Cr-rich CoCr alloy forms on top of the dots, whereas the sidewalls are coated with almost pure magnetic Co. The information can be stored by setting the magnetization of these "vertical nanomagnets" to be up or down on the sidewalls.

Figure 5.3-39 compares the magnetic contrast measured by magnetic force microscopy (MFM) in two arrays of dots of the same geometry, one coated on the top of the dots with a (Pt/Co) multilayer, the other coated on the sidewalls with Co. Clearly the stray fields are much more localized and intense in the second case. It should be pointed out that the sidewalls are not single-domain here, but this can be fixed by using materials with higher coercivity ($Co_{80}Cr_{20}$ alloys for instance). Another advantage of this approach is that two opposite sidewalls can be used per dot, which allows one to double the areal density for the same geometry.

Manipulating the Magnetization of Individual Dots.
Two techniques have been developed to manipulate the magnetization of each dot individually. The first one is a thermomagnetic technique, which consists of us-

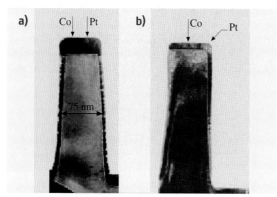

Fig. 5.3-37a,b Cross-sectional TEM views of dots prepared by electron-beam lithography followed by Si RIE and covering by (Pt/Co) multilayers. In (**a**), both Co and Pt were deposited at normal incidence. In (**b**), Co was deposited at normal incidence, whereas Pt was deposited at oblique incidence. The width of the dots in both cases is 75 nm. (From [3.123])

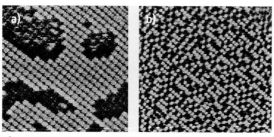

Fig. 5.3-38a,b 8 μm MFM images of domain patterns in arrays of (Pt/Co) dots deposited at normal incidence (**a**), and oblique incidence (**b**). (Courtesy of B. Dieny)

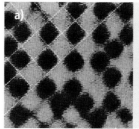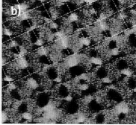

Fig. 5.3-39a,b MFM domain patterns in as-deposited arrays of 400×400 nm dots (**a**) coated on the top by a (Pt/Co) multilayer and (**b**) coated on the sidewall by 5 nm thick Co deposit. (Courtesy of B. Dieny)

ing a conductive atomic force microscope (AFM) tip to locally heat a given dot by passing a pulse of current between the tip and the dot [3.127]. The temperature rise leads to a decrease in the switching field of the magnetization of the addressed dot. Simultaneously, a static field of -200 Oe is applied to the whole sample. This field is large enough to switch the magnetization of the heated dot in the desired direction but low enough not to switch the magnetization of the other dots [3.123]. This is illustrated in Fig. 5.3-40. These experiments are preliminary, but impressive and highly encouraging. The main difficulty was irreproducibility in the contact resistance.

A second technique has also been developed to manipulate the magnetization of individual dots; this technique is based on the use of write heads developed by Headway Inc. The array of dots is mounted on a piezoelectric stage, which allows displacement of the sample over an area of $120\,\mu\mathrm{m} \times 120\,\mu\mathrm{m}$. The head is in contact with the surface of the sample. A write current of 30 mA was used in the writer. An example of controlled writing of a line of dots is shown in Fig. 5.3-41 [3.123].

Fig. 5.3-40 MFM images of an array of dots covered by a multilayer of composition Pt 20 nm/(Co 0.5 nm/Pt 1.8 nm). The dots are 400 nm in size, with an edge-to-edge spacing of 100 nm. A controlled switching of individual dots has been achieved by passing a pulse of current from a metallic AFM tip to the dot and simultaneously applying a uniform field of -200 Oe over the whole sample. (From [3.123])

Fig. 5.3-41 $4\,\mu\mathrm{m} \times 4\,\mu\mathrm{m}$ MFM image of a line written on an array of rectangular dots $100\,\mathrm{nm} \times 200\,\mathrm{nm}$ in size with an edge-to-edge spacing of 100 nm. The writing was performed using a commercial inductive write head provided by Headway Inc. (From [3.123])

5.3.5 Preparation Techniques

Various strategies have been developed for the manufacture of nanostructured materials. They follow essentially two approaches:

- The *top-down* manufacturing paradigm is that of the existing electronic technologies. It consists in downscaling the pattern size in lithographically structured materials.
- The *bottom-up* paradigm of nanosciences consists in the fabrication of atomically precise structures of increasing size, up to the mesoscopic scale.

Many preparation techniques are very specific to one particular type of material: for instance, many different chemical and physical techniques are used to produce porous silicon. Hence, we list and briefly describe below only the most common and generic methods for the synthesis and fabrication of nanostructured materials. For a more detailed, recent, comprehensive description and discussion of the following production techniques, see [3.5, 128].

5.3.5.1 Molecular-Beam Epitaxy

2-D Structures

An atomic or molecular beam of Ga, As, Al, and/or other elements is created thermally in a ultrahigh-vacuum chamber and directed at a specific, well-defined, atomically flat facet of a single-crystal substrate. A very low deposition rate, associated with a precisely controlled substrate temperature and various real-time techniques for monitoring the state of the surface, permit one to find growth conditions that preserve a coherent crystalline order or minimize the density of structural defects. Because of the atomic layer-by-layer deposition, MBE is the technique that allows the best control over the thickness of a 2-D film down to the subatomic level. This has motivated a huge number of research studies on the mechanism of MBE growth. MBE is used when the highest thickness accuracy, abrupt interfaces, the highest purity, and/or the lowest density of surface defects is needed. Its main applications are for semiconductor quantum wells, since the band gap of $Al_xGa_{1-x}As$ can be tuned continuously through the composition x of the ternary alloy without inducing too large a lattice mismatch (see e.g. [3.32]).

1-D Structures

Quantum wires are less commonly encountered, since their fabrication procedures are even more complicated: one method exploits the step-flow growth of vicinal surfaces [3.129].

Self-assembly in specific MBE-type situations leads to the formation of wire structures. Thus, straight atomic lines of Si have been grown on a β-SiC(100) surface [3.130].

0-D Structures

MBE was primarily designed for the growth of 2-D nanostructures. However, it has been observed that the combination of two lattice-mismatched materials can lead to the formation of self-assembled size-monodispersed clusters or dots during growth in the Stranski–Krastanov (SK) mode. The most studied system is that of InAs deposited on a GaAs(100) substrate [3.131]. In that case, the lattice mismatch is 7%, and the nanocluster size is $20\,\text{nm} \pm 10\%$. The average nanocluster size and density are tunable over a narrow range through the choice of growth parameters. Moreover, various cluster shapes are obtained with different substrate faces. More complicated systems have been produced, such as 2-D core–shell magnetic dots [3.132]. Finally, the use of a high-index substrate (such as GaAs(311)) can lead to a lateral ordering of the nanocluster distribution [3.133].

Clusters of various metals can be formed by use of a dewetting (Vollmer–Weber) type of epitaxial growth; this has been done mainly on highly oriented pyrolytic graphite (HOPG). Formation of 3-D islands has been reported for Cu, Ag, Au [3.134], Pt [3.135], Co [3.136], Mo, [3.137], and Pd [3.138].

In some specific situations, a 3-D stacking of quantum dot layers has been achieved [3.139].

5.3.5.2 Metal-Organic Chemical Vapor Deposition (MOCVD)

This type of growth takes place in a reaction chamber in which the substrate is heated to a very high temperature (500 °C to 1000 °C). A combination of metal-organic and metal hydride vapors (e.g. triethylgallium, $Ga(C_2H_5)_3$, and arsine, AsH_3, for GaAs formation) is then brought into contact with the substrate by use of a laminar flow of a carrier gas (typically N_2 or H_2). The desired atoms are deposited layer by layer through a chemical reaction taking place at the substrate surface. MOCVD can be adapted to large-scale, long, continuous runs. Today, it is thus the most conventional industrial manufacturing technique for semiconductor heterostructures.

5.3.5.3 Lithography

The lateral patterning in large-scale integrated electronic circuits is produced by projection photolithography. The resolution is limited not by the resists themselves but by optical diffraction during exposure, the minimum line spacing being of the order of λ/NA, where λ is the wavelength and NA the numerical aperture of the projection system. Today, the deep-UV technology is used; this is based on ArF excimer laser sources ($\lambda = 193\,\text{nm}$), the numerical aperture is of the order of 0.8, and image enhancement techniques such as phase-shift masks are used [3.140]. This technique achieves typical line spacings close to 100 nm. To achieve higher resolution, alternative radiation sources and/or exposure techniques have been used and are summarized in Table 5.3-3.

5.3.5.4 Nanocrystals in Matrices

Diffusion-Controlled Growth

The growth of clusters from ions dissolved in a glass can be kinetically controlled by tuning the temperature between the glass transition temperature and the melting temperature. Empirical methods for the growth of metal clusters for use in stained-glass windows have been known for centuries. This technique is now mainly used for the fabrication of color cutoff filters using II–VI compounds. The main compound used is CdS_xSe_{1-x}, and such filters are made by Schott in Germany, Corning in the USA, and Hoya in Japan. It is also used for photochromic glasses based on I–VII compounds (mainly AgBr).

Table 5.3-3 Radiation sources and exposure techniques for higher resolution

Radiation source	Energy
Deep UV	6.4 eV
X-rays	2 keV
Electrons	20 keV
Ions	100 keV

Exposure techniques	Comments
Contact printing	The resolution is the limited by the optical skin depth of the mask material.
Proximity printing	The resolution is limited by diffraction, with a minimum line spacing of $\sqrt{g\lambda}$, where g is the mask–surface distance. Proximity printing techniques have been used mainly in conjunction with X-ray radiation sources.
Nanoimprinting	A master stamp is used directly for hot embossing of a substrate, usually made of a polymer. This allows the stamp to produce many patterned substrates. Although there is no diffraction limit here, hot embossing has been applied mainly for low-resolution applications, in various fields such as miniaturized total-analysis Systems (µTAS), microfluidics, and microoptics. UV–nanoimprint lithography corresponds to contact printing with UV exposure.

Scanned-source techniques	
Electron and ion beams	Scanning electron beam lithography is the most widely used technique for nanostructuring surfaces beyond the optical resolution limit. The resist used is polymethylmethacrylate (PMMA).
Near-field scanning optical microscope (NSOM)	This is the scanned-source equivalent of contact and proximity printing.
Scanning tunneling microscope (STM)	In contrast to electron-beam lithography, the mechanism of STM lithography is generally an inelastic electron tunneling process, with energy quanta of an electron volt. STM can also be used for single-atom or single-molecule manipulations for the fabrication of individual nanostructures.

Sol–Gel Synthesis

Cold chemical synthesis of nanoporous glasses by the sol–gel technique allows the embedding of temperature-sensitive clusters such as organic nanocrystals. Moreover, it often permits one to achieve larger densities of semiconductor nanocrystals than in diffusion-controlled growth.

Zeolites

Zeolites constitute a group of hydrated crystalline aluminosilicates containing regularly shaped pores with sizes from 1 nm to several nanometers. This provides them with the ability to reversibly adsorb and desorb specific molecules. Chemical synthesis of semiconductors inside such pores permits the fabrication of regularly distributed, highly size-monodispersed nanocrystal arrays. This has been applied to CdS, PbI_2, and Ge.

5.3.5.5 Ex Situ Synthesis of Clusters

As an alternative to in situ growth, nanocrystals can be synthesized either in solution (e.g. by colloidal synthesis) or in the gas phase, prior to dispersion in a matrix or deposition or self-assembly onto a substrate.

Colloidal Synthesis

Although gold colloids were known and studied very early [3.2], the synthesis technique has been much studied and greatly improved during the past few decades. The two-phase approach permits the fabrication of thiol-covered noble-metal nanoparticles [3.141]. The thiol layer prevents coalescence of the metal particles, thus stabilizing the colloid and preserving the size distribution. Motivated by the self-ordering properties of layers of nanoparticles with a narrow size distribution,

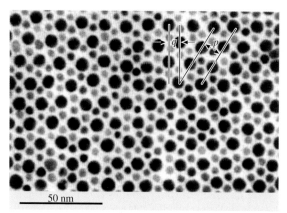

Fig. 5.3-42 TEM image showing a superlattice of thiol-passivated Au clusters with a bimodal size distribution (∼ 4.5 and 7.8 nm diameter). The 2-D supracrystallization shows a regular self-organized bimodal packing. (After 3.144)

been synthesized using a reverse-micelle approach and assembled as a hexagonal self-organized superlattice [3.146]. 3-D self-assembled superlattices of FePt nanoparticles have also been produced [3.121].

Ligand-stabilization techniques have led to macromolecular complexes consisting of a metal core containing a fixed number of atoms surrounded by a precise number of organic ligand molecules. Following the synthesis of $Au_{55}(P(C_6H_5)_3)_{12}Cl_3$ clusters [3.147], a number of such macromolecules of various sizes and compositions have been studied, such as giant ligand-stabilized clusters containing 561 Pd atoms [3.148].

Variations of the colloidal-synthesis and size selection techniques have also been developed in parallel for the preparation of semiconductor nanoparticles, particularly II–VI compounds such as CdSe [3.149], from which self-assembled superlattices can be grown [3.150]. GaAs nanocrystals have been produced similarly [3.151].

Gas-Phase Production

Several techniques have been applied for the production of "preformed" clusters in the gas phase, prior to their deposition onto a substrate. The low-energy cluster beam deposition technique was developed especially for the deposition of small clusters of transition metals (Ni, Fe, Co, etc.) [3.152]. An extensive review of this vast subject can be found in [3.153].

researchers have developed various size-selection techniques [3.142, 143]. As shown in Fig. 5.3-42, relatively complex self-organized structures are now achievable through self-assembly of size-selected thiol-capped nanoparticles obtained from colloidal synthesis [3.144], including mixtures of particles of different chemical nature (Au and Ag, [3.145]).

This technique has been extended to transition metals. 6 nm trioctylphosphine-coated Co particles have

References

3.1 M. José-Yacaman, L. Rendon, J. Arenas, M. C. S. Puche: Maya blue paint: An ancient nanostructured material, Science **273**, 223–225 (1996)

3.2 M. Faraday: Experimental relations of gold (and other metals) to light, Phil. Trans. R. Soc. **147**, 145–181 (1857)

3.3 R. Dingle, W. Wiegmann, C. H. Henry: Quantum states of confined carriers in very thin $Al_xGa_{1−x}As–GaAs–Al_xGa_{1−x}As$ heterostructures, Phys. Rev. Lett. **33**, 827–830 (1974)

3.4 B. Bhushan (Ed.): *Springer Handbook of Nanotechnology* (Springer, Berlin, Heidelberg 2004)

3.5 P. Moriarty: Nanostructured materials, Rep. Prog. Phys. **64**, 297–381 (2001)

3.6 M. Milun, P. Pervan, D. P. Woodruff: Quantum well structures in thin metal films: Simple model physics in reality?, Rep. Prog. Phys. **65**, 99–141 (2002)

3.7 S. A. Lindgren, S. Valden: Electron-energy-band determination by photoemission from overlayer states, Phys. Rev. Lett. **61**, 2894–2897 (1988)

3.8 T. Miller, A. Samsavar, G. E. Franklin, T.-C. Chiang: Quantum-well states in a metallic system: Ag on Au(111), Phys. Rev. Lett. **61**, 1404–1407 (1988)

3.9 S. Bjornholm, J. Borggren, O. Echt, K. Hansen, J. Pedersen, H. D. Rasmussen: The influence of shells, electron thermodynamics, and evaporation on the abundance spectra of large sodium metal clusters, Z. Phys. D **19**, 47 (1991)

3.10 W. Ekardt: Work function of small metal particles: Self-consistent spherical jellium-background model, Phys. Rev. B **29**, 1558–1564 (1984)

3.11 J. H. Cho, K. S. Kim, C. T. Chan, Z. Y. Zhang: Oscillatory energetics of flat Ag films on MgO(001), Phys. Rev. B **63**, 113408 (2001)

3.12 J. J. Paggel, T. Miller, T. C. Chiang: Quantum-well states as Fabry-Perot modes in a thin-film electron interferometer, Science **283**, 1709–1711 (1999)

3.13 M. S. Hybertsen: Mechanism for light emission from nanoscale silicon. In: *Porous Silicon Science and Technology*, ed. by J.-C. Vial, J. Derrien (Les éditions de physique, Les Ulis 1995) p. 67

3.14 L.T. Canham: Silicon quantum wire array fabrication by electrochemical and chemical dissolution of wafers, Appl. Phys. Lett. **57**, 1046–1048 (1990)

3.15 V. Lehmann, U. Gösele: Porous silicon formation: A quantum wire effect, Appl. Phys. Lett. **58**, 856–858 (1990)

3.16 A. I. Belogorokhov, V. A. Karavanskii, A. N. Obraztsov, V. Y. Timoshenko: Intense photoluminescence in porous gallium-phosphide, JETP Lett. **60**, 274–279 (1994)

3.17 A. Anedda, A. Serpi, V. A. Karavanski, I. M. Tiginyanu, V. M. Ichizli: Time-resolved blue and ultraviolet photoluminescence in porous GaP, Appl. Phys. Lett. **67**, 3316–3318 (1995)

3.18 J.S. Shor, L. Bemis, A.D. Kurtz, I. Grimberg, B.Z. Weiss, M. F. Macmillian, W. J. Choyke: Characterization of nanocrystallites in porous p-type 6H-SiC, J. Appl. Phys. **76**, 4045–4049 (1994)

3.19 D.J. Lockwood, Z. H. Lu, J.-M. Baribeau: Quantum confined luminescence in Si/SiO_2 superlattices, Phys. Rev. Lett. **76**, 539–541 (1996)

3.20 R.W. Collins, P.M. Fauchet, I. Shimizu, J.-C. Vial, T. Shimada, A. P. Alivisatos (Eds.): *Advances in Microcrystalline and Nanocrystalline Semiconductors*, Vol. 452 (Materials Research Society, Pittsburgh 1997)

3.21 N.M. Park, T.S. Kim, S.J. Park: Band gap engineering of amorphous silicon quantum dots for light-emitting diodes, Appl. Phys. Lett. **78**, 2575–2577 (2001)

3.22 L. Pavesi, L. Dal Negro, C. Mazzoleni, G. Franzo, F. Priolo: Optical gain in silicon nanocrystals, Nature **408**, 440–444 (2000)

3.23 P. Ball: Let there be light, Nature **409**, 974–976 (2001)

3.24 L. E. Brus: Electron–electron and electron–hole interactions in small semiconductor crystallites: The size dependence of the lowest excited electronic state, J. Chem. Phys. **80**, 4403–4409 (1984)

3.25 Y. Kayanuma: Quantum-size effects of interacting electrons and holes in semiconductor microcrystals with spherical shape, Phys. Rev. B **38**, 9797–9805 (1988)

3.26 Al. L. Efros, A. L. Efros: Interband absorption of light in a semiconductor sphere, Sov. Phys. Semicond. **16**, 772–775 (1982)

3.27 A. Shik: *Quantum Wells, Physics and Electronics of Two-Dimensional Systems* (World Scientific, Singapore 1997) Chap. 4.4

3.28 S. Glutsch, D.S. Chemla: Transition to one-dimensional behavior in the optical absorption of quantum-well wires, Phys. Rev. B **53**, 15902–15908 (1996)

3.29 D.J. Norris, A. Sacra, C.B. Murray, M.G. Bawendi: Measurement of the size-dependent hole spectrum in CdSe quantum dots, Phys. Rev. Lett. **72**, 2612–2615 (1994)

3.30 G. Banfi, V. Degiorgio: Neutron scattering investigation of the structure of semiconductor-doped glasses, J. Appl. Phys. **74**, 6925–6935 (1993)

3.31 M. H. Yukselici: Growth kinetics of CdSe nanoparticles in glass, J. Phys.: Condens. Matter **14**, 1153–1162 (2002)

3.32 C. Weisbuch, B. Vinter: *Quantum Semiconductor Structures* (Academic Press, Boston 1991) Chap. 3

3.33 B. Zhao, A. Yariv: Quantum well semiconductor lasers. In: *Semiconductor Lasers I: Fundamentals*, ed. by E. Kapon, P.L. Kelley (Academic Press, San Diego 1999) Chap. 1

3.34 E. Kapon: Quantum wire and quantum dot lasers. In: *Semiconductor Lasers I: Fundamentals*, ed. by E. Kapon, P.L. Kelley (Academic Press, San Diego 1999) Chap. 4

3.35 V.I. Klimov, A.A. Mikhailovsky, S. Xu, A. Malko, J.A. Hollingsworth, C.A. Leatherdale, H.J. Eisler, M.G. Bawendi: Optical gain and stimulated emission in nanocrystal quantum dots, Science **290**, 314–317 (2000)

3.36 H.J. Eisler, V.C. Sundar, M.G. Bawendi, M. Walsh, H.I. Smith, V. Klimov: Color-selective semiconductor nanocrystal laser, Appl. Phys. Lett. **80**, 4614–4616 (2002)

3.37 P.T. Guerreiro, S. Ten, N.F. Borrelli, J. Butty, G.E. Jabbour, N. Peyghambarian: PbS quantum-dot doped glasses as saturable absorbers for mode locking of a Cr:forsterite laser, Appl. Phys. Lett. **71**, 1595–1597 (1997)

3.38 K. Wundke, S. Potting, J. Auxier, A. Schulzgen, N. Peyghambarian, N. F. Borrelli: PbS quantum-dot-doped glasses for ultrashort-pulse generation, Appl. Phys. Lett. **76**, 10–12 (2000)

3.39 S. Schmitt-Rink, D.S. Chemla, D.A.B. Miller: Linear and nonlinear optical properties of semiconductor quantum wells, Adv. Phys. **38**, 89–188 (1989)

3.40 F. Devaux, F. Dorgeuille, A. Ougazzaden, F. Huet, M. Carre, A. Carenco, M. Henry, Y. Sorel, J.-F. Kerdiles, E. Jeanney: 20 Gbit/s operation of a high-efficiency InGaAsP/InGaAsP MQW electroabsorption modulator with 1.2-V drive voltage, IEEE Photonics Tech. Lett. **5**, 1288–1990 (1993)

3.41 D. Delprat, A. Ramdane, A. Ougazzaden, H. Nakajima, M. Carre: Integrated multiquantum well distributed feedback laser-electroabsorption modulator with a negative chirp for zero bias voltage, Electron. Lett. **33**, 53–55 (1997)

3.42 T.D. Krauss, F. W. Wise: Coherent acoustic phonons in a semiconductor quantum dot, Phys. Rev. Lett. **79**, 5102–5105 (1997)

3.43 D. Donovan, J. F. Andress: *Lattice Vibrations* (Chapman and Hall, London 1971)

3.44 J.F. DiTusa, K. Lin, M. Park, M.S. Isaacson, J.M. Parpia: Role of phonon dimensionality on electron–phonon scattering rates, Phys. Rev. Lett. **68**, 1156–1159 (1992)

3.45 S.H. Song, W. Pan, D.C. Tsui, Y.H. Xie, D. Monroe: Energy relaxation of two-dimensional carriers in strained $Ge/Si_{0.4}Ge_{0.6}$ and $Si/Si_{0.7}Ge_{0.3}$ quan-

3.45 ...tum wells: Evidence for two-dimensional acoustic phonons, Appl. Phys. Lett. **70**, 3422–3424 (1997)

3.46 V. Narayanamurti, H. L. Störmer, M. A. Chin, A. C. Gossard, W. Wiegmann: Selective transmission of high frequency phonons by a superlattice: The dielectric phonon filter, Phys. Rev. Lett. **43**, 2012–2016 (1979)

3.47 P. V. Santos, L. Ley, J. Mebert, O. Koblinger: Frequency gaps for acoustic phonons in a-Si:H/a-SiN$_x$:H superlattices, Phys. Rev. B **36**, 4858–4867 (1987)

3.48 M. Fujii, T. Nagareda, S. Hayashi, K. Yamamoto: Low-frequency Raman scattering from small silver particles embedded in SiO$_2$ thin films, Phys. Rev. B **44**, 6243–6248 (1991)

3.49 L. Saviot, B. Champagnon, E. Duval, A. I. Ekimov: Size-selective resonant Raman scattering in CdS doped glasses, Phys. Rev. B **57**, 341–346 (1998)

3.50 H. Lamb: On the vibrations of an elastic sphere, Proc. Math. Soc. London **13**, 187 (1882)

3.51 S. Rufo, M. Dutta, M. A. Stroscio: Acoustic modes in free and embedded quantum dots, J. Appl. Phys. **93**, 2900–2905 (2003)

3.52 N. Del Fatti, C. Voisin, F. Chevy, F. Vallee, C. Flytzanis: Coherent acoustic mode oscillation and damping in silver nanoparticles, J. Chem. Phys. **110**, 11484–11487 (1999)

3.53 M. Ikezawa, T. Okuno, Y. Masumoto, A. A. Lipovskii: Complementary detection of confined acoustic phonons in quantum dots by coherent phonon measurement and Raman scattering, Phys. Rev. B **64**, 201315 (2001)

3.54 A. Arbouet, C. Voisin, D. Christofilos, P. Langot, N. Del Fatti, F. Vallee, J. Lerme, G. Celep, E. Cottancin, M. Gaudry, M. Pellarin, M. Broyer, M. Maillard, M.-P. Pileni, M. A. Treguer: Electron–phonon scattering in metal clusters, Phys. Rev. Lett. **90**, 177401 (2003)

3.55 S. Schmitt-Rink, D. A. B. Miller, D. S. Chemla: Theory of the linear and nonlinear optical properties of semiconductor microcrystallites, Phys. Rev. B **35**, 8113–8125 (1987)

3.56 D. Kovalev, H. Heckler, M. Ben-Chorin, G. Polisski, M. Schwartzkopff, F. Koch: Breakdown of the k-conservation rule in Si nanocrystals, Phys. Rev. Lett. **81**, 2803–2806 (1998)

3.57 M. S. Hybertsen: Absorption and emission of light in nanoscale silicon structures, Phys. Rev. Lett. **72**, 1514–1517 (1994)

3.58 N. F. Mott: Conduction in non-crystalline materials III. Localized states in a pseudogap and near extremities of conduction and valence bands, Philos. Mag. **19**, 835–852 (1969)

3.59 B. F. Levine: Quantum-well infrared photodetectors, J. Appl. Phys. **74**, R1–81 (1993)

3.60 L. Esaki, R. Tsu: Superlattice and negative differential conductivity in semiconductors, IBM J. Res. Devel. **14**, 61–65 (1970)

3.61 L. Esaki, L. L. Chang: New transport phenomenon in a semiconductor superlattice, Phys. Rev. Lett. **33**, 495–498 (1974)

3.62 K. K. Choi, B. F. Levine, R. J. Malik, J. Walker, C. G. Bethea: Periodic negative conductance by sequential resonant tunneling through an expanding high-field superlattice domain, Phys. Rev. B **35**, 4172–4175 (1987)

3.63 N. F. Mott: *Conduction in Non-Crystalline Materials*, Oxford Science Publications, 2nd edn. (Clarendon, Oxford 1993)

3.64 P. Sheng, J. Klafter: Hopping conductivity in granular disordered systems, Phys. Rev. B **27**, 2583–2586 (1983)

3.65 R. P. Andres, T. Bein, M. Dorogi, S. Feng, J. I. Henderson, C. P. Kubiak, W. Mahoney, R. G. Osifchin, R. Reifenberger: Coulomb staircase at room temperature in a self-assembled molecular nanostructure, Science **272**, 1323–1325 (1996)

3.66 M. A. Kastner: The single-electron transistor, Rev. Mod. Phys. **64**, 849–858 (1992)

3.67 A. L. Efros, B. I. Shklovskii: Coulomb gap and low-temperature conductivity of disordered systems, J. Phys. C: Solid State Phys. **8**, L49–51 (1975)

3.68 M. Fujii, T. Kita, S. Hayashi, K. Yamamoto: Current-transport properties of Ag–SiO$_2$ and Au–SiO$_2$ composite films: Observation of single-electron tunnelling and random telegraph signals, J. Phys.: Condens. Matter **9**, 8669–8677 (1997)

3.69 P. E. Trudeau, A. Escorcia, A. A. Dhirani: Variable single electron charging energies and percolation effects in molecularly linked nanoparticle films, J. Chem. Phys. **119**, 5267–5273 (2003)

3.70 A. Taleb, F. Silly, A. O. Gusev, F. Charra, M.-P. Pileni: Electron transport properties of nanocrystals: Isolated, and "supra"-crystalline phases, Adv. Mater. **12**, 633–637 (2000)

3.71 J. Fick, A. Martucci, M. Guglielmi, J. Shell: Nonlinear optical properties of semiconductor-doped sol-gel waveguides, Fiber Integrated Opt. **19**, 43–56 (2000)

3.72 U. Kreibig, M. Vollmer: *Optical Properties of Metal Clusters* (Springer, Berlin, Heidelberg 1995) p. 50

3.73 K. Kolwas, S. Demianiuk, M. Kolwas: Optical excitation of radius-dependent plamon resonances in large metal clusters, J. Phys. B **29**, 4761–4770 (1996)

3.74 S. Link, Z. L. Wang, M. A. El-Sayed: Alloy formation on gold–silver nanoparticles and the dependence of the plasmon absorption on their composition, J. Phys. Chem. B **103**, 3529–3533 (1999)

3.75 F. Hache, D. Ricard, C. Flytzanis: Optical nonlinearities of small metal nanoperticles: Surface-mediated resonance and quantum size effects, J. Opt. Soc. Am. B **3**, 1647–1655 (1986)

3.76 R. Elghanian, J. J. Storhoff, R. C. Mucic, R. L. Letsinger, C. A. Mirkin: Selective colorimetric detection of polynucleotides based on the distance-dependent optical properties of gold nanoparticles, Science **277**, 1078–1081 (1997)

3.77 J. A. Creighton, D. G. Eadon: Ultraviolet-visible spectra of the colloidal metallic elements, J. Chem. Soc. Faraday Trans. **87**, 3881–3891 (1991)

3.78 P. E. Batson: Surface plasmon coupling in clusters of small spheres, Phys. Rev. Lett. **49**, 936–940 (1982)

3.79 J. R. Krenn, A. Dereux, J. C. Weeber, E. Bourillot, Y. Lacroute, J. P. Goudonnet, G. Schider, W. Gotschy, A. Leitner, F. R. Aussenegg, C. Girard: Squeezing the optical near-field zone by plasmon coupling of metallic nanoparticles, Phys. Rev. Lett. **82**, 2590–2593 (1999)

3.80 F. Silly, A. O. Gusev, A. Taleb, F. Charra, M. P. Pileni: Coupled plasmon modes in an ordered hexagonal monolayer of metal nanoparticles: A direct observation, Phys. Rev. Lett. **84**, 5840–5843 (2000)

3.81 S. A. Maier, M. L. Brongersma, P. G. Kik, S. Meltzer, A. A. G. Requicha, H. A. Atwater: Plasmonics – a route to nanoscale optical devices, Adv. Mater. **13**, 15 (2001)

3.82 R. Mayoral, J. Requena, J. S. Moya, C. Lopez, A. Cintas, H. Miguez, F. Meseguer, L. Vazquez, M. Holgado, A. Blanco: 3D long-range ordering in an SiO_2 submicrometer-sphere sintered superstructure, Adv. Mater. **9**, 257–260 (1997)

3.83 J. V. Sanders: Colour of precious opal, Nature **204**, 1151–1153 (1964)

3.84 E. Yablonovitch: Inhibited spontaneous emission in solid-state physics and electronics, Phys. Rev. Lett. **58**, 2059–2062 (1987)

3.85 S. John: Strong localization of photons in certain disordered dielectric superlattices, Phys. Rev. Lett. **58**, 2486–2489 (1987)

3.86 M. N. Baibich, J. M. Broto, A. Fert, F. Nguyen Van Dau, F. Petroff, P. Etienne, G. Creuzet, A. Friederich, J. Chazelas: Giant magnetoresistance of (001)Fe/(001)Cr magnetic superlattices, Phys. Rev. Lett. **61**, 2472 (1988)

3.87 P. Grünberg, R. Shreiber, Y. Pang, M. B. Brodsky, H. Sowers: Layered magnetic structures: Evidence for antiferromagnetic coupling of Fe layers across Cr interlayers, Phys. Rev. Lett. **57**, 2442 (1986)

3.88 R. L. Comstock: *Introduction to Magnetism and Magnetic Recording* (Wiley, New York 1999)

3.89 N. F. Mott: The electrical conductivity of transition metals, Proc. R. Soc. **153**, 699 (1935)

3.90 J. F. Gregg: Introduction to spin electronics. In: *Spin Electronics*, Lecture Notes in Physics, No. 569, ed. by M. Ziese, M. J. Thornton (Springer, Berlin, Heidelberg 2001) Chap. 1

3.91 G. Binasch, P. Grünberg, F. Saurenbach, W. Zinn: Enhanced magnetoresistance in layered magnetic structures with antiferromagnetic interlayer exchange, Phys. Rev. B **39**, 4828 (1989)

3.92 R. Schad, C. D. Potter, P. Beliën, G. Verbanck, V. V. Moshchalkov, Y. Bruynseraede: Giant magnetoresistance in Fe/Cr superlattices with very thin Fe layers, Appl. Phys. Lett. **64**, 3500 (1994)

3.93 A. Bartélémy, A. Fert, F. Petroff: Giant magnetoresistance in magnetic multilayers. In: *Handbook of Magnetic Materials*, Vol. 12, ed. by K. H. J. Buschow (Elsevier, Amsterdam 1999) Chap. 1, pp. 1–96

3.94 S. S. P. Parkin, D. Mauri: Spin engineering: Direct determination of the Ruderman–Kittel–Kasuya–Yosida far-field range function in ruthenium, Phys. Rev. B **44**, 7131 (1991)

3.95 A. Barthélémy, A. Fert, J-P. Contour, M. Bowen, V. Cros, J. M. De Teresa, A. Hamzic, J. C. Faini, J. M. George, J. Grollier, F. Montaigne, F. Pailloux, F. Petroff, C. Vouille: Magnetoresistance and spin electronics, J. Magn. Magn. Mater. **242–245**, 68 (2002)

3.96 D. H. Mosca, F. Petroff, A. Fert, P. A. Schroeder, W. P. Pratt, R. Loloee: Oscillatory interlayer coupling and giant magnetoresistance in Co/Cu multilayers, J. Magn. Magn. Mater. **94**, L1 (1991)

3.97 W. P. Pratt, S. F. Lee, J. M. Slaughter, R. Loloee, P. A. Schroeder, J. Bass: Perpendicular giant magnetoresistances of Ag/Co multilayers, Phys. Rev. Lett. **66**, 3060 (1991)

3.98 A. Fert, L. Piraux: Magnetic nanowires, J. Magn. Magn. Mater. **200**, 338 (1999) (Special issue)

3.99 M. A. M. Gijs, M. T. Johnson, A. Reinders, P. E. Huisman, R. J. M. van de Veerdonk, S. K. J. Lenczowski, R. M. Gansewinkel: Perpendicular giant magnetoresistance of Co/Cu multilayers deposited under an angle on grooved substrates, Appl. Phys. Lett. **66**, 1839 (1995)

3.100 B. Dieny, V. S. Speriosu, S. S. Parkin, B. A. Gurney, D. R. Wilhoit, D. Mauri: Giant magnetoresistance in soft ferromagnetic multilayers, Phys. Rev. B **43**, 1297 (1991)

3.101 J. Barnas, A. Fuss, R. E. Camley, P. Grünberg, W. Zinn: Novel magnetoresistance effect in layered magnetic structures: Theory and experiment, Phys. Rev. B **42**, 8110 (1990)

3.102 C. Tsang, D. E. Heim, R. E. Fontana, V. S. Speriosu, B. A. Gurney, M. L. Williams: Design, fabrication and testing of spin-valve read heads for high density recording, IEEE Trans. Magn. **30**, 3801 (1994)

3.103 W. F. Egelhoff Jr, P. J. Chen, C. J. Powell, D. Parks, G. Serpa, R. D. McMichael, D. Martien, A. E. Berkowitz: Specular electron scattering in metallic thin films, J. Vac. Sci. Technol. B **17**, 1702 (1999)

3.104 P. M. Tedrow, R. Meservey: Spin-dependent tunneling into ferromagnetic nickel, Phys. Rev. Lett. **26**, 192 (1971)

3.105 P. M. Tedrow, R. Meservey: Spin polarization of electrons tunneling from films of Fe, Co, Ni, and Gd, Phys. Rev. B **7**, 318 (1973)

3.106 T. Dimopoulos: Spin polarised tunnel transport in magnetic tunnel junctions: The role of metal/oxide interfaces in the tunnel process. Ph.D. Thesis (University Louis Pasteur, Strasbourg 2002)

3.107 T. Miyazaki, N. Tezuka: Giant magnetic tunneling effect in Fe/Al$_2$O$_3$/Fe junction, J. Magn. Magn. Mater. **139**, L231 (1995)

3.108 N. Tezuka, T. Miyazaki: Magnetic tunneling effect in Fe/Al$_2$O$_3$/Ni$_{1-x}$Fe$_x$ junctions, J. Appl. Phys. **79**, 6262 (1996)

3.109 J. S. Moodera, L. R. Kinder, T. M. Wong, R. Meservey: Large magnetoresistance at room temperature in ferromagnetic thin film tunnel junctions, Phys. Rev. Lett. **74**, 3273 (1995)

3.110 M. Jullière: Tunneling between ferromagnetic films, Phys. Lett. **54**, 225 (1975)

3.111 J. Nassar, M. Hehn, A. Vaures, F. Petroff, A. Fert: Magnetoresistance of ferromagnetic tunnel junctions with Al$_2$O$_3$ barriers formed by rf sputter etching in Ar/O$_2$ plasma, Appl. Phys. Lett. **73**, 698 (1998)

3.112 M. Viret, M. Drouet, J. Nassar, J. P. Contour, C. Fermon, A. Fert: Low-field colossal magnetoresistance in manganite tunnel spin valves, Europhys. Lett. **39**, 545 (1997)

3.113 J. Nassar, M. Viret, M. Drouet, J. P. Contour, C. Fermon, A. Fert: Low-field colossal magnetoresistance in manganite tunnel junctions, Mater. Res. Soc. Symp. Proc. **494**, 231 (1998)

3.114 M. Bowen, M. Bibes, A. Barthélémy, J.-P. Contour, A. Anane, Y. Lemaître, A. Fert: Nearly total spin polarization in La$_2$O$_3$Sr$_1$O$_3$MnO$_3$ from tunneling experiments, Appl. Phys. Lett. **82**, 233 (2003)

3.115 T. Obata, T. Manako, Y. Shimakawa, Y. Kubo: Tunneling magnetoresistance at up to 270 K in La$_{0.8}$Sr$_{0.2}$MnO$_3$/SrTiO$_3$/La$_{0.8}$Sr$_{0.2}$MnO$_3$ junctions with 1.6-nm-thick barriers, Appl. Phys. Lett. **74**, 290 (1999)

3.116 S. Gota, J. B. Moussy, M. Henriot, M. J. Guittet, M. Gautier-Soyer: Atomic-oxygen-assisted MBE growth of Fe$_2$O$_4$(111) on α-Al$_2$O$_3$(0001), Surf. Sci. **482-485**, 809–816 (2001)

3.117 S. Van Dijken, X. Jiang, S. S. P. Parkin: Giant magnetocurrent exceeding 3400% in magnetic tunnel transistors with spin-valve base layers, Appl. Phys. Lett. **83**, 951 (2003)

3.118 S. A. Wolf, A. Y. Chtchenlkanova, D. M. Treger: Spintronics: Spin based electronics. In: *Handbook of Nanoscience, Engineering and Technology*, Electrical Engineering Textbook Series, Vol. 27, ed. by W. A. Goddard III, D. W. Brenner, S. E. Lyshevski, G. J. Iafrate (CRC, Boca Raton 2003) Chap. 8

3.119 E. E. Fullerton, D. T. Margulies, M. E. Schabes, M. Carey, B. Gurney, A. Moser, M. Best, G. Zeltzer, K. Rubin, H. Rosen, M. Doerner: Antiferromagnetically coupled magnetic media layers for thermally stable high-density recording, Appl. Phys. Lett. **77**, 3806 (2000)

3.120 C. Fermon: Micro- and nanofabrication techniques. In: *Spin Electronics*, Lecture Notes in Physics, No. 569, ed. by M. Ziese, M. J. Thornton (Spinger, Berlin, Heidelberg 2001) Chap. 16

3.121 S. H. Sun, C. B. Murray, D. Weller, L. Folks, A. Moser: Monodisperse FePt nanoparticles and ferromagnetic FePt nanocrystal superlattices, Science **287**, 1989–1992 (2000)

3.122 J. Moritz, B. Dieny, J. P. Nozières, S. Landis, A. Lebib, Y. Chen: Domain structure in magnetic dots prepared by nanoimprinting and E-beam lithography, J. Appl. Phys. **91**, 7314 (2002)

3.123 J. Moritz, S. Landis, J. C. Toussaint, P. Bayle-Guillemaud, B. Rodmacq, G. Casali, A. Lebib, Y. Chen, J. P. Nozières, B. Dieny: Patterned media made from pre-etched wafers: A promising route toward ultrahigh-density magnetic recording, IEEE Trans. Magn. **38**, 1731 (2002)

3.124 S. Landis, B. Rodmacq, B. Dieny, B. Dal'Zotto, S. Tedesco, M. Heitzmann: Domain structure of magnetic layers deposited on patterned silicon, Appl. Phys. Lett. **75**, 2473 (1999)

3.125 A. Lebib, Y. Chen, F. Carcenac, E. Cambril, L. Manin, L. Couraud, H. Launois: Tri-layer systems for nanoimprint lithography with an improved process latitude, Microelectron. Eng. **53**, 175 (2000)

3.126 S. Landis, B. Rodmacq, B. Dieny: Magnetic properties of Co/Pt multilayers deposited on silicon dot arrays, Phys. Rev. B **62**, 12271 (2000)

3.127 X. G. Binning, M. Despont, U. Drechsler, W. Häberle, M. Lutwyche, P. Vettiger, H. J. Mamin, B. W. Chui, T. W. Kenny: Ultrahigh-density atomic force microscopy data storage with erase capability, Appl. Phys. Lett. **74**, 1329 (1999)

3.128 J. L. Marin, R. Riera, R. A. Rosas: Confined systems and nanostructured materials. In: *Nanomaterials*, Handbook of Advanced Electronic and Photonic Materials and Devices, Vol. 9, ed. by H. S. Nalwa (Elsevier, Amsterdam 2000) p. 55

3.129 J. M. Hartmann, M. Charleux, J. L. Rouviere, H. Mariette: Growth of CdTe/MnTe tilted and serpentine lattices on vicinal surfaces, Appl. Phys. Lett. **70**, 1113–1115 (1997)

3.130 P. Soukiassian, F. Semond, A. Mayne, G. Dujardin: Highly stable Si atomic line formation on the beta-SiC(100) surface, Phys. Rev. Lett. **79**, 2498–2501 (1997)

3.131 D. Leonard, K. Pond, P. M. Petroff: Critical layer thickness for self-assembled InAs islands on GaAs, Phys. Rev. B **50**, 11687–11692 (1994)

3.132 S. Rusponi, T. Cren, N. Weiss, M. Epple, P. Buluschek, L. Claude, H. Brune: The remarkable difference between surface and step atoms in the magnetic anisotropy of two-dimensional nanostructures, Nature Mater. **2**, 546–551 (2003)

3.133 R. Nötzel, T. Fukui, H. Hasegawa, J. Temmyo, T. Tamamura: Atomic-force microscopy study of strained InGaAs quantum disks self-organizing on GaAs(n11) substrates, Appl. Phys. Lett. **65**, 2854–2856 (1994)

3.134 E. Ganz, K. Sattler, J. Clarke: Scanning tunneling microscopy of the local atomic structure of two-dimensional gold and silver islands on graphite, Phys. Rev. Lett. **60**, 1856–1859 (1988)

3.135 G. W. Clark, L. L. Kesmodel: Ultrahigh vacuum scanning tunneling microscopy studies of platinum on graphite, J. Vac. Sci. Technol. B **11**, 131–136 (1993)

3.136 H. Xu, K. Y. S. Ng: Scanning tunneling microscopy investigation of Co cluster growth and induced surface morphology changes on highly oriented pyrolitic graphite, J. Vac. Sci. Technol. B **13**, 2160–2165 (1995)

3.137 H. Xu, K. Y. S. Ng: Surface process and interaction of Mo clusters on highly oriented pyrolytic graphite observed by scanning tunneling microscopy, J. Vac. Sci. Technol. B **15**, 186–191 (1997)

3.138 A. Piednoir, E. Perrot, S. Granjeaud, A. Humbert, C. Chapon, C. R. Henry: Atomic resolution on small three-dimensional metal clusters by STM, Surf. Sci. **391**, 19–26 (1997)

3.139 G. Springholz, V. Holy, M. Pinczolits, G. Bauer: Self-organized growth of three-dimensional quantum-dot crystals with fcc-like stacking and a tunable lattice constant, Science **282**, 734–737 (1998)

3.140 M. D. Levenson, N. S. Viswanathan, R. A. Simpson: Improving resolution in photolithography with a phase-shifting mask, IEEE Trans. Electron Devices **ED-29**, 1812–1836 (1982)

3.141 M. Brust, M. Walker, D. Bethell, D. J. Schiffrin, R. Whyman: Synthesis of thiol-derivatized gold nanoparticles in a 2-phase liquid–liquid system, J. Chem. Soc., Chem. Commun. **1994**, 801–802 (1994)

3.142 R. P. Andres, J. D. Bielefeld, J. I. Henderson, D. B. Janes, V. R. Kolagunta, C. P. Kubiak, W. J. Mahoney, R. G. Osifchin: Self-assembly of a two-dimensional superlattice of molecularly linked metal clusters, Science **273**, 1690–1693 (1996)

3.143 A. Taleb, C. Petit, M. P. Pileni: Synthesis of highly monodisperse silver nanoparticles from AOT reverse micelles: A way to 2D and 3D self-organization, Chem. Mater. **9**, 950–959 (1997)

3.144 C. J. Kiely, J. Fink, M. Brust, D. Bethell, D. J. Schiffrin: Spontaneous ordering of bimodal ensembles of nanoscopic gold clusters, Nature **396**, 444–446 (1998)

3.145 C. J. Kiely, J. Fink, J. G. Zheng, M. Brust, D. Bethell, D. J. Schiffrin: Ordered colloidal nanoalloys, Adv. Mater. **12**, 640 (2000)

3.146 C. Petit, A. Taleb, M.-P. Pileni: Self-organization of magnetic nanosized cobalt particles, Adv. Mater. **10**, 259–261 (1998)

3.147 G. Schmid, R. Pfeil, R. Boese, F. Bandermann, S. Meyer, S. G. Calis, J. W. A. Van Der Velden: Au55[P(C6H5)]12Cl6 ein Goldcluster ungewöhnlicher Größe, Chem. Ber. **114**, 3634–3642 (1981)

3.148 V. Oleshko, V. Volkov, R. Gijbels, W. Jacob, M. Vargaftik, I. Moiseev, G. Vantendeloo: High-resolution electron-microscopy and electron-energy-loss spectroscopy of giant palladium clusters, Z. Phys. **D34**, 283–291 (1995)

3.149 C. B. Murray, D. B. Norris, M. G. Bawendi: Synthesis and characterization of nearly monodisperse CdE (E=S, Se, Te) semiconductor nanocrystallites, J. Am. Chem. Soc. **115**, 8706–8715 (1993)

3.150 C. B. Murray, C. R. Kagan, M. G. Bawendi: Self-organization of CdSe nanocrystallites into 3-dimensional quantum-dot superlattices, Science **270**, 1335–1338 (1995)

3.151 M. A. Olshavsky, A. N. Goldstein, A. P. Alivisatos: Organometallic synthesis of GaAs crystallites exhibiting quantum confinement, J. Am. Chem. Soc. **112**, 9438–9439 (1990)

3.152 C. Binns: Nanoclusters deposited on surfaces, Surf. Sci. Rep. **44**, 1–49 (2001)

3.153 W. A. De Heer: The physics of simple metal clusters: Experimental aspects and simple models, Rev. Mod. Phys. **65**, 611 (1993)

Acknowledgements

2.1 The Elements
by Werner Martienssen

We thank Dr. G. Leichtfried, Plansee AG, A-6600 Reutte/Tirol for recently determined new data on the refractory metals Nb, Ta, and Mo, W.

4.1 Semiconductors
by Werner Martienssen

In selecting the "most important information" from the huge data collection in Landolt–Börnstein, the author found great help in the new *Semiconductors: Data Handbook* [1]. Again, the data in this Springer Handbook of Condensed Matter and Materials Data represent only a small fraction of the information given in *Semiconductors: Data Handbook*, which is about 700 pages long. I am much indebted to my colleague O. Madelung for kindly presenting me the manuscript of that Handbook prior to publication.

[1] O. Madelung (Ed.): *Semiconductors: Data Handbook*, 3rd Edn. (Springer, Berlin, Heidelberg 2004)

4.5 Ferroelectrics and Antiferroelectrics
by Toshio Mitsui

The author of this subchapter thanks the coauthors of LB III/36 for their helpful discussions and suggestions. Especially, he is much indebted to Prof. K. Deguchi for his kind support throughout the preparation of the manuscript.

1074

About the Authors

Wolf Assmus

Johann Wolfgang Goethe-University
Physics Department
Frankfurt am Main, Germany
assmus@physik.uni-frankfurt.de
http://www.rz.uni-frankfurt.de/piweb/kmlab/Leiter.html

Chapter 1.3

Dr. Wolf Assmus (Kucera Professor) is Professor of Physics at the University of Frankfurt and Dean of the Physics-Faculty. He is a solid state physicist, especially interested in materials research and crystal growth. His main research fields are: materials with high electronic correlation, quasicrystals, materials with extremely high melting temperatures, magnetism, and superconductivity.

Stefan Brühne

Johann Wolfgang Goethe-University
Physics Department
Frankfurt am Main, Germany
bruehne@physik.uni-frankfurt.de

Chapter 1.3

Dr. Stefan Brühne, née Mahne, a chemist by education in Germany and England, received his PhD in 1994 from Dortmund University, Germany, on giant cell crystal structures in the Al–Ta system. Following a post doc position at the Materials Department (Crystallography) at ETH Zurich he spent seven years in the ceramics industry. His main activity was R&D of glasses, frits and pigments for high-temperature applications, thereby establishing design of experiment (DoE) techniques. Since 2002, at the Institute of Physics at Frankfurt University he has been investigated X-ray structure determination of quasicrystalline, highly complex and disordered intermetallic materials.

Fabrice Charra

Commissariat à l'Énergie Atomique, Saclay
Département de Recherche sur l'État Condensé, les Atomes et les Molécules
Gif-sur-Yvette, France
fabrice.charra@cea.fr
http://www-drecam.cea.fr/spcsi/

Chapter 5.3

Fabrice Charra conducts research in the emerging field of nanophotonics, in the surface physics laboratory of CEA/Saclay. The emphasis of his work is on light emission and absorption form single nanoscale molecular systems. His area of expertise also extends to nonlinear optics, a domain to which he contributed several advances in the applications of organic materials.

Gianfranco Chiarotti

University of Rome "Tor Vergata"
Department of Physics
Roma, Italy
chiarotti@roma2.infn.it

Chapter 5.2

Gianfranco Chiarotti is Professor Emeritus, formerly Professor of General Physics, Fellow of the American Physical Society, fellow of the Italian National Academy (Accademia Nazionale dei Lincei). He was Chairman of the Physics Committee of the National Research Council (1988–1994), Chair Franqui at the University of Liège (1975), Assistant Professor at the University of Illinois (1955–1957), Editor of the journal Physics of Solid Surfaces, and Landolt–Börnstein Editor of Springer-Verlag from 1993 through 1996. He has worked in several fields of solid state physics, namely electronic properties of defects, modulation spectroscopy, optical properties of semiconductors, surface physics, and scanning tunnelling microscopy (STM) in organic materials.

Claus Fischer

Formerly Institute of Solid State and Materials Research (IFW)
Dresden, Germany
A_C.FischerDD@t-online.de

Chapter 4.2

Claus Fischer recieved his PhD from the Technical University Dresden (Since his retirement in 2000 he continues to work as a foreign scientist of IFW in the field of high-T_c superconductors.) His last position at IFW was head of the Department of Superconducting Materials. The main areas of research were growth of metallic single crystals in particular of magnetic materials, developments of hard magnetic materials, of materials for thick film components of microelectronics and of low-T_c and high-T_c superconducting wires and tapes. Many activities were performed in cooperation with industrial manufacturers.

Günter Fuchs

Leibniz Institute for Solid State and
Materials Research (IFW) Dresden
Magnetism and Superconductivity in the
Institute of Metallic Materials
Dresden, Germany
fuchs@ifw-dresden.de
http://www.ifw-dresden.de/imw/21/

Chapter 4.2

Dr. Günter Fuchs studied physics at the Technical University of Dresden, Germany, and received his PhD in 1980 on the pinning mechanism in superconducting NbTi alloys. Since 1969 he has been at the Institute of Solid State and Materials Research (IFW) in Dresden. His activities are in superconductivity (HTSC, MgB_2, intermetallic borocarbides) and the applications of superconductors. He received the PASREG Award for outstanding scientific achievements in the field of bulk cuprate superconductors in high magnetic fields in 2003.

Frank Goodwin

International Lead Zinc Research
Organization, Inc.
Research Triangle Parc, NC, USA
fgoodwin@ilzro.org
http://www.ilzro.org/Contactus.htm

Chapter 3.1

Frank Goodwin received his Sc.D. from the Massachusetts Institute of Technology in 1979 and is responsible for all materials science research at International Lead Zinc Research Organization, Inc. where he has conceived and managed numerous projects on lead and zinc-containing products. These have included lead in acoustics, cable sheathing, nuclear waste management and specialty applications, together with zinc in coatings, castings and wrought forms.

Susana Gota-Goldmann

Commissariat à l'Energie Atomique (CEA)
Direction de la Recherche Technologique
(DRT)
Fontenay aux Roses, France
susana.gota-goldmann@cea.fr

Chapter 5.3

Dr. Susana Gota-Goldmann received her PhD in Materials Science form the Université Pierre et Marie Curie (Paris V) in 1993. After her PhD, she was engaged as a researcher in the Materials Science Division of the CEA (Commissariat à l'Énergie Atomique, France). She has focused her scientific activity on the growth and characterisation of nanometric oxide layers with applications in spin electronics and photovoltaics. In parallel she has developed the use of synchrotron radiation techniques (X-ray absorption magnetic dicroism, photoemission, resonant reflectivity) for the study of oxide thin layers. Recently she has moved from fundamental to technological research. Dr. Gota-Goldmann is now working as a project manager at the scientific affairs direction of the Technology Research Division (CEA/DRT).

Sivaraman Guruswamy

University of Utah
Metallurgical Engineering
Salt Lake City, UT, USA
sguruswa@mines.utah.edu
http://www.mines.utah.edu/metallurgy/
MML

Chapter 3.1

Dr. Guruswamy is a Professor of Metallurgical Engineering at the University of Utah. He obtained his Ph.D. degree in Metallurgical Engineering from the Ohio State University in 1984. He has made significant contributions in several areas including magnetic materials development, deformation of compound semiconductors, and lead alloys. His current work focuses on magnetostrictive materials and hybrid thermionic/thermoelectric thermal diodes.

Gagik G. Gurzadyan

Technical University of Munich
Institute for Physical and Theoretical
Chemistry
Garching, Germany
gurzadyan@ch.tum.de
http://zentrum.phys.chemie.
tu-muenchen.de/gagik

Chapter 4.4

Gagik G. Gurzadyan, Ph.D., Dr. Sci., has extensive experience in nonlinear optics and crystals, laser photophysics and spectroscopy. He has authored several books including the Handbook of Nonlinear Optical Crystals published by Springer-Verlag. He worked in the Institute of Spectroscopy (USSR), CEA/Saclay (France), Max-Planck-Institute of Radiation Chemistry (Germany). At present he works at the Technical University of Munich with ultrafast lasers in the fields of nonlinear photochemistry of biomolecules and femtosecond spectroscopy.

Hideki Harada

High Tech Association Ltd.
Higashikaya, Fukaya, Saitama, Japan
khb16457@nifty.com
http://homepage1.nifty.com/JABM

Chapter 4.3

Dr. Hideki Harada is chief advisor of magnetic materials and their application and President of High Tech Association Ltd., Saitama, Japan. He is Chairman of the Japan Association of Bonded Magnet Industries (JABM) and received his Ph.D. in 1987 with a work on electrostatic ferrite materials. He worked in research and development of magnetic materials and cemented carbide tools at Hitachi Metals where he also was on the Board of Directors. He received the Japanese National Award for Industries Development Contribution.

Bernhard Holzapfel

Leibniz Institute for Solid State and Materials Research Dresden – Institute of Metallic Materials
Superconducting Materials
Dresden, Germany
B.Holzapfel@ifw-dresden.de
http://www.ifw-dresden.de/imw/26/

Chapter 4.2

Dr. Bernhard Holzapfel is head of the superconducting materials group at the Leibniz Institute for Solid State and Materials Research (IFW) Dresden, Germany. His main area of research is pulsed laser deposition of functional thin films and superconductivity. Currently he works on the development of HTSC high J_c coated conductors using ion beam assisted deposition or highly textured metal substrates. His work is supported by a number of national and European founded research projects.

Karl U. Kainer

GKSS Research Center Geesthacht
Institute for Materials Research
Geesthacht, Germany
karl.kainer@gkss.de
http://www.gkss.de

Chapter 3.1

Professor Kainer is director of Institute for Materials Research at GKSS-Research Center, Geesthacht and Professor of Materials Technology at the Technical University of Hamburg-Harburg. He obtained his Ph.D. in Materials Science at the Technical University of Clausthal in 1985 and his Habilitation in 1996. In 1988 he received the Japanese Government Research Award for Foreign Specialists. His current research activities are the development of new alloys and processes for magnesium materials.

Catrin Kammer

METALL – Intl. Journal for Metallurgy
Goslar, Germany
Kammer@metall-news.com
http://www.giesel-verlag.de

Chapter 3.1

Catrin Kammer received her Ph.D. in materials sciences from the Technical University Bergakademie Freiberg, Germany, in 1989. She has been working in the field of light metals and is author of several handbooks about aluminium and magnesium. She is working as author for the journal ALUMINIUM and is teaching in material sciences. Since 2001 she is editor-in-chief of the journal METALL, which deals with all non-ferrous metals.

Wolfram Knabl

Plansee AG
Technology Center
Reutte, Austria
wolfram.knabl@plansee.com
http://www.plansee.com

Chapter 3.1

Dr. Wolfram Knabl studied materials science at the Mining University of Leoben, Austria and received his Ph.D. at the Plansee AG focusing on the development of oxidation protective coatings for refractory metals. Between 1996 and 2002 he was responsible for the test laboratories at Plansee AG and since October 2002 he is working in the field of refractory metals, especially material and process development in the technology center of Plansee AG.

Alfred Koethe

Leibniz-Institut für Festkörper- und Werkstoffforschung
Institut für Metallische Werkstoffe (retired)
Dresden, Germany
alfred.koethe@web.de

Chapter 3.1

Dr. Alfred Koethe is physicist and professor of Materials Science. He retired in 2000 from his position as head of department in the Institute of Metallic Materials at the Leibniz Institute of Solid State and Materials Research in Dresden, Germany. His main research activities were in the fields of preparation and properties of ultrahigh-purity refractory metals and, especially, of steels (stainless steels, high strenght steels, thermomechanical treatment, microalloying, relations chemical composition/microstructure/properties).

Dieter Krause

Schott AG
Research and Technology-Development
Mainz, Germany
dieter.krause@schott.com

Chapter 3.4

Dieter Krause studied physics at the universities of Erlangen and Munich, Germany, where he received his Ph.D. for work on magnetism and metal physics. He was professor in Tehran, Iran, lecturer in Munich and Mainz, Germany. As scientist and director of Schott's corporate research and development centre he was involved in research on optical and mechanical properties of amorphous materials, thin films, and optical fibres. Now he is consultant, chief scientist, and the editor of the "Schott Series on Glass and Glass Ceramics – Science, Technology, and Applications" published by Springer.

Manfred D. Lechner

Universität Osnabrück
Institut für Chemie – Physikalische Chemie
Osnabrück, Germany
lechner@uni-osnabrueck.de
http://www.chemie.uni-osnabrueck.de/pc/index.html

Chapter 3.3

Professor Lechner has a PhD in chemistry from the University of Mainz, Germany. Since 1975 he is Professor of Physical Chemistry at the Institute of Chemistry of the University of Osnabrück, Germany. His scientific work concentrates on the physics and chemistry of polymers. In this area he is mainly working on the influence of high pressure on polymer systems, polymers for optical storage and waveguides as well as synthesis and properties of superabsorbers from renewable resources.

Gerhard Leichtfried

Plansee AG
Technology Center
Reutte, Austria
gerhard.leichtfried@plansee.com
http://www.plansee.com

Chapter 3.1

Dr. Gerhard Leichtfried received his Ph.D from the Montanuniversität Leoben and is qualified for lecturing in powder metallurgy. For 20 years he has been working in various senior positions for the Plansee Aktiengesellschaft, a company engaged in refractory metals, composite materials, cemented carbides and sintered iron and steels.

Werner Martienssen

Universität Frankfurt/Main
Physikalisches Institut
Frankfurt/Main, Germany
Martienssen@Physik.uni-frankfurt.de

Chapters 1.1, 1.2, 2.1, 4.1

Werner Martienssen studied physics and chemistry at the Universities of Würzburg and Göttingen, Germany. He obtained his Ph.D. in Physics with R.W. Pohl, Göttingen, and holds an honorary doctorate at the University of Dortmund. After a visiting-professorship at the Cornell University, Ithaca, USA in 1959 to 1960 he taught physics at the University of Stuttgart and since 1961 at the University of Frankfurt/Main. His main research fields are condensed matter physics, quantum optics and chaotic dynamics. Two of his former students and coworkers became Nobel-laureates in Physics, Gerd K. Binnig for the design of the scanning tunneling microscope in 1986 and Horst L. Störmer for the discovery of a new form of quantum-fluid with fractionally charged excitations in 1998. Werner Martienssen is a member of the Deutsche Akademie der Naturforscher Leopoldina, Halle and of the Akademie der Wissenschaften zu Göttingen. Since 1994 he is Editor-in-Chief of the data collection Landolt–Börnstein published by Springer, Heidelberg.

Toshio Mitsui

Osaka University
Takarazuka, Japan
t-mitsui@jttk.zaq.ne.jp

Chapter 4.5

Toshio Mitsui is an emeritus professor of Osaka University. He studied solid state physics and biophysics at Hokkaido University, Pennsylvania State University, Brookhaven National Laboratory, the Massachusetts Institute of Technology, Osaka University and Meiji University. He was the first to observe the ferroelectric domain structure in Rochelle salt with a polarization microscope. He proposed various theories on ferroelectric effects and biological molecular machines.

Manfred Müller

Dresden University of Technology
Institute of Materials Science
Dresden, Germany
m.mueller33@t-online.de

Chapter 4.3

Dr.-Ing. habil. Manfred Müller is a Professor emeritus of Special Materials at the Institute of Materials Science of the Dresden University of Technology. Before his retirement he was for many years head of department for special materials at the Central Institute for Solid State Physics and Materials Research of the Academy of Sciences in Dresden, Germany. His main field was the research and development of metallic materials with emphasis on special physical properties, such as soft and hard magnetic, electrical and thermoelastic properties. His last field of research was amorphous and nanocrystalline soft magnetic alloys. He is a member of the German Society of Materials Science (DGM) and was a member of the Advisory Board of DGM.

Sergei Pestov

Moscow State Academy of Fine Chemical Technology
Department of Inorganic Chemistry
Moscow, Russia
pestovsm@yandex.ru

Chapter 5.1

Dr. Pestov is a docent of the Inorganic Chemistry Department and a head of group on liquid crystals (LC) at the Moscow State Academy of Fine Chemical Technology. He earned his Ph.D. in physical chemistry in 1992. His research is focused on thermal analysis and thermodynamics of systems containing LC and physical properties of LC. He is an author of a Landolt–Börnstein volume and two books devoted to liquid crystals.

Günther Schlamp

Metallgesellschaft Ffm and Degussa Demetron (retired)
Steinbach/Ts, Germany

Chapter 3.1

Günther Schlamp received his Ph.D. from the Johann-Wolfgang-Goethe University of Frankfurt/Main, Germany, in Physical Chemistry. His industrial activities in research include the development and production of refractory material coatings, high purity materials and parts for electronics, and sputter targets for the reflection-enhancing coating of glas. He has contributed to several Handbooks with repoprts on properties and applications of noble metals and their alloys.

Barbara Schüpp-Niewa

Leibniz-Institute for Solid State and Materials Research Dresden
Institute for Metallic Materials
Dresden, Germany
b.schuepp@ifw-dresden.de
http://www.ifw-dresden.de

Chapter 4.2

Barbara Schüpp-Niewa studied chemistry in Gießen and Dortmund where she received her Ph.D. in 1999. Since 2000 she has been a scientist at the Leibniz-Institute for Solid State and Materials Research Dresden with a focus on crystal structure investigations of oxometalates with superconducting or exciting magnetic ground states. Her current research activities include coated conductors.

Roland Stickler

University of Vienna
Department of Chemistry
Vienna, Austria
roland.stickler@univie.ac.at

Chapter 3.1

Professor Stickler received his master and Dr. degree from the Technical University in Vienna. From 1958 to 1972 he was manager of physical metallurgy with the Westinghouse Research Laboratory in Pittsburgh, Pa. In 1972 he accepted a full professorship at the University of Vienna heading a materials science group in the Institute of Physical Chemistry, and from 1988 he was head of this institute until his retirement as professor emeritus in 1998. He was involved in research and engineering work on superalloys, semiconductor materials and high melting point materials, investigating the relationship between microstructure and mechanical behavior, in particular fatigue and fracture mechanics properties. He was leader of a successful project on brazing under microgravity conditions in the Spacelab-Mission. Further activities included the participation in European COST projects, in particular as chairman of actions on powder metallurgy and light metals. He has authored and coauthored more than 250 publications in scientific journals and proceedings.

Pancho Tzankov

Max Born Institute for Nonlinear Optics
and Short Pulse Spectroscopy
Berlin, Germany
tzankov@mbi-berlin.de
http://staff.mbi-berlin.de/tzankov/

Chapter 4.4

Pancho Tzankov studied laser physics at Sofia University, Bulgaria, and received his Ph.D. in physical chemistry from the Technical University of Munich, Germany. He is now a postdoctoral fellow at the Max Born Institute in Berlin, Germany. His research activities involve development of new nonlinear optical parametric sources of ultrashort pulses and their application for time-resolved spectroscopy.

Volkmar Vill

University of Hamburg
Department of Chemistry, Institute of
Organic Chemistry
Hamburg, Germany
vill@chemie.uni-hamburg.de
http://liqcryst.chemie.uni-hamburg.de/

Chapter 5.1

Professor Volkmar Vill received his Diploma in Chemistry in 1986, his Diploma in Physics in 1988 and his Ph.D. in Chemistry in 1990 from the University of Münster, Germany. In 1997 he earned his Habilitation in Organic Chemistry from the University of Hamburg where he is Professor of Organic Chemistry since 2002. He is the author of the LiqCryst – Database of Liquid Crystals and the Editor of the Handbook of Liquid Crystals, of Landolt–Börnstein, Organic Index, and Vol. VIII/5a, Physical Properties of Liquid Crystals.

Hans Warlimont

DSL Dresden Material-Innovation GmbH
Dresden, Germany
warlimont@ifw-dresden.de

Chapters 3.1, 3.2, 4.2, 4.3

Hans Warlimont is a physical metallurgist and has worked on numerous topics in several research institutions and industrial companies. Among them were the Max-Planck-Institute of Metals Research, Stuttgart, and Vacuumschmelze, Hanau. He was Scientific Director of the Leibniz-Institute of Solid State and Materials Research Dresden and Professor of Materials Science at Dresden University of Technology. Recently he has established DSL Dresden Material-Innovation GmbH to industrialise his invention of electroformed battery grids.

Detailed Contents

List of Abbreviations XV

Part 1 General Tables

1 The Fundamental Constants
Werner Martienssen 3
- 1.1 What are the Fundamental Constants and Who Takes Care of Them? 3
- 1.2 The CODATA Recommended Values of the Fundamental Constants 4
  - 1.2.1 The Most Frequently Used Fundamental Constants 4
  - 1.2.2 Detailed Lists of the Fundamental Constants in Different Fields of Application 5
  - 1.2.3 Constants from Atomic Physics and Particle Physics 7
- References 9

2 The International System of Units (SI), Physical Quantities, and Their Dimensions
Werner Martienssen 11
- 2.1 The International System of Units (SI) 11
- 2.2 Physical Quantities 12
- 2.3 The SI Base Units 13
  - 2.3.1 Unit of Length: the Meter 13
  - 2.3.2 Unit of Mass: the Kilogram 14
  - 2.3.3 Unit of Time: the Second 14
  - 2.3.4 Unit of Electric Current: the Ampere 14
  - 2.3.5 Unit of (Thermodynamic) Temperature: the Kelvin 14
  - 2.3.6 Unit of Amount of Substance: the Mole 14
  - 2.3.7 Unit of Luminous Intensity: the Candela 15
- 2.4 The SI Derived Units 16
- 2.5 Decimal Multiples and Submultiples of SI Units 19
- 2.6 Units Outside the SI 20
  - 2.6.1 Units Used with the SI 20
  - 2.6.2 Other Non-SI Units 20
- 2.7 Some Energy Equivalents 24
- References 25

3 Rudiments of Crystallography
Wolf Assmus, Stefan Brühne 27
- 3.1 Crystalline Materials 28
  - 3.1.1 Periodic Materials 28
  - 3.1.2 Aperiodic Materials 33
- 3.2 Disorder 38

		3.3	Amorphous Materials	39
		3.4	Methods for Investigating Crystallographic Structure	39
		References		41

Part 2 The Elements

1 The Elements
Werner Martienssen 45

	1.1	Introduction		45
		1.1.1	How to Use This Section	45
	1.2	Description of Properties Tabulated		46
		1.2.1	Parts A of the Tables	46
		1.2.2	Parts B of the Tables	46
		1.2.3	Parts C of the Tables	48
		1.2.4	Parts D of the Tables	49
	1.3	Sources		49
	1.4	Tables of the Elements in Different Orders		49
	1.5	Data		54
		1.5.1	Elements of the First Period	54
		1.5.2	Elements of the Main Groups and Subgroup I to IV	59
		1.5.3	Elements of the Main Groups and Subgroup V to VIII	98
		1.5.4	Elements of the Lanthanides Period	142
		1.5.5	Elements of the Actinides Period	151
	References			158

Part 3 Classes of Materials

1 Metals
Frank Goodwin, Sivaraman Guruswamy, Karl U. Kainer, Catrin Kammer, Wolfram Knabl, Alfred Koethe, Gerhard Leichtfried, Günther Schlamp, Roland Stickler, Hans Warlimont 161

	1.1	Magnesium and Magnesium Alloys		162
		1.1.1	Magnesium Alloys	163
		1.1.2	Melting and Casting Practices, Heat Treatment	168
		1.1.3	Joining	169
		1.1.4	Corrosion Behavior	169
		1.1.5	Recent Developments	170
	1.2	Aluminium and Aluminium Alloys		171
		1.2.1	Introduction	171
		1.2.2	Production of Aluminium	171
		1.2.3	Properties of Pure Al	172
		1.2.4	Aluminium Alloy Phase Diagrams	174
		1.2.5	Classification of Aluminium Alloys	179
		1.2.6	Structure and Basic Mechanical Properties of Wrought Work-Hardenable Aluminium Alloys	180

	1.2.7	Structure and Basic Mechanical Properties of Wrought Age-Hardenable Aluminium Alloys	182
	1.2.8	Structure and Basic Mechanical Properties of Aluminium Casting Alloys	184
	1.2.9	Technical Properties of Aluminium Alloys	186
	1.2.10	Thermal and Mechanical Treatment	194
	1.2.11	Corrosion Behavior of Aluminium	202
1.3	Titanium and Titanium Alloys		206
	1.3.1	Commercially Pure Grades of Ti and Low-Alloy Ti Materials	207
	1.3.2	Ti-Based Alloys	208
	1.3.3	Intermetallic Ti–Al Materials	209
	1.3.4	TiNi Shape-Memory Alloys	216
1.4	Zirconium and Zirconium Alloys		217
	1.4.1	Technically-Pure and Low-Alloy Zirconium Materials	217
	1.4.2	Zirconium Alloys in Nuclear Applications	218
	1.4.3	Zirconium-Based Bulk Glassy Alloys	218
1.5	Iron and Steels		221
	1.5.1	Phase Relations and Phase Transformations	222
	1.5.2	Carbon and Low-Alloy Steels	227
	1.5.3	High-Strength Low-Alloy Steels	240
	1.5.4	Stainless Steels	240
	1.5.5	Heat-Resistant Steels	257
	1.5.6	Tool Steels	262
	1.5.7	Cast Irons	268
1.6	Cobalt and Cobalt Alloys		272
	1.6.1	Co-Based Alloys	272
	1.6.2	Co-Based Hard-Facing Alloys and Related Materials	274
	1.6.3	Co-Based Heat-Resistant Alloys, Superalloys	274
	1.6.4	Co-Based Corrosion-Resistant Alloys	276
	1.6.5	Co-Based Surgical Implant Alloys	277
	1.6.6	Cemented Carbides	277
1.7	Nickel and Nickel Alloys		279
	1.7.1	Commercially Pure and Low-Alloy Nickels	279
	1.7.2	Highly Alloyed Ni-Based Materials	279
	1.7.3	Ni-Based Superalloys	284
	1.7.4	Ni Plating	288
1.8	Copper and Copper Alloys		296
	1.8.1	Unalloyed Coppers	296
	1.8.2	High Copper Alloys	297
	1.8.3	Brasses	298
	1.8.4	Bronzes	298
	1.8.5	Copper–Nickel and Copper–Nickel–Zinc Alloys	300
1.9	Refractory Metals and Alloys		303
	1.9.1	Physical Properties	306
	1.9.2	Chemical Properties	308
	1.9.3	Recrystallization Behavior	311
	1.9.4	Mechanical Properties	314

        1.10 Noble Metals and Noble Metal Alloys 329
              1.10.1 Silver and Silver Alloys 330
              1.10.2 Gold and Gold Alloys 347
              1.10.3 Platinum Group Metals and Alloys 363
              1.10.4 Rhodium, Iridium, Rhutenium, Osmium, and their Alloys 386
        1.11 Lead and Lead Alloys 407
              1.11.1 Pure Grades of Lead 407
              1.11.2 Pb–Sb Alloys 411
              1.11.3 Pb–Sn Alloys 414
              1.11.4 Pb–Ca Alloys 416
              1.11.5 Pb–Bi Alloys 419
              1.11.6 Pb–Ag Alloys 420
              1.11.7 Pb–Cu, Pb–Te, and Pb–Cu–Te Alloys 421
              1.11.8 Pb–As Alloys 421
              1.11.9 Lead Cable Sheathing Alloys 421
              1.11.10 Other Lead Alloys 421
        References 422

2 **Ceramics**
   Hans Warlimont 431
        2.1 Traditional Ceramics and Cements 432
              2.1.1 Traditional Ceramics 432
              2.1.2 Cements 432
        2.2 Silicate Ceramics 433
        2.3 Refractory Ceramics 437
        2.4 Oxide Ceramics 437
              2.4.1 Magnesium Oxide 444
              2.4.2 Alumina 445
              2.4.3 Al–O–N Ceramics 447
              2.4.4 Beryllium Oxide 447
              2.4.5 Zirconium Dioxide 447
              2.4.6 Titanium Dioxide, Titanates, etc. 450
        2.5 Non-Oxide Ceramics 451
              2.5.1 Non-Oxide High-Temperature Ceramics 451
              2.5.2 Borides 451
              2.5.3 Carbides 458
              2.5.4 Nitrides 467
              2.5.5 Silicides 473
        References 476

3 **Polymers**
   Manfred D. Lechner 477
        3.1 Structural Units of Polymers 480
        3.2 Abbreviations 482
        3.3 Tables and Figures 483
              3.3.1 Polyolefines 483
              3.3.2 Vinyl Polymers 489

	3.3.3	Fluoropolymers	492
	3.3.4	Polyacrylics and Polyacetals	497
	3.3.5	Polyamides	501
	3.3.6	Polyesters	503
	3.3.7	Polysulfones and Polysulfides	506
	3.3.8	Polyimides and Polyether Ketones	508
	3.3.9	Cellulose	509
	3.3.10	Polyurethanes	511
	3.3.11	Thermosets	512
	3.3.12	Polymer Blends	515
References			522

4 Glasses
Dieter Krause ... 523

4.1	Properties of Glasses – General Comments		526
4.2	Composition and Properties of Glasses		527
4.3	Flat Glass and Hollowware		528
	4.3.1	Flat Glass	528
	4.3.2	Container Glass	529
4.4	Technical Specialty Glasses		530
	4.4.1	Chemical Stability of Glasses	530
	4.4.2	Mechanical and Thermal Properties	533
	4.4.3	Electrical Properties	537
	4.4.4	Optical Properties	539
4.5	Optical Glasses		543
	4.5.1	Optical Properties	543
	4.5.2	Chemical Properties	549
	4.5.3	Mechanical Properties	550
	4.5.4	Thermal Properties	556
4.6	Vitreous Silica		556
	4.6.1	Properties of Synthetic Silica	556
	4.6.2	Gas Solubility and Molecular Diffusion	557
4.7	Glass-Ceramics		558
4.8	Glasses for Miscellaneous Applications		559
	4.8.1	Sealing Glasses	559
	4.8.2	Solder and Passivation Glasses	562
	4.8.3	Colored Glasses	565
	4.8.4	Infrared-Transmitting Glasses	568
References			572

Part 4 Functional Materials

1 Semiconductors
Werner Martienssen ... 575

1.1	Group IV Semiconductors and IV–IV Compounds	578
1.2	III–V Compounds	604

		1.2.1	Boron Compounds	604
		1.2.2	Aluminium Compounds	610
		1.2.3	Gallium Compounds	621
		1.2.4	Indium Compounds	638
	1.3	II–VI Compounds		652
		1.3.1	Beryllium Compounds	652
		1.3.2	Magnesium Compounds	655
		1.3.3	Oxides of Ca, Sr, and Ba	660
		1.3.4	Zinc Compounds	665
		1.3.5	Cadmium Compounds	676
		1.3.6	Mercury Compounds	686
	References			691

2 Superconductors
Claus Fischer, Günter Fuchs, Bernhard Holzapfel, Barbara Schüpp-Niewa, Hans Warlimont ... 695

	2.1	Metallic Superconductors		696
		2.1.1	Elements	696
		2.1.2	Practical Metallic Superconductors	704
	2.2	Non-Metallic Superconductors		711
		2.2.1	Oxide Superconductors	711
		2.2.2	Superconductors Based on the Y–Ba–Cu–O System	720
		2.2.3	Superconductors Based on the Bi–Sr–Ca–Cu–O System	736
		2.2.4	Carbides, Borides, Nitrides	744
	References			749

3 Magnetic Materials
Hideki Harada, Manfred Müller, Hans Warlimont ... 755

	3.1	Basic Magnetic Properties		755
		3.1.1	Atomic Moment	755
		3.1.2	Magnetocrystalline Anisotropy	756
		3.1.3	Magnetostriction	757
	3.2	Soft Magnetic Alloys		758
		3.2.1	Low Carbon Steels	758
		3.2.2	Fe-based Sintered and Composite Soft Magnetic Materials	759
		3.2.3	Iron–Silicon Alloys	763
		3.2.4	Nickel–Iron-Based Alloys	769
		3.2.5	Iron–Cobalt Alloys	772
		3.2.6	Amorphous Metallic Alloys	772
		3.2.7	Nanocrystalline Soft Magnetic Alloys	776
		3.2.8	Invar and Elinvar Alloys	780
	3.3	Hard Magnetic Alloys		794
		3.3.1	Fe–Co–Cr	795
		3.3.2	Fe–Co–V	797
		3.3.3	Fe–Ni–Al–Co, Alnico	798
		3.3.4	Fe–Nd–B	800

		3.3.5	Co–Sm	803
		3.3.6	Mn–Al–C	810
	3.4	Magnetic Oxides		811
		3.4.1	Soft Magnetic Ferrites	811
		3.4.2	Hard Magnetic Ferrites	813
	References			814

4 Dielectrics and Electrooptics
Gagik G. Gurzadyan, Pancho Tzankov 817

4.1	Dielectric Materials: Low-Frequency Properties	822
	4.1.1 General Dielectric Properties	822
	4.1.2 Static Dielectric Constant (Low-Frequency)	823
	4.1.3 Dissipation Factor	823
	4.1.4 Elasticity	823
	4.1.5 Piezoelectricity	824
4.2	Optical Materials: High-Frequency Properties	824
	4.2.1 Crystal Optics: General	824
	4.2.2 Photoelastic Effect	824
	4.2.3 Electrooptic Effect	825
	4.2.4 Nonlinear Optical Effects	825
4.3	Guidelines for Use of Tables	826
4.4	Tables of Numerical Data for Dielectrics and Electrooptics	828
	Isotropic Materials	828
	Cubic, Point Group $m3m$ (O_h) Materials	828
	Cubic, Point Group $\bar{4}3m$ (T_d) Materials	834
	Cubic, Point Group 23 (T) Materials	838
	Hexagonal, Point Group $\bar{6}m2$ (D_{3h}) Materials	838
	Hexagonal, Point Group $6mm$ (C_{6v}) Materials	840
	Hexagonal, Point Group 6 (C_6) Materials	842
	Trigonal, Point Group $\bar{3}m$ (D_{3d}) Materials	844
	Trigonal, Point Group 32 (D_3) Materials	844
	Trigonal, Point Group $3m$ (C_{3v}) Materials	848
	Tetragonal, Point Group $4mmm$ (D_{4h}) Materials	852
	Tetragonal, Point Group $4/m$ (C_{4h}) Materials	854
	Tetragonal, Point Group 422 (D_4) Materials	854
	Tetragonal, Point Group $\bar{4}2m$ (D_{2d}) Materials	856
	Tetragonal, Point Group $4mm$ (C_{4v}) Materials	864
	Orthorhombic, Point Group mmm (D_{2h}) Materials	866
	Orthorhombic, Point Group 222 (D_2) Materials	866
	Orthorhombic, Point Group $mm2$ (C_{2v}) Materials	872
	Monoclinic, Point Group 2 (C_2) Materials	884
References		890

5 Ferroelectrics and Antiferroelectrics
Toshio Mitsui 903

5.1	Definition of Ferroelectrics and Antiferroelectrics	903
5.2	Survey of Research on Ferroelectrics	904

5.3	Classification of Ferroelectrics		906
	5.3.1	The 72 Families of Ferroelectrics	909
5.4	Physical Properties of 43 Representative Ferroelectrics		912
	5.4.1	Inorganic Crystals Oxides [3.1, 2]	912
	5.4.2	Inorganic Crystals Other Than Oxides [3.3]	922
	5.4.3	Organic Crystals, Liquid Crystals, and Polymers [3.4]	930
References			936

Part 5 Special Structures

1 Liquid Crystals
Sergei Pestov, Volkmar Vill 941

1.1	Liquid Crystalline State		941
	1.1.1	Chemical Requirements	943
	1.1.2	Physical Properties of Liquid Crystals	943
	1.1.3	Applications of Liquid Crystals	944
	1.1.4	List of Abbreviations	944
	1.1.5	Conversion Factors	945
1.2	Physical Properties of the Most Common Liquid Crystalline Substances		946
1.3	Physical Properties of Some Liquid Crystalline Mixtures		975
References			977

2 The Physics of Solid Surfaces
Gianfranco Chiarotti 979

2.1	The Structure of Ideal Surfaces		979
	2.1.1	Diagrams of Surfaces [3.2]	979
	2.1.2	Crystallographic Formulas	986
2.2	Surface Reconstruction and Relaxation		986
	2.2.1	Definitions and Notation	986
	2.2.2	Metals	987
	2.2.3	Semiconductors	987
2.3	Electronic Structure of Surfaces		996
	2.3.1	Metals	997
	2.3.2	Semiconductors	1003
	2.3.3	Magnetic Surfaces	1007
2.4	Surface Phonons		1012
	2.4.1	Metals	1012
	2.4.2	Semiconductors and Insulators	1017
	2.4.3	Atom–Surface Potential	1019
2.5	The Space Charge Layer at the Surface of a Semiconductor		1020
	2.5.1	Definitions and Notation	1020
	2.5.2	Useful Formulas and Numerical Values	1022
	2.5.3	Surface Conductivity	1023
2.6	Most Frequently Used Acronyms		1026
References			1029

3 Mesoscopic and Nanostructured Materials
Fabrice Charra, Susana Gota-Goldmann 1031
- 3.1 Introduction and Survey 1031
  - 3.1.1 Historical Review 1031
  - 3.1.2 Definitions 1032
  - 3.1.3 Specific Properties 1034
  - 3.1.4 Organization of this Chapter 1034
- 3.2 Electronic Structure and Spectroscopy 1035
  - 3.2.1 Electronic Quantum Size Effects 1035
  - 3.2.2 Breakdown of the Momentum Conservation Rule 1036
  - 3.2.3 Excitons in Quantum-Confined Systems 1036
  - 3.2.4 Vibrational Modes and Electron–Phonon Coupling 1040
  - 3.2.5 Electron Transport Phenomena 1042
- 3.3 Electromagnetic Confinement 1044
  - 3.3.1 Nanoparticle-Doped Materials 1044
  - 3.3.2 Periodic Electromagnetic Lattices 1048
- 3.4 Magnetic Nanostructures 1048
  - 3.4.1 Spin Electronics 1049
  - 3.4.2 Ultrahigh-Density Storage Media in Hard Disk Drives 1060
- 3.5 Preparation Techniques 1063
  - 3.5.1 Molecular-Beam Epitaxy 1063
  - 3.5.2 Metal-Organic Chemical Vapor Deposition (MOCVD) 1064
  - 3.5.3 Lithography 1064
  - 3.5.4 Nanocrystals in Matrices 1064
  - 3.5.5 Ex Situ Synthesis of Clusters 1065
- **References** 1066

Acknowledgements 1073
About the Authors 1075
Detailed Contents 1081
Subject Index 1091

Subject Index

π-bonded chain geometry 993
π-bonded chain model
– diamond(111)2×1 1005
$\alpha = 0$ glass-ceramics 558
$(CH_3NHCH_2COOH)_3 \cdot CaCl_2$ family 932
5CB
– liquid crystals 948
8CB
– liquid crystals 948
8OCB
– liquid crystals 949

A

Abbe value
– glasses 543, 548
Abrikosov vortices 717
absorption and fluorescence spectra of CdSe 1039
absorption coefficient
– two-photon 826
absorption spectra of spherical particles 1046
Ac actinium 84
acceptor surface level 1023
acceptor surface state 1022
accumulation layer 1020
accuracy 4
acids
– liquid crystals 946
acoustic band 1012
acoustic surface wave 906
acronyms
– solid surface 1026
actinium Ac
– elements 84
adatom 995
adopted numerical values for selected quantities 23
Ag silver 65
Ag-based materials 344
age hardening 198
AISI (American Iron and Steel Institute) 221
Al aluminium 78
Al bronzes 298
alkali aluminium silicates
– electrical properties 434
– mechanical properties 434
– thermal properties 434

alkali halides
– surface phonon energy 1017
alkali–alkaline-earth silicate glasses 530
alkali–lead silicate glasses 530
alkaline-earth aluminium silicates
– electrical properties 436
– glasses 530
– mechanical properties 435
– thermal properties 436
allotropic and high-pressure modifications
– elements 46
alloy
– cast irons 270
– cobalt 272
– elinvar 780
– invar 780
– lead, battery grid 413
– lead–antimony 412
– lead–tin 415
– magnesium 163
– Ti_3Al-based 210
– TiAl-based 213
– titanium 206
– wear resistant 274
alloy systems 296
Allred 46
Alnico 798
Al–O–N ceramics
– dielectric properties 447
– optical properties 447
alum family 911
– ferroelectrics 911
alumina
– electrical properties 446
– mechanical properties 445
– properties 445
– thermal properties 446
aluminium Al
– aluminium alloys 171
– aluminium production 171
– chemical properties 172
– cold working 195
– corrosion behavior 204
– elements 78
– hot working 195
– mechanical properties 172
– mechanical treatment 195
– surface layers 204
– work hardening 195

aluminium alloy
– abrasion resistance 190
– aging 198
– behavior in magnetic fields 194
– binary Al-based systems 174
– classification of aluminium alloys 179
– coefficient of thermal expansion 192
– creep behavior 187
– elastic properties 194
– electrical conductivity 194
– hardness 186
– homogenization 198
– machinability 192
– mechanical properties 180, 182
– nuclear properties 194
– optical properties 194
– physical properties 192, 193
– sheet formability 190
– soft annealing 197
– specific heat 194
– stabilization 197
– stress-relieving 198
– structure 182
– technical property 186
– technological properties 190
– tensile strength 186
– thermal softening 195
– work-hardenable 180
aluminium antimonide
– crystal structure, mechanical and thermal properties 610
– electromagnetic and optical properties 619
– electronic properties 616
– transport properties 618
aluminium arsenide
– crystal structure, mechanical and thermal properties 610
– electromagnetic and optical properties 619
– electronic properties 616
– transport properties 618
aluminium casting alloys
– mechanical properties 184
– structure 184
aluminium compounds
– crystal structure, mechanical and thermal properties 610

- electromagnetic and optical properties 619
- electronic properties 614
- mechanical properties 610
- phonon dispersion curves 612
- thermal conductivity 618
- thermal properties 610
- transport properties 617
aluminium nitride
- crystal structure, mechanical and thermal properties 610
- electromagnetic and optical properties 619
- electronic properties 616
- transport properties 618
aluminium phase diagram
- aluminium alloy phase diagram 174
aluminium phosphide
- crystal structure, mechanical and thermal properties 610
- electromagnetic and optical properties 619
- electronic properties 616
- transport properties 618
aluminothermy 174
Al–Ni phase diagram 285
Am americium 151
americium Am
- elements 151
amorphous alloys
- cobalt–based 774
- iron–based 773
- nickel–based 774
amorphous materials 27, 39
amorphous metallic alloys 772
amount of substance
- definition 14
ampere
- SI base unit 14
amphiphilic compound 941
amphiphilic liquid crystal 942
anisotropic magnetoresistance 1058
annealing coefficient
- glasses 548
annealing of steel 223
antiferroelectric crystal 903
antiferroelectric hysteresis loop 904
antiferroelectric liquid crystal 911
antiferroelectrics
- definition 903
- dielectric properties 903
- elastic properties 903
- pyroelectric properties 903
antimony Sb

- elements 98
aperiodic crystals 27
aperiodic materials 33
apparent tilt angle 935
Ar argon 128
area of surface primitive cell
- crystallographic formulas 986
argon Ar
- elements 128
arsenic As
- elements 98
ARUPS 997
As arsenic 98
astatine At
- elements 118
ASTM 241, 242
ASTM (American Society for Testing and Materials) 330
ASW 906
At astatine 118
atomic moment 755
atomic number Z
- elements 45
atomic radius
- elements 46
atomic scattering 1019
atomic, ionic, and molecular properties
- elements 46
atomically clean crystalline surface 979
atom–surface potential
- surface phonons 1019–1020
Au gold 65
austenitizing 224

B

B boron 78
Ba barium 68
back-bond state 1006
bainite 223
$BaMnF_4$ family 922
band bending
- solid surfaces 1023
band gap see energy gap 592
band pass filters
- glasses 566
band structure
- aluminium compounds 614
- beryllium compounds 653
- boron compounds 606
- cadmium compounds 679
- group IV semiconductors and IV–IV compounds 589–592
- indium compounds 643

- magnesium compounds 657
- mercury compounds 688
- oxides of Ca, Sr, and Ba 662
- zinc compounds 668
barium Ba
- elements 68
barium oxide
- crystal structure, mechanical and thermal properties 660
- electromagnetic and optical properties 664
- electronic properties 661
- transport properties 663
barium titanate 915
base quantities 12
- ISO 13
base unit
- SI 13
basis
- crystal structure 28
$BaTiO_3$ 915
bcc positions
- surface diagrams 982
Be beryllium 68
becquerel
- SI unit of activity 19
benzene 946
berkelium Bk
- elements 151
beryllium Be
- elements 68
beryllium compounds 652
- crystal structure, mechanical and thermal properties 652
- electromagnetic and optical properties 655
- electronic properties 653
- mechanical and thermal properties 652
- optical properties 655
- thermal properties 652
- transport properties 655
beryllium oxide 447
- crystal structure, mechanical and thermal properties 652
- electrical properties 448
- electronic properties 653
- mechanical properties 447
- thermal properties 448
beryllium selenide
- crystal structure, mechanical and thermal properties 652
- electronic properties 653
beryllium sulfide
- crystal structure, mechanical and thermal properties 652

Subject Index

– electronic properties 653
beryllium telluride
– crystal structure, mechanical and thermal properties 652
– electronic properties 653
Bethe–Slater–Pauling relation 755, 756
Bh Bohrium 124
Bi bismuth 98
biaxial crystals 826
binding energy 998
– metal 999
Bioverit
– glasses 559
BIPM (Bureau International des Poids et Mesures) 3, 11, 12
bismuth Bi
– elements 98
Bi–Sr–Ca–Cu–O (BSCCO) 736
Bi–Sr–Ca–Cu–O
– coherence lengths 736
– London penetration depths 736
– maximum T_c 736
– structural data 736
– superconducting properties 741
– upper critical fields 736
BK 7
– glasses 537
Bk berkelium 151
blue phase 942
BNN
– ferroelectric material 920
Bohrium Bh
– elements 124
boiling temperature
– elements 47
Bondi 46
boracite-type family 911, 921
borides
– physical properties 452
Borofloat
– glasses 528, 529
boron antimonide
– crystal structure, mechanical, and thermal properties 604
– electromagnetic and optical properties 610
– electronic properties 607
– transport properties 608
boron arsenide
– crystal structure, mechanical, and thermal properties 604
– electromagnetic and optical properties 610
– electronic properties 606

– transport properties 608
boron B
– elements 78
boron compounds
– crystal structure, mechanical and thermal properties 604
– electromagnetic and properties 610
– electronic properties 606
– mechanical properties 604
– thermal properties 604
– transport properties 608
boron nitride
– crystal structure, mechanical, and thermal properties 604
– electromagnetic and optical properties 610
– electronic properties 606
– transport properties 608
boron phosphide
– crystal structure, mechanical, and thermal properties 604
– electromagnetic and optical properties 610
– electronic properties 606
– transport properties 608
borosilicate glasses 529, 530
Br bromine 118
Bragg equation 40
brasses 298
Bravais cell 979
– 2D lattice 980
Bravais lattice 32
– elements 47
breathing-mode acoustic oscillations 1040
Brillouin scattering 906
Brillouin zone 913
– aluminium compounds 614
– beryllium compounds 653
– boron compounds 606
– cadmium compounds 678
– gallium compounds 626
– group IV semiconductors and IV–IV compounds 589
– indium compounds 643
– magnesium compounds 657
– mercury compounds 688
– oxides of Ca, Sr, and Ba 661
– zinc compounds 668
Brillouin zone corner 921
bromine Br
– elements 118
bronzes 298
BSCCO
– films 738

– single crystal 739
– tapes 739
– wires 739
buckled dimer 991
bulk electron density 998
bulk glassy alloys 217, 218
bulk mobility 1026
bulk modulus
– elements 47

C

C carbon 88
Ca calcium 68
cadmium Cd
– elements 73
cadmium compounds
– crystal structure, mechanical and thermal properties 676
– electromagnetic and optical properties 683
– electronic properties 678
– mechanical and thermal properties 676
– optical properties 683
– thermal properties 676
– transport properties 682
cadmium oxide
– crystal structure, mechanical and thermal properties 676
– electromagnetic and optical properties 683
– electronic properties 678
– transport properties 682
cadmium selenide
– crystal structure, mechanical and thermal properties 676
– electromagnetic and optical properties 683
– electronic properties 678
– transport properties 682
cadmium sulfide
– crystal structure, mechanical and thermal properties 676
– electromagnetic and optical properties 683
– electronic properties 678
– transport properties 682
cadmium telluride
– crystal structure, mechanical and thermal properties 676
– electromagnetic and optical properties 683
– electronic properties 678
– transport properties 682
calamitic liquid crystal 941, 942

calcium Ca
– elements 68
calcium oxide
– crystal structure, mechanical and thermal properties 660
– electromagnetic and optical properties 664
– electronic properties 661
– transport properties 663
californium Cf
– elements 151
candela
– SI base unit 15
capacitor 903
capillary viscometer 943
carat 23
carbide
– cemented 277
– electrical properties 466
– mechanical properties 466
– physical properties 458
– thermal properties 466
carbon C
– elements 88
carbon equivalent (CE) 268
carbon fibers
– physical properties 477
carbon steels 230
carrier concentration n_i
– gallium compounds 631
– indium compounds 647
cast
– classification 268
cast iron 268
– grades 268
– mechanical properties 270
casting technology 170
catalysis 1020
Cd cadmium 73
Ce cerium 142
cell surface 943
cellulose 481, 509
– cellulose acetate (CA) 509, 510
– cellulose acetobutyrate (CAB) 509, 510
– cellulose propionate (CP) 509, 510
– ethylcellulose (EC) 509, 510
– polymers 509
– vulcanized fiber (VF) 509, 510
cement 432
cemented carbides 277
centering types
– crystal structure 28
centrosymmetric media 1007
Cerabone

– glasses 559
ceramic capacitor 906
ceramic thin film 914
– ferroelectrics 903
ceramics 345, 431
– Al–O–N 447
– applications 432
– non-oxide 451
– oxide 437
– properties 432
– refractory 437
– silicon 433
– technical 437
– traditional 432
Ceran®
– glasses 559
– linear thermal expansion 558
Ceravital
– glasses 559
cerium Ce
– elements 142
cesium Cs
– elements 59
Cf californium 151
CGPM (Conférence Générale des Poids et Mesures) 11, 12
CGS
– electromagnetic system 21
– electrostatic system 21
– Gaussian system 21
cgs definitions of magnetic susceptibility 48
chalcogenide glasses 568, 571
channel conductivity 1020
characterization of optical glasses 547
Charpy impact strength 478
– polymers 478
chemical disorder 38
chemical stability
– glasses 530, 531
– optical glasses 550
chemical symbols
– element 45
chemisorption 1020
chiral smectic C
– liquid crystals 944
chlorine Cl
– elements 118
cholesteric phase 942
cholesteryl (cholest-5-ene) substituted mesogens
– liquid crystals 968
cholesteryl compound 944
chromium Cr
– elements 114

CIP (current in the plane of layer) 1051
CIPM (Comité International des Poids et Mesures) 3, 11
Cl chlorine 118
clamped crystal 907, 921, 934
clamped dielectric constant 907
classifications of liquid crystals 942
climatic influences
– glasses 550
cluster boundaries 907
cluster formation 907
CM
– liquid crystals 970
Cm curium 151
CMOS (complementary metal-oxide-semiconductor) 1059
Co cobalt 135
$Co_{17}RE_2$ 805
Co_5RE 805
coating 943
cobalt
– alloys 272
– applications 273
– hard-facing alloy 274
– mechanical properties 277
– superalloys 274
– surgical implant alloys 277
cobalt Co
– elements 135
cobalt corrosion-resistant alloys 276
cobalt-based corrosion-resistant alloys 276
CODATA (Committee on Data for Science and Technology) 4
coefficient of expansion 478
– polymers 478
coefficient of thermal expansion
– glasses 526
coercive field 904
coherence length 920
coherent phonon and Raman spectra 1041
colloidal synthesis
– nanostructured materials 1065
colored glasses
– colorants 567
– glasses 565
columnar phase 941
commensurate reconstruction 986
commercially pure titanium (cp-Ti) 206
communication technology 912

compacted (vermicular) graphite (CG)   268
complex perovskite-type oxide   909
complex refractive index
– zinc compounds   674
composite medium   1045
– dielectric constant   1045
composite solder glasses
– glasses   563
composite structures
– crystallography   34
compound semiconductor   1003
compressibility   478
– polymers   478
compression modulus
– elements   47
condensed matter   27
– classification   28
conductivity
– frequency-dependent   823
conductivity tensor
– elements   47
conductor
– nanoparticle-based   1043
confined electronic systems
– nanostructured materials   1035
confinement effect
– nanostructured materials   1031
constants
– fundamental   3
container glasses   529
continuous distribution of states   1022
continuous-cooling-transformation (CCT) diagram   238
controlled rolling   240
Convention du Mètre   12
conventional system
– ISO   13
conversion factor   945
– density   945
– diamagnetic anisotropy   945
– dipole moment   945
– dynamic viscosity   945
– kinematic viscosity   945
– molar mass   945
– temperatures of phase transitions   945
– thermal conductivity   945
cooperative interactions   903
coordination number
– elements   46, 47
copolymers
– physical properties   477, 483
copper
– unalloyed   297

copper alloys   296, 297
copper Cu
– elements   65
copper–nickel   300
copper–nickel–zinc   300
Co–RE
– phase equilibria   805
corrosion   1020
– resistance   218
Coulomb blockade
– nanostructured materials   1031, 1044
coupled plasmon modes   1048
Co–Sm   803
CPP (current flows perpendicular to the plane of the layer)   1051, 1054
Cr chromium   114
creep modulus   478
– polymers   478
critical field   904
critical slowing-down   907, 908
critical temperature
– elements   47, 48
– Pb alloys   699
CrNi steels   252
Cronstedt, swedish mineralogist   1032
crystal axes   980
crystal morphology   30
crystal optics   824
crystal structure
– beryllium compounds   652
– cadmium compounds   676
– group IV semiconductors   578
– III–V compounds   610
– indium compounds   638
– IV–IV compound semiconductors   578
– magnesium compounds   655
– mercury compounds   686
– oxides of Ca, Sr, and Ba   660
– zinc compounds   665
crystal structure, mechanical and thermal properties
– III–V compounds   621, 638
– III–V semiconductors   604
– II–VI compounds   652, 655, 660, 665, 676, 686
crystal symmetry
– elements   47
– ferroelectrics   915
crystalline ferroelectric   911
crystalline materials
– definition   27
crystalline surface
– atomically clean   979

crystallization
– glasses   524
crystallographic formulas   986
crystallographic properties
– elements   46, 47
crystallographic space group   946–952, 955–961, 964, 966, 968–971
crystallographic structure
– methods to investigate   39
crystallography
– concepts and terms   27
– rudiments of   27
crystals
– biaxial   826
– cubic   826
– isotropic   826
– uniaxial   826
Cryston
– glasses   559
Cs cesium   59
CS2004
– liquid crystals   975
Cu copper   65
cubic
– dielectrics   828, 838
cubic $BaTiO_3$ 916
cubic boron nitride   451
cubic crystal   47
cubic system   986
Curie point   906
Curie temperature   906
– surface   1009
Curie–Weiss constant   931, 932
curium Cm
– elements   151
current/voltage characteristics of films   1044
Cu–Ni
– electrical conductivity   302
– thermal conductivity   303
Cu–Zn
– electrical conductivity   302
– thermal conductivity   303
Cu–Zn phase diagram   299
cyclohexane   946
cycloolefine copolymer (COC)   486, 487
cyclosilicate   433

D

D deuterium   54
damping constant
– optical mode frequency   916
dangling bonds (DBs)   991

DAS 991
data storage media
– nanostructured materials 1031
Db dubnium 105
de Broglie wavelength 1035
Debye length
– solid surfaces 1020
Debye temperature Θ_D
– boron compounds 605
– cadmium compounds 678
– gallium compounds 623
– group IV semiconductors 584
– indium compounds 640
– IV–IV compound semiconductors 584
– mercury compounds 687
– metal surfaces 1013
– oxides of Ca, Sr, and Ba 661
– solid surfaces 1014
– surface phonons 1012
– zinc compounds 667
decimal multiples of SI units 19
degree Celsius
– unit of temperature 14
density ϱ 478, 945–954, 956–971
– aluminium compounds 611
– beryllium compounds 653
– boron compounds 604
– cadmium compounds 676
– elements 47
– gallium compounds 621
– group IV semiconductors and IV–IV compounds 579
– indium compounds 638
– magnesium compounds 656
– mercury compounds 686
– oxides of Ca, Sr, and Ba 660
– polymers 478
– temperature dependence 943
– zinc compounds 665
density of electronic states
– magnesium compounds 658
density of electronic states see also density of states 658
density of phonon states see also density of states 658
density of states (DOS)
– nanostructure 1034
dentistry 330
depletion layer 1020
– solid surfaces 1024
derived quantities 12
– ISO 13
derived units
– SI 16
– special names and symbols 16

deuteration 925
deuterium D
– elements 54
devices
– optical 903
– piezoelectric 903
– pyroelectric 903
devil's staircase 932
devitrifying solder glasses 562
DFB (distributed feedback) 1038
DFG (difference frequency generation) 825
diamagnetic anisotropy 943, 945
diamond
– crystal structure, mechanical and thermal properties 578–588
– electromagnetic and optical properties 601
– electronic properties 589–594
– transport properties 595
diamond positions
– surface diagrams 983
diamond-like structure 979
Dicor
– glasses 559
dielectric
– constant (low-frequency) 823
– dissipation factor 823
– elasticity 823
– general properties 822
– lossy 823
– low-frequency materials 822
– stiffness constant 823
dielectric anisotropy 943, 946, 947
dielectric anomaly 922, 934
dielectric constant ε 907, 947–958, 960–966, 968–971, 973, 1045
– aluminium compounds 619
– beryllium compounds 655
– boron compounds 610
– cadmium compounds 683
– elements 46
– gallium compounds 635, 637
– group IV semiconductors and IV–IV compounds 601–603
– indium compounds 650
– magnesium compounds 660
– mercury compounds 691
– oxides of Ca, Sr, and Ba 664
– zinc compounds 672
dielectric dispersion 907
dielectric dissipation factor
– glasses 538
dielectric function 1001
– surface layer 1007
dielectric loss tan δ

– gallium compounds 635
dielectric material
– properties 826
dielectric polarization 825
dielectric properties
– glasses 538
dielectric strength
– glasses 539
dielectric tensor 827
dielectrics
– α-iodic acid, α-HIO$_3$ 868
– α-mercuric sulfide, α-HgS 846
– α-silicon carbide, SiC 840
– α-silicon dioxide, α-SiO$_2$ 846
– α-zinc sulfide, α-ZnS 842
– β-barium borate, β-BaB$_2$O$_4$ 848
– 2-cyclooctylamino-5-nitropyridine, C$_{13}$H$_{19}$N$_3$O$_2$ 874
– 3-methyl 4-nitropyridine 1-oxide, C$_6$N$_2$O$_3$H$_6$ 868
– 3-nitrobenzenamine, C$_6$H$_4$(NO$_2$)NH$_2$ 878
– 4-($N1N$-dimethylamino)-3-acetamidonitrobenzene 884
– ADA 856
– ADP 856
– aluminium oxide, α-Al$_2$O$_3$ 844
– aluminium phosphate AlPO$_4$ 844
– ammonium dideuterium phosphate, ND$_4$D$_2$PO$_4$ 856
– ammonium dihydrogen arsenate, NH$_4$H$_2$AsO$_4$ 856
– ammonium dihydrogen phosphate, NH$_4$H$_2$PO$_4$ 856
– ammonium Rochelle salt 870
– ammonium sulfate, (NH$_4$)$_2$SO$_4$ 872
– "banana" 872
– barium fluoride, BaF$_2$ 828
– barium formate, Ba(COOH)$_2$ 866
– barium magnesium fluoride, BaMgF$_4$ 872
– barium nitrite monohydrate, Ba(NO$_2$)$_2 \cdot$ H$_2$O 842
– barium sodium niobate, Ba$_2$NaNb$_5$O$_{15}$ 872
– Barium titanate, BaTiO$_3$ 864
– BBO 848
– berlinite 844
– beryllium oxide, BeO 840
– BGO 838
– BIBO 884
– bismuth germanium oxide, Bi$_{12}$GeO$_{20}$ 838

Subject Index

- bismuth silicon oxide, $Bi_{12}SiO_{20}$ 838
- bismuth triborate, BiB_3O_6 884
- BK7 Schott glass 828
- BMF 872
- BSO 838
- cadmium germanium arsenide, $CdGeAs_2$ 856
- cadmium germanium phosphide, $CdGeP_2$ 856
- cadmium selenide, CdSe 840
- cadmium sulfide, CdS 840
- cadmium telluride, CdTe 834
- calcite, $CaCO_3$ 844
- calcium fluoride, CaF_2 828
- calcium tartrate tetrahydrate, $Ca(C_4H_4O_6) \cdot 4H_2O$ 874
- CBO 868
- CDA 858
- cesium dideuterium arsenate, CsD_2AsO_4 856
- cesium dihydrogen arsenate, CsH_2AsO_4 858
- cesium lithium borate, $CsLiB_6O_{10}$ 858
- cesium triborate, CsB_3O_5 868
- cinnabar 846
- CLBO 858
- CNB 874, 875
- COANP 874
- copper bromide, CuBr 834
- copper chloride, CuCl 834
- copper gallium selenide, $CuGaSe_2$ 858
- copper gallium sulfide, $CuGaS_2$ 858
- copper iodide, CuI 834
- cubic $m3m$ (O_h) 828
- cubic $\bar{4}3m$ (T_d) 834
- cubic, 23 (T) 838
- D(+)-saccharose, $C_{12}H_{22}O_{11}$ 888
- DADP 856
- DAN 884
- DCDA 856
- deuterated L-arginine phosphate, $(ND_xH_{2-x})_2^+(CND)(CH_2)_3CH(ND_yH_{3-y})^+COO^- \cdot D_2PO_4^- \cdot D_2O$ 884
- diamond, C 830
- dipotassium tartrate hemihydrate, $K_2C_4H_4O_6 \cdot 0.5H_2O$ 886
- DKDA 858
- DKDP 858
- DKT 886
- DLAP 884, 886
- DRDA 860
- DRDP 860
- fluorite 828
- fluorspar 828
- forsterite 866
- gadolinium molybdate, $Gd_2(MoO_4)_3$ 876
- gallium antimonide, GaSb 834
- gallium arsenide, GaAs 834
- gallium nitride, GaN 840
- gallium phosphide, GaP 834
- gallium selenide, GaSe 838
- gallium sulfide, GaS 838
- germanium, Ge 830
- GMO 876
- greenockite 840
- halite 832
- hexagonal, 6 (C_6) 842
- hexagonal, $6mm$ (C_{6v}) 840
- hexagonal, $\bar{6}m2$ (D_{3h}) 838
- high-frequency (optical) properties 817
- Iceland spar 844
- indium antimonide, InSb 836
- indium arsenide, InAs 836
- indium phosphide, InP 836
- Irtran-3 828
- Irtran-6 834
- isotropic 828
- KB5 880
- KBBF 846
- KDA 860
- KDP 860
- KLINBO 864
- KTA 880
- KTP 880
- LBO 878
- L-CTT 874
- lead molybdate, $PbMoO_4$ 854
- lead titanate, $PbTiO_3$ 864
- LFM 878
- list of described substances 818
- lithium fluoride, LiF 830
- lithium formate monohydrate, $LiCOOH \cdot H_2O$ 876
- lithium gallium oxide, $LiGaO_2$ 876
- lithium iodate, α-$LiIO_3$ 842
- lithium metagallate 876
- lithium niobate (5% MgO-doped), $MgO:LiNbO_3$ 848
- lithium niobate, $LiNbO_3$ 848
- lithium sulfate monohydrate, $Li_2SO_4 \cdot H_2O$ 886
- lithium tantalate, $LiTaO_3$ 848
- lithium tetraborate, $Li_2B_4O_7$ 864
- lithium triborate, LiB_3O_5 878
- low-frequency properties 817
- magnesium fluoride, MgF_2 852
- magnesium oxide, MgO 830
- magnesium silicate, Mg_2SiO_4 866
- Maxwell's equations 824
- m-chloronitrobenzene, $ClC_6H_4NO_2$ 874, 875
- mNA 878
- m-nitroaniline 878
- MNMA 874
- monoclinic, 2 (C_2) 884
- N,2-dimethyl-4-nitrobenzenamine, $C_8H_{10}N_2O_2$ 874
- N-[2-(dimethylamino)-5-nitrophenyl]-acetamide 884
- nantokite 834
- numerical data 818, 828
- orthorhombic, 222 (D_2) 866
- orthorhombic, $mm2$ (C_{2v}) 872
- orthorhombic, mmm (D_{2h}) 866
- paratellurite 854
- physical properties 817
- PMMA (Plexiglas) 828
- POM 868
- potassium acid phthalate, $KH(C_8H_4O_4)$ 878
- potassium bromide, KBr 830
- potassium chloride, KCl 830
- potassium dideuterium arsenate, KD_2AsO_4 858
- potassium dideuterium phosphate, KD_2PO_4 858
- potassium dihydrogen arsenate, KH_2AsO_4 860
- potassium dihydrogen phosphate, KH_2PO_4 860
- potassium fluoroboratoberyllate, $KBe_2BO_3F_2$ 846
- potassium iodide, KI 830
- potassium lithium niobate, $K_3Li_2Nb_5O_{15}$ 864
- potassium niobate, $KNbO_3$ 880
- potassium pentaborate tetrahydrate, $KB_5O_8 \cdot 4H_2O$ 880
- potassium sodium tartrate tetrahydrate, $KNa(C_4H_4O_6) \cdot 4H_2O$ 870
- potassium titanate (titanyl) phosphate, $KTiOPO_4$ 880
- potassium titanyl arsenate, $KTiOAsO_4$ 880
- proustite 850
- pyragyrite 850
- quartz 846

- RDA 860
- RDP 862
- Rochelle salt 870
- rock salt 832
- RTP 882
- rubidium dideuterium arsenate, RbD$_2$AsO$_4$ 860
- rubidium dideuterium phosphate, RbD$_2$PO$_4$ 860
- rubidium dihydrogen arsenate, RbH$_2$AsO$_4$ 860
- rubidium dihydrogen phosphate, RbH$_2$PO$_4$ 862
- rubidium titanate (titanyl) phosphate, RbTiOPO$_4$ 882
- rutile 852
- sapphire 844
- Silicon dioxide, SiO$_2$ 828
- silicon, Si 830
- silver antimony sulfide, Ag$_3$SbS$_3$ 850
- silver arsenic sulfide, Ag$_3$AsS$_3$ 850
- silver gallium selenide, AgGaSe$_2$ 862
- silver thiogallate, AgGaS$_2$ 862
- sodium ammonium tartrate tetrahydrate, Na(NH$_4$)C$_4$H$_4$O$_6 \cdot$ 4H$_2$O 870
- sodium chlorate, NaClO$_3$ 838
- sodium chloride, NaCl 832
- sodium fluoride, NaF 832
- sodium nitrite, NaNO$_2$ 882
- strontium fluoride, SrF$_2$ 832
- strontium titanate, SrTiO$_3$ 832
- sucrose 888
- sylvine 830
- sylvite 830
- TAS 850
- tellurium dioxide, TeO$_2$ 854
- tellurium, Te 846
- tetragonal, $4/m$ (C_{4h}) 854
- tetragonal, $4/mmm$ (D_{4h}) 852
- tetragonal, 422 (D_4) 854
- tetragonal, $4mm$ (C_{4v}) 864
- tetragonal, $\bar{4}2m$ (D_{2d}) 856
- TGS 888
- thallium arsenic selenide, Tl$_3$AsSe$_3$ 850
- titanium dioxide, TiO$_2$ 852
- tourmaline, (Na,Ca)(Mg,Fe)$_3$B$_3$Al$_6$Si$_6$(O,OH,F)$_{31}$ 850
- triglycine sulfate, (CH$_2$NH$_2$COOH)$_3 \cdot$ H$_2$SO$_4$ 888
- trigonal, 32 (D_3) 844
- trigonal, $3m$ (C_{3v}) 848
- trigonal, $\bar{3}m$ (D_{3d}) 844
- urea, (NH$_2$)$_2$CO 862
- wurtzite 842
- YAG 832
- YAP 866
- YLF 854
- yttrium aluminate, YAlO$_3$ 866
- yttrium aluminium garnet, Y$_3$Al$_5$O$_{12}$ 832
- yttrium lithium fluoride, YLiF$_4$ 854
- yttrium vanadate, YVO$_4$ 852
- YVO 852
- zinc blende 842
- zinc germanium diphosphide, ZnGeP$_2$ 862
- zinc oxide, ZnO 840
- zinc selenide, ZnSe 836
- zinc telluride, ZnTe 836
- zincite 840

difference frequency generation (DFG) 825
differential conductance as function of voltage 1043
diffraction method 39
diffusion coefficient 946–949, 956–960, 970
diffusion-controlled growth
- nanostructured materials 1064

dimensionality
- nanostructured materials 1033

dimensions
- physical quantities 4, 13

dimer bond length 995
dioxide
- zirconium 448

dipole glass 906
dipole moment 945, 947–953, 955–963, 968–971
direct gap
- group IV semiconductors and IV–IV compounds 592

direct piezoelectric effect 824
director 943
discotic liquid crystal 941, 942, 972
- physical properties 972

disk-like molecule 942
disordered materials 38
dispersion
- glasses 543

dispersion curves
- electronic structure of surfaces 999

dispersion hardening 329
displacement of atoms
- surface phonons 1012

displacive disorder 38
displacive ferroelectrics 907
display lifetime 944
dissipation factor 826, 828
dissipative dispersion 907
dissociation energy of molecule
- elements 46

DOBAMBC (liquid crystal) 934
domain pattern
- perpendicular magnetization 1062

domain wall
- ferroelectric 907
- ferromagnetic 907

donor surface state 1022
doping
- chemical 576

DOS (density of states) 330, 1049
- nanostructured materials 1034

drain 1024
Drude-model metal 1045
drug delivery 944
DSP family 931
DTA (differential thermal analysis) 39
dubnium Db
- elements 105

ductile iron 269
Duran
- glasses 527, 531, 537

Dy dysprosium 142
dye dopant 944
dynamic viscosity 945–948, 955–959, 961, 962, 967, 969–971, 973
dysprosium Dy
- elements 142

E

E7
- liquid crystals 975

ECS 1011
EDX (energy-dispersive analysis of X-rays) 39
EELS 1013
effect of solute elements conductivity of Cu 297
effective masses
- aluminium compounds 617
- boron compounds 607
- cadmium compounds 681
- group IV semiconductors and IV–IV compounds 593
- indium compounds 646
- mercury compounds 689

- oxides of Ca, Sr, and Ba   661
- zinc compounds   670
effective masses m_n and m_p
- gallium compounds   629
einsteinium Es
- elements   151
elastic compliance   828
elastic compliance tensor   934
- elements   47
elastic constant c_{ik} 823
- aluminium compounds   611
- beryllium compounds   652
- cadmium compounds   676
- gallium compounds   621
- group IV semiconductors and IV–IV compounds   580
- indium compounds   638
- magnesium compounds   656
- mercury compounds   687
- oxides of Ca, Sr, and Ba   660
- zinc compounds   665
elastic modulus   478, see elastic constant 580
- elements   46, 47
- polymers   478
elastic stiffness   828
- elements   47
elastic tensor   827
elastooptic coefficient   825
elastooptic constant   828
elastooptic tensor   827
electric strength   478
- polymers   478
electrical conductivity
- aluminium compounds   618
- boron compounds   608
- elements   46
- group IV semiconductors and IV–IV compounds   595
electrical conductivity   see also electrical resistivity 659
electrical conductivity σ
- indium compounds   647
- magnesium compounds   659
- mercury compounds   690
- oxides of Ca, Sr, and Ba   663
electrical resistivity
- boron compounds   609
- elements   48
- gallium compounds   630
electrical steel   766
electroforming   288
electromagnetic and optical properties
- group IV semiconductors and IV–IV compounds   601

- III–V compounds   610, 619, 635, 650
- II–VI compounds   655, 660, 664, 672, 683, 691
electromagnetic concentration effect   1046
electromagnetic confinement
- nanostructured materials   1044
electromechanical coupling constant   912, 918
electron affinity
- elements   46
electron and hole mobilities
- aluminium compounds   618
electron density of states   1034
- cadmium compounds   680
electron diffraction   41
electron effective mass m_n 593, see effective mass 661
electron g-factor g_c
- gallium compounds   629
- indium compounds   646
electron microscope image
- magnetic tunneling junction   1057
electron microscopy   1049
electron mobility μ_n 663, see mobility μ 682
- elements   48
- gallium compounds   631
- indium compounds   648
- mercury compounds   690
- oxides of Ca, Sr, and Ba   663
electron transport phenomena
- nanostructured materials   1042
electron tunneling
- phonon-assisted   1043
electronegativity
- elements   46
electronic band gap
- elements   48
electronic conductivity σ
- zinc compounds   671
electronic configuration
- elements   46
electronic dispersion curves   1000
electronic ground state
- elements   46
electronic properties
- group IV semiconductors and IV–IV compounds   589–594
- III–V compounds   606, 614, 626, 643
- II–VI compounds   653, 657, 661, 668, 678, 688

electronic structure
- solid surfaces   996
electronic transport, general description
- aluminium compounds   617
- beryllium compounds   655
- boron compounds   608
- cadmium compounds   682
- gallium compounds   629
- group IV semiconductors and IV–IV compounds   595
- indium compounds   647
- magnesium compounds   659
- mercury compounds   689
- oxides of Ca, Sr, and Ba   663
- zinc compounds   670
electronic work function
- elements   48
electronic, electromagnetic, and optical properties
- elements   46
electron–phonon coupling   1040, 1041
electrooptic coefficients   826
electrooptic modulators
- nanostructured materials   1040
electrooptic tensor   827
electro-optical constants
- zinc compounds   675
electro-optical effect   944
electrostrictive constant   930
elemental semiconductor   1003
elements   45
- allotropic modifications   48
- atomic properties   46
- electromagnetic properties   48
- electronic properties   48
- high-pressure modifications   48
- ionic properties   46
- macroscopic properties   46
- materials data   46
- molecular properties   46
- optical properties   48
- ordered according to the Periodic table   52
- ordered by their atomic number   51
- ordered by their chemical symbol   50
- ordered by their name   49
elinvar alloys   786
- antiferromagnetic   789
elinvar-type alloys
- nonmagnetic   792
energy bands   see band structure 590

energy diagram for a MIM tunnel junction   1053
energy dispersion curve   1000
energy equivalents   24
energy equivalents in different units   24
energy exchange time τ_{e-ph} 1041
energy gap
– aluminium compounds   616
– beryllium compounds   654
– cadmium compounds   681
– gallium compounds   628
– group IV semiconductors   592, 593
– indium compounds   644
– IV–IV compound semiconductors   592, 593
– magnesium compounds   659
– mercury compounds   688
– oxides of Ca, Sr, and Ba   661
– zinc compounds   669
energy gaps
– boron compounds   607
energy shifts in the luminescence peaks   1037
energy-storage cell   1043
engineering critical current density   741
enthalpies of phase transitions   946–973
enthalpy change
– elements   47
enthalpy of combustion   477
– polymers   477
enthalpy of fusion   477
– polymers   477
entropy of fusion   477
– polymers   477
Er erbium   142
erbium Er
– elements   142
Es einsteinium   151
Eu europium   142
europium Eu
– elements   142
EXAFS (extended X-ray atomic fine-structure analysis)   39
excess carrier density   1023
excitation energy
– nanostructured materials   1038
exciton binding energy
– gallium compounds   628
– zinc compounds   669
exciton Bohr radii
– semiconductors   1037

exciton energy
– group IV semiconductors and IV–IV compounds   593
exciton peak energy
– indium compounds   645
exciton Rydberg series   1038
excitons
– nanostructured materials   1036
external forces   46
external-field dependence   46
extinction coefficient k
– gallium compounds   635
– zinc compounds   674

F

F fluorine   118
F2
– optical glasses   551
F5
– glasses   537
facets
– crystallography   27
Fahrenheit   48
families of ferroelectrics   909
fast displays   903
fcc positions
– surface diagrams   981
Fe iron   131
$Fe_{14}Nd_2B$
– commercial magnets   804
– magnetic materials   803
Fe–C(-X)
– carbide phases   224
Fe–Cr alloy   226
Fe–Mn alloys   226
Fe–Ni alloys   225
Fermi energy   998
Fermi function   1022
Fermi level pinning   1025
Fermi surface
– nanostructured materials   1055
Fermi surface shift in an electric field   1050
Fermi surfaces
– nanostructured materials   1049
Fermi wavelength
– nanostructured materials   1035
fermium Fm
– elements   151
ferrielectric triple hysteresis loop   936
ferrielectricity   927
ferrite
– applications   812

– hard magnetic   813
– MnZn   812
– NiZn   813
– soft magnetic   811
ferroelectric ceramics   906
ferroelectric hysteresis loop   904
ferroelectric liquid crystal   906, 911, 945, 967
– physical properties   967
ferroelectric mixtures
– liquid crystals   975
– physical properties   975
ferroelectric phase transition   906, 933
ferroelectric polymers   906
ferroelectric transducer   906
ferroelectrics
– classification   906
– definition   903
– dielectric properties   903
– displacive type   906
– elastic properties   903
– families   909, 911
– general properties   906
– indirect type   906
– inorganic crystals   903
– inorganic crystals other than oxides   922
– inorganic crystals oxides   912
– liquid crystals   903, 930
– order–disorder type   906
– organic crystals   903, 930
– phase transitions   903
– piezoelectric properties   903
– polymers   903, 930
– pyroelectric properties   903
– symbols and units   912
ferromagnetic surface   1008
Fe−Co−Cr   795
Fe−Co−V   797
Fe−Nd−B   800
– phase relations   800
Fe−Ni−Al−Co   798
Fe−Si alloys
– rapidly solidified   768
Fibonacci sequence   35
field-effect mobility   1024
– solid surfaces   1024
first Brillouin zone   *see* Brillouin zone 657
flat glasses   528
flat-band condition   1020
fluorinated three-ring LC   944
fluorine F
– elements   118

fluoropolymers 480, 496
– poly(ethylene-co-
  chlorotrifluoroethylene) (ECTFE)
  496, 497
– poly(ethylene-co-
  tetrafluoroethylene) (ETFE) 496,
  497
– poly(tetrafluoroethylene-co-
  hexafluoropropylene) (FEP) 496,
  497
– polychlorotrifluoroethylene
  (PCTFE) 496, 497
– polytetrafluoroethylene (PTFE)
  496, 497
flux flow (FF) 718
Fm fermium 151
formation curve
– glasses 524
formulas
– crystallographic 986
Fotoceram
– glasses 559
Fotoform
– glasses 559
Foturan
– glasses 559
Fourier map 926
four-ring system 964
Fr francium 59
francium Fr
– elements 59
Frantz–Keldysh effect 1040
free dielectric constant 907
frequency conversion 825
Fresnel reflectivity
– glasses 549
Friedel–Creagh–Kmetz rule 943
fundamental constants 3
– 2002 adjustment 4
– alpha particle 9
– atomic physics and particle physics
  7
– CODATA recommended values
  4
– electromagnetic constants 6
– electron 7
– meaning 4
– most frequently used 4
– neutron 8
– proton 8
– recommended values 3, 4
– thermodynamic constants 6
– units of measurement 3
– universal constants 5
– what are the fundamental
  constants? 3

fused silica
– glasses 534, 537

G

Ga gallium 78
GaAs positions
– surface diagrams 983
gadolinium Gd
– elements 142
gallium antimonide
– crystal structure, mechanical and
  thermal properties 621
– electromagnetic and optical
  properties 635
– electronic properties 626
– transport properties 631
gallium arsenide
– crystal structure, mechanical and
  thermal properties 621
– electromagnetic and optical
  properties 635
– electronic properties 626
– transport properties 631
gallium compounds
– crystal structure, mechanical and
  thermal properties 621
– electromagnetic and optical
  properties 635
– electronic properties 626
– mechanical and thermal properties
  621
– thermal conductivity 634
– thermal properties 621
– transport properties 629
gallium Ga
– elements 78
gallium nitride
– crystal structure, mechanical and
  thermal properties 621
– electromagnetic and optical
  properties 635
– electronic properties 626
– transport properties 631
gallium phosphide
– crystal structure, mechanical and
  thermal properties 621
– electromagnetic and optical
  properties 635
– electronic properties 626
– transport properties 631
gamma titanium aluminides 213
gas permeation 478
– polymers 478
Gd gadolinium 142
Ge germanium 88

germanium
– band structure 590
– crystal structure, mechanical and
  thermal properties 578–588
– electromagnetic and optical
  properties 601
– electronic properties 589–594
– transport properties 598
germanium Ge
– elements 88
g-factor
– cadmium compounds 681
g-factor, conduction electrons
– group IV semiconductors and
  IV–IV compounds 594
glass designation 544
glass formers 527
glass matrix 1040
glass number 8nnn 534, 537
glass number nnnn
– sealing glasses 563
glass structure
– sodium silicate glasses 524
glass temperature
– glasses 524
glass transition temperature 477
– polymers 477
glass-ceramics 525, 526, 558
– density 558
– elastic properties 558
– manufacturing process 558
glasses 523
– Abbe value 547
– abbreviating glass code 543
– acid attack 532
– acid classes 533
– alkali attack 532
– alkali classes 533
– alkali–alkaline-earth silicate 530
– alkali–lead silicate 530
– alkaline-earth aluminosilicate 530
– amorphous metals 523
– armor plate glasses 529
– automotive applications 529
– band pass filters 566
– Borofloat 528, 529
– borosilicate 529, 530
– borosilicate glasses 529
– brittleness 536
– chemical constants 553
– chemical properties 549
– chemical resistance 549
– chemical stability 530, 531
– chemical vapor deposition 523
– color code 554
– composition 527

- compound glasses 529
- container glasses 528, 529
- crack effects 536
- density 528
- dielectric properties 538
- Duran® 531
- elasticity 536
- electrical properties 537
- engineering material 523
- fire protecting glasses 529
- flat 528
- fracture toughness 537
- frozen-in melt 533
- halide glasses 568
- hydrolytic classes 533
- infrared transmitting glasses 568
- infrared-transmitting 571
- inhomogeneous 525
- internal transmission 554
- linear thermal expansion 536, 556
- long pass filters 566
- major groups 526
- manufacturers, preferred optical glasses 545
- melting range 533
- mixtures of oxide compounds 524
- neutral density filters 566
- optical 551
- optical characterization 547
- optical glasses 543
- optical properties 539, 543
- oxide glasses 568
- passivation glasses 562
- physical constants 553
- plate glasses 528
- properties 526
- quasi-solid melt 533
- refractive index 539
- Schott filter glasses 569
- sealing glasses 559
- short pass filters 566
- silicate based 526
- soda–lime type 528
- solder glasses 562
- strength 534
- stress behavior 535
- stress rate 535
- stress-induced birefringence 539
- supercooled melt 533
- surface (cleaning and etching) 533
- surface modification 532
- surface resistivity 538
- technical 530
- tensile strength 536
- thermal conductivity 556
- thermal strength 537
- transmittance 539
- viscosity 534
- vitreous silica 556
- volume resistivity 537
- wear-induced surface defects 535
glasses, colored
- nomenclature 566
- optical filter 566
glasses, sealing
- ceramic 562
- principal applications 561
- recommended material combinations 560
- special properties 561
glasses, solder and passivation
- composite 563
- properties 564
glassy state
- crystallography 39
GMO family
- ferroelectrics 920
GMR (giant magnetoresistance) 1049, 1050
- mechanism 1051
- thickness dependence 1051
GMR ratio 1050
gold
- alloys 347
- applications 347
- chemical properties 361
- electrical properties 356
- electrical resistivity 356
- intermetallic compounds 350
- magnetic properties 358
- mechanical properties 352
- optical properties 359
- phase diagrams 347
- production 347
- special alloys 361
- thermal properties 359
- thermochemical data 347
- thermoelectric properties 358
gold Au
- elements 65
golden mean 35
granular materials 1043
gray
- SI unit of absorbed dose 19
gray iron 269
gray tin
- band structure 590
- crystal structure, mechanical and thermal properties 578–588
- electromagnetic and optical properties 601
- electronic properties 589–594
- transport properties 599
Griffith flaw
- glasses 534
Group IV semiconductors 576, 578–603
- electron mobility 597
- hole mobility 597
Group IV semiconductors and IV–IV compounds
- crystal structure 578
- electromagnetic and optical properties 601
- electronic properties 589–594
- mechanical properties 578
- thermal properties 578
- transport properties 595–601
groups of elements (Periodic table) 45

H

H hydrogen 54
hafnium Hf
- elements 94
Haigh Push-Pull test 408
Hall coefficient
- elements 48
- group IV semiconductors and IV–IV compounds 595
Hall mobility
- group IV semiconductors and IV–IV compounds 596, 597
Hall mobility see also mobility μ 596
halogen-substituted benzene 946
hard disk drive 1060
- limits 1060
- technology 1060
hard ferrites
- magnetic properties 814
hard magnetic alloys 794
hardenability 237
hardmetals 277
hassium Hs
- elements 131
Hatfield steel 226
HATOF 1013
HCl family 924
He helium 54
heat capacities c_p, c_V
- boron compounds 605
- group IV semiconductors 584

– IV–IV compound semiconductors 584
heat capacity 477, 956–960, 970
– cadmium compounds 678
– gallium compounds 623
– indium compounds 640
– mercury compounds 687
– oxides of Ca, Sr, and Ba 661
– polymers 477
– zinc compounds 667
heat-resistant steels 258
HEIS (high-energy ion scattering) 989, 1013
helical structure 942
helium He
– elements 54
Hermann–Mauguin symbols 30
hertz
– SI unit of frequency 19
heterocycles 946
hexagonal $BaTiO_3$ 917
Hf hafnium 94
Hg mercury 73
high copper alloy 297
high-T_c superconductors
– lower critical 716
– upper critical 716
high-frequency dielectric constant ε see also dielectric constant ε 601
high-Ni alloys 281
high-pressure die casting (HPDC) 168
high-pressure modifications 48
high-strength low-alloy 240
hip implants 277
HMF (half-metallic ferromagnet) 1055
Ho holmium 142
hole effective mass m_p 594, see effective mass 661
hole mobility μ_p see mobility μ 682
– elements 48
– gallium compounds 631
– indium compounds 648
hollow-ware
– glasses 528
holmium Ho
– elements 142
holohedry
– crystallography 30
homeotropic alignment 943
Hooke's law 47, 823
hopping mechanism
– nanostructured materials 1043
host–guest effect 944
hot forming

– glasses 524
Hoya code
– glasses 544
HPDC (high-pressure die casting) 168
HRTEM (high-resolution transition electron microscopy) 39
Hs hassium 131
Hume-Rothery phase 333, 350
Hume-Rothery phases 296
hydrogen H
– elements 54
hydrostatic pressure 926
hyper-Raman scattering 915

I

I iodine 118
IBA 991, 1024
ICSU (International Council of the Scientific Unions) 4
ideal surface 979
III–V compound semiconductors 604
II–VI semiconductor compounds 652
image potential 996
image state 996, 997, 999
– effective mass 999
impact strength 478
– polymers 478
impurity elements 206
impurity scattering
– group IV semiconductors and IV–IV compounds 599
In indium 78
incommensurate phase 930
incommensurate phases 906
incommensurate reconstruction 986
index of refraction
– complex 674
indirect gap
– group IV semiconductors and IV–IV compounds 589
indium antimonide
– crystal structure, mechanical and thermal properties 638
– electromagnetic and optical properties 650
– electronic properties 643
– transport properties 647
indium arsenide
– crystal structure, mechanical and thermal properties 638
– electromagnetic and optical properties 650

– electronic properties 643
– transport properties 647
indium compounds
– crystal structure, mechanical and thermal properties 638
– electromagnetic and optical properties 650
– electronic properties 643
– mechanical and thermal properties 638
– optical properties 650
– thermal properties 638
– transport properties 647
indium In
– elements 78
indium nitride
– crystal structure, mechanical and thermal properties 638
– electromagnetic and optical properties 650
– electronic properties 643
– transport properties 647
indium phosphide
– crystal structure, mechanical and thermal properties 638
– electromagnetic and optical properties 650
– electronic properties 643
– transport properties 647
induced phase transition 904
inelastic neutron scattering 906
infrared-transmitting glasses 568, 571
inorganic ferroelectrics 903
inorganic ferroelectrics other than oxides 922
inosilicates 433
insulator
– surface phonon 1017
intercritical annealing 240
interface state density
– solid surface 1026
interlayer distance
– crystallographic formulas 986
International Annealed Copper Standard (IACS) 296
international system of units 11
international tables for crystallography 31
International Union of Pure and Applied Chemistry (IUPAC) 15
International Union of Pure and Applied Physics (IUPAP) 14
internuclear distance
– elements 46
intrinsic carrier concentration

– group IV semiconductors and
  IV–IV compounds 595, 596
intrinsic charge carrier concentration
– elements 48
intrinsic Debye length 1021
intrinsic Fermi level 1020
invar alloys
– Fe–Ni-based 782
– Fe–Pd base 785
– Fe–Pt-based 784
Invar effect 385
inversion center 30
inversion layer 1020
– solid surfaces 1024
inversion layer channel 1024
iodine I
– elements 118
ionic radius
– elements 46, 49
ionization energy
– elements 46
Ir iridium 135
iridium 393
– alloys 393
– applications 393
– chemical properties 398
– diffusion 398
– electrical properties 397
– lattice parameter 394
– magnetic properties 397
– mechanical properties 395
– optical properties 398
– phase diagram 393
– production 393
– thermal properties 398
– thermoelectrical properties 397
iridium Ir
– elements 135
iron and steels 221
iron Fe
– elements 131
iron miscibility gap 226
iron phase diagram 226
iron-carbon alloys 222
iron-cobalt alloys 772
iron–silicon alloys 763
Ising model 906
ISO (International Organization for
  Standardization) 12
isothermal transformation (IT)
  diagram 238
isotropic
– dielectrics 828
isotropic liquid 943
IV–IV compound semiconductors
  576, 578–603

– electron mobility 597
– hole mobility 597

J

JDOS 1006
jellium 997
jellium model 998
– work functions 999
jewellery 330
joint density of states 1005
jominy apparatus 238
Josephson vortices 717
joule
– SI unit of energy 19

K

K 50
– glasses 537
K potassium 59
K10
– optical glasses 551
K7
– optical glasses 551
katal
– SI unit of catalytic activity 19
KDP family 925
kelvin 48
– SI base unit 14
Kerr effect 825
– optical 826
Kerr ellipticity 1011
KH_2PO_4 family 925
kilogram
– SI base unit 14
kinematic viscosity 945, 947, 950,
  951, 955, 962, 964–967, 969, 975
Kleinman symmetry conditions 825
$KNbO_3$ 913
knee joint replacements 277
KNO_3 family 925
Knoop hardness
– optical glasses 550
Kr krypton 128
KRIPES 997
Kroll process 206
krypton Kr
– elements 128
$KTaO_3$ 913

L

LA
– phonon spectra 915
La lanthanum 84

Lamb theory of elastic vibrations
  1040
lamellar (flake) graphite (FG) 268
langbeinite-type family 911, 930
lanthanum La
– elements 84
LASF35
– optical glasses 551
LAT family 932
lattice concept
– crystallography 28
lattice constants see lattice
  parameters 656
lattice dynamics 1012
lattice parameter 980
– aluminium compounds 610
– beryllium compounds 652
– boron compounds 604
– cadmium compounds 676
– gallium compounds 621
– group IV semiconductors and
  IV–IV compounds 579
– indium compounds 638
– magnesium compounds 656
– mercury compounds 686
– oxides of Ca, Sr, and Ba 660
– zinc compounds 665
lattice scattering
– group IV semiconductors and
  IV–IV compounds 598
lattice vibration 906
lattices
– planes and directions 28
Laue images 41
lawrencium Lr
– elements 151
layer-structure family 909
LB (Langmuir–Blodgett) film 944
LC display 944
LC materials (LCMs) 944
LC–surface interaction 944
lead 407
– antimony 412
– arsenic alloys 421
– battery grid alloys 413
– bearing alloys 415
– bismuth 419
– cable sheathing alloys 421
– calcium–tin 417
– calcium–tin, battery grid 418
– coper alloys 421
– corrosion 408
– corrosion classification 411
– fusible alloys 420
– gamma-ray mass-absorption 412
– grades 407

– internal friction   408
– low-melting alloys   419
– mechanical properties   408
– quaternary eutectic alloy   420
– recrystallization   409
– silver alloys   420
– solder alloys   415
– solders   416
– tellurium alloys   421
– ternary alloys   413
– tin alloy   415
lead glasses   527
lead Pb
– elements   88
lead–antimony
– phase diagram   413
LEED (low-energy electron diffraction)   987, 1013
LEIS (low-energy ion scattering)   989, 1013
LF5
– optical glasses   551
Li lithium   59
$Li_2Ge_7O_{15}$ family   920
$LiBaO_3$ 919
light transmittance
– glasses   539
light-emitting diode   1043
$LiNbO_3$ family   909, 919
linear thermal expansion
– optical glasses   556
linear thermal expansion coefficient α
– aluminium compounds   611
– beryllium compounds   653
– boron compounds   605
– cadmium compounds   677
– gallium compounds   623
– group IV semiconductors   582, 583
– indium compounds   639
– IV–IV compound semiconductors   582, 583
– magnesium compounds   656
– mercury compounds   687
– oxides of Ca, Sr, and Ba   660
– zinc compounds   666
$LiNH_4C_4H_4O_8$ family   911
liquid crystal
– anisotropy   943
– refractive index   943
– rod-like   943
liquid crystal family   911, 934
liquid crystal ferroelectrics   903
liquid crystal material (LCM)
– degradation   944
liquid crystal salts   973

– physical properties   973
liquid crystal two-ring systems with bridges
– physical properties   955
liquid crystal two-ring systems without bridges
– physical properties   947
liquid crystalline acids
– physical properties   946
liquid crystalline compound   943
– mesogenic group   943
– side group   943
– terminal group   943
liquid crystalline mixtures
– physical properties   975
liquid crystals
– ferroelectric properties   905
liquid crystals (LCs)   941
list of described physical properties   822
lithium Li
– elements   59
lithium niobate   919
lithography
– nanostructured materials   1064
Lithosil^{TM} 551
LLF1
– optical glasses   551
local-field effect   1045
long pass filters
– glasses   566
longitudinal acoustic branch   915
long-range order   39, 927
– glasses   524
long-range order of molecules   943
losses
– dynamic eddy current   763
– hysteresis   763
low dielectric loss glasses   527
low-dimensional system
– nanostructured materials   1034
low-frequency dielectric constant   917
low-temperature annealing   298
Lr lawrencium   151
LSMO ($La_{0.7}Sr_{0.3}MnO_3$)   1056
Lu lutetium   142
lumen
– non-SI unit in photometry   15
luminescence
– nanostructured materials   1036
luminous flux
– photometry   15
luminous intensity I_v
– photometry   15
lutetium Lu

– elements   142
lyotropic liquid crystals   942

M

M. Faraday   1032
Macor
– glasses   559
magnesium alloys   163
– corrosion behavior   169
– heat treatments   169
– joining   169
– mechanical properties   168
– nominal composition   165
– solubility data   163
– tensile properties   167
– tensile property   166
magnesium compounds
– crystal structure, mechanical and thermal properties   655
– electromagnetic and optical properties   660
– electronic properties   657
– mechanical and thermal properties   655
– optical properties   660
– thermal properties   655
– transport properties   659
magnesium Mg
– casting practices   168
– elements   68
– magnesium alloys   162
– melting practices   168
magnesium oxide   444
– applications   444
– crystal structure, mechanical and thermal properties   655
– electrical properties   444
– electromagnetic and optical properties   660
– electronic properties   657
– mechanical properties   444
– thermal properties   444
– transport properties   659
magnesium selenide
– crystal structure, mechanical and thermal properties   655
– electromagnetic and optical properties   660
– electronic properties   657
– transport properties   659
magnesium silicate
– electrical properties   435, 436
– mechanical properties   435, 436
– thermal properties   435, 436
magnesium sulfide

- crystal structure, mechanical and thermal properties 655
- electromagnetic and optical properties 660
- electronic properties 657
- transport properties 659
magnesium telluride
- crystal structure, mechanical and thermal properties 655
- electromagnetic and optical properties 660
- electronic properties 657
- transport properties 659
magnet
- Mn−Al−C 811
magnetic domain 1058
magnetic dots
- nanostructured materials 1049
magnetic dots, arrays of 1061
magnetic field constant
- fundamental constant 14
magnetic layers 1048, 1050
- spin valve 1052
magnetic materials 755
- Co_5Sm based 806
- hard 794
- permanent 794
magnetic nanostructures 1031, 1048, 1050
- information storage 1048
- read heads 1048
- sensors 1048
magnetic oxides 811
magnetic periodic structures 33
magnetic reading head 1053, 1058
magnetic recording
- perpendicular discontinuous media 1060
magnetic sensors 1048, 1058
magnetic surface 1008
magnetic susceptibility 948–950, 956, 957
- elements 46, 48
magnetic tunnel junction 1054
- manganite-based 1057
magnetic tunneling junctions (MTJ) 1056
magnetization
- elements 48
magnetocrystalline anisotropy 756
magnetoelectronic devices 1058
magnetoresistance effect 1050
magnetoresistance of Fe/Cr multilayers 1051
magnetostriction 757

magnets
- 17/2 type 807
- 5/1 type 806
- $TM_{17}Sm_2$ 808
majority carriers 1023
malleable irons 270
manganese Mn
- elements 124
manipulating the dot magnetization 1062
manufacturing
- Fe−Nd−B magnets 801
manufacturing process
- glasses 525
martensite 223
martensitic transformation 225, 226
mass magnetic susceptibility
- elements 48
mass susceptibility
- elements 48
mass-production glasses 526
materials
- semiconductors 576
materials data
- elements 46
material-specific parameters 46
matrix composites 170
Maxwell–Garnett model 1045
Maya blue color 1032
MBBA
- liquid crystals 955
MBE (molecular-beam epitaxy) 991, 1032, 1049, 1063
- 0-D structures 1064
- 1-D structures 1064
- 2-D structures 1063
Md mendelevium 151
mechanical and thermal properties
- III–V compounds 621, 638
- II–VI compounds 652, 655, 660, 665, 676, 686
mechanical properties
- elements 46, 47
- group IV semiconductors 578–588
- III–V compounds 610
- III–V semiconductors 604
- IV–IV compound semiconductors 578–588
- optical glasses 550
- technical glasses 533
MEIS (medium-energy ion scattering) 989, 1013
meitnerium Mt
- elements 135
melt viscosity 478

- polymers 478
melting point T_m
- aluminium compounds 611
- beryllium compounds 653
- boron compounds 605
- cadmium compounds 677
- gallium compounds 622
- group IV semiconductors and IV–IV compounds 580
- indium compounds 639
- magnesium compounds 656
- mercury compounds 686
- oxides of Ca, Sr, and Ba 660
- zinc compounds 666
melting temperature 477
- elements 47, 48
- polymers 477
memory devices 903
mendelevium Md
- elements 151
mercury compounds
- crystal structure, mechanical and thermal properties 686
- electromagnetic and optical properties 691
- electronic properties 688
- mechanical and thermal properties 686
- thermal properties 686
- transport properties 689
mercury Hg
- elements 73
mercury oxide
- crystal structure, mechanical and thermal properties 686
- electromagnetic and optical properties 691
- electronic properties 688
- transport properties 689
mercury selenide
- crystal structure, mechanical and thermal properties 686
- electromagnetic and optical properties 691
- electronic properties 688
- transport properties 689
mercury sulfide
- crystal structure, mechanical and thermal properties 686
- electromagnetic and optical properties 691
- electronic properties 688
- transport properties 689
mercury telluride
- crystal structure, mechanical and thermal properties 686

– electromagnetic and optical properties 691
– electronic properties 688
– transport properties 689
mesogen 943
mesogenic group 943
mesophase 941
mesoscopic material
– conductivity 1044
– nanoparticle doped 1044
– waveguide applications 1045
mesoscopic materials 1031, 1033
– manufacturing 1033
mesoscopic system
– quantum size effect 1035
– thermodynamic stability 1035
metal
– nanoparticle 1040
– resonance state 999
– surface 987
– surface core level shifts (SCLS) 998
– surface Debye temperature 1013
– surface phonon 1012
– surface state 999
– work function 997
metal surface 987
– jellium model 998
metals 997
– vertical relaxation 989
meter
– SI base unit 13
metrologica, international journal 12
MFM image of a written line on an array of dots 1063
MFM image of arrays of dots 1063
MFM image of domain pattern 1062
– sidewalls 1062
Mg magnesium 68
MHPOBC (liquid crystal) 935, 967
microphase separation 941
Mie theory 1045
Miller delta 825
Miller indices 28
M–I–M (metal–insulator–metal) heterostructure 1053
minority carriers 1023
missing-row reconstruction 987
Mn manganese 124
Mn−Al−C 810
– phase relations 810
Mo molybdenum 114
Mo-based alloys 317
mobility μ

– aluminium compounds 618
– cadmium compounds 682
– group IV semiconductors and IV–IV compounds 597–599
– indium compounds 648
MOCVD (metal-organic chemical vapor deposition) 1064
modulated crystal structure 924
modulated structures
– crystallography 34
moduli see elastic constant 580
Mohs hardness 822, 826
– elements 47
molar enthalpy of sublimation
– elements 47
molar entropy
– elements 47
molar heat capacity
– elements 47
molar magnetic susceptibility
– elements 48
molar mass 945–972
– glasses 527
molar susceptibility
– elements 48
molar volume
– elements 47
mole
– definition 14
– SI base unit 14
mole fraction
– glasses 527
molecular architecture 477
molybdenum Mo
– elements 114
momentum-conservation rule 1036
monocrystalline material 47
monolithic alloys 170
monophilic liquid crystal 942
MOS devices 979
MOS field-effect transistor 1020
MOSFET
– electron and hole mobility 1025
– equilibrium condition 1024
– schematic drawing 1025
Mott–Wannier exciton 1037
MQW (multiple quantum well) 1038
MRAM (magnetic random access memories) 1058
MRAM cell
– schematic diagram 1059
Mt meitnerium 135
MTJ (magnetic tunnel junction) 1053, 1054
MTJ sensor

– operation principle 1059
multi-component alloys 219
multiphase (MP) alloys 276
multiple hysteresis loops 932
multiple quantum wells 1042
MVA-TFT (multidomain vertical-alignment thin-film transistor) 944

N

N nitrogen 98
N16B
– glasses 537
– sealing glasses 563
N-4
– liquid crystals 959
Na sodium 59
$NaNO_2$ family 924
nanocrystal 1038
nanoimprint lithography 1061
nanoimprinted single domain dots
– images 1061
nanoimprinting 1061
nanolithography 1031
nanomaterial 1032
nanometric multilayers 1049
nanoparticle 1031
– doped material 1045
– local-field 1045
nanopatterning 1061
nanophase materials 1032
nanoporous materials 1032, 1043
nanoscience 1032
nanostructured material 1031, 1032
– classification scheme 1034
– conductance 1043
– definition 1032
– electrical conductivity 1043
– manufacturing 1063
– preparation 1031, 1035
– zeolites 1065
nanostructures
– magnetic 1048
nanotechnology 1032
Nb niobium 105
N-BAF10
– optical glasses 551
N-BAF52
– optical glasses 551
N-BAK4
– optical glasses 551
N-BALF4
– optical glasses 551
N-BASF64
– optical glasses 551

Nb-based alloys 318
N-BK7
– glasses 548
– optical glasses 551
Nd neodynium 142
Nd–Fe–B
– physical properties 802
Ne neon 128
near field microscopy 1049
nematic mixtures
– liquid crystals 975
– physical properties 975
nematic phase
– liquid crystals 941
nematic–isotropic transition 945
nematic-phase director 943
Neoceram
– glasses 559
neodynium Nd
– elements 142
neon Ne
– elements 128
Neoparies
– glasses 559
neptunium Np
– elements 151
nesosilicate 433
neutral density filters
– glasses 566
neutron diffraction 41, 926
neutron scattering 915
neutron spectrometer scan 916
new rheocast process (NRC) 170
newton meter
– SI unit of moment of force 19
N-FK51
– glasses 548
– optical glasses 551
N-FK56
– optical glasses 551
$(NH_4)_2SO_4$ family 927
$(NH_4)_3H(SO_4)_2$ family 928
$(NH_4)HSO_4$ family 928
$(NH_4)LiSO_4$ family 928
Ni nickel 139
Ni superalloys 294
nickel
– alloys 279
– application 279
– carbides 285
– low-alloy 279
– mechanical properties 280
– plating 288
nickel Ni
– elements 139
nickel-based superalloys 284

nickel–iron alloys 769
nickel-silvers 300
NIMs (National Institutes for Metrology 4
niobium Nb
– elements 105
nitride
– electrical properties 467
– mechanical properties 467
– thermal properties 467
nitrides
– physical properties 468
nitrogen N
– elements 98
N-KF9
– optical glasses 551
N-KZFS2
– optical glasses 551
N-LAF2
– optical glasses 551
N-LAK33
– optical glasses 551
N-LASF31
– optical glasses 551
No nobelium 151
nobelium No
– elements 151
noble metals 329
– Ag 329
– alloys 329
– applications 330
– Au 329
– catalysts 330
– corrosion resistance 329
– hardness 329
– Ir 329
– optical reflectivity 330
– Os 329
– Pd 329
– Pt 329
– Rh 329
– Ru 329
– vapour pressure 330
NOL (nano-oxide layer) 1053
noncrystallographic diffraction symmetries 36
nondestructive testing 941
nonlinear field-dependent properties 46
nonlinear optical coefficients
– nanostructured materials 1039
nonlinear optical device 903, 925
nonlinear optical susceptibility 920
nonlinear susceptibility tensors 827
non-oxide ferroelectrics 905
non-SI units 11, 20

normalizing 224
Np neptunium 151
N-PK51
– optical glasses 551
N-PSK57
– optical glasses 551
N-SF1
– optical glasses 551
N-SF56
– optical glasses 551
N-SF6
– glasses 548
– optical glasses 551
N-SK16
– optical glasses 551
N-SSK2
– optical glasses 551
nuclear incoherent scattering 916
nuclear reactor 218

O

O oxygen 108
occupied electron shells
– elements 46
Ohara code
– glasses 544
one-dimensional liquid 943
one-electron potential 999
opal
– nanostructured materials 1032
opals 1048
OPO (optical parametric oscillation) 825
optical constants
– gallium compounds 635
– group IV semiconductors and IV–IV compounds 601–603
– indium compounds 651
optical constants n and k
– cadmium compounds 684
optical glasses 526, 543
– thermal properties 556
optical materials
– high-frequency properties 824
optical mode frequency 916
optical parametric oscillation (OPO) 825
optical parametric oscillator 920
optical phonon scattering
– group IV semiconductors and IV–IV compounds 598
optical phonon scattering *see also* phonon scattering 598
optical phonon softening 913
optical properties

– group IV semiconductors and IV–IV compounds   601
– III–V compounds   610, 619, 635, 650
– II–VI compounds   655, 660, 664, 672, 683, 691
optical second-harmonic generator   903
optical transparency range   826
optical visualization   941
optoelectronic devices
– nanostructured materials   1038
order parameter   946–952, 956–960, 962, 963
order parameter, S 943
order principle for mesogenic groups   946
organic ferroelectrics   903
orientational order   941
Os osmium   131
osmium Os
– alloys   402
– applications   402
– cathodes   404
– chemical properties   406
– electrical properties   404
– elements   131
– lattice parameter   402
– magnetic properties   405
– mechanical properties   404
– phase diagrams   402
– production   402
– thermal properties   405
– thermoelectric properties   404
outer-shell orbital radius
– elements   46
over-aging   201
oxidation states
– elements   46
oxide   437
– beryllium   447
– magnesium   444
– physical properties   438
oxide ceramics
– production of   444
oxide ferroelectrics   903, 905
oxide superconductors
– low-T_c oxide   711
oxides of Ca, Sr, and Ba
– crystal structure, mechanical and thermal properties   660
– electromagnetic and optical properties   664
– electronic properties   661
– mechanical and thermal properties   660

– thermal properties   660
– transport properties   663
oxygen O
– elements   108

P

P phosphorus   98
Pa protactinium   151
PAA
– liquid crystals   958
pair distribution function
– crystallography   40
palladium Pd
– alloys   364
– applications   364
– electrical properties   370
– elements   139
– lattice parameters   366
– magnetic properties   372
– mechanical properties   368
– phase diagrams   364
– production   364
– thermoelectric properties   370
parabolic band   1000
Parkes process   330
partially stabilized zirconia (PSZ)
– electrical properties   449
– mechanical properties   449
– thermal properties   449
particle in a box model   1035
particle intensity I_p
– radiometry   15
passivation glasses   562, 564
– glasses   565
– properties   564
pattern transfer by imprinting   1061
Pauling   46
Pb lead   88
PbHPO$_4$ family   927
PbTiO$_3$ 916
PbZrO$_3$ 918
Pb–Ca–Sn
– battery grid alloys   418
PCH-7
– liquid crystals   950
Pd palladium   139
pearlite   222
Pearson symbol
– elements   47
percolation density
– nanostructured materials   1033
periodes of elements (Periodic table)   45
Periodic table
– elements   45

Periodic table of the elements   53
permanent magnets
– Co–Sm   806
perovskite-type family   909, 911, 912
perovskite-type oxide   909
phase diagram
– Fe–C   222
– Fe–Cr   226
– Fe–Mn   225
– Fe–Ni   225
– Fe–Si   227
– Ti–Al   210
– Zr–Nb   218
– Zr–O   218
phase separation
– glasses   525
phase transition temperature   943
phase-matching angle   826
phase-matching condition   825
phonon confinement
– nanostructured materials   1042
phonon density of states
– gallium compounds   623, 625
– indium compounds   642
phonon dispersion
– surface phonons   1012
phonon dispersion curve
– aluminium compounds   612
– gallium compounds   623–625
– indium compounds   641
– surface phonons   1015–1026
phonon dispersion relation
– group IV semiconductors   585–587
– IV–IV compound semiconductors   585, 588
phonon energies
– nanostructured materials   1040
– solid surfaces   1013
phonon frequencies ν
– cadmium compounds   678
– gallium compounds   624
– group IV semiconductors   585
– indium compounds   640
– IV–IV compound semiconductors   585
– magnesium compounds   657
– mercury compounds   687
– oxides of Ca, Sr, and Ba   661
– zinc compounds   667
phonon frequencies ν see also phonon wavenumbers $\tilde{\nu}$ 605
phonon instability   921
phonon mode frequency   908
phonon scattering

- group IV semiconductors and
  IV–IV compounds 598
phonon wavenumbers $\tilde{v}$ *see also*
  phonon frequencies v 605
- aluminium compounds 612
- cadmium compounds 678
- gallium compounds 624
- indium compounds 640
- magnesium compounds 657
- mercury compounds 687
- zinc compounds 667
phonon wavenumbers $\tilde{v}$/frequencies v
- boron compounds 605
phonon–phonon coupling 907
phosphorus P
- elements 98
photochromic glasses
- nanostructured materials 1032
photoconductive crystal 922
photoelastic effect 824
photoemission spectra for Ag
  quantum wells 1036
photoluminescence spectra of CdSe 1039
photometry
- intensity measurements 15
photonic band-gap materials 1048
phyllosilicate 433
physical properties
- liquid crystals 943
physical quantities 12
- base 11, 12
- data 13
- definition 12
- derived 11, 12
- general tables 4
piezoelectric
- element 918
- material 917
- strain constant 918
- strain tensor 824
- tensor 827
piezoelectricity 824, 906
piezooptic coefficient 825
planar alignment 943
planar electromechanical coupling
  factor 918
Planck radiator 15
plasmon excitations
- nanostructured materials 1031
plasmon oscillation
- nanostructured materials 1032
plasmon peak
- metals 1045
plasmon resonance 1045

plastic crystal 941
platinum group metals (PGM) 363
- alloys 363
platinum Pt
- alloys 376
- applications 376
- catalysis 385
- chemical properties 385
- electrical properties 381
- elements 139
- magnetical properties 384
- mechanical properties 378
- optical properties 385
- phase diagrams 376
- production 376
- thermal properties 385
- thermoelectric properties 382
plutonium Pu
- elements 151
PLZT
- ceramic material 918
Pm promethium 142
Po polonium 108
Pockels effect 825
point groups
- crystallography 30
Poisson equation 1022
Poisson number
- elements 47
Poisson's ratio 478
- optical glasses 550
- polymers 478
polarization microscopy 943
polonium Po
- elements 108
poly(4-methylpentene-1) (PMP)
  488, 489
poly(ethylene-co-acrylic acid) (EAA)
  486, 487
poly-(ethylene-co-norbornene) 486, 487
poly(ethylene-co-vinyl acetate)
  (EVA) 486, 487
poly(vinyl chloride) 492, 495
- plastisized (60/40) (PVC-P2) 492, 495
- plastisized (75/25) (PVC-P1) 492, 495
- unplastisized (PVC-U) 492, 494–496
polyacetals 480, 497
- poly(oxymethylene) (POM-H) 497–500
- poly(oxymethylene-co-ethylene) (POM-R) 497, 498, 500
polyacrylics 480, 497

- Poly(methyl methacrylate)
  (PMMA) 497–499
polyamides 480, 501
  polyamide 11 (PA11) 501
- polyamide 12 (PA12) 501
- polyamide 6 (PA6) 501, 502
- polyamide 610 (PA610) 501, 502
- polyamide 66 (PA66) 501, 502
polybutene-1 (PB) 488, 489
polyesters 481, 503
- poly(butylene terephthalate) (PBT) 503–505
- poly(ethylene terephthalate) (PET) 503–505
- poly(phenylene ether) (PPE) 504–506
- polycarbonate (PC) 503, 504
polyether ketones 481, 508
- poly(ether ether ketone) (PEEK) 508, 509
polyethylene 483–486
- high density (HDPE) 483–486
- linear low density (LLDPE) 484–486
- low density (LDPE) 484–486
- medium density (MDPE) 484–486
- ultra high molecular weight
  (UHMWPE) 484–486
polyethylene ionomer (EIM) 486, 487
polyimides 481, 508
- poly(amide imide) (PAI) 508
- poly(ether imide) (PEI) 508, 509
- polyimide (PI) 508
polyisobutylene (PIB) 488, 489
polymer
- physical properties 483
polymer blend 515
- physical properties 477, 483
- poly(acrylonitrile-co-butadiene-co-acrylester) + polycarbonate (ASA + PC) 515–517
- poly(acrylonitrile-co-butadiene-co-styrene) + polyamide (ABS + PA) 515–517
- poly(acrylonitrile-co-butadiene-co-styrene) + polycarbonate (ABS + PC) 515–517
- poly(butylene terephthalate) + poly(acrylonitrile-co-butadiene-co-acrylester) (PBT + ASA) 517, 521
- poly(butylene terephthalate) + polystyrene (PBT + PS) 515, 519, 520

– poly(ethylene terephthalate) + polystyrene (PET + PS)   515, 519, 520
– poly(phenylene ether) + polyamide 66 (PPE + PA66)   517, 521
– poly(phenylene ether) + polystyrene (PPE + PS)   520, 521
– poly(styrene-co-butadiene) (PPE + SB)   517, 520, 521
– poly(vinyl chloride) + chlorinated polyethylene (PVC + PE-C)   515, 518
– poly(vinyl chloride) + poly(acrylonitrile-co-butadiene-co-acrylester) (PVC + ASA)   515, 518
– poly(vinyl chloride) + poly(vinyl chloride-co-acrylate) (PVC + VC/A)   515, 518
– polycarbonate + liquid crystal polymer (PC + LCP)   515, 519, 520
– polycarbonate + poly(butylene terephthalate) (PC + PBT)   515, 519, 520
– polycarbonate + poly(ethylene terephthalate) (PC + PET)   515, 519, 520
– polypropylene + ethylene/propylene/diene rubber (PP + EPDM)   515, 516
– polysulfone + poly(acrylonitrile-co-butadiene-co-styrene) (PSU + ABS)   517, 521
polymer family   911, 936
polymer ferroelectrics   903
polymer matrix   1040
polymers   477
– abbreviations   482
– Charpy impact strength   478
– coefficient of expansion   478
– compressibility   478
– creep modulus   478
– crystallinity   477
– density   478
– elastic modulus   478
– electric strength   478
– enthalpy of combustion   477
– enthalpy of fusion   477
– entropy of fusion   477
– ferroelectric properties   905
– gas permeation   478
– glass transition temperature   477
– heat capacity   477
– impact strength   478
– melt viscosity   478
– melting temperature   477
– physical properties   477
– physicochemical properties   477
– Poisson's ratio   478
– refractive index   478
– relative permittivity   478
– shear modulus   478
– shear rate   478
– Shore hardness   478
– sound velocity   478
– steam permeation   478
– stress   478
– stress at 50% strain (elongation)   478
– stress at fracture   478
– stress at yield   478
– structural units   479–481
– surface resistivity   478
– thermal conductivity   478
– Vicat softening temperature   477
– viscosity   478
– volume resistivity   478
polyolefines   480, 483–486
polypropylene (PP)   488, 489
polysulfides   481, 506
– poly(phenylene sulfide) (PPS)   506, 507
polysulfones   481, 506
– poly(ether sulfone) (PES)   506, 507
– polysulfone (PSU)   506, 507
polyurethanes   481, 511
– polyurethane (PUR)   511, 512
– thermoplastic polyurethane elastomer (TPU)   511, 512
polyvinylidene fluoride
– ferroelectrics   911
porous aluminium silicates
– electrical properties   436
– mechanical properties   436
– thermal properties   436
Portland cement
– ASTM types   433
– chemical composition   432
positional order   941
potassium K
– elements   59
potential barrier   1022
powder-composite materials   277
Powder-in-Tube (PIT)   739
power-law dependence of conductivity on film thickness   1042
Pr praseodynium   142
practical superconductors

– characteristic properties   705
praseodynium Pr
– elements   142
prefixes
– decimal multiples of units   19
primitive cell
– crystal structure   28
projected band structure   996
projected bond length   995
promethium Pm
– elements   142
property tensor   46
– independent components   827
protactinium Pa
– elements   151
proton distribution   926
pseudopotential calculation   1006
Pt platinum   139
p-type diamond
– Debye length   1021
Pu plutonium   151
pulsed infrared lasers   1040
pyrochlore-type family   909
pyroelectric coefficient   917
pyroelectric measurement   931
PZT
– piezoelectric material   917

Q

quantum confinement   1036
– nanostructured materials   1042
quantum dots
– nanostructured materials   1031, 1035
quantum size effect
– nanostructured materials   1031, 1035
quantum transport
– nanostructured materials   1053
quantum well
– coupled   1043
– nanostructured materials   1031, 1035
quantum wires
– nanostructured materials   1035
quantum-well superlattices   1033
quasicrystals   34
QWIP (quantum well infrared photodetector)   1042

R

Ra radium   68
radiant intensity I_e
– radiometry   15

radiation sources and exposure techniques in lithography   1065
radiometric and photometric quantities   16
radiometry
– intensity measurements   15
radium Ra
– elements   68
radon Rn
– elements   128
Raman scattering   906
Raman scattering spectroscopy   1040
Raman spectrum   908
RAS   997
Rayleigh mode   1012
Rb rubidium
– elements   59
Re rhenium
– elements   124
real and imaginary parts ε_1 and ε_2 of the dielectric constant see dielectric constant ε 602
– gallium compounds   637
reconstruction model   987
– solid surfaces   991
reconstruction of semiconductors   991
reconstruction of surface   986
– metals   987
recording media
– arrays of magnetic dots   1061
reduced surface state energy   1022
reduced wave vector   1012
reduced-dimensional material geometries   1033
references
– solid surfaces   1029
reflectance anisotropy spectroscopy (RAS)   1007
refractive index   478, 829, 946–972
– elements   48
– glasses   539, 543
– polymers   478
– Sellmeier dispersion formula   547
– temperature dependence   548
refractive index n 619
– boron compounds   610
– cadmium compounds   684
– gallium compounds   635
– group IV semiconductors and IV–IV compounds   601–603
– indium compounds   650
– mercury compounds   691
– zinc compounds   672
refractories

– boride-based   452
– carbide-based   458
– nitride-based   468
– oxide-based   438
– silicide-based   472
refractory ceramics   437
refractory metals   303
– alloys   303
   compositions   305
   dispersion-strengthened   304
– annealing   311
– chemical properties   308
– crack growth behavior   325
– creep elongation   316
– creep properties   327
– dynamic properties   318
– evaporation rate   307
– fatigue data   321
– flow stress   316
– fracture mechanics   322
– grain boundaries   314
– high-cycle fatigue properties   319
– linear thermal expansion   306
– low-cycle fatigue properties   320
– mechanical properties   314
– metal loss   308
– microplasticity   318
– oxidation behavior   308
– physical properties   306
– production routes   304
– recrystallization   311
– resistance against gaseous media   309
– resistance against metal melts   309
– specific electrical resistivity   307
– specific heat   307
– static mechanical properties   315
– stress–strain curves   320
– thermal conductivity   306
– thermomechanical treatment   314
– vapor pressure   307
– Young's modulus   307
refractory metals alloys
– application   306
– products   306
refractory production
– raw materials   444
relative permittivity   478
– polymers   478
relaxation of semiconductors   991
relaxation of surface   986
– metals   987
relaxor   906, 909, 918
remanent magnetization   922
remanent polarization   918

residual resistance ratio (RRR)   397
residual resistivity ratio (RRR)   338
resistivity
– gallium compounds   630
response of material   46
Rf rutherfordium   94
Rh rhodium
– elements   135
RHEED (reflection high-energy electron diffraction)   990
rhenium Re
– elements   124
rhodium
– alloys   386
– applications   386
– chemical properties   392
– electrical properties   390
– magnetic properties   391
– mechanical properties   387
– optical properties   392
– phase diagrams   386
– production   386
– thermal properties   392
– thermoelectrical properties   391
rhodium Rh
– elements   135
ribbon silicates   433
RIE (reactive ion etching)   1061
Rn radon   128
Rochelle salt   904
Rochelle salt family   932
rod-like molecule   942
RT (room temperature)   49
RTP (room temperaure and standard pressure)   49
Ru ruthenium   131
rubidium Rb
– elements   59
ruthenium Ru
– alloys   399
– applications   399
– chemical properties   402
– electrical properties   401
– elements   131
– lattice parameter   400
– magnetic properties   401
– mechanical properties   400
– optical properties   402
– phase diagrams   399
– production   399
– thermal properties   402
– thermoelectric properties   401
rutherfordium Rf
– elements   94
RW (weighted sound reduction)   409

S

S sulfur 108
SAE (Society of Automotive Engineers) 221
SAM (self-assembled monolayer) 944
samarium Sm
– elements 142
SAW (surface acoustic wave) 912
Sb antimony 98
SbSI family 922
Sc scandium 84
SC(NH$_2$)$_2$ family 930
scandium Sc
– elements 84
scattering
– nanoscale objects 1048
scattering losses of a waveguide 1045
Schoenflies symbol 30
– elements 47
Schott AG 523
Schott code
– glasses 544
Schott filter glasses
– glasses 569
Schott glasses 8nnn 540, 541
SDR 997
Se selenium 108
seaborgium Sg
– elements 114
sealing glasses 527
– glasses 559
second
– SI base unit 14
secondary hardening 263
second-harmonic generation (SHG) 825, 906
second-order elastic constants *see* elastic constant 580
second-order phase transition 907
selenium Se
– elements 108
Sellmeier dispersion formula
– glasses 547
Sellmeier equations 826
SEM (scanning electron microscopy) 39
semiconductor 1003
– covalent 992
– field-effect mobility 1024
– III–V compounds 1004
– intrinsic Debye length 1021

– nanocrystal 1040
– polar 992
– quantum confinement 1036
– reconstruction 1004
– reconstruction model 991
– surface 990
– surface core level shift 1003
– surface Debye temperature 1017
– surface phonon 1017
– surface shift 1004
semiconductor band bending 1020
semiconductor nanostructures 1035
semiconductor surface
– Fermi level pinning 1025
– ionization energy 1003
semiconductors
– aluminium compounds 610
– boron compounds 604
– cadmium compounds 676
– chemical doping 576
– gallium compounds 621
– group IV semiconductors and IV–IV compounds 578–603
– III–V compounds 576, 604
– II–VI compounds 576, 652
– indium compounds 638
– introduction 575
– IV–IV compounds 576
– magnesium compounds 655
– mercury compounds 686
– oxides of Ca, Sr, and Ba 660
– physical properties 577
– table of contents 575
– zinc compounds 665
semi-solid metal processing (SSMP) 170
sensor
– chemiresistor-type 1043
SF1
– optical glasses 551
SF11
– optical glasses 551
SF2
– optical glasses 551
SF6
– glasses 537, 548
– optical glasses 551
SF66
– optical glasses 551
SFG (sum frequency generation) 825
SFM (superfluorinated material) 975
Sg seaborgium 114

shape memory 298
– nickel 279
shape-memory alloys
– TiNi 216
shear modulus 478
– elements 47
– polymers 478
shear rate 478
– polymers 478
SHG (second-harmonic generation) 825, 906
Shore hardness 478
– polymers 478
short pass filters
– glasses 566
short-range order 39
– glasses 524
Shubnikov groups
– crystallography 33
SI (Système International d'Unités) 3, 11
SI (the International System of Units) 12
SI base unit 13
SI definitions of magnetic susceptibility 48
SI derived units 16, 17
– with special names 17, 18
SI prefixes 19
Si silicon 88
SI units
– base quantities 13
– base units 13
Si$_3$N$_4$ ceramics 451
Si$_3$N$_4$ powders 472
SiC ceramics 451
side group 943
sievert
– SI unit of dose equivalent 19
silica
– glasses 524
silicate 433
silicate based glasses 526
silicide 473
– physical properties 472
silicon
– electromagnetic and optical properties 601
– electronic properties 589–594
– transport properties 598
silicon carbide
– band structure 590
– crystal structure, mechanical and thermal properties 578–588
– electromagnetic and optical properties 601

– electronic properties 589–594
– transport properties 595
silicon nitride 467
silicon Si
– crystal structure, mechanical and thermal properties 578–588
– elements 88
silicon steels
– grain-oriented 765
– non-oriented 763
silicon technology 1036
silicon-based lasers 1036
silicon–germanium alloys
– band structure 590
– transport properties 601
silicon-germanium alloys
– crystal structure, mechanical and thermal properties 578–588
– electromagnetic and optical properties 601
silicon–silicon oxide interface 1025
silver 330
– alloys 330
– application 330
– chemical properties 344
– crystal structures 333
– diffusion 342
– electrical properties 338
– intermetallic phases 333
– magnetic properties 339
– mechanical properties 335
– optical properties 341
– phase diagrams 331
– production 330
– ternary alloys 345
– thermal properties 340
– thermodynamic data 331
– thermoelectric properties 339
silver Ag
– elements 65
simple perovskite-type oxide 909
single hysteresis loop 904
SiO_2
– glasses 524
SK51
– optical glasses 551
Sm samarium
– elements 142
smectic C* phase 934
smectic phase 942
Sn tin 88
SNR (signal-to-noise ratio) 1060
soda lime glasses 528, 529, 534
sodium Na

– elements 59
soft annealing 224
soft magnetic alloys 758
– nanocrystalline 776
soft magnetic materials
– composite 759
– sintered 759
soft-mode spectroscopy 906
solar cell 1043
solder alloy 345
solder glasses 562
sol–gel synthesis
– nanostructured materials 1065
solid material
– structure 27
solid material, structure 27
solid surface energy 944
solid-state polymorphism 943
sorosilicate 433
sound velocity 478, 947–949, 956–962, 968–971
– elements 47
– polymers 478
source 1024
sp^3-bonded crystal 991
space charge function
– solid surfaces 1022
space charge layer
– semiconductor surface 1020
– solid surfaces 1020
space groups
– crystallography 31
SPARPES 1011
speed of light
– fundamental constant 13
spheroidal (nodular) graphite (SG) 268
spheroidite 223
spin accumulation 1049
spin diffusion length 1050
spin electronics
– applications 1057
– nanostructured materials 1031, 1049
spin polarization 1055
– nanostructured materials 1049
spin valve multilayers 1052
spin valve read head
– schematic diagram 1058
spin valve sensor 1053
spin-asymmetric material 1049
spinel structure
– crystallography 33
spin-electronic switch 1055
spin–orbit splitting energy Δ_{so}
– aluminium compounds 617

– cadmium compounds 681
– gallium compounds 628
– group IV semiconductors and IV–IV compounds 593
– indium compounds 644
– mercury compounds 689
– zinc compounds 670
spintronics
– nanostructured materials 1031, 1049
SPLEED 1011
spontaneous electric polarization 903
spontaneous polarization 917, 945
Sr strontium 68
$Sr_2Nb_2O_7$ family 920
SRI (sound reduction index) 409
$SrNb_2O_7$ family 909
$SrTeO_3$ family 919
$SrTiO_3$ 913
stabilized zirconia (PSZ) 448
stacking faults
– crystallography 41
stain resistance
– optical glasses 550
stainless steels 240
– austenitic 252
– duplex 257
– ferritic 246
– martensitic 250
– martensitic-ferritic 250
standard electrode potential
– elements 46
standard entropy
– elements 47
standard temperature and pressure (STP) 46
Stark effect
– nanostructured materials 1040
static dielectric constant 826
– elements 48
static dielectric constant 828–889
STC (sound transmission classification) 409
steam permeation 478
– polymers 478
steel
– austenitic 259
– carbon 227
– ferritic 258
– ferritic austenitic 259
– hardening 237
– heat-resistant 258, 261
– high-strength low-alloy (HSLA) 240

– low-alloy carbon steel 227
– mechanical properties 237
– stainless 240
– tool 262
stibiotantalite family 909
stiffness constant see elastic constant 580
STM (scanning tunneling microscopy) 988, 992, 997
STM spectroscopy 1005
STN (supertwisted nematic) effect 944
STO ($SrTiO_3$) 1056
storage capacity of hard disks 1048
storage density evolution of hard disk drives 1048
storage media 1060
– arrays of nanometer-scale dots 1060
– limits 1060
– technology 1060
storing information on the sidewalls of the dots 1062
strain
– polymers 478
strength
– glasses 534
stress 478
– polymers 478
stress at 50% strain (elongation) 478
– polymers 478
stress at fracture 478
stress at yield 478
– polymers 478
stress birefringence
– glasses 539, 549
stress intensity factor
– glasses 534
strong-confinement regime
– nanostructured materials 1037, 1038
strontium oxide
– crystal structure, mechanical and thermal properties 660
– electromagnetic and optical properties 664
– electronic properties 661
– transport properties 663
strontium Sr
– elements 68
strontium titanate 913
structural parameters 990
structural phase transitions 906
structure

– diamond-like 979
structure type
– crystallography 33
Strukturbericht type 47
sublattice 903
sublattice polarization 904
submicrometer magnetic dots 1060
substituted mesogens (liquid crystals)
– physical properties 968
sulfur S
– elements 108
sum frequency generation (SFG) 825
superalloys
– Ni-based cast 288
– nickel 294
superconducting high-T_c
– crystal structure 712
superconducting oxides
– high-T_c chemical composition 712
superconductivity
– elements 48
superconductor 695
– borides 745
– borocarbides 746, 747
– carbides 745
– commercial Nb_3Sn 709
– critical temperature 699
– crystal structure 712
– Debye temperature 696
– device applications 719
– high-T_c cuprates 712, 713, 720
– industrial wire performance 719
– metallic 696
– Nb alloys 702
– non-metallic 712
– Pb alloys 696
– pinning 717
– practical metallic 704
– production Nb_3Sn 708
– Sommerfeld constant 696
– SQUIDs 720
– structural data 720
– thermodynamic properties 696
– Type I 695
– Type II 695
– V alloys 700
– vortex lines 717
– Y–Ba–Cu–O 723
supercooled liquid
– glasses 524
supercooled mesophase 945
superstructures

– crystallography 41
supertwisted nematic (STN) effect 944
Supremax
– glasses 527
surface
– Curie temperature 1009
– diagram 979
– ionization energy 1003
– magnetic 1008
– semiconductor 990
– structure of an ideal 979
surface band structure 996
surface Brillouin zone (SBZ) 996
surface conductivity
– solid surfaces 1024
surface core level shifts (SCLS) 998
– solid surfaces 1003, 1004
surface differential reflectivity (SDR) 1008
surface excess conductivity 1024
surface magnetization 1011
surface mobility 1024
surface of diamond 1004
surface phonon 1012
– dispersion 1019
– metal 1013
– mode 1012
surface plasmon
– absorption of nanoparticles 1046
– dispersion curve 1001
surface resistivity 478
– polymers 478
surface resonance 996
– phonons 1012
surface response
– dielectric theory 1007
surface state
– acceptor 1022
– band 1005
– donor 1022
– transitions 1005
surface state bands
– solid surfaces 1004
surface states 996
surface tension 948–950, 955–962, 968–971, 975
– elements 47
surface tension (γ_{LV}) 943
surfaces 979
surgical implant alloys
– cobalt-based 277
susceptibility

– magnetic   48
– mass   48
– molar   48
– nonlinear dielectric   825, 829
– second-order nonlinear dielectric   826
– third-order nonlinear dielectric   826
SV (spin valve)   1052
symmetry elements of point groups   30
synthesis of clusters
– gas-phase production   1066
– nanostructured materials   1065
synthetic silica
– glasses   557

T

T tritium   54
TA
– phonon spectra   915
Ta tantalum   105
Ta-based alloys   318
tailoring of the electronic wave function   1032
tantalum Ta
– elements   105
Tb terbium   142
TC 12 (Technical Committee 12 of ISO)   12
Tc technetium   124
Te tellurium   108
technetium Tc
– elements   124
technical ceramics   437
technical coppers   297
technical glasses   527, 530
technical specialty glasses   526
tellurium Te
– elements   108
TEM (transmission electron microscopy)   39
TEM image of a superlattice of Au clusters   1066
TEM views of single dots   1062
temper graphite (TG)   268
temperature dependence of carrier concentration
– group IV semiconductors and IV–IV compounds   596
temperature dependence of electrical conductivity
– indium compounds   647
temperature dependence of electronic mobilities

– indium compounds   649
temperature dependence of energy gap
– indium compounds   645
temperature dependence of linear thermal expansion coefficient
– cadmium compounds   677
– magnesium compounds   656
temperature dependence of the lattice parameters
– group IV semiconductors and IV–IV compounds   580–582
temperature dependence of thermal conductivity
– group IV semiconductors and IV–IV compounds   599, 600
– indium compounds   650
temperatures of phase transitions   946–976
tempering of steel   223
template synthesis   944
tensile strength
– elements   47
tensor
– elastooptic   826
– piezoelectric strain   826
terbium Tb
– elements   142
terminal group   943
ternary alloys   298
terne steel coatings   415
TFT (thin-film transistor)   944
TGS family   932
Th thorium   151
thallium Tl
– elements   78
thermal and thermodynamic properties
– elements   46
thermal conductivity κ 478, 945, 947–949, 956, 958
– aluminium compounds   618
– cadmium compounds   682
– elements   46, 47
– gallium compounds   634
– group IV semiconductors and IV–IV compounds   599
– indium compounds   650
– magnesium compounds   659
– mercury compounds   690
– oxides of Ca, Sr, and Ba   663
– polymers   478
– zinc compounds   670
thermal expansion
– glasses   526
thermal expansion coefficient, linear

– elements   47
thermal gap
– indium compounds   645
thermal properties
– group IV semiconductors   578–588
– III–V compounds   610, 621, 638
– III–V semiconductors   604
– II–VI compounds   652, 655, 660, 665, 676, 686
– IV–IV compound semiconductors   578–588
– technical glasses   533, 536
thermal vibrations
– surface phonons   1012
thermal work function
– elements   48
thermally activated flux flow (TAFF)   718
thermochromic material   944
thermodynamic properties
– elements   47
thermoelectric coefficient
– elements   48
thermoelectric power
– oxides of Ca, Sr, and Ba   664
thermography   941, 944
thermomechanical treatment (TMT)   314
thermosets   481, 512
– diallyl phthalate (DAP)   512, 514
– epoxy resin (EP)   514, 515
– melamine formaldehyde (MF)   512, 513
– phenol formaldehyde (PF)   512, 513
– polymers   512
– silicone resin (SI)   514, 515
– unsaturated polyester (UP)   512–514
– urea formaldehyde (UF)   512, 513
thermotropic liquid crystal   942
thin film   906
thin-film transistor (TFT)   944
thixomolding   170
thorium Th
– elements   151
three and four-ring systems
– liquid crystals   964
three-dimensional long-range order   941
three-ring system   964
three-wave interactions
– in crystals   825
thulium Tm
– elements   142

Ti titanium 94
time-temperature-transformation (TTT) diagram 238
tin Sn
– elements 88
titanates 450
titanium 206
– commercially pure grades 207
– creep behavior 208
– creep strength 210
– hardness 207
– high-temperature phase 206
– intermetallic materials 210
– phase transformation 206
– sponge 207
– superalloys 210
– titanium alloys 206
titanium alloys 209
– applications 209
– chemical composition 209
– chemical properties 213
– mechancal properties 213
– mechanical properties 209
– physical properties 213
– polycrystalline 213
– single crystalline 213
– thermal expansion coefficient 214
titanium dioxide
– mechanical properties 450
– thermal properties 450
titanium oxide
– phase diagram 206
titanium Ti
– elements 94
Tl thallium 78
Tm thulium 142
TMR (tunnel magnetoresistance) 1054, 1055
TN (twisted nematic)
– liquid crystals 944
TO 915
tool steels 262
torsional modulus
– optical glasses 550
total losses 763
transformation temperature
– glasses 524
transition range
– glasses 524
transition temperature
– glasses 525
transitions
– surface states 1005
transmission spectra

– colored glasses 566, 567
transmission window
– glasses 524
transmittance
– glasses 548
transmittance of glasses
– color code 549
transport properties
– group IV semiconductors and IV–IV compounds 595–601
– III–V compounds 608, 617, 629, 647
– II–VI compounds 655, 659, 663, 670, 682, 689
transverse acoustic branch 915
transverse optical branch 915
transverse optical mode 906
triple point of water 48
tritium T
– elements 54
truncated crystal 986
tungsten bronze-type family 909, 920
tungsten W
– elements 114
tunnel junction
– magnetic 1053
tunnel magnetoresistance
– function of field and temperature 1057
tunnel magnetoresistance as a function of magnetic field 1055
tunneling
– nanostructured materials 1053
tunneling mechanism
– nanostructured materials 1043
twisted nematic (TN) effect 944
two-dimensional liquid 942
two-photon absorption coefficient 829
two-ring systems with bridges
– liquid crystals 955
two-ring systems without bridges
– liquid crystals 947
Type II superconductors
– anisotropy coefficients 716
– coherence lengths 716
– high-T_c cuprate compounds 716
type metals 414

U

U uranium 151
ultrahigh density storage media 1049, 1060

unalloyed coppers 296
uniaxial crystals 826
Unified Numbering System for Metals and Alloys (UNS) 296
unit cell of Si(111) 7×7 995
units
– amount of substance 14
– atomic 21
– atomic units (a.u.) 22
– candela 15
– CGS units 21
– coherent set of 20
– crystallography 21
– electric current 14
– general tables 4
– length 13
– luminous intensity 15
– mass 14
– natural 21
– natural units (n.u.) 21
– non-SI 22
– non-SI units 20, 21
– other non-SI units 23
– temperature 14
– the international system of 11
– time 14
– used with the SI 20
– X-ray-related units 22
units of physical quantities
– fundamental constants 3
units outside the SI 20
UNS (Unified Numbering System) 221
UPS 998
uranium U
– elements 151
UTS – ultimate tensile strength 219

V

V vanadium 105
van der Waals attraction 1019
vanadium V
– elements 105
vertical nanomagnets 1062
vertical relaxation of metals 989
VFT (Vogel, Fulcher, Tammann) equation
– glasses 533
Vicat softening temperature 477
– polymers 477
Vickers hardness
– elements 47

vinylpolymers   480, 489–492
– poly(acrylonitrile-co-butadiene-co-styrene) (ABS)   492–494
– poly(acrylonitrile-co-styrene-co-acrylester) (ASA)   492–494
– poly(styrene-co-acrylnitrile) (SAN)   489, 490, 492
– poly(styrene-co-butadiene) (SB)   489–491
– poly(vinyl carbazole) (PVK)   492, 493
– polystyrene (PS)   489–491
VIP (viewing-independent panel)   975
viscosity   478, 948, 950–954, 963–965, 967, 968, 975
– dynamic   943
– elements   47
– glasses   524
– kinematic   943
– optical glasses   556
– polymers   478
– technical glasses   533
– temperature dependence   525
viscosity of glasses
– temperature dependence   534
vitreous silica
– electrical properties   557
– gas solubility   557
– glasses   526, 556
– molecular diffusion   557
– optical constants   557
vitreous solder glasses   562
Vitronit
– glasses   559
volume compressibility
– elements   47
volume magnetization
– elements   48
volume of primitive cell
– crystallographic formulas   986
volume resistivity   478
– polymers   478
volume–temperature dependence
– glasses   524
Vycor^{TM}
– glasses   527

W

W tungsten   114
wavelength dependence of refractive index n
– indium compounds   651

WDX (wavelength-dispersive analysis of X-rays)   39
weak-confinement regime nanostructured materials   1037, 1038
wear-induced surface defects
– glasses   535
Weibull distribution
– glasses   535
weight fraction
– glasses   527
Wood's metal   420
work function Φ
– metal   997
– solid surfaces   997
work hardening wrought copper alloys   300
wrought alloys   298
wrought magnesium alloys   164
wrought superalloys   284
wtppm (weight part per million)   407
Wyckoff position
– crystallography   32

X

Xe xenon   128
xenon Xe
– elements   128
X-ray diffraction   39
X-ray interferences
– crystallography   27

Y

Y yttrium   84
Yb ytterbium   142
Young's modulus   823
– elements   47
– optical glasses   550
YS – yield stress   219
ytterbium Yb
– elements   142
yttrium Y
– elements   84
Y–Ba–Cu–O
– critical current density   733
– crystal defects   728
– crystal structure   723
– electric resistivity   730
– grain boundaries   730
– hole concentration   733
– lattice parameters   726
– lower critical field   734
– oxygen content   724

– pinning   735
– substitutions   725
– superconducting properties   731
– thermal conductivity   730
– transition temperature   733
– upper critical field   734

Z

zeolites
– nanostructured materials   1031, 1065
Zerodur$^\circledR$
– glasses   558
– linear thermal expansion   558
zinc compounds
– crystal structure, mechanical and thermal properties   665
– effective hole mass   670
– electromagnetic and optical properties   672
– electronic properties   668
– mechanical and thermal properties   665
– optical properties   672
– thermal properties   665
– transport properties   670
zinc oxide
– crystal structure, mechanical and thermal properties   665
– electromagnetic and optical properties   672
– electronic properties   667
– transport properties   670
zinc selenide
– crystal structure, mechanical and thermal properties   665
– electromagnetic and optical properties   672
– electronic properties   667
– transport properties   670
zinc sulfide
– crystal structure, mechanical and thermal properties   665
– electromagnetic and optical properties   672
– electronic properties   667
– transport properties   670
zinc telluride
– crystal structure, mechanical and thermal properties   665
– electromagnetic and optical properties   672
– electronic properties   667
– transport properties   670

zinc Zn
– elements   73
zircaloy   219
– irradiation effect   219
zirconium
– alloys   217
– bulk glassy alloys   218
– bulk glassy behavior   220
– low alloy materials   217
– nuclear applications   218
– technically-pure materials   217
zirconium dioxide   448

zirconium Zr
– elements   94
ZLI-1132
– liquid crystals   975
Zn zinc   73
Zr zirconium   94

Most Frequently Used Fundamental Constants

CODATA Recommended Values of Fundamental Constants

Quantity	Symbol and relation	Numerical value	Unit	Relative standard uncertainty
Speed of light in vacuum	c	299 792 458	m/s	Fixed by definition
Magnetic constant	$\mu_0 = 4\pi \times 10^{-7}$	$12.566370614\ldots \times 10^{-7}$	N/A^2	Fixed by definition
Electric constant	$\varepsilon_0 = 1/(\mu_0 c^2)$	$8.854187817\ldots \times 10^{-12}$	F/m	Fixed by definition
Newtonian constant of gravitation	G	$6.6742(10) \times 10^{-11}$	m^3/(kg s^2)	1.5×10^{-4}
Planck constant	h	$4.13566743(35) \times 10^{-15}$	eV s	8.5×10^{-8}
Reduced Planck constant	$\hbar = h/2\pi$	$6.58211915(56) \times 10^{-16}$	eV s	8.5×10^{-8}
Elementary charge	e	$1.60217653(14) \times 10^{-19}$	C	8.5×10^{-8}
Fine-structure constant	$\alpha = (1/4\pi\varepsilon_0)(e^2/\hbar c)$	$7.297352568(24) \times 10^{-3}$		3.3×10^{-9}
Magnetic flux quantum	$\Phi_0 = h/2e$	$2.06783372(18) \times 10^{-15}$	Wb	8.5×10^{-8}
Conductance quantum	$G_0 = 2e^2/h$	$7.748091733(26) \times 10^{-5}$	S	3.3×10^{-9}
Rydberg constant	$R_\infty = \alpha^2 m_e c/2h$	10 973 731.568525(73)	1/m	6.6×10^{-12}
Electron mass	m_e	$9.1093826(16) \times 10^{-31}$	kg	1.7×10^{-7}
Proton mass	m_p	$1.67262171(29) \times 10^{-27}$	kg	1.7×10^{-7}
Proton–electron mass ratio	m_p/m_e	1836.15267261(85)		4.6×10^{-10}
Avogadro number	N_A, L	$6.0221415(10) \times 10^{23}$		1.7×10^{-7}
Faraday constant	$F = N_A e$	96 485.3383(83)	C	8.6×10^{-8}
Molar gas constant	R	8.314472(15)	J/K	1.7×10^{-6}
Boltzmann constant	$k = R/N_A$	$1.3806505(24) \times 10^{-23}$	J/K	1.8×10^{-6}
		$8.617343(15) \times 10^{-5}$	eV/K	1.8×10^{-6}
Josephson constant	$K_J = 2e/h$	$483\,597.879(41) \times 10^9$	Hz/V	8.5×10^{-8}
von Klitzing constant	$R_K = h/e^2 = \mu_0 c/2\alpha$	25 812.807449(86)	Ω	3.3×10^{-9}
Bohr magneton	$\mu_B = e\hbar/2m_e$	$927.400949(80) \times 10^{-26}$	J/T	8.6×10^{-8}
		$5.788381804(39) \times 10^{-5}$	eV/T	6.7×10^{-9}
Atomic mass constant	$u = (1/12)m(^{12}\mathrm{C})$ $= (1/N_A) \times 10^{-3}$ kg	$1.66053886(28) \times 10^{-27}$	kg	1.7×10^{-7}
Bohr radius	$a_0 = \alpha/4\pi R_\infty$ $= 4\pi\varepsilon_0 \hbar^2/m_e e^2$	$0.5291772108(18) \times 10^{-10}$	m	3.3×10^{-9}
Quantum of circulation	$h/2m_e$	$3.636947550(24) \times 10^{-4}$	m^2/s	6.7×10^{-9}